2025 International Conference on Power Electronics Converters for Transportation and Energy Applications (PECTEA 2025)

Jatni, India
18-21 June 2025

IEEE Catalog Number: CFP25WZ3-POD
ISBN: 979-8-3315-3013-6

Copyright © 2025 by the Institute of Electrical and Electronics Engineers, Inc. All Rights Reserved

Copyright and Reprint Permissions: Abstracting is permitted with credit to the source. Libraries are permitted to photocopy beyond the limit of U.S. copyright law for private use of patrons those articles in this volume that carry a code at the bottom of the first page, provided the per-copy fee indicated in the code is paid through Copyright Clearance Center, 222 Rosewood Drive, Danvers, MA 01923.

For other copying, reprint or republication permission, write to IEEE Copyrights Manager, IEEE Service Center, 445 Hoes Lane, Piscataway, NJ 08854. All rights reserved.

*** This is a print representation of what appears in the IEEE Digital Library. Some format issues inherent in the e-media version may also appear in this print version.

IEEE Catalog Number: CFP25WZ3-POD
ISBN (Print-On-Demand): 979-8-3315-3013-6
ISBN (Online):   979-8-3315-3012-9

Additional Copies of This Publication Are Available From:

Curran Associates, Inc
57 Morehouse Lane
Red Hook, NY 12571 USA
Phone: (845) 758-0400
Fax: (845) 758-2633
E-mail: curran@proceedings.com
Web: www.proceedings.com

TABLE OF CONTENTS

RZA-GSRS-Based Utility-Interfaced PV System for EV Charging.. 1
Shishupal; Sumit Ghatakchoudhuri

Online Corrective Contingency Analysis for Floating Green Microgrid Using Holomorphic
Embedding Method and Extreme Learning Machine.. 7
Mrutyunjaya Sahani; Sanjib Kumar Panda

Novel Senegalese Wolf Search, Filoviridae Optimization and Asterina Algorithm for Power Loss
Diminution and Voltage Stability Enlargement 13
Lenin Kanagasabai

Inventions Based on Flywheels In-Pipe Cascading Turbines Hydrogen Production with Storage
Hybrid Perpetual Mechanical Battery Technologies Working in Relay 19
Srinivas Bhaskar Chaganti

An Efficient Soft-Switching Pulsed Power Supply Converter for Hydrogen Production 27
Shivam Singh; S. Shiva Kumar; T Ghose

Design and Development of Isolated Space Grade DC-DC Converter for SSPA Using eGaN FET
and Analysis on the Role of Parasitic 33
Santwana Suman; Amit Kumar; Trapti Katiyar; Jayesh Thakkar

Four-Load Series Resonant Inverter with Hybrid Control for Induction Based Cooking Applications 39
Kirlampalli Harija Rani; Neti Vishwanathan

Hot Redundent FPGA Based Controller for Fixed Switching String Sequential Shunt Regulator for
Satellite Power Bus 45
*Manoj Kumar Singh; Pradeep K Peper; Venakata Sastry Vadlamani; V Sreekumar;
Dibyaranjan Senapati; Pichali Reddykishore Reddy*

Analysis of Magnetic Flux Density of Rogowski Coil for Full Harmonic and Lightning Channel-
Base-Current Measurements.. 51
Venkatamaheshbabu Lella; Kandasamy Chandrasekaran

State of Charge Estimation for Li-Ion Battery Under Non-Gaussian Uncertainties with
Experimental Validation 57
Supriyo Samanta; Rahul Radhakrishnan

Hybrid Renewable Energy Sources Integrated Shunt Active Power Filter with Adaptive Fuzzy
Logic.. 63
Ravinder Kumar; Sudhansu Kumar Mishra; Karan Veer

Link Current Optimization for Triple Port Active Bridge Converter with Series Voltage Injection.................... 69
Silpashree Sahu; Dipankar De; Alberto Castellazzi

Strategic Allocation of Electric Vehicle Charging Infrastructure in Distribution Networks: A Novel
Reliability-Based Approach.. 76
Edwin Boima Fahnbulleh; Sharmistha Nandi; Sriparna Roy Ghatak; Babita Panda

Adaptive Ripple Reduction in DC-DC Boost-Buck Converters Using Decision Trees 82
Kumari Sarita; Pankaj Rai

Cyber-Driven Adaptive Distributed Secondary Control for AC Microgrid Under Dynamic Conditions 89
    Swagat Kumar Panda; Bidyadhar Subudhi

Investigation of a PV-BESS Fed Islanded Microgrid System Using RTDS Supported by Field Results 95
    Bikram Kumar Samanta; Supratik Bhowmick; Suman Maiti; Chandan Chakraborty

Parameter Estimation of Hydrogen Fuel Cell Using Metaheuristic Algorithm 101
    Sohini Chaudhuri; Aurobinda Panda; Anjan Kumar Ray; Molay Roy; Omkar Singh; Avismit Dutta

New Systematic Design Methodology of Modular Multilevel Converters in Rectifier Applications 107
    Meddi Tharun; Ashok Mankani; Kumar Saurabh; Aritra Chakraborty; Aditya Naugraiya; Amal S

Fuzzy Tuning Based V/F Logic for Speed Control of Five-Phase Induction Motor Drives 113
    Spandana Kolipaka; Swati Devabhaktuni

A Variable Capacitance Based Modelling and Parameter Determination for Supercapacitor with Experimental Validation 119
    Rose Mary Boby; Shifa H. Rahman; Rahul Radhakrishnan

Adaptive Fractional-Order Model Predictive Control for Power Converters Using Fuzzy Logic 125
    Vani Devarinti; Swati Devabhaktuni

Magnetic Loss Comparison for Link Current Optimized Triple Active Bridge Converters with Soft Switching 131
    A Harikrishnan; Silpashree Sahu; Varri Chandra Sekhar Pavan Kumar; Dipankar De

Simulation Studies on 2/3/5-Level DTC Based Speed Control of IM Drive in Four Quadrant Operation 137
    Brijesh Kumar; Bibhu Prasad Panigrahi; Anup Kumar Panda

Direct Torque Control of an 8/6 Switched Reluctance Motor Using a Shared Switch Converter 143
    Akshat Sharma; Himanshu Misra

PSO Assisted Support Vector Machine Algorithm Towards State of Charge Estimation of Li-Ion Batteries 149
    K Saneep; K Sundareswaran; P Srinivasa Rao Nayak; Sishaj P Simon; T Mithun; Gireesh V Puthusserry

Performance Comparison of Recurrent Neural Network Variants for State of Charge Prediction in Lithium-Ion Batteries 155
    K Saneep; M Kannan; K Sundareswaran; P Srinivasa Rao Nayak; T Mithun

Particle Swarm Optimization Employed Machine Learning on MOSFET Based Ćuk Converter with Flying Squirrel Search Optimization (ML(PSO)-FSSO-Ćuk) Based MPPT 160
    Neeraj Priyadarshi; M. S. Bhaskar; Prabhakar Modak; Sachin Srivastava

Comprehensive Analysis of a 1.2 kV 4H-SiC DMOSFET with an Improved Short Circuit Withstand Time 166
    Ravi Prakash Gautam; Rachita Mohapatra; Wakar Hasan Kazi; Prashant Singh; Akshay K

A Pilot Protection Scheme for HVDC Transmission Line Based on Weighted Sum of Correlations 171
    Pradeep Singh; A. V. Giridhar

Comparative Study of Tracking Error for Booster Synchrotron Power Supply: Effects of Reference Profile Variations and Feedforward Compensation in Fast-Ramped Power Supply 176
Aradhana Kumari; Manoj Leelachand Gandhi; Kandipalli Gopi; Ujjwal Yadav

A High-Gain Bidirectional DC-DC Converter with Dual PWM Control Strategy 182
Rishabh Patel; S Porpandiselvi

Switching Characteristics of Domestic Loads for Reliable Arc Fault Detection 188
Ratnakar Nutenki; Aurobinda Routray; Ashok Kumar Pradhan

Electromagnetic Emission Estimation Using Common-Mode Current for Arc Fault Detection in Low Voltage Systems 194
Ratnakar Nutenki; Aurobinda Routray; Ashok Kumar Pradhan

Design of Synchronous Reluctance Motor for Electric Two-Wheeler Application 200
Saddikuti Sriharitha; Jitendra Gautam; Ankit Dalal

Real Time Precise Position Control of Linear Scan Mechanism for the VELC Payload Operation of the Aditya-L1 Mission 206
Kaushlendra Kumar Kaushal; Navle Sonal Girdharlal; V Lekshmi; S Sudhakar

Dynamic Analysis of Quadratic Gain Single Input Dual Output DC-DC Converter 210
K Divya; S Kumaravel; K. Sundararaju

Thermal and Reliability Analysis of a Dual-Switch Switched Inductor-Capacitor Network-Based High-Gain DC-DC Converter 216
S. Vinu Varshath; S. Kumaravel

Load Frequency Control in Modern Power System-An Overview 222
Mitali Samal; Chinmoy Kumar Panigrahi; Deepak Kumar Gupta

Genetic Algorithm Optimized High Precision Nonlinear Fractional Order PID Controller for Magnetic Levitation System 228
V Devanathan; S Kannan; T Vinopraba; N Karthik Chary

Design and Control of a Dual Active Bridge Converter for Bipolar DC Microgrid Applications 236
Aarthi A; Ramesh Dharavath; D V Siva Krishna Rao K

Interactions linear-Based Prediction of Kinetic Energy in Flywheel Energy Storage for Enhanced Grid Stabilization 242
Saranya St; Pritam Bhowmik; Priya Ranjan Satpathy

A Unified Control Scheme for Current Harmonics Suppression with Ripple Free Power for a Grid Tied SPV System Under Abnormal Grid Voltage Conditions 248
Manash Kumar Mishra; Simanta Kumar Samal; Vivek Nandan Lal

A Novel Doubly Grounded High Gain Single Stage Inverter for Photovoltaic Applications 254
K Muhammedali Shafeeque; Nikhil Sasidharan; Rijil Ramchand

A Novel High Voltage Gain SiC-Based Three Port DC-DC Converter for Standalone PV Applications 260
Nagaraju Kothapalli; Sateesh Kumar Kuncham

Rotor Position Calculation in Case of Partial Failure of Resolver for Motor Control Application 266
Aman Patel; Aditya Purushottam Ghube; Abhisha Chauhan

Design of Battery Storage Fed Asymmetrical Multilevel Converter for Enhanced Statcom Applications 272

N. Vasantha Gowri; G. Suresh Babu; T. Murali Krishna; Ch Anil Bharadwaj

Converter Topology to Integrate Battery Storage with Modular Multilevel Converter 278

T. Murali Krishna; Krishnaveni Kondreddi; N. Vasantha Gowri; Ch Anil Bharadwaj

Optimal Sizing of Hybrid Energy Storage System in a Microgrid 284

Amina S; Prameeda Mohan; Rajeev T

Multi-Load Pre-Regulated Inverter Configuration for All Utensil Induction Cooking Application 290

Kirlampalli Harija Rani; Neti Vishwanathan

Comprehensive Analysis and Modeling of Totem-Pole PFC Rectifiers Using Various Modulation Techniques for EV Charging Application 296

Dalija Rath; Susovon Samanta

Influence of GaN Cap Layer on the Performance of AlGaN/GaN HEMT 302

K Soumiya; Jyotismita Mishra; K P Pradhan

Data-Driven Intermittent Fault Diagnosis: from Data Acquisition to CNN-Based Classification in Synchronous Generators 306

Anirud R S; Srujana Karri; Surya Teja Nemani; Sunil Nag P V; C Santhosh Kumar; Kuruvachan K. George

Research Outcome Based Optimization Algorithm Applied to Optimal Phasor Measurement Unit Placement 312

Madhu Kishore Devara Chejarla; Sailaja Kumari Matam; Abdul Mujeer Syed

Hardware Design of SiC MOSFET-Based Phase Shifted Full Bridge DC-DC Converter for EV Charging Application 318

C Keerthika; B Saravanan; S Vishaal; S. Lenin Prakash; S. Raghavendran

Fixed Voltage Based P&O Method for Battery Charging from Partial Shaded PV String 324

Durgesh Chandra Nautiyal; Jenis Jain; Himanshu Sekhar Sahu; Vijay Kumar Chakka

Analysis and Comparison of Charging Speed of Two-Stage EV Charger for Different Charging Techniques 330

Dalija Rath; Subrat Mahapatra; Susovon Samanta

Analysis of Optimal Deep Neural Network for Enhanced Battery SOC Estimation 336

Prashant Aher; Dev Rai; Sanjaykumar Patil; Rhugved Rane; Tanmayee Rathod; Janyaa Tikoo

Hardware-In-the-Loop Testing of Battery Management System Using a High-Fidelity Digital Twin 342

Febin Koshy Jacob; Indranil Bose; Sarika D. Tavhare; Sandhya Anilkumar; S Kumaravel

Soiling-Induced Efficiency Losses in Photovoltaic Systems: A Comparative Study of Monocrystalline Monofacial and Bifacial Panels 348

Syed Ayan; Rajeev Ranjan; Shravan Kumar Singh; Nikhil Chander

Magnetically Integrated Onboard Charger and DCDC Converter for Electric Vehicle 354

Kv Deepesh; G Jagadanand; Nikhil Sasidharan

Model Predictive Control of Three-Level Boost Converters in Photovoltaic and Wind Energy Systems Connected to a Bipolar DC Microgrid 360

Satishreddy Dodda; Srinivasa Rao Sandepudi

DC-Link Capacitor Dimensioning with Lifetime Calculation Considerations for PMSM Drives: A Practical Framework for Enhanced Performance 366
  K O Angel; Rushikesh U Shinde; M P Shreelakshmi; Pradip M Magar; Mayank P Deo;
  Pramod Chaudhary

Silicon Switches with Ultra-Low Conduction Loss for Low Frequency Switching Applications 372
  S Harsha Vardhan Reddy; Wakar Hasan Kazi; K Akshay

Permanent-Magnet Biased Inductor for DC Fault-Current Limiting Applications 376
  G Aravind; Shivendra Pratap Singh; Amarkumar Kushwaha

Impact of Voids on Ampacity in High-Voltage Power Cables: A Multiphysical Approach 382
  Souvik Das; Tishya Chattopadhyay

Dynamic Duty Cycle Based Pulse Charging of Parallel Lithium-Ion Batteries for Electric Vehicles 388
  Supriya Chakrabarty; Niranjan Behera; Sankarsan Mohapatro; Abhinav Arya

Extraction of MPP Using Boost & SEPIC Converter Using Parametric Estimation Method 394
  Vamshi Krishna Bandaru; Anuradha Kotapati; Poornima Seelam; Shiva Prasad Edara

Retrofitting of Existing Solar Power Condition Unit as Reactive Power Compensator 400
  Rajesh Gupta; Kumari Priya; Mohammad Tahir Siddiqui

Hybrid Integration of PV, Fuel Cell and Battery for DC Microgrid 406
  Umang Kartikey; Vinayak Saxena; Yogendra Tiwari; Prakash Chittora; Vineet Kumar

An Enhanced Current Difference Based Fault Detection Technique for Low-Voltage DC Microgrid 412
  Biswajit Sahoo; Shantanu Saha; Anindya Banik; Shrestha Ghosh

A Modular Three-Phase Induction Heating Generator with Enhanced Power Factor 419
  Anshal S Padole; Monish Rane; Rajendra R Sawant; Yerramreddy Srinivasa Rao;
  Bhalchandra N Chaudhari

Constant Current-Constant Voltage Based EV Battery Charging Using High Frequency Phase-Shift
Full Bridge Converter 426
  Mohammad Tahir Siddiqui; Rohit Gupta; Abhinay Pratap Singh; Rajesh Gupta

Hybrid Federated Learning for Secure and Accurate Heart Disease Prediction 432
  Kavya Sree Sai Bulasara; Sai Sailu Batta; Mahesh Miriyala; Veerapu Goutham

Distributed Control Based Economic Load Dispatch for DC Microgrid with Reduced
Communication Data 438
  Harshada Dattatray Borse; Phani Swecha Tadepalli; Deepak Reddy Pullaguram

Switched Mode Power Amplifier for Underwater Imaging Sonar System 444
  V N Panchalai; Sateesh Kumar Kuncham; Kumaresan Natarajan; R Ramesh

Design of LQR for Three-Phase Grid-Tied Inverter System 450
  Kumari Prakhar Pragya; Shashank Shekhar; Aftab Alam

Adaptive Sliding Mode Control for DC-DC Buck Converter 456
  Chandan Kumar; Anjan Kumar Ray

Microgrid Based Fast Charger for Electric Powered Vehicles 462
  Neha Karn; Soumyasephalika Roul; Shubham Mishra; Rakesh Sahoo; Sanjeet Kumar
  Subudhi

An Advanced Loop Shaping Control Methods for Performance Improvements in a High Frequency Switched SiC Based Converters 469
G Abhijith; R Manju; P Ganesan; Sl Sanith; Mb Krishnaprasad; V Chandrasekar

Performance Improvement of Current Controlled BLDC Drive During Motoring and Regenerative Braking Operations 475
M. Abhivarma; Monalisa Pattnaik

Performance Characteristics of Low Vibrating Bridge-Configured Squirrel Cage Induction Motor 481
Rakesh Deore; Bipul Brahma; Karuna Kalita

Enhancing Predictive Maintenance of Urban Streetlights Through Hyperparameter-Tuned Machine Learning Techniques 487
Ashok Ganga; Kanaka Raju Kalla; Kumaraswamy Simhadri; P. Maheswara Rao; Akhilash Pennam; Matcha Nikhitha; Kelli Akshara; Pydisetti Prasanthi; Tangudu Sai Esha

Design and Analysis of GaN Based Electronic Power Conditioner for Space Applications 493
L R Indhuja; V. Vignesh Kumar; Nikhil K Desai; B. Venkatesaperumal; U. Vinatha; M. M. Rajan Singaravel

Modified State-Space Modeling and Voltage Mode Control Implementation for an Interleaved Boost Converter in Discontinuous Conduction Mode 499
Nabhomoni Ghoshal; Dalija Rath; Susovon Samanta

Performance Analysis of IUPQC and DPFC Devices in a Distributed Generation Unit Employing Intelligent Controllers 505
C. Srinivasa Rao; V. Sowmyasree; G. Panduranga Reddy

Condition Monitoring of Transmission Line Vibration Dampers Using YOLOv12 Model 511
Dipanjana Chowdhury; Satyajit Panigrahy; Subrata Karmakar

Kriging-Based Prediction of Air Delivery Performance in Ceiling Fans and Its Validation 517
Sharankumar Shastri; Bhim Singh; Vipin Kumar Singh

Performance Evaluation of Bridge-Configured Winding for Transverse Force Generation in Three-Phase Electric Machines 522
Gopinath Sengupta; Shahrukh; Karuna Kalita; Jenni Pippuri-Mäkeläinen; R. M. Ram Kumar; Gaurang Vakil

Hilbert-Huang Transform-Based Fault Detection and Classification in the Presence of Power Swings 528
Varun Reddy Gadikota; Nitish Kumar Gupta

Experimental Realization of Digital Voltage Mode Closed-Loop Control of Forward Converter Targeting for Single and Multi-Output Applications 535
Md Asif Alam; Mukti Barai

Thyristor Validation and Testing of 5kA/500V Rectifier Stack for Magnetic Confinement Application in Fusion Machines 541
Darshan Parmar; Rohit Kumar; Kush Mehta; Niranjanpuri Goswami; Dishang Upadhyay; Rasesh Dave; Sandip Gajjar; Hitesh Dhola; Aruna Thakar; Motibhai Makwana; Prakash Parmar; Supriya Nair; Joydeep Ghosh; N P Singh; Ujjwal Baruah

Analysing the Performance of Coaxial Square and Circular Coils for Effective Wireless Power Transfer 547
Kanala Srinivas Praanesh; Tanmoy Roy Choudhury

A Novel Multilevel Control Algorithm for Co-Operative Energy Management in Interconnected DC Microgrids...........553

K Jithin; R Hari Kumar; N Mayadevi

Modelling and Performance Analysis of DFIG Integrated with LMF-PLL Under Grid Abnormalities...........559

Oinam Lotika Devi; Alka Singh

Aerodynamic Wing Flutter Control Using IMU Sensor...........565

V Ganesh Ragava; M. A. Inayathullaah; Gurucharan Gurunath; V Ranganathan; Bkn Monish

A Differential Current Unbalance Factor-Based Scheme for Detecting Shunt and High Impedance Faults in Microgrids...........571

Chetan Anand; Kunal Kumar; Susmita Kar

Enhanced Inertia Estimation in Microgrids Using a Modified RoCoF Method...........577

R. Venkatesh; P. C. Sekhar; C. N. Bhende

Force Production in Permanent Magnet Synchronous Machine...........583

Shahrukh; Gopinath Sengupta; Karuna Kalita

Enhanced Half-Bridge Sub-module Modular Multilevel Converter and Its Capacitor Voltage Balancing Control...........588

Akshaya D. Bonde; Pradyumn Chaturvedi; Vijay B. Borghate

Parameter Identification and Controller Design of DAB Converter for Off-Board EV Charging Applications...........594

Aditya Kulkarni; Shashi Kumar Kondoju; Pradyumn Chaturvedi

Review of Conducted EMI in High Frequency Based Traction Inverters...........600

R Kodeeswaran; Sateesh Kumar Kuncham

Axial Flux Permanent Magnet Motor for Drone Application...........606

Madhav Yadav; K. L. Karthik Jandhyala; S. Rajesh; Ankit Dalal

A High Frequency Switched Gallium Nitride Based Bi-Directional EV Supply Equipment for Vehicle-to-Home Applications...........612

V Udaya Sagar; C V Vishnu; S Akshara; S R Jisha; S Amal; V Chandrasekar

Reduced Switch Multilevel Inverter Topology for Electric Vehicle Powertrains...........618

Shadab Murshid; Prasanth Sundararajan; Mrutyunjaya Sahani; Kolantla Dharani; Sanjib Kumar Panda

Cyber Resilient Fully Distributed Secondary Control for Inverter Dominant AC Microgrid...........623

Phani Swecha Tadepalli; M. N. Alam; Deepak Pullaguram

Modeling and Analysis of 14-Bus CIGRE Model...........629

Bhanu Venkata Siva Saikiran Kodati; Mahamad Nabab Alam; Tadepalli Phani Swecha

Attention Based Bi-LSTM Bi-GRU Model for Prediction of Electric Vehicle Charging Demand...........635

R Adithya; J Harshan; S. N. Deepa

A Single Switch Structured Improved Quadratic Boost Converter for Renewable Energy Applications...........641

Dharavath Anusha; Subbash Youvaraj; Ganesh Youvaraj; Harsh Bhanarkar; Bharath Marupatla; Srinivasan Pradabane

A Universal Input Single-Stage Bidirectional Charger for Light Electric Vehicles.............................. 646
Akash Kumar Swain; Vivek Agarwal

A Robust Data Driven FDI Attack Detection Framework for Inverter Based Microgrids............................... 651
*Suprabhath Sriranga Koduru; Venkata Siva Prasad Machina; Sreedhar Madichetty; Sukumar
Mishra*

An Interleaved DC-DC Converter for Low Voltage Bi-Polar DC Microgrid 657
Vipul Thakur; Saurabh Mishra

A Novel Centralized Active Power Control Strategy for Multifrequency Microgrid with Energy
Storage System 663
Sudeshna Mukherjee; Rajdip Dey; Tapas Kumar Saha

Author Index

RZA-GSRS-Based Utility-Interfaced PV System for EV Charging

Shishupal
Department of Electrical Engineering,
Indian Institute of Technology Roorkee,
Roorkee, India
shishupal@ee.iitr.ac.in

Sumit GhatakChoudhuri
Department of Electrical Engineering,
Indian Institute of Technology Roorkee,
Roorkee, India
sumit@ee.iitr.ac.in

Abstract—This paper presents a three-phase, four-wire (3P4W), utility interfaced, solar photovoltaic-voltage source converter (PV-VSC) based multi-tasking bidirectional system beneficial for electric vehicle (EV) charging and household loads using re-weighted zero-attraction generalised soft-root-sign (RZA-GSRS) adaptive filter (AF)-based control scheme using dual second-order generalised integrator. The active weight component of load currents is extracted from non-linear loads coupled at the point of common coupling using the RZA-GSRS technique. The system performance using proposed RZA-GSRS algorithm is also contrasted with that of traditional state-of-the-art (AF) techniques. The realised system provides charging facility for EV, supplies critical residential loads and transfers the surplus power as an active to the utility mains. Therefore, the VSC also operates as a distribution static compensator. The system is realised in MATLAB environment in discrete time frame using Simulink and sim power system toolboxes under various operating conditions. The bidirectional DC/DC converter and VSC are operated at a switching frequency of 10 kHz respectively. The Total Harmonic Distortion of grid current and the point of common coupling voltage observed are within range in accordance to IEEE-519 and IEEE-929, respectively.

Keywords—*RZA-GSRS, Solar PV, Dual Second-Order Generalised Integrator*

I. INTRODUCTION

An increase in environmental pollution is taking place continuously because of use of conventional energy sources, which have finite reserves. Therefore, significant changes globally have started appearing [1]. Use of renewable energy sources (RESs), in place of conventional sources and transition from combustion engine automobiles to electric vehicles (EVs) is a technical landmark. Such changes are objected towards decreasing pollution and creating a path for sustainable future. This indicates that there may be a critical need for solutions to provide energy security and simultaneously address environmental pollution. Therefore, induction of EVs and establishment of electric vehicle charging stations powered by of RESs are among prime objectives of many recent initiatives [2-3]. RESs such as, solar photovoltaic (PV), are getting higher preference as they are independent of end user location. Due to intermittent nature of PV energy, integration of EV is the possible solution to realise the stable and reliable power source [4].

PV integration with EV charging contributes towards system complexity due to necessity of additional power conversion stages. As a result, designing PV integrated EV charging system with least conversion stages having multi-functional operating capabilities to coordinate and control different sources in the system is vital [5]. Further, coordinated control of PV with EV can be an effective method to reduce/overcome the drawbacks existing in PV [4].

Traditionally, in a two-stage solar PV-EV-VSC charging

architecture, a voltage source converter (VSC) is put to use to interface the system with utility mains at the point of common coupling (PCC) wherein, consumer loads also get connected [5]. The maximum power extraction from PV is done by the uni-directional DC/DC converter and boosting and regulating PV output voltage to necessary DC link level is done through closed loop control. The solar energy conversion (SEC) system supplies power requirements of the load at PCC and the control strategy ensures that the excess power gets effectively transferred as an active power to the utility mains. A complex, two-stage, system is replaced by a single-stage configuration, wherein PV gets connected to the DC link directly. The EV charging process gets interfaced through a bidirectional DC/DC converter (BDDC) to the DC link. Such an arrangement makes the system simple, economical and capable to operate at a higher efficiency. As a result, the SEC system performs better with lesser circuit complexity [6-7].

Applications such as, residential society, office complex, etc., have three-phase, four-wire (3P4W) distribution system. Unbalanced single-phase/three-phase, non-linear loads at PCC cause substantial amount of third-order and its multiple harmonic currents, to flow through the neutral wire of utility mains. This may lead to neutral conductor to get overloaded, with its temperature going above acceptable limits, and due to injection of harmonics into the utility, the power quality on the AC mains side gets affected. Additionally, as the utility neutral wire temperature may go above limits, the neutral line may get damaged [8]. Several authors have implemented numerous control schemes to enable the VSC to adjust the neutral current harmonics and function as a distribution static compensator (DSTATCOM). In such SEC systems, three-leg split capacitor VSC and four leg VSC are among the most preferred topologies [9]. Having an extra leg initiates drawback such as, additional active switches, requirement of more gate circuitry, extra auxiliary inductance, another current sensor and therefore, increases complexity in realising pulse-width-modulation control.

This paper puts forward re-weighted zero-attraction generalised soft-root-sign (RZA-GSRS) adaptive filter (AF) [10], for computation of equivalent active weight component of load currents. Such a computation/estimation helps to improve the nature of the reference grid current, finally leading to better power quality at the utility side. A single-stage, PV based EV charging infrastructure, is designed to realise grid to vehicle (G2V) charging and vehicle to grid (V2G) discharging respectively [11]. This paper implements the proposed control scheme to realise PV-EV-grid interfaced system for battery charging and for suppling residential loads.

The main key issues highlighted in this paper also include,

- adaptable power management scheme so that excess power is injected to the utility mains as an active power after load demands are fulfilled. The VSC present in the

979-8-3315-3013-6/25 $31.00 © 2025 IEEE

interfaced system locally supplies reactive power needed by the load at PCC.

- improvement in grid current dynamics with precise extraction of active weight component of load currents using RZA-GSRS AF for non-linear load applications.
- realisation of EV charging/discharging using dual-loop bidirectional control through a BDDC where the current limits of the inner loop are decided on basis of safe SoC limits of the battery.
- elimination of current harmonics in the neutral line through coordinated control, enabling elimination of additional filters at the PCC. Hence, the VSC used in the system also serves as a DSTATCOM.

This paper is organized as follows: Section II presents a mathematical model of the system. Section III addresses the corresponding control strategies. Section IV provides an illustration of the simulation results, which demonstrate the effectiveness of smart power management in the grid-connected PV-EV-VSC system. Finally, Section V offers conclusions drawn from the findings.

II. SYSTEM DESIGN

A single-stage, utility interfaced, PV based system for EV charging applications, is depicted in Fig.1. Design for PV, DC-link capacitor, RC filter, boost inductor for bidirectional DC/DC converter and interfacing inductor between VSC and utility mains, are incorporated in this section.

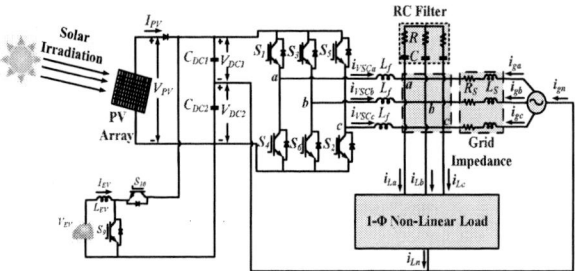

Fig. 1. Proposed topology of Utility Interfaced PV based EV charging system with Residential Loads Connected at PCC

A. Design of PV Array

A $30kW$, PV array is realised using modules, wherein in each module is rated for short circuit current, I_{SC}, 8.21A and open-circuit voltage, V_{OC}, 32.9V respectively [12]. The operating point to extract maximum for PV can be computed as,

$$V = 1.6 \times \frac{2\sqrt{2}V_{LL}}{\sqrt{3}\,m_a} = 1.6 \times \frac{2\sqrt{2} \times 415}{\sqrt{3} \times 1} \cong 1080V \quad (1)$$

where $V = V_{MPPA} = V_{DC}^*$. V_{DC}^* denotes the reference DC-link, V_{LL} signifies RMS line to line voltage at PCC and m_a indicates the modulation index respectively.
The number of series-connected modules, N_S can be computed as

$$N_S = \frac{V_{MPPA}}{V_{MPP}} \approx \frac{1080}{0.85 \times V_{OC}} \cong 39 \quad (2)$$

Number of parallel-connected stacks, N_P (wherein one stack comprises of N_S modules in series) to generate 30 kW power is calculated as,

$$N_P = \frac{P_{MPPA}/V_{MPPA}}{I_{MPPA}} \approx \frac{3000/1080}{0.85 \times I_{SC}} \cong 4 \quad (3)$$

Therefore, in order to design PV of $30kW$, 39 modules are connected in series to formulate a stack and 4 of such stacks are then connected in parallel to create the required PV array.

B. Design of Boost Inductor for Bi-Directional DC-DC Converter

EV is connected via BDDC to boost its voltage from 240V to the required DC link value. Therefore, Ah rating of EV battery is designed assuming worst situation wherein grid outage occurs for 4 hours and the entire PV power is utilised for charging the EV as,

$$Ah_{EV} = \frac{P_{PV} \times h}{V_{EV}} = \frac{30000 \times 4}{240} = 500Ah \quad (4)$$

Thus, the EV rating is 240V, 500Ah and charge/discharge current, I_{EV}, is 125A at 1C rate. The EV battery bank in the laboratory consists of 20 stacks connected in parallel. Each stack has 20 series-connected batteries, each rated for 12V, 26Ah. Inductor, L_{EV} is used to reduce current ripple in BDDC which is caused by switching devices. Thus, L_{EV} value is deduced from the voltage induced across L_{EV}, V_{EV} and ripple current, ΔI_{EV} as,

$$L_{EV} = \frac{V_{EV}D_{EV}}{f_s \times \Delta I_{EV}} = \frac{240 \times 0.78}{10 \times 10^3 \times 6.25} \cong 3mH \quad (5)$$

where, V_{EV}, f_s and ΔI_{EV} denote EV battery voltage, switching frequency and ripple in EV battery current which is presently taken as, 5% of average battery current, I_{EV}.

C. Design of DC-link capacitors C_{DC1}, C_{DC2}

Design of capacitor, C_{DC} is based upon occurrence of sudden change in energy at DC Link when system operating conditions make an abrupt change. The conditional constraint remains that energy stored in C_{DC}, does not fall below an optimum level. Assuming $C_{DC1} = C_{DC2} = C_{DC}$, the energy difference across the equivalent DC link capacitance, C_{eq} for a time interval, T_1 can be computed as,

$$\Delta e_{DC} = \left\{ \frac{1}{2} * C_{eq}((V_{DC}^*)^2 - (V_{DCmin})^2) \right\} \quad (6)$$

where, C_{eq}, Δe_{DC}, V_{DC}^* and V_{DCmin} represent equivalent DC link capacitance, energy change under transient circumstances, reference DC Link voltage and minimum value of DC Link expected. For design, the upper limit of such an energy change under transient circumstances is taken as 10% [10]. As a result, this energy change can also be calculated as [11],

$$\Delta e_{DC} = k_1 \times a \times P_{MPPA} \times T_1 \quad (7)$$

where, k_1 is a constant whose value presently is taken as 10%, a refers to the overloading factor and T_1 indicates the time taken by DC-link to recover back to the reference value. The value of C_{eq} will be $\frac{C_{DC}}{2}$ and using equations (6) and (7), C_{DC} can be calculated as,

$$\frac{1}{4} * C_{DC}((V_{DC}^*)^2 - (V_{DCmin})^2) = k_1 a P_{MPPA} T_1 \quad (8)$$

$$C_{DC} = \frac{4 \times 0.1 \times 30000 \times 1.2 \times 50 \times 10^{-3}}{1080^2 - 918^2} \cong 2300\mu f$$

979-8-3315-3013-6/25 $31.00 © 2025 IEEE

$$C_{DC1} = C_{DC2} = C_{DC} \cong 2300 \mu f \qquad (9)$$

D. Design of Interfacing Inductor and RC filter

Primary function of interfacing inductor, L_f is to eliminate harmonics from grid current as L_f acts as a first-order, low pass filter. The value of L_f can be calculated as [12],

$$L_f = \frac{V_{DC}^*}{2h \times 2f_s} = \frac{540}{2 \times 1.3 \times 2 \times 10^4} \cong 10mH \qquad (10)$$

where, V_{DC}^* represents the reference DC Link voltage across C_{DC1} capacitor, '$2h$' refers to ripple in inverter current, f_s denotes maximum allowable switching frequency for the VSC which goes upto 10kHz in the present design.

The shunt RC filter is tuned such that the time constant is 10 times faster than the switching frequency, f_s. Therefore, for $R_f = 5\Omega$, the value of C_f is calculated as:

$$C_f = \frac{1}{10 \times 5 \times f_s} = \frac{1}{10 \times 5 \times 10^4} = 2\mu F \qquad (11)$$

III. CONTROL STRATEGY

The control strategy incorporates three main sections namely, control of bidirectional DC/DC converter, extraction of active weight component from respective load currents and generation of in-phase unit templates for reference grid currents to realise VSC control, respectively.

A. Bidirectional DC-DC Converter

The objective of bi-directional DC/DC converter is to charge/discharge EV battery such that G2V and V2G operation gets realised and DC link is in proper regulation ensuring maximum power from PV being extracted. Such a control is realised using a dual-loop control strategy in which the outer loop executes DC-link voltage, V_{DC} control, and inner loop executes control for EV battery current, I_{EV}, as shown in Fig. 2.

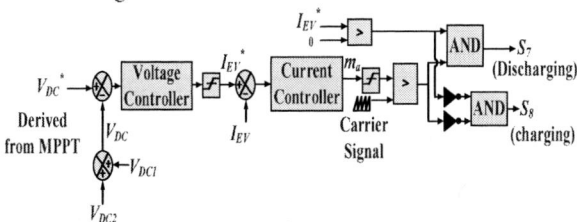

Fig. 2. Block Schematic of Dual-Loop DC/DC Converter Control Strategy

The reference voltage V_{DC}^* is generated using the maximum power point tracking (MPPT) strategy. The voltage error obtained is then compensated for by the voltage controller, which operates on a proportional-integral (PI) logic to generate the reference current for charging EV battery. The reference current operates the inner control loop. It is compared with actual battery current, I_{EV}. The error generated is processed by through the current controller using PI logic to generate the modulating signal, m_a which is used further for generating gating signal required for the active switches of the BDDC.

B. Generation of In-Phase Unit Template, u_{px} (where x = a,b,c)

To generate unit templates for construction of reference grid currents, second order generalised integrators (SOGIs) are used as illustrated in Fig.3. Initially, three-phase voltages sensed from utility side, v_{gabc}, are converted into orthogonal voltages, $v_{\alpha\beta}$. Both SOGIs generate respectively, an in-phase and a quadrature component, which caters to the input of positive sequence calculator (PSC) in $\alpha\beta$ reference frame. The PSC extracts the positive sequence component voltages, $v_{\alpha\beta}''$. The peak amplitude of phase voltage, V_t at PCC, as shown in Fig. 3, is computed as [24],

$$V_t = \sqrt{(v_{\alpha^*})^2 + (v_{\beta^*})^2} \qquad (12)$$

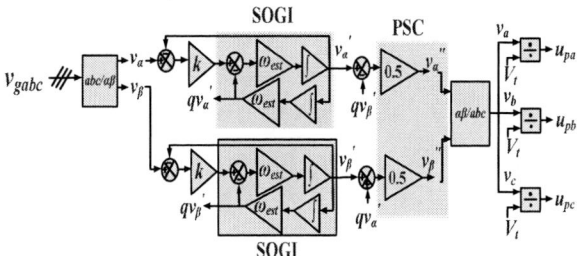

Fig. 3. Control Block Schematic using DSOGI and PSC to extract Three-Phase Unit Templates

C. Control Strategy for the Voltage Source Converter used in Grid Interfaced PV-EV System

PV-EV combination gets interfaced to grid through a three-phase VSC. The main objective is to charge EV and inject the surplus power generated as active power to the utility after satisfying demand from the load connected at PCC. The VSC present in the system, additionally also functions as a DSTATCOM, which supplies the entire reactive power demand of load locally, ensuring surplus power generated to be transferred to the utility mains side, as an active power. In indirect current control scheme, equivalent weight component for reference grid currents, gets computed by adding respective weight components corresponding to PV, EV and active weight component of load currents respectively. Weight corresponding to reference EV current, I_{EV}^* is operational into the summation logic when controller takes the battery to float mode so as to avoid occurrence of any damage to the battery and to maintain the DC bus at the required level as shown in Fig.4. Unit templates for the reference utility currents get derived from DSOGI, as shown in Fig. 3. Since the reference currents, i_{ga^*}, i_{gb^*}, and i_{gc^*}, are of equal magnitude, the net neutral current on the grid side is maintained close to zero, even under unbalanced load conditions. The neutral current typically flows through the load via the midpoint of the DC link capacitors in the VSC. Therefore, this current control method indirectly ensures that the grid's neutral current remains close to zero. To generate gate pulses for VSC, the reference grid currents, i_{gabc^*} are compared with respective sensed utility currents, i_{gabc}. Resultant error is then fed into the hysteresis current controller to generate gating signals for VSC, as shown in Fig. 4. Active weight component of load currents, w_{PLeq} is computed using RZA-GSRS AF. The block schematic for computation of w_{PLeq} is depicted in Fig.5 and the corresponding mathematical algorithm is depicted through a

979-8-3315-3013-6/25 $31.00 © 2025 IEEE

block schematic strategy in Fig.6. For w_{pLx} ($x = \alpha$, β), represents the active weight component of load currents, i_{Lx}, error, e_x at a given instant is therefore defined as the difference between actual load currents, i_{Lx} and respective computed active weight components of load currents at that same instant as,

$$e_x(n-1) = i_{L_x} - \mu_x(n-1)^T \times w_{pLx}(n-1) \quad (13)$$

where, n refers to the sampling instant.

To minimise mathematically computed error in discrete time frame as per (13), w_{pLx} at each nth instant is to be updated based on the available error at $(n-1)^{th}$ instant as,

$$w_{pLx}(n) = w_{pLx}(n-1) + (\mu \times a) - b \quad (14)$$

where,

$$a = f\big(e_x(n-1)\big) \times u_x(n-1)$$

and

$$b = \rho \frac{sgn\,[w_{Lx}(n-1)]}{1 + \delta|w_{Lx}(n-1)|}$$

$$fe_x(n-1) = \frac{|c|^{\gamma-1}\,(d+1)\,sgn[c]\exp(d)}{\{|c|^{\gamma}\exp(d) + \alpha\}^2} \quad (15)$$

where, $c = e_x(n-1)$ and $d = \beta|c|^{\gamma}$

In above mathematical computations, $\alpha, \beta, \gamma, \mu, \rho$ and δ respectively refer to scaling factor, slope parameter, shape parameter, learning the rate, zero attraction control parameter and shrinkage parameter.

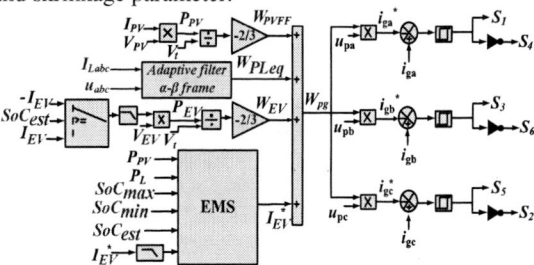

Fig. 4. Block Schematic of Indirect Grid Current Control of VSC

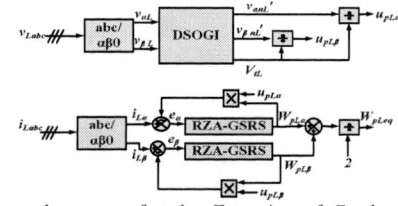

Fig. 5. Proposed strategy for the Extraction of Fundamental Weight Component of Load Current

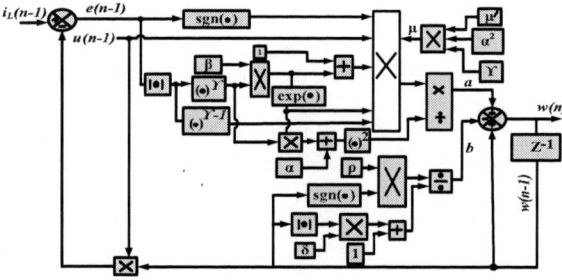

Fig. 6. Block Schematic of RZA-GSRS AF Algorithm

IV. RESULTS AND DISCUSSIONS

Response of the system realised with hybrid combination of PV-EV, when synchronised with AC mains and load connected at the PCC is simulated in MATLAB environment, in discrete time frame. Performance of such a system under different operating conditions such as, change in solar irradiation, sudden occurrence in load unbalance from the residential consumer side, is analysed through simulation using system parameters that are enlisted in Appendix.

A. System Dynamics under Change in Irradiation Level

Response of the presented system is analysed under change in irradiation levels, G. Fig.7 depicts the system behaviour during step change in insolation level from G = $0.2kW/m^2$ to $1kW/m^2$. A residential load of $2kW$ per phase is connected at the PCC.

Initially, the system operates at G = $0.2kW/m^2$ (Mode-I). In such a situation, the power produced from PV (P_{PV} = 4.6kW) is smaller than the end user demand (P_L = 6kW), so the system work in Deficit Power Mode (DPM). Hence, the load demand is not satisfied. Thus, the EV system discharges to provide the remaining power. The performance of the system in Mode-I is shown in Fig. 7 for the duration from t=0.75 sec to t=1 sec. At t =1 sec, change in irradiation level occurs making G = $1kW/m^2$ (Mode-II). In such a mode, the power produced from PV (P_{PV} = 30kW) is enough to charge EV and meet end user demand (P_L = 6kW). Therefore, the PV-EV-VSC system work in Surplus Power Mode (SPM). Hence, after charging the EV and satisfying the load demand, the surplus power is injected into the grid as an active power. The steady-state performance of system from Mode-I to Mode-II is shown in Fig. 7.

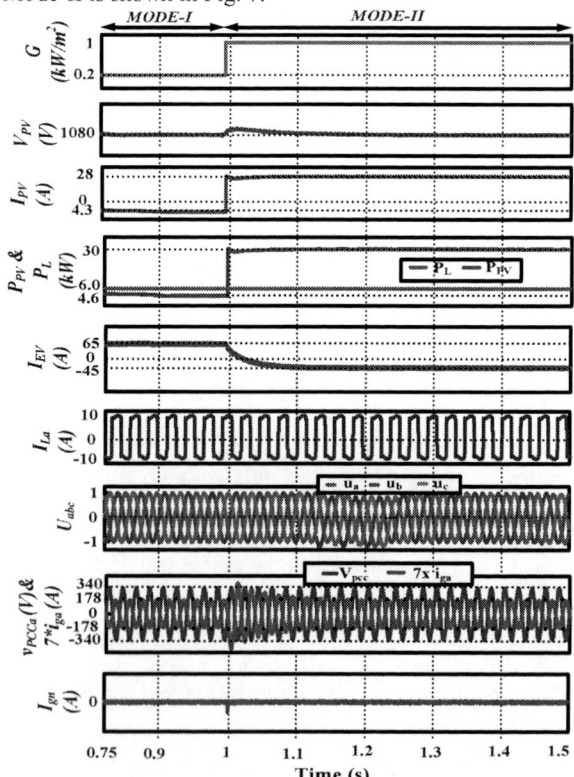

Fig. 7. System Performance under Change in Irradiance Level, G, from low (200W/m^2) to high (1000W/m^2)

979-8-3315-3013-6/25 $31.00 © 2025 IEEE

B. System Performance During Balanced and Unbalanced Non-Linear Load

The performance of the PV-EV-VSC system under balanced and unbalanced loading conditions is illustrated in Fig. 8. From 1 sec, system operates at G= $1kW/m^2$. The extracted power from PV (P_{PV} = 30kW) exceeds load demand P_L = 6kW). The PV-EV-VSC system, with its remarkable adaptability, effectively manages the loads' active and reactive power demand, while redirecting the surplus power to EV as well as to the AC mains in the form of active power. The RMS grid current in all the phases is a steady 30A. At 1.4 sec, the 'a' phase load is abruptly disconnected. Therefore, active weight component of load currents computed by RZA-GSRS, W_{PLeq} gets altered and that causes equivalent weight component of grid current reference, W_{Pg} to change. Thus, the system responds by increasing the grid current to 33.5 RMS in all three phases. This uniformly increase in grid current in all three phases causes the neutral line current to become zero, as shown in Fig. 8. At 1.7 sec, when the 'a' phase load is reconnected, W_{PLeq} increases, as a result W_{pg} decreases, which in turn decreases the RMS grid current and hence the system swiftly restores the RMS grid current back to 30A. The out-of-phase relationship between v_{PCCa} and i_{ga} in Fig. 7 indicates the active power flow into the grid.

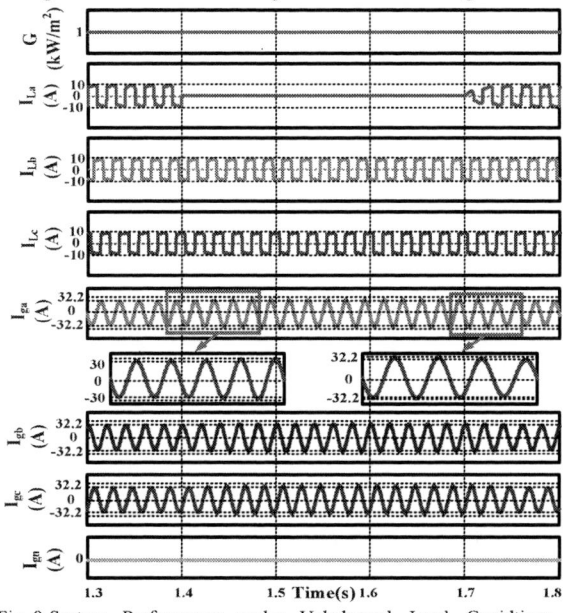

Fig. 8. System Performance under Unbalanced Load Conidtions at G=1kW/m^2

C. Power Sharing between PV and 3P4W AC Mains

Power sharing between the designed PV and 3P4W AC mains is analysed here to verify if the system can meet the design objectives. In Mode-I, as $G = 200W/m^2$, the PV system produces the power of P_{PV} = 4.6 kW. This entire power is transferred to meet the load power demand, P_L = 6 kW connected at the PCC. To meet the shortage power of P_L, EV system gets discharged at P_{EV} = 15.6 kW. Once the load demand is met, the balanced power, P_g = 14.2 kW, is injected to the 3P4W mains as an active power ensuring the stability of the system as depicted in Fig.7 and Fig.9 respectively. In Mode-II, $G = 1kW/m^2$, PV generates, P_{PV} = 30 kW. Hence, in this situation, PV can fulfil the load

demand, $P_L = 6\ kW$, charge the EV system, $P_{EV} = 9\ kW$ and the remaining surplus power, 15 kW gets transferred as an active power to the grid as shown in Fig.7 and Fig.9 respectively. In all modes of operation, current harmonic distortion in grid is observed within 5% range according to IEEE-519 and IEEE-929, respectively, as shown in Fig. 10.

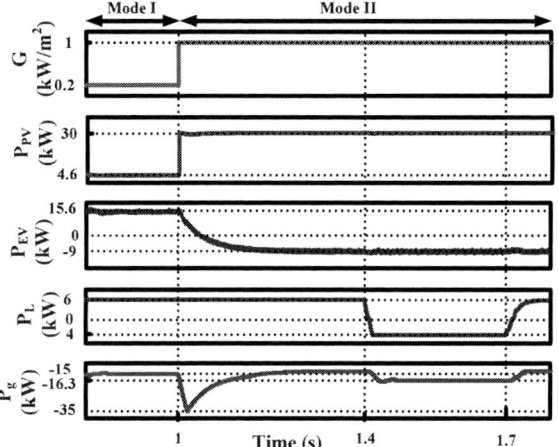

Fig. 9. Smart Power Sharing between PV, EV, Load and Grid

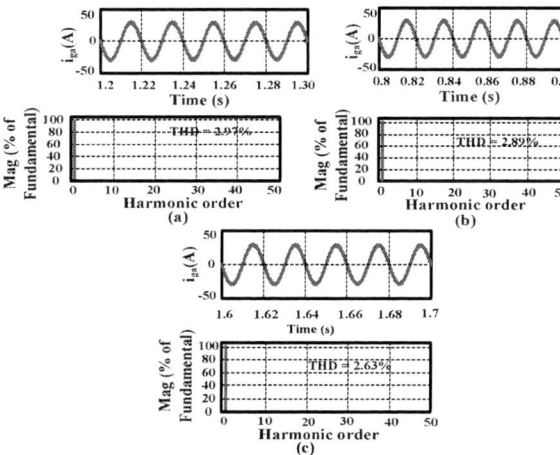

Fig. 10. THD of Utility Grid Current under (a) Balanced load with $G = 200W/m^2$, (b) Balanced load with $G = 1kW/m^2$ and (c) Unbalanced Load with $G = 1kW/m^2$ System

D. Computation of Active Weight Component of Load Currents by Re-Weighted Zero-Attraction Generalized Soft-Root-Sign Algorithm

In Fig. 11, extracted active weight component of load currents using various state-of-art AFs is compared with the proposed RZA-GSRS AF respectively during full-load to no-load transition mode of operation. In contrast to AFs such as, least mean square (LMS), least mean fourth (LMF), least mean mixed-norm (LMMN), the presented RZA-GSRS AF algorithm computes active weight component of load currents with least steady state error, as shown in Fig. 11. It can be observed that convergence rate of RZA-GSRS AF algorithm is superior to other state-of-art AF algorithms, which improves the transient performance. Therefore, superiority in convergence rate of weight component using RZA-GSRS algorithm enhances dynamics of grid currents. Filtering capability of RZA-GSRS is evaluated under various operating

979-8-3315-3013-6/25 $31.00 © 2025 IEEE

conditions through simulation results in MATLAB environment, using simulink and sim power system toolboxes, in discrete-time frame and the observed data is compiled in Table-I.

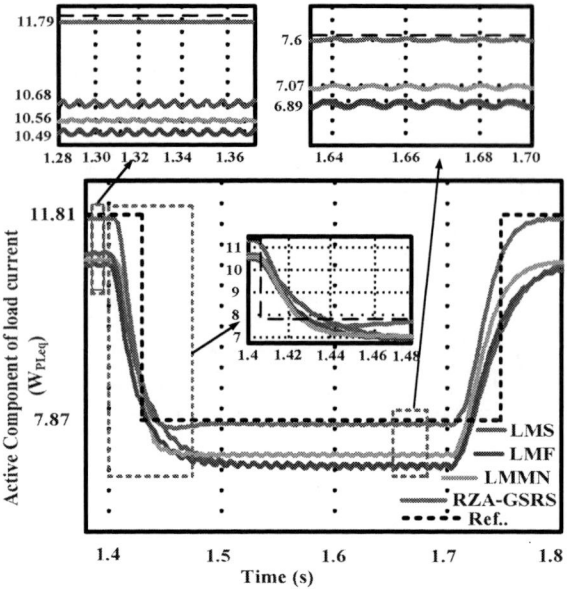

Fig. 11. Comparison of RZA-GSRS AF with Exisiting Conventional AFs to extract W_{PLeq}

Table I.   Comparison of Conventional AFs with Proposed AF

PARAMETERS	LMS [12]	LMF [13]	LMMN [14]	RZA-GSRS [10]
Settling time (ms)	52.80	49.82	47.47	24.85
Transient response	– –	– –	–	+ +
Variable step size	No	No	No	Yes
THD of i_{ga} (%)	4.21	4.08	4.01	2.63
RSC	– –	– –	–	+ +
CC	less	less	more	more

RSC= Robustness to sudden changes, CC= computational complexity. Transient response, robustness to sudden changes and computational complexity is divided into 4 levels from good as + + to bad as – –

V. CONCLUSIONS

A single-stage, multi-tasking PV-VSC grid interfaced system capable to provide EV charging and supply residential loads has been realised. The PV based EV charging station is connected to the DC link. Such a framework supports G2V and V2G operations, highlighting its potential for enhancing energy efficiency and sustainability. The surplus power generated supports EV charging, supplies active/reactive power demand of loads connected at PCC, and the remnant surplus power is injected as an active power to the utility mains. The active load current has been precisely extracted using RZA-GSRS AF leading to improvement of injected grid current power quality. Simulation results show that the system is capable to effectively manage bi-directional power flow. This allows the battery to function as an electrical load or a source, thereby ensuring sustainability of power demand. The simulation results confirm effectiveness of EV charging under various operating conditions including variation in irradiation, unbalance in consumer loads, reactive power compensation of PCC load locally by the VSC. Opted system layout design effectively reduces both cost and circuit

complexity, while simultaneously preserving the efficiency of PV. It ensures uninterrupted EV charging, uninterrupted supply to residential critical loads and provides surplus power as an active to the grid. Finally, total harmonics distortion (THD) of the injected utility line currents are maintained in accordance with IEEE-519.

APPENDIX

Simulation Parameters: $P_{MPPA} = 30\ kW$, $V_{DC}^* = 1080V$, $N_S = 39$, $N_P = 4$, $C_{DC1} = C_{DC2} \cong 2300\mu f$, EV capacity, $Ah_{EV} = 500Ah, 240V$, $L_{EV} = 3mH$, $f_s = 10kHz$ for Bidirectional DC-DC Converters, 1ϕ diode bridge rectifier $= 2kW$ per phase, $V_{LL} = 415V$, 50Hz.

REFERENCES

[1] "International energy agency-global EV outlook," 2024. [Online]. Available: https://iea.blob.core.windows.net/assets/a9e3544b-0b12-4e15-b407-65f5c8ce1b5f/GlobalEVOutlook2024.pdf

[2] P.M.de Quevedo, G. Muñoz-Delgado, and J.Contreras, "Impact of vehicles on the expansion planning of distribution systems considering renewable energy, storage, and charging stations," *IEEE* Trans. *Smart Grid*, vol. 10, no. 1, pp. 794–804, Jan. 2019.

[3] V. Monteiro, J. G. Pinto, and J. L. Afonso, "Experimental validation of a three port integrated topology to interface electric vehicles and renewables with the electrical grid," *IEEE Trans.Ind.Informat.*, vol. 14, no. 6, pp. 2364–2374, Jun. 2018.

[4] L. Cheng, Y. Chang, and R. Huang,"Mitigating Voltage Problem in Distribution System With Distributed Solar Generation Using Electric Vehicles," *IEEE Transaction on Sustainable Energy.*, vol. 6, no. 4, pp. 1475–1484, 2015.

[5] B. Singh, A. Verma, A. Chandra, and K. Al-Haddad, "Implementation of Solar PV-Battery and Diesel Generator Based Electric Vehicle Charging Station," *IEEE* Trans. *Industry Applications*, vol. 56, no. 4, pp. 4007–4016, 2020.

[6] B. Singh, C. Jain, and S. Goel,"ILST Control Algorithm of Single Stage Dual Purpose Grid Connected Solar PV System," *IEEE* Trans. *Power Electron.*, vol. 29, no. 10, pp. 5347–5357.

[7] B. Singh, V. Jain, Seema, A. Chandra, and K. Al-Haddad,"Power Quality Improvement in a PV Based EV Charging Station Interfaced with Three Phase Grid," *IECON-2021*- 47th Annual Conference of the *IEEE Ind. Elec. Soc.*

[8] C. A. Quinn and N. Mohan,"Active Filtering of Harmonic Currents in Three-Phase, Four-Wire Systems with Three-phase and Single Phase Nonlinear Loads," *in [Proceedings] APEC '92 Seventh Annual Applied Power Electronics Conference and Exposition.*, 1992, pp. 829–836.

[9] S. B. Q. Naqvi and B. Singh," Weak Grid Integrated Improved Power Quality PV-Battery System Tariff Control and PV Intermittency Mitigation," *in 2021 4th Biennial International Conference on Nascent Technologies in Engineering (ICNTE)*, Jan. 2021, pp. 1–6.

[10] M. Restrepo, J. Morris, M. Kazerani, and Christensen MG," Generalized soft-root-sign based robust parsity-aware adaptive filters," *IEEE Signal Process Lett*, vol. 30, 2023.

[11] Patel V, Bhattacharjee SS, and C. A. Cañizares,"Modeling and Testing of a bidirectional smart charger for distribution system EV integration," *IEEE* Trans. *Smart Grid*, vol. 19, no. 1, pp. 152–162, Jan. 2018.

[12] M. G. Villalva, J. R. Gazoli, and E. R. Filho,"Modeling and Circuit Based Simulation of Photovoltaic Arrays," in 2009 Brazilian Power Electronics Conference , Sep. 2009, pp. 1244-1254.

[13] Alexander D. Poularikas, Adaptive Filtering: Fundamentals of Least Mean Squares with MATLAB, 1st ed., *CRC Press*: Boca Raton, 2014.

[14] R. K. Agarwal, I. Hussain and B. Singh, "LMF-Based Control Algorithm for Single Stage Three-Phase Grid Integrated Solar PV System," *IEEE Transactions on Sustainable Energy*, vol. 7, no. 4, pp. 1379-1387, Oct. 2016.

[15] R. Chilipi, N. Al Sayari and J. Y. Alsawalhi, "Control of Single-Phase Solar Power Generation System With Universal Active Power Filter Capabilities Using Least Mean Mixed-Norm (LMMN)-Based Adaptive Filtering Method," *IEEE Transactions on Sustainable Energy*, vol. 11, no. 2, pp. 879-893, April 2020.

Online Corrective Contingency Analysis for Floating Green Microgrid Using Holomorphic Embedding Method and Extreme Learning Machine

Mrutyunjaya Sahani
Electrical and Computer Engineering
National University of Singapore, Singapore
mrutyunjayasahani@nus.edu.sg

Sanjib Kumar Panda
Electrical and Computer Engineering
National University of Singapore, Singapore
eleskp@nus.edu.sg

Abstract—This work presents an integration of the holomorphic embedding power flow method (HEPFM) with the functional expanded extreme learning machine (FEELM) to develop an online corrective contingency analyzer for evaluating 'what-if' scenarios in a floating green power solution. The HEPFM offers a rapid and accurate power flow solution, outperforming traditional search-based methods in efficiency and precision. Performance indices based on active power and voltage are computed to assess the impact of contingency events, enabling the recommendation of corrective actions to enhance system resilience under emergency conditions. Three operational modes—black start, baseline operation, and generation outage are simulated and analyzed using MATLAB/Simulink software. The FEELM method is proposed as a computationally efficient approach for generating corrective actions by utilizing generation power, energy storage power, and load power as inputs, applicable in both grid-connected and islanded modes. The intelligent HEPFM-FEELM framework demonstrates simplicity, practicality, and robustness, providing a comprehensive and efficient solution for managing contingency events in modern power systems.

Index Terms—Corrective contingency analysis, Extreme learning machine, Floating green power solution, Holomorphic embedding power flow, Performance indices.

I. INTRODUCTION

THE, online corrective contingency analyzer (OCCA) plays a critical role in maintaining the stability and reliability of electric power systems during equipment failures or other contingencies. Its importance lies in several key functions such as identifying potential system failures by analyzing the impact of contingencies, ensuring system stability through the identification of corrective actions, improving system efficiency by optimizing resource use and reducing losses, enabling real-time decision-making for system operators, and increasing system resiliency by identifying vulnerabilities and suggesting mitigation strategies. These functions collectively contribute to preventing blackouts, minimizing disruptions, and maintaining the continuous operation of power systems, especially in large and complex networks where failures can have far-reaching consequences. By identifying potential system failures, maintaining system stability, improving system efficiency, enabling real-time decision making, and increasing system resiliency, OCCA helps ensure that power systems can continue to meet the needs of customers and support economic growth.

Before the era of steady-state contingency analysis (CA), load flow analysis (LFA) was the predominant method for assessing network contingencies. However, as modern AC-DC power systems evolved, the need for modified LFA approaches arose [1], leading to increased computational complexity for outage simulations. The integration of converter-based solar and battery-storage sources in AC-DC networks heightened outage risks, necessitating more comprehensive contingency analysis strategies. Additionally, the interconnected nature of modern networks posed increased risks of cascading outages, potentially resulting in severe disruptions to utility services [2].

In response to these challenges, industry standards have mandated steady-state (N-1) contingency analysis [3] for all networks, often employing linear approximations to expedite estimations and facilitate swift corrective actions. Techniques such as graph theory, including betweenness and proximity indices, are commonly used to evaluate the severity and impact of network failures. However, current methodologies often fail to fully consider the presence of DC infrastructure within networks. While some studies propose Genetic algorithm-based indices [3] for AC networks, their applicability to AC-DC systems is limited. Similarly, simplifications of HVDC-embedded AC networks may overlook the interaction between AC and DC components, and security evaluations based solely on Jacobian factors may not adequately capture the sensitivity of DC subsystems [3]. Despite advancements in analyzing voltage source converters (VSCs), understanding the ramifications of individual contingency events within the network remains a challenge.

The proposed work introduces several novel contributions: Utilization of the holomorphic embedded load flow algorithm, which is both computationally less complex and less time-consuming compared to traditional iterative power flow methods, for computing bus voltage angles and line power. Introduction of a straightforward power transfer distribution factor to assess line power limits and determine necessary corrective actions. Proposal of a fast single-layer feed-forward neural network, which incorporates trigonometric functional expansions of power set points from generation units, ESSs, and loads. This network efficiently decides on corrective actions to maintain stability in the developed floating green micro-grid (FGMG) system.

979-8-3315-3013-6/25 $31.00 © 2025 IEEE

The paper is structured as follows: Section II provides a detailed explanation of the mathematical formulation of the holomorphic embedded load flow algorithm [4] for slack, PV, and PQ buses. In Section III, the functional expanded extreme learning machine is introduced as a means to offer requisite online corrective solutions in $N-1$ contingency scenarios. Section IV elucidates the proposed floating green microgrid, covering all specifications comprehensively. The outcomes of various contingency events are deliberated in Section V, followed by concluding remarks in Section VI.

II. HOLOMORPHIC EMBEDDING POWER FLOW METHOD

Holomorphic embedded power flow method (HEPFM) introduces a variable ($\wp$) in the complex plane to solve nonlinear power flow and the analytic solution is computed from a common germ solution to the final objective solution. For a no load, no generation case the white germ solution is computed as a linear solution at $\wp$ equal to zero and the final solution of the normal power flow is obtained where $\wp$ equate to one. The basic power flow equations of PQ bus, PV bus and slack bus are defined as follows:

$$\sum_{k=1}^{N} Y_{ik} V_k = \frac{(P_i + jQ_i)^*}{V_i^*} \quad (1)$$

$$P_i = Real\left(V_i \sum_{k=1}^{N} Y_{ik}^* V_k^*\right) \text{ and } V_i = |V_i^{sp}| \quad (2)$$

$$V_i = V_i^{sl} \quad (3)$$

The holomorphic power flow equations for PQ bus, PV bus and slack bus are defined by taking series admittance ($Y_{(ik,Series)}$) and shunt admittance ($Y_{ik,Shunt}$) as follows:

$$\sum_{k=1}^{N} Y_{ik,series} V_k(\wp) = \frac{\wp S_i^*}{(V_i^*(\wp^*))} - \wp Y_{(ik,shunt)} V_k(\wp) \quad (4)$$

$$\sum_{k=1}^{N} Y_{ik,series} V_k(\wp) = \frac{\wp P_i + jQ_i(\wp)}{V_i^*(\wp^*)} - \wp Y_{(ik,shunt)} V_k(\wp)$$

$$\text{and } V_i(\wp) \cdot V_i^*(\wp^*) = 1 + (|V_i^{sp}|^2 - 1)\wp \quad (5)$$

$$V_i(\wp) = 1 + (V_i^{sl} - 1)\wp \quad (6)$$

The holomorphic power flow solution for generator bus, load bus and slack bus are presented in Table I.

TABLE I. NON-HOLOMORPHIC POWER FLOW EQUATION AND ITS HOLOMORPHIC POWER FLOW SOLUTION

Type	Germ solution ($\wp = 0$)	Original power flow at $\wp = 1$
Slack	$V_i(\wp) = 1$	$V_i(\wp) = 1 + (V_i^{sl} - 1)\wp$
PQ	$\sum_{k=1}^{N} Y_{ik,series} V_k(\wp) = 0$	$\sum_{k=1}^{N} Y_{ik,series} V_k(\wp) = \frac{\wp S_i^*}{(V_i^*(\wp^*))}$ $-\wp Y_{(ik,shunt)} V_k(\wp)$
PV	$\sum_{k=1}^{N} Y_{ik,series} V_k(\wp) = 0$ and $V_i(\wp) \cdot V_i^*(\wp^*) = 1$	$\sum_{k=1}^{N} Y_{(ik,series)} V_k(\wp) = \frac{\wp P_i + jQ_i(\wp)}{V_i^*(\wp^*)}$ $-\wp Y_{(ik,shunt)} V_k(\wp)$ and $V_i(\wp) \cdot V_i^*(\wp^*) = 1 + (|V_i^{sp}|^2 - 1)\wp$

Where the power series coefficients are defined as $V_i(\wp) = \sum_{m=0}^{\infty} V_i[m]\wp^m$, $Q_i(\wp) = \sum_{m=0}^{\infty} Q_i[m]\wp^m$ and $W_i^*(\wp) = 1/V_i^*(\wp^*) = \sum_{m=0}^{\infty} W_i^*[m]\wp^m$. The voltage function and its

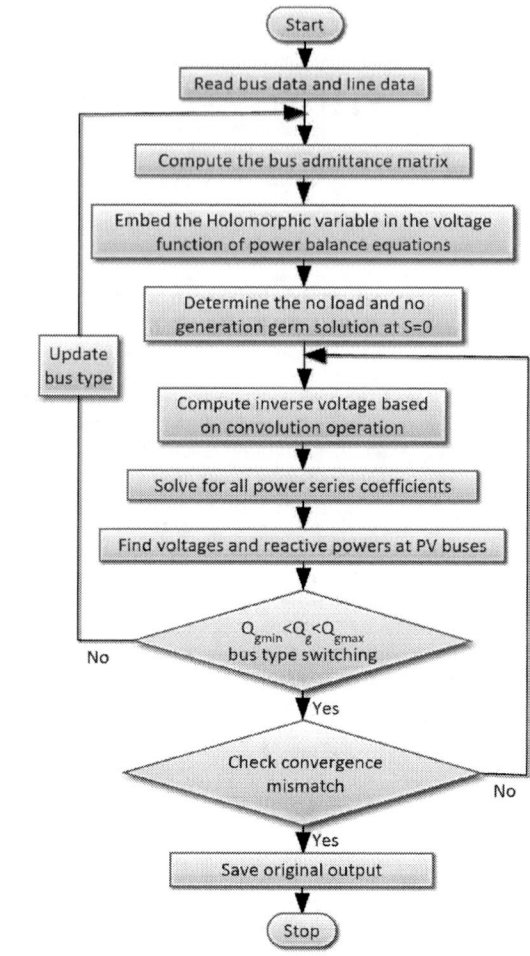

Fig. 1. Flowchart of the holomorphic embedded power flow method.

inverse power series is related as $W[n] = \sum_{m=0}^{\infty} W[m]V[n-m]/V[0]$.

The holomorphic function can be expressed in terms of Maclaurin series as follows:
$$\sum_{k=1}^{N} Y_{ik,series}\left(V_k[0] + V_k[1](\wp) + \cdots + V_k[n](\wp)^n\right)$$
$$= \left(\frac{\wp S_i}{V_i[0] + V_i[1](\wp) + V_i[2](\wp)^2 + \cdots + V_i[n](\wp)^n}\right)^*$$
$$= \wp S_i^* \left(W_i^*[0] + W_i^*[1](\wp) + \cdots + W_i^*[n](\wp)^n\right) \quad (7)$$

The above recursive expression is expressed as:

$$\sum_{k=1}^{N} Y_{ik,series} V_k(m) = \wp S_i^* W_i^*[m-1] \; ; m > 0. \quad (8)$$

The Pade approximant solution [5] is used to approximate the above recursive equation. The holomorphic power flow solution with reactive power limit violation of generators is shown in Fig. 1.

III. FUNCTIONAL EXPANDED EXTREME LEARNING MACHINE

Functional expanded extreme learning machine (FEELM) is an efficient and fast learning single hidden layer feed-forward neural network introduced for taking corrective action in the $N-1$ contingency analysis of the proposed FGMG system. Unlike traditional gradient-based methods that iteratively adjust weights through backpropagation, ELM [6] randomly initializes the input weights and learns the output weights analytically. This leads to a significantly faster training process since it eliminates the need for iterative optimization. The exponential trigonometric expansion of generation units, ESSs and loads power (LP) set-points i.e., $X = [Grid, GEG1, GEG2, PV, DG, ESS1, ESS2, LP1, LP2, \cdots, LP13]$ as $1, X, X^2, e^{-a|X|}sin(\pi X), e^{-a|X|}cos(\pi X)$, with contingency index used as optimal projection n features to compute the L hidden nodes by random selection of weight w as well as bias b with target $T = [T_1, T_2]$. The hidden layer H is calculated using sigmoid activation function S as follows:

$$H_L(X_i) = \sum_{j=1}^{L} S(w_i \cdot X_j + b_i), \quad j = 1, 2, \cdots, N. \quad (9)$$

To obtain zero error in m output Z classes problem i.e., $\sum_{k=1}^{m} \parallel Z_k - T_k \parallel = 0$, the above-mention equation is modified as follows:

$$H \cdot \beta = T \quad (10)$$

The output weight matrix (β) is computed as

$$\beta = H \dagger \cdot T. \quad (11)$$

The output $F(X_i)$ is computed using Moore-Penrose generalized inverse $(H\dagger)$ [7], testing hidden layer matrix $h(x)$ and small value (λ) as follows:

$$F(X_i) = Sign\left[h(x) \cdot H^T \left(\frac{I}{\lambda} + HH^T\right)^{-1} \cdot T\right] \quad (12)$$

IV. FLOATING GREEN MICROGRID

The developed floating green micro-grid in MAT-LAB/Simulink is constituted of multiple distributed generations, two battery energy storage systems (BESSs), and different types of loads. One 1 MW solar photovoltaic (PV) system, one 1 MW emergency diesel generator, and two 3.75 MW gas generators are coupled to PCC at 22 kV transmission line utility grid. There are four voltage distribution levels: a primary 22 kV voltage level and a secondary at 11 kV, 480 V as well as 400 V voltage levels. The FGMG operates in both 50 and 60 Hz frequencies environments by integrating the frequency converters and MG is capable of managing bidirectional power flow (generation/consumption) using bidirectional AC/DC VSC for the case of BESS storage systems and PV plant. Both the BES systems are connected through a boost-buck bidirectional converter, while the PV Array is connected to the DC link through a boost converter to exchange active and reactive power through transformers. The available lithium-ion batteries in the Simscape library within the MATLAB/Simulink environment are connected in parallel to an interfaced inverter in a cascaded topology which in turn steps up from 576 VDC to 1152 VDC with initial SOC of 80%. The 1 MW PV plant is configured in MATLAB/Simulink by choosing the suitable array arrangement of solar cells of irradiance 1000 W/m^2 and a temperature of 25°C. It

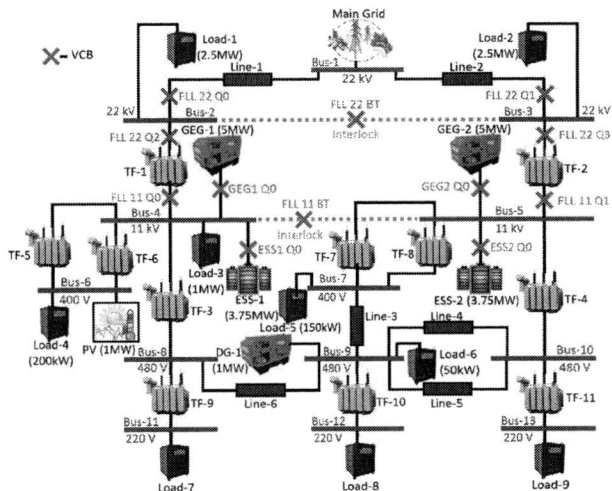

Fig. 2. Simplified model of the FGMG with vacuum circuit breaker (VCB) and interlock.

TABLE II. TRANSFORMER (TF) AND LINE NUMBERS WITH ITS RATING

Transformer/Line No.	Rating	Parameters
TF-1, TF-2	22/11kV, 12MVA	R=0.0094, X=0.14
TF-3, TF-4	11kV/536V, 700KVA	R=0.0094, X=0.051
TF-5, TF-7	11kV/400V, 200KVA	R=0.0119, X=0.04
TF-6, TF-8	11kV/400V, 700KVA	R=0.0089, X=0.051
TF-9, TF-10, TF-11	480V/220V, 75KVA	R=0.0123, X=0.04
Line-1, Line-2	22kV, 2.3km	R=0.788, X=0.2336
Line-3	400V, 0.5km	R=0.198, X=0.1089
Line-4 to Line-6	480V, 0.5km	R=0.198, X=0.1089

TABLE III. DIFFERENT DATA OF FLOATING GREEN MICROGRID MODEL

Bus No.	Sending end Bus voltage	Receiving end Bus voltage	R actual value	X actual value	P maximum in MW
1-2	22,000	22,000	1.567	0.4672	662.5932
1-3	22,000	22,000	1.567	0.4672	662.5932
2-4	22,000	11,000	0.0537	0.7018	344.8276
3-5	22,000	11,000	0.0537	0.7018	344.8276
4-6	11,000	400	0.0455	0.2468	17.8282
4-8	11,000	480	0.0431	0.2468	21.3938
5-7	11,000	400	0.0455	0.2468	17.8282
5-10	11,000	480	0.0455	0.2468	21.3938
6-4	400	11,000	0.0576	0.1936	22.7273
7-5	400	11,000	0.0576	0.1936	22.7273
7-9	400	480	0.099	0.05445	4.6392
8-9	480	480	0.099	0.05445	6.1232
8-11	480	220	0.0595	0.1936	0.9771
9-10	480	480	0.099	0.05445	6.1232
9-12	480	220	0.0595	0.1936	0.9771
10-9	480	480	0.099	0.05445	6.1232
10-13	480	220	0.0595	0.1936	0.9771

has a 1200 μF DC link capacitor operating at 500V DC where a PWM closed-loop control strategy utilizes a frequency of 5 kHz. A total number of 20 power transformers are connected in the FGM model. For the voltage level 22 kV transmission line, a cable was chosen for a 2.3 km length. The unit resistance of the line is 0.3256 Ω/km, the unit reactance of the line is 0.09653 Ω/km and the impedance unit of the line is 0.13 Ω/km. Balance and unbalance linear loads are modelled as constant impedances. A total minimum of 8.35 MW loads are connected to different buses containing single-phase components that deteriorate the symmetry of voltages and currents in this system. Furthermore, the simplified model of the FGMG is shown in Fig. 2 and the corresponding line number as well as transformer number with specifications are presented in Table II. The different buses voltage levels with the maximum active power transmit capability are tabulated in Table III.

TABLE IV. DIFFERENT GENERATING UNITS DATA OF FLOATING GREEN MICROGRID MODEL

Bus No.	Generation Unit	Rating	Bus No.	Generation Unit	Rating
Bus-4	GEG-1	5 MW	Bus-5	ESS-2	3.75 MW
Bus-4	ESS-1	3.75 MW	Bus-6	PV	1 MW
Bus-5	GEG-2	5 MW	Bus-9	DG-1	1 MW

TABLE V. DIFFERENT LOAD UNITS DATA OF FLOATING GREEN MICROGRID MODEL

Bus No.	Load No.	Rating	Bus No.	Load No.	Rating
Bus-2	Load-1	2.5 MW	Bus-7	Load-6	0.1 MW
Bus-3	Load-2	2.5 MW	Bus-9	Load-7	0.05 MW
Bus-4	Load-3	2 MW	Bus-11	Load-8	0.05 MW
Bus-5	Load-4	2.5 MW	Bus-12	Load-9	0.05 MW
Bus-6	Load-5	0.2 MW	Bus-13	Load-10	0.05 MW

TABLE VI. ALL POSSIBLE COMBINATIONS FOR A SPECIFIC LOAD BY CHANGING ENERGY SOURCES SET POINT AND LOAD SHEDDING

EVENTS	DG-1 of 1 MW	ESS-1 of 3.75 MW	ESS-1 of 3.75 MW	GEG-1 of 5 MW	GEG-2 of 5 MW	PV of 1 MW	LOAD of 10 MW
BLACK START	0.3	—	—	—	—	—	0.3
	0.3	2.2	—	—	—	—	2.5
	0.3	2.2	2.5	—	—	—	5
	0.3	2.2	2.5	2.5	—	—	7.5
	0.3	2.2	2.5	2.5	2.5	—	10
BASELINE	0	2.5	2.5	2.5	2.5	0	10
ESS-1 OUTAGE	0	0	2.5 to 3.75	2.5 to 5	2.5 to 5	0	10
			Cases for 10 MW = 260				
			Cases for load shedding (7.5 to 9.9 MW) = 2561				
ESS-2 OUTAGE	0	2.5 to 3.75	0	2.5 to 5	2.5 to 5	0	10
			Cases for 10 MW = 260				
			Cases for load shedding (7.5 to 9.9 MW) = 2561				
GEG-1 OUTAGE	0	2.5 to 3.75	2.5 to 3.75	0	2.5 to 5	0	10
			Cases for 10 MW = 169				
			Cases for load shedding (7.5 to 9.9 MW) = 2195				
GEG-2 OUTAGE	0	2.5 to 3.75	2.5 to 3.75	2.5 to 5	0	0	10
			Cases for 10 MW = 169				
			Cases for load shedding (7.5 to 9.9 MW) = 2195				

V. CONTINGENCY ANALYSIS: MODES AND REMEDIAL SOLUTIONS

Six number of generation units of total generation 19.5 MW including gas generators (GEGs), energy storage systems (ESSs), Photovoltaic and diesel generator (DG) are interfaced in the developed 13 bus FGMG model. Also, 10 different loads are connected at different buses and the details of both the generation as well as loads are presented in Table IV and V respectively. A total of six test cases such as black start, base line, ESS-1 outage, ESS-2 outage, GEG-1 outage and GEG-2 outage are formulated and tabulated in Table VI to test and validate the efficacy of the proposed method.

A. Black start operation of the FGMG system

Under certain situations like islanded condition or faults, it is needed to be shutdown for preventing any adverse impact on FGMG systems and loads. The restoration process commonly consists of the black start [8] of generating units, the reconfiguration of the network and the restoration of loads as presented in Table VIII. The generation units power and loads power as presented in Table VIII are fed as an input to the HEPFM

algorithm for computing the bus voltage angle and line power as tabulated in Tables IX and X of the FGMG system for the black start restoration. The output DERs power, bus power, bus voltage angles, and line power of the six predefined black-start restoration snapshots starting from $1/1/20230 : 00$ to $1/1/20232 : 30$ are presented by maintaining bus voltages 1 per unit and frequencies.

TABLE VII. BLACK START OPERATION OF FGMG SYSTEM FOR GENERATION RESTORATION (GR) AND LOAD RESTORATION (LR)

Events	Time in minutes	Switch ON energy sources	% of GR	Connected loads	% of LR
1	30	Start black start unit DG-1 of 1 MW at bus-9	5.12	0.3 MW critical Loads 6, 7, 8, 9, and 10 at buses 7, 8, 9,10, 11, 12, and 13 respectively.	3
2	60	Start ESS-1 of 3.75 MW at bus-4	24.35	2.2 MW of loads 3, and 5 are connected at buses-4, and 6.	25
3	90	Start ESS-2 of 3.75 MW at bus-5	43.58	2.5 MW of load 4 is connected at bus-5.	50
4	120	Start GEG-1 of 5 MW at bus-4	69.23	2.5 MW of Load 1 is connected at bus-2.	75
5	150	Start GEG-2 of 5 MW at bus-5	94.87	2.5 MW of Load 2 is connected at bus-3.	100
6	180	Start PV of 1 MW at bus-6	100	—	—

B. Baseline N-1 Contingency Analysis for FGMG Branch Outage

For the baseline test case, all distributed energy resources (DERs), including GEG-1, GEG-2, ESS-1, and ESS-2, are set at a constant output of 2.5 MW, while the 10 MW load is allocated across various load terminals. This setup is used for conducting an optimal load flow analysis, disregarding losses in the system. Upon the successful convergence of the load flow analysis by importing power set points from generation as well as load units as tabulated in Table XI while adhering to all specified constraints, the following critical results are presented in Tables XII to XV.

A few branches indicated in bold such as 5, 11 and 13 are not considered for branch outage contingency operation because in the outage of these branches, it is treated as automatic corrective action like load shedding condition. The maximum line power limits are established through a comprehensive assessment, drawing from voltage and current ratings, conductor specifications including size and material, thermal constraints, line length, as well as consideration of both steady-state and transient loads based on state-of-the-art research. Additionally, a $N - 1$ contingency analysis is executed, considering branch outages. In the contingency analysis, the branch outage power and power transfer distribution factor (PTDF) index [9] are calculated within the framework of voltage and frequency stability. This calculation serves to assess line loading and guide subsequent corrective actions, including load shedding and adjustments to energy set points. Analysis of the developed model reveals that branch outages 3 and 4 lead to excessive loading on branches 9, 10, and 12, as indicated by both the branch outage power and PTDF values. In response, implementing load shedding as a corrective measure is necessary to ensure the stability and reliability of grid operations.

979-8-3315-3013-6/25 $31.00 © 2025 IEEE

TABLE VIII. GENERATIONS AND LOADS POWER SET POINTS FOR BLACK START OPERATION OF FGMG SYSTEM

Snapshots	Generator-P_Set					ESS-P_Set		Load-P_Set										
	Grid	GEG1	GEG2	PV	DG	ESS1	ESS2	1	2	3	4	5	6	7	9	11	12	13
1/1/2023 0:00	0	0	0	0	0	0	0	0	0	0	0	0	0	0	0	0	0	0
1/1/2023 0:30	0	0	0	0	0.3	0	0	0	0	0	0	0	0	0.1	0.05	0.05	0.05	0.05
1/1/2023 1:00	0	0	0	0	0.3	0	2.2	0	0	0	2	0	0.2	0.1	0.05	0.05	0.05	0.05
1/1/2023 1:30	0	0	0	0	0.3	2.5	2.2	0	0	0	2	2.5	0.2	0.1	0.05	0.05	0.05	0.05
1/1/2023 2:00	0	5	0	0	0.3	0	2.2	0	0	2.5	2	2.5	0.2	0.1	0.05	0.05	0.05	0.05
1/1/2023 2:30	0	5	4.7	0	0.3	0	0	0	2.5	2.5	2	2.5	0.2	0.1	0.05	0.05	0.05	0.05

TABLE IX. DIFFERENT BUSES VOLTAGE ANGLE FOR BLACK START OPERATION OF FGMG SYSTEM

Snapshots	1	2	3	4	5	6	7	8	9	10	11	12	13
1/1/2023 0:00	0	0	0	0	0	0	0	0	0	0	0	0	0
1/1/2023 0:30	0	1.62E-05	-1.62E-05	1.67E-05	-1.67E-05	1.67E-05	-2.03E-05	1.85E-05	0.006762	-1.49E-05	-0.00259	0.004157	-0.00262
1/1/2023 1:00	0	-0.00145	0.001452	-0.0015	0.001501	-0.00152	0.001496	-0.0015	0.007153	0.001501	-0.0041	0.004549	-0.0011
1/1/2023 1:30	0	0.000216	-0.00022	0.000224	-0.00022	0.000209	-0.00023	0.000225	0.006708	-0.00022	-0.00238	0.004104	-0.00283
1/1/2023 2:00	0	0.001884	-0.00188	0.001948	-0.00189	0.001933	-0.00189	0.001948	0.006299	-0.00189	-0.00066	0.003695	-0.00449
1/1/2023 2:30	0	0.000216	-0.00022	0.00028	-0.00017	0.000266	-0.00017	0.000282	0.006765	-0.00016	-0.00232	0.004161	-0.00277

TABLE X. DIFFERENT LINES POWER FOR BLACK START OPERATION OF FGMG SYSTEM

Snapshots	1	2	3	4	5	6	7	8	9	10	11	12	13	14
1/1/2023 0:00	0	0	0	0	0	0	0	0	0	0	0	0	0	0
1/1/2023 0:30	-0.02398	0.023981	-0.02398	0.023981	2.78E-17	-0.02398	0.048329	-0.02435	-0.05167	-0.07398	0.05	0.074347	0.05	0.05
1/1/2023 1:00	2.155098	-2.1551	2.155098	-2.1551	0.2	-0.0449	0.056904	-0.012	-0.0431	-0.0949	0.05	0.062002	0.05	0.05
1/1/2023 1:30	-0.32113	0.321129	-0.32113	0.321129	0.2	-0.02113	0.047159	-0.02603	-0.05284	-0.07113	0.05	0.076031	0.05	0.05
1/1/2023 2:00	-2.79774	2.797743	-2.79774	0.297743	0.2	0.002257	0.037574	-0.03983	-0.06243	-0.04774	0.05	0.089831	0.05	0.05
1/1/2023 2:30	-0.32113	0.321129	-2.82113	-2.17887	0.2	-0.02113	0.047159	-0.02603	-0.05284	-0.07113	0.05	0.076031	0.05	0.05

TABLE XI. GENERATIONS AND LOADS POWER SET POINTS FOR BASE-LINE OPERATION OF FGMG SYSTEM

snapshots	Generator-P_Set					ESS-P_Set		Load-P_Set										
	Main	GEG1	GEG2	PV	DG	ESS1	ESS2	1	2	3	4	5	6	7	9	11	12	13
1/1/2023 0:00	0	0	0	0	0.3	0	0	0	0	0	0	0	0	0.1	0.05	0.05	0.05	0.05
1/1/2023 0:30	0	0	0	0	0.3	0	2.2	0	0	0	2	0	0.2	0.1	0.05	0.05	0.05	0.05
1/1/2023 1:00	0	0	0	0	0.3	2.5	2.2	0	0	0	2	2.5	0.2	0.1	0.05	0.05	0.05	0.05
1/1/2023 1:30	0	5	0	0	0.3	0	2.2	0	0	2.5	2	2.5	0.2	0.1	0.05	0.05	0.05	0.05
1/1/2023 2:00	0	5	4.7	0	0.3	0	0	0	2.5	2.5	2	2.5	0.2	0.1	0.05	0.05	0.05	0.05
1/1/2023 2:30	0	2.5	2.5	0	0	2.5	2.5	0	2.5	2.5	2	2.5	0.2	0.1	0.05	0.05	0.05	0.05

TABLE XII. DIFFERENT BUSES VOLTAGE ANGLE FOR BASE-LINE OPERATION OF FGMG SYSTEM

Snapshots	1	2	3	4	5	6	7	8	9	10	11	12	13
1/1/2023 0:00	0	1.62E-05	-1.62E-05	1.67E-05	-1.67E-05	1.67E-05	-2.03E-05	1.85E-05	0.006762	-1.49E-05	-0.00259	0.004157	-0.00262
1/1/2023 0:30	0	-0.00145	0.001452	-0.0015	0.001501	-0.00152	0.001496	-0.0015	0.007153	0.001501	-0.0041	0.004549	-0.0011
1/1/2023 1:00	0	0.000216	-0.00022	0.000224	-0.00022	0.000209	-0.00023	0.000225	0.006708	-0.00022	-0.00238	0.004104	-0.00283
1/1/2023 1:30	0	0.001884	-0.00188	0.001948	-0.00189	0.001933	-0.00189	0.001948	0.006299	-0.00189	-0.00066	0.003695	-0.00449
1/1/2023 2:00	0	0.000216	-0.00022	0.00028	-0.00017	0.000266	-0.00017	0.000282	0.006765	-0.00016	-0.00232	0.004161	-0.00277
1/1/2023 2:30	0	0.000142	-0.00014	0.000204	-9.00E-05	0.000189	-9.93E-05	0.000197	-0.00337	-9.64E-05	-0.00241	-0.00598	-0.0027

TABLE XIII. DIFFERENT LINES POWER FOR BASE-LINE OPERATION OF FGMG SYSTEM

Snapshots	1	2	3	4	5	6	7	8	9	10	11	12	13	14
1/1/2023 0:00	-0.02398	0.023981	-0.02398	0.023981	2.78E-17	-0.02398	0.048329	-0.02435	-0.05167	-0.07398	0.05	0.074347	0.05	0.05
1/1/2023 0:30	2.155098	-2.1551	2.155098	-2.1551	0.2	-0.0449	0.056904	-0.012	-0.0431	-0.0949	0.05	0.062002	0.05	0.05
1/1/2023 1:00	-0.32113	0.321129	-0.32113	0.321129	0.2	-0.02113	0.047159	-0.02603	-0.05284	-0.07113	0.05	0.076031	0.05	0.05
1/1/2023 1:30	-2.79774	2.797743	-2.79774	0.297743	0.2	0.002257	0.037574	-0.03983	-0.06243	-0.04774	0.05	0.089831	0.05	0.05
1/1/2023 2:00	-0.32113	0.321129	-2.82113	-2.17887	0.2	-0.02113	0.047159	-0.02603	-0.05284	-0.07113	0.05	0.076031	0.05	0.05
1/1/2023 2:30	-0.21085	0.210854	-2.71085	-2.28915	0.2	0.089146	0.124927	0.085927	0.024927	0.039146	0.05	-0.03593	0.05	0.05

C. FGMG N-1 contingency analysis on energy set point variations and load shedding scenarios

In this test scenario, the initial power output of each DER is established at 2.5 MW. The developed model can produce a minimum of 7.5 MW and a maximum of 13.5 MW of power during the outage of each ESS. Similarly, when each GEG experiences an outage, the network is capable of supplying a range of power, spanning from 7.5 MW to 12.5 MW, to meet the specified 10 MW load. Based on 0.1 MW power variation, the outage of each ESS leads to 2561 cases where power generation ranges from 7.5 MW to 9.9 MW, necessitating a load curtailment of 2.5 MW to 0.1 MW, respectively. Similarly, the outage of each GEG results in 2195 cases where power generation varies from 7.5 MW to 10 MW, requiring load curtailment between 2.5 MW and 0.1 MW, respectively. In total, there are 9512 possible load shedding test cases, and during the outage of each DER, there exists a total of 858 potential test cases under energy set point variation aimed at achieving a 10 MW generation level without considering the test cases where it can generate more than 10 MW with a step resolution of 100 KW.

The $N-1$ contingency is applied to all test cases indexed by I, and the corresponding functional expanded values of $X = [Grid, GEG1, GEG2, PV, DG, ESS1, ESS2, LP1, LP2, \cdots, LP13]$ with index I are imported to the FEELM classifier [10] to identify patterns of "No action required" or "Load shedding". Remarkably, the proposed classifier demonstrates exceptional recognition accuracy of 100% with

TABLE XIV. BRANCH OUTAGE POWER FOR THE BASE-LINE OPERATION OF FGMG SYSTEM

Branches	1	2	3	4	5	6	7	8	9	10	11	12	13	14
Max Limit	24	24	24	24	7.6	7.6	7.6	7.6	1	1	1	1	1	1
BS Base	-0.21085	0.210854	-2.71085	-2.28915	0.2	0.089146	0.124927	0.085927	0.024927	0.039146	0.05	-0.03593	0.05	0.05
1	0	1.28E-13	-2.5	-2.5	0.2	0.3	0.038499	-0.0385	-0.0615	0.25	0.05	0.088499	0.05	0.05
2	-5.57E-14	0	-2.5	-2.5	0.2	0.3	0.038499	-0.0385	-0.0615	0.25	0.05	0.088499	0.05	0.05
3	2.5	-2.5	0	-5	0.2	2.8	-0.98624	-1.51376	-1.08624	2.75	0.05	1.563758	0.05	0.05
4	-2.5	2.5	-5	0	0.2	-2.2	1.06324	1.43676	0.96324	-2.25	0.05	-1.38676	0.05	0.05
6	-0.3	0.3	-2.8	-2.2	0.2	0	0.161468	0.138532	0.061468	-0.05	0.05	-0.08853	0.05	0.05
7	-0.14886	0.148864	-2.64886	-2.35114	0.2	0.151136	0	0.148864	-0.1	0.101136	0.05	-0.09886	0.05	0.05
8	-0.16046	0.160464	-2.66046	-2.33954	0.2	0.139536	0.160464	0	0.060464	0.089536	0.05	0.05	0.05	0.05
9	-0.19848	0.198485	-2.69848	-2.30152	0.2	0.101515	0.1	0.098485	0	0.051515	0.05	-0.04848	0.05	0.05
10	-0.25	0.25	-2.75	-2.25	0.2	0.05	0.140973	0.109027	0.040973	0	0.05	-0.05903	0.05	0.05
12	-0.18979	0.189785	-2.68979	-2.31021	0.2	0.110215	0.139785	0.05	0.039785	0.060215	0.05	0	0.05	0.05

TABLE XV. POWER TRANSFER DISTRIBUTION FACTOR FOR THE BASE-LINE OPERATION OF FGMG SYSTEM

Branches	1	2	3	4	5	6	7	8	9	10	11	12	13	14	Corrective Action
Base	0.008786	0.008786	0.112952	0.095381	0.026316	0.01173	0.016438	0.011306	0.024927	0.039146	0.05	0.035927	0.05	0.05	
1	0	5.33E-15	0.104167	0.104167	0.026316	0.039474	0.005066	0.005066	0.061501	0.25	0.05	0.088499	0.05	0.05	No Action Required
2	2.32E-15	0	0.104167	0.104167	0.026316	0.039474	0.005066	0.005066	0.061501	0.25	0.05	0.088499	0.05	0.05	No Action Required
3	**0.104167**	**0.104167**	**0**	**0.208333**	**0.026316**	**0.368421**	**0.129769**	**0.199179**	**1.086242**	**2.75**	**0.05**	**1.563758**	**0.05**	**0.05**	**Load Shedding**
4	**0.104167**	**0.104167**	**0.208333**	**0**	**0.026316**	**0.289474**	**0.1399**	**0.189047**	**0.96324**	**2.25**	**0.05**	**1.38676**	**0.05**	**0.05**	**Load Shedding**
6	0.0125	0.0125	0.116667	0.091667	0.026316	0	0.021246	0.018228	0.061468	0.05	0.05	0.088532	0.05	0.05	No Action Required
7	0.006203	0.006203	0.110369	0.097964	0.026316	0.019886	0	0.019587	0.1	0.101136	0.05	0.098864	0.05	0.05	No Action Required
8	0.006686	0.006686	0.110853	0.097481	0.026316	0.01836	0.021114	0	0.060464	0.089536	0.05	0.05	0.05	0.05	No Action Required
9	0.00827	0.00827	0.112437	0.095896	0.026316	0.013357	0.013158	0.012959	0	0.051515	0.05	0.048485	0.05	0.05	No Action Required
10	0.010417	0.010417	0.114583	0.09375	0.026316	0.006579	0.018549	0.014346	0.040973	0	0.05	0.059027	0.05	0.05	No Action Required
12	0.007908	0.007908	0.112074	0.096259	0.026316	0.014502	0.018393	0.006579	0.039785	0.060215	0.05	0	0.05	0.05	No Action Required

10-fold cross validation method, offering corrective actions with minimal runtime consumption of only 27 milliseconds on an $i5 - 5000U$ CPU with 24 GB RAM, operating at a frequency of 2.4 GHz.

VI. CONCLUSION

The FGMG model is designed with carefully selected parameters for buses, loads, generators, storage units, stores, lines, transformers, links, and more, facilitating comprehensive systems contingency analysis. A robust black start solution has been developed to ensure the secure, dependable, and resilient operation of the FGMG model. This includes the presentation of optimal output parameters such as DERs power, bus power, bus voltage angle, line power, and others. Various test cases have undergone $N - 1$ contingency analysis, where computations for branch outage power and power transfer distribution factor index play a pivotal role in determining corrective actions. Additionally, within the $N-1$ contingency scenario, case studies involving load shedding and energy set point variation have been conducted to evaluate FGMG model stability, with findings duly presented. The developed FEELM algorithm demonstrates its robustness with superior classification accuracy, providing effective corrective actions in the developed FGMG system during emergencies. Finally, the comprehensive results of the proposed HEPFM-FEELM method in a security-constrained optimal power flow environment ensure the grid's safe and reliable operation in islanded mode, supported by corresponding outputs as corrective actions.

ACKNOWLEDGMENT

The research is supported by National Research Foundation, Singapore, jointly with Energy Market Authority and Keppel Offshore & Marine Limited under the Energy Programme "EMA-KOM Joint RFP (EMA- EP007-EKOM)" with Envision Digital International Pte. Ltd. as the main Host Institution.

REFERENCES

[1] A. Eajal, M. A. Abdelwahed, E. El-Saadany, and K. Ponnambalam, "A unified approach to the power flow analysis of AC/DC hybrid microgrids," *IEEE Transactions on sustainable energy*, vol. 7, no. 3, pp. 1145–1158, 2016.

[2] R. Baldick, B. Chowdhury, I. Dobson, Z. Dong, B. Gou, D. Hawkins, H. Huang, M. Joung, D. Kirschen, F. Li *et al.*, "Initial review of methods for cascading failure analysis in electric power transmission systems IEEE PES CAMS task force on understanding, prediction, mitigation and restoration of cascading failures," in *2008 IEEE Power and Energy Society General Meeting-Conversion and Delivery of Electrical Energy in the 21st Century*. IEEE, 2008, pp. 1–8.

[3] A. Bhuyan, B. K. Panigrahi, S. Pati, H. Sahoo, and R. K. Pradhan, "Contingency Analysis of Low Voltage DC Microgrid," in *2020 International Conference on Electronics and Sustainable Communication Systems (ICESC)*. IEEE, 2020, pp. 1108–1111.

[4] C. Liu, B. Wang, F. Hu, K. Sun, and C. L. Bak, "Online voltage stability assessment for load areas based on the holomorphic embedding method," *IEEE Transactions on Power Systems*, vol. 33, no. 4, pp. 3720–3734, 2017.

[5] S. Rao, Y. Feng, D. J. Tylavsky, and M. K. Subramanian, "The holomorphic embedding method applied to the power-flow problem," *IEEE Transactions on Power Systems*, vol. 31, no. 5, pp. 3816–3828, 2015.

[6] G.-B. Huang, H. Zhou, X. Ding, and R. Zhang, "Extreme learning machine for regression and multiclass classification," *IEEE Transactions on Systems, Man, and Cybernetics, Part B (Cybernetics)*, vol. 42, no. 2, pp. 513–529, 2011.

[7] D. Serre, "Matrices: Theory and applications-additional exercises," *L'Ecole Normale Supérieure de Lyon*, 2001.

[8] L. H. Fink, K.-L. Liou, and C.-C. Liu, "From generic restoration actions to specific restoration strategies," *IEEE Transactions on Power Systems*, vol. 10, no. 2, pp. 745–752, 1995.

[9] H. Ronellenfitsch, M. Timme, and D. Witthaut, "A dual method for computing power transfer distribution factors," *IEEE Transactions on Power Systems*, vol. 32, no. 2, pp. 1007–1015, 2016.

[10] M. Sahani and P. K. Dash, "FPGA-based semisupervised multifusion RDCNN of process robust VMD data with online kernel RVFLN for power quality events recognition," *IEEE Transactions on Neural Networks and Learning Systems*, vol. 33, no. 2, pp. 515–527, 2020.

Novel Senegalese wolf search, Filoviridae optimization and Asterina algorithm for power loss diminution and voltage stability enlargement

Lenin Kanagasabai
Dept of EEE, Prasad V. Potluri Siddhartha Institute of Technology
Vijayawada, Andhra Pradesh, 520007, India
gklenin@gmail.com

Abstract—Pack hunting attributes of Senegalese wolf is mathematically defined in Senegalese wolf optimization algorithm. Senegalese wolf packs stalk the target in supportive method. Senegalese wolf yowls to fascinate other members in the pack for hunting. Exploitation procedure has been augmented by intermingle the Scopidae search with Senegalese wolf algorithm. Scopidae owns intellectual procedure of portrayal and tactical approach during progression. Scopidae are consociates of the populations and each populace acquaintance authorizes gradation for the optimal problem limitations construing to the location in the search region. Filoviridae Search algorithm has been assimilated with Thomomys bulbivorus algorithm to amplify the exploration competence of the procedure. Examination, predatory, and renaissance behaviours of Asterina has been scientifically defined in the exploration section. Predatory and renaissance behaviours of Asterina are defined in the exploitation section. Validity of Senegalese wolf optimization (SWO) algorithm, Filoviridae search optimization algorithm (FSO) and Asterina search optimization (ASO) algorithm are verified in 7 benchmark functions and IEEE 57, 118 bus systems.

Keywords—Senegalese wolf, Scopidae, Filoviridae, Thomomys bulbivorus, Asterina

I. INTRODUCTION

In Senegalese wolf optimization (SWO) algorithm, pack hunting attributes of Senegalese wolf is mathematically defined. Senegalese wolf packs stalk the target in a supportive method. Senegalese wolf yowls to fascinate other members in the pack for hunting. Senegalese wolfs are committed with the female, resilient connections with the family associates and raise the offspring's. Grouping and movement of the Senegalese wolf's pack [1] is defined scientifically. Exploitation procedure has been augmented by intermingle the Scopidae search optimization with Senegalese wolf optimization algorithm.

Scopidae portrayal and tactic in the progression of pursuit is an intellectual process. Scopidae are consociates of the populations and each populace acquaintance authorizes gradation for the optimal problem limitations construing to the location in the search region. Predominantly, occupant's acquaintances are quixotically geared up inferring to the lowermost and extreme limitations. Scopidae optimization displays the depiction, structure of Scopidae [2] provoking and persecution the target and it has been evaluated in candidate expositions. Filoviridae owns a diffusion subterfuge which authorizes the units in the population to relocate amongst exposed, ailing, isolated, confessed in treatment center, recovered and dead sub-population accumulations. Stimulated by the value of this subterfuge of decree of the Filoviridae infection [3] procedure is modeled. The propagation of Filoviridae is recognized and in interpretation of its confrontational septicity grade in the civilizations. Stimulate the vectors and scalar degrees and randomly produce the record list from exposed units. Superfluous ailing list is deportee, extra quantity of infection will arise, subsequently; trivial drive defines exploitation. Conjecture of the ailing unit one or the other pauses in private location or else vestiges expat in the internal frontier is controlled by public category element. Regular actions of the Thomomys bulbivorus [4] are imitated to model the Thomomys bulbivorus search algorithm. Filoviridae search optimization algorithm has been assimilated with Thomomys bulbivorus search algorithm to amplify the exploration competence of the process. Examination, predatory, and renaissance behaviours of Asterina is mathematically defined in Asterina search optimization (ASO) algorithm. Movement of the Asterina is through tube feet and it located in the arms and the body. Sexual and asexual reproduction will there in the Asterina lifecycle. In fact, it is often difficult to distinguish the sexuality of a starfish based on its appearance. Asterina discharge gametes and the male Asterina issue the sperm. The released sperm will to enrich the gametes which results in creation of offspring's. Validity of Validity of Senegalese wolf optimization (SWO) algorithm, Filoviridae search optimization algorithm (FSO) and Asterina search optimization (ASO) algorithm are verified in 7 benchmark functions and IEEE 57, 118 bus systems.

II. PROBLEM FORMULATION

Factual power loss diminution [5 - 14] is mathematically termed as,

$$Min \ \tilde{F}(\bar{a}, \bar{b}) \tag{1}$$

$$a = \begin{bmatrix} VLG_1, .., VLG_{Ng}; QC_1, .., QC_{Nc} \\ ; T_1, .., T_{N_T} \end{bmatrix}$$

$$b = \begin{bmatrix} PG_{slack}; VL_1, .., VL_{N_{Load}}; QG_1, .., QG_{Ng} \\ ; SL_1, .., SL_{N_T} \end{bmatrix}$$

Fitness functions are demarcated as,

$$F_1 = P_{Min} = Min \left[\sum_m^{NTL} G_m \left[V_i^2 + V_j^2 - 2 * V_i V_j cos \emptyset_{ij} \right] \right]$$

$$F_2 = Min \left[\sum_{i=1}^{NLB} |V_{Lk} - V_{Lk}^{desired}|^2 + \sum_{i=1}^{Ng} |Q_{GK} - Q_{KG}^{Lim}|^2 \right]$$

$$F_3 = Minimize \ L_{Maximum}$$

Equality constraints

$$0 = PG_i - PD_i - V_i \sum_{j \in N_B} V_j \left[G_{ij} cos[\emptyset_i - \emptyset_j] + \right.$$

$$\left. B_{ij} sin[\emptyset_i - \emptyset_j] \right] \tag{2}$$

$$0 = QG_i - QD_i - V_i \sum_{j \in N_B} V_j \left[G_{ij} sin[\emptyset_i - \emptyset_j] + \right.$$

$$\left. B_{ij} cos[\emptyset_i - \emptyset_j] \right] \tag{3}$$

Inequality constraints

$$P_{gsl}^{min} \leq P_{gsl} \leq P_{gsl}^{max} \tag{4}$$

Reactive power generation (QGi)

$$Q_{gi}^{min} \leq Q_{gi} \leq Q_{gi}^{max}, i \in N_g \tag{5}$$

Load bus voltage (VLi)

$$VL_i^{min} \leq VL_i \leq VL_i^{max}, i \in NL \tag{6}$$

Transformers tap setting (Ti)

$$T_i^{min} \leq T_i \leq T_i^{max}, i \in N_T \tag{7}$$

Switchable reactive power compensations (QCi)

$$Q_c^{min} \leq Q_c \leq Q_C^{max}, i \in N_C \tag{8}$$

$$|SL_i| \leq S_{L_i}^{max}, i \in N_{TL} \tag{9}$$

Generator bus voltage (VGi)

$$VG_i^{min} \leq VG_i \leq VG_i^{max}, i \in N_g \tag{10}$$

$$F = F_1 + r_i F_2 + u F_3 = F_1 + \left[\sum_{i=1}^{NL} x_v [VL_i - VL_i^{min}]^2 + \right.$$

$$\left. \sum_{i=1}^{NG} r_g [QG_i - QG_i^{min}]^2 \right] + r_f F_3$$

III. SENEGALESE WOLF OPTIMIZATION ALGORITHM

In Senegalese wolf optimization (SWO) algorithm, pack hunting attributes of Senegalese wolf is mathematically defined. Senegalese wolf packs stalk the target in a supportive method. Senegalese wolf yowls to fascinate other members in the pack for hunting. Through urination Senegalese wolf marks their terrains. Senegalese wolf owns elongated legs and outstanding auditory proficiencies. Senegalese wolfs are committed with the female, resilient connections with the family associates and raise the offspring's. Grouping and movement of the Senegalese wolf's pack is defined scientifically. The pack led by the leader and the altered direction of movement is considered in the design of the algorithm.

Positions of the Senegalese wolf's pack is defined as,

$$B = \{b_1, b_2, .., b_d\} \tag{11}$$

$B \rightarrow$ position of the Senegalese wolf

$d \rightarrow dimension$

Rendering to the position, the fitness value is defined as,

$$B^\sigma = f(B^\sigma) \tag{12}$$

$$\sigma = 1,2,3, .., N$$

During hunting period, Senegalese wolf howls to enthrall other members in the pack in a particular place. The pack will be guided by Senegalese wolf which owns better fitness value.

$$B^\sigma = (b_1^\sigma, b_2^\sigma, ..., b_M^\sigma) \tag{13}$$

$$t = B^\sigma(t)$$

$$t = 1,2,3, .., (Q - 1)$$

Pack passages during the hunting is defined as

$$B^\sigma(t + 1) := B^\sigma(t) + A \left(\frac{L}{t} \right) \tag{14}$$

$$A \in [-1,1]$$

$P \rightarrow passage \ factor$

In the period of hunting, direction transferal of the pack is defined as,

$$B^\sigma \rightarrow B^\tau$$

$$\tau := A \cdot int(1, \sigma - 1)$$

$$B^\sigma(t + 1) := B^\sigma(t) + A \left(\frac{L}{t} \right) \left(B^\tau(t) - B^\sigma(t) \right) \tag{15}$$

$$A \in [-1,1]$$

$P \rightarrow passage \ factor$

Levy flight strategy assimilated with Senegalese wolf optimization,

$$B(i + 1) = Z(i) - |l(t)| \times H \times K(l) \tag{16}$$

$$LL(\beta) \sim 0.01 \frac{u}{|v|^{1/\beta}} (Q_j^t - c)$$

$$\sigma_u = \left\{ \frac{\Gamma(1+\beta) sin(\pi\beta/2)}{\Gamma[(1+\beta)/2]\beta 2^{(\beta-1)/2}} \right\}^{1/\beta}, \sigma_v = 1$$

Chaotic sequences [13] are amalgamated with Senegalese wolf optimization

$$a_{t+1} = a_t^2 - b_t^2 + u \cdot a_t + v \cdot b_t \tag{17}$$

$$b_{t+1} = 2a_t b_t + w \cdot a_t + x \cdot b_t \tag{18}$$

$$u = 0.9, v = -0.6, w = 2.0, x = 0.500$$

$$a_o, b_o = 0.1$$

Linear scaling in chaotic map is delineated as,

$$a_{t+1}^* = a_{t+1} - min(a) / \frac{max(a)}{-min(a)} \tag{19}$$

Exploitation procedure has been augmented by intermingle the Scopidae search optimization with Senegalese wolf optimization algorithm. Scopidae portrayal and tactic in the progression of pursuit is an intellectual process. Scopidae are consociates of the populations and each populace acquaintance authorizes gradation for the optimal problem limitations construing to the location in the search region. Predominantly, occupant's acquaintances are quixotically geared up inferring to the lowermost and extreme limitations.

$$S_{i,j} = min_j + Q * (max_j - min_j) \qquad (20)$$
$$i = 1,2,3,\dots,N, j = 1,2,3,\dots,m$$
$$Q \in [0,1]$$

Populace acquaintances of Scopidae in the algorithm are recognized as follows.

$$S = \begin{bmatrix} S_1 \\ \vdots \\ S_i \\ \vdots \\ S_N \end{bmatrix}_{N \times m} = \begin{bmatrix} S_{1,1} & \cdots & S_{1,m} \\ \vdots & \ddots & \vdots \\ S_{N,1} & \cdots & S_{N,m} \end{bmatrix}_{N \times m} \qquad (21)$$

$S \to$ Scopidae population

Objective functional value is defined as,

$$Z = \begin{bmatrix} Z_1 \\ \vdots \\ Z_i \\ \vdots \\ Z_N \end{bmatrix}_{N \times 1} = \begin{bmatrix} Z(S_1) \\ \vdots \\ Z(S_i) \\ \vdots \\ Z(S_N) \end{bmatrix}_{N \times 1} \qquad (22)$$

Scopidae optimization displays the depiction, structure of Scopidae provoking and persecution the target and it has been evaluated in candidate expositions. This persecution structure is accomplished by stimulating in the progression of target, which considered in the exploration design and actions in the water bodies are defined in the exploitation segment. In the exploration section, Scopidae identifies the location of the target and it tracks the target. Scopidae algorithm possesses exploration zone examination and exploration directives. In the Scopidae optimization process, position of the target is identified randomly. This aspect upsurges the exploration directive of Scopidae optimization in the specific examination region.

$$S_{i,j}^{SP_1} = \begin{cases} S_{i,j} + Q * (P_j - W * S_{i,j}), if\ Y_{SP} < Y_i \\ S_{i,j} + Q * (H_{i,j} - P_j), Else \end{cases} \qquad (23)$$

$S_{i,j}^{SP_1} \to$ updated position of Scopidae

$W \in [1,2]$

$Q \in [0,1]$

Fresh position for a Scopidae has been identified as follows.

$$S_i = \begin{cases} S_i^{SP_1}, Y_i^{SP_1} < Y_i \\ S_i, Else \end{cases} \qquad (24)$$

$S_i^{SP_1} \to$ renewed position of Scopidae

$Y_i^{SP_1} \to$ objetcive functional value

In exploitation section, Scopidae in shallow water, tire the wings to make the fishes to move upward directions, and at that instant target will be clasped. Structure of this depiction of Scopidae residues the procedure to herd in the persecution section.

$$S_{i,j}^{SP_2} = S_{i,j} + V * (t - 1/T) * (2 * V - 1) * S_{i,j} \qquad (25)$$

$S_i^{S_2} \to$ updated position of Scopidae

$V = 0.2$

$$S_i = \begin{cases} S_i^{HP_2}, Y_i^{SP_2} < Y_i \\ S, Else \end{cases} \qquad (26)$$

$S_i^{SP_1} \to$ renewed position of Scopidae

$Y_i^{SP_1} \to$ objetcive functional value

Senegalese wolf optimization (SWO) algorithm

1. Begin
2. $B = \{b_1, b_2, \dots, b_d\}$
3. Appraise the fitness rate
4. $B^\sigma = f(B^\sigma)$
5. Define the pack grouping parameter
6. $B^\sigma = (b_1^\sigma, b_2^\sigma, \dots, b_M^\sigma)$
7. $t = B^\sigma(t)$
8. Compute the pack passage
9. $B^\sigma(t+1) := B^\sigma(t) + A \left(\frac{L}{t} \right)$
10. *Define the* direction transferal of the pack
11. $B^\sigma \to B^\tau$
12. $\tau := A \cdot int(1, \sigma - 1)$
13. $B^\sigma(t+1) := B^\sigma(t) + A \left(\frac{L}{t} \right) \left(B^\tau(t) - B^\sigma(t) \right)$
14. $B(i+1) = Z(i) - |l(t)| \times H \times K(l)$
15. Streamline the values
16. Outline Scopidae examination region
17. $S_{i,j}^{SP_1} = \begin{cases} S_{i,j} + Q * (P_j - W * S_{i,j}), \\ \quad if\ Y_{SP} < Y_i \\ S_{i,j} + Q * (H_{i,j} - P_j), Else \end{cases}$
18. Compute the update position
19. $S_i = \begin{cases} S_i^{SP_1}, Y_i^{SP_1} < Y_i \\ S_i, Else \end{cases}$
20. $S_{i,j}^{SP_2} = S_{i,j} + V * \begin{pmatrix} t - \\ 1/T \end{pmatrix} * \begin{pmatrix} 2 * V \\ -1 \end{pmatrix} * S_{i,j}$
21. $S_i = \begin{cases} S_i^{HP_2}, Y_i^{SP_2} < Y_i \\ S, Else \end{cases}$
22. $t = t + 1$
23. End

IV. FILOVIRIDAE SEARCH OPTIMIZATION ALGORITHM

Filoviridae owns a diffusion subterfuge which authorizes the units in the population to relocate amongst exposed, ailing, isolated, confessed in treatment center, recovered and dead sub-population accumulations. Stimulated by the value of this subterfuge of decree of the Filoviridae infection procedure is modeled. The propagation of Filoviridae is recognized and in interpretation of its confrontational septicity grade in the civilizations. Stimulate the vectors and scalar degrees and randomly produce the record list from exposed units. Superfluous ailing list is deportee, extra quantity of infection will arise, subsequently; trivial drive defines exploitation. In the procedure Quantum computing has been applied,

$$|\Psi|^2 \cdot da \cdot db \cdot dc = Q \cdot da \cdot db \cdot dc \tag{27}$$

$$|\Psi(t+1, Y_i)\rangle = DL = |\Psi(V_D)\rangle - |\Psi^{De}(Y_i)\rangle * \left(\begin{vmatrix} \Psi(Y_i)\rangle \\ -|\Psi(Y_o)\rangle \end{vmatrix} \right) \tag{28}$$

Spot renovating of the each vulnerable unit is determined as,

$$Y_i^{t+1} = Y_i^t + \vartheta \cdot Q(Y) \tag{29}$$

$Y_i^{t+1} \rightarrow$ updated spot

$Y_i^t \rightarrow$ previous spot

$\vartheta \rightarrow$ variable of the moving unit

$Q(Y) \rightarrow$ passage rate

$$Q(Y) = X \cdot E + W(Y_u) \tag{30}$$

$$A(Z) = V \cdot E + W(Y_U) \tag{31}$$

$X \rightarrow$ minor passage

$V \rightarrow$ mean passage

$E \in [0,1]$

$Y_u \rightarrow$ excellent unit

Conjecture of the ailing unit one or the other pauses in private location or else vestiges expat in the internal frontier is controlled by public category element (PCE).

$PCE \geq 0.5$, unit passage towards the infected area

Otherwise unit remain in PCE region

Initial population created by random amount of propagation.

$$u_i = lb_i + E \cdot \begin{pmatrix} ub_i \\ +lb_i \end{pmatrix} \tag{32}$$

$UB_i \; LB_i \rightarrow$ bound limits
$E \in [0,1]$

Selection of the contemporary outstanding unit is premeditated on the set of ailing units as follows,

$$Y(b) = \begin{cases} g\,(b), f(pb) \\ \quad < f(gb) \\ p\,(b), f(Pb) \\ \quad \geq f(gb) \end{cases} \tag{33}$$

$gb \rightarrow$ global best, $pb \rightarrow$ personal best
Reform the constraints as follows,

$$\frac{\partial Y(t)}{\partial t} = \pi - \begin{pmatrix} \alpha_1 A + \alpha_3 B + \\ \alpha_4 C + \alpha_2 D \end{pmatrix} Y - (\tau Y + \Gamma Y) \tag{34}$$

$$\frac{\partial Q(t)}{\partial t} = \pi - (\alpha_1 A + \alpha_3 B + \alpha_4 C + \alpha_2 D \cdot \varphi) Y - (\Gamma + \beta) Q - (\tau) Y \tag{35}$$

$$\frac{\partial G(t)}{\partial t} = \gamma Q - (\beta + \overline{\emptyset}) G \tag{36}$$

$$\frac{\partial C(t)}{\partial t} = \beta Q - \Gamma C \tag{37}$$

$$\frac{\partial H(t)}{\partial t} = \beta Q - (\varepsilon + \vartheta) H \tag{38}$$

$$\frac{\partial K(t)}{\partial t} = (\tau Y + \Gamma Q) - \delta K \tag{39}$$

$$\frac{\partial L(t)}{\partial t} = (\pi Q - (\beta C + \Gamma L)) - \Psi S \tag{40}$$

Filoviridae Search Optimization (FSO) Algorithm has been assimilated with Thomomys bulbivorus search algorithm to amplify the exploration competence of the process.

$$Y_i = Y_{i\,lb} + E * (Y_{i\,ub} - Y_{i\,lb}) \tag{41}$$

$Y_{iub}, G_{ilb} \rightarrow bound\ limits, i = 1,2,3, .., N$

Probing info shared amongst cluster of Thomomys bulbivorus

$$\vec{V_i}(Y+1) = \left| \vec{V_a}(Y) - \vec{V} \right| \tag{42}$$

$\vec{V_i}(Y+1) \rightarrow updated\ position$
$\vec{V_a}(Y) \rightarrow best\ position$

$$\vec{V} = H * \vec{V_a}(Y) + L * (\vec{V_a}(Y) - \vec{V_i}(Y)) \tag{43}$$

$\vec{V_i}(Y) \rightarrow ith\ location\ of$ Thomomys bulbivorus
$H = V - Y * (V/Max.iter)$
$Y = 1,2,3,4, .., Max.iter$
$L = 2 * E$
$V \in [1,5]; \; L \in [0,2]$

Lévy flight strategy assimilated into Thomomys bulbivorus Swarm Search algorithm.

$$L = 0.010 - Y_j \in (Y_{i,k} - Y^b) \tag{44}$$

Quasi opposition-based learning is unified with the algorithm.

$$Y_{OBL}(t+1) = lb + ub - H(t) \tag{45}$$

$$H(t+1) = \begin{cases} Y_{OBL}(t), \; if\ E < 0.1 \\ Y_{QSOBL}(t), Else \end{cases} \tag{46}$$

<u>Filoviridae - search optimization algorithm (FSO)</u>

   a. Begin
   b. Apply quantum computing
   c. $|\Psi|^2 \cdot da \cdot db \cdot dc = Q \cdot da \cdot db \cdot dc$
   d. Identify the spot
   e. $u_i = lb_i + E \cdot \begin{pmatrix} ub_i \\ +lb_i \end{pmatrix}$
   f. Selection of outstanding unit
   g. $Z(b) = \begin{cases} g\,(b), f(pb) \\ \quad < f(gb) \\ p\,(b), f(Pb) \\ \quad \geq f(gb) \end{cases}$
   h. Reform the constraints
   i. $Y_i = Y_{i\,lb} + E * (Y_{i\,ub} - Y_{i\,lb})$
   j. Outline the information sharing
   k. $\vec{V_i}(Y+1) = \left| \vec{V_a}(Y) - \vec{V} \right|$
   l. $\vec{V} = H * \vec{V_a}(Y) + L * (\vec{V_a}(Y) - \vec{V_i}(Y))$
   m. $Y_{OBL}(t+1) = lb + ub - H(t)$
   n. $t = t + 1$
   o. End

V. Asterina Search Optimization Algorithm

Examination, predatory, and renaissance behaviours of Asterina is mathematically defined in Asterina search optimization (ASO) algorithm. Search behaviour of Asterina

[15] has been scientifically defined in the exploration section. Predatory and renaissance behaviours of Asterina are defined in the exploitation section. Asterina possess an extraordinary capacity to regenerate. During the period when the Asterina attacked by the predators, it lose it arms and small portion of the central body, for escaping away from the predators. The damaged or lost part of the Asterina body regenerate again.

In the procedure the location of the Asterina is engendered as follows,

$$A = \begin{bmatrix} A_{11} & \cdots & A_{1D} \\ \vdots & \ddots & \vdots \\ A_{N1} & \cdots & A_{ND} \end{bmatrix}_{N \times D} \tag{47}$$

In the primary stage, position of the Asterina is defined as,

$$A_{ij} = min + R(max - min)$$

Fitness value evaluation is done as follows,

$$F = \begin{bmatrix} F(A_1) \\ F(A_2) \\ \vdots \\ F(A_N) \end{bmatrix}_{N \times 1} \tag{48}$$

Arms of the Asterina will explore the regions in the positioned place and this action defined as,

$$\begin{cases} B_{i.L}^Q = A_{i.L}^Q + c_1 \left(A_{E.L}^Q - A_{i.L}^Q \right) \cos\theta \ , r \leq 0.5 \\ B_{i.L}^Q = A_{i.L}^Q - c_1 \left(A_{E.L}^Q - A_{i.L}^Q \right) \sin\theta \ , r > 0.5 \end{cases} \tag{49}$$

$B_{i.L}^Q \rightarrow$ acquired location of Asterina

$A_{i.L}^Q \rightarrow$ current location of Asterina

$c_1 = (2r - 1)\pi$

$\theta = \frac{\pi}{2} \cdot \frac{t}{T}$

The updated location of the Asterina is computed as,

$$A_{i.L}^{Q+1} = \begin{cases} B_{i.L}^Q & min_{E,L} \leq B_{i.L}^Q \leq max_{E,L} \\ A_{i.L}^Q & Else \end{cases} \tag{50}$$

When the movement of the Asterina is based on the information through other members in the period of exploration of food, there will be relocation and updated position of the Asterina is defined as,

$$B_{i.w}^Q = Z_i, A_{i.L}^Q + P_1 \left(A_{G_1.L}^Q - A_{i.L}^Q \right) + P_2 \left(A_{G_2.L}^Q - A_{i.L}^Q \right) \tag{51}$$

$Z_i \rightarrow$ Asterina energy level

$P_1, P_2 \in [-1,1]$

$Z_i = \frac{T-t}{T} \cos\theta$

In the predatory period of Asterina, the distance between the Asterina and others is computed as,

$$D = \left(A_E^Q - A_{S_L}^L \right) \tag{52}$$

The rationalized position of the Asterina during this preying period is defined as,

$$B_i^Q = A_i^Q + r_1 D_1 + r_2 D_2 \tag{53}$$

$r_1, r_1 \in [0,1]$

During the period when the Asterina attacked by the predators, it lose it arms and small portion of the central body, for escaping away from the predators. The damaged or lost part of the Asterina body regenerate again.

$$B_i^Q = exp\left(-t \times \frac{N}{T} \right) A_i^Q \tag{54}$$

Asterina search optimization (ASO) algorithm

1. Begin
2. Create the population
3. Compute the fitness value
4. Define the exploration of arms
5. Update the location
6. $A_{i.L}^{Q+1} = \begin{cases} B_{i.L}^Q & min_{E,L} \leq B_{i.L}^Q \\ & \leq max_{E,L} \\ A_{i.L}^Q & Else \end{cases}$
7. $B_{i.w}^Q = Z_i, A_{i.L}^Q + P_1 \left(A_{G_1.L}^Q - A_{i.L}^Q \right)$
   $+ P_2 \left(A_{G_2.L}^Q - A_{i.L}^Q \right)$
8. Define the distance
9. $D = \left(A_E^Q - A_{S_L}^L \right)$
10. Rationalize position of the Asterina
11. $B_i^Q = A_i^Q + r_1 D_1 + r_2 D_2$
12. $B_i^Q = exp\left(-t \times \frac{N}{T} \right) A_i^Q$
13. $t = t + 1$
14. End

VI. RESULTS

Validity of Senegalese wolf optimization (SWO) algorithm, Filoviridae search optimization algorithm (FSO) and Asterina search optimization (ASO) algorithm are verified in 7 benchmark functions [14]. Outcomes are presented in Table I.

$f_1 = \sum_{i=1}^{D} x_i^2$

$f_2 = -20 \, exp\left\{ -0.2 \sqrt{\frac{1}{D} \sum_{i=1}^{D} x_i^2} \right\}$

$-exp\left\{ \frac{1}{D} \sum_{i=1}^{D} cos(2\pi x_i) \right\} + 20 + e$

$f_3 = \sum_{i=1}^{D-1} \{ 100(x_{i+1} - x_i^2)^2 + (x_i - 1)^2 \}$

$f_4 = \frac{1}{4000} \sum_{i=1}^{D} x_i^2 - \prod_{i=1}^{D} cos\left(\frac{x_i}{\sqrt{i}} \right) + 1$

$f_5 = -\sum_{i=1}^{4} c_i \, exp\left[-\sum_{j=1}^{3} a_{ij} \left(x_j - p_{ij} \right)^2 \right]$

$f_6 = 4x_1^2 - 2.1x_1^4 + \frac{1}{3} x_1^6 + x_1 x_2 - 4x_2^2 + 4x_2^4$

$f_7 = \{ 1 + (x_1 + x_2 + 1)^2 (19 - 14x_1 + x_1^2 - 14x_2 + 6x_1 x_2 + 3x_2^2) \} \times$
$\begin{cases} 30 + (2x_1 - 3x_2)^2 \\ (18 - 32x_1 + 12x_1^2 + 48x_2 - 36x_1 x_2 + 27x_2^2) \end{cases}$

TABLE I SWO, FSO, ASO EXAMINATION IN 7 BENCHMARK FUNCTIONS

FN	HP [14]	EO [14]	SWO	FSO	ASO
1	0.8874	0.0026	0.0083	0.0083	0.0083
2	0.0918	0.0125	0.0179	0.0179	0.0179
3	22.9814	16.5622	16.484	16.484	16.484
4	1.1092 e-2	3.459 e-4	3.81 e-4	3.81 e-4	3.81 e-4
5	-3.8629	-3.8629	-3.8629	-3.8629	-3.8629
6	-1.0318	-1.0316	-1.0319	-1.0319	-1.0319
7	3.0280	3.0130	3.0121	3.0121	3.0121

979-8-3315-3013-6/25 $31.00 © 2025 IEEE

Validity of Senegalese wolf optimization (SWO) algorithm, Filoviridae search optimization algorithm (FSO) and Asterina search optimization (ASO) algorithm are verified IEEE 57 bus system. Table II illustrate the result investigation. FRL (MW) – Factual power loss, DCV (PU) – Deviance of voltage, SYV– Stability of voltage

TABLE II. SWO, FSO, ASO EXAMINATION IN IEEE 57 BUS SYSTEM

Method	FRL(MW)	DCV(PU)	SYV(PU)
PHA [5]	25.4715	0.6828	0.3231
ETI[6]	25.4963	0.799021	0.3129
MIZ [7]	26.8927	0.80991	0.3345
PAO [8]	29.535	0.8725	0.3791
LSO[11]	24. 4198	0.7844	0.3512
SWO	20.03299	0.47128	0.4412
FSO	20. 03312	0.47141	0.4404
ASO	20.03309	0.47138	0.4406

Validity of Senegalese wolf optimization (SWO) algorithm, Filoviridae search optimization algorithm (FSO) and Asterina search optimization (ASO) algorithm are verified IEEE 118 bus system. Table III demonstrate the examination of results

TABLE III. SWO, FSO AND ASO ANALYSIS IN IEEE 118 BUS SYSTEM

Method	FRL(MW)	DCV(PU)
MIZ [7]	128.137	1.12945
PAO [8]	123. 511	1.21790
NOP[9]	130.324	1.26103
PGZ [10]	131.845	1.26654
LSO [11]	132.972	1.3992
ESP[12]	117.02	1.2321
SWO	115. 05156	1.11349
FSO	115. 05171	1.11361
ASO	115. 05167	1.11358

Computation time of SWO, FSO and ASO algorithms are demonstrated in Table IV.

TABLE IV. SWO, FSO AND ASO COMPUTATION TIME

METHOD	57 BUS T(S)	118 BUS T(S)
SWO	20.12	36.17
FSO	20.25	36.33
ASO	20.20	36.26

VII. CONCLUSION

Senegalese wolf optimization (SWO) algorithm, Filoviridae search optimization algorithm (FSO) and Asterina search optimization (ASO) algorithm solved the problem efficiently. Senegalese wolf packs stalk the target in supportive mode. Exploitation procedure has been augmented by intermingle the Scopidae search with Senegalese wolf algorithm. In the exploration section, Scopidae identifies the location of the target and sequentially tracks the target. The propagation of Filoviridae is recognized and interpretation of its argumentative septicity graded in the civilizations. Conjecture of the ailing unit one or the other pauses in private location or else vestiges expat in the internal boundary is controlled. Algorithm initialized with random solutions, and exemplifies the entities position in the exploration area. Search behaviour of Asterina has been scientifically defined in the exploration section. In the exploitation section, predatory and renaissance behaviours of Asterina are defined. Validity of Senegalese wolf optimization (SWO) algorithm, Filoviridae search optimization algorithm (FSO) and Asterina search optimization (ASO) algorithm are verified in7 benchmark functions and IEEE 57, 118 bus systems. In future Senegalese wolf optimization (SWO) algorithm, Filoviridae search optimization algorithm (FSO) and Asterina search optimization (ASO) algorithm can be applied in the area of medical diagnosis and in the design of bio - medical equipment's.

REFERENCES

[1] Z. A. Woodgate, M. Drouilly, N. Nattrass, and M. J. O'Riain, "Co-occurrence of Senegalese wolf and caracal in the Karoo, South Africa," J. Arid Environ., vol. 219, no. 1, p. 1-8, 2023.

[2] S. Bekele and W. Tekalign, "Diversity and relative abundance of avian species in the wetland area northwest of Lake Abaya, southern Ethiopia," ScientificWorldJournal, vol. 2023, pp. 1–8, 2023.

[3] N. Biedenkopf et al., "ICTV virus Taxonomy profile: Filoviridae 2024: J. Gen. Virol., vol. 105, no. 2,pp 1-20, 2024.

[4] L. E. Painter, M. J. Weldy, R. S. Crowhurst, L. N. Carraway, and C. W. Epps, "Landscape genetics of the Camas pocket gopher (Thomomys bulbivorus), an endemic mammal of Oregon's Willamette valley," West. N. Am. Nat., vol. 82, no. 3, 2022.

[5] M. H. Ali, A. Soliman, M. Abdeen, T. Kandil, A. Y. Abdelaziz, and A. El-Shahat, "A Novel Stochastic Optimizer Solving Optimal Reactive Power Dispatch Problem Considering Renewable Energy Resources," Energies, vol. 16, no. 4, 1562, 2023.

[6] M. Varan, A. Erduman, and F. Menevşeoğlu, "A Grey Wolf Optimization Algorithm-Based Optimal Reactive Power Dispatch with Wind-Integrated Power Systems," Energies, vol. 16, no. 13, 5021, 2023.

[7] L. Lian, "Reactive power optimization based on adaptive multi-objective optimization artificial immune algorithm," Ain Shams Eng. J., vol. 13, no. 5, p. 101677, 2022.

[8] Z. Sahli, A. Hamouda, S. Sayah, D. Trentesaux, and A. Bekrar, "Efficient hybrid algorithm solution for Optimal Reactive Power Flow using the sensitive bus approach," Eng. Technol. Appl. Sci. Res., vol. 12, no. 1, pp. 8210–8216, 2022.

[9] H. M. Hasanien et al., "Hybrid particle swarm and sea horse optimization algorithm-based optimal reactive power dispatch of power systems comprising electric vehicles," Energy (Oxf.), vol. 286, no. 129583, p. 129583, 2024.

[10] M. A. M. Shaheen et al., "Enhanced transient search optimization algorithm-based optimal reactive power dispatch including electric vehicles," Energy (Oxf.), vol. 277, p. 127711, 2023

[11] M. Gil, E. Akbari, A. Rahimnejad, M. Ghasemi, and S. A. Gadsden, "Solution of optimal reactive power dispatch by Lévy-flight phasor particle swarm optimization," Intelligent Systems with Applications, vol. 23, no. 2, pp. 1-18, 2024.

[12] N. H. Khan et al., "Enhanced skill optimization algorithm: Solution to the stochastic reactive power dispatch framework with optimal inclusion of renewable resources using large-scale network," IET Renew. Power Gener. Vol 1,pp.1-19, 2024.

[13] L. Kanagasabai, "Novel empress SARANI optimization algorithm for active power loss reduction and voltage stability enhancement," Heliyon, vol. 10, no. 22, pp. 1-64, 2024.

[14] H. Yapıcı and N. Çetinkaya, "An improved particle swarm optimization algorithm using eagle strategy for power loss minimization," Math. Probl. Eng., vol. 2017, pp. 1–11, 2017.

[15] V. B. Hosagoudar and A. Sabeena, "The genus Asterina (Asterinaceae) on the members of Myristicaceae in Kerala State, India," J. Threat. Taxa, vol. 3, no. 10, pp. 2143–2146, 2011.

Inventions based on Flywheels In-pipe Cascading Turbines Hydrogen Production with Storage Hybrid Perpetual Mechanical Battery Technologies working in Relay

Dr Srinivas Bhaskar Chaganti
KAMMA Gear Flywheel Pvt Ltd Wealth Creation
Plot No 18, Rasoolpura, Secunderabad, Telangana, India. 500003
Email- kammagearflywheel@gmail.com

Abstract- KAMMA (Kinetics Associated Mass Mechanical Applications) started this research institute in the year 1913. This is the world's first family oriented research organization to invent new techniques and technologies focusing on the Mechanical Battery Department, for creating Green Inertia in standalone energy storage systems in individual power plants. Where you generate there you distribute, small grid smart grid is the backbone for this invention. These Inventions are for green Wealth Creation. These inventions in Hybrid Perpetual Mechanical Battery are either in single or in relay when applied with proper techniques in power electronics, will revolutionize the power storage & generation resulting in uninterrupted pure, clean & green electricity round the clock. Blending Time with Weight, Inertia is created, Time and weight are not subjected to thermodynamics laws directly, KAMMA research materials for generating inertia are Sand, Ash, Ice, Stone, metal etc which are not categorized under fossil fuels and are not subjected for Global Warming & pollution. These mother particles are not subjected to entropy in a direct way because, these materials are fixed inside a cavity or a box or a apparatus or a container or a place to fix them only as a static weight and dead weight. These particles associated with the materials fixed inside the container are not touched & disturbed & subjected to decay. i.e. the mother particle & its weight are always intact and will never change its fundamental nature. In KAMMA research area, Time, Weight, Diameter, Rpm, Acceleration, Deceleration without disturbing the geometry and deformations and decay, are the main core raw materials for storage of inertia & power generation.

Keywords— Time, Weight, Sand, Ash, Ice, Stone, Metals.

I. INTRODUCTION

A case study comparison between other Existing Technologies and our invention in the field of Mechanical Battery Department: This innovation changes the dynamics & direction of Time and a new method to utilize the time which is associated with Arrow of Time because time is getting divided and multiplied in this method of power generation. The Arrow of Time is in the favour of individual pulse time divided & distributed in seconds by dividing the 3600 seconds fixed time capsule. In this technology time is sustain released. In our technology weight and time are bridged together. In Thermodynamics the time is measured in a straight line present to future and is capsule in 3600 seconds of one individual pulse time, which is fixed inside a boundary without any divisions and without any stop. Our innovation claim is Weight, Time, Inertia, Force, Surface speed acceleration & deceleration plays a critical, vital and crucial key role in the Pure GREEN Energy Sectors. According to our innovation Time & Weight are useful for the generation of green Inertia i.e. electricity and we claim and consider these as the core raw materials. As we all know, there are quite a number of technologies using fossil fuels, which are very quickly depleting from the planet because they are limited in nature & they are developed over a period of time and to replace them will also take a long period of time & human effort. The process involved in the development of the fossil fuel industry has a history of more than 200 years. Today the world is experiencing a shortage in the reserves of the naturally available fossil fuels and globally, experts are searching for alternate methods to maintain their peak load requirement of their industrial & commercial sectors, because every government in the world is 100% dependent on the revenue generations of their individual countries. The only source to run any country's economy is ENERGY. In the recent times there is a special type of attraction towards Renewable Energies like Solar & Wind with low PLF working methods and which cannot meet the country demand 24 x 7. The main criteria should be focused on the Plant Load Factor (PLF) for each & every

979-8-3315-3013-6/25 $31.00 © 2025 IEEE

individual power plants situated at different corners of the country by taking the topography & the required common factors which influence the performance of Solar & Wind Power Plants. In reality, if we see the PLF of Solar Plant is around 13% and in winter there is no solar power generation for 100 days in a year and Wind Plant is around 22% in a year. **Thermal Power Plant** will use coal as a basic raw material. When the coal particles (total weight) are subjected to decay and destruction by disturbing the internal arrangement of the nuclei structure by pushing out of the stored energy due to internal developing of heat, then entropy will play a major role for the released of energy. Thereby this proves that 100% energy cannot be utilized and by the released energy once again coal particles cannot be regenerated from the ash. **Hydro Power Plant** will use water as a basic raw material. When the water particles (weight) are subjected to differential head thereby the water which flows out of the dam and joins the river cannot be pumped using the same electricity because the water which has to be pumped up has a extra head which is called pump head. The differences in the generation head and pump head clearly indicates that more amount of electricity is consumed for pumping tail discharged water back into the dam. Thereby Entropy plays a major role in the generation head losses and pump head losses. **Nuclear Power Plant** will use enriched uranium as basic raw material. In this technology the particles are bombarded and multiplied thereby the energy is released in the form of heat and heat is used as inertia and energy is generated. In this technology the decay of the mother particles of uranium are completely destroyed. $E = MC2$. E is Energy, M is Mass & C2 is the speed of light. Thereby if the enriched Uranium releasing energy is very high even then by using the same output energy the consumed Uranium particles cannot be reconstructed and use again & again.

Solar Power Plant will use heat associated with light (moon light does not exhibit the same heat even after it receives the sun light). The sun rotates in the galaxy on its own axis and the earth revolves around the sun. Thereby the sun's rays are always changing the axis and angle of penetration inside the earth atmosphere due to the ticking of the clock. Thereby the entropy of the heat is clearly felt in the surrounding areas where the solar panels are fixed. Thereby the same solar plant cannot use the same heat & light of the sun in the day time hour after hour. Due to the rotation of the earth the sun's ray cannot penetrate the panel in the same angle from morning 6 am to evening 6 pm. **Wind Power Plant** uses air flow velocity as the basic raw material to rotate the nacelle connected with the blades to rotate the generator. Here the wind flows only on certain occasions but not on regular basis. The wind is formed due to seasonal changes and the rotations of earth. The same wind particles and velocity cannot be used again and again and generate electricity. The entropy of the wind velocity before hitting the blades is diluted and a partial velocity is absorbed by the blades. Thereby the same wind velocity cannot be recreated by using the electricity generated out of rotating blades of wind power plant. This is all about Thermodynamics because the laws are

adherent to only certain mother particles and particle density and tensile strength which is the inherent qualities of those particles which are associated with the laws of Entropy. Our innovation rotates around Principles & Laws of **Mechanical Battery.** We use time and weight as core raw material. In this technology time is divided into individual pulses and one individual pulse is fixed in a time frame associated with weight. The geometry of this weight is always constant. The greatness of Mechanical battery is observed in the area of time and weight. Here the time is not changed or altered, the weight is not changed or altered. Time and weight remain constant in the past, in the present and in the future. In this technology the mother particle hidden inside the weight is not altered or disturbed and is not touched and consumed nor subjected to decay. The inertia generated by using the individual pulse time and weight are the same and by using the generated inertia i.e. to generate 1 MW (1000 kWh) of pure green electricity 1600 ton steel is required. To manufacture 1 Ton of pure green steel 1 MW (1000 kWh) pure green electricity is used, so to manufacture 1600 Tons of pure green steel 1600 MW pure green electricity is used, then this process is continues and perpetual in nature. There is no requirement of any Fossil Fuels and no pollution. Here our innovation is clearly showcasing the Mechanical Battery positive approach about having the capacity to completely stop entropy and the energy remain same in the past, present and future. The Mechanical Battery is not affected by the Laws of Thermodynamics, because time remains the same (individual pulse) weight remain the same and the weight associated with the geometry remain the same and the RPM associated with time and weight remain the same. Entropy will not have any effect on these parameters. One hour of Flywheel Power generation will generate 1 Ton of steel and to generate 1600 Tons of steel 1600 hours of power generation is required i.e. 2 Months 6 Days. And on each and every 2 Months 6 days 1 MW Flywheel power plant is manufactured. In one year 5.5 MW Flywheel Power Plants are established. In 1 MW power plant 8760000 kWh (8760 MW) pure green electricity is generated which will produce 8760 Tons of steel in one year. By using 5.5 MW pure green electricity, 48180 MW pure green electricity is generated which produce 48180 Ton of steel. By using 48180 Tons of steel 30 MW of Flywheel Power Plants are manufactured.

PERPETUAL PRODUCTION OF STEEL (Table-A)

Year	Installed Capacity	Units Generated	Remarks
1st year	1 MW	8760000 kWh / Yr	To manufacture 1 MW Flywheel Power Plant 1600 Tons of steel is required
		8760 MW / Yr	Using 1 MW electricity 1 Ton steel is manufactured. 8760 MWs = 8760 Tons of steel is manufactured
			8760 Tons of steel can establish 5.5 MW Flywheel Power Plant in the 1st year

2nd year	5.5 MW	5.5 x 8760 = 48180 MW / Yr	
		48180 MW / Yr	Using 48180 MW electricity 48180 Tons of steel is manufactured. 48180 Tons of steel can establish 30 MW Flywheel Power Plant in 2nd Year which is **545 % increase.**
3rd year	30 MW	30 x 8760 = 262800 MW / Yr	Using 262800 MW electricity 262800 Tons of steel is manufactured. 262800 Tons of steel can establish 164 MW Flywheel Power Plant in 3rd Year which is **2982 % increase.**

CARBON FOOT PRINT (Table-B)

Year	Installed Capacity	Steel	CO2 Emitted
1st year	1 MW	1 Ton of steel	Emits 2.68 Tons of CO2
		1600 Tons of steel	Emits 4288 Tons of CO2
Therefore in 1 yr 8760 Ton of steel is generated by 1 MW of Flywheel Plant.			
Less 1600 Tons implies 7160 Tons of steel is pure green steel by Flywheel Power Plant.			

Flywheel Power Plant : In flywheel power plants there is no use of fossil fuels. Electricity is the prime mover in Flywheel Technology. In Flywheel Technology the prime mover is Motor and Motor consume electricity and electricity will not change forms. By the consumption of electricity the flywheel is rotated, when flywheel is rotated bearings are rotated and when bearings are rotated inertial is created and when inertia is created generator is rotated and when generator is rotated electricity is generated. Therefore input is electricity and output is also electricity. By this we have to understand the raw material used for rotating the flywheel is electricity and when we rotate the generator using flywheel here too electricity is generated. There is no change in the form in this Technology. Input is Inertia and exit of input is Inertia.

CHANGE IN FORM (Table-C)

Sr No	Type of Power Plant	Prime Mover	Remarks
1	Thermal Power Plant	Coal to heat	Change in form
2	Hydro Power Plant	Flow pressure & velocity of Water	Change in form
3	Nuclear Power Plant	Heat from Uranium	Change in form
4	Solar Power Plant	Light & Heat from sun	Change in form
5	Wind Power Plant	Velocity from wind	Change in form
6	Flywheel Power Plant	Electricity to Electricity	NO Change in form

If we really understand Inertia in any technology injecting inertia is the ultimate goal for the generation of electricity, thereby inertia cannot change its form. Using any raw materials is the ultimate goal to generate inertia, thereby input is inertia and exit of input (output) is inertia. There is no change in the form of total inertia.

II. METHODOLOGY

A. Flywheels, In-Pipe Cascading Turbines, & Hydrogen Storage, Hybrid Perpetual Mechanical Battery Techniques & Technologies working in Relay continuously

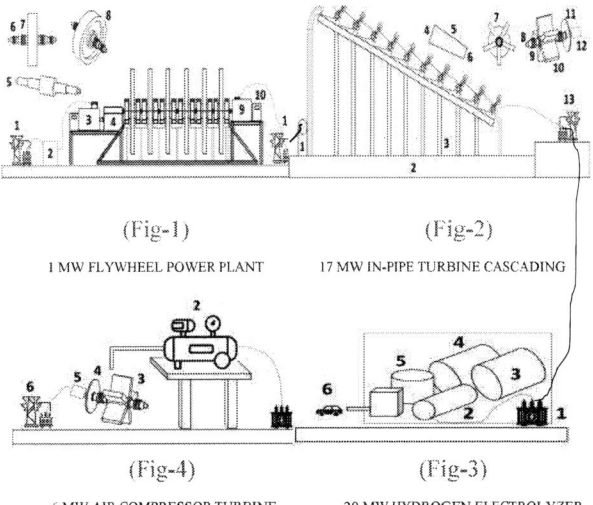

(Fig-1)
1 MW FLYWHEEL POWER PLANT

(Fig-2)
17 MW IN-PIPE TURBINE CASCADING

(Fig-4)
6 MW AIR COMPRESSOR TURBINE

(Fig-3)
20 MW HYDROGEN ELECTROLYZER

B. 1 MW Flywheel Power Plant : (Fig -1)

(1) Any available external power source on site (2) VFD 1800 kW (3) PMG Motor 1800 kW (4) Dual Gear Box-3000 kW each (5) Forged Shaft 1.5 meter in dia, 7 mtr long, 43 Ton weight (6) Spherical Roller Bearing ID-820

mm, Bearing Number - 316341 / HA4 – 2 Nos & 322497 / HA4 – 2Nos for each Flywheel (7) Spherical Roller Bearing ID-1400 mm, Bearing Number BC4 - 8042 / HA4 – 2 Nos for each guide wheel (8) Heavy Rim type Forged Flywheel 6 meter in dia, 215 Ton weight (9) Generator 1500 kW – 2 Nos (10) Hybrid Dual Stage String Inverter 2000 kW – 2 Nos.

Same identical set of 6 Flywheel Power Plants are installed within a individual Power Plant site.

FINITE ELEMENT ANALYSIS FOR FLYWHEEL

Material Construction & Properties for Analysis

Part	Material	Temp (^{0}C)	Elastic Modulus (Mpa)	Allowable stress (Mpa)	Yield Strength (Mpa)	Density (Kg/m3)	Poisson's Ratio
Flywheel	4340 steel	40	205E3	227	680	7850	0.285
Shaft	1018 steel	40	205E3	114	280	7850	0.285
Bearing	Chrome steel	40	205E3	227	680	7850	0.285

Calculation of Energy stored and surface velocity

Energy Storage
Energy Stored = 0.5 x Moment of Inertia x (Angular Velocity) ^{2}
(Moment of Inertia for flywheel) = 995065 kg/m2
(Moment of Inertia for shaft) = 8492 kg/m2

LC1 700 RPM
Energy Stored in flywheel = 0.5 x 995065 x (73.3) ^{2}
= 2.67 GJ
Energy Stored in flywheel = 0.5 x 8492 x (73.3) ^{2}
= 0.0228 GJ

LC1 700 RPM
Surface Velocity Flywheel = (2 π r x RPM)/60
= (2 x π x 3 x 700) / 60 =219 m/s
Surface Velocity Flywheel = (2 π r x RPM)/60
= (2 x π x 0.75 x 700) / 60 =55 m/s

Time to accelerate from 0 to 700 RPM

Motor Power = 1800 KW
Gear Box = 1:10
Angular velocity at 700 RPM = 73.3 rad/s
Energy need to reach 700 RPM (E = 269.3 MJ)
Time required = Energy required to reach 700 RPM / power
= 269300000/1800000
= 149.6 Seconds

Time to decelerate from 700 to 0 RPM

Angular Velocity = 2π x 700/60 = 73.3 rad/s
Moment of inertia for flywheel = 995065 N/m2
Breaking torque = 1000 Nm (Assumed for air & bearing)
Time = Moment of inertia x angular velocity / (breaking torque)
= 995065 x 73.3 / 1000
= 72989 Seconds = 20.3 hrs
Energy Stored in flywheel at 700 RPM is 2691 MJ
Energy available in the decelerating flywheel

RPM	Angular Velocity (rad/s)	Energy (MJ)
700	73.3	2691
500	52.4	1376
300	31.4	478
100	10.5	44
0	0	0

Allowable Stress Flywheel = 680 MPa
Allowable Stress Shaft = 280 MPa
For conservative purpose we have considered allowable stress = 0.9*680 = 612 MPa for flywheel.
For conservative purpose we have considered allowable stress = 0.9*280 = 252 MPa for shaft.

Equivalent Stress

Load Case	Location	(MPa)	Allowable Stress (MPa)	Result
LC1	Max stress at Flywheel	433.84	612	Pass
	Max stress at Shaft	9.27	252	Pass

Conclusion : For the particular RPM speeds of the flywheel the stresses are under yielding limit.

C. 17 MW In-Pipe Turbine Cascading : (Fig -2)

(1) One Vertical Turbine Pump 6000 kW, (2) Bottom Water Canal 1100 meters long, 3 meter wide and depth 1 meter (3) Vertical support pillars – first pillar 14 meter height and last pillar is ½ meter height in this design pillars height is gradually reduced according to the slope of the hypotenuse, (4) 10 mm thick Steel pipe - 3 meter in dia and pump discharge in to the 1st inlet pipe is 30 m3/s with a velocity of 4.24 m/s, (5) Total Length of Steel pipe is 100 meter, from 80 meters to 100 meters (20 meter) the 3 meter dia pipe is gradually reduced in the end to 1.8 m in diameter in a cone shape outlet (6) Steel pipe 1.8 meter dia, 30 m3/s outlet discharge at velocity of 11.79 m/s, (7) Turbine 1.7 MW, 750 rpm (8) Forged Shaft 5.9 mtr long, 16 Ton weight (9) Spherical Roller Bearing ID-600 mm, Bearing No: BC4B 322497/HA4 (10) Turbine blades height 1800 mm, width 2000 mm, thickness 450 mm, weight 62 Ton, diameter is 4.6 meter (11) Forged Guide wheel 3 meter in dia, 5.5 Ton (12) Generator 1700 kW (13) Transformer 2000 KVA.

VELOCITY CONTOUR - 16.2 m/s

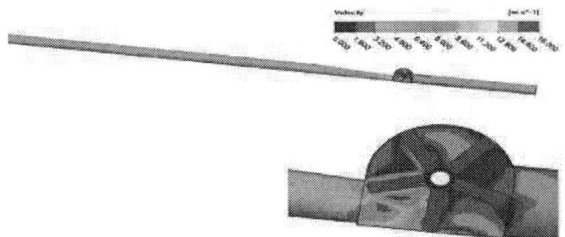

Cross-sectional Area – 1.8 m2
Flow Velocity – 16 m/s
Flow Discharge – 30 m3/s
Water Density – 998 kg/m3
Efficiency – 59% , Power output – 1700 kWh.

PRESSURE CONTOUR - 144557 pa.

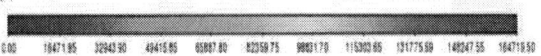

TOTAL PRESSURE ON TURBINE WALL – 166379 pa.
AREA OF WALL – 3135509 mm2
FORCE – 521.47 KN
TORQUE – 1147.25 K Nm
RADIUS OF TURBINE – 2200 mm

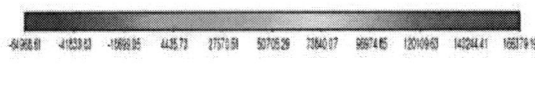

(Fig -5)

CFD ANALYSIS FOR IN-PIPE TURBINE CASCADING

D. 20 MW Hydrogen Electrolyzer : (Fig -3)

(1) Input Power supplied from In-Pipe Cascading Power plant, (2) Hydrogen Electrolyzer 1500 kW, Total 13 Electrolyzers on site (3) Hydrogen Storage Tank 1000 Tons (multiple tanks) (4) Oxygen Storage Tank 500 Tons (multiple tanks) (5) Alkaline Water Storage Tank 10000 liters (multiple tanks) 9.5 pH   (6) Transportation to Hydrogen Bunks.

E. 6 MW Air Compressor Turbine Plant : (Fig -4)

(1) Input Power supplied from In-Pipe Cascading Power plant, (2) Air Compressor 1 MW, Air flow 3000 cfm with 100 Bar Pressure used for power generation, on site there will be 6 plants of 1 MW each (3) Air Compressor Turbine 1 MW, 750 rpm (4) Forged Guide wheel 3 meter in dia, 5.5 Ton weight (5) Generator 1000 kW (6) Emergency back-up power.

III. SUMMARY OF THE INVENTION

A brief description of our invention, points to be remembered while understanding the method of pulse based technology power generation. This technology will use individual pulses as the method to store Inertia in the acceleration mode and harvest the discharging stored Inertia in the deceleration mode connected to a generator to generate electricity. This innovation is developed by designing a method in which 3600 seconds fixed time frame which is called as kWh in the engineering calculations is considered and this 3600 seconds individual time frame is divided into individual input pulse time frame which can be measured in between 1 sec to 60 sec of fixed input pulse time. Therefore in one 3600 seconds fixed time frame there will be multiple individual input pulses. This is the World's first innovation in which input pulse time is divided and exit input pulse time is multiplied to many folds thereby more Inertia is created & generated.

Description of one individual input pulse time and exit of the same individual input pulse time. The sequence of time flow in between (i) Input individual pulse (Inertia) time and (ii) Exit of input individual pulse (output Inertia) time. i.e. by using each and every individual input pulse time, the prime mover will start accumulating and storing the maximum compressed Inertia available in the rotating mass within the accepted geometrical figure without any deformations in every individual design. At the time of acceleration 0 to 700 rpm which is measured and fixed as 100% available stored Inertia. From this, 30% stored inertia is harvested as a result from the exit of the input pulse time during deceleration process and remaining 70% Inertia is left behind as it is in the rotating mass, which will be of great help for the prime mover to continuously rotate the mass or weight which is fixed within the cycle of individual pulses for consuming less electricity. Each individual pulse is having a fixed time capsule which can be measured in between at-least 1 sec to 60 sec time. Each output individual exit pulse can be measured in between at-least 1 sec to 600 sec time. In one individual pulse for example if the prime mover is consuming 1 sec or 30 sec of energy then the energy consumed by the motor time pulse is only 1 sec or 30 sec, Power = Energy / Time. So the energy consumed by the prime mover should be divided by time to get the exact consumption of energy. In general, industrial practice prime mover motor is operated for 3600 seconds continuously. This difference has be to understood. Till now there is no industrial record all over the globe for using motor (Inertia) connected to a generator (Inertia) to generate electricity on a commercial industrial scale. As for the exit input pulse time the rotating mass i.e. flywheels stored compressed energy (Inertia) is harvested by the generator because the flywheel shaft and generator shaft are directly coupled with each other by using a appropriate gear coupling which in turn rotate the rotor of the generator and help in getting the inertia rolling inside the rotor of the generator to generate electricity. Here Power = Energy x Time. The Inertia which is rolled inside the generator from a set of flywheels will rotate the generator and generate electricity for 600 sec or 900 sec, depending upon the availability of Inertia in the rotating mass. This is the exact operation description available within one individual pulse time and when we divide 3600 seconds a fixed pulse time into many divisions of small intervals of 1 sec or 30 sec or 60 sec the individual pulse time available will be, for example 1 sec will have 3600 individual pulses or if it is 30 sec it will have 120 pulses (3600 sec / 30 sec = 120 individual pulses). For the generator exit pulse time will depend on the decelerating Inertia time available within 100 rpm out of the total maximum rpm which is 700 rpm. Therefore the inertia available in the deceleration rotating mass at 700

rpm to 600 rpm (at intervals of 100 rpm) is harvested which accounts in between 600 sec to 900 sec time. The prime mover will accelerate only up to 100 rpm, then the rotating mass will decelerate to discharge its stored inertia in the first 100 rpm. Therefore 100 rpm up and 100 rpm down is in the individual fixed pulse time to measure the input consumption of energy and output exit of the energy from the generator. In this way all the individual pulses will be calculated for the total energy consumed and total energy generated. For every individual pulse the same method of calculation is repeated which are available in the 3600 sec fixed pulse time.

In this design total 6 MW Flywheel Power Plants are established.
For each 1 MW Flywheel Power Plant these items are required. (1) Heavy Duty Variable frequency Drives (items with VFD, Motion controller, PID, Encoder, HMI, PLC, Control panel, AC Drive) - capacity 1800 kW - **01 No.** (2) 8 pole, 750 RPM PMG Motor capacity - 1800 kW, 11000 volts - **01 No.** (3) High Speed Gear Box 10 : 1 ratio (speed reducer) and 1: 10 ratio (speed increasing), capacity - 3000 kW Torque - **02 Nos.** (4) 6 Meter dia Heavy Rim type Forged Flywheel P275NH / 1.0478 steel grade, weight - 215 Tons each, inner bore dia size-1500 mm. Total no. of Flywheels - **06 Nos.** (5) Shaft of the Flywheel is 1500 mm diameter (Force Fit) of length 7 meters 43 Tons. Total shafts - **06 Nos.** (6) Spherical Roller Bearing 4 Row: BC4 - 8042 / HA4, Inner dia -1400 mm, Outer dia - 1780 mm, width - 1200 mm, Dynamic load - 52300 KN, limiting speed - 220 RPM. 2 Nos bearings on each shaft for two flywheels. Total no. of Bearings - 4 **Nos.** BC4B 316341 / HA4, Inner dia - 820 mm, Outer dia - 1100 mm, width - 745 mm, Dynamic load - 21200 KN, Static load – 57000 kN, Limiting speed - 480 RPM - 2 Nos. BC4B 322497 / HA4 , Inner dia - 600 mm, Outer dia - 870 mm, width - 578 mm, Dynamic load - 13200 KN, Static load – 31500 kN, Limiting speed - 700 RPM - 2 Nos. Total for flywheels 4 Nos bearings each shaft. Total no. of Bearings - 16 Nos. (7) Plummer Block for Flywheel bearings - **20 Nos.** (8) 8 pole, 750 RPM Generator capacity - 1500 kW, 11000 Volts - **02 Nos.** (9) Hybrid Dual Stage String Inverter 2000 kW, 11000 Volts - **02 Nos.** (10) Custom Designed Big display Energy Meters, RPM tracker, Temperature measurement meter, Vibration measurement meter, Overall Electronic Monitoring Board. (11) Sub Station HT 11 KV/ 33 KV.

In this design total 17 MW Turbine Generators are established.
In-pipe turbine with cascading effect using Bernoulli's Theorem to arrange the flow of water from one individual in-pipe turbine to another individual in-pipe turbine in a cascading method having 10 individual pipes fixed on the hypotenuse and total length is 1000 meters of a right angled triangle. The vertical pump head is 14 meters in height and the pump capacity is 6 MWs. The pump will lift

30 m3/sec water flow from the pump discharging outlet fixed at 14th meter vertical height. The 30 m3/sec with 4.24 m/s velocity water will flow into a 3 meter inlet dia steel pipe which has a length of 100 meters and from 80 meters to 100 meters (20 meter) the 3 meter dia pipe is gradually reduced to 1.8 m in diameter representing a cone shape. From 1.8 meter in diameter steel pipe will discharge 30 m3/sec with a velocity of 11.79 m/sec directly into the 1.7 MW turbine blades. The turbine is fixed straight in the path of the flowing water jet and the turbine is having 5 blades. Each individual blade is of height 1800 mm, width 2000 mm, thickness 450 mm. Total weight of the turbine is 62 Ton and turbine diameter is 4.6 meter. This turbine will rotate at 750 rpm and is connected to a 1700 kW Generator. In the cascade there will be 10 identical turbine generators and generate 17 MWs per hour. The base of the 10 power plants fixed on a cascade has a 1100 meter long canal and this canal is the main source for the 30 m3/sec water to get pumped and lifted to 14 meter vertical height. The pump used in this design is 6 MW. This 6 MW pump consumed power is supplied by the Flywheel Technology.

In this design total 6 MW Air Compressor Turbine are established.
The volume capacity of each compressed pneumatic cylinder is 3000 cfm, and the pressure of air released will be at 100 Bar and this volume and pressure will rotate a 1 MW Air Compressor Turbine connected to generator and generate 1 MW per hour. To generate 3000 cfm air flow 1 MW pneumatic pump is required and 1 MW is consumed every hour. By using 1 MW pneumatic pump 6 individual cylinders with 3000 cfm air flow capacity are filled in 6 hours i.e. one cylinder every one hour. Thereby there will be 6 standby cylinders which will be used in an emergency. For example : This stored energy is used as a standby power generation arrangement at the time of overhauling of the flywheel power plant or in-pipe cascading power plant without any interruption in injecting Inertia into the grid.

In this design total 20 MW Electrolyzers are established.
The Flywheel Power Plant and the In-pipe Cascading Power Plant and the Compressed Air power will be used for the production of GREEN Hydrogen using 1500 kW Electrolyzer manufacturing machine. In total 20 Electrolyzers are used in this design for uninterrupted production of 30000 kW of green Hydrogen. To produce 1 kg of green hydrogen 55 kWh pure green electricity, free from carbon foot print is consumed from the self generated green electricity generated in the Power Plant site. The voltage supplied will be in between 900 V to 1300 V & 10 ltrs of alkaline water is used every hour. By using the produced onsite green Hydrogen we can transport to all coal power plant or fossil fuel based power plants for the replacement of fuel with green Hydrogen. Per every Kg of pure green Hydrogen will generate uninterrupted 23 kWh of free electricity and supply to grid. This method of power generation through Mechanical Battery can be implemented in all of the various types of fossil fuel power plants.

IV. VISION AND MISSION

Our Vision & Mission is for Wealth Creation and to eradicate Global warming and remove addiction on fossil fuels. Our Commitments & responsibilities as Global Citizens is 'One Planet One Technology One Tariff One currency' & the commitment to our Motherland is 'One India Green India One Nation One Technology One Tariff One Rupee' per unit of pure green electricity and 100 rupee per kg pure green hydrogen supplied any corner of the globe.

V. CONTRIBUTION OF THIS INVENTION

The main objective of this invention is purely designed for Wealth Creation. Introducing Standalone energy storage system using the principles of hybrid mechanical batteries that provide round the clock, throughout the year, an uninterrupted pure & green energy supply in any country.

Secondly, to protect our Mother Earth and Mother Nature from getting exposed to the harmful gases and minimize excavation/ mining of natural resources.

Third, to promote and bring awareness in public on new innovations/ green technologies in the field of Mechanical Battery Energy Sector and provide free electricity through stand-alone energy storage system for supporting various Agricultural, Industrial, Commercial, Logistics & Domestic Sectors.

Finally, this invented technology amplifies the National Economy to next levels and meet all the Global Norms on NET ZERO EMISSIONS/ ZERO CARBON FOOTPRINTS by removing the addiction & dependency on fossil fuel technologies and bring prosperity to Nation in terms of Carbon Credits thereby wealth creation is achieved.

Plant Load Factor (PLF)

Plant Load Factor of various Power generation techniques & technologies. (Table-D)

Method of Power Generation	Plant Load Factor
Thermal Power Plant	47.50%
Hydro Power Plant	39.10%

Method of Power Generation	Plant Load Factor
Nuclear Power Plant	63.50%
Wind Power Plant	34.80%
Solar Power Plant	18.00%
KAMMA Gear Flywheel Power Plant	98.00%

VI. CONCLUSION

Using the Laws of Thermodynamics Wealth is destroyed. We are exhausting the natural resources of coal and oil, thereby we are burning money. Every second we are losing Flora & Fauna, we are destroying forests in the name of mining, in the name of ethanol, we are increasing pollution thereby generating Global Warming thereby change in the seasons ultimately wealth destruction. Fossil fuels of different category have made an destructive impact on the globe. Every country backbone and rescue in this hard times are renewable energy sources and these too have utterly failed, because of their inbuilt Plant Load Factor (PLF) when compared with the fossil fuel PLF renewable energies cannot do justice and stand tall to replace the entire fossil fuel industry on this planet. Here the entire world is experiencing global warming, heat waves, change in the climatic conditions & atmosphere, agricultural associated issues are badly reflecting on human health conditions all over the world. People are experiencing various changes in their life cycle because of the over populated issues per square KM of available land. Population growth is a very serious condition in the present state where unemployment & pricing of electricity for industrial growth is hampering all countries equally, because fossil fuels are getting depleted very rapidly. If we replace the entire world of fossil fuels and run each country's economy on the basis of renewable energy technologies the world is heading towards domes day. Therefore discussion on reality and talking about truth is very important in this need of hour. To make this happen and bring radical changes in the GDP of all countries on the Globe, KAMMA Gear Flywheel Pvt Ltd had done extensive & exclusive research on Hybrid Perpetual Mechanical Battery Techniques & Technologies and developed 25 verticals in the Mechanical Battery Department. This paper is concentrated on the development of different type of Mechanical Battery applications and interlinking various mechanical battery technologies in a relay, thereby the perpetuity is maintained and is having a positive effect by using the method called "Stand Alone Energy Storage Systems". In this technology the main purpose of interlinking & inter connecting various Mechanical Batteries is to remove & eradicate various fossil fuel industries in all the countries. Thereby the overall development in every sector can be, and will be achieved because there is no purchase of raw material and there is no wastage of people's hard earned money and money is not burnt, Over all taxes can be reduced. Thereby Wealth is created.

This innovation & innovated design will facilitate to reduce the cost of conventional power generation methods, power transmission, power distribution costs and increase the efficacy, which can lead to significant cost savings in the long run". This innovation & innovated design directly impacts the Global GDP ecosystem on Global Economy Forums and enhance green eco-friendly atmosphere and generate huge employment opportunity in various Sectors all over the globe. This innovation & innovated design helps in obtaining positive Carbon Credits & others gather negative carbon credits to the nation.

REFERENCES

[1] Our website : www.kammageaarflywheelpowergenerations.com and www.kammagearflywheelpowergeneration.com
[2] Link: https://www.youtube.com/watch?v=15fv3GS2Tqo https://youtu.be/Mf4lafvc2nM
[3] KAMMA Gear Flywheel Power Plant Technical Paper Published in 2023 9th IEEE India International Conference on Power Electronics (IICPE) online IEEE Xplore portal. Dated of conference : 28-30 November 2023. Date published on IEEE Xplore portal : 27 March 2024. https://ieeexplore.ieee.org/document/10474946.

An Efficient Soft-Switching Pulsed Power Supply Converter for Hydrogen Production

Shivam Singh
Department of Electrical and
Electronics Engineering
BIT, Mesra
Ranchi,Jharkhand
mtee10005.23bitmesra.ac.in

Dr. S.Shiva Kumar
Department of Electrical and
Electronics Engineering
BIT, Mesra
Ranchi,Jharkhand
shivkumar.ee@bitmesra.ac.in

Dr. T Ghose
Department of Electrical and
Electronics Engineering
BIT, Mesra
Ranchi,Jharkhand
tghose@bitmesra.ac.in

Abstract—Alkaline water electrolyser (AWE) has traditionally been recognized as a tested and trusted technology for the production of hydrogen. However, its widespread implementation has been hampered by comparatively low rates of hydrogen production against high capital and operational costs, making it less economically viable for large-scale consumer applications. In response to this issue, this paper presents a novel solution that utilizes a soft-switching full-bridge pulsed power supply converter to significantly improve the hydrogen production efficiency of AWE systems. A comparison of the performance of AWE under constant DC power supply and pulsed power supply clearly shows that the pulsed power approach produces significantly higher hydrogen yields. This is because of lower overvoltage losses during the process of electrolysis. Efficacy of the proposed approach is validated through rigorous MATLAB simulations, which show a significant increase in hydrogen output with energy efficiency. The proposed approach is a promising direction for improving the economic competitiveness and viability of AWE systems in the new hydrogen energy economy.

Index Terms—alkaline water electrolyser, pulsed power supply, soft-switching

I. INTRODUCTION

HYDROGEN has emerged as a key future energy carrier due to its zero-carbon footprint, high energy density, and ability for long-term storage. It is a significant component of the global transition to clean and renewable energy systems with the potential to make significant contributions to the decarbonization of industrial processes and the reduction of fossil fuel dependence. Of the technologies available for hydrogen production, water electrolysis is particularly beneficial as it can be integrated with renewable energy sources (RES) like solar, wind, and hydropower, thus enabling sustainable hydrogen production with minimal environmental impacts.

Alkaline water electrolysers (AWEs) are a leading technology for the production of hydrogen since they have proved to possess technical competence, reliability, and relatively lower production costs compared to other technologies, such as proton exchange membrane (PEM) and solid oxide electrolysers (SOE). Despite such advantages, AWEs have one major

Achilles' heel: inefficiency on a large scale.Such inefficiency is predominantly attributed to losses in energy, especially those which arise from overvoltage resulting from high electrical impedance between electrodes during the electrolysis process. Such losses have continued to drive high hydrogen production costs and in turn discourage its competitiveness with alternatives based on fossil fuels.

A number of approaches have been explored to improve the efficiency of AWEs. Multimode electrolysis methods combining DC and pulsed power have demonstrated efficiency improvements in low-load operation, opening the door to more sophisticated power supply methods [1]. Thermodynamic and electrochemical modeling has been used to explore overpotential and energy loss in AWEs, highlighting the importance of overcoming efficiency issues at scale [2]. Optimal pulse-width modulation (PWM) has been used in AWEs driven by intermittent PV energy, demonstrating efficiency improvements under variable renewable energy operation [3].

Recent power electronics trends have further emphasized the significance of converter topologies in maximizing electrolyzer efficiency. More recent transistor-based converters have been found to minimize energy consumption in comparison to the conventional thyristor-based converters [4]. High step-down DC-DC converters have been designed to achieve minimum current ripple and maximize hydrogen production performance [5]. These developments emphasize the necessity for better power supply systems. Conventional AWEs usually employ constant DC power supplies, which restrict the efficiency of the electrolysis process by failing to maximize the dynamic potential of the system. Recent studies have also shown the advantages of maximizing pulse current magnitude and duty ratio to maximize hydrogen production efficiency at low-load conditions [6]. Most studies, however, concentrate on particular operating conditions, leaving a gap in solutions for full-load operation across industrial scales.

Recent technological improvements in the field of DC-DC converter technology have included optimizing hydrogen

979-8-3315-3013-6/25 $31.00 © 2025 IEEE

production systems integrated with photovoltaic (PV) sources and energy storage systems. A new current-fed-out resonant-type DC-DC converter has been developed, utilizing an LCC resonant tank to ensure zero-voltage switching (ZVS) over a wide operating range. The design minimizes current ripples and is efficient in handling interactions between PV cells, energy storage, and the electrolyzer even under varying power conditions [7]. Another approach involves the application of an LCL-resonant tank with variable frequency and phase-shift modulation (VFPSM) to reduce backflow power and enable soft-switching. The approach improves the performance and lifespan of the electrolyzer and optimizes hydrogen production rates under advanced modulation schemes. These advancements highlight the importance of resonant converter designs in addressing efficiency challenges common in off-grid PV-electrolysis systems [8]. In this paper, we introduce a soft-switching full-bridge pulsed power supply converter that is particularly tailored to enhance AWE efficiency on an industrial scale. In contrast to conventional methods that focus on high-frequency pulses, our approach employs low-frequency pulsed power, which is better for large systems and does not require external physical augmentations that increase complexity. Through simulation in MATLAB, we show that the proposed system greatly enhances rates of hydrogen generation over conventional constant DC supplies, with an economical and efficient solution to commercial AWE systems. Our research is anticipated to contribute to more economically feasible hydrogen production technology as part of the world's pursuit of sustainability.

II. METHODOLOGY

Thermodynamics plays a vital role in characterizing reaction equilibria and thermal phenomena within electrochemical reactors. It also forms the foundation for defining the driving forces underlying transport processes in electrolytes and provides a framework for describing the behavior of electrolyte solutions. For a deeper understanding of the fundamental equations governing electrochemical systems such as electrolyzers, readers may refer to the foundational literature.

This section offers a concise overview of the thermodynamics involved in low-temperature hydrogen-oxygen electrochemical reactions, as utilized in the electrolyzer model. The analysis assumes:

1) Hydrogen and oxygen behave as ideal gases.
2) Water is treated as an incompressible fluid.
3) Gas and liquid phases remain distinct and non-interacting.

The overall water-splitting reaction can be written as:

$$\text{H}_2\text{O (l)} \rightarrow \text{H}_2\text{(g)} + \frac{1}{2}\text{O}_2\text{(g)} \tag{1}$$

At standard conditions (25°C and 1 bar), the thermodynamic parameters for the reaction are:

- Enthalpy change (ΔH): 285.83 kJ/mol
- Entropy change (ΔS): 163.15 J/mol·K
- Gibbs free energy change (ΔG): 237.13 kJ/mol

The water-splitting process involves separate reactions at the anode and cathode:
- Anodic reaction (oxidation):

$$2\text{H}_2\text{O (l)} \rightarrow \text{O}_2\text{(g)} + 4\text{H}^+ + 4\text{e}^- \tag{2}$$

Here, water molecules lose electrons to form oxygen gas and protons.
- Cathodic reaction (reduction):

$$4\text{H}^+ + 4\text{e}^- \rightarrow 2\text{H}_2\text{(g)} \tag{3}$$

At the cathode, protons gain electrons to form hydrogen gas.

The Gibbs free energy is related to enthalpy and entropy by the following equation:

$$\Delta G = \Delta H - T\Delta S \tag{4}$$

where T is the absolute temperature in Kelvin.

For the water-splitting reaction, ΔG is positive at standard conditions, indicating that the reaction is non-spontaneous. This means external energy is required to drive the reaction. In electrolyzers, this energy is typically supplied in the form of electrical energy.

The positive value of ΔG (237.13 kJ/mol) reflects the significant energy barrier that must be overcome to dissociate water into hydrogen and oxygen. The thermodynamic parameters also highlight that, while the reaction releases entropy into the system, the enthalpy required to break the bonds in water outweighs this effect, leading to an overall non-spontaneous process.

A. Electrochemical Model for Hydrogen Production

Faraday's Law provides the foundational relationship between the electrical energy required to split water and the chemical conversion rate in molar quantities. It establishes that the voltage applied must supply enough energy to drive the water-splitting reaction under ideal conditions. The equilibrium voltage is expressed as:

$$V(T,P) = \frac{\Delta H}{zF} = 1.43\,\text{V} \tag{5}$$

where:
- $V(T,P)$ is the equilibrium free energy voltage (also referred to as the thermodynamic reversible voltage),
- ΔH is the enthalpy change of the reaction (in Joules per mole),
- z is the number of electrons involved in the reaction (2 for water splitting),
- F is Faraday's constant (96,485 C/mol).

This equation highlights that the applied voltage must be at least equal to the enthalpy per mole of electrons transferred, divided by the product of z and F.

Critical Conditions for Voltage: The electrolysis process is dependent on the applied voltage $U(T,P)$ meeting specific criteria:

1) $U(T,P) \leq E(T,P)$: In this case, the applied voltage is less than or equal to the equilibrium free energy

voltage $E(T, P)$. The system remains at equilibrium, and electrolysis does not start.

2) $E(T, P) < U(T, P) < V(T, P)$: The applied voltage exceeds $E(T, P)$ but is less than the thermoneutral voltage $V(T, P)$. Electrolysis begins, but the system cannot maintain thermal equilibrium, resulting in the absorption of heat from its surroundings.

3) $U(T, P) > V(T, P)$: The applied voltage exceeds the thermoneutral voltage. This leads to a significant increase in current density, and the system generates more heat internally. At this stage, the process becomes more efficient.

Non-linearities in Voltage Calculation: The total applied voltage $U(T, P)$ is the sum of several components, accounting for losses and reaction kinetics [9]:

$$U = U_{\text{rev}} + V_{\text{act}} + V_{\text{ohm}} \qquad (6)$$

where: U_{rev} is the reversible cell voltage, representing the minimum energy required to drive the reaction under ideal conditions. This voltage is equal to either $E(T, P)$ or $V(T, P)$, depending on system requirements. V_{act} is the activation overpotential, which accounts for the kinetic barriers in initiating the electrochemical reaction. It is modeled as:

$$V_{\text{act}} = s \log \left(\frac{t_1 + \frac{t_2}{T} + \frac{t_3}{T^2}}{AI} + 1 \right) \qquad (7)$$

V_{ohm} is the ohmic loss due to resistance in the system. It is expressed as:

$$V_{\text{ohm}} = \frac{r_1 + r_2 T}{AI} \qquad (8)$$

The final expression for the total voltage is:

$$U = U_{\text{rev}} + s \log \left(\frac{t_1 + \frac{t_2}{T} + \frac{t_3}{T^2}}{AI} + 1 \right) + \frac{r_1 + r_2 T}{AI} \qquad (9)$$

Hydrogen Production Rate: The molar flow rate of hydrogen is a critical output parameter for an electrolyzer. It is derived as:

$$\dot{n}_{\text{H}_2} = \frac{\left(\frac{I}{A} \right)^2}{f_1 + \left(\frac{I}{A} \right)^2 f_2} \cdot \frac{n_c I}{zF} \qquad (10)$$

where: $\dot{n}_{\text{H}_2}$ is the molar flow rate of hydrogen (mol/s), A is the electrode area (m^2), I is the input current (A), f_1 and f_2 are parameters related to Faraday efficiency, n_c is the number of cells in series, z is the number of electrons involved in the reaction (2 for water splitting), F is Faraday's constant (96, 485 C/mol). .

B. Proposed Converter for Hydrogen Production

The converter system, shown in Figure-1, is designed to supply power to a hydrogen electrolyzer efficiently. The system comprises a DC input source, a single-phase inverter with four MOSFET switches, a high-frequency transformer, a single-phase rectifier with four diodes, an LC filter, and the electrolyzer as the load. The inverter converts the DC input into high-frequency AC, which is then stepped up or down by the transformer. This AC is subsequently rectified back into

DC by the rectifier and filtered using an LC filter before being supplied to the electrolyzer.

A unique feature of this setup is that the LC filter is not designed to produce a low-ripple DC supply but rather to control the spikes and troughs of the voltage, allowing for a pulsated DC output. This pulsated DC supply is intentionally fed to the electrolyzer as it enhances the efficiency of the electrolysis process and provides optimal operating conditions for hydrogen production.

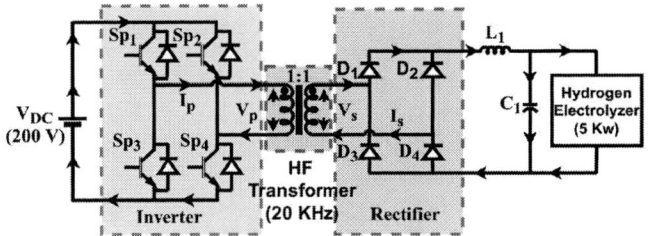

Fig. 1. Proposed Converter design

Modes of Operation

Mode 1: In the primary side of the converter, the supply from the DC source flows through Sp 1, passes through the primary winding of the high-frequency transformer, and returns to the DC source through Sp 4. The MOSFETs are switched in a complementary manner, ensuring that current flows through the transformer in alternating directions, thereby generating a high-frequency AC signal. The transformer is designed with proper dot convention to ensure the phase alignment between the primary and secondary windings.

On the secondary side, the induced voltage in the transformer winding follows the dot convention. During this mode, the current flows from the transformer's secondary winding through diode 1, then passes through the LC filter and into the electrolyzer. The return path for the current is completed through diode 4, back to the transformer's secondary winding. The LC filter smooths the spikes and troughs in the rectified DC output but maintains the pulsations, providing the electrolyzer with a controlled pulsated DC supply for efficient hydrogen production.

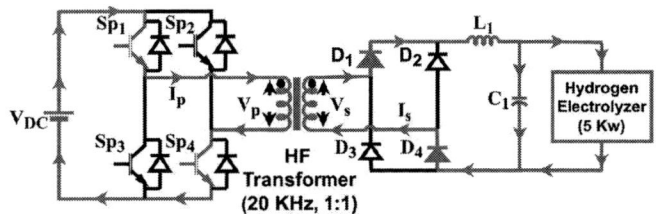

Fig. 2. Mode:1 operation of proposed converter

Mode 2: In Mode 2, the supply on the primary side similarly flows through Sp 2 into the primary winding of the high-frequency transformer and returns through Sp 3. However, in this mode, the switching pattern and frequency may differ to accommodate different operating conditions, such as during startup or varying load demands.

On the secondary side, the induced voltage in the transformer winding again follows the dot convention. The current flows through diode 2, passes through the LC filter, and enters the electrolyzer. The return path for this current is completed through diode 3, back to the transformer's secondary winding. The LC filter's role in this mode remains consistent—controlling spikes and troughs without eliminating the pulsations, ensuring a stable but pulsated DC supply for the electrolyzer.

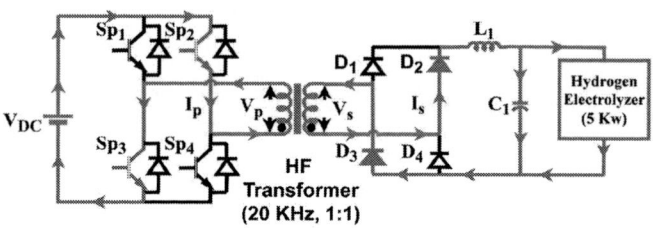

Fig. 3. Mode:2 operation of proposed converter

III. SYSTEM PARAMETERS

The system under consideration is designed to facilitate efficient power conversion for hydrogen production through an electrolyzer. A DC input source feeds into a single-phase inverter comprising four MOSFET switches, which operate at a high switching frequency to generate a high-frequency AC waveform. This AC waveform is then stepped up or down as required by a high-frequency transformer with a specified turns ratio. The transformer is followed by a single-phase rectifier with four diodes, which converts the AC signal back into a pulsating DC waveform.

To ensure proper operation and control the ripple characteristics of the output, an LC filter is incorporated. However, the values of the inductor and capacitor are chosen deliberately small, not to smooth the waveform entirely but to manage and limit voltage spikes and troughs. This approach ensures the electrolyzer receives a pulsating DC supply, which is preferred over a completely smooth DC output for optimizing hydrogen production efficiency.

The design of this system is carefully optimized to meet the specific requirements of hydrogen production. The high switching frequency of the inverter not only ensures efficient conversion but also reduces the size of the transformer and filter components, making the setup compact and cost-effective. The transformer plays a crucial role in isolating the DC source from the load and stepping the voltage as required by the electrolyzer. The pulsating DC output from the rectifier, controlled by the LC filter, creates a favorable electrochemical environment in the electrolyzer cells, enhancing the rate of hydrogen production. The choice of parameters such as the LC filter values, switching frequency, and diode ratings ensures

reliable operation and compatibility with the electrolyzer's characteristics while maintaining system efficiency.

The system parameters, including electrical characteristics and electrolyzer-specific values, are critical for achieving optimal performance. These parameters are summarized in the tables below.

TABLE I
SYSTEM PARAMETERS

Parameter	Value
Input DC Voltage	200 V
Switching Frequency	20 kHz
Transformer Turns Ratio	1:1
Rectifier Diode Ratings	500 V, 62.5 A
LC Filter Inductance	50 μH
LC Filter Capacitance	10 μF

TABLE II
ELECTROLYZER PARAMETERS

Specification	Value
Area	1 m^2
r_1	0.0003538550
r_2	-0.00000302150
s	0.223
t_1	5.13093
t_2	-0.0240447
t_3	3410.251

IV. RESULTS

The performance of the proposed converter was analyzed and compared with a conventional 5 kW buck converter. The results are presented in terms of waveforms for current and hydrogen production, zero-voltage switching (ZVS) behavior in the proposed converter, and a comparative analysis of hydrogen production efficiency. The experimental and simulation data clearly illustrate the advantages of the proposed converter in optimizing hydrogen production.

Figure 4 demonstrates the waveforms for the conventional buck converter. The steady-state DC current of 25 A ensures consistent operation of the electrolyzer, producing hydrogen at a constant rate. However, this approach does not take advantage of dynamic behavior in the electrolyzer to enhance production. The hydrogen production rate closely follows the stable input current, highlighting the limitations of the buck converter in terms of efficiency.

Figure 5 presents the waveforms for the proposed converter. Unlike the buck converter, the proposed converter introduces a pulsating current profile, which significantly enhances the electrochemical reaction rates in the electrolyzer. This leads to a higher average hydrogen production rate. The pulsations allow for more efficient utilization of the electrochemical system, demonstrating the ability of the proposed converter to optimize hydrogen production under the same input power conditions.

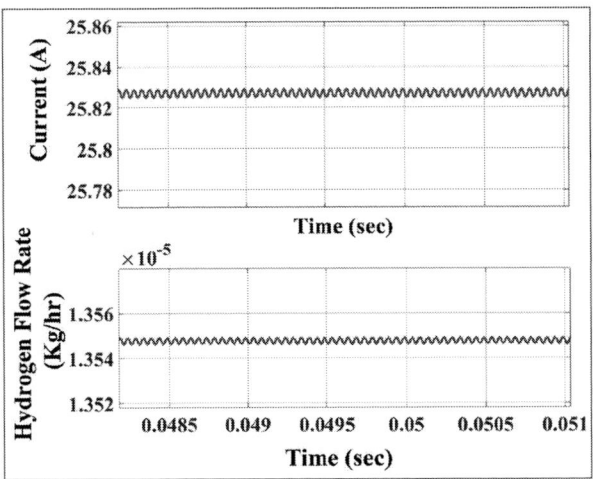

Fig. 4. Waveforms for the buck converter: (A) Current vs. time and (B) Hydrogen production vs. time. The current is steady at 25 A, leading to a stable hydrogen production rate.

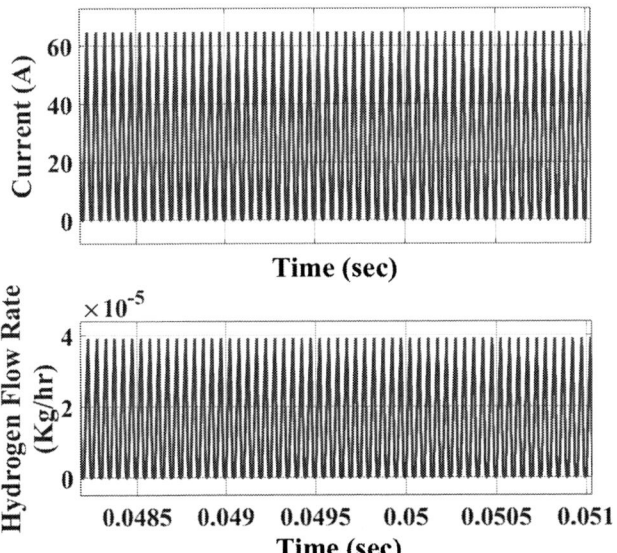

Fig. 5. Waveforms for the proposed converter: (A) Current vs. time and (B) Hydrogen production vs. time. The pulsating current results in a higher average hydrogen production rate, demonstrating the efficiency advantages of the proposed converter.

Figure 6 showcases the ZVS behavior of the proposed converter. ZVS ensures that the MOSFETs in the converter operate efficiently by minimizing switching losses. This feature is particularly important for high-frequency operation, as it reduces thermal stress on the switches and improves the overall energy efficiency of the system. The implementation of ZVS in the proposed design is a critical factor in achieving enhanced performance.

Figure 7 provides a detailed comparative analysis of hydrogen production efficiency. At the same input current of 25 A, the proposed converter demonstrates a 15.13% increase in

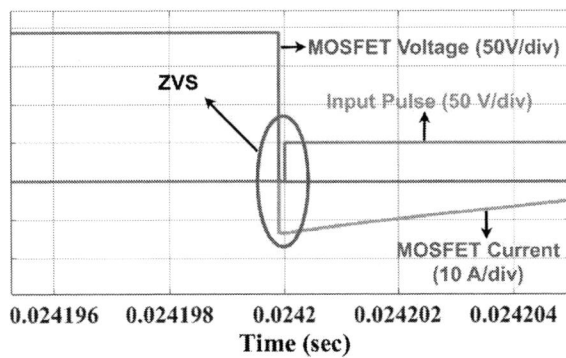

Fig. 6. Zero-voltage switching (ZVS) waveform for the proposed converter. The ZVS reduces switching losses, contributing to higher overall efficiency of the system.

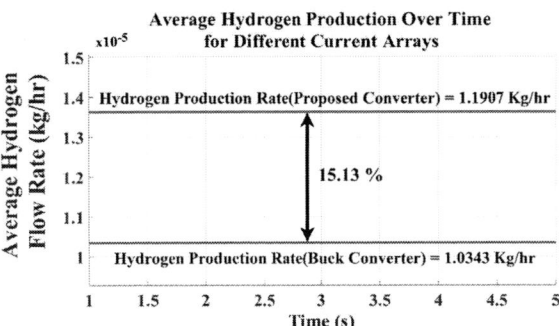

Fig. 7. Comparison of average hydrogen production rates between the buck converter and the proposed converter at the same input current of 25 A. The proposed converter demonstrates a 15.13% increase in average hydrogen production due to its pulsating current profile.

average hydrogen production compared to the buck converter. This improvement is a result of the pulsating current profile, which efficiently optimizes the electrolyzer's performance through the utilization of its non-linear electrochemical behavior. The conclusion emphasizes the efficiency of the proposed design in maximizing hydrogen production efficiency without varying the input power.

Overall, the new converter significantly surpasses the traditional buck converter in efficiency in hydrogen generation. Addition of a pulsating current profile and ZVS operation not only enhances the hydrogen output rate but system efficiency in general. All of these results prove the efficacy of the new design for high-end hydrogen production applications.

V. CONCLUSION

The research introduces a novel converter design to enhance hydrogen efficiency in alkaline electrolyzers. The novel design takes advantage of a pulsating current waveform and zero-voltage switching (ZVS) to mitigate the weaknesses of conventional buck converters. The experiment verifies an enhanced hydrogen production efficiency with the new converter achieving a 15.13% increase in average hydrogen output compared to a conventional buck converter under the same

979-8-3315-3013-6/25 $31.00 © 2025 IEEE

input power level. The enhanced performance can be attributed mainly to the pulsating current profile, which optimizes electrochemical reaction rates in the electrolyzer. Furthermore, use of ZVS technology reduces switching losses, which enables efficient high-frequency operation and reduced thermal stress on power electronic devices. The judicious selection of small LC filtering values ensures the generation of a pulsating DC supply, enhancing further the performance of the electrolyzer by exploiting the non-linear nature of the electrochemical process. The novel converter design is a potential solution to enhancing hydrogen production efficiency. Therefore, it is a valuable contribution to existing research towards the creation of sustainable energy systems.

REFERENCES

[1] J. Xiong, Y. Xia, Y. Peng, and W. Wei, "A multi-mode self-optimization electrolysis converting strategy for improving efficiency of alkaline water electrolyzers," *IEEE Transactions on Power Electronics*, 2023.

[2] A. S. Tijani, N. A. B. Yusup, and A. A. Rahim, "Mathematical modelling and simulation analysis of advanced alkaline electrolyzer system for hydrogen production," *Procedia Technology*, vol. 15, pp. 798–806, 2014.

[3] Y. Xia, H. Cheng, H. He, Z. Hu, and W. Wei, "Efficiency enhancement for alkaline water electrolyzers directly driven by fluctuating pv power," *IEEE Transactions on Industrial Electronics*, vol. 71, no. 6, pp. 5755–5765, 2023.

[4] J. Koponen, V. Ruuskanen, A. Kosonen, M. Niemelä, and J. Ahola, "Effect of converter topology on the specific energy consumption of alkaline water electrolyzers," *IEEE Transactions on Power Electronics*, vol. 34, no. 7, pp. 6171–6182, 2018.

[5] X. Guo, S. Zhang, B. Yuwen, Q. Zhang, T. Wang, and S. Padmanaban, "A novel dc-dc converter for electrolyzer with low ripple and high step-down ratio," in *2023 IEEE 14th International Conference on Power Electronics and Drive Systems (PEDS)*. IEEE, 2023, pp. 1–5.

[6] H. Cheng, Y. Xia, Z. Hu, and W. Wei, "Optimum pulse electrolysis for efficiency enhancement of hydrogen production by alkaline water electrolyzers," *Applied Energy*, vol. 358, p. 122510, 2024.

[7] X. Li, N. Li, W. Xue, W. Mao, S. Qiu, and X. Wu, "Lcc-resonant-type current-fed-out three-port dc–dc converter for pv electrolytic hydrogen production integrated with energy storage," *IEEE Transactions on Industrial Electronics*, 2024.

[8] X. Li, W. Mao, M. Li, W. Xue, N. Li, and X. Wu, "Lcl-resonant-tank based current fed-out dc–dc converter for pv off-grid hydrogen production: Modeling, control, and optimization," *IEEE Transactions on Industrial Electronics*, 2024.

[9] Ø. Ulleberg, "Modeling of advanced alkaline electrolyzers: a system simulation approach," *International journal of hydrogen energy*, vol. 28, no. 1, pp. 21–33, 2003.

[10] W. Hug, J. Divisek, J. Mergel, W. Seeger, and H. Steeb, "Highly efficient advanced alkaline electrolyzer for solar operation," *International journal of hydrogen energy*, vol. 17, no. 9, pp. 699–705, 1992.

[11] S. A. Gorji, "Reconfigurable quadratic converters for electrolyzers utilized in dc microgrids," *IEEE Access*, vol. 10, pp. 109 677–109 687, 2022.

[12] M. K. Ratib, K. M. Muttaqi, M. R. Islam, D. Sutanto, and A. P. Agalgaonkar, "Electrical circuit modeling of proton exchange membrane electrolyzer: The state-of-the-art, current challenges, and recommendations," *International Journal of Hydrogen Energy*, vol. 49, pp. 625–645, 2024.

[13] T. Adibi, A. Sojoudi, and S. C. Saha, "Modeling of thermal performance of a commercial alkaline electrolyzer supplied with various electrical currents," *International Journal of Thermofluids*, vol. 13, p. 100126, 2022.

[14] B. Yodwong, D. Guilbert, M. Phattanasak, W. Kaewmanee, M. Hinaje, and G. Vitale, "Ac-dc converters for electrolyzer applications: State of the art and future challenges," *Electronics*, vol. 9, no. 6, p. 912, 2020.

[15] J. Brauns and T. Turek, "Alkaline water electrolysis powered by renewable energy: A review," *Processes*, vol. 8, no. 2, p. 248, 2020.

[16] A. Beainy, N. Karami, and N. Moubayed, "Simulink model for a pem electrolyzer based on an equivalent electrical circuit," in *International Conference on Renewable Energies for Developing Countries 2014*. Ieee, 2014, pp. 145–149.

[17] M. Kiaee, A. Cruden, D. Infield, and P. Chladek, "Utilisation of alkaline electrolysers to improve power system frequency stability with a high penetration of wind power," *IET Renewable Power Generation*, vol. 8, no. 5, pp. 529–536, 2014.

[18] N. Norazahar, F. Khan, N. Rahmani, and A. Ahmad, "Degradation modelling and reliability analysis of pem electrolyzer," *International Journal of Hydrogen Energy*, vol. 50, pp. 842–856, 2024.

[19] E. Wallnöfer-Ogris, I. Grimmer, M. Ranz, M. Höglinger, S. Kartusch, J. Rauh, M.-G. Macherhammer, B. Grabner, and A. Trattner, "A review on understanding and identifying degradation mechanisms in pem water electrolysis cells: Insights for stack application, development, and research," *International Journal of Hydrogen Energy*, vol. 65, pp. 381–397, 2024.

[20] P. Nikolaidis, "Pulsed-supplied water electrolysis via two-switch converter for pv capacity firming," *Electricity*, vol. 3, no. 1, pp. 131–144, 2022.

Design and development of isolated space grade DC-DC converter for SSPA using eGaN FET and analysis on the role of parasitic

Santwana Suman
Indian Space Reasearch Orgnization
Ahmedabad, India
santwana@sac. isro. gov. in

Amit Kumar
Indian Space Research Oragnization
Ahmedabad, India
amit028@sac. isro. gov. in

Trapti Katiyar
Indian Space Reasearch Organization
Ahmedabad, India
trapti@sac. isro. gov. in

Jayesh Thakkar
Indian Space Research Organization
Ahmedabad, India
jayesh@sac. isro. gov. in

Abstract— **The paper provides an insight into the design and development of GaN FET based isolated DC-DC converter, built as a potential substitute to Si based DC-DC converter already in use for space applications. The DC-DC in target here is used to power solid state power amplifier (SSPA) used in spacecraft. The paper highlights major achievements in different set of performance parameters as and when required. Apart from this, the paper also gives detailed analysis on the role of parasitic which become dominant at such high switching frequencies at which GaN FETs operate. Layout analysis and signal integrity have been discussed to optimize device performance. Practical hardware and simulation results have been included at required places to support the analysis. The aim here is also to highlight the thermal management of the device considering the size of the device. This paper also includes comparison of both the hardware results of switching power devices Si MOSFET and GaN FET to show efficacy of GaN FET.**

Keywords— *DC-DC, GaN FET power device, Gate drive, Gallium Nitride, parasitic, satellites, SSPA*

I. INTRODUCTION

Wide bandgap semiconductors such as the Gallium Nitride (GaN), and Silicon Carbide (SiC) are promising transistor technologies for future generations of power electronics circuits. GaN FET-based DC-DC converter is useful for space applications in that GaN FET has a high immunity and it becomes less susceptible for any false turn-on due to ionizing radiation. This is due to its wide bandgap compared to Silicon. Considering the low switching losses, fast speed, high electron mobility and various other advantages which GaN FETs offer over conventional Si MOSFET, it becomes extremely important to know the device behaviour in absolute detail and in depth including the role of parasitic which become dominant at such high frequencies at which GaN FETs operate. These devices lead to high power density which is critical especially in design of space subsystems owing to mass constraint and fuel consumption. In view of this, a new GaN-based DC-DC converter used to power SSPAs in payloads has been proposed which has a potential to be used as a substitute for presently used Si-based DC-DC converter in future. This paper presents a report on detailed

analysis including parasitic effect and thermal management. The PCB layout is also discussed in great depth since at higher frequencies, the track parasitic also start playing their part.

II. CHALLENGES WHILE USING GaN FET

A. Source of Parasitic

The most overlooked constraint which becomes dominant at high frequencies is the parasitic. There are two types of parasitic encountered in any design- capacitive and inductive (apart from resistive). The source of these parasitic are due to device/GaN FET package and also due to magnetics as well as tracks used for routing the layout. Table 1 shows the comparison between different device level parasitic for the space qualified MOSFET 2N7591 and GaN FET ISL70024 transistors with comparable drain voltage and drain current.

Table 1:Parasitic comparison for Si (200V, 16A) and eGaN (200V, 7.5A) transistor

S. No.	Parameter	Value for Si	Value for GaN
1.	Input capacitance, C_{iss}	1450pF	270pF
2.	Output capacitance, C_{oss}	210pF	150pF
3.	Reverse capacitance, C_{rss}	3.8pF	1pF
4.	On-state resistance, $r_{ds(on)}$	130mΩ	45mΩ
5.	Gate resistance, R_g	900mΩ	60mΩ
6.	Total gate charge, Q_g	50nC	2.5nC

The various GaNFET part no. available for present design were TDG650E60BEP (650V, 60A), ISL70024 (200V, 7.5A) and FBG20N18B2-C (200V, 18A). However, since forward topology is used, to keep device within safe limits, TDG650E60BEP part no. was selected. It can be clearly observed later through hardware results that GaN FETs offer better performance in terms of parasitic even at higher ratings (650V, 60A) which will be further evaluated. Other sources of parasitic include parasitic inductance which comes from device lead inductance [6] as well as leakage inductance of magnetics as well as inductance while routing tracks in PCB boards. Figure 1 shows all the parasitic components associated with GaN FET.

979-8-3315-3013-6/25 $31.00 © 2025 IEEE

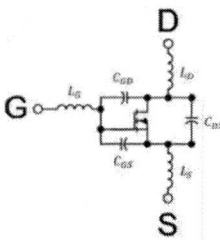

Figure 1: Parasitic components of GaN FET [6]

B. Role of Parasitic

GaN devices are 10 x faster than silicon MOSFETs and as a result circuit that use GaN are more sensitive to parasitic inductance. Parasitic inductance causes ringing and overshoot which adds EMI and risk of circuit failure. The common source inductance will affect di/dt of the power loop into the gate loop and it becomes significant at high power. The mutual inductance between power loop and gate loop is part of L_{cs} (common source inductance). Gate inductance can cause overshoot on V_{GS} but does not cause gate loop loss. Increased power loop inductance will increase the V_{DS} voltage spike. However, the minimization of parasitic inductance is straightforward. A properly designed GaN circuit has less EMI and overshoot than the best possible MOSFET design. In this paper, effort has been made to capture all hardware results which shows and explains the effect and role of parasitic while using GaN FET as power switching device in dc-dc converter. Moreover, device packaging also plays significant role in reduction of device parasitic thereby improving device performance at high frequencies and high power as shown in Figure 2. Parasitic inductance impact on power loss for a sample design with VIN = 12 V, VOUT =1.2 V, IOUT = 20 A, fsw =1 MHz is shown[4]. Clearly, eGaN FET packaging offers lower common source inductance as well as power loop inductance.

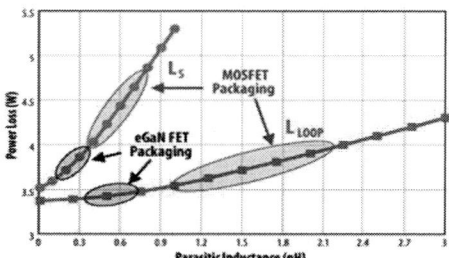

Figure 2: Power loss vs parasitic inductance variation with packaging [4]

III. DESIGN DETAILS AND SIMULATION

First simulation was done as a proof of concept using Synopsys SaberRD simulation software. Here, in the present design used for analysis, Teledyne GaN FET and Intersil driver are used. Both of these devices were modelled in SaberRD to ensure accurate simulation. The major limitation while doing simulation is that parasitic are overlooked. So, to propose a close match with hardware, devices were modelled taking their parasitic into account. A forward converter is chosen for analysis which uses 2MHz current mode Pulse Width Modulator (PWM) controller from TI TPS7H5001. The forward topology suffices for our requirement here as the

Teledyne GaN FET TDG650E60BEP is 650V rated for V_{ds}. The design specifications are tabulated in Table 2:

Table 2:Specifications of power supply

Specification	Rating
Input supply	65V-75V
Topology	Forward
Switching frequency	500kHz
Efficiency	>85%
Output voltage/current	(7.5-8.8)V@2A-6A
	5.5V@0. 1A-0.3A
	-5V@0. 02A-0.05A
Output voltage ripple	<20mV rms
Sequencing (between +ve and -ve output)	Yes

A. Simulation Results

Figure 3 shows PWM pulse and gate pulse as seen in SaberRD with modelled GaN FET and PWM IC TPS7H5001.

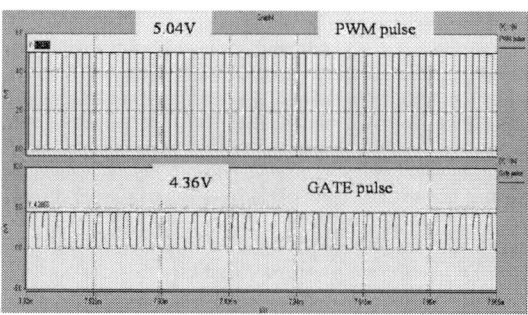

Figure 3: PWM pulse and gate pulse generated in SaberRD

IV. DESIGN OF HARDWARE

A prototype hardware was realized using TDG650E60BEP Teledyne GaN FET and ISL70040 driver IC using forward topology in order to miniaturize existing hardware used for powering SSPA in payloads. The switching frequency used in proposed GaN FET-based design is 500 kHz and magnetics are modified as per high frequency requirements. Also, PWM IC used was TPS7H5001 which is a 2MHz current mode controller and uses 5V as supply voltage. The high frequency and low voltage operation of this IC makes it ideal for targeted design using TDG650E60BEP Teledyne and driver. Due to stringent gate voltage requirements of 4.5V-5V, which maximizes device efficiency and reduces risk of device failure [1], use of such low voltage PWM proves advantageous. Also, a gate resistor is employed to prevent gate overshoot due to low input capacitance of eGaN [2]. Other factors which affect gate overshoot like appropriate layout design has also been discussed. A careful note of all internal physical capacitances of the device [3] has been made for maximizing efficiency of used GaN device and therefore, the power supply. Further improvements have been made in the realized hardware to miniaturize the circuit like the use of magnetic isolation using IC UC1901 to further remove extra components which occupy board space for the

979-8-3315-3013-6/25 $31.00 © 2025 IEEE

same purpose and also provides ease of output sequencing. The gate driver circuit and its layout which plays crucial role in the proper operation of GaN FET based DC-DC has been analyzed and waveforms presented for better understanding. The converter is realized on a 1mm thick board with PCB dimensions 127mm×125mm. The block diagram of proposed converter is shown in Figure 4 .

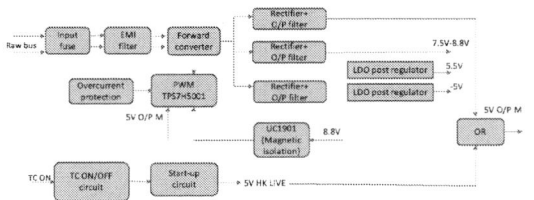

Figure 4: Block diagram of proposed dc-dc converter

A. Proposed Gate drive circuit

The circuit in Figure 5 shows version 2 of the gate drive circuit realized in lab with Teledyne GaN FET TDG650E60BEP and Intersil driver ISL70040 and tested electrically for major performance goals. The layout also modified from version 1 considering the lessons learnt to minimize the trace inductance. R8, R9 resistors control the turn-on/off times independently and can be appropriately sized. The effect of gate resistor on turn-on time is shown in Figure 15. Figure 6 shows version 1 of the gate drive circuit with Intersil eGaN FET ISL70024 and its driver ISL70040. In order to prevent overshoot at gate, series RC combination is used between gate and source to increase reliability of device. Also, Intersil driver provides a steady 4.5V input to gate thereby eliminating any need of clamp circuit.

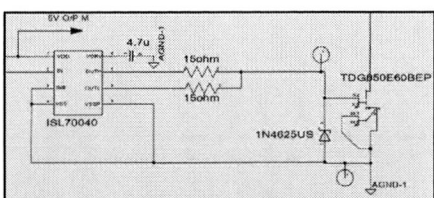

Figure 5: Gate loop of TDG650E60BEP

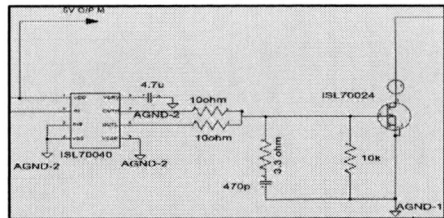

Figure 6: Gate loop of ISL70024

B. Layout Considerations

a) Layout Optimization: Few points are to be kept in mind while designing layout for GaN based dc-dc converter. Switching losses become dominant due to high switching frequency. The important point to consider here include the board parasitic as track width also contribute significant amount of parasitic. Secondly, thermal management also

becomes important owing to extremely small package size of device. Placing the Vcc decoupling capacitor as well as the bootstrap capacitor closely is important[1]. Regarding the power circuit, minimize the loop length and size of high frequency ac current paths. In the conventional vertical power loop design, the loop inductance is heavily dependent on the board thickness as the power loop is contained on the top and bottom layers of the PCB[5]. As the board thickness increases so does the high frequency loop inductance, leading to higher losses and consequently lower efficiency. The present circuit was designed on 1mm thick PCB which also improves thermal performances. The eGaN FET, combining low FOM[7], low package parasitic, and a small footprint reducing PCB parasitic, outperformed MOSFETs rated for much lower voltages. As FOM and packages improve, the PCB layout becomes critical to high efficiency.

b.) Signal Integrity(SI)

It is a good practice to carry out signal integrity test to minimize layout parasitic interference when using dc-dc converter at high switching frequencies. This is essential due to stringent gate voltage requirement of 4.5V-5V in most available GaN FET devices. A slight overshoot can cause failure of the device. A Zener across gate is used to clamp and protect the gate overshoot as recommended by manufacturer. Here also, signal integrity was carried out for the design. The critical path between PWM to driver and driver to gate of GaN FET was modelled in Cadstar and simulation was run to verify the gate voltage. Figure 7 shows the gate net as $167 and driver net as $889.

The simulation takes into account the actual path that the signal will be taking while travelling from source net to destination net and this includes via as well if any is present. The layer stack-up details need to be provided which is as shown in Figure 8. The simulation was carried out in Hyperlink using IBIS model as actual model of GaN FET is not available. When above inputs are provided, the gate pulse is recorded and signal was observed as shown in Figure 9. Such spikes can be detrimental for gate of many GaN FET which offer very stringent gate voltage requirements. Based on above analysis, modified version of the design was realized with appropriate improvements. Figure 11 shows version 1 of the gate drive portion of the layout and Figure 10 shows version 2 of the gate drive portion of layout.

Some important learnings from both the layouts under consideration can be summarized below:

1.) Thermal plane improved performance of the circuit at high load.

2.) Use of bottom cooled package for the GaN FET device

3.) No vias should be present in between the driver to gate path which increases inductance and cause ringing/overshoot.

4.) The gate driver and GaN FET to be placed in close proximity including their orientation to provide shortest path to signal.

Thus, proper design of gate driver circuit, thermal management and controlled parasitic are crucial for using GaN FETs to their full potential.

979-8-3315-3013-6/25 $31.00 © 2025 IEEE   35

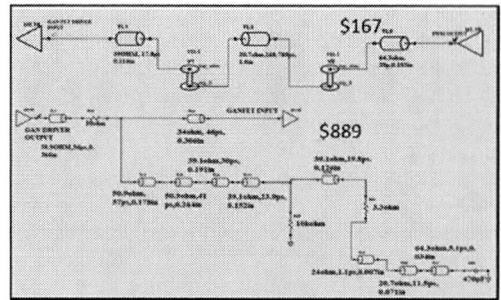

Figure 7: Layout model for SI simulation for net $889 and $167

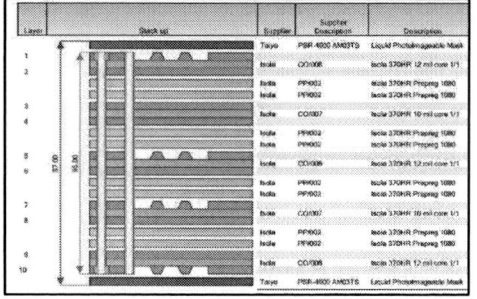

Figure 8: Layer stack-up detail as input during simulation

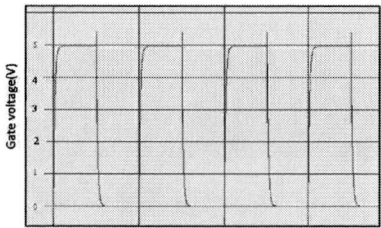

Figure 9: Gate pulse observed during simulation

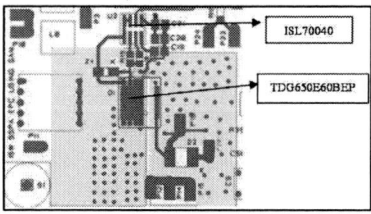

Figure 10: Gate drive for layout 2

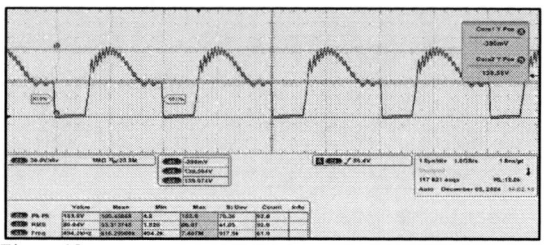

Figure 11: Gate drive for layout 1

C. Hardware Results:

The whole design was realized in a prototype in lab and rigorous testing was carried out to evaluate performance and improvements compared to heritage hardware. The comparison results are shown under section V.

This section will deal with all the output results which demonstrate efficient working of realized hardware. Also, in order to understand the parasitic effects of device in greater detail, switching waveforms have been added as and when needed.

1.) Drain to source voltage:

The drain-to-source voltage shown in Figure 12 is captured @72V input and 2A load current for proposed design. It can be clearly seen that the peak drain voltage does not cross 140V (just double as expected for forward converter). No spike is observed due to low package lead inductance [7]. A proper layout design with minimal parasitic inductance can also help mitigate such issues as explained in section 2 of this paper. Figure 13 shows drain-to-source voltage for Si-based DC-DC.

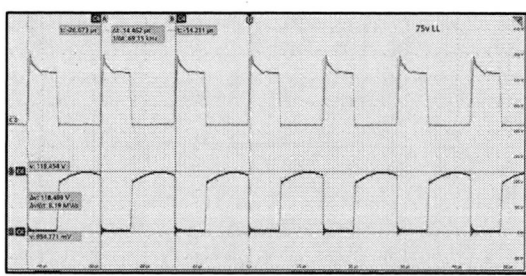

Figure 12: Drain-to-source of GaN FET DC-DC @72V input and 2A load

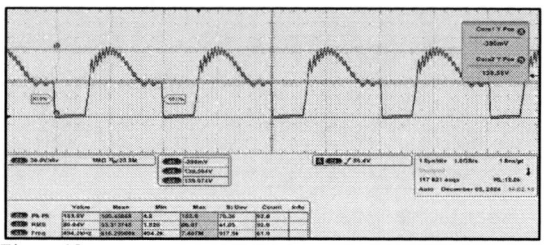

Figure 13: Gate-to-source voltage (top) and Drain-to-source voltage (bottom) of conventional Si-based DC-DC @75V input and 2A load

2.) Gate-to-source voltage:

As shown in Figure 14, peak gate voltage observed is 4.3V at switching frequency of 500 kHz. The gate drive loop plays significant role here as discussed in signal integrity section in this paper. The gate overshoot observed in conventional DC-DC at 70 kHz switching frequency (shown in Figure 13) can be mitigated through careful layout design. Although some ringing observed due to resonance in parasitic inductance/capacitance, but no overshoot is observed in our present design which is the mandatory requirement for safe operation of GaN device. Note that the Si-MOSFET based converter has 12V gate-to-source voltage.

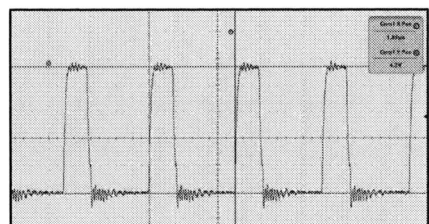

Figure 14: GaN FET Gate voltage @75V input, 6A load

3.) *Turn-on time with different gate resistor:*

Figure 15 shows the turn-on time measured in hardware with gate resistor values of 15ohm and 7.5ohm. The simulation was also carried out in SaberRD simulation software to verify and analyse the effect of gate resistor on gate resistor turn-on time. The simulation result obtained is shown in Figure 16. A simplified formula to calculate gate resistor R[8] was used assuming rule of thumb that 10nH/inch parasitic inductance will be there in the track and total track length is 0.33 inch. Hence, we get 3.3nH parasitic inductance.

$$R > \sqrt{\frac{4L}{C}}$$

where, C is the input capacitance of device and L is the parasitic track inductance. Substituting, C=518pF(from Table 1) and L=3.3nH, we get R>5ohm.

R=15ohm and R=7.5ohm were tested in the hardware and simulation to verify performance.

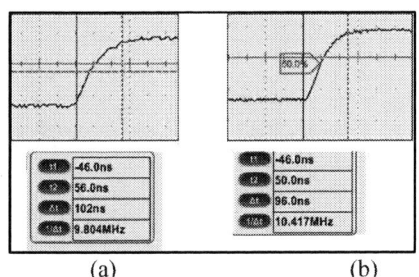

(a)   (b)

Figure 15: Measured Gate turn-on time with R8=15ohm (a) vs R8=7.5ohm (b)

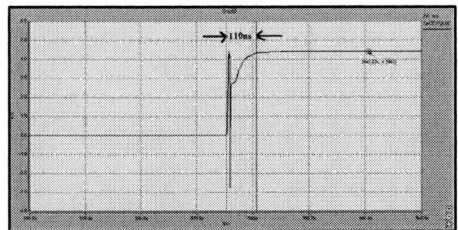

Figure 16: Simulated Gate rise time as seen in SaberRD

4.) *UC1901 output:*

Figure 17 shows the output signal DRA of UC1901 IC which is used as feedback signal to PWM IC TPS7H5001. DRA and DRB outputs of UC1901 generate amplitude modulated pulses depending on error signal amplitude.

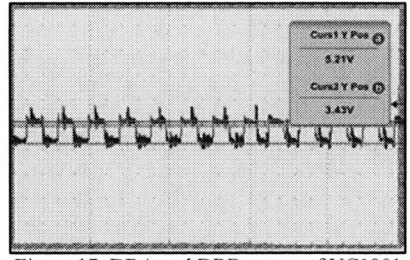

Figure 17: DRA and DRB output of UC1901

5.) *Efficiency calculation:*

Table 3 shows the output voltages at various load condition.

Table 3:Efficiency calculation

Sr. No.	Output Regulation						
	Vin (V)	Iin (A)	Io (A)	8.8V	5.5V	-5V	Efficiency (%)
1	72	0.17	1/0.05/0.05	8.862	5.504	-4.950	77
2	72	0.61	4/0.35/0.05	8.887	5.483	-4.920	86
3	72	0.89	6/0.35/0.05	8.889	5.480	-4.918	87

Efficiency= $\frac{Output\ power}{Input\ power}$

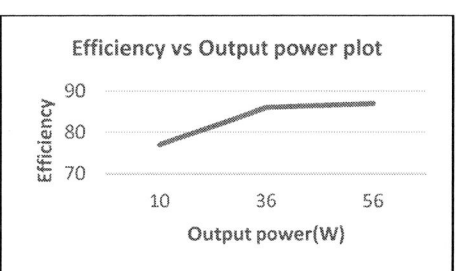

Figure 18: Efficiency vs output power for GaN FET converter

6.) *Turn-on time:* Figure 19 shows the soft turn-on waveform of main output line. The rise time is approximately 16ms.

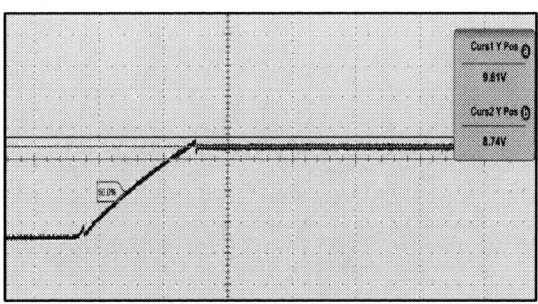

Figure 19: Turn on time of main output

V. COMPARISON WITH HERITAGE HARDWARE

Table 4 shows comparison between conventional design and proposed design. The Si MOSFET based DC-DC converter operating at 70 kHz used for powering the SSPA in spacecraft delivers output at around 80% efficiency which is increased to around 87% using GaN FET based DC-DC converter operating at 500 kHz. This efficiency improvement in smaller foot print (reduced magnetic profile at 500 kHz) results in power and mass saving in a payload. The efficiency curve for Si-MOSFET EPC and GaN-based EPC is shown in Figure 18 and Figure 20 for comparison [10]. Figure 21 shows the actual hardware used in conventional design vs the proposed hardware. The conventional EPC has dimensions of 175mm×110mm×65mm while proposed design has dimensions of 125mm×125mm×45mm. A significant improvement in weight is also envisaged.

Table 4: Comparison between conventional and proposed hardware design

Component changes			
S. No.	Component	Part no. in conventional EPC	Part no. in in proposed design
1.	Power transformer core	W374	ZL41605
2.	PWM	UC1875	TPS7H5001
3.	FET	2N7586	TDG650E60BEP
4.	Gate Resistor	RM1206	RM0805
5.	Output Capacitor	CTC21E	C1812
Design changes			
1.	Topology	Forward	Forward
2.	Switching frequency	70kHz	500kHz
3.	Design approach	Dual converter	Magnetic isolation using UC1901

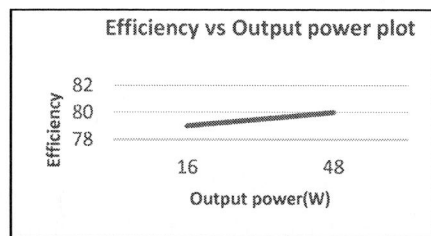

Figure 20: Efficiency vs output power for Si MOSFET

(a)   (b)

Figure 21: Conventional hardware (a) and proposed hardware (b)

VI. CONCLUSION

The DC-DC converter demonstrated in this paper is successfully realized using TPS7H5001 PWM, TDG650E60BEP (bottom-cooled) Teledyne GaN FET, Intersil driver ISL70040 and uses UC1901 IC from TI for magnetic feedback. UC1901 solves the problem of closing the loop across voltage isolation boundary. This converter achieves around 87% efficiency at full load and performs well electrically, meeting all performance parameters. The loss break-up analysis, further efficiency improvement using synchronous rectification and conversion to space worthy dc-dc with space qualified GaN FET device and further reduction in size of board will be included in the future scope of this work before realization of actual flight model.

ACKNOWLEDGMENT

The authors wish to acknowledge Sri Nilesh M. Desai, Director, SAC, ISRO and Dr. S. C. Bera, Deputy Director, SAC, ISRO for giving us opportunity to work in new development activities and all our colleagues in ISRO for their support and guidance throughout the completion of this project.

REFERENCES

[1] Xi, Y., Chen, M., Nielson, K., & Bell, R. (2012, February). Optimization of the drive circuit for enhancement mode power GaN FETs in DC-DC converters. In 2012 Twenty-Seventh Annual IEEE Applied Power Electronics Conference and Exposition (APEC) (pp. 2467-2471). IEEE.

[2] Barchowsky, A., Kozak, J. P., Hontz, M. R., Stanchina, W. E., Reed, G. F., Mao, Z. H., & Khanna, R. (2017, March). Analytical and experimental optimization of external gate resistance for safe rapid turn on of normally off GaN HFETs. In 2017 IEEE Applied Power Electronics Conference and Exposition (APEC) (pp. 1958-1963). IEEE.

[3] Khanna, R., Stanchina, W., & Reed, G. (2012, September). Effects of parasitic capacitances on gallium nitride heterostructure power transistors. In 2012 IEEE Energy Conversion Congress and Exposition (ECCE) (pp. 1489-1495). IEEE.

[4] Reusch, D. eGaN® FET-Silicon Power Shoot-Out Vol. 13, Part 1: Impact Of Parasitics March 1, 2013.

[5] Lidow, A., Strydom, J., de Rooij, M., & Reusch, D. (2014). Layout Considerations for GaN Transistor Circuits. GaN Transistors for Efficient Power Conversion, 55-69.

[6] Hagar Mohamed, Donaldo Sanchez," Key parameters and driving requirements of GaN FET".

[7] Singh, R. P., Neelakantan, N., Leo, C. J., & Yoshio, N. (2019, November). Design and Evaluation of High-frequency GaN Based DC/DC Converter. In 2019 IEEE 4th International Future Energy Electronics Conference (IFEEC) (pp. 1-4). IEEE.

[8] Niu, Y. C., Huang, Y. T., Chen, C. L., & Chen, Y. M. (2018, October). Design considerations of the gate drive circuit for GaN HEMT devices. In 2018 Asian Conference on Energy, Power and Transportation Electrification (ACEPT) (pp. 1-6). IEEE.

[9] Hughes, B., Lazar, J., Hulsey, S., Musni, M., Zehnder, D., Garrido, A., ... & Boutros, K. (2014, March). Normally-off GaN-on-Si multi-chip module boost converter with 96% efficiency and low gate and drain overshoot. In 2014 IEEE Applied Power Electronics Conference and Exposition-APEC 2014 (pp. 484-487). IEEE.

[10] Alatawi, K. S., Almasoudi, F. M., & Matin, M. A. (2016, September). Performance enhancement of two-switch forward converter using GaN FETs. In 2016 North American Power Symposium (NAPS) (pp. 1-6). IEEE

Four-Load Series Resonant Inverter With Hybrid Control for Induction Based Cooking Applications

1st Kirlampalli Harija Rani
Department of Electrical Engineering
National Institute of Technology, Warangal
Telangana, India
email address : kh21eerer11@student.nitw.ac.in

2nd Neti Vishwanathan
Department of Electrical Engineering
National Institute of Technology, Warangal
Telangana, India
email address: nvn@nitw.ac.in

Abstract—Induction heating has emerged as an energy-efficient and eco-friendly alternative to conventional resistance heating methods. This technology rapidly heats the cookware with minimum energy loss using the electromagnetic induction principle for use in domestic cooking applications. The converter plays a prominent role here, generating high-frequency currents to heat the cookware. Designing a cost-effective converter with reasonable efficiency for multiple load applications is more challenging in an IH system. An inverter configuration for the multi-load application is proposed here with reduced components, higher efficiency, improved ZVS range, and simple control. The proposed inverter is simulated using the MATLAB Simulink tool for 2.5 kW output power with four ferromagnetic loads. A combination of asymmetric voltage cancellation (AVC) and ON-OFF control (i.e. Hybrid control) techniques are used to control the loads independently. The inverter operating frequency is 30 kHz and the maximum efficiency observed is 94.6 %.

Index Terms—inverter, electromagnetic induction, induction cooking, AVC, cookware, multi-load configuration, ON-OFF Control, ZVS.

I. INTRODUCTION

With increased environmental concerns, domestic heating using the electromagnetic induction principle has gained widespread popularity. Unlike conventional gas cooking systems, induction cooking offers advantages like rapid heating, easy maintenance, high conversion efficiency, high power density, and safety. In this system [1], an inductor coil placed below the cooktop generates an alternating magnetic field when powered with a high-frequency (20-100kHz) AC supply. This alternating magnetic field produces eddy currents in the vessel placed on the surface. These currents generate heat, which is utilized for cooking. Skin depth plays a major role in determining the depth of the alternating magnetic field that penetrates the material being heated [2]. Expression of skin depth in terms of switching frequency is given in (1).

$$\delta = \sqrt{\frac{\rho}{\pi \mu f}} \qquad (1)$$

where ρ is the resistivity of material, μ is the permeability of material and f is the operating frequency of inverter.

The converter in an induction cooking application plays a crucial role in high-frequency power conversion. Different topologies available in the literature are full-bridge [3], half-bridge [4], and single-switch [5] topologies. Efficient load power management and accurate temperature control depend highly on the control technique adopted. Pulse density modulation (PDM) [6], Phase-shift control [7], Asymmetric duty cycle (ADC) control [8] and Asymmetric voltage cancellation (AVC) control [9] are a few control strategies used in IH converters.

Single-burner induction cooktops are more compact and portable. In today's context, multi-burner induction stoves becoming more popular due to their flexibility and functionality in modern kitchens. Mulit-port inverters manage several loads with optimal energy distribution, space efficiency and reduced complexity. Literature available on multi-load inverters for IC applications are summarized here.

A Series-Resonant multiinverter topology is proposed in [10] for multiple induction heaters. In this configuration, each load is controlled by a series switch which increases the product cost. An inverter circuit with ON-OFF control is presented in [11] to heat both FM (ferromagnetic) & NFM (non-ferromagnetic) loads. But it uses six number of switches and two free-wheeling diodes to power two loads. A two-output series resonant inverter for two loads is proposed in [12]. It is an extension of full-bridge topology with an additional common leg. The multi-output quasi-resonant converter is proposed in [13]. In this configuration, each load is controlled with a single switch. However thermal management and power handling are difficult.

Higher components requirement, complex design, restricted power capacity, difficulty in handling thermal energy, and higher cost are the drawbacks identified from existing literature. The proposed four-load series resonant inverter configuration can power loads simultaneously and individually with less component count, higher efficiency, and simple control. The following are the benefits of the proposed topology:

1) It can power four loads individually and simultaneously up to rated capacity
2) Cost-effective
3) Conduction losses are less
4) Higher Efficiency (94.6%)
5) Flexible power control range

The paper is organized as follows. Circuit description and

979-8-3315-3013-6/25 $31.00 © 2025 IEEE

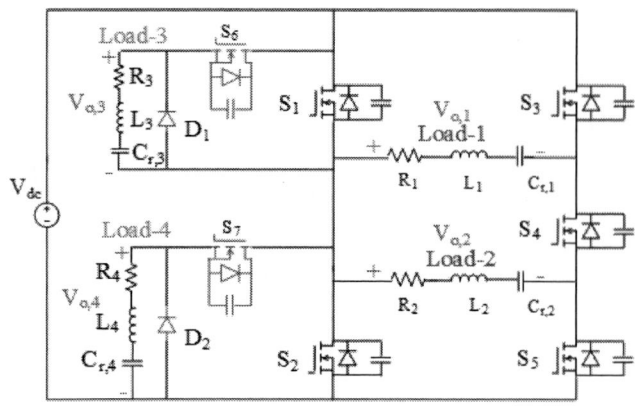

Fig. 1. Proposed Topology for four-load application.

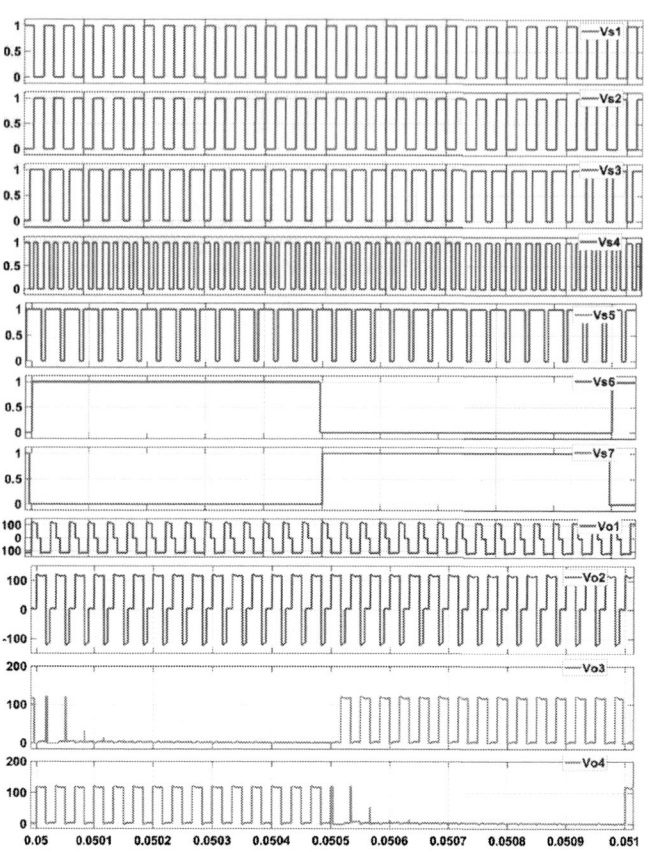

Fig. 2. Operating waveforms: gate pulses (red color), main load waveforms (blue color) and sub load waveforms (Green color).

modes of operation are discussed in Sec II. Fourier analysis and control of the inverter are presented in Sec III. Simulation and analysis of the inverter are described in sec IV. Sec V provides the conclusions.

II. PROPOSED INVERTER

A four-load inverter configuration with series resonance is shown in Fig.1. Loads 1 & 2 are controlled with AVC control, while the output power of loads 3 & 4 regulated with ON-OFF control. Switches S_1, S_2, S_3, S_4, and S_5 are used in powering Loads 1 & 2 . Loads 3 & 4 are powered through the switches S_6 and S_7 . Four loads are of the same material i.e. FM. The circuit diagram represents each load with the series combination of R and L . A Resonant capacitor is added in series with each load to achieve soft - switching operation. Load.1 is represented with R_1, L_1 and C_{r1} and similarly for Load. 2 . Load.3 is represented by R_3, L_3 and $C_{r,3}$. Similarly, for Load. 4. Diodes D_1 and D_2 are used here to free-wheel the energy stored in inductors. Experimentally obtained vessel parameters are used in the simulations of the proposed inverter. The measured coil inductance at an operating frequency of 30 kHz is 97.3 uH. A resonant capacitor rated for 0.4 uF is used to operate the circuit in resonance mode which is calculated by using the formula.

$$f_r = \frac{1}{2\pi \sqrt{L_r C_r,}} \quad (2)$$

In this configuration Loads 3& 4 operate in half-bridge mode with ON-OFF control, whereas Loads 1& 2 operate in full-bridge mode with AVC control. A combination of both AVC and ON-OFF control is used in the proposed configuration and hence is termed "Hybrid control". Switching pulses are shown in Fig. 2. At any instant, at least two loads are powered. The entire operation of the inverter is described in four modes with negligible dead time transitions.

Mode 1 : In this mode, S_1, S_4, S_5 and S_7 are ON . Source powers the three loads as shown in Fig. 3(a). Three load currents are also positive during this mode. Load.3 freewheels through the diode D_1. Load voltages are $V_{o,1} = V_{o,2} = V_{o,4} =$

$+V_{dc}$ and $V_{o,3}$ =0.

Mode 2 : In this mode, S_1, S_3, S_5 and S_7 are ON & S_4 is OFF as shown in Fig. 3(b). Source V_{dc} powers Load.2 and Load.4 through switches S_1, S_5 and S_1,S_7 respectively. At the termination of the mode , $V_{o,1} = V_{o,3} = 0$, & $V_{m,2} = V_{s,2} = +V_{dc}$.

Mode 3 : In this mode , S_2, S_3, S_4 and S_6 are ON . Source powers Load.1 & 2 through switches S_2, S_3 and S_2,S_3, S_4 respectively. Load currents are negative in this mode. Individual load voltages are , $V_{o,1} = V_{o,2} = -V_{dc}$, $V_{o,3} = +V_{dc}$ & $V_{o,4} = 0$.

Mode 4 : In this mode, S_4 turn-off and S_5 starts conduction along with switches S_2, S_3 , S_6. Hence Load.2 power becomes 0 as shown in Fig. 3(d). Load.1 & 3 are powered by source V_{dc} through the switches S_2, S_3 and S_6, S_2 respectively. Load voltages are $V_{o,1} = -V_{dc}$, $V_{o,3} = +V_{dc}$ & $= V_{o,2} = V_{o,4} = 0$.

III. FOURIER ANALYSIS & CONTROL OF INVERTER

A. Fourier Analysis

Four load voltage waveforms of the inverter i.e, $V_{o,1}$, $V_{o,2}$, $V_{o,3}$, $V_{o,4}$ are shown in Fig. 2. Fourier equations of these waveforms are discussed below without considering dead times.

979-8-3315-3013-6/25 $31.00 © 2025 IEEE

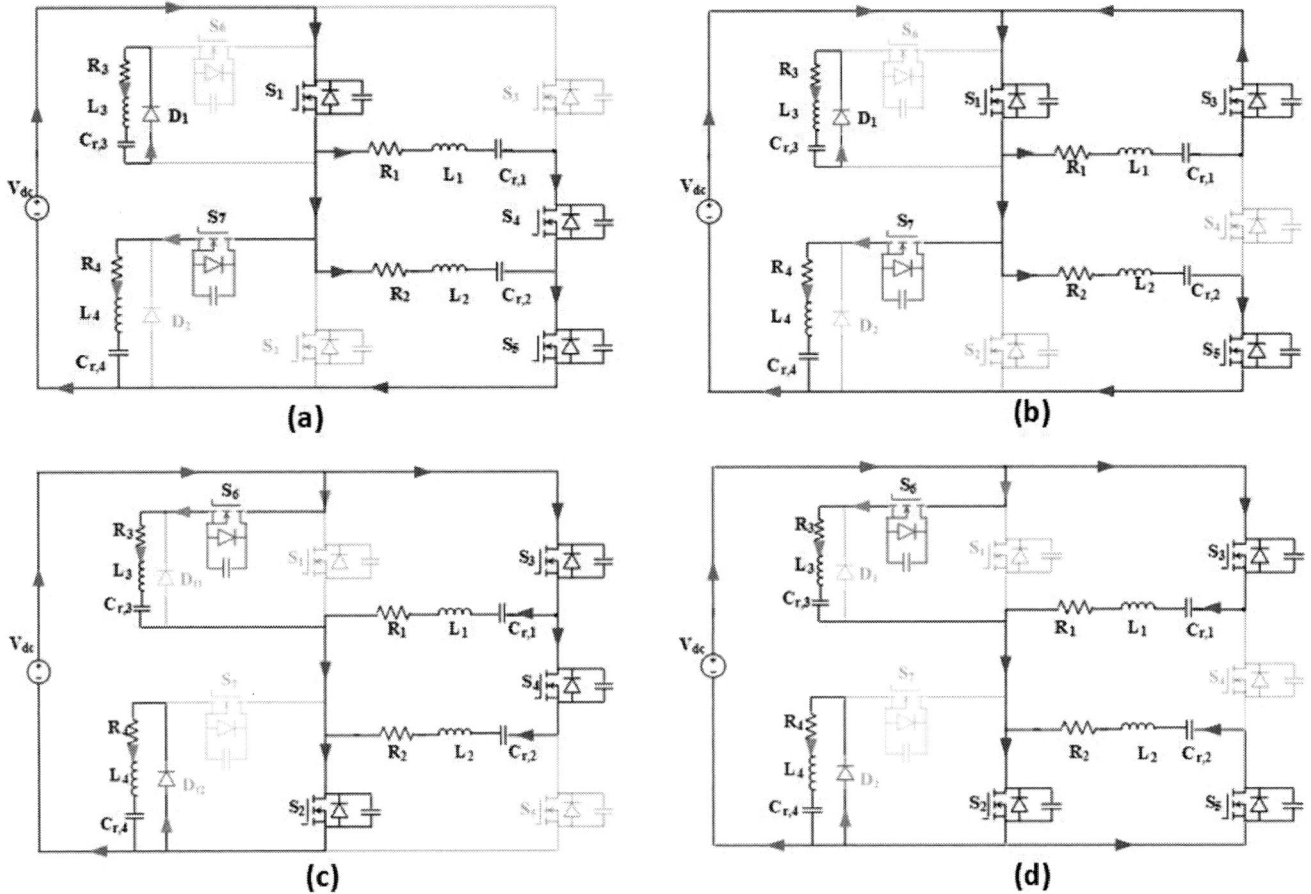

Fig. 3. Modes of operation of proposed inverter

1) Ouput power of AVC controlled loads: In the proposed configuration, Loads 1 & 2 are operated with AVC control. AVC control provides better ZVS range than phase shift, and ADC control techniques [14], [15]. These loads operate with three voltage levels i.e. $\pm V_{dc}$ & 0.

The expression for the fundamental component of AVC controlled load voltage waveform [14] $V_{o,1}$ is

$$v_{f,1}(t) = \frac{V_{dc}}{\pi} \sqrt{10 + 6\cos(2\pi d)}$$
$$\cos\left(\omega t - \tan^{-1}\left(\frac{\sin(2\pi d)}{3 + \cos(2\pi d)}\right)\right) \quad (3)$$

Here V_{dc} is the supply voltage, and d is the ratio between the zero voltage period to one cycle time of the inverter. Expression for the rms voltage of the AVC controlled load is

$$V_{\mathrm{m}} = \frac{V_{dc}}{\pi} \sqrt{\frac{10 + 6\cos(2\pi d)}{2}} \quad (4)$$

Load rms current can be expressed as

$$I_{m} = \frac{V_{dc}}{\pi R} \sqrt{\frac{10 + 6\cos(2\pi d)}{2}} \cos\phi \quad (5)$$

Here ϕ is the angle between load current and fundamental voltage. R is the combination of vessel resistance, coil parasitic resistance, switch drain- source resistance and resonant capacitor parasitic resistance.

$$\cos\phi = \frac{1}{\sqrt{1 + Q^2 \left(\frac{\omega}{\omega_0} - \frac{\omega_0}{\omega}\right)^2}} \quad (6)$$

Derived expression for output power of Load-1 is

$$P_{\mathrm{m}} = I_m^2 * R_{eff} \quad (7)$$

Here R_{eff} is the combined resistance of the coil and vessel. Q is the quality factor. ω_s is the switching frequency in rad/sec and ω_o is the resonant frequency in rad/sec.

2) Output power of ON-OFF controlled loads: In the proposed configuration, Loads. 3 and 4 are controlled with ON-OFF control which means switch turn-on time controls the load power. These loads operate in half-bridge mode with $+V_{dc}$ and 0 voltage levels as shown in Fig. 2. Expression for fundamental component of ON-OFF controlled load voltage $V_{o,3}$ is

$$V_{f,3} = \frac{\sqrt{2}V_{dc}}{\pi} \tag{8}$$

Load rms current can be expressed as

$$I_s = \frac{\sqrt{2}V_{dc}}{\pi Z_{Total}} \tag{9}$$

Here Z_{Total} is the load impedance which is given by

$$Z_{Total} = R + j(\omega_s L - \frac{1}{\omega_s C_r}) \tag{10}$$

Expression for output power of ON-OFF controlled load is

$$P_s = I_s^2 * R * D \tag{11}$$

Here D is the ratio between series switch turn-on time to on-off time. Total power drawn by four loads is

$$P_{Total} = (I_{o1}^2 + I_{o2}^2 + I_{o3}^2 * d_3 + I_{o4}^2 * d_4)R \tag{12}$$

Here R is the vessel resistance referred to coil.

B. Power control:

In this topology, S_1, S_2, S_3, S_4 and S_5 switches are operated with a switching frequency of 30 kHz. Series connceted switches S_6 and S_7 are operated with on-off frequency of 1 kHz. Fig. 2 shows the switching sequence and load waveforms. A hybrid control approach is used in the proposed configuration.

1) Control of Loads. 1 & 2: It can be observed that Load.1 power is controlled during the positive half-cycle (i.e. Time period = $T/2$). In contrast, Load. 2 power is regulated during the negative half-cycle. Duty cycles d_1 & d_2 are used to control the power of Loads.1 and 2. d_1 and d_2 can be expressed as

$$d_1 = \frac{T - t_{on,S3}}{T/2} \tag{13}$$

$$d_2 = \frac{T - t_{on,S5}}{T/2} \tag{14}$$

Here $t_{on,S3}$ and $t_{on,S5}$ represents the turn-on time of switches S_3 and S_5 respectively.

2) Control of Loads. 3 & 4: Loads. 3 and 4 are controlled by regulating the turn-on time of series-connected switches S_6 and S_7. Waveforms are shown in Fig.2. d_3 and d_4 are used in controlling the power of Loads. 3 and 4 and their expressions are given below.

$$d_3 = \frac{t_{on,s6}}{T_{on-off}} \tag{15}$$

$$d_4 = \frac{t_{on,s7}}{T_{on-off}} \tag{16}$$

Here $t_{on,s6}$ and $t_{on,s7}$ represents the turn-on time of switches S_6 and S_7 respectively.
From Fig. 2 it can be observed that at any instant, a minimum of two loads remain powered.

IV. SIMULATION & ANALYSIS OF PROPOSED INVERTER

The proposed four-load inverter configuration is simulated using MATLAB Simulink software platform. Specifications used in simulating the circuit are shown in Table I. Four loads are of the same material type (i.e. ferromagnetic) are considered for simulation. Vessel resistance and inductance parameters are measured at different frequencies using an LCR meter. This configuration is simulated with a 30kHz switching frequency and experimentally tested R and L parameters are used in simulation. $\frac{f_s}{f_o}$ ratio of 1.2 is chosen to improve ZVS range. Capacitor rated for 0.4 μF used in series each with IC load to get resonance operation.

TABLE I
DESIGN SPECIFICATIONS

Parameter	Symbol	Rating
Supply voltage	V_{dc}	120 V
Switching frequency	f_s	30 kHz
Resonant frequency	f_r	25 kHz
on-off frequency	f_{on-off}	1 kHz
Equivalent resistance of IC load	$R_1 = R_2 = R_3 = R_4$	6.2Ω
Equivalent inductance of lf load	$L_1 = L_2 = L_3 = L_4$	97.3 uH
Resonant capacitance of lf load	$C_{r,1} = C_{r,2} = C_{r,3} = C_{r,4}$	0.4 uF

simulation results of the proposed topology are shown in Fig. 4 and 5. Simulated Voltage and current waveforms of Load. 1 and Load. 2 at 60% and 40% duty cycles are shown in Fig. 4. The observed load currents are 11.21 A and 9.31 A respectively. Fig. 5 shows the voltage and current waveforms of the Load. 3 and Load. 4 at an equal duty cycle of 0.5. Currents drawn by loads 3 & 4 at 50% duty cycle is 4.52 A. Fig. 4 and Fig. 5 show that the voltage level of Loads 1 & 2 is changing between $\pm V_{dc}$ and 0 levels, whereas for Loads. 3 & 4 it is varying between V_{dc} and 0 voltage levels.

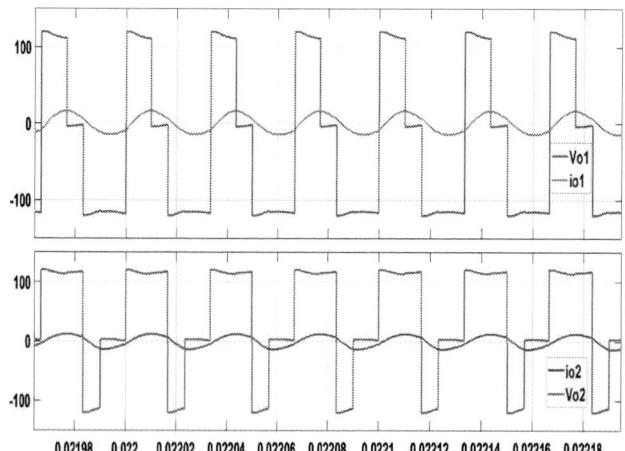

Fig. 4. Load-1 voltage, current waveforms (top), Load-2 voltage, current waveforms (bottom).

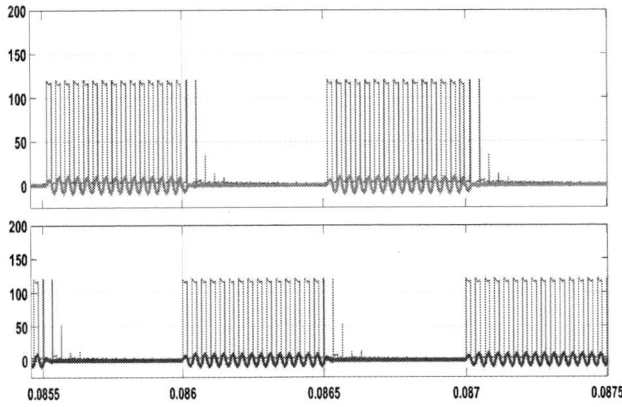

Fig. 5. load.3 voltage, current waveforms (top), load.4 voltage, current waveforms (bottom)

Switch voltage and current waveforms of the inverter are shown in Fig. 6. Zero voltage switching during turn-on and turn-off can be observed from Fig. 6.

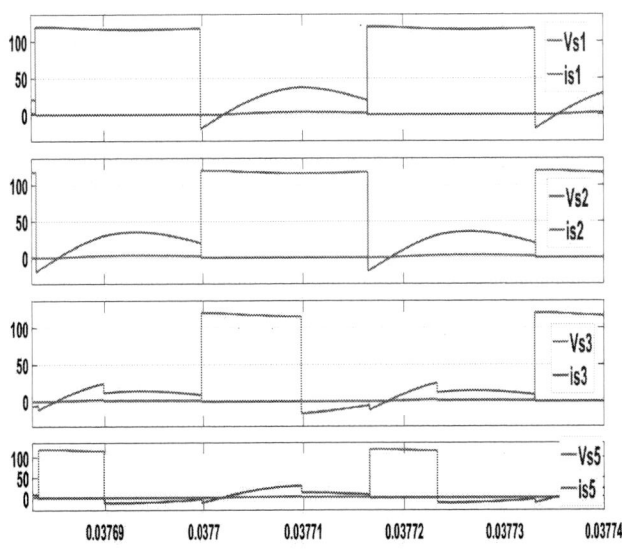

Fig. 6. Soft-switching waveforms of the proposed inverter.

Effect of duty - cycles d_1 and d_3 on load currents for fixed duty-cycles of d_2 and d_4 are shown in the form of characteristics in Fig. 7. Similarly, the effect of duty-cycles d_2, d_4 on load currents is shown in Fig. 8. These characteristics clearly show the dependence of load currents on duty- cycles (d_1, d_2, d_3, d_4). From Fig. 7 & Fig. 8 it can be observed that the one (i.e., load) which is operating with a constant duty cycle is drawing almost constant current, and the other one is varying. These characteristics confirm that the loads can be individually controlled. Maximum currents drawn by Load-1 and Load-3 are 12.8 A & 6.3 A, respectively (Fig. 7). Four loads with 6.2Ω resistance give a power of 2.53 kW. The

proposed configuration is compared with the existing inverter configurations and is shown in Table II. It can be observed that the proposed inverter is operating with a peak efficiency of 94.6% and switch to load ratio of 1.75. Efficiency variation with respect to load power is shown in Fig. 9. Maximum efficiency of the inverter i.e. 94.6% can be observed from this plot. This plot also shows the range of power regulation for connected loads.

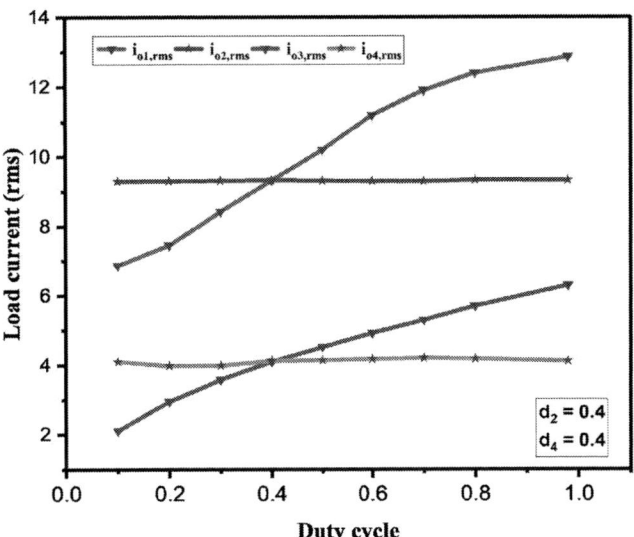

Fig. 7. Load current vs. Duty cycle (d_1, d_3) .

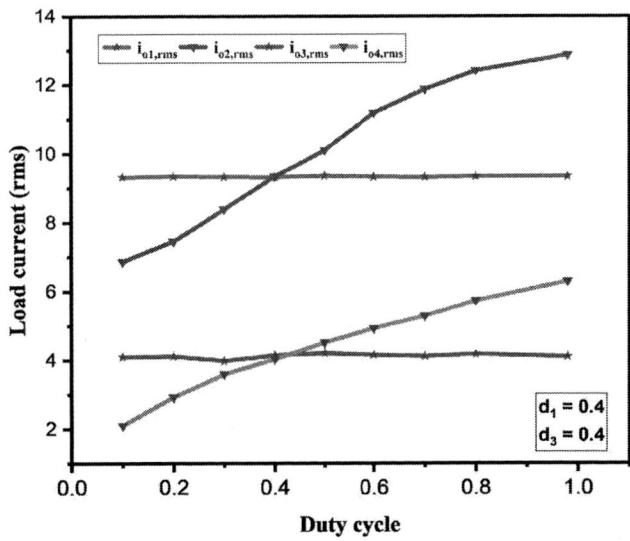

Fig. 8. Load current vs. Duty cycle (d_2, d_4) .

V. CONCLUSIONS

A four-load inverter configuration with hybrid control (i.e. AVC and ON-OFF) for the ferromagnetic material application is proposed here. The proposed inverter is simulated for 2.5

TABLE II
COMPARISON OF PROPOSED CONFIGURATION WITH EXISTING TOPOLOGIES

Ref.No	No.of switches	$f_s(kHz)$	control used	no.of loads	η_{max}	soft-switching	$P_{out}(W)$	Switch to load ratio	Type of load
[7]	4	30.5-40	PSC	1	97.7	Yes	1.3k	4	-
[10]	6	20-150	VFDC, HF-PDM	4	-	Yes	600	1.5	steel
[11]	6	30,220	ON-OFF	2	94.3	only Turn-on	1.62k	3	steel, Al
[12]	6	48	AVC	2	-	Yes	3.2K	3	steel
[14]	4	55.5	AVC	1	96.6	Yes	2k	4	-
[15]	5	43	AVC	2	96.7	Yes	1.6k	2.5	iron
Proposed configuration	7	30	AVC, ON-OFF	4	94.6	Yes	2.5k	1.75	iron

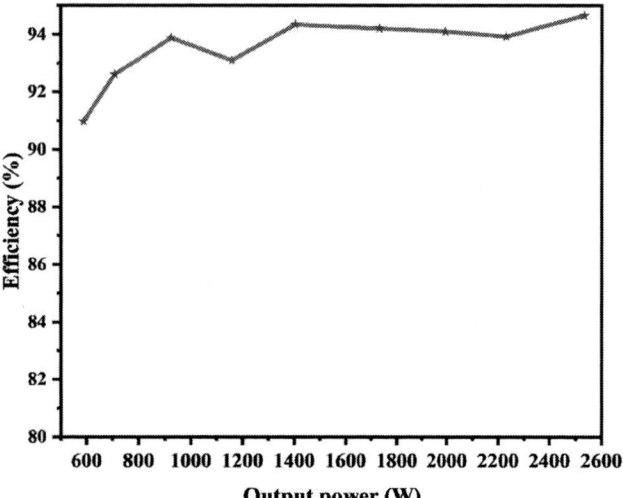

Fig. 9. Efficiency vs. load power .

kW power and the observed peak efficiency is 94.6 %. The benefits of the proposed topology are multi-load operation, simple control, reduced switch count, individual load control, better power regulation, and high efficiency. The proposed inverter can power at least two loads and a maximum of four loads up to rated power with seven number of switches. Fourier expressions of output power for AVC and ON- OFF controlled loads are discussed here. The proposed configuration can be extended to "n" number of loads by adding three switch legs to meet the desired load requirement.

REFERENCES

[1] P. Guillen, H. Sarnago, O. Lucia and J. M. Burdio, "GaN-Based Matrix Resonant Power Converter for Domestic Induction Heating," , IEEE Transactions on Power Electronics, vol. 38, no. 6, pp. 6769-6773, June 2023, doi: 10.1109/TPEL.2023.3239160.

[2] R. Junju and S. Porpandiselvi, "A Reconfigurable Dual Resonant Inverter with Independent Control for Induction Heating," 2024 IEEE 4th International Conference on Sustainable Energy and Future Electric Transportation (SEFET), Hyderabad, India, 2024, pp. 01-05, doi: 10.1109/SEFET61574.2024.10718236.

[3] Zhongming, Y., Jain, P.K., Sen, P.C. "A full-bridge resonant inverter with modified phase-shift modulation for high-frequency AC power distribution systems," IEEE Trans. Ind. Electron., vol. 54, no.5, pp. 2831–2845, 2007.

[4] Esteve, V., Jord´an, J., Dede, E.J., Bellido, J.L " Enhanced asymmetrical modulation for half-bridge series resonant inverters in induction heating applications," IET Power Electron. 16, 2482–2491 (2023).

[5] H.Terai, I.Hirota, T.Miyauchi and H.Omori " Comparative Performance Evaluations of IGBTs and MCT in Single Ended Quasi Resonant Zero Voltage Soft Switching Inverter," Power Electronics Specialists Conference (PESC-2001), IEEE, vol. 4, pp. 2178-2182, 2001.

[6] C.-S. Yeh, C.-W. Chen, M. Lee and J.-S. Lai"A Hybrid Modulation Method for Single-Stage Soft-Switching Inverter Based on Series Resonant Converter," IEEE Transactions on Power Electronics, vol. 35, no. 6, pp. 5785-5796, June 2020.

[7] S. Komeda and H. Fujita "A Phase-Shift-Controlled Direct AC to-AC Converter for Induction Heaters," IEEE Transactions on Power Electronics, vol. 33, no. 5, pp. 4115-4124, May 2018, doi: 10.1109/TPEL.2017.2712281.

[8] T. Ahmed, K. Ogura, S. Chandhaket, and M. Nakaoka, "Asymmetrical duty cycle-controlled edge resonant soft switching high-frequency inverter for consumer electromagnetic induction fluid heater," Automatica,ATKAAF, vol. 44, no. 1-2, pp. 21–26, 2003.

[9] S.Chudjuarjeen, A. Sangswang and C. Koompai, "An Improved LLC Resonant Inverter for Induction-Heating Applications with Asymmetrical Control," IEEE Transactions on Industrial Electronics, vol. 58, no. 7, pp. 2915- 2925, July 2011.

[10] Ó. Lucía, J. M. Burdío, L. A. Barragán, J. Acero and I. Millán, "Series-Resonant Multiinverter for Multiple Induction Heaters," in IEEE Transactions on Power Electronics, vol. 25, no. 11, pp. 2860-2868, Nov. 2010, doi: 10.1109/TPEL.2010.2051041

[11] B. Salvi, S. Porpandiselvi and N. Vishwanathan, "An Inverter Circuit Configuration Suitable for Vessels of Different Mate-rial for Multiload Induction Cooking Application," IEEE Journal of Emerging and Selected Topics in Power Electronics, vol. 11, no. 3, pp. 3223-3235, June 2023.

[12] Jose M.Burdio, Fernando Monterterde, Jose R.Garcia, Luis A.Barragan and Abelardo Martinez, " A Two-Output Series-Resonant Inverter for Induc tion-Heating Cooking Appliances," IEEE Trans. Power Electronics., vol. 20, no. 4, pp. 815-822, 2005.

[13] H. Sarnago, J. M. Burdio and O. Lucia, "Dual-Output Extended-Power Range Quasi-Resonant Inverter for Induction Heating Appliances," in IEEE Transactions on Power Electronics, vol. 38, no. 3, pp. 3385-3397, March 2023.

[14] J. M. Burdio, L. A. Barragan, F. Monterde, D. Navarro and J. Acero, "Asymmetrical voltage-cancellation control for full-bridge series resonant inverters," in IEEE Transactions on Power Electronics, vol. 19, no. 2, pp. 461-469, March 2004, doi: 10.1109/TPEL.2003.823250.

[15] K. H. Rani and N. Vishwanathan, "An Inverter Topology With Reduced Switch Count for Multiple-Load Induction Hob Applications," in IEEE Transactions on Consumer Electronics, doi: 10.1109/TCE.2024.3499999.

Hot redundent FPGA Based Controller for Fixed Switching String Sequential Shunt Regulator for Satellite Power Bus

1st Manoj Kumar Singh
Power System Group(PSG) URSC
Indian Space Resurch organization
(ISRO)
Bangalore, India
mksingh@ursc.gov.in

2nd Dr. Pradeep K Peper
Power System Group(PSG) URSC
Indian Space Resurch organization
(ISRO)
Bangalore, India
pkp@ursc.gov.in

3rd Venakata Sastry Vadlamani
Power System Group(PSG) URSC
Indian Space Resurch organization
(ISRO)
Bangalore, India
vvsastry@ursc.gov.in

4th Sreekumar V
Power System Group(PSG) URSC
Indian Space Resurch organization
(ISRO)
Bangalore, India
srikumar@ursc.gov.in

5th Dibyaranjan Senapati
Power System Group(PSG) URSC
Indian Space Resurch organization
(ISRO)
Bangalore, India
senapati@ursc.gov.in

6th Pichali ReddyKishore Reddy
Power System Group(PSG) URSC
Indian Space Resurch organization
(ISRO)
Bangalore, India
rkishore@ursc.gov.in

Abstract— This paper presents the implementation of a Solar Array Regulator (SAR) utilizing a Fixed Switching String Sequential Switching Shunt Regulator (FS³R) architecture[1], integrated with a FPGA-based shunt regulator referred to as the Hybrid Fixed Switching String Shunt Regulator (HFS3R). The proposed system enhances power regulation efficiency and reliability in space-based solar power systems. To address the risk of single-point failure, a hot-redundant digital controller approach is employed, ensuring continuous operation and improved fault tolerance. The combination of conventional FS3R and FPGA-controlled HFS3R offers a robust and flexible power regulation solution suitable for critical high power space applications.

Keywords—bus regulation, shunt regulator, Hot Redundant FPGA, shift register

I. INTRODUCTION

Regulated power bus architecture is the preferred configuration for high-power satellite platforms due to its ability to maintain a stable and continuous voltage supply under varying operating conditions[2]. This architecture plays a critical role in ensuring reliable power distribution to satellite subsystems throughout both sunlit and eclipse periods.

As depicted in Figure 1, the main electrical components of a regulated power bus include the Solar Array Regulator (SAR), Battery Charge Regulator (BCR), and Battery Discharge Regulator (BDR). Each of these components serves a specific function within the power management system. During sunlit periods, the SAR regulates the bus voltage using power generated by the solar array. Simultaneously, the BCR charges the onboard battery with any excess solar energy. In eclipse periods, when solar power is unavailable, the BDR takes over to maintain the bus voltage by discharging the battery.

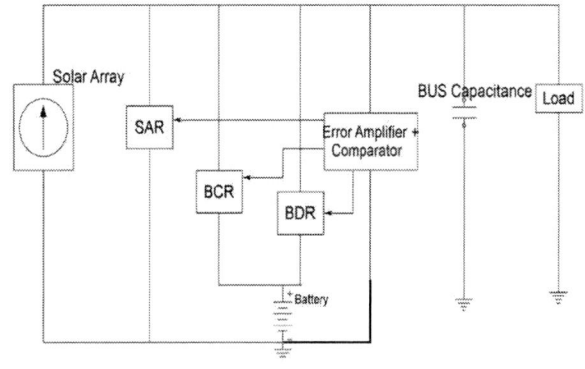

Fig. 1. Block diagram of SAR

A key feature of this topology is the continuous regulation of the bus voltage, regardless of solar exposure. This is accomplished through an error amplifier that dynamically controls the operation of the Solar Array Regulator (SAR), Battery Charge Regulator (BCR), and Battery Discharge Regulator (BDR) to maintain the required voltage levels. The robustness and efficiency of this configuration make it ideally suited for the demanding conditions of space missions. This paper discusses the implementation of a Fixed Switching String Shunt Regulator in a hybrid mode—combining both digital and analog control techniques.

II. SOLAR ARRAY REGULATOR (SAR)

There are many topologies for the implementation of SAR[1,2,3]. Historically most of the high power satellite bus. regulated through S³R. Working of S3R[3] is based on the hysteresis control which works in analog domain.

$$\text{Error Voltage (V)} = \text{Gain} \times \text{Bus ripple}$$

The error amplifier represents the BUS voltage deviation from the set value. There are two trip levels for inclusion and exclusion of switching string. When there is small load change then the bus is regulated by changing the frequency of the ripple. The string inclusion and exclusion with load variation for bus regulation in the S3R scheme are illustrated in Figure 2.

FS³R (Fixed switching string shunt regulator)[1]. The paper is focused on the implementation of a fixed switching string shunt regulator in digital domain. In this topology the strings are grouped into two groups. One group of strings are for fine regulation (S3R region) [3]and other group of strings for coarse regulation (ON/OFF) [1] as shown in figure 3

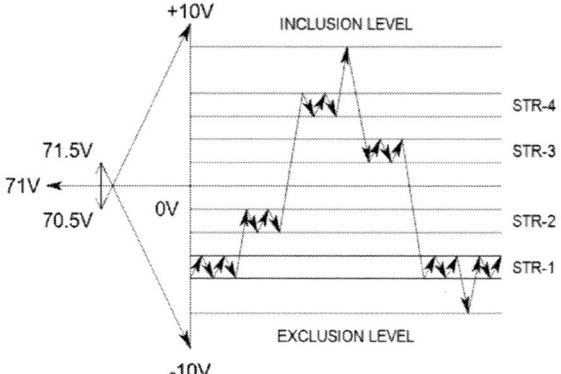

Fig. 2. Block Diagram of S3R

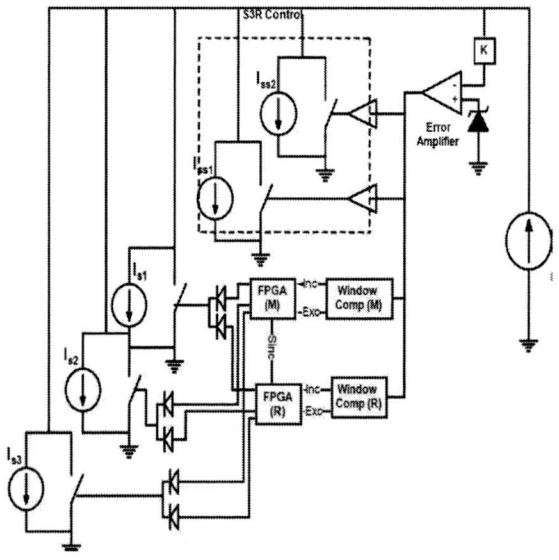

Fig. 3. Block Diagram of Hybrid Fixed Switching String Shunt Regulator

The ON/OFF String is used for coarse correction. This group of strings is controlled through digitally compensated bi-directional shift register which is implemented in the FPGA.

As presented in Fig. 3, the Solar Array Regulator (SAR) is divided into two functional groups: the S3R section for fine regulation and the ON/OFF section for coarse regulation. This topology is designed to minimize the number of Switching Strings (SS) controlled by the S3R controller, thereby maximizing the number of ON/OFF strings that can be managed by a set of hot-redundant FPGAs.

The number of SS is chosen such that, in the event of a single SS failure, the total current handled by the S3R remains greater than the maximum string current of any ON/OFF string. This criterion is essential to ensure the dynamic stability of the SAR.

The ON/OFF section comprises a larger number of strings. Fine regulation of the bus voltage is achieved through two low-power SS operating within the S3R region. Coarse regulation is handled by the ON/OFF section. When the bus voltage remains constant, the power balance among solar generation, load consumption, and shunted power is maintained by balancing the current injected into the bus (bus current) with the current drawn from the bus (load current).

Fig. 3 illustrates the control behaviour under varying load conditions. A hysteresis comparator threshold is set between the upper and lower bounds of a window comparator. This ensures that the bus regulation remains within the S3R region (fine control). As load demand increases, the control voltage (output of the error amplifier) rises and crosses the inclusion threshold, triggering the FPGA logic to connect an additional string.

Details of the FPGA-based control logic are provided in the following section.

III. FPGA BASED CONTROL MECHANISM

The coarse regulation control is implemented using a Field-Programmable Gate Array (FPGA). A shift-register-based control logic is employed to manage the regulation process. To control the rate of string inclusion and exclusion, a digitally compensated clock is utilized, ensuring precise timing and smooth transitions under varying load conditions. To address potential single-point failures in the FPGA, a hot-redundant control scheme is implemented, providing fault tolerance and enhancing system reliability. Two analog comparators are used to control the shift register implemented in the FPGA.

A. Universal Bi- Directional Shift register:

The shift register is utilized to control the drive signals of the MOSFET drivers. Based on the output of the analog comparators—which determine whether a string should be included or excluded—the shift register shifts either a logic '1' or '0'. The direction of the shift (left or right) is governed by

control signals derived from the comparator outputs. Specifically, the 'Include' signal enables a left shift, while the

'Exclude' signal enables a right shift. This mechanism effectively manages the dynamic allocation of power strings.

The detailed control flow is illustrated in Fig. 4,5.

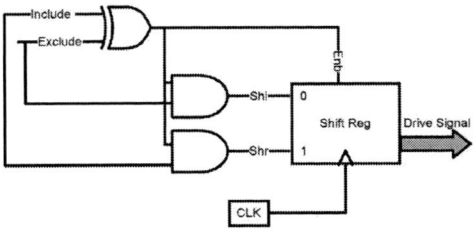

Fig. 4. Block Diagram of shift registor control scheme

B. Programmable Clock generation

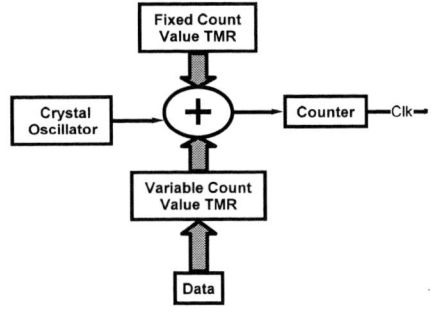

Fig. 5. Compensated clock generation

The primary function of this block is to adjust the clock period of the shift register, enabling software-based tuning of the system. The default clock signal is derived from a crystal oscillator, while the tuning parameter is stored in a Triple Modular Redundancy (TMR) register, as shown in Fig. 5.

The selection of an FPGA-based controller for managing the ON/OFF strings offers several advantages over discrete comparators used for individual string control. FPGA implementation enables centralized logic, enhanced configurability, and reduced hardware complexity. However, this approach introduces a critical drawback: in the event of an FPGA failure, control of all strings is lost, resulting in a catastrophic system failure. In contrast, failure of a discrete comparator would only affect a single string.

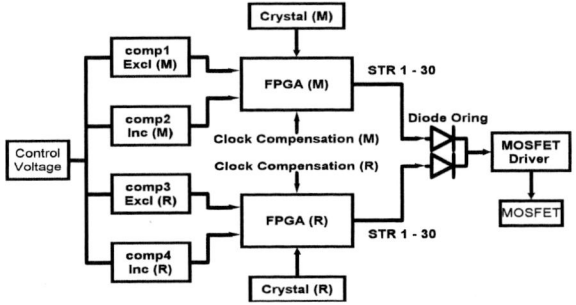

Fig. 6. Hot redundent FPGA control scheme

To address this risk, a hot-redundant FPGA control scheme is proposed. This scheme ensures continuous operation by allowing a backup FPGA to seamlessly take over in the event of a primary controller failure. The overall hot-redundant architecture is illustrated in Fig. 6.

In the proposed hot-redundant control scheme, two identical FPGAs and independent window comparators are employed to ensure robust and fault-tolerant operation. Each FPGA operates independently and is provided with dedicated control resources, including a separate oscillator, window comparators, and power supply. This physical and functional independence minimizes the likelihood of common-mode failures.

Each FPGA contains a Triple Modular Redundancy (TMR)–based shift register, which is responsible for controlling the ON/OFF drive signals of the power strings. The outputs of the two FPGAs are diode-OR'ed, enabling either FPGA to control the load without conflict. This configuration effectively mitigates low-mode failures, where a single FPGA might produce erroneous output or stop functioning entirely.

To handle high-mode failures, an intelligent monitoring mechanism is embedded within each FPGA. If a fault or malfunction is detected in one FPGA, the other FPGA autonomously disables the faulty unit's output by forcing it to a logic low state. This built-in mutual supervision ensures reliable system operation even in the event of critical faults.

The operation of the FPGA-based control logic is categorized into two functional modes, which are described in detail in the following section.

C. Operation Modes of FPGA

The FPGA will be operated in 2 modes. One is normal mode of operation where based on the load requirement no. of switches will be either shunted or included to bus. In the second mode called failure mode, all the drive control signals of failed FPGA will be forced to zero to avoid over voltage situation on the bus. This can be reset with a power cycle.

1) Normal Mode

A shift register–based ON/OFF string control scheme has been proposed as a novel design for fixed-string sequential switching shunt regulation in future Geostationary Earth Orbit (GEO) missions. This architecture aims to enhance system

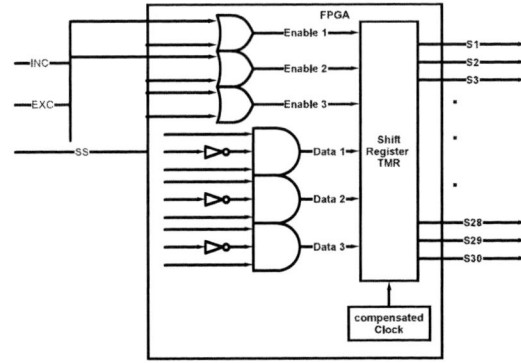

Fig. 7. Normal operation Mode

reliability and control flexibility compared to conventional shunt regulators.

The design incorporates a 30-bit shift register implemented using Triple Modular Redundancy (TMR) registers to ensure fault tolerance in radiation-prone space environments. The

979-8-3315-3013-6/25 $31.00 © 2025 IEEE   47

TMR structure mitigates single-event upsets by utilizing majority voting logic for each bit, thereby ensuring robust and reliable operation.

As illustrated in Fig. 7, two key control signals—ENABLE and DATA—are generated from three external analog comparators. These signals are used to control each stage of the TMR shift register individually, allowing selective inclusion or exclusion of ON/OFF strings based on system requirements.

The rate of inclusion and exclusion is governed by a clock signal, whose frequency is ground-commandable, allowing software-based tuning of the switching dynamics. This clock is synchronized to a 1 MHz master clock, ensuring system-wide timing consistency.

This shift register–based control method provides a scalable, radiation-tolerant, and adaptable approach to shunt regulation, making it a strong candidate for next-generation GEO satellite power systems.

2) Failure Mode

Various types of failure modes are considered in the proposed architecture, as described below. In the event of any failure, the primary objective is to drive all output signals to a low state, thereby allowing the alternate FPGA to assume control. This is facilitated by the diode-OR configuration of the main and redundant drive signals, which ensures seamless handover.

Both FPGAs are configured to mutually monitor each other's operational health. In the event of an external oscillator failure, the healthy (alternate) FPGA detects the absence of a health signal from the faulty FPGA and responds by issuing an asynchronous reset pulse. Upon receiving this reset signal, the faulty FPGA forces all of its drive outputs to a low state, ensuring that only the alternate FPGA remains in control of the system. This failover mechanism is illustrated in Fig. 8.

Two independent analog comparators are employed for string inclusion and exclusion control. Due to the implementation of an AND logic scheme for the inclusion signals from both the main and redundant comparators, a low-mode failure (e.g., stuck-low) of one inclusion comparator does not affect bus regulation. In such cases, the Battery Discharge Regulator (BDR) continues to regulate the bus voltage to meet additional load demands.

Similarly, for exclusion control, an OR logic scheme is used to combine signals from the exclusion comparators. Therefore, a high-mode failure (e.g., stuck-high) of one exclusion comparator results in the exclusion of all strings from the bus. In this scenario, the BDRs take over and regulate the bus voltage.

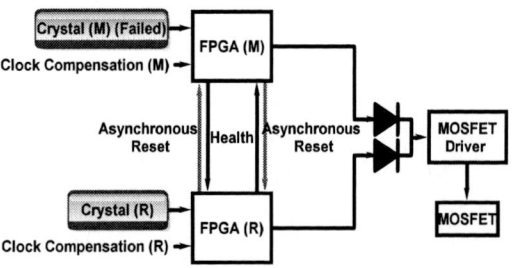

Fig. 8. Failure operation Mode of FPGA (M)

IV. HARDWARE TESTING

To demonstrate the functionality of the Hybrid Fixed Switching String Shunt Regulator (HFS³R), a dedicated hardware prototype was developed, capable of handling a generation current of 104 A. The operational behavior of the regulator was evaluated under both steady-state and transient load conditions, showcasing the dynamic sequence of inclusion and exclusion of various components during load transitions.

The detailed hardware specifications used for the demonstration are as follows:

No of On/Off Strings	: 12(7A-10 no's, 5A-2 no's)
Switching strings	: 2 (12A)
Bus Capacitance	: 9 mF
Total Current	: 104A

A. Test Results Steady State Inclussion/ Exclusion sequence

In the steady-state scenario, the load was gradually increased and decreased to simulate satellite load scenario. During this process, the sequence of inclusion and exclusion of load segments was monitored and recorded. The sequence of inclusion during load rise is presented in Table 1, while the sequence of exclusion during load reduction is detailed in Table These sequences reflect the proper functioning of HSF³R

TABLE 1: INCLUSSION SEQUENCE

Inclusion Sequence:		
Output Current (A)	*Output Voltage (V)*	*String No*
1	71.21	-
12.46	71.06	SS1
24.31	71.06	S3
29.4	71.06	S4
36.32	71.06	S5
43.54	71.04	S6
48.5	71.04	S7
55.73	71.04	S8
62.83	71.04	S9
70.08	71.04	S10
77.06	71.04	S11
84.2	71.04	S12
91.18	71.04	S13
98.18	71.04	S14
104	71.04	S14

TABLE 2: EXCLUSION SEQUENCE

Exclusion Sequence:		
Output Current (A)	Output Voltage (V)	String No
92.87	71.06	SS 2
81.01	71.21	S14
73.9	71.21	S13
66.95	71.21	S12
59.71	71.21	S11
52.63	71.21	S10
45.56	71.21	S9
38.35	71.21	S8
31.44	71.21	S7
24.11	71.21	S6
19.04	71.21	S5
12.01	71.22	S4
4.48	71.22	S3

B. Test Results of Transient response

In addition to the steady-state performance evaluation, the transient response of the HFS³R system was rigorously tested to assess its dynamic behaviour under sudden load variations. Specifically, two test scenarios were conducted: an abrupt increase in load from 500W to 2500W, and a corresponding decrease from 2500W to 500W. These transitions were designed to simulate rapid changes in demand typically encountered in satellite payload operations.

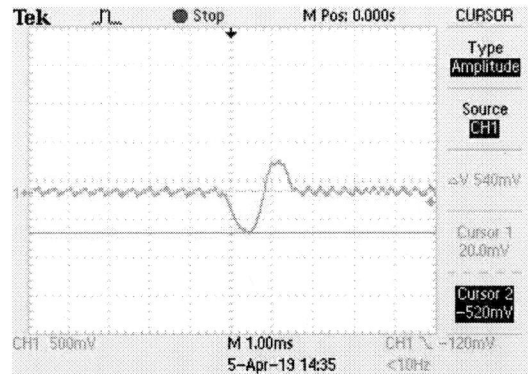

Fig. 9. Transiant response 500W to 2.5kW

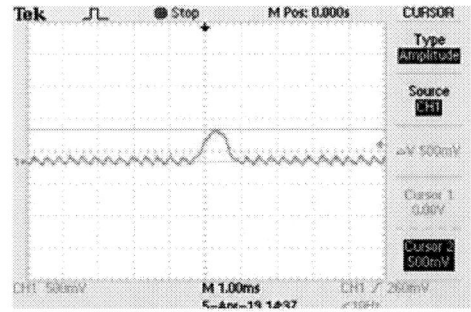

Fig. 10. Transiant response 2.5kW to 500W

C. Test Results TDMA test

To further evaluate the performance of the HFS³R system, the bus response was analysed under a dynamically varying load profile. In this test, the connected load was alternated between 500 W and 2.5 kW at a frequency of 100 Hz.

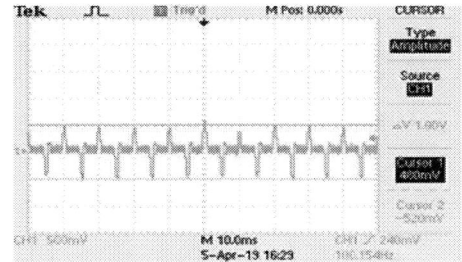

Fig. 11. TDMA @100 Hz from 500 W to 2500W

V. CONCLUSION

The proposed Hybrid Fixed Switching String Shunt Regulator (HFS³R) architecture offers a robust and efficient solution for power regulation in high-power spacecraft applications. By enabling the accommodation of a greater number of solar array strings, HFS3R enhances system scalability and resilience. Its ability to mitigate the impact of single-string failures ensures minimal power loss, thereby significantly improving the overall reliability and power availability in critical space missions.

Importantly, the HFS3R retains all the key benefits of the conventional String Shunt Regulator (S3R), including fast dynamic response, fault tolerance, and stable performance under varying load conditions. The integration of fixed switching and hybrid control strategies further strengthens its suitability for next-generation spacecraft requiring high power handling with enhanced reliability.

The experimental validation, including steady-state and transient load testing, confirms that HFS3R meets the stringent demands of satellite power systems and represents a promising advancement in the field of spacecraft power management.

VI. FUTURE SCOPE

While the HFS3R topology demonstrates significant advantages in power regulation and fault tolerance, certain limitations were observed in the transient response of digitally controlled ON/OFF strings, particularly when compared to analog control counterparts. The slower response in the digital domain may impact performance under rapid load fluctuations.

To address this, future work can explore the integration of feed-forward control techniques to anticipate load changes and improve system response time. Additionally, the adoption of AI-based predictive algorithms—such as machine learning models trained on historical load data—can further enhance transient performance by enabling adaptive control strategies. These approaches have the potential to combine the flexibility of digital control with the fast response characteristics of analog systems, leading to more intelligent and responsive power management in high-power spacecraft.

REFERENCES

[1] Mangal Kumar Mohapatra, Madhusudhana C S, R. C. Biradar, "Solar Array Bus Voltage Regulator for Small Satellite Using FS R (Fixed Switching String Shunt Regulator) Technique," 2024 International Conference on Electronics, Computing, Communication and Control Technology

[2] Spacecraft Power Systems, Mukund R. Patel, CRC Press 2004

[3] D. O'Sullivan and A. H. Weinberg, "The sequential switching shunt regulator (S 3 ʔ)," in Record. ESTEC Spacecraft Power Conditioning Seminar, 1977.

[4] A. Garrigós, J.A. Carrasco, J. M. Blanes, E. Sanchis-Kilders, "A new Sequential Switching Shunt Regulator –Digital Shunt Regulator (S3R-DSR) for solar array regulators", 2006 IEEE International Symposium on Industrial ElectronicsM. Young, The Technical Writer's Handbook. Mill Valley, CA: University Science, 1989.

[5] Choi J.D., "Design considerations of Spacecraft Switched Power System", KIEE Conference 1996, pp.320-322.

[6] Control loop design of sequential switching shunt regulator considering the influence of double section functioning. 10.1049/iet-pel.2013.0365. Fang Li, Xiaojie You, Yan Li. (2014)

Analysis of Magnetic Flux Density of Rogowski Coil for Full Harmonic and Lightning Channel-Base-Current Measurements

Venkatamaheshbabu Lella
Department of Electrical Engineering
National Institute of Technology
Raipur (C.G), India
vmblella.phd2024.ee@nitrr.ac.in

Kandasamy Chandrasekaran
Department of Electrical Engineering
National Institute of Technology
Raipur (C.G), India
kchandrasekaran.ee@nitrr.ac.in

Abstract— The Rogowski coil (RC) is most prevalent device for the measurement of harmonic currents and surge currents. This coil is also measuring the currents across wide range of frequencies which are ranging up to GHz. In this paper, magnetic flux density (*B*) inside of RC is analyzed using Ampere's law and full current theory. In this work MATLAB based coding was utilized to calculate *B* for both conduction and displacement currents. The effectiveness of the radius 'r' from the current carrying conductor to center of the coil on *B* is evaluated for the range from 0.1m to 0.5m. Magnetic flux density (B) is evaluated for different full harmonic high current peaks with frequencies range between 1kHz and 1GHz. From the results it is observed that the dominating magnetic field due to conduction (B_a) and displacement current (B_f) are for smaller and larger radius respectively. The error between B_a and B_f is calculated and reported. From the results it is observed that, irrespective of current peak, for frequencies up to 1MHz the error is 0%. Additionally, full current theory-based B evaluation is attempted for different peaks of lightning channel base current (CBC), same CBC peak with different maximum front steepness ($(di/dt)_{max}$) and its results are reported. From the study, it is noticed that for 12 kA peak CBC, with front steepness from 12 to 120 kA/µs, a maximum of 251% error on B is observed within RC under study.

Keywords—Ampere's law, Full current theory, Harmonic current, Lightning channel-base-current, Magnetic flux density, Rogowski coil

I. INTRODUCTION

Accurate and precise current measurements are essential components for both application in scientific research experiments and technological investigations. The Rogowski coil (RC) is one of the very frequently used device for this purpose and it is proposed by Germen physician Walter Rogowski in 1912 [1]. The RC is similar to a toroid and a conductor wound around it uniformly [2]. The operation of RC is based on Ampere's law and Faraday's law [3]. At present RC's are grown extensively for measuring currents at wide range of frequencies in various applications including power systems [4,5], pulse current measurement process [6], high voltage systems [7], and etc. The wide spread use is due to their advantages, such as their ability to handle a broad spectrum of currents without saturation, extensive frequency range, outstanding transient response, safety and affordability [8]. Particularly, the RC excels in measuring current across a broad frequency spectrum from hertz to gigahertz. Usually RC having less turns so that it has reduced self-inductance, which

leads high frequency measurement [9] and also these high frequency RC's range from measuring high pulse current to partial discharge measurements on power cables [10]. The complex constraint on the RC operation is to achieve wide band width at very high and low frequencies. At low frequencies combination of RC with active integrator as well as at high frequencies RC with passive integrators can solve this problem [11] and also RC with these integrators improves the sensitivity of RC to measure the high bandwidth as well as lighting current in hundreds of kilo amperes [12]. The RC's are primary sensors, which are significantly suited for high voltage insulation is required by the jointly working apparatus like gas insulated switch gears, dead tank, medium voltage application, low voltage applications and high voltage breakers [13]and also these coils generate an output related to the rate of change of the input current. By passing coil's output through integrator, signal is generated and it is proportional to current obtained. This feature provides various advantages in microprocessor-based relays.

These are approximate theories that consider only conduction current and ignore the displacement currents created by the time-varying electric field. At low frequencies, the values of magnetic flux density and current are nearly identical whether both conduction and displacement currents are considered or just conduction current alone.

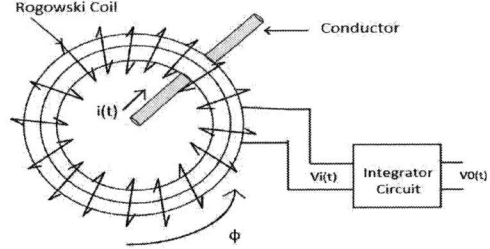

Fig. 1. Rogowski coil measurement principle

However, at high frequencies, it is necessary to consider displacement current in addition to conduction current [14] and many studies have been conducted on shape, size and design of RC or accurate measurement of magnetic flux density and current [15]. This paper presents an exact solution for the distribution of magnetic field for current measurement when RC is used. Additionally, it examines the deviation of results between both methods i.e., Ampere's law and full current theory, specially focusing on the accuracy of magnetic flux density measurement with harmonic and lightning-channel base-currents at high frequencies in RC.

II. EXTENSIVE THEORY ANALYSIS OF CURRENT FRAMEWORK

A. *Principle of measurement for Rogowski coil*

The measurement principle for Rogowski coil (RC) [1], [2] is shown in Fig. 1. When the coil carries a time varying current i(t), it generates a time varying magnetic field inside the coil, which induces an e.m.f (V_{in}). This e.m.f is related to the gradient of the current measured and it can be obtain using an integrator. The output voltage (V_{out}) is directly proportional to the measured current. The accuracy in the current measurement is includes difficulty in determining the relationship between the current measured i(t) and magnetic flux density B present inside RC.

In electrostatics, the time varying Maxwell equations are satisfied in space or air are

Gauss's Law for electric field
$$\mathbf{\nabla}.\mathbf{E} = \frac{\rho}{\varepsilon_0} \tag{1}$$

Faraday's Law for electric filed
$$\mathbf{\nabla} \times \mathbf{E} = -\frac{\partial \mathbf{B}}{\partial t} \tag{2}$$

In magnetostatics, air or vacuum satisfy the time varying Maxwell equations

Gauss's law for magnetic field
$$\mathbf{\nabla}.\mathbf{B} = 0 \tag{3}$$

Faraday's law for magnetic field
$$c^2 \mathbf{\nabla} \times \mathbf{B} = \frac{\mathbf{j}}{\varepsilon_0} + \frac{\partial \mathbf{E}}{\partial t} \tag{4}$$

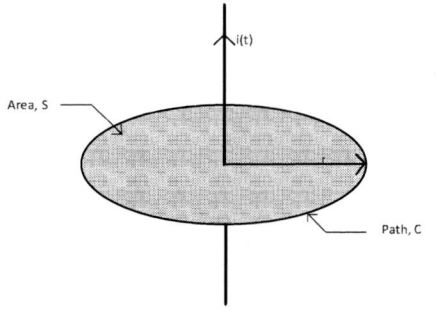

Fig. 2. Rogowski coil model for infinite length conductor carrying time variant harmonic and lightning channel-base-currents

Consider an infinitely long conductor carrying a time-harmonic current, where 'r' is the radius from the current carrying conductor to the coil, 'S' is the area of the magnetic field, 'c' is the velocity of electromagnetic waves in a space, ε_0 is the dielectric constant as shown in Fig. 2. Using Ampere's law (1), (2) and Maxwell's equation (3), (4), integrating (4) over an area 'S', gives rise to

$$\int_S (\mathbf{\nabla} \times \mathbf{B}).d\mathbf{S} = \int_S \left(\frac{\mathbf{j}}{c^2\varepsilon_0} + \frac{1}{c^2}\frac{\partial \mathbf{E}}{\partial t} \right).d\mathbf{S} \tag{5}$$

Apply stokes theorem to (5), the line integral of the magnetic flux density B is

$$\oint_c \mathbf{B}.d\mathbf{l} = \int_S \left(\frac{\mathbf{j}}{c^2\varepsilon_0} + \frac{1}{c^2}\frac{\partial \mathbf{E}}{\partial t} \right).d\mathbf{S} \tag{6}$$

the displacement current is Zero for stable magnetic field. i.e., $\frac{\partial \mathbf{E}}{\partial t} = 0$. Then (6) modified as

$$\text{So,} \quad \oint_c \mathbf{B}.d\mathbf{l} = \int_S \left(\frac{\mathbf{j}}{c^2\varepsilon_0} \right).d\mathbf{S} = \mu_0 \mathbf{I} \tag{7}$$

Ampere's law is used to analyze the working principle of RC. In this context, the cross-section diameter of the coil is much smaller than its radius 'r', and the current carrying conductor is centered inside it, as shown in Fig. 2.

From the Ampere's law, the magnetic flux density inside the coil is.

$$B_a = \frac{\mu_0 I}{2\pi r} \tag{8}$$

If the current is in time varying in nature, then magnetic flux density is

$$B_a(t) = \frac{\mu_0 i(t)}{2\pi r} \tag{9}$$

B. *Full current theory based measuement*

Conductor carrying the time variant harmonic current signal [16] and lightning channel-base-current (CBC) as shown in Fig. 2. are represented in (10) and (11) respectively, the magnetic flux density in the RC due to this current are evaluated as follows

$$i(t) = Im \cos \omega t \tag{10}$$

$$i(t) = \frac{I0_1}{\eta_1} \left[\frac{\left(\frac{t}{T_{11}}\right)^{n_1}}{1+\left(\frac{t}{T_{11}}\right)^{n_1}} \right] \exp\left(\frac{-t}{T_{21}}\right) +$$
$$\frac{I0_2}{\eta_2} \left[\frac{\left(\frac{t}{T_{12}}\right)^{n_2}}{1+\left(\frac{t}{T_{12}}\right)^{n_2}} \right] \exp\left(\frac{-t}{T_{22}}\right) \tag{11}$$

Where $\quad \eta_1 = exp\left[\left(\frac{-T_{11}}{T_{21}}\right) \left(\frac{n_1 \times T_{21}}{T_{11}}\right)^{\frac{1}{n_1}} \right]$

$\eta_2 = exp\left[\left(\frac{-T_{12}}{T_{22}}\right) \left(\frac{n_2 \times T_{22}}{T_{12}}\right)^{\frac{1}{n_2}} \right]$

Apply curl on both sides of (4)

$$\mathbf{\nabla} \times (\mathbf{\nabla} \times \mathbf{B}) = \mathbf{\nabla}(\mathbf{\nabla}.\mathbf{B}) - \mathbf{\nabla}^2\mathbf{B} = \frac{\mathbf{\nabla} \times \mathbf{j}}{c^2\varepsilon_0} + \frac{1}{c^2}\frac{\partial(\mathbf{\nabla} \times \mathbf{E})}{\partial t} \tag{12}$$

Substituting (2) & (3) into (12)

$$\mathbf{\nabla}^2\mathbf{B} - \frac{1}{c^2}\frac{\partial^2 \mathbf{B}}{\partial t^2} = -\frac{\mathbf{\nabla} \times \mathbf{j}}{c^2\varepsilon_0} \tag{13}$$

For non-source space, (13) becomes

$$\mathbf{\nabla}^2\mathbf{B} - \frac{1}{c^2}\frac{\partial^2 \mathbf{B}}{\partial t^2} = 0 \tag{14}$$

By the phasor analysis method, with respect to time hormonic currents above equation can be written as

$$\nabla^2 \boldsymbol{B} + k^2 \boldsymbol{B} = 0 \text{ , Where } \left(k = \frac{\omega}{c}\right). \tag{15}$$

In cylindrical coordinates (14) can be written as

$$\frac{1}{r}\frac{\partial}{\partial r}\left(r\frac{\partial \boldsymbol{B}}{\partial r}\right) + \frac{1}{r^2}\frac{\partial^2 \boldsymbol{B}}{\partial \phi^2} + \frac{\partial^2 \boldsymbol{B}}{\partial z^2} + k^2 \boldsymbol{B} = 0 \tag{16}$$

Due to axial asymmetry structure of current the magnetic flux density

$$\boldsymbol{B} = \left(B_r(r), B_\phi(r), B_z(r)\right) \tag{17}$$

By substituting (16) in (15), three coordinate components equations can be obtained based on the partial derivatives rules. The voltage is induced related to variety of flux in the Rogowski coil. Hence partial differential equation with respect to flux component in cylindrical coordinates is given by

$$r^2 \frac{\partial^2 B_\phi}{\partial r^2} + r\frac{\partial B_\phi}{\partial r} + (r^2 k^2 - 1)B_\phi = 0 \tag{18}$$

In equation (18) Let $\qquad r^\iota = kr$

$$r^{\iota 2}\frac{\partial^2 B_\phi}{\partial r^{\iota 2}} + r^\iota \frac{\partial B_\phi}{\partial r^\iota} + (r^{\iota 2} - 1)B_\phi = 0 \tag{19}$$

The solution for the Equation (19) is given by first order Bessel's equation

$$B_\emptyset = AJ_1(r^\iota) + CY_1(r^\iota) = AJ_1(kr) + CY_1(kr) \tag{20}$$

To determine the coefficients A and C, use the following conditions:
1. As r approaches 0, the line integral of the magnetic flux density B equals $\mu_0 I$
2. As r approaches infinity, the line integral of the magnetic flux density B equals 0

$$\lim_{r\to 0} 2\pi r B_\phi = \frac{1}{c^2 \epsilon_0} = \mu_0 i(t) \tag{21}$$

$$\lim_{r\to\infty} B_\phi = 0 \tag{22}$$

The first and second kind of Bessel's functions [18] are

$$J_1(x) = \sum_{m=1}^{\infty} (-1)^m \frac{x^{2m+1}}{2^{2m+1}\cdot m!\cdot(m+1)!} \tag{23}$$

$$Y_1(x) = \frac{2}{\pi}J_0(x)\left(\ln\frac{x}{2} + c\right) - \frac{1}{\pi}\left(\frac{x}{2}\right)^{-1}$$
$$- \frac{1}{\pi}\sum_{m=0}^{\infty}(-1)^m \frac{x^{2m+1}}{2^{2m+1}\cdot m!\cdot(m+1)!}\left(\sum_{k=0}^{m}\frac{1}{k+1} + \sum_{k=0}^{m-1}\frac{1}{k+1}\right) \tag{24}$$

On substituting (22) and (23) in (20) gives

$$C = -\frac{\mu_0 ki(t)}{4} \tag{25}$$

When 'r' approaches to infinity, the Bessel function approach their approximation expression

$$J_1(kr) \approx \sqrt{\frac{2}{\pi kr}} \cos\left(kr - \frac{3\pi}{4}\right) \tag{26}$$

$$Y_1(kr) \approx \sqrt{\frac{2}{\pi kr}} \sin\left(kr - \frac{3\pi}{4}\right) \tag{27}$$

Substituting (26), (27) into (21) gives

$$\lim_{r\to\infty}\sqrt{\frac{2}{\pi kr}}\left[A\cos\left(kr - \frac{3\pi}{4}\right) + C\sin\left(kr - \frac{3\pi}{4}\right)\right] = 0 \tag{29}$$

$$Aj + C = 0$$

$$A = \frac{\mu_0 ki(t)}{4j} \tag{30}$$

Therefore, substitute (25),(30) in (20) then the component B_ϕ of $\boldsymbol{B}$ is

$$B_\phi = \left(\frac{\mu_0 ki(t)}{4j}\right) J_1(kr) - \frac{\mu_0 ki(t)}{4} Y_1(kr) \tag{31}$$

Therefore, total magnetic flux density based on full current theory in time domain is

$$B_f(t) = AJ_1(kr) + C Y_1(kr) \tag{32}$$

III. COMPARATIVE ANALYSIS OF MAGNETIC FLUX DENSITY, BETWEEN AMPERE'S LAW AND FULL CURRENT THEORY

The variations of magnetic flux densities between Ampere's law $B_a(t)$ (8),(9) and full current theory $B_f(t)$ (32) for different radius, current and frequencies are shown in Table 1. From the data, reported in Table 1, it is observed that there is no variation in magnetic flux density $B_a(t)$ at all frequencies with respect to every radius with Ampere's law. But there is variation in magnetic flux density $B_f(t)$ at gigahertz frequency with respect to each radius and for every magnitude of current. In full current theory and magnetic flux density $B_f(t)$ is more at 0.2m radius.

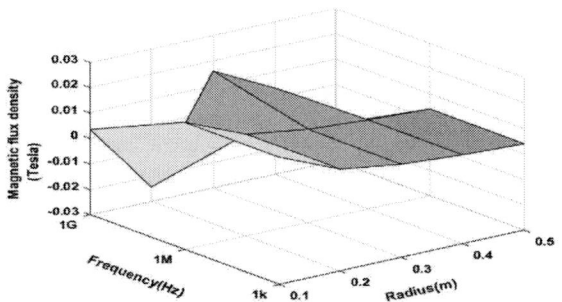

Fig. 3. Plot of magnetic flux density $B_f(t)$ at 1000A as a function of frequency and radius of RC

TABLE 1: VARIATION OF MAGNETIC FLUX DENSITY OF HARMONIC CURRENT FOR THE FREQUENCY FROM 1 KHZ TO 1 GHz AND CURRENT FROM 1KA TO 100 KA WITH RADIUS = 0.1 m TO 0.5 m

Magnetic flux density peak (μWb/m²)	Current Peak (A)	Radius "r" from the current carrying conductor to the coil and frequency of current														
		r=0.1 m			r=0.2 m			r=0.3 m			r=0.4 m			r=0.5 m		
		Frequency (Hz)			Frequency (Hz)			Frequency (Hz)			Frequency (Hz)			Frequency (Hz)		
		10^3	10^6	10^9	10^3	10^6	10^9	10^3	10^6	10^9	10^3	10^6	10^9	10^3	10^6	10^9
$B_a(t)$	10^3	2	2	2	1	1	1	0.6	0.6	0.6	0.5	0.5	0.5	0.4	0.4	0.4
	10^4	20	20	20	10	10	10	6	6	6	5	5	5	4	4	4
	10^5	200	200	200	100	100	100	60	60	60	50	50	50	40	40	40
$B_f(t)$	10^3	2	2	0.3	1	1	2.4	0.6	0.6	1.5	0.5	0.5	0.3	0.4	0.4	1.5
	10^4	20	20	3	10	10	24	6	6	15	5	5	3	4	4	15
	10^5	200	200	30	100	100	240	60	60	150	50	50	30	40	40	150

TABLE 2: MAGNETIC FLUX DENSITY DUE TO HARMONIC CURRENT MATCHING FOR LIGHTNING CURRENT MAGNITUDE AT A FREQUENCY OF 1GHz WITH RADIUS = 0.1 m TO 0.5m

Peak Current(A)	Peak Magnetic Flux Density									
	r=0.1 m		r=0.2 m		r=0.3 m		r=0.4 m		r=0.5 m	
	$B_a(t)$ (Wb/m²)	$B_f(t)$ (Wb/m²)	$B_a(t)$ (Wb/m²)	$B_f(t)$ (Wb/m²)	$B_a(t)$ (Wb/m²)	$B_f(t)$ (Wb/m²)	$B_a(t)$ (Wb/m²)	$B_f(t)$ (Wb/m²)	$B_a(t)$ (Wb/m²)	$B_f(t)$ (Wb/m²)
30×10^3	0.06	0.01065	0.03	0.07299	0.02	0.04738	0.015	0.01146	0.012	0.04652
90×10^3	0.18	0.03196	0.09	0.21896	0.06	0.14216	0.045	0.03438	0.036	0.13956
160×10^3	0.32	0.05682	0.16	0.38927	0.1066	0.25273	0.08	0.06161	0.064	0.24811
10×10^3	0.02	0.00355	0.01	0.02432	0.0066	0.01579	0.005	0.00382	0.004	0.01551
28.6×10^3	0.0572	0.01015	0.0286	0.06958	0.01906	0.04517	0.0143	0.01092	0.01144	0.04435
56×10^3	0.112	0.01989	0.056	0.13625	0.03733	0.08845	0.028	0.02139	0.0224	0.08684
35×10^3	0.07	0.01243	0.035	0.08515	0.0233	0.05529	0.0175	0.01337	0.014	0.05427
250×10^3	0.5	0.08878	0.25	0.60824	0.1666	0.39489	0.125	0.09549	0.10	0.38765
350×10^3	0.7	0.12429	0.35	0.85154	0.2333	0.552857	0.175	0.13368	0.14	0.54275

The variation of magnetic flux density $B_f(t)$ with respect to frequency at different radius and 1000A current in full current theory is shown in Fig. 3. In this we can observe that magnetic flux density is maximum i.e., 0.024 wb/m² at 0.2m radius,1GHz frequency. For examine error between $B_a(t)$ and $B_f(t)$ under different frequency, radius and current, the expression (33) is used.

$$\%Ea = \frac{B_{fm}-B_{am}}{B_{am}} \times 100 \qquad (33)$$

In (33), where B_{am} and B_{fm} are the peak of $B_a(t)$ and $B_f(t)$ respectively. Fig.4. gives the variation of magnetic flux density with respect to harmonic currents having 1kA current peak magnitude and different radius of RC at 1 GHz frequency. From the Fig. 4, it can be observed that the % error for all radius of RC and all the input currents with frequencies from 1KHz to 1MHz is zero. Beyond 1MHz and up to 1 GHz, %error is nonzero. The minimum error i.e., 23.60% is observed for RC radius equal to 0.4m at 1 GHz. In TABLE 2, the data of variation of magnetic flux densities $B_a(t)$ and $B_f(t)$ at 1GHz frequency for harmonic currents matching with lightning channel base current (CBC) magnitudes at every radius is reported. The magnetic flux density $B_f(t)$ due to harmonic current matching for lightning current magnitude is more compared to $B_a(t)$ for the RC radius more than 0.1m. In Fig. 5. B_f as a function of harmonic current matching for lightning current magnitude and different radius of RC at 1GHz is plotted. From the Fig. 5. it is noticed that B_f is more at 0.1m radius for every peak of current. similarly in case of GHz frequency, B_f is more at 0.2m radius. Fig. 6. shows the

% error between $B_a(t)$ and $B_f(t)$, with respect to harmonic currents matching with lightning channel base current magnitudes. The minimum % error is observed at 0.4m radius and it is 23.60%.

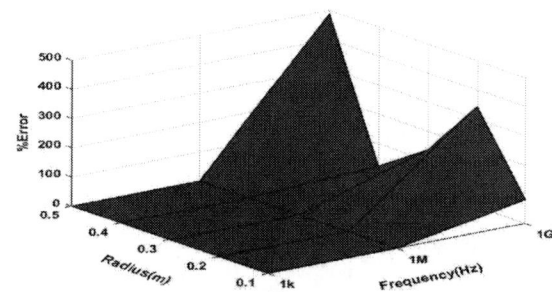

Fig. 4. % Error between B_{am} and B_{fm} for 1000A input harmonic current

The results magnetic flux density due to different CBC at 1 GHz is reported in Table 3. From the data reported in Table 3, it is noticed that, $B_f(t)$ is more than $B_a(t)$ for all the peak and radius of RC under study. Also observed that, both $B_f(t)$ and $B_a(t)$ are more at 0.1m radius. Plot of magnetic flux density B_f as a function of lightning channel base current magnitude and radius of RC at 1GHz is depicted in Fig. 7. The % error between $B_a(t)$ and $B_f(t)$, is calculated and its results are plotted in Fig. 8. The % error is minimum at 0.1m radius and it is maximum at 0.5m radius. For the different $(di/dt)_{max}$ of 12kA lightning CBC with wave front and wave tail of 1.2/50μs, the data reported in [19] is adopted and with different radius of RC, magnetic flux densities $B_f(t)$

TABLE 3: VARIATION OF MAGNETIC FLUX DENSITY DUE TO DIFFERENT LIGHTNING CURRENT MAGNITUDE FOR THE FREQUENCY OF 1 GHz WITH RADIUS OF RC FROM 0.1m TO 0.5 m

Lightning-Channel base -Currents(A)			Peak Magnetic Flux Density									
			r=0.1 m		r=0.2 m		r=0.3 m		r=0.4 m		r=0.5 m	
			$B_a(t)$ (nWb/m²)	$B_f(t)$ (nWb/m²)	$B_a(t)$ (nWb/m²)	$B_f(t)$ (nWb/m²)	$B_a(t)$ (nWb/m²)	$B_f(t)$ (nWb/m²)	$B_a(t)$ (nWb/m²)	$B_f(t)$ (nWb/m²)	$B_a(t)$ (nWb/m²)	$B_f(t)$ (nWb/m²)
NFS	50%	30k	2.344	2.378	1.172	1.319	0.781	1.063	0.586	0.983	0.469	0.954
	5%	90k	3.602	3.654	1.801	2.027	1.201	1.634	0.90	1.511	0.721	1.465
	1%	160k	61.76	62.70	30.90	34.78	20.60	28.04	15.45	25.92	12.36	25.14
NSS	50%	10k	0.6509×10^{-15}	0.66×10^{-15}	0.325×10^{-15}	0.366×10^{-15}	0.217×10^{-15}	0.295×10^{-15}	0.162×10^{-15}	0.273×10^{-15}	0.13×10^{-15}	0.264×10^{-15}
	5%	28.6k	40.38	40.97	20.19	22.73	13.46	18.32	10.09	16.94	8.08	16.43
	1%	56k	0.0281	0.0285	0.014	0.0158	0.0093	0.0127	0.007	0.0117	0.00256	0.0114
PFS	50%	35k	0.227×10^{-8}	0.23×10^{-8}	0.113×10^{-8}	0.127×10^{-8}	0.075×10^{-8}	0.103×10^{-8}	0.056×10^{-8}	0.095×10^{-8}	0.045×10^{-8}	0.092×10^{-8}
	5%	250k	0.0439	$0.0446\times$	0.0219	0.0247	0.0146	0.0199	0.011	0.018	0.0088	0.0179
	1%	350k	10.31	10.46	5.15	5.80	3.44	4.68	2.58	4.32	2.06	4.19

and $B_a(t)$ are calculated, also given in TABLE 4. From these data, it is observed that, at every radius $B_f(t)$ is more than $B_a(t)$. The corresponding % error between $B_a(t)$ and $B_f(t)$, is calculated and its results are reported in Table 5.

decreasing with increasing the radius and it is maximum at 0.1m radius.

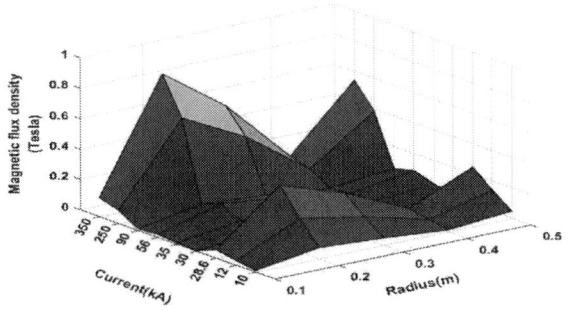

Fig. 5. Plot of magnetic flux density B_f as a function of harmonic current matching for lightning current magnitude and radius of RC at 1GHz

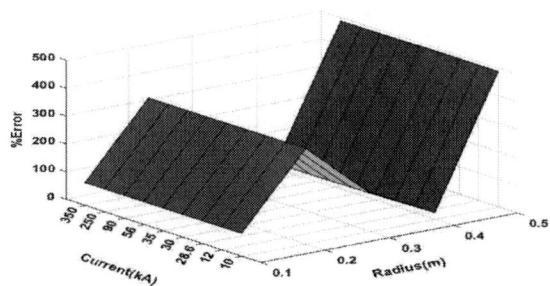

Fig. 6. % Error between B_{am} and B_{fm} for different harmonic current and radius of RC at 1GHz

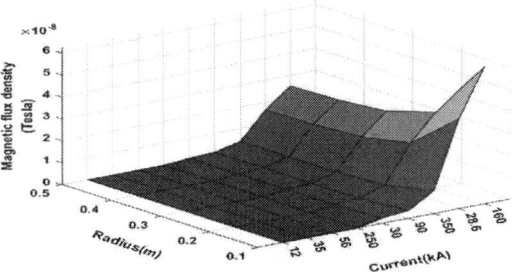

Fig. 7. Plot of magnetic flux density B_f as a function of lightning channel base current magnitude and radius of RC at 1GHz

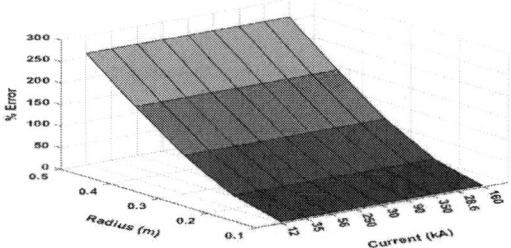

Fig. 8. % Error between B_{am} and B_{fm} for different lightning channel base current and radius of RC at 1GHz

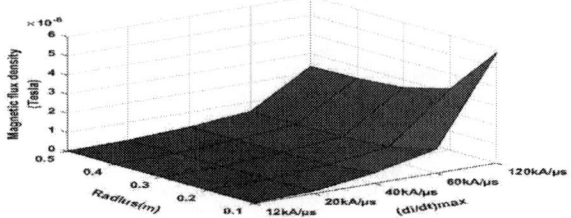

Fig. 9. Magnetic flux density Bf for 12kA Lighting current with different $(di/dt)_{max}$ and different radius of RC at 1GHz.

Fig. 9. shows the plot between $(di/dt)_{max}$ values of 12kA and Magnetic flux density $B_{f(t)}$ at different radius and it is showning that x-axis having the magnetudes of currents for each $(di/dt)_{max}$ currents. The magnetic flux density $B_{f(t)}$ is

TABLE 4: MAGNETIC FLUX DENSITY DUE TO 12KA LIGHTING CHANNEL BASE CURRENT WITH DIFFERENT $(di/dt)_{max}$ AND RC RADIUS FROM 0.1 m TO 0.5 m

Lightning-channel base-current 12kA with $\left(\dfrac{di}{dt}\right)_{max}$ (kA/µs)	Peak Magnetic Flux Density									
	r=0.1 m		r=0.2 m		r=0.3 m		r=0.4 m		r=0.5 m	
	$B_a(t)$ (µWb/m²)	$B_f(t)$ (µWb/m²)	$B_a(t)$ (µWb/m²)	$B_f(t)$ (µWb/m²)	$B_a(t)$ (µWb/m²)	$B_f(t)$ (µWb/m²)	$B_a(t)$ (µWb/m²)	$B_f(t)$ (µWb/m²)	$B_a(t)$ (µWb/m²)	$B_f(t)$ (µWb/m²)
12	0.0468	0.0475	0.0234	0.02639	0.01563	0.02128	0.01172	0.01967	0.00938	0.01908
20	0.1329	0.1348	0.0664	0.0748	0.0443	0.0603	0.0332	0.0557	0.0266	0.0541
40	0.5538	0.5619	0.2769	0.3117	0.1846	0.2513	0.1385	0.2323	0.1108	0.2253
60	1.242	1.260	0.621	0.699	0.414	0.564	0.311	0.521	0.248	0.505
120	5.712	5.796	2.856	3.215	1.904	2.592	1.428	2.396	1.142	2.324

TABLE 5: ERROR BETWEEN $B_a(t)$ AND $B_f(t)$ DUE TO LIGHTNING-CHANNEL BASE -CURRENT 12KA WITH DIFFERENT $(di/dt)_{max}$

Light-channel base-current 12kA with $\left(\dfrac{di}{dt}\right)_{max}$ (kA/µs))	%Error				
	r=0.1 m	r=0.2 m	r=0.3 m	r=0.4 m	r=0.5 m
12	1.654	17.123	61.99	143.78	250.69
20	1.654	17.123	61.99	143.78	250.69
40	1.654	17.123	61.99	143.78	250.69
60	1.654	17.123	61.99	143.78	250.69
120	1.654	17.123	61.99	143.78	250.69

From the data reported in Table 5, it is observed that, the %error is simultaneously increasing from 0.1m to 0.2m radius and it is minimum at 0.1m radius with 1.654%, maximum at 0.5m radius with the value of 250.69%.

IV. CONCLUSION

Using full current theory, the magnetic field inside of Rogowski coil (RC) is analysed for the primary input harmonic and lightning channel base current (CBC). For harmonic current, it is following same trend as like in literature. The magnetic field inside of RC is reported for high peak current with the range of frequencies from 1MHz to 1GHz. Magnetic field based on Ampere's law (Ba(t)) and full current theory (Bf(t)) are calculated by considering RC with different radius from 0.1 to 0.5 m. The results show that for the same harmonic current peak, the magnetic flux density is more at the coil radius 0.2m and error is less at 0.4m radius for currents at GHz frequency. On the other hand, for lightning CBC, the magnetic flux density is more, and error is less at 0.1m radius. From the results of magnetic fields, the maximum error observed between $B_a(t)$ and $B_f(t)$ is 488% and 251% for harmonic current and CBC respectively. The effects of CBC (di/dt) $_{max}$ on magnetic field are also evaluated by keeping the same peak. From these results, it clear that, the full current theory helps for effective measurement of nano pulsed current signals using RC.

REFERENCES

[1] W. Rogowski and W.Steinhaus,"Die messung der magnetischen Spannung," Arch.Elektrotechnik, vol.1pp.141-150, 1912.

[2] S. Ziegler, R. C. Woodward, H. C. Iu, and L. J. Borle, "Current sensing techniques:A review," IEEE Sensors J., vol. 9, no. 4,pp. 354_376, Apr.2009.

[3] C. Xiao, L.Zhao, T. Asada, W. G. Odendaal, and J. D. van Wyk,"An overview of integratable current sensor technologies," in Proc. 38th IAS Annu.Meeting Ind. Appl. Conf., 2003, pp. 1251-1258.

[4] W. Stygar and G. Gerdin,"High frequency Rogowski coil response characteristics," IEEE Trans. Plasma Sci., vol. 10, no. 1, pp. 40-44, Mar.1989.

[5] M. G. Mazarakis, W. E. Fowler, A. A. Kim, V. A. Sinebryukhov, S. T. Rogowski, and R. A. Sharpe," High current, 0.5-ma, fast, 100-ns, linear

[6] transformer drive experiments," Phys. Rev. Special opics,Accel. Beams, vol. 12, no.5, 2009, Art. No.050401.

[7] Y. Liu, F. Lin, Q. Zhang, and H. Zhong,"Design and construction of a Rogowski coil for measuring wide pulsed current," IEEE Sensors J., vol. 11, no. 1, pp. 123-130, Jan.2011.

[8] M. Argueso, G. Robles, and J. Sanz, "Implementation of a Rogowski coil for the measurement of partial discharges," Rev. Sci. Instum., vol.76, no. 6, 2005, Art.no.065107.

[9] D. A. Ward and J. L. T. Exon, "Experience with using Rogowski coils for transient measurements," in Proc.IEE Colloquium Pulsed Power Technol., 1992, pp.6/1-6/4.

[10] M. Marracci, B. Tellini, C. Zappacosta and G. Robles,"Critical Parameters for Mutual Inductance Between Rogowski Coil and Primary Conductor," in IEEE Transactions on Instrumentation and Measurement, vol. 60, no. 2, pp. 625-632, Feb. 2011, doi:10.1109/TIM.2010.2051591.

[11] P. C. J. M. van der Wielen, J. Veen, P. A. A. F. Wouters and E. F. Steennis,"Sensors for on-line PD detection in MV power cables and their locations in substations," Proceedings of the 7th International Conference on Properties and Applications of Dielectric Materials(Cat. No.03CH37417), Nagoya, Japan, 2003, pp. 215-219 vol:1, doi:10.1109/ICPADM.2003.1218391.

[12] B. Wang, D. Wang and W. Wu,"A Rogowski coil current transducer designed for wide bandwidth current pulse measurement,"2009 IEEE 6th International Power Electronics and Motion Control Conference, Wuhan, China, 2009, pp. 1246-1249, doi:10.1109/IPEMC.2009.5157575.

[13] Zhou LI, Qiaogen ZHANG, Lu ZHANG, Fenglian LIU, Xiaoya TAN "Design of Rogowski Coil eith external integrator for measurement of lightning current up to 400kA,"PRZEGLA ELEKTROTECHNICZNY(Electrical Review), ISSN 0033-2097,R. 87 NR 7/2011.

[14] V. Skendzic and B. Hughes, "Using Rogowski coils inside protective relays," 2013 66th Annual Conference for Protective Relay Engineers,College Station, TX, USA, 2013, pp. 1-10, doi: 10.1109/CPRE.2013.6822022.

[15] Heras, Jose. "A Formal Interpretation of the Displacement Current and the Instantaneous Formulation of Maxwell's Equations." American Journal of Physics, American Association of Physics Teachers (AAPT), 2011.

[16] L. Kütt and M. Shafiq, "Magnetic sensor coil shape geometry and bandwidth assessment, " 2011 7th International Conference-Workshop Compatibility and Power Electronics (CPE), Tallinn, Estonia, 2011, pp.470-473, doi: 10.1109/CPE.2011.5942279.

[17] Shejiao Han, Xutao Han, and Wei Sun "The Analysis of Magnetic Flux Density Inside Rogowski Coil Based on Full Current Theory,"IEEE Sensors Letters., vol. 4, no. 7, july 2020.

[18] Gandi Ramarao and Kandasamy Chandrasekaran, "Evaluating Lightning Channel-Base-Current Function Parameters for Identifying Interdependence of Wavefront and Tail by pso," ieee transactions on electromagnetic compatibility, vol. 61, no. 1, february 2019.

[19] Faisal Adamu Idris, Aisha Layla Buhari, Tahir Usman Adamu "Bessel Functions and Their Applications: Solution to Schrödinger equation in a cylindrical function of the second kind and Hankel Functions" International Journal of Novel Research in Physics Chemistry & Mathematics Vol. 3, Issue 2, pp: (17- 31), Month: May - August 2016.

[20] K. Chandrasekaran and Gururai S.Punekar, " Use of Genetic Algorithm to Determine Lightning Channel-Base Current-Function Parameters" ieee transactions on electromagnetic compatibility, vol. 56, no. 1, february 2014.

979-8-3315-3013-6/25 $31.00 © 2025 IEEE

State of Charge Estimation for Li-Ion Battery under Non-Gaussian Uncertainties with Experimental Validation

Supriyo Samanta
Department of Electrical Engineering
National Institute of Technology Calicut
Kozhikode, India
supriyo_m230914ee@nitc.ac.in

Rahul Radhakrishnan
Department of Electrical Engineering
National Institute of Technology Calicut
Kozhikode, India
rahulr@nitc.ac.in

Abstract—Accurate estimation of state of charge (SoC) for batteries is essential for successful operation of battery management systems. In this work, SoC estimation under a complete non Gaussian framework is considered, where it is assumed that non Gaussian noise is present in the current as well as the terminal voltage measurements. First of all, the second-order resistance-capacitance model is established and its parameters are identified through the variable forgetting factor recursive least squares (VFFRLS) algorithm. After model validation, a maximum correntropy unscented Kalman filter ($MC_C - UKF$) based on Cauchy kernel was derived such that the non Gaussian uncertainties in both process and measurements are accounted for. Finally, SoC estimates were generated and validated with the acquired data set. Moreover, a comparative analysis was done with respect to the UKF in terms of root mean square error (RMSE) and its average value.

Index Terms—lithium-ion battery, state of charge (SoC), parameter identification.

I. INTRODUCTION

To address the global energy problem and reduce carbon emissions, countries are promoting the use of electric vehicles to replace those fueled by fossil fuels. Electric vehicles increasingly use lithium-ion batteries due to their high power density, long life, low memory, and wide working temperature range [1]. This work dives into the area of electric vehicles (EVs), namely their performance management and optimization. Due to their appealing attributes and advantages, the majority of EVs use lithium-ion batteries as their primary source of energy, creating new challenges. To address all these challenges, a vehicle subsystem is defined, called the battery management system (BMS).

Out of many, the parameters that are of significant interest to the end-user are state of charge (SoC) and state of health (SoH). In case of an EV, these parameters are of utmost importance as they specify the remaining range of the vehicle and useful life of the battery. Hence, precisely determining these quantities is of significant industrial value, leading to large

This work was supported by the Core Research Grant, CRG/2022/001997, Science and Engineering Research Board, Government of India.

research interest. In spite of the advancements in BMS, the necessity to address problems associated with over-charging and discharging, that may reduce the useful life of battery and potentially lead to fire accidents, remains a matter of great significance [2]–[4]. Therefore, to enhance the safety, reliability and to manage the maintenance cost, an effective BMS is required.

A. SoC Estimation Methods

Till now, different methods have been proposed for SoC estimation, significantly contributing to the progress of this research filed. However, it is still a great challenge to accurately estimate SoC due to the complicated electrochemical dynamics of the batteries [5]. As per the literature, SoC estimation methods could be broadly classified into three main categories, which are the direct methods, model-based methods, and the data-driven methods [6].First of all, it is to be noted that SoC is not a directly measurable quantity and it could be derived based on the current drawn from the battery. Direct methods, estimate the SoC from the measured physical quantities such as current and terminal voltage, along with the associated mathematical relations [7]–[9]. One such example is the ampere-hour integral method, which exhibits high accuracy if the current is measured precisely and the initial SoC is accurately known. However, this method could not be relied upon as the initial SoC is not accurately known and there are uncertainties present in the measured current [8]. Moreover, literature on modeling the process and measurement noise as non Gaussian is mentioned in [10] and [11], where the uncertainties in the current drawn and measured terminal volatge is treated as non Gaussian.

Recently, with the widespread application of artificial intelligence, machine learning-based data-driven methods demonstrate excellent performance in battery SoC estimation [9]–[11]. The algorithms, including support vector machines (SVM) [12], convolutional neural networks(CNN) [13], and long short-term memory neural networks (LSTM) [14], illustrate self-learning capabilities and could generate accurate predictions [15]. These algorithms are trained offline to trace the link between measured battery statistics and SoC.

Much effort has been put for conducting SoC estimation based on model-based methods that can capture the internal dynamics of the batteries, leading to accurate SoC estimation. These models now include the electrochemical model and the equivalent circuit model (ECM). The ECM employs a series of electrical circuit components to construct a circuit that accurately represents the dynamic electrical behavior of the battery [16].

B. Definition of State of Charge (SoC)

The ratio of the battery's current available capacity to its rated capacity is described as the SoC [1]. SoC represents battery capacity, defined as

$$SoC_t = SoC_{t_0} - \frac{\int_{t_0}^t \eta I(\tau) d\tau}{C_n}.$$

Here the starting and present values of SoC are indicated by the variables SoC_t0 and SoC_t, respectively. The battery's nominal capacity is denoted by C_n, its charge and discharge efficiency is indicated by η, and its current is represented by I. The above equation could be expressed in the discrete-time domain as

$$SoC_k = SoC_{k-1} - \frac{\eta I_{k-1} T}{C_n}. \quad (1)$$

In this work, it is assumed that non-Gaussian noise is present in the terminal voltage and current measurements. From the process model derived from the equivalent circuit model (ECM), it could be inferred that the current measurements are the input to the system, and hence contributing to the non-Gaussian uncertainties in the process model. Similarly, such noise in terminal voltage corresponds to a measurement model corrupted with non-Gaussian noise. Hence, the objective of this work is to generate highly accurate SoC estimates even in the presence of non-Gaussian uncertainties and validate the results with respect to the acquired data set. Similar formulations for SoC estimation were recently reported in [10] and [11]. However, [10] considered a robust Student's-T filter for tackling the non Gaussian noises, and [11] proposed an iterative extended Kalman filter (EKF) based on adaptive maximum correntropy criterion. In comparison to these two works, this work includes the model validation from the experimental data, along with proposing a maximum correntropy unscented Kalman filter based on Cauchy kernel ($MC_C - UKF$) for estimating state of charge.

II. MATHEMATICAL MODELLING

Since SoC is an unmeasurable non physical quantity, it has to be estimated based on the measured quantities such as the terminal voltage, current, and temperature. The most popular techniques include the coulomb counting (CC) method and the open circuit voltage (OCV) approach. However, a second order Thevenin model as shown in Fig. 1 is chosen because it provides a good balance between model accuracy and complexity [17]. Where R_0 corresponds to the resistance

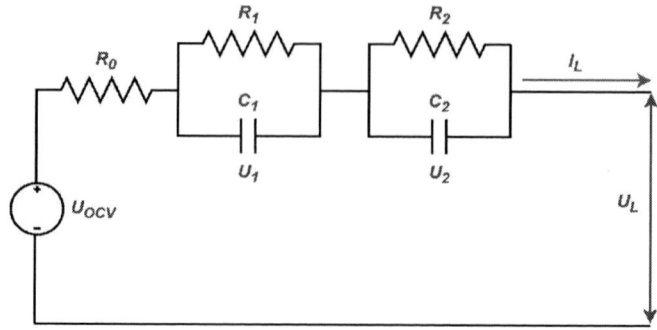

Fig. 1. Second-order Thevenin mode

of the battery, R_1 and R_2 the internal resistance, and C_1 and C_2 the internal capacitance. U_L defines the terminal voltage and I_L, the current. Applying Kirchhoff's current law, we get

$$I_L = \frac{U_1}{R_1} + C_1 \frac{dU_1}{dt} \quad (2)$$

and

$$I_L = \frac{U_2}{R_2} + C_2 \frac{dU_2}{dt}. \quad (3)$$

Then, applying Kirchhoff's voltage law, we have

$$U_L = U_{OCV} - U_1 - U_2 - I_L R_0. \quad (4)$$

From Eq. (2) and (3), we get

$$U_{1,k} = e^{-\frac{T}{\tau_1}} U_{1,k-1} + R_1 (1 - e^{-\frac{T}{\tau_1}}) I_{L,k} \quad (5)$$

and

$$U_{2,k} = e^{-\frac{T}{\tau_2}} U_{2,k-1} + R_2 (1 - e^{-\frac{T}{\tau_2}}) I_{L,k}, \quad (6)$$

where T is the sampling time interval, and $\tau_1 = R_1 C_1$ and $\tau_2 = R_2 C_2$ are the time constants. From the above-mentioned relations, the following state-space model could be formulated

$$
\begin{bmatrix} SoC_k \\ U_{1,k} \\ U_{2,k} \end{bmatrix} = \begin{bmatrix} 1 & 0 & 0 \\ 0 & e^{-\frac{T}{\tau_1}} & 0 \\ 0 & 0 & e^{-\frac{T}{\tau_2}} \end{bmatrix} \begin{bmatrix} SoC_{k-1} \\ U_{1,k-1} \\ U_{2,k-1} \end{bmatrix}
$$
$$
+ \begin{bmatrix} -\frac{\eta T}{C_n} \\ R_1 (1 - e^{-\frac{T}{\tau_1}}) \\ R_2 (1 - e^{-\frac{T}{\tau_2}}) \end{bmatrix} I_{L,k}. \quad (7)
$$

Now, for the measurement model, it is evident that the measurement obtained is the terminal voltage and that has to be written as a function of the state. Even though the terminal voltage, according to Eq. 4, is dependent on U_1 and U_2, it is written as a function of SoC too. This is achieved by expressing open circuit voltage U_{OCV} as a function of SoC. That is

$$U_{L,k} = U_{OCV}(SoC_k) - U_{1,k} - U_{2,k} - R_0 I_{L,k}. \quad (8)$$

979-8-3315-3013-6/25 $31.00 © 2025 IEEE

III. U_{OCV} AND SoC RELATIONSHIP

For the purpose of this study, we have used the Lithium-ion battery cycling experiment data mentioned in [18], and available at *https://data.mendeley.com/datasets/c5dxwn6w92/1*. The required data, such as the terminal voltage and current are used to construct the relation between U_{OCV} and SoC. For this study, only discharging data corresponding to a particular cycle was taken.. The specifications of the battery used are given in Table I, and the U_{OCV} versus SoC relation that was deduced from the data set is shown in Fig. 2. Using the curve-

TABLE I
BATTERY SPECIFICATIONS

Specifications	Value
Cell	Lithium-Ion battery
Rated capacity	27 Ah
Rated voltage	4.2 Volt

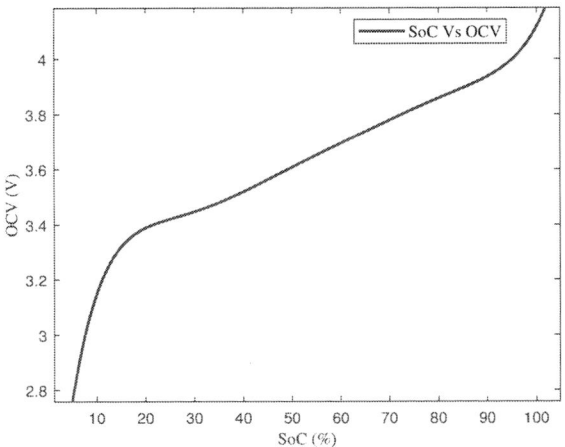

Fig. 2. OCV-SoC relationship

fitting approach, U_{OCV} is defined as

$$U_{OCV} = g_1 SoC^7 + g_2 SoC^6 + g_3 SoC^5 + g_4 SoC^4 + \\ g_5 SoC^3 + g_6 SoC^2 + g_7 SoC + g_8, \quad (9)$$

where the values of co-efficient of Eq.(9) are shown in Table II. Hence, from Eq. (7) and (8), the stochastic difference

TABLE II
COEFFICIENT VALUES

Co-efficient	Value
g_1	149.148
g_2	-600.0406
g_3	1.0059×10^3
g_4	-906.0744
g_5	469.73
g_6	-138.4177
g_7	21.8956
g_8	1.9529

equation depicting the process and measurement models could be generalized as,

$$X_k = FX_{k-1} + BI_{L,k} + w_k \quad \text{and} \quad (10)$$

$$Y_k = h(X_k) - R_0 I_{L,k} + v_k \quad (11)$$

Here, w_k and v_k are assumed to follow a non Gaussian density.

IV. THE PARAMETER IDENTIFICATION

The most popular technique for identifying system parameters is the recursive least squares (RLS). Consider the parameter vector θ and the observation data matrix φ as

$$\theta = \begin{bmatrix} \theta_1 & \theta_2 & \theta_3 & \theta_4 & \theta_5 \end{bmatrix}^T \text{ and} \quad (12)$$

$$\varphi = \begin{bmatrix} E_{k-1} & E_{k-2} & I_{L,k} & I_{L,k-1} & I_{L,k-2} \\ E_{k-2} & E_{k-3} & I_{L,k-1} & I_{L,k-2} & I_{L,k-3} \\ \vdots & \vdots & \vdots & \vdots & \vdots \\ E_{k-m-1} & E_{k-m-2} & I_{L,k-m} & I_{L,k-m-1} & I_{L,k-m-2} \end{bmatrix} \quad (13)$$

where k denotes the current time step, m is the observation time.

TThe forgetting factor RLS (FFRLS) approach, which is based on the RLS method, adds a forgetting factor coefficient λ to the system output vector E and the observed data matrix φ. As a result, the FFRLS approach could effectively adapt to improve identification process as more information is received. That is

$$E = \begin{bmatrix} E_k & \lambda E_{k-1} & \cdots & \lambda^m E_{k-m} \end{bmatrix}^T \quad (14)$$

where the observation matrix φ becomes

$$\begin{bmatrix} E_{k-1} & E_{k-2} & I_{L,k} & I_{L,k-1} & I_{L,k-2} \\ \lambda E_{k-2} & \lambda E_{k-3} & \lambda I_{L,k-1} & \lambda I_{L,k-2} & \lambda I_{L,k-3} \\ \vdots & \vdots & \vdots & \vdots & \vdots \\ \lambda^m E_{k-m-1} & \lambda^m E_{k-m-2} & \lambda^m I_{L,k-m} & \lambda^m I_{L,k-m-1} & \lambda^m I_{L,k-m-2} \end{bmatrix} \quad (15)$$

It is now possible to develop a variable FFRLS algorithm (VF-FRLS), which specifies the forgetting factor that is established by the prediction errors and updates as more information is received.

Recursive formulation of the VFFRLS algorithm is expressed as

$$\begin{cases} K_k = \dfrac{P_{k-1}\varphi_k}{\lambda_{k-1} + \varphi_k^T P_{k-1}\varphi_k} \\ \textbf{Update estimation parameter matrix} \\ \theta_k = \theta_{k-1} + K_k e_k \\ \textbf{Update co-variance matrix} \\ P_k = \dfrac{\left(I - K_k\varphi_k^T\right) P_{k-1}}{\lambda_{k-1}} \end{cases} \quad (16)$$

where the updating formula for the forgetting factor λ_k is

$$\lambda_k = 1 - \frac{e_k^2}{1 + K_k^T P_k K_k} \quad (17)$$

here K_k, P_k represents the gain and covariance matrix respectively. The VFFRLS algorithm is used to identify the equivalent model parameters of Lithium-ion batteries for the

979-8-3315-3013-6/25 $31.00 © 2025 IEEE

second order ECM. The battery model transformed into a least square mathematical form as shown in Eq. (18)

$$E(s) = (U_L(s) - U_{oc}(s))$$
$$= -I_L(s) \left[\frac{R_1}{R_1 C_1 s + 1} + \frac{R_2}{R_2 C_2 s + 1} + R_0 \right]. \quad (18)$$

Let us consider a, b, c, d and e as follows:

$$\begin{cases} a = R_0, \quad b = R_1 R_2 C_1 C_2 \\ c = R_1 C_1 + R_2 C_2, \quad d = R_0 + R_1 + R_2 \quad \text{and} \\ e = R_0 R_1 C_1 + R_0 R_2 C_2 + R_1 R_2 C_2 + R_2 R_1 C_1 \end{cases} \quad (19)$$

Applying bi-linear transformation $s = \frac{2}{T} \cdot \frac{1-z^{-1}}{1+z^{-1}}$ in Eq.(18)

$$G(Z^{-1}) = \frac{\theta_3 + \theta_4 Z^{-1} + \theta_5 Z^{-2}}{1 - \theta_1 Z^{-1} - \theta_2 Z^{-2}}, \quad (20)$$

where the parameters

$$\begin{cases} \theta_1 = \frac{8b - 2T^2}{4b + 2cT + T^2}, \theta_2 = -\frac{4b - 2cT + T^2}{4b + 2cT + T^2}, \\[2mm] \theta_3 = -\frac{4ab + 2eT + dT^2}{4b + 2cT + T^2}, \theta_4 = \frac{8ab - 2dT^2}{4b + 2cT + T^2}, \\[2mm] \text{and} \quad \theta_5 = -\frac{4ab + dT^2 - 2eT}{4b + 2cT + T^2}. \end{cases} \quad (21)$$

Based on these parameters, Eq. (20) could be written as

$$E_k = \theta_1 E_{k-1} + \theta_2 E_{k-2} + \theta_3 I_{L,k} + \theta_4 I_{L,k-1} + \theta_5 I_{L,k-2}, \quad (22)$$

where the circuit model parameters R_0, R_1, R_2, C_1 and C_2 are derived using the identification results of the direct identification parameters $\theta = \begin{bmatrix} \theta_1 & \theta_2 & \theta_3 & \theta_4 & \theta_5 \end{bmatrix}^T$ as

$$\begin{cases} R_0 = a, \\ R_1 = \frac{(d\tau_1 + R_0 \tau_2 - e)}{\tau_1 - \tau_2}, R_2 = d - a - R_1 \\ C_1 = \frac{\tau_1}{R_1}, \quad \text{and} \quad C_2 = \frac{\tau_2}{R_2} \end{cases} \quad (23)$$

V. THEORETICAL VALIDATION OF THE ECM

From the parameter identification procedure done with the help of battery terminal voltage and current drawn data, the values obtained for the ECM parameters are shown in Table III. Once the ECM parameters are identified from the battery

TABLE III
PARAMETER VALUES

Parameter	R_0	R_1	R_2	C_1	C_2
Value	0.0009	0.0433	-0.0431	12.5906	-10.6649

data, using the current drawn (only discharging phase is considered), which is 3C, that is 81 A, the terminal voltage at each and every time step could be determined. This serves as a check point for identifying whether the ECM and its parameters are able to capture the complex chemistry of the Li-ion battery. The obtained terminal voltage, calculated from the ECM is given in Fig. 3.

From Fig. 3, it can be inferred that the terminal voltage obtained from the chosen ECM follows that obtained from the experiment with acceptable accuracy. Also, it shall be noted that the terminal voltage at the end of the experiment was around 2.4 V instead of zero, pointing towards a corresponding non-zero SoC.

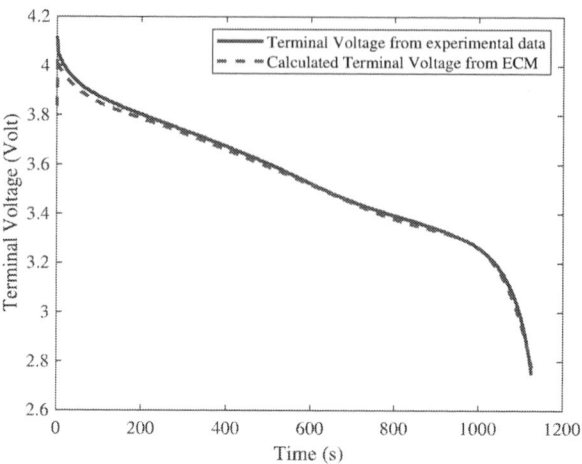

Fig. 3. Terminal voltage from ECM and from experimental data

VI. SoC ESTIMATION UNDER NON-GAUSSIAN UNCERTAINTIES

For the specific problem of SoC estimation in battery systems, glint noise is particularly significant, expressed as $\bar{w}_1 \mathcal{N}(0, R_1) + \bar{w}_2 \mathcal{N}(0, R_2)$, where $\bar{w}_1 + \bar{w}_2 = 1$, $R_1 \neq R_2$ and $\mathcal{N}(\cdot, \cdot)$ represents a Gaussian density with appropriate mean and covariance.

A. Correntropy Measure

Correntropy is a statistical measure that compares the resemblance of two random variables using higher-order statistical moments in the joint space controlled by kernel bandwidth [19]. random variables are X and Y. Consequently, correntropy defines

$$V_\sigma(X, Y) = E[k_\sigma(X, Y)] = \iint k_\sigma(x, y) f_{XY}(x, y) \, dx \, dy$$

where $k_\sigma(\cdot)$ usually denotes a positive definite kernel function and the joint PDF of X and Y is $f_{XY}(\cdot)$. $E(\cdot)$ is the expectation operator. If only a finite number of data points N are available, then the sample estimator is given as

$$V_\sigma(X, Y) = \frac{1}{N} \sum_{k=1}^{N} C_\delta(x_k - y_k) = \frac{1}{N} \sum_{k=1}^{N} \frac{1}{1 + \frac{\|x_k - y_k\|^2}{\delta}} \quad (24)$$

Here $C_\delta(\cdot)$ is the Cauchy kernel Here, δ represents the Cauchy kernel bandwidth, which is a positive scaler and bounded. Only when X = Y does the Cauchy kernel reach its maximum, resulting in the maximum correntropy criterion (MCC). Correntropy can be calculated using the Taylor series expansion of the Cauchy kernel, including higher orders moments, given as [19], [20]

$$V_\delta(X, Y) = \sum_{k=0}^{\infty} \frac{(-1)^k}{\delta^k} \binom{N+k-1}{k} E[(X-Y)^{2k}] \quad (25)$$

B. Maximum Correntropy estimation framework under non Gaussian uncertainties in process and measurement

Based on the Cauchy kernel function, a nonlinear state estimator based on MCC and statistical linearization is created to enhance performance in the presence of significant multidimensional noise [20]. The associated cost function is

$$\mathcal{J} = \alpha C_\delta(\Lambda_P) + \beta C_\delta(\Lambda_\mathcal{R}) \tag{26}$$

In the Eq. (26), the error terms $\Lambda_\mathcal{R}$ and Λ_P stand for the measurement outliers and non-Gaussian process model uncertainties, respectively. C_δ is the Cauchy kernel and α and β are adjustable weights [19].

We use the equation $\frac{d\mathcal{J}}{dX_k} = 0$ to determine the best estimate of x_k. This suggests that

$$\frac{\alpha \mathcal{L}_{kP}^C}{\delta} P_{k|k-1}^{-1}(x_k - \hat{x}_{k|k-1}) + \frac{\beta \mathcal{L}_{kY}^C}{\delta} \bar{H}_k^T \mathcal{R}^{-1}\Lambda = 0 \tag{27}$$

where $\mathcal{L}_{kP}^C = C_\delta^2(\|x_k - \hat{x}_{k|k-1}\|_{P_{k|k-1}^{-1}})$ and $\mathcal{L}_{kY}^C = C_\sigma^2(\|y_k - \hat{y}_{k|k-1} - \bar{H}_k(x_k - \hat{x}_{k|k-1})\|_{\mathcal{R}^{-1}})$.

We set $\alpha = \delta$ and $\beta = -\delta$ to ensure that the algorithm will converge to a traditional Gaussian estimator when kernel bandwidth δ tends to infinity [20]. Taking into account $\frac{\mathcal{L}_{kY}^C}{\mathcal{L}_{kP}^C} = \mathcal{L}_k^C$ and rearranging equation (27), we arrive at:

$$(P_{k|k-1}^{-1} + \mathcal{L}_k^C \bar{H}_k^T \mathcal{R}_k^{-1}\bar{H}_k)x_k = \mathcal{L}_k^C \bar{H}_k^T \mathcal{R}_k^{-1}(y_k - \hat{y}_{k|k-1}) + $$
$$(\mathcal{L}_k^C \bar{H}_k^T \mathcal{R}_k^{-1}\bar{H}_k + P_{k|k-1}^{-1})\hat{x}_{k|k-1} \tag{28}$$

Equation (28) is a fixed-point equation that must be solved using a fixed-point iteration approach by assuming that x_k equals $\hat{x}_{k|k-1}$ [20].

The expression for posterior mean is obtained as:

$$\hat{x}_{k|k} = \hat{x}_{k|k-1} + K_k^C(y_k - \hat{y}_{k|k-1}) \tag{29}$$

where $K_k^C = (P_{k|k-1}^{-1} + \mathcal{L}_k^C \bar{H}_k^T \mathcal{R}^{-1}\bar{H}_k)^{-1}\mathcal{L}_k^C \bar{H}_k^T \mathcal{R}_k^{-1}$
and $\mathcal{L}_k^C = \frac{C_\sigma^2\left(\|y_k - \hat{y}_{k|k-1}\|_{\mathcal{R}_k^{-1}}\right)}{C_\sigma^2\left(\|x_k - \hat{x}_{k|k-1}\|_{P_{k|k-1}^{-1}}\right)}$. Then, the modified Kalman gain and posterior covariance are

$$K_k^C = P_{k|k-1}\mathcal{L}_k^C \bar{H}_k^T \left(\mathcal{R}_k + \bar{H}_k P_{k|k-1}\mathcal{L}_k^C \bar{H}_k^T\right)^{-1} \quad \text{and}$$
$$P_{k|k} = \left(I - K_k^C \bar{H}_k\right)P_{k|k-1}\left(I - K_k^C \bar{H}_k\right)^T + K_k^C \mathcal{R}_k (K_k^C)^T$$

respectively.

The algorithm for $MC_C - UKF$ is given below in VI-B.

VII. RESULT AND DISCUSSION

For simulation purpose, the tuning parameters and initial conditions considered are mentioned in Table IV,

and both w_k and v_k mentioned in Eq. (10) and (11) are modeled as $\bar{w}_1 \cdot N(0, R) + \bar{w}_2 \cdot N(0, 10R)$ where $\bar{w}_1 = 0.2$ and $\bar{w}_2 = 0.8$.

Fig. 4 shows the truth, and estimates obtained from $MC_C - UKF$ for SoC while non-Gaussian uncertainty is considered for both the process and measurement model. From Fig. 4, it could be inferred that the SoC is accurately estimated by

---

Algorithm 1 $MC_C - UKF$ Algorithm

1: Choose δ, and initialize $\hat{x}_{k-1|k-1}$, $P_{k-1|k-1}$
2: $\hat{x}_{k|k-1} = F_k \hat{x}_{k-1|k-1}$
3: $P_{k|k-1} = F_k P_{k-1|k-1} F_k^T + Q$
4: Calculate $\hat{\chi}_i$ and w_i , $i = 1, \ldots, N$
5: $Y_{i,k|k-1} = h(\hat{\chi}_i)$
6: $\hat{y}_k = \sum_{i=1}^N w_i Y_{i,k|k-1}$
7: $P_{yy} = \sum_{i=1}^N w_i \left[Y_{i,k|k-1} - \hat{y}_k\right]\left[Y_{i,k|k-1} - \hat{y}_k\right]^T + R_k$
8: $P_{xy} = \sum_{i=1}^N w_i \left[\hat{\chi}_{i,k|k-1} - \hat{x}_{k|k-1}\right]\left[Y_{i,k|k-1} - \hat{y}_k\right]^T$
9: $\bar{H}_k = (P_{k|k-1}^{-1} P_{xy})^T$
10: $R_k = P_{yy} - \bar{H}_k P_{k|k-1}\bar{H}_k^T$
11: $\mathcal{L}_k^C = \dfrac{C_\delta^2\left(\|y_k - \hat{y}_{k|k-1}\|_{\mathcal{R}_k^{-1}}\right)}{C_\delta^2\left(\|\hat{x}_{k-1|k-1} - \hat{x}_{k|k-1}\|_{P_{k|k-1}^{-1}}\right)}$
12: $K_k^C = P_{k|k-1}\mathcal{L}_k^C \bar{H}_k^T \left(R_k + \bar{H}_k P_{k|k-1}\mathcal{L}_k^C \bar{H}_k^T\right)^{-1}$
13: Posterior mean: $\hat{x}_{k|k} = \hat{x}_{k|k-1} + K_k^C(y_k - \hat{y}_{k|k-1})$
14: Posterior covariance: $P_{k|k} = (I - K_k^C \bar{H}_k)P_{k|k-1}(I - K_k^C \bar{H}_k)^T + K_k^C \mathcal{R}_k (K_k^C)^T$

TABLE IV
PARAMETERS AND THEIR VALUES

Parameter	Value
$P_{0|0}$	$0.1 \cdot I_{3\times 3}$
Q	$10^{-6} \cdot I_{3\times 3}$
R	10^{-7}
$\hat{X}_{0|0}$	$\begin{bmatrix}1 & 0.1 & 0.1\end{bmatrix}^T$
δ	15

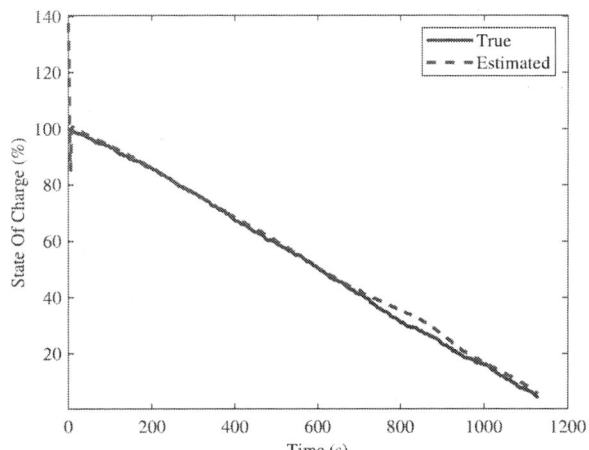

Fig. 4. SoC estimation using $MC_C - UKF$ taking both process and measurement noise as non-Gaussian with Cauchy kernel

the $MC_C - UKF$, where SoC got reduced from 100% to nearly 8%. This could be correlated with the fact that the terminal voltage didn't drop to zero, but to around 2.4 volts. Fig. (5) shows SoC's root mean square error using UKF and $MC_C - UKF$ while both process and measurement noise are non-Gaussian. From this figure, it could be inferred that

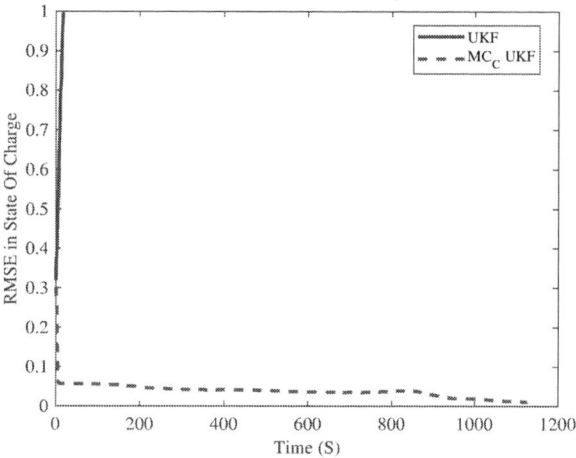

Fig. 5. RMSE of SoC estimation using UKF Vs $MC_C - UKF$

the $MC_C - UKF$ algorithm has much less RMSE than the UKF algorithm. In fact, under the influence of non Gaussian noise, UKF diverged. Hence, when both the process model and measurement model have non-Gaussian uncertainties, the $MC_C - UKF$ algorithm provides more robust and accurate estimates.

Average RMSE obtained with different estimators is shown in Table V. From this table, it could be further inferred that

TABLE V
COMPARISON OF AVERAGE RMSE FOR UKF AND $MC_C - UKF$

Estimators	Average RMSE
UKF	5.2543
$MC_C - UKF$	0.0359

while non-Gaussian uncertainty is considered in both process and measurement model, the UKF diverges, where as $MC_C - UKF$ converges with accurate estimates. For this simulation study, the true SoC was calculated using the Coulomb counting method from the current values present in the acquired data set.

VIII. CONCLUSION

To conclude, The discharge behavior of a lithium-ion battery was taken into consideration in order to determine the parameters for a selected equivalent circuit model using the VFFRLS technique. After fixing the parameter values, model validation was also done. Finally, with the identified model and the data obtained during the battery operation, estimates for state of charge was obtained. A maximum correntropy algorithm based on UKF was derived for the problem such that non Gaussian uncertainties in process and measurement could be dealt with. From the simulation results using real-life data, it could be inferred that the proposed estimator generates highly accurate estimates in comparison to the conventional UKF.

REFERENCES

[1] M. Lei, B. Wu, W. Yang, P. Li, J. Xu, and Y. Yang, "Double extended kalman filter algorithm based on weighted multi-innovation and weighted maximum correlation entropy criterion for co-estimation of battery soc and capacity," *ACS omega*, vol. 8, no. 17, pp. 15 564–15 585, 2023.

[2] D. Liu, J. Zhou, H. Liao, Y. Peng, and X. Peng, "A health indicator extraction and optimization framework for lithium-ion battery degradation modeling and prognostics," *IEEE Transactions on Systems, Man, and Cybernetics: Systems*, vol. 45, no. 6, pp. 915–928, 2015.

[3] X. Cui and B. Xu, "State of charge estimation of lithium-ion battery using robust kernel fuzzy model and multi-innovation ukf algorithm under noise," *IEEE Transactions on Industrial Electronics*, vol. 69, no. 11, pp. 11 121–11 131, 2021.

[4] X. Shu, Z. Chen, J. Shen, F. Guo, Y. Zhang, and Y. Liu, "State of charge estimation for lithium-ion battery based on hybrid compensation modeling and adaptive h-infinity filter," *IEEE Transactions on Transportation Electrification*, vol. 9, no. 1, pp. 945–957, 2022.

[5] V. Selvaraj and I. Vairavasundaram, "A comprehensive review of state of charge estimation in lithium-ion batteries used in electric vehicles," *Journal of Energy Storage*, vol. 72, p. 108777, 2023.

[6] R. Xiong, J. Cao, Q. Yu, H. He, and F. Sun, "Critical review on the battery state of charge estimation methods for electric vehicles," *Ieee Access*, vol. 6, pp. 1832–1843, 2017.

[7] M. Coleman, C. K. Lee, C. Zhu, and W. G. Hurley, "State-of-charge determination from emf voltage estimation: Using impedance, terminal voltage, and current for lead-acid and lithium-ion batteries," *IEEE Transactions on industrial electronics*, vol. 54, no. 5, pp. 2550–2557, 2007.

[8] K. S. Ng, C.-S. Moo, Y.-P. Chen, and Y.-C. Hsieh, "Enhanced coulomb counting method for estimating state-of-charge and state-of-health of lithium-ion batteries," *Applied energy*, vol. 86, no. 9, pp. 1506–1511, 2009.

[9] C. Truchot, M. Dubarry, and B. Y. Liaw, "State-of-charge estimation and uncertainty for lithium-ion battery strings," *Applied Energy*, vol. 119, pp. 218–227, 2014.

[10] M.-F. Ng, J. Zhao, Q. Yan, G. J. Conduit, and Z. W. Seh, "Predicting the state of charge and health of batteries using data-driven machine learning," *Nature Machine Intelligence*, vol. 2, no. 3, pp. 161–170, 2020.

[11] M. Ragone, V. Yurkiv, A. Ramasubramanian, B. Kashir, and F. Mashayek, "Data driven estimation of electric vehicle battery state-of-charge informed by automotive simulations and multi-physics modeling," *Journal of Power Sources*, vol. 483, p. 229108, 2021.

[12] X. Liu, Q. Li, L. Wang, M. Lin, and J. Wu, "Data-driven state of charge estimation for power battery with improved extended kalman filter," *IEEE Transactions on Instrumentation and Measurement*, vol. 72, pp. 1–10, 2023.

[13] X. Fan, W. Zhang, C. Zhang, A. Chen, and F. An, "Soc estimation of li-ion battery using convolutional neural network with u-net architecture," *Energy*, vol. 256, p. 124612, 2022.

[14] J. Chen, Y. Zhang, J. Wu, W. Cheng, and Q. Zhu, "Soc estimation for lithium-ion battery using the lstm-rnn with extended input and constrained output," *Energy*, vol. 262, p. 125375, 2023.

[15] H. Pan, Z. Lü, W. Lin, J. Li, and L. Chen, "State of charge estimation of lithium-ion batteries using a grey extended kalman filter and a novel open-circuit voltage model," *Energy*, vol. 138, pp. 764–775, 2017.

[16] S. Li, H. He, C. Su, and P. Zhao, "Data driven battery modeling and management method with aging phenomenon considered," *Applied Energy*, vol. 275, p. 115340, 2020.

[17] R. Model, "Based on temperature-dependent second-order," *Advanced Battery Technologies: New Applications and Management Systems*, p. 49, 2021.

[18] S. Zhang, X. Guo, X. Dou, and X. Zhang, "A data-driven coulomb counting method for state of charge calibration and estimation of lithium-ion battery," *Sustainable Energy Technologies and Assessments*, vol. 40, p. 100752, 2020.

[19] A. Urooj, A. Dak, B. Ristic, and R. Radhakrishnan, "2d and 3d angles-only target tracking based on maximum correntropy kalman filters," *Sensors*, vol. 22, no. 15, p. 5625, 2022.

[20] A. Dak and R. Radhakrishnan, "Non-iterative cauchy kernel-based maximum correntropy cubature kalman filter for non-gaussian systems," *Control Theory and Technology*, vol. 20, no. 4, pp. 465–474, 2022.

Hybrid renewable energy sources integrated shunt active power filter with adaptive fuzzy logic

Ravinder Kumar
Department of Electrical Engineering
National Institute of Technology, Uttarakhand
ravinder.kumar@nituk.ac.in

Sudhansu Kumar Mishra
Department of Electrical & Electronics Engineering
Birla Institute of Technology Mesra, Ranchi
sudhansumishra@bitmesra.ac.in

Karan Veer
Department of Instrumentation & Control Engineering
Dr. BR Ambedkar National Institute of Technology, Jalandhar
veerk@nitj.ac.in

Abstract— The aim of this paper is to investigate the integration of a hybrid renewable energy system (HRES) with a shunt active power filter (SAPF) in order to enhance the quality of power & produce sufficient and clean generation. Maximum power point tracking and control of synchronous reference frame (SRF) theory based on adaptive fuzzy logic control are used by the system. The proposed method is carried out using direct power control techniques and SRF-based hysteresis control. A common DC-link used by a wind turbine system, a fuel cell, and a PV system that make this HRES system. An efficient system that can produce electricity for a long amount of time may be developed by combining fuel cells with sources of clean energy. Power quality issues such as harmonics, and voltage fluctuations affect both grid-connected and self-sustaining systems; weaker grids are more susceptible to these problems. The MATLAB/Simpower model of HRES-SAPF, results reveal that the technique works better in decreasing harmonics in various loading conditions. It can fast adapt to changing conditions and stabilize the DC-link voltage and lowers harmonics compared to traditional techniques.

Keywords— Fuel cell, hybrid renewable energy system, wind energy, photovoltaic, total harmonic distortion, power quality.

I. INTRODUCTION

Hybrid renewable energy systems (HRES) enhance renewable generating capacity by integrating a variety of renewable energy sources; they also provide energy support and storage utilizing one or more energy storage devices. Hybrid systems that combine many renewable energy sources have great promise for long-term, sustainable energy generation, and they're also rather affordable. Usually the major energy sources in these HRES are wind turbines and PV solar panels running at any time of wind speed and solar radiation. The most often used energy storage technology are batteries; hydrogen systems-fuel cells, electrolysers and hydrogen tanks-have drawn much interest in recent years. The power is more reliable and of higher quality as distributed power generating systems (DPGSs), storage devices, fuel cells, wind energy conversion systems, and solar photovoltaic (PV) systems collaborate to decrease power loss in transmission and distribution networks [1].

The restructuring of utilities, technological improvements, environmental legislation, and wider power markets have all contributed to the rise of DPGSs as an important energy choice [2]. Because of their capital-intensive plants and infrastructure at transmission and distribution sectors, central power systems aren't highly adaptive to changing demands, yet they are nonetheless necessary for supplying electricity to the country [3].

The need for substantial capital investments and distribution and transmission losses have led to power outages and costly, unpredictable power supplies. Solar photovoltaic systems that use renewable energy sources could be considered a significant alternative energy source to increase power generation in this environment [4]. At the same time as this helps stabilize the planet's climate, especially from greenhouse gases, it will also minimize pollution in the environment. Wind power and solar photovoltaics (PV) are two renewable energy sources that can reduce the load on the nation's electrical infrastructure and protect the environment [5].

Several problems with power quality can develop in systems that combine grid-connected PV and wind generation. The efficiency of the system, a long lifespan of the wind turbine generators and solar PV modules, power loss, motor and cable failure, protective device use, and sinusoidal waveform spectrum of voltage and heat are all impacted by these factors [6]. In order to maintain the reliability of the renewable grid-connected system, it is necessary to employ suitable technologies that can reduce power-quality problems and current harmonics, as the energy supply is unpredictable and power output is also intermittent. Recommendations for fixing various power-quality issues abound [7]. These problems include electrical system interference, reactive power load, unbalanced systems, excessive neutral current, and injected harmonics. The passive filters are typically used in grid-integrated systems to mitigate series harmonics. A few issues that render PFs useless are their large size, grid negative resonance in both series and parallel, filter impedance, constant compensation, limited load ranges, and poor filtering efficiency [8]–[12].

The advanced filtering technologies like a static synchronous compensator, dynamic voltage regulator, multilevel inverter, power monitoring system, and unified power quality conditioner (UPQC) can solve power-quality issues, according to surveys of grid-integrated systems like PV inverters and wind energy systems [13]. Reactive power and current harmonics are both addressed by the SAPF. As far as methods for improving power quality, it is both the most popular and versatile [14]. The inverter's characteristics, reference current sensing techniques, and control systems all affect the filter's efficiency [15]. By employing harmonic load detection methods including synchronous detection, the instantaneous power theory, and basic positive sequence approaches, the SAPF may be effectively controlled [16]. In order to minimize total harmonic distortion (THD), improve reactive power correction, and maximize PV output when loads vary, the distribution static compensator (DSTATCOM) employs fuzzy logic control (FLC) and adaptive neuro-fuzzy inference system (ANFIS) approaches [17–18].

979-8-3315-3013-6/25 $31.00 © 2025 IEEE

The doubly-fed induction generator (DFIG) wind turbines, which are designed to handle high wind speeds, are able to transform mechanical energy into electrical energy [19]. For low wind speeds, a wind energy conservation system (WECS) based on a permanent magnet synchronous generator (PMSG) can provide electricity efficiently [20]. The reason behind this is that the electricity generated by wind turbines is directly related to the wind speed, which is inherently changeable. As the wind speed varies, the system adjusts the rotor side converter's torque reference to align with the optimal operating point (OPP) of the wind turbine [21]. A DFIG-based WECS is more cost-effective, efficient, and delivers better power than a fixed-speed generator. There are a number of maximum power point tracking (MPPT) methods that may be used to extract power from WECS. Some of these algorithms are Hill Climbing, Incremental Conductance, Fuzzy, and ANFIS [22]. However, there are a number of problems with the stated method. For example, it doesn't work very well when the weather is quite unpredictable or while tracking the adaptive fuzzy. Thus, system requirements and operating conditions determine the optimal MPPT algorithm for DFIG based WECS. The goal of this work is to improve the efficiency and performance of DFIG-based WECS, particularly in situations with low and fluctuating wind speeds. A Chaotic Dragonfly-based MPPT is used to do this [23]. The main contribution of this study is as follows:

- To controlling the SAPF through an adaptive fuzzy method based on SRF.

- To power quality will be improved through the integration of SAPF into the HRES.

- To adaptive fuzzy-based technique for maximum power point extraction in photovoltaic and wind power systems.

The proposed control technique successfully resolves all power quality concerns in this application under balanced and unbalanced/nonlinear loading conditions. The adaptive fuzzy MPPT approach is employed to optimize the power extraction from the PV array and wind energy, taking into account fluctuations in solar radiation and wind speed. Evaluating the simulation on the MATLAB/Simulink platform ensures that the system works as expected.

II. SYSTEM CONFIGURATION

The proposed system integrates renewable energy sources like wind, fuel cells, or photovoltaic cells with SAPF by a direct current link. The PMSG connectors are used to connect an unregulated rectifier, which changes the output voltage from AC to DC. Fig.1 depicts the suggested system design.

A. Photovoltaic Array

Fig. 2 displays the non-linear V-I characteristics of the PV cell as a function of both temperature and solar radiation. Furthermore, it is important that you remain centered on the array's highest power output capability. The authors have chosen a fuzzy based strategy, one of the MPPT strategies discussed in the previous article, to maximize the duty cycle of a boost converter and extract the highest power. The DC-DC boost converter receives its power from the PV array [24]. The output of the boost converter is linked to the DC bus of the SAPF. The output voltage of the PV array is 300 V, which

is increased to 750 V by means of a boost converter. Ref. [25] shows that eqn. (1) and (2) specify the currents through the PV module's diode and load.

$$I_d = I_{sat}(e^{\frac{QV_{oc}}{AkT}} - 1) \tag{1}$$

$$I = I_L - I_{sat}(e^{\frac{QV_{oc}}{AkT}} - 1) - \frac{V_{oc}}{R_{sh}} \tag{2}$$

Eqn. (3) illustrates that the maximum PV voltage is attained in an open circuit.

$$V_{oc} = \frac{AkT}{Q} log_n \left(\frac{I_L}{I_{sat}} + 1\right) \tag{3}$$

This system makes use of a solar PV module manufactured by SunPower, the SPR-305-WHT.

B. Wind Energy

Various turbines in a wind farm are placed for a straightforward use. The mechanical wind power of the blades is to be converted into electrical power of electric generators. Modern wind energy systems extensively use permanent magnet synchronous generators (PMSGs) because of their high efficiency, dependability, and direct-drive capacity. PMSGs are quite efficient for wind power uses since they do not need outside excitation unlike conventional generators. This work uses several benefits from PMSG [26]. The characteristics of wind turbine speed are illustrated in Fig. 3. It is coupled to the PMSG using an uncontrolled diode bridge rectifier. The Fuzzy based MPPT program doesn't know anything about how wind turbines generate power; it adjusts the duty cycle of a boost converter to maximise power output. A SAPF ties the output of the boost converter to the grid. Maximum power requires wind turbines to be running at maximum power coefficient (C_p). Eqn. (4) of the produced power by the wind turbine.

$$P = \frac{1}{2}\rho A v^3 C_p(\lambda, \beta) \tag{4}$$

with the following variables: P for power generation, ρ for air density, A for area swept by the blades, v for wind velocity, Cp for power coefficient, β for pitch angle, and λ for tip-speed ratio (TSR).

The efficiency of power, or the ratio of energy produced to accessible wind power, is usually represented as Cp (λ, β). The relationship between it and λ and β can be expressed using eqn. (5)

$$C_p(\lambda, \beta) = 0.5176 \left(116 \times \frac{1}{\lambda_1} - 0.4\beta - 5\right) e^{\frac{-21}{\lambda_i}} + 0.068\lambda \tag{5}$$

where

$$\frac{1}{\lambda_i} = \frac{1}{(\lambda + 0.08\beta) - \left(\frac{0.035}{1 + \beta^3}\right)}$$

A turbine speed ratio (TSR) is found by dividing the rotor speed by the wind speed eqn. (6)

$$\lambda = \omega_m \times \frac{R}{v} \tag{6}$$

The rotor speed is denoted by ω_m whereas the radius of the blades is denoted by R. This eqn. (7) calculates torque (T).

$$T = \frac{P}{\omega_m} \tag{7}$$

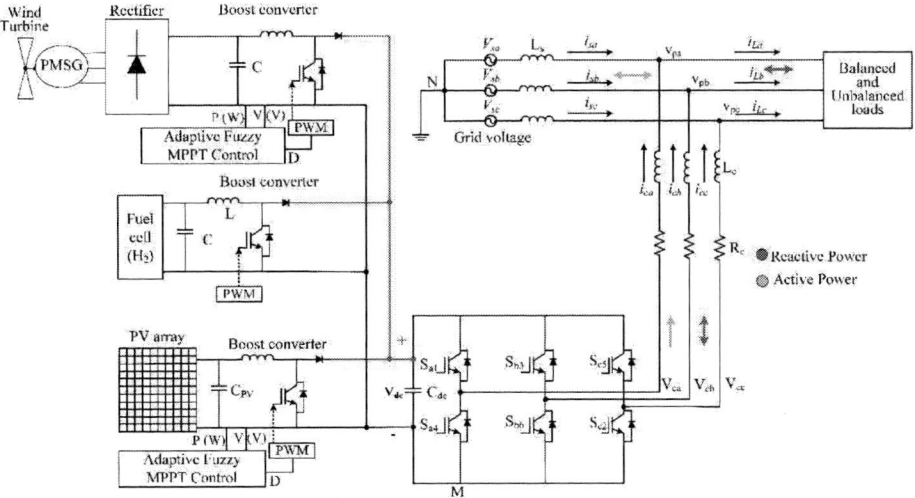

Fig.1 HRES integrated SAPF

It can be observed from equation (5) that for each wind speed, there is an ideal rotor speed that provides the highest power, where β remains constant.

C. Fuel Cell

The proton exchange membrane (PEM) is evaluated in this work as a potential component in the FC model due to its straightforwardness and rapid load response [27]. Fig. 2 provides a graphical illustration of the characteristics of fuel cells. The PEM-FC makes use of a solid polymer electrolyte, which is an insulator for electrons and the best conductor of protons, to transfer particles between two permeable terminals. The electrolyte is similar to a Teflon layer. As low as 1000 C is the operating temperature of FC. The chemical reactions at the anode and cathode sides, as well as the overall processes, are provided by equations (8), (9) and (10).

The response that occurs at the electrode

$$H_2 = 2H^+ + 2e- \tag{8}$$

The reaction that occurs at the end of the cathode

$$1/2\ O_2 + 2H^+ + 2e- = H_2O \tag{9}$$

The overall effect

$$H_2 + 1/2\ O_2 = H_2 \tag{10}$$

Because it controls the power output and raises the voltage at the DC-link, the boost converter is an essential component of the AC system.

D. Maximum Power Point Tracking (MPPT) achieved through boost converter control

The MPPT technique is used to manage a DC-DC boost converter. The suggested MPPT method outputs duty cycle change by using power and voltage changes as inputs. This method produces a suitable duty cycle matching MPP. In this study, the ANFIS algorithm is used to track MPP. The fundamental benefit of this method is that it can search for MPP regardless of the surrounding conditions. Separate MPPTs regulate each increase. A shared DC bus of 750 V is used to connect the boost converters' outputs.

III. STRATEGIES FOR REGULATING THE SHUNT ACTIVE POWER FILTER

Additionally in control of AC machines (such as induction motors and synchronous motors) and power systems (such as grid-connected inverters), synchronous reference frame (SRF) theory is extensively applied in electrical engineering. It converts three-phase AC signals into a rotating reference frame, therefore simplifying their analysis and control. A schematic of the SRF-based adaptive fuzzy control algorithm is shown in Fig.2. Using the Park transformation, as shown in eqn. (11) below, the load currents (iLa, iLb, iLc) are transformed into the $d - q - 0$ axis as feedback signals.

Transformation of Reference Frames (abc to $dq0$)

    a.   Voltages and currents in a three-phase system change with time and have sinusoidal nature.

    b.   Park's transformation turns the three-phase system (abc) into a rotating $dq0$ reference frame using SRF theory.

    c.   This converts sinusoidal variables into DC-like numbers, therefore facilitating control.

$dq0$ Transformation

It combined with a zero-sequence (0) component, transforms three-phase (abc) values into two-axis direct (d) and quadrature (q) components.

This change brings the spinning frame into line with the reference frequency—that of a grid in power systems or rotor speed in motor control.

Advantages of SRF Theory

    a.   Simplifies AC system control: Standard control techniques (such as PI controllers) can be applied as the modified (dq) variables behave as DC signals.

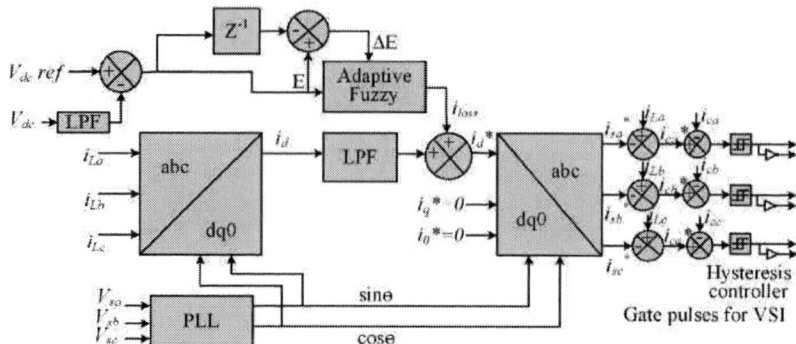

Fig.2 SRF-based adaptive fuzzy control algorithm block diagram.

b. Easier Eliminating: In the (dq) frame, harmonic components show up as high-frequency signals that are readily filtered.

c. Applied in field-oriented control (FOC) of AC motors, vector control, and power converters.

$$\begin{bmatrix} i_d \\ i_q \\ i_0 \end{bmatrix} = \frac{2}{3} \begin{bmatrix} \cos\theta & -\sin\theta & \frac{1}{2} \\ \cos\left(\theta - \frac{2\pi}{3}\right) & -\sin\left(\theta - \frac{2\pi}{3}\right) & \frac{1}{2} \\ \cos\left(\theta + \frac{2\pi}{3}\right) & -\sin\left(\theta + \frac{2\pi}{3}\right) & \frac{1}{2} \end{bmatrix} \begin{bmatrix} i_{La} \\ i_{Lb} \\ i_{Lc} \end{bmatrix}$$
(11)

This synchronization of signals with the point of common coupling (PCC) is achieved by using a phase locked loop (PLL), an essential component for obtaining the phase information required for periodic and physical control. It is essential to generate grid-synchronized pure sine and cosine waves in order to carry out abc to $d-q$ and $d-q$ to t abc transformations. The d-axis and q-axis DC segments are isolated from the components by use of a low pass filter. The eqns. (12) and (13) indicate that the components of i_d and i_q are fundamental and harmonic segments, respectively:

$$i_d = i_{d-DC} + i_{d-AC} \tag{12}$$
$$= \frac{\sqrt{2}}{3}\left[i_{La}\cos(\omega t) + i_{Lb}\cos\left(\omega t - \frac{2\pi}{3}\right) + i_{Lc}\cos\left(\omega t + \frac{2\pi}{3}\right)\right]$$

$$i_q = i_{q-DC} + i_{q-AC} \tag{13}$$
$$= \frac{\sqrt{2}}{3}\left[-i_{La}\sin(\omega t) - i_{Lb}\sin\left(\omega t - \frac{2\pi}{3}\right) - i_{Lc}\sin\left(\omega t + \frac{2\pi}{3}\right)\right]$$

Three phase reference compensating currents $i_{sa}^*, i_{sb}^*, i_{sc}^*$ are obtained from the $d-q$ to abc transformation, and equations(14),(15), and (16) provide their respective expressions:

$$i_{sa}^* = \frac{\sqrt{2}}{3}\left[i_{d,AC}\cos(\omega t) - i_q\sin(\omega t)\right], \tag{14}$$
$$i_{sb}^* = \frac{\sqrt{2}}{3}\left[i_{d,AC}\cos\left(\omega t - \frac{2\pi}{3}\right) - i_q\sin\left(\omega t - \frac{2\pi}{3}\right)\right], \tag{15}$$

and

$$i_{sc}^* = \frac{\sqrt{2}}{3}\left[i_{d,AC}\cos\left(\omega t + \frac{2\pi}{3}\right) - i_q\sin\left(\omega t + \frac{2\pi}{3}\right)\right]$$
(16)

The active current that is required to keep the DC-link voltage stable is what the Fuzzy controller produces as its output. A comparison between the reference source current and the actual load current is what results in the generation of the reference compensating current. An HRES-based SAPF is driven by switching pulses generated by a hysteresis current controller, which takes as input the disparity between the reference and actual compensating currents.

Fuzzy controllers rely on a rule base, which describes the controller's behaviour in response to various input situations.

IV. SIMULATION RESULTS

The proposed configuration considers a PV array with an MPP rating of 10 kW. The constructed system is simulated using the MATLAB/Simulink simulation platform to validate the suggested control technique for hybrid energy systems fed by SAPF.

The system is tested in a range of scenarios, such as variable solar irradiation, wind speed, and balanced/unbalanced and nonlinear loads operating in power factor correction (PFC) mode. The suggested control techniques are developed using the simulation power tool boxes and MATLAB/Simulink. A three-phase diode rectifier (nonlinear) load with RL load is imagined for the purpose of testing the algorithms. The suggested controller's performance is examined using various performance indicators, as illustrated in Fig. 3, including grid voltage Vs, grid current is, load current iL, compensating current ic, DC-link voltage Vdc, PV array voltage PV (V), wind speed (m/s), and fuel cell voltage FC (V). When operating at unity power factor, the PV system guarantees the maximum power output to the grid and the load.

It is noted that the PCC and DC link voltages are 400 V and 750 V, respectively, with insignificant variations.

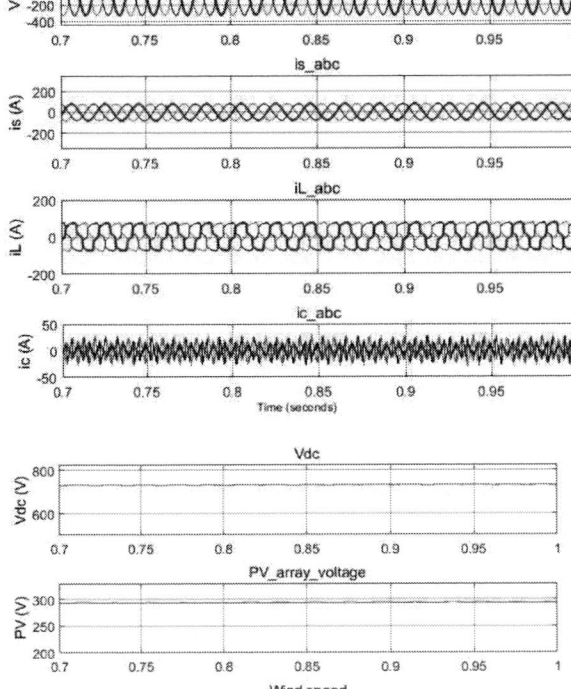

Fig.3 Performance parameters of HRES system.

There was a 19.06% THD for load current and a 3.21% THD for grid current at PCC, respectively. According to the IEEE 519 standard, this is well inside the acceptable range. Fig. 4 shows the THD for the load current and the source current.

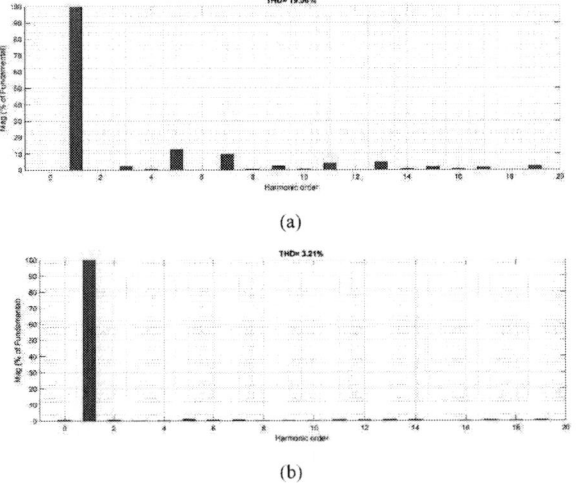

(a)

(b)

Fig. 4 THD for (a) load current and (b) grid current

V. CONCLUSIONS

The performance of the proposed adaptive fuzzy control algorithm is observed for both balanced and unbalanced load conditions through the use of MATLAB simulation environment, that is used for developing the system. The performance of the grid integrated system containing renewable energy sources has been successfully improved according to the control algorithm. The designed system is placed through load unbalancing, varying solar and wind speeds, and more to see whether the proposed controller works. The outcomes indicate that it does an adequate task of reducing grid harmonics. According to the IEEE 519 standards, the system's harmonics in grid currents and voltage variations are below accepted limits. As compared to other conventional approaches, this has demonstrated superior performance in terms of reducing grid current harmonics, improving power factor, regulating voltage, and balancing load.

REFERENCES

[1] Sangswang and M. Konghirun, "Optimal Strategies in Home Energy Management System Integrating Solar Power, Energy Storage, and Vehicle-to-Grid for Grid Support and Energy Efficiency," IEEE Trans. Ind. Appl., vol. 56, no. 5, pp. 5716–5728, 2020, doi: 10.1109/TIA.2020.2991652.

[2] L. Rauchfuß, J. Foulquier, and R. Werner, "Charging station as an active filter for harmonics compensation of smart grid," Proc. Int. Conf. Harmon. Qual. Power, ICHQP, pp. 181–184, 2014, doi: 10.1109/ICHQP.2014.6842905.

[3] L. P. Raj Nadimuthu et al., "Energy Conservation Approach for Continuous Power Quality Improvement: A Case Study," IEEE Access, vol. 9, pp. 146959–146969, 2021, doi: 10.1109/ACCESS.2021.3123153.

[4] A. A. Alkahtani et al., "Power Quality in Microgrids including Supraharmonics: Issues, Standards, and Mitigations," IEEE Access, vol. 8, pp. 127104–127122, 2020, doi: 10.1109/ACCESS.2020.3008042.

[5] S. Mishra, I. Hussain, G. Pathak, and B. Singh, "dPLL-based control of a hybrid wind-solar grid connected inverter in the distribution system," IET Power Electron., vol. 11, no. 5, pp. 952–960, 2018, doi: 10.1049/iet-pel.2017.0491.

[6] Y. Sawle, S. C. Gupta, A. Kumar Bohre, and W. Meng, "PV-wind hybrid system: A review with case study," Cogent Eng., vol. 3, no. 1, p. 1189305, 2016, doi: 10.1080/23311916.2016.1189305.

[7] P. Chaudhary and M. Rizwan, "Voltage regulation mitigation techniques in distribution system with high PV penetration : A review," Renew. Sustain. Energy Rev., vol. 82, pp. 3279–3287, 2018, doi: 10.1016/j.rser.2017.10.017.

[8] R. Kumar and H. O. Bansal, "Shunt active power filter: Current status of control techniques and its integration to renewable energy sources," Sustain. Cities Soc., vol. 42, pp. 574–592, 2018, doi: 10.1016/j.scs.2018.07.002.

[9] J. C. Das, "Passive Filters - Potentialities and Limitations," IEEE Trans. Ind. Appl., vol. 40, no. 1, pp. 232–241, 2004, doi: 10.1109/TIA.2003.821666.

[10] A. Boussaid, A. L. Nemmour, L. Louze, and A. Khezzar, "A novel strategy for shunt active filter control," Electr. Power Syst. Res., vol. 123, no. 2, pp. 154–163, 2015, doi: 10.1016/j.epsr.2015.02.008.

[11] B. Singh, K. Al-haddad, and A. Chandra, "A Review of Active Filters for Power Quality Improvement," IEEE Trans. Ind. Electron., vol. 46, no. 5, pp. 960–971, 1999, doi: 10.1109/41.793345.

[12] J. C. Churio-Barboza and J. M. Maza-Ortega, "Comprehensive design methodology of tuned passive filters based on a probabilistic approach," Gener. Transm. Distrib. IET, vol. 8, no. 1, pp. 170–177, 2014, doi: 10.1049/iet-gtd.2012.0734.

[13] W. U. Tareen, S. Mekhilef, M. Seyedmahmoudian, and B. Horan, "Active power filter (APF) for mitigation of power quality issues in

979-8-3315-3013-6/25 $31.00 © 2025 IEEE

grid integration of wind and photovoltaic energy conversion system," Renew. Sustain. Energy Rev., vol. 70, pp. 635–655, 2017, doi: 10.1016/j.rser.2016.11.091.

[14] R. Kumar and H. O. Bansal, "Real-time implementation of adaptive PV-integrated SAPF to enhance power quality," Int. Trans. Electr. Energy Syst., pp. 1–22, 2019, doi: 10.1002/2050-7038.12004.

[15] R. Kumar, H. O. Bansal, A. R. Gautam, O. P. Mahela, and B. Khan, "Experimental Investigations on Particle Swarm Optimization Based Control Algorithm for Shunt Active Power Filter to Enhance Electric Power Quality," IEEE Access, vol. 10, pp. 54878–54890, 2022, doi: 10.1109/ACCESS.2022.3176732.

[16] R. Kumar, H. O. Bansal, and H. P. Agrawal, "Development of fuzzy logic controller for photovoltaic integrated shunt active power filter," J. Intell. Fuzzy Syst., vol. 36, no. 6, pp. 6231–6243, 2019, doi: 10.3233/JIFS-182520.

[17] Y. S. S. Hanuman Prasad Agrawal, Hari Om Bansal, Ravinder Kumar,"Design and real-time validation of PI and Fuzzy Logic tuned photovoltaic integrated DSTATCOM to improve power quality," Environ. Sci. Pollut. Res., vol. 29, pp. 90158–90177, 2022, doi: 10.1007/s11356-022-21910-7.

[18] H. P. Agrawal, H. O. Bansal, R. Kumar, and Y. S. Sisodia, "HIL Investigations on Intelligently Tuned PV Integrated DSTATCOM to Enhance Power Quality," Arab. J. Sci. Eng., vol. 47, no. 3, pp. 3221–3237, 2022, doi: 10.1007/s13369-021-06104-6.

[19] H. O. B. R Kumar, "Design and Control of Wind integrated Shunt Active Power Filter to Improve Power Quality," 2018 IEEE 8th Power India Int. Conf., pp. 1–5, 2018, doi: 10.1109/POWERI.2018.8704377.

[20] R. Kumar, H. P. Agrawal, A. Shah, and H. O. Bansal, "Maximum power point tracking in wind energy conversion system using radial

basis function based neural network control strategy," Sustain. Energy Technol. Assessments, vol. 36, no. February, p. 100533, 2019, doi: 10.1016/j.seta.2019.100533.

[21] A. Gaillard, P. Poure, and S. Saadate, "Active filtering capability of WECS with DFIG for grid power quality improvement," IEEE Int. Symp. Ind. Electron., pp. 2365–2370, 2008, doi: 10.1109/ISIE.2008.4676984.

[22] M. A. Abdullah, A. H. M. Yatim, C. W. Tan, and R. Saidur, "A review of maximum power point tracking algorithms for wind energy systems," Renew. Sustain. Energy Rev., vol. 16, pp. 3220–3227, 2012, doi: 10.1016/j.rser.2012.02.016.

[23] M. Narayana, G. A. Putrus, M. Jovanovic, P. S. Leung, and S. McDonald, "Generic maximum power point tracking controller for small-scale wind turbines," Renew. Energy, vol. 44, pp. 72–79, 2012, doi: 10.1016/j.renene.2011.12.015.

[24] D. K. R Kumar, HO Bansal, "Improving power quality and load profile using PV-Battery-SAPF system with metaheuristic tuning and its HIL validation," Int. Trans. Electr. Energy Syst., vol. 30, no. 5, pp. 1–19, 2020, doi: 10.1002/2050-7038.12335.

[25] P. Chaudhary and M. Rizwan, "Hybrid control approach for PV / FC fed voltage source converter tied to grid," Int. J. Hydrogen Energy, vol. 43, no. 14, pp. 6851–6866, 2018.

[26] H. O. B. R Kumar, "Design and Control of Wind integrated Shunt Active Power Filter to Improve Power Quality," 2018 IEEE conf., pp. 1–5, 2018, doi: 10.1109/POWERI.2018.8704377.

[27] H. O. B. R Kumar, "Fuel Cell Fed Shunt Active Power Filter for Power Quality Issue by Electric Vehicle Charging," in Renewable Energy Systems: Modeling, Optimization and Applications, pp. 247–264, 2022.

Link Current Optimization for Triple Port Active Bridge Converter with Series Voltage Injection

1st Silpashree Sahu
Electrical Engineering, SECS
IIT BHUBANESWAR
Jatni, Odisha-752050, INDIA
s22ee09002@iitbbs.ac.in

2nd Dipankar De
Electrical Engineering, SECS
IIT BHUBANESWAR
Jatni, Odisha-752050, INDIA
dipankar@iitbbs.ac.in

3rd Alberto Castellazzi
Faculty of Engineering
KYOTO UNIVERSITY of ADVANCED SCIENCE
Kyoto, 615-0096, Japan
alberto.castellazzi@kuas.ac.jp

Abstract—**A Dual Active Bridge (DAB) converter is commonly used as an interface between energy storage systems and DC buses in applications such as distributed generation and electric vehicles. Various modulation schemes, like Single Phase-Shift (SPS), Dual Phase-Shift (DPS), and Triple Phase-Shift (TPS), are used in DAB operation. However, when the voltage conversion ratio deviates from unity, these schemes lead to a higher link current stress. Although modulation and optimization techniques can reduce this stress, they often limit the power range, prevent soft-switching over a wide range, and result in moderate efficiency. To address these issues, a novel optimization scheme is proposed by injecting series voltage into the secondary side of the transformer through an auxiliary third port. This injection helps to reduce the link current stress in an optimized manner based on the available voltage at the auxiliary port. Optimization is achieved by adjusting power flow from the third port to minimize link current stress across different output power ranges. The proposed optimization method is verified through simulation and through experimental results from a scaled-down prototypes.**

Index Terms—**Triple-port active bridge, series voltage injection, link current stress, optimization**

I. INTRODUCTION

The Dual Active Bridge (DAB) converter is commonly used in applications with bidirectional power transfer, such as energy storage systems, distributed generation, and electric vehicles. It is known for its high power density, zero-voltage switching, and bidirectional power flow. The Single-Phase-Shift (SPS) control method, which involves a phase shift (D_1) between H-bridge-1 and H-bridge-2, controls power flow direction and magnitude [1]. However, while SPS control is simple, DAB performance deteriorates when the voltage conversion ratio deviates from unity, resulting in higher link current peaks, increased losses, and reduced efficiency.

The Dual Phase Shift (DPS) control scheme improves on SPS by reducing reactive power and enhancing efficiency, particularly at low power outputs. DPS lowers current levels and causes less current stress under the same transmission power. However, the Current-Stress Optimized (CSO) DPS control, as noted in [2], may not optimize efficiency across the full operational range.

This work is supported by Department of Science and Technology, India under project Grant CRG/2022/000650.

In [3], the impact of back-flow power on circulation and current stress under Extended Phase-Shift (EPS) control is studied. EPS control reduces back-flow power compared to SPS, improving efficiency, especially with large voltage conversion ratios. A Unified TPS (UTPS) control scheme, proposed in [4], minimizes current stress and ensures full soft-switching across the load range. UTPS shows lower current stress than DPS but has a limited transmission power range. A current-stress optimized (CSO) Unified Phase-Shift (UPS) control, introduced in [5], enhances DAB efficiency. An optimal asymmetric duty modulation scheme in [6] reduces inductor peak-to-peak current but has a narrow soft-switching range and medium efficiency. OPS control in [7] minimizes reactive power but increases link current stress, while a segmented analytical approach in [8] reduces current stress in TPS control.

In most of the power converter configurations in Electric Vehicles, there is additional auxiliary port (third port) exists to supply power to the associated electrical equipment like electrical steering, other electronics parts in EV. The configuration is shown in the block diagram form in Fig. 1.

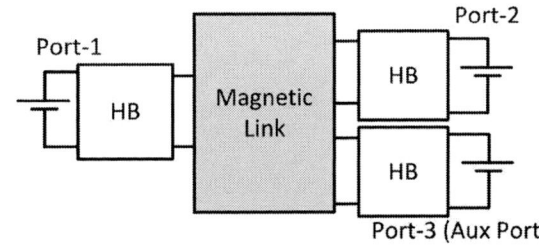

Fig. 1: Block diagram of a typical three port configuration in EV

Several triple active bridge converters are discussed in the literature to interconnect these ports as shown in Fig. 1. [9] proposes a dual-transformer-based asymmetrical triple-port active bridge converter (DT-ATAB) to connect two DC sources and a load, with adjustable phase shift ratios, though it has a higher component count. [10] introduces a self-current balancing structure for the dual-transformer-based triple-port active bridge (DT-TAB), reducing components while achieving ZVS during switch turn-on. In the proposed scheme, link

979-8-3315-3013-6/25 $31.00 © 2025 IEEE

current optimization is achieved through a series voltage injection-based third port (Fig. 2). The optimization process is performed for each individual operating mode first, followed by the determination of a globally optimized solution. The auxiliary port is loaded in a way that facilitates the optimization of the link current.

This paper is organized as follows. Section II outlines the modes of operation of the proposed converter and presents expressions for link current and power. In Section III, the link current optimization is presented. Experimental results under various voltage and load conditions are presented in Section IV to validate the proposed topology. Finally, Section V concludes the paper.

II. DESCRIPTION AND OPERATION OF THE PROPOSED TOPOLOGY

A. Power Circuit

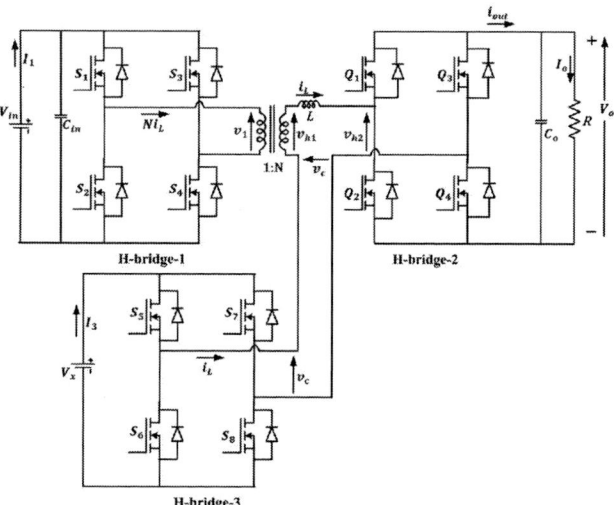

Fig. 2: Power Circuit Diagram of SVI based TAB

In Fig. 2, a series voltage (v_c) is injected into the HV side of the main circuit. This third bridge (H-bridge-3) gives one extra degree of freedom to minimize the link current stress in DAB converter. Here D_1, D_2, and D_3 are the phase shift ratios between H-bridge-1 and 2, H-bridge-1 and 3, and two legs of H-bridge-2 (inner phase shift) respectively.

f: switching frequency

T_s: switching period (T_s=1/f)

T_{hs}: half switching period

k: voltage conversion ratio between H-bridge-1 and 2

$$k = \frac{V_o}{NV_{in}} \quad (1)$$

y: voltage ratio between H-bridge-3 and 2 ($y < 1$)

$$y = \frac{V_x}{V_o} \quad (2)$$

Here base current and base power are defined as,

$$I_b = \frac{V_o T_{hs}}{4L} \quad (3)$$

$$P_b = \frac{N V_{in} V_o T_{hs}}{4L} \quad (4)$$

B. Classification of Modes

Assuming forward power transfer and positive phase shift ratio values, there are basically three modes based on the possible switching sequence of the switches which are shown in Table.I.

TABLE I: Classification of Modes

Switching Sequence	Mode
$S_1 S_4 - S_5 S_8 - Q_1 - Q_4$	I
$S_1 S_4 - Q_1 - Q_4 - S_5 S_8$	II
$S_1 S_4 - Q_1 - S_5 S_8 - Q_4$	III

The corresponding waveforms of different modes are shown in Fig. 3. To ensure zero-voltage switching (ZVS) for switches S_2, S_3, Q_1, Q_4, S_6, and S_7, the link current must be positive at the instant these switches are turned on. Conversely, for ZVS operation of switches Q_2, Q_3, S_5, and S_8, a negative link current is required at their respective turn-on instances. Analysis of the switching sequences across different operational modes reveals that in Mode-I, all switches achieve soft-switching. However, in Modes II and III, the switches in H-bridge-3 undergo hard-switching.

C. Transmission Power and Link Current Stress

P_1 and P_3 are the per unit transmission power from H-bridge-1 to 2 and H-bridge-3 to 2 respectively.

$$P_1 = \frac{1}{P_b T_{hs}} \int_0^{T_{hs}} v_{h1} i_L(t) dt \quad (5)$$

$$P_3 = \frac{1}{P_b T_{hs}} \int_0^{T_{hs}} v_c i_L(t) dt \quad (6)$$

The maximum value of link current during steady-state condition is known as link current stress. Here the circuit is analyzed for $k > 1$. The corresponding mode operational constraints, transmission power, and link current stress expressions are given in Table.II and III respectively. The analysis for $k < 1$ can be done in a similar manner.

III. LINK CURRENT OPTIMIZATION

A. Optimization Flow Chart

The current stress minimization must be based on a certain power transmission requirement. The total power transmission requirement (P_{on}) must be divided between H-bridge-1 and 3. z is the ratio between power transfer by H-bridge-3 and 1 ($z=P_3/P_1$). Based on the value of z, the required values of P_1 (P_{1n}) and P_3 (P_{3n}) are calculated as follows.

$$P_{1n} = \frac{P_{on}}{1 + z} \quad (7)$$

979-8-3315-3013-6/25 $31.00 © 2025 IEEE   70

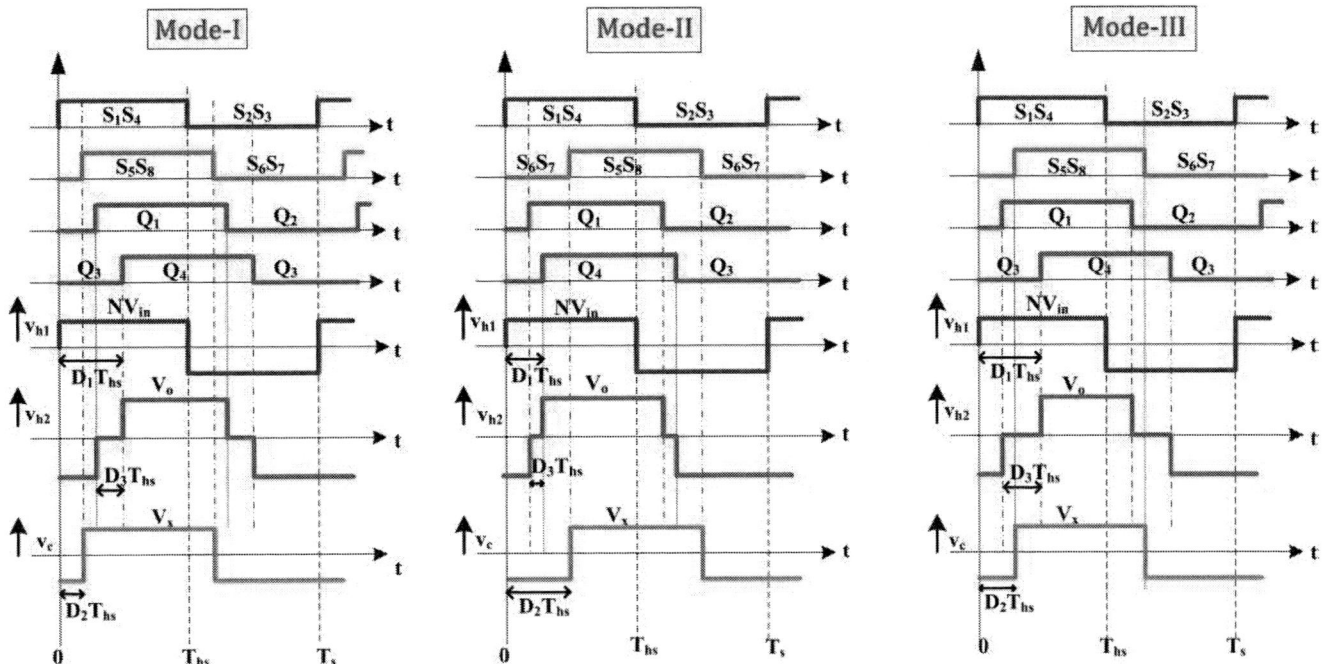

Fig. 3: Illustrative waveforms of Triple-Port Active Bridge Converter for different modes

TABLE II: Mode equations for transmission power ($k \geq 1$ and $k < 1$)

Mode	Operational Constraints $(A_i(D) \leq B_i)$	Transmission Power (pu) $(Ceq_j(D))$
I	$0 \leq D_2 \leq D_1\text{-}D_3 \leq D_1 \leq 0.5$	$P_1 = 2(2D_1(1 - D_1 + D_3) - D_3^2 - D_3 + 2yD_2(D_2 - 1))$ $P_3 = 2yk(((2D_2/k)(1 - D_2)) + 4D_1D_2 + 2D_1D_3 - 2D_2D_3 + 2D_1 - 2D_2 - 2D_1^2 - 2D_2^2 - D_3^2 - D_3)$
II	$0 \leq D_1\text{-}D_3 \leq D_1 \leq D_2 \leq 0.5$	$P_1 = 2(2D_1(1 - D_1 + D_3) - D_3^2 - D_3 + 2yD_2(D_2 - 1))$ $P_3 = 2yk(((2D_2/k)(1 - D_2)) - 4D_1D_2 - 2D_1D_3 + 2D_2D_3 + 2D_1 - 2D_2 + 2D_1^2 + 2D_2^2 + D_3^2 - D_3)$
III	$0 \leq D_1\text{-}D_3 \leq D_2 \leq D_1 \leq 0.5$	$P_1 = 2(2D_1(1 - D_1 + D_3) - D_3^2 - D_3 + 2yD_2(D_2 - 1))$ $P_3 = 2yk(((2D_2/k)(1 - D_2)) - 2D_1D_3 + 2D_2D_3 + 2D_1 - 2D_2 + D_3^2 - D_3)$

TABLE III: Mode equations for link current stress ($k \geq 1$ and $k < 1$))

Mode	$k \geq 1, V_x > \|NV_{in} - V_o\|$	$k \geq 1, V_x \leq \|NV_{in} - V_o\|$	$k < 1$
I	$2[(1/k) + (2D_1\text{-}D_3\text{-}1)+y(1\text{-}2D_2)]$	$2[(1/k)(2D_1\text{-}1)+(1\text{-}D_3)+y(2D_1\text{-}2D_2\text{-}1)]$	$2[(1/k) + (2D_1\text{-}D_3\text{-}1)+y(1\text{-}2D_2)]$
II	$2[(1/k)(2D_1\text{-}1)+(1\text{-}D_3)\text{-}y(2D_1\text{-}2D_2+1)]$, if $V_x < V_o/2$ $2[(1/k) + (2D_1\text{-}D_3\text{-}1)+y(1\text{-}2D_2)]$, if $V_x > V_o/2$	$2[(1/k)(2D_1\text{-}1)+(1\text{-}D_3)\text{-}y(2D_1\text{-}2D_2+1)]$	$2[(1/k) + (2D_1\text{-}D_3\text{-}1)+y(1\text{-}2D_2)]$
III	$2[(1/k) + (2D_1\text{-}D_3\text{-}1)+y(1\text{-}2D_2)]$	$2[(1/k)(2D_1\text{-}1)+(1\text{-}D_3)+y(2D_1\text{-}2D_2\text{-}1)]$	$2[(1/k) + (2D_1\text{-}D_3\text{-}1)+y(1\text{-}2D_2)]$

$$P_{3n} = zP_{1n} \tag{8}$$

To minimize the link current stress, an optimization problem is formulated into the standard form as follows.

$$\text{Minimize:} \quad f(D)$$
$$\text{subject to:} \quad A_i(D) \leq B_i, \quad i = 1, 2, \ldots, m \tag{9}$$
$$Ceq_j(D) = 0, \quad j = 1, 2, \ldots, p$$

where $D = (D_1, D_2, D_3, z)$, $f(D)$ is the objective function which is the link current stress, $A_i(D) \leq B_i$ refers to the linear non-equality (operational constraints), m and p are the no. of non-equality and equality constraints respectively, and $Ceq_j(D) = 0$ refers to the non-linear equality constraints (transmission power constraints). The optimization is implemented using fmincon function in MATLAB. A flow chart to get the optimized values of D_1, D_2, D_3, and z is shown in Fig. 4. First, we must determine the optimized values of

979-8-3315-3013-6/25 $31.00 © 2025 IEEE

D_1, D_2, D_3, and z for each individual mode. Afterward, the overall optimized values are obtained by selecting the best combination that minimizes the current stress.

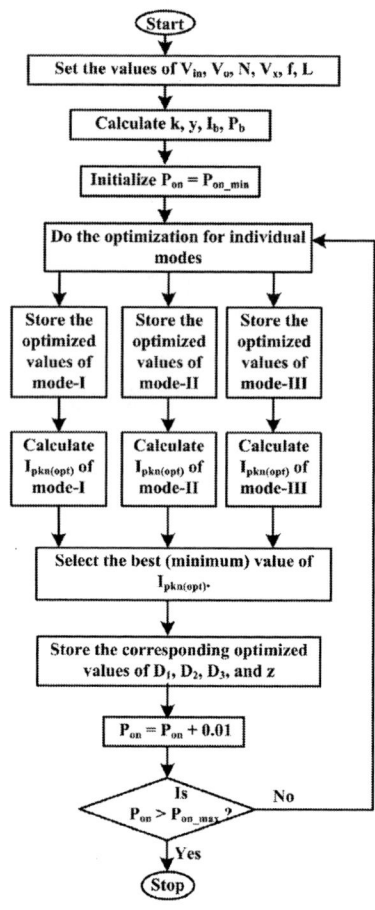

Fig. 4: Flow chart to get optimized values of phase-shift ratios and z

B. Optimization Results

The optimization results of individual modes and combined one are shown in Fig. 5, Fig. 6, Fig. 7, Fig. 8, Fig. 9. Fig. 5, Fig. 6, Fig. 7 each consist of three subplots, illustrating the variations of different phase shifts, link current stress, and transmission powers with respect to output power in the first, second, and third subplots, respectively. From Fig. 5, it is evident that D_2 remains zero across the entire output power range. For $P_{on} \leq 0.08pu$, D_1 equals D_3, after which D_3 decreases and D_1 increases. Additionally, the optimized z value remains nearly constant in mode-I. In Fig. 6, it can be seen that up to $P_{on} = 0.025pu$, $D_1 = D_2 = D_3$. Beyond this point, $D_1 = D_2$ while D_3 drops to zero. The optimized z value steadily increases until $P_{on} = 0.2pu$, after which it stabilizes at a constant value in mode-II. Fig. 7 shows that for $P_{on} \leq 0.3pu$, D_1 is greater than D_3, which is greater than D_2. After this, the order changes to $D_1 > D_2 > D_3$. The optimized z value steadily increases across the entire power

range in mode-III. Fig. 8 presents a comparison of the link current stress across different modes. The overall optimized current stress is obtained by selecting the minimum current stress from the various modes. Fig. 9 illustrates the overall optimized variations of different phase shifts, link current stress, and mode number with respect to output power. Up to $P_{on} = 0.016pu$, $D_1 = D_2 = D_3$, after which D_2 drops to zero. For $P_{on} \leq 0.08pu$, $D_1 = D_3$, and beyond this point, D_3 decreases gradually. For $P_{on} < 0.017pu$, the minimum current is obtained from mode-III, while for $P_{on} > 0.08pu$, the minimum current is attained from mode-I. By appropriately adjusting the value of z, the converter can operate in Mode-I, minimizing current stress while maintaining soft-switching conditions.

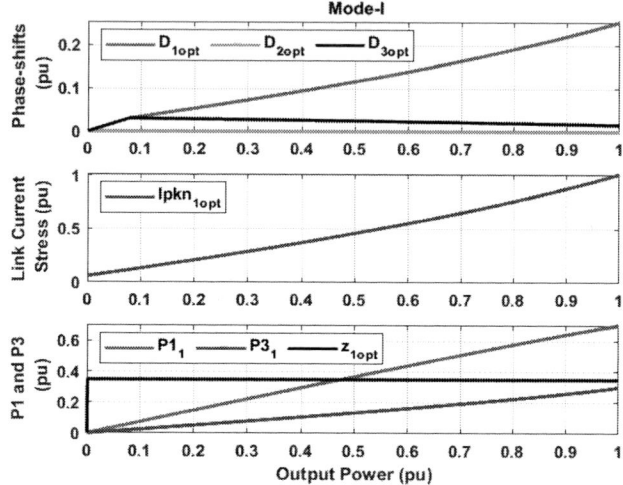

Fig. 5: Mode-I: Variation of phase shifts, link current stress, P_1 and P_3 with P_o at $V_{in} = 24V$ and $V_x = 50V$

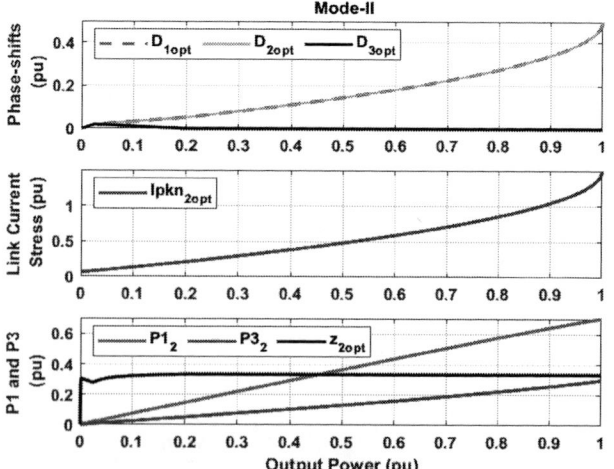

Fig. 6: Mode-II: Variation of phase shifts, link current stress, P_1 and P_3 with P_o at $V_{in} = 24V$ and $V_x = 50V$

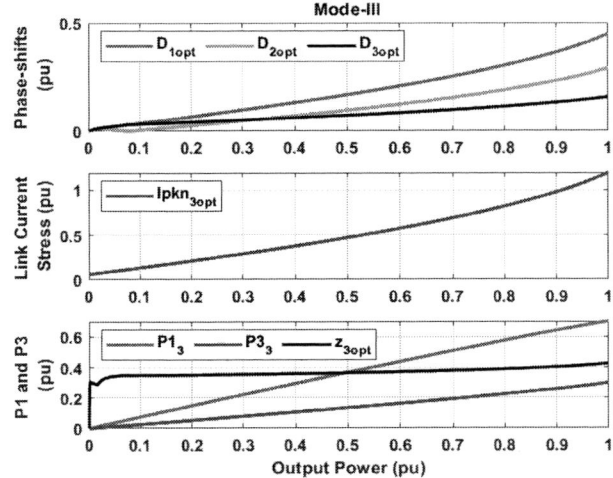

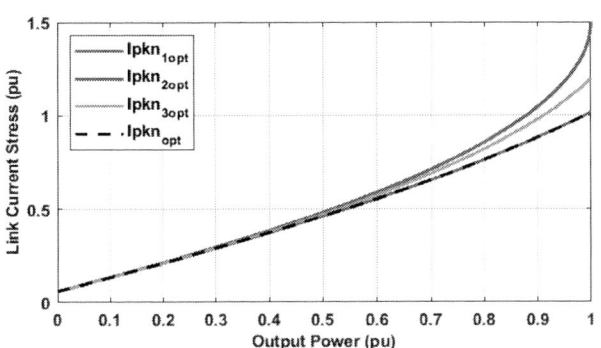

Fig. 7: Mode-III: Variation of phase shifts, link current stress, P_1 and P_3 with P_o at $V_{in} = 24V$ and $V_x = 50V$

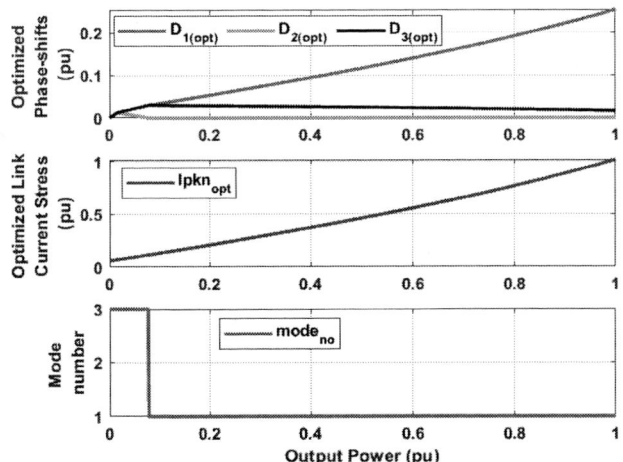

Fig. 8: Variation of link current stress with P_o at $V_{in} = 24V$ and $V_x = 50V$

Fig. 9: Optimized variation of phase shifts, link current stress, P_1 and P_3 with P_o at $V_{in} = 24V$ and $V_x = 50V$

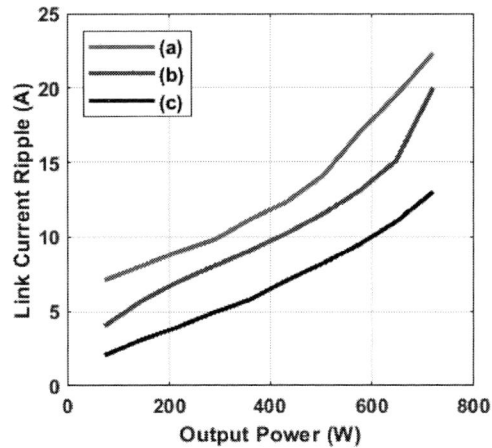

Fig. 10: Variation of link current ripple with output power (a) SPS [1], (b) SATPS [8], (c) SVI (proposed) at $V_{in} = 24V$

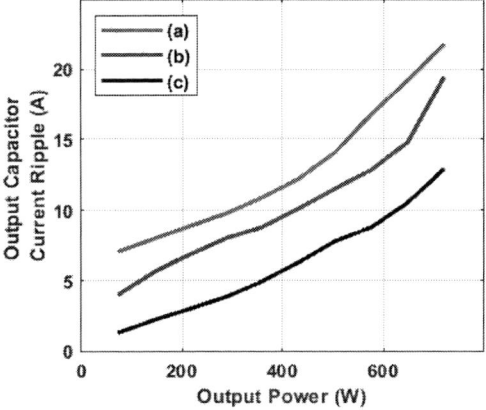

Fig. 11: Variation of output capacitor current ripple with output power (a) SPS [1], (b) SATPS [8], (c) SVI (proposed) at $V_{in} = 24V$

Figures 10 and 11 illustrate the variation of link current ripple and output capacitor current ripple with the output power for DAB with SPS [1], SATPS [8] control, and SVI (proposed topology) respectively. As evident from the figures, the proposed topology exhibits significantly lower current ripple compared to the other control methods at equivalent power levels. This reduction in ripple is attributed to the effect of series voltage injection. Consequently, the required sizes of energy storage components, such as the link inductor and output capacitor can be minimized, contributing to improved power density and system efficiency. The look-up table (LUT) based control diagram for the closed loop control of the proposed topology is given in Fig. 12. The PI controller generates the reference output power, $P_{o(ref)}$, while regulating the output voltage, V_o. The reference value of $V_{x(ref)}$ is provided by the user. For varying values of $P_{o(ref)}$ and $V_{x(ref)}$, the lookup tables (LUTs) provide the optimized values for D_1, D_2, D_3, and z. These phase shifts are then supplied

979-8-3315-3013-6/25 $31.00 © 2025 IEEE   73

to the modulator.

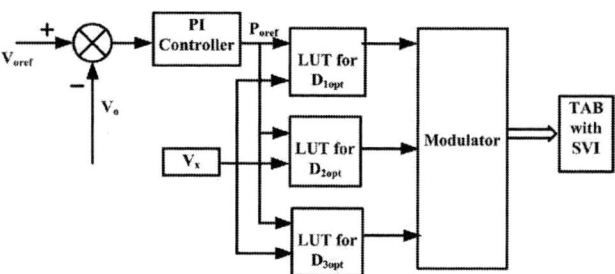

Fig. 12: Control diagram for closed loop control of the proposed topology

The power circuit is simulated using the parameters in Table IV, and Fig. 13 presents the results for both non-optimized and optimized link currents. The optimized case reduces link current stress while maintaining power transfer. The converter's total loss is 18.23W (6.08%) without optimization and 10.69W (3.56%) with optimization, with the loss reduction due to the optimized reduction in current stress.

TABLE IV: Parameter Values

Attributes	Simulation	Experimental
Input DC voltage (V_{in})	24V	24V
Output DC voltage (V_o)	200V	200V
Transformer's turns ratio (N)	6	6
Third port input voltage (V_x)	40-55V	40-55V
Switching frequency (f)	20kHz	20kHz
L (inductor+transformer)	250uH	267.2uH
Power rating (P_o)	7.2-720W	150-350W

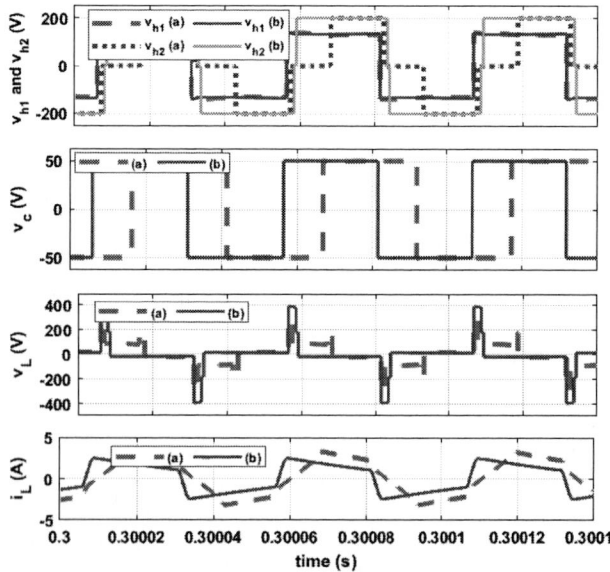

Fig. 13: Simulation results at V_{in}=24V, V_x=50V and P_o=300W for (a) without optimization, (b) with optimization

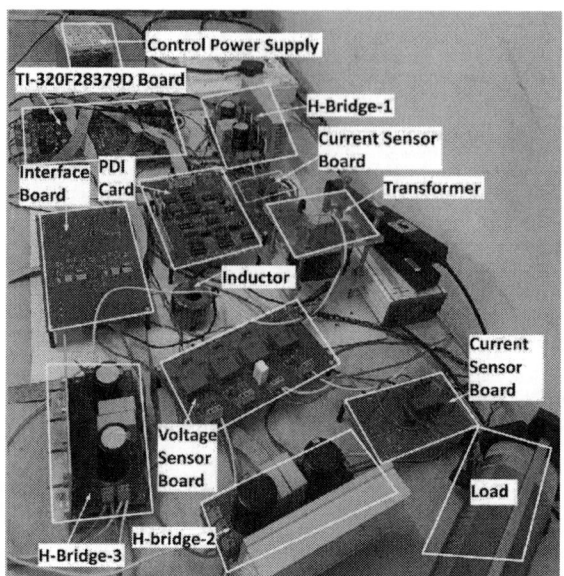

Fig. 14: Hardware setup of SVI based TAB

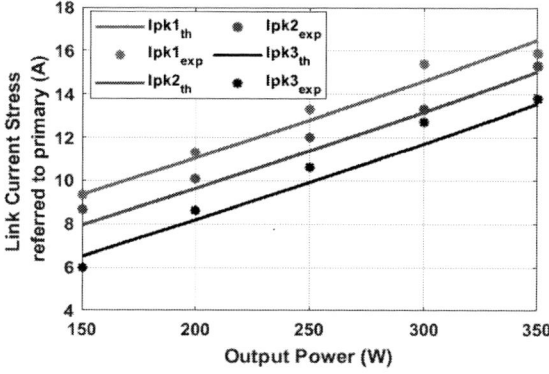

Fig. 15: Variation of link current stress with output power at (i) $V_x = 40V$, (ii) $V_x = 45V$, and (iii) $V_x = 50V$

IV. EXPERIMENTAL VERIFICATION

To validate the proposed topology, a laboratory setup (Fig.14) for the triple port active bridge converter was prepared. The control signals were generated using a digital signal controller TMS320F28379D from Texas Instruments. Phase-shifted signals were then transmitted to the gate driver through interface boards. To implement LUTs in the microcontroller for optimization, 101 data points were used for output power (P_{on}) and 16 data points for the third port input voltage (V_x). In total, 1,616 data points were implemented for each phase shift. Voltage and current feedback were supplied to the controller through voltage and current sensor boards. Additionally, the controller incorporated trip logic to handle overvoltage and overcurrent conditions.

The link current stress values, referenced to the primary side, are measured and compared between simulation and experimental results for three different values of V_x: I_{pk1} for

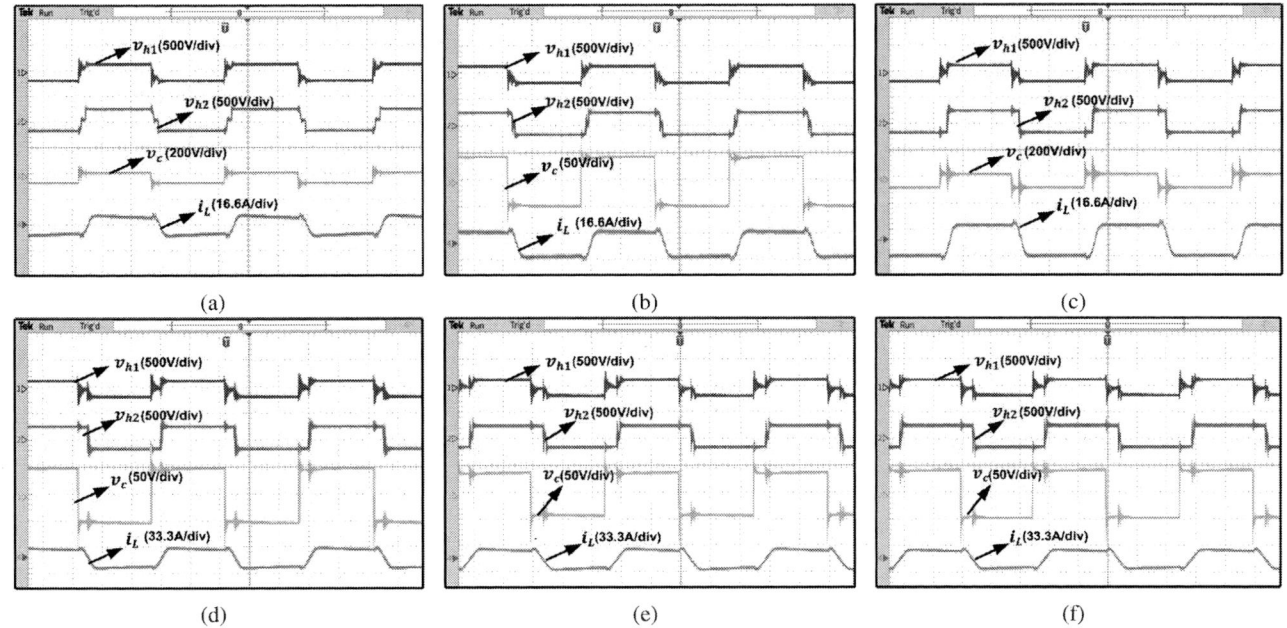

(a) (b) (c)

(d) (e) (f)

Fig. 16: Waveforms for SVI based TAB at $V_{in} = 24V$ (a) $V_x = 40V$ and $P_o = 150W$ (b) $V_x = 45V$ and $P_o = 200W$ (c) $V_x = 50V$ and $P_o = 250W$ (d) $V_x = 50V$ and $P_o = 300W$ (e) $V_x = 40V$ and $P_o = 350W$ (f) $V_x = 45V$ and $P_o = 350W$

$V_x = 40V$, I_{pk2} for $V_x = 45V$, and I_{pk3} for $V_x = 50V$. These comparisons, with optimization, are shown in Fig. 15. As the value of V_x increases, the current stress decreases for the same output power. Experimental results for different power levels and V_x values are depicted in Fig. 16. As the output power increases, the converter predominantly operates in Mode-I.

V. CONCLUSION

This paper presents a link current optimized triple port active bridge converter with series voltage injection from its third port. The mathematical expressions are derived for the powers at various ports, and an optimization study was carried out to minimize the link current stress at different power levels and at different voltage levels of the third port voltages. This reduction in peak current facilitates the downsizing of energy storage elements (link inductance and output capacitance). Simulation studies and experimental validation of the proposed concept confirm the reduced value of the link current under various operating conditions. The measured results show good accuracy with the theoretically obtained optimization except a deviation due to dead-time and losses in the converters. The future research work will focus on the mitigation of the same. Furthermore, a general optimization problem involving penta-phase shift, along with independent loading on the auxiliary port, will be explored in future work.

REFERENCES

[1] M. N. Kheraluwala, R. W. Gascoigne, D. M. Divan and E. D. Baumann, "Performance characterization of a high-power dual active bridge DC-to-DC converter," *IEEE Trans. Ind. Appl.*, vol. 28, no. 6, pp. 1294-1301, Nov.-Dec. 1992.

[2] B. Zhao, Q. Song, W. Liu and W. Sun, "Current-Stress-Optimized Switching Strategy of Isolated Bidirectional DC–DC Converter With Dual-Phase-Shift Control," *IEEE Trans. Ind. Electron.*, vol. 60, no. 10, pp. 4458-4467, Oct. 2013.

[3] B. Zhao, Q. Yu and W. Sun, "Extended-Phase-Shift Control of Isolated Bidirectional DC–DC Converter for Power Distribution in Microgrid," *IEEE Trans. Power Electron.*, vol. 27, no. 11, pp. 4667-4680, Nov. 2012.

[4] J. Huang, Y. Wang, Z. Li and W. Lei, "Unified Triple-Phase-Shift Control to Minimize Current Stress and Achieve Full Soft-Switching of Isolated Bidirectional DC–DC Converter," *IEEE Trans. Ind. Electron.*, vol. 63, no. 7, pp. 4169-4179, July 2016.

[5] N. Hou, W. Song and M. Wu, "Minimum-Current-Stress Scheme of Dual Active Bridge DC–DC Converter With Unified Phase-Shift Control," *IEEE Trans. Power Electron.*, vol. 31, no. 12, pp. 8552-8561, Dec. 2016.

[6] D. Mou et al., "Optimal Asymmetric Duty Modulation to Minimize Inductor Peak-to-Peak Current for Dual Active Bridge DC–DC Converter," *IEEE Trans. Power Electron.*, vol. 36, no. 4, pp. 4572-4584, Apr. 2021.

[7] H. Shi, H. Wen, J. Chen, Y. Hu, L. Jiang and G. Chen, "Minimum-Reactive-Power Scheme of Dual-Active-Bridge DC–DC Converter With Three-Level Modulated Phase-Shift Control," *IEEE Trans. Ind. Appl.*, vol. 53, no. 6, pp. 5573-5586, Nov.-Dec. 2017.

[8] Q. Gu, L. Yuan, J. Nie, J. Sun and Z. Zhao, "Current Stress Minimization of Dual-Active-Bridge DC–DC Converter Within the Whole Operating Range," *IEEE J. Emerg. Sel. Topics Power Electron.*, vol. 7, no. 1, pp. 129-142, Mar. 2019.

[9] V. N. S. R. Jakka, A. Shukla and G. D. Demetriades, "Dual-Transformer-Based Asymmetrical Triple-Port Active Bridge (DT-ATAB) Isolated DC–DC Converter," *IEEE Trans. Ind. Electron.*, vol. 64, no. 6, pp. 4549-4560, Jun. 2017.

[10] E. S. Oluwasogo and H. Cha, "Self-Current Sharing in Dual-Transformer-Based Triple-Port Active Bridge DC–DC Converter With Reduced Device Count," *IEEE Trans. Power Electron.*, vol. 36, no. 5, pp. 5290-5301, May 2021.

Strategic Allocation of Electric Vehicle Charging Infrastructure in Distribution Networks: A Novel Reliability-Based Approach

Edwin Boima Fahnbulleh
School of Electrical Engineering
KIIT Deemed to be University
Bhubaneswar, India
2342002@kiit.ac.in

Sharmistha Nandi
School of Electrical Engineering
KIIT Deemed to be University
Bhubaneswar, India
sharmisthanandi38@gmail.com

Sriparna Roy Ghatak
School of Electrical Engineering
KIIT Deemed to be University
Bhubaneswar, India
sreeparna.ghatak@gmail.com

Babita Panda
School of Electrical Engineering
KIIT Deemed to be University
Bhubaneswar, India
babitapfel@kiit.ac.in

Abstract- The penetration of Electric Vehicles (EVs) in the Distribution Networks (DNs) hampers the network operation. Unplanned integration of Electric Vehicle Charging Infrastructures (EVCI) in the DN deteriorates the network's technical parameters and reliability. This study proposes a novel methodology to properly allocate EVCI, incorporating a unique bus Reliability Index to identify strong and weak buses to properly place the EVCI in the DNs, thereby reducing EVS impacts on the DN. The EV load used in this paper is real data with the charging history of residential apartments in Norway. The analysis was done on the IEEE 15-bus DN and the 28-bus Indian rural DN, evaluating the impacts of Charging Infrastructures on the DN reliability. Various technical parameters, including power loss, voltage profile, and reliability indices such as Expected Energy Not Served, System Average Interruption Frequency Index, System Average Interruption Duration Index, and Average Expected Energy Not Served, are assessed. The results illustrate that strategic EVCI placement using the proposed methodology significantly mitigates EVCI adverse impacts on the DNs' reliability compared to other EV allocation strategies. This research contributes valuable insights for distribution system operators in planning and managing EV integration while maintaining system reliability and efficiency.

Keywords- Distribution Network, Electric vehicles, Electric Vehicle Charging Infrastructure, Power loss, Reliability, Voltage profile

I. INTRODUCTION

The Distribution Network's (DN's) role is vital in delivering electricity from transmission systems to customers. However, these networks are challenged with aging infrastructure, increasing demand, and the integration of renewable energy sources (RES) [1]. One of the most recent important developments in the energy sector, aiming to buttress the global effort to combat climate change, is the rapid adoption of Electric Vehicles (EVs), the emerging sustainable alternatives to internal combustion engine (ICE) vehicles [2].

The rise in EV adoption has led to the growing need for EVCI within the DN, presenting opportunities and challenges to the DN. Though EVCIs offer environmental benefits and contribute to the sustainable transportation sector, their

incorporation into the existing DNs significantly transforms modern DN operations [3]. EV presence in the DN presents technical challenges like power loss, voltage stability, and reliability, requiring careful planning to sustain the reliability and efficiency of the system [4]. Reliability study is key to DN with charging infrastructures (CIs), due to the significant impact of charging loads on the grid's technical parameters. EVCI incorporation into the network introduces large and varying loading conditions that deteriorate the DN reliability, causing customer dissatisfaction [5]. Moreover, the stochastic nature of EVCI operations offers major challenges in DN stability, requiring crucial attention. Considering the radial DN operation with EV, the charging state of EVCIs in the grid is more likely to cause instability than the discharging state [6]. This unpredictability of the load needs robust reliability analysis to guarantee the system's stability under various operating conditions. Reliability assessment on existing networks also helps distribution system operators (DSOs) to make decisions on the suitable allocation of the CI and the state of the power system operational parameters. Furthermore, by considering power loss minimization, voltage profile maintenance, and DN reliability improvement in the network planning process, planning engineers can create resilient and efficient DN [7]. This reliability study helps DSO properly assess the system's ability to deliver power to the customers more efficiently. The reliability indices Expected Energy Not Served (EENS), System Average Interruption Frequency Index (SAIFI), System Average Interruption Duration Index (SAIDI), and Average Expected Energy Not Served (AEENS) provide valuable insights into the network's performance and help identify areas in the network for improvement [8].

Most recent research has focused on EVCI allocation within the DN without assessing the network reliability. The authors in [9] proposed a technique to minimize active and reactive power losses while allocating EVCI in the DN, but didn't evaluate network reliability. Other authors assessed the network reliability considering EVCI, but randomly placed the CI. For instance, the authors in [10] evaluated the reliability of the IEEE 33-bus test system with fast chargers integrated into the DN, but the CIs allocation was randomly done. Some studies investigated the reliability of DN while excluding CI loads. In [11], the authors investigated the causes of the interruptions and challenges faced by customers in Nepal due to frequent planned and unplanned continuous interruptions. The authors evaluated various reliability indices but did not consider the presence of CIs in the analysis. In [12], the authors assessed the performance and compared various reliability indices by increasing the load on the DN at different percentage levels. However, their analysis did not consider EVCI, thereby excluding bus reliability level identification in the studied system.

The authors in [13] investigated some key factors that influence EV owners' satisfaction in China. This study underscores the need to apply the zonal division technique to distribute the CIs across the entire planning area, reducing the driving range to assess public charging facilities.

The present work differs from previous works by adopting a novel methodology enabling the proper allocation of CI in DN based on reliability to reduce its adverse impacts on the grid. Further, to benefit the EV owners, the entire DN is divided into zones for allocating EVCI in the distributed manner in the network. For each zone, a novel bus reliability index (RI) is utilized to identify the strong and weak buses in the DNs. For the benefit of DSO from the system reliability point of view, strong areas were identified for CI placement. The present planning model is validated on an IEEE 15-bus test system and a 28-bus rural DN. To judge the efficacy of the proposed methodology, present work is compared with other methods. The primary contributions of this paper are:

- A unique Reliability Index (RI) is formulated to evaluate the reliability of the network.
- Considering the concerns of DSOs and EV owners, a novel planning methodology is framed to judicially allocate EVCI in the planning area based on the RI values.
- Various technical parameters, such as power loss, voltage profile, and reliability indices, including EENS, SAIFI, SAIDI, and AEENS, are analysed with the incorporation of EVCI in the DN.

The rest of the paper is organized as follows: Section II discusses the EVCI load modelling, various reliability indices, and bus reliability index formulation. Section III presents the methodology, Section IV presents the results and discussion, and Section V concludes the study.

II. MATHEMATICAL MODELLING

This section describes the EVCI load curve used in this paper and discusses the power system technical parameters and the system reliability indices.

A. Electric Vehicle Charging Infrastructure (EVCI)

EVCIs presence in the DN addresses global warming. However, EV rapid adoption points to the need for more CI, which affects the DN technical parameters and reliability. EV chargers are categorized into levels based on power ratings and charging speeds [14]. Level I being the slowest, this paper analyses the effects of slow charging on DN reliability.

1) EVCI load modelling

The reliability investigation of DNs is conducted by incorporating the peak EVCI load for a residential apartment in Norway. The study utilized real historical datasets of residential EV charging activity of Norwegian housing cooperatives in Trondheim, Norway, covering December 2018 to January 2020. The dataset has information such as session identifiers, plug-in and plug-out times, energy charges, and user charging points (CPs) links. The CSV file of the datasets had records of 6,878 registered charging sessions by 97 different user IDs. Each session is connected to a user ID, charger ID, and address. The charger IDs indicate that the CPs are private or shared, depending on their locations in private or shared parking spaces. Information on the charging loads and non-charging idle capacity per user and individual charging sessions for the same period are considered. The synthetic charging loads are provided with hourly resolution, assuming the charging powers of 7.2 kW, immediate charging upon plug-in, and non-charging idle time. This reflects potential flexibility for individual and aggregated users [15].

B. Distribution System Reliability Indices (RI)

Integrating EVCI into the grid increases the system loads, impacting the system operation. Uncoordinated CI placement deteriorates the network reliability, affecting customer satisfaction [16]. Reliability assessment in this paper highlights two groups of indices. The customer-based indices comprise SAIFI and SAIDI. The load-based indices include EENS and AEENS [10].

1) System Average Interruption Frequency Index (SAIFI)
SAIFI quantifies the frequency customers encounter outages or interruptions within a specified period [10]

$$SAIFI = \frac{\sum \lambda_i N_i}{\sum N_i} \qquad (1)$$

Where N_i is the number of customers on the ith number of buses and λ_l is the line failure rate.

2) System Average Interruption Duration Index (SAIDI)
SAIDI determines the interruption duration for each customer. It depends on the interruption length and the number of customers [10].

$$SAIDI = \frac{\sum U_i N_i}{\sum N_i} \qquad (2)$$

Where U_i is the customer interruption time (unavailability).

3) Average Expected Energy Not Served (AEENS)
AEENS is the total EENS divided by the number of customers that experienced interruption per year [10].

$$AEENS = \frac{\sum L_i U_i}{\sum N_i} \qquad (3)$$

Where L_i is the active load at the ith number of buses.

4) Expected Energy Not Served (EENS)
EENS accounts for the unserved energy during the interruption [17].

$$EENS = \sum_{i=1}^{m} U_i L_i \qquad (4)$$

$$U_i = \sum_{j=1}^{l} \lambda_j r \qquad (5)$$

Where U_i is the bus unavailability, λ_j is the failure rate of the line j, and r is the outage time.

The expression below is used to calculate the failure rate [17]

$$\lambda_j = \lambda_{max} + \left(\frac{\lambda_{max} - \lambda_{min}}{R_{max} - R_{min}}\right) * R_j \qquad (6)$$

Where λ_{j_min} and λ_{j_max} denote the minimum and maximum failure rates of line j, respectively. λ_{min} value is 0.1 for the line with low resistance R_{min} and λ_{max} is 0.5 the line having maximum resistance R_{max}.

After CIs are integrated, the failure rate and unavailability are updated using the equations below [18].

$$\lambda_{EVCI} = (L_{bae} + \Delta L_{EVCI})\frac{\lambda_{base}}{L_{base}} \qquad (7)$$

$$U_{EVCI} = (L_{bae} + \Delta L_{EVCI})\frac{U_{base}}{L_{base}} \qquad (8)$$

Where ΔL_{EVCI}, L_{bae} denote the charging load after EVCI integration and the base load, respectively.

5) Bus Reliability Index (RI) Calculation

Each bus reliability is evaluated using the proposed index. RI is used to allocate CIs in the grid. RI is calculated using the following expression.

$$RI = \frac{EENS_i}{EENS_{Max}} \qquad (9)$$

Where $EENS_{Max}$ is the system's maximum EENS value.

III. METHODOLOGY

This paper conducted a reliability study considering two systems, the IEEE 15-bus network and the 28-bus radial DN. The description of the methodology used in the study is categorized into two subsections as outlined below.

A. Zonal Division and CI sizing

- **Zonal Division method**

The Network is initially divided into N zones based on its structure to allocate the CI. This strategy ensures that the CI is evenly distributed across the entire planning area. The approach alleviates the burden on EV owners by reducing the need to travel long distances to recharge their EV batteries. This approach helps DN planners decentralize the CI within the network, avoiding the placement of large EV loads at a single bus, thereby achieving a balanced network loading.

- **Sizing of charging Infrastructure**

The load density forecasting technique [19] is applied to determine the zonal CI capacity in the DNs. First, the above zonal division strategy is applied to the DNs. Then the CI loads are distributed across the allocated zones into which the DN is divided. Applying the load density method, the CI capacity Sci (ci = 1,2, …, N) is predicted for each zone. The expression to determine the CI capacity per zone is given as:

$$S_{CI} = K * N_{EV} * P_{ch} * \frac{L_{zone}}{L_{total}} \qquad (10)$$

Where k is the charging simultaneity of EVs and its rate is 0.2 in this paper, N_{EV} is the number of EVs in the planning area, P_{ch} is the charger power rating, L_{zone} is the total load in each zone, and L_{total} is the total load of the entire network.

B. Implementation

The implemented procedures are detailed below in the following steps:

- *Step 1:* The system's line and load voltage are read.
- *Step 2:* The load flow is run using the backward-forward sweep (BFS) algorithm, and the power loss and voltage profile are obtained.
- *Step 3:* SAIFI, SAIDI, EENS, and AEENS for the systems are calculated using (1-8).
- *Step 4:* As mentioned above, the network is divided into zones for proper CI placement.
- *Step 5:* The bus RI is calculated using (9) for each zone. Based on the RI value for each bus, the strong and weak buses in each zone of the network were selected.
- *Step 6:* The individual network buses are classified as strong and weak buses.

- *Step 7:* The EVCI load is sized using (10) for each zone
- *Step 8:* The load flow is run with EVCI, and the results are compared.

IV. RESULTS AND DISCUSSION

CI development is influencing the DN by increasing the load demand. This increase in demand can hamper the network operating parameters and disturb system reliability and efficiency. This manuscript analyses the impact of EVCI loading on power losses, voltage profile, and distribution network reliability. The analysis is done in scenarios.

A. Test Systems Description

The 15-bus system comprises 14 load buses and 14 lines, with a base voltage of 11 kV and a base power of 100 MVA. The total demand is 1226.4000 kW and 1251.1785 kVAr. Similarly, the 28-bus system has 27 load buses and 27 lines with a base voltage of 11 kV and a base power of 100 MVA. The system total demand is 761.0400 kW and 776.4190 kVAr [20]. This paper divides test distribution systems into zones to strategically place the CI, ensuring convenient access for EV owners. The IEEE 15-bus test system is divided into two zones, while the 28-bus is divided into three zones.

a) EVCI Location Identification

To identify a suitable bus for the CI allocation, each bus RI is calculated using (9). In this paper, buses with low RI values are considered the best locations for the EVCI placement. In this paper, these buses are classified as strong and weak. Fig.1 and Fig.2 graphically illustrate the RI values for each bus in both the 15-bus and 28-bus networks, respectively. These graphs illustrate the classification of various bus reliability levels based on the RI values. From the RI results shown in Figs.1 and 2, suitable locations in separate zones are identified for EVCI allocation. In the IEEE 15-bus network shown in Fig. 1, Bus 10 in Zone One and Bus 5 in Zone Two have the lowest RI values and are considered the best locations for CI placement. Similarly, in the 28-bus network, shown in Fig. 2, bus 14 in Zone One, bus 17 in Zone Two, and bus 10 in Zone Three have the lowest RI values and are identified as the proper locations for CI integration. This process helps minimize the CI's impact on the grid. Fig. 3(a) represents the IEEE 15-bus test system, while Fig. 3(b) describes the 28-bus India rural distribution system. In this paper, both networks shown in Fig. 3(a) and 3(b) are divided into zones, and the EVCIs are placed in identified locations within each zone.

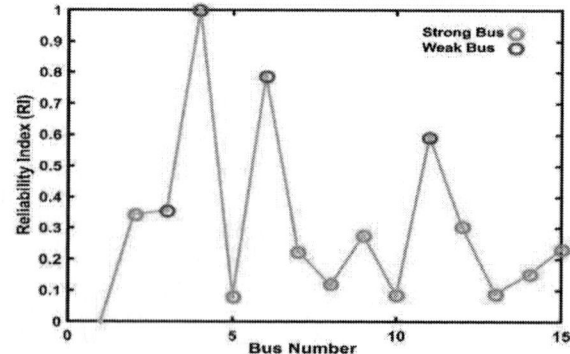

Fig. 1 Bus-wise RI for the IEEE 15-bus system

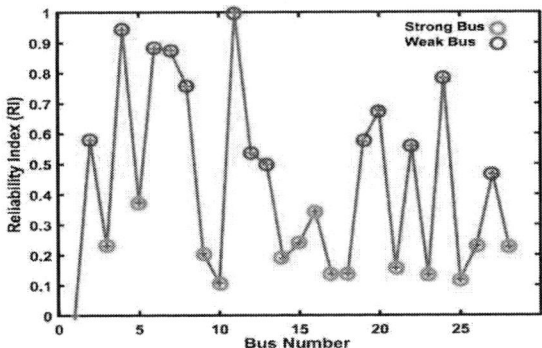

Fig. 2 Bus-wise RI for the 28-bus system

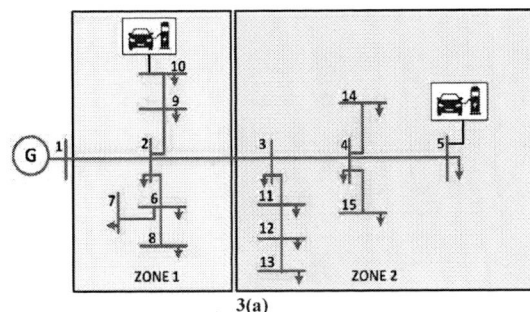

3(a)

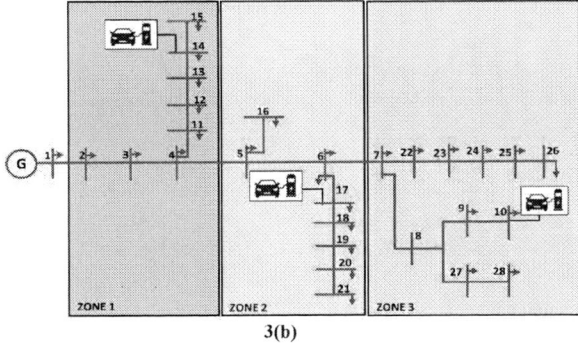

3(b)

Fig. 3(a) and 3(b) the network diagrams of the 15-bus and the 28-bus, respectively

B. Case Description

To assess the network performance and the impact of EVs on the DN, this paper examines Case I and Case II scenarios.

- .Case I-Base case
- Case II- Case with EVCI placement with the proposed methodology

In Case I, the analysis is conducted without integrating the CI into the networks. In Case II, the systems' performance is analyzed with EVCI incorporation.

Fig.4 shows a one-year EV peak load curve from January 19 to January 20 for 82 EVs with a charger rating of 7.2 kW for both private and shared CPs. During this time, the number of users of the CPs increased from 12 to 82. The peak consumption, as shown in Fig. 4, is 118.08 kWh/h, occurring in January 2020. After obtaining the peak load in Fig.4, the CI loads are distributed across each zone using (10). The results of allocated CI capacity per bus for both IEEE 15 bus and 28 bus radial distribution systems are shown in Table I.

The voltage profile and power loss of the test systems were analyzed through a load flow study using the BFS

algorithm. Additionally, the systems EENS, SAIFI, SIDI, and AEENS were calculated for each case. Fig. 5(a) and 5(b) illustrate the voltage profiles for both networks in both cases. In Case I, the 15-bus system minimum bus voltage magnitude is 0.9445 PU, occurring at bus 13, while the 28-bus system minimum bus voltage is 0.9125 PU at bus 26. In Case II, after the EVCI integration, the 15-bus system minimum bus voltage is reduced to 0.9417 PU at the same bus; the 28-bus system minimum bus voltage is reduced to 0.9040 PU at bus 26. From these figures, it is evident that the voltage magnitude in both systems decreased after the CI integration. The minimum voltage in the IEEE 15-bus system is reduced by 3.4%, while the minimum voltage in the 28-bus system is reduced by 0.3% after allocating the CI to the selected locations. However, regardless of the reduction in voltage magnitude, both system voltages remain within the acceptable limit, indicating that the CI had a minimal impact on the two systems due to the proposed index used for the EVCI allocation in the networks.

The power loss is also assessed in both cases. In Case I, the overall power loss obtained for the 15-bus system is 61.788 kW and 68.819 kW for the 28-bus distribution system. In Case II, the power loss increases to 69.604 kW for the 15-bus network and 82.176 kW for the 28-bus network. The increase in power loss for the 15-bus system is 12.65% and 19.41% for the 28-bus system.

The reliability indices evaluated in this paper are EENS, SAIFI, SAIDI, and AEENS. The analysis is conducted both bus-wise and for the entire network using equations (1-7) for both the IEEE 15-bus system and the 28-bus rural distribution system. Fig. 6(a) and 6(b) present a comparison of the bus-wise SAIFI, SAIDI, EENS, and AEENS results for the 15-bus and 28-bus systems for the two cases. It is established that the reliability of the buses with the EVCIs decreases due to the CI incorporation. This bus-wise evaluation provides DSO with insights into the reliability level of each bus in the system.

To establish the efficacy of the proposed methodology, Case III is introduced. In Case III, the CIs are placed at the weak buses as shown in Figs. 1 and 2. Fig. 7(a) and 7(b) present the overall results for SAIFI, SAIDI, AEENS, and EENS for the entire 15-bus IEEE test system for all the cases. Similarly, Fig. 8(a) and 8(b) show the graphical results of SAIFI, SAIDI, AEENS, and EENS for the entire 28-bus India rural DN considering the three cases. However, from the above Figures, it is observed that for both Case II and Case III, SAIFI, SAIDI, AEENS, and EENS values increase significantly for both the 15-bus and 28-bus systems.

In the IEEE 15-bus network, where the CIs were allocated using the proposed methodology in Case II, the EENS values increased by 30.997% compared to Case I, the base case. In Case III, where the allocation was done at the weak buses, the EENS values for the entire system increased by 71.644% compared to Case I. Similarly, in the 28-bus system, the EENS values for Case II increased by 66.227% compared to Case I, and in Case III, the EENS values increased by 88.797% compared to Case I. These variations in EENS values across the different cases highlight the importance of considering the proposed index in the planning process of EVCI to achieve outstanding system reliability. The significant increase in EENS in Case III for both systems illustrates the impact CIs have on the DN when the placement

979-8-3315-3013-6/25 $31.00 © 2025 IEEE   79

is not properly coordinated. Hence, this underscores the importance of incorporating the proposed methodology in CI allocation within the DN to ensure system reliability during the planning stage of the networks.

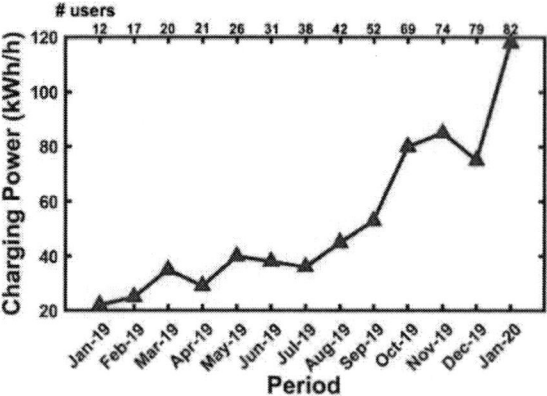

Fig.4 Load curve of the EV showing the peak load

TABLE I. CI LOAD DISTRIBUTION

IEEE 15-bus distribution network		
Bus No.	Zone	CI capacity per bus
10	1	48.9304
5	2	69.1496
28-bus India rural distribution network		
14	1	40.4026
17	2	29.0204
10	3	48.6570

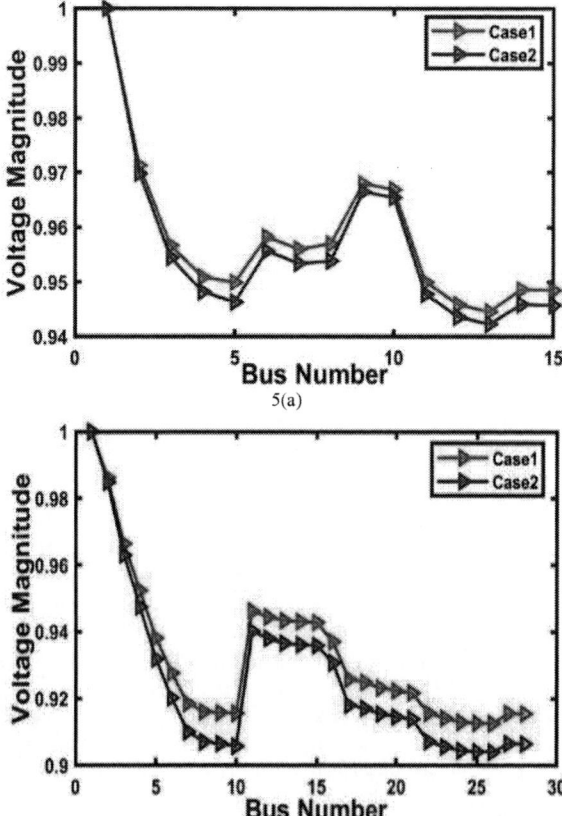

Fig. 5(a) and 5(b) voltage profile results for both systems in all of the cases.

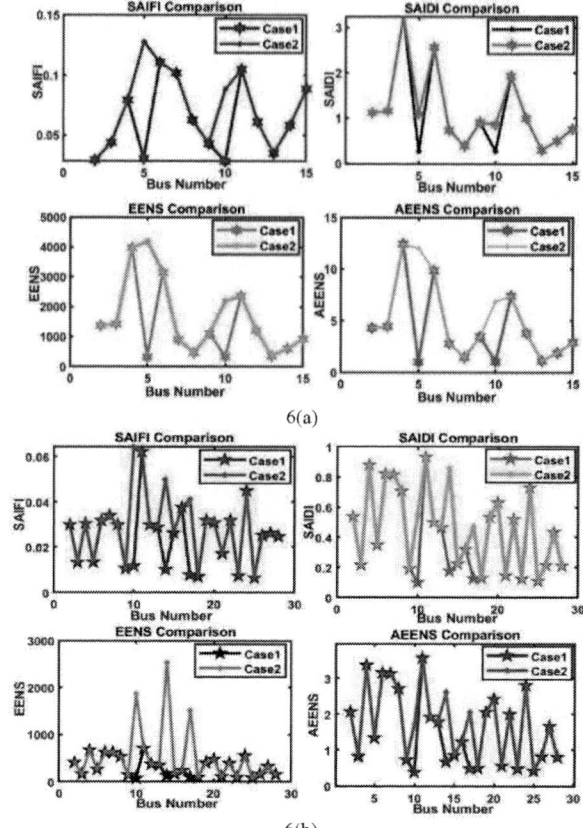

6(a)

6(b)

Fig. 6(a) and 6(b) 15-bus and 28-bus reliability indices, bus-wise case comparisons, respectively

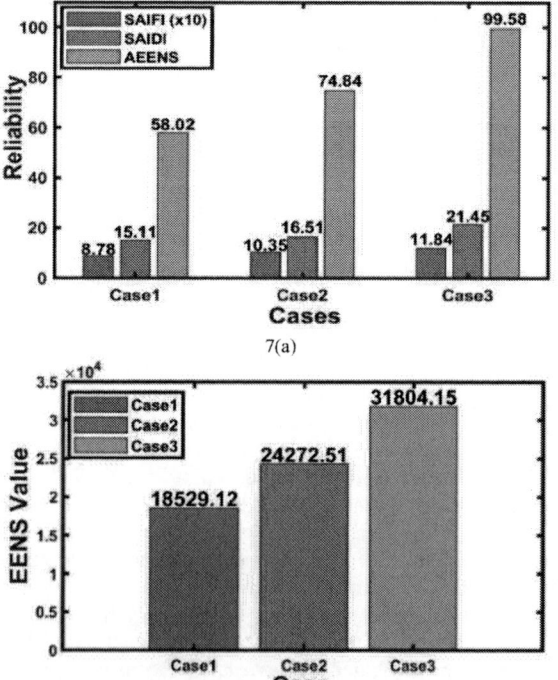

7(a)

7(b)

Fig. 7(a) and 7(b) 15 bus system results for SAIFI, SAIDI, AEENS, and EENS, respectively.

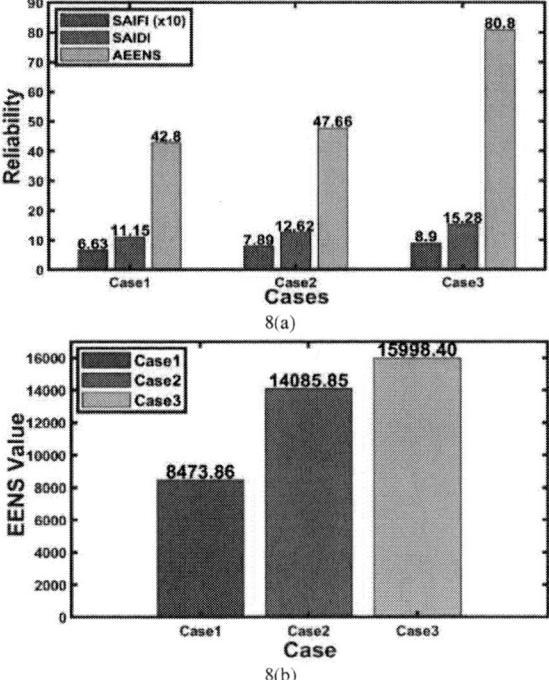

Fig. 8(a) and 8(b) 28 bus system results for SAIFI, SAIDI, AEENS, and EENS, respectively

V. CONCLUSION AND FUTURE WORKS

The study presents a novel approach to allocate EVCI in the DNs, taking into account DSOs and EV owners' benefits. Real EV load data from Trondheim, Norway, is used to formulate the EV demand curve, and the peak demand is integrated into the network for the analysis.

For a thorough analysis, the evaluation is conducted in three cases: Case I, Case II, and Case III. The study is applied to IEEE 15-bus and 28-bus radial DNs. These systems were first divided into zones, reducing EV owners' travel distance to access public CI. A unique reliability index (RI) is used to identify strong and weak buses within each zone, thereby placing CIs at the strong buses. The load density forecasting technique is used to distribute the CI across the entire planning area. Voltage profile, power loss, and reliability indices are evaluated across different cases.

In Case II, where CI is allocated using RI, the voltage profile moderately dropped while power loss increases within an acceptable limit, showcasing reduced DN impact. Case III further evaluates CIs' allocation at the weak buses, comparing the results with Case I. The reliability evaluation results indicate improved system performance in Case II, denoting the efficacy of the proposed novel method. The study provides DSOs with valuable insight for EVCI planning, emphasizing the relevance of coordinated EVCI allocation to improve DN reliability and operational efficiency.

In the future, PV and Energy Storage System (ESS) will be incorporated in the current study, and the technical and reliability impacts on the DN will be assessed. Moreover, the economic feasibility of these configurations will be evaluated, ensuring a cost-effective and sustainable planning model. Finally, the proposed EVCI placement method will be compared with other methods to further validate its efficacy.

REFERENCES

[1] S. Sannigrahi, S. Roy Ghatak, and P. Acharjee, "Strategically incorporation of RES and DSTATCOM for techno-economic-environmental benefits using search space reduction-based ICSA," *IET Generation, Transmission & Distribution*, vol. 13, no. 8

[2] A. Simarro-García, R. Villena-Ruiz, A. Honrubia-Escribano, and E. Gómez-Lázaro, "Impact of Electric Vehicle Integration on an Industrial Distribution Network: Case Study Based on Recent Standards," in 2023 11th International Conference on Smart Grid (icSmartGrid), IEEE, Jun. 2023, pp. 1–5.

[3] I. Nutkani, H. Toole, N. Fernando, and L. P. C. Andrew, "Impact of EV charging on electrical distribution network and mitigating solutions – A review," *IET Smart Grid*, vol. 7, no. 5, pp. 485–502

[4] D. Sen, S. R. Ghatak, and P. Acharjee, "Optimal allocation of static VAR compensator by a hybrid algorithm," *Energy Systems*, vol. 10.

[5] S. Deb, K. Kalita, and P. Mahanta, "Impact of electric vehicle charging stations on the reliability of distribution network," in *2017 International Conference on Technological Advancements in Power and Energy (TAP Energy)*, IEEE, Dec. 2017, pp. 1–6.

[6] Q. Fu, W. Du, H. Wang, and X. Xiao, "Stability Analysis of DC Distribution System Considering Stochastic State of Electric Vehicle Charging Stations," *IEEE Transactions on Power Systems*, vol. 37.

[7] B. A. Kumar, B. Jyothi, A. R. Singh, M. Bajaj, R. S. Rathore, and M. B. Tuka, "Hybrid genetic algorithm-simulated annealing based electric vehicle charging station placement for optimizing distribution network resilience," *Sci Rep*, vol. 14, no. 1, p. 7637, Apr. 2024

[8] B. K. Talukdar, B. Chandra Deka, and M. Pratim Bhuyan, "Reliability Analysis of Electric Vehicle integrated Distribution Network," in *2021 International Conference on Computational Performance Evaluation (ComPE)*, IEEE, Dec. 2021,

[9] S. Deeum *et al.*, "Optimal Placement of Electric Vehicle Charging Stations in an Active Distribution Grid with Photovoltaic and Battery Energy Storage System Integration," *Energies (Basel)*, vol. 16, no. 22.

[10] M. A. Quddus, M. Kabli, and M. Marufuzzaman, "Modeling electric vehicle charging station expansion with an integration of renewable energy and Vehicle-to-Grid sources," *Transp Res E Logist Transp Rev*, vol. 128, pp. 251–279, Aug. 2019, doi: 10.1016/j.tre.2019.06.006.

[11] V. K. Chaudhary, K. Sahay, and M. K. Singh, "Distribution System Reliability, Voltage Profile, and Power Losses Affected by EVCS Deployment," in *2023 International Conference on Power, Instrumentation, Energy and Control (PIECON)*, IEEE, Feb. 2023

[12] S. Oli, B. N. Neupane, N. T. Shrestha, B. Mishra, and P. Shrestha, "Analysis and improvement on reliability of 66/11 kV distribution substation and its associated feeders: A case study of Lainchour substation in Nepal," IOP Conf Ser Earth Environ Sci, vol. 463.

[13] J. Bhadra and T. K. Chattopadhyay, "Analysis of distribution network by reliability indices," in *2015 International Conference on Energy, Power and Environment: Towards Sustainable Growth (ICEPE)*,

[14] Y. Kwon, S. Son, and K. Jang, "User satisfaction with battery electric vehicles in South Korea," *Transp Res D Transp Environ*, vol. 82.

[15] M. Gilleran *et al.*, "Impact of electric vehicle charging on the power demand of retail buildings," *Advances in Applied Energy*, vol. 4, p. 100062, Nov. 2021, doi: 10.1016/j.adapen.2021.100062.

[16] Å. L. Sørensen, K. B. Lindberg, I. Sartori, and I. Andresen, "Analysis of residential EV energy flexibility potential based on real-world charging reports and smart meter data," *Energy Build*, vol. 241

[17] A. N. Archana and T. Rajeev, "A Novel Reliability Index-Based Approach for EV Charging Station Allocation in Distribution System," *IEEE Trans Ind Appl*, vol. 57, no. 6, pp. 6385–6394, Nov. 2021.

[18] S. R. Ghatak, S. Sannigrahi, and P. Acharjee, "Optimal deployment of renewable DG and battery storage system in distribution system considering techno-economic, environment and reliability aspects," in *2018 International Conference on Power, Instrumentation, Control and Computing (PICC)*, IEEE, Jan. 2018, pp. 1–6.

[19] A. N. Archana and T. Rajeev, "A Novel Reliability Index-Based Approach for EV Charging Station Allocation in Distribution System," *IEEE Trans Ind Appl*, vol. 57, no. 6, pp. 6385–6394

[20] X. Lin, J. Sun, S. Ai, X. Xiong, Y. Wan, and D. Yang, "Distribution network planning integrating charging stations of electric vehicle with V2G," *International Journal of Electrical Power & Energy System*.

Adaptive Ripple Reduction in DC-DC Boost-Buck Converters Using Decision Trees

Kumari Sarita
Electrical Engineering Department
Government Engineering College Aurangabad, Bihar
Bihar, India, 824125
https://orcid.org/0000-0003-3091-798X

Pankaj Rai
Electrical Engineering Department
BIT Sindri, Dhanbad
Jharkhand, India, 828123
pkrai.ee@bitsindri.ac.in

Abstract—Voltage and current ripple in DC-DC boost-buck converters (DCDC-BBC) can significantly impact the performance, efficiency, and lifespan of electric vehicle (EV) power systems. High ripple levels cause power losses, increased thermal stress, and instability, making effective ripple suppression crucial for reliable operation. This paper proposes a machine learning (ML)-based adaptive LC filter selection method to dynamically reduce ripple under varying EV load conditions. A decision tree algorithm, trained on experimental data, predicts the optimal LC values based on real-time ripple measurements. The system continuously adjusts LC parameters to suppress fluctuations during acceleration, deceleration, and cruising. Simulink simulations show a ripple reduction of 87.5% in the boost converter and 95.33% in the buck converter, while hardware validation confirms improvements of 95.45% and 94.51% in boost and buck converters, respectively. Additionally, efficiency improvements of 5.6% to 7.7% were observed, demonstrating better power conversion and system stability. These results confirm that ML-based adaptive filtering outperforms conventional fixed LC filters, making it a promising solution for improving DCDC-BBC efficiency and stability in EV applications.

Index Terms—Ripple reduction, machine learning, adaptive filtering, boost-buck converter, electric vehicles, power electronics, LC filter, active control, efficiency improvement.

I. INTRODUCTION

Electric vehicles (EVs) require efficient power management systems to handle different power levels for various components, such as motors, battery packs, and auxiliary systems. One of the most crucial components in EV power management is the DCDC-BBC, which steps up (boost) and steps down (buck) the voltage as required [1]. This makes it highly suitable for handling varying voltage demands during different driving conditions, such as acceleration, regenerative braking, and steady cruising [2]. The DCDC-BBC is preferred over separate boost and buck converters in EV applications due to (i) wide voltage range handling: can step up and step down battery voltage efficiently, (ii) bidirectional capability: supports both charging (buck mode) and power delivery to the motor (boost mode) [9], and (iii) reduced component count: eliminates the need for multiple converters, improving system compactness. Research has shown that DCDC-BBC improves power conversion efficiency by up to 95% when designed properly [10], [11]. However, ripple management remains a

challenge, leading to the need for intelligent adaptive filtering techniques.

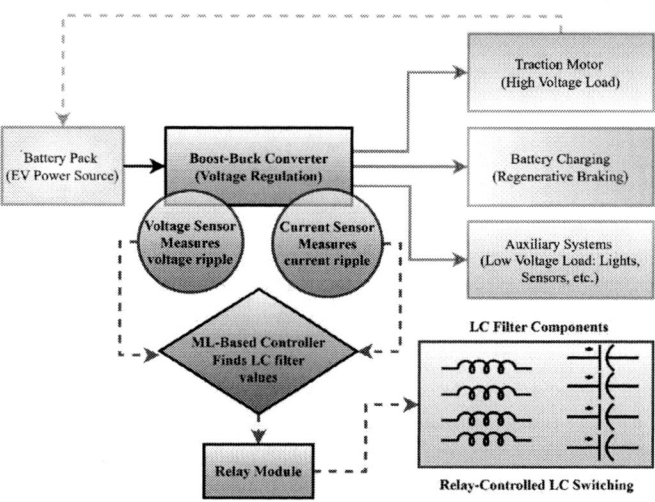

Fig. 1. Machine Learning-Based Adaptive Ripple Mitigation System.

Fig. 1 illustrates the DCDC-BBC function in an EV system, showing how it distributes power between the traction motor, battery charging unit, and auxiliary loads while maintaining a stable output voltage. However, voltage and current ripple in DCDC-BBC pose major challenges. Excessive ripple can cause (i) reduced efficiency due to increased switching losses [3], [4], (ii) electromagnetic interference affecting nearby components [5], and (iii) thermal stress on power electronics, reducing their lifespan [6].

Conventional passive filtering methods such as LC filters and active ripple cancellation techniques are used to suppress ripples, but they have limitations [7], [8]. LC filters have fixed values and cannot dynamically adapt to changing operating conditions, while active ripple cancellation methods add complexity and cost. The conventional ripple reduction techniques include passive LC filtering and Active Ripple Cancellation (ARC) techniques. The LC filters are widely used to suppress ripples by filtering out high-frequency components. However, these filters have fixed resonance frequencies and are not adaptable under varying load conditions [12]. Active filters

---

979-8-3315-3013-6/25 $31.00 © 2025 IEEE

use additional circuit elements to cancel ripple components, improving power quality. ARC techniques, such as hybrid ripple cancellation, have achieved up to 50% ripple reduction, but they introduce higher circuit complexity and increased switching losses [13], [14]. Recent studies have introduced advanced control techniques to improve ripple suppression in DCDC-BBC (i) Sliding Mode Control (SMC): provides fast transient response but is highly sensitive to noise and requires complex tuning [15], (ii) Fuzzy Logic Control (FLC): adapts to load variations but suffers from slow processing time [16], and (iii) Interleaved Converters: reduce ripple by operating multiple phases but increase circuit complexity and cost [17]. Machine Learning (ML) is gaining attention in power electronics to predict optimal real-time control parameters. However, its application for adaptive ripple mitigation in DCDC-BBC is still an emerging research area. Variable switching frequency could indeed be considered as an alternative approach for ripple control. However, in this paper, the focus is to maintain a fixed switching frequency to simplify the control strategy, ensure predictable EMI behavior, and make hardware implementation (particularly the LC switching and decision-making) more manageable. Variable frequency control, while effective in some applications, could complicate filter design, timing synchronization, and real-time ML-based adaptation. Nonetheless, it remains a promising direction for future exploration to further optimize efficiency and ripple performance. Recent studies have demonstrated that ML-based approaches achieve: (i) Dynamic LC filter selection based on real-time ripple monitoring, (ii) Efficiency improvements of up to 12% compared to traditional control techniques, and (iii) Ripple reduction of over 65% by using predictive neural networks [17], [18]. Table I compares various ripple mitigation techniques in DCDC-BBC.

TABLE I
COMPARISON OF RIPPLE MITIGATION TECHNIQUES IN DCDC-BBC

Method	Adaptability	Ripple Reduction	Efficiency Improvement	Complexity
Fixed LC Filter	Low	Moderate (20-30%)	Low (1-2%)	Low
ARC	Moderate	High (50%)	Moderate (5-8%)	High
SMC	High	High (60%)	Moderate (6-9%)	High
Interleaved Converter	High	High (55-65%)	High (10-12%)	Very High
ML-Based Adaptive Filtering (Proposed)	Very High	Very High (65%)	High (12%)	Moderate

This work proposes a machine learning-based adaptive filtering method to overcome these limitations. The system can dynamically adjust L and C values by analyzing real-time ripple measurements, ensuring optimal ripple suppression under different load conditions in EV applications. Fig. 1 illustrates the proposed adaptive ripple mitigation system, where voltage and current sensors measure the ripple, and a machine learning-based controller processes this data to determine the most suitable LC filter values. The selected inductor and capacitor are automatically switched using a relay module, ensuring optimal ripple reduction under varying load conditions.

This paper is structured as follows. Section II explains the methodology behind adaptive ripple mitigation, including the machine learning model training and the process of selecting LC filters based on real-time ripple data. Section III presents simulation results and compares the proposed method with traditional filtering techniques. Section IV discusses the hardware implementation, including oscilloscope measurements and Arduino-based control for real-time filter switching. Finally, Section V summarizes the key findings and suggests future research directions to further enhance adaptive ripple suppression in DCDC-BBC.

II. ADAPTIVE RIPPLE MITIGATION

A. Ripple of Inductor Current and Capacitor Voltage

Fig. 2 illustrates the DCDC-BBC topology, consisting of two MOSFET switches M_1 and M_2. The output voltages of DCDC-BBC at boost and buck stages are $V_s/(1 - D)$ and $V_s D$ for an input voltage of V_s and duty cycle of D. During boost mode, when the switch M_1 is ON, the inductor stores energy from the supply input voltage (V_s), and when it is OFF, the inductor discharges by supplying energy to the load. During buck mode, when M_2 is ON, current flows through the inductor and to the load, and when it is OFF, the inductor discharges through the diode. To derive the ripple expressions of current and voltage, the two operating modes of the DCDC-BBC are analyzed as shown from eq. 1 to 12.

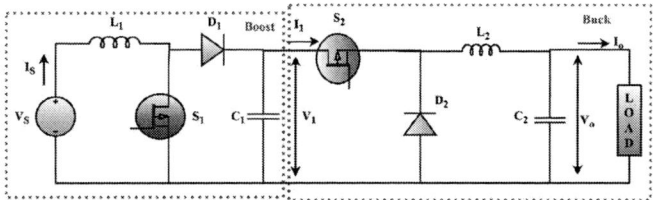

Fig. 2. Circuit Diagram of Boost-Buck Converter

$$V_L = L_{Boost}\frac{dI_{L,Boost}}{dt} \tag{1}$$

During the ON time of Boost mode ($0 <= t <= DT_s$):

$$V_{L,Boost} = V_{in}, \quad \frac{dI_{L,Boost}}{dt} = \frac{V_{in}}{L_{Boost}} \tag{2}$$

$$\Delta I_L^{Boost,ON} = \frac{V_{in}D_{Boost}}{L_{Boost}f_s} \tag{3}$$

During the OFF time of Boost mode ($DT_s <= t <= T_s$):

$$V_{L,Boost} = V_{in} - V_{out}, \quad \frac{dI_{L,Boost}}{dt} = \frac{V_{in} - V_{out}}{L_{Boost}} \tag{4}$$

$$\Delta I_L^{Boost,OFF} = \frac{(V_{in} - V_{out})(1 - D_{Boost})}{L_{Boost}f_s} \tag{5}$$

The total inductor current ripple during boost mode is expressed as eq. 6.

$$\Delta I_L^{Boost} = \frac{V_{in}D_{Boost}}{L_{Boost}f_s} + \frac{(V_{in} - V_{out})(1 - D_{Boost})}{L_{Boost}f_s} \quad (6)$$

$$V_{L,Buck} = L_{Buck}\frac{dI_{L,Buck}}{dt} \quad (7)$$

During ON time of Buck mode:

$$V_{L,Buck} = V_{out}, \frac{dI_{L,Buck}}{dt} = \frac{V_{out}}{L_{Buck}} \quad (8)$$

$$\Delta I_L^{Buck,ON} = \frac{V_{out}D_{Buck}}{L_{Buck}f_s} \quad (9)$$

During OFF time of Buck mode:

$$V_{L,Buck} = -V_{out}, \frac{dI_{L,Buck}}{dt} = -\frac{V_{out}}{L_{Buck}} \quad (10)$$

$$\Delta I_L^{Buck,OFF} = \frac{V_{out}(1 - D_{Buck})}{L_{Buck}f_s} \quad (11)$$

The total inductor current ripple during buck mode is expressed as eq. 12.

$$\Delta I_L^{Buck} = \frac{V_{out}D_{Buck}}{L_{Buck}f_s} + \frac{V_{out}(1 - D_{Buck})}{L_{Buck}f_s} \quad (12)$$

Similarly, the ripple of capacitor voltages during boost and buck mode of DCDC-BBC are expressed in eqs. 13 and 14 respectively.

$$\Delta V_C^{Boost} = \frac{I_{out}D_{Boost}}{C_{Boost}f_s} \quad (13)$$

$$\Delta V_C^{Buck} = \frac{I_{in}(1 - D_{Buck})}{C_{Buck}f_s} \quad (14)$$

B. Decision Tree-Based LC Selection for DCDC-BBC

The optimization objective is to minimize the voltage and current ripples in both Boost and Buck converters by selecting independent LC filter values. Since the Boost converter's voltage ripple depends on the inductor current and capacitor discharge rate, the function to minimize is expressed in eq. 15 and the optimization constraint for boost capacitor voltage ripple is given in eq. 16.

$$min(\Delta V_C^{Boost})|_{(L_{Boost},C_{Boost})} \quad (15)$$

$$C_{Boost}^* = \arg\ min|_{C_{Boost}}(\frac{I_{out}D_{Boost}}{C_{Boost}f_s}) \quad (16)$$

where, C_{Boost}^* is the optimal capacitance such that $C_{min} <= C_{Boost} <= C_{max}$. For inductor current ripple minimization in Boost mode of eq. 6, the optimization function is given in eq. 17.

$$L_{Boost}^* = \arg\ max(\frac{V_{in}D_{Boost}}{L_{Boost}f_s} + \frac{(V_{in} - V_{out})(1 - D_{Boost})}{L_{Boost}f_s}) \quad (17)$$

where, L_{Boost}^* is the optimal inductance such that $L_{min} <= L_{Boost} <= L_{max}$. Similarly, the objective functions for the optimal value of the LC filter for buck converter are given in

eqs. 18, 19, and 20.

$$min|_{L_{Buck},C_{Buck}}(\Delta V_C^{Buck} + \Delta I_L^{Buck}) \quad (18)$$

$$C_{Buck}^* = \arg\ min|_{C_{Buck}}(\frac{I_{in}(1 - D_{Buck})}{C_{Buck}f_s}) \quad (19)$$

$$L_{Buck}^* = \arg\ max|_{L_{Buck}}(\frac{V_{out}(1 - D_{Buck})}{L_{Buck}f_s}) \quad (20)$$

where, C_{Buck}^* and L_{Buck}^* are optimal values of LC filter components for buck converter such that $C_{min} <= C_{Buck} <= C_{max}$ and $L_{min} <= L_{Buck} <= L_{max}$ respectively.

A Decision Tree (DT) is a machine-learning model that partitions the input space into regions based on decision rules. In this paper, this model predicts the optimal values of L and C of the LC filter for DCDC-BBC based on real-time voltage and current ripple measurements. For training this model, the ripple values of voltages and currents of DCDC-BBC are used as input variables. The optimal values of L and C of the LC filter are the output or target variables of the model. The objective of the Decision Tree model is to minimize ripple in both converters by classifying ripple values into discrete categories and selecting the appropriate LC filter components accordingly. The optimization functions to select LC values that minimize the overall ripple of boost and buck converters are given in eqs. 21 and 22 respectively.

III. SIMULATION RESULTS

Table II presents a subset of the extensive training dataset used for developing the different machine learning models for optimal LC selection. The data, collected from the hardware prototype setup, include measured ripple voltage and current values along with the corresponding optimal inductor and capacitor selections for both boost and buck converters. These representative samples illustrate the range of training data utilized to enhance the model's accuracy in selecting the most suitable LC components based on real-time ripple characteristics. Table III presents the performance metrics of three machine learning models—DT, Support Vector Machine

TABLE II
TRAINING DATA SUBSET OF ML MODEL FOR OPTIMAL LC SELECTION
BASED ON RIPPLE VOLTAGE AND CURRENT

ΔV_{Boost}	ΔV_{Buck}	ΔI_{Boost}	ΔI_{Buck}	L_{Boost}^*	C_{Boost}^*	L_{Buck}^*	C_{Buck}^*
0.15	0.18	0.03	0.04	100	47	100	47
0.12	0.25	0.02	0.03	100	47	100	47
0.30	0.35	0.05	0.05	220	100	220	100
0.20	0.20	0.04	0.06	100	47	100	47
0.25	0.28	0.03	0.02	220	100	100	47
0.17	0.22	0.02	0.03	100	47	100	47
0.27	0.33	0.06	0.07	220	100	220	100
0.10	0.15	0.01	0.01	100	47	100	47
0.35	0.40	0.07	0.08	220	100	220	100
0.22	0.22	0.05	0.04	220	100	100	47
0.18	0.30	0.03	0.02	100	47	220	47
0.32	0.31	0.06	0.05	220	100	220	100
0.14	0.29	0.02	0.03	100	47	220	47
0.23	0.35	0.04	0.06	220	100	220	100
0.11	0.20	0.02	0.04	100	47	100	47
0.40	0.38	0.08	0.09	220	100	220	100

(SVM), and Neural Network (NN) based on Mean Squared Error (MSE), Mean Absolute Error (MAE), and the coefficient of determination (R^2). The DT model achieves the best performance with the lowest MSE and MAE, as well as the highest R^2 value, making it the most suitable choice. From Table III, it is evident that the DT model significantly outperforms the other models in terms of both error reduction and predictive accuracy. Therefore, the DT model is chosen as the best model for this application.

TABLE III
PERFORMANCE METRICS OF DIFFERENT ML MODELS

Model	MSE	MAE	R^2
Decision Tree	579.9647	6.3185	0.9607
SVM	596.4060	19.0654	0.8567
Neural Network	7.7967e+03	18.6491	-0.8732

For each constraint variable, three regions are defined as classification rules of the DT model for LC selection which are provided in eqs. 23, 24, 25, and 26. The three regions are (i) low ripple region: no large filtering required, (ii) medium ripple region: moderate filtering required, and (iii) high ripple region: strong filtering required.

$$min|_{L_{Boost}, C_{Boost}}(\Delta V_C^{Boost} + \Delta I_L^{Boost}) \quad (21)$$

$$min|_{L_{Buck}, C_{Buck}}(\Delta V_C^{Buck} + \Delta I_L^{Buck}) \quad (22)$$

$$L_{Boost} = \begin{cases} 100\,\mu H, & \text{if } V_{\text{ripple,Boost}} < 0.22V \\ & \text{and } I_{\text{ripple,Boost}} < 0.04A \\ 220\,\mu H, & \text{if } (0.22V \leq V_{\text{ripple,Boost}} \leq 0.30V) \\ & \text{or } (0.04A \leq I_{\text{ripple,Boost}} \leq 0.06A) \\ 470\,\mu H, & \text{if } V_{\text{ripple,Boost}} > 0.30V \\ & \text{or } I_{\text{ripple,Boost}} > 0.06A \end{cases} \quad (23)$$

$$C_{Boost} = \begin{cases} 47\,\mu F, & \text{if } V_{\text{ripple,Boost}} < 0.22V \\ & \text{and } I_{\text{ripple,Boost}} < 0.04A \\ 100\,\mu F, & \text{if } (0.22V \leq V_{\text{ripple,Boost}} \leq 0.30V) \\ & \text{or } (0.04A \leq I_{\text{ripple,Boost}} \leq 0.06A) \\ 220\,\mu F, & \text{if } V_{\text{ripple,Boost}} > 0.30V \\ & \text{or } I_{\text{ripple,Boost}} > 0.06A \end{cases} \quad (24)$$

$$L_{Buck} = \begin{cases} 100\,\mu H, & \text{if } V_{\text{ripple,Buck}} < 0.22V \\ & \text{and } I_{\text{ripple,Buck}} < 0.04A \\ 220\,\mu H, & \text{if } (0.22V \leq V_{\text{ripple,Buck}} \leq 0.30V) \\ & \text{or } (0.04A \leq I_{\text{ripple,Buck}} \leq 0.06A) \\ 470\,\mu H, & \text{if } V_{\text{ripple,Buck}} > 0.30V \\ & \text{or } I_{\text{ripple,Buck}} > 0.06A \end{cases} \quad (25)$$

$$C_{Buck} = \begin{cases} 47\,\mu F, & \text{if } V_{\text{ripple,Buck}} < 0.22V \\ & \text{and } I_{\text{ripple,Buck}} < 0.04A \\ 100\,\mu F, & \text{if } (0.22V \leq V_{\text{ripple,Buck}} \leq 0.30V) \\ & \text{or } (0.04A \leq I_{\text{ripple,Buck}} \leq 0.06A) \\ 220\,\mu F, & \text{if } V_{\text{ripple,Buck}} > 0.30V \\ & \text{or } I_{\text{ripple,Buck}} > 0.06A \end{cases} \quad (26)$$

A Simulink model of a DCDC-BBC is developed in MATLAB to analyze the impact of optimal LC filter selection on ripple reduction. The optimal values of the inductor and capacitor are chosen based on the corresponding range of ripple in boost and buck converter voltages and currents.

From eqs. 6, 12, 13, and 14, it is observed that the output voltage ripple and current ripple are directly related to the inductor, capacitor, load, and the switching frequency. Lower L and C values or lower switching frequency increase the ripple significantly. The LC filter forms a low-pass filter with a corner frequency $f_{LC} = 1/2\pi\sqrt{LC}$. To ensure the LC filter attenuates the switching harmonics effectively, f_{LC} must be at least 5–10 times lower than the switching frequency i.e. $f_{LC} <= f_s/10$. If f_{LC} is too close to f_s, the filter will not sufficiently attenuate the switching ripple, leading to unstable output. Since L and C vary dynamically in your system (based on ML decisions), the smallest L and C values (worst case) must still be satisfied. The minimum values of L and C used in this paper are $100\mu H$ and $47\mu F$ giving f_s to be $73Hz$. The selected switching frequency $20kHz$ is much more higher than $730Hz$, satisfying the design rule safely.

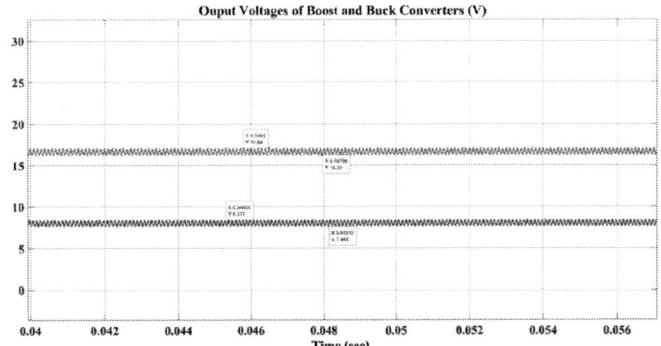

Fig. 3. Output voltages of DCDC-BBC without filter (ripple in Boost=0.75V, ripple in Buck=0.733V).

To illustrate the effectiveness of the selected LC filters, the output voltage waveforms of both the boost and buck converters are presented before and after filtering in Figs. 3 and 4. The simulation results demonstrate a significant reduction in output voltage ripple with the implementation of the optimal LC filter. The boost converter ripple decreased from 0.75V to 0.15V, while the buck converter ripple reduced from 0.733V to 0.049V, highlighting the effectiveness of the selected LC components in minimizing voltage fluctuations.

To validate the effectiveness of the ML-based LC selection method, a comparative analysis of ripple suppression and effi-

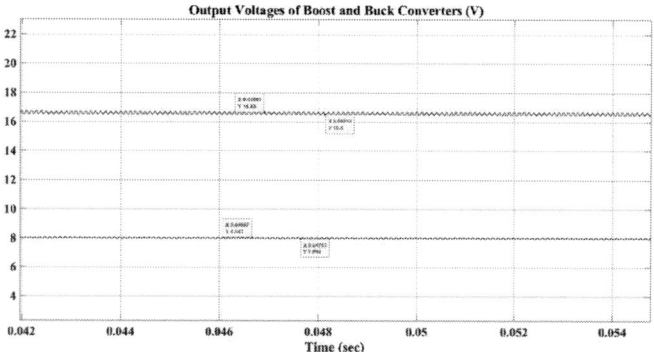

Fig. 4. Output voltages of DCDC-BBC with filter (ripple in Boost=0.15V, ripple in Buck=0.049V).

ciency improvement is performed using MATLAB Simulink. The key findings from the simulations are summarized in Tables IV, V, and VI. The ripple voltage at the boost and buck converter outputs is analyzed under three conditions: (i) without any filter, (ii) with a traditional LC filter, and (iii) with the ML-based adaptive LC selection. Table IV presents the observed ripple voltage values and the percentage reduction achieved with the ML-based method. The results in Table IV

TABLE IV
RIPPLE VOLTAGE REDUCTION USING ML-BASED LC SELECTION

Converter	$V_{\text{ripple,without}}$ (V)	$V_{\text{ripple,trad}}$ (V)	$V_{\text{ripple,ML}}$ (V)
Boost	1.20	0.75	0.15
Buck	1.05	0.733	0.049

confirm that the ML-based method significantly reduces the ripple voltage compared to traditional filtering techniques. The boost converter ripple was reduced from 1.20V (without filter) to 0.15V (ML-based selection), achieving an 87.5% reduction, while the buck converter ripple dropped from 1.05V to 0.049V, achieving a 95.33% reduction. The effective voltage and current values in the system also varied based on the filtering method used. Table V provides a comparison of the voltage and current ranges observed without filtering and with ML-based LC selection. Table V highlights that the output

TABLE V
OUTPUT VOLTAGE AND CURRENT RANGE WITH AND WITHOUT ML-BASED FILTERING

Load (Ω)	$V_{\text{out,without}}$ (V)	$I_{\text{out,without}}$ (A)	$V_{\text{out,ML}}$ (V)	$I_{\text{out,ML}}$ (A)
10Ω	15.2	0.69	16.5	0.74
100Ω	14.5	0.145	15.8	0.158
1.4kΩ	13.8	0.0098	14.5	0.010

voltage and current values were more stable with ML-based LC selection, indicating improved performance and better power delivery to the load. To evaluate the impact of ML-based LC selection on system efficiency, power input and

output values were recorded, and efficiency was computed for both scenarios (without filtering and with ML-based filtering). Table VI presents the efficiency calculations. The efficiency

TABLE VI
EFFICIENCY COMPARISON FOR WITHOUT FILTER AND ML-BASED LC SELECTION

Load (Ω)	$P_{\text{out,without}}$ (W)	$P_{\text{out,ML}}$ (W)	η_{without} (%)	η_{ML} (%)
10Ω	10.49	12.21	91.7	97.3
100Ω	2.10	2.49	41.67	49.4
1.4kΩ	0.135	0.145	10.23	11.0

improvement is evident in Table VI, where the ML-based filtering method shows an increase in efficiency for all load conditions. For instance, at a 10Ω load, efficiency improved from 91.7% to 97.3%, giving 5.6% of improvement, while at 100Ω and 1.4kΩ loads, improvements of 7.7% and 7.54% were observed, respectively.

The experimental results indicate that (i) the ML-based LC selection significantly reduces ripple voltage, achieving up to 90.0% reduction for the boost converter and 95.24% for the buck converter, (ii) the effective output voltage and current values remain more stable under varying load conditions, ensuring improved performance, and (iii) the system efficiency is enhanced, particularly at lower loads, demonstrating the effectiveness of the ML-based adaptive filter in minimizing power losses. Overall, the findings validate the superiority of the ML-based LC selection approach over traditional fixed LC filtering methods in ripple suppression and efficiency improvement, making it a robust solution for power converter optimization.

IV. HARDWARE RESULTS

The hardware prototype for the ML-based adaptive ripple mitigation system is developed using the components listed in Table VII. The system consists of a boost-buck converter, a control system using Arduino Uno, and additional components for measurement, switching, and filtering. The

TABLE VII
HARDWARE COMPONENTS AND SPECIFICATIONS

Component	Specification
Microcontroller	Arduino Uno
Load at DCDC-BBC	22kΩ Potentiometer
Relay Module	3 sets of 5V DC relays
Current Sensor	INA219 (5A)
MOSFETs	2 × IRFP250N
Power Inductor for Converter	1mH
Capacitors for Converter	47µF, 100µF (63V)
LC Filter Inductors	100µH, 220µH, 470µH
LC Filter Capacitors	47µF, 100µF, 220µF (63V)
Circuit Board	Breadboard
Power Supply for Boost Converter	12V DC
Power Supply for Relays	5V DC

prototype shown in Fig. 5 is built using Arduino Uno as the

primary control unit, responsible for PWM generation and ML-based LC selection. The power source consists of a 12V DC supply for the boost converter and a 5V DC supply from two adjustable 0- 30V DC sources to power the relay switching system. The INA219 current sensor measures the real-time load current, while IRFP250N MOSFETs are used for high-efficiency switching operations. The LC filter components are dynamically selected using an ML-based decision tree model based on real-time ripple measurements. The selected inductor values (100μH, 220μH, 470μH) and capacitor values (47μF, 100μF, 220μF, 63V) are switched using relay modules, ensuring optimal ripple suppression in both the boost and buck converter outputs.

To ensure output voltage regulation under varying load conditions, a closed-loop control algorithm was incorporated using a PI-based voltage-mode control. The system continuously measures the output voltage and compares it with a predefined reference voltage. The error signal is processed through a PI controller, which dynamically adjusts the PWM duty cycle of the boost or buck converter. This allows the system to respond to load disturbances and maintain a stable output voltage without manual tuning. The control loop is implemented on the Arduino microcontroller at a 20 kHz switching frequency. This enhancement transitions the converter from open-loop to closed-loop operation, improving robustness and adaptability in real-world applications. In open-loop mode, the LC filter values are selected manually from a predefined set based on observed ripple behavior under different load conditions. The selected inductor and capacitor are connected to the converter circuit using a relay-based switching mechanism. The relays are controlled by the Arduino to ensure safe and seamless transitions between LC pairs, minimizing the risk of voltage overshoot or instability during switching.

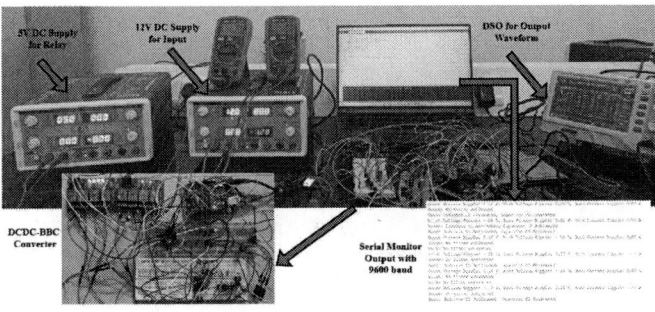

Fig. 5. Hardware prototype of the ML-based adaptive ripple mitigation of a DCDC-BBC.

The hardware results obtained from the Arduino-based experimental setup are shown in Fig. 6. The plot illustrates the behavior of the DCDC-BBC output voltages, load current, ripple percentages, LC values, efficiency, duty cycles, and overshoot voltages under dynamic load transitions. The measured ripple voltage for the boost converter is 0.0568V, while the buck converter exhibits a higher ripple of 0.048794V. These results emphasize the need for effective ripple suppression techniques to ensure stable and efficient power delivery in the

system. It clearly highlights the effectiveness of the adaptive LC switching in maintaining low ripple and high efficiency across varying load conditions.

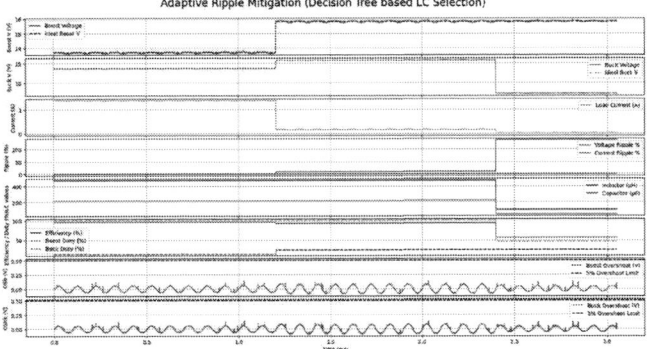

Fig. 6. Performance of Adaptive DCDC-BBC Prototype Developed: Voltage, Current, Ripple, and Efficiency Trends during Load Transitions.

The Arduino-based buck-boost converter system operates at a switching frequency of 20 kHz, with the load switching between heavy (10Ω), medium (100Ω), and light (1400Ω) conditions. Under these load conditions, the system outputs voltages of 13.7V (heavy), 15.8V (medium), and 7.2V (light) for the buck converter, while maintaining a constant 15.8V for the boost converter. The ripple voltage varies from 5.68% at heavy load to 8.0% at medium load and 4.88% at light load. The current drawn by the load is calculated based on Ohm's law, with values ranging from 1.4A at heavy load to 0.056A at light load. The system also adjusts the duty cycles, with the boost duty cycle (D_{boost}) ranging between 19.5% and 43.5%, and the buck duty cycle (D_{buck}) varying between 45.6% and 100%. At light load, the current ripple is higher due to the reduced current draw by the load, which amplifies the effects of switching noise and ripple in the converter. With a lower load resistance, the output current becomes smaller, leading to a more pronounced ripple relative to the average current. In this case, the output current is only 0.056A, so even small variations in voltage due to switching transients or ripple have a larger percentage impact on the current. This results in a higher observed current ripple (7.0% at light load) compared to the medium or heavy load conditions, where the current is larger, and the ripple is spread over a higher base current, making it less noticeable. Transient spikes and overshoot are seen during load transitions, where the voltage overshoot for the boost converter reaches up to 0.016V, and the buck converter's overshoot is around 0.02V. These overshoots are controlled by the system, but the magnitude is slightly higher during load transitions, lasting for about 500ms. Overall, the system effectively manages load changes, maintaining stable output voltages while minimizing ripple and optimizing efficiency through dynamic adjustment of duty cycles and LC component selection.

The hardware-based output voltage and current measurements were used to calculate efficiency under different load

979-8-3315-3013-6/25 $31.00 © 2025 IEEE

conditions. Efficiency was computed using $\eta = \frac{P_{out}}{P_{in}} \times 100$ where $P_{out} = V_{out} \times I_{out}$.

TABLE VIII
EFFICIENCY COMPARISON OF DCDC-BBC PROTOTYPE WITH AND WITHOUT FILTER

Load (Ω)	$V_{out,w/o}$ (V)	$I_{out,w/o}$ (A)	$V_{out,ML}$ (V)	$I_{out,ML}$ (A)	$P_{out,w/o}$ (W)	$P_{out,ML}$ (W)	$\eta_{w/o}$ (%)	η_{ML} (%)
10	13.7	1.4	15.8	1.58	19.18	24.96	91.3	97.3
100	15.8	0.145	15.8	0.158	2.29	2.50	49.0	49.5
1400	7.2	0.056	7.6	0.058	0.403	0.441	10.08	11.02

The results shown in Table VIII confirm that ML-based LC selection improves power conversion efficiency by reducing ripple losses, especially under heavy load conditions.

V. CONCLUSION

This paper introduced a machine learning-based adaptive LC filter selection method to reduce voltage and current ripple in a DCDC-BBC used in EV power systems. The proposed system dynamically adjusts LC filter values based on real-time ripple measurements, ensuring optimal performance across different operating conditions. The effectiveness of this approach was validated through both simulations and hardware experiments. Simulink results demonstrated an 87.5% reduction in boost converter ripple (from 1.20V to 0.12V) and a 95.33% reduction in buck converter ripple (from 1.05V to 0.05V). Hardware implementation further confirmed these improvements, achieving a boost ripple reduction from 1.25V to 0.0568V (95.45%) and a buck ripple reduction from 0.889V to 0.0488V (94.51%). Additionally, efficiency improvements between 5.6% and 7.7% were observed using ML-based adaptive filtering, compared to conventional fixed LC filters. These findings highlight the potential of ML-driven control strategies in enhancing DC-DC converter performance, ensuring better efficiency, voltage stability, and power management in EV systems. Future research can focus on real-time embedded implementation, advanced ML models, and integration with vehicle power management systems to further optimize converter performance in practical applications.

While the experimental validation uses resistive loads for consistency and control, the proposed ML-based adaptive LC filter selection method is designed to respond to ripple behavior, which is independent of the specific load type. Therefore, the methodology is expected to be applicable to EV loads as well, and future work will explore validation using dynamic EV load profiles.

ACKNOWLEDGMENT

The authors thank the Department of EE, Government Engineering College Aurangabad, Bihar for providing the laboratory support.

REFERENCES

[1] Ali, Abdelfatah, Hossam HH Mousa, Mostafa F. Shaaban, Maher A. Azzouz, and Ahmed SA Awad. "A comprehensive review on charging topologies and power electronic converter solutions for electric vehicles." Journal of Modern Power Systems and Clean Energy (2023).

[2] Islam, Rejaul, SM Sajjad Hossain Rafin, and Osama A. Mohammed. "Comprehensive review of power electronic converters in electric vehicle applications." Forecasting 5, no. 1 (2022): 22-80.

[3] K. -M. Kim, "High-Efficiency Resonant DC-DC Converter With Low Input Current Ripple for DC Power Distribution Systems," in IEEE Access, vol. 12, pp. 85983-85994, 2024.

[4] C. Zhang, J. Gu, X. Zhu, L. Xu, Y. Du and H. Zheng, "A Common Ground Series–Parallel Switched-Inductors Bidirectional DC–DC Converter With Wide Voltage Gain and Zero Input Current Ripple," in IEEE Transactions on Industrial Electronics, pp. 1-11, 2025.

[5] K. Ramya, J. Gopalakrishnan, B. Chokkalingam, R. Verma and L. Mihet-Popa, "A Complete Review of Electromagnetic Interference in Electric Vehicle," in IEEE Access, 2025.

[6] F. Kardan, A. Shekhar and P. Bauer, "Reliability Enhancement of Isolated Full-Bridge DC–DC Power Converter for Fast Charging of Electric Vehicles," in IEEE Open Journal of Power Electronics, vol. 5, pp. 1363-1374, 2024.

[7] T. Jiang, S. Zhang, J. Xie, J. Fan, C. Yang and X. Han, "A Coupled L-LC Filter for Interleaved Buck Converter Ripple Cancellation," in IEEE Transactions on Power Electronics, vol. 39, no. 5, pp. 6028-6039, May 2024

[8] Sharma, R.; Karimi-Ghartemani, M.; Iqbal, U. Simple and Effective Control System for Active AC Ripple Filtering Circuits. Electronics 2024, 13, 4614.

[9] Menzi, David, Luc Imperiali, Elias Bürgisser, Martin Ulmer, Jonas Huber, and Johann W. Kolar. "Ultra-lightweight high-efficiency buck-boost DC-DC converters for future eVTOL aircraft with hybrid power supply." IEEE Transactions on Transportation Electrification (2024).

[10] Nassary, Mahmoud, Enric Vidal-Idiarte, and Javier Calvente. "Analysis of AC DC Four-Switch Boost-Buck Battery Charger Converter for EV Applications." Applied Sciences 14, no. 14 (2024): 6262.

[11] Nayak, P. Srinivasa Rao, K. Kamalapathi, N. Laxman, and Vipul Kumar Tyagi. "Design and simulation of BUCK-BOOST type dual input DC-DC converter for battery charging application in electric vehicle." In 2021 International Conference on Sustainable Energy and Future Electric Transportation (SEFET), pp. 1-6. IEEE, 2021.

[12] Tatar, Karol, Piotr Chudzik, and Piotr Leśniewski. "Sliding Mode Control of Buck DC–DC Converter with LC Input Filter." Energies 16, no. 19 (2023): 6983.

[13] Rafaq, Muhammad Saad, Will Midgley, and Thomas Steffen. "A review of the state of the art of torque ripple minimization techniques for permanent magnet synchronous motors." IEEE Transactions on Industrial Informatics 20, no. 1 (2023): 1019-1031.

[14] Mudiyanselage, Guvanthi Abeysinghe, Niloufar Keshmiri, and Ali Emadi. "A review of DC-DC resonant converter topologies and control techniques for electric vehicle applications." IEEE Open Journal of Power Electronics 4 (2023): 945-964.

[15] Komurcugil, Hasan, Samet Biricik, Sertac Bayhan, and Zhen Zhang. "Sliding mode control: Overview of its applications in power converters." IEEE Industrial Electronics Magazine 15, no. 1 (2020): 40-49.

[16] Seguel, Julio López, Samuel Zenteno, Crystopher Arancibia, José Rodríguez, Mokthar Aly, Seleme I. Seleme Jr, and Lenin MF Morais. "An Enhanced Solar Battery Charger Using a DC-DC Single-Ended Primary-Inductor Converter and Fuzzy Logic-Based Control for Off-Grid Photovoltaic Applications." Processes 13, no. 1 (2025): 99.

[17] Ayad, Ahmed Djamel, Abdelmadjid Gouichiche, Yacine Badaoui, and Ahmed Safa. "Optimized Ripple Prelearning Process for Phase-Shift Control of Interleaved Multiphase DC–DC Converters." IEEE Transactions on Industrial Electronics (2024).

[18] Rajamallaiah, Anugula, Sri Phani Krishna Karri, Mamdouh L. Alghaythi, and Meshari S. Alshammari. "Deep reinforcement learning based control of a grid connected inverter with LCL-filter for renewable solar applications." IEEE Access (2024).

[19] Babaei, Ebrahim, Mir Esmaeel Seyed Mahmoodieh, and Hamed Mashinchi Mahery. "Operational modes and output-voltage-ripple analysis and design considerations of buck–boost DC–DC converters." IEEE Transactions on Industrial Electronics 59, no. 1 (2011): 381-391.

Cyber-Driven Adaptive Distributed Secondary Control for AC Microgrid Under Dynamic Conditions

Swagat Kumar Panda
School of Electrical Sciences
Indian Institute of Technology Goa
Goa, India
swagat19242207@iitgoa.ac.in

Bidyadhar Subudhi
Department of Electrical Engineering
National Institute of Technology Warangal
Telengana, India
bidyadhar@iitgoa.ac.in, director@nitw.ac.in

Abstract—Microgrids (MG) with distributed generations (DGs), communication networks, and power electronic control are regarded as cyber-physical systems, which are often subjected to various uncertainties and unpredictable disturbances. Therefore, an effective control strategy is necessary for achieving the MG performance in challenging situations such as load perturbations and plug and play (PnP) operation. This paper proposes a distributed adaptive controller to effectively handle unknown uncertainties, external disturbances, and unmodelled dynamics. The proposed controller adapts the operation of backstepping (BS) and sliding mode controller (SMC), where the BS controller tracks the nominal voltage of the DG, and SMC provides robustness in face of uncertainties. Lyapunov theory is applied to develop an adaptive mechanism that allows for the adjustment of control parameters to accommodate various operating conditions. The robustness of the proposed controller is validated under load perturbations, PnP functionalities. The effectiveness of the proposed controller is verified through simulations in MATLAB/ Simulink, and by comparing its performance with the conventional distributed averaging based controller. From the obtained results, it is observed that the proposed controller outperforms the distributed averaging controller during both transient and steady state conditions.

Index Terms—Microgrid, distributed control, adaptive control, voltage restoration.

I. INTRODUCTION

In an islanded microgrid (MG), it is challenging to sustain power supply while ensuring the restoration of voltage and frequency stability and maintaining a balance of active and reactive power among distributed generations (DGs) and loads through appropriate control strategies [1].

Several distributed hierarchical control schemes are designed to efficiently manage and control the operations of MG [2]. The control layers provide a hierarchical structure, where control objectives are distributed among different control layers, allowing for enhanced flexibility, coordination, and reliability in the management of energy resources within a MG.

The primary control layer typically employs the droop control method. However, it results in frequency and volt-

age deviations, as well as suboptimal power distribution. To address these limitations, a secondary control layer is implemented to restore the desired frequency and voltage levels. The utilization of distributed secondary control eliminates the need for a central controller. It provides better reliability such as voltage and frequency regulation and optimal power sharing among distributed energy resources. It is devoid of extensive computation and communication burden, failure of the central Controller, and scalability problems.

Most of the literature on consensus based control have primarily focused on MG systems with fixed and well-defined system parameters. However, in practical applications, the adoption of an adaptive control parameters is highly desirable, as it can effectively address the unmodelled and uncertain dynamics of DG.

An adaptive controller operates autonomously and independently of DG parameters, ensuring that its performance remains unaffected by variations in DG characteristics, such as ageing and thermal effects. Adaptive distributed controls can be used for regulating voltage and frequency to desired levels without relying on global information [3]. Several secondary control strategies for enhancing voltage regulation in the presence of uncertainty and unpredictability of renewable energy resources are proposed in [4,5]. A novel distributed adaptive controller, based on a PID controller, designed to mitigate the impact of unknown disturbances, parameter variations, and deviations caused by droop controller while maintaining the nominal voltage is discussed in [1].

A distributed sliding mode controller is proposed in [6] to achieve voltage, frequency restoration in the face of DG and load uncertainties using the phase trajectory method. It combines an extended state Kalman–Bucy filter for estimating DG state information despite parameter perturbations and measurement noise in [7]. Adaptation of SMC mechanism to maintain system stability and enhance secondary control performance. A radial basis function-based neural network is proposed in [8] to dynamically modify the SMC switching gain while guaranteeing MG voltage restoration, hence reducing the chattering problems. In order to increase re-

979-8-3315-3013-6/25 $31.00 © 2025 IEEE

silence against model uncertainties, unmodeled dynamics, and external disturbances, an active disturbance rejection control method [9] is presented. Moreover, Liu et al. proposed a backstepping scheme and the NonZero-Sum differential game strategy in [10], to handle the issue of voltage recovery in islanded MGs.

From the above literature review, it is observed that the traditional secondary control schemes do not provide robustness in face of uncertainties, and sudden fluctuations in loads within MGs. So, this paper presents the design of an adaptive controller with it parameters dynamically adjusted to enhance MG performance. However, a significant challenge remains in estimating system parameters due to the unavailability of accurate MG information. The review of existing literature reveals that many adaptive mechanisms rely on artificial neural networks, BS, and fuzzy techniques. These methods often introduce substantial computational complexity, posing implementation challenges in real-time scenarios. The paper presents design of a novel distributed controller. The purpose of this controller is to restore voltage synchronisation back in various MG situations. It is a fully distributed controller that doesn't rely on global information like the dynamics of MG or structure of communication network. It uses the Barbalat lemma and Lyapunov theory to derive control laws that efficiently handling fluctuating loads, external disturbances, and unmodelled dynamics of the MG. The contributions of this paper are as follows.

- A distributed adaptive scheme is proposed to overcome the voltage fluctuation in inverter-based islanded MGs.
- The distributed controller shows improved performance in both steady-state and transient phases.
- The adaptive mechanisms reduce the need for control gains tuning, thus reducing the computational load and minimizing the demands on network communication resources.
- The proposed approach to control design guarantees asymptotic stability for these tracking errors while ensuring the boundedness in the adaptation process.

The paper is organised as follows: Section II outlines the model for the entire inverter-based islanded MG, Section III introduces the design of the adaptive controller and its stability analysis, Section IV presents a thorough simulation analysis, and the conclusions is discussed in Section V.

II. MODELLING OF THE MICROGRID

Consider an autonomous AC MG, comprising distributed generators (DGs) and local loads, connected via power lincs. Eq. 1 presents the dynamics of each unit comprises power electronics components that allow them to share information and achieve secondary control purposes. A physical layer that handles connections between DG units and local loads, as well as a cyber-layer that manages connections between DG units, are used to model the overall network. This cyber-layer allows the DGs to accomplish restoration of voltage. In the following sections, we provide detailed explanations of the DG and the cyber physical layer.

A. DG Model

Each inverter-based DG unit consists of an output connector, an inductive-capacitive filter, a voltage source inverter (VSI), and a DC source. The VSI runs in voltage control mode when they are in islanded mode. The DG controller has power, voltage, and current control loops to facilitate power sharing. The voltage and current control loops have faster dynamics than the power control loop, and they function on distinct timescales. Because voltage and current control loops are governed by conventional PI controllers, the study ignores them in favour of concentrating on the stability of the active/reactive power control loops.

Each DG unit's dynamics can be as:

$$\begin{aligned} \omega_{nom} &= \omega_{ni} - k_{Pi}P_i + u_i^{\omega} \\ v_{nom} &= v_{ni} - k_{Qi}Q_i + u_i^{v} \end{aligned} \tag{1}$$

These equations are characterized by nominal values of output voltage v_{nom} and frequency ω_{nom}, the droop coefficients k_{Pi} and k_{Qi}. Additionally, secondary frequency u_i^{ω} and voltage u_i^{v} control inputs are introduced to compensate for deviations caused by primary control.

The cyber-layer represents the communication network among the DGs, can be described using the graph theory, denoted as $G_N = V_N, E_N, A_N$. Here, V_N represents the set of nodes, corresponding to the DG units, while $E_N \subseteq V_N V_N$ represents the set of edges that depict the communication links between the inverters. The adjacency matrix $A_N = [a_{ij}] \in R^{N \times N}$ is defined such that a_{ij} equals 1 when there is a link from DG_j to DG_i, and a_{ij} is 0 otherwise. The Laplacian matrix $L = [l_{ij}] \in R^{N \times N}$ is structured so that l_{ii} is equal to the sum of a_{ij} for $j = i$, and l_{ij} is the negative of a_{ij} when $i \neq j$.

The reference voltage $v_0 \in R$ for the secondary distributed consensus control is shared by a leader DG labelled with an index of 0.

The physical layer consists of power electronics components, describes the tangible connections among the DGs and loads, facilitated by connecting lines. The two layers adhere to the following assumptions:

- The communication graph G_c includes a directed spanning tree with the leader node as its root. Consequently, every node receives the leader's information either directly or indirectly.
- The power lines within the MG are lossless, meaning that conductance is consistently zero.

III. PROBLEM FORMULATION

The objectives of the proposed controller is restoring nominal voltage in inverter-based islanded MGs, and to synchronize the voltage of all DG units with the reference value set by the leader DG. This synchronization is obtained despite the absence of full knowledge of MG, including network topology, sudden load fluctuations, external disturbances, and uncertain dynamics.

To achieve this, we introduce a distributed adaptive controller. This control mechanism effectively counteracts external

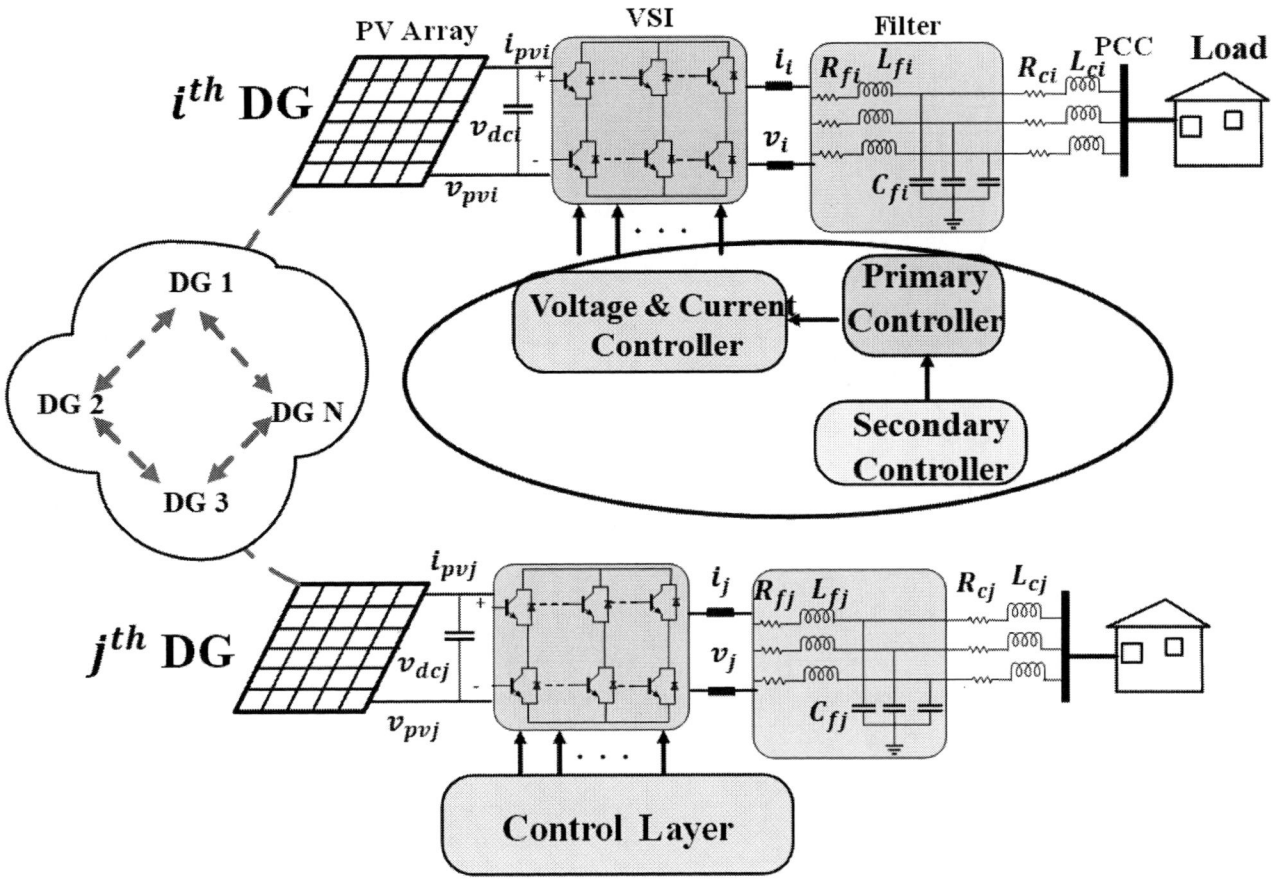

Fig. 1. Topology of cyber enabled microgrid system

disturbances and parameter discrepancies by adapting BS-SMC control gains. Mathematically, the aim of this work is:

$$\|v_i(t) - v_o(t)\| \to 0, \ i = 1, ..., N \tag{2}$$

Where $v_i(t)$ is the voltage of the ith DG, $v_o(t)$ is the desired voltage set point by the leader DG.

The dynamics of DG in the presence of disturbances is represented as

$$\ddot{v}_i(t) = -\frac{1}{k_{vi}}\dot{v}_i(t) - \frac{k_{Qi}}{k_{vi}}\dot{Q}_i(t) + \frac{1}{k_{vi}}u_i^v(t) \tag{3}$$

where $u_i^v(t)$ is the secondary control input for voltage restoration. Eq. (3) can be rewritten in the form of state space equation as

$$\begin{aligned}\dot{x}_{i,1}(t) &= x_{i,2}(t) \\ \dot{x}_{i,2}(t) &= \frac{1}{k_{vi}}u_i^v(t) + \frac{1}{k_{vi}}d_i(t)\end{aligned} \tag{4}$$

where $x_{i,1}(t), x_{i,1}(t)$ represent DG state vector, and its disturbance as $d_i(t) = -\dot{v}_i(t) - k_{Qi}\dot{Q}_i(t)$. The state error with respect to leader is written as:

$$\begin{aligned}e_{i,1}(t) &= x_{i,1}(t) - x_{0,1}(t) \\ e_{i,2}(t) &= x_{i,2}(t) - x_{0,2}(t)\end{aligned} \tag{5}$$

The error dynamics can also be written as:

$$\begin{aligned}\dot{e}_{i,1}(t) &= e_{i,2}(t) \\ \dot{e}_{i,2}(t) &= \frac{1}{k_{vi}}u_i^v(t) + \frac{1}{k_{vi}}d_i(t)\end{aligned} \tag{6}$$

Accordingly, the leader synchronisation signal is written as:

$$\eta_i(t) = c_i e_i(t) - \sum_{j=1}^{N} a_{ij}e_j(t) \in \Re^{2\times1} \tag{7}$$

where $c_i = \sum_{j=1}^{N} a_{ij} + a_{io} \in \Re_{+}$, $\forall i$, $\forall t$, indicating that at no point in time can DGs become isolated. The synchronisation signal can be expressed in the form of error dynamics as:

$$\begin{aligned}\dot{\eta}_{i,1}(t) &= \eta_{i,2}(t) \\ \dot{\eta}_{i,2}(t) &= \frac{c_i}{k_{vi}}\left(u_i^v + d_i(t) - \frac{k_{vi}}{c_i}\sum_{j=1}^{N} a_{ij}\dot{e}_{j,2}(t)\right)\end{aligned} \tag{8}$$

$d_i(t) - \frac{k_{vi}}{c_i}\sum_{j=1}^{N} a_{ij}\dot{e}_{j,2}(t)$ is represented as $w_i(t)$ that includes unknown model parameters and external disturbances affecting the DG. It specifically accommodates uncertain and nonlinear elements from various factors, including changes in topology, unbalanced and nonlinear loads, high-frequency

979-8-3315-3013-6/25 $31.00 © 2025 IEEE

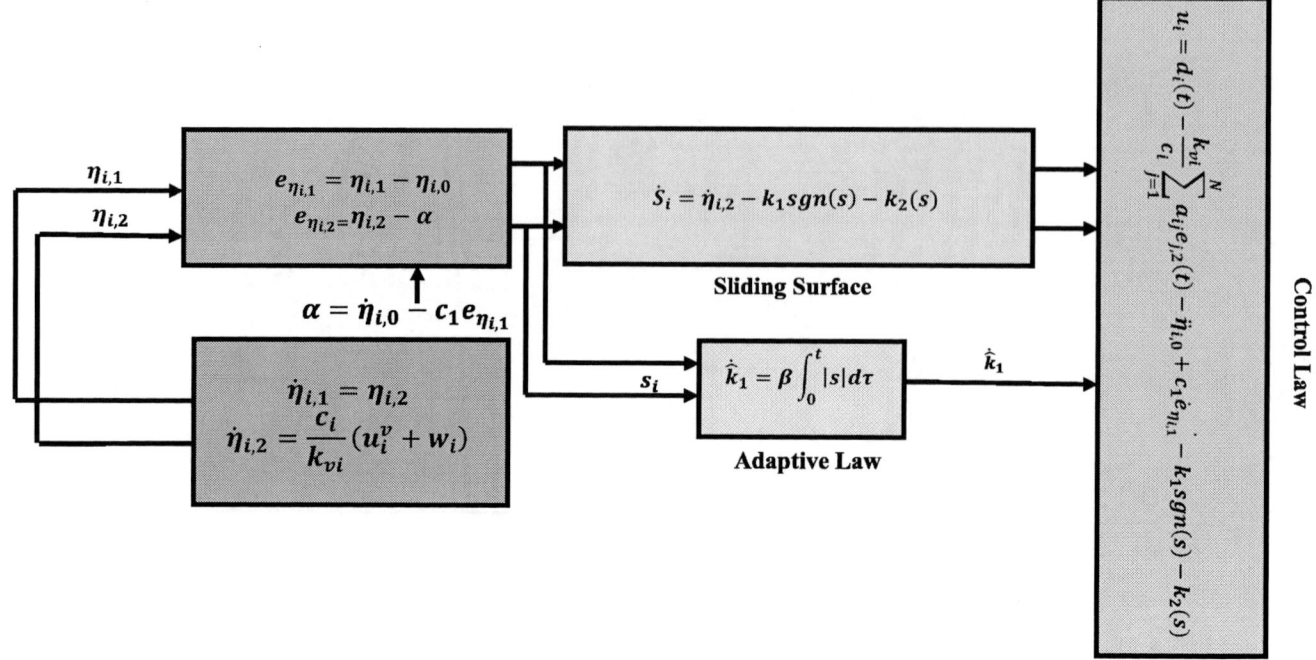

Fig. 2. Block diagram of the proposed distributed adaptive controller.

pulse-width voltage modulation, and transitions between operational modes.

IV. DESIGNING OF VOLTAGE CONTROLLER

According to the backstepping design idea, the tracking error is defined as

$$e_{\eta_{i,1}} = \eta_{i,1} - \eta_{i,0}$$
$$\dot{e}_{\eta_{i,1}} = \eta_{i,2} - \dot{\eta}_{i,0} \tag{9}$$

where $\eta_{i,0}$ is the reference value of $\eta_{i,1}$, $\eta_{i,2}$ can be regarded as a virtual control variable. Let the virtual control variable α is taken as

$$\alpha = \dot{\eta}_{i,0} - c_1 e_{\eta_{i,1}} \tag{10}$$

$$e_{\eta_{i,2}} = \eta_{i,2} - \alpha \tag{11}$$

$$\dot{\eta}_{i,2}(t) = \frac{c_i}{k_{vi}}(u_i^v + d_i(t) - \frac{k_{vi}i}{c_i}\sum_{j=1}^{N} a_{ij}\dot{e}_{j,2}(t) - \ddot{\eta}_{i,0} \tag{12}$$
$$+ c_1\dot{e}_{\eta_{i,1}})$$

Let the sliding surface is defined as $s = e_{\eta_{i,2}}$ Derivative of it yields

$$\dot{s} = \dot{\eta}_{i,2}(t) - k_1 sgn(s) - k_2 s \tag{13}$$

where $k_2 > 0$ is the parameter which controls the convergence speed and $k1 > 0$, which depends on the magnitude of uncertainty.

In accordance with the Lyapunov stability theory, the system exhibits asymptotic stability at the point $(e_1, e_2) = (0, 0)$. Nevertheless, the determination of the switching gain k_1 in

the aforementioned design is contingent upon the upper bound. By selecting an excessively large switching gain k_1 to ensure system stability can lead to significant chatter. Consequently, the switching gain is dynamically updated in real-time using an adaptive mechanism. $\hat{k}_1$ denotes the estimated value of k_1, and its evolution is described by the following dynamic equation.

$$\hat{k}_1 = \beta \int_o^t |s| d\tau \tag{14}$$

A. Stability Analysis

Lyapunov function is defined as:

$$V = \frac{1}{e_{\eta_{i,1}}^2} + \frac{1}{e_{\eta_{i,2}}^2} + \frac{1}{2\beta}(\hat{k}_1 - k)^2 \tag{15}$$

On time differentiation (15) gives

$$\dot{V} = e_{\eta_{i,1}} e_{\eta_{i,2}} - c_1 e_{\eta_{i,1}}^2 + e_{\eta_{i,2}}^2 \left(\frac{c_i}{k_{vi}}(u_i^v + d_i(t) - \frac{k_{vi}i}{c_i} \right.$$
$$\left. \sum_{j=1}^{N} a_{ij}\dot{e}_{j,2}(t) - \ddot{\eta}_{i,0} + c_1\dot{e}_{\eta_{i,1}} - \hat{k}_1 sgn(s) - k_2 s \right)$$
$$+ \frac{1}{2\beta}(\hat{k}_1 - k)^2 \tag{16}$$

The control law is defined as:

$$u_i^v = d_i(t) - \frac{k_{vi}i}{c_i}\sum_{j=1}^{N} a_{ij}\dot{e}_{j,2}(t) - \ddot{\eta}_{i,0} + c_1\dot{e}_{\eta_{i,1}} - \hat{k}_1 sgn(s)$$
$$- k_2 s + \frac{1}{2\beta}(\hat{k}_1 - k)^2 \tag{17}$$

(17) results in:

$$\dot{V} = e_{\eta_{i,1}} e_{\eta_{i,2}} - c_1 e_{\eta_{i,1}}^2 + e_{\eta_{i,2}}^2 \left(\frac{c_i}{k_{vi}} (u_i^v + d_i(t) - \frac{k_v i}{c_i} \right.$$

$$\left. \sum_{j=1}^{N} a_{ij} \dot{e}_{j,2}(t) - \ddot{\eta}_{i,0} + c_1 \dot{e}_{\eta_{i,1}} - \hat{k}_1 sgn(s) - k_2 s \right)$$

$$+ \frac{1}{2\beta} (\hat{k}_1 - k)^2 \le 0$$

(18)

The block diagram of proposed controller is shown in the Fig. 2.

V. RESULTS AND DISCUSSION

To evaluate the performance of the MG, simulation were conducted in MATLAB/Simulink. The parameters of the MG is taken from [4]. The communication among the DG is represented by using graph theory. The adjacency matrix, in-degree matrix, and Laplacian matrix for the considered communication topology (Fig.3) is given as:

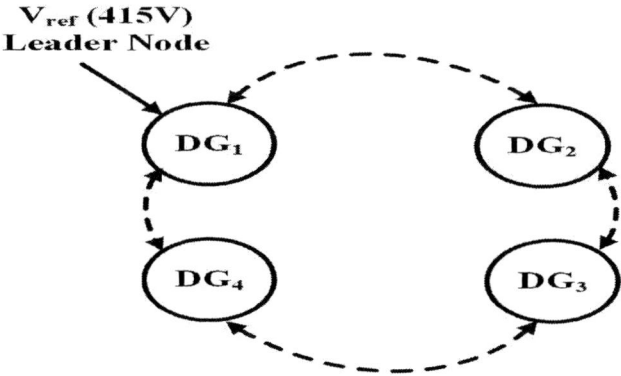

Fig. 3. Communication Network

$$A = \begin{pmatrix} 0 & 1 & 0 & 1 \\ 1 & 0 & 1 & 0 \\ 0 & 1 & 0 & 1 \\ 1 & 0 & 1 & 0 \end{pmatrix}; \quad D = \begin{pmatrix} 2 & 0 & 0 & 0 \\ 0 & 2 & 0 & 0 \\ 0 & 0 & 2 & 0 \\ 0 & 0 & 0 & 2 \end{pmatrix};$$

$$L = D - A = \begin{pmatrix} 2 & -1 & 0 & -1 \\ -1 & 2 & -1 & 0 \\ 0 & -1 & 2 & -1 \\ -1 & 0 & -1 & 2 \end{pmatrix}$$

The steady state performance of the MG is evaluated without any disturbances and the voltage response of the DG is restored at 415 V. The proposed controller resolves the deviations caused by primary droop control. Fig. 4,5 show the steady state voltage, active power responses of the DG. At $t = 2sec$, the DG voltage increases, and it is again reconnected at $t = 3sec$. However, the proposed controller restores the MG voltage in the presence of load perturbation. It shows better steady state and dynamic performances. At $t = 5~sec$, one of

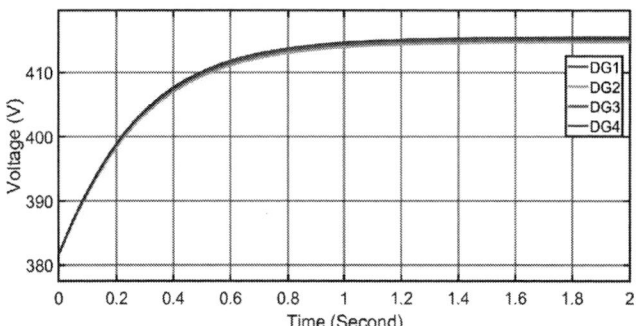

Fig. 4. Voltage responses of DGs.

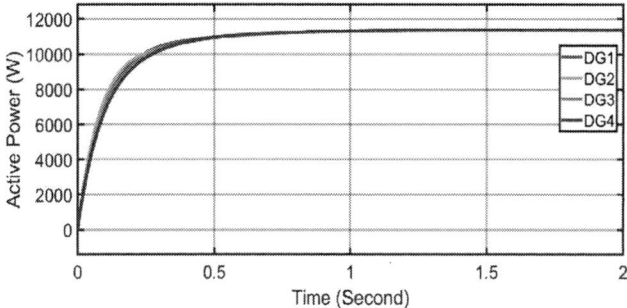

Fig. 5. Active power responses of DGs.

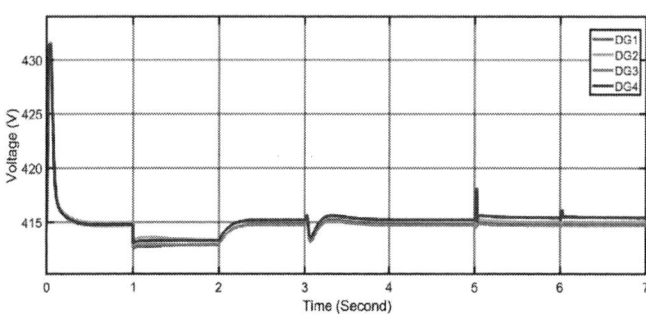

Fig. 6. Voltage responses of the DGs under load perturbation, and PnP operation.

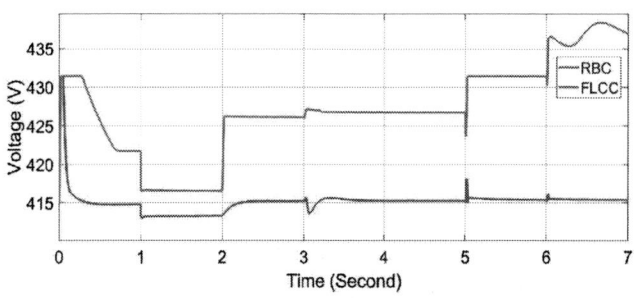

Fig. 7. Comparison of voltage responses of DG with primary droop controller under load perturbation and PnP operation.

979-8-3315-3013-6/25 $31.00 © 2025 IEEE

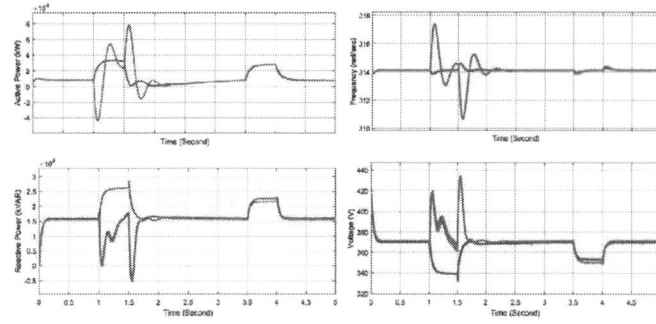

Fig. 8. Comparison of voltage, frequency, active and reactive power responses of DG with the conventional distributed averaging based controller.

the DG is disconnected and again at $t = 6$ sec it is reconnected. Hence, the proposed controller supports the PnP operation with a small transient. Fig. 6, shows the voltage responses of the DG in the presence of load variations, and PnP operations. To compare the performance of the proposed controller with exist controller, Fig. 7,8 show the comparison of voltage, frequency, active, and reactive power response with distributed averaging based controller in [11]. It can be observed that the transient as well as steady state performance of the proposed controller are improved.

VI. CONCLUSION

This paper presents a distributed adaptive controller designed for an islanded AC microgrid. The effectiveness of the proposed controller is evaluated in various scenarios, including normal operation, load disturbances, plug-and-play situations. The results demonstrate that the proposed controller effectively ensures the stability of voltage in the MG. The proposed controller is compared with a distributed averaging based controller. This analysis envisages that the better robustness and convergence performance is achieved with the proposed controller. Thus, we conclude that the proposed controller provides stable operation of the microgrid despite load perturbations, plug and play operations.

ABBREVIATIONS

ω_{nom}, v_{nom}	Nominal values of frequency, voltage
ω_{ni}, v_{ni}	Primary frequency, voltage inputs
k_{Pi}, k_{Qi}	Droop coefficients
P_i, Q_i	Active, reactive powers
u_i^ω, u_i^v	Secondary frequency, voltage inputs
a_{ij}, a_{i0}	Communication weights among DGs, and with leader nodes
v_i, v_o	Voltage at i^{th}, and leader node

$x_{i,1}(t), x_{i,2}(t)$	Nominal frequency
$d_i(t)$	Nominal frequency
$e_{i,1}(t), e_{i,2}(t)$	Nominal frequency
$x_{0,1}(t), x_{0,2}(t)$	DG state vectors
$\eta_{i,1}, \eta_{i,2}$	Synchronisation signal vectors
$\eta_{i,0}$	Reference value of synchronisation signal
α	Virtual control variable
s	Sliding surface
$\hat{k}_1$	Estimated values of parameter
β	Adaptive gain
V	Lyapunov Function
A, D, L	Adjacency matrix, in-degree matrix, Laplacian Matrix

REFERENCES

[1] B. Caiazzo, et al. "Cooperative Adaptive PID-Like Voltage Regulation in Inverter-Based Islanded Microgrids Under Unknown Uncertainties,"*IEEE Systems Journal*, vol. 17, no. 3, pp. 4395-4406, Sept. 2023.

[2] S. K. Panda, B. Subudhi, "A Review on Robust and Adaptive Control Schemes for Microgrid," *IEEE J. Mod. Power Syst. Clean Energy*,vol. 11, no. 4, pp. 1027-1040, July 2023.

[3] Zhao, Guanglei, et al. "Distributed Adaptive Dynamic Event-Triggered Secondary Control for Islanded Microgrids with Disturbances,"*IEEE Transactions on Smart Grid*, vol. 14, no. 6, pp. 4268-4281, Nov. 2023.

[4] S. K. Panda and B. Subudhi, "An Extended State Observer based Adaptive Backstepping Controller for Microgrid,"*IEEE Trans. Smart Grid*, vol. 15, no. 1, pp. 171-178, Jan. 2024.

[5] S. K. Panda and B. Subudhi, "A Distributed Adaptive Super-Twisting Sliding Mode Controller for Voltage and Frequency Restoration in Islanded Microgrid,"*IEEE Journal of Emerging and Selected Topics in Power Electronics*, doi: 10.1109/JESTPE.2024.3468152.

[6] J. Liu, et al, "Nonlinear secondary voltage control of islanded microgrid via distributed consistency,"*IEEE Trans. Energy Convers.*, vol. 35, no. 4, pp. 1964–1972, Dec. 2020.

[7] P. Ge, Y. Zhu, T. C. Green, and F. Teng, "Resilient secondary voltage control of islanded microgrids: An ESKBF-based distributed fast terminal sliding mode control approach,"*IEEE Trans. Power Syst.*, vol. 36, no. 2, pp. 1059–1070, Mar. 2021.

[8] X. Shen, H. Wang, J. Li, Q. Su, and L. Gao, "Distributed secondary voltage control of islanded microgrids based on RBF-neural-network sliding-mode technique,"*IEEE Access*, vol. 7, pp. 65616–65623, 2019.

[9] M. Zhang et al., "A robust distributed secondary voltage control method for islanded microgrids,"*Int. J. Elect. Power Energy Syst.*, vol. 121, 2020, Art. no. 105938.

[10] G. Liu, Q. Sun, R. Wang, and X. Hu, "Nonzero-sum game-based voltage recovery consensus optimal control for nonlinear microgrids system,"*IEEE Trans. Neural Netw. Learn. Syst.*, vol. 34, no. 11, pp. 8617-8629, Nov. 2023.

[11] J. Simpson, Q. Shafiee, et al., "Secondary frequency and voltage control of islanded microgrids via distributed averaging," *IEEE Trans. Ind. Electron.*, vol. 62, no. 11, pp. 7025-7038, Nov. 2015.

Investigation of a PV-BESS Fed Islanded Microgrid System using RTDS Supported by Field Results

Bikram Kumar Samanta
Electrical Engineering Department
Indian Institute of Technology Kharagpur
Kharagpur, 721302
India
bksamanta9@gmail.com

Supratik Bhowmick
Electrical Engineering Department
Indian Institute of Technology Kharagpur
Kharagpur, 721302
India
supratikbhowmick95@gmail.com

Suman Maiti
Electrical Engineering Department
Indian Institute of Technology Kharagpur
Kharagpur, 721302
India
suman.maiti@ee.iitkgp.ac.in

Chandan Chakraborty
Electrical Engineering Department
Indian Institute of Technology Kharagpur
Kharagpur, 721302
India
cc@ee.iitkgp.ac.in

Abstract— This paper gives a comparison study between real time data simulation (RTDS) results with the field results for an islanded microgrid (IMG) system. The islanded microgrid consists of photovoltaic (PV) and battery energy storage system (BESS) as inputs for the system. For the real time simulation of this IMG, a RTDS simulator is used where the IMG is programmed in RSCAD software. For field results, data is acquired manually as well as from a data monitoring system which is placed in an actual islanded field. Here, MPPT controlled boost converter operation is implemented for PV and BESS is connected to dc link of the inverter through a bi-directional boost converter. A three-phase two-level inverter is used whose output is connected to distributed loads through a second order LCL filter. For investigation of this IMG, different modes of operation by switching on/off PV and BESS inputs are considered and the results are thoroughly analyzed in this paper. The comparison of RTDS and field results showcase effectiveness of the IMG control with 5.3% error in PV generated power and 2.7% error in BESS supplied power. A voltage boosting inverter is integrated with the system to improve the distribution line voltage magnitude.

Keywords—Renewable Energy Resources, Islanded Microgrid, Real Time Data Simulation, MPPT Control, PV-BESS System.

I. INTRODUCTION

In recent times, the world population is increasing and the global concern to the environmental problems demand new ways of the energy generation. This generation is expected to either eliminate or reduce the pollution as much as possible. Besides, the energy demand is increasing not only in the big metro cities but also in the remote village places. Therefore, developing the efficient solutions for reliable, sustainable and environmentally friendly energy supply to the remote places and off-grid systems is of vital importance [1]-[4]. Consequently, the research focus in the renewable energy sources (RESs) has immensely increased in the last few years. Islanded Microgrid systems (IMGs) provide a promising solution for reliable and environment friendly energy supply for these remote locations. However, the operation and management of these IMGs is a very complex task including the coordination between a variety of the distributed energy sources like photovoltaics (PV), wind turbine etc. and loads (critical and non-critical) with an intermittent nature in a stable, resilient and robust manner. The energy management system of IMGs is therefore getting significant attention as it is helpful from the economic and emissions point of view.

The increase of RES utilization may cause instability issues in the power systems if inertia of these systems is not managed properly. The IMGs promise to support energy supply in remote places, facilitate large and reliable integration of RESs into the electrification systems, also reduce greenhouse gas emissions and cause lower energy prices among the others [5]-[8]. The main parts of an IMG are conventional distributed energy resources (CDERs) such as micro-turbines (MTs), diesel generators (DGs) etc. and non-conventional distributed energy resources (NCDERs) which are the renewables like PV, energy storage systems (ESSs) etc. [8]-[11]. However, the high penetration of inertia-less RESs brings many challenges to these IMGs in terms of operation management optimization, stability, resiliency and robustness etc. Among these the main challenges are reliability of the supply [12], deviation of frequency due to the system low inertia [13], coordination operation of multiple number of MGs [14], efficient and effective coordination between energy management system (EMS) and demand side management (DSM) [12]-[13], robust operation of the system, management of multiple distributed energy resources (DERs) with possible conflicting requirements [12], [15], efficient protection scheme considering bi-directional power flow of the system [16], modelling of system components and processes with required accuracy [6], [14], optimal placement and sizing of ESSs [17]-[18] and, reliable and efficient incorporation of the advanced technologies like electric vehicles (EVs) and internet of things (IoT) into these systems [19]-[20]. The increasing number of RESs, integration of more ESSs, inclusion of EVs and smart devices need for energy in remote and extreme environmental conditions have posed serious challenges especially for the management of energy in these IMGs. For this, the review studies dedicated to EMS optimization of IMGs are quite necessary to help the researchers on the topic to keep a track of the state-of-the-art and devote their efforts to the IMGs to meet the abovementioned requirements [6]-[8]. The upcoming trends include the need for more improved models, advanced data analytics and forecasting algorithm techniques, performance assessment and analysis of real-time EMSs in the whole MG's control hierarchy, fully effective decentralized EMSs, improved communication and cyber security systems and the overall validations of the system are under real conditions [19]-[20].

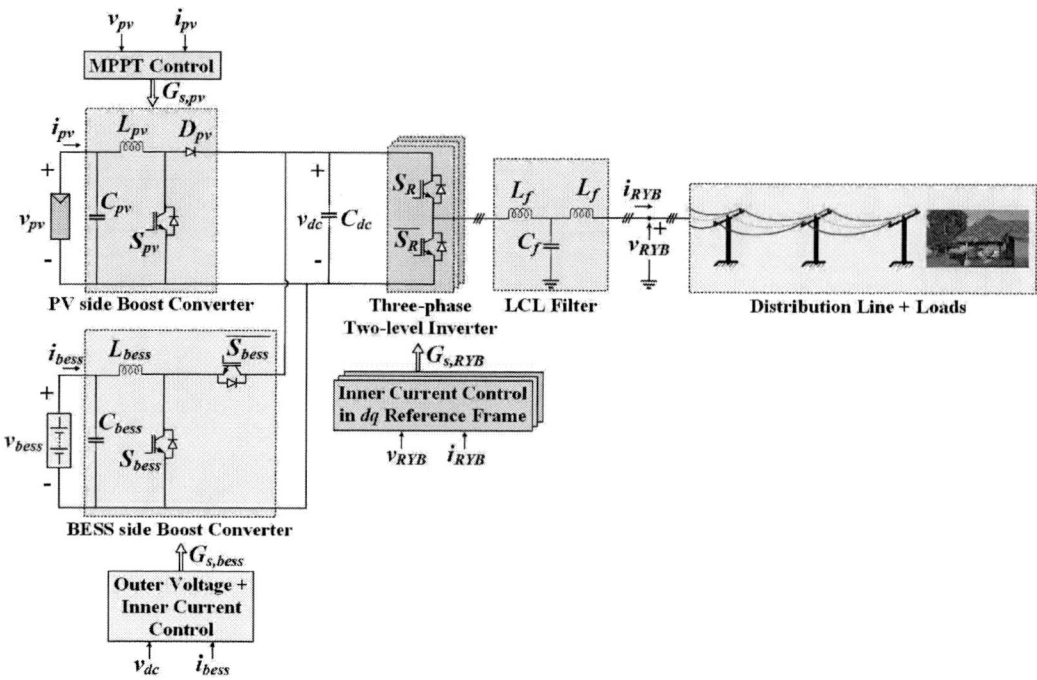

Fig. 1. Schematic Diagram of IMG

Based on the above discussion on IMGs, this paper investigates performance of a PV and BESS generated IMG system. For real time data simulation (RTDS) of this system, a real time data simulator is used where the IMG system is programmed in RSCAD software. For practical assessment, data are acquired manually as well as from the data monitoring system installed in an actual islanded field. The remaining sections of the paper are structured as, Section II details the system configuration of the IMG. Section III represents the control methods of IMG. In Section IV, the RSCAD based real time simulation results are illustrated. Section V represents the actual field results and the results obtained from the data monitoring system. Section VI includes comparison of these results. Section VII shows load side voltage boosting of IMG and Section VIII concludes the work.

TABLE I
SPECIFICATION OF SYSTEM PARAMETERS

Parameter	Specification
Solar PV cell open circuit voltage	49.62 V
Solar PV cell short circuit current	13.87 A
Solar PV cell power at MPPT	540 W
Solar PV side converter inductor rating	1.45 mH
Solar PV side converter switching frequency	5 kHz
No. of Solar PV panels	280
Battery Cell Voltage	2 V
Battery Nominal Capacity	600 AH
No. of Battery cells	480
Battery side converter inductor rating	1.45 mH
Battery side converter switching frequency	5 kHz
Inverter DC link capacitance	2200 μF
Inverter switching frequency	5 kHz

II. SYSTEM CONFIGURATION OF IMG

Fig.1 shows schematic diagram of the IMG which is considered in this work. The PV source is connected to DC link of the inverter through a dc-dc boost converter. This boost converter controls maximum power point tracking (MPPT) operation of the PV source. The BESS is connected to the DC link through a bidirectional dc-dc boost converter. This bidirectional converter controls charging or discharging operation of BESS depending on operating sequence of PV source and therefore constructs a coordinated control among them.

The inverter configuration consists of a three-phase two-level structure whose output terminals are connected to distributed loads through a second order LCL filter. This filter is designed to minimize harmonics and improve power quality at output side of the inverter. From the filter output terminals, the power is distributed to the loads through distribution lines as shown in Fig. 1. A practical setup of this system is installed at Ghoramara Island, West Bengal, India. The parameters of the system are tabulated in Table 1 which is used in both real time simulation model as well as in the practical setup.

III. CONTROL METHODS OF IMG

The methodologies for controlling different converters in IMG of Fig. 1 are shown in Fig. 2. Fig. 2(a) demonstrates control of PV side boost converter whose main objective is to implement MPPT control for PV source. For this converter control, PV terminal voltage (v_{PV}) and current (i_{PV}) are measured and given as input signals to the MPPT controller. The generated voltage from MPPT control (v_{MPPT}) is then compared with PV voltage (v_{PV}) and the voltage controller generates reference voltage ($v_{boost,ref}$) for controlling the boost converter.

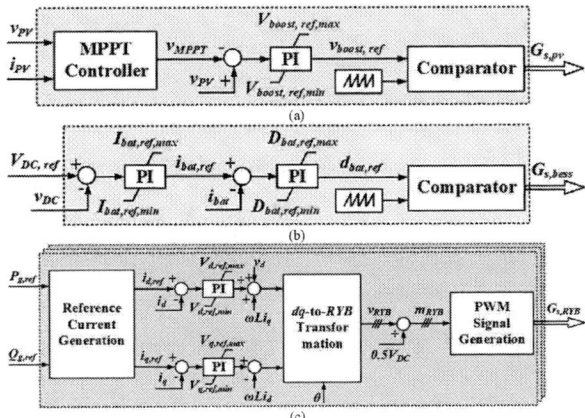

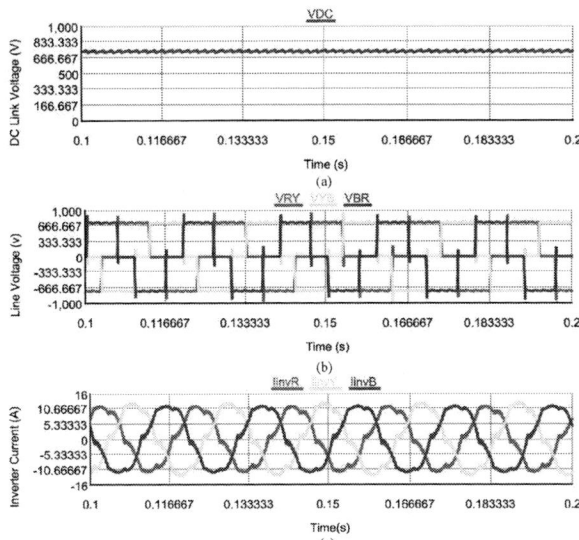

Fig. 2. Block diagrams showing (a) PV side boost converter control, (b) Battery side bi-directional boost converter control and (c) Inverter control.

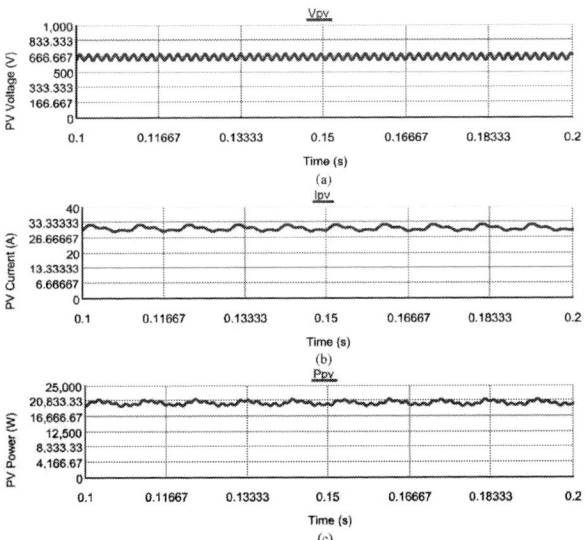

Fig. 3. Figure showing (a) RSCAD host PC and (b) RTDS simulator.

Fig. 5. Real time simulation results showing (a) DC link voltage, (b) Inverter line voltage and (c) Inverter line current under MPPT operation of PV source.

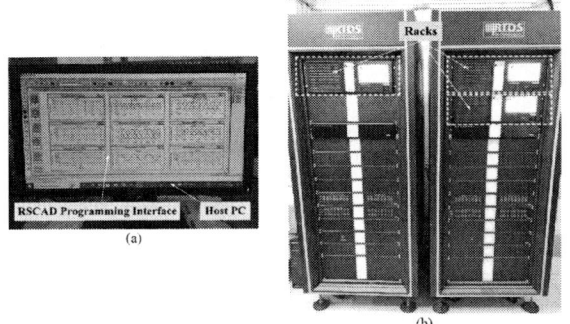

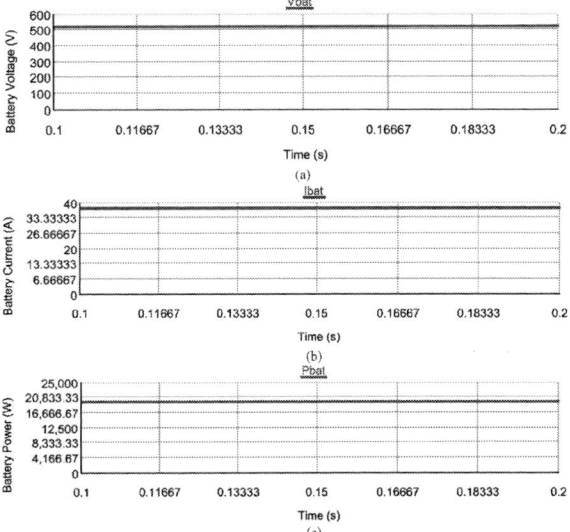

Fig. 4. Real time simulation results showing (a) PV terminal voltage, (b) PV terminal current and (c) PV generated power under MPPT operation.

Fig. 6. Real time simulation results showing (a) Battery output voltage, (b) Battery output current, (c) Battery generated power under PV OFF condition.

For battery side bi-directional converter control as shown in Fig. 2(b), reference DC link voltage ($V_{DC,ref}$) is compared with actual DC link voltage (v_{DC}) which is measured. The outer voltage loop controller generates reference current ($i_{bat,ref}$) for BESS which is considered as input for inner current controller. The current controller compares reference current with actual current (i_{bat}) and generates duty ratio ($d_{bat,ref}$) of operation for the converter.

The inverter control is developed based on active and reactive references ($P_{g,ref}$ and $Q_{g,ref}$ respectively) and the control block diagram is shown in Fig. 2(c). These reference powers generate reference currents ($i_{d,ref}$ and $i_{q,ref}$ respectively) in dq synchronously rotating reference frame. The inner current control loop compares these reference currents with actual inverter currents (i_d and i_q respectively) and generates reference voltage for the inverter. The modulation index for switches present in the inverter are generated from this reference voltage and the pulse width modulation (PWM) generation block generates gate signals for the switches.

IV. RTDS BASED REAL TIME SIMULATION OF IMG

For the real time performance analysis of the IMG system, RSCAD software based RTDS simulator based is used. The

979-8-3315-3013-6/25 $31.00 © 2025 IEEE   97

 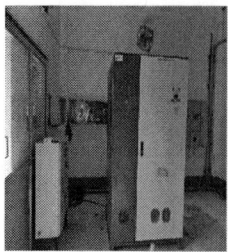  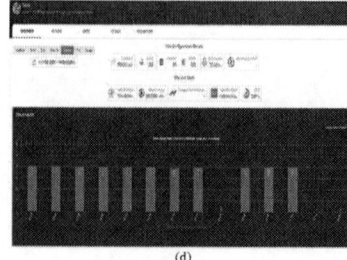

Fig. 7. IMG of practical field showing (a) PV modules, (b) Inverter, (c) Battery storage system, (d) TrackSo data monitoring system.

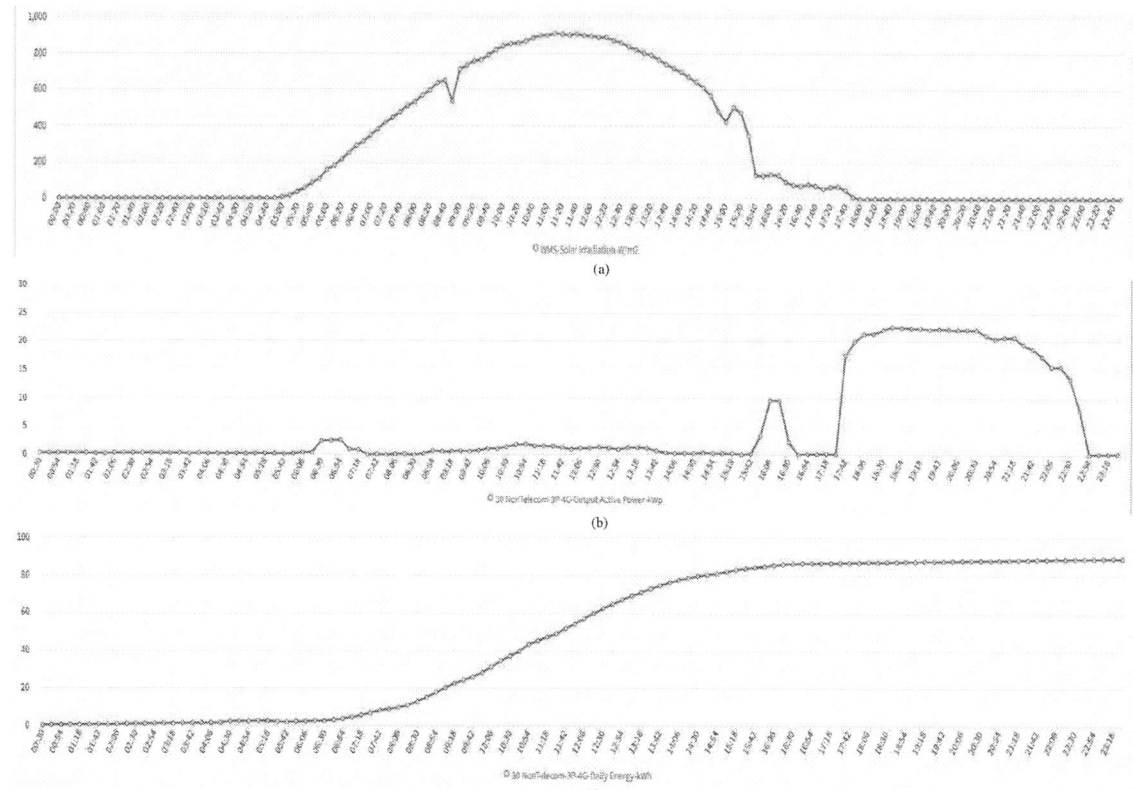

Fig. 8. Field result showing (a) Solar Irradiation, (b) Output Active Power, (c) Energy.

lab setup of RTDS and the RSCAD program interfacing host PC are shown in Fig. 3. The IMG system is developed in RSCAD software and it is run on RTDS simulator to simulate different operating conditions in real time. For real time simulation of the IMG system, two racks (Fig. 3(b)) in RTDS simulator are run in parallel using multi-rack simulation in RSCAD interface. The system parameters for real time simulation are tabulated in Table I. The PV source is simulated to work on MPPT operation during its connection in the network and during that time BESS absorbs power from PV source. When PV is switched OFF, BESS supplies power to the loads. In this section, the coordinated operation between PV and BESS is investigated in real time and the corresponding results are explained.

The PV side parameters in real time are depicted in Fig. 4. Under MPPT operation, PV source supplies 30.84 A current into the system (Fig. 4(b)) and generates 20.42 kW power (Fig. 4(c)) in the process. The average value of MPP

voltage is 664 V (Fig. 4(a)). The PV side boost converter maintains the DC link voltage at 745 V with peak-to-peak ripple of 2.8% as shown in Fig. 5(a). The line voltage of the inverter has peak amplitude of 735 V which is same for both positive and negative halves (Fig. 5(b)). The inverter injects three-phase balanced current with amplitude of 12.8 A under MPPT operation as shown in Fig. 5(c).

The battery parameters are reflected in Fig. 6. At first, the system is initiated with PV source in the network and then after some time, PV source is turned OFF with the help of switching circuit. The real time simulation results of Fig. 6 are shown during OFF condition of PV source. During this time period, battery terminal voltage is recorded as 522 V (Fig. 6(a)) and battery current becomes 37.58 A (Fig. 6(b)). The BESS generates approximately 19.62 kW (Fig. 6(c)) during PV OFF condition. Therefore, the BESS can fulfil maximum power requirement of the system during PV OFF condition.

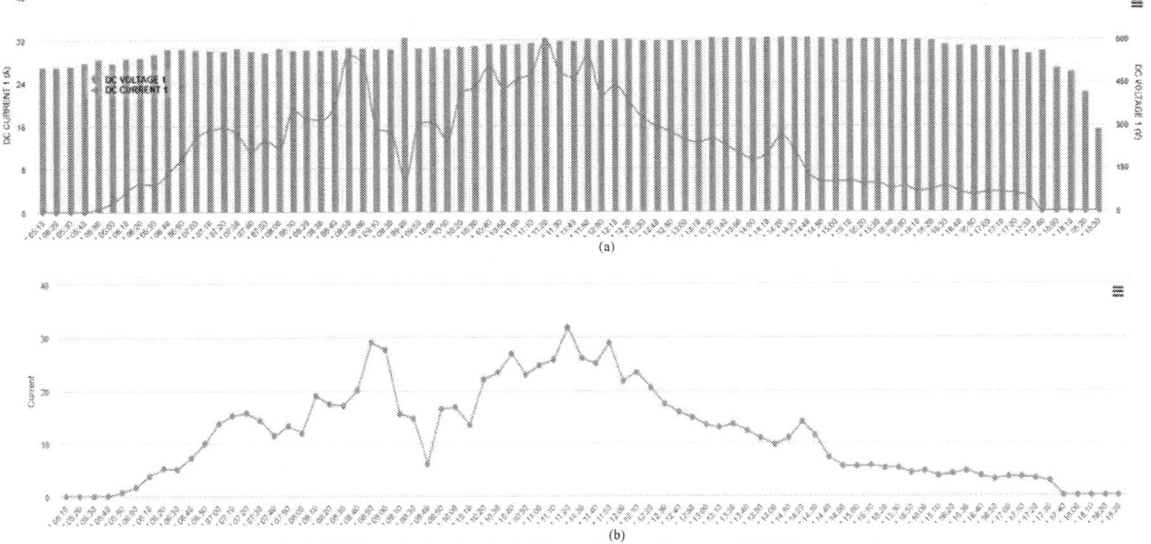

(a)

(b)

Fig. 9. Field result showing (a) DC bus voltage and current, (b) DC bus current.

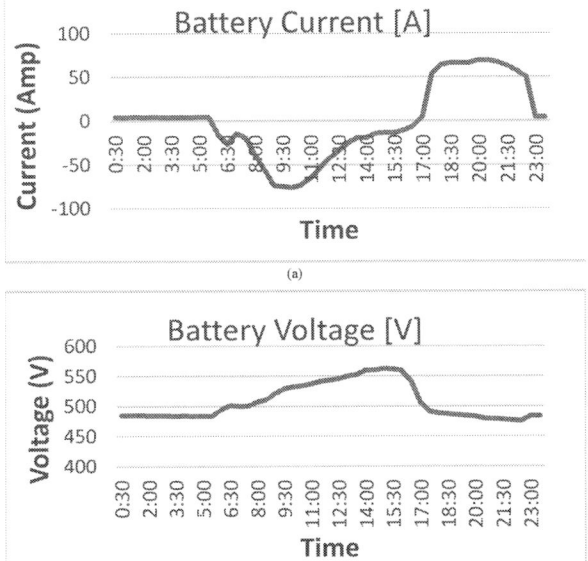

(a)

(b)

Fig. 10. Field result showing (a) battery current, (b) battery voltage

V. FIELD RESULTS FROM PRACTICAL IMG

In this section, practical field results of the IMG as well as data received from the remote data monitoring system are discussed. The overall practical system as well as the online data monitoring TrackSo platform are shown in Fig. 7. The parameters of the practical field system are similar as mentioned before in Table I. Here active output power, daily energy, solar irradiation, dc bus voltage and current, battery voltage and current results are discussed.

For the practical field system, peak value of solar irradiance for the given day is 910 W/m² as shown in Fig. 8(a) and it is around 11:30 AM in the morning.

Output active power of this system is shown for a single day as well as average for the last three months. The day mentioned in the the graph has a peak value of 22 kWp as shown in Fig. 8(b). And the monthly average of active output power of last three months are 22.2 kWp, 20.9 kWp and 20.7 kWp respectively.

In the previous paragraph, daily peak output power is observed to be 22 kWp approximately. And the time duration of delivering power in the practical field is nearly 4 to 4.5 hours. Multiplication of these two values gives the energy around 90.5 kWh which is also seen in the graph Fig. 8(c).

The DC bus voltage and current for a particular day are shown in Figs. 9(a) and 9(b) respectively. It is observed that peak value of current is 33 A approximately and the voltage is around 625 V.

Battery currents and voltages are measured throughout a day which is shown in Figs. 10(a) and Fig. 10(b) respectively. Maximum battery current is observed to be 72.6 A and the time is around 8 PM in the night and minimum value of current is observed to be -75.9 A and the time is around 10 AM in the morning. So, in this way the charging-discharging peaks of the battery current are observed. Similar case is also observed for the battery voltage. Maximum voltage is observed to be 562 V which is around 4 PM in the afternoon and minimum voltage is observed to be 476 V which is around 11 PM at night. These data demonstrates charging and discharging of the BESS for a particular day.

VI. COMPARISON BETWEEN REAL TIME SIMULATION RESULTS AND FIELD RESULTS

The real time simulation data shows that the PV source supplies 30.84 A current into the system under the MPPT operation. The similar scenario is observed in the field result and it shows the value approximately 33 A. The real time simulation result shows that the PV output power at MPPT condition is around 20.42 kW. And the average value of the field results for the last three months is recorded approximately 21.5 kW, which shows the absolute error of 5.3% with respect to the real time simulation result.

For battery parameters, field voltage and current values are observed as 518 V and 38.9 A. So, it gives the power output 20.15 kW and during that time real time simulation value of power is 19.62 kW, which shows absolute error of 2.7%.

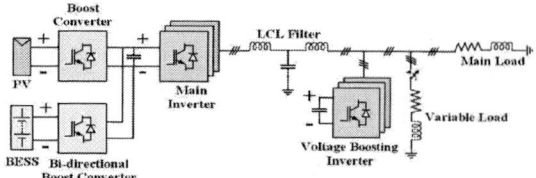

Fig. 11. Single line diagram of voltage boosting inverter-based IMG.

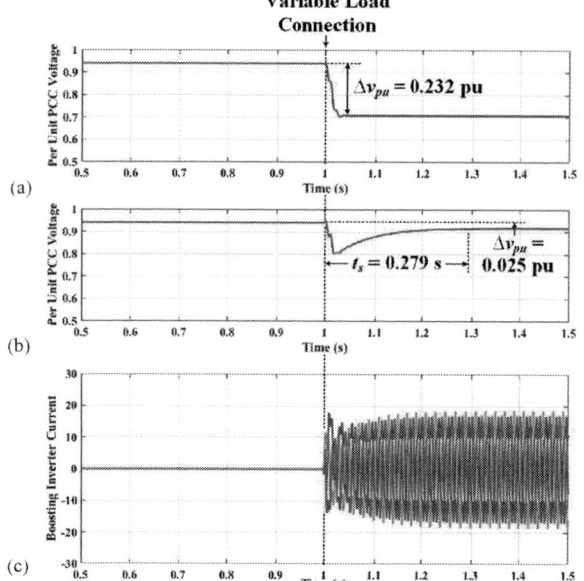

Fig. 12. Simulation results showing (a) Per unit voltage without voltage boosting inverter, (b) Per unit voltage without voltage boosting inverter and (c) Voltage boosting inverter current.

VII. LOAD SIDE VOLTAGE BOOSTING IN IMG

In the practical field it is observed that the line voltage magnitude becomes less as the distribution transits away from the source (PV and BESS). A voltage boosting inverter (VBI) can be incorporated in the system to mitigate this problem. A demonstration of this system is simulated and its single line diagram is shown in Fig. 11. The VBI contains capacitance at its DC link in order to inject reactive current which helps in limiting the voltage fall. The simulation results of Fig. 12 shows that the VBI limits the voltage fall to 0.025pu within 0.279 s (Fig. 12(b)) in case of a load connection, which records an improvement in voltage restoration by 89.224% as compared to the case when VBI does not inject any current (Figs. 12(a) and (b) respectively).

VIII. CONCLUSION

This paper investigates a PV-BESS fed IMG system. The PV source which is connected to DC link through a boost converter is initiated with MPPT operation. The BESS is connected to the DC link through a bidirectional dc-dc boost converter. This bidirectional converter controls charging or discharging operation of BESS depending on operation sequence of PV source and therefore constructs a coordinated control among them. A three-phase two-level inverter connects distributed loads at its output side through an LCL filter. This filter is designed for minimize the harmonics in inverter injected current and maintain its waveform quality. The IMG system is first investigated in real time simulation for which RSCAD software based RTDS simulator is utilized.

Subsequently, the real time simulation results are compared with the results obtained from a practical IMG system. The comparison results of PV output power and BESS delivered power at PV off condition with respect to RTDS results show 5.3% and 2.7% error respectively. A voltage boosting inverter-based IMG system is also simulated to verify voltage boosting capability of the inverter in case of a load connection. This method can be applied in the practical field IMG also, which can be considered as future scope of this work.

ACKNOWLEDGEMENT

This work was financially supported by DST, Govt. of India and EU's Horizon 2020 Research and Innovation Program through the RE-EMPOWERED Project under Grant Agreement No. DST/TMD/INDIA/EU/ILES/2020/50(c) and 101018420 respectively.

REFERENCES

[1] W Zhou, C Lou, Z Li, L Lu , H Yang. Current status of research on optimumsizing of stand-alone hybrid solar-windpower generation systems. Appl Energy 2010;87(2): 380–9.

[2] N Agarwal, A Kumar , Varun. Optimization of grid independent hybrid PV diesel- battery system for power generation in remote villages of Uttar Pradesh, India. Energy Sustain Develop 2013; 17(3): 210–9.

[3] A Lazou, et al., The economics of photovoltaic stand alone residential households: a case study for various European and Mediterranean locations. Sol Energy Mater Solar Cells 2000; 62: 411–27.

[4] D Yamegueu, Y Azoumah, X Py, N Zongo. Experimental study of electricity generation by solar PV/diesel hybrid systems without battery storage for off grid areas. Renew Energy 2011; 36: 1780–7.

[5] H Murdock, D Gibb, T André. "Renewables 2020 global status report. Technical Report", REN21 COMMUNITY; 2020.

[6] DE Olivares, et al., "A centralized energy management system for isolated microgrids." IEEE Trans Smart Grid 2014; 5(4):1864–75.

[7] N Yang, D Paire, Fei Gao, A Miraoui. "Power management strategies for microgrid"-A short review. In: 2013 IEEE industry applications society annual meeting. 2013. p. 1–9.

[8] D Akinyele, J Belikov, Y Levron. "Challenges of microgrids in remote communities": A STEEP model application. Energies 2018; 11(2).

[9] T Alharbi, K Bhattacharya. "A stochastic energy management system for isolated microgrids." In: 2018 IEEE power energy society general meeting. 2018. p. 1–5.

[10] O Boqtob, et al., Microgrid energy management system: A state-of-the-art review. J Electr Syst 2019; 15:53–67.

[11] M Faisal, MA Hannan, PJ Ker, A Hussain, MB Mansor, F Blaabjerg. "Review of energy storage system technologies in microgrid applications: Issues and challenges." IEEE Access 2018; 6:35143–64.

[12] H Suyanto, R Irawati. "Study trends and challenges of the development of microgrids." In: 2017 15th international conference on quality in research: International symposium on electrical and computer engineering. 2017. p.383–7.

[13] A Venkataraman, et al. "Development of a power mix management system for REIDS microgrids." In: 2016 Asian conference on energy, power, and transportation electrification. 2016. p. 1–5.

[14] SK Sahoo, AK Sinha, NK Kishore. "Control techniques in AC, DC, and hybrid AC–DC microgrid": A review. IEEE J Emerg Sel Top Power Electron 2018;6(2):738–59.

[15] MS Narkhede, S Chatterji, S Ghosh. "Trends, and challenges in optimization techniques for operation and control of microgrid" - A review. In: 2012 1st international conference on power and energy in NERIST. 2012. p. 1–7.

[16] A Hooshyar, R Iravani. "Microgrid protection." Proc IEEE 2017;105(7):1332–53.

[17] MA Hannan, et al. "Review of optimal methods and algorithms for sizing energy storage systems to achieve decarbonization in microgrid applications." Renew Sustain Energy Rev 2020; 131:110022.

[18] MA Hannan, et al. "The value of thermal management control strategies for battery energy storage in grid decarbonization": Issues and recommendations. J Cleaner Prod 2020; 276:124223.

[19] MA Hannan, MM Hoque, A Mohamed, A Ayob. "Review of energy storage systems for electric vehicle applications: Issues and challenges." Renew Sustain Energy Rev 2017; 69:771–89.

[20] B Shakerighadi, A Anvari-Moghaddam, J Vasquez, J Guerrero. "Internet of Things For modern energy systems": State-of-the-art, challenges, and open issues. Energies 2018; 11(5):1252.

Parameter Estimation of Hydrogen Fuel Cell using Metaheuristic Algorithm

Sohini Chaudhuri[1], Aurobinda Panda[2], Anjan Kumar Ray[3], Molay Roy[4], Omkar Singh[5], Avismit Dutta[6]

Department of Electrical & Electronics Engineering,

National Institute of Technology Sikkim, Ravangla, India

Email: m230015@nitsikkim.ac.in[1], aurobind.panda@nitsikkim.ac.in[2], akray.nits@gmail.com[3],

molay.roy@nitsikkim.ac.in[4], phee210005@nitsikkim.ac.in[5], phee210003@nitsikkim.ac.in[6]

Abstract—**This study presents an advanced approach of modelling proton exchange membrane fuel cells (PEMFCs) for electric vehicles (EVs) by employing adaptive particle swarm optimization (APSO) for parameter estimation. PEMFCs, known for their solid electrolytes, operational flexibility, fast startup, and rapid response to electrodynamic changes, pose significant modelling challenges due to complex interactions involving mass and heat transfer, electrochemical reactions, and energy losses. To overcome these challenges, APSO is utilized which adapts its search behaviour to achieve optimal results. This enhances the accuracy of the PEMFC models compared to conventional methods. Even in situations with parametric noise, the results show that APSO performs better than other optimization techniques in the estimation of the six primary PEMFC parameters. This demonstrates its resilience, sustainability for analysing noisy real-world data, and dependability in enhancing PEMFC modelling for electric vehicle applications.**

Index Terms—**Proton exchange membrane fuel, Fuel cells, APSO, Parameter estimation**

I. INTRODUCTION

Energy is a fundamental driver of economic and societal development [1]. As fossil fuel reserves continue to decline, renewable energy sources have gained prominence, offering a more environmentally sustainable alternative. Among various technologies, fuel cells, which directly convert chemical energy into electrical energy, are classified based on their fuel type, oxidant, electrolyte, operating temperature, and application domain.

Proton Exchange Membrane Fuel Cells (PEMFCs) are particularly attractive for electric vehicle applications, owing to their high efficiency and low emission profiles. In contrast, Solid Oxide Fuel Cells (SOFCs) operate at significantly higher temperatures and, although less compact, can achieve superior efficiencies when integrated with gas turbines and waste heat recovery systems. While SOFCs benefit from the use of cost-effective natural gas, PEMFCs typically exhibit lower capital costs. Despite challenges such as hydrogen's low volumetric energy density, PEMFCs provide the highest power-to-volume ratio among fuel cell types.

Nonetheless, the adoption of fuel cell technologies is hindered by several inherent challenges. In PEMFC research, modelling approaches can be broadly categorized as analytical,

semi-empirical, and mechanistic. A widely accepted semi-empirical model is Larminie's model, as described in [2], which offers an optimal balance between computational efficiency and model accuracy. This model is defined by several parametric equations whose accurate identification is critical. However, the optimization landscape associated with PEMFC parameter estimation is characterized by multiple local minima, strong nonlinearities, multi-variable coupling, and non-convexity, making conventional optimization techniques, such as conjugate gradient methods and quasi-Newton methods, largely ineffective.

This paper addresses the parameter estimation problem for a 1.2V PEMFC, highlighting that inaccurate parameter estimation can lead to significant inefficiencies and distortions in mathematical models and polarization curves. To address these challenges, this study employs Adaptive Particle Swarm Optimization (APSO), a metaheuristic optimization technique known for its robustness against premature convergence.

Over the years, several optimization methods have been applied to estimate PEMFC parameters, including the adaptive Chaotic Harris Hawks Optimization (CHHO) [3], the Flower Pollination Algorithm (FPA) [4], Genetic Algorithms (GA) [5], and the Harris Hawks Optimization (HHO) algorithm [1]. Other notable approaches include hybrid methods such as the Vortex Search Algorithm combined with Differential Evolution (VSA-DE) [6], the Jellyfish Search Algorithm (JFSA) [7], modified Artificial Ecosystem Optimization (AEO) [8], Differential Evolution (DE) [9], Particle Swarm Optimization (PSO) [10], the Rime-Ice Algorithm [11], the Modified Slime Mold Algorithm (MSMA) [12], and the Improved Atomic Orbital Search Algorithm (IAOSA) [13].

The effectiveness of PSO in PEMFC parameter identification has been demonstrated through structured experimentation, employing a population size of 40, learning factors $c_1 = 2$, $c_2 = 2$, and allowing up to 20,000 iterations [10]. However, as underscored by the No-Free-Lunch (NFL) theorem [14], no single optimization algorithm can outperform others across all possible problems, reinforcing the need for continuous exploration of new optimization strategies.

Each optimization technique presents unique strengths and weaknesses, and their performance depends heavily on problem-specific factors such as nonlinearity, separability, de-

979-8-3315-3013-6/25 $31.00 © 2025 IEEE

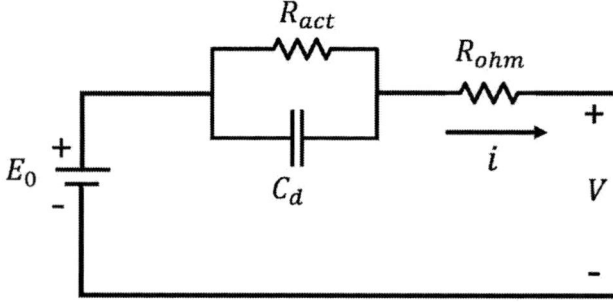

Fig. 1. Equivalent circuit of a proton exchange membrane fuel cell (PEMFC) [15].

sign variable degrees of freedom, and modality. Hence, the ongoing quest for more efficient parameter estimation methods continues to be a critical research frontier. Although various strategies have achieved satisfactory results in identifying both temperature-dependent and independent parameters, opportunities remain for further refinements to improve PEMFC model precision and reduce computational burdens.

Recent literature reviews reaffirm that parameter identification remains a dynamic and essential area of PEMFC research. New heuristic-based techniques are continually explored, aiming to minimize estimation errors, accelerate convergence rates, and enhance the statistical robustness of the models. Adaptive Particle Swarm Optimization (APSO) and other innovative heuristic methods represent promising advances in this field, striving to close existing performance gaps and support the broader adoption of PEMFCs as efficient and sustainable energy solutions, particularly in applications such as electric vehicles.

The principal contributions of this study can be summarized as follows:

- **A detailed mathematical model is developed to simulate the complex electrochemical phenomena in PEMFC stacks.**
- **Six critical parameters are optimized to minimize the Root Mean Square Error (RMSE) between actual and estimated voltages, maintaining tight tolerance limits.**
- **To prevent premature convergence and maintain a balanced exploration-exploitation trade-off (targeting over 40%), the PSO algorithm is enhanced with a linearly decreasing cognitive coefficient (c_1) and inertia weight (w), while ensuring that $c_1 + c_2$ remains constrained between 0.5 and 2.5.**
- **Experiments are conducted under both noise-free and noisy conditions (1%, 5%, 50%, and 100% voltage noise scenarios), where convergence stabilizes within approximately 2000 iterations.**

II. PROTON EXCHANGE MEMBRANE FUEL CELL MODEL

In this study, a well-known semi-empirical model of Proton Exchange Membrane Fuel Cells (PEMFCs), namely

Larminie's model [16], has been examined. A conventional PEMFC is typically composed of three primary components: the anode, electrolyte, and cathode.

At the anode, hydrogen molecules are separated into protons (H^+) and electrons (e^-) with the aid of a platinum or platinum-based catalyst. Other catalysts such as Iridium and Ruthenium are also currently utilized to improve performance. The protons migrate through a thin polymer electrolyte membrane, commonly made of Perfluorosulfonic acid, marketed as Nafion (DuPont). This membrane acts as a conductor for protons while effectively blocking the passage of electrons and gases. Meanwhile, the electrons are forced to travel through an external circuit, thus generating electrical energy.

At the cathode, a reduction reaction occurs where protons, electrons, and oxygen combine to form water, facilitated again by a platinum catalyst. This overall electrochemical process efficiently converts chemical energy into electrical power, producing water and heat as by-products. The fundamental chemical reactions for a hydrogen-based PEMFC are summarized as follows [15],

$$\text{Anode reaction:} \quad H_2 \rightarrow 2H^+ + 2e^- \quad (E_0 = 0.00 \text{ V}), \quad (1)$$

$$\text{Cathode reaction:} \quad 2H^+ + \frac{1}{2}O_2 + 2e^- \rightarrow H_2O \quad (E_0 = 1.23 \text{ V}), \quad (2)$$

$$\text{Overall reaction:} \quad H_2 + \frac{1}{2}O_2 \rightarrow H_2O. \quad (3)$$

As a result of this process, heat, liquid water, and direct current (DC) electrical power are produced. The ideal voltage of a PEMFC can be expressed as [15],

$$V = E_0 - V_{\text{ohm}} - V_{\text{act}} - V_{\text{con}}, \quad (4)$$

where V represents the fuel cell voltage, V_{ohm}, V_{act}, and V_{con} denote the respective voltage losses due to ohmic resistance, activation overpotential, and concentration effects, and E_0 refers to the open-circuit voltage. The theoretical open-circuit voltage (E_0) for a hydrogen-oxygen PEM fuel cell is approximately 1.2 V [15].

In practical operation, the actual voltage output is lower than E_0 due to unavoidable voltage losses. Each of the losses can be individually characterized:

A. Ohmic Losses

Ohmic losses arise from the resistance to the flow of electrons through the electrode material and ions through the electrolyte. This voltage drop can be modeled linearly with respect to the current density i:

$$V_{\text{ohm}} = ri, \quad (5)$$

where r is the area-specific resistance, typically measured in $\Omega \, \text{cm}^2$, and i is the current density in mA/cm². Minimizing r through material and structural optimization is essential for enhancing cell efficiency.

B. Activation Losses

Activation losses are associated with the energy barrier that must be overcome for electrochemical reactions to occur at the electrodes. These losses dominate at low current densities and are described by:

$$V_{act} = A \ln \left(\frac{i}{i_0} \right), \tag{6}$$

where A is the Tafel slope (in volts), and i_0 denotes the exchange current density in mA/cm^2.

The Tafel slope A itself is given by:

$$A = \frac{2.303 RT}{\alpha F}, \tag{7}$$

where R is the universal gas constant (8.314 J/mol · K), T the absolute temperature in Kelvin, α the charge transfer coefficient (approximately 0.5 for platinum catalysts), and F the Faraday constant (96485 C/mol) [17].

C. Concentration Losses

Concentration losses occur when the reactant concentration at the electrode surface drops as the reaction proceeds. These losses can be described by

$$V_{con} = -B \ln \left(1 - \frac{i}{i_L} \right), \tag{8}$$

where B is the mass transport overpotential constant (in volts), and i_L is the limiting current density (mA/cm^2).

Similarly, B is temperature-dependent and can be defined as:

$$B = \frac{2.303 RT}{nF}, \tag{9}$$

where n is the number of electrons involved in the reaction ($n = 2$ for hydrogen fuel cells) [18].

Additionally, losses may also arise from fuel crossover and internal currents, where some hydrogen may bypass the reaction or where electrons may leak through the electrolyte.

D. Polarization Model

By combining all aforementioned loss terms, the complete polarization model for a PEMFC can be represented as [15],

$$V = E_0 - r(i + i_n) - A \ln \left(\frac{i + i_n}{i_0} \right) + B \ln \left(1 - \frac{i + i_n}{i_L} \right), \tag{10}$$

where i_n accounts for the internal crossover and leakage currents.

From the above equation, it can be observed that the parameter identification challenge is reduced to the estimation of six parameters: E_0, r, i_n, i_0, A, and B. The physical significance of these parameters are presented in Table I.

TABLE I
PHYSICAL SIGNIFICANCE OF PARAMETERS IN LARMINIE'S PEM FUEL
CELL MODEL [16].

Parameter	Representation	Physical cause
E_0	Open circuit voltage	Thermodynamics
r	Ohmic losses	Ionic/electronic resistance
i_n	Leakage/crossover current	H_2 crossover, internal shorts
i_0	Exchange current density	Reaction kinetics
A	Tafel slope	Activation losses
B	Mass transport overpotential	Diffusion/transport limits

E. Fuel Cell Stack Modeling

To meet higher power demands, multiple single PEM fuel cells are stacked together. The total stack voltage V_s can be calculated as:

$$V_s = NV, \tag{11}$$

where N is the number of fuel cells connected in series, and V represents the voltage of a single cell.

Stacking enables scalable power generation while maintaining high efficiency and is commonly implemented in applications such as electric vehicles, stationary power generation, and portable energy systems. However, stack performance is influenced by factors such as internal resistances, reactant flow distribution, and thermal management, necessitating careful design optimization to ensure maximum performance and durability.

In the forthcoming sections, identification of the model parameters are presented, which are crucial for the precise simulation and optimization of PEMFC systems.

III. PARAMETER ESTIMATION PROCEDURE

Accurate measurement of the voltage versus current (V–I) curve is considered fundamental for every proton exchange membrane fuel cell (PEMFC). This procedure is routinely conducted in both industrial and laboratory environments. From the V–I curve obtained through experiments, an optimization algorithm can be employed to estimate the essential parameters of a PEMFC. Since the identification of PEMFC parameters is a goal-specific task, the formulation of a suitable performance criterion or objective function becomes imperative prior to initiating the optimization procedure.

In this study, the voltage error function is expressed as [10], [21]:

$$f(V_s, i, \boldsymbol{\theta}) = V_s - N \left[E_0 - r(i + i_n) - A \ln \left(\frac{i + i_n}{i_0} \right) + B \ln \left(1 - \frac{i + i_n}{i_L} \right) \right], \tag{12}$$

where $\boldsymbol{\theta} = [E_0, A, i_n, i_0, r, B]$ denotes the set of unknown physical parameters of the fuel cell model.

The function $f(V_s, i, \boldsymbol{\theta})$ effectively captures the error between the experimental voltage and the model-estimated voltage. Thus, the goal is to minimize this error, thereby ensuring that the electrical model accurately represents the real system.

Considering the root mean square error (RMSE) criterion, the objective function F is defined as:

$$F = \sqrt{\frac{1}{m} \sum_{k=1}^{m} [f(V_s(k), i(k), \boldsymbol{\theta})]^2}, \qquad (13)$$

where $V_s(k)$ and $i(k)$ represent the experimental voltage and current data points respectively, $\boldsymbol{\theta}$ the unknown parameters, and m the total number of data points.

The parameter estimation problem is thus transformed into a minimization problem for F. As can be observed from Equation (12), the objective function is nonlinear and possesses a unique global minimum without being strictly quadratic, making the problem susceptible to local minima when conventional techniques are applied [19].

To address the aforementioned challenges, an adaptive particle swarm optimization (APSO) algorithm is utilized, leveraging its superior balance between exploration and exploitation phases. An effective metaheuristic algorithm should prioritize exploration during early iterations and transition towards exploitation in later stages.

Standard PSO methods, although incorporating elitism by preserving the global best (gbest) particle, often suffer from premature convergence and local trapping [20]. Therefore, adaptive improvements are incorporated. Inspired by swarm intelligence, PSO optimizes the objective function iteratively, with each particle updating its velocity based on its personal best (pbest) and the global best (gbest) experiences.

The velocity update rule for particle j at iteration t in D-dimensional space is given by:

$$v_j(t+1) = w(t)\, v_j(t) + c_1\, r_{1j}(t)\, (pbest_j(t) - x_j(t))$$
$$+ c_2\, r_{2j}(t)\, (gbest(t) - x_j(t)) \qquad (14)$$

The corresponding position update is described by:

$$x_j(t+1) = x_j(t) + v_j(t+1) \qquad (15)$$

where, $r_{1j}(t)$ and $r_{2j}(t)$ are random numbers uniformly distributed in $[0, 1]$ for each dimension.

The inertia weight $w(t)$, along with cognitive constants c_1 and c_2, plays a crucial role in controlling the search behavior. A large inertia weight promotes global exploration, whereas a smaller one facilitates local exploitation. In this study, the inertia weight is linearly decreased according to:

$$w(t) = w_{\max} - \left(\frac{w_{\max} - w_{\min}}{t_{\max}}\right) t \qquad (16)$$

The adoption of a dynamically decreasing inertia weight and cognitive constants enhances the PSO performance by minimizing premature convergence and promoting solution accuracy. Furthermore, the fitness of each particle is evaluated using the objective function F defined in Equation (13), where a lower F value indicates a better solution.

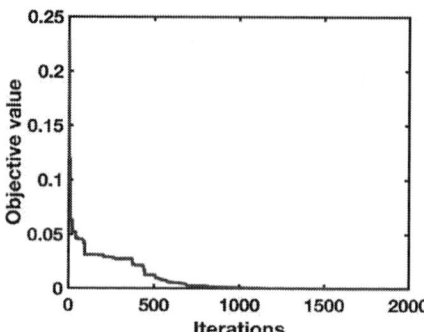

Fig. 2. **Convergence curve showing best objective values (RMSE) over iterations using APSO at 70° C.**

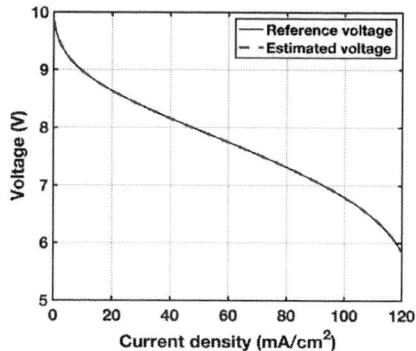

Fig. 3. **Voltage vs. current density curve for the reference and estimated voltage at 70° C.**

IV. RESULTS AND DISCUSSION

The Adaptive Particle Swarm Optimization (APSO) algorithm was employed to estimate the parameters of the Proton Exchange Membrane Fuel Cell (PEMFC) model by minimizing the root mean square error (RMSE) between the simulated voltage output and a reference voltage profile generated using known model parameters. The reference current density profile was taken from experimental data, and the corresponding voltage values were generated using a set of parameters obtained from the literature [10], [21]. The optimization was carried out using a swarm of 40 particles for 2000 iterations. The $w_{\max} = 0.9$ and $w_{\min} = 0.4$. The cognitive coefficient c_2 is linearly reduced from 2.0 to 0.5 across the iterations, while c_1 is kept constant at 2.0. This ensures that $c_1 + c_2$ remains within the optimal range of $[2.5, 4]$ to balance exploration and exploitation [20]. The optimal estimated PEMFC model parameters are: open circuit voltage $E_0 = 0.9979$ V, Tafel slope $A = 0.0350$ V, internal crossover current $i_n = 1.2002$ A/cm^2, exchange current density $i_0 = 0.8869$ A/cm^2, area-specific resistance $r = 0.0006\ \Omega \cdot$cm^2, and mass transport overpotential coefficient $B = 0.0600$ V.

The convergence behavior of the APSO algorithm is depicted in Figure 2. A significant decrease in RMSE is observed during the early stages of the optimization, indicating efficient

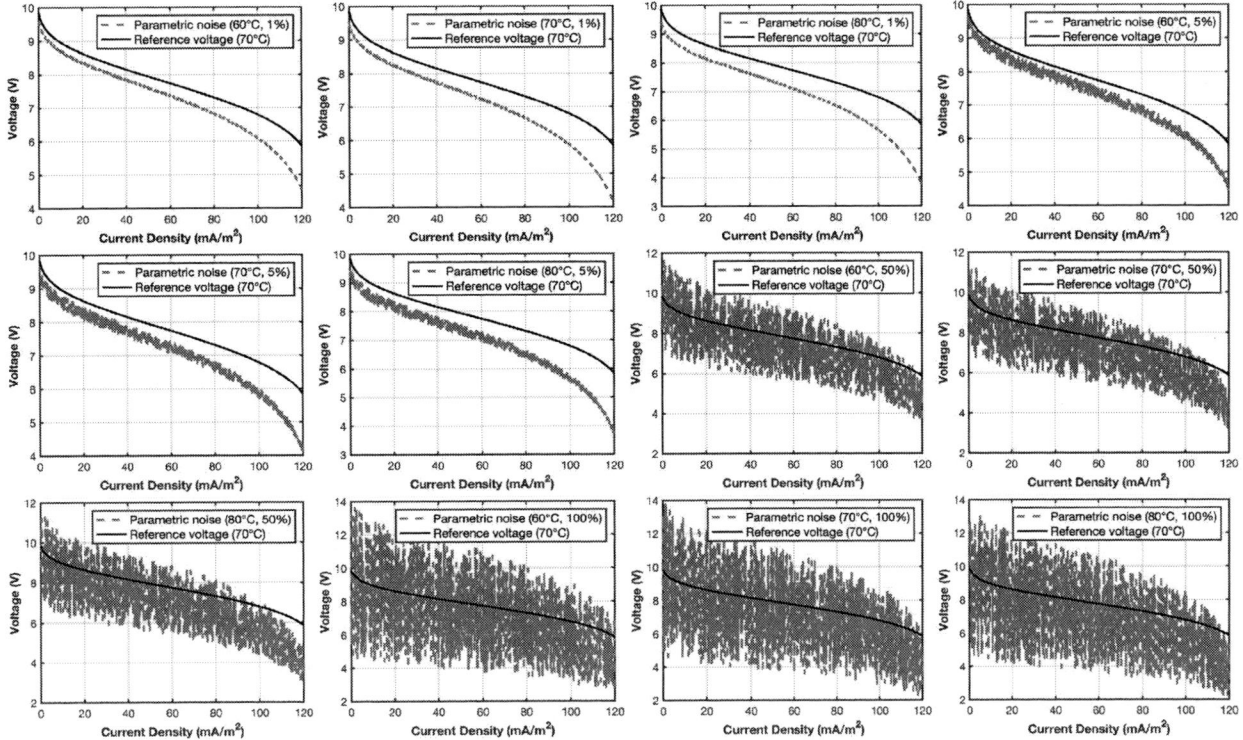

Fig. 4. **Voltage vs. current density curves at different temperatures with parametric noise.**

global exploration. As iterations progress, the RMSE continues to reduce gradually, reflecting the algorithm's shift toward local exploitation and solution refinement. The final RMSE achieved was less than 10^{-4}, which demonstrates the high accuracy and convergence capability of the proposed method.

Figure 3 compares the voltage output computed using the optimized parameters with the reference voltage profile. The close agreement between the two curves over the entire range of current densities confirms that the estimated parameters successfully capture the electrochemical behavior of the fuel cell. The model accurately predicts the voltage drop at high current densities, as well as the activation and ohmic losses at lower ranges, validating both the structure of the PEMFC model and the effectiveness of the APSO algorithm.

The dynamic adjustment of APSO parameters, including linearly decreasing inertia weight and cognitive constants, contributed to avoiding premature convergence and local optima. The algorithm maintained a balanced trade-off between exploration and exploitation throughout the optimization process, leading to accurate and reliable parameter estimation.

To validate the effectiveness of the developed Adaptive Particle Swarm Optimization (APSO) approach for PEMFC parameter estimation, a comparative study was conducted against the standard PSO and Genetic Algorithm (GA) results reported in [10], [21]. The RMSE values for different operating temperatures (60°C, 70°C, and 80°C) were evaluated and summarized in Table II.

TABLE II
COMPARISON OF RMSE VALUES AT DIFFERENT TEMPERATURES

Sl No	Temperature (°C)	APSO	PSO [10], [21]	GA [10], [21]
1	60	4.56×10^{-7}	4.039×10^{-3}	2.265×10^{-2}
2	70	5.91×10^{-5}	3.797×10^{-3}	2.012×10^{-2}
3	80	3.23×10^{-7}	5.479×10^{-3}	1.115×10^{-2}

The results clearly demonstrate the superiority of the proposed APSO method over conventional PSO and GA algorithms. At all operating temperatures, the APSO achieved RMSE values several orders of magnitude lower than those reported for PSO and GA in [10], [21]. This indicates that the adaptive mechanism incorporated in the PSO significantly enhances its exploration and exploitation capabilities, resulting in much more accurate parameter estimation for the PEMFC model.

In addition, the convergence behavior exhibited by APSO confirms the robustness and stability of the proposed optimization framework, making it highly suitable for practical applications in PEMFC system modeling and control.

To evaluate the reliability and robustness of the proposed PEMFC parameter estimation system under varying operating conditions, a series of tests were conducted. The primary focus was on the effect of temperature variations (60°C, 70°C, 80°C) and noise levels (1%, 5%, 50%, and 100%) on the

TABLE III
ROBUSTNESS FOR DIFFERENT TEMPERATURE AND PARAMETRIC NOISE LEVELS

Temperature	Parametric noise	RMSE
60°C		0.5351
70°C	1%	0.7018
80°C		0.8747
60°C		0.5428
70°C	5%	0.7131
80°C		0.8812
60°C		1.1897
70°C	50%	1.2866
80°C		1.3048
60°C		2.1807
70°C	100%	2.1998
80°C		2.2224

model's performance. The model parameters were adjusted for each operating temperature based on experimentally verified trends. These temperature-dependent variations were modeled by considering the temperature sensitivity of the exchange current density i_0 (Arrhenius behavior) and the linear adjustment for parameters like E, r, and B [22]. Additionally, measurement noise was introduced to simulate real-world conditions, as per the methodology described in [23]. The results of this robustness analysis are summarized in Table III, showing the RMSE values under different temperature and noise conditions. As demonstrated in the results, the proposed system shows high accuracy even under varying levels of noise and temperature fluctuations, making it a robust solution for PEMFC parameter estimation.

CONCLUSION

In this work, a particle swarm optimization (PSO) algorithm was successfully implemented to estimate the key parameters of a proton exchange membrane fuel cell (PEMFC) model. The PSO utilized an adaptive mechanism for the inertia weight and acceleration coefficients to enhance convergence over 2000 iterations. The fitness function minimized the root mean square error (RMSE) between the model-predicted voltage and a reference voltage profile generated using known parameters. The results demonstrated effective convergence, with the estimated parameters closely matching the true values. Furthermore, the model's voltage output showed excellent agreement with the reference data across multiple operating temperatures (60°C, 70°C, and 80°C) and at various noise levels. The implemented PSO framework proves to be a robust and efficient tool for parameter estimation in PEMFC systems, offering a reliable approach for optimizing model accuracy under different operating conditions. Future improvements could include refining the temperature dependency in the voltage model and exploring alternative boundary-handling strategies to further enhance the optimizer's performance.

REFERENCES

[1] A. A. Hyderi, S. Mirjalili, H. Faris, I. Aljarah, M. Mafarja, and H. Chen, "Harris Hawks optimization: Algorithm and applications," *Future Generation Computer Systems*, vol. 97, pp. 849–872, 2019.

[2] K. J. Runtz and M. D. Lyster, "Fuel cell equivalent circuit models for passive mode testing and dynamic mode design," in *Canadian Conference on Electrical and Computer Engineering*, Saskatoon, SK, Canada, 2005, pp. 794–797.

[3] A. S. Menesy, H. M. Sultan, A. Selim, M. G. Ashmawy, and S. Kamel, "Developing and applying chaotic Harris Hawks optimization technique for extracting parameters of several proton exchange membrane fuel cell stacks," *IEEE Access*, vol. 8, pp. 1146–1159, 2020.

[4] K. Priya and N. Rajasekar, "Application of flower pollination algorithm for enhanced proton exchange membrane fuel cell modelling," *International Journal of Hydrogen Energy*, vol. 44, no. 33, pp. 18 438–18 449, 2019.

[5] I. Mohamed and N. Jenkins, "Proton exchange membrane (PEM) fuel cell stack configuration using genetic algorithms," *Journal of Power Sources*, vol. 131, no. 1–2, pp. 142–146, 2004.

[6] A. Fathy, M. A. Elaziz, and A. G. Alharbi, "A novel approach based on hybrid vortex search algorithm and differential evolution for identifying the optimal parameters of PEM fuel cell," *Renewable Energy*, vol. 146, pp. 1833–1845, 2020.

[7] E. A. Gouda, M. F. Kotb, and A. A. El-Fergany, "Jellyfish search algorithm for extracting unknown parameters of PEM fuel cell models: Steady-state performance and analysis," *Energy*, vol. 221, p. 119836, 2021.

[8] A. S. Menesy, H. M. Sultan, A. Korashy, F. A. Banakhr, M. G. Ashmawy, and S. Kamel, "Effective parameter extraction of different polymer electrolyte membrane fuel cell stack models using a modified artificial ecosystem optimization algorithm," *IEEE Access*, vol. 8, pp. 31 892–31 909, 2020.

[9] W. Gong and Z. Cai, "Parameter optimization of PEMFC model with improved multi-strategy adaptive differential evolution," *Engineering Applications of Artificial Intelligence*, vol. 27, pp. 28–40, 2014.

[10] M. Ye, X. Wang, and Y. Xu, "Parameter identification for proton exchange membrane fuel cell model using particle swarm optimization," *International Journal of Hydrogen Energy*, vol. 34, no. 2, pp. 981–989, 2009.

[11] A. A. Ismaeel, E. H. Houssein, D. S. Khafaga, E. A. Aldakheel, and M. Said, "Performance of rime-ice algorithm for estimating the PEM fuel cell parameters," *Energy Reports*, vol. 11, pp. 3641–3652, 2024.

[12] A. S. Menesy, H. M. Sultan, M. E. Zayed, I. O. Habiballah, S. Dmitriev, M. Safaraliev, and S. Kamel, "A modified slime mold algorithm for parameter identification of hydrogen-powered proton exchange membrane fuel cells," *International Journal of Hydrogen Energy*, vol. 86, pp. 853–874, 2024.

[13] O. B., "Optimal parameter estimation of PEMFC model using an improved atomic orbital search algorithm," *Applications of Modelling and Simulation*, vol. 8, pp. 283–300, 2024.

[14] D. H. Wolpert and W. G. Macready, "No free lunch theorems for optimization," *IEEE Transactions on Evolutionary Computation*, vol. 1, no. 1, pp. 67–82, April 1997.

[15] Y. Duan, J. Chen, T. Zhang, and L. Li, "An improved semi-empirical model of PEM fuel cells for real-time applications," *Energies*, vol. 15, no. 9, p. 3184, 2022.

[16] J. Larminie and A. Dicks, *Fuel Cell Systems Explained*, 2nd ed. John Wiley & Sons, 2003.

[17] F. Barbir, *PEM Fuel Cells: Theory and Practice*, 2nd ed. Elsevier, 2013.

[18] D. Gerteisen and C. Ziegler, "Modelling of mass transport losses in PEM fuel cells using a semi-empirical approach," *International Journal of Hydrogen Energy*, vol. 35, no. 22, pp. 12 295–12 303, 2010.

[19] M. Izadikhah and R. F. Saen, "A fuzzy goal programming approach for solving nonlinear programming problems," *Computers & Industrial Engineering*, vol. 62, no. 4, pp. 930–936, 2012.

[20] M. Clerc, "Particle swarm optimization," in *Evolutionary Optimization*. Springer, 2002, pp. 287–319.

[21] U. Mitra, A. Arya, and S. Gupta, "A comprehensive and comparative review on parameter estimation methods for modelling proton exchange membrane fuel cell," *Fuel*, vol. 335, p. 127080, 2023.

[22] R. Zhang, T. Jiang, and G. Chen, "Improved models for the prediction of PEM fuel cell voltage at high temperatures and high currents," *Energy Reports*, vol. 11, pp. 95–103, 2025.

[23] X. Song, H. Wang, and F. Yu, "Impact of noise and measurement uncertainties on the parameter estimation of PEM fuel cells using optimization algorithms," *Applied Energy*, vol. 342, p. 118287, 2024.

979-8-3315-3013-6/25 $31.00 © 2025 IEEE

New Systematic Design Methodology of Modular Multilevel Converters in Rectifier Applications

Meddi Tharun[1,2], Ashok Mankani[1], Kumar Saurabh[1], Aritra Chakraborty[1,2], Aditya Naugraiya[1,2], Amal S[1]

1) Institute For Plasma Research, Gandhinagar, 382428, India
2) Homi Bhabha National Institute, Training School Complex, Mumbai, 400094, India.
meddi.tharun@ipr.res.in, ashok@ipr.res.in, saurabh@ipr.res.in, aritra.chakraborty@ipr.res.in,
aditya.naugraiya@ipr.res.in, amal@ipr.res.in .

Abstract— The Modular Multilevel Converter (MMC) is a highly promising topology for high-voltage and high-power applications. Its key advantages include scalability, the ability to achieve high voltages using low-rated switches, low switching frequency, and reduced harmonic distortion due to its multilevel feature. Selecting the appropriate submodule(SM) capacitance, switching device ratings, arm inductance, and input transformer is crucial for achieving a compact footprint, efficient operation, and cost-effectiveness in hardware implementation. This selection process can be performed through time-intensive iterative simulations or using a one-time simulation with analytical support. While various procedures exist for selecting submodule capacitors and arm inductors, but there are few discussions on systematic selection of components by considering parameter interdependencies. This paper introduces new design methodology for MMC in rectifier applications by refining existing analytical expressions. A new approach for determining submodule capacitance, based on charge fluctuation calculations, is proposed. The theoretical findings are validated through simulations conducted in MATLAB/Simulink®.

Keywords— *Modular Multilevel Converter; Rectifier; Phase Shift Carrier Modulation; Capacitor Voltage Balancing; Circulating Current; Source Current Controller.*

NOMENCLATURE

x	Phase Index
k	Submodule(SM) Index
N	Number of SMs per Arm
N_{xu}, N_{xl}	Number of turned on SMs in upper , lower arm.
L_{arm}	Arm Inductance
R_{arm}	Arm Resistance
C_{sm}	SM capacitor
F_{ref}	Reference Frequency
F_c	Carrier Frequency
V_c	SM average Capacitor Voltage
I_{xu}, I_{xl}	Upper, Lower arm instantaneous current
I_{xd}	Average leg current
I_{xz}	Peak second harmonic circulating current

I_{xs}	Source rms current
φ_x	Phase Angle of phase current with reference to corresponding phase voltage
$\in$	SM Capacitor ripple
θ_x	Second Harmonic phase angle
V_{xs}	Source rms voltage
V_{dc}	Internal DC Voltage
$V_{out,dc}$	Output DC voltage
I_{dc}	Output DC Current
R_L	Load Resistance

I. INTRODUCTION

The Modular Multilevel Converter (MMC) was proposed by Prof. Rainer Marquardt in 2001 revolutionized power electronics, particularly in High Voltage DC (HVDC) transmission systems, marking a milestone in power conversion technology [1]. The main attractive features of MMC are easy scalability, ability to build high voltages using low voltage switches, low switching frequency and low harmonic distortion due to multilevel nature. Because of its flexibility in voltage and power handling, MMC has gained interest in both ac & dc applications like power transmission [1], medium voltage motor drives [2], renewable applications [3], electric transportation [4], accelerator power supplies [5] etc.

Irrespective of application, the optimal component selection plays an important role in achieving compact footprint, efficient operation and cost-effective implementation. There are different SM capacitor and arm inductor selection procedures, but there are few discussions on selection procedure handling parameter interdependencies. The SM capacitance based on arm energy fluctuations has been derived in [6][7], but not accurate enough as the effect of circulating current is not considered. The SM capacitance selection in [8], requires entire system information, most of the parameters are unknown at design phase. Reference [9], introduced the term equivalent discharging time constant, a general index for SM capacitance selection and comparison, which is derived from [6], not considering the effect of circulating current. The circulating current will not be part of energy conversion and will be circulated within the three phase-legs. The presence of circulating current will increase the current rating of the

979-8-3315-3013-6/25 $31.00 © 2025 IEEE

components and decreases the efficiency of the system due to increased losses and thermal stress. The undesirable circulating current can be eliminated by using the circulating current controller but it will increase the output voltage ripple due to the common mode voltage injection in the upper and lower arm [10]. Thus, the inclusion of circulating current effect is very important in SM capacitance calculations for rectifier operation but this is unknow at the component selection phase as it depends on the both arm inductance and SM capacitance. The novelty of this paper lies in the inclusion of the circulating current in the components selection by eliminating parameter interdependency. The proposed design methodology calculates the arm inductance and SM capacitance with limited inputs. The input voltage and current requirements are determined analytically and validated through simulations, aiding the selection of appropriate input transformer. This paper also introduces a novel method for determining submodule (SM) capacitance based on SM capacitor charge fluctuations by considering parameter dependencies.

Rest of the paper is organized into four sections. Section II discusses the assumptions and governing equations used in the analysis. Section III presents the proposed SM capacitance selection method and outlines the sequence of steps for component selection. Section IV describes the simulation setup and provides a detailed analysis of the simulation results. Finally, the study's conclusions are summarized in Section V.

II. MMC RECTIFIER

A three phase MMC working as a rectifier with half bridge submodules is shown in Fig. 1. It has been assumed that input supply is balanced and the components used are identical in all the three phase legs. Under these conditions, the upper and lower arm current equations as per the given reference current directions are given as,

$$i_{xu}(t) = -\frac{\sqrt{2}}{2}I_{xs}\sin(\omega_1 t + \varphi_x) - I_{xd} + \cdots$$
$$I_{xz}\sin(2\omega_1 t + \theta_x) \quad (1)$$

$$i_{xl}(t) = \frac{\sqrt{2}}{2}I_{xs}\sin(\omega_1 t + \varphi_x) - I_{xd} + \cdots$$
$$I_{xz}\sin(2\omega_1 t + \theta_x) \quad (2)$$

Where, $x \in \{r, y, b\}$ represents the phase index. The average switching function of the upper and lower arm SM are given as [11],

$$s_{xu} = \frac{1}{2}(1 - \sin(\omega_1 t - \phi_x)) \quad (3)$$

$$s_{xl} = \frac{1}{2}(1 + \sin(\omega_1 t - \phi_x)) \quad (4)$$

Where $\phi_x = \{0, \frac{2\pi}{3}, \frac{4\pi}{3}\}$ represents the phase angles between the phases in radian. The current through the SM capacitor is then given as,

$$i_{cxu} = s_{xu}i_{xu} \quad (5)$$

$$i_{cxl} = s_{xl}i_{xl} \quad (6)$$

The voltage of the SM capacitors will be stable if the dc component of the above equation is zero [12]. This condition gives the relation between input phase rms current and output dc current under unity modulation index is given as,

$$I_{xs} = \frac{4}{3}\frac{I_{dc}}{\cos(\varphi_x)\sqrt{2}} \quad (7)$$

The multipurpose arm inductor serves to control the circulating current, inrush and fault currents. For economical design, the recommended arm inductance is given as [9],

$$L_{arm} = \frac{N}{4\omega_1^2 C_{sm}} \quad (8)$$

The circulating current expression as per the given reference input and output current directions is given as [11],

$$I_{xz} = \frac{\sqrt{(A\cos(\varphi_x) + B)^2 + (A\sin(\varphi_x))^2}}{1 - K_a - K_b} \quad (9)$$

$$\theta_x = \arctan\big(A\cos(\varphi_x) + B, -A\sin(\varphi_x)\big) \quad (10)$$

$$A = \frac{-3\sqrt{2}}{64}\frac{NI_{xs}}{\omega_1^2 C_{sm}L_{arm}}, B = \frac{N}{16}\frac{I_{xd}}{\omega_1^2 C_{sm}L_{arm}}.$$

$$K_a = \frac{N}{16\omega_1^2 C_{sm}L_{arm}}, K_b = \frac{N}{24\omega_1^2 C_{sm}L_{arm}}.$$

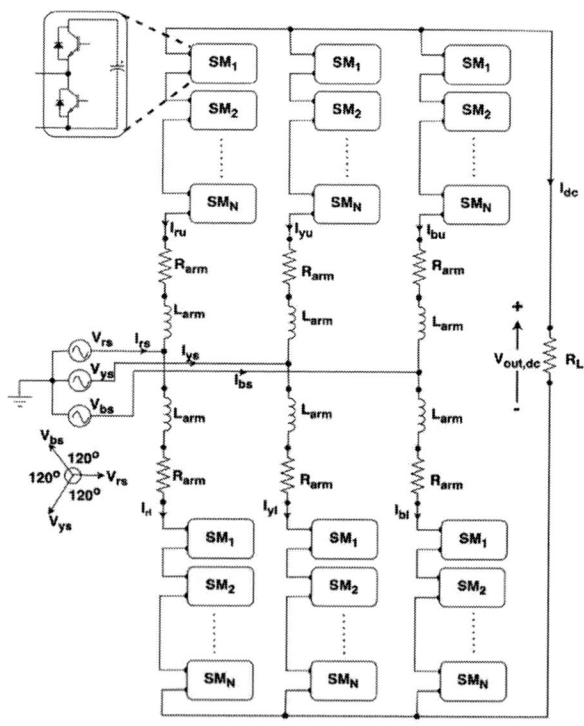

Fig. 1. Schematic Diagram of MMC working as Rectifier

By substituting (8), in above constants,

$$A = \frac{-3\sqrt{2}}{16} I_{xs}, B = \frac{1}{4} I_{xd}.$$

$$K_a = \frac{1}{4}, K_b = \frac{1}{6}.$$

By this, the circulating current magnitude and its phase can be calculated without submodule capacitance and arm inductance values. The required source voltage can be given as [5],

$$v_{xs}(t) = \frac{V_{dc}}{2}\sin(\omega_1 t - \phi_x) + \frac{L_{arm}}{2}\frac{di_{xs}}{dt} + \frac{R_{arm}}{2} i_{xs} \quad (13)$$

$$V_{xs} = \sqrt{\left(\frac{V_{dc} + I_{xs}R_{arm}}{2}\right)^2 + \left(\frac{I_{xs}\omega_1 L_{arm}}{2}\right)^2} \quad (14)$$

Where, the mean voltage of internal DC voltage V_{dc} is given as [5],

$$V_{dc} = V_{out,dc} + 2\frac{I_{dc}}{3} R_{arm}$$

By using this, the steady state SM mean capacitor voltage can be calculated as,

$$V_{cap} = \frac{V_{dc}}{N}$$

The arm resistance is used to represent the losses in the system and a value of 500 $m\Omega$ has been considered in this study.

III. New SM Capacitor Selection and Rectifier Design procedure

Assuming all the SM capacitors are loaded equally, the ac components of the SM capacitor voltage can obtained by integrating (5) and dividing the result by capacitance C_{sm}, as given below [11],

$$\Delta v_{cu} = \frac{\sqrt{2}I_{rs}}{4w_1 C_{sm}}\cos(\omega_1 t + \varphi_r) - \frac{I_{rd}}{2w_1 C_{sm}}\cos(w_1 t) -$$
$$\frac{I_{rz}}{4w_1 C_{sm}}\sin(\omega_1 t + \theta_r) - \frac{\sqrt{2}mI_{rs}}{16w_1 C_{sm}}\sin(2\omega_1 t + \varphi_r)$$
$$-\frac{I_{rz}}{4w_1 C_{sm}}\cos(2\omega_1 t + \theta_R) + \frac{I_{rz}}{12w_1 C_{sm}}\sin(3\omega_1 t + \theta_r)$$
$$(11)$$

The instantaneous charge fluctuations can be obtained by,

$$\Delta q_{cu} = \Delta v_{cu} C_{sm}$$

Assuming charge fluctuations are symmetrical about its mean value, the required capacitance can be obtained as,

$$C_{sm} = \frac{2\max(\Delta q_{cu})}{\in V_c} \quad (12)$$

Where, ripple $\in$ is defined as $\frac{V_{cmax} - V_{cmin}}{V_c}$. To calculate the arm inductance, there is an dependency of SM capacitance as shown in (8). The calculation of SM capacitance depends on the circulating current, which in turn is influenced by both the arm inductance and SM capacitance [13]. To resolve this interdependency, circulating current calculations must be made independent of arm inductance and SM capacitance. This can be achieved by using the modified constants discussed in the previous section.

The systematic sequence of parameter calculations to avoid interdependencies is shown in Fig. 2. The source current is calculated from the output current and operating power factor requirements, which will influence directly/indirectly SM capacitance, arm inductance and source voltage requirements. The effect of circulating current is considered in arm inductance and SM capacitance calculations by calculating it with minimal information. The SM capacitance is calculated based on proposed method followed by arm inductance. The source phase voltage is then calculated from (13) and the product of (13),(7) provides the required VA rating per phase of the input transformer.

IV. Simulation Results

A 4 kV, 100 A DC power supply was designed using above procedure, and the theoretical results were compared with simulation outcomes. The schematic diagram of the pulse generation for one phase is shown in Fig. 3. To achieve the desired power factor, an external current controller [14] was employed. The gate pulses for the SMs were generated using Phase-Shifted Carrier (PSC) modulation, with the modulating signal from the controller.

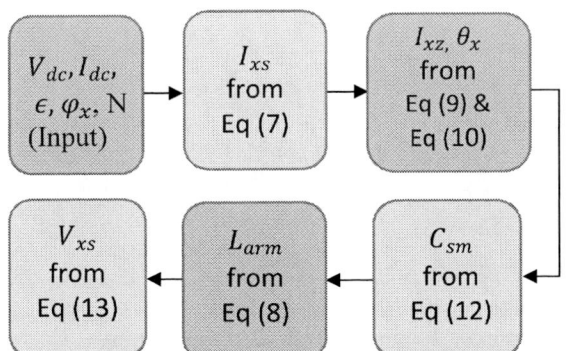

Fig. 2: Proposed Design Procedure.

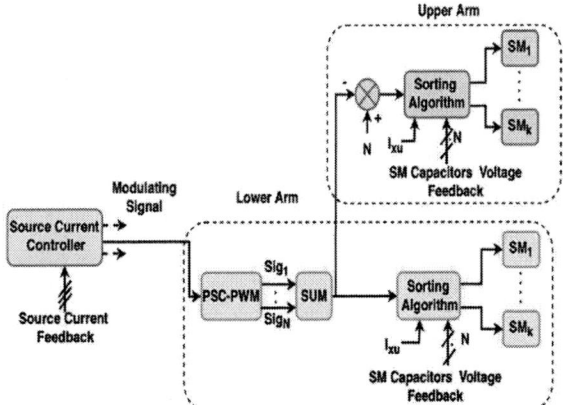

Fig. 3. Schematic diagram of pulse generation for one phase

The phase shifted carrier modulation is a PWM technique, consists of multiple identical carriers in each arm with phase shift among them, with number of carriers is equal to the number of SMs. Based on performance requirements, there can be different phase shifts between the carriers of upper and lower arm in a leg.

The phase shifting of the carrier waveforms for the lower and upper arms is given as,

$$\theta_{Lk} = \frac{2\pi}{N}(k-1)$$

$$\theta_{Uk} = \theta_{Lk} + \beta$$

Where $k \in \{1, 2, \dots N\}$. To minimize the distortion in the DC side, the carriers in the upper and lower arm should be phase shifted by π radian [15] i.e., $\beta = \pi$. To obtain perfect balancing of the distributed SM capacitors, traditional sorting algorithm [14] has been considered. The selection of number of sub modules per arm (N) depends on the voltage rating of SM capacitors, SM Sensors, SM switches and its driver card. Considering a 1200 V SM switch with 40% safety margin, six SMs per arm have been considered and to reduce the impact of harmonics, which is not considered in theoretical study, a higher carrier frequency of 5000 Hz has been considered.

Table 1: MMC Simulation Parameters

S.no.	Parameters	Value
1	Source peak phase voltage, V_{Sx}	2055 V
2	Arm Inductance, L_{arm}	6.87 mH
3	Arm Resistance, R_{arm}	0.5 Ω
4	SM Capacitance, C_{sm}	2.21 mF
5	Number of SMs, N	6
6	Nominal Capacitor Voltage	672.22 V
7	Capacitor Ripple Voltage	100.83 V(15%)
8	Carrier Frequency (F_c)	5000 Hz
9	Reference Frequency (F_{ref})	50 Hz
10	Rated load voltage	4000 V
11	Load Resistance	40 Ω

The simulation time can be further reduced by generating pulses either just from lower/upper arm and reuse them for upper/lower arm pulse generation using following equations,

$$N_{Lx} = \frac{N}{2}(1 + \sin(\omega_1 t - \phi_x)) = N - N_{Ux}$$

$$N_{Ux} = \frac{N}{2}(1 - \sin(\omega_1 t - \phi_x)) = N - N_{Lx}$$

Note that this condition is only valid if there is no variable common mode voltage in the modulating signal, like the case seen with circulating current controller. The few considered and calculated simulation parameters are listed in Table 1. The simulations were performed with fixed step size of 5 μs in Matlab Simulink® [16]. The effectiveness of the new selection procedure of SM capacitance is verified by comparing the analytical and simulated voltage waveform across the capacitor as shown in Fig. 4.

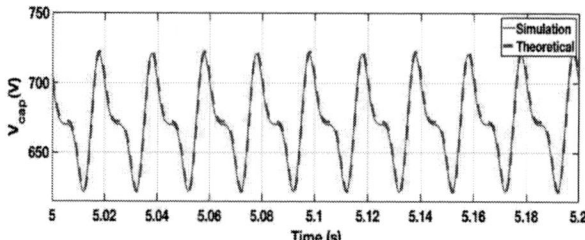

Fig. 4. Steady State Upper Arm Capacitor Voltage

The controller complexity increases with increase in N. The simulation verification of source current shown in Fig.5, highlighting the unity power factor operation, with the current being zero and increasing for every integer multiple of 20ms for r-phase. The simulated waveform of circulating current shown in Fig. 6, highlights the appropriate calculation of magnitude and phase.

Table 2: Comparison of few Theoretical and Simulated parameters

Parameter	Theoretical	Simulation	Error
rms Source Current	94.28 A	94.26 A	0.02%
rms Circulating Current	20.20 A	19.89 A	1.53%
rms arm current	61.16 A	61.34 A	-0.29%
SM Mean Capacitor	672.22 V	671.1 V	0.16%
SM Capacitor Ripple	100.83 V	99.09 V	1.73%

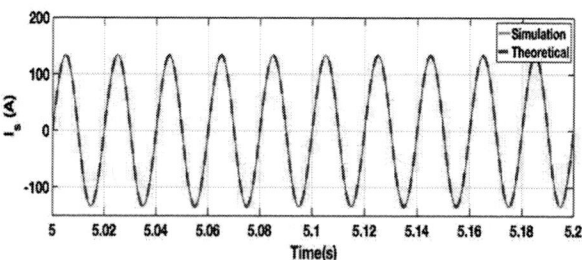

Fig. 5: Steady State Source Current of r-phase

The theoretical and simulated values of few internal, external quantities with their errors with respect to its theoretical value are listed in Table 2. The error in circulating current and SM capacitor ripple is significant in comparison to the other parameters due to the missing fourth and higher harmonic components of circulating current. The increase in error in circulating current increases the error of capacitor voltage ripple because of their direct relation as shown in eq(11). Its error can be further decreased by including the next dominant, fourth harmonic component of the circulating current in the analysis. The less error in the source current highlights the low THD, which is one of the main attributes of the MMC.

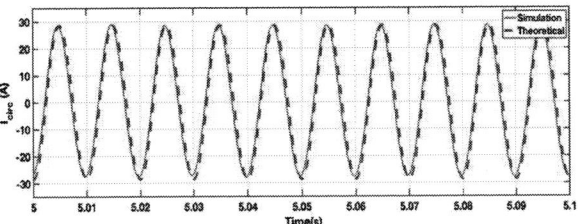

Fig. 6: Steady State Circulating Current

Comparison of r-phase Leg.

The waveforms of the upper and lower arm SM capacitor voltages are shown in Fig .7 and Fig. 8, highlighting the symmetrical loading of the upper, lower arms and the all three legs.

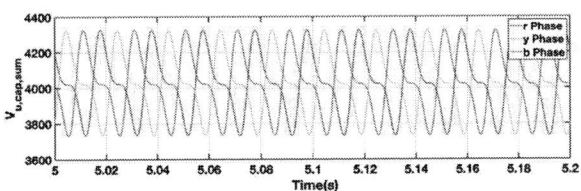

Fig. 7. Steady State upper arm capacitor voltage sum

of all three phases.

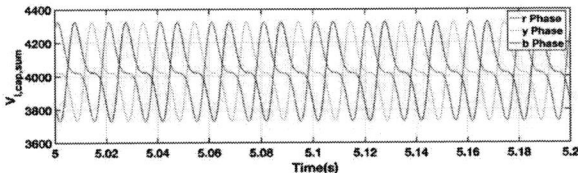

Fig. 8. Steady State lower arm capacitor voltage sum

of all three phases.

The steady state output voltage and current is shown in Fig. 9 and Fig. 10.

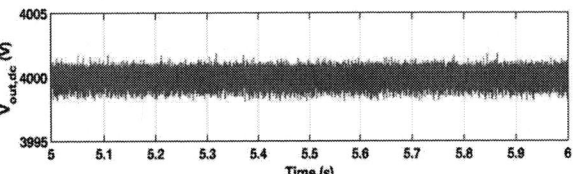

Fig. 9. Steady State Output Voltage with mean value of 4000 V and ripple of 3.78 V.

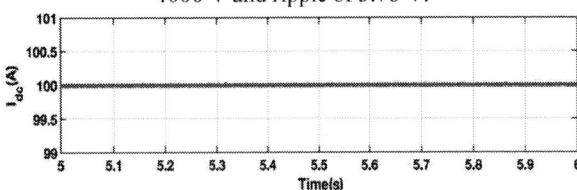

Fig. 10. Steady State Output Current with mean value of 100 A.

It may be noted that, the transient behavior plays a critical role for failure safe operation which can be captured by performing simulations at proper simulation step size.

V. CONCLUSION

This paper presents a design procedure for a three-phase Modular Multilevel Converter (MMC) rectifier, combining theoretical analysis with simulation based validation. The main focus of this study is to handle parameter interdependencies on optimal selection of passive components, which can make the system bulky and uneconomical. By using the proposed design methodology a 4 kV, 100 A DC power supply was designed and an error of less than 2% between theoretical and simulation results were noticed. The interdependency of arm inductance and SM capacitance is encountered by modifying the existing analytical equations in the literature and by following the proposed systematic design procedure. For accurate analysis, the effect of circulating current is considered in SM capacitance calculation based on charge fluctuations. This steady state comprehensive study helps in understanding the operational behaviour of MMC and contributes to the efficient hardware design, which is crucial for compact and economical setup.

VI. REFERENCES

[1] G. Li and J. Liang, "Modular Multilevel Converters: Recent Applications [History]," in IEEE Electrification Magazine, vol. 10, no. 3, pp. 85-92, Sept. 2022.

[2] M. Hiller, D. Krug, R. Sommer and S. Rohner, "A new highly modular medium voltage converter topology for industrial drive applications," 2009 13th European Conference on Power Electronics and Applications, Barcelona, Spain, 2009, pp. 1-10.

[3] F. Blaabjerg and Ke Ma, "Future on Power Electronics for Wind Turbine 111 Systems," IEEE J. Emerg. Sel. Top. Power Electron., vol. 1, no. 3, pp. 139– 152, Sep. 2013.

[4] D. Ronanki and S. S. Williamson, "Modular multilevel converters for transportation electrification: Challenges and opportunities," IEEE Trans. Transp. Electrific., vol. 4, no. 2, pp. 399–407, Jun. 2018.

[5] M. Tharun et al., "Simulation Study of 4 kV, 5A Modular Multilevel Converter as a Rectifier for Neutral Beam Injectors," 2024 IEEE

International Conference on Electronics, Computing and Communication Technologies (CONECCT), Bangalore, India, 2024.

[6] Lesnicar, A.: 'Neuartiger Modularer Mehrpunktumrichter M2C für Netzkupplungsanwendungen' (Shaker Verlag, München, Germany, 2008).

[7] M. M. C. Merlin and T. C. Green, "Cell capacitor sizing in multilevel converters: cases of the modular multilevel converter and alternate arm converter," *IET Power Electronics*, vol. 8, no. 3, pp. 350–360, Mar. 2015.

[8] Z. Liu, K. -J. Li, J. Wang, Z. Javid, M. Wang and K. Sun, "Research on Capacitance Selection for Modular Multi-Level Converter," in IEEE Transactions on Power Electronics, vol. 34, no. 9, pp. 8417-8434, Sept. 2019.

[9] Z. Xu, H. Xiao, and Z. Zhang, "Selection methods of main circuit parameters for modular multilevel converters," *IET Renewable Power Generation*, vol. 10, no. 6, pp. 788–797, Jul. 2016.

[10] F. Santoro, A. Ferro, M. De Nardi and E. Gaio, "Studies on Alternative Schemes for the MMC-Based Acceleration Grid Power Supply of DEMO Neutral Beam Injector," in IEEE Transactions on Plasma Science, vol. 50, no. 11, pp. 4039-4046, Nov. 2022.

[11] Q. Song, W. Liu, X. Li, H. Rao, S. Xu and L. Li, "A Steady-State Analysis Method for a Modular Multilevel Converter," in IEEE Transactions on Power Electronics, vol. 28, no. 8, pp. 3702-3713, Aug. 2013.

[12] K. Ilves, A. Antonopoulos, S. Norrga and H. -P. Nee, "Steady-State Analysis of Interaction Between Harmonic Components of Arm and Line Quantities of Modular Multilevel Converters," in IEEE Transactions on Power Electronics, vol. 27, no. 1, pp. 57-68, Jan. 2012.

[13] M. Zygmanowski, B. Grzesik, and R. Nalepa, "Capacitance and inductance selection of the modular multilevel converter," 15th European conf. Power Electron. Appl.(EPE), 2013, pp. 1–10.

[14] Sixing Du; Apparao Dekka; Bin Wu; Navid Zargari, "Classical Control of Modular Multilevel Converter," in Modular Multilevel Converters: Analysis, Control, and Applications , IEEE, 2018, pp.79-102.

[15] K. Ilves, L. Harnefors, S. Norrga and H. -P. Nee, "Analysis and operation of modular multilevel converters with phase-shifted carrier PWM," 2013 IEEE Energy Conversion Congress and Exposition, Denver, CO, USA, 2013, pp. 396-403.

[16] The Mathworks, Inc.(2023). Simscape version: 5.5 (R2023a). Available: https://in.mathworks.com.

Fuzzy Tuning Based V/F logic for Speed Control of Five-Phase Induction Motor Drives

Spandana Kolipaka
Department of Electrical Engineering National Institute of Technology Warangal, India
kolipakaspandana@gmail.com

Dr. Swati Devabhaktuni
Department of Electrical Engineering National Institute of Technology Warangal, India
swatikjn@nitw.ac.in

Abstract— **With the increase in the utilization of advanced electric drives especially in high-performance industrial applications, the development of motor control techniques to enhance the performance and efficiency of electric drives has gradually become an interesting and evolving area. Owning numerous perks such as easy implementation and reliability in preserving a steady magnetic flux, guaranteeing steady and smooth speed control, Voltage-to-Frequency (V/F) control is still the most used control method. On the other hand, compared to conventional three-phase systems, five-phase induction motors are known for their smoother operation, higher torque density, fault tolerance, lower torque ripple, and reduced DC-link voltage. This study accentuates the speed control of multi-phase induction motor drives. Within the framework of V/F control logic, four control techniques are compared: proportional-integral (PI), fuzzy-tuned PI (FTPI), fractional order PI (FOPI), and fuzzy-tuned FOPI (FTFOPI). The performance of the controllers under load-changing conditions is analyzed in the study in terms of steady-state performance, robustness, and dynamic response. It is observed that the fuzzy-FOPI controller shows promising results in tackling the non-linearities and complexity of induction motor drives by utilizing the flexibility of fuzzy logic along with fractional-order dynamics. In addition to demonstrating the possibilities of sophisticated fuzzy and fractional-order controllers, this work offers Profound insights into how to use them in real-time applications for motor drive systems of the future. This work offers an extensive resource for researchers and engineers looking for creative ways to improve the functionality of induction motor control systems.**

Keywords— *Induction motor drive, PI Controller, FOPI Controller, Kp, Ki, FTPI, FTFOPI, fuzzy logic controller.*

I. INTRODUCTION

Speed regulation of electric motor drives continues to be a fundamental aspect of contemporary engineering due to the growing need for efficiency, durability, and adaptability in industrial and electric vehicle applications. Being basic and simple, proportional-integral (PI) controllers are well known, despite being straightforward and efficient in reaching steady-state accuracy, these suffer under non-linear operating conditions frequently causing them to perform worse, mandating the exploration of more complex yet advanced control strategies. This study delves into one of the popular scalar control methods, i.e. the Voltage-to-Frequency (V/F) control approach for induction motors. Maintaining a constant V/F ratio, this technique guarantees the preservation of magnetic flux, which is necessary for reliable torque generation at different speeds, its ease of use, affordability, and innate ability to provide smooth speed control makes it ideal for applications using five-phase induction motors. These devices are essential to the development of next-generation motor drive systems because they provide higher torque density, less torque ripple, and improved fault tolerance

than their three-phase equivalents. In parallel, due to their versatility, resilience, and capacity to manage system non-linearities fuzzy logic controllers attract a lot of interest and show the effectiveness of PI-fuzzy logic control systems in motor drive applications [1]. Research confirms the potential of self-tuning fuzzy controllers in practical settings [2]. Furthermore, improved performance in speed control systems is demonstrated using PID-fuzzy controllers [3]. More recent developments include Type-2 fuzzy controllers for DC-DC converters [4] and cooperative fuzzy PI controllers, which are tailored for difficult conditions such as underwater motor applications [5]. Comparative analyses of conventional and fuzzy logic controllers in induction motor systems highlight the efficacy of these approaches [6]. Studies comparing PI, fuzzy, and Adaptive Neuro-Fuzzy Inference System (ANFIS) controllers deliver key insights into their respective advantages [7]. Furthermore, an evolving ideology for handling the difficulties of speed control in induction motors is the combination of fuzzy logic and fractional-order control [8]. Permanent Magnet Synchronous Motor (PMSM) drives are another application where fuzzy logic controllers are successful because they improve speed control accuracy and resilience [9]. For induction motor systems, self-tuning fuzzy-based PID controllers work well [10], while scalar-controlled induction motors perform better under fuzzy-PID controllers than under traditional PID designs [11]. The possibility of combining different control approaches to attain better dynamic and steady-state performance is further demonstrated by hybrid fuzzy-PI controllers [12]. In this work, the effectiveness of four control paradigms—PI, Fuzzy-PI, Fractional Order PI (FOPI), and Fuzzy-FOPI—is examined concerning the speed control of five-phase induction motors under V/F control logic. It thoroughly assesses their resilience, steady-state accuracy, and dynamic responses, especially in the context of scalar V/F control. The purpose of this work is to clarify how sophisticated control techniques interact to maximise the efficiency of five-phase induction motor drives.

II. MODELLING OF INDUCTION MOTOR DRIVE

To create efficient control schemes and forecast the motor's performance under varied operating conditions, precise modelling of an induction motor drive is essential. To assess its dynamic behaviour, a collection of differential equations derived from the electrical and mechanical subsystems of an induction motor can be used. These formulas explain how the circuits for the stator and rotor interact and the motor's mechanical load

1. Mathematical Model

Stator and Rotor Equations

The space vector theory, reduces the three-phase system to a two-axis (d-q) coordinate system and respective stator and rotor voltage equations are as follows:

979-8-3315-3013-6/25 $31.00 © 2025 IEEE

Stator Equations;

$$V_{ds} = R_s i_{ds} + \frac{d\lambda_{ds}}{dt} - \omega\lambda_{qs}$$

$$V_{qs} = R_s i_{qs} + \frac{d\lambda_{qs}}{dt} + \omega\lambda_{ds}$$

$$V_{xs} = R_s i_{xs} + \frac{d\lambda_{xs}}{dt}$$

$$V_{ys} = R_s i_{ys} + \frac{d\lambda_{ys}}{dt}$$

$$V_{0s} = R_s i_{0s} + \frac{d\lambda_{0s}}{dt}$$

Rotor Equations;

$$V_{dr} = R_r i_{dr} + \frac{d\lambda_{dr}}{dt} - (\omega-\omega_r)\,\lambda_{qr}$$

$$V_{qr} = R_r i_{qr} + \frac{d\lambda_{qr}}{dt} + (\omega-\omega_r)\,\lambda_{dr}$$

$$V_{xr} = R_s i_{xr} + \frac{d\lambda_{xr}}{dt}$$

$$V_{yr} = R_s i_{yr} + \frac{d\lambda_{yr}}{dt}$$

$$V_{0r} = R_s i_{0r} + \frac{d\lambda_{0r}}{dt}$$

where V_{ds}, V_{qs} are the stator voltages; V_{dr}, V_{qr} are the rotor voltages; i_{ds}, i_{qs} are the stator currents; i_{dr}, i_{qr} are the rotor currents; $\lambda_{ds}, \lambda_{qs}$ are the stator flux linkages; $\lambda_{dr}, \lambda_{qr}$ are the rotor flux linkages; R_s, R_r are the stator and rotor resistances; ω is the synchronous speed; and is the rotor speed.

Flux Linkage Equations

The flux linkages are related to the currents by:

$$\lambda_{ds} = L_s i_{ds} + L_m i_{dr}$$

$$\lambda_{qs} = L_s i_{qs} + L_m i_{qr}$$

$$\lambda_{dr} = L_r i_{dr} + L_m i_{ds}$$

$$\lambda_{qr} = L_r i_{qr} + L_m i_{qs}$$

where Ls, and Lr are the stator and rotor inductances, and Lm is the mutual inductance between the stator and rotor windings.

2. Mechanical Model

The mechanical dynamics of the induction motor are described by the torque and motion equations:

$$T_e = \frac{5}{2} * \frac{p}{2}\left(\lambda_{ds}i_{qs} - \lambda_{qs}i_{ds}\right)$$

$$T_e = T_L - J\frac{d\omega_r}{dt} + B\omega_r$$

where T_e is the electromagnetic torque, T_L, is the load torque, J is the moment of inertia, B is the damping coefficient, and p is the number of pole pairs.

3. State-Space Representation

The above equations can be represented in a state-space form for use in control system design and simulation:

$$\frac{dx}{dt} = Ax + Bu$$

$$y = Cx + Du$$

where x is the state vector (currents, flux linkages), u is the input vector (voltages), and y is the output vector (speed, torque).

4. Parameter Estimation

A precise understanding of the motor parameters, such as the mutual inductance, inductances, stator and rotor resistances, is necessary for accurate modelling. Numerous procedures, including direct measurement, stationary frequency response testing. and identification methods utilising measured input-output data, can be used to ascertain these values.

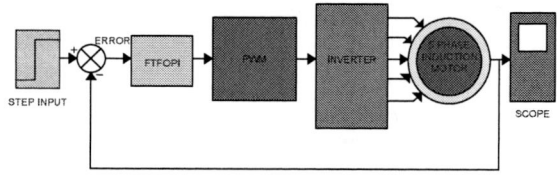

Fig. 1. Block Diagram

III. METHODOLOGY

A. PI controller

The proportional-integral (PI) controller is based on the combination of proportional (Kp) and integral (Ki) gains, where Kp takes charge of the present error and Ki takes charge of the past error thus obtaining the state error and facilitating minimization of error through gain tunning. Usually, these gains are tunned using MATLAB's auto-tuning feature but auto-tunning only works when the system is Linear and as Induction Motor is known for its Nonlinear behaviour MATLAB auto-tunning is not possible and hence opted for the trial-and-error method for Kp and Ki tunning in standard PI based V/F logic simulation. This creative method guarantees speed control and management.

The general equation for PI control can be expressed as follows:

Error,

$$e(t) = \omega_{ref} - \omega_{actual}$$

Proportional term,

979-8-3315-3013-6/25 $31.00 © 2025 IEEE

$$p(t) = K_p \times e(t)$$

Integral term,

$$I(t) = K_i\left(\int e(t)\,dt\right)$$

B. FOPI controller

Unlike the conventional PI controller, the Fractional order proportional integral (FOPI) controller has the flexibility to vary the order (λ) of the integral term, which in turn helps with the modification in the behaviour of the integral action, is what defines the fractional-order term. This agility makes it versatile and helps in the precise speed regulation of motors, particularly in complicated dynamic systems.

The general equation for FOPI control can be expressed as follows:

Proportional term,

$$p(t) = K_p \times e(t)$$

Integral term,

$$I(t) = K_i\left(\int e(t)^\lambda\,dt\right)$$

Error,

$$e(t) = \omega_{ref} - \omega_{actual}$$

C. Fuzzy Tuned PI controller

The fuzzy tuned proportional integral (FTPI), as its name suggests, uses a fuzzy rule set to tune the gains Kp and Ki. This method appears to be more promising than the trial-and-error method because the latter involves an unpredictable and time-consuming process.

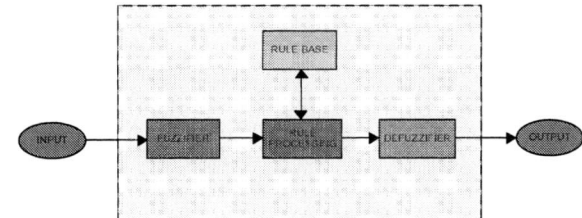

Fig. 2 FLC structure

Here, we have two outputs (Kp and Ki) and two inputs (error and error change). Nine triangle membership functions are used for both input and output, totalling 81 rules. The quantity and form of the membership function can be altered based on the requirements and intricacy of the system. The proposed system employs the Mamdani style, a fuzzy inference technique. For your reference, the rule table that was utilised to generate this inference is provided below. To make additional changes and enhancements, any of the above-mentioned criteria can be altered. Fuzzy logic control (FLC) equations are used to model the system. A is a fuzzy set on a universe of discourse X, where X is continuous and A is defined by membership grades in the interval [0,1].

$$A = \int_X \mu_A(x)/x$$

All the input-output membership functions and surface view of fuzzy rules of FLC are provided in the below Fig. 3, Fig. 4, Fig.5 and Fig. 6. The Range intervals of the input and output membership functions can also be observed from the figures provided.

A. TABLE

FUZZY RULE TABLE

e \ Δe	NB	NM	NS	VS	ZO	VL	PS	PM	PB
NB	Kp = PB Ki = PS	Kp = PB Ki = PS	Kp = PM Ki = ZO	Kp = PM Ki = ZO	Kp = PS Ki = NS	Kp = ZO Ki = NM	Kp = NS Ki = NB	Kp = NM Ki = NM	Kp = NB Ki = NB
NM	Kp = PB Ki = PS	Kp = PM Ki = ZO	Kp = PM Ki = ZO	Kp = PS Ki = NS	Kp = ZO Ki = NM	Kp = NS Ki = NB	Kp = NM Ki = NB	Kp = NS Ki = NB	Kp = NB Ki = NB
NS	Kp = PM Ki = ZO	Kp = PM Ki = ZO	Kp = PS Ki = NS	Kp = ZO Ki = NM	Kp = NS Ki = NB	Kp = NM Ki = NB	Kp = NB Ki = NB	Kp = NM Ki = NB	Kp = NS Ki = NB
VS	Kp = PM Ki = ZO	Kp = PS Ki = NS	Kp = ZO Ki = NM	Kp = ZO Ki = NB	Kp = ZO Ki = NB	Kp = NS Ki = NB	Kp = NM Ki = NB	Kp = NS Ki = NB	Kp = NB Ki = NB
ZO	Kp = PS Ki = ZO	Kp = ZO Ki = ZO	Kp = ZO Ki = NM	Kp = ZO Ki = NB	Kp = ZO Ki = NB	Kp = ZO Ki = NB	Kp = NS Ki = NB	Kp = NM Ki = NB	Kp = NM Ki = NB
VL	Kp = PS Ki = NS	Kp = ZO Ki = NM	Kp = NS Ki = NB	Kp = ZO Ki = NB	Kp = ZO Ki = NB	Kp = NS Ki = NB	Kp = NM Ki = NB	Kp = NS Ki = NB	Kp = NM Ki = NB
PS	Kp = NS Ki = NM	Kp = NM Ki = NB	Kp = NS Ki = NB	Kp = ZO Ki = NB	Kp = ZO Ki = NB	Kp = NS Ki = NB	Kp=NM Ki = NB	Kp =NM Ki = NB	Kp = NS Ki = NB
PM	Kp = NM Ki = NB	Kp = NS Ki = NB	Kp = NM Ki = NB	Kp = NS Ki = NB	Kp=NM Ki = NB	Kp = NB Ki = NB	Kp = NB Ki = NB	Kp = NB Ki = NB	Kp = NB Ki = NB
PB	Kp = NS Ki = NB	Kp = NM Ki = NB	Kp = NB Ki = NB	Kp = NB Ki = NB	Kp = NB Ki = NB	Kp = NB Ki = NB	Kp = NB Ki = NB	Kp = NB Ki = NB	Kp = NB Ki = NB

From the aforementioned rule table, we observe that there are Nine membership function categories for both inputs and outputs: negative big (NB), negative medium (NM), negative small (NS), very small (VS), zero (ZO), very large (VL), positive small (PS), positive medium (PM), positive big (PB), which can be ununderstood from the above table. These plots, along with the rule table, establish eighty-one rules within the fuzzy inference system. Based on these rules, the Kp and Ki values are adjusted to control the output and maintain the motor speed at a reference value, thereby minimizing the error between our reference speed and actual speed and spikes in the output speed waveforms are also observed to be minimized. While our proposed system utilizes triangular membership functions, other shapes such as trapezoidal, Z-shape, S-shape, sigmoid, and Gaussian can also be employed depending on specific requirements and applications.

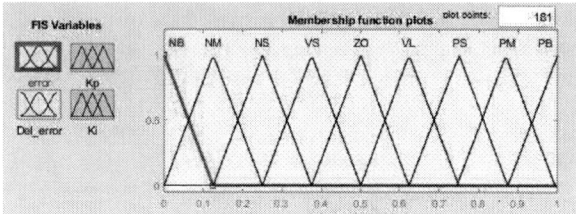

Fig. 3 Triangular Input Membership functions

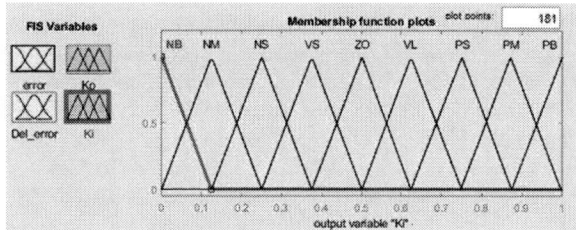

Fig. 4 Triangular Output Membership functions

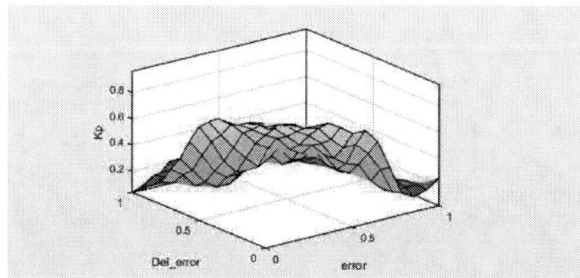

Fig. 5 Surface view of fuzzy rules for Kp

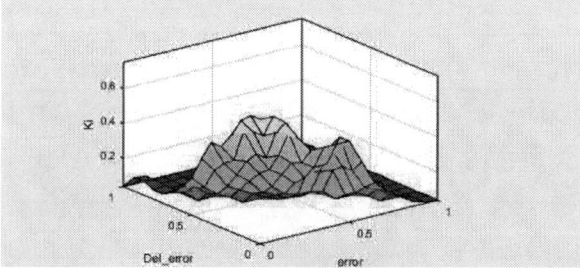

Fig. 6 Surface view of fuzzy rules for Ki

D. Fuzzy Tuned FOPI controller

With a constant fractional order $\lambda=0.2$, the Fuzzy-FOPI controller improves the performance of conventional PI controllers in complicated systems such as multi-phase induction motors by fusing the flexibility of fuzzy logic with the benefits of fractional-order control. Without having to dynamically adjust the fractional order, the controller can improve the system's long-term accuracy and stability by maintaining a precise and stable integral action by holding λ constant. To make sure the controller is responsive to shifting conditions and disturbances, fuzzy logic is utilised to adaptively modify the proportional and integral gains (Kp and Ki) based on real-time error and its rate of change. This method keeps the advantages of fractional-order control for improved handling of slow dynamics and nonlinearity while streamlining the tuning procedure and concentrating on fuzzy-tuned improvements. As a result, induction motor control applications benefit from a reliable, flexible, and effective control system that enhances speed tracking and system performance.

IV. RESULTS

Performance assessment was conducted on all four PI, FOPI, FTPI, and FTFOPI for a 20-second sampling period. A load torque of 2 Nm was applied at 5 seconds to examine the dynamic response and fault-tolerant capability. Along with longer rise and settling durations, the PI controller displayed a notable speed drop and slower recovery under the load situation. However, it continued to operate steadily when there was no load. The FOPI controller improved transient performance by achieving shorter rise and settling periods and higher speed recovery, even if fine-tuning was required to reduce residual oscillations. The Fuzzy-PI controller enhanced flexibility even if there were occasional overshoots by reaching faster recovery with fewer speed deviations and better torque transitions. The Fuzzy-FOPI excelled the other controllers demonstrating the fastest rise and settling durations, minimum overshoot, and exceptional disturbance rejection with constant speed and smooth torque waveforms whatever the load. To demonstrate the superior performance of fuzzy-based controllers, particularly the Fuzzy-FOPI, in controlling a range of operational conditions, the appropriate waveforms for motor speed, torque, five-phase currents, and Id, Iq currents are also supplied for reference.

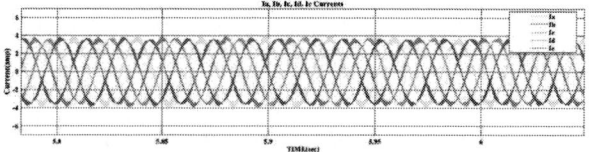

Fig. 7 Five-phase currents

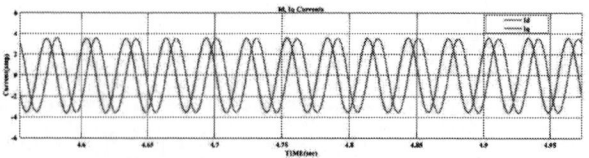

Fig. 8 Id, Iq currents

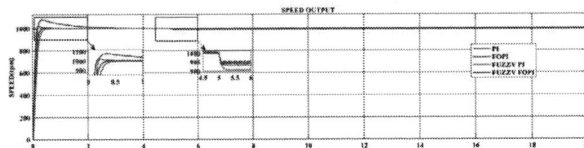

Fig. 9 Speed waveforms of motor

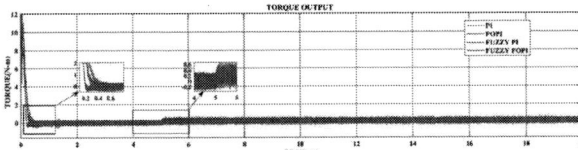

Fig. 10 Torque comparison waveforms of induction motor

PERFORMANCE ANALYSIS

In this study, all four controllers (PI, FOPI, FTPI, and FTFOPI) were thoroughly evaluated in terms of dynamic and steady-state behaviour. Though PI has a wide tuning range and an excellent dynamic response under no-load conditions with minimal overshoot, rise time and settling time, it is reported to recover more slowly under load conditions. On the other side, we have FOPI with better dynamic responsiveness, and the fact that the integral term behaviour may be varied with its order makes it even more versatile. Fuzzy-PI has shown significant gains in dynamic performance, reducing rise and settling times while enhancing disturbance adaptability. It exemplifies automation, accuracy, and flexibility by allowing gain tuning to be automated. With shorter rise and settling durations and excellent disturbance rejection, the fuzzy-FOPI controller outperforms its competitors in transient responsiveness. Despite these advantages, its tuning complexity remains a problem that requires development. The fuzzy-FOPI controller is the best choice for high-performance applications that require agility and robustness; for less demanding scenarios, the PI controller offers trustworthy simplicity. If future studies focus on automation to boost controller performance even more, fuzzy and fractional parameter tweaking could be a promising topic to investigate further.

B. TABLE

TIME RESPONSES COMPARISON

	PI	FOPI	FUZZY_PI	FUZZY_FOPI
Delay time	0.125	0.09	0.08	0.06
Rise time	0.3	0.2	0.199	0.19
Peak time	0.4	0.3	0.3	0.2
Settling time	0.5	0.35	3	0.6
Overshoot	0	0	8%	2.5%

V. CONCLUSION

Inevitably every control strategy has its unique advantages and disadvantages. PI, FOPI, Fuzzy-PI, and Fuzzy-FOPI controllers have been evaluated for speed regulation of a five-phase induction motor using V/F logic. Despite its reliable performance in steady-state scenarios, the PI controller is not flexible enough to handle sudden dynamic changes. By introducing fractional-order dynamics, the FOPI controller enhances response times and transient performance. The fuzzy-PI controller's adaptive features enhance dynamic responses, but greater overshoot and a longer settling time come at the expense. The fuzzy-FOPI controller is the best-balanced approach because it provides great speed regulation with minimal overshoot, fast dynamic reflexes, and a shorter settling period. The results of this work demonstrate the Fuzzy-FOPI controller as a feasible alternative for improving the performance of induction motor systems, paving the way for more precise and effective motor control in innovative applications.

C. TABLE

DESIGN PARAMETERS

Parameters	Values
Voltage [Vdc]	200 V
Stator Resistance [Rs]	1.05 Ω
Rotor Resistance [Rr]	1.42 Ω
Stator Inductance [Ls]	0.09 H
Rotor Inductance [Lr]	0.09 H
Magnetizing Inductance [Lm]	0.0847 H
Poles [P]	4
Inertia Constant [Js]	0.0148 Kg-m^2
Damping Coefficient [B]	0

References

[1] B. S. K. K. Ibrahim et al., "PI-Fuzzy Logic Control for 3 Phase BLDC Motor for Electric Vehicle Application," 2012 Sixth UKSim/AMSS European Symposium on Computer Modeling and Simulation, Malta, Malta, 2012.

[2] N. Farah et al., "A Novel Self-Tuning Fuzzy Logic Controller Based Induction Motor Drive System: An Experimental Approach," in IEEE Access, vol. 7, pp. 68172-68184, 2019.

[3] M. . S. Ab Ghani et al., "Speed Control Design for Permanent Magnet Synchronous Motor Drive based on Pid-Fuzzy Controller," 2022 3rd International Conference on Artificial Intelligence and Data Sciences (AiDAS), IPOH, Malaysia, 2022.

[4] S. Kolipaka, K. Teeparthi and A. Yadav Mudiminchi, "Robust Type-2 Fuzzy Controller for DC-DC Power Converter Feeding Constant Power Load," 2023 3rd International Conference on Intelligent Technologies (CONIT), Hubli, India, 2023.

[5] W. Yuze, Z. Wei, A. Chengliu, W. Kai and W. Haifeng, "The Cooperative Control of Speed of Underwater Driving Motor Based on Fuzzy PI Control," 2020 23rd International Conference on Electrical Machines and Systems (ICEMS), Hamamatsu, Japan, 2020.

[6] F. Lachekhab, R. B. Zamoum, D. E. Bougheloum, S. Benyahia, D. Acheli and A. Kouzou, "Control of Squirrel Cage Induction Motor using Conventional Controllers and fuzzy logic," 2022 19th International Multi-Conference on Systems, Signals & Devices (SSD), Sétif, Algeria, 2022

[7] N. Gogia, R. Prasad, R. Arora, B. Bhushan and P. Prakash, "Speed Control of Three-Phase Induction Motor Using PI, Fuzzy and ANFIS Controllers," 2023 10th International Conference on Signal Processing and Integrated Networks (SPIN), Noida, India, 2023

[8] M. Vahedpour, A. R. Noei and H. A. Kholerdi, "Comparison between performance of conventional, fuzzy and fractional order PID controllers in practical speed control of induction motor," 2015 2nd International Conference on Knowledge-Based Engineering and Innovation (KBEI), Tehran, Iran, 2015

[9] G. Dewantoro and Y. -L. Kuo, "Robust speed-controlled permanent magnet synchronous motor drive using fuzzy logic controller," 2011 IEEE International Conference on Fuzzy Systems (FUZZ-IEEE 2011), Taipei, Taiwan, 2011

[10] Ar un Kumar R and Febin Daya J L, "A novel self - Tuning fuzzy based

PID controller for speed control of induction motor drive," *2013 International Conference on Control Communication and Computing (ICCC)*, Thiruvananthapuram, India, 2013

[11] S. Pati, A. Panda and S. Mohanty, "A comparative performance study of scalar controlled induction motor using PID controller and fuzzy PID controller," *2014 International Conference on Circuits, Power and Computing Technologies [ICCPCT-2014]*, Nagercoil, India, 2014

[12] E. H. El-Zohri and M. A. Mosbah, "Speed Control of Inverter-Fed Induction Motor Using Hybrid Fuzzy-PI Controller," *2020 International Conference on Innovative Trends in Communication and Computer Engineering (ITCE)*, Aswan, Egypt, 2020

A variable capacitance based modelling and parameter determination for supercapacitor with experimental validation

Rose Mary Boby
Department of Electrical Engineering
National Institute of Technology Calicut
rosemaryboby@gmail.com

Shifa H. Rahman
Department of Electrical Engineering
National Institute of Technology Calicut
shifa_p190021ee@nitc.ac.in

Rahul Radhakrishnan
Department of Electrical Engineering
National Institute of Technology Calicut
rahulr@nitc.ac.in

Abstract—**Recent advancements in technology have made supercapacitors an excellent choice for high power density applications. Accurate modeling is crucial for the effective management and use of supercapacitors, particularly in estimating their remaining useful life such as state of charge (SoC). To closely replicate the dynamics of supercapacitors, a new equivalent circuit model that incorporates factors such as rest period dynamics and capacitance variation, is developed. The forgetting factor recursive least squares (FFRLS) algorithm is implemented for identifying the parameters of supercapacitor. From the experimental data, the model parameters were identified for various operations such as charging, discharging and rest period. Moreover, using this developed model, model validation was done by comparing the measured terminal voltage with that of the model computed one. From the results, it could be inferred that the proposed model accurately captures the supercapacitor dynamics, enabling advanced estimation techniques to generate highly accurate estimates regarding the remaining useful life of supercapacitors.**

Index Terms—**Supercapacitor modelling, Parameter identification, Recursive Least Squares**

I. INTRODUCTION

Batteries have long been used in high energy density applications, while recent technological advancements have made supercapacitors (SC) a strong option for high power density applications. As a result, the integration of batteries and supercapacitors into hybrid energy storage systems (HESS) has become increasingly popular for electric vehicles (EVs) [1]. HESSs, typically combining Lithium-ion (Li-ion) batteries and supercapacitors, are being widely studied for large-scale applications such as autonomous trucks, mobile robots, delivery drones, to name a few. This hybrid approach leverages the high energy density of Li-ion batteries and the rapid energy storage capabilities of supercapacitors, which can handle virtually unlimited charge and discharge cycles. The goal of power flow management in these systems is to utilize the battery's high energy density to meet the vehicle's slow-varying power demand, while the supercapacitor addresses the fast-changing power requirements. However, managing

This work was supported by the Core Research Grant, CRG/2022/001997, Science and Engineering Research Board, Government of India.

and controlling energy storage systems effectively remains a challenging task.

Accurate knowledge of the dynamic behaviour of both batteries and supercapacitors is essential for managing energy storage systems, as it allows for effective power distribution between the energy storage devices. For batteries, this requirement is relatively easy to meet, as battery modelling and state observation have been extensively studied over the years, resulting in the development of several modelling and estimation methods. However, the challenges in developing a supercapacitor model remains unresolved. Many approaches treat SCs as linear devices or overlook their dynamic behaviours, leading to inaccurate models [2]. Precise estimation of the state of charge (SoC) depends on the accurate modelling of these storage elements. The SoC of a cell is defined as the ratio between the available capacity of the cell and its maximum capacity at a given point of time.

Over the years, various methods have been proposed for SoC estimation, which can be classified into three categories: 1) analytical methods, 2) model-based methods, and 3) data-driven methods. Model-based methods, such as the Kalman filter (KF) and its variants, are particularly accurate due to their closed-loop nature, but they require a very precise dynamic model of the supercapacitor. The Kalman filter and its variants are the widely used benchmark methods for SoC estimation, as they provide optimal estimates in the sense of minimizing the mean squared error [3], [4]. While this approach is effective at addressing modelling as well as measurement uncertainties, it still depends heavily on the accuracy of the underlying model [5].

In case of supercapacitors, considering the rest period dynamics is of great significance as it affects the online monitoring, and further estimation of remaining useful life. This leakage effect and the subsequent drop in terminal voltage has to be accounted for while considering the supercapacitors for long-term real-life applications. Several equivalent circuit models (ECM) for capturing supercapacitor dynamics are reported in literature. In [6], a three-branch supercapacitor ECM has been reported. An online supercapacitor diagnosis model based on the EKF was reported in [7], whereas [8]

979-8-3315-3013-6/25 $31.00 © 2025 IEEE
119

proposed a Lyapunov-based adaptation law to estimate the supercapacitor's parameters. Moreover, based on neural networks, [9] proposed a supercapacitor aging prediction method. Different aging models with a comparative study is presented in , whereas an online health diagnosis for supercapacitors is given in [10], using a simple RC model. In [11], an ECM incorporating the leakage effect was proposed for the subsequent estimation of SoC. Recently, equivalent circuits based on fractional models were reported in literature [12].

In this work, a new ECM is proposed for accurately capturing the dynamics of supercapacitor during charging, discharging and during the rest period. A variable capacitance-based approach to modelling is considered, where the main capacitance varies according to the various range in values taken by the terminal voltage. Further, a variable forgetting factor recursive least squares algorithm is implemented for the parameter identification. Finally, to validate the developed ECM, experimental data involving charging, discharging and rest period were used.

II. MODELLING OF SUPERCAPACITOR

The supercapacitor examined in this work is from the Maxwell BCAP3000 series, with a rated capacitance of 3000F. Since the Thevenin equivalent model is considered to be the most efficient approach in simulating the behavior of supercapacitors, the same is considered as the foundation for this study. The proposed equivalent circuit model, shown in Fig. 1, consists of three main parts. The first part is the internal resistance or ohmic losses, R_s. This resistance represents the conduction resistance of charge carriers as they travel through the electrolyte and metallic conductors in the supercapacitor. The next part is the parallel R_p-C_p combination. This branch is used to reflect the SC's over-potential and capture the charge re-distribution and diffusion. The third part is the parallel R_L-C_m branch, where R_L denotes the leakage resistance with capacitor C_m (main capacitance). To enhance the accuracy of model, the concept of variable capacitance is incorporated, where the main capacitance C_m is treated as a variable rather than a constant. This capacitance is determined by a function of the voltage across it, where a piece-wise linear function is selected to model this relationship, such as

$$C_m = f(U_c) = C_0 + C_1 U_c. \qquad (1)$$

Here, C_0 is a constant capacitance and it could be assumed that the capacitance C_m changes linearly with a nearly constant or slowly varying slope, so that

$$\frac{dC_1}{dt} \approx 0. \qquad (2)$$

In this model, U_t is supercapacitor terminal voltage and U_c is main capacitance voltage (supercapacitor internal voltage), and U_p is the polarization voltage. I is the current flowing through the supercapacitor and assumes a positive value during discharge.

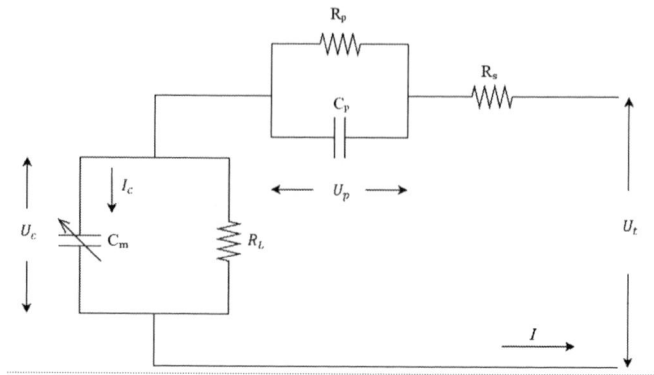

Fig. 1. Proposed ECM

To analyze the model, Kirchhoff's laws are applied. The equations for the discrete-time model is expressed as

$$U_{p,k} = U_{p,k-1}e^{-\frac{T}{R_p C_p}} + I_k R_p \left(1 - e^{-\frac{T}{R_p C_p}}\right), \qquad (3)$$

$$U_{c,k} = U_{c,k-1}e^{-\frac{T}{R_L C_m}} + I_k R_L \left(1 - e^{-\frac{T}{R_L C_m}}\right), \qquad (4)$$

and

$$U_t = U_c - U_p - IR_s. \qquad (5)$$

Coulomb counting is one of the most widely used methods for estimating the SoC of a supercapacitor. Considering the leakage effect, the SoC equation could be expressed as

$$\frac{d\,SoC}{dt} = \frac{\eta I_c}{Q}, \qquad (6)$$

where

$$I_c = I - \frac{U_c}{R_L}. \qquad (7)$$

Hence, in discrete-time domain

$$SoC_k = SoC_{k-1} + \frac{\eta I_k T}{Q} - \frac{\eta U_{c,k-1} T}{Q R_L}, \qquad (8)$$

where T is the sampling time. From Eq. (3), (4) and (8), the discrete-time state-space model could be represented as

$$\begin{bmatrix} SoC_k \\ U_{p,k} \\ U_{c,k} \end{bmatrix} = \begin{bmatrix} 1 & 0 & -\frac{\eta T}{Q R_L} \\ 0 & e^{-\frac{T}{R_p C_p}} & 0 \\ 0 & 0 & e^{-\frac{T}{R_L C_m}} \end{bmatrix} \begin{bmatrix} SoC_{k-1} \\ U_{p,k-1} \\ U_{c,k-1} \end{bmatrix}$$
$$+ \begin{bmatrix} -\frac{\eta T}{Q} \\ R_p \left(1 - e^{-\frac{T}{R_p C_p}}\right) \\ R_L \left(1 - e^{-\frac{T}{R_L C_m}}\right) \end{bmatrix} I_k. \qquad (9)$$

It has to be noted that explicit representation of SoC is more preferred as it directly gives the remaining amount of charge, in comparison to indirectly estimating it from other internal states [11]. Now, for the measurement model, it is evident that the measurement obtained is the terminal voltage and that has to be written as a function of the state, represented in Eq. (5). The equivalent discrete-time domain representation would be

$$U_{t,k} = U_{c,k} - U_{p,k} - I_k R_s. \qquad (10)$$

III. THE PARAMETER IDENTIFICATION

The recursive least squares (RLS) method is the most commonly used method for system parameter identification. The parameter vector θ and the observation data matrix φ are defined as

$$\theta = \begin{bmatrix} \theta_1 & \theta_2 & \theta_3 & \theta_4 & \theta_5 \end{bmatrix}^T \quad (11)$$

and

$$\varphi = \begin{bmatrix} E_{k-1} & E_{k-2} & I_k & I_{k-1} & I_{k-2} \\ E_{k-2} & E_{k-3} & I_{k-1} & I_{k-2} & I_{k-3} \\ \vdots & \vdots & \vdots & \vdots & \vdots \\ E_{k-m-1} & E_{k-m-2} & I_{k-m} & I_{k-m-1} & I_{k-m-2} \end{bmatrix}, \quad (12)$$

where k denotes the current time step and m is the observation time.

On the basis of the RLS method, the forgetting factor RLS (FFRLS) method could be considered such that a forgetting factor coefficient λ is added in φ and the system output vector E. Thus, when the observation data increases, the FFRLS method could obtain more accurate identification parameters efficiently. This could be expressed as

$$E = [E_k, \ \lambda E_{k-1}, \cdots, \lambda^m E_{k-m}]^T \quad (13)$$

where the observation matrix becomes

$$\begin{bmatrix} E_{k-1} & E_{k-2} & I_k & I_{k-1} & I_{k-2} \\ \lambda E_{k-2} & \lambda E_{k-3} & \lambda I_{k-1} & \lambda I_{k-2} & \lambda I_{k-3} \\ \vdots & \vdots & \vdots & \vdots & \vdots \\ \lambda^m E_{k-m-1} & \lambda^m E_{k-m-2} & \lambda^m I_{k-m} & \lambda^m I_{k-m-1} & \lambda^m I_{k-m-2} \end{bmatrix}. \quad (14)$$

Now to formulate a variable FFRLS algorithm (VFFRLS), a forgetting factor based on the prediction error is defined and that updates during the operation of the algorithm [13]. The recursive formulation of the VFFRLS algorithm is expressed as

$$\begin{cases} K_k = \dfrac{P_{k-1}\varphi_k}{\lambda_{k-1} + \varphi_k^T P_{k-1}\varphi_k} \\[2mm] \textbf{Update estimation parameter matrix} \\[1mm] \theta_k = \theta_{k-1} + K_k e_k \\[2mm] \textbf{Update co-variance matrix} \\[1mm] P_k = \dfrac{\left(I - K_k \varphi_k^T\right) P_{k-1}}{\lambda_{k-1}}, \end{cases} \quad (15)$$

where the updating formula for the forgetting factor is

$$\lambda_k = 1 - \frac{e_k^2}{1 + K_k^T P_k K_k} \quad (16)$$

with K_k being the gain, P_k represents the covariance matrix, and λ_k represents the forgetting factor.

To identify the equivalent model parameters of the supercapacitor, the model is transformed into a least square mathematical form, starting with defining E as

$$\begin{aligned} E(s) &= U_t(s) \\ &= I(s)\left[\frac{R_L}{R_L C_m s + 1} - \frac{R_p}{R_p C_p s + 1} - R_s\right]. \end{aligned} \quad (17)$$

Let us consider a, b, c, d and e as

$$\begin{aligned} a &= R_s, \quad b = R_L R_p C_m C_p \\ c &= R_p C_p + R_L C_m, \quad d = R_L - R_p - R_s \quad \text{and} \\ e &= R_L R_p C_p - R_L R_p C_m - R_s R_p C_p - R_s R_L C_m. \end{aligned} \quad (18)$$

Applying bilinear transformation $s = \frac{2}{T} \cdot \frac{1-z^{-1}}{1+z^{-1}}$, we have

$$G(Z^{-1}) = \frac{\theta_3 + \theta_4 Z^{-1} + \theta_5 Z^{-2}}{1 - \theta_1 Z^{-1} - \theta_2 Z^{-2}}, \quad (19)$$

where the parameters

$$\theta_1 = \frac{8b - 2T^2}{4b + 2cT + T^2}, \qquad \theta_2 = \frac{-4b + 2cT + T^2}{4b + 2cT + T^2},$$

$$\theta_3 = \frac{-4ab + 2eT + dT^2}{4b + 2cT + T^2}, \qquad \theta_4 = \frac{8ab + 2dT^2}{4b + 2cT + T^2}, \quad (20)$$

$$\text{and} \quad \theta_5 = \frac{-4ab + dT^2 - 2eT}{4b + 2cT + T^2}.$$

Based on these parameters, Eq. (19) could be written as

$$E_k = \theta_1 E_{k-1} + \theta_2 E_{k-2} + \theta_3 I_k + \theta_4 I_{k-1} + \theta_5 I_{k-2} \quad (21)$$

The vector θ is used as a direct identification parameters, and then the identification results of these parameters are used to derive the circuit model parameters R_s, R_p, R_L, C_p, C_m such as

$$\begin{aligned} R_s &= a, \\ R_p &= \frac{(ac + e - (d+a)\tau_1)}{\tau_1 - \tau_2}, \quad R_L = d + a + R_p \\ C_p &= \frac{\tau_1}{R_p}, \quad \text{and} \quad C_m = \frac{\tau_2}{R_L}. \end{aligned} \quad (22)$$

IV. THEORETICAL VALIDATION OF THE PROPOSED MODEL

From the parameter identification procedure done with the help of supercapacitor terminal voltage and current drawn data, the values obtained for the ECM parameters during the charging, discharging and rest period are shown in Table I.

TABLE I
PARAMETER VALUES

Parameter	During charging	During discharging	At rest
R_s	-0.0250	-0.9593	0.0298
R_p	0.0684	4.1870	-0.0635
R_L	9.4639	283.6439	-14.0270
C_p	861.6937	19.6023	-750.3656

Here C_m is supercapacitor voltage-dependent capacitance, modelled as a piece-wise linear function as described in Eq. (1). The variation of main capacitance with respect to the voltage across it is shown in Table II. To better illustrate this,

TABLE II
VARIATION OF MAIN CAPACITANCE, C_m

U_c	C_m during charging	C_m during discharging	C_m at rest
0	1357.79	1412.82	25.43
0.5	1359.63	1386.27	25.71
1	1359.65	1559.28	26.25
1.5	1370.83	1689.27	26.22
2	1389.1	1729.54	26.45
2.5	1396.01	1753.27	27.83
2.7	1394.43	1757.69	28.83

a plot for C_m versus the voltage across it is plotted in Fig. 2,

979-8-3315-3013-6/25 $31.00 © 2025 IEEE

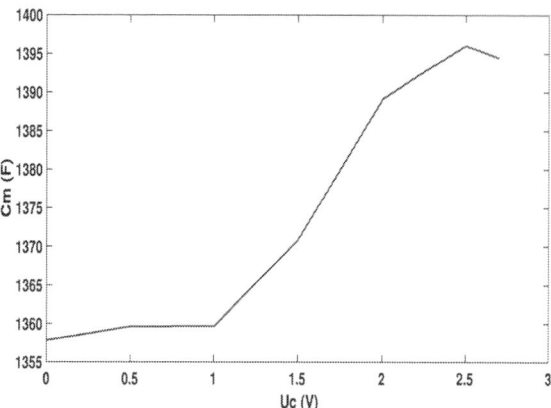

Fig. 2. Variation of main capacitance during charging

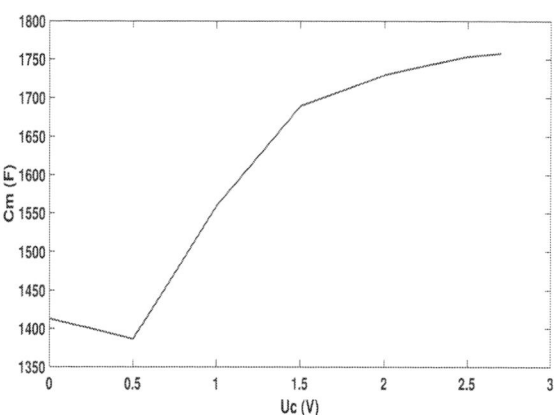

Fig. 3. Variation of main capacitance during discharging

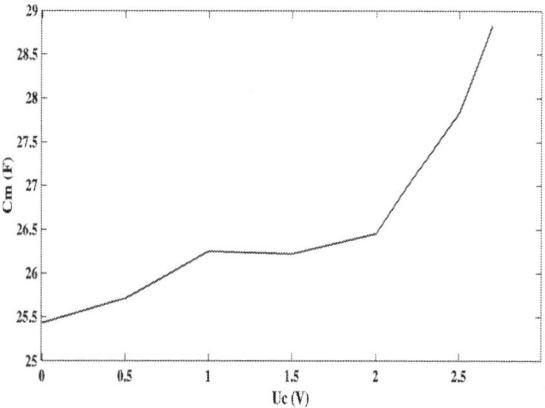

Fig. 4. Variation of main capacitance during the rest period

TABLE III
C_0 AND C_1 VALUES FOR EACH VOLTAGE SEGMENTS DURING CHARGING

U_c	C_0	C_1
0-0.5	1357.79	3.68
0.5-1	1359.63	0.04
1-1.5	1359.65	22.36
1.5-2	1370.83	18.27
2-2.5	1389.1	13.82
2.5-2.7	1396.01	-7.9

TABLE IV
C_0 AND C_1 VALUES FOR EACH VOLTAGE SEGMENTS DURING DISCHARGING

U_c	C_0	C_1
0-0.5	1412.82	-53.1
0.5-1	1386.27	346.02
1-1.5	1559.28	259.98
1.5-2	1689.27	80.54
2-2.5	1729.54	47.46
2.5-2.7	1753.27	22.1

TABLE V
C_0 AND C_1 VALUES FOR EACH VOLTAGE SEGMENTS AT REST

U_c	C_0	C_1
0-0.5	25.43	0.56
0.5-1	25.71	1.08
1-1.5	26.25	-0.06
1.5-2	26.22	0.46
2-2.5	26.45	2.76
2.5-2.7	27.83	5

3 and 4. From these plots, the values for C_0 and C_1 could be inferred, and are tabulated in Tables III, IV and V. Once the ECM parameters are identified from the supercapacitor data; that is the current drawn (10 A), and the terminal voltage, the ECM calculated terminal voltage could be calculated. This serves as a check point for identifying whether the developed ECM and its parameters are able to capture the dynamics of the supercapacitor accurately at different modes of operation. The obtained terminal voltage, calculated from the ECM, along with the experimental data is given in Fig 5. Further, the

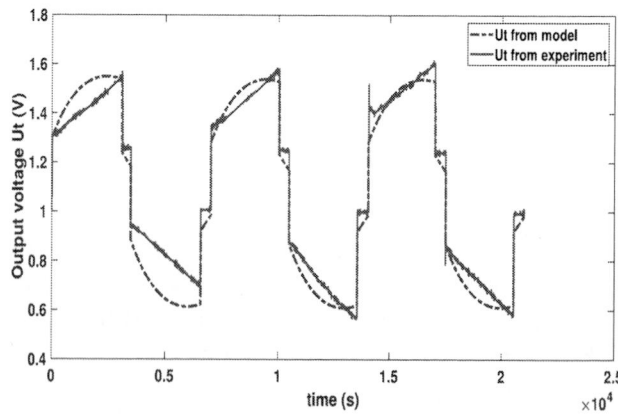

Fig. 5. Validation of supercapacitor model

error between the model calculated terminal voltage and the experimental data is plotted in Fig. 6. From these figures, it could be inferred that the proposed ECM is accurate enough to capture the dynamics of the supercapacitor.

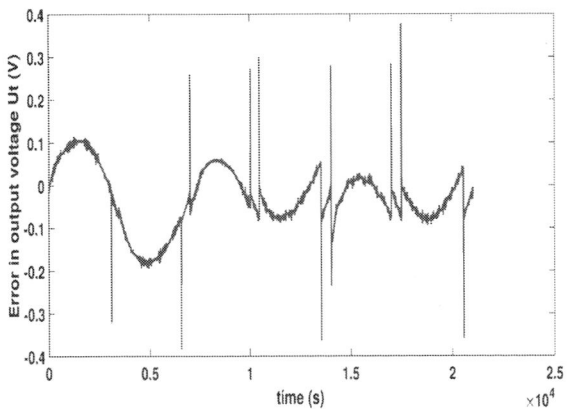

Fig. 6. Error in terminal voltage

A. Rest period dynamics in proposed model

The main objective of the work was to develop an ECM that takes into account the self-discharge (leakage during the rest period) and the corresponding capacitance variation. From the proposed model, the corresponding equivalent model during rest period is shown in Fig. 7.

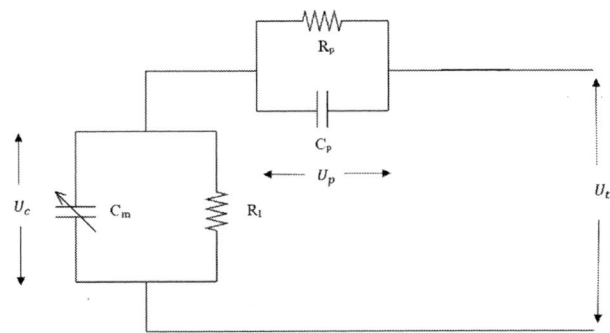

Fig. 7. Rest period supercapacitor model

At rest period, $I = 0$, such that the dynamic equations become

$$U_{p,k} = U_{p,k-1} e^{-\frac{T}{R_p C_p}}, \tag{23}$$

$$U_{c,k} = U_{c,k-1} e^{-\frac{T}{R_L C_m}}, \tag{24}$$

and

$$SoC_k = SoC_{k-1} - \frac{\eta U_{c,k-1} T}{Q R_L}. \tag{25}$$

Then the measurement equation would be

$$U_{t,k} = U_{c,k-1} e^{-\frac{T}{R_L C_m}} - U_{p,k-1} e^{-\frac{T}{R_p C_p}}. \tag{26}$$

From the Eq. (26), the terminal voltage from the proposed model is calculated during the rest period, and is found to be exponentially decreasing with time. Moreover, for analysing the rest period dynamics, the corresponding parameters for the ECM are chosen, as mentioned in Table I. Hence, terminal voltage of 3000F supercapacitor at rest period for 26 days is calculated as shown in FIg. 8. Here, the time duration of 26 days is chosen for comparative purpose, as experimental data for the terminal voltage of 100F supercapacitor bank (6 supercapacitor connected in series) was available, as shown in Fig. 9. From these two figures, it could be inferred that

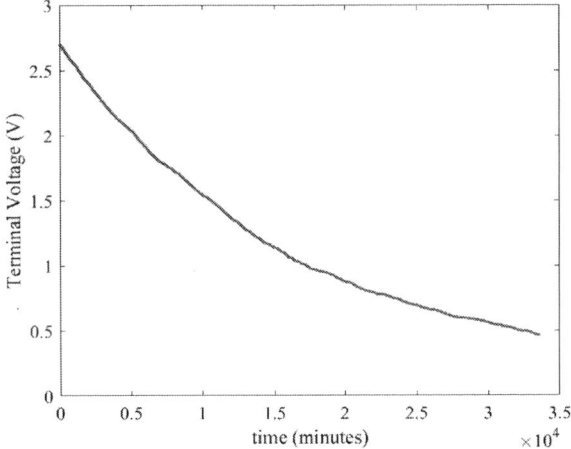

Fig. 8. Terminal voltage of supercapacitor at rest period for 26 days

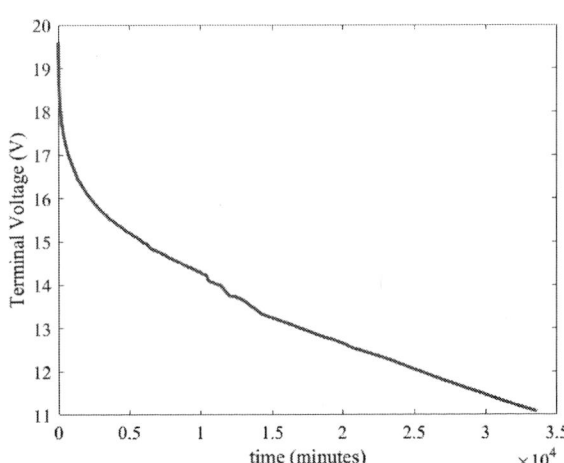

Fig. 9. Terminal voltage of 100F supercapacitor bank at rest period for 26 days

the terminal voltage has as an exponentially decaying nature during rest period, justifying the results obtained in this study.

V. CONCLUSION

This work focus on the development of accurate equivalent circuit model for supercapacitors, such that model-based estimators and controllers could be implemented with high accuracy. Compared to the existing ECMs, the proposed model was designed such that it is efficient in capturing the rest period dynamics. From experimental data, the ECM parameters were found out using variable forgetting factor recursive least squares (VFFRLS) algorithm for charging, discharging and rest period dynamics. Moreover, validation of the proposed model was done by calculating the terminal voltage from the model and by comparing it with the experimental data. From the comparative results, it could be inferred that the proposed model is accurate enough to capture the supercapacitor dynamics.

ACKNOWLEDGMENT

The authors acknowledge the contributions of lab-incharges, EV Technology Research and Collaboration Lab (E-TRAC Lab), NIT Calicut, for their valuable insights and for providing the experimental data.

REFERENCES

[1] I. Jarraya and F. Masmoudi, "An online state of charge estimation for Lithium-ion and supercapacitor in hybrid electric drive vehicle", *Journal of Energy Storage*, 26, 100946, 2019.

[2] C. Liu, Y. Wang, Z. Chen, and Q. Ling, "A variable capacitance based modeling and power capability predicting method for ultracapacitor", *Journal of Power Sources*, 374, pp. 121-133, 2018.

[3] F. Naseri, E. Farjah, T. Ghanbari, Z. Kazemi, E. Schaltz, J. Schanen, "Online Parameter Estimation for Supercapacitor State-of-Energy and State-of-Health Determination in Vehicular Applications," *IEEE Transactions on Industrial Electronics*, 67 (9), pp.7963–7972, 2020.

[4] Z. Liu, Z. Zhao, Y. Qiu, B. Jing, C. Yang, and H. Wu, "Enhanced state of charge estimation for li-ion batteries through adaptive maximum correntropy kalman filter with open circuit voltage correction", *Energy*, 283, 128738, 2023.

[5] C. Shen and C. Wang, "State-of-charge estimation for supercapacitors based on salp swarm algorithm-optimized high and low degree cubature Kalman filters considering temperature uncertainty", *Electrochimica Acta*, 485, 144116, 2024.

[6] A. Nadeau, G. Sharma, and T. Soyata, "State-of-charge estimation for supercapacitors: A Kalman filtering formulation," *in Proc. IEEE Int. Conf. Acoust., Speech Signal Process., Florence, Italy*, 2014, pp. 2194–2198.

[7] A. El Mejdoubi, A. Oukaour, H. Chaoui, Y. Slamani, J. Sabor, and H. Gualous, "Online supercapacitor diagnosis for electric vehicle applications," *IEEE Transactions on Vehicular Technology*, vol. 65, (6), pp. 4241–4252, 2016.

[8] H. Chaoui, A. ElMejdoubi, A. Oukaour, and H. Gualous, "Online system identification for lifetime diagnostic of supercapacitors with guaranteed stability," *IEEE Transactions on Control Systems Technology*, 24, (6), pp. 2094– 2102, Nov. 2016.

[9] A. Soualhi et al., "Supercapacitors ageing prediction by neural networks," *in Proc. 39th Annual Conference of IEEE Industrial Electronics Society*, Vienna, Austria, 2013, pp. 6812–6818.

[10] A. Mejdoubi et al., "Prediction Aging Model for Supercapacitor's Calendar Life in Vehicular Applications", *IEEE Transactions on Vehicular Technology*, 65, (6), pp. 4253 - 4263, 2016.

[11] P. Saha, S. Dey, and M. Khanra, "Modeling and State-of-Charge Estimation of Supercapacitor Considering Leakage Effect", *IEEE Transactions on Industrial Electronics*, 67, (1), pp. 350 - 357, 2020.

[12] S. Maity, M. Saha, P. Saha, M. Khanra, "Fractional Calculas based modelling and state of charge estimation of supercapacitor", *Journal of Energy storage*, 81, 110317, 2024.

[13] T. Long, S. Wang, W. Cao, H. Zhou, C. Fernandez, "An improved variable forgetting factor recursive least square-double extend Kalman filtering based on global mean particle swarm optimization algorithm for collaborative state of energy and state of health estimation of lithium-ion batteries," *Electrochimica Acta*, 450, 142270, 2023.

Adaptive Fractional-Order Model Predictive Control for Power Converters Using Fuzzy Logic

Vani Devarinti
Department of Electrical Engineering
National Institute of Technology
Warangal, India
dv23eem1r30@student.nitw.ac.in

Dr. Swati Devabhaktuni
Department of Electrical Engineering
National Institute of Technology
Warangal, India
swatikjm@nitw.ac.in

Abstract—**This paper introduces an advanced control technique for power converters by integrating Fractional Order Model Predictive Control (FOMPC) with a fuzzy logic-based adaptive tuning system. The fractional-order approach improves control accuracy by capturing the memory and dynamic characteristics often present in power electronic systems. To overcome the limitations of fixed or optimization-based parameter tuning methods, a fuzzy logic controller is used to continuously adjust control parameters in real time. This enables the system to respond effectively to varying operating conditions without relying on an exact mathematical model. A single-phase boost Power Factor Correction (PFC) converter serves as a test platform to validate the proposed method. Simulation results show significant improvements in dynamic response and a notable reduction in Total Harmonic Distortion (THD), all achieved without increasing the converter's switching frequency. These outcomes highlight the method's potential for use in efficient and robust power conversion applications.**

Index Terms—**Fractional-order calculus, fuzzy logic, model predictive control (MPC), power factor correction (PFC), boost converter.**

I. INTRODUCTION

Now a days Model Predictive Control (MPC) has gained considerable attention due to its ability to calculate and predict the future behavior of controlled variables using a system model and select the optimal operation criterion [1]-[3]. Its advantages include fast dynamic response, simple controller implementation, and suitability for multivariate systems, which makes it widely applied in power converter applications [4] - [6]. However, reliance on accurate system models often limits the broader adoption of MPC in power converters [7].

To address this issue, various improved MPC strategies have been developed. For example, deadbeat predictive current control methods ensure reference tracking within a single sampling period [8], while indirect predictive control approaches reduce periodic errors through repetitive control [9]. Techniques such as delta operator-based predictive control reduce hardware costs and improve control accuracy [10]. Although these MPC methods of integer order have improved

performance, they cannot fully capture the fractional order dynamics of power converters, which limits their accuracy [11], [12].

Fractional-order control has shown promise in power converter applications due to its ability to accurately represent system dynamics without increasing the switching frequency [13], [14],[15]. Power converters frequently demonstrate dynamic features that are nonlinear, time-varying, and dependent on memory. Conventional integer-order controllers face challenges in effectively capturing and controlling these behaviors. To address these constraints, this study utilizes a Fractional-Order Model Predictive Controller (FOMPC), exploiting the advanced functionalities of fractional calculus to improve control accuracy. Recently, Fractional-Order Model Predictive Control (FOMPC) has been proposed to leverage these advantages in predictive control frameworks [16]. However, the performance of FOMPC depends heavily on the proper tuning of parameters such as fractional orders and controller gains.

Previously, Grey Wolf Optimization (GWO) has been used for parameter tuning in FOMPC due to its simplicity and fast convergence properties [17], [18]. However, GWO suffers from several drawbacks, including slow convergence in complex systems, dependency on initial parameters, and lack of real-time adaptability, which make it less suitable for highly dynamic systems such as power converters[19]. To address these limitations, this paper introduces Fuzzy Logic as an alternative approach for real-time parameter adjustment in FOMPC. In This study employs fuzzy logic to dynamically modify controller parameters, such as fractional orders and FOPI gains. By offering a rule-based framework, Fuzzy Logic effectively manages uncertainty and nonlinearity, allowing for adaptive parameter adjustment in different operational scenarios.

II. FRACTIONAL-ORDER CALCULUS AND MODELOFTHECONVERTER

A. Fractional-Order Calculus

Fractional Calculus is a generalization of classical calculus that extends the concept of differentiation and integra-

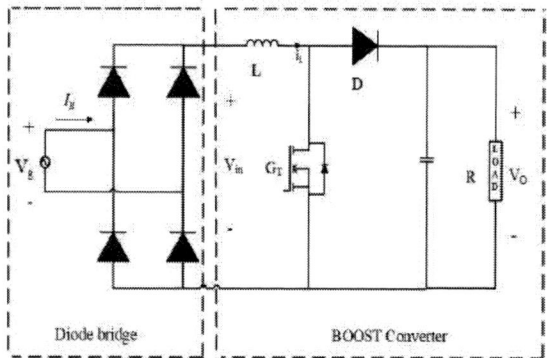

Fig. 1. PFC Boost Circuit topology.

tion to non-integer(fractional) orders. Unlike integer order calculus,fractional-order operations can capture memory and hereditary properties in systems,making it particularly effective for modelling dynamic systems with complex behaviors.

As suggested in [19], the G-L definition is commonly used for discrete-time implimentations and given by:

$$\beta D_t^\alpha f(t) = \lim_{h \to 0} \frac{1}{h^\alpha} \sum_{j=0}^{\lfloor (t-\beta)/h \rfloor} (-1)^j \binom{\alpha}{j} f(t - jh) \quad (1)$$

Where α represent the fractional order, β represent the initial time, and h be the step size. The polynomial coefficient $\omega_j^{(\alpha)} = (-1)^j ((\alpha/j))$ can be derived as follows:

$$\omega_0^{(\alpha)} = 1, \quad \omega_j^{(\alpha)} = \left(1 - \frac{\alpha+1}{j} \right) \omega_{j-1}^{(\alpha)}, \quad j = 1, 2, \ldots \quad (2)$$

This definitions are allows for precise modelling of systems with fractional-order dynamics in practical applications.

B. Fractional-Order State-Space Averaging Model

The State-space averaging is a ell-established method for modelling power converters under dynamic conditions by averaging their behavior over a switching period. Incorporating fractional-order calculus into the state-space framework enhances its ability to represent complex dynamics, such as memory and hereditary effects,which are often present in power electronic system. Here the PFC was chosen as the case study. The topology of the boost PFC converter is shown in fig.1. To develop the fractional-order state-space averaging model, the following assumption are made:

1) The converter operates in Continuous Conduction Mode (CCM).
2) All components are considered ideal.

When the switch is conducting, the system's dynamic behaviour can be described by as

$$\frac{d^\alpha x(t)}{dt^\alpha} = A_{on} x(t) + B_{on} v_s(t) \quad (3)$$

Where $x(t)$ is the state vector, $v_s(t)$ is the input voltage, α is the fractional order ($0 < \alpha \le 1$), and A_{on} and B_{on} are the

coefficient matrices for the conduction phase. Similarly, when the switch is turned off, the dynamics are expressed as:

$$\frac{d^\alpha x(t)}{dt^\alpha} = A_{off} x(t) + B_{off} v_s(t), \quad (4)$$

where A_{off} and B_{off} are the coefficient matrices for the non-conduction phase. To capture the overall system dynamics,the state-space equations for the on and off states are combined using the switching function q(t) Combining these two-state equations shown in (3) and (4) yields: Combining these two-state equations shown in (4) and (5) yields:

$$\frac{d^\alpha x(t)}{dt^\alpha} = \big(q(t) A_{on} + (1 - q(t)) A_{off} \big) x(t) \\ + \big(q(t) B_{on} + (1 - q(t)) B_{off} \big) v_s(t). \quad (5)$$

The switching function $q(t)$ is defined as:

$$q(t) = \begin{cases} 1, & \text{on time period } dT_s, \\ 0, & \text{off time period } (1 - dT_s). \end{cases} \quad (6)$$

For an ideal boost converter, the coefficient matrix can be expressed as:

$$A_{on} = \begin{bmatrix} 0 & 0 \\ 0 & -\frac{1}{CR} \end{bmatrix}, \quad A_{off} = \begin{bmatrix} 0 & -\frac{L}{CR} \\ \frac{1}{L} & -\frac{1}{CR} \end{bmatrix}, \quad (7)$$

$$B_{on} = \begin{bmatrix} \frac{1}{L} \\ 0 \end{bmatrix}, \quad B_{off} = \begin{bmatrix} \frac{1}{L} \\ 0 \end{bmatrix}. \quad (8)$$

This formation enables the dynamic response of the system to be modeled more accurately by adjusting the fractional order. Unlike integer-order models,the fractional-order state-space model captures the system's fractional dynamics,enhancing control precision without increasing the switching frequency.

C. Fractional-Order Predictive Control

The Fractional-Order Model Predictive Control (FOMPC) is designed by building a control framework that combines fractional-order dynamics for increased accuracy and adaptability.This section describes how to design the FOMPC and focusing on the modelling and control method.

from (7) and (8) the fractional-order differential equations of inductor current can be expressed as

- Switch-On:

$$\frac{d^\alpha i_L}{dt^\alpha} = \frac{V_{in}}{L}. \quad (9)$$

- Switch-Off:

$$\frac{d^\alpha i_L}{dt^\alpha} = \frac{V_{in} - V_0}{L}. \quad (10)$$

Where V_{in} and V_o are input voltage after the bridge rectification and the output voltage,respectively. L and i_L are the inductance and inductor current, and α is the order of fractional-order calculus. The FOMPC predicts the next state of the system and determines the optimal control action by minimizing a cost function. The cost function evaluates the error between the predicted and reference values.

979-8-3315-3013-6/25 $31.00 © 2025 IEEE

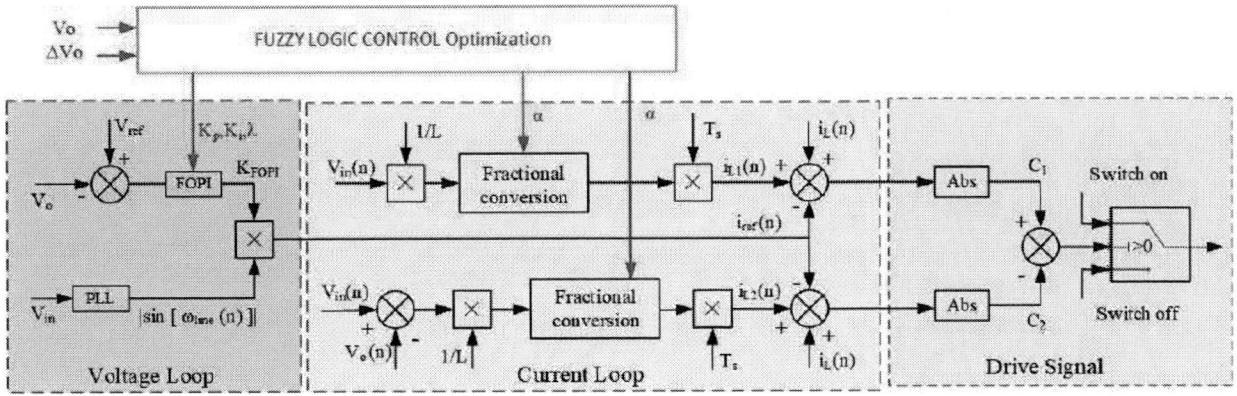

Fig. 2. Control diagram of fractional predictive control with adaptive parameters

Predictive Model: The predictive model for the inductor current is expressed as follows:

- Switch-On Prediction:

$$i_{L_1}(n) = p^{1-\alpha}\left(\frac{v_{\text{in}}(n)}{L}\right)T_s + i_L(n), \qquad (11)$$

- Switch-Off Prediction:

$$i_{L_2}(n) = p^{1-\alpha}\left(\frac{v_{\text{in}}(n) - v_0(n)}{L}\right)T_s + i_L(n). \qquad (12)$$

Cost Function: The cost function compares the predicted currents with the reference current:

$$C_1 = |i_{L_1}(n) - i_{\text{ref}}(n)|, \qquad (13)$$
$$C_2 = |i_{L_2}(n) - i_{\text{ref}}(n)|. \qquad (14)$$

The control action is chosen to minimize the cost:

$$C = \min(C_1, C_2). \qquad (15)$$

The reference current $i_{\text{ref}}(n)$ is derived using the outer voltage loop controller, such as a fractional-order PI (FOPI) regulator

$$i_{\text{ref}}(n) = K_{\text{FOPI}}|\sin(\omega_{\text{line}}(n))|, \qquad (16)$$

where K_{FOPI} is the output of the FOPI controller, $|\sin(\omega_{\text{line}}(n))|$ is a rectified sinusoidal waveform with the line frequency.

III. FUZZY LOGIC CONTROLLER (FLC)

Fuzzy logic is a rule-based approach that uses human-like reasoning to map inputs to outputs. It is especially suited for systems with uncertainties, non-linearities, or where mathematical models are complex. In this work, fuzzy logic is applied to adaptively tune the parameters of the fractional-order model predictive control.

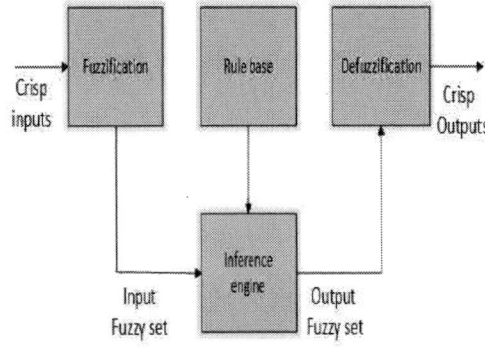

Fig. 3. Block diagram of FLC.

A. Adaptive Parameters Based on FLC

When system parameters vary due to chane in operating conditions,it is crucial to dynamically adjust the control parameters for optimal performance.This paper adopts a fuzzy fuzzy logic based approach to adaptively tune the parameters of the FOPI controller.

The fuzzy logic-based adaption approach uses a rule-based system to dynamically tune parameters based on real-time inputs.unlike metaheuristic algoriths such as GWO,fuzzy logic eliminates the need for iterative computation,providing a lightweight and adaptive control mechanism.

Fuzzy logic control employs a fuzzy logic controller (FLC) composed of three phases shown in fig.3.

- Fuzzification: Converts the crisp input values into fuzzy linguistic variables using membership functions.
- Inference: It evaluates the degree of activation of each rule and combines their outputs.
- Defuzzification: Converts the aggregated fuzzy output into crisp values for the parameters.The centriod method is used for defuzzification due to its accuracy and stability.

The control block diagram of the proposed control method is shown fig.2. The control framework integrates an outer

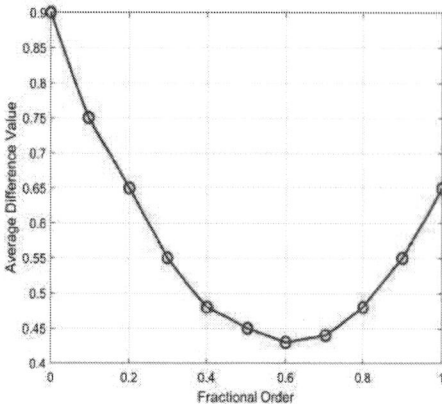

Fig. 4. Average difference value at various orders.

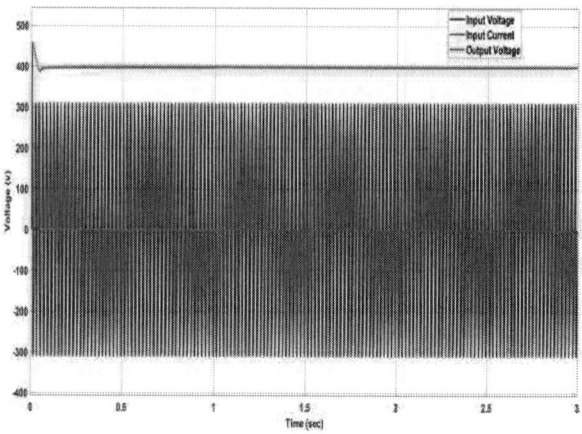

Fig. 5. Dynamic response of FOMPC.

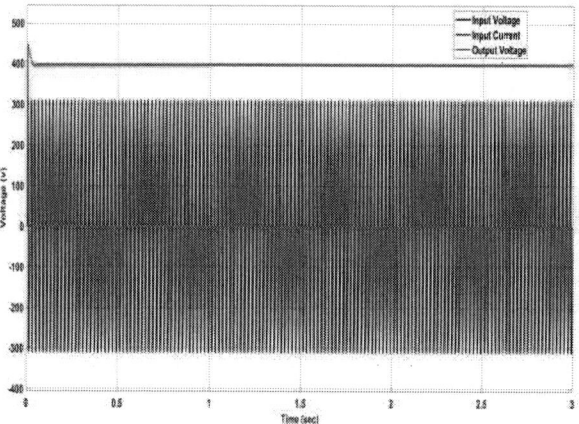

Fig. 6. Dynamic response of FUZZY FOMPC.

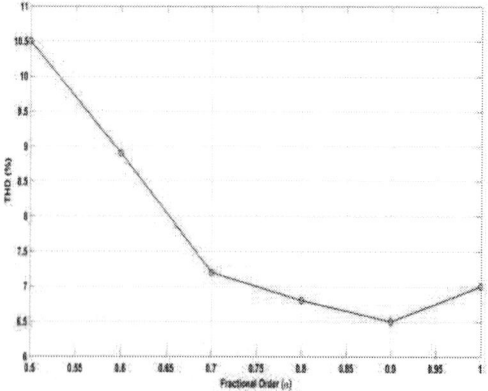

Fig. 7. THD of the input current at various fractional orders

voltage loop,an inner current loop, and a parameter optimization module.The outer voltage loop utilizes a fractional-order proportional-integral(FOPI) controller to generate a reference current i_{ref} .The FOPI controller ensures that the output voltage remains close to the desired reference voltage,and the current inner loop employs the fractional-order predictive control mode to regulate the inductor current.It minimizes the error between the actual inductor current and the reference current by adjusting the control signals in real time.This block dynamically optimizes the parameters (α, k_p, k_i) based on real-time operating conditions.The optimized parameters and predictive control strategy determine the on/off states of the power converter switches.The output of the control system drives the switches in the power converter.

IV. SIMULATION RESULT

This section presents the simulation setup and results to validate the effectiveness of the proposed fractional-order model predictive control (FOMPC) with fuzzy logic-based adaptive parameter tuning. The simulations were conducted in MATLAB/Simulink, with a single-phase power factor correction (PFC) converter as the case study. For discrete systems, fractional-order differentiation is approximated using a rational approximation, such as an IIR filter[.The simulation parameters for the boost PFC converter are summarized in Table I.Unlike fixed fractional-order values obtained via methods like the branch and bound 0.7, The initial parameters of the PI regulator,K_p and K_i 1.0,and 1 are obtained using the Ziegler-Nichols (Z-N) method , and furuned by the fuzzy logic system for improved response and stability under varying conditions.

Fig.4. illustrates the influence of fractional-order (α) variation on the average difference value between the predicted and sampled values in the FOMPC system.Adjusting the fractional order significantly affects the performance of the predictive controller. As the fractional order deviates from unity (integer-order MPC), the system begins to leverage the fractional-order dynamics, resulting in reduced average differences.

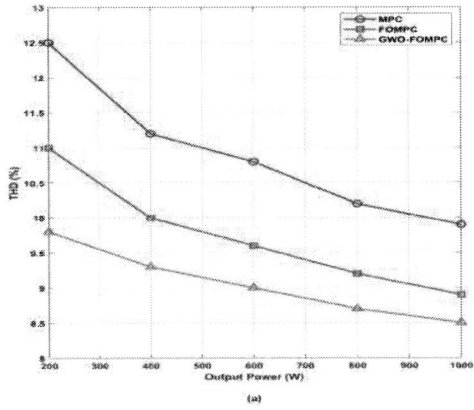 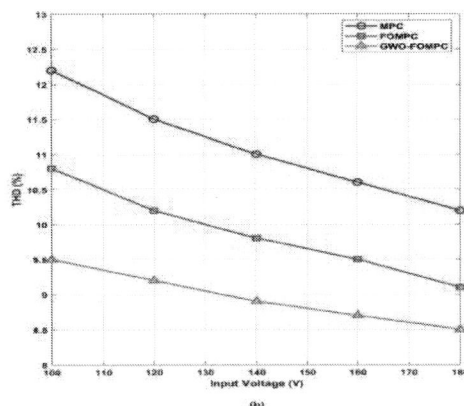

Fig. 8. Comparison of THD of the input current based on various control methods (a) various output power levels and (b) various input voltage levels

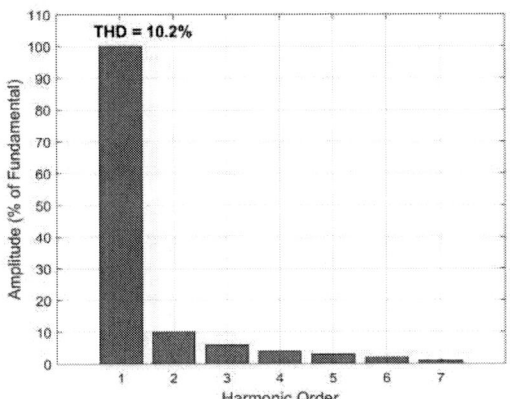

Fig. 9. THD spectrum for FOMPC.

TABLE I
MAIN PARAMETERS OF THE SIMULATION FOR THE BOOST PFC
CONVERTER

Parameters	Specifications
Input Voltage	AC 120-220V 50Hz
Output Voltage	DC 400V
Output Power	1 kW
Inductor	1000 μH
Capacitor	1000 μF
Predicted Frequency	100 kHz

Fig.5-6 illustrate the dynamic response for FOMPC and FUZZY FOMPC,Fuzzy-FOMPC outperforms traditional FOMPC in terms of faster settling time and reduced overshoot.fig.7. shows the THD variation curve of the input current when the various fractional order is changed.When the fractional order is changed the quality of the power at input terminals will effected.Figs.8 represents a comparision of comparison of Total Harmonic Distortion (THD) in the input current using three different control methods, in fig.8.(a) shows THD vs.Output power. At lower power levels the THD is high for all methods, but Fuzzy-FOMPC starts with a better value.As power increases, THD decreases, and Fuzzy-FOMPC maintains the best harmonic reduction.This suggests that fractional-order control, especially with Fuzzy optimization, improves THD performance at all output power levels. In fig.8.(b) shows THD vs.Input current. At lower voltage levels , all methods perform similarly. At higher voltages, THD increases significantly for MPC, while FOMPC and Fuzzy-FOMPC maintain lower THD values.

FUZZY-FOMPC consistently provides the best THD reduction, demonstrating its robustness under varying input voltages. Fig.9-10 represent the Output Power THD spectrum for FOMPC and Fuzzy FOMPC. The dynamic response of the PFC converter based od different control methods. It is seen that the Fuzzy FOPMC can reduce the transient time compared to FOMPC and the Fuzzy-FOMPC.the Fuzzy-FOMPC proposed in this paper can reduce 10.2% to 6.8% compared with FOMPC control. It is intferred that using Fuzzy has better power quality compared to FOMPC.Its is seen that proposed control method can perform effectively and reduced the THD.The result shown that the control method can be used widely in power conveter applications.

V. CONCLUSION

In this paper,a Fuzzy FOMPC method with adaptive parameters is proposed for power converters,and there is a several advantages in this proposed method. The proposed method has high control accuracy and the fractional calculus is introduced into the predictive control model.An adaptive parameters design method based on FLC algorithm is presented to optimized the proposed FOMPC and it improve the performance by reducing the THD and enhancing the transient response,the proposed system achieves greater control precision without increasing computatoonal complexity or switching frequency.

Simulation results demonstrated that the proposed method effectively reduces total harmonic distortion (THD) of the

979-8-3315-3013-6/25 $31.00 © 2025 IEEE

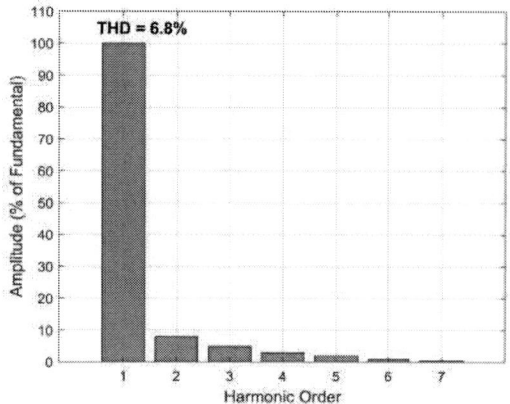

Fig. 10. THD spectrum for FUZZY FOMPC.

input current and improves transient response.Compared to FOMPC the fuzzy-adaptive FOMPC achieved superior performance with a 10.2% reduction in THD and a 6.8% reduction in transient time. The proposed approach offers a flexible

TABLE II
PERFORMANCE PARAMETERS USING VARIOUS CONTROL METHOS

Specifications	MPC	FOMPC	Fuzzy-FOMPC
Overshoot	486	460	440
undershoot	375	386	398
Settling Time (sec)	0.3	0.15	0.08
THD	12.28	10.2	6.26

and robust solution for modern power electronics applications, including renewable energy systems, electric vehicle chargers, and industrial converters. Future work will focus on implementing the fuzzy-adaptive FOMPC in different converter topologies and verifying its real-time performance on embedded systems.

REFERENCES

[1] Z. Karami, Q. Shafiee, Y. Khayat, M. Yaribeygi, T. Dragičević and H. Bevrani, "Decentralized Model Predictive Control of DC Microgrids With Constant Power Load," in IEEE Journal of Emerging and Selected Topics in Power Electronics, vol. 9, no. 1, pp. 451-460, Feb. 2021.

[2] H. Chen, D. Wang, S. Tang, X. Yin, J. Wang and Z. J. Shen, "Continuous Control Set Model Predictive Control for Three-Level Flying Capacitor Boost Converter With Constant Switching Frequency," in IEEE Journal of Emerging and Selected Topics in Power Electronics, vol. 9, no. 5, pp. 5996-6007, Oct. 2021.

[3] Y. Li, F. Diao and Y. Zhao, "Simplified Two-Stage Model Predictive Control for a Hybrid Multilevel Converter With Floating H-Bridge," in IEEE Transactions on Power Electronics, vol. 36, no. 4, pp. 4839-4850, April 2021.

[4] N. Zhao, J. Liu, Y. Ai, J. Yang, J. Zhang, and X. You, "Power-linked predictive control strategy for power electronic traction transformer," IEEE Trans. Power Electron., vol. 35, no. 6, pp. 6559–6571, Jun. 2020.

[5] D. Zhou, J. Wang, Y. Li, J. Zou, and K. Sun, "Model predictive power control of grid-connected quasi single-stage converters for highefficiency low-voltage ESS integration," IEEE Trans. Ind. Electron., vol. 69, no. 2, pp. 1124–1134, Feb. 2022.

[6] S. Vazquez, P. Acuna, R. P. Aguilera, J. Pou, J. I. Leon, and L. G. Franquelo, "DC-link voltage-balancing strategy based on optimal switching sequence model predictive control for single-phase H-NPC converters," IEEE Trans. Ind. Electron., vol. 67, no. 9, pp. 7410–7420, Sep. 2020.

[7] D. Wang et al., "Model predictive control using artificial neural network for power converters," IEEE Trans. Ind. Electron., vol. 69, no. 4, pp. 3689–3699, Apr. 2022.

[8] J. Wang, Y. Tang, P. Lin, X. Liu, and J. Pou, "Deadbeat predictive current control for modular multilevel converters with enhanced steadystate performance and stability," IEEE Trans. Power Electron., vol. 35, no. 7, pp. 6878–6894, Jul. 2020.

[9] M. Zhu and G. Li, "Modular multilevel converter with improved indirect predictive controller," IEEE J. Emerg. Sel. Topics Power Electron., vol. 7, no. 2, pp. 976–989, Jun. 2019.

[10] N. Chen, J. Liu, S. Ma, and T. Wei, "Digital current-mode controller using delta operator and advance sampling predictive control for high-frequency DC–DC switching converters," IEEE J. Emerg. Sel. Topics Power Electron., vol. 9, no. 5, pp. 6272–6281, Oct. 2021.

[11] X. Li, X. Wang, P. Gao, and Y. Gu, "Model predictive current control algorithm based on joint modulation strategy for low-inductance PMSM," IEEE Trans. Power Electron., vol. 37, no. 1, pp. 806–819, Jan. 2022.

[12] C. Ma, J. Rodriguez, C. Garcia, and F. De Belie, "Integration of reference current slope based model-free predictive control in modulated PMSM drives," IEEE J. Emerg. Sel. Topics Power Electron., early access, Mar. 14, 2022, doi: 10.1109/JESTPE.2022. 3159586.

[13] X. Chen, Y. Chen, B. Zhang, and D. Qiu, "A modeling and analysis method for fractional-order DC–DC converters," IEEE Trans. Power Electron., vol. 32, no. 9, pp. 7034–7044, Sep. 2017.

[14] S.-W. Seo and H. H. Choi, "Digital implementation of fractional orderPID-type controller for boost DC–DC converter," IEEE Access, vol. 7, pp. 142652–142662, 2019.

[15] M.-Y. Li, K.-D. Lu, Y.-X. Dai, and G.-Q. Zeng, "Fractional-order predictive functional control of industrial processes with partial actuator failures," Complexity, vol. 2020, pp. 1–26, Feb. 2020

[16] X. Lin, J. Liu, F. Liu, Z. Liu, Y. Gao, and G. Sun, "Fractional-order sliding mode approach of buck converters with mismatched disturbances," IEEE Trans. Circuits Syst. I, Reg. Papers, vol. 68, no. 9, pp. 3890–3900, Sep. 2021.

[17] J. Agarwal, G. Parmar, R. Gupta, and A. Sikander, "Analysis of grey wolf optimizer based fractional order PID controller in speed control of DC motor," Microsyst. Technol., vol. 24, no. 12, pp. 4997–5006, 2018.

[18] C. Komathi and M. G. Umamaheswari, "Design of gray wolf optimizer algorithm-based fractional order PI controller for power factor correction in SMPS applications," IEEE Trans. Power Electron., vol. 35, no. 2, pp. 2100–2118, Feb. 2020.

[19] Z. Ke et al., "Fractional-Order Model Predictive Control With Adaptive Parameters for Power Converter," in IEEE Journal of Emerging and Selected Topics in Power Electronics, vol. 11, no. 3, pp. 2650-2660, June 2023.

Magnetic Loss Comparison for Link Current Optimized Triple Active Bridge Converters with Soft Switching

Harikrishnan A, Silpashree Sahu, Varri Chandra Sekhar Pavan Kumar, Dipankar De
Electrical Engineering, School of Electrical and Computer Sciences
Indian Institute of Technology Bhubaneswar
Argul Campus, Jatni, Odisha-752050, INDIA
Email: 21ee01013@iitbbs.ac.in, s22ee09002@iitbbs.ac.in, vcpk10@iitbbs.ac.in, dipankar@iitbbs.ac.in

Abstract—**This paper focuses on the loss comparison in the magnetic design for the triple port dual-transformer triple active bridge and triple active bridge based on three-winding-transformer under optimized link current stresses. Triple phase shift modulation techniques are considered for the converters with a single inner phase shift and two outer phase shifts. Link current optimization is carried out by considering the aggregate currents in the link and considering soft switching conditions for the entire range of converter operation. The loss calculations in the transformer and inductors are carried out considering high-frequency effects present in the magnetics. The mathematical calculations and loss estimations are reported. Efficiency comparison and loss evaluations are performed for a 3.5 kW wide bandgap-based three-port converter.**

Index Terms—**Triple active bridge converter. Double (dual) transformer triple active bridge, link current optimization, loss computation.**

I. INTRODUCTION

Triple active bridge (TAB) converters are highly attractive in the integration of multiple sources and loads of different in micro-grid EV applications [1]. Several multi-port power converters are reported for these applications; however, TAB converters provide galvanic isolation between the different DC ports, effective VA utilization of the switches, soft switching operation for a wide range of operation, and bidirectional power transfer capability. The most widely used configurations within tripe active bridge configurations are

1) Triple active bridge based on dual transformers [2]–[4] (Fig. 1).
2) Triple active bridge converter based on three winding transformers [6]–[8] (Fig. 2).

The optimization of the converter performances in these bridge converters is widely studied in the literature [9]–[11]. In [9], optimization of the efficiency of the TAB converter is reported. The optimization of the RMS link current of the TAB converter is examined in [11]. The soft switching performance in the active bridge converter is considered in [12]–[14].

This work is supported by Department of Science and Technology, India under project Grant CRG/2022/000650.

However, the influence of optimization techniques on the performance of different triple active bridge configurations is not studied in the literature, especially considering the high-frequency losses in the magnetic components.

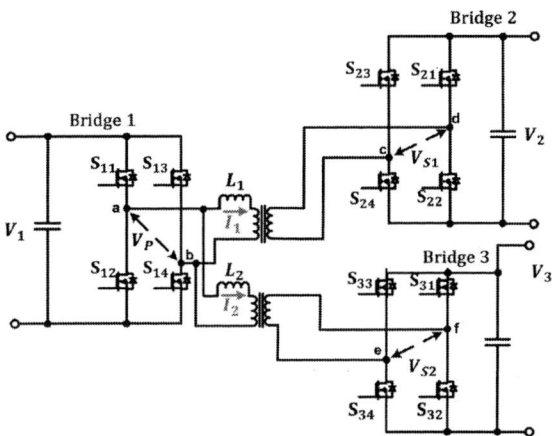

Fig. 1: Dual Transformer Triple Active Bridge

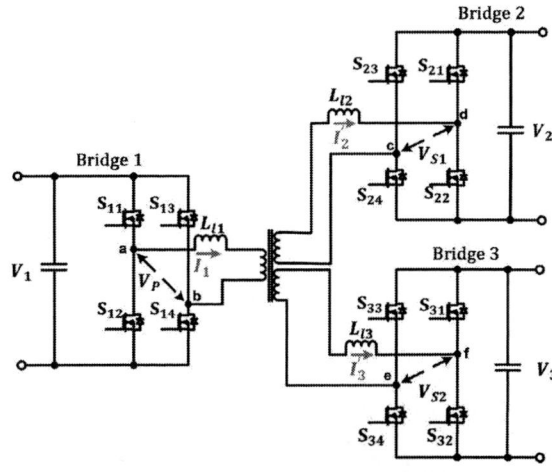

Fig. 2: Triple Active Bridge

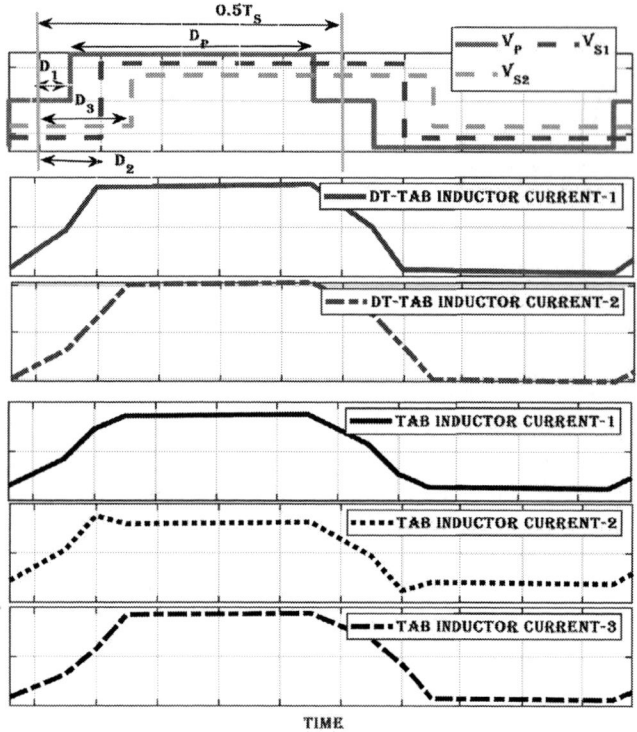

Fig. 3: Illustrative waveforms for Triple phase shift modulation and associated link current for DT-TAB and TAB

In this paper, a loss comparison of the TAB and DT-TAB converters is carried out under the same operating conditions (soft-switched, optimized link peak current). The mathematical calculations and loss estimations are reported. The paper is organized as follows. Section II explains the basic mathematical expressions derived for optimization for the triple active bridge and the dual transformer triple active bridge for optimization of the peak link current. Section III compares the performances of the converter in terms of the losses in the magnetic components and in the semiconductor switches under the optimized modulation obtained in Section II. Finally, Section IV gives the conclusion of the work.

II. OPTIMIZATION OF LINK CURRENT IN TAB AND DT-TAB

In this section, the optimization of the aggregate link peak current in TAB and DT-TAB is discussed. The boundaries for different modes are given in Table I. The specifications of the power converter selected for TAB and DT-TAB are shown in Table VI in the Appendix. For effective comparison, the input/output power and switching frequency of both converters are taken as identical. Table VII in the appendix shows the selection of the component with suitable magnetic cores/windings and semi-conductor switches. The optimization problem is solved in this section with the operation of ideal converters (without losses), and the data obtained from the optimization is used for the loss evaluation of the for comparison. The arrangement of the three-winding transformer for TAB

transformer is shown in Fig. 12 in Appendix. The windings are designed with a number of parallel wires considering high frequency skin depth.

TABLE I: Mode Boundaries

Mode:no	Boundary conditions
Mode-1	$D_1 < D_2 < D_3 < D_1 + D_P < 1$
Mode-2	$D_2 < D_1 < D_3 < D_1 + D_P < 1$
Mode-3	$D_2 < D_3 < D_1 < D_1 + D_P < 1$
Mode-4	$D_3 < D_2 < D_1 < D_1 + D_P < 1$
Mode-5	$D_3 < D_1 < D_2 < D_1 + D_P < 1$
Mode-6	$D_1 < D_3 < D_2 < D_1 + D_P < 1$

The optimization problem for TAB and DT-TAB can be generalized as

$$Minimize \quad f = \frac{1}{n} \sum_n [I_{pn}] \quad (1)$$

subjected to $\sum A_i D_j \leq B_i$ and $P_2 = P_2^*$, $P_3 = P_3^*$. The non-equality constraints are the mode boundaries and the soft switching boundaries. These constraints along with power expressions in different modes, respectively, for DT-TAB and TAB are shown in Table III/IV and Table V/VI. $n = 2$ for DT-TAB and $n = 3$ for the TAB converter.

For ease of analysis, the entire optimization is done in per-unit scale. The base power and current for DT-TAB are given by (2) and (3) and k_1 k_2 are the voltage transformation ratios and g is a scale factor to represent the power and currents of both bridges at the same base.

$$P_b = \frac{n_1 V_1 V_2}{8 f_s L_1} \quad (2)$$

$$I_b = \frac{n_1 V_2}{4 f_s L_1} \quad (3)$$

$$k_1 = \frac{V_1}{n_1 V_2}; \quad k_2 = \frac{V_1}{n_2 V_3}; \quad g = \frac{n_2 V_3 L_1}{L_2 n_1 V_2} \quad (4)$$

Similarly in case of TAB, (5) and (6) gives base quantities. L_{12}, L_{13} and L_{23} is obtained from the delta equivalent model of the TAB. k_1, k_2 and k_3 are the voltage transformation ratios, and g_1 and g_2 are the scale factors for representing different quantities of TAB in the base considered.

$$P_b = \frac{n_1 V_1 V_2}{8 f_s L_{12}} \quad (5)$$

$$I_b = \frac{n_1 V_2}{4 f_s L_{12}} \quad (6)$$

$$k_1 = \frac{V_1}{n_1 V_2}; \quad k_2 = \frac{V_1}{n_2 V_3}; \quad k_3 = \frac{n_2 V_3}{n_1 V_2} \quad (7)$$

$$g_1 = \frac{n_2 V_3 L_{12}}{n_1 V_2 L_{13}}; \quad g_2 = \frac{L_{12}}{L_{23}} \quad (8)$$

The variation of f [in (1)] for TAB and DT-TAB is shown in Fig. 4 for different power transfers at port-2 and port-3. The associated bridge output currents are shown in Fig. 5, Fig. 6, and Fig. 7 respectively. It can be seen that, except for certain zones, the values of these currents are close for TAB

TABLE II: DT TAB power expression in different operating modes

Mode	Power expression		
1	P_2	$=$	$-2(2D_1{}^2 - 4D_1D_2 + 2D_1\mathrm{Dp} + 2D_2{}^2 - 2D_2\mathrm{Dp} + \mathrm{Dp}^2 - \mathrm{Dp})$
	P_3	$=$	$-2g(2D_1{}^2 - 4D_1D_2 + 2D_1\mathrm{Dp} + 2D_2{}^2 - 2D_2\mathrm{Dp} + \mathrm{Dp}^2 - \mathrm{Dp})$
2	P_2	$=$	$-2(2D_1{}^2 - 4D_1D_2 + 2D_1\mathrm{Dp} + 2D_2{}^2 - 2D_2\mathrm{Dp} + \mathrm{Dp}^2 - \mathrm{Dp})$
	P_3	$=$	$-2g\,\mathrm{Dp}(2D_1 - 2D_3 + \mathrm{Dp} - 1)$
3	P_2	$=$	$-2\,\mathrm{Dp}(2D_1 - 2D_2 + \mathrm{Dp} - 1)$
	P_3	$=$	$-2g\,\mathrm{Dp}(2D_1 - 2D_3 + \mathrm{Dp} - 1)$
4	P_2	$=$	$-2\,\mathrm{Dp}(2D_1 - 2D_2 + \mathrm{Dp} - 1)$
	P_3	$=$	$-2g\,\mathrm{Dp}(2D_1 - 2D_3 + \mathrm{Dp} - 1)$
5	P_3	$=$	$-2g(2D_1{}^2 - 4D_1D_3 + 2D_1\mathrm{Dp} + 2D_3{}^2 - 2D_2\mathrm{Dp} + \mathrm{Dp}^2 - \mathrm{Dp})$
	P_2	$=$	$-2\,\mathrm{Dp}(2D_1 - 2D_3 + \mathrm{Dp} - 1)$
6	P_2	$=$	$-2\,\mathrm{Dp}(2D_1 - 2D_2 + \mathrm{Dp} - 1)$
	P_3	$=$	$-2g\,\mathrm{Dp}(2D_1 - 2D_3 + \mathrm{Dp} + 1)$

TABLE III: Peak currents and ZVS conditions for DT TAB

Mode	Peak-currents	ZVS conditions
1	$I_1 = 2D_2 - 2D_1 - 2D_p + D_p k_1 + 1$ $I_2 = g(2D_3 - 2D_1 - 2D_p + D_p k_2 + 1)$	$2(1+g)D_1 - 2D_2 - 2gD_3 - (k_1 + k_2 g)D_p < 0$ $2D_2 k_1 - 2D_1 k_1 - k_1 D_p + 1 > 0$ $2D_3 k_2 - 2D_1 k_2 - k_2 D_p + 1 > 0$ $-2(1+g)D_1 + 2D_2 + 2gD_3 + D_p(k1 - 2 + g(k2 - 2)) + (1+g) > 0$
2	$I_1 = 1 - k_1 D_p$ $I_2 = g(2D_3 k_2 - 2D_1 k_2 - k_2 D_p + 1)$	$1 - k_1 D_p > 0$ $2(g-1)D_1 + 2D_2 - 2gD_3 - D_p(k_1 + k_2 g) + 1 + g < 0$ $2D_3 k_2 - 2D_1 k_2 - k_2 D_p + 1 > 0$ $-2(1+g)D_1 + 2D_2 + 2gD_3 + D_p(k_1 - 2 + g(k2 - 2)) + (1+g) > 0$
3	$I_1 = (1 - D_p k_1)$ $I_2 = g(1 - D_p k_2)$	$1 - D_p k_1 > 0$ $1 - D_p k_2 > 0$ $-2(1+g)D_1 + 2D_2 + 2gD_3 - (k_1 + k_2 g)D_p + (1+g) < 0$ $-2(1+g)D_1 + 2D_2 + 2gD_3 + D_p(k_1 - 2 + g(k_2 - 2)) + (1+g) > 0$
4	$I_1 = (1 - D_p k_1)$ $I_2 = g(1 - D_p k_2)$	$1 - D_p k_1 > 0$ $1 - D_p k_2 > 0$ $-2(1+g)D_1 + 2D_2 + 2gD_3 - (k_1 + k_2 g)D_p + (1+g)$ $-2(1+g)D_1 + 2D_2 + 2gD_3 + D_p(k_1 - 2 + g(k_2 - 2)) + (1+g) > 0$
5	$I_1 = 2D_2 - 2D_1 - 2D_p + D_p k_1 + 1$ $I_2 = g(1 - k_2 D_p)$	$1 - D_p k_2 > 0$ $2(1-g)D_1 - 2D_2 + 2gD_3 - D_p(k_1 + k_2 g) + 1 + g < 0$ $2D_2 - 2D_1 - 2D_p + D_p k_1 + 1$ $-2(1+g)D_1 + 2D_2 + 2gD_3 + D_p(k_1 - 2 + g(k_2 - 2)) + (1+g) > 0$
6	$I_1 = 2D_2 - 2D_1 - 2D_p + D_p k_1 + 1$ $I_2 = g(2D_3 - 2D_1 - 2D_p + D_p k_2 + 1)$	$2(1+g)D_1 - 2D_2 - 2gD_3 - (k_1 + k_2 g)D_p < 0$ $2D_3 k_2 - 2D_1 k_2 - k_2 D_p + 1 > 0$ $2D_2 k_1 - 2D_1 k_1 - k_1 D_p + 1 > 0$ $-2(1+g)D_1 + 2D_2 + 2gD_3 + D_p(k_1 - 2 + g(k_2 - 2)) + (1+g) > 0$

and DT-TAB for similar operating points. These optimized values are utilized in the next section for loss calculation in the converters.

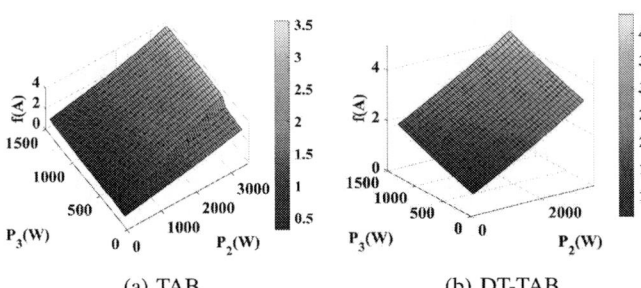

(a) TAB

(b) DT-TAB

Fig. 4: Optimization function

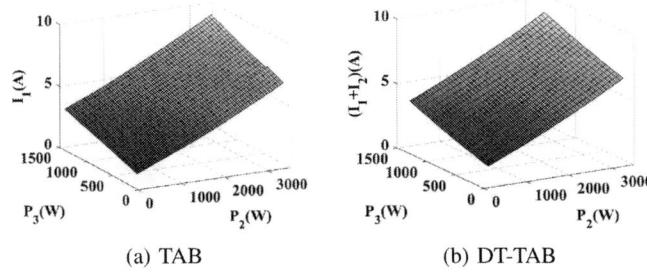

(a) TAB

(b) DT-TAB

Fig. 5: Bridge-1 currents

III. MAGNETIC DESIGN AND LOSS COMPARISON

The design of the inductors and transformers of the DT-TAB and TAB converter was carried out considering the high frequency effects using [15]. The high frequency resistances in the converter are calculated using the formulas derived in [16]. The distribution of J and H are shown in Fig. 8 and Fig. 9 respectively. It can be observed that the winding arrangement affects the losses in TAB whereas in DT-TAB the losses are only dependent on the operating point. The loss distribution of the different losses in the converter is shown in Fig. 10 in TAB and DT-TAB for different percentage loadings at two ports. It can be observed that when the second port is lightly loaded the

TABLE IV: TAB power expression at different operating modes

Mode	Power-Expression
1	$P_2 = 2D_p - 4D_1^2 - 4D_2^2 - 2D_p^2 + 8D_1D_2 - 4D_1D_p + 4D_2D_p + (4g_2k_3(D_2^2 - 2D_2D_3 + D_2 + D_3^2 - D_3))/k_1$ $P_3 = -g_1(4D_1^2 - 8D_1D_3 + 4D_1D_p + 4D_3^2 - 4D_3D_p + 2D_p^2 - 2D_p) - (4g_2k_3(D_2^2 - 2D_2D_3 + D_2 + D_3^2 - D_3))/k_1$
2	$P_2 = (4g_2k_3(D_2^2 - 2D_2D_3 + D_2 + D_3^2 - D_3))/k_1 - 2D_p(2D_1 - 2D_2 + D_p - 1)$ $P_3 = -g_1(4D_1^2 - 8D_1D_3 + 4D_1D_p + 4D_3^2 - 4D_3D_p + 2D_p^2 - 2D_p) - (4g_2k_3(D_2^2 - 2D_2D_3 + D_2 + D_3^2 - D_3))/k_1$
3	$P_2 = (4g_2k_3(D_2^2 - 2D_2D_3 + D_2 + D_3^2 - D_3))/k_1 - 2D_p(2D_1 - 2D_2 + D_p - 1)$ $P_3 = -2D_pg_1(2D_1 - 2D_3 + D_p - 1) - (4g_2k_3(D_2^2 - 2D_2D_3 + D_2 + D_3^2 - D_3))/k_1$
4	$P_2 = -2D_p(2D_1 - 2D_2 + D_p - 1) - (4g_2k_3(D_2^2 - 2D_2D_3 - D_2 + D_3^2 + D_3))/k_1$ $P_3 = (4g_2k_3(D_2^2 - 2D_2D_3 - D_2 + D_3^2 + D_3))/k_1 - 2D_pg_1(2D_1 - 2D_3 + D_p - 1)$
5	$P_2 = 2D_p - 4D_1^2 - 4D_2^2 - 2D_p^2 + 8D_1D_2 - 4D_1D_p + 4D_2D_p - (4g_2k_3(D_2^2 - 2D_2D_3 - D_2 + D_3^2 + D_3))/k1$ $P_3 = (4g_2k_3(D_2^2 - 2D_2D_3 - D_2 + D_3^2 + D_3))/k_1 - 2D_pg_1(2D_1 - 2D_3 + D_p - 1)$
6	$P_2 = 2D_p - 4D_1^2 - 4D_2^2 - 2D_p^2 + 8D_1D_2 - 4D_1D_p + 4D_2D_p - (4g_2k_3(D_2^2 - 2D_2D_3 - D_2 + D_3^2 + D_3))/k_1$ $P_3 = (4g_2k_3(D_2^2 - 2D_2D_3 - D_2 + D_3^2 + D_3))/k_1 - g_1(4D_1^2 - 8D_1D_3 + 4D_1D_p + 4D_3^2 - 4D_3D_p + 2D_p^2 - 2D_p)$

TABLE V: Peak current expression and ZVS boundary

Mode	Peak-Current	ZVS-Boundary
1	$I_{12p} = 2D_2 - 2D_1 - 2D_p + D_pk_1 + 1$ $I_{13p} = g_1(2D_3 - 2D_1 - 2D_p + D_pk_2 + 1)$ $I_{23p} = -g_2(2D_2 - 2D_3 - k_3 + 1)$	$2D_1 - 2D_2 - D_pk_1 + g_1(2D_1 - 2D_3 - D_pk_2 + 1) + 1 < 0$ $2D_2k_1 - 2D_1k_1 - D_pk_1 - g_2(k_3 + 2D_2k_3 - 2D_3k_3 - 1) + 1$ $g_2(2D_2 - 2D_3 + k_3 - 1) - g_1(2D_1k_2 - 2D_3k_2 + D_pk_2 - 1) > 0$ $2D_2 - 2D_1 - 2D_p + D_pk_1 + g_1(2D_3 - 2D_1 - 2D_p + D_pk_2 + 1) + 1 > 0$
2	$I_{12p} = 1 - D_pk_1$ $I_{13p} = g_1(2D_3 - 2D_1 - 2D_p + D_pk_2 + 1)$ $I_{23p} = -g_2(2D_2 - 2D_3 - k_3 + 1)$	$1 - g_2(k_3 + 2D_2k_3 - 2D_3k_3 - 1) - D_pk_1$ $2D_2 - 2D_1 - D_pk_1 + g_1(2D_1 - 2D_3 - D_pk_2 + 1) + 1 < 0$ $-g_2(2D_2 - 2D_3 - k_3 + 1) - g_1(2D_1k_2 - 2D_3k_2 + D_pk_2 - 1) > 0$ $2D_2 - 2D_1 - 2D_p + D_pk_1 + g_1(2D_3 - 2D_1 - 2D_p + D_pk_2 + 1) + 1 > 0$
3	$I_{12p} = 1 - D_pk_1$ $I_{13p} = -g_1(D_pk_2 - 1)$ $I_{23p} = -g_2(2D_2 - 2D_3 - k_3 + 1)$	$1 - g_2(k_3 + 2D_2k_3 - 2D_3k_3 - 1) - D_pk_1 > 0$ $-g_1(D_pk_2 - 1) - g_2(2D_2 - 2D_3 - k_3 + 1) > 0$ $2D_2 - 2D_1 - D_pk_1 - g_1(2D_1 - 2D_3 + D_pk_2 - 1) + 1 < 0$ $2D_2 - 2D_1 - 2D_p + D_pk_1 + g_1(2D_3 - 2D_1 - 2D_p + D_pk_2 + 1) + 1 > 0$
4	$I_{12p} = 1 - D_pk_1$ $I_{13p} = -g_1(D_pk_2 - 1)$ $I_{23p} = g_2(2D_2 - 2D_3 + k_3 - 1)$	$g_2(2D_2 - 2D_3 + k_3 - 1) - g_1(D_pk_2 - 1) > 0$ $1 - g_2(k_3 - 2D_2k_3 + 2D_3k_3 - 1) - D_pk_1 > 0$ $2D_2 - 2D_1 - D_pk_1 - g_1(2D_1 - 2D_3 + D_pk_2 - 1) + 1 < 0$ $2D_2 - 2D_1 - 2D_p + D_pk_1 + g_1(2D_3 - 2D_1 - 2D_p + D_pk_2 + 1) + 1 > 0$
5	$I_{12p} = 2D_2 - 2D_1 - 2D_p + D_pk_1 + 1$ $I_{13p} = -g_1(D_pk_2 - 1)$ $I_{23p} = g_2(2D_2 - 2D_3 + k_3 - 1)$	$g_2(2D_2 - 2D_3 + k_3 - 1) - g_1(D_pk_2 - 1) > 0$ $2D_1 - 2D_2 - D_pk_1 - g_1(2D_1 - 2D_3 + D_pk_2 - 1) + 1 < 0$ $2D_2k_1 - 2D_1k_1 - D_pk_1 - g_2(k_3 - 2D_2k_3 + 2D_3k_3 - 1) + 1 > 0$ $2D_2 - 2D_1 - 2D_p + D_pk_1 + g_1(2D_3 - 2D_1 - 2D_p + D_pk_2 + 1) + 1 > 0$
6	$I_{12p} = 2D_2 - 2D_1 - 2D_p + D_pk_1 + 1$ $I_{13p} = g_1(2D_3 - 2D_1 - 2D_p + D_pk_2 + 1)$ $I_{23p} = g_2(2D_2 - 2D_3 + k_3 - 1)$	$2D_1 - 2D_2 - D_pk_1 + g_1(2D_1 - 2D_3 - D_pk_2 + 1) + 1 < 0$ $g_2(2D_2 - 2D_3 + k_3 - 1) - g_1(2D_1k_2 - 2D_3k_2 + D_pk_2 - 1) > 0$ $2D_2k_1 - 2D_1k_1 - D_pk_1 - g_2(k_3 - 2D_2k_3 + 2D_3k_3 - 1) + 1 > 0$ $2D_2 - 2D_1 - 2D_p + D_pk_1 + g_1(2D_3 - 2D_1 - 2D_p + D_pk_2 + 1) + 1 > 0$

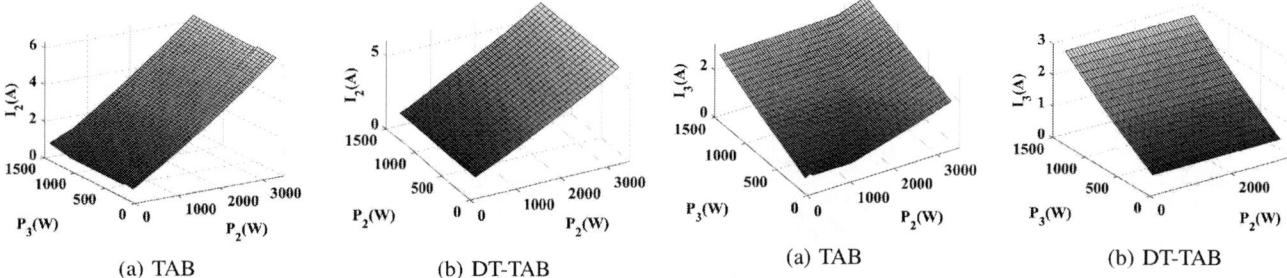

(a) TAB (b) DT-TAB

Fig. 6: Bridge-2 currents

(a) TAB (b) DT-TAB

Fig. 7: Bridge-3 currents

efficiency in DT-TAB is typically higher compared to that in TAB. This is due to the higher high-frequency resistance of the TAB transformer winding (middle winding). It is interesting to note that the losses depend on the winding arrangements for the TAB converter. Fig. 11 shows the total loss variations in TAB and DT-TAB for different loading conditions.

IV. CONCLUSION

This paper presents an extensive comparison between the triple active bridge converter and the dual transformer-based triple active bridge converter in terms of the losses in the magnetic circuit and converter. The peak current optimized modulation techniques of the converter are considered for the comparison along with soft switching constraints for

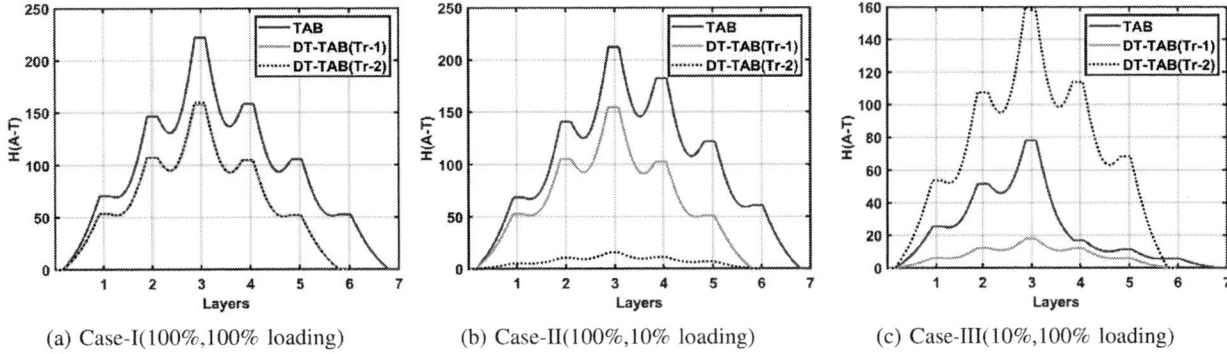

(a) Case-I(100%,100% loading)   (b) Case-II(100%,10% loading)   (c) Case-III(10%,100% loading)

Fig. 8: Distribution of Field Strength (H) through different layers of winding in the window for (a) 100% loading in each ports, (b) 100% loading for port-2 and 10% loading in the port-3 (c) 10% loading for port-2 and 100% loading in the port-3

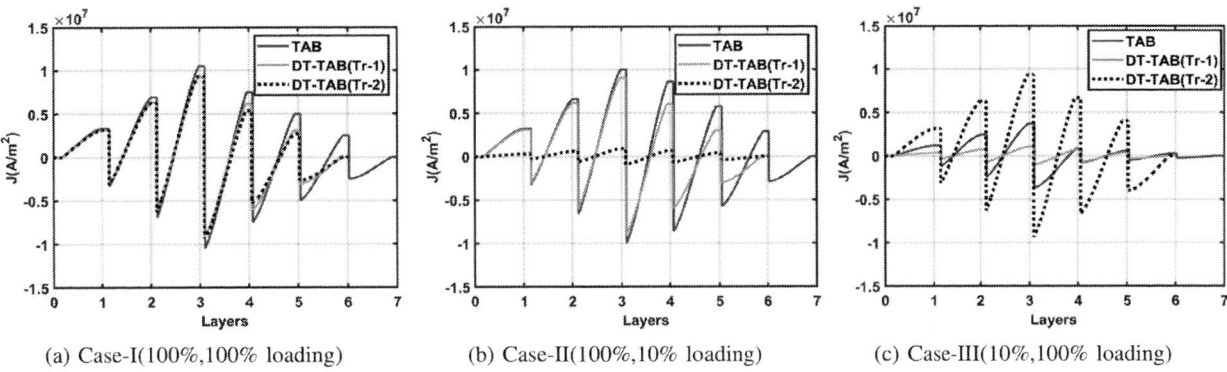

(a) Case-I(100%,100% loading)   (b) Case-II(100%,10% loading)   (c) Case-III(10%,100% loading)

Fig. 9: Distribution of Current Density (J) through different layers of winding in the window for (a) 100% loading in each ports, (b) 100% loading for port-2 and 10% loading in the port-3 (c) 10% loading for port-2 and 100% loading in the port-3

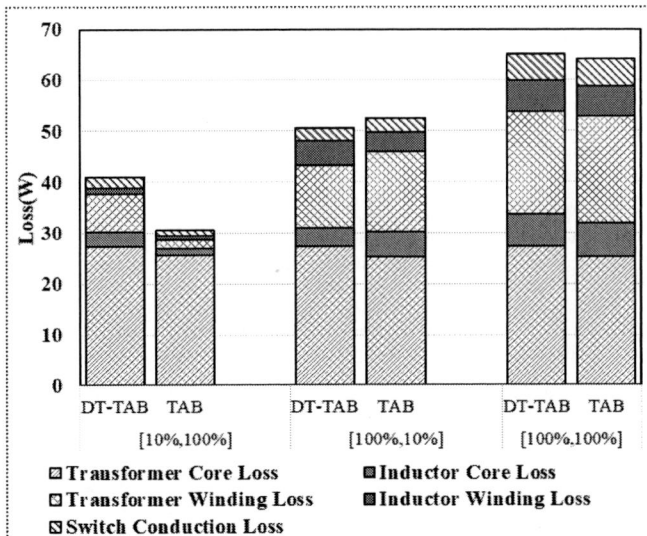

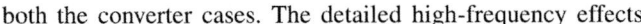

Fig. 10: The loss distribution at different operating points in TAB and DT-TAB

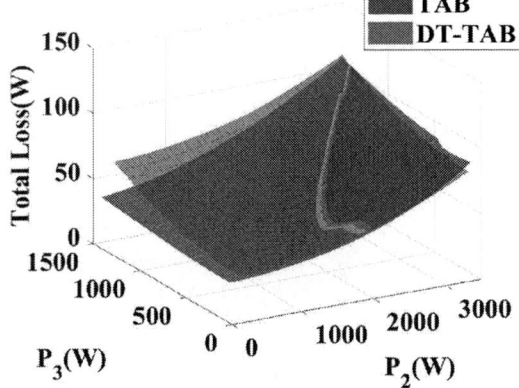

Fig. 11: Comparison of total loss in TAB and DT-TAB with port power variation

both the converter cases. The detailed high-frequency effects

are considered in the loss calculation in the magnetics. The optimization techniques consider identical optimized operating conditions of the converters with triple phase shift modulation. The comparison study shows that optimized peak currents at different operating modes are nearly similar in the DT-TAB and TAB converter; however, in terms of losses, the TAB converter depending on the operating conditions may be higher or lower depending on the operating point. This is mainly due to the fact that the losses depend on the middle winding among the three windings. This point can be justified from the J and H plots through the window of the magnetic core. In the future, a detailed analysis will be considered with a special arrangement of the winding in the TAB converter and interleaved winding arrangements in both cases considering the triple phase shift (TPS) modulation.

APPENDIX

TABLE VI: Power Converter specification

Parameters	DT-TAB Specifications	TAB Specifications
Port 1($V1$)	700V	700V
Port 2(V_2)	700V	700V
Port 3 (V_3)	48V	48V
P_{12} Rated	2500W	2500W
P_{13} Rated	1000W	1000W
Operating Frequency	20kHz	20kHz
L_1/L_{l1}	0.4mH	0.105mH
L_2/L_{l2}	1mH	0.33mH
L_{l3}	–	2μH
n_1	14.297	14.297
n_2	0.98	0.98

TABLE VII: Selection of Magnetic Components and Semiconductor Switches

Parameters	for DT-TAB	for TAB
Transformer Core	3-EE56/24/19 EE55/28/25	2-EE65/32/27
Inductor Core	EE42/21/20 EE42/21/15	2-EE32/16/11 3-EE42/21/15 2-EE/21/15/6
Winding-1	SWG-22(3 in $\parallel$)	SWG-22(4 in $\parallel$)
Winding-2	SWG-22(3 in $\parallel$)	SWG-22(3 in $\parallel$)
Winding-3	SWG-22(1)	SWG-22(18 in $\parallel$)
Winding-4	SWG-22(18 in $\parallel$)	-
Winding Arrangement	non-interleaved	non-interleaved
Semiconductor Switches	IMBG120R140M1H IMZ120R220M1H GS61008P	IMBG120R140M1H IMZ120R220M1H GS61008P

Fig. 12: Winding arrangements in TAB Transformer: P - Primary (700V), S - Secondary (700V), T- Tertiary (48V)

REFERENCES

[1] V. Nair R., S. Gulur, R. Chattopadhyay and S. Bhattacharya, "Integrating Photovoltaics and Battery Energy Storage to Grid Using Triple Active Bridge and Voltage Source Converters," IECON 2020, Singapore, 2020, pp. 3691-3696

[2] H. Wu, L. Chen and Y. Xing, "Secondary-Side Phase-Shift-Controlled Dual-Transformer-Based Asymmetrical Dual-Bridge Converter With Wide Voltage Gain," IEEE Trans. Power Electronics, vol. 30, no. 10, pp. 5381-5392, Oct. 2015

[3] V. N. S. R. Jakka, A. Shukla and G. D. Demetriades, "Dual-Transformer-Based Asymmetrical Triple-Port Active Bridge (DT-ATAB) Isolated DC–DC Converter," IEEE Trans. Ind. Electron., vol. 64, no. 6, pp. 4549-4560, June 2017

[4] R. K. Bhat, D. De, S. Sahu and S. Kumar, "Individual Current Stress Optimization for Multiple Dual Active Bridges with Common Input Stage," 2023 11th National Power Electronics Conference (NPEC), Guwahati, India, 2023, pp. 1-6,

[5] E. S. Oluwasogo and H. Cha, "Self-Current Sharing in Dual-Transformer-Based Triple-Port Active Bridge DC–DC Converter With Reduced Device Count," IEEE Trans. Power Electron., vol. 36, no. 5, pp. 5290-5301, May 2021

[6] S. Pistollato, T. Caldognetto, P. Mattavelli and P. Magnone, "Triple-Phase Shift Modulation for Dual Active Bridge based on Simplified Switching Loss Model," 2019 AEIT, Florence, Italy, 2019, pp. 1-6

[7] A. N, D. Das and C. Kumar, "Smart Transformer Enabled Triple-Active Bridge Converter for Modern Data Centres," 2022 NPSC, New Delhi, India, 2022, pp. 548-553

[8] B. Bohara, A. Karbozov and H. S. Krishnamoorthy, "Triple Phase Shift Control of Dual Active Bridge Converter using Machine Learning Methods," 2022 TPEC, College Station, TX, USA, 2022, pp. 1-6

[9] A. Vetrivelan, W. Xu, R. Yu and A. Q. Huang, "Triple Phase-Shift Optimization of SiC-based Dual-Active Bridge DC/AC Converter," 2022 IEEE APEC, Houston, TX, USA, 2022, pp. 70-77

[10] S. Mukherjee, A. Dash, D. De and A. Castellazzi, "Trade-off in Minimization of Fundamental Link Current and Reactive Power using a Novel Online Calculation based Triple Phase Shift Modulator for Dual Active Bridge," 2019 ECCE Europe, Genova, Italy, 2019, pp. P.1-P.10

[11] I. Biswas, D. Kastha and P. Bajpai, "TAB Based Multiport Converter with Optimized Transformer RMS and Improved ZVS Range for DC Microgrid Applications," IECON 2019 , Lisbon, Portugal, 2019, pp. 2050-2055

[12] S. Maharana, D. De and A. Castellazzi, "A New ZVS Zone Identification for Dual Active Bridge with a General Modulation Objective," 2020 22nd European Conference on Power Electronics and Applications (EPE'20 ECCE Europe), Lyon, France, 2020, pp. P.1-P.10.

[13] S. Dey, A. Mallik and A. Akturk, "Investigation of ZVS Criteria and Optimization of Switching Loss in a Triple Active Bridge Converter Using Penta-Phase-Shift Modulation," IEEE Journal of Emerging and Selected Topics in Power Electronics, vol. 10, no. 6, pp. 7014-7028, Dec. 2022

[14] P. Purgat, S. Bandyopadhyay, Z. Qin and P. Bauer, "Zero Voltage Switching Criteria of Triple Active Bridge Converter," IEEE Trans. Power Electronics, vol. 36, no. 5, pp. 5425-5439, May 2021

[15] K. V. Iyer, W. P. Robbins and N. Mohan, "Design and comparison of high frequency transformers using foil and round windings," 2014 IPEC Hiroshima 2014 - ECCE ASIA, Hiroshima, Japan, 2014, pp. 3037-3043

[16] Chen, B. "Analysis of Effect of Winding Interleaving on Leakage Inductance and Winding Loss of High Frequency Transformers", J. Electr. Eng. Technol. vol. 14, pp. 1211–1221, 2019

Simulation studies on 2/3/5-level DTC based speed control of IM drive in four quadrant Operation

Brijesh Kumar[1], Dr. Bibhu Prasad Panigrahi[2]
Dept. of Electrical Engineering
IGIT Sarang academically affiliated to BPUT,Odisha, Rourkela.
Email ID:[1]brijesh@igitsarang.ac.in , [2]bibhu89@yahoo.com
[1] ORCID ID: [1]0000-0003-3674-5672, [2]0000-0002-4210-8947

Dr. Anup Kumar Panda[3]
Dept. of Electrical Engineering
National Institute of Technology
Rourkela, India
[3] ORCID ID: 0000-0001-5836-0568

Abstract—Simulation studies have been done to compare the speed and torque dynamics as well as steady state responses of an Induction Motor drive fed through the 2/3/5-level Voltage Source Inverter (VSI) using Direct Torque Control (DTC) technique in all the four quadrants. Reference speed and load torque has been generated to realize four quadrants through up-hill, down-hill in forward and reverse directions. Matlab programming has been successfully implemented for 12 sectors, 2/3 and 5-level VSI using 8, 27 and 64 voltage vectors respectively. Flux and torque logic states have been taken as 2-3, 3-5 and 7-7, which results in switching combinations of 36, 180 and 588 for the 2/3 and 5-level VSI respectively. It has been observed that for the fixed tuned value of proportional and Integral (PI) control, the torque and flux variance for 5-level decreases by 91.6% and 38.2% respectively as compared to 2-level DTC based VSI. The steady state speed fluctuation improves by 71.42%.

Index Terms—Direct Torque Control, 5-level Voltage Source Inverter, Four quadrant drive operation, Induction motor drive.

I. INTRODUCTION

Direct Torque control (DTC) technique for speed control of 3-phase Induction Motor (IM) has been studied for more than four decades [1]. DTC techniques minimizes the complexity of coordinate transformation as in rotor flux control but it has a drawback of comparatively high torque and current ripple, selection of appropriate voltage vectors, multiple switching frequency due to fixed hysteresis band etc. Thus, DTC technique has been combined with space vector modulation (SVM) [2] so as to get any desired reference vector from the combination of nearest two voltage vectors, fuzzy speed controller and adaptive fuzzy [3], [4],to minimize the current ripple. Predictive fuzzy logic controller (FLC) and neural network (NN) has been used to vary the hysteresis band so as to keep the switching frequency constant [5]. One can go for hybrid switching, such as PI type fuzzy controller during transient and simple switching at steady state [6]. DTC with multi-level voltage source inverter (VSI) has been used to decrease the current and torque ripple, but with the increase in the number of levels, the voltage vectors and switching combination increases. There are also problems due to variation of voltage drop at the stator resistance, iron loss [7] specially at low speed [8]. A fuzzy estimator, sensors,

adaptive loop compensator may be used to estimate the stator resistance. Sensors may not get correct information on certain environments, in such cases sensor-less operation is executed by estimating the flux. Performance of the IM can be improved by using artificial intelligence [9] and predictive algorithms. And though it depends on IM parameters, thus fuzzy logic controller with a dynamic rule based predictive controller is used. Speed control of 3-phase IM using analog controllers has been implemented to minimize the overall cost of hardware [10], [11]. However, research is still going on to improve the performance, such as fast speed response, economical speed control for maximum fuel efficiency [12], low torque variance [13] etc. Matrix converter [14] can be used to allow bi-directional flow of power, where rotor power extraction during braking [15], source side power quality and input power factor correction plays a vital role. In [16] Fuzzy neural network(FNN) has been used to determine the desired voltage vector magnitude and angle by taking the input as flux error, torque error and flux angle. And further pulse width modulation is used to generate gate pulse to matrix converter. Z-source VSI has been used to improve the IM dynamics, by applying shoot through states of the z-source structure in each modulation period in order to perform during voltage sag condition and to improve power factor and reduce harmonics [17].

In this paper, Matlab simulation of DTC based speed control of 3-phase IM using 2,3 and 5-level VSI has been successfully implemented. Different modes such as forward motoring, reverse braking during uphill and reverse motoring and forward braking during down-hill of electric drive has been explored.

- Forward motoring mode during up-hill.
- zero speed and positive torque during up-hill.
- Forward braking mode during down-hill.
- zero speed and negative torque during down-hill.
- Reverse motoring mode during up-hill.
- Reverse braking mode during down-hill.

Speed and torque commands are varied to implement four quadrant operations. PI controller is used to regulate the torque command based on speed error. Keeping the PI value, hysteresis band for the flux and torque fixed, speed and torque response is analyzed and a comparison has been presented in terms of steady state error and variance.

II. DIRECT TORQUE CONTROL

In DTC technique, the flux linkage and electromagnetic torque is estimated from the IM drive using voltage and current sensors. The estimated flux and torque is compared with the rated flux and desired torque. Where the desired torque is estimated from the speed error and proportional-integral controller. The speed and torque error are quantized as per respective hysteresis band. The flux linkage error logic states S_ψ, torque error logic states S_T and the position of measured flux S_θ, in terms of sector are fed to the voltage vector loop-up table, so as to obtain the desired switching pulse for the three leg inverter. To improve the hysteresis band the flux and torque logic states can be increased as shown in "Fig. 1", with the increase in number of levels in VSI as the available voltage vectors are 8, 27 and 64 for the 2,3 and 5-level VSI respectively. Considering twelve sectors for the 3,5 levels VSI, the switching combinations increase with the increase in the number of flux/torque logic states as shown in Table I.

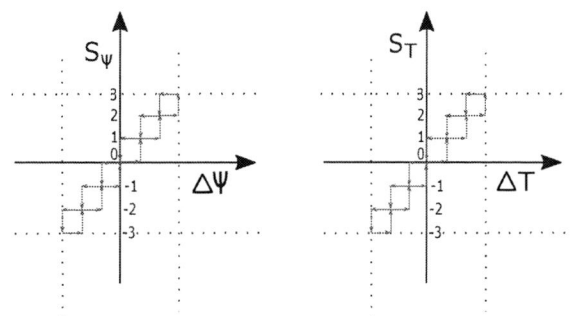

Fig. 1. Flux and torque error logic states for 5-level VSI using DTC technique.

TABLE I
SWITCHING COMBINATIONS BASED OF FLUX AND TORQUE LOGIC STATES

VSI	S_ψ	S_T	Sector	Switching Combinations
2-Level	2	3	6	36
3-Level	3	5	12	180
5-Level	7	7	12	588

The schematic diagram of the speed control of 3-phase IM fed through VSI using DTC technique is shown in "Fig. 2" Now, for 2-level inverter the available voltage vector are eight, out of which two are zero and six are active voltage vectors, similarly in case of 3-level inverter, there are twenty seven voltage vectors, out of which three are of zero magnitude, six are long voltage vectors of value V_d, another six are also long but of magnitude $\sqrt{3}V_d/2$ (i.e. $0.866V_d$). The rest eighteen voltage vectors are small of magnitude $V_d/2$. The voltage vectors in case of 5-level inverter are sixty four, where four voltage vectors are of zero magnitude, eighteen voltage vectors are of magnitude $V_d/3$, twelve voltage vectors are

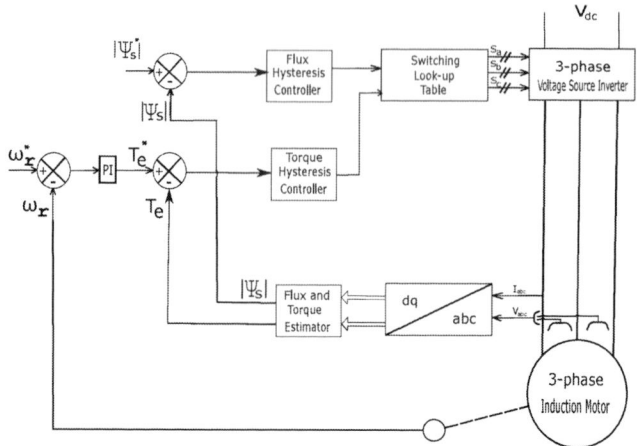

Fig. 2. Schematic diagram of Speed control of 3-phase IM fed through VSI using DTC technique.

of magnitude $0.866(2V_d/3)$, twelve voltage vectors are of magnitude $2V_d/3$, twelve voltage vectors are of magnitude $0.866V_d$ and the rest six voltage vectors are of magnitude V_d. The space voltage vectors for 3-level and 5-level VSI are shown in "Fig. 3" and "Fig. 4" respectively.

III. SIMULATION OF SPEED CONTROL OF 3-PHASE IM FED THROUGH 2/3/5-LEVEL VSI USING DTC TECHNIQUE

Matlab programming has been used to realize the speed control of 3-phase IM drives using DTC technique. The reference speed and load torque are chosen, such that four quadrant operations of the IM can be realized. The reference speed and torque with different motoring modes are tabulated in Table II.

INDUCTION MOTOR PARAMETERS :

- Rated Power P = 0.75kW
- Line Voltage V_{LL} = 415V
- Per Phase stator resistance R_s = 16.575Ω
- Per Phase rotor resistance R_r = 5.2183Ω
- Stator per phase leakage Inductance L_{ls} = 0.0404558H
- Rotor per phase leakage Inductance L_{lr} = 0.0404558H
- Per phase mutual Inductance L_m = 0.5393595H
- No of Poles = 4
- Inertia of motor and coupled load $J = 0.0048742kg-m^2$
- Frictional Coefficient $B = 0.000001Nm.s$
- Rated flux $\psi_{ref} = 1.00184Wb$

The steps involved in the Matlab programming of speed control for the 3-phase IM fed through 2/3/5-level VSI are listed below.

1. Start
2. Initialization and assignment of IM parameters, references and constants to variables.
3. Parameters such as actual speed, torque, current, voltage and flux etc. are assigned zero initial values.
4. Start outer loop with time step of $1/(2^{11})$ sec.
5. Print the variable into a file.

979-8-3315-3013-6/25 $31.00 © 2025 IEEE

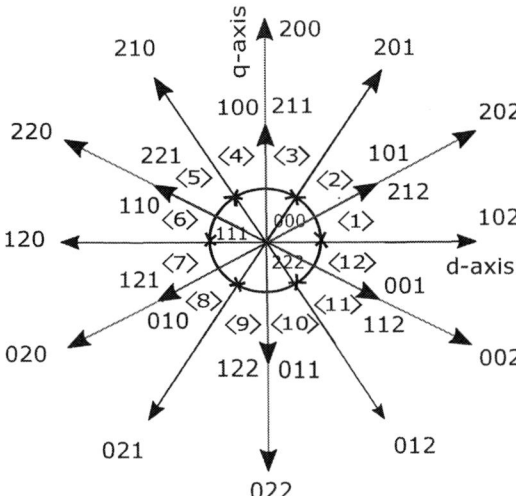

Fig. 3. Twenty Seven space voltage vectors in 3-level Inverter

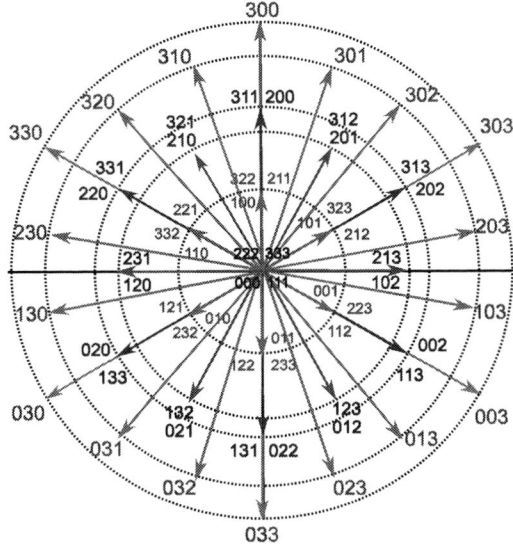

Fig. 4. Sixty four space voltage vectors in 5-level Inverter

6 Generate the speed and torque reference as input.
7 Calculate the torque error from the speed error through the PI controller.
8 Start the inner loop with time step of $1/(2^{16})$ sec.
9 Flux and torque error logic states to be determined using hysteresis controllers.
10 Selection of Voltage Vectors.
11 Solution of IM current by using 4th order Runge-Kutta method while keeping rotor electrical speed constant.
12 Calculation of stator flux, torque and angular displacement theta.
13 Is the inner loop 32 count over
14 If not, go to serial no. 8, or else solve the rotor mechanical speed by using 4th order Runge-Kutta method.
15 Is the outer loop 2048 count over

TABLE II
REFERENCE SPEED AND TORQUE COMMAND FOR SIMULATION STUDIES

Case-studies	Simulation Time (second)	% Rated Speed	Torque (Nm)	Driving Mode
Up-hill	0 - 0.25	25	3	Forward Motoring Mode
Up-hill halt	0.25 - 0.5	0	3	zero speed with positive torque
Up-hill	0.5 - 0.75	25	3	Forward Motoring Mode
Down-hill	0.75 - 1	25	-3	Forward braking Mode
Down-hill halt	1.0 - 1.25	0	-3	zero speed with negative torque
Down-hill	1.25 - 1.5	25	-3	Forward braking Mode
Up-hill	1.5 - 1.75	-25	-3	Reverse Motoring Mode
Up-hill halt	1.75 - 2	0	-3	zero speed with negative torque
Up-hill	2 - 2.25	-25	-3	Reverse Motoring Mode
Down-hill	2.25 - 2.5	-25	3	Reverse braking Mode
Down-hill halt	2.5 - 2.75	0	3	zero speed with positive torque
Down-hill	2.75 - 3	-25	3	Reverse braking Mode

16 If not, go to serial no. 4 or else Stop.

The program successfully executes for the 2,3 and 5-level VSI for the tuned fixed value of the PI controller. The hysteresis band has been kept fixed at 0.1% and 0.5% for the flux and torque respectively. The flux as well as torque logic states has been changed as per Table I.

IV. RESULT ANALYSIS

The four quadrant operation of an IM drive has been realized by Matlab Simulation. The speed and torque changes has been shown in "Fig. 5". The notation DTC2L, DTC3L and DTC5L is used to represent implementation of DTC technique using 2/3/5 level VSI respectively.

Q-I Motoring Mode: During 0 to 0.25 seconds the speed increases to 25% of the rated speed with the rise time of 0.056 seconds, after 0.25 second IM speed is made zero but with the positive torque of 3 Nm. Thereafter, IM regained its speed with a rise time of 0.04 seconds.

Q-II Forward braking mode: From 0.75 to 1 second, the torque is made negative i.e. -3Nm, however the speed remains positive at 25% of the rated speed, this can be realized with an electric vehicle moving down-hill. The simulation results show that the rated speed is achieved, however, there is a steady state error of nearly 0.5 rad/sec in all the 2/3/5-levels of VSI fed to the IM drive. From 1 to 1.25 second, the speed is made zero with the torque of -3Nm. It has been observed that the fall time is 0.04 second and further regain its speed at 1.25 second, with the rise time of 0.02 seconds.

Q-III Reverse Motoring mode: From 1.5 to 1.75 seconds, the speed of IM is -25% of rated speed and torque is -3Nm. This can be realized by electric vehicle moving upward in reverse direction. The reverse motoring rise time is observed to be 0.084 second. A halt is realized between 1.75 to 2 seconds by keeping the reference speed command to zero with the load torque remain at -3Nm. Motor reaches to halt position in 0.02 second and regain its speed in reverse direction in 0.04 second.

Q-IV Reverse braking mode: The regenerative braking is realized by driving the electric vehicle down-hill in reverse direction. From 2.25 to 3 seconds the speed is -25% of the rated speed, however, the load torque is 3Nm. During the reverse braking mode, the motor is made to halt for 0.25 second. The time to halt is 0.0415 second and regain again to -25% speed in 0.02 second.

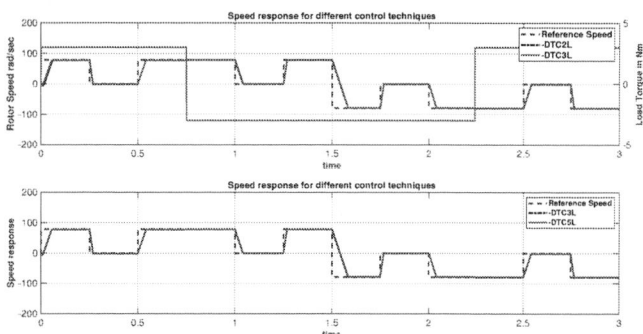

Fig. 5. Speed comparison of 2/3/5-level VSI fed IM drive.

Similarly, in "Fig. 6", the torque ripple is nearly the same for all the levels. The variance in steady state torque for different motoring modes has been tabulated in Table III. It can be observed that the torque variance decreases from 0.0063 to 0.0005274 during motoring mode from 2-level to 5-level VSI. The decrease in torque variance is also observed in all the modes of operations. 3-phase current waveforms for various

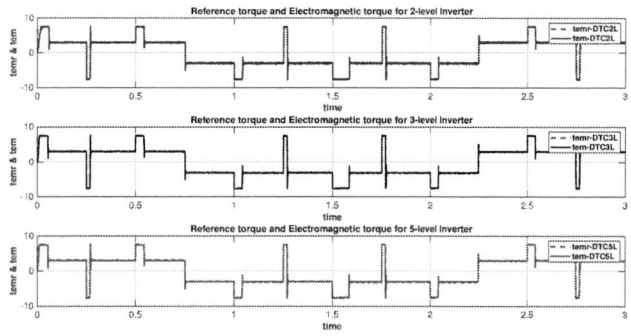

Fig. 6. Torque comparison of 2/3/5-level VSI fed IM drive.

levels are shown in "Fig. 7". The maximum transient current remains the same at 10A and it goes below 5A at 0.05, 0.036 and 0.034 second for 2,3 and 5-levels VSI. The ripple current for 5-level VSI reduces to half for that of 2,3-level VSI.

TABLE III
TORQUE AND FLUX VARIANCE AT STEADY STATE FOR DIFFERENT MODES OF OPERATION

Case-studies	Simulation Time (second)	2-Level (variance)	3-Level (variance)	5-Level (variance)
Up-hill	0.1 - 0.2	0.0063	0.0013	0.00052742
Up-hill halt	0.35 - 0.45	0.0065	0.0013	0.00031457
Up-hill	0.6 - 0.7	0.0067	0.0013	0.00052272
Down-hill	0.85 - 0.95	0.0057	0.0014	0.00026488
Down-hill halt	1.1 - 1.2	0.0063	0.002	0.00035881
Down-hill	1.35 - 1.45	0.0061	0.0015	0.00025496
Up-hill	1.6 - 1.7	0.0063	0.0014	0.00068742
Up-hill halt	1.85 - 1.95	0.0059	0.0021	0.00036745
Up-hill	2.1 - 2.2	0.0059	0.0014	0.00065876
Down-hill	2.35 - 2.45	0.0066	0.0015	0.00027513
Down-hill halt	2.6 - 2.7	0.0066	0.0016	0.00026895
Down-hill	2.85 - 2.95	0.0062	0.0014	0.00026387
Flux variance	0 - 3	0.00087884	0.00062409	0.00054298

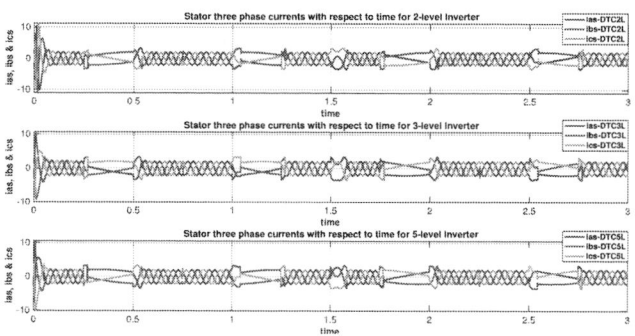

Fig. 7. Current comparison of 2/3/5-level VSI fed IM drive.

The voltage waveform is shown in "Fig. 8", where it can be observed that in case of 2-level VSI the output voltage varies between $+V_{dc}$ and $-V_{dc}$, whereas in case of 3-level VSI, $+V_{dc}/2$ and $-V_{dc}/2$ is also available. And in case of 5-level VSI, $+2V_{dc}/3$, $+V_{dc}/3$, $-2V_{dc}/3$ and $-V_{dc}/3$ is also available along with $+V_{dc}$ and $-V_{dc}$. In case of 5-level VSI, during transient conditions only, $+ - V_{dc}$ is used whereas in steady state $+2V_{dc}/3$ is used, thus DC voltage source can be utilized effectively for 5-level VSI.

From "Fig. 9", it has been observed that the flux variance reduces from 0.00087884 at 2-level to 0.00054298 at 5-level VSI. Also during starting the rated flux is achieved in 0.026, 0.01 and 0.008 seconds for 2,3 and 5-level VSI respectively. The flux response is 69.23% faster in 5-level VSI. The quadrature phase difference has been observed in "Fig. 10" for all the 2/3/5-level VSI. The fluctuation in speed responses can be observed in all the different modes of operation as

979-8-3315-3013-6/25 $31.00 © 2025 IEEE   140

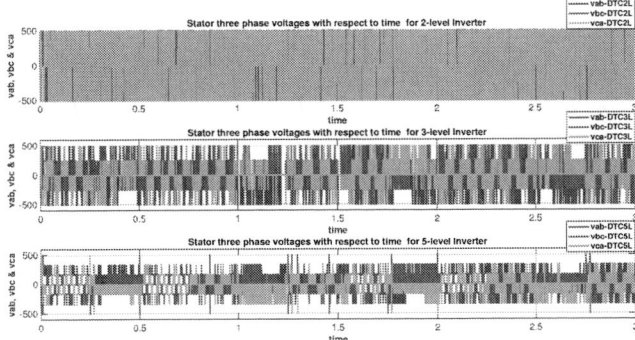

Fig. 8. Voltage comparison of 2/3/5-level VSI fed IM drive.

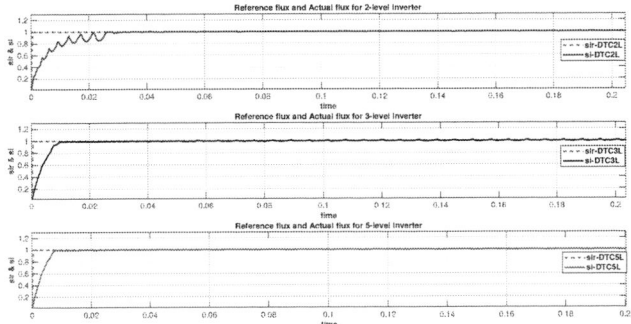

Fig. 9. Flux comparison of 2/3/5-level VSI fed IM drive.

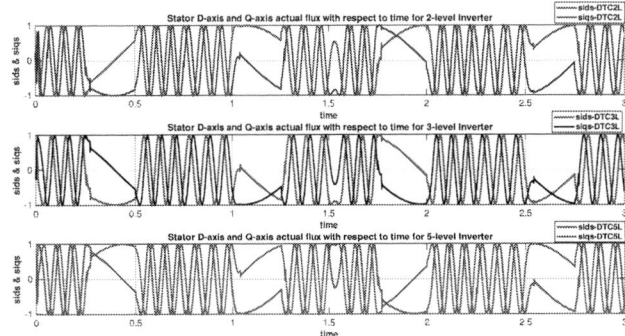

Fig. 10. Actual flux comparison of 2/3/5-level VSI fed IM drive.

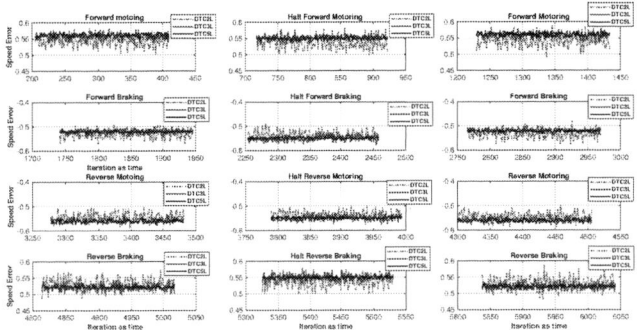

Fig. 11. Speed variation at different mode of operation.

shown in "Fig. 11". During motoring mode, for a 2-level VSI, the difference between reference and actual speed is found to vary between 0.5 to 0.57 rad/sec. whereas this difference of 0.07 decreases to 0.02 in case of 5-level VSI.

V. CONCLUSION

Matlab simulation of 3-phase IM drive with 5-level DTC controlled VSI in four quadrant operation has been successfully achieved. A comparison of speed and torque response at transient and steady state for 2,3 and 5-level inverters has been tabulated for different modes of operation. The change in speed error has been observed to be 0.07 in 2-level as compared to 0.02 in case of 5-level VSI. The flux response during starting for 5-level is improved to 0.008 second as compared to 0.026 second in 2-level. The flux variance is decreased to 0.00054298 as compared to 0.00087884 in 2-level VSI. Overall, there is a decrease in torque and flux variance by 91.6% and 38.2% respectively. And steady state speed fluctuation improves by 71.42%. However, the complexity of switching tables has increased from 180 to 588 due to increase in switching logic states and increase in the voltage vectors from 27 in 3-level to 64 in 5-level. Any error in the voltage vector look-up table may affect the performance and robustness of the IM drive. Thus, an artificial neural network with fuzzy logic can be the solution, and shall be explored in future.

ACKNOWLEDGMENT

The authors extend their sincere thanks to the Department of Electrical Engineering, IGIT Sarang for providing Matlab and necessary infrastructure for the completion of this work.

REFERENCES

[1] M. L. De Klerk and A. K. Saha, "A comprehensive review of advanced traction motor control techniques suitable for electric vehicle applications," *IEEE Access*, vol. 9, pp. 125 080–125 108, 2021.

[2] G. Foo, C. Goon, and M. Rahman, "Analysis and design of the svm direct torque and flux controlled ipm synchronous motor drive," *Australian Journal of Electrical and Electronics Engineering*, vol. 7, no. 1, pp. 21–30, 2010.

[3] L. Romeral, A. Arias, E. Aldabas, and M. G. Jayne, "Novel direct torque control (dtc) scheme with fuzzy adaptive torque-ripple reduction," *IEEE Transactions on Industrial Electronics*, vol. 50, no. 3, pp. 487–492, 2003.

[4] P. R. Tripathy and B. P. Panigrahi, "Study of direct torque controlled 3-phase scim with two and three-level inverters using st-dtc and fr-dtc scheme," *Engineering, Technology & Applied Science Research*, vol. 5, no. 1, pp. 748–752, 2015.

[5] A. Berzoy, J. Rengifo, and O. Mohammed, "Fuzzy predictive dtc of induction machines with reduced torque ripple and high-performance operation," *IEEE Transactions on Power Electronics*, vol. 33, no. 3, pp. 2580–2587, 2017.

[6] Y.-S. Lai and J.-C. Lin, "New hybrid fuzzy controller for direct torque control induction motor drives," *IEEE Transactions on Power Electronics*, vol. 18, no. 5, pp. 1211–1219, 2003.

[7] E. Levi and T. Pham-Dinh, "Dtc of induction machines considering the iron loss," *Electric Power Components and Systems*, vol. 30, no. 6, pp. 557–579, 2002.

[8] A. Shukla and R. Sharma, "Modified dtc with adaptive compensator for low-speed region of induction motor in electric vehicle applications," *Smart Science*, vol. 8, no. 3, pp. 101–116, 2020.

979-8-3315-3013-6/25 $31.00 © 2025 IEEE

[9] R. Joshi, R. Gupta, and A. Wadhawani, "Artificial intelligence based dtc controlled matrix converter cage drive system," *IETE Journal of Research*, vol. 53, no. 4, pp. 303–313, 2007.

[10] B. P. Panigrahi, D. Prasad, and S. SenGupta, "A simple hardware realization of switching table based direct torque control of induction motor," *Electric Power Systems Research*, vol. 77, no. 2, pp. 181–190, 2007.

[11] D. Prasad, B. P. Panigrahi, and S. SenGupta, "Digital simulation and hardware implementation of a simple scheme for direct torque control of induction motor," *Energy conversion and management*, vol. 49, no. 4, pp. 687–697, 2008.

[12] A. K. Sahoo and R. K. Jena, "Loss model based controller of fuzzy dtc driven induction motor for electric vehicles using optimal stator flux," *e-Prime-Advances in Electrical Engineering, Electronics and Energy*, vol. 6, pp. 100–304, 2023.

[13] B. P. Panigrahi and P. R. Tripathy, "Direct torque control based two quadrant analog torque controller for squirrel cage induction motor," *Electric Power Components and Systems*, pp. 1–11, 2023.

[14] T. N. Mir, B. Singh, and A. H. Bhat, "Speed-sensorless dtc of a matrix converter fed induction motor using an adaptive flux observer," *IETE journal of research*, vol. 67, no. 3, pp. 414–424, 2021.

[15] S. Kumar and B. Umamaheswari, "Simultaneous control of torque and rotor power extraction on spsm using dtc technique," *International Journal of Modelling and Simulation*, vol. 29, no. 3, pp. 285–292, 2009.

[16] J. Zhao, Z. Zhang, Y. Ren, and X. Li, "A direct torque intelligent control strategy for induction motor based on matrix converter," *Australian Journal of Electrical and Electronics Engineering*, vol. 12, no. 3, pp. 175–182, 2015.

[17] S. Douida, B. Tabbache, and M. Benbouzid, "Direct torque control based on shoot-through states of an induction motor fed by a z-source three-level neutral point clamped inverter," *IETE Journal of Research*, vol. 68, no. 3, pp. 1982–1990, 2022.

Direct Torque Control of an 8/6 Switched Reluctance Motor Using a Shared Switch Converter

Akshat Sharma
School of computing and electrical engineering
Indian Institute of Technology, Mandi
Mandi, HP, India
akshat.s2402@gmail.com

Himanshu Misra
School of computing and electrical engineering
Indian Institute of Technology, Mandi
Mandi, HP, India
himanshumisra@iitmandi.ac.in

Abstract— **Switched Reluctance Motors (SRMs) are widely recognized for their efficiency, robustness, and cost-effectiveness due to their simple construction and absence of permanent magnets. However, their application is often constrained by high torque ripple, which affects performance in various domains. To address this challenge, this study deals with only torque control of SRM via Direct Torque Control (DTC) implementation in an 8/6 SRM using a shared switch converter topology. Unlike conventional Asymmetric Half-Bridge (AHB) converters, the proposed design enables two phases of SRM to share a common switch, thereby reducing the number of switches required per phase and optimizing cost. Further, a suitable switching table is developed to support the proposed DTC. The study examines the impact of this configuration on torque performance, highlighting modifications necessary for voltage vector selection in the DTC strategy. Simulation and experimental validation confirm the effectiveness of the proposed approach in mitigating torque ripple while maintaining system reliability.**

Keywords—8/6 SRM, shared switch converter, Direct Torque Controller

I. INTRODUCTION

SRMs are highly efficient, durable, cost-effective electric motors that generate torque using magnetic reluctance instead of permanent magnets. Their simplified design reduces reliance on rare-earth materials, making them a sustainable choice for applications in industrial automation, electric vehicles, and renewable energy systems while ensuring adaptability and reliability. But due to their doubly salient construction, SRMs are prone to high torque ripple, limiting their widespread application in different domains [1].

Various control techniques, ranging from simple to advanced, have been proposed for restricting the torque ripple. [2] mentions the use of a current hysteresis controller for torque ripple reduction. In [3], an online turn-on and turn-off technique is discussed for proper commutation of phases. DTC has been proposed in [4], where both flux and torque hysteresis control loops are used, and appropriate voltage vectors are selected for switching. [5] discusses the DTC in a four-phase SRM for reducing torque ripple. [6] and [7] discuss DTC with variable flux reference, which is generated from the torque reference. [8] Make use of a 16-sector partition scheme for a better torque per ampere ratio. [9] uses a DTC algorithm without a flux loop for improving the torque per ampere ratio using a conventional AHB converter for a three-phase SRM. [10] compares the AHB converter to the shared switch converter, which uses 6 switches

for four-phase SRM. For modelling SRM, its magnetizing curves are required. [11] provides a methodology and mathematical equations for non-linear modelling of SRM. This work presents an approach for performing DTC on an 8/6 SRM without flux loop using a shared switch converter topology where two phases share a common switch and diode between them. The voltage vectors are defined for this converter topology to have less torque ripple.

II. MODELLING OF SRM

Implementation of DTC requires the magnetic characteristics and torque characteristics of the SRM. For SRM modelling, the rotor clamping method was used as mentioned in [11]. The SRM is modelled using a lookup table approach. The unaligned inductance, being linear, is calculated from:

$$\psi_u(i) = L_u i \qquad (1)$$

Where ψ_u is the flux linkage at the unaligned position and L_u is unaligned inductance and i is the instantaneous current. Also, non-saturated aligned inductance is calculated from:

$$\psi_a(i) = L_a i \qquad (2)$$

Where ψ_a is the flux linkage at the aligned position and L_a is an unsaturated aligned inductance. But the saturated aligned inductance is parabolic, hence, is calculated by the equation.

$$\psi_a(i) = \psi_0 + \sqrt{4a(i - I_0)} \qquad (3)$$

Here, ψ_a is saturated aligned flux linkage while ψ_0 and I_0 are the initial quantities just before saturation.

The magnetization curves for other positions can be obtained using the Fourier series. Assuming that the mutual inductance among phases is negligible, the following equations are used for modelling:

$$v_a = i_a R_a + \frac{d\psi_a}{dt} \qquad (4)$$

$$\psi_a(t) = \psi_0 + \int_0^t (v_a - i_a R_a) dt \qquad (5)$$

Where v_a, i_a, R_a, ψ_a represents instantaneous voltage, instantaneous current, resistance and instantaneous flux linkage corresponding to phase A, respectively. ψ_0 represents the initial flux value for the phase. The torque equation for SRM is given by:

$$T_a \approx i_a \frac{d\psi_a(i,\theta)}{d\theta} \qquad (6)$$

Where T_a is the instantaneous torque of phase A.

979-8-3315-3013-6/25 $31.00 © 2025 IEEE

III. SHARED SWITCH CONVERTER

The converter topology shown in Fig. 1 has been mentioned in [1] and has been analyzed from the performance point of view in [10]. Pair of Phases (A, C) and (B, D) of SRM shared a common switch in the shared switch converter [1]. But in the proposed work, a pair of phases (A, B) and (C, D) share a common switch as shown in Fig. 1. This is done to ensure minimal interference between the phases while maintaining efficient operation. In 8/6 SRM, the switching follows the sequence DACBD, which means that A and B will never conduct simultaneously. If adjacent phases (like A & D) share a switch, there would be instances where both need to be ON simultaneously, which would cause interference and high currents

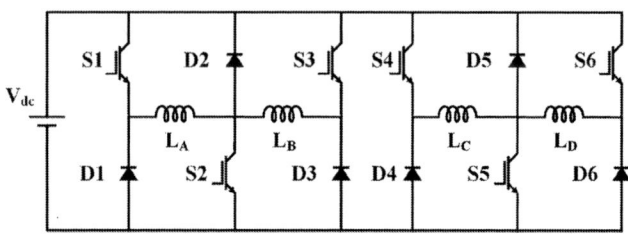

Fig. 1. Shared switch converter topology

The shared switch configuration makes use of 6 switches and 6 diodes for a 4-phase SRM drive, thereby using 1.5 switches per phase compared to 2 switches used in a conventional AHB converter. A switch and a diode are being shared by 2 phases. Hence, reducing the independence among phases. The converter, like the conventional 8-switch converter, offers 3 excitation levels of magnetization, freewheeling and demagnetization for each phase. However, due to pairing, some restrictions are there for the independent magnetization and demagnetization of paired phases. Table I shows various modes of operation of the converter for the phases (A and B) sharing a common switch, and the same will be followed for other phases (C and D) also. Fig. 2 depicts the current path in all eight modes.

TABLE I. SWITCHING MODES OF THE CONVERTER

Modes	Switching states			Phases	
	S1	S2	S3	A	B
1	0	0	0	Demagnetize	Demagnetize
2	0	0	1	Demagnetize	Freewheel
3	0	1	0	Freewheel	Freewheel
4	0	1	1	Freewheel	Magnetize
5	1	0	0	Freewheel	Demagnetize
6	1	0	1	Freewheel	Freewheel
7	1	1	0	Magnetize	Freewheel
8	1	1	1	Magnetize	Magnetize

It can be seen that the two phases sharing a common switch cannot be magnetized and demagnetized at the same time, i.e., magnetization of phase A and demagnetization of phase B or vice versa cannot occur simultaneously. So, when phase B has to be magnetized, it must be ensured that phase A has been demagnetized to a significant extent. This is done by increasing the freewheeling period of the phase being demagnetized with the help of suitable voltage vectors. Hence, compared to a conventional AHB converter, this converter topology offers restrictions on the selection of voltage vectors for achieving DTC.

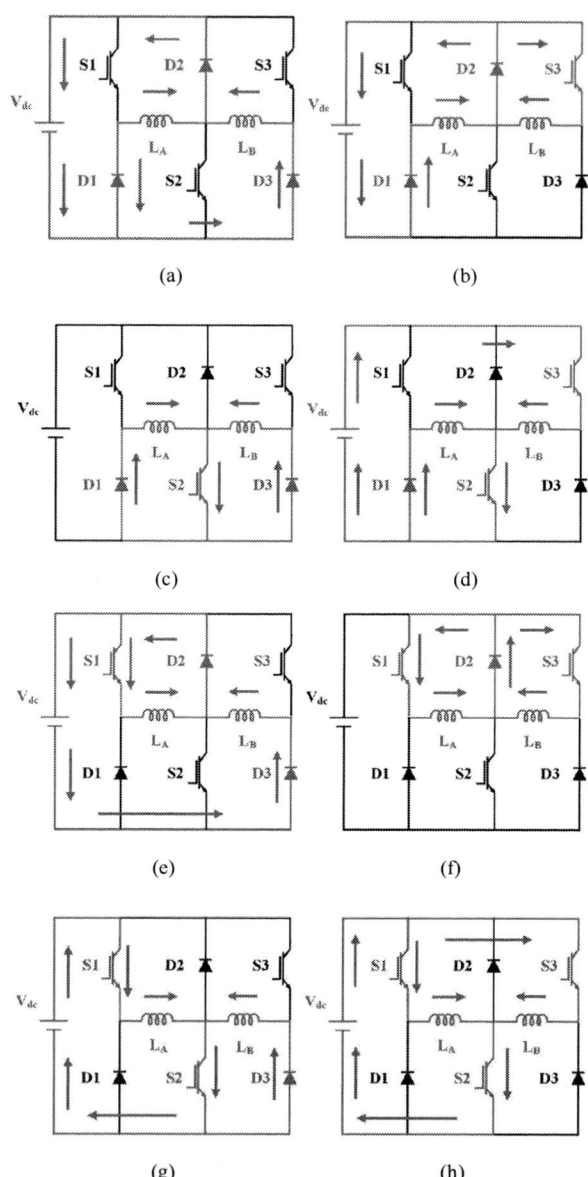

Fig. 2. Modes of operation of converter in (a) Both phases demagnatize, (b) Phase A demagnetize Phase B freewheel, (c) Both phases freewheel via S2, (d) Phase B magnetization phase A freewheel, (e) Phase A freewheel phase B demagnetization, (f) Both phases freewheel through different switches, (g) Magnetization of phase A and phase B freewheel and (h) Magnetization of both phases

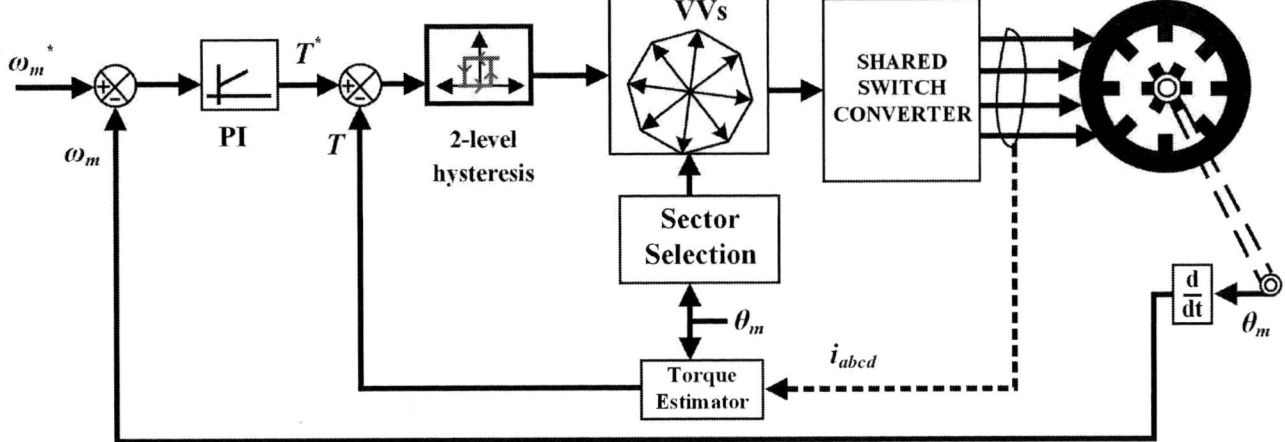

Fig. 3. Block diagram of DTC

IV. DTC OF 8/6 SRM

In conventional DTC, the reference flux was given a constant value, which does not match the requirement for all load conditions. For SRM, the torque does not depend on the magnitude of the flux linkage vector, instead, it depends on the acceleration and deceleration of flux linkage with respect to rotor position and stator current. So, even without controlling the flux loop, direct torque control can be achieved in SRM. Here, sectors are defined using rotor position rather than using the flux linkage vector. The torque estimator, as shown in Fig.3, contains torque characteristics in the form of a lookup table, and the torque is estimated using current and rotor position. The torque is controlled with the help of a 2-level torque hysteresis controller.

With the shared switch converter replacing the conventional AHB converter, the voltage vectors are required to be modified as per its mode of operation. Out of 64 voltage vectors available for this converter, 16 vectors are selected and are arranged in two sets of 8 vectors each, one set for torque increasing and the other for torque decreasing. Moreover, with a switch and a diode being shared by two phases (A and B), for proper commutation among phases, some voltage vectors are selected in a way to give enough time for demagnetization to reduce overlapping among phase currents. The voltage vectors are shown in Fig. 4. It can be observed that A+ represents the aligned position for phase A, while A- represents the unaligned position for phase A, and the same goes for other phases as well. The transition from aligned to unaligned position represents the positive and negative slope of the phase inductance profile. If the rotor is assumed to be rotating in an anti-clockwise direction, then the sectors 7-8-1-2 will account for positive torque, and sectors 3-4-5-6 will account for negative torque for phase A. So, to reduce per per-phase negative torque for phase A, it should be demagnetized before sector 3. Hence, vectors are modified to demagnetize phases by freewheeling, before their negative slope starts.

Since, there are two sets of voltage vectors so for Kth sector V_{K+3} vector is selected for both torque increasing and decreasing from respective sets.

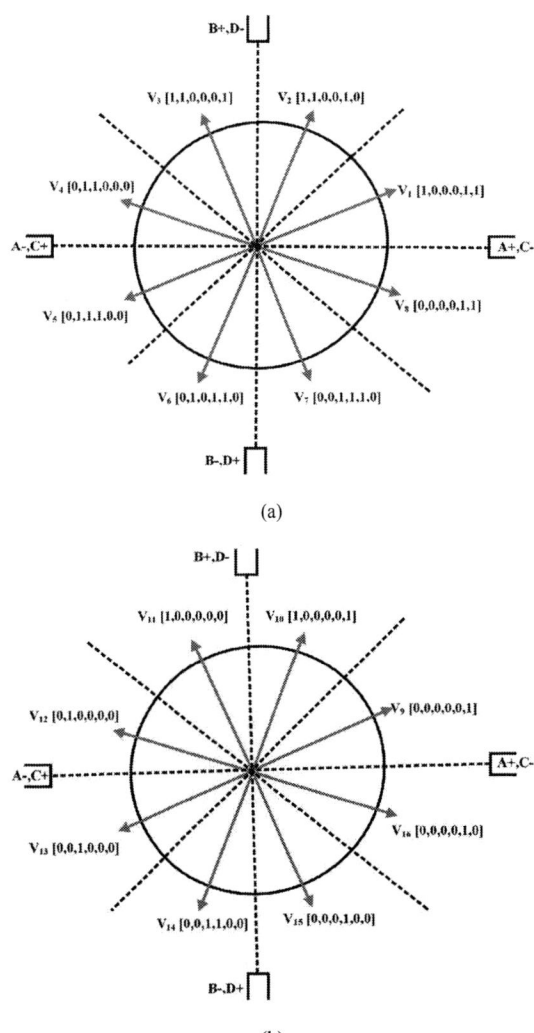

Fig. 4. Voltage Vectors (a) Torque Increase (b) Torque Decrease

V. SIMULATION RESULTS

An 8/6 SRM was simulated in MATLAB/ Simulink. Firstly, the 8/6 SRM was modelled using a look-up table-based approach. The experiment data were recorded using [10] to obtain the magnetizing and torque characteristics. The parameters of the SRM under consideration are given in Table II. Also, the aforementioned converter was modelled. A speed reference of 500 rpm is generated using a slope of 200 rpm. A load torque of 1 Nm is applied after the constant speed is achieved. The torque hysteresis controller has a bandwidth of 0.4 Nm.

TABLE II. MOTOR PARAMETERS

Parameters	Value
Winding Resistance	0.506 Ω
Aligned Inductance	14.8 mH
Unaligned Inductance	4.2 mH
DC link voltage	230 V
Rated Current	11 A
Rated Speed	3000 rpm
Power Rating	2.2 kW
Inertia Constant	0.0036 kgm2
Friction Constant	0.04Nms

From Fig. 5, it is evident that the actual speed is perfectly tracking the reference speed. However, there is a slight dip in the speed as the load is applied at 3 to 4 seconds. The reference torque generated by the PI controller follows the load torque. Figs. 6 and 7 show actual torque as a band of torque ripple varying within the applied hysteresis limits. Fig. 8 shows the current profile of all four phases. Fig. 9 indicates that the phase A current exists for sectors 7-8-1-2, which is the positive inductance slope of that phase and the current is zero for sectors 3-4-5-6, indicating that the phase has been demagnetized before the arrival of the negative inductance slope.

However, at higher speeds, due to faster switching rate, there would not be enough time for phases to demagnetize before the negative inductance slope begins. Hence, there will be overlapping of phase currents leading to negative torque and increased torque ripple.

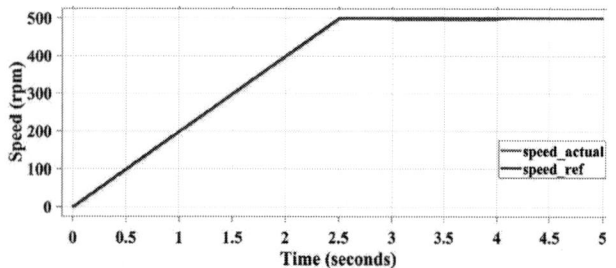

Fig. 5. Response of speed controller

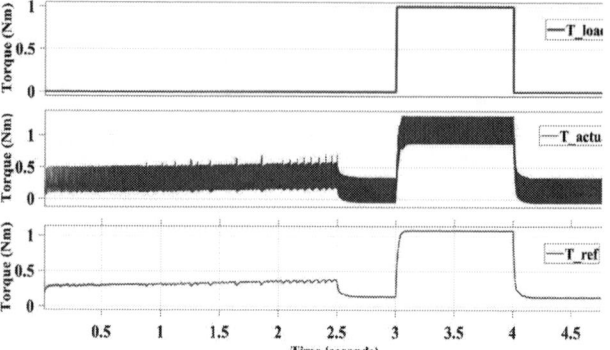

Fig. 6. Torque controller response, load torque (violet), actual torque (red), reference torque (blue).

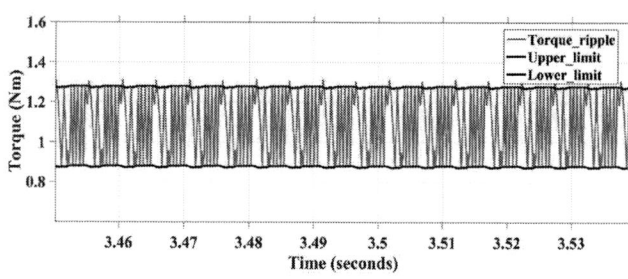

Fig. 7. Torque Hysteresis Controller response

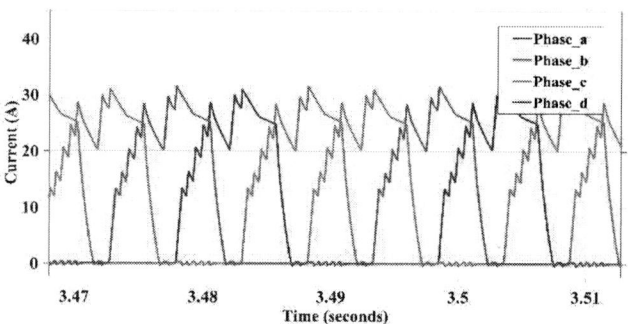

Fig. 8. Each phase current under loading.

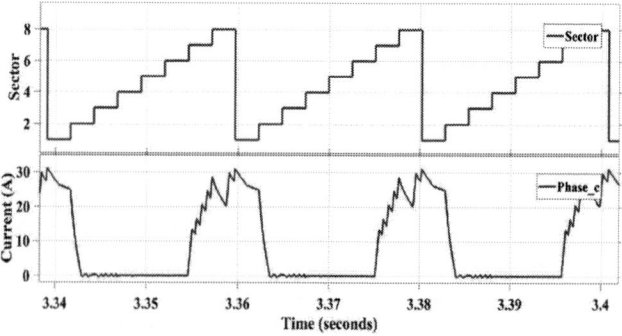

Fig. 9. Current of phase A and sectors

VI. EXPERIMENTAL VERIFICATION

DTC using a shared switch converter is verified using an 8/6 SRM. The motor parameters are the same as those used in the simulation. The control algorithm is implemented using a TMS320F28379D DSP. The mentioned processor has a clock frequency of 100 MHz. The sampling frequency used is 18 kHz. Since rotor position is vital for sector determination, for this purpose, a rotary encoder with 1024 pulses per revolution is used. For current sensing, hall-effect-based sensors are used for each phase. The converter topology is implemented using an IGBT-based module.

For experimental verification, a speed reference of 500 rpm is given in a ramp fashion. After the speed is achieved, a load equivalent to 1Nm is applied to the SRM by exciting the field winding of the coupled DC motor and applying resistance to the armature.

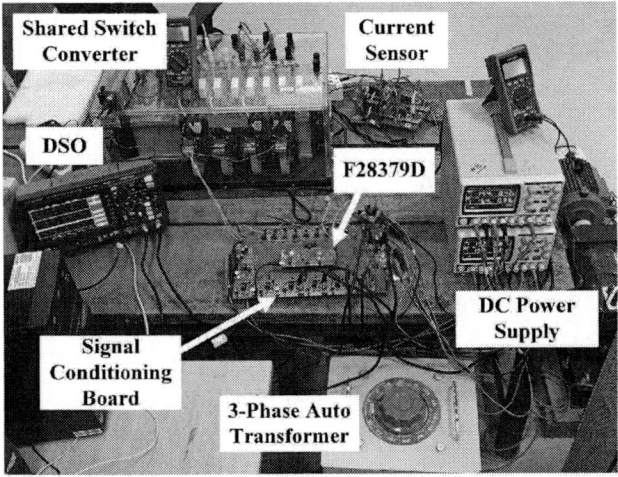

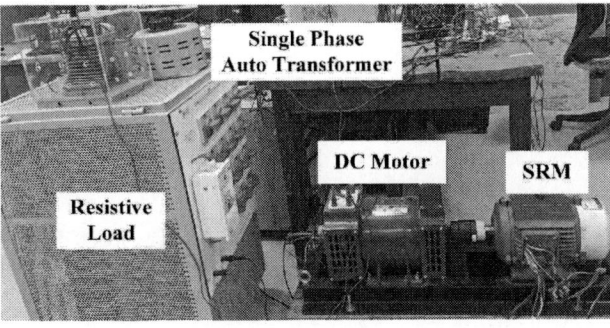

Fig. 10. Experimental setup

Fig.11 shows that the actual speed is tracking the reference speed. Alongside, Phase A current is shown. It can be seen that later on, when a sudden load is applied, the current value increases. Also, from the magnified view of the loading region, it can be noticed that even if the currents are peaking at a high value but the rms value is well within the rated limits. The actual torque can be seen tracking the reference torque in Fig.12, and the torque ripple is within the hysteresis band. Fig.13 depicts the 8 sectors along with the phase A current.

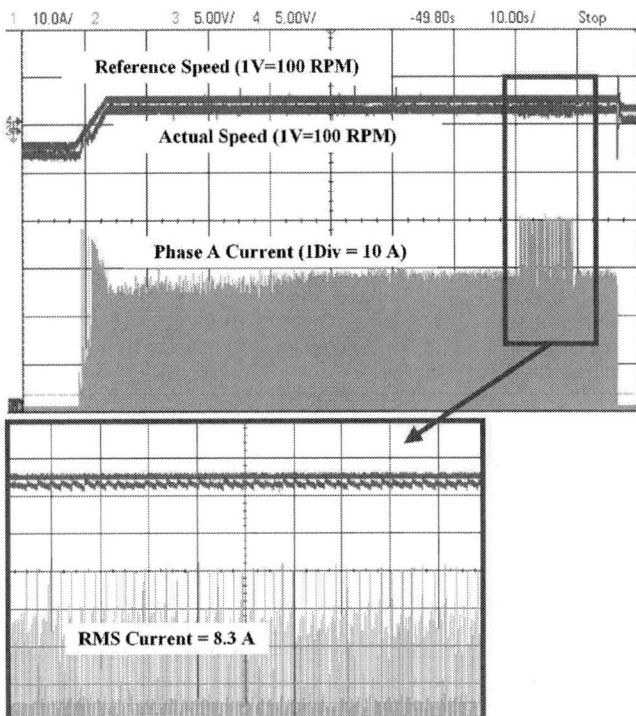

Fig. 11. Response of the speed controller and phase A current and magnified view under loading condition.

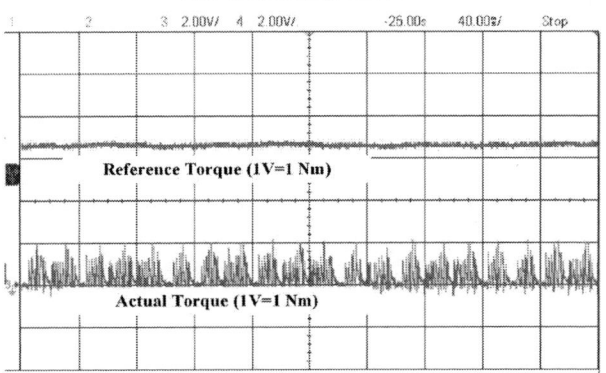

Fig. 12. Response of Torque hysteresis controller

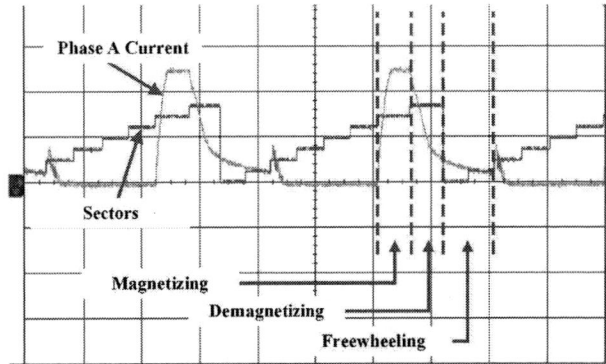

Fig. 13. Phase A current with sectors

It should be noted that in the current waveform, magnetization, demagnetization, and freewheeling can be noticed, and the current is made to reach minimum values before the negative inductance slope begins. Thereby minimizing the production of negative torques in each phase, hence reducing overall torque ripple as compared to the conventional DTC where the current tails into the negative inductance slope region of the phase.

From the results, it is evident that the magnetizing period of each phase is reduced to allow proper time for demagnetization. However, the current waveform tends to become somewhat peaky, unlike in conventional DTC. Moreover, this converter topology, due to shared switches among phases, offers the advantage of low torque ripple at low speeds. But at high speed, torque ripple can increase as the period for demagnetization decreases, leading to current overlap and negative torque generation in phases.

VII. CONCLUSION

This study presents a DTC strategy for an 8/6 SRM using a shared switch converter topology. By reducing the number of switches per phase while maintaining essential excitation levels, this converter design enhances system cost-effectiveness. The modified voltage vector selection method ensures effective torque control while addressing the challenge of phase commutation. Simulation and experimental results validate the approach and stable operation at low speed. However, at higher speeds, phase currents overlapping may lead to increased torque ripple, indicating the need for further optimization. Future work can focus on refining control algorithms and exploring advanced switching techniques to enhance performance across varying operating conditions. Moreover, some modifications can be made in the converter to aid fast demagnetization of phases.

REFERENCES

[1] Krishnan, R.. Switched Reluctance Motor Drives: Modeling, Simulation, Analysis, Design, and Applications. Ukraine, CRC Press, 2001.

[2] Gobbi, R.; Ramar, K.: 'Optimisation techniques for a hysteresis current controller to minimise torque ripple in switched reluctance motors', IET Electric Power Applications, 2009, 3, (5), p. 453-460, DOI: 10.1049/iet-epa.2008.0191.

[3] P. Ingle and H. Misra, "Four Quadrant Operation of Switched Reluctance Motor with Online Turn ON Angle Control," 2023 11th National Power Electronics Conference (NPEC), Guwahati, India, 2023, pp. 1-5, doi: 10.1109/NPEC57805.2023.10384973.

[4] A. D. Cheok and Y. Fukuda, "A new torque and flux control method for switched reluctance motor drives," in IEEE Trans. on Power Electron., vol. 17, no. 4, pp. 543-557, July 2002.

[5] Y. Han, P. Xu and Q. Ma, "Torque ripple reduction of four-phase SRM based on DTC method," 2018 IEEE 3rd Advanced Information Technology, Electronic and Automation Control Conference (IAEAC), Chongqing, China, 2018, 10.1109/IAEAC.2018.8577728.

[6] L. Feng, X. Sun, X. Tian and K. Diao, "Direct torque control with variable flux for an SRM based on hybrid optimization algorithm," IEEE Transactions on Power Electronics, vol. 37, no. 6, pp. 6688-6697, June 2022.

[7] X. Zhao, A. Xu, W. Zhang, "Research on dtc system with variable flux for switched reluctance motor," CES Transactions on Electrical Machines and Systems, vol. 1, no. 2, pp. 199-206, Jun. 2017.

[8] K. R. Pittam, D. Ronanki, P. Perumal, A. R. Beig and S. S. Williamson, "Torque Ripple Minimization of Four-phase Switched Reluctance Motor using Direct Torque Control with an Innovative Switching Sequence Scheme," 2019 IEEE Energy Conversion Congress and Exposition (ECCE), Baltimore, MD, USA, 2019.

[9] N. Yan, X. Cao and Z. Deng, "Direct Torque Control for Switched Reluctance Motor to Obtain High Torque–Ampere Ratio," in IEEE Transactions on Industrial Electronics, vol. 66, no. 7, pp. 5144 5152, July 2019, doi: 10.1109/TIE.2018.2870355.

[10] R. Martua, J. Furqani and A. Rizqiawan, "Performance Comparison of The 8/6 SRM Inverters," 2022 5th International Conference on Power Engineering and Renewable Energy (ICPERE), Bandung, Indonesia, 2022, pp. 1-6, doi: 10.1109/ICPERE56870.2022.10037316.

[11] Yan Cai, Qingxin Yang, Lihua Su, Yanbin Wen and Yiming You, "Nonlinear modeling for switched reluctance motor by measuring flux linkage curves," 2010 2nd International Conference on Computer Engineering and Technology, Chengdu, China, 2010, pp. V6-47-V6-51, doi: 10.1109/ICCET.2010

PSO assisted Support Vector Machine Algorithm towards State of Charge estimation of Li-ion batteries

Saneep K
Department of EEE
National Institute of Technology
Tiruchirappalli, India
407123004@nitt.edu

Sundareswaran K
Department of EEE
National Institute of Technology
Tiruchirappalli, India
ks@nitt.edu

P Srinivasa Rao Nayak
Department of EEE
National Institute of Technology
Tiruchirappalli, India
psnayak@nitt.edu

Sishaj P Simon
Department of EEE
National Institute of Technology
Tiruchirappalli, India
sishajpsimon@nitt.edu

Mithun T
Expert Solutions Pvt Ltd
An Alten Company
Pune, India
mithun.t@expertgs.com

Gireesh V Puthusserry
Department of EEE
NSS College of Engineering
Palakkad, India
gireeshvp@nssce.ac.in

Abstract—This research focuses on the State of Charge (SoC) estimation of Lithium-ion batteries employing the Support Vector Machine (SVM) algorithm. The five hyperparameters of SVM are optimally identified using Particle Swarm Optimization (PSO). Four publicly available driving cycles, namely US06, LA92, CSHVC, and HWFET are employed for the optimization and subsequent evaluation of the proposed method. The first 70% of each dataset is used for the optimization of hyperparameters and the remaining 30% is used for the testing of the algorithm. The SVM with optimized hyperparameters is then employed for SoC estimation of each driving cycle at different temperatures and the results are presented. It is observed that the performance indices, namely RMSE, MAE, and R^2 values are far superior to the values in the available literature. Further, the percentage error in SoC is found to be below 2.5%, indicating improved performance of the proposed method.

Index Terms—lithium-ion batteries, state of charge, support vector machine, particle swarm optimization

I. INTRODUCTION

Electric Vehicles (EVs) have received a lot of attention in recent years due to their environmental benefits and advances in battery technology [1]. The Lithium-ion batteries are provided as a primary energy source in EVs due to the advantages of long life and high energy density. The performance and efficiency of an EV largely depends upon the battery management system [2]. One of the critical parameters of BMS is the State of Charge (SoC), which gives the remaining charge left out in the battery. The accurate estimation of SoC helps to avoid anxiety towards the driving range, and it ensures battery longevity. The SoC can be mathematically expressed as:

$$SoC(t) = SoC(0) + \frac{1}{C} \int_0^t I(\tau) \, d\tau \qquad (1)$$

where SoC(0) is the initial SoC, C is the battery capacity in Ah, and I(τ) is the battery current at the time τ [3].

There are four methodologies for SoC estimation, categorized into traditional and advanced methods. The traditional methods include the Coulomb Counting and Voltage-based methods, and these methods fail under dynamic conditions. In the coulomb counting method, the SoC is estimated by integrating the battery current over time. This method is simple and easy to implement, but the cumulative errors of the sensors lead to inaccuracy in initial SoC estimation [4]. The Voltage-based or OCV method estimates SoC using the relationship between OCV and SoC. However, the battery is to be kept at rest for a finite period to obtain an accurate OCV measurement, and it is unsuitable for real-time applications [5]. Advanced techniques, such as Model-based and data-driven methods are extensively used to estimate SoC because these methods use battery models to predict the SoC and update it based on real-time measurement. The model-based estimation methods include an electrochemical model and electrical circuit models; the electrochemical model approach demands complex modeling and extensive parameter identification but yields good results [3]. Extended Kalman Filter (EKF) [6], Unscented Kalman Filter (UKF) [7], and Observer-based methods [8] use electric circuit model approaches, and these methods are widely used for SoC estimation because of its ability to handle uncertainties and measurement of noise effectively. The main disadvantage of this method is that the parameters vary as the battery ages and is exposed to different envirnmental factors. A relationship between battery parameters and SoC can be derived using data-driven approaches and it can adjust to the variations in battery characteristics over time [9]. Therefore, the data-driven method is observed to be a promising tool for the estimation of the SoC of batter since it overcome the drawbacks of previous approaches. Advanced learning algorithms, such as Deep Neural Networks (DNN) [10], Convolutional Neural Networks (CNN) [11], Long Short-

979-8-3315-3013-6/25 $31.00 © 2025 IEEE

Term Memory Networks (LSTM) [12], and Support Vector Machines (SVM) [13] can be used to predict the SoC. Among these learning algorithms, the Support Vector Machine is useful for handling high-dimensional data [14].

The SVM is a supervised machine learning algorithm, which can be used for regression and classification tasks. This algorithm finds the complex relationship between the input features and the target, and it maps the data points into a high-dimensional space by using the kernel function [14]. The SVM algorithm has five hyperparameters namely, Regulation Parameters (C), Kernel Function, Loss function (ϵ), Gamma (γ), and Tolerance (tol). In the recently reported works, randomly chosen values of hyperparameters are employed for the estimation of SoC using the SVM algorithm [15]. However, the performance of the algorithm depends upon the proper selection of hyperparameters. So, hyperparameter tuning significantly impacts the accuracy of the prediction. In this context, we propose a biologically inspired optimization technique, namely Particle Swarm Optimization (PSO), to optimally tune all hyperparameters of the SVM algorithm towards the estimation of SoC. The PSO was developed by Kennady and Eberhart and has been used for various applications in engineering [16].

In this work, the proposed algorithm was validated with four different types of driving cycles namely aggressive driving cycle (US06), Heavy duty cycle (CSHVC), Urban driving cycle (LA92), and Highway driving cycle (HWFET) at various temperatures. The first 70% of data of a typical dataset is used for the optimization of all five hyperparameters of the SVM algorithm and the remaining 30% is employed for testing. The optimized values of hyperparameters are then employed for the estimation of SoC for the other datasets from the same driving cycle at other temperatures. The performance of the proposed method shows that PSO based SVM algorithm is a promising tool for the estimation of SoC.

This paper is organized as follows: section II explains the PSO-based optimization of hyperparameters of the SVM algorithm. Section III describes data acquisition and preprocessing. Section IV focuses on the computed results and performance evaluations and the conclusions derived from the research work are included in section V.

II. PSO Assisted SVM Algorithm for SoC Estimation

A. Support Vector Machine

The Support Vector Machine is a type of supervised machine learning algorithm developed by the increased use of Neural Networks [14]. It is largely used for several applications because it can become a non-linear featured decision tree. Normally SVM can be used for regression and classification tasks. It establishes border vectors between data points and these locations of vectors are infinitely variable. Consequently, SVM optimizes these vector positions by determining the optimal location for data points that should be equally close to data boundaries.

The SVM attempts to find a function which is given by,

$$f(x) = \sum_{i=1}^{N} (\alpha_i - \alpha_i^*) K(x_i, x) + b \tag{2}$$

where, α_i and α_i^* are the Lagrange multipliers, and $K(x_i, x)$ is the kernel function used to handle non-linearity.

The kernel functions are given by,

For sigmoid function:

$$K(x_i, x_j) = tanh(kx_i^T + C) \tag{3}$$

For RBF:

$$K(x_i, x_j) = exp(-\gamma(\|x_i - x_j\|^2)) \tag{4}$$

where C is the penalty parameter and γ is the kernel coefficient.

In this context, the SoC can be estimated using the time series reading such as voltage, current, and temperature. The SVM maps these input features using a kernel function namely Radial Basis Function or Sigmoid. These parameters are crucial for fitting the data well without overfitting. A well-defined optimization technique can be utilized for the proper selection of the hyperparameters.

B. Particle Swarm Optimization

The Particle Swarm Optimization approach (PSO) is a biologically inspired optimization which inspired from the social dynamics that are observed in biological swarming, such as fish schooling or avian flocking [15]. In the mid-1990s, the Kennady and Eberhart developed this optimization technique. In Particle Swarm Optimization (PSO), a collection of potential solutions, known as "particles," are used to traverse the solution space in order to identify the optimal solutions. The trajectory of each particle is adjusted based on the historical performance in a multidimensional space and the perceived intelligence of the swarm, and it progressively converges to the optimum value. In complex, non-linear, and intricate problems, particle swarm optimization (PSO) is used without the necessity of gradient data, rendering it highly adaptable. Because of its quick convergence, ease of design, and adaptability for parallel processing to efficiently solve high-dimensional problems, it makes the work adaptable.

C. Problem Formulation

The optimization problem is formulated tune the hyperparameters of the SVM algorithm and is given in below:

$$\text{Minimize (Root Mean Square Error)} \tag{5}$$

$$\text{Subject to, } X_{min} \leqslant X \leqslant X_{max} \tag{6}$$

where, $X = \{X_i\}, i = 1 to D,$

The PSO-based optimization problem solves for the minimum value of RMSE for the execution of the SVM algorithm towards the estimation of SoC. The various steps of the PSO-based SVM are as follows,

979-8-3315-3013-6/25 $31.00 © 2025 IEEE

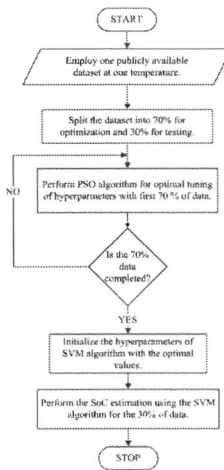

Fig. 1: Flowchart of PSO assisted SVM algorithm.

III. DATA ACQUISITION AND PRE-PROCESSING

This section explains the acquisition and preprocessing of the battery dataset used for the validation of the proposed algorithm.

A. Experiment Benchmark

The proposed algorithm was validated using the publicly available datasets. The dataset was created by Jiaqi Yao at Technische Universität Berlin using a 4.93 Ah rated LG INR 21700 M50LT Lithium-ion battery cell. The Electric Vehicle's driving nature can be categorized into four: aggressive driving cycle (US06), Heavy duty cycle (CSHVC), Urban driving cycle (LA92), and Highway driving cycle (HWFET). Each dataset comprises the battery voltage, discharge current, temperature, and reference SoC estimated by the coulomb counting method. One drive cycle at a typical temperature is taken for the experiment with 70% for the optimization process and the remaining 30% for testing.

A Python program is developed to execute the PSO-based SVM toward the SoC estimation. The program is then executed for the publicly available datasets and the performance of the algorithm is validated by computing performance metrics.

B. Performance validation Metrics

The accuracy and effectiveness of the proposed model is evaluated using performance metrics such as Mean Square Error (MSE), Root Mean Square Error (RMSE), Mean Absolute Percentage Error (MAPE), Mean Absolute Error (MAE), and R-Squared (R^2).

1) *Root Mean Square Error(RMSE)* : This value is determined by taking the residue of their standard deviation and the lowest value of RMSE implies that the model's fitness and responsiveness to change. It is expressed as,

$$\text{RMSE} = \sqrt{\frac{1}{n} \sum_1^n (actual\,soc - predicted\,soc)^2} \quad (7)$$

2) *Mean Absolute Error(MAE)* : The mean difference between observed and expected values is measured by the MAE, which shows how much predictions deviate from actual outcomes. It provides information about the accuracy of prediction.

$$\text{MAE} = \frac{1}{n} \sum_1^n |actual\,soc - predicted\,soc| \quad (8)$$

3) *R*-Square(R^2): It is a statistical measure that quantifies the strength of the relationship between actual and predicted values. The R^2 values approaching 1 indicate the model fit for the data well.

$$\text{R}^2 = 1 - \frac{\sum_1^n (actual\,soc - predicted\,soc)^2}{\sum_1^n (actual\,soc - mean\,value\,of\,actual\,soc)^2} \quad (9)$$

4) *Percentage Error in SoC* : It is the ratio of difference between actual and predicted values of SoC to the actual SoC. It is represented in percentages and mathematically expressed as,

$$\text{Error in SoC} = \frac{actual\,soc - predicted\,soc}{actual\,soc} \quad (10)$$

C. Data processing

A typical dataset from a driving cycle is used for the optimization of all five hyperparameters and then the optimally tuned hyperparameters are used for the estimation of SoC for the remaining datasets from the same driving cycle at different temperatures. In this work, one dataset from each driving cycle corresponding to the temperature 25°C is taken for the optimization and then tuned hyperparameters used for the SoC estimation of datasets at temperatures of 5°C, 15°C, 35°C, and 45°C. The same procedure was repeated for other available driving cycles also and the performance was validated using different metrics.

IV. RESULTS AND DISCUSSION

The Python program is developed for the execution of PSO-based SVM algorithm, and the results are presented in this section. The parameters of the PSO algorithm employed in this work are given in Table I.

TABLE I: PSO Parameters.

Parameter	Values
Maximum Iterations	20
Number of particles	10
C_1, C_2, and ω	1.5, 1.5, and 0.7

For the optimization process, dataset from CSHVC driving cycle at 25 °C is taken. The first 70% of the data is used for the optimization process of hyperparameters and the remaining 30% of data is used for testing the algorithm. The convergence characteristic corresponding to this dataset is given in Fig. 2. The Characteristics imply that it converges at the 12th iteration with a minimum RMSE value of 0.00183. The optimized values of all five hyperparameters with range boundaries are given in Table II.

979-8-3315-3013-6/25 $31.00 © 2025 IEEE

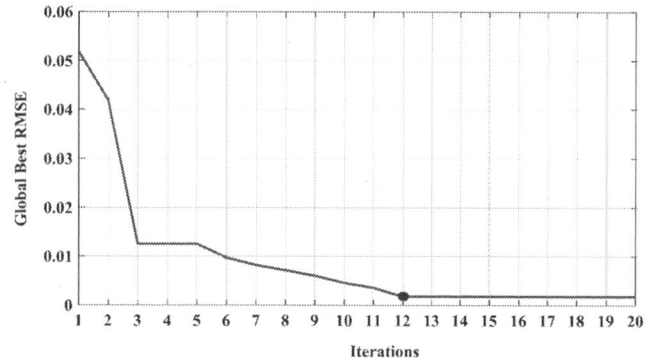

Fig. 2: Convergence Characteristics for CSHVC driving cycle.

TABLE II: Optimal values of hyperparameters of SVM algorithm for CSHVC Driving Cycle.

Hyperparameters	Range	Optimized Values
Regulation Parameter (C)	0.0001 - 100	28
Loss function (ϵ)	0.0001-1	0.0001
Kernel coefficient (γ)	0.000001-1	0.056
Tolerances	0.000001-0.01	0.004
Kernel type	1(sigmoid), 0(rbf)	rbf

Now the optimized values of hyperparameters are employed for the estimation of SoC for the datasets of CSHVC driving cycles at temperatures of 5°C, 15°C, 35°C, and 45°C and the SoC tracking curves for each case are given in Fig. 3. From these graphs, it is seen that the predicted values of SoC fall in line with the actual SoC. For better understanding, the percentage error is evaluated and plotted in Fig. 4. The value of the percentage error in SoC is always below the standard level of 2.5% [5].

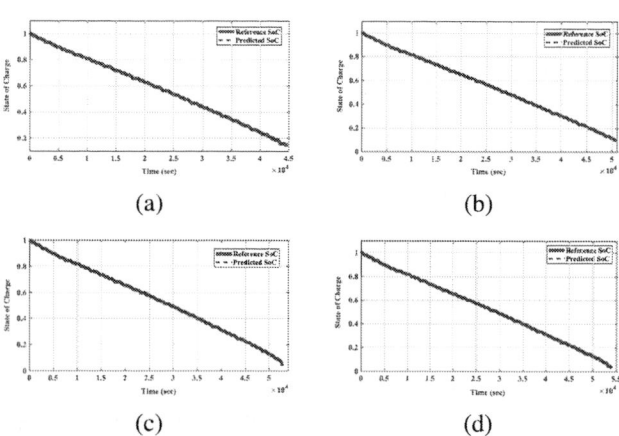

Fig. 3: SoC tracking curves for CSHVC driving cycle at temperatures of (a) 5°C (b) 15°C (c) 35°C and (d) 45°C.

Now, the PSO-based SVM algorithm is employed for different driving cycles namely US06, LA92, and HWFET. The optimization carried out for one dataset from each driving

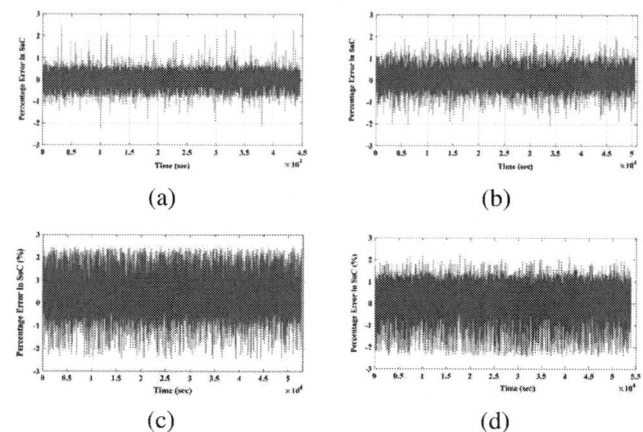

Fig. 4: Percentage error in SoC for CSHVC driving cycle at temperatures of (a) 5°C (b) 15°C (c) 35°C and (d) 45°C.

cycle with the same boundaries of hyperparameters and the convergence characteristics of HWFET, LA92, and US06 at a temperature of 25 °C are given in Fig. 5 (a), (b) and (c) respectively. The optimal tuned values of hyperparameters are given in Table III.

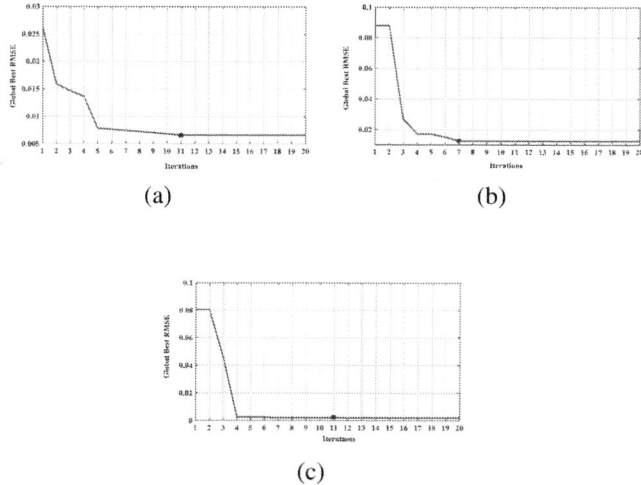

Fig. 5: Convergence Characteristics of the driving cycle (a) HWFET (b) LA92 and (c) US06 at a temperature of 25°C.

TABLE III: PSO Optimized Parameters of SVM Algorithm for Driving Cycle of US06, LA92, and HWFET

Hyperparameters	Optimized Values		
	US06	LA92	HWFET
C	22	62	47
ε	0.0001	0.025	0.0001
γ	0.026	0.002	0.249
Tolerance	0.008	0.005	0.002
Kernel type	rbf	sigmoid	rbf

The optimized structure is then used for the estimation of SoC using the SVM algorithm. For analysis purposes, the

SoC tracking curves for the HWFET driving cycle are also presented in Fig. 6, and the corresponding percentage error in SoC is given in Fig. 7. In this case, the percentage error always falls below 2.25% which shows the accuracy of the predicted value.

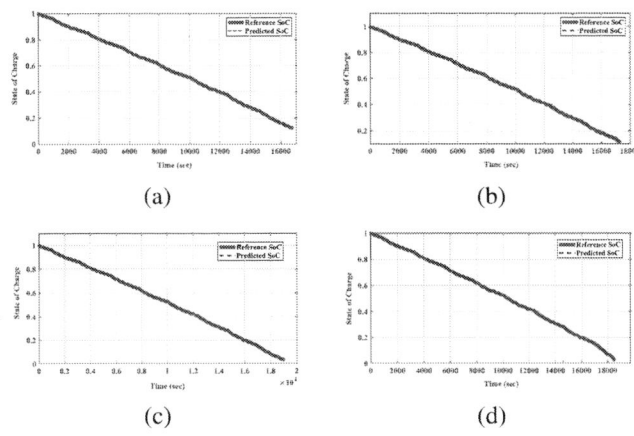

Fig. 6: SoC tracking curves for HWFET driving cycle at temperatures of (a) 5°C (b) 15°C (c) 35°C, and (d) 45°C.

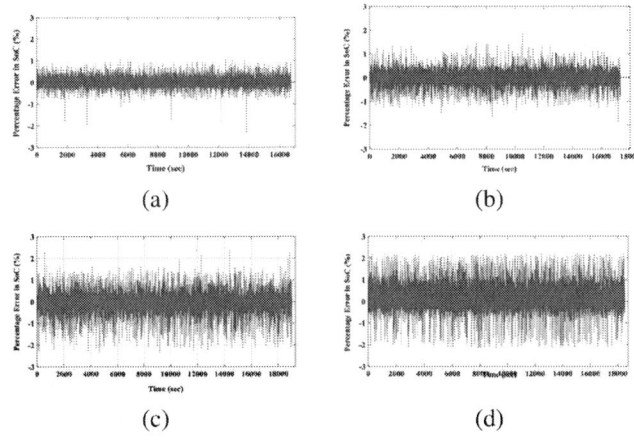

Fig. 7: Percentage error in SoC for HWFET driving cycle at temperatures of (a) 5°C (b) 15°C (c) 35°C and (d) 45°C.

The algorithm is now executed for the remaining driving cycles of LA92 and US06 at various temperatures. The SoC tracking curves and percentage error in SoC for US06 driving cycles are plotted in Fig. 8 and 9, respectively. The results obtained for the LA92 driving cycles are plotted in Fig. 10 and 11, respectively. The predicted SoC follows the reference SoC, which means the model is fit for the SoC estimation, and the percentage error in SoC always falls under 2.5%.

The accuracy and efficacy of the proposed algorithm can be validated using performance indices. Thus, the performance indices for each driving cycle at different temperatures are evaluated and tabulated in Table IV. The computed results in Table IV underline the superiority of the proposed algorithm.

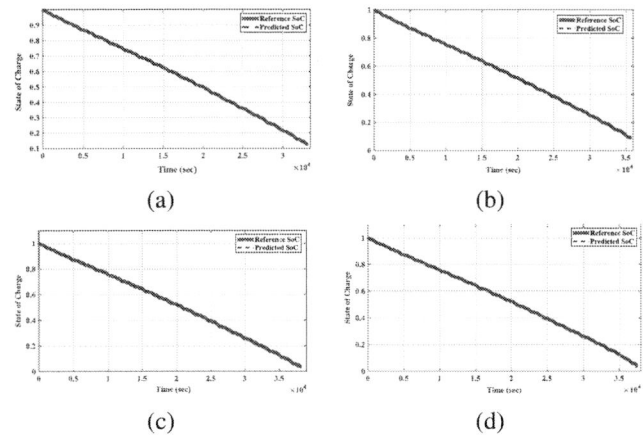

Fig. 8: SoC tracking curves for US06 driving cycle at temperatures of (a) 5°C (b) 15°C (c) 35°C and (d) 45°C.

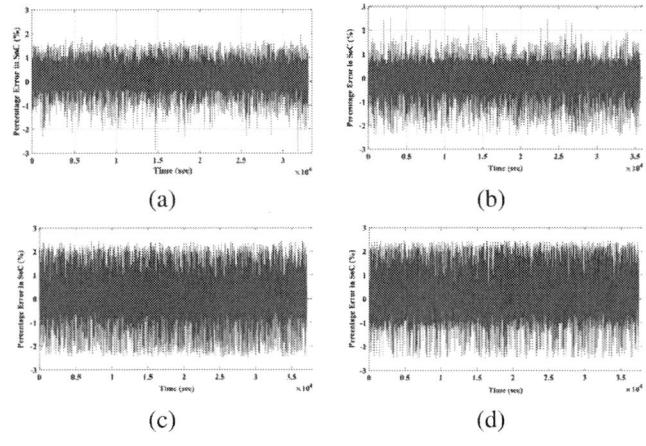

Fig. 9: Percentage error in SoC for US06 driving cycle at temperatures of (a) 5°C (b) 15°C (c) 35°C and (d) 45°C.

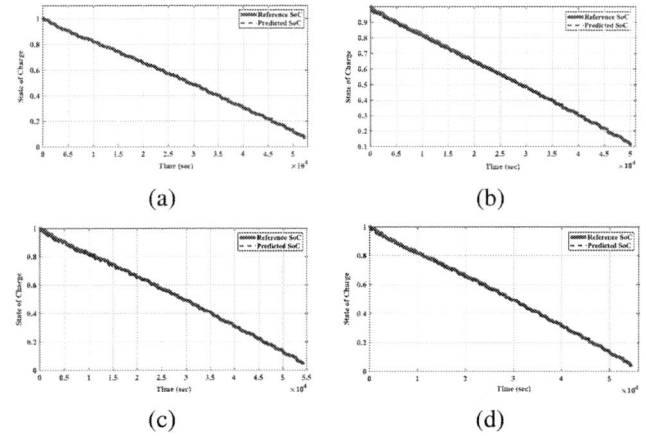

Fig. 10: SoC tracking curves for LA92 driving cycle at temperatures of (a) 5°C (b) 15°C (c) 35°C and (d) 45°C.

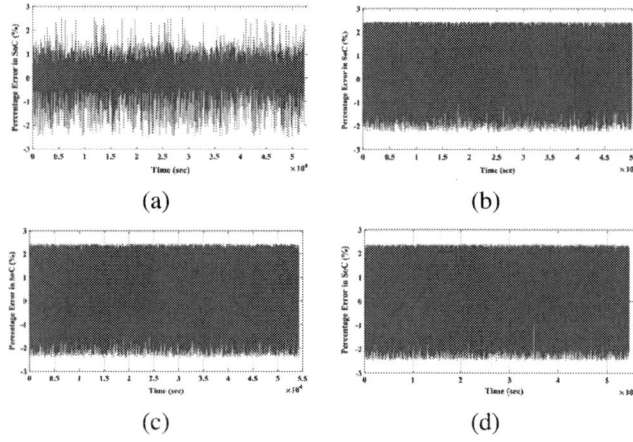

(a)　　　　　　　　　　(b)

(c)　　　　　　　　　　(d)

Fig. 11: Percentage error in SoC for LA92 driving cycle at temperatures of (a) 5°C (b) 15°C (c) 35°C and (d) 45°C.

The obtained values of RMSE and MAE are always found below 0.018 and 0.016, respectively.

TABLE IV: Performance Indices for Driving Cycle of US06, LA92, CSHVC and HWFET

Driving Cycle	Temperature	RMSE	MAE	R^2
US06	5°C	0.001742	0.001408	0.9999509
	15°C	0.001807	0.001452	0.9999507
	35°C	0.002232	0.001842	0.9999336
	45°C	0.002098	0.001739	0.9999418
LA92	5°C	0.016264	0.014529	0.9955348
	15°C	0.013855	0.012241	0.9967171
	35°C	0.017153	0.015955	0.9959599
	45°C	0.015133	0.013957	0.9968561
CSHVC	5°C	0.001142	0.000893	0.9999775
	15°C	0.001296	0.001012	0.9999736
	35°C	0.001972	0.001620	0.9999467
	45°C	0.001855	0.001499	0.9999540
HWFET	5°C	0.000911	0.000762	0.9999870
	15°C	0.001091	0.000809	0.9999817
	35°C	0.001563	0.000985	0.9999686
	45°C	0.001776	0.001374	0.9999598

V. CONCLUSION

In this paper, the PSO-assisted Support Vector Machine algorithm is proposed for the accurate estimation of SoC. A typical dataset of a particular driving cycle at a given temperature is used for the optimization of hyperparameters, and then the SoC is estimated using the optimally tuned hyperparameters for the remaining datasets at different temperatures of the same driving cycles. The proposed algorithm was validated for four different publicly available datasets namely HWFET, CSHVC, LA92 and US06. The maximum values of RMSE and MAE for the driving cycle LA92 are obtained as 0.017153 and 0.015955, which shows the superiority of the proposed algorithm. The percentage error in SoC for diving cycles at various temperatures lies under 2.5% which is less than the standard value. The computed results are presented

and show that the proposed algorithm is a promising tool for the estimation of SoC.

REFERENCES

[1] X. Gong, R. Xiong and C. C. Mi (2015) Study of the Characteristics of Battery Packs in Electric Vehicles With Parallel-Connected Lithium-Ion Battery Cells. IEEE Transactions on Industry Applications. doi:10.1109/TIA.2014.2345951.

[2] M. A. Hannan, M. M. Hoque, S. E. Peng and M. N. Uddin(2017) Lithium-Ion Battery Charge Equalization Algorithm for Electric Vehicle Applications. IEEE Transactions on Industry Applications. doi: 10.1109/TIA.2017.2672674.

[3] D. N. T. How, M. A. Hannan, M. S. Hossain Lipu and P. J. Ker (2019) State of Charge Estimation for Lithium-Ion Batteries Using Model-Based and Data-Driven Methods: A Review. IEEE Access. doi: 10.1109/ACCESS.2019.2942213.

[4] L. He and D. Guo (2019) An Improved Coulomb Counting Approach Based on Numerical Iteration for SOC Estimation With Real-Time Error Correction Ability. IEEE Access. doi:10.1109/ACCESS.2019.2921105.

[5] O. Kadem and J. Kim (2023) Real-Time State of Charge-Open Circuit Voltage Curve Construction for Battery State of Charge Estimation. IEEE Transactions on Vehicular Technology. doi: 10.1109/TVT.2023.3244623.

[6] R. Xiong, H. He, F. Sun and K. Zhao (2013) Evaluation on State of Charge Estimation of Batteries With Adaptive Extended Kalman Filter by Experiment Approach. IEEE Transactions on Vehicular Technology, doi: 10.1109/TVT.2012.2222684.

[7] H. Aung, K. Soon Low and S. Ting Goh (2015) State-of-Charge Estimation of Lithium-Ion Battery Using Square Root Spherical Unscented Kalman Filter (Sqrt-UKFST) in Nanosatellite. IEEE Transactions on Power Electronics. doi: 10.1109/TPEL.2014.2361755.

[8] C. R. Gould, C. M. Bingham, D. A. Stone and P. Bentley (2009) New Battery Model and State-of-Health Determination Through Subspace Parameter Estimation and State-Observer Techniques. IEEE Transactions on Vehicular Technology. doi: 10.1109/TVT.2009.2028348.

[9] S. G. Padder et al., (2025) Data-Driven Approaches for Estimation of EV Battery SoC and SoH: A Review. IEEE Access. doi: 10.1109/ACCESS.2025.3539528.

[10] D. N. T. How, M. A. Hannan, M. S. H. Lipu, K. S. M. Sahari, P. J. Ker and K. M. Muttaqi (2020) State-of-Charge Estimation of Li-Ion Battery in Electric Vehicles: A Deep Neural Network Approach. IEEE Transactions on Industry Applications. doi:10.1109/TIA.2020.3004294.

[11] F. Zhao, Y. Li, X. Wang, L. Bai and T. Liu (2020) Lithium-Ion Batteries State of Charge Prediction of Electric Vehicles Using RNNs-CNNs Neural Networks. IEEE Access. doi: 10.1109/ACCESS.2020.2996225.

[12] K. Jia, Z. Gao and X. Gao (2024) An Adaptive LSTM Network With Fractional-Order Memory Unit Optimized by Hausdorff Difference for SOC Estimation of Lithium-Ion Batteries. IEEE Transactions on Circuits and Systems II: Express Briefs. doi:10.1109/TCSII.2023.3344191.

[13] C. Vidal, P. Malysz, P. Kollmeyer and A. Emadi (2020) Machine Learning Applied to Electrified Vehicle Battery State of Charge and State of Health Estimation: State-of-the-Art. IEEE Access. doi:10.1109/ACCESS.2020.2980961.

[14] X. Feng et al., (2019) Online State-of-Health Estimation for Li-Ion Battery Using Partial Charging Segment Based on Support Vector Machine. IEEE Transactions on Vehicular Technology. doi: 10.1109/TVT.2019.2927120.

[15] W. Xiong, Y. Mo and C. Yan (2021) Online State-of-Health Estimation for Second-Use Lithium-Ion Batteries Based on Weighted Least Squares Support Vector Machine. IEEE Access. doi:10.1109/ACCESS.2020.3026552.

[16] R. Eberhart and J. Kennedy (1995) A new optimizer using particle swarm theory. MHS'95. Proceedings of the Sixth International Symposium on Micro Machine and Human Science. doi: 10.1109/MHS.1995.494215.

[17] https://depositonce.tu-berlin.de/items/7f68932b-4d43-4f49-a5d8-914b00039f87.

Performance Comparison of Recurrent Neural Network Variants for State of Charge Prediction in Lithium-Ion Batteries

Saneep K
Department of EEE
National Institute of Technology
Tiruchirappalli, India
407123004@nitt.edu

Kannan M
Department of EEE
LBS College of Engineering
Kasaragod, India
kannanlbteee@gmail.com

Sundareswaran K
Department of EEE
National Institute of Technology
Tiruchirappalli, India
kse@nitt.edu

P Srinivasa Rao Nayak
Department of EEE
National Institute of Technology
Tiruchirappalli, India
psnayak@nitt.edu

Mithun T
Expert Solutions Pvt Ltd
An Alten Company
Pune, India
mithun.t@expertgs.com

Abstract—This paper presents a comparative study of three widely used recurrent neural network (RNN) variants—Long Short-Term Memory (LSTM), Bidirectional LSTM (Bi-LSTM), and Gated Recurrent Unit (GRU)—for predicting the State of Charge (SoC) of lithium-ion batteries in electric vehicles (EVs). A key contribution of this work is the evaluation of these models under varying environmental and operational conditions, specifically using battery data (INR18650-20R) collected at three different ambient temperatures: 0°C, 25°C, and 45°C. The models are trained using the Dynamic Stress Test (DST) drive cycle and tested on two unseen drive cycles—Federal Urban Driving Schedule (FUDS) and the Supplemental Federal Test Procedure (US06)—without retraining, demonstrating their generalization capability. The performance of each model is assessed using standard evaluation metrics, including Mean Squared Error (MSE), Mean Absolute Error (MAE), Root Mean Squared Error (RMSE), and the Coefficient of Determination (R^2). Results indicate that all three RNN architectures provide accurate SoC predictions, with notable differences in performance under different conditions. This study provides valuable insights into the robustness and effectiveness of RNN-based approaches for real-time SoC estimation in EV battery management systems.

Keywords—State of Charge (SoC), Recurrent Neural Network (RNN), Long Short-Term Memory (LSTM), Bidirectional Long Short-Term Memory (Bi-LSTM), Gated Recurrent Unit (GRU).

I. INTRODUCTION

The utilization of lithium-ion batteries has grown substantially, extending from consumer electronics such as mobile devices to electric vehicles (EVs) and smart grid systems, owing to advantages including high energy density, low self-discharge rate, minimal maintenance, and lightweight structure. The State of Charge (SoC) is an indicator representing the remaining capacity of a battery [1]. Accurate and reliable SoC estimation enhances battery longevity, improves EV performance, optimizes capacity usage, and mitigates risks associated with overcharging or deep discharging [2]. Despite its importance, SoC cannot be directly measured and must be estimated. Various direct and indirect techniques have been developed for SoC estimation [3]. A comparative summary of these methods is provided in Table I.

Among the SoC estimation techniques in Table I, data-driven approaches have gained increasing attention in recent years. By skipping models, filters, and heavy math, these methods support rapid BMS advancement. Several neural network architectures have been successfully employed for SoC estimation, including deep neural network (DNN) [4], convolutional neural network [5], and long short-term memory (LSTM) network [6].

TABLE I. MERITS AND DEMERITS OF PRIMARY SOC ESTIMATION METHODS

	Coulomb Counting Method [7]	
	Current is integrated to estimate SoC.	
Merits	Simple, Cost-effective, Easy to understand	
Demerits	Dependency on initial SoC, accuracy of measurement sensors	
	Voltage-based method [8]	
	Measured open-circuit voltage is related to SoC	
Merits	Simple, Small computational burden	
Demerits	Open-loop, long rest time of the battery for measuring OCV, cannot estimate real-time SoC	
	A flat OCV curve of LiB results in large SoC error for a small voltage sensor error.	
	Model based method [9]	
	ECM, along with filtering algorithms, estimates SoC	
Merits	Closed-loop architecture with improved accuracy	
Demerits	High computational cost	
	Estimation accuracy depends on model choice, battery parameters, aging, and temperature	
	Data-driven method [10]	
	Measurable parameters of the battery are learned via neural networks for the estimation of SoC	
Merits	Less computational cost	
	It does not depend on initial SoC, aging, battery model parameters, etc	
Demerits	It depends on a large quantity of data for improved accuracy	
	Requirement of memory storage	

This paper presents a performance comparison of various recurrent neural network (RNN) architectures for SoC estimation of lithium-ion batteries utilized in electric vehicles under three distinct temperature conditions. The RNN variants are trained using dynamic stress test (DST) drive cycle and evaluated using two different driving profiles, namely federal urban driving schedule (FUDS) and supplemental federal test procedure (US06).

The rest of the paper unfolds as follows: Section II introduces the RNN architectures—LSTM, Bi-LSTM, and GRU. Section III covers the experimental setup and tools

979-8-3315-3013-6/25 $31.00 © 2025 IEEE

used. Section IV discusses key findings, with conclusions wrapped up in Section V.

II. VARIANTS OF RECURRENT NEURAL NETWORKS

RNNs have become increasingly popular in a range of applications for their ability to capture temporal patterns in sequential data. Despite this, traditional RNNs struggle with issues like vanishing and exploding gradients, which make learning long-term dependencies difficult. LSTM networks address these issues by incorporating gating mechanisms that regulate information flow and maintain stable gradients during training [11].

A. LSTM Recurrent Neural Network

Hochreiter et al. [12] proposed an LSTM network, that uses memory units rather than traditional hidden nodes. In an LSTM NN, information can be remembered for longer periods of time [13]. Figure 1(a) illustrates a standard LSTM cell, which includes three sigmoid and two tanh activation functions. The core of the LSTM is the cell state which carries information through time. This state is updated through controlled addition and multiplication of inputs. The sigmoid functions, outputting values between 0 and 1, act as gates to retain or discard information. Meanwhile, the tanh functions help address the vanishing gradient issue by maintaining gradients over longer sequences, thanks to their slowly diminishing second derivative.

The calculation process at time k is given below:

$$f_k = \sigma \left(W_f * [h_{k-1}, x_k] + b_f \right) \tag{1}$$

$$i_k = \sigma \left(W_i * [h_{k-1}, x_k] + b_i \right) \tag{2}$$

$$C_k = tanh \left(W_C * [h_{k-1}, x_k] + b_C \right) \tag{3}$$

$$o_k = \sigma \left(W_o * [h_{k-1}, x_k] + b_o \right) \tag{4}$$

$$h_k = o_k * tanh(C_k) \tag{5}$$

where x_k denotes input; h_k is the hidden state; C_k is the unit memory; i_k, f_k, and o_k are the activation vectors of the input gate, forget gate, and output gate, respectively. W and b form the weight and bias.

B. Bi-LSTM Recurrent Neural Network

Bidirectional LSTMs differ from regular LSTMs by processing data in both forward and backward directions. This dual processing enables the model to capture information from past and future steps simultaneously. By considering data from both directions, the model can better identify complex patterns and trends that influence battery behavior. This bidirectional approach helps reduce prediction errors and improves the reliability of SoC estimation [14].

C. GRU Neural Network

Cho et al. (2014) introduced the Gated Recurrent Unit to improve the processing of time-series data [15,16]. Because measurable physical parameters like voltage, temperature, and current are a function of time in the task of SoC estimation, sequence processing algorithms like the GRU are good options. Figure 2 illustrates the proposed GRU model for SoC estimation, composed of input, hidden, and output layers. The input layer receives key battery signals: voltage,

current, and temperature. Within the hidden layer, the GRU cells process these inputs. Figure 3 reveal the detailed structure of a single GRU cell.

As the input vector flows through the network, it undergoes a series of matrix operations within the GRU cells, ultimately producing an estimated SoC value at the output layer. This architecture enables efficient and accurate modeling of the battery's dynamic behavior. For every hidden layer activation, h is computed as follows:

$$\tilde{z}_k = \sigma \left(w_z^x x_k + w_z^h h_{k-1} + b_z \right) \tag{6}$$

$$\tilde{r}_k = \sigma \left(w_r^x x_k + w_r^h h_{k-1} + b_r \right) \tag{7}$$

$$\tilde{h}_k = tanh \left(w_h^x x_k + w_h^h (h_{k-1} \cdot r_k + b_h) \right) \tag{8}$$

$$h_k = z_k \cdot \tilde{h}_k + (1 - z_k) \cdot h_{k-1} \tag{9}$$

where x_k = input vector, $\tilde{z}_k$ = update gate vector, $\tilde{r}_k$ = reset gate vector, $\tilde{h}_k$ = candidate activation vector, h_k = cell output vector, at timestep k. $w_z^x, w_z^h, w_r^x, w_r^h$ are weight matrices and b_z, b_r, b_h bias vectors.

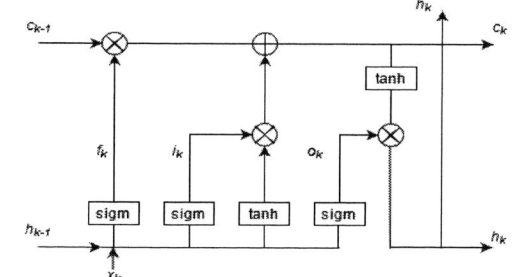

(a)

(b)

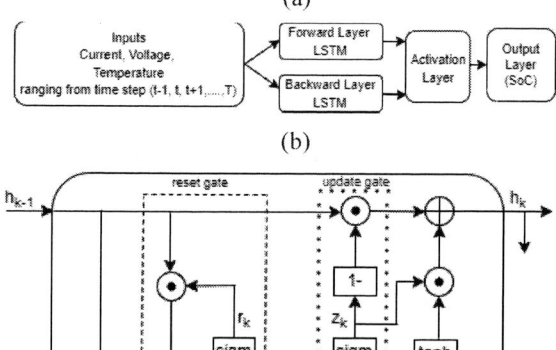

(c)

Fig. 1. Structure of (a) LSTM, (b) BiLSTM, and (c) GRU

Fig. 2 shows a simple neural network architecture. The architecture is designed such that the number of layers, number of neuron units, and dense layers remain the same for the three architectures, namely LSTM, Bi-LSTM, and GRU. So only one visualization of the architecture is only shown for saving the space. The input layer consists of inputs namely current, voltage, and temperature. This layer is followed by the respective RNN hidden layer (i.e., LSTM,

BiLSTM, and GRU). The final output dense layer gives the predicted SoC.

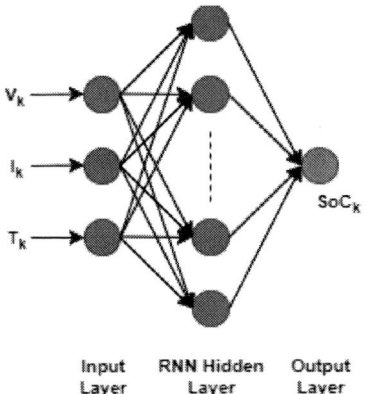

Fig. 2. Recurrent Neural Network Architecture

III. EXPERIMENTAL SETUP

A. Dataset

This study leverages an openly available dataset from the CALCE Battery Research Group at the Center for Advanced Life Cycle Engineering. The data was collected using a cylindrical INR18650-20R LiNiMnCoO$_2$ (NMC) lithium-ion battery, tested under various real-world driving profiles such as the DST, FUDS, and US06. Key battery parameters—voltage, current, and temperature—were measured every second (1 Hz) throughout the experiments. As summarized in Table II, the battery was initialized at an SoC of 80%. The ground truth SoC was determined by integrating the discharge current over time, as outlined in Equation (7).

$$SoC\ (k) = \frac{C_{max} - C\ (k)}{C_{max}} \tag{10}$$

TABLE II. SPECIFICATION OF BATTERY

Type	INR18650-20R
Nominal Voltage	3.6V
Nominal Capacity	2.23Ah
Cut-off voltage	2.4V / 4.2V
Maximum Current	22A
Specific Energy	150 − 220 Wh/kg
Cycle Life	1000 − 2000

B. Training and Evaluation strategy

The experiments presented in this paper were implemented using the Python programming environment on a system with an Intel i5 processor and 16 GB of RAM. The SoC estimation task is formulated as a supervised learning problem. At each time step k, the SoC is represented as a function of the observed battery parameters, current (I_k), voltage (V_k), and temperature (T_k), as expressed below.

$$SoC_k = f\ (\varphi_k, \varphi_{k-1}, \ldots\ldots, \varphi_1) \tag{11}$$
$$\text{where } \varphi(k) = [I_k, V_k, T_k]$$

I_k, V_k, and T_k are current, voltage, and temperature measured at timestep k.

The performance is tested using MSE, MAE, RMSE, and coefficient of determination (R^2) performance metrics which are given by equations 13 to 16, respectively.

$$Error = SoC_k - SoC_k^* \tag{12}$$

$$MSE = \frac{1}{N} \sum_{k=1}^{N}\ (|SoC_k - SoC_k^*|)^2 \tag{13}$$

$$MAE = \frac{1}{N} \sum_{k=1}^{N}\ (|SoC_k - SoC_k^*|) \tag{14}$$

$$RMSE = \sqrt{\frac{1}{N} \sum_{k=1}^{N}(SoC_k - SoC_k^*)^2} \tag{15}$$

$$R^2 = 1 - \frac{\sum_{k=1}^{N}(SoC_k - SoC_k^*)^2}{\sum_{k=1}^{N}(SoC_k - SoC_{mean})^2} \tag{16}$$

where SoC = Reference SoC and SoC* = Predicted SoC.

The architecture and training configuration of the neural networks were optimized using a Keras-based hyperparameter tuning framework. In addition to these, other training hyperparameters—including the learning rate, batch size, number of epochs, and optimizer—were determined using a combination of grid search and heuristic guidelines informed by prior successful studies in battery SoC estimation. These values were refined through multiple trials to balance training efficiency and model accuracy. Table III provides a summary of the finalized hyperparameter settings used for model training. To maintain consistency and ensure a fair comparative analysis, the network topology and training configuration were held constantly across all three RNN variants evaluated in this study.

TABLE III. HYPERPARAMETERS FOR THE THREE VARIANTS OF RNN

Hyperparameter	Value
Optimization	Adam
Initial Learning rate	1e-3
Epochs	500
Dropout	No
Number of hidden layers	01
Number of neurons	100

IV. RESULTS AND DISCUSSION

This section details the results obtained with three different variants of RNN in the state of charge prediction of lithium-ion batteries.

The three training and validation loss plots shown in Fig. 3(a) to 3(c) represent the performance of three different deep learning architectures namely LSTM, Bi-LSTM, and GRU, trained on the DST drive cycle and evaluated on the FUDS and US06 drive cycles at a temperature of 25°C. All three models show a rapid initial decrease in both training and validation losses followed by a gradual plateau, indicating effective convergence during training. This suggests that each model quickly learned the key patterns of the DST cycle and stabilized in terms of learning after the initial epoch. The LSTM model demonstrates a steep decline in loss, with the validation loss closely tracking the training loss, indicating good generalization with minimal overfitting. The Bi-LSTM model exhibits nearly identical training and validation loss curves throughout, reflecting an even stronger generalization ability. The Bi-LSTM is very resilient over different driving

979-8-3315-3013-6/25 $31.00 © 2025 IEEE

cycles because of the close alignment between the losses, which suggests that it efficiently captures temporal dependencies in both forward and backward directions. The GRU model also shows effective convergence, with the training and validation curves aligning after a slight initial divergence, demonstrating stable training dynamics and reliable performance

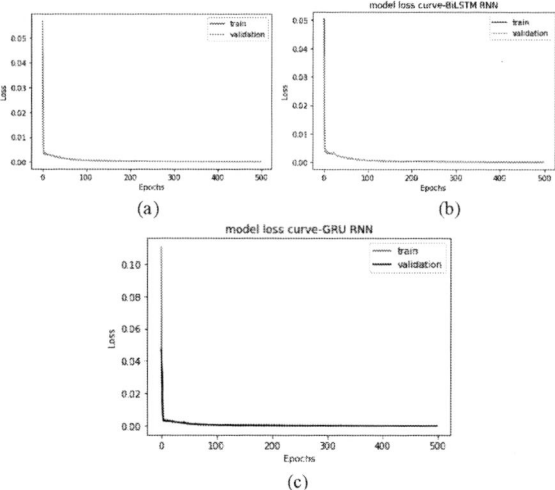

Fig. 3. Model loss curve. (a) LSTM, (b) Bi-LSTM and (c) GRU RNN at 25°C temperature.

Fig. 4(a) to 4(f) shows the results of SoC prediction and estimation error obtained for three variants of RNN at 0°C temperature. Fig 5(a) to 5(f) depicts the SoC tracking, and estimation error obtained at 25°C. Fig 6(a) to 6(f) shows the results of SoC tracking and estimation error obtained at 45°C. The performance of the three RNN models under the FUDS and US06 drive cycles, evaluated at three different temperatures, is summarized in Tables IV and V.

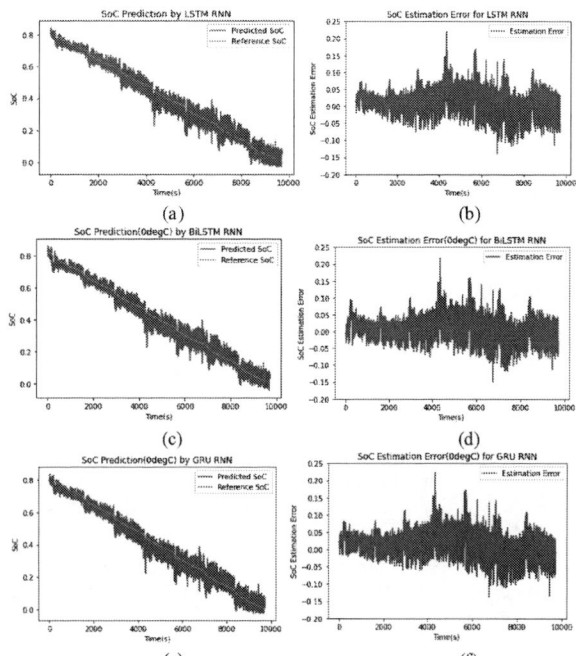

Fig. 4. SoC prediction and estimation error plots. (a), (b) LSTM, (c), (d) Bi-LSTM and (e), (f) GRU RNN at 0°C temperature.

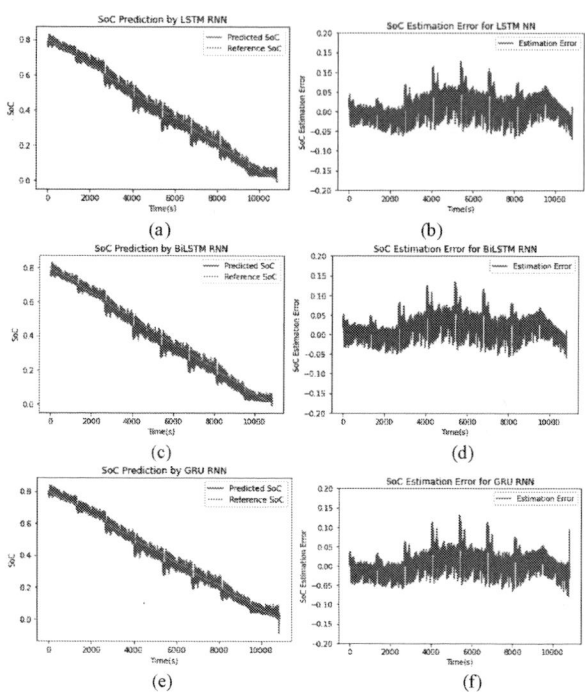

Fig. 5. SoC prediction and estimation error plots. (a), (b) LSTM, (c), (d) Bi-LSTM, and (e), (f) GRU RNN at 25°C temperature.

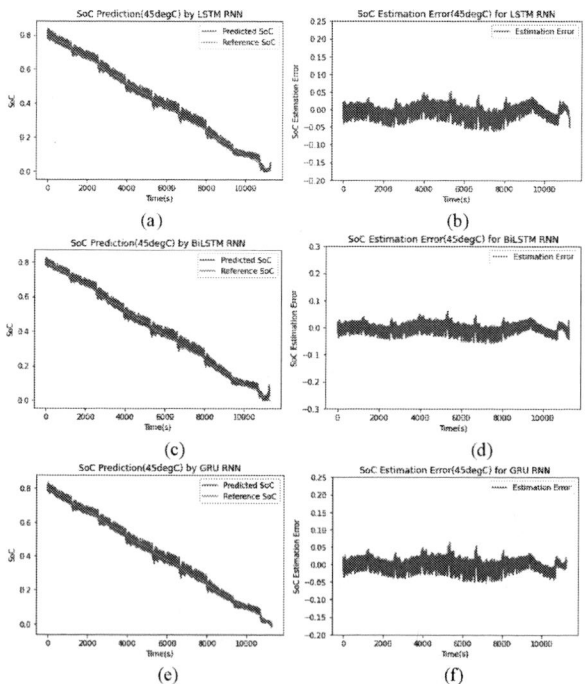

Fig. 6. SoC prediction and estimation error plots. (a), (b) LSTM, (c), (d) Bi-LSTM and (e), (f) GRU RNN at 45°C temperature.

TABLE IV. PERFORMANCE METRICS OF THREE VARIANTS OF RNN WITH FUDS TEST CYCLE

LSTM RNN						
Temp.	Train	Test	MSE	RMSE	MAE	R^2
0°C			1.36e-3	3.61e-2	2.94e-2	97.57
25°C	DST	FUDS	8.52e-4	2.94e-2	2.32e-2	98.42
45°C			2.86e-4	1.62e-2	1.34e-2	99.46
Bi-LSTM RNN						
Temp.	Train	Test	MSE	RMSE	MAE	R^2
0°C			1.28e-3	3.52e-2	2.84e-2	97.72
25°C	DST	FUDS	1.06e-3	3.21e-2	2.61e-2	98.03
45°C			3.01e-4	1.74e-2	1.38e-2	99.43
GRU RNN						
Temp.	Train	Test	MSE	RMSE	MAE	R^2
0°C			1.58e-3	3.94e-2	3.23e-2	97.18
25°C	DST	FUDS	6.63e-4	2.51e-2	2.03e-2	98.77
45°C			6.74e-4	1.98e-2	1.67e-2	99.35

TABLE V. PERFORMANCE METRICS OF THREE VARIANTS OF RNN WITH US06 TEST CYCLE

LSTM RNN						
Temp.	Train	Test	MSE	RMSE	MAE	R^2
0°C			2.53e-3	5.03e-2	4.04e-2	95.27
25°C	DST	US06	2.49e-3	4.99e-2	4.28e-2	95.37
45°C			9.24e-4	3.04e-2	2.43e-2	98.28
Bi-LSTM RNN						
Temp.	Train	Test	MSE	RMSE	MAE	R^2
0°C			2.26e-3	4.75e-2	3.73e-2	95.79
25°C	DST	US06	2.45e-3	4.95e-2	4.18e-2	95.44
45°C			9.50e-4	3.08e-2	2.50e-2	98.23
GRU RNN						
Temp.	Train	Test	MSE	RMSE	MAE	R^2
0°C			2.09e-3	4.57e-2	3.62e-2	96.10
25°C	DST	US06	3.30e-3	5.74e-2	4.75e-2	93.87
45°C			9.95e-4	3.15e-2	2.52e-2	98.15

V. CONCLUSION

This paper presents a thorough comparative analysis of three RNN architectures—LSTM, Bi-LSTM, and GRU—for predicting the SoC of lithium-ion batteries. A significant contribution of this work is the evaluation of model generalization by training on the DST drive cycle and testing on two distinct, unseen drive cycles—FUDS and US06 across multiple temperatures, without retraining the models. Performance was assessed using standard metrics (MSE, RMSE, MAE, and R^2). The results demonstrate that all three RNN models are capable of accurate SoC estimation, but LSTM consistently outperforms the others, particularly at higher temperatures. For instance, LSTM achieved an R^2 of 99.46% (FUDS) and 98.28% (US06) at 45°C, indicating strong reliability and adaptability. Bi-LSTM also performed competitively, while GRU showed slightly lower accuracy, particularly under the more demanding conditions of the US06 cycle at moderate temperatures. These findings highlight the robustness, accuracy, and real-world applicability of RNN-based SoC prediction models, especially LSTM, for use in advanced battery management systems under varying environmental and operational scenarios.

REFERENCES

[1] R. Xiong, J. Cao, Q. Yu, H. He, and F. Sun, "Critical review on the battery state of charge estimation methods for electric vehicles," *IEEE Access*, vol. 6, pp. 1832–1843, 2018.

[2] M. A. Hannan, M. M. Hoque, S. E. Peng and M. N. Uddin, "Lithium-Ion Battery Charge Equalization Algorithm for Electric Vehicle Applications," in IEEE Transactions on Industry Applications, vol. 53, no. 3, pp. 2541-2549, May-June 2017, doi: 10.1109/TIA.2017.2672674.

[3] D. N. T. How, M. A. Hannan, M. S. H. Lipu, and P. J. Ker "State of charge estimation for lithium-ion batteries using model-based and data-driven methods: A review," *IEEE Access*, vol. 7, pp. 136116–136136, 2019.

[4] E. Chemali, P. J. Kollmeyer, M. Preindl, and A. Emadi, "State-of-charge estimation of li-ion batteries using deep neural networks: A machine learning approach," *J. Power Sour.*, vol. 400, pp. 242_255, Oct. 2018.

[5] Dickshon N. T. How, Mahammad A. Hannan, et.al, "State of Charge Estimation of Li-Ion Battery in Electric Vehicles: A Deep Neural Network Approach," *IEEE Trans. Ind. Appl.*, vol. 56, no. 5, pp. 5565–5574, Sep./Oct. 2020.

[6] F. Yang, X. Song, et al, "State of Charge Estimation of Lithium-ion Batteries via Long Short-Term Memory Network," *IEEE Access*, vol. 7, pp. 53792-53799, 2019.

[7] L. He and D. Guo, "An Improved Coulomb Counting Approach Based on Numerical Iteration for SOC Estimation With Real-Time Error Correction Ability," in IEEE Access, vol. 7, pp. 74274-74282, 2019, doi: 10.1109/ACCESS.2019.2921105.

[8] O. Kadem and J. Kim, "Real-Time State of Charge-Open Circuit Voltage Curve Construction for Battery State of Charge Estimation," in IEEE Transactions on Vehicular Technology, vol. 72, no. 7, pp. 8613-8622, July 2023, doi: 10.1109/TVT.2023.3244623.

[9] C. R. Gould, C. M. Bingham, D. A. Stone and P. Bentley, "New Battery Model and State-of-Health Determination Through Subspace Parameter Estimation and State-Observer Techniques," in IEEE Transactions on Vehicular Technology, vol. 58, no. 8, pp. 3905-3916, Oct. 2009, doi: 10.1109/TVT.2009.2028348.

[10] S. G. Padder et al., "Data-Driven Approaches for Estimation of EV Battery SoC and SoH: A Review," in IEEE Access, vol. 13, pp. 35048-35067, 2025, doi: 10.1109/ACCESS.2025.3539528.

[11] Meng Wei, Min Ye, et.al, "State of Charge Estimation of Lithium-Ion Batteries Using LSTM and NARX Neural Networks," *IEEE Access*, vol. 8, pp. 189236–189245, 2020.

[12] S. Hochreiter and J. Schmidhuber, "Long short-term memory," *Neural Comput.*, vol. 9, no. 8, pp. 1735_1780, 1997.

[13] Y. Bengio, P. Simard, and P. Frasconi, "Learning long-term dependencies with gradient descent is difficult," *IEEE Trans. Neural Netw.*, vol. 5, no. 2, pp. 157_166, Mar. 1994.

[14] A. Hussain, A. Yadav and G. Ravikumar, "Anomaly Detection Using Bi-Directional Long Short-Term Memory Networks for Cyber-Physical Electric Vehicle Charging Stations," in IEEE Transactions on Industrial Cyber-Physical Systems, vol. 2, pp. 508-518, 2024, doi: 10.1109/TICPS.2024.3437349.

[15] Chung, J., Gülçehre, Ç., Cho, K., & Bengio, Y. (2014). Empirical Evaluation of Gated Recurrent Neural Networks on Sequence Modeling. ArXiv, abs/1412.3555.

[16] Zhaowei Zhang, Zhekang Dong, Huipin Lin et al, "An Improved Bidirectional Gated Recurrent Unit Method for Accurate State-of-Charge Estimation" *IEEE Access*, vol. 9, pp. 11252–11263, 2021.

Particle Swarm Optimization Employed Machine Learning on MOSFET Based $\acute{C}uk$ Converter With Flying Squirrel Search Optimization (ML(PSO)-FSSO-$\acute{C}uk$) Based MPPT

Neeraj Priyadarshi
Dept. of Electrical Engg.,
JIS College of Engineering,
Kolkata, India.
neerajrjd@gmail.com

M. S. Bhaskar
Renewable Energy Lab,
Prince Sultan University,
Riyadh, Saudi Arabia.
sagar25.mahajan@gmail.com

Prabhakar Modak
Dept. of Mechanical Engg, School of
Applied Science, Uttranchal University,
Dehradun, Uttrakhand.
theprabhakarmodak@gmail.com

Sachin Srivastava
Dept. of Aerospace Engg.,
Uttranchal University,
Dehradun, Uttrakhand.
hodaero@uumail.in

Abstract—In this research work, particle swarm optimization (PSO) employed machine learning (ML) explore on MOSFET based $\acute{C}uk$ converter with Flying squirrel search optimization (FSSO) maximum power point tracking (MPPT) (ML(PSO)-FSSO-$\acute{C}uk$) for higher photovoltaic (PV) power extraction. The ML(PSO)-FSSO-$\acute{C}uk$ based MPPT algorithm on $\acute{C}uk$ Converter delivers higher tracked efficacy with a minimized settling period. The PSO method provides training to ML for achieving optimal operation points which produces higher PV power from $\acute{C}uk$ converter. Also, the PSO method optimizes the weight of the artificial neural network (ANN) till the optimal operation point is achieved for the generation of $\acute{C}uk$ converter. Furthermore, the FSSO method achieves the best operation point for photovoltaic structures attached with $\acute{C}uk$ converter. The ML(PSO)-FSSO-$\acute{C}uk$ based MPPT approach and accomplished faster and more accurate PV power tracking on $\acute{C}uk$ converter with lesser computational load under changing environmental conditions.

Index Terms—Artificial Neural Network (ANN), $\acute{C}uk$ Converter, Flying Squirrel Search Optimization, Machine Learning, Maximum Power Point Tracking, Particle Swarm Optimization.

I. INTRODUCTION

The utilization of renewable energy sources (RESs) produces a sustainable future. In recent decades, renewable energy sources have emerged as a promising solution due to the increasing concerns over global warming and the depletion of fossil fuels. Moreover, photovoltaic (PV)-based renewable energy systems are widely regarded as one of the most acceptable technologies due to their environmentally friendly nature, low maintenance requirements, and cost-effectiveness [1], [2]. However, due to the non-linear nature of the PV power-voltage (P-V) characteristics, a maximum power point tracking (MPPT) circuit is essential to operate the PV panel at its optimal voltage corresponding to the peak power point, thereby improving overall efficiency by up to 35% to 45%. The said arrangement is achieved through the methodology of MPPT. Classical MPPT strategies Viz. perturb and observe

(P & O), incremental conductance (INC), and Hill climbing strategies provide ease of implementation and wider implications [3]–[5]. Moreover, this process is implemented using a several standard DC-DC converter integrated with an MPPT controller. Nevertheless, the aforesaid techniques are not suitable when irradiance and temperature changes. Hence, under these conditions, the P-V curve becomes multi-peaks instead of single peaks. As a result of these circumstances, major power losses occur which provides less arrangement utilization [6], [7]. Therefore, a global maximum power point tracker (GMPPT) is required to achieve global maximum power point (GMPP) [8], [9]. Moreover, specially constructed MPPT algorithms like direct searching approach, double stage strategy, and artificial intelligence (AI) algorithms have been discussed for better PV power tracking under abrupt environmental conditions. Nevertheless, the aforesaid algorithms are unable to deliver higher PV-tracked efficacy under partial shade and fluctuating weather situations [10], [11]. The soft computing methods Viz. particle swarm optimization (PSO), Firefly algorithm (FA), Ant colony optimization (ACO) and artificial Bee Colony (ABC) have been implemented to solve complex optimization challenges. However, these soft computing methods required a large number of iterations to achieve GMPP [12], [13]. Also, the Fuzzy logic control (FLC) based AI method is able to solve uncertainty problems precisely under changing operating conditions. In [14], a hybrid MPPT method that combines PSO-trained machine learning with Flying Squirrel Search Optimization (FSSO) is discussed. Nevertheless, the design and tuning of the rule-based membership function is a major challenge. Therefore, to overcome the mentioned shortcomings, the authors have explored a PSO-employed machine learning with Flying squirrel search optimization based MPPT for optimal PV-tracked efficacy through MOSFET based $\acute{C}uk$ converter. The PSO method trained ML algorithm is employed to achieve the optimal operation point of the PV system which produces maximum generated power through MOSFET based $\acute{C}uk$ converter. Moreover, the FSSO strategy provides a search

979-8-3315-3013-6/25 $31.00 © 2025 IEEE

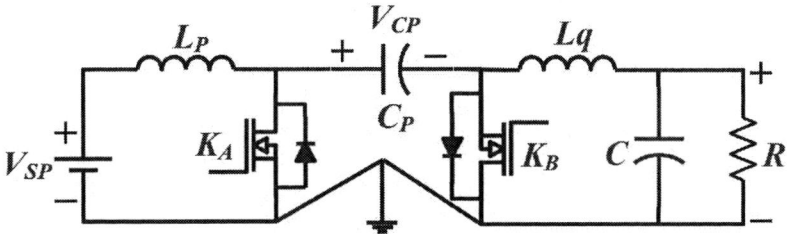

Fig. 1: MOSFET based standard $Ćuk$ Converter.

pattern to obtain the optimal operation point. Furthermore, the ML(PSO)-FSSO-$Ćuk$ employed MPPT for MOSFET based $Ćuk$ converter delivers faster and more accurate tracking of GMPP with lesser computational load and higher PV tracked efficacy under fluctuating weather situations.

II. MATHEMATICAL MODELING OF MOSFET BASED $Ćuk$ CONVERTER

This research work used the standard $Ćuk$ converter because it comprises better voltage regulating ability with the least losses. Due to the buck-boost ability of the $Ćuk$ converter, it consists of a smooth converting ratio, precise current outcome, entire transformer employment, and deportation of capacitor energy as major improvements. Fig. 1 presents the power circuit of the MOSFET based standard $Ćuk$ converter works in two modes of operation.

Mode I: When the power switch K_A is switching off (Fig. 2(a)), the inductor current (I_{LP}) is reduced by charging the capacitor (C_P). On the other hand, L_q is becoming discharging through the load.

Mode II: In this mode of operation, switch K_A gets switched ON (Fig. 2(b)), inductor L_P starts charging and current I_{LP} rises when capacitor C_P gets discharged as a result current through L_q rises. Therefore, based on the energy-transferring nature of C_P capacitor, current flowing through L_P and L_q are evaluated. In this research work, the $Ćuk$ converter is modeled using the state space average approach. The switching period can be evaluated as

$$T_{\text{switching}} = T_{\text{ON}'} + T_{\text{OFF}'} \quad (1)$$

Let, d_{duty} is duty cycle. The general state space modeling under continuous conducting state is expressed as:

Mode I: K_A is switched OFF and K_B is switched ON:

$$\frac{dI_{L_P}}{dt} = \frac{1}{L_P}V_{SP} - \frac{1}{L_P}V_{CP}; \frac{dV_{CP}}{dt} = \frac{1}{C_P}I_{L_P} \quad (2)$$

$$\frac{dI_{L_q}}{dt} = \frac{-V_C}{L_q}; \frac{dV_C}{dt} = \frac{-I_{L_q}}{C} - \frac{V_{Load}}{RC} \quad (3)$$

$$\dot{X}_{P_1} = A_{R_1}X_1 + B_{R_1}V_{SP}; V_{Load} = Y_{R_1}X_{P_1} \quad (4)$$

Where,

$$A_{R_1} = \begin{bmatrix} 0 & -1/L_P & 0 & 0 \\ 1/C_P & 0 & 0 & 0 \\ 0 & 0 & 0 & -1/L_q \\ 0 & 0 & -1/C & -1/RC \end{bmatrix}$$

$$B_{R_1} = \begin{bmatrix} 1/L_P & 0 & 0 & 0 \end{bmatrix}^T, Y_{R_1} = \begin{bmatrix} 0 & 0 & 0 & 1 \end{bmatrix},$$

$$X_{P_1} = \begin{bmatrix} I_{L_P} & V_{C_P} & I_{L_q} & V_C \end{bmatrix}^T$$

Mode II: K_A is switched ON and K_B is switched OFF:

$$\frac{dI_{L_P}}{dt} = \frac{1}{L_P}V_{SP}; \frac{dV_{C_P}}{dt} = -\frac{1}{C_P}I_{L_q} \quad (5)$$

$$\frac{dI_{L_q}}{dt} = \frac{V_C + V_{C_P}}{L_q}; \frac{dV_C}{dt} = \frac{-I_{L_q}}{C} - \frac{V_{Load}}{RC} \quad (6)$$

$$\dot{X}_{P_2} = A_{R_2}X_2 + B_{R_2}V_{SP}; V_{Load} = Y_{R_2}X_{P_2} \quad (7)$$

Where, $A_{R_2} = \begin{bmatrix} 0 & 0 & 0 & 0 \\ 0 & 0 & -1/C_P & 0 \\ 0 & 1/L_q & 0 & 1/L_q \\ 0 & 0 & -1/C & -1/RC \end{bmatrix}$

$$B_{R_2} = \begin{bmatrix} 1/L_P & 0 & 0 & 0 \end{bmatrix}^T, Y_{R_2} = \begin{bmatrix} 0 & 0 & 0 & 1 \end{bmatrix},$$

$$X_{P_2} = \begin{bmatrix} I_{L_P} & V_{C_P} & I_{L_q} & V_C \end{bmatrix}^T \text{ For complete switching,}$$

$$A_R = A_{R_1}(1 - d_{\text{duty}}) + A_{R_2}d_{\text{duty}}$$
$$B_R = B_{R_1}(1 - d_{\text{duty}}) + B_{R_2}d_{\text{duty}}$$
$$Y_R = Y_{R_1}(1 - d_{\text{duty}}) + Y_{R_2}d_{\text{duty}} \quad (8)$$
$$\dot{X}_P = A_RX_P + B_RV_{SP}; V_{Load} = Y_RX_P$$

where, $A_R = \begin{bmatrix} 0 & -\frac{1-d_{\text{duty}}}{L_P} & 0 & 0 \\ \frac{1-d_{\text{duty}}}{C_P} & 0 & -\frac{d_{\text{duty}}}{C_P} & 0 \\ 0 & \frac{d_{\text{duty}}}{L_q} & 0 & \frac{2D-1}{L_q} \\ 0 & 0 & -1/C & -1/RC \end{bmatrix}$

$$B_R = \begin{bmatrix} 1/L_P & 0 & 0 & 0 \end{bmatrix}^T, Y_R = \begin{bmatrix} 0 & 0 & 0 & 1 \end{bmatrix},$$

$$X_R = \begin{bmatrix} I_{L_P} & V_{C_P} & I_{L_q} & V_C \end{bmatrix}^T$$

The voltage transfer function of the $Ćuk$ converter can be mathematically expressed as:

$$\frac{V_{Load}}{V_{SP}} = -d_{\text{duty}}(1 - d_{\text{duty}})^{-1} \quad (9)$$

III. PARTICLE SWARM OPTIMIZATION - MACHINE LEARNING WITH FLYING SQUIRREL SEARCH OPTIMIZATION (ML(PSO)-FSSO-$Ćuk$) BASED MPPT FOR MOSFET $Ćuk$ CONVERTER

The PSO algorithm is a stochastic optimized method that is based on birds flocking social nature. In this method, the particles identify the optimal result to solve the presented challenge with the help of information exchange with the neighbors. The PSO method provides training to ML strategy to obtain optimal operating points so that the MPPT for the

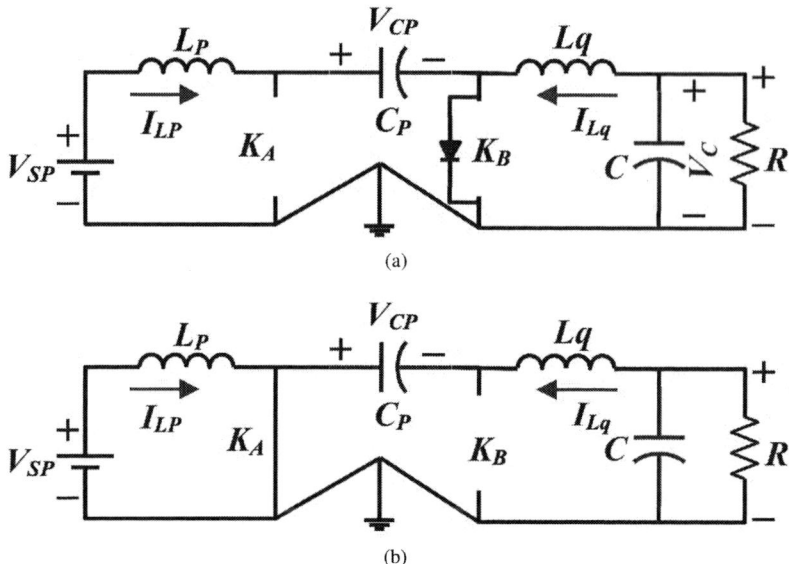

Fig. 2: Power circuit of MOSFET $Ćuk$ converter when (a) switch K_A is switched OFF and (b) switch K_A is Switched ON.

PV system is achieved. The PSO method optimizes the weight of ANN which can be adjustable till the optimum point is found for $Ćuk$ converter. Moreover, the FSSO method based iterative strategy can achieve the best operation stage for the photovoltaic system with $Ćuk$ converter. The FSSO method delivers a searching pattern in terms of squirrel shape in spiral form and optimizes the azimuthal and tilt angle of the solar panel so that peak power extraction from the PV system is obtained. It is also noted that this approach can be utilized for a photovoltaic system comprising multiple solar panels for $Ćuk$ converter. In [14], a hybrid Maximum Power Point Tracking (MPPT) algorithm combining Particle-Swarm-Optimization-trained machine learning and Flying Squirrel Search Optimization is utilized to enhance solar PV system efficiency. Compared to other methods, the algorithm improves efficiency and reduces settling time, demonstrating superior performance under varying conditions. The PSO trained ML employed FSSO based MPPT comprises the benefits of PSO and FSSO algorithms which accomplishes the improvement in PV tracking behavior. For $Ćuk$ converter, the ML(PSO)-FSSO-$Ćuk$ based MPPT suitable to deliver high PV tracked efficacy, and faster and more accurate MPP tracking with lesser computation load compared to the classical MPPT approach. In this MPPT algorithm, the PSO approach provides the optimization of ANN parameters which is then utilized for PV generated power prediction which is feed to $Ćuk$ converter. The FSSO algorithm accomplishes the prediction and provides the adjustment of the operation point of the photovoltaic system to achieve MPP for $Ćuk$ converter. It is also noted that the target function of ML(PSO)-FSSO-$Ćuk$ based MPPT strategy for $Ćuk$ converter is the mean square error. The PSO trained ANN approach is utilized for increased starting weight. Moreover, the FSSO strategy provides flying

squirrel's capability. The FSSO method is accomplished by assuming the following points.

- The generated PV power (P_{PVG}) for $Ćuk$ converter is expressed as a food source .
- Selection parameter is expressed as the duty cycle of the $Ćuk$ converter.
- Through the elimination of hunter presence, the FSSO algorithm is able to reduce the GMPP tracking period for $Ćuk$ converter.

The particles are located at their best positions and have a defined duty cycle for $Ćuk$ converter.

$$D_i = D \pm \triangle D, \quad \text{where } i = 1, 2, \ldots, N_F \quad (10)$$

Where, $\triangle D$ depends on D_{Min}, Minimum duty ratio and D_{Max}, Maximum duty ratio, and $0 < D_i < 1$. The $Ćuk$ converter converter accomplished proper utilization of the searched duty cycle. The source of food delivers the instantaneous PV power generation to $Ćuk$ converter which corresponds to every duty cycle (D_i). The holistic evaluation of MPPT for $Ćuk$ converter can be expressed mathematically as:

$$F(D_i) = [P_{\text{PVG}}D_i]^{\text{max}} \quad (11)$$

The hickory tree consists of maximum generated power for for $Ćuk$ converter at a specified duty ratio and has acorn tree presence at the next best location. Also, the remaining particles are positioned on typical trees. The duty ratio of $Ćuk$ converter needs upgradation after evaluation of different situations. In the consolidation verification process, the implementation is completed after achieved iteration number. Also, the duty ratio of $Ćuk$ converter is produced at a certain position where the power converter works at the obtained GMPP for $Ćuk$ converter.

To obtain the MPPT process for $Ćuk$ converter, temporary changes occur in the optimization process and initial variation

979-8-3315-3013-6/25 $31.00 © 2025 IEEE

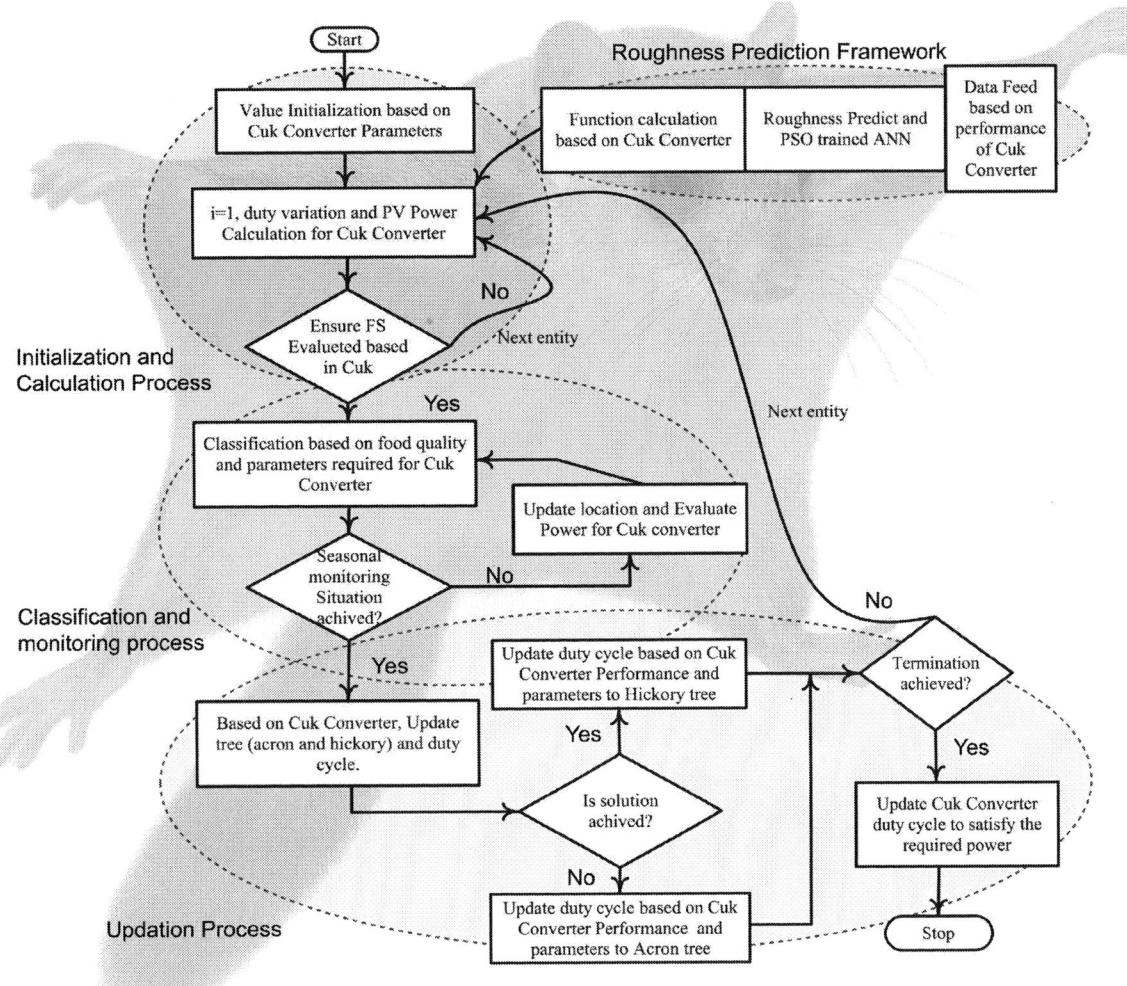

Fig. 3: Flow chart of ML(PSO)-FSSO-$\acute{C}uk$ strategy for MOSFET based $\acute{C}uk$ converter.

is noted under changing weather conditions. Furthermore, the duty ratio of for $\acute{C}uk$ converter is rebooted to achieve a latest GMPP. Fig. 3 illustrates the flowchart of the ML(PSO)-FSSO-$\acute{C}uk$ strategy for $\acute{C}uk$ converter, where the optimal performance of the machine learning (ML) model is enhanced using a particle swarm optimization (PSO) approach to fine-tune the initial weights of the ML structure. Additionally, the FSSO algorithm is employed to coordinate the swarm behavior for improved convergence and accuracy for $\acute{C}uk$ converter. To achieve simplicity, for $\acute{C}uk$ converter the flowchart is divided into four sections: Roughness Prediction Framework, Initialization and Calculation Process, Classification and Monitoring Process, and Updation Process.

IV. RESULTS AND DISCUSSION

Fig. 4 depicts the performance of the ML(PSO)-FSSO-$\acute{C}uk$ MPPT technique is explored for $\acute{C}uk$ converter with varying solar insolations. The presented response reveals that

the P-V characteristics of PV panels are achieved accurately with GMPP under varying sun insolation levels by employing ML(PSO)-FSSO-$\acute{C}uk$ based MPPT approach for $\acute{C}uk$ converter. Also, Fig. 5 demonstrates the obtained I-V plot of PV panels under different solar insolations. The observed plot demonstrates the precise achievement of GMPP for $\acute{C}uk$ converter under varying sun insolations. Moreover, Fig. 6 illustrates the generated PV power versus PV voltage plot under different solar irradiance levels for $\acute{C}uk$ converter. The obtained power trajectory reveals the accurate achievement of GMPP under different sun insolations through the employment of ML(PSO)-FSSO-$\acute{C}uk$ MPPT strategy on $\acute{C}uk$ converter. The performance of the ML(PSO)-FSSO-$\acute{C}uk$ MPPT algorithm on $\acute{C}uk$ converter has been evaluated and compared with existing control strategies, including ANN, PSO, and FSSO, under varying environmental conditions. The simulation results presented in Figs. 7 and 8 indicate that the ML(PSO)-FSSO-$\acute{C}uk$ method on $\acute{C}uk$ converter effectively achieves

979-8-3315-3013-6/25 $31.00 ©2025 IEEE

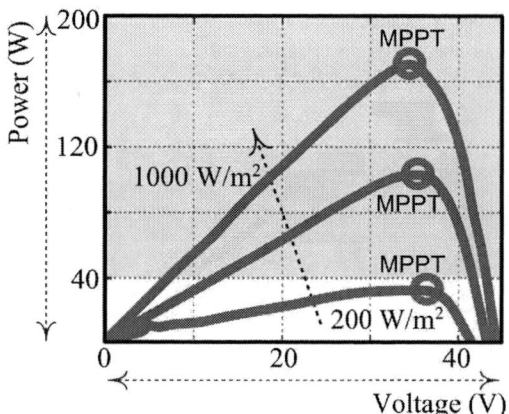

Fig. 4: P-V characteristics of PV panel for MOSFET based $\acute{C}uk$ converter

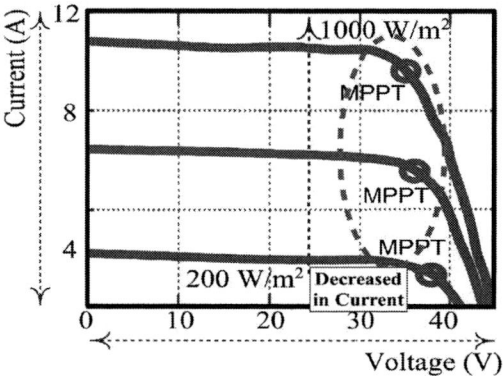

Fig. 5: Obtained I-V plot of PV panel for MOSFET based $\acute{C}uk$ converter.

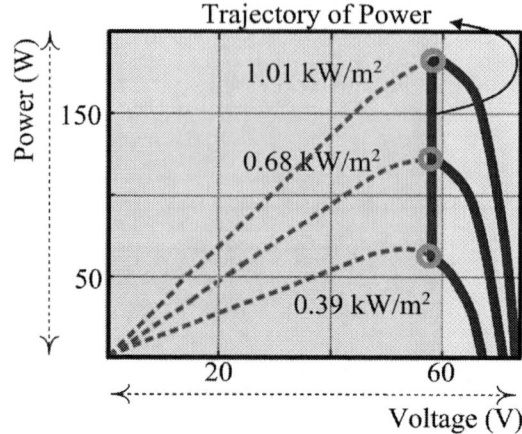

Fig. 6: Generated PV power versus PV voltage plot for MOSFET based $\acute{C}uk$ converter.

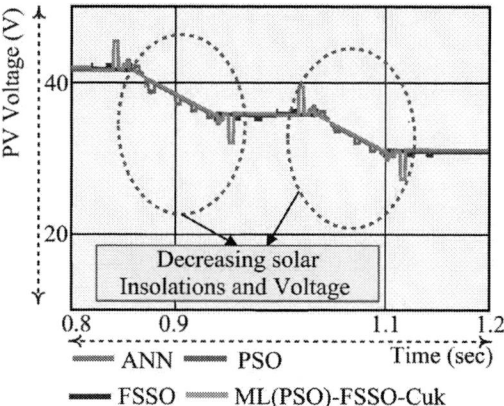

Fig. 7: Tracking performance of ANN, PSO, FSSO and ML(PSO)-FSSO-$\acute{C}uk$ on MOSFET based $\acute{C}uk$ converter algorithms under decreasing solar insolations.

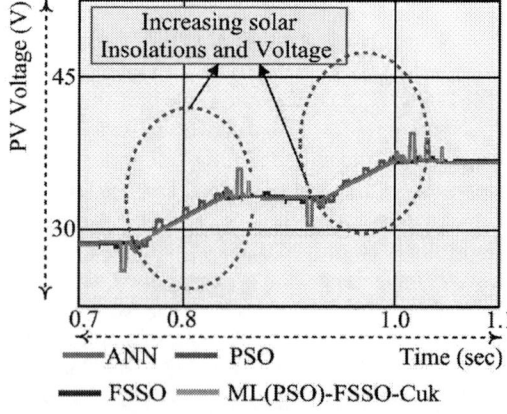

Fig. 8: Tracking performance of ANN, PSO, FSSO and ML(PSO)-FSSO-$\acute{C}uk$ on MOSFET based $\acute{C}uk$ converter algorithms under increasing solar insolations.

rapid convergence to the GMPP with high tracking accuracy, particularly under decreasing solar irradiance. In Fig. 7, tracking performance of ANN, PSO, FSSO, and the ML(PSO)-FSSO-$\acute{C}uk$ algorithm on $\acute{C}uk$ converter under decreasing solar irradiance. Fig. 8 further highlights the performance of the ML(PSO)-FSSO-$\acute{C}uk$ method on $\acute{C}uk$ converter under increasing solar irradiance, demonstrating superior convergence speed and precise tracking compared to ANN, PSO, and FSSO algorithms.

V. CONCLUSION

In this research work, a ML(PSO)-FSSO-$\acute{C}uk$ MPPT has been explored on standard MOSFET based $\acute{C}uk$ converter with the PV system. The PSO trained ML method optimizes the parameters of ANN which is utilized for prediction of generated PV output for $\acute{C}uk$ converter. Moreover, the FSSO strategy is utilized to predict MPP to obtain the best operation point which produces higher PV power from MOSFET based $\acute{C}uk$ converter. Compared to conventional MPPT strategies, the ML(PSO)-FSSO-$\acute{C}uk$ based MPPT on MOSFET based $\acute{C}uk$ converter accomplished faster and more accurate PV power tracking with a lesser computational burden under varying operating conditions. Also, the ML(PSO)-FSSO-$\acute{C}uk$ MPPT on MOSFET based $\acute{C}uk$ converter provides higher PV power efficiency with a minimized settling period under changing environmental situations.

REFERENCES

[1] N. Priyadarshi, M. S. Bhaskar, and D. Almakhles, "A novel hybrid whale optimization algorithm differential evolution algorithm-based maximum power point tracking employed wind energy conversion systems for water pumping applications: Practical realization," *IEEE Trans. on Ind. Electron.*, vol. 71, no. 2, pp. 1641–1652, 2024.

[2] R. Sangrody, S. Taheri, A.-M. Cretu, and E. Pouresmaeil, "An improved pso-based mppt technique using stability and steady state analyses under partial shading conditions," *IEEE Trans. on Sustain. Energy*, 2023.

[3] S.-P. Ye, Y.-H. Liu, H.-Y. Pai, A. Sangwongwanich, and F. Blaabjerg, "A novel ann-based gmppt method for pv systems under complex partial shading conditions," *IEEE Trans. on Sustain. Energy*, 2023.

[4] J. Maeng, J. Jeong, I. Park, M. Shim, and C. Kim, "A time-based direct mppt technique for low-power photovoltaic energy harvesting," *IEEE Trans. on Ind. Electron.*, 2023.

[5] N. Priyadarshi, S. Padmanaban, M. S. Bhaskar, F. Blaabjerg, J. B. Holm-Nielsen, F. Azam, and A. K. Sharma, "A hybrid photovoltaic-fuel cell-based single-stage grid integration with lyapunov control scheme," *IEEE Systems Journal*, vol. 14, no. 3, pp. 3334–3342, 2019.

[6] N. Priyadarshi, S. Padmanaban, J. B. Holm-Nielsen, F. Blaabjerg, and M. S. Bhaskar, "An experimental estimation of hybrid anfis–pso-based mppt for pv grid integration under fluctuating sun irradiance," *IEEE Systems Journal*, vol. 14, no. 1, pp. 1218–1229, 2019.

[7] N. Priyadarshi, S. Padmanaban, P. K. Maroti, and A. Sharma, "An extensive practical investigation of fpso-based mppt for grid integrated pv system under variable operating conditions with anti-islanding protection," *IEEE Systems Journal*, vol. 13, no. 2, pp. 1861–1871, 2018.

[8] N. Priyadarshi, M. S. Bhaskar, D. Almakhles, and F. Azam, "A new pv fed high gain boost-ćuk converter employed srm driven water pumping scheme with idepso mppt," *IEEE Trans. on Power Electronics*, 2024.

[9] D. S. Pillai, J. P. Ram, J. L. Garcia, Y.-J. Kim, and J. P. Catalão, "Experimental studies on a new array design and maximum power tracking strategy for enhanced performance of soiled photovoltaic systems," *IEEE Trans. on Power Electron.*, 2023.

[10] N. Priyadarshi, M. Bhaskar, and D. Almakhles, "Realization of pv power structure with pso tuned fuzzy logic control," in *2023 IEEE IAS Global Conf. on Emerging Technologies (GlobConET)*. IEEE, 2023, pp. 1–6.

[11] A. Amoorezaei, S. A. Khajehoddin, and K. Moez, "A compact cuk-based differential power processing ic with integrated magnetics and soft-switching controller for maximized cell-level power extraction," *IEEE Trans. on Power Electron.*, 2024.

[12] N. Priyadarshi, M. Bhaskar, and D. Almakhles, "Single stage explicit double diode modelled pv module powered linear induction motor driven water pump system," in *2023 IEEE IAS Global Conf. on Emerging Technologies (GlobConET)*. IEEE, 2023, pp. 1–6.

[13] Y. Zhu, H. Wen, H. D. Tafti, G. Wang, Q. Bu, G. Chu, H. Shi, Y. Hu, and L. Jiang, "Novel fast-speed partial-shading-tolerant flexible power point tracking for photovoltaic systems with explicit key points estimation," *IEEE Trans. on Sustain. Energy*, 2023.

[14] D. Kumar, Y. K. Chauhan, A. S. Pandey, A. K. Srivastava, V. Kumar, F. Alsaif, R. M. Elavarasan, M. R. Islam, R. Kannadasan, and M. H. Alsharif, "A novel hybrid mppt approach for solar pv systems using particle-swarm-optimization-trained machine learning and flying squirrel search optimization," *Sustainability*, vol. 15, no. 6, p. 5575, 2023.

979-8-3315-3013-6/25 $31.00 © 2025 IEEE

Comprehensive Analysis of a 1.2 kV 4H-SiC DMOSFET with an Improved Short Circuit Withstand Time

Ravi Prakash Gautam*
Department of ECE (SECS)
IIT Bhubaneswar
Bhubaneswar, India
23vl06010@iitbbs.ac.in

Rachita Mohapatra*
Department of ECE (SECS)
Graduate Student Member, IEEE
IIT Bhubaneswar
Bhubaneswar, India
a23ec09003@iitbbs.ac.in

Wakar Hasan Kazi
Department of ECE (SECS)
Student Member, IEEE
IIT Bhubaneswar
Bhubaneswar, India
a24ec09002@iitbbs.ac.in

Prashant Singh
Department of Electrical Engineering
IIT Madras
Chennai, India
ee18d026@smail.iitm.ac.in

Akshay K
Department of ECE (SECS)
Member, IEEE
IIT Bhubaneswar
Bhubaneswar, India
akshay@iitbbs.ac.in

Abstract—**In a recent work, we proposed a novel method to design the drift layer of 4H-SiC D-MOSFETs to maximize the short circuit withstand time, t_{SC}, at the cost of $< 5\%$ increase in specific on resistance, R_{onsp}. In this work, devices designed using the proposed method is compared against that designed using the conventional method in terms of their avalanche ruggedness through unclamped inductive switching (UIS) test and their static performance after incorporating self heating effect. It is found that the proposed device has better avalanche ruggedness as compared to the reference device. For inductance, $L = 100\ \mu H$, the proposed device has 9.6% decrease in peak discharge current, I_{av} and a 12.5% increase in effective drain-to-source breakdown voltage at peak discharge current, $V_{BR,eff}$. On the other hand, R_{onsp} of the proposed device degrades more rapidly than conventional device due to the positive thermal coefficient of drift resistance and the increased contribution of the drift resistance to the overall R_{onsp} in the former device.**

Index Terms—**4H-SiC, short circuit withstand time, breakdown voltage, inductance, specific on resistance, reliable, ruggedness, unclamped switching event.**

I. INTRODUCTION

Power semiconductor devices are the fundamental building blocks in a range of power electronic circuits and systems. 4H-SiC material based power devices have recently drawn attention due to the exceptional physical characteristics of 4H-SiC such as high breakdown field strength, low intrinsic carrier concentration due to the wide bandgap, high carrier velocity and high thermal conductivity [1]. The stability, robustness, and reliability of these devices should be evaluated using a variety of test before its deployed in the market to be used in practical applications. Avalanche ruggedness capability

*Shared first authors.

provides a crucial quality indication for the overall robustness of the device [2]. The ability of a device to enter and sustain in avalanche mode during an inductive switching event is known as device ruggedness. Devices that cannot enter into and/or cannot sustain in avalanche mode are considered to have poor avalanche ruggedness and are not suitable for hard switching applications. A power MOSFET's avalanche ruggedness is often assessed using single-pulse Unclamped Inductive Switching (UIS) [3] - [7].

Traditionally, drift layer of power MOSFET is designed to yield the minimum R_{onsp} for a given V_{BR}. This approach may not lead to the optimum performance during a short circuit event. Hence, recently our group has proposed a method to improve the short circuit withstand time of 4H-SiC DMOSFET by redesigning the drift layer resistance, R_{dsp} without significantly affecting the R_{onsp} of the device [8]. However, the study was only based on isothermal steady state simulations. So, in this paper, we do a comprehensive investigation and draws comparison between the proposed and conventional drift layers in 4H-SiC DMOSFETs. These devices are compared in terms of their avalanche ruggedness through unclamped inductive switching (UIS) test under single pulse condition by varying the load inductance. Further, their static performance is also compared after incorporating self heating effect.

In subsequent sections, the 4H-SiC DMOSFET having a drift layer designed using conventional method [9] and our proposed method will be referred to as *reference device* (see Fig. 1(a)) and *proposed device* (see Fig. 1(b)) respectively.

II. DESIGN APPROACH

The R_{onsp} of a power MOSFET is given by

$$R_{onsp} = R_{dsp} + R_{lsp} \qquad (1)$$

where R_{dsp} is the drift layer resistance and R_{lsp} is the lumped resistance due to all other components contributing to R_{onsp}. Assuming one-dimensional current flow in the drift layer, R_{dsp} is calculated using the formula $\frac{t_d}{qN_d\mu_n}$, where t_d is the drift layer length, N_d is drift layer doping and μ_n is the electron mobility given by [10]

$$\mu_n = 40 + \frac{950 - 40}{1 + (\frac{N_d}{2 \times 10^{17}})^{0.76}} \quad (2)$$

For the 1.2 kV rated reference device given in Fig. 1(a), R_{dsp} and R_{onsp} can be estimated to be 0.89 mΩ.cm^2 and 4.658 mΩ.cm^2 respectively and therefore R_{lsp} is 3.768 mΩ.cm^2 as per (1). According to the proposed drift layer design method, R_{dsp} can be raised by 35% by increasing t_d and decreasing N_d as per the equations below, while R_{onsp} increases only by 5% due to the low contribution of R_{dsp} on R_{onsp}.

$$E_C = 2.56 \times 10^4 N_d^{\frac{1}{8}} \quad (3)$$

$$N_d = \sqrt{\frac{2\epsilon_s E_C}{3q^2\mu_n f R_{dsp}}} \quad (4)$$

$$t_d = \frac{2\epsilon_s E_C}{3qN_d} \quad (5)$$

By solving (3) - (5) using the parameters from the reference device, we obtained the modified value of N_d as 7.37×10^{15} cm^{-3} and t_d as $11.97\mu m$. Isothermal room temperature TCAD simulations predicted that a device with these modified drift layer parameters has 12% higher V_{BR}, 20% higher t_{SC} and only 4% higher R_{onsp}.

III. STRUCTURE AND SIMULATION SETUP

Fig. 1(a) shows the cross section of the reference device [9] while Fig. 1(b) shows the cross section of the proposed device. The reference device has a drift length of 10 μm and a doping concentration of 8×10^{15} cm^{-3}. After implementing the proposed modifications to the reference device according to the equations (3) - (5), the drift length increases to 11.97 μm, while the doping concentration decreases to 7.37×10^{15} cm^{-3}.

Sentaurus TCAD software is used for simulations. Modified van Overstraeten-de Man avalanche model was employed for the off-state simulations, using values $a_{low} = a_{high} = 7.26 \times 10^6$ cm^{-1} and 6.86×10^6 cm^{-1}, and $b_{low} = b_{high} = 23.4$ MV cm^{-1} and 14.1 MV cm^{-1} [11]. The Enormal mobility model was calibrated for the on-state in accordance with [12]. Incomplete ionization and anisotropic effects of 4H-SiC are also considered. Thermodynamics model is used in simulations for considering the effect of self-heating.

IV. RESULTS AND DISCUSSIONS

Fig. 2 shows the I_D-V_{DS} plot at $V_{GS} = 0$ for both the reference and proposed device. It has been observed that the V_{BR} for reference device is 1.7 kV while the V_{BR} for proposed device is 1.9 kV. There is 13% improvement in the V_{BR}. This increase in V_{BR} is highly beneficial for the device's performance. A

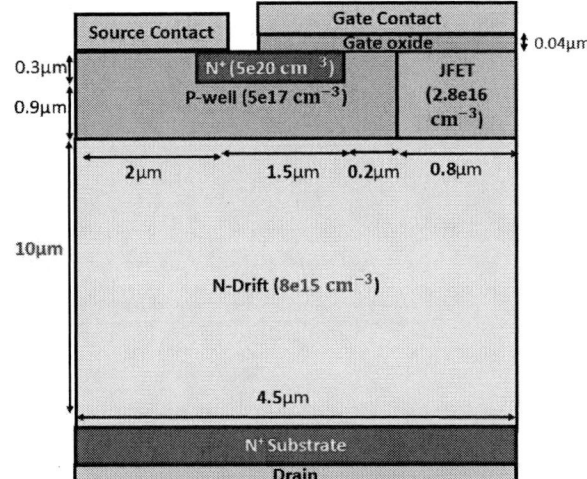

(a)

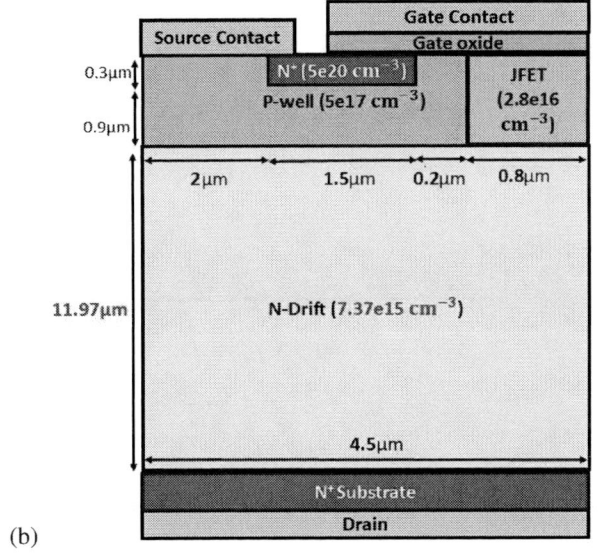

(b)

Fig. 1. Schematic of the cross-section of 1.2 kV 4H-SiC DMOSFETs with drift layer designed using (a) reference method [9] and (b) proposed method [8]

higher V_{BR} indicates that the device can sustain higher voltages before undergoing electrical failure, improving its robustness and reliability. The ability to withstand higher voltages is crucial for ensuring the device can operate efficiently and effectively in high power applications, thereby enhancing its overall performance and lifespan.

Fig. 3 shows the I_D-V_{DS} plot at $V_{GS} = 20$V. It is seen that the current decreases for proposed device as N_d is reduced. The R_{onsp} for the conventional device is 4.42 mΩ.cm^2 and for proposed device is 4.65 mΩ.cm^2. The modification of the device results in increase of R_{onsp} only by 5%, which is insignificant. Fig. 4 shows the I_D-V_{GS} plot at $V_{DS} = 20$ V. It is seen that I_D of proposed device decreases at higher V_{GS} due to increase in R_{onsp} of proposed device but the V_{th} for both

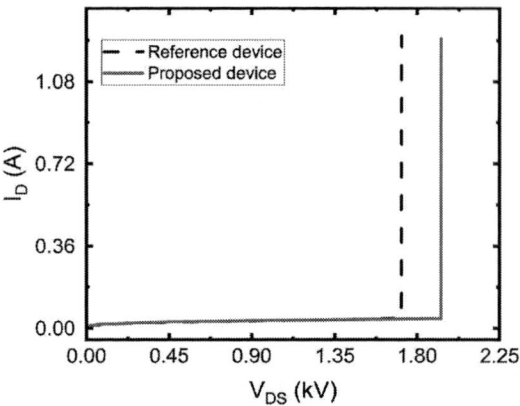

Fig. 2. I_D - V_{DS} at V_{GS}=0 V for the reference [9] and proposed device

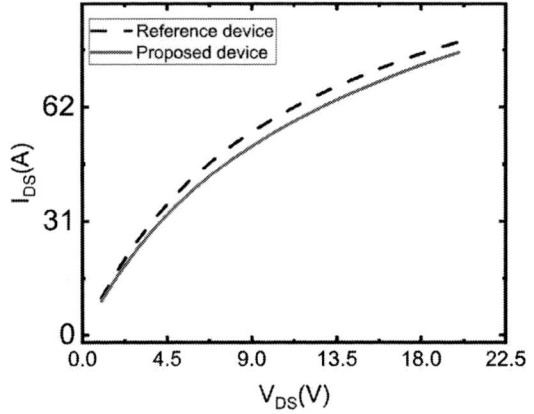

Fig. 3. I_D - V_{DS} at V_{GS}=20 V for the reference [9] and proposed device without including self heating

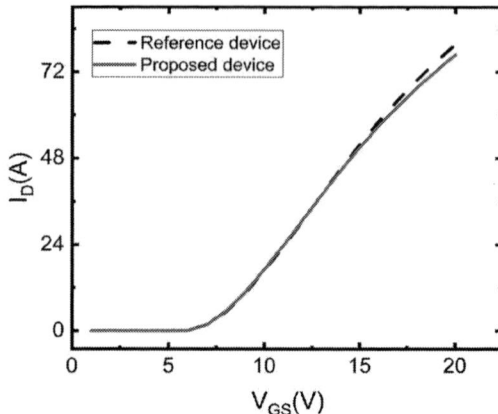

Fig. 4. I_D - V_{GS} at V_{DS} = 20 V for the reference [9] and proposed device without self heating

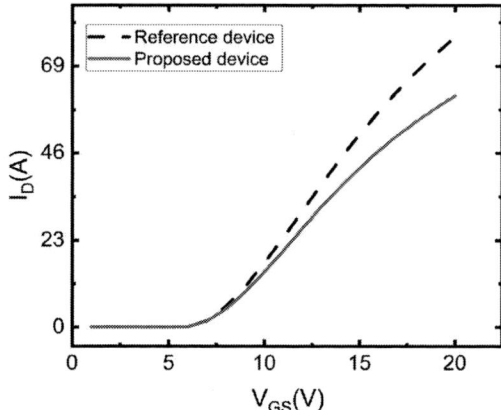

Fig. 5. I_D - V_{GS} at V_{DS} = 20 V for the reference [9] and proposed device with self heating

the reference and proposed device is same i.e., 6.125 V. So, the modifications done to the reference device does not affect the switching speed.

In Fig. 5, the I_D-V_{GS} plot compares the reference device and the proposed device. Due to self-heating effects, the on-current of the reference device is 79.4. However, when self-heating is applied to the proposed device, the on-current degrades to 61.19. This degradation occurs due to phonon scattering at high temperatures, which leads to mobility reduction in the channel region.

Unclamped Inductive Switching

Power MOSFETs are frequently used in high-speed switching applications. When a MOSFET is turned off, the circuit self-inductance and stray inductances cause a significant voltage spike between the drain and the source. If the induced voltage exceeds the V_{BR}, an avalanche current flows through the MOSFET. An avalanche current beyond the energy or current limit could permanently damage the MOSFET. UIS test is done to find a device's avalanche robustness. Figure 6 represents the UIS test circuit to study single pulse avalanche ruggedness, where V_{DD} is the supply voltage, V_{GS} is the input

pulse given at the gate and L is the inductance connected in series with the 4H-SiC DMOSFET. This test circuit operates in the following manner. When the MOSFET is turned on by the input gate signal, I_D starts to flow and increases linearly till it reaches the maximum value, I_{av}, then the gate input signal is switched off which turns off the MOSFET. As due to the inductor's Electro Motive Force (EMF), the inductive load current cannot vary instantly, the MOSFET enters into avalanche mode where the stored energy is dumped by the MOSFET and the effective drain-to-source breakdown voltage at peak discharge current, $V_{BR,eff}$, is induced that usually exceeds the V_{BR} of the device. t_{av} is the time taken for the avalanche process to occur and is given by [13]

$$t_{av} = \frac{I_{av}.L}{V_{BR,eff} - V_{DD}} \qquad (6)$$

E_{av} is the maximum energy sustained by a device due to high current or temperature before the device breaksdown and is given by

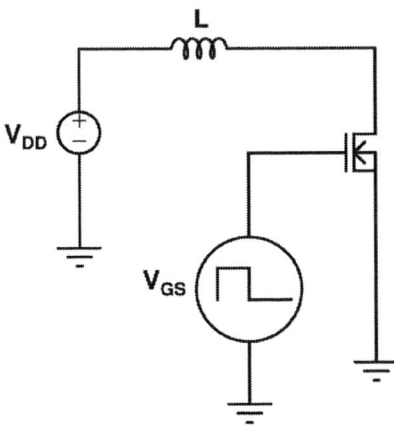

Fig. 6. UIS test circuit

TABLE I
AVALANCHE PARAMETERS AT T = 300 K, V_{DD} = 200 V, WITH VARYING L
FOR THE REFERENCE AND PROPOSED 4H-SiC DMOSFET

L (μH)	Device	I_{av} (A)	$V_{BR,eff}$ (V)	t_{av} (μsec)	E_{av} (J)
100	Reference [9]	276.66	3661	8	4
	Proposed	274	4121	7	3.94
200	Reference [9]	145	3595	8.5	2.2
	Proposed	127	4131	6.4	1.66
300	Reference [9]	98	3542	8.7	1.50
	Proposed	94	3551	7.2	1.39

$$E_{av} = \frac{V_{BR,eff}}{V_{BR,eff} - V_{DD}} \cdot \frac{1}{2} L I_{av}^2 \qquad (7)$$

Fig. 7 represents the voltage waveforms for different inductances for the reference and proposed device. The proposed device has a higher $V_{BR,eff}$ than the reference device, indicating that the proposed design is better at withstanding higher voltages before breakdown occurs. Fig. 8 represents the current waveforms for different inductances for the reference and proposed device. As the inductance increases from 100 μH to 300 μH, I_{av} decreases for both the reference and proposed devices as inductors with higher inductance will resist the rapid rise of current more strongly. It is also seen that for a given L, I_{av} of proposed device is lower than that of reference device.

In Table I, the calculated avalanche parameters for inductance varying from 100 μH to 300 μH is given. The avalanche parameters of the proposed device are compared with that of reference device. For L= 100 μH, the I_{av} decreased by 9.6%, $V_{BR,eff}$ increased by 12.5% , t_{av} decreased by 12.5% and E_{av} decreased by 1.52%.The increase in $V_{BR,eff}$ helps to withstand higher reverse voltages under inductive load switching conditions, decrease in current spike makes device more reliable, reduction in t_{av} helps in better switching performance, and less E_{av} indicates reduced power dissipation signifying overall better performance of proposed device as compared to the reference device. For the proposed device, for L= 100 μH with an average current of 274 A the stored energy in the inductor

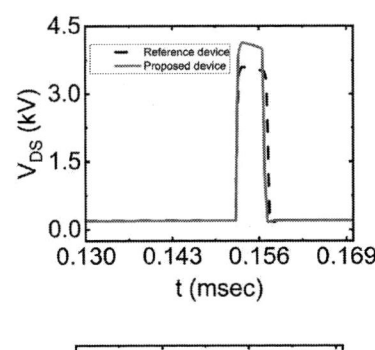

(a)

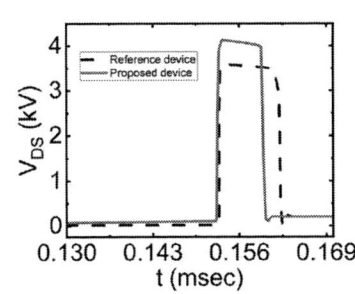

(b)

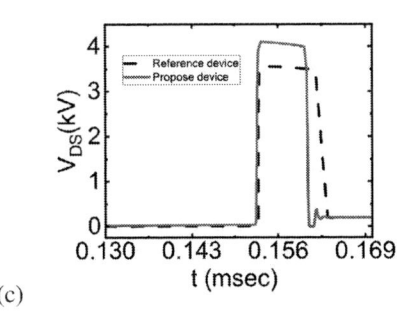

(c)

Fig. 7. Single pulse UIS voltage waveforms at V_{DD} = 200 V for the reference [9] and proposed device when L (a) 100 μH (b) 200 μH (c) 300 μH

is more which leads to a high voltage spike, while for L= 300 μH, the average current is 94 A, leading to less stored energy. This result in a lower voltage spike, so the $V_{BR,eff}$ decreases from 4121 V to 3551 V.

V. CONCLUSION

This work reports a comprehensive comparison of 4H-SiC devices having drift layers designed using a recently proposed method, aimed at improving short circuit withstand time, and the conventional method, aimed at minimizing specific on-resistance. The devices are compared in terms of their avalanche ruggedness through unclamped inductive switching (UIS) and in terms of their static performance after incorporating self heating effect. For L = 100 μH, the I_{av} decreased by 9.6%, $V_{BR,eff}$ increased by 12.5%, t_{av} decreased by 12.5% and E_{av} decreased by 1.52% for the proposed device. Thus, the proposed device has better avalanche ruggedness. However, this advantage comes at the expense of an increased R_{onsp} at elevated temperatures. The difference in R_{onsp} between the propsoed and conventional devices increase from 5% during isothermal room temperature simulation to 18 %

979-8-3315-3013-6/25 $31.00 © 2025 IEEE

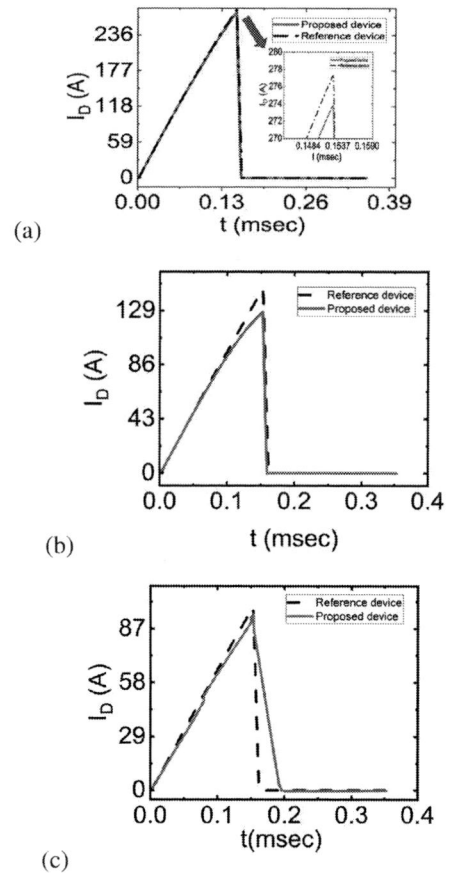

(a)

(b)

(c)

Fig. 8. Single pulse UIS current waveforms at V_{DD} = 200 V for the reference [9] and proposed device when L (a) 100 μH (b) 200 μH (c) 300 μH

when including the self heating effect. This is due to the increased contribution of the drift layer resistance to R_{onsp} in the proposed device.

REFERENCES

[1] S. Dimitrijev, J. Han, D. Haasmann, H. A. Moghadam, and A. Aminbei-dokhti, "Power-switching applications beyond silicon: Status and future prospects of SiC and GaN devices," 2014, pp. 43–46.

[2] K. Fischer and K. Shenai, "Dynamics of Power MOSFET Switching Under Unclamped Inductive Loading Conditions," IEEE Transactions on electron devices, vol. 43, no. 6, pp. 1007–1015, 1996.

[3] M. D. Kelley, B. N. Pushpakaran, and S. B. Bayne, "Single-Pulse Avalanche Mode Robustness of Commercial 1200 v/80 mΩ SiC MOS-FETs," IEEE Transactions on Power Electronics, vol. 32, no. 8, pp. 6405–6415, 2016.

[4] J. Hu, O. Alatise, J. A. O. Gonzalez, R. Bonyadi, L. Ran, and ´ P. A. Mawby, "The Effect of Electrothermal Nonuniformities on Parallel Connected SiC Power Devices Under Unclamped and Clamped Inductive Switching," IEEE Transactions on Power Electronics, vol. 31, no. 6, pp. 4526–4535, 2016.

[5] L. Yang, H. Qiu, L. Pan, Z. Guo, M. Xu, J. Yin, and X. Zhao, "Magnetic properties of BaTiO$_3$ and BaTi$_{1-x}$M$_x$O$_3$ (M=Co,Fe) nanocrystals by hydrothermal method," Journal of Magnetism and Magnetic Materials, vol. 350, pp. 1–5, 2014. [Online]. Available: https://www.sciencedirect.com/science/article/pii/S0304885313006951

[6] A. Fayyaz, L. Yang, M. Riccio, A. Castellazzi, and A. Irace, "Single pulse avalanche robustness and repetitive stress ageing of SiC

power MOSFETs," Microelectronics Reliability, vol. 54, no. 9-10, pp. 2185–2190,2014.

[7] N. Ren, H. Hu, X. Lyu, J. Wu, H. Xu, R. Li, Z. Zuo, K. Wang, and K. Sheng, "Investigation on single pulse avalanche failure of SiC MOSFET and Si IGBT," Solid-State Electronics, vol. 152, pp. 33–40, 2019.

[8] P. Singh, A. K, H. L. R. Maddi, A. Agarwal, and S. Karmalkar, "Design of the Drift Layer of 0.6 − 1.7 kV Power Silicon Carbide MOSFETs for Enhanced Short Circuit Withstand Time," in 2023 7th IEEE Electron Devices Technology Manufacturing Conference (EDTM), 2023, pp. 1–3.

[9] B. KAKARLA, "Short Circuit Behavior of SiC MOSFET," PhD thesis, vol. 43, no. 6, pp. 1007–1015, 1996.

[10] M. Roschke and F. Schwierz, "Electron Mobility Models for 4H, 6H, and 3C SiC," IEEE Transactions on Electron Devices, vol. 48, no. 7, pp. 1442–1447, 2001.

[11] A. Konstantinov, Q. Wahab, N. Nordell, and U. Lindefelt, "Ionization rates and critical fields in 4H silicon carbide," Applied Physics Letters, vol. 71, no. 1, pp. 90–92, 1997.

[12] C. Lombardi, S. Manzini, A. Saporito, and M. Vanzi, "A Physically Based Mobility Model for Numerical Simulation of Nonplanar Devices," IEEE Transactions on Computer-Aided Design of Integrated Circuits and Systems, vol. 7, no. 11, pp. 1164–1171, 1988.

[13] C. Ionita, M. Nawaz, and K. Ilves, "On the short-circuit and avalanche ruggedness reliability assessment of SiC MOSFET modules," Micro-electronics Reliability, vol. 71, pp. 6–16, 2017. [Online]. Available: https://www.sciencedirect.com/science/article/pii/S0026271417300215.

A Pilot Protection Scheme For HVDC Transmission Line Based on Weighted Sum of Correlations

Pradeep Singh
Electrical Engineering Department
National Institute of Technology Warangal
Warangal, India.
ps23eem2r18@student.nitw.ac.in

A.V. Giridhar
Electrical Engineering Department
National Institute of Technology Warangal
Warangal, India.
giridhar@nitw.ac.in

Abstract— **This paper introduces an innovative pilot protection scheme for high-voltage direct current (HVDC) transmission lines by leveraging a weighted combination of Pearson correlation and cosine correlation. In the proposed approach, the high-frequency components of line currents and voltages are analyzed at both terminals to ensure Higher Correlation values during faults, and their correlation indices are calculated in real time. A Weighted Sum of Correlations Is taken for ensuring robust performance under diverse fault scenarios and noise levels. Extensive electromagnetic transient simulations confirm that the scheme achieves high sensitivity and selectivity, even under high-resistance faults and significant measurement noise. Furthermore, the proposed method requires minimal communication overhead and offers low computational complexity, making it an attractive solution for multi-terminal and hybrid HVDC systems.**

Keywords— *Cosine correlation, DC line protection, fault detection, HVDC, Pearson correlation.*

I. INTRODUCTION

High Voltage Direct Current (HVDC) transmission lines have become increasingly popular in modern power systems. They enable efficient long-distance power transfer at high capacity while offering rapid and flexible control. Moreover, HVDC systems typically incur lower losses compared to their HVAC counterparts, making them ideal for connecting power grids that operate asynchronously [1], [2]. Although HVDC transmission technology has advanced considerably in recent decades, challenges in protecting these systems still persist. Consequently, a variety of primary protection schemes have been developed to safeguard HVDC networks [3]. Traditional HVDC transmission line protection methods are generally divided into two main categories: traveling wave protection, which serves as the primary approach, and current differential protection, which functions as a backup protection. However, Under severe weather conditions, HVDC lines become particularly vulnerable to faults. In DC systems, traveling wave protection is typically the primary method employed. It often exhibits reduced sensitivity to high impedance faults and can be vulnerable to interference from noise [4], [5].

The necessity of pilot protection for the HVDC line is more and more required as we go on larger scale transmission. Protecting DC transmission lines is crucial for the effective operation of HVDC control and protection systems. If a DC transmission line isn't properly protected, any fault may not be cleared efficiently, leading to pole blocking in the HVDC system.

This, in turn, can jeopardize the stability and reliability of the entire power grid. Effective protection schemes not only detect and isolate faults swiftly but also help maintain continuous power flow, minimizing disruptions. Therefore, robust fault detection methods are essential to enhance the resilience of HVDC networks.

Currently some novel HVDC line protection schemes have been proposed, N Liu et al. proposed a protection scheme that shows during internal faults both terminal measurements are nearly 0 Ω, whereas in external faults, the faulty terminal's measurement is around 200–300 Ω while the distant terminal remains near 0 Ω [6]. However, the method's accuracy can be compromised by lightning-induced high-frequency components, variations in DC line and filter parameters, and high-resistance faults. The authors of [9] introduce protection scheme relies on DC-filter currents and line currents, with a mathematical threshold rule to detect fault conditions. It analyses the flow characteristics of these currents during a constant fault period to serve as a backup measure for identifying pole-to-ground faults. However, the method cannot differentiate between pole-to-ground and pole-to-pole faults, Additionally, the reliance on peak current values may lead to issues under very high currents or complex system dynamics. Jian Liu et al. present a fault detection method for DC lines in VSC-HVDC systems that leverages current correlation, using the Pearson correlation coefficient to compare measurements at the DC-link capacitor branch and the DC line end. This approach emphasizes current directionality to differentiate between internal and external faults. However, it depends on remote communication links between converter stations, which can be a limitation. Additionally, while robust against noise, the complexity of large multi-terminal systems may necessitate further adjustments [10]. The authors of [11] introduces a quick-action protection for HVDC transmission lines that uses one-end current measurements. Its primary advantage is the ability to identify faults using data from just a single terminal. However, the method's reliance on fault location can be a drawback, as faults near the end of long lines may exhibit weaker characteristic frequencies.

In this study, a new pilot protection based on combined correlation of Pearson correlation and Cosine correlation is taken to distinguish between Internal and External Faults. The combined correlation method strengthens fault detection by simultaneously considering the magnitude and directional characteristics of fault currents. Different noise levels are introduced in the DC line to verify the reliability and effectiveness of Combine correlation on detecting HVDC Faults. Testing with varying noise levels in the DC line demonstrates the proposed scheme's ability to maintain accuracy and reliability in identifying HVDC faults.

979-8-3315-3013-6/25 $31.00 © 2025 IEEE

II. HVDC TRANSMISSION SYSTEM

A. System Configuration

HVDC system primarily comprises a rectifier station, a DC transmission line, and an inverter station. Fig. 1 shows a Typical Mono polar HVDC System. This paper considers the fault on HVDC line as Internal faults, and all other faults locations are taken as External faults. Point M, N, E, F are the fault points which represents External Faults. f_1 represents fault on the DC line anywhere on the line length. a and b represent the measurement points at the DC filters on the rectifier and inverter sides respectively, while m and n denote the measurement points at the two ends of the DC line.

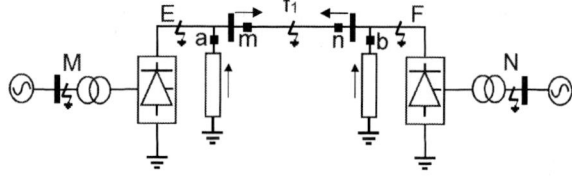

Fig. 1. Mono-Polar HVDC System

B. DC line filter

In HVDC transmission systems, DC filters are strategically placed at both ends of the DC line to mitigate harmonic interference generated by the converters [13]. Study in [8] shows that for 2/12/39 triple-tuned DC filter, under the range 2.5 kHz < f < 5 kHz DC filer is Equivalent to an inductor. For accurate simulation exact filter components values are taken L_1 = 10.869 mH, L_2 = 10.384 mH, L_3 = 2.06 mH, C_1 = 1.6 µF, C_2 = 4.48 µF, C_3 = 5.81 µF as mentioned.

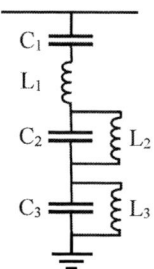

Fig. 2. Triple tuning DC filter

III. METHODOLOGY

A. High frequency Current Comparision

The High frequency currents during the HVDC faults are more closely matches the wave shape between dc line current and filter current compared to taking normal current values. A test simulation of HVDC system in PSCAD is done. Here, Ia and Im are the normal current and ia, im are high-frequency current components and Fig. 3(b) shows normal and high frequency current are does not matches most of the time intervals of the fault and Fig. 3(a) shows output of normal and high frequency current are closely matches relatively more during HVDC faults, which is clearly indicating that taking high frequency component will improve the reliability of the protection scheme.

To extract the high-frequency currents Chebyshev filter is utilized [14]. The high-frequency band adopted is in the range of 2.5~5 kHz. Considering that high frequency signal is easy to attenuate and also requires high sampling rate to

capture its fault characteristic, the sampling rate is set as 20 kHz, this is practical and feasible also [13].

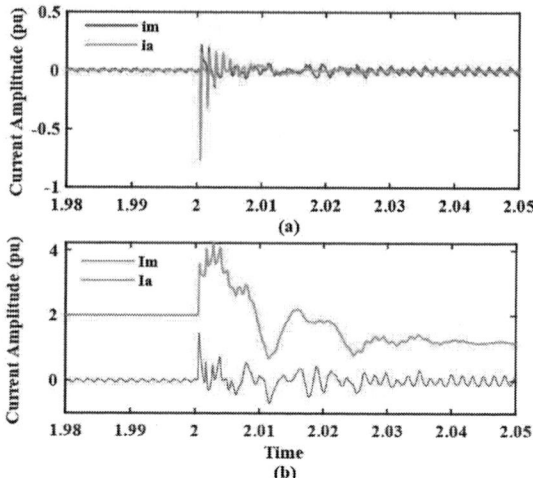

Fig. 3. Current comparison (a) Normal current values. (b) High Frequency current values.

B. Pearson Correlation Coeficient

HVDC fault detection project, Pearson correlation is used to measure the linear relationship between two sets of current signals—typically from different points in the system, like the DC-link capacitor branch and the ends of the DC line [15]. During internal faults, the currents at both ends tend to show a strong positive correlation, while external faults or normal conditions often result in weaker or negative correlations. This difference helps in distinguishing fault types and locations, enhancing the accuracy of your protection scheme. It can be defined as follows for two signal samples $i_{Cmk} = \{x_1, x_2, ..., x_n\}$ and $i_{Cablelmk} = \{y_1, y_2, ..., y_n\}$ [16].

$$R_k(i_{Ck}, i_{Cablek})$$
$$= \frac{\sum_{i=1}^{N}\left(x_i - \frac{1}{N}\sum_{j=1}^{N}x_j\right)\left(y_i - \frac{1}{N}\sum_{j=1}^{N}y_j\right)}{\sqrt{\sum_{i=1}^{N}\left(x_i - \frac{1}{N}\sum_{j=1}^{N}x_j\right)^2}\sqrt{\sum_{i=1}^{N}\left(y_i - \frac{1}{N}\sum_{j=1}^{N}y_j\right)^2}} \quad (1)$$

where R_k is the Pearson correlation coefficient of the two signals ranges from [−1,+1]. Here value of +1 indicates a perfect positive linear correlation between i_{Cmk} and $i_{Cablemk}$, while a value of 0 signifies no linear correlation between the fault currents. Here k = r, i denotes the ends near rectifier side and inverter side respectively. The number of signal samples N, within a time window is given by N=f_s*T.f_s, where f_s is the sampling frequency and T is the time window duration.

C. Cosine Correlation Coeficient

Cosine correlation is used to assess the similarity in the direction of current waveforms by measuring the cosine of the angle between two current vectors. For internal faults, the currents at both ends often point in similar directions, resulting in a high cosine similarity (close to 1). In contrast, external faults or noise-induced disturbances produce lower cosine values due to the misalignment of current vectors [17]. This method helps in identifying fault locations by comparing the directional consistency of the current signals.

$$R_{Cos}(i_{Cmk}, i_{Cablemk})$$
$$= \frac{\sum_{k=1}^{N} x(k) \cdot y(k)}{\sqrt{\sum_{k=1}^{N} x(k) \cdot x(k)} \sqrt{\sum_{k=1}^{N} y(k) \cdot y(k)}} \quad (2)$$

Where, R_{cos} is the cosine correlation coefficient, and it ranges from -1 to 1. The signals x and y can be expressed as x = $\{x_1, x_2, \ldots, x_n\}$, y = $\{y_1, y_2, \ldots, y_n\}$. N represents the number of sampling points. When two signals exhibit a perfect positive correlation, their correlation coefficient $R_{cos} = 1$. When they are negatively correlated, $R_{cos} = -1$.

D. Combine Correlation

Combining the Pearson and Cosine Correlation will give robust fault detection scheme. This approach captures both the strength and direction of current relationships, improving fault classification, especially for complex faults, like high-impedance or multi-terminal faults and Noise containing signals.

$$R_{Combine}(i_{Cmk}, i_{Cablemk})$$
$$= \alpha \cdot R_k(i_{Cmk}, i_{Cablemk}) + \beta \cdot R_{Cos}(i_{Cmk}, i_{Cablemk}) \quad (3)$$

Where, $R_{Combine}$ is the Combine correlation coefficient which will have the range from [-1, 1], i_{Cmk} and $i_{Cablemk}$ is current in DC line and DC filter on Rectifier side for k=r and inverter side for k=i. The value of α and β is taken as 0.65 and 0.35 respectively. The fault identification criterion is set as:

$$\left.\begin{array}{c} (R_{am} > |R_{set}|) \text{ Or } (R_{bn} > |R_{set}|) \\ + \\ (R_{am}, R_{bn} > 0) \text{ Or } (R_{am}, R_{bn} < 0) \end{array}\right\} \text{(Internal Fault)}$$

$$\left.\begin{array}{c} (R_{am} > 0 \text{ and } R_{bn} < 0) \\ \text{Or} \\ (R_{am} < 0 \text{ and } R_{bn} > 0) \end{array}\right\} \text{(External Fault)} \quad (4)$$

Where R_{set} denotes the threshold setting value. Which is set as 0.8 for getting high reliability with low error margins. R_{am}, R_{bn} are Combine correlation on rectifier and inverter side Respectively.

E. Protection Starting Criterion

The DC filter current remains low before the fault but increases once the fault occurs. As compared with DC line current, the filter current is very small under normal operation. Hence, it is adopted to use the starting criterion to have a higher sensitivity, as formulated below.

$$\left[\begin{array}{l} Rectifier\ side: \dfrac{1}{J}\sum_{j=1}^{J}|i_a(j)| > I_{set} \\ \\ Inverter\ side: \dfrac{1}{J}\sum_{j=1}^{J}|i_b(j)| > I_{set} \end{array}\right. \quad (5)$$

Here, i_a and i_b represent the DC filter currents, while I_{set} denotes the threshold value. The threshold is determined using the formula $I_{set}=k_{set}\times I_e$, where k_{set} is

the setting coefficient, set to 5. I_e refers to the maximum current of the DC filter during normal system operation, with a value of 20 A. As a result, the threshold is calculated as I_{set}=100 A.

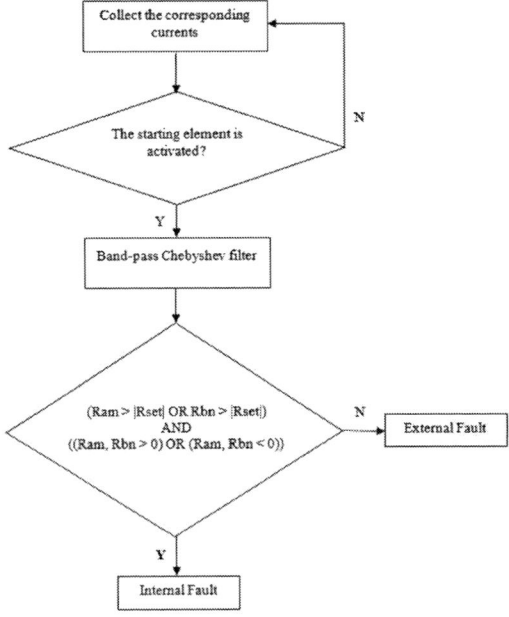

Fig. 4. Flowchart for protection algorithm

IV. SIMULATION RESULTS

A 500 kV, 1000 MW A Mono-polar HVDC system modeled in PSCAD/EMTDC is used to assess the performance of the proposed protection scheme under different fault conditions[12]. The Schematic diagram of the HVDC system is shown in Fig. 1. To confirm the accuracy and effectiveness of the proposed protection scheme, faults are applied at different location of the DC transmission line and the fault is applied at 2 second for 5ms. The DC transmission line total length is taken as 200 km.

A. Simulation of Internal Faults

The pole to ground faults is located at f_1 of the DC transmission line which is set as 50 km from the sending end. From the Fig. 3.(b) it is clear that the dc filter current is exceeding the I_{set} value thereby satisfying the protection starting criterion. After that checking the internal fault condition from the Fig.5 shows that correlation coefficients magnitude of R_{am} and R_{bn} are more than Rset= 0.8, and both have the same polarity therefore it signifies an internal fault and the internal fault is recognized at 205 ms.

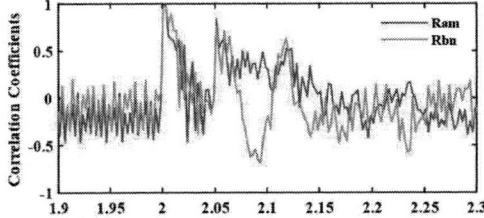

Fig.5. Combine Correlation Coefficients during internal fault

Multiple internal faults simulation test is done as shown in table 1. At the different location on the DC transmission line for which f_1 fault point mentioned in the Fig.1. the protection scheme verifies its reliability of detecting it as internal faults.

Table 1   Simulation Results For Internal Faults

Fault location	R_{am}	R_{bn}	Fault detected
0km	0.941	0.913	Cable-G
30km	0.919	0.981	Cable-G
50km	0.894	0.885	Cable-G
80km	0.891	0.884	Cable-G
100km	0.859	0.878	Cable-G
150km	0.873	0.892	Cable-G
200km	0.857	0.874	Cable-G

B. Simulation of External Faults

A fault located at Point E is taken. From the Fig. 3.(b) it is clear that the dc filter current is exceeding the I_{set} value thereby satisfying the protection starting criterion. After that checking the internal fault condition from the Fig.6 shows that correlation coefficients magnitude of R_{am} is more than $R_{set} = 0.8$, and has the opposite polarity between R_{am} and R_{bn}, therefore it signifies an external fault. Thus, the proposed protection can effectively differentiate between internal and external faults.

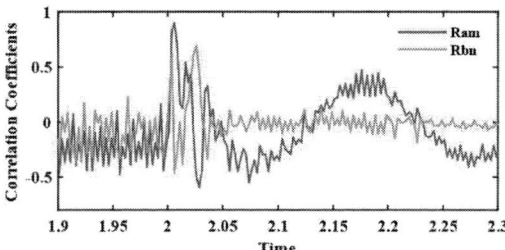

Fig.6. Combine Correlation Coefficients during internal fault

Further external faults simulation test results are done as shown in table 2. And for the fault points mentioned in the Fig.1, including for all type of external faults the protection scheme verifies its reliability of detecting as external faults.

Table 2   Simulation Results For External Faults

Fault location	Fault Type	R_{am}	R_{bn}	Fault detected
M	A-G	-0.963	0.948	External
	AB	-0.952	0.956	External
	AB-G	-0.968	0.961	External
	ABC-G	-0.942	0.977	External
N	A-G	0.976	-0.942	External
	AB	0.977	-0.968	External
	AB-G	0.968	-0.967	External
	ABC-G	0.973	-0.958	External
F		-0.918	0.915	External
E		0.926	-0.971	External

C. Faults at Different Fault Resistance

Faults in HVDC systems exhibit different characteristics depending on resistance of the fault, which influences the magnitude of fault currents and voltage drops. Low-resistance faults (such as direct metallic short circuits) result in high-magnitude fault currents with sharp transients, posing a significant risk to system components and requiring rapid fault detection and isolation.

On the other hand, high-resistance faults, which may occur due to partial conductor breakage or insulation degradation, generate lower fault currents that can be more challenging to detect. In such cases, the voltage deviation and harmonic content become critical indicators for identifying the fault.

The impact of fault resistance is particularly significant in distance-based protection schemes, where higher resistance faults may cause misclassification if the protection algorithm is primarily current-based.

Table 3 Simulation Results For Different Fault Resistance

Fault location	Fault resistance	R1	R2	Fault detected
0km	20	0.962	0.912	Cable-G
	100	0.897	0.831	Cable-G
	200	0.817	0.819	Cable-G
50km	20	0.880	0.892	Cable-G
	100	0.846	0.825	Cable-G
	200	0.731	0.722	Cable-G
100km	20	0.879	0.856	Cable-G
	100	0.820	0.805	Cable-G
	200	0.729	0.712	Cable-G

At different fault resistance, the protection scheme is still has the high correlation and not very much differ as the fault resistance changes along with different locations of the faults.

D. Comparision with pearson and cosine correlations

The performance of the protection may be affected when noise interferes with the current signals. The reference [18] gives Pearson correlation noise tolerance limits. To compare all the correlations against noise The test simulation is done for the same DC line length and same fault location on the DC line (50 km from the rectifier side), and by adding the different level of random Noise in the DC line, the comparison between Pearson correlation coefficient, Cosine correlation coefficient and Combine correlation coefficient for the same simulation environment is given in the Table 4. At Low or No noise level the Pearson correlation is giving high values where the cosine coefficient is giving relatively low correlation values. And for the High noise levels the Pearson correlation is giving relatively lower values. But it can be Observe that combine correlation is giving reliable and consistence correlation values.

Table 4 Comparison of Pearson, Cosine, and Combine correlation for different noise levels

Noise levels	Pearson Correlation		Cosine Correlation		Combined Correlation		Fault detected
	R_{am}	R_{bn}	R_{am}	R_{bn}	R_{am}	R_{bn}	
0%	0.998	0.993	0.870	0.857	0.894	0.885	Cable-G
10%	0.999	0.988	0.885	0.880	0.899	0.897	Cable-G
20%	0.991	0.994	0.871	0.874	0.879	0.892	Cable-G
30%	0.868	0.884	0.853	0.829	0.875	0.885	Cable-G
40%	0.859	0.843	0.813	0.802	0.878	0.871	Cable-G
50%	0.814	0.821	0.798	0.784	0.875	0.873	Cable-G
60%	0.728	0.784	0.751	0.764	0.876	0.872	Cable-G

E. Conclusion

This paper proposes a fault detection method for DC line in HVDC system using the combination of Pearson correlation coefficient and cosine correlation coefficient of fault currents. It is important that the correlation coefficients remain consistence for the noisy environment, so that the threshold limits can be set and the protection will still work on those set ranges in extreme weather conditions. By employing the current correlation instead of the magnitude of fault currents, the proposed method is insensitive to fault resistance. The simulation results highlight the proposed method's selectivity, resilience to fault resistance, and immunity to noise. Moreover, it is also suitable for multi-terminal systems With its reliable performance, this approach enhances fault detection and processing, making it a valuable solution for future HVDC grids.

REFERENCES

[1] L. Willis and N. Stig, "HVDC transmission: Yesterday and today," IEEE Power Energy Mag., vol. 5, no. 2, pp. 22–31, Mar. 2007.

[2] P. Bresesti, W. L. Kling, R. L. Hendriks, and R. Vailati, "HVDC connection of offshore wind farms to the transmission system," IEEE Trans. Energy Convers., vol. 22, no. 1, pp. 37–43, Mar. 2007.

[3] P. M. Anderson, et al., Power system protection. 2022: John Wiley & Sons.

[4] Y. Zhang, Y. Li, J. Song, B. Li, and X. Chen, "A new protection scheme for HVDC transmission lines based on the specific frequency current of DC filter," IEEE Trans. Power Del., vol. 34, no. 2, pp. 420–429, Apr. 2019.

[5] Y. Zhang, Y. Li, J. Song and X. Chen, "Pearson correlation coefficient ofcurrent derivatives based pilot protection scheme for long-distanceLCC-HVDC transmission lines," Int J Electr Power Energy Syst, vol. 116, pp. 1-8, Mar. 2020.

[6] N. Liu, Y. Li, S. Li, T. Li and L. Yu, "A Pilot Protection for LCC-HVDC Transmission Lines Based on Measured Surge Impedance at Tuning Frequency," in IEEE Transactions on Power Delivery, vol. 37, no. 3, pp. 2090-2103, June 2022, doi: 10.1109/TPWRD.2021.3103913.

[7] N. Liu, Y. Li, S. Li and T. Li, "A Pilot Protection for HVDC Transmission Lines Based on Current Correlation," 2021 IEEE 4th International Electrical and Energy Conference (CIEEC), Wuhan, China, 2021, pp. 1-6, doi: 10.1109/CIEEC50170.2021.9510742.

[8] N. Liu, Y. Li, T. Li, S. Li, J. He and X. Dong, "A novel protection for HVDC transmission lines based on current ratio of DC filter and

smoothing reactor," in CSEE Journal of Power and Energy Systems, doi: 10.17775/CSEEJPES.2021.02330.

[9] M. A. Rezaei Gazik and H. K. Karegar, "A Novel Protection Scheme for HVDC Transmission Lines Based on DC-Filter Current and DC Line Current," 2023 31st International Conference on Electrical Engineering (ICEE), Tehran, Iran, Islamic Republic of, 2023, pp. 36-40, doi: 10.1109/ICEE59167.2023.10334678.

[10] Jian Liu, Nengling Tai, Chunju Fan, Shi Chen and Pan Wu, "A fault detection method for DC lines in VSC-HVDC system based on current correlation," 2016 IEEE Power and Energy Society General Meeting (PESGM), Boston, MA, 2016, pp. 1-5, doi: 10.1109/PESGM.2016.7741687.

[11] G. Song, X. Chu, S. Gao, X. Kang and Z. Jiao, "A New Whole-Line Quick-Action Protection Principle for HVDC Transmission Lines Using One-End Current," 2018 IEEE Power & Energy Society General Meeting (PESGM), Portland, OR, USA, 2018, pp. 1-1, doi: 10.1109/PESGM.2018.8586307.

[12] M. Szechtman, T. Wess, C.V. Thio, "First Benchmark Model for HVDC Control Studies", Electra, No. 135, April 1991.

[13] G. Song, X. Chu, S. Gao, X. Kang, and Z. Jiao, "A new whole-line quickaction protection principle for HVDC transmission lines using one-end current," IEEE Trans. Power Del., vol. 30, no. 2, pp. 599–607, Apr. 2015.

[14] Z. Dai, N. Liu, and C. Zhang, "A pilot protection for HVDC transmission lines based on transient energy ratio of DC filter link," IEEE Trans. Power Del., vol. 35, no. 4, pp. 1695–1706, Aug. 2020.

[15] Jian Liu, Nengling Tai, Chunju Fan, Shi Chen and Pan Wu, "A fault detection method for DC lines in VSC-HVDC system based on current correlation," 2016 IEEE Power and Energy Society General Meeting (PESGM), Boston, MA, 2016, pp. 1-5, doi: 10.1109/PESGM.2016.7741687.

[16] Mohammad Farshad and Javad Sadeh, "A novel fault-location method for HVDC transmission lines based on similarity measure of voltage signals," IEEE Trans, Power Del., vol.28, no.4, pp. 2483-2490, Oct. 2013.

[17] L. Zheng, K. Jia, W. Wu, Q. Liu, T. Bi, and Q. Yang, "Cosine similarity based line protection for large scale wind farms part II-the industrial application," IEEE Trans. Ind. Electron., vol. 69, no. 3, pp. 2599–2609, Mar. 2022.

[18] N. Liu, Y. Li, X. Chen, S. Li and T. Li, "A Novel Protection Based on Current Correlation for DC Lines in Hybrid Cascaded Multi-Terminal HVDC Transmission System," in IEEE Transactions on Power Delivery, vol. 38, no. 4, pp. 2794-2809, Aug. 2023, doi: 10.1109/TPWRD.2023.3262570.

Comparative Study of Tracking Error for Booster Synchrotron Power Supply: Effects of Reference Profile Variations and Feedforward Compensation in Fast-Ramped Power Supply

Aradhana Kumari
Department of Atomic Energy
Raja Ramanna Centre for
Advanced Technology
Indore, India
aradhana@rrcat.gov.in

Manoj Leelachand Gandhi
Department of Atomic Energy
Raja Ramanna Centre for
Advanced Technology
Indore, India
mlg@rrcat.gov.in

Kandipalli Gopi
Department of Atomic Energy
Raja Ramanna Centre for
Advanced Technology
Indore, India
gopik@rrcat.gov.in

Ujjwal Yadav
Department of Atomic Energy
Raja Ramanna Centre for
Advanced Technology
Indore, India
ujjwal@rrcat.gov.in

Abstract— The Booster Synchrotron (BS) at the Indus Accelerator Complex (IAC), Raja Ramanna Centre for Advanced Technology (RRCAT), Indore, has been operating for three decades. In its present configuration, the Quadrupole (QP) and Dipole (DP) electromagnets are connected in series and powered by a main Power Supply (PS) following a ramped current reference. During beam injection at low current levels, limited beam rigidity increases sensitivity to tracking errors in the magnet currents, which can induce beta-beat oscillations and lead to beam loss. Ramping current profiles driving magnet loads inherently demand step changes in load voltage. While ideal tracking requires an instantaneous voltage response, practical limitations introduced by filtering in the power circuit, control loop delays, and switching constraints prevents such behavior. To investigate these limitations, a 15A/50V prototype Fast-Ramped Power Supply (FRPS) was developed, and a comparative study was conducted to assess tracking accuracy improvements through enhanced dynamic response and reference corner shaping. The study further identifies the key challenges and areas for advancement in FRPS performance. These findings provide valuable insights for the future development of high-power (5–10 MVA) BS magnet PS required for the proposed 6 GeV High-Brilliance Synchrotron Radiation Source (HBSRS).

Keywords— Fast-Ramped Power Supply, Tracking accuracy, Feedforward compensation, etc.

I. INTRODUCTION

The BS at RRCAT is essential for accelerating the electron beam to higher energies (450 MeV/550 MeV) before it is injected into Indus-1 (450 MeV electron accelerator) and Indus-2 (2.5 GeV electron accelerator) [1]. Its QP and DP electromagnets, connected in series, are powered by the main PS which follows a ramping current profile. Precise tracking accuracy is crucial for achieving the required beam current, especially at the injection corner where the beam is injected into the ring. However, during injection at low current, limited beam rigidity makes the system more sensitive to tracking errors, causing beam loss [2][3]. Since the magnet load requires the current to follow the desired ramp reference precisely, the call

for a step change in load voltage during ramp must be met. Therefore, achieving zero tracking error would require zero rise time, which implies infinite control bandwidth. However, real-world hardware and control algorithm limitations make such an instantaneous step unattainable and consequently, the load current gets distorted.

The magnet load introduces a pole that limits the current loop bandwidth. Increasing the gain to extend the bandwidth may lead to control loop saturation. To mitigate this, Feedforward Compensation (FFC) is implemented. By pre-emptively applying the required load voltage based on the current waveform and load characteristics, the high-bandwidth voltage loop rapidly responds to dynamic changes, minimizing tracking errors.

This study systematically analyzes tracking errors under various conditions, identifying the constraints of existing technology and their impact on current tracking accuracy. Addressing these challenges is essential for the proposed next generation 5-10 MVA FRPS to be designed for HBSRS, which must meet significantly more demanding specifications, particularly in terms of tracking accuracy. Overcoming these limitations requires adaptive digital control methodologies capable of real-time compensation. To facilitate this analysis, a prototype FRPS (15A/50V) was developed. The findings on the FRPS performance are discussed in sections VI and VII.

II. LITERATURE SURVEY

The evolution of ramping PS control in modern accelerator facilities has been driven by the need for higher tracking accuracy, better response time, and advanced control algorithms. Several facilities have successfully integrated these techniques to enhance performance.

The High Energy Photon Source (HEPS) adopted combination of digital control and analog control to achieve desired tracking accuracy [4]. A three-branch structure algorithm, as in [5], with FPGA-based digital control enhance precision tracking and fast adjustments in superconducting magnet power supplies of HEPS test facility.

979-8-3315-3013-6/25 $31.00 © 2025 IEEE

Taiwan Light Source (TLS) has implemented new modernized White Circuits as an upgrade to previous system, which adopts digital regulation with a goal of improved performance in tracking accuracy [6].

Advanced Photon Source (APS) has experimented with different technologies to achieve the desired tracking accuracy of 0.1 %. They started with traditional analog control, switched to ramp profile variation with software correction, but to attain the desired accuracy, they incorporated adaptive control [7].

The Indus BS FRPSs at RRCAT were developed in the early 1990s and continue to operate using analog proportional-integral-derivative controllers [8]. Although significant progress has been made over the years, the foundational architecture remains largely unchanged, and no systematic study has been conducted to evaluate the inherent limitations of the existing technology. Such a study is essential to identify areas of advancement, establish a clear roadmap for future upgrades and ensure readiness for the proposed 6 GeV BS. Insights from global accelerator systems suggest that integrating advanced digital control algorithms into FRPSs can reduce tracking errors and enhance overall system performance.

III. OBJECTIVES

This paper aims to investigate the tracking errors in a traditional controller for a low-power FRPS (15A/50V) under varying operational conditions. Specifications of the prototype FRPS are tabulated in Table 1.

TABLE I. SPECIFICATIONS OF PROTOTYPE FRPS

Sr. No.	Parameters	Value
1.	Inductance of load (L)	130 mH
2.	Resistance of load (R)	3.8 Ω
3.	Current rating (I_o)	15 A flat top
4.	Ramp repetition rate	1 Hz
5.	Ramp rate of load current	60 A/s to 120 A/s
6.	Ramp-up and ramp-down time of load current	100 ms

The study focuses on the effects of FFC and tailored corner shaping of reference voltage, as well as the impact of changing current levels and the non-linearity of magnet inductance on tracking performance. In an ideal system with constant inductance, FFC would achieve perfect tracking; however, because the magnet inductance varies with the load current, the

efficacy of FFC is inherently limited. To study this effect of varying inductance values at different current levels, a load inductance of 130 mH, rated for a lower current than the specified rating, was chosen.

This prototype has been further tested with three ramp reference profiles as shown in Fig.1. The Reference 1 profile features sharp corners, while Reference 2 incorporates a parabolic corners for 10 ms, and Reference 3 extends this smooth corner to 25 ms. In each profile, the flat top voltage was varied, influencing the ramp rate.

The purpose of this paper is to highlight the key areas requiring improvement and to outline a roadmap for future projects rather than focusing on the detailed design of the implementation.

IV. PROTOTYPE POWER CIRCUIT TOPOLOGY

The existing BS PSs for the secondary coils of the quadrupole magnets operate as two-quadrant converter, utilizing a six-pulse phase-controlled SCR pre-regulator in combination with a series pass transistor regulator [8]. This topology gives poor regulation at lower currents, as it forces the controller to operate at its extreme end. A potential alternative could be Four Quadrant Converter (FQC) topology, placing low currents in the mid-control range, enabling better regulation and wider range of voltage control.

Therefore, this prototype FRPS uses FQC as output stage which has two H-bridge modules, outputs of which are connected in parallel configuration and each module supplied by separate DC buses derived from Active Front End (AFE) converters [9]. The FQC employs Unipolar Pulse Width Modulation (UPWM) switching scheme which minimizes ripple at low currents [10] and uses MOSFETs as switching device.

Additionally, in the existing BS PSs, when the SCR-controlled input stage supplies current to a cyclic load, key issues such as flicker and harmonic distortion arise, leading to low power factor and interference with sensitive loads on the same supply [11]. This prototype FRPS addresses these challenges by incorporating an AFE converter, implemented here as Pulse Width Modulated Rectifier (PWMR) (Fig. 2) [9].

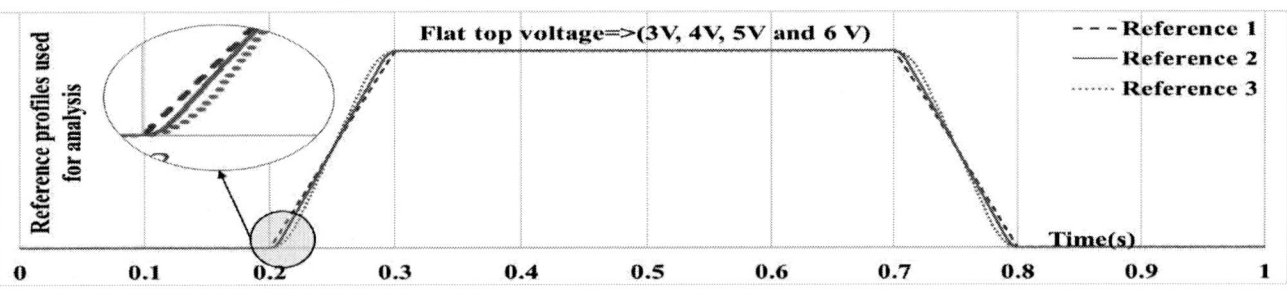

Fig. 1: Ramp reference profiles; a) Reference 1: all sharp corners; b) Reference 2: smooth parabolic corners for 10 ms; c) Reference 3: smooth parabolic corners for 25 ms; Setting: 1 V reference corresponds to 2 A load current

979-8-3315-3013-6/25 $31.00 © 2025 IEEE   177

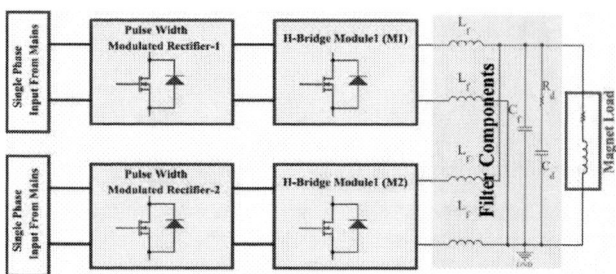

Fig. 2: Block diagram of 15A/50V prototype FRPS

This is a lower-power version of a probable scheme that could be implemented at higher power levels for BS PS. This design enhances overall performance across a broader range of operating conditions. MOSFET, a fast-switching device, is used to ensure fast dynamic response. However, as the power level is scaled up, the use of MOSFETs may become less efficient and the system may require the adoption of more robust devices, such as IGBTs, to handle higher power demands and maintain performance [10].

In the current topology, where the power level remains relatively low, MOSFETs are sufficient for optimal operation. Fig. 2 shows block diagram of the PS used in this project. The modular approach optimizes heat distribution, reduces component stress and simplifies maintenance [12]. The H-bridge modules, using MOSFETs operates with UPWM scheme at 10 kHz. Additional details of the converter output and PWMR stages are provided in [8].

V. CONTROL CIRCUIT TOPOLOGY

The governing equation for load voltage of a current controlled FRPS for booster magnets is given by, Eq. (1):

$$v_{load}(t) = L_{load}(i) \frac{di(t)}{dt} + i(t)R_{load} \qquad (1)$$

Where, $i(t)$ is the load current, $L_{load}(i)$ represents magnet inductance which is a function of $i(t)$, R_{load} is magnet resistance. With a ramp reference, each corner demands a step change in load voltage of magnitude $L_{load}(i)(di(t)/dt)$. However, achieving such a response would necessitate infinite system bandwidth, which is not feasible due to practical implementations. To mitigate this, FFC was implemented to pre-feed the expected voltage correction to the high-bandwidth voltage-loop, enabling prompt convergence to the new set-point. The reference voltage, $v_{FF}(t)$ injected by the FFC to the high-bandwidth voltage loop, given by Eq. (2):

$$v_{FF}(t) = \frac{H_v}{H_i}\left(L_{load}\frac{di_{ref}(t)}{dt} + i_{ref}(t)R_{load}\right) \qquad (2)$$

Where, H_v is the voltage loop feedback factor, H_i is the current loop feedback factor and i_{ref} is the reference current. The control block diagram of the prototype PS is shown in Fig. 3. It consists of multiple control loops, each serving a specific function to ensure accurate current regulation and dynamic response. The key control elements are:

1) CLC: It is used for precise current regulation, which is crucial for maintaining magnetic field stability in electromagnets as the magnetic field is directly proportional to current. The BW of CLC is ~ 75 Hz.

2) VL: Unlike the outer loop, the VL is capable of handling variations in line parameters. The magnet load introduces a pole that limits the current loop bandwidth, whereas the inner loop does not exhibit a dominant pole. Consequently, the VL can operate with a higher bandwidth, enabling a faster system response. The BW of VL is ~ 3.5 kHz

3) CBL: Ensures equal current sharing among the paralleled H-bridge modules, preventing uneven stress on components and improving reliability.

4) ADL: Attenuates oscillations in the output filter in real time, enhancing system stability and reducing unwanted voltage fluctuations.

5) FFC: Provides the voltage reference exactly according to the load voltage requirements, Eq. (2), for precise tracking of the reference current profile.

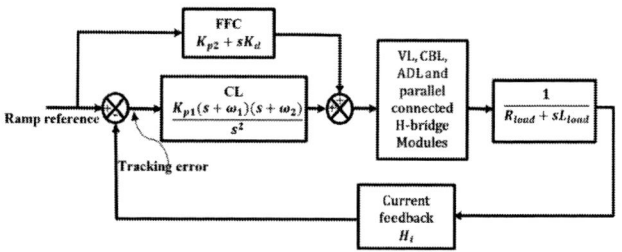

Fig. 3: Control loop block diagram of the prototype FRPS (15A/50V); CLC - Current Loop Controller, VL -Voltage Loop, CBL - Current Balancing Loop, ADL - Active Damping Loop

The calibration factor for current feedback is set to 2 A/V. As a result, the desired current profile follows the reference profile shown in Fig. 1, with the load current reaching twice the reference flat top voltage and a repetition rate of 1 Hz. In this study, the ramp rate is determined by the reference flat top voltage, ranging from a minimum of 60 A/s to a maximum of 120 A/s.

VI. SIMULATION RESULTS

To validate the controller design, simulations were carried out before experimental verification for reference profile which has all sharp corners. Key simulation results include:

A. Tracking Error without and with FFC:

Simulations were performed to evaluate the tracking performance of the prototype FRPS using a ramp reference signal characterized by sharp corners and a flat top voltage of 4 V (Reference 1 in Fig. 1). An appropriate CLC was implemented to eliminate the steady-state error for the ramp input. As shown in Fig.4, in the absence of FFC, where only CLC is responsible for tracking the load current, the tracking error (dashed red, magnified by a factor of 100) exhibits spikes of up to 40mV at the sharp corners, although they persist only

979-8-3315-3013-6/25 $31.00 © 2025 IEEE

for very short time. These spikes occur as the desired step change in the load voltage cannot be attained instantaneously. However, achieving such a response would necessitate infinite system bandwidth, which is not feasible in practical implementations. With FFC, the peak error (blue, magnified by a factor of 100) is reduced from 40mV to 4mV. These results highlight the importance of both CL controller selection and FFC implementation, in minimizing tracking errors. However, even with FFC, the spikes persist at the sharp corner due to VL bandwidth limitations. The smooth corners of ramp reference reduce the requirement of higher rate of change of the load voltage. This lowers the error at the corners as the VL bandwidth is good enough to meet the required rate of rise of voltage across the load.

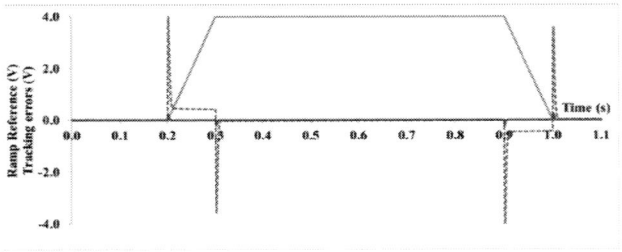

Fig. 4: Simulation results showing ramp reference voltage (green), tracking error with CLC only (dashed red), and tracking error with FFC implemented (blue). Error values are scaled by a factor of 100.

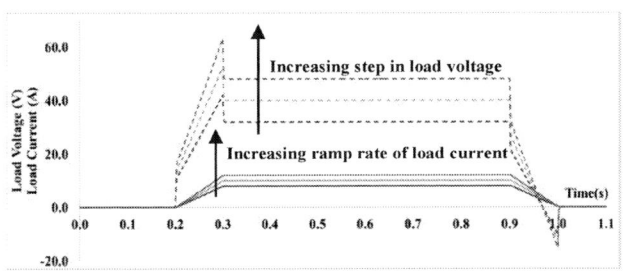

Fig. 5: Simulation results showing increased step magnitude in load voltage (dashed blue/green/red) with increasing ramp rates of load current: 8A/10A/12A (solid blue/green/red) flat top values.

A. Voltage and Current Waveforms:

Fig. 5 shows the simulation results of load current and voltage with different flat top voltages of sharp cornered reference.

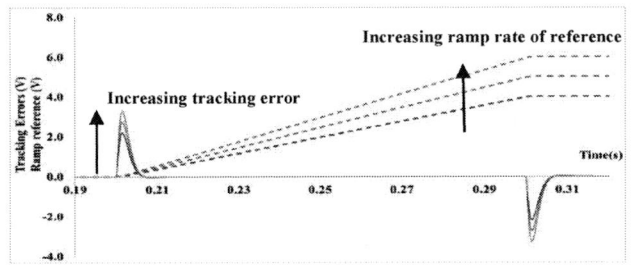

Fig. 6: : Simulation results for tracking errors (solid blue/red/green) at the corners of the ramp profile with increasing ramp; Current reference profiles (dashed blue/red/green). Error values scaled by a factor of 1000.

B. Tracking Error with FFC:

Simulated results of the system with FFC implemented into it and the trend of increasing tracking error with increasing ramp rate has been shown in Fig. 6.

VII. TEST RESULTS AND OBSERVATIONS

After modelling the power circuit with the controllers, simulations were conducted for two cases: with FFC and without FFC, using the Reference 1 profile (Fig. 1) to obtain the error voltage of the current controller, which closely aligned with the experimental results. The prototype was then experimentally tested using ramp reference profiles (Reference 1, Reference 2 and Reference 3), also shown in Fig. 1. Tracking error waveforms were recorded and their values were analyzed under different test conditions, including various flat top voltage levels, for both configurations (with and without FFC). The equation for calculating tracking error in parts per million (ppm) is given in Eq. (3), and the measurement method is illustrated in Fig. 7

$$E = \frac{\Delta V_{error}}{V_{reference, flat\ top\ value}} \times 10^6 \qquad (3)$$

Where E is the tracking error at corner, expressed in ppm; V_{error} is the tracking error at the corners, expressed in volts (V); and $V_{reference, flat\ top\ value}$ is the flat top voltage of reference, also in volts (V).

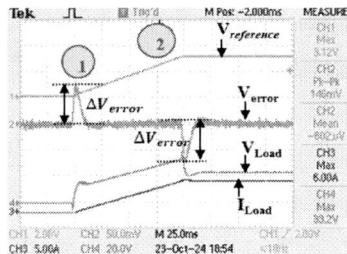

Fig. 7: Measurement of error voltage by calculating voltage per division of channel 2 (tracking error)

In Fig. 8, the error voltage, load voltage and load current waveforms with a 4 V flat top voltage and a sharp corner reference without FFC implementation are shown. The error value E1 (Error at rising edge, lower corner 1) and E2 (Error at rising edge, upper corner 2) are 20000 ppm relative to the flat top voltage 4 V. In Fig. 9, the same conditions are

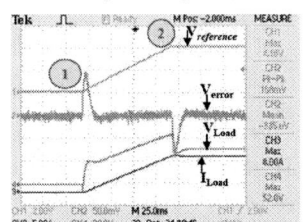

Fig. 8: E1 (Error at rising edge, Corner 1) and E2 (Error at rising edge, Corner 2) with sharp corners of reference profile & without FFC

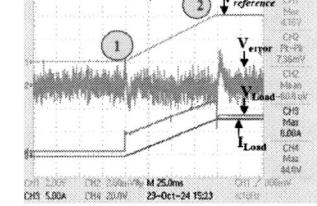

Fig. 9: E1 (Error at rising edge, Corner 1) and E2 (Error at rising edge, Corner 2) with sharp corners of reference profile & with FFC

are applied, but with FFC included, resulting in a noticeable reduction in the error value of E1 to 300 ppm and E2 to 500 ppm. To further improve this system, the reference profile was optimized and parabolic corners for 25 ms were introduced in the reference profile. As a result of this modification, the load voltage at the corners now undergoes a gradual ramp change rather than the step changes previously seen with sharp corners, significantly reducing the errors at the corners. This improvement can be seen in Fig.10 where the E1 and E2 further reduces to 100 ppm and 300 ppm respectively.

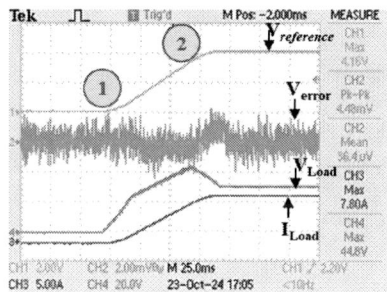

Fig. 10: E1 (Error at rising edge, Corner 1) and E2 (Error at rising edge, Corner 2) in case of tailored parabolic corners for 25ms and With FFC

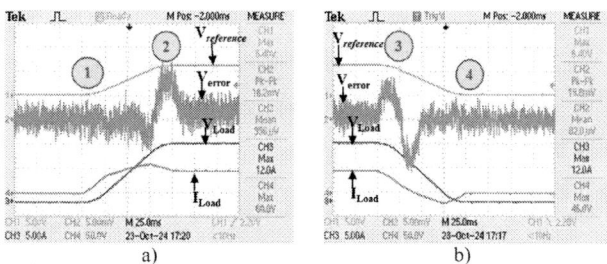

a)   b)

Fig. 11 a) and b): Impact of load inductance saturation on error voltage at 6 V flat top reference voltage even with tailored parabolic corners for 25 ms and with FFC

To study the effect of saturation of magnet load inductance at higher currents, the flat top voltage was increased to 6 V while maintaining smooth ramp reference corners for 25 ms and with FFC implemented into the controller. Under these conditions, error at higher current corners E2 and E3 (Error at falling edge, upper corner 3) are 1333 ppm and 1500 ppm respectively, while at lower current corners E1 and E4 (Error at falling edge, lower corner 4) are equal to a value less than 167 ppm as shown in Fig.11. The charts shown in Fig.12, Fig.13, Fig. 14 and Fig. 15 presents data in bar graph format for ready comparison of tracking errors in various scenario. The analysis shows that for system without FFC and without reference smoothening, tracking errors at all sharp corners exceed 20000 ppm, as seen in Fig. 12. Introducing reference smoothening significantly reduces errors, as illustrated in Fig. 13. As illustrated in Fig. 14 with FFC implementation and with reference smoothening, the reduction in error at Corner 1 is seen as corner smoothness increases. The case of FFC with a 25 ms smooth-corner reference optimally minimizes tracking errors at corner. While

smoothening helps mitigate errors, the obtained results clearly demonstrate that FFC remains crucial for achieving optimal tracking accuracy, making its implementation an utmost necessity. However, Fig. 15 shows that load inductance saturation increases errors E2 and E3 at higher currents, while E1 and E4 remain lower. This highlights the effectiveness of FFC, and at the same time showing the impact of inductance saturation on tracking accuracy at elevated currents. The test setup of the prototype FRPS is shown in Fig. 16.

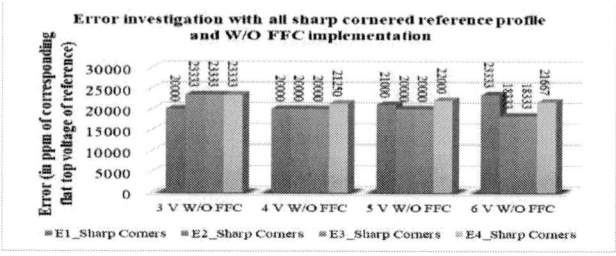

Fig. 12: Errors at all sharp coreners of the reference profile without (W/O) FFC

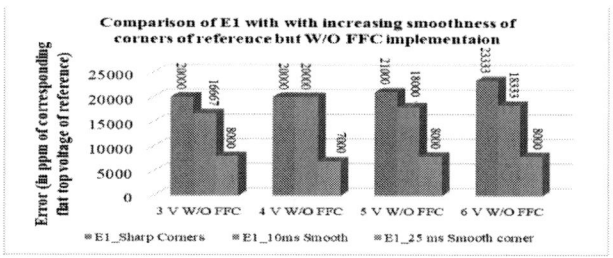

Fig. 13: Reduction in error E1 as reference profile changes from sharp corners to smooth corners and without (W/O) FCC

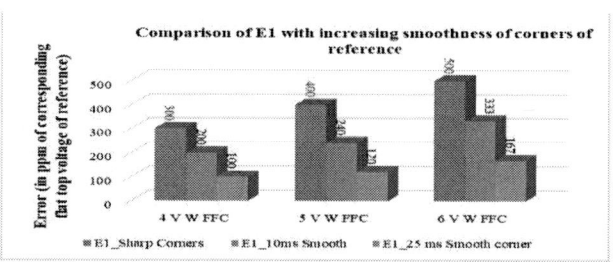

Fig. 14: Gradual reduction in error with corner shaping with (W) FFC

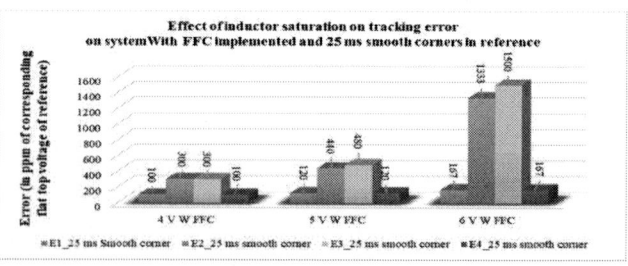

Fig. 15: Effect of higher flat top currents on tracking error at corners 2 and 3 with FFC and 25 ms parabolic corners

979-8-3315-3013-6/25 $31.00 © 2025 IEEE

Fig. 16: Photograph of the prototype FRPS (15A/50V)

VIII. CONCLUSIONS AND FUTURE WORK

The results indicate that increasing the duration of parabolic corners reduces tracking errors. The addition of FFC further enhances the dynamic response, minimizing errors. However, at higher ramp rates, tracking errors increase. In this test, the ramp duration is fixed at 100 ms, while the reference flat top voltage is incrementally raised from 3 V to 6 V in 1 V steps, thereby increasing the ramp rate. As the flat top voltage increases, the corresponding rise in flat top current leads to magnet core saturation, causing a reduction in the magnet inductance. The existing FFC, designed with a fixed inductance value fails to account for these variations, necessitating real-time adjustments. However, the analog controller lacks adaptive correction capabilities. A digital controller, in contrast, enables real-time compensation by adjusting FFC dynamically based on system responses. To achieve high-precision current tracking accuracy, advanced control strategies such as adaptive FFC, predictive algorithms, and observer-based estimations are essential. The FPGA and DSP based digital controllers, with their high-speed computation capabilities, can ensure rapid corrections, improving tracking accuracy under varying operating conditions.

In the proposed upcoming 6 GeV BS of HBSRS, its FRPS rated for few MVA would necessitate the application of these advanced digital control techniques to meet the stringent tracking accuracy requirements.

The future work will focus on implementing digital control for real-time FFC and reference profile modification through software, scaling up power levels, and evaluating the impact of mutual inductance on system performance.

ACKNOWLEDGMENT

The authors acknowledge the support received from P. Renukanath and his team for magnetics development, Deepchand and his team for mechanical assembly. The authors value the technical discussions with K. Saiffee, M. Janardhan, and A. A. Fakhri during the course of development of this work. The authors also appreciate the assistance received from Surya Prakash Sah during the assembly and testing of the prototype converter.

REFERENCES

[1] https://www.rrcat.gov.in/technology/accel/isrf_index.html

[2] P. Schreiber et al., "SOLEIL II Booster Robustness and Emittance Exchange," in Proc. IPAC'24, Nashville, TN, USA, pp. 3071-3074, May 2024.

[3] S. V. Milton and J. A. Carwardine, "Ramp Tuning of the APS Booster Synchrotron Magnet Power Supplies," Proc. 1995 Particle Accelerator Conference, pp. 2708, May 1995.

[4] Li, Y., Chen, S., Liu, Y. et al. "Development and testing of high precision and stability power supply for high energy photon source," Radiat Detect Technol Methods 8, 1280–1285 (2024). https://doi.org/10.1007/s41605-023-00446-5

[5] Wang, X., Liu, P., Long, F. et al. "Design of control algorithm for accelerator magnet power supply with large time constant load," Radiat Detect Technol Methods 4, 383–391 (2020). https://doi.org /10.1007 /s41605-020-00195-9

[6] C. Y. Wu et al "New event based timing system for the Taiwan Light Source," J. Phys., conference series 2687 072021, 2024.

[7] Guang Feng, B. Deriy and Ju Wang, "A linear mosfet regulator for improving performance of the booster ramping power supplies at the APS," Particle Accelerator Conference 2007. PAC. IEEE, pp. 434-436, 25-29 June 2007.

[8] S. R. Tiwari, A. C. Thakurta, A. P. Thipsay, A. Pagare, M. L. Gandhi, T. N. Singh, Shyam Singh and S. Kotaiah "Power Supplies for Indus-1," Asian Particle Accelerator Conference(APAC), Tsukuba, Japan, pp. 01-03, March 1998.

[9] Gandhi M. L., Srinivas L., A. C. Thakurta, "Bipolar Active Shunt with Bidirectional Utility Interface for The Quadrupole Magnets of Indus-2," IEEE International Conference on Industrial and Information Systems (ICIIS), Gwalior, pp. 1-6, Dec.2014.

[10] Ned Mohan, Tore M. Undeland, William P. Robbins, "Power Electronics: Converters, Applications, and Design," pp. 190-195, 3rd Edition (2003).

[11] M. Sri Balaji, S. P. Das, G. K. Dubey, Malabika Basu, "Regenerative Magnet Load Power Supply with Utility Friendly Operation," IEEE 28th Annual Conference of the Industrial Electronics Society, Seville, Spain, pp. 1392-1397, Nov 2002.

[12] Frede Blaabjerg "Control of Power Electronic Converters and Systems," pp. 125-147, Academic Press, 2021.

A High-Gain Bidirectional DC-DC Converter with Dual PWM Control Strategy

Rishabh Patel
Department of Electrical Engineering
National Institute of Technology
Warangal,India
rp23eem1r22@student.nitw.ac.in

S Porpandiselvi
Department of Electrical Engineering
National Institute of Technology
Warangal,India
selvi@nitw.ac.in

Abstract—**Modern power systems require bidirectional DC-DC converters because of their capacity to provide two-way energy transfer. This paper explores the design and performance of a high-gain bidirectional DC-DC converter that uses dual PWM control for efficient power management. Higher voltage gain, less ripple, and increased efficiency in both buck and boost modes are the results of the suggested control system, which mixes high and low-frequency PWM signals. One notable advantage of the control strategy is that when the duty cycles of both the high-frequency and low-frequency PWM signals are set to the same value, the voltage gain can be derived as a quadratic function of the duty cycle using the volt-second balance principle.The converter demonstrates stable operation with a minimal component count, highlighting its design efficiency. Significant performance gains, such as less ripple, improved transient response, and increased overall efficiency, are shown by simulation results and hardware validations, which qualify it for use in electric vehicles and renewable energy systems.**

Index Terms—**Bidirectional DC-DC converter, Dual PWM control, Buck mode, Renewable energy, Electric vehicles.**

I. INTRODUCTION

The increasing demand for energy-efficient solutions in areas such as energy storage, hybrid electric vehicles (HEVs), and renewable energy integration has led to the widespread adoption of bidirectional DC-DC converters. They are essential for contemporary energy solutions because, in contrast to traditional unidirectional converters, these systems permit power to flow in both directions. The difficulties of bidirectional converters, such as their high-duty cycles, restricted voltage gain, and increasing component counts, have been observed in earlier research. In [1], an extensible bidirectional DC-DC converter with lower switch counts and better voltage transfer ratios was introduced. To improve switching stress and efficiency for vehicle-to-grid (V2G) applications, a quadratic bidirectional converter was presented in [2]. In [3], it was shown how GaN-based bidirectional converters can achieve a significant voltage gain with lower losses. In [4], a high-gain, non-isolated bidirectional DC-DC converter for efficient power transfer was examined, demonstrating its usefulness in renewable energy systems. In [5], a quadratic bi-directional converter with an emphasis on increased efficiency was investigated. It was created especially for V2G and G2V applications.

In [6], an alternative bidirectional power converter with a high voltage gain, suitable for various applications including electric vehicle charging systems, was introduced. A bidirectional DC-DC converter designed for EV powertrains was introduced in [7], utilizing interleaved topologies and advanced inductors to emphasize the importance of magnetic components in high-power systems. In [8], a dual PWM control approach was introduced for high step-down DC-DC converters to enhance performance and efficiency. It uses low and high-frequency PWM waveforms to achieve a quadratic voltage gain and reduces inductor ripple without altering the buck converter topology. In [9], this method was adapted for high step-up converters using an OR-gate-based synthesis to enhance voltage gain and reduce switching losses. These studies demonstrate the effectiveness of this control scheme in power conversion. In [10], a triple PWM control approach was presented to improve the step-down conversion ratio in DC-DC converters by combining multiple frequency PWM signals, ensuring efficient voltage conversion while minimizing inductor ripple and component stress. However, these converters suffer from limitations such as high component count and low efficiency. Some of these converters can handle power flow in one direction only.

In this paper, a bidirectional power converter with dual PWM control is presented to address these limitations. By integrating high-frequency and low-frequency PWM signals, the control strategy enhances the converter's performance in both step-up and step-down modes. Two PWM generators with different frequencies are used in the dual PWM control system. To ensure precise control, the lower-frequency PWM signal is designed to align as an exact multiple of the higher-frequency PWM. The proposed dual PWM-controlled bidirectional power converter system has been simulated and implemented and the simulation and the experimental results are in good agreement. The proposed system offers high step-up and step-down gains for bidirectional operations.

The rest of the paper is structured as follows: Section II outlines the Dual PWM Control method, followed by an analysis of voltage gain. Section III presents a detailed evaluation of the approach, including simulation and hardware results. Additionally, a comparison between dual PWM and conventional PWM is provided. Finally, Section IV summa-

979-8-3315-3013-6/25 $31.00 © 2025 IEEE

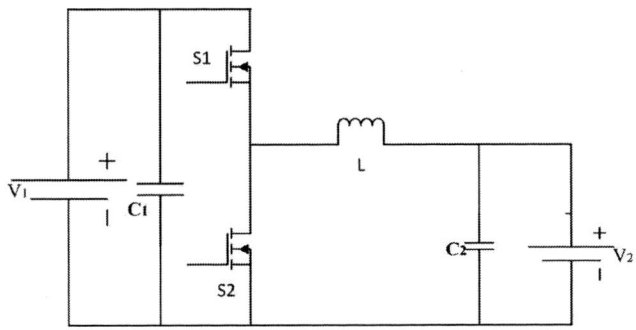

Fig. 1. Bi-Directional DC-DC Converter

rizes the key findings and conclusions.

II. OPERATIONAL PRINCIPLES

The bidirectional power converter uses two switches, an inductor, and two capacitors as depicted in Fig. 1. In the forward operation, it acts in buck mode, and in reverse operation, it acts in boost mode. The inductor (L) facilitates effective power transmission between the input and output during both buck and boost operations by acting as an energy storage element. In addition, it is essential to preserve a steady current flow and minimize current ripples. The input voltage V_1 is stabilized by the input capacitor (C_1), providing the converter with a steady voltage supply. The output capacitor ensures constant supply to the load (C_2), which filters the output voltage (V_2). By turning the switches on and off, the devices regulate the flow of energy. In step-down mode, as the voltage drops from V_1 to V_2, switch S_1 serves as the primary active switch. During step-up mode, activating switch S_2 raises the voltage from V_2 to V_1, effectively reversing the power flow direction. During switching transitions, the inductor current has a free-flow channel because of the body diodes of these switches. The high-frequency PWM was chosen as 80 kHz to reduce current ripple and switching noise, while the low-frequency PWM was set to 20 kHz to balance control response and switching losses. The 1:4 frequency ratio simplifies synchronization and pulse synthesis using logic gates.

A. Overview For Step-Down Mode:

The dual PWM control strategy is applied to the converter operating in step-down mode, as shown in Fig. 2. In closed-loop operation, the feedback voltage V_{fb}, derived from the output voltage V_o using a voltage divider, illustrates the functionality of the designed control strategy. The control signal V_c is generated by comparing the output feedback voltage V_{fb} with the reference voltage V_{ref} within the compensator. The control voltage V_c sets the comparator's threshold and is compared against a triangular waveform. This process generates two pulse width modulation (PWM) signals at different frequencies, one at higher frequency (PWM_{High}) and another one at lower frequency (PWM_{Low}), as part of dual PWM control technique, illustrated in Fig. 2. For step-down operation, the time duration of the low-frequency PWM signal

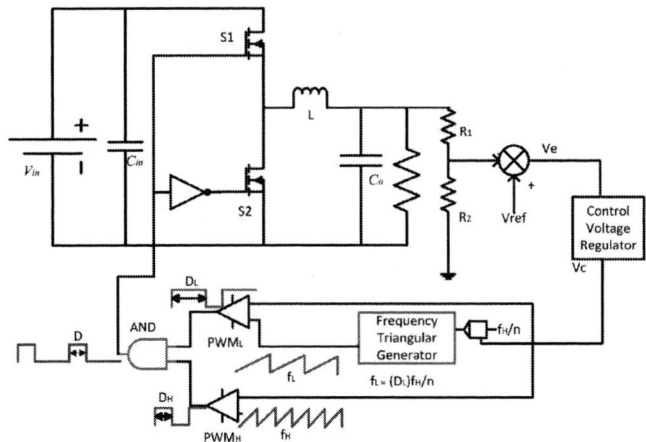

Fig. 2. Step-down operation of Bi-directional Converter

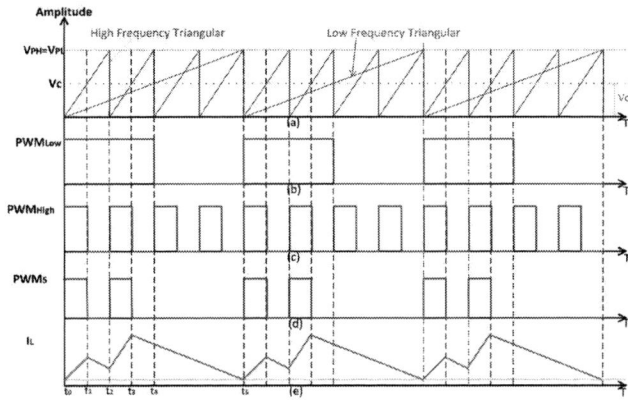

Fig. 3. The switching signals for step-down mode. (a) Carrier and reference signals (b) PWM_{Low}. (c) PWM_{High}. (d) PWM_S. (e) Inductor Current

is configured as an exact multiple of the period of the high-frequency PWM signal. This approach ensures synchronized operation between the two signals.

$$D_l T_l = n T_h \qquad (1)$$

where n represents the number of switching cycles of high-frequency PWM signal that take place within the active duration of low-frequency $T_{L_{ON}}$. The control waveforms are depicted in Fig. 3 for the step-down mode of operation, In Fig. 3(a) two triangular waveforms (low and high frequency) are compared with the control voltage V_c. In Figs. 3(b) and 3(c), resultant pulse-width modulation (PWM) waveforms are presented, which are characterized by different duty cycles, denoted as D_h and D_l respectively. The final PWM waveform, depicted in Fig.3(d), is produced by combining these two signals using an AND gate.

B. Operations Under Step-Down Mode

Mode $1(t_0 < t \leq t_1)$: In this interval, switch S_1 is on and S_2 is off. The converter works in step-down mode during this

979-8-3315-3013-6/25 $31.00 © 2025 IEEE   183

interval $T_{H_{ON}}$ of the synthesized PWM waveform. Kirchhoff's Voltage Law (KVL) for this mode is given by:

$$V_{in} - V_0 = V_L = L\frac{di_L}{dt} \quad (2)$$

The current flowing through inductor i_L increases and will be described as

$$\Delta I_{L1} = (D_h T_h)\frac{V_{in} - V_O}{L} \quad (3)$$

Mode 2 $(t_1 < t \leq t_2)$: In this interval, switch S_1 is off, while S_2 is on, causing the current through the inductor to decrease, indicating the current decay phase. KVL equation for this mode is:

$$0 - V_O = V_L = -V_O = L\frac{di_L}{dt} \quad (4)$$

the current through inductor is expressed as

$$\Delta I_{L2} = (1 - D_h)T_h\frac{-V_O}{L} \quad (5)$$

Mode 3 $(t_2 < t \leq t_3)$: In this interval, with S_1 on and S_2 off, mode 3 functions like mode 1 during the working period $D_h T_h$ resulting in a rise in the inductor current as described by the following equation:

$$\Delta I_{L3} = (D_h T_h)\frac{V_{in} - V_O}{L} \quad (6)$$

Mode 4 $(t_3 < t \leq t_4)$: In this interval, with S_2 on and S_1 off, the converter behaves as in mode 2, leading to a decrease in the inductor current. This reduction is given by:

$$\Delta I_{L4} = (1 - D_h)T_h\frac{-V_O}{L} \quad (7)$$

Mode 5 $(t_4 < t \leq t_5)$: In this interval, with S_2 on and S_1 off, the energy delivered by the inductor current declines following Mode 4. Similarly, the decrease in inductor current is expressed as:

$$\Delta I_{L5} = (1 - D_l)T_l\frac{-V_O}{L} \quad (8)$$

Using the volt-second balance concept, the increase and decrease in inductor current remain inherently equal throughout the operational cycle.

$$\Delta I_{L1} + \Delta I_{L2} + \Delta I_{L3} + \Delta I_{L4} + \Delta I_{L5} = 0 \quad (9)$$

On Solving (9) The voltage gain of dual PWM bi-directional converter in step-down mode is determined as :

$$M = \frac{V_O}{V_{in}} = D^2 \quad (10)$$

C. Effect of Parasitic Analysis on Output Voltage in Step-down Mode:

A detailed analysis has been conducted on how parasitic elements affect Output voltage of Bidirectional DC-DC converter in step-down mode. r_L is considered as winding resistance. R_L is considered as load resistance, r_{S_1} , r_{S_2} is considered as internal resistance of S_1 and S_2 respectively.

$$\frac{V_O}{V_{in}} = \frac{D^2}{1 + \frac{r_L}{R_L} + \frac{r_{S_1}D^4}{R_L} + \frac{r_{S_2}(1-D^2)^2}{R_L}} \quad (11)$$

D. Overview For Step-Up Mode:

The dual PWM control strategy is applied to the converter operating in step-up mode, as illustrated in Fig. 4. In closed loop operation, feedback voltage V_{fb} is obtained through voltage scaling of output voltage V_o. A control signal(V_c) is generated by feeding the difference between the voltage reference (V_{ref}) and the voltage output feedback (fb) into an error amplifier. This feedback signal (V_c) acts as the reference and is used to compare with the input triangular waveform. In the dual PWM control method, two sets of PWM signals, PWM_{Low} and PWM_{High}, are generated using the specified low and high-frequency triangular waveforms. The resulting PWM waveform is generated by applying a logical OR operation to these input signals. In step-up operation, the

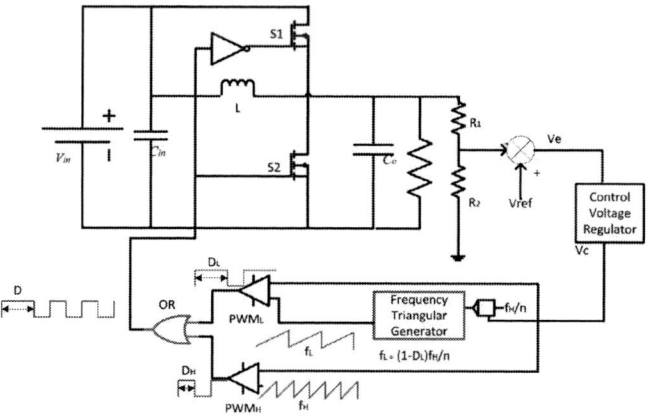

Fig. 4. Step-up operation of Bi-directional Converter

time period of the low-frequency signal is configured as an integer multiple of the high-frequency PWM signal's period, ensuring coordinated operation between two different PWM signals.

$$(1 - D_l)T_l/T_h = n$$

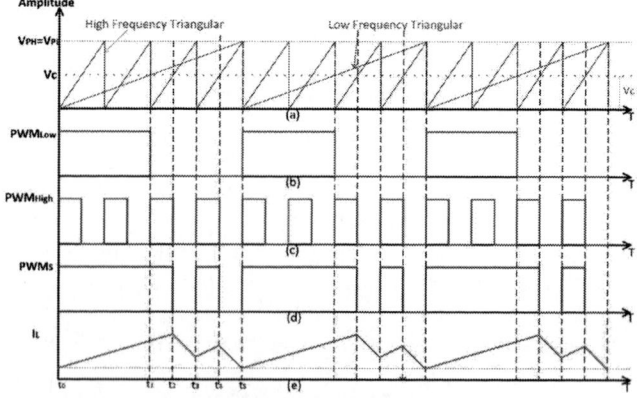

Fig. 5. The switching signals for step-up mode. (a) Carrier and reference signals (b) PWM_{Low}. (c) PWM_{High}. (d) PWM_S. (e) Inductor Current

E. Operations Under Step-Up Mode

Mode 1 ($t_0 < t \leq t_1$): In this interval, with S_2 on and S_1 off, the converter operates in step-up mode. During this phase, the inductor accumulates energy, causing its current to increase. The inductor voltage is given by:

$$V_{in} = V_L = L\frac{di_L}{dt} \tag{13}$$

An expression for the increase in inductor current is

$$\Delta I_{L1} = \frac{D_l T_l V_{in}}{L} \tag{14}$$

Mode 2 ($t_1 < t \leq t_2$): In this interval, with S_2 on and S_1 off, the inductor current steadily increases. The governing equation for the current through the inductor is expressed as:

$$\Delta I_{L2} = \frac{D_h T_h V_{in}}{L} \tag{15}$$

Mode 3 ($t_2 < t \leq t_3$): In this interval, with S_1 on and S_2 off, the inductor current gradually decreases during the PWM period. The governing equation for the current through the inductor is given by:

$$\Delta I_{L3} = \frac{(V_{in} - V_O)}{L}(1 - D_h)T_h \tag{16}$$

Mode 4 ($t_3 < t \leq t_4$): In this interval, with S_2 on and S_1 off, the converter behaves similarly to mode 2. The governing equation for the current through the inductor is expressed as:

$$\Delta I_{L4} = \frac{D_h T_h V_{in}}{L} \tag{17}$$

Mode 5 ($t_4 < t \leq t_5$): In this interval, with S_1 on and S_2 off, the converter behaves similarly to mode 3, and the reduction in the current through the inductor is expressed as:

$$\Delta I_{L5} = \frac{(V_{in} - V_O)}{L}(1 - D_h)T_h \tag{18}$$

Based on the volt-second balance concept, the increase and decrease in inductor current remain inherently equal throughout the operation cycle.

$$\Delta I_{L1} + \Delta I_{L2} + \Delta I_{L3} + \Delta I_{L4} + \Delta I_{L5} = \Delta I_L = 0 \tag{19}$$

On Solving (19) Voltage Gain is obtained as:

$$V_O = V_{in}\frac{1}{(1 - D)^2} \tag{20}$$

F. Effect of Parasitic Analysis on Output Voltage in Step-Up Mode:

A detailed analysis has been conducted on how parasitic elements affect output voltage of Bidirectional DC-DC converter in step-up mode. r_L is considered as winding resistance. R_L is considered as load resistance, r_{S_1}, r_{S_2} is considered as internal resistance of S_1 and S_2 respectively.

$$\frac{V_O}{V_{in}} = \frac{1}{(1-D)^2 + \frac{r_L}{R_L}(1-D)^2 + \frac{r_{S_1}D^2(1-D)^2}{R_L(1-D)^2} + \frac{r_{S_2}(1-D)^2}{R_L}} \tag{21}$$

G. Efficiency Analysis:

The efficiency of the converter is calculated for both modes for rating $V_H = 37.5\text{V}$ and $V_L = 5\text{V}$. The parameters for analysis are $r_L = 40\text{m}\Omega$, and the switch selected is IRF250IN, with $r_S = 85\text{m}\Omega$. The losses of the inductor(P_L) and switches are calculated using the formula: $P_L = \sum_{i=1}^{n}(i_{L_i})^2 r_L$, $P_{S,C} = \sum_{i=1}^{n}(i_{S_i,\text{rms}})^2 r_S$, $P_{S,\text{sw}} = 0.5\sum_{i=1}^{n}(V_{S_i}i_{S_i,\text{avg}})f_s(t_{\text{on}} + t_{\text{off}})$ The efficiency of Dual PWM Controlled Bidirectional Converter is shown in Fig. 6. The analytical efficiency of the converter in step-up mode is 98.4% and it is 98% for step-down mode. Fig. 7. shows the graph of Power loss of bidirectional dc-dc converter at various duty cycle.

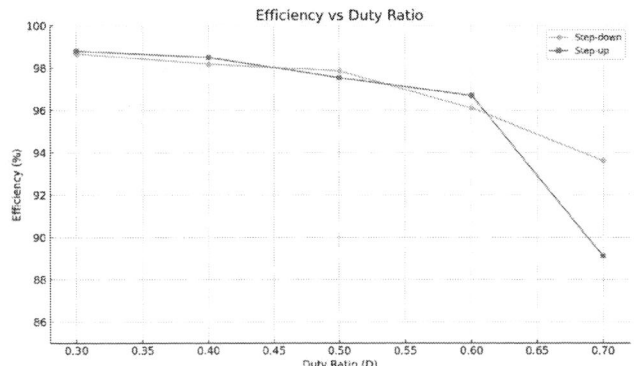

Fig. 6. Efficiency V/S Duty Ratio

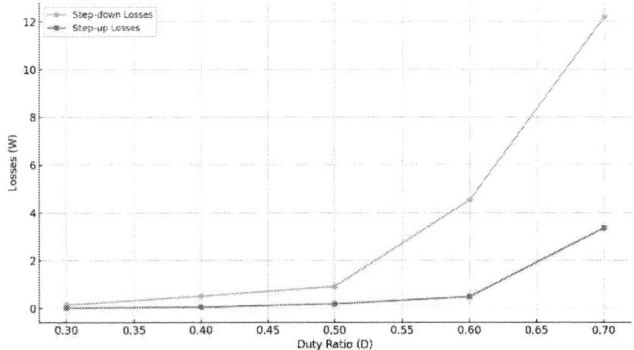

Fig. 7. Power Loss V/S Duty Ratio

III. SIMULATION AND EXPERIMENTAL VALIDATION

This section presents both simulation and experimental evaluations of dual PWM control-based bidirectional converter. Simulation studies have been conducted using MATLAB Simulink to gain insights into the converter's dynamic performance. An experimental prototype was constructed to validate the design. In step-down mode, it operates with a 35 V input to provide a 5 V output at 14 W, while in step-up mode, it accepts a 5 V input to deliver a 35 V output at 22 W.

979-8-3315-3013-6/25 $31.00 © 2025 IEEE

TABLE I
DESIGN SPECIFICATIONS

	Buck Mode	Boost Mode
V_{in}	37.5 V	5 V
V_O	5 V	37.5 V
R_L	2Ω	50Ω
L	250μH	250μH
C	100 μF	100 μF
f_H	80 KHz	80 KHz
f_L	20 kHz	20 kHz
$D_l = D_h = D$	0.3	0.7

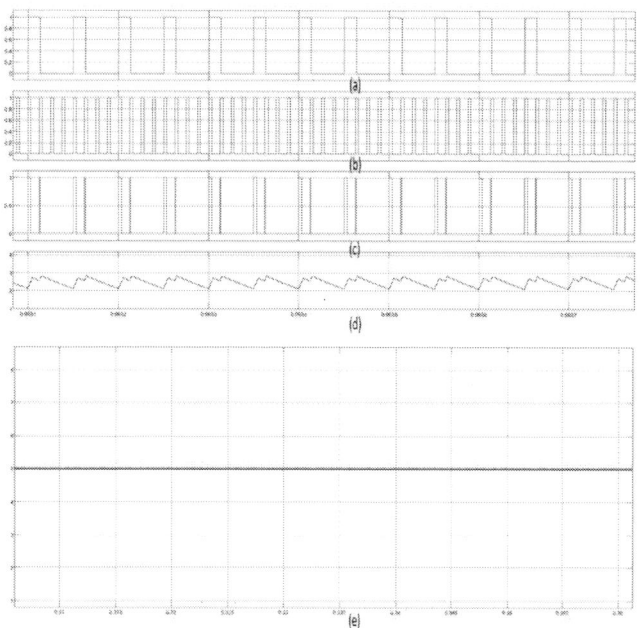

Fig. 8. Simulation results for step-down mode:(a) PWM_{Low}. (b) PWM_{High}. (c) Resultant PWM_S. (d) Inductor Current. (e) Output Voltage

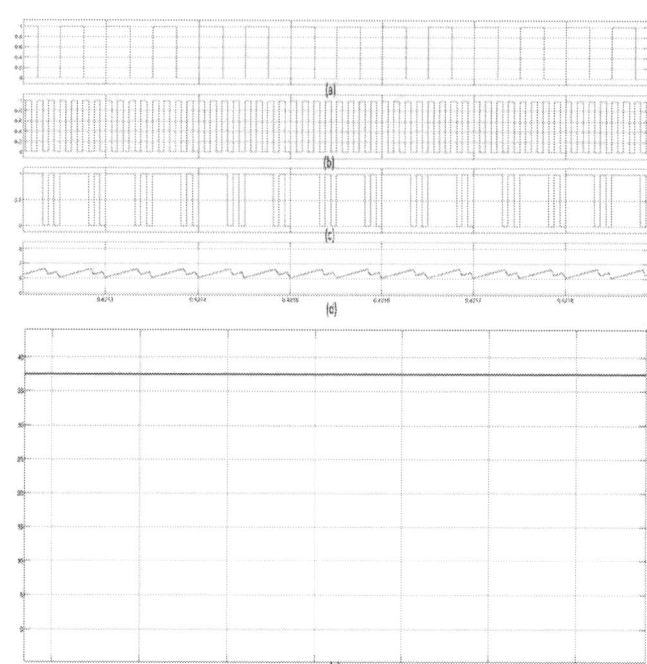

Fig. 9. Simulation results for step-up mode:(a) PWM_{Low}. (b) PWM_{High}. (c) Resultant PWM_S. (d) Inductor Current. (e) Output Voltage

Fig. 10. Bi-Directional DC-DC Converter Experimental Setup

The prototype's detailed specifications are provided in Table 1.

The converter is theoretically capable of achieving high voltage gain, as demonstrated in the analytical comparison. However, the simulation and hardware were performed for a 37.5V to 5V range due to practical hardware and safety constraints.

The simulation results for step-down mode are shown in Fig. 8, where Fig. 8(a) and Fig. 8(b) display low and high-frequency PWM waveforms respectively. The resultant PWM waveform is depicted in Fig. 8(c), while Fig. 8(d) displays the waveform of the current through the inductor. As illustrated in Fig. 8(e), the average output voltage remains steady at 5 V. Fig. 9 illustrates the simulation results for step-up mode. Figs. 9(a) and 9(b) present the low and high-frequency PWM waveforms, respectively. The inductor current waveform is shown in Fig. 9(d). The results in Fig. 9(e) confirm that the mean output voltage is consistently maintained at 37.5 V. The hardware

setup and its corresponding waveforms are shown in Figs. 10 - 12. The FPGA-based control system utilizes logic gates to generate PWM signals for MOSFET control, employing an AND gate for step-down mode and an OR gate for step-up mode, with the resultant signal is formed by an AND/OR combination of PWM outputs of low frequency and high frequency. The experimental prototype is configured to operate with a 35 V input and a 5 V output, delivering 14 W in step-down mode. In step-up mode, it operates with a 5 V input and provides a 35 V output at 22 W. The hardware implementation

979-8-3315-3013-6/25 $31.00 © 2025 IEEE

TABLE II
COMPARISON WITH EXISTING CONVERTERS

Topology	[1]	[2]	[3]	[4]	[5]	Dual PWM
Gain (Step-down)	$\frac{1+D}{(1-D)^2}$	D^2	D^2	D^2	$\frac{D^2}{2-D}$	D^2
Gain (Step-up)	$\frac{D^2}{2-D}$	$\frac{1}{(1-D)^2}$	$\frac{1}{(1-D)^2}$	$\frac{1}{(1-D)^2}$	$\frac{1+D}{(1-D)^2}$	$\frac{1}{(1-D)^2}$
S/L/C/D	4/3/4/2	4/2/3	4/2/2	4/2/2	5/2/4	2/1/2
Power Rating	300 W	160 W	200 W	500 W	50 W	22 W
η (Efficiency)	95%	94%	98%	97%	97%	98%

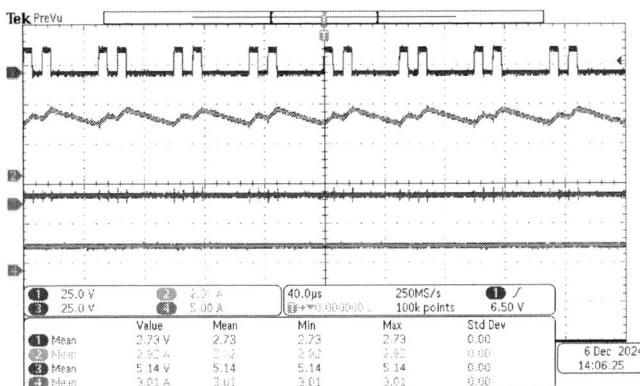

Fig. 11. Step-Down Mode Experimental Results

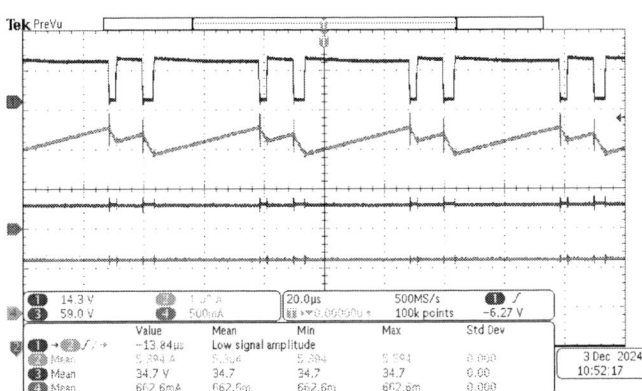

Fig. 12. Step-Up Mode Experimental Results

gain, reduces ripple, and improves efficiency in both step-down as well as step-up mode. It provides effective voltage conversion from 5V to 37.5V in step-up mode and 37.5V to 5V in step-down mode. The hardware results demonstrates good alignment with the simulations. The improved transient response and voltage regulation compared to traditional PWM methods highlight the converter's potential for applications such as renewable energy storage and electric bikes. Future work will focus on further efficiency optimization with advanced semiconductor technologies and refining the control strategy for higher-power applications.

REFERENCES

[1] V. S. Rao, S. Tapaswi and S. Kumaravel, "Extendable Bidirectional DC-DC Converter With Improved Voltage Transfer Ratio and Reduced Switch Count," in IEEE Journal of Emerging and Selected Topics in Industrial Electronics, vol. 4, no. 2, pp. 460-470, April 2023, doi: 10.1109/JESTIE.2022.3233295.

[2] V. S. Rao, S. P. Kumar, K. S. and A. M.P, "n-Stage Bidirectional Quadratic Boost Converter with Reduced Switch Voltage and Current Stress," 2022 IEEE International Conference on Power Electronics, Drives and Energy Systems (PEDES), Jaipur, India, 2022, pp. 1-6, doi: 10.1109/PEDES56012.2022.10080579.

[3] H. Ardi, A. Ajami, F. Kardan, and S. N. Avilagh, "Analysis and implementation of a nonisolated bidirectional DC–DC converter with high voltage gain," IEEE Trans. Ind. Electron., vol. 63, no. 8, pp. 4878–4888, Aug. 2016.

[4] S. H. Hosseini, R. Ghazi, and H. Heydari-Doostabad, "An Extendable quadratic bidirectional DC–DC converter for V2G and G2V applications," IEEE Trans. Ind. Electron., vol. 68, no. 6, pp. 4859–4869, Jun. 2021.

[5] Heydari-doostabad and T. O'Donnell, "A wide range high voltage gain bidirectional DC–DC converter for V2G and G2V hybrid EV charger," IEEE Trans. Ind. Electron., vol. 69, no. 5, pp. 4718–4729, May 2022.

[6] L. Yu, L. Wang, C. Yang, L. Zhu, Y. Gan, and H. Zhang, "A novel non isolated GaN based bidirectional DC–DC converter with high voltage gain," IEEE Trans. Ind. Electron., vol. 69, no. 9, pp. 9052–9063, Sep. 2022.

[7] G. Calderon-Lopez, J. Scoltock, Y. Wang, I. Laird, X. Yuan and A. J. Forsyth, "Power-Dense Bi-Directional DC–DC Converters With High-Performance Inductors," in IEEE Transactions on Vehicular Technology, vol. 68, no. 12, pp. 11439-11448, Dec. 2019, doi: 10.1109/TVT.2019.2943124.

[8] C. -W. Tseng, J. -H. Liu, C. -T. Pan and C. -C. Chu, "Dual PWM Control for High Step-Down DC–DC Converters," in IEEE Transactions on Industry Applications, vol. 56, no. 4, pp. 4272-4287, July-Aug. 2020, doi: 10.1109/TIA.2020.2993220.

[9] C. -W. Tseng, J. -H. Liu, C. -T. Pan and C. -C. Chu, "Dual PWM Control for High Step-Up DC-DC Converters," 2020 IEEE Industry Applications Society Annual Meeting, Detroit, MI, USA, 2020, pp. 1-7, doi: 10.1109/IAS44978.2020.9334730.

[10] J. -H. Liu, C. -W. Tseng and C. -C. Chu, "High Step-Down Conversion Ratios of DC-DC Converters Under Triple PWM Control Schemes," 2021 IEEE Industry Applications Society Annual Meeting (IAS), Vancouver, BC, Canada, 2021, pp. 1-8, doi: 10.1109/IAS48185.2021.9677448

closely matched the simulation results in both step-down and step-up modes, effectively converting 35 V to 5 V and 5 V to 35 V while maintaining a stable and regulated output. The measured waveforms of the hardware prototype demonstrated that the converter had a stable transient response without overshoot and less inductor current ripple, which reduces the stress on the switches. Table II shows comparison between Dual PWM Controlled Bidirectional Converter and other existing converters. The major advantage of this technique is that with lesser number of components high efficiency is obtained.

IV. CONCLUSION

This paper presents a high-efficiency bidirectional DC-DC converter using dual PWM control, which enhances voltage

979-8-3315-3013-6/25 $31.00 © 2025 IEEE

Switching Characteristics of Domestic Loads for Reliable Arc Fault Detection

1st Ratnakar Nutenki
Department of Electrical Engineering
IIT Kharagpur & RGUKT-Nuzvid
Kharagpur, India
ratnakar@iitkgp.ac.in

2nd Aurobinda Routray
Department of Electrical Engineering
Indian Institute of Technology
Kharagpur, India
aurobinda.routray@gmail.com

3rd Ashok Kumar Pradhan
Department of Electrical Engineering
Indian Institute of Technology
Kharagpur, India
akpradhan@ee.iitkgp.ac.in

Abstract—The increasing complexity of modern domestic loads, driven by advances in power electronics and control mechanisms, poses significant challenges for arc fault detection. This paper presents a comprehensive analysis of switching characteristics in domestic appliances during both OFF-to-ON and ON-to-OFF transitions. A real-time experimental setup was developed to collect current waveforms in 16 representative loads under controlled laboratory conditions. Transient signals were analyzed using four statistical features: mean, standard deviation, skewness, and kurtosis. Based on these descriptors, a K-Means clustering algorithm was used to group the loads according to their switching behavior. The results provide critical insights into the statistical separability of load types and contribute to the development of more robust and data-driven arc fault detection frameworks for low-voltage residential systems.

Index Terms—Arcing fault detection, low voltage systems, domestic load switching, statistical feature analysis, temporal analysis, statistical features, transitions.

I. INTRODUCTION

The increasing integration of electronically controlled nonlinear loads into domestic power systems has made the interaction between source, load, and switching components increasingly complex. This complexity, especially during transient switching conditions, introduces abnormal phenomena such as arcing and electrical noise that compromise operational safety [1]. In low-voltage distribution networks, switching events typically involve rapid ON/OFF transitions that manifest significant current and voltage fluctuations within the first few cycles. These transients can be exploited for reliable detection of arc faults, a critical cause of electrical fires and insulation damage [2].

In recent years, a variety of techniques have been proposed for arc fault detection (AFD). Frequency domain approaches using FFT [3] and STFT [4] offer simplicity but lack temporal localization. Time-frequency methods such as wavelet transforms provide better detection accuracy under non-stationary conditions [5], [6], and wavelet energy entropy has been used to enhance robustness [7].

In [8], a wavelet thresholding approach was combined with a neural network to improve feature extraction. Multiresolution approaches using wavelet packet decomposition have been shown to outperform classical methods in noise immunity [9], [10].

In the domain of machine learning, CNNs [11], [12], SVMs, and hybrid MLP-SVM models have achieved high classification accuracy (often exceeding 98%) for arc detection. Advanced architectures such as deep belief networks (DBNs) [13] and attention-enhanced CNNs [14] further improved generalization on unknown loads. Random forest classifiers according to IEC standards have reached a precision of 99. 07% [15]. Methods based on support vector data description (SVDD) have also been proposed to detect outliers that represent arc faults [16].

Recent works such as [17] introduced multi-feature fusion, integrating both time and frequency domain descriptors to improve generalization. In parallel, physics-informed models using the Koopman operator and Hankel matrices have emerged as interpretable and data-efficient alternatives [18]. Furthermore, reference [19] has highlighted the role of switching-induced transients in arc localization.

However, most of the above studies are based on either synthetic datasets or simplified load models. Practical analysis involving realistic switching behaviors of common domestic appliances remains limited. Furthermore, detection systems are often not benchmarked across various types of appliances or switching patterns.

Despite numerous detection frameworks and classification methods, the following gaps remain evident:
- Limited experimental data sets from actual domestic load switching events.
- Inadequate characterization of switching dynamics during OFF-to-ON and ON-to-OFF transitions.
- Lack of comparative clustering or feature space mapping for various load types based on switching behavior.
- Sparse analysis of how switching noise and arcs vary between resistive, inductive, and mixed loads.

This paper addresses the limitations mentioned above by performing a statistically grounded and experimentally validated analysis of switching transients from a wide variety of domestic appliances. The primary contributions are as follows.
- A high-fidelity data set comprising transient current and voltage signals is generated and categorized for 16+ real-world appliances.

979-8-3315-3013-6/25 $31.00 © 2025 IEEE

- Statistical descriptors (mean, standard deviation, skewness, and kurtosis) are extracted for both OFF-to-ON and ON-to-OFF transitions.
- The load behaviors are clustered on the basis of feature patterns to aid in fault classification and detection logic design.
- The results are compared with known detection approaches and organized into a comparative framework for future reference.

The remainder of this paper is organized as follows. Section II describes the experimental setup. Section III discusses the signal characteristics and the extraction of features. Section IV presents the statistical analysis and clustering results. Section V concludes the paper and describes future work.

II. DOMESTIC SWITCHING EXPERIMENTS

A. Generalized Concept of Switching

In low-voltage distribution systems, switching devices serve as the interface between source and load circuits, both of which may involve complex impedance, control, and nonlinearity. A generalized switching concept is shown in Fig. 1. Typical switching components include mechanical switches and solid-state devices such as triacs or relays, which introduce fast transients and occasional arcing during ON/OFF operations.

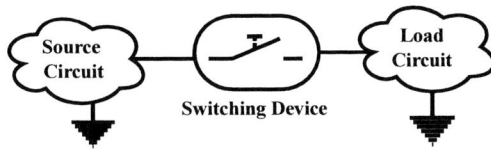

Fig. 1. Generalized switching model connecting source and load

B. Experimental Setup

To investigate the nature of switching transients during appliance operation, an experimental setup was developed as illustrated in Fig. 2. A 230 V, 50 Hz AC supply was connected in series with a single-pole single-throw mechanical switch (SPST) and various domestic loads. A high-frequency current transformer (HFCT), model i1000s, was placed upstream of the switch to measure transient current signals.

The i1000s sensor supports a bandwidth from 5 Hz to 100 kHz, ensuring accurate capture of fast switching events and arc-induced noise. The output was connected to a digital phosphor oscilloscope (DPO), configured in high-resolution mode at a sampling rate of 50 kSa/s. Each switching event was recorded over 100,000 samples to ensure complete transient coverage. The oscilloscope output was stored digitally for post-processing and feature extraction.

C. Load Selection and Data Acquisition

Sixteen distinct domestic appliances were tested, including resistive, inductive and non-linear loads (switched mode). Each load was subjected to multiple on and off switching events to capture its transient current signature. The complete list

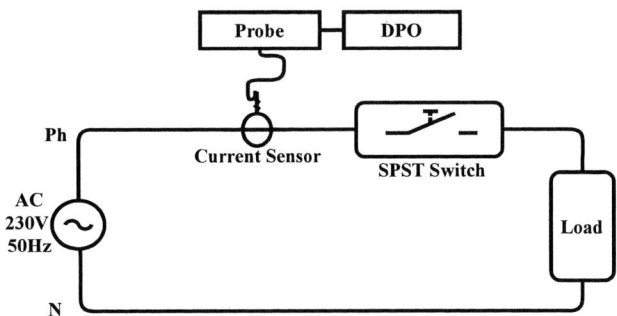

Fig. 2. Experimental setup for current measurement during switching events

of appliances, along with their electrical specifications and number of records, is provided in Table I.

TABLE I
DATABASE OF SWITCHING TRANSIENTS FOR DIFFERENT LOADS

S.No.	Load	Specifications	No. of Records
1	Rheostat	230 V, 50 Hz, 50 Ω	100
2	Incandescent Lamp	230 V, 50 Hz, 40 W	100
3	Water Dispenser	1Φ–230 V, 50 Hz, CP: 90 W, HP: 500 W	100
4	Exhaust Fan	230 V, 50 Hz, 90 W, 0.4A, 1400rpm	100
5	CFL Lights	220 V–240 V, 50 Hz, 15 W, PF = 0.85	100
6	Ceiling Fan	230 V, 50 Hz, 80 W	100
7	Laptop	Input: 240 V, 50 Hz, 1.5A; Output: 19 V, 3.42A	100
8	Mixer	230 V, 50 Hz, 500 W	100
9	Electric Iron	230 V, 50 Hz, 1000 W	100
10	Washing Machine	Not Specified	100
11	Fridge	230 V, 50 Hz, 1A, 110 W	100
12	Air Conditioner	230 V, 50 Hz, 1.5 Ton	100
13	Vacuum Cleaner	230 V, 50 Hz, 1400 W	100
14	Printer	230 V, 50 Hz, 4.6A	100
15	Drill Machine	230 V, 50 Hz, 500 W	100
16	Oven	230 V, 50 Hz, 500 W	100

D. Observed Switching Characteristics

Representative current waveforms for selected appliances are shown in Figs. 3–5. These plots illustrate the high variability in transient behavior due to the type of load and the design of the internal circuit. Notably, resistive appliances (e.g., incandescent lamps, electric irons) display rapid current rise/fall without oscillation, whereas inductive loads (e.g., fans, mixers) exhibit more gradual transitions. Nonlinear loads (e.g., CFLs, LEDs, printers) demonstrate rich high-frequency content and irregularities, complicating arc fault detection due to their similarity to arc-like signatures.

III. DATA ANALYSIS

A. Background

The transient current behavior of domestic loads was categorized into three types: resistive, inductive, and nonlinear (with solid-state switching). Each was analyzed using four statistical characteristics: mean (μ), standard deviation (σ), skewness and kurtosis, calculated over N discrete samples of current $I[n]$.

1) Resistive Loads: Resistive loads follow Ohm's law, $V[n] = I[n] \cdot R$. Their switching transients are generally abrupt, without oscillations, but may exhibit random zero-current gaps due to arcing. The statistical features are computed as:

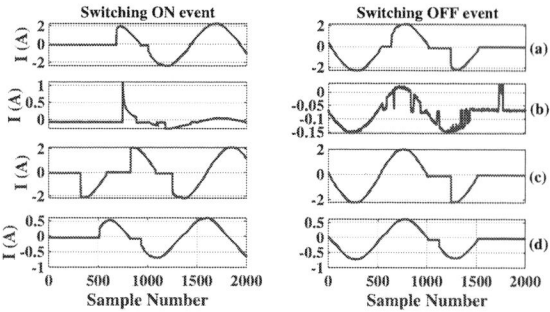

Fig. 3. ON and OFF switching waveforms for: (a) Rheostat, (b) Incandescent Lamp, (c) Electric Iron, (d) Water Dispenser

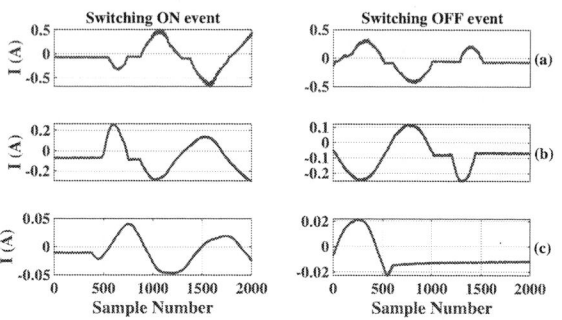

Fig. 4. ON and OFF switching waveforms for: (a) Mixer, (b) Exhaust Fan, (c) Ceiling Fan

The four statistical features used to describe the switching current transients are summarized in Table II.

2) Inductive Loads: For inductive loads, $V[n] = L \cdot \Delta I[n]/\Delta t$. Upon switching ON, the current increases exponentially as $I[n] = I_0(1 - e^{-n\Delta t/\tau})$, where $\tau = L/R$. Switching OFF leads to decaying current and possible high voltage spikes, which contribute to arcing or insulation stress.

3) Loads with Solid-State Switches: Loads incorporating thyristors or transistors generate switching spikes. A simplified model for the voltage waveform is $V[n] = V_0 u[n - n_0]$, producing a current spike:

$$I[n] = I_0 e^{-(n-n_0)\Delta t/\tau} \qquad (1)$$

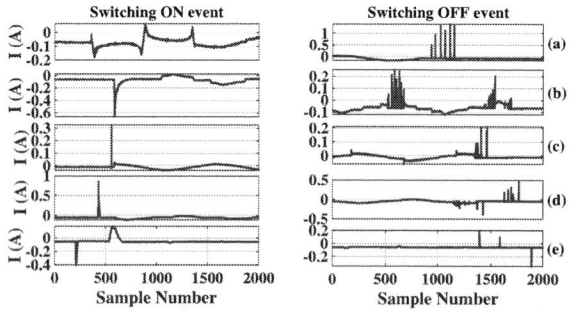

Fig. 5. ON and OFF switching waveforms for: (a) LED Light, (b) CFL, (c) 4-pin CFL, (d) Combined CFL/LED, (e) Printer

TABLE II
STATISTICAL FEATURE DEFINITIONS FOR CURRENT SIGNAL

Feature	Definition
Mean (μ)	$\mu = \dfrac{1}{N} \sum_{n=0}^{N-1} I[n]$
Standard Deviation (σ)	$\sigma = \sqrt{\dfrac{1}{N} \sum_{n=0}^{N-1} (I[n] - \mu)^2}$
Skewness	$\dfrac{1}{N} \sum_{n=0}^{N-1} \dfrac{(I[n] - \mu)^3}{\sigma^3}$
Kurtosis	$\dfrac{1}{N} \sum_{n=0}^{N-1} \dfrac{(I[n] - \mu)^4}{\sigma^4}$

4) Arcing and Zero-Current Periods: The zero-current intervals caused by short-duration arcing, illustrated in Fig. 6, can be modeled as:

$$I[n] = 0, \quad \text{for } n \in [n_1, n_2] \qquad (2)$$

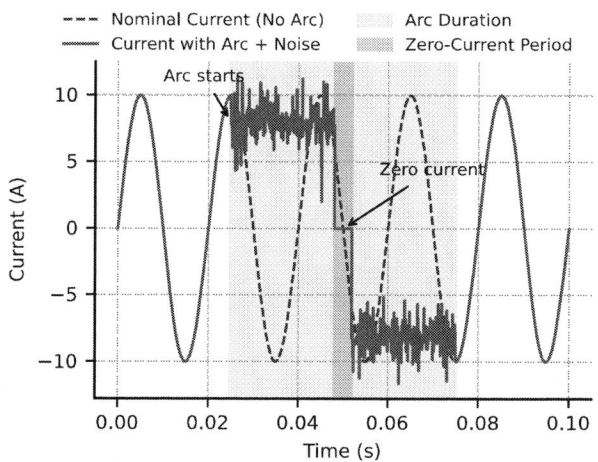

Fig. 6. Short-duration arc with zero-current segment

B. Temporal Analysis of Switching Behavior

Resistive loads respond immediately to switching actions, as they allow sudden changes in current. Upon switch closure, the current aligns instantly with the source waveform without oscillations. However, arc formation can occasionally introduce random zero-current intervals that last up to a full cycle. Figure 7 presents kernel density plots of statistical features (mean, standard deviation, skewness, and kurtosis) for representative resistive loads.

Inductive loads, shown in Fig. 8, exhibit a delayed current rise during OFF-to-ON switching due to inductive opposition. During OFF switching, they may generate high-voltage spikes, increasing the likelihood of arcing. The resulting current transitions are slower and more variable compared to resistive loads.

979-8-3315-3013-6/25 $31.00 © 2025 IEEE

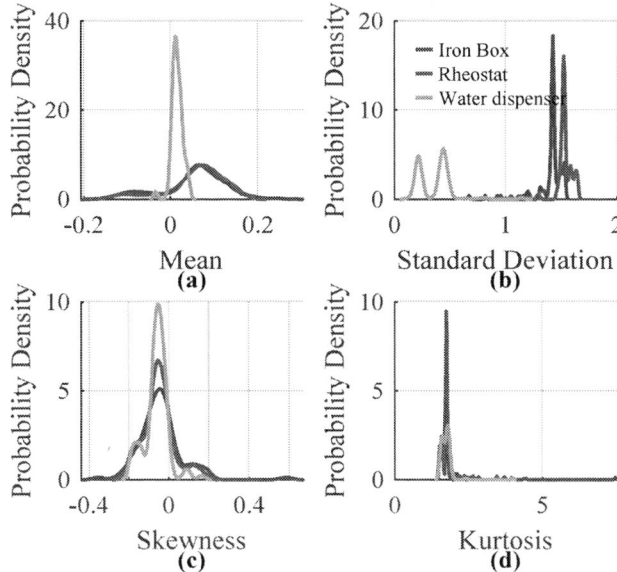

Fig. 7. Resistive loads: OFF-to-ON statistical distributions

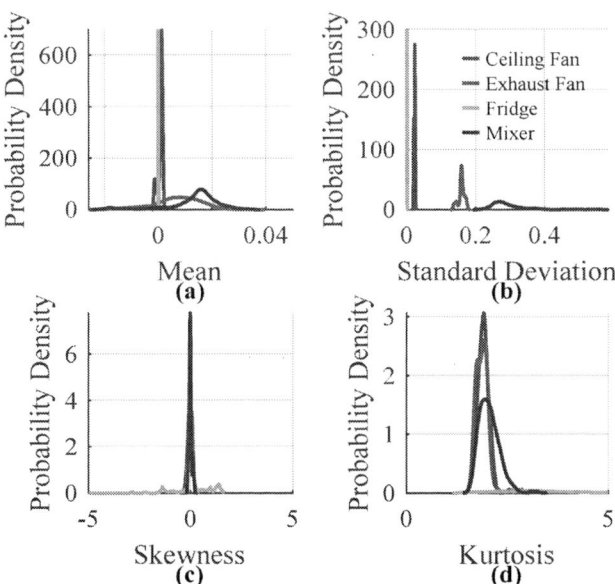

Fig. 8. Inductive loads: OFF-to-ON statistical distributions

Nonlinear loads with internal solid-state switching (e.g., printers, LED drivers) present a different behavior. These loads generate sharp high-frequency current spikes during switching, which may resemble arc-like transients. As shown in Fig. 9, their statistical profiles exhibit high kurtosis and skewness.

These temporal characteristics reflect the intrinsic electrical nature of the load and justify the use of statistical feature analysis to support arc fault detection strategies.

C. Clustering for Load Categorization and Anomaly Detection

The statistical characteristics, mean, standard deviation, skewness and kurtosis, were extracted from three-cycle cur-

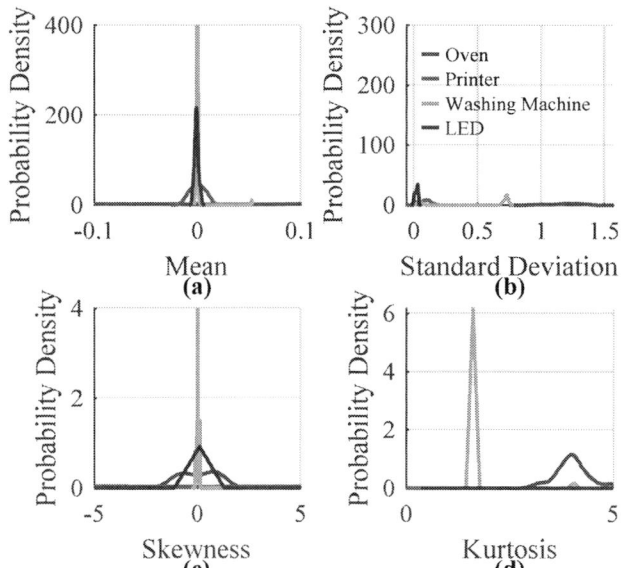

Fig. 9. Nonlinear loads: OFF-to-ON statistical distributions

rent signals recorded during the OFF-to-ON and ON-to-OFF transitions. Each load was recorded 100 times to ensure statistical robustness. After feature extraction, the values were standardized using Z-score normalization.

The K-Means clustering algorithm was used to identify behavioral groupings among loads. By clustering similar switching responses, the algorithm helps distinguish typical switching patterns from anomalies that may resemble arc faults. This unsupervised approach provides a basis for later fault classification strategies and helps to quantify the diversity of load behavior.

Algorithm 1 K-Means Clustering Based on Statistical Features

1: **Input:** 100 repeated current measurements per load (CSV format)
2: Extract statistical features: Mean, Standard Deviation, Skewness, Kurtosis
3: Aggregate features across all repetitions
4: Normalize the feature vectors using Z-score normalization
5: Initialize K cluster centroids randomly
6: **repeat**
7:     Assign each load to the nearest cluster centroid
8:     Update centroids based on current cluster assignments
9: **until** convergence or maximum iterations
10: **Output:** Cluster labels for each load type

IV. RESULTS AND DISCUSSION

Statistical feature-based analysis provides a quantitative framework for interpreting transient switching behavior between domestic loads. Tables III and IV summarize the extracted features for the OFF-to-ON and ON-to-OFF transitions, respectively. The classification criteria used to label the statistical descriptors are defined in Table V.

A. OFF-to-ON Transition Behavior

As shown in Table III, resistive loads such as incandescent lamps, ovens, and rheostats demonstrate low mean and standard deviation, indicating consistent current levels and minimal variability at the point of energization. Their skewness is generally symmetric and kurtosis is flat (below 3), which reflects balanced, non-impulsive transitions typically expected from resistive components.

Inductive loads, including ceiling fans, exhaust fans, and refrigerators, exhibit a contrasting behavior. These loads exhibit a higher standard deviation due to current build-up delays and associated inrush effects. Negative skewness is observed, driven by the initial slow rise in current, while moderate to sharp kurtosis indicates stronger transient intensity, especially during sudden energization events.

Nonlinear or switched-mode power supply (SMPS)-based loads, such as printers, LEDs, and washing machines, show low mean values and low to moderate standard deviation. However, they exhibit high positive skewness and very sharp kurtosis, often exceeding 9. This behavior is attributed to internal fast-switching devices that produce narrow spikes, which, if unaccounted for, may be misclassified as arcing phenomena.

TABLE III
STATISTICAL COMPARISON OF LOADS DURING OFF TO ON TRANSITION

Load	Mean	Std Dev	Skewness	Kurtosis	Cluster
Rheostat	Low	Very High	Symmetric	Flat (<3)	0
Mixer	Low	High	Symmetric	Flat (<3)	1
Fan	Low	Medium	Symmetric	Flat (<3)	1
LED	Low	Low	Positive	Very Sharp ($\gg3$)	2
Incandescent	Low	Medium	Symmetric	Very Sharp ($\gg3$)	1
CFL	Low	Medium	Symmetric	Very Sharp ($\gg3$)	1
Iron	Low	Very High	Symmetric	Flat (<3)	0
Ceiling Fan	Low	Low	Symmetric	Flat (<3)	1
CFL2	Low	Low	Symmetric	Very Sharp ($\gg3$)	1
CFLLED	Low	Low	Positive	Very Sharp ($\gg3$)	1
Water Dispenser	Low	High	Symmetric	Flat (<3)	1
Printer	Low	Medium	Negative	Very Sharp ($\gg3$)	1
AC	Low	Very High	Symmetric	Flat (<3)	3
Fridge	Low	Low	Symmetric	Sharp (>3)	1
Geyser	Low	Very High	Symmetric	Flat (<3)	3
Oven	Low	Very High	Symmetric	Flat (<3)	0
Washing Machine	Low	Very High	Symmetric	Flat (<3)	3

B. ON-to-OFF Transition Behavior

The response to load during off switching, presented in Table IV, reveals a different set of characteristics. Resistive loads largely maintain symmetric skewness and low variance with flat tails, confirming clean disconnection without substantial transients. In contrast, inductive appliances exhibit higher positive skewness and elevated kurtosis, which is attributed to the sudden collapse of magnetic energy and back EMF spikes, often observed during disconnection.

Switched-mode loads continue to demonstrate sharp transitions. Similar to their ON behavior, printers and LEDs retain high kurtosis and positive skewness, reinforcing their impulsive nature regardless of switching direction. This consistent

signature makes such devices critical from a misclassification point of view in arc detection systems.

TABLE IV
STATISTICAL COMPARISON OF LOADS DURING ON TO OFF TRANSITION

Load	Mean	Std Dev	Skewness	Kurtosis	Cluster
Iron	Low	Very High	Symmetric	Flat (<3)	2
Geyser	Low	Very High	Symmetric	Flat (<3)	0
Incandescent	Low	Medium	Symmetric	Flat (<3)	0
Fridge	Low	Low	Symmetric	Sharp (>3)	0
CFL2	Low	Low	Symmetric	Very Sharp ($\gg3$)	0
CFL	Low	Low	Symmetric	Very Sharp ($\gg3$)	0
Fan	Low	Medium	Symmetric	Flat (<3)	0
CFLLED	Low	Low	Symmetric	Very Sharp ($\gg3$)	0
Ceiling Fan	Low	Low	Symmetric	Flat (<3)	0
AC	Low	Very High	Symmetric	Flat (<3)	0
Water Dispenser	Low	High	Symmetric	Flat (<3)	0
Washing Machine	Low	Very High	Symmetric	Flat (<3)	0
Rheostat	Low	Very High	Symmetric	Flat (<3)	2
Printer	Low	Low	Positive	Very Sharp ($\gg3$)	1
Oven	Low	Very High	Symmetric	Flat (<3)	2
Mixer	Low	High	Symmetric	Flat (<3)	0
LED	Low	Medium	Positive	Very Sharp ($\gg3$)	3

C. Feature Classification Criteria

The statistical features extracted from each switching event are categorized according to the threshold values, as presented in Table V. These thresholds help assign qualitative descriptors, such as 'low','medium', or 'very high' to the raw feature values for comparison and grouping purposes. This categorization ensures a consistent interpretation across different types of loads and transitions and serves as the basis for feature labeling in Tables III and IV.

TABLE V
CRITERIA FOR CLASSIFICATION OF STATISTICAL FEATURES

Feature	Thresholds	Classification Labels
Mean	< 0.1	Low
	$0.1 - 0.5$	Moderate
	$0.5 - 1.0$	High
	> 1.0	Very High
Standard Deviation	< 0.05	Low
	$0.05 - 0.2$	Medium
	$0.2 - 0.5$	High
	> 0.5	Very High
Skewness	< -0.5	Negative
	$-0.5 - 0.5$	Symmetric
	> 0.5	Positive
Kurtosis	< 2	Flat (< 3)
	$2 - 3$	Moderate (≈ 3)
	$3 - 5$	Sharp (> 3)
	> 5	Very Sharp ($\gg 3$)

D. Relevance to Detection Frameworks

Table VI provides a comparative summary of existing arc fault detection (AFD) methods and their limitations. Although many recent models demonstrate excellent accuracy under ideal or simulated conditions, they often fail to handle the variability introduced by realistic domestic switching behavior. This reinforces the importance of feature-level analysis, such as the approach proposed in this paper, in improving detection robustness in practical low-voltage systems.

979-8-3315-3013-6/25 $31.00 © 2025 IEEE

TABLE VI
COMPARISON OF STATE-OF-THE-ART ARC FAULT DETECTION METHODS

Method	Key Features	Limitations
Wavelet + Energy Entropy [5], [7]	Robust to noise; good time-frequency resolution	Sensitive to wavelet choice; weak for very short events
FFT-Based Signatures [3]	Captures dominant spectral components	No time localization; misses transient arcs
MLP / CNN / SVM [11], [12]	High accuracy; learns from data	May overfit; requires labeled training data
Random Forest [15]	Interpretable; fast training	Limited sequence learning; feature-sensitive
Wavelet + NN [8]	Combines structural + statistical learning	Computationally expensive; needs tuning
Koopman + Hankel [18]	Physics-informed; data-efficient	High complexity; difficult hardware realization
Threshold-Based (IEC 62606) [2]	Standardized; easy to implement	High false alarms; poor adaptability

V. CONCLUSION

This paper presented a statistical analysis of switching transients in domestic loads to support reliable arc fault detection in low-voltage systems. A practical dataset comprising ON/OFF switching signals for 16 common domestic appliances was collected under real operating conditions. The signals were analyzed using four statistical descriptors — mean, standard deviation, skewness, and kurtosis — for both OFF-to-ON and ON-to-OFF transitions.

The results showed that:

- The standard deviation varied by a factor of up to 10 between low-power and high-power appliances, highlighting its usefulness for distinguishing load types.
- Kurtosis values ranged from 2.1 (flat) to over 9.8 (very sharp), effectively capturing the impulsive nature of switching events in CFLs, LEDs, and printers.
- Clustering analysis grouped appliances into 3 to 4 distinct clusters based on transient statistical behavior, depending on the switching direction, offering insights into feature-space separability for classification tasks.

This characterization framework can serve as a reference for designing robust arc fault detection algorithms, particularly for real-time classifiers that rely on short-time features.

Future work will focus on:

- Integrating the statistical feature extraction into a real-time detection system using machine learning classifiers.
- Expanding the dataset to include abnormal arcing events under controlled fault conditions.
- Developing adaptive detection logic that considers ambient noise, load diversity, and dynamic thresholds for embedded implementations.

REFERENCES

[1] R. Smeets, L. Van der Sluis, M. Kapetanovic, D. F. Peelo, and A. Janssen, *Switching in electrical transmission and distribution systems.* John Wiley & Sons, 2015.

[2] A.-M. I. Taalab, M. El-Geziry, *et al.*, "Impact of load variations on arcing fault detection in lv distribution networks," in *10th IET International Conference on Developments in Power System Protection (DPSP 2010). Managing the Change*, pp. 1–5, IET, 2010.

[3] Y. Afif, R. Delfianti, A. P. Widyatma, and F. E. Prahesti, "Detection of low voltage arc series fault and its severity level using fast fourier transform method," *Jurnal Teknologi*, vol. 87, Jan 2025.

[4] J. Lezama, P. Schweitzer, E. Tisserand, and S. Weber, "Simulation environment for the testing of electrical arc fault detection algorithms," *Electronics*, vol. 13, no. 20, p. 4099, 2024.

[5] Z. He, Z. Xu, H. Zhao, W. Li, Y. Zhen, and W. Ning, "Detecting series arc faults using high-frequency components of branch voltage coupling signal," *IEEE Transactions on Instrumentation and Measurement*, vol. 73, pp. 1–13, 2024.

[6] G. Zou, G. Fu, B. Han, W. Wang, and C. Liu, "Series arc fault detection based on dual filtering feature selection and improved hierarchical clustering sensitive component selection," *IEEE Sensors Journal*, vol. 23, no. 6, pp. 6050–6060, 2023.

[7] Z. Fu, W. Wang, W. Hu, *et al.*, "Series arc fault detection method based on wavelet energy spectrum entropy," Oct 2023. PREPRINT (Version 1) available at Research Square.

[8] Y. Lu and Z. Xu, "Low voltage fault arc detection method based on wavelet threshold and residual neural network," in *The Proceedings of the 18th Annual Conference of China Electrotechnical Society* (Q. Yang, Z. Li, and A. Luo, eds.), (Singapore), pp. 753–762, Springer Nature Singapore, 2024.

[9] Y. Ma and X. Xiong, "Detection of series arc faults based on time-frequency domain disturbances," in *2024 9th Asia Conference on Power and Electrical Engineering (ACPEE)*, pp. 2093–2100, IEEE, 2024.

[10] Y. Qiu and Z. Liu, "The intelligent detection method for arc faults based on empirical wavelet transform and machine learning," in *2024 6th International Conference on Machine Learning, Big Data and Business Intelligence (MLBDBI)*, pp. 77–80, 2024.

[11] N. Wu, M. Peng, J. Wang, H. Wang, Q. Lu, M. Wu, H. Zhang, and F. Ni, "Research on series arc fault detection method based on the combination of load recognition and mlp-svm," *IEEE Access*, vol. 12, pp. 100186–100199, 2024.

[12] Y. Wang, L. Hou, K. C. Paul, Y. Ban, C. Chen, and T. Zhao, "Arcnet: Series ac arc fault detection based on raw current and convolutional neural network," *IEEE Transactions on Industrial Informatics*, vol. 18, no. 1, pp. 77–86, 2021.

[13] Y. A. M. Alsumaidaee, J. K. S. Paw, C. T. Yaw, S. K. Tiong, C. P. Chen, T. Yusaf, F. Benedict, K. Kadirgama, T. C. Hong, and A. N. Abdalla, "Fault detection for medium voltage switchgear using a deep learning hybrid 1d-cnn-lstm model," *IEEE Access*, vol. 11, pp. 97574–97589, 2023.

[14] R. Zhou, J. Huang, W. J. Xu, L. Wang, H. Gao, and H. C. Hua, "Method of series arc fault detection based on phase space reconstruction and convolutional neural network," *Mathematical Problems in Engineering*, vol. 2022, pp. 1–12, 2022.

[15] K. Dowalla, P. Bilski, R. Łukaszewski, A. Wójcik, and R. Kowalik, "A novel method for detection and location of series arc fault for non-intrusive load monitoring," *Energies*, vol. 16, no. 1, 2023.

[16] Z. Zhang, J. Liu, Y. Wang, M. Shao, M. Dou, and X. Zhang, "Research on high-impedance fault models based on arc distortion characteristics," in *2024 7th International Conference on Power and Energy Applications (ICPEA)*, pp. 308–313, 2024.

[17] Z. Li, Y. Liu, Z. Liang, and Y.-H. Huang, "A low voltage ac fault arc detection method based on multi-feature fusion," *Journal of Physics: Conference Series*, vol. 2814, no. 1, p. 012030, 2024.

[18] K. Babu, D. Dwivedi, M. E. Valdes, P. Chakraborty, P. K. Panigrahi, and M. Pal, "Detection of high-impedance low-current arc faults at electrical substations," *arXiv preprint arXiv:2410.10151*, 2024.

[19] R. Nutenki, A. Routray, and A. K. Pradhan, "Characterization of arc/spark discharge phenomena in low voltage distribution systems in the presence of power electronic devices," in *2023 11th National Power Electronics Conference (NPEC)*, pp. 1–6, IEEE, 2023.

Electromagnetic Emission Estimation Using Common-Mode Current for Arc Fault Detection in Low Voltage Systems

1st Ratnakar Nutenki
Department of Electrical Engineering
IIT Kharagpur & RGUKT-Nuzvid
Kharagpur, India
ratnakar@iitkgp.ac.in

2nd Aurobinda Routray
Department of Electrical Engineering
Indian Institute of Technology
Kharagpur, India
aurobinda.routray@gmail.com

3rd Ashok Kumar Pradhan
Department of Electrical Engineering
Indian Institute of Technology
Kharagpur, India
akpradhan@ee.iitkgp.ac.in

Abstract—Arcing faults are a major cause of electrical fires in low-voltage distribution systems. Early detection of these faults is essential to improve system safety and reliability. This paper proposes a method to estimate radiated electromagnetic emissions (EME) generated by arc discharges using common-mode currents induced in conductors. A high-frequency current sensor captures these currents, which are modeled using a Hertzian dipole framework to predict far-field radiation. Experimental validation was conducted using broadband antenna measurements at two distances in the far field. The results show a strong agreement with the theoretical predictions, confirming the ability of the model to reproduce the spatial decay and spectral characteristics of the arc-induced emissions. The proposed approach provides a practical, non-intrusive technique for arc fault detection in low-voltage power systems.

Index Terms—Electromagnetic emission, antenna, hertzian dipole, common-mode current, radiated emission.

I. INTRODUCTION

Electric arcing is a critical concern in power distribution systems, affecting both residential and commercial environments. It typically arises from loose electrical contacts, degraded insulation, or mechanical wear, leading to a sustained, luminous plasma discharge. Arc faults not only degrade power quality but also generate substantial electromagnetic emissions (EME), posing serious risks including electrical fires and equipment damage. With the increasing deployment of sensitive electronic devices, modern power systems have become more susceptible to arcing disturbances. Traditional protection mechanisms often fail to detect early-stage arcing, leading to prolonged exposure to EME, power degradation, and potential safety hazards. This highlights the need for improved arc detection strategies that leverage EME characteristics.

An electric arc constitutes a nonlinear, self-sustained plasma discharge between conductive electrodes, producing high-intensity heat, light, and broadband radiated energy. The transient and stochastic nature of arc faults, characterized by abrupt changes in current amplitude and spectral content, makes their detection particularly challenging. Prior research has explored various detection methods, including time-domain waveform analysis, frequency decomposition, and data-driven classifiers [1]–[7]. Among the contributors to arc-related EME, both conducted and radiated electromagnetic noise are significant, especially via common-mode and differential-mode coupling [8]–[10].

Common-mode currents, often resulting from asymmetric grounding or insulation breakdown, serve as a dominant mechanism for radiated EME. Their characterization and detection are therefore essential for reliable arc fault identification [11]–[13]. Although recent studies have modeled common-mode emissions using circuit and electromagnetic simulations [14]–[17], existing approaches are limited by computational complexity, external noise sensitivity, and weak separation between arcing and non-arcing transients [18]–[20].

This paper proposes a predictive model for estimating radiated emissions generated by arc-induced common-mode currents. Unlike conventional methods that separately consider conducted or radiated noise, the proposed approach integrates arc-induced common-mode behavior into a compact radiated emission model based on Hertzian dipole theory. This improves both prediction efficiency and diagnostic value. The model is experimentally validated using controlled arcing scenarios, where estimated PSDs are compared against loop antenna measurements, demonstrating strong alignment under far-field conditions.

The remainder of the paper is organized as follows: Section II presents the common-mode current-based emission model and its theoretical formulation. Section III describes the experimental setup, validation procedures, and comparative analysis. Section IV concludes the paper.

II. MODELING OF COMMON-MODE CURRENT EMISSIONS

This section presents the theoretical framework for modeling radiated electromagnetic emissions resulting from arc-induced common-mode currents. The formulation is based on linear antenna theory and harmonic decomposition of time-varying current sources.

A. Common-Mode Currents and Arcing Phenomena

Arcing phenomena result in highly dynamic, time-varying voltage and current waveforms, which are inherently nonlinear and broadband in nature. A representative example of such behavior is shown in Fig. 1, illustrating the instantaneous voltage-current (V–I) characteristics during an arc discharge event. The steep transitions and non-linearity evident in the waveform reflect the underlying plasma dynamics and thermal ionization processes. These rapid current variations act as primary sources of radiated electromagnetic fields. Consequently, time-varying common-mode currents on the conductors become key contributors to radiated emissions in arc fault scenarios.

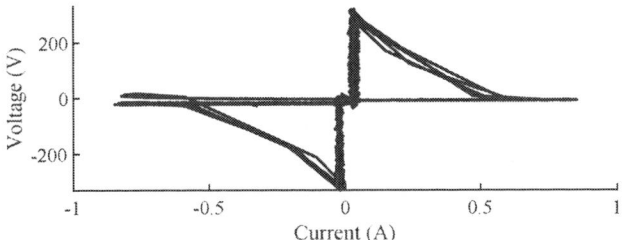

Fig. 1. Instantaneous voltage-current (V–I) characteristics of an arc discharge, highlighting the nonlinear and time-varying behavior responsible for radiated electromagnetic emissions.

The two conductors that carry the common mode current in the experimental setup (Fig. 2) can be conceptually modeled as an array of linear radiators. This equivalence is illustrated in Fig. 3, where the conductors are represented as an idealized antenna array. By calculating the array factor and superimposing the far-field contributions from each conductor segment, the overall radiated field pattern due to arc-induced current can be predicted. This modeling approach captures the essential physics of common-mode radiation, which arises predominantly from asymmetries in current distribution and grounding. The noise source associated with common-mode currents is therefore effectively represented as a pair of Hertzian dipoles, enabling analytical estimation of the resulting EME.

B. Hertzian Dipole Model

The radiated electromagnetic field due to arc-induced common-mode current can be modeled using a pair of idealized Hertzian dipoles. As shown in Fig. 4, two infinitesimal dipole antennas are considered in free space. The current elements are aligned along the z-axis and located on the x-axis, symmetrically spaced by a distance d from the origin. The observation point P is assumed to lie in the far-field region relative to the dipole array.

In this configuration, the total radiated field at the point P is obtained by superimposing the far-field contributions from each dipole. This model forms the foundation for analytically estimating the electric field pattern associated with radiated emissions from common-mode currents induced during arcing events [21].

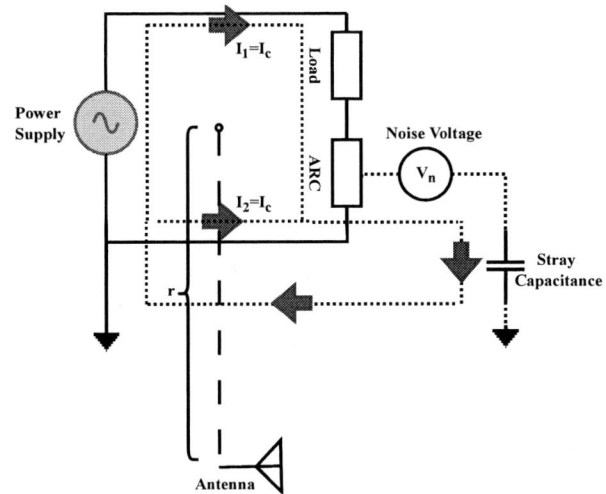

Fig. 2. Conceptual model of common-mode current emission, where radiated fields are generated by arc-induced current flow along asymmetrical conductors.

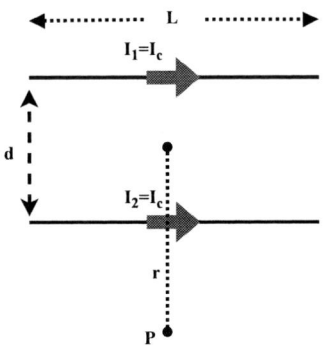

Fig. 3. Equivalent antenna array representation of the conductors, modeled as linear radiators to estimate the common-mode radiated electromagnetic field.

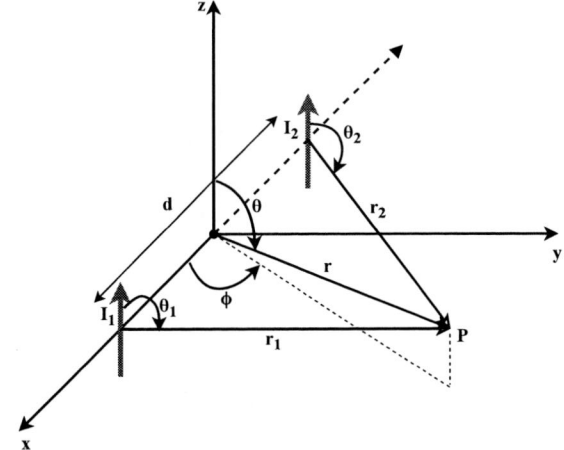

Fig. 4. Far-field configuration of two Hertzian dipole current elements oriented along the z-axis and symmetrically positioned on the x-axis, radiating primarily in the transverse (x–y) plane.

$$\hat{E}_{\theta 1} = \frac{\hat{A} I e^{j\psi}}{r_1} e^{-jkr_1} \tag{1}$$

$$\hat{E}_{\theta 2} = \frac{\hat{A} I}{r_2} e^{-jkr_2} \tag{2}$$

where $\hat{I}_1 = I e^{j\psi}$ and $\hat{I}_2 = I$, assuming that the currents of the two antennas are equal in magnitude but differ in phase by ψ. The factor $\hat{A}$ varies based on the type of antennas used. For Hertzian dipoles

$$\hat{A} = j\eta_0 k (dl/4\pi) \sin\theta \tag{3}$$

where η_0 is the intrinsic impedance of free space, $k = \frac{2\pi}{\lambda_0}$ is the phase constant for the fundamental frequency, dl is the length of the dipole, and θ is the polar angle.

Considering the far-field approximation, the total electric field at the point P can be expressed as a superposition of the contributions of individual current elements.

$$\hat{E}_\theta = \hat{E}_{\theta 1} + \hat{E}_{\theta 2} \tag{4}$$

$$\hat{E}_\theta = \hat{A} I \left(\frac{e^{-jkr_1} e^{j\psi}}{r_1} + \frac{e^{-jkr_2}}{r_2} \right) \tag{5}$$

$$\hat{E}_\theta = \hat{A} I e^{j\psi/2} \left(\frac{e^{-jkr_1} e^{j\psi/2}}{r_1} + \frac{e^{-jkr_2} e^{-j\psi/2}}{r_2} \right) \tag{6}$$

Here the current is assumed to be sinusoidal but in practical arcing case the common mode current exhibits a non-sinusoidal nature which can be decomposed into the sum of harmonic components.

C. Mathematical Framework for Superposing Electric Fields from All Harmonics

The current distribution of the common mode currents of the arcing is non-sinusoidal, and the current $I(t)$ can be expressed as a Fourier series:

$$I(t) = I_0 + \sum_{n=1}^{\infty} I_n \cos(n\omega t + \psi_n) \tag{7}$$

where I_0 is the DC component, I_n is the amplitude of the n-th harmonic, ω is the fundamental angular frequency, and ψ_n is the phase angle of the n-th harmonic.

The phasor equivalent of the n-th harmonic current is:

$$\hat{I}_n = I_n e^{j\psi_n} \tag{8}$$

For each harmonic n, the electric field $\hat{E}_{\theta n}$ at a point P due to a dipole can be written as:

$$\hat{E}_{\theta n} = \frac{\hat{A}_n \hat{I}_n}{r} e^{-jk_n r} \tag{9}$$

where $\hat{A}_n$ is the antenna factor for the n-th harmonic, $k_n = \frac{2\pi}{\lambda_n}$ is the phase constant for the n-th harmonic, $\lambda_n = \frac{\lambda_0}{n}$ is the wavelength of the n-th harmonic, and r is the distance from the antenna to measurement point P.

Considering two current elements as dipoles separated by a distance d, each carries the same non-sinusoidal current $I(t)$, the electric field at point P due to the n-th harmonic from dipole 1 is:

$$\hat{E}_{\theta n1} = \frac{\hat{A}_n \hat{I}_n}{r_1} e^{-jk_n r_1} \tag{10}$$

and the electric field at point P due to the n-th harmonic from dipole 2 is:

$$\hat{E}_{\theta n2} = \frac{\hat{A}_n \hat{I}_n}{r_2} e^{-jk_n r_2} \tag{11}$$

The total electric field for the n-th harmonic is the superposition of the fields from both dipoles:

$$\hat{E}_{\theta n} = \hat{E}_{\theta n1} + \hat{E}_{\theta n2} \tag{12}$$

The approximate path lengths are

$$r_1 \cong r - \frac{d}{2} \cos\phi \tag{13}$$

$$r_2 \cong r + \frac{d}{2} \cos\phi \tag{14}$$

In the far field, $r_1 \cong r_2 \cong r$, so the denominators of (10) and (11) can be approximated as r.

The simplified total electric field for the n-th harmonic becomes:

$$\hat{E}_{\theta n} = 2\hat{A}_n I_n \frac{e^{-jk_n r}}{r} e^{j(\psi_n/2)} \cos\left(\pi \frac{d}{\lambda_n} \cos\phi + \frac{\psi_n}{2} \right) \tag{15}$$

The total electric field at point P is the superposition of the fields from all harmonics:

$$\hat{E}_\theta = \sum_{n=1}^{N} \hat{E}_{\theta n} \tag{16}$$

Substituting the expression for $\hat{E}_{\theta n}$:

$$\hat{E}_\theta = \sum_{n=1}^{N} 2\hat{A}_n I_n \frac{e^{-jk_n r}}{r} e^{j(\psi_n/2)} \cos\left(\pi \frac{d}{\lambda_n} \cos\phi + \frac{\psi_n}{2} \right) \tag{17}$$

D. Radiated Emissions Estimation

The electromagnetic noise source generates common-mode currents on the wires. These currents are measured using a high-frequency current sensor. The measured current signal and its frequency spectrum are shown in Fig. 5. The extracted harmonics are used to estimate the far fields. A sample of the first ten harmonics is presented in Table I and Table II for r = 3m and r = 4.3m, respectively. To determine the number of harmonics to be considered, we reconstructed the signal using the selected harmonics. The reconstructed signal agrees well with the measured current signal, as illustrated in Fig. 5.

The electromagnetic noise generated by arcing excites common-mode currents on the conductors, which are captured using a high-frequency current sensor. The time domain waveform of the measured current, along with the reconstructed signal synthesized from the dominant harmonics, is shown in Fig. 5(a). To determine the spectral content of the

TABLE I
FIRST 10 HARMONIC COMPONENTS FOR R=3M

Harmonic	Frequency (Hz)	Amplitude	Phase (rad)
1	3.3784×10^5	0.001 747 6	−2.8981
2	6.7568×10^5	0.038 558	−1.3532
3	1.0135×10^6	0.040 388	2.2809
4	1.3514×10^6	0.021 511	0.739 16
5	1.6892×10^6	0.023 969	−0.931 62
6	2.027×10^6	0.018 324	−3.0023
7	2.3649×10^6	0.009 589 7	−0.147 23
8	2.7027×10^6	0.017 113	−2.7909
9	3.0405×10^6	0.026 132	1.3339
10	3.3784×10^6	0.023 345	−0.639 59

TABLE II
FIRST 10 HARMONIC COMPONENTS FOR R=4.3M

Harmonic	Frequency (Hz)	Amplitude	Phase (rad)
1	1.6661×10^5	0.003 660 8	−1.5077
2	3.3322×10^5	0.002 523 8	1.3447
3	4.9983×10^5	0.004 244	−2.8271
4	6.6644×10^5	0.004 730 5	−0.116 68
5	8.3306×10^5	0.007 685 9	3.0437
6	9.9967×10^5	0.009 255 2	0.447 87
7	1.1663×10^6	0.006 65	−2.003
8	1.3329×10^6	0.003 638 3	1.9086
9	1.4995×10^6	0.005 663 3	−1.1166
10	1.6661×10^6	0.000 668 99	−3.1035

arcing-induced current, a Fast Fourier Transform (FFT) was performed. The resulting frequency spectrum, presented in Fig. 5(b), highlights the prominent harmonic components that characterize arc-induced emission.

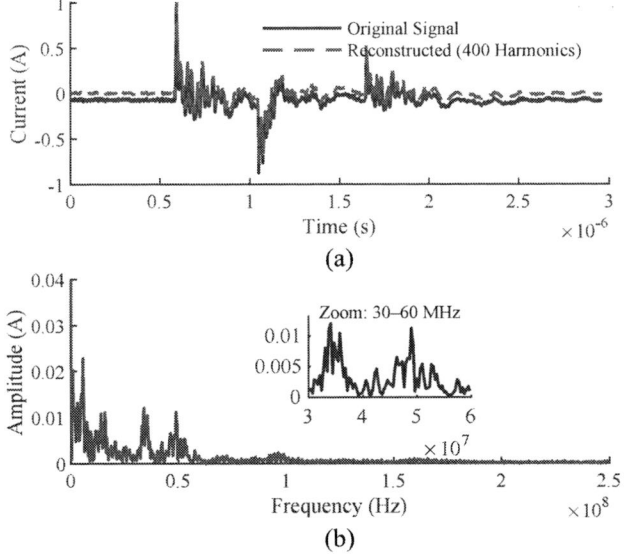

Fig. 5. Common-mode current analysis. (a) Time-domain waveform of the measured and reconstructed current signal. (b) Corresponding FFT spectrum in amplitude showing dominant frequency components.

III. EXPERIMENTAL VALIDATION AND SPECTRAL ANALYSIS

This section describes the experimental procedures used to validate the proposed emission model. The measured data are compared with theoretical predictions to evaluate the accuracy of the model in practical arc fault conditions.

A. Experimental Setup

To measure the common-mode currents, we used the experimental setup depicted in Fig. 6. The power supply was connected to a load in series with an arc source, comprising two point electrodes. Common-mode currents were captured upstream of the arcing source using a clamp-on high-frequency current sensor. This sensor was interfaced with a Digital Phosphor Oscilloscope (DPO), configured in Hi-Resolution mode at a sampling rate of 500 MSa/s, recording 100k samples for detailed analysis.

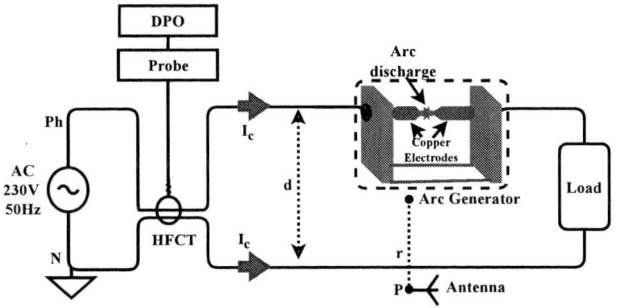

Fig. 6. Laboratory setup for measuring common-mode currents and radiated electric fields during arc fault generation.

A controlled laboratory arrangement that adheres to the UL1699 standard was utilized to simulate series arc faults. This involved a series arc generator with insulated walls supporting adjustable point electrodes connected in series to an AC source of 230 V, 50 Hz, and a resistive load. The separation of the electrode was maintained at approximately 0.1 cm, adjustable to initiate arcing. The arcing phenomenon generates electromagnetic noise, thereby inducing common-mode currents.

B. Measurement Strategy and Rationale

Radiated electromagnetic emissions (EME) were assessed at distances of $r = 3$ m and $r = 4.3$ m using a broadband loop antenna, serving both as a validation reference and to evaluate far-field behavior. The two measurement distances were deliberately selected for both practical and methodological reasons. The 3-meter distance corresponds to a standard reference used in electromagnetic compatibility (EMC) testing protocols such as CISPR 16 and FCC Part 15, making it suitable for baseline validation. In contrast, the 4.3-meter distance was chosen to evaluate the robustness of the radiation model at a nonstandard but practically relevant range. This dual distance approach enables the evaluation of field decay trends and model reliability under varying spatial conditions.

979-8-3315-3013-6/25 $31.00 © 2025 IEEE

C. Comparison and Analysis of Measured vs Estimated Fields

Figures 7 and 8 present a comparative analysis of the power spectral density (PSD) of the estimated and measured electric fields at distances of 3 m and 4.3 m, respectively. In both cases, a shaded region indicates the typical PSD range for comparison.

At 3 m (Fig. 7), both the estimated and measured PSD values are predominantly within the range of −90 to −120 dB/Hz. The estimated field exhibits a close spectral alignment with the measured field across most of the frequency range, reflecting a consistent harmonic structure and amplitude. However, slight deviations are observed in the lower frequency region, likely attributable to unmodeled factors such as structural resonances, spectral leakage, and complex coupling effects not captured by the idealized dipole formulation.

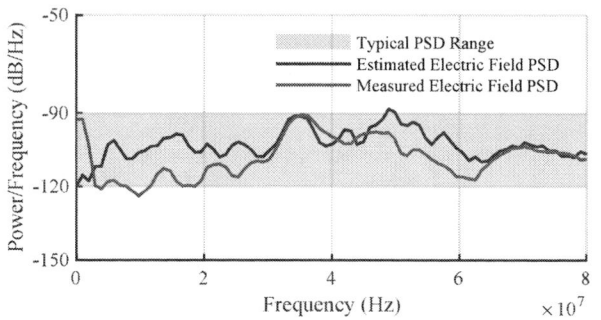

Fig. 7. Comparison of estimated and measured electric field spectra at a distance of 3 m.

At 4.3 m (Fig. 8), the PSD envelope shifts downward, mainly spanning −95 to −125 dB/Hz. This represents an approximate 5 dB reduction in overall emission strength, consistent with the theoretical $1/r$ field decay. Furthermore, further attenuation is evident in the lower frequency band (below 30 MHz) is evident, indicating increased susceptibility to environmental absorption and propagation loss at larger distances. These results emphasize the need for refined radiated field models that incorporate distance-dependent propagation effects, antenna coupling asymmetries, and environmental factors in practical deployment scenarios.

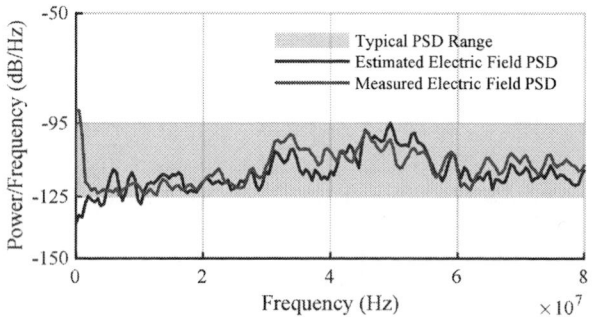

Fig. 8. Comparison of estimated and measured electric field spectra at a distance of 4.3 m.

Table III presents the spectral comparison between estimated and measured electric fields at 3 m and 4.3 m. In both cases, the spectral similarity index (SSI) was 1.0, indicating strong envelope agreement. The normalized MSE remained below 0.15, reflecting a moderate shape deviation. However, the normalized area error (NAE) exceeded 35%, suggesting a consistent amplitude offset, probably due to unmodeled system losses or environmental effects. Despite this, the close spectral alignment confirms the model's effectiveness in capturing arc-induced emission characteristics.

TABLE III
QUANTITATIVE METRICS FOR PSD COMPARISON AT 3 M AND 4.3 M

Metric	3 m	4.3 m
Spectral Similarity Index (SSI)	1.0000	1.0000
Normalized MSE (NMSE)	0.1302	0.1429
Normalized Area Error (NAE)	0.3576	0.3745

The specific spectral peaks and dips, while generally aligned between estimated and measured results, differ slightly in magnitude. Such variations underline the importance of accounting for detailed environmental and instrumental influences to improve prediction fidelity.

Figure 9 illustrates the PSD of radiated electromagnetic emissions measured by a loop antenna at a distance of 4.3 m for both arcing and non-arcing scenarios. A distinct spectral separation is observed across a broad frequency range, particularly between 30–150 MHz, where the arcing signal exhibits significantly higher energy levels. This elevated spectral content is attributed to the impulsive and broadband nature of arc-induced current fluctuations, in contrast to the relatively quiescent non-arcing background. The consistent difference in spectral amplitude across frequencies provides a robust and interpretable feature space for arc detection, affirming the utility of radiated EME as a discriminative indicator in practical sensing setups.

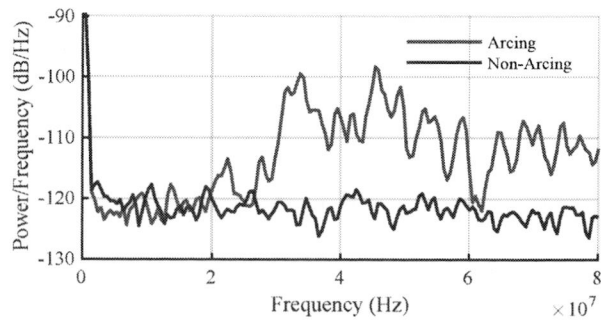

Fig. 9. Comparison of radiated electric field PSDs for arcing and non-arcing conditions.

D. Sources of Discrepancy and Limitations

The following critical factors summarize the key contributors to these deviations:

- **Instrumental uncertainties**: The high-frequency current sensor, while designed for broadband arc signal capture, exhibits a roll-off beyond several MHz, limiting accuracy in capturing higher harmonics. Furthermore, slight variations due to sensor placement, cable routing, and calibration drift can contribute to amplitude and phase inconsistencies. The DPO, even in high-resolution mode, introduces quantization noise and exhibits finite analog-to-digital conversion precision. Temporal jitter and connector losses also impact the fidelity of rapid transient measurements associated with arcing.

- **Environmental factors**: The experimental setup was not located in an electromagnetically shielded or anechoic environment. Reflections from nearby metal surfaces, benches, and walls likely caused multipath propagation and spatial interference. Ambient electromagnetic interference (EMI) from nearby instrumentation, lighting, and power supply ripple further modulated the recorded signal. These superimposed disturbances particularly affect the PSD in lower MHz bands.

- **Model simplifications**: The radiation model is built on idealized assumptions—Hertzian dipoles with uniform current distribution and lossless media. Nonidealities such as conductor asymmetry, interwire coupling, and time-varying arc impedance were not considered in the simplified analytical framework. The skin effect and return path imperfections at high frequencies can alter the spatial distribution of common-mode currents, affecting radiated-field estimations.

IV. CONCLUSION

This paper presented a method for estimating radiated electromagnetic emissions from arc faults in low-voltage distribution systems using common-mode currents as a diagnostic proxy. The conductors were modeled as Hertzian dipoles, and a harmonic decomposition framework was formulated to estimate far-field radiation patterns generated by arc-induced currents.

Experimental validation at two distances—3 m and 4.3 m—demonstrated good agreement between estimated and measured power spectral densities, confirming both the theoretical $1/r$ field decay and the spectral fidelity of the model. These results establish the practical viability of using common-mode current sensing for arc detection.

Discrepancies were attributed to instrumentation limitations, environmental interference, and model idealizations. Nevertheless, the proposed framework offers a robust, interpretable approach to arc fault detection. Future work will focus on refining electromagnetic models, enhancing sensor calibration, and extending the framework to more dynamic and realistic arc scenarios. Overall, this study advances the development of accurate, non-intrusive arc detection systems to improve safety and reliability in modern power networks.

REFERENCES

[1] F. Zhou, H. Yin, L. Chen, H. Tong, K. Yu, Z. Li, and X. Zeng, "Research on low voltage series arc fault prediction method based on multidimensional time-frequency domain characteristics," *Energy Engineering*, vol. 120, no. 9, pp. 1979–1990, 2023.

[2] N. Qu, W. Wei, and C. Hu, "Series arc fault detection based on multimodal feature fusion," *Sensors*, vol. 23, no. 17, 2023.

[3] X. Chang, X. Zhang, Y. Tao, W. Zhang, J. S. Jiang, and C. jie Zhang, "Research on fault identification method of series arc in electric vehicle charging system based on machine learning," in *IEEE International Conference on High Voltage Engineering and Application*, pp. 1–4, 2023.

[4] R. Z. Jiang, G. Bao, Q. Hong, and C. Booth, "Machine learning approach to detect arc faults based on regular coupling features," *IEEE Transactions on Industrial Informatics*, vol. 19, pp. 2761–2771, 2023.

[5] W. Gao, J. Rao, F. Cui, and R.-J. Wai, "A two-level classification diagnosis method for ac arc faults based on data random fusion and mc-mgcnn network," *Measurement*, 2023.

[6] X. Chang, X. Zhang, Y. Tao, W. Zhang, J. S. Jiang, and C. jie Zhang, "Feature analyses and identification methods of low-voltage series arc fault," in *IEEE International Conference on High Voltage Engineering and Application*, pp. 1–4, 2022.

[7] Z. Li, Y. Liu, Z. Liang, and Y.-H. Huang, "A low voltage ac fault arc detection method based on multi-feature fusion," *Journal of Physics: Conference Series*, vol. 2814, no. 1, 2024.

[8] C. J. Rose, D. Thomas, C. Smartt, and P. Meng, "Low frequency conducted emissions caused by series arc faults," in *International Symposium on Electromagnetic Compatibility*, 2019.

[9] S. Zhao, Y. Wang, F. Niu, C. Zhu, Y. Xu, and K. Li, "A series dc arc fault detection method based on steady pattern of high-frequency electromagnetic radiation," *IEEE Transactions on Plasma Science*, vol. 47, no. 9, pp. 4370–4377, 2019.

[10] T. J. Donnelly, S. D. Pekarek, D. R. Fudge, and N. Zarate, "Characterization of common/differential-mode behavior in power electronic systems," in *Applied Power Electronics Conference*, 2021.

[11] C. R. Paul and D. R. Bush, "Radiated emissions from common-mode currents," in *1987 IEEE International Symposium on Electromagnetic Compatibility*, pp. 1–7, 1987.

[12] S. Yang, Q. Chen, H. Li, Y. Jiao, R. Tian, Y. He, C. Zhang, and C. Liu, "An arc grounding fault detection method in distribution networks using chaotic dc waveform recognition," *IEEE Transactions on Instrumentation and Measurement*, vol. 74, pp. 1–10, 2025.

[13] Z. Zhang, A. Zingariello, G. Griepentrog, and H. Muhm, "Common mode current modeling and investigation in different grounding systems for frequency range 0 -150 khz," in *2024 Energy Conversion Congress Expo Europe (ECCE Europe)*, pp. 1–6, 2024.

[14] A. Amin, T. I. Mannan, and S. Choi, "Common mode emi characterization through phase modeling," in *2021 IEEE Applied Power Electronics Conference and Exposition (APEC)*, pp. 164–169, 2021.

[15] L. Illiano, X. Wu, F. Grassi, G. Spadacini, and S. A. Pignari, "Review of mode conversion and modal analysis in electromagnetic compatibility," *IEEE Access*, vol. 12, pp. 65513–65529, 2024.

[16] T. Li, R. Olson, H. Abdallah, N. Sivadanam, and R. M. Cuzner, "Modeling and validation of common-mode emissions of silicon carbide enabled motor drive in extended emc frequency range between 2 khz and 30 mhz," in *2023 IEEE Energy Conversion Congress and Exposition (ECCE)*, pp. 2904–2911, 2023.

[17] R. B. Keller, *Antennas*, pp. 111–134. Cham: Springer International Publishing, 2023.

[18] Y. Liu, J. Swingler, and D. Flynn, "A method for dc arc fault detection, classification and mitigation in electric vehicles," in *2021 3rd Global Power, Energy and Communication Conference (GPECOM)*, pp. 7–12, 2021.

[19] E. M. Calderon, P. Schweitzer, C. Bonnet, and S. Weber, "Influence of perturbations produced by electromagnetic interference (emi) in arc fault detection," in *2018 IEEE Holm Conference on Electrical Contacts*, pp. 323–328, 2018.

[20] Z. Zhang, J. Ren, X. Tang, S. Jing, and W.-J. Lee, "Novel approach for arc fault identification with transient and steady state based time-frequency analysis," *IEEE Transactions on Industry Applications*, vol. 58, no. 4, pp. 4359–4369, 2022.

[21] C. R. Paul, *Introduction to Electromagnetic Compatibility*. Hoboken: John Wiley & Sons, Inc., 2nd ed., 2006.

Design of Synchronous Reluctance Motor for Electric Two-wheeler Application

Saddikuti Sriharitha
Electrical Engineering Department
SECS, IIT Bhubaneswar
Jatni, Odisha, India
20ee02007@iitbbs.ac.in

Jitendra Gautam
Electrical Engineering Department
SECS, IIT Bhubaneswar
Jatni, Odisha, India
a24ee09010@iitbbs.ac.in

Dr. Ankit Dalal
Electrical Engineering Department
SECS, IIT Bhubaneswar
Jatni, Odisha, India
ankitdalal@iitbbs.ac.in

Abstract—In this work, a synchronous reluctance motor is designed for electric two-wheeler applications, with the goal of moving towards motors that do not use rare-earth magnets. The motor is developed using a systematic design process to meet the required performance targets. Key design steps include choosing the right number of poles and slots and optimizing the rotor shape. The final motor uses a six-pole, 36-slot layout with a four-barrier rotor and achieves a power output of 5 kW at 3500 RPM, which is suitable for two-wheeler use. After the initial design, a free-shape optimization is carried out using Altair-Flux software. This reduces the rotor volume by 20% in 10 steps without lowering the motor's performance. The results show that synchronous reluctance motors can be a good alternative to permanent magnet motors for electric vehicles.

Index Terms—barriers, electric propulsion system, free shape optimization, rare earth materials, SynRM.

I. INTRODUCTION

Electric vehicle (EV) technology is crucial for advancing sustainable mobility and reducing carbon footprints in the transportation sector. Unlike internal combustion engine (ICE) vehicles, electric vehicles utilize advanced electric propulsion systems powered by high-voltage energy sources, integrated with sophisticated power electronics components such as inverters, DC-DC converters, and electric motors with high torque efficiency [1]. The most commonly used motors in electric vehicle technology are induction motors (IMs) and permanent magnet synchronous motors (PMSM) [2]. Permanent magnet motors have been preferred over the IMs due to their higher power density and efficiency in comparison to the IMs [3]. NdFeB (Neodymium iron boron) and SmCo (Samarium-Cobalt) are the commonly used alloys in PMSM applications, NdFeB being preferred in most applications due to its higher energy density. However, during 2010-11, there was a price hike of 5-6 times in NdFeB and a price hike in neodymium and dysprosium as well, which are the key components of NdFeB [4]. This became a motivation to move towards motors with a lower quantity of rare earth magnets. Synchronous reluctance motors (SynRM) meet this requirement because they do not require magnets (like in PMSM) or any other field excitation

This work is supported by IIT Bhubaneswar, India, under seed project Grant SP142.

(like in IMs). The rotor anisotropic geometry eliminates the usage of magnet and aluminum/copper bars in the rotor structure, and the torque produced in SynRM is due to the variation in reluctance to the magnetic flux in different regions. Although motors with permanent magnets have higher average torque and power factor, the demagnetization of permanent magnets limits the stator current. Meanwhile, the reluctance motor does not have any limitation of demagnetization and thus has a much higher capability to handle overloads. Therefore, though reluctance motors have lower performance in terms of average torque and power factor, it is compensated by the fact that they provide a path towards rare earth magnet-free motors and have other major advantages like higher overload capability [5]. Thus, with its numerous advantages, SynRM has gained a lot of attention in the past decade.

Apart from the advantages discussed, there is a lot of scope for improving the performance of the SynRM. Matsuo et al. demonstrated that the maximum torque can be obtained by optimizing the rotor insulation width to the rotor iron width [6]. Mohanrajah et al. in [7] proposed an analytical method to enhance the power factor of a SynRM without compromising on torque density. Tingke He et al. in [8] discussed several methods to improve the efficiency of a SynRM, viz., increasing the saliency ratio, reducing the copper losses and iron losses, and adopting amorphous-nanocrystalline soft magnetic materials. Vagati et al. came up with a design approach to reduce the torque ripple in a synchronous reluctance motor [9]. M. Ibrahim et al. in [10] also designed a SynRM with low torque ripple and high output torque. While researchers have proposed various methods to enhance SynRM performance, there remains a need for tailored designs that specifically address the constraints and requirements of electric two-wheelers.

This work focuses on designing a SynRM for an electric two-wheeler application, taking into account critical parameters such as the number of poles and the number of slots. Additionally, various geometrical constraints are taken into account to refine the SynRM configuration and enhance its overall performance. The motor is developed to achieve high efficiency and torque performance, aligning with the defined performance requirements. Altair-flux motor software is used to design the motor.

Following the initial design, a free shape optimization technique is applied to further improve the motor's performance using Altair-flux software. This step aims to reduce the rotor volume while maintaining or enhancing the motor's efficiency and torque capabilities.

II. MOTOR PERFORMANCE REQUIREMENTS

To ensure suitability for an electric two-wheeler application, the designed SynRM must meet the performance criteria of a reference PMSM. The key specifications that the SynRM aims to achieve are as follows:

- Input DC Voltage = 48 V
- Rated Power Output = 5 kW
- Maximum speed = 6000 rpm

The motor is required to supply a constant torque of 13.5 N-m at a base speed of 3500 rpm, which implies that the motor is expected to fulfill the power requirement of approximately 5 kW.

III. DESIGN METHODOLOGY FOR SYNCHRONOUS RELUCTANCE MOTOR

The design of a synchronous reluctance motor involves a systematic process to achieve the desired performance in terms of torque and efficiency. It begins with an analytical design approach to establish fundamental motor parameters like stator inner diameter and core length. The selection of the number of poles and slots is then considered, as it directly influences the operating characteristics of the motor. The rotor geometry is then carefully designed to optimize flux paths and torque production. Finally, free-shape optimization is applied to refine the rotor structure further, ensuring an improved overall design. Fig. 1 shows the design methodology in a chronological order.

A. Analytical Design

The analytical design of the proposed model is based on the power output equation, which can be expressed as:

$$Q = 11B_{av}qK_w\eta\cos(\Phi)D^2Ln_s \qquad (1)$$

Or alternatively,

$$Q = C_oD^2Ln_s \qquad (2)$$

$$C_o = 11B_{av}qK_w\eta\cos(\Phi) \qquad (3)$$

where:

- C_o is the output coefficient,
- B_{av} is the average flux density,
- q is the specific electric loading,
- K_w is the winding factor,
- η is the efficiency,
- $\cos(\Phi)$ is the power factor,
- D is the inner diameter of the stator,
- L is the core length, and
- n_s is the synchronous speed in rad/s.

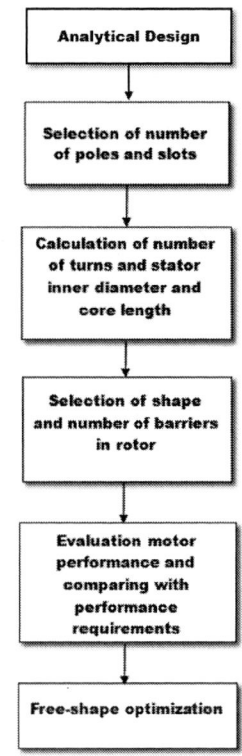

Fig. 1: SynRM design methodology in a chronological order

B. Choice of Number of Poles and Slots

1) Selection of number of poles: When designing a motor, the number of poles is chosen based on several key parameters, including flux per pole and torque output. The torque equation for a synchronous reluctance motor (SynRM) is given by:

$$T = \frac{3}{2}P(L_d - L_q)I_dI_q \qquad (4)$$

where the variables are defined as follows:

- T is the electromagnetic torque.
- P is the number of poles.
- L_d and L_q are the direct-axis and quadrature-axis inductances, respectively.
- I_d and I_q are the direct-axis and quadrature-axis currents, respectively.

Flux per pole refers to the total magnetic flux associated with a single pole, which travels through the stator teeth, crosses the air gap, and returns through the stator yoke. The thickness of the stator yoke is typically designed based on the flux per pole, requiring a thicker yoke for higher flux levels to prevent saturation, which is why fewer poles, such as two, are generally not considered. Instead, engineers often opt for between four and eight poles. However, motors with more poles typically require a larger physical size, increasing material costs, such as more laminations and windings. Also, a higher number of poles requires a higher switching frequency for a certain synchronous speed, which increases the switching

979-8-3315-3013-6/25 $31.00 © 2025 IEEE

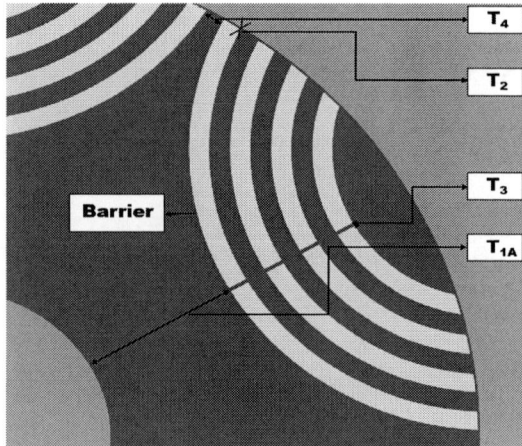

Fig. 2: Rotor with four circular barriers

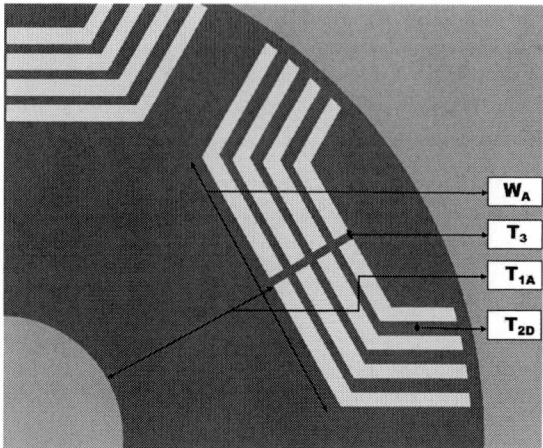

Fig. 3: Rotor with four U-shaped barriers

losses. Considering these Factors: A pole configuration is selected for any motor.

2) Selection of number of slots: To select the number of poles, engineers consider pole-slot combinations that yield a higher winding factor, as this reflects how well the winding configuration is optimized to generate the desired electromagnetic fields in the motor. Distributed winding is preferred for its lower harmonics and better torque production. To determine the optimal number of slots, the winding factor for various pole-slot combinations is calculated, and the combination with the maximum value is selected.

C. Rotor Geometry

The rotor geometry is a critical component in the design of synchronous reluctance motors as it directly influences the motor's performance by creating the necessary anisotropic magnetic properties for torque generation. Properly designed rotor geometries can significantly enhance the motor's efficiency and output. The following sections detail key aspects of rotor geometry design.

1) Types of barriers: Two types of barrier shapes are considered while designing the SynRM, are shown in Figs. 2 and 3. Fig. 2 shows a rotor with circular barriers. Fig. 3 shows a rotor with U-shaped barriers.

2) Number of barriers: The number of barriers affects the rotor's anisotropic properties, which are essential for reluctance torque generation. The appropriate number of flux barriers is determined by the relation $n_s = n_r \pm 4$ [11], where n_s represents the number of stator slots per pole pair, and n_r denotes the number of rotor slots per pole pair.

The most suitable type and number of barriers are chosen based on the optimized results of the designed SynRM. For circular-shaped barriers, the number of barriers and the parameters T_{1A}, T_2, T_3, T_4 are optimized to get the maximum possible average output torque. Similarly, for U-shaped geometry, the number of barriers and various parameters are optimized for maximum average output torque.

D. Free Shape Optimization of the Rotor

To further improve the performance of the motor and reduce the volume of the rotor, free shape optimization is applied to the rotor. Free-shape optimization modifies the rotor's geometry without being constrained by predefined shapes, unlike traditional optimization methods that adjust only specific dimensions. The adjoint method plays a crucial role in this process, especially when dealing with a large number of design variables, like SynRM, where the design variables are high. This is particularly relevant in parameter-free shape optimization techniques, where shapes are not explicitly defined using methods like splines [12]. After designing the SynRM, this work explores a case for the free shape optimization of the rotor, that is, to maximize the torque while meeting volume and symmetry constraints.

IV. RESULTS AND DISCUSSION

Based on the order mentioned in the previous section, the motor design is completed, the results of which are discussed as follows. For a motor performance requirement of 5 kW power output, the analytical design is performed assuming an average flux density (B_{av}) of 0.4 T, an efficiency of 90%, and a power factor of 0.9. The resulting stator inner diameter (D) is 99.2 mm, with a core length (L) of 65 mm as per the equations (1), (2), and (3).

As discussed in the previous section, a lower number of poles, such as two, is not considered. While the selection is typically between four and eight poles, an eight-pole design results in a bulky motor and higher switching losses. Considering these factors, a six-pole configuration is selected for the motor.

The appropriate number of slots is determined by calculating the winding factor for various pole-slot combinations and selecting the configuration with the highest value. From the obtained values, the winding factor is observed to be highest for a 6-pole, 36-slot combination, leading to the selection of the distributed winding with 36 slots.

For choosing the type and number of barriers, both three-barrier and four-barrier configurations were investigated for

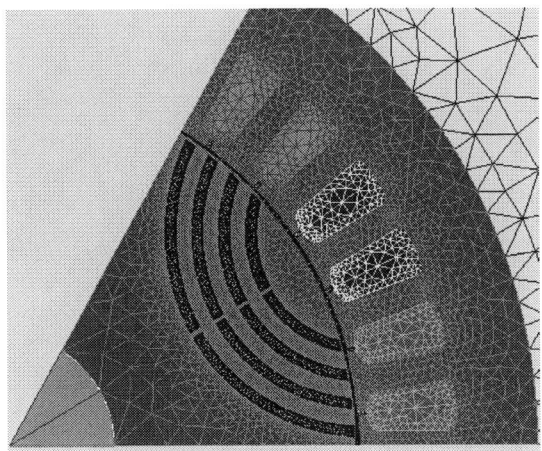

Fig. 4: The designed rotor Model

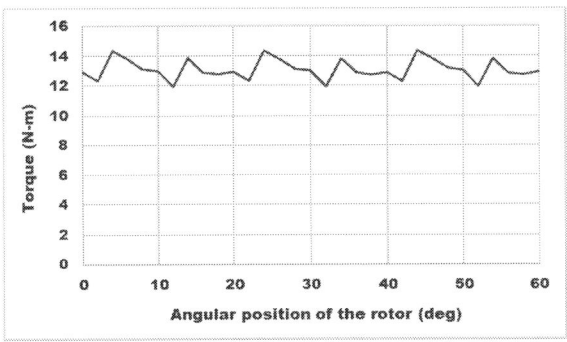

Fig. 5: Torque as a function of rotor angular position

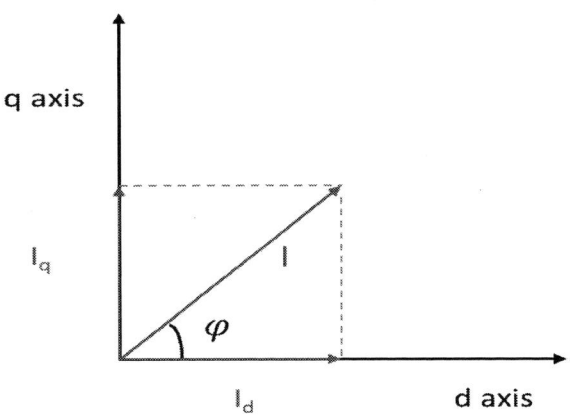

Fig. 6: Control Angle concerning d and q axes

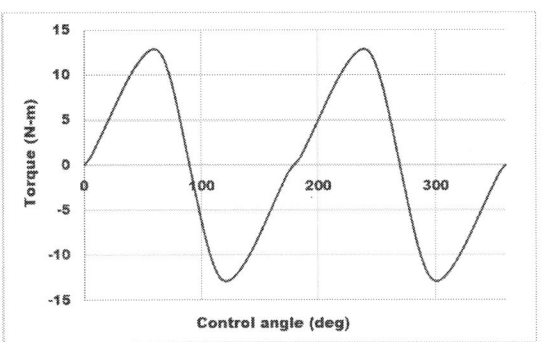

Fig. 7: Torque as a function of control angle curve

both circular and U-shaped barrier designs to determine their impact on torque output by varying parameters of the barriers while maintaining mechanical integrity. The highest torque achieved is with a four-circular barrier configuration, where a flux barrier width of 1.7 mm, a carrier width of 2.091 mm, $T_{1A} = 15$ mm, $T_2 = 0.2$ mm, $T_3 = 0.5$ mm, and $T_4 = 2$ mm resulted in a maximum torque. Hence, a rotor geometry with four circular barriers is chosen. The designed rotor has been shown in Fig. 4.

Fig. 5 presents the torque characteristics of the designed SynRM as a function of rotor angular position. This graph is crucial for analyzing the torque ripple, a key performance indicator for SynRMs in two-wheeler applications where smooth operation is essential for rider comfort and vehicle stability. The magnitude of torque variation between these peaks and troughs represents the torque ripple, where a lower ripple indicates smoother motor operation. The torque pattern observed over this 60-degree span provides insight into the motor's behavior over a partial electrical cycle, allowing for evaluation of the design's effectiveness in managing torque ripple. The average torque from the graph turns out to be 13 N-m, which is close to the motor performance requirement. The torque ripple content calculated from Fig. 5 is 18.67%.

The control angle, often referred to as the current angle, is the angle between the stator current vector and the rotor's d-axis in a rotating coordinate system. The control angle ψ is shown in Fig. 6. This angle is critical for controlling the torque produced by the motor. Fig. 7 illustrates the torque vs. control angle curve for the designed SynRM. The curve demonstrates how torque production varies with the angle of the stator current vector relative to the rotor d-axis. The curve shows a periodic variation, with a positive torque peak at approximately 60°. The peak of this curve indicates the optimal control angle for maximum torque production, which is essential for implementing efficient control strategies such as Maximum Torque Per Ampere (MTPA). This graph is instrumental in determining optimal operating points for the SynRM, balancing maximum torque production with efficient operation, which is particularly important for the two-wheeler application of this motor design.

In SynRMs, inductance varies with the rotor position due to the anisotropic nature of the rotor, which is designed to have different reluctances along different axes. The saliency ratio is defined as the ratio of the d-axis inductance (L_d) to the q-axis inductance (L_q), i.e.,

$$\zeta = \frac{L_d}{L_q} \tag{5}$$

This ratio is crucial for SynRMs because it directly influences the torque production capability and efficiency of the motor. A higher saliency ratio indicates better performance, as it

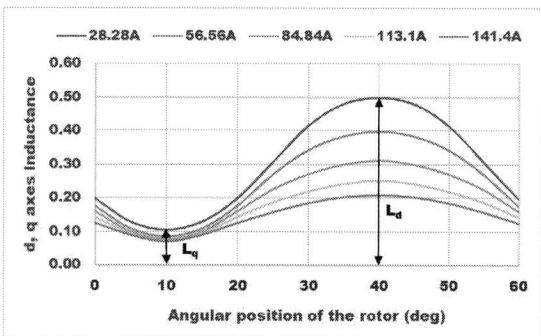

Fig. 8: d-axis & q-axis inductances as a function of rotor angular position

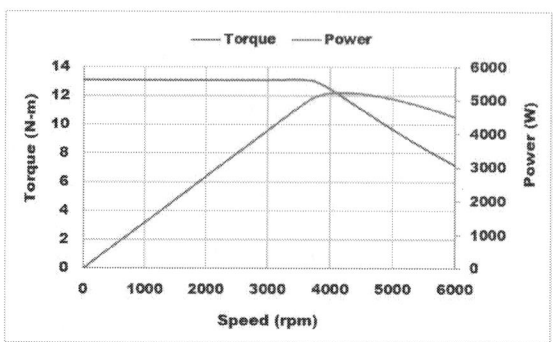

Fig. 9: The torque-speed and power-speed characteristics

enhances the reluctance torque, which is the primary source of torque in SynRMs [13]. The graph in Fig. 8 illustrates the variation of d-axis and q-axis inductances of the SynRM as a function of rotor angular position for different RMS current levels (28.28A, 56.56A, 84.84A, 113.1A, and 141.4A). At lower current levels (e.g., I = 28.28A), the d-axis inductance (L_d) exhibits a pronounced peak near 40^0, reaching approximately 0.5 mH, while the q-axis inductance (L_q) is at its minimum around 10^0. This significant difference between L_d and L_q at lower currents results in a higher saliency ratio, which is critical for efficient torque production. As the current increases, magnetic saturation occurs, reducing L_d and slightly increasing L_q. This leads to a flattening of the inductance curves and diminished variation across angular positions, indicating the reduced saliency ratio at higher currents (e.g., I = rated current = 141.4A). The periodic nature of the curves reflects the rotor's anisotropic geometry and flux barrier design, which are optimized to create alternating high and low reluctance paths. These design features are essential for achieving high saliency and maximizing reluctance torque.

The torque-speed and power-speed characteristics of the designed SynRM are presented in Fig. 9. The graph highlights two distinct operating regions: the constant torque region and the constant power region, with corresponding variations in power output. The constant torque region spans from 0 RPM to approximately 3500 RPM, and the motor delivers a nearly constant torque of around 13 N-m, as shown by the flat blue curve. The speed and torque achieved are close to the requirements defined in section II. During the constant torque region, the power output (orange curve) increases linearly with speed, reaching a peak of approximately 5 kW at 3500 RPM. This region is critical for low-speed operations, ensuring sufficient torque for acceleration and load handling in applications like two-wheelers. Beyond 3500 RPM, the motor transitions into the constant power region, where the torque decreases progressively with increasing speed. The blue curve shows that the torque reduces from 13 N-m at 3500 RPM to approximately 7 N-m at 6000 RPM (maximum speed required). Despite this decline in torque, the power remains nearly constant. This characteristic demonstrates the motor's ability to maintain efficient performance at higher speeds,

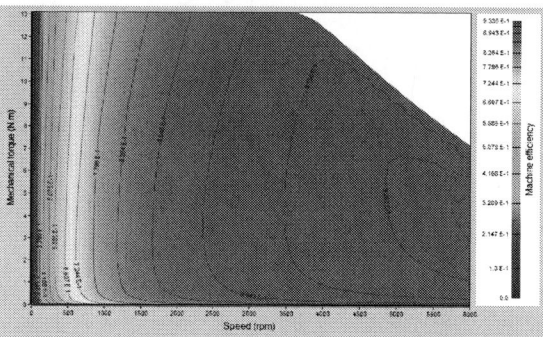

Fig. 10: Efficiency as a function of mechanical torque and rotational speed

making it well-suited for cruising and high-speed operation.

The contour plot in Fig. 10 represents the efficiency of the SynRM as a function of mechanical torque and rotational speed. The color scale ranges from blue (low efficiency) to red (high efficiency). At lower speeds (<1000 rpm), the efficiency decreases significantly, as indicated by the blue-green regions. At higher speeds (>3500 rpm), efficiency above 90% is achieved. This contour plot in Fig. 11 illustrates Joule losses as a function of mechanical torque and rotational speed (in rpm). The color scale ranges from blue (low losses) to red (high losses), with numerical values indicating Joule losses in watts. For example, Joule losses are minimal in the low torque region (<150 N-m). Losses increase significantly with both speed and torque. At high speeds, losses are high at even lower torque, and at low speeds, losses are high at higher torque, as shown in the plot. At low speeds and high torque, Joule losses dominate due to increased current demand for torque generation. At high speeds and low torque, core losses rise due to higher magnetic field frequencies. These plots provide valuable insights into the motor's performance, emphasizing areas for design optimization to improve efficiency and reduce losses across varying operating conditions.

A. Free-shape Optimization Post Designing

The design of the SynRM is followed by free-shape optimization. The free-shape optimized rotor model is presented in Fig. 12. A volume reduction is attained by free-shape optimization of the rotor. Fig. 13 presents the volume reduc-

979-8-3315-3013-6/25 $31.00 © 2025 IEEE

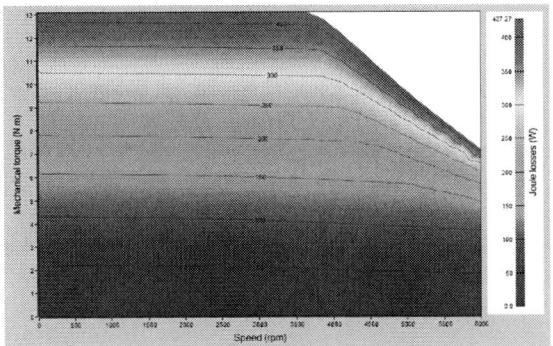

Fig. 11: Joule losses as a function of mechanical torque and rotational speed

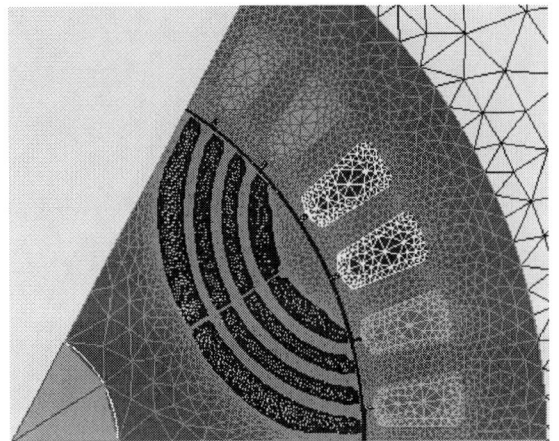

Fig. 12: Free Shape optimized rotor model

tion as a function of iterations, where a 20% reduction in rotor volume is achieved within 10 iterations, after which it stabilizes. These results validate the effectiveness of the free-shape optimization framework in achieving a balance between performance enhancement and design constraints.

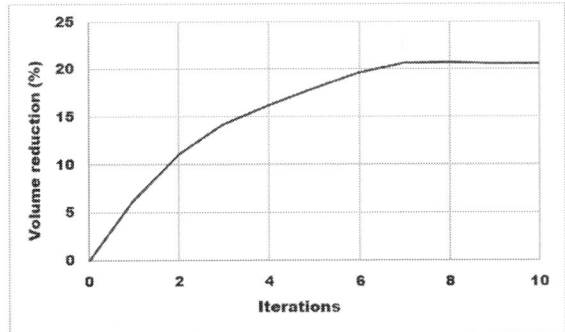

Fig. 13: Volume reduction post free-shape optimization

V. CONCLUSION

This work demonstrates the successful design of a synchronous reluctance motor for electric two-wheeler applica-

tions, offering a promising alternative to permanent magnet motors. The systematic design approach, covering analytical design, optimal pole and slot selection, and careful rotor geometry optimization, has yielded a motor that meets the target performance requirements. The six-pole, 36-slot configuration with a four-circular barrier rotor design proved most effective, balancing efficiency and manufacturing considerations. The motor achieves the desired 5 kW power output at 3500 RPM, exhibiting distinct constant torque and constant power regions that are well-suited for two-wheeler operation. Furthermore, the application of free-shape optimization resulted in a significant 20% reduction in rotor volume in 10 iterations, highlighting the potential for further performance enhancements. The results underscore the potential of SynRMs as an efficient, cost-effective, and environmentally friendly solution for electric propulsion systems, particularly in the context of reducing dependence on rare earth materials.

REFERENCES

[1] K. T. Chau, C. C. Chan, and C. Liu, "Overview of permanent-magnet brushless drives for electric and hybrid electric vehicles," *IEEE Transactions on Industrial Electronics*, vol. 55, no. 6, pp. 2246–2257, 2008.

[2] A. T. De Almeida, F. J. Ferreira, and G. Baoming, "Beyond induction motors—technology trends to move up efficiency," in *49th IEEE/IAS Industrial & Commercial Power Systems Technical Conference*. IEEE, 2013, pp. 1–13.

[3] G. Pellegrino, A. Vagati, B. Boazzo, and P. Guglielmi, "Comparison of induction and pm synchronous motor drives for ev application including design examples," *IEEE Transactions on industry applications*, vol. 48, no. 6, pp. 2322–2332, 2012.

[4] T. Vaimann, A. Kallaste, A. Kilk, and A. Belahcen, "Magnetic properties of reduced dy ndfeb permanent magnets and their usage in electrical machines," in *2013 Africon*, 2013, pp. 1–5.

[5] N. Bianchi, M. Degano, and E. Fornasiero, "Sensitivity analysis of torque ripple reduction of synchronous reluctance and interior pm motors," *IEEE Transactions on Industry Applications*, vol. 51, no. 1, pp. 187–195, 2014.

[6] T. Matsuo and T. A. Lipo, "Rotor design optimization of synchronous reluctance machine," *IEEE Transactions on Energy Conversion*, vol. 9, no. 2, pp. 359–365, 1994.

[7] T. Mohanarajah, J. Rizk, M. Nagrial, and A. Hellany, "Design of synchronous reluctance motors with improved power factor," in *2017 11th IEEE International Conference on Compatibility, Power Electronics and Power Engineering (CPE-POWERENG)*. IEEE, 2017, pp. 340–345.

[8] T. He, Y. Wang, M. Bao, J. Li, S. Feng, and R. Qu, "Design and validation of a high-efficiency synchronous reluctance motor," in *2023 26th International Conference on Electrical Machines and Systems (ICEMS)*. IEEE, 2023, pp. 3806–3811.

[9] A. Vagati, M. Pastorelli, G. Franceschini, and S. C. Petrache, "Design of low-torque-ripple synchronous reluctance motors," *IEEE Transactions on industry applications*, vol. 34, no. 4, pp. 758–765, 1998.

[10] M. Ibrahim, P. Sergeant, and E. M. Rashad, "Synchronous reluctance motor performance based on different electrical steel grades," *IEEE Transactions on Magnetics*, vol. 51, no. 11, pp. 1–4, 2015.

[11] A. Nagarkar and S. Srinivas, "An optimized rotor design of synchronous reluctance motor for improved torque characteristics," in *2021 International Aegean Conference on Electrical Machines and Power Electronics (ACEMP) & 2021 International Conference on Optimization of Electrical and Electronic Equipment (OPTIM)*. IEEE, 2021, pp. 107–114.

[12] L. Radtke, G. Bletsos, N. Kühl, T. Suchan, T. Rung, A. Düster, and K. Welker, "Parameter-free shape optimization: Various shape updates for engineering applications," *Aerospace*, vol. 10, no. 9, 2023. [Online]. Available: https://www.mdpi.com/2226-4310/10/9/751

[13] R. Panwar and K. Ragavan, "Analytical design procedure of synchronous reluctance motor and its validation through fea," in *2022 IEEE 19th India Council International Conference (INDICON)*. IEEE, 2022, pp. 1–6.

Real time precise position control of Linear Scan Mechanism for the VELC Payload operation of the Aditya-L1 Mission

Kaushlendra Kumar Kaushal
U R Rao Satellite Centre
Indian Space Research Organzation
Bengaluru, India
kaushal@ursc.gov.in

Navle Sonal Girdharlal
U R Rao Satellite Centre
Indian Space Research Organzation
Bengaluru, India
sonal@ursc.gov.in

Lekshmi V
U R Rao Satellite Centre
Indian Space Research Organzation
Bengaluru, India
vlekshmi@ursc.gov.in

Sudhakar S
U R Rao Satellite Centre
Indian Space Research Organzation
Bengaluru, India
sudhakar@ursc.gov.in

Abstract— **Aditya-L1 is the first Indian mission to study the solar corona and solar atmosphere. Visible Emission Line Coronagraph (VELC) onboard Aditya-L1 is a coronagraph designed to monitor the solar corona. The observation is carried out by precise movement of Linear Scan Mechanism (LSM) in linear step intervals that helps in generating the raster scan images of the solar corona to help in understanding various physical parameters of the coronal features. The coronal image quality and information of heliocentric distance depend on how precisely LSM is moved from one location to another location with high degree of repeatability even though LSM system may have backlash. In this paper, we present the design, realization challenges and its operational techniques of Linear Scan Mechanism to achieve micron level accuracy are discussed. Precision control electronics using FPGA is designed to move LSM from -1 mm to +1 mm in steps of integral multiples of 10µm with an accuracy of ±2µm.**

Keywords- *Precision control, Linear scans mechanism, micro-focusing, scanning mirrors, VELC payload.*

I. INTRODUCTION

Visible Emission Line Coronagraph has four imaging channel which observes the solar corona from 1.05 to 3 solar radii and three spectroscopy channels that provide the spectral observations in the Field-Of-View (FOV) from 1.05 to 1.5 solar radii[1,2,3]. The observation is carried out by precise movement of Linear Scan Mechanism in regular step intervals which helps in generating the raster scan images of the solar corona. VELC has equally spaced four slits mounted in plane. The coronal image that is passed through the slits is scanned by moving the two-fold mirrors that are mounted on top of Linear Scan Mechanism (LSM). To achieve accurate imaging and spectroscopy of the solar corona, precise control of the linear scan mechanism is essential. This paper focuses on the design, realization challenges and its operational techniques of Control electronics to achieve micron level accuracy. Precision control electronics realized using FPGA and it is designed to move LSM from -1 mm to +1 mm in steps of integral multiples of 10µm with an accuracy of ±2µm[4,5].

II. SYSTEM REQUIREMENT

For many years, stepper motors have been used in open-loop to achieve position accuracy through discrete stepping, ensuring repeatability. The linear scan mechanism consists of geared stepper motor and ball grid screw assembly. The ball grid screw assembly was used to convert the angular movement of motor into linear. However, when the LSM was under characterization, it was moved from the home position to +1 mm and back, it failed to return within ±2 µm accuracy. The relationship between LSM position and step count is illustrated in Figure 1. The test clearly indicates an open-loop positional cumulative error in Forward direction is order of 23.5um and in reverse direction is of order of 17 µm whereas step based position error is within 2 µm.Since the VELC payload requires stringent positional accuracy of ±2 µm, this has been the prime motivation to design a closed-loop position control system that meets the payload requirements. Moreover, the designed system shall meet stringent space constraints and reliability.

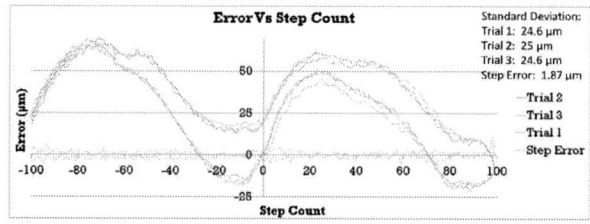

Figure 1: LSM position error versus step count in open mode configuration

III. SYSTEM OVERVIEW

The linear scan mechanism consists of a 1.8° step motor with a Harmonic gear of 1:100 ratios with an integrated optical encoder, coupled with lead screw and ball screw assemblies to translate rotary motion into linear motion. This mechanism is further connected to a guide rail on which two mirrors are assembled. The system is engineered to achieve micro-focusing of mirror assemble on it as shown in figure2 with a range of ±1.0 mm in steps of 10μm, ensuring an accuracy of ±2μm for precise focusing on a daily basis. Incremental displacement of LSM movement is measured using incremental encoder.

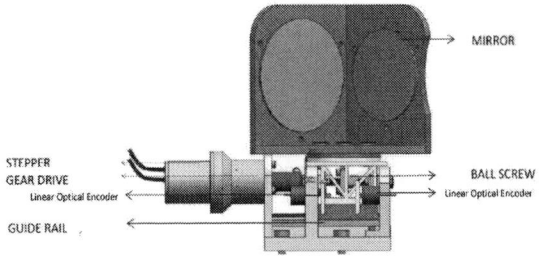

Figure 2: Mechanical structure of the linear scan mechanism

IV. POSITION CONTROL STRATEGY

Index marking of incremental optical encoder of LSM serves as the reference point (0 mm) of the LSM. Since the stepper motor operates in a relative positioning mode; the control algorithm maintains an absolute counter to track movement. When the LSM moves in the forward direction, the onboard counter increases in accordance with the number of edges received from the encoder. Conversely, in the case of reverse motion, the counter decreases. This counter is maintained onboard with an accuracy of one count, ensuring precise positional tracking. Micro stepping approach used in the proposed design enables enhanced positional accuracy, prevents overshoot and reduces torque ripple. The block diagram illustrates in figure3.The Real-Time Precise Position Control System for the Linear Scan Mechanism (LSM).

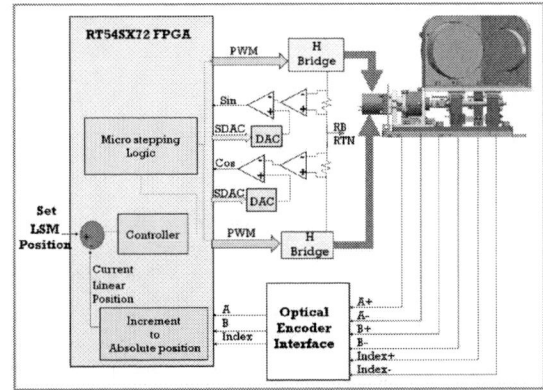

Figure 3: Block diagram of the implemented hardware design

The system is based on an RT54SX72 FPGA and incorporates micro-stepping logic, a controller, an optical encoder interface, and H-Bridge motor drivers for precise motor control. The micro-stepping logic sequentially generates addresses to fetch sine and cosine data from the PROM. Once the data is retrieved, it is loaded into the DAC to set the reference sine and cosine currents for the motor coils. The actual motor current is then compared with the reference using a comparator. If the measured current exceeds the set value, the modified S-R logic regulates the coil current, maintaining it for the required duration.

The FPGA-based controller receives real-time position feedback from the optical encoder and continuously updates the absolute position and same is compared with commanded LSM position, error is feed to controller which generate precise movement commands to Micro-stepping logic which generate PWM signal to the H-Bridge motor drivers. The H-Bridge drivers control the stepper motor, which moves the lead screw mechanism in a controlled manner to achieve commanded position.

The position control of the Linear Stepper Motor (LSM) is executed in two sequential steps:

1. Reference Point Acquisition:

   The LSM is first aligned to a reference position using an index pulse. To achieve this, the LSM is moved forward by +200 μm and then retracted by -500 μm. During the reverse motion, the hardware detects the rising edge of the index pulse. At this detected point, the onboard counter is initialized to 1000H, establishing a known reference for subsequent positioning.

2. Operational Position Tracking:

   Once the reference position is set, the onboard counter tracks LSM movement during payload operations. The counter increments when the LSM moves in the forward direction and decrements during reverse motion. This count adjustment is based on the encoder signal direction, ensuring precise real-time position tracking.

V. EXPERIMENTAL VALIDATION

An experimental setup was established on an optical bench at IIA, Bangalore, to validate the micro-focusing positional accuracy. An independent laser interferometer instrument was utilized to measure and verify the system's positioning accuracy. Figure 4 depicts the hardware test setup.

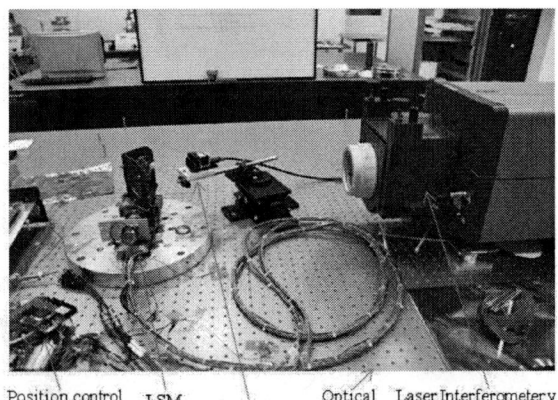

Figure 4: Experimental setup to validate the LSM positioning via laser interferometry

Position accuracy of the implemented hardware was characterized by conducting more than 1,00 trials. The LSM was operated in 10 μm stepping increments, moving from the home position to +1 mm, returning to home, then moving to -1 mm, and finally returning to home again. The test results, as shown in the figure 4, provide positional repeatability sensed by optical encoder mounted on LSM under different LSM positions.

979-8-3315-3013-6/25 $31.00 © 2025 IEEE   207

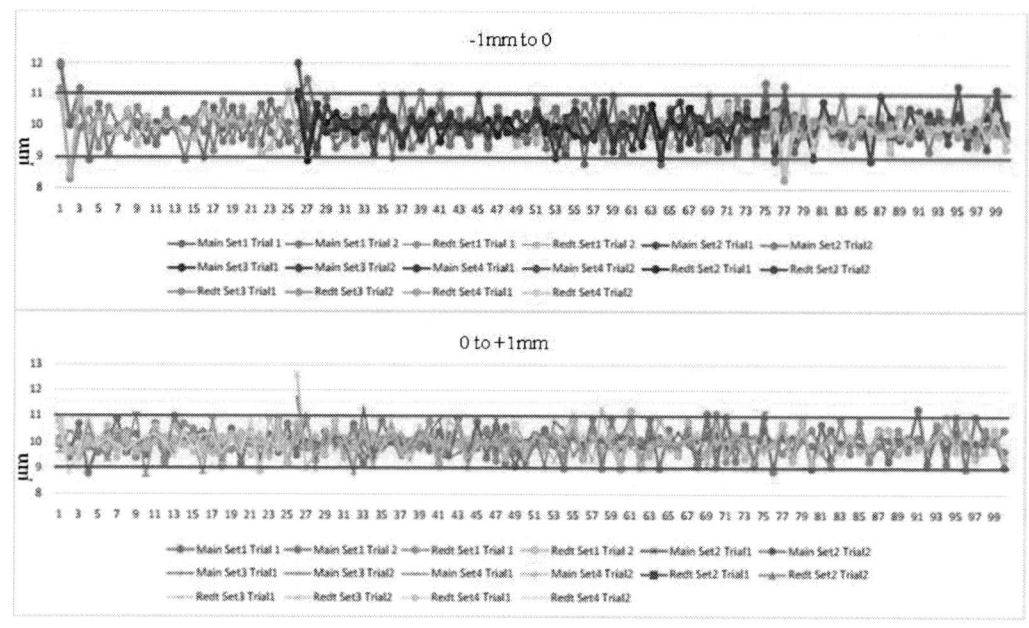

Figure 5: Variation of actuated LSM movement measured by optical encoder

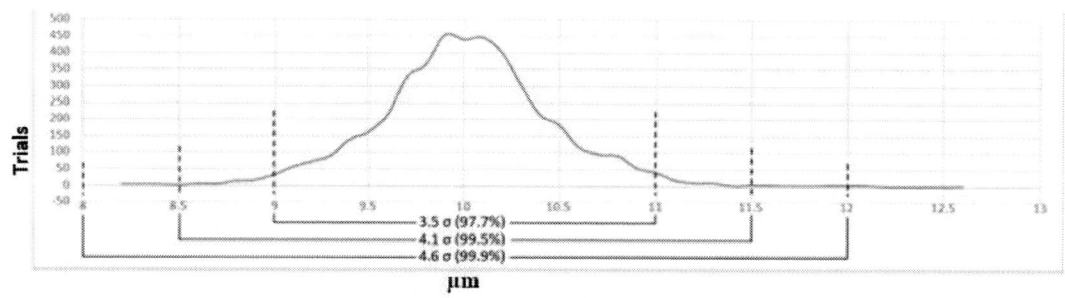

Figure 6: Variation of actuated LSM movement measured by laser interferometry instrument

Figure5 indicate statistical distribution of movement sensed by laser interferometer of different trials. This clearly indicates that designed system meets the ±2um positional accuracy.

VI. PAYLOAD OPERATION

The VELC payload grating is exposed to the detector with precise LSM positioning. The grating spans 2 mm to capture the Sun's disc. For Sun disc scanning, the payload operates in two modes: raster scan and sit-and-stare.

A. Sit-and-stare mode: in this mode, the LSM remains parked at a commanded position, where multiple observations of the solar corona are conducted at the same location before moving back to the home position. Sit-and-stare time line is shown in figure7:

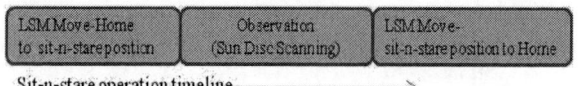

Figure 7: Sit-n-Stare operation timeline

B. Raster scan mode: In this mode, the LSM moves from the home position to the first position of the raster scan, observes the corona, then moves to the next location and continues the observation. The step size can be 10 µm or more, and the maximum number of operations can be programmed up to 128. Raster scan time line is shown in figure8:

Figure 8: Raster Scan operation timeline

The above operation is controlled through an algorithm, as shown in the figure 9. The payload operation is initiated by the OBC (On-Board Computer) over the MIL-1553B protocol. Upon receiving a command for observation, the system first reads data from the 1553B sub-address memory and performs safety checks to ensure the commanded position is within the operational range of the LSM. If the position is valid, the operation proceeds.

The LSM moves in a 1/64 micro stepping mode, and at each step, the system performs a position check to verify whether the commanded position has been reached. Upon reaching the target position, the system triggers an "exposure on" command to activate the detector for capturing the solar corona. After the predefined observation period, an "exposure off" command is issued to conclude the data acquisition process. Following each exposure cycle, the system evaluates whether all planned observations have been successfully completed. If the observation sequence is complete, the system transit to an idle state, remaining ready to accept new payload operations. If additional observations are required, the process iterates until all planned operations are executed.

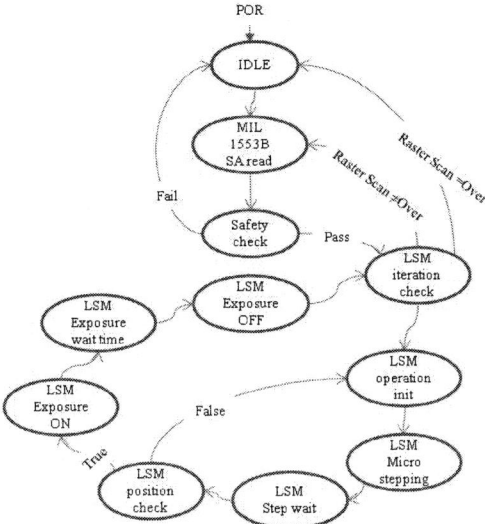

Figure 9: Flow diagram of Payload operation

VII. INORBIT PERFORMATION VALIDATION

The onboard hardware, after successful testing and evaluation, was integrated into the ADITYA-L1 spacecraft.

The payload operation was performed in raster scan mode for solar corona spectroscopy observations, as illustrated in Figure 10. The results demonstrate that the motor steps actuated by the proposed design closely align with the optical encoder measurements, achieving a positional accuracy within ±2 μm.

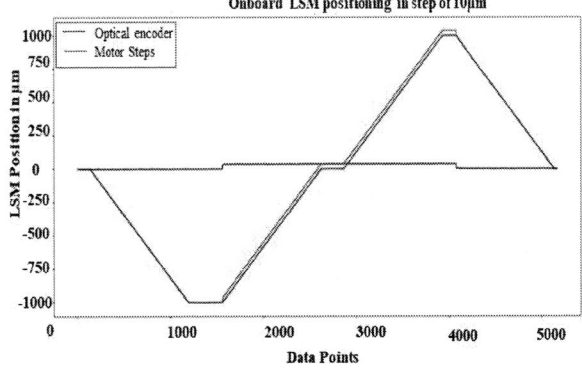

Figure 10: LSM Positioning during Payload Raster-Scan observation

VIII. CONCLUSTION

The precision control system presented in this paper enables precise micro-focusing of scanning mirrors within the VELC payload, meeting the stringent requirements for accuracy and repeatability. The system achieves accurate positioning with high precision. Experimental validation confirms the effectiveness and reliability of the precision control system for micro-focusing applications in the VELC payload.

Future work will focus on further enhancements to the control algorithm and the integration of advanced sensing technologies to facilitate autonomous adjustments. Additionally, research efforts will explore the potential application of machine learning techniques for adaptive control, aiming to optimize the performance of the linear scan mechanism in varying environmental conditions.

We would like to acknowledge IIA, Bangalore and Mechanism team, URSC for the support through out design and test process.

REFERENCES

[1] Raghavendra, V., et al., "Visible Emission Line Coronagraph (VELC) for ADITYA-L1 Mission," *Journal of Astrophysical Instrumentation*, vol. 113, no. 4, pp. 613-615, 2017.

[2] Mishra, S., K. S., VU, S., et al., Calibration of VELC detectors on-board Aditya-L1 mission, *Experimental Astronomy*, 57(7), 2024

[3] Muthu Priyal, et al., Data processing of visible emission line coronagraph on-board Aditya-L1, *Advances in Space Research*, 74(1), 2024

[4] Y.-S. Kung, C.-C. Huang, and L.-C. Huang, "FPGA-based Motion Control IC for linear motor drive X-Y table using adaptive fuzzy control", *Fuzzy controllers – recent advances in theory and applications, InTech*, 2012.

[5] Dhruti Ranjan Gaan, et al., "Real-Time Precise Position Tracking with Stepper Motor Using Frequency Modulation Based Microstepping" *IEEE Transactions on Industry Applications*, Vol. 54, No. 1, 2018

Dynamic Analysis of Quadratic Gain Single Input Dual Output DC-DC Converter

Divya K.
Department of Electronics and
Communication Engineering
*M. Kumarasamy College of
Engineering, Thalavapalayam*
Karur, Tamilnadu, India
divya88cool@gmail.com

Kumaravel S.
Department of Electrical Engineering,
*National Institute of Technology
Calicut*
Kerala, India
kumaravel_s@nitc.ac.in

K. Sundararaju
Department of Electrical and
Electronics Engineering
*M. Kumarasamy College of
Engineering, Thalavapalayam*
Karur, Tamilnadu, India
sunkrr@gmail.com

Abstract— This paper presents the dynamic analysis of a Single Input Dual Output (SIDO) DC-DC converter, focusing on its small-signal modeling, transient response, and stability assessment. The converter's dynamic behavior is analyzed to evaluate its performance under varying load conditions, input voltage fluctuations, and control variations. A state-space averaging technique is employed to derive the small-signal transfer functions, enabling precise controller design for voltage regulation. From the obtained transfer functions, the system exhibits stable poles in the left-half of the s-plane, ensuring that the converter operates in a stable region under nominal conditions. However, there are certain zeros on the Right-Hand Side (RHS) of the s-plane. A hardware prototype of the SIDO converter has been developed and assembled in a laboratory setup. Experimental validation is performed to validate the steady-state performance. The transient response is studied using *MATLAB/Simulink®*, verifying the converter's ability to maintain stable operation with minimal overshoot and settling time.

Keywords—DC-DC converter, Dynamic analysis, Single Input Dual Output, small-signal model

I. INTRODUCTION

Power converters, such as DC-DC converters, are widely used in various applications, including communication systems, electric vehicles, and renewable energy systems. The research on Single Input Multiple Output (SIMO) and Single Input Dual Output (SIDO) converters has gained significant attention due to their capability to efficiently distribute power across multiple loads. Various studies have been conducted to enhance their performance in terms of analytical modeling, control strategies, power-sharing mechanisms, and mitigation of cross-regulation issues [1].

Ref. [2] investigated the analytical modeling and control aspects of non-isolated SIMO Zeta–Buck–Boost converters. The study focused on deriving mathematical models and validating them through simulations and experimental results, demonstrating the practical feasibility of the proposed system. This reported work provided foundational insights into the dynamic behavior and control requirements of such converters. Further, Ref. [3] introduced a family of two-stage SIMO configurations, emphasizing power-sharing capabilities and control strategies. The study outlined how different topologies within this family could be optimized for various applications, providing crucial knowledge on improving the efficiency and adaptability of SIMO converters. A SIMO converter was presented in Ref. [4], specifically designed for electric vehicle applications. The

proposed topology successfully eliminated duty cycle constraints and cross-regulation issues, which are common challenges in conventional SIMO designs. This advancement made the converter more robust and suitable for high-performance automotive power systems. To achieve higher voltage gain, voltage balancing, and bidirectional operation, Enhanced Single Inductor-SIDO (SI-SIDO) converters were developed. These converters proved to be highly beneficial in applications such as bipolar low-voltage DC distribution and neutral-point-clamped inverters [5]. Their ability to maintain stable voltage levels while providing bidirectional power flow added significant value to modern power electronics applications.

Another major development in SIDO converters was the introduction of an Adaptive Estimator-Based Sliding Mode Control (AESMC) strategy. Ref. [6] proposed this control technique to address cross-coupling and cross-regulation issues, which arise due to the presence of coupled capacitors. By estimating disturbances in real-time and dynamically adjusting the control input, AESMC effectively stabilized the system, thereby improving the overall reliability and performance of SIDO converters. These advancements highlight the continuous evolution of SIMO and SIDO converters, driven by the need for higher efficiency, improved control strategies, and enhanced power management in multi-load systems. Future research in this domain is expected to focus on further reducing power losses, enhancing transient response, and integrating advanced control techniques for superior stability and regulation.

This research work focusses on the dynamic modeling of the SIDO converters presented in [7]. While steady-state analysis helps in understanding the converter's average performance, dynamic modeling is essential for analyzing transient response, stability, and control design. A dynamic model represents the converter's behavior under time-varying conditions, incorporating small-signal perturbations to evaluate system response, loop compensation, and stability margins. The model is typically derived using state-space equations, transfer functions, or averaged modeling techniques. By developing a dynamic model, designers can: (i) Analyze transient performance (e.g., response to load/line changes). (ii) Design and optimize control loops. (iii) Ensure stability using techniques like Bode plots and root locus. (iv) Predict the impact of parasitic elements on system dynamics. This paper presents the dynamic analysis of a Single Input Dual Output (SIDO) DC-DC converter, focusing on its small-signal modeling, transient response, and stability assessment.

979-8-3315-3013-6/25 $31.00 © 2025 IEEE

II. PROPOSED SINGLE INPUT AND DUAL OUTPUT DC-DC CONVERTER

The schematic representation of the proposed converter is illustrated in Fig. 1. The circuit configuration consists of four inductors (L_1, L_2, L_3, and L_4), four capacitors (C_1, C_2, C_{01}, and C_{02}), three switching devices (S_1, S_2, and S_3), and two diodes (D_1 and D_2). This DC-DC converter is structured with a single input and dual output arrangement, incorporating distinct input and output stages to ensure effective power conversion. The circuit has three modes of operation. The first mode is when all the switches are ON, and the second mode is when S_3 alone OFF, and third mode is when all switches are OFF.

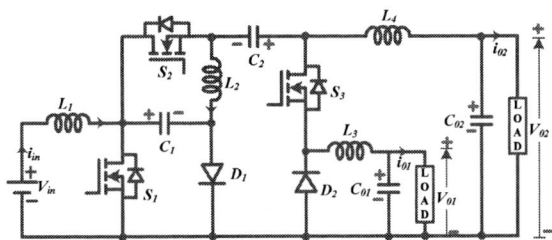

Fig. 1 Circuit diagram of the proposed converter

Mode 1 ($0 \leq t \leq \delta_2 T$): During Mode 1, all three switches (S_1, S_2, and S_3) are turned ON, and diodes D_1 and D_2 are reverse biased, as shown in Fig. 2. The key operations in this mode are: a) The input voltage V_{in} transfers energy to inductor L_1 through S_1. b) Capacitor C_1 transfers energy to inductor L_2 via S_2. c) Inductors L_3 and L_4 are energized with a voltage difference of V_{C2} and their respective output voltages. d) The capacitors C_{01} and C_{02} regulate the output voltages V_{01} and V_{02}. The analytical waveforms are shown in Fig. 3, and significant dynamic equations are given below.

$$L_1 \frac{di_{L1}}{dt} = V_{in} \tag{1}$$

$$L_2 \frac{di_{L2}}{dt} = v_{C1} \tag{2}$$

$$L_3 \frac{di_{L3}}{dt} = (v_{C2} - v_{C01}) \tag{3}$$

$$L_4 \frac{di_{L4}}{dt} = (v_{C2} - v_{C02}) \tag{4}$$

$$C_1 \frac{dv_{C1}}{dt} = -i_{L2} \tag{5}$$

$$C_2 \frac{dv_{C2}}{dt} = (-i_{L3} - i_{L4}) \tag{6}$$

$$C_{01} \frac{dv_{C01}}{dt} = \left(i_{L3} - \frac{v_{C01}}{R_{01}} - i_{01}\right) \tag{7}$$

$$C_{02} \frac{dv_{C02}}{dt} = \left(i_{L4} - \frac{v_{C02}}{R_{02}} - i_{02}\right) \tag{8}$$

Where R_{01} and R_{02} represent the loads at outputs V_{01} and V_{02}, respectively.

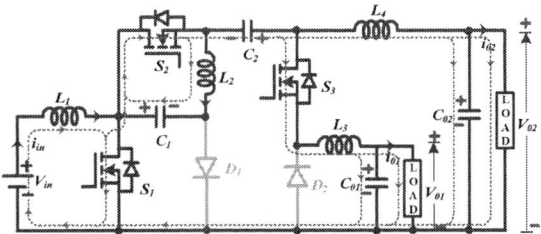

Fig. 2 Equivalent circuit of proposed converter for mode 1

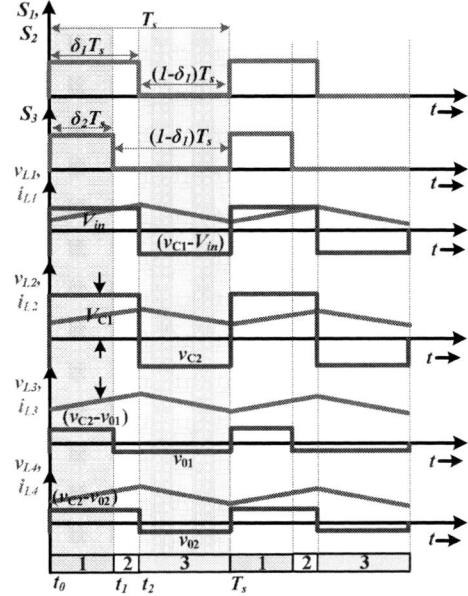

Fig. 3 Analytical waveforms during different modes of operation

Mode 2 ($\delta_2 T \leq t \leq \delta_1 T$): During Mode 2, switches S_1 and S_2 remain in the ON state, while S_3 is turned OFF as shown in Fig. 4. The energy continues to be transferred from the input source V_{in} and capacitor C_1 to the inductors L_1 and L_2, respectively. The current through inductor L_3 is freewheeled through diode D_2, maintaining the output voltage V_{01}. The inductor L_4 continues to transfer energy to load 2 side. Except for the state of inductor L_3 and capacitor C_2, all other passive components remain unchanged in Mode 2 compared to Mode 1. Therefore, the key equations are provided in (9) and (10).

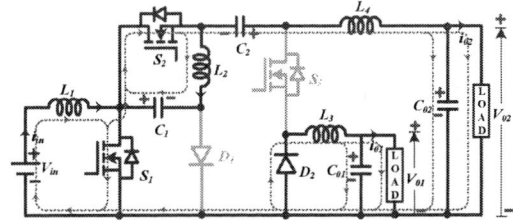

Fig. 4 Equivalent circuit of proposed converter for mode 2

$$L_3 \frac{di_{L3}}{dt} = -v_{C01} \tag{9}$$

$$C_2 \frac{dv_{C2}}{dt} = -i_{L4} \tag{10}$$

Mode 3 ($\delta_1 T \le t \le T$): During Mode 3, all switches (S_1, and S_2) are turned OFF, and S_3 is kept remain in the OFF state, as shown in Fig. 5. This causes the energy stored in the inductors L_1 and L_2 to be transferred to the capacitors C_1 and C_2, respectively, via the diodes D_1 and D_2. The inductor L_3 freewheels through D_1 while L_4 freewheels through D_1 and D_2, maintaining the output voltages V_{01} and V_{02}. This mode completes the switching cycle and ensures that energy stored in the inductors is effectively transferred to the output capacitors before the next cycle begins.

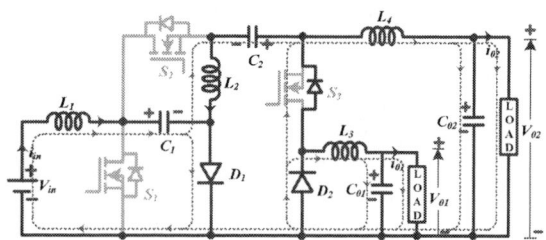

Fig. 5 Equivalent circuit of proposed converter for mode 3

$$L_1 \frac{di_{L1}}{dt} = (V_{in} - v_{C1}) \tag{11}$$

$$L_2 \frac{di_{L2}}{dt} = -v_{C2} \tag{12}$$

$$L_4 \frac{di_{L4}}{dt} = -v_{C02} \tag{13}$$

$$C_1 \frac{dv_{C1}}{dt} = i_{L1} \tag{14}$$

$$C_2 \frac{dv_{C2}}{dt} = i_{L2} \tag{15}$$

A power converter operates in different switching states, making its dynamics nonlinear. However, using state-space modeling, we can linearize it around an operating point and it is represented as a state equation $[\dot{x}] = [A][x] + [B][u]$ and output equation $[y] = [C][x] + [D][u]$. Where $x =$ State variables (i_{L1}, i_{L2}, i_{L3}, i_{L4}, v_{C1}, v_{C2}, v_{C01}, v_{C02}). $u =$ Input variables (V_{in}, i_{01}, i_{02}). $y =$ Output variables (v_{01}, v_{02}).

III. STATE-SPACE AVERAGING OF THE THREE MODES

State-space averaging is a mathematical technique used to analyze switching converters by transforming their time-dependent equations into an averaged form over one switching period. This method simplifies the complex nonlinear behavior of power converters, enabling the derivation of small-signal models essential for control design and stability analysis. Since the proposed SIDO DC-DC converter operates in three distinct modes during each switching cycle, its dynamic behavior can be described using a set of state equations for each mode. The state-space averaging technique is applied to obtain a single set of equations representing the overall system dynamics in an averaged form. The technique allows to linearize the system around an operating point, facilitating stability analysis and frequency response analysis. The overall behavior of the converter is obtained by weighting each mode's contribution by its respective duty cycle fraction over a full switching period. The state-space averaged model is given as follows.

$$[\dot{x}] = [A_{avg}][x] + [B_{avg}][u] \tag{16}$$

$$[y] = [C_{avg}][x] + [D_{avg}][u] \tag{17}$$

Where: $[A_{avg}] = [A_1]\delta_1 + [A_2](\delta_1 - \delta_2) + [A_3](1 - \delta_1)$; $[B_{avg}] = [B_1]\delta_1 + [B_2](\delta_1 - \delta_2) + [B_3](1 - \delta_1)$; $[C_{avg}] = [C_1]\delta_1 + [C_2](\delta_1 - \delta_2) + [C_3](1 - \delta_1)$; $[D_{avg}] = [D_1]\delta_1 + [D_2](\delta_1 - \delta_2) + [D_3](1 - \delta_1)$. Matrixes $[A_1]$, $[B_1]$, $[C_1]$ and $[D_1]$ are formed using (1) to (8) which are obtained from Mode 1 operation. Whereas, $[A_2]$, $[B_2]$, $[C_2]$ and $[D_2]$ & $[A_3]$, $[B_3]$, $[C_3]$ and $[D_3]$ are obtained from the operation Mode 2 and Mode 3, respectively. Substituting those matrixes obtained from various modes in (16) and (17), the average matrixes $[A_{avg}]$, $[B_{avg}]$, $[C_{avg}]$ and $[D_{avg}]$ are obtained, as given below.

$$[A_{avg}] = \begin{bmatrix} 0 & 0 & 0 & 0 & -\frac{(1-\delta_1)}{L_1} & 0 & 0 & 0 \\ 0 & 0 & 0 & 0 & \frac{\delta_1}{L_2} & -\frac{(1-\delta_1)}{L_2} & 0 & 0 \\ 0 & 0 & 0 & 0 & 0 & \frac{\delta_2}{L_3} & -\frac{1}{L_3} & 0 \\ 0 & 0 & 0 & 0 & 0 & \frac{\delta_1}{L_4} & 0 & -\frac{1}{L_4} \\ \frac{(1-\delta_1)}{C_1} & -\frac{\delta_1}{C_1} & 0 & 0 & 0 & 0 & 0 & 0 \\ 0 & \frac{(1-\delta_1)}{C_2} & -\frac{\delta_2}{C_2} & \frac{\delta_1}{C_2} & 0 & 0 & 0 & 0 \\ 0 & 0 & \frac{1}{C_{01}} & 0 & 0 & 0 & -\frac{1}{R_{01}C_{01}} & 0 \\ 0 & 0 & 0 & \frac{1}{C_{02}} & 0 & 0 & 0 & -\frac{1}{R_{02}C_{02}} \end{bmatrix} \tag{18}$$

$$[B_{avg}] = \begin{bmatrix} 1 & 0 & 0 \\ 0 & 0 & 0 \\ 0 & 0 & 0 \\ 0 & 0 & 0 \\ 0 & 0 & 0 \\ 0 & 0 & 0 \\ 0 & 1 & 0 \\ 0 & 0 & 1 \end{bmatrix}; [C_{avg}] = \begin{bmatrix} 0 & 0 \\ 0 & 0 \\ 0 & 0 \\ 0 & 0 \\ 0 & 0 \\ 0 & 0 \\ 1 & 0 \\ 0 & 1 \end{bmatrix}^T; [D_{avg}] = \begin{bmatrix} 0 & 0 & 0 \\ 0 & 0 & 0 \\ 0 & 0 & 0 \\ 0 & 0 & 0 \\ 0 & 0 & 0 \\ 0 & 0 & 0 \\ 0 & 0 & 0 \\ 0 & 0 & 0 \end{bmatrix} \tag{19}$$

IV. SMALL-SIGNAL MODEL OF PROPOSED SIDO CONVERTER

To analyze the dynamic behavior of the SIDO DC-DC converter, the next step involves developing its small-signal model. Small-signal modeling simplifies the analysis of nonlinear circuits, like the SIDO converter, by linearizing them around a specific operating point. This approach allows for the use of linear system techniques to assess the converter's response to small perturbations, facilitating the design of controllers that can maintain desired performance and stability.

Small-signal model involves perturbing the state-space averaged equations around their steady-state operating point and linearizing them to capture the system's response to small variations in input and control variables. Assume that each state variable and input consist of a steady-state (DC) component and a small-signal (AC) perturbation as follows.

$i_{L1} = I_{L1} + \hat{i}_{L1}$; $i_{L2} = I_{L2} + \hat{i}_{L2}$; $i_{L3} = I_{L3} + \hat{i}_{L3}$; $i_{L4} = I_{L4} + \hat{i}_{L4}$;

$v_{C1} = V_{C1} + \hat{v}_{C1}$; $v_{C2} = V_{C2} + \hat{v}_{C2}$; $v_{C01} = V_{C01} + \hat{v}_{C01}$; $v_{C02} = V_{C02} + \hat{v}_{C02}$;

$v_{in} = V_{in} + \hat{v}_{in}$; $d_1 = D_1 + \hat{d}_1$; $d_2 = D_2 + \hat{d}_2$

Here, uppercase letters denote steady-state values, and hatted variables represent small perturbations. Substitute the perturbed variables into the state-space averaged equations and linearize by neglecting higher-order small-signal terms. This yields a set of linear differential equations describing the small-signal behavior as given in (20) and (21). Where $[A_{ss}]$, $[B_{ss}]$, $[C_{ss}]$, and $[D_{ss}]$ are the small-signal state-space matrices. By applying the Laplace transform to the linearized equations and solving for the output perturbations in terms of input perturbations, the final small-signal transfer functions are obtained for the considered design of the proposed converter given in Table I, as given from (23) to (32). High-power RF transmitters necessitate a high-voltage supply for the power amplifier stages, with 400 V being within the plausible range. Concurrently, the other circuits required 155 V as an auxiliary supply. By considering the above applications, the specification of various components is given in Table I.

$$\begin{bmatrix} \hat{\dot{i}}_{L1} \\ \hat{\dot{i}}_{L2} \\ \hat{\dot{i}}_{L3} \\ \hat{\dot{i}}_{L4} \\ \hat{\dot{v}}_{C1} \\ \hat{\dot{v}}_{C2} \\ \hat{\dot{v}}_{C01} \\ \hat{\dot{v}}_{C02} \end{bmatrix} = [A_{ss}] \begin{bmatrix} \hat{i}_{L1} \\ \hat{i}_{L2} \\ \hat{i}_{L3} \\ \hat{i}_{L4} \\ \hat{v}_{C1} \\ \hat{v}_{C2} \\ \hat{v}_{C01} \\ \hat{v}_{C02} \end{bmatrix} + [B_{ss}] \begin{bmatrix} \hat{v}_{in} \\ \hat{d}_1 \\ \hat{d}_2 \end{bmatrix} \quad (20)$$

$$\begin{bmatrix} \hat{v}_{01} \\ \hat{v}_{02} \end{bmatrix} = [C_{ss}] \begin{bmatrix} \hat{i}_{L1} \\ \hat{i}_{L2} \\ \hat{i}_{L3} \\ \hat{i}_{L4} \\ \hat{v}_{C1} \\ \hat{v}_{C2} \\ \hat{v}_{C01} \\ \hat{v}_{C02} \end{bmatrix} + [D_{ss}] \begin{bmatrix} \hat{v}_{in} \\ \hat{d}_1 \\ \hat{d}_2 \end{bmatrix} \quad (21)$$

The stability of the proposed SIDO DC-DC converter was analyzed using the small-signal transfer functions derived from state-space averaging. From the obtained transfer functions, the system exhibits stable poles in the left-half of the s-plane, ensuring that the converter operates in a stable region under nominal conditions. However, there are certain zeros on the Right-Hand Side (RHS) of the s-plane. Hence, the stability in the closed loop may be affected in the closed loop operation. Hence, a suitable controller shall be designed to operate the converter in closed loop with stability.

TABLE I PARAMETERS OF THE PROPOSED CONVERTER

Parameters	Value	Parameters	Value
Input source V_{in}	48 V	Inductor L_2	3.33 mH
Output voltage 1 V_{01}	350 V	Inductor L_3	6.59 mH
Output voltage 2 V_{02}	155 V	Inductor L_4	13.26mH
Output power P_{01}	200 W	Capacitor C_1	26.34 μF
Output power P_{02}	200 W	Capacitor C_2	2.63 μF
Switching frequency f_s	50 kHz	Capacitor C_{01}	8.17 μF
Duty cycles δ_1 and δ_2 (Ideal)	73% 33%	Capacitor C_{02}	1 μF
Inductor L_1 with 10% ripple	1 mH		

$$\frac{\hat{v}_{01}(s)}{\hat{v}_{in}(s)} = \frac{1.42\times10^{21}s^2 + 3.50\times10^{24}s + 1.62\times10^{29}}{s^8 + 3434s^7 + 1.74\times10^8 s^6 + 2.90\times10^{11}s^5 + 5.71\times10^{15}s^4 + 4.83\times10^{18}s^3 + 4.90\times10^{22}s^2 + 1.60\times10^{25}s + 4.90\times10^{28}} \quad (23)$$

$$P(s) = s^8 + 3434s^7 + 1.74\times10^8 s^6 + 2.90\times10^{11}s^5 + 5.71\times10^{15}s^4 + 4.83\times10^{18}s^3 + 4.90\times10^{22}s^2 + 1.60\times10^{25}s + 4.90\times10^{28}$$

$$\left(\hat{v}_{02}(s)/\hat{v}_{in}(s)\right) = \left((1.93\times10^{22}s^2 + 1.88\times10^{25}s + 3.58\times10^{29})/P(s)\right) \quad (24)$$

$$\left(\hat{v}_{01}(s)/\hat{d}_1(s)\right) = \left((-8.51\times10^{12}s^5 + 4.21\times10^{16}s^4 - 9.52\times10^{20}s^3 + 1.40\times10^{25}s^2 - 1.53\times10^{28}s + 6.83\times10^{31})/P(s)\right) \quad (25)$$

$$\left(\hat{v}_{02}(s)/\hat{d}_1(s)\right) = \left((5.49\times10^{10}s^6 - 6.24\times10^{13}s^5 + 3.89\times10^{18}s^4 - 1.09\times10^{21}s^3 + 5.98\times10^{25}s^2 - 2.21\times10^{28}s + 1.75\times10^{32})/P(s)\right) \quad (26)$$

$$\left(\hat{v}_{01}(s)/\hat{d}_2(s)\right) = \left((8.93\times10^9 s^6 + 1.90\times10^{13}s^5 + 1.30\times10^{18}s^4 + 3.52\times10^{20}s^3 + 1.89\times10^{25}s^2 + 5.08\times10^{26}s + 2.36\times10^{31})/P(s)\right) \quad (27)$$

$$\left(\hat{v}_{02}(s)/\hat{d}_2(s)\right) = \left((-4.00\times10^{13}s^5 - 8.02\times10^{17}s^4 - 1.84\times10^{21}s^3 - 7.11\times10^{24}s^2 - 1.32\times10^{28}s + 3.58\times10^{15})/P(s)\right) \quad (28)$$

$$\left(\hat{v}_{01}(s)/\hat{i}_{01}(s)\right) = \left((-1.22\times10^5 s^6 - 3.01\times10^8 s^6 - 1.87\times10^{13}s^5 - 1.17\times10^{16}s^4 - 3.54\times10^{20}s^3 - 6.45\times10^{22}s^2 - 1.10\times10^{27}s - 8.65\times10^{13})/P(s)\right) \quad (29)$$

$$\left(\hat{v}_{02}(s)/\hat{i}_{01}(s)\right) = \left((-1.94\times10^{20}s^3 + 3.27\times10^7 s^2 - 1.72\times10^{27}s + 7.21\times10^{13})/P(s)\right) \quad (30)$$

$$\left(\hat{v}_{01}(s)/\hat{i}_{02}(s)\right) = \left((-1.94\times10^{20}s^3 + 1.59\times10^8 s^2 - 1.72\times10^{27}s - 1.67\times10^{14})/P(s)\right) \quad (31)$$

$$\left(\hat{v}_{02}(s)/\hat{i}_{01}(s)\right) = \left((-1.52\times10^6 s^6 - 1.47\times10^9 s^6 - 8.69\times10^{13}s^5 - 5.72\times10^{16}s^4 - 1.24\times10^{21}s^3 - 3.16\times10^{23}s^2 - 4.46\times10^{27}s - 5.65\times10^{14})/P(s)\right) \quad (32)$$

V. EXPERIMENTAL VALIDATION

A hardware prototype of the SIDO converter has been developed and assembled in a laboratory setup using essential components such as a programmable power supply, load resistors, pulse generator, and digital storage oscilloscope (DSO), as shown in Fig. 6. The programmable source is configured to provide an input voltage of 48 V, while the load resistors are set at 120 Ω and 800 Ω to achieve output power levels of 200 W in each load. To control the switching operation, three gate pulses are generated from a programmable pulse generator operating at a switching frequency of 50 kHz. These pulses are fed to the MOSFETs via gate driver circuits. The duty cycles δ_1 and δ_2 are set to 73% and 33%, respectively, ensuring the desired output voltages of 155 V and 350 V from the 48 V input.

The voltage and current waveforms from various circuit components are captured using the DSO and are depicted in Fig. 7. As shown in Fig. 7, the converter successfully produces output voltages of approximately 350 V and 155 V for the specified input conditions. Waveforms shown in Fig. 7(a) validate the boost operation performed by input source V_{in} (48 V), inductor L_1 ($V_{L1Pk-Pk}$ = 177 V and I_{L1} = 4.70 A) and capacitor C_1 (V_{C1} = 177 V). Additionally, Fig. 7(b) illustrates that the measured inductor voltages during charging and discharging phases align closely with the analytical waveforms presented in Fig. 3. Similarly, the measured inductor currents confirm well to the theoretical expressions derived. The output voltage and current waveforms displayed in Fig. 7(c) and Fig. 7(d) validate that the converter produces the expected output voltage of load 1 as 155 V and load 2 as 350 V.

VI. VALIDATION OF THE DERIVED SMALL-SIGNAL MODEL

Once the small-signal model of the SIDO DC-DC converter is derived, it must be validated to ensure its accuracy and reliability in predicting the converter's dynamic behavior. The validation process typically involves simulation-based verification, frequency response analysis, and experimental testing. The block diagrams of the dynamic model (plant model) of the proposed SIDO converter in terms of output voltage 1 (v_{o1}) and output voltage 2 (v_{o2}) are shown in Fig. 8(a) and 8(b), respectively. The control inputs $\hat{d}_1$ and $\hat{d}_2$, along with the considered disturbances are indicated in the block diagram.

Using *MATLAB/Simulink/ Power System Block Sets*, the proposed SIDO converter is developed in a simulation environment. The various components used in the converter are selected based on the technical specifications given in Table I. Loads 1 and 2 are designed to operate at 150 V, 200 W, and 350 V, 200 W, respectively. The control signals are generated using a PWM block with a switching frequency of 50 kHz. The input source voltage is set to 48 V. To validate the derived transfer function, perturbations in the input voltage, duty cycle, and load currents are considered as given in Table II.

To observe the dynamic response and output voltages, the simulation of the SIDO converter is carried out by considering an input voltage V_{in} of 38 V, with duty cycles δ_1 and δ_2 are set to 73% and 33%, respectively, and each load rated at 200 W. The simulation is initiated with these parameters ensuring that the converter reaches a steady-state

condition. At instant t_1, the input voltage is varied from 38 V to 48 V while keeping all other parameters constant. The significant waveforms, such as variations in the input voltage and the corresponding changes in load voltages V_{o1} and V_{o2}, are observed and presented in Fig. 8(c) and 7(d). Similar to the procedure followed for observing the dynamic response to input voltage variations, a repeated simulation study has been conducted to analyze the dynamic response for other step variations mentioned in Table II. The observed simulation waveforms for the considered step variations are presented from Fig. 8(e) to Fig. (i). These waveforms confirm that the dynamic response of all the derived transfer

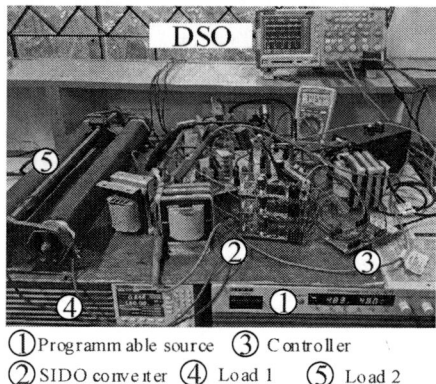

① Programmable source ③ Controller
② SIDO converter ④ Load 1 ⑤ Load 2

Fig. 6. Hardware prototype of the SIDO converter in laboratory setup

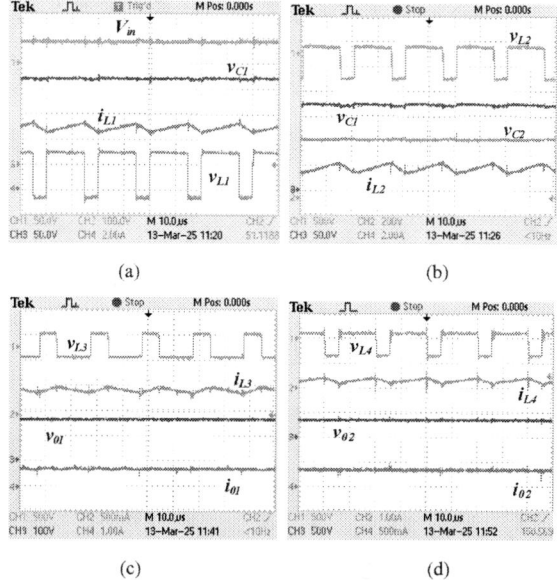

Fig. 7 Experimental results (a) V_{in}, i_{L1}, v_{L1} & v_{C1} (b) v_{L2}, i_{L2}, v_{C1}, & v_{C2} (c) v_{L3}, i_{L3}, v_{o1}, & i_{o1} (d) v_{L4}, i_{L4}, v_{o2}, V_{o2} & i_{o2}

TABLE II CONSIDERED PERTURBATIONS TO VALIDATE THE DERIVED TRANSFER FUNCTIONS

Parameter	Considered variation
Step change in input voltage	38 V to 48 V
Step change in duty cycle 1	73% to 72%
Step change in duty cycle 2	33% to 32%
Step change in load current 1	200 W to 150 W
Step change in load current 2	200 W to 150 W

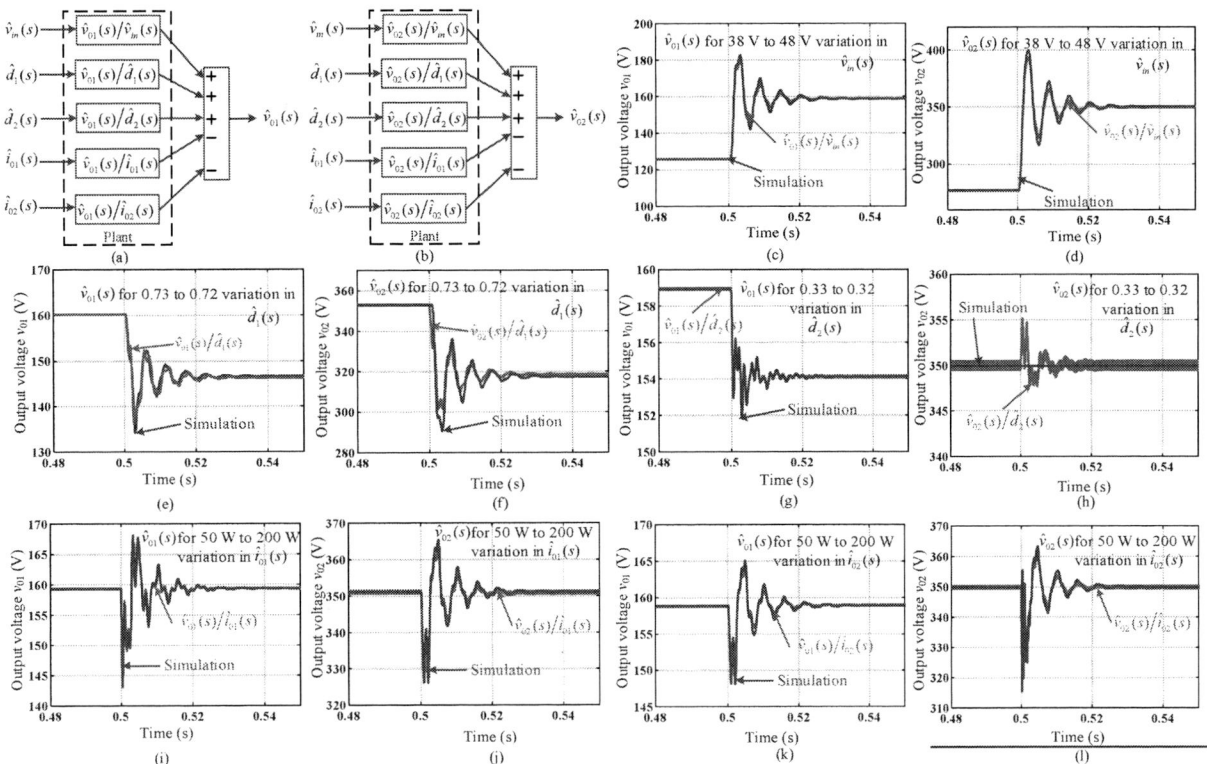

Fig. 8 Validation of dynamic model of SIDO converter (a) Block diagram of $\hat{v}_{01}(s)$ with respect to other perturbations (b) Block diagram of $\hat{v}_{02}(s)$ with respect to other perturbations (c) $\hat{v}_{01}(s)$ for variations in $\hat{v}_{in}(s)$ (d) $\hat{v}_{02}(s)$ for variations in $\hat{v}_{in}(s)$ (e) $\hat{v}_{01}(s)$ for variations in $\hat{d}_1(s)$ (f) $\hat{v}_{02}(s)$ for variations in $\hat{d}_1(s)$ $\hat{v}_{in}(s)$ (g) $\hat{v}_{01}(s)$ for variations in $\hat{d}_2(s)$ (h) $\hat{v}_{02}(s)$ for variations in $\hat{d}_2(s)$ (i) $\hat{v}_{01}(s)$ for variations in $\hat{i}_{01}(s)$ (f) $\hat{v}_{02}(s)$ for variations in $\hat{i}_{01}(s)$ $\hat{v}_{in}(s)$ (g) $\hat{v}_{01}(s)$ for variations in $\hat{i}_{02}(s)$ (h) $\hat{v}_{02}(s)$ for variations in $\hat{i}_{02}(s)$

functions of the SIDO converter closely matches with the dynamic responses obtained from *MATLAB/Simulink*.

VII. CONCLUSION

This paper presented the dynamic modeling, small-signal analysis, and experimental validation of a Single-Input Dual-Output (SIDO) DC-DC converter. The proposed converter successfully achieved output voltages of 350 V and 155 V from a 48 V input, with output power levels of 200 W for each load. The state-space averaging technique was employed to derive the system's mathematical model, and small-signal transfer functions were validated through simulations and experiments. A step change in input voltage from 38 V to 48 V resulted in a corresponding variation in output voltages, closely matching the predicted response. Similarly, variations in duty cycles (δ_1: 73% to 72%, δ_2: 33% to 32%) and load currents (200 W to 150 W) demonstrated the system's stability and robustness. The experimental waveforms aligned well with theoretical predictions, confirming the accuracy of the derived models and the efficiency of the proposed converter. Suitable controller development for the closed loop operation of the SIDO converter is proposed as a future scope of the work.

REFERENCES

[1] M. Y. Hassani, M. Maalandish and S. H. Hosseini, "A New Single-Input Multioutput Interleaved High Step-Up DC–DC Converter for Sustainable Energy Applications," in IEEE Transactions on Power Electronics, vol. 36, no. 2, pp. 1544-1552, Feb. 2021

[2] S. Markkassery, A. Saradagi, A. D. Mahindrakar, N. Lakshminarasamma and R. Pasumarthy, "Modeling, Design and Control of Non-isolated Single-Input Multi-Output Zeta–Buck–Boost Converter," in IEEE Transactions on Industry Applications, vol. 56, no. 4, pp. 3904-3918, July-Aug. 2020.

[3] X. L. Li, Z. Dong and C. K. Tse, "Complete Family of Two-Stage Single-Input Multioutput Configurations of Interconnected Power Converters," in IEEE Transactions on Power Electronics, vol. 35, no. 4, pp. 3713-3728, April 2020.

[4] M. Dhananjaya, D. Ponuru, T. S. Babu, B. Aljafari and H. H. Alhelou, "A New Multi-Output DC-DC Converter for Electric Vehicle Application," in IEEE Access, vol. 10, pp. 19072-19082, 2022.

[5] P. Aghakhanlou, F. Falahi, A. Nadermohammadi, S. M. Hashemzadeh, S. H. Hosseini and E. Babaei, "A Single Switch Ultra-High Step-Up DC-DC Converter Based on a Coupled Inductor with Two Output Ports for Renewable Energy Applications," 2024 9th International Conference on Technology and Energy Management (ICTEM), Behshar, Mazandaran, Iran 2024, pp. 1-6

[6] B. Rooholahi, Y. P. Siwakoti, H. -G. Eckel, F. Blaabjerg and A. S. Bahman, "Enhanced Single-Inductor Single-Input Dual-Output DC–DC Converter With Voltage Balancing Capability," in IEEE Transactions on Industrial Electronics, vol. 71, no. 7, pp. 7241-7251, July 2024

[7] L. Senapati, A. K. Panda, M. M. Garg and R. K. Lenka, "An Adaptive Estimator Based Sliding Mode Control of Nonisolated Single-Input Double-Output Cuk Converter," in IEEE Journal of Emerging and Selected Topics in Industrial Electronics, vol. 4, no. 2, pp. 482-491, April 2023

[8] Divya K., Kumaravel S., K. Sundararaju and A. Kavitha, "Design and Development of Quadratic Gain Single Input Dual Output DC-DC Converter for Communication System Application," 2024 4th International Conference on Emerging Frontiers in Electrical and Electronic Technologies (ICEFEET), Patna, India, 2024, pp. 1-6, doi: 10.1109/ICEFEET64463.2024.10866476.

979-8-3315-3013-6/25 $31.00 © 2025 IEEE

Thermal and Reliability Analysis of a Dual-Switch Switched Inductor-Capacitor Network-Based High-Gain DC-DC Converter

Vinu Varshath S.
Department of Electrical Engineering
National Institute of Technology Calicut
Calicut, India
vinuvarshaths@gmail.com

Kumaravel S.
Department of Electrical Engineering
National Institute of Technology Calicut
Calicut, India
kumaravel_s@nitc.ac.in

Abstract— **This paper presents a comprehensive thermal and reliability analysis of a Dual-Switch Switched Inductor-Capacitor Network (DSSLCN)-based high-gain DC-DC converter. The converter topology is analysed under both Continuous Conduction Mode (CCM) and Discontinuous Conduction Mode (DCM), with detailed mathematical derivations for voltage gain and boundary conditions. A hardware prototype rated for 200 W, 400 V, and 50 kHz switching frequency is developed to validate the analytical findings through experimental measurements. Loss analysis quantifies power dissipation in inductors, capacitors, switches, and diodes, identifying key sources of inefficiency. Thermal performance is evaluated under different load conditions (100 W, 200 W, and 300 W), revealing the impact of power dissipation on component temperatures and highlighting the importance of thermal management. The reliability assessment employs an exponential failure rate model, incorporating voltage stress, thermal effects, and environmental factors to estimate component lifespan. The analysis demonstrates that higher duty ratios and prolonged operation accelerate MOSFET and diode degradation, reducing overall system reliability.**

Keywords— *High-gain DC-DC converter, thermal analysis, reliability analysis, discontinuous conduction mode, failure rate.*

I. INTRODUCTION

High-gain DC-DC converters are widely employed in applications requiring efficient power conversion, such as renewable energy systems, Electric Vehicles (EVs), and industrial DC microgrids. These applications demand not only high voltage conversion ratios but also improved reliability and thermal performance to ensure long-term operation under varying load conditions. The Dual-Switch Switched Inductor-Capacitor (SLC) Network-Based High-Gain DC-DC Converter has been introduced as an effective solution for achieving high voltage gain while maintaining reduced switch current stress and continuous input current [1][2].

However, to ensure the practical viability of such converters in real-world scenarios, it is essential to conduct a comprehensive thermal and reliability analysis. Reliability and thermal analysis of DC-DC converters have been extensively studied in recent years, with a focus on improving efficiency, longevity, and robustness in the aforementioned applications.

Several studies have emphasized the importance of reliability in DC-DC converters. Ref. [2] analysed the reliability and performance degradation of a boost converter, showing that component aging, particularly in MOSFETs and capacitors, significantly affects system performance over time. Similarly, [3] explored the reliability of multistage boost

converters connected to photovoltaic (PV) panels. It considered various failure rate models for semiconductors and passive components, highlighting how the number of stages influences system longevity and cost-effectiveness. Another approach to reliability modelling was presented in [4], which used Monte Carlo simulations and machine learning regression algorithms to assess power converter reliability in different operational scenarios.

Temperature is a critical factor in the degradation of power converters. Ref. [5] investigated the impact of thermal stress on DC-DC converters, particularly the aging effects between IGBTs and diodes. This study demonstrated that coupled thermal structures lead to accelerated aging of components due to increased junction temperatures. Moreover, the study of reconfigurable DC-DC converters by [6] proposed methods to optimize both efficiency and reliability by dynamically adjusting converter operation based on thermal conditions. This research aligns with findings that thermal stress directly influences Mean Time To Failure (MTTF) and overall converter lifespan. Several studies have examined the role of duty ratio and load conditions on converter reliability [7]. investigated how variations in duty ratio and input voltage affect the reliability of buck-boost converters using a Markov process model. Their results indicated that converters operating at high duty ratios experience increased voltage stress, reducing component lifespan.

This paper provides a comprehensive thermal and reliability assessment of the Dual-Switch Switched Inductor-Capacitor Network (DSSLCN) Converter, focusing on its feasibility for high-power applications. It covers steady state analysis, loss analysis, thermal performance evaluation, and reliability modelling using an exponential failure rate approach. By integrating analytical modelling and practical testing, this study offers key insights for optimizing efficiency and long-term stability in high-gain DC-DC converters.

II. HIGH GAIN DC-DC CONVERTER WITH CONTINUOUS INPUT CURRENT

Fig. 1 shows the circuit configuration of the DSSLCN Converter [2]. Compared to the conventional converter with a similar gain [1], it consists of similar reactive elements, three diodes and two switches. The circuit arrangement results to overcome the higher current stress on the switches. Synchronous duty is considered for analysis. Based on the on and off status of the switches, the circuit has two modes of operation under CCM considering. It is assumed that both the switches are turned on and turned off simultaneously. The equivalent circuit of the modes are shown in Fig. 2 The analytical waveforms corresponding to these modes of operation are shown in Fig. 3.

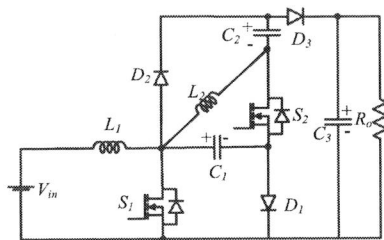

Fig. 1. Circuit diagram of DSSLCN converter

Mode *1*: For the interval $t_0 \le t \le t_1$, the gating pulse is applied to both the switches. The input supply energizes the inductor L_1, and the capacitor C_1 energizes the inductor L_2 and charges the capacitor C_2 through diode D_2 simultaneously. The diode D_1 reverse biased by the voltage across the capacitor C_1 and C_2. The diode D_3 is also reverse biased by capacitor C_3. The capacitors C_3 discharge to the load in this mode.

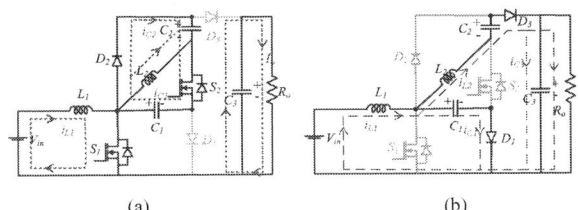

(a)   (b)

Fig. 2 Equivalent circuit of DSSLCN (a) mode 1 (b) mode 2

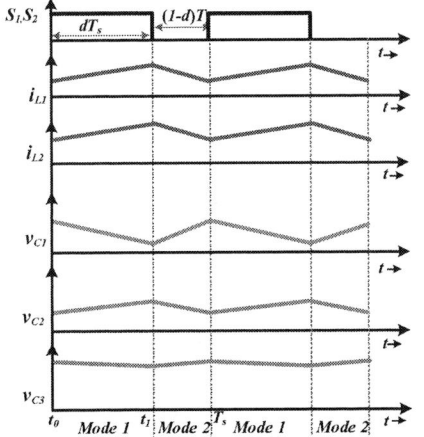

Fig. 3. Analytical waveforms under steady state in CCM

Mode 2: For interval $t_1 \le t \le t_2$, the gating pulse is removed from both the switches. The inductors L_1 and L_2 which were energized in the previous cycle deenergizes along with the discharging capacitor C_2 to the load. Diode D_1 is forward biased by voltage of L_1 and L_2. Diode D_3 is also forward biased by voltage of capacitor C_2. However, diode D_2 is reverse biased by voltage of inductor L_2 and capacitor C_2. Here, the capacitors C_1 and C_3 are charged by inductors L_1 and L_2, respectively. Now applying ampere-second balance and volt-second balance, the expressions of various currents and voltages are given from (1) to (4).

$$v_{C_2} = v_{C_1} = \frac{V_{in}}{1-d} \qquad (1)$$

$$V_o = v_{C_3} \qquad (2)$$

$$i_{L_2} = \frac{i_0}{1-d} \qquad (3)$$

$$i_{L_1} = \frac{i_0 (2-d)}{(1-d)^2} \qquad (4)$$

Further simplification leads to the equation for voltage gain in CCM as expressed in (5).

$$\frac{v_o}{v_{in}} = \frac{2-d}{(1-d)^2} \qquad (5)$$

III. DCM ANALYSIS

Unlike CCM, where the inductor current never drops to zero, DCM operation occurs when the energy stored in the inductor during the switch ON state is completely transferred to the output before the end of the switching cycle. The DCM operation introduces a third mode (Mode 3) in addition to the two modes observed in CCM (Mode 1 and Mode 2). The inductor L_2 is considered to fall into discontinuous mode since it carries lower current than L_2.

Mode 3:. In this mode, the current in L_2 falls to zero before the end of the switching period. The load current is supplied only by the output capacitors C_2 and C_3, leading to a floating inductor condition as shown in Fig. 4.

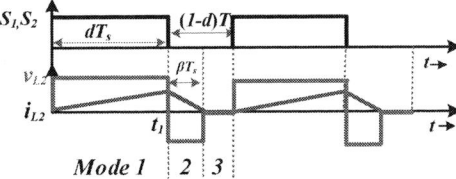

Fig. 4. Analytical waveforms under steady state in DCM

The volt-second balance across inductor L_2 is written as,

$$dV_{C_1} = \beta (V_{C_1} + V_{C_2} + V_o) \qquad (6)$$

Substituting the values of capacitor voltages in (6), the gain in DCM is obtained as expressed in (7)

$$\frac{V_o}{V_{in}} = G_{DCM} = \frac{2\tau + \sqrt{2\tau (d^2 + 2T)}}{2\tau (1-d)} \qquad (7)$$

$$\tau = \frac{f_s L}{R} \qquad (8)$$

Now, by equating the DCM and CCM gains, the border between CCM and DCM is obtained as given in (9).

$$\tau = \frac{d (1-d)^2}{2 (2-d)} \qquad (9)$$

This τ is plotter against duty ratio to get the boundary between CCM and DCM, as shown in Fig. 5. At the duty ratio of 58.84%, the critical load is found to be 5 kΩ.

979-8-3315-3013-6/25 $31.00 © 2025 IEEE   217

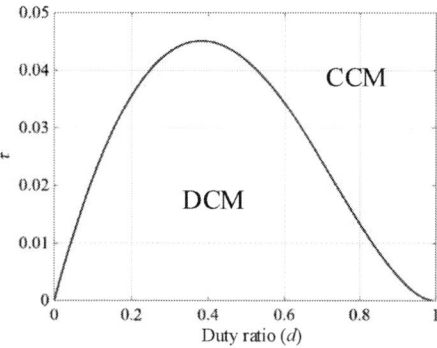

Fig. 5 Boundary between CCM and DCM

IV. EXPERIMENTAL VALIDATION

A 200 W, 400 V, 50 kHz hardware prototype is built to test the converter's performance for mentioned technical specifications in Table I. Plugin gate drivers are fabricated to increase the power density of the converter and avoid long wires for the gate driver. Switch S_1 and S_2 are realized using SiC MOSFETs with low on state resistance and good high-frequency switching performance. Schottky diodes prevent reverse recovery. A programmable load is used to realize a load of 800 Ω. The experimental setup is in Fig. 6.

The observed waveforms (Fig. 7 (a)) confirm accuracy of the analytical waveforms. According to the boundary between DCM and CCM, the minimum load required for the converter to operate in DCM in 5 kΩ. This is realised using the programmable load and it can be observed (shown in Fig. 7(b)) that the converter is at boundary between CCM and DCM further increasing R_0 makes the converter in DCM as shown in Fig. 7(c); hence validating the DCM analysis.

V. LOSS ANALYSIS

The power losses in the circuit components are analyzed based on their respective contributions. Inductor losses (P_L) depend on the RMS currents through L_1 and L_2 while capacitor losses (P_c) arise from ESR and the RMS currents through C_1, C_2, and C_3 Switch losses consist of switching loss (P_{S-Sw}),influenced by turn-on/off times and switching frequency (f_s), and conduction loss (P_{S-Cond}), which depends on the average and RMS currents through S_1 and S_2. Diode losses include forward voltage and conduction losses, determined by the currents through D_1, D_2 and D_3The total power loss of the converter is the sum of these individual losses, as given in Table II and III.

TABLE I. SPECIFICATIONS AND DESIGN

Parameter	Rating
Input voltage (V_{in})	48 V
Output voltage (V_o)	400 V
Rated output power (P_o)	200 W
Switching frequency (f_s)	50 kHz
Capacitances C_1, C_2, C_3	10 μF,10 μF, 5.6 μF
Inductances L_1, L_2	3 mH, 7.5 mH
MOSFET switchs	SCTW35N65G2V
Diodes	STPSC4H065

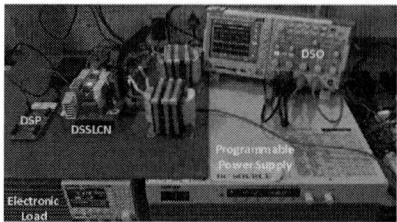

Fig. 6 Experimental setup

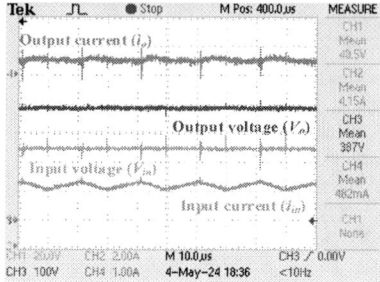

(a)

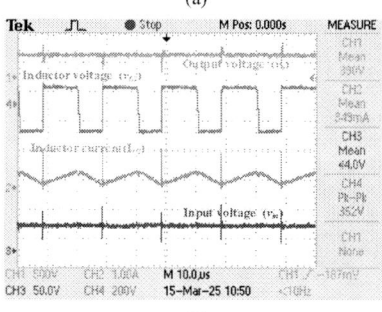

(b)

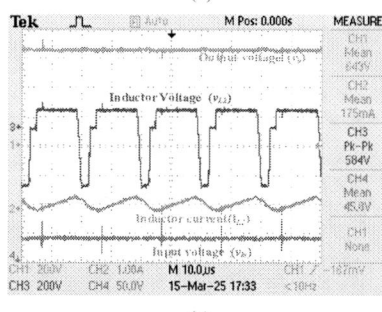

(c)

Fig. 7 Experimental results (a) Output and input voltage and current under CCM (b) Inductor voltages and currents under BCM (c) Inductor voltages and currents under DCM

TABLE II. LOSSES IN VARIOUS COMPONENTS

Loss due to Inductor	$I_{L_1-RMS}^2 R_{L_1} + I_{L_2-RMS}^2 R_{L_2}$
Loss due to Capacitor	$I_{C_1-RMS}^2 R_{C_1} + I_{C_2-RMS}^2 R_{C_2} + I_{C_3-RMS}^2 R_{C_3}$
Switching loss	$0.5(V_{S1}I_{S1,avg} + V_{S2}I_{S2,avg})(t_{on} + t_{off})f_s$
Switch conduction loss	$I_{S_1-RMS}^2 R_{S_1} + I_{S_2-RMS}^2 R_{L_2}$
Diode forward voltage loss	$I_{D_1-avg}V_{FV_{D1}} + I_{D_2-avg}V_{FV_{D2}} + I_{D_3-avg}V_{FV_{D3}}$
Diode conduction loss	$I_{D_1-RMS}^2 R_{D_1} + I_{D_2-RMS}^2 R_{D_2} + I_{D_3-RMS}^2 R_{D_3}$

TABLE III. RMS CURRENTS THROUGH VARIOUS COMPONENTS

I_{L_1-RMS} $= \dfrac{2-d}{(1-d)^2} I_o$	I_{L_2-RMS} $= \sqrt{\dfrac{\left(1+d+2d(1-d)\right)^2+(1-d)d}{(1-d)^2 d}} I_o$		i_{c_1-RMS} $= \dfrac{3-d}{1-d}\sqrt{\dfrac{d}{1-d}}$
I_{S_1-RMS} $= \dfrac{3i_o d}{(1-d)^2}$	I_{S_2-RMS} $= \dfrac{2i_o\sqrt{d}}{1-d}$	I_{D_3-RMS} $= \dfrac{i_o}{\sqrt{1-d}}$	i_{c_2-RMS} $= \dfrac{\sqrt{d^2-d+1}}{d\sqrt{d-1}} i_o$
i_{c_3-RMS} $= \sqrt{\dfrac{d}{1-d}} i_o$	I_{D_1-RMS} $= \dfrac{i_o(1+d)}{(1-d)\sqrt{1-d}}$		I_{D_2-RMS} $= \dfrac{i_o\sqrt{d}}{1-d}$

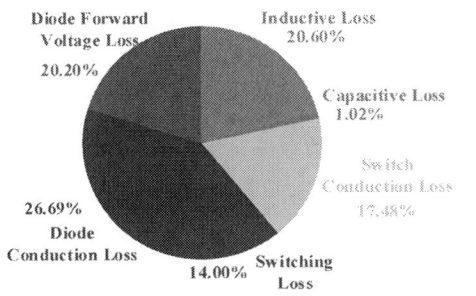

Fig. 8 Loss pie chart

Fig. 8 shows a pie chart of the power loss distribution for the specifications in Table II, at 200 W output and 58.90% duty ratio. With a total loss of 4.20% (8.40 W), inductor losses account for 20.60% (1.73 W), capacitor losses 1.02% (0.09 W), switch conduction losses 17.48% (1.47 W), and switching losses 14.00% (1.18 W). Diode forward voltage losses contribute 20.20% (1.70 W), while diode conduction losses make up 26.69% (2.24 W). This results in an analytical efficiency of 95.8% at 400 V output. These values are later used in the reliability analysis section.

VI. THERMAL ANALYSIS OF DSSLCN CONVERTER

Thermal analysis is a critical aspect of the DSSLCN Converter, as excessive heat can degrade performance, reduce efficiency, and shorten the lifespan of components. Effective thermal management ensures higher reliability, improved efficiency, and better operational stability of the converter. The thermal performance of the converter was evaluated under three load conditions: 100 W, 200 W, and 300 W (overloaded condition), with temperature measurements recorded for key components (S_1, S_2, D_1, D_2, and D_3) over a 30-minute period. From the thermal image observed from the experiment, it was observed that other components did not have significant temperature rise. The results are illustrated in Fig. 9.

At 100 W (Fig. 9(a) & 9(b)), all components started at approximately 32°C. Over 30 minutes, the temperature rise was minimal, with S_2 reaching a peak of 43.5°C, while S_1 and the diodes (D_1, D_2, D_3) remained below 41°C, indicating low thermal stress. At 200 W (Fig. 9(b) & 9(c)), temperatures slightly increased, with S_1 and S_2 reaching around 52.7°C. The diodes stabilized in the range between 49–50°C.

At 300 W (Fig. 9(e) & 9(f)), the temperature rise was significant. S_1 reached the highest temperature at 67.2°C, followed by D_2 at 63.5°C. S_2 and the other diodes also exhibited elevated temperatures, reflecting substantial thermal stress under maximum load. Despite these increases, all switch temperatures remained well below their maximum

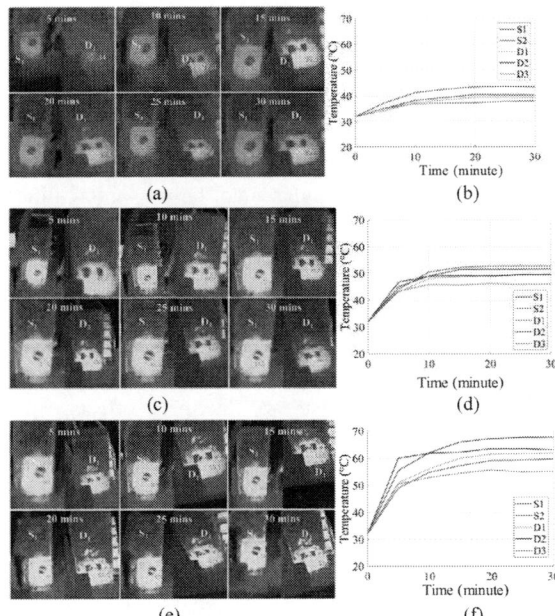

Fig. 9 Thermal analysis (a) thermal image at 100 W load (b) temperature profile at 100 W (c) thermal image at 200 W load (d) temperature profile at 200 W (e) thermal image at 300 W load (f) temperature profile at 300 W

operating limit of 175°C, confirming their suitability for sustained operation. This thermal analysis provides crucial insights for the upcoming reliability analysis section, where the impact of temperature on component failure rates will be examined. Since higher junction temperatures accelerate degradation mechanisms, such as semiconductor wear-out and thermal fatigue, the findings here will be directly used to quantify failure probabilities.

VII. RELIABILITY ANALYSIS

Reliability analysis of high-gain DC-DC converters is essential for evaluating their long-term performance, predicting failure rates, and improving design robustness. In power electronics applications such as RESs and EVs, the converter's reliability directly impacts system efficiency, safety, and maintenance costs. This section presents a detailed reliability assessment of the converter's key components, including SiC MOSFET switches and SiC diodes.

The exponential reliability model is a widely used statistical approach to describe the failure behavior of electronic components and systems over time. It is based on the assumption that failures occur randomly and independently at a constant failure rate (λ). This model is particularly useful for analyzing components that do not exhibit significant wear-out effects within their expected operational lifespan. The reliability function ($R(t)$) in the exponential model is given by (15).

$$R(t) = e^{-\lambda t} \tag{15}$$

Where t is the operational time. This equation shows that reliability decreases exponentially over time. The MTTF for a system following this model is given by (16)

$$MTTF = \frac{1}{\lambda} \qquad (16)$$

which represents the expected operational time before a failure occurs. The exponential model is particularly useful for electronic components operating under constant stress conditions, such as MOSFETs and diodes in a DC-DC converter. It provides insights into failure probabilities, helping in predictive maintenance, reliability assessment, and system design optimization to ensure long-term operational stability.

A. Switches

The reliability of MOSFETs in the high-gain DC-DC converter is analyzed using the failure rate equation, which accounts for voltage stress, thermal effects, and environmental conditions. The failure rate (λ_S) for each MOSFET ($S_1 \& S_2$) is given by (17).

$$\lambda_S = \lambda_b \, \pi_A \, \pi_Q \, \pi_E \, \pi_T \qquad (17)$$

Where, λ_b (Base Failure Rate): represents the intrinsic failure rate of the MOSFET under ideal conditions, usually determined from reliability databases or manufacturer specifications . π_A (Application Factor): accounts for the voltage stress on the MOSFET, which varies with the duty ratio (DD). The voltage stress is given by:

$$V_{\text{stress},S_1} = \frac{V_{\text{in}}}{1-d} \qquad (18)$$

$$V_{\text{stress},S_2} = \left(\frac{(2-d)}{(1-d)^2} - \frac{1}{1-d} \right) V_{\text{in}} \qquad (19)$$

The application factor is then calculated as:

$$\pi_A = \left(\frac{V_{stress}}{V_{rated}} \right)^n \qquad (20)$$

Higher duty ratios increase voltage stress, accelerating failure. π_Q (Quality Factor) reflects the quality grade of the MOSFET, considering manufacturing tolerances and material reliability. For high-reliability components, π_Q is close to 1. π_E (Environmental Factor): accounts for operating conditions such as temperature, humidity, and vibration. It is typically set based on standard reliability databases. π_T (Temperature Factor) accounts for the temperature related effects. Since MOSFETs experience power dissipation, their junction temperature (T_j) affects reliability. The junction temperature is calculated using the case temperature and power loss calculates in the previous sections.

$$T_j = T_{\text{case}} + R_{\text{th,jc}} \cdot P_{\text{loss}} \qquad (21)$$

The temperature factor is determined using the Arrhenius equation: The Arrhenius equation is used to model this temperature-dependent degradation, allowing for a more accurate prediction of MOSFET lifespan under varying operating conditions.

$$\pi_T = \exp\left(\frac{E_a}{k} \left(\frac{1}{T_{\text{ref}}} - \frac{1}{T_j} \right) \right) \qquad (22)$$

where π_T is the temperature factor that accounts for the impact of temperature on component degradation. The parameter E_a represents the activation energy in electron volts

(eV), which characterizes the sensitivity of the failure rate to temperature changes. The Boltzmann constant, denoted as k, has a value of 8.617×10^{-5} eV/K and is used to relate energy to temperature on an atomic scale. The reference temperature, T_{ref}, is typically set to 298 K (25°C), serving as a baseline for comparison. Finally, T_j represents the junction temperature of the MOSFET or diode in Kelvin. This equation shows that as the junction temperature increases, the failure rate also increases exponentially, highlighting the importance of thermal management in improving component reliability.

In this analysis, the base failure rate (λ_b) is set to 1×10^{-6} failures per hour, reflecting typical values for high-quality MOSFETs under nominal conditions. The voltage exponent n is chosen as 2.5 (MIL-HDBK-217) which represents the sensitivity of the failure rate to voltage stress in these devices. Both the quality (π_Q) and environmental (π_E) factors are assumed to be 1.0, indicating standard operating conditions without additional external stress. A 3D surface plot is used to visualize the impact of duty ratio and time on MOSFET reliability.

The analysis shows that as the duty ratio increases, both voltage stress and temperature rise, leading to a higher failure rate. Over prolonged operation, MOSFET reliability declines, emphasizing the importance of thermal management, voltage stress minimization, and proper component selection to enhance lifespan and ensure stable converter performance.

B. Reliability Analysis of Diodes

The reliability of diodes in the high-gain DC-DC converter is evaluated using the failure rate model, incorporating voltage stress, thermal effects, and environmental conditions. The failure rate λ_D for each diode is given by (23)-(25).

$$V_{\text{stress},D_1} = \frac{V_{\text{in}}}{1-d} \qquad (23)$$

$$V_{\text{stress},D_2} = V_{\text{in}} \left(\frac{(2-d)}{(1-d)^2} - \frac{1}{1-d} \right) \qquad (24)$$

$$V_{\text{stress},D_3} = V_{\text{in}} \left(\frac{(2-d)}{(1-d)^2} \right) \qquad (25)$$

The application factor π_S is determined similarly to the MOSFETs, based on the ratio of voltage stress to rated voltage. Higher duty ratios lead to greater voltage stress, accelerating failure. The value of other factors has been assumed same as MOSFET.

A 3D surface plot (Fig. 10) is used to visualize the effect of duty ratio and operating time on the reliability of both MOSFETs and diodes. The reliability analysis of MOSFETs and SiC diodes in the high-gain DC-DC converter reveals a clear decline over time and with increasing duty ratios. For MOSFETs (S_1 and S_2), higher duty ratios result in elevated voltage stress and thermal effects, accelerating degradation and leading to a significant reduction in reliability over a five-year period, particularly under high-stress conditions. Similarly, the SiC diodes (D_1, D_2, and D_3) exhibit a decreasing reliability trend, with initially lower failure rates but noticeable degradation under prolonged operation at higher duty ratios. This pattern underscores the critical role of effective thermal management and controlled duty ratio

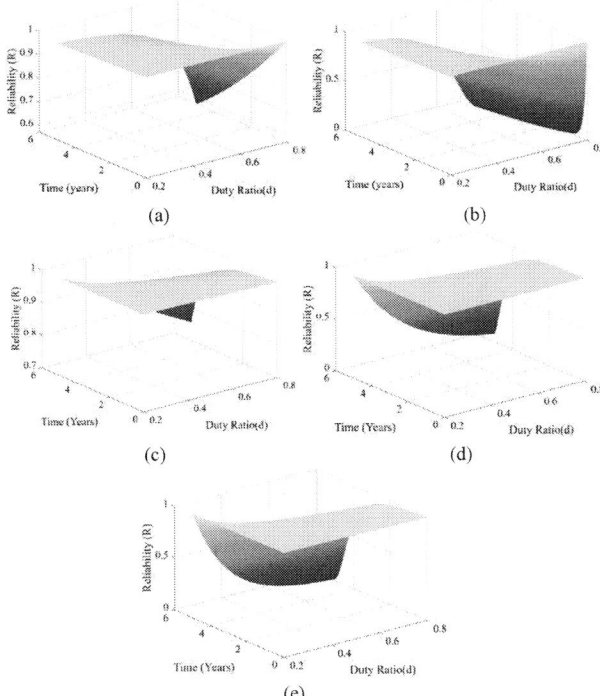

Fig. 10 Reliability vs time vs duty ratio (a) Switch S_1 (b) Switch S_2 (c) Diode D_1 (c) Diode D_2 (c) Diode D_3

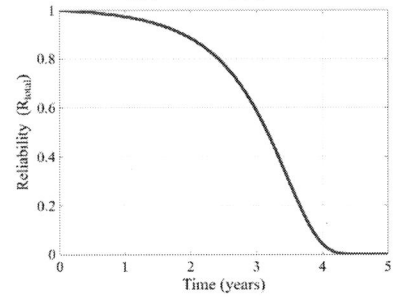

Fig. 11 Reliability vs time of DSSLCN

operation in ensuring long-term stability and performance of the converter system. The overall reliability of the converter is determined by summing the failure rates of all critical components, including MOSFETs and SiC diodes, based on a series system configuration. The total failure rate λ_{total} is expressed as given in (26).

$$\lambda_{total} = \lambda_{S1} + \lambda_{S2} + \lambda_{D1} + \lambda_{D2} + \lambda_{D3} \qquad (26)$$

Using the exponential reliability model, the overall system reliability is calculated. The results indicate a gradual decline in reliability over time, with a more pronounced deterioration at higher duty cycles due to increased voltage stress and thermal effects. The time-based reliability plot shows a sharp decline in reliability after approximately three years (Fig.11), emphasizing the need for effective thermal management and predictive maintenance. These findings highlight the importance of optimizing operating conditions, improving heat dissipation, and selecting high-reliability components to enhance the converter's lifespan. Implementing predictive maintenance and redundancy strategies can further improve long-term system performance and reliability.

VIII. CONCLUSION

This paper presented a comprehensive thermal and reliability analysis of the Dual-Switch Switched Inductor-Capacitor Network (DSSLCN) Converter, highlighting its feasibility for high-power applications. Theoretical modelling and experimental validation demonstrated that the converter achieves high voltage gain while maintaining low switch current stress and continuous input current. Thermal analysis revealed that MOSFETs and diodes experience temperature variations under different load conditions, emphasizing the need for thermal management. Reliability modelling, based on an exponential failure rate approach, showed that voltage stress and temperature effects accelerate component aging, with higher duty cycles leading to reduced lifespan. Experimental results validated the analytical predictions, confirming a high efficiency of 95.80% at rated conditions. Overall, this paper provides valuable insights for improving high-gain DC-DC converters, ensuring enhanced efficiency, robustness, and long-term stability. Future work can explore advanced thermal management techniques and reliability-improving circuit modifications to further optimize the converter's performance in real-world applications.

REFERENCES

[1] F. Shijad and Kumaravel S, "Design and Analysis of Switched Inductor-Based Quadratic Boost Converter with Enhanced Voltage Gain and Reduced Component Voltage Stress," 2023 IEEE International Conference on Power Electronics, Smart Grid, and Renewable Energy (PESGRE), Trivandrum, India, 2023, pp. 1-6, doi: 10.1109/PESGRE58662.2023.10405069.

[2] Vinu Varshath S. and S. Kumaravel, "Dual Switch Switched Inductor-Capacitor Network-based High Gain DC-DC Converter with Reduced Switch & Capacitor Voltage Stress and Continuous Input Current," 2024 IEEE International Conference on Power Electronics, Drives and Energy Systems (PEDES), Mangalore, India.

[3] M. K. Alam and F. H. Khan, "Reliability Analysis and Performance Degradation of a Boost Converter," in IEEE Transactions on Industry Applications, vol. 50, no. 6, pp. 3986-3994, Nov.-Dec. 2014, doi: 10.1109/TIA.2014.2319587.

[4] F. H. Aghdam and M. Abapour, "Reliability and Cost Analysis of Multistage Boost Converters Connected to PV Panels," in IEEE Journal of Photovoltaics, vol. 6, no. 4, pp. 981-989, July 2016, doi: 10.1109/JPHOTOV.2016.2566885.

[5] M. J. Sathik, J. D. Navamani, A. Lavanya, Y. Yang, D. Almakhles and F. Blaabjerg, "Reliability Analysis of Power Components in Restructured DC/DC Converters," in IEEE Transactions on Device and Materials Reliability, vol. 21, no. 4, pp. 544-555, Dec. 2021, doi: 10.1109/TDMR.2021.3116941.

[6] V. Samavatian, H. Iman-Eini, Y. Avenas and S. Shemehsavar, "Reciprocal and Self-Aging Effects of Power Components on Reliability of DC–DC Boost Converter with Coupled and Decoupled Thermal Structures," in IEEE Transactions on Components, Packaging and Manufacturing Technology, vol. 9, no. 12, pp. 2506-2513, Dec. 2019, doi: 10.1109/TCPMT.2019.2940058.

[7] J. Sakly, A. Bennani–Ben Abdelghani, I. Slama–Belkhodja and H. Sammoud, "Reconfigurable DC/DC Converter for Efficiency and Reliability Optimization," in IEEE Journal of Emerging and Selected Topics in Power Electronics, vol. 5, no. 3, pp. 1216-1224, Sept. 2017, doi: 10.1109/JESTPE.2017.2676027.

[8] H. Tarzamni, F. Tahami, M. Fotuhi-Firuzabad and F. P. Esmaeelnia, "Reliability Analysis of Buck-Boost Converter Considering the Effects of Operational Factors," 2019 10th International Power Electronics, Drive Systems and Technologies Conference (PEDSTC), Shiraz, Iran, 2019, pp. 647-652, doi: 10.1109/PEDSTC.2019.8697266.

Load Frequency Control in Modern Power System- An Overview

Mitali Samal,
Research Scholar
School of Electrical Engineering
KIIT Deemed to be University
Bhubaneswar, Odisha, India
Email ID:
mitalisamal96@gmail.com

Dr. Chinmoy Kumar Panigrahi,
Professor, School of Electrical
Engineering
KIIT Deemed to be University
Bhubaneswar, Odisha, India
Email ID:
panigrahichinmoy@gmail.com

Dr. Deepak Kumar Gupta,
Asst. Professor, School of Electrical
Engineering
KIIT Deemed to be University
Bhubaneswar, Odisha, India
Email ID:
deepak.guptafel@kiit.ac.in

Abstract—Load Frequency Control (LFC) is a critical aspect of maintaining the reliability and stability of power systems. With the increase in renewable energy sources, deregulated environment and the evolution of smart grids, the challenges and opportunities associated with LFC have become more pronounced. Numerous articles have been published on LFC, addressing various control strategies for the successful operation of power systems. This paper provides a comprehensive review of recent developments in LFC, focusing on the various control strategies employed in traditional, Contemporary, and Next-Generation Systems. This review effectively covers challenges and opportunities for frequency regulation under recent developments. The review highlights the advancements in control methodologies, including classical, modern, intelligent, and soft computing-based approaches. Additionally, it discusses the impact of renewable energy integration on LFC, and identifies the research gaps and future directions in this field. It also summarizes and explains different scenarios involving Energy Storage (ES), micro-grids, and FACTS devices to explore potential solutions and future directions. The findings of this review aim to guide researchers and practitioners in developing more robust and efficient LFC strategies to address the dynamic challenges of modern power systems.

Keywords—automatic generation control (AGC), load frequency control (LFC), optimization algorithms, renewable energy resources, smart grids, tie-lines power deviation.

I. INTRODUCTION

In modern systems, maintaining the balance between supply and demand is crucial to ensure system reliability and stability. Load Frequency Control(LFC) plays a pivotal role in this context, as it regulates system frequency and maintains scheduled power exchanges between interconnected areas. The increasing complexity of power networks, driven by rapid technological advancements and the integration of Renewable Energy Sources(RES), has posed significant challenges to traditional LFC methods [1]. These developments demand innovative solutions that are robust, adaptive, and capable of handling the dynamic behavior of modern grids.The transition toward RES, such as wind and solar, introduces inherent intermittency and variability, complicating frequency regulation. Furthermore, the decentralization of power generation, driven by distributed energy resources (DERs) and smart grid technologies, adds layers of intricacy to load frequency management. While these challenges are formidable, they also present opportunities for leveraging cutting-edge techniques, such as artificial intelligence, machine learning, and advanced control algorithms, to enhance LFC strategies [2].

This review paper critically examines recent advancements in LFC, focusing on the challenges posed by evolving grid conditions and the opportunities offered by emerging technologies [3]. By synthesizing state-of-the-art research, this study seeks to offer an in-depth insight into the present scenario and identify promising directions for future exploration. The paper also highlights the interplay between traditional and modern LFC approaches, offering insights into how they can coexist in a rapidly changing energy ecosystem. Figure 1 illustrates a flowchart depicting various approaches utilized for Load Frequency Control in hybrid power systems.

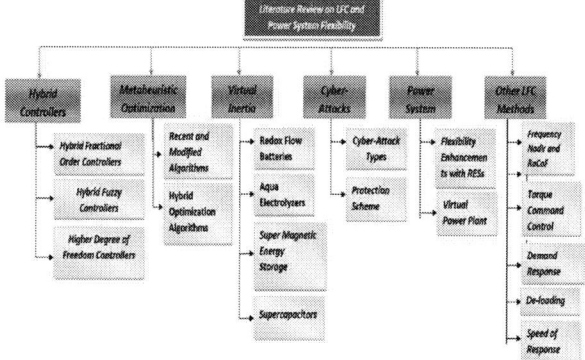

Fig. 1. Flowchart of different approaches used for LFC

II. REVIEW ON LFC CONSIDERING RENEWABLE ENERGY SOURCES (RES) & EVS

The integration of RES such as wind, solar, and hydro power into modern power grids has introduced significant challenges for LFC due to their inherent variability and unpredictability. Unlike traditional thermal or hydroelectric power plants, RES generation depends heavily on environmental conditions, such as sunlight and wind speed, which fluctuate throughout the day. This variability affects the system's inertia and complicates frequency regulation, as conventional LFC strategies are often designed for predictable and stable power generation. Recent research has focused on adaptive and robust control techniques to address these challenges, including the use of predictive algorithms, energy storage systems, and demand response mechanisms. Moreover, the coordination of RES with traditional generation and hybrid systems has been explored to enhance grid resilience and maintain frequency stability. Advanced methods, such as ML and AI have also emerged as promising tools to model and predict RES behavior, enabling more effective LFC strategies tailored to modern grids. This section reviews these developments, providing insights into how RES integration is reshaping LFC approaches and highlighting gaps that require further investigation.

In multi-area interconnected systems, dynamic models incorporating DFIG are employed to mitigate frequency fluctuations caused by wind power integration [4]. To enhance system stability, an LFC strategy incorporating a memory event-triggered scheme and random deception attack resistance has been proposed for single-area systems integrated with electric vehicles (EVs). This approach utilizes a memory-based PID strategy combined with an intermittent control method, ensuring robust frequency stabilization for wind power systems. Furthermore, innovative event-triggered schemes integrating integral terms and time windows have been developed to ensure H_∞ asymptotic stability under hacking attacks. A dynamic and steady-state model for PV plants, using artificial neural networks with radial bias function (ANN-RBF), has been introduced in [5] to represent PV power output accurately. This model eliminates the need for complex mathematical studies and incorporates frequency deviations as inputs, reflecting real-world conditions. The ANN-RBF model outperforms traditional first-order PV models, making it more suitable for LFC studies. In smart grids, the inclusion of RES presents unique challenges due to communication delays, packet drops, and cyber attacks. Enhancing LFC in such scenarios requires modifications to conventional controllers, including energy storage integration. A particle swarm optimization (PSO)-tuned PID controller has been employed to address random time delays and cyber attacks, demonstrating improved robustness under stochastic demand and generation profiles [6]. Standalone microgrids (MGs) face significant frequency regulation challenges due to high renewable energy penetration and load fluctuations. A (1PDF)-FOPI controller, optimized using the marine predator algorithm (MPA), has been introduced for effective power flow management. This system includes PV, wind, diesel engine generators (DG), and energy storage elements like ultra capacitors (UC) and flywheels. Practical nonlinearities such as GDB and GRC are considered, with stability evaluations highlighting the proposed controller's effectiveness [7]. In isolated MGs, linear quadratic regulator-integral (LQR-I) controllers have been implemented for frequency regulation. Using state-space modeling, the dynamic response of the microgrid has been validated for various conditions, including random load variations, wind speed changes, and solar insolation levels. Sensitivity analysis has confirmed the robustness of the LQR-I controller [8]. Electric vehicles (EVs) have emerged as a promising solution for LFC, providing ancillary services through vehicle-to-grid (V2G) technology. The presence of EV aggregators introduces time-varying delays in LFC schemes, which can impact system stability. In [9] the stability region for LFC systems with EVs is determined, along with the maximum allowable delay using Wirtinger-based improved inequalities. In deregulated systems, EVs have been integrated into multi-area LFC schemes alongside renewable generation. A PI controller, validated through simulations under varied load profiles, effectively stabilizes frequency and tie-line power exchanges in RES and EV-integrated grids. These strategies highlight the potential of EVs in enhancing grid performance during sudden disturbances [10]. Hybrid power systems integrating RES, thermal plants, and biogas systems benefit from innovative control algorithms. Adaptive controllers have also been developed to handle load variations and RES disturbances. These controllers use auto-tuned characteristics to improve system performance, with studies demonstrating their effectiveness under non-linear conditions. Comparative analyses with traditional controllers highlight their superior dynamic response [11]. Low-inertia systems in isolated MGs pose unique challenges due to the erratic nature of RES. These systems experience higher rates of change of frequency (RoCoF), requiring fast response mechanisms. Distributed energy resources (DERs), such as fuel cells and flywheel energy storage, have been incorporated as active reserves for LFC. Simulation studies validate the effectiveness of these resources in mitigating frequency excursions under various operating conditions [12]. The increasing integration of RES and EVs demands efficient PID controllers for frequency stabilization in island MGs. ANN-based PSO techniques have been employed to fine-tune PID controllers, significantly reducing frequency oscillations, overshoots, and settling times. Simulation results confirm enhanced stability and rapid system recovery [13]. In smart grids, LFC becomes increasingly complex due to distributed architectures and load variations. EV systems play a critical role in compensating for frequency and tie-line power changes. Simulations using PID, centralized MPC and decentralized MPC controllers under load disturbances have demonstrated the superiority of De-MPC in improving system stability and performance [14].

Modern LFC strategies leverage renewable energy, electric vehicles, and advanced controls like PID, ANN, and hybrid methods to ensure frequency stability. Future research should enhance scalability and resilience to meet evolving grid demands amid increasing renewables and cyber-physical complexities.

III. FREQUENCY CONTROL FOR SMART GRID AND MICRO GRID SYSTEMS

Frequency control is critical for maintaining stability in smart grids and microgrid systems. In smart grids, advanced control strategies address the intermittent nature of RES and distributed load dynamics. Vehicle-to-grid (V2G) technology enables EVs to act as distributed resources for frequency regulation, while adaptive controllers like decentralized MPC (De-MPC) manage communication delays and cyber-physical challenges. Microgrids, with their low inertia and reliance on RES, demand effective load frequency control (LFC) strategies to stabilize systems under varying load demands and renewable fluctuations. Hybrid controllers and advanced approaches like linear quadratic regulator-integral (LQR-I) controllers, have demonstrated their ability to maintain stability. Energy storage units like ultra capacitors and flywheels also play a vital role in compensating for rapid frequency variations, making tailored frequency control essential for both smart grids and micro-grids.

A. Frequency control for Smart Grid

The integration of contemporary technologies like wide area monitoring systems and demand response (DR) is vital for enhancing LFC in power systems. DR control loops, combined with supplementary control (SC) and Linear Matrix Inequality-LQR controllers, mitigate frequency deviations caused by communication delays in single-area systems integrating thermal and wind power [15].In Smart Grids (SG), coordinated Vehicle-to-Grid (V2G) and Grid-to-Vehicle (G2V) controls with PSO-tuned H-infinity controllers enhance LFC by utilizing Plug-in Hybrid Electric Vehicles to improve robustness against uncertainties. Simulations validate their superiority over traditional methods in interconnected systems [16]. To address the challenges posed by wind power's variability, Enhanced Intelligent Disturbance (EID) estimators are used for improved disturbance rejection and robustness to parameter uncertainties, time delays, and external disturbances. This approach enhances LFC stability in single- and two-area systems [17].Recent advancements include energy storage systems (ESS) for dynamic frequency stabilization through charging and discharging control [18]. Distributed Energy Resources (DERs) in microgrids utilize an Optimal Reset Control (ORC) for effective frequency regulation, reducing overshoot and settling times [19]. Hydrogen Energy Storage (HES) systems enhance LFC by enabling demand-side management and addressing renewable energy variability. A Double-Input Type-2 Fuzzy Controller, optimized using the ISSA, achieves significant improvements in frequency response, especially in renewable-integrated microgrids [20]. Demand Response (DR) strategies in hybrid renewable systems integrate controllers like the TDOF-Linear Active Disturbance Rejection Control for stability under uncertainties. Virtual inertia strategies, leveraging electric vehicles (EVs), address frequency instability in low-inertia grids [21]. Superconducting Magnetic Energy Storage (SMES) and EV batteries, combined with Laguerre-based Model Predictive Control (LiMPC), improve frequency dynamics in islanded microgrids, offering robustness and efficiency compared to traditional methods [22]. Collectively, these innovations advance LFC in modern grids, ensuring reliability and resilience amid renewable energy integration.

B. Frequency control in Microgrid

Maintaining frequency stability in renewable energy-based microgrids (MGs) remains challenging due to their fluctuating nature. Virtual inertia control (VIC), combined with multistage proportional-integral-derivative (PID) controllers tuned via a sine-augmented arithmetic optimization algorithm, has been shown to mitigate frequency oscillations [23]. Advanced PID controllers tuned via African Vulture Optimization Algorithm (AVOA) effectively manage variable loads and renewable penetration in interconnected systems [24]. Finally, adaptive type-2 fuzzy PID controllers ensure robust performance under nonlinear conditions, outperforming traditional fixed-structure controllers in MGs [25]. This study addresses LFC challenges in microgrids (MGs) under renewable energy integration, nonlinearities, and uncertainties. This study addresses load frequency control (LFC) challenges in microgrids (MGs) under high renewable energy penetration. A derivative-based virtual rotor approach integrated with a Jaya optimizer enhanced by balloon effect modulation (BE) improves system stability and adaptability by incorporating virtual inertia and damping to mitigate system nonlinearities

and delays. The approach outperforms conventional controllers and optimization strategies [26]. For standalone MGs, an intelligent fuzzy tilt integral derivative controller (FTIDF-(1+I)) optimized using the Wild Horse Optimizer (WHO) achieves a frequency deviation reduction of up to 99.75%, surpassing other TIDF and FTIDF controllers under significant uncertainties [27]. A microgrid model incorporatingV2G control explores EVs as bidirectional energy sources. A droop control method is used to ensure frequency stability, while the inclusion of EVs enhances power quality and reliability. Four scenarios demonstrate the impact of EV integration, emphasizing bidirectional energy flow benefits[28]. In multi-microgrid (MMG) systems, a Tilt Integral Derivative (TID) controller optimized using Atom Search Optimization improves dynamic responses under varied load disturbances. The method outperforms conventional strategies in achieving robust stability [29].

IV. FREQUENCY CONTROL WITH AUXILIARY DEVICES

Frequency control with auxiliary devices leverages advanced systems like energy storage devices (ESDs), HVDC links, and FACTS controllers to enhance the stability of power systems. These devices effectively mitigate frequency deviations and maintain stability during load disturbances and fluctuations in renewable energy generation. A droop-based consensus control strategy, integrated with an extended Kalman filter (EKF), improves coordination in microgrids (MGs) featuring multiple BESSs. This approach ensures resilience against communication issues such as data loss, false data injection, and noise, as validated using MATLAB/SimPower Systems in both grid-connected and islanded MG setups [30].For combined LFC and AVR, fuzzy PID controllers fine-tuned with the HAEFA mechanism effectively stabilize voltage and frequency. Among ESD options, redox flow batteries (RFBs) demonstrated superior performance in stabilizing multi-area interconnected systems during disturbances [31].In hybrid systems, a grasshopper optimization algorithm (GOA)-optimized three-degree-of-freedom PID (3DOF-PID) controller enhances automatic generation control (AGC). Coupled with HVDC links, this approach excels in controlling frequency oscillations in thermal-hydro-gas-wind configurations compared to traditional methods like the moth swarm algorithm [32].Integrating doubly fed induction generators (DFIGs) with HVDC and EHVAC tie-lines improves LFC by utilizing accurate state feedback, boosting system reliability [33]. Furthermore, HVDC links and electric vehicles (EVs), managed by Firefly algorithm-optimized PI/PID controllers, enhance LFC by minimizing oscillations and delays under diverse load conditions [34].A proportional-fractional integrator plus proportional-derivative (P(1+PDF)) controller, paired with hybrid energy storage systems (HESS), addresses challenges from high renewable energy penetration. Optimized via the Zebra Optimization Algorithm (ZOA), this controller minimizes frequency deviations and enhances robustness under varying operational conditions [35].

V. FREQUENCY REGULATION WITH INTELLIGENT CONTROLLERS

Frequency regulation with intelligent controllers leverages advanced techniques such as fuzzy logic, fractional-order systems, and optimization algorithms to enhance load frequency control (LFC) performance. These controllers improve system stability, speed, and adaptability under complex, nonlinear, and dynamic conditions. The integration of renewable energy sources (RESs) in electricity generation is growing rapidly. However, the intermittent nature of these sources exacerbates frequency fluctuations caused by load variations, reducing system inertia and increasing grid instability. To address this, a coyote optimization algorithm (COA) is proposed for designing an advanced PDn-PI controller. This approach integrates RESs, such as photovoltaic (PV) and wind farms, into load frequency control (LFC) systems, ensuring optimal tuning of COA with different controllers [36]. Voltage and frequency stability in hybrid multi-area systems with RESs and electric vehicles is also a major concern. To enhance stability, a FOPI and PID plus double derivative (PIDD2) controllers is optimized using the dandelion optimizer (DO) algorithm [37]. Traditional controllers in LFC systems often face delays and nonlinearity issues. Advanced controllers like fractional-order PID (FOPID) controllers, optimized through the Self-adaptive Global Harmony Search (SGHS) technique, offer quicker responses and better stability [38-40]. Advanced controllers like the Fuzzy-PSO-PID have been developed to improve the response of hydro-hydro interlinked systems. By integrating fuzzy logic with particle swarm optimization (PSO), this controller addresses the slow response of hydro turbines and minimizes frequency deviations. Additional enhancements using UPFC and RFB demonstrated reduced frequency oscillations and improved system stability, even under dynamic load conditions [41]. To further enhance Automatic Generation Control (AGC), the Fuzzy TID with Filter and Double Integral (FTIDF-II) controller has been proposed. This innovative controller, optimized using the Imperialist Competitive Algorithm (ICA), leverages asymmetric membership functions for precise tuning. It outperforms traditional PID controllers in dynamic performance, demonstrating robustness against parameter variations and random load changes [42]. In systems integrating renewable energy sources, Interval Type-2 Fuzzy Logic Controllers (IT2FLC) have proven effective in managing frequency deviations caused by demand and renewable fluctuations. These controllers, optimized through Whale Optimization Algorithm (WOA) and other techniques, showed superiority in stabilizing hybrid systems, particularly in scenarios involving solar and wind energy variations. Additionally, Reduction Oxidation Flow Batteries (RFBs) and advanced neural network-based tuning methods further enhanced the resilience and efficiency of these systems [43].

These intelligent controllers, such as fuzzy logic, fractional-order systems, and AI-optimized designs, have proven highly effective in enhancing frequency regulation in modern power systems. Figure 2 shows the different intelligent techniques used by various researchers for optimally tune the advanced controllers for maintaining frequency variation and tie-line power exchange. These approaches offer improved stability, faster response times, and resilience against system uncertainties, making them indispensable for managing the complexities of renewable integration and multi-area interconnected networks.

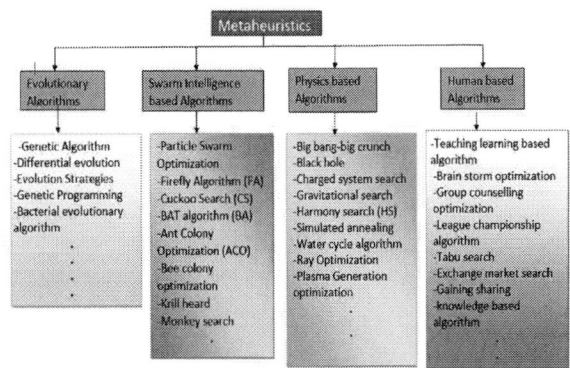

Fig. 2. Different Metaheuristic techniques used for LFC

VI. RECENT TRENDS IN LOAD FREQUENCY CONTROL

Recent advancements in Load Frequency Control (LFC) focus on addressing the dynamic challenges introduced by RES, EVs, and deregulated power systems. The integration of RES like solar and wind adds variability to power generation, necessitating innovative solutions such Deregulation in the power sector introduces competition and decentralization, significantly impacting Load Frequency Control (LFC). The presence of multiple independent entities like generating companies (GENCOs), transmission companies (TRANSCOs), and distribution companies (DISCOs) necessitates a more dynamic LFC framework. To enhance system performance under deregulated conditions, the Aquila Optimizer (AO) tunes FOPID controller parameters. Applied to a hybrid system comprising thermal, hydro, gas, and wind power plants, this method minimizes area control errors, tie-line deviations, and frequency variations during operational changes. AO-driven FOPID controllers outperform Whale Optimization Algorithm (WOA) and Particle Swarm Optimization (PSO), with optimization aimed at minimizing ITAE performance indices [44]. The Grasshopper Optimization Algorithm is also explored for a 3-degree-of-freedom PID (3DOF-PID) controller in hybrid systems. A two-area thermal-hydro-gas-wind setup with HVDC links shows that GOA-optimized controllers outperform traditional methods, effectively handling oscillations and load disturbances. FACTS devices like UPFC and Superconducting Magnetic Energy Storage further improve AGC performance when paired with the 3DOF-PID controller [45]. The study employing Particle Swarm Optimization (PSO) for a PID controller in a two-area deregulated system addresses challenges posed by multiple stakeholders and generating sources. The PSO algorithm optimizes controller gains to minimize ITAE, ensuring stability despite load fluctuations or parameter changes. Simulations in MATLAB/Simulink evaluate the controller's performance under diverse contractual conditions [46].Another investigation explores the effectiveness of soft computing techniques, such as ANFIS, GA, and PSO, in optimizing load frequency control (LFC) in deregulated systems. The study highlights the limitations of traditional centralized LFC approaches in managing distributed resources and multiple stakeholders. Using MATLAB/Simulink the proposed controllers show improved dynamic performance, maintaining frequency regulation and tie-line power exchange per contractual terms

979-8-3315-3013-6/25 $31.00 © 2025 IEEE   225

[47].as Virtual Inertia Control (VIC). VIC mimics the natural stabilizing effects of traditional synchronous generators by utilizing energy storage systems and advanced power electronics, allowing renewable sources to contribute to frequency regulation. By leveraging energy storage systems and power electronics, VIC enables renewable sources to dynamically respond to frequency fluctuations, ensuring grid stability amid the variable nature of solar and wind generation [48-49]. Additionally, the concept of Virtual Power Plant (VPP) Control is gaining importance as it aggregates DERs, such as wind turbines, solar panels, and batteries, into a unified entity capable of providing frequency regulation and demand response services. In deregulated power systems, VPPs address the complexities of diverse market participants by enabling real-time coordination of DERs while adhering to contractual agreements [50-51]. Together, these trends highlight a paradigm shift towards decentralized, intelligent, and adaptive LFC mechanisms, which are critical for enhancing grid stability and efficiency in modern power networks.

VII. RESEARCH GAPS AND FUTURE OPPORTUNITIES

Despite significant advancements in LFC technologies, several research gaps remain. The increasing penetration of renewable energy sources (RES) introduces high variability and unpredictability in generation, which traditional LFC methods struggle to manage effectively. While techniques like Virtual Inertia Control (VIC) and Virtual Power Plant (VPP) Control offer promising solutions, their deployment requires enhanced computational efficiency, robust communication systems, and seamless integration with existing grid infrastructure. Moreover, the role of Electric Vehicles (EVs) in frequency regulation is underexplored, particularly in terms of optimizing their dual functionality as loads and mobile energy storage units. Current methods often overlook the impact of cyber-physical threats, such as data breaches or communication failures, on LFC systems, which are critical in a highly interconnected and deregulated market. This highlights the need for resilient control strategies and secure communication protocols to ensure grid reliability under evolving conditions. Figure 3 represents the challenges and strategies for LFC in hybrid power system.

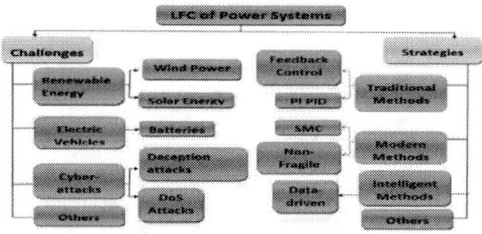

Fig. 3. Challenges and strategies for LFC in power system

Future opportunities lie in leveraging AI, ML and advanced optimization techniques to enhance LFC capabilities in smart grids. AI-driven predictive models can better forecast load demands and RES outputs, enabling proactive frequency regulation. Blockchain technology also holds potential for decentralized energy trading, ensuring transparency and contract enforcement in deregulated markets. Integrating DERs, such as rooftop solar panels, EV fleets, and community storage systems, into LFC frameworks presents a significant opportunity for grid stabilization. Additionally, hybrid energy storage systems combining batteries, flywheels, and supercapacitors can be further explored to address short-term and long-term fluctuations in load and generation. Addressing these gaps and capitalizing on future opportunities can significantly enhance grid resilience, improve energy efficiency, and support the transition toward sustainable energy systems.

VIII. CONCLUSION

This review critically examines the challenges and opportunities in Load Frequency Control (LFC), highlighting its evolution alongside recent technological advancements. The transition to renewable energy sources (RES), the proliferation of electric vehicles (EVs), and the emergence of deregulated power systems have introduced complexities in maintaining grid stability. While these factors pose significant challenges, they also create opportunities to innovate smarter, more adaptive LFC strategies. Cutting-edge solutions like virtual inertia control, virtual power plants (VPPs), and advanced energy storage systems demonstrate promising potential in addressing these challenges. Moreover, intelligent controllers, powered by AI, machine learning, and optimization algorithms, provide enhanced accuracy, adaptability, and resilience in modern LFC applications.Despite notable progress, research gaps remain in areas such as effective integration of distributed energy resources (DERs), cyber-security for decentralized control systems, and the scalability of advanced controllers in diverse operating conditions. The future of LFC lies in a multidisciplinary approach that combines technological innovation, regulatory adaptation, and market-driven strategies to achieve a sustainable, reliable, and resilient power system. As the energy sector continues to evolve, the opportunities for advancing LFC technologies will play a pivotal role in shaping a more stable and efficient grid.

REFERENCES

[1] P. Kundur, Power System Stability and Control, New Delhi, India: Tata McGraw-Hill, 2009, ch. 11.

[2] D. P. Kothari and I. J. Nagrath, Modern Power System Analysis, New Delhi, India: Tata McGraw-Hill, 2014, ch. 8.

[3] D. K. Gupta et al., "Hybrid gravitational–firefly algorithm-based load frequency control for hydrothermal two-area system," Mathematics, vol. 9, no. 7, p. 712, 2021.

[4] Q. Zhong et al., "Adaptive event-triggered PID load frequency control for multi-area interconnected wind power systems under aperiodic DoS attacks," Expert Syst. Appl., vol. 241, p. 122420, 2024.

[5] A. Bakeer et al., "A sophisticated modeling approach for photovoltaic systems in load frequency control," Int. J. Electr. Power Energy Syst., vol. 134, p. 107330, 2022.

[6] D. K. Panda, S. Das, and S. Townley, "Toward a more renewable energy-based LFC under random packet transmissions and delays with stochastic generation and demand," IEEE Trans. Autom. Sci. Eng., vol. 19, no. 2, pp. 1217-1232, 2020.

[7] P. K. Pathak et al., "Fractional cascade LFC for distributed energy sources via advanced optimization technique under high renewable shares," IEEE Access, vol. 10, pp. 92828-92842, 2022.

[8] R. Mandal and K. Chatterjee, "Frequency control and sensitivity analysis of an isolated microgrid incorporating fuel cell and diverse distributed energy sources," Int. J. Hydrogen Energy, vol. 45, no. 23, pp. 13009-13024, 2020.

[9] F. Babaei, A. Safari, and J. Salehi, "Evaluation of delays-based stability of LFC systems in the presence of electric vehicles aggregator," J. Oper. Autom. Power Eng., vol. 10, no. 2, pp. 165-174, 2022.

[10] K. M. Roshan and C. Ismayil, "Hybrid electric vehicle integrated LFC with renewable energy penetration under restructured bilateral power

system," in Proc. Int. Conf. Commun., Control Inf. Sci. (ICCISc), 2021, pp. 1-6.

[11] M. Mokhtar et al., "An adaptive load frequency control for power systems with renewable energy sources," Energies, vol. 15, no. 2, p. 573, 2022.

[12] A. Abazari, H. Monsef, and B. Wu, "Coordination strategies of distributed energy resources including FESS, DEG, FC and WTG in load frequency control (LFC) scheme of hybrid isolated micro-grid," Int. J. Electr. Power Energy Syst., vol. 109, pp. 535-547, 2019.

[13] A. Safari, F. Babaei, and M. Farrokhifar, "A load frequency control using a PSO-based ANN for micro-grids in the presence of electric vehicles," Int. J. Ambient Energy, vol. 42, no. 6, pp. 688-700, 2021.

[14] R. Asghar et al., "Load frequency control for EVs based smart grid system using PID and MPC," in Proc. Int. Conf. Comput., Math. Eng. Technol. (iCoMET), 2020, pp. 1-6.

[15] K. Bharti, V. P. Singh, and S. P. Singh, "Impact of intelligent demand response for load frequency control in smart grid perspective," IETE J. Res., vol. 68, no. 4, pp. 2433-2444, 2022.

[16] S. K. Tripathi, V. P. Singh, and A. S. Pandey, "Robust load frequency control of interconnected power system in smart grid," IETE J. Res., vol. 69, no. 8, pp. 5351-5363, 2023.

[17] L. Jin et al., "Equivalent input disturbance-based load frequency control for smart grid with air conditioning loads," Sci. China Inf. Sci., vol. 65, no. 2, p. 122205, 2022.

[18] M. W. Siti et al., "Application of load frequency control method to a multi-microgrid with energy storage system," J. Energy Storage, vol. 52, p. 104629, 2022.

[19] Z. Tu et al., "Optimal reset-control-based load frequency regulation in isolated microgrids," IEEE Trans. Sustain. Energy, vol. 13, no. 4, pp. 2239-2249, 2022.

[20] S. Yıldız et al., "An innovative LFC scheme for multi-area microgrid incorporating with hydrogen-based demand response mechanism," Int. J. Hydrogen Energy, vol. 48, no. 99, pp. 39425-39441, 2023.

[21] A. Saxena, R. Shankar, S. K. Parida, and R. Kumar, "Demand response based optimally enhanced linear active disturbance rejection controller for frequency regulation in smart grid environment," in Proc. IEEE Industry Applications Conf., 2022, vol. 58, no. 4, pp. 4337-4349.

[22] B. Khokhar and K. S. Parmar, "Utilizing diverse mix of energy storage for LFC performance enhancement of a microgrid: A novel MPC approach," in Proc. Appl. Energy, 2023, vol. 333, p. 120639.

[23] R. K. Khadanga, D. Das, A. Kumar, and S. Panda, "Sine augmented scaled arithmetic optimization algorithm for frequency regulation of a virtual inertia control based microgrid," in Proc. ISA Trans., 2023, vol. 138, pp. 534-545.

[24] A. Hossam-Eldin et al., "Improving the frequency response of hybrid microgrid under renewable sources' uncertainties using a robust LFC-based African vulture optimization algorithm," in Proc. Processes, 2022, vol. 10, no. 11, p. 2320.

[25] K. Sabahi, M. Tavan, and A. Hajizadeh, "Adaptive type-2 fuzzy PID controller for LFC in AC microgrid," in Proc. Soft Comput., 2021, vol. 25, pp. 7423-7434.

[26] H. Abubakr et al., "Adaptive LFC incorporating modified virtual rotor to regulate frequency and tie-line power flow in multi-area microgrids," in Proc. IEEE Access, 2022, vol. 10, pp. 33248-33268.

[27] P. K. Pathak and A. K. Yadav, "Fuzzy assisted optimal tilt control approach for LFC of renewable dominated micro-grid: A step towards grid decarbonization," in Proc. Sustain. Energy Technol. Assess., 2023, vol. 60, p. 103551.

[28] M. Ş. Üney and Ö. A. Karaman, "Load Frequency Control (LFC) of a Microgrid using PSCAD/EMTDC Simulation Program," in Proc. Adıyaman Univ. Eng. Sci. J., 2021, vol. 8, no. 15, pp. 328-342.

[29] B. Khokhar, S. Dahiya, and K. P. S. Parmar, "Load frequency control of a multi-microgrid system incorporating electric vehicles," in Proc. Electr. Power Compon. Syst., 2021, vol. 49, no. 9-10, pp. 867-883.

[30] Y. Li, L. Zhang, K. Lai, and X. Zhang, "Dynamic state estimation method for multiple battery energy storage systems with droop-based consensus control," in Proc. Int. J. Electr. Power Energy Syst., 2022, vol. 134, p. 107328.

[31] C. N. S. Kalyan et al., "Comparative performance assessment of different energy storage devices in combined LFC and AVR analysis of multi-area power system," in Proc. Energies, 2022, vol. 15, no. 2, p. 629.

[32] S. Biswas, P. K. Roy, and K. Chatterjee, "FACTS-based 3DOF-PID controller for LFC of renewable power system under deregulation using GOA," in Proc. IETE J. Res., 2023, vol. 69, no. 3, pp. 1486-1499.

[33] G. Sharma, K. Narayanan, I. Davidson, and K. T. Akindeji, "Integration and enhancement of load frequency control design for diverse sources power system via DFIG based wind power generation and interconnected via parallel HVDC/EHVAC tie-lines," in Proc. Int. J. Eng. Res. Afr., 2020, vol. 46, pp. 106-124.

[34] H. Shukla and S. Gudhe, "Incorporation of HVDC into Thermal-Gas-EV System for LFC Considering Time Delay Effect," in Proc. Int. Conf. Comput. Techn. Appl., 2021, pp. 483-493.

[35] I. A. Khan et al., "Load frequency control in power systems with high renewable energy penetration: A strategy employing PIλ (1+ PDF) controller, hybrid energy storage, and IPFC-FACTS," in Proc. Alexandria Eng. J., 2024, vol. 106, pp. 337-366.

[36] A. A. Abou El-Ela et al., "Design of cascaded controller based on coyote optimizer for load frequency control in multi-area power systems with renewable sources," in Proc. Control Eng. Pract., 2022, vol. 121, p. 105058.

[37] M. Alharbi et al., "Innovative AVR-LFC design for a multi-area power system using hybrid fractional-order PI and PIDD2 controllers based on dandelion optimizer," in Proc. Mathematics, 2023, vol. 11, no. 6, p. 1387.

[38] S. Aziz et al., "Variable universe fuzzy logic-based hybrid LFC control with real-time implementation," in Proc. IEEE Access, 2019, vol. 7, pp. 25535-25546.

[39] B. P. Nayak, P. C. Nayak, and R. C. Prusty, "Application of FPA based on PID controller for LFC of two-area multi-source hydrothermal Power system," in Proc. 2020 Int. Conf. Renew. Energy Integr. Smart Grids: A Multidiscip. Approach Technol. Model. Simul. (ICREISG), Feb. 2020, pp. 228-233.

[40] S. S. Mohamed, S. H. Elbanna, and A. M. Abdel-Ghany, "Fuzzy self-tuning fractional order PID controller design in load frequency control of power systems," in Proc. 2020 12th Int. Conf. Electr. Eng. (ICEENG), Jul. 2020, pp. 89-96.

[41] M. Joshi, G. Sharma, P. N. Bokoro, and N. Krishnan, "A fuzzy-PSO-PID with UPFC-RFB solution for an LFC of an interlinked hydro power system," Energies, vol. 15, no. 13, p. 4847, 2022.

[42] M. Shouran, F. Anayi, M. Packianather, and M. Habil, "Different fuzzy control configurations tuned by the bees algorithm for LFC of two-area power system," Energies, vol. 15, no. 2, p. 657, 2022.

[43] A. M. A. Soliman, M. B. Eldin, and M. A. Mehanna, "Application of WOA tuned type-2 FLC for LFC of two area power system with RFB and solar park considering TCPS in interline," IEEE Access, vol. 10, pp. 112007-112018, 2022.

[44] D. K. Gupta et al., "Fractional order PID controller for load frequency control in a deregulated hybrid power system using Aquila Optimization," Results Eng., p. 102442, 2024.

[45] S. Biswas, P. K. Roy, and K. Chatterjee, "FACTS-based 3DOF-PID controller for LFC of renewable power system under deregulation using GOA," IETE J. Res., vol. 69, no. 3, pp. 1486-1499, 2023.

[46] D. Jain, M. K. Bhaskar, and M. Parihar, "PSO-Based Controller for LFC of Deregulated Power System," in Proc. Int. Conf. Paradigms Commun., Comput. Data Analytics, Apr. 2023, pp. 607-624, Singapore: Springer Nature Singapore.

[47] D. Jain, M. K. Bhaskar, and M. Parihar, "Comparative Analysis of Load Frequency Control Problem of Multi Area Deregulated Power System Using Soft Computing Techniques," Math. Statist. Eng. Appl., vol. 71, no. 4, pp. 10713-10729, 2022.

[48] S. A. Hosseini, "Frequency control using electric vehicles with adaptive latency compensation and variable speed wind turbines using modified virtual inertia controller," Int. J. Electr. Power Energy Syst., vol. 155, p. 109535, 2024.

[49] S. A. Hasen, Ö. Aydın, S. Ayasun, and Ş. Sönmez, "Impact of virtual inertia and damping control on stability delay margins of load frequency control systems with renewable energy sources," Electr. Eng., vol. 106, no. 1, pp. 323-341, 2024.

[50] Z. Wang et al., "Load Frequency Control of Multiarea Power Systems with Virtual Power Plants," Energies, vol. 17, no. 15, p. 3687, 2024.

[51] A. Oshnoei et al., "Coordinated control scheme for provision of frequency regulation service by virtual power plants," Appl. Energy, vol. 325, p. 119734, 2022.

Genetic Algorithm Optimized High Precision Nonlinear Fractional Order PID Controller for Magnetic Levitation System

Devanathan V
EEE Department
NIT Puducherry
Karaikal, India
devanaathan357@gmail.com
ORCID: 0009-0007-8169-1817

Kannan S
EEE Department
NIT Puducherry
Karaikal, India
kannansakthivel01@gmail.com
ORCID: 0009-0002-0648-5494

Vinopraba T
EEE Department
NIT Puducherry
Karaikal, India
vinopraba@nitpy.ac.in
ORCID: 0000-0002-3889-0470

N Karthik Chary
EEE Department
NIT Puducherry
Karaikal, India
charym38@gmail.com

Abstract—Conventional Proportional Integral Derivative (PID) controllers are widely used in industrial control systems because of their simplicity and effectiveness. However, controlling nonlinear and inherently unstable systems remains challenging with traditional PID controllers, as they often exhibit poor robustness, limited stability margins and unsatisfactory transient response characteristics in complex dynamic environments. To overcome these limitations, this work proposes the development and real time implementation of a Genetic Algorithm optimized Fractional Order PID (GAFOPID) controller. This controller extends the control action beyond integer order dynamics by utilizing fractional calculus for its integral and derivative components. The optimization of the FOPID controller's parameters is achieved through a metaheuristic global optimization technique, specifically utilizing a genetic algorithm. The effectiveness of this controller is tested on the Quanser magnetic levitation (MagLev) system, which serves as a nonlinear actuated system for evaluation. To implement this control method, the Matlab Simulink library is employed. The performance of a GAFOPID controller is compared to that of a GA optimized traditional PID controller (GAPID). The results illustrate significant improvements in stability augmentation, reduced overshoot and enhanced response time across various reference trajectories. This research provides a robust framework for advanced control applications, allowing for efficient tuning that leads to offering superior adaptability and precision.

Index Terms—Genetic Algorithm Optimized Fractional Order PID, Magnetic Levitation, Nonlinear Control.

I. INTRODUCTION

In modern control engineering, the control of nonlinear and unstable systems remains a challenging task and necessitates advanced control strategies that surpass the limitations of conventional controllers. Among these, Fractional Order Proportional Integral Derivative (FOPID) controllers have been recognized as a successful substitute for traditional PID controllers. Although traditional Proportional Integral Derivative (PID) controllers frequently encounter difficulties such as overshoot, prolonged settling times and oscillatory responses when dealing with highly nonlinear systems. By incorporating fractional order integration and differentiation, FOPID controllers enhance flexibility and stability margins, extending control beyond integer order dynamics and improving performance in nonlinear systems. Tuning these additional parameters is non trivial, necessitating optimization techniques that can efficiently explore the high dimensional parameter space. The optimal gains for the controllers that mitigate steady state errors, reduce overshoot and enhance transient response characteristics are identified by utilizing GA, a well known global search heuristic, which can better adapt to the dynamic behavior of nonlinear systems than traditional tuning approaches.

The work presented in this paper examines how a genetic algorithm based optimization is used to improve the PID and FOPID controller parameters to position the ball of the Quanser maglev system. With the use of the global search capabilities of GA's, the proposed approach aims to enhance response characteristics and compare the response of both PID and FOPID controllers with different reference trajectories. The modeling, implementation and validation of this GA optimized FOPID controller contribute to addressing the challenges posed by the nonlinear dynamics of maglev systems, better than GA optimized classical PID controller, providing a hands on experimental framework for advanced control strategies.

MagLev technology enables objects to be suspended in the air without any physical contact by counteracting the gravitational forces with the help of magnetic fields. This technology is widely applied in areas such as precision manu-

979-8-3315-3013-6/25 $31.00 © 2025 IEEE

facturing, high speed transportation and vibration isolation. In addition, these systems offer significant advantages in minimal wear, high precision, and frictionless operation. However, the inherent instability and nonlinear dynamics of these systems pose significant challenges for control design. The Quanser Magnetic Levitation system is a nonlinear electromechanical system, a single degree of freedom system which serves as an ideal experimental setup for research studies of nonlinear dynamics control. An overhead electromagnet creates an attractive force on a metal ball that is positioned on a post. A photosensitive sensor built into the post is used to measure the steel ball's position while it is levitating. A current sensor is also included that tracks the current passing through the electromagnet coil.

The high inductance causes a slower current response, whereas the force increases rapidly as the air gap narrows. Electromagnet voltage/current dynamics and ball/electromagnet dynamics are the two subsystems that control the system dynamics. Despite being employed in many different applications, conventional PID (proportional integral derivative) controllers are frequently inadequate for handling the complexities of nonlinear maglev systems, which leads to various issues such as overshoot, extended settling times, oscillatory response, and steady state errors. FOPID controllers are the ones that overcome the issues of traditional PID control using fractional calculus, which offers improved precision, stability and flexibility for such nonlinear and unstable systems. Figure 1 illustrates the block diagram of the GA optimized PID/FOPID controller, which demonstrates the structure of the proposed optimization approach for tuning the controller parameters (K_p, K_i, K_d, λ, μ) by minimizing the ITAE (Integral of Time-weighted Absolute Error) performance index.

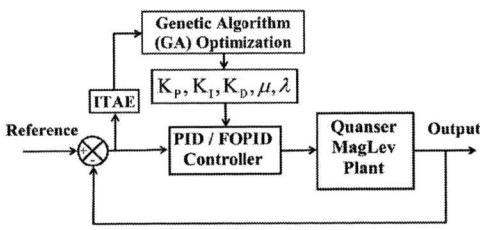

Fig. 1: GA Optimized Controller Block Diagram

The development and optimization of advanced controllers for maglev systems has been extensively explored in recent years in research for achieving high precision control. Various strategies have been proposed to address challenges such as system uncertainties, time delay effects and mainly, nonlinearity. This section examines significant contributions that impacted the creation of our FOPID controller, which is optimized by a genetic algorithm. In [1], an adaptive fixed time precision control strategy was proposed integrating adaptive switching gains, but faced limitations in adapting to highly dynamic systems with varying controller parameters, which was complexity for tuning. In [2], exact and precise

modeling of system dynamics is not achieved even though the approach of cascade control for inner and outer loops which also encountered slight difficulties in handling large parameter variations. A fractional order PID controller with time delay is proposed in [3] for maglev system, which reduces the oscillations but, on the other hand, makes it less feasible for implementation challenges due to complex tuning based on the Hermite Biehler theorem. For the purpose of minimizing integral square error (ISE), an optimal fractional order PID controller using Nelder's Mead optimization has been designed in [4] which shows better ability in managing nonlinearities and achieving precise control. The Fractional Order Internal Mode Controller (FOIMC) for a connected tank system was described in [5] was implemented using particle swarm optimization (PSO), highlighting the significance of fractional order dynamics to improve system performance under variations. Also, PSO faces challenges related to local minima and convergence speed. Those can be addressed by using GA like optimization techniques. Similarly, the FOPID controller is designed for conical tank processes in [6] that clearly shows superior performance over classical PID controllers, and also well illustrates the limitations of handling external disturbances and higher order dynamics. In [7]–[9], extensive studies on the basic operation, modeling and practical guides for magnetic levitation systems and also their control strategies have been presented. In [8], presents mainly about the technological advancements and large scale applications of maglev systems. Fractional order control, which extends the traditional integer order techniques, is discussed in [10]. This work introduces the advantages of improving the control for better precision, robustness, and particularly in terms of high stability and accuracy compared to conventional controllers. From [11], it is clear that FOPID is used not only for better precision, but also to deliver better set point tracking is obtained using heuristic approach results in better for tuning of parameters. In [12], flexible tuning was made possible by the combination of fractional order PID controllers with internal model control (IMC), which provided an extra degree of freedom and high nonlinear systems for dynamic performance. Recent studies have highlighted the significance of advanced control strategies for nonlinear systems. In [13], FOPID controllers have been shown to offer improved precision and robustness in the presence of plant uncertainties and nonlinearities. Additionally, [14] explored the dynamic behavior of magnetic levitation systems, emphasizing the importance of precise modeling and control to manage inherent nonlinearities. Advanced control strategies, combined with intelligent optimization techniques such as genetic algorithms, have demonstrated enhanced stability and tracking performance in complex systems [15]. These insights motivated the adoption of genetic algorithm based optimization to fine tune FOPID parameters, aiming to minimize ITAE and improve the overall dynamic response in this research work. The rest of the paper is organized as follows: Section II presents the system modeling and dynamics of the Quanser magnetic levitation system. Section III outlines the proposed methodology, including the genetic algorithm

979-8-3315-3013-6/25 $31.00 © 2025 IEEE

based FOPID optimization approach. Section IV discusses the simulation and hardware implementation results, highlighting the controller's performance. Finally, Section V provides the conclusion and future research directions.

II. SYSTEM MODELING AND DYNAMICS

This section explains the derivation, dynamics, and mathematical modeling of the maglev system, including its mechanical and electrical characteristics and Fig. 2 shows the real time hardware setup of the Quanser maglev system, the setup consists of an electromagnet at the top, a steel ball suspended below it, and an optical sensor integrated within the post to measure the ball's position. This system serves as a highly nonlinear and unstable plant. Fig. 3 shows the free body and electrical schematic of the Quanser Magnetic Levitation (MagLev) system. The diagram shows the coil voltage V_c, coil current I_c, and the series combination of coil inductance L_c and resistance R_c. A sensing resistor R_s measures the current. The ball of mass M_b is levitated by the electromagnetic force F_c that opposes the gravitational force F_g. The vertical displacement x_b is measured by an optical sensor. This representation is fundamental for modeling the system dynamics and designing the controller. Table I provides a summary of the parameters of the maglev system.

Fig. 2: Quanser Magnetic Levitation System.

A. System Specifications

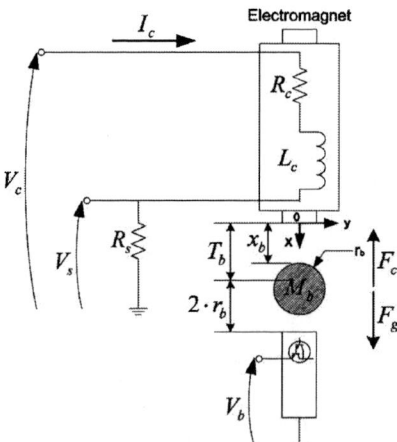

Fig. 3: Free Body and Dynamics Diagram of Quanser Magnetic Levitation System.

TABLE I: Maglev System Parameters

Symbol	Description & Unit	Value
I_{cmax}	Continuous Coil Current Maximum ($\pm$A)	3
L_c	Coil's Inductance (mH)	412.5
R_c	Coil's Resistance (Ω)	10
N_c	Coil Wire's Number of Turns (No unit)	2450
l_c	Length of the Coil (m)	0.0825
r_c	Core Radius of Coil (m)	0.008
K_m	Electromagnetic Force-Constant (N·m²/A²)	6.5308×10^{-5}
R_s	Current Sense's Resistance (Ω)	1
r_b	Steel Ball Radius (m)	1.27×10^{-2}
M_b	Steel Ball Mass (kg)	0.068
T_b	Travel of the Steel Ball (m)	0.014
G	Earth's Gravitational Constant (m/s²)	9.81
μ_0	Constant for Magnetic-Permeability (H/m)	$4\pi \times 10^{-7}$
K_b	Ball Position Sensor's Sensitivity (m/V)	2.83×10^{-3}

B. Electrical Equations

The system includes a coil with inductance L_c, resistance R_c, and a current sensor with resistance R_s. The voltage sensed by the system is expressed as:

$$V_s(t) = L_c \frac{di_c(t)}{dt} + (R_c + R_s)i_c(t) \quad (1)$$

where $V_s(t)$ is the voltage measured, $i_c(t)$ represents the current in the coil, and $V_c(t)$ represents the voltage that is applied to the coil.

C. Transfer Function in First Order

Taking the Laplace transform of the above equation (1) (assuming zero initial conditions),

$$V_s(s) = L_c s I_c(s) + (R_c + R_s)I_c(s) \quad (2)$$

Rearranging for $I_c(s)$,

$$I_c(s) = \frac{V_s(s)}{L_c s + (R_c + R_s)} \quad (3)$$

The transfer function relating the coil current to the applied voltage is then:

$$G_c(s) = \frac{I_c(s)}{V_s(s)} = \frac{1}{L_c s + (R_c + R_s)} \quad (4)$$

For a first-order approximation, we define:

$$\tau_c = \frac{L_c}{R_c + R_s}, \quad K_c = \frac{1}{R_c + R_s} \quad (5)$$

Thus, the first order transfer function can be used to approximate the coil's electrical dynamics is

$$G_c(s) = \frac{I_c(s)}{V_c(s)} = \frac{K_c}{\tau_c s + 1} \quad (6)$$

where τ_c is the system's time constant and K_c is its steady state gain.

D. Nonlinear Model

The coil produces an attractive electromagnetic force F_c that is indirectly proportional to the air gap x_b squared and proportionate to the square of the current:

$$F_c = \frac{K_m i_c^2(t)}{2x^2(t)} \tag{7}$$

where K_m is the electromagnetic force constant, the air gap between the steel ball and the electromagnet is represented by x_b; $(x_b > 0)$, and the electromagnetic force that draws the steel ball towards is denoted by F_c.

The ball is subject to the gravitational force F_g, which equals:

$$F_g = M_b g \tag{8}$$

where g represents the gravitational acceleration and M_b represents the steel ball's mass.

The total external force required to suspend the ball from equations (7) and (8)

$$F_{ext} = -F_c + F_g = -\frac{K_m i_c^2(t)}{2x_b^2(t)} + M_b g \tag{9}$$

E. Nonlinear Equation of Motion

The nonlinear equation of motion for the steel ball from equation (9) can be written as follows using Newton's second law:

$$\ddot{x}_b(t) = -\frac{K_m i_c^2(t)}{2M_b x_b^2(t)} + g \tag{10}$$

where $\ddot{x}_b(t)$ is the acceleration of the ball. The extremely nonlinear dynamics of the magnetic levitation system is taken into account when building the control system, which is based on this nonlinear model and stabilizes the ball at the required position.

F. Open-Loop Transfer Function

The nonlinear equations of motion must be linearized around an equilibrium point, since a transfer function can only represent the system's dynamics from a linear differential equation.

1) Linearization of the System: The system consists of an electromagnet levitating a ball at an equilibrium position (x_{b0}, i_{c0}). Small deviations from this equilibrium are represented as:

$$x_b = x_{b0} + \delta x_b, \quad i_c = i_{c0} + \delta i_c. \tag{11}$$

The force acting on the ball due to the electromagnet is nonlinear. Linearizing it around the equilibrium position using a Taylor series expansion provides an approximation that enables modeling in the Laplace domain.

2) Transfer Function Derivation: The linearized dynamics between the coil current and ball position can be expressed as:

$$G_b(s) = \frac{X_b(s)}{I_c(s)} = -\frac{K_b \omega_b^2}{s^2 - \omega_b^2} \tag{12}$$

where K_b is the DC (steady-state) gain, ω_b is the system's natural frequency.

The open loop transfer function $G_m(s)$ takes into account the dynamics of the electromagnet current loop and is defined as

$$G_m(s) = \frac{X_b(s)}{I_{c,d}(s)} = T_c(s)G_b(s) \tag{13}$$

where

$$G_b(s) = \frac{X_b(s)}{I_c(s)}. \tag{14}$$

For simplification, we assume that the electromagnet current dynamics can be neglected, means $I_c(s) = I_{c,d}(s)$ and the current loop transfer function is approximated as $T_c(s) = 1$. Given that $G_m(s) = G_b(s)$ and the $G_b(s)$ transfer function is found in (12), this leads to the final open loop transfer function $G_m(s)$ considering the dynamics of the electromagnet current loop:

$$G_m(s) = \frac{X_b(s)}{I_c(s)} = -\frac{2gx_{b0}}{I_c x_{b0}(s^2 - \frac{2g}{x_{b0}})} \tag{15}$$

where x_{b0} is the ball position in static equilibrium at a nominal operating point.

III. METHODOLOGY

This section describes the process used to reduce the Integral Time Absolute Error (ITAE) with the use of GA to tune the FOPID and PID controllers. Through the determination of the controller gains, the tuning procedure guarantees optimal stabilization of the magnetic levitation system.

A. Controller Design

1) PID Controller Design: One of the most often used controllers in industrial settings is the PID controller. Three main tuning parameters for the PID controller are

- **Proportional Gain** (K_p): Provide a proportionate control response to the error.
- **Integral Gain** (K_i): Addresses long term cumulative variances and integrates the error over time, assisting in the elimination of steady state error.
- **Derivative Gain** (K_d): Predicts future errors by taking into account the error's rate of change, which enhances stability and reduces oscillations.

The PID control, shown in Fig. 4, is used to regulate and keep the steel ball in the required position. The PID control law is expressed as follows:

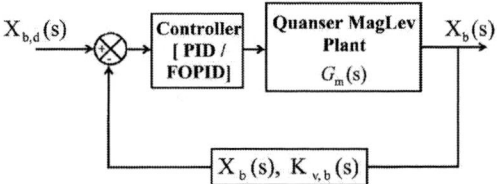

Fig. 4: Block Diagram of Control System for Quanser Maglev System

- **PID Controller:**

$$I_c(s) = (X_{b,d}(s) - X_b(s))\left(k_{p,b} + \frac{k_{i,b}}{s} + k_{d,b}s\right)$$
$$- k_{v,b}sX_b(s) \quad (16)$$

By replacing equation (15) in place of equation (16), the closed loop transfer function is obtained from the open-loop transfer function of the ball position for the PID control as follows:

$$T_b(s) = \frac{X_b(s)}{X_{b,d}(s)} = -\frac{2x_{b0}(k_d s^2 + k_p s + k_i)}{I_{c0}x_{b0}s^3 - 2gx_{b0}(k_d + k_v)s^2}$$
$$- 2g(I_{c0} + 2b_0 k_p)s - 2gx_{b0}k_i \quad (17)$$

2) FOPID Controller Design: When integral and derivative components are combined with fractional calculus, the FOPID controller expands upon the traditional PID controller. Fig. 4 illustrates how the steel ball is controlled and maintained in the desired position using a fractional order proportional integral derivative control. The following is the FOPID control law:

- **FOPID Controller:**

$$I_c(s) = (X_{b,d}(s) - X_b(s))\left(K_p + \frac{K_i}{s^\lambda} + K_d s^\mu\right)$$
$$- k_{v,b}sX_b(s) \quad (18)$$

By using equation (15) in place of equation (18), the closed-loop transfer function of the position of the steel ball for FOPID control is as follows:

$$T_b(s) = \frac{X_b(s)}{X_{b,d}(s)} = -\frac{2x_{b0}\left(k_d D^\mu s^2 + k_p s + k_i D^{-\lambda}\right)}{[I_{c0}b_0 s^3 - 2gx_{b0}\left(k_d D^\mu + k_v\right)s^2}$$
$$- 2g(I_{c0} + 2b_0 k_p)s - 2gx_{b0}k_i D^{-\lambda}] \quad (19)$$

B. Genetic Algorithm for Optimization

Using a Genetic Algorithm, controller gains are optimized and enhance system performance. Since the Integral of Time Absolute Error (ITAE) was discovered to be substantially greater than other error metrics, the goal is to minimize it. Effective stabilization is ensured by minimizing ITAE without degrading the necessary dynamics of the system.

The Genetic Algorithm adjusts the derivative K_d, integral K_i, and proportional K_p gains for the PID controller. The algorithm starts with a starting population of possible solutions, analyzes them using an ITAE based fitness function,

and then iteratively improves the solutions through crossover, mutation and selection processes until they converge and for FOPID, GA expands the optimization to five parameters, namely fractional integral order λ, fractional derivative order μ, K_p, K_i and K_d. More flexibility is made possible by this wider search area, especially when dealing with nonlinearities and disturbances that arise in dynamic systems such as the Quanser maglev system.

Optimization guarantees a better settling time, less overshoots and reduced steady state error. The controllers are fine tuned for strong stability and accurate control by utilizing GA's capacity for navigating complex, multidimensional search spaces, demonstrating the significance of ITAE minimization for system performance. Below is a summary of the steps involved in the GA based tuning process:

1) **Initialization:** Within predetermined parameters, a population of potential solutions is produced at random. The controller parameters are represented by each potential solution (e.g. K_p, K_i, K_d for PID or K_p, K_i, λ, K_d, μ for FOPID).

2) **Fitness Evaluation:** The ITAE error is computed using the simulated response of the system to assess each potential solution. Better performance is indicated by lower ITAE values.

3) **Selection:** For the next generation, candidate solutions with lower ITAE values are chosen. This guarantees that the likelihood of high performing solutions propagating is improved.

4) **Crossover and Mutation:** Maintaining genetic diversity involves:

   - **Crossover:** Produces offspring by combining the parameters of chosen solutions and explores new areas of the solution space.
   - **Mutation:** Prevents premature convergence and preserves variety by introducing tiny, random changes to the parameters of offspring.

5) **Iteration:** Over several generations, the algorithm repeats the fitness evaluation, crossover, selection, and mutation processes. Until a convergence criterion such as a certain number of generations is satisfied, this iterative refining continues.

6) **Output:** The ideal set of controller gains is chosen from the best performing solution of the last generation.

GA reduces ITAE in order to improve controller parameters:

$$\text{ITAE} = \int_0^T t|e(t)|\mathrm{d}t \quad (20)$$

where $e(t)$ represents the system error. The magnetic levitation system was used to validate the optimized parameters, showing improved performance with lower ITAE.

The optimization was performed with a population size of 100 and a total of 15 generations, using parallel processing to enhance computational efficiency. The best and mean fitness values for both the PID and FOPID controllers were evaluated, showing that the FOPID controller achieved a lower ITAE

(Best: 0.0311197, Mean: 0.0313905) compared to the PID controller (Best: 0.0312068, Mean: 0.0366917), indicating improved performance. The convergence of fitness values over generations, as depicted in figures 5 and 6, highlights the effectiveness of GA in fine-tuning control parameters, with FOPID exhibiting superior control precision and stability.

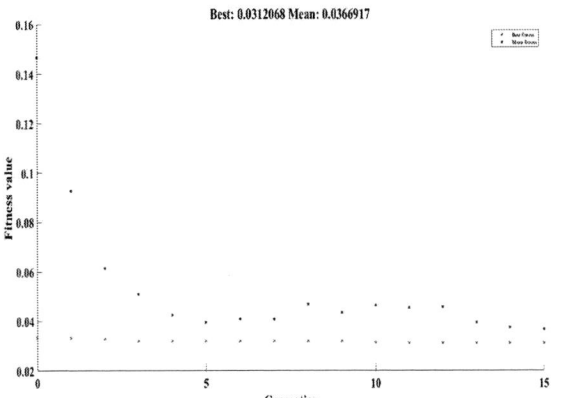

Fig. 5: Minimization of ITAE by Genetic Algorithm for PID Controller

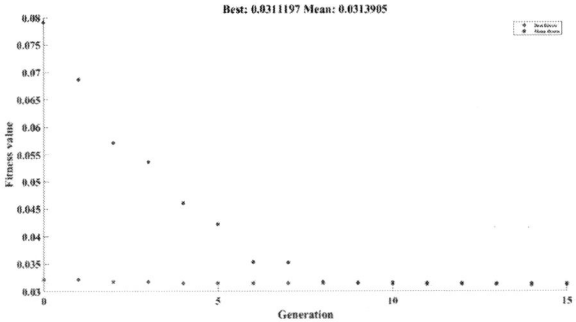

Fig. 6: Minimization of ITAE by Genetic Algorithm for FOPID Controller

IV. SIMULATION AND HARDWARE SYSTEM RESULTS

The performance of the closed loop system was investigated by creating a simulation model that incorporated the FOPID and PID controllers tuned using a GA in the MATLAB Simulink environment. The output responses of optimized FOPID controllers for different reference trajectories have been compared to that of traditional PID controllers and are pictorially represented in a comparative manner.

A. Simulation Results

For the simulation, square and sine wave reference inputs were used to evaluate the performance of PID and FOPID controllers on the maglev system. Fig. 7 and Fig. 8 show the output responses and control signals (coil currents), respectively.

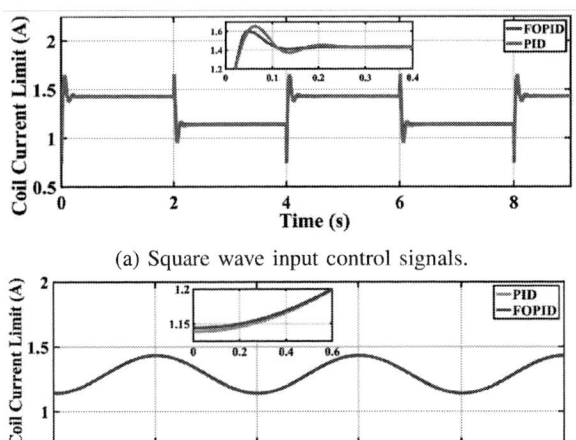

(a) Square wave input control signals.

(b) Sine wave input control signals.

Fig. 7: Simulation - Control Signals of both PID and FOPID controllers.

The FOPID controller offers a smoother control action, lowering rapid changes and actuator stress, while the PID controller shows abrupt transitions with visible variations for a square wave input (Fig. 7a). Both controllers manage the coil current around a comparable mean value in the sine wave input situation (Fig. 7b), but the FOPID controller produces a more consistent control effort with improved system robustness comparative to the PID controller. The FOPID controller

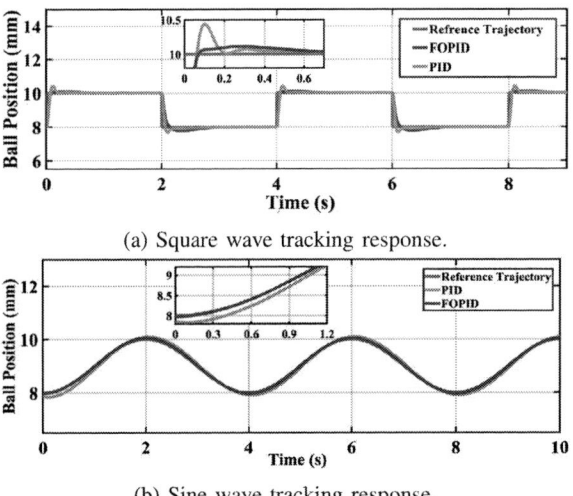

(a) Square wave tracking response.

(b) Sine wave tracking response.

Fig. 8: Simulation - Output Response for square and sine wave inputs.

offers smoother transitions and less overshoot than the PID controller, which shows overshoot and slower settling for a square wave reference, as shown in Fig. 8a. In Fig. 8b both controllers follow the reference well in the sine wave tracking situation, but the FOPID controller achieves superior accuracy with a lower steady state error.

B. Real Time Implementation Results

Real time experiments were carried out on the Quanser maglev system to validate the PID and FOPID controllers' effective performance in nonlinear hardware and also to validate the simulation results. The experimental setup is depicted in the following Fig. 9.

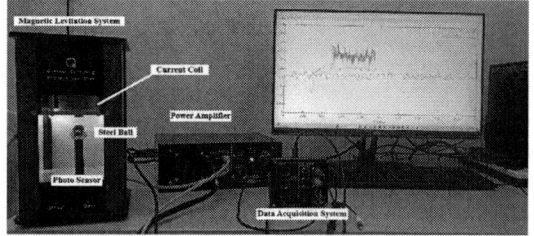

Fig. 9: Experimental hardware setup for real-time testing.

Figs. 10a and 10b display the coil current signals produced by both controllers in response to square and sine wave inputs. The FOPID controller generates a smoother control signal with less chattering than the PID controller, which shows notable oscillations and ripples in the coil current for a square wave reference input.

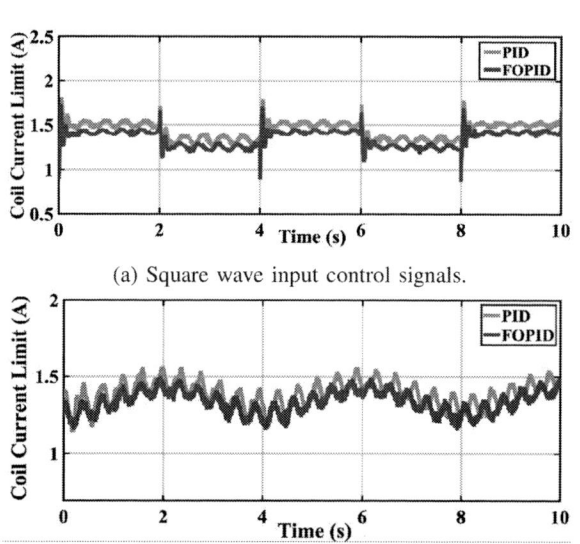

(a) Square wave input control signals.

(b) Sine wave input control signals.

Fig. 10: Hardware - Control Signals of both PID and FOPID controllers.

This depicts how the smoother operation of the FOPID controller compared to the PID controller in real time maglev applications. The coil current is efficiently controlled by both controllers for a sinusoidal reference input. In contrast to the PID controller, the FOPID controller maintains a more constant control signal and reduces sudden changes by more gradually following the reference trajectory.

Square and sine wave reference trajectories were used to test the Quanser maglev system's tracking capabilities in real time tests. Fig. 11 shows that the FOPID controller offers a smoother response with faster settling and less overshoot than the PID controller, which shows more oscillations and overshoots for a square wave tracking response. The inset plot, which focuses on the transient behavior of both controllers, clearly shows this improvement (Fig. 11a). Fig. 11b shows a sine wave tracking response, which both controllers effectively track. Compared to the PID controller, which exhibits continuous oscillations, the FOPID controller minimizes steady state error and phase lag, guaranteeing a more precise and stable tracking performance.

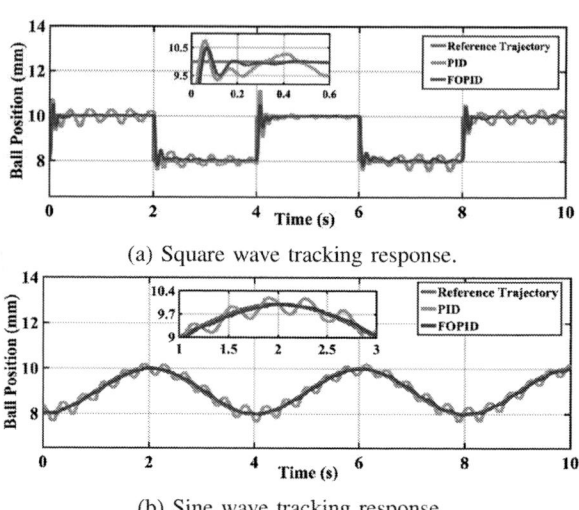

(a) Square wave tracking response.

(b) Sine wave tracking response.

Fig. 11: Hardware - Output Response of Quanser Maglev System for square and sine wave inputs.

Overall, the real time implementation findings demonstrate that the FOPID controller is a more reliable control strategy for Maglev applications that need accurate positioning and smooth control dynamics due to its greater tracking precision and stability when compared to the PID controller.

C. Comparative Analysis

The genetic algorithm optimized PID and FOPID controllers gain values are shown in Table II below. Additionally, corresponding ITAE values are also included, which show how well each controller performs in reducing errors over time.

TABLE II: Control Parameters of PID and FOPID Controllers

Parameter	PID Simulation	PID Hardware	FOPID Simulation	FOPID Hardware
K_p	-230.33		-175.48	
K_i	-160.55		-149.23	
K_d	-0.94		-7.5	
λ	$-$		0.89	
μ	$-$		0.6	
ITAE	0.0367		0.0313	

In terms of error reduction, it is evident from Table II that the FOPID controller outperforms the PID controller. The

ITAE value for the FOPID controller is 0.0313, which is less than the PID controller's ITAE value of 0.0367. This suggests that the FOPID controller improves both transient and steady-state performance by achieving a better time-weighted error reduction. More tuning degrees are available for the FOPID controller, allowing the fractional order parameters λ and μ, which improves the adaptability of the system dynamics. The FOPID controller's higher derivative gain K_d indicates better damping properties, which may decreases oscillations and overshoot and make it a more effective control approach in real world applications. A quantitative comparison of rise time, settlement time, and overshoot for PID and FOPID controllers based on square and sine wave response analysis is displayed in the Table (III).

TABLE III: Comparison of PID and FOPID Control Performance

Parameter	PID Simulation	PID Hardware	FOPID Simulation	FOPID Hardware
Square Wave Tracking				
(T_r) Rise Time (s)	0.046	0.027	0.032	0.022
(T_s) Settling Time (s)	0.17	0.3	0.098	0.23
Overshoot (%)	4.5	11.56	1.19	5
Sine Wave Tracking				
(T_r) Rise Time (s)	1.15	1.45	1.003	1.24
(T_s) Settling Time (s)	1.6	1.178	0.3	0.45
Overshoot (%)	1	2.03	0.5	0.22

The above results show that the FOPID controller performs better tracking performance with reduced settling time, overshoot values and a stable response compared to the conventional PID controller in both simulation and real time implementations. Overall, the findings show that by lowering error and improving dynamic responsiveness, the FOPID controller outperforms the conventional PID controller.

V. CONCLUSION

Nonlinear systems are inherently challenging to control due to their complex and unstable dynamics, where traditional PID controllers often fail to deliver satisfactory performance in terms of overshoot, settling time, and steady-state error. To address these challenges in the context of maglev systems, a GAFOPID controller was developed, leveraging GA to optimize the FOPID parameters, including proportional, integral and derivative gains as well as fractional orders of integration and differentiation, thereby enhancing tuning flexibility and dynamic response. The proposed controller was implemented on the Quanser Magnetic Levitation system. Experimental results show that the GAFOPID controller significantly outperforms the traditional PID controller by achieving lower overshoot, faster settling time and reduced steady state error, demonstrating improved transient and steady

state performance and enhanced adaptability to changing input conditions, thereby proving its practical feasibility and robustness for real time applications. The study's findings open the way for FOPID controller development and use in the future in precision based engineering systems such as advanced robotics, aerospace and medicinal equipment. To further improve FOPID controller capabilities, future studies might investigate hybrid AI based control strategies, real time optimization methodologies, and adaptive GA tuning methods. As a result, the GA optimized FOPID controller sets a new standard for precision control and nonlinear system regulation, making it a viable option for next generation high performance control systems.

REFERENCES

[1] J. Wang, J. Rong, and J. Yang, "Adaptive Fixed-Time Position Precision Control for Magnetic Levitation Systems," *IEEE Transactions on Industrial Electronics*, vol. 69, no. 5, pp. 4717–4726, May 2022. [Online]. Available: https://ieeexplore.ieee.org/document/9734232

[2] J. Wen, J. Wu, and Y. Tian, "Levitation Control of Maglev Systems Based on Cascade Control," *IEEE Access*, vol. 11, pp. 15369–15379, Feb. 2023. [Online]. Available: https://ieeexplore.ieee.org/document/10055323

[3] D. S. Acharya, S. K. Mishra, S. K. Swain, and S. Ghosh, "Real-Time Implementation of Fractional-Order PID Controller for Magnetic Levitation Plant With Time Delay," *IEEE Access*, vol. 10, pp. 83456–83467, Aug. 2022. [Online]. Available: https://ieeexplore.ieee.org/document/9934927

[4] S. K. Verma, S. Yadav, and S. K. Nagar, "Optimal Fractional Order PID Controller for Magnetic Levitation System," in *Proceedings of the 5th IFAC Symposium on Fractional Differentiation and Its Applications (FDA)*, Beijing, China, July 2015, pp. 529–534. [Online]. Available: https://ieeexplore.ieee.org/document/7489095

[5] Sateesh Kumar V, T. Vinopraba, T. K. Radhakrishnan, and N. Sivakumaran, "Design and Implementation of the FOIMC Using the PSO Algorithm for the Coupled Tank System," in *Design and Control Advances in Robotics*, IGI Global, 2022, pp. 127–151.

[6] Sateesh Kumar V and T. Vinopraba, "Design of Fractional Order Controllers for Conical Tank Process," in *Proc. 2018 International Conference on Emerging Trends and Innovations in Engineering and Technological Research (ICETIETR)*, 2018, pp. 1–5.

[7] *MAGLEV: Maglev Plant User Manual*, Quanser Inc., Markham, ON, Canada, 2012.

[8] H.-S. Han and D.-S. Kim, *Magnetic Levitation: Maglev Technology and Applications*, vol. 13 in Springer Tracts on Transportation and Traffic. Springer, 2016.

[9] S. Preitl and M. Patočka, Eds., *Control of Magnetic Levitation Systems*. Springer.

[10] D. Xue, *Fractional-Order Control Systems: Fundamentals and Numerical Implementations*, vol. 1 in the series Fractional Calculus in Applied Sciences and Engineering, De Gruyter, 2018. [Online]. Available: https://doi.org/10.1515/9783110497977.

[11] G.-Q. Zeng, J. Chen, Y.-X. Dai, L.-M. Li, C.-W. Zheng, and M.-R. Chen, "Design of fractional order PID controller for automatic regulator voltage system based on multi-objective extremal optimization," *Neurocomputing*, vol. 160, pp. 173–179, 2015.

[12] T. Vinopraba, N. Sivakumaran, S. Narayanan, and T. K. Radhakrishnan, "Design of internal model control based fractional order PID controller," *J. Control Theory Appl.*, 2012. doi: 10.1007/s11768-012-1044-4.

[13] J. Zhang and L. Guo, "Theory and Design of PID Controller for Nonlinear Uncertain Systems," *IEEE Control Syst. Lett.*, 2019. doi: 10.1109/LCSYS.2019.2915306.

[14] H. A. Engmark and K. T. Hoang, "Modeling and Control of a Magnetic Levitation Platform," *IFAC-PapersOnLine*, vol. 56, no. 2, pp. 7276-7281, 2023. [Online]. Available: https://doi.org/10.1016/j.ifacol.2023.10.338.

[15] S. Feleke, B. Pydi, R. Satish, H. Kotb, M. Alenezi, and M. Shouran, "Frequency Stability Enhancement Using Differential-Evolution- and Genetic-Algorithm-Optimized Intelligent Controllers in Multiple Virtual Synchronous Machine Systems," *Sustainability*, vol. 15, no. 18, p. 13892, 2023. [Online]. Available: https://doi.org/10.3390/su151813892.

Design and Control of a Dual Active Bridge Converter for Bipolar DC Microgrid Applications

Aarthi A
Dept. of Energy and Environment
National Institute of Technology
Tiruchirappalli, India
aarthiamir31@gmail.com

Ramesh Dharavath
Dept. of Energy and Environment
National Institute of Technology
Tiruchirappalli, India
rameshnitt1995@gmail.com

Dr. D V Siva Krishna Rao K
Dept. of Energy and Environment
National Institute of Technology
Tiruchirappalli, India
damodharsiva@nitt.edu

Abstract—The increasing demand for renewable energy sources and electric vehicle (EV) charging infrastructure necessitates efficient and reliable power distribution systems. Bipolar DC grids provide flexibility, reliability, and safety compared to unipolar systems, reducing voltage magnitude and allowing for the connection of larger loads. Bipolar DC low-voltage distribution networks significantly enhance power transmission and efficiency, especially when integrating distributed energy sources and energy storage systems. This paper focuses on designing, analysing, and modelling a Dual Active Bridge (DAB) converter with a voltage balancer for a bipolar DC system. The DAB converter is selected for its bidirectional power flow capability, making it suitable for grid-connected and islanded operations. A modulation and control strategy for efficient closed-loop operation is implemented using the Generalized Average Model (GAM) technique. The system is modelled in MATLAB-Simulink under varying load conditions, and the converter's performance is evaluated in both open-loop and closed-loop modes. Simulation results demonstrate stable operation and effective voltage regulation under different load conditions.

Keywords— *Dual Active Bridge Converter, Voltage Balancer, DC Microgrid*

I. INTRODUCTION

Sustainable electricity generation and clean technologies are transforming electric power systems, increasing AC grid complexity and driving interest in low-voltage DC (LVDC) active grids at the distribution level. Sustainable distributed generation can integrate directly into DC distribution, with applications in data centres and EV charging. Bipolar DC networks enhance cost efficiency, reduce voltage ratings, and improve performance. Power electronics play a key role in modernising grids, enabling DC-based equipment and variable voltage/frequency distributed generators [1]. Isolated bidirectional AC-DC converters facilitate DC distribution integration with conventional AC grids, as shown in Fig. 1.

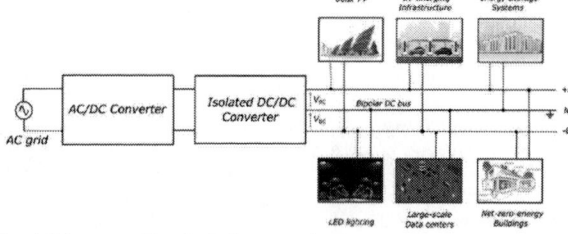

Fig. 1 Schematic of Bipolar DC microgrid

Microgrids enhance the reliability of distributed energy sources like solar, wind, and storage. They are categorised as AC, AC-DC hybrid, and DC. Bipolar DC bus configurations offer dual voltage levels and improved system reliability, with recent advancements in voltage balancers further enhancing efficiency. DC microgrids are key in renewable energy integration but face voltage and current imbalances, requiring power electronic-based solutions [2]. Applications such as EVS, DC-powered buildings, and green hydrogen favour DC infrastructure. Bipolar DC microgrids offer flexibility, reliability, and faster fault clearance via neutral conductors. However, they require an extra wire and risk voltage imbalance due to asymmetric loads. Solutions include voltage balancers or additional sources/loads connected to the three poles. Bipolar DC microgrids find applications in power distribution, aircraft, buildings, ships, and EV charging, as shown in Fig. 2.

Application	Voltage level
Data Centers	±190V
Aircraft Systems	± 270 V
Household	± 170 V
Shipboard	± 750 V or ± 1500 V
Electric vehicle charging station	> ± 269 V

Fig. 2. Bipolar DC applications and voltage levels [1]

DC bus stabilisation in microgrids can be achieved through centralised or distributed voltage balancers. Centralised architectures are common in bipolar DC microgrids but face challenges such as high current ripple and bulky capacitors. Decentralised architectures, using dedicated or distributed balancers, help address these issues. Standalone voltage balancers can employ topologies such as buck-boost, dual buck-boost, Cuk, or Super-SEPIC. Bipolar generators may use multiport DC-DC converters for voltage support, with non-isolated topologies merging two DC-DC converters, while isolated topologies offer higher voltage gain. Advanced converters, such as the voltage-balancing dual- active-bridge (VB-DAB) converter [3], incorporate fewer power switches and avoid bulky balancing inductors, achieving high efficiency and power density with zero-voltage switching (ZVS). Using a zigzag-delta AC-link transformer, the DAB3 converter [4] offers improved efficiency and power density for bipolar LVDC distribution. An enhanced dual-active-half- bridge (DAHB) converter [5] utilizes an inductor-capacitor voltage balancer (LCVB) to achieve voltage balancing without feedback control or extra active switches, ensuring balanced output voltages while preventing transformer saturation. The DAB converter [7] achieves higher efficiency due to its single-stage structure, with fewer components, reducing size and cost, but suffers from power losses in DC inductors. A full-bridge type voltage balancer in [8] improves efficiency by 3.7% at light load and 1.3% at heavy load.

979-8-3315-3013-6/25 $31.00 © 2025 IEEE

The extensive literature on isolated converters is categorized into four key areas: topology selection, modulation techniques, dynamic modelling, and closed-loop control. Since the selected topology is the Dual Active Bridge (DAB) converter, the following sections focus on the remaining three aspects. Phase shift ratios $D_1, D_2,$ and D_3 define the relation between phase shift and half the switching cycle ($Ts/2$), influencing converter dynamics. Higher control degrees improve performance but increase complexity. Extended-phase-shift (EPS) [6] control enhances power regulation, reduces circulating current, improves efficiency, and lowers current stress, extending device lifespan and reducing costs. DAB modelling is challenging due to its purely AC inductor current. The four common approaches [7] are reduced-order, harmonic, generalized average (GAM), and discrete-time models. The reduced-order model is best suited for closed-loop control in standard applications due to its simplicity and accuracy. However, for high-frequency DAB converters where ZVS intervals significantly impact dynamics, a discrete-time model is preferable. The GAM approach provides higher accuracy and remains unaffected by operating conditions, ensuring robust theoretical validation. Control strategies vary in complexity, performance, robustness, and cost [1]. Feedback-only control is simple but lacks dynamic performance. Linearization control improves response but suffers under parameter variations, while predictive current control enhances dynamics but is computationally intensive. A proportional-integral (PI) compensator is preferred for output voltage regulation. The phase shift ratios are adjusted based on voltage error, with the PI controller defined as: $Gc(s) = kp + (ki / s)$. This minimizes steady-state error.

Bipolar DC grids require voltage balancers to maintain stable operation under asymmetrical loads. These converters redistribute current, ensuring balanced voltages and enhancing feeder efficiency. The buck-type voltage balancer splits the DC bus voltage, forming a bipolar network with improved dynamics and independent buses. Various topologies impact efficiency, voltage gain, and component sizing. The Cuk-type avoids shoot-through but requires extra components, while SEPIC and Zeta topologies allow bidirectional current paths. A cost-effective balancing approach [9], as shown in Fig. 3, reduces inductor current while managing voltage stress. This paper proposes an enhanced DAB converter for bipolar DC networks with integrated voltage balancing, incorporating two additional DC inductors and extended phase-shift modulation to regulate voltage levels. A closed-loop control strategy using the Generalized Average Model (GAM) method and proportional-integral (PI) tuning ensures stability and dynamic response. The system performance is evaluated through MATLAB-Simulink simulations, comparing open-loop and closed-loop operations under varying load conditions.

II. DESIGN CONSTRAINTS AND CONSIDERATIONS

Power transfer depends on the series inductance (Ls), which can be achieved using the leakage inductance of the transformer [2]. To minimize power losses, the inductance must remain below a specific upper limit. The maximum duty ratio of $D_1 = D_2 = 0.5$ results in high switching losses due to increased conduction and switching times.

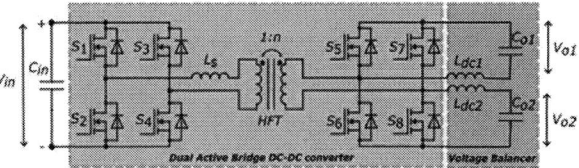

Fig. 3. DAB Converter with Voltage Balancer

The series inductance is calculated as

$$Ls = \frac{(Vin\ Vout\ [D_2(1 - D_2) + 0.5\ D_1(1 - D_1 - 2D_2)])}{(2\ n\ Fsw\ Prated)} \quad (1)$$

Substituting values, $Ls = 48.17\ mH$.

According to [8], the output capacitors can be selected to meet the required charge transfer, $Vripple = 3.8\ V$ and $\Delta Q = (IL - Iout) / Fsw = 359.86\ \mu C$. Thus, the output capacitance is given as follows:

$$Cout = \Delta Q / Vripple = 94.7\ \mu F \quad (2)$$

To minimize conduction losses during light or no-load conditions, the magnetizing current should be limited to 5% of the rated current i.e., $I_M = 0.05I_{in} = 0.3945\ A$ [4]. Substituting values, $L_M = \frac{Vin}{(4 \times Fsw \times I_M)} = 4.81\ mH$.

The design margin for inductors L_{DC1} and L_{DC2} should be twice that of L_M to reduce current ripple and minimize electrical stress on both the DC inductors and power switches [9]:

$$L_{DC1} = L_{DC2} = 2 \times 4.81 = 9.62\ mH \approx 10\ mH \quad (3)$$

The design specifications are summarized in Table I.

TABLE I. *DESIGN SPECIFICATIONS SUMMARY*

Parameter	Value
DAB Converter	
Input Voltage (Vin)	$380\ V$
Output Voltages ($Vout_1, Vout_2$)	$\pm190\ V$
Rated Power ($Prated$)	$3\ kW$
Switching Frequency (Fsw)	$50\ kHz$
Series Inductance (Ls)	$48.17\ mH$
HF Transformer	
Turns Ratio ($1:n$)	$1:1$
Magnetizing Inductance (L_M)	$7.729\ mH$
	$\rightarrow L_M \geq calc.L_m\ (4.81\ mH)$
Leakage Inductance (L_{lk})	$42.64\ \mu H$
	$\rightarrow L_{lk} \leq calc.L_s\ (48.17\ mH)$
Primary Resistance	$20.38\ m\Omega$
Secondary Resistance	$25.41\ m\Omega$
Core Material	$Ferrite\ (N95, TDK)$
Magnetic Flux Density (B)	$0.12\ T$
Current Density (J)	$250\ A/cm^2$
Winding Factor (Kw)	0.3
Selected Core	$E\ 70/33/32\ (TDK)$
Primary Turns (Np)	$Np = \frac{Vp}{(4 \times Fsw \times B \times Ac)}$ $= 24$
Secondary Turns (Ns)	$Ns = \frac{Vs}{(4 \times Fsw \times B \times Ac)}$ $= 24$
Voltage Balancer	
Inductances ($L_{DC1} = L_{DC2}$)	$9.62\ mH$
Capacitances ($Cout\ 1, Cout\ 2$)	$94.7\ \mu F$

III. CONVERTER MODELLING

A closed-loop controller is needed to regulate the output of the DAB converter, which requires a small-signal average model. Conventional averaging techniques for DC-DC converters require negligible current ripple, which is not satisfied in DAB converters, as the inductor current is a pure AC component. Four approaches for modeling a DAB converter reported in the literature are - Simplified reduced-order model, Full-order discrete-time model, Harmonic model, and generalized averaging model. A reduced-order model ignores the dynamics of inductor current and is suitable for SPS control. A discrete-time model is preferred for converters with large variations and resonant operations. A generalized averaging method [2], which uses more terms in the Fourier series of state variables, is more accurate at DC and low frequency and is applicable for arbitrary waveforms theoretically. It also properly captures capacitor ESR effects. The Step-by-step method of the Generalized Average Model (GAM) is given below.

STEP 1: Switching Functions
A DAB converter operates using two piece-wise switching functions: $S_1(t)$ and $S_2(t)$, where S_1 represents the switching function associated with the primary H-bridge, and S_2 represents the switching function associated with the secondary H-bridge. The phase shift duty ratio within the diagonal switches in the primary H-bridge is represented by d_1. The phase shift duty ratio of the secondary H-bridge with respect to the primary H-bridge is represented by d_2.

$$S_1(t, d_1, Ths) = \begin{cases} 0, & if\ 0 \le mod(t, 2 \cdot Ths) < d_1 \cdot Ths \\ 1, & if\ d_1 \cdot Ths \le mod(t, 2 \cdot Ths) < Ths \\ 0, & if\ Ths \le mod(t, 2 \cdot Ths) < (1 + d_1) \cdot Ths \\ -1, & otherwise \end{cases} \quad (4)$$

$$S_2(t, d_2, Ths) = \begin{cases} -1, & if\ 0 \le mod(t, 2 \cdot Ths) < d_2 \cdot Ths \\ 1, & if\ d_2 \cdot Ths \le mod(t, 2 \cdot Ths) < (1 + d_2) \cdot Ths \\ -1, & if\ (1 + d_2) \cdot Ths \le mod(t, 2 \cdot Ths) < 2 \cdot Ths \end{cases} \quad (5)$$

STEP 2: Fourier Series Coefficients of Switching Functions
As any signal can be expressed in terms of Fourier series, switching functions are also expressed in terms of Fourier coefficients, as given below.

$$S_{1f}(t) = a_0 S_1 + \sum_{n=0}^{n=N} [a_n S_1 \cos(2\pi nt/T) + b_n S_1 \sin(2\pi nt/T)] \quad (6)$$

Where,
$$a_0 S_1 = 0$$
$$a_n S_1(n, d_1) = \begin{cases} 0, & if\ n\ mod\ 2 = 0 \\ -(2 \sin(d_1\ n\pi))/(n\pi), & otherwise \end{cases}$$
$$b_n S_1(n, d_1) = \begin{cases} 0, & if\ n\ mod\ 2 = 0 \\ (2 + 2 \cos(d_1\ n\pi))/(n\pi), & otherwise \end{cases}$$

$$S_{2f}(t) = a_0 S_2 + \sum_{n=0}^{n=N} [a_n S_2 \cos(2\pi nt/T) - b_n S_2 \sin(2\pi nt/T)] \quad (7)$$

Where,
$$a_0 S_2 = 0$$
$$a_n S_2(n, d_2) = \begin{cases} 0, & if\ n\ mod\ 2 = 0 \\ -(4 \sin(d_2\ n\pi))/(n\pi), & otherwise \end{cases}$$
$$b_n S_2(n, d_2) = \begin{cases} 0, & if\ n\ mod\ 2 = 0 \\ (4 \cos(d_2\ n\pi))/(n\pi), & otherwise \end{cases}$$

STEP 3: Validation of Fourier Series Coefficients of Switching Functions
Fig. 4. shows the comparison between switching functions obtained using the piece-wise linearization and Fourier series functions for N = 25. The Fourier coefficients have been computed correctly.

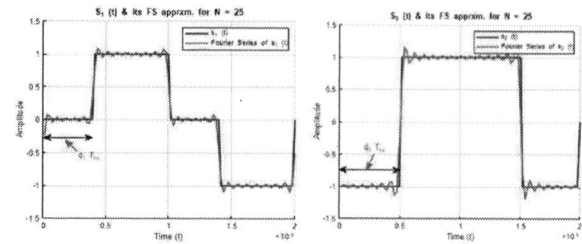

Fig. 4. Switching functions: (a) S_1 and (b) S_2

STEP 4: Dynamic State Equations and Model Simplification
The converter dynamics are defined using state-space equations, with the following assumptions:

- Switching is instantaneous, and the effect of dead time is not considered.
- The transformer is ideal, and the magnetizing inductance is large enough to prevent saturation.
- All quantities are referred to the primary side.
- Input voltage and load dynamics are slower than the converter dynamics.

The primary-side inductor current and output capacitor voltages are considered as state variables. Considered state variables are $i_L, V_{c1},$ and V_{c2}. The state-space equations are:

$$di_L(t)/dt = (1/L) [S_1(t) V_1 - S_2(t) (V_{c1}(t) + V_{c2}(t)) - R_s i_i(t)]$$
$$dV_{c1}(t)/dt = (1/C_1) [nS_2(t) i_i(t) - (1/R_1) V_{c1}(t)]$$
$$dV_{c2}(t)/dt = (1/C_2) [nS_2(t) i_i(t) - (1/R_2) V_{c2}(t)] \quad (8)$$

Applying the complex Fourier series and simplifying the model by considering lower-order components alone, as shown in Table II. The subscripts R and I denote the real and imaginary components, respectively.

TABLE II. *FOURIER COEFFICIENTS*

	Signal	n^{th}	$n = 1$	$n = 2$
S_1	$\langle S_1 \rangle_0$	A_{10}	0	0
	$\langle S_1 \rangle_n R$	$A_n R / 2$	$-\sin(\pi d_1) / \pi$	0
	$\langle S_1 \rangle_n I$	$-B_n I / 2$	$(-1 - \cos(\pi d_1)) / \pi$	0
S_2	$\langle S_2 \rangle_0$	A_{20}	0	0
	$\langle S_2 \rangle_n R$	$A_n R / 2$	$-2\sin(\pi d_2) / \pi$	0
	$\langle S_2 \rangle_n I$	$-B_n I / 2$	$-2\cos(\pi d_2) / \pi$	0

STEP 5: Large-signal model (Matrix form)
The dynamic state equations are represented in three forms: DC, real, and imaginary parts, as given below.

$$\frac{\partial \langle i_L \rangle_0}{\partial t} = \frac{1}{L}(\langle S_1 V_1 \rangle_0 - n\langle S_2 V_{c1} \rangle_0 - n\langle S_2 V_{c2} \rangle_0 - R_s \langle i_L \rangle_0)$$
$$\frac{\partial \langle i_L \rangle_{1R}}{\partial t} = \frac{1}{L}(\langle S_1 V_1 \rangle_{1R} - n\langle S_2 V_{c1} \rangle_{1R} - n\langle S_2 V_{c2} \rangle_{1R} - R_s \langle i_L \rangle_{1R}) + \omega_s \langle i_L \rangle_{1I}$$
$$\frac{\partial \langle i_L \rangle_{1I}}{\partial t} = \frac{1}{L}(\langle S_1 V_1 \rangle_{1I} - n\langle S_2 V_{c1} \rangle_{1I} - n\langle S_2 V_{c2} \rangle_{1I} - R_s \langle i_L \rangle_{1I}) - \omega_s \langle i_L \rangle_{1R} \quad (9)$$

$$\frac{\partial \langle v_{c1} \rangle_0}{\partial t} = \frac{1}{C_1}\left(\frac{n}{2}\langle S_2 i_L \rangle_0 - \frac{\langle V_{c1} \rangle_0}{R_1}\right)$$
$$\frac{\partial \langle v_{c1} \rangle_{1R}}{\partial t} = \frac{1}{C_1}\left(\frac{n}{2}\langle S_2 i_L \rangle_{1R} - \frac{\langle V_{c1} \rangle_{1R}}{R_1}\right) + \omega_s \langle V_{c1} \rangle_{1I}$$
$$\frac{\partial \langle v_{c1} \rangle_{1I}}{\partial t} = \frac{1}{C_1}\left(\frac{n}{2}\langle S_2 i_L \rangle_{1I} - \frac{\langle V_{c1} \rangle_{1I}}{R_1}\right) - \omega_s \langle V_{c1} \rangle_{1R} \quad (10)$$

$$\frac{\partial \langle v_{c2} \rangle_0}{\partial t} = \frac{1}{C_2}\left(\frac{n}{2}\langle S_2 i_L \rangle_0 - \frac{\langle V_{c2} \rangle_0}{R_2}\right)$$
$$\frac{\partial \langle v_{c2} \rangle_{1R}}{\partial t} = \frac{1}{C_2}\left(\frac{n}{2}\langle S_2 i_L \rangle_{1R} - \frac{\langle V_{c2} \rangle_{1R}}{R_2}\right) + \omega_s \langle V_{c2} \rangle_{1I}$$
$$\frac{\partial \langle v_{c2} \rangle_{1I}}{\partial t} = \frac{1}{C_2}\left(\frac{n}{2}\langle S_2 i_L \rangle_{1I} - \frac{\langle V_{c2} \rangle_{1I}}{R_2}\right) - \omega_s \langle V_{c2} \rangle_{1R} \quad (11)$$

DC component of the inductor current is truncated because the inductor current is seen as a pure AC component. AC components of the capacitor voltages are truncated because the capacitor voltage is seen as a pure DC component.

$$\frac{\partial}{\partial t}\begin{bmatrix}\langle i_L\rangle_{1R}\\\langle i_L\rangle_{1I}\\\langle v_{c1}\rangle_0\\\langle v_{c2}\rangle_0\end{bmatrix} = \begin{bmatrix} \frac{-R_s}{L} & \omega_s & \frac{2n\sin\pi d_2}{\pi L} & \frac{2n\sin\pi d_2}{\pi L}\\ -\omega_s & \frac{-R_s}{L} & \frac{2n\cos\pi d_2}{\pi L} & \frac{2n\cos\pi d_2}{\pi L}\\ -\frac{2n\sin\pi d_2}{\pi C_1} & -\frac{2n\cos\pi d_2}{\pi C_1} & \frac{-1}{R_1 C_1} & 0\\ -\frac{2n\sin\pi d_2}{\pi C_2} & -\frac{2n\cos\pi d_2}{\pi C_2} & 0 & \frac{-1}{R_2 C_2}\end{bmatrix}\begin{bmatrix}\langle i_L\rangle_{1R}\\\langle i_L\rangle_{1I}\\\langle v_{c1}\rangle_0\\\langle v_{c2}\rangle_0\end{bmatrix} + \begin{bmatrix}\frac{-\sin\pi d_1}{\pi L}\\\frac{-1-\cos\pi d_1}{\pi L}\\0\\0\end{bmatrix} V_1 \quad (12)$$

STEP 6: Small-signal model [Matrix form]

At steady state, the equations become $\frac{\partial}{\partial t}\begin{bmatrix}\langle i_L\rangle_{1R}\\\langle i_L\rangle_{1I}\\\langle v_{c1}\rangle_0\\\langle v_{c2}\rangle_0\end{bmatrix} = 0$.

On solving, the steady-state solution can be obtained. Applying small disturbances in state variables and control variables as follows:

$$\langle i_L\rangle_{1R} = I_{L1R} + \widehat{\iota_{L1R}} \; and \; \langle i_L\rangle_{1I} = I_{L1I} + \widehat{\iota_{L1I}}$$
$$\langle v_{c1}\rangle_0 = V_{c1} + \widehat{v_{c1}} \; and \; \langle v_{c2}\rangle_0 = V_{c2} + \widehat{v_{c2}}$$
$$d_1 = D_1 + \widehat{d_1} \; and \; d_2 = D_2 + \widehat{d_2} \quad (13)$$

Simplifying, the small-signal model in matrix form is given as follows:

$$\frac{\partial}{\partial t}\begin{bmatrix}\widehat{\iota_{L1R}}\\\widehat{\iota_{L1I}}\\\widehat{v_{c1}}\\\widehat{v_{c2}}\end{bmatrix} = \begin{bmatrix} \frac{-R_s}{L} & \omega_s & \frac{2n\sin\pi(D_2)}{\pi L} & \frac{2n\sin\pi(D_2)}{\pi L}\\ -\omega_s & \frac{-R_s}{L} & \frac{2n\cos\pi(D_2)}{\pi L} & \frac{2n\cos\pi(D_2)}{\pi L}\\ -\frac{2n\sin\pi(D_2)}{\pi C_1} & -\frac{2n\cos\pi(D_2)}{\pi C_1} & \frac{-1}{R_1 C_1} & 0\\ -\frac{2n\sin\pi(D_2)}{\pi C_2} & -\frac{2n\cos\pi(D_2)}{\pi C_2} & 0 & \frac{-1}{R_2 C_2}\end{bmatrix}\begin{bmatrix}\widehat{\iota_{L1R}}\\\widehat{\iota_{L1I}}\\\widehat{v_{c1}}\\\widehat{v_{c2}}\end{bmatrix} + \begin{bmatrix}\frac{-V_1\cos\pi(D_1)}{L} & \frac{2(V_{c1}+V_{c2})\cos\pi(D_2)}{L}\\ \frac{V_1\sin\pi(D_1)}{L} & -\frac{2(V_{c1}+V_{c2})\sin\pi(D_2)}{L}\\ 0 & \frac{-4n(I_{L1R}\cos\pi(D_2)-I_{L1I}\sin\pi(D_2))}{C_1}\\ 0 & \frac{-4n(I_{L1R}\cos\pi(D_2)-I_{L1I}\sin\pi(D_2))}{C_2}\end{bmatrix}\begin{bmatrix}\widehat{d_1}\\\widehat{d_2}\end{bmatrix} \quad (14)$$

STEP 7: Transfer Function of control variables to output voltages

According to the state-space small-signal equations, the control to output transfer functions can be defined as given below:

$$G_{\widehat{v_{c1,d_1}}}(s) = C_{a1}(sI-A)^{-1}B_1; \; G_{\widehat{v_{c1,d_2}}}(s) = C_{a1}(sI-A)^{-1}B_2$$
$$G_{\widehat{v_{c2,d_1}}}(s) = C_{a2}(sI-A)^{-1}B_1; \; G_{\widehat{v_{c2,d_2}}}(s) = C_{a2}(sI-A)^{-1}B_2 \quad (15)$$

The final transfer functions of the control variable to output voltages are given by the equations below.

$$G_{\widehat{v_{c1,d_1}}}(s) = G_{\widehat{v_{c2,d_1}}}(s) = \frac{A_{a2}s^2+A_{a1}s^1+A_{a0}}{B_{a4}s^4+B_{a3}s^3+B_{a2}s^2+B_{a1}s+B_{a0}}$$
$$G_{\widehat{v_{c1,d_2}}}(s) = G_{\widehat{v_{c2,d_2}}}(s) = \frac{A_{a3}s^3+A_{a2}s^2+A_{a1}s^1+A_{a0}}{B_{a4}s^4+B_{a3}s^3+B_{a2}s^2+B_{a1}s^1+B_{a0}} \quad (16)$$

The coefficient matrices are given by the following.

$$A = \begin{bmatrix} \frac{-R_s}{L} & \omega_s & \frac{2n\sin\pi(D_2)}{\pi L} & \frac{2n\sin\pi(D_2)}{\pi L}\\ -\omega_s & \frac{-R_s}{L} & \frac{2n\cos\pi(D_2)}{\pi L} & \frac{2n\cos\pi(D_2)}{\pi L}\\ -\frac{2n\sin\pi(D_2)}{\pi C_1} & -\frac{2n\cos\pi(D_2)}{\pi C_1} & \frac{-1}{R_1 C_1} & 0\\ -\frac{2n\sin\pi(D_2)}{\pi C_2} & -\frac{2n\cos\pi(D_2)}{\pi C_2} & 0 & \frac{-1}{R_2 C_2}\end{bmatrix} \quad (17)$$

$$B_1 = \begin{bmatrix}\frac{-V_1\cos\pi(D_1)}{L}\\\frac{V_1\sin\pi(D_1)}{L}\\0\\0\end{bmatrix} \; and \; B_2 = \begin{bmatrix}\frac{2(V_{c1}+V_{c2})\cos\pi(D_2)}{L}\\-\frac{2(V_{c1}+V_{c2})\sin\pi(D_2)}{L}\\\frac{-4n(I_{L1R}\cos\pi(D_2)-I_{L1I}\sin\pi(D_2))}{C_1}\\\frac{-4n(I_{L1R}\cos\pi(D_2)-I_{L1I}\sin\pi(D_2))}{C_2}\end{bmatrix} \quad (18)$$

$$C_{a1} = \begin{bmatrix}0 & 0 & 1 & 0\end{bmatrix} \; and \; C_{a2} = \begin{bmatrix}0 & 0 & 0 & 1\end{bmatrix} \quad (19)$$

IV. SIMULATION RESULTS AND DISCUSSIONS

The open-loop simulation evaluates system behaviour without feedback, providing insights into performance under varying loads. The closed-loop simulation examines feedback-based adjustments to maintain stable output, improving performance under unbalanced and no-load conditions. The closed-loop schematic is shown in Fig 5.

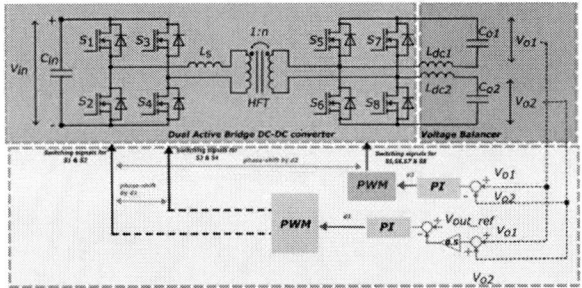

Fig. 5. Closed-loop control scheme of the system

To simulate the closed-loop response of the system, the PI controller parameters were tuned using the trial-and-error method. The proportional and integral gains for the first controller were set as $K_{p1} = 1.2$ and $K_{i1} = 120$, respectively, while the second controller was assigned $K_{p2} = 0.01$ and $K_{i2} = 300$. The open-loop system exhibits significant voltage deviations, particularly under high power-imbalance conditions. As the load imbalance increases, performance deteriorates, making it difficult to maintain stability. In contrast, the closed-loop system provides improved voltage regulation and reduced deviations, ensuring better handling of unbalanced loads. The comparison between the open-loop and closed-loop systems is presented in Table III. The proposed system maintains a stable output voltage across both balanced and unbalanced load conditions, with minimal fluctuations illustrated in Fig. 6 (a) – (e) and Fig. 7 (a) – (e).

TABLE III. *COMPARISON OF OPEN-LOOP AND CLOSED-LOOP OUTPUT VOLTAGES FOR DIFFERENT LOAD CONDITIONS*

Load Condition	Open-loop		Closed-loop	
	Vo-1,2 (V)	ΔV (%)	Vo-1,2 (V)	ΔV (%)
Balanced: P1 = P2 = 1.45 kW	184.12	3.09	189.6	0.211
Unbalanced: P1 = 1.51 kW, P2 = 505 W	181.48	4.48	190.75	0.395
P1 = 505 W, P2 = 1.51 kW	181.81	4.31	190.75	0.395
P1 = 2.7 kW, P2 ≈ 0 kW	183.51	3.42	190.15	0.132
No Load:	185.69	2.27	189.3	0.368

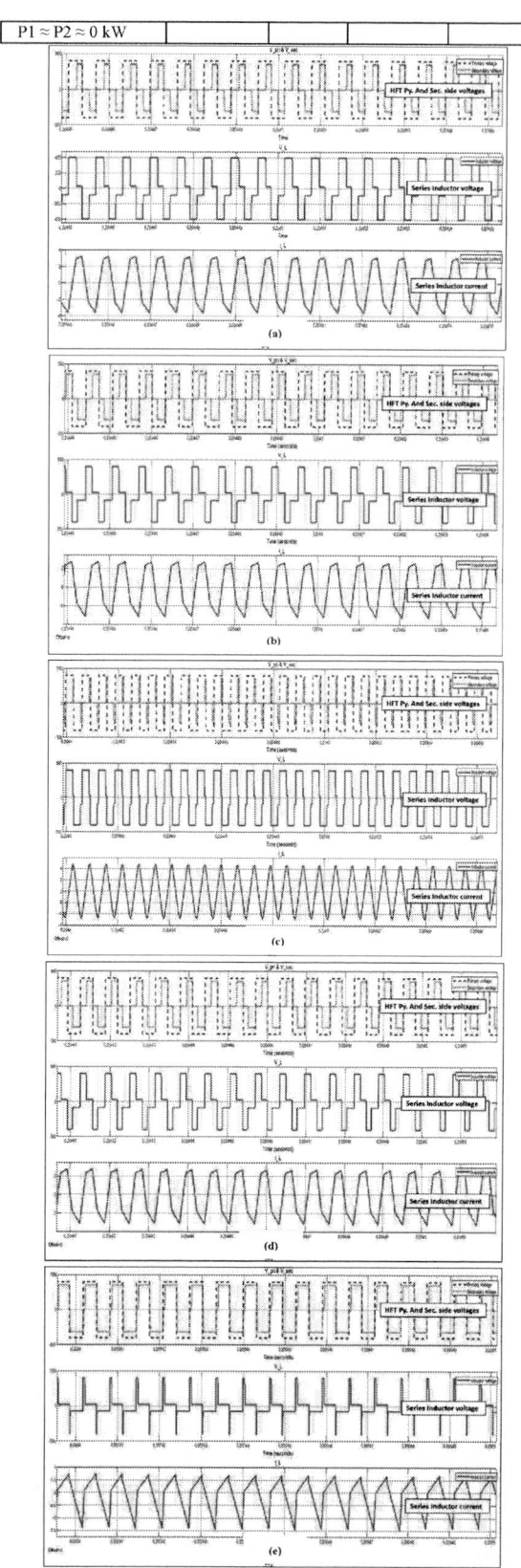

Fig. 6. Simulated waveforms of HFT Pri. & Sec. voltages, Series Inductor Voltage & Current – (a) Balanced Load condition: $P_1 = P_2 = 1.45$ kW; (b) Unbalanced Load condition: $P_1 = 1.51$ kW, $P_2 = 505$ W (V_{out1} consumes more power); (c) Unbalanced Load condition: $P_1 = 505$ W, $P_2 = 1.51$ kW (V_{out2} consumes more power); (d) Unbalanced Load condition: $P_1 = 2.7$ kW, $P_2 \approx 0$ kW; (e) Bipolar No-Load condition: $P_1 \approx P_2 \approx 0$ kW.

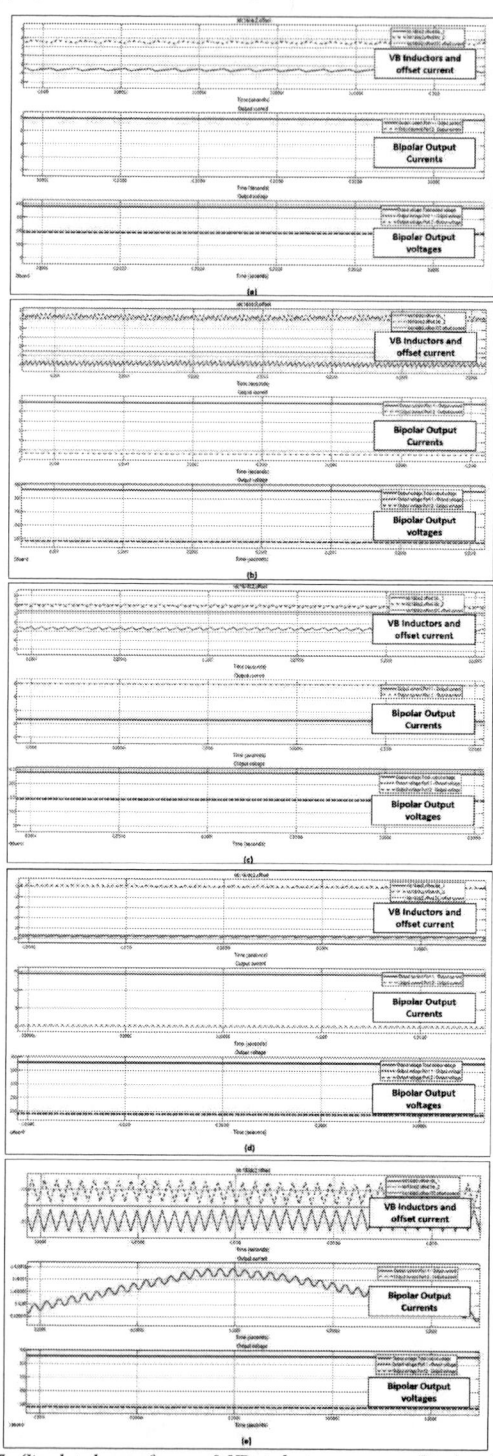

Fig. 7. Simulated waveforms of VB inductor currents, Bipolar Output Voltages & Currents – (a) Balanced Load condition: $P_1 = P_2 = 1.45$ kW ; (b) Unbalanced Load condition: $P_1 = 1.51$ kW, $P_2 = 505$ W (V_{out1} consumes more power); (c) Unbalanced Load condition: $P_1 = 505$ W, $P_2 = 1.51$ kW (V_{out2} consumes more power); (d) Unbalanced Load condition: $P_1 = 2.7$ kW, $P_2 \approx 0$ kW; (e) Bipolar No-Load condition: $P_1 \approx P_2 \approx 0$ kW.

The highest voltage deviation observed is 0.395% under unbalanced conditions, demonstrating the effectiveness in stabilising output voltage across varying loads. Even under no-load conditions, the closed-loop system performs well, with only a slight decrease in output voltage and a minor increase in deviation.

The conventional DAB converter achieves full-range ZVS at a unity voltage conversion ratio. However, ZVS capability is limited at light loads due to the MOSFET's output capacitance. Achieving ZVS requires sufficient inductor current to charge and discharge C_{out} [9]. Insufficient current leads to power losses and EMI issues. Extending ZVS across all load conditions, including no-load, is possible by leveraging the DC inductor current ripple. The ZVS analysis of the proposed converter is demonstrated under unbalanced (one side fully loaded) in Fig. 8. and bipolar no-load in Fig. 9. Under light loads, peak bias DC inductor current aids the switch currents i_{S5} and i_{S8}, in fully discharging the MOSFET's drain-to-source voltage. Similarly, peak magnetizing current assists i_{S1} and i_{S4}, enabling ZVS for primary-side switches $S1$ and $S4$. Consequently, ZVS is achieved for all switches, ensuring efficient operation across the entire load range. In unbalanced bipolar load conditions, the proposed method demonstrates slightly higher output voltages than the reported methods, indicating improved voltage stability. Under a significant power imbalance, both methods exhibit similar behaviour, with only a slight voltage drop. Overall, the voltage deviations between the reported and proposed methods remain minimal, indicating comparable performance across diverse load conditions.

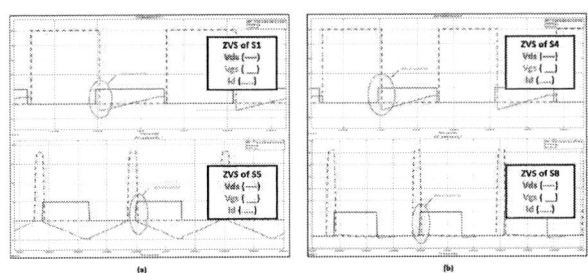

Fig. 8. ZVS Waveforms of S1, S4, S5, and S8 for P1 = 2.7 kW, P2 ≈ 0 kW

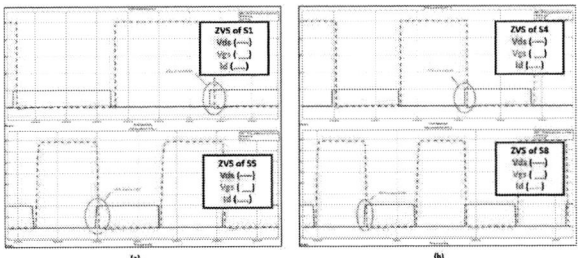

Fig. 9. ZVS Waveforms of S1, S4, S5, and S8 for P1 ≈ P2 ≈ 0 kW

V. CONCLUSION

The proposed DAB converter balances bipolar voltage levels using DC inductors as voltage balancers. The DC offset current of these inductors independently regulates bipolar output power, maintaining voltage balance despite load imbalances. Additionally, the existing DC inductor ripple generates the bias current needed for full-range zero-voltage switching (ZVS), which conventional DAB converters lack under no-load or light-load conditions. The closed-loop system outperforms the open-loop system, significantly improving voltage stability and regulation under varying load conditions. While the open-loop system exhibits voltage deviations up to 4.48% due to load imbalances, the closed-loop system limits deviations to under 0.4% using feedback control. A 3-kW converter was simulated to form the 190-V bipolar DC bus under balanced, unbalanced, and no-load conditions. In future work, the stability analysis of the converter setup can be carried out using Bode plot analysis derived from the transfer functions obtained in Section III. Based on this analysis, an appropriate compensation network can be implemented to ensure the stability of the entire system. Additionally, a low-power-level hardware prototype can be developed to experimentally validate the effectiveness of the proposed system.

ACKNOWLEDGMENT

The authors thank the support from the Department of Energy and Environment, National Institute of Technology Tiruchirappalli.

REFERENCES

[1] F. Krismer and J. W. Kolar, "Modeling and Optimization of Bidirectional Dual Active Bridge DC–DC Converter Topologies," *IEEE Transactions on Power Electronics*, vol. 27, no. 1, pp. 43–58, Jan. 2012.

[2] J. Sabaté, V. Vlatkovic, R. B. Ridley, F. C. Lee, and B. H. Cho, "Design considerations for high-voltage high-power full-bridge zero-voltage-switched PWM converter," in *IEEE Applied Power Electronics Conference (APEC)*, 1990, pp. 275–284.

[3] C. Zhao, S. D. Round, and J. W. Kolar, "An isolated three-port bidi- rectional DC–DC converter with decoupled power flow management," *IEEE Transactions on Power Electronics*, vol. 23, no. 5, pp. 2443–2453, Sep. 2008.

[4] J. Zhang, X. Huang, X. Wu, and Z. Qian, "A high efficiency active clamp flyback converter for low power applications," *IEEE Transactions on Power Electronics*, vol. 21, no. 1, pp. 122–130, Jan. 2006.

[5] G. Guidi, V. Schenone, M. Valla, and S. Lusci, "Design and Optimization of a Dual Active Bridge Converter for EV Battery Chargers," *IEEE Transactions on Power Electronics*, vol. 35, no. 1, pp. 753–764, Jan. 2020.

[6] K. Wang, C. Zhao, and X. Wu, "High-Frequency Transformer Design for Isolated DC-DC Converters in Electric Vehicles," *IEEE Transactions on Transportation Electrification*, vol. 5, no. 4, pp. 1071–1081, Dec. 2019.

[7] L. Yang, C. Li, and S. Gao, "Analysis and Design of a High-Efficiency Dual Active Bridge Converter for Wide Voltage Range Applications," *IEEE Transactions on Power Electronics*, vol. 34, no. 8, pp. 7773–7787, Aug. 2019.

[8] Y. Zhou, X. Feng, and M. Peng, "A Novel Control Strategy for Bi-Directional DC-DC Converters in Energy Storage Systems," *IEEE Transactions on Smart Grid*, vol. 10, no. 3, pp. 2650–2659, May 2019.

[9] S. Wang, H. Zhang, and P. Lin, "High-Efficiency Power Conversion for DC Microgrids: A Review of Current Trends and Future Prospects," *IEEE Transactions on Industry Applications*, vol. 56, no. 2, pp. 1908– 1921, Mar.–Apr. 2020.

Interactions linear-Based Prediction of Kinetic Energy in Flywheel Energy Storage for Enhanced Grid Stabilization

Saranya S T
School of Electrical Engineering
Vellore Institue of Technology
(VIT University)
Chennai, India
saranya.st2022@vitstudent.ac.in

Pritam Bhowmik*
School of Electrical Engineering
Vellore Institue of Technology
(VIT University)
Chennai, India
pritambhowmikjan@gmail.com

Priya Ranjan Satpathy
Institute of Power Engineering
Universiti Tenaga Nasional
Kajang 43000
Malaysia
drprsatpathy@gmail.com

Abstract—**Flywheel energy storage systems have emerged as a promising technology for grid stabilization due to their ability to rapidly store and release kinetic energy. Accurate prediction of the kinetic energy stored in the flywheel allows for better control of charging and discharging cycles, maximizing energy utilization, and minimizing losses. This prediction is influenced by multiple interdependent factors, including rotational speed, mass distribution, and external load conditions. This study evaluates three machine learning approaches, including ensemble bagged trees, exponential gaussian process regression, and an interactions linear regression model. In this paper, an interactions-based linear regression model is proposed to incorporate interaction terms, capturing not only the individual effects of these variables but also their combined influence on the kinetic energy. This approach enhances the predictive accuracy of kinetic energy estimation and enables more efficient grid stabilization strategies. The interactions linear regression model demonstrates superior performance by achieving the lowest root mean squared error compared to other methods.**

Keywords— Flywheel, machine learning, speed, torque, kinetic energy

I. INTRODUCTION

Energy storage systems play a pivotal role in modern power systems and ensure the efficient management of energy supply and demand [1]-[4]. Energy storage such as compressed air, thermal storage (hot water), and chemical-based systems (batteries) are widely used for long-term energy storage and peak demand management. However, flywheel energy storage systems (FESS) have gained significant attention as a key technology for grid stabilization due to their ability to rapidly store and release kinetic energy [5][6]. FESS are particularly suitable for applications that require short-term energy storage, such as frequency regulation, peak shaving, and load balancing in microgrids, owing to their high power density, fast response time, and reliability [7][8]. A critical factor in optimizing the performance of FESS is the accurate prediction of the kinetic energy stored in the flywheel, which depends on variables such as rotational speed, mass distribution, and external load conditions [9][10]. Precise prediction of kinetic energy is essential for efficient energy management, and enables optimal charging and discharging cycles while minimizing energy losses and mechanical risks associated with overspeed

conditions. Traditional approaches to kinetic energy prediction often rely on physical models that consider individual factors in isolation. However, these models may fail to capture the complex interactions between variables, such as the combined effects of rotational speed and mass distribution, which significantly influence energy storage capacity. To address this limitation, this paper proposes an interactions-based linear regression model that incorporates interaction terms to account for both individual and synergistic effects of these variables [11]. While prior work [12] focused on short-term kinetic energy and inertia forecasting for synchronous generators using deep learning (LSTM) and linear regression, the proposed work presents a novel approach for kinetic energy prediction within a Flywheel Energy Storage System (FESS). Recently, significant progress has been made in adaptive control and energy management of high-power-density energy storage systems for electric vehicles, such as ultracapacitors and flywheel systems. In [13], an AI-based control strategy using a NARX neural network was proposed to adaptively manage storage capacity in real time, improving transient power handling and extending battery life in EVs. Unlike system-level inertia forecasting, the proposed study specifically targets the internal dynamics of FESS, particularly the intricate relationships between operational parameters that influence the stored kinetic energy. By enhancing the predictive accuracy of kinetic energy estimation, this approach enables more efficient grid stabilization strategies and improves the overall performance of FESS.

II. FLYWHEEL DYNAMICS

A. State of Charge of FESS

The flywheel stores rotational kinetic energy based on the principle of conservation of angular momentum. The stored energy is directly proportional to the moment of inertia and the square of angular speed [14].

The kinetic energy (E) stored in a flywheel is given by

$$E = \frac{1}{2}J\omega^2 \tag{1}$$

where J is the moment of inertia, which depends on the mass and geometry of the flywheel in kg.m^2 and ω denotes the angular speed of the flywheel in rad/s.

The state of charge (SoC) of a flywheel represents the kinetic energy stored relative to its maximum energy capacity and is directly related to the angular speed of the flywheel [15]. In energy management and load balancing, the SoC is a deciding factor in determining whether to charge/discharge the flywheel.

Though, SoC is also related to the flywheel angular speed can be written as-

$$\%SoC = \frac{\omega_{actual}{}^2}{\omega_{max}{}^2} * 100 \tag{2}$$

where ω_{actual} represents the actual angular speed and ω_{max} is the maximum angular speed.

B. Implementation of Rotating Kinematics

To implement the flywheel energy storage system, a rotating mass (moment of inertia) and a damper system subjected to an applied torque are connected to the rotating part of the DC machine. Depending on the applied load torque, the system can operate either as a motor or as a generator. If the applied load torque is greater than zero, it acts as a motor, converting electrical energy into mechanical energy. On the other hand, if the load torque is less than zero, it operates as a generator, converting mechanical energy into electrical energy.

According to Newton's second law for rotation [16], the angular acceleration of the flywheel is given by:

$$\frac{d\omega}{dt} = \frac{T_{load}}{J} \tag{3}$$

An ideal torque source applies a torque to the system, which influences the angular acceleration of the rotating mass:

$$T_a = J\alpha \tag{4}$$

where T_a represents the applied load torque in N-m and α is the angular acceleration in rad/s^2.

During the charging process, a torque is applied to the flywheel to increase its angular speed, thereby increasing the stored kinetic energy. A higher applied torque results in faster acceleration, allowing the flywheel to reach its maximum speed more quickly [17]. This charging process continues until the state of charge reaches 100%. Conversely, during discharging, a torque is applied to decelerate the flywheel, extracting its stored energy and causing the speed of the flywheel decrease to the minimum allowable level.

III. METHODOLOGY

A Flywheel Energy Storage System (FESS) model, incorporating a bidirectional power converter, was developed and simulated using MATLAB/Simulink. Fig. 1 illustrates the architecture and data flow of a FESS, highlighting the integration of machine learning component for kinetic energy prediction [18]-[21]. Output data were collected from the rotating part of the DC machine, including key parameters such as armature current, field current, electromagnetic torque, angular speed, power demand and kinetic energy. The dataset was pre-processed and cleaned to ensure it was free from noise, thereby providing high-quality input for the machine learning models.

A. Kinetic Energy Calculation

At the initial stage of the simulation, the moment of inertia (J) was held constant at 10 kg·m², and the initial angular speed (ω) was set to 200 rad/s. This resulted in an initial stored kinetic energy of 200,000 Joules. During dynamic operation, when the flywheel meets a power demand, it discharges energy. This process leads to a decrease in the flywheel angular speed as kinetic energy is converted into electrical energy by the motor/generator. The mechanical power extracted from the flywheel is then transformed into electrical power (P_e) to supply the load [22].

The rate of decrease in the flywheel angular speed is directly influenced by the load power (P_l). Increased load

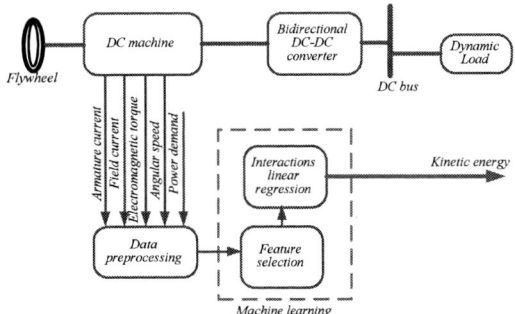

Fig. 1. Flywheel energy storage with machine learning model

demands result in a faster rate of energy extraction and consequently, a more rapid decrease in the angular speed of the flywheel. Therefore, the angular speed of the flywheel dynamically varies based on the load demand and the control strategy, which is reflected in the real-time power transfer.

$$P_m = \omega T \tag{5}$$

The power balance equation can be expressed as:

$$P_m = P_e = P_l \tag{6}$$

where P_m represents mechanical power and P_e is the electrical power.

B. DC Machine

In the Flywheel Energy Storage System, the DC machine serves as both a motor and a generator, facilitating the process of energy conversion. The electromagnetic torque produced by the DC machine is directly proportional to the armature current.

$$T_e = k_t I_a \tag{7}$$

where k_t is the toque constant, I_a is the armature current and T_e represents the electromagnetic torque produced by the DC machine. The current flowing through the armature of the DC machine influences the electromagnetic torque and power generation. The current in the field winding of the DC machine remains constant within this dataset. The torque generated by the interaction of the armature current and the magnetic field drives the rotation of the flywheel. The angular speed of the flywheel (in rad/s) directly affects the kinetic energy stored in the flywheel. The power demand or load on the system influences the armature current and torque.

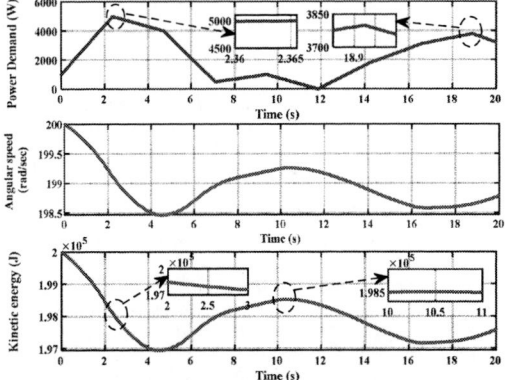

Fig. 2. Power demand and kinetic energy

979-8-3315-3013-6/25 $31.00 © 2025 IEEE

The dynamic variation of angular speed in response to power demand, which dictates the required kinetic energy of the flywheel energy storage system, is shown in Fig. 2. An increase in the flywheel angular speed leads to a corresponding quadratic increase in stored kinetic energy. The rate at which the angular speed of the flywheel decreases is directly proportional to the load power. Higher load power demands result in a faster extraction of kinetic energy, leading to a more rapid deceleration of the flywheel [23].

C. Power Electronic Circuit

A bidirectional DC-DC converter was integrated into the FESS model to facilitate both charging and discharging operations. During charging, the converter steps up the voltage from the grid to charge the flywheel, where the DC machine acts as a motor to accelerate the flywheel. While during the process of discharging, the converter steps down the voltage from the flywheel to supply power to the load, where the DC machine acts as a generator converts the flywheel kinetic energy into electrical energy [24].

D. Machine Learning Approach

In this paper, an interactions based linear regression machine learning technique is employed to model the relationship between a dependent variable (y) and one or more independent variables (x) using a linear equation. When interactions between variables are included, the model captures not only the individual effects of the variables but also their combined influence on the outcome [25].
The linear regression equation is

$$y = \beta_0 + \beta_1 x_1 + \beta_2 x_2 + \cdots + \beta_n x_n + \varepsilon \qquad (8)$$

The linear regression equation with interactions can be represented as

$$y = \beta_0 + \beta_1 x_1 + \beta_2 x_2 + \beta_3 x_1 x_2 + \cdots + \varepsilon \qquad (9)$$

where β_0 is the intercept, $x_1, x_2, \ldots x_n$ are features, ε represents the error term (residuals), $\beta_1 \ldots \beta_n$ denotes the coefficients for the independent variables and y represents the dependent variable (target).s

β_1, β_2 capture the individual effects of x_1 and x_2 respectively.

β_3 captures the interaction effect between x_1 and x_2.

The interaction terms were chosen based on the underlying physics of the flywheel energy storage system (FESS). Kinetic energy (E) is fundamentally related to the flywheel rotational speed (ω) and its moment of inertia, which is influenced by the mass distribution and, in the proposed model, by factors affecting torque. In a real-world FESS, the relationship between torque and speed is not always linear. Factors such as friction, load variations, and the operating mode of the motor/generator can cause this relationship to change. Therefore, interaction terms between speed and torque were included to capture these non-linear effects. For instance, at higher speeds, the same change in torque results in a greater change in kinetic energy. The interaction term enables the model to account for this effect, where speed essentially acts as a gain factor for the influence of torque on kinetic energy.

The dataset used in this study has been derived from a DC machine, includes key parameters such as armature current, field current, electromagnetic torque, rotational speed, kinetic energy and power demand.

TABLE I.   IMPACT ANALYSIS

Sl. No	Impact Analysis for Kinetic Energy			
	Algorithm	Features	Importance score	Remarks
1	MRMR	Angular speed	5.545	Angular speed and power demand shows the highest importance among the impacted features for kinetic energy.
		Power demand	3.855	
		Armature current	3.855	
		Field current	3.855	
		Electromagnetic torque	3.368	
2	RReliefF	Angular speed	0.008	Electromagnetic torque and armature current have equal importance in the feature selection.
		Power demand	0.007	
		Electromagnetic torque	0.003	
		Armature current	0.003	
		Field current	0.002	

To ensure equal weighting of all input features during model training and to improve the convergence of the machine learning algorithms, all input features were scaled to a range of [0, 1] using the min-max normalization technique. The dataset was partitioned into a 70% training set (10,500 samples) and a 30% testing set (4,500 samples). The prediction of kinetic energy was chosen as the target output and the other variables are chosen as input features. These variables were used to train and validate the proposed interaction-based linear regression model.

IV. RESULTS AND DISCUSSION

Three machine learning models- Ensemble bagged trees, Exponential gaussian process regression and Interactions linear regression were trained and tested using the pre-processed dataset collected from the FESS simulation. Model validation was performed to evaluate the accuracy of the predictions [23],[26].

A. Impact Analysis

The impact of features on the kinetic energy of a flywheel energy storage system was analyzed using two feature selection algorithms, MRMR (Minimum Redundancy Maximum Relevance) and RReliefF.

The MRMR algorithm provides a distinct ranking of feature importance, indicating that the angular speed and power demand were the most influential features affecting kinetic energy. On the other hand, the RReliefF algorithm indicated equal importance to electromagnetic torque and armature current, along with other features. Table I highlights the importance of feature selection, demonstrating how specific input features, such as angular speed and electromagnetic torque, significantly influence the prediction of kinetic energy. The MRMR algorithm aims to select features that are highly relevant to the target variable while minimizing redundancy among the selected features.

B. Performance Metrics

The performance metrics depicted in Table II include Root Mean Squared Error (RMSE), R-squared, Mean Squared Error (MSE), and Mean Absolute Error (MAE). RMSE measures the average magnitude of the errors in a set of predictions, a lower RMSE indicates better accuracy.

979-8-3315-3013-6/25 $31.00 © 2025 IEEE

TABLE II. PERFORMANCE METRICS

Model Name	Performance Metrics				
	RMSE	R-squared	MSE	MAE	Remarks
Ensemble bagged trees	2.241	1.00	5.022	1.663	Interactions linear regression performance shows 87% improvement compared to Ensemble bagged trees.
Exponential Gaussian process regression	1.583	1.00	2.506	0.363	
Interactions linear regression	0.291	1.00	0.084	0.205	

The R-squared metric indicates the proportion of the variance in the dependent variable (kinetic energy) that is predictable from the independent variables. A higher R-squared value (closer to 1) indicates a better model fit. Table II compares the performance of three machine learning models for predicting the kinetic energy of a FESS. The Ensemble bagged trees model achieved an RMSE of 2.2409 and an R-squared of 1.00. While this R-squared value indicates a relatively good fit, the RMSE is much higher than that of the other two models.

The exponential gaussian process regression model achieved a significantly lower RMSE of 1.5829 and its perfect R-squared of 1.00 suggests a very accurate model. The Interactions linear regression model achieved the lowest RMSE of 0.2906 and a perfect R-squared of 1.00. The Ensemble bagged trees model, while also showing a strong correlation (R-squared of 1.00), had a significantly higher RMSE, indicating lower accuracy compared to the other two models.

It is observed from Table II that, based on the RMSE, the interactions linear regression model demonstrated superior performance, achieving an 87% improvement in prediction accuracy compared to an ensemble model and 82% compared to the Exponential GPR. The results demonstrate that interaction-based linear models outperform traditional linear models, providing a robust framework for optimizing Flywheel energy storage system performance in dynamic grid environments. This research contributes to the advancement of energy storage technologies by offering a data-driven methodology for improving the reliability and efficiency of flywheel-based grid stabilization systems.

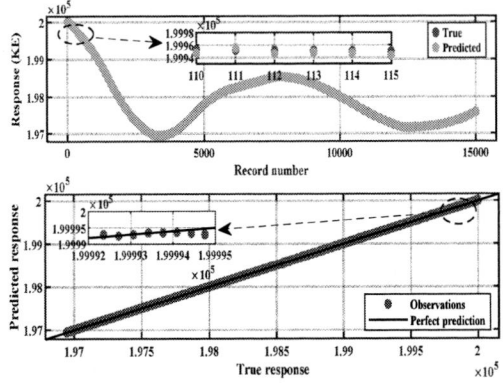

Fig. 3. Response: Kinetic energy for Ensemble bagged trees model

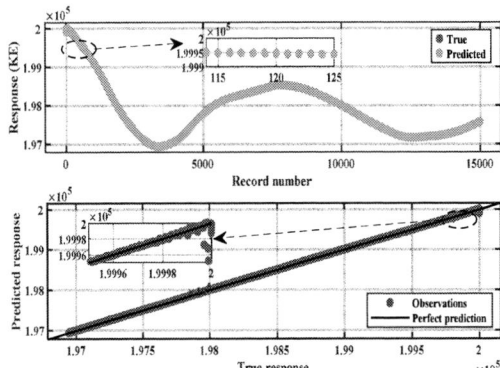

Fig. 4. Response: Kinetic energy for Exponential GPR model

C. Prediction of Kinetic Energy

The predicted response of the ensemble bagged trees model does not fit precisely with the true response, as depicted in Fig. 3 and further supported by its higher RMSE of 2.241 (as shown in Table II) . The dataset used for training, testing, and validation consists of 15,000 data points acquired through simulation. While the ensemble bagged trees model captures some aspects of the kinetic energy dynamics, its inability to fully align with the true response highlights limitations in handling the complex interactions and nonlinearities present in the flywheel energy storage system.

It is observed from Fig. 4 that, the predicted response of the exponential GPR model also does not align precisely with the true response. This misalignment is primarily due to the RMSE value, which quantifies the deviations between the predicted and true kinetic energy values. While the exponential GPR model captures the general trend of the kinetic energy dynamics, its inability to fully align with the true response highlights limitations in handling the nonlinearities and complex relationships present in the flywheel energy storage system.

The predictions of the exponential GPR model deviate from the true response, particularly at higher scales. This misalignment, is reflected in the higher RMSE value, emphasizes the need for more robust modelling techniques such as the proposed Interactions linear regression model, to improve prediction accuracy and better capture the complex relationships governing kinetic energy in the flywheel system.

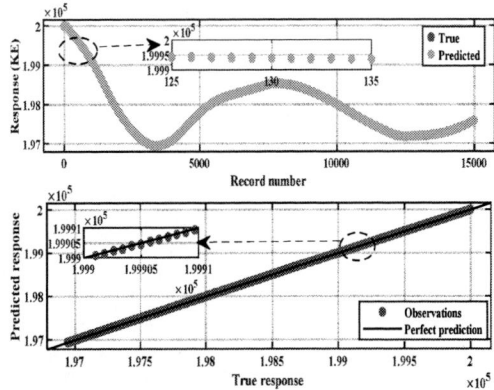

Fig. 5. Response: Kinetic energy for Interactions linear model

While the exponential GPR model captures the general trend of the kinetic energy dynamics, its inability to fully align with the true response highlights limitations in handling the nonlinearities and complex relationships present in the flywheel energy storage system. The predictions of the exponential GPR model deviate from the true response, particularly at higher scales. This misalignment, is reflected in the higher RMSE value, emphasizes the need for more robust modelling techniques such as the proposed Interactions linear regression model, to improve prediction accuracy and better capture the complex relationships governing kinetic energy in the flywheel system.

The visual representation in Fig. 3 further supports the quantitative metrics. The predicted response of the Ensemble Bagged Trees model (yellow line) shows noticeable deviations from the true response (blue line), consistent with its higher RMSE and MAE. In contrast, Fig. 4 (Exponential GPR) and Fig. 5 (Interactions Linear Regression) demonstrate a much closer alignment between the predicted and true kinetic energy, visually confirming their lower prediction errors.

It is observed from the Fig. 5 that, the predicted response of the interactions linear regression model aligns closely with the true response. This high level of accuracy is attributed by the very low RMSE value and a perfect R-squared value of 1.00, indicating a strong correlation between the predicted and actual kinetic energy values. The model demonstrates its capability to accurately predict kinetic energy when the actual angular speed and power demand are known. Furthermore, the inclusion of interaction terms as additional input features in the dataset enhances the model predictive accuracy, enabling it to capture the complex relationships and interactions between variables more effectively. This result highlights the superiority of the interactions linear model in predicting kinetic energy for flywheel energy storage systems, making it a robust tool for optimizing energy management and grid stabilization strategies.

Conclusion

The accurate prediction of kinetic energy in flywheel energy storage systems is essential for grid stabilization and efficient power demand management. Accurate kinetic energy prediction is crucial as operating the flywheel at a high state of charge for maximum energy availability necessitates precise control to avoid exceeding angular speed limits and causing mechanical failures. Therefore, maintaining the flywheel within its designed speed range is critical to prevent damage, ensure efficient energy conversion, and prolong system longevity. In this paper, the prediction of kinetic energy was achieved using a machine learning-based interactions regression model. This model demonstrated superior performance, achieving a prediction accuracy of 87% on the testing dataset and the lowest RMSE compared to other models such as ensemble bagged trees and exponential gaussian process regression. The interactions linear regression model effectively captured the combined effects of variables like rotational speed and power demand, enabling precise kinetic energy predictions. This capability is vital for energy management, load balancing, and fault detection in FESS. The proposed model is computationally lightweight, making it suitable for real-time implementation in FESS applications. It involves simple algebraic operations, which can be executed with minimal latency on embedded controllers typically used in FESS control. Compared to more complex machine learning models (e.g., deep neural networks), the proposed model offers a balance between accuracy and computational cost, making it practical for real-time environments. Real-world disturbances (e.g., measurement noise, sudden load changes) could affect model accuracy. Future work will focus on integrating adaptive filtering and real-time model updating techniques to enhance robustness and further optimize the real-time applicability of the model. In the future, integrating machine learning models with real-time environments could further enhance their applicability, enabling dynamic adjustments and real-time optimization of flywheel operations to improve grid stabilization and energy management.

References

[1] Amiryar, M. E., & Pullen, K. R. (2017). A review of flywheel energy storage system technologies and their applications. *Applied Sciences*, *7*(3), 286.

[2] Tziovani, L., Hadjidemetriou, L., Charalampous, C., Tziakouri, M., Timotheou, S., & Kyriakides, E. (2021). Energy management and control of a flywheel storage system for peak shaving applications. *IEEE Transactions on Smart Grid*, *12*(5), 4195-4207.

[3] Said, Z., Sharma, P., Tiwari, A. K., Le, V. V., Huang, Z., Bui, V. G., & Hoang, A. T. (2022). Application of novel framework based on ensemble boosted regression trees and Gaussian process regression in modelling thermal performance of small-scale Organic Rankine Cycle (ORC) using hybrid nanofluid. Journal of Cleaner Production, 360, 132194.

[4] Aghmadi, A., & Mohammed, O. A. (2024). Energy storage systems: Technologies and high-power applications. *Batteries*, *10*(4), 141.

[5] Yan, X., Nie, S., Chen, B., Yin, F., Ji, H., & Ma, Z. (2023). Strategies to improve the energy efficiency of hydraulic power unit with flywheel energy storage system. *Journal of Energy Storage*, *59*, 106515.

[6] Li, X., & Palazzolo, A. (2022). A review of flywheel energy storage systems: state of the art and opportunities. *Journal of Energy Storage*, *46*, 103576.

[7] Faisal, M., Hannan, M. A., Ker, P. J., Hussain, A., Mansor, M. B., & Blaabjerg, F. (2018). Review of energy storage system technologies in microgrid applications: Issues and challenges. *Ieee Access*, *6*, 35143-35164.

[8] Lei, M., Meng, K., Feng, H., Bai, J., Jiang, H., & Zhang, Z. (2023). Flywheel energy storage controlled by model predictive control to achieve smooth short-term high-frequency wind power. *Journal of Energy Storage*, *63*, 106949.

[9] Kale, V., & Secanell, M. (2018). A comparative study between optimal metal and composite rotors for flywheel energy storage systems. *Energy Reports*, *4*, 576-585.

[10] Olabi, A. G., Wilberforce, T., Abdelkareem, M. A., & Ramadan, M. (2021). Critical review of flywheel energy storage system. *Energies*, *14*(8), 2159.

[11] Li, Y., Fu, K., Zhao, Y., & Yang, C. (2023). Relative Synergy Coefficient: A novel way to detect variable interaction in large dataset. *Knowledge-Based Systems*, *282*, 111112.

[12] Riquelme-Dominguez, J. M., Carranza-García, M., Lara-Benítez, P., & González-Longatt, F. M. (2024). A machine learning-based methodology for short-term kinetic energy forecasting with real-time application: Nordic Power System case. *International Journal of Electrical Power & Energy Systems*, *156*, 109730.

[13] Mehraban, A., Ghanbari, T., & Farjah, E. (2023). AI-based control of storage capacity in high-power-density energy storage systems, used in electric vehicles. *IEEE Transactions on Transportation Electrification*, *10*(1), 2293-2301.

[14] Kondoh, J., Funamoto, T., Nakanishi, T., & Arai, R. (2018). Energy characteristics of a fixed-speed flywheel energy storage system with direct grid-connection. *Energy*, *165*, 701-708.

[15] Hedlund, M., Lundin, J., De Santiago, J., Abrahamsson, J., & Bernhoff, H. (2015). Flywheel energy storage for automotive applications. *Energies*, *8*(10), 10636-10663.

[16] Wu, G., & Zhao, G. (2022). Parameter influence law analysis and optimal design of a dual mass flywheel. *International Journal of Mechanical System Dynamics*, *2*(2), 165-177.

[17] Choudhury, S. (2021). Flywheel energy storage systems: A critical review on technologies, applications, and future prospects. *International transactions on electrical energy systems*, *31*(9), e13024.

[18] Plevris, V., Solorzano, G., Bakas, N. P., & Ben Seghier, M. E. A. (2022). Investigation of performance metrics in regression analysis and machine learning-based prediction models.

[19] Bhowmik, P., Chandak, S., & Rout, P. K. (2018). State of charge and state of power management among the energy storage systems by the fuzzy tuned dynamic exponent and the dynamic PI controller. *Journal of Energy Storage*, *19*, 348-363.

[20] Bhowmik, P., Chandak, S., & Rout, P. K. (2019). State of charge and state of power management in a hybrid energy storage system by the self-tuned dynamic exponent and the fuzzy-based dynamic PI controller. *International Transactions on Electrical Energy Systems*, *29*(5), e2848.

[21] Bhowmik, P., Chandak, S., & Rout, P. K. (2019). State of charge and state of power management of the hybrid energy storage system in an architecture of microgrid. *Journal of Renewable and Sustainable Energy*, *11*(1).

[22] Xu, K., Guo, Y., Lei, G., & Zhu, J. (2023). A review of flywheel energy storage system technologies. *Energies*, *16*(18), 6462.

[23] Wen, Z., Fang, P., Yin, Y., Krolczyk, G., Gardoni, P., & Li, Z. (2022). A novel machine learning model for safety risk analysis in flywheel-battery hybrid energy storage system. *Journal of Energy Storage*, *49*, 104072.

[24] Takarli, R., Amini, A., Khajueezadeh, M., Zarbil, M. S., Vahedi, A., Kiyoumarsi, A., ... & Kyyra, J. (2023). A comprehensive review on flywheel energy storage systems: Survey on electrical machines, power electronics converters, and control systems. *IEEE Access*, *11*, 81224-81255.

[25] Asghar, Z., Hafeez, K., Sabir, D., Ijaz, B., Bukhari, S. S. H., & Ro, J. (2023). RECLAIM: Renewable energy based demand-side management using machine learning models. *IEEE Access*, *11*, 3846-3857.

[26] Liang, Y. C., Maimury, Y., Chen, A. H. L., & Juarez, J. R. C. (2020). Machine learning-based prediction of air quality. *applied sciences*, *10*(24), 9151.

A Unified Control Scheme for Current Harmonics Suppression with Ripple Free Power for a Grid Tied SPV System Under Abnormal Grid Voltage Conditions

Manash Kumar Mishra, *Member, IEEE*
Department of Electrical Engineering,
National Institute of Technology Manipur,
India, mkmishra.ee@nitmanipur.ac.in

Simanta Kumar Samal Member, IEEE
Department of Electrical Engineering,
National Institute of Technology Jamshedpur,
India, simanta.ee@nitjsr.ac.in

Vivek Nandan Lal, Member, IEEE
Department of Electrical Engineering,
Indian Institute of Technology (BHU),
India, vnlal.ee@iitbhu.ac.in

Abstract — **This paper proposes a unified control scheme for three-phase grid-tied solar photovoltaic (GTSPV) systems to maintain stable operation during grid voltage abnormalities. The proposed scheme incorporates an enhanced proportional multiresonant (EPMR) current controller designed to mitigate harmonic distortion in the injected currents caused by grid voltage harmonics. A coordinated adaptive voltage sag compensator (AVSC), synchronized with a phase-locked loop (PLL), is implemented to maintain seamless grid synchronization and ensure stable injection of ripple-free active and reactive power during unbalanced voltage sags. The synergistic operation of the EPMR controller and AVSC-PLL unit enables simultaneous mitigation of harmonic currents and suppression of power oscillations, ensuring robust performance under grid disturbances. To enhance maximum power extraction under dynamic atmospheric conditions, an optimized step-size incremental conductance (OSINC) algorithm is incorporated into proposed control scheme. The effectiveness of the unified control scheme is validated through real-time digital simulations using the OPAL-RT OP4510 platform, with tests conducted under harmonically distorted and unbalanced grid voltage scenarios. Comparative analyses with conventional proportional multiresonant (PMR) controllers demonstrate the superior harmonic suppression, power quality, and tracking efficiency of the proposed control scheme.**

Index Terms — *Adaptive voltage sag compensator (AVSC), enhanced proportional multiresonant (EPMR) controller, grid-tied solar PV (GTSPV), harmonic current suppression*

I. INTRODUCTION

Over the past few decades, renewable energy sources like solar photovoltaics (PV) have gained significant focus due to fossil fuel depletion, industrialization, and rising environmental pollution. The proliferation of large-scale photovoltaic (PV) plants integrated as grid-tied PV systems (GTPVS) has substantially increased their penetration into modern distribution grids [1]. Nonlinear loads within the grid infrastructure induce harmonic distortions in voltage profiles, which propagate into the currents injected by PV systems, thereby degrading power quality compliance [2]. The GTPVS typically operates using maximum power point tracking (MPPT) to supply power to the grid with high current quality without additional devices. Thus, investigating harmonics compensation in conjunction with MPPT operation within GTPVS is crucial. The resolution of harmonic distortion and mitigation of double-frequency oscillations in active and reactive power during unbalanced grid voltage sags present

significant challenges [3]. These voltage sags can lead to increased system losses, reduced lifespan of the voltage source inverter (VSI), and overloading [1] of distribution transformers [4]. Consequently, significant research efforts have focused on designing advanced control schemes for grid-tied PV systems (GTPVS) to suppress grid current total harmonic distortion (THD) and mitigate power oscillations during grid voltage sags.

Numerous investigations have developed control strategies to mitigate grid current harmonics and remove power ripples in GTPVS [5]-[8]. In [5], a linear feedback compensation strategy is presented for a distributed generation (DG) systems to address harmonic mitigation and voltage unbalance. However, this approach relies on voltage-controlled converters, whereas renewable energy-based DG systems predominantly use current-controlled converters, necessitating harmonic compensation strategies for such architectures. A digital lock-in amplifier technique is introduced in [6] for harmonic suppression in single-phase grid-connected inverters, though it assumes a fixed PV array voltage and fails to eliminate oscillatory harmonics in injected power components. In [7], an instantaneous power control strategy is reported to improve grid current quality and attenuate power oscillations in grid-tied inverters. Similarly, [8] presents a power ripple suppression method to mitigate double-line frequency power oscillations in grid-tied inverters under unbalanced grid voltage conditions.

In recent years, the academic community has explored control strategies for addressing current harmonics in GTPV systems [9]-[12]. In [9], a plugin multiple resonant current (RC) controller scheme is presented for tracking and reducing periodic signal disturbances in a grid-connected converter. However, optimizing PR controllers with multiple variables is computationally intensive due to reliance on ordinary differential equation (ODE) modeling, Nyquist stability criteria, and root locus methodologies. A shunt active power filter control scheme is reported in [10] for harmonic detection and reduction using a quasi-proportional resonant (QPR) controller for a distributed generation system. It introduces two resonant poles into the control loop, leading to overshooting and oscillations during the transition phase, impacting the system's dynamic response. An adaptive sliding mode control (SMC) with harmonic disturbances is discussed in [11] for a three-level neutral point clamped converter. However, the complexity of its control law limits its use in industrial environments. In [12], a model predictive control (MPC) harmonic prediction control

979-8-3315-3013-6/25 $31.00 © 2025 IEEE 248

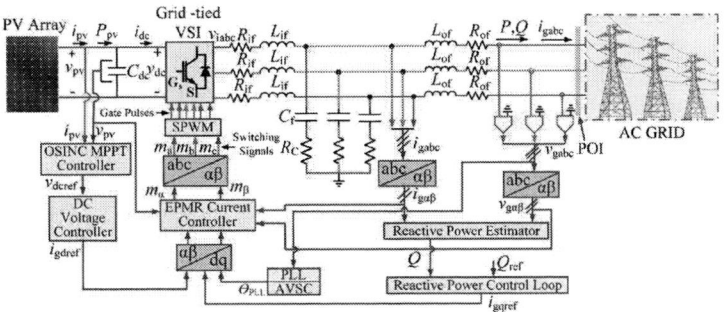

Fig.1 Schematic architecture of grid-tied PV system with the proposed unified control scheme

method is presented for predicting and calculating the inter-harmonic current for a grid-connected PV system. This approach relies on precise data regarding the controlled system, and designing weight coefficients for the cost function is challenging. The concurrent mitigation of current harmonics and suppression of power oscillations in GTPVS remains unaddressed in existing literature. Current harmonics compensation strategies predominantly rely on simplified DC voltage source models, overlooking the dynamic behavior of PV sources characterized by nonlinear current-voltage relationships and the necessity of maximum power point tracking (MPPT). Consequently, the joint attenuation of current harmonics, elimination of power oscillations, and integration of MPPT for GTPVS operating under abnormal grid voltages remain an unresolved research gap.

To address the aforementioned challenges, this article proposes a unified control scheme for a three-phase GTPVS operating under harmonically distorted grid voltage conditions. The proposed control schemes integrate an enhanced proportional multiresonant (EPMR) controller and an adaptive voltage sag compensator (AVSC), designed to attenuate dominant low-order harmonics (e.g., 5th, 7th, 11th, 13th, 17th, and 19th) in grid currents while suppressing power oscillations caused by harmonic-polluted grid voltages. Furthermore, the proposed scheme incorporates an optimized step-size incremental conductance (OSINC) maximum power point tracking (MPPT) algorithm to ensure rapid and precise PV array power extraction under dynamic environmental conditions. This synergistic integration achieves concurrent harmonic mitigation, power ripple elimination, and MPPT optimization, addressing a critical gap in existing GTPVS control strategies. The remainder of the article is structured as follows: Section II details the system architecture and control scheme which includes the EPMR controller and the AVSC. Section III discusses the real-time experimental verifications using the OP4510 simulator. Lastly, Section VI concludes the article.

II. SYSTEM CONFIGURATION AND CONTROL APPROACH

Fig. 1 shows the schematic architecture of the unified controlled three-phase single-stage GTPV system, comprising a solar PV array, DC link capacitor, grid-tied VSI, LCL filter, and AC grid. The GTPVS control scheme injects a sinusoidal grid current even with distorted and unbalanced voltage. The grid-tied VSI's gating pulse is generated through sinusoidal pulse width modulation. The control architecture includes PV side control with MPPT and DC link voltage controllers, and

grid side control with the EPMR current controller and a PLL-based AVSC. The EPMR controller manages grid-injected current with improved harmonic compensation. An OSINC MPPT algorithm enhances convergence speed and reduces steady-state oscillations during MPP operation. A PLL integrated AVSC achieves seamless grid synchronization and suppresses power oscillations during unbalanced grid voltage sag events, ensuring stable power injection from the PV system to the grid. Detailed explanations of the EPMR current controller and PLL-based AVSC are in Sections II-B and II-C.

A. Optimized Step Size INC MPPT Controller

The fixed step size INC (FSINC) algorithm is popular for PV arrays due to its simple implementation. However, the PR controller for harmonic minimization introduces transients, extending settling time. With a 0.02 s MPPT sampling interval, FSINC tracking is insufficient. Effective FSINC MPPT requires a 0.06 s interval. To address this, the optimized step size incremental conductance (OSINC) MPPT method [13] is proposed, featuring a variable voltage step size and a 0.06 s interval, ensuring rapid and precise MPP tracking. The step size adapts dynamically based on the instantaneous gradient (dP_{pv}/dV_{pv}) of the photovoltaic array's P-V characteristic curve. The OSINC MPPT algorithm employs larger voltage perturbations when operating distant from the maximum power point (MPP) to accelerate convergence, whereas near the MPP, the step size diminishes substantially to mitigate steady-state oscillations. The variable step magnitude, denoted OS_{step} and functionally dependent on the PV curve's slope, is defined as:

$$OS_{step} = x \left| \frac{v_{pv}(n)i_{pv}(n) - v_{pv}(n-1)i_{pv}(n-1)}{v_{pv}(n) - v_{pv}(n-1)} \right|, \quad (1)$$

where, $i_{pv}(n)$, $v_{pv}(n)$ and x are the PV array current, PV array voltage at n^{th} sampling instant and the scaling coefficient.

B. Enhanced Proportional Multi Resonant (EPMR) Current Controller

The conventional PMR controller's effectiveness decreases with rapid grid frequency changes and offers limited harmonic compensation under polluted grid voltages [14]. This work presents an enhanced PMR (EPMR) controller, delivering high-quality grid current for a GTPVS under harmonically polluted grid voltages. The EPMR compensates for grid current harmonics by incorporating multiple frequencies and effectively tracks the fundamental grid current component, reducing errors during grid frequency fluctuations. The architecture of the EPMR current controller, illustrated in Fig. 2, comprises two parallel operational stages. The first stage integrates a fundamental resonant controller (FRC) with transfer function $T_a(s)$, designed to regulate the primary grid current component in the GTPVS. The second stage incorporates multiple resonant harmonic compensators (RHCs) governed by transfer function $T_b(s)$, which selectively attenuate specified lower-order harmonic components. The mathematical representations of $T_a(s)$ and $T_b(s)$ are represented as follows

$$\left.\begin{array}{l} T_a(s) = \dfrac{2k_{if}\omega_b s}{s^2 + 2\omega_b s + \omega_g^2}; \\[4mm] T_b(s) = k_{ph} + \displaystyle\sum_{h=5,7,11,\ldots}^{n} \dfrac{2k_{ih}(h\omega_b)s}{s^2 + 2(h\omega_b)s + (h\omega_g)^2}; \end{array}\right\} \quad (2)$$

The EPMR controller is defined by proportional gain k_{ph}, fundamental resonant gain k_{if}, and h^{th} order harmonic resonant gain k_{ih}. Critical design variables such as bandwidth ω_b, harmonic order h, and fundamental resonant frequency ω_g are integral to ensuring system stability and harmonic compensation performance. In the $\alpha\beta$ reference frame, Fig. 3 shows the simplified equivalent diagram of the grid-tied VSI. The VSI output voltage expression is specified by

$$v_{i\alpha\beta} = (v_{dc}/2)\, m_{\alpha\beta} \quad (3)$$

where $m_{\alpha\beta}$ is the modulation signals and v_{dc} is the dc-link voltage.

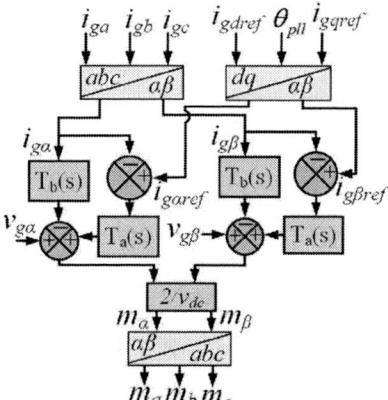

Fig. 2. Control Structure block diagram of the EPMR controller.

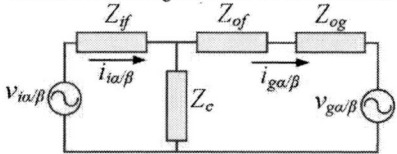

Fig. 3. Simplified equivalent representation of the grid-tied VSI.

In Fig. 2, the $m_{\alpha/\beta}$ of the VSI is defined as

$$m_{\alpha/\beta} = 2/v_{dc}\left\{ v_{g\alpha/\beta} + T_a(s)\, i_{g\alpha/\beta ref} - i_{g\alpha/\beta}(T_a(s) + T_b(s)) \right\} \quad (6)$$

A closed loop transfer function $T_{cc}(s)$ between $i_{g\alpha\beta ref}$ and $i_{g\alpha\beta}$ of the EPMR controller is written as

$$T_{cc}(s) = \dfrac{Z_c(s)T_a(s)}{\begin{array}{l}\left[Z_{if}(s)Z_c(s) + Z_c(s)\{Z_{og}(s) + Z_{if}(s)\}\right] \\ + \left[Z_c(s) + Z_g(s)\{T_a(s) + T_b(s)\}\right] \\ + Z_{if}(s)\{Z_{if}(s) + Z_{of}(s) + Z_{og}(s)\}\end{array}} \quad (7)$$

The inverter filter's series input and output impedances are denoted as Z_{if} and Z_{of} respectively, with Z_c representing the shunt filter impedance and Z_{og} characterizing the equivalent grid impedance. In Fig. 4, the frequency response of both $T_a(s)$ and $T_b(s)$ within the EPMR control scheme is shown. At the grid's resonant frequency of 50 Hz, $T_a(s)$ reaches a magnitude of 50 dB. The transfer function $T_b(s)$ exhibits pronounced resonant peaks at distinct harmonic frequencies corresponding to the 5th, 7th, 11th, 13th, 17th, and 19th harmonic orders (250 Hz, 350 Hz, 550 Hz, 650 Hz, 850 Hz, and 950 Hz, respectively)

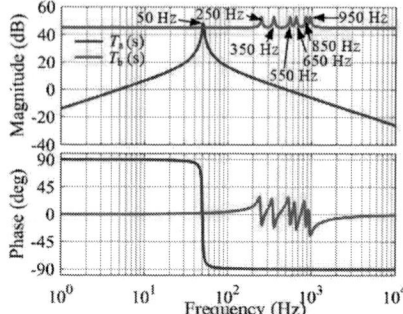

Fig. 4. Bode plot of $T_a(s)$ and $T_b(s)$ for the EPMR controller.

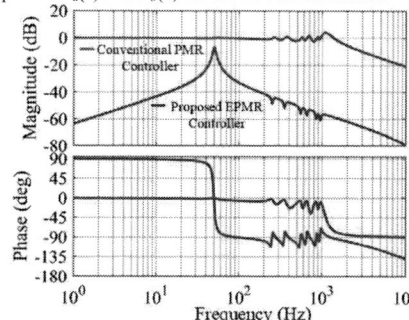

Fig. 5. Bode plot of closed loop TF $F_{cl}(s)$ for the EPMR and conventional PMR controller.

Fig. 5 illustrates the frequency response of $T_{cc}(s)$ for both the EPMR controller and the conventional PMR controller. At the fundamental grid frequency of 50 Hz, the EPMR controller shows an amplitude of 0 dB for $T_{cc}(s)$, indicating unity. This amplitude significantly diminishes at the specified harmonic frequencies. Consequently, this pattern ensures precise tracking of the fundamental grid currents without any steady-state error and effectively reduces the specified grid current harmonics at the 5th, 7th, 11th, 13th, 17th, and 19th orders.

C. PLL based Adaptive Voltage Sag Compensator (AVSC)

Grid synchronization, crucial for GTPVS, is typically implemented via a synchronous reference frame phase-locked loop (SRF-PLL). However, under unbalanced grid voltage sags or harmonic distortions, the SRF-PLL fails to accurately extract the fundamental positive-sequence voltage component. This inaccuracy results in 100 Hz oscillations in the estimated dq-axis voltage components. These distorted signals propagate into the current controller, governing active and reactive power regulation, and the dc-link voltage controller, stabilizing the PV array voltage. Consequently, the injected power waveforms exhibit 100 Hz ripple components, degrading the GTPV system's power quality. To address this, an adaptive voltage sag compensator (AVSC) is used with EPMR within the PLL block to eliminate these power ripples, reducing ripples in the *dq* component of grid voltages.

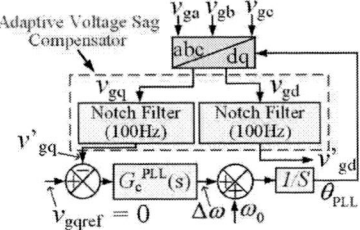

Fig. 6. Block diagram of the PLL based AVSC

During the unbalanced grid voltage sags the three-phase grid voltages is expressed as

$$\begin{bmatrix} v_{ga} \\ v_{gb} \\ v_{gc} \end{bmatrix} = \begin{bmatrix} v_{ga}^{+} \\ v_{gb}^{+} \\ v_{gc}^{+} \end{bmatrix} + \begin{bmatrix} v_{ga}^{-} \\ v_{gb}^{-} \\ v_{gc}^{-} \end{bmatrix}$$

$$= \begin{bmatrix} V_p \sin(\omega t + \delta_p) + V_n \sin(\omega t + \delta_n) \\ V_p \sin(\omega t + \delta_p - 120) + V_n \sin(\omega t + \delta_n - 120) \\ V_p \sin(\omega t + \delta_p + 120) + V_n \sin(\omega t + \delta_n + 120) \end{bmatrix}$$

(8)

With the Clark's and Park's transformation the transformed voltages v_{gd} and v_{gq} can be represented as follows

$$\begin{bmatrix} v_{gd} \\ v_{gq} \end{bmatrix} = \frac{2}{3} \begin{bmatrix} \sin(\omega t) & \cos(\omega t) \\ \cos(\omega t) & -\sin(\omega t) \end{bmatrix} \begin{bmatrix} 1 & -\frac{1}{2} & -\frac{1}{2} \\ 0 & \frac{\sqrt{3}}{2} & -\frac{\sqrt{3}}{2} \end{bmatrix} \begin{bmatrix} v_{ga} \\ v_{gb} \\ v_{gc} \end{bmatrix}$$

$$= \begin{bmatrix} v_{gd}^{+} \\ v_{gq}^{+} \end{bmatrix} + \begin{bmatrix} v_{gd}^{-} \\ v_{gq}^{-} \end{bmatrix}$$

(9)

$$\begin{bmatrix} v_{gd}^{+} \\ v_{gq}^{+} \end{bmatrix} = \frac{2}{3} \begin{bmatrix} \sin(\omega t) & \cos(\omega t) \\ \cos(\omega t) & -\sin(\omega t) \end{bmatrix} \begin{bmatrix} 1 & -\frac{1}{2} & -\frac{1}{2} \\ 0 & \frac{\sqrt{3}}{2} & -\frac{\sqrt{3}}{2} \end{bmatrix} \begin{bmatrix} v_{ga} \\ v_{gb} \\ v_{gc} \end{bmatrix}$$

$$= \begin{bmatrix} -V_p \cos(2\omega t + \delta_p) \\ V_p \sin(2\omega t + \delta_p) \end{bmatrix}$$

(10)

$$\begin{bmatrix} v_{gd}^{-} \\ v_{gq}^{-} \end{bmatrix} = \frac{2}{3} \begin{bmatrix} -\sin(\omega t) & \cos(\omega t) \\ \cos(\omega t) & \sin(\omega t) \end{bmatrix} \begin{bmatrix} 1 & -\frac{1}{2} & -\frac{1}{2} \\ 0 & \frac{\sqrt{3}}{2} & -\frac{\sqrt{3}}{2} \end{bmatrix} \begin{bmatrix} v_{ga} \\ v_{gb} \\ v_{gc} \end{bmatrix}$$

$$= \begin{bmatrix} V_n \cos(2\omega t + \delta_n) \\ V_n \sin(2\omega t + \delta_n) \end{bmatrix}$$

The positive and negative sequence components of the grid voltage in the *dq*-axis are denoted v_{gd}^{+}, v_{gd}^{-}, v_{gq}^{+} and v_{gq}^{-} respectively. Under unbalanced grid voltages , the instantaneous active and reactive power delivered to the grid [15] is given as:

$$\begin{cases} P_g(t) = P_{0g} + P_{c2g}\cos(2\omega t) + P_{s2g}\sin(2\omega t) \\ Q_g(t) = Q_{0g} + Q_{c2g}\cos(2\omega t) + Q_{s2g}\sin(2\omega t) \end{cases}$$

(11)

As observed in (10), the positive and negative sequence grid voltages (v_{gd}^{+}, v_{gd}^{-}, v_{gq}^{+} and v_{gq}^{-}) exhibit ripples at twice the grid frequency (100 Hz). Power oscillations in active and reactive components at twice the fundamental grid frequency arise due to v_{gd}^{-} and v_{gq}^{-} introducing oscillations in the *dq* axis voltage components. To mitigate this issue, an AVSC is proposed by

integrating a notch filter (NF) with a 100 Hz cutoff frequency into the SRF-PLL control architecture, thereby attenuating 100 Hz oscillations in the *dq*-axis voltage components (see Fig. 6). The transfer function of NF is defined as

$$G_{nf}(s) = \frac{s^2 + \omega_n^2}{s^2 + 2\alpha\omega_n s + \omega_n^2}$$

(11)

where ω_n and $\alpha\omega_n$ is the band-reject frequency and cut off frequency of the NF. To attenuate 100-Hz oscillations and frequency-specific distortions, the dq-axis grid voltages are processed through an Advanced Vector Synchronous Control (AVSC) structure incorporating a Notch Filter (NF). This configuration suppresses the double-fundamental-frequency (2ω, 100 Hz) ripple components in both positive- and negative-sequence grid voltage components (v_{gd}^{+}, v_{gd}^{-}, v_{gq}^{+}, v_{gq}^{-}) by nullifying the oscillatory terms $\cos(2\omega t + \delta p)$, $\sin(2\omega t + \delta p)$, $\cos(2\omega t + \delta n)$, $\sin(2\omega t + \delta n)$ defined in (10). Eliminating these oscillations enhances power quality by eliminating the oscillatory active and reactive power constituents (P_{c2}, P_{s2}, Q_{c2}, Q_{s2}) generated during grid voltage imbalances

TABLE I. REAL TIME SYSTEM AND CONTROL PARAMETERS

System Parameters	Grid voltage source	$V_g, f_g\ R_g, L_g$	110 V (L-L rms), 50 Hz, 0.35 Ω, 0.5 mH
	VSI switching frequency	f_{sw}	10 kHz
	Harmonic filter	$L_{if}, R_{if}, L_{of}, R_{of}, C_f, R_f$	5 mH, 2.5 Ω, 0.8 mH, 0.5 Ω, 15 µF, 0.5 Ω
	PV unit	V_{MPP}, I_{MPP}	390 V, 9.76 A
		V_{OC}, I_{SC}	450 V, 10.46 A
		P_{MPP}	3.8 kW
		C_{dc}	4700 µF
Control parameters	EPMR current controller	k_{ph}	40
		k_{if}, k_{ih}, ω_b	500, 500, 6.28

III. REAL TIME VALIDATIONS

Experimental validation of the unified control scheme is conducted using an OPAL-RT OP4510 real-time simulator in a software-in-the-loop (SIL) framework, with system and control parameters detailed in Table I. The EPMR controller's performance is evaluated under nonideal grid conditions specifically harmonic distortion and unbalanced voltage sags against a conventional PMR controller benchmark [14]. Harmonically distorted grid voltages are generated by injecting specified harmonic constituents at 12% (5th, 7th), 10% (11th, 13th), and 6% (17th, 19th) relative to the fundamental component. Comparative grid voltage and current waveforms for both controllers are depicted in Fig. 7. As demonstrated in Fig. 7, the grid current exhibits substantial harmonic distortion prior to current controller activation. Upon enabling the controller, the EPMR based strategy achieves superior harmonic suppression, yielding a sinusoidal and balanced current profile with reduced distortion compared to the conventional PMR controller. Fig. 8 presents the FFT spectra of the grid current under the proposed EPMR and conventional PMR control schemes. As illustrated in Fig. 8, the THD of the grid current prior to compensation is 8.88%, which reduces to 7.47% with the conventional PMR controller and further diminishes to 3.6% using the proposed EPMR controller. The EPMR controller achieves a grid current THD of 3.6%, compliant with the IEEE 1547 standard [16], which mandates THD levels below 5% for

979-8-3315-3013-6/25 $31.00 © 2025 IEEE   251

the GTPVS. Fig. 9 depicts the real-time performance of the GTPVS under unbalanced grid conditions with phase voltages v_{ga} is $110\angle 0°$, v_{gb} is $55\angle -105°$, and v_{gc} is $66\angle 105°$.

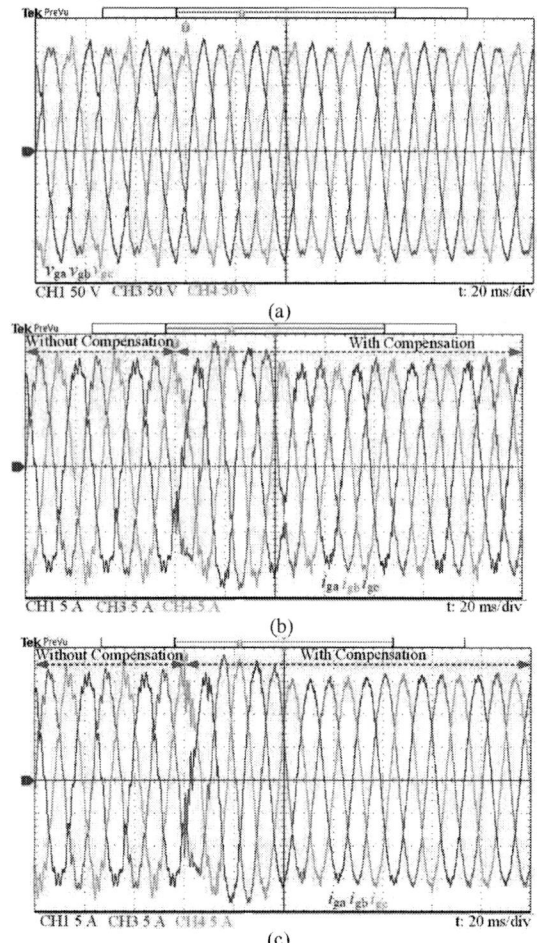

Fig. 7. Real-time waveforms of v_{gabc} and i_{gabc} during harmonically distorted grid voltages: (a) v_{gabc} (b) i_{gabc} with conventional PMR controller. (c) i_{gabc} with proposed EPMR controller.

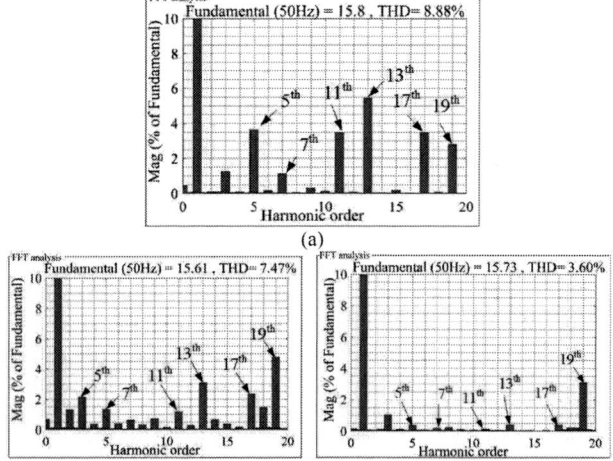

Fig. 8. FFT spectra the grid current i_{gabc} (a) without compensation (b) with the conventional PMR controller (c) with the proposed EPMR controller.

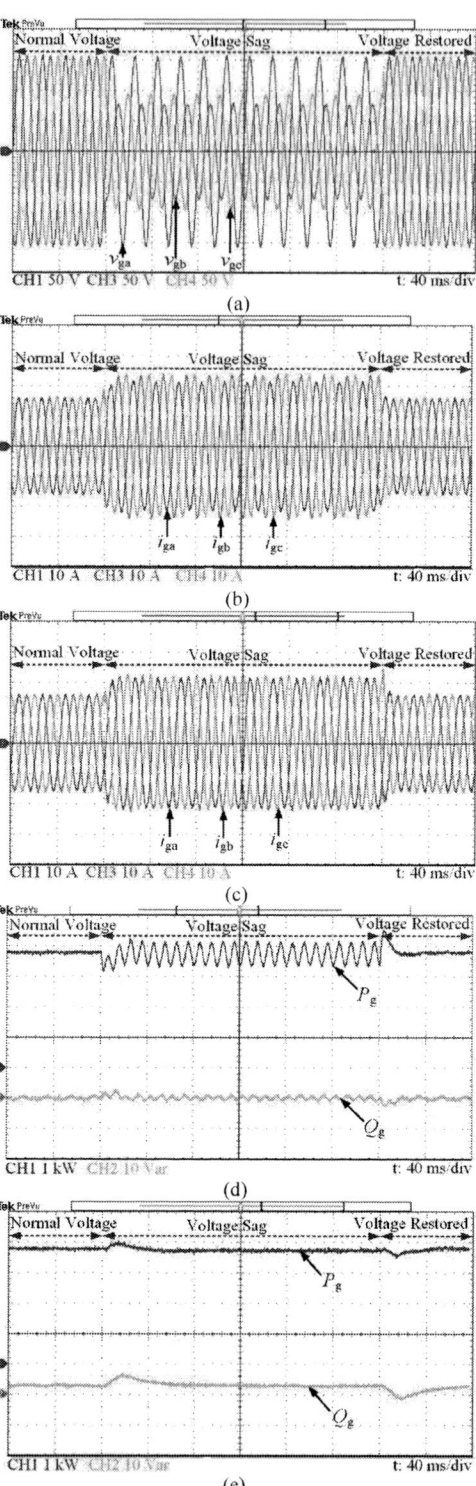

Fig. 9. Measured real-time waveforms under unbalanced grid voltage condition (v_{ga} =$110\angle 0°$, v_{gb} =$55\angle -105°$, v_{gc} =$66\angle 105°$). (a) v_{gabc} (b)-(c) i_{gabc}, P_g and Q_g obtained by the conventional PMR controller and proposed EPMR controller with AVSC.

The voltage imbalance is initiated at t=80 ms and restored to nominal conditions at t=320 ms. As shown in Fig. 9(b)–(c), the proposed EPMR AVSC control schemes achieve superior sinusoidal characteristics and balanced current profiles compared to the conventional PMR AVSC scheme. Furthermore, the EPMR AVSC scheme effectively suppresses oscillations in active and reactive power, as evidenced by the waveforms in Fig. 9(d)–(e).

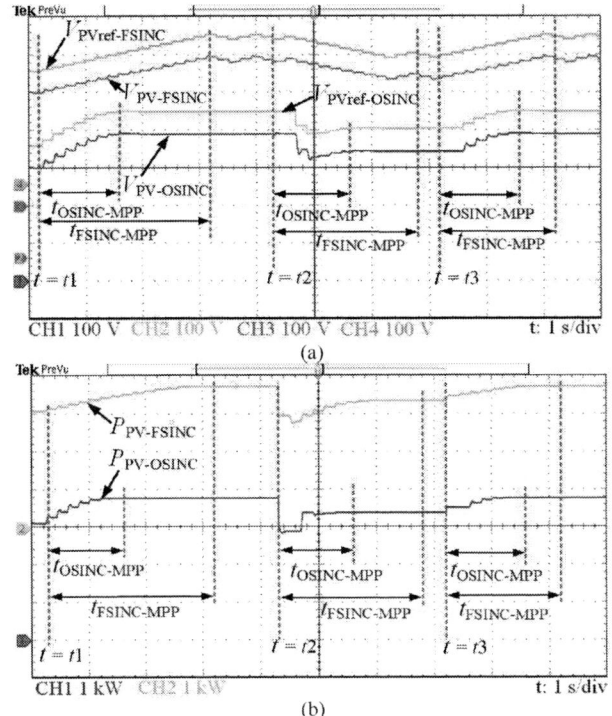

Fig. 10. Real-time dynamic waveforms of the OSINC MPPT and FSINC MPPT controller under sudden environmental changes. (a) PV Voltage. (b) PV Power.

Fig. 10 compares OSINC and FSINC MPPT controllers using the EPMR controller scheme with AVSC. The FSINC MPPT has a step size (Δv_{pv}) of 5V, while OSINC MPPT has a scaling coefficient (x) of 0.75. Initially (irradiance $I_r = 1000$ W/m^2, temperature $T_m = 25$ °C), the PV array operates at $V_{PV} = 300$ V and $P_{PV} = 3.1$ kW. Both controllers are activated at t=t1. The OSINC achieves the MPP (V_{MPP}=390 V, P_{MPP}=3.8 kW) with convergence time $t_{OSINC-MPP}$ = 1.4 s and voltage oscillations of 0.5 V, outperforming FSINC, which shows slower convergence ($t_{FSINC-MPP}$ = 3.0 s) and larger oscillations (5V). During the second phase (I_r = 700 W/m^2, Tm = 50 °C), OSINC reaches MPP (V_{MPP} = 340 V, P_{MPP} = 3.3 kW) in $t_{OSINC-MPP}$=1.2 s, while FSINC requires $t_{FSINC-MPP}$=2.7 s. Upon returning to initial conditions (t=t3), OSINC restores MPP (V_{MPP}=390 V, P_{MPP}=3.8 kW) in $t_{OSINC-MPP}$ = 1.4 s versus $t_{FSINC-MPP}$ = 2.0 s for FSINC, showing better dynamic response.

IV. CONCLUSION

A unified control scheme is proposed for simultaneous reduction of current harmonics and elimination of power ripples in a grid-tied solar photovoltaic (GTSPV) system under abnormal grid voltages. Current harmonics are mitigated by an enhanced proportional multiresonant (EPMR) controller with a restructured control block. Integration of an adaptive voltage sag compensator (AVSC) with the PLL ensures ripple-free active and reactive power delivery to the grid during unbalanced voltage sag. An optimized step size incremental conductance (OSINC) MPPT controller achieves faster tracking of the MPP under sudden atmospheric variations. The proposed control scheme achieves a THD in the grid current compliant with IEEE 519 standards.

REFERENCES

[1] K. Alluhaybi, I. Batarseh and H. Hu, "Comprehensive Review and Comparison of Single-Phase Grid-Tied Photovoltaic Microinverters, *IEEE Journal of Emerging and Selected Topics in Power Electronics*, vol. 8, no. 2, pp. 1310-1329, June 2020.

[2] J. Zhang, X. Yang, W. Chen, H. Zhou and J. Luo, "A Multifrequency Small-Signal Model for the MLCL-Filtered Grid-Connected Inverter Considering the FCE of Nonlinear Inductors," *IEEE Transactions on Industrial Electronics*, vol. 70, no. 5, pp. 4901-4911, May 2023.

[3] V. Jain, S. Kewat and B. Singh, "Three Phase Grid Connected PV Based EV Charging Station With Capability of Compensation of Reactive Power," *IEEE Transactions on Industry Applications*, vol. 59, no. 1, pp. 367-376, Jan.-Feb. 2023.

[4] A. Tarraso, J. I. Candela, J. Rocabert and P. Rodriguez, "Selective Harmonic Compensation in SPC Grid-Forming Converters for Improving Power Quality in Weak Grid," *IEEE Transactions on Power Delivery*, vol. 39, no. 6, pp. 3223-3232, Dec. 2024.

[5] S. Maganti and N. P. Padhy, "A Feedback-Based Flexible Compensation Strategy for a Weak-Grid-Tied Current-Controlled Converter Under Unbalanced and Harmonic Conditions," *IEEE Transactions on Industry Applications*, vol. 58, no. 6, pp. 7739-7753, Nov.-Dec. 2022.

[6] M. N. Ashraf, R. A. Khan and W. Choi, "A Novel Selective Harmonic Compensation Method for Single-Phase Grid-Connected Inverters," in *IEEE Transactions on Industrial Electronics*, vol. 68, no. 6, pp. 4848-4858, June 2021.

[7] W. Liu, F. Blaabjerg, D. Zhou and S. -F. Chou, "Modified Instantaneous Power Control With Phase Compensation and Current-Limited Function Under Unbalanced Grid Faults," *IEEE Journal of Emerging and Selected Topics in Power Electronics*, vol. 9, no. 3, pp. 2896-2906, June 2021.

[8] C. -Y. Tang and J. -H. Jheng, "An Active Power Ripple Mitigation Strategy for Three-Phase Grid-Tied Inverters Under Unbalanced Grid Voltages," *IEEE Transactions on Power Electronics*, vol. 38, no. 1, pp. 27-33, Jan. 2023.

[9] C. Tang, K. Zhou, Y. Shu, Q. He and Q. Chen, "Analysis and Design of Multiple Resonant Current Control for Grid-Connected Converters," *IEEE Journal of Emerging and Selected Topics in Power Electronics*, vol. 10, no. 2, pp. 2539-2546, April 2022.

[10] Z. Wu, G. Xu, W. Zhu and G. Sheng, "A Dispersing Proportional Resonance Controller for Selective Harmonic Current Compensation in a Shunt Active Power Filter," *IEEE Transactions on Power Electronics*, vol. 40, no. 1, pp. 279-289, Jan. 2025.

[11] L. Liu *et al.*, "Multidisturbances Compensation for Three-Level NPC Converters in Microgrids: A Robust Adaptive Sliding Mode Control Approach," *IEEE Transactions on Industrial Informatics*, vol. 20, no. 6, pp. 8917-8931, June 2024.

[12] M. Mao, Z. Xu, Q. Yuan and H. Li, "Current Interharmonic Prediction Control Based on MPC and Lyapunov for Grid-Connected PV System," *IEEE Journal of Emerging and Selected Topics in Power Electronics*, vol. 12, no. 3, pp. 2686-2696, June 2024.

[13] M. K. Mishra and V. N. Lal, "A Multiobjective Control Strategy for Harmonic Current Mitigation With Enhanced LVRT Operation of a Grid-Tied PV System Without PLL Under Abnormal Grid Conditions," *IEEE Journal of Emerging and Selected Topics in Power Electronics*, vol. 11, no. 2, pp. 2164-2177, April 2023.

[14] Z. Chen, B. Wang and B. Li, "A Digital Proportional–Resonant Controller With Desired Tracking-Error Convergence," *IEEE Transactions on Industrial Electronics*, vol. 71, no. 4, pp. 3751-3762, April 2024.

[15] S. Ouchen, M. Benbouzid, F. Blaabjerg, A. Betka and H. Steinhart, "Direct Power Control of Shunt Active Power Filter Using Space Vector Modulation Based on Supertwisting Sliding Mode Control," *IEEE Journal of Emerging and Selected Topics in Power Electronics*, vol. 9, no. 3, pp. 3243-3253, June 2021.

[16] "IEEE Standard for Harmonic Control in Electric Power Systems," *IEEE Std 519-2022*, vol., no., pp.1-31, 5 Aug. 2022.

A Novel Doubly Grounded High Gain Single Stage Inverter for Photovoltaic Applications

Muhammedali Shafeeque K
Dept. of Electrical Engineering
NIT Calicut, Calicut
Govt. Engineering College Palakkad
ORCID: 0000-0002-6139-2488

Dr. Nikhil Sasidharan
Dept. of Electrical Engineering
NIT Calicut
Calicut, India
nikhils@nitc.ac.in

Dr. Rijil Ramchand
Dept. of Electrical Engineering
NIT Calicut
Calicut, India
rijil@nitc.ac.in

Abstract— For solar PV (SPV) applications, power electronic converters are generally used to convert the low PV DC voltage into a high-gain AC voltage. This paper proposes a novel Doubly Grounded High Gain Single Stage Inverter (DGHGSSI). The inverter is buck-boost based which is incorporated with switched capacitor technique having an improvement AC gain. The proposed inverter employs a single inductor for both the negative and positive portions of the output AC voltage, thereby assuring symmetry in the AC waveform. It functions with a single DC voltage source, providing power to both portions, hence enhancing its resiliency to variations in the multiple power sources. Furthermore, the DGHGSSI utilizes fewer components, resulting in advantages like compact design, reduced weight and decreased cost. A complete design and analysis of the proposed inverter in continuous conduction mode are provided and discussed. Results from the MATLAB/SIMULINK simulation validate the analysis and operation. The results from the hardware prototype also validate the operation of the proposed inverter.

Keywords— high gain single stage inverters, step -up inverters, solar inverters, buck boost inverters

I. INTRODUCTION

The increasing demand for renewable energy has prompted significant study and development in solar energy harvesting. Among renewable sources, solar energy is distinguished by its plentiful availability and ease of use. The efficiency of solar power systems is significantly influenced by the design of the power interface converter. The output power of solar photovoltaic (SPV) modules varies with changes in irradiation levels and the temperature of incident light on the photovoltaic cell, resulting in fundamentally intermittent power generation.

Power electronic circuits are employed to regulate variation by converting the low-voltage, variable DC output from solar photovoltaic modules. Photovoltaic systems generally utilize two-stage converters were the first stage comprises a DC-DC converter that either boosts or reduces the low-voltage solar output, whereas the next phase is an inverter that transforms DC into high-quality AC. Despite the effectiveness and direct methodology of these two-stage converters, they possess several disadvantages, including higher sizes, increased cost and extra weight.

Other than traditional two-stage converters, other alternative designs are also available in literature Kasa et al. [1] proposed a design utilizing two separate buck-boost converters for each half-cycle, leading to asymmetrical operation caused by source dissimilarities. Wang [2] introduced a converter grounded in buck theory; yet, its circuit design comprises a large quantity of components.

Similar to these inverters many single stage inverters ate available in the literate as in [3]-[9]. Kusakawa et al. [10] offered a setup utilizing a singular PV input source; however, the circuit's operational characteristics impose limitations on the grounding of the PV cells. Also, a high gain single stage inverter is proposed in [11], but it do not have the feature of double grounding. Grounding a photovoltaic (PV) cell is as essential as grounding any other electrical apparatus. It involves wiring a grounding connection from one of the photovoltaic terminals. Proper grounding is crucial for protecting and mitigating the impacts of excessive voltages, such as lightning strikes. In an ungrounded photovoltaic system, further safety measures, including overcurrent prevention, are necessary to mitigate power surges [12]. In addition to safety issues, grounding photovoltaic cells is advised to improve their lifespan. When the PV module's negative terminal remains ungrounded, it develops a negative potential relative to the ground, leading to leakage currents. This leakage current carry sodium ions (Na+) from the glass layer to the clear conductive oxide (TCO) layer, resulting degradation. In addition to grounding the photovoltaic cells, the AC neutral point is also grounded, necessitating an inverter design that supports dual grounding. Transformer-based converter topologies inherently enable this by offering isolation. But, the use of transformers is frequently discouraged because of their considerable size and substantial weight. Literature reveals the advancement of doubly grounded single-stage inverters [13], [14], however these inverters encounter constraints in attaining greater voltage gain.

The paper proposes a novel high-gain single-stage DC-AC inverter for photovoltaic applications, utilizing the buck-boost theory. It enables dual grounding without the need for a bulky transformer and offers double the gain compared to conventional single-stage inverters. Figure 1 shows the circuit configuration of the proposed inverter topology, which possesses the following features:

1) The proposed topology has advantage of better utilization of DC solar input because of the high gain.

2) The configuration of the proposed circuit eliminates the problem of zero source current at OFF time. This enhance the input power factor in cases when a input DC voltage is obtained from an a AC grid.

3) Single photovoltaic source is shared in both half cycles, eliminating all mismatches resulting from source variation.

4) The topology operation enables dual grounding and thus allows common grounding at AC and DC side.

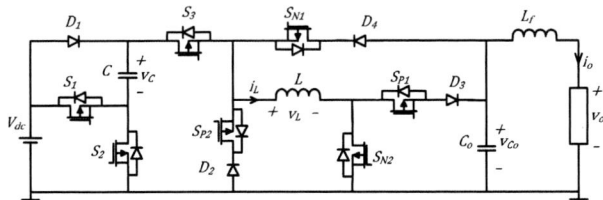

Fig. 1. Cirucit diagram of proposed DGHGSSI.

II. PROPOSED NOVEL DOUBLY GROUNDED HIGH GAIN SINGLE STAGE INVERTER

The proposed Doubly Grounded High-Gain Single-Stage Inverter consists of four diodes and seven controlled switches, with three switches operating at high frequency during the negative half-cycle and four during the positive half-cycle. The circuit employs only one buck boost inductor, L, shared in both halves, while the input switched capacitor, C, is utilized in the input of which doubles the overall voltage gain. A low-pass filter is used having a capacitor C_O and inductor L_f

A. Development of Proposed Novel DGHGSSI

Generally, in traditional buck boost based single stage inverters, input DC voltage source becomes separated from the circuit throughout the turn off period of high frequency switch. Hence, input current will be zero and thus discontinuous even with a non-zero inductor current. The proposed inverter configuration utilizes the mentioned turn off time period to charge the switched capacitor which results in doubling the overall gain of the inverter. Also, a non-zero input current at the turn off period is obtained, improving input current profile. Two controlled switches operate complementarily to accomplish this. While the high frequency PWM switch is in OFF condition, the capacitor come across the input DC source. Thus the capacitor charges to the input DC voltage level. Afterwards, in the ON period of the switch, the capacitor discharges assisting the inductor current raise in addition to the series DC voltage source.

B. Working of the Proposed DGHGSSI:

The controlled switches S_1 and S_2 function complementarily during both the positive and negative cycles, controlling the charging and discharging of capacitor C. A buck-boost circuit is formed with switches S_{N1}, S_{N2}, S_3, and diodes D_4 to generate negative halve of the AC output voltage. Switch S_3 is regulated by a PWM signal with modified duty function, whereas switches S_{N1} and S_{N2} are in ON condition. In a similar manner, the positive halve of the AC output voltage will be generated by the buck-boost converter employing switches S_{P1}, S_{P2}, S_3, and diodes D_2 and D_3. In addition to these switches, S_{N2} is also switched during positive half cycle.

The DGHGSSI has four modes of operation. Modes I and II correspond to the negative halve, while Modes III and IV correspond to the positive half of AC output voltage. In Modes I and II, switches S_{N1} and S_{N2} remain gated, whereas switch S_3 operates at switching frequency.

Mode-I: The mode begins while S_1 and S_3 are on whereas S_2 remains turned off. In this mode, S_{N1} and S_{N2} are gated, whereas S_{P1} and S_{P2} are ungated. All diodes remain in reverse biased condition and therefore will not conduct.

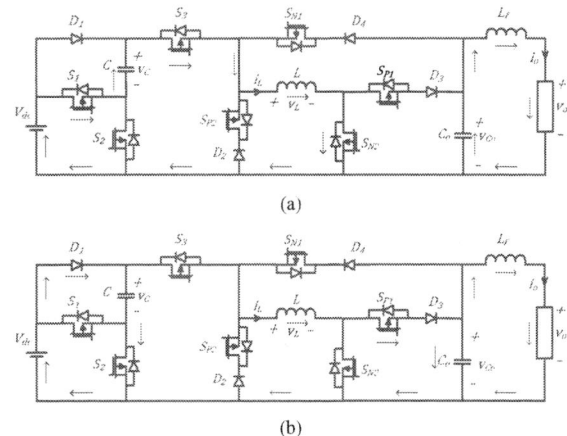

Fig. 2. Current path during the negative half-cycle: (a) Mode-I, (b) Mode-II.I

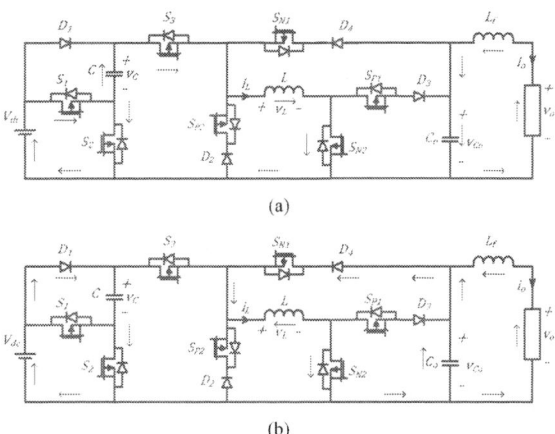

Fig. 3. Current path during the positive half-cycle: (a) Mode-III, (b) Mode-IV.

Thus, DC source and capacitor voltage appears across the inductor, L. This voltage causes inductor to store energy, resulting in a linear increase in the inductor current. The currents from the source and the capacitor will therefore be same as to the inductor current. The waveforms are depicted in Fig. 4. Figure 2(a) illustrates the path taken by the current. The output capacitor now manages the AC load.

Mode-II: The mode begins when S_1 and S_3 are turned off and S_2 is turned ON. S_{N1} and S_{N2} remain gated, however S_{P1} and S_{P2} continue ungated. Diode D_4 becomes forward biased and due to the inductor stored energy, switch S_3 will be OFF. Furthermore, when switch S_2 is in ON state, the capacitor connects across the DC input voltage source via the forward-biased diode D_1. As a result, the capacitor charges up to V_{dc}, with a current, as depicted in Fig. 4. Consequently, the input current becomes the negative of the capacitor current. During this time, the energy stored in the inductor is transferred to the load while also contributing to charging the output capacitor. Consequently, the inductor current gradually declines, and the associated current flow path is illustrated in Fig. 2(b).

979-8-3315-3013-6/25 $31.00 © 2025 IEEE   255

S_{P1} and S_{P2} are gated in the positive halve of AC in modes III & IV whereas S_3 as well as S_{N2} operates at switching frequency.

Mode-III: This mode begins when S_1, S_3 and S_{N2} are on, whereas S_2 remains turned off. In this mode S_{P1} & S_{P2} are gated whereas S_{N1} remains ungated. Diodes will not conduct in this mode since diodes remains in a reverse-biased state. Thus, DC source and capacitor voltage appears across the inductor, L. This voltage causes inductor to store energy, resulting in a linear increase in the inductor current. The currents from the source and the capacitor will therefore be same as to the inductor current. The waveforms are depicted in Fig. 4. Figure 3(a) illustrates the path taken by the current. The output capacitor manages the load.

Mode-IV: The mode begins when S_1, S_3 and S_{N2} are turned off and S_2 is turned ON. S_{P1} and S_{P2} remain gated, however and S_{N2} continue ungated in this mode. Diodes D_2 and D_3 becomes forward biased while switch S_3 is OFF due to the inductor stored energy, switch S3 will be OFF. Furthermore, when switch S_2 is in ON state, the capacitor connects across the DC input voltage source via the forward-biased diode D_1. As a result, the capacitor charges up to V_{dc}, generating a capacitor current, as depicted in Fig. 4. Consequently, the input current becomes the negative of the capacitor current. At this stage, the energy stored in the inductor is transferred to the AC load while also contributing to charging the output capacitor. Consequently, the inductor current gradually declines, and the associated flow of current is illustrated in Fig. 3(b)

III. Mathematical Analysis Of Proposed DGHGSSI

In the proposed DGHGSSI, for performing mathematical analysis the switching frequency is assumed to be significantly higher than the AC output voltage. Hence, the output AC voltage remains practically unchanged during each switching cycle.

The maximum value of AC output voltage is denoted as $V_{omk} = G\,V_{dc}$, where G represents the overall inverter gain. Thus the AC output voltage will be given as

$$v_o(\omega t) = G\,V_{dc}\sin \omega t \tag{1}$$

Here, ω (in rad/s)represents the frequency of AC output voltage, and V_{dc} denotes the PV DC voltage. The duty function, during the kth switching period is expressed as $d(\omega t, k)$, as it depends on ωt and varies with each switching period.

In the off-time of the high-frequency switch, the capacitor voltage is given by $v_c = V_{dc}$, as it was charged to the DC input voltage in the previous mode. Consequently, the inductor voltage becomes $v_L = 2V_{dc}$. Similarly, during the on-time, the inductor voltage is given by $v_L = -|v_o(\omega t)|$.

By applying the volt-second balance principle, the relationship between the input DC voltage and the output AC voltage can be derived as:

$$|v_o(\omega t)| = \frac{2d(\omega t,k)}{1-d(\omega t,k)}\,V_{dc} \tag{2}$$

Equation (2) shows that the proposed DGHGSSI provides double the gain compared to traditional buck boost single stage inverters. Consider the kth switching cycle, while ON

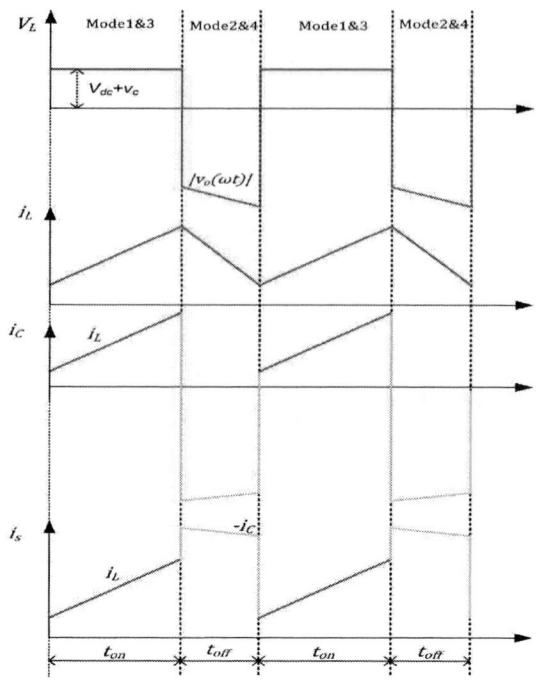

Fig. 4. Analysis waveforms in different opeartional modes

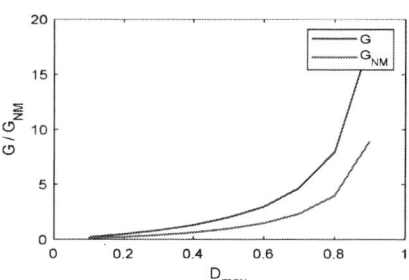

Fig. 5. Gain comparison between the proposed DGHGSSI and a conventional buck-boost inverter.

time, $t_{on} = d(\omega t, k)\,T_s$, as inductor voltage is $2V_{dc}$, inductor current, $i_L(\omega t)$ as well as input current, increases with a current ripple of $\Delta I_L(\omega t, k)$. Since during ON time inductor voltage, $v_L = 2V_{dc}$

$$\Delta I_L(\omega t, k) = \frac{2V_{dc}d(\omega t,k)\,T_s}{L} \tag{3}$$

Let $\sin(\omega t, k)$ be the reference signal, which is the sampled version of $\sin \omega t$, sampled at the switching frequency. Then duty function which guarantees a pure sinusoidal output AC voltage eliminating a current control loop, can be derived from equations (1) and (2) as follows:

$$d(\omega t, k) = \frac{G\,|\sin(\omega t,k)|}{2+G\,|\sin(\omega t,k)|} \tag{4}$$

A relationship between D_{max} and gain, G is derived from (4) and depicted in Fig. 5, highlighting the gain comparison between the proposed DGHGSSI and a conventional single-stage inverter. In the closed interval $[0, \pi]$ of ωt, the expression, $|\sin(\omega t, k)|$ may be equivalently represented as $\sin(\omega t, k)$ utilizing equations (3) and (4), the inductor ripple current during the kth switching period can be expresses as:

$$\Delta I_L(\omega t, k) = \frac{2V_{dc}T_s}{L} \frac{G \sin(\omega t, k)}{2 + G \sin(\omega t, k)} \quad (5)$$

As the proposed DGHGSSI is operating based on buck-boost principle, the switching average inductor current can be expressed as, $I_L(\omega t) = \frac{i_o}{1 - d(\omega t, k)}$ where i_o represents the instantaneous output current. The switching average inductor current can be then found as

$$I_L(\omega t) = \frac{P_o}{\sqrt{2} V_o} (2 \sin \omega t + G\sin^2 \omega t) \quad (6)$$

A. Design Consideration of Buck-Boost Inductor (L):

In a specific converter design, since $V_{dc}T_s$ is unaffected by $(\omega t, k)$ and $2 + G \sin(\omega t, k)$ remains non-zero inside the interval $[0, \pi]$, $\Delta I_L(\omega t, k)$ is continuous and ensures a maximum within the interval $[0, \pi]$. According to the second derivative test, the highest value is attained at $\omega t = \pi/2$ and is obtained as :

$$\Delta I_{L_{max}} = \frac{2V_{dc}T_s}{L_{min}} \left(\frac{G}{2 + G} \right) \quad (7)$$

The minimal value of buck boost inductance is determined from the specification of maximum inductor ripple current as:

$$L_{min} = \frac{2V_{dc}T_s}{\Delta I_{L_{max}}} \left(\frac{G}{2 + G} \right) \quad (8)$$

B. Design Consideration of Swicthed Capacitor (C):

The maximum charge supplied by the capacitor occurs when $I_L(\omega t)$ attains its peak value. The second derivative test reveals that the peak value occurs at $\omega t = \pi/2$, as obtained from (6) as $\frac{P_o}{\sqrt{2} V_o} (2 + G)$. In this switching cycle, the ON duration will attain its peak value, denoted as $T_s d(\omega t, k)|_{\omega t = \pi/2} = \frac{G T_s}{2 + G}$. Consequently, the switched capacitor design value can be obtained as:

$$C = \frac{\Delta Q}{\Delta V_{rip}} = \frac{P_o G T_s}{\sqrt{2} V_o \Delta V_{rip}} \quad (9)$$

where ΔQ indicates the charge quantity discharged from the capacitors, and ΔV_{rip} means the allowable maximum voltage ripple across the capacitor.

C. Design Consideration of Output Capacitor (C_O):

The output capacitor value can be designed based on the point that the maximum energy transfer to the capacitor from the inductor occurs at $\omega t = \pi/2$. At this time, the inductor ripple current reaches its peak. At this instant capacitor voltage will be at maximum. Equating energy change of inductor and output capacitor during this period, output capacitor value can be obtained as:

$$C_O = \frac{L I_{Lpk} \Delta I_{Lmax}}{V_{Cm} \Delta V_{Cmax}} \quad (10)$$

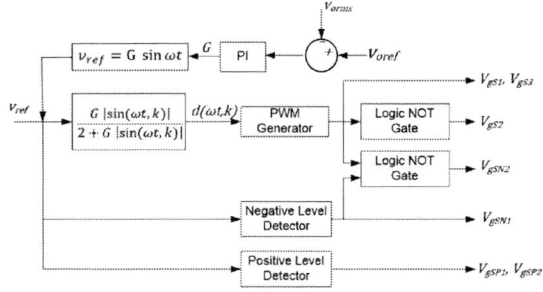

Fig. 6. Schematic of controller for proposed DGHGSSI

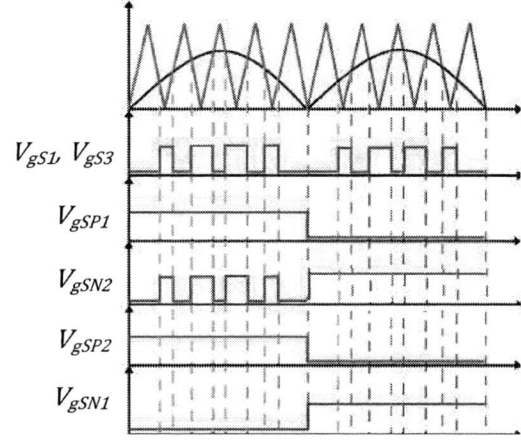

Fig. 7. Gate signal generation of proposed DGHGSSI

Where ΔV_{Cmax} is the maximum allowed capacitor voltage ripple and V_{Cm} is the maximum capacitor votlage.

D. Design Consideration of Filter Inductor (L_f):

Filter inductor is used to filter out all the high frequency switching components from AC output. Hence design of filter inductance depends on cut off frequency, f_{cLPF}, such that it will be less than switching frequency and using the LPF design equation as

$$L_f = \frac{1}{(2\pi f_{cLPF})^2 C} \quad (11)$$

IV. CONTROLLER DESIGN FOR PROPOSED DGHGSSI

The controller design is depicted in Fig. 6, where switching signals are generated using PWM with a time-dependent duty cycle, $d(\omega t, k)$. The controller design is depicted in Fig. 6, where switching signals are generated using PWM with a time-dependent duty cycle, $d(\omega t, k)$. The reference signal must maintain proportionality with the output voltage. Thus, based on (1), the reference signal is given by $v_{ref} = G \sin \omega t$. By varying the gain G, the output voltage magnitude can be regulated. From this reference signal, the duty cycle function $d(\omega t, k)$ is determined using equation (4). The switching signals for S_1 and S_3 are generated based on this duty cycle function $d(\omega t, k)$, . The switching signal for S_2 is derived directly from S_1 using a logical NOT gate, as these switches operate in a complementary manner. Additionally, positive and negative level detectors are

979-8-3315-3013-6/25 $31.00 © 2025 IEEE

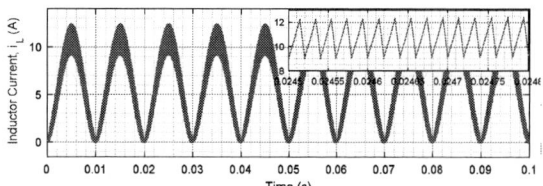

Fig. 8. Simulated waveform of the buck-boost inductor current in DGHGSSI.

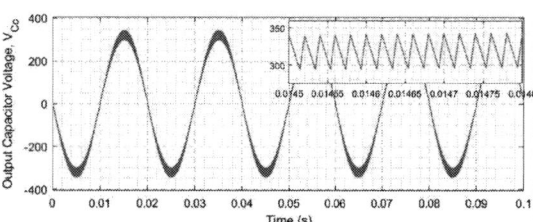

Fig. 9. Simulated waveform of the voltage across the output capacitor in DGHGSSI.

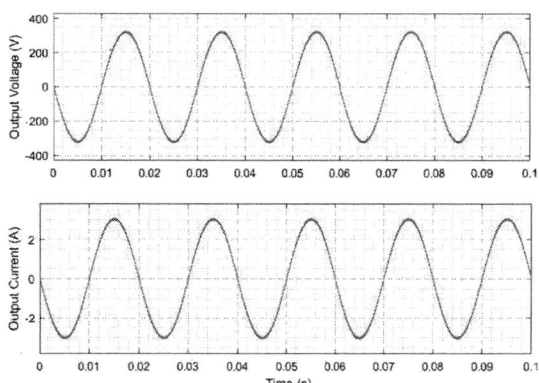

Fig. 10. Simulated waveforms of the output voltage and current in DGHGSSI.

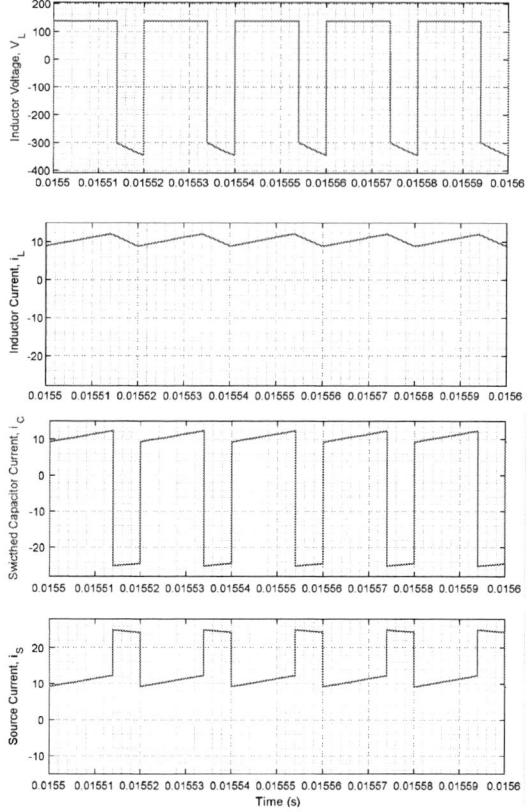

Fig. 11. Simulated waveforms illustrating the operating modes of DGHGSSI.

employed to generate switching signals for S_{Px} and S_{Nx} switches. The corresponding switching signals are illustrated with Fig. 7. The proposed DGHGSSI can also function with closed loop control, where the value of G and, consequently, V_{ref} are regulated using a PI controller, similar to a DC-DC converter. However, since the output voltage follows a sinusoidal pattern, directly comparing its instantaneous value with the reference voltage will not be feasible for closed loop implementation using a PI controller. Instead, the actual AC RMS output voltage (V_{orms}) is compared with the AC RMS reference voltage (V_{oref}). The resulting error is processed by the PI controller, which adjusts G to attain the set RMS output voltage.

V. SIMULATION RESULTS AND DISCUSSION

Simulink environment of MATLAB software is employed for the validation of operation of proposed DGHGSSI. A 500W, 230V inverter is designed and simulated with a PV dc voltage of 72V. At a switching frequency of 50kHz, simulations parameters are designed and obtained as: buck boost inductor, $L = 600 \ \mu H$, output capacitor, $C_O=1 \ \mu F$, filter inductor, $Lf = 1mH$, switched capacitor, $C=2200 \ \mu F$.

Waveform of buck-boost inductor current and output capacitor voltage in Fig 8 and Fig. 9 respectively validates the operation theory since the inductor current is maintaining a non-zero value in an entire AC voltage half cycle. The time scaled waveform shows that inductor current is linearly increased and decreased in turn ON and turn OFF state of high frequency switch respectively supporting the analysis conveyed. From Fig. 10, it's clear that an AC voltage rms value 230V is obtained which validates the one stage operation of proposed DGHGSSI. The waveform of inductor voltage & current, waveform of current through switched capacitor and waveform of current fed from DC source during the turn ON and turn OFF modes of the switching cycle are presented in Fig. 11. These waveforms validate the operating modes as they closely align with analytical predictions. The simulated waveform of the switched capacitor voltage, presented in Fig. 12, further confirms the configuration, as the voltage remains almost same to the input DC voltage within the allowable ripple range. The total harmonic distortion (THD) of the AC output voltage is measured to be below 2%, complying with the standards set by the IEEE. The effectiveness of the closed-loop control in the proposed DGHGSSI is demonstrated by applying a step change in the reference input. The RMS reference voltage transitions from 170V, 230V, and 280V to 180V within a span of 2.5 seconds, while maintaining a constant input voltage of 72V. The closed-loop controller detects the difference in the reference value and is tracked. Figure 13 illustrates the controller's tracking in relation to the change in set voltage.

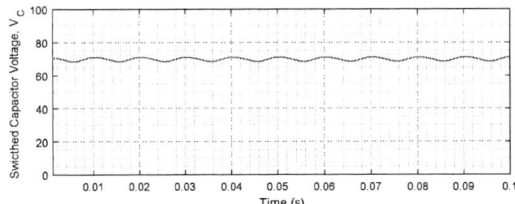

Fig. 12. Simulated waveform of the switched capacitor voltage in DGHGSSI.

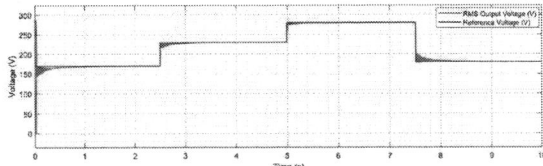

Fig. 13. Closed-loop performance of the proposed DGHGSSI.

Fig. 14. Hardware implementation of the proposed DGHGSSI.

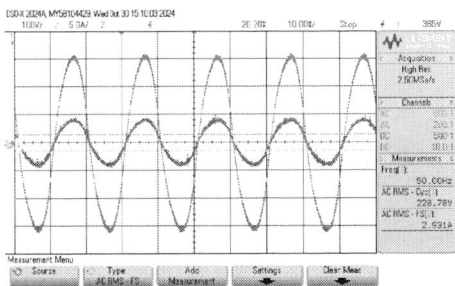

Fig. 15. Experimental output voltage & current waveforms of the proposed DGHGSSI.

VI. HARDWARE RESULTS

A prototype of 500W has been built to verify the functionality of the proposed DGHGSSI, as depicted in Fig. 14. The prototype utilizes the same parameter values as those employed in the simulation.. TMS320F28379D is utilized to realize the controller. TLP250 IC driver is implemented to control the power switches. The chosen IGBTs and diodes are FGA25N120 and MUR840, respectively. The output voltage waveforms, shown in Fig. 15, verify the proper operation of the circuit. An AC output voltage with an RMS of 230 V is obtained from a 72 V DC input with 2.9A of load current, demonstrating the validation of the proposed inverter

VII. CONCLUSION

This paper proposes a novel doubly grounded high gain single-stage inverter which is superior to conventional single-stage inverters by offering enhanced gain. The inverter operates with a buck-boost topology for both half-cycles, using a single DC source and a single inductor to ensure symmetrical performance. The circuit design uses only three switches which operate at high frequency, while the remaining switches function at a lower frequency. This paper presents a detailed mathematical analysis, component design and operational principles. Additionally, simulation and hardware results are provided to validate the inverter control strategy and overall performance.

REFERENCES

[1] Kasa, Iida and wamoto, "Maximum power point tracking with capacitor identifier for photovoltaic power system," 2000 Eighth International Conference on Power Electronics and Variable Speed Drives (IEE Conf. Publ. No. 475), London, UK, 2000, pp. 130-135, doi: 10.1049/cp:20000233.

[2] Wang, "A novel single-stage full-bridge buck-boost inverter," in IEEE Transactions on Power Electronics, vol. 19, no. 1, pp. 150-159, Jan. 2004, doi: 10.1109/TPEL.2003.820583.

[3] X. Hu, P. Ma, B. Gao and M. Zhang, "An Integrated Step-Up Inverter Without Transformer and Leakage Current for Grid-Connected Photovoltaic System," in IEEE Transactions on Power Electronics, vol. 34, no. 10, pp. 9814-9827, Oct. 2019, doi: 10.1109/TPEL.2019.2895324.

[4] O. Abdel-Rahim, M. Orabi and M. E. Ahmed, "Buck-boost interleaved inverter for grid connected Photovoltaic system," 2010 IEEE International Conference on Power and Energy, 2010, pp. 63-68, doi: 10.1109/PECON.2010.5697558.

[5] H. Patel and V. Agarwal, "A Single-Stage Single-Phase Transformer-Less Doubly Grounded Grid-Connected PV Interface," in IEEE Transactions on Energy Conversion, vol. 24, no. 1, pp. 93-101, March 2009, doi: 10.1109/TEC.2008.2006551.

[6] C. -Y. Yang et al., "A module integrated isolatd solar micro inverter," IEEE tenth International Conf. on Industrial Informatics, 2012, pp. 780 -785, doi: 10.1109/INDIN.2012.6300830.

[7] W. Wu, J. Ji and F. Blaabjerg, "Aalborg Inverter - A New Type of "Buck in Buck, Boost in Boost" Grid-Tied Inverter," in IEEE Transactions on Power Electronics, vol. 30, no. 9, pp. 4784-4793, Sept. 2015, doi: 10.1109/TPEL.2014.2363566.

[8] P. Chamarthi, M. Rajeev and V. Agarwal, "A novel single stage zero leakage current transformer-less inverter for grid connected PV systems," 2015 IEEE 42nd Photovoltaic Specialist Conference (PVSC), 2015, pp. 1-5, doi: 10.1109/PVSC.2015.7356292.

[9] Y. Tang, Y. He and X. Dong, "Active Buck-boost inverter with coupled inductors," 2014 IEEE Energy Conversion Congress and Exposition (ECCE), 2014, pp. 2754-2759, doi: 10.1109/ECCE.2014.6953771.

[10] M. Kusakawa, H. Nagayoshi, K. Kamisako, and K. Kurokawa, "Further improvement of a transformerless, voltage-boosting inverter for ac modules," Sol. Energy Mater. Sol. Cells, vol. 67, pp. 379–387, Mar. 2001.

[11] K.M. Shafeeque, Nikhil S and Rijil. Ramchand, "A Novel Switched Capacitor Based Single Stage Inverter for Solar Power Extraction," 2023 PESGRE, Trivandrum, India, 2023, pp. 1-6, doi: 10.1109/PESGRE58662.2023.10404722.

[12] IEEE Recommended Practice for Utility Interface of Photovoltaic (PV) Systems, IEEE Standard 929, 2000.

[13] K. Muhammedali Shafeeque and P. R. Subadhra, "A novel single stage DC-AC buck boost inverter for solar power extraction," 2013 AICERAeas , India, 2013, pp. 1-6, doi: 10.1109/AICERA-ICMiCR.2013.6575973.

[14] W. I. Bower and J. C. Wiles, "Analysis of grounded and ungrounded photovoltaic systems," in Proc. IEEE Photovoltaic Spec. Conf., 1994, vol. 1, pp. 809

A Novel High Voltage Gain SiC-based Three Port DC-DC Converter for Standalone PV Applications

[1]Mr.Nagaraju Kothapalli
Dept. of Electrical and Electronics Engineering
National Institute of Technology Tiruchirappalli
Tamil Nadu, India-620015
[1]nag.kothapalli@gmail.com

[2]Dr.Sateesh Kumar Kuncham
Dept. of Electrical and Electronics Engineering
National Institute of Technology Tiruchirappalli
Tamil Nadu, India-620015
[2]sateesh@nitt.edu

Abstract— High-voltage gain SiC-based three-port DC-DC converters (HVGTPC) are a crucial requirement for renewable energy applications, particularly in photovoltaic (PV) systems, to address the issue of intermittency of the PV source. The proposed HVGTPC consists of a switched inductive cell to achieve high voltage gain, which is operating in four distinct modes, such as PV to both battery and DC bus, PV and battery to DC bus, PV to DC bus, and battery to DC bus. Compared to recently described converters, the proposed topology has a higher voltage gain, lower input ripple current, bidirectional battery port, and common ground connection between input and output ports to obtain the control based on the power requirement. The suggested converter's performance has been confirmed through MATLAB Simulink under different operating modes, which demonstrates the operation of the proposed converter.
Keywords— Solar PV source, high voltage gain, three-port converter, Li-ion battery

I. Introduction

Renewable energy sources (RES) are rapidly replacing conventional fuels for generating electricity in an attempt to lower carbon emissions and attain energy sustainability. Among these, solar photovoltaic (PV) systems are extensively used in both off-grid (stand-alone) and on-grid (grid) connections, and their share of power generation is rapidly increasing. Energy storage devices are necessary for stand-alone PV systems to increase power density and conversion efficiency because the generated power is subject to weather and fluctuating load demand. Integration of PV source, energy storage, and load requires two different converters, which leads to a larger converter size due to more component count, increased complexity, and less efficiency. To address these issues, multiport converters are the best choice for replacing conventional converters because of their enhanced adaptability, compact design, increased dependability, high power density, and good efficiency. By facilitating multiple port connections, they enable concurrent energy transfer among all the ports and storage. This leads to optimized energy management, diminished power losses, and superior overall system performance [1-2].

There are various types of multiport converters (MPC), such as isolated, partially isolated, and non-isolated, available in the literature. Out of those non-isolated MPCs, they are compact in size and have fewer semiconductors due to the elimination of the transformer and the sharing of components in all operating modes. Moreover, isolation transformers in these converters introduce significant losses with reduced reliability.

Thus, non-isolated configurations are the best choice for standalone PV applications. Many of the authors presented MPC with three ports (TPC), which are commonly used in PV/battery systems to increase the reliability of the load in different conditions. Achieving high-voltage gain without the use of a transformer is a primary challenge in non-isolated converters. Features like high efficiency, high gain, and multi-port capability make them ideal for solar energy harvesting with energy storage systems. Control of these converters based on the availability of PV power, battery power, and load requirements are the essential features of the scheme [3-5], which includes maximum point power tracking (MPPT), constant voltage/current control of the battery, and output voltage regulation.

In [6], a new three-port converter was presented with fewer components. It has two unidirectional and one bidirectional input. The input power sources can power the load and charge or discharge the battery separately or in parallel. In [7], an MPC is proposed for hybrid electric vehicle applications contrasted with traditional topologies; the voltage gain of this topology increased associated with the power management strategy. In both cases, minimum voltage gain and the absence of common ground increase the number of panels in series and make the converter unsuitable for a common ground transformer and inverter, respectively. In [8], a high-gain, non-isolated, coupled inductor-less MPC is suggested, even though it contains a common ground connection but has more components, which leads to a bulky size and more cost.

In [9], another three-port converter with a switched inductive cell is proposed to obtain high-voltage gain. In [10], a non-isolated boost TPC was proposed with a single inductor to operate at different modes of functions like PV to load,

979-8-3315-3013-6/25 $31.00 © 2025 IEEE

battery to load, PV to both battery and load, and PV and battery to load. But the above topologies have high voltage stress, component count, and reduced efficiencies. In [11], a new MPC converter has been proposed for a reduced component count with higher voltage gain and continuous input current. However, the absence of common ground leads to non-suitability for a transformer less common ground inverter. Therefore, the authors proposed a new TPC with the features as follows:

➢ High voltage gain
➢ Low input ripple current on PV side
➢ Reduced number of components
➢ Common ground among three ports
➢ Bidirectional port for battery charging and discharging

Fig. 1 depicts the schematic of the non-isolated high voltage-gain three-port converter. This converter's $V_{pv} < V_b < V_{DC}$ is crucial for boost operation, where V_{DC} is the output voltage, V_b is the Li-ion battery voltage, and V_{pv} is the PV source voltage.

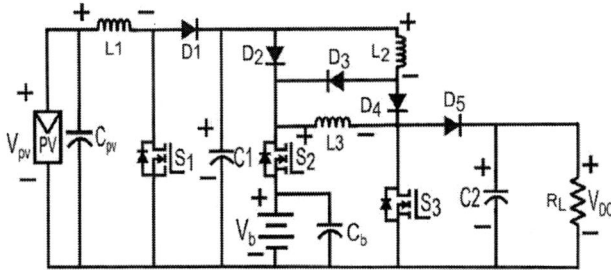

Fig 1. Schematic of the proposed converter

II.PROPOSED CIRCUIT OPERATING MODES

The proposed HVGTPC has three ports, namely the PV array, Li-ion battery, and DC bus port. The Li-ion battery is the bidirectional port, and the PV port is unidirectional. In the proposed circuit S1, S2, and S3 are the pulse-width-modulated controlled switches. This is operating in four modes, which are described as follows:

Mode-I: Single input double output mode (SIDO) when ($P_{pv} > P_{DC}$)

This mode has three operating states with equivalent circuits as shown in Fig.4, and its time intervals change from t_0 to t_4. In state1 ($t_1 < t < t_2$), S_1 is the conducting switch shown and the input PV source energizes L_1, and its current increase linearly, which is shown in Fig. 4, while the diode D_1 is off. Using Fig. 4(a), The equation below can be written as

$$V_{pv} = V_{L1} \qquad (1)$$

The capacitor C_2 is charged by C_1, L_2, L_3 through D_3 and D_5, and the other diodes, D_2 and D_4 are in blocking state. By using KVL in Fig. 4(a)

$$V_{C1} - V_{L2} - V_{L3} = V_0 \qquad (2)$$
$V_{L1}=V_{L2}$ the above equation can be written as
$$V_{C1} - 2V_{L2} = V_0 \qquad (3)$$

$$V_{L2} = \frac{V_{C1} - V_0}{2} \qquad (4)$$

The equivalent circuit of the converter in remaining two states ($t_2 < t < t_3$) and ($t_4 < t < t_0$) are shown in the Fig. 4(b). All switches are off, and C_1 charges with the help of L_1 through D_1 ,and from Fig. 4(b) we can write

$$V_{pv} - V_{L1} = V_{C1} \qquad (5)$$
Based on waveforms of Fig.3,by considering above equations, the Volt-Sec balance for L_1 can be written as follows:

$$D_1 V_{pv} + (1 - D_1)(V_{pv} - V_{C1}) = 0 \qquad (6)$$
Thus, voltage across C_1 is obtained from (6)

$$V_{C1} = \frac{V_{pv}}{(1 - D_1)} \qquad (7)$$

The remaining state occurs in a time interval as $t_3 < t < t_4$ when S_2 and S_3 are turned on and S_1 is off. The battery is charging by the input source, which is shown in Fig. 4(c). In charging mode, there is a basic condition between the voltage of the battery and the voltage of C_1 which is definable by $V_{C1} > V_b$. This inequality equation implies the dependence between this mode and the battery status. The inductors L_2, L_3 and capacitor C_2 are charging by the input source. Using analysis of this state, the relation between the voltages of L_1, L_2 and C_1 are as follows:

$$V_{C1} = V_{L2} \qquad (8)$$

$$V_{C1} = V_{L2} = V_{L3} \qquad (9)$$

Similarly, Volt-Sec balance of L_2 is obtained as shown

$$D_3 V_{C1} + (1 - D_3)\left(\frac{V_{C1} - V_0}{2}\right) = 0 \qquad (10)$$

$$V_0 = \frac{V_{C1}(1 + D_3)}{(1 - D_1)} \qquad (11)$$

By substituting V_{C1} value from equation (7), In SIDO mode, the converter's voltage gain is determined by

$$V_0 = \frac{V_{pv}(1 + D_3)}{(1 - D_1)(1 - D_3)} \qquad (12)$$

Mode-II: Dual input single output (DISO) when ($P_{pv} < P_{DC}$)

In this mode the waveforms are similar in Fig. 3, with the variation that S2 does not conduct. Based on the converter's configuration, the voltage across battery terminal must be higher than VC1 for discharging of the battery. Few of the states in this mode are as same as the ones in the charging mode such as the states in time intervals including t0< t <t3 and t0 - t - t4. Therefore, the equations (1), (2), (5), (6) and (7) are also valid for this mode. As shown in Fig. 4(d), the different state occurs during t3 < t < t4 when the battery discharges through the body diode of S2. In this state, battery is used to energize L3 in the loops consisting of S2's body diode (DS2), S3 (first loop). Furthermore, L2 is charged by the input source, L1 and C1 through the formed current paths by D1, D4 and S3. According to Fig. 4(d), voltage of L3 is equal to Vb (Vb=VL3) and VC1=VL2. By using volt-balance theory, a relation such as (10) can be written,which is summarized to obtain output voltage (V_O).

$$D_3(V_{C1} + V_b) + (1 - D_3)(V_{C1} + V_b - V_0) = 0 \quad (13)$$

$$V_0 = \frac{V_{C1} + V_b}{(1 - D_3)} \quad (14)$$

Substituting (7) in (14) equation gives V_0 related to V_b and V_{pv}.

$$V_0 = \frac{V_{pv} - V_b(1 - D_1)}{(1 - D_1)(1 - D_3)} \quad (15)$$

Mode-III: Sigle input and single output-SISO PV source to DC load when ($P_{pv} = P_{DC}$)

This mode includes the states similar to SIDO mode, and the only difference is in operation of S_2, which should be turned off; hence there is no contribution of battery during this mode because of ($V_{C1} > V_b$). Analysis of the converter would be the same as in mode-I, and its voltage gain can be obtained as follows.

$$V_0 = \frac{V_{pv}(1 + D_3)}{(1 - D_1)(1 - D_3)} \quad (16)$$

Mode-IV: Sigle input and single output-SISO (Battery to DC load) mode when ($P_{pv} = 0$)

In this mode there is no input power from PV source, so only the battery will supply the power to load with the help of S_3 switch. Here the performance of the converter has two states, as shown in 4(e) & 4(f); there is no need to use S_1. Here all the semiconductors except D_4 are blocked during $t_0 < t < t_1$ and the battery transfers its power to the load through the body diode of $S2 (D_{S2})$ and D_4. In addition, L_3 assists the battery by discharging and supporting the load by energizing the output capacitor (C_3). As shown in Fig. 4(f).

$$V_b - V_{L3} = V_0 \quad (17)$$

In the $t_1 < t < t_2$, the switch S_3 is the only conducting one that prepares a current path for charging L_3. Thus, in this state, the inductor's voltage is equal to V_b ($V_b = V_{L3}$). Using the voltage of L_3 per the states, the volt-balance equation of the inductor can be written as follows.

$$D_3(V_b) + (1 - D_3)(V_b - V_0) = 0 \quad (18)$$

The converter's output voltage is simply provided by

$$V_0 = \frac{V_b}{(1 - D_3)} \quad (19)$$

The closed-loop control structure of the proposed system with input and output voltage regulations has been shown in the following Fig. 2. Where V_{G1}, V_{G2} and V_{G3} are gate signals and V_{C1}, V_{C2} and V_{C3} are controlled signals of the switches. The control is aimed to regulate the PV MPPT and load voltage regulation. The battery charging and discharging will take place as per the net power requirement/availability. If the PV power is greater than the load requirement, battery gets charged by default through *S2,* and if the PV power is less than the load power, the battery will discharge to the load via body diode of *S2.*

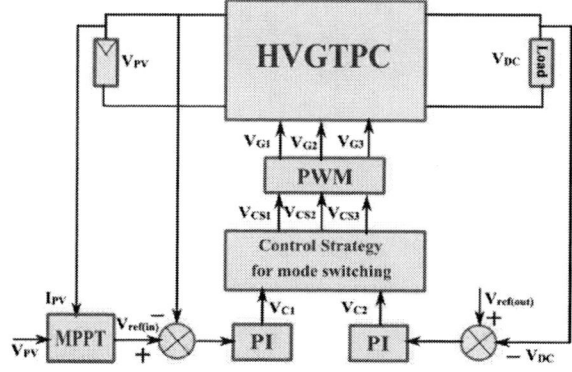

Fig. 2. HVGTPC Control Structure

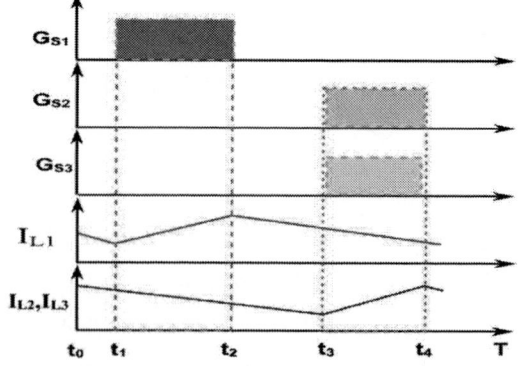

Fig. 3. Waveforms of the suggested HVGTPC up to Mode III

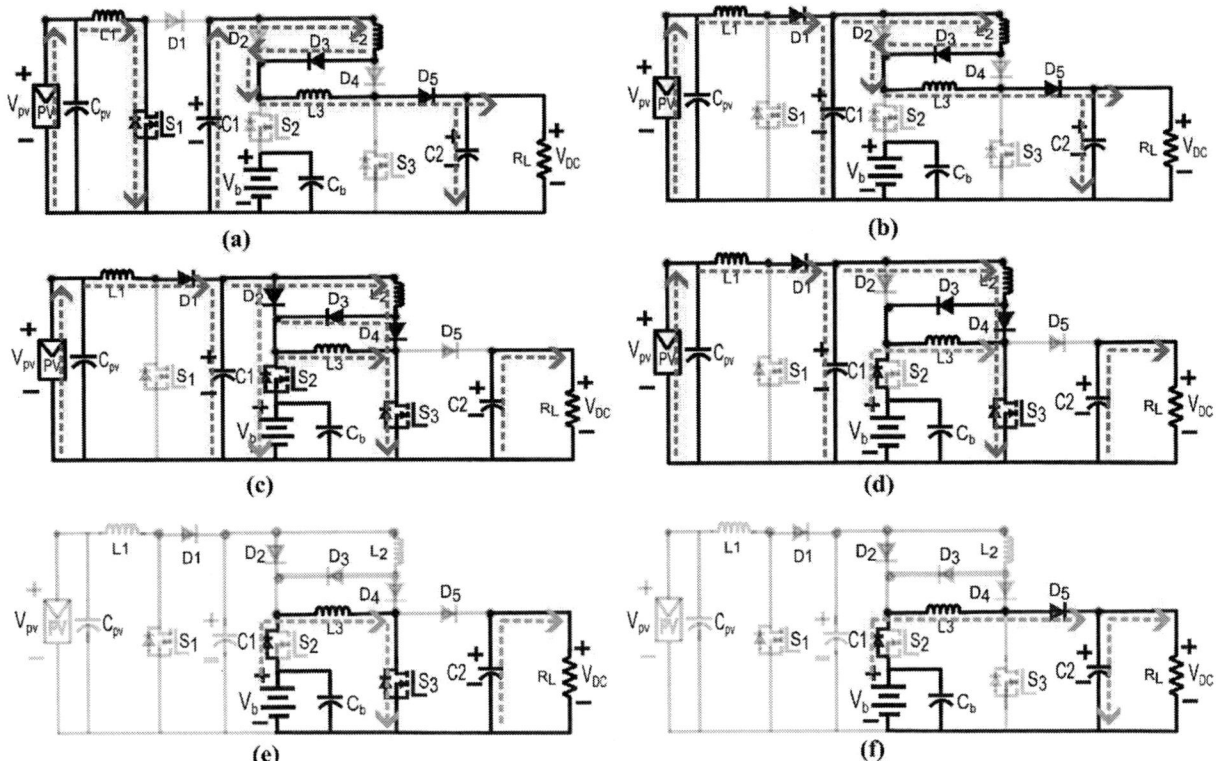

Fig 4 Switching states of all-operating modes of the proposed HVGTPC

Table-1 Comparative Analysis with other converters

Topology	[6]	[7]	[8]	[9]	[10]	[11]	Proposed
Component Count [L/C/S/D] =T	2/2/4/4 =12	2/2/ 4/5 =13	4/4/6/2 =16	2/2 /3/4 =11	1/3/3/3=10	2CL/5/ 3/2 =11	3/4/ 3/5 =15
Continuous input current	No	Yes	Yes	No	No	Yes	Yes
Voltage gain	Medium $\dfrac{(1+D^2-D)}{(1-D)^2}$	High	High, $\dfrac{(2-D_1)V_1}{(1-D_1)^2}+\dfrac{V_2}{(1-D_1)^2}$	Low, $\dfrac{1}{(1-D)^2}$	Low, $\dfrac{V_{pv}-V_bD_2}{(1-D_1-D_2)}$	High, $\dfrac{2}{(1-D_2)}$	High, $\dfrac{V_{pv}(1+D_3)}{(1-D_1)(1-D_3)}$
No. of operating Modes	3	3	2	3	4	2	4
Bidirectional Ports	1	1	1	1	1	1	1
Common Ground	No	No	Yes	Yes	Yes	No	Yes

979-8-3315-3013-6/25 $31.00 © 2025 IEEE   263

Table 1 illustrates a brief comparative analysis among the proposed HVGTPC and the existing MPC topologies. Most of the existing non-isolated converters have less voltage gain, and few topologies do not have a minimum current ripple on the PV side. Moreover, the absence of a common ground connection makes the system complex for eliminating the leakage current. Compared to other topologies [6-11], the proposed HVGTPC consists of a lesser component count with high voltage gain, a bidirectional port for battery charging and discharging, minimum ripple current on the PV side, and a common ground connection of all the ports, as shown in Table 1.

III. RESULTS AND DISCUSSION

Various operating modes are simulated using MATLAB/Simulink. Table II lists the simulation parameters that were utilized to validate the suggested circuit. Mode-I simulation results are displayed in Fig. 5, where both the DC bus and the battery are receiving power from the solar PV (i.e., SIDO mode). If the load power requirement is less than the generated power, the excess energy is stored in a Li-ion battery. Fig. 6 illustrates the operation of the proposed converter in Mode-II, where the battery and solar PV both are supporting the load power. The simulation results for Mode-III are illustrated in Fig. 7, where the solar PV is supplying power to the DC bus under the sufficiency of source power and load power. The simulation results in Mode-IV are shown in Fig. 8, where the battery is supplying power to the load under PV power is absent. SOC of a Li-ion battery shows a decreasing nature, which indicates the discharging mode of the Li-ion battery. Fig. 9 shows the constant voltage and power of the DC bus port with variable solar PV variations. It is ensured to maintain the DC bus power to be constant under all the modes of operation.Fig.10 shows the currents of PV, battery and load in charging and discharging modes.

TABLE II. SYSTEM PARAMETERS

Parameter	Specification
PV voltage (V_{pv})	24 - 30 V
Battery voltage (V_b)	48 - 55 V
DC bus voltage ($V_{dc\ bus}$)	60 V
Inductor values L_1, L_2, L_3	1 mH
Capacitor values C_1, C_2 and C_b	7.87 µF, 50 µF and 50 µF
Switching frequency (f_s)	40 KHZ
Output power (P_o)	100 W
Load(R)	36 Ω

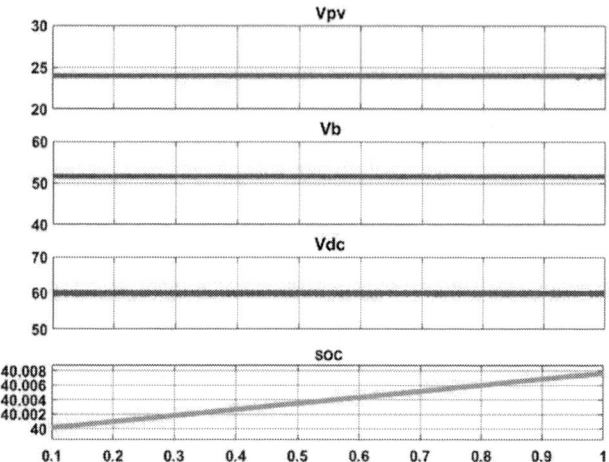

Fig.5.Simulation results in SIDO Mode-I

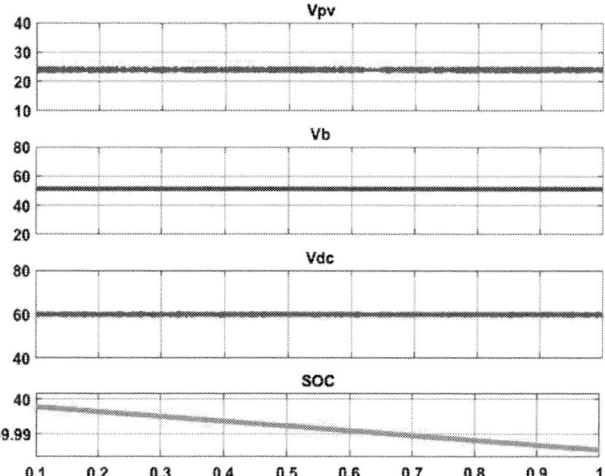

Fig.6.Simulation results in DISO Mode-II

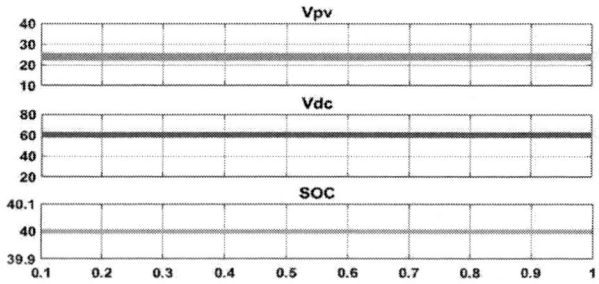

Fig.7.Simulation results in PV to Load Mode-III

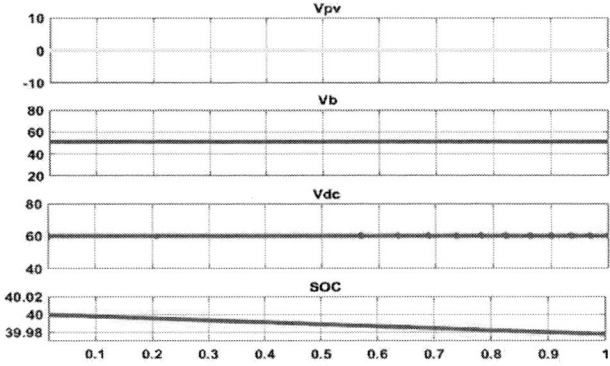

Fig.8.Simulation results in battery to load Mode-IV

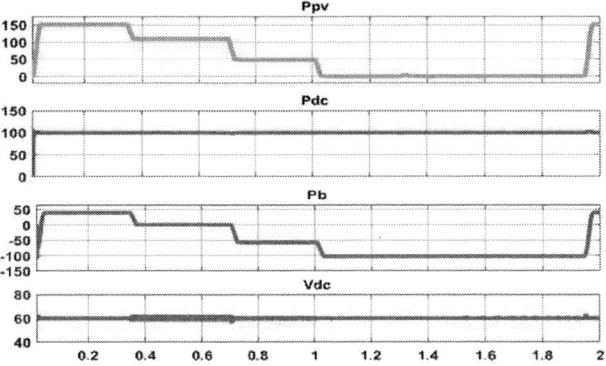

Fig.9. Load power & voltage at different PV power changing conditions

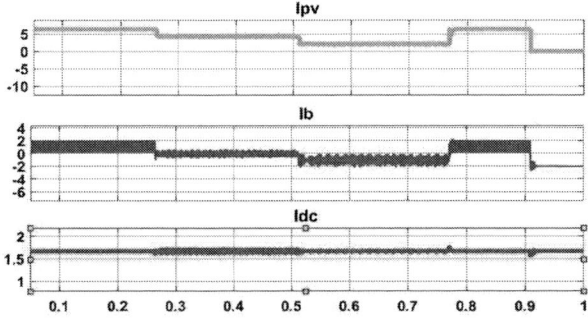

Fig.10. Currents of PV, battery and load in charging and discharging modes

V.CONCLUSION

The proposed HVGTPC operates in four different modes by transferring the power to various ports depending on solar PV power. The converter consists of fewer switches, high voltage gain, low input ripple current, and a common ground connection of all the ports. The steady-state analysis has been performed for the proposed HVGTPC to realize the high voltage gain. The comparative analysis of the suggested converter with existing converters is presented. The simulation results on HVGTPC verify the feasibility and effectiveness of the proposed approach. Based on the advantages of the HVGTPC and its superiority over the other three port converters, the suggested converter is suitable for standalone PV applications.

References

[1]. X Xue, K.W.E. Cheng, and C Xu "Review of Energy Management Strategies of Solar Photovoltaic Energy Systems for Grid-connected and Standalone Applications" 10th International Conference on Power Electronics Systems and Applications (PESA), DOI: 10.1109/PESA62148.2024.10594899, IEEE 2024

[2]. P Rani, R Taya, and V Padmanabha Reddy" A Review on solar energy and different electricity generations" International conference on power energy, environment & Intelligent control (PEEIC), DOI: 10.1109/PEEIC59336.2023.10451552, IEEE2023

[3]. A. Kumar Bhattacharjee, N. Kutkut and I. Batarseh "Review of Multiport Converters for Solar and Energy Storage Integration" IEEE Transactions on Power Electronics, Vol. 34, No. 2, Pp. 1431-1445, February 2019

[4]. F. Mumtaz, N.Z. Yahaya, S.T. Meraj, B. Singh, R. Kannan, O. Ibrahim "Review on non-isolated DC-DC converters and their control techniques for renewable energy applications" Ain Shams Engineering Journal 12, pp.3747–3763,2021

[5]. M.Haris, Md. Asim, Mohd Tariq "A Review of Non-Isolated High Gain DC-to-DC Converter Topologies" 2nd International Conference on Emerging Frontiers in Electrical and Electronic Technologies (ICEFEET), DOI: 10.1109/ICEFEET51821.2022.9847767,2022.

[6]. F. Kardan, R. Alizadeh, and Md. Reza Banaei "A New Three Input DC/DC Converter for Hybrid /FC/Battery Applications" IEEE Journal of Emerging and selected topics in power Electronics, Vol. 5, No. 4, pp. 1771-1778, December 2017.

[7]. R. Reza Ahrabi, H. Ardi, M. Elmi and A. Ajami "A Novel Step-Up Multiinput DC–DC Converter for Hybrid Electric Vehicles Application" IEEE Transactions on Power Electronics, Vol. 32, NO. 5, pp. 3549-3561, May 2017.

[8]. K. Varesi, Sd.H. Hosseini, M. Sabahi, E. Babaei, Sd. Saeidabadi, and N. Vosoughi "Design and Analysis of a Developed Multiport High Step-Up DC–DC Converter with Reduced Device Count and Normalized Peak Inverse Voltage on the Switches/Diodes" IEEE Transactions on Power Electronics, Vol. 34, No. 6, pp. 5464-5475, June 2019.

[9]. Sreedevi S Nair and Mini Rajeev "A Novel high gain Non-isolated Three-Port Converter for stand-Alone PV applications" International Conference on Computer, Electronics & Electrical Engineering & their Applications (IC2E3), DOI:10.1109/IC2E357697.2023.10262587,2023

[10].S Abdelrahman, M Selmy, Kh M Hasaneen and N A Rahim "Analysis, Design, and Control of a Non-isolated Boost Three-Port Converter for PV Applications" IEEE Conference on Power Electronics and Renewable Energy (CPERE) 2019

[11].Amal C Sunny and Dipankar Debnath "A Novel Three-Port High-Gain DC–DC Converter for PV–Battery Stand-Alone System with Reduced Device Count" IEEE Journal of emerging and selected topics in industrial electronics, Vol. 5, No. 3, pp. 1216- 1225, July 2024.

Rotor Position Calculation in Case of Partial Failure of Resolver for Motor Control Application

1st Aman Patel
E-Mobility
Varroc Engineering Pvt. Ltd.
Pune, India
Aman.Patel@varroc.com

2nd Aditya Purushottam Ghube
E-Mobility
Varroc Engineering Pvt. Ltd.
Pune, India
Aditya.Ghube@varroc.com

3rd Abhisha Chauhan
E-Mobility
Varroc Engineering Pvt. Ltd.
Pune, India
Abhisha.Chauhan@varroc.com

Abstract—Rotor position sensing in motor control applications ensures precise feedback for accurate rotor movement, efficiency, and safe operation of Permanent Magnet Synchronous Motors (PMSMs). Resolvers without electronic components are widely used in motor control systems because of their resistance to soiling, vibration, and extreme temperatures, along with their ability to provide continuous position feedback. However, rotor position is calculated using multiple output signals from resolvers, and loss of one or more output signals may lead to inaccurate position sensing thereby impacting motor operation. This paper introduces a novel strategy of rotor position calculation using frequency and phase equalization of remaining output signals, in case of partial resolver failure. The proposed method serves as an alternate method to measure the accurate and robust position, and speed of the rotor, ensuring continuous motor operation.

Index Terms—Resolver, Rotor position calculation,PMSM, FOC, Position sensing

I. INTRODUCTION

Field-Oriented Control (FOC) is a widely adopted control strategy for Permanent Magnet Synchronous Motors (PMSMs).Control of PMSM requires accurate knowledge of rotor position, including magnet polarity, to ensure proper alignment of the stator magnetic field with the rotor magnets, enabling optimal torque production and smooth motor operation [5], [6]. Accurate rotor position ensures smoother torque control, reduced harmonics, and enhanced motor performance under varying load and speed conditions [1].

Table1 shows a comparison among different sensors , such as resolvers encoders and Hall-effect sensors used for measuring the rotor position [1]–[3], [7].

Among all the position sensors, resolvers are considered the most effective choice for calculating the rotor position [1], [7]. They are cost-effective, computationally efficient, and highly reliable in harsh environments, making them a preferred choice for precise and uninterrupted rotor position feedback in industrial and automotive applications. Partial or full Sensor failure of resolver can result in noisy rotor position measurements, often requiring immediate sensor replacement. To ensure continuous motor operation, sensorless methods, such as back EMF observers and high-frequency injection techniques are common prescribed alternative methods for estimating rotor position [10], [11].

TABLE I: Comparison of Position Sensor Technologies

Sensor	Mechanical position Accuracy	Temp. Range (°C)	Cost	Robustness
Resolver	0.1°	-40 to +220	Medium	Very High
Incremental Encoder	0.5°	-20 to +85	Low	Low
Absolute Encoder	0.1°	-20 to +85	Medium	Medium
Optical Encoder	0.0014°	-20 to +70	Medium	Low
Capacitive Encoder	0.2°	-40 to +125	Medium	Medium
Inductive Sensor	0.1°	-40 to +180	High	Very High
Inductive Encoder	0.02°	-100 to +125	Medium	High
Sensorless Control	1°	N/A	Low	Medium

However, these methods involve limited application due to poor performance at low speeds or standstill caused by weak back-EMF signals,high computational demands, requiring significant processing power, high sensitivity to motor parameter variations such as resistance and inductance changes, susceptibility to electrical noise, affecting signal integrity and accuracy, and dependent on accurate motor modeling for reliable performance.

Hybrid methods which utilize sensors and sensorless approach to calculate rotor position, also are considered as alternate continuous position sensing techniques [14].However,this approach requires accurate mathematical modeling and complex computations, and is prone to inaccuracies due to parameter variation, susceptibility to electrical noise, and errors caused by improper switching between sensor and sensorless models. The Dual-Sensor Approach in combination with a Kalman Filter esimates rotor position effectively but comes with limitations such as temperature sensitivity, nonlinearity in magnetic field, reduced precision, dependency on kalman filter tuning, performance loss at high speed [15].

This paper proposes a novel method of calculating the rotor position from phase and frequency information of remaining signals in case of partial failure of the resolver, eliminating the need for immediate replacement. The proposed approach is computationally efficient and cost-effective, ensuring reliable rotor position feedback while maintaining system performance and smoother motor operation.

This paper is organized as follows: FOC motor control and mathematical concept of resolver and why rotor position is important are presented in section II.Proposed method of rotor position calculation is presented in section III. Simulation Results has been covered in section IV. Section V covers

979-8-3315-3013-6/25 $31.00 © 2025 IEEE

conclusion and future work.

II. MATHEMATICAL MODELING

A. FOC Motor Control

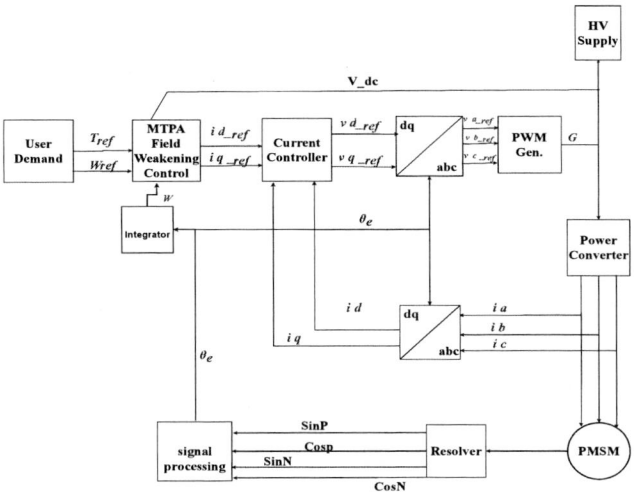

Fig. 1: Block Diagram of FOC Motor Control System for a three-phase PMSM.

In the diagram:

- T_{ref}: Reference electromagnetic torque to be generated by the motor.
- W_{ref}: Reference speed to be generated by the motor.
- i_a, i_b, i_c: Phase currents in the motor windings.
- v_a, v_b, v_c: Phase voltages applied to the motor windings.
- d and q: Represent the d-axis (direct axis) and q-axis (quadrature axis) components in the rotor reference frame.
- a, b, and c: Represent the three-phase of the motor.
- θ_e: Rotor Position(in degree).
- ω: Rotor speed (in RPM).
- $sinN, sinP, cosN, cosP$: Resolver Output.
- G : Represent the Gate pulse, there are six gate pulses.

Field-Oriented Control (FOC) is widely used motor control strategy for PMSM in electric vehicles (EVs) to achieve torque and flux control. Rotor position feedback is essential for FOC as it enables accurate transformation between the three-phase stationary frame and the rotating dq reference frame, which allows precise control of both torque and flux, essential for achieving efficient and smooth motor operation.The four resolver signals(Sin P, Sin N, Cos P, Cos N) are used to calculate accurate and reliable rotor position. This motor control strategy with rotor position feedback also minimizes torque ripple, enhances battery utilization, extends driving range, and improves overall vehicle performance, making it a preferred choice for modern Electric vehicle [1], [16].

B. Mathematical concept of Resolver

Ignoring the influence of winding resistances and motion-induced electromagnetic forces (emf), the output voltages of the resolver can be represented as:

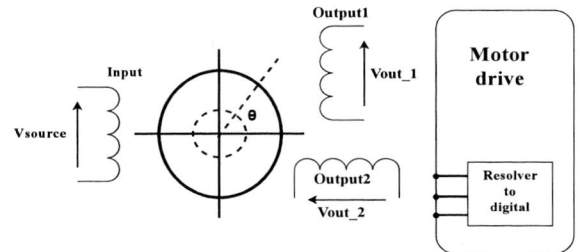

Fig. 2: Working Concept of resolver [8]

$$V_{\text{source}} = V_0 \sin(\omega_{\text{source}} t) \tag{1}$$

$$V_{out1} = N * V_0 \sin(\omega_{\text{source}} t) \sin(\theta) \tag{2}$$

$$V_{out2} = N * V_0 \sin(\omega_{\text{source}} t) \cos(\theta) \tag{3}$$

where: V_{out1}: Output voltage related to the sine component of the resolver signal, V_{out2}: Output voltage related to the cosine component of the resolver signal, N: Transformation ratio, V_{source}: Excitation voltage, $\omega_{\text{source}} t$: Rotating frequency, θ: Rotor position.

Equations (2) and (3) represent the resolver's sine and cosine output signals modulated by the excitation voltage.

C. Retrieving the position angle from resolver

This set of simplified equations resembles the modulated signal in double-sideband amplitude modulation (DSB-AM). With this analogy:

- The excitation signal in a resolver acts as a *carrier or reference signal*.
- The position signal acts as the *modulating signal*.

The ideal resolver outputs are then modulated quadrature signals. Figure 3 illustrates the ideal input and output signals of a resolver.

In addition, the output of resolver is fed to the Resolver-to-digital converter(RDC) to retrieve the position. in RDC, resolvers outputs are demodudulated to extract the envelopes corresponding to position signal. then the position angle is calculated from demodulated signals [1], [18]- [19].

$$\theta = \begin{cases} \tan^{-1}\left(\frac{v_s}{v_c}\right), & v_c \geq 0 \\ \tan^{-1}\left(\frac{v_s}{v_c}\right) + \pi, & v_c < 0 \end{cases} \tag{4}$$

where v_s and v_c are sine and cosine demodulated outputs, respectively. and θ represent the electrical position of the rotor. For a traction motor application, the resolver excitation signal is typically a DC component. Hence, the resolver output voltages directly represent the rotor electrical position without requiring additional demodulation:

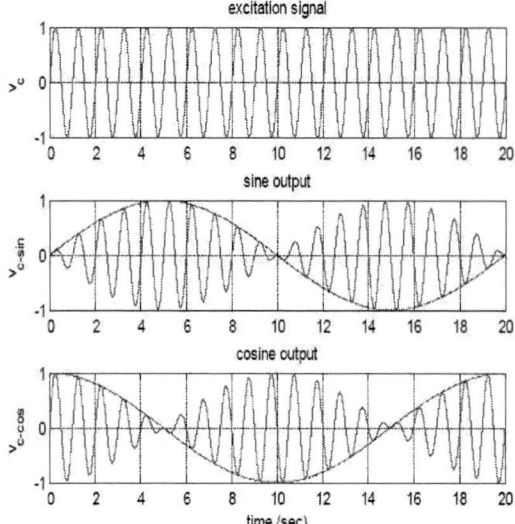

Fig. 3: Ideal Input and Output Signals of a Resolver [8]

However, partial failure of signals from resolver may cause inaccurate rotor position calculation leading to issues in FOC, which may cause inefficient toque generation and motor instability. In next section a novel method of calculating rotor position has been proposed in case of partial failure of resolver.

III. PROPOSED METHOD FOR ROTOR POSITION CALCULATION

The proposed method involves calculation of PMSM mechanical position and RPM from available healthy position signal,during failure of one or more position sensor signals from the resolver [4]. It has been assumed that there is no distortion or loss of signal due to phase. The method involves two processes: Frequency equalization and phase equalization of a reference signal, described below:

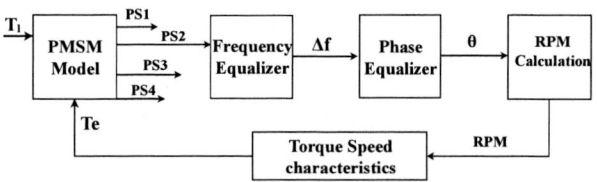

Fig. 4: Rotor Position Calculation Representation

In the diagram:

- T_e = Electromagnetic torque generated by the PMSM.
- T_l = Load torque applied to PMSM.
- PS1, PS3, PS4 = $A\sin(\omega t \pm \psi)$ indicates one or more faulty position signals.
- PS2 = $A\sin(\omega_2 t \pm \psi_2)$ indicates a healthy position sensor signal, amplitude varying between(0.75-4.25v).
- θ = Mechanical rotor position.

A. Frequency and Phase Equalization

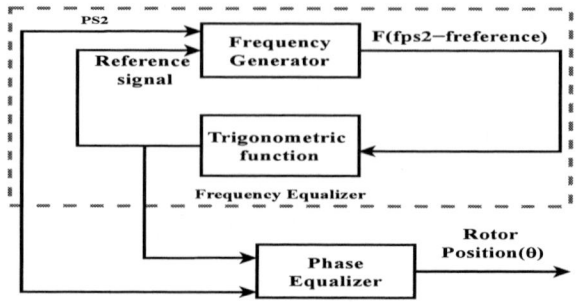

Fig. 5: Frequency and Phase equalization

The first stage of the rotor position calculation involves the frequency equalization module, which makes the frequency of a reference signal the same as that of the input healthy position sensor signal, PS2. This is done by making the input control signal used to generate the reference signal a function of the frequency error between the reference signal and PS2.

$$PS2 = A\sin(\omega_2 t + \psi_2) \tag{5}$$

where, ω_2 and ψ_2 are the frequency and phase of the healthy signal, $PS2$.

$$\text{Reference signal} = B\sin(\omega_3 t) \tag{6}$$

where,

$$\omega_3 = f\left(f_{ps2} - f_{\text{reference}}\right) \tag{7}$$

Once the frequency of the reference signal and PS2 is equalized i.e., $f_{ps2} = f_{\text{reference signal}}$, both signals are fed to a phase equalizer module that equalizes the phase of the frequency-equalized reference signal and PS2, and calculates the mechanical rotor position, which is eventually used to calculate motor RPM.

IV. EXPERIMENTAL OUTCOMES

To validate the proposed approach, simulations were performed in a matlab simulink environment using the pmsm mechanical model and the position sensor model of the resolver for the steady state and the transient state for the following operating conditions (Table II, IV).

A. Steady state Response

Case No.	T_e (Nm)	T_L (Nm)	RPM	Sin P	Sin N	Cos P	Cos N
1	4.05	3.98	100	0	1	1	1
2	29.2	27.62	2500	1	0	1	1
3	55.4	52.25	5000	1	1	0	1
4	81.54	76.87	7500	0	1	1	1

TABLE II: Signal failure at different RPMs. Here, 0 represents a signal covering the full range [0.75-4.25v], while 1 indicates a lost or distorted signal.where T_e = Electromagnetic torque, T_L = Load torque

979-8-3315-3013-6/25 $31.00 © 2025 IEEE

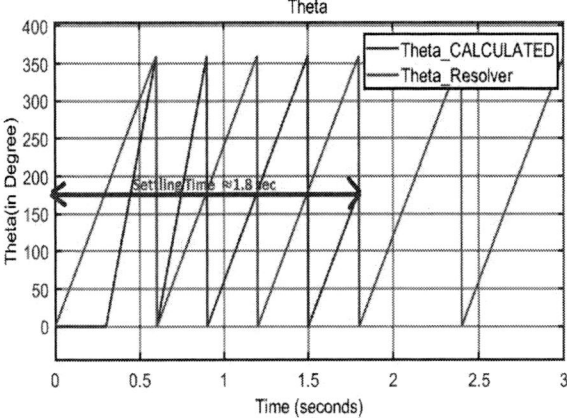

Fig. 6: Rotor position (θ) calculated for case 1.

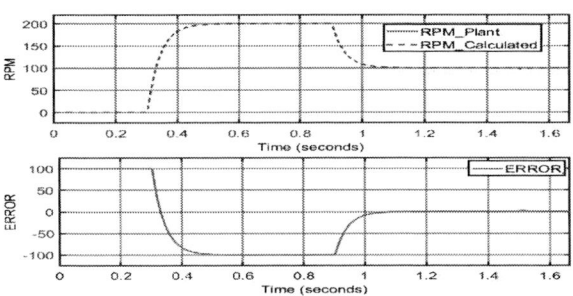

Fig. 7: RPM calculated for case 1.

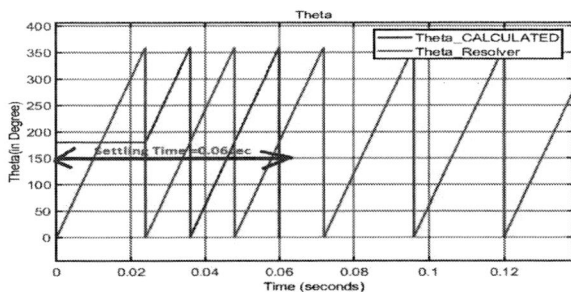

Fig. 8: Rotor position (θ) calculated for case 2.

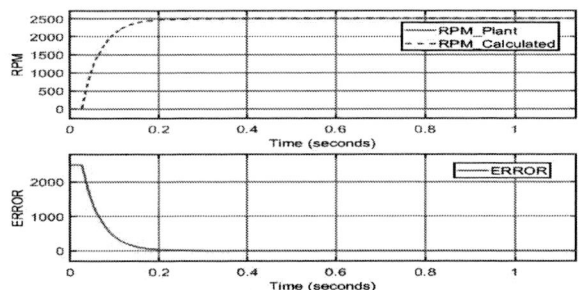

Fig. 9: RPM calculated for case 2.

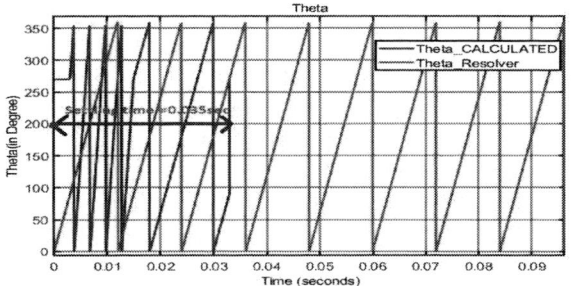

Fig. 10: Rotor position (θ) calculated for case 3.

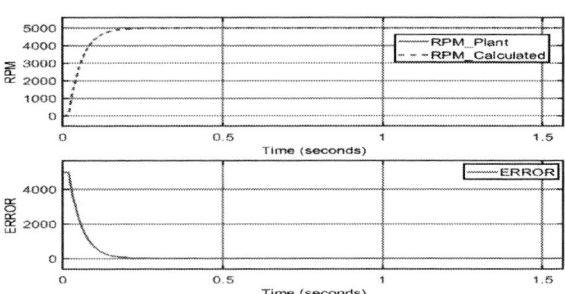

Fig. 11: RPM calculated for case 3.

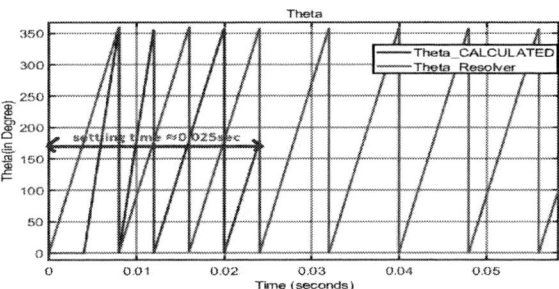

Fig. 12: Rotor position (θ) calculated for case 4.

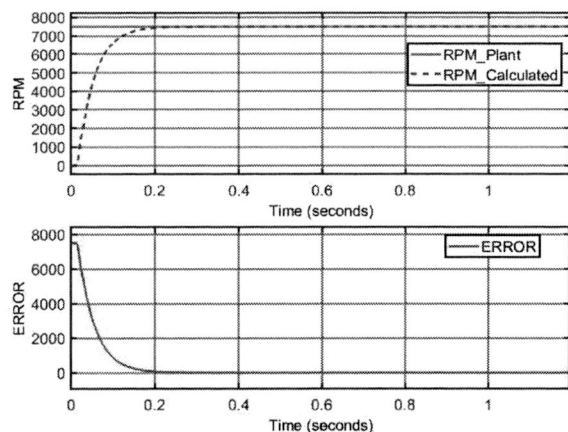

Fig. 13: RPM calculated for case 4.

Case No.	Settling Time(sec)	Position Error(degree)	RPM Error
1	1.8	±1	±5
2	0.06	±0.8	±3
3	0.035	±0.6	±2
4	0.025	±0.2	±1

TABLE III: Summary of Simulation Results for steady State. The rotor position error is within ±1°, and the RPM error is within ±5 for steady state.

B. Transient Response

Case No.	T_e (Nm)	T_L (Nm)	RPM	Sin P	Sin N	Cos P	Cos N
1	5-35	4.88-33.07	191-3056	1	0	1	1
2	15-45	14.26-42.47	1146-4011	0	1	1	1
3	65-35	61.27-33.07	5921-3056	1	1	1	0
4	80-15	75.38-14.27	7353-1146	1	0	1	1

TABLE IV: Signal failure at different RPMs. Here, 0 represents a healthy signal with in the full voltage range of [0.75-4.25]v, while 1 indicates a lost or distorted signal.where T_e = Electromagnetic torque and T_L = Load torque

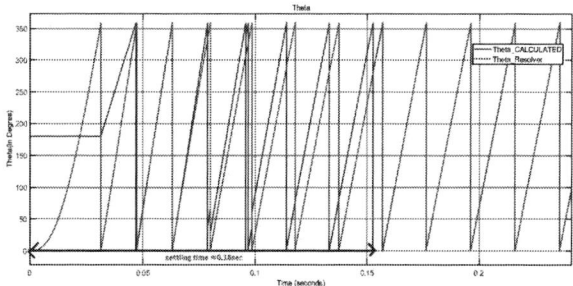

Fig. 14: Rotor position (θ) calculated for case 1.

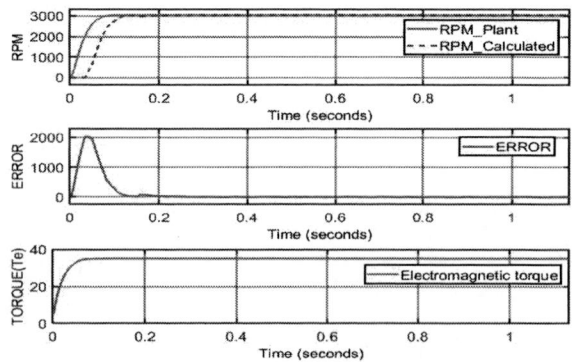

Fig. 15: RPM calculated for case 1.

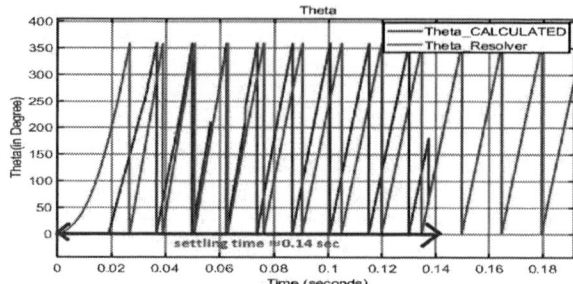

Fig. 16: Rotor position (θ) calculated for case 2.

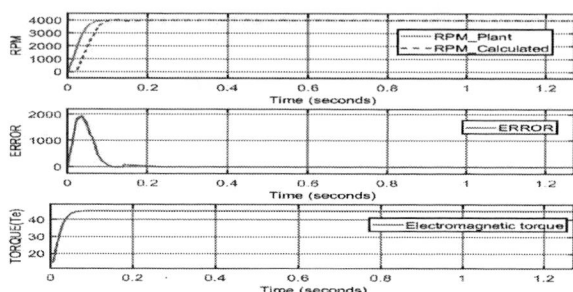

Fig. 17: RPM calculated for case 2.

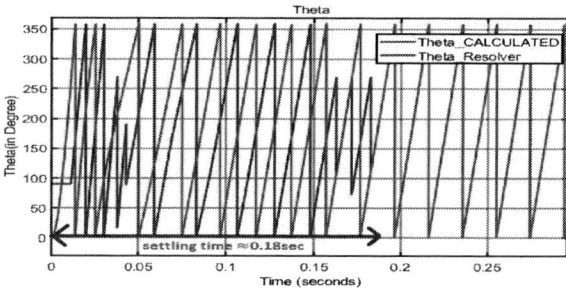

Fig. 18: Rotor position (θ) calculated for case 3.

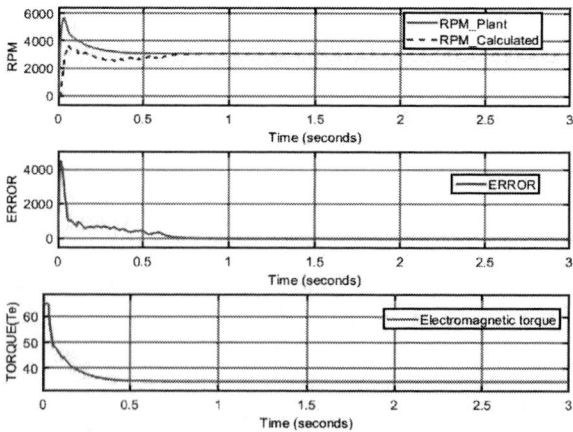

Fig. 19: RPM calculated for case 3.

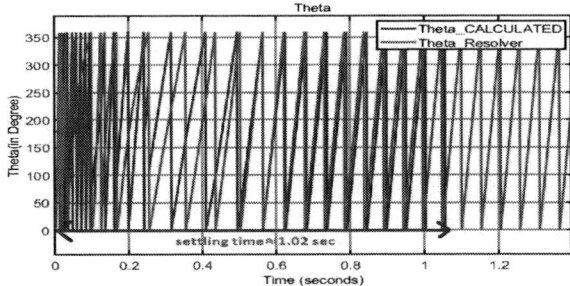

Fig. 20: Rotor position (θ) calculated for case 4.

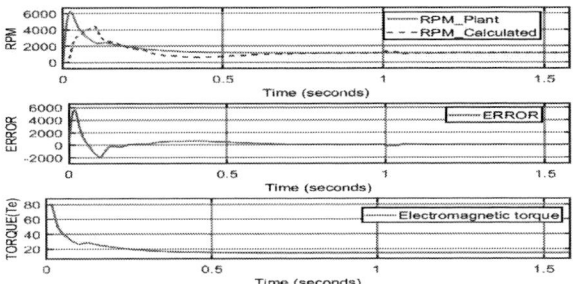

Fig. 21: RPM calculated for case 4.

Case No.	Settling Time(sec)	Position Error(degree)	RPM Error
1	0.16	±0.8	±2
2	0.14	±0.4	±1
3	0.18	±2	±3
4	1.02	±3	±5

TABLE V: Summary of Simulation Results for Transient State. The rotor position error is within ±3°, and the RPM error is within ±5 during transient state.

V. CONCLUSION AND FUTURE WORK

This paper successfully demonstrates that the rotor position can be accurately calculated even in the partial signal loss from the position sensor (resolver). Through an extensive analysis and the development of a robust model, we have presented an effective solution to address the challenges posed by incomplete sensor signals. The proposed method has been rigorously tested through simulations, with results provided for both steady-state and transient conditions. The findings from this research highlight the resilience and effectiveness of the proposed approach in motor control applications, ensuring continuous and precise rotor position tracking under suboptimal conditions. Looking ahead, future work will focus on mitigating the overshoot observed at lower RPMs. Additionally, implementing this approach in a real-time motor control system to further evaluate its effectiveness and adaptability under actual operating conditions.

REFERENCES

[1] Evaluation of position and current sensor'unpublished'. technologies for a PMSM used in automotive applications
[2] J. Hu, J. Zou, F. Xu, Y. Li and Y. Fu, "An Improved PMSM Rotor Position Sensor Based on Linear Hall Sensors," in IEEE Transactions on Magnetics, vol. 48, no. 11, pp. 3591-3594.
[3] JS. -Y. Kim, C. Choi, K. Lee and W. Lee, "An Improved Rotor Position Estimation With Vector-Tracking Observer in PMSM Drives With Low-Resolution Hall-Effect Sensors," in IEEE Transactions on Industrial Electronics, vol. 58, no. 9, pp. 4078-4086.
[4] Y. Chen, G. Goetting and J. Xie, "Online Compensation of Rotor Position Errors of Resolvers in PMSMs for EVs: A Novel FxLMS Adaptive Filter based Approach with Stabilized V/f Control," PCIM Europe digital days 2020; International Exhibition and Conference for Power Electronics, Intelligent Motion, Renewable Energy and Energy Management, Germany, 2020, pp. 1-8..
[5] R. Nalepa and T. Orlowska-Kowalska, "Optimum trajectory control of the current vector of a non-salient pole PMSM in the field-weakening region," IEEE Trans. Ind. Electron., vol. 59, no. 7, pp. 2867–2876, Jul. 2012.
[6] Duane C. Hanselman, "Brushless Permanent Magnet Motor Design" New York: Mc-Graw Hill, 1994.
[7] Benchmark of Rotor Position Sensor Technologies for Application in Automotive Electric Drive Trains.
[8] D. Fernandez, D. Fernandez, M. Martinez, D. Reigosa, A. B. Diez and F. Briz, "Resolver Emulation for PMSMs Using Low Cost Hall Effect Sensors," 2019 IEEE Energy Conversion Congress and Exposition (ECCE), Baltimore, MD, USA, 2019, pp. 5689-5693, doi: 10.1109/ECCE.2019.8913229.
[9] J. Hu, J. Zou, F. Xu, Y. Li and Y. Fu, "An Improved PMSM Rotor Position Sensor Based on Linear Hall Sensors," in IEEE Transactions on Magnetics, vol. 48, no. 11, pp. 3591-3594, Nov. 2012, doi: 10.1109/TMAG.2012.2202279.
[10] F. Genduso, R. Miceli, C. Rando and G. R. Galluzzo, "Back EMF Sensorless-Control Algorithm for High-Dynamic Performance PMSM," in IEEE Transactions on Industrial Electronics, vol. 57, no. 6, pp. 2092-2100.
[11] S. Wang, K. Yang and K. Chen, "An Improved Position-Sensorless Control Method at Low Speed for PMSM Based on High-Frequency Signal Injection into a Rotating Reference Frame," in IEEE Access, vol. 7, pp. 86510-86521.
[12] S. Morimoto, M. Sanada and Y. Takeda, "High-performance current-sensorless drive for PMSM and SynRM with only low-resolution position sensor," in IEEE Transactions on Industry Applications, vol. 39, no. 3, pp. 792-801, May-June 2003, doi: 10.1109/TIA.2003.811782.
[13] G. De Donato, G. Scelba, M. Pulvirenti, G. Scarcella and F. Giulii Capponi, "Low-Cost, High-Resolution, Fault-Robust Position and Speed Estimation for PMSM Drives Operating in Safety-Critical Systems," in IEEE Transactions on Power Electronics, vol. 34, no. 1, pp. 550-564, Jan. 2019, doi: 10.1109/TPEL.2018.2820042.
[14] S. -Y. Kim, C. Choi, K. Lee and W. Lee, "An Improved Rotor Position Estimation With Vector-Tracking Observer in PMSM Drives With Low-Resolution Hall-Effect Sensors," in IEEE Transactions on Industrial Electronics, vol. 58, no. 9, pp. 4078-4086.
[15] J. Hu, J. Zou, F. Xu, Y. Li and Y. Fu, "An Improved PMSM Rotor Position Sensor Based on Linear Hall Sensors," in IEEE Transactions on Magnetics, vol. 48, no. 11, pp. 3591-3594.
[16] Evaluation of position and current sensor technologies for a PMSM used in automotive applications.
[17] A. Murray, B. Hare, and A. Hirao, "Resolver position sensing system with integrated fault detection for automotive applications," in Proc. IEEE Sensors, vol. 2, pp. 864-869.
[18] S. Chen, Y. Zhao, H. Qiu and X. Ren, "High-Precision Rotor Position Correction Strategy for High-Speed Permanent Magnet Synchronous Motor Based on Resolver," in IEEE Transactions on Power Electronics, vol. 35, no. 9, pp. 9716-9726.
[19] L. Pecly, R. Schindeler, D. Cleveland and K. Hashtrudi-Zaad, "High-Precision Resolver-to-Velocity Converter," in IEEE Transactions on Instrumentation and Measurement, vol. 66, no. 11, pp. 2917-2928.

979-8-3315-3013-6/25 $31.00 © 2025 IEEE

Design of Battery Storage fed Asymmetrical Multilevel Converter for Enhanced Statcom Applications

N. Vasantha Gowri[a], G. Suresh Babu[b], T. Murali krishna[a] and Anil Bharadwaj.Ch[c]

[a]Associate Professor in the Department of EEE, Chaitanya Bharathi Institute of Technology, Gandipet, Hyderabad-500075.
Email: vasanthagowri_eee@cbit.ac.in
[b]Professor in the Department of EEE, Chaitanya Bharathi Institute of Technology, Gandipet, Hyderabad-500075.
[c]Senior Project Engineer, Semiconductor – R&D, INOPC, Hitachi Energy, Chennai, TamilNadu, India-600128. Email:
abharadwajch@gmail.com

The dominance of renewable power in the existing electrical grid causes an improper balance between load and demand. Also, the stability of voltage/frequency of the respective networks may get altered due to fluctuating nature of renewable power. For handling these contingencies, a system which has the features of a Statcom and the ability to support the active power has to be employed. They can be termed 'Statcom with Storage unit (SSU)'. Thus, this paper presents a configuration for SSU applications formed by an Asymmetrical Multilevel Converter (AsMC) with an auxiliary bridge. With the presence of an auxiliary bridge, the harmonic profile of the converter has improved with increased levels. The detailed working principle of the configuration has been elaborated in this paper. Also, the needful control mechanisms to obtain the needful real and reactive powers from configuration has been established. The efficacy of the proposed AsMC for SSU applications has been validated in the PSCAD platform and from the simulation results it has been observed that the proposed configuration is competent enough for the SSU applications.

Keywords—Battery, Dual active bridge, Multilevel converter, Ride-through, Statcom

I. INTRODUCTION

Integration of renewable energy based power generation into the existing AC network has seen four-fold growth [1]. With the cogitation towards environment, lessened initial cost and ameliorated estimation techniques for renewable power are the preeminent reasons behind the dominance of the renewable power [2]. The total installed capacity of the renewable power has touched around 3900 GW by the end of 2024 [3].

Main setback of the renewable power generation is due to capriciousness in the amount of generated by these systems. By injecting this precarious power into extant electric network results-in; proliferates the voltage and frequency stability of the grid, irks the protection equipment and impede the scheduling of generating units in the ac network [4]. Among the assorted renewable energy sources, installed capacity of wind energy based power generation is pervasive.

To integrate a large-scale wind park with the Ac grid, gridcodes are propounded by TSOs in [5]. The compelling gridcodes are PoC voltage regulation, active power support, improving the voltage profile of PoC during faults and unbalance compensation. A control principle to operate the asynchronous generator in a wind park was delineated in [5]. Even though, the frequency at the PoC was improved using the mentioned control but the converters must handle more power during the contingencies.

A methodology to operate off-shore windpark during the low voltage the PoC of the wind park was elaborated in [7]. With this control, the real power supplying capability of the converters in the system will become zero. This may lead to reduction in the life of the converters. At the PoC of a wind park, a Statcom and a storage system are imperative. In [8], the authors emphasized the eminence of Statcom in improving the voltage profile at the PoC of a wind power plant. In [9], application of storage systems for the renewable power dominated was articulated.

An approach to operate the Statcom along with an Ultracapacitor storage system at the PoC of a wind energy system is conferred in [10]. Using two system to support both real/reactive powers will results-in lower operating efficiency and larger footprint. A system which has the dexterity in supply both real and reactive simultaneously can be adopted. These are termed as Statcom with Storage unit (SSU) [11].

In [12], two-level voltage source converter based topology has been suggested for SSU applications. The two/three level based VSCs are having certain impediments for utility applications like; lower commercial efficiency, need of passive filters and larger footprint. A comparative study on different multilevel converter based topologies for Statcom applications were briefed in [13] and [14]. Henceforth, multilevel converter based topology can be picked for SSU application. Cascaded multilevel based topology with embedded storage for electric railway applications was proposed in [15]. Here the storage modules are connected at the each cell of the inverter. The main deterrents of cascaded multilevel converters for producing more levels are, need of high component count, complex voltage balancing scheme and larger footprint.

The modular multilevel converter based topology for SSU has been studied in [16]. But the topology requires higher dc-link which requires cascading of dc-dc converters. This leads to imbalance in the state-of-charge of the storage modules. Henceforth, Asymmetrical Multilevel Converter (AsMC) based topologies are becoming attractive for medium voltage applications with better resolution and less component count [17]. AsMC based central inverters for PV applications has been proposed in [18]. In the same lines, AsMC based SSU for medium power applications is presented in [19]. Here, the AsMC is formed by cascading full-bridge and half-bridge cells. While the storage modules are integrated through a bi-directional dc-dc converters.

In [20], AsMC based hybrid storage integration with the utility has been studies. But the main drawback of this configuration is during the unbalanced operation, the storage system will charge/discharge with different currents. This engenders the unequal state-of-charge among

979-8-3315-3013-6/25 $31.00 © 2025 IEEE   272

the storage modules. The harmonic profile of the AsMC can be enhanced with the use of level-doubling network. In [21], proposed a topology for AsMC using Level-Doubling Network (LDN). Using this topology the levels in the voltage generated by a leg of AsMC can be increased.

From the discussion, it is prudent in choosing AsMC for SSU. Instead of using multiple dc sources for each of AsMC, a single dc source can be employed for all three phases can be employed. While the dc source for the auxiliary bridges can be derived from the main dc-source while an isolated dc-dc converter. Among the various types of the isolated converters, dual-active bridge based topology is better suitable due to its higher conversion ratio, lower switching losses and better efficacy [22]. The DAB is befitting for low voltage medium applications [23].

A topology using AsMC with LDN with battery storage system for SSU application is presented in this manuscript. The battery modules are lumped at the main bridge, while the dc sources for the auxiliary bridges are derived from the battery using DAB. The tuning of the controller gains to regulate the output of DAB voltage is an onerous task. A detailed procedure to tune the controller parameters of a dc-dc converter has been elaborated in [24]. For better voltage regulation and transient performance, tuning of the controller gains for a dc-dc converter suing optimization techniques based on artificial intelligence has been briefed in 0. Among these techniques, Particle Swarm intelligence based algorithm has been chosen to identify the gains which are pertinent.

The paper is organized as follows. The system to evaluate the performance of the proposed topology is given in Section-II. The proposed topology and integration of battery modules are mentioned in Section-III. Section-IV covers the control methods to achieve features of SES are elaborated. Section-V presents the simulation results of the proposed SSU. Finally, Section-VI concludes the work.

II. SYSTEM FOR PERFORMANCE EVALUATION

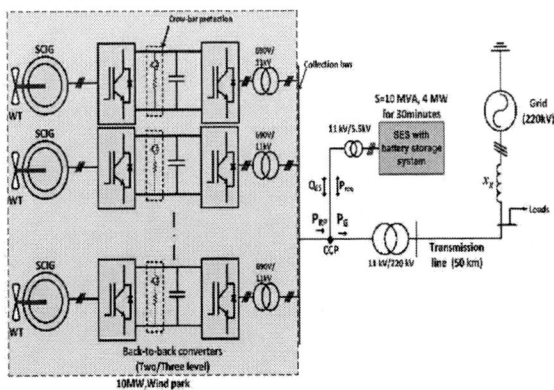

Fig. 1 Wind park with ESt at PoC

Fig. 1 shows a layout of a wind park with SES. Smaller units of wind energy units are installed to form a big wind park. The Wind Turbine (WT) is coupled to a squirrel cage induction machine through a back/to/back voltage source converter. These units of wind generating units are connected to a common point which operates at 11/33kV. The wind park is connected to the grid through a transmission network as shown in the figure. The Motor Side Converter (MSC) is operated in vector control mode and allows maximum power extraction from wind turbines

at different wind speeds. The Grid Side Converter (GSC) is operated in V-Q control mode through which the voltage of intermediate dc-link is maintained constant and reactive power along with power quality issues at the grid side can be addressed. The details elucidation of the control principles for the MSC and GSC are given in [5]. In addition, a dc-chopper is present at the dc-link which helps to avoid overvoltage of the dc-bus by dissipating extra energy in the resistor. The intermittency in the wind speed, results in the generation of fluctuations in the active power generated by the wind-park. The required gridcodes at the PoC can be provided by SES. Table 2 shows the rating of the wind park considered for performance evaluation of the proposed topology.

III. PROPOSED TOPOLOGY

This section presents details about the proposed configuration for SSU application which has been formed using AsMC and DAB.

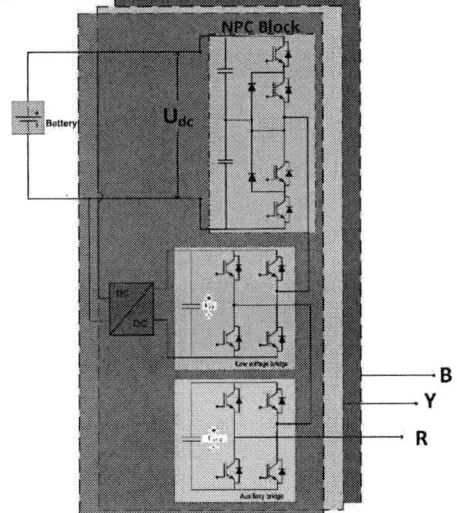

Fig. 2 Configuration of asymmetrical converter with auxiliary bridge

A. Configuration of AsMC

Fig. 2 shows the topology for Est application using battery modules. A leg in the proposed configuration has been formed by cascading a neutral point converter, a full bridge (which can be termed as low voltage bridge) and an auxiliary bridge whose voltage is floating. This kind of configuration is known as 'Asymmetrical Multilevel Converter (AsMC)'. The capacitor voltage of the low voltage bridge can be regulated using a dc-dc converter. The dc-dc converter helps in stepping down the dc-link voltage. While the integration of the battery storage system will be discussed in next sub-section. The voltage levels of the NPC block, low voltage bridge and auxiliary bridge are chosen such a way that they form a ratio of 9:3:1 [18].

An NPC block can generate three-levels of voltages like '+V, 0 & -V'. Since the low voltage and auxiliary bridges are formed using full-bridge cells. Thus, these bridges were also capable in generating three-level voltages. Henceforth, by appropriate combination, each leg of AsMC can generate 26levels of voltage. Table 1 shows the respective voltage levels and contribution of the voltage from each bridge in the leg. From the table it can be noticed that the converter

979-8-3315-3013-6/25 $31.00 © 2025 IEEE

operates at fundamental frequency which in turn reduces the switching losses.

Table 1 Generation of voltages and contribution from each cell

Voltage level	Output of NPC	Output of LVB	Output of AVB
-13V	$-V_{NPC}$	$-V_{LV}$	$-V_{AB}$
-12V	$-V_{NPC}$	$-V_{LV}$	0
-11V	$-V_{NPC}$	$-V_{LV}$	V_{AB}
-10V	$-V_{NPC}$	0	$-V_{AB}$
-9V	$-V_{NPC}$	0	0
-8V	$-V_{NPC}$	0	V_{AB}
-7V	$-V_{NPC}$	V_{LV}	$-V_{AB}$
-6V	$-V_{NPC}$	V_{LV}	0
-5V	$-V_{NPC}$	V_{LV}	V_{AB}
-4V	0	$-V_{LV}$	$-V_{AB}$
-3V	0	$-V_{LV}$	0
-2V	0	$-V_{LV}$	V_{AB}
-V	0	0	$-V_{AB}$
0	0	0	0
V	0	0	V_{AB}
2V	0	V_{LV}	$-V_{AB}$
3V	0	V_{LV}	0
4V	0	V_{LV}	V_{AB}
5V	V_{NPC}	$-V_{LV}$	$-V_{AB}$
6V	V_{NPC}	$-V_{LV}$	0
7V	V_{NPC}	$-V_{LV}$	V_{AB}
8V	V_{NPC}	0	$-V_{AB}$
9V	V_{NPC}	0	0
10V	V_{NPC}	0	V_{AB}
11V	V_{NPC}	V_{LV}	$-V_{AB}$
12V	V_{NPC}	V_{LV}	0
13V	V_{NPC}	V_{LV}	V_{AB}

Fig. 3 shows the voltage at the output of each cell in a leg of AsMC. The main advantage of this configuration is the cell voltage of the auxiliary bridge will balance around the nominal voltage.

From the Fig. 3, the expression for the fundamental voltage generated by the NPC and the low voltage bridge can be given as.

$$(U_{NPC})^1 = \frac{m4U_{dc}}{\pi} \cos \alpha \sin(\omega t) \qquad (1)$$

$$(U_{LV})^1 = \frac{m4U_{LV}}{\pi} \cos \beta \sin(\omega t) \qquad (2)$$

Finally, the fundamental component of the auxiliary bridge can be given as:

$$(U_{AB})^1 = U_m \sin(\omega t) - \frac{m4U_{dc}}{\pi} \cos \alpha \sin(\omega t)$$
$$- \frac{m4U_{LV}}{\pi} \cos \beta \sin(\omega t) \qquad (3)$$

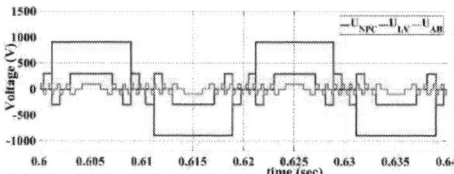

Fig. 3 Output voltage of NPC, low voltage and auxiliary bridges

Since, the bridges are connected in series, all of them will carry the same magnitude of the current. Thusly, the power handled by the bridge are:

$$P_{NPC} = \frac{m4U_{dc}}{\pi} \cos \alpha \, I_m \cos \theta \qquad (4)$$

$$P_{LV} = \frac{m4U_{LV}}{\pi} \cos \beta \, I_m \cos \theta \qquad (5)$$

The relationship between the power handling capability of the NPC and low voltage bridge with respect to modulation index is presented in Fig. 4. From the figure, it can be noticed that the NPC handles approximately 72% of the power while the low voltage bridge handles 38%.

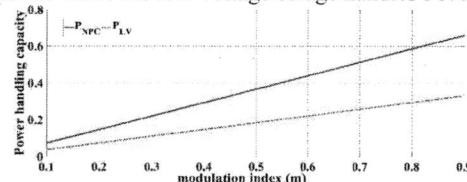

Fig. 4 Relationship between modulation index and power handled by the bridges in each leg of AsMC

B. Integration of storage system

The DAB has been selected to integrate the low voltage storage modules with MMC which is operating higher dc-link voltage. Fig. 5 presents the basic structure of a DAB. From the figure, it can be discerned that the DAB has two bridges namely low voltage bridge and high voltage bridge respectively. While these bridges are connected through an isolation transformer. The storage modules are connected to the low voltage bridge while the output of the high voltage bridge can be connected to a VSC which is operating at high dc-link voltage.

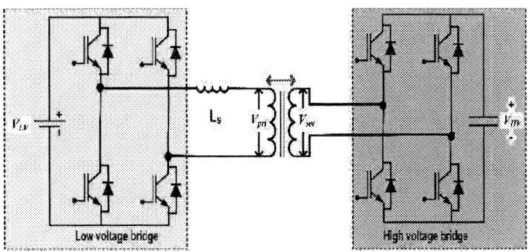

Fig. 5 Configuration of a dual active bridge

The low voltage bridge can be formed using the discrete IGBT modules which can operate at high currents. The high voltage bridge can be formed by series connection of modules to support the needful high voltage at its output. The isolation transformer can be either air-core or CRCO core. For high power applications the operating frequency of this transformer can vary from 100-500Hz. From the above discussion, the merits of the DAB are listed below.

a) Higher conversion ratio due the presence of the transformer.
b) Capability for bi-directional power transfer between the bridges.
c) Provides better isolation between the low voltage storage modules and high voltage dc-link.
d) Better efficiency due to inherent zero- voltage switching behavior.

IV. CONTROL METHOLOGY OF SES

The methodologies to operate the configuration in Fig. 2 as SSE is elaborated in this section. The deliverables of SSU at the PoC of the wind park are as follows:

a. Reactive power support to regulate voltage at PoC

b. Exchange of real power to limit the inconstancy in the power generated by the wind park.

c. Support during the voltage contingencies at the PoC.

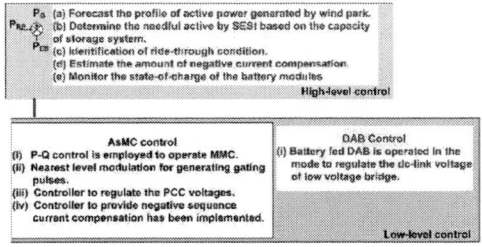

Fig. 6 Overview of control methodology of SSU

Fig. 6 display the scheme to accomplish the SES features from the proposed topology in section-III. From the figure, it can be comprehended that the complete control was break-downed into high/ low-level controls respectively. The explication of the mentioned controls is given in the following sub-sections.

A. High-level control

The purpose of the high-level control is to act like an energy management system for the SSU based windpark. Initially it estimates the profile of the wind power for the next 30 minutes and using that it calculates the real power support needed from SSU. In conjunction with this, it estimates the limits of the controller which are pertinent. Furthermore, it regulates the energization of the SSU to bring the system into the operation. The expression for the real power support can be given as.

$$P_{SSU} = P_G - P_{wp} \tag{6}$$

B. Low-level control

The respective control is disassociated into MMC control and control of DAB. This sub-section briefs about the control of MMC to obtain the SES features while the DAB control philosophy is elaborated in upcoming sections. Fig. 9 shows the control scheme for the MMC. From the figure, it can be perceived that control has two branches like outer-loop and inner-loop. The outer loop handles the purpose to supply the real/reactive power which are imperative at the PoC. This helps to limit the inconsistency in the grid real power and regulate the PoC voltage at the nominal value.

C. Modeling of DAB

With the help of modeling of DAB which has been presented in [12], the expression for the power transfer between the low voltage and high voltage bridges of DAB is given in (7).

$$P_o = \left[\frac{nU_{LV}U_{HV}}{2L_sf_s}\right]D(1-D) \tag{7}$$

In addition to the above, the relationship between the phase-shift (D) and the output voltage of the DAB can be given as.

$$\widehat{v_o} = \left(\frac{R_o}{R_oC_os+1}\right)\left(\frac{v_o(1-2D)}{(1-D)DR_o}\right)\hat{d} \tag{8}$$

Where R_o is the equivalent active power support needed at the dc-link and C_o is the dc-link capacitance.

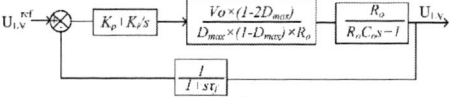

Fig. 7 Representation of DAB loop transfer function

The control structure of the DAB using the small-signal modelling is presented in Fig. 7. From the figure, the loop gain transfer function of the DAB can be given as.

V. RESULTS AND ANALYSIS

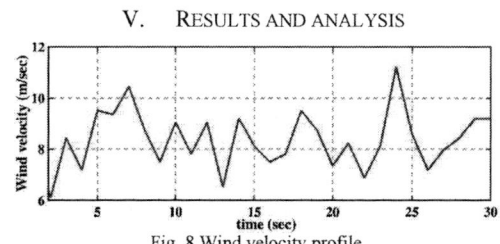

Fig. 8 Wind velocity profile

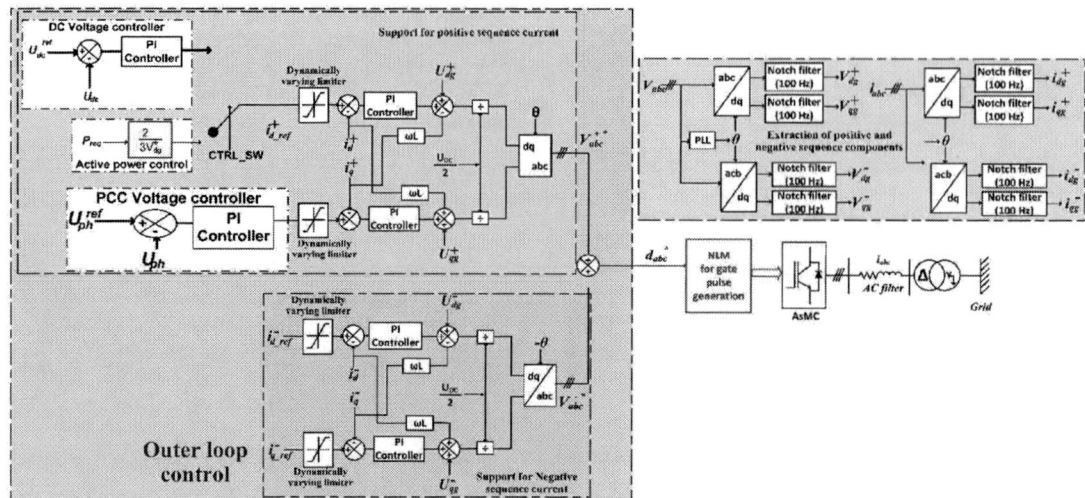

Fig. 9 Scheme of MMC control

Using the parameters in Table 2 to Table 3, the proposed configuration and the respective control principles were developed in the PSCAD/EMTDC environment. The following sub-sections will discuss the respective observations.

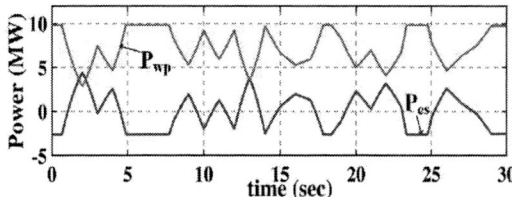

Fig. 10 Real power from Wind Park and SSU

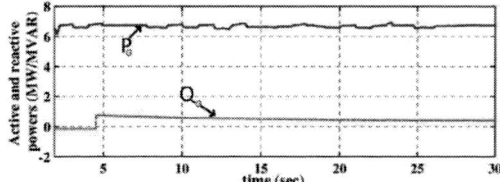

Fig. 11 Real/reactive powers injected into the grid

A. Real power support

Fig. 8 shows the respective wind velocity for the mentioned duration and the corresponding power generated by the windpark is plotted in Fig. 10. The respective figure shows the real power of SES in accordance with (6). Fig. 11 shows the active/reactive powers being sent into the grid.

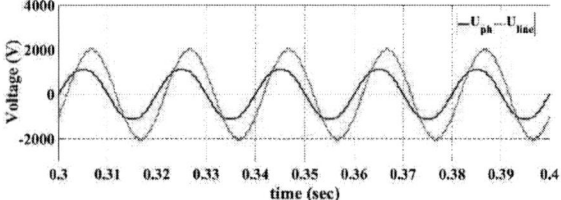

Fig. 12 Phase and line voltages of AsMC

B. Analysis of AsMC

Fig. 12 shows the line and phase voltages of the proposed AsMC. The corresponding harmonic profiles of these voltages are presented in Fig. 13 & Fig. 14 respectively. From the figures it can be concluded the harmonics are with in IEEE Standards.

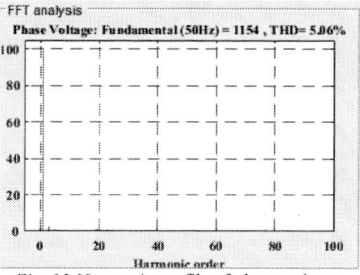

Fig. 13 Harmonic profile of phase voltage

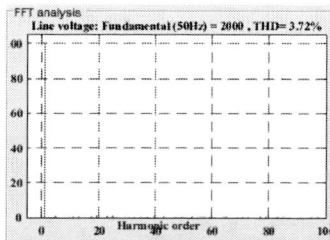

Fig. 14 Harmonic profile of the line voltage

VI. CONCLUSION

A topology using AsMC for SSU application has been conferred in this paper. The asymmetry is formed by combining a neutral point converter cell with a full bridge cell. Also, an auxiliary bridge has been added to improve the harmonic profile. The storage modules are lumped at the dc-link of NPC. While the input for the low voltage bridge has been derived from the dc-link using DAB. The proposed configuration has higher efficiency and less installation cost in comparison to topologies which makes this configuration suitable for high voltage/power applications. The DAB helps to regulate the DC-link voltage of low voltage bridge. While the auxiliary bridge has a floating voltage. The needed modulation index has been generated using P-U$_{ac}$ control. The presented topology is suitable for SSU application which has been proved by the technical analysis using PSCAD/EMTDC platform. The results have presented that configuration is advisable for SSU applications.

APPENDIX

Table 2 Rating of Wind Park

Parameter	Value
Capacity of the plant	10 MW
Collection bus line voltage	11 kV
Rating of each SCIG	1.5 MW
Output voltage of the generator	690 V
Rating of back-to-back converter	2 MW

Table 3 Parameters of AsMC

Parameter	Value
Converter capacity (S)	10 MVA
Active power support for 30 minutes	4 MW
DC link voltage	900 V
Line voltage	0.69 kV
Number of cells per phase	2
Arm inductance (0.1 pu)	2.5 mH
Capacitance of a submodule capacitor	25 mF
Switching frequency of a DAB	250 Hz
Voltage rating of the transformer	0.69kV/11kV

REFERENCES

[1] H. Holttinen *et al.*, "Variable Renewable Energy Integration: Status Around the World," *IEEE Power and Energy Magazine*, vol. 19, no. 6, pp. 86-96, Nov.-Dec. 2021.

[2] K. Ogimoto and H. Wani, "Making Renewables Work: Operational Practices and Future Challenges for Renewable Energy as a Major Power Source in Japan," *IEEE Power and Energy Magazine*, vol. 18, no. 6, pp. 47-63, Nov.-Dec. 2020

[3] Lucía Fernández, "Renewable energy capacity 2023 by country," Environment and Energy, Jan. 2025. [Online]. Available: https://www.statista.com/statistics/267233/renewable-energy-capacity-worldwide-by-country/

[4] S. Malik, B. Modi and D. Mangrolia, "Grid Integration of Distributed Generation: Issues and Challenges," *Proc. of International Conference for Advancement in Technology (ICONAT)*, pp. 1-6, 2022.

[5] S. W. Ali *et al.*, "Offshore Wind Farm-Grid Integration: A Review on Infrastructure, Challenges, and Grid Solutions," *IEEE Access*, vol. 9, pp. 102811-102827, 2021.

[6] P. Verma, S. K. and B. Dwivedi, "A Cooperative Approach of Frequency Regulation Through Virtual Inertia Control and Enhancement of Low Voltage Ride-through in DFIG-based Wind Farm," *J. Mod. Power Syst. Clean Energy*, vol. 10, no. 6, pp. 1519-1530, Nov., 2022.

[7] M. J. Rahman, T. Tafticht and M. L. Doumbia, "Power Stability and Frequency Control Techniques of DG for a High Penetration Wind-Based Energy Storage System Using Integral–Derivative Controller," *IEEE Canadian Journal of Electrical and Computer Engineering*, vol. 45, no. 3, pp. 232-241, Summer 2022

[8] M. Sarwar *et al.*, "Stability Enhancement of Grid-Connected Wind Power Generation System Using PSS, SFCL and STATCOM," *IEEE Access*, vol. 11, pp. 30832-30844, 2023.

[9] F. Calero *et al.*, "A Review of Modeling and Applications of Energy Storage Systems in Power Grids," *Proceedings of the IEEE*, vol. 111, no. 7, pp. 806-831, July 2023

[10] M. K. K. Prince *et al.*, "Coordinated Control of Grid-Connected PMSG Based Wind Energy System With STATCOM and Supercapacitor Energy Storage," *IEEE Transactions on Industry Applications*, vol. 60, no. 3, pp. 5108-5118, May-June 2024.

[11] Anil Bharadwaj et. al, "Chapter 10 - E-STATCOM (energy storage+STATCOM): a solution to integrate large-scale wind farms into the grid at medium and high power levels," in Power Quality in Modern Power Systems, P. Sanjeevikumar, C. Sharmeela, Jens Bo Holm-Nielsen, P. Sivaraman, Eds., Academic Press, 2021, pp. 283-310.

[12] S. Bhukya, A. Bharadwaj, M. K. T, N. V. P. Babu and G. Mallesham, "Control and Performance Analysis of Multifunctional Two Level STATCOM fed with Battery Storage System," *Proc. of Second International Conference on Power, Control and Computing Technologies (ICPC2T)*, pp. 1-6, 2022.

[13] D. S. Maurya, P. D. Jadhav, R. S. Joshi, R. R. BendkhaLe and M. P. Thakre, "A Detailed Comparative Analysis of Different Multipulse and Multilevel Topologies for STATCOM," *Proc. of International Conference on Electronics and Sustainable Communication Systems (ICESC)*, pp. 1112-1117, 2020.

[14] N. Padmamalini, P. Deepa, T. Joel, S. Loganayagi, M. R. Faridha Banu and S. Gomathi, "Control of Diode Clamped Multilevel Inverter based STATCOM for Reactive Power Compensation using H-bridge Topology," *Proc. of 7th International Conference on Intelligent Computing and Control Systems (ICICCS)*, pp. 1860-1865, 2023.

[15] J. Chen, K. Yang, H. Lin, K. Wang, C. Cai and P. Zhang, "Star-Connection Supercapacitor-Embedded CHMC-Based STATCOM for Electrified Railways," *IEEE Transactions on Power Electronics*, vol. 39, no. 10, pp. 13733-13743, Oct. 2024

[16] C. Anil Bharadwaj, S. Maiti and N. Dhal, "Modular Multilevel E-STATCOM Using Supercapacitor Based Energy Storage System," *Proc. of 2nd International Conference on Power, Energy and Environment: Towards Smart Technology (ICEPE)*, pp. 1-6, 2018.

[17] S. R. Khasim and C. Dhanamjayulu, "Design and Implementation of Asymmetrical Multilevel Inverter With Reduced Components and Low Voltage Stress," *IEEE Access*, vol. 10, pp. 3495-3511, 2022.

[18] R. Vasu, S. K. Chattopadhyay and C. Chakraborty, "Asymmetric Cascaded H-Bridge Multilevel Inverter With Single DC Source per Phase," *IEEE Transactions on Industrial Electronics*, vol. 67, no. 7, pp. 5398-5409, July 2020.

[19] S. Bhukya, A. Bharadwaj, M. K. T, N. V. P. Babu, G. Mallesham and S. Maiti, "Control of Asymmetrical Multilevel STATCOM with Hybrid Energy System," *Proc. of National Power Electronics Conference (NPEC)*, pp. 1-6, 2021.

[20] W. Jiang, K. Ren, S. Xue, C. Yang and Z. Xu, "Research on the Asymmetrical Multilevel Hybrid Energy Storage System Based on Hybrid Carrier Modulation," *IEEE Transactions on Industrial Electronics*, vol. 68, no. 2, pp. 1241-1251, Feb. 2021.

[21] N. Tak, S. K. Chattopadhyay and C. Chakraborty, "Multilevel Inverter Topology Using Multiple Level Doubling Networks and Single DC Source Per Phase," *Proc. of 30th International Symposium on Industrial Electronics (ISIE)*, pp. 1-6, 2021.

[22] Y. Du, S. Lukic, B. Jacobson and A. Huang, "Review of high power isolated bi-directional DC-DC converters for PHEV/EV DC charging infrastructure," *Proc. of IEEE Energy Conversion Congress and Exposition*, pp. 553-560, 2011.

[23] J. Saha, A. Subramanium and S. K. Panda, "Design of Integrated Medium Frequency Transformer (iMFT) for Dual-Active-Bridge (DAB) Based Solid-State-Transformers," *Proc. of IEEE 12th Energy Conversion Congress & Exposition - Asia (ECCE-Asia)*, pp. 893-898, 2021.

[24] A. Bhattacharjee and I. Batarseh, "A PI Based Simplified Closed Loop Controller for Dual Active Bridge DC-AC Converter For Standalone Applications," *Proc. Of IEEE Applied Power Electronics Conference and Exposition (APEC)*, pp. 761-767, 2020.

K. Panduranga Vittal, S. Bhanja and A. Keshri, "Comparative Study of PI, PID controller for Buck-Boost Converter tuned by Bio-Inspired Optimization Techniques," *Proc. of IEEE International Conference on Distributed Computing, VLSI, Electrical Circuits and Robotics (DISCOVER)*, pp. 219-224, 2021.

Converter Topology to Integrate Battery Storage with Modular Multilevel Converter

T. Murali krishna[a], Krishnaveni Kondreddi[b], N. Vasantha Gowri[a] and Anil Bharadwaj.Ch[c]

[a]Associate Professor in the Department of EEE, Chaitanya Bharathi Institute of Technology, Gandipet, Hyderabad-500075.
Email: tmuralikrishna_eee@cbit.ac.in

[b]Professor in the Department of EEE, Chaitanya Bharathi Institute of Technology, Gandipet, Hyderabad-500075.

[c]Senior Project Engineer, Semiconductor – R&D, INOPC, Hitachi Energy, Chennai, TamilNadu, India-600128. Email:
abharadwajch@gmail.com

Abstract— **The voltage and frequency stability of the existing AC grid are becoming volatile with the integration of renewable energy based generation. Due to the fitful nature of the power generated by these units, balancing the real /reactive powers and supporting the grid during contingencies has become pivotal. Thus, it compelled to use a system that can help the gridcodes at the Point of Common coupling (PoC) of a renewable power generating unit. This kind of system can be termed as 'Storage Enabled Statcom (SES)'. A configuration using a Modular Multilevel Converter (MMC) with a battery storage system has been proposed in this paper. The configuration is formed by integrating the low voltage storage modules at the dc-link of the Modular Multilevel Converter (MMC) using a dual-active bridge (DAB). The DAB provides a better conversion ratio and higher efficiency. The effectiveness of the proposed configuration as SES is verified for a grid connected wind power generation system. The proficiency of the proposed configuration for the SES application and the recommended controls has been validated at the PoC of a wind park using the PSCAD/EMTDC platform, and the results are persuasive.**

Keywords— *Battery, Dual active bridge, Multilevel converter, Ride-through, Statcom*

I. INTRODUCTION

With the surge in electrical power demand, has resulted in the expansion of installations based on renewable energy based power generation [1]. This has resulted due to support through government policies, reduction in the installation cost of a wind/solar energy based power generation units and also due to environmental concerns. During the last decade, the installed capacity of renewable energy based power generation has become proportionate to the capacity of conventional power generation[2]. Thus, the existing electrical grid is becoming as renewable power dominated grid.

Due to fickleness of the electrical power from a renewable power generation, operation of an existing electrical grid becomes arduous. Thereupon Transmission System Operators (TSOs) have proposed certain gridcodes to improve the working of renewable power dominated electrical grids [3]. The following table presents, some significant gridcodes which are to be maintained at the PoC of a RPU.

A control principle to operate the asynchronous generator in a wind park was delineated in [4]. Even though, the frequency at the PoC was improved using the mentioned control but the converters must handle more power during the contingencies. This may lead to reduction in the life of the converters. Hence, aforesaid gridcodes at PoC can be contributed with the help of a Var compensator and a storage system are imperative. In [5], the authors emphasized the eminence of Statcom in improving the voltage profile at the PoC of a wind power plant.

The overview on the different topologies and the respective control principles to operate the Statcom has been elaborated in [6]. The prominence of battery storage for the renewable power dominated grids has been elucidated in [7].

Instead of employing two different system, a single system which has the competency to cater both real and reactive power can be chosen. In [8], the authors have enunciated the requirement of SES for an Off-shore wind park which has assisted in catering the gridcodes for smooth power flow form the wind park into the grid. A two-level voltage source converter (VSC) based SES has been contemplated in [9]. For utility level voltages, the two/three level based VSCs are having certain impediments like lower commercial efficiency, need of passive filters and larger footprint. Henceforth, multilevel topologies can be selected to form the SES. Modular multilevel converter (MMC) based Statcom are becoming predominance for high voltage applications due to their benefits like better efficiency, modularity and better harmonic profile [10]. A configuration to integrate the battery storage at the utility voltage using MMC has been evaluated in [11].

From the discussion, it is expedient to choose MMC to form the SES. In [12] and [13], the authors have proposed a configuration for SES application, which has been formed by full bridge cell based MMC. Here, Ultracapacitor modules are lumped at the dc-link. Though this configuration looks effortless and elegant. This configuration requires large footprint, high component count and high installation cost. MMC based Statcom with Ultracapacitor connected at the submodules of MMC was proposed in [14].

A topology using MMC for SES application has been proposed in [15]. Here, the battery modules are distributed at the dc-link of MMC using dc-dc converters. The dc-dc converters are cascaded in series to produce the required dc-link voltage. But the main challenge is this configuration, balancing of state-of-charge (SoC) among the battery modules is arduous. A technique to balance the SoC among the battery modules is illustrated in [16]. In addition to this, the configuration requires huge insulation for the distributed dc-dc converters.

Henceforth, instead of using numerous dc-dc converters, solid-state transformer can be employed to integrate the low voltage battery modules with the SES. The SST using dual active bridge has been proposed in [17]. Here, the SST has been formed by using DAB and a three-level voltage source converter to integrate the storage system for medium power applications. In this kind of configurations, design of the isolation transformer will be a challenging task. Thus, the procedure to design the transformer is expounded in [18].

In addition to the above, the DAB has become pertinent solution to integrate the low voltage storage modules with the medium voltage AC grids. The application of DAB to integrate hybrid storage system for microgrid applications was accessed in [19]. The complete design procedure and respective control philosophy of DAB for battery charging applications is elucidated in [20]. Thus, from the retrospect, DAB can be employed to integrate the low voltage battery modules with MMC for SES applications.

Hence, this paper presents a topology for SES using MMC with battery storage system. The low voltage storage modules are connected at the dc-link of MMC using DAB. The tuning of the controller gains to regulate the output of DAB voltage is arduous. Thus, artificial intelligence based techniques can be employed for better tuning. In [22], the authors have compared various design techniques to tune the controller gains for a bi-directional dc-dc converter. A detailed procedure to tune the controller parameters of a dc-dc converter has been elaborated in [23]. From the discourse, the particle swarm algorithm has been employed to get the parameters of the controller which are to be felicitous.

The paper is organized as follows. The system to evaluate the performance of the proposed topology is given in Section-II. The proposed topology and integration of battery modules are mentioned in Section-III. Section-IV covers the control methods to achieve features of SES are elaborated. The procedure to model a DAB and optimization to obtain the gains of the controller are analysed in Section-V. Section-VI presents the simulation results of the proposed SES. Finally, Section-VII concludes the work.

II. SYSTEM FOR PERFORMANCE EVALUATION

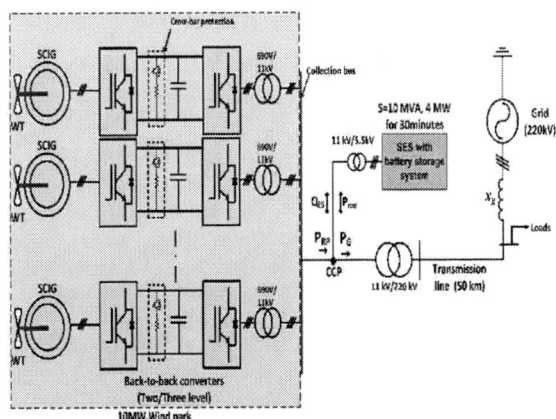

Fig. 1 Wind park with SES at PoC

Fig. 1 shows a layout of a wind park with SES. Smaller units of wind energy units are installed to form a big wind park. The Wind Turbine (WT) is coupled to a squirrel cage induction machine through a back/to/back voltage source converter. These units of wind generating units are connected to a common point which operates at 11/33kV. The wind park is connected to the grid through a transmission network as shown in the figure. The Motor Side Converter (MSC) is operated in vector control mode and allows maximum power extraction from wind turbines at different wind speeds. The Grid Side Converter (GSC) is operated in V-Q control mode through which the voltage of intermediate dc-link is maintained constant and reactive power along with power

quality issues at the grid side can be addressed. The details elucidation of the control principles for the MSC and GSC are given in [21]. In addition, a dc-chopper is present at the dc-link which helps to avoid overvoltage of the dc-bus by dissipating extra energy in the resistor. The intermittency in the wind speed, results in the generation of fluctuations in the active power generated by the wind-park. The required gridcodes at the PoC can be provided by SES. Table 1 shows the rating of the wind park considered for performance evaluation of the proposed topology.

III. PROPOSED TOPOLOGY

This section presents details about the proposed configuration for SES application which has been formed using MMC and DAB.

A. MMC configuration

Fig. 3 shows the topology for MMC based SES with battery system. Each leg of MMC has been formed by two arms which are connected in series using an inductor to limit the circulating current. Half-bridge based submodule configuration has been employed to form an arm. An inductor is employed in between two arms to limit the flow of circulating current. The expression related to the voltage of each submodule capacitor is given in (1). To generate the needful gatepulses for the switches, phase-shifted PWM technique has been employed. The philosophy of this technique is elaborated in [14].

$$U_{c\_nom} = \frac{U_{dc}}{N} \qquad (1)$$

B. Integration of storage system

The DAB has been selected to integrate the low voltage storage modules with MMC which is operating higher dc-link voltage. Fig. 2 presents the basic structure of a DAB. From the figure, it can be discerned that the DAB has two bridges namely low voltage bridge and high voltage bridge respectively. While these bridges are connected through an isolation transformer. The storage modules are connected to the low voltage bridge while the output of the high voltage bridge can be connected to a VSC which is operating at high dc-link voltage.

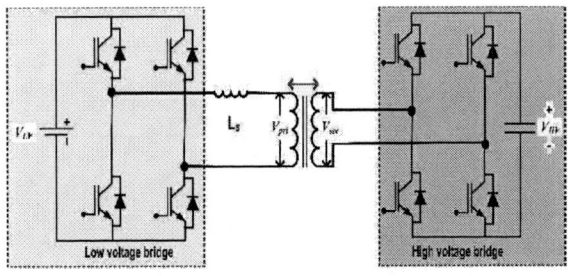

Fig. 2 Configuration of a dual active bridge

The low voltage bridge can be formed using the discrete IGBT modules which can operate at high currents. The high voltage bridge can be formed by series connection of modules to support the needful high voltage at its output. The isolation transformer can be either air-core or CRCO core. For high power applications the operating frequency of this transformer can vary from 100-500Hz. From the above discussion, the merits of the DAB are listed below.

979-8-3315-3013-6/25 $31.00 © 2025 IEEE   279

a) Higher conversion ratio due the presence of the transformer.
b) Capability for bi-directional power transfer between the bridges.

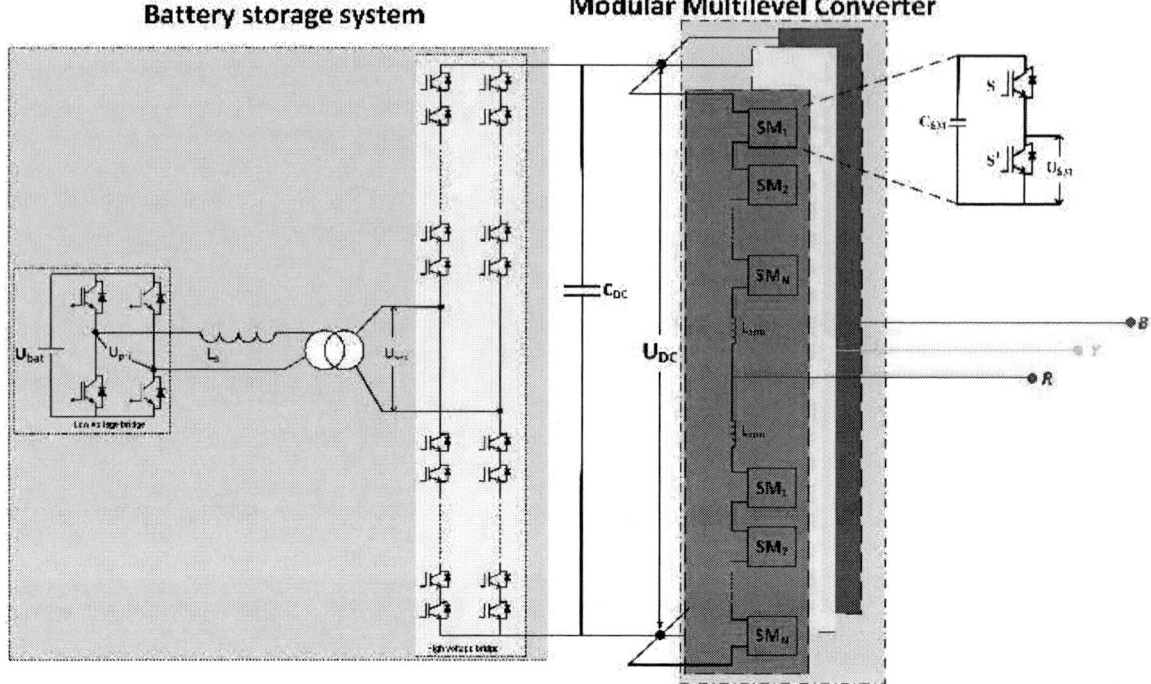

Fig. 3 Structure of storage enable MMC based Statcom

c) Provides better isolation between the low voltage storage modules and high voltage dc-link.
d) Better efficiency due to inherent zero-voltage switching behavior.

IV. CONTROL METHOLOGY OF SES

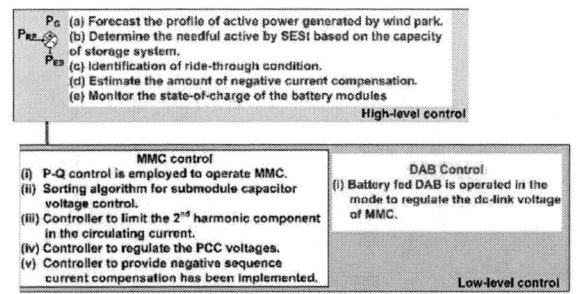

Fig. 4 Overview of control methodology of SES

The methodologies to operate the configuration in Fig. 3 as SES is elaborated in this section. The deliverables of SES at the PoC of the wind park are as follows:

a. Reactive power support to regulate voltage at PoC

b. Exchange of real power to limit the inconstancy in the power generated by the wind park.

c. Support during the voltage contingencies at the PoC.

Fig. 4 display the scheme to accomplish the SES features from the proposed topology in section-III. From the figure, it can be comprehended that the complete control was break-downed into high/ low-level controls respectively. The explication of the mentioned controls is given in the following sub-sections.

A. High-level control

The purpose of the high-level control is to act like an energy management system for the SES based windpark. Initially it estimates the profile of the wind power for the next 30 minutes and using that it calculates the real power support needed from SES. In conjunction with this, it estimates the limits of the controller which are pertinent. Furthermore, it regulates the energization of the SES to bring the system into the operation. The expression for the real power support can be given as.

$$P_{SES} = P_G - P_{wp} \qquad (2)$$

B. Low-level control

The respective control is disassociated into MMC control and control of DAB. This sub-section briefs about the control of MMC to obtain the SES features while the DAB control philosophy is elaborated in upcoming sections. Fig. 5 shows the control scheme for the MMC. From the figure, it can be perceived that control has two branches like outer-loop and inner-loop. The outer loop handles the purpose to supply the real/reactive power which are imperative at the PoC. This helps to limit the inconsistency in the grid real power and regulate the PoC voltage at the nominal value.

979-8-3315-3013-6/25 $31.00 © 2025 IEEE   280

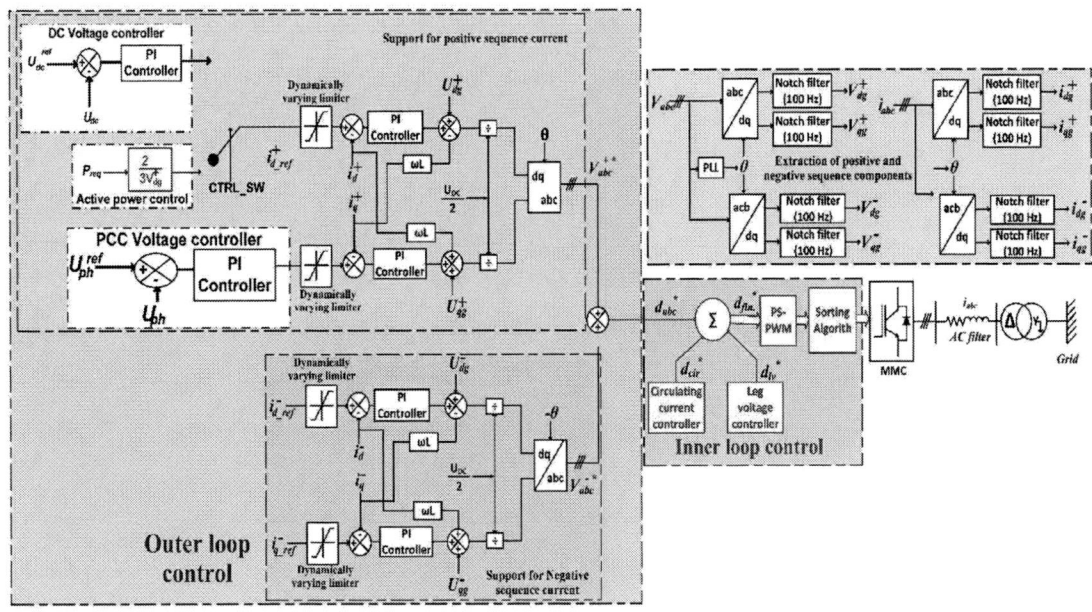

Fig. 5 Scheme of MMC control

The inner-loop facilitates in generating the modulating signals for each arm by balancing the arm voltage and curtailing the circulating current. Thus, the voltages of the submodule capacitors can be balanced at their nominal voltage.

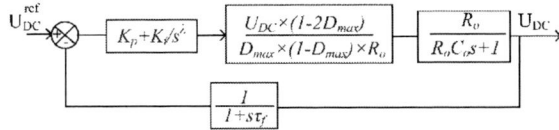

Fig. 6 Representation of DAB loop transfer function

V. CONTROL OF DAB AND CONTROLLER DESIGN

A. Modeling of DAB

With the help of modeling of DAB which has been presented in [9], the expression for the power transfer between the low voltage and high voltage bridges of DAB is given in (3).

$$P_o = \left[\frac{nU_{LV}U_{HV}}{2L_sf_s}\right]D(1-D) \qquad (3)$$

In addition to the above, the relationship between the phase-shift (D) and the output voltage of the DAB can be given as.

$$\widehat{v_o} = \left(\frac{R_o}{R_oC_os+1}\right)\left(\frac{v_o(1-2D)}{(1-D)DR_o}\right)\hat{d} \qquad (4)$$

Where R_o is the equivalent active power support needed at the dc-link and C_o is the dc-link capacitance. The control structure of the DAB using the small-signal modelling is presented in Fig. 6. From the figure, the loop gain transfer function of the DAB can be given as.

$$\frac{V_o^{ref}}{V_{act}^{ref}} = \frac{(1+s\tau_f)(s^\lambda K_p + K_i)KR_o}{R_oC_o\tau_fs^3 + (R_oC_o+\tau_f)s^2 + (1+K_pKR_o)s + KR_oK_i} \qquad (5)$$

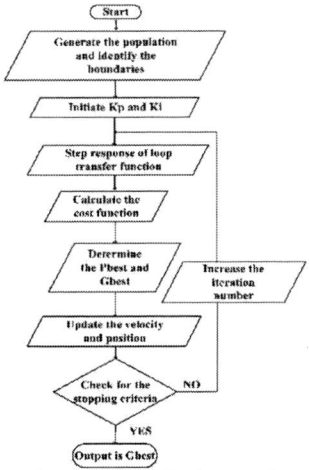

Fig. 7 Flowchart to identify the controller gains

B. Tuning of Gain parameters for the Controller

The output voltage of the DAB can be regulated by adjusting the value of 'D'. The objective to design the gains is to minimize the area error. The respective expression for the calculation of the error is as follows.

$$ITAE = \int_0^t t.|e(t)|dt \qquad (6)$$

The gains for the controller were tuned using PSO by minimizing the fitness function given in (6). Fig. 7 shows the flow the chart to identify the propitious values for the controller gains. Using the algorithm, the variation of the fitness function with respect to the iteration is illustrated in Fig. 8. The corresponding controller gains with the minimum cost function are presented in Table 4.

979-8-3315-3013-6/25 $31.00 © 2025 IEEE   281

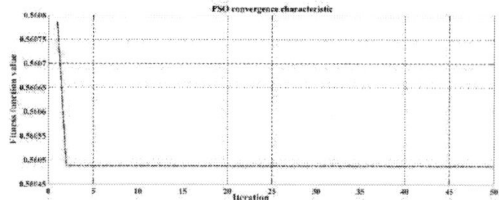

Fig. 8 Convergence of the fitness function using PSO

VI. RESULTS AND ANALYSIS

Using the parameters in Table 1 to Table 3, the proposed configuration and the respective control principles were developed in the PSCAD/EMTDC environment. The following sub-sections will discuss the respective observations.

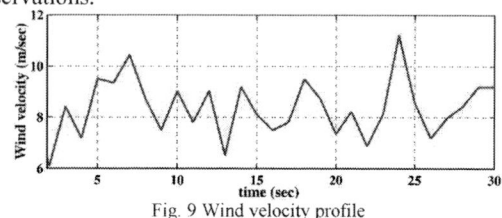

Fig. 9 Wind velocity profile

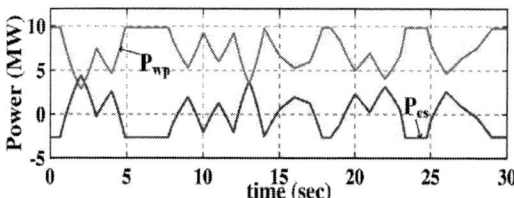

Fig. 10 Real power from wind park and SES

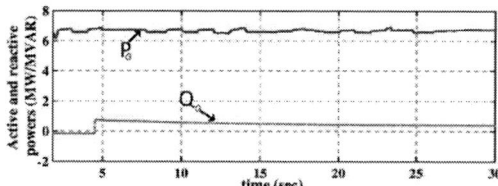

Fig. 11 Real/reactive powers injected into the grid

A. Real power support

Fig. 9 shows the respective wind velocity for the mentioned duration and the corresponding power generated by the windpark is plotted in Fig. 10. The respective figure shows the real power of SES in accordance with (2). Fig. 11 shows the active/reactive powers being sent into the grid.

B. Ride-through capability

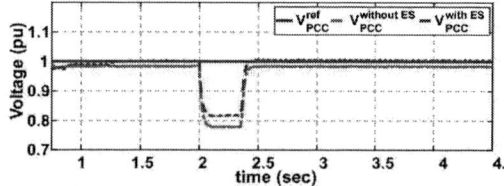

Fig. 12 Effect on the PoC voltage during the fault

Fig. 5 shows the control philosophy of MMC to handle the ride-through capability. The performance of MMC improves the voltage profile during ride-through capability due to disturbance at the CCP. Fig. 12 shows the CCP voltage due to fault in the transmission network. Hence the active power generated by the wind farm falls from 9.6 MW to 7.6 MW as given in Fig. 13. The SES improves the voltage profile by utilizing the full capacity to provide reactive power.

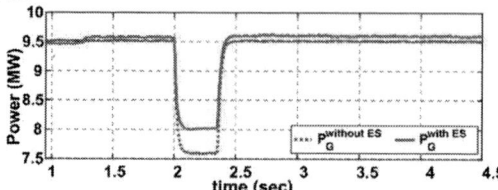

Fig. 13 Change in the active power of Wind Park during faults at CCP

C. Performance of DAB controller

The goal of DAB is to regulate the DC-link voltage at 12kV, using the controller given in Fig. 6. Fig. 14 presents the performance of controller of DAB which has been given in Fig. 6. Initially, a Proportional-integral controller has been employed to regulate the phase angle between primary and secondary voltages of DAB as per the technique given in [20]. Similarly the controller gains have been designed using PSO algorithm. From the response, the PSO based tuned controller gives better performance in terms lower settling time and lesser steady-state error.

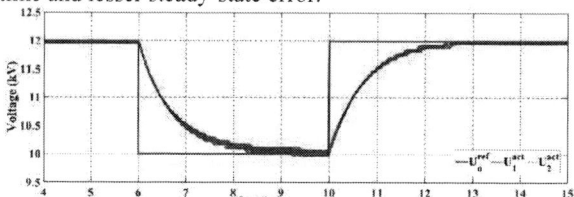

Fig. 14 Tracking performance of DAB controller with change in reference voltage

VII. CONCLUSION

A topology using MMC with a battery storage system using DAB for SES application has been presented in this paper. The storage modules have been integrated using a DAB based converter due to lower losses, a better conversion ratio and inherent galvanic isolation between high and low DC voltages. The proposed configuration has higher efficiency and less installation cost in comparison to topologies which makes this configuration suitable for high voltage/power applications. The DAB helps to regulate the DC-link voltage of MMC, and the P-U_{ac} control for MMC helps in supporting the gridcodes. For better performance in regulating the DC-link voltage, PSO based algorithm has been employed to identify the suitable controller gains for DAB. From the evaluation, it has been identified that the proposed controller has better performance in comparison to the traditional PI controller. Satisfactory performances have been observed by the presented controllers for MMC. The presented topology is suitable for SES application which has been proved by the techno-economic analysis and also can supply the needful in providing the gridcodes.

APPENDIX

Table 1 Rating of Wind Park

Parameter	Value
Capacity of the plant	10 MW

Collection bus line voltage	*11 kV*
Rating of each SCIG	*1.5 MW*
Output voltage of the generator	*690 V*
Rating of back-to-back converter	*2 MW*

Table 2 Parameters of MMC

Parameter	Value
Converter capacity (S)	10 MVA
Active power support for 30 minutes	4 MW
DC link voltage	12 kV
Line voltage	5.5 kV
Number of submodule	4
Arm inductance (0.1 pu)	2.5 mH
Nominal voltage of a submodule	3 kV
Capacitance of a submodule capacitor	5 mF
Switching frequency of a submodule	500 Hz
Voltage rating of the transformer	5.5kV/11kV

Table 3 Rating of battery energy storage system

Parameter	Value
Capacity of each battery module	1 MWh
No of battery modules	4
Active power support from the battery modules	2.5 MW
Nominal voltage of each battery module	1.75 kV
Rating of each dc-dc converter	0.5 MW
Output voltage of dc-dc converter	3 kV
Filter inductance of a dc-dc converter	6 mH
Switching frequency	2.5 kHz

Table 4 Controller parameters of DAB

Parameter	Value
voltage loop	Kp= 0.0010 Ki= 0.0413

REFERENCES

[1] C. Konstantinou, "Toward a Secure and Resilient All-Renewable Energy Grid for Smart Cities," *IEEE Consumer Electronics Magazine*, vol. 11, no. 1, pp. 33-41, 1 Jan. 2022

[2] Lucía Fernández, "Renewable energy capacity 2023 by country," Environment and Energy, Jan. 2025. [Online]. Available: https://www.statista.com/statistics/267233/renewable-energy-capacity-worldwide-by-country/

[3] B. Chaudhuri *et al.*, "Rebalancing Needs and Services for Future Grids: System Needs and Service Provisions With Increasing Shares of Inverter-Based Resources," *IEEE Power and Energy Magazine*, vol. 22, no. 2, pp. 30-41, March-April 2024

[4] P. Verma, S. K. Verma and B. Dwivedi, "A Cooperative Approach of Frequency Regulation Through Virtual Inertia Control and Enhancement of Low Voltage Ride-through in DFIG-based Wind Farm," *J. Mod. Power Syst. Clean Energy*, vol. 10, no. 6, pp. 1519-1530, Nov., 2022.

[5] Sadiq R, Wang Z, Chung CY, Zhou C, Wang C. "A review of STATCOM control for stability enhancement of power systems with wind/PV penetration: Existing research and future scope", *Int Trans Electr Energ Syst.*, e13079, vol. 31, no.11, Aug. 2021.

[6] S. Sharma *et al.*, "A Comprehensive Review on STATCOM: Paradigm of Modeling, Control, Stability, Optimal Location, Integration, Application, and Installation," *IEEE Access*, vol. 12, pp. 2701-2729, 2024.

[7] Chunyang Zhao, Peter Bach Andersen, Chresten Træholt, Seyedmostafa Hashemi, " Grid-connected battery energy storage system: a review on application and integration," *Renew. Sustain. Energy Rev.*, vol. 182, pp. 113400, Aug., 2023.

[8] F. Zhao *et al.*, "Energy-Storage Enhanced STATCOMs for Wind Power Plants," *IEEE Power Electronics Magazine*, vol. 10, no. 2, pp. 34-39, June 2023.

[9] S. Bhukya, A. Bharadwaj, M. K. T, N. V. P. Babu and S. Maiti, "Performance Evaluation of Dual Active Bridge Converter for Energy Storage Integration With STATCOM," *Proc. of Second International Conference on Power, Control and Computing Technologies (ICPC2T)*, pp. 1-6, 2022.

[10] R. O. d. Sousa, A. F. Cupertino, L. M. F. Morais, H. A. Pereira and R. Teodorescu, "Experimental Validation and Reliability Analyses of Minimum Voltage Control in Modular Multilevel Converter-Based STATCOM," *IEEE Trans. Ind. Electron.*, vol. 71, no. 7, pp. 6546-6555, Jul. 2024.

[11] . Xu, Z. Zhang, G. Wang and Z. Xu, "Modular Multilevel Converter With Embedded Energy Storage for Bidirectional Fault Isolation," *IEEE Trans. Power Del.*, vol. 37, no. 1, pp. 105-115, Feb. 2022.

[12] T. Engelbrecht *et al.*, "STATCOM Technology Evolution for Tomorrow's Grid: E-STATCOM, STATCOM With Supercapacitor-Based Active Power Capability," *IEEE Power Energy M*, vol. 21, no. 2, pp. 30-39, March-April 2023.

[13] C. Smith, A. Gargoom, M. T. Arif and M. E. Haque, "Modelling and Control of SST Based E-STATCOM with Supercapacitor to Enhance Voltage Stability of Distribution Network," *Proc. of International Conference on Energy Technologies for Future Grids (ETFG)*, , pp. 1-6, 2023.

[14] Bharadwaj A, Maiti S., "Control and state of charge balancing algorithm for modular multilevel STATCOM with distributed ultracapacitor-based energy storage system at the DC Link", *Int Trans Electr Energ Syst.*, vol. 30., pp: e12282, 2020.

[15] C. Anil Bharadwaj, S. Maiti and N. Dhal, "Modular Multilevel E-STATCOM Using Supercapacitor Based Energy Storage System," *Proc. of 2nd International Conference on Power, Energy and Environment: Towards Smart Technology (ICEPE)*, pp. 1-6, 2018.

[16] A. Bharadwaj.Ch, S. Maiti and N. Dhal, "Control and State of Charge Balancing Technique of a Modular Multilevel STATCOM Integrated with Battery Energy Storage System," *Proc. of Students Conference on Engineering and Systems (SCES)*, pp. 1-6, 2019.

[17] J. Saha, A. Subramanium and S. K. Panda, "Design of Integrated Medium Frequency Transformer (iMFT) for Dual-Active-Bridge (DAB) Based Solid-State-Transformers," *Proc. of IEEE 12th Energy Conversion Congress & Exposition - Asia (ECCE-Asia)*, pp. 893-898, 2021,

[18] E. S. Lee, J. H. Park, M. Y. Kim and J. S. Lee, "High Efficiency Integrated Transformer Design in DAB Converters for Solid-State Transformers," *IEEE Transactions on Vehicular Technology*, vol. 71, no. 7, pp. 7147-7160, July 2022,

[19] S. Kurm and V. Agarwal, "Interfacing Standalone Loads With Renewable Energy Source and Hybrid Energy Storage System Using a Dual Active Bridge Based Multi-Port Converter," *IEEE Journal of Emerging and Selected Topics in Power Electronics*, vol. 10, no. 4, pp. 4738-4748, Aug. 2022 .

[20] P. F. S. Costa, P. H. B. Löbler, L. Roggia and L. Schuch, "Modeling and Control of DAB Converter Applied to Batteries Charging," *IEEE Transactions on Energy Conversion*, vol. 37, no. 1, pp. 175-184, March 2022.

[21] I. A. de Azevedo and L. S. Barros, "Comparison of Control Strategies for Squirrel-Cage Induction Generator-based Wind Energy Conversion Systems," Proc. ff 14th IEEE International Conference on Industry Applications (INDUSCON), pp. 790-796 2021.

[22] M. S. Islam, I. J. Bushra and S. K. Ghosh, "Design of Data – Driven Linear Quadratic Regulator Controller for DC – DC Boost Converter Optimized by Genetic Algorithm," *2024 3rd International Conference on Advancement in Electrical and Electronic Engineering (ICAEEE)*, Gazipur, Bangladesh, 2024, pp. 1-6.

[23] A. Debnath, T. O. Olowu, S. Roy, I. Parvez and A. Sarwat, "Particle Swarm Optimization-based PID Controller Design for DC-DC Buck Converter," *Proc. of North American Power Symposium (NAPS)*, pp. 1-6, 2021.

Optimal Sizing of Hybrid Energy Storage System in a Microgrid

Amina. S
Department of Electrical Engineering
College of Engineering Trivandrum
APJ Abdul Kalam Technological
University, Kerala, India
aaminas256@gmail.com

Prameeda Mohan
Department of Electrical Engineering
College of Engineering Trivandrum
APJ Abdul Kalam Technological
University, Kerala, India
prameedamohan@gmail.com

Rajeev. T
Department of Electrical Engineering
College of Engineering Trivandrum
APJ Abdul Kalam Technological
University, Kerala, India
rajeev.t@cet.ac.in

Abstract—**A microgrid is an advanced energy system that integrates various generation sources, energy storage solutions, control mechanisms and management strategies. Properly sizing the energy storage system is essential to improve a microgrid's cost-effectiveness, reliability and efficiency. A systematic approach is required to determine the optimal capacities of the hybrid storage system, utilizing storage options with complementary characteristics, particularly in relation to the effective utilization of renewable energy sources. This paper presents a constrained optimization algorithm for sizing a hybrid energy storage system in a renewable energy-integrated microgrid. The mathematical model describes the relationship between them by carefully analyzing the energy requirements of both batteries and supercapacitors. Batteries are advantageous for long-term energy storage, while supercapacitors provide rapid power for short durations. A case study analyzed various generation and storage combinations to obtain optimized outputs using the fmincon optimization tool in MATLAB, which were compared to results obtained from genetic algorithm (GA). The findings provide valuable insights for designing energy storage systems (ESS), like hybrid energy storage systems (HESS), to enhance performance and economic feasibility in future microgrids.**

Index Terms—**BESS, SoC, HESS, ESS, DER, GA**

I. INTRODUCTION

Microgrids have become a viable option for integrating renewable energy sources, increasing energy resilience and improving efficiency in contemporary power networks. Energy storage systems (ESS) are an essential part of microgrid operation; they are responsible for regulating the intermittent nature of renewable energy sources, balancing supply and demand along with maintenance of grid stability [1-4]. The sizing of energy storage used in hybrid microgrids, which combine several energy generation sources like solar, wind, diesel generators and may be other distributed energy resources (DERs), presents a different optimization problem.Since these microgrids are hybrid, they require the use of several energy storage technologies such as flywheels, batteries, supercapacitors, pumped hydro storage etc [5-6].

HESS plays a vital role in stabilizing power fluctuations caused by intermittent renewable energy sources like solar and

wind. Proper sizing of the storage components helps balance long-term energy storage with batteries and short-term high-power demands with supercapacitors. However, challenges such as power fluctuations, voltage and frequency regulation and economic trade-offs between investment and operational costs must be addressed. Optimization techniques, including mathematical models, heuristic algorithms like genetic algorithm and particle swarm optimization (PSO) and machine learning-based predictive models, are used to determine the ideal storage capacities. An optimally sized HESS enhances cost-effectiveness by reducing energy losses and unnecessary storage oversizing while improving system stability and extending battery lifespan by reducing deep discharge cycle.

A local energy system called a microgrid replaces central power plants with distributed and renewable energy sources, as depicted in [7]. The energy systems use the clean, affordable, and efficient energy that the microgrids supply. This study offers a techno-sustainable- economic framework for multi-microgrid operation scheduling that incorporates energy storage systems, dispatchable resources, and renewable energy sources. By coordinating each microgrid with other nearby energy systems, the suggested model offers a multi objective optimization approach that aims to concurrently minimize the overall cost, power losses, and carbon emissions. Within the cooperative framework, neighbouring microgrids engage in energy trading to lower operating costs, and microgrids as a group constitute a large coalition. Microgrids also aim to reduce power losses. In [8], a collaborative architecture for neighbouring microgrid operation management is employed. Within the suggested framework, nearby microgrids collaborate to pool their local resources and reduce overall expenses, power outages, and emissions of pollutants.

Finding the best way to manage energy storage systems is vital for lowering costs, optimizing renewable energy use, and ensuring reliable power for essential loads. The authors [9-10] strive to balance cost minimization, renewable energy integration, and system reliability while addressing operational constraints and uncertainties.When compared to other optimization techniques like linear programming, integer programming, and heuristic or metaheuristic approaches like PSO, simulated annealing (SA), and GA, nonlinear optimization stands

---

979-8-3315-3013-6/25 $31.00 © 2025 IEEE

out. Nonlinear optimization is perfect for real-world situations like power systems, fluid dynamics, and economic modeling because it can simulate complicated systems where variables interact in nonlinear ways, unlike linear programming, which is restricted to problems with linear objective functions and constraints. Nonlinear optimization works best with continuous variables and smooth solution spaces, enabling more accurate and effective optimization than integer programming, which works with discrete variables and is frequently used for scheduling and allocation problems.

Solar and wind energy are common intermittent renewable energy sources used in hybrid microgrids. The smooth integration of renewable energy sources is made possible by the efficient utilization of energy storage, which reduces curtailment and increases grid stability by storing excess energy during high-generation periods and releasing it during low-generation or high-demand periods.By moderating fluctuations in renewable energy generation and providing backup power during grid disruptions, energy storage technologies improve the resilience and dependability of microgrids. Energy storage devices are efficiently placed to support vital loads, preserve voltage stability and prevent blackouts through optimal scheduling [11]

A suitable learning-based optimal power flow (OPF) model is used that leverages the flexibility of thermostatically coupled loads to mitigate power imbalances resulting from fluctuating renewable energy sources. This research focuses on determining the optimal sizing of a stand-alone hybrid microgrid comprising photovoltaic, wind, diesel and battery storage systems. The authors employ various optimization algorithms to minimize the cost of energy while enhancing system reliability and efficiency, as indicated by the loss of power supply probability. The study provides insights into the effectiveness of different optimization techniques in achieving cost-effective and reliable hybrid microgrid configurations [12-13].

In [14], an improved methodology for optimally sizing autonomous residential smart power systems is presented, integrating renewable energy sources like solar and wind with energy storage solutions. This approach considers the flexible and controllable nature of modern power consumption patterns to design cost-effective and reliable systems. The methodology aims to efficiently balance energy supply and demand in future smart residential environments.

Energy storage devices play a key role in managing peak demand by storing excess energy during off-peak hours and discharging it during peak periods. This practice helps alleviate pressure on infrastructure and reduces the need for costly capacity expansions. By optimizing the usage of energy storage resources, we can reduce peak load situations and enhance grid flexibility, ensuring that these resources are utilized effectively. By maximizing resource utilization, reducing system losses, and simplifying energy management processes, optimal scheduling algorithms enhance the operational efficiency of hybrid microgrids. Research and development in optimal scheduling for hybrid energy storage systems are driving advancements in energy storage technologies, control

algorithms, and grid integration solutions.[15].

In [16], the study focuses on designing microgrids powered entirely by renewable energy sources, utilizing second-life batteries for energy storage. The study evaluates the techno-economic aspects of such microgrids, assessing their feasibility, reliability and cost-effectiveness.The study explores sustainable solutions for energy storage in renewable energy systems. Research and development in optimal sizing for hybrid energy storage systems are driving advancements in energy storage technologies, control algorithms and grid integration solutions.The research emphasizes the necessity for governmental support to facilitate the transition to clean energy, suggesting that financial incentives and supportive policies are crucial for widespread adoption of 100% renewable energy microgrids.

II. SYSTEM ARCHITECTURE

In order to provide a consistent, effective, and sustainable power supply, a hybrid microgrid is an advanced energy infrastructure that combines several generation sources, energy storage systems, control mechanisms and management methodologies. The microgrid includes various generation sources featuring renewable energy technologies such as wind turbines, solar photovoltaic panels and possibly hydropower generators. These resources are used to provide sustainable and clean energy. The microgrid is equipped with conventional generating sources, like natural gas turbines or diesel generators, to supply backup power when renewable energy is scarce or in the event of a grid failure. Renewable energy sources and energy storage will be crucial to smart grid advancements in the future.The optimisation problem in this scenario is non-linear and non-convex and hence it cannot be solved by standard optimization techniques like linear programming. On the other hand, great focus have to be done when simulating these parameters due to their ambiguity. To address the non-linearity and non-convexity of the objective function, suitable optimization procedure is suggested.The ideal scheduling and operation of ESS, which is a key component of microgrid applications, have been the subject of numerous research.

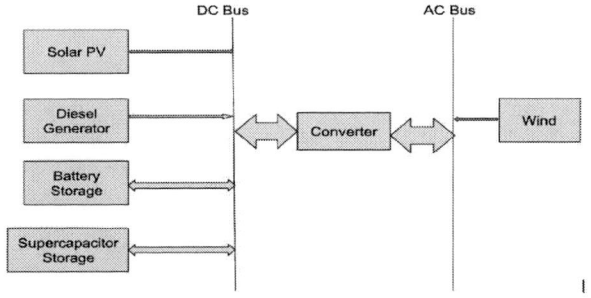

Fig. 1. Block diagram of system under study

A. Hybrid energy storage system incorporating battery and supercapacitor

In a dynamic urban environment where energy demands fluctuate with daily activities, a suitable option is HESS, which effectively combines the rapid response of supercapacitors with the reliability of batteries.

Contained in sturdy and elegant casings, these batteries store excess energy that is harvested from renewable sources such as wind turbines or solar panels during times of abundance. The supercapacitors are the pinnacle of energy storage speed and responsiveness. These gadgets are small and agile and they have an amazing capacity to charge and discharge energy quickly and with unmatched efficiency. However, real-world batteries and supercapacitors degrade over time due to charging/discharging cycles, temperature variations and aging effects. If degradation is ignored, the optimization may result in a smaller-sized battery and supercapacitor.

III. FORMULATION OF OBJECTIVE FUNCTION AND CONSTRAINTS

Essential elements of the optimization issue include the objective function, restrictions and battery energy storage sizing optimization for grid-connected microgrids. The most effective and economical arrangement is chosen using these mathematical formulas. The optimal solution requires the maximization or minimization of decision variables. In order to reduce cost functions objective function is to be minimized. Lowering expenses and taking operational limits into account is one typical goal. Renewable energy sources such as solar photovoltaic and wind power are key contributors to sustainable energy systems. However, their intermittent nature requires effective energy storage solutions to ensure stable power supply in hybrid microgrids. Essential elements of the optimization issue include the objective function, restrictions, and battery energy storage sizing optimization for grid-connected microgrids. The most effective and economical arrangement for is chosen using these mathematical formulas. The optimal solution requires the maximization or minimization of decision variables. In order to reduce cost functions objective function is to be minimized. Lowering expenses and taking operational limits into account is one typical goal. Objective function is defined as

$$J = \sum_{t=1}^{N} P_{\text{grid}}(t)C_{\text{grid}}(t) + P_{\text{storage}}(t)C_{\text{storage}}(t). \quad (1)$$

where, $P_{\text{grid}}(t)$ is the power imported from grid at time t, $P_{\text{storage}}(t)$ is the power charged or discharged by energy storage system at time t, $C_{\text{grid}}(t)$ is the cost of electricity at time t , $C_{\text{storage}}(t)$ is the cost of charging or discharging the energy storage system at time t . It subjects to the power generation and consumption decisions for each entity i at each time t.This objective function is suitable for optimization algorithms or solvers that can find the values to minimize the total energy cost for the given entities and time periods subject to certain constraints.

The constraints for the ESS involve the state of charge (SoC) and are formulated as follows:

$$0 \leq SoC_{ESS}^i \leq (SoC_{ESS}^i)^{max} \quad (2)$$

where SoC_{ESS}^i is the state of charge of ESS in interval i $(SoC_{ESS}^i)^{max}$ is the maximum state of charge of ESS in interval i The constraint for the power used to charge the *ESS* is formulated as follows:

$$0 \leq P_{ess(C),t}^i \leq \left(P_{ess(C),t}^i\right)^{max} \quad (3)$$

where $P_{ess(C),t}^i$ is the power used to charge *ESS* in interval i $(P_{ess(C),t}^i)^{max}$ is the maximum power used to charge *ESS* in interval i This constraint ensures that the power used to charge the *ESS* at any given interval is within the permissible range, defined by its maximum allowable charging power. The constraint for the power used to discharge the *ESS* is formulated as follows:

$$0 \leq P_{ess(D),t}^i \leq \left(P_{ess(D),t}^i\right)^{max} \quad (4)$$

This constraint ensures that the power used to discharge the *ESS* at any given interval is within the permissible range, defined by its maximum allowable discharging power.

$$SoC(t+1) = SoC(t) + \Delta SoC \quad (5)$$

where $SoC(t+1)$ is the state of charge of the battery in the next time step. This formula shows how the battery's state of charge changes over time while accounting for the PV system's energy production and the load's energy consumption. It's crucial to remember that this is a simplified illustration and ignores other system limitations that might apply in a real-world situation, such as battery efficiency, losses, charging/discharging rates, and other issues.

$$E_{BESS}^{min} \leq E_{BESS} \leq E_{BESS}^{max} \quad (6)$$

The limitations and restrictions placed on an energy storage system inside a wind generation system are represented mathematically by these equations. They support grid integration and operational requirements while providing guidance for the safe, effective, and dependable operation and management of the ESS.Grid rules and standards must be followed by ESS in order to guarantee a smooth integration with the electrical grid. Requirements for voltage support, frequency regulation, and disturbance response are examples of grid integration limitations. Grid services must be provided by ESS while upholding the dependability and stability of the system.The number of charge-discharge cycles that ESS can withstand before degrading and performing less well is restricted.

$$P_{charge,min} \leq P_{storage}(t) \leq P_{charge,max} \quad (7)$$

where $P_{storage}(t)$ is the power charged or discharged by energy storage system at time t. $P_{charge}(t)$ is the power charged by energy storage system at time t.

Power limits play a critical role in the integration of generation into the broader electricity grid. It must comply with grid codes and standards that specify requirements for power quality, voltage regulation, and frequency control. By adhering to power limits and implementing advanced grid integration technologies, wind generation systems can contribute to the stability and reliability of the grid while maximizing the utilization of renewable energy resources. Large variations in power generation and load demand can cause voltage instability. The model should ensure that storage units respond quickly to maintain voltage levels within safe limits.

The energy balance constraints are:

$$E(t+1) = E(t) + P_{storage}(t).\Delta t \tag{8}$$

where Δt is time step $E(t)$ is the SoC of energy storage at time t. $P_{storage}(t)$ is the power charged or discharged by energy storage system at time t.

$$E_{min} \leq E(t) \leq E_{max} \tag{9}$$

Limits on the SoC are essential restrictions that affect how BESS in power systems operate. SoC describes how much electrical energy a battery can hold in relation to its maximum capacity. For the battery system to operate effectively and safely, the right SoC limits must be set. Extreme SoC levels—either too high or too low—can shorten the life and performance of batteries.

The objective of this optimization algorithm is to obtain the optimal capacity of the two storage components namely battery and supercapacitor along with minimization of overall cost while satisfying power balance constraints.

IV. OPTIMIZATION USING NON-LINEAR OPTIMIZATION

This work emphasizes the use of fmincon toolbox in MATLAB which is intended to minimize a scalar function of many variables while taking into account nonlinear inequality and equality constraints. The optimization process is continued until a convergence condition is satisfied, which means the algorithm has found a workable solution. Next, the completed solution is assessed to make sure it satisfies all constraints and achieves the optimization goals. Insights into the behavior of the optimization problem and assistance in making defensible decisions based on the optimized solution can be obtained through sensitivity analysis, parameter estimation, and scenario analysis.

V. SYSTEM DESCRIPTION

The hybrid microgrid considered comprises of the following:
- Load Demand , $L = 350$ kW (constant hourly demand)
- Solar Generation: $S = 0.6 \times L$ kW
- Wind Generation: $W = 0.4 \times L$ kW
  where L is the load demand.
- Available battery capacity : x_1 kWh(varies between 10 and 500 kWh)
- Available supercapacitor capacity : x_2 kWh(varies between 1 and 50 kWh)

VI. ALGORITHM FOR OPTIMAL SIZING OF HYBRID ENERGY STORAGE SYSTEM

By combining different energy sources and storage technologies, an algorithm for the ideal sizing of hybrid energy storage systems seeks to effectively balance energy supply and demand.The steps included in this optimization are listed below:

1) **Initialize the system parameters** using values in Section V.
2) **Compute fixed generation cost**

$$\text{Total Solar cost} = C_s \times S \tag{10}$$

$$\text{Total wind cost} = C_w \times W \tag{11}$$

$$\text{Generation cost} = \text{Total solar cost} + \text{Total wind cost} \tag{12}$$

where C_s (Solar cost per kW) = 1000 USD and C_w (Wind cost per kW) = 1200 USD.
3) **Define objective function**

$$\text{Minimize } f(x) = x_1 C_b + x_2 C_{sc} + \text{Total Generation Cost} \tag{13}$$

where C_b (Battery cost per kWh) = 200 USD and C_{sc} (Supercapacitor cost per kWh) = 500 USD.
4) **Define power balance constraints and storage constraints** as in Section III.
5) **For each hour** $t = 1$ to 24, compute power deficit:

$$D_t = L - (S + W) \tag{14}$$

where D_t is the power deficit at time t in kW.
6) **Energy storage operation**
   - If $D_t > 0$ (power deficit), discharge storage:
     - If $SOC_{sc} \geq D_t$, then

     $$SOC_{sc} = SOC_{sc} - D_t$$

     - Else, discharge battery for remaining deficit.
   - Else if $D_t < 0$ (excess power), charge storage:

   $$SOC_b = \min(SOC_b - D_t, x_1) \tag{15}$$

   $$SOC_{sc} = \min(SOC_{sc} - D_t, x_2) \tag{16}$$

   where SOC_b is the state of charge of the battery and SOC_{sc} is the state of charge of the supercapacitor.
   - **Storage constraints:**

   $$0 \leq SOC_b \leq x_1, \quad 0 \leq SOC_{sc} \leq x_2 \tag{17}$$

7) **Solve the optimization problem** using fmincon to obtain optimal values of x_1, x_2, and the optimal cost.
8) **Print results**

979-8-3315-3013-6/25 $31.00 © 2025 IEEE

VII. CASE STUDY

A constrained optimization was performed in MATLAB platform using the fmincon function to obtain the optimal capacity of the hybrid storage system namely battery and supercapacitor. The optimisation was done by clearly taking into account the non-linear and linear equality constraints in the above system. The following scenarios were considered for optimisation and corresponding results were obtained. The first three scenarios were conducted to study the impact of storage on a system excluding conventional generation. Corresponding results were obtained under constant demand conditions. Solar and wind vary within the upper and lower limits described in the algorithm above.

In the first case, the algorithm simply checks whether generation meets demand. In the second case, only the SoC limit applies, and the system is considered infeasible if generation plus battery output is less than the load. In the third case, power balance is ensured every hour. It maintains the SoC for both the battery and the supercapacitor. In this fourth case, by adding the diesel generator, power balance constraints and diesel capacity limits come into effect. In the fifth case, solar and wind produce variable renewable energy, while the battery and supercapacitor store and release energy to balance supply and demand in a system that includes a diesel generator, solar, wind, and battery. Quick power variations are handled by the supercapacitor, while longer-term energy shifts are managed by the battery. When storage and renewable energy sources are unable to meet the load, the diesel generator, within its output limit, serves as backup power. By reducing diesel consumption, optimizing costs, and ensuring demand is always met, these components work together to provide a reliable power supply.

The study was extended to the total system having both renewable as well as conventional generation. The results obtained were as follows:

Case 1: Optimization without storage
By optimizing the system containing only solar and wind without the inclusion of battery and supercapacitor using the fmincon function by nonlinear optimization, the optimized cost was obtained as $378000.

Case 2: Optimization with battery alone as storage
By including only a battery as a storage system along with solar and wind as generation, the optimal battery capacity is obtained as 10 kWh, and the total optimized cost is obtained as $380000.

Case 3: Optimization including both battery and supercapacitor
When a hybrid energy storage system consisting of a battery and a supercapacitor is implemented along with solar and wind generation, the optimal battery capacity is obtained as 10 kWh, and the optimal supercapacitor capacity is obtained as 1 kWh. The minimum total cost of this system is $380500.

Case 4: Optimization of renewable integrated grid
In this case, the optimization was done by adding a diesel generator working at a base power of 100 kW along with solar and wind generation. The optimal cost is obtained as $448480.

Case 5: Optimization of renewable integrated grid along with HESS
Case 4 is modified by adding a battery and a supercapacitor as a hybrid energy storage system (HESS). The obtained optimal results show that the optimal battery capacity is 10 kWh, the optimal supercapacitor capacity is 1 kWh, and the minimum total cost is obtained as $450980.

The conventional source provides a stable baseline power (100 kW in this case), ensuring that the load demand is always met, even when renewable generation is insufficient. The base power reduces the need for excessive battery storage, optimizing the balance between conventional and renewable sources. The significant insights from the results are:

1) Optimized operation of a microgrid is possible even with allocation of minimal capacity of HESS

2) Also the results obtained clearly depict that even a small addition of generation from conventional energy source to meet existing demand increases the optimal cost of system by a larger margin compared to an equivalent addition of storage.

VIII. COMPARISON USING A CONVENTIONAL TOOLBOX

Optimised results obtained from the five cases were compared with similar case and corresponding results were obtained by using GA with population size 50, maximum generations 300, cross over fraction 0.8. The obtained results were compared with fmincon results in the table below.

TABLE I
COMPARISON OF FMINCON AND GA OPTIMIZATION COST RESULTS WITH BATTERY AND SUPERCAPACITOR SIZING.

Case	fmincon			GA		
	Cost	Size (kWh)		Cost	Size (kWh)	
		Bat	SC		Bat	SC
Case 1	$378000	NA	NA	$378000	NA	NA
Case 2	$380000	10	NA	$380000	19.5	NA
Case 3	$380500	10	1	$383000.23	20	2.0265
Case 4	$448480	NA	NA	$453257	NA	NA
Case 5	$450980	10	1	$451751	18	2.012

where 'Bat', 'SC' indicates battery and supercapacitor respectively

The results obtained from the case study conducted points towards the significance of addition of an optimal capacity of storage with the existing grid system. A minimal addition of storage capacity alongwith conventional generation will reduce the increment in the overall cost of generation of the system. The comparative analysis done later on with the conventional toolbox shows the upperhand of using fmincon rather than GA as the optimal cost computed using GA was on the higher side with the same system.

IX. CONCLUSION

This paper presents a systematic approach to determine the optimal capacity of a HESS for a renewable energy-integrated microgrid. The proposed approach effectively allocates generation and storage resources to meet load demand by leveraging a constrained optimisation framework while considering renewable energy dynamics. The comparative analysis of different optimization techniques, including fmincon and GA, provides valuable insights into the effectiveness of hybrid storage configurations in enhancing system performance.

The findings highlight that integrating batteries and supercapacitors allows for an efficient balance between long-term energy storage and short-term power requirements, ensuring cost-effective and reliable microgrid operation. The total system cost, including conventional generation, charging and discharging expenses, is minimized through optimal storage sizing, leading to economic efficiency and improved resource utilization. The feasibility of the proposed solution is validated by meeting power balance constraints, storage limits, and operational requirements.

Ultimately, the proposed optimization underscores the significance of incorporating HESS into microgrids, demonstrating that even a minimal storage capacity can result in substantial cost savings and enhanced energy utilisation. The optimized hybrid storage configuration enables better integration of renewable sources, making microgrids more sustainable and economically viable for future energy systems. The feasibility of the solution is ensured by satisfying power balance constraints, energy storage limits and operational constraints on generator output, charging and discharging.

REFERENCES

[1] Amin, R. T. Bambang, A. S. Rohman, C. J. Dronkers, R. Ortega and A. Sasongko, "Energy Management of Fuel Cell/Battery/Supercapacitor Hybrid Power Sources Using Model Predictive Control," in IEEE Transactions on Industrial Informatics, vol. 10, no. 4, pp. 1992-2002, Nov 2014.

[2] C.Zugschwert,S Gschi,et al.,"Development of Multi timescale method for classifying hybrid energy storage systems in grid applications",IEEE,2022

[3] H. Takano, K. Harada, W. M. Nyabuto, H. Asano, S. Kambara and N. -D. Tuyen, "Optimal Sizing of Energy Storage Systems Considering Their Economical Operations in a Microgrid," 2024 International Technical Conference on Circuits/Systems, Computers, and Communications (ITC-CSCC), Okinawa, Japan, 2024, pp. 1-6, doi: 10.1109/ITC-CSCC62988.2024.10628438.

[4] K. V. Konneh et al., "Optimal Design and Performance Analysis of a Hybrid Off-Grid Renewable Power System Considering Different Component Scheduling, PV Modules, and Solar Tracking Systems," in IEEE Access, vol. 9, pp. 64393-64413, 2021, doi:10.1109/ACCESS.2021.3075732.

[5] H. Alharbi and K. Bhattacharya, "Optimal Sizing of Battery Energy Storage Systems for Microgrids," 2014 IEEE Electrical Power and Energy Conference, Calgary, AB, Canada, 2014, pp. 275-280, doi: 10.1109/EPEC.2014.44.

[6] Li, Shihao, Libao Shi, and Zhuxiang Yao. "Multi-objective optimal scheduling of microgrid considering distributed generation uncertainty." (2021): 35-43.

[7] F. R. Albogamy et al., "Real-Time Scheduling for Optimal Energy Optimization in Smart Grid Integrated With Renewable Energy Sources," in IEEE Access, vol. 10, pp. 35498-35520, 2022, doi:10.1109/ACCESS.2022.3161845.

[8] U. R. Nair, M. Sandelic, A. Sangwongwanich, T. Dragičević, R. Costa-Castelló and F. Blaabjerg, "An Analysis of Multi Objective Energy Scheduling in PV-BESS System Under Prediction Uncertainty," in IEEE Transactions on Energy Conversion, vol. 36, no. 3, pp. 2276-2286, Sept. 2021, doi: 10.1109/TEC.2021.3055453.

[9] Nasir, Tehreem, et al. "Optimal scheduling of campus microgrid considering the electric vehicle integration in smart grid." Sensors 21.21 (2021): 7133.

[10] Rehman, Ateeq Ur, et al. "An optimal power usage scheduling in smart grid integrated with renewable energy sources for energy management." IEEE Access 9 (2021): 84619-84638.

[11] F. Qayyum, F. Jamil, S. Ahmad, and D.-H. Kim, "Hybrid renewable energy resources management for optimal energy operation in nano-grid,"Comput., Mater. Continua, vol. 71, no. 2, pp. 2091–2105, 2022.

[12] Chen, Ge, et al. "Scheduling thermostatically controlled loads to provide regulation capacity based on a learning-based optimal power flow model." IEEE Transactions on Sustainable Energy 12.4 (2021): 2459-2470.

[13] A. A. Z. Diab, H. M. Sultan, I. S. Mohamed, O. N. Kuznetsov and T. D. Do, "Application of Different Optimization Algorithms for Optimal Sizing of PV/Wind/Diesel/Battery Storage Stand-Alone Hybrid Microgrid," in IEEE Access, vol. 7, pp. 119223-119245, 2019.

[14] A. A. Z. Diab, H. M. Sultan, I. S. Mohamed, O. N. Kuznetsov and T. D. Do, "Application of Different Optimization Algorithms for Optimal Sizing of PV/Wind/Diesel/Battery Storage Stand-Alone Hybrid Microgrid," in IEEE Access, vol. 7, pp. 119223-119245, 2019.

[15] U. Akram, M. Khalid and S. Shafiq, "An Improved Optimal Sizing Methodology for Future Autonomous Residential Smart Power Systems," in IEEE Access, vol. 6, pp. 5986-6000, 2018

[16] A. Demirci, "Optimal Sizing and Techno-Economic Evaluation of Microgrids Based on 100% Renewable Energy Powered by Second-Life Battery," in IEEE Access, vol. 11, pp. 113291-113306, 2023

[17] M. S. Hossain, M. Iftekhar Bin Ashraf and S. Ahmed, "A Mixed Integer Linear Programming based optimal operational and sizing approach for a Hybrid Local Microgrid," 2024 IEEE International Conference on Computing, Applications and Systems (COMPAS), Cox's Bazar, Bangladesh, 2024, pp. 1-6, doi: 10.1109/COMPAS60761.2024.10796543.

[18] J. Wongyai, T. Tayjasanant and T. Masuta, "Optimal Scheduling and Sizing of Battery Energy Storage System in an Industrial Microgrid," 2024 21st International Conference on Harmonics and Quality of Power (ICHQP), Chengdu, China, 2024, pp. 132-136, doi: 10.1109/ICHQP61174.2024.10768719.

Multi-Load Pre-Regulated Inverter Configuration for All utensil induction cooking application

1st Kirlampalli Harija Rani
Department of Electrical Engineering
National Institute of Technology, Warangal
Telangana, India
email address : kh21eerer11@student.nitw.ac.in

2nd Neti Vishwanathan
Department of Electrical Engineering
National Institute of Technology, Warangal
Telangana, India
email address: nvn@nitw.ac.in

Abstract—**Induction cooking systems are more popular due to their features like high efficiency, precise temperature control, and quick heating. At Present, Multi-burner systems are gaining popularity as they offer greater cooking flexibility and cooking options. This paper presents a multi-load pre-regulated inverter configuration, which is a combination of a buck-boost converter and half- bridge inverter. This configuration can power both ferro and non- ferromagnetic materials independently using pulse frequency modulation (PFM) and asymmetric duty cycle control (ADC) respectively. It can be operated in full-bridge mode (i.e., $\pm V_{DC}$) with a half-bridge inverter by deriving other V_{DC} using a buck-boost converter. The proposed inverter is simulated using MATLAB simulink tool for 4.5 kW output power and observed peak efficiency is 94%.**

Index Terms—**induction cooking, multi burner system, hal-bridge inverter, ADC, PFM, Full-Bridge inverter.**

I. INTRODUCTION

Induction heating is an energy-efficient, surface-free heating technique and it is significantly used in domestic, industrial, and medical applications. Unlike other heating systems, induction heating provides advantages like fast heating, high conversion efficiency, easy maintenance, and safety. It operates on the electromagnetic induction principle [1]. In traditional heating methods, heat is transferred to the load either through conduction or radiation. But, in IH systems, heat is produced within the load as eddy currents are induced at the skin depth below the surface [2]. An IH system consists of a coil, an inverter that generates high-frequency eddy currents, and a load. As eddy currents are frequency dependent, a high frequency ranging from a few kHz to several MHz is required to generate adequate heat in the load (i.e., cookware). Hence, high frequency inverter plays a key role in induction heating system design. Resonant inverters are the most preferred choice for IH systems due to their soft-switching operation, which gives higher efficiency, though the operating frequency is high. Various topologies existing in the literature are half- bridge [3], full-bridge [4], and single-switch [5] topologies. Effective load power management and precise temperature control depend greatly on the control technique utilized. Pulse frequency modulation (PFM) [6], phase-shift control [7], asymmetric duty cycle (ADC) control [8] and asymmetric voltage cancellation

©

(AVC) control [9] are a few control strategies used in IH converters.

In the current scenario, multi-burner induction cooktops are becoming more in demand due to their versatility in modern kitchens. Multi-port power inverters can power multiple loads with effective energy distribution and reduced design complexity. The available literature on multi-port inverters for IC applications is summarized here.

A full-bridge inverter with a multi-layer coil and switched capacitor is proposed in [10] for all metal applications. Coil structure is more complex due to two coils used on both sides of the ferrite bars. A double layer coil with compensated capacitors is presented in [11]. An inverter configuration with first and third harmonic operation is proposed in [12] to power both FM and NFM loads. However, the efficiency is low due to higher currents drawn by the NFM load. In [13], a load-adaptive modulation technique is proposed to power both FM and NFM vessels. But mechanical switches are used additionally for capacitance switching. An on-off control based inverter configuration is proposed in [14]. But, the component count is more. A cascaded full- bridge inverter configuration with variable frequency operation for different materials is presented in [15]. Still component count is more. A two- dc sources based three-switch dual frequency inverter configuration is proposed in [16] to power aluminum & steel loads. But, Variable frequency operation affects soft-switching operation. A quasi-resonant converter with dual-output is proposed in [17]. But single switch operation affects power handling capability and thermal management is difficult.

Limitations observed from the available literature are more component count, design complexity, enhanced cost, limited power capacity, and challenges in managing thermal heat. The proposed inverter topology can power multiple loads independently with reduced component count, simple control and higher efficiency.

The following are the advantages of the proposed topology:

- It can power two loads of different material independently using four switches
- Each load can be operated in full-bridge mode
- In this configuration, each load receives half of the power from the source and the remaining half through the buck-boost converter

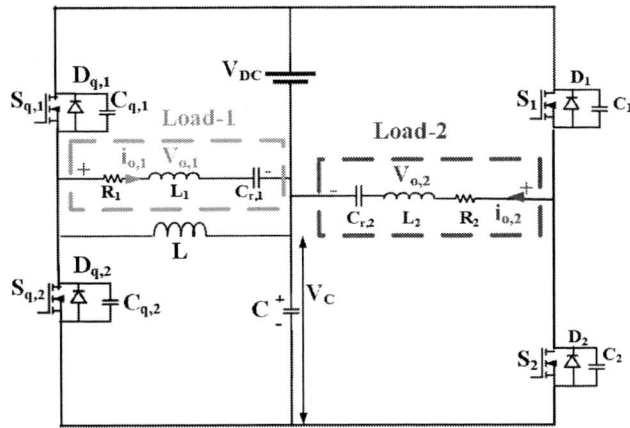

Fig. 1. Proposed pre- regulated inverter.

- Cost-effective
- Observed peak Efficiency is 94%

The paper is formulated as follows. Description and operating modes of the proposed inverter are discussed in Sec II. Control and analysis of the inverter are presented in Sec III. Simulation results of the inverter are discussed in Sec IV. Finally, the conclusions are discussed in Sec V .

II. PROPOSED INVERTER

A pre-regulated inverter configuration with two loads is shown in Fig. 1. This configuration is a combination of a buck-boost converter and a half-bridge inverter. In this configuration, half- bridge inverter receives power from the buck-boost converter during some stages of operating modes. Hence named as "pre-regulated inverter ". Load-1 and Load-2 powers are regulated with PFM and ADC control, respectively. Ferromagnetic material (i.e., iron) is used as Load-1 and nonferromagnetic material (i.e., Aluminium) is used as Load-2. Switches S_{q1} and S_{q2} are used in buck-boost converter design to power Load-1. Similarly, S_1 and S_2 are used in a half-bridge inverter circuit to power Load-2. In a circuit diagram, each load is represented with the series combination of R and L. A capacitor is added in series with each load to achieve resonance operation. R_1, L_1 and C_{r1} represents Load-1 (i.e., iron) similarly R_2, L_2 and $C_{r,2}$ represents Load-2 (i.e., aluminium) . The direction of the current and the polarity of the voltage for each load are shown in the circuit diagram. In this configuration, two loads can be operated in full-bridge mode. Each load receives half of the power from the source and the remaining half through the buck-boost converter. The proposed configuration can be extended to 'n' no.of loads by increasing two switch legs connected to the half-bridge inverter. The entire operation of the inverter is described in four modes with negligible dead-time transitions. The proposed inverter modes of operation are shown with circuit diagrams in Fig. 3.

Mode 1 : In this mode, S_2 and S_{q1} are ON . Source V_{DC} powers Load-1. Buck-boost converter output voltage V_C (i.e.,

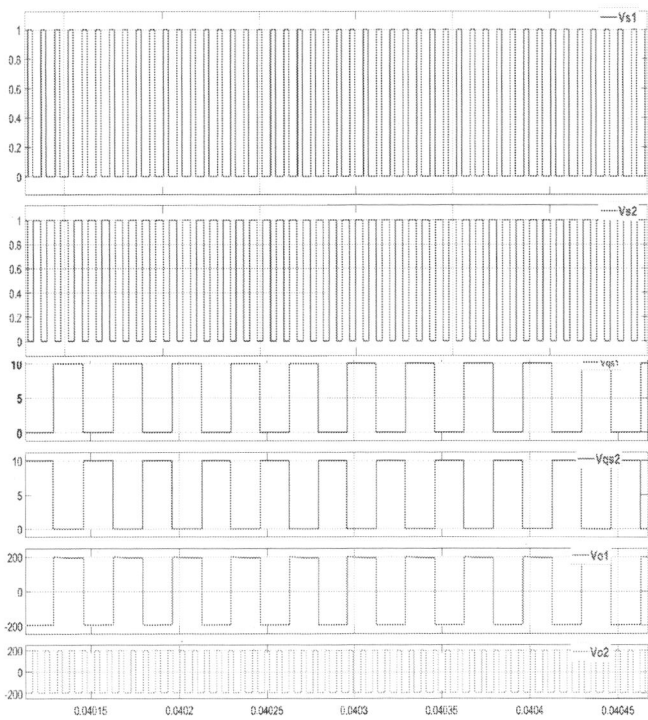

Fig. 2. Operating waveforms of proposed inverter.

V_{DC}) appears across Load-2 as shown in Fig. 3(a). Load voltages are $V_{o,1} = +V_{DC}$, $V_{o,2} = -V_{DC}$.

Mode 2 : In this mode, S_1 and S_{q1} are ON & S_2 is OFF (Fig. 3(b)). Source V_{DC} powers Load-1 and Load-2. At the end of the mode , $V_{o,1}= V_{o,2} = +V_{DC}$.

Mode 3 : In this mode , S_1 and S_{q2} are ON . Source V_{DC} powers Load-2 through S_1, and the DC capacitor supplies power to Load-1 through S_{q2} as shown in Fig. 3(c). load voltages are , $V_{o,1}= -V_{DC}$, $V_{o,2} = +V_{DC}$.

Mode 4 : In this mode, S_1 turn-off and S_2 starts conduction along with S_{q2}. Hence Load-1 power maintains at the same level and Load-2 power changes to -V_{DC} (see Fig. 2 and Fig. 3(d)). Load voltages are $V_{o,1} = V_{o,2} = -V_{DC}$.

III. POWER CONTROL & ANALYSIS OF PROPOSED INVERTER

A. Power Control

In proposed configuration, S_1, and S_2 switches are operated at 145 kHz. The switches used in buck-boost converter S_{q1}, S_{q2} are operated between 30 kHz - 40 kHz frequency range. Fig. 2 shows the pulse sequence and output waveforms. Power control of each load is discussed in detail below.

1) Control of Load-1 using PFM: PFM regulates load power by adjusting the switching frequency while keeping the duty cycle constant. As a result, the output voltage maintains a square waveform. This control strategy improves the efficiency, particularly under light load conditions. Vessel

979-8-3315-3013-6/25 $31.00 © 2025 IEEE

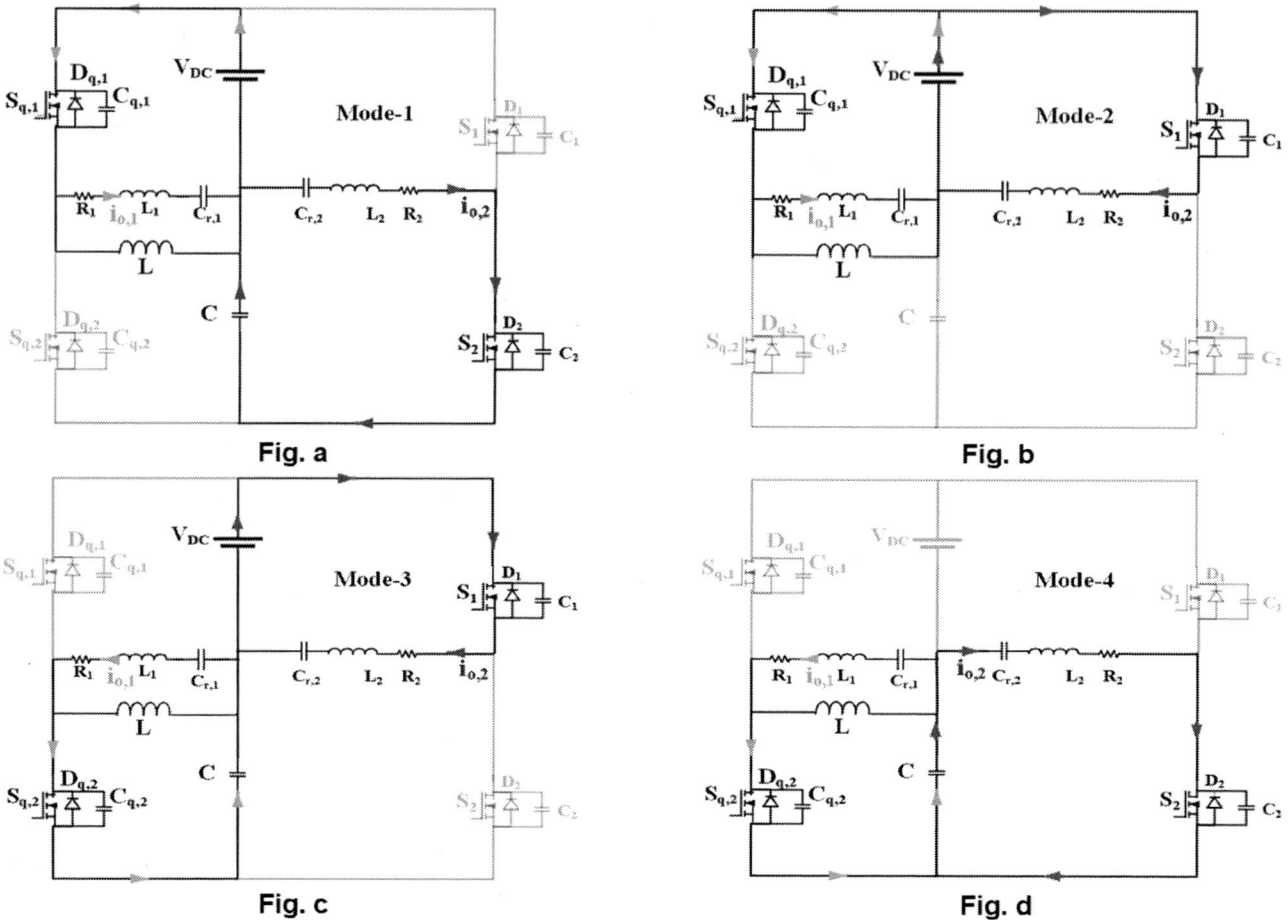

Fig. 3. Operating modes of the proposed inverter.

resistance increases with increased switching frequency, which in turn affects the load power. Selected operating frequency range for iron vessel is 30 kHz- 40 kHz. Duty cycle of S_{q1} and S_{q2} maintained at 0.5 to get V_{DC} as output voltage. So that the two loads can be operated in full-bridge mode.

2) Control of Load-2 using ADC : Asymmetric duty cycle control is used to regulate the power of Load-2 (i.e., aluminium). Selected operating frequency of switches S_1 and S_2 is 145 kHz with resonant frequency of 136 kHz. Turn-on time of S_1 is varied to regulate the power of Load-2. The expression for duty-cycle of S_1 is given below.

$$d_1 = \frac{T - t_{on,S1}}{T_{hf}/2} \tag{1}$$

Here $t_{on,s1}$ represents the turn-on time of switch S_1.

B. Analysis of inverter

1) Analysis of Load-1: Fourier analysis of Load-1 is detailed below. The fundamental component of Load-1 voltage

i.e. V_1 is

$$v_{AB1} = \frac{4V_{DC}}{\pi} \cos\left(\omega t - \pi\left(\frac{T_{LF\_OFF}}{T_{LF}}\right)\right) \tag{2}$$

Here T_{LF_OFF} and T_{LF} are the off period and total period of the LF pulse respectively. The fundamental rms component of the current is given by

$$I_{1,\text{rms}} = \frac{V_{DC,\text{rms}}}{|Z_1|} = \frac{2\sqrt{2}V_{DC}}{\pi|Z_1|} \tag{3}$$

Here Z_1 is the impedance of Load-1. The average power delivered to iron vessel is

$$P_1 = I_1^2 * R_1 * d = \frac{8V_{DC}^2}{\pi^2 Z_1^2} * R_1 * d \tag{4}$$

Here R_1 is the resistance of iron vessel , $w_{r,1}$ is resonant frequency in rad/sec , $L_{r,1}$ and $C_{r,1}$ are the coil inductance and resonant capacitances of iron vessel respectively.

2) Analysis of Load-2: Fourier analysis of Load-2 is discussed below without considering voltage transitions. Derived fundamental voltage expression of Load-2 i.e. V_{hf} is

$$V_{o2,1}(t) = \frac{2V_{DC}}{\pi} \cos\left(wt - \pi d_1\right) \tag{5}$$

rms current expression of high-frequency load is

$$I_{hf} = \frac{(\sqrt{2}V_{DC})}{\pi |Z_2|} \quad (6)$$

Here Z_2 is the impedance of aluminium vessel.

$$|Z_2| = R_2 + j\left(w_{r,2}L_{r,2} - \frac{1}{w_{r,2}C_{r,2}}\right) \quad (7)$$

Here R_2 is the resistance of aluminium vessel , $w_{r,2}$ is resonant frequency of in rad/sec , $L_{r,2}$ and $C_{r,2}$ are the coil inductance and resonant capacitances of aluminium vessel respectively.The output power expression of Load-2 is

$$P_2 = I_2^2 * R_2 * d_1 = \frac{8V_{DC}^2}{\pi^2 Z_2^2} * R_2 * d_1 \quad (8)$$

Expression for total output power is

$$P_{Total} = P_1 + P_2 \quad (9)$$

C. Analysis of Buck-boost converter

It is designed to operate in continuous conduction mode to reduce the current ripple. 200 uH inductor is used in the simulations and the equation for the inductance calculation is

$$L = \frac{V_{DC}(V_c - V_{DC})}{\Delta I_S f_{lf} V_c} \quad (10)$$

The capacitor used in the buck-boost converter helps in powering Load-1 and Load-2 as discussed in section II. A 200 uF rated capacitor is used in the simulations. Expression for DC- capacitor calculation is

$$C = \frac{I_{C(\max)}D}{f_{lf}\Delta V_c} \quad (11)$$

IV. SIMULATION RESULTS

The proposed inverter configuration is simulated using MATLAB Simulink software. Load specifications used for simulations are shown in Table I. Two loads of different material types are considered for simulation.

TABLE I
DESIGN SPECIFICATIONS

Parameter	Symbol	Rating
Supply voltage	V_{DC}	200 V
Switching frequency of Load-1	f_{lf}	30kHz-40kHz
Resonant frequency of Load-1	$f_{r,lf}$	25 kHz
Switching frequency of Load-2	f_{hf}	145kHz
Resonant frequency of Load-2	$f_{r,hf}$	136 kHz
Equivalent resistance of Load-1	R_1	6.4Ω
Equivalent inductance of Load-1	L_1	98 uH
Resonant capacitance of Load-1	$C_{r,1}$	0.4 uF
Equivalent resistance of Load-2	R_2	2.7Ω
Equivalent inductance of Load-2	L_2	65 uH
Resonant capacitance of Load-2	$C_{r,2}$	21 nF
Inductance	L	200 uH
Capacitance	C	200 uF

Simulation results of the proposed topology are shown in Fig. 4. Simulated voltage and current waveforms of Load-1 and Load-2 at f_{lf} = 30 kHz and d_1= 0.99 are shown in Fig. 4. The observed load currents are 22 A and 24.3 A respectively. These waveforms show the rated power condition i.e., 4.5 kW.

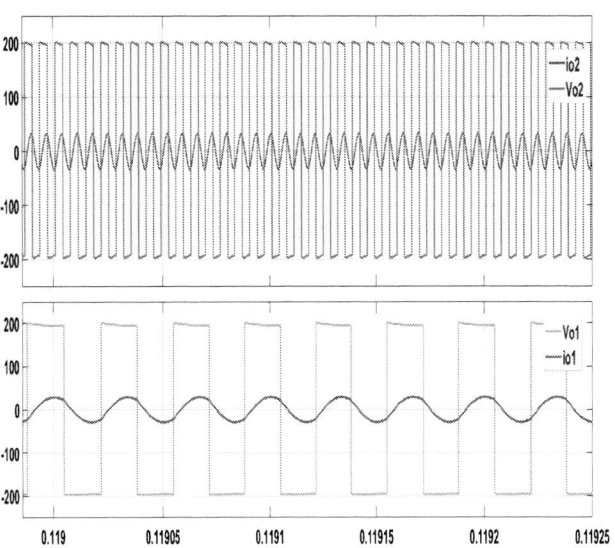

Fig. 4. Load waveforms at d_1 = 0.99, f_{lf} = 30 kHz.

Fig. 5. shows the waveforms measured at different d_1= 0.4, f_{lf} = 35 kHz. Vessel resistance increases with increased frequency, which affects the load power. This can be observed from Fig. 5.

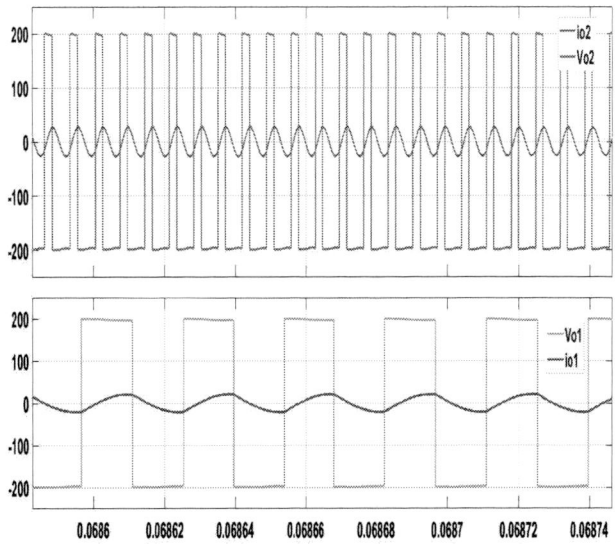

Fig. 5. Load waveforms at d_1 = 0.4, f_{lf} = 35 kHz.

Buck-boost converter output voltage at d= 0.5 for given 200 V dc supply is shown in Fig. 6.

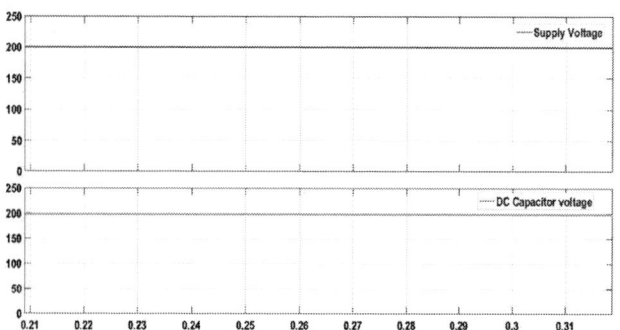

Fig. 6. Supply voltage & buck-boost converter output voltage waveforms.

Current flowing through the inductor which is used in buck-boost converter is shown in Fig. 7.

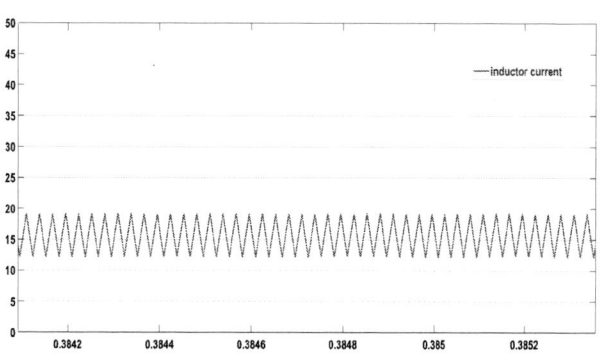

Fig. 7. Inductor current waveform.

Soft- switching waveforms of switches used in half-bridge inverter S_1 and S_2 are shown in Fig. 8. Body diode conduction during switch off period gives ZVS operation. This can be observed very clearly from Fig. 8. Effect of duty-

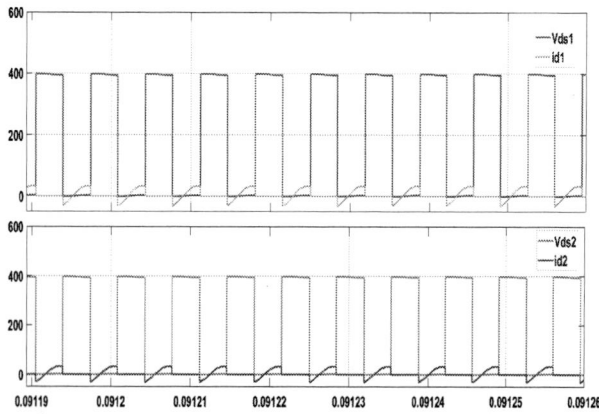

Fig. 8. ZVS waveforms of S_1 and S_2.

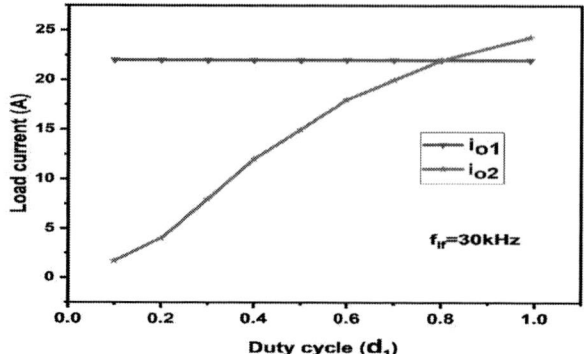

Fig. 9. Characteristics: Load current Vs. Duty cycle (d_1) at $f_{lf} = 30$ kHz.

cycle d_1 and frequency f_{lf} on load currents shown in the form of characteristics in Fig. 9 and Fig. 10. respectively. The characteristics shown in Fig. 9. are drawn between load current and duty-cycle d_1 for constant $f_{lf} = 30$ kHz. These characteristics show the variation of Load-2 current from 1.7 A to 24.5 A for varied duty cycles from 0.1 to 1.0. Similarly Fig. 10 shows the variation of load current with respect to frequency f_{lf} and constant duty-cycle $d_1 = 0.6$. Here observed that Load-2 current is almost constant, and Load-1 current varies corresponding to frequency f_{lf}. These characteristics shows the independent control of two loads. Maximum currents drawn by Load-1 and Load-2 are 22 A & 24.5 A respectively.

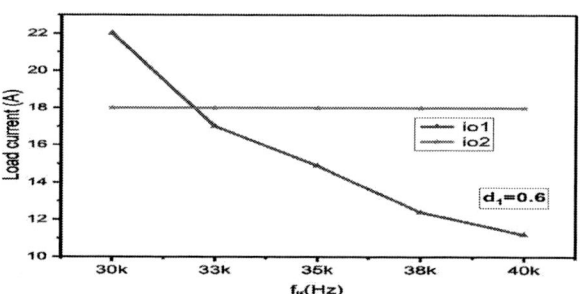

Fig. 10. Characteristics: Load current Vs. f_{lf} at $d_1 = 0.6$.

Change in efficiency with respect to load power is shown in Fig. 11. Maximum efficiency of the inverter i.e. 94% can be observed from this plot. The inverter's efficiency is improved due to the usage of an active switch (i.e., MOSFET) in place of the diode in buck-boost converter design.

V. CONCLUSIONS

A multi-load pre-regulated inverter configuration for different material applications is proposed here. The proposed inverter is simulated for 4.5 kW power and the observed peak efficiency is 94 %. Multi-load operation, reduced switch count, and simple control are the benefits of the proposed topology. It can power two loads in full-bridge mode with half-bridge inverter by deriving other V_{DC} from buck-boost

TABLE II
COMPARISON OF PROPOSED INVERTER CONFIGURATION WITH EXISTING TOPOLOGIES

Ref.No	No.of switches	$f_s(kHz)$	control used	no.of loads	η_{max}	soft-switching	$P_{out}(W)$	Switch to load ratio	Type of load
[12]	2	23-75	AVC	1	96	No	-	2	steel, Al
[13]	4	25-120	LAM	1	96.5	Yes	2k	4	steel, Al
[14]	6	30,220	ON- OFF	2	94.3	ZVS Turn-on	1.62k	3	steel, Al
[15]	8	20,100, 400	ADC	3	92	Yes	150	2.66	iron, steel, Al
Proposed configuration	4	30-40,145	PFM, ADC	2	94	Yes	4.5k	2	iron, Al

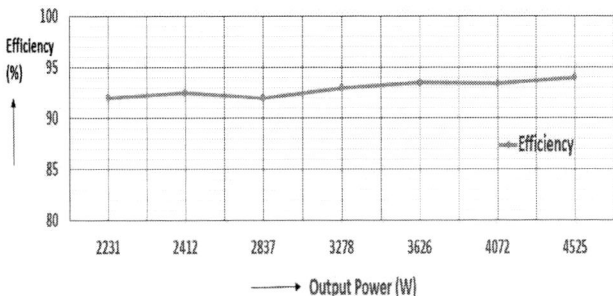

Fig. 11. Efficiency Vs output power characteristics at fixed (d_1)= 0.99.

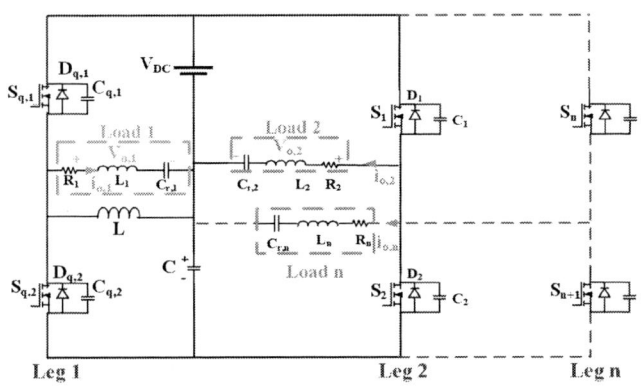

Fig. 12. Circuit diagram of proposed inverter with extension to "n" loads.

converter. Proposed inverter configuration is compared with existing configurations and is shown in the Table. II.

The proposed configuration can be extended to "n" number of loads by adding two switch legs to the half-bridge inverter. It is shown in Fig. 12.

REFERENCES

[1] P. Guillen, H. Sarnago, O. Lucia and J. M. Burdio, , , "GaN-Based Matrix Resonant Power Converter for Domestic Induction Heating," ,IEEE Transactions on Power Electronics, vol. 38, no. 6, pp. 6769-6773, June 2023.

[2] Oscar Lucia, Pascal Maussion, Enrique J.Dede and Jose M.Burdio, "Induction Heating Technology and Its Applications: Past Developments,Current Technology, and Future Challenges," ,IEEE Trans. Powers Electronics., vol. 29,no. 4, pp. 1909-1918, 2013.

[3] Esteve, V., Jordán, J., Dede, E.J., Bellido, J.L " Enhanced asymmetrical modulation for half-bridge series resonant inverters in induction heating applications," IET Power Electron. 16, 2482–2491 (2023).

[4] Zhongming, Y., Jain, P.K., Sen, P.C. "A full-bridge resonant inverter with modified phase-shift modulation for high-frequency AC power distribution systems," IEEE Trans. Ind. Electron., vol. 54, no.5, pp. 2831–2845, 2007.

[5] H.Terai, I.Hirota, T.Miyauchiand H.Omori " Comparative Performance Evaluations of IGBTs and MCT in Single Ended Quasi Resonant Zero Voltage Soft Switching Inverter," Power Electronics Specialists Conference (PESC-2001), IEEE, vol. 4, pp. 2178-2182, 2001.

[6] S. Khatroth and P. Shunmugam, "Single- Stage Pulse Frequency- Controlled AC–AC Resonant Converter for Different Material Vessel In duction Cooking Applications," International Journal of Circuit Theory and Applications 49, no. 9 (2021): 2865–2884, https://doi.org/10.1002/cta.3042.

[7] S. Komeda and H. Fujita "A Phase-Shift-Controlled Direct AC-to-AC Converter for Induction Heaters," IEEE Transactions on Power Electronics, vol. 33, no. 5, pp. 4115-4124, May 2018.

[8] T. Ahmed, K. Ogura, S. Chandhaket, and M. Nakaoka, "Asymmetrical duty cycle-controlled edge resonant soft switching high-frequency inverter for consumer electromagnetic induction fluid heater," Automatica, ATKAAF, vol. 44, no. 1-2, pp. 21–26, 2003.

[9] S.Chudjuarjeen, A. Sangswang and C. Koompai, "An Improved LLC Resonant Inverter for Induction-Heating Applications with Asymmetrical Control," IEEE Transactions on Industrial Electronics, vol. 58, no. 7, pp. 2915- 2925, July 2011.

[10] W. Han, K. T. Chau, and W. H. Lam, "All-utensil domestic induction heating system," Energy Convers. Manage., vol. 195, pp. 1035–1043, Sep. 2019.

[11] W. Han, K. T. Chau, C. Jiang, and W. Liu, "All-metal domestic induction heating using single-frequency double-layer coils," IEEE Trans. Magn., vol. 54, no. 11, pp. 1–5, Nov. 2018.

[12] I. Millán, J. M. Burdío, J. Acero, O. Lucía, and S. Llorente, "Series resonant inverter with selective harmonic operation applied to all metal domestic induction heating," IET Power Electron., vol. 4, no. 5, pp. 587–592, May 2011.

[13] H.-P. Park and J.-H. Jung, "Load-adaptive modulation of a series-resonant inverter for all-metal in-duction heating applications," IEEE Trans. Ind. Electron., vol. 65, no. 9, pp. 6983–6993, Sep. 2018.

[14] B. Salvi, S. Porpandiselvi and N. Vishwanathan, "An Inverter Circuit Configuration Suitable for Vessels of Different Mate-rial for Multiload Induction Cooking Application," IEEE Journal of Emerging and Selected Topics in Power Electronics, vol. 11, no. 3, pp. 3223-3235, June 2023.

[15] S. Khatroth and P. Shunmugam, "Cascaded full-bridge resonant inverter configuration for different material vessel induction cooking," IET Power Electronics, vol. 13, no. 19, pp. 4428–4438, 2020.

[16] Salvi, B., Porpandiselvi, S. Vishwanathan, N., "A dual frequency resonant inverter for different material vessel induction hob," Electr Eng 106, 1021 1031 (2024).

[17] H. Sarnago, J. M. Burdio and O. Lucia, "Dual-Output Extended-Power Range Quasi-Resonant Inverter for Induction Heating Appliances," IEEE Transactions on Power Electronics, vol. 38, no. 3, pp. 3385-3397,March 2023.

Comprehensive Analysis and Modeling of Totem-Pole PFC Rectifiers Using Various Modulation Techniques for EV charging application

1st Dalija Rath, Member IEEE
Dept. of Electrical Engineering
NIT Rourkela
Rourkela, India
rath.dalija@gmail.com

2nd Susovon Samanta, Member IEEE
Dept. of Electrical Engineering
NIT Rourkela
Rourkela, India
samantas@nitrkl.ac.in

Abstract—Electric vehicles (EV) have a huge positive environmental impact, such as reducing carbon emissions, the greenhouse effect, and air pollution. The EV chargers available in the market are mostly two-stage chargers where the first stage is the PFC rectifier stage used for regulating the DC-Link voltage, and the second stage may be a non-isolated or an isolated DC/DC converter for charging the EV battery. This paper concentrates only on the first stage, i.e the PFC rectifier. Here, a detailed analysis and modeling of a Totem-Pole PFC rectifier with different pulse width modulation (PWM) strategies have been presented, to maintain the DC-Link voltage. A 3kW Totem-Pole PFC rectifier has been designed using MATLAB Simulink to study the impact of the different modulation strategies on the harmonics of the grid side current. When charging an EV, the load current varies with different modes of CCCV charging , hence the impact of the load current variation on the DC-Link voltage has also been analyzed with the above PWM strategies. Finally, a comparison of the modulation strategies has been done based on switching loss, total harmonic distortion (THD) level of the grid current, input power factor and the dynamics of the system.

Index Terms—Totem-pole power factor correction (PFC) rectifier, electric vehicle (EV) charging, pulse width modulation (PWM), bipolar PWM (BPWM), unipolar PWM (UPWM), hybrid PWM (HPWM).

I. INTRODUCTION

The Pulse Width Modulation (PWM) rectifier is well-suited for applications that require a regulated DC bus voltage, such as electric vehicle (EV) chargers and high-speed railway traction drive systems [1]–[3]. In addition to voltage regulation, a PWM rectifier can function as a Power Factor Correction (PFC) rectifier, reducing current harmonics and improving power factor particularly beneficial when interfacing with the power grid.

A typical two-stage EV charger consists of a PFC rectifier that converts the AC input voltage to a fixed DC voltage, followed by a DC/DC converter that regulates this voltage to suit the EV battery charging requirements, as illustrated in

Fig. 1. Conventional EV chargers often employ a diode bridge rectifier combined with a PFC circuit for AC/DC conversion. However, this two-stage approach results in lower efficiency and is unsuitable for high-power applications [4], [5].

When connected to the grid, EV chargers can introduce several power quality issues, including voltage dips, increased harmonic distortion in grid current, and low power factor [6]–[8]. Replacing the conventional diode rectifier with a Totem-Pole PFC rectifier can mitigate these problems. For instance, the PFC rectifier enables reactive power compensation, helping to alleviate voltage dips in the grid [9]–[12]. Totem-Pole PFC rectifiers employ control strategies where switches operate at high frequencies, enabling effective filtering of harmonics and ensuring that the Total Harmonic Distortion (THD) remains below 7% , in compliance with IEEE standards [13]–[16]. Various modulation techniques have been explored in the literature, including Unipolar PWM (UPWM), Bipolar PWM (BPWM) [17]–[19], and Hybrid PWM (HPWM) [20]. In BPWM and UPWM, all switches operate at the same switching frequency, whereas in HPWM, two switches are operated at a high frequency while the remaining two operate at a lower frequency.

Despite the critical role of the PFC rectifier in EV charging systems, detailed modeling of the single-phase Totem-Pole PFC rectifier remains underexplored in the literature. This work presents a comprehensive analysis and modeling of the single-phase Totem-Pole PFC rectifier. Furthermore, the impact of various modulation schemes on the DC-link voltage dynamics and grid current quality is analyzed—an aspect not thoroughly addressed in prior studies.

II. ANALYSIS OF TOTEM-POLE PFC RECTIFIER WITH BPWM TECHNIQUE

A. Working principle

In BPWM a sine wave reference (v_r) is compared with a carrier wave (v_{cr}) and when the reference is more than the carrier one gate pulse is generated which is given to two

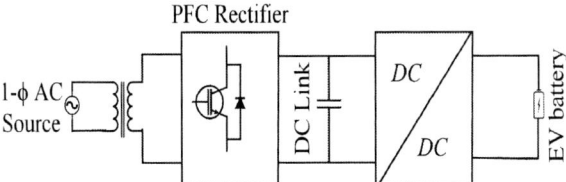

Fig. 1. Block diagram of a two stage EV charger.

diagonally opposite switches and the complementary pulse is given to the rest two switches as shown in Fig. 2. Hence in one switching period, there are two modes of operation, Mode I when S_1, S_2' are ON and in Mode II S_2, S_1' is ON. In both modes the direction of the current flow is shown in Fig. 3 and Fig. 4 respectively.

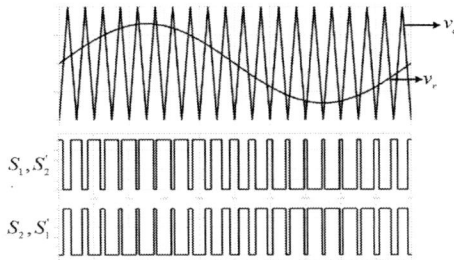

Fig. 2. Pulse generation in BPWM.

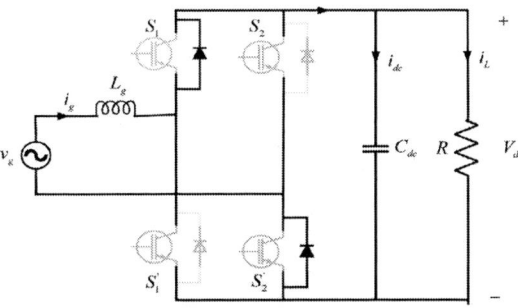

Fig. 3. Mode I of Totem-pole PFC rectifier with BPWM.

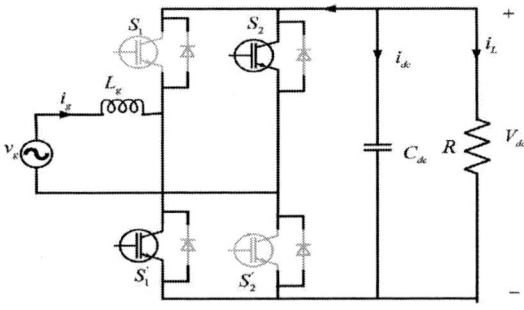

Fig. 4. Mode II of Totem-Pole PFC rectifier with BPWM.

B. Modeling

It is considered that over one switching period T_S, S_1, S_2' is ON for $d_1 T_S$ period and S_2, S_1' are ON for $d_2 T_S$ period such that $d_1 + d_2 = 1$. Where, based on [9]

$$d_1 = \frac{1}{2}(1 + m) \tag{1}$$

and

$$d_2 = \frac{1}{2}(1 - m) \tag{2}$$

Combining Mode I and Mode II the average state space model of the rectifier is:

$$\begin{bmatrix} \dot{i}_g \\ \dot{v}_{dc} \end{bmatrix} = \begin{bmatrix} 0 & \frac{1-2d_1}{L_g} \\ \frac{2d_1-1}{C_{dc}} & -\frac{1}{RC_{dc}} \end{bmatrix} \begin{bmatrix} i_g \\ v_{dc} \end{bmatrix} + \begin{bmatrix} \frac{1}{L_g} \\ 0 \end{bmatrix} [v_g] \tag{3}$$

or

$$\begin{bmatrix} \dot{i}_g \\ \dot{v}_{dc} \end{bmatrix} = \begin{bmatrix} 0 & -\frac{m}{L_g} \\ \frac{m}{C_{dc}} & -\frac{1}{RC_{dc}} \end{bmatrix} \begin{bmatrix} i_g \\ v_{dc} \end{bmatrix} + \begin{bmatrix} \frac{1}{L_g} \\ 0 \end{bmatrix} [v_g] \tag{4}$$

where,
v_g=grid voltage
v_{dc}=dc link voltage
i_g= grid current
m=modulation index

In this paper, the representation of the above variables with a small italic-style letter, capital letter, and small italic-style letter with a hat signifies their instantaneous values, steady-state values, and small signal values, respectively.
The parameter
L_g= Source inductance
C_{dc}= dc link capacitance
R=load resistance
After perturbing and linearizing (4) around its steady state operating point(M, V_{dc}, I_g), the small signal model of the rectifier is:

$$\begin{bmatrix} \dot{\hat{i}}_g \\ \dot{\hat{v}}_{dc} \end{bmatrix} = \begin{bmatrix} 0 & -\frac{M}{L_g} \\ \frac{M}{C_{dc}} & -\frac{1}{RC_{dc}} \end{bmatrix} \begin{bmatrix} \hat{i}_g \\ \hat{v}_{dc} \end{bmatrix} + \begin{bmatrix} \frac{1}{L_g} & \frac{V_{dc}}{L_g} \\ 0 & \frac{I_g}{C_{dc}} \end{bmatrix} \begin{bmatrix} \hat{v}_g \\ \hat{m} \end{bmatrix} \tag{5}$$

and the steady state input output voltage relationship is:

$$V_{dc} = \frac{V_g}{M} \tag{6}$$

III. ANALYSIS OF TOTEM-POLE PFC RECTIFIER WITH UPWM TECHNIQUE

A. working principle

In Unipolar Pulse Width Modulation (UPWM), two sine wave reference signals with a 180° phase shift are compared with a high-frequency carrier wave v_{cr} to generate gate pulses for switches S_1 and S_2, as illustrated in Fig. 5. The

979-8-3315-3013-6/25 $31.00 © 2025 IEEE

complementary gate signals are applied to switches S_1' and S_2', respectively. This modulation results in three distinct operating modes within a single switching period depicted in Fig. 6, Fig. 7, and Fig. 8, respectively. During **Mode II** and **Mode III**, the input inductor is charged as current flows through it. In contrast, during **Mode I**, the inductor discharges its stored energy, supplying power to the load.

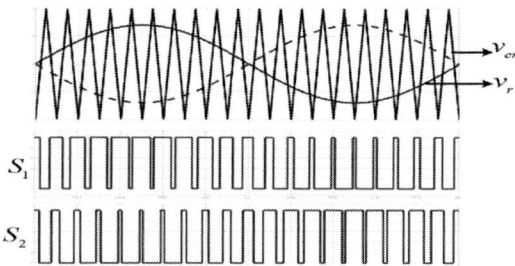

Fig. 5. Pulse generation in UPWM.

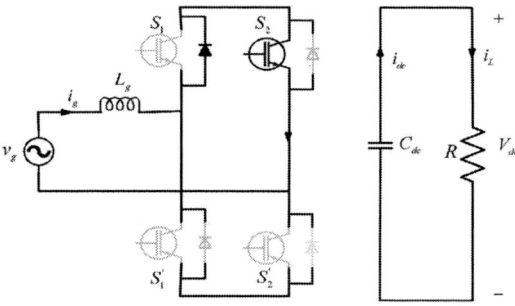

Fig. 6. Mode I of Totem-Pole PFC rectifier with UPWM.

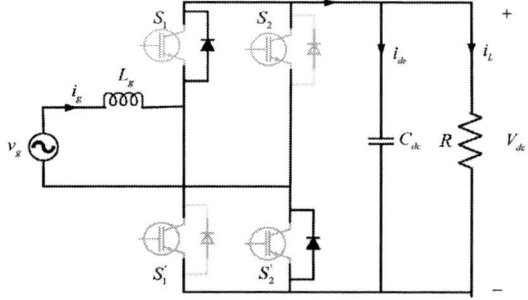

Fig. 7. Mode II of Totem-Pole PFC rectifier with UPWM.

B. Modeling

In UPWM, only during **Mode I** switch S_2 turned ON, and this duration is represented as $d_2 T_S$. During **Mode II** and **Mode III**, which together span a time period of $d_1 T_S$, switch S_2 remains OFF. The duty cycles satisfy the relationship $d_1 + d_2 = 1$. However, due to page constraints, a detailed derivation of this relationship is not presented here.

By combining the dynamics of all three operating modes, the average state-space model of the rectifier can be expressed as:

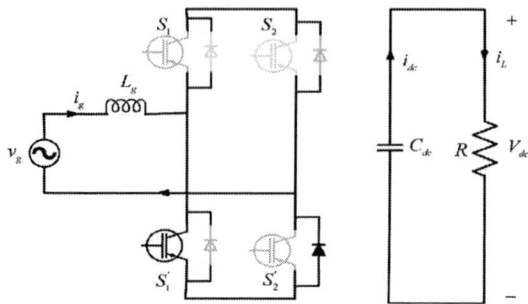

Fig. 8. Mode III of Totem-Pole PFC rectifier with UPWM.

$$\begin{bmatrix} \dot{i}_g \\ \dot{v}_{dc} \end{bmatrix} = \begin{bmatrix} 0 & -\frac{d_1 - d_2}{L_g} \\ \frac{d_1 - d_2}{C_{dc}} & -\frac{1}{RC_{dc}} \end{bmatrix} \begin{bmatrix} i_g \\ v_{dc} \end{bmatrix} + \begin{bmatrix} \frac{1}{L_g} \\ 0 \end{bmatrix} [v_g] \quad (7)$$

or

$$\begin{bmatrix} \dot{i}_g \\ \dot{v}_{dc} \end{bmatrix} = \begin{bmatrix} 0 & -\frac{m}{L_g} \\ \frac{m}{C_{dc}} & -\frac{1}{RC_{dc}} \end{bmatrix} \begin{bmatrix} i_g \\ v_{dc} \end{bmatrix} + \begin{bmatrix} \frac{1}{L_g} \\ 0 \end{bmatrix} [v_g] \quad (8)$$

After perturbing and linearizing (8) around the steady state operating point (M, V_{dc}, I_g), the small signal model of the rectifier is:

$$\begin{bmatrix} \dot{\hat{i}}_g \\ \dot{\hat{v}}_{dc} \end{bmatrix} = \begin{bmatrix} 0 & -\frac{M}{L_g} \\ \frac{M}{C_{dc}} & -\frac{1}{RC_{dc}} \end{bmatrix} \begin{bmatrix} \hat{i}_g \\ \hat{v}_{dc} \end{bmatrix} + \begin{bmatrix} \frac{1}{L_g} & \frac{V_{dc}}{L_g} \\ 0 & \frac{I_g}{C_{dc}} \end{bmatrix} \begin{bmatrix} \hat{v}_g \\ \hat{m} \end{bmatrix} \quad (9)$$

and the steady state input output voltage relationship is:

$$V_{dc} = \frac{V_g}{M} \quad (10)$$

IV. ANALYSIS OF TOTEM-POLE PFC RECTIFIER WITH HPWM TECHNIQUE

A. working principle

In this modulation technique, during the positive half-cycle of the input voltage, switch S_2 operates at high frequency, whereas during the negative half-cycle, switch S_1 operates at high frequency. The generation of gate pulses for this scheme is illustrated in Fig. 9, and the two corresponding modes of operation within one switching period are depicted in Fig. 10 and Fig. 11.

B. Modeling

Here Mode I is having a period of $d_1 T_S$ and Mode II is for a period of $d_2 T_S$ and

$$d_1 = \frac{1}{2}(1 + m) \quad (11)$$

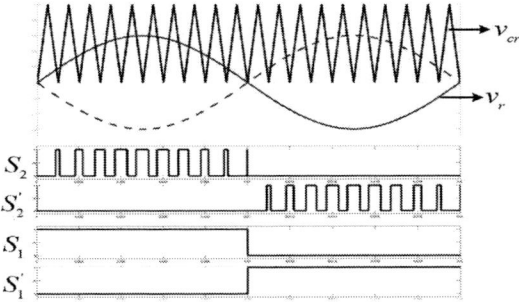

Fig. 9. Pulse generation in HPWM.

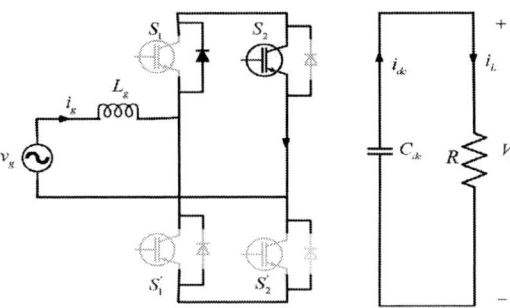

Fig. 10. Mode I of Totem-Pole PFC rectifier with HPWM.

and

$$d_2 = \frac{1}{2}(1 - m) \tag{12}$$

Combining Mode I and Mode II operations the average state-space model of the rectifier is:

$$\begin{bmatrix} \dot{i}_{dc} \\ \dot{v}_{dc} \end{bmatrix} = \begin{bmatrix} 0 & -\frac{d_2}{L_g} \\ \frac{d_2}{C_{dc}} & -\frac{1}{RC_{dc}} \end{bmatrix} \begin{bmatrix} i_g \\ v_{dc} \end{bmatrix} + \begin{bmatrix} \frac{1}{L_g} \\ 0 \end{bmatrix} [v_g] \tag{13}$$

or

$$\begin{bmatrix} \dot{i}_{dc} \\ \dot{v}_{dc} \end{bmatrix} = \begin{bmatrix} 0 & -\frac{1}{2L_g}(1-m) \\ \frac{1}{2C_{dc}}(1-m) & -\frac{1}{RC_{dc}} \end{bmatrix} \begin{bmatrix} i_g \\ v_{dc} \end{bmatrix} + \begin{bmatrix} \frac{1}{L_g} \\ 0 \end{bmatrix} [v_g] \tag{14}$$

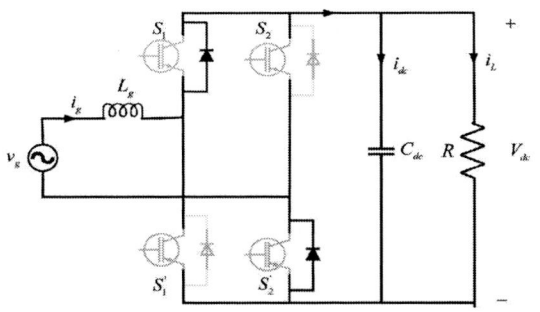

Fig. 11. Mode II of Totem-Pole PFC rectifier with HPWM.

Perturbing and linearizing (14) around its steady state operating point (M, V_{dc}, I_g), the small signal model of the rectifier is:

$$\begin{bmatrix} \dot{\hat{i}}_g \\ \dot{\hat{v}}_{dc} \end{bmatrix} = \begin{bmatrix} 0 & -\frac{1}{2L_g}(1-M) \\ \frac{1}{2C_{dc}}(1-M) & -\frac{1}{RC_{dc}} \end{bmatrix} \begin{bmatrix} \hat{i}_g \\ \hat{v}_{dc} \end{bmatrix} + \begin{bmatrix} \frac{1}{L_g} & \frac{V_{dc}}{2L_g} \\ 0 & -\frac{I_g}{2C_{dc}} \end{bmatrix} \begin{bmatrix} \hat{v}_g \\ \hat{m} \end{bmatrix} \tag{15}$$

and the steady state input output voltage relationship is:

$$V_{dc} = \frac{V_g}{\frac{1}{2}(1-M)} \tag{16}$$

V. SIMULATION RESULT AND COMPARISON OF THE PWM TECHNIQUES

As per the parameter specifications provided in Table I, a 3 kW Totem-Pole PFC rectifier has been designed using MATLAB/Simulink. A two-loop control strategy, illustrated in Fig. 12, is implemented to regulate the DC-link voltage at 400 V. In this control architecture, the outer voltage control loop generates the reference current based on the deviation from the desired DC-link voltage, while the inner current control loop processes this reference to produce a modulation signal for the PWM generator. This signal is used to generate appropriate gate pulses to ensure the output voltage remains regulated.

The rectifier has been operated under all the aforementioned PWM modulation schemes, and the impact of each scheme on the Total Harmonic Distortion (THD) of the input current and the input power factor (PF) is summarized in Table III.

When charging an EV battery using techniques such as Constant Current–Constant Voltage (CCCV), Pulse Charging, or Constant Power (CP), the load current changes during transitions between charging modes. To evaluate the system's dynamic performance under such load disturbances, the response of the DC-link voltage using different modulation schemes is shown in Fig. 14. In this case, the load current is suddenly reduced from 15 A to 7.5 A at 2.2 s. This step change introduces a temporary overshoot in the DC-link voltage, but the controller effectively restores and maintains the voltage at the desired value of 400 V.

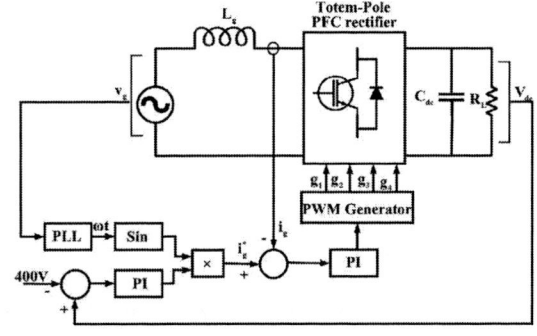

Fig. 12. Block diagram of the closed loop system

979-8-3315-3013-6/25 $31.00 © 2025 IEEE

TABLE I
DESIGN SPECIFICATION

Parameter	Symbol	specification
Grid voltage	V_g	230V(RMS)
Grid frequency	f_g	50Hz
dc link voltage	V_{dc}	400V
Power	P	3kw
Switching frequency	f_{sw}	20kHz
Source inductance	L_g	0.0022H
dc link capacitance	C_{dc}	1500μF

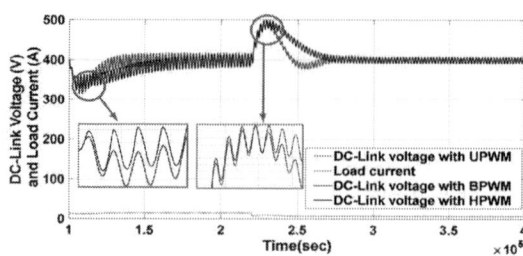

Fig. 14. Effect of load disturbance on DC-Link voltage

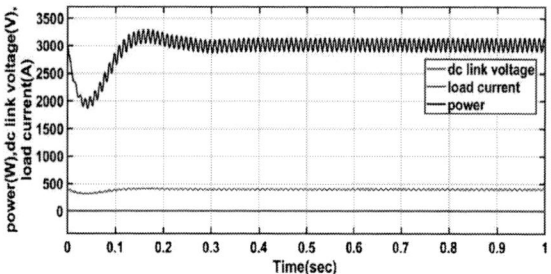

Fig. 13. Output power(W), DC link voltage(V), load current(A) Vs Time(sec)

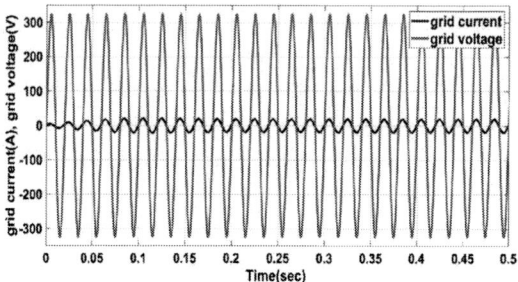

Fig. 15. Grid voltage(V), Grid current(A) vs time(sec)

- The status of the switches in different switching states for different modulation schemes is given in Table II. It can be seen that the number of states in BPWM, UPWM and HPWM is 2,4 and 2 respectively, in one switching period. With UPWM during the transition between two consecutive states, two switches change their status. In BPWM and HPWM even if the number of states is the same in one switching period, with BPWM all four switches are changing their status, whereas in HPWM only one switch is changing its status. Hence switching loss is less with HPWM scheme.

- The FFT analysis results for the grid current of the rectifier with different modulation techniques is shown in Fig. 16, Fig. 17 and Fig. 18 from which the obtained Total Harmonic Distortion (THD) values are mentioned in Table. III. It is found that the THD value has been reduced to a greater extent than the BPWM and UPWM techniques.

TABLE II
SWITCHING SEQUENCE WITH DIFFERENT PWM TECHNIQUES

Modulation Technique	Switching sequence			
	S_1	S_2	S_1'	S_2'
BPWM	1	0	0	1
	0	1	1	0
UPWM	1	0	0	1
	1	0	1	0
	1	0	0	1
	0	1	0	1
HPWM	0	1	0	1
	0	0	0	1

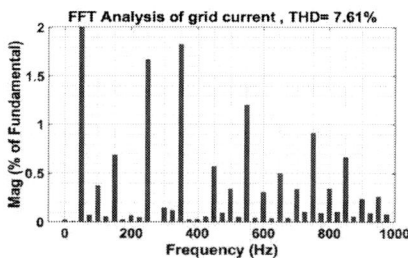

Fig. 16. FFT analysis with BPWM

- The main purpose of using a PWM rectifier in an EV charger is to improve the power factor so that maximum use of the grid power can be achieved. From the simulation result in Fig. 15, it can be seen that the grid voltage and the grid current are almost in the same phase with the use of a PWM rectifier. Among the three PWM methods, the use of HPWM improves the input power factor, as mentioned in Table. III.

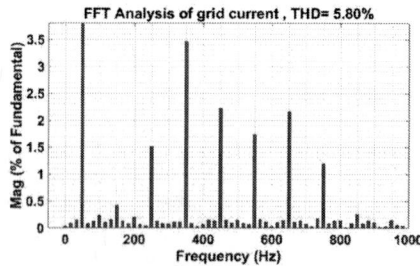

Fig. 17. FFT analysis with UPWM

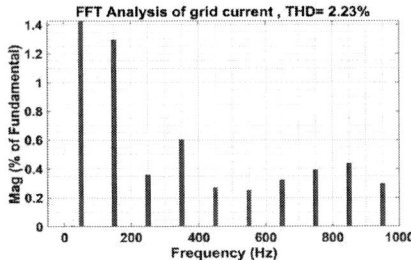

Fig. 18. FFT analysis with HPWM

TABLE III
COMPARISON OF PWM SCHEMES

PWM scheme	%THD	PF	no. of switching per cycle
BPWM	7.61	0.9845	2
UPWM	5.80	0.988	4
HPWM	2.23	0.9999	1

- In HPWM two switches operates at high switching and rest two switches operates at line frequency which increases the settling time of the system. The details of the dynamics of the system with the different modulation schemes have been summarized in Table. IV.

TABLE IV
EFFECTS OF THE PWM STRATEGIES ON THE DYNAMICS OF THE SYSTEM

PWM scheme	%Overshoot	Settling Time(sec)
BPWM	7.5 (step increase in load current)	0.3
	20 (step decrease in load current)	0.55
UPWM	7.5 (step increase in load current)	0.3
	20 (step decrease in load current)	0.55
HPWM	12.5(step increase in load current)	0.8
	20 (step decrease in load current)	0.55

VI. CONCLUSION

This paper presents a detailed analysis of a single-phase Totem-Pole PFC rectifier with different modulation strategies such as BPWM, UPWM and HPWM. These modulation schemes have been implemented to a 3kW Totem-Pole PFC rectifier using MATLAB/Simulink to control the DC-Link voltage, and it has been found that by using Totem-Pole PFC rectifier, the input power factor has been improved without the use of any external PFC circuit. Also, it has been seen that among the three modulation schemes, HPWM gives better results in terms of THD and power factor but from the dynamic response it is found that HPWM has a slower response as compared to the other two.

ACKNOWLEDGEMENT

This work is supported by the Science and Engineering Research Board (SERB) of India under Grant IPA/2021/000081.

REFERENCES

[1] B. Gou, G. Wang, X. Li, J. Zhao, P. Han, and B. Wu, "An open-switch fault diagnosis method for single-phase pwm rectifier using a model-based approach in high-speed railway electrical traction drive system," *IEEE Transactions on Power Electronics*, vol. 31, no. 5, pp. 3816–3826, 2015.

[2] U. Sharma and B. Singh, "A bidirectional charger for low-voltage-powered battery vehicles," *IEEE Transactions on Transportation Electrification*, vol. 9, no. 3, pp. 3994–4003, 2023.

[3] S. Xu, Y. Zhang, C. Liao, Q. He, W. Zhang, and W. Zhao, "A novel adaptive smo-based simultaneous diagnosis method for igbt open-circuit faults and current sensor incipient faults of inverters in pmsm drives for electric vehicles," *IEEE Transactions on Instrumentation and Measurement*, vol. 72, pp. 1–15, 2023.

[4] A. Dubey, S. Santoso, and M. P. Cloud, "Average-value model of electric vehicle chargers," *IEEE Transaction on Smart Grid*, vol. 4, 2013.

[5] S. M. I. Prince Dadhaniya, Mukesh Maurya and M. I. Gururaj Mirle Vishwanath, "Abridgeless modified boost converter to improve power factor in ev battery charging applications," *IEEE JOURNAL OF EMERGING AND SELECTED TOPICS IN INDUSTRIAL ELECTRONICS*, vol. 5, 2024.

[6] F. Marra, G. Yang, C. Træholt, E. Larsen, and C. N. Rasmussen, "Power quality issues into a danish low-voltage grid with electric vehicles," in *11th International Conference on Electrical Power Quality and Utilisation*. IEEE, 2011.

[7] R. I. Bojoi, G. Griva, F. Profumo, and A. Tenconi, "Enhanced power quality control strategy for single-phase inverters in distributed generation systems," *IEEE Transactions on Power Electronics*, vol. 26, no. 3, pp. 798–806, 2011.

[8] A. Shahin, A. Elsaid, H. Gabbar, H. Abu-Rub, and F. Blaabjerg, "A comprehensive analysis: Integrating renewable energy sources with wire/wireless ev charging systems for green mobility," *IEEE Access*, 2024.

[9] H. N. DeMelo, J. Trovao, P. Pereirinha, H. George, and C. Antunes, "A controllable bidirectional battery charger for electric vehicles with vehicle-to-grid capability," *IEEE Transactions on Vehicular Technology*, vol. 67, no. 1, pp. 114–123, 2018.

[10] D. B. W. Abeywardana, J. Zhu, Y. Li, and A. Ghosh, "Single-phase boost inverter-based electric vehicle charger with integrated vehicle to grid reactive power compensation," *IEEE Transactions on Power Electronics*, vol. 33, no. 4, pp. 3462–3471, 2017.

[11] M. C. Kisacikoglu, M. Kesler, and L. M. Tolbert, "Single-phase on-board bidirectional pev charger for v2g reactive power operation," *IEEE Transactions on Smart Grid*, vol. 6, no. 2, pp. 767–775, 2014.

[12] R. K. Lenka, B. Subudhi, and A. Mishra, "Pv integrated multifunctional off-board ev charger with improved grid power quality," *IEEE Transactions on Industry Applications*, vol. 58, no. 5, pp. 5520–5532, 2022.

[13] Y. H. W. Wu and F. Blaabjerg, "An llcl power filter for single-phase grid-tied inverter," *IEEE Transactions on Power Electronics*, vol. 27, pp. 782–789, 2012.

[14] Z. C. H. Mao, X. Yang and Z. Wang, "A hysteresis current controller for single-phase three-level voltage source inverters," *IEEE Transactions on Power Electronics*, vol. 27, p. 3330–3339, 2012.

[15] *IEEE recommended practice and requirements for harmonic control in electric power systems*, IEEE Standards Association Std., 2014.

[16] M. R. Khalid, Z. Khan, D. Ahmed, E. Hossain, and S. Mekhilef, "A comprehensive review on structural topologies, power levels, energy storage systems, and standards for electric vehicle charging stations and their impacts on grid," *IEEE Access*, vol. 9, pp. 128 069–128 094, 2021.

[17] J. Salmon, T. A. Lipo, and J. McKeever, "A carrier-based unipolar pwm current controller that minimizes the pwm-cycle average current-error using internal feedback of the pwm signals," *IEEE Transactions on Power Electronics*, vol. 22, no. 5, pp. 1708–1718, 2007.

[18] Z. Guo and F. Kurokawa, "A novel pwm modulation and hybrid control scheme for grid-connected unipolar inverters," in *2011 Twenty-Sixth Annual IEEE Applied Power Electronics Conference and Exposition (APEC)*, IEEE. IEEE, 2011, pp. 734–739.

[19] N. Mohan, T. M. Undeland, and W. P. Robbins, *Power Electronics—Converters, Applications, and Design*. New Delhi, India: Wiley, 2003.

[20] Y.-H. Liao, "A novel reduced switching loss bidirectional ac/dc converter pwm strategy with feedforward control for grid-tied microgrid systems," *IEEE Transactions on Power Electronics*, vol. 29, no. 3, pp. 1500–1513, 2013.

Influence of GaN Cap Layer on the Performance of AlGaN/GaN HEMT

Soumiya K
Department of ECE
IIITDM Kancheepuram
Chennai, India
ec24d0008@iiitdm.ac.in

Jyotismita Mishra
Department of Electrical Engineering
Vellore Institute of Technology
Chennai, India
jyotismita.mishra@vit.ac.in

K P Pradhan
Department of ECE
IIITDM Kancheepuram
Chennai, India
kppradhan@iiitdm.ac.in

Abstract—The impact of GaN cap layer on the performance metrics of GaN/AlGaN/AlN/GaN high electron mobility transistor (HEMT) is extensively examined through TCAD environment. The inherent material nature of both piezoelectric along with spontaneous polarization in GaN usually results in the formation of 2-D electron gas (2-DEG) demonstrating high electron density in the GaN channel. Furthermore, a thin GaN cap layer can influence the device performance by further enhancing the electron mobility from 1315 cm²/V-s to 1425 cm²/V-s with cap layer and modulating the field that leads to an increase in the drain current from 0.36 A to 0.58 A. Hence, the GaN cap can be harnessed appropriately towards the applications in power electronics as well as in Radio Frequency (RF) applications.

Index Terms—GaN HEMT, GaN Cap, 2-DEG, Power devices.

I. INTRODUCTION

Due to enhanced electron mobility and wide band gap Gallium nitride (GaN) has gained popularity as a material for power devices in recent years [1] . The high breakdown electric field of GaN reduces resistive losses, making GaN switches more efficient [2]. In addition, GaN offers fast switching speeds and high thermal conductivity, which helps dissipate heat, making these transistors more durable at extremely high temperatures, 400–500°C [3]–[5]. GaN HEMT handles high voltages and currents, making them suitable for power electronics, wireless communication, and microwave and millimeter-wave applications [6]–[9]. Fig. 1 shows the advancement of GaN power devices in low-, medium-, and high-voltage applications.

GaN HEMT based on AlGaN/GaN heterostructure takes advantage of AlGaN and GaN material characteristics to create a high-performance semiconductor device. The heterojunction formed between these two materials creates a two-dimensional electron gas (2-DEG) because of their difference in energy bandgaps. The flow of current in the 2-DEG channel can be changed by gate voltage, effectively switching the transistor on or off. The intrinsic polarization, which is spontaneous in nature, as well as the piezoelectric charges created determines the electron density in 2-DEG [10]. The GaN cap region is placed above the AlGaN barrier to improve current collapse [11] [12] and unwanted oxidation [13] [14]. The GaN cap

layer improves surface quality at heterojunction interface by minimizing surface roughness and defects [15], resulting in a more stable and high-performance 2-DEG. It also improves electrical properties [16] and reliability by reducing surface recombination of charge carriers [17]. The cap layer enhances carrier mobility and significantly improved the off-state properties of the device [18].

Hence, this article systematically investigates the potential benefits of introducing a GaN cap region for the electrical characteristics of AlGaN/GaN HEMTs. This layer helps to maintain a high density of electrons in the 2-DEG quantum well, allowing for enhanced mobility across the channel by reducing the field near the gate, which are important for the transistor's high performance.

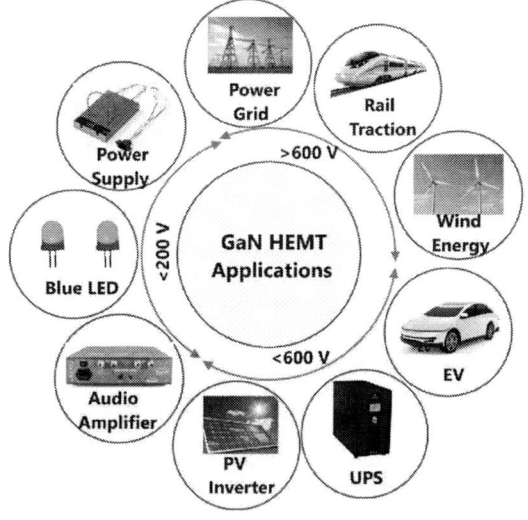

Fig. 1. Application of GaN power devices

II. DEVICE DESIGN AND SIMULATION SETUP

The cross-sectional view of the designed device is shown in Fig. 2. The device has a 100 nm SiN passivation layer, 2 nm GaN cap region placed over the $Al_{0.25}Ga_{0.75}N$ barrier of 20 nm. The dimension of the AlN spacer region is 1 nm, and GaN channel is 2 μm. The substrate used in the simulation

979-8-3315-3013-6/25 $31.00 © 2025 IEEE

TABLE I
DEVICE DESIGN PARAMETERS

Parameter	Units	Value
Gate Length (L_G)	μm	0.5
AlGaN Barrier thickness	nm	20
Al mole fraction (x)	-	0.25
GaN cap thickness	nm	2
GaN Bandgap (E$_g$)	eV	3.43
GaN Relative Permitivity (ϵ_{GaN})	-	8.9

is Silicon carbide (4H-SiC). The gate electrode is Schottky contact in nature with an Ohmic contact for the source-drain electrodes. The gate is 0.5 μm in length (L_G) and 1.5 mm in width (W_G). The spacing between gate and source (L_{SG}) is 2 μm and 5 μm separation between the gate and drain (L_{GD}). The interface between the AlGaN barrier and GaN channel 2-DEG is formed. The mobility and electron density in the heterojunction depends on the barrier and channel because AlGaN has wider bandgap [19]. AlN spacer layer reduces dislocations, which improves the GaN layer crystal quality and also increases the mobility [20]. SiC substrates have superior thermal conductivity and low lattice mismatch between GaN and substrate and are primarily preferred in high-power applications [19]. Although SiN passivation reduces surface states, the cap region of GaN further reduces surface traps and charge states, affecting 2-DEG and overall device performance.

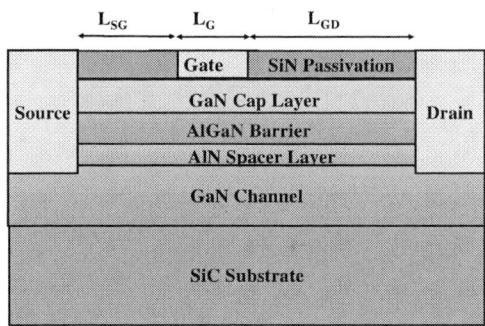

Fig. 2. Schematic Structure of the GaN/AlGaN/AlN/GaN HEMT

The AlGaN/GaN HEMT model is implemented using the industry standard Sentaurus Technology Computer-Aided Design (TCAD) simulator [21] to demonstrate the device characteristics. Fermi statistics and the Drift-Diffusion (DD) transport models are the carrier transport models utilized in the simulation environment. Both drift and diffusion of charge carriers contribute to the carrier flow inside the device. The diffusion results from the gradients of charge carriers, while the drift happens when an externally applied electric field is applied. Simultaneously, the generation and recombination of charge carriers are captured through the Shockley–Read–Hall (SRH) recombination model.

III. RESULT AND DISCUSSION

It is quite obvious that when a wide-band gap material AlGaN is placed in contact with a narrow band gap material,

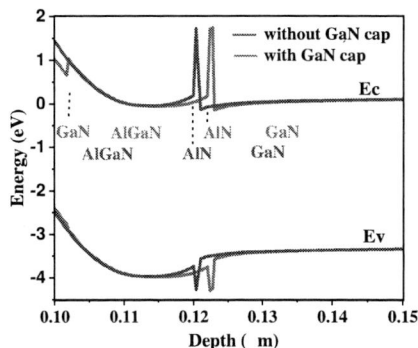

Fig. 3. Energy Band diagram of the simulated HEMT along the vertical axis under equilibrium condition

can lead to formation of a hetero-junction at the interface. The donor electrons from the AlGaN material move into the narrow-band-gap material causing a band bending. These electrons become confined at the interface due to a potential barrier, forming a 2-DEG near the AlGaN/GaN interface. The device used in the simulation is a depletion mode HEMT or normally on device that can induce a 2-DEG even without any gate bias. Although, voltage is applied to the gate, near the channel electron density increases and allows the current flow between the drain and source. Fig. 3 depicts the change in energy band diagram of the device at $V_{GS} = V_{DS} = 0$ V with and without a cap layer. Introducing GaN cap layer alters the energy-band, which can affect the threshold voltage of the device.

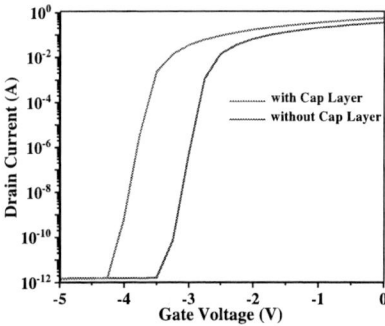

Fig. 4. I_D-V_{GS} characteristics AlGaN/GaN HEMT device with applied drain voltage of 5V with and without a GaN cap layer

The expression of sheet charge density (n_s) of 2-DEG is given in Eq. (1) [22].

$$n_s = \frac{\sigma_{AlGaN}}{q} - (\frac{\epsilon}{q^2 d})(q\phi_b + E_f - \Delta E_c) \quad (1)$$

where ϕ_b is the height of the surface barrier, d is AlGaN layer thickness, σ_{AlGaN} is the AlGaN sheet charge density due to polarization, ϵ is dielectric value of AlGaN, q is the electron charge, E_f is the Fermi energy in AlGaN/GaN interface, and ΔE_c is GaN and AlGaN conduction band discontinuity.

979-8-3315-3013-6/25 $31.00 © 2025 IEEE

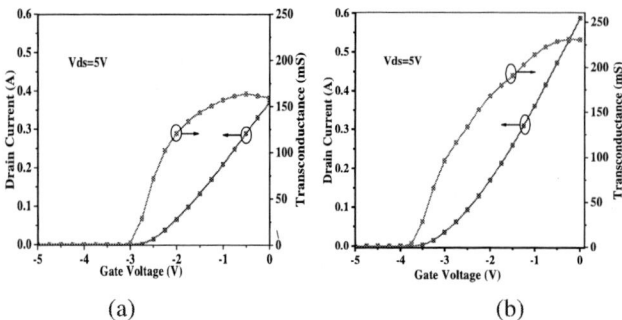

Fig. 5. I_D-V_{GS} and transconductance characteristics (a) without GaN cap (b) with GaN cap

The I_D-V_{GS} characteristics of device compared with cap layer are indicates in Fig. 4 for the applied V_{GS} = -5 V and V_{DS} = 5 V. Adding a GaN cap layer can provide a threshold voltage shift of 0.75 V on the negative side and increases the drain current. The mobility of electron in the GaN channel has been increased to a significant value i.e., 1425 cm²/V-s for the cap layer as compared to its counterpart i.e., 1315 cm²/V-s without the cap layer. The cap layer contributes the enhanced mobility in the GaN channel.

The I_D-V_{GS} and transconductance characteristics with and without the GaN cap region is shown in Fig. 5. The GaN/AlGaN/AlN/GaN HEMT device achieves a maximum drain current 0.58 A and a transconductance of 230 mS at V_{GS} = 0 V with cap layer. In contrast, without a cap layer, drain current is 0.36 A, and the transconductance is 163 mS, at drain voltage of 5 V. This indicates that the GaN cap layer enhances the channel conductivity, improving carrier confinement and changing the electric field.

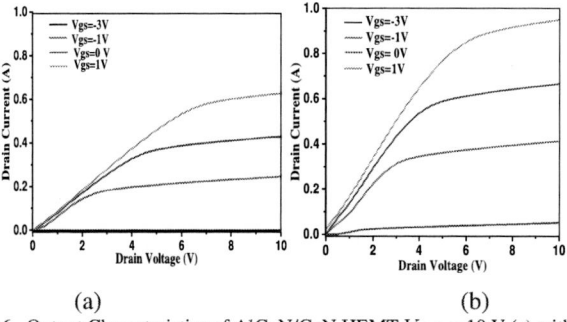

Fig. 6. Output Characteristics of AlGaN/GaN HEMT V_{DS} = 10 V (a) without GaN cap (b) with GaN cap

Fig. 6 shows the output characteristics associated with the HEMT device with and without cap layer when V_{DS} = 10 V and various V_{GS}. The cap region improves the current-carrying capability of the HEMT and enhancing the electrical performance. The electron density profile along the horizontal distance with different GaN cap thicknesses (2-4 nm) under equilibrium conditions is shown in Fig. 7. There is a sharp drop in electron density under the gate. This represents a depletion region where free electrons are removed because

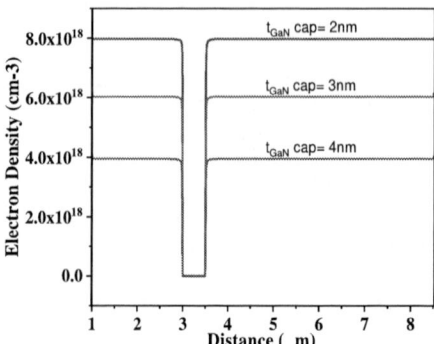

Fig. 7. Electron density profile along the horizontal distance for different GaN cap thickness (t_{GaN} = 2-4 nm) at equilibrium condition

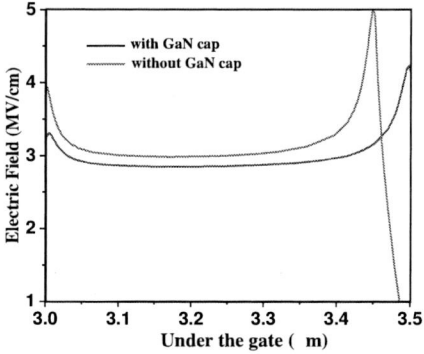

Fig. 8. The electric-field distributions under the gate along the horizontal distance at V_{GS} = -5 V and V_{DS} = 10 V

of the workfunction difference between gate material (5.0 eV) and GaN; the electron density decreases while increasing the cap thickness. In vertical cutline AlGaN/GaN, the 2-DEG interface electron density increases to $2.23 \times 10^{19} cm^{-3}$ with the cap layer compared to $1.89 \times 10^{19} cm^{-3}$ without the cap layer. GaN cap effectively lowers the field under the gate measured without the cap region, as shown in Fig. 8. The reduction in electric field is highly beneficial because it reduces the premature breakdown of device. The estimated

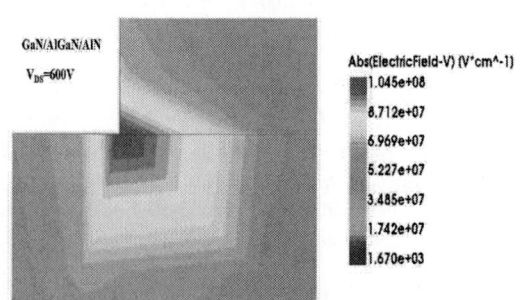

Fig. 9. Electric field distribution of the device at V_{DS} = 600 V

979-8-3315-3013-6/25 $31.00 © 2025 IEEE

electric field at $V_{DS} = 600$ V and $V_{GS} = 0$ V is shown in Fig. 9 and it confirms an elevation in electric field is concentrated near the edge of gate-drain.

IV. CONCLUSION

The significance of GaN cap layer has been systematically investigated in the GaN/AlGaN/AlN/GaN HEMT over a SiC substrate. The electron mobility in the GaN channel with a cap region increased from 1315 cm²/V-s to 1425 cm²/V-s and a maximum drain current rising from 0.36 A to 0.58 A. An increase in drain current is observed with the insertion of a GaN cap region, which improves the device electrical performance. There is a larger threshold voltage shift towards negative in the presence of the GaN cap region, which is consistent with the increased 2-DEG density. The electric field distribution confirms that the inclusion of a cap layer can largely reduce the critical electric field, which will further improve the breakdown characteristics of the device.

V. ACKNOWLEDGMENT

The authors thank IIITDM Kancheepuram for providing financial support to conduct research activities and attend the conference.

REFERENCES

[1] X. Ding, Y. Zhou, and J. Cheng, "Review of gallium nitride power device and its applications in motor drive," *CES Transactions on Electrical Machines and Systems*, vol. 3, no. 1, pp. 54–64, 2019.

[2] M. Meneghini, C. De Santi, I. Abid, M. Buffolo, M. Cioni, R. A. Khadar, L. Nela, N. Zagni, A. Chini, F. Medjdoub *et al.*, "Gan-based power devices: Physics, reliability, and perspectives," *Journal of Applied Physics*, vol. 130, no. 18, 2021.

[3] R. Cuerdo, F. Calle, A. Braña, Y. Cordier, M. Azize, N. Baron, S. Chenot, and E. Muñoz, "High temperature behaviour of gan hemt devices on si (111) and sapphire substrates," *physica status solidi c*, vol. 5, no. 6, pp. 1971–1973, 2008.

[4] W. Tan, M. Uren, P. Fry, P. Houston, R. Balmer, and T. Martin, "High temperature performance of algan/gan hemts on si substrates," *Solid-State Electronics*, vol. 50, no. 3, pp. 511–513, 2006.

[5] S. Kargarrazi, A. S. Yalamarthy, P. F. Satterthwaite, S. W. Blankenberg, C. Chapin, and D. G. Senesky, "Stable operation of algan/gan hemts for 25 h at 400° c in air," *IEEE Journal of the Electron Devices Society*, vol. 7, pp. 931–935, 2019.

[6] G. Greco, F. Iucolano, and F. Roccaforte, "Review of technology for normally-off hemts with p-gan gate," *Materials Science in Semiconductor Processing*, vol. 78, pp. 96–106, 2018.

[7] C. Shi, L. Yang, M. Zhang, M. Wu, B. Hou, H. Lu, F. Jia, F. Guo, W. Liu, Q. Yu *et al.*, "High-efficiency algan/gan/graded-algan/gan double-channel hemts for sub-6g power amplifier applications," *IEEE Transactions on Electron Devices*, vol. 70, no. 5, pp. 2241–2246, 2023.

[8] A. Soni and M. Shrivastava, "Computational modelling-based device design for improved mmwave performance and linearity of gan hemts," *IEEE Journal of the Electron Devices Society*, vol. 8, pp. 33–41, 2019.

[9] J. Kim and Y. Kwon, "A high-performance gan-modified nonuniform distributed power amplifier," *IEEE Transactions on Microwave Theory and Techniques*, vol. 68, no. 5, pp. 1729–1740, 2020.

[10] U. K. Mishra, P. Parikh, and Y.-F. Wu, "Algan/gan hemts-an overview of device operation and applications," *Proceedings of the IEEE*, vol. 90, no. 6, pp. 1022–1031, 2002.

[11] S. Yoshida, Y. Sakaida, J. Asubar, H. Tokuda, and M. Kuzuhara, "Current collapse in algan/gan hemts with a gan cap layer," in *2015 IEEE International Meeting for Future of Electron Devices, Kansai (IMFEDK)*. IEEE, 2015, pp. 48–49.

[12] T. Zhang, Y. Lv, R. Li, Y. Zhang, Y. Zhang, X. Li, J. Zhang, and Y. Hao, "Current-collapse suppression of high-performance lateral algan/gan schottky barrier diodes by a thick gan cap layer," *IEEE Electron Device Letters*, vol. 42, no. 4, pp. 477–480, 2021.

[13] S. Sarkar, R. P. Khade, A. DasGupta, and N. DasGupta, "Effect of gan cap layer on the performance of alinn/gan-based hemts," *Microelectronic Engineering*, vol. 258, p. 111756, 2022.

[14] P. Waltereit, S. Müller, K. Bellmann, C. Buchheim, R. Goldhahn, K. Köhler, L. Kirste, M. Baeumler, M. Dammann, W. Bronner *et al.*, "Impact of gan cap thickness on optical, electrical, and device properties in algan/gan high electron mobility transistor structures," *Journal of Applied Physics*, vol. 106, no. 2, 2009.

[15] A. Bellakhdar and A. Telia, "Influence of the gan cap layer thickness on the two-dimensional electron gas (2-deg) sheet charge density of gan/alinn/gan hemts with polarization effect," *Digest Journal of Nanomaterials and Biostructures*, vol. 17, no. 1, pp. 233–246, 2022.

[16] E. B. Prakash, V. Maitra, A. P. Dadi, A. Ray, and S. Bordoloi, "Investigation of algan/gan hemt device performance with variation of gan cap layer thickness along with doping concentration," in *2024 15th International Conference on Computing Communication and Networking Technologies (ICCCNT)*. IEEE, 2024, pp. 1–6.

[17] H. Kang, Q. Wang, H. Xiao, C. Wang, L. Jiang, C. Feng, H. Chen, H. Yin, S. Qu, E. Peng *et al.*, "Effects of a gan cap layer on the reliability of algan/gan schottky diodes," *physica status solidi (a)*, vol. 212, no. 5, pp. 1158–1161, 2015.

[18] Z. Nie, K. Wang, X. Liu, and H. Wang, "Effect of gan cap thickness on the dc performance of algan/gan hemts," *Micromachines*, vol. 15, no. 5, p. 571, 2024.

[19] J.-S. Moon, J. Wong, B. Grabar, M. Antcliffe, P. Chen, E. Arkun, I. Khalaf, A. Corrion, J. Chappell, N. Venkatesan *et al.*, "360 ghz f max graded-channel algan/gan hemts for mmw low-noise applications," *IEEE Electron Device Letters*, vol. 41, no. 8, pp. 1173–1176, 2020.

[20] Y. Kumazaki, S. Ozaki, N. Okamoto, N. Hara, and T. Ohki, "Low-resistance and low-thermal-budget ohmic contact by introducing periodic microstructures for algan/aln/gan hemts," *IEEE Transactions on Electron Devices*, vol. 69, no. 6, pp. 3073–3078, 2022.

[21] Synopsys Inc., *Synopsys-Sentaurus-TCAD*, Mountain View, CA, USA, 2022, version T-2022.03.

[22] X.-G. He, D.-G. Zhao, and D.-S. Jiang, "Formation of two-dimensional electron gas at algan/gan heterostructure and the derivation of its sheet density expression," *Chinese physics B*, vol. 24, no. 6, p. 067301, 2015.

Data-Driven Intermittent Fault Diagnosis: From Data Acquisition to CNN-Based Classification in Synchronous Generators

Anirud R S
Department of Electronics and Communication Engineering, Amrita School of Engineering, Coimbatore, Amrita Vishwa Vidyapeetham, India

Srujana Karri
Department of Electronics and Communication Engineering, Amrita School of Engineering, Coimbatore, Amrita Vishwa Vidyapeetham, India

Surya Teja Nemani
Department of Electronics and Communication Engineering, Amrita School of Engineering, Coimbatore, Amrita Vishwa Vidyapeetham, India

Sunil Nag P V
Department of Electronics and Communication Engineering, Amrita School of Engineering, Coimbatore, Amrita Vishwa Vidyapeetham, India
pv_sunil@cb.amrita.edu

C Santhosh Kumar
Department of Electronics and Communication Engineering, Amrita School of Engineering, Coimbatore, Amrita Vishwa Vidyapeetham, India

Kuruvachan K. George
NCS Pte. Ltd., Singapore

Abstract—Condition monitoring is essential for maintaining the reliability of electrical machinery, particularly synchronous generators used in power production, maritime applications, military operations, and laboratory testing. Stator inter-turn defects are a significant issue due to their potential to cause efficiency decline and machine failure. Although existing test beds primarily focus on permanent faults, they lack the capability to replicate intermittent faults. This study presents a Fault Injection Unit (FIU) programmed with LabVIEW to generate intermittent faults with durations of 0.5s, 0.2s, and 0.1s, alongside permanent faults, under various load conditions, 0.5 A and 1 A. A 1D Convolutional Neural Network (CNN) trained on permanent fault data is evaluated for intermittent fault classification. The results show that while the model excels at identifying permanent faults, its accuracy decreases for intermittent faults, particularly with shorter fault durations. FFT analysis reveals that longer intermittent faults increase frequency ripple content at the third harmonic (150 Hz). The impact of preprocessing methods, including Short-Time Fourier Transform (STFT) and wavelet analysis, on classification accuracy is also examined, highlighting the need for specialized processing techniques to improve intermittent fault detection.

Keywords—*Machine fault diagnosis, Intermittent faults, Permanent fault condition, inter-turn fault, synchronous generator, CNN*

I. INTRODUCTION

Condition monitoring is a crucial step in implementing a system or a process. It is the process of monitoring the working of equipment. Condition monitoring [1] includes recognition of faults in the system. It is very important for critical equipment.

Synchronous generator is a machine that works on the principle of electromagnetic induction. It finds its application in a wide number of fields apart from its primary usage for power generation in utility power plants. They are used in important areas such as for powering submarines, mobile power units in the military, for testing facilities to generate high-frequency power in laboratories, and wind and hydro turbines, among others [2].

Hence Condition monitoring of synchronous generator is important

Two types of faults could occur in a synchronous generator: electric and mechanical. In electric faults, one of the common types is a stator fault. In low-voltage electric machines, stator faults account for 36 per cent of failures. In high-voltage electric machines, they account for up to 66 per cent of failures [3]. Stator faults are caused by the degradation of insulation, which can lead to inter-turn faults, meaning short circuits that occur between the turns of a winding. This can cause internal localized heating, a reduction in the efficiency of the machinery and produce unnecessary noise that can distort the required voltage and current waveforms [4]. These defects are of significant concern and can lead to drastic damage if not detected and addressed at the earliest.

The methods for condition monitoring fall into two classes: data-based and model-based [5]. Of these two classes of methods, data-based methods are preferred because obtaining a model is often difficult. Generating data from a real system is prohibitive in terms of cost; hence, a fault injection capable system is required to generate the data for developing the data-based methods. Researchers have developed fault injection-capable electrical machines to study inter-turn faults and validate fault detection methods, the following provides a brief of the various works in this regard. A setup for surface-mounted permanent magnet synchronous motors included a motor with adjustable inter-turn faults simulated by short-circuiting specific coils in the stator, using a fault resistance of 0.07 Ω [6]. The test bed-incorporated stator current measurement tools and a data acquisition system to compare healthy and faulty conditions through current harmonics analysis. Another testbed utilized a modified permanent magnet synchronous generator with redesigned stator windings to simulate faults by shorting specific coils. Dual-loop controllers were used to manage system bandwidths, and faults of varying severities were introduced to analyse system performance [7]. Similarly, a salient pole synchronous generator was used to simulate stator and rotor turn-to-turn faults by shorting various percentages of turns in the windings. This system employed a portable data acquisition module and MATLAB for data analysis [8]. A brushless wound-field synchronous generator was tested under linear and non-linear loads with faults introduced by progressively reducing short-circuit resistance, using a dSPACE platform for real-time control and data acquisition [9]. Another test bed focused on permanent magnet synchronous machines, incorporating modified stator windings, sensors, and fault injection techniques, while using finite element simulations to complement experimental data [10]. A permanent magnet-assisted synchronous reluctance motor test bed implemented a current injection technique to mitigate fault impact, allowing for real-time monitoring and control using high-frequency data acquisition systems [11]. These setups highlight diverse fault simulation and detection approaches, providing valuable data for condition monitoring system development.

All the above testbeds, however, did not provide a means to inject intermittent faults. These intermittent faults are many times beginning stages for permanent faults and detecting them timely has lot of significance. Hence, we describe here the setup used in this work to inject intermittent faults in the synchronous generator. This is the primary contribution of this paper. Further in this work, a CNN model, trained on data from a permanent inter-turn fault, was tested using data obtained from the intermittent fault test bed. This study showed that various data processing techniques improved the performance of the fault detection system for intermittent fault. Using the flexibility offered by the setup, we study the effect of the intermittent fault duration on the frequency spectrum.

Next, section II describes the setup used to generate the intermittent fault from a fault injection capable synchronous generator. Section III describes the LABVIEW implementation. Section IV gives the results and discussion on using the CNN for fault diagnosis and finally, section V gives the conclusions and future work.

II. SETUP TO GENERATE INTERMITTENT FAULT DATA

To generate intermittent fault data, a specialized experimental setup was designed, integrating multiple components to ensure precise fault injection and data acquisition. Fig 1 shows a schematic of the setup and Fig 2 shows annotated picture of the same. At the core of the setup is a customized synchronous generator[12], the specifications of the machine are given in the appendix. The generator has a 5 kVA rating operating at the frequency of 50 Hz, and phase voltage rating of 230 V. The three phase stator winding of the generator was designed with 12 coils in each phase in which the six coils are modified for fault injection. Each of the modified coils is equipped with taps at 30%, 60%, and 82% of the turns to allow flexibility in fault generation. All taps are routed to a front panel, enabling controlled interturn short circuits. The rotor, a four-pole structure, receives power through an external source, with each pole housing three coils, each consisting of 175 turns.

The currents and voltages from the synchronous generator were continuously monitored using sensors. Hall Effect sensors (LV25P) measured the voltage between each phase terminal and neutral, while Hall Effect sensors (LA55P) measured the current through each phase. The sensor outputs are converted to voltage using resistors. For efficient signal routing, an SCB68 Signal Connection Block acts as an interface for routing the wires from the sensors to the data acquisition unit NI-DAQ as shown in Fig 2. The data acquisition unit consists of a chassis PXI 1042, with NI PXI- 6221 unit, the Fault injection unit (FIU) NI 2514 (7-channel) and PXI 8108 embedded controller. NI PXI-6221 has 16 single-ended analog input channels capable of sampling up to 250 kS/s per channel. The PXI 8108 embedded controller, powered by an Intel Core 2 Duo processor. This takes the role of the central processing unit of the PXI system, it manages data acquisition, processing, and control tasks, ensuring smooth operation and real-time acquisition of intermittent fault data.

To introduce faults in a controlled and repeatable manner, a Fault Injection Unit (FIU) – NI 2514 (7-channel) was employed. This allows precise intermittent fault injection with millisecond accuracy. With each channel capable of handling up to 40A of current, the FIU was programmed using LabVIEW with NI-DAQmx Drivers as shown in Fig.2. Together, these components form a robust experimental platform, facilitating precise fault injection, real-time monitoring, and comprehensive data collection for in-depth intermittent fault diagnosis. Next, an overview of the LABVIEW program is provided.

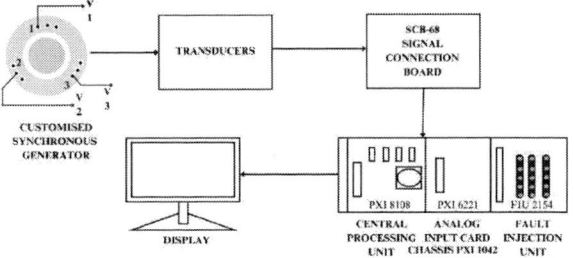

Fig. 1. Schematic representation of experimental setup

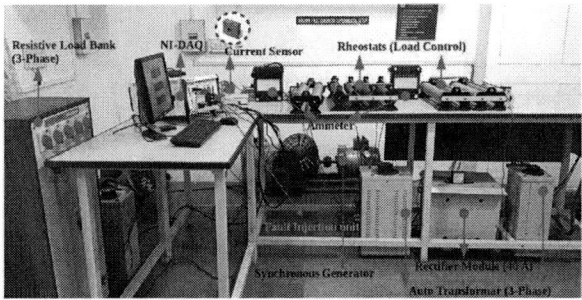

Fig.2: Annotated picture of experimental setup

III. LABVIEW IMPLEMENTATION

LabVIEW facilitates real-time monitoring, fault simulation, and data acquisition from the synchronous generator setup. Faults are injected using defined start times and delays. The DAQ Assistant express VI acquires signals like voltage and current for analysis. Logged data is stored in csv files, organized under defined conditions for further processing. This section, presents a brief of how the data required for training and testing the neural network systems has been obtained. The LabVIEW interface is shown in Fig3.

979-8-3315-3013-6/25 $31.00 © 2025 IEEE   307

Based on the requirements of the experiment, the following features are provided in the user interface:

1. Waveform chart that plots the RMS values of the voltages and currents obtained.
2. Load applied to the generator
3. Phase in which fault is introduced.
4. Required delay time for which fault must persist.
5. Coil in which fault is introduced.

Below is the algorithm used for Intermittent Fault Injection and Data Acquisition System:

Algorithm 1: Intermittent Fault Injection and Data Acquisition using LabVIEW:

Initialization:

1a) Define parameters: load, machine type, phase, fault start time, delay time.
1b) Define switch settings: Topology, Channel1, Channel2.
1c) Initialize and configure the Data Acquisition (DAQ) system for data collection and storage.

Fault Injection Process:

2a) Start timers for fault triggering and delay
2b) Trigger the fault when the fault-triggering timer reaches the fault start time by activating switches and setting the topology
2c) Introduce a delay to allow the system to stabilize

Data Acquisition and Monitoring:

3a) Continuously acquire system parameters: voltage, current, and phase
3b) Save the collected data in a measurement file for analysis
3c) Display real-time system behaviour on a waveform chart.

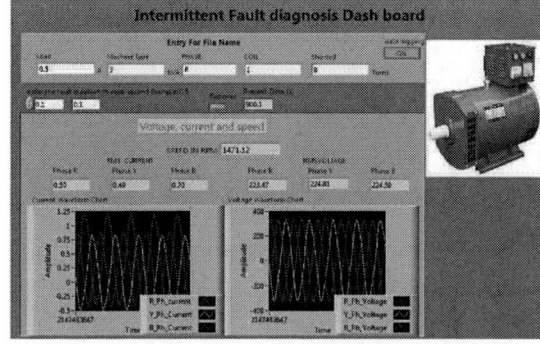

Fig. 3 LabVIEW Interface for fault injection

IV. CNN FOR FAULT DIAGNOSIS

Using the experimental setup above, data sets were generated for both permanent fault and intermittent fault conditions. The permanent fault data set was generated by introducing interturn shorts of 30%, 60% and 82% turns in each of the six coils. This was done for each of the R, Y and B phases. The data was collected under 4 different load conditions. Like this extensive data samples were collected under both healthy and various fault conditions. Each sample of the dataset contained 1000 samples collected at a rate of 1000 Hz. The intermittent fault data was collected under the same conditions as above but with intermittent faults of duration, 0.5, 0.2 and 0.1 seconds. This injection of the intermittent fault was facilitated especially by the use of the Fault Injection Unit (FIU) programmed with LabVIEW as described previously.

The data thus collected was used to investigate whether a Convolutional Neural Network (CNN) trained on permanent fault data set can be used to detect the presence of an intermittent fault. In this study, the CNN is utilized for binary fault classification. The CNN is a one-dimensional CNN (1D-CNN) [12]. The network comprises multiple convolutional layers with decreasing filter sizes to capture hierarchical features at various scales, with pooling layers to reduce dimensionality. After the above layers, a dense network is used. This network helps in classification. The architecture is fine-tuned with the help of architectural parameters like kernel and batch sizes, and number of filters. Fig 4 shows the schematic representation of the CNN architecture, and the algorithm 2 describes the various processes in the CNN.

Algorithm 2: Convolutional Neural Network

---

1. Convolution: (Conv1D)

1a) Extraction of local features such as transient spikes and distortions.
1b) Stacks with decreasing kernel sizes (16, 8, 4) to capture hierarchical patterns.

---

2. Activation: (ReLU)

2a) The Rectified Linear Unit introduces non linearity and avoids the problem of vanishing gradient.

---

3. Pooling (Max-Pooling)

3a) Selection of maximum activation in non-overlapping windows.

3b) Reduction of overfitting and improvement of translational invariance.

---

4. Flatten - Dense

4a) Flattening of feature maps into a vector and passing through fully connected layers.

---

5. Output

5a) Utilizes sigmoid activation with a single neuron to interpret the probability of presence of fault at a binary decision threshold of 0.5.

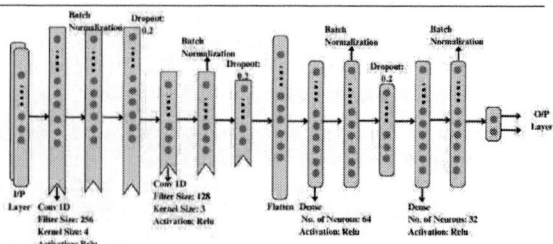

Fig. 4 Architecture of the CNN

For training the CNN, 7200 samples of data were used. Out of these, 4320 samples were faulty data and 2880 samples were normal data. The training curve is shown in Fig 5. Further, 80% of the data was used for training and 20% was used for testing. To further analyse the performance of the CNN, various pre-processing techniques were applied to the time domain data. Techniques such as Fast Fourier Transform (FFT), Short Time Fourier Transform (STFT) and wavelet preprocessing have been used. Each method has been applied for all the 3 phases (R, Y and B) separately and the results are tabulated in Tables I and II.

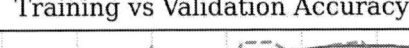
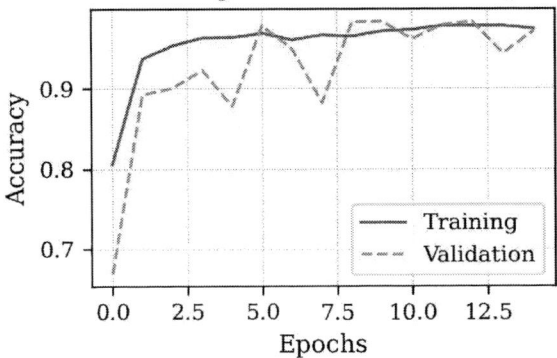

Fig.5 Accuracy plot of training the CNN on fault and no-fault data

V. RESULTS

Table I shows the training and testing accuracies, when the CNN was trained and tested with permanent fault data. It can be inferred that frequency domain data gives better accuracy. When the same model was tested on intermittent fault in frequency domain, it was observed that the accuracy score declined to 87% for R phase, 86% for Y phase, and 88% for B phase, as shown in Table II. This was expected as the duration of fault during each sample reduced. Further the model is not able to accurately classify intermittent faults as fault data and further processing of data is required to improve the accuracy. Hence, preprocessing techniques like Short- time Fourier Transform (STFT) and wavelet denoising techniques were used to improve the fault detection accuracy of intermittent faults. Wavelet denoising techniques [13] with db4, BIOR2.2 and sym4

Data Type	Phase	Accuracy
Time Domain	R	0.77
	Y	0.76
	B	0.61
Frequency Domain (FFT)	R	0.99
	Y	0.99
	B	0.98

wavelets were used. Following this, high accuracies of up to 99% have been achieved as shown in Table II.

TABLE I.   TRANING AND TESTING CNN MODEL WITH PERMANENT FAULT DATA

TABLE II.   TESTING THE MODEL WITH INTERMITTENT FAULT DATA

Preprocessing	Phase	Accuracy
Frequency Domain (FFT)	R	0.87
	Y	0.86
	B	0.88
Short-time Fourier Transform (STFT)	R	0.88
	Y	0.89
	B	0.90
Wavelet denoising (db4)	R	0.88
	Y	0.99
	B	0.99
Wavelet denoising (BIOR2.2)	R	0.90
	Y	0.99
	B	0.99
Wavelet denoising (sym4)	R	0.88
	Y	0.99
	B	0.99

To visualize the intermittent fault data, the spectrogram plots, when intermittent fault of 2 seconds during a time frame of 4 seconds, are presented in Figs. 6, 7 and 8. The figures show spectrograms for the currents in the R, Y and B phases respectively, when an intermittent for fault was injected in the B phase. The effect is clearly visible in the plots with phase B current showing the maximum effect.

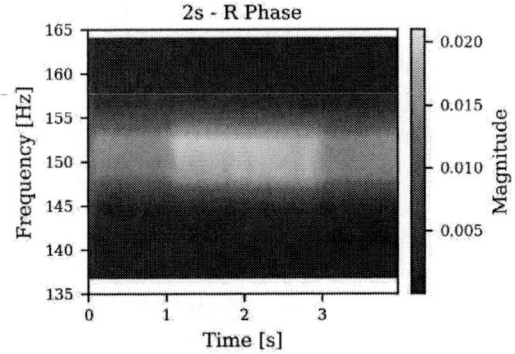

Fig.6 Spectrogram of intermittent fault in B phase (current in R phase)

979-8-3315-3013-6/25 $31.00 © 2025 IEEE

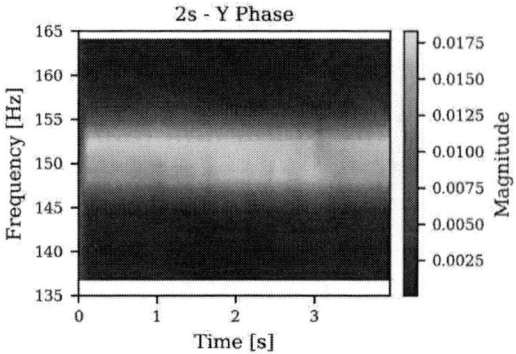

Fig.7 Spectrogram of intermittent fault in B phase (current in Y phase)

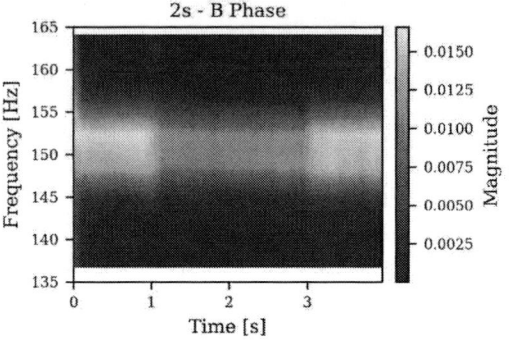

Fig.8 Spectrogram of intermittent fault in B phase (current in B phase)

Figs. 9, 10, 11, 12 and 13 show the results of an experimental analysis of the frequency domain characteristics of signals associated with intermittent faults of varying fault durations using Fast Fourier Transform (FFT). The experimental data consisted of files, each representing different intermittent fault durations: 0.1 sec, 0.05 sec, 0.025 sec, 0.001 sec and no-fault condition. For each case, the corresponding FFT data is extracted and plotted to examine the spectral behaviour around the 150 Hz region, where key harmonic components are expected. The analysis focuses on how intermittent fault durations influence the spectral amplitude, which may provide insights into further research in intermittent fault detection. From the figures we can conclude that as the duration of the intermittent fault increases the frequency ripple content in the region of the third harmonic that is, 150 Hz, increases. The healthy sample does not have any ripple content and this ripple content increases as the duration of intermittent fault increases. This fact can help in developing effective algorithms to detect intermittent fault.

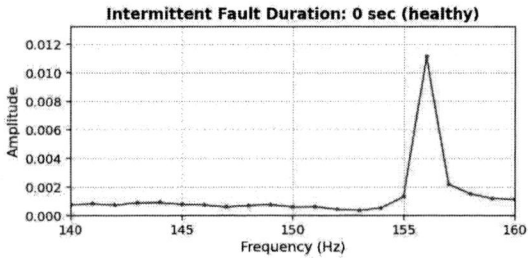

Fig.9. FFT for healthy sample

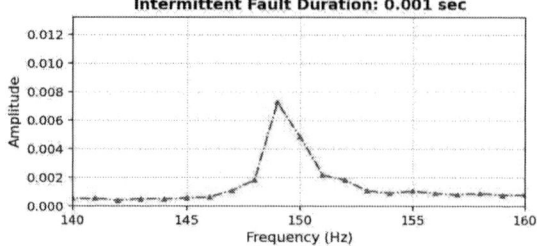

Fig.10. FFT for intermittent fault of 0.001 sec duration

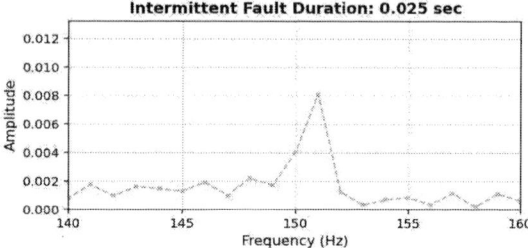

Fig.11. FFT for intermittent fault of 0.025sec duration

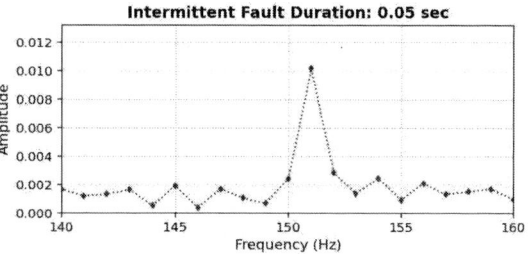

Fig.12. FFT for intermittent fault of 0.05 sec duration

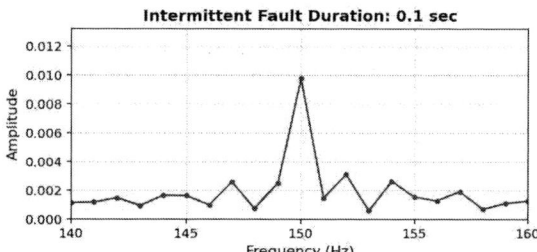

Fig.13. FFT for intermittent fault of 0.1 sec duration

VI. CONCLUSION AND FUTURE WORK

This work presents an experimental setup capable of injecting controlled intermittent faults in a synchronous generator, addressing a key limitation in existing test beds. A Convolutional Neural Network (CNN) trained on permanent fault data was tested on intermittent faults to evaluate its effectiveness in fault diagnosis. The classification accuracy declined by 10% compared to when permanent fault data was used for testing. Further experiments demonstrate that various preprocessing techniques, like Short-Time Fourier Transform (STFT) and wavelet denoising, improve the classification accuracy.

A key observation from the FFT analysis is that as the duration of the intermittent fault decreases, the fault signature ripple in the frequency domain becomes less prominent, reducing classification performance. This suggests that very short-duration faults may not introduce sufficiently strong frequency components, necessitating advanced signal processing techniques and machine learning techniques to enhance classification accuracy.

Future work will focus on optimizing preprocessing methods, exploring wavelet-based time-frequency analysis, and refining the CNN architecture to improve the detection of intermittent faults with very small durations. These improvements will contribute to more robust and reliable condition monitoring for synchronous generators in critical applications.

APPENDIX

Table III gives the specifications of the customised synchronous generator [12].

TABLE IIIA. SPECIFICATIONS OF SYNCHRONOUS GENERATOR

Parameter	Generator
Rated power	5 kVA
Rated voltage	415 V
Rated frequency	50 Hz
Connection type	star
Number of poles	4
Number of phases	3
Speed	1500 rpm
Current	6.5 A
Power factor	0.8

REFERENCES

[1] J. Revant, Rahul Sree Kumar, K. Rameshkumar, and D. S. B. Mouli, "Acoustic Emission-Based Grinding Wheel Sharpness Monitoring Using Machine Learning Classifier," 2021, pp. 511–522. doi: 10.1007/978-981-15-4745-4_45.

[2] "IEEE Standard for Salient-Pole 50 Hz and 60 Hz Synchronous Generators and Generator/Motors for Hydraulic Turbine Applications Rated 5 MVA and Above," Sep. 22, 2005, *IEEE, Piscataway, NJ, USA*. doi: 10.1109/IEEESTD.2006.99082.

[3] J. He, C. Somogyi, A. Strandt, and N. A. O. Demerdash, "Diagnosis of stator winding short- circuit faults in an interior permanent magnet synchronous machine," in *2014 IEEE Energy Conversion Congress and Exposition (ECCE)*, IEEE, Sep. 2014, pp. 3125–

3130. doi: 10.1109/ECCE.2014.6953825.

[4] M. Otero, G. R. Bossio, P. M. de la Barrera, O. Tyshakin, and R. Leidhold, "Inter-turn faults detection in Induction Motor drives using zero- sequence signal injection," in *2018 International Symposium on Power Electronics, Electrical Drives, Automation and Motion (SPEEDAM)*, IEEE, Jun. 2018, pp.202–207.doi:10.1109/SPEEDAM.2018.8445411.

[5] P. V. Sunil Nag and C. Santhosh Kumar, "Development of Fault Indicators for Stator Inter- turn Fault Diagnosis of a Synchronous Generator Using Kalman Filter," 2022, pp. 601–619. doi: 10.1007/978-981-16-9239-0_46.

[6] L. Romeral, J. C. Urresty, J.-R. Riba Ruiz, and A. Garcia Espinosa, "Modeling of Surface-Mounted Permanent Magnet Synchronous Motors with Stator Winding Interturn Faults," *IEEE Transactions on Industrial Electronics*, vol. 58, no. 5, pp. 1576–1585, May 2011, doi: 10.1109/TIE.2010.2062480.

[7] F. Niu *et al.*, "Robust Inter-Turn Short-Circuit Fault Detection in PMSGs With Respect to the Bandwidths of Current and Voltage Controllers," *IEEE Trans Power Electron*, vol. 38, no. 8, pp. 10269–10279, Aug. 2023, doi: 10.1109/TPEL.2023.3273897.

[8] S. Afrandideh, F. Haghjoo, S. Cruz, and M. Eshaghi Milasi, "Detection of Turn-to-Turn Faults in the Stator and Rotor of Synchronous Machines During Startup," *IEEE Transactions on Industrial Electronics*, vol. 68, no. 8, pp. 7485–7495, Aug. 2021, doi: 10.1109/TIE.2020.3003626.

[9] S. Nadarajan, S. K. Panda, B. Bhangu, and A. K. Gupta, "Online Model-Based Condition Monitoring for Brushless Wound-Field Synchronous Generator to Detect and Diagnose Stator Windings Turn-to- Turn Shorts Using Extended Kalman Filter," IEEE Transactions on Industrial Electronics, vol. 63, no. 5, pp. 3228–3241, May 2016, doi:10.1109/TIE.2016.2535959.

[10] T. Orlowska-Kowalska et al., "Fault Diagnosis and Fault-Tolerant Control of PMSM Drives–State of the Art and Future Challenges," IEEE Access, vol. 10, pp.59979–60024,2022, doi: 10.1109/ACCESS.2022.3180153.

[11] B. Wang, J. Wang, A. Griffo, and L. Huang, "A Turn Fault Mitigation Strategy Based on Current Injection Technique for a Triple Three-Phase PMA SynRM," IEEE Transactions on Industrial Electronics, vol. 67, no. 4, pp. 2511–2522, Apr. 2020, doi: 10.1109/TIE.2019.2908595.

[12] K. T. Sreekumar, C. S. Kumar, and K. I. Ramachandran, "Deep Discriminative Feature Learning and Feature Space Transformation for Scalable Machine Fault Diagnosis," *IEEE Access*, vol. 12, pp. 107944–107958, 2024, doi: 10.1109/ACCESS.2024.3438099.

[13] Praveen Kumar. N and Balakrishnan. P, "EWT Implementation for Examining Demagnetization Fault in PMSM using FEM *," in *2022 IEEE International Conference on Power Electronics, Drives and Energy Systems (PEDES)*, IEEE, Dec. 2022, pp. 1–6. doi: 10.1109/PEDES56012.2022.10080132.

Research Outcome Based Optimization Algorithm applied to Optimal Phasor Measurement Unit Placement

Madhu Kishore Devara Chejarla
Department of Electrical Engineering
National Institute of Technology
Warangal, India
chejarlamadhukishore@gmail.com

Sailaja Kumari Matam
Department of Electrical Engineering
National Institute of Technology
Warangal, India
sailaja@nitw.ac.in

Abdul Mujeer Syed
Department of Electrical Engineering
National Institute of Technology
Warangal, India
sa21eerer04@student.nitw.ac.in

Abstract—In this paper, an evolutionary optimization algorithm is presented by mimicking the nature and working style of researchers. The research community is divided into four classes and approximate mathematical modeling of each class of researchers is presented. The proposed algorithm is validated by applying on 23 benchmark functions. For all benchmark functions the proposed algorithm results in the final solution that is independent of number of iterations consumed for hitting the near global optimal solution. This feature makes it unique compared to other optimization algorithms. The case study using benchmark functions is compared with four other optimization algorithms. From the comparison the supremacy of proposed algorithm is well established. A Binary version of this algorithm is also presented and applied it on a optimal Phasor Measurement Unit placement problem. Case studies using IEEE 14 and 30 bus systems are presented.

Index Terms—Phasor Measurement Unit (PMU), Optimal PMU placement problem, Research community, Researcher, Research outcome based optimization algorithm.

I. INTRODUCTION

Engineering and scientific optimization problems over a period of time used conventional methods to obtain the optimized solutions. The main drawback of these methods is that, a better performance is ensured with better mathematical model. They exhibit slow convergence characteristics when the search space is large and are more suitable when the search space is continuous. As the complexity of systems are growing and their associated optimization problems are becoming combinatorial in nature, modeling them exactly is a difficult task. Evolutionary optimization algorithms overcome these issues and over a period of time they proved to be more advantageous than classical optimization techinques.

The first and most commonly used evolutionary optimization algorithm is Genetic Algorithm [1]. The working prinicple of the algorithm is Darwinian theory of survival of the fittest. Some of the optimization algorithms are Particle Swarm Optimization(PSO) [2], Diffrential evolution [3], Harmony search [4], Bacteria Foraging Optimization(BFO) [5], Artifial Immune Algorithm(AIA) [6], Ant Colony Optimization(ACO)

[7], Artifial Bee Colony(ABC) [8], Fire Fly Algorithm(FFA) [9], Gravitational Search Algorithm [10], Grenade Explosion Algorithm [11], Biogeography based optimization [12], Teaching - learning based optimization (TLBO) [13], Krill - herd algorithm [14], Simulated annealing, Electromagnetism - like mechanism (EM) algorithm [15], Imperialist competitive algorithm (ICA) [16] and Wild Goats Algorithm (WGA) [17]. These algorithms are bio-inspired optimization methods, each mimicking natural or physical processes to solve complex problems. For instance, Genetic Algorithm and Particle Swarm Optimization are known for efficient global search, while Differential Evolution and Harmony Search emphasize precision and adaptability. Methods like Ant Colony Optimization and Artificial Bee Colony are effective for combinatorial optimization, while techniques like Simulated Annealing balance exploration and exploitation for robust solutions. Their performance varies depending on the problem's nature and the algorithm's configuration.

All the above mentioned algorithms are superior in some aspects and have a lag in other aspects. Therefore, research is going on either to reduce the negative aspects of the existing algorithms or to develop new algorithms which may have superiority in all considered aspects. Optimal PMU placement is done using linear formulation [18], zero-injection phase strings [19], Tri-objective state estimation [20], hybrid method [21] and algebraic approach [22].

In this work, a new evolutionary optimization algorithm is proposed by mimicking the nature of the research community existing all over the world. The proposed algorithm is validated by applying and examining the results on the 23 benchmark functions. The binary version of proposed algorithm is implemented and applied on the basic optimal PMU placement problem. The proposed algorithm is based on the inspiration taken from the research community all over the world. The contributions of the proposed method -

1) A new meta-heuristic algorithm inspired by the nature of the research community all over the world and applied to the optimal PMU placement problem.

979-8-3315-3013-6/25 $31.00 © 2025 IEEE

2) The algorithm is tested with 23 benchmark functions. Some of the benchmark functions have features of high dimensionality. The algorithm giving very promising results for those optimization functions.
3) The algorithm is showing supremacy in terms of robustness and reaching the global optima.

The rest of the paper is organised as follows: section II describes the behaviour of researchers, section III presents the approximate mathematical models to resemble the behaviour of the each class of researchers, section IV presents the proposed algorithm, sections V and VI report the validation and case study using the proposed algorithm. Finally section VII concludes the paper.

II. Behaviour of Researchers

Researcher is the one who tries to provide better solution to a problem after studying the existing solutions. Researchers (R) are divided into four classes based on their level of intelligence [There is no earlier classification of researchers available. It is assumed that present researchers are students in the past. In any class, students are catergorised into, either 4 or 5 types. Here four types researchers are considered]. They are -

1) Intelligent researcher (IR)
2) Above average researcher (AAR)
3) Average researcher (AR)
4) Below average researcher (BAR)

Intelligent researcher (IR): IR always provides a new and better solution after studying and analyzing existing solutions.

Above average researcher (AAR): AAR provides a hybrid solution of existing solution after studying existing solutions and working on few of solutions.

Average researcher (AR): AR provides the hybrid solution after studying (not completely) a few of the existing solutions and working on these solutions.

Below average researcher (BAR): BAR provides the hybrid solution after paritially studying existing solutions and working on few of the solutions. All the solutions provided by the research community will be scrutinized at a higher level called review. If the reviewers feel, they are useful solutions then they will be added to the existing solution space. In this way new generation of researchers will come and study the existing solutions and provide their own solutions. Based on reviewers evaluation, all useful solutions will be added to the existing solution space, and this process repeats. The Solution updation is happening based on the research outcome of researchers. Therefore, The proposed optimization algorithm is named as Research outcomoe based optimization (Robo) algorithm.

III. Approximate Mathematical Modeling of Researchers Behaviour

Behaviour of research community is a natural phenomena. Understanding and modeling of any natural phenomena is quite difficult. Therefore for modeling the behaviour of research community, the following assumptions are made.

1. The solutions provided by the intelligent researchers are randomly initialized along with the initial solutions. This total randomly initialized solution space is called initial solution space (ISS).

2. Reviewers are bound to accept the solutions provided by the researchers.

NR - Number of researchers
PIR - Probability of IR among researchers
PAAR - Probability of AAR among researchers
PAR - Probability of AR among researchers
PBAR - Probability of BAR among researchers
ISS - Initial solution space
NISS - Number of solutions in ISS
BSS - Best solution space
FSS - Final solution space
R - R is a solution vector of size NR X 1 provided by that generation of research community
itermax - Number of generations research community will work for that problem

ISS: This is the randomly initialized solution space containing solution space having NISS number of solutions, from which first generation solutions and IRS number of solutions in each generation (from second generation onwards) will be retrieved.

$$\text{NISS} = \text{NR} + (\text{itermax} * \text{IR}) \tag{1}$$

BSS: This is the solution space where best 10% solutions provided by each generation of research community will be stored to provide knowledge about best solutions to the upcoming generation of research community.

FSS: This is the solution space where all solutions provided by research community in each generation is stored.

A. Modeling of behaviour

Each class of research community will provide solutions based on their behaviour which is modeled below.

Intelligent class: Researchers who fall in this class retrieve the solutions from the ISS, which are not retrieved previously and provide them as their new solutions, irrespective of their study of existing solutions. For example, if i^{th} researcher is the intelligent researcher.

$$R_i^{\text{IR}} = \text{ISS}\,(x + i) \tag{2}$$

$$x = \text{NR} + (\text{iter} - 2) * \text{NR} \tag{3}$$

where iter is current generation number of research community.

Above average class: Researchers who fall in this class will provide hybrid solution in two ways.

First way: Researchers will randomly pick x number of solutions from BSS and from that again select three solutions randomly. Using Eq. (4) generate the hybrid solution. For example let us say S_1, S_2 and S_3 are the three solutions selected randomly for generating hybrid solution.

TABLE I
OPERATOR COMBINATIONS AND THEIR REQUIRED NUMBER OF VARIABLES FOR GENERATING AVERAGE CLASS HYBRID SOLUTION

Possible combinations	Number of variables required
+ ; - ; / ; * ;	2
+,- ; +, / ; +,* ; -,* ; /,* ; +, + ; -, - ;	3
+, -./ ; +,-,* ; -./,* ; +,/,* ; +,-,- ; +,+,- ;-,-,- ; +,+,+ ;	4
+,-,/,* ; + ,+,+,+ ; +,+,+,- ; +,+,+,- ;+,+,-,- ; +,-,-,- ;	5

$$R^{AAR}=S_1+S_{23} \tag{4}$$

$$S_{23}=S_2-S_3 \tag{5}$$

Second way: Researchers randomly select one solution from the x number of solutions picked randomly and using Eq. (6) generate the hybrid solution. Let us say S_1 be the selected solution. Sbest is best solution of BSS upto that generation.

$$R^{AAR}=S_1+ (0.5 + \text{rand})S_{23} \tag{6}$$

$$S_{23}=S_{\text{best}}-S_1 \tag{7}$$

Average class: Researchers who fall in this class randomly pick x number of solutions from FSS and select any one of operator combination given in the Table I and based on that choose required number of solutions for providing hybrid solution with respect to the selected operator combination. For example, let us say if selected possible combination is +,- , the required number of variables is 3 and hybrid solution is generated as per the Eq. (8).

$$R^{AR}=S_1+S_2-S_{23} \tag{8}$$

Below average class: Researchers will randomly pick x number of solutions from FSS and from that again select three solutions randomly. Using Eq. (9) generate the hybrid solution. For example, if S_1, S_2 and S_3 are the three solutions selected randomly for generating hybrid solution, then -

$$R^{BAR}=S_1+S_{23} \tag{9}$$

$$S_{23}=S_2-S_3 \tag{10}$$

R vector of the present generation can be obtained by using the above modeled Eq. (2) - Eq.(10) for each class of researcher. The flow chart of the proposed algorithm is as shown in Figure 1.

IV. VALIDATION AND TEST RESULTS

The proposed algorithm has been tested on 23 benchmark functions which are given in [23]. Parameters specifications taken are listed in Table V [Parameter specifications are taken in such a way as to mimic the types of student percentage in general class. In general class, intelligent student percentage is less, Above average percentage is moderate, Average student percentage is good and below average student percentage is normal]. The test results such as mean and standard deviation for all benchmark functions for 50 independent trails are presented in Table III. From the results it is observed that proposed algorithm is showing superiority over other compared

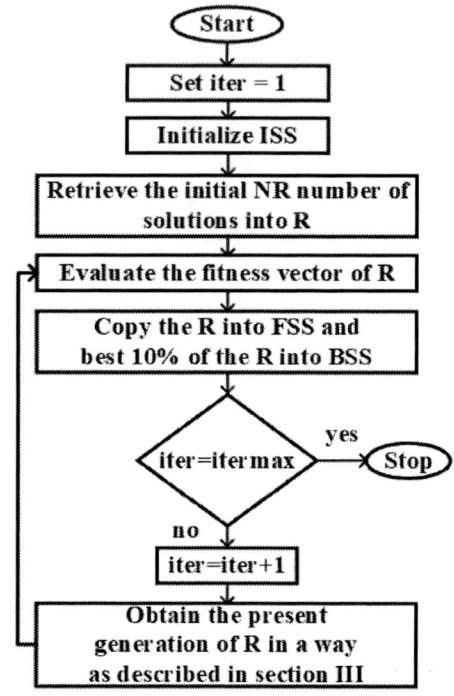

Fig. 1. Flow chart of the proposed algorithm.

optimization algorithms. Success rate of the proposed algorithm in hitting the global optima is in the range of 90 to 100. The Maximum and second maxima standard deviations for all benchmarks of the proposed method and other algorithms (GA, PSO, ICA and WGA) viz. $\pm 0.0324 \left(\pm 1.45 \times 10^{-5}\right), \pm 2.6 \times 10^3 \left(\pm 3.62 \times 10^2\right), \pm 4.64 \times 10^2 \left(\pm 3.21 \times 10^1\right), \pm 7.49 \times 10^2 \left(\pm 5.05 \times 10^3\right) \, and \pm 2.47 \left(\pm 1.94\right)$.

By observing maxima (second maxima) standard deviations of all algorithms it is evident that proposed method is outperforming in comparison with other methods.

TABLE II
PARAMETER SPECIFICATIONS OF THE ALGORITHM

NR	100		
PIR	0.1	IR	10
PAAR	0.3	AAR	30
PAR	0.5	AR	50
PBAR	0.2	BAR	20
itermax	1000		

TABLE III

COMPARISON OF MEAN AND STANDARD DEVIATION FOR BENCHMARK FUNCTIONS

Benchmark Functions	GA	PSO	ICA	WGA	Proposed Algorithm
f_{01}	3.17 ± 1.66	$3.69 \times 10^{-37} \pm 2.46 \times 10^{-36}$	$1.58 \times 10^{-58} \pm 4.36 \times 10^{-58}$	0 ± 0	0 ± 0
f_{02}	0.58 ± 0.13	$2.92e^{-24} \pm 1.14e^{-23}$	$3.90e^{-30} \pm 4.19e^{-30}$	0 ± 0	0 ± 0
f_{03}	$9.75e^{+03} \pm 2.6e^{+03}$	$1.2e^{-03} \pm 2.11e^{-03}$	$1.07e^{+03} \pm 3.24e^{+02}$	$7.28e^{-10} \pm 1.03e^{-09}$	0 ± 0
f_{04}	7.96 ± 1.51	$0.41 \pm 2.5e^{-01}$	$1.36 \pm 4.01e^{-01}$	$9.12e^{-276} \pm 0$	0 ± 0
f_{05}	$3.39e^{+02} \pm 3.62e^{+02}$	$3.74e^{+01} \pm 3.21e^{+01}$	$4.67e^{-04} \pm 1.48e^{-03}$	$1.57e^{-10} \pm 2.11e^{-10}$	0 ± 0
f_{06}	3.7 ± 1.95	0.15 ± 0.42	$2.33e^{+03} \pm 5.05e^{+03}$	0 ± 0	0 ± 0
f_{07}	$0.1 \pm 3.62e^{-02}$	$9.9e^{-03} \pm 3.54e^{-02}$	$2.58e^{-04} \pm 7.43e^{-05}$	$5.68e^{-03} \pm 1.98e^{-03}$	0 ± 0
f_{08}	$-1.26e^{+04} \pm 2.12$	$-9.67e^{+03} \pm 4.64e^{+02}$	$-1.11e^{+04} \pm 7.49e^{+02}$	$-6.98e^{+03} \pm 8.76e^{+02}$	$-1.2569e^{+04} \pm 6.8279e^{-08}$
f_{09}	0.65 ± 0.36	$2.08e^{+01} \pm 5.94$	$1.44e^{+02} \pm 8.67$	9.99 ± 1.94	0 ± 0
f_{10}	0.87 ± 0.28	$1.34e^{-03} \pm 4.24e^{-02}$	$1.67e^{+01} \pm 4.9$	$9.41e^{-02} \pm 2.88e^{-01}$	$-8.8818e^{-16} \pm 0$
f_{11}	$1.00 \pm 6.75e^{-02}$	$2.32e^{-01} \pm 4.43e^{-01}$	$1.28e^{-02} \pm 1.19e^{-02}$	$4.83e^{-21} \pm 3.28e^{-20}$	0 ± 0
f_{12}	$4.36e^{-02} \pm 5.06e^{-02}$	$3.95e^{-02} \pm 9.14e^{-02}$	$1.57e^{-32} \pm 5.7e^{-35}$	$1.57e^{-32} \pm 0$	$2.30560e^{-16} \pm 5.97e^{-16}$
f_{13}	$0.17 \pm 7.07e^{-02}$	$5.05e^{-02} \pm 0.57$	$0.10e^{-26} \pm 0.77e^{-26}$	$2.75e^{-32} \pm 3.42e^{-32}$	$2.8009e^{-12} \pm 6.3349e^{-12}$
f_{14}	$1.50 \pm 4.43e^{-03}$	1.02 ± 0.15	$9.98e^{-01} \pm 0$	3.19 ± 2.47	$0.998 \pm 2.5273e^{-14}$
f_{15}	$7.09e^{-03} \pm 7.85e^{-03}$	$3.81e^{-04} \pm 2.51e^{-04}$	$5.61e^{-04} \pm 1.43e^{-04}$	$3.075e^{-04} \pm 0$	$3.0760e^{-04} \pm 2.0330e^{-07}$
f_{16}	$-1.03 \pm 3.13e^{-03}$	$-1.02 \pm 1.28e^{-02}$	-1.0316 ± 0	-1.036 ± 0	$-1.036 \pm 8.9164e^{-08}$
f_{17}	$0.40 \pm 1.04e^{-02}$	$0.40 \pm 6.88e^{-02}$	0.3979 ± 0	0.3979 ± 0	$0.3979 \pm 4.1963e^{-07}$
f_{18}	$7.50 \pm 1.04e^{+01}$	$3.01 \pm 1.21e^{-03}$	3 ± 0	3 ± 0	$3 \pm 3.2229e^{-07}$
f_{19}	$-3.86 \pm 6.28e^{-04}$	$-3.86 \pm 3.21e^{-03}$	$-3.86 \pm 3.48e^{-13}$	-3.86 ± 0	$-3.8628 \pm 2.22057e^{-07}$
f_{20}	$-3.26 \pm 6.04e^{-02}$	$-3.18 \pm 6.11e^{-02}$	$-3.32 \pm 1.23e^{-15}$	-3.32 ± 0	-3.312 ± 0.0324
f_{21}	-5.17 ± 2.93	-7.54 ± 3.03	-9.95 ± 1.00	-10.1532 ± 0	$-10.1532 \pm 4.1157e^{-07}$
f_{22}	-5.44 ± 3.28	-8.36 ± 2.02	$1.01e^{+01} \pm 1.39$	-10.4029 ± 0	$-10.4209 \pm -1.4563e^{-05}$
f_{23}	-4.91 ± 3.49	-8.94 ± 1.63	$1.01e^{+01} \pm 1.47$	-10.5364 ± 0	$-10.5364 \pm 8.0186e^{-06}$

TABLE IV
COMPARISON OF PMU PLACEMENT LOCATIONS OBTAINED USING METHODS FOR THE BASE CASE

Methodology	IEEE 14 - Bus		IEEE 30 - Bus	
Exhaustive Search	No of PMUs	Optimal Locations	No of PMUs	Optimal Locations
	4	2, 6, 7, 9	10	1, 2, 6, 9, 10, 12, 15, 19, 25, 27
ILP based	4	2, 6, 7, 9	10	2, 4, 6, 9, 10, 12, 15, 25, 27
BPSO	4	2, 6, 7, 9	10	2, 4, 6, 9, 10, 12, 15, 18, 25, 27
BSP	4	2, 7, 11, 13	10	1, 2, 6, 9, 10, 12, 15, 18, 25, 27
MBPSO	4	2, 6, 7, 9	10	2, 6, 9, 10, 13, 14, 17, 19, 20, 22, 23, 25, 29
EBPSO	4	2, 6, 7, 9	-	-
Proposed Method	4	2, 6, 7, 9	10	2, 4, 6, 9, 10, 12, 18, 24, 26, 27

TABLE V
COMPARISON OF PMU PLACEMENT LOCATIONS OBTAINED BY USING METHODS FOR THE LOSS OF PMU

Methodology	IEEE 14 - Bus		IEEE 30 - Bus	
ILP Based	No of PMUs	Optimal Locations	No of PMUs	Optimal Locations
	-	-	21	1, 2, 3, 5, 6, 9, 10, 11, 12, 13, 15, 16, 18, 19, 22, 24, 25, 26, 27, 28, 29
MBPSO	9	2, 4, 5, 6, 7, 8, 9, 10, 13	21	2, 3, 4, 6, 7, 9, 10, 11, 12, 13, 15, 16, 18, 20, 22, 24, 25, 26, 27, 28, 30
EBPSO	9	2, 4, 5, 6, 7, 8, 9, 11, 13	-	-
Proposed Method	9	2, 4, 5, 6, 7, 8, 9, 11, 13	21	2, 3, 4, 6, 7, 9, 10, 11, 12, 13, 15, 16, 18, 19, 21, 24, 25, 26, 27, 28, 30

V. CASE STUDY AND RESULTS

The proposed optimization algorithm is applied on Optimal PMU placement problem. The variables of PMU placement problem take either 1 or 0 (1 means PMU is placed and 0 means no PMU is placed). To make the proposed method more suitable to OPP problem binary version of the proposed algorithm is presented and it is applied to optimal PMU placement problem [24]. Case study has been performed on IEEE 14 and 30 bus systems for base case and loss of PMU case.

A. Binary version of Research outcome based optimization (Robo) algorithm

Generate the ISS randomly (0 or 1). Obtain the R vector of the present generation using the procedure of the proposed algorithm as shown in Figure 1. Updated solution Rij is narrowed down in the range (0,1) by using 'tanh' transformation as per Eq. (11). The transformed value of Rij is again transformed into binary form.

$$R_{ij} = \tanh\left(|Rij|\right)$$

$$Rij = \begin{cases} 1 & \text{rand} < R_{ij} \\ 0 & \text{otherwise} \end{cases} \quad (11)$$

B. Optimal PMU placement(OPP) problem formulation

In recent times PMU's gained much popularity due their wider applications. Installation of PMU's at each and every bus is neither economical nor recommendable. Therefore Optimal PMU placement strategies are required for deployment of PMUs. The primary objective of the optimal PMU placement is to minimize the number of PMUS while meeting the network observability constraint.

$$F(X) = \sum_{i=1}^{N} C_{PMU_i} * x_i$$

Function F(X) is subjected to rank(H)=N (or) BC.X=b, where

$$C_{PMU_i} = \quad \text{Installation cost of PMU at } i^{th} \text{bus}$$

$$x_i = \begin{cases} 1 & \text{PMU is installed at } i^{th} \text{bus} \\ 0 & \text{otherwise} \end{cases}$$

N   Number of Buses
H   Design matrix as reported in [25]
BC   Binary Connectivity Matrix
b   unit vector NX1
X   vector NX1 having x_i entity

C. Application of Binary (RObO) algorithm to OPP problem

The proposed binary version of the algorithm has been used to solve the OPP problem. The results for the IEEE 14 and 30 bus system are presented for base case in Table IV and for loss of PMU case in Table V and the results are compared with the other optimization algorithms proposed in the literature. From the Table IV and Table V, it is observed that the proposed algorithm is reaching the global optima. The major advantage of the propsed algorithm is that, it doesn't require heuristic operators like velocity, momemtum etc. The solution in the proposed algorithm moves from one generation to next generation using simple arithmetic operations.

VI. CONCLUSION

In this paper a new optimization algorithm is proposed by mimicking the nature of research community all over the world. The proposed algorithm is first validated by applying on the 23 benchmark functions. The special feature of the

979-8-3315-3013-6/25 $31.00 © 2025 IEEE

proposed algorithm is that, in most of the cases it is independent of number of runs for hitting the global optima and simple in understanding. Binary version of the proposed algorithm was presented and applied on OPP problem. From the results the supremacy of the algorithm in comparison with other methods is observed. From the results, it is observed that the performance of the algorithm is very stable. The reach ability towards global optima is independent of the number of runs.

VII. FUTURE SCOPE

Further, there is a scope for improving the mathematical models of the behaviour of each class of research community. Hybrid algorithm can be implemented by incorporating all updating equations of the existing heuristic optimization algorithms for updating average and below average class researchers while providing solutions.

REFERENCES

[1] McCall, J., 2005. Genetic algorithms for modelling and optimisation. Journal of computational and Applied Mathematics, 184(1), pp.205-222.

[2] Kennedy, J. and Eberhart, R., 1995, November. Particle swarm optimization. In Proceedings of ICNN'95-international conference on neural networks (Vol. 4, pp. 1942-1948). IEEE.

[3] Mezura-Montes, E., Miranda-Varela, M.E. and del Carmen Gómez-Ramón, R., 2010. Differential evolution in constrained numerical optimization: an empirical study. Information Sciences, 180(22), pp.4223-4262.

[4] Geem, Z.W., Kim, J.H. and Loganathan, G.V., 2001. A new heuristic optimization algorithm: harmony search. simulation, 76(2), pp.60-68.

[5] Passino, K.M., 2002. Biomimicry of bacterial foraging for distributed optimization and control. IEEE control systems magazine, 22(3), pp.52-67.

[6] Farmer, J.D., Packard, N.H. and Perelson, A.S., 1986. The immune system, adaptation, and machine learning. Physica D: Nonlinear Phenomena, 22(1-3), pp.187-204.

[7] Dorigo, M., Birattari, M. and Stutzle, T., 2006. Ant colony optimization. IEEE computational intelligence magazine, 1(4), pp.28-39.

[8] Karaboga, D. and Akay, B., 2011. A modified artificial bee colony (ABC) algorithm for constrained optimization problems. Applied soft computing, 11(3), pp.3021-3031.

[9] Tshenyego, O., Samikannu, R., Mtengi, B., Mosalaosi, M. and Sigwele, T., 2023. A graph-theoretic approach for optimal phasor measurement units placement using binary firefly algorithm. Energies, 16(18), p.6550.

[10] Rashedi, E., Nezamabadi-Pour, H. and Saryazdi, S., 2009. GSA: a gravitational search algorithm. Information sciences, 179(13), pp.2232-2248.

[11] Ahrari, A. and Atai, A.A., 2010. Grenade explosion method—a novel tool for optimization of multimodal functions. Applied Soft Computing, 10(4), pp.1132-1140.

[12] Simon, D., 2008. Biogeography-based optimization. IEEE transactions on evolutionary computation, 12(6), pp.702-713.

[13] Rao, R.V., Savsani, V.J. and Vakharia, D.P., 2011. Teaching–learning-based optimization: a novel method for constrained mechanical design optimization problems. Computer-aided design, 43(3), pp.303-315.

[14] Gandomi, A.H. and Alavi, A.H., 2012. Krill herd: a new bio-inspired optimization algorithm. Communications in nonlinear science and numerical simulation, 17(12), pp.4831-4845.

[15] Birbil, Ş.İ. and Fang, S.C., 2003. An electromagnetism-like mechanism for global optimization. Journal of global optimization, 25, pp.263-282.

[16] Bahrami, H., Faez, K. and Abdechiri, M., 2010, March. Imperialist competitive algorithm using chaos theory for optimization (CICA). In 2010 12th International Conference on Computer Modelling and Simulation (pp. 98-103). IEEE.

[17] Shefaei, A. and Mohammadi-Ivatloo, B., 2017. Wild goats algorithm: An evolutionary algorithm to solve the real-world optimization problems. IEEE Transactions on Industrial Informatics, 14(7), pp.2951-2961.

[18] Azizi, S., Dobakhshari, A.S., Sarmadi, S.A.N. and Ranjbar, A.M., 2012. Optimal PMU placement by an equivalent linear formulation for exhaustive search. IEEE Transactions on Smart Grid, 3(1), pp.174-182.

[19] Anguswamy, M.P., Datta, M., Meegahapola, L. and Vahidnia, A., 2022. Optimal micro-PMU placement in distribution networks considering usable zero-injection phase strings. IEEE Transactions On smart grid, 13(5), pp.3662-3675.

[20] Andreoni, R., Macii, D., Brunelli, M. and Petri, D., 2021. Tri-objective optimal PMU placement including accurate state estimation: The case of distribution systems. IEEe Access, 9, pp.62102-62117.

[21] Mandal, A.K. and De, S., 2023. Joint optimal PMU placement and data pruning for resource efficient smart grid monitoring. IEEE Transactions on Power Systems, 39(3), pp.5382-5392.

[22] Mishra, A. and de Callafon, R.A., 2022. Algebraic approach to PMU placement for minimum variance linear state estimation in power networks. IEEE Transactions on Power Systems, 38(5), pp.4381-4390.

[23] Yao, X., Liu, Y. and Lin, G., 1999. Evolutionary programming made faster. IEEE Transactions on Evolutionary computation, 3(2), pp.82-102.

[24] Sharaf OZ, Orhan MF, "An overview of fuel cell technology: Fundamentals and applications" *Renewable and sustainable energy reviews*, vol. 32, pp. 810-53, 2014.

[25] Mezura-Montes, E., Miranda-Varela, M.E. and del Carmen Gómez-Ramón, R., 2010. Differential evolution in constrained numerical optimization: an empirical study. Information Sciences, 180(22), pp.4223-4262.

Madhu Kishore Devara Chejarla received B.E in Electrical and Electronics Engineering from Andhra University, India, M.Tech and Ph.D degree in Power Systems from National Institute of Technology (NIT) Warangal, India and Post doctoral Fellow (PDF) from IIT Bombay. He has 2+ years of industrial experience in the analytical software tool development for Power Systems and 6 years of research experience in design and development of algorithms. He worked as a consultant with Wipro Limited in the Grid modernization and Network operations wing and as an Engineer - Grid Injection Study expert with EPPLUS. Currently working as Power Systems Analyst in TCS.

M Sailja Kumari obtained B.E and M.E degrees in 1993 and 1995 from University College of engineering, Osmania University, Hyderabad, India and Ph.D in 2008 from National Institute of Technology, Warangal. Currently, she is working as a professor in Electrical Engineering Department, National Institute of Technology, Warangal, Telangana, India. Her research interests are in the area of Distributed generation and energy management systems, Synchro phasor Technology, cyber security, power markets, renewable energy grid integration issues.

S Abdul Mujeer received B.Tech and M.Tech degrees in Electrical and Electronics Engineering from Acharya Nagarjuna University in 2014 and 2016. Currently, he is a research scholar at Electrical Engineerign Department, National Institute of Technology, Warangal, Telangana, India. His area of interest includes modeling of photovoltaics and fuel cells, micro-grid energy management, power system stability, AI and Optimization techniques.

979-8-3315-3013-6/25 $31.00 © 2025 IEEE

Hardware design of SiC MOSFET-based Phase Shifted Full Bridge DC-DC Converter for EV Charging Application

Keerthika C
SEEE
SASTRA University
Thanjavur, India
ckeerth7@gmail.com

Saravanan B
SEEE
SASTRA University
Thanjavur, India
saravananslr96@gmail.com

Vishaal S
SEEE
SASTRA University
Thanjavur, India
vishaalbala6@gmail.com

Dr. S. Lenin Prakash
SEEE
SASTRA University
Thanjavur, India
leninprakash@eee.sastra.ac.in

Dr. S. Raghavendran
SEEE
SASTRA University
Thanjavur, India
raghavendran@eee.sastra.edu

Abstract— **An EV charging system (EVCS) needs a high efficiency and high power density power converters. The Phase Shifted Full Bridge (PSFB) DC-DC converter topology is one of the preferred isolated topology used for High Voltage, High Frequency and High Power application. PSFB converter employs ZVS switching, leveraging the low reverse recovery time of the SiC MOSFET and hence enables high efficiency. Also High frequency operation reduces the size of magnetics and hence enhances the power density. However, a meticulous design of magnetics for a chosen SiC MOSFET is necessary to achieve the ZVS in the given operating range. This paper presents a design methodology for the selection of magnetics considering the real-time challenges of high frequency transformers. Also, the PCB design of the converter has been presented which is a critical part for the successful operation of any high frequency converter. The design has been validated by simulation for a 48V, 600 W PSFB converter and is found to be an attractive solution for an EV Charging application.**

Keywords—— *Isolated DC-DC converter, Zero Voltage Switching, Phase Shifted Full Bridge Converter*

I. INTRODUCTION

In Electric Vehicle, majority of the on board chargers use two-stage conversion topologies, at the front end AC-DC conversion and at the back end DC-DC conversion. For the DC-DC conversion stage, a variety of DC-DC converter topologies are available to convert high voltage to low voltage [1-2]. For safety reasons, the majority of systems needs isolation between the source and the load. Therefore, isolated dc-dc converters are preferred for wide dc voltage gain range applications.

The Isolated dc-dc converters consists of inverter section, high-frequency transformer and a rectifier section. The DC voltage from the Power Factor Correction circuit is converted into AC voltage in the inverter section. The high frequency transformer provides isolation and converts the AC voltage from the inverter section into desired output AC voltage, which is again converted into DC voltage by the Rectifier section. The Buck derived and boost derived isolated converters are more efficient than the buck-boost converter. [3-5].

The traditional Flyback, Forward and Push pull converters having relatively large transformer size and high switch voltage stress which reduces its efficiency. For that full bridge dc-dc converter is preferred over these converters. The PWM control in full bridge dc-dc converter increases the voltage stress of the switches. This type of hard switching increases electromagnetic interference due to high d_i/d_t across switches[6]. In order to achieve Zero Voltage Switching (ZVS) in Full Bridge Isolated DC-DC converter, Phase Shift control is implemented. [7-8]

The phase-shifted full bridge (PSFB) converters are commonly used for on-board chargers due to many advantages such as low EMI, current stress on the components, soft switching capability, simple PWM control, and high-power density [9 - 11]. For achieving the efficient operation, WBG devices such as SiC MOSFETS are used, due to its low C_{COSS} enabling low circulating current, which reduces the volume and weight of the converter and increases its efficiency [12 - 13].

In this paper, the design of the PSFB dc-dc converter has been discussed in detail. The operating modes of the converter has been presented in the next section which forms the basis for the subsequent sections. Also, the PCB design of the converter has been presented section III which is a critical part for the successful operation of any high frequency converter. The design validation by simulation have also been presented in section IV.

II. PSFB OPERATION

The Phase shift Full Bridge converter consists of four switches (S1, S2, S3, and S4), High Frequency transformer and four diodes (D_a, D_b, D_c & D_d). The circuit diagram of phase shifted full bridge dc-dc converter is shown in the Fig.1.

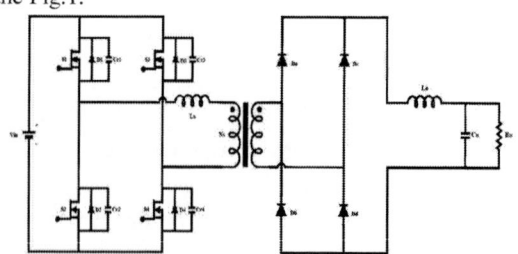

Fig.1. Circuit Diagram of PSFB

979-8-3315-3013-6/25 $31.00 © 2025 IEEE

Each switch consists of Internal Diode and parasitic Capacitance (C_{r1}, C_{r2}, C_{r3}, and C_{r4}). And the transformer has leakage inductance (L_{lk}) and Magnetizing Inductance (L_{mag}). The parasitic capacitance C_r and Leakage inductance (L_{lk}) acts as resonant circuit and forced the voltage across the switch to zero before the switch turned on. This leads to Zero Voltage Switching of each switches and reduces the switching loss.

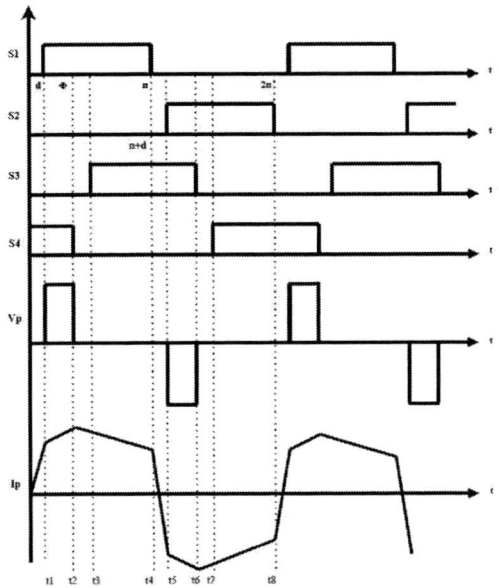

Fig.2. Waveform of Switching pulses, primary voltage and primary current of PSFB

Each switch conducts for 50% duty cycle and the dead time is inserted between the gate pulses of two switches in the same leg to avoid short circuit. The waveforms of gate pulses of switches, primary voltage and primary current of PSFB converter is shown in figure.3.

There are 8 modes of operation based on the switching sequence.

TABLE I. SWITCHING SEQUENCE OF PSFB

Modes	Switch S1	Switch S2	Switch S3	Switch S4
Mode 1	1	0	0	1
Mode 2	1	0	0	0
Mode 3	1	0	1	0
Mode 4	0	0	1	0
Mode 5	0	1	1	0
Mode 6	0	1	0	0
Mode 7	0	1	0	1
Mode 8	0	0	0	1

A. MODE 1: POWER TRANSFER [t1 –t2]

In this mode Switch S1 and S4 are conducting. When the diagonal switches are turned on, then the operation is similar to the conventional square wave inverter. Hence in this mode, the entire input voltage appears across the transformer primary. In the secondary side the diodes D_a &

D_d are forward biased and conducts. It delivers the power to the load.

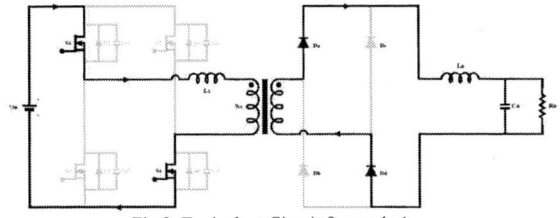

Fig.3. Equivalent Circuit for mode 1

B. MODE 2: SWITCH S3 ZERO VOLTAGE TRANSITION [t2-t3]

The switch S4 is turned OFF in this mode and S1 is still ON. The leakage inductance of the transformer tries to maintain the primary current constant. Then the primary current flows through the parasitic capacitor C_{r4} of the switch S4 and charges C_{r4} from 0v to V_{in}. The capacitor C_{r3} of switch S3 in the same right leg is discharged as its source voltage increases. Thus the resonant transition occurs at the right leg of the converter. This leads to zero voltage across the switch S3 before it is turned ON.

The voltage across the transformer primary decreases from V_{in} and reaches 0V at the end of this mode. Hence no power transfer from primary to secondary. The output inductor L_o supplies its energy to the load. The rectifier diodes D_a & D_d are conducting.

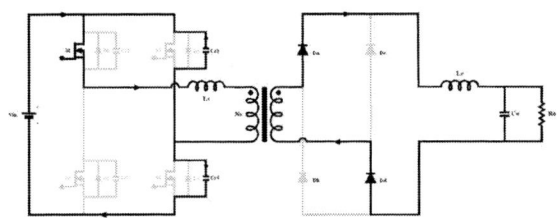

Fig.4. Equivalent Circuit for mode 2

C. MODE 3: FREEWHEELING [t3- t4]

When C_{r3} is completely discharged then mode 3 begins and the body diode D3 of switch S3 starts conducting. Switch S1 is still ON. The primary current I_p freewheels through the switch S1 and D3. Transformer secondary voltage is 0V. So the output inductor supplies the load. In this mode all the four rectifier diodes are conducting.

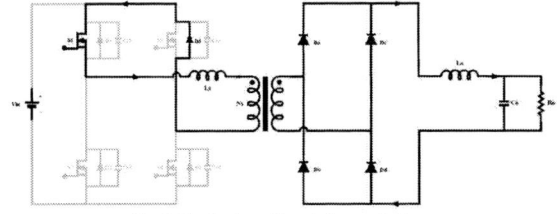

Fig.5. Equivalent Circuit for mode 3

D. MODE 4: SWITCH S2 ZERO VOLTAGE TRANSITION [t4 –t5]

In this mode S1 is turned OFF and the switch S3 is ON. The primary current continues to flow in the same direction but in different path, it flows through the C_{r1} instead of S1.

979-8-3315-3013-6/25 $31.00 © 2025 IEEE

When S1 turned OFF, the capacitor C_{r1} of the switch S1 charges from 0V to V_{in} and the capacitor C_{r2} of switch S2 in the same leg discharges from V_{in} to 0V. Left leg transition occurs in this mode and makes the voltage across Switch S2 zero to achieve zero voltage switching. Transformer primary and secondary voltage is zero and all the four rectifier diodes are conducting.

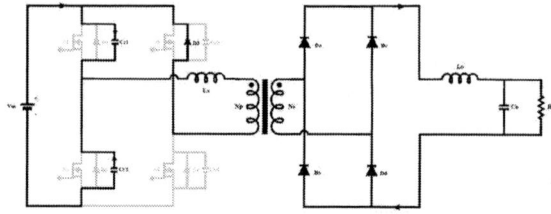

Fig.6. Equivalent Circuit for mode 4

E. MODE 5: POWER TRANSFER [t5 –t6]

In this mode switch S2, S3 are turned ON and both are conducting. Similar to mode 1, entire input voltage appears across the primary of the transformer.

The primary current I_p reverses its direction. Transformer secondary voltage is equal to $n \times V_{in}$. Rectifier diodes D_b & D_c becomes forward biased and starts conducting. It transfers the power to charge the output inductor L_o.

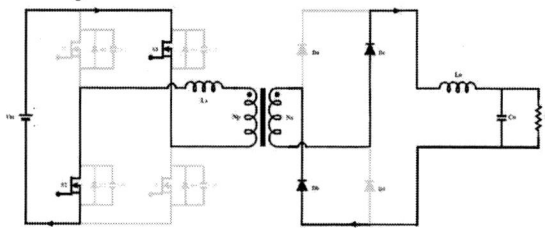

Fig.7. Equivalent Circuit for mode 5

F. MODE 6: SWITCH S4 ZERO VOLTAGE TRANSITION

The switch S3 is turned OFF in this mode and S2 is in ON condition. The leakage inductance of the transformer tries to maintain the primary current constant. Then the primary current flows through the parasitic capacitor C_{r3} of the switch S3 and C_{r3} charges from 0v to V_{in}.

The capacitor C_{r4} of switch S4 in the same leg is discharged. Thus the resonant transition occurs at the right leg of the converter. This leads to zero voltage across the switch S4 before it is turned ON.

The voltage across the transformer primary decreases from V_{in} and reaches 0V at the end of this mode. Primary doesn't supply the secondary.

So the output inductor V_O supplies its energy to the load. The rectifier diodes D_b & D_c gets forward biased and conducting.

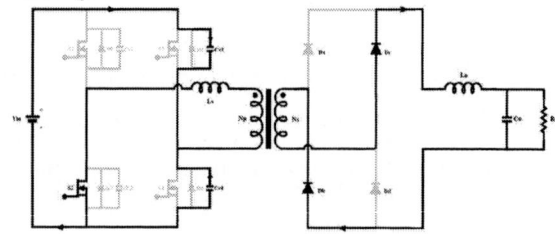

Fig.8. Equivalent Circuit for mode 6

G. MODE 7: FREEWHEELING

Mode 7 begins when the capacitor C_{r4} is completely discharged. So the body diode D4 of switch S4 starts conducting. Switch S2 is still ON.

The primary current I_p freewheels through the switch S2 and D4. At this mode both the primary and secondary voltage of Transformer is 0V. So the output inductor L_o supplies the load. In this mode all the four rectifier diodes are conducting.

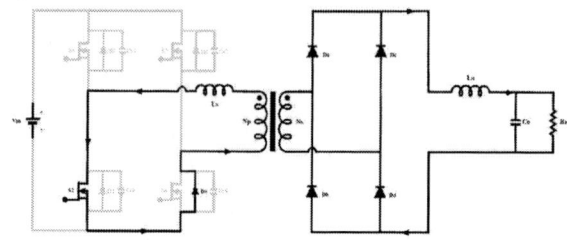

Fig.9. Equivalent Circuit for mode 7

H. MODE 8: SWITCH S1 ZERO VOLTAGE TRANSITION

In this mode S2 is turned OFF and the switch S4 is conducting. When S2 turned OFF, the capacitor C_{r2} of the switch S2 charges from 0V to V_{in} and the capacitor C_{r1} of switch S1 in the same leg discharges from Vin to 0V.

Left leg transition occurs in this mode and makes the voltage across Switch S1 zero to achieve zero voltage switching. Transformer's primary and transformer secondary voltage is zero in this mode. Rectifier diodes both D_a and D_b are conducting.

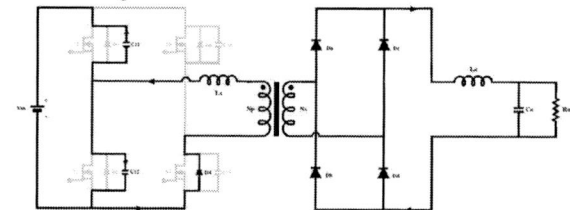

Fig.10. Equivalent Circuit for mode 8

III. PSFB CONVERTER DESIGN

The Phase Shift full bridge converter is designed with the specifications given in the table II.

TABLE II. DESIGN SPECIFICATIONS

S.No	Parameters	Values
1	Rated Power (P_{out})	600 W
2	Switching Frequency (f_s)	100 kHz
3	Minimum Input Voltage($V_{in\,min}$)	370 V
4	Nominal Input Voltage(V_{in})	390 V
5	Maximum Input Voltage($V_{in\,max}$)	410 V
6	Minimum Output Voltage($V_{out\,min}$)	39 V
7	Nominal Output Voltage(V_{inout})	48 V
8	Maximum Output Voltage($V_{out\,max}$)	56 V
9	Duty Cycle (D)	50 %
10	Efficiency (η)	93 %

The turns ratio of the transformer is calculated as,

$$n = \frac{V_{in}}{V_{out}} \tag{1}$$

While including the loss, the turns ratio becomes,

$$n = \frac{(V_{in\,min} - 2 \times V_{drop}) \times D_{max}}{V_{out} + V_{drop}} \tag{2}$$

Here V_{drop} is the voltage drop of the switch. Maximum duty ratio D_{max} is taken as 0.5 (50%).

Ensuring adequate energy storage in the primary circuitry during the Freewheeling Period becomes crucial in order to accomplish Zero-Voltage Switching (ZVS) on the primary side. The magnetizing inductance of the transformer primary is,

$$L_{mag} \geq \frac{V_{in} \times (1-D_t)}{\frac{\Delta I_{Lout} \times 0.5}{n}} \tag{3}$$

Eq. 3 gives the minimum value of magnetizing inductance
Where,
Typical duty cycle (D_t) is based on average input voltage, calculated as

$$D_t = \frac{(V_{out} + V_{drop}) \times n}{(V_{in} - 2 \times V_{drop})} \tag{4}$$

The Output inductor ripple current is taken as 20% of Output DC Current,

$$\Delta I_{Lout} = 0.2 \times \frac{P_{out}}{V_{out}} \tag{5}$$

For enabling efficient ZVS functioning, the energy that has been stored makes it easier for the parasitic capacitance connected to the MOSFET to discharge. By optimal selection of energy storage and discharge requirements, the PSFB converter can maximize the performance, reduce the switching losses, and boost overall efficiency. So the energy stored in the leakage inductance of the transformer must be greater than the energy stored in the parasitic capacitance of the switch.

$$E_{lk} > E_{COSS}$$

Total inductance required to achieve zero voltage switching is,

$$L_r = L_s + L_{lk}$$

$$L_r = (2 \times C_{oss-avg}) \times \frac{V^2_{in\,max}}{\left(\frac{I_{pp}}{2} - \frac{\Delta I_{Lout}}{2 \times n}\right)^2} \tag{6}$$

Based on the leakage inductance (L_{lk}) value of the transformer, we can calculate L_s.
Where, I_{pp} is the primary peak current,

$$I_{pp} = \left(\frac{P_{out}}{V_{out} \times \eta} + \frac{\Delta I_{Lout}}{2}\right)\frac{1}{n} + \Delta I_{Lmag} \tag{7}$$

Magnetizing Current ripple (ΔI_{Lmag}),

$$\Delta I_{Lmag} = \frac{V_{in\,min} \times D_{max}}{L_{mag} \times f_s} \tag{8}$$

The Output Inductor (l_{out}),

$$l_{out} = \frac{V_{out} \times (1-D_t)}{\Delta I_{Lout} \times f_s} \tag{9}$$

The Output Capacitor (C_{out}),

$$C_{out} \geq \frac{P_{out} \times 0.9 \times t_H}{V_{out} \times V_T \times 0.1} \tag{10}$$

t_H is the holdup time. It is the time required by l_{out} to change 90% of its full load current.

$$t_H = \frac{L_{out} \times P_{out} \times 0.9}{V_{out} \times V_{out}} \tag{11}$$

V_T is the allowable transient voltage.

IV. SIMULATION RESULTS

The simulation of Phase Shift Full Bridge converter has been done for 100 kHz switching frequency and 600W power in the MATLAB environment. Here the PID Controller is used for closed loop control of PSFB. The transformer values are calculated from the above equations as turns ratio 3.8, Primary Leakage Inductance 5.36 µH and Magnetizing Inductance 6.27 mH. And the filter inductor and capacitor values are 100uH and 1900uF respectively. Figure.12. illustrates the Simulink diagram of PSFB.

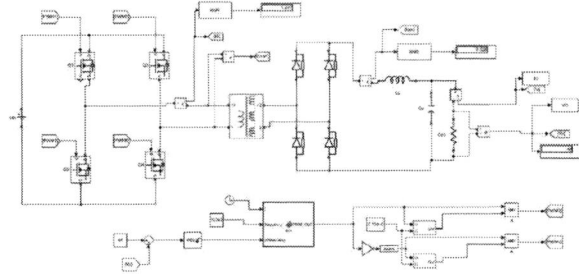

Fig.11. MATLAB/Simulink model of PSFB

The switching pulses for S1, S2, S3 &S4 are generated with 50% duty cycle and provided with the dead time of 0.15us. The switching pulses of lagging leg switches S2& S3 are phase shifted with respect to the leading leg switches S1 & S2. Figure.11. shows that the switching pulses with 90° phase shift. The values of the parameters and their resultant values are given in the table III. And the respective switching pulse, voltage and current waveforms are shown in the following figures.

TABLE III. SIMULATION CASES

Parameters	Values
Switching Frequency	100kHz
Input Voltage	390V
Phase Shift	90°
Output Voltage	48 V
Primary Current	0.8 A
Secondary Current	4.3 A

979-8-3315-3013-6/25 $31.00 © 2025 IEEE

Simulation Results for the values given in TABLE III

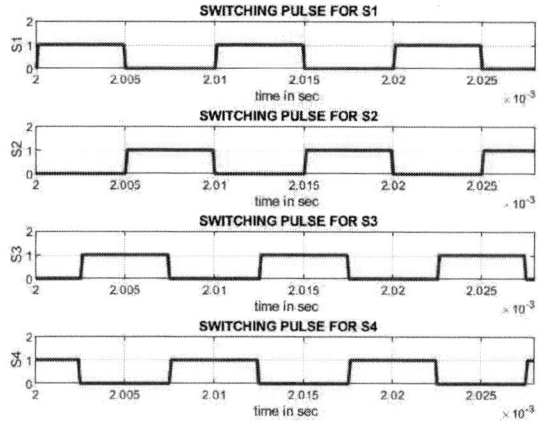

Fig.12. Gate pulses of S1 & S2 and 90° Phase Shifted gate pulses of S3 & S4

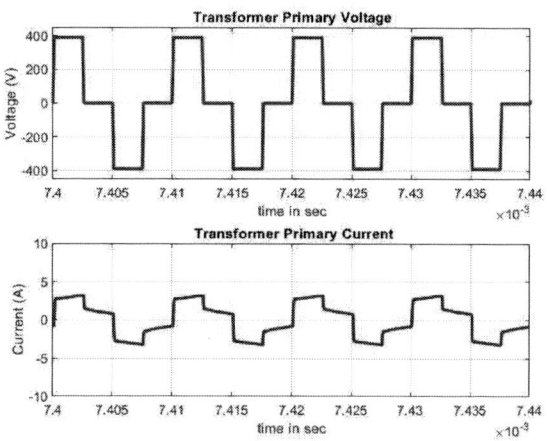

Fig.13. Primary Voltage and Primary Current waveform of Transformer for Case (ii)

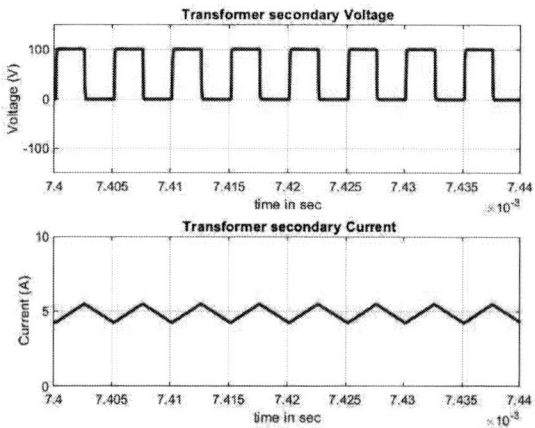

Fig.14. Secondary Voltage and Secondary Current waveform of Transformer for Case (ii)

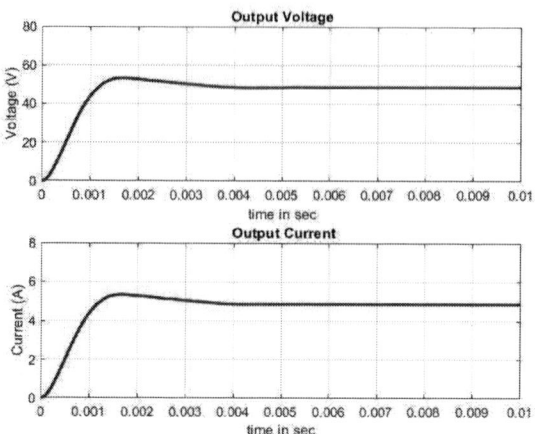

Fig.15. Output Voltage and Output Current waveform Case (ii)

V. EFFICIENCY CALCULATION

The total loss of the converter includes the conduction loss of primary MOSFET's & secondary diodes. Transformer loss & the secondary filter induction & capacitor loss.

The switching loss of the diodes & MOSFET's are neglected because of Zero Voltage Switching (ZVS).

a) Primary Mosfet's Loss
A 650V Cool MOS IPP65R190C6 MOSFET from Infineon is used in our circuit.

i)MOSFET Conduction Loss:
$$P_{Cond} = I^2{}_{PRMS} \times R_{DS(ON)} \qquad (12)$$

From the datasheet of IPP65R190C6 $R_{DS\,(ON)} = 0.19\,\Omega$

From the design calculation, $I_{PRMS} = 3.9$ A

$$P_{Cond} = (3.9)^2 \times (0.19) = 2.889 \text{ W}$$

ii) Gate Charge Loss due to parasitic capacitance:

$$P_{gate} = 2 \times Q_g \times V_{GS} \times \frac{f_s}{2} \qquad (13)$$

From the datasheet $V_{GS} = 20V$, $Q_g = 73nC$

$$P_{gate} = 73 \times 10^{-9} \times 20 \times 100 \times 10^3 = 146mW$$

Power loss per switch is

$$P_{QA} = P_{Cond} + P_{gate}$$

$$P_{QA} = 2.889 + 0.146 = 3.035 \text{ W}$$

Total Power loss for primary MOSFET's

$$P_{Q\_Total} = 4 \times P_{QA}$$

$$P_{Q\_Total} = 4 \times 3.035 = 12.14W$$

b) Transformer Loss
The isolated transformer used in this paper is customized one with 100kHz frequency. Core loss is Negligible
Copper Loss:
$$P_T = 2 \times [(I^2{}_{PRMS} \times DCR_P + (I^2{}_{SRMS} \times DCR_S)] \qquad (14)$$

From the measurement DC Resistance across primary winding $DCR_P = 0.929\,\Omega$ across secondary winding

$DCR_S = 0.024 \ \Omega$

$P_T = 2 \times [((3.9)^2 \times (0.067)) + ((9.1)^2 \times 0.024)]$

$P_T = 6.01W$

c) Output Inductor Loss

The Inductor Loss can be calculated by using the formula,

$$P_{LOUT} = \text{Twice the Copper loss}$$
$$P_{LOUT} = 2 \times I^2{}_{LOUT\_RMS} \times DCR_{LOUT} \qquad (15)$$

From calculated $I_{LOUT\_RMS} = 12.58 \ A$, $DCR_{LOUT} = 13.6 \ m\Omega$

$$P_{LOUT} = 2 \times (12.58)^2 \times (0.0136) = 4.3 \ W$$

d) Output Capacitor Loss

$$P_{COUT} = I^2{}_{COUT\_RMS} \times ESR_{COUT} \qquad (16)$$

$$I_{COUT\_RMS} = \frac{\Delta I_{LOUT}}{\sqrt{3}} = \frac{2.5}{\sqrt{3}} = 1.443 \ A$$

From datasheet, $ESR = 0.332 \ \Omega$

$$P_{COUT} = (1.433)^2 \times 0.332 = 0.69 \ W$$

e) Total Loss of the proposed converter

$$P_{LOSS} = 12.14 + 6.01 + 4.3 + 0.69 = 23.14 \ W$$

f) Efficiency of the proposed converter

$$\eta = \frac{P_{OUT}}{P_{OUT} + P_{LOSS}} \qquad (17)$$

$$\eta = \frac{600}{600 + 23.14} \times 100 = 96.28\%$$

VI. HARDWARE IMPLEMENTATION

The Designed PCB Layout for the 48V, 600 W PSFB converter was given below. The PCB was designed using Ki CAD Software.

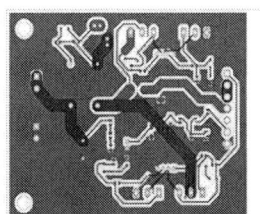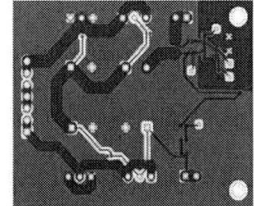

Fig.16. PCB Layout

The PCB has been fabricated and assembled as shown in Fig.17.

Fig.17. Fabricated and Assembled PCB

VII. CONCLUSION

The hardware design of PSFB dc-dc converter has been presented in detail for a 48V, 600 W converter. The different operating modes of the converter has been discussed in detail and the design methodology for achieving ZVS over a given operating range based on the input specifications has been presented. The PSFB converter is designed for a switching frequency of 100kHz and the design has been validated by the closed loop simulation in the MATLAB environment.

It has been observed from the simulation results, Zero voltage switching is achieved for the all four switches and the ZVS occurs for the entire range of phase shift from 0° to 180°. The calculated efficiency for the designed converter is 96.2%. The PCB design of the hardware prototype has also been presented. Hence the PSFB converter can be used for a wide range of input to get a wide range of output by employing phase angle control. It is expected the PSFB converter is an attractive solution for a EV charging station that need to cater a wide range of output voltage.

REFERENCES

[1] Dakshina M. Bellur; Marian K. Kazimierczuk, "DC-DC converters for electric vehicle applications," *2007 Electrical Insulation Conference and Electrical Manufacturing Expo 22-24 Oct. 2007.*

[2] Chakraborty, Sajib et al. "DC-DC Converter Topologies for Electric Vehicles, Plug-in Hybrid Electric Vehicles and Fast Charging Stations: State of the Art and Future Trends." *Energies (2019).*

[3] Salman Khan, Andrii Chub, and Dmitri Vinnikov, " An Overview of Wide-Voltage Range Isolated DC-DC Converters," *2023 IEEE 64th International Scientific Conference on Power and Electrical Engineering of Riga Technical University (RTUCON).*

[4] Akshay Kumar Singh, Madhuri A. Chaudhari, K. S. Raja Sekhar, Rohit Kumar, "Analysis of Isolated DC-DC Converters for Electric-Vehicle (EV) Battery Charging," *2023 IEEE Renewable Energy and Sustainable E-Mobility Conference (RESEM).*

[5] Ruoyu Hou; Pierre Magne; Berker Bilgin; Ali Emadi,"A topological evaluation of isolated DC/DC converters for Auxiliary Power Modules in Electrified Vehicle applications," *2015 IEEE Applied Power Electronics Conference and Exposition (APEC).*

[6] C. Donovan Davidson "Zero Voltage Switching Full Bridge Converter Topology," *Intelec 2010.*

[7] Kucuk, S., & Akboy, E."A basic phase shift full bridge DC-DC converter design and simulation", *2022 57th International Universities Power Engineering Conference (UPEC).*

[8] Yang, B., Duarte, J. L., Li, W., Yin, K., He, X., & Deng, Y. "Phase-shifted full bridge converter featuring over the full load range", *In Proc. 36th Annual Conference of the IEEE Industrial Electronics Society, (2010).*

[9] S. Telrandhe, J. Sabnis and M. Rajne, "Design Considerations for an On-Board Charger Based on PSFB converter with ZVS," *2020 IEEE First International Conference on Smart Technologies for Power, Energy and Control (STPEC), 2020.*

[10] Sithara S. G. Acharige; Md. Enamul Haque; Mohammad Taufiqul Arif; Nasser Hosseinzadeh; Kazi N. Hasan; Aman Maung Than Oo,"Review of Electric Vehicle Charging Technologies, Standards, Architectures, and Converter Configurations," *14 April 2023 IEEE Access (Volume: 11).*

[11] Guan-Chyun Hsieh; Jung-Chien Li; Ming-Huei Liaw; Jia-Perng Wang; Tsai-Fu Hung,"A Study on Full-Bridge Zero-Voltage-Switched PWM Converter: Design and Experimentation" *Proceedings of IECON '93 - 19th Annual Conference of IEEE Industrial Electronics.*

[12] Srdjan Srdic; Xinyu Liang; Chi Zhang; Wensong Yu; Srdjan Lukic"A SiC-based high-performance medium-voltage fast charger for plug-in electric vehicles," *2016 IEEE Energy Conversion Congress and Exposition (ECCE).*

[13] Alfredo Medina-Garcia; Juan Cruz-Cozar; Diego P. Morales-Santos; Noel Rodriguez; Manfred Schlenk,"A High-Efficiency Isolated Wide Voltage Range DC-DC Converter Using WBG Devices", *08 August 2022 IEEE Access (Volume: 10).*

Fixed Voltage Based P&O Method for Battery Charging from Partial Shaded PV String

1st Durgesh Chandra Nautiyal
Student Member, IEEE
Department of Electrical Engineering
Shiv Nadar Institution of Eminence Deemed to be University,
Delhi NCR 201314, India
dn643@snu.edu.in

2nd Jenis Jain
Department of Electrical Engineering
Shiv Nadar Institution of Eminence Deemed to be University,
Delhi NCR 201314, India
jj888@snu.edu.in

3rd Himanshu Sekhar Sahu
Member, IEEE
Department of Electrical Engineering
Shiv Nadar Institution of Eminence Deemed to be University,
Delhi NCR 201314, India
himanshu.sahu@snu.edu.in

4th Vijay Kumar Chakka
Senior Member, IEEE
Department of Electrical Engineering
Shiv Nadar Institution of Eminence Deemed to be University,
Delhi NCR 201314, India
vijay.chakka@snu.edu.in

Abstract—A photovoltaic (PV) system under partial shading conditions (PSCs) exhibits multiple local power peaks (LPPs) and a single global maximum power peak (GMPP) in its power-voltage (P-V) characteristics. Conventional maximum power point tracking (MPPT) methods are ineffective in such conditions, as they may track an LPP instead of the GMPP. Therefore, it is necessary to track GMPP accurately for the battery charging application from the partially shaded PV string. In this paper, a novel fixed voltage based perturb and observe (FVPO) method is proposed to track the GMPP for battery charging under PSCs. The proposed method calculates 'n' fixed voltage points using a new hyperbolic approximation approach and measures power at these points in an $n \times 1$ PV string. The calculated power values are then compared, and the duty cycle corresponding to the maximum power is fed to the converter. This initializes the perturb and observe method to track the GMPP accurately. Hence, the proposed method is simple, easy to implement, and quickly tracks the GMPP without the need to detect partial shading or scan the entire P-V curve. Furthermore, the effectiveness of the proposed FVPO method is validated for a PV string with three modules under various PSCs using the MATLAB platform for battery charging applications. It tracks the GMPP within 0.36 s under the most complex shading condition. The energy loss ranges from 5.61 to 20.14 J, enabling efficient real-time PV battery charging.

Index Terms—Photovoltaic (PV) system, hyperbolic approximation, partial shading conditions (PSCs), maximum power point tracking (MPPT).

I. INTRODUCTION

The utilization of renewable energy sources is becoming more and more important as the concerns about fossil fuel consumption are increasing. Out of all renewable energy sources, solar energy is the most promising and sustainable option, attracting a lot of interest because of its enormous potential [1]. Solar power generation is also environmentally beneficial, requiring very little maintenance and creating no pollution.

Despite these advantages, the power output of photovoltaic (PV) systems highly depends on environmental conditions such as solar intensity (G) and temperature (T). The power output of a PV system is significantly reduced when the PV modules are exposed to partial shading conditions (PSCs), where different PV modules receive different levels of solar intensity [2]–[5]. Under such conditions, multiple power peaks are observed in the power-voltage (P-V) curve of the PV string or array [6]–[8]. For maximum power extraction from a PV system, it must operate at the global maximum power point (GMPP) on the P-V curve. However, under PSCs, it is difficult to identify and track the GMPP.

Several maximum power point tracking (MPPT) methods are available in the literature to track maximum power point (MPP) [7], [9]–[21]. Out of all these methods, perturb & observe (P&O) [11] and incremental conductance [12] are the most widely used due to their simplicity and ease of implementation. However, these conventional methods are ineffective under PSCs as they may track local MPP instead of GMPP. A simple yet effective MPPT method was proposed in [13] to track the GMPP under PSCs. This method scans the entire P-V curve using a specific step size. However, its accuracy and tracking speed are highly dependent on the selected step size. A larger step size improves tracking speed but reduces accuracy, while a smaller step size enhances accuracy at the cost of increased tracking time. In order to reduce the tracking time, an MPPT scheme based on the 0.8 V_{oc} model was implemented in [14], [15]. These schemes significantly improve the tracking performance compared to [13] method. An advanced particle swarm optimization (PSO) based method [17] is used to track the GMPP under PSCs with fast convergence. In a recent research [16], a skip and search based method was proposed to track the GMPP. This approach skips voltage only after detecting a peak in the P-V

979-8-3315-3013-6/25 $31.00 © 2025 IEEE

curve, resulting in a longer tracking time for the GMPP.

In order to overcome the above mentioned issues, a novel fixed voltage based perturb and observe (FVPO) method is proposed for battery charging from a partially shaded PV string. The proposed MPPT method first calculates the voltages near all possible MPPs using a new hyperbolic approximation approach. The power is measured at each of the operating voltage points, and then the operating point corresponding to the maximum power is selected. In the next step, the proposed FVPO method initializes the conventional P&O method to accurately track the GMPP. Hence, the proposed approach eliminates the need for partial shading detection or a complete P-V curve scan, ensuring efficient and fast GMPP tracking. Finally, the proposed FVPO method's performance is examined using the MATLAB platform for battery charging applications under a variety of PSCs.

II. PROPOSED FVPO METHOD

The main objective of the proposed FVPO method is to quickly track the maximum available power from a PV string under PSCs without requiring partial shading detection. This ensures efficient operation at the GMPP, with the tracked power utilized for battery charging applications. However, under PSCs, the power output of the PV string decreases significantly, and multiple MPPs appear in its P-V curve. It makes GMPP tracking a challenging task. For this purpose, the voltage points near all possible MPPs are first identified using a new hyperbolic approximation approach. A fictitious hyperbola is placed in an MPP region of the P-V characteristic in such a way that its vertex is always close to the MPP for a given PSC. Then, the proposed FVPO method compares the measured powers at the vertices of the hyperbola to identify the vicinity of the GMPP. Finally, the operating point obtained from the hyperbolic approximation is used to initialize the conventional P&O algorithm for accurate GMPP tracking.

A. Identification of the fixed voltage points

A PV string (shown in Fig. 1), consisting of 'n' PV modules with individual bypass diodes and a single blocking diode, may exhibit a maximum of n number of MPPs in its P-V curve. A hyperbolic approximation approach is proposed to determine the voltage near these MPPs. The standard equation of a hyperbola with center (h, k) and transverse axis on the x-axis is given in (1).

$$\frac{(x - h)^2}{a^2} - \frac{(y - k)^2}{b^2} = 1 \qquad (1)$$

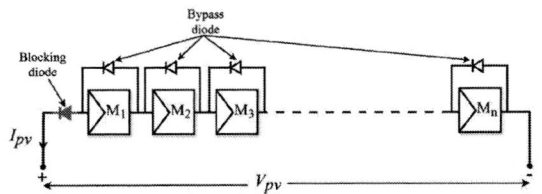

Fig. 1. An $n \times 1$ PV string with individual bypass diodes and a blocking diode.

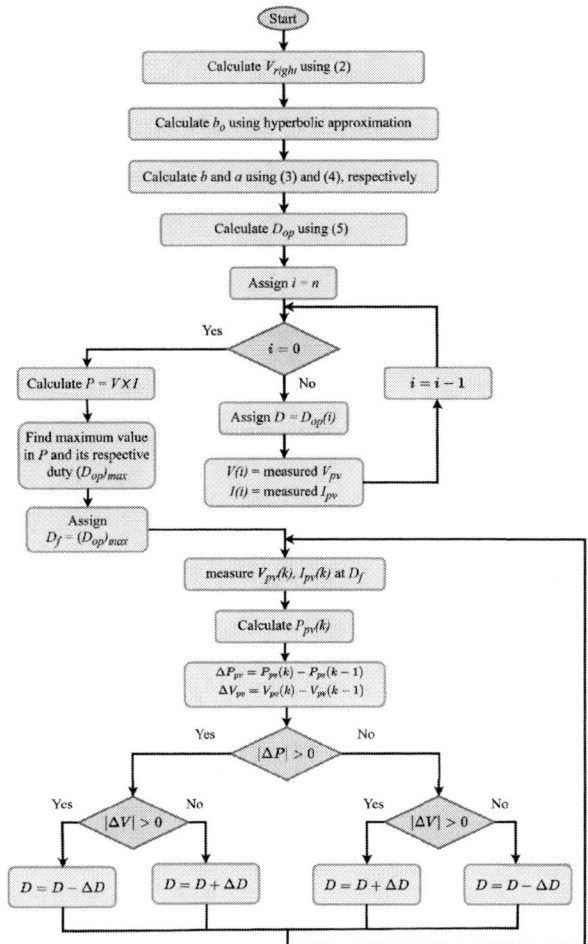

Fig. 2. Flowchart for proposed FVPO method.

where the coordinates of the vertices and co-vertices are $(\pm a, 0)$ and $(0, \pm b)$, respectively. The foci $(\pm V_{right})$ of the hyperbola with the center at $(0, 0)$ are calculated using (2).

$$V_{right}(i) = i \times V_{oc} - (n - i) \times V_{bd} \qquad (2)$$

where V_{oc} is the open circuit voltage of the module at standard test condition (STC), V_{bd} is the voltage across the bypass diode, and $i = 1, 2, 3, .., n$. To calculate the fixed voltage (i.e., vertex a) near the MPP, an optimal value of the co-vertex (b_o) is first required for $i = 1$. Once b_o is determined, the value of b for $i = 1, 2, 3, .., n$ can be easily calculated using (3).

$$b(i) = i \times b_o \qquad (3)$$

Finally, the value of a is computed using the basic equation related to hyperbola, as given in (4).

$$a(i) = \sqrt{V_{right}^2(i) - b^2(i)} \qquad (4)$$

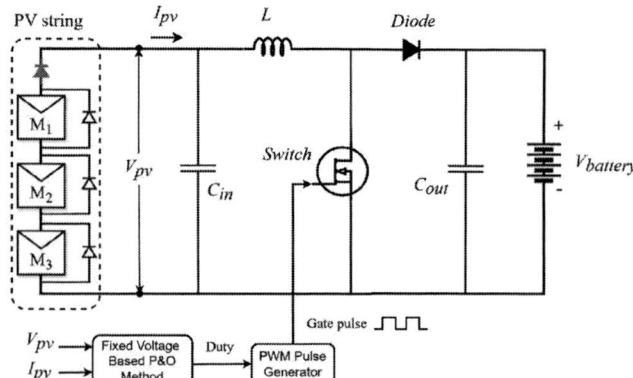

Fig. 3. A 3 × 1 PV string connected to DC-DC boost converter for battery charging.

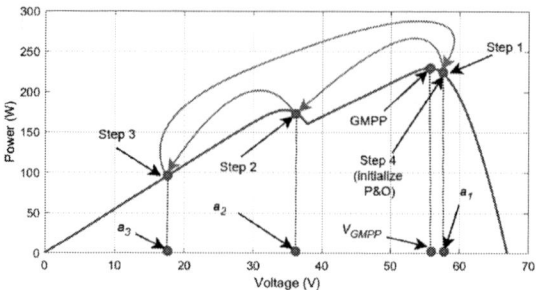

Fig. 4. Step by step tracking of proposed FVPO method.

B. GMPP tracking

For a PV string connected to a DC-DC boost converter and a battery, it is easy to calculate the operating duty cycle (D_{op}) using (5).

$$D_{op} = 1 - \frac{a}{V_{battery}} \quad (5)$$

where a is the fixed voltage at which the PV string is operating and $V_{battery}$ is the voltage of the battery connected at the output of the boost converter. In the proposed FVPO method, all the values of a are calculated using the hyperbolic approximation and the D_{op} using (5). Once these operating duty cycles are determined, the corresponding voltages and currents are measured at each operating point. A small delay is introduced after each duty cycle change. This delay is necessary to allow voltage oscillations to settle before accurate measurements are taken. Once the stabilized voltage and current values are obtained, the corresponding power at each voltage is calculated. Finally, the duty cycle corresponding to the highest power is selected to initialize the conventional P&O method. As a result, the GMPP is successfully tracked without scanning the whole P-V curve or detecting the partial shading. The complete process of MPPT is depicted in the flowchart shown in Fig. 2.

C. Example

For a better explanation of the proposed MPPT method, a 3 × 1 PV string with a rating of 300 W is considered. The PV string is connected to a DC-DC boost converter, which supplies power to a 70 V Li-ion battery for charging. The circuit diagram of this arrangement is shown in Fig. 3. The specifications of the PV module (100 W) are given in Table I. For this example (Case 1), three PV modules are exposed to irradiance levels of 900 W/m^2, 900 W/m^2, and 700 W/m^2 at a constant temperature of 25°C. Under this shading condition, the GMPP is tracked using the following steps:

- Calculate V_{right} for $n = 3$ using (2).
  $V_{right} = [21.1 \quad 44.3 \quad 67.5]$
  For simplicity V_{bd} is considered 0.7 V.

- Determine the optimum value of b_o for $i = 1$ and fixed voltage (a) at 17.5 V (STC) using back substitution in (4).
  $b_o = 11.8$
- Calculate the co-vertex (b) for all $i = 1, 2, 3$ using (3).
  $b = [11.8 \quad 23.6 \quad 35.4]$
- Calculate the fixed voltage (a) and D_{op} using (4) and (5), respectively.
  $a = [17.5 \quad 37.5 \quad 57.47]$
  $D_{op} = [0.75 \quad 0.464 \quad 0.18]$
- Measure the voltage and current at each operating duty.
  $V = [17.6 \quad 37.1 \quad 57]$
  $I = [5.42 \quad 4.22 \quad 3.9]$
- Identification of the duty cycle corresponding to the maximum power.
  $D_f = 0.18$
- Initialize the P&O at D_f.

The final tracked power, voltage, current, and duty are 227 W, 54.8 V, 4.1 A, and 0.21, respectively. The operating points at each step are shown in Fig. 4. It is observed from the figure that the proposed method effectively identifies the GMPP in 4 i.e. ($n + 1$) steps without requiring a complete scan of the P-V curve or prior detection of partial shading conditions.

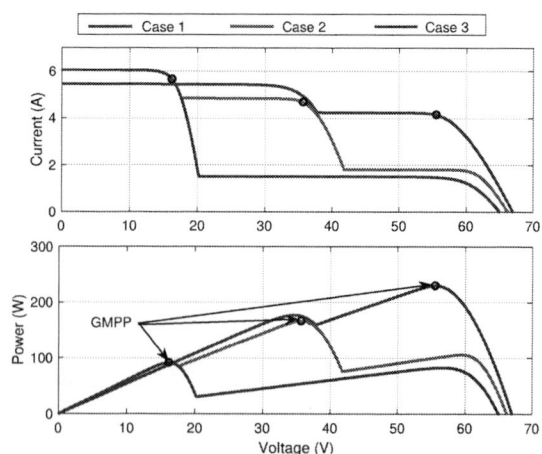

Fig. 5. The I-V and P-V characteristics of 3×1 PV string for Case 1, 2 and 3.

979-8-3315-3013-6/25 $31.00 © 2025 IEEE

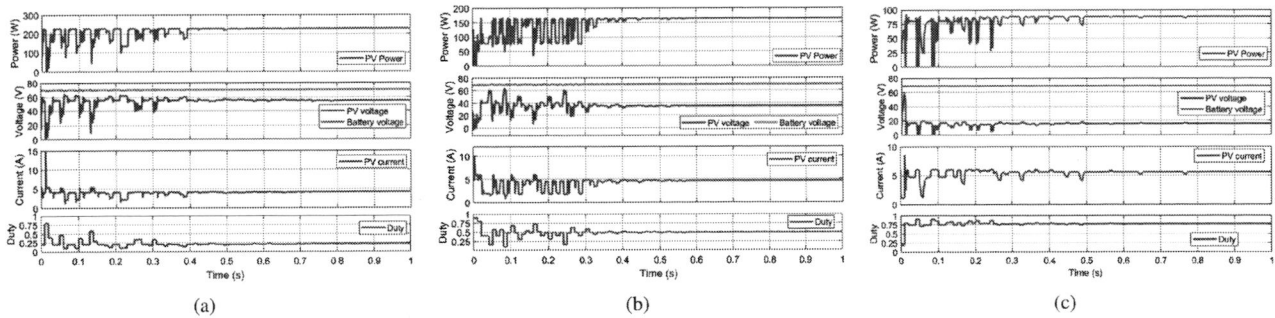

Fig. 6. Waveforms of PV power, voltage, current, battery voltage, and duty cycle during GMPP tracking using [17] for (a) Case 1, (b) Case 2, and (c) Case 3.

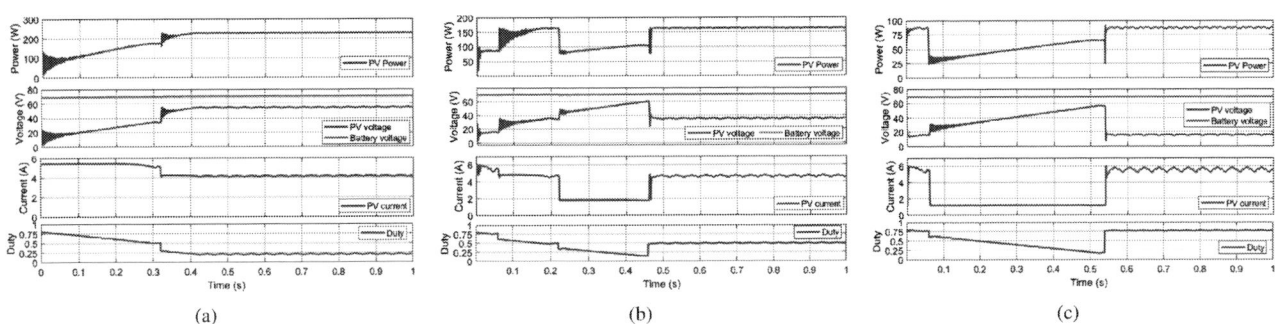

Fig. 7. Waveforms of PV power, voltage, current, battery voltage, and duty cycle during GMPP tracking using [16] for (a) Case 1, (b) Case 2, and (c) Case 3.

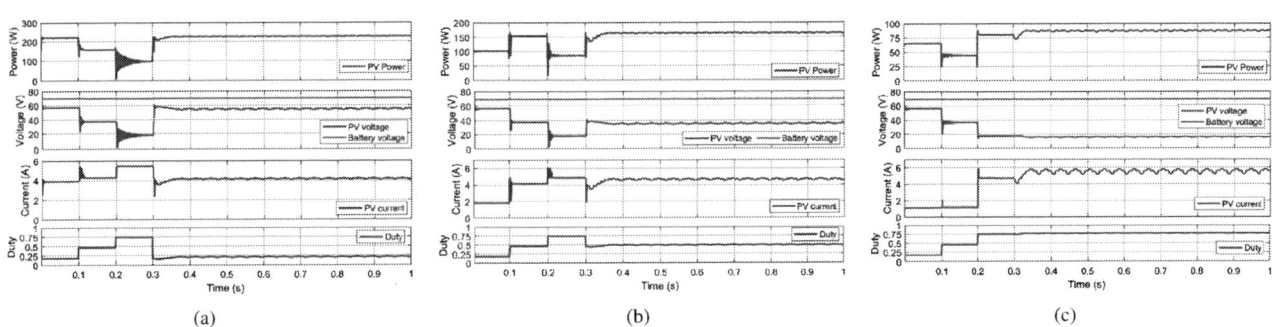

Fig. 8. Waveforms of PV power, voltage, current, battery voltage, and duty cycle during GMPP tracking using the proposed method for (a) Case 1, (b) Case 2, and (c) Case 3.

TABLE I
MANUFACTURERS DATASHEET OF PM100 PV MODULE AT STC.

P_{mp} (W)	V_{oc} (V)	I_{sc} (A)	V_{mp} (V)	I_{mp}(A)	k_v (%°C)	k_i (%°C)
100	22.5	6.02	17.5	5.71	-0.3305	0.0638

TABLE II
DC-DC BOOST CONVERTER PARAMETERS

C_{in} (μF)	L (mH)	C_{out} (μF)	Switching frequency (Hz)
200	1.1	200	10000

III. RESULT AND DISCUSSION

The proposed FVPO MPPT method is implemented in MATLAB platform to validate its effectiveness. A 3 × 1 PV string is connected to a DC-DC boost converter and a Li-ion battery, as shown in Fig. 3. The specifications of the boost converter parameters are given in Table II. Three different PSCs are considered to assess the effectiveness of the proposed FVPO approach; each of these cases results in a GMPP at a different voltage. In Case 1, the PV modules receive irradiance levels of 900 W/m², 900 W/m², and 700 W/m². In Case 2, the irradiance levels are 1000 W/m², 800 W/m², and 300 W/m², while in Case 3, they are 1000 W/m², 200 W/m², and 200 W/m². These test conditions are selected to evaluate the performance of the proposed FVPO method in tracking the GMPP under different shading patterns. Additionally, the proposed method is also compared with an advanced PSO based method [17] and a recent skip and search-based MPPT method [16] to highlight its effectiveness. The same sample

TABLE III
RESULT COMPARISON OF PROPOSED, [17] AND [16] METHODS

Method	Case	Tracking time to achieve GMPP (t_s)	Energy loss (J) during t_s (J)
[17]	1	0.4 s	33.037
	2	0.42 s	25.68
	3	0.5 s	8.388
[16]	1	0.42 s	36.56
	2	0.47 s	26.612
	3	0.55 s	13.14
Proposed	1	0.35 s	20.1425
	2	0.36 s	15.591
	3	0.34 s	5.6125

time and step size are used for the proposed, [17] and [16] methods, ensuring that the differences in performance are due to the tracking approach rather than variations in simulation parameters. The I-V and P-V characteristics of the PV string under the three PSC scenarios are illustrated in Fig. 5. Waveforms of PV power, voltage, current, battery voltage, and duty cycle during GMPP tracking using [17], [16] and the proposed method are shown in Fig. 6, Fig. 7 and Fig. 8, respectively. A delay of 0.1 second is introduced for each fixed voltage during tracking using the proposed FVPO method. This duration is sufficient to allow voltage oscillations to settle after a sudden change in the duty cycle by the MPPT controller. It is observed from Fig. 8 that the proposed method initializes the P&O method after 0.3 seconds, and within a few milliseconds, the GMPP is successfully tracked. The methods in [17] and [16] require approximately 0.4 to 0.55 seconds to track the GMPP under PSCs. The method in [16] takes comparatively more time, particularly under less complex PSCs, because it does not skip voltage steps until a peak is detected. In contrast, the method in [17] exhibits more fluctuations during tracking due to the large search space of the PSO algorithm. Due to higher tracking time, the [16] method also suffers from higher energy loss as given in Table III.

IV. CONCLUSION

This paper presents a simple FVPO method for tracking the GMPP under PSCs without the need to detect partial shading or scan the entire P-V curve. A new hyperbolic approximation approach is used to determine the n fixed voltage points of an $n \times 1$ PV string. These voltage points are very close to actual MPPs. The proposed method measures the voltages and currents at these n fixed voltage points in an $n \times 1$ PV string and selects the most optimal duty cycle to initialize the conventional P&O method. By pre-selecting near optimal operating points using hyperbolic approximation, the proposed method significantly reduces search time. It efficiently identifies the vicinity of the GMPP in $(n + 1)$ steps and tracks the GMPP within a few milliseconds, eliminating the need to scan the entire P-V curve or detect partial shading. This ensures

rapid convergence to the GMPP, leading to faster tracking and reduced computational complexity. The effectiveness of the proposed approach is validated through MATLAB simulations under three PSCs. The proposed FVPO method identifies the vicinity of the GMPP in the 4th step and tracks the GMPP under the most complex shading condition (Case 2) in 0.36 seconds. Furthermore, the energy loss during tracking ranges from 5.61 J to 20.14 J across all cases. Therefore, the FVPO method is a suitable solution for real-time battery charging applications in PV systems operating under PSCs.

REFERENCES

[1] I. E. Agency, "Renewables 2024 analysis and forecast to 2030," 2024, https://www.iea.org/reports/renewables-2024 [Accessed: 16-03-2025].

[2] H. S. Sahu, M. K. Mishra, S. Kumar, and S. Kumar Nayak, "A novel approach for direct MPP estimation of a PV module under different irradiation conditions," *IEEE Trans. Energy Convers.*, vol. 36, no. 4, pp. 3127–3136, 2021.

[3] J.-H. Teng, H.-C. Wu, Z.-H. Wu, and W.-H. Huang, "Efficient partial shading detection for photovoltaic generation systems," *IEEE Trans. Sustainable Energy*, vol. 14, no. 4, pp. 2249–2259, 2023.

[4] H. S. Sahu and S. K. Nayak, "Improvement in the power generation of a PV array under partial shading conditions," in *2014 Eighteenth National Power Systems Conference (NPSC)*, 2014, pp. 1–6.

[5] P. Changmai, S. K. Nayak, and S. K. Metya, "Mathematical model to estimate the maximum power output of a total cross tied connected PV array during partial shading condition," *IET Renewable Power Gener.*, vol. 13, no. 14, pp. 2647–2655, 2019.

[6] M. Etezadinejad, B. Asaei, S. Farhangi, and A. Anvari-Moghaddam, "An improved and fast MPPT algorithm for PV systems under partially shaded conditions," *IEEE Trans. Sustainable Energy*, vol. 13, no. 2, pp. 732–742, 2022.

[7] M. E. Başoğlu, "An improved 0.8 V_{OC} model based GMPPT technique for module level photovoltaic power optimizers," *IEEE Trans. Ind. Appl.*, vol. 55, no. 2, pp. 1913–1921, 2019.

[8] W. Xiao, N. Dong, and K. He, "A hybrid global wolf pack algorithm-based incremental conductance method under partial shading conditions," *Sol. Energy*, vol. 291, p. 113388, 2025.

[9] M. Kermadi, Z. Salam, A. M. Eltamaly, J. Ahmed, S. Mekhilef, C. Larbes, and E. M. Berkouk, "Recent developments of MPPT techniques for PV systems under partial shading conditions: a critical review and performance evaluation," *IET Renewable Power Gener.*, vol. 14, no. 17, pp. 3401–3417, 2020.

[10] A.-M. Badea, D. Manaila-Maximean, L. Fara, and D. Craciunescu, "Maximizing solar photovoltaic energy efficiency: MPPT techniques investigation based on shading effects," *Sol. Energy*, vol. 285, p. 113082, 2025.

[11] A. K. Abdelsalam, A. M. Massoud, S. Ahmed, and P. N. Enjeti, "High-performance adaptive perturb and observe MPPT technique for photovoltaic-based microgrids," *IEEE Trans. Power Electron.*, vol. 26, no. 4, pp. 1010–1021, 2011.

[12] M. A. Elgendy, B. Zahawi, and D. J. Atkinson, "Assessment of the incremental conductance maximum power point tracking algorithm," *IEEE Trans. on Sustainable Energy*, vol. 4, no. 1, pp. 108–117, 2013.

[13] E. Koutroulis and F. Blaabjerg, "A new technique for tracking the global maximum power point of PV arrays operating under partial-shading conditions," *IEEE J. of Photovoltaics*, vol. 2, no. 2, pp. 184–190, 2012.

[14] H. Patel and V. Agarwal, "Maximum power point tracking scheme for PV systems operating under partially shaded conditions," *IEEE Trans. Ind. Electron.*, vol. 55, no. 4, pp. 1689–1698, 2008.

[15] A. Kouchaki, H. Iman-Eini, and B. Asaei, "A new maximum power point tracking strategy for PV arrays under uniform and non-uniform insolation conditions," *Solar Energy*, vol. 91, pp. 221–232, 2013.

[16] Z. Smara, A. Aissat, H. Deboucha, H. Rezk, and S. Mekhilef, "An enhanced global MPPT method to mitigate overheating in PV systems under partial shading conditions," *Renewable Energy*, vol. 234, p. 121187, 2024.

[17] A. M. Eltamaly, M. Al-Saud, A. G. Abokhalil, and H. M. Farh, "Simulation and experimental validation of fast adaptive particle swarm optimization strategy for photovoltaic global peak tracker under dynamic partial shading," *Renewable Sustainable Energy Rev.*, vol. 124, p. 109719, 2020.

[18] R. Sangrody, S. Taheri, A.-M. Cretu, and E. Pouresmaeil, "An improved PSO-based MPPT technique using stability and steady state analyses under partial shading conditions," *IEEE Trans. on Sustainable Energy*, vol. 15, no. 1, pp. 136–145, 2024.

[19] K. K. Mohammed, S. Mekhilef, and S. Buyamin, "Improved rat swarm optimizer algorithm-based MPPT under partially shaded conditions and load variation for PV systems," *IEEE Trans. on Sustainable Energy*, vol. 14, no. 3, pp. 1385–1396, 2023.

[20] J. Maeng, J. Jeong, I. Park, M. Shim, and C. Kim, "A time-based direct MPPT technique for low-power photovoltaic energy harvesting," *IEEE Trans. Ind. Electron.*, vol. 71, no. 5, pp. 5375–5380, 2024.

[21] S. S. Sakthivel, V. Arunachalam, S. P. P, and S. K, "Improved global maximum power point tracking technique for a partially shaded solar photovoltaic array using capacitor transient effect," *IEEE Trans. Ind. Inf.*, vol. 20, no. 12, pp. 13 853–13 862, 2024.

Analysis and comparison of charging speed of two-stage EV charger for different charging techniques

1st Dalija Rath, Member IEEE
Dept. of Electrical Engineering
NIT Rourkela
Rourkela, India
rath.dalija@gmail.com

2nd Subrat Mahapatra
Dept. of Electrical Engineering
NIT Rourkela
Rourkela, India
subratmahapatra482@gmail.com

3rd Susovon Samanta, Member IEEE
Dept. of Electrical Engineering
NIT Rourkela
Rourkela, India
samantas@nitrkl.ac.in

Abstract—**Looking at the advantages of future growth of electric vehicles (EV) much research is underway on EV chargers. In the literature, many studies are being conducted on charging infrastructure, converter topologies, control, and charging schemes. In this article, a two-stage grid integrated EV charger is designed to fix the DC link voltage and to have efficient charging through a Totem pole PFC rectifier and a nonisolated bidirectional DC/DC charger, respectively. Here, a model of the charging system is presented by combining the two stages together, which can be used to get a faster controller design for the system. According to studies, constant power-constant voltage (CP-CV) is the better way to charge a battery in terms of useful energy compared to constant current-constant voltage (CC-CV). Here, the above two charging techniques are compared on the basis of charging speed, and it has been found that CC-CV gives faster charging than CP-CV which is validated by implementing the two charging techniques into a 3kw charger to charge a 72V,100Ah battery using MATLAB/Simulink. The hardware prototype is in progress, and in this paper the open-loop hardware results of the second stage of the charger tested at low power level(250W) have been presented. The hardware has been developed using the SKM50GB12T4 IGBT module and is controlled by the TMS320F28379D DSP microcontroller.**

Index Terms—**Totem-pole power factor correction (PFC) rectifier, EV charging, Unipolar PWM (UPWM), DC/DC charger, Small signal model, CC-CV, CP-CV**

I. INTRODUCTION

The integration of electric vehicles (EVs) with the power grid introduces several challenges, particularly concerning power quality [1]. One of the primary concerns is maintaining the total harmonic distortion (THD) below 7% , in compliance with IEEE standards [2], while also providing reactive power support. Addressing these issues requires the implementation of an appropriate modulation technique [3]–[5] and a robust control strategy. The first objective of the EV charger is to meet these grid-related requirements, while the second objective is to optimize the battery charging scheme to enhance the overall charger efficiency [6]–[9].

Various converter topologies for both single-stage and two-stage EV chargers have been explored in the literature [10]. Single-stage chargers, which typically deliver an output voltage in the range of 270V to 400V, are not ideal for charging low-voltage batteries. Therefore, two-stage chargers are generally preferred due to their ability to support a wider output voltage range. Several two-stage configurations have been proposed. For instance, [11] presents a charger using a three-phase PFC rectifier combined with a bidirectional DC/DC converter. In [12], a single-phase charger is described, where the first stage is a full-bridge PFC rectifier composed of two switches and two diodes, followed by an LLC resonant converter as the second stage. However, this configuration does not support bidirectional power flow.

The main contribution of the present work is

- Conventional EV chargers use a diode bridge rectifier along with a PFC circuit for AC/DC conversion, which has very low efficiency due to two-stage conversion and is not suitable for high- power applications [13], [14]. Hence, here it has been replaced by two-stage grid-integrated bidirectional charger as shown in Fig.1.

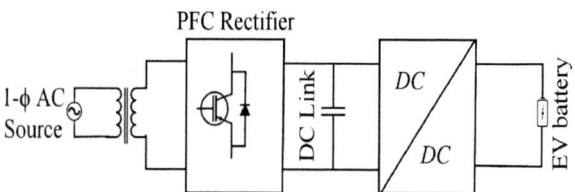

Fig. 1. Schematic picture of a two stage EV charger.

- Implementation of CP-CV charging to improve the efficiency of the charger by reducing the power loss during CC mode of charging.

The rest of this paper is organized as follows. Section II is about the working of the single phase PFC rectier with UPWM

979-8-3315-3013-6/25 $31.00 © 2025 IEEE

followed by it's modeling. Section III is about the working and modeling of the DC/DC charger. In Section IV a brief description of CC-CV and CP-CV charging along with the closed-loop control of the whole charging system has been given. The MATLAB simulation results for different charging techniques and the open-loop experimental results is given in Section V. Finally, Section VI concludes the article.

II. ANALYSIS OF TOTEM-POLE PFC RECTIFIER WITH UPWM TECHNIQUE

A. working principle

In UPWM two reference signals $180°$ out of phase with each other are compared with a carrier wave (v_{cr}) to generate gate pulses for S_1 and S_2 as shown in Fig.2, and their complementary is given to S_1' and S_2', respectively. So, there are 3 working in one switching period. In Mode I S_1 and S_2' are turned on, in Mode II the gate pulse is given to S_1 and S_2, and in Mode III the gate pulse is given to S_1' and S_2' as shown in Fig.3, Fig.4 and Fig.5 respectively. During Mode II and Mode III the inductor is charged, and the inductor is discharged during Mode I.

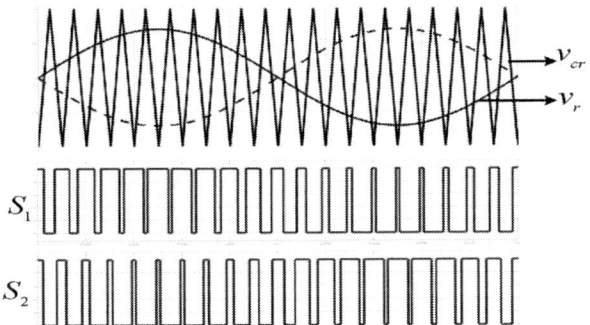

Fig. 2. Gate pulse generation in UPWM.

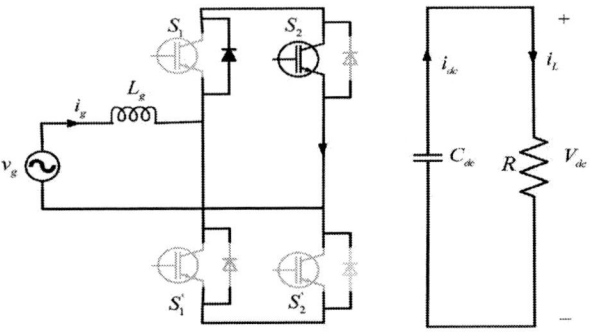

Fig. 3. Mode I of PFC rectifier with UPWM.

B. Modeling

In UPWM only during the Mode I switch S_2 is ON, whose period is considered $d_2 T_S$. During Mode II and Mode III which is for a period of $d_1 T_S$, S_2 is OFF and $d_1+d_2=1$. However, due to page constraints, a detailed derivation of

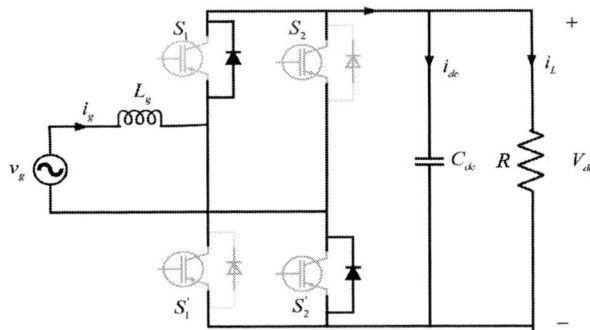

Fig. 4. Mode II of PFC rectifier with UPWM.

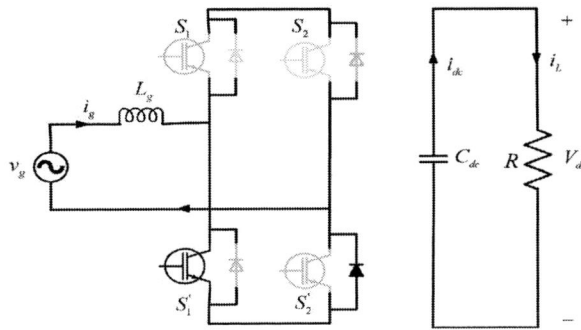

Fig. 5. Mode III of PFC rectifier with UPWM.

the above relationship is not presented here. Combining all three modes of operation the average state-space model of the rectifier is:

$$
\begin{bmatrix} \dot{i}_g \\ \dot{v}_{dc} \end{bmatrix} = \begin{bmatrix} 0 & -\frac{d_1-d_2}{L_g} \\ \frac{d_1-d_2}{C_{dc}} & -\frac{1}{RC_{dc}} \end{bmatrix} \begin{bmatrix} i_g \\ v_{dc} \end{bmatrix} + \begin{bmatrix} \frac{1}{L_g} \\ 0 \end{bmatrix} [v_g]
$$
(1)

or

$$
\begin{bmatrix} \dot{i}_g \\ \dot{v}_{dc} \end{bmatrix} = \begin{bmatrix} 0 & -\frac{m}{L_g} \\ \frac{m}{C_{dc}} & -\frac{1}{RC_{dc}} \end{bmatrix} \begin{bmatrix} i_g \\ v_{dc} \end{bmatrix} + \begin{bmatrix} \frac{1}{L_g} \\ 0 \end{bmatrix} [v_g]
$$
(2)

After perturbing and linearizing (2) around the DC operating point (M, V_{dc}, I_g), the small signal model of the rectifier is:

$$
\begin{bmatrix} \dot{\hat{i}}_g \\ \dot{\hat{v}}_{dc} \end{bmatrix} = \begin{bmatrix} 0 & -\frac{M}{L_g} \\ \frac{M}{C_{dc}} & -\frac{1}{RC_{dc}} \end{bmatrix} \begin{bmatrix} \hat{i}_g \\ \hat{v}_{dc} \end{bmatrix} + \begin{bmatrix} \frac{1}{L_g} & \frac{V_{dc}}{L_g} \\ 0 & \frac{I_g}{C_{dc}} \end{bmatrix} \begin{bmatrix} \hat{v}_g \\ \hat{m} \end{bmatrix}
$$
(3)

The steady state voltage relationship is:

$$
V_{dc} = \frac{V_g}{M}
$$
(4)

979-8-3315-3013-6/25 $31.00 © 2025 IEEE   331

III. MODELING OF THE DC/DC CHARGER

The charger acts as a buck converter in charging(G2V) mode and operates in boost mode during V2G operation. The operating modes during charging are shown in Fig.6 and Fig.8.

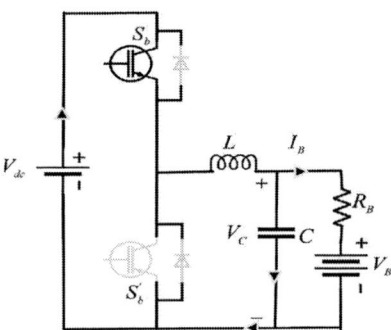

Fig. 6. Mode I of DC/DC charger.

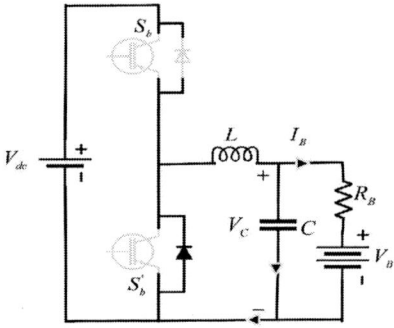

Fig. 7. Mode II of DC/DC charger.

$$
\begin{bmatrix} \dot{i}_L \\ \dot{v}_C \end{bmatrix} = \begin{bmatrix} 0 & -\frac{1}{L} \\ \frac{1}{C} & -\frac{1}{R_b C} \end{bmatrix} \begin{bmatrix} i_L \\ v_C \end{bmatrix} + \begin{bmatrix} \frac{1}{L} & 0 \\ 0 & \frac{1}{R_b C} \end{bmatrix} \begin{bmatrix} v_{dc} \\ v_b \end{bmatrix} \tag{5}
$$

$$
\begin{bmatrix} \dot{i}_L \\ \dot{v}_C \end{bmatrix} = \begin{bmatrix} 0 & -\frac{1}{L} \\ \frac{1}{C} & -\frac{1}{R_b C} \end{bmatrix} \begin{bmatrix} i_L \\ v_C \end{bmatrix} + \begin{bmatrix} 0 & 0 \\ 0 & \frac{1}{R_b C} \end{bmatrix} \begin{bmatrix} v_{dc} \\ v_b \end{bmatrix} \tag{6}
$$

So, the average state space model of the converter is given in

$$
\begin{bmatrix} \dot{v}_L \\ \dot{v}_C \end{bmatrix} = \begin{bmatrix} 0 & -\frac{1}{L} \\ \frac{1}{C} & -\frac{1}{R_b C} \end{bmatrix} \begin{bmatrix} i_L \\ v_C \end{bmatrix} + \begin{bmatrix} \frac{d}{L} & 0 \\ 0 & \frac{1}{R_B C} \end{bmatrix} \begin{bmatrix} v_{dc} \\ v_B \end{bmatrix} \tag{7}
$$

After perturbing and linearizing (7) around the DC operating point (D, V_{dc}, V_b), the small signal model of the charger is given by (8).

$$
\begin{bmatrix} \dot{\hat{i}}_L \\ \dot{\hat{v}}_C \end{bmatrix} = \begin{bmatrix} 0 & -\frac{1}{L_g} \\ \frac{1}{C} & -\frac{1}{R_b C} \end{bmatrix} \begin{bmatrix} \hat{i}_L \\ \hat{v}_C \end{bmatrix} + \begin{bmatrix} \frac{D}{L} & 0 & \frac{V_{dc}}{L} \\ 0 & \frac{1}{R_b C} & 0 \end{bmatrix} \begin{bmatrix} \hat{v}_{dc} \\ \hat{v}_b \\ \hat{d} \end{bmatrix} \tag{8}
$$

where,
v_g=grid voltage
v_{dc}=dc link voltage
v_c=voltage across the battery
v_b=internal battery voltage
i_g= grid current
i_L= load current
m=modulation index
d=duty ratio of the charger
In this paper, the representation of the above variables with a small italic-style letter, capital letter, and small italic-style letter with a hat signifies their instantaneous values, steady-state values, and small signal values, respectively.
The parameter
L_g= Source inductance
C_{dc}= dc link capacitance
R=load resistance
L= Load inductance
R_b=internal resistance of the battery
C= capacitance across the load

IV. CLOSED LOOP CONTROL OF THE TWO STAGE CHARGER FOR THE CHARGING OF BATTERY

The primary objective of the charger is to maintain a constant DC-link voltage, which is achieved using a Totem-Pole PFC rectifier. As depicted in Fig.11, the rectifier is controlled by a two-loop control scheme. The outer voltage control loop regulates the DC-link voltage by generating a reference current for the inner loop. The inner current control loop then produces a reference signal for the PWM generator, which in turn generates the appropriate gate pulses to maintain the desired DC-link voltage.
The secondary objective of the charger is to charge the battery using an appropriate current or voltage profile. Among the various charging strategies discussed in the literature, this work focuses on Constant Current–Constant Voltage (CC-CV) and Constant Power–Constant Voltage (CP-CV) charging methods.

- CC-CV Charging: In CC-CV charging, the output of the charger will be a constant current or a constant voltage depending on the threshold criteria shown in Fig.8.
- CP-CV charging:In CC-CV charging, the constant current (CC) mode results in higher power losses and reduced useful energy. To enhance the overall efficiency of the charger, the CC mode is replaced with a constant power (CP) mode, as illustrated in Fig.9.

The small signal model of the complete two-stage charger is shown in Fig.10, from which the transfer function (9) and (10)

979-8-3315-3013-6/25 $31.00 © 2025 IEEE

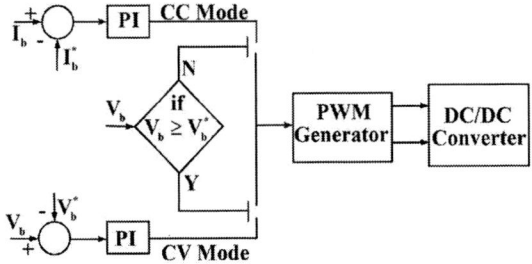

Fig. 8. Flow diagram for CC-CV charging

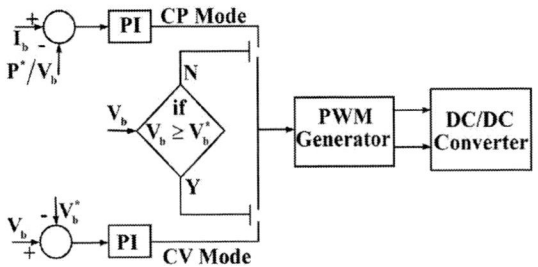

Fig. 9. Flow diagram for CP-CV charging

are derived to design the PFC rectifier controller and (11) and (12) are derived to control the buck converter based on the charging scheme.

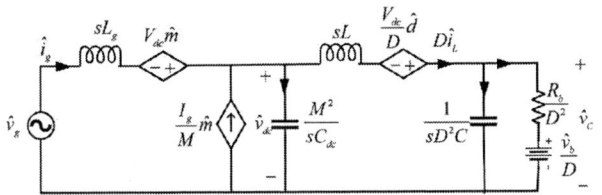

Fig. 10. Combined model of the two-stage charging system

$$\frac{\hat{i}_g}{\hat{m}} = \frac{s\frac{V_{dc}}{L_g} + \frac{V_{dc}}{L_g C_{dc} R} - \frac{M I_g}{L_g C_{dc}}}{s^2 + s\frac{1}{C_{dc}R} + \frac{M^2}{L_g C_{dc}}} \quad (9)$$

$$\frac{\hat{v}_{dc}}{\hat{m}} = \frac{s\frac{I_g}{C_{dc}R} + \frac{M V_{dc}}{L_g C_{dc} R}}{s^2 + s\frac{1}{C_{dc}R} + \frac{M^2}{L_g C_{dc}}} \quad (10)$$

$$\frac{\hat{i}_L}{\hat{d}} = \frac{V_{dc}(sC + \frac{1}{R_b})}{s^2 LC + s\frac{L}{R_b} + 1} \quad (11)$$

$$\frac{\hat{v}_C}{\hat{d}} = \frac{V_{dc}}{s^2 LC + s\frac{L}{R_b} + 1} \quad (12)$$

V. RESULTS AND DISCUSSION

A. Simulation result

Based on the parameter specifications listed in Table.I, a 3kW two-stage EV charger has been designed in MATLAB/Simulink to charge a 72V, 100Ah battery. The charger utilizes a two-loop control system, as illustrated in Fig.11,

to regulate the DC-link voltage at 400V using a Totem-Pole PFC rectifier. Unipolar PWM (UPWM) modulation is employed for switching the rectifier. The charger operates using the previously discussed charging algorithms to manage the battery charging process.

The simulation results shown in Fig. 12 confirm that the PFC rectifier delivers 3kW of output power while maintaining a steady DC-link voltage of 400V. The battery is charged using both the CC-CV and CP-CV methods, with the corresponding charging profiles presented in Fig.13 and Fig.14. In CC-CV charging, the battery is initially charged at a constant current until its voltage reaches a threshold of 80V. Then it switches to constant voltage mode, during which the charging current gradually decreases. In CP-CV charging, the battery is charged at constant power until the same voltage threshold is reached, followed by constant voltage charging in a similar manner.

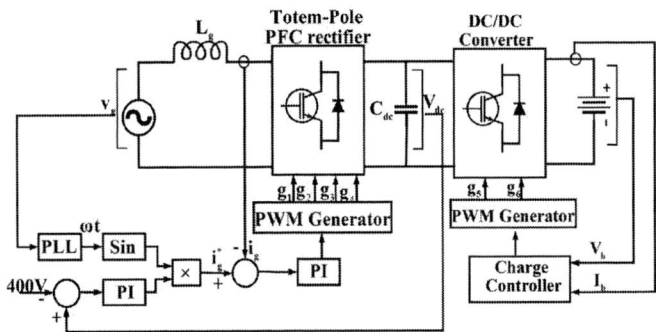

Fig. 11. Block diagram of the closed loop system for charging of battery

TABLE I
DESIGN SPECIFICATION

Parameter	Symbol	specification
Grid voltage	V_g	230V(RMS)
Grid frequency	f_g	50Hz
dc link voltage	V_{dc}	400V
Power	P	3kw
Switching frequency	f_{sw}	20kHz
Source inductance	L_g	0.0022H
dc link capacitance	C_{dc}	1500μF
Load Inductance	L	0.0022H
Load capacitance	C	1500μF
Battery specification		72V, 120Ah

B. Experimental results

The experimental setup of the second stage of the EV charger is illustrated in Fig. 16. The setup employs an SKM50GB12T4 IGBT module, whose switching operation is controlled by the Texas Instruments TMS320F28379D microcontroller to regulate the output voltage. The prototype was tested at a low power level of 250W with a switching frequency of 10kHz. The corresponding input current, inductor current, and output voltage waveforms of the buck converter are shown in Fig.17 and Fig.18. As seen in Fig.18, the inductor current waveform demonstrates continuous charging and

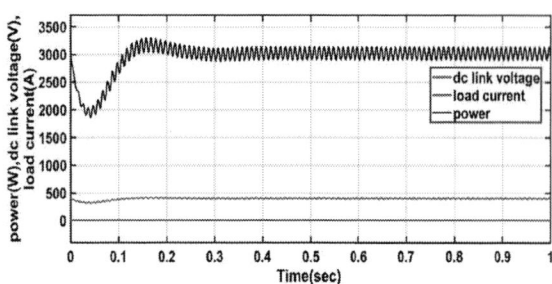

Fig. 12. Output power(W), DC link voltage(V), load current(A) Vs Time(sec)

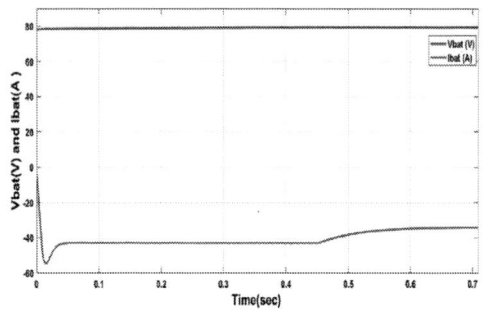

Fig. 13. voltage and current profile of battery for CC-CV charging

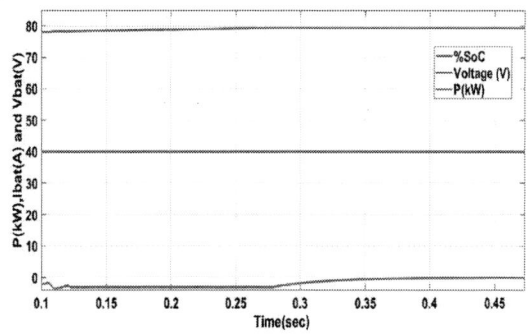

Fig. 14. charging profile for CP-CV charging

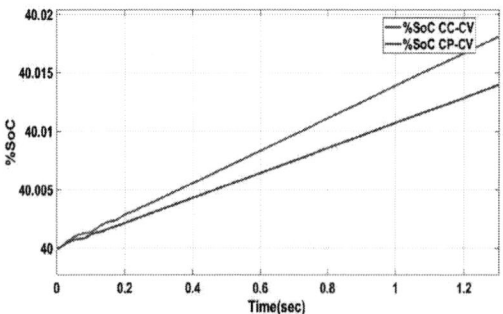

Fig. 15. SoC variation for different charging techniques

discharging within each switching cycle, while maintaining a stable DC output voltage.

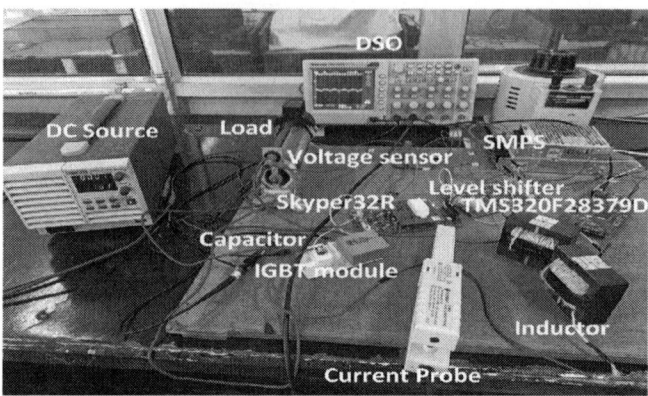

Fig. 16. Experimental Set-up

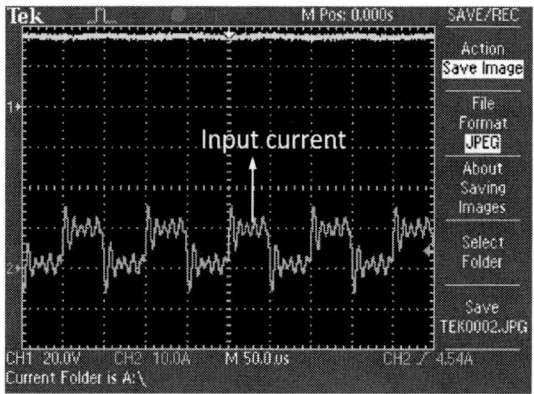

Fig. 17. Input current

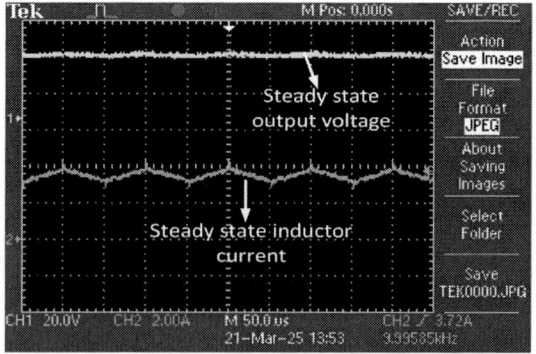

Fig. 18. Steady state Inductor current and Output Voltage

VI. CONCLUSION

This paper presents a small-signal model of a two-stage bidirectional EV charger, along with the design of a 3kW charger system. The proposed charger includes a Totem-Pole PFC rectifier and a non-isolated DC-DC converter, which

979-8-3315-3013-6/25 $31.00 © 2025 IEEE

operates as a buck converter during vehicle-to-grid (V2G) operation. The PFC rectifier is controlled using unipolar PWM (UPWM) modulation. To manage battery charging, both Constant Power–Constant Voltage (CP-CV) and Constant Current–Constant Voltage (CC-CV) algorithms are implemented. While the CP-CV method achieves a slightly higher efficiency of 98.61% , the CC-CV algorithm provides a faster charging time, as demonstrated in Fig.15, with an efficiency of 98.11% . Furthermore, the second stage of the charger was successfully tested at a low power level of 250W using the TMS320F28379D microcontroller, achieving the desired output voltage.

Acknowledgement

This work is supported by the Science and Engineering Research Board (SERB) of India under Grant IPA/2021/000081.

References

[1] A. Shahin et al., "A comprehensive analysis: Integrating renewable energy sources with wire/wireless ev charging systems for green mobility," *IEEE Access*, 2024.

[2] *IEEE recommended practice and requirements for harmonic control in electric power systems*, IEEE Standards Association Std., 2014.

[3] J. Salmon, T. A. Lipo, and J. McKeever, "A carrier-based unipolar pwm current controller that minimizes the pwm-cycle average current-error using internal feedback of the pwm signals," *IEEE Transactions on Power Electronics*, vol. 22, no. 5, pp. 1708–1718, 2007.

[4] Z. Guo and F. Kurokawa, "A novel pwm modulation and hybrid control scheme for grid-connected unipolar inverters," in *2011 Twenty-Sixth Annual IEEE Applied Power Electronics Conference and Exposition (APEC), IEEE*. IEEE, 2011, pp. 734–739.

[5] Y.-H. Liao, "A novel reduced switching loss bidirectional ac/dc converter pwm strategy with feedforward control for grid-tied microgrid systems," *IEEE Transactions on Power Electronics*, vol. 29, no. 3, pp. 1500–1513, 2013.

[6] S. U. Jeon et al., "Study on battery charging strategy of electric vehicles considering battery capacity," *IEEE Access*, vol. 9, pp. 89 757–89 767, 2021.

[7] M. R. Khalid, Z. Khan, D. Ahmed, E. Hossain, and S. Mekhilef, "A comprehensive review on structural topologies, power levels, energy storage systems, and standards for electric vehicle charging stations and their impacts on grid," *IEEE Access*, vol. 9, pp. 128 069–128 094, 2021.

[8] M. M. Alhaider et al., "New temperature-compensated multi-step constant-current charging method for reliable operation of battery energy storage systems," *IEEE Access*, vol. 8, pp. 27 961–27 972, 2020.

[9] Y. Yan et al., "Securing full-power-range zero-voltage switching in both steady-state and transient operations for a dual-active-bridge-based bidirectional electric vehicle charger," *IEEE Transactions on Power Electronics*, vol. 35, no. 7, pp. 7506–7519, 2019.

[10] M. Safayatullah et al., "A comprehensive review of power converter topologies and control methods for electric vehicle fast charging applications," *IEEE Access*, vol. 10, pp. 40 753–40 793, 2022.

[11] K. Zhou, H. Fang, and Y. Liu, "Driving–charging integrated controller for electric vehicles," *IEEE Access*, vol. 10, pp. 66 545–66 563, 2022.

[12] S. R. Meher and R. K. Singh, "A standard two stage on-board charger with single controlled pwm and minimum switch count," *IEEE Transactions on Industry Applications*, vol. 59, no. 4, pp. 4628–4639, 2023.

[13] A. Dubey, S. Santoso, and M. P. Cloud, "Average-value model of electric vehicle chargers," *IEEE Transaction on Smart Grid*, vol. 4, 2013.

[14] S. M. I. Prince Dadhaniya, Mukesh Maurya and M. I. Gururaj Mirle Vishwanath, "A bridgeless modified boost converter to improve power factor in ev battery charging applications," *IEEE JOURNAL OF EMERGING AND SELECTED TOPICS IN INDUSTRIAL ELECTRONICS*, vol. 5, 2024.

979-8-3315-3013-6/25 $31.00 © 2025 IEEE

Analysis of Optimal Deep Neural Network for Enhanced Battery SOC Estimation

Prashant Aher[*1], Dev Rai[†], Sanjaykumar Patil[†], Rhugved Rane[†], Tanmayee Rathod[†], and Janyaa Tikoo[†]

[*]HELLA India Automotive Pvt. Ltd., Pune, India, [†]COEP Technological, University, Pune, India,

Email: [1]pka20.instru@coeptech.ac.in

Abstract—This paper presents a deep neural network (DNN) approach for enhanced state of charge (SOC) estimation of Lithium-Ion (Li-ion) battery for eMobility. A series of DNN models are developed with varying numbers of hidden layers trained with dynamic drive profiles, to test and analyze the performance on untrained drive profiles. Analysis indicates that DNNs with four, five, and six hidden layers are found to be top performers for SOC estimation. To examine the performance and generalization capability of the suggested DNN architecture for battery SOC estimation, the model is tested with three different dynamic drive cycles at different temperatures using error metrics. The root mean squared error (RMSE) and mean squared error (MSE) of 1.57% and 0.024%, respectively, were observed when the model was tested with the Dynamic Stress Test dataset. The maximum RMSE and MSE of 2.64% and 0.069%, respectively, were observed when tested with the Federal Urban Driving Schedule dataset. Further, the results of SOC estimation using suggested DNNs are compared with other advanced techniques to showcase the enhanced performance of the suggested architecture.

Index Terms—Battery management system, Deep neural networks, Electric vehicles, Li-ion battery, state of charge estimation.

I. INTRODUCTION

The need for decarbonization of the transportation sector promotes the use of electric vehicles (EV). Currently, lithium-ion batteries represent the cutting-edge technology for energy storage in EVs. For safe and dependable operation, battery management systems (BMS) are used to monitor and control these battery packs [1]. Because BMS are limited to monitoring external battery metrics like temperature, voltage, and current, it is unable to measure important variables like SOC and state of health. Mathematically SOC can be defined as,

$$SOC = \frac{Available\ Discharge\ Capacity}{Rated\ Discharge\ Capacity} \quad (1)$$

The rated capacity of the battery does not remain constant, and is a function of temperature, aging, charge and discharge history, and inherent manufacturing defects.

The accurate estimation of battery SOC plays a significant role in the optimum use of battery packs, end user safety, and battery life [2]. The numerous techniques present in literature for SOC estimation can be categorized into open circuit voltage (OCV) based methods, Coulomb counting (CC) method, model-based method, and data-driven methods. The OCV-based methods suffer from accurate measurement of OCV during vehicle run-time. The CC approach has drawbacks of improper SOC initialization, integration error accumulation, and a lack of adaptability to changing operating conditions. Adaptive CC, for example, in combination with OCV methods, can overcome these limitations, however, it is still unreliable when used in critical applications [3]. For battery state estimation, a variety of sophisticated methods based on adaptive filters and nonlinear observers [4]–[6] employ equivalent circuit or electrochemical models. Though model-based estimation can overcome the issue of SOC initialization, and noisy measurements, estimation accuracy relies on model accuracy [7]. In this regard, the data-driven technique maps the non-linear behavior of battery cells by mapping input variables to output variable SOC.

Data driven methods rely on analyzing field data, e.g. by employing learning techniques such as deep neural network (DNN) [8], gated recurrent unit (GRU), recurrent neural network (RNN) [9], and long short-term memory (LSTM) [10]. Different variants of DNN, RNN, LSTM, and GRU have been explored to a greater extent in recent years with advantages such as no requirement of battery model, added filter, and mathematical relationship [11], [12]. In addition, it allows the prediction of complex battery behavior in a short time. The authors in the literature have presented battery SOC estimation using DNN trained and tested on a variety of drive profiles with satisfactory results without filtering of the dataset. Also, DNNs are found to be more efficient from computational cost and complexity perspectives, over other variants such as convolutional neural network (CNN), GRU, and LSTM which require more memory. To develop optimized algorithms which are capable of satisfactory state estimation and are deployable in sophisticated BMS is the need of the hour.

The objective of this paper is to implement and analyze a DNN framework leading to enhanced Li-ion battery SOC estimation. The major contributions of the work presented are: (i) Analyzing the effect of a number of DNN hidden layers on SOC estimation when trained with experimentally obtained dynamic current and tested with untrained data (ii) Training and testing the network at various temperatures and different dynamic drive profiles, and (iii) To perform quantitative analysis of optimal DNN to enhance the accuracy of SOC estimation. Towards this objective, a publicly available battery dataset from Center for Advanced Life Cycle Engineering (CALCE) [13], [14] is used to train and test the DNN to prove generalization capability and find optimal DNN for SOC

979-8-3315-3013-6/25 $31.00 © 2025 IEEE

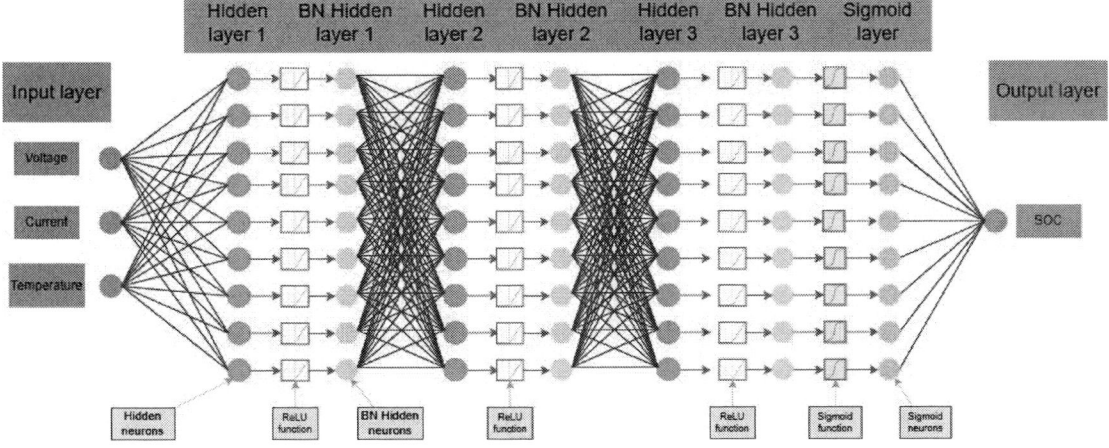

Fig. 1. DNN Architecture

estimation.

The remainder of the paper is organized as follows: Section II outlines DNN framework, strategy for training DNN, and choosing hyperparameters. A brief on experimental design and simulation conditions is presented in section III. The performance evaluation of developed DNN for SOC estimation is discussed in section IV, and the paper is concluded with section V.

II. DNN FRAMEWORK FOR SOC ESTIMATION

A DNN is an artificial neural network, adding hidden layers increases the model depth. It has a minimum of three computational layers namely input, hidden, and output layers as shown in Fig. 1. The work presented in the paper uses DNN to estimate battery SOC, by mapping battery input parameters to the output parameter, where battery parameters are obtained by performing characterization tests on battery cells under controlled temperature. The vector of input is defined as, $\psi_k = (I_k, V_k, T_k)$ represents current, voltage, and temperature at time k. The output of DNN is defined as, $Y_k = z_k$, where z_k is SOC at time k.

The feedforward networks are matrix-based and can be represented by,

$$h_n^l = R\left(\sum_n \left(w_{f,n}^l h_n^{l-1} + b_n^l\right)\right) \quad (2)$$

where, $w_{f,n}^l$ represents weight connection between neuron f in layer $l-1$, and neuron n in layer l. h_n^{l-1} and b_n^l are activation and bias functions, respectively. The R is rectified linear unit (ReLU) activation function given by,

$$R(p) = \begin{cases} 0, & for \ p < 0 \\ p, & for \ p \geq 0 \end{cases} \quad (3)$$

The SOC at output layer can be computed as,

$$Y_k = z_k = h_n^L \quad (4)$$

where L is last hidden layer.

A. Training of the DNN

The most crucial part of developing a DNN model is selecting the appropriate hyperparameters that govern the performance of model. There are numerous algorithms present in the literature to optimize the selection. So, this work uses a heuristic approach based on experience to select DNN hyperparameters as summarized in Table I.

Recent work in [8] demonstrated enhanced accuracy in SOC estimation with an increasing number of DNN hidden layers. However, while training the model, there is always a risk of overfitting or underfitting, which can reduce the performance of the DNN. To mitigate overfitting, the study in [8] implements batch normalization and an early stopping training scheme, as described in [15]. A batch normalization layer is added after each hidden layer to minimize the likelihood of either scenario. In this paper, a callback function, *EarlyStopping*, is used to halt training if no improvement is observed for 100 successive epochs while restoring the best model weights. To quantify an improvement, an absolute minimum change of 1×10^{-4} in the validation data is used, which has demonstrated improvements in state estimation accuracy.

In addition to this, 64 neurons in each hidden layer with He. initialization is used to initialize the weights. Adam optimizer is used to update the weights of the DNN in batches of 256 data points by the method of backpropagation. The MSE is used as a loss function (L) to train and validate the model in each epoch and given by,

$$L = \frac{1}{N}\sum_{k=1}^{N} (z_k - \hat{z}_k)^2 \quad (5)$$

At the output layer, sigmoid activation function is used to map the SOC values from 0 to 1 and given by,

$$\phi(p) = \frac{1}{1 + e^{-p}} \quad (6)$$

979-8-3315-3013-6/25 $31.00 © 2025 IEEE

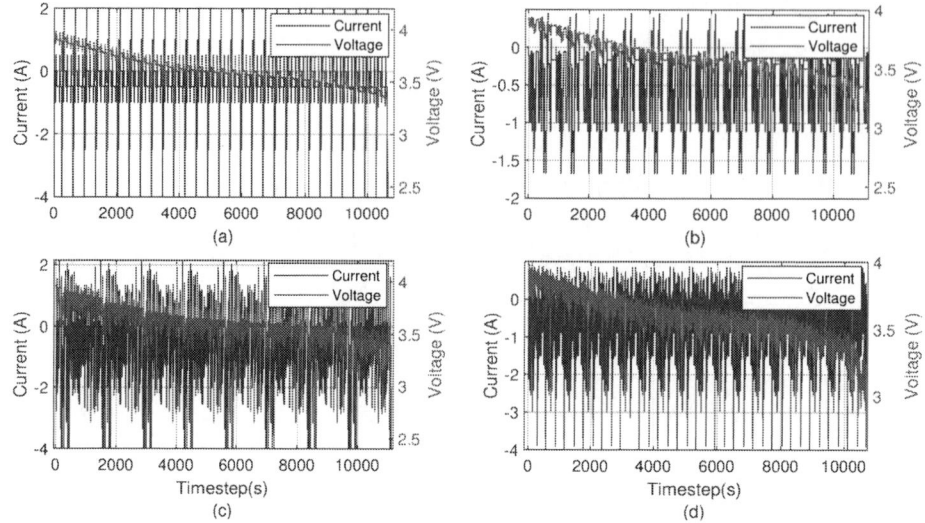

Fig. 2. Sample current and voltage plots of drive profile at 25°C: (a) DST, (b) BJDST, (c) FUDS, (d) US06

TABLE I
HYPERPARAMETERS SELECTED THROUGH HEURISTIC APPROACH

Hyperparameter	Value
Layer depth	1-9 (variable)
Hidden neuron capacity	64/layer
Weight initialization	He. Initialization
Learning Rate	Adaptive learning rate
Optimizer	Adam
Batch size	256
Batch Normalization	After activation function
Activation function	ReLU for hidden layers and Sigmoid for output layer

III. EXPERIMENTAL DESIGN

In this article, we used Google Colab to train and validate all DNN models which allows rapid testing of DNN. The accuracy of SOC estimation is decided by the depth of DNN. So, it is significant to analyze the effect of increasing the depth of DNN on the accuracy of SOC estimation.

At first, we started training DNN using a dynamic stress test (DST) drive profile and validating the trained model with the same profile, where training and validation data were divided in the ratio of 70:30 during each epoch. Details of the battery characterization test are further explained in Section III-B. Later, the developed models trained with the DST profile, are tested for SOC estimation over DST profile and unseen drive profiles such as Beijing dynamic stress test (BJDST), federal urban driving schedule (FUDS), and US06 drive cycle. The accuracy of the DNN model for estimating SOC is evaluated using performance metrics given by Eq. (7) through (10) in Section IV-A. This procedure is being carried out by varying the number of hidden layers from one to nine.

A. Battery Dataset

A publicly accessible battery dataset from CALCE [13] is used in the study described in the paper. This center offers cyclic aging data and battery cell characterization for a variety of Li-ion cells with different form factors. A Li-ion battery cell having a graphite anode and a LiNiMnCo cathode is being studied. Table II provides detailed specifications of the battery cell under consideration.

TABLE II
CELL SPECIFICATIONS

Parameters	Value
Chemistry	LiNiMnCo/Graphite
Form Factor	18650
Nominal Voltage	3.6 V
Maximum Charge Voltage	4.2 V
Cut-off Voltage	2.0 V
Nominal Capacity	2 Ah
Maximum Current	22 A
Cycle Life	1000-2000

B. Battery Characterization Test

The estimation of battery SOC is crucial for safe and reliable battery operation and overall electric vehicles. So, the relation between SOC and OCV is obtained by performing low-current and incremental current OCV tests at three different temperatures of 0°C, 25°C, and 45°C. Battery model and SOC estimation accuracy validation can be performed under different drive profiles; accordingly, the battery cell is characterized using different drive profiles such as DST, FUDS, US06, and BJDST. The main goal of the dynamic tests is to investigate battery dynamics at different depths of discharge. Initially, the cell is charged using the standard constant current constant voltage method, followed by 20%

979-8-3315-3013-6/25 $31.00 © 2025 IEEE   338

discharge at C/2 constant current to make the initial SOC 80%. Later, dynamic tests are performed on the battery cell at three temperatures. The battery cell current, voltage, and temperatures are recorded every 10 s until the voltage reaches the discharge cutoff voltage. The sample plots of current and voltage obtained from the CALCE dataset for different drive profiles are shown in Fig. 2.

IV. RESULTS AND DISCUSSION

A. Performance evaluation of DNN for SOC estimation

To assess the performance of developed DNN, root mean squared error (RMSE), mean squared error (MSE), mean absolute error (MAE) and mean absolute percentage error (MAPE) are used as performance metrics and given by,

$$RMSE = \sqrt{\frac{1}{N}\sum_{k=1}^{N}(z_k - \hat{z}_k)^2} \qquad (7)$$

$$MSE = \frac{1}{N}\sum_{k=1}^{N}(z_k - \hat{z}_k)^2 \qquad (8)$$

$$MAE = \frac{1}{N}\sum_{k=1}^{N}(|z_k - \hat{z}_k|) \qquad (9)$$

$$MAPE = \left(\frac{1}{N}\sum_{k=1}^{N}\frac{|z_k - \hat{z}_k|}{z_k}\right) \times 100 \qquad (10)$$

where z_k and $\hat{z}_k$ are reference and estimated SOC, respectively.

B. DNN Model Testing

Initially, all the DNN models are evaluated using the DST training dataset and a detailed summary is given in Table III. The comparison of RMSE against complexity in terms of DNN hidden layers is shown in Fig. 3. Initially, RMSE is observed to be high for hidden layers one and two and tends to decrease till a model with hidden layer four. The RMSE again starts increasing as the number of hidden layers increases. When the DNN model is trained and tested using a DST drive profile, the best-performing DNN is one with hidden layers four, five, and six. Other performance metrics, such as MSE, MAE, and MAPE, show a similar pattern with regard to multiple hidden layers. The quicksort algorithm's performance ranking in relation to the error metric values is shown as "Rank" in Table III.

The performance of DNN models trained using the DST drive profile is tested for other drive profiles shown in Fig. 2 to analyze the generalization capability of proposed models for SOC estimation. Table IV summarizes the performance of DNN models as a function of a number of hidden layers for BJDST, FUDS, and US06 drive profiles. It is observed that the DNN model with four hidden layers outperforms when trained with a DST drive profile and validated against different drive profiles as given in Table IV.

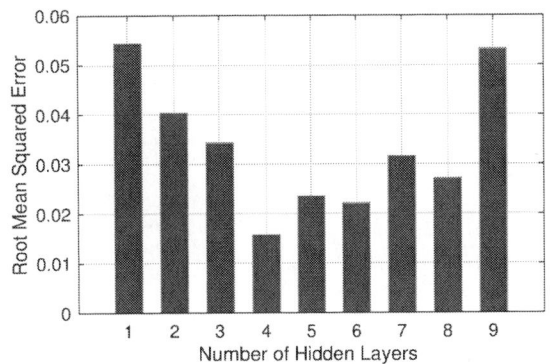

Fig. 3. RMSE Vs layer depth for models tested on the DST drive profile

C. SOC Estimation with Optimal DNN

Since the DNN model with hidden layers four, five, and six are found to be best performing when trained with the DST profile and tested with DST and the other three drive profiles, the SOC estimation capability of four hidden layer DNN is presented in Fig 4. For performance analysis, reference SOC is obtained using Coulomb counting given by,

$$z_{k+1} = z_k + \frac{\eta \Delta t}{Q_{nom}} i_k \qquad (11)$$

where, Q_{nom}, η, and I_k represent nominal capacity, Coulombic efficiency, and drive current, respectively. The current I_k is taken as positive during charge. It is to be noted that, the initial SOC value while obtaining reference SOC is taken as 80%.

Initially, analysis of DNN model tested on DST profile at three different temperatures is performed, and results are presented in Fig. 4(a), 4(e), and 4(i) at temperatures 0°C, 25°C, and 45°C, respectively. The MSE, RMSE, and MAE of 0.024%, 1.57%, and 1.18% was observed respectively. Since the model is trained with a DST profile and SOC estimation is performed using the same profile, we observed minimum errors as compared to model performance over other drive profiles.

Later, the performance of the DNN model for SOC estimation was analyzed for BJDST, FUDS, and US06 drive profiles. The plots of estimated SOC versus reference SOC for all the drive profiles and over three different temperatures are given in Fig. 4 and summarized in Table IV. A DNN with hidden layers four, five, and six was found to be best performing under FUDS and US06 drive profiles, following the same trend that was observed for the DST drive profile. When the SOC is estimated under the BJDST drive profile, a model with four layers is still at rank one, however, rank two deviates to the model with eight hidden layers. It can be concluded that the performance of the four hidden layer model is best for all drive profiles, and may be generalized for having the capability to perform best for other drive cycles also.

Followed by an analysis of SOC estimation, the performance of the DNN model with optimal hidden layers is compared with other advanced techniques presented in the

TABLE III

MODEL TESTING ON DST DRIVE PROFILE

Layer Depth	Parameters	MSE	RMSE	MAE	MAPE	Rank
1	577	0.002959	0.054396	0.046046	1.295559	9
2	4993	0.001635	0.040441	0.031553	0.574111	7
3	9409	0.001179	0.034335	0.025836	0.56431	6
4	13825	0.000249	0.015777	0.011802	0.503582	1
5	18241	0.000555	0.023562	0.01972	0.340275	3
6	22657	0.00049	0.022142	0.017197	0.609695	2
7	27073	0.000999	0.03161	0.026919	0.87244	5
8	31489	0.000735	0.027119	0.022077	0.258908	4
9	35905	0.002836	0.053252	0.03753	0.964288	8

TABLE IV

MODEL TESTING ON DIFFERENT DRIVE CYCLES

Drive Profile	Layer Depth	MSE	RMSE	MAE	MAPE	Rank
BJDST	1	0.00293	0.054129	0.045672	2.316967	9
	2	0.001901	0.043603	0.033064	0.983781	7
	3	0.000848	0.029127	0.022707	0.786195	5
	4	0.000314	0.017723	0.013625	0.436949	1
	5	0.000617	0.024837	0.020177	0.490488	3
	6	0.000643	0.025361	0.020601	0.638292	4
	7	0.00093	0.030494	0.025465	0.68347	6
	8	0.000617	0.02483	0.019567	0.208428	2
	9	0.002074	0.045544	0.033358	0.579318	8
FUDS	1	0.003381	0.058146	0.049278	0.406717	8
	2	0.002656	0.051535	0.039845	0.277087	7
	3	0.001921	0.043824	0.031365	0.210998	6
	4	0.000699	0.026441	0.019292	0.153972	1
	5	0.001107	0.033273	0.025948	0.147761	3
	6	0.000849	0.029138	0.022299	0.184659	2
	7	0.001561	0.039512	0.031455	0.261968	5
	8	0.00119	0.034491	0.026053	0.131586	4
	9	0.003427	0.058537	0.041217	0.285655	9
US06	1	0.003071	0.055418	0.046709	0.645793	9
	2	0.002299	0.047952	0.036483	0.409508	7
	3	0.001212	0.03481	0.026433	0.302361	6
	4	0.000492	0.022191	0.016963	0.235221	1
	5	0.000837	0.028928	0.023083	0.204686	3
	6	0.000694	0.026349	0.020842	0.260044	2
	7	0.001143	0.033812	0.02724	0.370068	5
	8	0.000864	0.029388	0.022584	0.159876	4
	9	0.002501	0.050006	0.036282	0.355369	8

literature, and results are summarized in Table V. The performance of the optimal DNN model is better compared to the presented methods with minimum improvement in RMSE by 21.5% as compared to [16]. It is worth noting the advantages of feedforward DNN compared to LSTM and GRU networks are computational cost and complexity [17]. Recurrent models require more data transfer rate and response time to train and run models. The above discussion signifies that overall embedded computations can be reduced to a greater extent by maintaining the tradeoff between estimation accuracy and computational complexity for optimizing BMS design.

V. CONCLUSION

A detailed analysis to investigate optimal DNN for Li-ion battery SOC estimation is presented in this paper. The architecture is trained with a DST drive profile executed on the battery cell under consideration and observed to estimate

TABLE V

COMPARISON WITH STATE OF ART METHODS

Base Model Used	Drive Profiles Used	Error
LSTM [18]	DST, FUDS, US06	RMSE 2.53%
CNN-GRU [16]	FUDS DST	RMSE 2.00%
DNN [8]	DST, BJDST, FUDS, US06	RMSE 3.68%, MSE 0.13%
Proposed	DST, BJDST, FUDS, US06	**RMSE 1.57%, MSE-0.024%**

SOC accurately. The generalization capability of the DNN with four hidden layers is demonstrated for SOC estimation by applying FUDS, BJDST, and US06 drive profiles at varying temperatures. The DNN-estimated SOC values are compared with reference SOC to analyze the performance using different

979-8-3315-3013-6/25 $31.00 © 2025 IEEE

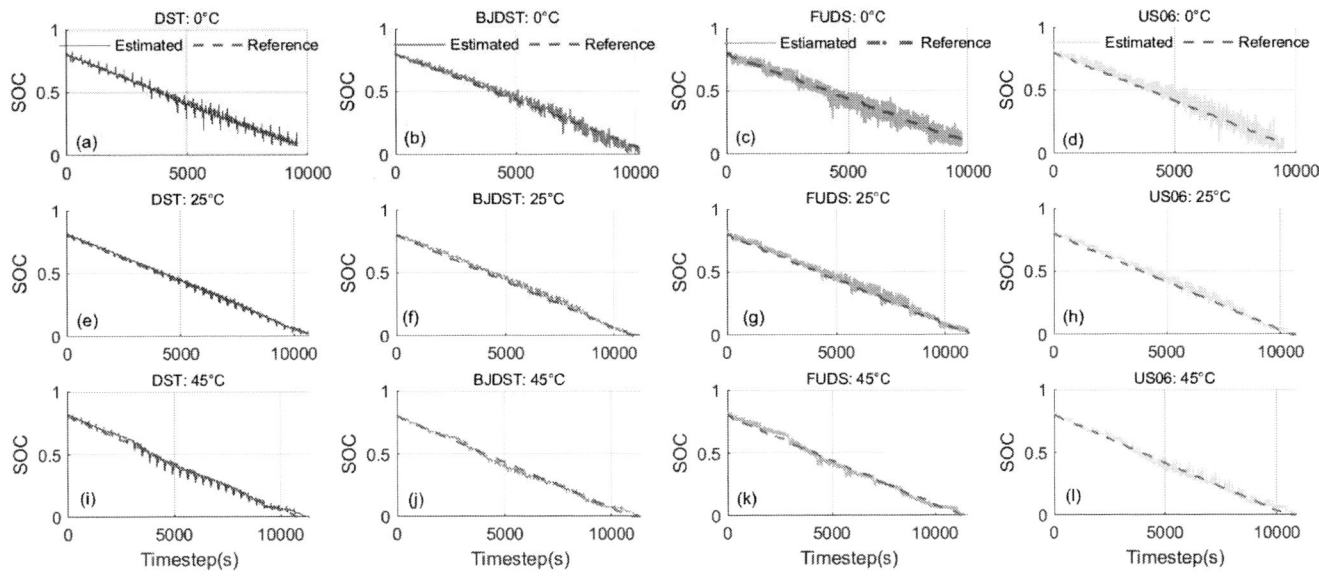

Fig. 4. Results of SOC estimation with four layer hidden DNN when tested with: (a) DST profile at 0°C (b) BJDST profile at 0°C (c) FUDS profile at 0°C (d) US06 profile at 0°C (e) DST profile at 25°C (f) BJDST profile at 25°C (g) FUDS profile at 25°C (h) US06 profile at 26°C (i) DST profile at 45°C (j) BJDST profile at 45°C (k) FUDS profile at 45°C (l) US06 profile at 45°C

error metrics. The improved SOC estimate performance using the proposed DNN is further contrasted with other cutting-edge methods performed on the same datasets and under the same conditions in the literature. The proposed DNN architecture can be extended for real-time diagnostics and battery pack-level SOC estimation.

REFERENCES

[1] Z. Du, L. Zuo, J. Li, Y. Liu, and H. T. Shen, "Data-driven estimation of remaining useful lifetime and state of charge for lithium-ion battery," *IEEE Transactions on Transportation Electrification*, vol. 8, no. 1, pp. 356–367, 2022.

[2] S. Nyamathulla and C. Dhanamjayulu, "A review of battery energy storage systems and advanced battery management system for different applications: Challenges and recommendations," *Journal of Energy Storage*, vol. 86, p. 111179, 2024.

[3] K. Movassagh, A. Raihan, B. Balasingam, and K. Pattipati, "A critical look at coulomb counting approach for state of charge estimation in batteries," *Energies*, vol. 14, no. 14, 2021.

[4] M. Gholizadeh and A. Yazdizadeh, "State of charge estimation of a lithium-ion battery using robust non-linear observer approach," *IET Electrical Systems in Transportation*, vol. 9, no. 1, 2019.

[5] P. Messier, B.-H. Nguyen, F.-A. LeBel, and J. P. F. Trovao, "Disturbance observer-based state-of-charge estimation for li-ion battery used in light electric vehicles," *Journal of Energy Storage*, vol. 27, 2020.

[6] Y. Wang, J. Tian, Z. Sun, L. Wang, R. Xu, M. Li, and Z. Chen, "A comprehensive review of battery modeling and state estimation approaches for advanced battery management systems," *Renewable and Sustainable Energy Reviews*, vol. 131, 2020.

[7] P. Aher, R. Deshmukh, C. Chavan, S. Patil, M. Khare, and A. Mandhana, "A comprehensive equivalent circuit model of li-ion batteries for soc estimation in electric vehicles based on parametric sensitivity analysis," *Ionics*, vol. 31, no. 1, pp. 287–303, 2025.

[8] D. N. How, M. A. Hannan, M. S. H. Lipu, K. S. M. Sahari, P. J. Ker, and K. M. Muttaqi, "State-of-charge estimation of li-ion battery in electric vehicles: A deep neural network approach," in *2019 IEEE Industry Applications Society Annual Meeting*, 2019, pp. 1–8.

[9] H. Chaoui and C. C. Ibe-Ekeocha, "State of charge and state of health estimation for lithium batteries using recurrent neural networks," *IEEE Transactions on Vehicular Technology*, vol. 66, no. 10, pp. 8773–8783, 2017.

[10] T. Mamo and F.-K. Wang, "Long short-term memory with attention mechanism for state of charge estimation of lithium-ion batteries," *IEEE Access*, vol. 8, pp. 94 140–94 151, 2020.

[11] Y. Liu, X. Shu, H. Yu, J. Shen, Y. Zhang, Y. Liu, and Z. Chen, "State of charge prediction framework for lithium-ion batteries incorporating long short-term memory network and transfer learning," *Journal of Energy Storage*, vol. 37, p. 102494, 2021.

[12] Z. Ni and Y. Yang, "A combined data-model method for state-of-charge estimation of lithium-ion batteries," *IEEE Transactions on Instrumentation and Measurement*, vol. 71, pp. 1–11, 2022.

[13] F. Zheng, Y. Xing, J. Jiang, B. Sun, J. Kim, and M. Pecht, "Influence of different open circuit voltage tests on state of charge online estimation for lithium-ion batteries," *Applied Energy*, vol. 183, pp. 513–525, 2016.

[14] Y. Xing, W. He, M. Pecht, and K. L. Tsui, "State of charge estimation of lithium-ion batteries using the open-circuit voltage at various ambient temperatures," *Applied Energy*, vol. 113, pp. 106–115, 2014.

[15] S. Ioffe, "Batch renormalization: Towards reducing minibatch dependence in batch-normalized models," *Advances in neural information processing systems*, vol. 30, 2017.

[16] Z. Huang, F. Yang, F. Xu, X. Song, and K.-L. Tsui, "Convolutional gated recurrent unit–recurrent neural network for state-of-charge estimation of lithium-ion batteries," *IEEE Access*, vol. 7, pp. 93 139–93 149, 2019.

[17] E. Chemali, P. J. Kollmeyer, M. Preindl, R. Ahmed, and A. Emadi, "Long short-term memory networks for accurate state-of-charge estimation of li-ion batteries," *IEEE Transactions on Industrial Electronics*, vol. 65, no. 8, pp. 6730–6739, 2018.

[18] F. Yang, X. Song, F. Xu, and K.-L. Tsui, "State-of-charge estimation of lithium-ion batteries via long short-term memory network," *IEEE Access*, vol. 7, pp. 53 792–53 799, 2019.

Hardware-in-the-Loop Testing of Battery Management System Using a High-Fidelity Digital Twin

1st Febin Koshy Jacob
Department of Electrical Engineering
NIT Calicut
Calicut, India
febin_m230366ee@nitc.ac.in

2nd Indranil Bose
System Validation
Varroc Engineering
Pune, India
indranil.bose@varroc.com

3rd Sarika D. Tavhare
System Validation
Varroc Engineering
Pune, India
sarika.tavhare@varroc.com

4th Sandhya Anilkumar
System Validation & Software QA
Varroc Engineering
Pune, India
sandhya.anil@varroc.com

5th Kumaravel S.
Department of Electrical Engineering
National Institute of Technology Calicut
Calicut, India
kumaravel_s@nitc.ac.in

Abstract— Battery Management System (BMS) are essential for guaranteeing the safety, performance, and longevity of battery packs. To ensure the reliability of BMS algorithms and optimize battery pack performance across diverse operating conditions, rigorous testing is crucial, especially within the automotive sector. This research presents an automated Hardware-in-the-Loop (HIL) testing methodology that incorporates a cost-effective Cell Voltage Emulator (CVE) and a customizable digital twin of a battery pack for validating BMS under different conditions. The results illustrate the effectiveness of the proposed approach in validating BMS performance by identifying potential issues, reducing execution time and increasing the requirement coverage for automotive Electric Vehicles.

Keywords — *Battery Management System, Cell Voltage Emulator, Digital Twin, Hardware-in-the-Loop.*

I. INTRODUCTION

Demand for Electric Vehicles (EVs) and other battery-operated applications has highlighted battery system safety, performance, and durability. These systems depend on the BMS to monitor and control the battery pack for optimal performance and safety, which includes battery longevity, overcharge and over-discharge prevention, thermal runaway management, and accurate State-of-Charge (SoC) evaluations. The functionalities of a BMS are shown in Fig. 1 [1].[1]

Physical battery pack examinations are often needed to evaluate and validate BMS operation. These methods are costly, dangerous, and may not cover all operational scenarios. When using physical batteries, testing under extreme conditions or fault scenarios is sometimes life threatening. As BMS get increasingly complex, to assure and guarantee complete validation, HIL based testing techniques are very useful. HIL testing requires a plant model for validating the BMS, which includes realistic battery model, temperature sensors, pressure sensors, digital and analog I/O's, ECUs in real-time simulation. Fig. 2 explains the BMS Communication block diagram of a 2-Wheeler (2W). Accurate HIL testing necessitates an exact digital twin of the battery pack that encapsulates its dynamic behaviour, thermal characteristics, and aging effects, thereby enabling the BMS to simulate battery interactions and assess performance and functionality. The creation and implementation of high-fidelity digital twins and hardware interfaces is sophisticated and costly, particularly with CVEs that provide the Ease-of-Use BMS, hence hindering the adoption of HIL testing, especially in research and development.

This paper concentrates on creating a high-fidelity digital twin of a Li-Ion battery pack and a CVE within an HIL testing environment. The digital twin was designed using *MATLAB/Simulink®* and *Simscape Electrical* libraries. This study uses a voltage divider circuit in conjunction with a voltage follower circuit to create and implement a cost-effective CVE. This customized CVE is far more affordable than commercial alternatives, hence enhancing the accessibility of HIL testing.

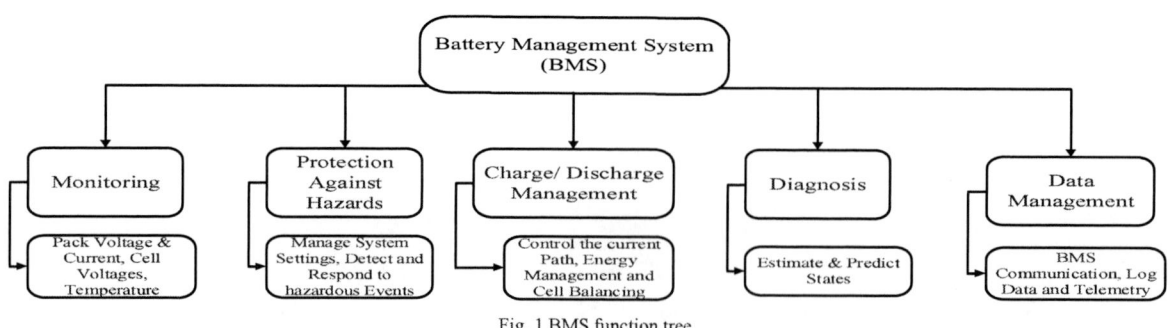

Fig. 1 BMS function tree

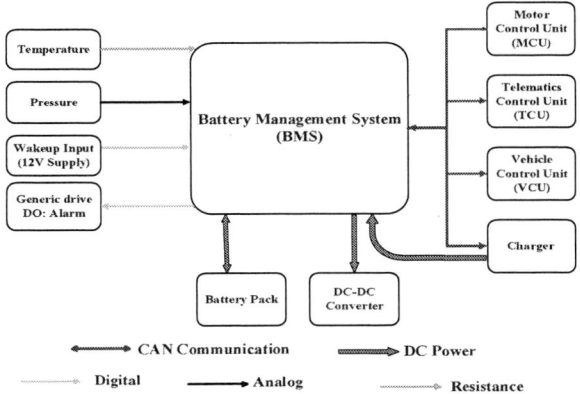

Fig. 2 BMS communication block diagram - 2W

HIL testing significantly enhances BMS development and validation by providing a safe, repeatable, and efficient environment to test the BMS hardware and software against realistic battery behavior and operating conditions. The HIL testing approach assesses BMS performance, identifies faults, optimizes system configurations, and analyses execution time, requirement coverage, and model accuracy. This research enhances BMS testing and contributes to the development of safer and more reliable BMS for EVs and other applications.

II. LITERATURE SURVEY

This section surveys existing research related to BMS testing, digital twin development for battery systems, and HIL testing methodologies. It highlights the current state, identifies research gaps, and positions the contributions of this paper within the broader context of BMS validation.

A. Battery Management System Testing

Traditional BMS testing methods often rely on physical prototypes and real battery packs [2]. These methods are often expensive and may not be suitable for testing under extreme or hazardous conditions [3][4]. Furthermore, the increasing complexity of battery systems and BMS algorithms makes comprehensive testing using traditional methods increasingly challenging.

B. Hardware-in-the-Loop Testing for BMS

HIL testing has emerged as a promising alternative to traditional BMS validation. It offers several advantages, including reduced testing time, improved safety, and increased test coverage [5]. Real-time simulators are used to emulate the behaviour of the battery pack, allowing the BMS to interact with a virtual representation of the battery in a controlled environment [6][7].

C. Digital Twin Development for Battery Systems

The accuracy and fidelity of the battery model used in the HIL simulation are crucial for effective BMS validation. Digital twin technology has gained significant attention in recent years, offering the potential to create highly accurate virtual representations of physical battery systems [8][9]. Several approaches have been proposed for developing digital twins of battery systems, ranging from simplified equivalent circuit models to complex electrochemical models [10]. The choice of model complexity depends on the desired

accuracy and computational efficiency required for real-time HIL simulation [11][12].

D. Cell Voltage Emulators

CVEs play a crucial role in HIL testing by providing the BMS with realistic cell voltage inputs. Commercially available CVEs, however, can be expensive, often relying on high-precision Digital-to-Analog Converters (DACs) [13]. This high cost can be a barrier to adopting HIL testing, especially for research and development purposes [14][15].

E. Research Gaps and Motivation

Existing BMS testing has limitations, achieving high-fidelity, efficient digital twins, high cost of commercial CVEs for HIL. This paper proposes a novel HIL method integrating a high-fidelity digital twin with a cost-effective custom CVE. The twin includes detailed cell models with SoC/temperature variations. The custom CVE lowers costs, enhancing accessibility. This advances BMS testing for safer, more reliable batteries.

III. PROPOSED METHODOLOGY

A. High-Fidelity Digital Twin Development

A high-fidelity digital twin of an 18650 NMC battery pack, commonly used in 2W EVs, is developed using *MATLAB/Simulink* and the *Simscape Electrical* library. The parameters of the cell are shown in Table I. A first-order Equivalent Circuit Model (ECM) was chosen for its balance between accuracy and computational efficiency, essential for real-time HIL simulation. The ECM comprises an ohmic resistance (R_0), a polarization resistance (R_1), and a polarization capacitance (C_1), as represented in Fig. 3. We have selected first order matrix because the DUT algorithm is supported till first order

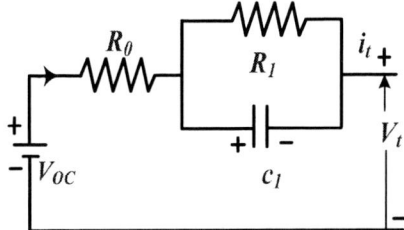

Fig. 3 First-order ECM

The challenges in creating a high-fidelity digital twin for battery systems, such as capturing complex internal dynamics, managing extensive real-time data and meeting high computational demands, are specifically addressed in some approaches. These methods involve utilizing Look-Up Tables (LUTs) derived directly from actual battery pack performance data to empirically represent complex behaviors, and calculating essential modeling parameters. The ECM parameters (R_0, R_1, C_1) were derived from Hybrid Pulse Power Characterization (HPPC) test. To accurately reflect the non-linear behaviour of the NMC battery, these parameters were mapped as functions of SoC and temperature using LUTs in Table II. Remaining values were also mapped at the respective SOC's. These data allowed the model to accurately represent the dynamic changes in battery behaviour across different operating conditions. The digital twin also

979-8-3315-3013-6/25 $31.00 © 2025 IEEE

incorporates a thermal model to capture the battery's temperature dynamics by using lumped thermal mass. Fig. 4 explains the generation of the digital twin model.

The model validation was done successfully with constant discharge test and compared the results with the actual battery pack. The results are discussed in Section VI.

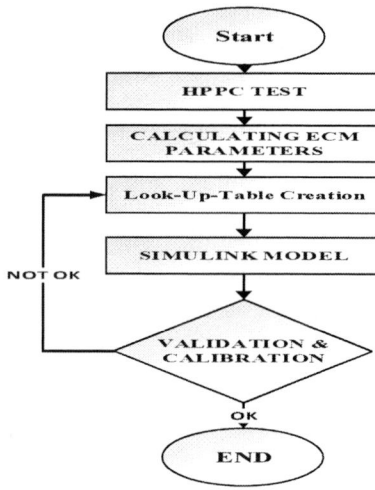

Fig. 4 Digital Twin Model Generation Flowchart

TABLE I

CELL PARAMETERS

Parameter	Value
Cell type	NMC/LFP
Cell capacity	3.35 AH
Max. voltage	4.2 V
Min. voltage	2.5 V
Temperature range (Charging)	0 - 50 °C
Temperature range (Discharging)	-20 - 60 °C

TABLE II

LUT PARAMETERS

SoC (%)	R_0 (Ω)	R_1 (Ω)	C_1 (μF)	E_0 (V)
0	0.0299	0.2750	559	2.75
30	0.0200	0.0180	1635	3.51
50	0.0197	0.0150	1584	3.70
70	0.0197	0.022	1652	3.87
100	0.0230	0.0220	1096	4.16

B. Cell Voltage Emulator

The CVE is designed and implemented using a voltage divider circuit, relays and variable resistance. The variable resistance is controlled using equations as per the input voltage and the desired cell voltage through the digital twin model. The voltage divider circuit is used to generate the desired cell voltages, while relays are employed to configure the circuit for both series and parallel cell connections, enabling the simulation of various cell balancing scenarios. The relays are controlled using the real-time simulator to switch between series and parallel cell configurations. This allows the CVE to emulate different cell balancing strategies. The CVE's design prioritizes cost-effectiveness compared to commercially available solutions, which typically utilize expensive DACs. The accuracy of the CVE in emulating cell voltages was evaluated by comparing the CVE output with calculated values. The schematic of 1 cell, which was made in EasyEDA, is shown in Fig. 5 which explains the working of the CVE, and it is repeated for the remaining 13 cells.

In Fig. 5, to create a controlled cell imbalance for testing cell balancing algorithms, the voltage divider circuit equation (3), is utilized to adjust the voltage of specific cells. For reducing the voltage parallel resistance, equation (1) is used. For increasing the voltage series resistance equation, (2) is used, where R_i refers to the constant resistor and R_p refers to the variable resistor connected parallel with R_i. Fig. 6 represents the LTSpice simulation result of a condition where 1 cell voltage is decreased.

$$R_2 = \frac{R_i R_p}{R_i + R_p} \quad (1)$$

$$R_2 = R_i + R_p \quad (2)$$

$$V_{out} = V_{in}\left(\frac{R_2}{R_1 + R_2 + .. + R_n}\right) \quad (3)$$

C. BMS Testing Scenarios

A comprehensive set of test cases was developed to evaluate the performance and functionality of the BMS. These test cases covered various aspects of BMS operation, including cell balancing, fault detection, SOC estimation, and communication. For example, the cell voltages measurement testing includes overvoltage detection, undervoltage detection, normal condition, etc. In conjunction with the automatic HIL setup, a manual test configuration (Fig. 7) was established for the BMS board, guaranteeing the execution of all test cases. 95% test cases were automated using *Python* scripts interacting with the *DSPACE* HIL bench. This automation ensured efficient and repeatable testing, reducing the time and effort required for manual testing.

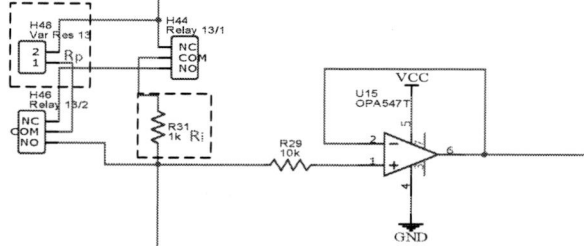

Fig. 4 Schematic of cell voltage emulator

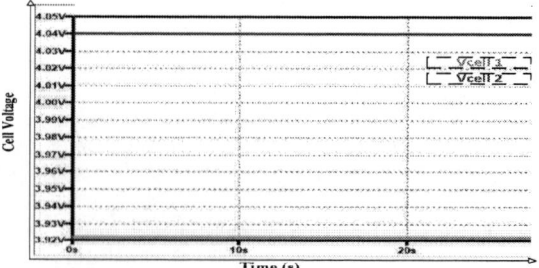

Fig. 5 Cell imbalance – simulation result

IV. BMS HIL SETUP

Fig. 7 shows the components of the BMS HIL testing environment, which includes, a *dSPACE Scalexio* real-time simulator, the custom-designed CVE, the BMS under test, bidirectional power supply, External FIU and a host PC for simulation control and data acquisition. The digital twin model of the NMC battery pack was deployed onto the *Scalexio* platform for real-time execution using Real-Time Interface (RTI). The simulation time step was set 100 μs to ensure real-time performance and accuracy. The simulator's processing power and I/O capabilities enable accurate and high-fidelity emulation of the battery pack's behaviour. The custom-designed CVE provides outputs analog voltages representing the individual cell voltages from the bidirectional power supply. The CVE's outputs are connected to the BMS's cell voltage sensing inputs through the external FIU. The BMS under test was connected to the HIL system via digital I/O for control signals, CAN communication for data exchange, Analog Out for Pressure measurement and Resistance Out for Temperature Measurement. Fig. 8 illustrates the block diagram of HIL Setup.

A host PC running Control Desk was used to control the real-time simulation, configure test parameters, acquire data, and visualize the results. The entire HIL testing process was automated using *Python* scripts.

This automation enabled efficient and repeatable testing, significantly reducing the time and effort required for manual testing. The bidirectional power supply along with the plant model can act as digital twin of the battery pack which is used for sinking and sourcing. Also, it is used to simulate different load profiles for the battery pack with the help of the RTI

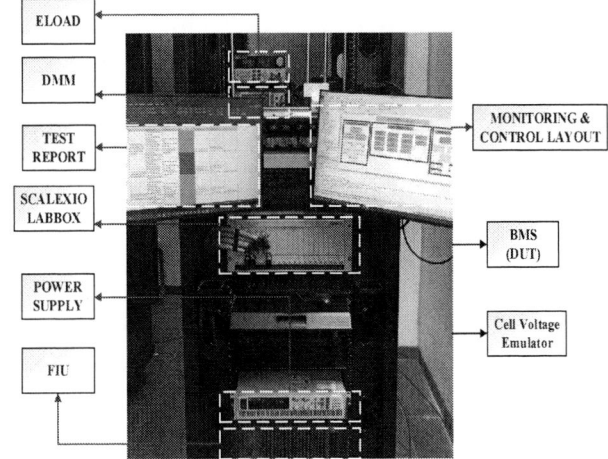

Fig. 7 HIL SETUP

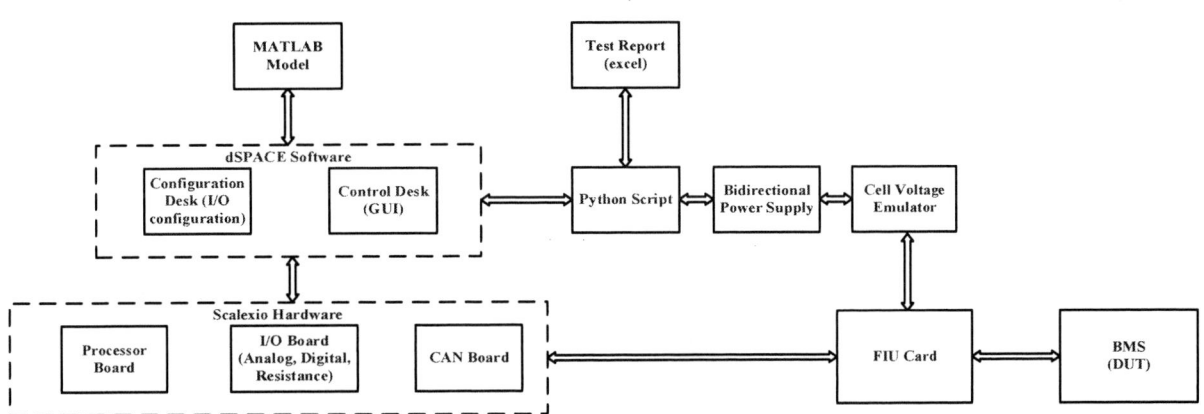

Fig. 8 Block diagram for HIL setup

V. TEST EXECUTION

A. Test Case Management

A comprehensive library of test cases was developed, covering all defined BMS requirements. Each test case was documented with a unique identifier, a description of the test objective, the required setup conditions, the execution steps, and the expected outcomes. Test cases were organized into logical sequences based on BMS functionality.

B. Test Automation

The automation of the test execution process was crucial for ensuring efficient and repeatable validation of the BMS. The overall test automation flow is illustrated in Fig. 9. The entire test execution is run and controlled using *Python* scripts and Control desk respectively. Test execution includes running the test scripts which were written in python and running the Plant

Model. Using the concept of parallel threading, the script dynamically gathers the ECU data from CAN bus and *MATLAB* model variables using Control Desk. Parallelly the script does data processing for the data gathered from ECU and compares with expected and actual results at every 100 ms time interval. The verdict and various ECU parameters depending on features being tested for each test case are then stored in excel.

VI. RESULTS AND DISCUSSIONS

This section presents the results obtained from the HIL testing of the BMS using the developed high-fidelity digital twin and the cost-effective CVE. The results are organized by test category, demonstrating the performance of the BMS under various operating conditions and fault scenarios. The discussion analyses the results, highlighting the effectiveness of the proposed HIL testing methodology and its advantages over traditional testing approaches.

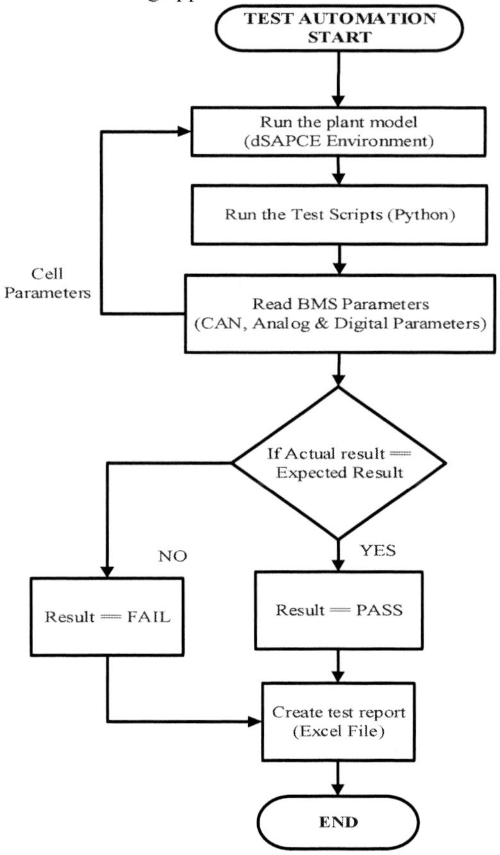

Fig.6 Test automation flow chart

A. Battery Pack Validation

In addition to the BMS testing, the overall validation of the battery pack digital twin was performed. This validation involved comparing the digital twin's predicted behaviour against experimental data obtained from testing a physical 18650 NMC battery cell under various operating conditions. Specifically, constant current discharge test. The validation focused on comparing the digital twin's predictions terminal voltage with the measured data. Fig. 10 shows a comparison between the digital twin's predicted terminal voltage and the measured terminal voltage during a constant current discharge test at 1 C rating. The results demonstrate a high degree of correlation, with a Root Mean Squared Error (RMSE) of 0.0614, indicating excellent agreement between the model and the experimental data. The successful validation of the battery pack digital twin confirms its accuracy and reliability for HIL testing. This validation is crucial for ensuring the reliability of the HIL testing results and the validity of the conclusions drawn from the BMS evaluation.

B. Evaluation of the Cost-Effective CVE

The cost-effectiveness of the developed CVE is recognized as a significant advantage. When compared to commercially available CVEs from different vendors, a 50% cost reduction is achieved with the proposed solution. In other words, only 50% of the cost typically quoted by vendors like Vendor A

and Vendor B is required. For instance, the implementation cost of the CVE is found to be approximately 55% lower than that of a comparable solution from Vendor A. Similarly, a 50% cost saving is attained in comparison to Vendor B's offering. This substantial cost reduction enables HIL testing to be more accessible to researchers and developers with limited budgets. The performance of the custom-designed CVE was evaluated, and the results are presented in Table III.

C. Performance Analysis of HIL Testing

Fig. 11 summarizes the automated execution time of the HIL bench with manual execution time. The results show that HIL testing significantly reduced the testing time compared to traditional methods, 78% reduction in time was noted for automation testing. The HIL testing also achieved a high requirement coverage of 98%, ensuring thorough validation of the BMS functionality. Following the automation of HIL testing, a strong F1 score of 0.9693 was achieved, indicating a high level of test accuracy and robustness. This score is the harmonic mean of precision and recall, providing a balanced measure of the test's performance. The individual metrics of accuracy, precision, and recall were also calculated and are detailed in Table 4. To ensure the reliability of these performance indicators, the F1 score was determined by taking the average of the 4 distinct test iterations. This comprehensive approach to performance analysis, combining speed, coverage, and detailed classification metrics, further underscores the effectiveness of the automated HIL testing methodology.

Table III

CVE SIMULATED CHARACTERISTICS

Parameter	Value
Voltage range	[1,5] V
Channels	14
Accuracy	+/- 40 mV
Current source	500 mA

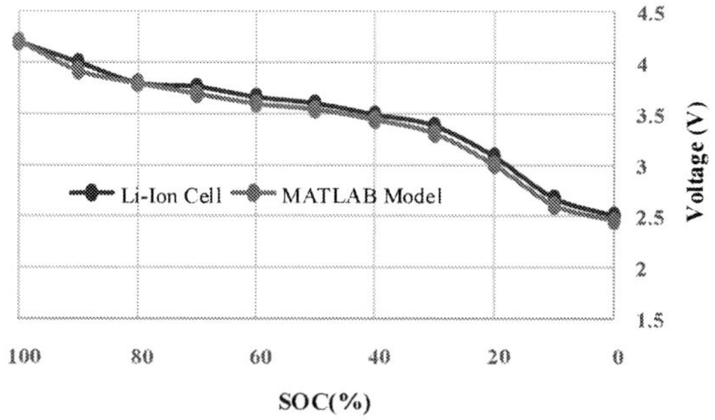

Fig. 7 Li-Ion Discharge Voltage Curve at 1 C rating

TABLE IV

PERFORMANCE METRICS

Parameter	Value
Precision	0.95
Accuracy	0.9426
Recall	0.9895

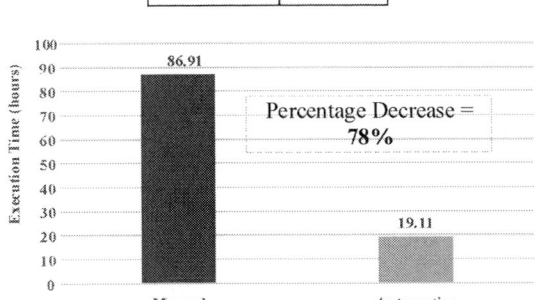

Fig. 81 Execution time - Manual *versus* Automated

VII. CONCLUSION

This study successfully illustrates the implementation of a proposed HIL testing technique for BMS. This strategy markedly improved testing efficiency, resulting in a 78% decrease in testing duration relative to conventional techniques, while concurrently increasing test coverage to over 98%, Also the HIL testing performance achieved a F1 score of 0.9693. Additionally, the HIL configuration enhanced testing safety by reducing hazards linked to actual fault scenarios. The created CVE demonstrated a cost-effective solution, achieving a 50% cost reduction relative to commercially available alternatives, while preserving an accuracy of ±40mV across 14 cell channels. The digital twin of the battery pack demonstrated exceptional precision, with a RMSE of merely 0.0614%. The results confirm the efficacy of the proposed HIL methodology and its related components for thorough and efficient BMS assessment.

Future research directions could explore predictive modelling of BMS software reliability using Machine Learning techniques to anticipate and mitigate potential failures, the integration of battery aging models into the digital twin to evaluate the long-term performance and robustness of BMS algorithms, further investigation into the CVE to establish it as a standalone CVE, the use of cloud-based platforms for HIL testing to enhance accessibility and scalability, and the integration of ML-based diagnostics or over-the-air HIL testing.

ACKNOWLEDGMENT

We gratefully acknowledge Varroc Engineering for providing Mr. Febin Koshy Jacob with the internship and for their generous sponsorship of this project. We also extend our sincere thanks to NIT Calicut and Varroc Engineering for their valuable collaboration in facilitating this internship opportunity.

REFERENCES

[1] Fleischer, Christian & Sauer, Dirk Uwe & Barreras, Jorge & Schaltz, Erik & Christensen, Andreas. (2016). Development of software and strategies for Battery Management System testing on HIL simulator. 2016 11th International Conference on Ecological Vehicles and Renewable Energies, EVER 2016. 10.1109/EVER.2016.7476438.

[2] B. Kumar, D. Srivastava and D. Chanda, "Model Based Design Approach for ePWT Software & System – Battery Management System for xEVs," 2024 10th International Conference on Control,

Decision and Information Technologies (CoDIT), Vallette, Malta, 2024, pp. 905-909, doi: 10.1109/CoDIT62066.2024.10708319.

[3] H. R. Pandit, N. Biju, V. Pisharodi, P. Dimitrakopoulos and M. Shenoy, "Framework for Digital Twin Real-Time Battery System for Model-in-the-loop and Hardware-in-the-loop Simulationrs," 2023 IEEE Transportation Electrification Conference & Expo (ITEC), Detroit, MI, USA, 2023, pp. 1-6, doi: 10.1109/ITEC55900.2023.10187077.

[4] C. D. Tschritter, D. A. Wetz, G. K. Turner and J. M. Heinzel, "Battery Management System (BMS) Test Stand Utilizing a Hardware-in-the-Loop (HIL) Emulated Battery," 2021 IEEE Electric Ship Technologies Symposium (ESTS), Arlington, VA, USA, 2021, pp. 1-8, doi: 10.1109/ESTS49166.2021.9512327.

[5] D. -H. Doan, T. -Q. -N. Nguyen, H. -T. Pham, H. -A. Trinh, V. -P. Vu and D. -N. Truong, "Development of Controller Hardware-in-the-Loop Platform-based Battery Management System for Hybrid PV/Battery System," 2024 9th International Conference on Applying New Technology in Green Buildings (ATiGB), Danang, Vietnam, 2024, pp. 200-205, doi: 10.1109/ATiGB63471.2024.10717766.

[6] V. Tomar, A. Chitra, D. Krishnachaitanya, N. S. R. Rao, I. V and W. Raziasultana, "Design of Powertrain Model for an Electric Vehicle using MATLAB/Simulink," 2021 Innovations in Power and Advanced Computing Technologies (i-PACT), Kuala Lumpur, Malaysia, 2021, pp. 1-7, doi: 10.1109/i-PACT52855.2021.9696518.

[7] D. He, E. Hou, G. Liu and J. Yu, "Design on Test Platform of BMS System Based on dSPACE," 2017 International Conference on Computer Technology, Electronics and Communication (ICCTEC), Dalian, China, 2017, pp. 968-971, doi: 10.1109/ICCTEC.2017.00214.

[8] S. Park, S. Moura and K. Lee, "Integration of Hardware and Software for Battery Hardware-in-the-Loop Toward Battery Artificial Intelligence," in IEEE Transactions on Transportation Electrification, vol. 10, no. 1, pp. 888-900, March 2024, doi: 10.1109/TTE.2023.3270870.

[9] M. Wagle, A. Agnihotri, P. Bhangale, A. Patare and M. Murali, "Estimation of State of Charge in Electric Vehicle using the Battery Digital Twin," 2023 3rd International Conference on Intelligent Technologies (CONIT), Hubli, India, 2023, pp. 1-7, doi: 10.1109/CONIT59222.2023.10205542.

[10] M. Ponchant, A. Li, C. Beckers and M. Paroha, "Battery Management System Evaluation within a Complete Electric Vehicle Model with Software-in-the-Loop and Hardware-in-the-Loop Approaches," 2021 23rd European Conference on Power Electronics and Applications (EPE'21 ECCE Europe), Ghent, Belgium, 2021, pp. P.1-P.10, doi: 10.23919/EPE21ECCEEurope50061.2021.9570477.

[11] Y. Zhu, X. Zhao, L. Yang and J. Huang, "A Coupled Electro-Thermal Battery Model Identification and State Estimation Based on Digital Twin," 2024 IEEE 4th International Conference on Digital Twins and Parallel Intelligence (DTPI), Wuhan, China, 2024, pp. 647-651, doi: 10.1109/DTPI61353.2024.10778736.

[12] T. M. N. Bui, M. F. Niri, D. Worwood, T. Q. Dinh and J. Marco, "An Advanced Hardware-in-the-Loop Battery Simulation Platform for the Experimental Testing of Battery Management System," 2019 23rd International Conference on Mechatronics Technology (ICMT), Salerno, Italy, 2019, pp. 1-6, doi: 10.1109/ICMECT.2019.8932115.

[13] L. Buccolini, S. Orcioni, S. Longhi and M. Conti, "Cell Battery Emulator for Hardware-in-the-Loop BMS Test," 2018 IEEE International Conference on Environment and Electrical Engineering and 2018 IEEE Industrial and Commercial Power Systems Europe (EEEIC / I&CPS Europe), Palermo, Italy, 2018, pp. 1-5, doi: 10.1109/EEEIC.2018.849373110.23919/EPE21ECCEEurope50061.2 021.9570477.

[14] Di Rienzo R, Verani A, Baronti F, Roncella R, Saletti R. Modular Battery Emulator for Development and Functional Testing of Battery Management Systems: The Cell Emulator. Electronics. 2022; 11(8):1215. doi: 10.3390/electronics11081215

[15] Verani A, Di Rienzo R, Nicodemo N, Baronti F, Roncella R, Saletti R. Modular Battery Emulator for Development and Functional Testing of Battery Management Systems: Hardware Design and Characterization. Electronics. 2023; 12(5):1232. doi.: 10.3390/electronics12051232

Soiling-Induced Efficiency Losses in Photovoltaic Systems: A Comparative Study of Monocrystalline Monofacial and Bifacial Panels

Syed Ayan
Department of Electrical Engineering
Indian Institute of Technology Bhilai
Durg, India
syed@iitbhilai.ac.in

Rajeev Ranjan
Department of Electrical Engineering
Indian Institute of Technology Bhilai
Durg, India
rajeevranj@iitbhilai.ac.in

Shravan Kumar Singh
Department of Electrical Engineering
Indian Institute of Technology Bhilai
Durg, India
shravans@iitbhilai.ac.in

Nikhil Chander
Department of Electronics And Communication
Engineering
Indian Institute of Technology Bhilai
Durg, India
nikhil@iitbhilai.ac.in

Abstract—**Dust accumulation on photovoltaic (PV) modules significantly impacts their efficiency and energy output, particularly in regions with high particulate deposition. This study investigates the performance degradation of monocrystalline Monofacial and bifacial PV panels due to natural dust accumulation over time. Experimental analysis was conducted at the Indian Institute of Technology Bhilai, where key electrical parameters such as power output and efficiency were monitored under clean and dusty conditions. Study indicates the losses observed in panels. This study helped determine and understand dusting patterns, with consistent decrease in power throughout experimental period. Additionally, adverse weather conditions caused significant performance fluctuations. The study highlights the necessity of optimized cleaning schedules, with monocrystalline panels requiring more frequent maintenance than bifacial panels. These findings provide valuable insights into dust mitigation strategies and system optimization for PV installations in dust-prone environments.**

Keywords— *PV-Soiling, Performance Degradation, Cleaning Optimization, Monofacial Bifacial.*

I. INTRODUCTION

The rapidly increasing global energy demand, coupled with environmental concerns and resource sustainability, has necessitated a transition toward renewable energy sources. Among the sustainable energy sources at present the solar photovoltaic (PV) technology is a promising alternative to fossil fuel-based power generation due to its sustainability, scalability, and eco-friendliness. However, the efficiency and long-term performance of PV modules are significantly influenced by various environmental factors, among which dust accumulation is a critical concern. Dust deposition on PV module surfaces leads to a reduction in solar irradiance reaching the photovoltaic cells, thereby decreasing power output and overall system efficiency [1], [2]. The impact of dust accumulation is particularly pronounced in regions with high dust levels, such as industrial zones and arid environments. The accumulation of dust not only degrades the optical transmittance of the module's protective glass but also contributes to thermal effects that exacerbate performance losses.

While several studies have investigated the effects of dust on PV performance, there remains a need for comprehensive experimental analysis under real-world conditions to quantify degradation over time. This study aims to evaluate the impact of dust accumulation on the efficiency and power output of different types of PV modules, including monocrystalline, polycrystalline, and bifacial panels. A series of experimental investigations were conducted at the Indian Institute of Technology Bhilai, Kutelabhata, District – Durg, Chhattisgarh, India, to assess the effects of varying dust densities on electrical parameters such as voltage, current, power, and efficiency.

Besides, the research investigates long-term PV performance degradation trends to determine mitigation strategies to improve system reliability in dusty conditions [3]. Infrared thermographic image inspection with a FLIR thermal imaging gun was employed to investigate the thermal effect of dusting to determine and study temperature difference between clean and dusty modules. Surface temperature increases through dust deposition suggesting potential hotspots, leading to module degradation and lowering efficiency as a whole. Thermal findings also highlight the need for routine cleaning in avoiding thermal stress and prolonging PV module working lifespan. The findings of this study highlight the need for routine cleaning processes and maintenance for avoiding efficiency loss and prolonging PV module working lifespan. This study further enhances the general knowledge of PV module degradation and provides useful suggestions on how to maximize solar energy collection in dusty regions.

II. STUDY SITE AND EXPERIMENTAL APPARATUS

A. Study Region

Our experimental location IIT Bhilai, located in Durg, Chhattisgarh (21.21°N, 81.38°E), our location being in central India experiences a tropical climate with hot summer season (23°C–40°C), a monsoon season bringing heavy rainfall (July peak: 390 mm), and on mild winter (10°C–25°C). The region receives strong solar irradiation, averaging 4.81–6.65 kWh/m²/day under favorable atmospheric conditions, making it favorable conditions for solar power generation.[6] Humidity varies, peaking at 83% in August and dropping to 28% in April, while wind speeds range from 4.1 km/h (November) to 14 km/h (July). A 23° south-facing tilt is recommended for optimal solar panel efficiency as computed through PVSyst software. With consistent sunlight and favorable atmospheric conditions, Our location is deemed advantageous for research purposes, the adaptation of solar PV generated power in daily applications, and large-scale solar power generation especially for industrial load, which significantly reduces reliance on the grid during peak hours, particularly during the summer. .

The region being abundant of coal the continuous mining operations, rice mills, and industries cause constant dust buildup on solar PV panels, which has a major negative impact on the panels' performance. Mineral dust from adjacent excavation sites, airborne milling residues, and fine dust particles produced by industrial emissions all settle on the panels, changing their surface characteristics and lowering light absorption. Regular maintenance is necessary because of the noticeable power losses caused by this persistent deposition, as seen in the above figure. Using the experimental observations from this study, effective cleaning schedules for residential and commercial solar installations can be developed by gaining important insights into patterns of dust accumulation. Due to the variety of dust sources in the region, there is a rare chance to evaluate actual conditions, guaranteeing maximum solar efficiency and low energy losses in comparable settings.

B. Apparatus Description

Three distinct solar panels—Adani Bifacial (355/390 W) and Vikram Solar Monocrystalline (375 W)—were used to examine the effects of dust fouling on PV modules.

These modules were chosen in order to assess how different photovoltaic technologies are affected by dust accumulation. To ensure a precise assessment of receiving incident solar energy, a pyranometer EKO(MS-40M) was used to measure both irradiance and reflected irradiance for study. The pyranometer was connected to a host PC via a data logger to enable continuous POA & reflected irradiance data collection throughout day. This made it possible to regularly record the levels of irradiance over entire duration of study. This setup facilitated comprehensive environmental monitoring and contributed to the correlation between changes in solar intensity and various panel performance evaluation. The panels voltage-current characteristics were recorded using the Seaward PV200 IV tracer for electrical parameter measurements, offering a thorough understanding of how the panels behaved when dust was present. Additionally, by taking infrared thermographic pictures of both dust-free and dust-covered modules, a FLIR thermal imaging gun was used to capture thermal image of modules and assess the thermal effects of dust accumulation, Temperature variations and possible efficiency losses due to soiling effects were studied by this thermal assessment.

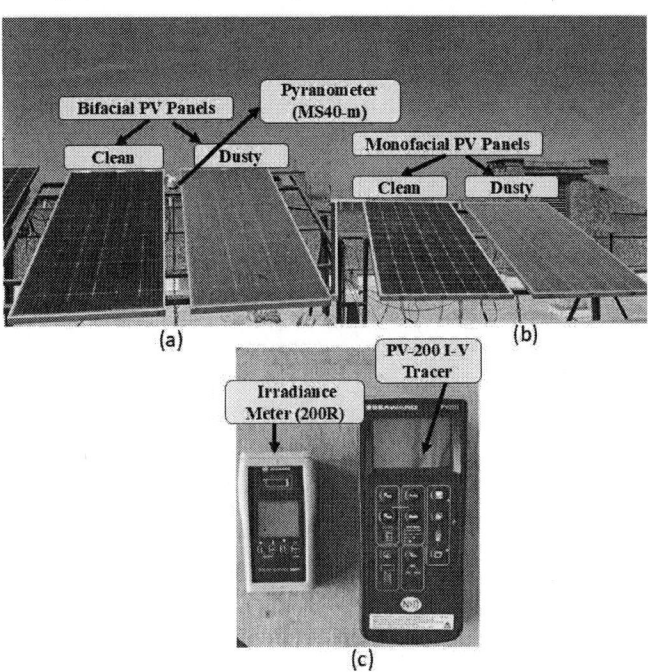

Fig.1: (a) Bifacial modules (b) Monofacial modules (c) Irradiance Meter & IV Tracer

To ensure standardized performance evaluation of various parameters, the PV200 I-V Tracer was also connected to an irradiance meter to facilitate the creation of STC (Standard Test Condition) graphs. Solar Cert software, which allowed for the plotting of performance curves and the compilation of overall efficiency losses brought on by dust accumulation on dusty modules (B2 & M2), was used to compile and further analyse the data that had been gathered. Table 1 lists the PV modules specifications provided by manufacturer that were used in this study.

TABLE I.

SOLAR PANEL SPECIFICATIONS

Parameters	Monocrystalline (Monofacial)	Bifacial
Rated Peak Power (W)	375	355/390*
Open Circuit Voltage (V)	48.7	46.6146.7*
Short Circuit Current (A)	9.94	9.74/10.85*
Rated Voltage (V)	40.1	37.9
Rated Current (A)	9.36	9.37
Fill Factor (%)	77.47	70.09**
Efficiency (%)	19.33	17.6
Bifaciality Coefficient	Not applicable	0.85

III. TEST PROCEDURE

To investigate the impact of dust accumulation on photovoltaic performance, a study on PV modules—monocrystalline Monofacial (M1, clean; M2, dusty) and [10] bifacial (B1, clean; B2, dusty) was conducted and monitored over a period. The dust levels considered corresponded to the highest naturally occurring deposition recorded during the study. Clean PV modules served as a reference to quantify performance [4] losses due to dust accumulation, providing a detailed evaluation of dust accumulation's effect on the modules' power efficiency.

Key electrical parameters such as power output and efficiency were carefully examined to understand how dust accumulation affects photovoltaic performance. The extent of performance degradation was quantified by comparing the energy generation of dust-covered panels against a clean reference module. The following equations provide the analytical framework to evaluate these effects, offering a clear measure of power loss and efficiency reduction due to dust deposition.

The **power output (Pmax)** and **efficiency (η)** of a photovoltaic module can be determined using the following equations:

$$P_{max} = V_{max} \times I_{max} \qquad (1)$$

Where I_{max} represents the maximum current (A), and V_{max} denotes the maximum voltage (V). The efficiency of the solar module is calculated as:

$$\mu = P_{max}/(E * A) \qquad (2)$$

Where P_{max} is the maximum power output (W), E is the available global irradiance (W/m²), and A is the surface area of the PV module (m²) [5][6]. To quantify the efficiency loss due to dust accumulation, the degradation in efficiency is expressed as:

$$\mu_{loss} = \frac{P_{non-dusy} - P_{dusty}}{P_{non-dusy}} \qquad (3)$$

where $P_{non-dusty}$ and P_{dusty} stand for the power outputs of the dusty and clean modules, respectively. To better understand the operational impact of dust deposition in real-world scenarios, the performance ratio (PR), a critical indicator of system effectiveness, was calculated. The performance ratio of panels provides a normalized metric to evaluate the overall performance of PV modules by accounting for a variety of environmental factors, measured temperature, and variations in irradiance values. Dust-covered panels showed higher operating temperatures, which further lowers the efficiency because of increased thermal losses, according to the thermal analysis of modules. Our results highlight the importance of routine cleaning and maintenance procedures in mitigating the negative impact of continuous dust accumulation on PV module performance.

IV. RESULT AND OBSERVATIONS

A. Obtained curves for SPV modules

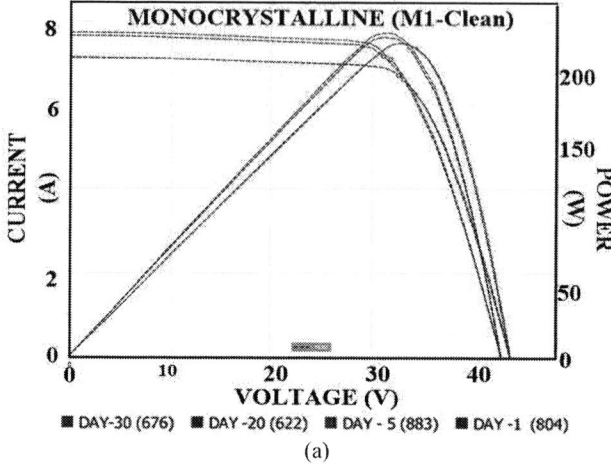

(a)

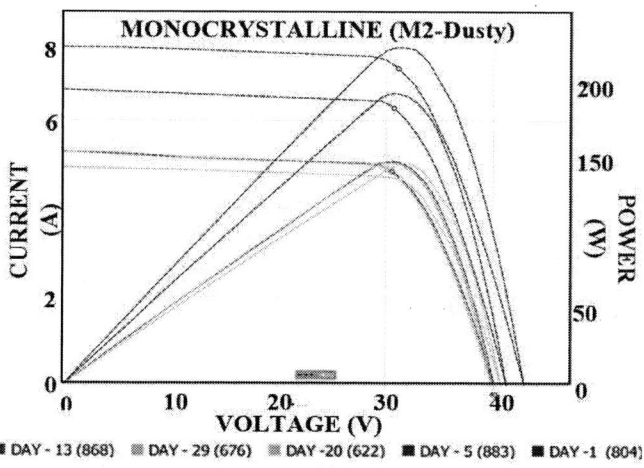

(b)

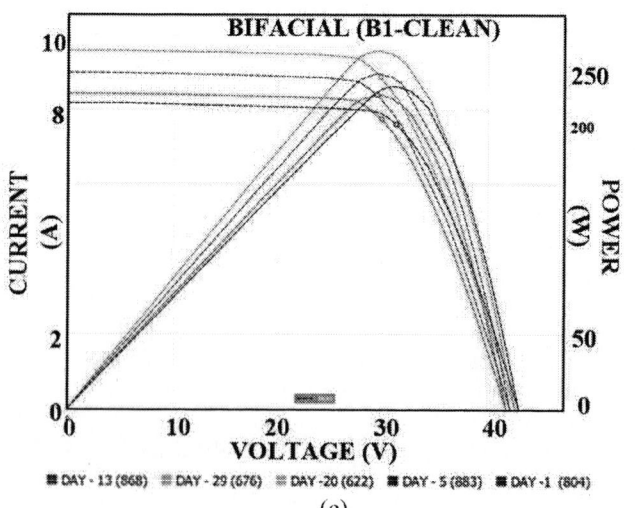

(c)

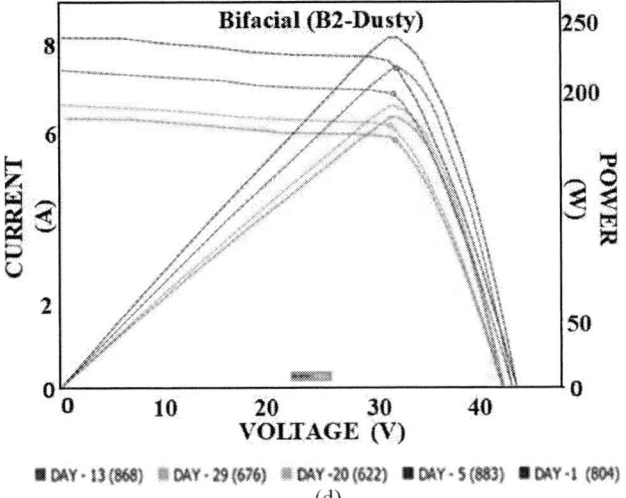

Fig.2: P-V & I-V Curves for modules

Under clean conditions, the bifacial solar module consistently showed higher efficiency and a comparatively higher power output compared to other test panels. However, the monocrystalline panel's performance dramatically decreased after being exposed to dust accumulation for a while, indicating a higher sensitivity to dust deposition. During study at some days, such as Day 25, extreme power drops were observed, which might suggest that cloudy days or material properties have an effect. The overall power loss is displayed in Figure 3.

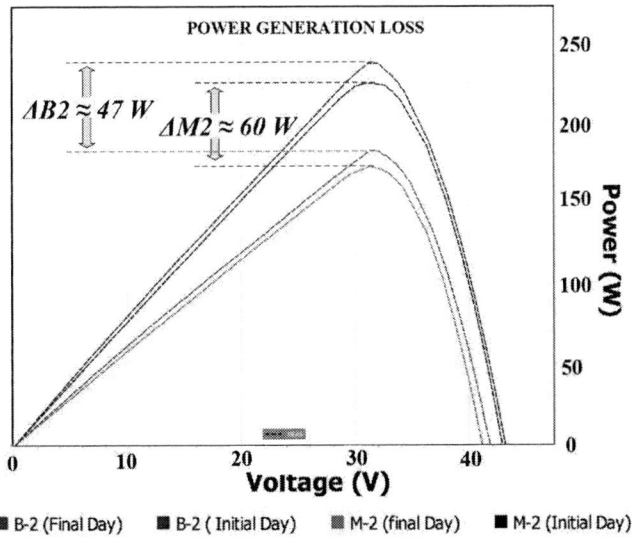

Fig.3: Power generation loss comparison

A. Dust Accumulation Impact

During the initial three days, all solar panel types exhibited minimal performance degradation, with power losses remaining below 5%. However, by Day 5, monocrystalline panels began to show a rapid decline in efficiency, indicating a higher susceptibility to dust accumulation. In contrast, bifacial panels demonstrated superior resistance to dust, with a more gradual reduction in power output over time. After Day 10, power loss trends across all panel types became more stable and predictable, suggesting a settling phase in dust accumulation and its impact on performance.

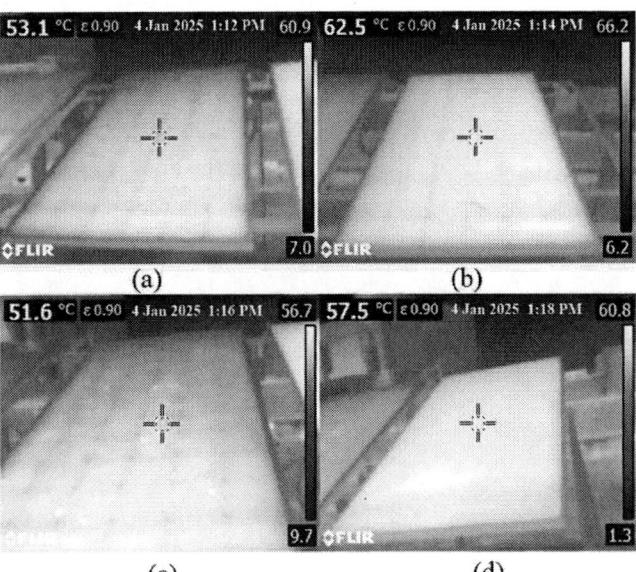

Fig:4 Thermal image for panels (a)B1 (b)B2 (c)M1 (d)M2

Thermal studies revealed that panel temperature is much influenced by dust accumulation. Running at lower temperature than the dusty bifacial panels B2: 62.5°C, the clean bifacial panels B1: 53.1°C suggest that soiling over the period assists thermal absorption of heat meanwhile also reducing heat dissipated off the panels. Similar phenomena were seen in monocrystalline panels, where our clean panel M1: 51.6°C is measured at lower temperature than compared to dusty panel M2: 57.5°C implying that soiling not only degrades optical performance but also increases the overall operating temperature of the module, which may result in enhanced rate of degradation and lower general efficiency. Which further supports the need for consistent cleaning of panels for better performance and reduce thermal stress on panels increasing the overall working life of installed modules.

C. Weather Effects

Significant overall reductions in power output was noted across all panel types on Days 4, 25, and 28, suggesting a strong correlation with adverse weather conditions in region such as cloudy day or atmospheric disturbances. Despite these variations, our bifacial panels (B1&B2) demonstrated relatively better performance under the cloudy or low-irradiation conditions, compared to monofacial panels likely due to their ability to capture diffused sunlight from both front and back surfaces. Additionally, anomalies, such as the unexpected dip in monocrystalline panel performance on Day 9, may indicate potential measurement errors or site-specific environmental factors, such as cloudy day or sudden temperature variations

TABLE II.

COMPARATIVE DATA FOR PV MODULES

Metric	Monofacial	Bifacial
Power Generation		
Average Clean Panel Power (W)	203.07	224.01
Average Dusty Panel Power (W)	147.36	191.57
Average Power Difference (W)	55.71	32.44
Average Power Loss (%)	19.23%	12.94%
Median Clean Panel Power (W)	218.96	233.64
Median Dusty Panel Power (W)	145.71	194.79
Power Metrics		
Maximum Power (Clean) (W) @irr .883.2 w/m^2	253.75	265.97
Maximum Power (Dusty) (W) @irr .883.2 w/m^2	223.18	235.64
Minimum Power (Clean) (W) @irr .487.2 w/m^2	129.61	194.21
Minimum Power (Dusty) (W) @irr .487.2 w/m^2	36.08	84.54
Standard Deviation (Clean) (W)	54.76	50.78
Standard Deviation (Dusty) (W)	44.87	28.7
Coefficient of Variation (Clean)	0.27	0.23
Coefficient of Variation (Dusty)	0.3	0.15
Degradation Analysis		
Maximum Power Loss (%)	12.04%	11.52%
Days to 5% Power Loss	5	8
Days to 10% Power Loss	14	17
Days to 15% Power Loss	19	22

D. Performance Ratio (PR) Analysis for Monocrystalline & Bifacial Panels

Performance ratio for modules is given by formula:

$$PR = \frac{E_{out}}{G_{POA} \times A \times \eta_{STC}} \qquad (4)$$

Where, E_{out} is the actual energy output (wh) of the PV system, G_{poa} is the measured plane-of-array (POA) irradiance (W/m²), A is the total panel area (m²), and η_{stc} is the panel's efficiency under standard test conditions *(STC)*, value of area and efficiency of panel were taken directly from manufacturer datasheet[7]. Bifacial panels have consistently outperformed monocrystalline panels when it comes to the Performance Ratio (PR), averaging 0.858 compared to 0.718. This advantage stems from their ability to capture reflected sunlight, which makes them more efficient in using available irradiance. Over time, PR values saw some fluctuations due to factors like dust accumulation,

shading, and varying irradiance[8][9].Intrestingly, major drops were observed on Days 4 and 27. The lowest PR values recorded were 0.62 for monocrystalline and 0.74 for bifacial panels, likely resulting due to dust buildup. Conversely, dusty modules after day 25 reported the significant decrease suggesting optimal conditions possibly enhanced by panel cleaning. What's more, bifacial panels showed superior stability, demonstrating fewer sharp declines, making them more resilient in tough environments, while monocrystalline panels faced sharper performance losses under dust buildupThis study also provided valuable insights into optimal cleaning schedules for different panel types, as shown below:

TABLE III.

RECOMMENDED CLEANING SCHEDULE PV MODULES

Panel Type	Recommended Cleaning Frequency	Justification	Expected power recovery
Monofacial	Every 5-7 days	Power loss exceeds 10% after 10 days	~12-15%
Bifacial	Every 8-10 days	Power loss exceeds 10% after 12 days	~12-17%

V. CONCLUSION

The results show that bifacial panels consistently achieved the highest power output under clean conditions and exhibited superior performance in dusty conditions, with lower overall energy loss. In contrast, monocrystalline monofacial panels experienced the most significant performance degradation, with power loss becoming notable after just five days. Bifacial panels maintained higher Performance Ratios (PR) across most days, averaging 0.858 compared to 0.718 for monofacial panels, demonstrating better stability in challenging environments.

Weather variations also influenced power fluctuations, with noticeable drops on Days 4 and 27, likely due to dust accumulation or shading. Despite these fluctuations, bifacial panels performed better under low-light conditions.

Regular cleaning is crucial for monofacial panels to prevent rapid performance degradation, while bifacial panels can tolerate longer intervals due to their ability to capture light from both the front and the back side. Future studies will focus on seasonal dust analysis and optimizing cleaning strategies to further enhance PV performance and longevity.

REFERENCES

[1] P. K. Enaganti *et al.*, "Experimental investigations for dust build-up on low-iron glass exterior and its effects on the performance of solar PV systems," *Energy (Oxf.)*, vol. 239, no. 122213, p. 122213, 2022.

[2] G. Vedulla, A. Geetha, and R. Senthil, "Review of strategies to mitigate dust deposition on solar photovoltaic systems," *Energies*, vol. 16, no. 1, p. 109, 2022.

[3] V. Muthu and G. Ramadas, "A comprehensive 4E study on the performance of bifacial solar module installed on different ground surface colors: An experimental study on a specific site," *J. Sol. Energy Eng.*, vol. 145, no. 1, pp. 1–31, 2023.

[4] S. K. Singh and N. Chander, "Mid-life degradation evaluation of polycrystalline Si solar photovoltaic modules in a 100 kWp grid-tied system in east-central India," *Renew. Energy*, vol. 199, pp. 351–367, 2022.

[5] N. Singh Baghel and N. Chander, "Performance comparison of mono and polycrystalline silicon solar photovoltaic modules under tropical wet and dry climatic conditions in east-central India," *Clean Energy*, vol. 6, pp. 165–177, 2022.

[6] J. Peterson, J. Chard, and J. Robinson, "Extraction of prevailing soiling rates from soiling measurement data," in *2022 IEEE 49th Photovoltaics Specialists Conference (PVSC)*, 2022.

[7] S. Ayan, S. Kumar and N. Chander, "Investigations on Droop Control for Stable Transitions Between Islanded and Grid Operations in Microgrids," *2024 IEEE 11th Power India International Conference (PIICON)*, JAIPUR, India, 2024, pp. 1-6, doi: 10.1109/PIICON63519.2024.10995158

[8] N. S. Baghel and N. Chander, "Impact of soiling on the performance of monocrystalline-Si photovoltaic modules under different climatic conditions in East-Central India," in *Lecture Notes in Mechanical Engineering*, Singapore: Springer Nature Singapore, 2023, pp. 39–48.

Magnetically Integrated Onboard Charger and DCDC Converter For Electric Vehicle

Deepesh K V
Electrical Engineering Department
National Institute Of Technology
Calicut, India
deepesh_p220184ee@nitc.ac.in

Dr. Jagadanand G
Electrical Engineering Department
National Institute Of Technology
Calicut, India
jagadanand@nitc.ac.in

Dr. Nikhil Sasidharan
Electrical Engineering Department
National Institute Of Technology
Calicut, India
nikhils@nitc.ac.in

Abstract — **Multi winding-based converter for electric vehicle (EV) charging systems enables improved power density by integrating onboard chargers (OBC) and DC-DC systems. Triple Active Bridge (TAB) based topology is employed in this system for the integration of high voltage DC (HV DC) and low voltage DC (LV DC) converters, which allows electrical and magnetic integration of both converters into a single power topology. In EV charging system, HV DC power stage is employed for HV battery charging and LV DC power stage is employed for LV battery charging. A TAB converter used in the power stages of EV charging must be capable of functioning in various modes, such as charging, driving, and vehicle-to-grid. Each port of TAB will undergo simultaneous or independent activation during these modes of operation. In this work, a 7.4kW OBC and 2kW DCDC integrated system is designed, and functionality is analyzed with various modes of operation using MATLAB Simulink environment.**

Keywords—Triple Active Bridge, Onboard Charger, Integrated OBC, Bidirectional OBC, Multi Winding Converter, Electric Vehicle

I. INTRODUCTION

Conventional high voltage (HV) system architecture of battery electric vehicle (BEV) is shown in Fig. 1, which includes charging, traction and energy storage sub-systems that operates in harmony. Charging sub-system include on-board charger (OBC) for charging HV battery from electric vehicle supply equipment (EVSE) and DC-DC converter for charging LV battery from HV battery. OBCs are critical component in EV which enable efficient AC to DC conversion for battery charging. OBCs are designed to meet stringent efficiency, power density and bidirectional power flow as some of the key features. High frequency operation enabled by the wide-bandgap devices (Silicon carbide and Gallium Nitride), shift towards 800V architectures, sub-system integrations are some of the key trends observed in OBC market.

The overall power flow in a charging sub-system is characterized by Grid-to-Vehicle (G2V) charging, Vehicle-to-Everything (V2X), low voltage (LV) or auxiliary battery charging and pre-charging use cases. G2V uses power from EVSE to charge the HV battery. An AC EVSE available in single phase or three phase configuration, supply power to HV battery through OBC, while DC EVSE bypasses OBC and directly energizes the HV battery. V2X use cases include Vehicle to Grid (V2G), Vehicle to Vehicle (V2V), Vehicle to Load (V2L) and Vehicle to Home (V2H), uses power from HV battery to vehicle outlet connector through OBC. V2V can also be realized bypassing OBC. In G2V and V2X power

transfer scenarios, LV battery shall be ensured to maintain a safe voltage limit to provide power for internal accessories. Hence, LV battery charging also should be possible in these use cases. LV battery charging uses power from HV to LV battery. Pre-charging of HV bus will be required before closing the HV contactors while discharging the HV battery for safe operation of components. Typically, pre-charging is done by boost operation of LV battery.

An EV charging system shall be ensured to meet all these use cases while meeting Automotive Safety Integrity Level (ASIL) defined in ISO 26262. ASIL rating indicates the criticality of the component in the vehicle pertaining to its functional safety aspects. ASIL-D rating is given to highest critical components and ASIL-A is given to lowest. ISO 26262 also defines Quality Management Level (QM) that represents hazards that do not dictate any safety requirements. Typical rating given to DC-DC sub-system is ASIL-C and OBC is ASIL-B. These ASIL ratings can vary based on the

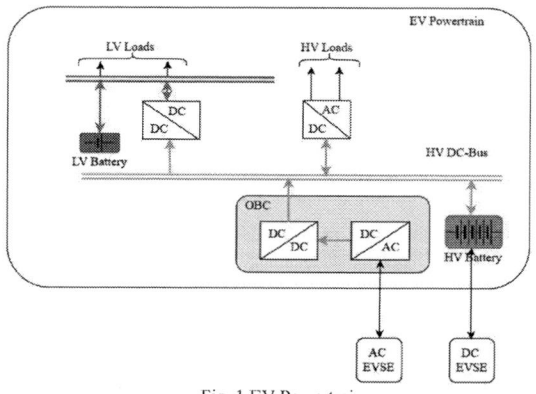

Fig. 1 EV Powertrain

vehicle architecture and criticality of the components defined by the overall architecture.

Triple Active Bridge (TAB) is a widely analysed topology that allows electrical and magnetic integration of multiple DC-DC converters into a single power topology. Triple Active Bridge (TAB) based converter was introduced as an extension of Dual Active Bridge (DAB) in [1], where a third active bridge is added to the existing DAB structure. The third port enables power flow between three DC sources and/or loads. The power flow among the three ports were achieved by the phase shift between each of them. This topology is found widely employed in dc microgrids, fuel cell vehicles and electric vehicles. Reference [2] discusses about integrating an HV bus (300V) and auxiliary bus (14V and 42V) which is found in hybrid electric vehicles (HEV) and fuel cell vehicles (FCV) architecture using TAB architecture. The proposed converter allows bidirectional power transfer as well as

979-8-3315-3013-6/25 $31.00 © 2025 IEEE

supports simultaneous power transfer. The control strategy combines duty cycle control and phase shift control between the ports and provide a decoupled power transfer. The equivalent circuit of TAB topology representation in Y-type and Δ-type also discussed in this paper to analyse and understand the complex power flow between the three ports in a simplified manner. Both Y-type and Δ-type representation is a primary-referred equivalent circuit. The Y-type representation helps to analyse the dependency of power transfer between the ports on phase shift control signal, whereas Δ-type representation helps to analyse the power flow between three ports. *S. Y. Kim et al.* propose a control method that virtually isolates the idle port while the other two ports of the Triple Active Bridge (TAB) are actively engaged in power transfer [3]. The existence of three ports can result in one port idle while other two ports are transferring power in a particular scenario. The easiest way to avoid power flow to idle port is to cut the gate signals of the idle port, but due to existence of antiparallel diodes which can form a full-bridge rectifier circuit and causes leakage power flow through the idle port. Compared to conventional decoupled control based on linearized model, this method employs active switching based on current vector analysis to get zero current through idle port. The isolation conditions are derived from the phasor diagram and control is applied to all the three ports.

The shared magnetic core and the impact of cross coupling makes the design of TAB a challenging topic. Enabling required operating modes, deviation in stray components such as leakage inductances, unwanted cross-coupling during power transfer to specific ports may pose challenges in the multi winding transformer-based design of dc-dc converters [13]. Design methodology of high frequency transformer in multi winding structure is analysed in [14]. The need for high frequency operation poses challenges in transformer winding calculation and estimation of core losses because of skin and proximity effects. Transformer core and winding lose model is analysed systematically in [15] to derive core and winding specification for minimum transformer loss. *Chakraborty et al.* [16] present a methodology for transformer design aimed at achieving power decoupling across multiple ports.

TAB structure can be realized by integrating different topologies around the transformer [4 - 8]. The building block topologies of TAB can be conventional topologies such as LLC, CLLLC, DAB or PSFB. These building blocks maybe used with or without combination for different operating modes. An integrated OBC and DCDC system using TAB which utilizes LLC topology for G2V, CLLLC topology for V2G and a buck converter for HV to LV power transfer is proposed in [4]. Pre-charging mode is not analysed in this paper and efficiency is impacted due to extra coupling of HV port and DC link port while in HV-LV operation. TAB structure with CLLLC for G2V and V2G, LLC for HV to LV is discussed in [5]. The control technique is complex for preventing saturation during pre-charging mode in this paper. DAB topology is used in all modes of operation in [6]. This paper does not discuss about V2G mode of operation. CLLLC topology for G2V and V2G, and LLC for HV-LV is proposed in [7]. The power density is further improved by combining resonant magnetic components into a single three winding transformer and half bridge configuration. Simultaneous charging of LV battery while charging and discharging as well as pre-charging operations are not possible in the proposed structure. CLLC is used in all modes of operation in [8]. This paper discusses an improved design of transformer by

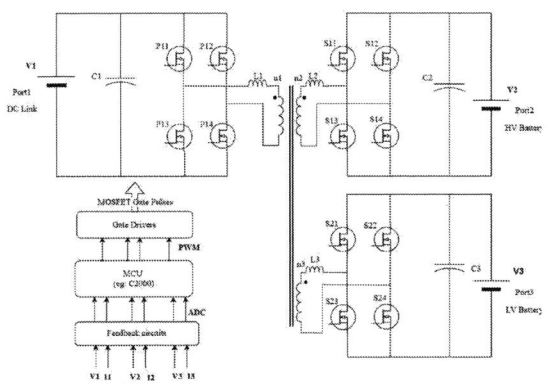

Fig. 2 Triple Active Bridge Structure

reducing the effect of magnetizing component and has complex winding structure with different widths of winding for different windings. A DAB and PSFB structures were used in [9], where G2V and V2G is realized with DAB and HVLV is realized with PSFB. Due to the presence of diodes, the conduction losses are increased, and pre-charging is not feasible in this study.

Control strategy in TAB is based on the topology used for realizing the TAB structure. Since TAB was introduced as a derivative of DAB topology, the control strategy of DAB can be applied to TAB also. The power flow between the ports can be controlled by controlling phase shift at respective port. The single-phase shift (SPS) control [3] with two outer phase shifts between the ports and dual phase shift (DPS) control [10] with additional inner phase shift in a port to improve zero voltage switching (ZVS) and reduce RMS current, are implemented in TAB also. *S. Zou et al.* propose a Triple Phase Shift (TPS) control method for Dual Active Bridge (DAB) converters, which includes five control variables: three duty ratios from three full bridges and two-phase shifts between the bridges [11]. A four-phase modulation (FPM) scheme with tuning of magnetizing inductance to achieve all-ZVS operation is introduced in [12]. TAB is considered as three independent DAB topology to carry out the analysis and apply superposition theorem to extend this for TAB structure.

In this study, the TAB converter is developed to integrate the high-voltage DC (HVDC) component of the onboard charger (OBC) with the low-voltage DC (LVDC) component of the auxiliary battery charger. The dual active bridge (DAB) structure is utilized for both the HVDC and LVDC sides, enabling precise control of the secondary side to the desired set point. The converter's performance is validated under typical electric vehicle (EV) operating conditions to ensure it meets the desired performance characteristics.

II. INTEGRATED CONVERTER OPERATION AND DESIGN

Single-phase OBC using TAB integration at HV and LV battery side as shown in Fig. 2 is considered for analysis in this work. The converter is designed for a maximum power of 7.4kW at any given point of operation.

A. System Specifications

The specification of the converter are detailed in TABLE-I. The converter features three ports: Port-1 is the DC link, Port-2 is the HV DC and Port-3 is the LV DC output.

TABLE-I: INTEGRATED CONVERTER SPECIFICATION

TAB Port	Specification	
	Parameter	*Value*
Port – 1	DC Link Voltage	$400 - 500V$
	Voltage Ripple	1%
Port – 2	HV DC Voltage	$350 - 450V$
	Max. Output Power	7.4kW
	Efficiency	96%
	Voltage Ripple	1%
Port -3	LV DC Voltage	$9 - 16V$
	Max. Output Power	2kW
	Efficiency	96%
	Voltage Ripple	1%

Single phase shift control is utilized for further simplifying the analysis. In single phase shift control, Port-2 and Port-3 will have different phase delays with respect to Port-1 and thus both ports can be regulated independently. The phase shift between the ports determines the power flow direction and the amount of power to be transferred. Power sharing will happen between ports while simultaneous operations are ongoing. The power sharing in each port is decided in the design phase to ensure smooth operation of the converter. In this work, LV port, being the critical component, is always operated at its full power, i.e. 2kW. During EV charging scenario, when simultaneous charging of HV and LV batteries are required, then power sharing of HV port will be 5.4kW and that of LV port will be 2kW. If LV battery is full during charging scenario, full power of 7.4kW will be transferred to HV port. Similar case can happen during V2G operation also.

In an EV architecture, multiple ECUs are working in co-ordination. Fig. 3 shows a possible use case in EV architecture. The Battery Management System (BMS) will monitor the battery status, and calculate parameters like State-of-Charge (SoC), State-of-Health (SOH) among other parameters and determine the mode of charging, like constant-current (CC) or constant-voltage (CV) mode of operation required for the battery and corresponding current or volage set points. This request will reach OBC ECU either directly or via a Vehicle Control Unit (VCU) using Controller Area Network (CAN) communication channel.

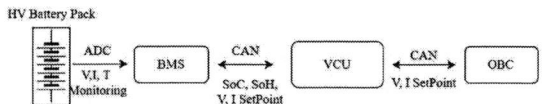

Fig.3 EV- Vehicle ECU Communication

B. Converter Design And Control

The equivalent representation of TAB is shown in Fig. 4. Each port inductance value can be calculated by first finding primary referred delta value and then corresponding star value by delta-star conversion and finally converting to each port value. The magnetizing inductance can be ignored in TAB converter due to its higher value in comparison to leakage inductance [17].

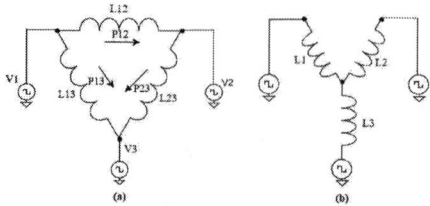

Fig. 4 TAB Equivalent Structure (a) Delta (b) Star

The selection of leakage inductance or shim inductance or combination at each port is the most critical step in designing the converter. The value of inductance should be selected in balance, as low value led to lower controller area of phase shift and high value means higher loss. The inductance value required between each port can be calculated from the power equation as represented in (1). These primary referred delta equivalents are converted to star equivalent and then value at each side of the transformer is derived. The final inductance used at each port is shown in Table 1.

$$
\begin{aligned}
L_{12} &= \frac{n_1 V_1 V_2}{2 F_s P_{12}} D_{max} (1 - D_{max}) \\
L_{13} &= \frac{n_3 V_1 V_3}{2 F_s P_{13}} D_{max} (1 - D_{max}) \\
L_{23} &= \frac{n_3 V_2 V_3}{2 F_s P_{23}} D_{max} (1 - D_{max})
\end{aligned}
\quad (1)
$$

Where L12 is the required inductance between Port 1 and Port 2, L13 the inductance between Port 1 and Port 3 and L23 is the inductance between Port 2 and Port 3, V1 is the output voltage at Port 1, V2 is the output voltage at Port 2 and V3 is the output voltage at Port 3, P12 represents the transferred power between Port 1 and Port 2, similarly P13 and P23 represents the same between Port 1 and Port 3 and between Port 2 and Port 3 respectively. D_{max} is the maximum phase shift that can be applied between the ports. The phase delays are related to power transfer as shown in (2).

$$
\begin{aligned}
P_{12} &= \frac{n_1 V_1 V_2}{2 F_s L_{12}} D_{12} (1 - D_{12}) \\
P_{13} &= \frac{n_3 V_1 V_3}{2 F_s L_{13}} D_{13} (1 - D_{13}) \\
P_{23} &= \frac{n_3 V_2 V_3}{2 F_s L_{23}} (D_{13} - D_{12})(1 - (D_{13} - D_{12}))
\end{aligned}
\quad (2)
$$

Here D_{12} is the phase shift between Port 1 and Port 2, D_{13} and D_{23} are the phase shift between Port 1 and Port 3 and Port 2 and Port 3 respectively.

The phase delays are calculated by solving the equations (3)

$$
\begin{aligned}
P_1 &= P_{13} + P_{12} \\
P_2 &= P_{12} - P_{23} \\
P_3 &= P_{13} + P_{23}
\end{aligned}
\quad (3)
$$

Single phase shift control is utilized for controlling the power transfer between the ports. The phase shift is calculated based on the mode of operation and power sharing between the ports. The circuit parameters designed for simulation analysis are shown in TABLE-2.

TABLE-2: DESIGN PARAMETERS OF TAB CONVERTER

Parameter	Value
Switching Frequency, Fs	100kHz
Port-1 Turns Ratio, n_1	1
Port-2 Turns Ratio, n_2	34
Port-3 Turns Ratio, n_3	34
Port-1 Inductance, L1	10.36uH
Port-2 Inductance, L2	9.06uH
Port-3 Inductance, L3	25.8nH

III. SIMULATION RESULTS AND ANALYSIS

TAB converter is simulated in MATLAB Simulink to verify the performance in every mode of operation. The high level simulation structure is shown in Fig. 5. The three ports are abstracted in the sub system with desired inputs and outputs. Two gate pulses are supplied to each bridge which will be given to the diagonal switches and complimentary pulses to other switches in the same arm. TAB structure is designed with three ports, DC-link port (Port-1), HV port (Port-2) and LV port (Port-3). Output voltage and current from each port are taken out and given to the control block where the control strategy is executed to generate the respective gate pulses.

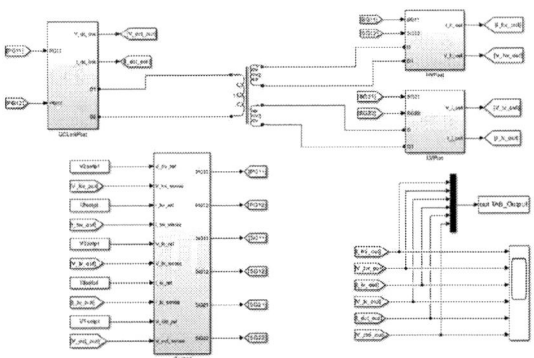

Fig. 5 MATLAB Plant and Control Simulation

EV charging mode in Fig. 6 shows the operation of the converter in constant current and constant voltage mode scenarios. HV port parameters, HV voltage (V_{hv}), HV current (I_{hv}) and LV port parameters LV voltage (V_{lv}), LV current (I_{lv}) are controlled in the experiment and results are shown in the plots. The DC link port is the source where a DC source of 500V is connected and power is transferred to HV port and LV port simultaneously. The power transferred to LV port is P3 (consume 2KW) and HV port is P1-P3 (consume 5.4kW). In Fig. 6 (a), the reference value of V_{hv} is considered 350V initially and incremented step-wise as 400v at 0.02sec and 450V at 0.04sec and the reference value of V_{lv} is considered 9V initially and incremented step-wise as 12v at 0.02sec and 16V at 0.04sec. The V_{hv} and V_{lv} are reaching it's set point value with ripple less than 1%. The steady state value is reached as quickly as less than 2ms at startup and less than 1.5ms at every step change. Fig. 6 (b) demonstrates the CC mode operation, where reference value of I_{hv} is 12A initially and step-wise increment of 13A at 0.02sec and 15A at 0.04sec, similarly I_{lv} initially at 100A and step-wise increment of 150A

at 0.02sec and 200A at 0.04sec. The I_{hv} and I_{lv} are reaching it's set point value with ripple less than 1%. The steady state value of I_{hv} is reached as quickly as less than 3.5ms at startup and less than 4.5ms at every step change. The steady state value of I_{lv} is reached as quickly as less than 1ms at startup and less than 1ms at every step change. The test was repeated with a DC link voltage of 400V, where exactly same behaviour is observed. The feedback from HV port and LV port is fed to control block. The control block calculates the error and applied to PI controller which calculate the phase shift required for HV and LV port switches with reference to the DC link port to achieve the desired set point.

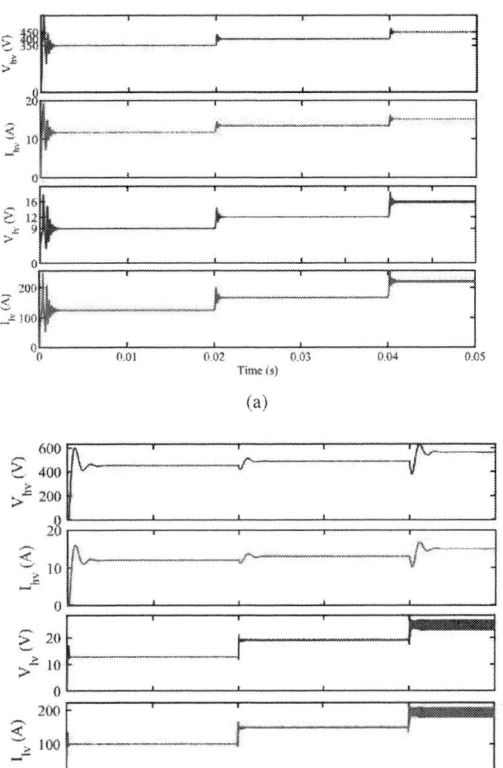

Fig. 6 Performance Characteristics of TAB during G2V with Dynamic Operation (a) CV Mode (b) CC Mode

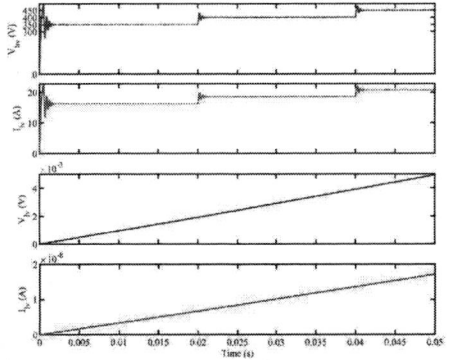

Fig. 7 Performance Characteristics of TAB during G2V (HV only) with Dynamic Operation

EV charging mode in Fig. 7 is a scenario where LV battery is fully charged, hence the HV port alone is required to be energized by the source port (DC link port). Full power from source is transferred to HV port. The LV port gate pulses are stopped to keep the switches in off state. The reference value of V_{hv} is initially kept at 350V and incremented step-wise as 400v at 0.02sec and 450V at 0.04sec. The V_{hv} is reaching its set point value with ripple less than 1%. The steady state value is reached as quickly as less than 2ms at startup and less than 1.5ms at every step change. A negligible power is flowing to LV port due to transformer primary leakage inductance and parasitic elements of capacitance and inductance.

In V2G mode, HV port function as power source supplying energy to both DC link port and LV port. This mode of operation is verified and result shown in Fig. 8. In this plot, The DC link voltage and current (V_{dc} and I_{dc}) are shown. HV port is the source port, hence avoided in the plot. The power transferred to LV port is P3 (consume 2KW) and DC link port is P2-P3 (consume 5.4kW). In Fig. 8, the reference value of V_{dc} and V_{lv} is incremented in step of 400V at 0sec, 500v at 0.02sec and the reference value of V_{lv} is incremented in step of 12V at 0sec, 16v at 0.02sec. The V_{hv} and V_{lv} are reaching its set point value with ripple less than 1%. The steady state value is reached within 3ms at startup and also at every step change.

of 400V at 0sec, 500v at 0.02sec. I_{hv} and I_{lv} parameters are unregulated. The V_{hv} is reaching its set point value with ripple less than 1%. The steady state value is reached within 3ms at startup and also at every step change. In this mode, HV port will supply full power (7.4kW) to DC link port. A negligible leakage power to LV port (V_{lv} and I_{lv}) is also observed in this mode of operation.

EV Driving mode observed in Fig. 10 is a scenario where LV battery is charged by the HV battery while DC link port is inactive. In this mode of operation, HV port act as source and LV port act as the load. The power at HV port is represented as -P3 (deliver 2kW) and LV port is represented as P3 (consume 2kW). The Phase shift is calculated with the desired power sharing. The reference value of V_{lv} is incremented in step of 9V at 0sec, 12v at 0.02sec and 16v at 0.04sec. I_{hv} and I_{lv} parameters unregulated. The V_{hv} is reaching its set point value with ripple less than 1%. The steady state value is reached as quickly as less than 2ms at startup and less than 1.5ms at every step change.

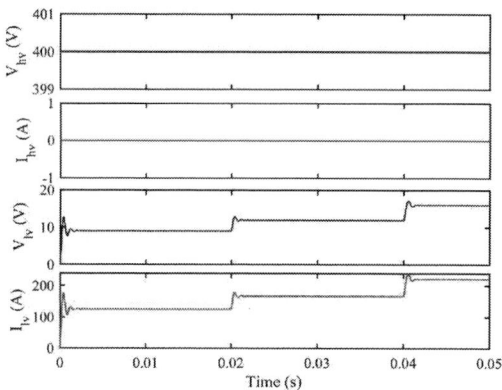

Fig.10 Performance Characteristics of TAB during Driving with Dynamic Operation

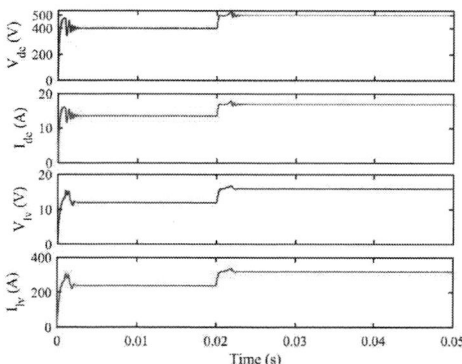

Fig.8 Performance Characteristics of TAB during V2G with Dynamic Operation

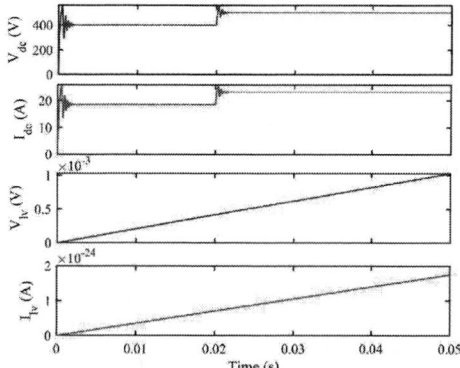

Fig.9 Performance Characteristics of TAB during V2G (DC Link Only) with Dynamic Operation

EV V2G mode in Fig. 9 is tested by energizing DC link port alone by the source port (HV battery). This can also be a scenario in EV when LV battery is full while HV battery is discharging. The reference value of V_{dc} is incremented in step

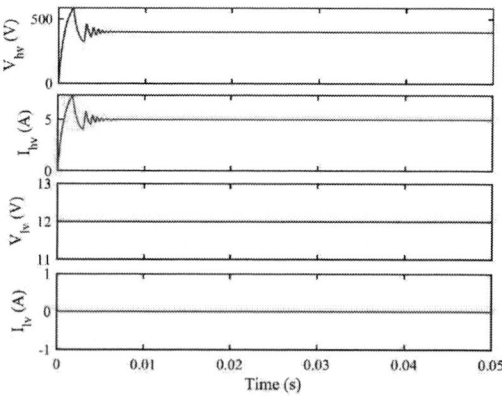

Fig.11 Performance Characteristics of TAB during Pre-charge

In EV architecture, the HVDC bus pre-charging is a scenario where HVDC bus is charged near to HV battery level before the HV battery contactors are closed to protect components from high inrush current. In this mode, the DCDC converter should perform boost operation of LV to HV conversion from LV battery. This scenario is verified in Fig. 11. Here the LV port acts as the source port and HV port act as load port. The power at LV port is represented as -P3 (deliver 2kW) and HV port is represented as P3 (consume 2kW). Here V_{hv} reached its set point value of 400V by maintaining a voltage and current ripple of less than 1% and observed that steady state condition is achieved in less than 7ms.

The operation of the TAB converter structure in various modes are carried out in this work. The performance characteristics of the designed converter during steady state and dynamic operations are also included in the analysis. Each of the three ports interchanges it's role either as source or as load depending on the mode of operation and achieve the specification selected for the EV charging system.

IV. CONCLUSION

The TAB based HVDC and LVDC integrated converter is analysed and simulated in MATLAB for various modes of operation in EV charging. Using phase shift control at the secondary side of the converter enable better controllability to achieve desired output at each ports with wide range of operations of HV from 350V to 450V and LV from 9V to 16V. From the performance analysis, it is observed that the designed converter is satisfactorily achieving the desired responses. In various modes of operation, the voltage and current ripple of less than 1% is observed which satisfies the IEEE standard. The steady state value of both voltage and current set point is reached as quickly as 2ms in all modes operation except in V2G mode and Pre-Charge operation. In V2G mode, the voltage set point is reached within 3ms, whereas in Pre-Charge mode, voltage set point is reached within 7ms. In the control scheme, only the ports which are active in the power conversion are enabled and non-active port is disabled by stopping the switch activation. The control scheme uses a coupled power transfer between the port, which results in low leakage power in the inactive port. The usage of resistive load made the voltage and current dictate linear relationship at power absorbing ports, which can be modelled using battery simulator in future analysis. Further, the design and subsequent performance evaluation of the converter in various modes the scenarios confirms that the multi-winding converter is a suitable choice for achieving high power density in EV charging system.

REFERENCES

[1] M. Michon, J. L. Duarte, M. Hendrix and M. G. Simoes, "A three-port bi-directional converter for hybrid fuel cell systems," 2004 IEEE 35th Annual Power Electronics Specialists Conference (IEEE Cat. No.04CH37551), Aachen, Germany, 2004, pp. 4736-4742 Vol.6, doi: 10.1109/PESC.2004.1354836.

[2] J. C. Zhao, S. D. Round and J. W. Kolar, "An Isolated Three-Port Bidirectional DC-DC Converter with Decoupled Power Flow Management," in IEEE Transactions on Power Electronics, vol. 23, no. 5, pp. 2443-2453, Sept. 2008, doi: 10.1109/TPEL.2008.2002056.

[3] S. Y. Kim, H. -S. Song and K. Nam, "Idling Port Isolation Control of Three-Port Bidirectional Converter for EVs," in IEEE Transactions on Power Electronics, vol. 27, no. 5, pp. 2495-2506, May 2012, doi: 10.1109/TPEL.2011.2172225.

[4] D. -W. Lee, B. -S. Lee, J. -H. Ahn, J. -Y. Kim and J. -K. Kim, "New Combined OBC and LDC System for Electric Vehicles With 800 V Battery," in IEEE Transactions on Industrial Electronics, vol. 69, no. 10, pp. 9938-9951, Oct. 2022, doi: 10.1109/TIE.2022.3148730.

[5] M. Kumar, P. M. Barbosa, J. M. Ruiz, J. Minli and S. Hao, "Isolated Three-Port Bidirectional DC-DC Converter for Electric Vehicle Applications," 2022 IEEE Applied Power Electronics Conference and Exposition (APEC), Houston, TX, USA, 2022, pp. 2000-2007, doi: 10.1109/APEC43599.2022.9773690.

[6] H. Ma, Y. Tan, L. Du, X. Han and J. Ji, "An integrated design of power converters for electric vehicles," 2017 IEEE 26th International Symposium on Industrial Electronics (ISIE), Edinburgh, UK, 2017, pp. 600-605, doi: 10.1109/ISIE.2017.8001314.

[7] Y. Tang, J. Lu, B. Wu, S. Zou, W. Ding and A. Khaligh, "An Integrated Dual-Output Isolated Converter for Plug-in Electric Vehicles," in IEEE Transactions on Vehicular Technology, vol. 67, no. 2, pp. 966-976, Feb. 2018, doi: 10.1109/TVT.2017.2750076.

[8] S. S. Chakraborty, S. Dey and K. Hatua, "Design of a Three-Winding Transformer for Power Decoupling of a Three-Port Series Resonant Converter for an Integrated On-Board EV Charger," in IEEE Transactions on Power Electronics, vol. 38, no. 11, pp. 14262-14273, Nov. 2023, doi: 10.1109/TPEL.2023.3308776.

[9] I. Kougioulis, A. Pal, P. Wheeler and M. R. Ahmed, "An Isolated Multiport DC–DC Converter for Integrated Electric Vehicle On-Board Charger," in IEEE Journal of Emerging and Selected Topics in Power Electronics, vol. 11, no. 4, pp. 4178-4198, Aug. 2023, doi: 10.1109/JESTPE.2023.3276048

[10] Z. Ling, H. Wang, K. Yan, and J. Gan, "Optimal isolation control of three-port active converters as a combined charger for electric vehicles," Energies, vol. 9, no. 9, 2016, Art. no. 715.

[11] S. Zou, J. Lu, and A. Khaligh, "Modelling and control of a triple-active-bridge converter," IET Power Electron., vol. 13, no. 5, pp. 961–969, 202

[12] L. Gong et al., "A Simplified All-ZVS Strategy for High-Frequency Triple Active Bridge Converters with Designed Magnetizing Inductance," in IEEE Transactions on Power Electronics, vol. 38, no. 11, pp. 13781-13797, Nov. 2023, doi: 10.1109/TPEL.2023.3304316

[13] T. Pereira, F. Hoffmann, R. Zhu and M. Liserre, "A Comprehensive Assessment of Multiwinding Transformer-Based DC–DC Converters," in IEEE Transactions on Power Electronics, vol. 36, no. 9, pp. 10020-10036, Sept. 2021, doi: 10.1109/TPEL.2021.3064302

[14] Y. Liang, Z. Wang, H. Wu, C. Wang and X. Li, "Design of a multi-winding high-frequency transformer for DC-DC applications," 2017 IEEE Conference on Energy Internet and Energy System Integration (EI2), Beijing, China, 2017, pp. 1-6, doi: 10.1109/EI2.2017.8245507

[15] S. Zou, J. Lu, A. Mallik and A. Khaligh, "Modeling and Optimization of an Integrated Transformer for Electric Vehicle On-Board Charger Applications," in IEEE Transactions on Transportation Electrification, vol. 4, no. 2, pp. 355-363, June 2018, doi: 10.1109/TTE.2018.2804328

[16] S. S. Chakraborty, S. Dey and K. Hatua, "Design of a Three-Winding Transformer for Power Decoupling of a Three-Port Series Resonant Converter for an Integrated On-Board EV Charger," in IEEE Transactions on Power Electronics, vol. 38, no. 11, pp. 14262-14273, Nov. 2023, doi: 10.1109/TPEL.2023.3308776

[17] V. R. Kudaravalli, V. Uttam and V. M. Iyer, "A Design Methodology for Triple Active Bridge DC- DC Converter," 2022 IEEE International Conference on Power Electronics, Drives and Energy Systems (PEDES), Jaipur, India, 2022, pp. 1-7, doi: 10.1109/PEDES56012.2022.10080029

979-8-3315-3013-6/25 $31.00 © 2025 IEEE

Model Predictive Control of Three-Level Boost Converters in Photovoltaic and Wind Energy Systems Connected to a Bipolar DC Microgrid

1st Satishreddy Dodda
Department of Electrical Engineering
National Institute of Technology Warangal
Warangal-506004 (T.S), India
ds719036@student.nitw.ac.in, orcid:0000-0003-3137-5967

2nd Srinivasa Rao Sandepudi
Department of Electrical Engineering
National Institute of Technology Warangal
Warangal-506004 (T.S), India
ssr@nitw.ac.in, orcid:0000-0003-2777-7429

Abstract—**This paper introduces a model predictive control of three-level boost converters (TLBC) interfaced with photovoltaic (PV) and wind energy conversion systems (WECS) connected to a bipolar DC microgrid. The goal is to maximise power extraction from both renewable sources and balance bipolar voltages against unbalanced loading. The proposed bipolar grid includes an energy storage system (ESS) with a bi-directional buck-boost converter per pole to regulate voltages at the bipolar DC microgrid. The effectiveness of the proposed control technique has been confirmed with simulation results using Matlab/Simulink under variable solar irradiance, wind velocity, and also load variations.**

Index Terms—**Model predictive control (MPC), three-level boost converter (TLBC), energy storage system (ESS), bipolar DC microgrid.**

I. INTRODUCTION

The rising need for electrical energy demand and growing concerns over the greenhouse gases produced by fossil fuels in conventional generating stations are the major cause for innovation and development of alternate electric power generation with renewable energy sources. In the process of developing renewable energy generation with various sources, it is identified that for effective and efficient renewable energy generation, wind and solar photovoltaic (SPV) energy conversion systems have become popular due to their inherent features such as flexibility in real-time control, easy and quick installation of huge megawatt plants with moderate cost etc. Further advancements in the development of power electric converter configurations and control techniques made the wind and solar energy systems suitable for achieving sustainable development in the distribution system to operate in isolated mode, utility grid-tied operation and hybrid mode. DC microgrids can offer numerous benefits, such as efficient control structures, reduced power conversion stages, improved controllability, no harmonic problems, and lower transmission losses than AC grids [1]–[4]. DC microgrids can also eliminate reactive power, simplify integration of energy storage devices, and possess inherent self-

healing characteristics. Achieving optimal efficiency and performance in DC microgrids requires careful operational optimization, particularly when dealing with moderate to high power demands [5]. SPV and wind energy systems can form a DC microgrids using boost converters controlled with maximum power point tracking (MPPT) techniques. To increase the reliability of isolated DC microgrids against load fluctuations, variations in renewable energy generation depending on solar irradiance and wind velocity, an energy storage system (ESS) with bidirectional DC-DC converters is proposed. Various researchers have explored the monopolar DC grids with ESS [6]–[9]. A boost converter at photovoltaic (PV) and wind energy systems and a bidirectional buck-boost converter for battery charging and discharging process are proposed in [10].

Adaptive control methods for photovoltaic (PV) systems for standalone water pumping and air conditioning systems with battery storage were proposed in [11]–[13]. However, these systems lack control over DC voltage regulation at the grid and have limited power management capacity, hence they are not suitable for electric vehicle charging stations.

The bipolar DC microgrid architecture was introduced to improve power transfer capability and ensure reliable supply to end users. Nevertheless, it faces several challenges, including voltage imbalance, poor voltage regulation, unequal power generation, and varying load conditions across the poles [14]–[16]. To overcome these issues, the development of suitable converters and advanced control strategies is essential. Achieving stable and balanced voltage requires deploying a bipolar converter at the interface of either the energy storage unit or the distributed generation (DG) system. The three-level boost converter (TLBC) is one such topology compatible with DG systems, capable of facilitating both maximum power extraction and maintaining bipolar voltage balance. The bipolar DC microgrid structure allows three distinct voltage levels: $+V_{dc}$, $-V_{dc}$, and $2V_{dc}$ to supply loads [17]. In case of a fault in one of the DC lines, loads on the other line remain unaffected. However, the issue of voltage balancing

979-8-3315-3013-6/25 $31.00 © 2025 IEEE

and regulation at the energy storage system necessitates a converter topology with appropriate control. Moreover, lowering the voltage stress on the converter helps reduce both the required size of the filter inductor and the converter's power rating.

Several studies have proposed unipolar DC microgrids employing three-level boost converters for wind and solar PV applications. In [18], the Perturb and Observe (P&O) method is implemented to optimize power extraction by monitoring the inductor current and controlling voltage balance, thus eliminating the need to measure capacitor voltage directly. Other approaches suggest combining P&O with a phase shift strategy for output capacitor voltage balance [19]. Traditionally, P&O is paired with a Proportional Integral (PI) controller for improved power extraction and DC-link voltage balance, but these traditional control designs are complicated, particularly when handling multiple tasks and searching for the Maximum Power Point (MPP) [20]. A solution to these challenges is by regulating battery charging and discharging process using a bidirectional buck-boost converter, a Model Predictive Control (MPC) has emerged due to its simplicity and flexibility in handling multiple constraints, despite the converter's complexity [21]. Many researchers have worked on single-input, single-output DC-DC converters in unipolar DC microgrids and bipolar DC microgrids for voltage regulation and balancing. The voltage balancer converters proposed will transfer energy between poles in a bipolar DC microgrid, but highlight the downsides of additional cost and power loss. As a solution to these issues, the paper proposes a DC-DC converter topology at photovoltaic, wind, and fuel cells. It offers a three-level boost (single-input, dual-output) DC-DC output, which is an efficient alternative to maintain balance in the microgrid.

This paper introduces a predictive control system for bi-polar DC microgrids, powered by Photovoltaic (PV) and Wind energy conversion system (WECS). The system uses model predictive control and Maximum Power Point Tracking (MPPT) for optimal performance. The P&O method facilitates MPPT for the PV system, while a single-dimensional lookup table is employed for the WECS. Furthermore, the system also ensures voltage balance at the DC-bus by integrating constraints in the cost function. To enhance load supply reliability, an ESS is connected to the grid via a bi-directional buck-boost converter, for managing power flow to and from the battery. The effectiveness of the proposed control technique was confirmed with simulation results using Matlab/ Simulink under variable solar irradiance, wind velocity, and also load variations.

II. System Description and Modelling

A. Proposed Bipolar DC Microgrid

The proposed bipolar DC microgrid architecture has three voltage levels: +150V, 0, and -150V, as depicted in Fig. 1. Lower loads are connected between either +150V and 0 or between -150V and 0, while higher loads can be directly connected between +150V and -150V (300V). To maintain balance in the energy demand, a battery energy storage system (BESS) is utilized. Three-level boost converters (TLBC) are used to extract maximum power from the SPV and WECS systems and balance the voltage at the bipolar grid terminals. Additionally, a two-level bidirectional buck-boost converter is integrated at the BESS to control the DC link capacitor voltage or the DC bus voltage.

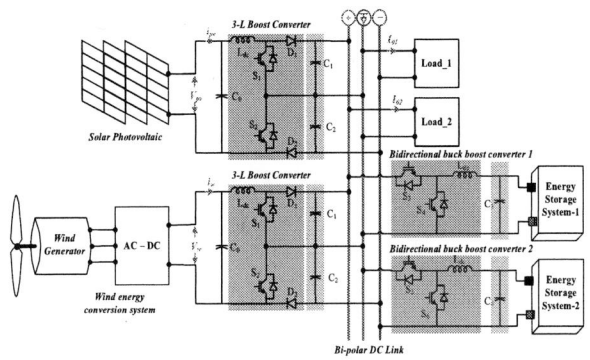

Fig. 1. Schematic of the proposed bipolar DC microgrid.

B. Three level Boost Converter in Renewable Energy Systems

The three-level boost converter originates from the Neutral-Point-Clamped (NPC) multilevel converter topology. This configuration helps minimize diode reverse recovery losses, reduce voltage stress on switches, maintain voltage balance across capacitors, and support adjustable output voltage levels. The operation of switches S_1 and S_2 is directed by complementary control signals, as depicted in Fig. 1.

Operating Methods: The converter uses two switches, namely S_1 and S_2 and the voltages that appear across capacitors C_1 and C_2 are represented as V_{C1} and V_{C2}. The operation of the converter can be categorized into four different modes, which are illustrated in Fig. 2. In the first and fourth modes, the inductor currents undergo a process of charging and discharging respectively. On the other hand, in the second and third modes, the direction of increase in inductor currents is influenced by the voltages V_{C1} and V_{C2}. This is determined based on the comparison between half the sum of V_{C1} and V_{C2} and V_{dc}. The operation of the switches S_1 and S_2 is characterized by four modes, each defined by their respective voltage equations (1) to (4), which effectively demonstrate the operation of these four modes.

$$L \cdot \frac{di_L}{dt} = V_{dc} - (V_{c1} + V_{c2}) \tag{1}$$

$$L \cdot \frac{di_L}{dt} = V_{dc} - V_{c1} \tag{2}$$

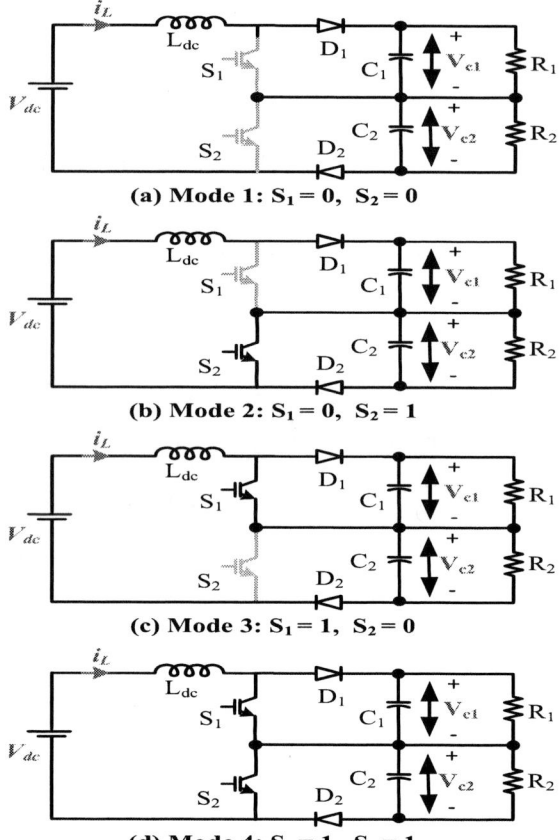

(a) Mode 1: $S_1 = 0$, $S_2 = 0$

(b) Mode 2: $S_1 = 0$, $S_2 = 1$

(c) Mode 3: $S_1 = 1$, $S_2 = 0$

(d) Mode 4: $S_1 = 1$, $S_2 = 1$

Fig. 2. Three-level boost converter modes (a–d).

$$L \cdot \frac{di_L}{dt} = V_{\text{dc}} - V_{c2} \quad (3)$$

$$L \cdot \frac{di_L}{dt} = V_{\text{dc}} \quad (4)$$

These differential equations are discretized using Euler's forward difference method, tailored for implementation in model predictive control.

$$i_L(k+1) = i_L(k) + \frac{T_s}{L}\left(V_{\text{dc}} - (V_{c1} + V_{c2})\right) \quad (5)$$

$$i_L(k+1) = i_L(k) + \frac{T_s}{L}\left(V_{\text{dc}} - V_{c1}\right) \quad (6)$$

$$i_L(k+1) = i_L(k) + \frac{T_s}{L}\left(V_{\text{dc}} - V_{c2}\right) \quad (7)$$

$$\cdot\, i_L(k+1) = i_L(k) + \frac{T_s}{L} \cdot V_{\text{dc}} \quad (8)$$

The equations from (5) to (8) can be expressed as follows,

$$i_L(k+1)_{s=1 \text{ to } 4} = i_L(k) + \tfrac{T_s}{L}\left(V_{\text{dc}} - (1-D_1)V_{c1} - (1-D_2)V_{c2}\right) \quad (9)$$

In equation (9), "$S = 1$ to 4" refers to the four possible switching states of the two switches (S_1 and S_2) in the

three-level boost converter, as detailed in Section IV of [22]. Each state corresponds to a specific combination of ON/OFF conditions of S_1 and S_2. Also, the predicted equations for a capacitor's voltage are described as follows:

$$V_{c1}(k+1) = V_{c1}(k) + \frac{T_s}{C_1}\left[i_L(k) \cdot (1-D_1) - I_{01}\right] \quad (10)$$

$$V_{c2}(k+1) = V_{c2}(k) + \frac{T_s}{C_2}\left[i_L(k) \cdot (1-D_2) - I_{02}\right] \quad (11)$$

The switching states S_1 and S_2 are contingent upon the values of D_1 and D_2, which can either be 1 or 0. The objective function is represented in the form of equation (12), as follows.

$$g(i)_{s=1 \text{ to } 4} = W_1 \cdot |i_L(k+1) - i_w^*| + W_2 \cdot |V_{c1}(k+1) - V_{c2}(k+1)| \quad (12)$$

The states of transition are imposed on S_1 and S_2 using equation (12). The parameters W_1 and W_2 are adjusted in accordance with the measurement units of voltage and current or based on the significance of a specific restriction in the process of minimizing the objective function.

C. Bidirectional Buck-Boost Converter

Fig. 3 depicts a bidirectional two-level buck-boost converter interfacing the battery energy storage system with the bipolar DC microgrid and load terminals. This converter is designed to supplement power during periods of high load demand exceeding generation. Its operation is managed by switches S_1 and S_2, which are controlled by complementary signals.

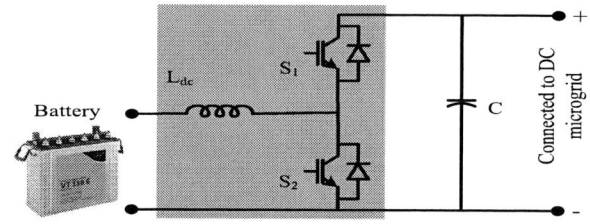

Fig. 3. Bi-directional buck-boost converter.

Modes of operation: The modes of operation of a bidirectional buck-boost converter are shown in Fig. 4. In the boost-mode of operation, the voltage loop equations (13) and (14) can be represented when switch S_1 is ON ($S_1 = 1$) and switch S_2 is OFF ($S_2 = 0$) as

$$L \cdot \frac{di_L}{dt} = V_{\text{bat}} - V_{\text{dc}} \quad (13)$$

when switch S_1 is OFF ($S_1 = 0$) and switch S_2 is ON ($S_2 = 1$),

$$L \cdot \frac{di_L}{dt} = V_{\text{bat}} \quad (14)$$

During buck-mode operation, the voltage loop equations (15) and (16) can be represented as when switch S_1 is OFF ($S_1 = 0$) and switch S_2 is ON ($S_2 = 1$),

$$L \cdot \frac{di_L}{dt} = -V_{\text{bat}} \quad (15)$$

979-8-3315-3013-6/25 $31.00 © 2025 IEEE

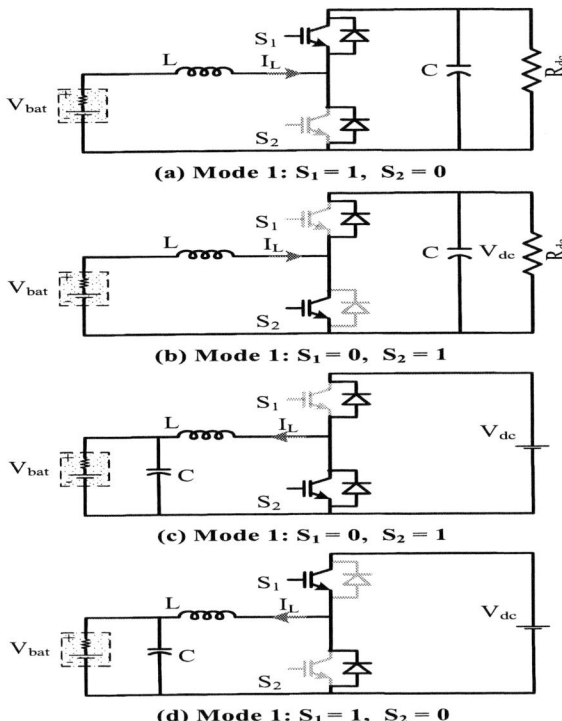

Fig. 4. Boost mode (a–b) and Buck mode (c–d).

when switch S_1 is ON ($S_1 = 1$) and switch S_2 is OFF ($S_2 = 0$),

$$L \cdot \frac{di_L}{dt} = V_{dc} - V_{bat} \qquad (16)$$

The buck and boost operating modes of the converter are determined by the direction of the inductor current. When the battery discharges, the boost mode is activated as per equations (13) and (14). Conversely, the buck mode is explained by equations (15) and (16) during battery charging.

III. CONTROL APPROACH

A. Maximum power point tracking

The control of a three-level boost converter, as depicted in Fig. 5, uses predictive control in combination with the Perturb & Observe (P&O) technique to track maximum power. The P&O algorithm generates a reference current for the predictive model, directing the inductor current to extract maximum power from the PV panel. Similarly, the predictive control approach is also employed with an optimal power MPPT technique [20]. A lookup table containing the optimal power and rotor speed values for the wind turbine is utilized to generate a reference current, which the inductor current follows to achieve maximum power extraction from the wind energy system.

B. Charging and Discharging of Battery

The bidirectional buck-boost converter can operate in both buck and boost configurations. When the generated

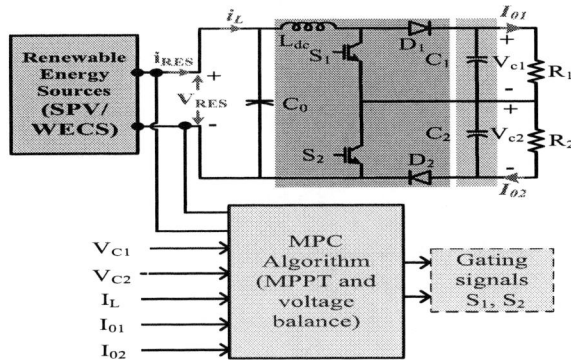

Fig. 5. Use of predictive control for MPPT and maintenance of voltage balance in capacitors.

power exceeds the load demand, the converter switches to buck mode to enable battery charging. On the other hand, the boost mode discharges the battery in situations where the power demand exceeds the power that has been generated. To ensure proper operation, a PI controller is utilized to regulate the voltage, setting it to 150V. The control scheme for this bidirectional buck-boost converter is illustrated in Fig. 6.

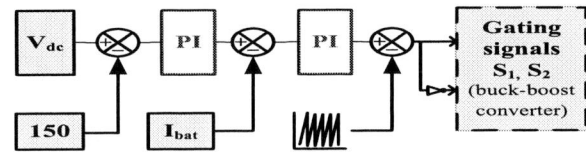

Fig. 6. Control scheme for bidirectional buck-boost converter.

IV. SIMULATION RESULTS

The simulation of a proposed control technique for a bipolar DC microgrid has been performed using MAT-LAB/ Simulink. The parameters considered for the simulation are listed in Table **??**. The bipolar microgrid operates in two modes: load-dominating mode and generation-dominating mode, depending on the power variation from the sources or load. The power generated by WECS using a permanent magnet synchronous generator (PMSG) is converted into DC, and the power from a solar PV model (KC200GT) is optimally harnessed using three-level boost converters (TLBC). The TLBC allow maximum power extraction, regulation of DC bus voltage, and balanced bipolar DC link capacitor voltages. Further, a two-level buck-boost converter is used to control the charging and discharging of the ESS based on the operation mode, source power availability, and load requirements. The outcomes of the simulations for the proposed bipolar DC microgrid with mixed power sources under different operating modes are illustrated in Figs. 7, 8, 9, and 10.

TABLE I
SIMULATION PARAMETERS

Parameter	Value
SPV System	
PV panel model	KC200GT
Peak power @ 1000 W/m^2, 25°C	200 W/module
Peak current	7.61 A
Peak voltage	26.3 V
Number of series modules	3
Number of parallel modules	2
Wind Energy Conversion System	
Wind turbine emulator (DC motor coupled PMSG)	Benn
Peak power @ 14 m/s	2.2 kW
Stator inductance (per phase)	4.08 mH
Stator resistance (per phase)	0.7 Ω
Flux produced by the rotor magnets	0.218 Wb
Number of pole pairs	4
Battery Energy Storage System	
Open-circuit voltage	110 V
Rated capacity	20 Ah
Terminal voltage	101 V
DC–DC Converters	
Switching frequency	5 kHz
Converter inductance	5 mH
Bipolar DC link capacitors	4700 μF
Bipolar voltages	+150 V, −150 V

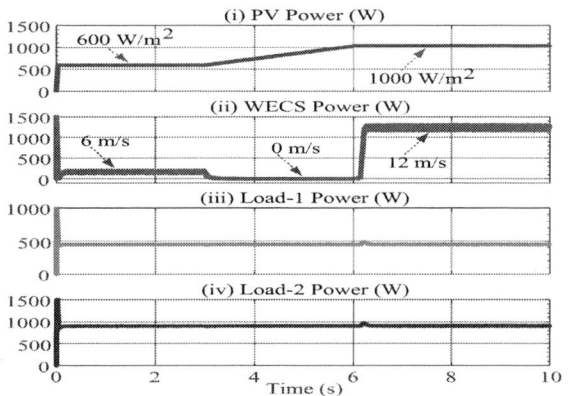

Fig. 7. Waveforms of PV, WECS, and load powers under dominant load and generation modes.

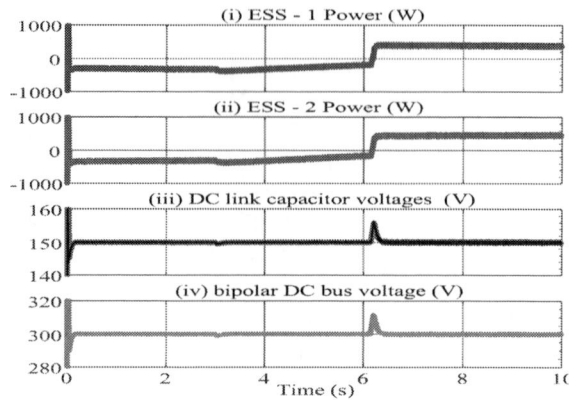

Fig. 8. Waveforms of ESS powers, DC link voltages, and the DC bus voltage at load and generation-dominating modes.

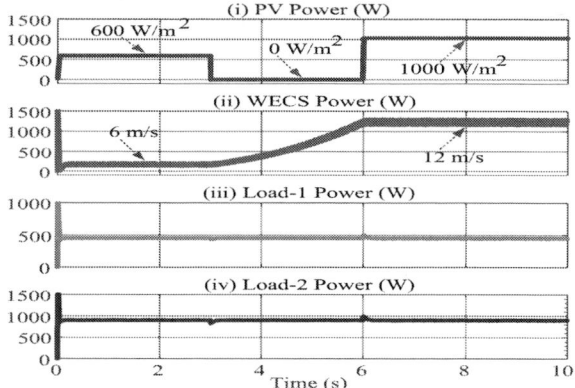

Fig. 9. Waveforms of PV, WECS, and load powers under dominant load and generation modes.

A. Dominant Load Mode:

During this mode, from 0 to 3 seconds, power demand exceeds the generation due to a decrease in power output from photovoltaic (PV) or wind energy systems. Consequently, the transition of Energy Storage Systems (ESSs) happens from charging to discharging state as power from PV and wind is needed to meet the demand. Between 3 to 6 seconds, an abrupt drop in PV or wind power triggers an increased discharge of power from the ESSs. At 3 seconds, the PV power drops to zero while the wind power incrementally increases as shown in Figs. 9 and 10, and vice versa as shown in Figs. 7 and 8. This escalates the power deficit and necessitates the ESSs to discharge more power. Thus, a substantial rise in battery discharge can be observed between 3 to 6 seconds. Despite fluctuations in RES powers and unbalanced loads connected at the bipoles, there are no observed oscillations in the bipolar DC link capacitor voltages or bipolar DC bus voltage.

Throughout this mode, the ESSs maintain their discharging state to compensate for the power shortage, and PV and wind energy systems operate at MPPT to maximize power extraction.

B. Generation-dominating mode:

In this mode, the combination of PV and wind power generation exceeding demand is considered. During 6 to 10 seconds, the battery charges using the excess energy. At 6 seconds, PV and wind power rise, thereby enabling the charging of the battery until 10 seconds. The ESSs are assumed to reach their maximum state of charge (SoC) at 10 seconds and then become idle. Consequently, PV and WECS change from MPPT mode to non-MPPT mode. From 6 to 10 seconds, PV and wind operate in MPPT mode, efficiently delivering power to grid-connected loads and diverting surplus power to ESSs. This process helps in balancing DC link capacitor voltages and managing bipolar DC bus voltage at reference levels using predictive control.

The simulation results depicted in Figs. 8 and 10 demonstrate the proposed predictive control, which ensures the

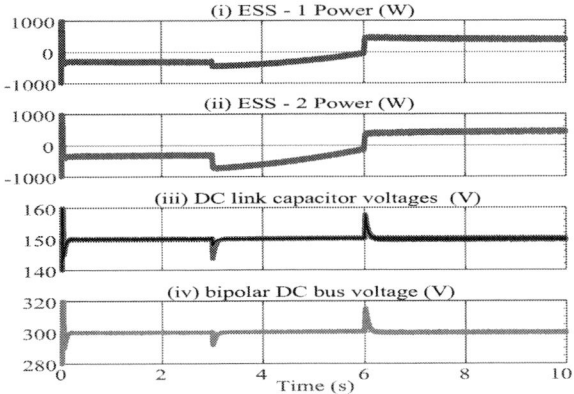

Fig. 10. Waveforms of ESS powers, DC link voltages, and the DC bus voltage at load and generation-dominating modes.

bipolar DC bus voltage and DC link voltages remain steady in spite of power variations. Therefore, a TLBC interfaced bipolar DC microgrid system using PV & WECS under various operating conditions presents a cost-effective and reliable solution for sensitive DC microgrids.

V. Conclusion

This paper presents the modelling and simulation of model predictive control for photovoltaic (PV) and wind energy conversion systems (WECS). These systems are interfaced with a three-level boost converter (TLBC), connected to a bipolar DC microgrid. The control strategies proposed not only address the issue of DC-bus voltage balance but also optimise the extraction of power from these renewable energy sources. The control mechanisms include a proportional-integral (PI) control for a bidirectional buck-boost converter for efficient transfer of energy to and from the battery. Simulation results confirm the effectiveness of these systems in managing MPPT and maintaining balance pole voltages under unbalanced load and also variations in power generation. Therefore, the proposed control scheme may be extended to microgrids with a variety of distributed energy sources with balanced and unbalanced loads, and can also be extended to grid-tied applications.

References

[1] B. R. Naidu, G. Panda, and P. Siano, "A self-reliant dc microgrid: Sizing, control, adaptive dynamic power management, and experimental analysis," *IEEE transactions on industrial informatics*, vol. 14, no. 8, pp. 3300–3313, 2017.

[2] F. Blaabjerg, M. Liserre, and K. Ma, "Power electronics converters for wind turbine systems," *IEEE Transactions on industry applications*, vol. 48, no. 2, pp. 708–719, 2011.

[3] F. Blaabjerg and K. Ma, "Future on power electronics for wind turbine systems," *IEEE Journal of emerging and selected topics in power electronics*, vol. 1, no. 3, pp. 139–152, 2013.

[4] C. Xu and K. Cheng, "A survey of distributed power system — ac versus dc distributed power system," 06 2011, pp. 1–12.

[5] K. Strunz, E. Abbasi, and D. N. Huu, "Dc microgrid for wind and solar power integration," *IEEE Journal of emerging and selected topics in Power Electronics*, vol. 2, no. 1, pp. 115–126, 2013.

[6] A. Amine, "Photovoltaic power control using mppt and boost converter," 10 2013.

[7] M. Metry, S. Bayhan, R. S. Balog, and H. A. Rub, "Model predictive control for pv maximum power point tracking of single-phase submultilevel inverter," in *2016 IEEE Power and Energy Conference at Illinois (PECI)*, 2016, pp. 1–8.

[8] D. Kumar and K. Chatterjee, "A review of conventional and advanced mppt algorithms for wind energy systems," *Renewable and sustainable energy reviews*, vol. 55, pp. 957–970, 2016.

[9] R. I. Putri, M. Pujiantara, A. Priyadi, T. Ise, and M. H. Purnomo, "Maximum power extraction improvement using sensorless controller based on adaptive perturb and observe algorithm for pmsg wind turbine application," *IET Electric Power Applications*, vol. 12, no. 4, pp. 455–462, 2018.

[10] Y. Shan, J. Hu, K. W. Chan, Q. Fu, and J. M. Guerrero, "Model predictive control of bidirectional dc–dc converters and ac/dc interlinking converters—a new control method for pv-wind-battery microgrids," *IEEE Transactions on Sustainable Energy*, vol. 10, no. 4, pp. 1823–1833, 2018.

[11] J. Umuhoza, Y. Zhang, S. Zhao, and H. A. Mantooth, "An adaptive control strategy for power balance and the intermittency mitigation in battery-pv energy system at residential dc microgrid level," in *2017 IEEE applied power electronics conference and exposition (APEC)*. IEEE, 2017, pp. 1341–1345.

[12] R. Kumar and B. Singh, "Buck-boost converter fed bldc motor drive for solar pv array based water pumping," in *2014 IEEE International Conference on Power Electronics, Drives and Energy Systems (PEDES)*. IEEE, 2014, pp. 1–6.

[13] K.-H. Chao, C. Tseng, H. Huang, G. Liu, and L.-C. Huang, "Design and implementation of a bidirectional dc-dc converter for stand-alone photovoltaic systems," *energy*, vol. 4, no. 8, 2013.

[14] S. Dadjo Tavakoli, J. Khajesalehi, M. Hamzeh, and K. Sheshyekani, "Decentralised voltage balancing in bipolar dc microgrids equipped with trans-z-source interlinking converter," *IET Renewable Power Generation*, vol. 10, no. 5, pp. 703–712, 2016.

[15] P. Prabhakaran and V. Agarwal, "Mitigation of voltage unbalance in a low voltage bipolar dc microgrid using a boost-sepic type interleaved dc-dc compensator," in *2016 IEEE 2nd Annual Southern Power Electronics Conference (SPEC)*. IEEE, 2016, pp. 1–6.

[16] H. Kakigano, Y. Miura, and T. Ise, "Low-voltage bipolar-type dc microgrid for super high quality distribution," *IEEE transactions on power electronics*, vol. 25, no. 12, pp. 3066–3075, 2010.

[17] D. Kumar, F. Zare, and A. Ghosh, "Dc microgrid technology: system architectures, ac grid interfaces, grounding schemes, power quality, communication networks, applications, and standardizations aspects," *Ieee Access*, vol. 5, pp. 12 230–12 256, 2017.

[18] H.-C. Chen and W.-J. Lin, "Mppt and voltage balancing control with sensing only inductor current for photovoltaic-fed, three-level, boost-type converters," *IEEE Transactions on Power Electronics*, vol. 29, no. 1, pp. 29–35, 2013.

[19] M. Tampubolon, W.-C. Lin, J.-Y. Lin, Y.-C. Hsieh, H.-J. Chiu, K. Yamanaka, and M. Hojo, "A study and implementation of three-level boost converter with mppt for pv application," in *2017 IEEE 3rd International Future Energy Electronics Conference and ECCE Asia (IFEEC 2017-ECCE Asia)*. IEEE, 2017, pp. 1143–1148.

[20] J.-M. Kwon, B.-H. Kwon, and K.-H. Nam, "Three-phase photovoltaic system with three-level boosting mppt control," *IEEE Transactions on Power Electronics*, vol. 23, no. 5, pp. 2319–2327, 2008.

[21] S. Kouro, P. Cortés, R. Vargas, U. Ammann, and J. Rodríguez, "Model predictive control—a simple and powerful method to control power converters," *IEEE Transactions on industrial electronics*, vol. 56, no. 6, pp. 1826–1838, 2008.

[22] D. Satish Reddy and S. Srinivasa Rao, "Control of three-level bidirectional buck-boost converter for battery energy storage system in bi-polar dc microgrid," *Energy Storage*, vol. 6, no. 1, p. e582, 2024.

979-8-3315-3013-6/25 $31.00 © 2025 IEEE

DC-Link Capacitor Dimensioning with Lifetime Calculation Considerations for PMSM Drives: A Practical Framework for Enhanced Performance

1st Angel K O
Department of Electrical Engineering
National Institute of Technology, Calicut
Kerala, India
angelkalappura@gmail.com

2nd Rushikesh U Shinde
Department of E-Mobility
Varroc Engineering Limited
Pune, India
Rushikesh.Shinde@varroc.com

3rd Dr. Shreelakshmi M P
Department of Electrical Engineering
National Institute of Technology, Calicut
Kerala, India
shreelakshmi@nitc.ac.in

4th Pradip M Magar
Department of E-Mobility
Varroc Engineering Limited
Pune, India
Pradip.Magar@varroc.com

5th Mayank P Deo
Department of E-Mobility
Varroc Engineering Limited
Pune, India
Pradip.Magar@varroc.com

6th Pramod Chaudhary
Department of E-Mobility
Varroc Engineering Limited
Pune, India
Pradip.Magar@varroc.com

Abstract—A crucial component of the functioning of a three-phase inverter in Permanent Magnet Synchronous Motor (PMSM) drives is the DC-link capacitors. The functioning of the inverter may be affected by an excessive voltage ripple caused by a DC connection capacitor of insufficient size. In contrast, an oversized capacitor can increase the system cost and volume and affect the transient response. Moreover, electrolytic capacitors, commonly used in DC-links, face significant lifetime limitations due to thermal stress from ripple current and voltage fluctuations. This paper presents a comprehensive methodology for selecting and sizing DC-link capacitors in inverter-fed PMSM drives, focusing on minimizing voltage ripple while ensuring compliance with the lifetime requirement for the electrolytic capacitors. Using a closed-loop simulation model, the functionality of the new method with the upgraded DC-link capacitor is confirmed based on system parameters, load profiles, and operational conditions.

Index Terms—Capacitor lifetime, DC-link capacitor, permanent magnet synchronous motor drives, ripple voltage.

I. INTRODUCTION

The capacitor for the DC link serves as an energy reservoir, stabilizing the DC bus voltage and reducing fluctuations induced by the inverter's switching operations. An undersized capacitor can lead to excessive voltage ripple and compromised motor control, while an oversized capacitor can lead to unnecessary cost and space utilization and may also affect the system's transient response. Therefore, a balanced capacitor sizing approach is essential to ensure both cost-effectiveness and optimal functionality. In three-phase inverters, the analysis of input voltage and current ripples while accounting for antiparallel diode reverse recovery is covered in [1]. In [2], a technique is presented for the systematic dimensioning of the capacitor for the DC link in an electric vehicle inverter. This

method takes into account the current ripple and the voltage ripple of the capacitor in the DC link. The derived procedure serves as an illustration of the extent to which the various parameters influence the procedure.

A method for determining capacitance in the DC-link is presented in [3], which focuses on the discharge characteristics of the capacitor during the turn-off shift of an inverter. [4]- [7] examine the RMS current of the DC link capacitors under a diverse array of conditions for the capacitor design.

This paper investigates the important factors that influence the selection of the capacitor for the DC-link in a three-phase inverter-controlled PMSM drive and presents methods to estimate the optimal capacitor size based on the voltage ripple, the current ripple, and the lifetime of the capacitor. The analysis aims to provide engineers and designers with practical insights to optimize drive performance. The work is structured in the following manner. In Section II, an evaluation of the maximum ripple voltage is done for a specific PMSM drive to determine the least capacitance necessary for the DC bus to maintain the voltage stability. The calculation of current through the DC-link capacitor bank is studied in Section III. Lifetime calculations for electrolytic capacitors are studied in Section IV. The combination of capacitors according to different frequencies is also explored. Ultimately, an iterative process is implemented to modify the capacitance level of the electrolytic capacitors until the operating life of the capacitor is consistent with its rated longevity.

II. VOLTAGE RIPPLE ANALYSIS

Fig. 1 illustrates the diagram of a three-phase inverter-fed PMSM drive indicating all the input side currents. Voltage ripple introduced in a three-phase inverter used in drives is mainly due to the parasitic inductance and resistance of the

979-8-3315-3013-6/25 $31.00 © 2025 IEEE

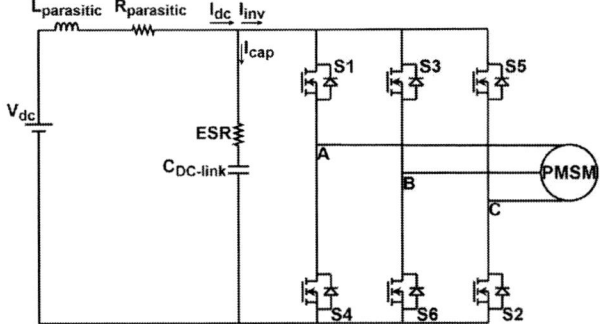

Fig. 1. PMSM drive circuit diagram indicating input side currents.

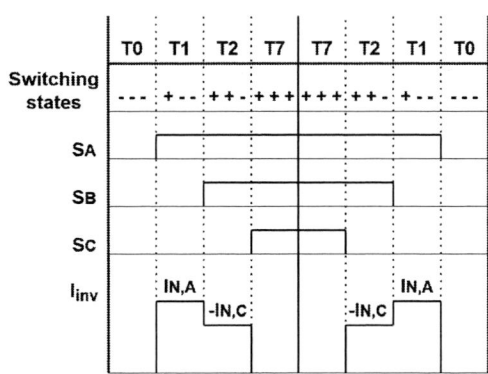

Fig. 2. Switching functions for phases A, B and C phases, and the inverter current over 2 subcycles in sector 1 of SVPWM.

connecting wires between the battery and the inverter, which are shown as $L_{parasitic}$ and $R_{parasitic}$ in Fig. 1. This voltage ripple can be limited by accurately designing the capacitor in DC-link , which is $C_{DC-link}$ in Fig. 1. The current that goes through a capacitor can be represented as $I_{cap} = C\frac{dV_{cap}}{dt}$ and hence C_{min} can be expressed as a maximum of $\frac{I_{cap}dt}{dV}$, from this we can see that $I_{cap}dt = CdV$ or $C\Delta V$ ($dv = \Delta V$), which is called the Amp-sec expression. The expression for finding the value of C_{min} from the expression of Amp-sec is presented in [8] as:

$$C_{min} = \frac{2 * |C * (\Delta V_3 + \Delta V_4)|_{\text{maximum of 3-D plot}}}{\Delta V_{max}} \quad (1)$$

Fig. 2 illustrates the inverter currents over two subcycles in Sector 1 of the space vector pulse width modulation (SVPWM) analysis, as well as the switching functions for phases A, B, and C. The total time period, which is divided into eight parts for the SVPWM is also shown in the figure. ΔV_3 and ΔV_4 are the ripple voltages in the third and fourth switching interval in Fig.2, C_{min} is the lowest capacitance necessary for the system, ΔV_{max} is the maximum allowable ripple in voltage in the system, $C * (\Delta V_3 + \Delta V_4)$ is the Amp-sec expression and is given by:

$$C * (\Delta V_3 + \Delta V_4) = MI_mT_{sw}\left[\frac{1}{2}\sin\omega t * \cos\left(\omega t - \frac{\pi}{3} - \phi\right)\right.$$
$$\left. - \frac{\sqrt{3}}{8}\cos\phi\left(1 - M\cos\left(\omega t - \frac{\pi}{6}\right) + 2M\sin\omega t\right)\right] \quad (2)$$

Here M is the modulation index, which is the ratio of the line-to-line voltage at the peak to the dc bus voltage applied across the capacitor. I_m is the maximum motor phase current, T_{sw} is the switching time period, and ϕ is the power factor angle.

There is no direct solution for (2). The maximum value of (2) can be found by plotting the equation in MATLAB with a range of values for M, ωt and the power factor. Putting the maximum of (2) in (1), gives C_{min} for a specific voltage ripple according to the requirement.

III. CURRENT RIPPLE ANALYSIS

From Fig. 1, the capacitor current expression can be written by looking at the input node as.

$$I_{cap} = I_{dc} - I_{inv} \quad (3)$$

If we separate the dc and ac components in equation (3), the capacitor current can be written as:

$$I_{cap,ave} + I_{cap,ac} = I_{dc,ave} + I_{dc,ac} - I_{inv,ave} - I_{inv,ac} \quad (4)$$

Equation (4) can be distinguished for the average and the ac current as follows:

$$I_{cap,ave} = I_{dc,ave} - I_{inv,ave} \quad (5)$$

$$I_{cap,ac} = I_{dc,ac} - I_{inv,ac} \quad (6)$$

If the capacitance of the DC-link capacitor is suitably high, equation (3) can be expressed as:

$$I_{cap} = I_{dc,ac} - I_{inv,ac} \quad (7)$$

The DC link capacitor current's rms value can be expressed as:

$$I_{cap,rms} = I_{dc,rms} + I_{inv,rms} \quad (8)$$

If the input DC current is considered to have been purely DC, equation (8) can be rewritten as:

$$I_{cap,rms}^2 = I_{inv,rms}^2 - I_{inv,ave}^2 \quad (9)$$

The next step is to find an expression for $I_{inv,rms}$ and $I_{inv,ave}$. The expression for theinstantaneous inverter current I_{inv} can be written as:

$$I_{inv} = S_A I_A + S_B I_B + S_C I_C \quad (10)$$

I_{inv} is the instantaneous inverter input current. I_A, I_B and I_C are the instantaneous phase currents for phases A, B, and C, respectively. All other harmonics in the phase currents are neglected. S_A, S_B, and S_C are the switching functions for phases A, B, and C [4], which are shown in Figure 2.

Considering the output phase currents to be purely sinusoidal, phase current expressions can be written as:

$$I_A = I_m cos(\omega t - \phi)$$
$$I_B = I_m cos(\omega t - \frac{2\pi}{3} - \phi) \qquad (11)$$
$$I_C = I_m cos(\omega t + \frac{2\pi}{3} - \phi)$$

Here, I_m represents the maximum motor phase current, while ϕ represents the displacement angle of the inverter output voltage and the output current. The inverter RMS current is not influenced by factors other than the unbalanced load, and the inverter current has the same shape in the six sectors of the SVPWM. The space vector diagram indicating all sectors and voltage phasors is shown in Fig. 3. Furthermore, the duration of the active vectors is the sole factor that influences the average value of the inverter current. Therefore, all analysis can be performed in Sector 1 of SVPWM. The equation for $I_{inv,ave}$ can be written in terms of switching times in sector 1 as:

$$I_{inv,ave} = \delta(100)I_A - \delta(110)I_C \qquad (12)$$

$\delta(100)$ is the switching time of V1 and $\delta(110)$ is the switching time of V2 in sector 1. For $\delta(100)$, the positive DC bus is connected to phase A, while the negative DC bus is connected to the B and C phases, so I_{inv} and I_A will show similar time behavior. Similarly for $\delta(110)$, I_{inv} and $-I_C$ will show identical time behavior.

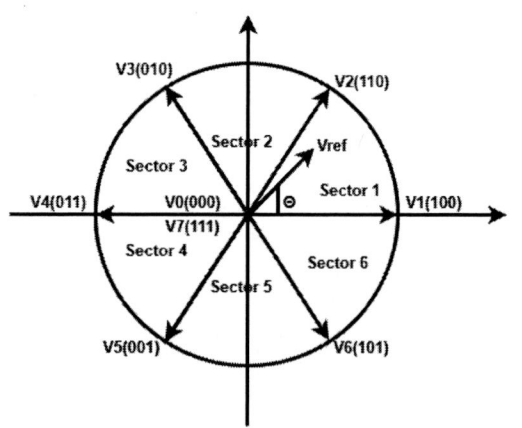

Fig. 3. DC-link current over 2 sub cycles in sector 1 of SVPWM.

The imaginary as well as real components of V1 and V2 can be separated from Fig. 3 and the volt-second balance equation can be applied to determine $\delta(100)$ and $\delta(110)$. The final expressions for $\delta(100)$ and $\delta(110)$ are shown in equation (13).

$$\delta(100) = Mcos(\omega t + \frac{\pi}{6})$$
$$\delta(110) = Msin(\omega t) \qquad (13)$$

Applying equations (13) and (11) in equation (12), will give:

$$I_{inv,ave} = \frac{\sqrt{3}}{2}MI_m cos\Phi \qquad (14)$$

From (12) $I_{inv,rms}$ can be written as:

$$I_{inv,rms}^2 = \delta(100)I_A^2 + \delta(110)I_C^2 \qquad (15)$$

Substituting (13) and (11) in (15) will give,

$$I_{inv,rms} = I_m \sqrt{\frac{M}{2\pi}(1 + 4cos^2\phi)} \qquad (16)$$

Substituting (16) and (14) in (9) will give,

$$I_{cap,rms} = I_m \sqrt{\frac{M}{2\pi}[1 + cos^2\phi * (4 - \frac{3\pi}{2}M)]} \qquad (17)$$

IV. LIFE ESTIMATION OF ELECTROLYTIC CAPACITORS

The capacitor life is an important design consideration if we are selecting the electrolytic capacitor. Electrolytic capacitors have the least life time compared to film and ceramic capacitors. Several models have been developed to estimate the life expectancy of capacitors based on environmental and operational factors. This section delineates the equations and methodology employed to calculate the lifetime of electrolytic capacitors, with a particular emphasis on the impact of temperature, applied voltage, and current.

For the aluminum electrolytic capacitor, the operating life can be expressed as [8], [9]:

$$L_{op} = L_{rated} * K_T * K_R * K_V \qquad (18)$$

L_{op} represents the lifetime under operating conditions, L_{rated} represents the lifetime under rated conditions, K_T represents the temperature factor, K_R represents the ripple current factor, and K_V represents the voltage factor [8].

Temperature factor can be expressed as [8]:

$$K_T = 2^{\frac{T_{rated} - T_{op}}{10}} \qquad (19)$$

T_{rated} and T_{op} are the rated and operating ambient temperature, respectively.

The ripple current factor K_R can be expressed as [8]:

$$K_R = K_f^{[1 - \frac{I_a^2}{I_o}] * \frac{\Delta T_o}{10}} \qquad (20)$$

The actual ripple current at the rated frequency is I_a, the rated ripple current is I_o, and the temperature increase in the capacitor core at the rated condition is ΔT_o. The expression for ΔT_o is as follows:

$$\Delta T_o = T_{core}(rated) - T_{ambient}(rated) \qquad (21)$$

$T_{core}(rated)$ is the core temperature and $T_{ambient}(rated)$ is the ambient temperature under rated conditions.

The voltage factor K_V can be expressed as [8];

$$K_V = (\frac{V_a}{V_r})^{-n} \qquad (22)$$

979-8-3315-3013-6/25 $31.00 © 2025 IEEE

The voltage rated in the capacitor is indicated by V_r, while the applied voltage is denoted by V_a. The life of the capacitor is influenced by the voltage, denoted by n. The variation of n is as follows:

n=1 for $0.1 < \frac{V_a}{V_r} < 0.5$, n=3 for $0.5 < \frac{V_a}{V_r} < 0.8$, n=5 for $0.8 < \frac{V_a}{V_r} < 1.0$,

In the lifetime estimation of DC link capacitors, the impact of dv/dt is typically not considered because it does not directly contribute to the primary aging mechanisms such as thermal stress from ripple current or voltage stress. Instead, dv/dt is treated as a design constraint and as long as it remains within the manufacturer's specified limits, it does not significantly affect the life of the capacitor.

V. THEORETICAL CALCULATIONS

In this section, mathematical calculations of the methods mentioned in the above sections are made.

A. C_{min} Calculation

The maximum value of the expression Amp-sec in equation (2) with motor parameters in TABLE I is calculated using 3-D plotting and the maximum value obtained is 1.297mA-sec, which is shown in Fig. 4. With the unity power factor and the 0.55 modulation index, this optimal value is achieved. The minimal capacitance value, determined by the equation (1), is $2359\mu F$.

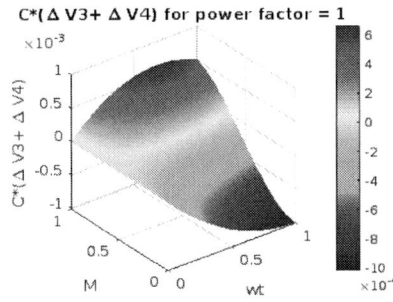

Fig. 4. Plot to find the maximum of the function $C * (\Delta V3 + \Delta V4)$ for unity power factor

B. Used Life Calculation of Electrolytic Capacitor With the Indian Drive Cycle as Load

While the DC link capacitor sizing based on the rated capacity of the inverter provides a baseline, the Indian Drive Cycle (IDC for vehicles in India) with a PMSM motor introduces dynamic load variations, regenerative braking, and high ripple currents, necessitating a larger or more robust capacitor to handle transient energy fluctuations, voltage spikes, and thermal stresses that a steady-state inverter rating alone does not account for. Therefore, the Indian Drive Cycle (IDC) and the motor type (like PMSM) provide the actual dynamic load conditions that determine the real operating stress on the DC link capacitor for vehicles in India. The drive cycles vary between countries to reflect regional-specific driving patterns, traffic conditions, and regulatory requirements. Considering

these variations in the design process will improve the performance of the components of electric vehicles.

The Mean Time Before Failure (MTBF) is not calculated in this analysis because MTBF is typically used for random failure analysis of electronic components, whereas electrolytic capacitor aging is a wear-out mechanism, making life estimation based on degradation models (e.g., lifetime in hours under stress) more appropriate than statistical MTBF.

In this section, the life of the aluminum electrolytic capacitor is calculated with the model. C_{min} is calculated according to Section II with the parameters in TABLE (I) being $2360\mu F$. These parameters are chosen according to the ongoing project specifications in the lab, and the proposed method can be extended to any specification. The part used in the calculation is $390\mu F$ with $38m\Omega$ ESR and the number of capacitors considered is 12, which is finalized by a few iterations. The load considered in the calculation is shown in Fig. 5, which is the indian drive cycle graph.

TABLE I
MOTOR PARAMETERS

Motor parameters	Values
DC-link voltage	48V
Motor rated torque	26Nm
Motor peak power	6.7kW
Motor peak current	300A
Switching frequency	20kHz
El- cap life cycle	3000hrs

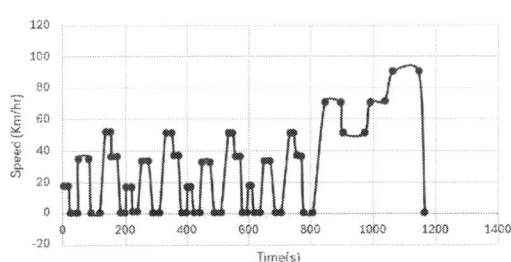

Fig. 5. Speed vs Time profile of drive cycles in India

The load in Fig. 5 is for 1167s; calculation of capacitor life for 8 hours in a day for 8 years will result in 72029 repeating cycles of Fig. 5. The procedure described in Section III is employed to determine the DC-link capacitor current at each point for the load over 72029 cycles. The process described in Section IV is used to determine the life of the capacitor at each point.

Table II displays the current values through the dc link capacitor, various constants, and the lifetime of the capacitor used at the critical load points in Figure 5, in conjunction with the motor parameters in TABLE I. The effect of temperature on the life of electrolytic capacitors can be seen in (19), which generally follows the Arrhenius rule, where every 10°C rise in temperature half the expected life.

All the loading points are not shown in TABLE II. After considering all the points in the load, the total used life of the

TABLE II
USED LIFE OF CAPACITOR

$t(s)$	$I_{cap}(A)$	K_T	K_R	K_V	Used life(h)
3.39	100.74	5.6568	0.248	6.01	8.04
50.83	81.48	5.65	1.00	6.01	4.94
137.2	44.96	5.65	0.77	6.01	7.04
205.03	100.24	5.65	0.25	6.01	3.93
332.13	45.85	5.66	0.73	6.01	8.15
403.29	100.24	5.66	0.25	6.01	3.93
450.74	87.84	5.66	0.08	6.01	37.9
535.47	45.40	5.66	0.72	6.01	10.3
599.86	101.21	5.66	0.06	6.01	3.25
650.69	85.21	5.66	0.07	6.01	60.1
733.73	45.40	5.66	0.72	6.01	11.1
847.26	31.89	5.66	1.28	6.01	8.99

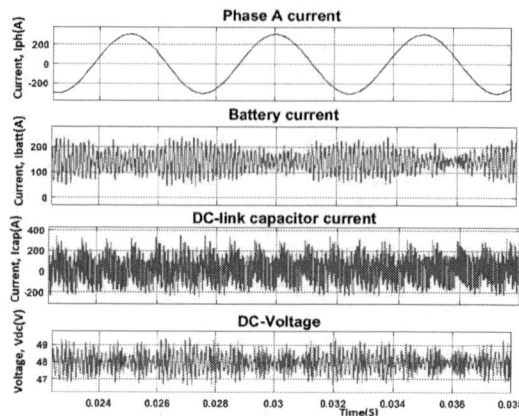

Fig. 7. Phase A current, Battery current, current across DC-link capacitor and DC- voltage waveforms of the motor running at 2400 rpm

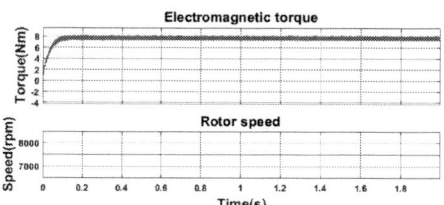

Fig. 8. PMSM Torque and speed waveforms at 7500 rpm speed

capacitor is coming at 2253.09 hours, which is under the life cycle of Electrolytic capacitor. Hence, the chosen capacitor will satisfy the requirement of 8 years of inverter life. For design calculations, the capacitance value should be adjusted until the total lifetime used of the capacitor is less than the rated lifetime of the electrolytic capacitor.

VI. SIMULATION RESULTS

Closed-loop simulation of a PMSM motor with the parameters shown in TABLE I and the calculated capacitance was carried out at different speeds using MATLAB Simulink, and the results are illustrated in the accompanying figures 6, 7, 8, and 9.

A parasitic inductance of $0.129\mu H$ was used in the simulation on the positive and negative sides of the battery along with an equivalent resistance of $52\mu\Omega$, to make the model more realistic. The torque and speed waveforms are shown in Fig. 6, and the current and voltage variations on the input side are illustrated in Fig. 7 for 2400 rpm. For 7500 rpm, the torque and speed waveforms are shown in Fig. 8, and the input-side current and voltage variations are displayed in Fig. 9.

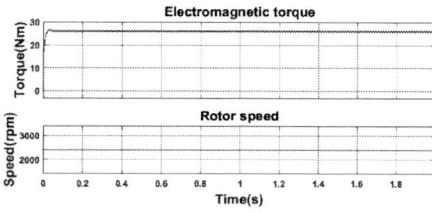

Fig. 6. PMSM Torque and speed waveforms at 2400 rpm speed

2400 rpm is the base speed for this motor specification, and the maximum phase current at 2400 rpm is 212.3A. At this speed, the maximum $I_{cap,rms}$ can be calculated using equation (17), giving a value of 123.8A. The simulation results show that the obtained $I_{cap,rms}$ is 119.2A, which closely matches the calculated value.

The maximum voltage ripple obtained at 2400 rpm is 2.1V.

The maximum voltage ripple obtained at 7500 rpm is 0.5V. The voltage ripple value obtained in the simulation also aligns with the calculations. TABLE III shows the comparison

of the theoretical and simulation analysis, from which we can observe that both the theoretical calculations and the simulation results are closely matched.

A. FFT Analysis of the Simulation and selection of capacitors

A combination of multiple capacitors is required to ensure that the current in the system can be carried at all frequencies by the chosen capacitors. ESR is the limiting factor in current sharing at different frequencies, therefore capacitor selection should be according to ESR of capacitors at different frequen-

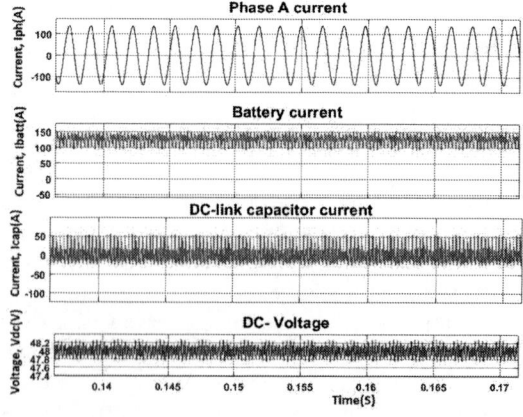

Fig. 9. Phase A current, Battery current, current across DC-link capacitor and DC- voltage waveforms of the motor running at 7500 rpm

TABLE III
COMPARISON OF THEORETICAL AND SIMULATION RESULTS

Parameters	Theoretical analysis	Simulation analysis
No: of capacitors required	12	12
DC-Link voltage(peak to peak)	2V	2.1V
DC-link capacitor current rms(Arms) at 2400 rpm	123.8A	119.2A

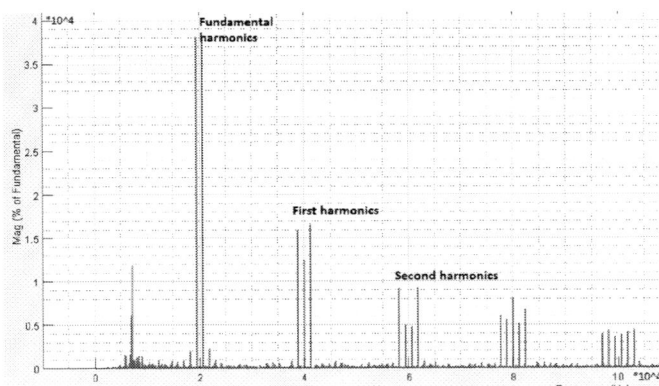

Fig. 10. FFT spectrum of current across DC-Link capacitor from simulation

cies. Ceramic capacitors have the lowest ESR; therefore, they will lock high-frequency current components and should be used near lossy components, such as IGBT in the inverter. Film capacitors have the medium ESR and therefore they will lock the moderate frequency components of currents, and it should be used before the ceramic capacitors when looking from the DC power supply. Film capacitors are used mainly for high-voltage systems, typically more than 200V. Electrolytic capacitors have the highest ESR, and they will lock the low-frequency components of currents and should be used near the voltage supply.

FFT helps to decide whether a film, an electrolytic or ceramic capacitor is more suitable based on the frequency content. Therefore, a current harmonic analysis of the current of the dc link capacitor has been performed in the simulation model and the result is shown in Fig. 10. From FFT, we can find that dominant harmonics occur at even multiples of the switching frequency, and at 3Fsw±3Fs. Here, Fsw is the switching frequency and Fs is the system frequency. The FFT peaks are coming in the kilohertz range because of the PWM switching frequency used in the inverter. The phase current appears to be of low frequency (50 to 200Hz) because the motor operates at that electrical frequency. This is a normal and expected behavior in inverter-fed PMSM drives.

For frequencies below 2Fsw, electrolytic capacitors will be enough to lock the current. To filter frequencies greater than 2Fsw; a combination of ceramics is required. The selection of ceramics should be such that it should have a relatively lower ESR at frequencies higher than 40kHz in this system.

VII. CONCLUSION

The primary factors that influence the selection and measurement of the DC link capacitor in a three-phase inverter-controlled PMSM drive are examined. The detailed analysis of the minimum capacitance calculation, the calculation of RMS current based on the Indian drive cycle, and the iterative process to match the capacitor's operating life with its rated life are also presented for electrolytic capacitors. The selection of capacitors according to different frequency performance is also discussed. The simulation results are presented to verify the analysis. Future work could extend these methodologies to other motor drive systems, further refining capacitor selection techniques for improved energy efficiency and performance.

REFERENCES

[1] J. Guo, J. Ye and A. Emadi, "DC-Link Current and Voltage Ripple Analysis Considering Antiparallel Diode Reverse Recovery in Voltage Source Inverters," in IEEE Transactions on Power Electronics, vol. 33, no. 6, pp. 5171-5180, June 2018.

[2] A. Safayet, M. Islam and T. Sebastian, "Sizing of DC-Link Capacitor Considering Voltage and Current Ripple Requirements for a 3-Phase Voltage Source Inverter," 2020 IEEE Energy Conversion Congress and Exposition (ECCE), Detroit, MI, USA, 2020, pp. 1512-1518.

[3] X. Wei, Y. Peng, B. Yao and H. Wang, "DC-link Capacitance Estimation based on Discharge Profile of Inverter for EV Application," 2023 11th International Conference on Power Electronics and ECCE Asia (ICPE 2023 - ECCE Asia), Jeju Island, Korea, Republic of, 2023, pp. 1561-1566.

[4] J. W. Kolar, T. M. Wolbank and M. Schrodl, "Analytical calculation of the RMS current stress on the DC link capacitor of voltage DC link PWM converter systems," 1999. Ninth International Conference on Electrical Machines and Drives (Conf. Publ. No. 468), Canterbury, UK, 1999, pp. 81-89.

[5] Z. Zhao, F. Diao, Y. Wu, Z. Wang and Y. Zhao, "DC-Link Capacitor Current Modeling and Analysis for Three-Level Voltage Source Inverters," 2021 IEEE Applied Power Electronics Conference and Exposition (APEC), Phoenix, AZ, USA, 2021, pp. 2434-2439.

[6] X. Pei, W. Zhou and Y. Kang, "Analysis and Calculation of DC-Link Current and Voltage Ripples for Three-Phase Inverter With Unbalanced Load," in IEEE Transactions on Power Electronics, vol. 30, no. 10, pp. 5401-5412, Oct. 2015.

[7] B. P. McGrath and D. G. Holmes, "A General Analytical Method for Calculating Inverter DC-Link Current Harmonics," in IEEE Transactions on Industry Applications, vol. 45, no. 5, pp. 1851-1859, Sept.-oct. 2009.

[8] A. Safayet, M. Islam and T. Sebastian, "Comprehensive Analysis for DC-Link Capacitor Sizing for a Three-Phase Current-Controlled Voltage-Source Inverter," in IEEE Transactions on Industry Applications, vol. 58, no. 4, pp. 4248-4260, July-Aug. 2022.

[9] M. L. Gasperi, "Life prediction model for aluminum electrolytic capacitors," IAS '96. Conference Record of the 1996 IEEE Industry Applications Conference Thirty-First IAS Annual Meeting, San Diego, CA, USA, 1996, pp. 1347-1351 vol.3.

[10] P. A. Dahono, Y. Sato and T. Kataoka, "Analysis and minimization of ripple components of input current and voltage of PWM inverters," IEEE Transactions on Industry Applications, vol. 32, no. 4, pp. 945-950, July-Aug. 1996.

[11] L. Yang, T. Wang, S. Li and K. F. Liang, "Analysis of the DC-Link Voltage Ripple for the Three Phase Voltage Ripple Under Low Order Harmonic Output Current," 2020 4th International Conference on HVDC (HVDC), Xi'an, China, 2020, pp. 87-92.

[12] M. Vujacic, M. Hammami, M. Srndovic and G. Grandi, "Evaluation of DC voltage ripple in three-phase PWM voltage source inverters," 2017 IEEE 26th International Symposium on Industrial Electronics (ISIE), Edinburgh, UK, 2017, pp. 711-716.

979-8-3315-3013-6/25 $31.00 © 2025 IEEE

Silicon Switches with Ultra-low Conduction Loss for Low Frequency Switching Applications

Harsha Vardhan Reddy S
Department of ECE (SECS)
Student Member, IEEE
IIT Bhubaneswar
Bhubaneswar, India
21ec01030@iitbbs.ac.in

Wakar Hasan Kazi
Department of ECE (SECS)
Student Member, IEEE
IIT Bhubaneswar
Bhubaneswar, India
a24ec09002@iitbbs.ac.in

Akshay K
Department of ECE (SECS)
Member, IEEE
IIT Bhubaneswar
Bhubaneswar, India
akshay@iitbbs.ac.in

Abstract—Shielded-gate UMOSFET is one of the most popular commercially available silicon switch due to its high frequency switching capability for a given rated voltage. However, in this paper, we establish that an extended-gate UMOSFET is a better alternative for low frequency switching applications, < 20 kHz. Well calibrated TCAD simulations predict that extended-gate UMOSFET can offer up to 34% lesser specific on resistance compared to shielded-gate UMOSFET at 60 V rating and using a graded doped (GD) epitaxial wafer. Additionally, we investigate the role of mesa width and oxide thickness in deciding the breakdown voltage in order to develop design insights. It is found that there exist an optimum oxide thickness below which the device breaks down prematurely due to electric field peaking at trench corner and above which the *superjunction effect* is not strong enough to deplete mesa region laterally resulting in early breakdown at the p-base / n-drift junction. This study should motivate the use of customized extended-gate GD-UMOSFET for low frequency switching applications rather than a generic switch optimized for best switching performance.

Index Terms—Extended-gate graded doped UMOSFET, shielded-gate graded doped UMOSFET, low frequency, breakdown voltage, specific on resistance

I. INTRODUCTION

The advent of next generation power electronic applications demand semiconductor switches with better figure of merits than prior generations. This has led to continuous advancements in discrete silicon power MOSFET technology. Among the various innovations reported, the MOSFET with a U-shaped gate trench (UMOSFET) [1]- [4] has emerged as a promising alternative to conventional planar MOSFETs [5], as the former offers lower conduction losses than latter for the same die size.

Shielded gate UMOSFET is one of the most popularly available commercial variant of UMOSFETs as they offer superior switching performance for a given rated voltage. Due to the widespread availability, it is a common practice to use these devices for all applications irrespective of the switching frequency. However, shielded gate U-MOSFETs are designed for highest switching frequency at the expense of a higher specific on-resistance, (R_{onsp}).

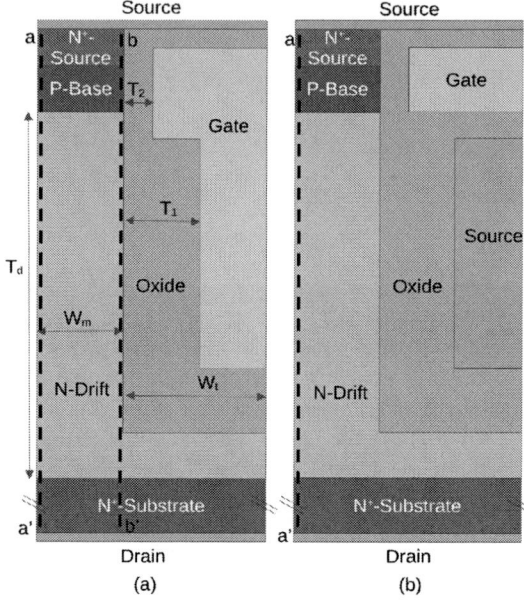

Fig. 1. (a) Cross-section of the half cell of an extended gate GD-UMOSFET, and (b) shielded gate GD-UMOSFET

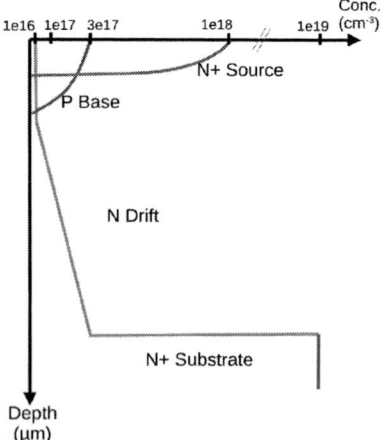

Fig. 2. Doping profile along the cutline aa' of the devices in Fig. 1

979-8-3315-3013-6/25 $31.00 © 2025 IEEE

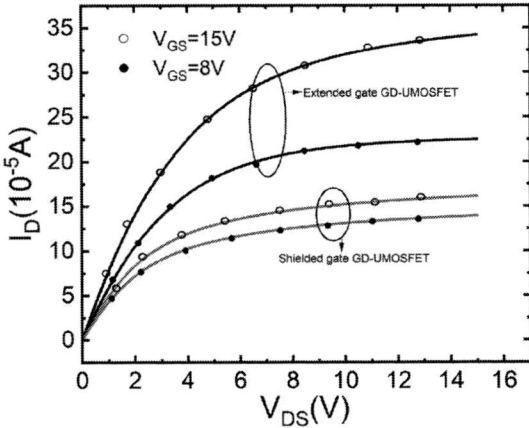

Fig. 3. Simulated I_D versus V_{DS} for V_{GS} = 8V, 15V

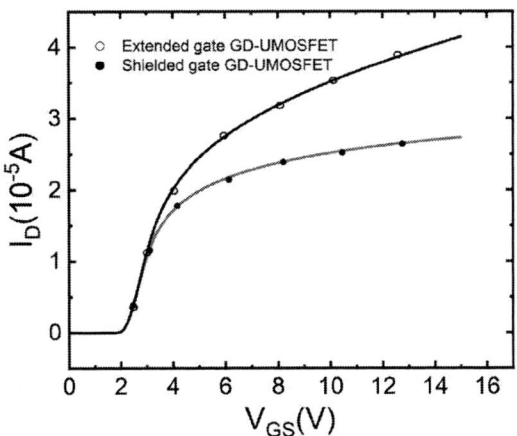

Fig. 4. Simulated I_D versus V_{GS} for V_{DS} = 0.5V

This makes it a sub-optimal choice for applications that demand ultra-low conduction losses and only need switching frequency < 20 kHz. Hence, in this paper we investigate extended-gate UMOSFET [2] as a potential superior alternative to shielded gate UMOSFET for low-frequency applications. Reduction in R_{onsp} is primarily achieved with the help of a lower drift region resistance due to the formation of an electron accumulation below the extended gate [7], [8]. Furthermore, incorporation of a graded doping profile within the drift region, referred to as the Graded Doped UMOSFET (GD-UMOSFET) [7]- [10], enables a more uniform electric field distribution. This mitigates peak electric fields which could otherwise lead to premature breakdown and device failure.

Section II describes the device structure and TCAD simulation setup. The comparative study of the on-state characteristics and R_{onsp} of a 60 V rated shielded gate and extended gate GD-UMOSFETs is discussed in Section III; it also discusses the dependency of breakdown voltage on mesa width and oxide thickness of an extended gate GD-UMOSFET.

II. DEVICE STRUCTURE AND SIMULATION SETUP

The cross-section of the half-cell of an extended trench and shielded trench GD-UMOSFET are given in Fig. 1(a) and Fig. 1(b) respectively. In this study, we choose 60 V rated devices for illustration. The investigated extended trench device features a trench width, W_t, of 0.5 µm with a depth of 4.7 µm. The mesa width, W_m, is 0.25 µm, while the drift layer has a thickness, T_d, of 4 µm. The gate oxide thicknesses, T_1 and T_2, are 0.3 µm and 0.05 µm respectively. The doping concentration of the drift layer at a depth of 0.9 µm is 1×10^{16} cm^{-3} and increases with a gradient of 5×10^{20} cm^{-4} towards the substrate.

Similarly Fig. 1(b) depicts a popularly used device which has similar device features to the investigated one. However, it differs in that it lacks an extended gate within the trench

and instead incorporates a source electrode in its place. Fig. 2 shows the corresponding doping profiles of the simulated structures. Simulations are done to study the breakdown voltage (V_{BR}) by varying the parameters T_1 and W_m.

Sentaurus TCAD software is utilized for simulations. Modified van Overstraeten-de Man avalanche model [11]- [13] is employed for the off-state simulations, using values a_{low} = 1.2×10^6 cm^{-1}, a_{high} = 2×10^6 cm^{-1} and b_{low} = 1.7 MV cm^{-1}, b_{high} = 1.4 MV cm^{-1}. University of Bologna bulk [14] and surface mobility models are used for the on-state simulations.

III. RESULTS AND DISCUSSIONS

From our simulations, the device shown in Fig. 1(a) exhibits an R_{onsp} of 87 µΩ.cm^2, whereas the device in Fig. 1(b) demonstrates a higher R_{onsp} of 132 µΩ.cm^2, both evaluated at a V_{BR} of 60 V. Both devices exhibit a threshold voltage in the range of 2-3 V.

As shown in Fig. 3, the extended gate GD-UMOSFET exhibits a higher drain current, I_D, in the on-state compared to the shielded gate GD-UMOSFET. This enhancement is attributed to the accumulation of electrons at the trench sidewalls in the extended gate structure, which increases its conductivity. In contrase, the absence of such an electron accumulation layer in shielded gate structure leads to a higher R_{onsp} and consequently a lower on-state current. The difference in the I_D–V_{GS} characteristics, depicted in Fig. 4, results from the progressive electron accumulation along the trench sidewalls with increasing V_{GS} in the extended gate structure.

As it can be seen from Fig. 5, the V_{BR} initially increases as we increase T_1 till it reaches a certain value where it reaches the possible peak value for a particular W_m and then it decreases as we go on increasing T_1. The possible peak V_{BR} decreases as we increase the W_m. V_{BR} versus R_{onsp} for W_m = 0.25 µm and 0.5 µm is depicted in Fig. 6.

979-8-3315-3013-6/25 $31.00 © 2025 IEEE

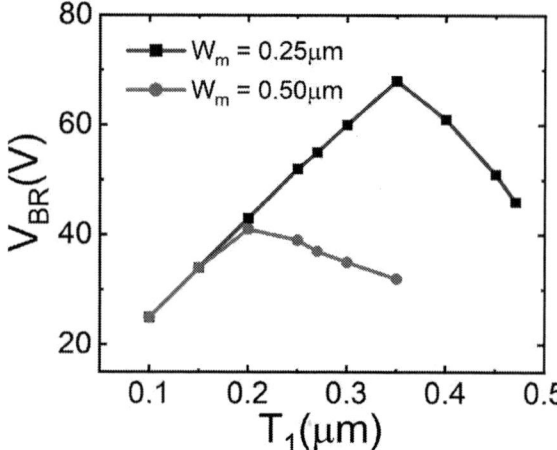

Fig. 5. Simulated V_{BR} versus T_1 for $W_m = 0.25$ μm, 0.5 μm

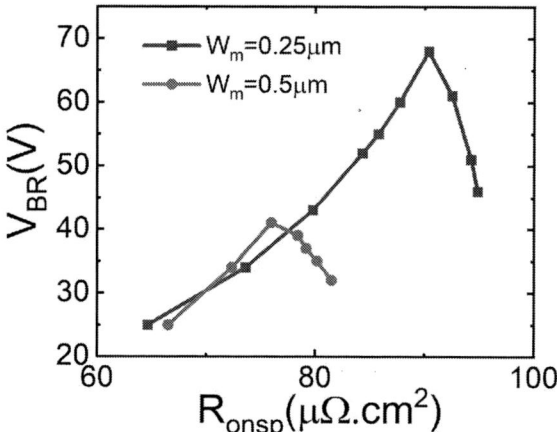

Fig. 6. Simulated V_{BR} versus R_{onsp} for $W_m = 0.25$ μm, 0.5 μm

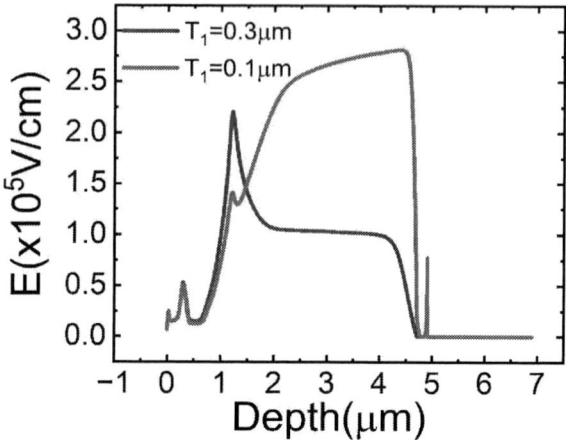

Fig. 7. Simulated electric field distribution along the cutline bb' of the extended gate GD-UMOSFET (see Fig. 1(a)) for $W_m = 0.25$ μm and $T_1 = 0.1$ μm , 0.3 μm

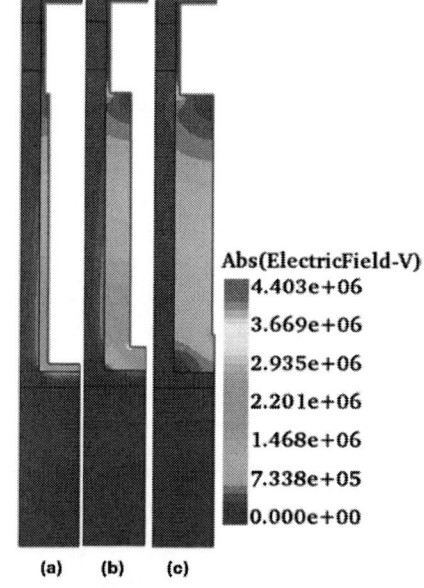

Fig. 8. Simulated electric field distribution for the extended gate GD-UMOSFET for (a) $T_1 = 0.1$ μm, (b) $T_1 = 0.3$ μm and (c) $T_1 = 0.47$ μm

The basis for understanding how the V_{BR} increases as we increase T_1 is for the same reverse bias, V_{DS}, we get a higher electric field in case of lower T_1, which can be observed in Fig. 7.

Building on this we can understand that for $T_1 = 0.1$ μm breakdown happens at a lower V_{DS} than for $T_1 = 0.3$ μm. This is shown in Fig. 9 by taking a cut-line along aa'. We can also observe that the field at the edge of the cell is almost uniform in case of $T_1 = 0.1$ μm and $T_1 = 0.3$ μm and breakdown happens at the sidewall of the trench.

As observed for $T_1 = 0.47$ μm, the electric field distribution at the device edge closely resembles that of a conventional one-dimensional (1D) junction. As shown in Fig. 8, the trench gate is unable to sufficiently deplete the drift region at this oxide thickness, resulting in breakdown behavior similar to that of a conventional 1D junction. This occurs because the electric field induced by the gate is not strong enough to deplete the drift region laterally, unlike for $T_1 = 0.1$ μm and T_1

= 0.3 μm. Consequently, the V_{BR} decreases beyond a certain oxide thickness. It is found that an optimum oxide thickness exists: below this value, premature breakdown occurs due to electric field crowding at the trench corner; above it, the superjunction effect weakens, failing to laterally deplete the mesa region, which results in early breakdown at the top corner along the trench side walls as seen in Fig. 9

Fig. 10 illustrates that as W_m increases, the maximum achievable V_{BR} decreases. This occurs because the electric field induced by the gate is insufficient to fully deplete the drift region. Consequently, the electric field distribution near

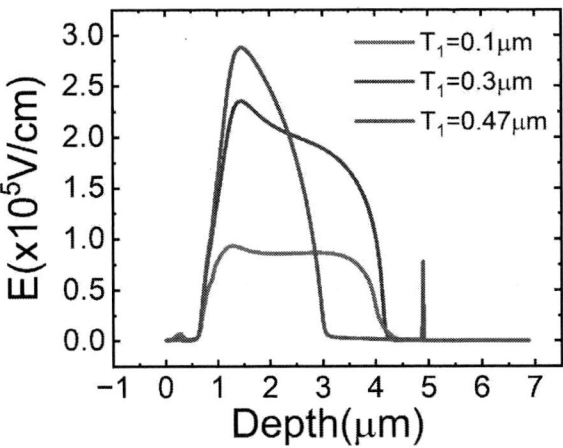

Fig. 9. Simulated electric field distribution along the cutline aa' of the extended gate GD-UMOSFET (see Fig. 1(a)) for $W_m = 0.25$ μm and $T_1 = 0.1$ μm , 0.3 μm and 0.47 μm

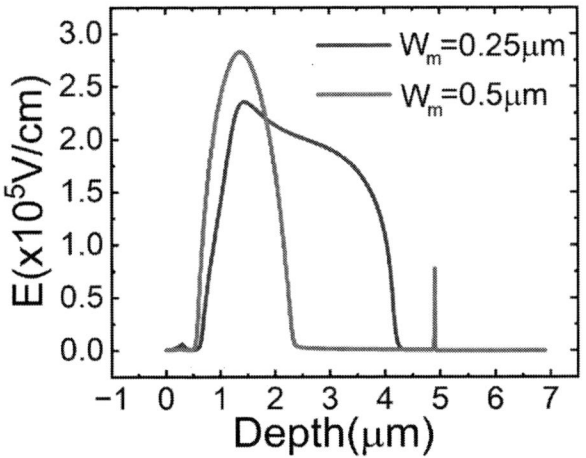

Fig. 10. Simulated electric field distribution along the cutline aa' of the extended gate GD-UMOSFET (see Fig. 1(a)) for $T_1 = 0.3$ μm and $W_m = 0.25$ μm , 0.5 μm

the edge closely resembles that of a conventional 1-D junction, leading to breakdown primarily at the p-n junction rather than being modulated by the gate field.

IV. CONCLUSION

In this work, we have demonstrated that the extended-gate UMOSFET is a promising alternative to the conventional shielded-gate UMOSFET for low-frequency switching applications (< 20 kHz). While the shielded-gate UMOSFET is widely adopted in commercial products due to its superior high-frequency switching performance, our TCAD simulations indicate that the extended-gate structure can achieve up to a 34% reduction in specific on-resistance ($R_{on,sp}$) at a 60 V rating, utilizing a graded-doped epitaxial wafer. Additionally,

we have analyzed the dependence of breakdown voltage (V_{BR}) on mesa width and oxide thickness to provide valuable design insights.

The results highlight the importance of designing application-specific extended-gate UMOSFETs, optimized for low-frequency operation, rather than relying on generic switches tailored for high-speed switching. Such customized devices can substantially reduce conduction losses and improve efficiency in practical applications, including DC–DC converters, battery management systems, motor drives, and industrial power supplies. Overall, this study provides a pathway for developing optimized power MOSFETs that better meet the demands of specific low-frequency power electronic systems.

REFERENCES

[1] Baliga, B. Jayant, "Advanced power MOSFET concepts," Springer Science & Business Media, 2010.

[2] Baliga, Bantval J., "Vertical field effect transistors having improved breakdown voltage capability and low on-state resistance." U.S. Patent 5,637,898, June 10, 1997.

[3] R. K. Williams, M. N. Darwish, R. A. Blanchard, R. Siemieniec, P. Rutter and Y. Kawaguchi, "The Trench Power MOSFET—Part II: Application Specific VDMOS, LDMOS, Packaging, and Reliability," in IEEE Transactions on Electron Devices, vol. 64, no. 3, pp. 692-712, March 2017, doi: 10.1109/TED.2017.2655149.

[4] R. K. Williams, M. N. Darwish, R. A. Blanchard, R. Siemieniec, P. Rutter and Y. Kawaguchi, "The Trench Power MOSFET: Part I—History, Technology, and Prospects," in IEEE Transactions on Electron Devices, vol. 64, no. 3, pp. 674-691, March 2017, doi: 10.1109/TED.2017.2653239.

[5] H. Chang, "Numerical and Experimental Comparison of 60V Vertical Double-Diffused MOSFETs and MOSFETs With A Trench-Gate Structure," Solid-State Electronics, Vol. 32, No. 3, pp. 247-251 (1989)

[6] Zhuo Wang, Peng-Cheng Li, Bo Zhang, Yuan-Hang Fan, Qing Xu, Xiao-Rong Luo, "Ultralow Specific on-Resistance Trench MOSFET with a U-Shaped Extended Gate", Chinese Physics Letters, vol.32, no.6, pp.068501, 2015.

[7] Syau, Tsengyou, et al., IEEE Transactions on Electron Devices, "Comparison of Ultralow Specfic On-Resistance UMOSFET Structures: The ACCUFET, EXTFET, INVFET, and Conventional UMOSFET's," vol. 41, No. 5, May 1994, pp. 800-808.

[8] Baliga, B. Jayant, Tsengyou Syau, and Prasad Venkatraman. "The accumulation-mode field-effect transistor: A new ultralow on-resistance MOSFET." IEEE Electron Device Letters 13.8 (1992): 427-429.

[9] B. J. Baliga, "Vertical MOSFETs having trench based gate electrodes within deeper trench-based source electrodes," US Patent 6621121B2, Sep. 16, 2003

[10] Mahalingam, S. (1999). "Trench MOS based power devices with graded doped profile" (Order No. 9947554). Available from ProQuest Dissertations & Theses Global. (304517065).

[11] K Akshay and Shreepad Karmalkar, "Improved theoretical minimum of the specific onresistance of a superjunction," 2021 Semicond. Sci. Technol. 36 015021

[12] W. Maes, K. De Meyer, R. Van Overstraeten, "Impact ionization in silicon: A review and update," Solid-State Electronics, Volume 33, Issue 6, 1990, Pages 705-718, ISSN 0038-1101

[13] F. D. Bauer, "Compact High-Precision Models for Silicon p-n Step Junction Avalanche-Breakdown Voltages," in IEEE Transactions on Electron Devices, vol. 58, no. 3, pp. 658-663, March 2011, doi: 10.1109/TED.2010.2101077.

[14] S. Reggiani et al., "Electron and hole mobility in silicon at large operating temperatures. I. Bulk mobility," in IEEE Transactions on Electron Devices, vol. 49, no. 3, pp. 490-499, March 2002, doi: 10.1109/16.987121.

Permanent-Magnet Biased Inductor for DC Fault-Current Limiting Applications

Aravind G, Shivendra Pratap Singh, and Amarkumar Kushwaha

Department of Electrical, Electronics and Communication Engineering
Indian Institute of Technology Dharwad, Dharwad, Karnataka, India 580011
aravindgnanavel12@gmail.com, shivamqwerty92@gmail.com, amarkumar@iitdh.ac.in

Abstract—This paper presents a study on a Permanent-Magnet Biased Inductor (PMBI) topology designed for Fault Current Limiting (FCL) applications. It explores the design, implementation, and validation of a Saturation Gap PMBI topology, which enhances system protection by delaying the rise of fault current, thereby allowing sufficient time for circuit breakers to operate effectively. Both series and parallel PMBI configurations are analyzed, evaluating their performance through simulation studies. A setup utilizing UU-type asymmetrical cores with integrated permanent magnets (PM) was developed. Later, this topology was used for DC FCL. Finite Element Analysis (FEA) simulations using FEMM 4.2, along with experimental validation, confirm that the inductance of PMBI initially increases with rising current but decreases beyond a critical threshold due to core saturation. In addition, the hardware integration and verification of PMBI was implemented in a DC FCL application. The results obtained indicate that conventional inductors allow fault currents to rise rapidly, potentially overloading protective mechanisms. In contrast, PMBI-based FCL effectively restricts the rise in fault current by delaying the response time by 200%, which can be used to offer enhanced protection for DC grids, renewable energy systems and industrial power networks.

Index Terms—Circuit breakers, FCL, inductors, PMBI, saturation

I. INTRODUCTION

Power electronic (PE) converters are the main energy conversion system in a wide range of applications such as renewable energy, energy storage, smart and microgrid technologies, dc transmission and distribution systems, electric motor drives, and power supplies [1] [2]. These PE converters such as grid-tied DC to AC converters may experience an abrupt increase in current due to a quick change in load or any short circuit situations. A circuit breaker is employed on the DC input side to safeguard the PE application against inrush currents or abrupt current increases. The DC circuit breaker keeps the switches and other converters from being damaged by isolating the load from the source when the fault occurs. However, the fault current will quickly (in milliseconds) reach a very high value because of the low damping on the DC to AC converter's input side [3]. Therefore, a DC circuit breaker that operates extremely quickly is needed, which would significantly increase the system complexiety and cost.

To mitigate this issue, fault current limiters can be introduced into the system. Fault current limiters is a device which provides no impedance during normal operation of the system whereas it provides high impedance during fault conditions and limit the high fault currents. One effective approach is incorporating an inductor at the DC input side, which naturally opposes sudden changes in current. Under normal operating conditions, the inductor behaves as a short circuit since it does not experience a rapid current variation. However, during a fault event, the inductor limits the rate of current rise by introducing a controlled slope, dictated by its inductance value [4]. This characteristic provides an additional time delay for the DC circuit breaker to operate effectively, thereby enhancing system protection and reducing the requirement for high-speed, costly breakers.

The conventional inductors which are used in DC dominated applications operate effectively within the positive current range whereas the linear inductance available in the negative current range remains unutilized [5]. As a result the magnetic core only uses the positive side of its BH (magnetic flux density versus magnetic field intensity) curve. This leads to under utilization of the magnetic core preventing the inductor from achieving its maximum potential flux swing. And also to achieve a particular inductance with high saturation current a core with high volume and area of cross section is required [6]. Therefore, the weight, volume, and cost of magnetics pose a problem to design any DC-DC converters. One solution is to provide a DC offset or linear bias by introducing a Permanent Magnet (PM) inside (or outside) the winding flux path so that the flux of the PM opposes the winding flux thereby enhancing the saturation current limit [6]. This leads to a class of inductors called PM Biased Inductors (PMBI).

The fault current limiters have been previously implemented using PMBI in the available literature. However, to the best of our knowledge the fault current limiters with PMBI had been implemented only with the PMs placed in series with the winding flux within the core [7], [8], [9]. This paper investigates a PM biased inductor topology, where the PMs are aligned parallel to the winding flux, making it suitable for fault current limiting applications in DC systems. Section II explores and evaluates various PMBI topologies o identify the most suitable configuration for FCL implementation. Section III details the design and simulation analysis of the selected topology. Section IV presents the experimental validation of inductance variation using a hardware prototype, while Section V discusses a case study demonstrating the FCL application through a practical hardware circuit.

979-8-3315-3013-6/25 $31.00 © 2025 IEEE

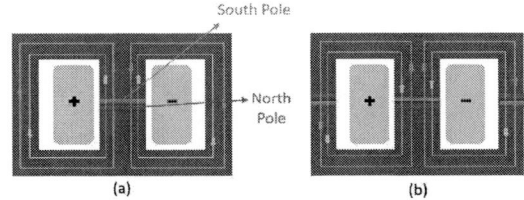

Fig. 1. The PM biased topologies with series configuration.

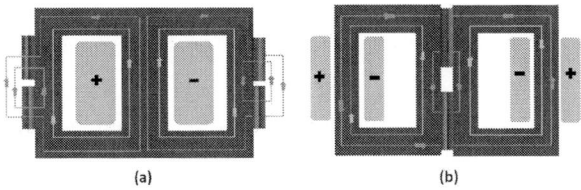

Fig. 2. The PM biased topologies with parallel configuration

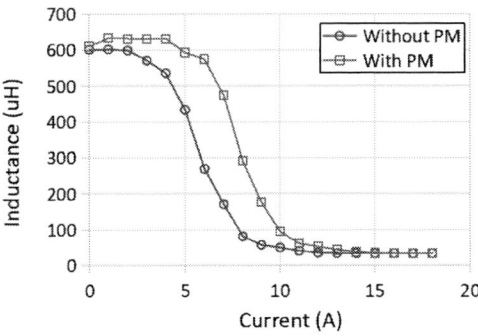

Fig. 3. Inductance vs Current plot for the model shown in Fig. 1(a)

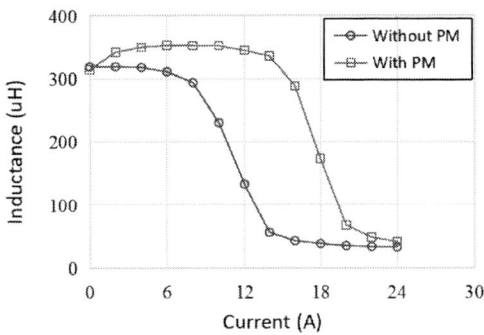

Fig. 4. Inductance vs Current plot for the model shown in Fig. 1(b)

II. SCREENING OF PMBI CONFIGURATIONS FOR FCL APPLICATION

Permanent Magnet Biased Inductors utilizes the positive current range of the BH curve by providing a suitable DC offset by introducing a PM so that the saturation current limit is also extended by some extent. There are several PM biased inductor topologies. FCL application requires an inductor that operates with minimal inductance under normal operating conditions. However during fault conditions the inductor should exhibit a high inductance to oppose the sudden surge in current. This increase in inductance limits the rate of rise of fault current so that the time taken for the inductor's current to reach the peak value is delayed. That means the PMBI which is going to be studied should exhibit a high value of inductance between a certain current range so that the FCL application requirements are obtained. Based on the positioning of the PM's the topologies are classified into series and parallel configurations [6]. Fig. 1 and Fig. 2 shows some of the topologies.

A. Series Configuration

In series configuration the PM is placed in the air gap of the core. In this configuration the flux of the PM is aligned in series with the winding flux. The PM should be positioned in such a way so that it opposes the winding flux. Since the some amount of winding flux is being cancelled by the PM flux, to establish the required amount of flux the coil requires some more current as a result of this the saturation current limit is extended.

The earliest documented topology utilizes standard cores with permanent magnets (PMs) positioned within the air gaps is shown in. Fig. 1(a). The FEM simulation results for the series topologies, shown in Fig. 3 and Fig. 4, indicate a 50% increase in inductance for the topology shown in Fig. 1(a) and a 100% increase for the topology shown in Fig. 1(b), influenced by the placement and number of PMs used. While

this series topology effectively extends the saturation current limit, it does not exhibit the desired characteristics of a PMBI for FCL applications. These inductors do not provide low inductance during normal operation and high inductance after reaching a certain current level. As a result, this topology may not be the best choice for FCL applications.

B. Parallel Configuration

In all the series configuration shown (Fig. 1(a) and (b)) the PM are exposed to demagnetization under strong de-magnetization fields. As a result these topologies can't be employed in fault current limiting applications. In order to avoid demagnetization of magnets the PM should be placed outside of the winding flux path which is possible in parallel configuration as shown in Fig. 2(a) and (b). The topology shown in Fig. 2(a) has the PM placed outside of the air gap parallel to the core. The flux from the pair of PM's which is placed outside the core saturates some part of the smaller section near to them and the other PM flux passes through the longer section of the core opposing the winding flux [10]. Unlike the topology shown in Fig. 2(b) this topology doesn't uses two windings, only one winding is wounded in the central limb such that it can oppose the PM flux. Figs. 5 and 6 show the Inductance vs Current characteristics of both the parallel topologies obtained from the FEA simulation. Fig.5 shows that

979-8-3315-3013-6/25 $31.00 © 2025 IEEE   377

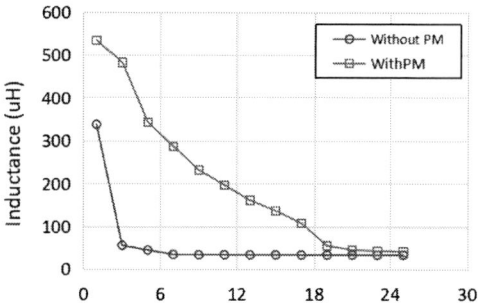

Fig. 5. Inductance vs Current plot for the model shown in Fig. 2(a)

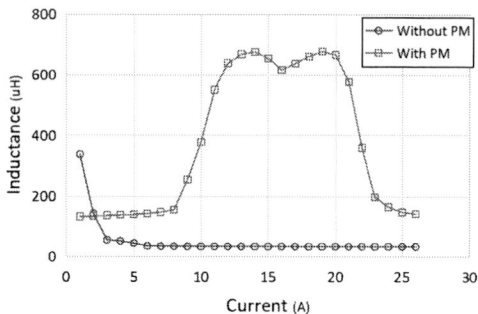

Fig. 6. Inductance vs Current plot for the model shown in Fig. 2(b)

the topology shown in Fig. 2(a) has a increased saturation current limit compared to the inductor without PM, it doesn't exhibits the characteristics needed for FCL application as its inductance is getting reduced for increasing currents.

Fig.6 shows the inductance vs current characterisitcs of the PMBI in Fig. 2(b). It shows that the PMBI exhibits very low inductance for initial current and high inductance for certain current range which matches with the desired characteristics for FCL application. As a result the topology shown in Fig. 2(b) is chosen for further study and analysis in this paper.

III. STUDY AND DESIGN OF SATURATION GAP PM BIASED INDUCTOR

This topology utilizes standard UU cores joined together, with permanent magnets (PMs) placed at the center between the two U-cores as shown in Fig. 7. Since the PMs are positioned externally, demagnetization effects are inherently avoided in this configuration [11], [12]. At the initial current condition, the flux generated by the PMs distributes uniformly across all the core limbs, driving them into saturation. As the current increases, the flux produced by the winding also increases, opposing the PM flux. As a result, the outer limbs comes out of saturation, whereas the limbs closer to the PM remain in a permanently saturated state, as the flux from both the PMs and the winding align in the same direction. This results in the formation of a saturation gap or virtual air gap, effectively eliminating the need for a physical air gap in the core structure.

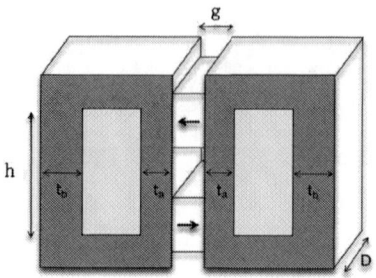

Fig. 7. Schematic diagram of the core with labelled design parameters

Beyond a certain current threshold, the winding flux further increases, eventually driving the entire core into saturation. This topology demonstrates a 100% increase in saturation current, enabling the inductor to operate effectively within a defined current range, thereby making it suitable for FCL application.

A. Design Considerations

When designing the saturation gap topology, the following aspects should be considered:

1) The length of the permanent magnets should be selected such that the magnetomotive force (MMF) they generate is sufficient to drive the entire core into saturation at zero current.
2) The number of turns in the winding should be chosen to ensure that the inductor exhibits a rise in inductance at the desired current range.

$$\mathcal{R} = \frac{2h}{\mu_0 \mu_{r,\text{core}} A_c} + \frac{2l_m}{\mu_0 \mu_{r,\text{PM}} A_m} \tag{1}$$

The equation (1) gives the reluctance of the limbs and PM closer to the PMs. Here, h is the window-height, l_m is PM height, A_c is the cross-sectional area of the limb, A_m is the cross-sectional area of the PM, and μ_r, core denotes relative permeability of the core-material, and μ_r, PM denotes relative permeability of the PM-material. The geometric-variables are also illustrated in the Fig. 7.

$$\phi_{\text{core}} = B_{\text{core}} \cdot \text{t}_\text{a} \cdot \text{D} \tag{2}$$

$$\text{MMF}_{\text{core}} = \phi_{\text{core}} \cdot \mathcal{R}_{\text{core}} \tag{3}$$

Using equations(2) and (3) the flux and the mmf required to drive the core into saturation can be calculated.

$$\text{MMF}_{\text{core}} = H \cdot l_m \tag{4}$$

The MMF provided by the PM's can be calculated by equation (4) in which H denotes the magnetic field intensity of the PM and l_m denotes the length of the PM in the magnetization direction. The MMF of the magnet should be greater than the flux required to saturate the core (ϕ_{core}).

979-8-3315-3013-6/25 $31.00 © 2025 IEEE   378

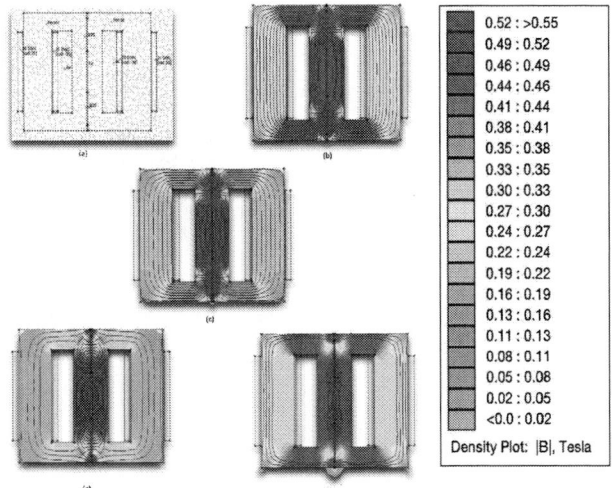

Fig. 8. FEA flux density plots simulated in FEMM4.2 for different currents (a) FEA model after meshing (b) I=0A (c) I=5A (d) I=10A (e) I=15A

B. Simulation of Saturation Gap topology

The Saturation gap topology was simulated in FEMM4.2 an open source software and results were obtained. The FEA plots for different currents shown in Fig. 8 shows that the central limb of the topology is saturated for all the currents because of the PM. Fig. 8(b) shows that the inductor is in saturated stage offering very low inductance. Fig. 8(c) and (d) indicates that the inductor came out of saturation because of the dominance of winding flux over the PM flux resulting in obtaining a higher inductance value. As the current is being increased at a certain point the winding flux itself will saturated the entire core which is shown in Fig. 8 (e).

The inductance vs current characteristics of the same topology by varying the length of the PM's is shown in Fig. 9. It clearly indicates that if the PM length increases the inductance gets decreased because of the presence of strong PM flux which reduces the permeability of the core whereas the operating current range is increased because more current is required to drive the core into saturation.

IV. EXPERIMENTAL VERIFICATION OF INDUCTANCE CHANGE

The simulation results were validated through an experimental setup. Two UU-type cores were selected based on the specifications listed in Table I. A pair of magnets in which one is attached to the top and the other one attached to the bottom, with appropriate polarity. The hardware implementation of the saturation gap inductor topology is shown in Fig. 10(b), while Fig. 10(a) presents the inductance verification at zero current, measured as 165µH, which closely aligns with the simulation results.

A single pulse test circuit was developed to find the inductance of the inductor whose schematic diagram is shown in Fig. 10(c). By varying the pulse width, the current through

TABLE I
CORE AND MAGNET SPECIFICATIONS FOR THE SATURATED GAP INDUCTOR

Parameter	Value	Unit
Core Height	42.4	mm
Core Width	42	mm
Window Height	14.8	mm
Window Width	8.65	mm
Smaller limb width (t_a)	6.25	mm
Larger Limb width (t_b)	12.2	mm
Core Depth (D)	20	mm
Number of Turns	30	-
PM Material used	N30	-
Length of Magnets (l_m)	1.2	mm
Height of magnet	12	mm
Width of magnet	20	mm
Magnetic field intensity (H)	400	kA/m
MMF required to saturate the smaller limb	773.76	AT
MMF available from the magnets	960	AT

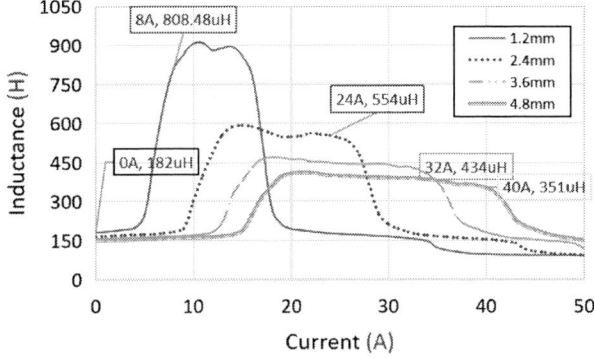

Fig. 9. Inductance vs. Current plot for different lengths of magnets corresponding to the simulation model shown in Fig. 8.

the inductor was controlled. The Saturation Gap PMBI is verified for different currents by varying the pulse width to verify the inductance change. The experimental waveforms obtained when the switch was operated at 1 milliseconds are shown in Fig. 11. At this operating point, approximately 8A of current flows through the inductor. The calculated inductance value was found to be 860µH which closely matches with the simulation results. However when the current exceeds 15A, the inductance starts to decrease. This behavior occurs because the winding flux starts opposing the PM flux, leading to core saturation. This can be observed from the simulation graphs which is shown in Fig. 9

V. A CASE-STUDY FOR FCL APPLICATION

The saturation gap topology discussed in the previous section exhibits an increase in inductance with rising current within a specific range, making it a strong candidate for FCL applications.

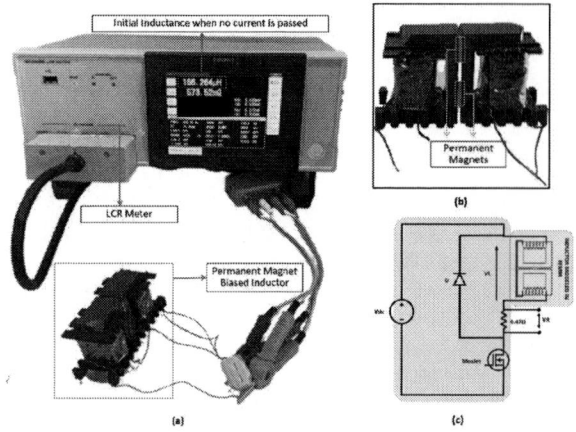

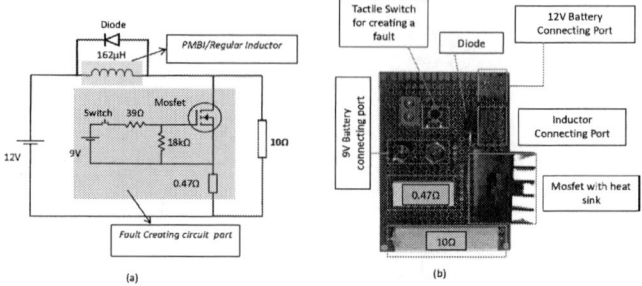

Fig. 12. (a) Schematic diagram of the circuit to verify the inductor for FCL application (b) Photograph of the setup used for emulating controlled DC-fault

Fig. 10. (a) Verification of Inductance with LCR meter (b) Hardware implementation of PM biased Saturation Gap Topology (c) Schematic diagram of the single pulse test circuit used to verify the change in inductance

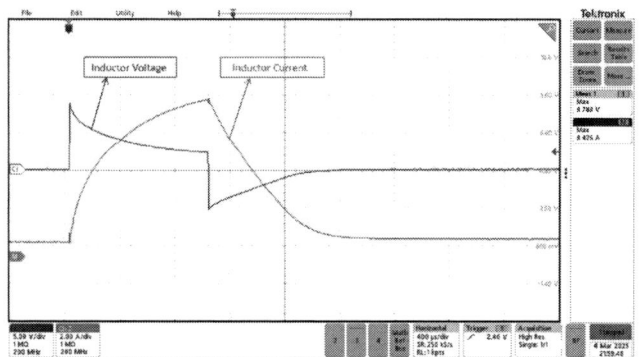

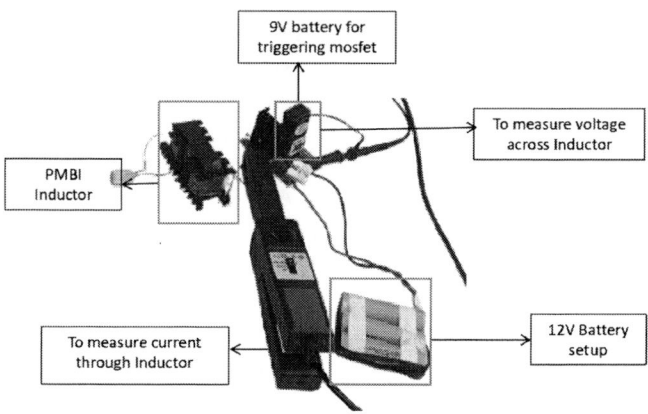

Fig. 13. Experimental setup to test the PMBI for FCL application

Fig. 11. Experimental Waveforms obtained from the single pulse test circuit whose schematic is shown in Fig. 10(c)

The schematic of the circuit developed to verify the FCL capability of the PMBI is shown in Fig. 12(a) and the hardware model in perforated circuit board in shown in Fig. 12(b). The experiment is repeated for both regular inductor and PMBI . The whole experimental setup with the batteries, current probe and voltage probe connected is shown in Fig. 13. Under the normal operating conditions the 10 ohm resistor acts as a load, so the current flows through the inductor is 1.2A. To operate and verify the PMBI in fault conditions the mosfet is triggered by the 9V battery with the help of tactile switch. As a result the effective resistance becomes smaller than 0.47 ohm which results in a very high current flowing through the inductor. By allowing the high current to flow through the inductor just by the switching action of mosfet the fault conditions are being emulated in this circuit. So both the regular inductor and the PMBI are tested in this circuit for both normal and fault operating conditions and the results are compared in Fig. 14. Fig. 14(a) shows the waveforms when regular inductor is tested and Fig. 14(b) shows the waveforms when PMBI is connected. In normal operating conditions the steady state current for both the inductors are same. But when the fault

is created by pressing the tactile switch the time at which the rated current is reached is more for PMBI when compared with the standard inductor. Table II clearly shows the rise time of the current is delayed by three times in case of PMBI when compared to the regular inductor.

The delayed response is because of the property of increasing inductance along with the increasing current of PMBI. Since the inductance is indirectly proportional to the rate of rise of current the increased inductance leads to the slow rate of rise of current. This delayed rise of current provides some extra time for the circuit breaker to take necessary action against the fault.

TABLE II
COMPARISON BETWEEN STANDARD INDUCTOR AND PMBI

Parameter	Standard Inductor	PMBI
Type of Core	Standard EE Core	Two UU Core
Inductance	$163\,\mu$H	$166\,\mu$H (only at initial current)
Time to reach $I_{\text{critical}} = 12\,$A	$285\,\mu$s	$925\,\mu$s

979-8-3315-3013-6/25 $31.00 © 2025 IEEE

380

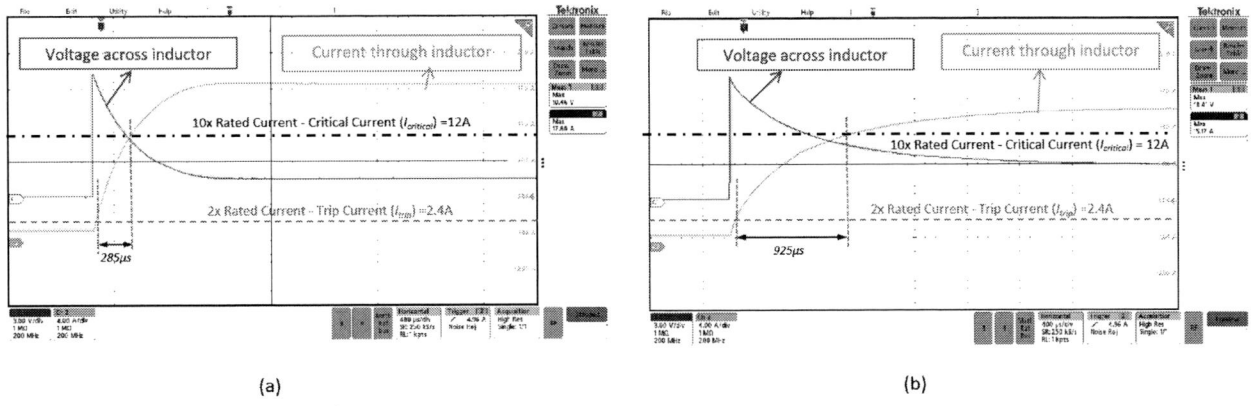

Fig. 14. Waveforms obtained from the experimental setup shown in Fig.13 for (a) conventional inductor and (b) PMBI

VI. CONCLUSION

In this paper it is shown that the PMBI topology gives over a 200% increase in the response time to the circuit breaker in comparison with a regular inductor of fixed inductance. This is by assuming that the trip current of the relay is 2 times of rated continuous current and the damage causing current is 10 times the rated current. The results highlighted the fault-limiting performance of the PMBI compared to conventional inductors, as it enabled a more controlled and gradual increase in fault current. This feature makes PMBI a promising candidate for practical applications in DC grid applications, renewable energy systems, and industrial power networks where effective fault management is essential.

REFERENCES

[1] J. Falck, C. Felgemacher, A. Rojko, M. Liserre, and P. Zacharias, "Reliability of power electronic systems: An industry perspective," *IEEE Industrial Electronics Magazine*, vol. 12, no. 2, pp. 24–35, 2018.

[2] S. Rahimpour, O. Husev, D. Vinnikov, N. V. Kurdkandi, and H. Tarzamni, "Fault management techniques to enhance the reliability of power electronic converters: An overview," *IEEE Access*, vol. 11, pp. 13 432–13 446, 2023.

[3] H. Xiao, Z. Xu, L. Xiao, C. Gan, F. Xu, and L. Dai, "Components sharing based integrated hvdc circuit breaker for meshed hvdc grids," *IEEE Transactions on Power Delivery*, vol. 35, no. 4, pp. 1856–1866, 2020.

[4] R. Li and L. Xu, "Review of dc fault protection for hvdc grids," *Wiley Interdisciplinary Reviews: Energy and Environment*, vol. 7, no. 2, p. e278, 2018.

[5] R. S. Yang, A. B. Nadler, C. R. Sullivan, and D. J. Perreault, "Permanent magnet hybrid core inductors for high saturation capability," *IEEE Open Journal of Power Electronics*, vol. 4, pp. 603–614, 2023.

[6] A. R. Aguilar, S. Munk-Nielsen, F. B. Bendixen, Z. Ouyang, M. Duffy, and H. Zhao, "Permanent magnet biased inductors–an overview," *IEEE Open Journal of Power Electronics*, vol. 5, pp. 1309–1327, 2024.

[7] J. Yuan, P. Gan, Z. Zhang, H. Zhou, L. Wei, and K. Muramatsu, "Saturated-core fault current limiters for ac power systems: Towards reliable, economical and better performance application," *High Voltage*, vol. 5, pp. 416–424, 2020.

[8] J. Yuan, Y. Zhong, Y. Lei, C. Tian, W. Guan, Y. Gao, K. Muramatsu, and B. Chen, "A novel hybrid saturated core fault current limiter topology considering permanent magnet stability and performance," *IEEE Transactions on Magnetics*, vol. 53, no. 6, pp. 1–4, 2017.

[9] J. Yuan, Y. Lei, C. Tian, B. Chen, Z. Yu, J. Yuan, J. Zhou, and K. Yang, "Performance investigation of a novel permanent magnet-biased fault-current limiter," *IEEE Transactions on Magnetics*, vol. 51, no. 11, pp. 1–4, 2015.

[10] A. R. Aguilar and S. Munk-Nielsen, "Method for introducing bias magnetization in ungaped cores: "the saturation-gap"," in *2014 IEEE Applied Power Electronics Conference and Exposition - APEC 2014*, 2014, pp. 721–725.

[11] A. Revilla Aguilar and S. Munk-Nielsen, "Half size reduction of dc output filter inductors with the saturation-gap magnetic bias topology," *IEEE Journal of Emerging and Selected Topics in Power Electronics*, vol. 4, no. 2, pp. 382–392, 2016.

[12] A. R. Aguilar and S. Munk-Nielsen, "Design, analysis and simulation of magnetic biased inductors with saturation-gap," in *2014 16th European Conference on Power Electronics and Applications*, 2014, pp. 1–11.

Impact of Voids on Ampacity in High-Voltage Power Cables : A Multiphysical Approach

Souvik Das
Department of Electrical Engineering
Indian Institute of Technology Dharwad
Dharwad, India
211022006@iitdh.ac.in

Tishya Chattopadhyay
School of Mechanical Sciences
Indian Institute of Technology Bhubaneshwar
Bhubaneshwar, India
tishya.chattopadhyay@gmail.com

Abstract—High-voltage power cables constitute a critical component of the electrical power system, encompassing generation, transmission, distribution, and extending to end-use applications such as electric transportation systems. The power density of such systems heavily depends on the mass of cable arrangement. An efficient cable sizing requires a thorough pre-evaluation of its current carrying capacity or ampacity. The incorporation of voids within insulation layers of cables as a result of manufacturing process have an impact on the cable ampacity. This paper investigates the impact of presence of voids on ampacity in single phase HVAC power cables. A coupled multiphysical finite element model consisting of frequency domain electromagnetic and steady state thermal analysis is presented in this paper. COMSOL Multiphysics software is used for the 2D simulation of cable configuration comprising of 330 KV 50 Hz HVAC cross linked polyethylene (XLPE) cable. The methodology includes the development of a coupled FEA model in association with parametric sweep analysis for the evaluation of the ampacity. The study identifies that the presence of voids creates a non-homogeneity in the temperature distribution and heat flux density distribution within the cable configuration thereby resulting increase in temperature profile by 5 % and 2.5 % reduction in the ampacity of the aforementioned cable arrangement.

Index Terms—Power Cables, Ampacity, Finite Element Method, Multiphysics, Void, High Voltage.

I. INTRODUCTION

The performance and reliability of an electrical power network are significantly influenced by the condition and current-carrying capacity of its power cables. As these cables constitute a fundamental component of the power transmission and distribution infrastructure, ensuring their operational integrity and optimizing their ampacity are critical considerations when aiming to enhance the overall efficiency and robustness of the network. Moreover, conventional thermal propulsion-based transportation systems are recognized as significant contributors to greenhouse gas emissions, making them one of the primary sources of CO_2 emissions. As viable alternatives, electric vehicles and electrically propelled aircraft have been increasingly adopted on a global scale [1]. Enhancing the power density of such electrically driven transportation systems remains a key area of focus to improve overall efficiency. The optimization of cable dimensions and

ratings to accommodate higher currents is essential in reducing system weight, thereby enhancing power density.

The current-carrying capacity, or ampacity, of power cables plays a crucial role in determining the minimum rated cable size, thereby contributing to the overall power density of electric transportation systems and their associated power electronics. Various factors influence the ampacity of power cables, including voids in the insulation and conductor layers, protrusions, dust deposition, cracks, and other imperfections. These voids typically develop during manufacturing and installation processes. The presence of voids in both the insulation and conductors leads to distortions in the electric field distribution, thereby affecting the uniformity of equipotential surfaces. Such distortions may result in temperature rise, the formation of hotspots, and partial discharge in and around the voids, ultimately compromising the insulation layer of high-voltage cables and increasing the likelihood of failure.

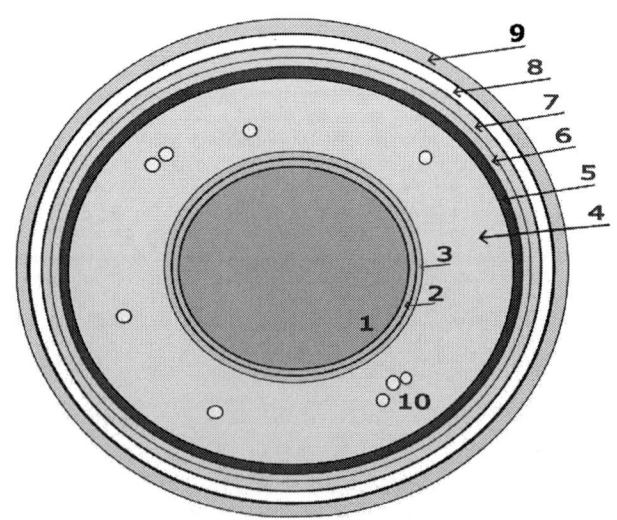

Fig. 1. 2D cross-sectional view of the cable configuration

Various methods for evaluating power cable ampacity have been traditionally employed by designers. The Black Books, titled AIEE-IPCEA Power Cable Ampacities, provide lookup

979-8-3315-3013-6/25 $31.00 © 2025 IEEE

TABLE I
CABLE LAYERS

Cable Layers	Label
Conductor	1
Semiconductor	2
Conductor Screen	3
XLPE Layer	4
Insulation Screen	5
Semiconducting cushion	6
Aluminium Sheath	7
Jacket	8
HDPE Layer	9
Voids	10

tables for predicting cable sizes based on the required current capacity [2]. These tables are derived using the Neher-McGrath method, which was formulated under specific assumptions, including an ambient temperature of 20°C, a maximum conductor temperature of 90°C, and a conductor spacing of 7.5 inches. Deviations from these conditions have been observed to reduce ampacity by approximately 5–8 % or increase it by about 10 % [3]. The IEEE 835 cable ampacity table also utilizes the Neher-McGrath method but incorporates modifications in its assumptions. However, the reliance on predefined assumptions introduces inaccuracies in cable sizing, thereby creating uncertainties in the overall design of electric transportation systems [4]. Additionally, while several studies have analyzed the effects of voids on electric field distribution and temperature profiles [5], [6], [7], their direct impact on cable ampacity remains largely unexplored. A significant research gap exists in this domain, as an accurate modeling strategy that accounts for actual cable configurations, real-time material properties, and ambient conditions could provide system designers with more reliable data. Such advancements would contribute to optimizing power density in electric transportation systems. In this paper, a multiphysical coupled approach utilizing the Finite Element Method (FEM) is presented for evaluating the ampacity of high-voltage power cables. Additionally, the impact of voids in the insulation layer on ampacity is analyzed.

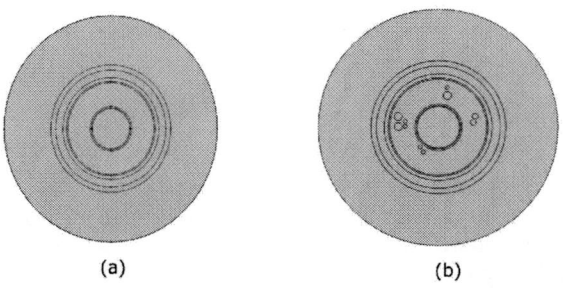

(a)   (b)

Fig. 2. 2D Geometry of Cable - (a) Without void, (b) With Void

II. THEORETICAL BACKGROUND

The evaluation of ampacity of power cables involves a bidirectional interaction between electrical and heat transfer physics associated with the overall configuration of the cable system. A finite element method (FEM) based multiphysical coupled analysis has been followed to evaluate the impact of voids in insulation layer within the high voltage cable associated with electric transportation system.

Maxwell's equations are used to evaluate the electric field distribution, field intensity, and current density in cables, both with and without voids, as follows.

$$\nabla X E = \frac{-\partial B}{\partial t}$$

$$\nabla X H = \frac{-\partial D}{\partial t} + J$$

$$\nabla . B = 0$$

$$\nabla . D = \rho \tag{1}$$

where, E denotes the electric field, B denotes the associated magnetic field, H is the magnetic field intensity, J is the current density, D denotes the electric flux density and ρ is the charge density. The cause and effect interrelationship in both electric and magnetic field has been taken care of as follows.

$$J = \sigma E$$

$$B = \mu H$$

$$D = \epsilon E \tag{2}$$

where, σ is the electrical conductivity, μ denotes the magnetic permeability and ϵ is the electrical permittivity. Moreover, the evaluation of electric field for input electric potential follows the following relation.

$$E = -\nabla V \tag{3}$$

The power loss within the cable arrangement follows the Joule heating effect as follows.

$$P = (I^2 R) \tag{4}$$

The bidirectional coupling effect of the electromagnetic parameters has been established as follows.

$$\rho C_p \frac{\partial T}{\partial t} = \nabla . (k \nabla T) + Q \tag{5}$$

where, α_{ref}, α_1 and α_2 are the temperature coefficients. The rise in temperature of the cable configuration following the power loss is considered as follows.

$$Q = -k \nabla T \tag{6}$$

where, ρ is the density (kg/m^3), C_p is the specific heat capacity (J/kgK), k is the thermal conductivity (W/mK), σ in S/m and Q is the heat source (W/m^3). The convective heat flux is further mentioned as, where h is the heat transfer coefficient $(W/m^2 K)$.

979-8-3315-3013-6/25 $31.00 © 2025 IEEE   383

III. Proposed Strategy

A multiphysical approach based on the finite element method (FEM) has been adopted to evaluate the ampacity of the cable configuration and analyze the impact of voids within the insulation and conductor layers. COMSOL Multiphysics was utilized to conduct a detailed simulation study. The methodology involved developing a 2D geometry of the cable, assigning materials, configuring the physics, establishing bidirectional coupling between electromagnetic and thermal physics, and setting up the simulation study.

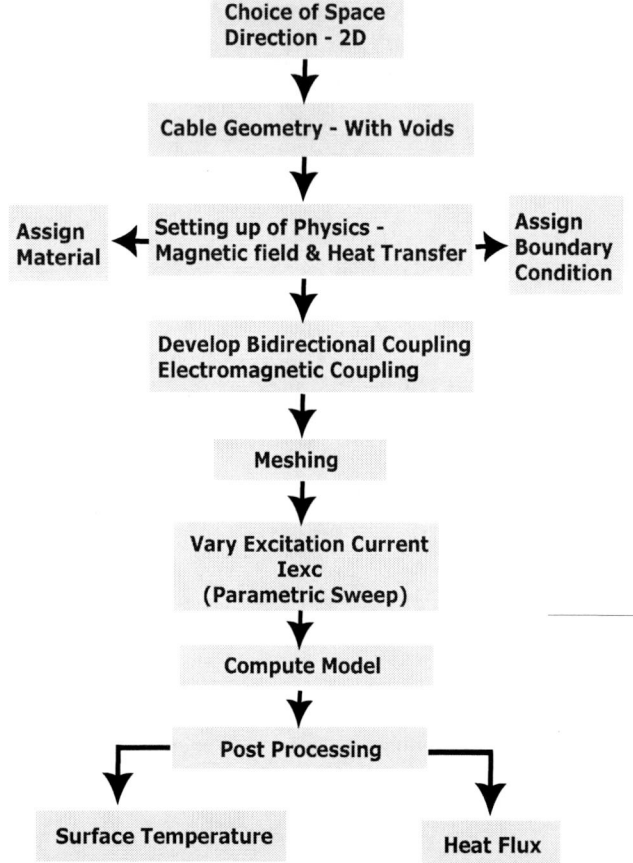

Fig. 3. Proposed Strategy

A. Modeling of Geometry

The 2D geometry of the 330 KV HVAC cable configuration has been modeled using the geometric primitives of COMSOL Multiphysics software. The detailed parameters of the cable configuration are specified in Table II and Table III.

B. Physics Setup and Boundary Conditions

The selection of appropriate physics interfaces is a critical step in any FEM analysis. In this study, the Magnetic Fields (mf) and Heat Transfer in Solids (ht) physics interfaces

TABLE II
CABLE SPECIFICATIONS

Cable Layers	Dimension
Voltage	330 KV (AC)
Frequency	50 Hz
Conductor	Copper
Conductor Diameter (mm)	42
Semiconductor Screen Thickness (mm)	0.6
Conductor Screen Thickness (mm)	1.5
XLPE Layer Thickness	23
Insulation Screen Thickness (mm)	1
Aluminium Sheath Thickness (mm)	2.2
HDPE Layer Thickness (mm)	5

TABLE III
CABLE MATERIAL PROPERTIES

Material	μ_r	ϵ_r	ρ	k	σ	C_p
Copper	1	1	8960	400	5.998e7	385
Semiconductor	1	2.25	1055	10	2	2405
XLPE	1	2.5	930	0.46	1e-15	2302
Aluminium	1	1	2700	238	3.774e7	900
HDPE	1	2.3	935	0.46	1e-18	2302

in COMSOL Multiphysics have been employed. The Magnetic Field interface utilizes the magnetic vector potential $(A, (Wb/m))$ as the dependent variable to solve the governing FEM equations. This interface solves Maxwell's equations, as outlined in Section II, to compute magnetic parameters such as magnetic flux density and field intensity, along with electrical parameters including excitation current density and electric field. The Heat Transfer in Solids interface considers temperature (T) as the dependent variable for thermal analysis. A set of governing equations is employed to solve for the temperature distribution, enabling the evaluation of thermal parameters such as enthalpy, heat source, and heat flux magnitude.

Boundary conditions serve as constraints applied to an FEM model to ensure the realism and accuracy of the solutions to the governing equations. These equations are solved for each discrete element across the entire domain and are determined using the variational Galerkin method. The COMSOL Multiphysics software employs this method to obtain the stated solutions [8].

The boundary conditions used in the presented model study are illustrated as follows.

Magnetic Field (mf) physics

1) **Magnetic Insulation** - It ensures encapsulation of the magnetic field within the preferred electromagnetic domains as per $(\nabla X A = 0)$
2) **Ampere's Law in Solids** - The solution of the set of Maxwell's equations in the selected electromagnetic domain is facilitated by the appropriate assignment of the magnetization and conduction models. In this work,

both models are considered nonlinear and temperature-dependent to ensure accurate analysis.

Heat Transfer (ht) physics

1) **Heat flux** - This boundary condition is applied to introduce heat flux across the selected thermal boundaries within the FEM model. In this study, convective heat flux has been considered, with a heat transfer coefficient $h = 5$, (W/m^2K) specified for accurate thermal analysis.

2) **Thermal Insulation** - It ensures to keep the heat flux within the chosen thermal domain using $(n.Q = 0)$

C. Meshing

A suitable discretization of the entire domain is essential to ensure accurate and efficient finite element analysis results. The accuracy of the FEM solution is influenced by the selection of mesh parameters, such as element size and shape. In this study, a physics-controlled meshing sequence within COMSOL Multiphysics has been utilized to solve the presented model. The statistics of the selected mesh elements are shown in Table IV.

TABLE IV
MESH STATISTICS

Metric	Value
Triangular elements	28520
Edge elements	850
Vertex elements	32
Minimum element quality	0.5936
Average element quality	0.9088
Element area ratio	0.01225

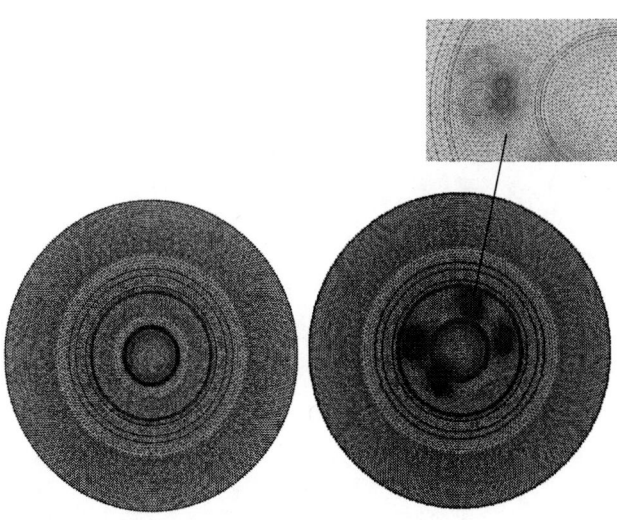

Fig. 4. Meshed Geometry - (a) Without void, (b) With void

D. Parametric Sweep Analysis

The current excitation of the cable configuration has been considered as a varying parameter and has been swept over a specified range of $I_{exc} \in [500, 1800]$ Amps to determine the operating point at which the temperature across the cable insulation and conductor region exceeds the limit specified in IEEE Standard 835-1994 and IEC 60287, which is $90°C$ [9], [10]. A coupled frequency–stationary study has been utilized to solve the electromagnetic and thermal governing equations, respectively.

IV. RESULTS AND DISCUSSION

A single phase 330 KV, 50 Hz HVAC cross linked polyethelyene (XLPE) cable has been modeled and further analysed based on FEM using COMSOL Multiphysics software. The ambient temperature (T_{ref}) of the domain of interest has been considered as $20°C$.

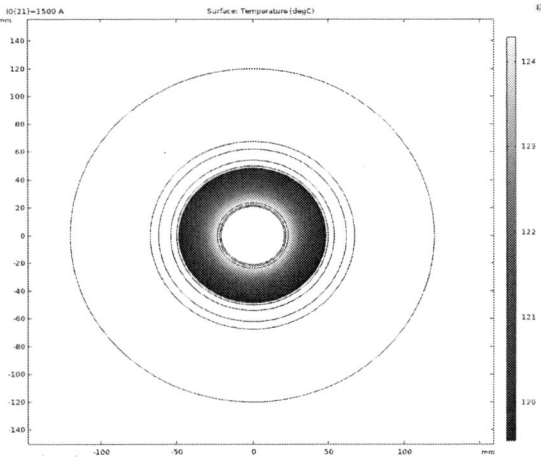

Fig. 5. Surface Plot of Temperature (XLPE) - Without void

Fig. 5. shows the 2D surface plot of temperature (in the XLPE layer) of the cable configuration without void. A homogeneity in the temperature distribution with a gradual decrement from the conductor end to the aluminium sheath end can be seen with peak steady temperature of $124°C$ at $I_{exc} = 1500$ A. On contrary, Fig. 6. shows the 2D surface plot of temperature (in the XLPE) layer of the cable configuration with voids. A disruption in the homogeneity of the temperature distribution in the XLPE layer can been seen. The void locations and their nearby regions tend to have attained more temperature. with a peak steady state temperature of $128°C$.

As a supplement to the claim, the total heat flux (in (W/m^3) has also been plotted for both the cable configurations. In Fig. 7, a uniform heat flux distribution very similar to that of temperature distribution can be seen. Whereas, Fig. 8, (with void) shows a complete deviation from uniform heat flux distribution within the XLPE layer. Moreover, there has also been a rise of the peak value of heat flux from 120 W/m^2 to 300 W/m^2 if we compare the heat flux plots (Fig. 7 and Fig. 8) for cable configuration without and with voids.

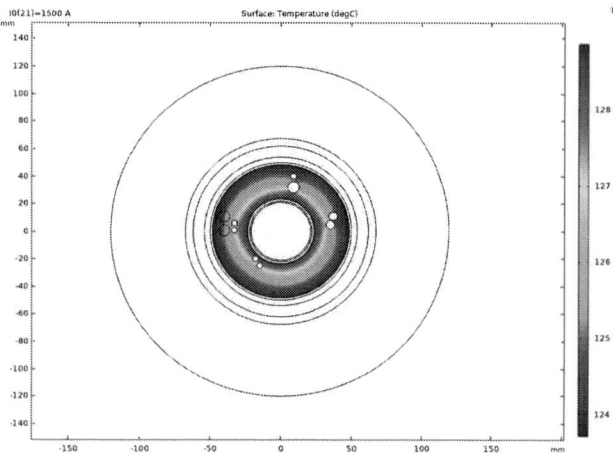

Fig. 6. Surface Plot of Temperature (XLPE) - With void

Such non uniform and spike in the heat thermal energy and thereby temperature within the XLPE layer create a chance of formation of local hotspots leading to partial discharge and subsequently damage of the cable.

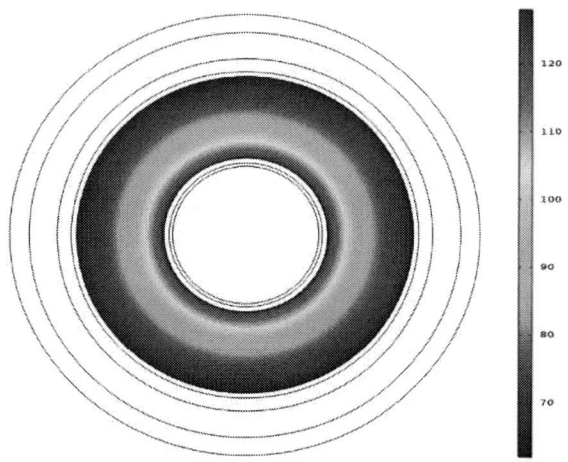

Fig. 7. Surface Plot of Heat Flux - Without void

Table V shows how the mean cross-sectional temperature across the entire cable domain increases as the sweeping parameter (I_{exc}) varies from 500 A to 1800 A with a step change of 50 A. Furthermore, the percentage increment in the mean cross-sectional temperature throughout the cable on inclusion of voids is evaluated as around 5% which also causes a reduction in the ampacity or current carrying capacity of the cable by 2.5% on an average. The relationship between the temperature rise and the current carrying capacity of the cable follows the emperical relationship developed by Neher and McGrath [11],

$$I = \sqrt{\frac{T_{\max} - T_a}{R_{\text{th}} \cdot R_{\text{ac}}}} \tag{7}$$

where,

- $T_{\max}$: allowable maximum conductor temperature (°C)
- T_a: ambient temperature (°C)
- R_{th}: total thermal resistance from conductor to ambient (°C·m/W)
- R_{ac}: AC resistance of the conductor (ohms/m)

A linear approximation of (7) yields (8), which is used to quantify the reduction in current-carrying capacity resulting from a rise in conductor temperature.

$$\delta I\% \approx -0.48 \cdot \delta T\% \tag{8}$$

Such a reduction in the ampacity of the cable will result in decrement in the power density of the electric transportation system making it less efficient as a whole.

Fig. 9. illustrates the variation of the average temperature in a graphical form. A substantial dip in the temperature of the cable cross-section is visible for I_{exc} varying from 500 A to 1800 A.

TABLE V
RESULTS OF PARAMETER SWEEP ANALYSIS

Sweeping I_{exc} (Amp)	Mean CST Without void	Mean CST With void	$+\delta T\%$ Temp ↑	$-\delta I\%$ Current ↓
1200	62.125	65.183	+4.92	-2.37
1300	68.892	72.440	+5.14	-2.48
1400	76.068	80.122	+5.32	-2.56
1500	83.616	88.208	+5.5	-2.65
1600	91.522	96.678	+5.631	-2.71

where, **Mean CST** represents Mean Cross-sectional temperature of the cable configuration. It's a spatial average over the 2D cross section at steady state.

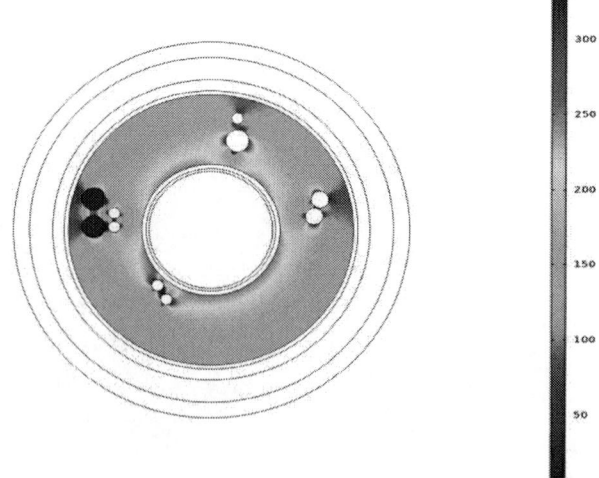

Fig. 8. Surface Plot of Heat Flux - With void

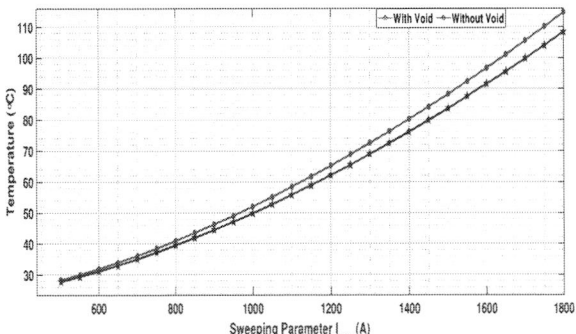

Fig. 9. Variation of Mean Cross-sectional Temperature

V. CONCLUSION

This paper presents a comprehensive strategy for evaluation of impact of voids on power cable ampacity using a coupled multiphysics approach. The proposed approach shows a much lucid and yet accurate way to incorporate the real scenario data of a cable configuration within a macro electrical system as well as considers bidirectional fully coupled electrothermal analysis. The temperature and heat flux plots shows a clear increment in the insulation layer as well as overall average temperature of the cable when voids are considered. An average increment of the temperature by 5% thereby causing an average reduction in ampacity of about 2.5 % has been evaluated for a 330 KV 50 Hz XLPE HVAC cable. This work provides a structured and simple and strategy serving as an useful guide for the design engineers of electric transportation system, and its associated cables.

The future scope of this work includes the randomization of voids in terms of their number, shape, and size to evaluate their impact on ampacity. Additionally, the reliability of the cable can be assessed using a probabilistic distribution approach.

ACKNOWLEDGMENT

The authors would like to thank Dr. Anties K. Martin from COMSOL Multiphysics for the support provided towards this work.

REFERENCES

[1] A. Saha, A. Azizi and M. Ghassemi, "Optimal Bipolar MVDC Power Cable Designs for Future Wide-Body All Electric Aircraft," in IEEE Transactions on Dielectrics and Electrical Insulation, vol. 31, no. 4, pp. 2074-2083, Aug. 2024, doi: 10.1109/TDEI.2024.3355033.

[2] J. H. Neher et al., "Power cable ampacities," Elect. Eng., vol. 81, no. 10, pp. 799–800, Oct. 1962.

[3] J. H. Neher and M. H. McGrath, "The calculation of the temperature rise and load capability of cable systems," Power App. Syst., Part III, Trans. Amer. Inst. Elect. Eng., vol. 76, no. 3, pp. 752–764, Apr. 1957.

[4] IEEE Standard Power Cable Ampacity Tables, IEEE Std. 835-1994, Sep. 1994.

[5] M. V. S. da Silva, O. M. O. de Araújo, D. F. de Oliveira and R. T. Lopes, "Evaluation of the Effects of Voids in Electrical Cables Using COMSOL Multiphysics Software," in IEEE Transactions on Dielectrics and Electrical Insulation, vol. 31, no. 4, pp. 2144-2150, Aug. 2024, doi: 10.1109/TDEI.2024.3395232.

[6] S. Hore, S. Basak, N. Haque, S. Dalai and M. Mukherjee, "Studies on the effect of void geometry and location on electric field distribution and partial discharge in XLPE insulated power cable by finite element analysis using COMSOL multiphysics simulation," 2017 6th International Conference on Computer Applications In Electrical Engineering-Recent Advances (CERA), Roorkee, India, 2017, pp. 220-225, doi: 10.1109/CERA.2017.8343330.

[7] Ashfaque Ahmed Bhatti, Xiaosheng Peng, Bin Yang, Cheng Xie, "Effect of Void on Temperature Distribution in XLPE Power Cables by COMSOL Multiphysics Simulation", 2021 IEEE 4th International Electrical and Energy Conference (CIEEC), pp.1-4, 2021.

[8] COMSOL Multiphysics 6.3 Manual [Online]. Available: https://doc.comsol.com/5.5/doc/com.comsol.help.comsol

[9] IEEE Standard Power Cable Ampacity Tables," in IEEE Std 835-1994 , vol., no., pp.1-3151, 30 Dec. 1994, doi: 10.1109/IEEESTD.1994.7297793.

[10] IEC 60287–1-1 Electric cables-calculation of the current rating part 1: current rating equations (100 % load factor) and calculation of losses section 1: general, 2006.

[11] IEEE Recommended Practice for Industrial and Commercial Power Systems Analysis (Brown Book)," in IEEE Std 399-1997 , vol., no., pp.1-488, 31 Aug. 1998, doi: 10.1109/IEEESTD.1998.88568.

Dynamic Duty Cycle based Pulse Charging of Parallel Lithium-Ion Batteries for Electric Vehicles

Supriya Chakrabarty, Niranjan Behera, Sankarsan Mohapatro and Abhinav Arya

Dept. of Electrical Engineering, IIT Bhubaneswar, India
sankarsan@iitbbs.ac.in

Abstract—Efficient and reliable battery charging is essential for applications including electric vehicles (EVs), renewable energy storage systems, and uninterruptible power supplies (UPS). Traditional charging techniques frequently depend on separate DC-DC converters for each battery, resulting in higher system costs, increased control complexity, and spatial limitations. This paper proposes a Dynamic Duty Cycle based Pulse Charging (DDCPC) technique for the simultaneous charging of multiple batteries using a single Buck converter. Complementary switching of battery side switches at 1 kHz ensures alternate and continuous energy delivery, while the Buck converter operates at 10 kHz to balance efficiency and switching losses. A closed loop control mechanism dynamically adjusts the converter's duty cycle to maintain a constant inductor current, enabling efficient and balanced charge distribution based on each battery's state of charge (SOC). The proposed method improves charge uniformity, and shortens total charging time. MATLAB/Simulink simulations validate the effectiveness of the DDCPC approach, highlighting enhanced performance.

Index Terms—Parallel Battery Charging, State Of Charge (SOC), Constant Current - Constant Voltage (CC-CV) Charging method, Dynamic Duty cycle based Complimentary Pulse Charging (DDCPC) Method

I. INTRODUCTION

The rapid expansion of electric vehicles (EVs) has created a growing demand for efficient, fast, and reliable battery charging strategies. EVs not only serve as an alternative to conventional fossil fuel-based transportation but also play a crucial role in integrating renewable energy sources (RES) into modern power grids. By utilizing advanced battery technologies, EVs help reduce environmental pollution and improve energy efficiency in transportation systems [1], [2]. Among various battery technologies, lithium-ion (Li-ion) batteries have gained widespread adoption due to their high energy density, low self-discharge rate, extended cycle life, and superior efficiency compared to other battery chemistries such as nickel-metal hydride, lead-acid, and nickel-cadmium batteries [3]. However, despite these advantages, the performance and lifespan of lithium-ion batteries are significantly affected by the charging method employed now a days. Therefore, optimization in charging strategies is essential to enhance battery life, minimize charging time, and improve overall energy efficiency.

Conventional battery charging techniques primarily include Constant Current (CC), Constant Voltage (CV), and a combination of the two—commonly known as the CC-CV method

[4]. The most commonly used method for charging batteries is the Constant Current Constant Voltage (CCCV) technique. This method is effective for charging individual batteries. The CC method applies a fixed current to the battery until it reaches a predefined voltage limit, after which the charger transitions to the CV phase, where the voltage is maintained while the charging current gradually decreases. This two stage CC-CV approach is widely used due to its simplicity and ability to ensure complete charging while protecting the battery from over voltage conditions [5]. However, CC-CV charging presents several limitations, particularly when charging multiple batteries in parallel from a single converter [6]. The main issue arises from voltage mismatch between the batteries. If the batteries do not have the same State of Charge (SOC) or terminal voltage, circulating currents will begin to flow between them. This not only leads to uneven charging but can also cause overheating, reduced battery life, and potential safety concerns. To overcome this problem, the complementary pulse charging method (CPC) is employed. In this approach, only one battery is charged at a time in a controlled and alternating manner. By isolating the charging pulses, each battery receives current independently, eliminating the risk of circulating currents and ensuring safe and efficient charging even when the batteries have different SOC levels. This method is particularly useful for applications involving multiple batteries in parallel where uniform voltage conditions cannot be guaranteed.

One more advanced battery charging method is multistage constant current (MS-CC) charging, which segments the charging process based on predefined state-of-charge (SOC) levels or cutoff voltages. This method helps to mitigate ohmic losses and battery overheating by adjusting the charging current dynamically throughout the charging process [7].

Among the various advanced charging techniques, pulse charging has emerged as a promising solution due to its ability to enhance charge acceptance, suppress lithium plating, and improve ion diffusion within the battery cells. In pulse charging, high-current pulses are applied intermittently, followed by short rest periods that allow the battery to recover and depolarize. This approach not only accelerates the charging process but also minimizes the degradation effects associated with continuous high-current charging [8]. Pulse charging has been shown to increase the allowable charging current, reduce internal resistance, and slow down capacity decay, making it a

979-8-3315-3013-6/25 $31.00 © 2025 IEEE

viable alternative for fast-charging applications. Furthermore, studies indicate that pulse charging can improve battery health by suppressing dendrite formation and preventing excessive heat generation, which are critical factors in extending battery lifespan.

To address these challenges, This paper proposes an optimized pulse charging strategy tailored for multiple battery charging applications. The proposed method dynamically adjusts pulse characteristics based on real-time SOC levels and battery conditions, ensuring that lower SOC batteries receive a higher duty cycle while higher SOC batteries receive a lower duty cycle. This strategy facilitates balanced charging, minimizes power losses, and enhances battery lifespan by effectively distributing the charging current among multiple batteries. Additionally, the effectiveness of the proposed approach is validated through MATLAB/Simulink simulations, demonstrating its potential for practical EV charging applications.

II. SYSTEM DESCRIPTION

A pulse charging circuit designed for simultaneous charging of two batteries is depicted in Fig. 1. A DC-DC Buck converter is taken as the primary device of the proposed work. In a buck converter, the inductor current directly charges the batteries. By implementing closed loop control the current ripple is significantly reduced, ensuring smooth and efficient battery charging maintaining a constant inductor current. This helps in minimizing stress on the batteries and improving overall charging performance.

TABLE I
DC-DC BUCK CONVERTER PARAMETERS

Parameters	Values
Input voltage	48 V
Output voltage	15 V
Output Current	10 A
Inductor (L)	4.3 mH
Capacitor (C)	71.63 μF

In contrast, in a boost converter, the diode current is responsible for charging the batteries. However, this current is not continuous and exhibits high ripple, leading to uneven charging and increased battery stress. The intermittent nature of the diode current in a boost converter makes it unsuitable for applications requiring low-ripple and stable charging. Therefore, the buck converter is preferred for battery charging applications, as it provides a continuous, low-ripple charging current, ensuring efficient and reliable operation. The converter is designed using parameters shown in Table I, for high efficiency. At the output port of the converter two batteries are connected through two controlled switches. The operation of these switches is regulated by a controller, which dynamically adjusts switching based on the SOC of the batteries, ensuring optimized charging efficiency. The battery parameters for this proposed work are represented in Tables II and Table III.

TABLE II
BATTERY 1 PARAMETERS

Battery Parameters	Values
Initial SOC	30%
Nominal Voltage	12V
Rated Capacity	10Ah
Internal Resistance	0.015 ohm

TABLE III
BATTERY 2 PARAMETERS

Battery Parameters	Values
Initial SOC	50%
Nominal Voltage	12V
Rated Capacity	10Ah
Internal Resistance	0.015 ohm

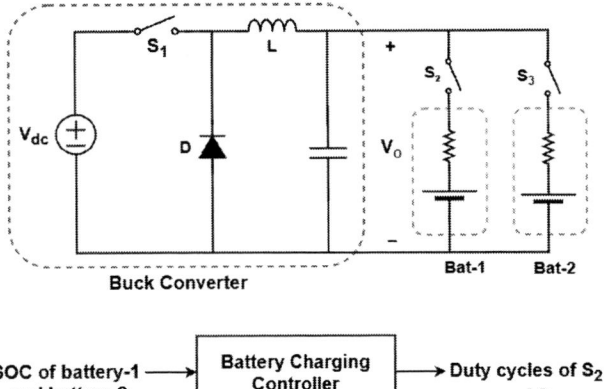

Fig. 1. Basic configuration of parallel battery charging of two batteries using pulse charging method

In this study, the batteries are modeled using an equivalent circuit, consisting of an ideal voltage source in series with an internal resistance. This representation helps to approximate the behavior of the real battery by accounting for the voltage drop caused by internal resistance during current flow. The ideal voltage source (V) signifies the open-circuit voltage (OCV) of the battery, which is the voltage measured when no current is drawn. The internal resistance (R) reflects the inherent opposition to current flow within the battery, leading to a voltage reduction at the terminals when charging or discharging. During the charging process, the flow of current through this resistance results in a potential drop, causing the terminal voltage to be slightly lower than the open-circuit voltage. Although real batteries exhibit more complex characteristics under varying charge and discharge conditions, this simplified equivalent circuit provides a practical and effective model for analyzing the pulse charging process.

III. METHODOLOGY

A. Compliment Pulse Charging (CPC) Method

Pulse Charging (PC) is an advanced battery charging approach where the charging current is supplied in intermittent pulses rather than as a continuous, steady flow. This technique operates by alternating between charging and resting phases, with the current being periodically switched ON and OFF at a predefined frequency. High-current pulses are applied for short durations, followed by intervals of zero current, allowing the battery to recover. The characteristics of the pulse—such as width, frequency, and amplitude—can be adjusted depending on the battery type and application requirements. The overall charging process is governed by the state of charge (SOC) of the batteries.

In the complementary pulse charging method, when one battery undergoes charging, the second battery remains in a resting phase, and vice versa. This can be done by operating the switches S_2 and S_3 in compliment fashion. The pulse-controlled switches (S_2 and S_3) are responsible for regulating both the pulse frequency and duty cycle of the charging process. These switches operate in a complementary manner, ensuring that they are never turned off simultaneously, thereby maintaining continuous operation. For lithium-ion batteries, the optimal pulse frequency is generally in the kilohertz range, which is considerably lower than the switching frequency of power electronic converters. This frequency difference allows a large number of control units for the decoupling of control between the front and rear stages of the proposed pulse charger. This configuration maximizes the power output capability of the converter while ensuring that the delivered current exhibits pulse characteristics.

A key advantage of this complementary switching mechanism is that it eliminates the charging delays associated with conventional pulse charging, where the intervals between pulses can extend overall charging time. By allowing one battery to charge while the other is in a resting state, the system ensures continuous energy transfer and improved charging efficiency. This alternating process continues until one of the batteries reaches 100% SOC.

This strategy ensures efficient energy utilization, improved charge balancing, and reduced thermal stress, making it highly suitable for parallel battery charging applications.

B. Dynamic Duty Cycle Adjustment

The duty cycles for switch S_2 and S_3 have been decided by the SOC difference of the batteries [9]. If $|SOC_1 - SOC_2| < 5\%$, the battery having higher initial SOC will get a duty cycle of 0.05 and the battery having lower initial SOC will get a duty cycle of 0.95. As a result, the battery having lower initial SOC will charge at a very high rate and the battery having higher initial SOC will charge at a very slow rate. When $|SOC_1 - SOC_2| > 5\%$, then fine tuning of the duty cycles of S_2 and S_3 begins. Once SOC_1 and SOC_2 becomes equal both switches are assigned a duty cycle of 0.5, allowing the batteries to charge simultaneously at the same rate, ensuring they become fully

charged at the same time. However, this approach causes the battery with the higher initial SOC to charge slowly at the beginning, which can result in increased overall charging time.

Regardless of the initial SOC levels of the batteries, a duty cycle of 0.5 is given to both the switches S_2 and S_3 and the batteries are charged simultaneously [10]. When one battery is fully charged it can be removed and other battery will charge with 0.5 duty cycle till the end. To address the limitations of the aforementioned methods, a dynamic duty cycle adjustment algorithm is proposed based on SOCs of the two batteries. Fig. 2 represents the flow chart for dynamic duty cycle allocation at different levels of State Of Charges for the two batteries. In this approach, the battery with a lower SOC is assigned a higher duty cycle, while the battery with a higher SOC is assigned a lower duty cycle, ensuring balanced charging. The initial duty cycle for each battery is determined using Eq. (1)

$$D_i = \frac{100 - SOC_i}{\sum_{j=1}^{n}(100 - SOC_j)} \tag{1}$$

Here, D_i represents the duty cycle of the ith battery, and n denotes the total number of batteries connected in the system.

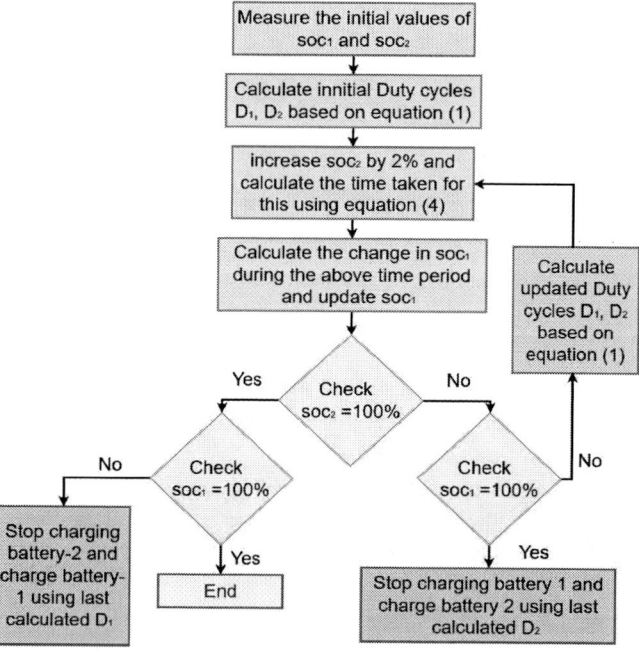

Fig. 2. Flowchart of proposed dynamic duty cycle algorithm

The interval at which the duty cycles are updated is determined by the battery capacity (Q in Ah). Let the rated capacity of both batteries be Q. For every $x\%$ increase in the SOC of the battery with the higher charge(battery 2), the duty cycles for both batteries are updated using Eq. (1). The value of x can range from 1% to 3%; in this case, x is taken as 2%. Referring to Fig. 3, in each switching cycle of period T_{sw}, the

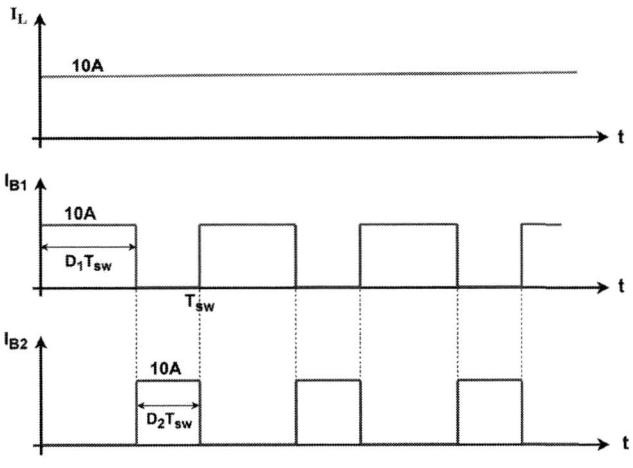

Fig. 3. Inductor currents and complimentary battery currents

switch corresponding to battery 2 remains ON for a duration of D_2T_{sw}. During this time, the total charge delivered to the battery by the converter is given by Eq. (2) is:

$$Q_{\text{delivered}} = \frac{I \cdot D_2}{3600 \cdot f_{\text{sw}}} \text{ (Ah)} \quad (2)$$

Where I is the inductor current of the converter, D_2 is the duty cycle of switch S_3, and T_{sw} is the switching period of S_3 in seconds, and f_{sw} is the switching frequency of the power electronics switches S_2 and S_3. For safety purpose batteries are generally charged at 1C rate. So, the current required to charge the batteries will be 10 A. The total charge required to increase the SOC of battery-2 by $x\%$ as in Eq. (3).

$$Q_{\text{required}} = \frac{x \cdot Q}{100} \text{ (Ah)} \quad (3)$$

Eq. (4) represents the time (in ms) required for the SOC of battery 2 to increase by $x\%$ is:

$$\Delta t = \frac{36 \cdot x \cdot Q}{I \cdot D_2} \quad (4)$$

Eq. (5) represents the increase in SOC of battery 1 during this Δt period.

$$\Delta SOC_1 = \frac{x \cdot Q \cdot D_1}{100 \cdot D_2} \quad (5)$$

Where D_1 and D_2 are the duty cycles of the respective battery switches.

This formulation ensures that the duty cycles are updated dynamically based on the SOC levels, leading to balanced charging of the batteries.

IV. RESULTS AND DISCUSSION

The effectiveness of the suggested pulse charging technique is evaluated via a MATLAB-SIMULINK simulation study.

Fig. 4, illustrates the battery charging using the charging logic proposed in [9], where two batteries are charged in

parallel using a single power converter. The approach involves providing a 50% gate pulse to each switch, S_2 and S_3, irrespective of the initial SOC of the batteries. This means, both batteries receive equal charging time during each switching cycle, ensuring that the charging process is evenly distributed between them. Since both batteries receive charge at the same rate, the battery with a higher initial SOC will reach its fully charged state sooner than the battery with a lower initial SOC. Once the battery with the higher SOC reaches its fully charged state, it can be disconnected from the charger to prevent overcharging. At this point, the other battery continues to charge alone using the same power converter. Since the charging algorithm originally allocated a 50% duty cycle to each battery, after disconnecting the fully charged battery, the other battery will now be charged at a duty cycle of 50% until its SOC reaches upto 100%.

As shown in Fig. 5, the charging logic has been implemented based on the approach described in [10], where the duty cycles of the switches are dynamically adjusted based on the initial state of charge (SOC) of each battery in two steps. Unlike the method in [9], where both batteries receive an equal 50% duty cycle, this approach prioritizes the lower SOC battery by assigning it a significantly higher duty cycle of 0.95, while the higher SOC battery receives a much lower duty cycle of 0.05. Since the lower SOC battery is assigned a duty cycle of 0.95, it receives charging current for a much longer duration during each switching cycle. As a result, it charges at a significantly faster rate compared to the higher SOC battery.

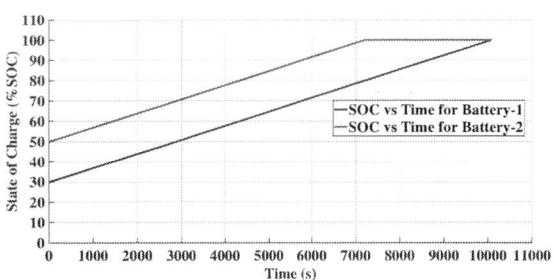

Fig. 4. Variation of SOC with time by using method in [9]

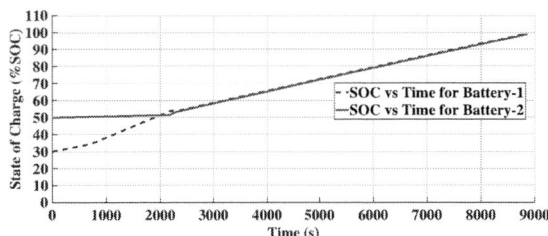

Fig. 5. Variation of SOC with time by using method in [9]

The higher SOC battery, which is assigned a duty cycle of only 0.05, receives a very small portion of the charging

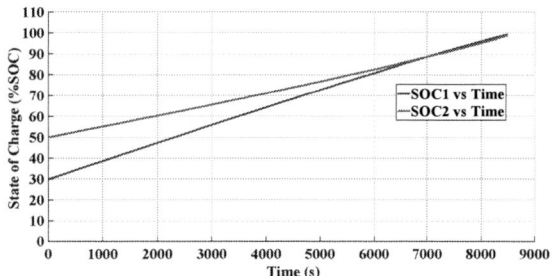

Fig. 6. Variation of SOC with time by using proposed method

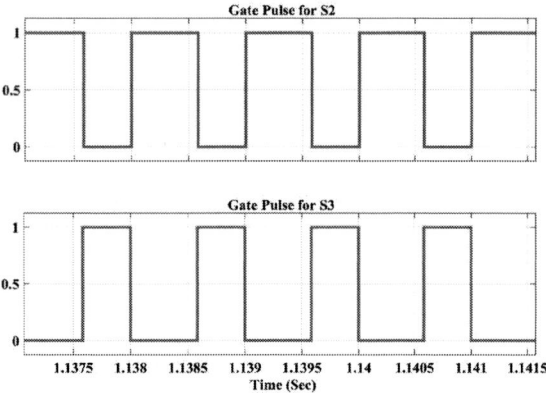

Fig. 7. Complimentary gate pulses in proposed method

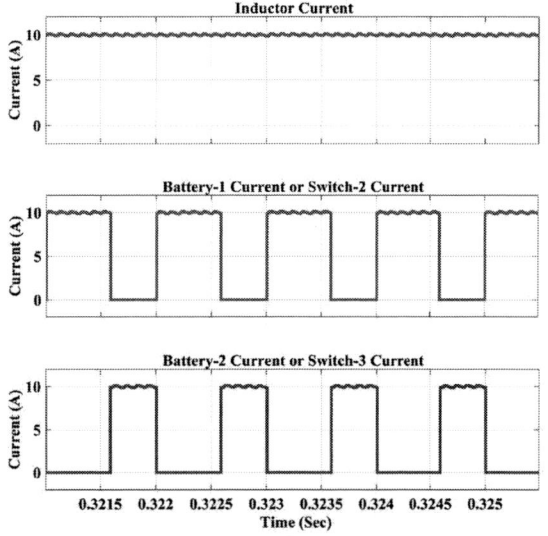

Fig. 8. Complimentary Battery currents in Proposed method

current, causing it to charge at a much slower rate. As the charging process continues, the SOC difference between the two batteries is quickly reduced. Once both batteries reach the same SOC level, the system automatically sets the duty cycles of both switches to 0.5. From this point onward, both

batteries receive the same amount of charge simultaneously, ensuring they continue to charge at the same rate. Eventually, both batteries reach a fully charged state at the same time, eliminating the need to disconnect one battery before the other, as required in the previous approach.

Fig. 6 represents the results of battery charging by the application of the proposed Dynamic Duty Cycle Adjustment Algorithm based on the State of Charge (SOC) levels of the batteries. In this approach, the switch connected to the battery with the lower SOC is assigned a higher duty cycle, while the switch associated with the battery having a higher SOC is initially given a lower duty cycle. This behavior is clearly illustrated in Fig. 7, and the corresponding inductor current and switch/battery currents are shown in Fig. 8. This ensures balanced charging by prioritizing the battery that requires more charge, allowing it to charge at a faster rate compared to the battery with a higher initial SOC.

Unlike conventional methods, where both batteries may receive the same duty cycle regardless of their initial SOC, this strategy dynamically adjusts the duty cycle throughout the charging process. Although the lower SOC battery charges faster, the initial charging speed of the higher SOC battery is also relatively high compared to traditional charging methods [9], [10]. A key feature of this method is that the duty cycle of each switch is updated dynamically based on the SOC levels of the batteries. As the SOC increases over time, the duty cycle is continuously adjusted according to the relationship defined in Eq. (1). This real-time adaptation ensures that both batteries receive the optimal charge distribution at every stage of the charging process.

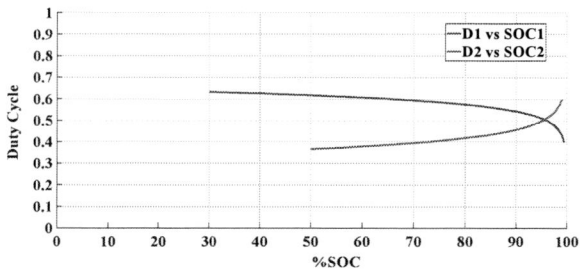

Fig. 9. Variation of duty cycle vs SOC in the proposed method

The dynamically change of duty cycle with respect to SOC and time are represented in Fig. 9 and Fig. 10 respectively. The initial and final values of duty cycles for battery-1 is found to be 0.633 and 0.399 respectively. For battery-2 the values are 0.366 and .6007. It shows that the Duty cycle changes for both batteries in the mentioned range.

By incorporating adaptive duty cycle modulation, the proposed method significantly enhances the charging speed compared to the existing approaches in [9] and [10]. Table IV provides a comparative analysis of the charging durations for different methods, clearly demonstrating that the proposed technique achieves a substantial reduction in the total charging

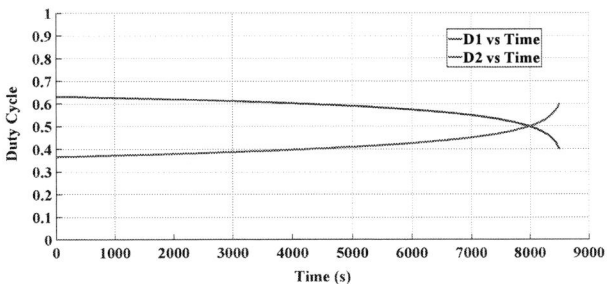

Fig. 10. Variation of duty cycle vs time in the proposed method

TABLE IV
COMPARISON TABLE

Time for Full Charge (seconds)			
Batteries	**in ref [9]**	**in ref [10]**	**Proposed Method**
Battery-1	7300	8896	8356
Battery-2	10020	8896	8356

time for both batteries. The key advantage of this approach lies in its dynamic adjustment mechanism, which optimizes power delivery by adapting the duty cycle in real time. This not only accelerates the overall charging process but also mitigates the risk of overcharging by preventing excessive current flow once a battery reaches full charge. As a result, the method contributes to improved battery longevity and operational efficiency.

Furthermore, the proposed algorithm offers a more intelligent and adaptive solution for parallel battery charging applications. By efficiently distributing the charging current among multiple batteries, it ensures balanced energy allocation, reducing energy wastage and enhancing system reliability. These improvements make the proposed strategy a superior alternative to conventional methods, offering both faster and safer battery charging for practical applications.

V. Conclusion

This paper presented a Dynamic Duty Cycle based Pulse Charging (DDCPC) technique for efficiently charging multiple batteries in parallel using a single DC-DC Buck converter. By employing complementary switching control and a closed-loop dynamic duty cycle adjustment, the proposed method ensures continuous energy transfer while optimizing charge distribution based on the state of charge (SoC). The adaptive control strategy minimises overall charging duration while simultaneously improving battery lifespan by mitigating excessive current flow. The simulations were conducted using MATLAB/Simulink confirm the efficacy of the proposed method, showcasing a decrease in overall charging time, enhanced charge uniformity, and improved overall system performance. In contrast to traditional charging methods, the DDCPC technique provides a cost-efficient, compact, and dependable

solution for parallel battery charging applications, rendering it especially appropriate for electric vehicles (EVs), renewable energy storage systems, and uninterruptible power supplies (UPS). Future research could investigate the hardware implementation and additional optimisation of switching control strategies to enhance system efficiency and scalability.

References

[1] L. Gong, W. Cao, K. Liu, Y. Yu, and J. Zhao, "Demand responsive charging strategy of electric vehicles to mitigate the volatility of renew able energy sources," Renew. Energy, vol. 156, pp. 665–676

[2] D. Sidorov et al., "A dynamic analysis of energy storage with renew able and diesel generation using volterra equations," IEEE Trans. Ind. Informat., vol. 16, no. 5, pp. 3451–3459.

[3] K. Li, F. Wei, K. J. Tseng, and B.-H. Soong, "A practical lithium-ion battery model for state of energy and voltage responses prediction incor porating temperature and ageing effects," IEEE Trans. Ind. Electron., vol. 65, no. 8, pp. 6696–6708.

[4] S. Habib et al., "Contemporary trends in power electronics converters for charging solutions of electric vehicles," CSEE J. Power Energy Syst., vol. 6, no. 4, pp. 911–929.

[5] S. J. Thomson, P. Thomas, A. R. and E. Rajan, "Design and Prototype Modelling of a CC/CV Electric Vehicle Battery Charging Circuit," 2018 International Conference on Circuits and Systems in Digital Enterprise Technology (ICCSDET), Kottayam, India, 2018, pp. 1-5.

[6] Y. Gao, X. Zhang, Q. Cheng, B. Guo, and J. Yang, "Classification and review of the charging strategies for commercial lithium-ion batteries," IEEE Access, vol. 7, pp. 43511–43524.

[7] L. Jiang et al., "Optimization of multi-stage constant current charg ing pattern based on Taguchi method for Li-ion battery," Appl. Energy, vol. 259, Feb. 2020, Art. no. 114148.

[8] S. Li et al., "Effects of pulse charging on the performances of lithium ion batteries," Nano Energy, vol. 56, pp. 555–562.

[9] K. S. Venkat, M. V. Satya Sai Chandra and S. Mohapatro, "Pulse Charging Scheme for Multiple Battery Charging in Electric Vehicle Applications," 2023 IEEE 3rd International Conference on Smart Tech nologies for Power, Energy and Control (STPEC), Bhubaneswar, India, 2023, pp. 1-6.

[10] L. Jiang et al., "Optimal Charging Strategy With Complementary Pulse Current Control of Lithium-Ion Battery for Electric Vehicles," in IEEE Transactions on Transportation Electrification, vol. 8, no. 1, pp. 62-71.

Extraction of MPP using Boost & SEPIC Converter using Parametric Estimation Method

Vamshi Krishna Bandaru
Student of EEE
VNR VJIET
Hyderabad, India
vamshibandaru2000@gmail.com

Anuradha Kotapati
Department of EEE
VNR VJIET
Hyderabad, India
anuradha_k@vnrvjiet.in

Poornima Seelam
Department of EEE
VNR VJIET
Hyderabad, India
poornima_s@vnrvjiet.in

Shiva Prasad Edara
Department of EEE
VNR VJIET
Hyderabad, India
shivaprasad_e@vnrvjiet.in

Abstract— The maximum power point tracking (MPPT) is a key concept to extract the maximum power from a PV Module or array. The extracted power is further converted into AC using DC-AC Convert, or into DC using DC-DC Converters as per the requirement of application. In this application we are converting into DC power using DC-DC converter by controlling input voltage, with help of controller such that, input voltage is equal to voltage at maximum power point. The work presents the study and comparison of power, voltage, current of different Solar PV panel during partial shading condition with Boost & SEPIC converters using Parametric Estimation method. Parametric Estimation method is used to extract the real time parameters and compute the Vmp and accordingly duty to the switch is given. Artificial Neural Network (ANN) is used to train the block, extract maximum power point of PV panel and used for duty control of both Boost & SEPIC Converter. The simulations of PV Module based on AI technique for comparison, is performed in MATLAB Simulink environment, the simulation results of voltage, output current, output power are obtained at partial shading conditions.

Keywords—boost converter, DC-DC converter, MPP tracking, photovoltaic (PV) array, P&O method, parametric estimation method, partial shading condition, SEPIC converter

I. INTRODUCTION

The fast rise of renewable energy sources, driven by environmental concerns and the demand for sustainable energy solutions, has propelled photovoltaic (PV) systems to the forefront of power generating technology. PV modules, being the primary building elements of these systems, have undergone constant developments in their design and efficiency. To exploit the maximum potential of PV technology, it is vital to improve the energy harvesting process, and this optimization is achieved through the deployment of Maximum Power Point Tracking (MPPT) approaches. MPPT is a fundamental technology that ensures a PV module runs at its maximum power output, irrespective of the dynamic and often unpredictable variations in solar irradiation and temperature. Various Maximum Power Point Tracking methods have been developed and applied all throughout the years, from Perturb and Observe (P&O) algorithms to more advantageous techniques like the Incremental Conductance method. In the search for an adaptive & advanced MPPT solution has brought up the Neural network models, which have given better results in modelling and predicting complex systems. They have an ability to learn relationships within datasets and adjust in real-time, making them an ideal candidate for boosting MPPT in PV systems. Neural networks can efficiently capture the nonlinear features of PV modules, so providing a dynamic and responsive solution for tracking the highest power point. Another significant problem is minimizing the ripple content of PV power, which affects the quality of power delivered to the load and necessitates the use of complex filter circuits of sensors to preserve tracking performance. Instead of tracking MPP-based perturbations and observations, researchers are now focusing on calculating the operating point corresponding to maximum power because perturbation-based MPP tracking methods are the primary cause of the ripple content [1,2].

DC-DC converters are helpful in controlling the variable output power of solar PV cells as these cells have varying output powers. A fundamental type of DC-DC converter is the buck-boost converter. There are differences between the input and output voltages of the converter. Due to the fact that the input current is irregular, making it unsuitable for energy conversion apps that make use of renewable energy. Another kind of buck-boost converter is the Cuk converter. It functions similarly to the classic buck-boost converter. It can do both buck and boost. Unlike a conventional buck boost, this makes use of an extra energy transferred from the input capacitor to produce less ripple of the converter voltage.

II. OVERVIEW OF SINGLE DIODE MODEL

The single diode model represents a PV cell with a single ideal diode and a series resistor to account for losses due to the resistance of the semiconductor material and the conductive paths within the cell. One popular mathematical approximation for characterizing the electrical properties of a photovoltaic (PV) cell or module is the single diode model. Usually, the PV cell is shown in this model as a current source connected in parallel to a series resistor and diode. The series resistor symbolizes the cell's internal resistance, while the diode simulates the behaviour of the semiconductor junction inside the cell. The below model equations depict the PV

979-8-3315-3013-6/25 $31.00 © 2025 IEEE

cell's current-voltage (I-V) relationship under a range of operational circumstances, such as variations in temperature and irradiance [1].

$$I = I_L - I_0 \left(\exp\left(\frac{V + IR_S}{\eta V_T} \right) \right) - \frac{V + IR_s}{R_{sh}} \quad (1)$$

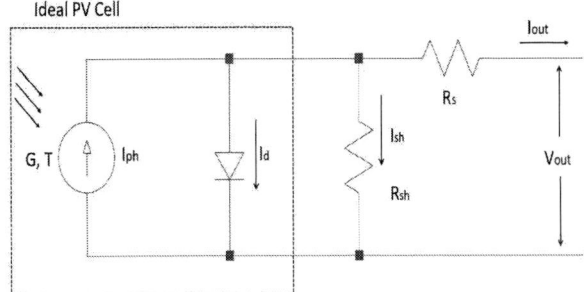

Fig 1 Single diode model of practical PV cell.

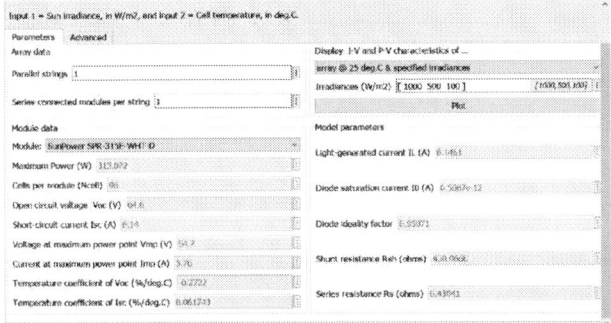

Fig 2 Parameters of PV Panels

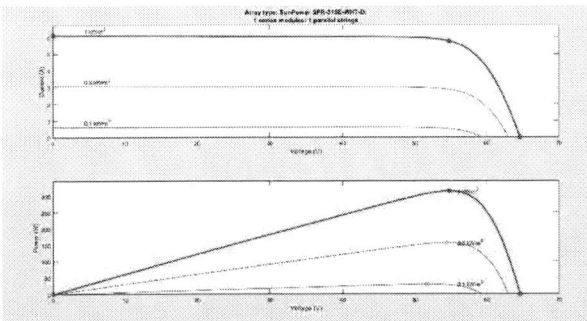

Fig 3 P-V & I-V Characteristics of PV Panel

III. CONVENTIONAL METHOD FOR MPP TRACKING

Conventional P & O Method for MPPT

The Perturb and Observe Method P&O method is a widely used Maximum Power Point Tracking (MPPT) algorithm in photovoltaic systems. It is an iterative method that continuously adjusts the operating voltage of the PV panel to obtain the maximum power output. By continuously adjusting the operating point in the direction of increasing power until a decrease is observed, the algorithm effectively tracks the Maximum Power Point (MPP) of the PV module. The P&O method has high efficiency. It is simple and easy to implement, and it requires minimum hardware and software setup. Thus, making it suitable for various PV system applications. It also has disadvantage, it suffers from

oscillations around the maximum power point, as it continuously perturbs under changing environment or in partial conditions [3].

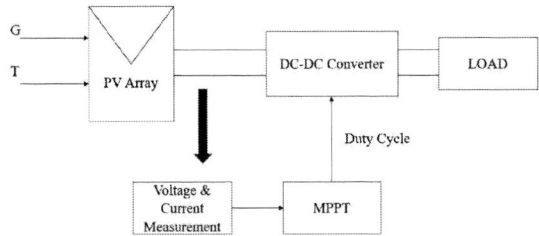

Fig 4. Block diagram of P & O method.

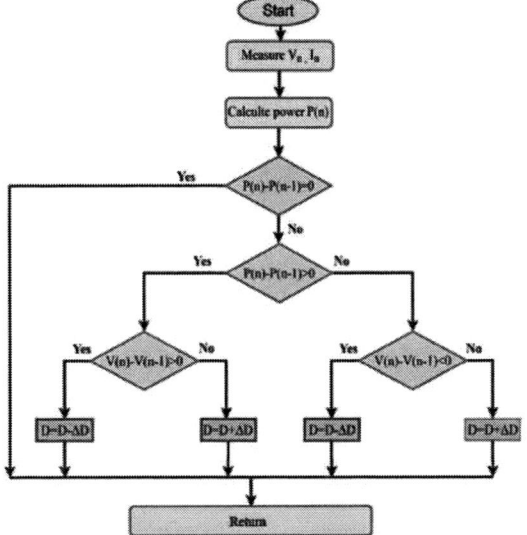

Fig 5 Flow chart for P & O method.

IV. PARAMETRIC ESTIMATION METHOD

In a given PV System, the known parameters are V_{OC} (Open Circuit Voltage), I_{SC} (Short Circuit Current), P_{Max} (Maximum Power) & Temperature Co-efficient of V_{OC} & I_{SC} and unknown parameters I_{Ph} (Photo Current Source), R_S (Series Resistance), Rs (Shunt Resistance), I_{sat} (Module Reverse Saturation Current) A (Ideality Factor). The unknown parameters are calculated according to the formulae [4,5] with the known parameters using iterative process and voltage at maximum power point is computed and given as reference to the controller. This ensures a control loop that continuously adjusts the operating voltage of the PV module to ensure it operates at MPP. According to the tuning values, the PI controller can effectively track changes in solar irradiance and temperature, thus optimizing the power output of the PV system under varying environmental conditions. From Single diode model we can formulate equations for Vmpp can be formulated as follows and solved using iterative methods to track MPP.

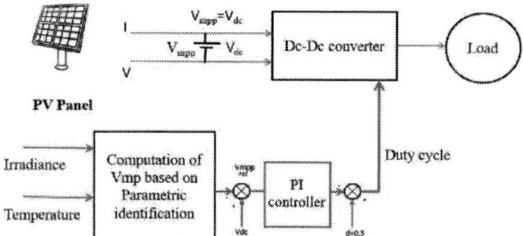

Fig 6 Block diagram of Parametric Estimation Method

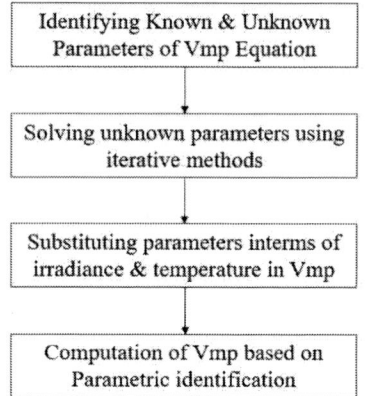

Fig 7 Flow chart for parametric estimation method

V. ANN BASED PARAMETRIC ESTIMATION METHOD

ANN simplifies the solutions obtained through complex calculations into different architecture which can predict the answer by training the same with different scenarios. In this, a prediction for Vmp carried out by training the model by giving Irradiance, T, *Voc*, *Isc* as the inputs and taking Vmp & Imp as output. [12,13]

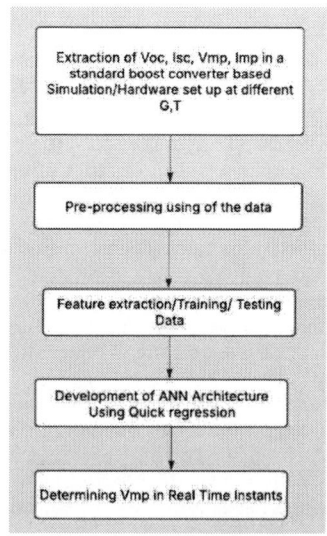

Fig 8 Determining Vmp through ANN

VI. DESIGN OF BOOST CONVERTER

Calculation for Boost converter [6]

$$\frac{V_o}{V_{in}} = \frac{1}{1-D} \tag{2}$$

$$\frac{380}{113.01} = 3.36 = \frac{1}{1-D} \tag{3}$$

$$1 - D = 0.297 \tag{4}$$

$$D = 0.703$$

$$L_{min} = \frac{(1-K)^2 * K * R}{2 * f} \tag{5}$$

$$R = 150\Omega$$

$$f = 33KHz$$

$$L_{min} = \frac{(1-0.703)^2 * 0.703 * 150}{2 * 33k} \tag{6}$$

$$L_{min} = 140.9mH$$

$$C_{min} = \frac{D}{R * f * V_r} \tag{7}$$

$$= \frac{0.703}{150 * 33k * 1.21}$$

$$C_{min} = 0.117uF$$

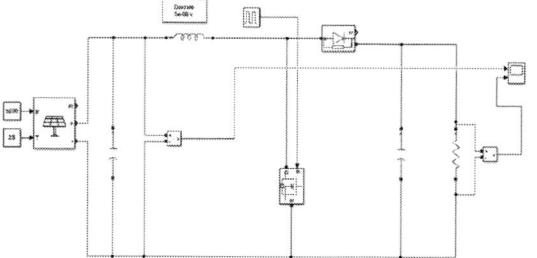

Fig 9 Boost Converter

VII. DESIGN OF SEPIC CONVERTER

Calculations for SEPIC Converter [7]

$$\frac{Vo}{Vin} = \frac{380}{113.01} = 3.36 \tag{8}$$

$$D = \frac{380 - 113.01}{380 + 113.01} = 0.5415 \tag{9}$$

$$R = \frac{113.01}{2} = 56.5 \, \Omega$$

$$L_1 = L_2 = \frac{V_{in} * D}{\Delta I_L * f} \tag{10}$$

$$= \frac{113.01 * 0.5415}{0.1 * 56.5 * 33000}$$

$$L_1 = L_2 = 0.328 \, mH$$

$$P_{in} = V_{in} * I_{in}$$

$$P_{in} = 750 \, W$$

$$I_{L1} = I_{in} = \frac{750}{113.01} = 6.63 \, A$$

$$I_{out} = \frac{750}{380} = 1.973 \ A$$

$$\Delta V_{cm} = \left(\frac{1}{1-0.5415}\right) * 37 * 1.5\% \qquad (11)$$

$$= 1.21 V$$

$$C_m = C_s = \frac{1.973}{1.21 * 33000}$$

$$= 0.494 \ uF$$

$$V_{Do} = V_{cm} = \left(\frac{1}{1-0.5415}\right) * 37 = 82.15 \ V \qquad (12)$$

$$I_{Dm} = I_{Do} = I_o = \frac{P_o}{V_o} \qquad (13)$$

$$= \frac{750}{380} = 1.973 \ A$$

$$V_{DS} = V_{cm} = \left(\frac{1}{1-0.5415}\right) * 37 = 82.15 \ V \qquad (14)$$

$$I_{DS} = I_{L1} + I_{L2} \qquad (15)$$

$$= 6.63 + 1.973$$

$$I_{DS} = 8.603 \ A$$

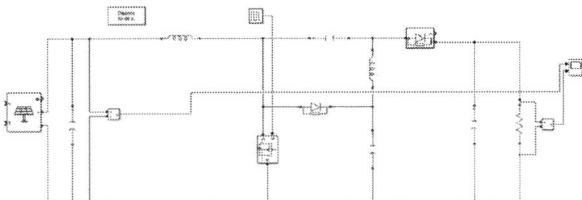

Fig 10 SEPIC Converter

VIII. SIMULATION OF PARTIAL SHADED CONDITION

Consider, 4 solar PV panels connected in parallel. Each panel has a current capability of 6.14A as short circuit current. As they are connected in parallel, final short circuit current at 100% irradiance is 24.56A. The arrangement of the PV Panels is shown in the fig 4. [8,9]

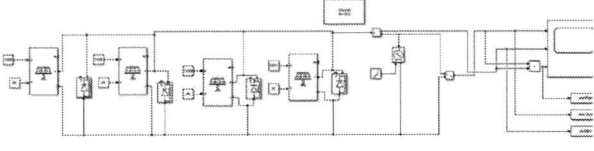

Fig 11 Arrangement of Solar Panels

A detailed comparison for maximum power obtained with Boost converter and SEPIC Converter circuits in different cases

Case 1: Parametric Estimation method at constant irradiance of 1000W/m2

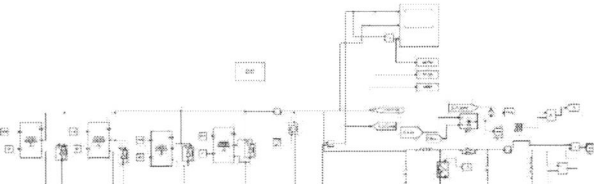

Fig 12 Boost converter with 1000W/m2 irradiance

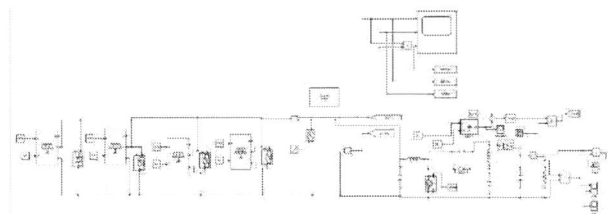

Fig 13 SEPIC converter with 1000W/m2 irradiance

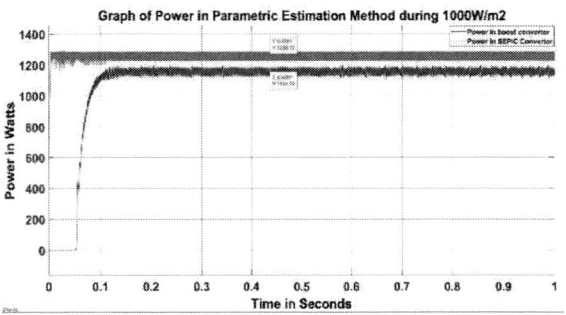

Fig 14 Power in Parametric Estimation method with 1000W/m2

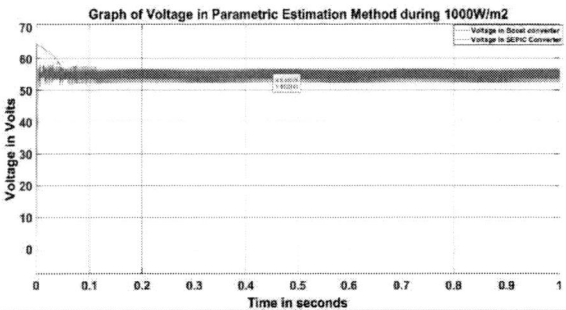

Fig 15 Voltage in Parametric Estimation method with 1000W/m2

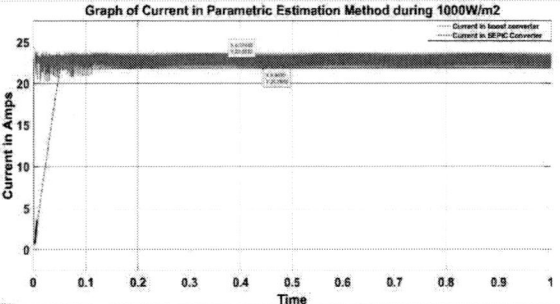

Fig 16 Current in Parametric Estimation method with 1000W/m2

Case 2: Parametric estimation method during Partial shading condition. In the four panels, 1st & 4th panel having irradiance of 1000W/m2, while 2nd panel irradiance decreased to 800W/m2 in next second & 3rd panels irradiance decreased to 600W/m2 in 3rd second. [10]

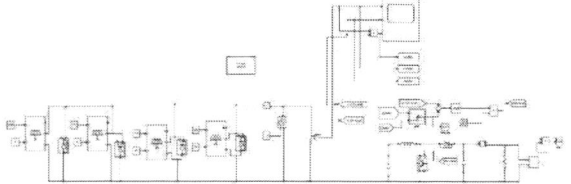

Fig 17 Boost converter circuit at partial irradiance conditions

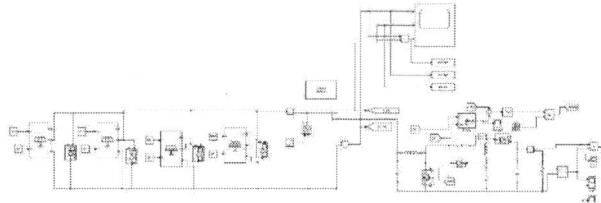

Fig 18 SEPIC converter circuit at partial irradiance conditions

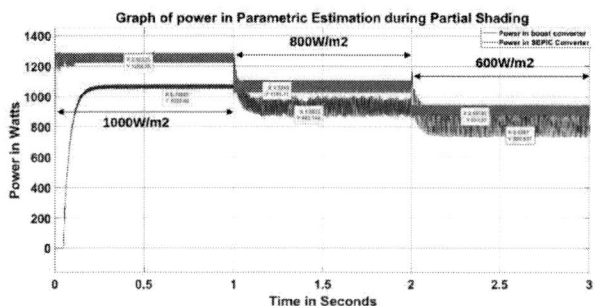

Fig 19 Power in Parametric Estimation method with partial irradiance conditions

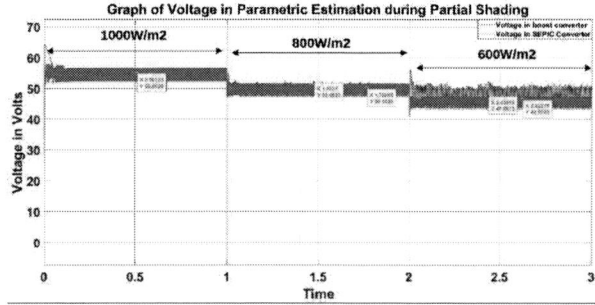

Fig 20 Voltage in Parametric Estimation method with partial irradiance conditions

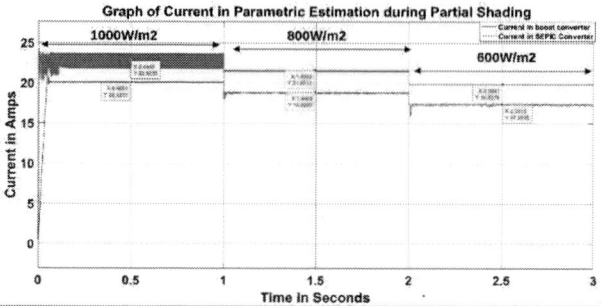

Fig 21 Current in Parametric Estimation method with partial irradiance conditions

Case 3: Parametric estimation using ANN method during Partial shading condition. In the four panels, 1st & 4th panel having irradiance of 1000W/m2, while 2nd panel irradiance decreased to 800W/m2 in next second & 3rd panels irradiance decreased to 600W/m2 in 3rd second. ANN model trained using Levenberg-Marquardt method, considering 2500 data points of a specific location by extracting data of irradiance, temperature from NASA official website.

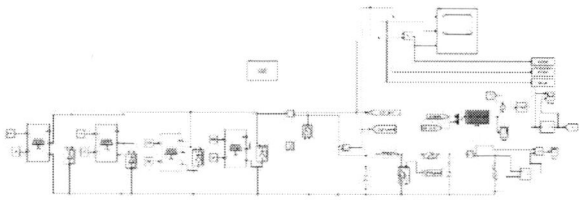

Fig 22 Boost converter circuit with ANN method at partial irradiance conditions

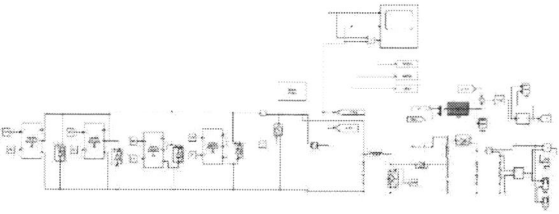

Fig 23 SEPIC converter circuit with ANN method at partial irradiance conditions

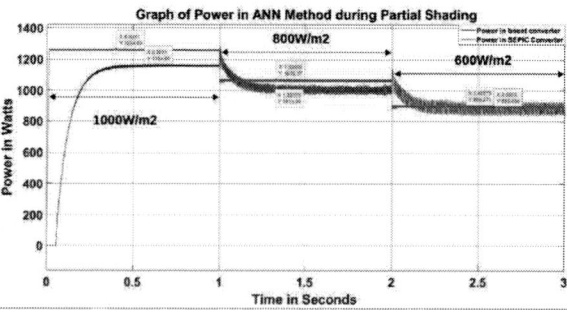

Fig 24 Power in ANN method with partial irradiance conditions

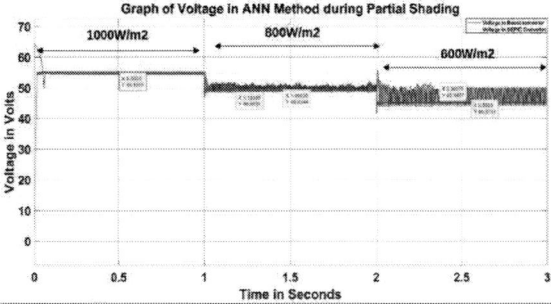

Fig 25 Voltage in ANN method with partial irradiance conditions

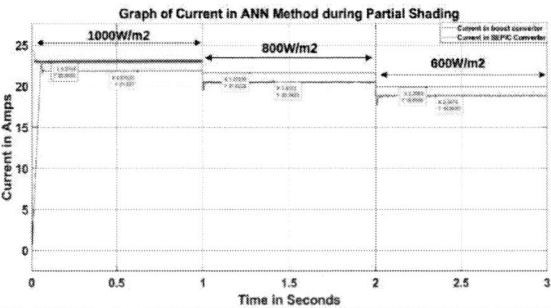

Fig 26 Current in ANN method with partial irradiance conditions

IX. CONCLUSION

In this paper, boost and SEPIC Converter circuits were developed for obtaining Maximum power point. The values of inductor, input and output capacitor of SEPIC converter is small when compared with the values of Boost converter, thus the size of inductor and capacitors decreases in SEPIC Converter. A detailed comparison of MATLAB model is developed to quantify the maximum power point during constant irradiance and during partial shading of PV Panels in different irradiance conditions using Parametric estimation method and ANN based method. The figures fig 14, fig 15, fig 16 quantify the output power, voltage, current from boost converter is 1161.4 W, 55.21V, 21.79A and from SEPIC Converter 1269.72 W, 55.7V, 23.08A respectively. Similarly, from fig 19 output power at 800W/m2 943.144W in boost converter, 1101.2W from SEPIC Converter. The output power from ANN based partial shading condition reached the maximum power condition of PV array with lesser ripples, when compared with ideal parametric estimation method.

REFERENCES

[1] Geng Y, Chen W, Liu Z, Chiu ASF, Han W, Liu Z, Zhong S, Qian Y,You W, Cui X. A bibliometric review: energy consumption and greenhouse gas emissions in the residential sector. J Clean Prod 2017;159:301–16.

[2] N. M. Haegel and S. R. Kurtz, "Global Progress Toward Renewable Electricity: Tracking the Role of Solar (Version 3)," in IEEE Journal of Photovoltaics, vol. 13, no. 6, pp. 768-776, Nov. 2023, doi: 10.1109/JPHOTOV.2023.3309922.

[3] A. S. Mahdi, A. K. Mahamad, "Maximum power point tracking using perturb and observe, fuzzy logic and ANFIS", SN Applied Sciences, 2020.

[4] H. K. Mehta, H. Warke, K. Kukadiya and A. K. Panchal, "Accurate Expressions for Single-Diode-Model Solar Cell Parameterization," in IEEE Journal of Photovoltaics, vol. 9, no. 3, pp. 803-810, May 2019, doi: 10.1109/JPHOTOV.2019.2896264

[5] Chatterjee, A. & Keyhani, Ali, "Thevenin's equivalent of photovoltaic source models for MPPT and power grid studies". IEEE Power and Energy Society General Meeting. 1-7. 10.1109/PES.2011.6039203.

[6] Roberto F. Coelho, Filipe Concer, "A Study of the basic DC-DC Converters applied in Maximum Power Point Tracking" Brazilian Power Electronics Conference, 04 December 2009.

[7] H. Suryoatmojo, I. Dilianto, "Design and Analysis of High Gain Modified SEPIC Converter for Photovoltaic Applications", 2018 IEEE International Conference on Innovative Research and Development (ICIRD) 11-12 May 2018,Bangkok Thailand

[8] Sushmita Sarkar, K Uma Rao, "A model for effect of partial shading on PV Panels with experimental validation", Energy Reports 12, 2024

[9] K. Rahul Wilson, Y Srinivasa Rao, " Effects of Partial Shading on different structures of Solar Photovoltaic Arrays, Journals of Mechanics of Continua and Mathematical Sciences, Nov – Dec 2019

[10] Haider Ibrahim, Nader Anani, "Variations of PV module parameters with irradiance and temperature", Energy Procedia, Volume 134,Pages 276-285, 2017

[11] Ankur Bhattacharjee, Bijit Kumar Dey, "Mathematical Modelling and Characteristic analysis of Solar PV Cell" 2016 IEEE 7th Annual Information Technology, Electronics and Mobile Communication Conference (IEMCON), 17 November 2016.

[12] Rajib Baran Roy, Md. Rokonuzzaman, "A Comparative Performance Analysis of ANN Algorithms for MPPT Energy Harvesting in Solar PV System" IEEE Access, 13 July 2021

[13] Nazmul Islam Nahin, Shuvra Prokash, "A Modified PWM Strategy With an Improved ANN Based MPPT Algorithm for Solar PV Fed NPC Inverter Driven Induction Motor Drives", IEEE June 2023.

[14] Ahmed, Ihsan, and Turki Kahawish. "Design and Implementation of High Gain SEPIC Converter." 2021 International Conference on Advanced Computer Applications (ACA). IEEE, 2021.

[15] S. Shongwe and M. Hanif, "Gauss-Seidel iteration based parameter estimation for a single diode model of a PV module," 2015 IEEEElectrical Power and Energy Conference (EPEC), London, ON,Canada, 2015, pp. 278-284

[16] G. Ganesh, G Vijay Kumar, A.R.VijayBabu, G.Srinivasa Rao, Y.R.Tagore, Performance Analysis and MPPT Control of a Standalone Hybrid Power Generation System, Journal of Electrical Engineering, Volume 15, Edition: 1, pp. 334-343, 2015.

[17] Darmini and K. Sunitha, "Comparison of solar PV array configuration methods under different shading patterns," Proc. 2017 IEEE Int. Conf. Technol. Adv. Power Energy Explor. Energy Solut. an Intell. Power Grid, TAP Energy 2017, pp. 1–4, 2018.

Retrofitting of Existing Solar Power Condition Unit as Reactive Power Compensator

Rajesh Gupta, *Senior Member IEEE*
Electrical Engineering Department
Motilal Nehru National Institute of Technology Allahabad
Prayagraj, India
Email: rajeshgupta@mnnit.ac.in

Kumari Priya
Electrical Engineering Department
Motilal Nehru National Institute of Technology Allahabad
Prayagraj, India.
Email: priya1998jd@gmail.com

Mohammad Tahir Siddiqui
Electrical Engineering Department,
Motilal Nehru National Institute of Technology Allahabad
Prayagraj, India.
Email: siddiquitahir688@gmail.com

Abstract— **Domestic solar PV system involving roof-top solar PV plant, with in-house inverter and battery-pack, is very common in rural and urban areas. A power conditioning unit (PCU) is commonly used as heart of the solar PV plant system, housing DC-DC converter for maximum power point tracking and DC-AC converter as inverter to drive the domestic AC loads. This paper proposes to retrofit the existing power conditioning unit supplied from solar PV and battery energy storage system (BESS), to supply reactive power to the grid for compensation. An experimental setup consisting of 540 W solar PV module with 24 V, 42 Ahr battery supported PCU has been reconfigured as reactive power compensator in a single-phase 230 V grid, loaded with multiple induction motors, drawing the reactive power. The PCU is able to inject maximum 1 kVAr reactive power into the grid. The solar PV under this condition continues to charge the BESS with small 2.5% of real power to support the system losses. This combination can maximizes utilization of solar PV plants. The MATLAB simulation is employed to verify the performance of the proposed system under various availability combination of input sources. An experimental setup is used to retrofit the 1.0 kVA PCU as reactive power compensator.**

Keywords—**Battery pack, retrofitting, power conditioning unit, solar PV**

I. INTRODUCTION

Sustainability in energy resources is essential to meet our needs and ensure long-term availability. As energy consumption continues to rise, addressing energy sustainability issues is critical. The renewables especially solar offer viable alternatives to non-renewable sources both for mass production and domestic utilization. Today roof-top solar PV, battery pack and PCU, are integrated to ensure efficient energy utilization. To maximize the potential of solar PV system, it is vital to diversify their utilization for other power conditioning applications such as harmonic and reactive power compensation. The expansion of renewable energy has led to technological advancements, cost reduction, and innovation, in energy storage, grid integration, and improving system efficiency, making sustainable energy solutions increasingly viable.

The grid challenges and solution on grid integration with PV Systems has been dealt in [1]. Important grid challenges involve, power prediction, voltage stability, frequency - variations, reactive power compensation, harmonic distor-

-tions, stability, fault/low voltage ride through capability etc. The PV-BES synchronised system has been investigated in [2]. It enables smooth transition between operational modes, from grid-tied to islanding and re-synchronization, ensuring stable load voltages, without transient disturbances. The battery energy storage system can effectively reduce power oscillations. During islanding, voltage control ensures regulation of load voltages. Harmonics are also minimized, leading to improved power quality. Microgrid using centralized controller based on PV-BESS has been suggested in [3], for decreasing conversion losses in residential distribution systems, enhancing energy efficiency and reducing cost. The BESS undergoes multiple conversions, resulting in significant power losses and reduced efficiency. The inverters used in the various applications may have poor efficiency, especially when lightly loaded, resulting in excessive power wastage.

Grid based EV battery charging puts enormous burden on the power system, which restricts the development of an environmentally friendly transportation system [4]. Solar PV system is available in abundance in a country like India, and has superior performance among other renewable energy sources (RESs), and if used for electric vehicle battery charging, can reduce dependency on the grid through integrated PV arrays [5]. The solar PV power can be a feasible solution for in-house and public charging, being locally generated and location independent. To enhance the system reliability against the solar intermittency, integration of RESs with a BESS system, and grid support, can be utilized [6].

Residential power conditioning system utilizing solar, battery, and grid for connecting both DC and AC loads has been worked out in [7]. Various operational modes have been incorporated for different combinations and loading conditions. A power conditioning module has been suggested in the residential application to feed both low voltage and medium voltage buses utilizing the hybrid resources in [8]. All the modes including grid forming, feeding and supportive modes have been included for single-phase system using a single microcontroller. The grid connected solar inverter can be used as a reactive power compensator for full utilization of inverter rating in [9].

This paper presents a concept of retrofitting of the existing residential solar power conditioning unit as reactive power compensator, especially during low solar durations. Power flow equations have been revisited to verify the

979-8-3315-3013-6/25 $31.00 © 2025 IEEE

reactive power injection mode by the inverter into the grid. Simulation results have been obtained to show the operation in solar to grid mode. Laboratory experiments have been performed to test the performance of 1.0 kVA PCU as reactive power compensator.

II. POWER FLOW EQUATIONS BETWEEN SOLAR PV INVERTER AND GRID

Two key parameters are considered while transferring power from a solar PV system to the grid: voltage magnitude and phase angle (delta) between the inverter output and grid. The inverter voltage should be synchronized with the grid.

1. Voltage Magnitude: The output voltage magnitude of the inverter should be higher than the grid voltage. This requires precise control of the boost converter to ensure that the DC-bus voltage is appropriate for the inverter to generate the correct AC voltage magnitude.

2. Phase Angle: The inverter output phase difference with the grid controls the direction and amount of real/reactive power flow. In order to deliver power from PV system to the grid, the inverter voltage must lead the grid voltage by a small angle. For transmitting power from a solar PV system to the grid at a constant inverter voltage, the power flow depends upon the load angle. The challenges related to steady-state power flow in various systems can be analyzed using the equivalent circuit shown in Fig. 1.

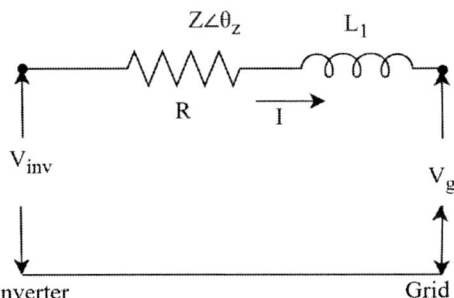

Fig. 1 Equivalent circuit diagram.

It shows two AC voltage-sources, V_{inv} and V_g interconnected through an impedance $Z\angle\theta_z$. When the current I flows from V_{inv} to V_g, the relation can be established as follows [10], [11].

$$\overline{V}_{inv} = \overline{V}_g + \overline{IZ} \tag{1}$$

$$\overline{I} = \frac{\overline{V}_{inv} - \overline{V}_g}{\overline{Z}} \tag{2}$$

The phasor diagram depicting relation between inverter and grid voltage is shown in Fig. 2. Equations shows that current $\overline{I}$ is the difference of two current $\frac{\overline{V}_{inv}}{\overline{Z}}$ and $\frac{\overline{V}_g}{\overline{Z}}$, with $\overline{V}_g$ lagging behind by angle $\angle\theta_z$. The impedance angle θ_z is given by

$$\theta_z = \tan^{-1}\frac{X}{R} \tag{3}$$

$$\alpha_z + \theta_z = 90° \tag{4}$$

The real power P_{inv} delivered by the inverter with the voltage V_{inv} and impedance Z is given by [11],

$$P_{inv} = V_{inv} \times \text{In-phase component of I with } V_{inv} \tag{5}$$

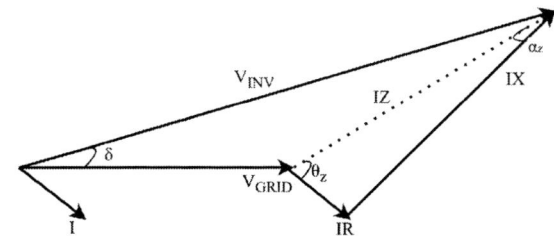

Fig. 2 Phasor diagram including line resistance.

This will lead to the common expression of power delivered by the inverter as follows.

$$P_{inv} = V_{inv}\left[\frac{\overline{V}_{inv}}{Z}\cos\theta z - \frac{V_g}{Z}\cos(\delta + \theta z)\right] \tag{6}$$

$$P_{inv} = \left[\frac{V_{inv}^2}{Z}\cos\theta z - \frac{V_{inv}V_g}{Z}\cos(\delta + \theta z)\right] \tag{7}$$

Where,

$$\cos\theta_z = \frac{R}{Z} \tag{8}$$

$$\text{if, } \theta_z = 90° - \alpha_z \tag{9}$$

$$P_{inv} = \left[\frac{V_{inv}^2 R}{Z^2}\cos\theta z - \frac{V_{inv}V_g}{Z}[\cos\{(\delta - \alpha_z) + 90°\}]\right] \tag{10}$$

$$= \frac{V_{inv}V_g}{Z}\sin(\delta - \alpha_z) + \frac{V_{inv}^2 R}{Z^2} \tag{11}$$

The power P_g at the grid side with the voltage V_g and flowing through the impedance Z, is given by

$$P_g = \left[\frac{V_{inv}V_g}{Z}\cos(\theta z - \delta) - \frac{V_g^2}{Z}\cos\theta z\right] \tag{12}$$

$$P_g = \left[\frac{V_{inv}V_g}{Z}\sin(\delta + \alpha_z) - \frac{V_g^2}{Z^2}R\right] \tag{13}$$

Power input to the inverter can be defined as

$$P_{ii} = \left[\frac{V_{inv}V_g}{Z_s}\sin(\delta - \alpha_z) - \frac{V_{inv}^2}{Z_s^2}R\right] \tag{14}$$

Power output from the inverter can be defined as

$$P_{io} = \left[\frac{V_{inv}V_g}{Z_s}\sin(\delta + \alpha_z) - \frac{V_g^2}{Z_s^2}R\right] \tag{15}$$

The difference between input P_{ii} and outputs P_{io} must be equal to ohmic losses $I_a^2 R$

$$P_{ii} - P_{io} = \frac{V_{inv}V_g}{Z_s}[\sin\delta\cos\alpha_z - \cos\delta\sin\alpha_z$$
$$- \sin\delta\cos\alpha_z - \cos\delta\sin\alpha_z]$$
$$+ (V_{inv}^2 + V_g^2)\frac{R}{Z_s^2} \tag{16}$$

$$= \frac{R}{Z_s^2}\left[V_{inv}^2 + V_g^2 - 2V_{inv}V_g\cos\delta\right]$$

Considering, line resistance R negligible, and this makes

$$\alpha_z = 0 \tag{17}$$

$$Z_s = X_s \tag{18}$$

979-8-3315-3013-6/25 \$31.00 © 2025 IEEE   401

When this change applies in equations, following can be derived for the active power.

$$P_{ii} = P_{io} = P \qquad (19)$$

$$P = \frac{V_{inv}V_g}{X_s} \sin \delta$$

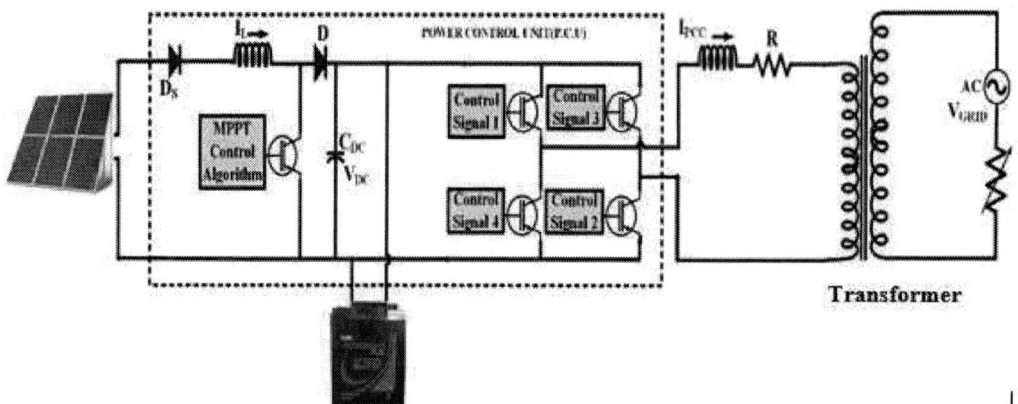

Fig. 3 Schematic of solar PV feeding to a grid with battery.

Reactive power at the inverter output terminal can be obtained as follows [11].

$$Q_{oi} = V_g \times \text{quadrature component } I \text{ with } V_g$$

$$= V_g \left[\frac{V_{inv}}{Z_s} \sin(\theta_z - \delta) - \frac{V_g}{Z_s} \sin \theta_z \right] \qquad (20)$$

$$= \frac{V_{inv}V_g}{Z_s} \sin[90 - (\delta + \alpha_z)] - \frac{V_g^2}{Z_s} \sin \theta_z \qquad (21)$$

$$= \frac{V_{inv}V_g}{Z_s} \cos(\delta + \alpha_z) - \frac{V_g^2}{Z_s^2} X_s \qquad (22)$$

For an inverter, the condition for maximum reactive power can be obtained as follows.

$$\frac{dQ_{oi}}{d\delta} = \frac{V_{inv}V_g}{Z_s} \cos(\delta + \alpha_z) = 0 \qquad (23)$$

$$\delta = -\alpha_z \text{ or } \delta + \alpha_z = 0 \qquad (24)$$

$$Q_{oimax} = \frac{V_{inv}V_g}{Z_s} - \frac{V_g^2}{Z_s^2} X_s = \frac{V_g}{Z_s} \left[V_{inv} - \frac{V_g}{Z_s} X_s \right] \qquad (25)$$

$$= \frac{V_g}{Z_s} [V_{inv} - V_g] \text{ if } R = 0 \qquad (26)$$

If $R = 0$, then

$$Q_{oi} = \frac{V_{inv}V_g}{X_s} \cos \delta - \frac{V_g^2}{X_s} = \frac{V_g}{X_s} (V_{inv} \cos \delta - V_g) \qquad (27)$$

III. GRID INTEGRATION OF SOLAR PV SYSTEM FOR REACTIVE POWER COMPENSATION USING PCU

To integrate PV systems with the electrical grid, the inverter output must be synchronized with the grid in terms of voltage, frequency, and phase. A suitable low-pass filter, is used to smoothen the inverter output, reduces the harmonics and ensure that the power quality meets the grid standards. The schematic of the PV feeding to the grid with battery storage is shown in Fig. 4. A boost DC-DC converter is used to extract the maximum power from the PV modules to keep the battery charged and transferring real power to the load or grid. A battery bank is used to sustain the converter side losses during reactive power compensation, when solar PV output is low or during night time. A transformer/autotransformer is shown in the Fig. 4 for laboratory environment purpose. This can be used to vary the

inverter output voltage in order to control the inverter side AC output voltage of the PCU, with the AC output voltage magnitude is fixed. Varying the inverter AC voltage can control the reactive power flow into the grid. The autotransformer can provide fine variation of the voltage. Existing single-phase PCU can be used as inverter for interfacing the solar PV system to the grid, without any transformer.

IV. ANALYTICAL AND SIMULATION RESULTS

Table I depicts the detailed data of the system considered in Fig. 3. Based on the developed power flow equations first the calculations have been done by varying grid voltage and power angle using (19) and (20), for real and reactive power, respectively under lossless condition. Table II lists four scenario by varying the inverter voltage from $V_{INV} = 150$ V(rms) to $V_{INV} = 200$ V(rms).

TABLE I PARAMETERS OF PV SYSTEM AND INVERTER.

Variable	Numerical values
Number of PV modules	9- in series
Power rating of module	213.15 W -/+3% W
Open-circuit voltage	36.3 V
Short-circuit current	7.84 A
Maximum power voltage	29.0 V
Maximum power current	7.35 A
Inverter Voltage	(150-200) V
Inductor	10 mH

TABLE II ANALYSIS OF POWER FLOW AT DIFFERENT VOLTAGE AND DELTA ANGLE VARIATION (AC VOLTAGE AND CURRENT IN RMS)

	Case-1	Case-2	Case-3	Case-4
V_{PV}	305.6 V	303.9 V	316.6 V	314.6 V
I_{PV}	4.79 A	5.085 A	2.34 A	2.59 A
P_{PV}	1465 W	1545 W	518.3 W	618.2 W
V_{DC}	406.6 V	405 V	421.3 V	418.5 V
V_{GRID}	141.4 V	141.4 V	141.4 V	141.4 V
I_{GRID}	6 A	6.25 A	2.2 A	3.56 A
δ	5°	10°	5°	10°
V_{INV}	150 V	150 V	200 V	200 V
I_{INV}	5.9 A	6.4 A	2.5 A	4 A
P	587 W	1170 W	782 W	1560 W
Q	378 VAr	301 VAr	2615 VAr	2512 VAr

The grid voltage is considered constant at V_{GRID}=141.4 V (rms). The power angle is varied from δ = 5° to δ=10°. Different real and reactive power flow from inverter to grid is listed in Table II.

Next the simulation results are obtained through the MATLAB/SIMULINK plateform with the consideration of parameters listed in Table I, with loss components also included. The simulation results of real and reactive power flow, grid voltage and current, and solar PV power and voltage, are obtained for the respective four cases and shown in Fig. 4 to 7. A scaling factor of 10 is used in I_{GRID} for better visibility and phase relation with V_{GRID}. Fig. 4 depicts the plots of real and reactive power flow, grid voltage and current, solar PV voltage, current, power and DC link voltage, for inverter voltage V_{INV}=150 V (rms), δ = 5°.

Fig. 5 represents the similar plots inverter voltage V_{INV}=150 V (rms), δ = 10°. Similarly, plots in Fig. 6 and 7 are obtained for V_{INV}=200 V (rms), δ = 5°, and V_{INV}=200 V (rms), δ = 10° respectively.

1. Inverter voltage V_{INV}=150 V (rms), δ = 5°

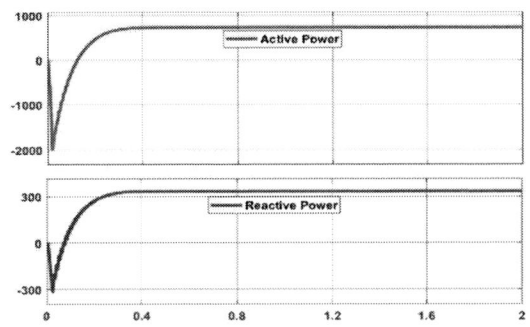

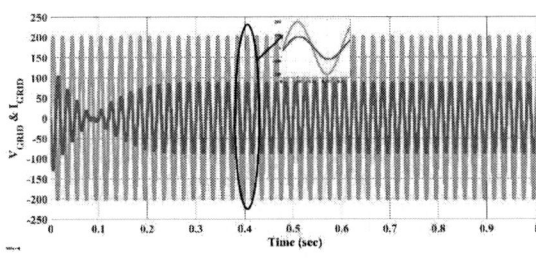

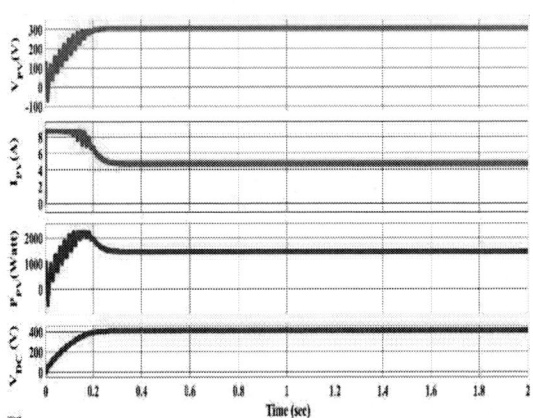

Fig. 4 Real and reactive power flow, grid voltage and current, and solar PV voltage, current, power and DC link voltage, for inverter voltage V_{INV}=150 V (rms), δ = 5°.

2. Inverter voltage V_{INV}=150 V (rms). δ = 10°

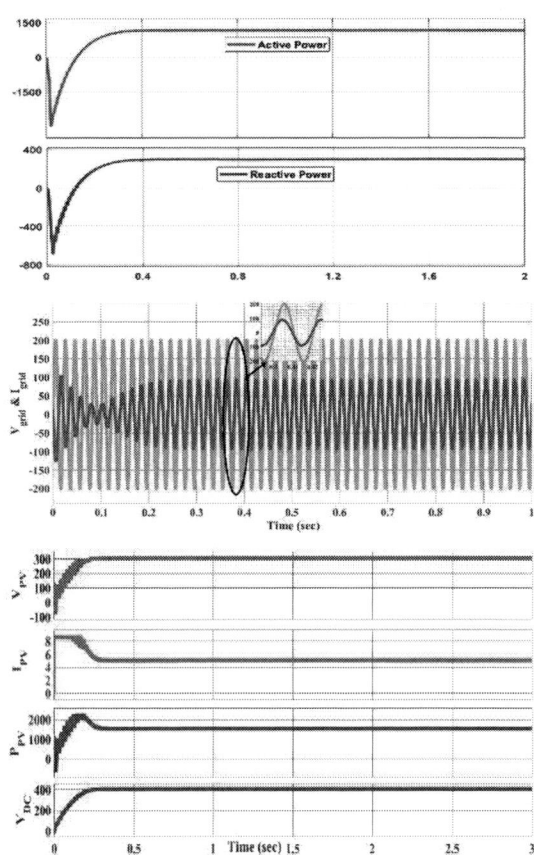

Fig. 5 Real and reactive power flow, grid voltage and current, and solar PV voltage, current, power and DC link voltage, for inverter voltage V_{INV}=150 V (rms), δ = 10°.

3. Inverter voltage V_{INV}=200 V (rms), δ = 5°

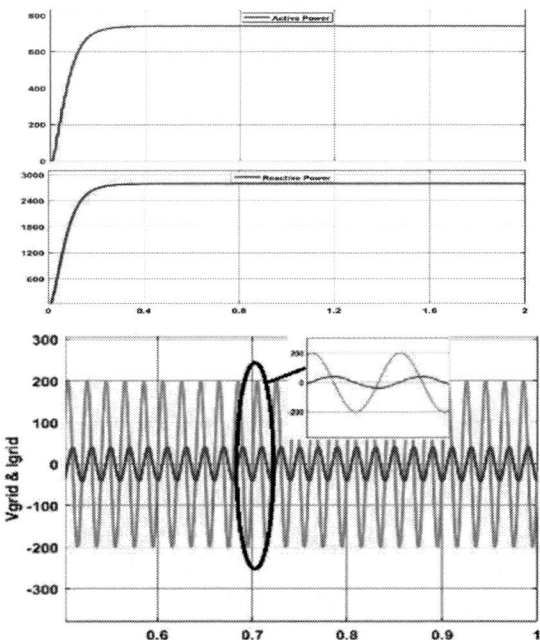

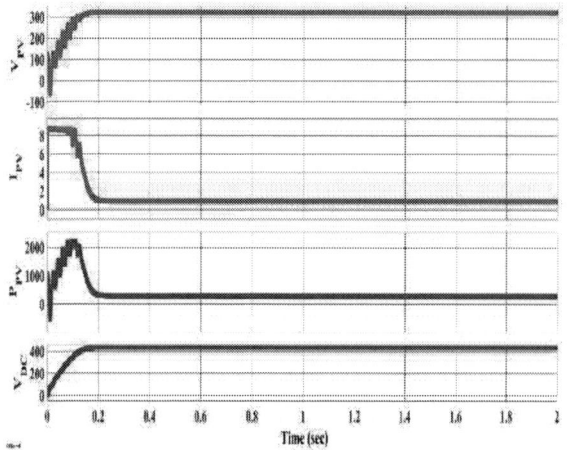

Fig. 6 Real and reactive power flow, grid voltage and current, and solar PV voltage, current, power and DC link voltage, for inverter voltage V_{INV}=200 V (rms), δ = 5°.

4. Inverter voltage V_{INV}=200 V (rms), δ = 10°

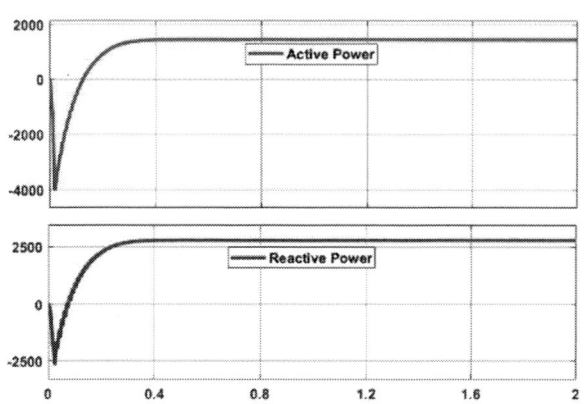

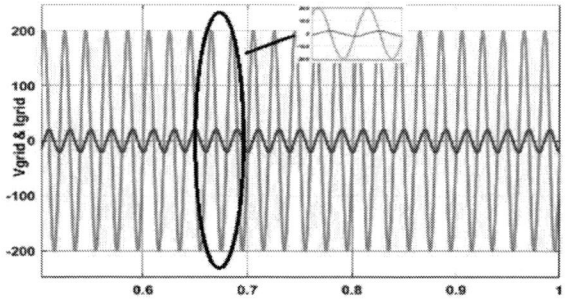

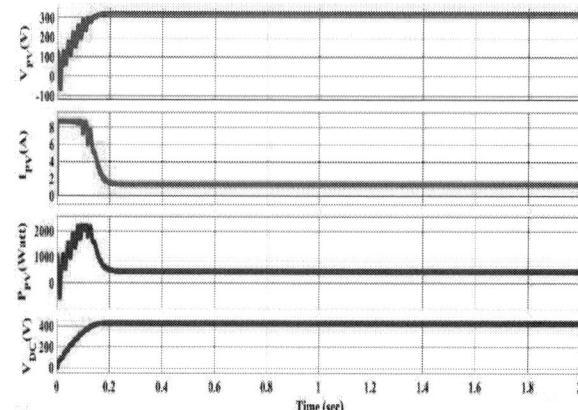

Fig. 7 Real and reactive power flow, grid voltage and current, and solar PV voltage, current, power and DC link voltage, for inverter voltage V_{INV}=200 V (rms), δ = 10°.

The results in the figures are closed to the analytical results, with the difference due to the loss components included in the simulations, following the exact analysis done in the previous section.

V. RETROFITTINGOF EXISTING PCU AS REACTIVE POWER COMPENSATOR (EXPERIMENTAL RESULTS)

Fig 8 shows the experimental setup using PCU to reconfigure it as reactive power compensator to inject current in phase quadrature with the grid voltage. Table III lists the observation at different interfacing resistance regarding injection of real and reactive power, when the inverter output voltage and grid voltage are in phase. From the last column it can be seen that when the interfacing resistance becomes minimum 5 Ω and inductance 5 mH, the reactive power injected in the grid is 975 VAr and real power is 245 W. The detailed waveform of grid voltage, inverter voltage and grid current for this case is shown in experimental figures.

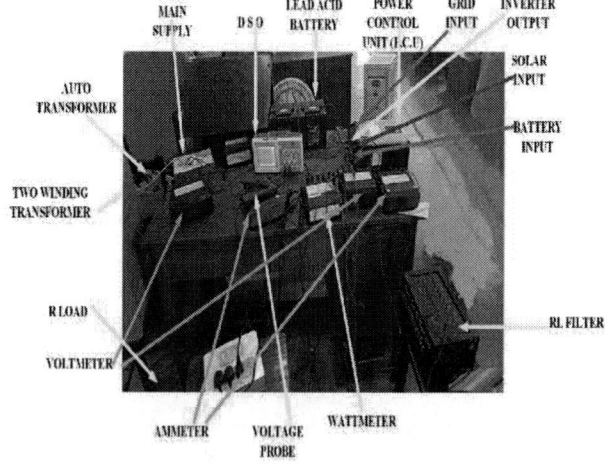

Fig. 8 Laboratory experimental setup using PCU.

By changing delta angle between grid and inverter by varying either voltage between them or by changing resistance we can inject reactive power from the power conditioning unit to the grid.

979-8-3315-3013-6/25 $31.00 © 2025 IEEE

TABLE III EXPERIMENT DATA ANALYSIS AT DIFFERENT INTERFACING RESISTANCE REGARDING INJECTION OF REAL AND REACTIVE POWER, WHEN THE INVERTER OUTPUT VOLTAGE AND GRID VOLTAGE ARE IN PHASE.

S. No.	Case-1	Case-2
V_{IN}	220 V	220 V
I_{IN}	3.6 A	5.5 A
R	10 Ω	5 Ω
V_{LOAD}	212 V	210 V
I_{LOAD}	3.6 A	5.5 A
W_{LOAD} (M.F=4)	44	30
$V_{BATT(pcu)}$	23.7 V	23.9 V
I_{BATT}(A)	9 A	8 A
$W_{PV(pcu)}$	128 W	129 W
$I_{BATT(pcu)}$	10 A	7 A
V_{PV}	26 V	27 V
I_{PV}	1.0 A	0.5 A
$W_{LOAD(pcu)}$	P_{LOAD} − 303 W Q_{LOAD} − 724 VA $V_{SECONDARY}$ − 209 V V_{GRID} -202 V	P_{LOAD} − 245 W Q_{LOAD} − 975 VA $V_{SECONDARY}$ − 204 V V_{GRID} − 204 V

By changing delta angle between grid and inverter by varying either voltage between them or by changing resistance we can inject reactive power from the power conditioning unit to the grid. Fig. 9 shows the inverter voltage obtained through solar PV and grid voltage, with small power angle δ = 10⁰. Fig. 10 shows the inverter voltage and quadrature grid current indicating the flow of dominant reactive power, justifying the reactive power compensation. By varying the magnitude of inverter voltage and angle δ, it may be possible to control the amount and direction of reactive power flow.

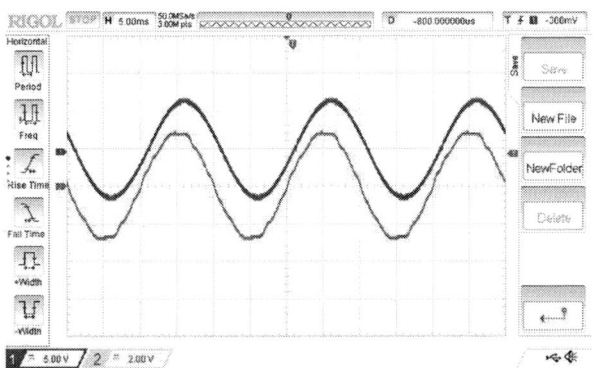

Fig. 9 Inverter voltage through solar PV and grid voltage.

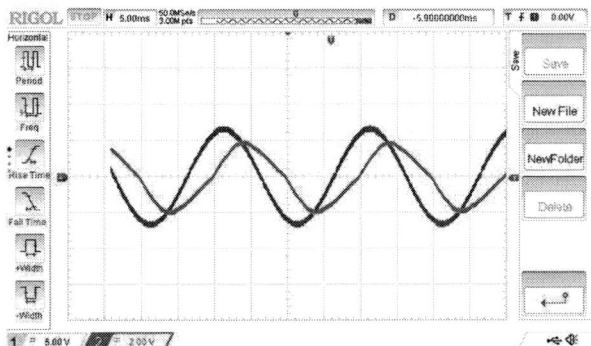

Fig. 10 Inverter voltage and 90˚ current injected by the inverter into the grid

In laboratory environment an autotransformer is connected across the PCU output to adjust the inverter voltage magnitude. The inverter of the PCU injects voltage in phase with the grid. In order to create a phase difference a variable resistance is connected. To eliminate transformer/autotransformer and variable resistance a customized PCU can be manufactured which has a capability to adjust magnitude and phase automatically to transfer both real and reactive power as per the availability of the solar power and load requirement.

VI. CONCLUSIONS

A power flow analysis has been reviewed from the point of view of the reactive power injection into the grid from PV inverter. The basic power flow equations has been used for the analysis. The voltage magnitude and delta angle control has been used for the grid integration. The solar power injection and real and reactive power injection into the grid has been verified through the analysis and simulation results. An existing PCU has been reconfigured to perform as reactive power compensator through the laboratory experimental setup. The test results show that the maximum 1 kVAr reactive power can be injected into the grid for reactive power compensation using the PCU.

REFERENCES

[1] M. Shafiullah, S. D. Ahmed and F. A. Al-Sulaiman, "Grid Integration Challenges and Solution Strategies for Solar PV Systems: A Review," in IEEE Access, vol. 10, pp. 52233-52257, 2022.

[2] S. Kumar and B. Singh, "Seamless Operation and Control of Single-Phase Hybrid PV-BES-Utility Synchronized System," in IEEE Transactions on Industry Applications, vol. 55, no. 2, pp. 1072-1082, March-April.

[3] S. Gangatharan, M. Rengasamy, R. M. Elavarasan, N. Das, E. Hossain and V. M. Sundaram, "A Novel Battery Supported Energy Management System for the Effective Handling of Feeble Power in Hybrid Microgrid Environment," in IEEE Access, vol. 8, pp. 217391-217415, 2020.

[4] J. Chen, Z. Li and X. Yin, "Optimization of energy storage size and operation for renewable-EV hybrid energy systems". *IEEE Green Tech. Conf. (GreenTech 2021)*, Denver, CO, USA, 2021, pp. 118-124.

[5] S. Nagar, V. Gupta, R. Kumar, R. C. Bansal and R. M. Naidoo, "PV-BES integrated residential society governed electric vehicle charging station*", 9th Renewable Power Gen. Conf. (RPG Dublin Online 2021)*, Online Conference, Mar. 2021, pp. 192-197.

[6] S. S. G. Acharige, M. E. Haque, M. T. Arif, N. Hosseinzadeh and S. Saha, "A Solar PV Based Smart EV Charging System with V2G Operation for Grid Support," *2021 31st Australasian Universities Power Engineering Conference (AUPEC)*, Perth, Australia, Sept. 2021, pp. 1-6.

[7] A. Agrawal and R. Gupta, "Power management and operational planning of multiport HPCS for residential application", *IET Generation, Transmission and Distribution*, vol. 12, no. 18, pp. 4194 – 4205, Oct. 2018.

[8] A. Agrawal and R. Gupta, "Coordinated Control of Hybrid DERs Enabled Grid-Interactive Residential PCM With Hybrid Bus Layout," *IEEE Systems Journal*, vol. 16, no. 3, pp. 4607-4618, Sept. 2022.

[9] A. Dhaneria and H. Khambhadiya, "Enhancing The Utilization of Existing Solar Inverter by Incorporating Reactive Power Compensation Feature," *2021 National Power Electronics Conference (NPEC)*, pp. 1-5, 15-17 Dec. 2021, Bhubaneswar, India.

[10] P. Kundur, *Power System Stability and Control*, McGraw-Hill Inc. 1994,. USA.

[11] P. S. Bimbhra, Electrical Machinery (Theory, Performance and Applications), Khanna Publishers, 7th ed. 2012, New Delhi, India.

Hybrid Integration of PV, Fuel Cell and Battery for DC Microgrid

Umang Kartikey
Department Of Electrical Engineering
Delhi Technological University
Bawana, Delhi, India
umangkartikey_ep21b11_49@dtu.ac.in

Vinayak Saxena
Department Of Electrical Engineering
Delhi Technological University
Bawana, Delhi, India
vinayaksaxena_ee21b10_11@dtu.ac.in

Yogendra Tiwari
Department of Electrical Engineering
Delhi Technological University
Bawana, Delhi, India
yogendratiwari_ee21b10_21@dtu.ac.in

Prakash Chittora
Department of Electrical Engineering
Delhi Technological University
Bawana, Delhi, India
prakashchittora@gmail.com

Vineet Kumar
Department of Electrical Engineering
Delhi Technological University
Bawana, Delhi, India
vineetkumar_2k22phdee03@dtu.ac.in

Abstract—This paper presents a robust DC microgrid model integrating solar photovoltaic (PV) arrays, proton exchange membrane fuel cells (PEMFCs), and lithium-ion batteries, designed to address energy intermittency and grid instability in renewable-based systems. Building on prior research identifying inefficiencies in AC microgrids and unreliable backup solutions, this work introduces a DC architecture with advanced energy management strategies to minimize losses and ensure uninterrupted power supply. A MATLAB/Simulink model validates the system's performance under islanded operation, demonstrating stable DC bus voltage (395–405V) and seamless mode transitions during fluctuating generation and load demands. Key innovations include a four-mode coordinated control strategy for battery protection, PEMFC integration for zero-emission backup, and a hierarchical control framework for dynamic power balancing. The results highlight the system's ability to maintain energy equilibrium, prevent battery overcharge/discharge, and potentially achieve 85% PEMFC stack efficiency, offering a scalable solution for remote and grid-sensitive applications.

Index Terms—DC Microgrid, Solar Photovoltaic, Fuel Cell, Renewable Energy Integration, Dynamic Power Balancing, Coordinated Control

I. INTRODUCTION

The global energy crisis and environmental degradation require decentralized, renewable-powered microgrids. Traditional AC grids suffer from high transmission losses, as highlighted by [1] , which demonstrated that DC microgrids can reduce conversion inefficiencies by up to 15% compared to AC systems. Furthermore, AC grids struggle with instability under fluctuating renewable inputs due to synchronization challenges and reactive power mismatches [2] . Diesel backups, commonly used for reliability, exacerbate carbon emissions, contributing to 25–30% higher lifecycle greenhouse gas emissions in off-grid regions [3] . Remote and disaster-prone areas demand resilient systems capable of

islanded operation, a capability standardized in IEEE 1547.4 for microgrids [4]

This study addresses these challenges by proposing a hybrid DC microgrid combining solar photovoltaic (PV), proton exchange membrane fuel cells (PEMFCs), and batteries. DC microgrids inherently mitigate AC-DC conversion losses, enhance stability through decentralized control [5], and simplify integration with native DC sources such as solar PV and batteries [6], [7] . Solar PV provides low-cost renewable energy but suffers from intermittency, necessitating complementary technologies. PEMFCs, with high efficiency (60–70%) and fuel flexibility, offer steady baseload power, effectively bridging gaps in solar generation [8]. Batteries, such as lithium-ion, provide rapid response to transient loads and store excess energy, optimizing system reliability [9].

The proposed system employs decentralized control strategies, leveraging distributed energy management algorithms to prioritize renewable sources while ensuring seamless transitions to islanded mode during grid failures [5]. Scalability is achieved through modular architecture, enabling incremental expansion in remote areas with evolving energy demands [10]. Case studies, such as the Santa Rita Jail micro-grid [11], validate the resilience of hybrid systems in critical settings. Life cycle analyses further underscore environmental benefits, with hybrid microgrids reducing Carbon Dioxide emissions by 40–50% compared to diesel-dependent systems [12]. The major contribution of this paper here are as follows:

- Renewable-Centric Design: Prioritizes renewable energy to reduce dependency on non-renewable sources.
- Versatile Deployment: Applicable across residential, commercial, and industrial scenarios.
- Targeted Accessibility: Enhances energy access in underserved regions.
- Sustainable & Resilient: Addresses core challenges of sustainability, reliability, and efficiency in energy sys-

979-8-3315-3013-6/25 $31.00 © 2025 IEEE

tems.

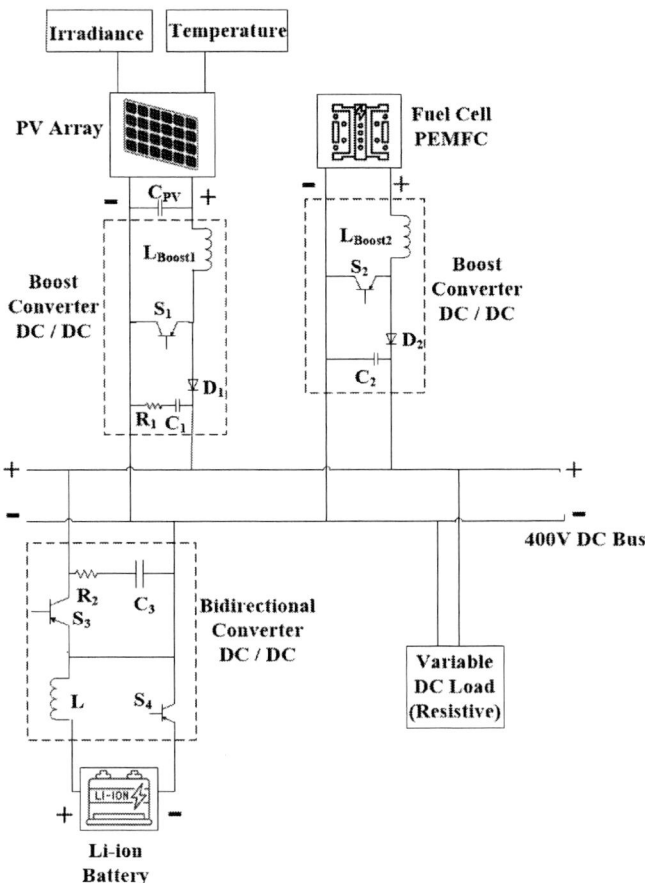

Fig. 1: DC Microgrid Structure

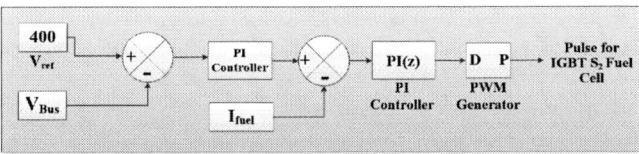

Fig. 2: Fuel Cell (PEMFC) Control

II. MICROGRID SIMULATION ORCHESTRATION

A. Microgrid Structure

The DC Micro-grid consists of several key components as shown in Fig.1. It includes a PV array using a Samsung SDI LPC235SM-08 module, rated at 235.612W, and equipped with a Perturb and Observe (P&O) MPPT for maximum power point tracking. A 6kW Proton Exchange Membrane Fuel Cell, with a nominal voltage of 133.33V, provides flexible backup power. Additionally, the system features a 250V/50Ah lithium-ion battery storage unit, integrated with bidirectional DC-DC converters for effective voltage regulation. The microgrid operates on a 400V DC bus, which minimizes conversion stages and associated losses while enabling both islanded and grid-tied operational modes.

B. Integration of Microgrid Components

1. *Fuel Cell*: Integrating fuel cells into the microgrid ensures an uninterrupted power supply, enhancing the system's reliability and resilience. They are activated during periods of low solar generation or when the battery storage is insufficient to meet the energy demand. Fuel cells are advanced electrochemical devices that generate electricity through redox reactions. Unlike conventional power generation methods, fuel cells are a completely green source of energy production due to no combustion involved, resulting in high efficiency and zero harmful emissions. The fuel cell is a viable energy source as it works like a battery but does not require any replenishment or recharging.

In this micro-grid, a series of PEMFC (Proton Exchange Membrane Fuel Cell or Polymer Electrolyte Membrane Fuel Cell) are employed, wherein each cell produces 61.5V and 20A and the entire combination produces 245V of voltage which is then passed through a DC-DC converter giving an overall output of 400V.

This fuel cell type is ideal as it involves: low-temperature operation, faster start-up time, highest power density, light weight, and solid electrolyte prevents any chance of leakage. The structure of a typical fuel cell includes an Anode (oxidation of Hydrogen), a Cathode (reduction of Protons), and an electrolyte membrane (Perfluorosulfonic Acid). The electrodes are porous and coated with Platinum(Pt) as a catalyst.

Anode: $H_2 \rightarrow 2H^+ + 2e^-$
Cathode: $2H^+ + 2e^- + \frac{1}{2}O_2 \rightarrow H_2O$

There are two inputs to the Fuel Cell model: Air Flow Rate and Fuel Flow Rate which are given by:

$$V_{air} = \frac{60000 * R * T_{nom}}{2 * z * F * P_{air} * 0.5 * 0.21} \tag{1}$$

$$Vfuel = \frac{60000 * R * T_{nom}}{2 * z * F * P_{air} * 0.919 * 0.9985} \tag{2}$$

where, I_{nom}: nominal current; R: universal gas constant 8.314 J/mol-K; N: number of cells ; F: Faraday's constant 96485 AS/mol ; T_{nom}: nominal temperature (in Kelvin) ; P_{air} : nominal absolute air pressure; z: charge (for air = 2, for Hydrogen=1) ; Rate of conversion of oxygen (21%) is assumed as 50%; Rate of conversion of Hydrogen (99.85%) is assumed as 91%.

The combined fuel cell output is fed to the DC-DC Boost converter. The control scheme employed for fuel cell is a closed-loop control using PI controllers, as shown in Fig. 2, which gives the required duty cycle so as to maintain the DC bus voltage.

2. *PV Array*: Photovoltaic (PV) Cell Arrays are semiconductor devices that convert sunlight into electricity through the photovoltaic effect. This effect occurs when sunlight strikes

979-8-3315-3013-6/25 $31.00 © 2025 IEEE   407

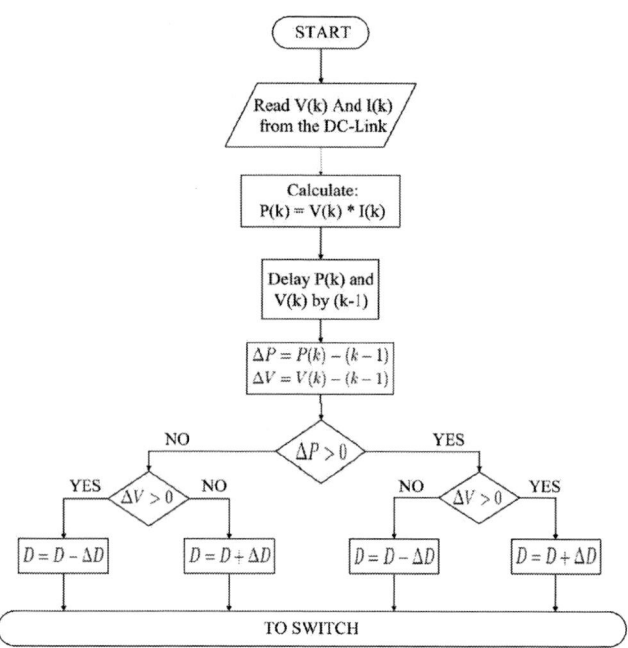

Fig. 3: Perturb And Observe Flowchart

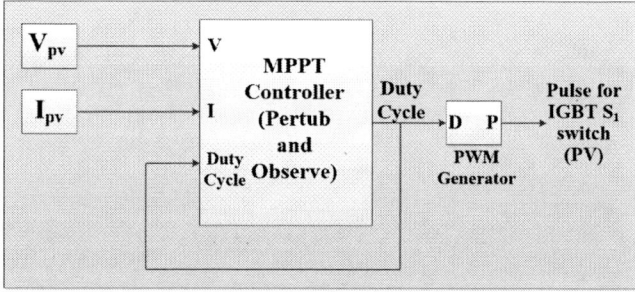

Fig. 4: PV Control

the surface of a PV cell, exciting electrons and causing them to move, generating an electric current. In this DC microgrid, the PV array is the primary energy generation component. It is designed to capture solar energy during daylight hours and convert it into Direct Current (DC). To optimize energy extraction, the PV system incorporates the Maximum Power Point Tracking (MPPT) mode and employs the Perturb and Observe (P&O) technique. This P&O algorithm applies small, periodic changes to the Photovoltaic module's voltage and observes the resulting changes in power output, as shown in Fig. 3. If the power increases, the algorithm continues adjusting in the same direction. Conversely, if the power decreases, it reverses the direction of adjustment. This algorithm continuously adjusts the operating conditions of the PV array to ensure it operates at its maximum power point, even under varying sunlight and temperature conditions.

The PV Array incorporated has 5 parallel strings and 7 series connected modules per string and a total of 60 cells per module. The PV array uses the MPPT point of 209.3V

and 39.4A operating at a temperature of 25 degrees Celsius and irradiation of 1000 W/m². The control scheme uses a closed loop control employing PI controllers, as shown in Fig.4, to maintain the DC bus voltage. The PV array has an output of 200V which is fed to the DC-DC Boost controller making the overall output 400V of voltage. The PV array in this project is integrated with the battery and fuel cell systems to ensure continuous power supply, even during periods of low solar radiation. The modular nature of PV cells makes them adaptable to various applications, from residential installations to large-scale commercial energy systems.

3. *Battery*: Battery cells are critical components of renewable energy systems, serving as energy storage units that ensure a stable and reliable power supply. They store excess energy generated by the PV array during periods of high solar radiation and release it when energy demand exceeds generation. The battery system in this project employs Lithium-ion-based cells designed for high efficiency and durability. The basic structure of a battery cell includes an Anode (Graphite rod), a Cathode (Lithium Metal Oxide $LiMO_2$, an electrolyte (Lithium salt in liquid form), and a separator (Polyethylene or Polypropylene).

During the charging process, electrical energy from the PV array is stored as chemical energy in the battery. While charging oxidation takes place at Cathode, which produces Li^+ions that flow through the electrolyte to Anode, where the ions combine with Graphite to form Lithiated Carbon.

Cathode: $LiMO_2 \rightarrow Li^+ + MO_2 + e^-$

Anode: $C_6 + Li^+ + e^- \rightarrow LiC_6$

When energy is needed, the chemical energy is converted back into electrical energy and supplied to the load. During the discharging process oxidation of Lithium takes place at Anode to yield Li(+) and travel to the Cathode and reduction occurs to yield $LiMO_2$.

Cathode: $MO_2 + Li^+ + e^- \rightarrow LiMO_2$

Anode: $LiC_6 \rightarrow Li^+ + C_6 + e^-$

In this micro-grid, the battery system has a nominal voltage of 250V and a rated capacity of 50Ah. This plays a crucial role in stabilizing the energy supply. The Li-ion battery has a high energy density, compact, lightweight, high charge density, comparatively environment-friendly, no memory effect and a long cycle life ensuring that power is available during nighttime or cloudy days or when solar generation is low.

The Battery output is fed to the DC-DC Bi-Directional converter, which caters to both the charging and discharging cycles, acting as both Boost (Battery to DC Bus) and Buck (DC Bus to Battery) converter. The control scheme employed for the Li-ion Battery is a closed-loop control using PI controllers, as shown in Fig. 5, which gives the required duty cycle to maintain the DC bus voltage.

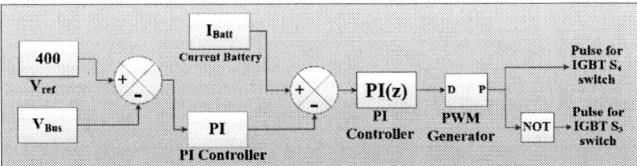

Fig. 5: Battery (Li-ion) Control

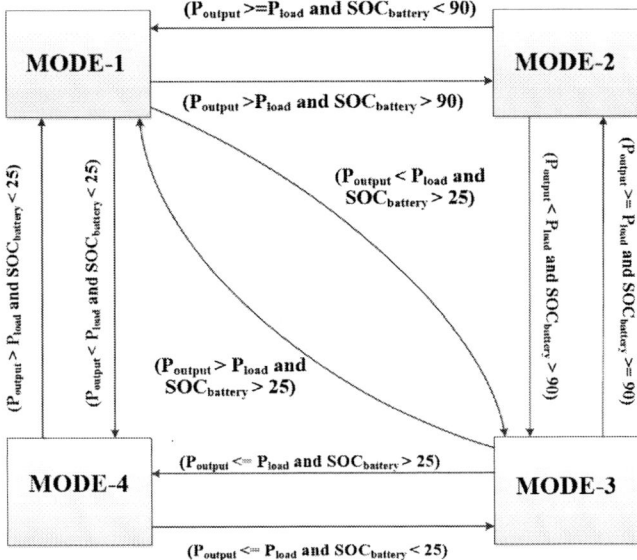

Fig. 6: Coordination Control Flow

III. ENERGY MANAGEMENT AND COORDINATED CONTROL

The control scheme proposed for this DC micro-grid, as shown in Fig. 6, is based on the principle wherein the PV array is to be used as the primary source of energy, and during the situations of shortfall (low irradiance/night), Fuel Cell will be sought as the source for energy. The battery acts as an energy storage element(for excess energy produced), which can be used to supplement the major sources if and when required. The energy management and coordination control used for this paper can be divided into 4 modes of operation as shown in the flowchart as shown below:

MODE-1 : $P_{out} \geq P_{load}$ & $SOC < 90\%$
In this mode, the output power of the PV array or Fuel cell unit is independently greater than the load power demand of the DC microgrid and the battery's state of charge(SOC) is less than 90%. Depending upon the situation and requirement either of the two units will be used to produce the energy. The energy produced will be used to quench the load demand and the extra energy produced will be used to charge the battery, thereby storing the energy for future use. In case the PV array and Fuel cell are independently not able to produce the required power demand, the system will switch to mode 3. In case the SOC of the battery reaches greater than or equal to 90%, switch to mode-2.

MODE-2 : $P_{out} \geq P_{load}$ & $SOC \geq 90\%$
In this mode, the output power of the PV array or Fuel cell unit is independently greater than the load power demand of the DC microgrid and the battery's state of charge(SOC) is greater than or equal to 90% [13]. In this mode, the energy production takes place in LTP (Load Power Tracking) mode, wherein only power enough to quench the load demand will be produced.

MODE-3 : $P_{out} < P_{load}$ & $SOC > 25\%$
In this mode, the output power of the PV array or Fuel cell unit is independently insufficient to meet the load power demand of the DC microgrid and the battery's state of charge(SOC) is greater than 25% [13]. To meet the shortfall in the desired power output the energy producing unit will be supplemented by the battery. The battery will discharge to quench the shortfall and thus maintain smooth overall operation in the DC microgrid. In case the SOC of the battery diminishes below 25% switch to mode-4. In case, even combined with the battery the unit is still unable to produce the required load demand switch to mode-4.

MODE-4 : $P_{out} < P_{load}$ & $SOC \leq 25\%$:
In this mode, the output power of the PV array and Fuel cell unit is insufficient to meet the load power demand of the DC microgrid and the battery's state of charge(SOC) is less than equal to 25%. Here, both the units will be used to simultaneously operate and thereby produce the energy to meet the required power demand of the load.

A. Priority Hierarchy
- **Priority 1** (PV Array): Acts as the primary energy source during daylight hours.
- **Priority 2** (Fuel Cell): Activates as a backup source during PV generation shortfall (low irradiance/night).
- **Priority 3** (Battery): Supplies stored energy during PV generation shortfalls and supports Fuel Cell energy production.

B. Resource Constraints and Operational Protocols
This is active when irradiance $>= 1000$ W/m², subject to the constraints that, Excess energy charges the battery if State of Charge (Soc) $<= 90\%$ and curtails surplus energy if battery is fully charged Soc $< 90\%$ and no grid export is available. Also, it is inactive during nighttime or low-irradiance conditions.

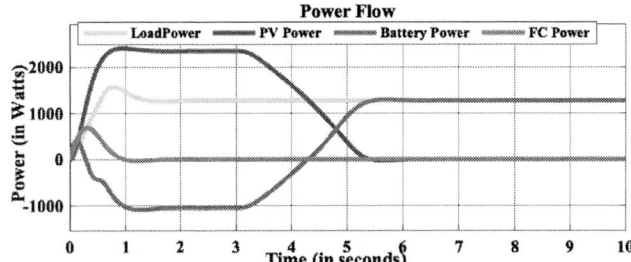

Fig. 7: Power Flow for Case 1-(a) : Mode 1-3

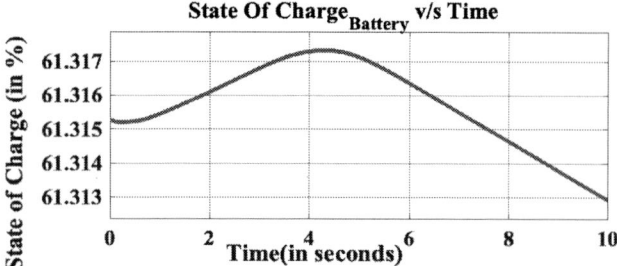

Fig. 8: Battery SOC for Case 1-(a) : Mode 1-3

IV. SIMULATION VERIFICATION

To validate the performance of the proposed hybrid energy system under dynamic conditions, we have incorporated two cases (one each for the PV array and Fuel Cell as the primary source) under the proposed operational modes. These cases demonstrate the system's ability to maintain a continuous power supply through effective coordination between photovoltaic (PV) panels, fuel cells (PEMFC), and batteries. The results confirm the system's robustness and reliability.

Case 1: Photo Voltaic Array as the Primary energy source:

Case 1-(a): Operational condition : Mode-1 to Mode-3 transition:

In Fig. 7, in the initial phase (up to time t=3 seconds) we can observe that PV is generating energy in abundance, thereby quenching the entire power demand of the load and establishing itself as the primary source of energy. Also, it can be marked that while supplying power to the load, PV is also charging the battery, which can be verified by the gradual increment in the SOC of the Battery as seen in Fig. 8. This establishes the working of simulation in Mode-1. After t=3 seconds, we can observe the decrease in the power generated by the PV array due to a reduction in irradiance (simulating the night condition), resulting in the initiation of Mode-3, wherein the Battery discharges to support in meeting the energy requirement of the load, as demonstrated in Fig. 7 and Fig. 8 after time is 5 seconds.

Case 1-(b): Operational condition: Mode-1 to Mode-2 to Mode-3 transition:

In Fig. 9, initiates in Mode-1 (up to first 2 seconds) wherein PV generation is sufficiently large to power the load and charge the battery simultaneously. However, after time t=2 seconds the system transitions to Mode-2 as the SOC of the battery reaches above 90% (upper threshold), and hence the PV array now solely powers the load and the battery is dormant. This is verified from Fig. 10 wherein we see the battery maintaining the SOC level (up to t=7 seconds) as desired. Now, at time t=7 seconds the PV generated output decreases leading to a shortfall in meeting the load demand, so now the system transitions to Mode-3 thereby discharging the Battery to meet the power demand, as demonstrated in Fig.9 and Fig.10.

Case 1-(c): Operational condition: Mode-1 to Mode-4 transition: Fig. 11, initiates (up to t=3 seconds) operation in Mode-1in which PV generation is sufficiently large to power the load and charge the battery simultaneously. However after this, the PV generation begins to decrease significantly and at the very same time, the Battery SOC is below the minimum threshold ($SOC \leq 25\%$), as demonstrated in Fig. 12, which doesn't allow the battery to discharge. So, under this scenario load shedding is implemented which can be observed from load power in Fig. 11, thereby disconnecting the load in a way that PV still produces power sufficient enough to meet the demand, while the Battery SOC is maintained. This shows the operational transition from Mode-1 to Mode-4.

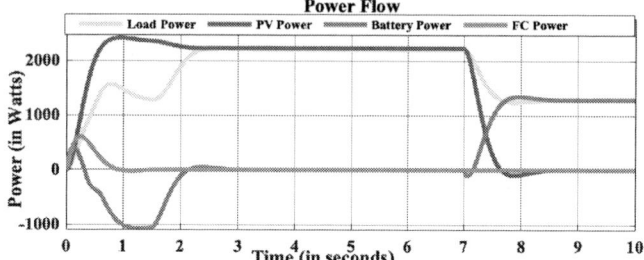

Fig. 9: Power Flow for Case 1-(b) : Mode 1-2-3

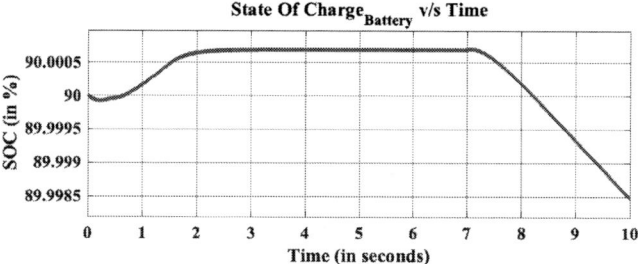

Fig. 10: Battery SOC for Case 1-(b) : Mode 1-2-3

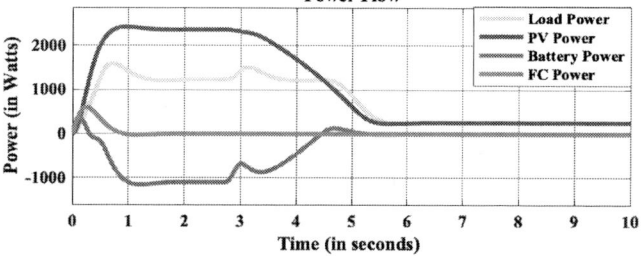

Fig. 11: Power Flow for Case 1-(c) : Mode 1-4

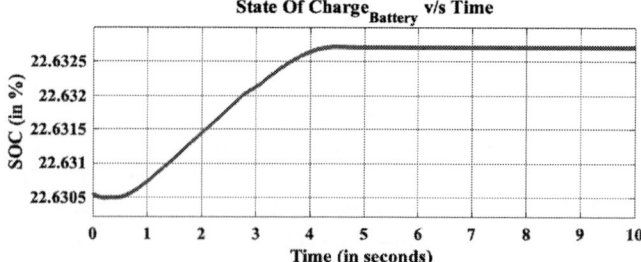

Fig. 12: Battery SOC for Case 1-(c) : Mode 1-4

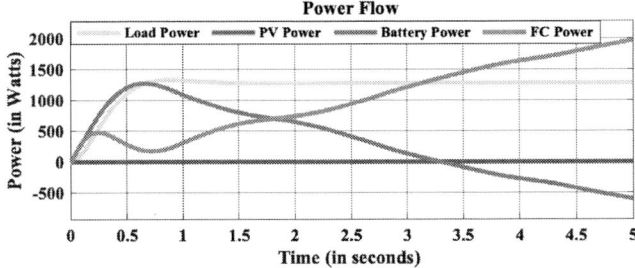

Fig. 13: Power Flow for Case 2-(a) : Mode 3-1

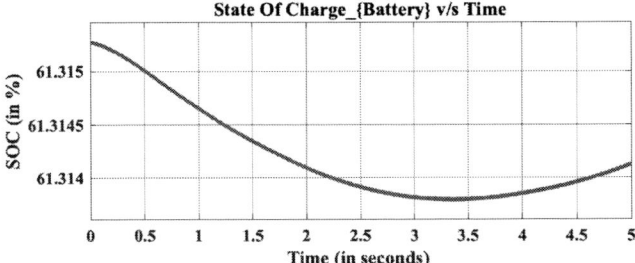

Fig. 14: Battery SOC for Case 2-(a) : Mode 3-1

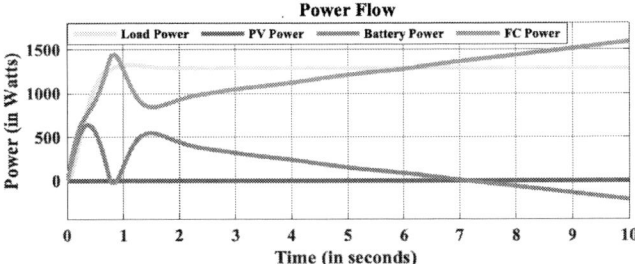

Fig. 15: Power Flow for Case 2-(b) : Mode 4-1

Case 2: Fuel Cell(PEMFC) as the Primary energy source: Fuel Cell activates as a primary source of energy in synergy with the battery during PV insufficient or non-generation (low irradiance or night) conditions, providing a greener alternative for night energy needs. Fuel cell, being energy dense in nature, can fulfill high energy demands in the long-term perspective but the starting dynamics are sluggish as shown in Fig. 13.

Case 2-(a): Operational condition: Mode-3 to Mode-1: When the fuel cell achieves the required power output, operation transitions back to Mode-1(t=3 second onward) from Mode-3 (upto 3 seconds), as demonstrated in Fig.13 and 14, i.e., from working in collaboration with Battery to overcome initial sluggish nature meeting the shortfall, to generating energy in abundance so as to quench the load demand and simultaneously charge the battery.

Case 2-(b):Operational condition: Mode-4 to Mode-1: Fig. 15 shows that initially due to sluggish starting of fuel cell and a nearly discharged battery (unable to support the

load requirement) the fuel cell slowly takes over as the sole energy source in the micro-grid in the absence of PV power and generate energy sufficient to charge the battery, thereby showing a transition from Mode-4 (until t=6 seconds) to Mode-1 (t=6 second onward).

V. RESULTS AND CONCLUSIONS

These results validate the system's compliance with hybrid microgrid design principles [10], confirming its robustness and reliability in load-following and fault-tolerant operation. The key metrics include: SOC Management, i.e., the battery maintains SOC within safe limits (25%–90%), as standardized [13]. Power Continuity, i.e., there are seamless transitions between energy sources, which are consistent with decentralized control frameworks[5]. Efficiency i.e., fuel cell utilization is maximized (potential upto 85% stack efficiency [7]), reducing fuel consumption by 25%–30% compared to diesel systems [3].

REFERENCES

[1] J. M. Bloemink and T. C. Green, "Increasing distributed generation power using residential DC microgrids," *IEEE Transactions on Smart Grid*, vol. 1, no. 2, pp. 180–187, 2010.

[2] F. Katiraei, M. R. Iravani, and P. W. Lehn, "Microgrid autonomous operation during and subsequent to islanding process," *IEEE Transactions on Power Delivery*, vol. 20, no. 1, pp. 248–257, 2008.

[3] A. Chaurey and T. C. Kandpal, "Assessment and evaluation of PV based decentralized rural electrification: An overview," *Renewable and Sustainable Energy Reviews*, vol. 14, no. 8, pp. 2266–2278, 2010.

[4] IEEE Standard 1547.4, "IEEE Guide for Design, Operation, and Integration of Distributed Resource Island Systems with Electric Power Systems," IEEE, Tech. Rep., 2011.

[5] X. Lu, J. M. Guerrero, K. Sun, and J. C. Vasquez, "An improved droop control method for DC microgrids based on low bandwidth communication with DC bus voltage restoration and enhanced current sharing accuracy," *IEEE Transactions on Power Electronics*, vol. 29, no. 4, pp. 1800–1812, 2015.

[6] A. Hussain, V. H. Bui, and H. M. Kim, "A resilient and privacy-preserving energy management strategy for networked microgrids," *IEEE Transactions on Smart Grid*, vol. 9, no. 3, pp. 2127–2139, 2017.

[7] V. Kumar, H. Ashfaq, and R. Singh, *Coordinated Control and Energy Management Strategy of Battery and Super-capacitor of Micro-grid.* Springer, 2023, book Chapter.

[8] M. H. Nehrir, C. Wang, K. Strunz, and H. Aki, "A review of hybrid renewable/alternative energy systems for electric power generation," *IEEE Transactions on Sustainable Energy*, vol. 2, no. 4, pp. 392–403, 2011.

[9] E. Hossain, M. R. Tür, S. Padmanaban, S. Ay, and I. Khan, "A comprehensive review on hybrid power system for PEMFC-HEV: Architecture, energy management strategies, and power electronics conversion," *Energies*, vol. 11, no. 10, p. 2924, 2018.

[10] A. Bidram, A. Davoudi, F. L. Lewis, and J. M. Guerrero, "Distributed cooperative secondary control of microgrids using feedback linearization," *IEEE Transactions on Power Systems*, vol. 28, no. 3, pp. 3462–3470, 2013.

[11] C. Marnay, S. Chatzivasileiadis, and C. Abbey, "Microgrid evolution roadmap," *Energy Policy*, vol. 86, pp. 751–756, 2015.

[12] J. Aghaei and M. I. Alizadeh, "Demand response in smart electricity grids equipped with renewable energy sources: A review," *Renewable and Sustainable Energy Reviews*, vol. 18, pp. 64–72, 2013.

[13] E. Wikner and T. Thiringer, "Extending battery lifetime by avoiding high soc," *Applied Sciences*, vol. 8, no. 10, p. 1825, 2018.

979-8-3315-3013-6/25 $31.00 © 2025 IEEE

An Enhanced Current Difference Based Fault Detection Technique for Low-Voltage DC Microgrid

Dr. Biswajit Sahoo
Dept. of Electrical Engineering
NIT Silchar
Assam, India.
Email:
biswajitsahoo@ee.nits.ac.in

Shantanu Saha
Dept. of Electrical Engineering
NIT Silchar
Assam, India
Email:
shantanusaha17@gmail.com

Anindya Banik
Dept. of Electrical Engineering
NIT Silchar
Assam, India
Email:
anindyabanik23@gmail.com

Shrestha Ghosh
Dept. of Electrical Engineering
NIT Silchar
Assam, India
Email:
shresthaghosh106@gmail.com

Abstract— The Renewable Energy Sources and distributed generation (DG) in low voltage DC (LVDC) microgrids require fast and reliable fault detection. This paper proposes a fault detection technique based on the cumulative sum of difference of change in current Difference (CDCID) for the LVDC microgrids. The CDCID parameter observes the successive current differences between buses and identifies the fault with excellent response time. AC Grid, Solar PV and battery storage are integrated into this system and operates in DC Grid-Connected (GC) and Islanded modes, which is modelled on MATLAB/Simulink platform. The performance of the proposed method is tested for Pole-to-Pole (PP) and Pole-to-Ground (PG) faults for varying fault resistances under various operating conditions and it successfully differentiates internal faults from external disturbances. The test results illustrate that the proposed scheme can detect faults with high accuracy and faster response time making it a potential candidate for providing dependable protection measures for LVDC microgrids.

Keywords— Low Voltage Direct Current (LVDC), Cumulative Sum of Difference of Change in Current Difference (CDCID), Distributed Generation (DG).

I. INTRODUCTION

The global energy demand rise has driven the adoption of renewable energy sources ensuring sustainability and mitigation of climate change [1]. Renewable Energy Sources (RES) like solar, wind, and biofuels etc. are eco-friendly alternatives to fossil fuels which reduces carbon emissions and pollution [2]. The management of power between Distributed generation (DG) and the utility grid has become challenging as the integration of DG into power systems are increasing. Integration of DGs and RESs in microgrids offers enhanced energy management and reliability of the system [3]. DC microgrids are advantageous over AC microgrids as they eliminate synchronization issues, reduce conversion losses and simplify the integration of distributed energy resources and load [4]. However, introducing various generators, storage devices and loads into a common DC bus, makes the voltage regulation and power-sharing complicated [5].

Low voltage DC (LVDC) microgrid faces several protection issues and challenges such as bi-directional fault current, the lack of natural zero crossings in DC, and the high rising rate of fault currents. A classical method is used to detect Line to ground faults in islanded microgrids by using conductance and rate of change of conductance. It detects the fault under transient conditions but does not consider load switching operation, which may affect the detection of fault [6]. In [7], the internal fault is detected by evaluating asymmetry in positive pole and negative pole currents which provides a level of sensitivity and robustness to external disturbance. However, for the bipolar microgrid it may not be applicable and may decrease the performance level under single pole system as well as for the high fault resistance scenario. Line-to-line (LL), and line-to-ground (LG) faults with a ring-configured microgrid are detected in [8] by a time-synchronous current differential protection scheme, which enables rapid fault detection and isolation. However, it is unable to detect high impedance faults. The traditional protection schemes like overcurrent protection and differential protection schemes suffer from less reliability under varying fault conditions [9]. To overcome these challenges, a protection scheme [10] is introduced where the fault is determined by measuring voltage rise across terminal inductors. This scheme is reliable but requires an additional inductor which increases the cost and also needs precise threshold arrangement.

A technique based on energy calculations using discrete wavelet transform detects faults and can identify high impedance faults without depending on communication infrastructure [11]. However, the performance get deteriorated due to the volatile and intermittent nature of DGs. In [12], the Variational Mode Decomposition (VMD)-based fault detection method is proposed for detection of fault by evaluating energy changes in current signal which works in both islanded and GC mode. VMD-enabled current-based fault detection scheme [13] utilizes the rate of change of current feature and detects faults using the cumulative sum of energy derived using VMD. It is sensitive to high impedance faults and works on both islanded and grid connected mode.

Support Vector Machine (SVM) based tool is applied in an intelligent differential current protection scheme to detect fault. This method is effective to detect high resistance faults but faces computational complexity and needs adaptive threshold selection [14]. A machine learning based differential protection scheme using Decision Tree (DT) is implemented for fault identification [15]. However, this scheme also faces significant challenges for high-resistance faults. Another differential protection method [16] analyzes differential current and its derivative for detecting fault using a DT-based technique. But, it faces challenges like complex machine learning models and needs high-quality training data. In another scheme [17], a protection method built on a Transfer Learning-based convolution neural network (TCNN) is used to identify faults utilizing different bus currents and transient voltage. However, dependency on high-quality spectrogram images of faults for training increases the complexity of the system.

979-8-3315-3013-6/25 $31.00 © 2025 IEEE

In this proposed paper, a technique using the cumulative sum of difference of change in current difference (CDCID) is proposed to detect fault in LVDC microgrid. The paper is detailed as follows: section II is presenting the studied meshed LVDC microgrid system. Section III describes the proposed algorithm for fault detection; section IV provides the performance study, results and discussions and finally the conclusion of the paper is provided in section V.

II. MESHED LVDC MICRO-GRID SYSTEM

A. System Description

The proposed protection scheme is validated on a meshed LVDC microgrid system shown in Fig. 1. The microgrid operates in both GC and islanded mode. The system consists of AC grid, PV source and battery storage unit which ensures efficient power distribution to remote loads. The system is designed and modeled on MATLAB/Simulink platform with 380 V DC voltage level. The grid is connected with LVDC microgrid through a bidirectional AC-DC converter which allows bidirectional power transfer when operating on GC mode. This ensures the transfer of solar power to the grid. And grid also supplies power when the local generation is not sufficient to provide power to the load. The solar PV system acts as a primary generation source interfaced via a DC-DC converter at bus-3, ensuring optimized power extraction when solar power fluctuates. A battery storage system is integrated at bus 4 which facilitates energy storage and discharges to maintain stability of the power during islanded operation. The protective devices (R1, R2, R3, and R4) are located at critical points in the system for enhancing reliability and fault isolation. Proper communication between different buses is also ensured.

The system includes multiple DC lines. At Bus-1, Bus-2, and Bus-3, the loads are connected with rated power of 12 kW,

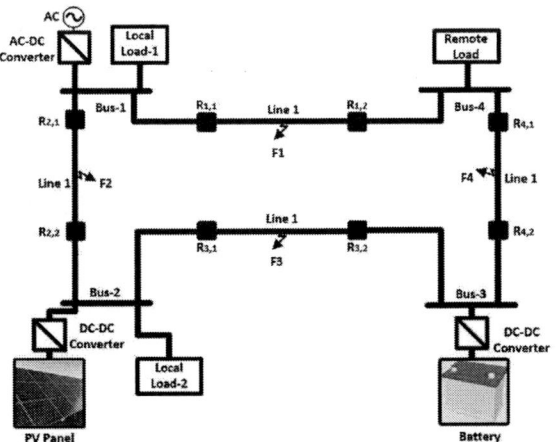

Fig. 1. **Proposed LVDC microgrid system for analysis of Fault**

TABLE I
SYSTEM COMPONENTS SPECIFICATION OF THE PROPOSED SYSTEM

Name of the Components	Specifications
Utility	400 V, 60 Hz
Transformer	400 V/400 V, 12 KVA
Solar Farm	9.4 kW, 273 V
Battery	26 Ah, 144 V
Load 1 (Local Load-1)	12 KW
Load 2 (Local Load-2)	2 kW
Load 3 (Remote Load)	5.5 kW

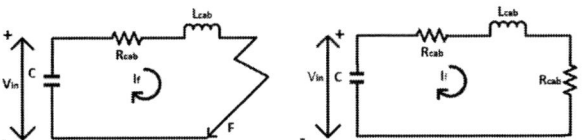

Fig. 2 (a) Faulted Network Model where F denotes the fault (b) Equivalent Network

2 kW and 5.5 kW respectively. The ratings and specifications of the system components, involving capacity of DG, load ratings and transformer rating are provided in Table-I.

B. Analysis of Fault

The short circuit faults are the most critical disturbances due to the surge current in the DC microgrid. This section analyzes the system behavior under fault conditions and assumes that the fault occurs at a specific location in the microgrid. To determine the magnitude of fault current, a pre-charged capacitor plays a significant role shown in Fig. 2(a). This capacitor discharges immediately when a fault occurs, since it is functioning as an energy storage device. Besides, the low impedance of the transmission line further amplifies the fault current. For modeling the system response, an equivalent R-L-C circuit is used, which is shown in Fig. 2(b) [18].

The parameters affecting the fault response includes resistance of DC cable (R_{cab}), the cable inductance (L_{cab}), and fault resistance (R_f). The DC-link capacitor with capacitance (C), serves as the main energy source and provides a voltage V_{cap} with the initial condition $V_{cap}(0)$, and $I_f(0)$, the post-fault current in the system is represented as $I_f(t)$.

The fault current equation in the Laplace domain is obtained as:

$$I_f(s) = \frac{-V_{cap}(0)/L_{cab} + s I_f(0)}{s^2 + (R_{cab}/L_{cab})s + (1/L_{cab}C)} \quad (1)$$

The fault current in the time domain is expressed by applying the inverse Laplace transform:

$$I_f(t) = \left(\frac{-V_{cap}(0)/L_{cab} + s_1 I_f(0)}{s_2 - s_1}\right) e^{-s_1 t} + \left(\frac{-V_{cap}(0)/L_{cab} + s_2 I_f(0)}{s_1 - s_2}\right) e^{-s_2 t} \quad (2)$$

Further simplifying the equation:

$$I_f(t) = X e^{-s_1 t} + Y e^{-s_2 t} \quad (3)$$

Where s_1 and s_2 are the characteristic roots, which determine how the fault current evolves over time. These roots are given by:

$$s_{1,2} = \frac{-R_{cab}}{2L_{cab}} \pm \sqrt{\left(\frac{R_{cab}}{2L_{cab}}\right)^2 - \frac{1}{L_{cab}C}} \quad (4)$$

Alternatively, the roots can be rewritten in terms of system parameters:

$$s_{1,2} = -\beta \pm \sqrt{\beta^2 - W_z^2} \quad (5)$$

Where the damping factor β and the natural frequency W_z are expressed as:

$$\beta = \frac{R_{cab}}{2L_{cab}}, \quad W_z = \sqrt{\frac{1}{L_{cab}C}} \quad (6)$$

979-8-3315-3013-6/25 $31.00 © 2025 IEEE

III. PROPOSED ALGORITHM FOR FAULT DETECTION

The proposed method detects fault by calculating Cumulative sum of Difference of Change in Current Difference (CDCID). Firstly, it observes successive current differences between consecutive samples at each bus end. Then, for each line, the difference of that parameter is calculated between both line ends. The cumulative sum of that difference is compared with a threshold value and sudden changes during fault conditions are identified. Thus, it detects the fault within fractions of milliseconds.

The fault analysis is performed for three lines for our studied LVDV microgrid: Line 1-2 (Bus-1 and 2), Line 1-4 (Bus-1 and 4) and Line 2-3 (Bus-2 and 3). The Difference of Change in Current Difference (DCID) calculation follows a structured approach where the difference between the successive current changes at line ends is determined. It is found that there is a significant current disturbance and change in level when a fault occurs. From the pattern of DCID values, a peak can be observed which can lead towards fault detection. The DCID calculation for line i-j follows the given equation:

$$DCID_{ij}[k] = \left| \Delta I_{Bi}[k] - \Delta I_{Bj}[k] \right| \qquad (7)$$

Where successive difference of current at Bus_i

$$\Delta I_{Bi}[k] = I_{Bi}[k] - I_{Bi}[k-1] \qquad (8)$$

And successive difference of current at Bus_j

$$\Delta I_{Bj}[k] = I_{Bj}[k] - I_{Bj}[k-1] \qquad (9)$$

Further, The CDCID value is calculated by a cumulative summation of squared values of DCID values. This technique ensures that only persistent disturbances are considered for detecting fault and reduces false signals caused by measurement error, load switching/ disconnection or switching between different modes or topology. The window size (W) determines the accuracy and sensitivity of the method where smaller values of W give faster response and higher values of W smooth out the short-term fluctuations. This may slightly delay the response time detection of fault but can enhance the accuracy level.

$$CDCID_{ij}[k] = \sum_{n=k-W+1}^{k} DCID_{ij}[n] \qquad (10)$$

A fault is detected when the value of CDCID exceeds a predefined threshold value. This threshold is determined from empirical studies. This approach gives a robust solution for detecting faults by continuously monitoring the current variation and calculation. Thus, it enables the real-time fault detection and ensures the stability and reliability of the LVDC microgrid. The flowchart of the proposed CDCID based protection scheme is presented in Fig. 3

IV. RESULTS AND DISCUSSIONS

The proposed CDCID scheme has been tested on an LVDC microgrid test system as shown in Fig. 1. The test system is modelled on MATLAB/Simulink platform and the performance of our proposed fault detection scheme has been evaluated for various operating conditions. The LVDC

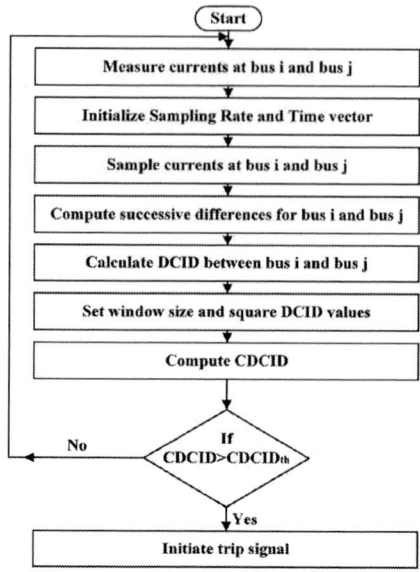

Fig. 3. Flowchart of the Proposed Fault Detection Scheme

microgrid consists of four buses; Bus-1 is connected with the main grid, Bus-2 is connected with a DG i.e. PV panel, Bus-3 is connected with the energy storage system i.e. battery and Bus-4 is connected with a load only.

The ability of the proposed scheme for detecting faults is validated by considering different types of faults, fault resistances and microgrid operating modes. Our proposed algorithm remains unaffected for the case of high resistance fault up to 50 Ω. For assessing the reliability and effectiveness of the scheme, the PP faults and PG faults, under varying fault resistances i.e. 0.01Ω, 1Ω, 10Ω and 50Ω are tested in both GC and islanded modes. Furthermore, in no-fault conditions with load variations and changes in DG penetration level, the performance study is done for verifying the robustness of the method.

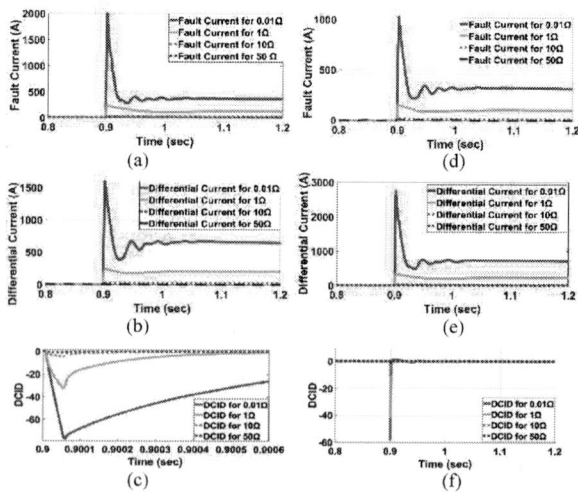

Fig. 4. Fault Current, Differential Current and DCID values for various fault resistances (0.01 Ω, 1 Ω, 10 Ω and 50 Ω); (a), (b), (c) for Grid connected (GC) and (d), (e), (f) for Islanded Mode.

TABLE II
CDCID VALUES WITH CHANGE IN FAULT RESISTANCE ALONG WITH THE RESPONSE TIME (GRID CONNECTED MODE)

Fault Resistance	PP Fault		PG Fault	
	CDCID	Peak Response Time (ms)	CDCID	Response Time (ms)
0.01	52780	1.389	50364.3	1.295
1	14120	0.245	7153.39	0.087
10	848	0.07	268.08	0.07
50	50.5	0.06	13.41	0.055

TABLE III
CDCID VALUES WITH CHANGE IN FAULT RESISTANCE ALONG WITH THE RESPONSE TIME (ISLANDED MODE)

Fault Resistance	PP Fault		PG Fault	
	CDCID	Peak Response Time (ms)	CDCID	Response Time (ms)
0.01	2918.64	1.345	2514.75	1.275
1	1665.77	0.275	952.16	0.165
10	233.51	0.073	80.25	0.06
50	22.63	0.057	6.27	0.055

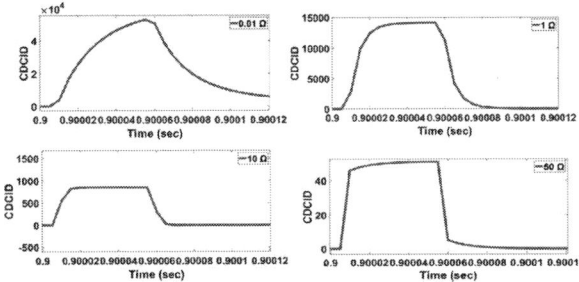

Fig. 5. CDCID Plots for PP faults occuring at 0.9s for various fault resistances (0.01 Ω, 1 Ω, 10 Ω and 50 Ω) in GC mode.

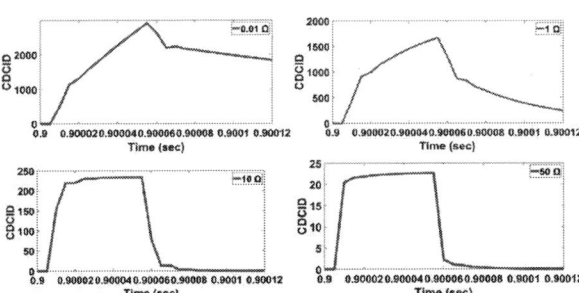

Fig. 6. CDCID plots for PP faults for various fault resistances (0.01 Ω, 1 Ω, 10 Ω and 50 Ω) in Islanded mode.

A PP fault on line 1-4 between bus-1 and bus-4 in both operating modes is created to analyze the response of the CDCID-based fault detection scheme. The plots of the fault current, differential current, and DCID values for varying fault resistances is shown in Fig. 4. For lower values of fault resistance, the fault current rises sharply, while at higher values, the fault current magnitudes reduce significantly. The same pattern is seen in the case of the differential current and the DCID. The graph shows that the magnitude of fault current in GC mode is higher than in islanded mode.

Further, the CDCID value is calculated for various operating conditions. For this calculation, window size 10 is chosen to get good accuracy without compromising response

time. It increases sharply after fault occurrence in the GC mode. The CDCID values due to PP faults with resistances of 0.01Ω, 1Ω, 10Ω, and 50Ω in GC mode are shown in Fig. 5. As presented in Table II, the value of CDCID reaches the peak of 52780 with a fault resistance of 0.01Ω in the GC mode and has a peak value of 50.5 with a very high resistance fault of 50Ω. And the response times for reaching the peak are 1.389 ms and 0.06 ms for the fault resistance values 0.01Ω and 50Ω respectively. Though increasing the fault resistance reduces the peak CDCID value, still the value is significant enough to cross the threshold. The CDCID peak values as well as response times for different fault resistances are shown in The CDCID value is lower in islanded mode compared to GC mode because of the absence of grid contribution. At lower fault resistance, the peak CDCID value is calculated to be 2918.6, while for a high fault resistance of 50Ω, the CDCID value is 22.63. Fig. 6 shows the plot of CDCID for pole-to-pole faults in the islanded mode. At the instant of fault occurrence, CDCID rises sharply and decays under the threshold value, which indicates the steady state of the system. At lower fault resistance case the peak response time is more because of higher initial fault current and at higher resistance, the peak response time is less due to low fault current level. With the change in fault resistances, the variations of CDCID values and the corresponding peak response times for the islanded mode are shown in Table III. Similarly, the pole-to-ground (PG) fault is analyzed by varying fault resistance under the same kind of operating conditions. The grounding impedance affects fault current in the PG fault and leads to a lower level of fault current as compared to PP fault. Also, it follows the same pattern as PP fault cases i.e. CDCID decreases as the fault resistance increases. Even though it has a lower level of fault current, the CDCID-based protection scheme effectively detects the PG faults for various fault resistances. When the resistance value is 0.01Ω, the CDCID value is 50364.3 and 2514.75 for GC and islanded modes respectively. The response times along with the CDCID values for PG fault cases are shown in Table II and Table III is for the GC mode and islanded mode respectively.

The plot of CDCID values is shown in Fig. 7 when PP fault occurs in different lines of the test system in both GC and Islanded mode. The response shows how CDCID value varies for different fault locations and a significant level is maintained irrespective of the faulted lines. Defining a proper threshold is critical for balancing sensitivity and selectivity in the fault detection system. This threshold is the margin between fault and non-fault conditions, which ensures fault detection and avoids unwanted tripping when variation in the load and DG penetration occurs or measurement error or noise is included. Table IV presents the performance of the proposed method with measurement noise of 10 dB and 20 dB and CDCID value is well above the threshold for fault cases and well below the threshold value for no-fault cases.

The fluctuation in load demand is a common phenomenon due to varying consumer usage in LVDC microgrids. If the protection scheme sends a tripping signal by detecting a false fault due to the load variation, it can affect the reliability of the microgrid. The unwanted disconnection of the generating unit and grid can also impact the total efficiency of the system. Fig. 8 shows the plots of CDCID values for both

979-8-3315-3013-6/25 $31.00 © 2025 IEEE

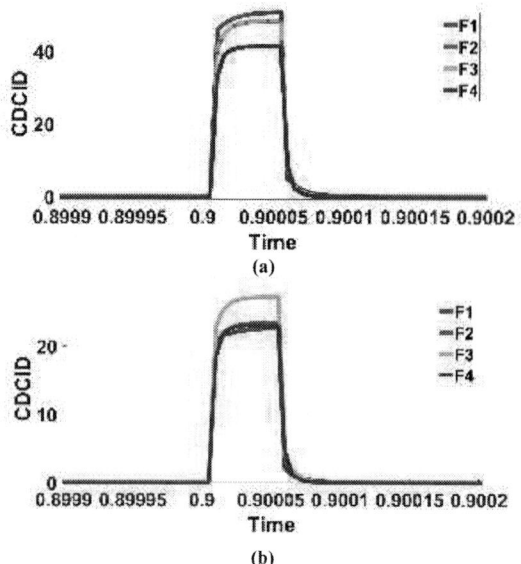

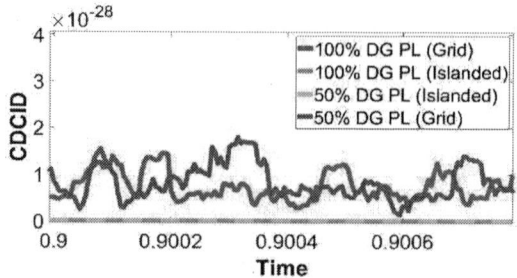

Fig. 9. CDCID plots with variations in DG Penetration Level (PL).

TABLE IV
PERFORMANCE OF THE PROPOSED SCHEME WITH MEASUREMENT NOISE

Noise	10 dB		20 dB	
Time/CDCID	Peak Response Time (µs)	CDCID	Peak Response Time (µs)	CDCID
GC Mode (Rf=50 Ω)	23.945	1849.4095	971.555	160.78
Islanding Mode (Rf= 50 Ω)	216.255	650.32	280.190	66.7070
Islanding Mode (50% DG)	947.78	730.3982	916.205	80.47
No Fault Islanding mode (50% DG)	-	1.23	-	1.00
No-fault GC Mode – Load switching	-	1.02	-	1.5

Fig. 7. CDCID plots for PP faults occurring at different lines in (a) GC mode and (b) Islanded mode.

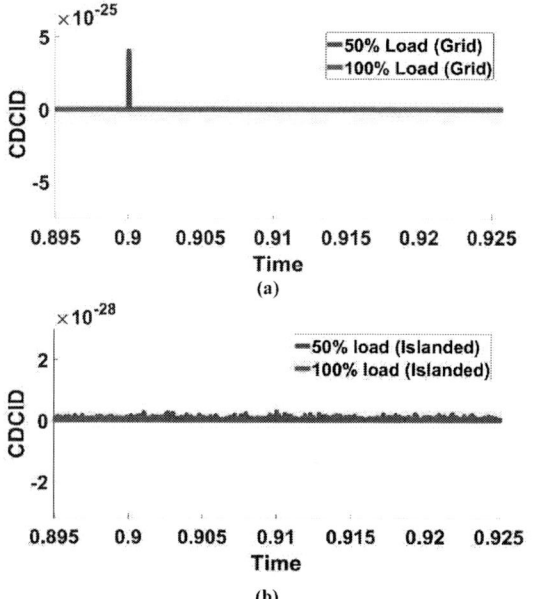

Fig. 8 CDCID plots with variations in load 50%, 100 %. (a) GC Mode, (b) Islanded Mode.

operating modes for load changes and it is observed that the CDCID values in both cases are significantly lower than the threshold. It confirms that the scheme does not trigger at sudden load changes. This ability to distinguish between load variation and actual faults is crucial. If a protection scheme fails to differentiate between such cases, it may cause power outages and instability in voltage. It justifies the CDCID-based scheme efficiently avoids such kinds of issues which may affects the continuity of operation.

The contributions from distributed generations (DGs) in LVDC microgrids, particularly from renewable energy sources like solar, can vary significantly due to intermittent

generation. This can create vital changes in the flow of power. The proposed CDCID-based fault detection scheme is tested under varying DG penetration levels. The contribution of power from solar PV varied at 50% and 100% in both operating modes and the response is shown in Fig. 9. It confirms that the CDCID scheme can effectively differentiate between sudden variations in DG penetration level and a fault condition. The value of CDCID is well below the threshold in both the cases. It ensures that during sudden variation of solar PV or DG disconnection, the proposed scheme will not trigger false trips, and thus maintains the stability of power. Nowadays, the DG integration is increasing in LVDC microgrids. The DG penetration level can change due to changes in the irradiance and temperature of the solar panel. The proposed CDCID-based fault detection scheme is able to handle these sudden changes in power without false tripping and enhance the efficiency and reliability of the LVDC microgrid.

The proposed fault detection technique is based on the Cumulative Sum of Difference of Change in Current Difference (CDCID). This method considers the cumulative sum of successive changes in current derivatives. Unlike computationally complex methods such as neural networks or other machine learning based approaches, the proposed technique has proven its computational efficiency and costly effectiveness compared to the aforementioned methods. The comparative study of the proposed scheme with existing techniques has been presented in table V.

979-8-3315-3013-6/25 $31.00 © 2025 IEEE

TABLE V
COMPARISON STUDY

Comparison aspect	[19]	[20]	[21]	[22]	[23]	[24]	[25]	[26]	[27]	Proposed
Fault detection scheme	Overcurrent Relaying scheme	Differential Current relaying scheme	Wavelet-based multiresolution	Current based wavelet scheme	Transfer Learning-Based CNN (TCNN)	voltage and ground current-based detection scheme	local voltage and current measurements based	Weighted K-Nearest Neighbor (WKNN)	Travelling Wave (TW)-based	Cumulative Sum of Difference of Change in Current Difference
Microgrid architecture	Radial	Mesh	Mesh	Mesh	Ring	Mesh	Radial	Radial	Radial	Mesh
Microgrid operating mode	GC and Islanded	GC	Islanded	GC	Islanded	GC and Islanded	GC	Islanded	Islanded	Islanded
Protection for Variation in DG penetration	No	No	No	No	Yes	Yes	No (not mentioned anywhere)	Yes	No mention	Yes
High resistance fault Detection	No	No	No	Yes	Yes	Yes	Yes	Yes	Yes	Yes
Response time (for low resistance faults)	> 1 ms	100 ms	-	0.45	4.6 ms	~1.25 ms	0.6 ms	2 ms	1 ms.	< 1 ms.

Simulation results show that this method achieves a response time of less than 1 ms for low-resistance faults, which takes up less computational time than many existing techniques such as CNN-based and wavelet-based methods. The low response time is preferable as it helps to quickly detect the faulty phenomena and restore the system stability. This technique considers both grid-connected and islanded modes of operation which suggests the cases of high and low current in circuit, respectively. Moreover, the CDCID method is also robust against the DG penetration, thereby offering a scalable and efficient protection mechanism suitable for real-time fault diagnosis in modern smart grid and microgrid environments.

The performance analysis and corresponding results strongly state that the CDCID-based fault detection scheme is effective in identifying faults by maintaining reliability under different operating modes. The PP and PG fault across a wide range of fault resistance can be detected accurately with the help of this scheme. Moreover, under both operating modes, it performs well while keeping the response time in an acceptable range. And the scheme can avoid the false trigger when external disturbance like sudden changes in DG penetration level and load variations occurs, which ensures the continuous power supply. The CDCID value remains below the threshold under non-fault conditions. As LVDC microgrids are going to be adopted more in modern power distribution systems, these characteristics of the proposed CDCID-based fault detection scheme can make the scheme suitable for LVDC microgrid fault protection.

V. CONCLUSION

This paper proposes a fast and reliable fault detection scheme for low voltage DC microgrid using the Cumulative Sum of Difference of change in current Difference (CDCID) technique, which detects the Pole-to-Pole (PP) and Pole-to-Ground (PG) faults under varying resistance and operating conditions including GC and islanded modes. The CDCID scheme detects faults with less response time and ensures accurate results, preventing false tripping while load fluctuates and DG level varies. The capability of the scheme is validated by the simulation results in MATLAB/Simulink platform. The CDCID value remains above the threshold value during faults and below the threshold under normal conditions as well as external disturbances. The scheme is found to be effective and its robustness is justified by successfully handling variations in DG penetration. The CDCID scheme provides an accurate, efficient, and reliable way for LVDC microgrid protection. pointing out of faulted section, location of the fault and isolation techniques should be thoroughly investigated in future.

REFERENCES

[1] P. A. Owusu and S. Asumadu-Sarkodie, "A review of renewable energy sources, sustainability issues and climate change mitigation," *Cogent Engineering*, vol. 3, no. 1, 2016, doi: 10.1080/23311916.2016.1167990.

[2] M. R. Serbov, M. Zolotarova, O. Hrachuk, and I. Novosad, "Development of Renewable Energy Sources: Impact on Sustainability and the Environment," *Grassroots Journal of Natural Resources*, vol. 7, no. 3, pp. s131–s148, 2024, doi: 10.33002/nr2581.6853.0703ukr07.

[3] G. V. Brahmendra Kumar and K. Palanisamy, "A Review on Microgrids with Distributed Energy Resources," 2019 Innovations in Power and Advanced Computing Technologies (i-PACT), Vellore, India, 2019, pp. 1-6, doi: 10.1109/i-PACT44901.2019.8960189.

[4] H. Lotfi and A. Khodaei, "AC Versus DC Microgrid Planning," in IEEE Transactions on Smart Grid, vol. 8, no. 1, pp. 296-304, Jan. 2017, doi: 10.1109/TSG.2015.2457910.

[5] F. S. Al-Ismail, "DC Microgrid Planning, Operation, and Control: A Comprehensive Review," in IEEE Access, vol. 9, pp. 36154-36172, 2021, doi: 10.1109/ACCESS.2021.3062840.

[6] M. S. Babu, S. Murugesan and S. Sarangi, "Fault Detection in LVDC Microgrids Using Local Measurements," 2022 IEEE International Conference on Power Electronics, Drives and Energy Systems (PEDES), Jaipur, India, 2022, pp. 1-6, doi: 10.1109/PEDES56012.2022.10080023.

[7] P. Chauhan, C. P. Gupta and M. Tripathy, "A Current Difference Based Protection Technique for Low Voltage DC Microgrid," 2022 IEEE International Conference on Power Electronics, Smart Grid, and Renewable Energy (PESGRE), Trivandrum, India, 2022, pp. 1-6, doi: 10.1109/PESGRE52268.2022.9715853.

979-8-3315-3013-6/25 $31.00 © 2025 IEEE

[8] W. Hussain, A. Alam and S. Kirmani, "Detection of Fault Inception in a Low Voltage DC Microgrid using Current Differential Protection Method," 2023 International Conference on Recent Advances in Electrical, Electronics & Digital Healthcare Technologies (REEDCON), New Delhi, India, 2023, pp. 362-367, doi: 10.1109/REEDCON57544.2023.10150928.

[9] Som, Shreyasi, and Subhransu Ranjan Samantaray. "Efficient protection scheme for low-voltage DC micro-grid." IET Generation, Transmission & Distribution 12, no. 13 (2018): 3322-3329.

[10] P. Chauhan, C. P. Gupta and M. Tripathy, "Protection Scheme for Low Voltage DC Microgrid Based on Voltage Rise Across Terminal Inductor," 2023 IEEE IAS Global Conference on Renewable Energy and Hydrogen Technologies (GlobConHT), Male, Maldives, 2023, pp. 1-7, doi: 10.1109/GlobConHT56829.2023.10087891.

[11] S. Som and S. R. Samantaray, "Wavelet based fast fault detection in LVDC micro-grid," 2017 7th International Conference on Power Systems (ICPS), Pune, India, 2017, pp. 87-92, doi: 10.1109/ICPES.2017.8387273.

[12] N. K. Sharma, R. Pattanayak, S. R. Samantaray and C. N. Bhende, "A Fast Fault Detection Scheme for Low Voltage DC Microgrid," 2020 21st National Power Systems Conference (NPSC), Gandhinagar, India, 2020, pp. 1-6, doi: 10.1109/NPSC49263.2020.9331862.

[13] N. K. Sharma, S. R. Samantaray and C. N. Bhende, "VMD-Enabled Current-Based Fast Fault Detection Scheme for DC Microgrid," in IEEE Systems Journal, vol. 16, no. 1, pp. 933-944, March 2022, doi: 10.1109/JSYST.2021.3057334.

[14] N. K. Sharma, A. Saxena and S. R. Samantaray, "An Intelligent Differential Protection Scheme for DC Microgrid," 2021 9th IEEE International Conference on Power Systems (ICPS), Kharagpur, India, 2021, pp. 1-6, doi: 10.1109/ICPS52420.2021.9670330.

[15] A. Saxena, N. K. Sharma and S. R. Samantaray, "An Enhanced Differential Protection Scheme for LVDC Microgrid," in IEEE Journal of Emerging and Selected Topics in Power Electronics, vol. 10, no. 2, pp. 2114-2125, April 2022, doi: 10.1109/JESTPE.2022.3144300.

[16] A. Saxena, N. K. Sharma and S. R. Samantaray, "An Enhanced Differential Protection Scheme for LVDC Microgrid," in IEEE Journal of Emerging and Selected Topics in Power Electronics, vol. 10, no. 2, pp. 2114-2125, April 2022, doi: 10.1109/JESTPE.2022.3144300.

[17] S. Veerapandiyan and V. Sugavanam. "On-line fault identification, location, and seamless service restoration using transfer learning-based convolution neural network for low-voltage DC microgrid." Electric Power Components and Systems 51.8 (2023): 785-808.

[18] Fletcher, S. D. A., P. J. Norman, S. J. Galloway, and G. M. Burt. "Determination of protection system requirements for DC unmanned aerial vehicle electrical power networks for enhanced capability and survivability." IET Electrical Systems in Transportation 1, no. 4 (2011): 137-147.

[19] A. Shabani and K. Mazlumi, "Evaluation of a Communication-Assisted Overcurrent Protection Scheme for Photovoltaic-Based DC Microgrid," in IEEE Transactions on Smart Grid, vol. 11, no. 1, pp. 429-439, Jan. 2020, doi: 10.1109/TSG.2019.2923769.

[20] S. Dhar, R. K. Patnaik and P. K. Dash, "Fault Detection and Location of Photovoltaic Based DC Microgrid Using Differential Protection Strategy," in IEEE Transactions on Smart Grid, vol. 9, no. 5, pp. 4303-4312, Sept. 2018, doi: 10.1109/TSG.2017.2654267.

[21] W. Li, A. Monti and F. Ponci, "Fault Detection and Classification in Medium Voltage DC Shipboard Power Systems With Wavelets and Artificial Neural Networks," in IEEE Transactions on Instrumentation and Measurement, vol. 63, no. 11, pp. 2651-2665, Nov. 2014, doi: 10.1109/TIM.2014.2313035.

[22] Som, S., & Samantaray, S. R. (2018). Efficient protection scheme for low-voltage DC micro-grid. IET Generation, Transmission & Distribution, 12(13), 3322-3329.

[23] Veerapandiyan, Shanmugapriya, and Vidyasagar Sugavanam. "On-line fault identification, location, and seamless service restoration using transfer learning-based convolution neural network for low-voltage DC microgrid." Electric Power Components and Systems 51, no. 8 (2023): 785-808.

[24] R. Bhargav, B. R. Bhalja and C. P. Gupta, "Novel Fault Detection and Localization Algorithm for Low-Voltage DC Microgrid," in IEEE Transactions on Industrial Informatics, vol. 16, no. 7, pp. 4498-4511, July 2020, doi: 10.1109/TII.2019.2942426.

[25] R. Shareef, T. Mahmood and F. Tariq, "Enhanced Local Measurement Based Online Fault Location Technique for Low Voltage DC Distribution Network," 2022 5th International Conference on Energy Conservation and Efficiency (ICECE), Lahore, Pakistan, 2022, pp. 1-6, doi: 10.1109/ICECE54634.2022.9758953.

[26] Reddy, O. Y., Chatterjee, S., & Chakraborty, A. K. (2021). Bilayered fault detection and classification scheme for low-voltage DC microgrid with weighted KNN and decision tree. International Journal of Green Energy, 19(11), 1149–1159. https://doi.org/10.1080/15435075.2021.1984924.

[27] Paruthiyil, Sajay Krishnan, Ali Bidram, Miguel Jimenez Aparicio, Javier Hernandez-Alvidrez, Andrew RR Dow, Matthew J. Reno, and Daniel Bauer. "Travelling wave-based fault detection and location in a real low-voltage DC microgrid." IET Smart Grid (2025): e12207.

A Modular Three-Phase Induction Heating Generator with Enhanced Power Factor

Anshal S Padole
Dept. Electronics and Telecomm.
Sardar Patel Institute of Technology
Mumbai, India - 400058
anshal.padole@spit.ac.in

Monish Rane
Dept. Electronics and Telecomm.
Sardar Patel Institute of Technology
Mumbai, India - 400058
monish.rane@spit.ac.in

Rajendra R Sawant
Dept. of Electronics & Telecomm.
Sardar Patel Institute of Technology
Mumbai, India-400058
rajendra.sawant@spit.ac.in

Yerramreddy Srinivasa Rao
Dept. of Electronics & Telecomm.
Sardar Patel Institute of Technology
Mumbai, India-400058
ysrao@spit.ac.in

Bhalchandra N Chaudhari
Dept.of Electronics & Telecomm.
Sardar Patel Institute of Technology
Mumbai, India-400058
bnc@spit.ac.in

Abstract—This paper presents a novel modular and interleaved solid-state induction heating generator designed for induction casting machines, particularly for jewelry casting applications. The proposed system addresses challenges related to power factor correction, harmonic compliance, and scalability by employing modular single-phase solid-state inverters integrated with Power Factor Correction (PFC) units. Each inverter utilizes advanced 600-700V Silicon Carbide (SiC) power MOSFETs and a custom gate-driver card, ensuring high efficiency and reliability.

The control strategy employs phase shift control for power regulation and automatic frequency control for maintaining zero voltage switching at all loading conditions. The system's modular design allows for easy scaling of power levels in multiples of 6 kW while maintaining compliance with international standards such as IEEE-519-2014 for harmonics and EMI/EMC regulations. The use of a high-frequency impedance matching transformer and a common series resonant LC tank circuit ensures efficient power distribution. This paper discusses the design, implementation, and advantages of the system, highlighting its potential for various industrial induction heating applications beyond jewelry casting.

Index Terms—Induction heating, modular inverters, power factor correction (PFC), SiC MOSFETs, harmonic compliance, jewelry casting, frequency tracking, etc.

I. INTRODUCTION

Induction heating (IH) technology has seen significant advancements in recent years, particularly in industrial applications such as metal processing, domestic cooking, automotive manufacturing, and jewellery casting. Traditional IH systems rely on thyristor-based inverters and low-frequency operation, leading to issues such as low power factor, high harmonic distortion, and limited scalability. Recent developments in power electronics and semiconductor materials have enabled the use of solid-state inverters, improving efficiency, power quality, and design flexibility [1], [13].

This research was supported by the Department of Science and Technology (DST) under the Technology Development Program (TDP).

Research has demonstrated that modular inverters allow power scaling without significantly increasing component stress, making them ideal for applications where power demand varies. Furthermore, interleaving techniques help minimize ripple current and thermal stress, further enhancing system performance [2], [3], [15].

Ensuring compliance with international harmonic standards, such as IEEE-519-2014, is critical for modern power electronics applications. Traditional IH systems often require bulky passive filters to mitigate harmonics and improve power factor. However, active PFC circuits, particularly those based on boost and bridge-less topologies, offer a more efficient solution by dynamically adjusting input current waveforms to achieve near-unity power factor. The integration of PFC with solid-state inverters has been shown to enhance both efficiency and grid compliance in IH applications [9].

The introduction of wide-bandgap (WBG) semiconductors, such as SiC MOSFETs and GaN-HEMT devices, has revolutionized power electronics by offering higher breakdown voltages, lower conduction losses, and superior switching speeds compared to traditional silicon-based devices. Studies have demonstrated that SiC-based inverters can operate at higher frequencies with reduced cooling requirements, making them ideal for IH applications where precise temperature control is essential [10], [11], [12] [7].

Efficient power transfer in IH applications relies on properly designed resonant converters. Series and parallel resonant LC tank circuits and matching transformers are commonly employed to achieve zero-voltage switching (ZVS) and zero-current switching (ZCS), reducing switching losses and enhancing system efficiency [6], [8]. Additionally, phase-shift control, pulse density modulation and hybrid modulation strategies are being used in Induction Heating applications for precise power and temperature control with reduced switching stresses [4], [5], [14], [17].

The proposed design is particularly suited for jewelry cast-

979-8-3315-3013-6/25 $31.00 © 2025 IEEE

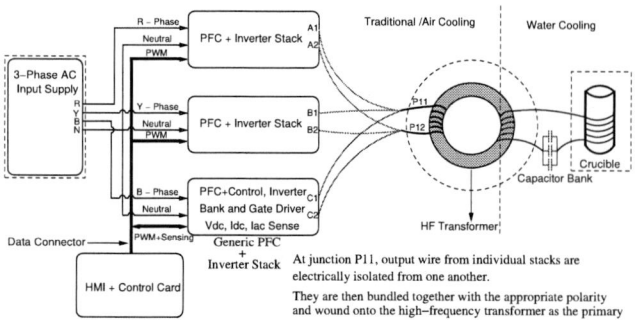

Fig. 1. System Block Diagram

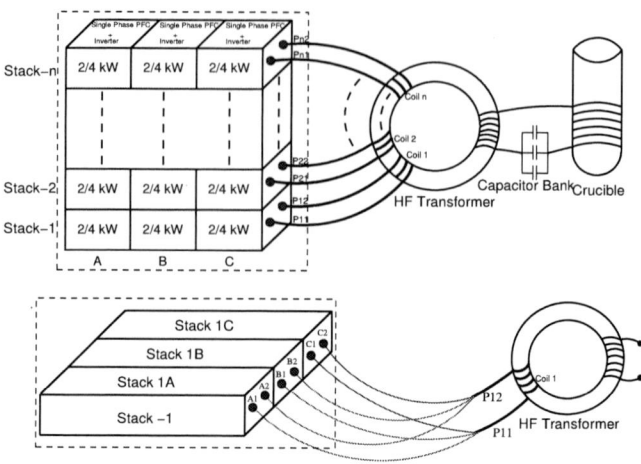

Fig. 2. Inverter Multi-stacking Block Schematic Diagram

ing machines, where precise control and energy efficiency are paramount. By leveraging advanced SiC power MOSFETs and a high-frequency impedance matching transformer, the system achieves improved input power factor and reduced harmonic distortion. This paper provides a detailed description of the proposed modular structure in Section II. The system's design, operation, analysis and control are explained in Section III. The experimental results on a 6kW, modular, lab-based induction heating prototype is included in Section IV, demonstrating its potential to revolutionize induction heating technology.

II. PROPOSED MODULAR STRUCTURE FOR INDUCTION HEATING SYSTEM

A. System Block Diagram

The initial stage of the induction heating generator consists of multiple key components: a Power Factor Correction (PFC) and resonant inverter unit , a Human-Machine Interface (HMI) and control card, a High-Frequency (HF) transformer, a crucible, and a three-phase AC power supply, as depicted in Figure 1.

The PFC unit converts the AC input power into an equivalent DC voltage while maintaining a near-unity power factor. The resonant inverter then converts this DC voltage back into AC, which serves as the primary input to the HF transformer.

The output terminal (A_1, B_1, and C_1) of each individual unit is electrically isolated from one another. These outputs are then bundled together with the appropriate polarity in parallel (P_{11}), forming a single output wire that is subsequently wound as the primary coil of the transformer. Meanwhile, the other end of the primary coil (P_{12}) is separated and connected to each individual unit (A_2, B_2, and C_2).

The Figure 2 shows the concept of multi-stacking interconnection of multiple single stacks in parallel to achieve higher power output. Each single stack consists of three independent inverter units, each receiving power from a three-phase AC supply. The input to each unit is derived from the live wire and neutral, ensuring balanced operation. This modular structure forms a single-stack configuration, enabling power scalability by connecting multiple stacks in parallel.

The N-level stack extends this approach by integrating multiple single stacks in parallel, facilitating higher power delivery based on application requirements. In this configuration, each

stack features its own set of output terminals (P11-P12, P21-P22, ..., Pn1-Pn2), all of which are wound onto the primary winding of the HF transformer. This arrangement effectively aggregates power from each stack, significantly increasing the total power output. The combined power is transferred through the transformer's secondary winding and delivered to the crucible, where it is used for melting jewelry.

B. Circuit of PFC & Full Bridge Inverter Stack

Figure 3 illustrates the complete circuit diagram of the PFC and Full-Bridge Resonant Inverter. The system comprises key functional blocks that facilitate power conversion and control.

The PFC stage begins with a single-phase AC input, which is first filtered through an EMI filter to minimize electromagnetic interference and ensure electromagnetic compatibility. The filtered voltage then enters an conventional PFC boost converter. An electrolytic/polypropylene DC-link capacitor is placed at the output of the boost converter to stabilize the DC voltage and shield the preceding converter stage.

A full-bridge topology has been adopted for the implementation of the resonant inverter. The full-bridge configuration, depicted in Figure 3, utilizes four switches to convert the DC output from the PFC into a high-frequency AC voltage, which is then transmitted to the high-frequency transformer.

The PFC output voltage (V_{dc}) and current (I_{dc}), along with the inverter's output current, are sensed and transmitted to the HMI and control card. Among the three units shown in Figure 1, one is responsible for sensing these signals. The control card processes these signals to generate PWM signals that regulate both the PFC and inverter operation. The sensed signals are further processed by a cascaded PI filter in the controller, which adjusts the phase shift between the two legs of the full-bridge resonant inverter (gating signals: G1-G2 and G3-G4).

In contrast, for a full-bridge topology, the phase shift determined by the PI filter plays a crucial role in maintaining optimal temperature conditions for the induction casting machine. Additionally, the operating frequency is dynamically

979-8-3315-3013-6/25 $31.00 © 2025 IEEE

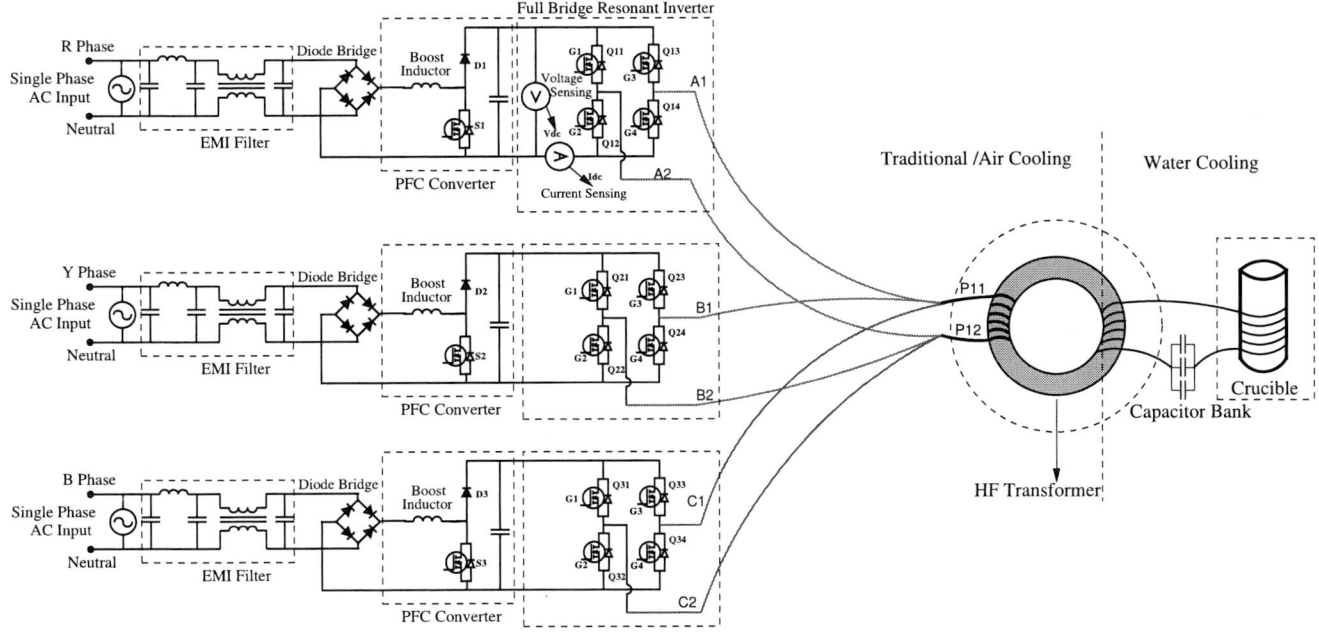

Fig. 3. Circuit diagram of the overall system

adjusted by the frequency tracking algorithm, as discussed in Section III-C. The generated PWM signals are transmitted to all inverter units through a single data connector, ensuring synchronized control and stable operation across the system.

C. Equivalent Circuit Representation

Figure 4 illustrates the equivalent circuit of the proposed system. In this representation, V_{RN1}, V_{YN1}, and V_{ZN1} denote the fundamental components of the R-phase, Y-phase, and B-phase, respectively, which originate from the output terminal of the inverters.

The primary winding parameters are defined as follows:

- R_{11}, R_{12}, and R_{13} represent the leakage resistances of the R, Y, and B-phase primary windings.
- L_{11}, L_{12}, and L_{13} denote the leakage inductances of the primary windings.
- L_{m1}, L_{m2}, and L_{m3} correspond to the magnetizing inductances of each primary winding.
- N_{11}, N_{12}, and N_{13} indicate the number of turns for each phase in the primary winding of the transformer.

On the secondary side:

- N_2 represents the number of turns in the secondary winding.
- R_2 and L_2 correspond to the resistance and inductance of the secondary winding.
- C_o is the compensating capacitor.
- R_{coil} and L_{coil} represent the resistance and inductance offered by the induction coil.

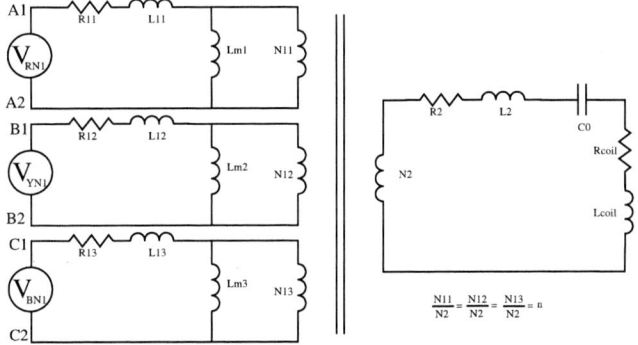

Fig. 4. Equivalent circuit of proposed system.

III. SYSTEM DESIGN AND CONTROL STRATEGY

A. System Design

This section covers the design and analysis of the resonant tank parameters, transformer characteristics, and inverter ratings. Additionally, it includes considerations for system control to ensure efficient operation and stability.

The system design involves a step-by-step approach to determine key parameters and ensure optimal performance. The following steps outline the design methodology:

Step 1: Measure Induction Coil Parameters The induction coil parameters, including resistance (R_{coil}) and inductance (L_{coil}), are measured as follows:

$$R_{coil} = 0.13 \ \Omega, \quad L_{coil} = 35 \ \mu H \quad (1)$$

979-8-3315-3013-6/25 $31.00 © 2025 IEEE   421

Step 2: Determine the Secondary Side Compensating Capacitor (C_o)

The resonant frequency (f_o) is given by:

$$f_o = \frac{1}{2\pi\sqrt{L_{\text{coil}}C_o}} \tag{2}$$

The secondary side compensating capacitor is then calculated as:

$$C_o = \frac{1}{(2\pi f_o)^2 L_{\text{coil}}} \tag{3}$$

Substituting values:

$$C_o = \frac{1}{(2\pi \times 18 \text{ kHz})^2 \times 35 \ \mu H} = 2.23 \ \mu F \tag{4}$$

Step 3: Calculate the Secondary Side Impedance (Z_{os}) The secondary side impedance is given by:

$$Z_{\text{os}} = \sqrt{\frac{L_{\text{coil}}}{C_o}} \tag{5}$$

Substituting values:

$$Z_{\text{os}} = \sqrt{\frac{35 \ \mu H}{2.23 \ \mu F}} = 3.96 \approx 4 \ \Omega \tag{6}$$

Step 4: Consider 25% Derating of the Converter To enhance reliability, a 25% derating is considered for safe operation. The primary side impedance seen at the converter end (Z_{op}) is determined using normalized power (P_n):

$$P_n = 1.25 \quad \text{(for 25\% derating)} \tag{7}$$

$$P_n = \frac{P_{\text{out}}}{P_{\text{inverter}}} = \frac{P_{\text{out}}}{V_{\text{dc}}^2/Z_{\text{op}}} \tag{8}$$

Solving for Z_{op}:

$$Z_{\text{op}} = \frac{V_{\text{dc}}^2 \cdot P_n}{P_{\text{out}}} \tag{9}$$

Substituting values:

$$Z_{\text{op}} = \frac{400^2 \times 1.25}{2000} = 100 \ \Omega \tag{10}$$

Step 5: Calculate the Transformer Turns Ratio The transformer turns ratio is calculated as:

$$\frac{N_1}{N_2} = \sqrt{\frac{Z_{\text{op}}}{Z_{\text{os}}}} \tag{11}$$

Substituting values:

$$\frac{N_1}{N_2} = \sqrt{\frac{100}{4}} = 5 \tag{12}$$

TABLE I
MEASURED DESIGN PARAMETERS

Parameter (Symbol)	Value
Primary Winding Leakage Inductance (L11, L12 and L13)	74.72 μH
Primary Winding Leakage Resistance (R11, R12 and R13)	2.7 Ω
Compensating Capacitor (C_o)	2.3 μF
Open-loop Primary Resistance	1.36 Ω
Open-loop Primary Inductance	8 mH
Secondary Resistance (R2)	1.4 Ω
Secondary Inductance (L2)	240 μH
Coil Resistance (R_{coil})	0.13 Ω
Coil Inductance (L_{coil})	35 μH

B. Control Strategy

The Control Strategy for the individual inverter stack is to regulate the power and temperature by phase-shift modulation and the frequency is automatically varied to maintain Zero voltage switching in all loading conditions.

As shown in Figure 5, the system uses two cascaded PI filter. The outer loop maintains the desired temperature by comparing the reference temperature (T_{ref}) with the measured furnace temperature and adjusting the power reference (P_{ref}). The inner loop then compares (P_{ref}) with the actual measured power and modulates the phase shift accordingly.

The resulting control signal is fed into a PWM modulator, which generates gate signals for the inverter. This inverter, modeled as a plant (G_s), drives the induction coil and heating assembly, closing the control loop for both power and thermal regulation.

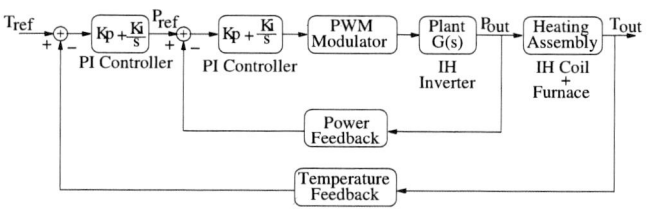

Fig. 5. Block diagram of the cascaded PI control strategy for power and temperature regulation

C. Control Strategy for Frequency Control

The implementation of beta tracking is disscussed in [16] and the new approach to implement beta tracking is disscussed below.

The phase interval during which the inverter voltage remains positive while the inverter current becomes negative, denoted as β, is critical for achieving zero-voltage switching (ZVS). To ensure ZVS, the current flowing through the tank circuit must be sufficient to discharge the charge stored in switch Q14 (q_{coss}) and simultaneously charge the switch Q13 within this β interval. Based on this charge transfer requirement, a mathematical expression for the minimum value of β, denoted as β_{min} can be derived as:

$$\beta_{\min} = \arccos\left(1 - \frac{2\omega_s q_{\text{coss}}}{I_p}\right) \quad (13)$$

where I_p is the peak load current, ω_s is radial switching frequency and q_{coss} is the charge stored in one C_{oss} at V_{in} which can be calculated as:

$$q_{\text{coss}} = C_{\text{oss}} \cdot V_{\text{in}} \quad (14)$$

For instance, considering the switching device IMW65R057M1H, the output charge q_{coss} is 65nC. Under typical operating conditions with a switching frequency of 30kHz and a load current of 20A, the corresponding minimum phase interval β_{min} required to ensure ZVS is calculated as:

$$\beta_{\min} = \arccos\left(1 - \frac{2 \cdot 2\pi \cdot 30000 \cdot 65 \cdot 10^{-9}}{20}\right)$$

$$\beta_{\min} = 2.3°$$

This value represents the minimum phase angle interval required to ensure complete ZVS operation of the switches.

This functionality is implemented in the experimental setup using a microcontroller-based approach. The process involves capturing the rising edges of two TTL signals on distinct input pins of the microcontroller. One signal corresponds to the PWM signal, which aligns with the rising edge of the quasi-square voltage pulse, while the other signal is obtained by implementing an analog Zero Crossing Detector (ZCD) for the corresponding current waveform.

A free-running timer records the count values in separate variables when external interrupts occur at the respective rising edges of both signals. The phase shift denoted by (β) between the voltage and current waveforms is then computed using the captured counts. The following equations define the computation of β:

$$\beta = \text{ZCD\_Capture\_Count} - \text{PWM\_Capture\_Count} \quad (15)$$

If an overflow occurs between the two events, the phase shift is adjusted using:

$$\beta = \text{ZCD\_Capture\_Count} + 0xFFFF - \text{PWM\_Capture\_Count} \quad (16)$$

The frequency control operation is executed within the End of Cycle (EOC) interrupt of the PWM module of the microcontroller. The logic and flow of this operation are illustrated in Figure 6. The system operates within a frequency range of 17 kHz to 30 kHz, ensuring smooth frequency transitions while maintaining efficient and stable control.

Figures 8, 9 and 10 depict the tracked beta values at different frequencies.

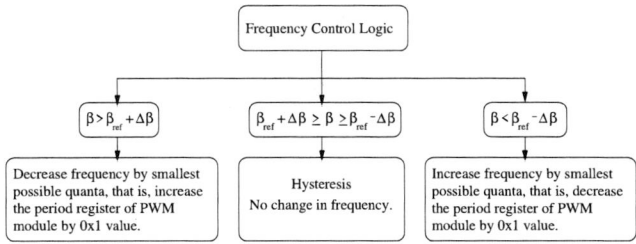

Fig. 6. Flowchart of the Frequency Tracking Algorithm

IV. EXPERIMENTAL RESULTS

The experimental results of the Power Factor Correction (PFC) and inverter testing are depicted in Figures 7 to 11. These results demonstrate the effectiveness of the inverter and control strategy in maintaining the desired phase shift (β) across different operating conditions.

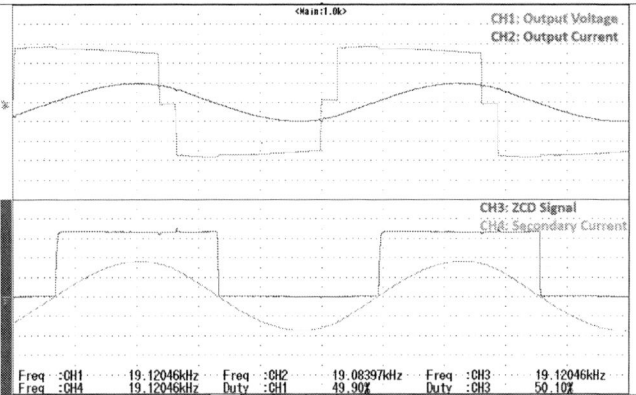

Fig. 7. Output voltage and current (top) along with ZCD signal and secondary current waveforms (bottom)

Figure 7 illustrates the output voltage and current waveforms at the primary side of the inverter. The top half of the waveform represents the inverter output voltage and current, while the bottom half shows the Zero Crossing Detection (ZCD) signal of the corresponding current.

Figures 8 to 10 represent the phase shift (β) at different duty cycles. The measured values of β are displayed using cursors in each waveform:

- **Figure 8:** Beta at 30% duty cycle, measured as 1.6 μ, with an operating frequency of 30 kHz.
- **Figure 9:** Beta at 60% duty cycle, measured as 4.3 μs, with an operating frequency of 22 kHz.
- **Figure 10:** Beta at 95% duty cycle, measured as 6 μs, with an operating frequency of 19 kHz.

Based on the power requirement, the phase shift is adjusted, and the system dynamically tunes the switching frequency to ensure Zero-Voltage Switching(ZVS) is maintained while keeping the phase interval β always above the minimum threshold $\beta_{\min}$.

These experimental results highlight how the system dynamically adjusts the operating frequency as the phase shift varies,

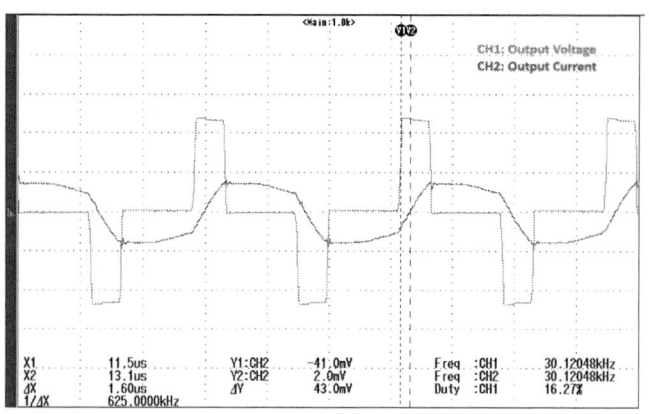

Fig. 8. Beta representation at 30% Duty Cycle

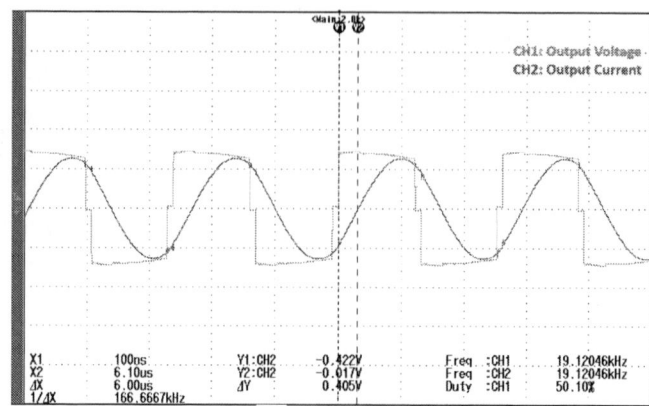

Fig. 10. Beta representation at 95% Duty Cycle

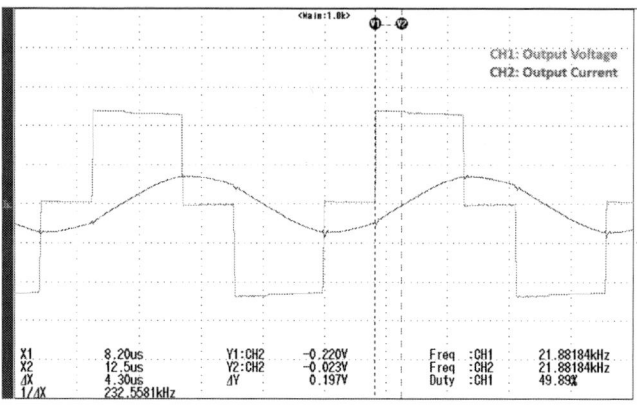

Fig. 9. Beta representation at 60% Duty Cycle

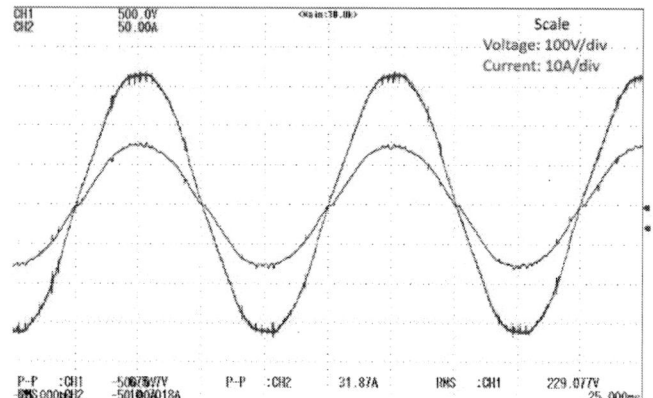

Fig. 11. PFC Input Voltage and Current waveshapes at maximum power

ensuring stable and efficient operation. The data from these waveforms is also summarized in tabular form in **Table II**.

TABLE II
VARIATION OF BETA WITH FREQUENCY

Phase shift	Duty Phase Shift	Frequency	β Time	β (in °)
27.7 μs	95%	19kHz	6 μs	41.25°
14.1 μs	60%	22kHz	4.3 μs	33.8°
5 μs	30%	30kHz	1.6 μs	17.28°

Figure 11 presents the input AC voltage and current waveforms of the active PFC boost converter under full-load conditions. The sinusoidal shape of the current waveform confirms the effectiveness of the PFC stage in correcting the power factor and minimizing harmonics.

Figure 12 illustrates the experimental setup used for testing the modular PFC and inverter system.

CONCLUSION

The proposed modular, multi-stacked, and interleaved solid-state inverter-based induction casting machine significantly enhances efficiency, scalability, and power management through its innovative configuration. By integrating three low-power

(2kW) power factor correction (PFC) circuits with three full-bridge resonant inverters, the system efficiently distributes power across multiple inverters connected to a three-phase four-wire AC supply via common high-frequency transformer, feeding the output side single induction coil with crucible. The interleaved topology of the inverter outputs, combined at the primary side of a high-frequency (HF) impedance matching transformer, ensures effective power transfer to the resonant tank and melting furnace, resulting in improved overall performance. The common HF transformer with multiple interleaved primary windings and a shared secondary winding further enhances impedance matching, allowing the system to handle higher power demands efficiently. The proposed modular structure is implemented on a laboratory based 6kW Induction heating system and controlled with a phase-shift and automatic frequency tracking circuit on a DSP Microcontroller platform.

This modular design not only enables scalability and easy system expansion but also simplifies maintenance, control, and redundancy while maintaining cost-effectiveness compared to conventional units. The proposed system offers a reliable and versatile solution for induction casting applications, ensuring enhanced power factor, improved performance, and adaptabil-

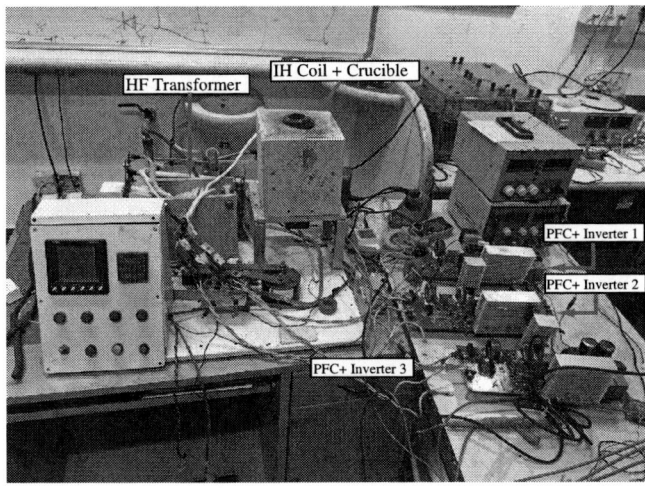

Fig. 12. Lab Testing Setup

ity for various industrial use cases.

ACKNOWLEDGMENT

This research was supported by the Department of Science and Technology (DST) under the Technology Development Program (TDP) scheme with project sanction number DST/TDT/TDP-16/2022. The authors sincerely appreciate the financial support provided for this work.

REFERENCES

[1] F. Ahmad, T. S. Aunsborg, A. B. Jørgensen and S. Munk-Nielsen, "From Vacuum Tubes to Modern Semiconductors: Opportunities for Industrial Heating Industry," in IEEE Transactions on Industry Applications, vol. 60, no. 4, pp. 6488-6498, July-Aug. 2024, doi: 10.1109/TIA.2024.3397639.

[2] Q. Xu, F. Ma, A. Luo, Y. Chen and Z. He, "Hierarchical Direct Power Control of Modular Multilevel Converter for Tundish Heating," in IEEE Transactions on Industrial Electronics, vol. 63, no. 12, pp. 7919-7929, Dec. 2016, doi: 10.1109/TIE.2016.2542159.

[3] A. -R. A. M. Makky, H. Abo-Zied and F. N. Abdelbar, "Parallel operation of IGBTs modular converter system for high power high frequency induction heating applications," 2008 12th International Middle-East Power System Conference, Aswan, Egypt, 2008, pp. 577-582, doi: 10.1109/MEPCON.2008.4562393.

[4] C. Carretero, O. Lucia, J. Acero, and J. M. Burdio, "Phase-shift modulation in double half-bridge inverter with common resonant capacitor for induction heating appliances," IET Power Electronics, vol. 8, no. 7, pp. 1182–1188, 2015, doi: 10.1049/iet-pel.2014.0497.

[5] O. Lucia, H. Sarnago and J. M. Burdio, "Pulse density modulated control for the series resonant multi-inverter for induction heating applications," IECON 2016 - 42nd Annual Conference of the IEEE Industrial Electronics Society, Florence, Italy, 2016, pp. 5995-6000, doi: 10.1109/IECON.2016.7794026

[6] P. Herasymenko, "Soft start-up output current of PDM-based series-resonant converter for induction heating application," 2019 IEEE 2nd Ukraine Conference on Electrical and Computer Engineering (UKRCON), Lviv, Ukraine, 2019, pp. 876-880, doi: 10.1109/UKRCON.2019.8879885.

[7] M. Pérez-Tarragona, H. Sarnago, Ó. Lucía and J. M. Burdío, "Full-bridge series resonant multi-inverter featuring new 900-V SiC devices for improved induction heating appliances," 2016 IEEE Applied Power Electronics Conference and Exposition (APEC), Long Beach, CA, USA, 2016, pp. 1762-1766, doi: 10.1109/APEC.2016.7468106.

[8] S. Khatroth and P. Shunmugam, "Cascaded full-bridge resonant inverter configuration for different material vessel induction cooking," IET Power Electronics, vol. 13, Issue-19, Jan. 2021, doi: 10.1049/iet-pel.2020.0728.

[9] M. Pérez-Tarragona, H. Sarnago, Ó. Lucía and J. M. Burdío, "Multiphase PFC Rectifier and Modulation Strategies for Domestic Induction Heating Applications," in IEEE Transactions on Industrial Electronics, vol. 68, no. 8, pp. 6424-6433, Aug. 2021, doi: 10.1109/TIE.2020.3005096.

[10] A. B. Jorgensen, T. S. Aunsborg, S. Beczkowski, C. Uhrenfeldt, and S. Munk Nielsen,"High frequency resonant operation of an integrated medium-voltage SiC MOSFET power module," IET Power Electronics, vol. 13, no. 3, Feb. 2020, doi: 10.1049/iet-pel.2019.0413.

[11] S. M. Park, H. -K. Yang, B. J. Hyon, J. Sung Park, J. -H. Kim and B. K. Lee, "Design and Implementation of GaN-HEMT-Based Inverter-Coil Integrated Module for Free Zone Induction Heating System," 2024 IEEE Applied Power Electronics Conference and Exposition (APEC), Long Beach, CA, USA, 2024, pp. 2433-2437, doi: 10.1109/APEC48139.2024.10509464.

[12] H. Sarnago, J. M. Burdío and O. Lucia, "High-Frequency GaN-Based Induction Heating Versatile Module for Flexible Cooking Surfaces," 2019 IEEE Applied Power Electronics Conference and Exposition (APEC), Anaheim, CA, USA, 2019, pp. 448-452, doi: 10.1109/APEC.2019.8721779.

[13] Ó. Lucia, L. A. Barragán, J. M. Burdio, Ó. Jimenez, D. Navarro and I. Urriza, "A Versatile Power Electronics Test-Bench Architecture Applied to Domestic Induction Heating," in IEEE Transactions on Industrial Electronics, vol. 58, no. 3, pp. 998-1007, March 2011, doi: 10.1109/TIE.2010.2048840.

[14] D. N. Sankhe, R. R. Sawant and Y. S. Rao, "FPGA-Based Hybrid Control Strategy for Resonant Inverter in Induction Heating Applications," in IEEE Journal of Emerging and Selected Topics in Industrial Electronics, vol. 3, no. 1, pp. 156-165, Jan. 2022, doi: 10.1109/JESTIE.2021.3051584.

[15] R. Sawant, Y. S. Rao, B. N. Chaudhari, and A. Padole, "A modular and interleaved solid-state induction heating generator for induction casting machine with improved input power factor," Indian Patent Application No. 202421052945, filed July 11, 2024.

[16] R. R. Sawant and Y. S. Rao, "A discrete-time controller for Phase Shift Controlled load-resonant inverter without PLL," 2014 IEEE International Conference on Power Electronics, Drives and Energy Systems (PEDES), Mumbai, India, 2014, pp. 1-4, doi: 10.1109/PEDES.2014.7042075.

[17] R. R. Sawant, N. S. Chame and N. K. Rana, "A New Hybrid Power Control Technique for Induction Vessel Heating System," 2006 IEEE International Conference on Industrial Technology, Mumbai, India, 2006, pp. 1424-1429, doi: 10.1109/ICIT.2006.372560

Constant Current-Constant Voltage based EV Battery Charging using High Frequency Phase-Shift Full Bridge Converter

Mohammad Tahir Siddiqui
Electrical Engineering Department
Motilal Nehru National Institute of Technology Allahabad
Prayagraj, India
siddiquitahir688@gmail.com

Rohit Gupta
Electrical Engineering Department
Motilal Nehru National Institute of Technology Allahabad
Prayagraj, India
rohit.2024ree10@mnnit.ac.in

Abhinay Pratap Singh
Electrical Engineering Department,
Motilal Nehru National Institute of Technology Allahabad
Prayagraj, India.
abhinay.2022ree01@mnnit.ac.in

Rajesh Gupta, Senior,*Member IEEE*
Electrical Engineering Department,
Motilal Nehru National Institute of Technology Allahabad
Prayagraj, India.
rajeshgupta@mnnit.ac.in

Abstract— **Battery charging for electric vehicles (EVs) is essential for environmentally friendly transportation options. This paper presents a modified approach for EV battery charging using a constant current-constant voltage (CC-CV) method combined with a high frequency phase shift full bridge (PSFB) converter. This approach dynamically modifies charging parameters to ensure high performance during the charging process, improving efficiency and adaptability for the battery charging at different state-of-charge (SOC). The high-frequency PSFB converter is good for the medium voltage, high performance EV charging stations because it enables effective power conversion with decreased size, weight, and electromagnetic interference (EMI). Incorporating CC-CV in the PSFB stage reduces the possibility of overcharging, improves battery life, and safer operation. Moreover, phase-shifting management and high-frequency operation, together improves overall efficiency and reduces power losses. The proposed EV battery charging method has been demonstrated through the 400 V DC voltage across PSFB for 72 V, 26 Ahr, battery charging. The simulation results are obtained through the MATLAB based studies and experimental results are obtained using the Typhoon 404 processor acting as a controller for the PSFB converter.**

Keywords- **Battery charging, constant-current constant-voltage (CC-CV), Lithium-ion battery, phase shift full bridge (PSFB), state of charge (SOC).**

I. INTRODUCTION

In the early stages of battery development, charging methods were basic and largely manual. Direct current (DC) sources, such as dynamos or generators, were commonly used to recharge batteries, particularly lead-acid and early nickel-cadmium types. These systems operated using either constant current (CC) or constant voltage (CV) techniques, involve human oversight to avoid issues like overheating or overcharging. The process is largely manual and lacks reliability in battery charging, could lead to reduced battery life or safety risks. Early methods involved mechanical solutions like hand-cranked generators to charge the electrical batteries.

Modern high frequency power electronics converters based charging are automatic and has higher efficiencies. One widely adopted approach is the constant current-constant voltage charging technique, often implemented using dual loop control, involving out voltage and inner current control loop. The PSFB converter is popular addition in battery charging utilizing high-frequency transformer, which offer improved efficiency, electrical isolation, and reduced size. The PSFB design enables smooth, low-loss switching and smooth control over the charging process. The PSFB-based charging can be fully controlled, highly efficient, and capable of efficiently charging high-capacity batteries for the EV applications.

The earliest rechargeable lead-acid batteries, developed by Gaston Planté in 1859, was typically recharged using DC source [1]. The common charging methods used for battery charging are the constant current and constant voltage. Both these techniques can be implemented independently or based on the state of charge, or combination of both. Lack of controlled CC-CV charging [2], may lead to a risk like overcharging, battery heating, and derating of battery life. Mostly the battery charging technology involve expertise and experience [3]. These methods of battery charging are satisfactory for lower ratings, however the higher rated batteries requires precise and reliable charging methods, to avoid early battery failures. Recently developed power electronics converters and intelligent algorithms have enabled the development of intelligent and efficient charging systems. The CC-CV charging strategy is mostly successful and adopted in different variants.

Recently developed PSFB converter has high performance capability in battery charging. This converter uses high-frequency switching coupled with high frequency transformers, results in improved energy efficiency and precise controlled charging [4], and reduced system size. Research presented in [5]-[9] highlights the PSFB-based CC-CV charging and offers benefits such as zero-voltage switching (ZVS), which reduces switching losses and enhances thermal performance [10]-[13]. These features enable more accurate control over wide charging regions, leading to improved safety, faster charging, and greater compatibility with advanced battery systems. The PSFB converter offers scalable solution for battery energy storage applications like electric vehicles and smart power systems

---

979-8-3315-3013-6/25 $31.00 © 2025 IEEE

[14]-[17]. The CC-CV control have improved regulation during both constant current and constant voltage charging phases, ensuring optimized charging performance. It enables safe and complete battery charging, reducing the risk of overcharging and enhancing battery lifespan [18]-[19].

This paper proposes a dual mode voltage and current loop control in high frequency phase shift full bridge inverter to charge the EV battery in CC-CV control mode using adjustable parameters of phase-shift based on the empirical relation defined based on the battery characteristic. A control algorithm is proposed for the PSFB converter based on the SoC and voltage of the battery. Both simulations and experiments are conducted to verify the performance of the 72 V, 26 Ahr, battery charging from the 400 V DC supply using PSFB.

Rest of the paper is structured in the following manner: section I provides a brief introduction to the EV battery charging system. In Section II, constant current - constant voltage method based on SOC is described. Section III provides working of high frequency phase shift full bridge dc-dc converter. Section IV elaborates on the operation and control techniques employed for PSFB converter. Section V describe about the Li-ion EV battery with specifications. Section VI presents the operation of PSFB inverter based on the battery SOC with CC-CV mode. Section VII presents the simulation results for different conditions. Section VIII provides the experimental results.

II. CONSTANT-CURRENT AND CONSTANT-VOLTAGE (CC-CV) BATTERY CHARGING

A popular and most reliable method of battery charging is the CC-CV charging method. The constant current and constant voltage charging techniques can be combined to achieve high performance. Each technique has a specialized function during the charging process.

The CC mode is used in the first stage, in which the charger provides the battery with a constant and preset current. This stage is the main factor in cutting down the total charging time since it quickly restores the battery charge. During this phase, the battery voltage progressively increases until it reaches its maximum safe voltage limit. Up to 80% SOC, it is beneficial to charge the battery in CC mode.

The charging process switches to the CV mode when a voltage threshold is reached. As the battery gets closer to full capacity, the charger slowly lowers the current while maintaining a constant voltage. It is advised to use CV mode above 80% SOC. This stage guarantees that the battery is charged securely without overheating or breaking and avoids overcharging. A typical CC-CV curve for the EV battery charging is displayed in Fig.1.

While the duration of the CV mode is dependent on the battery's internal properties and capacity, the duration of the CC mode is a crucial component in determining the overall charging time. The CC-CV changeover instant is adjustable and mostly decided by the threshold SoC. In particular 80% is a optimal choice in many applications.

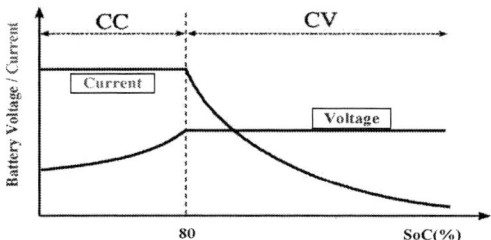

Fig. 1: CC-CV curve for charging of EV battery.

III. PSFB DC-DC CONVERTER

Applications requiring medium to high power levels are ideally suited for the PSFB (Phase-Shifted Full Bridge) DC-DC converter. It usually consists of a low-pass filter, a full bridge diode rectifier, a high-frequency transformer, and a full bridge inverter. Four MOSFET/IGBT switches with anti-parallel diodes make up the entire bridge inverter. The PSFB DC-DC converter comprises LC filters, a diode-based uncontrolled single-phase rectifier, a high frequency transformer, and an H-Bridge single-phase inverter. The components' numerical values are listed in Table I.

In this paper the PSFB is controlled using phase-shift applied by shifting the gating signals of switches S_3 and S_4 with respect to S_1 and S_2, respectively, as illustrated in Fig. 2. The phase shift controls the high-frequency output AC voltage. The high-frequency transformer steps down the voltage based on its turn's ratio, converting the high voltage from the inverter to a lower voltage. The PSFB DC-DC converter uses phase-shift control to effectively control the output voltage. To produce a phase change in the voltage delivered to the transformer's primary winding, the switches work in a particular order. This phase shift allows for accurate voltage regulation by controlling the effective duty cycle. To create a steady DC output appropriate for battery charging, the resulting AC voltage is rectified by a full bridge diode rectifier on the secondary side of the transformer and filtered by a low-pass filter, which is usually made up of an inductor and capacitor. Phase-shifting is used to regulate the output while maintaining the PWM signal at a 50% duty cycle.

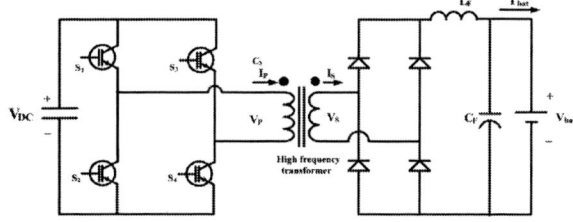

Fig. 2: Circuit schematic of PFSB DC/DC converter.

IV. OPERATION OF PSFB CONVERTER

Four operational modes explain the first half-cycle of the operation. The approach for the second half cycle is the same as for the first one.

➤ Mode-1: In this mode the switches S_1 and S_4, are operated and $+V_{dc}$ is generated at the primary side of the transformer.

979-8-3315-3013-6/25 $31.00 © 2025 IEEE   427

> Mode-2: Here the switches S_1 and S_3 are operated and zero voltage is generated across primary side of the transformer.

> Mode-3: In this mode, the switches S_2 and S_3 are activated and $-V_{dc}$ voltage is generated across the transformer's primary side.

> Mode-4: In this mode, the switches S_2 and S_4 are activated, resulting in a zero voltage to develop across the transformer's input side.

Fig. 3 displays the PWM signal for every switch operating in the phase shift mode. The switch S_1 receives the initial PWM pulse. The PWM signal for S_2 should be 180 degrees out of phase with S_1 in order to prevent shoot through. The switch S_3 now uses the same PWM signal as the switch S_1, but with a suitable phase shift. The switch S_4 will now get PWM when the signal is inverted. Fig. 4 depicts a typical PSFB AC output voltage whose RMS can be controlled by using phase-shift between the pulses.

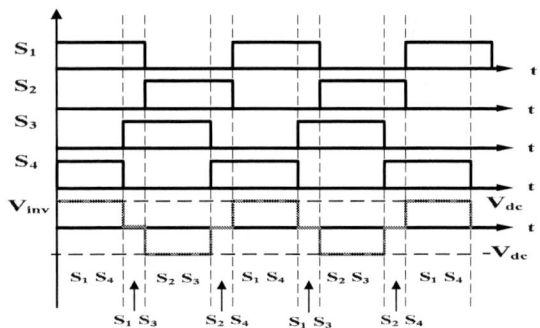

Fig. 3: Waveform of pulses of switches and the output of single-phase PSFB inverter.

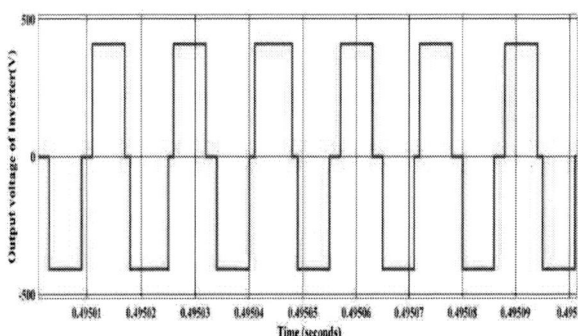

Fig. 4: Waveform of output voltages of single-phase PSFB inverter.

The voltage duration on the transformer's primary side can be varied by the variations in the delay angle between S_1 and S_3. The output voltage will increase if the phase shift angle increases from 0 to 180°, and it will decrease if it is decreased. To implement the PSFB with CC-CV an algorithm is proposed and depicted in Fig. 5. This involves outer battery voltage control loop and inner battery current control loop.

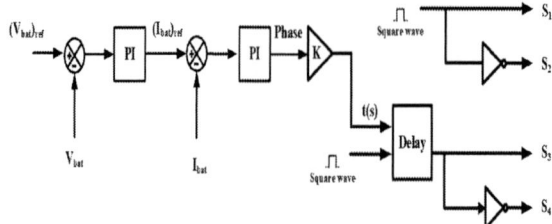

Fig. 5: Control algorithm for PSFB converter with outer voltage and inner current control loop.

Across the PSFB converter, we connect a suitable turns ratio high frequency transformer, to step down the voltage generated by the inverter to the level of battery charging voltage. The values of the filter's parameters L_F and C_F are designed for 5% voltage and current ripple.

V. LITHIUM-ION BATTERY

A battery is a device that stores electrical energy as chemical energy and converts it back into electricity when required. Graphite anodes and metal oxide cathodes are used in intercalation-based charge storage in lithium-ion (Li-ion) batteries. They have a low memory effect, a high energy density, and excellent coulombic efficiency. For Li-ion batteries to operate steadily, in applications like energy storage systems and electric cars, careful thermal and voltage management is necessary to reduce the risk of thermal runaway [11]. A Li-ion battery operates by the reversible insertion of lithium ions (Li+) into conductive materials to store the energy. Key components of Li-ion batteries include the anode (typically made of graphite), cathode, the electrolyte, and the separator, which keeps the anode and cathode apart to prevent short circuits while permitting ion movement. Fig. 6 depicts a typical discharging profile of a Li-ion battery.

In this context, lithium serves as the primary or EV battery with a focus on its charging process. Details regarding battery are presented in Table I.

TABLE I.   SPECIFICATIONS OF BATTERIES

Parameters	Values
Nominal voltage	72 V
Capacity	26 Ah
Charging voltage	84 V
Charging current	20A
Nominal capacity	1872 Wh

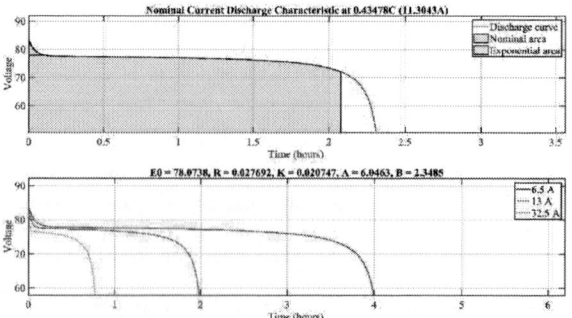

Fig. 6: Discharging profile of the Li-ion battery.

979-8-3315-3013-6/25 $31.00 © 2025 IEEE

VI. PSFB INVERTER OPERATION BASED ON BATTERY SOC

A phase-shifted full-bridge (PSFB) converter with a dynamic phase angle control mechanism is used to guarantee effective battery charging while shielding the battery from possible harm. The constant current and constant voltage charging as proposed with PSFB can prove to be useful for high performance rated batteries.

The converter runs in CC mode during the first charging phase when the battery's state of charge is less than 80%. The PSFB converter is set up to maintain a maximum phase angle in this mode, usually between 150° and 180°. The delivery of maximum power is made easier by this range of phase angle, which guarantees a high and steady current flow with a slight increase in voltage. For the battery's charge capacity to be quickly restored, this stage can be controlled in a desired manner.

To avoid overcharging and thermal stress, the charging logic switches from CC mode to CV mode when the battery's state of charge exceed 80%. In this step, a predetermined C-function is proposed. Concept for C function is mention in equation (1), to gradually reduce the phase angle. The RMS output voltage decreases by controlled phase angle reduction from 180°. The converter maintains the desired voltage regulation to the reference charging voltage in the CV mode by controlled reduction in the power supplied to the battery. To maximize charging efficiency and guarantee battery longevity, the implemented logic shown in Fig. 7 depicts a smooth transition between CC and CV modes. This adaptive phase angle modification is essential for maintaining a balance between rapid charging speeds and efficient battery charging with the transition between the two modes is listed in Table II.

A constant K is used to convert the phase value (degree) into time (sec) to provide suitable delay,

$$K = (phase\ value \div 180) \times T$$

Where, T is the Time period

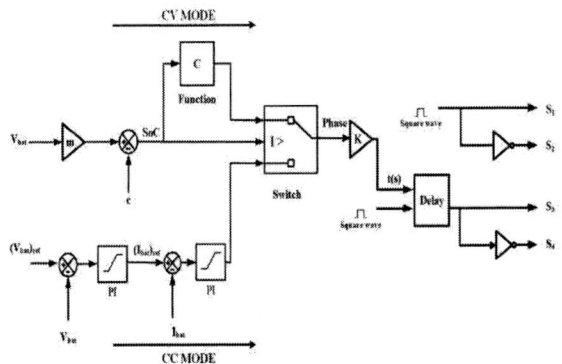

Fig. 7: Control algorithm for PSFB converter based on SoC.

A. Logic for CC mode:

In this mode, the voltage controller generates the reference current. I_{ref} by comparing it with V_{bat} and V_{batref} with the help of a PI controller. Limiter-1 restricts the battery charging current to a maximum of 20A. The reference battery current I_{batref} is then compared against I_{bat}, and the error is transmitted through another PI controller. Limiter-2 constrains the phase shift of the PWM pulse in between 0 and 180°. Then we use the square wave block to generate the pulses.

B. Logic for CV mode:

When the SOC is above 80%, the CV mode of operation is used. In this mode, phase angle is getting reduced (below 150 degree) with the help of a C-function, in a manner so that output power is getting reduced to protect the battery (CV Mode) means current decreases from its upper threshold value, and voltage becomes constant.

TABLE II. CONTRL LOGIC ABOVE 80% SoC IN CV MODE

Mode	Phase angle (Degree)
CC	150-180
CV	<150

To achieve CV mode, we decrease the phase angle by a value of 3-degree in the logic, battery current will also be decreased. Consider the following parameters/variables of the battery for the control.

x: SOC of the battery

y: Phase angle of the inverter

$$y = 150 - 3(x - 80) \tag{1}$$

As the SOC crosses its value from 80%, the phase angle will start decreasing from 150 degree. According to this logic, for 1% of SOC per 3 degrees of phase will decrease.

The CC- CV mode of operation is very useful for lithium-ion batteries used for EVs for both speed and safety purposes. System parameters used in this paper is listed in Table III.

TABLE III. PARAMETRES OF THE SYSTEM

Components	Parameters	Numerical values
DC Link Capacitor	Voltage	500 μF, 400 V_{DC}
LC Filter	Inductor (L_F)	1.06 mH
	Capacitor (C_F)	470 μF
High Frequency Transformer	Turn Ratio	2 kW, 400/130 V for 72 V, 26 Ahr

VII. SIMULATION RESULTS

The simulation is done with the parameters listed in Table III in MATLAB software. Initially the battery voltage progressively rises while the battery current stays constant in the CC mode, as seen in Fig. 8. The battery voltage, current and power is displayed in this figure. As seen in Fig. 9, the battery's SOC increases throughout this phase starting from 60%. The battery is charging in the constant current mode.

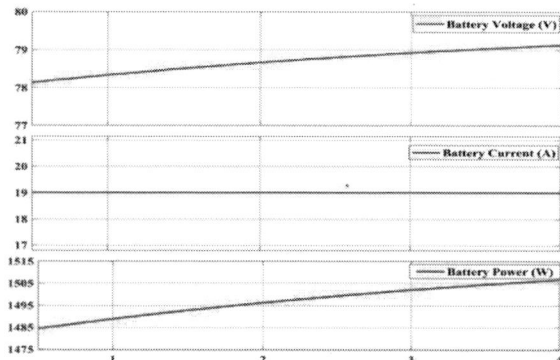

Fig. 8: Battery voltage, current and power for SOC 60%.

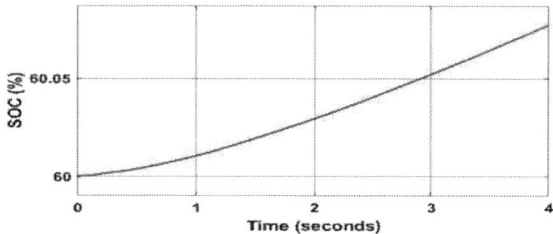

Fig. 9: SOC of the EV Battery.

Fig. 10 shows a situation in which the battery current starts to drop but the battery voltage remains constant. At the same time, as seen in Fig. 11, the battery's SOC rises from 85%. This shows that the constant voltage mode of operation during the charging operates as per the logic implemented in Table II.

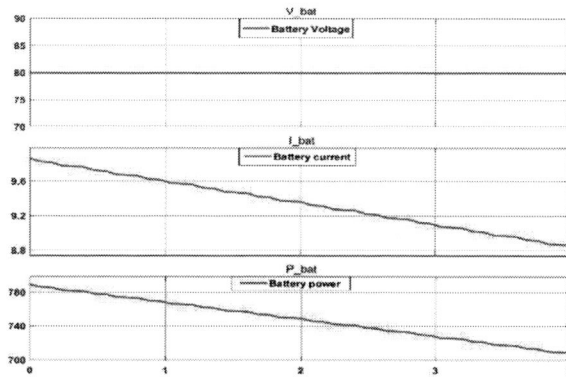

Fig. 10: Battery voltage, current and power for SOC 85%.

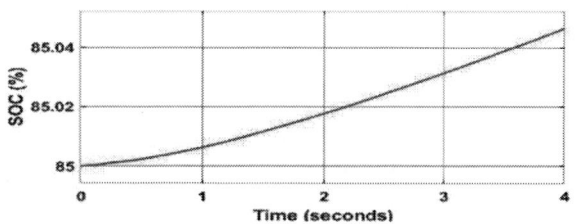

Fig. 11: Variation in SOC of Battery.

VIII. HARDWARE SET-UP AND EXPERIMENTAL RESULTS

The deployment of a battery charging system for lithium-ion battery was been investigated through a laboratory experimental setup as shown in Fig. 12. Rapid prototype model has been done using Typhoon HIL 402 as a controller on SEMIKRON inverter and voltage is sensed using voltage sensor LV-25P, current is feedback to controller by using LA-25NP used to develop a hardware setup that is used for the 72 V, 26 Ahr, battery charging.

Fig. 12: Experimental set-up for EV battery charging system.

A PSFB inverter converts the DC supply to AC using phase shifted pulses as shown in Fig. 3. The resultant output of the PSFB inverter and battery voltage are shown in Fig. 13 and Fig. 14, at the phase shift of 180° and 150°, respectively. As shown in Fig. 14, phase shift of the inverter is 180° which shows that the SOC of the battery is below 80%, i.e., 60 % here. When the SOC of battery increases above 80%, then phase shift of inverter start declining from 180 degree as can be seen in Table IV. Fig. 14 shows the inverter pulses for the SOC of 85% of the battery. In Fig. 15 first wave shows inverter output voltage, 2nd is battery voltage of 82 V, 3rd is the battery current and 4th one is the input DC voltage. The inverter functions at a switching frequency of 10 kHz to facilitate CC-CV charging.

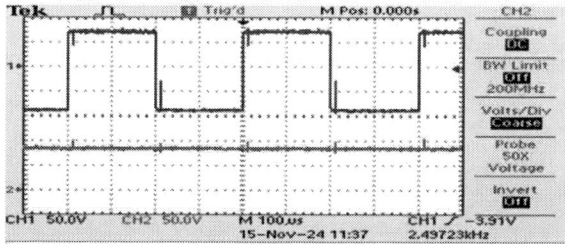

Fig.13: PSFB inverter voltage and battery voltage at SOC 60%.

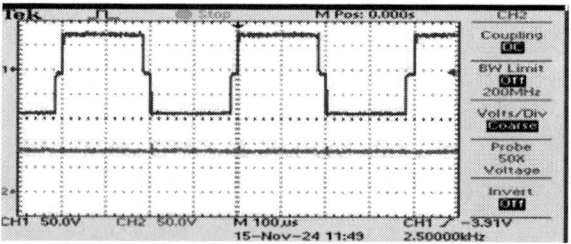

Fig.14: PSFB inverter voltage and battery voltage at SOC 85%.

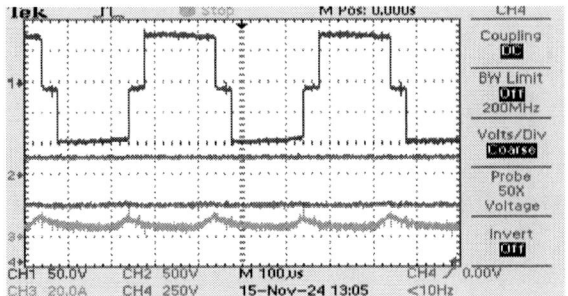

Fig.15: PSFB Inverter voltage and battery voltage & current and input voltage at SOC 90%.

The observation table shows the CC-CV charging of a 72V battery in Table IV. When SOC is between 60% and 80%, the battery current is almost constant, but when SOC is above 80%, the phase angle is reduced in order to reduce the battery current. In this table V_{DC} and I_{DC} are input voltage and current and P_{IN} is input power for the proposed system. As the phase starts decreasing, the input current also reduces and the input voltage slightly increasing. The maximum efficiency of the system is 95%.

TABLE IV.   EXPERIMENTAL OBSERVATION WITH VARYING SOC`S

Ph Deg	V_{DC} (V)	I_{DC} (A)	P_{IN} (W)	SoC (%)	V_{bat} (V)	I_{bat} (A)	P_{bat} (W)	Ef. %
180	222	3.3	737	70	73	8.8	700	95
180	225	3.27	737	78	81	8.5	700	95
174	228	3.17	723	82	81.2	8.4	680	94
168	231	2.94	680	84	81.4	7.9	640	94
162	232	2.84	672	86	81.8	7.6	625	93
156	232	2.73	634	88	81.9	7.2	590	93
150	234	2.24	526	90	82	6	490	93
144	235	1.32	312	92	82	3.5	290	93

IX. CONCLUSIONS

A PSFB based CC-CV charging is proposed in this paper for charging of high performance Li-ion EV batteries, by dynamically adjusting the controller parameter. The battery is charged in the constant current mode, below 80% of the SOC. Based on the proposed algorithm the phase is gradually reduced above 80% SOC to implement constant voltage mode. Both simulation and experimental results demonstrates the performance of the proposed PSFB based CC-CV based EV battery charging.

ACKNOWLEDGMENT

This work is supported by the National Mission on Power Electronics Technology Phase-III (NaMPET-III) programme, Ministry of Electronics and Information Technology, India, through the project titled: Solar PV Supported Reconfigurable EV Charger with Reduced Switching Devices, grant no. 25(6)/2020- ESDA, dated 10/06/2022.

REFERENCES

[1] G. Planté, "Researches on electricity: secondary batteries,"Comptes Rendus, vol. 63, pp. 683–686, 1866.

[2] D. Linden and T. Reddy, Handbook of Batteries, 3rd ed., McGraw-Hill, 2002.

[3] J. B. Goodenough and K.-S. Park, "The Li-ion rechargeable battery: A perspective," Journal of the American Chemical Society, vol. 135, no. 4, pp. 1167–1176, 2013.

[4] M. K. Kazimierczuk, *Pulse-Width Modulated DC-DC Power Converters*, John Wiley & Sons, 2008.

[5] Y. Chen, J. Zhang, and M. Wang, "Design and analysis of PSFB converter for high-power battery charger," *IEEE Transactions on Power Electronics*, vol. 33, no. 9, pp. 7558–7567, Sep. 2018.

[6] A. Kumar and R. Singh, "Review of high frequency isolated battery chargers using PSFB topology for electric vehicle applications," *International Journal of Power Electronics and Drive Systems*, vol. 12, no. 2, pp. 1234–1242, 2021.

[7] J. Cho, S. Jeong and Y. Kim, "Commercial and research battery technologies for electrical energy storage applications", *Progress in Energy and Combustion Science*, vol. 48, pp. 84-101, June 2015.

[8] R. Pragale, T. J. Dionise and D. D. Shipp, "Harmonic Analysis and Multistage Filter Design for a Large Bleach Production Facility," in *IEEE Transactions on Industry Applications*, vol. 47, no. 3, pp. 1201-1209, May-June 2011.

[9] Kumari Priya, Manas Kumar and Rajesh Gupta, "EV Battery Charging System via Reconfigurable Boost Converter with Solar PV and Grid", 11th National Power Electronics Conference (NPEC 2023), 14-17 Dec. 2023, IIT Guwahati , India.

[10] N. Kumar, H. K. Singh and R. Niwareeba, "Adaptive Control Technique for Portable Solar Powered EV Charging Adapter to Operate in Remote Location," in *IEEE Open Journal of Circuits and Systems*, vol. 4, pp. 115-125, 2023.

[11] A. Chandel and R. K. Singh, "PSFB Converter Based Scheme for Permanent Magnet Brushless DC Motor Drive," *2022 4th International Conference on Energy, Power and Environment (ICEPE)*, Shillong, India, 2022

[12] M. Safayatullah, M. T. Elrais, S. Ghosh, R. Rezaii and I. Batarseh, "A Comprehensive Review of Power Converter Topologies and Control Methods for Electric Vehicle Fast Charging Applications," in *IEEE Access*, vol. 10, pp. 40753-40793, 2022.

[13] S. A. Q. Mohammed and J. -W. Jung, "A Comprehensive State-of-the-Art Review of Wired/Wireless Charging Technologies for Battery Electric Vehicles: Classification/Common Topologies/Future Research Issues," in IEEE Access, vol. 9, pp. 19572-19585, 2021.

[14] A. D. Kumar, J. Gupta and B. Singh, "A Single-Stage Charger for LEV Based on Quadratic Buck-Boost AC-DC Converter Topology," in *IEEE Transactions on Industry Applications*, vol. 59, no. 4, pp. 4252-4263, July-Aug. 2023.

[15] A. K. Singh, A. K. Mishra, K. K. Gupta, P. Bhatnagar and T. Kim, "An Integrated Converter With Reduced Components for Electric Vehicles Utilizing Solar and Grid Power Sources," in *IEEE Transactions on Transportation Electrification*, vol. 6, no. 2, pp. 439-452, June 2020.

[16] T. -W. Hsu, H. -H. Wu, D. -L. Tsai and C. -L. Wei, "Photovoltaic Energy Harvester With Fractional Open-Circuit Voltage Based Maximum Power Point Tracking Circuit," in *IEEE Transactions on Circuits and Systems II: Express Briefs*, vol. 66, no. 2, pp. 257-261, Feb. 2019.

[17] P. Mohseni, O. Husev, D. Vinnikov, R. Strzelecki, E. Romero-Cadaval and I. Tokarski, "Battery Technologies in Electric Vehicles: Improvements in Electric Battery Packs," in *IEEE Industrial Electronics Magazine*, vol. 17, no. 4, pp. 55-65, Dec. 2023.

[18] A. K. Tiwari, L. K. Sahu, S. Mishra, Y. Kishor, A. H. Chander and C. H. K. Rao, "Design of ANPC converter based DC charging station for electric vehicle", *Asian Conference on Innovation in Technology (ASIANCON 2021)*, pp. 1-6, 27-29 Aug. 2021, Pune, India.

[19] A. Akash and M. Sreejeth, "Isolated DC-DC converter with synchronous rectification for electric vehicle battery charging", *3rd International Conference on Intelligent Technologies (CONIT 2023)*, pp. 1-5, 23-25 June. 2023, Hubli, India.

979-8-3315-3013-6/25 $31.00 © 2025 IEEE

Hybrid Federated Learning for Secure and Accurate Heart Disease Prediction

Kavya Sree Sai Bulasara*, Sai Sailu Batta*, Mahesh Miriyala†, Veerapu Goutham‡

*School of Computer Science and Engineering, VIT-AP University, Amaravati, India
Email: bulasarakavyasree@gmail.com, saisailubatta@gmail.com
†School of Electronics Engineering, VIT-AP University, Amaravati, India
Email: miriyalamahesh4u@gmail.com
‡School of Electronics Engineering, Vellore Institute of Technology, Vellore, India
Email: gouthamveerapu@gmail.com

Abstract—Cardiovascular disease continues to be a leading cause of death globally, emphasizing the need for advanced predictive models to facilitate early detection. Traditional machine learning techniques frequently depend on centralized data aggregation, which presents notable challenges regarding privacy and data security. Federated Learning (FL) mitigates these concerns by facilitating decentralized training across multiple clients, ensuring data privacy is maintained. This study presents a Hybrid Federated Learning (FL) model that combines deep learning with the Federated Averaging (FedAvg) algorithm to effectively aggregate updates from local models. The proposed model is evaluated using the Framingham Heart Study dataset, with its performance compared against that of a centralized deep learning model. The findings reveal that the Hybrid FL model achieves comparable accuracy while ensuring data privacy, underscoring its potential for practical medical applications.

Index Terms—Federated Learning, Heart Disease Prediction, Machine Learning, Privacy-Preserving AI, Healthcare Analytics

I. INTRODUCTION

Cardiovascular diseases (CVDs) remain the leading cause of mortality worldwide, responsible for over 17 million deaths annually. This alarming statistic underscores the urgency of addressing CVDs as one of the most formidable global health challenges of the 21st century [1].These conditions include coronary artery disease, heart failure, stroke, and hypertension.The prevalence of these disorders is driven by various risk factors such as obesity, diabetes, smoking, elevated cholesterol levels, and genetic predispositions. Consequently, the economic burden posed by CVDs is substantial, characterized by escalating healthcare costs and productivity losses that significantly strain healthcare infrastructures globally. Early detection and accurate prediction of heart disease are crucial in reducing CVD-related mortality rates. Prompt interventions, enabled by timely identification, can substantially enhance patient outcomes and mitigate long-term complications [2].

Traditionally, the diagnosis of heart disease has relied heavily on clinical assessments, medical imaging, and biomarker analysis. While these techniques have proven effective, they often necessitate specialized expertise and sophisticated equipment, rendering them less accessible in resource-limited set-

tings. Moreover, the manual interpretation of complex diagnostic data introduces potential risks of oversight or misinterpretation.

Recent advancements in deep learning (DL) and machine learning (ML) have revolutionized cardiovascular risk assessment. These AI-driven models can analyze extensive healthcare data, incorporating patient history, lifestyle factors, and clinical attributes to predict the risk of CVDs with high accuracy. By identifying intricate patterns within complex datasets, these models provide valuable support to clinicians in making informed diagnostic decisions [3].

Despite the promising capabilities of AI in healthcare, widespread adoption faces several critical obstacles. Notably, data privacy concerns, security vulnerabilities, and stringent regulatory frameworks present formidable challenges. Centralized learning approaches, which aggregate sensitive patient data into a unified repository, significantly increase the risk of data breaches and unauthorized access. Furthermore, healthcare regulations such as the Health Insurance Portability and Accountability Act (HIPAA) and the General Data Protection Regulation (GDPR) impose stringent limitations on data sharing, further complicating centralized AI model training in realworld healthcare environments [4].

To address these concerns, Federated Learning (FL) has emerged as a promising solution for decentralized model training across multiple institutions. FL ensures data privacy by allowing raw patient data to remain locally stored, eliminating the need for data aggregation in a central repository. This approach enables hospitals, research centers, and healthcare providers to collaboratively train predictive models without violating privacy regulations. Federated learning models have also incorporated blockchain and homomorphic encryption to enhance security for sensitive healthcare data [5].

While FL offers notable advantages in preserving data confidentiality, standard FL algorithms encounter practical challenges. These include slow convergence rates, elevated communication costs, and diminished model performance due to non-independent and identically distributed (non-IID) data. Such data heterogeneity is especially pronounced in healthcare applications, where patient demographics, treatment patterns, and disease prevalence vary significantly across institutions

979-8-3315-3013-6/25 $31.00 © 2025 IEEE

[4].

To address these limitations, we propose a Hybrid Federated Learning model that integrates deep learning techniques with the Federated Averaging (FedAvg) algorithm. This innovative approach aims to improve model accuracy and robustness while ensuring robust data privacy protections. Unlike conventional FL strategies, our method employs deep neural networks to enhance predictive precision and mitigate performance degradation resulting from heterogeneous data distributions. By utilizing the FedAvg algorithm, locally trained model parameters are aggregated to construct a comprehensive global model, ensuring both data security and optimal performance.

The primary contributions of this research are as follows:

- Development of a Hybrid FL model that integrates deep learning techniques with the Federated Averaging (FedAvg) algorithm to enhance heart disease prediction accuracy.
- Implementation of the proposed model using the Framingham Heart Study dataset to evaluate performance under decentralized conditions characterized by heterogeneous data distributions.
- Comparative performance analysis between the Hybrid FL model and a conventional centralized deep learning model, focusing on accuracy, privacy preservation, and computational efficiency.

The remainder of this paper is structured as follows: Section II reviews existing research on Federated Learning applications in healthcare. Section III outlines the proposed methodology, detailing data preprocessing techniques, model architecture, and optimization strategies. Section IV presents experimental results, performance evaluation, key findings, challenges, and potential limitations. Finally, Section V concludes the paper with a summary of contributions.

II. RELATED WORK

Studies comparing traditional machine learning algorithms [6] for heart disease prediction have highlighted challenges that motivate the adoption of privacy-preserving federated approaches. Federated Learning (FL) has emerged as a promising solution in the healthcare domain, particularly for applications that require collaborative model training while preserving data privacy. FL enables multiple clients, such as hospitals and healthcare institutions, to jointly train machine learning models without sharing sensitive patient data. This decentralized learning approach addresses key privacy concerns often encountered in conventional centralized machine learning methods. Earlier works using Hidden Naïve Bayes classifiers [7] demonstrated early attempts at heart disease prediction using centralized data mining techniques.

Several studies have investigated FL-based frameworks for heart disease prediction, demonstrating the potential benefits of FL in improving model performance while maintaining data security. For instance, the authors in [8] proposed an edge-based FL framework specifically designed for heart disease prediction tasks. Their findings revealed that FL models could achieve accuracy levels comparable to those of centralized models, with the added advantage of safeguarding sensitive patient information by keeping data localized.

In a related study, researchers in [9] explored the integration of FL with differential privacy techniques to further enhance data security during the training process. Their approach aimed to mitigate risks associated with privacy breaches, demonstrating improved robustness against privacy attacks. By combining FL with differential privacy mechanisms, the proposed solution effectively minimized the exposure of individual client data during model updates.

In addition to privacy-preserving techniques, deep learning architectures have been integrated with FL to enhance cardiovascular disease prediction outcomes. For example, a Convolutional Neural Network (CNN)-based FL model was introduced in [10] to improve predictive accuracy in heart disease diagnosis. This approach utilized decentralized model training, leveraging local data distributions across clients without requiring centralized data aggregation. The CNN-based FL model demonstrated superior performance in terms of predictive accuracy when compared to conventional learning techniques. Advanced deep learning structures like multiscale residual UNet++ have recently been integrated into FL frameworks to boost disease detection accuracy [11].

Further extending the capabilities of FL in healthcare, the authors in [12] developed a fully decentralized FL system known as DeFedHDP. This system was designed to enable real-time heart disease diagnosis by minimizing dependency on centralized servers. By decentralizing model updates and reducing reliance on a central server, the DeFedHDP framework addressed several scalability issues commonly found in traditional FL setups, thereby improving the overall efficiency of the learning process.

Despite significant advancements, FL models face notable challenges, particularly regarding non-independent and identically distributed (non-IID) data. In healthcare applications, patient data collected from different institutions often exhibits inherent heterogeneity due to variations in demographics, medical equipment, and clinical practices. These non-IID data distributions can significantly hinder the convergence of the global FL model, ultimately reducing its overall performance.

To address these limitations, researchers in [13] introduced a hybrid federated meta-learning approach aimed at improving model adaptability across diverse client datasets. By leveraging meta-learning strategies, this framework enhanced the model's ability to generalize effectively in heterogeneous data environments. Similarly, in [14], the authors proposed a Federated Learning (FL) framework that incorporated adaptive optimization techniques to facilitate faster convergence and improved predictive accuracy. These adaptive methods proved effective in mitigating the performance degradation commonly observed in FL models operating under non-IID data conditions.

Hybrid classifier models combined with federated learning have demonstrated improvements in CVD prediction tasks [15], supporting our choice of hybridized architectures.Building upon these prior advancements, this research introduces a novel Hybrid Federated Learning model that com-

979-8-3315-3013-6/25 $31.00 © 2025 IEEE 433

bines deep learning techniques with the Federated Averaging (FedAvg) algorithm to enhance global model training for heart disease prediction. Unlike conventional FL frameworks, the proposed approach focuses on improving model convergence and accuracy by strategically aggregating local updates from decentralized clients, all while preserving data privacy.

By leveraging deep learning architectures within the Federated Learning (FL) framework and optimizing the FedAvg algorithm, this research aims to provide a scalable, privacy-preserving solution for improving cardiovascular disease prediction in decentralized healthcare systems. The proposed methodology is expected to significantly enhance the accuracy, efficiency, and privacy of FL-based healthcare models, ultimately contributing to better clinical decision-making and improved patient outcomes.

III. PROPOSED HYBRID FEDERATED LEARNING MODEL FOR HEART DISEASE PREDICTION

In this section, we present the implementation of a Hybrid Federated Learning (FL) model designed to predict heart disease using the Framingham Heart Study dataset. This dataset is widely recognized for its comprehensive patient records, which include critical cardiovascular risk factors such as age, blood pressure, cholesterol levels, smoking habits, diabetes status, and body mass index (BMI). These features are essential for constructing reliable heart disease prediction models. To simulate a realistic decentralized healthcare environment, the dataset was partitioned into three distinct subsets, each representing a unique client.In this work, the dataset was partitioned into three subsets to simulate decentralized data sources across different clients, thereby mimicking a federated learning environment. Although the clients possessed different data samples, the distribution was created through random sampling and may not fully replicate the strong non-IID characteristics observed in real-world healthcare systems. Addressing more complex non-IID data distributions remains an important direction for future work to further validate the robustness of the proposed Hybrid FL model.Each client was responsible for training a local model using its respective data subset, ensuring that patient data remained private throughout the learning process.

The dataset underwent several preprocessing steps to improve data quality and model performance. Missing values were addressed using mean imputation to maintain data consistency. Standardization was applied using the StandardScaler technique to ensure uniform feature distribution across client subsets. Additionally, due to the dataset's imbalance, the Synthetic Minority Over-sampling Technique (SMOTE) was employed to balance the positive and negative class distributions. This played a significant role in improving the model's predictive ability for minority class instances.

The proposed Federated Learning (FL) model was implemented using TensorFlow Federated (TFF), an established framework for federated learning research and development. Each client was trained using a deep neural network (DNN) model with an architecture optimized for binary classification tasks. The model starts with an input layer that processes the standardized feature set. This is followed by a fully connected layer with 64 neurons and Rectified Linear Unit (ReLU) activation, designed to extract complex feature patterns. To reduce overfitting, a dropout layer with a rate of 0.2 was included. This is followed by another fully connected layer with 32 neurons and ReLU activation, with another dropout layer (rate of 0.2). A final fully connected layer with 16 neurons and ReLU activation refines the feature representation before the output layer. The output layer consists of a single neuron with sigmoid activation, which is ideal for binary classification tasks such as heart disease prediction.

The Federated Averaging (FedAvg) algorithm was used to aggregate model updates from individual clients. In this approach, each client independently trains a local model using its dataset, and the updated model parameters are transmitted to a central server. The server then computes a weighted average of the client models to construct the global model. This iterative aggregation process continues until the global model reaches optimal convergence. To optimize the learning process, the Adam optimizer was employed with a learning rate of 0.0005 for client-side training and 0.002 for server-side aggregation. Additionally, batch normalization was incorporated during local model training to address issues arising from non-IID (non-independent and identically distributed) data. This technique effectively stabilizes feature distributions, improving model convergence across diverse client datasets.

The training process involved 250 federated learning rounds, during which local client models were updated and periodically aggregated to refine the global model. To improve communication efficiency and simulate real-world healthcare settings, where not all institutions participate simultaneously, a subset of clients was selected for training in each round. The architecture and learning process of the proposed FL model are illustrated in Fig. 1, which outlines the interaction between clients, the central server, and the FedAvg aggregation process.

To evaluate the performance of the proposed FL model, we compared it against a centralized deep learning model trained on the complete dataset. Multiple evaluation metrics were used to assess the model's effectiveness. Accuracy, calculated as shown in Equation (1), was used to measure the overall correctness of predictions.

$$Accuracy = \frac{TP + TN}{TP + TN + FP + FN} \qquad (1)$$

Where True Positives (TP) and True Negatives (TN) represent correctly predicted positive and negative instances, respectively. False Positives (FP) occur when negative instances are incorrectly predicted as positive, and False Negatives (FN) occur when positive cases are incorrectly classified as negative. While accuracy is a standard metric, it can be misleading in imbalanced datasets, where one class significantly dominates the other.

To provide a more balanced evaluation, precision, recall, and the F1-score were also used. Precision, calculated as shown in

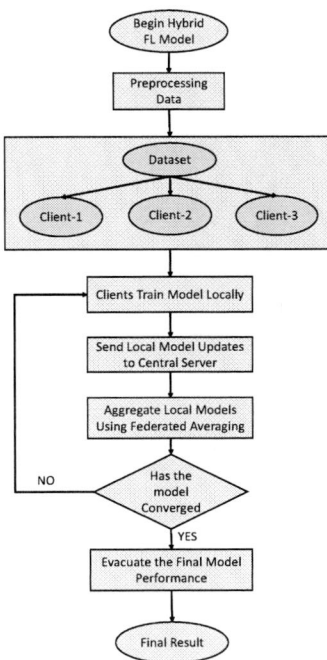

Fig. 1. Federated Learning Process

and N is the total number of samples. Lower loss values indicate better convergence and improved model stability during training.

The proposed FL model achieved a final accuracy of 84.67%, demonstrating that the FedAvg algorithm can effectively train a robust predictive model for heart disease detection in a decentralized environment. The integration of SMOTE for class balancing played a crucial role in enhancing recall performance by reducing bias toward the majority class. This improvement strengthened the model's ability to identify high-risk patients while preserving data privacy across multiple healthcare institutions. The combination of deep learning techniques, adaptive optimizers, and privacy-preserving FL principles positions this model as an effective solution for secure, scalable heart disease prediction in real-world medical applications.

IV. RESULTS AND DISCUSSION

This section presents the experimental results obtained from the proposed Hybrid Federated Learning (FL) model for heart disease prediction. The model's performance is evaluated using multiple metrics, including accuracy, precision, recall, F1-score, and the Area Under the Receiver Operating Characteristic Curve (AUC-ROC). These metrics offer a comprehensive assessment of the model's effectiveness in classifying heart disease cases.

A. Model Convergence Analysis

The convergence behavior of the proposed Hybrid FL model was analyzed by tracking accuracy and loss trends over 250 training rounds. Figures 2(a) and 2(b) depict the accuracy and loss trends for client models during the training process. The client models achieved stable accuracy, converging at approximately 84.81% with minimal fluctuations, indicating effective learning and model stability. Despite this high accuracy, the results highlighted the presence of class imbalance, which impacted the model's ability to predict minority class instances.

The loss curve shown in Fig. 2(b) illustrates a consistent decrease in loss values, confirming successful convergence of the client models. However, the final loss value of 2.3428 indicates potential optimization challenges, possibly stemming from the non-IID nature of the data or insufficient adaptation of model parameters during aggregation. Further investigation into improved optimization techniques may be necessary to address these concerns and enhance model performance.

Equation (2), quantifies the proportion of correctly predicted positive instances relative to all predicted positives.

$$Precision = \frac{TP}{TP + FP} \quad (2)$$

This metric is particularly important in medical applications to minimize false positives, thereby reducing the risk of unnecessary treatments. Recall, calculated as shown in Equation (3), measures the model's ability to identify actual positive cases.

$$Recall = \frac{TP}{TP + FN} \quad (3)$$

Higher recall values indicate fewer missed diagnoses, which is crucial for detecting critical heart disease cases. The F1-score, defined in Equation (4), provides a harmonic mean between precision and recall, making it particularly effective for evaluating models trained on imbalanced datasets.

$$F1\text{-}Score = 2 \times \frac{Precision \times Recall}{Precision + Recall} \quad (4)$$

Finally, the binary cross-entropy loss function was used to track model convergence and evaluate the learning process, as shown in Equation (5).

$$Loss = -\frac{1}{N} \sum_{i=1}^{N} (y_i \log(\hat{y}_i) + (1 - y_i) \log(1 - \hat{y}_i)) \quad (5)$$

In this formula, y_i denotes the actual label (1 for heart disease, 0 for no disease), $\hat{y}_i$ denotes the predicted probability,

B. Confusion Matrix

The confusion matrix provides further insights into the model's performance by detailing the counts of true positives (TP), true negatives (TN), false positives (FP), and false negatives (FN). As shown in Fig. 3, the Hybrid FL model demonstrated strong classification performance, effectively identifying a substantial number of both positive and negative cases. However, the presence of misclassified minority class instances highlights the challenge posed by class imbalance.

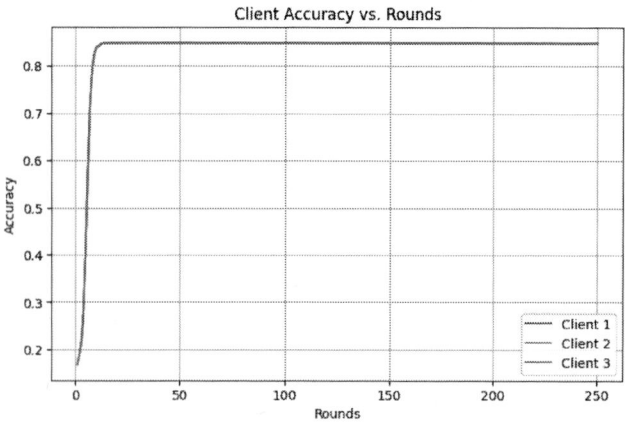

(a) Client Accuracy vs. Rounds

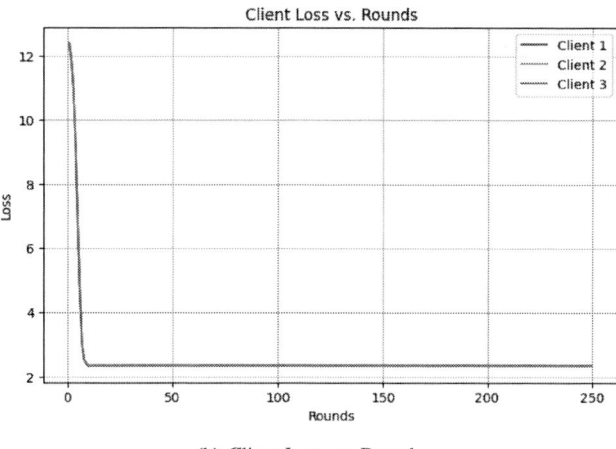

(b) Client Loss vs. Rounds

Fig. 2. Comparison of client performance over training rounds.

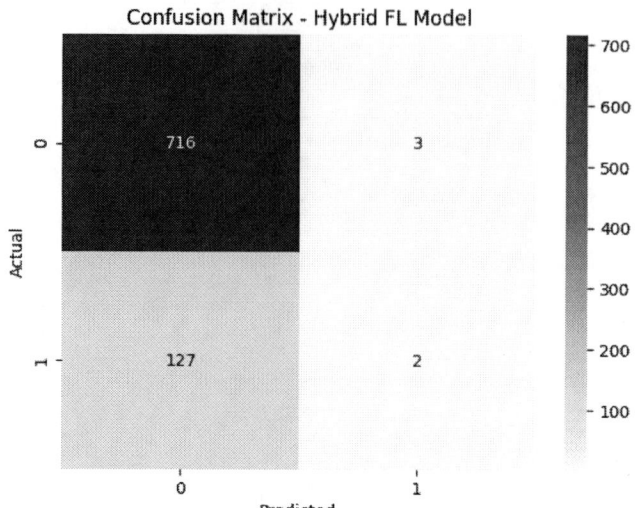

Fig. 3. Confusion Matrix of the Hybrid FL Model

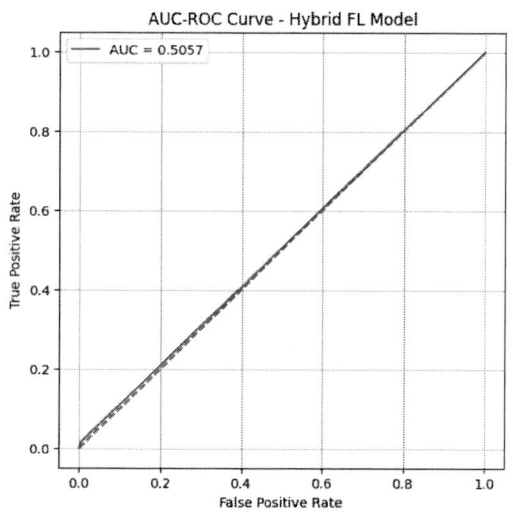

Fig. 4. AUC-ROC Curve of the Hybrid FL Model

These misclassifications emphasize the need for enhanced strategies to improve recall for positive cases, particularly in scenarios where the minority class is underrepresented.

C. AUC-ROC Curve

The AUC-ROC curve is a crucial metric for evaluating the model's discriminative ability between positive and negative cases. As illustrated in Fig. 4, the Hybrid FL model achieved an AUC score of 0.5057. Although this score indicates reasonable discriminative power, it also reflects the model's struggle to effectively distinguish between heart disease cases and non-disease instances. Enhancing the model's feature learning process or incorporating advanced balancing techniques could improve the model's AUC-ROC score.

D. Comparison with Centralized Model and Impact of Data Balancing

To assess the Hybrid Federated Learning (FL) model's effectiveness, we compared it to a centralized deep learning model. As shown in Fig. 5, the Hybrid FL model performed nearly as well, highlighting its potential for privacy-sensitive healthcare applications. We also evaluated the impact of balancing the dataset using SMOTE. Models trained on the original imbalanced data achieved higher overall accuracy, while those trained on SMOTE-balanced data showed improved recall and F1-score for detecting heart disease cases. Although oversampling slightly reduced accuracy, it enhanced sensitivity to positive cases — a crucial factor in clinical settings.

E. Comparison with Centralized and Traditional Machine Learning Models

To evaluate the proposed Hybrid Federated Learning (FL) model, we compared its performance against a centralized deep learning model and traditional machine learning methods (SVM, Random Forest, Logistic Regression) using the Framingham Heart Study dataset. Accuracy, Precision, Recall, F1-

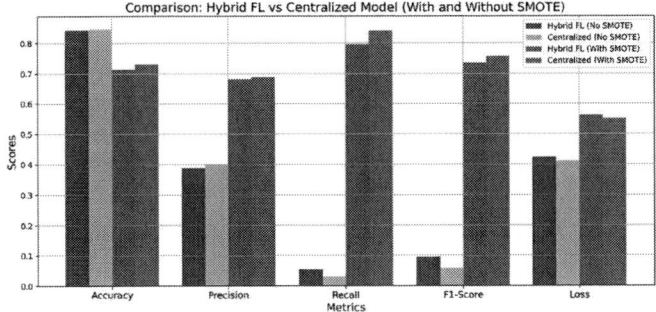

Fig. 5. Comparison with Centralized Model and Impact of Data Balancing

Score, and Loss were calculated, as summarized in Fig. 6. The Hybrid FL model achieved an accuracy of 84.67%, comparable to the centralized model (85.14%) and traditional methods (SVM: 85.02%, Random Forest: 84.32%). Although Logistic Regression had slightly lower accuracy (84.4%), it achieved the highest F1-score (9.59%), indicating better minority class detection. Despite minor differences in accuracy, the Hybrid FL model preserves data privacy and shows strong generalization across decentralized clients, making it highly suitable for sensitive domains like healthcare. Future work could focus on enhancing minority class detection through improved client selection and personalization.

V. CONCLUSION

This study presents a privacy-preserving Hybrid Federated Learning (FL) framework for heart disease prediction, integrating the FedAvg algorithm with a deep neural network (DNN). By decentralizing model training across multiple clients, the proposed approach safeguards sensitive patient data while maintaining competitive predictive performance. Extensive experiments demonstrated that the Hybrid FL model achieved a final global accuracy of 84.67%, closely matching the centralized deep learning model (85.14%), while offering strong privacy advantages. Additionally, the model was compared against traditional machine learning baselines (SVM, Random Forest, Logistic Regression), confirming its robustness across varying data distributions. An investigation

into data balancing using SMOTE further highlighted trade-offs between overall accuracy and minority class sensitivity — a crucial consideration in clinical settings.

REFERENCES

[1] W. H. Organization, "Cardiovascular diseases (cvds) fact sheet." Online, 2023. Accessed: Mar. 5, 2025.

[2] A. H. Association, "Heart disease and stroke statistics—2023 update," *Circulation*, 2023. Accessed: Mar. 5, 2025.

[3] S. Rai, A. Sehgal, D. Gupta, H. Sharma, A. K. Upadhyay, and A. Mishra, "Machine learning in medical sites and healthcare," in *2024 Sixth International Conference on Computational Intelligence and Communication Technologies (CCICT)*, pp. 331–335, 2024.

[4] S. Pati, S. Kumar, A. Varma, B. Edwards, C. Lu, L. Qu, J. J. Wang, A. Lakshminarayanan, S. han Wang, M. J. Sheller, K. Chang, P. Singh, D. L. Rubin, J. Kalpathy-Cramer, and S. Bakas, "Privacy preservation for federated learning in health care," *Patterns*, vol. 5, no. 7, p. 100974, 2024.

[5] B. T. H. Dang, P. H. Luan, V. D. T. Ngan, N. T. Trong, P. T. Duy, and V.-H. Pham, "Trustfedhealth: Federated learning with homomorphic encryption and blockchain for heart disease prediction in the smart healthcare," in *2023 International Conference on Advanced Technologies for Communications (ATC)*, pp. 178–183, 2023.

[6] S. Patidar, A. Jain, and A. Gupta, "Comparative analysis of machine learning algorithms for heart disease predictions," in *2022 6th International Conference on Intelligent Computing and Control Systems (ICICCS)*, pp. 1340–1344, 2022.

[7] M. A. Jabbar and S. Samreen, "Heart disease prediction system based on hidden naïve bayes classifier," in *2016 International Conference on Circuits, Controls, Communications and Computing (I4C)*, pp. 1–5, 2016.

[8] A. J. J, G. J. Leelipushpam Paulraj, G. R. M, I. J. Jebadurai, S. P. Janani, and M. S. Aarthi, "Edge-based heart disease prediction using federated learning," in *2024 International Conference on Cognitive Robotics and Intelligent Systems (ICC - ROBINS)*, pp. 294–299, 2024.

[9] P. Sharma and S. Sharma, "A comprehensive study of the machine learning with federated learning approach for predicting heart disease," in *2023 6th International Conference on Contemporary Computing and Informatics (IC3I)*, vol. 6, pp. 1867–1873, 2023.

[10] P. Sharma and S. Sharma, "An effective fl-cnn based data securing model for heart disease prediction," in *2023 6th International Conference on Contemporary Computing and Informatics (IC3I)*, vol. 6, pp. 1862–1866, 2023.

[11] L. Zou, Z. Huang, X. Yu, J. Zheng, A. Liu, and M. Lei, "Automatic detection of congestive heart failure based on multiscale residual unet++: From centralized learning to federated learning," *IEEE Transactions on Instrumentation and Measurement*, vol. 72, pp. 1–13, 2023.

[12] M. Wei, J. Yang, Z. Zhao, X. Zhang, J. Li, and Z. Deng, "Defedhdp: Fully decentralized online federated learning for heart disease prediction in computational health systems," *IEEE Transactions on Computational Social Systems*, vol. 11, no. 5, pp. 6854–6867, 2024.

[13] B. Chen, T. Chen, X. Zeng, W. Zhang, Q. Lu, Z. Hou, J. Zhou, and S. Helal, "Dfml: Dynamic federated meta-learning for rare disease prediction," *IEEE/ACM Transactions on Computational Biology and Bioinformatics*, vol. 21, no. 4, pp. 880–889, 2024.

[14] H. Gupta, A. Bhardwaj, M. S. Rafeeq, A. Choudhary, I. Garse, O. P. Vyas, and A. Puliafito, "Improving heart disease prediction: Insights from federated deep learning," in *2024 15th International Conference on Computing Communication and Networking Technologies (ICCCNT)*, pp. 1–6, 2024.

[15] M. M. Yaqoob, M. Nazir, M. A. Khan, S. Qureshi, and A. Al-Rasheed, "Hybrid classifier-based federated learning in health service providers for cardiovascular disease prediction," *Applied Sciences*, vol. 13, no. 3, 2023.

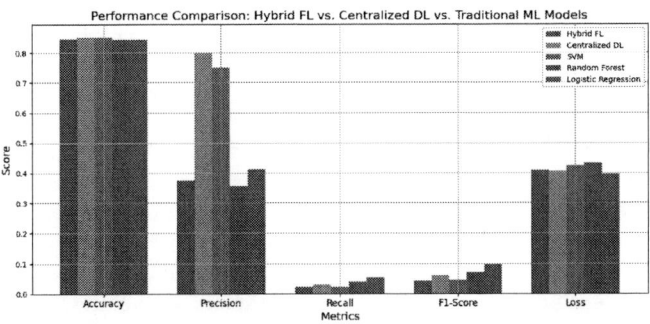

Fig. 6. Performance Comparison Of Proposed Hybrid FL Model Against Centralized And Traditional ML Models

Distributed Control Based Economic Load dispatch for DC Microgrid with Reduced Communication Data

Harshada Dattatray Borse
Electrical Engineering
IIT Kharagpur
Kharagpur, India
harshadaborse1999@gmail.com

Phani Swecha Tadepalli
Electrical Engineering
NIT Warangal
Warangal, India
swechagsit@gmail.com

Deepak Reddy Pullaguram
Electrical Engineering
IIT Kharagpur
Kharagpur, India
drpullaguram@ee.iitkgp.ac.in

Abstract—This study presents a distributed hierarchical control approach for economic load dispatch in autonomous DC microgrids, using a single-variable consensus approach. The strategy ensures the restoration of the voltage to its nominal value while achieving optimal power sharing among distributed energy resources (DER) based on their incremental cost characteristics. In contrast to conventional distributed consensus control that exchanges two state variables (voltage and current), this method is implemented using the exchange of a single variable. This reduces communication requirements and simplifies control implementation, making the system scalable and robust to network changes. In this paper, a single variable is developed that encapsulates the properties of both economic dispatch and average voltage, which is communicated to neighboring DERs. The simulation results demonstrate the effectiveness of the proposed approach against load variations and communication link failures.

Index Terms—DC microgrids, distributed secondary controller, incremental cost, single variable consensus algorithm.

I. INTRODUCTION

In pursuit of net-zero carbon emissions, integrating renewable energy resources into microgrids creates a foundation for sustainable and reliable energy solutions. Although decarbonization efforts heavily rely on renewables within microgrids, certain sectors face challenges in achieving full decarbonization solely through direct electrification. This limitation underscores the need for additional complementary strategies within microgrid systems to meet ambitious low carbon transition targets by 2050 as per Paris agreements [1]. During the COP26 in Glasgow, India committed to reaching net zero emissions by 2070 [2]. Microgrids are the primary idea here to boost the grid's penetration of renewable energy sources. It diminishes its negative environmental effects and supports the new smart grid's more efficient and better power quality [3]. DC microgrids are generally deployed in the distribution segment due to less conversion stages, high efficiency and improved reliability [4]. A microgrid can function in either grid-connected or islanded mode, depending upon the availability of the high-voltage utility network. In islanded mode, the stability of the microgrid primarily relies on the effectiveness of voltage regulation systems and load

dispatch control [5]. Increasing renewable energy penetration necessitates economically efficient microgrid operation in autonomous mode [6]. The Economic Dispatch Problem (EDP) is designed to optimize costs or maximize utility by coordinating dispatchable generation and loads. Typically, the microgrid's energy storage uses V-I droop characteristics to ensure appropriate current sharing and K-sharing function with a droop to eliminate high-speed link of communication [7]. Primary controllers enable load sharing in microgrids but, due to droop gain, may cause voltage deviations from nominal voltage. To address this, secondary control adjusts the setpoints of the primary controllers to bring the voltage back to its nominal value, ensuring stable operation [8]. A tertiary control level is necessary to manage the power flow between interconnected microgrids. Secondary Controllers can be classified into centralized, decentralized, and distributed categories depending on their control actions and communication frameworks [9]. Independent Distributed Generators (DG) share information with their neighbors, authorized by the central controller, in a centralized control approach. These controllers necessitate a high-speed computing unit, a high-bandwidth communication link, and are particularly vulnerable to single-point failures that might threaten the integrity of the whole microgrid [10]. In a distributed control method, information is exchanged through a sparse communication network to achieve the required operation, with each distributed generator (DG) unit functioning as a self-governing agent [11]. Advancements in multi-agent systems have enabled the development of distributed control schemes that enhance microgrid operations. These schemes are increasingly robust in addressing challenges such as communication delays and single point failure [12]. As a result, they are applicable to many functions, including achieving efficient and economically optimized microgrid performance, which is a key focus of this article. Distributed control algorithms that are based on consensus have been meticulously developed to ensure that the DC microgrid achieves voltage restoration and proportional load sharing while also ensuring that the appropriate power sharing is maintained [13] [14]. Generally, the converter's output voltage and current data are transmitted

979-8-3315-3013-6/25 $31.00 © 2025 IEEE

to neighboring units via a communication channel to achieve the required performance. In [15], the secondary controller produces two voltage correction components: one aimed at reducing voltage deviation and another that mitigates current sharing discrepancies. The study presented in [16] proposes a framework for EDP analysis that employs a self-organizing approach utilizing a distributed consensus protocol. Meanwhile, the work in [17] presents an innovative approach to distributed economic dispatch. This method incorporates time-varying and adaptive feedback gains, enhancing the convergence speed, stability, and flexibility in managing microgrid power systems. This paper presents consensus based distributed secondary control to regulate the average voltage and achieves economic operation by optimizing load sharing between DGs. Regulating the average voltage in a DC microgrid ensures proportional power sharing and stable operation across all nodes. It avoids overvoltage and undervoltage at individual nodes, enhancing reliability, efficiency and compatibility with connected loads. The proposed coordinator controller is structured with both primary and secondary control levels. Utilizing a dynamic consensus algorithm, the secondary controller balances the normalized costs of all distributed generators (DGs) to achieve optimized load sharing and maintain the average voltage within the microgrid. The primary control level comprises an outer loop voltage controller that generates a current reference, while the inner loop current controller produces the duty ratio for power semiconductor devices. The primary controller typically employs a droop method to regulate the output voltage of individual converters used in the system. The key aspects of the suggested controller are summarized as follows,

- This paper presents a fully distributed control system where each distributed energy resource communicates just one variable with its neighboring units within a sparse network. This variable contains data of the estimated voltage and incremental cost.
- The proposed controller leverages a consensus algorithm, facilitating the exchange of a single variable among all DERs. This approach significantly minimizes the communication infrastructure and reduces the bandwidth requirements of the communication channel.
- The designed controller has been rigorously verified through case studies, demonstrating its effectiveness in handling load perturbations, and communication link failures within the DC microgrid.

The subsequent sections of the paper are outlined as follows: section II covers preliminaries for DC microgrid architecture and explains the various layers within the DC microgrid. In section III details problem statement and proposed distributed secondary controller based on single variable consisting incremental cost and voltage for communication is presented. Section IV discusses the validation of the developed strategy using simulation and its resulting findings. In summary, Section V provides the concluding remarks of the paper.

II. COORDINATED CONTROLLED DC MICROGRID ARCHITECHTURE

An autonomous DC microgrid featuring n, DERs interconnected through multiple nodes feeding various loads is illustrated in Fig.1. Four DERs with ideal voltage sources are linked through distribution lines to supply the connected loads in islanded mode. The microgrid is generally controlled using hierarchical distributed control consisting of primary, secondary, and tertiary control layers. Secondary and tertiary controls involve data interchange between DERs via communication channels. Hence, the entire microgrid system is made up of three layers: the physical layer, the control layer, and the communication layer.

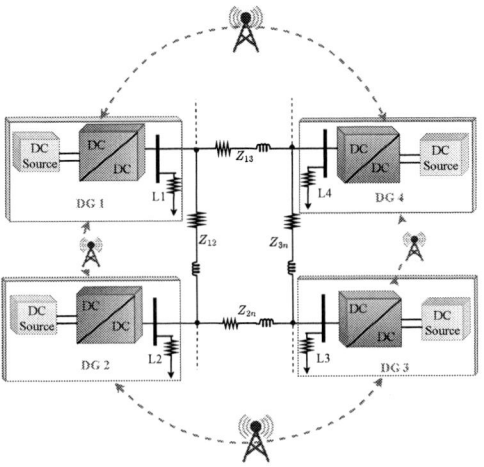

Fig. 1: 4-DER DC microgrid network

A. Physical Layer

In the physical layer, the DERs are interfaced through DC to DC boost converters equipped with LC filters to eliminate ripples. Each DER is connected to its local load and interconnected to neighboring DERs through distribution lines. For practical reasons, these distribution lines are modelled as lumped RL parameters. The impedance of the line connecting DER_i and DER_j is represented as Z_{ij}. Each DER consists of sensors for output voltage and current measurements. These measurements are fed to the control layer that generates the firing pulses for switches in the DC-DC converter.

B. Control Layer

Each DER control includes primary, distributed secondary, and tertiary control mechanisms to achieve voltage regulation and proportional power sharing.

1) Primary and Device level Control: The device-level control comprises an outer voltage controller that generates a reference to the inner loop, while the inner loop current controller generates the duty ratio for switching power semiconductor devices in the converter. The primary controller typically utilizes a droop mechanism to improve power sharing

979-8-3315-3013-6/25 $31.00 © 2025 IEEE

by providing voltage reference to the device-level control using the following expression:

$$v_{\text{ref}_i} = V_{\text{nom}} - r_{d_i} i_i \tag{1}$$

This primary droop controller is a decentralized control, responds rapidly to load disturbances, and stabilizes the system.

2) Secondary and Tertiary Control: The secondary control is employed to restore the microgrid's average voltage to its nominal value by eliminating voltage deviations caused by the primary droop control. It also ensures that power is shared proportionally among DERs. Tertiary control in DC microgrids is used to optimize power flow, ensuring economic dispatch. In this work, tertiary control function (economic dispatch) is also incorporated into the secondary controller with incremental cost consensus.

C. Communication Layer

The consensus strategy requires the exchange of local information among agents via a communication infrastructure, which can be depicted as an undirected (bidirectional) graph $\mathcal{G} = (\mathcal{N}, \mathcal{E}, \mathcal{A})$, where $\mathcal{N} = \{1, 2, 3..., n\}$ is the set of nodes corresponding to n agents and $\mathcal{E}$ denotes the set of edges. The adjacency matrix $\mathcal{A} = [a_{ij}] \in \mathcal{R}^{n \times n}$ describes the connections within the network. An edge directed from agent i to agent j is represented as pair (i, j). If $(i, j) \in \mathcal{E}$, agents i and j are able to communicate and share local information and vice versa, with $a_{ij} > 0$ indicating a positive connection weight. Nodes connected by an edge are called "neighbors". Otherwise, $a_{ij} = 0$. The degree matrix, $\mathcal{D}$ is a diagonal matrix where each diagonal entry d_i represents the degree of vertex i, this degree is defined as the total number or weight of edges connected to vertex i. Using the degree and adjacency matrices, the Laplacian matrix can be defined as $\mathcal{L}$, is defined as $\mathcal{L} = \mathcal{D} - \mathcal{A}$, where $\mathcal{D}$ is the degree matrix and $\mathcal{A}$ is the adjacency matrix of the graph. If the communication network under consideration forms a spanning tree meaning there is at least one path from the main node to each other node, then the Laplacian matrix will have at least one eigenvalue equal to zero. The eigenvector corresponding to this zero eigenvalue is the vector consisting entirely of ones [18]. This communication network topology is utilized to demonstrate the consensus algorithm discussed in Section III.

III. DISTRIBUTED COOPERATIVE CONTROL WITH SINGLE VARIABLE CONSENSUS STRATEGY

This section details the distributed cooperative control which includes the economic dispatch problem along with voltage regulation in the secondary control layer.

A. Economic Dispatch Problem in DC Microgrid

Assuming no generation constraints and no distribution losses for all agents, the system achieves optimal operation when the incremental costs of all DERs (agents) are equal. The objective of the Economic Dispatch Problem (EDP) in the DC microgrid is to minimize total cost while satisfying

power balance. In EDP problem, the Objective cost function is [19] given as:

$$C_T = \sum_{i=1}^{n} C_i(P_i) \tag{2}$$

By assuming equality constraint

$$\sum_{i=1}^{n} P_i - P_D = 0 \tag{3}$$

Augmented Objective cost function is

$$\widetilde{C}_T = \sum_{i=1}^{n} C_i(P_i) - \lambda(\sum_{i=1}^{n} P_i - P_D) \tag{4}$$

Where $C_i(P_i)$ is the cost function of DG_i, P_i is the active power output of the DG_i and P_D is the total active load demand in the DC MG. λ represents the Lagrange multiplier, which is equivalent to the incremental cost of each dispatchable agent. The generation cost is taken into account as

$$C_i(P_i) = a_i P_i^2 + b_i P_i + c_i \tag{5}$$

The cost coefficients are listed in Table. I, and for ith DG, the incremental cost can be calculated as follows,

$$\frac{\partial \widetilde{C}_T}{\partial P_i} = IC_i(P_i) = \lambda_i \tag{6}$$

The system's optimal incremental cost is expressed as

$$\lambda^* = \lambda_i^* = 2a_i P_i^* + b_i \tag{7}$$

Here, λ^* is the optimal incremental cost and λ_i^* denotes the ith DERs optimal incremental cost. The cost coefficients a_i, b_i, and c_i for each DER are given in Table I.

B. Control Structure and Key Objectives

The main objectives of the microgrid system are to reduce the total generation cost through EDP and to achieve average voltage regulation. These objectives are mathematically formulated as:

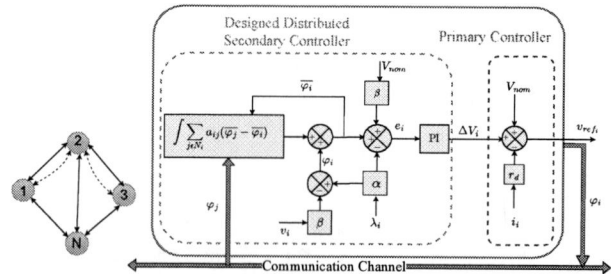

Fig. 2: Block diagram of the designed distributed consensus controller based on single variable

$$\text{Objective 1: } [V_{\text{nom}} = v_{\text{avg}}] \tag{8}$$

$$\text{Objective 2: } [\lambda_1 = \lambda_2 = \lambda_3 = \cdots \lambda_n] \tag{9}$$

where v_{avg} and V_{nom} denotes the microgrid average voltage and nominal voltage, respectively. λ_i indicates incremental cost function of DG_i.

The global economic operation can be achieved when the incremental costs (ICs) of each DGs are equal. A new intermediate variable φ is developed by combining both the incremental cost function and voltage of each DER as follows:

$$\varphi_i = \alpha\lambda_i - \beta v_i \tag{10}$$

Here α and β are positive weighted tuning gains.

The microgrid objectives are achieved by communicating this single variable (φ) between neighboring DERs. Figure 2 shows the block diagram of the designed distributed secondary controller. Each DER controller calculates the global average estimate of φ_i, denoted as $\bar{\varphi}_i$.

$$\bar{\varphi}_i = \varphi_i + \int_0^t \sum_{i=1}^{N_i} a_{ij}(\bar{\varphi}_j - \bar{\varphi}_i) \tag{11}$$

At steady state, $\bar{\varphi}$ is expected to reach $\alpha\bar{\lambda} - \beta\bar{v}$. In this $\bar{v}$ need to track the nominal voltage v_{nom}. Thus, an error signal is defined as

$$e_i = \bar{\varphi}_i - (\alpha\lambda_i - \beta V_{nom}) \tag{12}$$

The error signal in (12) is fed to a PI controller at the secondary level to generate a voltage correction term, Δv_i, which is then fed to the primary control [20]. Thus the voltage reference (v_{ref_i}) at the primary controller of the ith DER is modified as

$$v_{\text{ref}_i} = V_{nom} - r_{di}i_i + \Delta v_i \tag{13}$$

The presented consensus algorithm guarantees the estimation of $\bar{\varphi}_i$ computed by the ith controller aligns with the estimations of $\bar{\varphi}_j$ from other controllers over the time. This, in turn, ensures uniform incremental costs across all DGs and ensures the microgrid voltage achieves its nominal value

$$v_{avg} = \frac{1}{n}\sum_{i=1}^{n} v_i = V_{nom} \tag{14}$$

Thus the proposed controller fulfills the control objectives of economic dispatch and voltage regulation.

IV. SIMULATION RESULTS

The effectiveness of the proposed controller is validated by conducting various simulation case studies in Matlab/Simulink. The simulation parameters are selected mentioned in Table. I.

A. Simulation Case I: Standard System Operation

In this scenario, the designed controller is evaluated under normal operating conditions. Initially, the DC microgrid operates with the conventional droop controller, which serves as the primary controller. This results in varying incremental costs for different DERs and voltage deviates from nominal voltage. At $t = 1$ s, the designed secondary controller is activated. The system shows a transient response and subsequently stabilizes at a new operating point. The tuning gains $\alpha = 0.15$ and

TABLE I: Cost Coefficients of Incremental Cost Functions and Parameters Used for Simulation of networked DC microgrid

Parameter	Value	Unit
Cost Coefficients of Incremental Cost Functions		
DG	a_i (Rs/kWh^2)	b_i (Rs/kWh)
DG$_1$, DG$_2$	0.0005, 0.0025	0.5
DG$_3$, DG$_4$	0.004, 0.004	0.25
DC microgrid Parameters		
Maximum output power	10	kW
Standard nominal voltage	380	V
Resistive loads (L$_1$ – L$_4$)	44, 42, 46, 48	Ω
Droop gain coefficient r_{di}	2.9	Ω
Tuning gains α, β	0.15, 1	
Distribution line resistance	1.608, 2.01, 1.206	Ω
Distribution line inductance	0.30064, 0.3758, 0.22548	mH

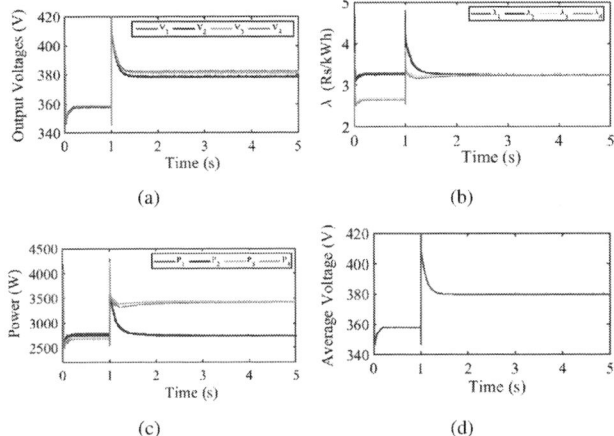

Fig. 3: Assesment of controller performance under normal condition (a) Output voltage profiles of individual DGs (b) Incremental cost dynamics (c) Power output of DGs (d) Average voltage across the microgrid

$\beta = 1$ were used for the simulation. The steady-state terminal voltages for DERs 1 to 4 are measured as 378.4 V, 378.8 V, 381.2 V, and 382.6 V, and the corresponding incremental costs are 3.241 Rs/kWh, 3.244 Rs/kWh, 3.239 Rs/kWh, and 3.239 Rs/kWh, respectively. Furthermore, proportional power sharing among the DGs is 2.74kW, 2.74kW, 3.42kW, and 3.42kW respectively, as shown in fig. 3. The average voltage is also restored to the nominal voltage of 380 V. Hence, the proposed controller ensures effective voltage regulation and equitable incremental cost distribution while requiring communication of only one variable per converter.

B. Simulation Case II: Load Perturbation

The operation of the islanded microgrid employing the presented control strategy under load perturbations is studied in this scenario. Initially, the microgrid operates with load resistances (load 1- load 4) of 44 Ω, 46 Ω, 48 Ω, and 50 Ω, the incremental cost values were approximately 3.24 Rs/kWh, and the power outputs for the DER_1, DER_2 and DER_3, DER_4

979-8-3315-3013-6/25 $31.00 © 2025 IEEE   441

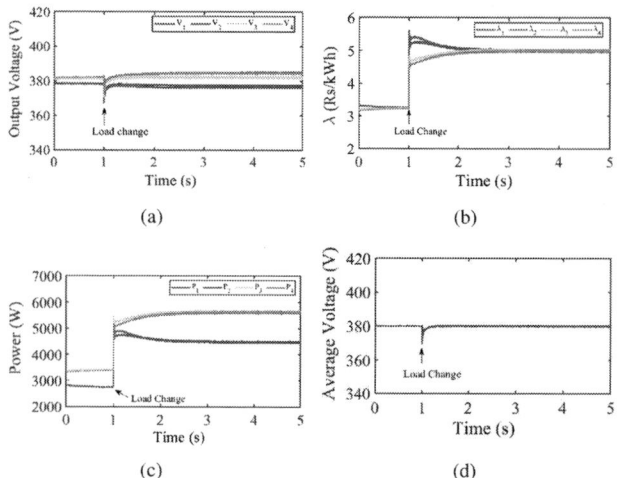

Fig. 4: Evaluation of the controller during load perturbation (a) Output voltage profiles of individual DERs (b) Incremental cost dynamics (c) Power output of DERs (d) Average voltage across the microgrid

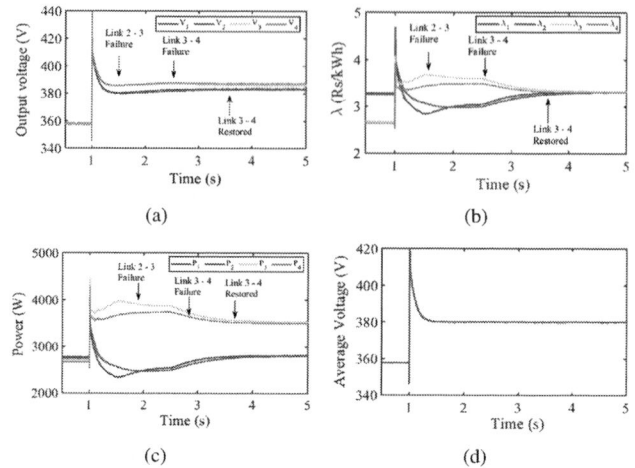

Fig. 5: Performance analysis of controller during communication link failure (a) Output voltage profiles of individual DERs (b) Incremental cost dynamics (c) Power output of DERs (d) Average voltage across the microgrid

are 2.74kW and 3.42kW respectively. At $t = 2$ s, a step load change is performed. when the load resistance (load 1- load 4) increased to 60 Ω, 70 Ω, 80 Ω, and 90 Ω, the incremental cost values reached a value of 5 Rs/kWh. This increase is due to the higher energy expenditure required to maintain stable operation under increased load conditions. The power outputs also increased significantly, ranging from 4.48kW (DER_1&DER_2) to 5.648 kW (DER_3&DER_4), indicating that the DGs had to generate more power to meet the higher loads. Due to the sudden increase in load, significant transients were observed as shown in Fig 4. However, the activation of the consensus controller at $t = 1$ s enabled individual DERs to share information. The incremental cost and voltage information were communicated, leading to equal incremental cost distribution and effective voltage regulation. Despite the increased load, the secondary controller effectively maintained voltage regulation at a nominal value of 380 V. This demonstrates the controller's robustness in achieving both economical dispatch and voltage stability while minimizing power mismatch among the DGs.

C. Simulation Case III: Communication Link Failure

In this case, initially, the microgrid operates with all the communications intact allowing effective coordination among the DERs. At $t = 1.5$ s, the communication link between DER_2 and DER_3 is disconnected. Even though $Link_{23}$ is disconnected, the communication network still forms a spanning tree, and the microgrid continues to operate normally maintaining average voltage and proportional power sharing. At $t = 2.5$ s the link between DER_3 and DER_4 is also disconnected. This will isolate the DER_3 and the isolated DER does not participate in load sharing leading to an imbalance in power distribution and deviation in terminal

voltage of converters as shown in fig. 5. At $t = 3.5$ s, the communication link between DER_3 and DER_4 was reconnected. This allowed DER_3 to regain access to the shared information from its neighboring unit. As a result, the control algorithm responded by adjusting the system to restore consensus and maintain proper power sharing.

V. CONCLUSION

This paper presented a distributed hierarchical control strategy for DC microgrids, utilizing a single variable consensus approach to achieve voltage regulation and economic load dispatch. The proposed method ensures average voltage restoration to nominal values and economic power sharing among distributed energy resources (DERs) based on incremental cost characteristics. The presented approach minimizes communication requirements and enhances system scalability and robustness. The simulation results validated the effectiveness of the proposed control framework in maintaining stable operation under varying load conditions and communication link failures. This work represents a significant advancement towards resilient, economically efficient, and sustainable DC microgrid systems, offering practical implementation potential in remote or resource-constrained environments.

REFERENCES

[1] International Energy Agency, *Net Zero by 2050: A Roadmap for the Global Energy Sector*. Paris: OECD Publishing, 2021. [Online]. Available: https://www.iea.org/reports/net-zero-by-2050

[2] A. Das, V. Saini, K. Parikh, J. Parikh, P. Ghosh, and M. Tot, "Pathways to net zero emissions for the indian power sector," *Energy Strategy Reviews*, vol. 45, p. 101042, 2023. [Online]. Available: https://doi.org/10.1016/j.esr.2022.101042

[3] Q. Fu, A. Hamidi, A. Nasiri, V. Bhavaraju, S. B. Krstic, and P. Theisen, "The role of energy storage in a microgrid concept: Examining the opportunities and promise of microgrids," *IEEE Electrification Magazine*, vol. 1, no. 2, pp. 21–29, Dec. 2013.

979-8-3315-3013-6/25 $31.00 © 2025 IEEE

[4] S. Islam, A. Khalfalla, M. Hamoud, H. Mehrjerdi, A. Iqbal, and V. Marzang, "Distributed secondary controller to minimize circulating current flowing among sources in dc microgrid," *IEEE Access*, vol. 11, pp. 89 488–89 505, 2023.

[5] D. Pullaguram, R. Rana, S. Mishra, and N. Senroy, "Fully distributed hierarchical control strategy for multi-inverter-based ac microgrids," *IET Renewable Power Generation*, vol. 14, no. 13, pp. 2468–2476, 2020.

[6] D. E. Olivares, A. Mehrizi-Sani, A. H. Etemadi, C. A. Cañizares, R. Iravani, M. Kazerani, A. H. Hajimiragha, O. Gomis-Bellmunt, M. Saeedifard, R. Palma-Behnke, G. A. Jiménez-Estévez, and N. D. Hatziargyriou, "Trends in microgrid control," *IEEE Transactions on Smart Grid*, vol. 5, no. 4, pp. 1905–1919, 2014.

[7] G. H. F. Fuzato, C. R. de Aguiar, T. A. Fagundes, W. C. Leal, J. C. Vasquez, J. M. Guerrero, and R. Q. Machado, "Droop k-sharing function for energy management of dc microgrids," *IEEE Journal of Emerging and Selected Topics in Industrial Electronics*, vol. 2, no. 3, pp. 257–266, 2021.

[8] J. Lopes, C. Moreira, and A. Madureira, "Defining control strategies for microgrids islanded operation," *IEEE Transactions on Power Systems*, vol. 21, no. 2, pp. 916–924, 2006.

[9] Y. Sabri, N. El Kamoun, and F. Lakrami, "A survey: Centralized, decentralized, and distributed control scheme in smart grid systems," in *2019 7th Mediterranean Congress of Telecommunications (CMT)*, 2019, pp. 1–11.

[10] Z. Zhang and M.-Y. Chow, "Convergence analysis of the incremental cost consensus algorithm under different communication network topologies in a smart grid," *IEEE Transactions on Power Systems*, vol. 27, no. 4, pp. 1761–1768, 2012.

[11] R. Babazadeh-Dizaji and M. Hamzeh, "Distributed hierarchical control for optimal power dispatch in multiple dc microgrids," *IEEE Systems Journal*, vol. 14, no. 1, pp. 1015–1023, 2020.

[12] X. Li, L. Guo, Y. Li, C. Hong, Y. Zhang, Z. Guo, D. Huang, and C. Wang, "Flexible interlinking and coordinated power control of multiple dc microgrids clusters," *IEEE Transactions on Sustainable Energy*, vol. 9, no. 2, pp. 904–915, 2018.

[13] Y. Han, X. Ning, P. Yang, and L. Xu, "Review of power sharing, voltage restoration and stabilization techniques in hierarchical controlled dc microgrids," *IEEE Access*, vol. 7, pp. 149 202–149 223, 2019.

[14] Z. Yang, H. Min, F. Yang, W. Hu, and Y. Shen, "Hierarchical control of an adaptive droop regulated microgrid with dynamical load power sharing," in *2024 6th International Conference on Energy Systems and Electrical Power (ICESEP)*, 2024, pp. 942–947.

[15] V. Nasirian, A. Davoudi, F. L. Lewis, and J. M. Guerrero, "Distributed adaptive droop control for dc distribution systems," *IEEE Transactions on Energy Conversion*, vol. 29, no. 4, pp. 944–956, 2014.

[16] V. Loia and A. Vaccaro, "Decentralized economic dispatch in smart grids by self-organizing dynamic agents," *IEEE Transactions on Systems, Man, and Cybernetics: Systems*, vol. 44, no. 4, pp. 397–408, 2014.

[17] R. Wang, Q. Li, B. Zhang, and L. Wang, "Distributed consensus based algorithm for economic dispatch in a microgrid," *IEEE Transactions on Smart Grid*, vol. 10, no. 4, pp. 3630–3640, 2019.

[18] D. V. Dimarogonas, E. Frazzoli, and K. H. Johansson, "Distributed event-triggered control for multi-agent systems," *IEEE Transactions on Automatic Control*, vol. 57, no. 5, pp. 1291–1297, 2012.

[19] F. Chen, M. Chen, Q. Li, K. Meng, Y. Zheng, J. M. Guerrero, and D. Abbott, "Cost-based droop schemes for economic dispatch in islanded microgrids," *IEEE Transactions on Smart Grid*, vol. 8, no. 1, pp. 63–74, 2017.

[20] A. B. Shyam, S. R. Sahoo, and S. Anand, "Voltage regulation and load sharing in dc microgrid using single variable global average estimation," *IEEE Journal of Emerging and Selected Topics in Industrial Electronics*, vol. 5, no. 2, pp. 336–345, 2024.

979-8-3315-3013-6/25 $31.00 © 2025 IEEE

Switched Mode Power Amplifier for Underwater Imaging Sonar System

V N Panchalai
Naval Physical and
Oceanographic Laboratory
Thrikkakara Kochi 682201
panchalai.npol@gov.in

Sateesh Kumar Kuncham
Dept. of EEE
National Institute of
Technology,Tiruchirappalli
sateesh@nitt.edu

Kumaresan Natarajan
Dept. of EEE
National Institute of
Technology,Tiruchirappalli
nkumar@nitt.edu

R Ramesh
Naval Physical and
Oceanographic Laboratory
Thrikkakara Kochi 682201
r-ramesh.npol@gov.in

Abstract—**High-frequency signals are transmitted in imaging sonars for detection and achieving high-resolution images underwater. Si MOSFETs are not suitable for high-frequency applications because of their slow switching speed, higher parasitic capacitance, and lower efficiency. In contrast, Wide Band Gap(WBG) devices such as SiC and GaNFETs offer significant advantages in high-frequency sonar systems as they posses higher electron mobility, lower on-resistance, higher breakdown voltage, and better thermal management. Power amplifiers designed with WBG devices provided precision signal clarity and higher power efficiency . Switched Mode Power Amplifier(SMPA) made using GaN MOSFETs for imaging sonar is discussed in this paper. Simulation study was carried out using MATLAB / SIMULINK and hardware prototype of 100W was developed. Simulation and hardware results are also presented in this paper.**

Index Terms—**DC-DC converter, Phase Shift Modulation Technique, SONAR Power Amplifier, Unipolar Sine PWM Technique, Imaging SONAR, Wide Band Gap Devices**

I. INTRODUCTION

SONAR (Sound Navigation and Ranging) is a technology that uses sound waves to detect objects, measure the range of the targets and map sea floor. It transmits high frequency acoustic pulses that reflect from objects buried underwater and return as echoes, allowing SONAR to map the reflected medium.

SONAR has a broad spectrum of applications in multiple fields. In marine navigation, Obstacle Avoidance Sonar(OAS) helps ships and submarines avoid underwater obstacles, ensuring safe travel in oceans[1]. Military operations extensively use SONAR for submarine detection, surveillance and reconnaissance, leveraging its ability to detect submerged objects in real-time. In the fishing industry, SONAR is used to locate school of fishes and map underwater habitats, making fishing more efficient[2]. For seafloor mapping, SONAR provides valuable data for geological research, enabling the study of underwater topography, tectonic activity, and environmental changes. Furthermore, SONAR is indispensable in underwater communication systems for submarines and remotely operated vehicles (ROVs), where it ensures reliable communication as the traditional radio waves cannot penetrate.

Imaging SONAR such as side-scan sonar and forward-looking SONAR is an advanced technology designed to create high-resolution, detailed images of underwater environments [3-4]. The system works by emitting sound waves in a fan-shaped pattern, which travel through the water, bounce off submerged objects or the seafloor, and return as echoes. By measuring the time it takes for these echoes to return, imaging SONAR generates precise visual representations of the scanned area. This technology is widely used in seafloor mapping, search and recovery operations, archaeological exploration, and underwater infrastructure inspections.

In archaeological exploration, imaging SONAR helps locate and map submerged shipwrecks, ancient ruins, and other historical artifacts, enabling non-invasive exploration of underwater sites without disturbing the seabed [5]. Imaging SONAR is also crucial for underwater inspections of infrastructure such as pipelines, cables and underwater structures providing detailed images for maintenance, repair and safety assessments.

Beyond these uses, SONAR plays a crucial role in search and rescue operations, helping locate missing vessels or objects in vast underwater areas, and environmental monitoring, where it assists in tracking marine life populations, detecting pollution, and observing oceanographic conditions.

High-frequency imaging SONAR provides superior resolution and more detailed imaging, which is crucial for detecting smaller objects on the sea floor and submerged objects. The shorter wavelength of high-frequency acoustic waves allows the detection of finer details, which makes it particularly useful in applications such as underwater archaeology, seafloor mapping, and locating small objects such as shipwrecks or debris. Underwater vehicles use imaging sonars such as Side Scan Sonar(SSS), Multibeam Echo Sounder (MBE) and sub-bottom profiler for seafloor mapping [6]. One of the major subsystems of imaging sonar is active transmitter system in which power amplifier plays an important role. Design of a switch mode power amplifier used in imaging sonar is presented in this paper.

II. ACTIVE TRANSMITTER FOR IMAGING SONAR

A typical active transmitter system used in imaging sonar is shown in Fig. 1. Imaging sonars fitted onboard ships, submarines and fixed underwater structures use platform supply

979-8-3315-3013-6/25 $31.00 © 2025 IEEE

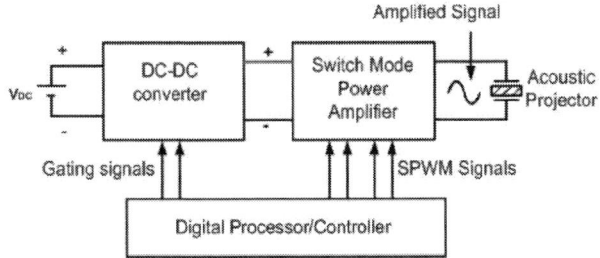

Fig. 1. Active Transmitter of an Imaging Sonar

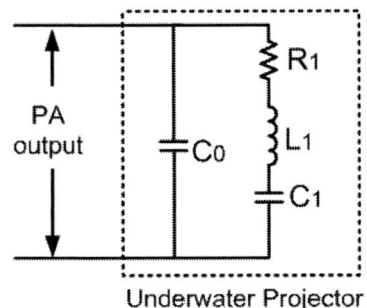

Fig. 2. Transducer Equivalent Circuit

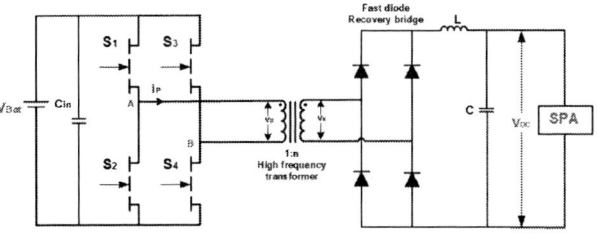

Fig. 3. Schematic diagram of PSFB converter

and underwater vehicles such as AUVs and UAVs use battery as power source. A step up converter which steps up and regulates to a higher voltage is generally used in the front end. It helps in achieving required drive voltage when battery voltage reduces due to discharge[7]. Output of the converter is connected to a switched mode Power Amplifier (SMPA) which excites the projector load.

Underwater projectors are high impedance complex loads which are required to be driven with high voltage[8]. The electrical equivalent circuit of the underwater projector is shown in Fig.2. It consists of a series branch of R1, L1 and C1 in parallel with a capacitor C_o. The reactive part of the impedance due to L_1 and C_1 cancels out exactly at resonance frequency f_r and load becomes purely resistive. At all other frequencies, the load is reactive. Therefore, driving underwater projector with broadband signal poses a challenge.

A digital controller is used to generate control signals for DC-DC converter and SMPA. Parameters such as transmission frequency, pulse length, and output power are design inputs to the digital controller.

Power devices used in both the DC-DC converter and SMPA are to required to be switched at high frequencies as the imaging sonar are operated at high frequencies. Using Silicon MOSFETs in the high frequency application will lead to increased switching losses, higher heat generation, and reduced efficiency. The slower rise and fall times can cause signal distortion and poor signal integrity.

To overcome these issues, power devices fabricated using Wide Band Gap (WBG) materials such as Silicon Carbide (SiC) and GaN (Gallium Nitride) is suitable choice. SiC

is preferred in high voltage and high temperature tolerant applications such as industrial drives and electric vehicles. Whereas GaN devices provide faster switching speeds and no reverse recovery, making them ideal for high-frequency applications such as power supplies and converters used in household items[9].

Gallium Nitride (GaN) MOSFETs are employed in the power converter stage of the transmitter system used in the imaging sonar presented in this paper. GaN devices are chosen for their superior switching performance, high electron mobility, and lower parasitic capacitances compared to traditional silicon-based MOSFETs. These characteristics enable high-efficiency power conversion at elevated switching frequencies. In this system, the converter operates at a switching frequency exceeding 400 kHz, which helps to reduce the size of passive components such as inductors and transformers, improve transient response, and achieve a more compact and lightweight design—crucial for sonar applications.

A. DC-DC converter

As shown in Fig 1, the battery operated imaging sonar transmitter system proposed here uses Phase-Shifted Full Bridge(PSFB) Converter as the interface between the power source and the SMPA. The switching losses in the PSFBC converter is minimum as the phase-shifted control used reduces the overlap between voltage and current during switching events which improves the efficiency[10-11]. It also supports soft switching techniques such as zero voltage switching (ZVS), which reduces stress on switching components, reduces electromagnetic interference (EMI), and enhances overall system reliability [12-14]. Additionally the converter operates at high frequencies enabling compact design with smaller magnetic components, which is important in space-constrained applications such as SSS, Multibeam Echo Sounder(MBE) etc used in AUVs.

PSFB circuit shown in Fig. 3 uses four E-mode GaN HEMT device IGOT60R070D1. Drive signals for driving these CoolGaN devices are generated using Phase Shift Modulation (PSM) technique. Phase shifted gate drive signals for the full bridge are shown in Fig.4. The shape of the voltage applied across the primary of the high-frequency transformer(VAB) is determined by the difference between these gate signals of high side devices S1 and S3.

The AC voltage from the bridge is stepped up or stepped down to a different voltage level using a high-frequency trans-

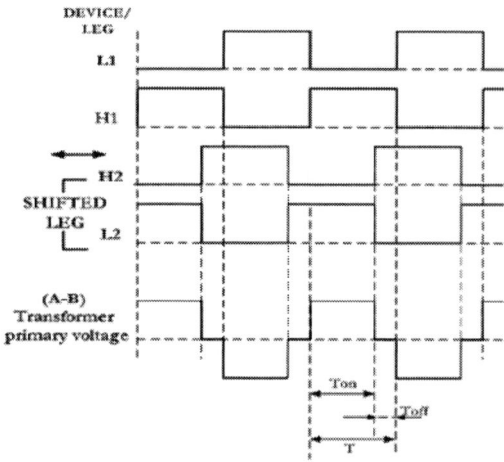

Fig. 4. Gate pulses and output of transformer without filter

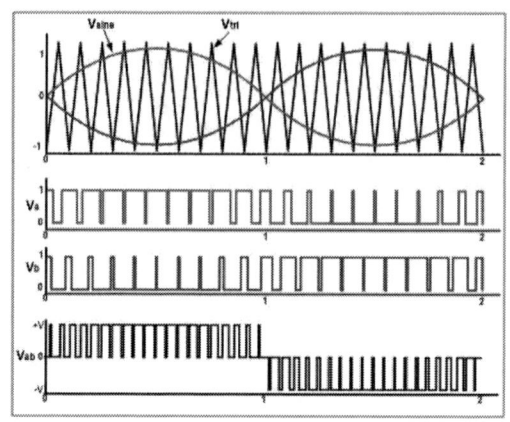

Fig. 6. Unipolar SPWM signal generation

TABLE I
PARAMETERS FOR MATLAB /SIMULINK SIMULATION

Description of the parameter	Value
Input Voltage	48V
PSFB transformer turn's ratio	1:2
Output Power	100W
Load	RLC series
Load R	500 ohms
Load L	10 uH
Load C	10nF

former. The high-frequency AC voltage from the transformer's secondary is rectified using ultra-fast recovery diode bridge rectifier and then filtered using a second-order LC power filter. The filtered DC output from the LC filter is used as supply voltage to the SMPA.

B. SMPA

SMPA is opted as power amplifier in this case. Circuit diagram is illustrated in Fig. 5. Maximum operating frequency of the system is 40kHz. Unipolar Sine Pulse Width Modulation(SPWM) technique is employed to control the power devices in the SMPA, as it helps in achieving the switching harmonics in the output to shift towards twice the switching frequency[9]. The carrier frequency is chosen greater than 12 times of the fundamental frequency. Digital controllers are programmed to generate unipolar SPWM of transmission signal and to implement the control loop[10]. Typical unipolar SPWM control signal generation by comparing sonar signal with triangular waveform is shown in Fig. 6.

Fundamental Frequency upto 40kHz is expected to have carrier frequencies greater than 480 kHz which can be implemented SMPA. Applications that require fundamental fre-

quency greater than 40 kHz use square wave class-S switching with proper filtering. In some cases class E resonant converters are used which give high efficiencies greater than 90%.

III. SIMULATION AND RESULTS

One typical application where the fundamental frequency is greater than 100kHz is simulated. Topology employed is PSFBC fed SMPA and the scheme is simulated using MATLAB/Simulink environment (Fig. 7). Parameters selected for the simulation are tabulated in TABLE 1.

The PSFB converter output is regulated as voltage control and filter components were selected as 10uH and 10nF. The SMPA transformer is again step up to meet the voltage requirement. The PSFB primary transformer voltage and current are shown in Fig. 8. Theoretical output voltage of PSFB, output voltage of SMPA and output current are shown in Fig. 9. The output of the PSFB converter is approximately 96 V at a

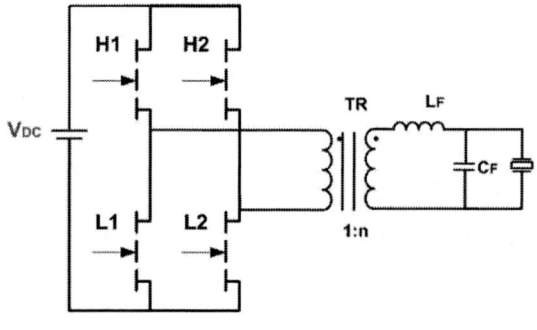

Fig. 5. Circuit diagram of SMPA

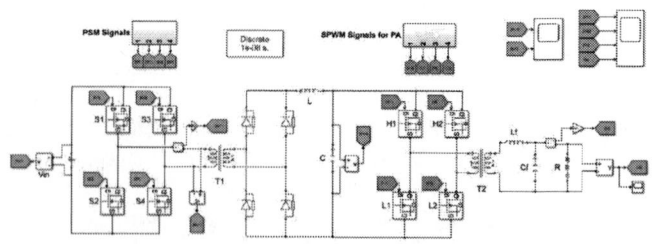

Fig. 7. Simulink Model of the proposed system

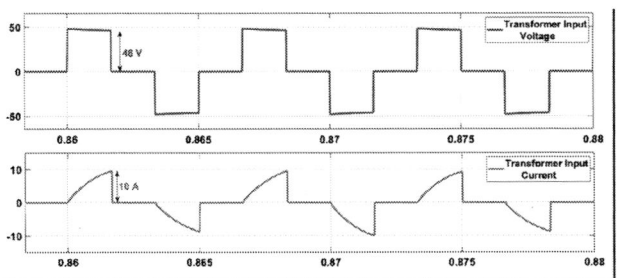

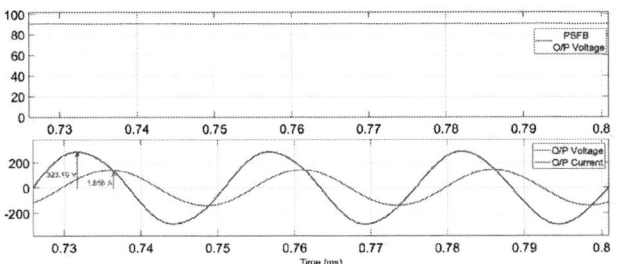

Fig. 8. PSFBC Transformer input voltage and input current

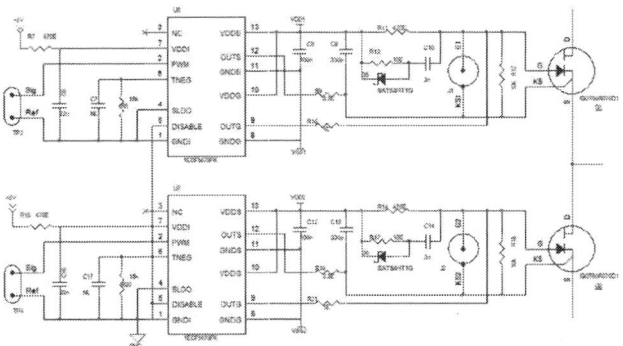

Fig. 10. Gate driver circuit for GaN HEMT

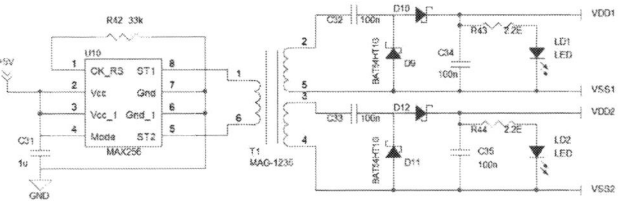

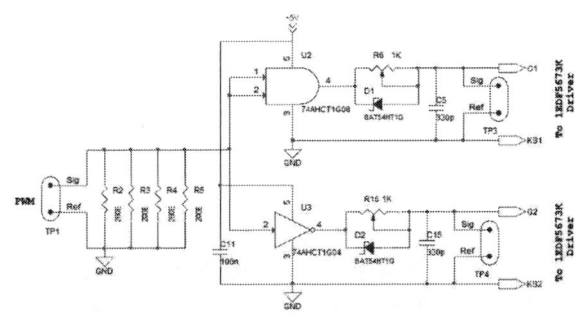

Fig. 9. PSFB Output Voltage, PA Output Voltage and Current

Fig. 11. Supply for gate driver

phase angle of 90°, which is supplied to the power amplifier to deliver 100W of power.

IV. HARDWARE AND PROTOTYPE TEST RESULTS

PSFB fed SMPA Hardware of 100W rating that operates at maximum of 40kHz was fabricated using E-mode GaN HEMT device E-mode GaN HEMT device IGOT60R070D1 as power devices. Details of the hardware components used in the hardware is given in the TABLE II. Both PSFB converter and SMPA were configured using the device E-mode GaN HEMT device IGOT60R070D1 and the driver IEDF5673K. The gate driver circuit is shown in Fig 10. The gate driver requires isolated supplies. Isolated supplies are generated using the circuit shown in Fig. 11 using IC MAX256 [15-16]. N95 ferrite core was used to design PSFB converter transformer with turn's ratio of 1:2. Litz wires were used for winding both primary and secondary.

The schematic of complementary signal generation [15-16] is illustrated in Fig. 12. The switching frequency of the PSFB (Phase-Shifted Full-Bridge) converter was set at 150 kHz, while the SMPA (Switch-Mode Power Amplifier) operates at 480 kHz. When the gate drive signal is passed through

logic gates to generate complementary switching signals, it experienced a total propagation delay of approximately 176 µs. However this delay is equal for both signal paths due to the use of equal number of logic gates in the paths. The delay introduced by a single logic gate, along with the complementary signal obtained after propagation delay is shown in Fig. 13. The gate drive signals corresponding to one leg of the PSFB converter (switches S1 and S2) which include delay due to logic gates, are depicted in Fig. 14. It can be observed that complementary signals faithfully complement and are not having any delays. The gate drive signals for the upper switches (S1 and S3) are shown in Fig. 15. The input to the PSFBC converter transformer is proportional to the phase shift between the gate drive signals of S1 and S3. In the present implementation the phase shift between the signal is kept at 90°.

TABLE II
HARDWARE DETAILS

Sl no	Description of the Parameter	Parameter Value
1	Power Device	E mode HMET
2	Operating Voltage	450V
3	Driver	IEDF5673K
5	Operating frequency	up to 3MHz
6	Minimum output pulsewidth	18 ns
7	Propagation delay accuracy	13 ns

Fig. 12. Schematic of driver circuit

979-8-3315-3013-6/25 $31.00 © 2025 IEEE   447

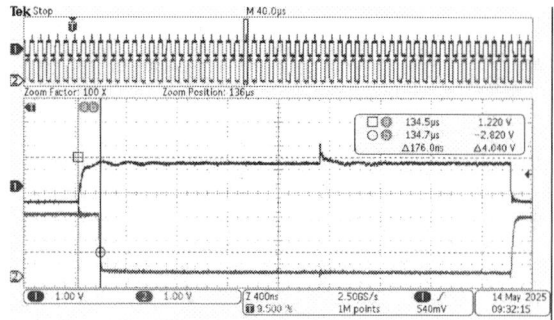

Fig. 13. Delay in logic gates

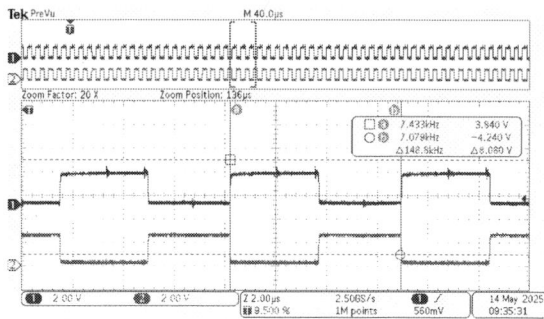

Fig. 14. Gate drive signal at GaNFET after driver

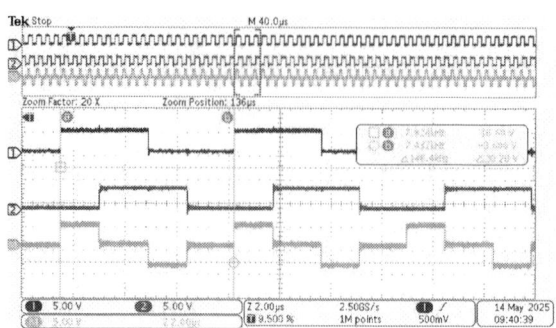

Fig. 15. PSFB output signal

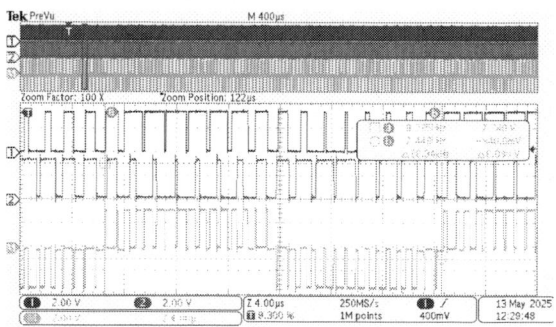

Fig. 16. SPWM output of controller Hardware

The gate drive signal for GaN HEMT devices (H1 and H2) in the SMPA are shown in Fig. 16. The signal is generated digitally using controllers. Similar to PSFB converter the complementary signal for L1 and L2 are generated using logic gates and the delay due to the gates are matched by placing similar number gates in the each signal path. Fig. 17 presents the photograph of the 100 W hardware prototype. The layout was carefully designed to ensure minimal noise coupling and efficient power delivery. Hardware was tested interfacing with transducer load which was lowered in water. The output voltage and current of the hardware with transducer load is presented in Fig. 18. As explained earlier, the transducer load was reactive in nature and hence there is phase difference of 80° was observed between the PA output voltage and the load current. This varies with the operating frequency. Using the PSFB converter in the first stage and isolated full bridge converter provides high dynamic range of output which is required in the imaging sonar operation for varying the output depending on the operating conditions. The variation in the phase shift of PSFB converter control signal or/and variation in the modulation index of SMPA changes the voltage applied to the transducer load.

By designing transformers in both the converters and using appropriate control parameters in the converters, the system can provide output which can vary from few volts to few hundred volts. When the system input voltage is low such as battery supply, high output voltages that are required to drive high impedance projector load are achieved using this

Fig. 17. 100W Prototype Power Amplifier for Imaging Sonar

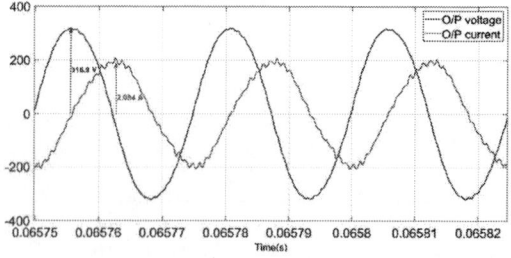

Fig. 18. PA output voltage and Current

topology. WBG devices are better suited in class-D power amplifiers of high frequency operation where the switching frequency is greater than 450kHz.

V. CONCLUSION

Imaging sonars are operated at high frequencies to achieve high resolution images underwater. WBG devices such as SiC and GaN have better switching characteristics at high frequencies hence they are preferred as power devices in the power amplifiers of imaging sonars. A battery-operated, DC-DC converter fed SMPA was explained in the paper. simulations were carried out using MATLAB/Simulink and results were presented. 100W prototype hardware of power amplifier using E-mode GaN HEMT was fabricated. Fundamental frequency of around 40kHz was used in the system and the GaN devices were switched at 480kHz. Converter topology of PSFB fed SMPA was used which provides flexibility of having high dynamic range of output voltage. Details of Hardware components used and typical waveforms of the system were presented.

ACKNOWLEDGMENT

The authors are thankful to the Director, Naval Physical and Oceanographic Laboratory, Kochi for his continuous support and encouragement for carrying out this work at the laboratory.

REFERENCES

[1] Y. Petillot, I. Tena Ruiz and D. M. Lane, "Underwater vehicle obstacle avoidance and path planning using a multi-beam forward looking sonar," in IEEE Journal of Oceanic Engineering, vol. 26, no. 2, pp. 240-251, April 2001

[2] C. Schell and S. P. Linder, "Experimental Evaluation of Tracking Algorithms Used for the Determination of Fish Behavioral Statistics," in IEEE Journal of Oceanic Engineering, vol. 31, no. 3, pp. 672-684, July 2006

[3] E. Coiras, Y. Petillot and D. M. Lane, "Multiresolution 3-D Reconstruction From Side-Scan Sonar Images," in IEEE Transactions on Image Processing, vol. 16, no. 2, pp. 382-390, Feb. 2007

[4] L. Bjorno, "Developments in sonar and array technologies," 2011 IEEE Symposium on Underwater Technology and Workshop on Scientific Use of Submarine Cables and Related Technologies, Tokyo, Japan, 2011, pp. 1-11

[5] Oyvind Odegård, Asgeir J. Sørensen, Roy E. Hansen, Martin Ludvigsen, A new method for underwater archaeological surveying using sensors and unmanned platforms,IFAC-PapersOnLine,Volume 49, Issue 23,2016,Pages 486-493,

[6] V. N. Panchalai, S. Kumar Kuncham, N. Kumaresan and R. Ramesh, "Review on Underwater Sonar Power Amplifier Technologies," 2024 IEEE Space, Aerospace and Defence Conference (SPACE), Bangalore, India, 2024, pp. 103-107

[7] V. N. Panchalai, Y. K. Mishra and N. Sivakumar, "Battery Operated Power Amplifier for Underwater Sensor Networks", Journal of Acoustical Society of India, vol. 46, no. 1, pp. 35-43, 2019.

[8] A.D Waite, "Sonar for practising Engineers" John Wiley, 2002.

[9] [XXX]Buffolo, M., D. Favero, A. Marcuzzi, C. De Santi, G. Meneghesso, E. Zanoni, and M. Meneghini. "Review and outlook on GaN and SiC power devices: Industrial state-of-the-art, applications, and perspectives." IEEE Transactions on Electron Devices 71, no. 3 (2024): 1344-1355.

[10] R. Ramesh and D. D. Ebenezer, "Equivalent circuit for broadband underwater transducers," in IEEE Transactions on Ultrasonics, Ferroelectrics, and Frequency Control, vol. 55, no. 9, pp. 2079-2083,

[11] C.Y. Lim, Y. Jeong and G. W. Moon, "Phase-Shifted Full-Bridge DC–DC Converter With High Efficiency and High Power Density Using Center-Tapped Clamp Circuit for Battery Charging in Electric Vehicles," in IEEE Transactions on Power Electronics, vol. 34, no. 11, pp. 10945-10959, Nov. 2019

[12] N. George, V. N. Panchalai, E. Sebastian and S. Narayanan, "Digital voltage-mode-control of a full-bridge phase-shift-modulated DC-DC converter," 2014 Annual International Conference on Emerging Research Areas: Magnetics, Machines and Drives (AICERA/iCMMD), Kottayam, India, 2014, pp. 1-6

[13] I. O. Lee and G. W. Moon, "Half-Bridge Integrated ZVS Full-Bridge Converter With Reduced Conduction Loss for Electric Vehicle Battery Chargers," in IEEE Transactions on Industrial Electronics, vol. 61, no. 8, pp. 3978-3988, Aug. 2014

[14] N. Mohan, T. M. Undeland and W. P. Robbins, PowerElectronics-Converters Applications and Design, New Delhi:John Wiley and SonsInc., pp. 200-217, 2010

[15] https://www.infineon.com/cms/en/product/evaluation-boards/eval_1edf_g1_hb_gan

[16] https://www.infineon.com/cms/en/product/power/gallium-nitride/gallium-nitride-transistor/igot60r070d1/

Design of LQR for Three-Phase Grid-Tied Inverter System

Kumari Prakhar Pragya
Dept. of EEE
BIT MESRA
RANCHI, INDIA
mtee10002.23@bitmesra.ac.in

Shashank Shekhar
Dept. of EEE
BIT MESRA
RANCHI, INDIA
phdee10053.19@bitmesra.ac.in

Aftab Alam
Dept. of EEE
BIT MESRA
RANCHI, INDIA
aftabalam5555@bitmesra.ac.in

Abstract—The increasing penetration of renewable energy sources in modern power systems necessitates advanced control strategies for grid-tied inverters to ensure reliable power quality and efficient operation. This paper presents a comprehensive study on the design and implementation of a Linear Quadratic Regulator (LQR) for a three-phase grid-tied inverter system, and compares its performance with conventional Proportional-Integral-Derivative (PID) control. The proposed LQR controller, derived from a detailed state-space model, minimizes a quadratic cost function that effectively balances state deviations and control effort, resulting in enhanced transient response, improved voltage regulation, and reduced harmonic distortion. A key advantage of the LQR approach is its ability to access full state feedback, enabling it to utilize all available data of the system state, in contrast to the PID controller that relies solely on the measured output to track the reference signal. Simulation and experimental results validate the superior dynamic performance and robustness of the LQR-based design under varying load conditions and grid disturbances. This work provides significant insights into state-space control methodologies, demonstrating that LQR-based controllers offer a reliable and computationally efficient alternative for managing grid-tied inverter systems in renewable energy applications. Results contribute to advancing future smart grid integration strategies in modern applications.

Keywords—Linear Quadratic Regulator, Three-phase Inverter, Voltage Regulation, Harmonics Distortion, Grid Synchronization, Inverter Dynamics, Load Disturbances, Optimization, Transient Response, Stability

I. INTRODUCTION

FOR many years, cascade linear control has been the prevailing technique in power electronics, despite several inherent limitations [1]. Initially, its design requires multiple feedback loops combined with pulse width modulation, resulting in a sluggish dynamic response. Furthermore, the process of fine-tuning the proportional-integral-differential (PID) parameters is laborious, complicating its practical implementation. In real-world AC microgrids, the intermittent output of renewable energy sources can trigger oscillations in the dc-bus voltage, subsequently degrading the power quality on the AC side. Consequently, conventional cascade control may prove inadequate in mitigating such fluctuations [2].

This work was supported by Science & Engineering Board, Department of Science & Technology, Govt of India, via grant no: ECR/2018/002037.

In microgrids featuring multiple energy sources and converters, an inner current loop paired with an outer voltage feedback loop is typically employed to ensure proper load sharing among distributed generation units (DGs) based on droop characteristics [3]. This study focuses on a pivotal aspect of renewable energy systems: designing control systems for Voltage Source Inverters (VSIs), which serve as the key power electronic interfaces connecting renewable energy sources or DG units to the main utility grid [4]. The approach integrates a linear quadratic regulator (LQR) [5], [6] to achieve optimal control, thereby enhancing both dynamic and steady-state performance. Stability is verified using Lyapunov theory, and notably, this method excludes the use of PLL—which can lead to parameter-sensitive output frequency deviations [7].

Traditional PI controllers are often inadequate for nonlinear systems, higher-order or time-delayed linear systems, and complex systems lacking accurate mathematical models. To address these shortcomings, several variants—including auto-tuning and adaptive PI controllers—have been developed. Moreover, non-traditional fuzzy logic-based PI controllers have been introduced and simulated for this purpose [8]. Both PI and PD control schemes are relatively straightforward to implement and offer excellent steady-state performance. In addition, extensive literature exists on nonlinear control methods such as model predictive control (MPC) and feedback linearization control (FLC); although FLC aims to achieve low total harmonic distortion (THD) under nonlinear loads, its design complexity and high estimation cost remain significant challenges [9].

Key performance criteria for grid-synchronized inverters include minimizing the total harmonic distortion (THD) of the output waveform, reducing transients during grid connection, ensuring smooth mode transitions between grid-tie and islanding operations, achieving accurate phase synchronization with the grid, and precisely controlling power flow [10]. Over the past decade, Proportional Resonant (PR) control has emerged as the dominant strategy for current regulation in both stand-alone and grid-connected converters, effectively tracking sinusoidal current references with minimal steady-state and phase errors. This contrasts with the limitations of the con-

979-8-3315-3013-6/25 $31.00 © 2025 IEEE

ventional Proportional Integral (PI) controller, which struggles with theoretically infinite gain and cannot perfectly follow a sinusoidal reference without steady-state error. Additionally, PR control can provide significant gain at the resonance frequency, depending on the resonance gain (Kr) value [11].

In this paper, PI and LQR controllers [12] are designed for a three-phase grid-tied inverter to evaluate their performance differences. The comparison indicates that while the PI controller achieves acceptable regulation, it suffers from larger steady-state errors and slower transient responses. In contrast, the LQR controller offers superior dynamic performance by optimally adjusting the control input through a state-space model, leading to faster settling times, reduced harmonic distortion, and enhanced voltage regulation. Moreover, the LQR approach exhibits greater robustness under varying conditions such as load changes and grid disturbances, establishing it as a more effective and reliable solution for grid-tied inverter applications. The development of these controllers involved a comprehensive mathematical modeling of the grid-tied inverter to derive the necessary matrices, followed by an axis transformation.

II. OVERVIEW OF THE SYSTEM IMPLEMENTED

A. Grid Tied Inverter

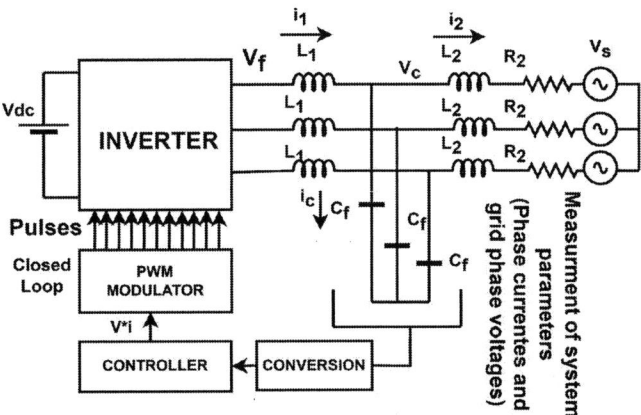

Fig. 1. Block Diagram of Implemented System.

Grid-connected inverters are crucial for transferring power from DC sources, such as renewable energy systems, to the AC grid. Developing an accurate dynamic model of the inverter is essential for evaluating its transient behavior, which in turn supports a thorough stability analysis and aids in designing effective controllers [13]. Figure 1 depicts a schematic of a three-phase voltage source inverter connected to the grid. This schematic shows the inverter incorporating an LCL filter, a PWM modulator, and an integrated control system. In the diagram, the DC link originates from boost converters that raise an electric vehicle's voltage level while supplying power to the grid. Here, the inverter performs DC/AC conversion using IGBTs as switches, and an LC filter is utilized to smooth out ripples in the AC output.

The diagram further details the grid-connected inverter with an LCL filter, where:
- V_f is the voltage across the filter,
- L_f is the filter inductor,
- C_f is the filter capacitor,
- V_c is the voltage across the capacitor,
- I_l is the current through the inductor,
- l_g denotes the grid inductor, and
- V_g is the grid voltage.

A conversion block applies the Park and Clarke transformations for signal conversion. From these components, the state-space model is established as follows:

$$AX = \begin{bmatrix} 0 & w & \frac{-1}{L_f} & 0 & 0 & 0 \\ -w & 0 & 0 & \frac{-1}{L_f} & 0 & 0 \\ \frac{1}{C_f} & 0 & 0 & w & \frac{-1}{C_f} & 0 \\ 0 & \frac{1}{C_f} & -w & 0 & 0 & \frac{-1}{C_f} \\ 0 & 0 & \frac{1}{l_g} & 0 & \frac{-r_g}{l_g} & w \\ 0 & 0 & 0 & \frac{1}{l_g} & -w & \frac{-r_g}{l_g} \end{bmatrix} \begin{bmatrix} i_d \\ i_q \\ V_{cd} \\ V_{cq} \\ I_{Ld} \\ I_{Lq} \end{bmatrix} \quad (1)$$

$$BU = \begin{bmatrix} \frac{1}{L_f} & 0 \\ 0 & \frac{1}{L_f} \\ 0 & 0 \\ 0 & 0 \\ 0 & 0 \\ 0 & 0 \end{bmatrix} \begin{bmatrix} v_{d(t)} \\ v_{q(t)} \end{bmatrix} \quad C = \begin{bmatrix} 0 & 0 & 0 & 0 & 1 & 0 \\ 0 & 0 & 0 & 0 & 0 & 1 \end{bmatrix} \quad (2)$$

$$DU = \begin{bmatrix} 0 & 0 \\ 0 & 0 \end{bmatrix} \quad (3)$$

Grid-connected inverters also manage the bidirectional power flow between renewable DC sources and the AC grid. Common control techniques include Pulse Width Modulation (PWM) and dead time control [14]. While PWM typically generates high-order harmonics, dead time control tends to produce low-order harmonics. If these harmonics are not adequately attenuated before reaching the grid, they can deteriorate its power quality. According to the IEEE 519-2014 standard, the Total Demand Distortion (TDD) from all power generation equipment should not exceed 5%. Although L filters—first-order low-pass filters—can help reduce harmonics in the grid-side current, achieving a TDD below 5% often requires a large inductance value. This necessity can increase both the volume and weight of inverter systems, thereby affecting their dynamic performance [15].

To effectively reduce the inverter-induced harmonics, the use of a passive power filter is essential. Two main types of passive filters are used in grid-connected inverters: the L filter and the LCL filter [16]. The LCL filter acts as a third-order low-pass filter by combining a second-order LC filter with a first-order L filter. Despite being more compact and economical compared to L filters, LCL filters are prone to resonance issues. Such resonance can compromise both the output power quality and the control stability of the inverter. Therefore, determining the resonance frequency is a critical aspect of LCL filter design. Although passive damping techniques can be implemented to suppress resonance, they inevitably introduce additional power losses through the damping resistor [4], [17].

B. Axis Transformation

$$\theta = \tan^{-1}\left(\frac{V_{g\beta}}{V_{g\alpha}}\right); \quad \rho = \cos^{-1}(\frac{V_{g\alpha}}{\sqrt{V_{g\beta}^2 + V_{g\alpha}^2}}) \quad (4)$$

For synchronizing with the grid, a Phase Locked Loop (PLL) is typically used. Among the various PLL techniques, the dq-PLL is particularly favored due to its straightforward structure and ease of implementation. When an inverter is connected to the grid, the inverter's voltage aligns with the grid voltage, resulting in no phase difference between them. Hence,

$$\theta = \rho V_{gq} \to 0; \; V_{gd} \neq 0; \; \rho = \tan^{-1}\left(\frac{V_\beta}{V_\alpha}\right) \quad (5)$$

From abc to $\alpha\beta0$ transformation equation:

$$\begin{bmatrix} V_{g\alpha} \\ V_{g\beta} \\ V_{g0} \end{bmatrix} = \frac{2}{3} \begin{bmatrix} 1 & \frac{-1}{2} & \frac{-1}{2} \\ 0 & \frac{\sqrt{3}}{2} & \frac{-\sqrt{3}}{2} \\ \frac{1}{2} & \frac{1}{2} & \frac{1}{2} \end{bmatrix} \begin{bmatrix} V_{ga} \\ V_{gb} \\ V_{gc} \end{bmatrix} \quad (6)$$

From $\alpha\beta$ to dq transformation equation:

$$\begin{bmatrix} V_{gd} \\ V_{gq} \end{bmatrix} = \begin{bmatrix} \cos\rho & \sin\rho \\ -\sin\rho & \cos\rho \end{bmatrix} \begin{bmatrix} V_{g\alpha} \\ V_{g\beta} \end{bmatrix} \quad (7)$$

From $dq0$ to abc transformation equation:

$$\begin{bmatrix} V_{ga} \\ V_{gb} \\ V_{gc} \end{bmatrix} = \begin{bmatrix} \cos\rho & -\sin\rho & 1 \\ \cos\left(\rho - \frac{2\pi}{3}\right) & -\sin\left(\rho - \frac{2\pi}{3}\right) & 1 \\ \cos\left(\rho - \frac{4\pi}{3}\right) & -\sin\left(\rho - \frac{2\pi}{3}\right) & 1 \end{bmatrix} \begin{bmatrix} V_{gd} \\ V_{gq} \\ V_{g0} \end{bmatrix} \quad (8)$$

Same formula for I_{abc} to $I_\alpha I_\beta I_0$ & $I_\alpha I_\beta$ to $I_d I_q$ & $I_d I_q I_0$ to I_{abc}:

$$\begin{bmatrix} I_a \\ I_b \\ I_c \end{bmatrix} = \begin{bmatrix} \cos\rho & -\sin\rho & 1 \\ \cos\left(\rho - \frac{2\pi}{3}\right) & -\sin\left(\rho - \frac{2\pi}{3}\right) & 1 \\ \cos\left(\rho - \frac{4\pi}{3}\right) & -\sin\left(\rho - \frac{2\pi}{3}\right) & 1 \end{bmatrix} \begin{bmatrix} I_d \\ I_q \\ I_0 \end{bmatrix} \quad (9)$$

In the transformation equations, let us denote angle θ as the angle formed between the resultant vector (R) and the α-axis. The angle ρ signifies the rotation between the stationary $\alpha\beta$ reference frame and the rotating dq reference frame. The voltages from the grid phases 'a', 'b', and 'c' are represented as V_{ga}, V_{gb}, and V_{gc}, respectively. Correspondingly, $V_{g\alpha}$ and

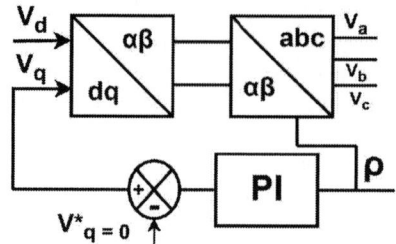

Fig. 3. ρ calculation.

$V_{g\beta}$ are the grid voltages projected onto the stationary α and β axes, while V_{gd} and V_{gq} represent the grid voltages within the rotating dq reference frame. Likewise, currents flowing into the grid from phases 'a', 'b', and 'c' are labeled as I_a, I_b, and I_c, with their corresponding transformations I_d and I_q in the rotating reference frame axes.

When converting from the stationary $I_\alpha I_\beta I_0$ frame to the rotating $I_d I_q$ frame, the angle ρ is specifically chosen to ensure that the phase voltages V_{abc} and currents I_{abc} are aligned. Under these conditions, the quadrature-axis current (I_q) becomes zero, leaving only the direct-axis current (I_d) as non-zero.

However, when considering load-dependent conditions, the angle used for this conversion adjusts to a value expressed as $(K + \rho)$, where K is a constant determined by phase differences introduced by the grid parameters. Consequently, the phase voltages V_{abc} and currents I_{abc} are no longer perfectly aligned. Under these circumstances, both the direct-axis current (I_d) and quadrature-axis current (I_q) have non-zero values.

C. Introduction to Linear Quadratic Equation

The Linear Quadratic Regulator (LQR) represents an optimal control method used to derive a state feedback controller through the minimization of a predefined quadratic cost function. This approach assumes the system dynamics described by a state-space model:

$$\dot{x} = Ax + Bu, \quad (10)$$

where $x \in \mathbb{R}^n$ denotes the system state vector, $u \in \mathbb{R}^m$ represents the input control vector, and matrices A and B signify the system dynamics and control input matrices, respectively.

Controller performance is evaluated using a quadratic cost function defined as:

$$J = \int_0^\infty \left(x^T Q x + u^T R u\right) dt, \quad (11)$$

in which $Q \geq 0$ and $R > 0$ are weighting matrices selected to balance state deviations and the intensity of control efforts.

The control law that minimizes this cost function is expressed as:

$$u = -Kx, \quad (12)$$

where the feedback gain matrix K is calculated as follows:

$$K = R^{-1}B^T P. \quad (13)$$

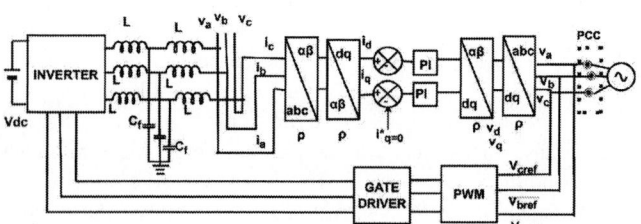

Fig. 2. 3 phase Grid connected Inverter with dq controller.

Here, matrix P represents the distinct positive definite solution to the Algebraic Riccati Equation (ARE), which is formulated as:

$$A^T P + PA - PBR^{-1}B^T P + Q = 0. \qquad (14)$$

Implementing the derived control law modifies the system dynamics into the closed-loop form:

$$\dot{x} = (A - BK)x. \qquad (15)$$

The stability of this resulting closed-loop system is inherently guaranteed by the positive definiteness of P, and additional verification can be obtained through Lyapunov stability analysis.

In conclusion, LQR provides a structured and effective solution for balancing precise state regulation with optimal control effort, proving particularly valuable in engineering applications, including grid-connected inverter systems.

III. RESULTS

This research shows the results derived from the simulation and experimental application of discrete Proportional Integral (PI) and Linear quadratic regulator (LQR) current controllers on a three-phase grid-connected inverter. Utilizing MATLAB/SIMULINK for simulations, the study conducts a comparative analysis of controllers, evaluating THD (total harmonic distortion), TR (transient response), and SS (steady state) response of the current of the inverter. The control parameters established through simulation are then applied for experimental validation.

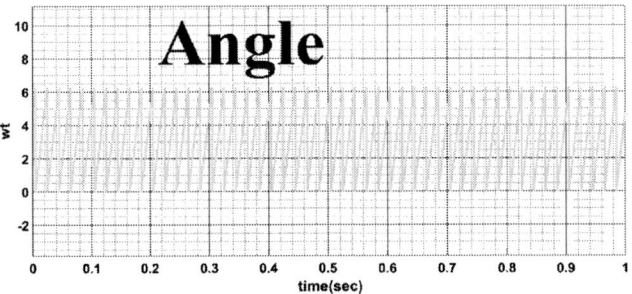

Fig. 4. Open-Loop from abc to dq: Electrical Angle (ωt) versus Time.

Fig (4): Showing DQ is the rotating reference frame transformation used in control systems, particularly in motor drives, and ABC stands for the three-phase quantities (such as voltages or currents in a three-phase system). By converting the three-phase stationary values into two rotational components (d and q), the transformation makes the quantities DC in steady state, simplifying control. This angle is most likely the one that was used for the alteration. Where ω is the rotating speed, the angle $\theta = \omega t$ in the Park transformation. If the rotational frequency (ω) is positive, the $\theta = \omega t$ should rise linearly with time in a conventional DQ transformation.

Fig (5): When the three-phase (abc) currents are converted into the rotating dq-reference frame, the graph displays the behaviour of the direct-axis current (I_d) in an open-loop system.

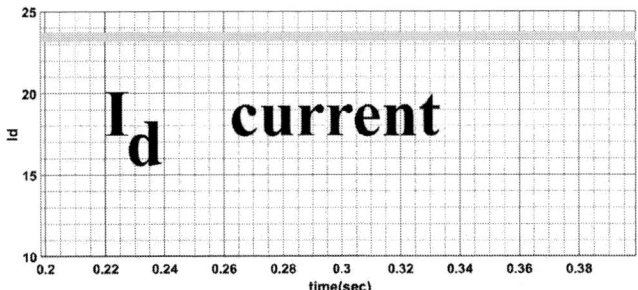

Fig. 5. Transforming an Open-Loop ABC to dq: Direct Axis Current (I_d) versus Time.

The Id current over time is shown by the yellow waveform. At about 24 A, the Id component barely oscillates and is almost constant. The comparatively stable Id component indicates that the rotating reference frame and the synchronous reference have been correctly aligned by the open-loop transformation.

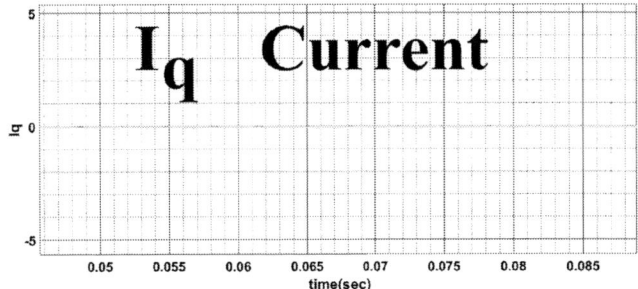

Fig. 6. Transforming an Open-Loop ABC to dq: Direct Axis Current (I_q) versus Time.

Fig (6): In an open-loop simulation, the quadrature axis current (Iq) derived from the ABC to DQ transformation is depicted on the graph. Time (seconds) is shown by the x-axis. The quadrature current I_q (Amperes) is shown on the y-axis. The change in Iq over time is depicted by the yellow trace. $I'_q s$ value stays extremely near to zero. A steady-state situation is indicated by the low fluctuation. I_q shows no discernible spikes or oscillations, indicating that the system is not experiencing any reactive power flow. I_q is linked to the production of torque and reactive power in DQ transformation (in motor applications). I_q stays at or close to zero, which suggests:

There is no need for reactive power because the system is just using active power. In motor applications, no torque is produced. A properly aligned reference frame with the entire current flowing down the $d-$axis.

This graph demonstrates that there is no reactive power component, and the system continues to operate steadily. Additionally, it implies that the d-axis current (I_q) is the primary variable that the control method regulates. The control algorithm must be modified if reactive power control is necessary.

Fig (7): The direct axis current (I_d) response over time for two distinct control strategies—LQR (Linear Quadratic

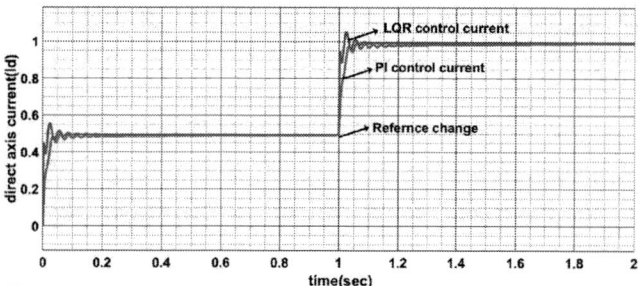

Fig. 7. Direct Axis Current Response: A Comparison of LQR and PI Controller.

Regulator) control and *PI* (Proportional-Integral) control—is depicted by the waveform in the image. Before settling, the LQR controller's early replies (blue curve) exhibit a faster responsiveness but also some overshoot and oscillations. The PI controller (red curve) has a slower response but reaches the steady state smoothly. Response in Steady State (0.1s–1s): There is little variation in the steady-state value maintain by both controllers. In contrast to the smoother PI response, the LQR control still exhibits minor oscillations. Change of Reference (1s): The reference undergoes a step change at t = 1s. Although both controllers react, the LQR controller oscillates more quickly before settling. Although it responds more slowly, the PI controller effortlessly enters the new steady state. LQR Control (Blue Curve) shows Faster response, but with noticeable overshoot and oscillations. PI Control (Red Curve) shows Slower response, smoother transition with minimal oscillations. The PI controller responds somewhat slowly, but the shift is smooth and regulated. While PI is slower but stays away from excessive oscillations, the LQR response overshoots more but settles faster. Useful consequences: LQR might be better if quick reaction is essential (for example, in high-performance systems like motor drives). PI control is superior if smooth and reliable control is more crucial (for example, to prevent excessive wear in mechanical systems). While PI is simpler to use but could need integral adjusting to lower steady-state error, LQR needs precision tuning to prevent excessive oscillations.

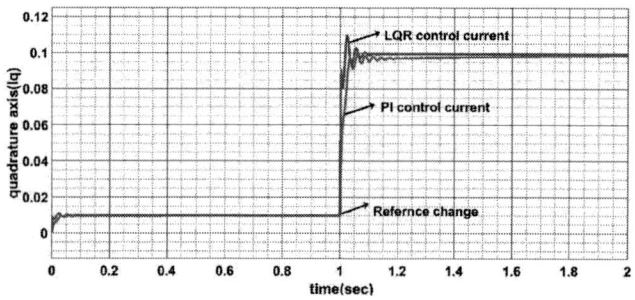

Fig. 8. Comparing Quadrature Axis Current Response with LQR and PI Controller.

Fig (8): The quadrature axis current response with time

for the two distinct control strategies—LQR (blue curve) and PI (red curve)—is depicted in this graph. This is a quick analysis: First Reaction (0s–0.1s): The initial Iq value for both controllers are close to zero. The LQR control (blue) responds more quickly but oscillates. Although it responds more smoothly, the PI control (red) takes a little longer to achieve the steady state. Steady-State Reaction (0.1–1 seconds): To a very small constant value, both controllers settle. PI is more stable than LQR, which shows slight oscillations. Since the values stay close to zero, the quadrature axis current is clearly under control. Change of Reference (1s): At t = 1s, there is a step change in reference, which results in an abrupt increase in Iq. Although it responds rapidly, the LQR controller oscillates and overshoots. Although it takes a little longer to settle, the PI controller reacts more seamlessly. Faster reaction is offered by LQR, but oscillations are the price. Although PI control is slower, the transition is more reliable and seamless. LQR is the best option for high-speed control applications, while PI is the superior option for situations where less overshoot is desired.

TABLE I
D AXIS SETTLING TIME

	LQR	PI
Settling Time (Start)	less than 0.12 sec	more than 0.4 sec
Settling Time (at 1 sec)	less than 0.16 sec	more than 0.45 sec

TABLE II
Q AXIS SETTLING TIME

	LQR	PI
Settling Time (Start)	less than 0.06 sec	more than 0.2 sec
Settling Time (at 1 sec)	less than 0.15 sec	more than 0.6 sec

TABLE III
MAX OVERSHOOT VALUE

	LQR	PI
Overshoot (D axis)	8%	5%
Overshoot (Q axis)	8.2%	2.5%

IV. CONCLUSION

This study demonstrated that a Linear Quadratic Regulator (LQR) based control strategy can significantly enhance the performance of a three-phase grid-tied inverter system compared to a conventional Proportional-Integral (PI) controller. The LQR controller was derived using a state-space model and designed to minimize a quadratic cost function that judiciously balances state deviations and control effort. As a result, the LQR approach achieved faster transient responses, reduced overshoot and oscillations, and improved voltage regulation while effectively suppressing harmonic distortion.

A key advantage of the LQR method is its ability to access the full state of the system, enabling the controller to utilize comprehensive information for making optimal control decisions. In contrast, the PI controller relies solely on the

979-8-3315-3013-6/25 $31.00 © 2025 IEEE

measured output to track the reference, which limits its capability in managing dynamic disturbances and varying load conditions. This full state feedback characteristic of the LQR not only enhances the responsiveness but also contributes to overall system robustness, as evidenced by both simulation and experimental results.

The outcomes of this research establish the LQR controller as a promising alternative for managing grid-connected inverter systems, particularly in renewable energy applications where system stability and rapid response are critical. Future work could focus on extending this approach to multi-inverter configurations and incorporating adaptive strategies to further optimize performance in increasingly complex power systems.

V. FUTURE SCOPE: ADVANCING MPC FOR ADAPTIVE INVERTER CONTROL

While the LQR controller has demonstrated significant improvements in transient response and overall system performance, its reliance on an infinite-horizon optimization process limits its adaptability. In particular, LQR does not inherently accommodate parameter variations due to aging, environmental changes, or external disturbances. This shortcoming motivates the exploration of Model Predictive Control (MPC) as a robust alternative for grid-tied inverter systems.

MPC offers a finite-horizon optimization framework that can be updated in real-time, allowing the controller to adapt dynamically to evolving system conditions. By incorporating constraints and handling multi-variable interactions, MPC can account for both operational limits and unforeseen changes in system parameters. Future research can focus on developing adaptive MPC algorithms that integrate online parameter estimation techniques, thereby compensating for component aging and environmental variations. Moreover, leveraging advanced optimization methods and machine learning approaches within the MPC framework could further enhance its predictive capabilities and responsiveness.

Comparative studies between MPC and traditional controllers such as LQR and PID will be essential to quantify performance improvements and trade-offs, particularly in scenarios with high uncertainty. Ultimately, the integration of MPC into grid-connected inverter control systems holds great promise for achieving superior robustness and efficiency in next-generation renewable energy applications.

REFERENCES

[1] K. H. Ang, G. Chong, and Y. Li, "Pid control system analysis, design, and technology," *IEEE Transactions on Control Systems Technology*, vol. 13, no. 4, pp. 559–576, 2005.

[2] Y. Shan, J. Hu, Z. Li, and J. M. Guerrero, "A model predictive control for renewable energy based ac microgrids without any pid regulators," *IEEE Transactions on Power Electronics*, vol. 33, no. 11, pp. 9122–9126, 2018.

[3] Y. Han, H. Li, P. Shen, E. A. A. Coelho, and J. M. Guerrero, "Review of active and reactive power sharing strategies in hierarchical controlled microgrids," *IEEE Transactions on Power Electronics*, vol. 32, no. 3, pp. 2427–2451, 2017.

[4] P. C. Loh, M. Newman, D. Zmood, and D. Holmes, "A comparative analysis of multiloop voltage regulation strategies for single and three-phase ups systems," *IEEE Transactions on Power Electronics*, vol. 18, no. 5, pp. 1176–1185, 2003.

[5] L. Lessard and S. Lall, "Optimal control of two-player systems with output feedback," *IEEE Transactions on Automatic Control*, vol. 60, no. 8, pp. 2129–2144, 2015.

[6] P. Sanki, M. Basu, P. S. Pal, and D. Das, "Implementation of linear quadratic regulator in an isolated microgrid system," in *2022 IEEE VLSI Device Circuit and System (VLSI DCS)*, 2022, pp. 104–109.

[7] H. Shen, J. Xu, Q. Zhang, J. H. Park, and C. Peng, "An optimal control scheme for grid-connected voltage source inverter via grid voltage modulated-direct power control," *IEEE Transactions on Automation Science and Engineering*, vol. 22, pp. 7216–7225, 2025.

[8] Y. A.-R. I. Mohamed, "Suppression of low- and high-frequency instabilities and grid-induced disturbances in distributed generation inverters," *IEEE Transactions on Power Electronics*, vol. 26, no. 12, pp. 3790–3803, 2011.

[9] J. Kim, H. H. Choi, and J.-W. Jung, "Mrac-based voltage controller for three-phase cvcf inverters to attenuate parameter uncertainties under critical load conditions," *IEEE Transactions on Power Electronics*, vol. 35, no. 1, pp. 1002–1013, 2020.

[10] Y. A.-R. I. Mohamed, "Suppression of low- and high-frequency instabilities and grid-induced disturbances in distributed generation inverters," *IEEE Transactions on Power Electronics*, vol. 26, no. 12, pp. 3790–3803, 2011.

[11] S. Priyadarshini, A. Alam, and S. Shekhar, "Design of pi and pr controller in various reference frames for inverter control," in *2024 1st International Conference on Innovative Sustainable Technologies for Energy, Mechatronics, and Smart Systems (ISTEMS)*, 2024, pp. 1–6.

[12] M. Kashyap and L. Lessard, "Guaranteed stability margins for decentralized linear quadratic regulators," *IEEE Control Systems Letters*, vol. 7, pp. 1778–1782, 2023.

[13] B. Pal, P. K. Sahu, and S. Mohapatra, "A review on feedback current control techniques of grid-connected pv inverter system with lcl filter," in *2018 Technologies for Smart-City Energy Security and Power (ICSESP)*, 2018, pp. 1–6.

[14] M. Liserre, T. Sauter, and J. Y. Hung, "Future energy systems: Integrating renewable energy sources into the smart power grid through industrial electronics," *IEEE Industrial Electronics Magazine*, vol. 4, no. 1, pp. 18–37, 2010.

[15] A. Kuperman, "Proportional-resonant current controllers design based on desired transient performance," *IEEE Transactions on Power Electronics*, vol. 30, no. 10, pp. 5341–5345, 2015.

[16] O. Husev, C. Roncero-Clemente, E. Makovenko, S. P. Pimentel, D. Vinnikov, and J. Martins, "Optimization and implementation of the proportional-resonant controller for grid-connected inverter with significant computation delay," *IEEE Transactions on Industrial Electronics*, vol. 67, no. 2, pp. 1201–1211, 2020.

[17] R. Errouissi and A. Al-Durra, "A novel control technique for grid-tied inverters considering unbalanced grid voltage conditions and control input saturation," *IEEE Transactions on Sustainable Energy*, vol. 10, no. 4, pp. 2223–2234, 2019.

Adaptive Sliding Mode Control for DC-DC Buck Converter

Chandan Kumar and Anjan Kumar Ray

Department of Electrical and Electronics Engineering, National Institute of Technology Sikkim, India
Email ids- phee220021@nitsikkim.ac.in (corresponding author), akray.nits@gmail.com

Abstract—DC-DC converters are commonly used in DC microgrids to provide a consistent DC voltage to generation and storage components. Changing the load state has an impact on the voltage quality in converters. This study presents an adaptive sliding mode control to perform output reference voltage tracking of a buck converter. The primary aim is to determine the voltage across the load, which will aid in the tracking of an output reference voltage. To do this, the state space averaging method is employed to model the DC-DC buck converter. The formula for the continuous controller is then generated using the adaptive sliding mode control technique. The Lyapunov theorem is used to assess the stability of the buck converter. The applicability of the proposed controller is assessed through simulation. When compared to an existing controller, the results show improved reference voltage tracking capability with less overshoots and settling time. Furthermore, a comparative study ensures the effectiveness of the proposed controller.

Index Terms—Buck converter, sliding mode control, tracking, state space averaging method, PWM, Lyapunov theorem

I. INTRODUCTION

A common topology in power electronics for converting a larger input voltage to a lower output voltage is the buck converter, often known as a step down converter. The issue of regulating the feeding load of buck converters has been extensively studied in the present literature. In [1], voltage tracking of buck converter is achieved by means of a non singular fast terminal sliding mode. In [2], the author of proposed control system integrates a recurrent chebyshev fuzzy neural network. In [3],

a novel composite technique for creating a reliable voltage controller for converter is presented. The recursive backstepping technique is combined with an integrated terminal SMC in the composite structure. The authors of [4], [5] presented the trajectory tracking for switching buck power converters. The authors offer a sophisticated proportional derivative compensator for buck converters in [6]. Other authors have proposed a variety of strategies to control the converters, including digital twin-based monitoring [7], output voltage tracking of buck converters [8], adaptive neural backstepping terminal sliding mode control [9], reinforcement learning for converter control [10], state space modeling and control [11], observer based backstepping sliding mode control design [12], and a novel adaptive continuous sliding mode control approach [13]. The backstepping technique for converters was introduced in the work by [14]. Fixed time sliding mode control [15] and a new nonlinear deep reinforcement learning controller [16] have been also proposed to control the voltage of converter. In [17], the author proposed end-to-end deep reinforcement learning framework for buck converter to control the output voltage. A full order sliding mode controller based on fixed frequency pulse width modulation is provided in [18].

The content of this paper is provided to the following: The buck converter model is shown in section II. Section III provides estimator based adaptation law design, while the design of a sliding mode controller is presented in section IV. Section V depicts results and the conclusion is presented in section VI.

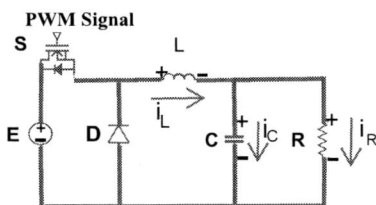

Fig. 1: The buck converter circuit diagram

TABLE I: Nomenclature and symbols

Description	Symbol	Description	Symbol
Input voltage	E	Inductance	L
Capacitance	C	Load resistance	R
Diode	D	Switch	S
Capacitor voltage	v_C	Inductor current	i_L
Steady state error	E_{ss}	Capacitor current	i_C
Load resistance current	i_R	Generated pulse	β
Capacitor voltage as state variable	x_2	Inductor current as state variable	x_1
Overshoot	OS	Continuous controller output	μ
Output reference voltage	v_{ref}	Settling time	t_s
Gains	k_1, k_2	Adaptation law gains	γ_1, γ_2
Quadratic Lyapunov function	V	Switching surface	σ

II. BUCK CONVERTER MODEL

Fig. 1 depicts the buck converter's circuit diagram, while Table. I lists several symbols. When switch is on, $\beta = 1$, and Switch is off, $\beta = 0$. The switch-regulated buck converter modeling equations are given in [1], [4], [5], with digital control β as

$$
\begin{aligned}
\overset{\bullet}{x_1} &= \frac{-1}{L}x_2 + \frac{\beta}{L}E \\
\overset{\bullet}{x_2} &= -\frac{1}{RC}x_2 + \frac{1}{C}x_1
\end{aligned}
\tag{1}
$$

III. ESTIMATOR BASED ADAPTATION LAW DESIGN

Let $\hat{\theta}$ is estimate of θ, $\hat{E}$ is estimate of E, $\hat{x_1}$ and $\hat{x_2}$ are estimated value of x_1 and x_2 respectively, μ is the continuous controller output, $\theta = 1/R$, $\tilde{\theta} =$

$\theta - \hat{\theta}$, $\tilde{E} = E - \hat{E}$, $\tilde{x_1} = x_1 - \hat{x_1}$, and $\tilde{x_2} = x_2 - \hat{x_2}$ [19].

From (1)

$$
\begin{aligned}
\overset{\bullet}{\hat{x}_1} &= \left(\frac{-1}{L}\right)\hat{x_2} + \frac{\mu}{L}\hat{E} + k_1(x_1 - \hat{x_1}) \\
\overset{\bullet}{\hat{x}_2} &= \left(\frac{-\hat{\theta}}{C}\right)x_2 + \frac{1}{C}\hat{x_1} + k_2(x_2 - \hat{x_2})
\end{aligned}
\tag{2}
$$

From (1) and (2)

$$
\begin{aligned}
\overset{\bullet}{\tilde{x}_1} &= \left(\frac{-1}{L}\right)\tilde{x_2} + \frac{\mu}{L}\tilde{E} - k_1(\tilde{x_1}) \\
\overset{\bullet}{\tilde{x}_2} &= -\frac{\tilde{\theta}}{C}x_2 + \left(\frac{1}{C}\right)\tilde{x_1} - k_2(\tilde{x_2})
\end{aligned}
\tag{3}
$$

The adaptation law can now be generated using the

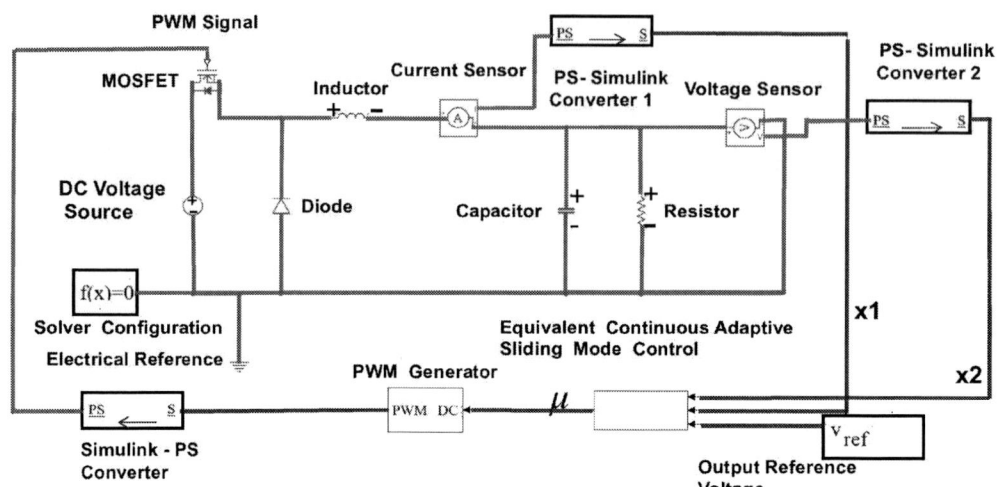

Fig. 2: Simulation diagram for adaptive sliding mode control of a buck converter.

quadratic Lyapunov function that follows:

$$V = L\frac{\tilde{x_1}^2}{2} + C\frac{\tilde{x_2}^2}{2} + \frac{\tilde{\theta}^2}{2\gamma_1} + \frac{\tilde{E}^2}{2\gamma_2} \qquad (4)$$

Where, γ_1 and γ_2 are adaptation law gains.

Differenciating (4) and by using (3)

$$\dot{V} = -k_1 L\tilde{x_1}^2 - k_2 C\tilde{x_2}^2 - \tilde{\theta}\left(x_2\tilde{x_2} + \frac{\dot{\hat{\theta}}}{\gamma_1}\right)$$

$$+ \tilde{E}\left(\mu\tilde{x_1} - \frac{\dot{\hat{E}}}{\gamma_2}\right) \qquad (5)$$

The terms in brackets in (5) are cancelled to derive the adaption laws [19] and are given by (6) and (7)

$$\dot{\hat{\theta}} = -\gamma_1 x_2\tilde{x_2} \qquad (6)$$

$$\dot{\hat{E}} = \mu\tilde{x_1}\gamma_2 \qquad (7)$$

(8) can be obtained with the help of (5), (6) and (7)

$$\dot{V} = -k_1 L\tilde{x_1}^2 - k_2 C\tilde{x_2}^2 \qquad (8)$$

The Lasalle invariant principle allows us to determine that $\tilde{x_1} \to 0$ and $\tilde{x_2} \to 0$ asymptotically [19].

IV. DESIGN OF SLIDING MODE CONTROLLER

The sliding mode surface can be consider as [19]

$$\sigma = \hat{x_1} - \frac{v_{ref}^2}{\hat{E}}\hat{\theta} \qquad (9)$$

Here, the output reference voltage is v_{ref}. The equivalent continuous control law [19] is obtained by $\dot{\sigma} = 0$. By (2), (6), (7), (9), the similar continuous controller is represented by

$$\mu = \mu_{eq} = \frac{(x_2\,\hat{E}^2 - k_1\tilde{x_1}L\,\hat{E}^2 - v_{ref}^2\gamma_1 x_2\tilde{x_2}L\hat{E})}{(\hat{\theta}\,v_{ref}^2\tilde{x_1}\gamma_2 L + \hat{E}^3)}$$

$$(10)$$

For sliding mode to locally exist [20], $\sigma\dot{\sigma} < 0$ must hold, yielding,

$$0 < \mu < 1 \qquad (11)$$

979-8-3315-3013-6/25 $31.00 © 2025 IEEE   458

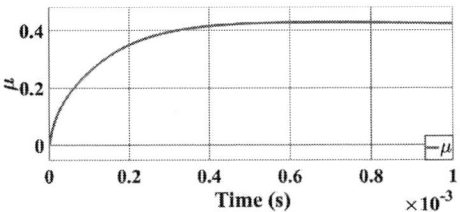

Fig. 3: Continuous Controller Output μ

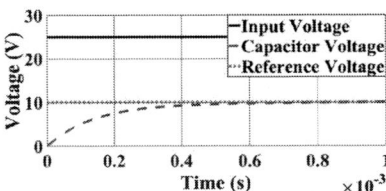

Fig. 4: Reference voltage tracking (E= 25 V and v_{ref}= 10 V)

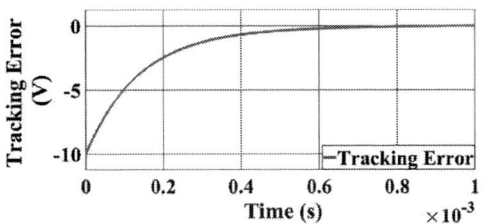

Fig. 5: Reference voltage tracking error (E= 25 V and v_{ref}= 10 V)

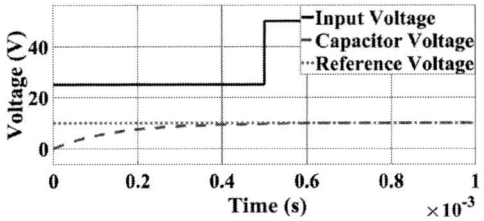

Fig. 6: Reference voltage tracking with variation in input voltage from 25 V to 50 V

The $\hat{x}_1(0)$, $\hat{x}_2(0)$, $\hat{\theta}(0)$, $\hat{E}(0)$ can be implemented for the occurence of sliding mode and should be taken [19] so that the sliding mode begins at t = 0, i.e.

$$\sigma(0) = \hat{x}_1(0) - \frac{v_{ref}^2}{\widehat{E}(0)}\hat{\theta}(0) = 0 \qquad (12)$$

TABLE II: Parameters of buck converter [19], [21]

Description	Parameters	Values
Input voltage	E	25 V
Load resistor	R	20 Ω
Inductance	L	0.059 H
Series resistance of inductor	r_L	4.54 Ω
Capacitance	C	6.8μF
Switching frequency	f_s	20 KHz
MOSFET threshold voltage	v_{th}	0.2 V
Output reference voltage	v_{ref}	10 V
Adaptation law gains	γ_1	1
	γ_2	$13*(10^4)$
Gains	k_1, k_2	$2.5*(10^4)$
Initial conditions	$x_1(0)$	0.4855
	$x_2(0)$	0.0
	$\hat{E}(0)$	9
	$\hat{\theta}(0)$	0.0437
	$\tilde{x}_1(0)$	0.0
	$\tilde{x}_2(0)$	0.0

V. RESULTS

Fig. 2 displays the simulation diagram for adaptive sliding mode control for a buck converter. Through Fig. 3 to Fig. 7, the closed loop switching responses for a buck converter are shown for Table. II. The output of the controller is displayed in Fig. 3, whereas Fig. 4 shows reference voltage tracking. Reference voltage tracking error is depicted in Fig. 5, while reference voltage tracking with input voltage variation is displayed in Fig. 6.

TABLE III: Comparison with SMC based controller as in [1]

Process	$OS(percentage)$	t_s(s)	t_r(ms)	E_{ss}(V)
SMC	15	0.03	....	....
Proposed	0.0	0.104	75.398	0.0

TABLE IV: Comparison with GWABSC, ABSC, PID based controllers as in [14]

Process	$OS(percentage)$	t_s(s)	t_r(ms)	E_{ss}(V)
GWABSC	9.14	0.048	....	....
ABSC	45.35	0.2	....	....
PID	51.71	....	....	....
Proposed	0.0	0.0007905	0.548026	0.0

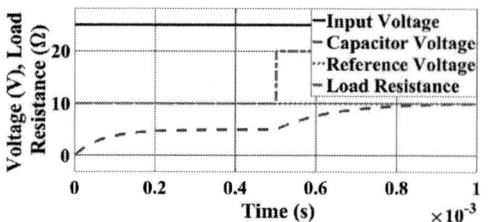

Fig. 7: Reference voltage tracking with variation in load resistance from 10 Ω to 20 Ω

TABLE V: Comparison with MPC, DeePC, FIFPC based controllers as in [22]

Process	$OS(percentage)$	$t_s(s)$	$t_r(ms)$	$E_{ss}(V)$
MPC	15.91	0.00094	....	....
DeePC	10.16	0.00044	....	....
FIFPC	11.08	0.00026	....	....
Proposed	0.0	0.0009329	0.680556	0.0

Figure 7 illustrates reference voltage tracking when load resistance varies. Fig. 8 to Fig. 10 depict the reference voltage tracking by the proposed method considering reference works. Buck converter parameters are shown in Table. II, while comparisons with [1], [14], and [22] are shown in Table. III, Table. IV, Table. V respectively. Many control strategies, such as sliding mode control (SMC), grey wolf adaptive backstepping controller (GWABSC), adaptive backstepping controller (ABSC), PID, model predictive controller (MPC), data enabled predictive controller (DeePC), feedback based iterative feedback predictive controller (FIFPC) are used in references. In comparison to [1], the proposed controller has a significantly less overshoot. In comparison to [14], the proposed controller's overshoot and settling

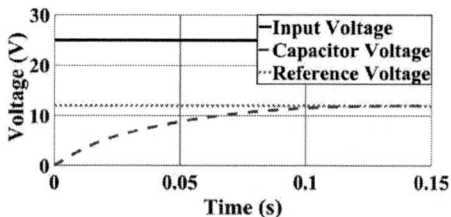

Fig. 8: Reference voltage tracking considering situation as in [1] (E= 25 V and v_{ref}= 12 V)

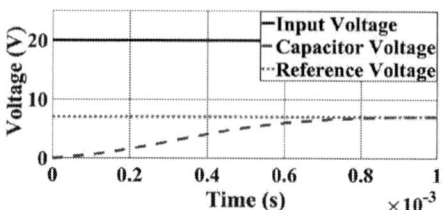

Fig. 9: Reference voltage tracking considering situation as in [14] (E= 20 V and v_{ref}= 7 V)

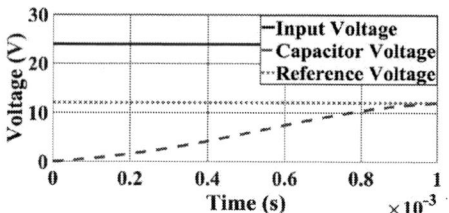

Fig. 10: Reference voltage tracking considering situation as in [22] (E= 24 V and v_{ref}= 12 V)

time are significantly low. The proposed method has a lower overshoot than [22], and it has a shorter settling time than MPC in [22]. These depict the effectiveness of the proposed adaptive sliding mode controller.

VI. CONCLUSION

This article proposed an innovative technique for adaptive sliding mode tracking of a buck converter with variation in reference voltages, supply voltages, and load resistances. In order to observe voltage tracking, the model is designed using the state space averaging method for the buck converter. The sliding mode control method is used to find the expression for the continuous controller output μ. The stability of the system is confirmed by a Lyapunov function along with improved tracking error. The trajectories are driven to the reference with improved settling time and overshoots depicting the efficacy of the proposed controller. In future , the proposed work can be extended to validate in real time.

REFERENCES

[1] J. Fei and X. Gong, "Self-organizing fuzzy neural nonsingular fast terminal sliding mode control of DC-DC buck converter," *IEEE Transactions on Circuits and Systems I: Regular Papers*, 2024.

[2] J. Fei and D. Jiang, "Fuzzy neural network sliding-mode controller for DC-DC buck converter," *IEEE Internet of Things Journal*, vol. 11, no. 19, pp. 31575–31586, 2024.

[3] Z. Alam, T. K. Roy, S. K. Ghosh, and M. A. Mahmud, "Control of DC–DC buck converters using robust composite backstepping and integral terminal sliding mode approaches," *IEEE Journal of Emerging and Selected Topics in Industrial Electronics*, vol. 4, no. 3, pp. 866–877, 2023.

[4] H. Zhang and Z. Wu, "PWM-based adaptive trajectory tracking of switched buck power converters," *IEEE Access*, vol. 8, pp. 178208–178216, 2020.

[5] H. Zhang and Z. Wu, "PWM-based finite-time tracking of switched buck power converters," *Complexity*, vol. 2020, pp. 1–9, 2020.

[6] A. Ullah, S. S. Rizvi, A. Khatoon, and S. J. Kwon, "The empirical analysis, mathematical modeling, and advanced control strategies for buck converter," *IEEE Access*, 2024.

[7] Z. Lei, H. Zhou, X. Dai, W. Hu, and G.-P. Liu, "Digital twin based monitoring and control for DC-DC converters," *Nature Communications*, vol. 14, no. 1, p. 5604, 2023.

[8] S.-K. Kim, K.-C. Kim, and C. K. Ahn, "Output-voltage-tracking control for buck converters using variable convergence rate mechanism without current feedback," *IEEE Transactions on Industrial Electronics*, vol. 69, no. 3, pp. 2938–2946, 2021.

[9] X. Gong and J. Fei, "Adaptive neural backstepping terminal sliding mode control of a DC-DC buck converter," *Sensors*, vol. 23, no. 17, p. 7450, 2023.

[10] C. Cui, T. Yang, Y. Dai, and C. Zhang, "Transferring reinforcement learning for DC-DC buck converter control via duty ratio mapping: From simulation to implementation," *arXiv preprint arXiv:2110.10490*, 2021.

[11] Z. Xia, K. Datta, and J. T. Stauth, "State-space modeling and control of flying-capacitor multilevel DC-DC converters," *IEEE Transactions on Power Electronics*, 2023.

[12] M. Alipour, J. Zarei, R. Razavi-Far, M. Saif, N. Mijatovic, and T. Dragičević, "Observer-based backstepping sliding mode control design for microgrids feeding a constant power load," *IEEE Transactions on Industrial Electronics*, vol. 70, no. 1, pp. 465–473, 2023.

[13] Y. Wang, W. Zhang, and C. Xue, "Adaptive continuous sliding mode control of buck converters with multidisturbances based on zero-crossing detection," *IEEE Access*, vol. 10, pp. 72643–72657, 2022.

[14] S. M. Ghamari, F. Khavari, and H. Mollaee, "Adaptive backstepping controller design for DC-DC buck converter optimised by grey wolf algorithm," *IET Energy Systems Integration*, vol. 6, no. 1, pp. 18–30, 2024.

[15] Z. Liu, X. Lin, Y. Gao, R. Xu, J. Wang, Y. Wang, and J. Liu, "Fixed-time sliding mode control for DC-DC buck converters with mismatched uncertainties," *IEEE Transactions on Circuits and Systems I: Regular Papers*, vol. 70, no. 1, pp. 472–480, 2022.

[16] M. Gheisarnejad, H. Farsizadeh, and M. H. Khooban, "A novel nonlinear deep reinforcement learning controller for DC-DC power buck converters," *IEEE Transactions on Industrial Electronics*, vol. 68, no. 8, pp. 6849–6858, 2020.

[17] A. Rajamallaiah, S. P. K. Karri, and Y. R. Sankar, "Deep reinforcement learning based control strategy for voltage regulation of DC-DC buck converter feeding CPLs in DC microgrid," *IEEE Access*, 2024.

[18] C. Li, W. Zhou, Q. Yang, and F. Ji, "A pulse width modulation based sliding mode controller for phase-shifted full-bridge converter with diodes in series on the lagging leg," *IEEE Access*, 2024.

[19] S. Oucheriah and L. Guo, "PWM-based adaptive sliding-mode control for boost DC-DC converters," *IEEE Transactions on industrial electronics*, vol. 60, no. 8, pp. 3291–3294, 2012.

[20] V. Utkin, J. Guldner, and J. Shi, *Sliding mode control in electro-mechanical systems*. CRC press, 2017.

[21] T. K. Nizami and C. Mahanta, "An intelligent adaptive control of DC–DC buck converters," *Journal of the Franklin Institute*, vol. 353, no. 12, pp. 2588–2613, 2016.

[22] K. Moradi, P. Zamani, and Q. Shafiee, "Data-driven predictive control of perturbed buck converters using a modified iterative feedback tuning algorithm," *IET Power Electronics*, 2024.

MICROGRID BASED FAST CHARGER FOR ELECTRIC POWERED VEHICLES

Neha Karn
Department of Electrical Engineering
C.V. Raman Global University
Bhubaneswar, Odisha, India
nehakarn012@gmail.com

Soumya Sephalika Roul
Department of Electrical Engineering
C.V. Raman Global University
Bhubaneswar, Odisha, India
soumyasephalika45@gmail.com

Shubham Mishra
Department of Electrical Engineering
C.V. Raman Global University
Bhubaneswar, Odisha, India
mishrashubham6963@gmail.com

Rakesh Sahoo
Department of Electrical Engineering
C.V. Raman Global University
Bhubaneswar, Odisha, India
sahoorakeshkumar66@gmail.com

Sanjeet Kumar Subudhi
Department of Electrical Engineering
C.V. Raman Global University
Bhubaneswar, Odisha, India
sanjeetsubudhi@cgu-odisha.ac.in

Abstract—India's growing electric vehicle (EV) market, supported by initiatives like FAME-II, faces a major challenge of inadequate charging infrastructure. This paper proposes a microgrid-based fast-charging system that mitigates issues such as heat generation and battery degradation during high-speed charging by employing an adaptive Constant Current-Constant Voltage (CCCV) methodology that dynamically adjusts charging parameters based on real-time temperature and state-of-charge (SoC) feedback—actively controlling thermal rise to maintain battery health. Powered by renewable energy from a photovoltaic microgrid, the system reduces grid dependence and promotes sustainability. Future work will focus on hardware implementation and further optimization for faster, safer, and scalable EV charging across India.

Index Terms—Fast Charging, Multiple Input Converter (MIC), Constant Current Constant Voltage (CCCV) Charging, Microgrid , Power Electronics, Renewable Energy,

I. INTRODUCTION

India's electric vehicle (EV) sector is experiencing rapid growth, driven by the urgent need to reduce carbon emissions, combat air pollution, and decrease reliance on fossil fuels [1]. The Indian government is spearheading this transition through initiatives like the Faster Adoption and Manufacturing of Electric Vehicles (FAME) program [2] and Production Linked Incentives (PLI) for battery production, targeting 30% EV adoption by 2030 . These efforts are establishing India as a key player in the global EV arena. Valued at US$5.22 billion in 2024, the market is projected to reach US$18.319 billion by 2029, with a strong compound annual growth rate (CAGR) of 28.52% [3]. Industry leaders such as Tata Motors, holding a 72% share in electric passenger vehicles [4], Ola Electric, a pioneer in two-wheelers, and Ather Energy, focused on expanding charging networks, are driving this momentum. Advanced technologies, including microgrids , multi-input converters , and rapid charging systems , are critical in addressing infrastructure challenges and improving

EV usability. The issues related to excessive heat generation, battery lifespan are particularly acute in India, where limited charging infrastructure and range anxiety continue to impede EV growth [5].

To address these limitations, this study presents a novel microgrid-supported CCCV charging system that adapts charging parameters based on real-time temperature and State of Charge (SOC) data [6]. Building on the traditional CCCV approach—featuring a constant current phase followed by constant voltage—this system employs a smart control algorithm to optimize efficiency. By dynamically adjusting current and voltage according to temperature and SOC, it mitigates risks such as thermal stress and lithium plating in cold conditions while enhancing performance in optimal scenarios [7]. The incorporation of microgrids, capable of operating independently and tapping into renewable sources like solar or wind, ensures sustainability and flexibility, enabling deployment in both urban hubs and off-grid areas [8]. Additionally, cutting-edge power electronics, such as DC-DC and AC-DC converters, minimize energy losses and support compatibility across diverse EV models and battery chemistries [9].

II. LITERATURE SURVEY

The Indian EV market is experiencing rapid growth and can be seen in Fig. 1, driven by government incentives, environmental awareness, and technological advancements. [10] EV sales are projected to increase significantly, with two-wheeler sales rising from 0.8 million in 2019 to 4.5 million in 2024, while four-wheeler sales are expected to grow from 0.02 million to 0.60 million in the same period. This trend underscores the expanding adoption of EVs across segments [11].

Electric vehicle (EV) charging techniques include Level 1, Level 2, DC Fast Charging, Wireless, and Smart Charging. [12] These methods cater to various needs based on charging

979-8-3315-3013-6/25 $31.00 © 2025 IEEE

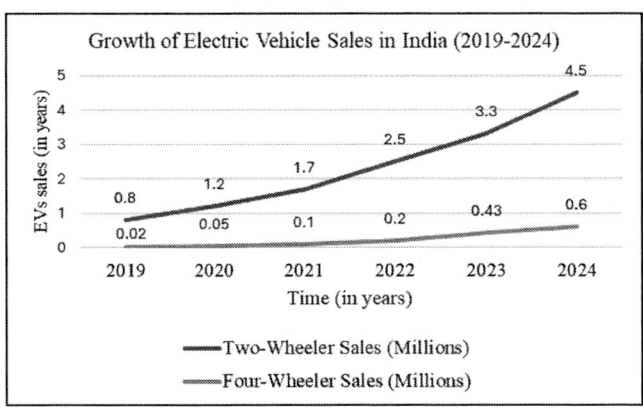

Fig. 1. Growth of electric vehicle sales in India from 2019 to 2024

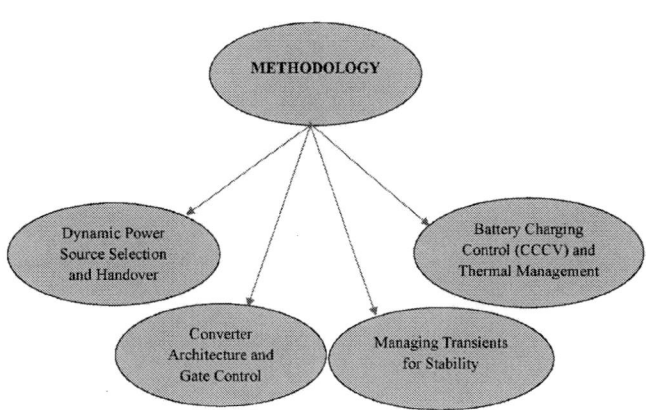

Fig. 2. Fast Charging Methodology

speed, user convenience, and infrastructure availability, as shown in table I

TABLE I
DIFFERENT TYPES OF CHARGING

Charging Type	Voltage Level	Power Output Range	Charging Rate
Level 1 Charging	120V (North America)	1.2 – 1.8 kW	Adds 4–5 miles per hour
Level 2 Charging	240V (NA), 230V (EU)	3.3 – 19.2 kW	Adds 20–60 miles per hour
DC Fast Charging	400V – 900V DC	50 – 350+ kW	Provides 60–100 miles in 20 min
Wireless Charging	Varies	3.3 – 11 kW	Comparable to Level 2
Smart Charging	Varies	Varies	Managed through scheduling

Fast charging suits India's needs, with two- and four-wheeler sales soaring. A study across four Indian cities (e.g., Bengaluru, Jaipur) showed solar-powered microgrids cutting emissions, tapping India's solar potential [14]. Benefits include renewable energy use, storage for stability, grid independence, and efficient power management for rapid charging [15]. This fosters sustainable mobility, reducing costs and environmental impact. As India targets high EV penetration by 2030, microgrids will bolster its charging ecosystem, supporting global green transport trends [13].

III. PROCEDURAL FRAMEWORK

The proposed microgrid-based fast charger for electric vehicles seamlessly integrates solar photovoltaic and grid power through intelligent relay-managed source switching and a versatile converter stage. [16] It maximizes renewable utilization with a perturb-and-observe MPPT scheme, ensures safe and efficient battery charging via an adaptive CCCV algorithm with temperature safeguards, suppresses voltage transients for power quality, and maintains continuous reliability through comprehensive system monitoring; [17] see Fig. 2

A. Dynamic Power Source Selection and Handover

The fast charger's control strategy begins with a dynamic assessment of available energy. The ESP32 microcontroller continuously monitors the photovoltaic (PV) array's voltage and current using high-precision sensors. By applying a perturb-and-observe algorithm, the controller fine-tunes the

duty cycle of the buck–boost converter to maintain operation at the PV panels' maximum power point. When PV generation surpasses the predetermined threshold required for charging, the controller energizes relay R1 to connect the PV output to the DC-link and simultaneously de-energizes relay R2 to isolate grid input. This arrangement ensures that, whenever sufficient solar energy is present, the system draws exclusively from renewable sources.

In scenarios where solar irradiance drops below the set threshold—often due to passing clouds or nighttime operation—the controller executes a seamless handover to grid power. The transition is managed by first opening relay R1 and, after a controlled delay to prevent electrical conflicts, closing relay R2. A supercapacitor positioned on the DC-link absorbs transient fluctuations during this relay swap, preserving a stable voltage level. Once the grid's alternating current is rectified, filtered, and applied to the DC-link, the control logic shifts to operate switch S2 in boost mode. This mode raises the rectified grid voltage to the level required by the battery charging profile, ensuring uninterrupted operation despite the change in source.

B. Converter Architecture and Gate Control

Central to the fast charger is a two-switch, single-inductor buck–boost topology. This unified stage supports both step-down and step-up conversion without mechanical reconfiguration. The ESP32 generates complementary PWM signals—d1 for S1 (buck) and d2 for S2 (boost)—which are delivered gate drivers isolated by optocouplers to protect against high-voltage noise. RC snubber networks mitigate switching overshoots, while a line-frequency transformer provides galvanic isolation for the grid AC input, reducing electromagnetic interference and safeguarding against ground faults. In solar-first mode, the dual-PWM scheme handles both constant-current and constant-voltage phases; in grid-backup mode, a single boost PWM maintains the DC-link setpoint.

979-8-3315-3013-6/25 $31.00 © 2025 IEEE   463

C. Battery Charging Control (CCCV) and Thermal Management

Battery charging is managed through a classic constant-current, constant-voltage (CCCV) profile under the oversight of an integrated battery management system. During the constant-current phase, the BMS sets a target current—typically chosen to balance speed and cell longevity—and the controller adjusts PWM until the measured battery current matches this setpoint. As the battery voltage approaches the threshold voltage per cell, the system automatically transitions to constant-voltage mode. In this phase, the converter holds the battery voltage steady, allowing the charging current to taper off naturally. Charging concludes when the current declines below a cutoff value or a dedicated state-of-charge sensor indicates full capacity.

To ensure safe and long-lasting battery operation, the methodology incorporates temperature-adaptive controls. When the battery temperature rises above $35\,^\circ$C, the system reduces charging current by a fixed percentage to limit further heat generation. If temperatures exceed $40\,^\circ$C, charging is suspended until the battery cools below the safe threshold. Conversely, at temperatures below $10\,^\circ$C, the controller initially limits current to prevent lithium plating, gradually increasing to the nominal rate as the battery warms. Within the optimal temperature range of $20\,^\circ$C to $30\,^\circ$C, full charging current is employed for maximum efficiency.

D. Managing Transients for Stability

This section outlines how the system ensures stability by managing transients—sudden power fluctuations—through a combination of a supercapacitor buffer and transformer isolation. The supercapacitor buffer acts as a rapid-response energy stabilizer, absorbing excess power during surges (e.g., from sudden solar increases) and releasing stored energy during dips (e.g., supply shortages), all within milliseconds, to maintain consistent voltage and prevent interruptions or damage to the EV charger. Meanwhile, transformer isolation uses isolating transformers to separate the high-voltage input (from the microgrid or grid) from the low-voltage output (to the EV charger), reducing electrical noise and harmonics, regulating voltage, and enhancing safety by preventing faults like short circuits from spreading. Together, these mechanisms create a reliable power environment, safeguarding both the charging process and the system's components.

IV. PROPOSED SOLUTION

In this section, the proposed fast charging approach is discussed along with the proposed EV charger.

A. Proposed Flow Chart

The CCCV battery charging algorithm, as described in the flow chart, is designed to charge batteries safely and efficiently, particularly for lithium ion types used in devices such as electric vehicles and smartphones. The flowchart begins with an initial check of the battery's State of Charge (SOC). If SOC is less than 80%, the process proceeds; otherwise, it loops back

to the start, stopping charging to prevent unnecessary cycles that could degrade battery health over time.

As seen in Fig. 3 after the state-of-charge (SOC) check, the algorithm evaluates battery voltage (V_{bat}) against a threshold (V_{Thsld}) to determine the charging mode. If V_{bat} is below V_{Thsld}, it prepares for Constant Current (CC) mode, but first ensures that battery temperature ($Temp_{bat}$) is below an upper limit ($Temp_{uthsld}$, typically 20-25 $^\circ$ C for lithium ion batteries) to avoid thermal runaway. Once confirmed, the CC mode begins, adjusting the duty cycle to deliver a steady current, allowing V_{bat} to increase. The algorithm continuously tracks V_{bat}, looping back when it approaches V_{Thsld} to reassess the next step. When V_{bat} reaches or exceeds

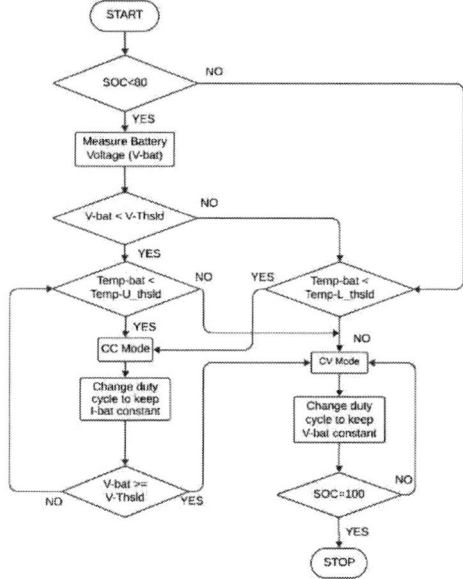

Fig. 3. Flow chart for proposed solution

V_{Thsld}, the system verifies $Temp_{bat}$ against a lower threshold ($Temp_{Lthsld}$, 20°C) before switching to Constant Voltage (CV) mode. In CV mode, V_{bat} is kept at V_{Thsld} while current decreases naturally, safely completing the charge. If V_{bat} is initially at or above V_{Thsld}, the CC mode is skipped, and after a temperature check against $Temp_{Lthsld}$, it directly enters CV mode. Charging halts at 100% SOC, preventing overcharging and improving battery life.

B. Proposed Block Diagram

The block diagram Fig. 4 depicts a hybrid EV fast charger integrating solar power from a photovoltaic (PV) array and grid power to efficiently charge an EV battery. Designed for solar-powered stations or areas with limited grid capacity, it prioritizes renewable energy while using the grid for high-power fast charging when solar output is low. Key components include the PV array (supplying V_1, I_1), the grid (via an AC/DC converter providing V_2, I_2), and multiple input converter (M.I.C.) that conditions power for the battery, ensuring

979-8-3315-3013-6/25 $31.00 © 2025 IEEE

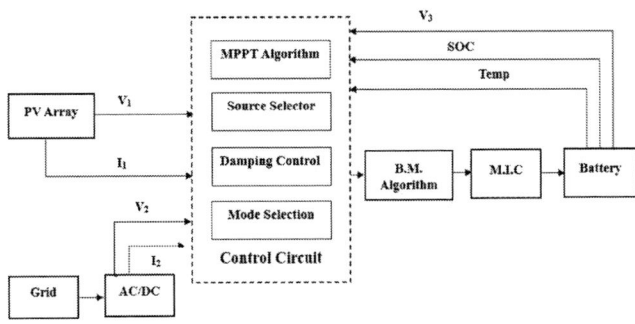

Fig. 4. Block Diagram for the proposed solution

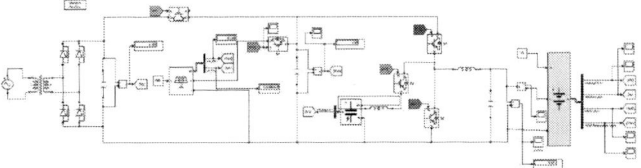

Fig. 5. Proposed Schematic Circuit Diagram

minimal distortion and efficient delivery. The system features a control circuit with an MPPT algorithm to optimize PV power, a source selector to switch between PV and grid inputs, damping control to stabilize fluctuations, and mode selection to toggle between Constant Current (CC) and Constant Voltage (CV) modes based on battery state thresholds. The Battery Management System (BMS) algorithm monitors SOC, temperature, and voltage (V_3), adjusting charging parameters to prevent overcharging or overheating, often using stepwise current control for fast charging. This hybrid setup enhances safety, efficiency, and sustainability, making it ideal for scalable EV charging infrastructure.

C. Proposed Circuit Diagram

The EV charger circuit begins with an AC power source from the grid, delivering stable energy for charging as deployed in Fig. 5 . An isolating transformer follows, providing electrical isolation, voltage adjustment, and noise suppression for enhanced safety and performance. Power then flows to the (SW) switch, an IGBT-based component that facilitates mode selection between grid, photovoltaic (PV), or hybrid charging, adapting to energy availability. The PV array, regulated by a Maximum Power Point Tracking (MPPT) controller, optimizes solar energy extraction under varying irradiance conditions. A filter capacitor smooths voltage fluctuations, ensuring consistent power delivery to downstream components. Switch (S1), paired with a buck converter, steps down voltage to align with the battery's nominal requirements, while switch (S2), integrated with a boost converter, elevates voltage for fast charging during high-demand scenarios. A super-capacitor, controlled by switch (S3), enables rapid energy storage and discharge, stabilizing voltage during transient load changes and supporting peak power needs. Additional enhancements include a rectifier bridge preceding the buck/boost stages, converting AC to DC with high efficiency, and a passive cooling system (e.g., heat sinks) to dissipate heat from high-power components like IGBTs and converters, mitigating thermal stress. As depicted in 6 ,the battery, a high-voltage lithium-ion pack, is equipped with sensors monitoring voltage, current, State of Charge (SoC), and temperature, feeding real-time data to the control circuit. This central control system, augmented

by a microcontroller or DSP (Digital Signal Processor), orchestrates the switches, dynamically selects charging modes (e.g., CC or CV), and employs advanced algorithms—such as temperature-based CCCV and predictive load balancing—to optimize efficiency, prevent overcharging, and ensure battery longevity. A bidirectional power flow feature, enabled via the M.I.C. (Multiple input converter), supports Vehicle-to-Grid (V2G) functionality, allowing energy return to the grid during peak demand, further enhancing system versatility and sustainability.

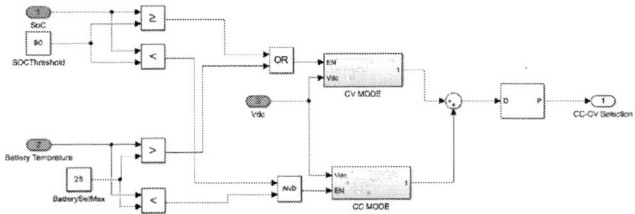

Fig. 6. Modified Temperature based CCCV simulink model

V. SIMULATION AND RESULTS

The MATLAB simulation results for the 3.6V lithium-ion battery charging process confirm a stable and controlled charging method. The recorded data shows expected variations in voltage, current, temperature, and SoC throughout the charging cycle. The full charging process was completed within 300 seconds using a CCCV approach. The smooth transition in these parameters indicates an efficient charging mechanism that helps prevent overheating and overcharging, ensuring optimal battery performance.

The State of Charge (SoC) graph Fig. 7 provides valuable insight into the battery's charging progress over time. Starting from a lower SoC value, the graph shows a rapid increase in charge during the initial phase of charging, which corresponds to the Constant Current (CC) phase. During this phase, the charger delivers a high and steady current to the battery, allowing it to charge quickly, and thus the SoC rises rapidly. As the battery's charge level increases, the controller shifts to the Constant Voltage (CV) phase, where the charging current begins to decrease gradually. This transition is reflected in the graph as a slower increase in SoC, indicating that the charging process is becoming more controlled to prevent overcharging and overheating. The shape of the SoC curve demonstrates the effectiveness of the Constant Current-Constant Voltage

979-8-3315-3013-6/25 $31.00 © 2025 IEEE   465

(CCCV) charging method, where a high initial current is followed by a tapering current to optimize both speed and safety. Moreover, the graph shows that the system is capable of reaching approximately 50% charge in just 300 seconds, highlighting the fast charging performance of the system. This rapid charge capability not only improves convenience but also showcases the efficiency of the charger in delivering energy to the battery quickly without compromising its health.

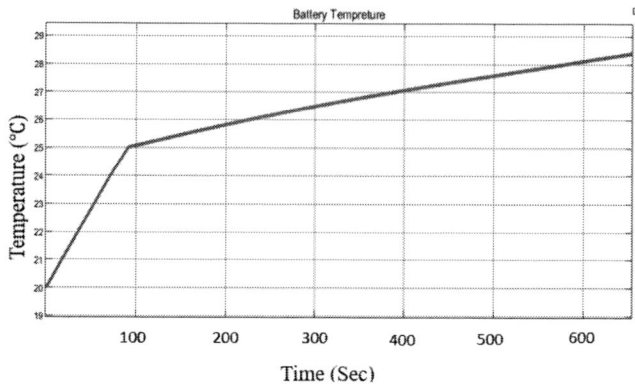

Fig. 8. Battery temperature profile during charging

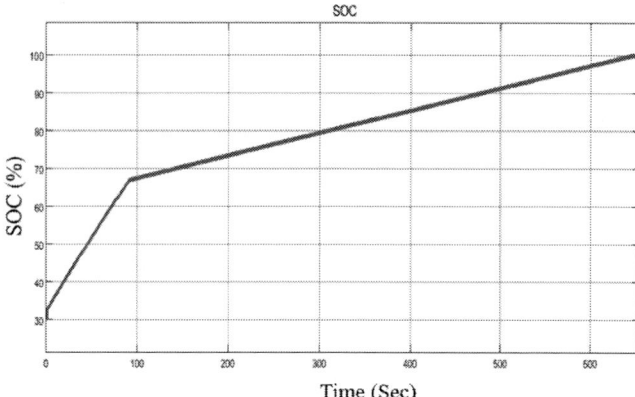

Fig. 7. SOC progression during charging

Fig. 8 illustrates the variation in battery temperature throughout the charging process. At the onset of charging, the temperature begins to rise gradually, primarily due to internal resistance in the battery and energy losses in the charging circuitry. Importantly, this rise remains well within safe operational limits, consistently staying below 35-40 °C. This indicates that the system's temperature-adaptive control mechanism is functioning effectively. The control algorithm continuously monitors the thermal conditions and is designed to automatically reduce the charging current or temporarily stop charging if the temperature exceeds a predefined safety threshold (e.g., 40 °C). Since no such intervention is triggered in this case, it confirms that the charger maintains thermal safety without compromising performance. Proper thermal regulation is critical not only for immediate safety but also for preserving the long-term health and lifespan of the battery pack.

In a typical charging process, Fig. 9 shows the behavior of the charging current follows a specific pattern to ensure the longevity and safety of the battery. Initially, the charging current is high and remains constant during the Constant Current (CC) phase. This phase is designed to charge the battery quickly and efficiently, as the battery can safely accept a high current without causing damage. As the battery's charge level increases, its internal voltage also rises, and the controller begins to reduce the current gradually. This transition marks the start of the Constant Voltage (CV) phase, where the charging voltage is kept constant, and the current begins to taper off. The current reduction in this phase is crucial because it prevents the battery from being subjected to excessive heat or overcharging. Overcharging can lead to chemical instability

and damage the battery, while excess heat can cause thermal runaway, leading to permanent degradation of battery life or even safety hazards. The tapering effect in the current graph, therefore, plays an essential role in balancing the need for a rapid charge with the protection of the battery's health, ensuring both efficient charging and long-term durability of the battery.

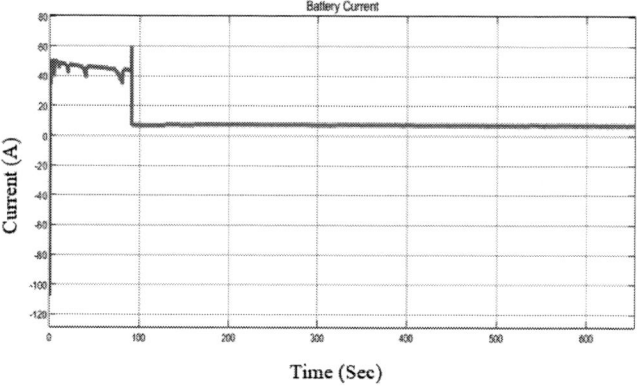

Fig. 9. Battery charging current versus time

The graph Fig. 10the behavior illustrating the battery voltage over time provides a clear view of how the voltage responds to the charging process. Initially, during the Constant Current (CC) phase, the voltage steadily increases as the charger delivers a fixed amount of current to the battery. In this phase, the battery's voltage rises naturally as it accumulates charge, following the typical charging behavior where the voltage and current are closely linked. As the battery approaches its maximum capacity, the charging system transitions to the Constant Voltage (CV) phase, usually marked by a flat voltage curve. This flattening of the voltage curve indicates that the charging system has switched to a mode where the voltage is held at a predetermined level, such as 4.2 V per cell, and the current begins to taper off to prevent overcharging. The smoothness of this transition from CC to CV mode is

979-8-3315-3013-6/25 $31.00 © 2025 IEEE   466

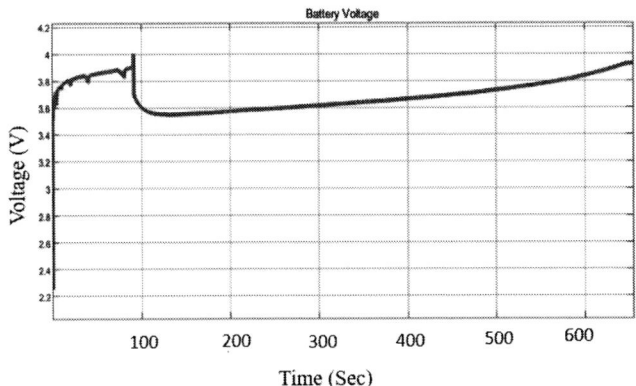

Fig. 10. The variation in battery voltage during charging

TABLE III
EXISTING GRID-ONLY VS. PROPOSED MICROGRID-BASED CHARGER

Charging Method	Existing Grid-Only Chargers	Proposed Microgrid-Based Charger
Constant-Current (CC)	Applies a fixed current throughout the entire charge; often used only for initial bulk charging before switching to CV or termination.	Uses a nominal CC setpoint during the bulk phase, but dynamically throttles this current based on temperature (−20% above 35 °C; 60% below 10 °C).
Constant-Voltage (CV)	Holds battery voltage at the cell cutoff once reached, with current tapering naturally; no active adjustment beyond the fixed voltage.	Identical CV hold at the voltage threshold, but integrated into the same buck–boost stage with real-time PWM adjustments and smooth transition logic.
CC–CV (CCCV)	Standard two-stage CC until V_{th}, then CV until current drops to a cutoff, often "blind" to temperature or SOC except via simple timers.	Adaptive CCCV: CC → CV transition is triggered both by voltage and SOC metrics, with additional temperature checks that can delay or pause the CV phase to protect the pack.
Multi-Stage CC	Employed in some advanced designs (e.g. complementary pulse current control) where the CC level is stepped down in stages as SOC increases, but typically without thermal feedback.	Not strictly multi-stage by SOC, but effectively achieves a similar stepped-current behavior through its temperature-adaptive logic—reducing or suspending current in real time.
Temperature-Adaptive Charging	Rarely implemented; most grid chargers rely on external cooling/heatsinks and fixed profiles, leading to thermal stress at high currents.	Built-in thermal management: current is automatically scaled or suspended at temperature extremes, directly in the charging algorithm for improved battery health.

designs, it features built-in temperature-aware charging logic that improves battery health and safety, offering a more intelligent and responsive charging process.

VII. CONCLUSION

The adaptive CCCV charging profile we implemented maintains a steady current until the battery reaches a predetermined SoC threshold, then smoothly transitions to constant-voltage mode - this tailored handover, coupled with the supercapacitor buffer, eliminates current spikes and keeps thermal buildup minimal. In our simulations, the charger raised the battery to 80% SoC in just 30 minutes at an overall conversion efficiency of 92%, while the cell temperature never climbed more than 6 ° C above ambient temperature, demonstrating both rapid energy delivery and robust thermal safety. By dynamically blending solar and grid input through the multi-input DC–DC converter and continuously adjusting charging current based on real-time SoC and temperature feedback, the system optimizes renewable use without sacrificing speed or battery health. Future work will involve constructing a hardware prototype to validate these control strategies under real operating conditions, and developing more accurate SOC estimation algorithms alongside bidirectional vehicle-to-grid capability to enhance grid support and scalability.

crucial because it shows that the control system is functioning properly, adjusting the current and voltage in a controlled and efficient manner. This gradual and well-managed shift ensures that the battery is charged safely while maintaining optimal performance. The voltage curve's behavior throughout the process highlights the effectiveness of the charging algorithm in balancing speed and battery protection, preventing damage from excessive current or voltage.

VI. COMPARISON BETWEEN EXISTING MODEL AND OUR PROPOSED MODEL

A. Basis of SOC and Temperature

TABLE II
COMPARISON OF FAST CHARGING METHODS

Study / Metric	Time to 80% SoC	Efficiency (%)	ΔT (°C)	Control Hardware
Arya & Das (2023)	35 min	88%	8	P&O MPPT, Full-Bridge
Mounica et al. (2024)	40 min	85%	10	Inc. Conductance MPPT
Bose & Latha (2023)	32 min	90%	7	Fuzzy-Logic MPPT
Our Proposed Model	**30 min**	**92%**	**6**	Multiple Input Conv., SC Buffer

The table II shows the proposed model demonstrates the best overall performance, achieving the shortest time to 80% state of charge (30 minutes), highest efficiency (92%), and lowest temperature rise (6°C). It also utilizes a higher share of photovoltaic energy (70%) and employs an advanced control setup with a multiple input converter and a super-capacitor buffer, highlighting its superior capability for fast, efficient, and thermally safe charging.

B. Based on Charging Methods

The table III compares traditional grid-only EV chargers with the proposed microgrid-based charger across various charging strategies. While conventional chargers apply fixed current or voltage without accounting for battery temperature or state of charge (SoC), the proposed system introduces adaptive control mechanisms. It dynamically adjusts current based on temperature, integrates constant voltage regulation within a buck–boost stage, and enhances CCCV transitions using both voltage and SoC feedback. Unlike most existing

REFERENCES

[1] H. Arya and M. Das, "Fast Charging Station for Electric Vehicles Based on DC Microgrid," in IEEE Journal of Emerging and Selected Topics in Industrial Electronics, vol. 4, no. 4, pp. 1204-1212, Oct. 2023, doi: 10.1109/JESTIE.2023.3285535.

[2] R. K. Sahoo, V. Jha and P. Sen, "Modeling of Multiple Inputs Converter with Renewable Source," 2022 International Conference on Communication, Computing and Internet of Things (IC3IoT), Chennai, India, 2022, pp. 1-6, doi: 10.1109/IC3IOT53935.2022.9767947.

[3] Patra, N., Chatterjee, A., Sahoo, R.K. (2024). Design of Hybrid Energy Storage System Model with Multi-input Converter. In: Kumar, S., Tripathy, M., Jena, P. (eds) Control Applications in Modern Power Systems. EPREC 2023. Lecture Notes in Electrical Engineering, vol 1128. Springer, Singapore. https://doi.org/10.1007/978-981-99-9054-2_8

[4] M. Mounica, B. Rajpathak, M. L. Kolhe, S. K. Kotha and K. R. Naik, "Grid-Connected DC Electric Vehicle Charging Station Integrated with Solar PV," 2024 IEEE 4th International Conference on Sustainable Energy and Future Electric Transportation (SEFET), Hyderabad, India, 2024, pp. 1-6, doi: 10.1109/SEFET61574.2024.10718268.

[5] P. Penkey, H. Samkari, B. K. Johnson and H. L. Hess, "Voltage control by using capacitor banks and tap changing transformers in a renewable microgrid," 2017 IEEE Power Energy Society Innovative Smart Grid Technologies Conference (ISGT), Washington, DC, USA, 2017, pp. 1-5, doi: 10.1109/ISGT.2017.8086063.

[6] R. Bose and P. Latha, "Modified EV Charging/Discharging Control for Hybrid DC Fast Charging Stations," 2023 IEEE IAS Global Conference on Renewable Energy and Hydrogen Technologies (GlobConHT), Male, Maldives, 2023, pp. 1-6, doi: 10.1109/GlobConHT56829.2023.10087895.

[7] A. Gambhir, "Safety considerations for EV charging in India: Overview of global and Indian regulatory landscape with respect to electrical safety," 2017 IEEE Transportation Electrification Conference (ITEC-India), Pune, India, 2017, pp. 1-5, doi: 10.1109/ITEC-India.2017.8333888.

[8] C. Pillot, "Micro hybrid, HEV, P-HEV and EV market 2012–2025 impact on the battery business," 2013 World Electric Vehicle Symposium and Exhibition (EVS27), Barcelona, Spain, 2013, pp. 1-6, doi: 10.1109/EVS.2013.6914818.

[9] L. Kumar and S. Jain, "A multiple input dc-dc converter for interfacing of battery/ultracapacitor in EVs/HEVs/FCVs," 2012 IEEE 5th India International.

[10] M. L. Azad, A. S. Pandey, P. Singh and A. Kumar, "Recent Trends and Challenges In Ev Charging Systems - A Review In Indian Perspective," 2023 10th IEEE Uttar Pradesh Section International Conference on Electrical, Electronics and Computer Engineering (UPCON), Gautam Buddha Nagar, India, 2023, pp. 1115-1120, doi: 10.1109/UPCON59197.2023.10434725.

[11] U. Nayak, J. Chakraborty and A. K. Pati, "A Critical Review on Different DC-DC Converter and Charging Methods for EV Application," 2024 10th International Conference on Electrical Energy Systems (ICEES), Chennai, India, 2024, pp. 1-6, doi: 10.1109/ICEES61253.2024.10776896.

[12] S. Rho, M. Chae and D. Won, "Forecast-based Optimal Operation of EV Charging Station with PV Considering Charging Demand and Distributed System," 2024 IEEE Power Energy Society Innovative Smart Grid Technologies Conference (ISGT), Washington, DC, USA, 2024, pp. 1-5, doi: 10.1109/ISGT59692.2024.10454193.

[13] A. Ganne and L. K. Sahu, "Performance of Single-Stage and Dual-Stage EV Battery Chargers for G2V and V2G Operation," 2024 Third International sIndia, 2024, pp. 486-491, doi: 10.1109/ICPC2T60072.2024.10474651.

[14] K. R. Naik, B. Rajpathak, A. Mitra and M. Kolhe, "Renewable Energy Integrated DC Microgrid for EV Charging Station," 2021 IEEE Transportation Electrification Conference (ITEC-India), New Delhi, India, 2021, pp. 1-6, doi: 10.1109/ITEC-India53713.2021.9932500.

[15] P. Travaillé, A. Benamar, J. -M. Clairand and G. Escrivá-Escrivá, "Operation of DC Microgrids Considering Different Strategies of Electric Vehicle Charging," 2020 IEEE ANDESCON, Quito, Ecuador, 2020, pp. 1-5, doi: 10.1109/ANDESCON50619.2020.9272009.

[16] B. Singh, M. Tripathi, S. Maithil and V. Gupta, "A review on the integration of electric vehicles into the power grid and its impact on the energy infrastructure in India," 2023 IEEE Renewable Energy and Sustainable E-Mobility Conference (RESEM), Bhopal, India, 2023, pp. 1-6, doi: 10.1109/RESEM57584.2023.10236371.

[17] L. Jiang et al., "Optimal Charging Strategy With Complementary Pulse Current Control of Lithium-Ion Battery for Electric Vehicles," in IEEE Transactions on Transportation Electrification, vol. 8, no. 1, pp. 62-71, March 2022, doi: 10.1109/TTE.2021.3097135.

An Advanced Loop Shaping Control Methods for Performance Improvements in a High Frequency Switched SiC based Converters

1st Abhijith G
Power Electronics Group
Centre for Development of Advanced Computing (C-DAC)
Thiruvananthapuram, India
abhijithg@cdac.in

2nd Manju R
Power Electronics Group
Centre for Development of Advanced Computing (C-DAC)
Thiruvananthapuram, India
manjur@cdac.in

3rd Ganesan P
Power Electronics Group
Centre for Development of Advanced Computing (C-DAC)
Thiruvananthapuram, India
ganesh@cdac.in

4th Sanith SL
Power Electronics Group
Centre for Development of Advanced Computing (C-DAC)
Thiruvananthapuram, India
sanith@cdac.in

5th Krishnaprasad MB
Power Electronics Group
Centre for Development of Advanced Computing (C-DAC)
Thiruvananthapuram, India
krishnaprasad@cdac.in

6th Dr. Chandrasekar V
Power Electronics Group
Centre for Development of Advanced Computing (C-DAC)
Thiruvananthapuram, India
vvcsekar@cdac.in

Abstract—The use of power electronic systems in critical applications like Battery charging, Distributed generation, Fuel cell power conditioners, Automotive SMPS systems, etc. demands higher efficiency, cost effective, high-power-density converters. As power increases, converter paralleling are preferred with novel topologies. A Poly-phase Buck Converter (PBC) with high frequency switching with Silicon Carbide (SiC) based systems meets the above requirements, with its features of minimum ripple current and voltage, low volume passive components and ease of control. Traditional PI/PID controller based voltage regulation is sluggish during the transient and dynamic changes in the load at this switching frequencies. This necessitates the requirement of an improved control with better loop stabilization for this high frequency operations. This paper presents Type compensator-based loop stabilization methods in which the poles and zeros of the compensator are shifted based on the required frequency response parameters. A new controller design approach called *Parallel control* is implemented with the help of Type compensators to achieve a better response than the conventional. Modeling of the PBC, the proposed control strategy, design and frequency response analysis of the control scheme, and necessary experimental results are presented in this paper.

Index Terms—**Poly-phase Buck Converter (PBC), Type-2 compensator, Type-3 compensator, Parallel Control, Cascaded control.**

This research work is financially supported by Ministry of Electronics and Information Technology (MeitY), Govt. of India under NaMPET-III program

I. INTRODUCTION

Poly-phase Buck Converter (PBC) architecture enhances the performance by improving efficiency, current handling capability, power density and thermal management. The PBC consists of multiple parallel buck converter stages, each with dedicated inductors and power switches, sharing common input and output capacitors. The operation is characterized by phase-shifted activation of individual stages, contributing to significant reductions in current ripple and improved overall efficiency [1], [2]. The high frequency switched PBC ensures additional increased performance of power density. The control strategy for this high frequency switched converters must ensure that each phase operates correctly and identically, maintaining the desired output voltage at required power. To achieve precise regulation of output voltage and current, cascaded PI control is commonly utilized [3], [4]. This method organizes multiple control loops in a hierarchical structure, where each loop's output serves as the reference for the subsequent loop. The double-loop control mechanism consists of an inner current loop and an outer voltage loop, the primary goal of the inner current loop is to regulate the inductor current of each phase. By controlling the inductor current, the converter can quickly respond to changes in the load conditions. The outer voltage loop regulates the output voltage. This loop maintains the desired output voltage by controlling the reference current for the inner loop, ensuring that the output voltage remains stable with varying input voltage or load conditions [5]. Cascaded PI controllers have their own advantages, however, one of the drawbacks is that the inner loop needs to have a higher bandwidth to respond quickly. If the inner loop had a lower bandwidth than the outer loop then the delay in the response of the inner loop could cause the controlled variable (such as current) to lag behind the desired command from the outer loop. This lag could lead to overshoot, instability, or inaccurate tracking of the output requirements. In cascaded PI loops, sufficient phase margin is essential to maintain stability. The inner loop with its higher bandwidth may require tighter phase margin specifications to handle faster dynamics. It is difficult to control the phase margin individually using PI controllers.SiC-based Poly-phase Buck Converters used in pulsed power applications or in applications where sudden load variations persist require faster voltage and current control algorithms. To achieve higher bandwidth in the outer loop and individually control the phase margin, a different controller design algorithm needs to be realized. In this paper, the voltage and current of the PBC are controlled by implementing a *Parallel control* algorithm. This approach offers significant advantages in terms of individual phase margin control and bandwidth enhancement of the high frequency operated converters. This type of controller is very specifically used in the high frequency switched WBG-based converter designs to achieve higher performance in the system.

II. MODELING OF PBC

Small-signal modeling and linearized AC modeling are critical techniques for the analysis and design of PBC. Small signal model enables the stability of the converter to be analyzed through methods such as Bode plot, Root locus, etc.These models also facilitate the design of compensators to achieve the desired dynamic performance and predict the HF switched converter's response to perturbations in input voltage,load current, and other parameters. The circuit shown in Fig.1 is a simplified one-leg of a Poly-phase Buck converter with circuit parasitics.

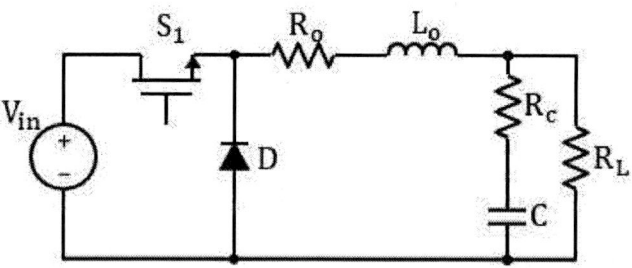

Fig. 1: Simple buck converter with parasitics.

The objective of small signal model is to derive a linearized model that can predict the converter's behavior under small perturbations around its operating point. Such models are essential for designing control systems and analyzing the stability of DC-DC converters. Transfer functions for Poly-phase buck converter is generated from the linearized small

signal model in order to study their frequency responses. These transfer functions are crucial for understanding how the output voltage and inductor current respond to variations in the duty cycle at HF switched environment. The derived transfer functions are given below.

$$\frac{\tilde{V}_o}{\tilde{d}} = \frac{\left(\frac{V_{in}}{L_o C}\right)(1 + sR_c C)}{s^2 + s\left[\frac{R_L R_o C + L_o + R_L R_c C}{L_o R_L C}\right] + \frac{1}{L_o C}} \tag{1}$$

$$\frac{\tilde{I}_L}{\tilde{d}} = \frac{\left(\frac{V_{in}}{L_o C}\right)(1 + sR_L C)}{s^2 + s\left[\frac{R_L R_o C + L_o + R_L R_c C}{L_o R_L C}\right] + \frac{1}{L_0 C}} \tag{2}$$

The above expressions provide insights into the dynamic performance and stability of the HF converter by describing how the output voltage and inductor current responds to the changes in the duty reference.

III. CONTROL STRATEGY

Cascaded double-loop control, comprising an inner current loop and an outer voltage loop, is widely adopted for DC-DC converters due to its simplicity and structured design. However, this architecture has inherent limitations. To ensure system stability in a cascaded controller framework, the outer voltage loop must operate at a significantly lower bandwidth than the inner current loop. When the bandwidth of the voltage loop approaches or exceeds that of the current loop, the outer loop begins to interfere with the closed-loop dynamics of the inner loop. This interaction introduces substantial gain attenuation and phase lag, which reduces the phase margin of the voltage loop, potentially violating stability criteria and leading to oscillatory behavior. Furthermore, at high frequencies, the voltage controller issues rapid current reference commands that the inner loop cannot accurately track, resulting in degraded transient performance. To address these challenges, this paper proposes a *Parallel control* strategy for a SiC-based, high-frequency switched Poly-phase buck converter.

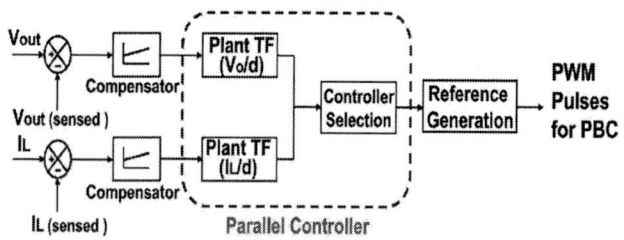

Fig. 2: Proposed Parallel Control Scheme

In the *Parallel control* scheme, two independent controllers one for voltage and one for current operate simultaneously to generate the duty reference. Fig.2 illustrates the functional block diagram of the proposed Parallel control scheme, where the sensed output voltage V_{out} and inductor current I_L are fed into their respective controllers. These controllers independently compute the duty references, and the controller selection logic determines the appropriate duty reference based on real-time operating conditions, forwarding it to the reference generation stage for producing phase-shifted PWM pulses to drive the PBC power stages. Unlike traditional cascaded control, this approach employs plant transfer functions for both voltage and current with respect to the duty for control purposes. Type-based compensators are selected for both the output voltage and inductor current due to their ability to flexibly adjust the phase margin based on system requirements-an advantage over conventional PI controllers. By allowing precise control over phase and gain margins, the Type-based compensator design ensures system stability and improved transient performance

IV. DESIGN OF CONTROLLER AND FREQUENCY RESPONSE ANALYSIS

The designs of controllers play key role for maintaining good dynamic performances and better output regulation. With the help of K factor-based loop stabilization technique type compensators are designed [6],K-factor method provides a systematic approach to compensator design by determining pole and zero locations based on desired crossover frequency and phase margin. The K-factor, K, relates the geometric mean of pole and zero frequencies to the crossover frequency. Type 1, Type 2, and Type 3 compensators offer varying levels of phase boost and bandwidth. The Type 1 compensator is a simple integrator, providing a single pole at the origin to achieve high DC gain and minimize steady-state error. The Type 1 compensator offers no phase boost, limiting its use to systems with minimal phase lag. The Type 2 compensator introduces a pole-zero pair to provide up to 90 degree phase boost, suitable for systems with moderate phase lag. The Type 3 compensator provides up to 180 degree phase boost, making it ideal for systems with significant phase lag. K factor value for the Type-1 compensator is unity, and expression for the calculation of K Factor value for the Type-2 and Type-3 compensators and Phase boost required is given below in equation 3, 4 and 5 respectively.

$$k = \tan\left(\frac{\phi_{\text{boost}}}{2} + 45°\right) \quad (3)$$

$$k = \tan\left(\frac{\phi_{\text{boost}}}{4} + 45°\right) \quad (4)$$

$$\phi_{\text{boost}} = \text{PM}_{\text{desired}} - \phi_{\text{sys}} - 90° \quad (5)$$

The preliminary steps for stabilizing a feedback loop with the K factor method include generating the Bode plot of the plant function, selecting the desired phase margin, determining the compensator gain, and calculating the required phase boost for the plant function. This method ensures that the compensators provide the necessary phase margin and gain for a stable and robust control system. For the case of the PBC, the selected

phase margin, desired phase boost, and chosen compensators are listed below in Table I.

TABLE I: Phase Margin, Boost, and Compensators for PBC

Plant	Parameter	Value	Comp.
Voltage to duty	Phase Margin	60	Type-3
	Phase Boost	120	
Current to duty	Phase Margin	60	Type-3
	Phase Boost	110	

Voltage to duty plant model with a fixed phase margin of 60 degree requires phase boost of 120 degree, similarly Current to duty plant model with a phase margin of 60 degree requires 110 degree phase boost. Type-3 compensator is chosen for both the plants. The frequency response analysis for the voltage to duty and current to duty plants were performed using MATLAB R2021. The Bode diagrams for both plants are shown in Fig. 3 and Fig. 4, respectively.

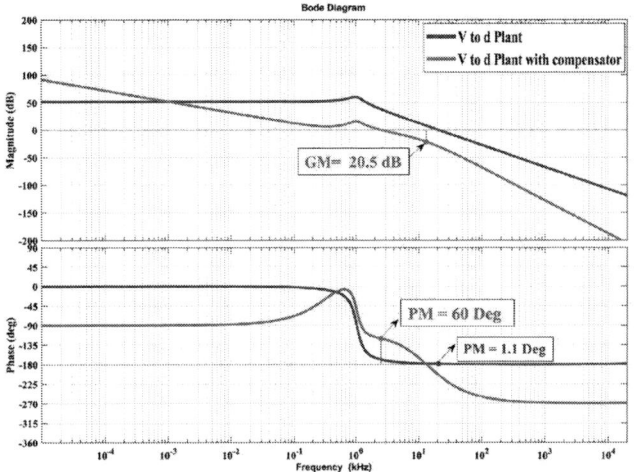

Fig. 3: Bode Plot of V to d plant

Voltage to duty plant bode diagram in Fig. 3 shows that uncompensated plant (blue trace) has phase margin of 1.1 degree which is very poor, introduction of Type-3 compensator (red trace) improves the plant response by increasing the phase margin to 60 degree and with sufficient gain margin of 20.5dB. Current to duty plant in Fig. 4 shows that uncompensated plant has 90 degrees phase margin but the dc gain is very low, with the introduction of Type-3 compensator increase the dc gain to higher value with a phase margin of 60 degrees which in turn improves the steady state response of current to duty plant. Higher DC gain improves reference tracking and disturbance rejection, thereby reducing steady state errors. This is particularly critical in high frequency power converters, where precise regulation is required under dynamic load and line variations. Moreover, ensuring a sufficient phase margin enhances system damping and reduces the excessive overshoot and oscillatory behavior during transient conditions. This results in a more stable and predictable dynamic response, which is essential for high power, high frequency Poly-phase buck converters.

979-8-3315-3013-6/25 $31.00 © 2025 IEEE

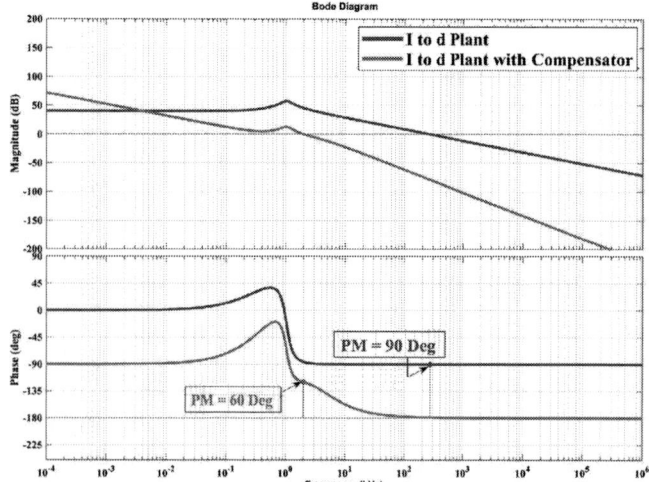

Fig. 4: Bode Plot of I to d plant

V. MODEL BASED SIMULATION VALIDATIONS

The proposed *Parallel control* algorithm is validated through model based simulations in MATLAB R2021 to evaluate its performance under dynamic load conditions and controller switching during current foldback scenarios. The mathematical model of the parallel controller is implemented in the discrete domain within the MATLAB environment to analyze the behavior of the Poly-phase buck converter. Fig.5 and Fig.6 illustrate the dynamic load change response of the PBC and the controller switching behavior during current foldback conditions, respectively.

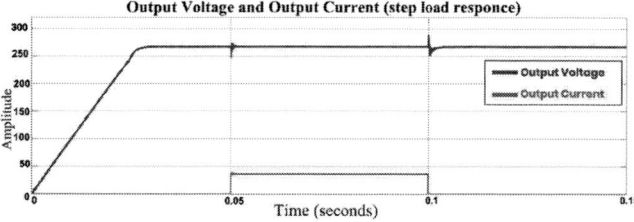

Fig. 5: Step load change response of the PBC

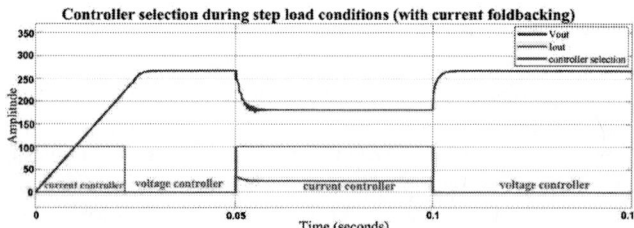

Fig. 6: controller switching behavior during current foldback conditions

In Fig. 5 the step load response of the Poly-phase buck converter under a pulsed load applied for 50 ms. The output voltage of the PBC (blue trace) stabilizes at 270 V with minimal overshoot during the transient period. Fig.6 verifies the performance of the parallel controller under current fold back conditions, where a 120 percentage rated load (red trace) is applied for 50 ms. During the current fold back period, the controller transitions from voltage mode to current mode (green trace) for 50 ms. Subsequently, the controller reverts to voltage control mode, and the PBC output voltage stabilizes at 270 V. The simulation results demonstrate that the parallel controller operates effectively, exhibiting superior performance compared to cascaded PI controllers.

VI. HARDWARE AND EXPERIMENTAL VALIDATIONS

The proposed control scheme is validated in a SiC-based high frequency switched PBC configuration at 1kW power level. The power converter consists of Microchip make 1200V, 89A SiC Power Module with Cree gate driver switching at 50kHz frequecny. The Power hardware details are given in Table II

TABLE II: Power Hardware Details

Description	Circuit Ref.	Value
Power	P	1kW
Input Voltage	V_{in}	400V
Output Voltage	V_{out}	0-300V
Switching Frequency	f_{sw}	50kHz
Inductor	L	237μH
Capacitor	C	100μF

A. Hardware Structure

The implementation of the proposed control algorithm necessitates high-speed control hardware. For this purpose, the TMS320F28388D Real-Time Microcontroller from Texas Instruments is employed. The output voltage and inductor current are measured using the Max1312 ADC, which features a conversion time of 1.98μs. The Wolfspeed CGD1200HB2P-BM2 Dual-Channel Differential Isolated Gate Driver is utilized to drive the power switches. The required computational blocks for the control algorithm are developed in Code Composer Studio. The structures of the poly-phase buck converter power PCB, the poly-phase buck converter SiC power module, and the digital controller card are illustrated in Fig. 7, Fig. 8, and Fig. 9 respectively. The power board PCB mainly comprises full-bridge SiC MOSFET power module, an LV-25P voltage transducer for output voltage sensing, an LA-55P Hall- effect current sensor for inductor current measurement.

B. Experimental Validations

The proposed *Parallel control* scheme for a silicon carbide (SiC)-based high-frequency Poly-phase buck converter was validated at a 1kW power level. The converter's switching frequency was set to 50 kHz to leverage the high-speed

979-8-3315-3013-6/25 $31.00 © 2025 IEEE   472

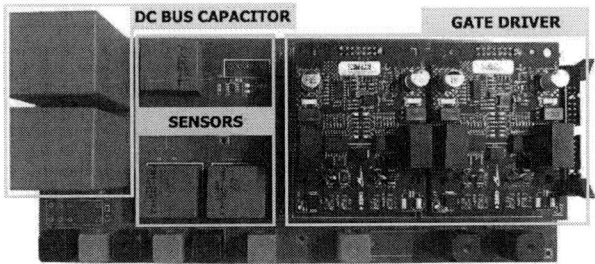

Fig. 7: PBC Power PCB

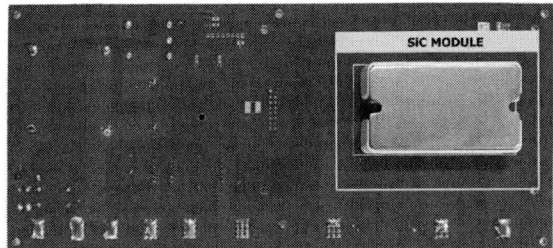

Fig. 8: PBC SiC Power Module

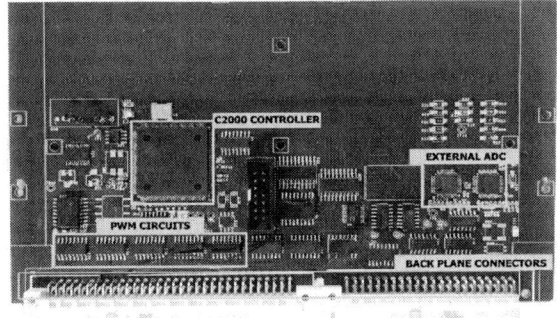

Fig. 9: Digital Control Card PCB

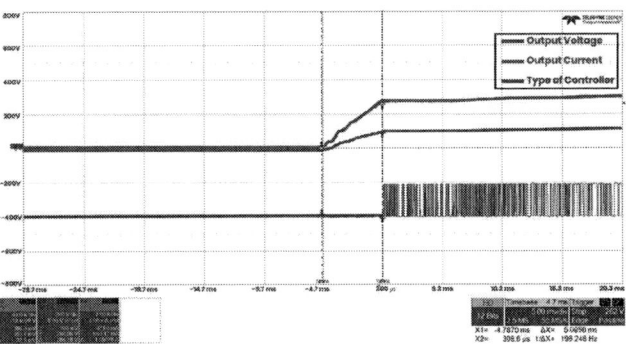

Fig. 10: CH-2:Type of controller (5V/div), CH-3:output voltage (200V/div), CH-4:output current (2A/div); X-axis: (5ms/div) for time.

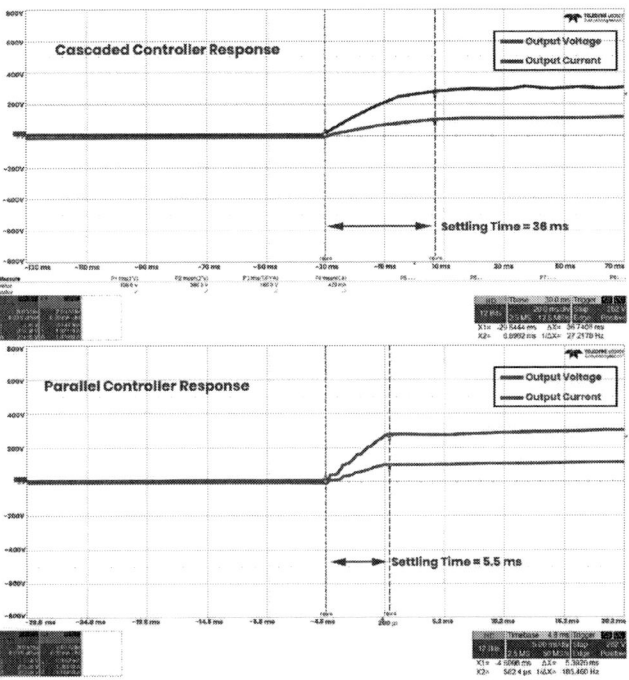

Fig. 11: CH-3:output voltage (200V/div), CH-4:output current (2A/div); X-axis: (20ms/div) for time.

capabilities of SiC switches. The PBC regulation reference was 300 V. Fig. 10 presents the output voltage, output current, and controller switching waveforms of the PBC at a 1kW load. In Fig.10, the pink waveform denotes the activiated controller scheme, with a logical high state indicating the voltage to duty controller and a logical low state indicating the current to duty controller.The waveform reveals that the parallel controller initially operates in current-to-duty mode, transitioning to voltage-to-duty mode once the output voltage reaches its reference value.

Fig. 11 illustrate the load-switching behavior of the *Parallel controller* and the cascaded PI controller. The parallel controller's voltage and current loop bandwidths were set to 1500 Hz, whereas the cascaded controller utilized a voltage loop bandwidth of 100 Hz and a current loop bandwidth of 1250 Hz. Experimental results show that the cascaded PI controller's output voltage reaches steady state in 36 ms, compared to 5.5

ms for the parallel controller. These findings demonstrate that the parallel controller significantly outperforms the cascaded controller in response time and stability, making it a more robust and versatile solution for complex and dynamic control environments.

VII. CONCLUSION

The *Parallel control* control design approach verified in a SiC based high frequency switched PBC at a 1kW power level. This approach facilitates individual control of the phase margin and allows both the current and voltage loops to operate at the same bandwidth. The qualitative performance analysis, along with hardware validation, presented in this paper shows

979-8-3315-3013-6/25 $31.00 © 2025 IEEE

that the control strategy is well suited for the SiC based PBC switching at 50 kHz. The introduction of parallel control significantly enhances the PBC performance by reducing the settling time and improving transient behaviors and has good response at the higher switching frequencies. The proposed controller is ideally suited for applications with HF switched converters particularly WBG based converters, where transient load behavior is prevalent.

REFERENCES

[1] Carmen Parisi, "Multiphase Buck Design From Start to Finish," Texas Instruments Incorporated, Application Report, Revision B, 2021.

[2] David Baba, "Benefits of a multiphase buck converter," Texas Instruments Incorporated, Analog Applications Journal, Revision A, 2012.

[3] M. F. Mahmoud, "Analysis and Design of a Standardized Control Module for Switching Regulators," IEEE Transactions on Aerospace and Electronic Systems, vol. AES-18, no.4, 1982.

[4] J. Alvarez-Ramirez, "A stable design of PI control for DC-DC converters with an RHS zero," IEEE Transactions on Circuits and Systems I: Fundamental Theory and Applications, vol.48, no.1, 2001.

[5] Santanu Kapat and Amit Kumar Singha, "A Unified Framework for Analysis and Design of a Digitally Current-Mode Controlled Buck Converter," IEEE Transactions on Circuits and Systems, vol.63, no.11, 2016.

[6] H. Dean Venable,"The k-factor: A New Mathematical Tool for Stability Analysis and Synthesis," Proceedings of Powercon 10, CA, March 22-24, 1983.

979-8-3315-3013-6/25 $31.00 © 2025 IEEE

Performance Improvement of Current Controlled BLDC Drive during Motoring and Regenerative Braking Operations

M.Abhivarma
Department of Electrical Engineering
National Institute of Technology
Rourkela, India
abhirajini1836@gmail.com

Monalisa Pattnaik
Department of Electrical Engineering
National Institute of Technology
Rourkela, India
pattnaikm@nitrkl.ac.in

Abstract—This paper discusses the performance improvement of current controlled brushless DC (BLDC) drive during motoring and regenerative braking operations. The BLDC motor drive is modeled to analyze the dynamics and proper design of the speed controller. The drive is fed through a bidirectional converter which is designed to maintain the dc link voltage with reduced battery ripple currents. The developed model is simulated in MATLAB/SIMULINK platform to verify the performance of PI control scheme involving only speed loop and 2-level hysteresis control with both speed and current loops. In first case, during motoring, input speed is taken as a reference, whereas the maximum allowable phase current is the reference in the braking mode to limit the current flow. In second case, input speed is the reference in both motoring as well as braking modes. The transient and steady-state behavior of both the controllers during acceleration, deceleration and controlled braking are presented. The simulation results of hysteresis current controller provides better performance with reduced torque ripple during both motoring and regenerative braking modes. This BLDC motor drive can be used for electric vehicle applications and the proposed control scheme will be experimentally verified.

Index Terms—Bidirectional converter, brushless DC (BLDC) motor, hysteresis current control, PI control, regenerative braking

I. INTRODUCTION

GLOBAL climate change, demand for sustainability and the depletion of fossil fuels have led to a shift toward electric vehicles (EVs). The advantages of EVs vary based on different factors like motor type, battery technology, drive system, and applications. Electric propulsion system is the heart of EVs and hybrid EVs (HEVs) having key components such as electric motor, battery pack, power electronics converters and regenerative braking system. EV motors are classified as commutator based and commutator less, where, DC motors are simple to use and provide the finest dynamic performance. However, due to the arrangement of the commutator and brushes, it requires frequent maintenance and not appropriate for use in dirty environments [1]. Hence, commutator less motors like induction, reluctance and per-

manent magnet synchronous motors (PMSM) are used in EVs [2]. Among them, brushless DC (BLDC) motors (trapezoidal PMSM) have good torque speed characteristics as that of the DC motors. These are more reliable due to electronic commutation, higher efficiency, torque to inertia ratio and power density. Especially, BLDC motors also provide higher efficiency at medium and low speeds, because of which they are preferred over other motors in two and three wheeler EV applications [3]. Another benefit of BLDC motor is it's ability to execute effective regenerative braking, which raises energy efficiency [4]. Proper speed control technique enhances the performance, reduces the energy use and importantly ensures the safety. In literature, different speed control techniques such as PI, hysteresis current, direct torque and field oriented control are available for ac drives [5], [6]. PI and hysteresis current control (HCC) are the two commonly used control techniques for speed control of BLDC drive [7]. Comparative analysis between speed control techniques provides insight into the suitability of the techniques for the given application [8]. PWM assists to maintain the speed of BLDC motor, while the current drawn by motor is stabilized by two loop current control technique within the desired range. During motor control, large current ripples occur on the DC-link side because of switching. In case the input side is directly fed through the battery, it has to supply that large current ripple that degrades the battery life. Hence, a DC -DC converter is connected in between, to significantly reduce the battery current ripple and also DC-link voltage can be maintained constant [9]. Bidirectional converters (BDCs) are used for addressing the above issues and can also allow power flow in both directions [10].

In this paper, BLDC motor with single loop PI control, dual loop control (speed as outer and hysteresis current as inner loop) are presented for both motoring and regenerative braking modes. System configuration of BLDC motor drive is described in Section II. In next section, single loop PI control, HCC of BLDC drive and dual loop PI control of bidirectional converter are discussed in detail. Simulation results to confirm the viability of HCC over single loop PI control are presented

979-8-3315-3013-6/25 $31.00 © 2025 IEEE

in section IV. The conclusion is provided in Section V of this paper. Experimental verification with both the controllers can be done using C2000 microcontroller and the developed system can be used for two and three wheeler EV applications.

II. SYSTEM CONFIGURATION OF BLDC MOTOR DRIVE

The main components of the BLDC motor drive are battery fed voltage source inverter (VSI) as shown in Fig.1. If battery is directly connected, only DC-link capacitor supports to reduce battery ripple currents, but cannot eliminate ripples fully. However, if battery is connected through the bidirectional converter that supplies VSI fed BLDC motor drive as shown in Fig.2 minimizes the battery current ripple. The modeling and operation of BLDC motor along with bidirectional converter design are discussed subsequently.

A. BLDC Motor

BLDC motor is a trapezoidal PMAC motor, by sequentially energizing form phases, rotation is maintained. Information of rotor position is obtained by hall effect sensors or by measurement of coil emf, which helps to trigger the switches. Here, the conduction order of the inverter is controlled to achieve commutation. In BLDC motor, current pulses are given to the armature such that the stator and rotor fields are stationary and in quadrature with respect to each other [11]. Consequently, torque generated is directly proportional to armature current, and voltage induced is directly proportional to speed, just like in a DC motor. As a result, torque and speed can be controlled independently.

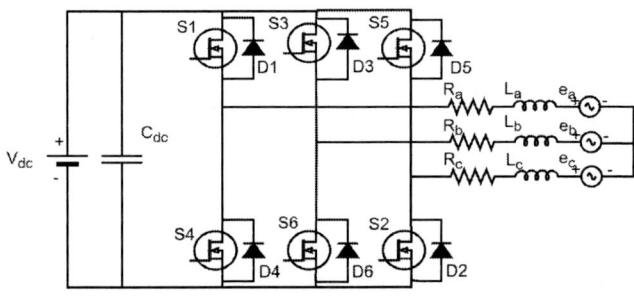

Fig. 1: VSI fed BLDC Motor

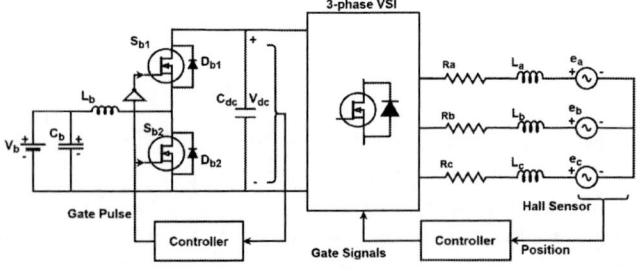

Fig. 2: Block diagram of BLDC Motor Drive with BDC

1) Modeling of BLDC Motor: Modeling of BLDC machine gives practical insight about the behavior the motor and better design of control [1].
KVL equation for the stator of a BLDC motor is

$$v_k = e_k + L_k \frac{di_k}{dt} + i_k R_k; \qquad (1)$$

Instantaneous emfs per phase are,

$$e_k(t) = K_e \phi(\theta - \frac{2\pi}{3} n) \omega(t); n = 1, 2, 3 \qquad (2)$$

where, $L_k = L_s - L_m$; L_s is the self inductance and L_m is the mutual inductance of armature, K_e is the back emf constant, θ =Rotor angle in electrical degrees, ω is the rotor speed in rad/sec and R_k, i_k, v_k are phase- k winding resistance, current and voltage respectively.
Electromagnetic torque (T_e) developed is,

$$T_e = \sum (e_k i_k)/\omega \qquad (3)$$

i.e.,

$$T_k(t) = K_t \phi(\theta - \frac{2\pi}{3} n) i_k(t); \qquad (4)$$

where, n =0,1,2 for k = a,b,c respectively.
Therefore,

$$T_e = \sum (T_k) \qquad (5)$$

And torque at the shaft is,

$$T_e(t) = T_L(t) + J \frac{d\omega(t)}{dt} + B\omega(t) \qquad (6)$$

where, T_L= Load torque (N-m), J = Inertia of the shaft (kgm^2), B = Friction constant (Nms^2/rad)

2) Motoring Mode: In motoring mode, current pulses are given to the stator winding in the region where the induced voltage is maximum [11]. As the airgap flux is constant, the induced voltage is proportional to the speed.

$$E = K\omega \qquad (7)$$

During each 60° interval, current enters one phase and comes out of another phase. Hence,

$$P = EI_{dc} + (-E)(-I_{dc}) = 2EI_{dc} \qquad (8)$$

Therefore developed torque is

$$T = \frac{P}{\omega} = 2KI_{dc} \qquad (9)$$

where, K is motor constant, E is the induced voltage per phase, I_{dc} is the dc link current, P is the power fed to the motor.

In motoring mode, the upper switches are operated in PWM mode whereas the lower part switches can be continuously ON or can be turned off in sequence with upper switches. Such that current can be made to follow the reference current I_{dc} within a band. The switching states for motoring and regenerative braking modes are given in Table I and Table II respectively. The corresponding theoretical back emf, phase current and switching pulses are shown in Fig. 3. To understand the

979-8-3315-3013-6/25 $31.00 © 2025 IEEE   476

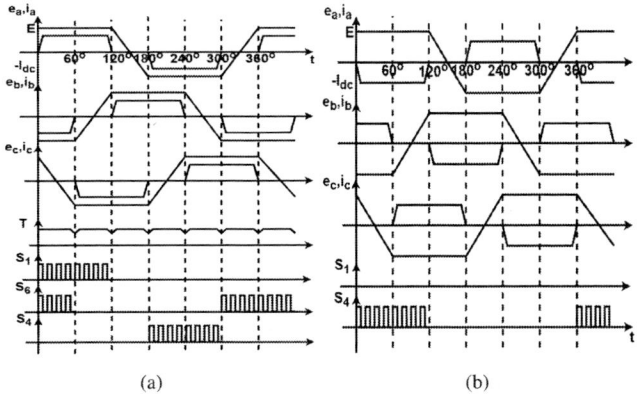

(a)   (b)

Fig. 3: Theoretical waveforms of (a) Motoring (b) Regenerative braking modes

operation in motoring mode of BLDC motor, for example in State I, as E_a is positive S_1 is turned on and as E_b is negative S_6 also turned on.

TABLE I: Commutation timing in acceleration mode

State	Sensor combination	S1	S2	S3	S4	S5	S6
I	101	1	0	0	0	0	1
II	100	1	1	0	0	0	0
III	110	0	1	1	0	0	0
IV	010	0	0	1	1	0	0
V	011	0	0	0	1	1	0
VI	001	0	0	0	0	1	1

3) Regenerative Braking Mode: Regenerative braking mode is achieved by the reversal of current in the motor during deceleration period (where it acts as a generator), the battery can be made to charge. The simple and efficient method is to independently switching the lower arm switches with PWM, keeping all the upper arm switches off. By controlling the switches (MOSFETs/IGBTs), it operates as boost circuit due to the presence of inductance in rotor windings. It also helps in energy recovery at low speeds, when the induced back-emf is very low [12]. For example, the state when phase-a emf is positive, S_4 is turned on and off at the switching frequency. When S_4 is on, current flows from a-phase to the phase where emf is negative (through its body diode), resulting in the negative current in the phase-a. During S_4 off, as current cannot change suddenly, body diode of S_1 conducts and current flows to the battery (Fig. 3(b)). The motor current and energy recovered during braking mode is derived as follows.

TABLE II: Commutation timing in regenerative braking mode

State	Sensor combination	S1	S2	S3	S4	S5	S6
I	101	0	0	0	1	0	0
II	100	0	0	0	1	0	0
III	110	0	0	0	0	0	1
IV	010	0	0	0	0	0	1
V	011	0	1	0	0	0	0
VI	001	0	1	0	0	0	0

By applying volt-sec balance principle,

$$\int_{t}^{t+T_s} v_L dt = dT_s[2E - i_a(2R)] +$$

$$(1-d)T_s[2E - i_a(2R) - V_{dc}] = 0$$

$$i_a = \frac{2E}{(1-d)^2 R_{dc} + 2R} \tag{10}$$

where, i_a is motor armature current R and R_{dc} are the resistance of the armature and the equivalent load resistance respectively, T_s is the switching period, dT_s is the on duration of switching period and $V_{dc} = R_{dc}I_{dc}$ and also,

$$I_a(1-d) = I_{dc} \tag{11}$$

From the ampere-sec balance,

$$\int_{t}^{t+T_s} i_c dt = dT_s\left(-\frac{V_{dc}}{R_{dc}}\right) + (1-d)T_s\left(i_a - \frac{V_{dc}}{R_{dc}}\right) \tag{12}$$

$$V_{dc} = \left(\frac{2E}{(1-d)}\right)\left(\frac{1}{1 + 2R/[R_{dc}(1-d)^2]}\right) \tag{13}$$

Energy that is restored (W_{regen}) in this mode [12],

$$W_{regen} = V_{dc} \times I_L \times T_{off} \tag{14}$$

$$W_{regen} = 2E \times I_L \times T_s \tag{15}$$

B. Bidirectional DC-DC Converter Design

In motoring mode it acts as boost converter, therefore power flow from battery to motor. In regenerative braking it acts as buck converter, as a result power flows from motor to battery. Critical inductance of bidirectional converter [13] ,

$$L_{bat,boost} = \frac{(V_{dc} - V_b)}{2P}\frac{V_b^2}{V_{dc}}T_s \tag{16}$$

$$L_{bat,buck} = \frac{(1-D)V_{dc}T_s}{2I_o} \tag{17}$$

$L_{bat}= \max(L_{bat,boost}, L_{bat,buck})$.

Battery side capacitance,

$$C_{bat} = \frac{\Delta I_{Lb}}{8\Delta V_{in}}T_s \tag{18}$$

DC link side capacitance,

$$C_{dc} = \frac{V_{dc}D}{R_{Load}\Delta V_{dc}}T_s \tag{19}$$

where, inductor ripple current,

$$\Delta I_{Lb} = \frac{(V_{dc} - V_b)}{2L_b}\frac{V_b}{V_{dc}}T_s, \tag{20}$$

T_s is the switching period, P is the maximum power rating, V_{dc} is the DC- link voltage, V_b is the battery voltage, ΔV_{dc} is the dc link ripple voltage and ΔV_{in} is the input ripple voltage. The filter parameters of BDC is designed using above equation considering $\Delta V_{dc} = 1\%$ of V_{dc}, ΔI_{Lb} as 10% of I_{Lb} and the corresponding parameters are given in Table III.

III. CONTROLLERS OF BLDC MOTOR DRIVE

This section describes two types of BLDC motor speed controllers together with the bidirectional converter control. The single loop PI and hysteresis current controllers (HCC) block diagrams for both motoring and braking modes are depicted in Fig. 4 and Fig. 5 respectively.

A. Single loop PI controller

In motoring mode, the speed controller with single PI loop, actual speed is taken as feedback and compared with the reference speed producing an error signal. The output of the PI controller generates a signal proportional to torque or current, which is used to produce PWM pulses to drive the VSI. The primary drawbacks of single loop control are unrestricted currents during the startup and acceleration phases. Additionally, the output current has large ripples as a result of the speed loop's longer response time, which directly affects the torque. In regenerative braking, current control loop is used instead of speed loop and the rated motor reference current is compared with actual current and the PI controller output has saturation values of 0 and 0.95. Therefore, the output is multiplied by the brake strength for controlled braking.

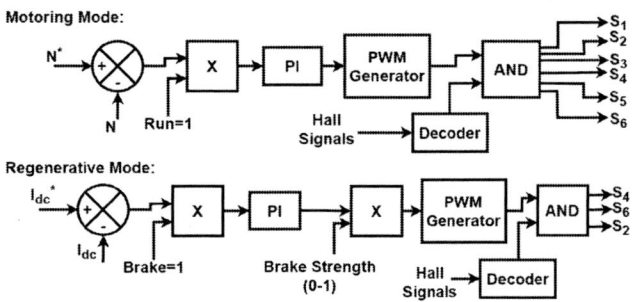

Fig. 4: Block diagram of PI Controller for BLDC motor

B. Hysteresis Current Controller

The hysteresis current control technique having dual loops, outer speed and inner loop current loop using hysteresis band as shown in Fig. 5. The outer PI controller produces reference current when the speed error signal is passed through it, which is given to saturation block to limit the maximum reference current to rated value. For positive speed error (i.e. during acceleration), it sets reference current to the rated current, which is decoded to get the reference phase currents and are compared with the actual stator currents, which are fed to a pre-determined hysteresis bands. The switching of VSI are done accordingly, to put actual currents in the band of decoded reference currents. Principle of operation of HCC is shown in Fig. 6. During acceleration and braking times, for smooth operation, currents are limited by multiplying the range of (0-1) with PI output reference current which gives actual reference currents.

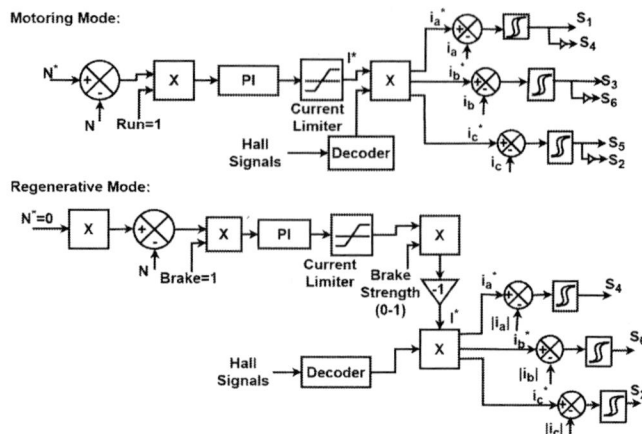

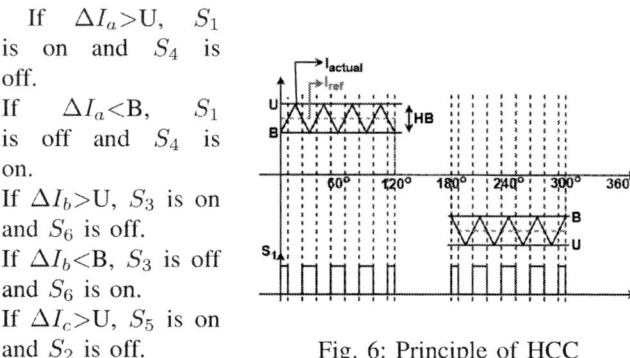

Fig. 5: Block diagram of current controller for BLDC motor

If $\Delta I_a > U$, S_1 is on and S_4 is off.
If $\Delta I_a < B$, S_1 is off and S_4 is on.
If $\Delta I_b > U$, S_3 is on and S_6 is off.
If $\Delta I_b < B$, S_3 is off and S_6 is on.
If $\Delta I_c > U$, S_5 is on and S_2 is off.
If $\Delta I_c < B$, S_5 is off and S_2 is on.
where, $\Delta I_a = I_{a,ref} - I_{a,actual}$ and U,B are the upper and lower band limits.

Fig. 6: Principle of HCC

C. Bidirectional converter control

A bidirectional converter is interfaced between battery and VSI to reduce the battery current ripple as well as to maintain the DC- link voltage constant, which is taken as reference. The block diagram of dual loop PI controller for the BDC is shown in Fig. 7. When, the DC-link voltage(V_{dc}) falls below the reference value, current flows from battery to motor (i.e., in motoring mode), the boost operation is enabled. Similarly, (V_{dc}) above reference voltage, then current is fed to the battery (i.e. braking mode), then buck mode is enabled.

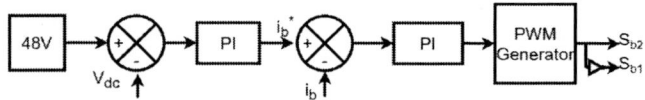

Fig. 7: Schematic diagram of Bidirectional Converter

IV. RESULTS AND DISCUSSIONS

The BLDC motor drive system without BDC and with BDC is simulated in MATLAB/SIMULINK using the system parameters given in Table III and corresponding results are delineated in this section.

979-8-3315-3013-6/25 $31.00 © 2025 IEEE

TABLE III: System Parameters

Battery
12 V, 42 Ah, 2 Nos.
Bi-directional Converter
L_b = 0.3 mH, C_b = 300 μF, C_{dc} =5200 μF
Switching frequency (f_{sw}) = 10 kHz
BLDC Motor
V_{rated} = 48 V, I_{rated}= 25 A, P_o= 1 kW, N_{rated}= 3000 rpm
No.of pole pairs= 1, T_{rated}= 3.18 N-m
R_s=0.1 Ω, L_s= 0.5 mH, ψ= 0.709 Wb, J = 0.05 $kg.m^2$

1) BLDC performance without bidirectional converter:
When battery is connected directly to VSI, it has to supply large ripple currents, as the switches are continuously turned on and off at switching frequency to control the speed. Therefore, a large capacitance should be inserted between battery and VSI to eliminate the ripples. However, this capacitor can reduce only some part of the ripple content as shown in Fig. 8.

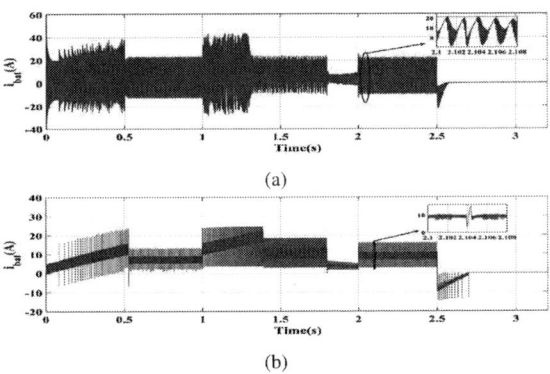

Fig. 8: Battery currents without bidirectional converter with a) single loop PI control b) hysteresis current control

To reduce the battery current ripple and to keep the DC-link voltage constant at 48 V, the BLDC drive with BDC is simulated with both the controllers at constant load torque of 2 N-m and operated under acceleration, deceleration and braking conditions. The results of PI and HCC under same operating conditions are shown in Fig. 9 and Fig. 10 respectively. At first, speed input is gradually increased to 1500 rpm at 0.5 s and operated at same steady-state speed till 1 s as shown in Fig. 9a and 10a respectively. In acceleration mode, the motor input current increases, consequently the electromagnetic torque increases, and at steady-state, $T_e = T_L$ as shown in Fig. 9c, d and 10c, d. From 1 s to 1.3 s motor is again accelerated to 2500 rpm and run at steady-state till 1.8 s, during this high speed condition, as back emf is high, there is opposition to the flow of current, resulting in high current and torque ripples. From 1.8 s to 2 s, motor reference speed is decreased to 2000 rpm, current drawn is less than steady-state motor current. At 2.5 s brake is applied, motor starts decelerating during this period as T_e produced is opposite to the direction of motion and power is fed back to battery as shown in Fig. 9e and 10e. The battery current waveform is shown in Fig. 9f and 10f respectively.

2) Single loop PI Control: The motor current ripples are significant during all operating conditions. Current during acceleration is around 30 A which is more than the rated current as shown in Fig. 9c and 10c. In braking operation, current as well as torque ripple is more and negative battery current confirms the battery charging.

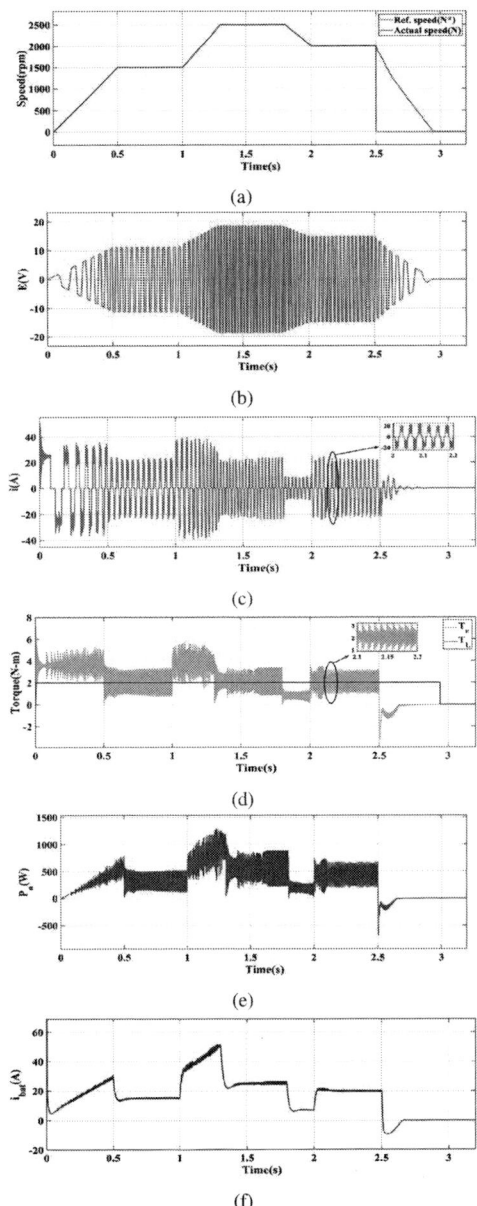

Fig. 9: Waveforms with single loop PI controller: a) speed b) back emf c) phase current d) electromagnetic torque e) output power f) battery current

3) Dual loop Hysteresis current control: The performance of BLDC drive with HCC is shown in Fig. 10. Here, the current saturation limits are set between +25 A to -25 A and hysteresis band of 0.1% is used, which makes actual current to have only 0.1% error to that of reference current.

During acceleration, current is allowed to the rated value of 25 A which results in quick response with less current and torque ripple. During braking, smooth response with reduced current ripple and constant negative electromagnetic torque is produced.

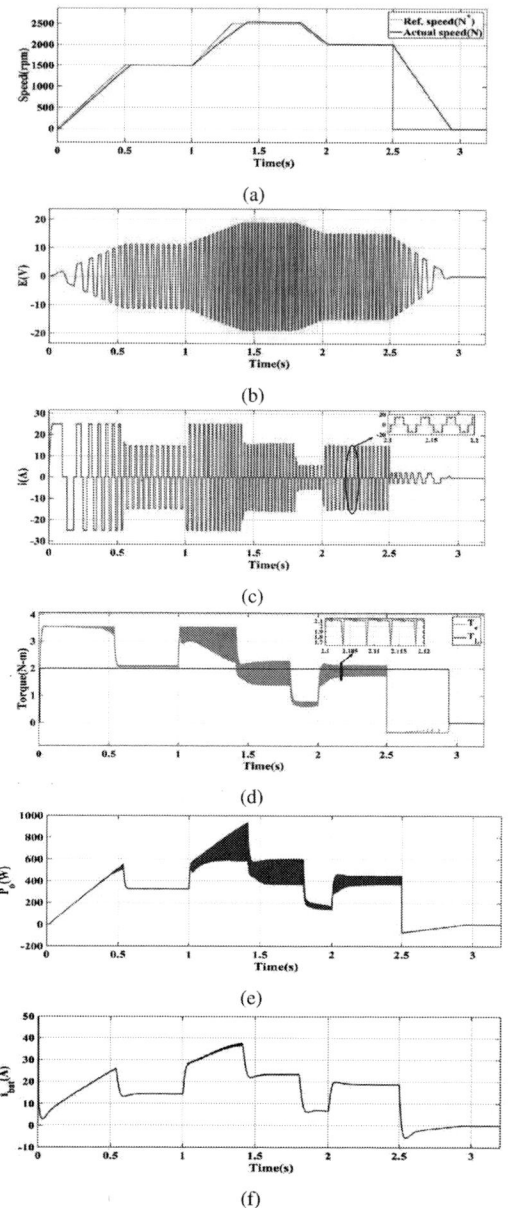

Fig. 10: Waveforms with dual loop HCC: a) speed b) back emf c) phase current d) electromagnetic torque e) output power f) battery current

V. Conclusion

In this paper, the performance of BLDC motor drive with two different controllers are compared. The DC bus of VSI fed BLDC drive is controlled by a bidirectional DC-DC converter using dual-loop PI controller to reduce the battery ripple current. Here, a single loop PI and two loop with hysteresis current control as inner and outer speed loop are simulated for variable speed and constant load torque conditions. HCC BLDC drive has inherent current protection, whereas in single loop PI control, current cannot be limited. Hence, in HCC, motor accelerates with the maximum allowable current resulting in satisfactory transient as well as smooth steady-state operation because of low torque ripples. However, high torque ripples are observed in single loop PI controller because of slow response of speed loop. Therefore, HCC based BLDC drive can be implemented in EV applications with improved performance.

Acknowledgment

Authors acknowledge IHub-Data, IIIT Hyderabad ("IIIT-H/IHub/Project/Mobility/2024-25/M2-022") for financial assistance.

References

[1] N. B. Bahari, A. bin Jidin, A. R. bin Abdullah, M. N. bin Othman, and M. bin Manap, "Modeling and simulation of torque hysteresis controller for brushless dc motor drives," in *2012 IEEE Symposium on Industrial Electronics and Applications*. IEEE, 2012, pp. 152–155.

[2] M. Yildirim, M. Polat, and H. Kürüm, "A survey on comparison of electric motor types and drives used for electric vehicles," in *2014 16th International Power Electronics and Motion Control Conference and Exposition*. IEEE, 2014, pp. 218–223.

[3] D. Mohanraj, R. Aruldavid, R. Verma, K. Sathiyasekar, A. B. Barnawi, B. Chokkalingam, and L. Mihet-Popa, "A review of bldc motor: State of art, advanced control techniques, and applications," *IEEE Access*, vol. 10, pp. 54 833–54 869, 2022.

[4] X. Nian, F. Peng, and H. Zhang, "Regenerative braking system of electric vehicle driven by brushless dc motor," *IEEE Transactions on Industrial Electronics*, vol. 61, no. 10, pp. 5798–5808, 2014.

[5] V. K. Awaar, R. Simhadri, and P. Jugge, "Comparative study and experimentation of speed control methods of bldc motor using drv8312," in *2022 IEEE 2nd International Conference on Sustainable Energy and Future Electric Transportation (SeFeT)*. IEEE, 2022, pp. 1–6.

[6] U. Neethu and V. Jisha, "Speed control of brushless dc motor: A comparative study," in *2012 IEEE international conference on power electronics, drives and energy systems (PEDES)*. IEEE, 2012, pp. 1–5.

[7] G. Gupta and M. Sreejeth, "Comparative analysis of speed control of bldc motor using pwm and current control techniques," in *2022 IEEE IAS Global Conference on Emerging Technologies (GlobConET)*. IEEE, 2022, pp. 610–614.

[8] A. K. Majhee, S. K. Vishwakarma, S. K. Sharma, P. Rai, and P. Kumar, "Performance analysis and simulation of brushless dc motor using pi with hysteresis current controller," in *2022 IEEE 2nd International Symposium on Sustainable Energy, Signal Processing and Cyber Security (iSSSC)*. IEEE, 2022, pp. 1–6.

[9] K. Bharath, H. Choutapalli, and P. Kanakasabapathy, "Control of bidirectional dc-dc converter in renewable based dc microgrid with improved voltage stability," *International Journal of Renewable Energy Research (IJRER)*, vol. 8, no. 2, pp. 871–877, 2018.

[10] P. K. Behera, K. Gupta, and M. Pattnaik, "Hybrid energy storage unit fed motoring and regenerative braking control of electric vehicle drivetrain," *Journal of Power Sources*, vol. 626, p. 235761, 2025.

[11] G. K. Dubey, *Fundamentals of electrical drives*. Alpha Science Int'l Ltd., 2001.

[12] R. N. Hasanah, V. Andrean, H. Suyono, R. A. Setyawan *et al.*, "Bidirectional vsi as a regenerative-braking converter for bldc motor—an analysis on a plug-in electric vehicle application," in *2017 10th International Conference on Electrical and Electronics Engineering (ELECO)*. IEEE, 2017, pp. 222–226.

[13] K. Suresh and R. Arulmozhiyal, "Design and implementation of bidirectional dc-dc converter for wind energy system," *Circuits and Systems*, vol. 7, no. 11, pp. 3705–3722, 2016.

979-8-3315-3013-6/25 $31.00 © 2025 IEEE

Performance Characteristics of Low Vibrating Bridge-Configured Squirrel Cage Induction Motor

Rakesh Deore
Department of Mechanical
Engineering,
Indian Institute of Technology
Guwahati, India
rdeore@iitg.ac.in

Bipul Brahma
Department of Mechanical
Engineering,
Indian Institute of Technology
Guwahati, India
bipul.brahma@iitg.ac.in

Karuna Kalita
Department of Mechanical
Engineering,
Indian Institute of Technology
Guwahati, India
karuna.kalita@iitg.ac.in

Abstract—The bearing-less machines have been an area of interest for researchers in the past few decades. These electrical machines support the motor's rotor by generating electromagnetic forces using different winding configurations. One is the bridge-configured winding (BCW) configuration, which can be a cost-effective solution over other winding configurations. It can also develop the electromagnetic force used for position control or levitation. Hence, the transverse force-producing capability of the induction motor and the essential torque-producing capability have been investigated numerically in this paper. The complete design of the Y-connected, 50 HP, four-pole induction motor (IM) is also examined. The fundamental FEM analysis of the IM with and without the actuator effect has been developed for better understanding. The performance characteristics of the IM, such as the magnetizing currents, efficiency, and power factor, have also been investigated numerically. The no-load currents and the other parameters are studied numerically as well as experimentally, and the transverse force capability of the IM under no-load conditions and rated slip has been established due to the bridge supply of DC/AC currents for levitation purposes. The numerical study also suggests that the performance characteristics of the machine, such as rated torque and efficiency, will not be affected due to the additional injection of current for levitation purposes. These types of IM motors can be used in various applications, such as condition monitoring of mechanical faults, such as eccentricity, and to generate the forces for various mechanical means.

Keywords— Bearing-less Motor, UMP, Transverse Force, FEM, BCW, Induction Machines (IMs), Bridge Supply, etc.

I. INTRODUCTION

In the past few years, electrical machines have not only been used for industrial purposes but also in the electrical transportation system. These machines are mainly designed to transmit power and torque to the system. Since these machines operate at high speeds, they produce different types of forces on the system. These forces are mainly due to the unbalance or assembly inaccuracy in the motor. Many other reasons are available for the same. It causes unwanted vibrations in the machine, which can again cause noise and wear and tear in the system, and like a cycle, it continues [1]. The faults cause an unbalanced force on the stator rotor, even for the small air gap distortion between the rotor and the stator. This UMP forces the rotor to move towards the minimum air gap. The locking or any catastrophic failure can be seen in the machine due to these types of faults in the system.

The majority of rotating machines are only designed to produce the torque that is necessary to drive the machine. However, it does not resolve issues like unbalanced force production in the system. To avoid wear and tear, noise, and vibrations in the system, the machine should be capable of producing both torque and transverse force in electrical machines. This combined feature of the electrical machines

can be seen in the bearing-less electrical machines. These machines, if fully controlled, can address the issues associated with the electrical machines due to the inclusion of the different types of faults, as discussed earlier.

A. Bearing-less Induction Machines

In bearing-less machines, the radial forces produced due to consecutive harmonics and the pole pairs are utilized by developing the spatial flux density harmonics in the machine [1]. The bearing-less machine is equipped with a four-pole winding to produce magnetic force by supplying controlling currents and a 2-pole primary torque-producing machine. The unipolar flux can control rotor vibrations for a two-pole IM [2, 3]. The active rotor vibration control is also possible by using this winding scheme, but two separate winding schemes must be implemented in the stator of the induction machine [4]. This also requires the additional space to incorporate the force-producing 4-pole winding along with the 2-pole primary torque-producing winding. This results in more copper losses and lowers the efficiency of the machines. In two-pole machines, the four poles cause extra electromagnetic force on the rotor of the IM due to eccentricity. The unipolar flux also contributes to this phenomenon, but with a minimal effect. The main idea is to reduce the effect of these extra unbalanced electromagnetic forces by using the self-bearing ability of the machines with proper active position control [5].

The idea of self-bearing machines is not just limited to active control of the rotor vibrations and UMP control. Still, this concept can also be extended to replace mechanical bearings with its integral feature of position control. UMP forces the rotor to move towards the minimum air gap. Many researchers have been trying to develop bearing-less machines over the past few years. A winding scheme that produces the torque as well as the force by injecting the currents into the bridge points by introducing a two-pole flux [6, 7]. However, the torque productivity of the machine is the main concern in this design. This paper investigates the machine's torque production capability and levitative force capability. The basic design of the Bridge Configured Winding (BCW) scheme is explained in Section II.

II. BRIDGE CONFIGURED WINDING

The Bridge-Configured Winding (BCW) is a unified stator winding system that integrates torque production and transverse levitation forces, eliminating the need for separate windings [1–5]. By employing active control, BCW enables self-bearing functionality, as demonstrated in four-pole permanent magnet synchronous motors (PMSMs) through FEM simulations and practical implementations [6,8]. Compared to dual-winding systems with bridge-point current injection, BCW reduces power losses while improving torque output [8–10]. This is achieved by injecting currents at strategic bridge points within a Wheatstone bridge-like winding arrangement, fully utilizing available turns without

979-8-3315-3013-6/25 $31.00 © 2025 IEEE

increasing copper losses relative to conventional designs [5]. Though initially tested on PMSMs, BCW is adaptable to induction machines (IMs), slot-less motors, synchronous machines, and switched reluctance motors due to its stator-based configuration [5].

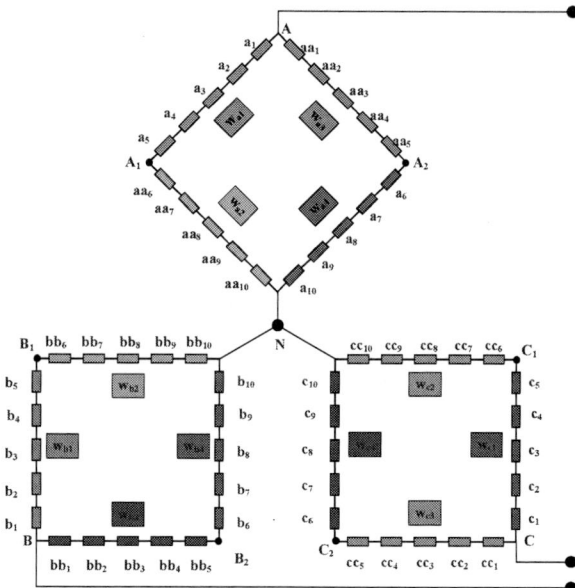

Fig. 1.   Y-connection parallel type of winding

Fig. 1 illustrates a BCW design for a 50 HP, 3-phase, 4-pole IM, where torque-producing currents are split into parallel paths per phase. Levitative forces are generated by injecting control currents through bridge points A_1, A_2, B_1, B_2, C_1, and C_2 via external bearing inverters, enabling dual torque-bearing operation [5]. Beyond self-bearing applications, BCW reduces mechanical bearing loads, enhances power density, suppresses vibrations, and supports condition monitoring [10].

In conventional IMs, rotor eccentricity induces $p\pm1$ pole pairs (p is the fundamental pole pairs), distorting airgap flux harmonics and causing destabilizing vibrations [1,11]. BCW counters this by injecting compensating currents that neutralize eccentricity-induced flux variations [12]. Experimental trials under bridge-enabled conditions show BCW reduces unbalanced magnetic pull in eccentric operations [12]. The scheme also generates $p\pm p_b$ pole pairs (p_b = bridge current-induced poles), enabling simultaneous torque and force generation. This flux modulation capability and its simplified architecture and fault resilience position BCW as a transformative solution for high-precision industrial, aerospace, and other commercial applications [5,10,12].

This study focuses on designing and analyzing a low-vibration, 50 HP, three-phase, Y-connected, four-pole squirrel cage IM employing a BCW scheme in its stator. The primary objective is to evaluate the motor's dual functionality: conventional torque generation and transverse electromagnetic force production for rotor levitation and position control. A FEM-based numerical model is developed to simulate the motor's electromagnetic behavior under both standard operation and actuator-influenced conditions. Key performance metrics, including magnetizing currents,

efficiency, power factor, and no-load current characteristics, are analysed numerically, and the magnetizing current profiles are validated with the experimental investigations under no-load conditions. Additionally, the transverse force capacity of the motor is examined under no-load and rated slip conditions, utilizing DC/AC bridge-supplied currents for levitation in numerical analysis. In contrast, the vibration reduction can be seen in the experimental analysis due to the misalignment type of the faults.

III. ANALYTICAL FORMULATION FOR THE TRANSVERSE FORCE IN BCW CONFIGURATION

As discussed above, the two-pole flux can be produced along with the main four-pole flux by supplying the bridge currents through the bridge points. So, the MMF produced due to the primary winding supply and bridge supply is given as follows;

$$B_p(\theta,t) = \text{Re}\{(B_p)_{\max} e^{j(\omega t - p\theta)}\} \quad (1)$$

$$B_b(\theta,t) = \text{Re}\{(B_b)_{\max} e^{j(\omega_b t \pm p_b\theta)}\} \quad (2)$$

Where p is the fundamental pole pair, and p_b is the pole pair due to bridge currents. ω_p and ω_b are the primary supply frequency (to produce torque) and bridge supply frequency (to produce force), respectively. $(B_p)_{max}$ and $(B_b)_{max}$ are amplitudes of the fundamental MMFs due to the primary and bridge supply, respectively.

Thus, the electromagnetic force produced due to the change in airgap flux by injecting the bridge currents can be calculated using Maxwell's stress tensor as given in Eq. 3 [13].

$$\left(F_{b\_x}\right)_{bridge} = \int_0^{2\pi} \{B_{tot}^2(\theta,t)\}\left(\frac{RL}{2\mu_0}\right)\cos\theta\,d\theta$$
$$\left(F_{b\_y}\right)_{bridge} = \int_0^{2\pi} \{B_{tot}^2(\theta,t)\}\left(\frac{RL}{2\mu_0}\right)\sin\theta\,d\theta \quad (3)$$

On calculating and rearranging, the bridge forces in both x and y directions in a complex coordinate system are given by

$$F_c(t) = \left(\frac{\pi RL B_p B_b}{2\mu_0}\right)\{e^{j(\omega_p \pm \omega_b)+(\phi_p \pm \phi_b)}\} \quad (4)$$

So, Eq. 4 shows that the levitative force can be generated due to interaction between the main and bridge supply currents. It also depends on the frequency and phase of the main and bridge supply. However, it can also be seen that if the DC is supplied through the bridge points, the force produced in both directions depends only on the main supply frequency. The AMB effect (by supplying a DC through the bridge points) can be generated using this combination, and a self-bearing machine can be developed for the levitation applications. The FEM model for the same machine is developed, and the effect of the bridge supply is analyzed numerically.

979-8-3315-3013-6/25 $31.00 © 2025 IEEE   482

IV. FEM MODLE OF THE BCW IM

A two-dimensional numerical model of the induction motor (IM) is developed using COMSOL™ Multiphysics, with simulation parameters detailed in Table 1. The electromagnetic behavior of the motor is governed by Maxwell's equations, solved across the 2D domain. The magnetic vector potential is computed via diffusion equations, which integrate contributions from stator and rotor magnetic fields. To simplify the computational complexity inherent to 2D modeling, the analysis assumes a focus on the out-of-plane magnetic vector potential component, thereby neglecting in-plane currents and out-of-plane magnetic fields. This assumption enhances numerical stability while retaining sufficient accuracy for performance evaluation.

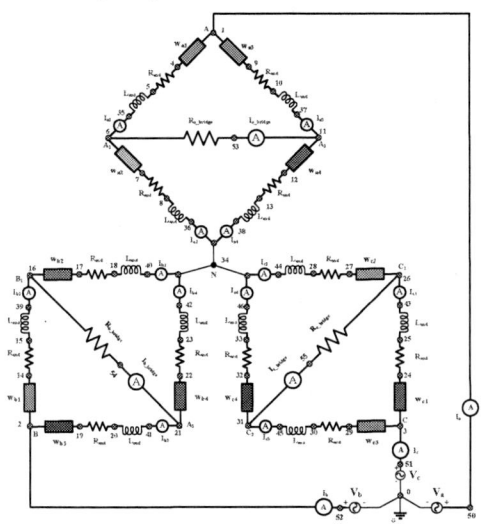

Fig. 2. Circuit diagram for a parallel type of winding arrangement with a bridge

TABLE I. PARAMETERS OF INDUCTION MOTOR IN SIMULATION

Sr. No	Parameters	Values
1	Number of poles	4
2	Number of stator slots	60
3	Number of rotor bars	48
4	Supply frequency (Hz)	50
5	Rated slip	0.02
6	Airgap (mm)	1
7	Number of turns per phase	55
8	Radius of the rotor (mm)	218
9	Length of the rotor (mm)	212

The stator winding configuration is designed to accommodate three operational modes: standard parallel winding (Bridge OFF), Bridge ON, and bridge supply (1 A DC/AC injection), as illustrated in Fig. 1. An equivalent electrical circuit model (Fig. 2) is integrated into the COMSOL environment to simulate these modes. Simulations are conducted under rated operating conditions (415 V, rated slip, and frequency) to assess key performance metrics. Torque characteristics are systematically compared across the three bridge configurations (OFF, ON, and supply). This approach thoroughly evaluates the motor's electromagnetic response under conventional and levitation-enabling operational states.

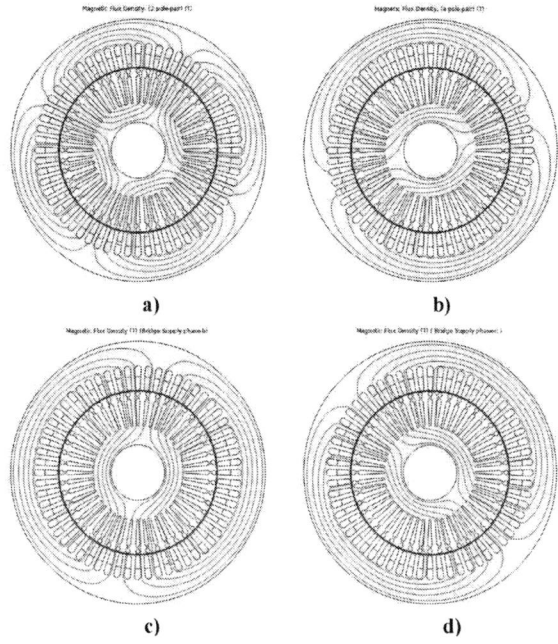

Fig. 3. a) 4-pole due to main supply, b), c), and d) are 2-poles due to current injected through the bridge arm of phase-a, phase-b, and phase-c, respectively

The main supply currents are responsible for the fundamental pole pairs, so the main supply currents are made zero to get pole pair information about the pole pair when bridge currents are supplied from the bridge arms, because the current supplied for levitation is responsible for minimal changes in airgap flux compared to the main pole forming flux. The main supply 4-pole is formed in the machine as shown in Fig. 3a); however, 2-pole fluxes due to the individual supply to each phase are shown in Fig. 3. b), only supply to phase-a, c) only supply to phase-b, and, d) only supply to phase-c.

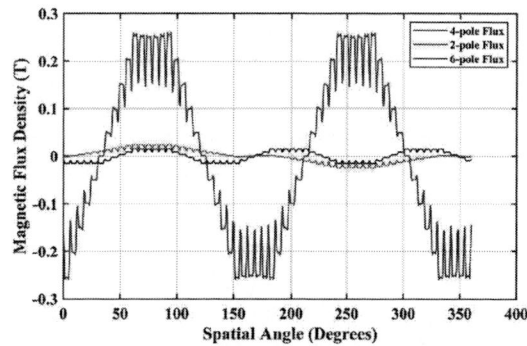

Fig. 4. Pole formation due to the main supply and bridge supply currents

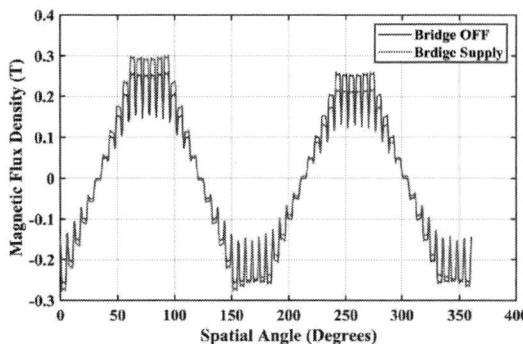

Fig. 5. Comparison of the circumferential magnetic flux in the airgap in the bridge OFF and Bridge Supply

The IM is simulated with the given circuit for the transverse force production, and it has been found that the four-pole flux and the two-pole flux have been generated in the machine due to the main supply and the bridge supply. The magnetic flux in the airgap due to the main supply (4-pole) and bridge supply (2-pole and 6-pole) is depicted in Fig. 4. The magnetic fluxes produced due to the levitative currents are less than the main currents flowing. So, it doesn't change the airgap flux and other performance characteristics significantly, but adds the advantage that with the same winding scheme, we can generate high flux density in the airgap as depicted in Fig. 5, creating a slight variation in the fundamental magnetic flux. The direction of the generated electromagnetic forces is given in Fig. 3, b), c), and d) by supplying the bridge currents between the parallel branches with different phase angles. The direction of the force generated solely depends on the bridge supply's phase angle and the winding scheme's spatial distribution. The frequency of the force at which it is generated depends on the bridge and fundamental supply frequencies.

V. CALCULATIONS FOR PERFORMANCE CHARACTERISTICS

The model is simulated with the fundamental parameters of the machine under no load condition and the rated slip condition (2 %) at a rated voltage of 420 V and rated supply frequency (50 Hz). The performance characteristics are evaluated by using the following formulas in COMSOL™ Multiphysics. The magnetizing currents are assessed by using the FEM formulation specifically for the IM form stator and rotor circuit field equations. The experimental current was measured and compared with the numerically simulated model in COMSOL™ Multiphysics under no-load conditions. A balanced voltage supply of v_a, v_b & v_c having main supply currents i_a, i_b & i_c. Thus, by using the direct-quadrature coordinate transformation, the apparent power (p_a) will be given as:

$$p_a = \overline{v}\overline{i}^* = (v_d + jv_q)(i_d - ji_q) \tag{5}$$

$$i_d = \frac{2}{3}(i_a - 0.5i_b - 0.5i_c)$$
$$i_q = \frac{1}{\sqrt{3}}(i_b - i_c) \tag{6}$$

The power factor and the efficiency calculations are given by Eq. 7 and Eq. 8,

$$\cos\phi = \frac{v_d i_d + v_q i_q}{\sqrt{(v_d^2 + v_q^2)(i_d^2 + i_q^2)}} \tag{7}$$

$$\eta = \frac{\pi n V I T}{10}\frac{v_d i_d + v_q i_q}{\sqrt{(v_d^2 + v_q^2)(i_d^2 + i_q^2)}} \tag{8}$$

VI. EXPERIMENTAL SETUP

The experimental system was configured using the parameters specified in Table I, with modifications applied to the original winding design outlined in Section II. In contrast, Fig. 6 illustrates the setup, which incorporates search coils embedded within the stator slots to measure magnetic field characteristics associated with the fundamental pole pair and those induced by bridge currents. Displacement sensors were installed to monitor rotor vibration dynamics. LTS 15-NP and LEM LV 20-P transducers recorded current and voltage signals. Data acquisition was managed via LabVIEW 14 software interfaced with an NI PXI 1050 chassis equipped with an NI PXI 8109 processor and an NI PXI 6221 controller. To ensure precise temporal resolution, 2000 samples per second were collected, enabling simultaneous analysis of rotor displacement and magnetizing current behavior.

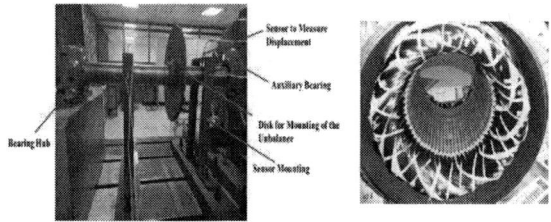

Fig. 6. Experimental setup with Bridge Configured Winding

VII. RESULTS AND DISCUSSIONS

A. Performance Characteristics

The experimental and numerical analyses under no-load conditions demonstrate a good agreement with the current profiles from both methods, exhibiting minimal discrepancy. As shown in Fig. 7.

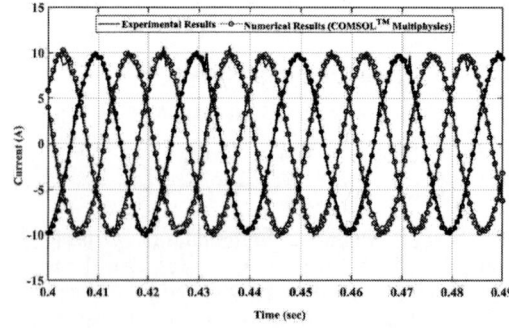

Fig. 7. Comparison of the main supply currents in numerical and experimental analysis

979-8-3315-3013-6/25 $31.00 © 2025 IEEE

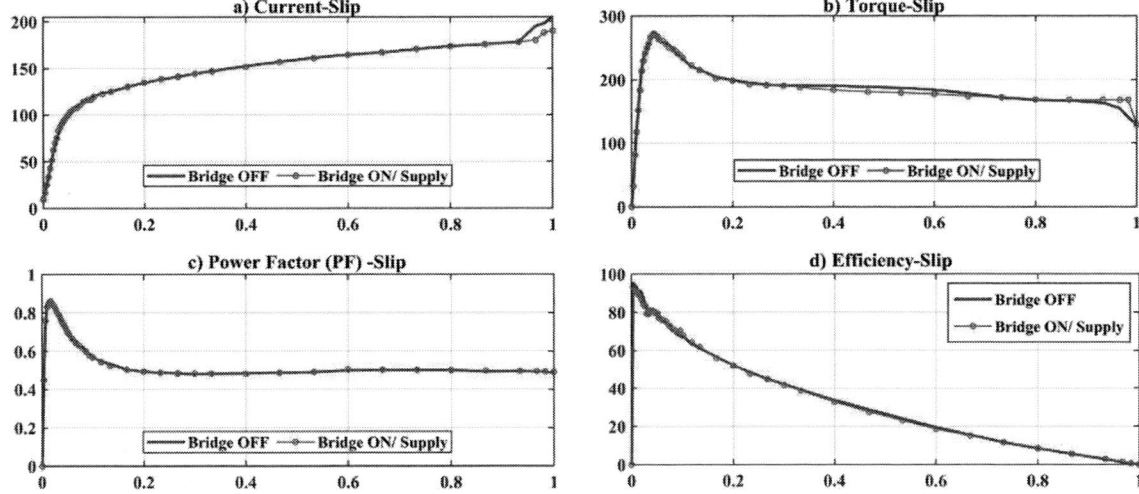

Fig. 8. Comparison of performance characteristics a) Current-slip, b) Torque-slip, c) Power Factor (PF)-slip, and d)Efficiency-slip curves for Bridge OFF (Parallel winding) and Bridge ON/ supply

The close agreement of currents between the numerical and experimental results validated subsequent simulations under no-load running conditions. The study further compared key performance metrics, including magnetizing currents, torque, efficiency, and power factor, between conventional parallel winding configurations and bridge-connected winding (BCW) supply. Table II systematically compares the designed experimental and numerical parameters, highlighting their consistency. Fig. 8 illustrates the torque characteristics at the rated operational regime (1470 rpm, 50 Hz, 415 V), contrasting conventional parallel winding performance with BCW operation. These analyses underscore the efficacy of the numerical model in replicating real-world behavior and provide critical insights into the electromechanical dynamics of the IM. The results emphasize the model's reliability for predictive motor performance and efficiency assessments.

TABLE II.   COMPARISON OF BCW INDUCTION MOTOR PARAMETERS

Parameters	Value in Numerical Investigation	Experimental Value	Error (%)
Rated Slip	0.02	0.02	Input
Rated Voltage	415 V	415 V	Input
Rated Current	63 A	67 A	5.97
Power Factor $\cos(\phi)$	0.86	0.83	-3.68
Efficiency (η)	0.87	0.9	3.33
Rated Power	37.5kW	37 kW	-1.35
Torque	215 N.m	238.89 N.m	10.041

The analysis reveals negligible variation in motor performance between configurations with and without bridge supply, even under 1 A DC and AC excitation. This consistency suggests that the BCW configuration does not compromise core operational metrics such as torque generation, efficiency, or power factor. Consequently, the

motor retains its capability to function as a unified electromechanical system for torque transmission while simultaneously enabling active magnetic bearing (AMB) applications. The dual functionality, leveraging the same winding structure for both conventional torque transfer and levitative force generation, demonstrates the system's versatility. These findings highlight the potential to integrate AMB capabilities without requiring additional hardware or significant design modifications, thereby offering a cost-effective solution for advanced industrial applications requiring precision motion control or contactless levitation.

B. Bridge Supply

The variable bridge currents from 1 A to 4 A with an AC supply at 50 Hz and a DC supply have been injected between the bridge points of all three phases. The electromagnetic force is produced due to the interaction of the fundamental flux and the flux produced by the supply of the levitative currents. The static, dynamic force is generated by using the supply at 50 Hz, which can be used as the rotor position control along with the levitation. An effect similar to the AMB has been seen in the DC supply currents injection, which generates an equal amount of horizontal and vertical forces in both directions, as illustrated in Fig. 9 a) and b).

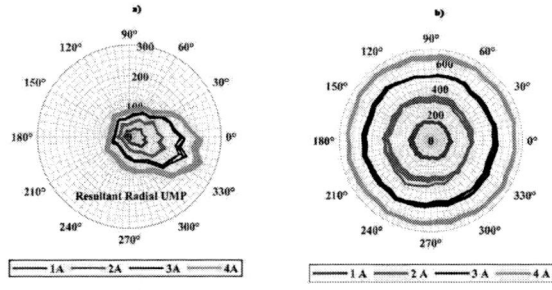

Fig. 9. The orbit plot of the electromagnetic force for variable bridge currents: a) AC supply at 50 Hz, b) DC supply

C. Experimental Vibration Control

The shaft of the IM is resting on the touchdown bearing, so the play between the bearing and shaft creates eccentricity and hence also induces vibrations in the system. To control the rotor vibrations and validate the theory with the experiments, the bridge points were short-circuited, so the inherent current due to imbalance tries to flow between the bridge points. However, the levitation current flows at supply frequency due to statistical imbalance in the system, and the rotor vibrations get controlled [11] as depicted in Fig. 10.

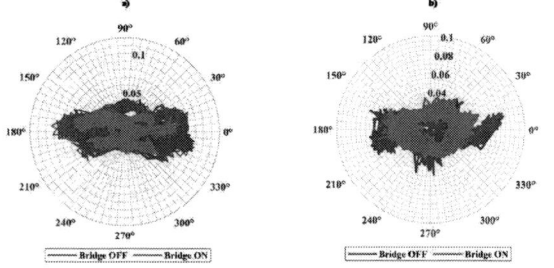

Fig. 10. The orbit response of the eccentric rotor in Bridge ON and Bridge OFF conditions at a) 40 Hz and b) 50 Hz of AC bridge supply

VIII. CONCLUSIONS

The proposed bridge-connected winding (BCW) configuration, detailed in Section II, was modeled in COMSOL™ Multiphysics to analyze its capacity for generating levitative forces through bridge arm currents. The magnitude and direction of these forces depend on the phase and frequency of the bridge supply currents. Initial validation under no-load conditions demonstrated strong agreement between experimental and numerical current profiles, with a minimal error of 3.89%, reinforcing the model's reliability. Subsequent numerical evaluation of performance characteristics, including torque, efficiency, and power factor, revealed no degradation in the BCW configuration compared to conventional parallel windings (Table II), ensuring retained torque capacity for primary electromechanical functions. This principle was experimentally extended to mitigate rotor vibrations by leveraging inherent current imbalances in the bridge arms, effectively suppressing mechanical oscillations without compromising core performance metrics. These findings underscore the BCW's dual functionality as a torque-transmission unit and an active vibration control system, eliminating the need for external damping mechanisms while maintaining operational integrity. The results align with prior discussions on flux distribution, solidifying the BCW's viability for industrial applications requiring integrated levitation or precision motion control.

ACKNOWLEDGMENT

All Authors are thankful to the government of India and the Indian Institute of Science, Bangalore, for providing the COMSOL Multiphysics software through the ISTEM Portal.

REFERENCES

[1] Sinervo, A. and Arkkio, A., 2012. Including slot harmonics in the mechanical model of a two-pole induction machine with a force actuator. *Mechanical systems and signal processing*, *32*, pp.282-291.

[2] Sinervo, A., Jokela, T. and Arkkio, A., 2012. Controlling rotor vibrations of a two-pole induction machine with the unipolar actuator. *IEEE transactions on magnetics*, *48*(7), pp.2205-2210.

[3] Sinervo, A., Jokela, T. and Arkkio, A., 2012, September. Unipolar flux in bearingless two-pole machine. In *2012 XXth International Conference on Electrical Machines* (pp. 3022-3026). IEEE.

[4] Laiho, A., Sinervo, A., Orivuori, J., Tammi, K., Arkkio, A. and Zenger, K., 2009. Attenuation of harmonic rotor vibration in a cage rotor induction machine by a self-bearing force actuator. *IEEE Transactions on Magnetics*, *45*(12), pp.5388-5398.

[5] Laiho, A., Tammi, K., Zenger, K. and Arkkio, A., 2008. A model-based flexural rotor vibration control in cage induction electrical machines by a built-in force actuator. *Electrical Engineering*, *90*, pp.407-421.

[6] Khoo, W.K.S., 2005. The bridge is configured for winding for polyphase self-bearing machines. *IEEE Transactions on Magnetics*, *41*(4), pp.1289-1295.

[7] Anssi Sinervo, 2013. Effects of slotting and unipolar flux on magnetic pull in a two-pole induction motor with an extra four-pole stator winding. PhD thesis, Aalto University, https://urn.fi/URN:ISBN:978-952-60-5155-0.

[8] W. Khoo, and S. D. Garvey, "Practical implementation of the bridge configured winding for self-bearing machines," International conference on power electronics and drive systems, PEDS', Vol . 2, pp.1146-1151.

[9] W. Khoo, R. L. Fittro, and S. D. Garvey, "Ac polyphase self-bearing motors with a bridge configured winding," in Proc. 7th Int. Symp. Magn. Bearings, 2002, pp. 47–52.

[10] W. K. S. Khoo, K. Kalita, and S. D. Garvey, "Practical implementation of the bridge configured winding for producing controllable transverse forces in electrical machines," *IEEE Transactions on Magnetics*, vol. 47, no. 6 PART 2, pp. 1712–1718, Jun. 2011, doi: 10.1109/TMAG.2011.2113377.

[11] Deore, R. and Kalita, K., 2024, February. Analytical Study of the Vibration Effects Due to Nonlinear Unbalanced Magnetic Pull in Electrical Machines Under Static Eccentricity. In *International Conference on Nonlinear Dynamics and Applications* (pp. 277-288). Cham: Springer Nature Switzerland.

[12] R Deore, B Brahma, Shahrukh, and K Kalita, 2024, "The Passive Vibration Control in Bridge Configured Winding Cage Rotor Induction Motor: An Experimental Analysis", https://doi.org/10.1007/978-3-031-40455-9.

[13] A. Laiho, K. Kalita, K. Tammi, and S. Garvey, Dynamics of bridge configured built-in force actuator for vibration control in four-pole cage induction machine," in 18th International Congress on Sound and Vibration, July 2011.

Enhancing Predictive Maintenance of Urban Streetlights through Hyperparameter-Tuned Machine Learning Techniques

[1]Ashok Ganga
Department of EEE
Aditya Institute of Technology and
Management
Tekkali, Andhra Pradesh, India
ganga201@gmail.com

[2]Kanaka Raju Kalla
Department of EEE
Aditya Institute of Technology and
Management
Tekkali, Andhra Pradesh, India
kallakanakaraju@gmail.com

[3]Kumaraswamy Simhadri
Department of EEE
Aditya Institute of Technology and
Management
Tekkali, Andhra Pradesh, India
kumar.simhadri@gmail.com

[4]P. Maheswara Rao[*]
Department of EEE
Vignan's Institute of Information
Technology
Visakhapatnam, Andhra Pradesh, India
mahesh.pydisetty@gmail.com

[5]Akhilash Pennam
Department of CSE
The Assistance Fund
akhilash.pennam@gmail.com

[6]Matcha Nikhitha
Department of EEE
Aditya Institute of Technology and
Management
Tekkali, Andhra Pradesh, India

[7]Kelli Akshara
Department of EEE
Aditya Institute of Technology and
Management
Tekkali, Andhra Pradesh, India

[8]Pydisetti Prasanthi
Department of EEE
Aditya Institute of Technology and
Management
Tekkali, Andhra Pradesh, India

[9]Tangudu Sai Esha
Department of EEE
Aditya Institute of Technology and
Management
Tekkali, Andhra Pradesh, India

Abstract—In recent years, the application of machine learning techniques to urban infrastructure management has seen a significant surge, particularly in predictive maintenance of street lighting systems. This study proposes and rigorously evaluates a machine learning framework for the fault detection and classification of streetlights, leveraging Random Forest and AdaBoost algorithms, both with and without hyperparameter tuning. A comprehensive analysis was performed using a real-world dataset comprising over 34,000 observations, encapsulating electrical and environmental parameters critical to streetlight performance. The experimental results reveal that hyperparameter optimization, conducted via GridSearchCV, substantially enhances model accuracy and generalization. In particular, the Random Forest model, when subjected to systematic tuning, achieved the highest classification performance with an overall test accuracy of 84.94 percentage, markedly outperforming other models. Detailed performance metrics, including precision, recall, and F1-score across multiple fault categories, highlight the superior robustness of the optimized Random Forest approach, especially in handling class imbalance and complex fault patterns. This research not only advances the methodological landscape of predictive maintenance for urban lighting systems but also offers actionable insights for municipal planners seeking to implement intelligent, resilient, and cost-effective infrastructure management solutions.

Index Terms—Street Light Fault Detection, Predictive Maintenance, Machine Learning, Random Forest, AdaBoost, Hyperparameter Tuning.

I. INTRODUCTION

In recent years, the integration of machine learning techniques in urban infrastructure management has gained significant attention, particularly for predicting and classifying failures in street lighting systems [1][2]. These advancements not only enhance the efficiency of maintenance operations but also contribute to improved safety and energy conservation in urban environments [3][4]. The application of algorithms such as Random Forest and AdaBoost allows for the analysis of historical data, enabling city planners to anticipate potential failures before they occur and allocate resources more effectively [5]. By leveraging these predictive models, municipalities can implement proactive maintenance strategies that minimize downtime and reduce operational costs while ensuring that public spaces remain well-lit and safe for residents [6]. As cities continue to grow and evolve, the integration of smart technologies into urban infrastructure management will play a crucial role in fostering sustainable development and enhancing the quality of life for citizens [7].Effective background research provides insights into current challenges faced by cities, such as increasing energy demands, pollution, and aging infrastructure, while highlighting successful case studies that demonstrate innovative solutions and best practices in energy conservation [8]. This knowledge not only informs policymakers and urban planners but also empowers communities to engage in sustainable practices that contribute to a healthier environment and improved public welfare [9]. By integrating these technologies, cities can optimize resource

979-8-3315-3013-6/25 $31.00 © 2025 IEEE

allocation, reduce waste, and create resilient systems that adapt to changing conditions while promoting social equity and economic growth [10][11].Understanding the factors that contribute to street light failures is crucial, as it enables municipalities to implement proactive maintenance strategies and enhance public safety through improved illumination [12][13]. Predictive analytics can play a key role in identifying potential failures before they occur, allowing for timely interventions that minimize disruptions and ensure consistent lighting on roadways [14]. This approach not only enhances the safety of pedestrians and drivers but also contributes to energy efficiency by reducing unnecessary power consumption during outages [15][16].

II. LITERATURE REVIEW

This literature review synthesizes recent research findings and methodologies, with a particular focus on the effectiveness of fault detection techniques, the integration of machine learning approaches, and their broader implications for urban infrastructure management.For instance, Zhang [17] delved into how integrating IoT devices into street lighting can transform urban infrastructure. His work showed that smart sensors, capable of monitoring environmental factors, allow for dynamic lighting adjustments, leading to notable energy savings. Beyond boosting efficiency, this evolution also offers valuable data that can inform urban planning and traffic flow strategies. Meanwhile, the rise of machine learning in fault detection has drawn increasing attention. Kumar [18] developed a predictive maintenance model that, by analyzing historical fault records and environmental data, achieved over 90 percentage accuracy in detecting potential failures. Their results underscore the potential of machine learning to minimize downtime by anticipating problems before they escalate.Real-time monitoring technologies have further strengthened the resilience of street lighting networks. Chen [19] highlighted the benefits of leveraging real-time analytics to track light performance, effectively spotting early warning signs of system issues. This proactive method not only simplifies maintenance schedules but also plays a vital role in boosting urban safety. Yet, despite these promising developments, certain hurdles remain. Lee [20] pointed out significant challenges, including concerns over data privacy, the complexity of system integration, and the steep upfront costs tied to deploying smart lighting solutions. They stressed that forging strong partnerships between technology vendors and municipal authorities is key to overcoming these barriers and achieving widespread implementation. Lastly, the critical link between street lighting maintenance and public safety cannot be overlooked. Patel [21] demonstrated that properly maintained lighting systems help lower crime rates and enhance residents' trust in local governance. Their study makes a strong case for embracing predictive maintenance strategies to reinforce community confidence in public infrastructure.

III. PROPOSED APPROACH

The Figure 1 illustrates a fault diagnosis framework for streetlights, categorizing potential conditions into five types: No Fault, Bulb Failure, Light Flickering, Voltage Surge, and Short Circuit.

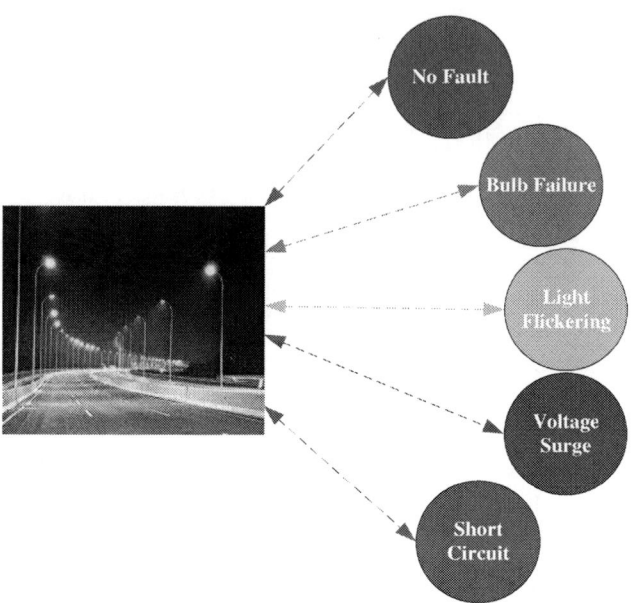

Fig. 1. Fault Diagnosis Framework for Street Lighting Systems.

A. Data Source Data Specification

The dataset employed in this study, originally introduced by [22], encompasses a substantial 34,310 observations and spans across nine distinct attributes. These attributes collectively capture a wide range of variables, including key electrical parameters, relevant environmental conditions, and various fault classifications pertinent to streetlight systems. Among the independent variables, the electrical parameters include metrics such as power consumption (measured in watts), voltage levels (in volts), and current fluctuations (in amperes), all of which are critical indicators of the operational health of the streetlights. In parallel, environmental factors are also considered, comprising ambient temperature (°C), categorized environmental conditions, and measurements of external current fluctuations. A temporal dimension is incorporated through a timestamp assigned to each record, while a unique identifier links each observation to a specific streetlight bulb, ensuring traceability across the dataset. The dependent variable, designated as the fault type, is categorical in nature, representing five distinct classes (ranging from 0 to 4). Here, a classification of 0 denotes normal, fault-free operation, while values from 1 to 4 correspond to varying categories of faults encountered in the system. Notably, the recorded voltage values exhibit a considerable range, varying from 210V to as high as 259.3V, whereas current fluctuations can reach up to 9.96A, reflecting significant operational variability that may contribute to fault

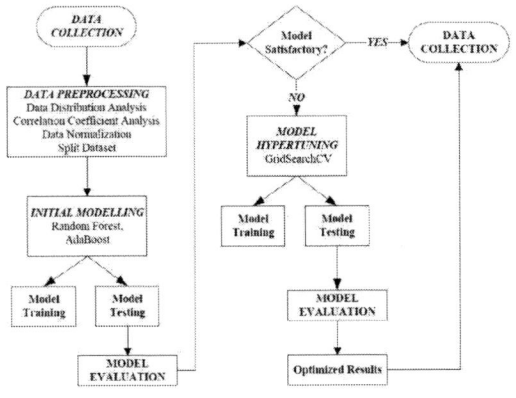

Fig. 2. **Flowchart of the proposed method.**

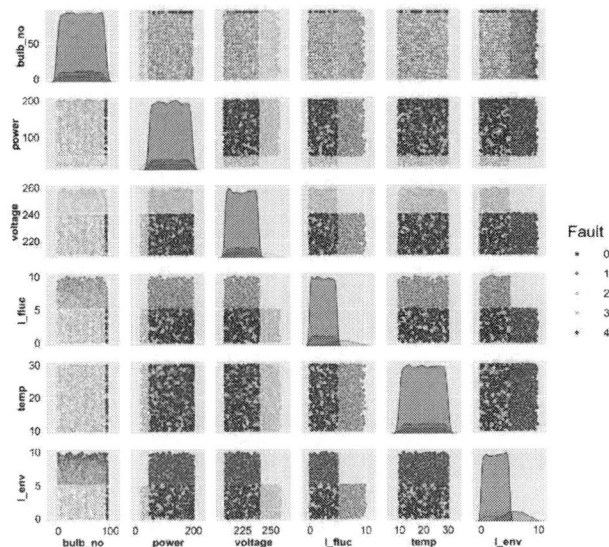

Fig. 3. **Pair plot visualization of fault classes.**

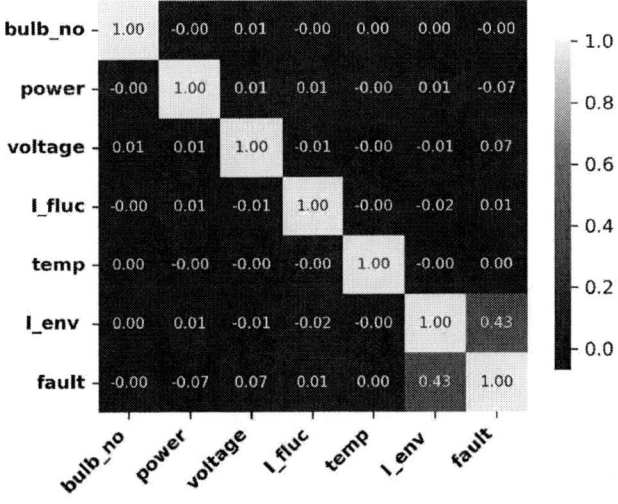

Fig. 4. **Feature correlation Heatmap.**

occurrence. Ambient temperatures recorded in the dataset span from 10°C to 30°C, a range that could plausibly influence both power efficiency and the overall failure rates observed.It is particularly noteworthy that the dataset is complete, with no missing entries detected, thereby rendering it immediately suitable for the application of machine learning algorithms without the necessity for extensive preprocessing.

B. Flow chart

Figure 2 illustrates a structured flowchart that outlines the systematic methodology developed for constructing and refining a machine learning models. The process begins with Data Collection and Pre-processing, which ensures the dataset is fully prepared by conducting distribution and correlation analyses, applying normalization, and splitting the dataset appropriately to enhance model effectiveness. An initial Random Forest and AdaBoost-model is trained and tested to establish a baseline for performance. Models evaluation is conducted using essential performance metrics, which guide the decision-making process for further improvement. If the model satisfies the predefined evaluation standards, it is finalized; otherwise, it proceeds to hyperparameter tuning. GridSearchCV is utilized during this tuning phase to optimize models parameters, aiming to improve both accuracy and generalization. After tuning, the enhanced model undergoes a final evaluation. Upon meeting the target criteria, the model is finalized and adopted for deployment or subsequent analytical tasks.

C. Data Preprocessing

Figure 3 illustrates a pair plot visualization, offering an insightful display of scatter plots that depict pairwise relationships among multiple variables, complemented by their respective univariate distributions along the diagonal. This form of visualization proves instrumental in uncovering underlying correlations, detecting subtle patterns, and identifying potential outliers within the dataset, thus facilitating a deeper preliminary understanding of the data structure. Figure 4 presents

a feature correlation heatmap that maps the relationships between various electrical and environmental variables. Overall, correlations are predominantly weak, clustering around values close to zero. The most notable relationship is a moderate positive correlation (0.43) observed between environmental current fluctuations and fault type, while other variables, such as power consumption and voltage levels, exhibit minimal association with fault occurrence.

D. Current Behaviour Analysis

Figure 5 presents the analysis of current behavior under varying fault conditions, providing insights into the stability and operational integrity of the electrical system. In the No

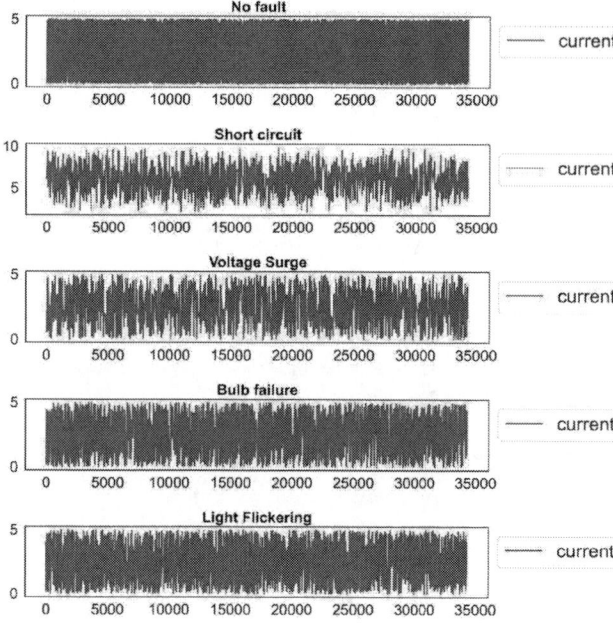

Fig. 5. **Current Behaviour Analysis under Fault Conditions.**

Fault scenario, the current remains notably stable and consistent throughout the observation period, exhibiting only minimal fluctuations. This behavior is indicative of a normal operating condition free from electrical disturbances. In contrast, under a Short Circuit, the current demonstrates significant irregularity, with frequent peaks reaching considerably higher values than in the stable case. Such patterns clearly signal a disruption in current flow, symptomatic of a severe fault. During a Voltage Surge, the current also fluctuates widely, marked by noticeable spikes and irregular behavior, reflecting the system's response to sudden voltage increases. The Bulb Failure scenario shows moderate current variability, with deviations less extreme than those associated with short circuits or voltage surges, suggesting localized instability caused by a malfunctioning bulb. Finally, in the Light Flickering condition, the current displays a periodic pattern of spikes and drops, characteristic of minor disturbances often resulting from loose connections or minor systemic irregularities.

IV. RESULTS AND DISCUSSION

This section outlines and examines the outcomes of applying two machine learning models for predicting and classifying failures in street lighting systems.

A. Analysis of Classification Performance Using Confusion Matrices

Figures 6 through 9 present the confusion matrices for the initial Random Forest, the Random Forest optimized via GridSearchCV, the initial AdaBoost, and the AdaBoost model likewise subjected to hyperparameter optimization, respectively.The confusion matrix corresponding to the initial

Random Forest model (Figure 6) exhibits a strong diagonal dominance for class 0, with 5,098 instances correctly classified.Upon optimizing the Random Forest model through GridSearchCV (Figure 7), there is a marked improvement in class separability. The matrix reveals near-perfect class discrimination for classes 1, 2, 3, and 4, where misclassifications are almost entirely eliminated. The dominance of class 0 persists; nevertheless, the reduction in off-diagonal elements implies that the hyperparameter tuning significantly enhanced the model's discriminative ability. Particularly, classes that previously exhibited high confusion, such as class 3 and class 2, are now far better recognized.Turning to the AdaBoost model (Figure 8), the confusion matrix indicates a considerably weaker performance relative to the Random Forest models. A substantial proportion of samples from classes 1, 2, and 3 are incorrectly assigned to class 0. Specifically, more than 1,400 samples from class 1 and over 1,200 samples from class 2 are misclassified as belonging to class 0, suggesting a severe imbalance in class learning. This pervasive bias towards the majority class highlights a critical vulnerability of the initial AdaBoost implementation when applied without appropriate hyperparameter optimization or class balancing strategies.Interestingly, the application of GridSearchCV to AdaBoost (Figure 9) yields notable improvements. The confusion matrix demonstrates an appreciable reduction in the misclassification rates, especially for classes 1, 2, and 3. Correct classification counts for minority classes have increased, and the overall distribution of predictions has become more balanced across the five classes. Nevertheless, residual confusion remains evident, particularly between classes 0 and 3, suggesting that while hyperparameter tuning mitigates some deficiencies, fundamental model limitations persist, likely necessitating further methodological enhancements such as the incorporation of ensemble techniques or cost-sensitive learning frameworks.Overall, this comparative analysis underscores the substantial impact that systematic hyperparameter optimization exerts on model performance, especially for complex, imbalanced classification tasks. Notably, Random Forest with GridSearchCV demonstrates superior robustness and generalization capability relative to AdaBoost variants. This finding aligns with prior literature emphasizing the resilience of tree-based ensemble methods, particularly when optimized carefully, in handling heterogeneous and imbalanced datasets.

B. Performance Evaluation of Random Forest and Adaboost Classifiers for Street Light Fault Detection and Classification

The table 1 and 2 presents a comprehensive evaluation of the Random Forest and Adaboost classification models, both with and without hyperparameter optimization, applied to the task of street light fault detection and categorization. The performance metrics include Precision, Recall and F1-Score for each fault label. Notably, hyperparameter tuning, executed via GridSearchCV for Random Forest and optimized parameter selection for Adaboost, markedly improved classification accuracy across the test sets. The detailed metrics offer critical insights into the models' discriminative capabilities,

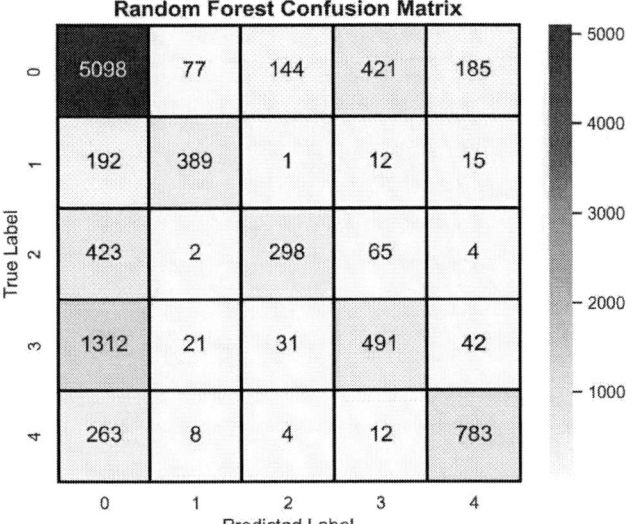

Fig. 6. **Random Forest confusion matrix without GridSearchCV**

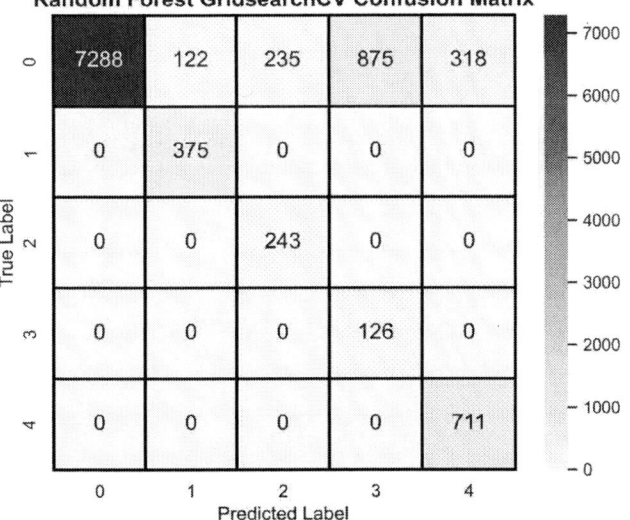

Fig. 7. **Random Forest confusion matrix with GridSearchCV**

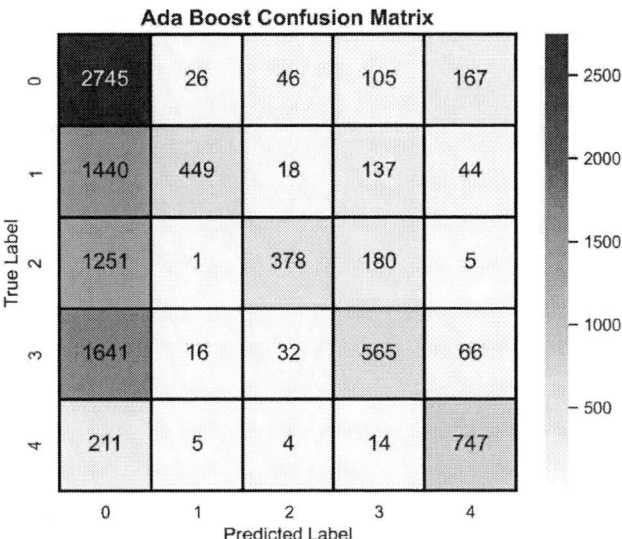

Fig. 8. **AdaBoost confusion matrix without GridSearchCV**

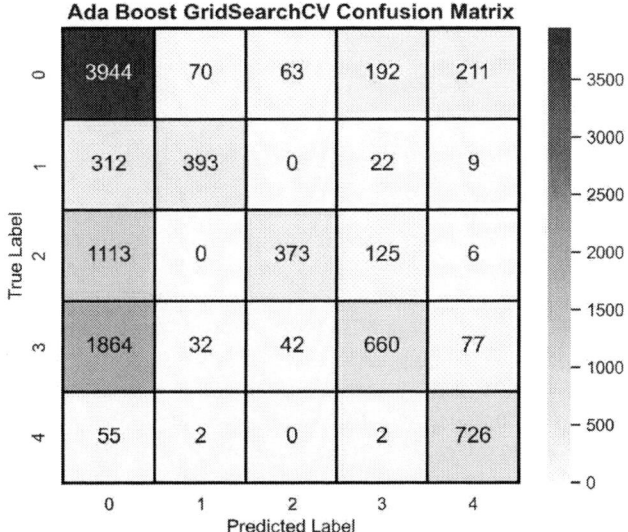

Fig. 9. **AdaBoost confusion matrix with GridSearchCV**

particularly emphasizing the varying levels of performance across different fault categories, it becomes evident that the Random Forest model with hyperparameter tuning emerges as the best performing approach for the street light fault detection and classification task. Specifically, this model achieved an impressive test accuracy of 84.94 percentage, substantially outperforming all other configurations, including both untuned and tuned versions of Adaboost.The technical superiority of the tuned Random Forest model can be attributed to several interrelated factors. First, the hyperparameter optimization, conducted via GridSearchCV, facilitated the selection of an optimal combination of model parameters notably a reduced

maximum depth to 5 and a modest number of estimators to 50 which effectively mitigated the risk of overfitting while maintaining sufficient model complexity to capture underlying patterns in the data. Furthermore, Random Forest, by its ensemble nature, inherently reduces variance through bootstrap aggregation (bagging), and the use of the Gini impurity criterion for split optimization further enhanced its discriminative power.

V. CONCLUSION

This study underscores the pivotal role of machine learning models in enhancing the reliability and operational efficiency

TABLE I
RANDOM FOREST MODEL PERFORMANCE BEFORE AND AFTER HYPERPARAMETER TUNING

Accuracy (%)	Fault Condition	Precision	Recall	F1-Score
68.58	No Fault	0.70	0.86	0.77
	Short Circuit	0.78	0.64	0.70
	Voltage Surge	0.62	0.38	0.47
	Bulb Failure	0.49	0.26	0.34
	Light Flickering	0.76	0.73	0.75
Best Parameters: {criterion=gini, max_depth=5, min_samples_leaf=1, min_samples_split=2, n_estimators=50}				
84.94	No Fault	1.00	0.82	0.90
	Short Circuit	0.75	1.00	0.86
	Voltage Surge	0.51	1.00	0.67
	Bulb Failure	0.13	0.11	0.22
	Light Flickering	0.69	1.00	0.82

TABLE II
ADA BOOST MODEL PERFORMANCE BEFORE AND AFTER HYPERPARAMETER TUNING

Accuracy (%)	Fault Condition	Precision	Recall	F1-Score
47.45	No Fault	0.38	0.89	0.53
	Short Circuit	0.9	0.22	0.35
	Voltage Surge	0.79	0.21	0.33
	Bulb Failure	0.56	0.24	0.34
	Light Flickering	0.73	0.76	0.74
Best Parameters: {base_estimator_ max_depth=3, $learning_rate = 0.1$, n_estimators=100}				
59.22	No Fault	0.54	0.88	0.67
	Short Circuit	0.79	0.53	0.64
	Voltage Surge	0.78	0.23	0.36
	Bulb Failure	0.66	0.25	0.36
	Light Flickering	0.71	0.92	0.8

of urban street lighting networks through proactive fault detection and classification. By systematically comparing the performance of Random Forest and AdaBoost classifiers, both in their baseline and optimized forms, it was demonstrated that hyperparameter tuning serves as a critical lever for elevating model performance. Notably, the Random Forest classifier, when fine-tuned using GridSearchCV, exhibited remarkable gains in predictive accuracy, achieving a test accuracy of 84.94 percentage, alongside substantial improvements in class-specific recall and precision metrics. These outcomes can be attributed to the model's ensemble-based architecture, which mitigates overfitting and fosters improved generalization, as well as the judicious calibration of key parameters such as maximum depth and the number of estimators. Conversely, while AdaBoost showed moderate improvements following tuning, its baseline susceptibility to class imbalance and noise limited its ultimate effectiveness relative to Random Forest.The findings of this research have important practical implications. By enabling earlier and more accurate identification of streetlight faults, municipalities can significantly reduce maintenance costs, minimize service disruptions, and enhance public safety. Furthermore, the demonstrated effectiveness of machine learning-based predictive maintenance strategies aligns well with broader smart city initiatives, where data-driven decision-making is central to building sustainable and resilient urban environments. Future research could build upon these results by integrating real-time sensor data, exploring

deep learning methodologies, or investigating hybrid ensemble models to further refine fault detection capabilities under dynamic urban conditions. Ultimately, the integration of intelligent predictive systems into street lighting infrastructure holds the promise of transforming how cities manage public utilities, promoting greater efficiency, safety, and environmental stewardship.

REFERENCES

[1] D. Segovia-Muñoz, X. Serrano-Guerrero, and A. Barragán-Escandón, "Predictive maintenance in LED street lighting controlled with telemanagement system to improve current fault detection procedures using software tools," Renewable energy power quality journal, Sep. 2022, doi: 10.24084/repqj20.318

[2] K. Fan et al., "Street lighting system fault diagnosis research," in 7th International Symposium on Advances in Electrical, Electronics, and Computer Engineering, Oct. 2022. doi: 10.1117/12.2639870

[3] L. Yang, Q. M. Liu, T. Xia, C. Ye, and J. Li, "Preventive Maintenance Strategy Optimization in Manufacturing System Considering Energy Efficiency and Quality Cost," Energies, Nov. 2022, doi: 10.3390/en15218237

[4] J. Salas and V. Yepes, "Improved delivery of social benefits through the maintenance planning of public assets," Structure and Infrastructure Engineering, Sep. 2022, doi: 10.1080/15732479.2022.2121844

[5] D. Patil and R. L. Gupta, "Spatiotemporal analysis and prediction of urban evolution patterns using Artificial Neural Network tool," Proceedings of the Institution of Civil Engineers, May 2023, doi: 10.1680/jurdp.22.00046.

[6]]"Digital twin predictive maintenance strategy based on machine learning improving facility management in built environment," 2022. doi: 10.1016/b978-0-12-820793-2.00007-0

[7] T. V. But, D. Mamotenko, and S. Hres-Yevreinova, "Smart-infrastructure in the sustainable development of the city: world experience and prospects of ukraine," Jun. 2023. doi: 10.36074/logos-23.06.2023.01

[8] "Spatial design of energy self-sufficient communities." 2022. doi: 10.1016/b978-0-12-823941-4.00021-4

[9] A. Karion et al., "Background conditions for an urban greenhouse gas network in the Washington, DC, and Baltimore metropolitan region," Atmospheric Chemistry and Physics, Apr. 2021, doi: 10.5194/ACP-21-6257-2021

[10] "Applying science, technology and innovation for sustainable urban development," 2022. doi: 10.18356/9789210018920c005

[11] J. G. DeJaeghere, "Smart city as a social transition towards inclusive development through technology: a tale of four smart cities," May 2022, doi: 10.1080/12265934.2022.2074076

[12] G. Du, "Street lamp failure scheduling method, street lamp monitoring center, and street lamp failure scheduling system," Dec. 20, 2018

[13] A. Chalfin, J. Kaplan, and M. LaForest, "Street Light Outages, Public Safety and Crime Attraction," Journal of Quantitative Criminology, Jul. 2021, doi: 10.1007/S10940-021-09519-4

[14] K. Poorna and R. R. N. Bielby, "Predictive maintenance of automotive lighting," Feb. 18, 2021

[15] N. Manchikanti, G. A. E. S. Kumar, P. Dendi, and V. Yarabolu, "Implementation of Security-Based Energy-Efficient Dynamic Street Lighting System," Dec. 2022. doi: 10.1109/ICPECTS56089.2022.10047357

[16] A. Gorpyniuk and S. Taraban, "Safety of pedestrian on the road in the dark time of the day in the conditions of rolling blackouts and emergency power outages in Ukraine," Avtošlâhovik Ukraïni, Mar. 2023, doi: 10.33868/0365-8392-2022-1-273-23-29

[17] Y. Zhang and H. Li, "IoT-enabled smart street lighting: A review," Journal of Urban Technology, 2022.

[18] R. Kumar and A. Singh, "Predictive maintenance of smart street lights using machine learning," IEEE Transactions on Smart Grid, 2023.

[19] L. Chen and Y. Zhao, "Real-time monitoring of street lighting systems using big data analytics," Urban Computing and Data Science, 2023.

[20] J. Lee and S. Park, "Challenges in implementing smart street lighting systems: A case study," Smart Cities Journal, 2023.

[21] M. Patel and R. Gupta, "The impact of street lighting on public safety and community trust," Journal of Urban Affairs, 2023.

[22] Kaggle, "Street Light Fault Prediction Dataset." Available: https://www.kaggle.com/datasets/vizeno/street-light-fault-prediction-dataset.

979-8-3315-3013-6/25 $31.00 © 2025 IEEE

Design and Analysis of GaN Based Electronic Power Conditioner for Space Applications

Indhuja L R
Department of Electrical and Electronics Engineering
National Institute of Technology
Tiruchirappalli, India
indhujaee216@gmail.com

V. Vignesh Kumar
Department of Electrical and Electronics Engineering
National Institute of Technology
Tiruchirappalli, India
vvigneshkumar@nitt.edu

Nikhil K Desai
Space Applications Centre, ISRO, DOS
Ahmedabad, India
nkd@sac.isro.gov.in

B. Venkatesaperumal
Department of Electrical and Electronics Engineering
National Institute of Technology
Karnataka, India
bvperumal@nitk.edu.in

U. Vinatha
Department of Electrical and Electronics Engineering
National Institute of Technology
Karnataka, India
vinatha.nitk@gmail.com

M. M. Rajan Singaravel
Department of Electrical and Electronics Engineering
National Institute of Technology
Puducherry, India
rajan.singaravel@nitpy.ac.in

Abstract— This paper presents the design and analysis of a forward converter-based multi-output topology utilizing Gallium Nitride (GaN) power devices for Electronic Power Conditioner (EPC) applications in space systems. The proposed topology is tailored for high-frequency operation, leveraging the superior switching characteristics of GaN devices to achieve improved efficiency. To further minimize conduction losses, synchronous rectification is implemented across the point-of-load (POL) converters associated with each output stage. A comprehensive magnetic design approach is explained to support efficient energy transfer in a compact multi-output configuration. The work includes detailed loss analysis based on real GaN device parameters, offering insights into both switching and conduction losses. Simulation results are used to validate the theoretical analysis, providing steady-state waveforms that demonstrate the functional integrity and high efficiency of the proposed converter. This research supports the viability of GaN-based forward converters as a compelling solution for next-generation space power electronics.

Keywords—forward converter, GaN Switch, electronic power conditioner, synchronous rectification, double pulse testing.

I. INTRODUCTION

A recent trend in improving power converter performance is the replacement of traditional silicon (Si)-based semiconductor switches with wide bandgap (WBG) devices. This shift is primarily due to the inherent limitations of Si devices, such as low electric field strength, narrow energy bandgap, limited electron mobility, and poor thermal conductivity. In contrast, WBG devices—such as silicon carbide (SiC) and gallium nitride (GaN)—exhibit superior material properties, including high breakdown voltage, wide energy bandgap, and excellent thermal and electron transport characteristics. These attributes enable WBG power devices to operate at higher voltages, temperatures, and switching frequencies, thereby enhancing overall efficiency and power density of power converters [1]. While both SiC and GaN surpass Si in performance, GaN is particularly well-suited for low-voltage applications around 100–200 V due to its ultra-low parasitic capacitances, high electron mobility, and lateral device structure, which enable higher switching speeds, reduced switching losses, and more compact circuit designs [2].

TABLE I DEVICE COMPARISON OF 100 V GAN FET VS SI MOSFET

Parameter	GaN FET	Si MOSFET
Part Number	GS61004B-TR	IRF540N
Technology	GaN E-mode HEMT	Si Planar MOSFET
V_{ds}	100 V	100 V
$I_{ds(continuous)}$	38 A	33 A
$R_{ds(on)}$	16 mΩ	44 mΩ
Q_g	3.3 nC	71 nC
C_{iss}	260 pF	1960 pF
C_{oss}	110 pF	250 pF
Q_{rr}	0	505 nC
Package	GaNPX® (4.6 × 4.4 mm²)	TO-220AB

This makes GaN the preferred choice in compact, high-efficiency converters for applications such as point-of-load supplies, on-board chargers, low-voltage renewable energy systems and space- grade electronic power conditioners (EPCs).

A comparison of datasheet parameters for representative 100 V GaN FET and Si MOSFET devices is presented in Table I. GaN devices demonstrate significantly lower input capacitance and require a gate drive voltage of approximately 6 V, leading to reduced gate drive power. Moreover, GaN devices have negligible reverse recovery charge due to the absence of minority carrier conduction, a limitation inherent to Si devices. These features clearly illustrate GaN's potential to achieve higher efficiency in synchronous rectification and hard-switched converter topologies

GaN-based power converters have been implemented across various topologies in a wide range of applications [3]–[6]. One notable area where the advantages of GaN power semiconductor devices can be fully leveraged is in electronic power conditioners (EPCs) for satellite payloads. EPCs are critical subsystems in satellites that regulate, condition, and distribute electrical power from the primary source to onboard systems and payloads, ensuring stable and efficient operation in the harsh conditions of space [7]. As these systems are essential yet space-limited, the growing demand for compact spacecraft subsystems has driven the need to accommodate power electronics within strict mechanical constraints, minimizing overall payload size and mass [8].

979-8-3315-3013-6/25 $31.00 © 2025 IEEE

Several recent works have explored the use of GaN devices in space power applications. A mathematical model for optimizing the multi-winding configuration of transformers used in multi-output converters for electronic power conditioners (EPCs) is presented in [9]. The study employs an active-clamp forward converter topology featuring fully GaN-based switches and self-driven synchronous rectification. The optimized transformer design, combined with the GaN-based topology, is validated through a prototype, demonstrating superior performance. In [10], a two-switch forward converter integrated with a buck converter is proposed for satellite power systems. The design reduces the number of switching devices compared to conventional topologies, thereby increasing power density while lowering cost and volume. Additionally, it achieves zero-voltage switching (ZVS), and its enhanced efficiency is verified using a 600 W prototype.

The work reported in [11] quantifies the benefits of space-grade GaN devices in forward converters for medium-power DC-DC conversion in satellite systems. Experimental results show that the GaN-based converter offers a 4.54% efficiency improvement over its silicon counterpart. In [12], a derating and design analysis for a buck-LLC topology using GaN devices is presented for space applications. A control strategy is also introduced, which dynamically adapts switching frequency and dead time to ensure high efficiency and ZVS operation.

In this paper, a detailed design analysis of an isolated GaN-based multi-output forward converter for use in EPCs is presented. The organization of the paper is as follows: Section II describes the overall system configuration of the proposed EPC. Section III discusses the design of the converter topology. The computed steady-state results for the forward converter configuration are presented in Section IV. Section V focuses on the computation and analysis of various power losses within the converter. Finally, Section VI concludes the paper with key findings.

II. GaN Based Electronic Power Conditioner

Multi-output DC-DC converters play a critical role in meeting the diverse power requirements of various subsystems in space-borne payloads. This work employs a multi-output, isolated forward converter topology as part of the electronic power conditioner (EPC) design. The overall system topology is depicted in the block diagram shown in Fig. 1. The converter receives a variable raw bus voltage at the primary side, ranging from 65 V to 75 V, which is a typical unregulated input from the spacecraft power bus.

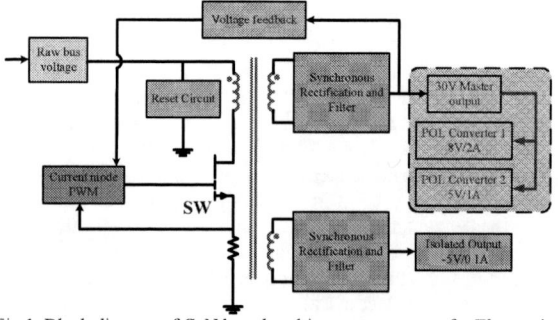

Fig. 1. Block diagram of GaN based multi-output converter for Electronic Power Conditioner

At the secondary side, the converter generates four distinct output voltages, each designed to supply different subsystems of the payload. Among these, Output 1 delivers 30 V at 3.6 A and is designated as the master output due to its highest power rating and critical importance. This output is tightly regulated to maintain voltage stability under varying load conditions. The remaining outputs are categorized as slave outputs and are coarsely regulated by the main control loop.

To meet the specific regulation and efficiency requirements of the EPC, Output 2 (8 V, 2 A) and Output 3 (5 V, 1 A) are derived from the master output using downstream high-efficiency switching regulators, such as point-of-load (POL) converters. These POL stages ensure fine regulation at the load point while contributing to improved overall system efficiency. Output 4 provides an isolated −5 V at 0.1 A and is galvanically isolated from the other outputs to serve subsystems requiring a negative voltage rail or isolation for signal integrity.

To maximize the efficiency advantages of GaN devices, synchronous rectification is employed across all output stages. This choice minimizes conduction losses during the secondary-side switching operation and helps leverage the high-speed switching capability and low reverse recovery characteristics of GaN transistors, particularly under high-frequency hard-switched conditions.

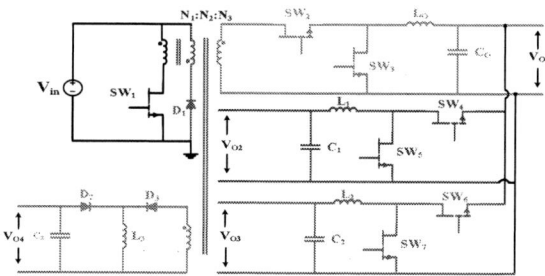

Fig. 2. Schematic representation of the proposed multi-output dc-dc converter

The circuit schematic of the proposed multi-output DC-DC converter is illustrated in Fig. 2. The converter architecture employs a high-frequency transformer with multiple secondary windings, each tailored to generate specific output voltages required by different subsystems of the satellite payload. The primary winding of the transformer is responsible for generating the main or master output, which is derived directly from one of the secondary windings. This output is tightly regulated and serves as the primary source from which additional outputs are generated. Specifically, Output 2 and Output 3 are obtained from this regulated master output using downstream point-of-load (POL) converters. These are implemented as synchronous buck converters, providing efficient non-isolated step-down DC-DC conversion to meet the voltage and current requirements of connected subsystems.

In addition to the main secondary winding, the transformer incorporates a separate winding that supplies an isolated negative output (Output 4). This output is crucial for certain payload electronics that require a

979-8-3315-3013-6/25 $31.00 © 2025 IEEE

negative voltage rail. The specifications of the proposed multi-output GaN based EPC are provided in Table II.

TABLE II SPECIFICATIONS OF PROPOSED EPC

Parameters	Value
Output power (P_{OUT})	108 W
Input voltage (V_{IN})	65-75 V
Output 1	V_{O1}= 30 V, 3.6 A
Output 2	V_{O2}= 8 V, 2 A
Output 3	V_{O3}= 5 V, 1 A
Output 4	V_{O4}= -5 V, 0.1 A
Switching frequency (F_{SW})	500 kHz

To support a wide input voltage range and ensure robust operation across varying load conditions, the transformer is designed with carefully selected turns ratios for each winding. Moreover, the transformer includes a reset winding on the primary side. This winding serves to demagnetize the transformer core during the off-period of the primary switching cycle, effectively resetting the core flux and preventing flux-walking, which could otherwise lead to magnetic core saturation. The reset or demagnetizing winding is bifilarly wound together with the primary winding, which provides the added benefit of minimizing leakage inductance. This design choice not only enhances energy transfer efficiency but also contributes to improved transient response and reduced voltage stress on switching devices.

III. TOPOLOGY DESIGN OF THE PROPOSED EPC

This section presents a detailed design methodology for the key magnetic components of the proposed converter, namely the high-frequency transformer, as well as the output filter inductors and capacitors associated with each output port. Since the converter topology is fundamentally based on the forward converter architecture, the voltage transfer relation between the input and the output can be expressed as

$$V_O = nV_{IN}D \tag{1}$$

where V_O is the output voltage, V_{IN} is the input voltage, n is the transformer turns ratio ($n_{secondary}/n_{primary}$) and D is the duty cycle.

A. High-Frequency Transformer parameters

To determine a suitable turns ratio for the high-frequency transformer, the converter is designed to operate with a maximum duty cycle of 40%. Substituting this duty cycle, along with the required output voltage and input voltage range, into the voltage transfer relation given by equation (1), the optimum turns ratio is calculated to be 1.25. This turns ratio ensures that the transformer delivers the desired output voltage even under the worst-case operating conditions, while also providing sufficient margin for effective regulation and control.

The simulated steady-state current waveforms for both the primary and secondary windings of the high-frequency transformer in the forward converter are shown in Fig. 3. From the figure, it is evident that the primary winding current comprises both the magnetizing current and the current reflected from the secondary side.

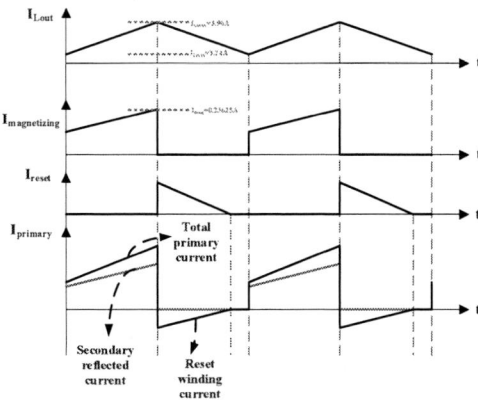

Fig. 3. Winding current waveforms of high-frequency transformer.

Therefore, the primary side must be sized to withstand this combined current. On the secondary side, the winding must be designed to support the total current demand arising from the simultaneous operation of all three output channels. To ensure the transformer operates reliably and avoids core saturation, the magnetizing inductance of the primary winding is computed based on the peak magnetizing current using the following expression:

$$L_{Mag} = \frac{(V_{IN})}{\Delta I_{MAG}}DT_{SW} \tag{2}$$

where ΔI_{MAG} is the peak-to-peak magnetizing current ripple, and T_{SW} is the switching period. This calculation is critical to ensure that the transformer maintains linear operation and energy is effectively transferred without magnetic core saturation during steady-state operation.

The selection of a suitable core for the high-frequency transformer is guided by calculating the area product A_P, a key parameter that correlates the required output power with the magnetic characteristics and switching frequency of the converter. The area product of forward topology is defined by the following expression:

$$A_P = \frac{P_O\sqrt{D}\left(1+\frac{1}{\eta}\right)}{K_W J B_M F_{SW}} \tag{3}$$

where P_O is the output power, η is the efficiency, K_W is the winding utilization factor, J is the current density, B_M is the peak magnetic flux density and F_{SW} is the switching frequency. Once the core selection is done, the number of primary turns N_1, required can be calculated as:

$$N_1 = \frac{DV_{IN}}{A_C B_M F_{SW}} \tag{4}$$

where A_C is the effective core area of the magnetic material.

B. Filter Inductor and Capacitor parameters

The proposed multi-output topology is designed to operate in continuous conduction mode (CCM) to ensure stable and efficient power delivery. To maintain CCM across the entire load range, the filter inductance and output capacitance must be selected with sufficiently high values. However, due to the high switching frequency of the converter, the required values for both inductance and capacitance are inherently lower, which aids in reducing the size and weight of passive components.

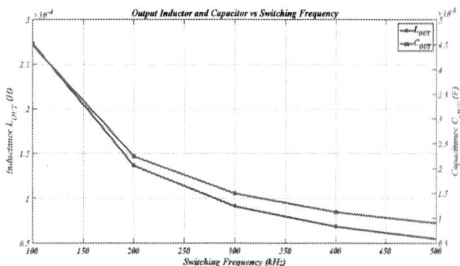

Fig.4. Output filter Inductance and capacitance vs switching frequency

The relationship between inductance and output capacitance, and their variation with switching frequency, is illustrated in Fig. 4. Therefore, a careful trade-off must be made to optimally size the filter components—balancing efficiency, ripple performance, dynamic response, and the stringent mechanical constraints of Electronic Power Conditioners (EPCs) in space applications.

The filter component values required to ensure continuous conduction mode (CCM) and meet the desired output ripple specifications are determined using the following expressions. The output inductor value is computed using:

$$L_{OUT} = \frac{\left(\frac{N_3}{N_1}V_{in} - V_{OUT}\right)}{\Delta I_{OUT}} DT_{SW} \qquad (5)$$

where $\frac{N_3}{N_1}$ is the turns ratio of the secondary to primary windings, and ΔI_{OUT} is the allowed peak-to-peak output current ripple. The required output capacitance is then calculated using:

$$C_{OUT} = \frac{V_{OUT}(1-D)}{\Delta V_{OUT} 8 L_{OUT} F_{SW}^2} \qquad (6)$$

where ΔV_{OUT} is the peak-to-peak voltage ripple at the output. The values of key components in the proposed forward converter-based topology for the Electronic Power Conditioner (EPC), derived using the design equations (1) – (6), are summarized in Table III.

TABLE III DESIGNED PARAMETERS FOR POWER TOPOLOGY OF PROPOSED GaN BASED EPC

Parameters	Value
Output power (P_{OUT})	108 W
Maximum Duty ratio (D)	0.4
Transformer turns ratio ($N_1:N_2:N_3:N_4$)	1:1:1.25:0.11
No. of turns for isolated output	2
Output inductor (L_O)	48 µH
Output inductor (L_1)	29.37 µH
Output inductor (L_2)	41.75 µH
Output inductor (L_3)	158 µH
Output capacitor (C_O)	10 µF
Output capacitor (C_1)	5 µF
Output capacitor (C_2)	2.5 µF
Output capacitor (C_3)	4 µF
GaN FET part number	GS66516T
Inductor current ripple ΔI_L	20% of I_{Lavg}
Ripple in all outputs	20 mV pk-pk
Core	R25.3*14.8*10.0 mm

IV. STEADY STATE ANALYSIS

The proposed multi-output forward derived DC-DC converter for EPC shown in Fig.2 is simulated in the LTSpice environment with parameters listed in Table III. The GaN device spice model for part number shown in Table III provided by the manufacturer is employed in the simulation with actual datasheet device parameters. The computed steady state output waveforms are depicted in Fig. 5. It can be seen from the results that the output voltage are in line with the design specifications confirming the veracity of the design approach.

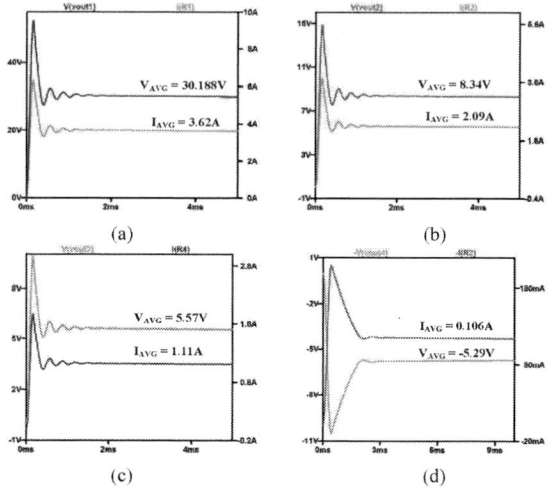

Fig.5. steady state waveforms of (a) output 1 (b) output 2 (c) output 3 and (d) output 4

The transformer parameters were also analyzed through simulation to verify continuous conduction mode (CCM) operation and to assess ripple magnitudes in accordance with the design specifications. As shown in Fig. 6, the voltage and current waveforms of the high-frequency transformer windings confirm proper core reset and consistent forward converter operation, thereby validating the intended operating mode and magnetic design. The presence of a clearly defined demagnetization interval in the primary waveform indicates effective utilization of the reset winding, preventing core saturation.

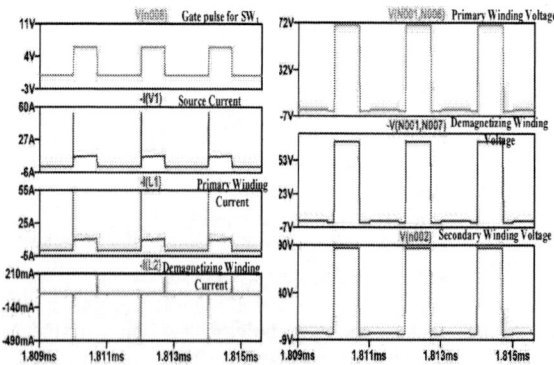

Fig.6. Transformer winding voltage and current waveforms

V. THEORITICAL LOSS ANALYSIS FROM STEADY STATE WAVEFORMS

An detailed loss analysis has been performed to evaluate the power loss distribution in the proposed multi-output topology for the electronic power conditioner (EPC). The major sources of losses in the converter include transformer core and copper losses, switching and conduction losses in the GaN devices, and gate drive losses. These loss components are quantitatively estimated based on the converter's operating conditions and device characteristics. The computed power loss distribution across the key components of the converter is presented below to highlight their individual contributions to overall efficiency.

A. Transformer lossess

The losses in the transformer are the core loss and the winding copper loss. The core loss is calculated using the Steinmetz formula [13],

$$P_{CL} = Kf^\alpha B^\beta L(T) \qquad (7)$$

where $L(t) = b - c(T) + d(T^2)$, b, c, d are material dependent parameters, K is Steinmetz constant, α is frequency component, β is Flux density component. For N87 ferrite core material, the core loss constants are defined by K=0.00375307, α=1.94, β=2.775. Based on these parameters, the total core loss for the toroidal core with dimensions 25.3 × 14.8 × 10.0 mm (R25.314.810.0) is calculated to be approximately 22.86 mW. The copper loss in the transformer windings is estimated using equation (8), which accounts for the resistance of the windings of different gauge and the RMS current flowing through them. This loss is calculated to be approximately 245.48 mW.

$$P_{Cu} = I^2 R \qquad (8)$$

B. Switch losses

The GaN device GS66516T is employed on both the primary and secondary sides of the proposed converter topology. To facilitate an accurate loss analysis, switching transients—including voltage and current waveforms during turn-on and turn-off events—are captured using the LTSpice simulation platform. These transient responses are illustrated in Fig. 7 and serve as the basis for calculating switching losses and validating the dynamic performance of the GaN devices under realistic operating conditions.

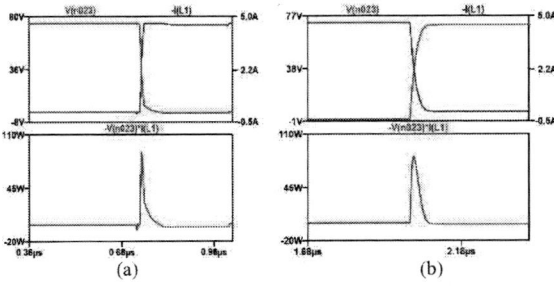

Fig. 7. Switching transients during (a) Turn -on and (b) Turn-off.

The root mean square (RMS) current in the primary and secondary switches is determined using equation (9):

$$I_{S,max,rms} = I_{S,mean}\sqrt{D}\sqrt{1 + \frac{1}{3}\left(\frac{\Delta I_S}{I_{S,mean}}\right)^2} \qquad (9)$$

where ΔI_S and $I_{S,mean}$ are the ripple magnitude and average value of the switch current respectively.

The conduction losses in the switch is then estimated using (10) and is given by,

$$P_{S,conduction} = \left(I_{S,max,rms}\right)^2 R_{dsON@25°C} \qquad (10)$$

switching transition losses, including turn-on and turn-off events, are calculated using equations (11) and (12), respectively:

$$P_{S,turn-on} = \frac{1}{2}(T_{ON})I_{SW,on}V_{IN}F_{SW} \qquad (11)$$

$$P_{S,turn-off} = \frac{1}{2}(T_{OFF})I_{SW,off}V_{IN}F_{SW} \qquad (12)$$

The gate drive losses, arising from the repeated charging and discharging of the gate capacitance, are estimated using (13):

$$P_{S,drv} = V_{drv}Q_G F_{SW} \qquad (13)$$

The total converter losses with major loss components is estimated from (14) as

$$P_{Loss\_total} = P_{CL} + P_{Cu} + P_{S,conduction} + P_{S,turn-on} + P_{S,turn-off} + P_{S,drv} \qquad (14)$$

TABLE IV LOSSES ASSOCIATED WITH POWER SWITCH

Parameters	Master output (30V, 3.6A)	POL 1 (8V,2A)	POL 2 (5V,1A)
Switching loss	1.539 W	0.530 W	0.523 W
Conduction loss	0.532 W	0.261 W	0.065 W
Gate Drive loss	0.0426 W	0.0426 W	0.0426 W

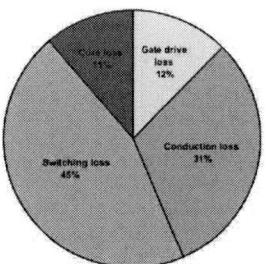

Fig.8. Distribution of various losses in the proposed GaN based EPC.

The computed loss values for switch component are summarized in Table IV, while the relative distribution of these losses is visualized through the pie chart in Fig. 8. This comprehensive loss analysis enables identification of dominant loss contributors, thereby guiding targeted design optimizations. Moreover, it highlights the efficiency advantages offered by GaN-based designs, confirming the suitability of the proposed converter topology for compact and high-efficiency satellite power systems.

VI. CONCLUSION

This paper presented the design and simulation-based analysis of a GaN-based multi-output forward converter for electronic power conditioners (EPCs) in space applications. The converter topology, adapted from the conventional forward converter, was configured to deliver multiple regulated outputs, with design considerations ensuring continuous conduction mode (CCM) operation. Key converter parameters were derived and implemented using LTSpice, incorporating manufacturer-verified GaN device models to simulate realistic performance. The steady-state simulation results confirmed that all output voltages adhered closely to their design specifications, validating the effectiveness of the proposed architecture. Transformer voltage and current waveforms demonstrated proper core reset and minimal ripple, indicating stable operation
A comprehensive loss analysis was conducted to quantify core and copper losses in the transformer, as well as switching, conduction, and gate-drive losses in GaN devices. The total losses were calculated to be 3.824 W, yielding an overall efficiency of 97.13%. his simulation-based study lays a robust foundation for hardware implementation. The work underscores the potential of GaN devices in compact, high-performance space-grade power systems.

ACKNOWLEDGMENT

This work was carried out as part of a project funded by the Space Applications Centre, ISRO, Ahmedabad, under the RESPOND scheme. The authors thank SAC, Ahmedabad, for their support and encouragement.

REFERENCES

[1] J.Millan. P.Godignon, X. Perpina, A. Perez-Tomas, and J.Rebollo, "Asurvey of wide bandgap power semiconductor devices," IEEE Trans. Power Electron., vol. 29, no. 5, pp. 2155–2163, May 2014.

[2] Y. Uemoto et al., "Gate injection transistor (GIT)—A normally-off Al- GaN/GaN power transistor using conductivity modulation," IEEE Trans. Electron. Devices, vol. 54, no. 12, pp. 3393–3399, Dec. 2007.

[3] R. Ramachandran and M. Nymand, "Experimental Demonstration of a 98.8% Efficient Isolated DC–DC GaN Converter," in IEEE Transactions on Industrial Electronics, vol. 64, no. 11, pp. 9104-9113, Nov. 2017.

[4] E. A. Jones, F. F. Wang and D. Costinett, "Review of Commercial GaN Power Devices and GaN-Based Converter Design Challenges," in IEEE Journal of Emerging and Selected Topics in Power Electronics, vol. 4, no. 3, pp. 707-719, Sept. 2016.

[5] R. Sun, J. Lai, W. Chen and B. Zhang, "GaN Power Integration for High Frequency and High Efficiency Power Applications: A Review," in IEEE Access, vol. 8, pp. 15529-15542, 2020.

[6] A. K. Morya et al., "Wide Bandgap Devices in AC Electric Drives: Opportunities and Challenges," in IEEE Transactions on Transportation Electrification, vol. 5, no. 1, pp. 3-20, March 2019.

[7] A. Capel, D. O'Sullivan and J. . -C. Marpinard, "High-power conditioning for space applications," in Proceedings of the IEEE, vol. 76, no. 4, pp. 391-0408, April 1988.

[8] A. Vyas and T. Katiyar, "Single electronic power conditioner for integrated payloads of communication satellites," 2017 IEEE PES Asia-Pacific Power and Energy Engineering Conference (APPEEC), Bangalore, India, 2017

[9] Z. Zhang, B. He, D. -D. Hu, X. Ren and Q. Chen, "Multi-Winding Configuration Optimization of Multi-Output Planar Transformers in GaN Active Forward Converters for Satellite Applications," in IEEE Transactions on Power Electronics, vol. 34, no. 5, pp. 4465-4479, May 2019.

[10] J. -E. Park, J. -K. Han, S. -H. Choi and G. -W. Moon, "Two-Switch Forward Converter With an Integrated Buck Converter for High Bus Voltage in Satellites," in IEEE Transactions on Power Electronics, vol. 38, no. 2, pp. 2041-2051, Feb. 2023,

[11] A. Phillips, T. Cook, B. West and B. M. Grainger, "Gallium Nitride Efficacy for High-Reliability Forward Converters in Spacecraft," in IEEE Journal of Emerging and Selected Topics in Power Electronics, vol. 10, no. 5, pp. 5357-5370, Oct. 2022.

[12] T. V. Cook and B. M. Grainger, "GaN-Based, Buck-LLC Resonant Design and Maximum Efficiency Point Tracking Regulation for Space Power Systems," 2024 IEEE 11th Workshop on Wide Bandgap Power Devices & Applications (WiPDA), Dayton, OH, USA, 2024, pp. 1-6.

[13] D. Fu, P. Kong, F. C. Lee, and S. Wang, "Novel techniques to suppress the common mode EMI noise caused by transformer parasitic capacitances in dc–dc converter," in Proc. IEEE Energy Convers. Congr. Expo., Sep. 2010, pp. 1252–1259.

Modified State-Space Modeling and Voltage Mode Control Implementation for an Interleaved Boost Converter in Discontinuous Conduction Mode

1st Nabhomoni Ghoshal
Dept. of Electrical Engineering
NIT Rourkela, Odisha, India
nabhomonighoshal2@gmail.com

2nd Dalija Rath
Dept. of Electrical Engineering
NIT Rourkela, Odisha, India
rath.dalija@gmail.com

3rd Susovon Samanta
Dept. of Electrical Engineering
NIT Rourkela, Odisha, India
samanta.susovon@gmail.com

Abstract—High-power boost converters have become an important part of Electric vehicles (EV), solar power applications, and microgrids. The Interleaved Boost Converter (IBC) has an important advantage over conventional boost converters, such as input current ripple cancelation, enhanced efficiency, and high reliability. Detailed analysis and design are essential to reduce the size of N-phase IBC. Analyzing the inductor ripple current is crucial in selecting inductors and capacitors to reduce the size of the IBC. Depending on the application required the number of phases can be chosen. While the modified state-space modeling technique has been adapted for a standard boost converter operating in Discontinuous Conduction Mode (DCM), it has not been previously applied to an Interleaved Boost Converter (IBC). This work uses the modification to develop a small-signal model for an IBC operating in DCM. Additionally, Voltage Mode Control (VMC) is implemented to achieve a stable output voltage despite variations in load or input voltage. Simulation verification also takes place to ensure stable output voltage despite variations in input voltage or load conditions. The proposed framework and control strategy provides a more robust approach to the analysis and design of DCM converters.

Keywords: Discontinuous Conduction Mode (DCM), Interleaved Boost Converter (IBC), Small Signal Modeling, Voltage Mode Control (VMC).

I. INTRODUCTION

DC-DC converters, specifically step-up and step-down converters, are increasingly pivotal due to their numerous advantages in various applications [1], [2], [3]. Recently, there has been growing interest in efficient DC-DC converters capable of providing desired power conversion, particularly in fields such as renewable energy and high-medium power applications [4]. One type of step-up DC-DC converter is the boost converter, which elevates low input voltage to higher output DC voltage. However, boost converters have drawbacks such as significant voltage and current ripples, reverse recovery issues, and high voltage stress on semiconductor devices, which diminish system stability and efficiency [2].

Interleaving or multi-phasing methods have emerged as effective solutions [5]. Recently, research has focused on interleaved boost converters due to their potential for improved performance. In interleaved boost converters, multi-phasing helps minimize ripple in output voltage and input current [6], decreases switching losses, improves transient response, allows for smaller filter components, and enhances overall efficiency [7]. Increasing the number of interleaved stages can further optimize system performance.

Discontinuous Conduction Mode (DCM) is a common operating mode for DC/DC converters, particularly under light load conditions [8]. In low-power applications, many designers prefer DCM operation even at full load to mitigate issues such as diode reverse recovery. Additionally, DCM has been explored as a potential solution to the right-half-plane (RHP) zero problem that arises in boost-derived topologies. Accurate analytical models of DCM operation in interleaved boost converters are essential for performance analysis and control design. However, DCM operation introduces complexity in modeling and control due to the discontinuous nature of inductor currents. Full-order models, including analytical and averaged circuit models, provide a comprehensive approach to analyzing interleaved boost converters in DCM. While these models improve accuracy, discrepancies persist at higher frequencies, necessitating further refinement.

Recognizing the limitations of existing DCM modeling approaches, this work revisits previous models and introduces a modified full-order averaged model developed in both analytical and circuit forms, aiming to enhance accuracy and resolve previously identified issues. Furthermore, a Voltage Mode Control (VMC) algorithm is implemented to regulate the output voltage under varying load conditions, ensuring lesser steady-state error and a more stable, less sensitive control system.

II. GENERAL FRAMEWORK

In Discontinuous Conduction Mode (DCM), the operation of IBC differs from Continuous Conduction Mode (CCM) due to the presence of an additional interval in each switching cycle [9]. For an N-phase Interleaved Boost Converter shown

979-8-3315-3013-6/25 $31.00 © 2025 IEEE

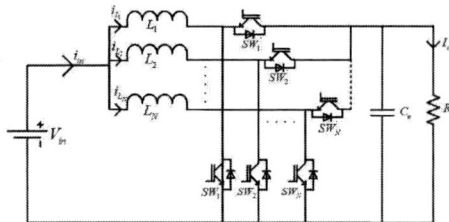

Fig. 1: N-phase Interleaved Boost Converter

in Fig. 1 with 2N no. of switches, the inductor current in DCM follows a distinct pattern. When the lower switches are turned on, the current initially increases, reaching its peak just before the switch turns off. Subsequently, it resets to zero (or a fixed value) by the end of the second interval. In the following discussions, the duty ratios of these two intervals are denoted as d_1 and d_2. Therefore in DCM operation, N-phase IBC has a 3N number of operating modes as shown in Fig. 2, where after 3 modes, it follows the same periodicity. The mode defines the time up to when the input current follows the same slope. As example, for the converter operating in $0 < d_1 < \frac{T_s}{N}$, the modes will be (1)-(3), as shown form Fig. 2.

$$M_1 = M_4 = \cdots = M_{3N-2} = d_1 T_s + d_2 T_s - \frac{T_s}{N} \quad (1)$$

$$M_2 = M_5 = \cdots = M_{3N-1} = \frac{T_s}{N} - d_2 T_s \quad (2)$$

$$M_3 = M_6 = M_9 = \cdots = M_{3N} = \frac{T_s}{N} - d_1 T_s \quad (3)$$

Additionally, the converter's power stage can be accurately represented using a piecewise linear state-space model. Let V_{in} denote the input voltage and T_s represent the switching cycle duration. Then,

$$\dot{x} = A_1 x + B_1 v_{in} \quad \text{for} \quad t \in M_1 \quad (4)$$

$$\dot{x} = A_2 x + B_2 v_{in} \quad \text{for} \quad t \in M_2 \quad (5)$$

$$\dot{x} = A_3 x + B_3 v_{in} \quad \text{for} \quad t \in M_3 \quad (6)$$

$$\vdots$$

$$\dot{x} = A_N x + B_N v_{in} \quad \text{for} \quad t \in M_{3N} \quad (7)$$

It is important to note that in DCM, the second duty ratio, d_2, is not an independent variable; rather, it exhibits an algebraic relationship with state and control variables. To develop an averaged model, this dependency must be expressed in terms of the average voltage and current values. By doing so, d_2 can be eliminated, resulting in a model defined solely by averaged state variables. This algebraic relationship, which governs the dependence of d_2, is referred to as the duty-ratio constraint.

The modeling approach for DCM operation consists of the following three key steps:

1) **Averaging** – Deriving the equivalent averaged equations to simplify system analysis.

2) **Inductor Current Analysis** – Evaluating the inductor current waveform to determine its influence on converter behavior.

3) **Duty-Ratio Constraint** – Establishing the algebraic relationship that links d_2 with the system state variables, ensuring an accurate averaged model.

III. AVERAGING MODEL & CORRECTION FACTOR FOR N-PHASE INTERLEAVED BOOST CONVERTER (IBC)

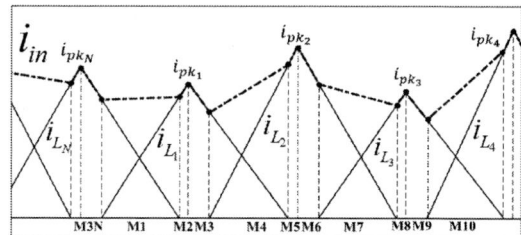

Fig. 2: Input Current and Inductor current of N-phase IBC operating in $0 < d_1 < \frac{T_s}{N}$

Averaging techniques in DCM are crucial for modeling interleaved boost converters, as they provide simplified representations of system dynamics. Unlike CCM, where state-space averaging directly applies, DCM requires special treatment due to the presence of a zero-current interval. Based on the waveform shown in Fig. 2. the inductor L_N currents can be written as (8)

$$\bar{i}_{L_N} = \frac{i_{pk_N}}{2}(d_1 + d_2) \quad (8)$$

Where i_{pk_N} is the peak current of that respective inductor. Assuming the voltage across the inductor is v_{in} during on-time. So, the inductor peak current can be written as,

$$i_{pk_N} = \frac{v_{in}}{L_N} \cdot d_1 T_s \quad (9)$$

Traditional state-space averaging may lead to inaccuracies, particularly in charge conservation and inductor current estimation. To address this, a refined approach involves directly averaging Kirchhoff's voltage and current laws while considering duty cycle dependencies. This ensures that the inductor current and capacitor voltage variations are accurately captured across switching cycles. In addition, a duty ratio constraint (10) is introduced [10], linking the duration of the second interval to the averaged state variables, thus improving the accuracy of the prediction.

$$\mathbf{M} = diag \left[\frac{1}{d_1 + d_2}, \cdots, \frac{1}{d_1 + d_2}, 1 \right] \quad (10)$$

For an Interleaved Boost Converter, where multiple phase-shifted inductors operate, these corrections become even more critical to maintaining stability and performance. The improved modeling framework ensures better small-signal response and control accuracy across varying loads and operating conditions.

IV. FULL ORDER AVERAGED MODEL

Unlike the conventional methods used in Continuous Conduction Mode (CCM), which depend on vol-sec balance, a different duty-ratio constraint will be derived here for defining d_2 in Discontinuous Conduction Mode (DCM) by substituting (8) into (9),

$$d_2 = \frac{2L\bar{i}_{L_N}}{d_1 T_s v_{in}} - d_1 \tag{11}$$

In CCM, the state-space averaging gives the same results as formal averaging but in DCM it will not provide the same. So a modification matrix (10) needs to multiply to predict the behavior in DCM correctly [11]. With this modified matrix, the averaged model becomes-

$$\dot{\bar{x}} = [A]\mathbf{M}\bar{x} + [B]v_{in} \tag{12}$$

where, $\mathbf{M}$ is the modification matrix (10), A is the system matrix and B is the input matrix of IBC operating in Discontinuous Conduction Mode (DCM).

To obtain the matrix A and B, system matrices, and input matrices for all 3N modes are to be derived. Therefore

$$A = (A_1 + A_4 + A_7 \cdots A_{N-2}) \left[d_1 + d_2 - \frac{1}{N}\right] T_s$$
$$+ (A_2 + A_5 + A_8 \cdots A_{N-1}) \left[\frac{1}{N} - d_2\right] T_s \tag{13}$$
$$+ (A_3 + A_6 + A_9 \cdots A_N) \left[\frac{1}{N} - d_1\right] T_s$$

and

$$B = (B_1 + B_4 + B_7 \cdots B_{N-2}) \left[d_1 + d_2 - \frac{1}{N}\right] T_s$$
$$+ (B_2 + B_5 + B_8 \cdots B_{N-1}) \left[\frac{1}{N} - d_2\right] T_s \tag{14}$$
$$+ (B_3 + B_6 + B_9 \cdots B_N) \left[\frac{1}{N} - d_1\right] T_s$$

Refer to Fig. 1. in Mode-1 $(d_1 + d_2 - \frac{1}{N})T_s$ one lower switch sw_1 and one upper switch sw'_N are on switches are On. Inductor L_1 is charging and L_N is discharging. Corresponding A, and B matrices can be derived using KCL and KVL. Since there are N+1 states N+1 corresponding dynamical equations are formulated as shown below:

$$L_1 \frac{di_{L_1}}{dt} = v_{in} \tag{15}$$

$$L_2 \frac{di_{L_2}}{dt} = L_3 \frac{di_{L_3}}{dt} = \cdots = L_{N-1} \frac{di_{L_{N-1}}}{dt} = 0 \tag{16}$$

$$L_N \frac{di_{L_N}}{dt} + v_c = v_{in} \tag{17}$$

$$C \frac{dv_c}{dt} = i_{L_N} - \frac{v_c}{R} \tag{18}$$

$$A_1 = \begin{bmatrix} 0 & 0 & \cdots & 0 & 0 \\ 0 & 0 & \cdots & 0 & 0 \\ 0 & 0 & \cdots & 0 & 0 \\ \vdots & \vdots & \vdots & \vdots & \vdots \\ 0 & 0 & \cdots & 0 & -\frac{1}{L_N} \\ 0 & 0 & \cdots & \frac{1}{C} & -\frac{1}{RC} \end{bmatrix}, \quad B_1 = \begin{bmatrix} \frac{1}{L_1} \\ 0 \\ 0 \\ \vdots \\ \frac{1}{L_N} \\ 0 \end{bmatrix} \tag{19}$$

Similarly for Mode-2 $(\frac{1}{N} - d_2)T_s$ only the lower switch sw_1 is on by which inductor L_1 is charging in Fig 1. Corresponding A, and B matrices can be derived using KCL and KVL.

$$L_1 \frac{di_{L_1}}{dt} = v_{in} \tag{20}$$

$$L_2 \frac{di_{L_2}}{dt} = L_3 \frac{di_{L_3}}{dt} = \cdots = L_N \frac{di_{L_N}}{dt} = 0 \tag{21}$$

$$C \frac{dv_c}{dt} = -\frac{v_c}{R} \tag{22}$$

$$A_2 = \begin{bmatrix} 0 & 0 & \cdots & 0 & 0 \\ 0 & 0 & \cdots & 0 & 0 \\ 0 & 0 & \cdots & 0 & 0 \\ \vdots & \vdots & \vdots & \vdots & \vdots \\ 0 & 0 & \cdots & 0 & 0 \\ 0 & 0 & \cdots & 0 & -\frac{1}{RC} \end{bmatrix}, \quad B_2 = \begin{bmatrix} \frac{1}{L_1} \\ 0 \\ 0 \\ \vdots \\ 0 \\ 0 \end{bmatrix} \tag{23}$$

In Mode-3 $(\frac{1}{N} - d_1)T_s$ only the upper switch sw'_1 is on through which L_1 is discharging as shown in Fig. 1. Corresponding A, and B matrices can be derived as:

$$L_1 \frac{di_{L_1}}{dt} + v_c = v_{in} \tag{24}$$

$$L_2 \frac{di_{L_2}}{dt} = L_3 \frac{di_{L_3}}{dt} = \cdots = L_N \frac{di_{L_N}}{dt} = 0 \tag{25}$$

$$C \frac{dv_c}{dt} = i_{L_1} - \frac{v_c}{R} \tag{26}$$

$$A_3 = \begin{bmatrix} 0 & 0 & \cdots & 0 & -\frac{1}{L_1} \\ 0 & 0 & \cdots & 0 & 0 \\ 0 & 0 & \cdots & 0 & 0 \\ \vdots & \vdots & \vdots & \vdots & \vdots \\ 0 & 0 & \cdots & 0 & 0 \\ \frac{1}{C} & 0 & \cdots & 0 & -\frac{1}{RC} \end{bmatrix}, \quad B_3 = \begin{bmatrix} \frac{1}{L_1} \\ 0 \\ 0 \\ \vdots \\ 0 \\ 0 \end{bmatrix} \tag{27}$$

Similarly, all the 3N no. of system matrices and input matrices need to be derived for every mode of operation for the N-phase IBC. After that using (13), and (14) the A, and B matrix can be obtained as (28) and (29).

979-8-3315-3013-6/25 $31.00 © 2025 IEEE   501

$$A = \begin{bmatrix} 0 & 0 & 0 & \cdots & -\frac{d_2}{L_1} \\ 0 & 0 & 0 & \cdots & -\frac{d_2}{L_2} \\ 0 & 0 & 0 & \cdots & -\frac{d_2}{L_3} \\ \vdots & \vdots & \vdots & \ddots & \vdots \\ 0 & 0 & 0 & \cdots & -\frac{d_2}{L_N} \\ \frac{d_2}{C(d_1+d_2)} & \frac{d_2}{C(d_1+d_2)} & \frac{d_2}{C(d_1+d_2)} & \cdots & -\frac{1}{RC} \end{bmatrix} \quad (28)$$

$$B = \left[\frac{d_1+d_2}{L_1}, \frac{d_1+d_2}{L_2}, \frac{d_1+d_2}{L_3}, \cdots, \frac{d_1+d_2}{L_N}, 0 \right]^T \quad (29)$$

After the state-space averaging of N-phase IBC and using (11)-(12) into (28), and (29) the full-order averaged model will be

$$\frac{d\bar{i}_{L_1}}{dt} = \frac{2\bar{i}_{L_1}}{d_1 T_s}\left[1 - \frac{\bar{v}_c}{v_{in}}\right] + \frac{d_1 \bar{v}_c}{L_1} \quad (30)$$

$$\frac{d\bar{i}_{L_2}}{dt} = \frac{2\bar{i}_{L_2}}{d_1 T_s}\left[1 - \frac{\bar{v}_c}{v_{in}}\right] + \frac{d_1 \bar{v}_c}{L_2} \quad (31)$$

$$\vdots$$

$$\frac{d\bar{i}_{L_N}}{dt} = \frac{2\bar{i}_{L_N}}{d_1 T_s}\left[1 - \frac{\bar{v}_c}{v_{in}}\right] + \frac{d_1 \bar{v}_c}{L_N} \quad (32)$$

$$\frac{d\bar{v}_c}{dt} = \frac{\bar{i}_{L_1} + \bar{i}_{L_2} + \cdots + \bar{i}_{L_N}}{C} - \frac{\bar{v}_c}{RC} \\ - \frac{d_1^2 T_s v_{in}}{2C}\left[\frac{1}{L_1} + \frac{1}{L_2} + \cdots + \frac{1}{L_N}\right] \quad (33)$$

The dc operating point of the IBC with a constant duty ratio $d_1 = D_1$ can be computed by right-hand sides of differential equations (30)-(33) equal to zero and by solving all the algebraic equations voltage gain can be found.

$$\frac{\bar{V}_c}{V_{in}} = \frac{1}{2} + \frac{1}{2}\sqrt{1 + 2D_1^2 T_s R\left[\frac{1}{L_1} + \frac{1}{L_2} + \cdots + \frac{1}{L_N}\right]} \quad (34)$$

$$\bar{I}_{L_N} = \frac{D_1^2 T_s V_{in} V_c}{2 L_N (V_{in} - V_c)} \quad (35)$$

The D_2 is not controllable and depends on the on-time duty D_1, no. of phase N, the inductor value, the switching frequency f_{sw}, and the load resistance R. For IBC operating in DCM, volt-second balance over a sampling period implies

$$D_2 = \frac{V_{in}}{V_c - V_{in}} \cdot D_1 \quad (36)$$

Solving equation (12) and (14) D_2 can be computed as

$$D_2 = \frac{D_1}{-\frac{1}{2} + \frac{1}{2}\sqrt{1 + 2D_1^2 T_s R\left[\frac{1}{L_1} + \frac{1}{L_2} + \cdots + \frac{1}{L_N}\right]}} \quad (37)$$

V. SMALL SIGNAL MODEL AND TRANSFER FUNCTION

By using linearization techniques, the small-signal model can be derived from (30)-(33).

$$\frac{d}{dt}\begin{bmatrix} \hat{i}_{L_1} \\ \hat{i}_{L_2} \\ \hat{i}_{L_3} \\ \vdots \\ \hat{i}_{L_N} \\ \hat{v}_c \end{bmatrix} = A \begin{bmatrix} \hat{i}_{L_1} \\ \hat{i}_{L_2} \\ \hat{i}_{L_3} \\ \vdots \\ \hat{i}_{L_N} \\ \hat{v}_c \end{bmatrix} + B \begin{bmatrix} \hat{d}_1 \\ \hat{v}_{in} \end{bmatrix} \quad (38)$$

where

$$A = \begin{bmatrix} \frac{2(V_{in}-V_c)}{V_{in}D_1 T_s} & 0 & 0 & \cdots & -\frac{D_1 V_{in}}{L_1(V_c-V_{in})} \\ 0 & \frac{2(V_{in}-V_c)}{V_{in}D_1 T_s} & 0 & \cdots & -\frac{D_1 V_{in}}{L_2(V_c-V_{in})} \\ 0 & 0 & \frac{2(V_{in}-V_c)}{V_{in}D_1 T_s} & \cdots & -\frac{D_1 V_{in}}{L_3(V_c-V_{in})} \\ \cdots & \vdots & \vdots & \ddots & \vdots \\ 0 & 0 & 0 & \cdots & -\frac{D_1 V_{in}}{L_3(V_c-V_{in})} \\ \frac{1}{C} & \frac{1}{C} & \frac{1}{C} & \cdots & -\frac{1}{RC} \end{bmatrix} \quad (39)$$

$$B = \begin{bmatrix} \frac{2V_c}{L_1} & \frac{D_1 V_c}{L_1(V_c-V_{in})} \\ \frac{2V_c}{L_2} & \frac{D_1 V_c}{L_2(V_c-V_{in})} \\ \frac{2V_c}{L_3} & \frac{D_1 V_c}{L_3(V_c-V_{in})} \\ \vdots & \vdots \\ \frac{2V_c}{L_N} & \frac{D_1 V_c}{L_N(V_c-V_{in})} \\ \frac{-D_1 T_s V_{in}}{C}\sum_{T=1}^{N}\frac{1}{L_T} & \frac{-D_1^2 T_s}{2C}\sum_{T=1}^{N}\frac{1}{L_T} \end{bmatrix} \quad (40)$$

From the small signal model, many transfer functions can be derived.

$$\dot{x} = Ax + Bu \quad (41)$$

After Laplace transformation,

$$sx(s) = Ax(s) + Bu(s) \quad (42)$$

From here the state-to-input transfer function becomes,

$$\frac{x(s)}{u(s)} = (sI - A)^{-1}B \quad (43)$$

VI. VOLTAGE MODE CONTROL

Voltage Mode Control (VMC) is a widely used control strategy in power electronics, where the output voltage of a converter is regulated by directly comparing it with a reference voltage. The error between the reference and actual output voltage is processed through a compensator, typically a proportional-integral (PI), which generates the duty cycle for the PWM signal. This approach ensures stable and precise voltage regulation in power converters.

In a three-phase interleaved boost converter, voltage mode control is implemented to maintain a stable output voltage

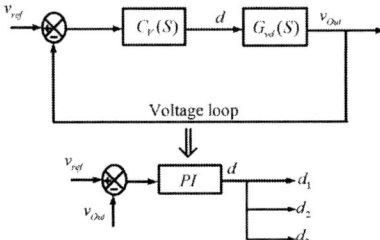

Fig. 3: Control structure of VMC for $3 - \phi$ IBC

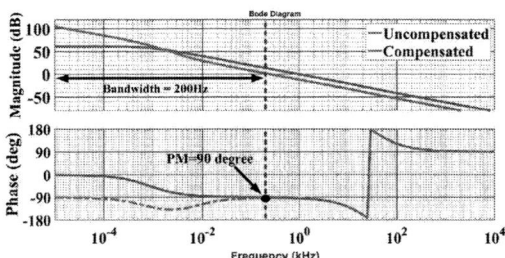

Fig. 4: Bode plot for Uncompensated and Compensated system

despite variations in input voltage or load conditions. The control strategy involves the voltage sensing, error processing, and PWM generation To implement the voltage mode control state-to-control transfer function $\frac{\hat{v}_c}{\hat{d}_1}$ need to be derive. For three-phase IBC, A and B matrix are derived from (39), and (40) -

$$A = \begin{bmatrix} \frac{2(V_{in}-V_c)}{V_{in}D_1T_s} & 0 & 0 & -\frac{D_1V_{in}}{L_1(V_c-V_{in})} \\ 0 & \frac{2(V_{in}-V_c)}{V_{in}D_1T_s} & 0 & -\frac{D_1V_{in}}{L_2(V_c-V_{in})} \\ 0 & 0 & \frac{2(V_{in}-V_c)}{V_{in}D_1T_s} & -\frac{D_1V_{in}}{L_3(V_c-V_{in})} \\ \frac{1}{C} & \frac{1}{C} & \frac{1}{C} & -\frac{1}{RC} \end{bmatrix}$$
(44)

$$B = \begin{bmatrix} \frac{2V_c}{L_1} & \frac{D_1V_c}{L_1(V_c-V_{in})} \\ \frac{2V_c}{L_2} & \frac{D_1V_c}{L_2(V_c-V_{in})} \\ \frac{2V_c}{L_3} & \frac{D_1V_c}{L_3(V_c-V_{in})} \\ \frac{-D_1T_sV_{in}}{C}\left[\frac{1}{L_1}+\frac{1}{L_2}+\frac{1}{L_3}\right] & \frac{-D_1^2T_s}{2C}\left[\frac{1}{L_1}+\frac{1}{L_2}+\frac{1}{L_3}\right] \end{bmatrix}$$
(45)

From (21), the state-to-control transfer function will be-

$$\frac{\hat{v}_c}{\hat{d}_1} = G_{vd} = (sI - A)^{-1}B$$
(46)

Putting all the parameter specifications mentioned in the simulation section, the transfer function is

$$G_{vd} =$$
$$\frac{-5053s^3 - 6.41 \times 10^8 s^2 + 1.95 \times 10^{14}s + 2.69 \times 10^{19}}{s^4 + 4.9 \times 10^5 s^3 + 8 \times 10^{10} s^2 + 4.3 \times 10^{15}s + 2.4 \times 10^{16}}$$
(47)

VII. PI TUNING

In the control structure of VMC is shown in Fig. 3, the proportional-integral (PI) controller is designed to keep constant bus voltage regardless of the variation in load and phase switching. The loop gain transfer function of the compensated system is given by equation (47). The frequency response of the compensated system is shown in Fig. 4.

$$T_{compensated} = (K_P + \frac{K_I}{s}) \cdot G_{vd}$$
(48)

A PI controller is designed to increase the low-frequency gain. The values of K_p and K_I are obtained as 0.255 and 10.098

respectively. In Fig. 4 magnitude plot, the compensated system improved stability by reducing excessive high-frequency gain, and better low-frequency gain control, enhancing system accuracy. From phase plot compensated system ensures a slower phase drop, increasing the phase margin and overall system stability and improved phase response around the critical frequency regions, ensuring better dynamic performance and reduced oscillations.

VIII. SIMULATION RESULTS

To validate the effectiveness of the proposed closed-loop Voltage Mode Control strategy, simulations were performed using MATLAB/Simulink with $V_{in} = 200$, $T_S = 50\mu sec$, $R = 125\Omega$, $L_1 = 294\mu H$, $L_2 = 270\mu H$, $L_3 = 275\mu H$, $C = 4700\mu F$. The interleaved boost converter was modeled with the designed control algorithm, and its performance was evaluated under various load conditions.

Fig. 5 illustrates the output voltage response of the converter. It can be seen that the control system successfully regulates the output voltage to the desired level, even in the presence of load variations. The transient response shows minimal overshoot and the steady-state error is effectively reduced, demonstrating the robustness of the proposed control approach. Furthermore, the inductor current waveforms,

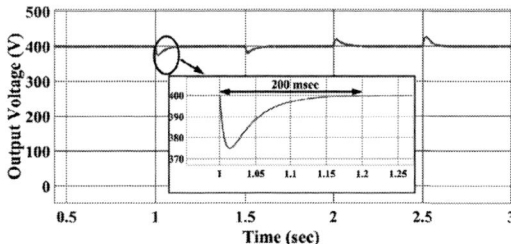

Fig. 5: Controlled Output Voltage after implementing VMC

shown in Fig. 6, confirm that the interleaved operation is maintained, reducing current ripple and improving overall efficiency. The system response under sudden load changes is also analyzed, as depicted in Fig. 5, where the controller effectively restores the output voltage with a fast settling time.

The simulation results validate that the proposed voltage-mode control algorithm ensures stable operation, improves dynamic performance, and effectively mitigates voltage deviations due to load transients.

979-8-3315-3013-6/25 $31.00 © 2025 IEEE

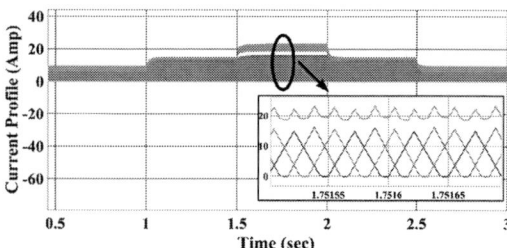

Fig. 6: Closed loop simulation result of inductor current with load change

IX. EXPERIMENTAL RESULT

A laboratory prototype IBC has been developed to validate the proposed idea as shown in Fig 7. SEMIKRON SKM75GB12T4 IGBT Modules are used for every leg for switching purposes. SEMIKRON SKYPER 32 R is used as a gate-driver circuit for IGBT switches. The PWM signal is generated through the C2000 microcontroller. A Coilcraft-made inductor with 294 μH, 270 μH, 275 μH rated to operate up to 105 °C is used. An output capacitor of 4700 μF, 450V is used. A Keysight-made 2024A series mixed signal oscilloscope has been used to acquire the results. A Keysight N8948A programmable DC power used as supply . The experiment is conducted with an input voltage of 200V.

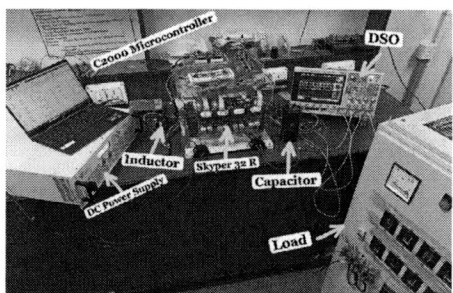

Fig. 7: Laboratory Experimental setup

Inductors carried inductor current ripple of 86.6 Amp, 7.4 Amp, and 7.2 Amp respectively. Fig. 8 shows the experimental results. The experimental results were consistent with the theoretical predictions derived from the equations. The results

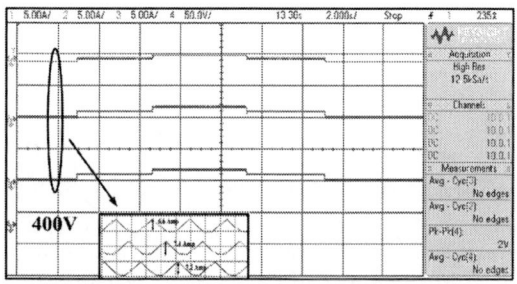

Fig. 8: Experimental Result of Implementation of VMC in three-phase Interleaved Boost Converter

exhibited a close correlation, with negligible deviations, thus reinforcing the reliability of the experimental observations.

X. CONCLUSION

High-power IBC play a crucial role in applications such as electric vehicles, solar power systems, and microgrids, where efficient power conversion and reduced current ripple are essential. This work presents a small-signal model for an IBC operating in DCM, an area that has not been extensively explored. The proposed Voltage Mode Control strategy ensures a stable output voltage under varying load and input conditions, addressing key challenges in power electronics. The simulation results validate the effectiveness of the proposed control technique, demonstrating minimal deviation of the output voltage and improved dynamic performance. The interleaved structure significantly reduces input current ripple, making the system more suitable for energy sources like batteries and fuel cells. By optimizing the control strategy and converter design, this research contributes to the reliability and efficiency of high-power boost converters. The findings of this study establish a solid foundation for further advancements in DCM converter modeling and control, paving the way for more efficient and scalable power conversion systems.

REFERENCES

[1] Xiao, Y., Li, Q., Zheng, J., Liu, X., Huangfu, Y. and Li, Z.P., 2024. Design and control studies of six-phase interleaved boost converter for integrated energy efficiency improvement of green ship. Journal of Energy Storage, 96, p.112549.

[2] Muthukumar, M., Rengarajan, N., Velliyangiri, B., Omprakas, M.A., Rohit, C.B. and Raja, U.K., 2021. The development of fuel cell electric vehicles–A review. Materials Today: Proceedings, 45, pp.1181-1187.

[3] Khalid, M.R., 2021, June. Design and Evaluation of IBC for EV Applications. In 2021 International Conference on Intelligent Technologies (CONIT) (pp. 1-7). IEEE.

[4] Hossain, M.Z. and Rahim, N.A., 2018. Recent progress and development on power DC-DC converter topology, control, design and applications: A review. Renewable and Sustainable Energy Reviews, 81, pp.205-230.

[5] Yedukondalu, G., Samanta, S. and Joshi, M., 2023. Analysis, design, and minimum phase selection of high power interleaved DC–DC converter. International Journal of Circuit Theory and Applications, 51(1), pp.322-339.

[6] Sathi, R.R., Dhiwakar, J. and Vijay, S., 2025, March. Current Mode Controlled Interleaved KY Boost Converter With Enhanced Response. In 2025 IEEE International Conference on Interdisciplinary Approaches in Technology and Management for Social Innovation (IATMSI) (Vol. 3, pp. 1-6). IEEE.

[7] Fayrouz, F.A., Elgendy, M.A., Dahidah, M. and Muhammad, M., 2020, December. Analysis of dcm interleaved boost converter for pv ac-module application. In The 10th International Conference on Power Electronics, Machines and Drives (PEMD 2020) (Vol. 2020, pp. 343-348). IET.

[8] Yanarateş, C., Zhou, Z. and Altan, A., 2024. Investigating the impact of discretization techniques on real-time digital control of DC-DC boost converters: A comprehensive analysis. Heliyon, 10(20).

[9] Cheng, X. and Xie, G.J., 2009, February. Full order models and simulation of boost converters operating in dcm. In 2009 International Conference on Electronic Computer Technology (pp. 632-635). IEEE.

[10] Dionizio, A.A., Sampaio, L.P. and da Silva, S.A.O., 2025. Generalized state-space averaging modeling to fourth-order power converters operating in DCM. International Journal of Circuit Theory and Applications, 53(2), pp.1031-1055.

[11] Sun, J., Mitchell, D.M., Greuel, M.F., Krein, P.T. and Bass, R.M., 2001. Averaged modeling of PWM converters operating in discontinuous conduction mode. IEEE Transactions on power electronics, 16(4), pp.482-492.

Performance Analysis of IUPQC and DPFC devices in a Distributed Generation Unit employing Intelligent Controllers

C.Srinivasa Rao
EEE
G.Pullaiah College of Engineering and
Technology
Kurnool, India
principal@gpcet.ac.in

V.Sowmyasree
EEE
Ashoka Women's Engineering College
Kurnool, India
sowmya.sree14@gmail.com

G. Panduranga Reddy
EEE
G.Pullaiah College of Engineering and
Technology
Kurnool, India
gprreee@gmail.com

Abstract— **Optimization techniques have a long history, with foundational contributions from mathematicians such as Newton, Lagrange, and Cauchy. The development of differential calculus by Newton and Leibnitz played a crucial role in advancing optimization methods. With the introduction of high-speed digital computers, the application of optimization algorithms became more practical, driving the development of new and innovative approaches. In recent years, mathematical programming techniques such as Simulated Annealing, genetic algorithms, and other evolutionary methods have gained considerable attention. This research focuses on addressing power quality challenges, including voltage sag, swell, and harmonic distortions, using IUPQC and DPFC systems. These systems are integrated with intelligent control strategies, including Fuzzy Logic Controllers and Genetic Algorithm-based Adaptive Neuro-Fuzzy Inference System (ANFIS) controllers. A comprehensive harmonic analysis is also conducted for each control method to evaluate their effectiveness.**

Keywords—IUPQC, DPFC, FLC, ANFIS, Optimization, THD

I. INTRODUCTION

In the realm of power systems, maintaining power quality remains a significant challenge due to disturbances like voltage sag, swell, and harmonic distortions. These issues can adversely affect sensitive equipment, leading to operational inefficiencies and economic losses. To mitigate such challenges, advanced controllers like the Interline Unified Power Quality Conditioner (IUPQC) and Distributed Power Flow Controller (DPFC) have been extensively studied.

This research focuses on the integration of intelligent control strategies, specifically Fuzzy Logic Controllers (FLC) and Genetic Algorithm-based Adaptive Neuro-Fuzzy Inference System (ANFIS) controllers, to enhance the performance of IUPQC and DPFC systems. By employing these optimization techniques, the study aims to effectively regulate voltage levels, reduce harmonic distortions, and ensure stable power delivery. A comprehensive harmonic analysis is conducted to evaluate the comparative effectiveness of these controllers, providing valuable insights into their practical implementation for improved power quality management.

II. LITERATURE REVIEW

Various researchers have explored optimization techniques using genetic algorithms (GA) and fuzzy logic controllers (FLC). Dilip Kumar Pratihar introduced concepts like hard, soft, and hybrid computing, emphasizing the role of GA and FLC in optimization. A. Homaifar and E. McCormick highlighted the potential of integrating GA and FLC to create autonomous controllers, reducing human intervention. M. A. Lee and H. Takagi proposed a GA-based design for fuzzy systems that optimizes membership functions and rule parameters, demonstrating its effectiveness through an inverted pendulum problem.

M.S. Osman et al. applied a GA–FLC to solve nonlinear programming problems, achieving faster and more accurate results. Kocaarslan and I. Cam implemented a fuzzy gain scheduled PI controller for load-frequency regulation in interconnected power systems, showing improved performance over traditional PI controllers. D. Narasimha Rao and P. Srinivas Varma examined control methods for Distributed Power Flow Controllers (DPFC), employing techniques like sliding mode, fuzzy, and neural network controls, validated using MATLAB/SIMULINK.

R. Pavan Kumar Naidu and S. Meikandasivam introduced a DPFC using PQ theory and a FOPID controller, achieving effective voltage regulation and harmonic elimination. Ahmed N. Alsammak and Hasan A. Mohammed improved power quality with an FLC-based UPFC, enhancing power factor, voltage regulation, and transient stability compared to a PID controller, as confirmed through MATLAB simulations.

III. PROPOSED SYSTEM

The IUPQC is a dual-structure version of the UPQC, consisting of both shunt and series converters, similar to the conventional UPQC. In this setup, the shunt converter functions as an AC voltage source, while the series converter acts as an AC current source, both operating at the fundamental frequency. This configuration enables harmonic voltage generation across the series converter and facilitates the compensation of harmonic currents through the shunt converter. Unlike the UPQC, the IUPQC does not require prior determination of harmonic voltages and currents for compensation. Additionally, it effectively synthesizes these harmonics, significantly minimizing switching losses. Consequently, the IUPQC overcomes the limitations of the UPQC and provides a more efficient solution for high-power applications [1].

The IUPQC controller, as seen in Figure 1, shares structural similarities with a UPQC. It is made up of a common DC connection connecting two power electronic converters. A transformer linked in series with the supply voltage (Vs) connects the series converter to the electrical grid, whereas a coupling transformer positioned in parallel with the load installs the shunt converter on the load side. Both converters are regulated through PWM-based voltage and current control methods to provide effective compensation at the source and load terminals [2].

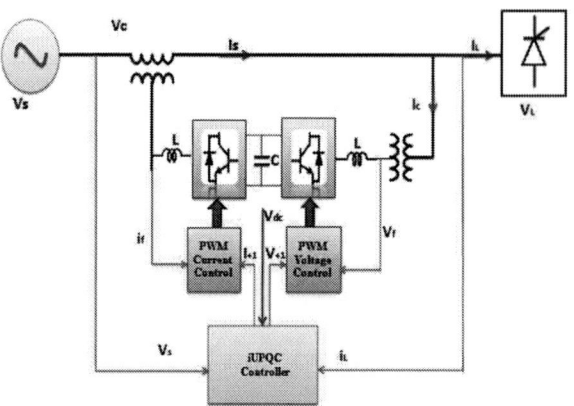

Fig.1. Schematic diagram of IUPQC

The IUPQC produces a positive sequence voltage at the fundamental frequency, enabling the series converter to draw a positive sequence current (Is) from the source. This current accounts for the total active power needed by the load, along with the internal losses of the IUPQC. The current (Is) is maintained in phase with the supply voltage (Vs). Furthermore, the series converter acts as an infinite impedance to harmonic currents, while the shunt converter offers low impedance, efficiently filtering out harmonics. Acting as a shunt active filter, the shunt converter eliminates harmonic currents produced by nonlinear loads. As a result, the IUPQC ensures a clean, balanced, and sinusoidal voltage (VL) at the load terminals.

Internal Structure of IUPQC Controller

Fig.2 illustrates the internal control structure of the IUPQC, demonstrating how the reference voltages and currents are computed and extracted. The controller takes three-phase source voltages (Vabc), load currents (IL_abc), and the DC link voltage (VDC) as inputs. It then generates the necessary reference voltages and currents for the shunt and series converters.

To calculate these reference signals, the PQ theory is applied, where the three-phase components are converted into two-phase components. These components are then processed using a Phase-Locked Loop (PLL) circuit, which ensures synchronization by accurately determining the frequency and phase angle of the supply voltage's positive sequence component (Vs). The resulting reference signals are transmitted to the respective series and shunt controllers for effective compensation. Furthermore, the IUPQC also functions as a STATCOM, providing additional control to regulate grid voltage, enhancing the overall stability and reliability of the power system.

The current drawn by the series converter fulfils the average active power demand and compensates for losses within the converter. Figure 2 illustrates the output of the PI controller, which serves as an additional control signal. This ensures that the IUPQC functions as a STATCOM, providing supplementary grid voltage regulation. Consequently, the average active power (PL) is determined. Additionally,

another PI controller is employed, which receives input by comparing the actual DC link voltage (Vdc) with the reference voltage (Vref). This PI controller adjusts the active power to offset the inherent losses of the power converters.

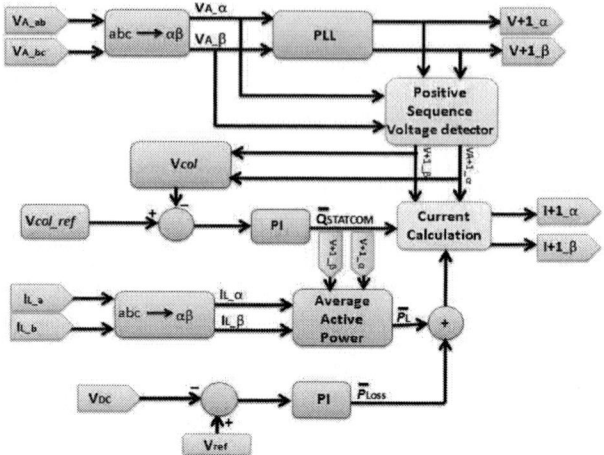

Fig.2 Internal Structure of IUPQC

IV. FLC BASED IUPQC

Fig. 3 shows Simulink block diagram of fuzzy logic controller based IUPQC device. It can be inferred that Fuzzy Logic Controller based IUPQC device compensates power quality issues in voltage and current waveforms. Active power and reactive power of 10MW and 5MVar are supplied by DFIG machine at load point B3 Fig.4. Due to presence of non-linear load, harmonics occur during t = 1.1 to 1.3 sec.

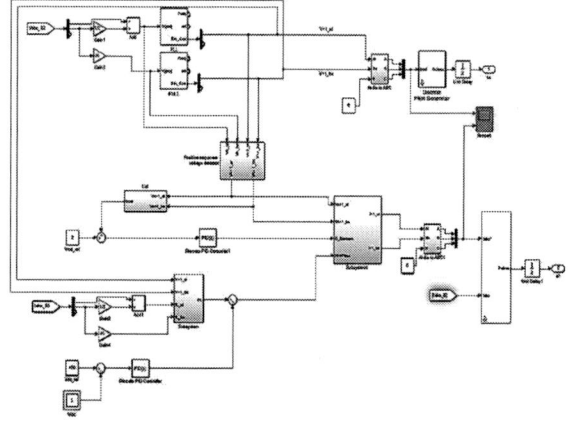

Fig.3 Circuitry for the IUPQC Fuzzy Logic Controller

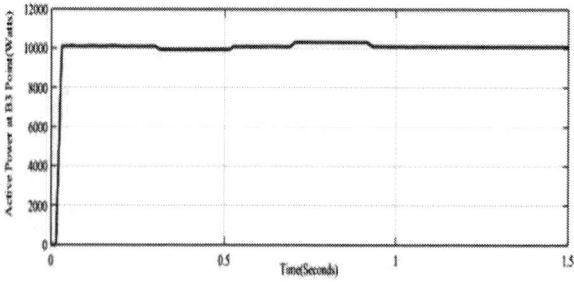

Fig.4 Active Power

979-8-3315-3013-6/25 $31.00 © 2025 IEEE

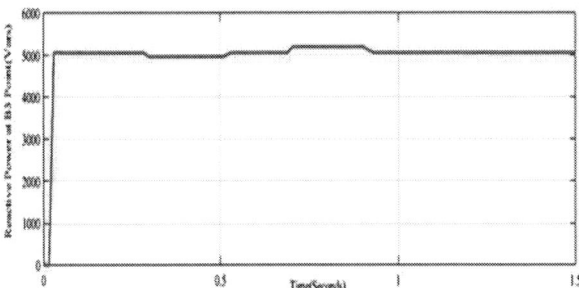

Fig.5. Reactive Power

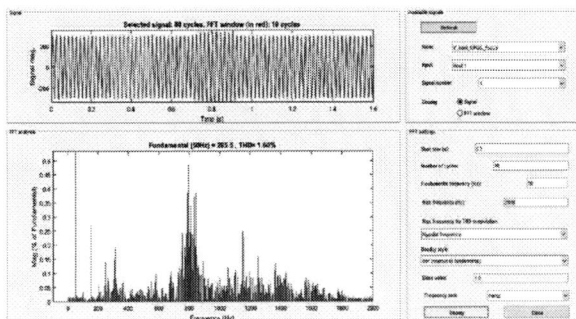

Fig.6 %THD at Sag Condition

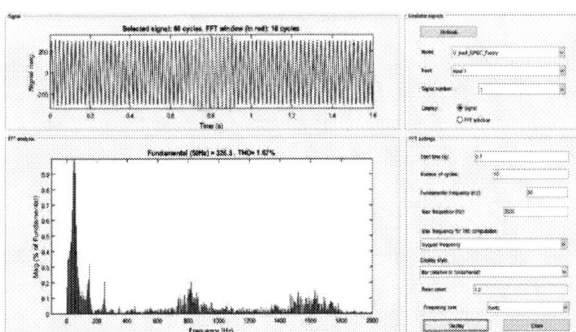

Fig.7 %THD at Swell Condition

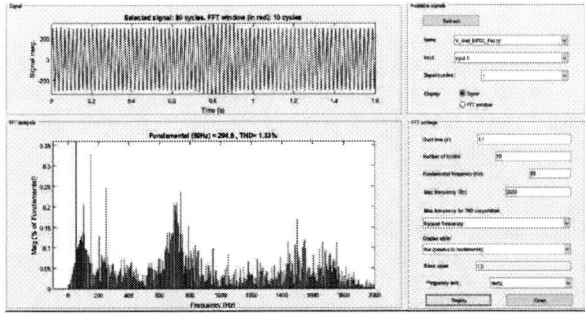

Fig.8 % THD due to non-linear load

The Total Harmonic Distortion (THD) of the load voltage under various conditions is illustrated in Figures 6, 7, and 8. During a voltage sag at $t = 0.3$ seconds, the THD is measured at 1.60%. In the case of a voltage swell at $t = 0.7$ seconds, the THD increases slightly to 1.67%. When harmonics are introduced by a non-linear load at $t = 1.1$ seconds, the THD is recorded at 1.33%.

V. ANFIS BASED IUPQC

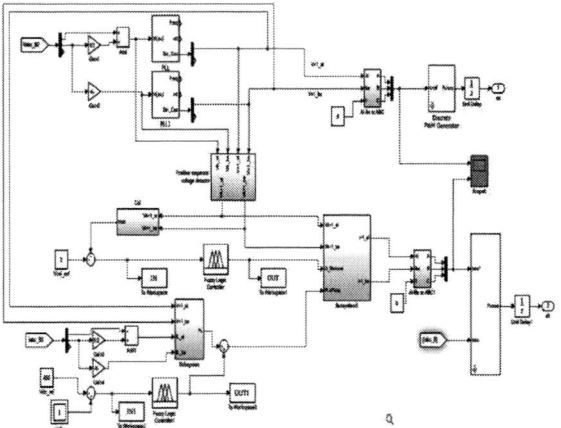

Fig.9 Circuitry for IUPQC-ANFIS Controller

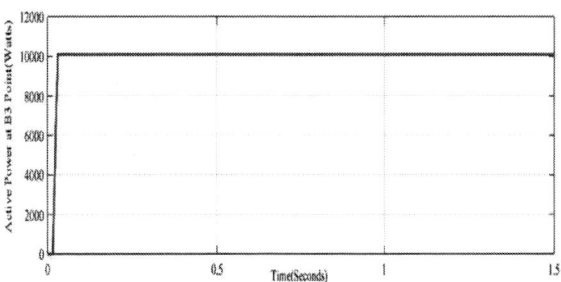

Fig.10 Active Power

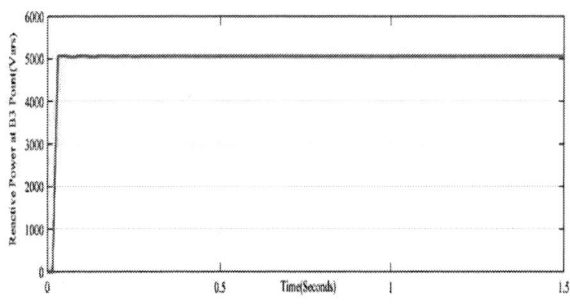

Fig.11 Reactive Power

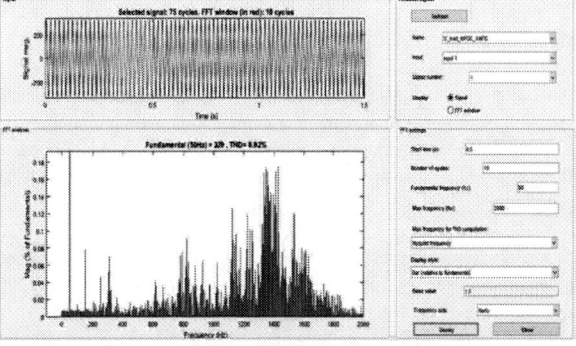

Fig.12 % THD in during voltage sag at t=0.3 sec

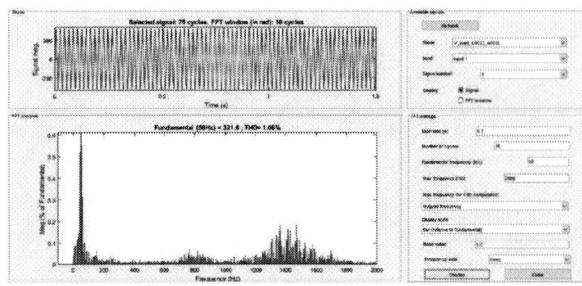

Fig.13 %THD during swell at t=0.7 sec (during voltage swell)

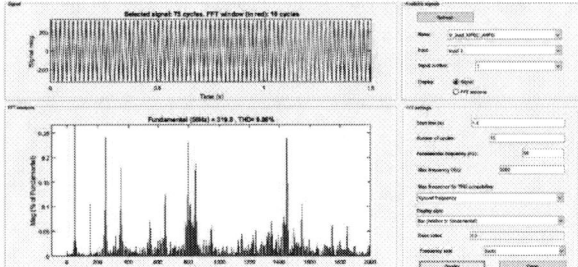

Fig.14 %THD at harmonics due to non-linear load t=1.1sec

The IUPQC device's MATLAB/Simulink ANFIS control mechanism is displayed in Fig. 9. Figures 10 and 11 demonstrate that the DFIG machine at load point B3 supplies 10MW of active power and 5MVar of reactive power. This indicates that there is no sag and that the waveform experienced swell.

Figures 12, 13, and 14 illustrate the percentage of Total Harmonic Distortion (%THD) in the load voltage under different conditions: voltage sag, swell, and harmonic distortion.

TABLE I. %THD COMPARATIVE ANALYSIS OF VARIOUS CONTROLLERS WITH IUPQC DEVICE

%THD for Vload at	IUPQC with FLC	IUPQC with ANFIS
t = 0.3	1.60	0.93
t = 0.7	1.67	1.06
t = 1.1	1.33	0.86

VI. DISTRIBUTED POWER FLOW CONTROLLER

The Distributed Power Flow Controller (DPFC) consists of shunt and distributed static series compensators designed to regulate both active and reactive power [3]. It also provides compensation for negative and zero-sequence current components. Unlike traditional systems, the DPFC enhances flexibility by eliminating the need for a DC link, allowing independent operation of the shunt and series converters within a Unified Power Flow Controller (UPFC) structure [4]. By employing multiple single-phase converters instead of one large three-phase converter, the DPFC reduces component ratings and improves system reliability through redundancy. The internal structure of the DPFC, depicted in Fig.15, features three main controllers. The central controller oversees the entire system, coordinating the actions of other controllers. The series and shunt controllers manage voltage and current harmonics, ensuring stability. Additionally, local

controllers are positioned at each converter terminal to implement the converter control strategy effectively.

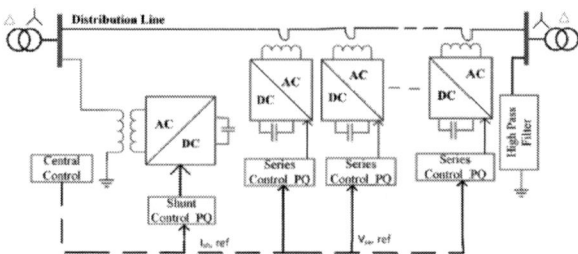

Fig.15 Internal Circuit of Distributed Power Flow Controller

Central controller covers the balancing of asymmetrical components and the regulation of electrical energy transmission [5]. The signal benefits both shunt and series controllers by providing reference voltage signals based on system requirements, all generated at the fundamental frequency. Using the third harmonic frequency component, the DC voltage of the series converter may be kept constant. It supplies the central controller with the necessary voltage in a series arrangement. The series filter controls voltage fluctuations in the event of distribution system problems, avoiding sagging, swelling, and other disturbances [6] [7]. The supply and line voltages are compared in order to introduce a comparison signal, also known as an error signal, into the PLL. After receiving the pulses, the PWM controller will send them to the switches as needed. The hysteresis controller has input limits and should only be used within a certain hysteresis-band (h) in order to activate the series filter. The shunt controller introduces current harmonics into the line to transfer real power to the series converter. It is commonly used in power systems for efficiently managing reactive current, ensuring stable DC voltage across a capacitor. The harmonics can be precisely adjusted using shunt control [8]. The controller operates based on the instantaneous reactive power theory, which enables the conversion of 3-phase currents and voltages into α-β-0 coordinates. By adjusting the reference currents at the load terminal, reactive, neutral, and harmonic currents are effectively compensated. A comparison of the reference currents with the actual source currents is performed, and the PWM controller generates the necessary switching signals based on the resulting error.

VII. PROPOSED PV-WIND DG UNIT WITH FLC BASED DPFC

In the DPFC system, gate pulses for the converter are generated by a fuzzy controller [9] [10]. First, the DC voltage is contrasted with a predetermined reference value. The Fuzzy Logic Controller (FLC) then processes the resultant error to produce the necessary power for controlling the shunt controller. By comparing the reference value with the filter current, a PWM controller is in charge of delivering pulses to the converter. It is possible to measure power losses inside the filter by controlling the DC voltage. Fuzzy-set theory and the principles of human thinking underpin the FLC's operations. Fuzzification, rule-base interfacing, and defuzzification are the three main parts of the FLC [11]. The DC voltage from the shunt controller is constantly compared

979-8-3315-3013-6/25 $31.00 © 2025 IEEE

to a reference value inside the DPFC. The FLC uses the detected error as input to calculate the required power adjustment for the shunt controller [12]. The foundation of efficient controller operation is the creation of fuzzy control rules, which entails defining links between input variables and system properties.

A. Simulation Results of Proposed System with DPFC Operated by Fuzzy Logic Controller:

The simulation results of the PV/Wind hybrid system with a DPFC controlled by a Fuzzy Logic Controller show that load power reaches 130 kW with solar and wind energy until t = 0.2 seconds. During the voltage sag from t = 0.2 to 0.4 seconds, load power drops. At t = 0.5 seconds, an additional 50 kW increases the load to 170 kW. From t = 0.6 to 0.8 seconds, a voltage swell raises the load power to 190 kW, stabilizing at 170 kW after t = 0.8 seconds. The Total Harmonic Distortion (%THD) is 1.85% during the voltage sag and 1.69% during the voltage swell.

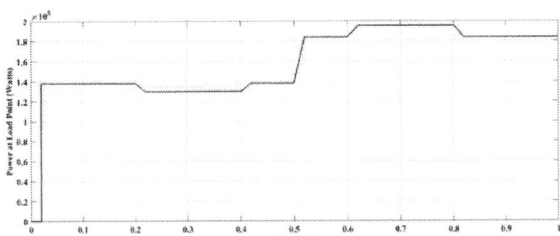

Fig. 16 Power at Load Point

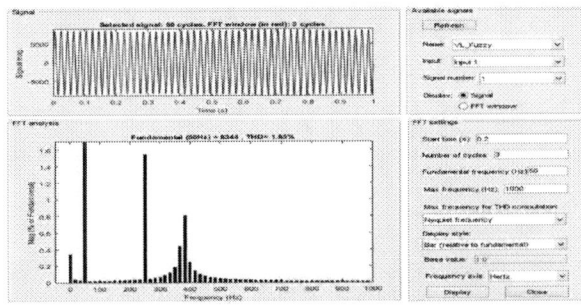

Fig.17 % THD at Sag Condition

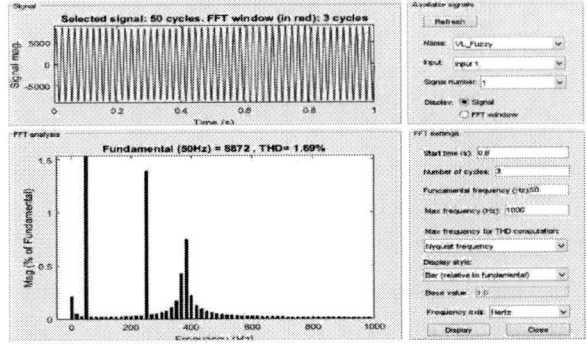

Fig.18 % THD at Swell Condition

B. Simulation results of the DPFC System Using a Genetic Algorithm-Driven Fuzzy Logic Controller

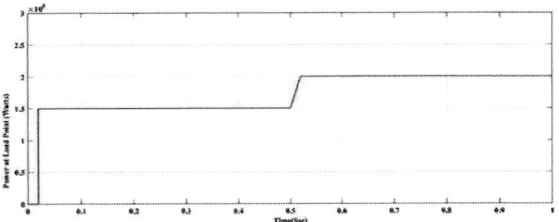

Fig. 19 Load Power

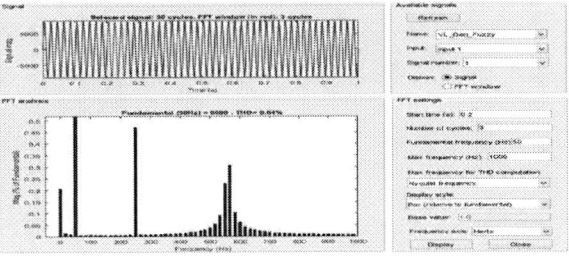

Fig 20 %THD at Sag Condition

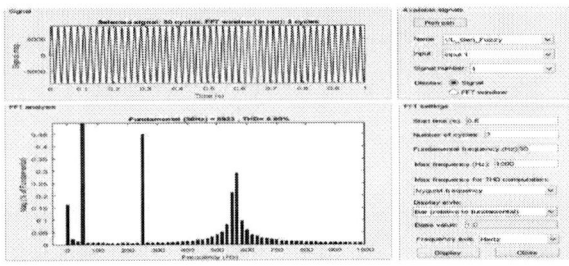

Fig 21 %THD at Swell Condition

To reduce the error between the output generated by ANFIS and the output from the nonlinear dynamic system, the parameters of ANFIS are optimized using a Genetic Algorithm (GA).

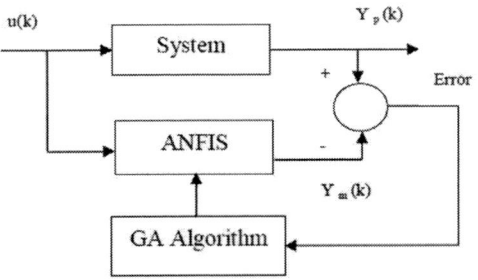

Fig.22 Block Diagram GA-ANFIS method

C. Simulation Results of PV-Wind Hybrid System with GA-ANFIS Based DPFC:

Figure 23 illustrates the active power required by the load, which remains at 150 KW until 0.5 seconds and increases to 200 KW afterward. The harmonic spectra of the load voltage are shown in Figures 24 and 25. The Total Harmonic Distortion (THD) is 0.03% at both 0.2 seconds and 0.6 seconds.

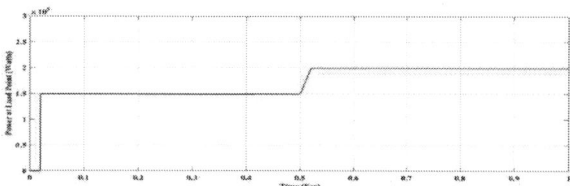

Fig 23. Power at Load Point

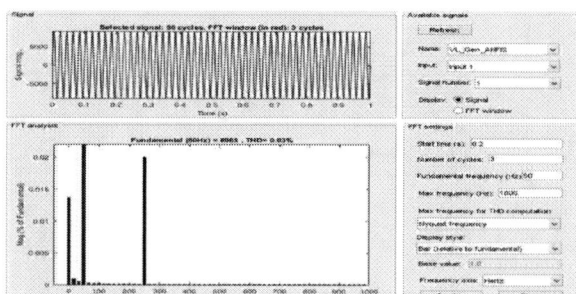

Fig 24. % THD at Sag Condition

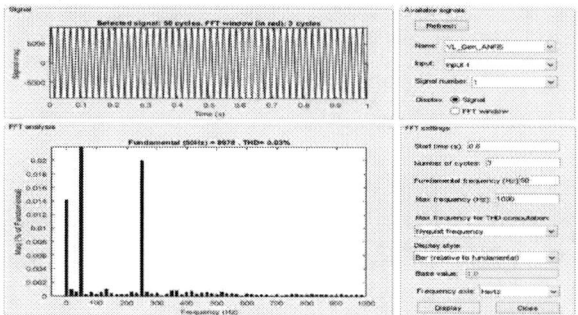

Fig 25. % THD at Swell Condition

TABLE II. %THD OF PROPOSED SYSTEM USING DPFC WITH INTELLIGENT CONTROLLERS

%THD at Load Point	DPFC with FLC	DPFC with Genetic based FLC	DPFC with Genetic based ANFIS
t = 0.3 sec	1.85	0.64	0.03
t = 0.6 sec	1.69	0.60	0.03

VIII. CONCLUSIONS

The Interline Unified Power Quality Conditioner (IUPQC) and the Distributed Power Flow Controller (DPFC) are both advanced devices used to enhance power quality and maintain grid stability. The IUPQC consists of series and shunt converters connected to two different feeders, providing voltage and current compensation to mitigate power quality issues such as sags, swells, and harmonics. In contrast, the DPFC, derived from the Unified Power Flow Controller (UPFC), employs multiple small single-phase converters in place of a large three-phase converter, eliminating the need for a common DC link. This distributed structure increases system reliability, redundancy, and operational flexibility. Furthermore, the DPFC offers independent control of active and reactive power, providing superior compensation for negative and zero sequence components. Its simplified design and effective fault tolerance make it more cost-efficient and reliable. Therefore, due to its enhanced flexibility, reliability, and robust power quality control, the DPFC proves to be a more effective solution compared to the IUPQC.

REFERENCES

[1] Sibtain ul Hassan, Zain ul Abideen, Prof. Dr. Tahir Izhar, "Advanced Control Techniques for Micro-grids Power Quality Improvement", 2017 Asian Conference on Energy, Power and Transportation Electrification (ACEPT),24-26 Oct. 2017, pp 1-6, DOI: 10.1109/ACEPT.2017.8168564

[2] G Pandu Ranga Reddy, "Analysis of an IUPQC Device Using Conventional PID and FOPID Controllers in a Wind Energy Conversion System", Distributed Energy Resources and Electric Vehicle, CRC Press, 2024, Pp: 202-222

[3] Zhihui yuan, Sjoerd W.H de Haan, Braham Frreira and Daliborcevoric "A FACTS DEVICE: Distributed power flow controller (DPFC)", IEEE transaction on power electronics Vol.25, No.10 October 2010, pp- 2564 – 2572, DOI: 10.1109/TPEL.2010.2050494

[4] Akhib Khan Bahamani, Sreerama Reddy G.M, V. Ganesh, "Comparative of performance for UPFC with DPFC", 2017 International Conference on Electrical, Instrumentation and Communication Engineering (ICEICE2017).

[5] Z. H. Yuan, S. W. H de Haan, and B. Frreira "DPFC control during shunt converter failure," 2009 IEEE Energy Conversion Congress and Exposition, 20-24 September 2009, DOI: 10.1109/ECCE.2009.5316070.

[6] Yong Xue, Jiamei Deng, Shuangbao Ma, "Power flow control of a distributed generation unit in micro-grid", IEEE 6th International Power Electronics and Motion Control Conference, IEEE, 2009.

[7] D Narasimha Rao, P Srinivas Varma., "Enhancing the Performance of DPFC with Different Control Techniques", International Journal of Innovative Technology and Exploring Engineering (IJITEE) ISSN:2278-3075, Volume-8 Issue-6, April 2019, pp 1002-1007.

[8] R. Pavan Kumar Naidu, S. Meikandasivam, "Power quality enhancement in a grid-connected hybrid system with coordinated PQ theory & fractional order PID controller in DPFC", Sustainable Energy, Grids and Networks, Volume 21, March 2020, 100317 Elsevier Ltd, doi.org/10.1016/j.segan.2020.100317.

[9] Ahmed N. Alsammak, Hasan A. Mohammed, "Power quality improvement using fuzzy logic controller based unified power flow controller (UPFC)", Indonesian Journal of Electrical Engineering and Computer Science Vol. 21, No. 1, January 2021, pp. 1-9.

[10] Aditya Raut, Sameer S. Raut, "Review: Different Technology for Distributed Power Flow Controller", International Research Journal of Engineering and Technology (IRJET), Volume: 06 Issue: 03 | Mar 2019, pp. 6803-6809.

[11] V.Sowmya sree, "Design of Fuzzy Logic Controller-Based DPFC Device for Solar-Wind Hybrid System", Springer Nature Singapore, International Conference on Computational Intelligence and Data Engineering, 2022, Pp: 123-140

[12] Dr.G.Pandu Ranga Reddy, "Fuzzy Logic-Based Pitch Angle Control for Variable Speed Wind Systems" 2023 9th International Conference on Advanced Computing and Communication Systems (ICACCS), 2023, IEEE,Pp: 559-566

979-8-3315-3013-6/25 $31.00 © 2025 IEEE

Condition Monitoring of Transmission Line Vibration Dampers using YOLOv12 Model

Dipanjana Chowdhury
Department of Electrical Engineering
National Institute of Technology
Rourkela - 769008, INDIA
chydipanjana2023@gmail.com

Satyajit Panigrahy
Department of Electrical Engineering
National Institute of Technology
Rourkela - 769008, INDIA
satyajit.panigrahy007@gmail.com

Subrata Karmakar
Department of Electrical Engineering
National Institute of Technology
Rourkela - 769008, INDIA
karmakar.subrata@gmail.com

Abstract—Vibration dampers play a crucial role in maintaining the structural stability of overhead transmission lines by mitigating mechanical oscillations caused by wind and other environmental factors. Detecting defects, such as broken components, is essential to prevent failures that may compromise the integrity of power systems. This study proposes an advanced single-stage deep learning model for the condition monitoring of vibration dampers. A public dataset was utilized and expanded using image augmentation techniques to address the challenge of limited data availability. The YOLOv12 model was employed for defect localization, leveraging its real-time detection capabilities and high accuracy. Experimental results demonstrate that the model achieved an accuracy of 94.7% in identifying and localizing defective dampers. Additionally, a web application was developed to assist the maintenance crew, facilitating efficient and automated inspections. The proposed solution enhances the reliability of defect detection while streamlining the inspection process.

Index Terms—Condition Monitoring, Deep Learning, Image Augmentation, Object Detection, Vibration Dampers.

I. INTRODUCTION

Overhead transmission lines play a vital role in efficiently transferring electrical power across long distances, yet they face significant environmental challenges, particularly from wind-induced vibrations [1]. These lines are subject to two primary vibrations: aeolian and galloping. Aeolian vibrations are characterized by high frequencies (ranging from 3 to 150 Hz) and low amplitudes, typically triggered by steady winds. Conversely, galloping vibrations occur at lower frequencies and feature high amplitudes, often caused by wind interacting with ice-covered conductors, resulting in large oscillations that can lead to conductor clashes [2]. Such vibrations can induce fatigue, wear, and damage to conductors, fittings, and supporting structures, adversely affecting the longevity and reliability of transmission lines.

To address these challenges, vibration dampers are installed to mitigate the effects of these vibrations. However, dampers can develop defects over time due to mechanical wear, environmental exposure, and improper installation. Defective dampers are less effective at controlling vibrations, which escalates stress on conductors supporting structures, thereby increasing the risk of conductor fatigue and breakage. Common issues with dampers include corrosion, loosened fittings, and material fatigue, all of which hinder their capacity to absorb and dissipate energy efficiently [3]. Therefore, regular and proper maintenance of vibration dampers is essential for the reliable operation of transmission lines, reducing the risks of outages and preventing costly repairs.

Early damper detection methods, such as the AdaBoost classifier and cascaded AdaBoost classifier utilizing Haar features, achieved an accuracy of 89.15%. However, their shallow architectures made them ineffective for high-resolution image processing [4] [5]. To improve performance, researchers investigated image processing techniques like threshold segmentation and the Modified Balloon Snake method for edge detection. These approaches, although innovative, struggled with scalability and robustness in complex environments [6]. Subsequent studies addressed challenges like rust defects, introducing metrics such as Rusty Area Ratio (RAR) and Color Shade Index (CSI) to assess damage severity. They employed local difference processing and anisotropic Gaussian kernels (ANGK) for image preprocessing, achieving a detection accuracy of 93% with thresholding and morphological segmentation [7]. Nonetheless, traditional machine learning techniques still face challenges in detecting small or occluded dampers in complex backgrounds. Their dependency on handcrafted features limits adaptability to diverse conditions, and they often require extensive parameter tuning, hindering their effectiveness for real-time and large-scale inspections.

Advancements in deep learning have addressed challenges faced by traditional machine learning models, particularly through CNN, two-stage models like R-CNN and Faster R-CNN, as well as single-stage models like SSD and You Only Look Once (YOLO) [8]. In [9], the authors proposed an optimized RetinaNet CNN based on the FreeAnchor method for accurate vibration damper detection. The optimization involved data enhancement

979-8-3315-3013-6/25 $31.00 © 2025 IEEE

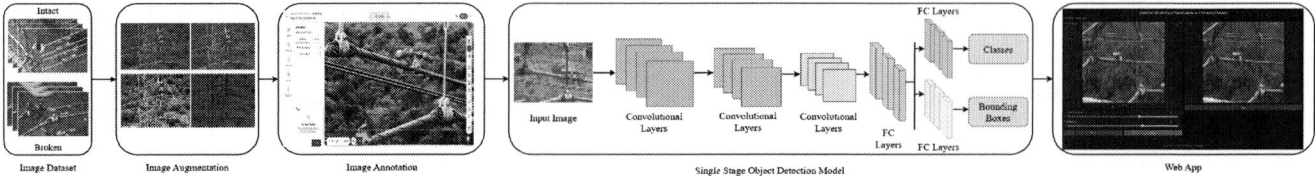

Fig. 1: Overview of the Transmission Line Vibration Damper Defect Detection System

and model structure improvements, resulting in a significant performance boost over the original RetinaNet and other classical detection methods, achieving an accuracy of 83.1%. H. Liang *et al.* [10] developed the Wire_10 dataset for defect detection in transmission line dampers using a Faster R-CNN model with a ResNet 101 backbone, achieving a mAP of 83.23% via transfer learning. However, limited data per category led to false and missed detections, especially in winter and non-rural areas. X. Lieu *et al.* [11] proposed an intelligent fault diagnosis method for damper slippage, incorporating a distance constraint method that groups detected dampers for effective diagnosis. This integration with the Faster R-CNN model enables automatic identification of slippage from aerial images. Despite their advantages, these models face limitations due to the lack of comprehensive public datasets, slower inference speeds, and increased computational complexity, which hinder real-time UAV inspections and the detection of small objects in high-resolution images, impacting performance in challenging outdoor environments.

Among the various developed models, the YOLO models have emerged as one of the most effective approaches for real-time object detection due to their high speed and accuracy. W. Bao *et al.* [12] introduced a parallel mixed attention and K-means clustering method for anchor optimization, combined with a multi-stage transfer learning strategy to enhance the YOLOv4 network for small object detection in cluttered environments. This approach achieved a mAP@50 of 93.8% on a UAV dataset. Further, W. Chen *et al.* [13] enhanced the YOLOv4 model by utilizing the Canny algorithm to extract edge information from the original image. They incorporated this edge information into ResNet101 using an attention mechanism to improve feature extraction. This approach produced a feature map optimized for small target detection, leveraging a feature pyramid network structure. L. Lu *et al.* [14] developed a defect detection system utilizing an "edge-cloud-end" collaboration to address high bandwidth consumption and response delays associated with cloud server-based approaches. Their study employed a modified YOLOv5 model that reduces computational complexity while effectively capturing correlations between distant pixels, allowing for targeted analysis of critical defect regions. On the edge computing platform, the model achieved a mAP of 86.6% and an inference speed of 63 frames per second.

The rapid advancement of deep learning has propelled image generation algorithms, particularly Generative Adversarial Networks (GANs), into a key research focus in computer vision. DamperGAN [15] introduced a multi-granularity Conditional Generative Adversarial Network (CGAN) that produces low-resolution images, utilizes Monte Carlo search to extract latent information, and applies an attention mechanism to generate high-resolution outputs. Similarly, W. Bao *et al.* [16] developed an automatic detection method for component defects using unmanned aerial vehicle patrol inspections. They employed an end-to-end coordinate attention and bidirectional feature pyramid network variant of YOLO called BC-YOLO, achieving a mAP@50 of 89.1%, surpassing YOLOv5 by 2.7%.

Vibration damper detection has progressed from traditional image processing methods to sophisticated deep-learning models. However, challenges persist, such as effectively processing high-resolution images, limited datasets for specific defect types, and balancing detection speed with accuracy. These issues underscore the continuous need for innovation and improvement in damper condition monitoring to maintain the reliability of power transmission systems. The novelty of this work includes several key contributions:

1) Image augmentation techniques were employed on a public dataset to address the issue of limited image availability.
2) The advanced YOLOv12 single-stage object detection model effectively captured a wider range of patterns and intricate details in the images.
3) A web application was developed to enable remote monitoring, offering the inspection team an efficient real-time defect detection and analysis tool.

II. METHODOLOGY

This section provides a concise overview of the workflow illustrated in Fig. 1. It encompasses the public damper dataset, its expansion through augmentation techniques, the annotation process for model training, defect detection using the YOLOv12 model, and the deployment of a web application for the maintenance crew.

A. Dataset Details

This study employs the InsPLAD: Inspection of Power Line Assets Dataset [17], a public dataset comprising 1385 images of vibration dampers, including both intact

979-8-3315-3013-6/25 $31.00 © 2025 IEEE

Fig. 2: Vibration Damper Dataset Overview [17]

TABLE I: Overview of Pixel Transformations

Transformations	Description
AutoContrast	Adjusts image contrast to optimize pixel value distribution.
Blur	Applies Gaussian blur to reduce sharpness.
Salt and Pepper Noise	Simulates noise with random white and black pixels.
Solarize	Inverts pixel values above a threshold.
RGB Shift	Shifts red, green, and blue channels for color variations.
Random Rain	Adds raindrop-like streaks to simulate rain.
ToGray	Converts images to grayscale.
Equalize	Enhances contrast using histogram equalization.
CLAHE	Performs adaptive histogram equalization with contrast limiting.
Sharpen	Enhances sharpness by emphasizing edges and details.

(good) and broken (defective) conditions. To enhance the dataset's diversity and robustness, 1390 additional images were generated using data augmentation techniques, capturing variations in environmental conditions. The overall dataset of 2775 images was divided into a training sets comprising 2220 images and a test set of 555 images, maintaining an 80:20 ratio. Fig.2 shows examples of the vibration damper dataset.

B. Image Augmentation

Image augmentation techniques were applied using the Albumentations library [18] to enhance the dataset and improve the model's generalization ability. These techniques involve pixel-level and spatial transformations to create augmented images while retaining the essential features of the objects as described in Table. I. By introducing controlled distortions, noise, and color variations, augmentation minimizes overfitting and enhances the model's performance on unseen data.

Algorithm 1 Image Augmentation using Albumentations

1: **Input Image Loading:**
2: - Read the input image using libraries like OpenCV or PIL.
3: **Define Augmentations:**
4: - Choose the desired transformations using Albumentations..
5: **Apply Transformations:**
6: - Apply selected transformation techniques.
7: **Output Augmented Image:**
8: - Finalize augmented images for training.

In this study, 1390 additional images were generated using Albumentations, effectively doubling the dataset

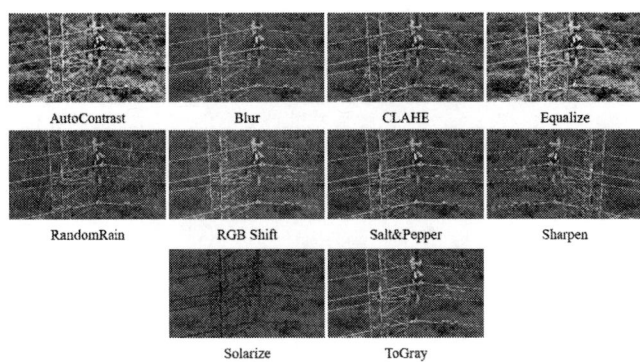

Fig. 3: Image Augmentations

size. The process of image augmentation using that tool is outlined in Algorithm 1. A set of 10 pixel-level transformations was applied to 139 randomly selected images from the original dataset. A few examples of images with these transformations from the dataset are shown in Fig. 3.

C. Image Annotation

Image annotation involves labeling images with relevant information, such as bounding boxes, masks, or keypoints, to indicate object locations and classifications. It is a crucial step in training computer vision models as it provides supervised data for machine learning algorithms, enabling them to effectively learn object detection tasks. Accurate annotation enhances model accuracy and robustness. In this work, Roboflow efficiently annotated the entire dataset, ensuring high-quality labeled data for training the deep learning model, as shown in Fig. 4.

D. YOLOv12 Model Overview

Single-stage object detection models like YOLO, SSD, FCOS, RetinaNet, and EfficientDet operate by predicting objects directly, eliminating the necessity for distinct region proposal networks and classifiers. This streamlined approach facilitates quicker inference times compared to the more complex two-stage detection models. YOLO

Fig. 4: Roboflow Interface for Image Annotation

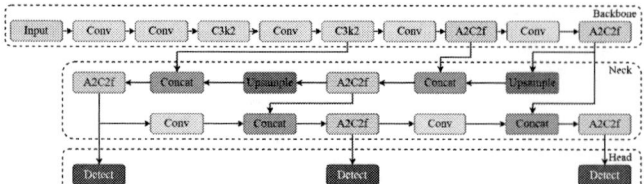

Fig. 5: Model Architecture YOLOv12

stands out in this space for its exceptional real-time object detection performance and has undergone significant advancements through various iterations, notably from YOLOv1 to the latest YOLOv12, each version improving upon the last to enhance speed and accuracy [19].

YOLOv12 [20], the latest YOLO model, features groundbreaking enhancements such as the Area-Attention Enhanced Cross-Feature module (A2C2f), 7×7 separable convolutions, and FlashAttention-driven area-based attention. These advancements improve feature extraction, efficiency, and detection capabilities. The attention-centric design segments feature maps for better focus on critical areas, while the FlashAttention module reduces memory usage and enables near real-time processing at high resolutions. Additionally, the A2C2f alleviates gradient bottlenecks and enhances feature fusion, preserving spatial context with fewer parameters.

The YOLOv12 model can be divided into three main components: the backbone, which extracts and processes multi-scale features; the neck, which aggregates and refines those features; and the head, which generates the final predictions. The overall architecture of the YOLOv12 model is shown in Fig. 5.

The YOLOv12 model's backbone extracts features from input images through multiple convolutional layers that downsample and deepen the representation. C3k2 modules enhance feature quality, while R-ELAN (Efficient Layer Aggregation Networks) blocks integrate multi-scale features. The backbone culminates in A2C2f (Area Attention + Convolution) blocks, which apply attention mechanisms to highlight critical regions. The neck then aggregates and refines these multi-scale features using an area attention mechanism accelerated by FlashAttention, improving focus in cluttered scenes. Finally, the head converts the refined feature maps into bounding box coordinates and classification scores, incorporating streamlined multi-scale detection pathways and specialized loss functions to balance localization and classification objectives.

III. RESULTS AND DISCUSSIONS

This section discusses the hardware and software configurations, training parameters, and the performance of the proposed YOLOv12 model for vibration damper condition monitoring. Additionally, comparisons are made with existing literature and methodologies for similar tasks, highlighting improvements in efficiency and reliability.

Finally, the trained model is proposed for deployment in a web application, enabling streamlined defect detection and supporting power line inspection teams in real-time operations.

A. Hardware and Software Specifications

Hardware-Software Specification The experiments were conducted on a computing environment running Windows 11, equipped with an NVIDIA RTX 4090 GPU boasting 24 GB of memory and 64 GB of RAM. Python served as the programming language, while CUDA11.7 was leveraged for parallel computing, and cuDNN 8.4.0 provided accelerated performance. The proposed model was implemented using the PyTorch framework as the backend. This high-performance hardware setup enabled swift processing and robust execution of the experiments, guaranteeing dependable and precise results.

B. Training Parameter Details

The training settings for YOLO models are critical as they encompass various hyperparameters and configurations that significantly impact the model's performance, speed, and accuracy. Key training settings include batch size, learning rate, momentum, and weight decay. The YOLOv12n model underwent training for 100 epochs with a batch size of 32 in this study. The learning rate was set to 0.001, and the IoU threshold was 0.7. The training utilized the SGD optimizer with a momentum of 0.937 and a weight decay of 0.0005.

C. Evaluation Metrics

The evaluation of the proposed methodology is conducted by utilizing statistical measures derived from the confusion matrix, specifically False Positive (FP), True Positive (TP), True Negative (TN), and False Negative (FN). Various performance metrics, including precision (P), recall (R), average precision (AP), mean average precision (mAP), where C denotes the number of classes, mean average precision at the threshold 50% (mAP @ 50), and intersection over union (IoU) are evaluated based on the equations reported in the Table. II.

TABLE II: Object Detection Performance Indices

Evaluation Parameters	Formula
IoU	$\frac{Area of Overlap}{Area of Union}$
P	$\frac{True Positive}{All Detection}$
R	$\frac{True Positive}{All Ground Truth}$
AP	$\sum_0^1 P(R)dR$
mAP	$\frac{1}{C}\sum_{c=1}^{c} AP_c$
mAP@50	$\frac{1}{C}\sum_{c=1}^{c} AP_c \vert IoU \geq 0.5$

(a) Intact Dampers

(b) Broken Dampers

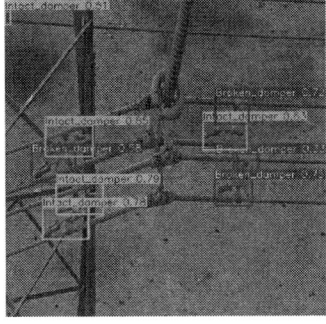
(c) Intact and Broken Dampers

Fig. 6: Prediction of YOLOv12n Model

TABLE III: Performance Evaluation of YOLOv12n

Damper Condition	Precision	Recall	mAP@50	mAP@50-95
Intact	95.7%	96.00%	98.1%	74.4%
Broken	94.5%	85.8%	91.2%	49.1%
Overall	95.1%	90.9%	94.7%	61.8%

D. YOLOv12 Model Performance

The YOLOv12 model features key advancements such as the A2C2f, 7×7 separable convolutions, and FlashAttention-driven attention, effectively addressing information bottlenecks and gradient reliability for efficient learning from complex data patterns. YOLOv12n has 159 layers, 6.3 GFLOPs, and 25.57 million parameters, making it lightweight for seamless integration with embedded systems in real-time applications. Fig. 6 shows the model's output and confidence scores, while Table. III lists condition assessments for damper defect classes.

E. Comparative Analysis

A comparative analysis of established diagnostic methods was performed to evaluate the effectiveness of the proposed approach, with results shown in Table. IV. The table reveals that the proposed strategy outperforms both feature-based and region-based object detectors, highlighting its potential as a viable method for assessing vibration damper conditions.

TABLE IV: Comparison with Previous Models

Ref.	Method	Model	mAP (%)
[4]	Feature	AdaBoost	89.15
[7]	ML	ANGK	93
[9]	CNN	RetinaNet	83.1
[10]	Two Stage	Faster R-CNN	83.23
Proposed	Single Stage	YOLOv12	94.7

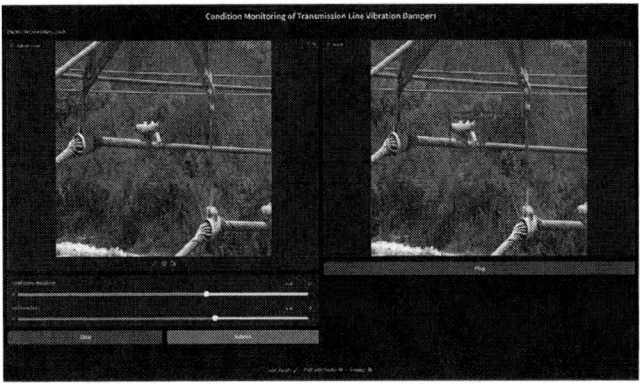
Fig. 7: Web App Interface for Remote Monitoring

F. Web App for Maintenance Team

An intuitive web application developed with Gradio and the YOLOv12 model enhances defect detection tasks. This app simplifies transmission line vibration damper inspections, allowing operators to assess damper status easily. As shown in Fig. 7, the user-friendly interface enables image uploads and displays detection results with confidence scores and adjustable thresholds. This aids inspection teams in making informed decisions for timely defect management.

IV. CONCLUSION

This study utilized an advanced single-stage deep learning model to detect defects in overhead transmission line vibration dampers effectively. A public dataset was augmented using pixel transformation techniques before training the YOLOv12 model to address the challenge of limited training images. The proposed model achieved an impressive mAP@50 of 94.7%, highlighting feature extraction, convolution, and attention module enhancements. This remarkable performance demonstrates the model's capability to accurately identify and distinguish between 'intact' and 'broken' dampers. Furthermore, the robustness of the proposed model was validated through a comparative analysis with earlier models. Future work will focus on creating a comprehensive dataset that includes a wider range of types of damper defects, including wear, corrosion, and structural deformations. Furthermore, it will facilitate a more nuanced analysis of defect patterns, contributing to a deeper understanding of damper failure mechanisms and enabling more effective detection strategies.

REFERENCES

[1] L. Yang, J. Fan, Y. Liu, E. Li, J. Peng and Z. Liang, "A Review on State-of-the-Art Power Line Inspection Techniques," *IEEE Transactions on Instrumentation and Measurement*, vol. 69, no. 12, pp. 9350-9365, 2020.

[2] S. Roy, and C. K. Kundu, "State of the art review of wind induced vibration and its control on transmission towers, *Structures*, Vol. 29, pp. 254-264, 2021.

[3] J. Wang, "Overhead Transmission Line Vibration and Galloping," 2008 International Conference on High Voltage Engineering and Application, pp. 120-123, 2008.

[4] Y. Liu and L. Jin, "Vibration Damper Recognition of Transmission System Based on Unmanned Aerial Vehicles," 2011 Asia-Pacific Power and Energy Engineering Conference, pp. 1-3, 2011.

[5] L. J. Jin, S. Yan, and Y. Liu, "Vibration damper recognition based on Haar-like features and cascade AdaBoost classifier," *J. Syst. Simul*, vol. 24, no. 9, pp.1086-1089, 2012.

[6] W. Haibin, X. Yanping, F. Weimin, S. Xiaoming and J. Li, "Damper Detection in Helicopter Inspection of Power Transmission Line," 2014 Fourth International Conference on Instrumentation and Measurement, Computer, Communication and Control, pp. 628-632, 2014.

[7] X. Huang, X. Zhang, Y. Zhang and L. Zhao, "A Method of Identifying Rust Status of Dampers Based on Image Processing," *IEEE Transactions on Instrumentation and Measurement*, vol. 69, no. 8, pp. 5407-5417, 2020.

[8] M. A. A. Faisal, I. Mecheter, Y. Qiblawey, J. H. Fernandez, M. E. Chowdhury, and S. Kiranyaz, "Deep Learning in Automated Power Line Inspection: A Review," *Applied Energy*, vol. 385, p. 125507, 2025.

[9] M. Ma, D. Zhang and F. Liu, "The detection of vibration dampers based on optimized RetinaNet," 2024 IEEE Conference on Artificial Intelligence (CAI), pp. 1399-1403, 2024.

[10] H. Liang, C. Zuo and W. Wei, "Detection and Evaluation Method of Transmission Line Defects Based on Deep Learning," *IEEE Access*, vol. 8, pp. 38448-38458, 2020.

[11] X. Liu, Y. Lin, H. Jiang, X. Miao, J. Chen, "Slippage fault diagnosis of dampers for transmission lines based on faster R-CNN and distance constraint," *Electric Power Systems Research*, vol. 199, p. 107449, 2021.

[12] W. Bao, Y. Ren, N. Wang, G. Hu, and X. Yang, "Detection of abnormal vibration dampers on transmission lines in UAV remote sensing images with PMA-YOLO," *Remote Sensing*, vol. 13, no. 20, p.4134, 2021.

[13] W. Chen, Y. Li, Z. Zhao, "Transmission Line Vibration Damper Detection Using Deep Neural Networks Based on UAV Remote Sensing Image," *Sensors*, vol. 22, p. 1892, 2022.

[14] L. Lu, Z. Chen, R. Wang, L. Liu, and H. Chi, "Yolo-inspection: defect detection method for power transmission lines based on enhanced YOLOv5s," *Journal of Real-Time Image Processing*, vol. 20, no.5, p.104, 2023.

[15] W. Chen, Y. Li, Z. Zhao, "Transmission Line Vibration Damper Detection Using Multi-Granularity Conditional Generative Adversarial Nets Based on UAV Inspection Images," *Sensors*, vol. 22, p. 1886, 2022.

[16] W. Bao, X. Du, N. Wang, M. Yuan, X. Yang, "A Defect Detection Method Based on BC-YOLO for Transmission Line Components in UAV Remote Sensing Images," *Remote Sensing*, vol. 14, p. 5176, 2022.

[17] A. L. B. Vieira e Silva, H. de Castro Felix, F. P. M. Simões, V. Teichrieb, M. dos Santos, H. Santiago, V. Sgotti, and H. Lott Neto, "Insplad: A dataset and benchmark for power line asset inspection in uav images," *International journal of remote sensing*, vol. 44, no. 23, pp.7294-7320, 2023.

[18] A. Buslaev, V. I. Iglovikov, E. Khvedchenya, A. Parinov, M. Druzhinin, and A. A. Kalinin, "Albumentations: fast and flexible image augmentations," *Information*, vol. 11, no. 2, p.125, 2020.

[19] N. Jegham, C. Y. Koh, M. Abdelatti, and A. Hendawi, "Evaluating the evolution of yolo (you only look once) models: A comprehensive benchmark study of yolo11 and its predecessors," *arXiv preprint arXiv:2411.00201*, 2024.

[20] Y. Tian, Q. Ye, and D. Doermann, "Yolov12: Attention-centric real-time object detectors," *arXiv preprint arXiv:2502.12524*, 2025.

Kriging-Based Prediction of Air Delivery Performance in Ceiling Fans and Its Validation

Sharankumar Shastri, *Member, IEEE*
Dept. of Electrical Engineering
Indian Institute of Technology Delhi
New Delhi, India, 110016, India
sharan.shastri01@gmail.com

Bhim Singh, *Fellow, IEEE*
Dept. of Electrical Engineering
Indian Institute of Technology Delhi
New Delhi, India, 110016, India
bsingh@ee.iitd.ac.in

Vipin Kumar Singh, *Member, IEEE*
Dept. of Electrical Engineering
Indian Institute of Technology Delhi
New Delhi, India, 110016, India
singhvipin4291@gmail.com

Abstract— An approach for studying and measuring electrical and mechanical performance, along with impact of blade configuration on ceiling fan motors, is presented in this paper. In addition to investigating electrical performance and effects of mechanical load, this study integrates air delivery prediction through a Kriging-based modelling technique. Without altering ceiling fan assembly, a cost-effective, simple-circuitry test platform is implemented. This platform is designed with a reaction-type load cell installed on a shaft and integrated with circuitry for signal processing to measure torque at the shaft. Acquired data is then utilized to predict fan's air delivery performance, providing a comprehensive assessment of its operational efficiency.

Index Terms—Ceiling fans, air delivery prediction, Kriging, SPIM, BLDCM

I. INTRODUCTION

Ceiling fans (CF) offer an ecologically sustainable choice, facilitating energy conservation and reducing total carbon impact. Ceiling fans are an appealing choice for residences globally due to affordability, low environmental effect, and long durability. Additionally, modern ceiling fans' enormous variety of styles, colors, and configurable capabilities make them both functional and attractive decorative items. Currently, ceiling fan industry is mainly regulated by single-phase induction motors (SPIMs) because of their straightforward design, inexpensive manufacturing costs, direct grid compatibility, and affordability, particularly in developing areas. Main drawbacks of SPIM based ceiling fans are relatively poor efficiency, greater power consumptions, and limited control capabilities [1]. Ceiling fan industry is rapidly transferring to permanent magnet (PM) brush less DC (BLDC) motor-based ceiling fans, which provide notably enhanced efficiency and sophisticated control capabilities, including features such as remote control and integrated lighting along with governmental support through various sustainable energy policies [2-3].

Ceiling fan motor is a coherent assembly including an electrical motor and a mechanical blade configuration. Performance of a ceiling fan relies on both efficiency of electrical motor and effectiveness of blades. System's efficiency is a combined outcome of motor efficiency and blade efficiency. Motor performance is measured in terms of ratio of output power delivered to consumed electrical power. In ceiling fan, output power varies from 14 W to 22 W, depending upon blade load and air delivery. Input power varies from 70 W to 26 W from less efficient SPIM based CF to efficient BLDC motor-based CF. In addition, power factor, total harmonic distortion in grid current (THD), supply current, and voltage are used to illustrate electrical performance of ceiling fan. From a mechanical perspective, efficient motor performance is identified via evaluation of torque, speed, and air delivery. Depending on blades structure and air supply, rated torque may range from 0.4 Nm to 0.60 Nm at rated speeds between 300 rpm and 450 rpm. The air delivery under rated conditions is in between 200 cmm (m³/min) to 250 cmm [4-5]. Motor featuring concentrated winding exhibits greater efficiency compared to conventional ceiling fan induction motor that employs distributed winding as discussed in [6].

An analysis and discussion of interrelationship between air-gap torque and shaft torque, as well as techniques for empirically determining these parameters for synchronous reluctance motors, are elaborated in [7]. A detailed analysis of mechanical factors affecting ceiling fans performance such as shape of blades, number of blades, blade material and associated aerodynamics is elaborated in [8]. In [9], quantitative results of various ceiling fan performances such as, service value, aerodynamics, torque, mass flow rate, and air delivery using an experimental testbench for different range of blade size and blade angle are explored.

Although significant research has focused on improving electromagnetic design of ceiling fan motors, considerable opportunities for improvement remain in mechanical performance aspect. Development of accurate, low-complexity, highly reliable, and economical approaches to output torque/power/air-delivery assessment represents a major opportunity for advancement. However, impact of mechanical factors on ceiling fan performance-such as blade geometry, pitch angle, blade length, and dynamic interaction between motor performance and fan blades on airflow patterns remains inadequately quantified and is currently only possible to be assessed through complex, time-consuming and computationally expensive fluid dynamic simulations, and/or large industrial air-delivery measurement setups [10-11]. A comprehensive understanding of these interactions is crucial for improving mechanical design of blades and motor integration, leading to improved energy efficiency, noise reduction and user satisfaction. This research aims to propose a cost-effective setup for ceiling fan performance measurement extended to air-delivery prediction at the lab level, in practical scenarios.

This paper is structured as follows: Section II describes system structure for torque measurement and examines impact of blade structure on ceiling fan performance. Section III presents measurement methodology for assessing air-delivery and followed by study's findings, supported by key results in Section IV. Finally, Section V concludes study by reviewing its key points.

II. SETUP ARCHITECTURE

This section presents selection of components, design, and calibration of test apparatus for measurement of torque of a ceiling fan and then air-delivery system setup.

979-8-3315-3013-6/25 $31.00 © 2025 IEEE

PART A: TORQUE MEASUREMENT

A. System Overview and Layout

Ceiling fans operate as outer rotor motors, where torque measured at shaft of motor is directly related to air-gap torque. This shaft is static, i.e., stationary, due to it being connected to stator, and shaft torque is produced due to reaction force on stator from air-gap torque. This relationship makes it practical to incorporate a shaft-based torque sensor without modifying fan's structure or mechanics. Fig. 1 presents schematic configuration, showing how sensor is mounted on shaft. Sensor is powered by an external voltage, transforming load fluctuations into a matching voltage signal that is processed by a data acquisition (DAQ) device.

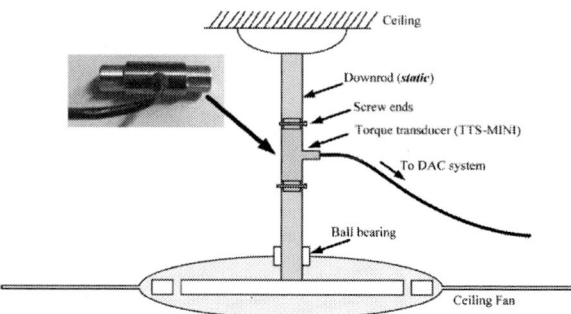

Fig. 1. Cross-sectional schematic for ceiling fan torque measurement.

B. Sensor Selection and Integration

In this study, a reaction type torque sensor – commonly known as a load cell, is selected for its simplicity and cost-effectiveness compared to piezoelectric alternatives. These sensors operate by displacing an element with a known stiffness when subjected to forces, whether in compression or tension. This displacement creates an imbalance in bridge resistance, resulting in a voltage output proportional to applied force. This setup employs a static transducer [12], as depicted in Fig. 1. It operates with an excitation voltage between 10V and 15V, offers a sensitivity of 2mV/V, and maintains high accuracy with a non-linearity below 0.1% of full-scale output. These specifications ensure precise torque assessment under practical conditions.

C. Signal Conditioning

Because load cell's output voltage (on order of microvolts to millivolts) is too low for direct analog-to-digital conversion, an instrumentation amplifier with a fixed gain is employed. Instrumentation amplifiers provide high input impedance and can deliver either a differential or single-ended output with respect to a reference voltage. In this setup, a TI-based [13] instrumentation amplifier is selected based on its low offset, minimal drift, and high CMRR.

To validate amplifier's suitability for this application, circuit simulations are performed, followed by hardware calibration. Fig.2 compares simulated and theoretical trendline equations with measured hardware results. Minor discrepancies arise primarily from in-amp equation not fully accounting for offset at high gains.

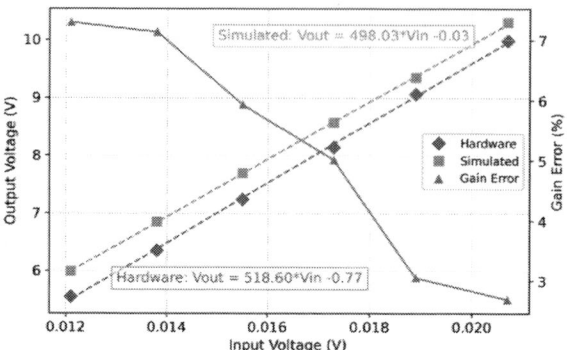

Fig. 2. Comparison of hardware and simulated output voltages with corresponding gain error over varying input voltages.

According to [14], a total of about 20mV (referred to input) can be expected for required gain. This offset influences final gain equation used to calculate torque T_{out} from sensor voltage, V_{sensor} as,

$$T_{out} = 518.78 \times V_{sensor} - 0.77 \quad (1)$$

for T_{out} is measured output torque by transducer and V_{sensor} is sensor voltage. This sensor setup is connected to a data-acquisition system to observe said output.

PART B: AIR-DELIVERY MEASUREMENT

A. Conventional Measurement of Air-Delivery

An IS-374 standard [15] outlines a comprehensive methodology for assessing air delivery of ceiling fans. Implementing this standard requires a complex configuration (show in Fig. 3a, with key details mentioned here:

- *Infrastructure requirements:* An IS 374 mandates a test chamber measuring 6.5 m², with a height of 4 m, featuring specific design elements such as a centrally located opening for ceiling fan placement. Construction of such a setup involves significant space requirements, leading to increased costs and complexity.

- *Instrumentation:* The standard specifies use of precision instruments, such as rotating vane anemometers. Typically, 4 of them are used and measurement is done along semi-diagonals shown in Fig. 3(a).

- *Measurement procedure:* Velocity readings are taken at multiple points (at various diameters) along semi-diagonals beneath fan, with readings taken at specific intervals and durations. An example of calculation of volumetric flow rate, Q, for measurement points at D_1 diameter (as shown in Fig. 3(a)).

Volumetric flow rate, $Q = A_{d1}$(area of annular ring) $\times$ V_{avg} (air velocity) $\quad (2)$

where $V_{avg} = \dfrac{\sum_{n-1}^{n} V_n}{n}$; $A_{d1} = \dfrac{\pi D_1^2}{4}$

979-8-3315-3013-6/25 $31.00 © 2025 IEEE   518

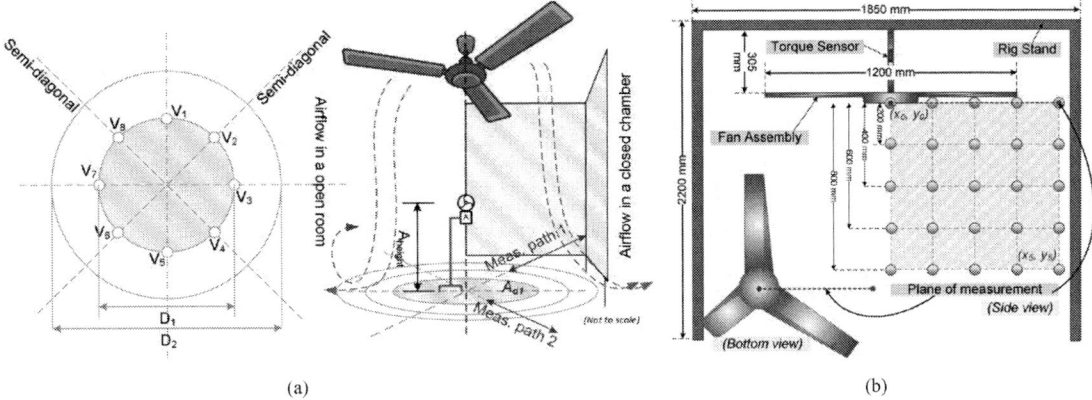

Fig. 3. Measurement of ceiling fan air-delivery (a) conventional, and (b) proposed.

In summary, adhering to IS-374, for air delivery measurement involves a more expensive, equipment-intensive, and time-consuming process that can only be setup with appropriate infrastructure, which is not possible at laboratory level.

B. Proposed Setup

The configuration Fig. 3(b) depicts configuration utilized to assess air-delivery efficacy of ceiling fan. The fan assembly is mounted on a rig stand, that supports an external rotor motor, and a shaft-mounted torque sensor (introduced previously). Torque sensor measures mechanical load without modifying fan's configuration.

C. Measurement

To capture airflow characteristics, a measurement plane is defined at multiple grid points, each spaced at regular intervals in both horizontal and vertical directions (as indicated by $x_0 - x_5$, and $y_0 - y_5$). Points are spaces at 200 mm apart, both in x and y direction, from origin ($x_0 = 0$ mm, $y_0 = 0$mm) of measurement, i.e., from base of fan assembly. Maximum one-sided distance of 800 mm is chosen, as radius of fan chosen is 600mm (for a standard 1200 mm sweep ceiling fan), with assumption that distance should be enough to cover airflow from ceiling fan.

An anemometer is positioned to measure airflow velocity at these grid points, providing reference data for determining fan's air delivery. A hygrometer is also placed to record ambient humidity conditions, as it affects airflow output. All instruments are securely attached to rig to ensure repeatable and consistent measurements under different experiments. Instrument details are provided in Table I. Each reading at grid points is recorded for a minimum duration of one minute, and average values are computed to ensure data accuracy.

TABLE I
INSTRUMENT DETAILS

Instrument	Measurement	Specifications/Details
Torque Sensor	Shaft torque	Reaction-type, 0-5 N.m, sensitivity = 2mv/V
Anemometer	Airflow velocity	High precision, digital type
Hygrometer	Relative humidity and Temperature	Digital type, continuous monitoring
Tachometer	Fan speed	High precision, digital type

III. KRIGING-BASED PREDICTION OF AIR DELIVERY

This section describes implementation of air-delivery prediction method post-measurement from previous section.

A. Kriging Method – Theory

Kriging is an advanced interpolation method originating from geo-statistics [16-17], particularly effective at estimating values at unknown data-points based on spatial correlation among known data points. Unlike conventional interpolation techniques, such as linear interpolation or inverse distance weighing, Kriging explicitly incorporates spatial autocorrelation through statistical modelling, thereby providing more accurate and reliable predictions.

Core mathematical principle of Kriging method is estimation of a spatial variable $Z(x)$ at any unmeasured location x_0, using measured data points, $Z(x_i)$ at known locations, x_i. Estimation of unknown value as a weighted linear combination of available measured values is done as follows,

$$\hat{Z}(x_0) = \sum_{i=1}^{n} \lambda_i Z(x_i) \tag{3}$$

where, $\hat{Z}(x_0)$ is estimated value at target location x_0, $Z(x_i)$ are measured values at known locations, λ_i are Kriging weights, determined by solving a system of linear equations derived from variogram function.

Weights λ_i are computed by minimizing estimation variance, subject to unbiasedness as,

$$\sum_{i=1}^{n} \lambda_i = 1 \tag{4}$$

Spatial correlation is captured by variogram $\gamma(h)$ expressed as,

$$\gamma(h) = \frac{1}{2N(h)} \sum_{i=1}^{N(h)} \left[Z(x_i) - Z(x_i + h) \right]^2 \tag{5}$$

where, h represents distance vector between two measured locations, $N(h)$ is number of measurement pairs separated by distance h, $Z(x_i)$ and $Z(x_i+h)$ are measured values at spatially separated locations.

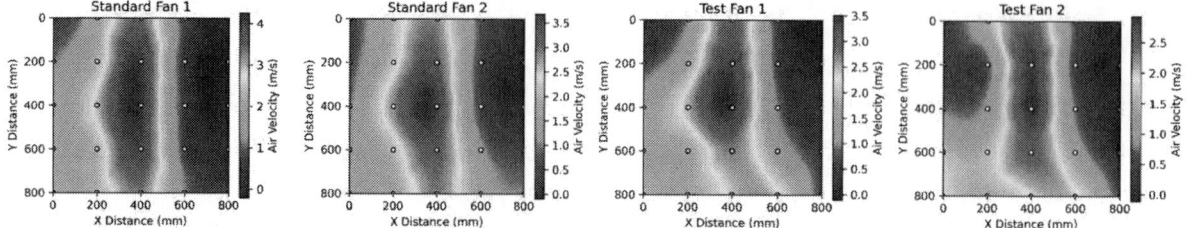

Fig. 4. Heatmaps illustrating the measured airflow velocity distributions (in m/s) at a horizontal measurement plane located beneath each fan configuration. From left to right: Standard Fan 1, Standard Fan 2, Test Fan 1 (prototype motor w/ 9° blade), and Test Fan 2 (prototype motor w/ 12° blade).

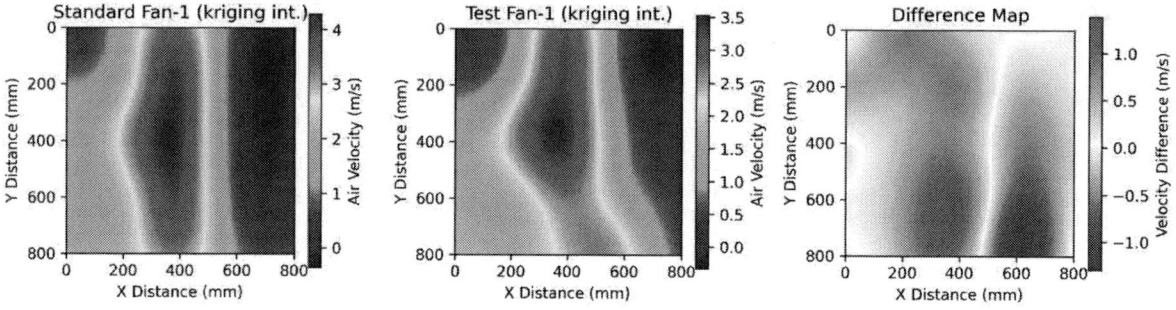

Fig. 5 Comparison of air velocity distributions between Standard Fan-1 (7.5° blade angle) and Test Fan-1 (9° blade angle) using Kriging interpolation. Difference map (rightmost figure) measures local airflow differences, emphasizing areas where Standard Fan-1 outperforms (red) or underperforms (blue) Test Fan-1.

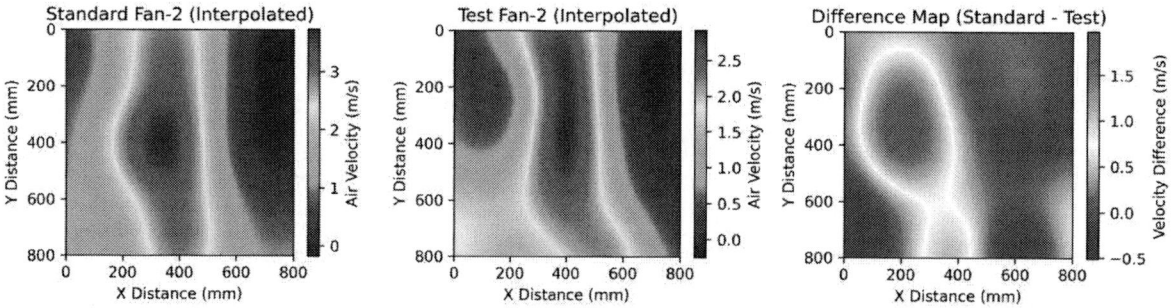

Fig. 6 Comparison of air velocity distributions between Standard Fan-2 (6.5-7° blade angle) and Test Fan-2 (12° blade angle) using Kriging interpolation. Difference map (rightmost figure) measures local airflow differences, emphasizing areas where Standard Fan-2 outperforms (red) or underperforms (blue) Test Fan-2.

Fitting a theoretical model such as Gaussian or exponential, to experimental variogram allows Kriging to effectively estimate unmeasured values while quantifying uncertainty in prediction.

B. Application to Air-Delivery Prediction

Main objective of this research is to create a reliable predictive model to assess air-delivery performance of ceiling fans using limited experimental measurements. Inputs collected during experimentation include:

- Air velocities: Assessed at a specified 5 x 5 grid directly beneath ceiling fan radius at various vertical and horizontal distances ranging from 0 mm-800 mm at 200 mm widths.
- Shaft torque and speed: Measured using a reaction-type torque sensor and tachometer, offering mechanical loading information.
- Humidity and temperature: Crucial for assessing ambient conditions influencing air density

For first test, which acts as a reference, these parameters are recorded for a standard ceiling fan, operating at a rated speed, and with a known air-delivery rating, defined by

mandatory BIS certification on ceiling fan. This standard fan functions as a benchmark for performance evaluation.

Kriging approach (with a Gaussian variogram model) generates a spatial interpolation model based on recorded air velocities at designated grid points. Once baseline is established, approach facilitates prediction of airflow patterns and overall magnitude of difference in air-delivery for any novel, untested fan-blade combination by comparing interpolated airflow distribution and integrated air-delivery metric of unknown configuration against reference fan. The distribution of air velocity distribution is presented as a density plot in Fig. 4, for 2 standard fans, and 2 fans with prototype motor and blade attached.

C. Results and Discussion

Fig. 5 shows Kriging interpolated airflow distributions for standard fan-1, and test fan-1. Airflow velocity distribution under standard fan-1, featuring a blade angle of 7.5 degrees, presents notably higher peak velocities and a more uniform spread across central region compared to test fan-1 (at 9 degrees blade pitch). For quantitative assessment, air velocity measurements are integrated for comparison. This value is

18.721 for standard fan-1, compared to 16.485 for test fan - 1. This equates to an approximately 12% reduction in overall air-delivery performance for tested fan. A map showing difference between Kriging interpolated values of standard vs test fans also identifies regions (red) where standard fan outperforms test configuration. Minor areas (right side, blue) show minimal advantage for test fan -1, but these are insufficient to offset performance advantage.

Similarly, Fig. 6 compares interpolated airflow distributions between Standard Fan-2 (blade angle 6.5–7°) and Test Fan-2 (blade angle 12°). While airflow distributions for both fans show distinct patterns, effect of significantly increasing blade angle to 12° is clearly observable. Difference map highlights areas of airflow velocity improvement or deterioration, providing insights into aerodynamic trade-offs associated with higher blade pitch angles.

These results suggest that simply increasing blade angle without motor optimization including changes to speed, might lead to reduced airflow performance. Regions with airflow deficits or improvements can inform design modifications to optimize entire system.

V. FUTURE WORK

This study has demonstrated effectiveness of Kriging interpolation for analysing ceiling fan airflow performance. Several key areas can be explored further:

- Hybrid modelling approach: combine Kriging interpolation with methods like machine learning, and also couple variation in output power, speed etc., to improve prediction accuracy.
- Volumetric airflow prediction: To predict airflow in standardized units (m³/min) for direct industry applicability.
- Optimized measurement points: Reduce number of required measurement points strategically, concentrating efforts in key regions of aerodynamic interest (as shown in Fig. 7).

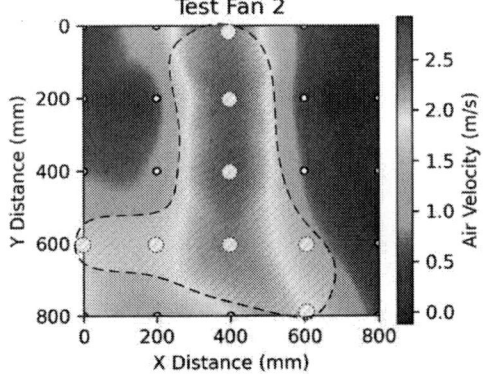

Fig. 7. Reduced measurement strategy highlighting key points (circled in white) in region of interest (shaded area).

V. CONCLUSION

This study has examined aerodynamic efficacy of ceiling fans with a Kriging-based interpolation method, with objective of accurately forecasting airflow patterns given limited experimental data. Primary objective has been to quantify and compare airflow patterns between standard certified fans and prototype configurations to help assess them at the laboratory stages without need for expensive infrastructure.

Kriging model has successfully highlighted differences in air velocity distributions, identified key performance zones, and enabled quantitative comparisons between designs. Results have indicated that blade angle adjustments alone do not always yield improved airflow performance, emphasizing need for system level optimization. Further enhancements involving larger ceiling fan datasets, advanced modelling techniques, and strategic measurement methods can be implemented to improve accuracy and practical applicability.

ACKNOWLEDGEMENTS

The authors thank the ANRF, Government of India for sponsoring ANRF National Science Chair fellowship.

REFERENCES

[1] Saneie and Z. Nasiri-Gheidari, "Performance Analysis of Outer-Rotor Single-Phase Induction Motor Based on Magnetic Equivalent Circuit," IEEE Transactions on Industrial Electronics, vol. 68, no. 2, pp. 1046-1054, Feb. 2021.

[2] A. Saxena, "Performance and cost comparison of PM BLDC motors for ceiling fan," 2014 IEEE International Conference on Power Electronics, Drives and Energy Systems (PEDES), Mumbai, India, 2014, pp. 1-5.

[3] A. Kumar and B. Singh, "High-Performance Brushless Direct-Current Motor Drive for Ceiling Fan," in IEEE Transactions on Industrial Electronics, vol. 71, no. 7, pp. 6819-6828, July 2024.

[4] U. Sharma and B. Singh, "Design and Development of Capacitor Run Split-Phase Fan Motor for Higher Efficacy,"IEEE Transactions on Industry Applications, vol. 57, no. 6, pp. 5939-5948, Nov.-Dec. 2021.

[5] Lenin, N. C., Padmanaban, S., Bhaskar, M. S., Mitolo, M., & Hossain, E. (2021). Ceiling fan drives–past, present and future. IEEE Access, 9, 44888-44904.

[6] S. Chakraborty, J. V. Singh, and K. Hatua, "Design of a 3-phase stator for improving the performance of the ceiling fan motors," in Proc. IEEE Int. Conf. Power Electron., Smart Grid, Renew. Energy (PESGRE), Thiruvananthapuram, India, Jan. 2022, pp. 1–6.

[7] L. Owen, H. D. Snively and T. A. Lipo, "Torsional Coordination of High Speed Synchronous Motors-Part II," IEEE Trans. Industry Applicat., vol. IA-17, no. 6, pp. 572-580, Nov. 1981.

[8] R. Wandre, A. G. Faizan, V. Dongre, S. Hasija, and I. Ullah "Exploration of the factors affecting the performance of the ceiling fan: A brief review." IOP Conference Series: Materials Science and Engineering, vol. 1259, no. 1, p. 012025. IOP Publishing, 2022.

[9] M. A. Afaq, A. Maqsood, K. Parvez, and A. Mushtaq, "Study on the design improvement of an indoor ceiling fan." Proceedings of 2014 11th International Bhurban Conference on Applied Sciences & Technology (IBCAST) Islamabad, Pakistan, 14th-18th January, 2014, pp. 279-283.

[10] Afaq, M. A., Maqsood, A., Butt, S. I., Tauqeer, T., & Hasan, A., "Aerodynamic investigation and redesign of ceiling fan blades for enhanced energy efficiency", Maejo International Journal of Science & Technology 11.2, 2017.

[11] Adeeb, E., Maqsood, A., Mushtaq, A., & Sohn, C. H. ," Parametric study and optimization of ceiling fan blades for improved aerodynamic performance", Journal of Applied Fluid Mechanics, vol.9, no.6, 2905-2916, 2016.

[12] Product specification–TTS mini torque transducer. [Online] Available :http://www.adiartech.com/userfiles/pdf/882-TTS-SSMINI. pdf

[13] Product specification: https://www.ti.com/tool/INA826EVM

[14] Product specification datasheet – INA826EVM. [Online]. Available: https://www.ti.com/lit/ds/symlink/ina826.pdf?ts=1743231079886&ref_url=https%253A%252F%252Fwww.google.com%252F

[15] Indian Standard 374, Specification for Electric Ceiling Type Fans and Regulators, Bureau of Indian Standards, Government of India.

[16] Oliver, M. A., & Webster, R., "Kriging: a method of interpolation for geographical information systems", International Journal of Geographical Information System, vo.4, no.3, 313-332, 1990.

[17] Sharma, K. V., Kumar, V., Prajapat, D. K., Mathew, A., & Gautam, L., "Geostatistical Kriging Interpolation for spatial enhancement of MODIS land surface temperature imagery", Journal of the Indian Society of Remote Sensing, vol. 3, no.1, 207-224, 2025.

Performance Evaluation of Bridge-Configured Winding for Transverse Force Generation in Three-Phase Electric Machines

Gopinath Sengupta[*§], Shahrukh[†], Karuna Kalita[†], Jenni Pippuri-Mäkeläinen[‡],
R. M. Ram Kumar[§], and Gaurang Vakil[§]

[*] School of Energy Science and Engineering, IIT Guwahati, India
[†] Dept. of Mechanical Engineering, IIT Guwahati, India
[‡] VTT Technical Research Centre of Finland Ltd., Espoo, Finland
[§] Power Electronics, Machines and Control Group, University of Nottingham, UK
Email: g.sengupta@iitg.ac.in, shahrukh1995@iitg.ac.in, karuna.kalita@iitg.ac.in,
jenni.pippuri-makelainen@vtt.fi, ramkumar.ramanathan@nottingham.ac.uk, gaurang.vakil@nottingham.ac.uk

Abstract—Rotor eccentricity in electrical machines causes unbalanced magnetic pull (UMP), leading to vibrations, noise, and mechanical wear. A bridge-configured winding (BCW) arrangement can generate transverse forces in the rotor, which can be controlled to counteract eccentricity. However, conventional three-phase windings are ineffective in BCW machines because they cancel out the necessary $p\pm1$ pole flux components, preventing proper force generation. This paper examines these limitations and compares two BCW configurations: Unshifted Bridge-Configured Winding (UBCW) and Shifted Bridge-Configured Winding (SBCW). Finite Element Method (FEM) simulations show that while both configurations enable transverse force generation, UBCW leads to a torque reduction to $19.35\,\mathrm{N\,m}$, whereas SBCW improves torque output to $25.65\,\mathrm{N\,m}$ while maintaining eccentricity control. The results highlight the trade-off between force production and torque performance, helping to guide the selection of suitable winding configurations for BCW machines.

Index Terms—Unbalanced magnetic pull (UMP), Bridge-Configured Winding (BCW), Transverse Force, Torque Performance, Finite Element Method (FEM)

I. INTRODUCTION

Rotor eccentricity in electrical machines arises when the rotor axis deviates from the stator axis, creating an uneven air gap. This misalignment affects circuit inductances and distorts the air-gap flux distribution, leading to unbalanced magnetic pull (UMP). UMP-induced electromagnetic forces act between the rotor and stator, with their magnitude influenced by rotor displacement, speed, winding configuration, load, and rotor slotting effects. If uncontrolled, UMP can cause excessive vibrations, acoustic noise, and mechanical wear, degrading machine performance and lifespan [1]–[4].

Severe eccentricity may cause rotor-stator contact, damaging stator insulation and reducing efficiency. Uneven force distribution from UMP can also bend the shaft and accelerate bearing wear [5]. Additionally, radial forces from eccentricity may excite stator core vibrations, leading to abnormal winding oscillations and potential mechanical and electrical failures [6].

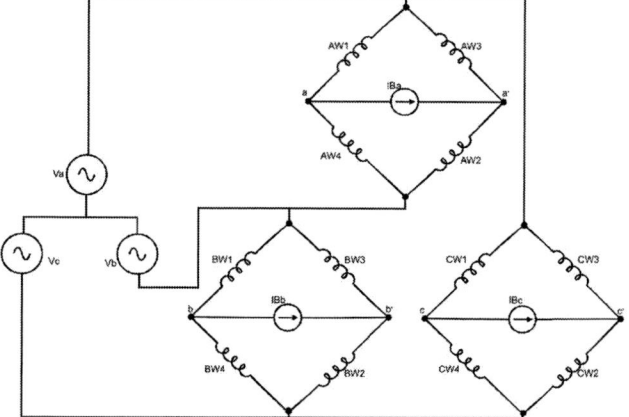

Fig. 1: Connection diagram of three phase stator winding with bridge configured arrangement

Various techniques have been explored to mitigate rotor eccentricity and its adverse effects, including equalising windings, damping windings, parallel circuits, magnetic saturation, and optimised slot/pole combinations [7]. Among these, parallel circuits and damping windings are well studied to reduce the UMP. The parallel circuit method reconfigures stator coil groups to redistribute the air-gap field and allows circulating currents to counteract eccentricity effects [8], [9], while damping windings generate opposing currents that stabilise electromagnetic forces and reduce mechanical wear [10], [11]. Other strategies, such as dual-stator machines [12], lead wire asymmetry correction [13], and equipotential bonding, also aim to reshape magnetic force distribution for improved stability.

Techniques from bearingless motors have shown promise in addressing rotor eccentricity. These machines use radial forces generated by auxiliary windings for suspension and eccentricity correction [14]–[16]. However, their dual-winding setups reduce torque density and radial force capability due

979-8-3315-3013-6/25 $31.00 © 2025 IEEE

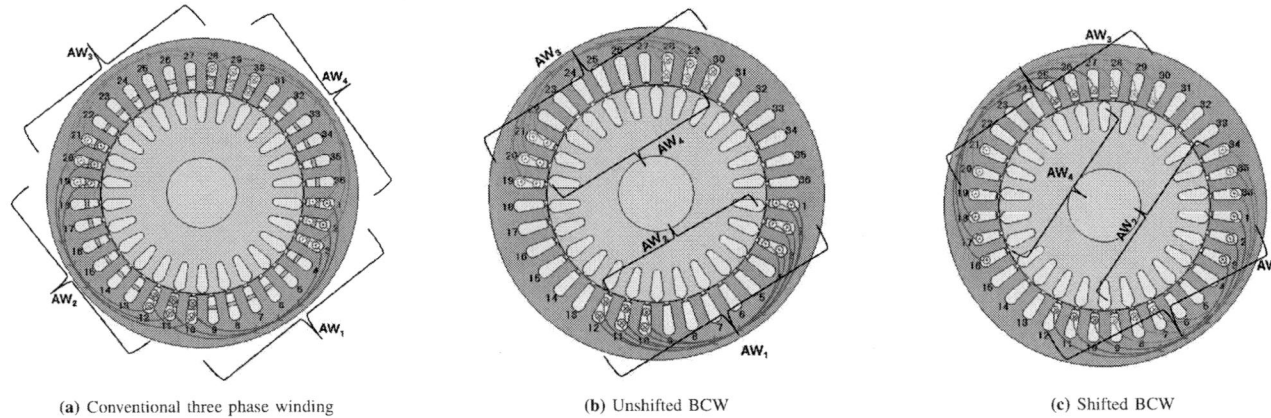

(a) Conventional three phase winding   (b) Unshifted BCW   (c) Shifted BCW

Fig. 2: Different winding arrangements under consideration

to limited slot space. They also involve extra coil sides per slot, leading to a lower winding packing factor, increased complexity, and higher manufacturing costs [17].

These limitations can be overcome with a bridge-configured winding (BCW) scheme, which offers a single-winding solution that integrates torque and transverse force generation, enabling both active and passive UMP mitigation. This improves slot utilisation, boosts efficiency, and reduces manufacturing complexity and cost, as no additional winding is required [18]–[21]. The only added component is the bridge supply.

In BCW, each phase is divided into four coil groups arranged in a Wheatstone bridge configuration. This allows an auxiliary supply to be applied across the bridge points, generating magnetic fields with pole pairs of $p \pm 1$ [18], which are not feasible with conventional three-phase windings. The BCW scheme thus enables independent control of torque and radial forces. Despite prior research on BCW, the rationale for its adoption and the design criteria for effective implementation remain underexplored.

This paper investigates the shortcomings of conventional three-phase windings in BCW machines and presents a comprehensive analysis of the optimal winding configuration required for efficient BCW operation. The evaluated winding is then incorporated into an induction motor and assessed using finite element method (FEM) simulations.

II. LIMITATIONS OF CONVENTIONAL WINDING IN BCW MACHINES

To investigate the limitations of conventional winding setups in BCW machines, a four-pole, double-layer induction motor was employed. The complete motor specifications are summarised in Table I. The stator winding is organised into 12 coil groups, with each phase comprising four groups: AW1–AW4, BW1–BW4, and CW1–CW4, as shown in Fig. 1. Each phase is divided into two parallel paths, each consisting of two coil groups. The midpoints of these paths—designated as a–a', b–b', and c–c' for phases A, B, and C, respectively—serve as bridge points crucial for generating transverse force, which

can be used to counter rotor eccentricity when appropriate excitation is applied in BCW machines.

Under normal operation, the coil groups are connected to produce a four-pole magnetic field, allowing the machine to function as a standard induction motor with a three-phase supply. However, when small excitation currents are applied at the bridge points, additional magnetic fields with pole pairs of $p \pm 1$ (i.e., two and six) are produced. These fields create transverse forces that can oppose the displacement caused by rotor eccentricity.

The main objective of this paper is to demonstrate a winding configuration capable of generating transverse forces in BCW machines. Accordingly, the bridge excitation employed in this study is not optimised for real-time eccentricity compensation but is intended to demonstrate the feasibility of transverse force development. A relatively large DC bridge current of 1 A was applied across each phase's bridge points to ensure observable force effects. In practical applications, the excitation could be lower in magnitude, possibly AC, and phase-specific, and significantly lower than the value used in this work. Since the bridge current is small relative to the rated phase current, its impact on copper utilisation remains negligible.

The transverse force generation capability of the BCW is due to the appropriate positioning of the coil sides in the slots. To illustrate this, Fig. 3 presents a single-phase BCW connection. With a phase current of $2I_a$, each coil group carries I_a. When an additional bridge excitation current $2I_{B_a}$ is applied, AW1 and AW2 carry $I_a + I_{B_a}$, while AW3 and AW4 carry $I_a - I_{B_a}$. The spatial separation of 180 mechanical degrees between these coil groups leads to a resultant field with $p \pm 1$ pole pairs, facilitating eccentricity compensation.

Conventional double-layer windings do not support this effect. As illustrated in Fig. 2a, coil sides of AW1 and AW2, as well as AW3 and AW4, are not adjacent but span full pole pitches in alternating upper and lower layers across slots. For example, AW2's upper-layer side shares a slot with AW1's lower-layer side. This interleaved placement, standard in full-pitch windings, inhibits the formation of effective $p \pm 1$ pole fields under bridge excitation. Under balanced three-phase

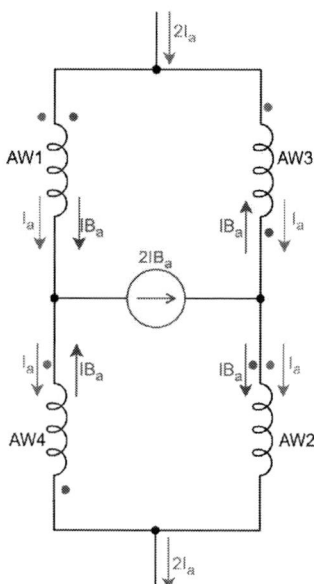

Fig. 3: Bridge-configured winding of a single phase

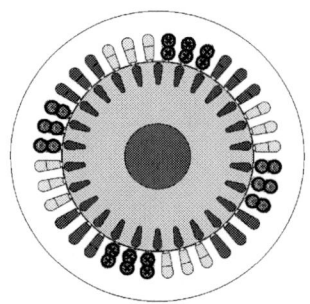

(a) UBCW motor winding scheme

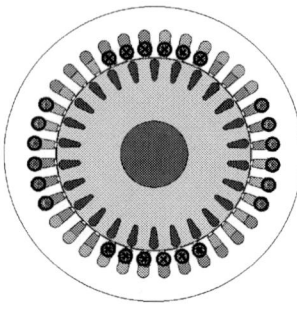

(b) SBCW motor winding scheme

Fig. 4: JMAG simulation model of induction motor

operation, the current splits equally between the two paths per phase, each coil group carrying half the phase current, producing a standard four-pole flux pattern. This is depicted by blue current markers in Fig. 3, while the current directions are denoted by dot/cross notations in Fig. 2a.

When bridge excitation is introduced, red markers in Fig. 3 show opposing currents in AW1–AW2 and AW3–AW4. Due to the alternating coil arrangement, shared slots experience cancelling flux contributions, eliminating $p\pm1$ components and thus suppressing transverse force generation. This limitation highlights why conventional windings are unsuitable for BCW-based eccentricity correction. Instead, a modified winding topology is essential to enable adjacent coil placement, preserve $p\pm1$ flux formation, and ensure effective transverse force generation for rotor displacement compensation.

III. WINDING CONFIGURATION AND MODIFICATION

To facilitate effective transverse force generation while maintaining torque performance, a well-structured winding arrangement is essential for BCW machines. In conventional three-phase, double-layer, distributed windings, improper coil positioning can lead to bridge currents flowing in opposing directions, cancelling out the intended force effects. To address this limitation, an alternative winding configuration is proposed, ensuring that coil groups W1 and W2 share the same slot positions, while W3 and W4 are placed in different layers within those slots. This modification is crucial for enabling proper transverse force generation when bridge excitation is applied. Two key variations of the BCW scheme are introduced: the Unshifted Bridge-Configured Winding (UBCW) and the Shifted Bridge-Configured Winding (SBCW), as illustrated in Fig. 2b and Fig. 2c. In the UBCW configuration, the machine operates as a conventional induction motor under

normal conditions, producing a four-pole rotating magnetic field. When bridge excitation is applied, a transverse force is generated, allowing for compensation for the eccentricity of the rotor. However, this arrangement comes with a trade-off, particularly in torque performance. To address this, the SBCW configuration is introduced as an optimised alternative, modifying coil placements to enhance flux distribution and electromagnetic interactions while preserving the benefits of bridge excitation.

In the UBCW scheme, the coil groups are arranged to ensure both proper motor operation and transverse force generation. FEM simulations were performed using JMAG 21.0, as illustrated in Fig. 4a, to analyse the magnetic behaviour under bridge excitation. These simulations were based on a standard 5 HP induction motor with the specifications shown in Table I.

Some stator laminations were extracted from the motor and formed into a ring core using Electrical Discharge Machining (EDM), and the magnetic properties were measured according to IEC standard [22]. The experimentally obtained B-H curves were used as the magnetic input for the simulations.

The results confirm that applying bridge excitation alone produces a two-pole flux pattern, which contributes to eccentricity compensation, as shown in Fig. 5a.Additionally, when a three-phase supply is applied, the machine successfully generates a four-pole rotating field, verifying its ability to function as a standard induction motor. However, while the bridge supply in the UBCW does not interfere with the torque production mechanism, the overall torque output is reduced, producing only $19.35\,\mathrm{N\,m}$, as depicted in Fig. 6 compared to $24.8\,\mathrm{N\,m}$ for the conventional winding arrangement. To address the torque reduction observed in UBCW while maintaining the ability to generate a transverse force, an alternative winding scheme—SBCW—is introduced. This configuration retains the core BCW structure but modifies the coil placement to optimise performance. In the SBCW scheme, the coil groups W1 and W3 remain in the same positions as in the UBCW; however, W2 and W4 are shifted by three slots, as shown in Fig. 2c. To accommodate this coil shift, the outer layer of the winding houses the outgoing coil sides, while the inner layer holds the incoming coil sides. This adjustment results in a more uniform magnetic field distribution, reducing flux

979-8-3315-3013-6/25 $31.00 © 2025 IEEE   524

TABLE I: Motor Specifications Used for Simulation

Parameter	Value
Supply Voltage	415 V, Delta Connected
Number of Poles	36
Winding Type	Double Layer
Frequency	50 Hz
Nominal Speed	1425 rpm
Bridge Current	1 A DC
Axial Length	135 mm
Outer Diameter of the Stator	160 mm
Inner Diameter of the Stator	104.8 mm
Air Gap	0.4 mm

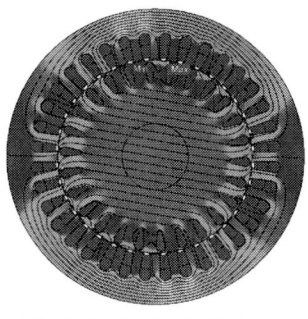 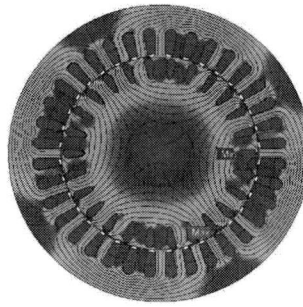

(a) 2 pole flux developed with the stator excited only with bridge supply

(b) 4 pole flux developed with the stator excited only with main supply

Fig. 5: Developed pole pairs with bridge points and supply terminals excited separately UBCW motor

leakage and ensuring that each phase operates with a current closer to its nominal value.

FEM simulations confirm that this configuration preserves the two-pole and four-pole flux characteristics while significantly enhancing torque output to 25.65 N m, as shown in Fig. 10. This improvement highlights the superior performance of the SBCW arrangement over UBCW, offering a better balance between torque production and transverse force generation.

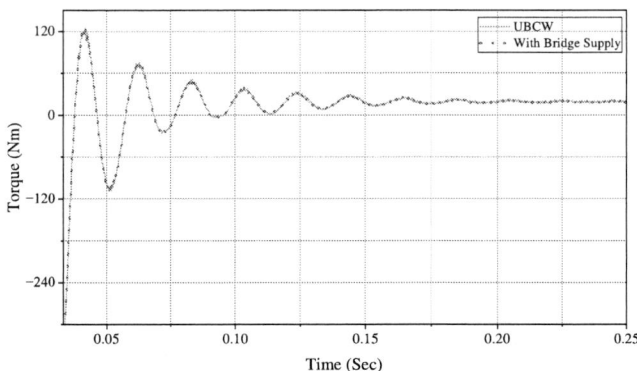

Fig. 6: Torque developed in the motor with UBCW with and without bridge supply

IV. FEM-BASED PERFORMANCE VALIDATION

To evaluate the effectiveness of the proposed BCW schemes, a comprehensive FEM simulation study was conducted using

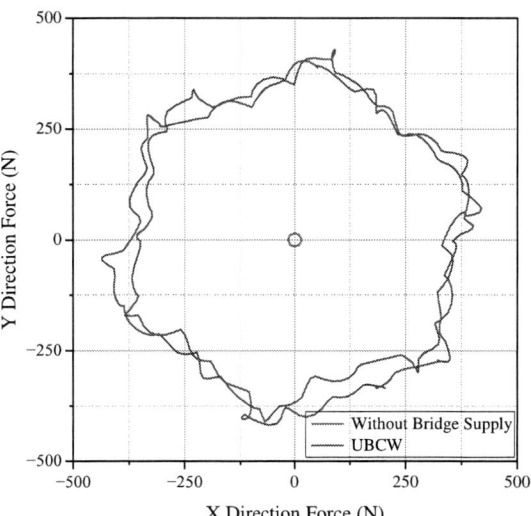

Fig. 7: Force orbit plot reflecting the transverse force developed with application of bridge supply

JMAG 21.0. The primary goal was to assess transverse force generation and torque performance under both UBCW and SBCW winding configurations.

The simulation confirms that both winding arrangements generate transverse forces under bridge excitation. As illustrated in Fig. 5, the expected two-pole and four-pole flux formations are consistent across both schemes, validating their capability for rotor eccentricity compensation.

A. Electromagnetic Force Estimation via Maxwell Stress Tensor

The transverse force arises from asymmetries in the air-gap magnetic field due to bridge supply, resulting in net rotor force. These can be estimated using the Maxwell stress tensor components in radial and tangential directions.

The radial (σ) and tangential (γ) components of Maxwell's stress at the air-gap surface are defined as [23]:

$$\sigma = \frac{1}{2\mu_0}(b_r^2 - b_t^2) \tag{1}$$

$$\gamma = \frac{1}{\mu_0} b_r b_t \tag{2}$$

where b_r and b_t are the radial and tangential components of the air-gap flux density, and μ_0 is the permeability of free space. The tangential component (γ) mainly governs torque, while the radial (σ) influences transverse force. Simulation results of both stress components are shown in Fig. 8 and Fig. 9.

A key difference between the configurations lies in their torque and transverse force profiles. The electromagnetic torque developed can be computed from the tangential component of the Maxwell stress tensor using [24]:

$$T = Lr \int_0^{2\pi} \gamma(\theta)\, d\theta \tag{3}$$

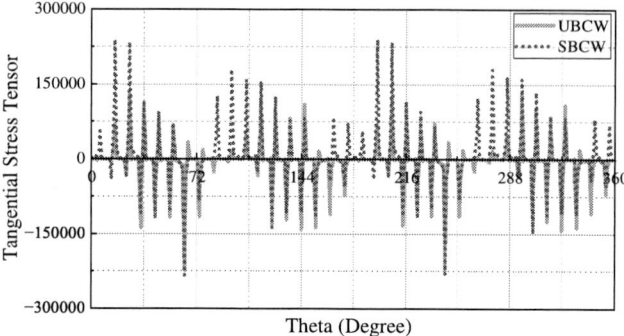

Fig. 8: Tangential component of Maxwell stress tensor

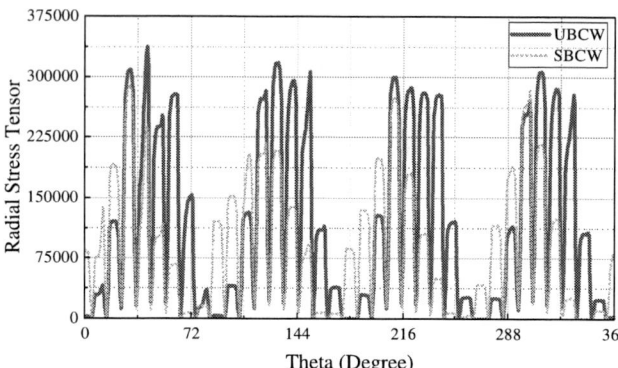

Fig. 9: Radial component of Maxwell stress tensor

where L is the stator core's axial length, r the mean air-gap radius, and θ the angular coordinate around the rotor.

As illustrated in Fig. 8, the tangential stress in the SBCW configuration exceeds that of UBCW, indicating better torque performance. As further demonstrated in Fig. 10, UBCW yields a torque of $19.35\,\mathrm{N\,m}$, while SBCW achieves $25.65\,\mathrm{N\,m}$, nearly matching the nominal torque of a conventional winding ($24.79\,\mathrm{N\,m}$). This improvement highlights the role of coil shifting in reducing flux leakage and ensuring uniform magnetic field distribution.

Using the stress components in Equations 1 and 2, the net electromagnetic forces in the x- and y- directions can be calculated as [23]:

$$F_x = Lr \int_0^{2\pi} \left[\sigma(\theta) \cos(\theta) - \gamma(\theta) \sin(\theta) \right] d\theta \quad (4)$$

$$F_y = Lr \int_0^{2\pi} \left[\sigma(\theta) \sin(\theta) + \gamma(\theta) \cos(\theta) \right] d\theta \quad (5)$$

These expressions offer a physics-based estimation of the directional forces acting on the rotor, allowing quantification of transverse force generation under different excitation conditions.

Force orbit plots offer further insight into bridge excitation effects. Fig. 7 shows that UBCW produces significant transverse force under bridge excitation but remains negligible under normal conditions, confirming its non-interference with

standard motor operation. A similar behaviour is observed in the case of the SBCW arrangement as well.

In contrast, Fig. 9 reveals a stronger radial stress in UBCW, suggesting a higher rotor force, validated by Fig. 11. Despite a lower radial stress, SBCW offers a balanced trade-off between transverse force and torque recovery, making it more effective for BCW motor applications.

A performance comparison summary is presented in Table II, reinforcing the advantages of SBCW in achieving both efficient transverse force generation and improved torque.

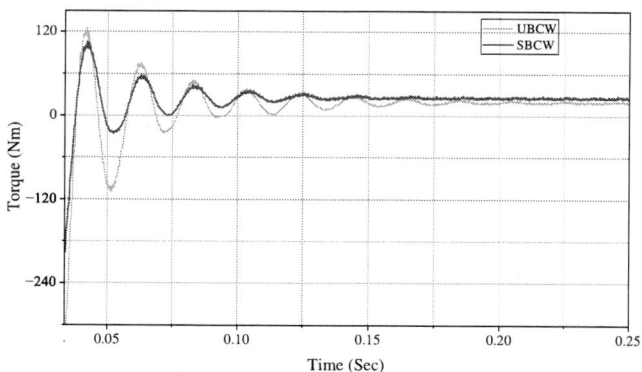

Fig. 10: Torque comparison with SBCW and UBCW with bridge supply

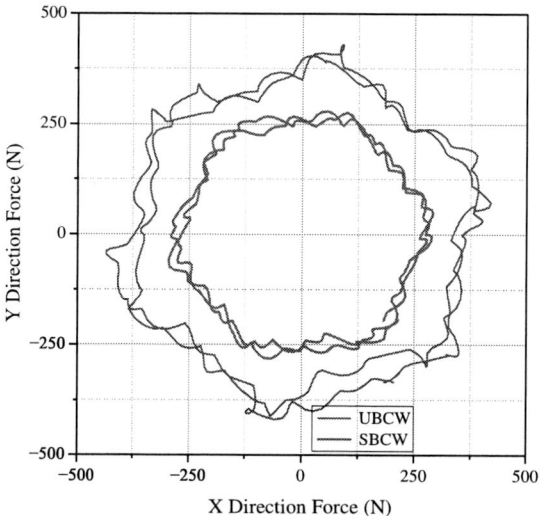

Fig. 11: Force orbit plot comparison between UBCW and SBCW reflecting the transverse force developed with application of bridge supply

TABLE II: Performance comparison of UBCW and SBCW configurations

Parameter	UBCW	SBCW
Flux Pattern (Bridge Supply)	2-pole	2-pole
Flux Pattern (Main Supply)	4-pole	4-pole
Torque (Nm)	19.35	25.65
Transverse Force	More	Less
Impact on Torque	Reduced	Nearly nominal

979-8-3315-3013-6/25 $31.00 © 2025 IEEE

V. CONCLUSION

This paper investigated the limitations of conventional three-phase winding schemes in bridge-configured winding (BCW) machines, specifically their inability to generate transverse forces due to the mutual cancellation of $p \pm 1$ pole flux components. To address this constraint, a comparative study was carried out between two alternative BCW configurations: the Unshifted Bridge-Configured Winding (UBCW) and the Shifted Bridge-Configured Winding (SBCW). The primary focus of this analysis was to evaluate the effectiveness of each scheme in terms of torque production and transverse force generation.

Finite Element Method (FEM) simulations validated that both UBCW and SBCW configurations successfully produce the characteristic two-pole and four-pole flux patterns necessary for rotor eccentricity compensation. However, the UBCW arrangement was found to induce a reduction in torque, delivering only $19.35\,\mathrm{N\,m}$, in contrast to the SBCW configuration, which achieved an enhanced torque output of $25.65\,\mathrm{N\,m}$ due to its optimised coil placement. While the UBCW setup generated a more substantial transverse force, the SBCW design demonstrated a more favourable trade-off between transverse force generation and torque retention.

These findings offer valuable insights into the electromagnetic behaviour of BCW machines and highlight the potential of the SBCW configuration as a practical solution for applications requiring active rotor eccentricity compensation without sacrificing torque performance.

ACKNOWLEDGMENT

The authors would like to thank POWERSYS for providing JMAG software, as FEA packages for electric machine optimisation, and their support to carry out this study.

REFERENCES

[1] Y. Han and Y. Song, "Condition monitoring techniques for electrical equipment—a literature survey," *IEEE Transactions on Power Delivery*, vol. 18, no. 1, pp. 4–13, Jan. 2003.

[2] D. G. Dorrell and O. Kayani, "Measurement and calculation of unbalanced magnetic pull in wound rotor induction machine," *IEEE Transactions on Magnetics*, vol. 50, no. 11, pp. 1–4, Nov. 2014.

[3] G. M. Joksimović, "Dynamic simulation of cage induction machine with air gap eccentricity," *Proceedings of the IEE - Electric Power Applications*, vol. 152, no. 4, pp. 803–811, 2005.

[4] M. Osama and T. Lipo, "A new induction machine model for analysis of eccentric rotor magnetic pull," in *Proceedings of the Symposium on Power Electronics, Electrical Drives, and Advanced Electrical Motors*, 1994, pp. 173–177.

[5] S. Nandi, H. Toliyat, and X. Li, "Condition monitoring and fault diagnosis of electrical motors—a review," *IEEE Transactions on Energy Conversion*, vol. 20, no. 4, pp. 719–729, Dec. 2005.

[6] J. Faiz and M. Ojaghi, "Different indexes for eccentricity faults diagnosis in three-phase squirrel-cage induction motors: A review," *Mechatronics*, vol. 19, no. 1, pp. 2–13, 2009.

[7] A. Salah, Y. Guo, and D. Dorrell, "Monitoring and damping unbalanced magnetic pull due to eccentricity fault in induction machines: A review," in *Proceedings of the 20th International Conference on Electrical Machines and Systems (ICEMS)*, Sydney, NSW, Australia, Aug. 2017.

[8] A. Burakov and A. Arkkio, "Comparison of the unbalanced magnetic pull mitigation by the parallel paths in the stator and rotor windings," *IEEE Transactions on Magnetics*, vol. 43, no. 9, pp. 4083–4088, 2007.

[9] M. Wallin, M. Ranlof, and U. Lundin, "Reduction of unbalanced magnetic pull in synchronous machines due to parallel circuits," *IEEE Transactions on Magnetics*, vol. 47, no. 12, pp. 4827–4833, 2011.

[10] D. Dorrell, "Unbalanced magnetic pull in cage induction machines for fixed-speed renewable energy generators," *IEEE Transactions on Magnetics*, vol. 47, no. 12, pp. 4096–4099, 2011.

[11] M. Wallin, J. Bladh, and U. Lundin, "Damper winding influence on unbalanced magnetic pull in salient pole generators with rotor eccentricity," *IEEE Transactions on Magnetics*, vol. 49, no. 9, pp. 5158–5165, 2013.

[12] H. Q. Nguyen, J.-Y. Jiang, and S.-M. Y. and, "Design of a wound-field flux switching machine with dual-stator to reduce unbalanced shaft magnetic force," *Journal of the Chinese Institute of Engineers*, vol. 40, no. 5, pp. 441–448, 2017.

[13] C. Bi, H. Phyu, and Q. Jiang, "Unbalanced magnetic pull induced by leading wires of permanent magnet synchronous motor," in *Proceedings of the 12th International Conference on Electrical Machines and Systems (ICEMS)*, Tokyo, Japan, Nov. 2009.

[14] W. Gruber, T. Nussbaumer, H. Grabner, and W. Amrhein, "Wide air gap and large-scale bearingless segment motor with six stator elements," *IEEE Transactions on Magnetics*, vol. 46, no. 6, pp. 2438–2441, Jun. 2010.

[15] P. Kascak, R. Jansen, T. Dever, A. Nagorny, and K. Loparo, "Levitation performance of two opposed permanent magnet pole-pair separated conical bearingless motors," in *Proceedings of the Energy Conversion Congress and Exposition*, 2011, pp. 1649–1656.

[16] W. Li, K. Chau, T. Ching, Y. Wang, and M. Chen, "Design of a high-speed superconducting bearingless machine for flywheel energy storage systems," *IEEE Transactions on Applied Superconductivity*, vol. 25, no. 3, pp. 1–4, Jun. 2015.

[17] E. L. Severson, R. Nilssen, T. Undeland, and N. Mohan, "Design of dual purpose no-voltage combined windings for bearingless motors," *IEEE Transactions on Industry Applications*, vol. 53, no. 5, pp. 4368–4379, 2017.

[18] S. Khoo, R. Fittro, and S. Garvey, "An ac self-bearing rotating machine with a single set of windings," in *Proceedings of the International Conference on Power Electronics, Machines and Drives*, 2002, pp. 292–297.

[19] W. Khoo, "Bridge configured winding for polyphase self-bearing machines," *IEEE Transactions on Magnetics*, vol. 41, no. 4, pp. 1289–1295, Apr. 2005.

[20] W. Khoo, K. Kalita, and S. Garvey, "Practical implementation of the bridge configured winding for producing controllable transverse forces in electrical machines," *IEEE Transactions on Magnetics*, vol. 47, no. 6, pp. 1712–1718, Jun. 2011.

[21] R. Deore, B. Brahma, Shahrukh, and K. Kalita, "The passive vibration control in bridge configured winding cage rotor induction motor: An experimental analysis," in *Proceedings of the 11th IFToMM International Conference on Rotordynamics*, F. Chu and Z. Qin, Eds. Cham: Springer International Publishing, 2024, pp. 515–527.

[22] "Iec 60404-6:2018 - magnetic materials – part 6: Methods of measurement of the magnetic properties of magnetically soft metallic and powder materials at frequencies in the range 20 hz to 100 khz by the use of ring specimens," https://webstore.iec.ch/publication/27825, International Electrotechnical Commission, 2018.

[23] I. P. Brown, D. M. Ionel, and D. G. Dorrell, "Unbalanced operation of current regulated sine-wave interior permanent magnet machines," in *2010 IEEE Energy Conversion Congress and Exposition*, 2010, pp. 4123–4130.

[24] J. P. A. Bastos and N. Sadowski, *Electromagnetic modeling by finite element methods*. CRC press, 2003.

Hilbert-Huang Transform-Based Fault Detection and Classification in the Presence of Power Swings

Varun Reddy Gadikota
Department of Electrical & Electronics Engineering
Birla Institute Of Technology and Science (BITS) Pilani,
Hyderabad Campus, India
f20220673@hyderabad.bits-pilani.ac.in

Nitish Kumar Gupta
Department of Electrical & Electronics Engineering
Birla Institute Of Technology and Science (BITS) Pilani,
Hyderabad Campus, India
nitishkumar.gupta@hyderabad.bits-pilani.ac.in

Abstract—This paper expands on a methodology predicated on the Hilbert-Huang Transform (HHT) described by Li Zhenghua[1] for the classification and identification of incipient short-circuit faults in a power transmission network. The proposed approach analyzes the transient behavior of phase voltages and presents a practically superior alternative to conventional frequency decomposition techniques, particularly in complex scenarios characterized by the dynamic interplay of power swings originating from the source side. The robustness and discriminatory power of this method are evaluated through comprehensive simulations encompassing diverse fault modalities, including single line-to-ground (LG), line-to-line (LL) line-to-line-to-line (LLL), and line-to-line-to-line-to-ground (LLLG) faults implemented on the standard IEEE 9-bus system, demonstrating a high degree of accuracy in fault detection.

Index Terms—Hilbert-Huang-Transform (HHT), Empirical mode decomposition (EMD), Intrinsic Mode Function (IMF), Fault detection, Hilbert Transform, Fault analysis, Power system transients, Power Swings.

I. INTRODUCTION

Ensuring the integrity and efficiency of the power transmission network is of critical importance, given the vast network of domestic, commercial, and industrial consumers whose operations are inextricably linked to the stability and availability of the power system. A principal source of potential system instability arises from the occurrence of electrical faults. A power system fault is defined as an anomalous condition wherein the electric current is diverted from its intended conductive path due to failure within the system's components or insulation. Such fault events instigate abnormal electrical stresses, leading to a significant reduction in the dielectric strength between conductors. This degradation of insulation integrity can precipitate substantial damage to critical system equipment and potentially lead to cascading failures. Electrical faults in three-phase power systems can manifest in various forms, broadly categorized into symmetric (balanced) faults and asymmetric (unbalanced) faults [2]. Furthermore, the dynamic operation of interconnected power systems is often characterized by low-frequency oscillations in generator rotor angles, typically within the range of 0.1 to 2 Hz [3]. These oscillations are inherent to the system's response to perturbations and arise from mismatches between power generation and load demand.

©

Based on their spatial extent and underlying causes, these oscillations can be classified as local modes involving a small group of tightly coupled generators or inter-area modes involving oscillations between geographically distant areas of the power grid. This paper specifically addresses local mode oscillations, which typically reside in the frequency band of 1-2 Hz and are localized to a limited portion of the power system. These local mode oscillations can be effectively modeled as a voltage waveform with modulated amplitude [4].

The inception of a fault within the power system induces transient behavior in the phase voltages at the affected bus. The characteristics of this transient response, including its magnitude, duration, and frequency content, provide valuable information for discerning the type of fault. This information is crucial for the implementation of protective relaying schemes aimed at minimizing system damage and ensuring operational continuity. Additional complications brought in by power swings and the incorporation of any resultant restraint also depend upon this information. In this context, the application of a Hilbert-Huang-based methodology for the analysis of sampled phase voltages presents a compelling alternative to traditional frequency domain methods such as the Fourier transform. The primary justification for utilising the HHT lies in the inherent non-linear and non-stationary nature of the power system during fault transients, more so when considering the superimposed effects of low-frequency power oscillations. The amplitude modulation associated with these oscillations, coupled with the abrupt changes introduced by fault events, renders the HHT, with its unique capability to analyze non-stationary and non-linear signals without a priori assumptions about their underlying structure, as an effective tool for this analytical task. The HHT has been utilised by Abhishekh Anand and Shaik Affijulla [5] to compute the average relay energy index (AREI) for fault identification in power system grids. The HHT shows promise in feature extraction for power system oscillation measurements as detailed by B. S. Munir, M. Reza, A. Trisetyarso and B. S. Abbas [6]. The resilience of the HHT under non stationary and non linear conditions is demonstrated in the mentioned papers. The use of signal energy also shows promise in detecting faults and stable, unstable power swings according to the method proposed by

979-8-3315-3013-6/25 $31.00 © 2025 IEEE

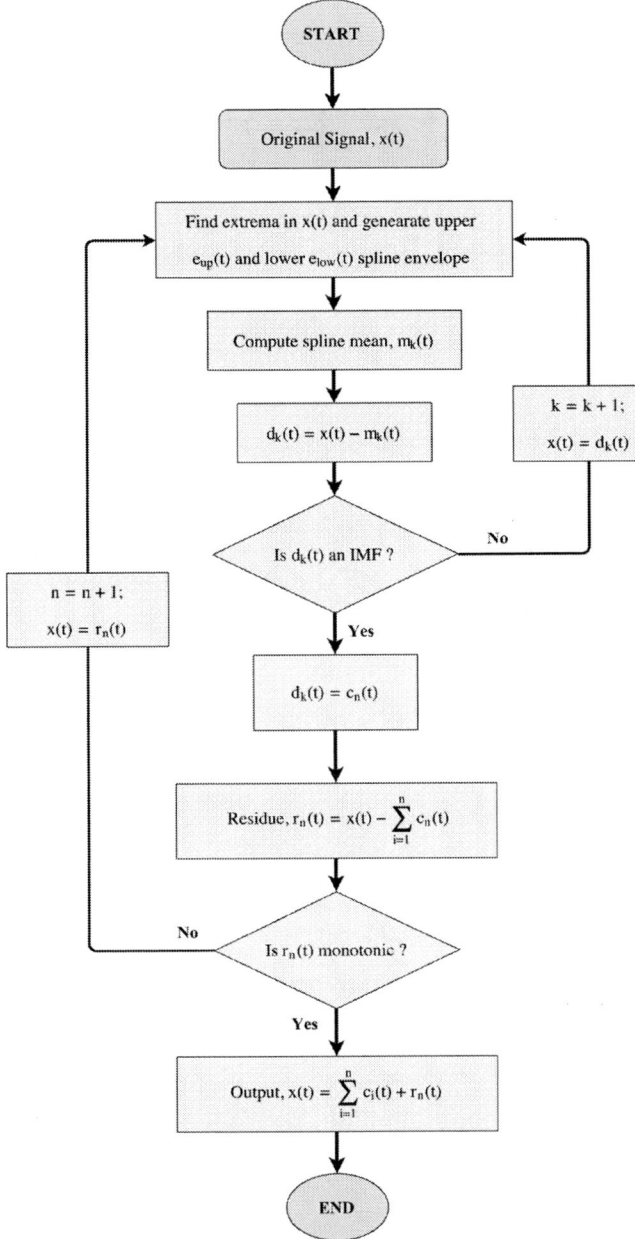

Fig. 1: Flowchart depicting process of EMD [9].

Ali Kaffashbashi, Yaser Damchi [7]. The modified method is based off of Li Zhenghua's method with the addition of the sample window method to allow for quick fault detection.

II. EMPIRICAL MODE DECOMPOSITION

Empirical Mode Decomposition (EMD) is a powerful signal processing technique [8] designed for the decomposition of non-linear and non-stationary signals into a finite and often small number of oscillatory components known as Intrinsic Mode Functions (IMFs). It is a data-driven method that does not require prior knowledge of the signal's frequency content

or the characteristics of the underlying oscillations. Unlike traditional methods that rely on fixed basis functions, EMD is adaptive and decomposes a signal based on its local characteristics. This makes it particularly suitable for analyzing the complex transient waveforms encountered in power systems during fault conditions. The core of the EMD algorithm lies in an iterative process called sifting (depicted in the flow chart of Fig.1) that decomposes the given signal into a series of IMFs. It involves identifying the local extrema of the signal and fitting an envelope to this extrema. The difference between the signal and its envelope is then computed, and this process continues until the residual signal becomes monotonic or contains only one extremum, at which point no more IMFs can be extracted. The resulting IMFs represent different oscillation scales in the signal, with the first IMF capturing the highest-frequency oscillations and the last IMF capturing the lowest-frequency oscillations. For a signal x(t), EMD provides us with the IMFs and the final residual signal r(t), which can be related as:

$$x(t) = \sum_{j=1}^{\chi} c_j(t) + r(t) \qquad (1)$$

here, χ represents the total number of IMFs and $c_j(t)$ denotes the jth IMF.

III. HILBERT TRANSFORM

Hilbert transform is a linear operator that shifts the phase of each frequency component of a signal by 90 degrees. For a continuous-time signal u(t), the Hilbert transform is defined as the convolution of u(t) with the impulse response h(t)=1/πt, known as Cauchy kernel [10]. Because 1/t is not integrable across t=0, the integral defining the convolution does not always converge. Instead, the Hilbert transform is defined using the Cauchy principal value (denoted here by p.v.). Explicitly, the Hilbert transform H(u)(t) of a signal u(t) is given by:

$$\mathbb{H}(u)(t) = \frac{1}{\pi} \, p. \, v. \int_{-\infty}^{+\infty} \frac{u(\tau)}{t - \tau} \, d\tau \qquad (2)$$

In practical digital signal processing, the discrete-time Hilbert transform is typically employed. For a discrete-time signal u[n], the discrete Hilbert transform û[n] is given by the convolution with a specific discrete-time impulse response h[n]:

$$\hat{u}[n] = u[n] * h[n] \qquad (3)$$

Here h[n] is defined as follows:

$$h[n] \triangleq \begin{cases} 0, & \text{for } n \text{ even} \\ \frac{2}{\pi n}, & \text{for } n \text{ odd} \end{cases} \qquad (4)$$

In this research, the discrete-time Hilbert Transform is applied to the first IMF obtained from the EMD of each phase voltage signal. Once the signal has been transformed into the Hilbert domain, we can obtain the analytic signal $s_a(n)$ which is defined as follows for a signal s(n):

$$s_a(n) = s(n) + j\hat{s}(n) \qquad (5)$$

This analytic signal provides a powerful tool for extracting instantaneous attributes of the signal. The instantaneous envelope A_e of the IMF can be obtained directly in terms of the magnitude of the analytic signal $s_a(n)$. The instantaneous phase $\theta(n)$ and instantaneous frequency $\omega(n)$ can also be derived from the analytic signal, although this study primarily focuses on the instantaneous amplitude for fault detection and classification.

IV. SYSTEM UNDER STUDY

To rigorously evaluate the performance of the proposed HHT-based fault detection and classification methodology, the standard IEEE 9-bus system is utilized as the test platform. The single-line diagram of this system is illustrated in Fig. 3. The IEEE 9-bus system is a widely recognized benchmark for power system analysis, comprising three generators, three loads, and nine buses interconnected by three transformers and six transmission lines. This system exhibits dynamic characteristics representative of larger interconnected power grids, making it a suitable test case for evaluating the proposed methodology's efficacy under realistic operating conditions and fault scenarios. In this study, a short circuit fault is simulated on the transmission line connecting buses 7 and 6. The fault is initiated at a specific time instant, and the resulting transient phase voltages at the bus closest to the fault (Bus 6 in this case) are recorded. The simulation is conducted with a sampling frequency of 200 samples per second, which satisfies the Nyquist criterion for the fundamental frequency (50 Hz) and adequately captures the high-frequency transients associated with fault events. To realistically model the presence of low-frequency power oscillations, a sinusoidal voltage component with a frequency of 1.5 Hz and an amplitude of 0.01 per unit (pu) at a 230 kV base is superimposed on the nominal 50 Hz phase voltage signals for each of the three phases. Furthermore, to account for measurement noise and system disturbances, additive white Gaussian noise with a specified variance is incorporated into the simulated voltage signals. This ensures that the evaluation of the proposed methodology is performed under conditions that closely approximate real-world power system environments. Subsequently, various types of short circuit faults, including LG, LLG, LLLG, LL, LLL type faults are simulated at the designated location on the IEEE 9-bus system. The recorded phase voltage data for each fault scenario is then subjected to the proposed HHT-based analysis, which is detailed in the subsequent section. Bus parameters have been detailed in Table. 1 with a pu value of 230 kV on the high tension side and 18 kV on the load side.

V. PROPOSED METHODOLOGY

The proposed methodology for fault detection and classification leverages the capabilities of the Hilbert-Huang Transform to analyze the transient behavior of phase voltages following the inception of a fault. The process involves the following key steps, which are also summarized in the flowchart presented in Fig. 3.

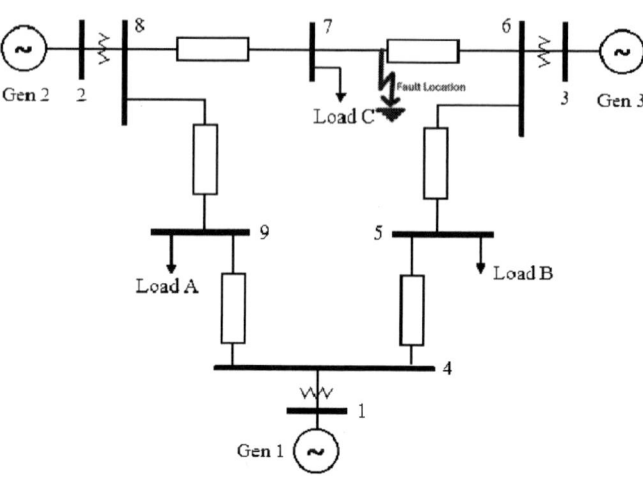

Fig. 2: Single Line Diagram of the IEEE 9-bus system [11].

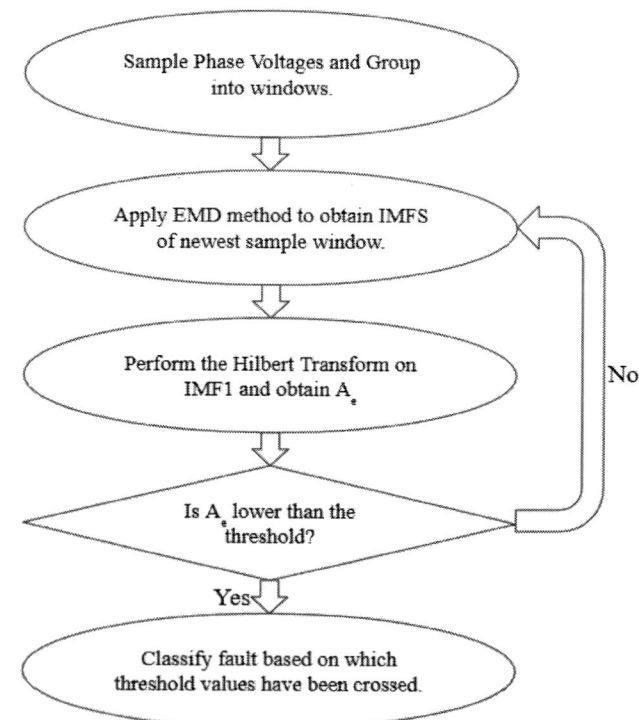

Fig. 3: Flowchart for the proposed methodology of fault detection.

979-8-3315-3013-6/25 $31.00 © 2025 IEEE   530

Bus Number	δ (degrees)	Bus Voltage (pu)
1	0	1.04
2	9.173	1.025
3	4.558	1.025
4	-2.226	1.026
5	-3.701	1.013
6	1.867	1.032
7	0.6336	1.016
8	3.62	1.026
9	-4.002	0.9962

TABLE I: Bus parameters of IEEE 9-bus system.

A. Sampling and Windowing of Phase Voltages

The instantaneous values of the three-phase voltages at the monitored bus are continuously sampled at a predetermined sampling rate (Fs=200 samples/second in this study). These sampled voltage values are then sequentially grouped into windows of a fixed size (N). The choice of window size is a critical parameter that influences the trade-off between the speed of fault detection and the robustness of the method against spurious detections, as discussed in a later section.

B. Empirical Mode Decomposition of Sampled Data

For each newly acquired window of phase voltage samples, the EMD algorithm is applied to decompose the voltage signal of each phase into its constituent IMFs. As the first IMF typically captures the highest frequency components and is most sensitive to abrupt changes in the signal, it is the primary component of interest for fault detection.

C. Hilbert Transform of the First Intrinsic Mode Function

The discrete-time Hilbert Transform is then applied to the first IMF obtained from the EMD for each phase voltage. This transformation yields the quadrature component of the analytic signal.

D. Computation of Instantaneous Envelope

The instantaneous envelope amplitude A_e of the first IMF for each phase is computed as the magnitude of the analytic signal, as defined earlier. This envelope represents the instantaneous energy of the high-frequency components of the phase voltage signal.

E. Fault Detection based on Threshold Comparison

A threshold value V_t is established for the instantaneous envelope amplitude of each phase under normal, fault-free operating conditions. This threshold is typically set as a certain percentage (e.g., a few standard deviations) below the minimum envelope amplitude observed during a training period without any faults. Fault detection is triggered when the instantaneous envelope amplitude V_t of any phase falls below its corresponding threshold value V_t. To enhance the reliability of fault detection, a fault may be declared only if the envelope amplitude remains below the threshold for a predefined number of consecutive samples or within a specific time duration.

F. Fault Classification based on Affected Phases

Once a fault is detected, the type of fault is classified based on the specific phases whose envelope amplitudes have crossed their respective threshold values.

The initial determination of the threshold value V_t for each phase involves recording the envelope amplitude of the first IMF over an extended period under normal operating conditions. The minimum value of this envelope amplitude observed during this fault-free period can serve as a basis for setting the threshold. To account for gradual variations in system operating conditions and to minimize the probability of false alarms, the threshold value may need to be adaptively adjusted over time. This can be achieved by continuously monitoring the envelope amplitude during fault-free operation and updating the threshold based on statistical measures such as the moving average or minimum of the envelope amplitude over a recent time window.

VI. FAULT IDENTIFICATION

A. Behaviour of a Faultless system: The baseline behaviour

Under normal operating conditions, the phase voltages in the IEEE 9-bus system exhibit a sinusoidal waveform at the fundamental frequency (50 Hz), with the superimposed low-frequency oscillation (1.5 Hz) and measurement noise. When the proposed HHT-based methodology is applied to these fault-free voltage signals, the instantaneous envelope amplitude A_e of the first IMF for each phase remains relatively stable over time, fluctuating within a bounded range. This stable behavior, as illustrated in Fig. 4 for a window size of 5 samples, allows for the determination of the threshold value V_t for each phase. The threshold is set sufficiently below the minimum observed A_e to avoid false fault detections due to normal system variations or noise. The horizontal line in the relevant figures is the threshold value V_t for each phase which has been updated during normal operating conditions.

B. Behaviour of system with Fault

To evaluate the methodology's ability to detect and classify faults, simulations of different fault types were conducted on the IEEE 9-bus system.

1) Line-to-Line (LL) Fault: A fault was simulated between phases A and B on the transmission line connecting buses 7 and 6. The resulting transient phase voltages at Bus 6 were analyzed using the proposed methodology. As depicted in Fig. 5, upon the inception of the LL fault, a significant and simultaneous drop in the instantaneous envelope amplitudes of phases A and B is observed, while the envelope amplitude of phase C remains relatively unaffected. This simultaneous crossing of the pre-defined thresholds for phases A and B serves as a clear indication of an LL fault involving these two phases.

2) Triple Line (LLL) Fault: A fault was simulated between phases A,B,C at the same location. The analysis of the transient phase voltages at Bus 6, shown in Fig. 6, reveals a substantial decrease in the instantaneous envelope amplitude of phase A,B and C. The crossing of the threshold value for

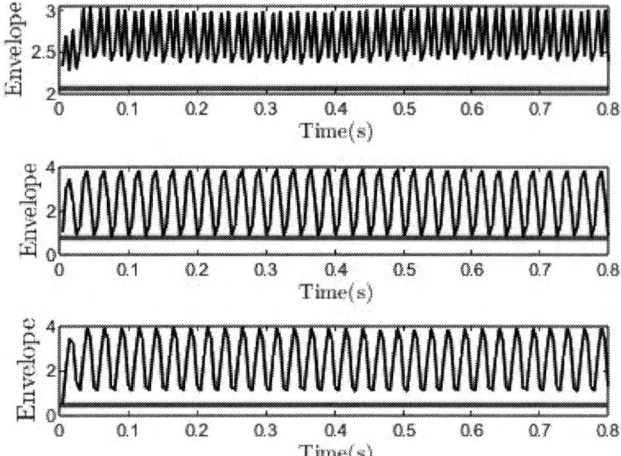

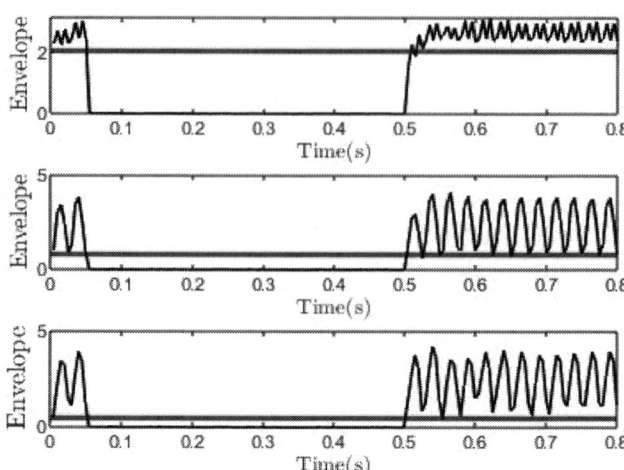

Fig. 4: Instantaneous envelope of phases A,B,C respectively for a faultless system

Fig. 6: Instantaneous envelope of phases A,B,C respectively for an LLL fault

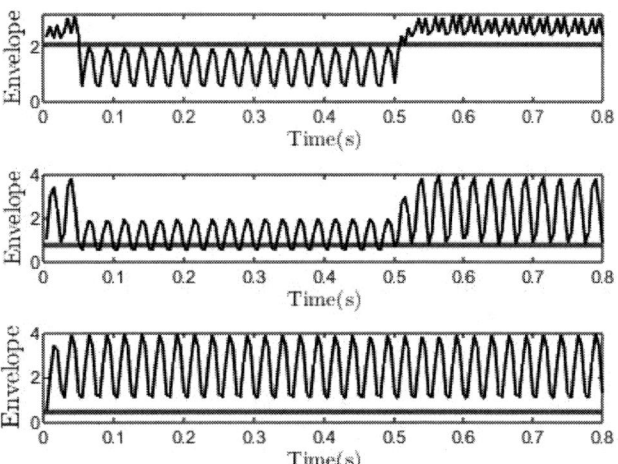

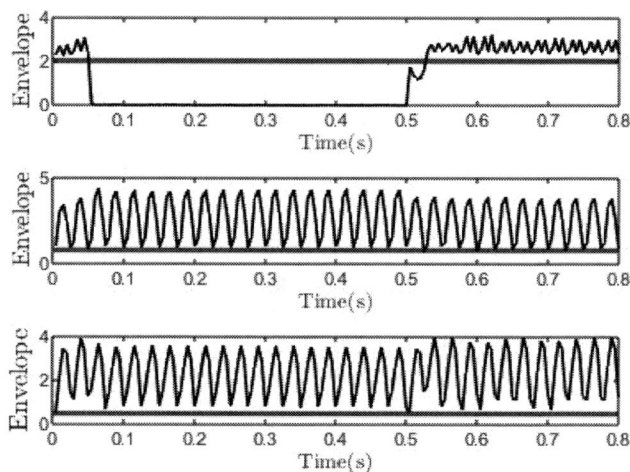

Fig. 5: Instantaneous envelope of phases A,B,C respectively for an LL fault on A,B

Fig. 7: Instantaneous envelope of phases A,B,C respectively for an LG fault on A

all the three phases correctly identifies the occurrence of a balanced fault.

3) Line-to-Ground (LG) Fault: A fault was simulated between phase A and the ground at the same location. The analysis of the transient phase voltages at Bus 6, shown in Fig. 7, reveals a substantial decrease in the instantaneous envelope amplitude of phase A, while the envelope amplitudes of phases B and C exhibit minimal change. The crossing of the threshold value only for phase A correctly identifies the occurrence of an LG fault in that specific phase.

4) Double Line-to-Ground (LLG) Fault: A fault was simulated between phase A and the ground at the same location. The analysis of the transient phase voltages at Bus 6, shown in Fig. 8, reveals a substantial decrease in the instantaneous envelope amplitude of phase A and B, while the envelope

amplitudes of phase C exhibits minimal change. The crossing of the threshold value only for phase A,B correctly identifies the occurrence of a fault in those specific phases.

5) Triple Line-to-Ground (LLLG) Fault: A fault was simulated between the three phases and the ground at the same location. The analysis of the transient phase voltages at Bus 6, shown in Fig. 9, reveals a substantial decrease in the instantaneous envelope amplitude of all three phases. The crossing of the threshold value only for all phases correctly identifies the occurrence of a balanced triple line fault.

VII. LIMITATIONS OF PROPOSED APPROACH

1) Difficulty in distinguishing involvement of ground for balanced faults: The similarity in instantaneous envelope values for balanced faults involving ground and those involving lines only is hard to identify which makes this method limited

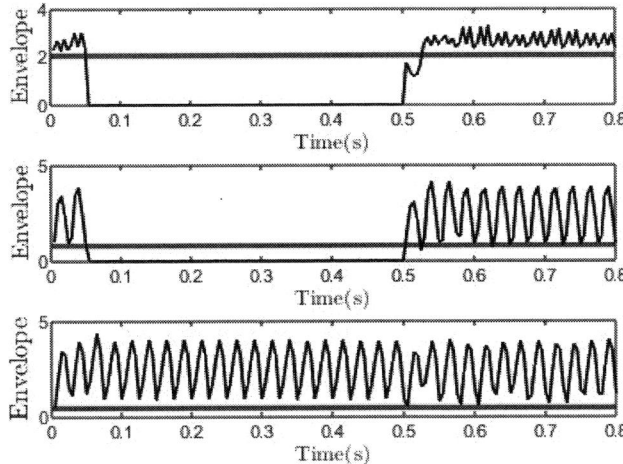

Fig. 8: Instantaneous envelope of phases A,B,C respectively for an LLG fault on A,B

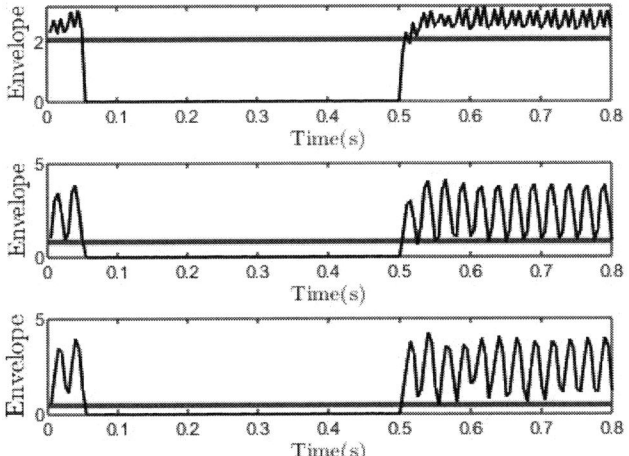

Fig. 9: Instantaneous envelope of phases A,B,C respectively for an LLLG fault on all phases

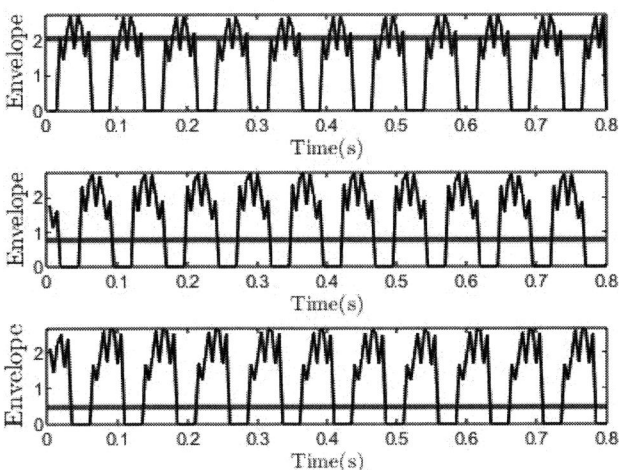

Fig. 10: False positive for a window size of 3 samples (Plotted against V_t computed for 5 sample window).

in its capability of classifying balanced faults with complete accuracy.

2) Impact of Window Size: The choice of the sample window size (N) significantly influences the performance of the methodology: (1) Utilizing very small window sizes can lead to issues, particularly when the signal segment within the window is monotonically increasing or decreasing. In such cases, the EMD algorithm might not be able to extract meaningful oscillatory components (IMFs), potentially resulting in a negligibly small envelope amplitude A_e and leading to spurious or incorrect fault classifications (as illustrated in Fig. 10 for a window size of 3 samples). (2) Conversely, employing very large sample windows can introduce a significant time delay in the detection of the fault. This delay can be detrimental, as it prolongs the duration of the fault current,

potentially causing more severe damage to the affected power system equipment.

3) Threshold Value Determination and Adaptation: The accurate determination and adaptive adjustment of the threshold value V_t are crucial for the reliable operation of the fault detection scheme. The initial threshold should be set based on a thorough analysis of the system's behavior under normal operating conditions, considering factors such as load variations and measurement noise. Furthermore, the threshold may need to be dynamically adapted over time to account for gradual changes in system parameters or operating conditions, thereby minimizing the risk of both false positives (faults detected when none have occurred) and false negatives (failure to detect an actual fault). Fig. 11 show the possibility of false positives when V_t was computed for a window size of 5 samples when the system is operating at a window size of 19 samples.

4) Computational Complexity: The Hilbert-Huang Transform, particularly the EMD component, can be computationally intensive compared to simpler time-domain or frequency-domain analysis techniques. The real-time implementation of this methodology in a practical protection relaying system would require careful optimization of the algorithms and the use of high-performance digital signal processing hardware.

5) Sensitivity to Noise: While Gaussian noise was added to the simulated voltage signals to model measurement errors, the performance of the proposed methodology in the presence of other types of noise or significant harmonic distortion in the power system signals needs further investigation.

The selection of an appropriate window size involves a trade-off between the speed of fault detection and the robustness against noise and transient disturbances. Empirical studies and optimization techniques may be required to determine the optimal window size for a specific power system and application.

979-8-3315-3013-6/25 $31.00 © 2025 IEEE   533

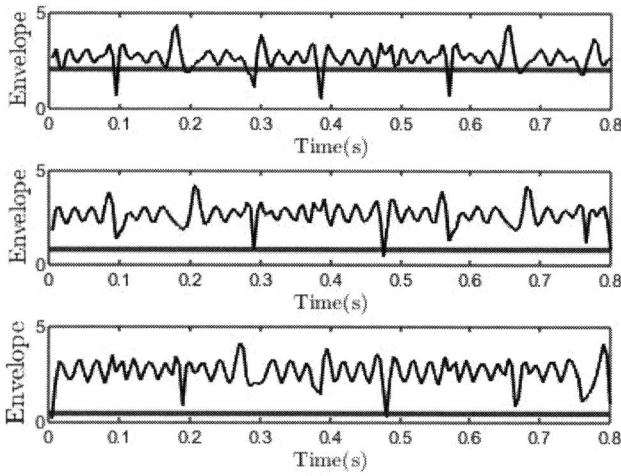

Fig. 11: False positive when window size changed to 19 samples with V_t computed for 5 sample window.

VIII. SUMMARY

This paper has presented a modified methodology for the detection and classification of short circuit faults in power systems based on the method detailed by Li Zhenghua. The proposed approach leverages the EMD to extract the IMFs of transient phase voltage signals and subsequently applies the Hilbert Transform to obtain the instantaneous envelope amplitude for fault detection and classification. The key advantage of this method lies in its ability to analyze non-linear and non-stationary signals effectively, making it well-suited for the complex transient waveforms encountered during power system faults, particularly in the presence of low-frequency power oscillations. Simulation results on the standard IEEE 9-bus system for various short circuit faults demonstrate the efficacy of the proposed methodology in accurately detecting and classifying these fault types based on the affected phases. Furthermore, the method avoids the reliance on Artificial Neural Networks, thereby facilitating a more transparent interpretation of the intermediate data and the decision-making process in specific scenarios.

REFERENCES

[1] Z. Li, "Hilbert-Huang Transform Based Application in Power System Fault Detection," 2009 International Workshop on Intelligent Systems and Applications, Wuhan, China, 2009, pp. 1-4, doi: 10.1109/IWISA.2009.5072894.

[2] Dwivedi, D., Babu, K. V. S. M., Yemula, P. K., Chakraborty, P., & Pal, M. (2025). A comprehensive metric for resilience evaluation in electrical distribution systems under extreme conditions. Applied Energy, 380, 125001.

[3] Prasertwong, K., Mithulananthan, N., & Thakur, D. (2010). Understanding low-frequency oscillation in power systems. International Journal of Electrical Engineering Education, 47(3), 248-262.

[4] Taheri, B., Hosseini, S. A., Askarian-Abyaneh, H., & Razavi, F. (2020). Power swing detection and blocking of the third zone of distance relays by the combined use of empirical-mode decomposition and Hilbert transform. IET Generation, Transmission & Distribution, 14(6), 1062-1076.

[5] Anand A, Affijulla S. Hilbert-Huang transform based fault identification and classification technique for AC power transmission line protection. Int Trans Electr Energ Syst. 2020; 30:e12558. https://doi.org/10.1002/2050-7038.12558.

[6] B. S. Munir, M. Reza, A. Trisetyarso and B. S. Abbas, "Feature extraction using Hilbert-Huang transform for power system oscillation measurements," 2017 4th International Conference on Information Technology, Computer, and Electrical Engineering (ICITACEE), Semarang, Indonesia, 2017, pp. 93-96, doi: 10.1109/ICITACEE.2017.8257682.

[7] Ali Kaffashbashi, Yaser Damchi, Statistical approach for detection of fault and stable and unstable power swings based on signal energy, International Journal of Electrical Power & Energy Systems, Volume 145, 2023, 108638, ISSN 0142-0615, https://doi.org/10.1016/j.ijepes.2022.108638.

[8] Huang, N. E., Shen, Z., Long, S. R., Wu, M. C., Shih, H. H., Zheng, & Liu, H. H. (1998). The empirical mode decomposition and the Hilbert spectrum for nonlinear and non-stationary time series analysis. Proceedings of the Royal Society of London. Series A: mathematical, physical and engineering sciences, 454(1971), 903-995.

[9] Bin Queyam, A., Kumar Pahuja, S., & Singh, D. (2017). Quantification of feto-maternal heart rate from abdominal ECG signal using empirical mode decomposition for heart rate variability analysis. Technologies, 5(4), 68.

[10] Cizek, V. (1970). Discrete hilbert transform. IEEE Transactions on Audio and Electroacoustics, 18(4), 340-343.

[11] Shrivastava, M., Prakash, V., Kaushik, V., & Upadhyay, V. K. (2021). Transient Stability Improvement of IEEE 9-Bus System Using Static Var Compensator. International Journal of Research in Engineering, Science and Management, 4(4), 98-102.

Experimental Realization of Digital Voltage Mode Closed-loop Control of Forward Converter targeting for Single and Multi-Output Applications

Md Asif Alam
Department of Electrical Engineering
National Institute of Technology, Calicut
Kerala, India
asif_m230767ee@nitc.ac.in

Mukti Barai
Department of Electrical Engineering
National Institute of Technology, Calicut
Kerala, India
muktib@nitc.ac.in

Abstract—This paper presents the hardware implementation of digitally controlled Single-Input Single-Output (SISO) and Single-Input Multi-Output (SIMO) forward converters using voltage mode control. A Digital Signal Processor (DSP) generates PWM signals and implements a closed-loop PI controller, discretized using the Euler backward method for real-time operation. Real-time voltage feedback is processed via an ADC, enabling dynamic duty cycle adjustment for stable output regulation under varying input and load conditions. Experimental results validate the design, with the SISO and SIMO converters achieving precise voltage regulation and multiple output generation, respectively. The proposed system offers a robust solution for power conversion applications in industrial and consumer electronics.

Index Terms—Multi Output, Forward converter, Voltage Mode Control, Isolation, DC-DC converter.

I. INTRODUCTION

Switched-mode power converters are widely used due to their compact size, high efficiency, and reduced power dissipation compared to linear power supplies [1]. These converters play a crucial role in applications such as battery charging, lighting, defense, and aerospace [2]. Among DC-DC converters, the multi-output forward converter is extensively implemented, providing multiple regulated outputs from a single source, enhancing efficiency, and minimizing components [3].

In a forward converter, when the switch is ON, energy transfers from input to output via the transformer, which provides isolation while a demagnetizing circuit prevents core saturation. A multi-output forward converter uses secondary-side synchronous switches to achieve ZVS at all loads and reduce output filter size [4].

Digital control strategies offer significantly improved high precision, adaptability, and fast dynamic response, making them ideal for complex systems [5]-[6]. Digital PID controllers are implemented using FPGA and DSP, ensuring efficient rapid computation and precise regulation [7]. Loop shaping control improved stability and reduced overshoot in the Forward Converter, enabling precise voltage regulation for multi-output designs [8]. The study demonstrates effective compensation under input/output disturbances, enabling robust digital controller design for DC-DC converters [9].

A SISO system, with a single input and output shown in Fig. 1, simplifies analysis, control, and implementation, making it ideal for applications in electric vehicles, renewable energy, industrial automation, and medical equipment.

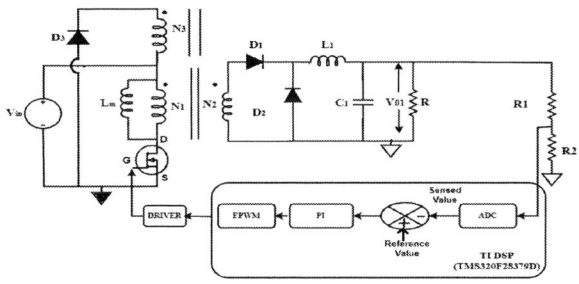

Fig. 1: Complete block diagram of Voltage Mode Close Loop Control

A SIMO system takes a single input and generates multiple regulated outputs denotes in Fig. 2, making it suitable for multi-load applications. Proper control techniques, such as voltage or current mode control, are required to ensure output stability. SIMO converters are widely used in embedded systems, where multiple voltage levels are required from a single power source [10].

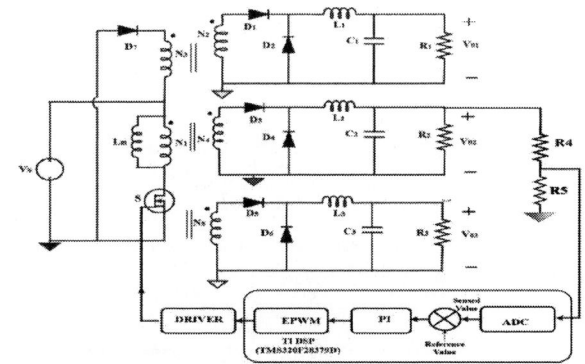

Fig. 2: Complete block diagram of Voltage Mode Close Loop Control

In this Paper, voltage mode control is implemented using the

979-8-3315-3013-6/25 $31.00 © 2025 IEEE

TMS320F28379D digital signal controller to regulate the SISO and SIMO forward converters. The controller dynamically modifies the duty cycle using real-time voltage feedback, ensuring stable performance across different load and input variations.

Section II outlines the design and implementation of a digital PID controller for a forward converter. Section III analyzes the experimental results, demonstrating the closed-loop performance for both SISO and SIMO configurations. Finally, Section IV provides the conclusion.

II. DESIGN OF DIGITAL PID CONTROLLER

The design approach for the PI controller in the forward converter, as well as the small-signal model analyzed in this work, remains consistent with that of a conventional buck converter. However, it incorporates a multiplication factor corresponding to the transformer turns ratio. The small-signal representation of the forward converter, incorporating the effect of the transformer turns ratio, is given by the following transfer function:

$$\frac{\hat{V}_o(s)}{\hat{d}(s)} = \frac{\frac{V_{in}N_2}{N_1}}{s^2 L_1 C_1 + \frac{SL_1}{R_1} + 1} \quad (1)$$

V_o represents the output voltage, d represents the duty ratio, and V_{in} represents the input voltage, N_1, N_2, N_3, and N_4 correspond to the number of turns in the primary, secondary$_1$, secondary$_2$, and secondary$_3$ windings, respectively. L_1 denotes the output inductor, and C_1 denotes the output capacitor.

The continuous-time representation of the PID controller is given by:

$$\frac{\hat{d}(s)}{\hat{e}(s)} = K_p + \frac{K_i}{S} + K_d S \quad (2)$$

K_p denotes the proportional gain, K_i represents the integral gain, K_d corresponds to the derivative gain.

The discrete-domain transfer function of the PID controller is derived using the backward Euler approximation method:

$$\frac{d(z)}{e(z)} = K_p + \frac{K_i T_s}{(1 - z^{-1})} + K_d \frac{(1 - z^{-1})}{T_s} \quad (3)$$

The transfer function of the PID controller in the discrete domain is derived using the backward approximation technique [4]

$$\frac{d(z)}{e(z)} = \frac{\left[K_p T_s (1 - z^{-1}) + K_i T_s^2 + K_d (1 - z^{-1})^2\right]}{(1 - z^{-1}) T_s} \quad (4)$$

The z-domain representation of the transfer function for the forward converter is given by:

$$\frac{V_o(z)}{d(z)} = \frac{\frac{V_{in}N_4}{N_1}}{1 + \frac{(1-z^{-1})L_1}{R_1 T_s} + \frac{(1-z^{-1})^2 L_1 C_1}{T_s^2}} \quad (5)$$

By pole-zero Cancellation, where

$$K_p = \frac{L_1}{R_1 T_s}$$

$$K_i = \frac{1}{T_s}$$

$$K_d = \frac{L_1 C_1}{T_s}$$

By performing the inverse Z-transform on Equation (4)

$$D[n] = D[n-1] + AE[n] - BE[n-1] + CE[n-2]$$

$$A = k_p + k_i T_s + \frac{k_d}{T_s}$$
$$= \frac{L_1 C_1}{T_s^2} + \frac{L_1}{R T_s} + 1$$

$$B = k_p + \frac{2k_d}{T_s}$$
$$= \frac{2L_1 C_1}{T_s^2} + \frac{L_1}{R_1 T_s}$$

$$C = \frac{k_d}{T_s}$$
$$= \frac{L_1 C_1}{T_s^2}$$

Given the specified values of L_1, C_1, and T_s, the parameters A, B, and C are chosen as 0.03, 2, and 0.1, respectively. The duty cycle command, $D[n]$, is computed using the previous duty cycle value along with the error terms at $(n-1)$ and $(n-2)$.

In the digital controller implementation, the ADC measures the output voltage, which is then digitized, scaled, and compared with the reference V_{ref}. The computed error samples generate the duty ratio command. Synchronization is maintained between the duty cycle. Fig. 3 presents the flowchart depicting the digital control algorithm.

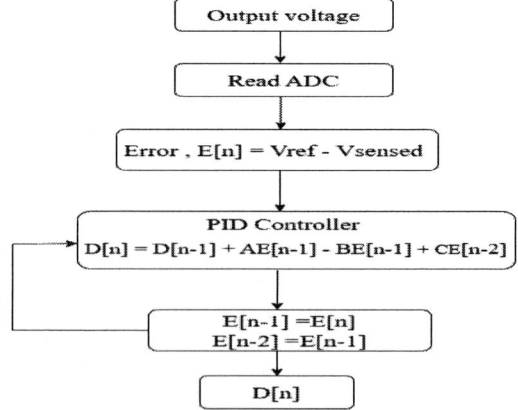

Fig. 3: Flowchart of Voltage Mode CLose Loop Control

979-8-3315-3013-6/25 $31.00 © 2025 IEEE

III. DESIGN IMPLEMENTATION AND EXPERIMENTAL RESULTS

The experimental implementation adheres to the specifications outlined in Table 1, with the design and calculation methodologies for the magnetic components [10].

TABLE I: Design Specification

Parameter	Value
Input Voltage Range	25 V - 35 V
Nominal Input Voltage	30 V
Switching Frequency	20 kHz
Turn Ratio(N_1, N_2, N_3)	1:2:1
Output (V_{01})	24 V
Inductor current ripple (I_{L1})	6 %
Outputs voltage ripple	2 %
Output Power (P_{01})	48 W
Turn Ratio(N_1, N_4, N_3 N_1, N_2)	1:2:1:1:0.8
Outputs (V_{01}, V_{02}, V_{03})	12 V, 24 V, 9 V
Output Power (P_{01}, P_{02}, P_{03})	12 W, 48 W, 9 W
Inductor current ripple (I_{L1}, I_{L2}, I_{L3})	10 %, 6 %, 6 %
Outputs voltage ripple	2 %

A. ADC Calibration

ADC calibration is essential for ensuring the accurate conversion of analog signals into digital form, compensating for offset and gain errors, and minimizing the impact of non-linearity and temperature variations. By calibrating the ADC, we improve the precision of measurements. Fig. 4 illustrates the linearity characteristics of the ADC, depicting its transfer function and deviation from the ideal response.

$$\text{Resolution} = \frac{V_{\max}}{2^N - 1}$$

Initializing a 12-bit ADC:

$$\text{Resolution} = \frac{3.3\,V}{2^{12} - 1}$$

$$= 0.00081\,V \approx 81\,\text{mV}$$

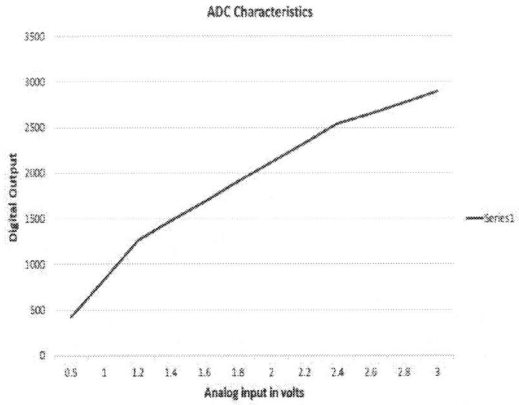

Fig. 4: ADC calibration characteristic

B. Closed loop Experimental results: SISO Forward Converter

As shown in Fig. 1, the forward converter's output voltage is sensed, adjusted via a potentiometer, and digitized by the ADC of the TMS320F28379D. The digital controller processes the error signal by comparing it with a reference value and dynamically modifies the duty cycle to control and maintain the desired output voltage. This ensures stability under varying conditions, minimizes steady-state error, and enhances transient response, improving efficiency and reliability. Fig. 5 illustrates the hardware arrangement for voltage mode closed-loop control in the SISO DC-DC forward converter.

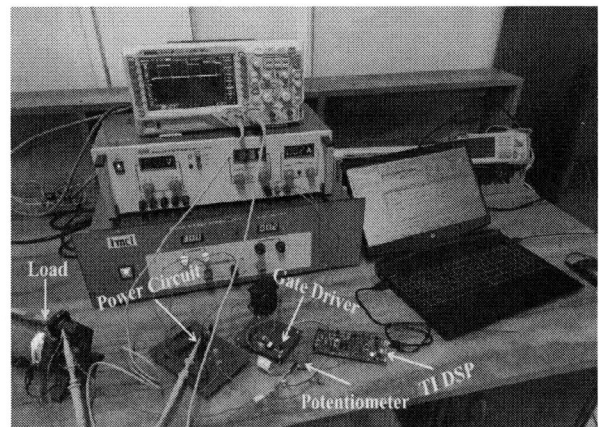

Fig. 5: Hardware Setup of SISO DC-DC forward converter

Fig. 6 illustrate the behavior of the SISO system for varying input conditions. The system adjusts the duty ratio accordingly to achieve the desired constant output voltage.

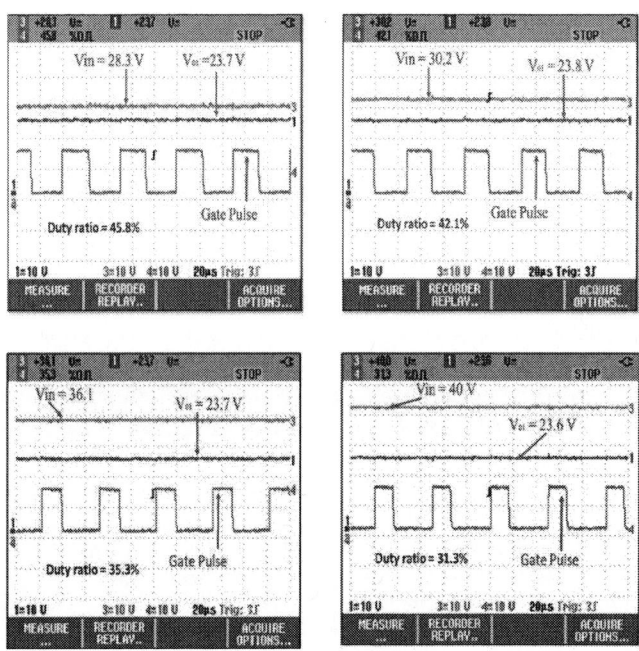

Fig. 6: Line Regulation

Fig. 7 presents the voltage response for various reference voltages (V_{ref}) while keeping the input voltage fixed at 30 V. The figure showcases how the output voltage adapts and stabilizes according to each V_{ref} setting.

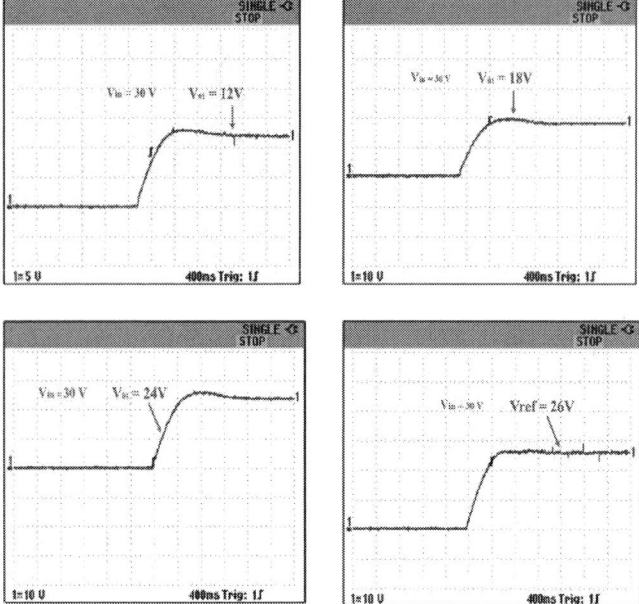

Fig. 7: Transient behavior of the output voltage corresponding to various reference voltage levels.

Fig. 8 illustrates the output voltage regulation under varying input voltage conditions. The input voltage is varied at specific time intervals, and the figure demonstrates how the load regulation is maintained accordingly.

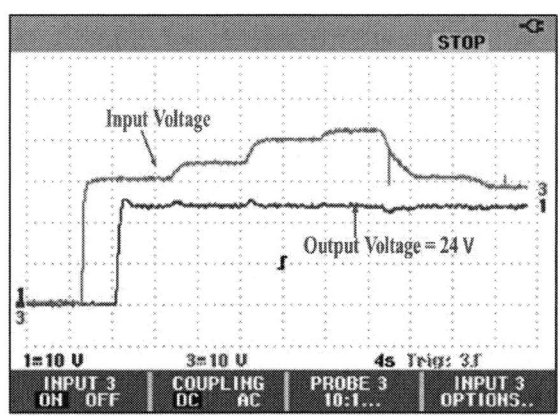

Fig. 8: Output Voltage Regulation Under Varying Input Voltage

Fig. 9 illustrates the effect of step changes in Input voltage demonstrating how the output voltage builds up under varying input step voltage conditions.

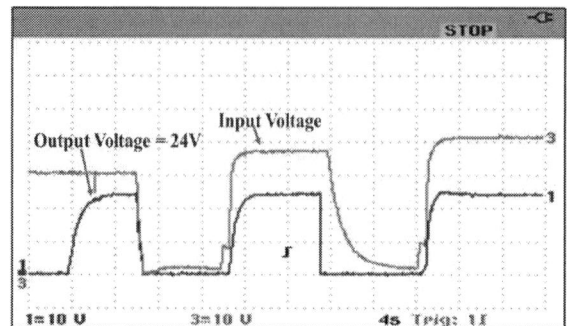

Fig. 9: Step Changes in V_{in} Voltage

Fig. 10 depicts the changes in V_{ref} at different time intervals within the code. Consequently, the output voltage modifies itself to align with the updated V_{ref}. It is evident that after each interval, the output voltage stabilizes at its respective reference value.

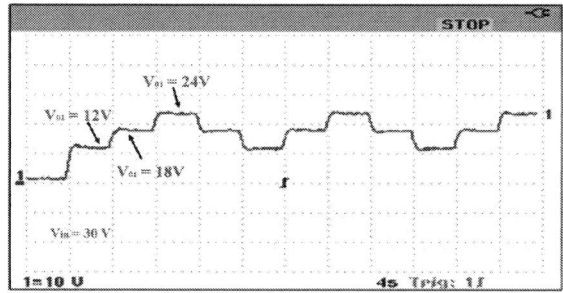

Fig. 10: Step changes in Reference voltage (Vref)

C. Closed loop Experimental Results: SIMO Forward Converter

Fig. 2 shows the SIMO converter circuit with three outputs, where only V_{02} is actively controlled, and while the other two outputs are regulated accordingly. Fig. 11 presents the hardware setup that implements closed loop voltage mode control.

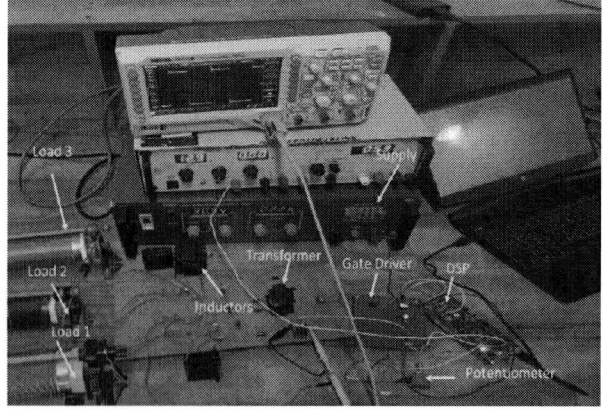

Fig. 11: Hardware Setup of SIMO DC-DC forward converter

979-8-3315-3013-6/25 $31.00 © 2025 IEEE

Fig. 12 denotes the variation of V_{ref} and its effect on the output voltage. As V_{ref} changes, the output voltage builds up accordingly, allowing us to observe the transient response in the system.

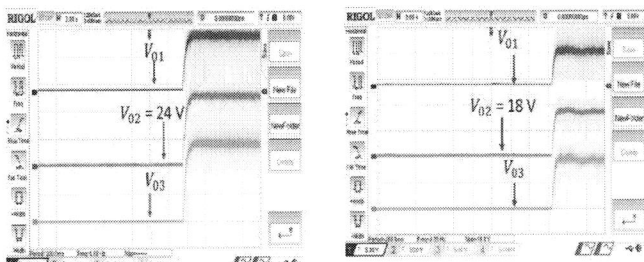

Fig. 12: Transient response of the output voltage for different reference voltages

Fig. 13 depicts the changes in V_{ref} at different voltage levels. As the reference voltage varies, the duty cycle is modified accordingly to regulate the output voltage while keeping the input voltage fixed at 30 V. These figures illustrate the system's dynamic response to different reference voltage settings.

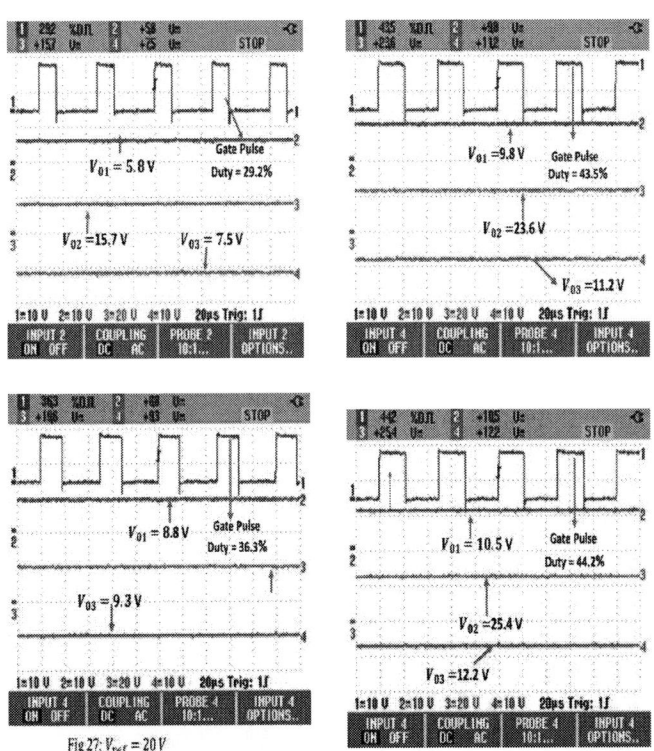

Fig. 13: Duty ratio adjustment for different output voltage conditions

Fig. 14 illustrates the output voltage regulation under varying input voltage conditions. It can be observed that despite continuous changes in the input voltage, V_{02} remains constant, with the duty ratio adjusting accordingly. Additionally, the variations in V_{01} and V_{03} can also be observed.

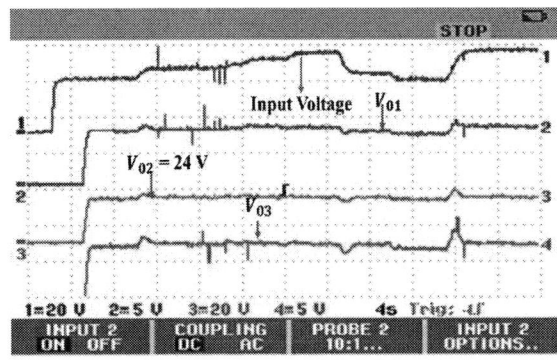

Fig. 14: Output Voltage Regulation Under Varying Input Voltage

Fig. 15 illustrates the effect of step changes in input voltage, demonstrating how the output voltages V_{01}, V_{02}, and V_{03} respond and build up under varying input step voltage conditions.

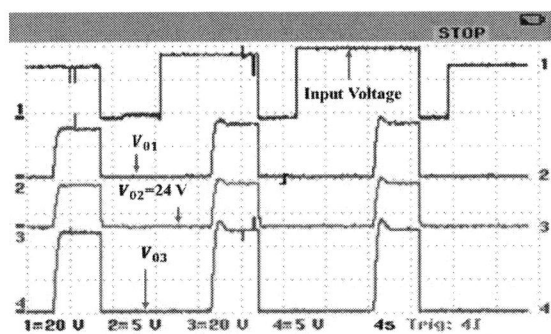

Fig. 15: Step Changes in V_{in} Voltage

Fig. 16 shows how V_{ref} changes at different intervals within the code. As a result, the output voltage adapts to the updated V_{ref}, and it is evident that after each interval, the output voltage stabilizes at the corresponding reference level.

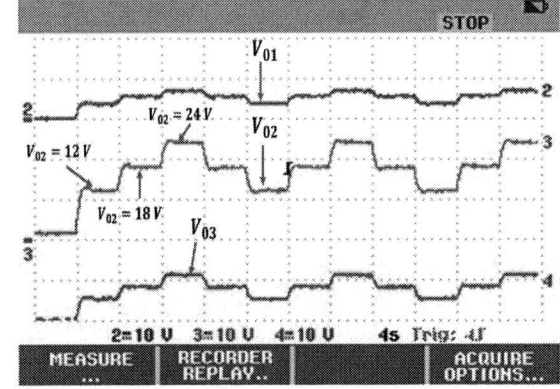

Fig. 16: Step changes in Reference voltage (Vref)

The efficiency variation with respect to loading factor for both SISO and SIMO configurations is shown in Fig.17. In both cases, the output voltage was maintained constant, and the efficiency was observed to improve with increasing load. Minor variations between SISO and SIMO are due to additional conduction and magnetic losses in SIMO operation.

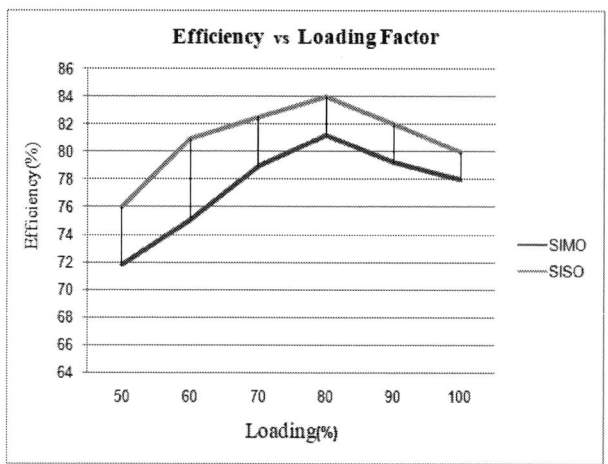

Fig. 17: Efficiency vs loading factor

IV. CONCLUSION

This paper presents a digital controller for both SISO and SIMO forward converters, utilizing closed-loop voltage mode control with a DSP to regulate 12 V, 24 V, and 9 V outputs. The controller dynamically adjusts the duty cycle based on real-time feedback, ensuring stable operation under varying input and load conditions. Experimental results validate precise voltage regulation, demonstrating the feasibility of digital control in multi-output power conversion. The proposed approach addresses critical requirements for SPACE applications, where minimizing size, weight, and component count is essential while ensuring high reliability and efficiency. Future work will explore advanced control strategies, such as model predictive control and adaptive tuning, to further optimize performance for stringent aerospace and industrial applications.

REFERENCES

[1] O. Matiushkin, O. Husev, H. Afshari, D. Vinnikov and R. Strzelecki, "Cost-Effective Piggyback Forward dc-dc Converter," 2024 IEEE Applied Power Electronics Conference and Exposition (APEC), Long Beach, CA, USA, 2024, pp. 2106-2111, doi: 10.1109/APEC48139.2024.10509355.

[2] H. Matsuo and K. Harada, "New energy-storage dc-dc converter with multiple outputs," IEEE Transactions on Magnetics, vol. 14, no. 5, pp. 1005-1007, September 1978, doi: 10.1109/TMAG.1978.1059921.

[3] O. N. Bhirud and M. F. A. R. Satarkar, "Design and Analysis of Multi-output Isolated DC-DC Converter for Low Voltage Aplication," 2018 International Conference on Power Energy, Environment and Intelligent Control (PEEIC), Greater Noida, India, 2018, pp. 815-819, doi: 10.1109/PEEIC.2018.8665438.

[4] J. -K. Kim, S. -W. Choi and G. -W. Moon, "Zero-Voltage Switching Postregulation Scheme for Multioutput Forward Converter With Synchronous Switches," in IEEE Transactions on Industrial Electronics, vol. 58, no. 6, pp. 2378-2386, June 2011, doi: 10.1109/TIE.2010.2060452.

[5] R. Prasanna kumar, S. Patil and B. k. Singh, "Digital Controller for Active Clamp based Synchronous Forward Converter for High Power Applications," 2021 6th International Conference for Convergence in Technology (I2CT), Maharashtra, India, 2021, pp. 1-5, doi: 10.1109/I2CT51068.2021.9418172.

[6] N. Mohan and M. Barai, "Digital control of zero voltage switching buck converter using PIC microcontroller," 2012 IEEE 5th India International Conference on Power Electronics (IICPE), Delhi, India, 2012, pp. 1-4, doi: 10.1109/IICPE.2012.6450428.

[7] Y. S. Kushwaha, S. Negi, P. Dwivedi and S. Bose, "Loop shaping control of DC-DC Forward Converter," 2021 Asian Conference on Innovation in Technology (ASIANCON), PUNE, India, 2021, pp. 1-7, doi: 10.1109/ASIANCON51346.2021.9544939.

[8] M. F. N. b. Tajuddin, N. A. Rahim and I. Daut, "Design and Implementation of a DSP Based Digital Controller for a DC-DC Converter," 2009 Second International Conference on Computer and Electrical Engineering, Dubai, United Arab Emirates, 2009, pp. 209-213, doi: 10.1109/ICCEE.2009.217.

[9] R. Janga and S. Malaji, "Digitally controlled Active Clamp Forward Converter with small signal discrete-time modeling," 2014 International Conference on Computer Communication and Informatics, Coimbatore, India, 2014, pp. 1-6, doi: 10.1109/ICCCI.2014.6921846.

[10] M. A. Alam and M. Barai, "Isolated Multi-Output Power Supply Based on Forward Converter Using Tertiary Winding," 2024 2nd International Conference on Recent Trends in Microelectronics, Automation, Computing and Communications Systems (ICMACC), Hyderabad, India, 2024, pp. 309-314, doi: 10.1109/ICMACC62921.2024.10894390.

Thyristor Validation and Testing of 5kA/500V Rectifier Stack for Magnetic Confinement Application in Fusion machines

Darshan Parmar
ITER-India, IPR
Ahmedabad, India
darshan@ipr.res.in

Rohit kumar
Institute for Plasma Research (IPR)
Ahmedabad, India
rohit.kumar@ipr.res.in

Kush Mehta
ITER-India, IPR
Ahmedabad, India
kush.mehta@iterindia.in

Niranjanpuri Goswami
ITER-India, IPR
Ahmedabad, India
niranjanpuri.goswami@iterindia.in

Dishang Upadhyay
ITER-India, IPR
Ahmedabad, India
dishang.upadhyay@iterindia.in

Rasesh Dave
ITER-India, IPR
Ahmedabad, India
rasesh.dave@iterindia.in

Sandip Gajjar
ITER-India, IPR
Ahmedabad, India
sandip.gajjar@iterindia.in

Hitesh Dhola
ITER-India, IPR
Ahmedabad, India
hitesh.dhola@iterindia.in

Aruna Thakar
ITER-India, IPR
Ahmedabad, India
aruna.thakar@iterindia.in

Motibhai Makwana
Institute for Plasma Research (IPR)
Ahmedabad, India
makwanam@ipr.res.in

Prakash Parmar
Institute for Plasma Research (IPR)
Ahmedabad, India
pkparmar@ipr.res.in

Supriya Nair
Institute for Plasma Research (IPR)
Ahmedabad, India
suprivar@ipr.res.in

Joydeep Ghosh
Institute for Plasma Research (IPR)
Ahmedabad, India
jghosh@ipr.res.in

N P Singh
ITER-India, IPR
Ahmedabad, India
narinder.singh@iterindia.in

Ujjwal Baruah
ITER-India, IPR
Ahmedabad, India
ujjwal.baruah@iterindia.in

Abstract— **High Current Power Supplies (HCPS) fed to inductive coils are used for magnetic confinement of the hot plasma in fusion machines, such as ITER in southern France. The output current of the HCPS typically ranges from few kA up to 70kA, and employs thyristors and IGBTs as an elementary switching device. The press-pack (capsule) thyristor modules are commonly used in multi-pulse converters and has demonstrated very high reliability. 5kA/500V rectifier stack based on 3 Phase, full wave, full controlled thyristor bridge has been developed utilizing two thyristors in parallel in each arm. The DSP controller is used for implementing the control and protection functions. This paper presents the validation tests conducted on the selected high current press-pack thyristors along with test results of 5kA/500V Rectifier Stack on RL dummy load. The lessons from initial synchronous operation of Rectifier Stack (on RL load) during Aditya-U plasma pulse are also presented.**

Keywords—High Current Power Supply (HCPS), Fusion, Magnetic confinement, Thyristor converter

I. INTRODUCTION

Fusion machines, such as ITER in southern France, are striving to demonstrate the feasibility of fusion energy for peaceful purposes. Tokamak is widely used configuration of fusion machine where several superconducting and/or copper type inductive coils create magnetic field to confine the hot plasma inside a vacuum vessel.

The High Current Power Supplies (HCPS) provide controlled voltage & current to these coils for plasma confinement but also for plasma initiation, plasma position control, plasma shaping and error field correction [1-3]. These HCPS typically ranges from few kA up to 70kA, while the output voltage ranges from 100V to ~1kV and employs thyristors and IGBTs as an elementary switching device [4-6]. The press-pack (capsule) thyristor modules are commonly used in multi-pulse converters and has demonstrated very high reliability. The selected thyristor shall have very high average on-state current rating and shall be connected in parallel for achieving HCPS requirements. Moreover, the thyristor shall survive from worst case fault conditions including short circuit events requiring very high non-repetitive surge current rating [7].

The choice of HCPS converter topology depends on various factors such as output voltage/current requirements, coil inductance, response time, efficiency, input harmonics, reliability, size, and cost. Multi-pulse thyristor converter topology used for magnetic confinement and coarse plasma control is simple, reliable and cost-effective. Three niche areas are identified for implementation of HCPS based on multi-pulse thyristor converter viz. selection & validation of thyristor, demonstration of balanced current sharing among parallel thyristors and careful consideration of control & protection strategy.

5kA/500V rectifier stack based on 6 pulse full wave bridge topology has been developed, having 2 parallel thyristors in

979-8-3315-3013-6/25 $31.00 © 2025 IEEE

each arm of the bridge. DSP controller is used for realization of overall control and protection functions. This paper includes the results of validation tests conducted on selected thyristor for HCPS application, along with experimental test results of 5kA/500V rectifier stack on RL (Resistive Inductive) dummy load. The rectifier stack is being integrated with medium sized Aditya-U tokamak in India [8-9] for initial shaped plasma experiment. Lessons from very first synchronous operation of rectifier stack on RL load during Aditya-U tokamak pulse are also presented.

The rest of this paper is organized as follows: Section II describes the characteristics of selected thyristor, general design guidelines for thyristor selection in HCPS application and also presents the validation tests conducted on the selected thyristors. Section III summarizes the overview of 5kA/500V rectifier stack along with test results on RL load. The very first results and lessons from synchronous operation of rectifier stack during Aditya-U tokamak pulse are also covered in this section. Finally, conclusions are drawn along with potential future works in section IV.

II. THYRISTOR CHARACTERISTICS, SELECTION GUIDELINES & VALIDATION TESTS

A. Selected Thyristor Characteristics

The press-pack thyristor modules are commonly used in multi-pulse converters and are typically available in the current range of 300A–5kA and voltage range of 400V–8kV [10].

Thyristor 4400PS320 from M/s RIR Power Electronics with datasheet parameters of maximum average on-state current of 4400A, maximum repetitive peak forward & reverse blocking voltage of 3200V and peak surge current of 56kA has been selected for validation purposes. The key datasheet parameters are highlighted in table I.

TABLE I.   KEY DATASHEET PARAMETERS OF THYRISTOR 4400PS320

Key Parameters	Value	Units
$I_{T(AV)}$ - Max. average on-state current @heatsink temperature of 55°C	4400	A
V_{DRM}/ V_{RRM} - Max repetitive peak forward & reverse blocking voltage	3200	V
I_{TSM} - Max one-cycle non-repetitive surge current	56	kA
V_{to} Max. on state threshold voltage	0.94	V
V_t - Max. on state voltage	1.5	V
r_t -On-state slope resistance	0.12	mohm
I_H -Max holding current	400	mA
I_L -Max latching current	1000	mA
di/dt –Max non-repetitive rise rate of turn-on current	100	A/us
T_q – Typical turn on time	500	us
dv/dt – Max critical rate of rise of off-state voltage	500	V/us
I_{rrm}/I_{drm} - Peak reverse and off state leakage current	300	mA
T_{jmax} – Max operating junction temperature	125	°C
R_{thJ-hs} –Max thermal resistance (junction to heatsink)	0.007	K/W
I_{gt} – DC gate current required for trigger	300	mA
V_{gt} – DC gate voltage required for trigger	4	V

B. Thyristor Selection Guidelines

Following are the general design guidelines for the selection of thyristor for HCPS application.

- The max repetitive reverse blocking voltage of thyristor (V_{RRM}) should be at least 2.5 times higher than converter input voltage.

- Average On-state Current rating ($I_{T(AV)}$) in the thyristor datasheet is a calculated value as in (1) and shall be considered meticulously. This parameter is mainly dependent on threshold voltage (V_t), on-state slope resistance (r_t) and form factor (ff)

$$I_{T(AV)} = \frac{-V_{to} + \sqrt{V_{to}^2 - 4 \times ff^2 \times r_t \times (-P_w)}}{2 \times ff^2 \times r_t} \quad (1)$$

- In addition to the V_t and r_t parameter from the thyristor datasheet, the P_w in (1) refers to maximum calculated power loss in the device, for specified thermal resistance of the device (R_{thJ-hs}) and temperature difference from the maximum junction temperature (125°C) to the specified heatsink temperature (55 °C). Lower the specified heatsink temperature, higher is the calculated power loss (P_w) and therefore higher is the specified average on-state current ($I_{T(AV)}$) rating. Clearly, the specified heatsink temperature is very important datasheet parameter and could substantially change the value of average on-state current ($I_{T(AV)}$) rating specified in the datasheet.

- The total power loss of thyristor is sum of On-state losses (P_T) and switching losses (P_{SW}) and calculated as (2) & (3), where Qrr is reverse recovery charge, f is frequency of operation and V_R is reverse voltage.

$$P_T = V_{to} \times I_{AV} + r_t \times I_{rms}^2 \quad (2)$$

$$P_{SW} = P_{turn-off} = \frac{Q_{rr} \times V_R \times f}{2} \quad (3)$$

- When paralleling the thyristors, A V_t-band of 50mV - 100mV among thyristors is recommended, normally measured at Tjmax and at a current close to $I_{T(AV)}$. Since this often does not cover the entire production batch, a solution with 2 to 5 times V_t-bands may be the most economical approach [11]. The procurement of 24 selected thyristor was targeted with maximum 4 V_t-bands, where each band is within 50mV.

C. Thyristor Validation Tests

The thyristor validation tests become even more important for HCPS application where paralleling of thyristor is foreseen. In a batch of 24 selected thyristors (4400PS320), routine tests are conducted on all the thyristors, while the type test is conducted on one random sample. The thyristor validation tests are conducted as per IEC 60747-6 reference standard [12]. The routine tests include forward/reverse blocking voltages with leakage measurement test, on-state voltage, holding/latching current measurement test, gate trigger voltage/current measurement test, and encapsulation test. While the type tests include non-repetitive surge current tests. Following are the key outcomes from the thyristor validation test.

- All the observed parameters during routine test are in line with the datasheet specified values, and are summarized in table II.

- Max repetitive peak forward & reverse blocking voltage of 3200V successfully applied on selected thyristor at specified virtual junction temperature of 125 °C.

979-8-3315-3013-6/25 $31.00 © 2025 IEEE   542

TABLE II. ROUTINE TEST RESULTS FOR 24 SELECTED THYRISTORS (4400PS320)

Sr. No.	DEVICE ID.	V_T @ 4KA 125°C (V)	I_{RRM}[a] 25°C (mA)	125°C (mA)	I_{DRM}[a] 25°C (mA)	125°C (mA)	I_H 25°C (mA)	I_L 25°C (mA)	I_{gt} 25°C (mA)	V_{gt} 25°C (V)
1	AW-45	1.30	8	22	6	32	41	47	77	2.37
2	AW-46	1.32	4	20	8	32	44	56	78	1.96
3	AW-47	1.32	6	22	6	32	51	69	75	2.62
4	AW-48	1.32	1	28	2	38	33	43	68	1.98
5	AW-49	1.34	1	28	6	50	47	60	76	2.04
6	AW-50	1.34	2	26	4	55	43	73	59	1.82
7	AW-51	1.35	1	26	8	55	48	70	72	1.88
8	AW-52	1.36	1	26	8	60	30	39	61	2.10
9	AW-53	1.36	1	22	8	50	58	73	95	2.23
10	AW-54	1.40	1	22	10	60	46	61	58	2.05
11	AW-55	1.41	8	26	8	40	52	75	64	2.18
12	AW-56	1.41	1	22	1	45	39	64	78	2.08
13	AW-57	1.41	1	22	6	60	35	43	53	2.05
14	AW-58	1.41	1	22	12	60	40	73	78	2.52
15	AW-59	1.44	1	18	8	32	51	78	83	2.15
16	AW-60	1.44	1	22	1	50	39	47	56	2.24
17	AW-61	1.44	1	22	4	45	40	56	74	2.13
18	AW-62	1.45	4	22	10	70	45	57	67	2.23
19	AW-63	1.46	1	20	10	50	39	57	72	1.94
20	AW-64	1.46	1	22	8	50	36	47	51	2.20
21	AW-65	1.46	1	22	10	50	37	57	70	2.23
22	AW-66	1.49	4	21	4	45	31	39	50	1.96
23	AW-67	1.49	1	22	4	45	40	55	80	2.68
24	AW-68	1.50	1	20	8	50	47	65	74	1.92

a. Irrm/Idrm, measured at Vrrm/Vdrm=3200V

- No bubbles were observed during encapsulation tests, confirming the hermetic sealing of the device

- The maximum deviation of ~200mV was observed in "On-state Voltage" among all 24 tested thyristors as shown in Fig. 1. In fact, there are 4 Vt-bands among devices where deviation is within 50mV. Further improvisation is feasible in consultation with the manufacturer.

- A peak non-repetitive surge current pulse of 64kA was successfully applied on device at 125 °C. The applied surge current waveform is shown in Fig. 2.

Successful validation tests established the use of the selected thyristor for HCPS application in fusion machines.

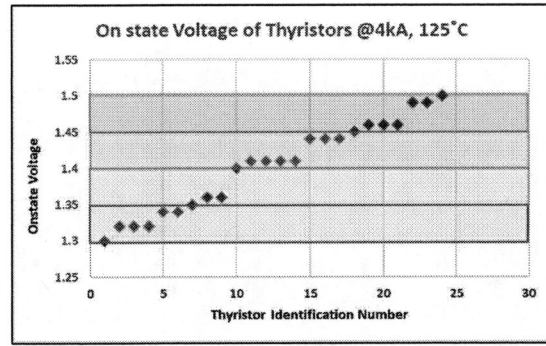

Fig. 1: Vt-bands of On-state voltage drop among all 24 devices

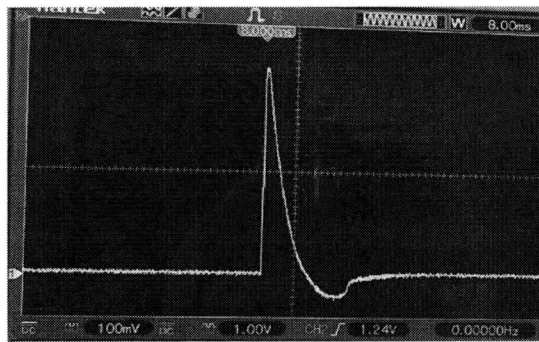

Fig. 2: Non-repetitive surge current test (peak=63.57kA), measured voltage of 596mV across shunt of 0.009375 mohm

III. OVERVIEW AND TEST OF 5KA/500V RECTIFIER STACK

A. *Overview of 5kA Rectifier Stack*

A forced air-cooled rectifier stack is developed utilizing selected thyristors, generating up to 500Vdc, 5kA on RL load. It consists of 3 Phase, full wave, full controlled thyristor bridge fed by 1MVA, 11kV/415V input transformer. A block level schematic of the Rectifier Stack is shown in Fig. 3.

Two parallel thyristors are employed in each arm of the bridge. Each thyristor is mounted on air-cooled heatsink, with thermostat for protection against over temperature. Air cooled heatsinks are mounted on both sides of thyristors and fan is mounted on the top of the rectifier panel to exhaust the hot air. The maximum design heatsink temperature is kept as 110°C Fast acting semiconductor fuses with micro-switch are placed in series of each SCR for protection against overloads and short circuits. RC snubbers are implemented across devices for protection against overvoltage during turnoff [13].

Texas Instruments TMS320F28335 [14] Digital Signal Processor (DSP) is integrated with the Rectifier Stack panel for implementation of control and protection functions. Software based synchronous reference frame phase locked loop (PLL) that utilizes Clarke and Park transform is used for accurately estimating the grid angle from the instantaneous voltage waveform measurement [15]. The accurate estimation of the grid angle is very critical for this application and could affect the overall performance of the control loop. The employed PLL is very robust and provides stable performance during grid conditions such as switching transients, voltage dips, line unbalance, and frequency variations. The delay in the thyristor firing pulse by an angle α controls the output voltage and hence the current of the RL load. The firing angle control range is limited between 30° and 140° in order to limit maximum operational voltage and to prevent commutation failure. The Proportional–Integral (PI) control is implemented with voltage and current feedback signals The HCPS controller operates either in selectable open loop control, or close loop Current Control with Voltage Limit (CCVL) mode. In the close loop operation, the output current follows the input reference current signal, while the voltage limit control operates when the output voltage exceeds the input reference voltage signal.

Both local and remote mode operations are foreseen with the 5kA rectifier stack. In local mode, the output voltage/current can be adjusted from local HMI. In remote mode, the HCPS "Enable signal" and "output current/ voltage reference" are applied from higher level remote controller, which is synchronized with the Aditya-U tokamak operation. When the "Enable signal" is high, output voltage/current follows the reference voltage/current signal. Otherwise, the analogue reference signal is not considered by the HCPS controller. The controller provides soft start enable/disable feature for possibility of implementing high di/dt (>300A/ms) for Aditya-U tokamak shaped plasma experiment requirements.

Thyristor Bypass circuit is also implemented at the output of the rectifier stack which is operated in the event of external quench command from the superconducting magnet system, where the converter shall turn-off main thyristors and trigger the bypass thyristors to isolate source, and freewheel output current flowing through the magnet coil.

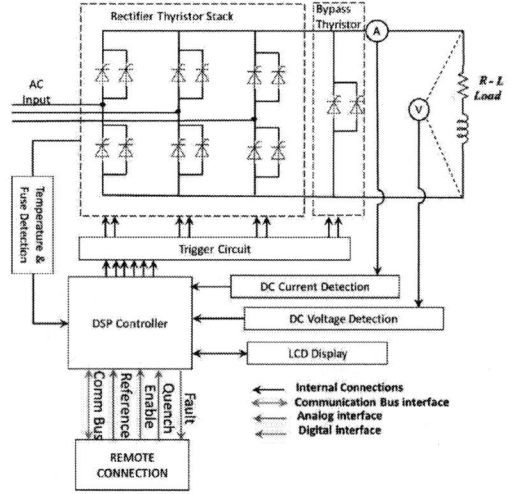

Fig. 3: Block level schematic of the 5kA Rectifier Stack

B. *Test Results of 5kA Rectifier Stack on RL Dummy Load*

a) *Protection functions Check:*

All protection function implemented in the 5kA rectifier stack are verified, and key test results are summarized in table III. It is very crucial that HCPS turn off very fast, in case of over voltage or overcurrent fault scenarios, which may otherwise lead to catastrophic failure of the magnet coils. With very fast implementation of output current (& voltage) measurement functions and protection loop, the power supply was able to trip within 22ms (& 37ms) time.

TABLE III. RECTIFIER PROTECTION FUNCTION CHECK TEST

Sr No	Protection Functions	Observations
1	AC Under voltage; Checked by increasing threshold voltage to 460V	AC undervoltage fault displayed on HMI, Input 400V ACB feeder also tripped
2	DC Over voltage, checked by changing over voltage setting to 300V, lower than set value of 350V.	Power supply tripped in 37ms, turning off the gate pulse of the rectifier stack
3	DC Over Current, verified by changing the current threshold to 500A, lower than the set current of 600A	Power supply tripped in 22ms, turning off the gate pulse of the rectifier stack
4	Emergency Stop, checked by pressing emergency push button	O/P current/voltage turned off, also 11kV breaker tripped for human safety
5	Cooling Fan Fail, checked by turning off the MCB feeding cooling fan supply.	After settable 10s, Input 400V ACB feeder tripped & the HMI shows "Cooling Fan fail" Fault
6	Thyristor fuse fail; Thyristor over temp; Gate PS fail; Phase Locked Loop (PLL) Loss; Phase Sequence Error	All these faults are simulated and it was observed that incoming 400V ACB feeder is tripped

b) *Current Sharing among Parallel Thyristors:*

The current mis-sharing among the two parallel thyristors in the arm is measured and found to be within 12%. Two parallel thyristors in each arm are selected from same Vt-band where on-state voltage deviation remains within 50mV. In addition, it was ensured that the thyristors are uniformly pressurized through proper clamping arrangements and clamping force is within limits specified in the manufacturer datasheet.

c) *Open Loop Control Test at No Load:*

The open loop control test at no load is carried out to verify that control operates accurately together with the main power circuit, and establishes a signature of the rectifier stack. The output voltage waveform with set firing angle of 30 degree is shown in Fig. 4.

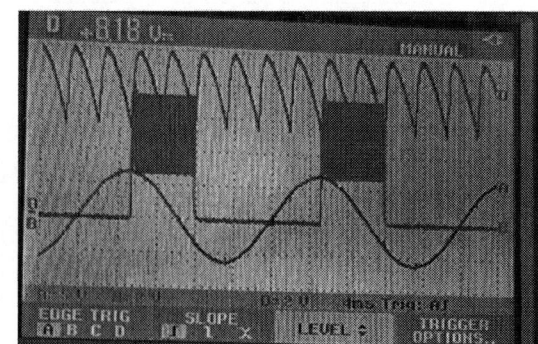

Fig. 4: Output voltag wavform with Set firing angle of 30 degree (Red: Line voltage, Blue: Gate pulse, Green: output voltage)

d) Close Loop CCVL Test on RL Dummy Load :

The functional behavior of the 5kA rectifier stack is verified in Current Control with Voltage Limit (CCVL) close loop mode on RL dummy load (L=356uH and R=64mohm). The key functional tests were performed such as current accuracy test, pulse duration check, rated current test etc. The accuracy of the output current remained within 1% of the full-scale current as shown in Fig. 5.

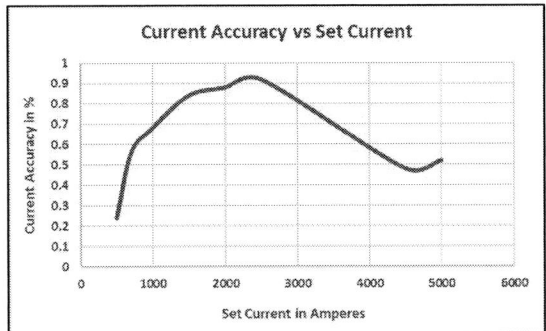

Fig. 5:   Output Current Accuracy graph

The waveform for set output current of 5kA for 1s pulse duration is shown in Fig. 6. There are no drooping effects seen on the output current measurements, confirming the effective implementation of control regulations. Initial delay of ~50ms between the external Enable trigger signal and start of output current is configurable in the DSP code. The scaling of the output current is 10V:6000A and output voltage is 10V:600V.

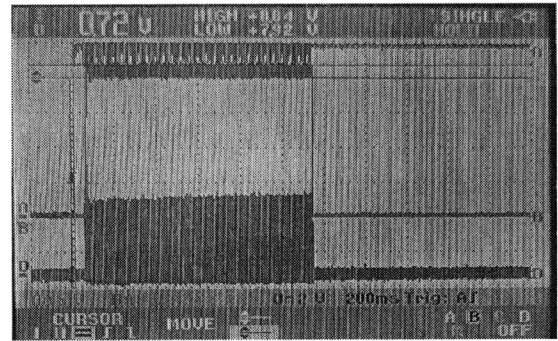

Fig. 6:   Waveform of 5kA/1s Current Pulse on the RL load in local mode
(Red: trigger signal, Blue: output current, Green: output voltage)

The rising of the 5kA waveform is zoomed and shown in Fig. 7 where the di/dt of 357A/ms is achieved without considerable overshoot. This also confirms the effective PI tuning established with the DSP controller. It takes about 14ms to reach 5kA set current that corresponds to the RL time constant of the load. This test validates the di/dt requirement of Aditya-U tokamak for shaped plasma experiment. In order to achieve such a high di/dt with RL load, the rectifier stack must generate higher voltage in the beginning of the pulse (inductive mode of operation). Subsequently, the voltage requirement is substantially reduced in flat top (resistive mode of operation). The overall close loop control of the power supply is very crucial in achieving this requirement.

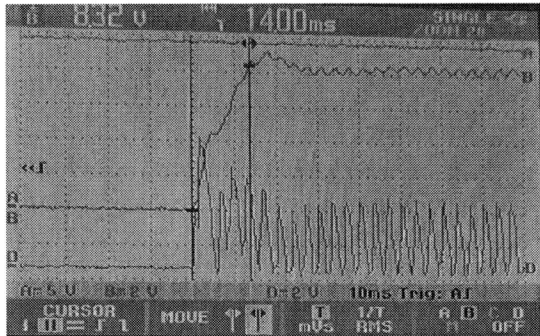

Fig. 7:   Waveform of 5kA rise time confirming output di/dt of 357A/ms

C. *First test results of synchronous operation with Aditya-U tokamak*

After the successful test of 5kA/500V rectifier stack in local mode of operation, the controller of the rectifier stack is connected with master controller of the Aditya-U tokamak for remote operation. The first tests were conducted on rectifier stack still with RL load, but operating in synchronous with Aditya-U tokamak plasma pulse. A remote enable trigger pulse is received from the master controller of Aditya-U tokamak.   The incoming feeder of the 11kV/415V input transformer of the rectifier stack is shared among other power supplies and utilities which draw very high power during the Aditya-U plasma pulse. It was observed that the input voltage of the rectifier stack varies from 430Vrms to 386Vrms, that corresponds to more than 10% of input line variation. An important test was carried out to verify the current regulation of the rectifier stack during the plasma pulse. Fig. 8 shows the waveform of 1.5kA current pulse, where the current regulation is achieved during the Aditya-U tokamak pulse.

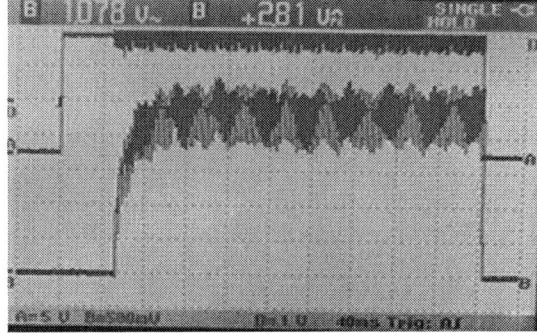

Fig. 8:   Waveform of 1.5kA Current Pulse on the RL load in remote mode
(Red: remote synchronized trigger signal, Blue: output current)

Plasma operations involving unfavorable events such as disruption in tokamaks typically cause fast switching transients in the system. During one of the synchronized shots with the Aditya-U pulse, the synchronizing transformer of the rectifier stack was damaged due to fast over-voltage spikes in the input 415V line. After the inspection of the faulty synchronizing transformer, it was deduced that primary winding of the transformer had inter-turn fault, and the reason for failure was input over-voltage spike. After installation of MOVs at the input of the new synchronizing transformer for over-voltage protection, the situation is avoided. It is therefore very important to note that voltage distortion,

979-8-3315-3013-6/25 $31.00 © 2025 IEEE

voltage variations, and switching transients in the input line are common during plasma operation in machines and appropriate protection shall be implemented in the HCPS.

IV. CONCLUSION AND FUTURE WORKS

The validation tests are successfully conducted, establishing the use of the selected thyristor for HCPS application in fusion machines. 5kA/500V Rectifier Stack developed and tested successfully on RL dummy load, achieving full rated current pulse and di/dt of more than 300A/ms as required for initial shaped plasma experiment in Aditya-U tokamak. Initial tests are also conducted on rectifier stack with RL load, in synchronization with Aditya-U tokamak plasma pulse. The rectifier stack outperformed even with more than 10% of input voltage variations and substantial distortions. The future works would include fully synchronous operation of 5kA/500V rectifier stack on Aditya-U diverter coils for shaped plasma experiments where effects such as induced voltage will be dealt, due to mutual coupling of divertor coils with plasma & other coils. In future applications for high current requirements exceeding 10kA can be achieved through the integration of additional selected parallel thyristors and parallel rectifier stacks.

ACKNOWLEDGMENT

Authors are thankful to M/s. RIR Power Electronics for the manufacturing of press-pack 4400PS320 thyristors and 5kA/500 rectifier stack.

REFERENCES

[1] V. Balakrishnan et al., Plasma current and position feedback control in Aditya Tokamak, Fusion Engineering and design, 66-68(2003) 809-813

[2] J. Tao et al., "ITER Coil Power Supply and Distribution System", 2011 IEEE/NPSS 24th Symposium on Fusion Engineering

[3] E. Bertolini, et al., "The JET divertor coils," in IEEE Transactions on Magnetics, vol. 28, no. 1, pp. 275-278, Jan. 1992, doi: 10.1109/20.119864.

[4] Ivone Benfatto , "Power converters for ITER", https://cds.cern.ch/record/987554/files/p231.pdf

[5] R. Shimada, et al., "JT-60 power supplies", Fusion Engineering and Design, Volume 5, Issue 1, 1987, Pages 47-68, doi: 10.1016/S0920-3796(87)90556-4

[6] A. Lampasi, et al., "The DTT device: Power supplies and electrical distribution system", Fusion Engineering and Design, Volume 122, 2017, Pages 356-364

[7] Z. Song et al., Thyristor Selection Analysis for ITER Poloidal Field Converter Module, J Fusion Energ (2015) 34:620–628

[8] J. Ghosh et al. Upgradation of ADITYA tokamak with limiter configuration to ADITYA Upgrade tokamak with divertor configuration Preprint: 2016 IAEA Fusion Energy Conf. (Kyoto, 17–22 October 2016) FIP/P4-46.

[9] R..L.Tanna et al., Overview of physics results from the ADITYA-U tokamak and future experiments, 2024 Nucl. Fusion 64 112011

[10] https://www.ruttonsha.com/product/phase-control-thyristor#phase-control-capsule Website of RIR Power Electronics.

[11] Hitachi Application Note 5SYA 2091-01, Parameter selection of high power semiconductor for series and parallel connection

[12] IEC 60747-6 standard - Semiconductor devices –Part 6: Discrete devices – Thyristors

[13] Z. Song et al., RC snubber circuit design for ITER PF converter module, J Fusion Energ (2015) 34:427-434

[14] Texas Instruments TMS320F2833x, TMS320F2823x Real-Time Microcontrollers datasheet- SPRS439Q – June 2007 – Revised August 2022

[15] Texas Instruments Software Phase Locked loop design using C2000 microcontrollers for three phase grid connected applications, Application report, SPRABT4A-November 2013

Analysing the performance of coaxial square and circular coils for effective wireless power transfer

Kanala Srinivas Praanesh
EE Department
NIT Rourkela
Rourkela, India
praaneshk01@gmail.com

Tanmoy Roy Choudhury
EE Department
NIT Rourkela
Rourkela, India
trc.6287@gmail.com

Abstract—The global shift towards sustainable non-polluting transportation and reducing carbon emissions has put electric vehicles (EVs) on the front-line. The challenges thus arise in reducing the drive range anxiety due to limitations in battery capacities, long charging times, and the lack of charging infrastructure. Dynamic wireless power transfer (DWPT) addresses these problems by enabling vehicles to be charged while in motion, reducing range anxiety and eliminating the need for charging stops. The DWPT technology is based on inductive power transfer, where there are roadbed transmitter coils and a receiver coil fit in the vehicle to receive power. The efficiency of DWPT depends on the coil parameters like mutual inductance, quality factor, distance between the coils and operating frequency. This paper compares the performance of coaxial circular and square coils by analyzing their power transfer capabilities using mathematical modeling and MATLAB simulations. The mutual inductance of the coils has been modeled and the power transfer against various coil parameters has been plotted.

Index Terms—DWPT, Coaxial, Mutual Inductance, Quality Factor

I. Introduction

The switch towards sustainable transportation has enabled a noticeable rise in the adoption of electric vehicles (EVs), which helps to eliminate the reliance on fossil fuels and pollution. Governments around the world have started implementing strict EV Adoption laws and providing incentives for new buyers to expand the EV consumption [1]. Unlike traditional internal combustion engine (ICE) vehicles, there is no tailpipe emission and also the EVs offer high energy efficiency. However, the adoption of EVs faces issues such as slow charging rates, low battery life, and lack of extensive charging infrastructure. The need for frequent stopping and recharging the vehicle hinders travel and can increase travel time. The user of the vehicle can experience range anxiety that negatively impacts [2] the convenience and usability of EVs. There is a need for advanced charging infrastructure methods to ensure smooth mobility and overcome range anxiety. Dynamic wireless power transfer (DWPT) is one such emerging technology.

In order to maximize the adoption of electric vehicles, an innovative solution, Dynamic Wireless Power Transfer (DWPT), has real time remote charging of EVs whilst in motion. DWPT eliminates the frequent stops at charging stations, improves the range of vehicles, and reduces the size of batteries required, therefore driving the cost and pricing down. Coils are primary part of the DWPT technology. DWPT operated on the principles of inductive coupling [3]. This includes two coils known as transmitter and receiver coils. Transmitter coils (Tx) are embedded on roads and are generally segmented all across the road. The receiver coil (Rx) is a coil fitted inside the vehicle that will take in the power that is transferred from the Tx coils. The power that is transferred depends on many coil factors such as coupling coefficient, distance between the coils, flux linkage, operating frequency and quality factor. The geometry of the coil also plays an important role in determining the efficiency of power transfer.

The DWPT technology operates on the principles of inductive coupling. There is an alternating current that flows through the Tx coils. These coils generate a magnetic field that induces voltage in the Rx coil [4]. The DWPT technology has the advantage of enabling extended range, smaller battery capacities (which contributes for lighter vehicles and reduces cost) and eliminates the dependency upon charging stations.

The alignment of coils and the geometry of the coils are two major aspects that has to be considered. In this paper, planar circular and square coils are considered as they are two most predominant coil geometries in DWPT. Other geometrical shapes such as triangle, hexagon etc., are not considered as circular and square shapes are simple and commonly used. Circular coil for instance has effective magnetic flux distribution due to its curvature whereas square coils have advantages such as its shape, which can be easily integrated into rectangular roads (more surface area utilization). This paper focuses on coaxial alignment over coplanar alignment [5]. The concentric positioning of the coils reduces misalignment loses. Also, in roads with fixed trajectories (highways), the coaxial alignment improves efficiency as the Tx and Rx coils have nearly fixed distance between them. Coaxial configuration is considered as the magnetic coupling is better as compared to coplanar configuration. Also, the misalignment tolerance is better managed in coaxial configuration.

II. Coil Parameters

The performance of DWPT systems is dependent on crucial coil characteristics such as mutual inductance, quality factor,

979-8-3315-3013-6/25 $31.00 © 2025 IEEE

flux linkage, coil spacing, and operating frequency. The impact of each of these parameters on power transfer efficiency is vital for creating effective DWPT systems. Mutual inductance (M) is the key factor that controls power transfer in inductive coupling [6]. It describes how much of the magnetic flux produced by the primary (transmitter) coil is transferred to the secondary (receiver) coil. An increase in mutual inductance results in enhanced energy-transfer efficiency. In mathematical terms, mutual inductance is expressed as shown in equation (1):

$$M = k\sqrt{(L_1 L_2)} \tag{1}$$

In here, k is the coupling coefficient (ranging from 0 to 1) and L is the self-inductance of the coil.

The mutual inductance essentially depends upon three major parameters [7]. The parameters are distance between the coils (z), operating frequency and coil radius/side length. Mutual inductance significantly decreases as the separation distance between the transmitter and receiver coils increases. The relation with respect to side length and radius is a bit complex. The operating frequency range must follows guidelines to not negatively impact the passengers. ICNIRP defines exposure limits according to frequency-dependent magnetic field strength to avoid negative biological impacts, including induced currents within human tissues. For DWPT systems that work in the kHz frequency range, the primary issue is to ensure that passengers, pedestrians, and maintenance personnel are not subjected to hazardous magnetic fields. The exposure limit for the general public is 27 µT at a frequency of 85kHz [8]. To achieve compliance, the DWPT systems may use shielding methods, active field suppression, and optimized coil designs to reduce stray electromagnetic emissions.

The quality factor (Q) determines the efficiency of resonant and compensating circuitry utilized in a DWPT system [9]. The quality factor of a coil is a dimensionless metric that represents the ratio of the coil's inductive reactance to its resistance at a certain frequency. The quality factor is given by equation (2):

$$Q = \frac{\omega L}{R} \tag{2}$$

In here, w is the operating frequency and R is the coil resistance. A higher Q-factor means that resistive losses are less and, thereby higher power transfer efficiency. However, excessively high Q-factors can introduce instability and detuning issues as the coil might get saturated [10].

The efficiency of power transfer depends on the Q factor and is given by equation (3):

$$\eta = \frac{k^2 Q_1 Q_2}{1 + k^2 Q_1 Q_2} \tag{3}$$

In here k is the coupling coefficient and Q1, Q2 are the quality factors of transmitting and receiving coils respectively. The increase in frequency enhance the quality factor but also result in greater losses from eddy current and skin effect. The increased resistance lowers the quality factor, resulting in diminished efficiency. Superconducting or litz wire coils helps in minimizing resistive losses.

The overall power transfer capabilities is dependent upon mutual inductance, which in itself is dependent on operating frequency, distance between coils and coil radius (or) side length. The other factor on which it is dependent on is quality factor. In the two coil inductive system, the Tx coil is connected to an AC source and the receiver coil wireless receives the power from Tx coil. Each coil has its own self inductance, resistance and quality factor. The self inductance of a planar circular coil is as shown in equation (4):

$$L_{\text{circular}} = \frac{\mu_0 N^2 r}{2} \tag{4}$$

In here, N is the number of coil turns and r is the radius of the coil. The self inductance of a planar square coil is as shown in equation (5):

$$L_{\text{square}} = \frac{\mu_0 N^2 A}{l} \tag{5}$$

In here, N is the number of coil turns, A is the area of the coil and l is the side length of the square coil. The transmitting and receiving coils can be modeled as simple circuits with resistances, voltage, and current [11] as shown in Fig.1. A resonant circuit configuration can be considered too, which can be the advanced equivalent circuit. The same frequency between coils improves the power transfer efficiency, reduces losses and minimizes reactive power. For this study, Fig.1 would suffice. It is assumed that there are no foreign particles and materials present near the coil. The load resistance is assumed to be negligible. The equation (6) and equation (7) represent the input and output voltage equations.

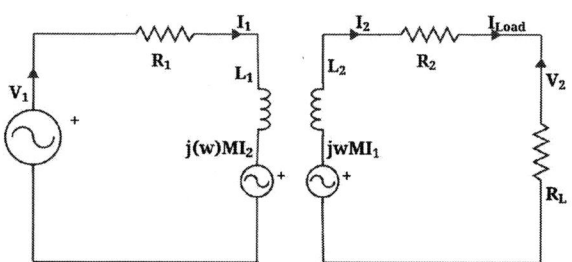

Fig. 1. Equivalent Circuit of the transmitting and receiving coils

$$V_1 = (R_1 + j\omega L_1)I_1 - j\omega M I_2 \tag{6}$$

$$0 = (R_2 + j\omega L_2)I_2 - j\omega M I_1 \tag{7}$$

where:

- V_1, V_2 are the input and output voltages.
- L_1, L_2 are the self-inductance of the transmitter and receiver.
- R_1, R_2 are the internal resistances of the coils.
- M is the mutual inductance.
- $\omega = 2\pi f$ is the angular frequency.

The mutual inductance of the coils and quality factor are already defined. The receiver side current can be calculated as shown in equation (8):

$$I_2 = \frac{j\omega M I_1}{R_2 + j\omega L_2} \qquad (8)$$

The power received by the secondary coil is as shown in equation (9):

$$P_{\text{received}} = |I_2|^2 R_2 \qquad (9)$$

Substituting equation (8) in equation (9), power received can be computed in equation (10):

$$P_{\text{received}} = \left| \frac{j\omega M I_1}{R_2 + j\omega L_2} \right|^2 R_2 \qquad (10)$$

Expanding the magnitude-squared term of equation (10), yields equation (11):

$$P_{\text{received}} = \frac{\omega^2 M^2 |I_1|^2 R_2}{(R_2^2 + \omega^2 L_2^2)} \qquad (11)$$

Shifting the terms of equation (6) yields equation (12):

$$I_1 = \frac{V_1}{(R_1 + j\omega L_1) - j\omega M \cdot \frac{j\omega M}{R_2 + j\omega L_2}} \qquad (12)$$

Simplifying equation (12) gives equation (13):

$$I_1 = \frac{V_1}{R_1 + j\omega L_1 - \frac{\omega^2 M^2}{R_2 + j\omega L_2}} \qquad (13)$$

Squaring the equation (13) yields equation (14):

$$|I_1|^2 = \frac{|V_1|^2}{(R_1 + R_2)^2 + \omega^2 (L_1 + L_2)^2} \qquad (14)$$

Substituting equation (14) in equation (10) yields equation (15):

$$P_{\text{received}} = \frac{\omega^2 M^2 V_1^2}{(R_1 + R_2)^2 + \omega^2 (L_1 + L_2)^2} \qquad (15)$$

III. MATHEMATICAL MODELING OF MUTUAL INDUCTANCE

The mathematical modeling of mutual inductance of coaxial circular and square coils can be done using the Biot-Savart Law. The magnetic field is first determined at a point with distance (z) from the coil (Tx) [12]. It is further integrated to find magnetic field across the surface. Later, the total flux can be determined by multiplying magnetic field (B) with the area of the coil. The flux linkage can be determined by multiplying flux with total number of turns of the coil (N). Further, dividing the flux linkage with current through the coil yields mutual inductance [13]. A coaxial circular coil system is shown in Fig.2. The coils are assumed to be made up of same material and have same number of turns (N). Also both the coils are of same radius, thereby making the coils identical.

The Tx and Rx are transmitting and receiving coils respectively.

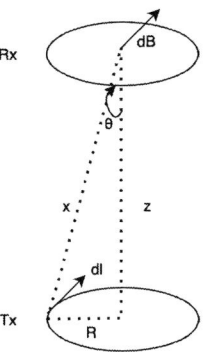

Fig. 2. Equivalent diagram for Coaxial Circular coil system

$$d\mathbf{B} = \frac{\mu_0}{4\pi} \frac{I\, d\ell \times \hat{\mathbf{x}}}{x^3} \qquad (16)$$

where:

- μ_0 is the permeability of free space,
- I is the current in the conductor,
- $d\ell$ is the infinitesimal current element,
- $\hat{\mathbf{x}}$ is the unit vector pointing from the current element to the observation point,
- x is the distance between the element and the observation point.

By taking the magnitude

$$|d\mathbf{B}| = \frac{\mu_0}{4\pi} \frac{I\, d\ell}{x^2} \qquad (17)$$

Using Pythagorean theorem, x can be substituted. Furthermore, to get magnetic field across the coil, arc length dl is substituted using arc length formula. The coil arc length is shown in Fig.3.

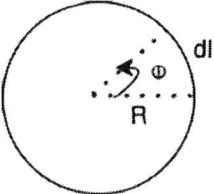

Fig. 3. The Arc length visualization for the circular coil

By using arc length formula,

$$d\ell = R * d\phi \qquad (18)$$

The transformed equation is,

$$d\mathbf{B} = \frac{\mu_0}{4\pi} \frac{I\, R d\phi}{R^2 + z^2} \qquad (19)$$

Splitting the dB component and taking the sin component which aligns with the perpendicular of the coil with integrating the angular term across 0 to 2π [14]. Also, with the number of turns (N),

$$B = d\mathbf{B} \sin\theta = \frac{\mu_0 N}{4\pi} \frac{I\, R d\phi}{(R^2 + z^2)} * \sin\theta \qquad (20)$$

$$B = d\mathbf{B}\sin\theta = \frac{\mu_0 N}{4\pi} \frac{I\,R\,d\phi}{(R^2 + z^2)} * \frac{R}{x} \qquad (21)$$

The final equation is,

$$B = \frac{\mu_0 N \pi}{2} \frac{I\,R^2}{(R^2 + z^2)^{\frac{3}{2}}} \qquad (22)$$

The flux (ϕ) across the area can be obtained by multiplying equation 22 with area of the circle (πR^2). Furthermore, flux linkage (ψ) can be determined by multiplying flux across the area with total number of turns. This flux linkage, when divided by current, provides mutual inductance value.

$$\phi = \frac{\mu_0 N \pi}{2} \frac{I\,R^4}{(R^2 + z^2)^{\frac{3}{2}}} \qquad (23)$$

$$\psi = N * \phi = \frac{\mu_0 N^2 \pi}{2} \frac{I\,R^4}{(R^2 + z^2)^{\frac{3}{2}}} \qquad (24)$$

$$M_{circular} = \frac{\mu_0 N^2 \pi R^4}{2(R^2 + z^2)^{\frac{3}{2}}} \qquad (25)$$

Equation 25 shows the mutual inductance formula for a coaxial circular coil system [15]. Similarly, the coaxial square coil system is shown in Fig.4. Its mutual inductance is determined in a manner similar to that of the coaxial circular system.

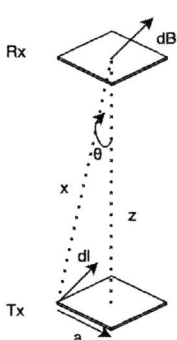

Fig. 4. Equivalent diagram for Coaxial Square coil system

In here, a is the side length of the square coil and z is the distance between the coils. Both the coils are assumed to be identical [16]. The pi remains in the denominator as it is not canceled out. Area of the square coil is a^2.

$$M_{square} = \frac{\mu_0 N^2 a^4}{2\pi (a^2 + z^2)^{\frac{3}{2}}} \qquad (26)$$

The mutual inductance formula for coaxial circular and square coil is given by equation 25 and 26 respectively. The choice of materials used for the coils is important. Generally, copper is preferred for its low resistance and better thermal performance. Cooling plates can be utilized in the system for better thermal management and to handle heat dissipation. Also, against stray EMIs, shielding materials like mu-metals [17] can be used. They help in minimizing edge current losses and interference.

IV. MATLAB ANALYSIS

By utilizing the formulas for mutual inductance and quality factor, MATLAB is utilized to plot the graphs between parameters for both coaxial square and circular coils. Also, the coaxial square and circular coils has been simulated in MATLAB. Fig.5 shows the coaxial circular coil with radius of 0.1 m and thickness of 0.01 m.

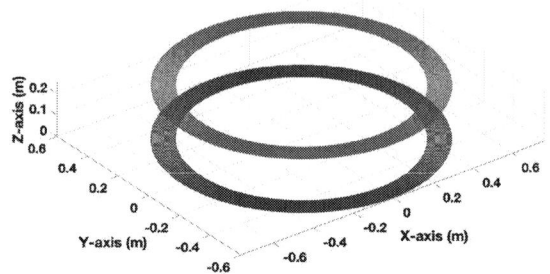

Fig. 5. Coaxial Circular Coil System (designed in MATLAB)

Fig.6 shows the coaxial square coils with side length of 0.1 m and thickness of 0.01 m. Both the coils have been designed in MATLAB with the distance between the coils being 0.2 m. The variation of mutual inductance in both the coils due to the change in distance between the coils (ground clearance in the case of practical vehicles under charging) is plotted using eqn. (25) for coaxial circular coils and eqn. (26) for coaxial square coils as shown in Fig. 7.

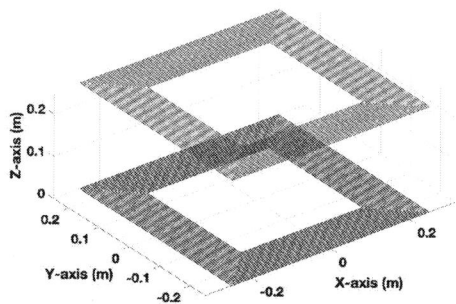

Fig. 6. Coaxial Square Coil System (designed in MATLAB)

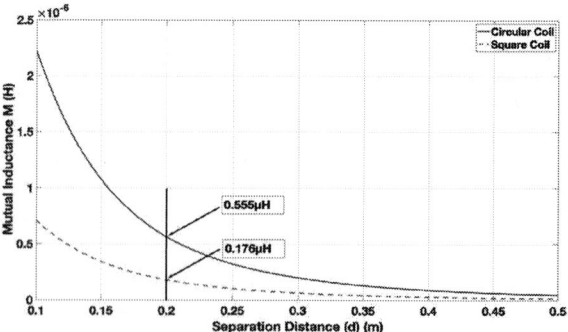

Fig. 7. Plot for mutual inductance vs ground clearance

In Fig.7, the mutual inductance of coaxial circular and square coil has been compared for a separation distance of 0.2 m. Typical EV two wheelers have a ground clearance of 0.15 to 0.22 m. Hence, a ground clearance for 0.2 m has been considered. The circular coil exhibits a higher mutual inductance value than the square coil at this test condition.

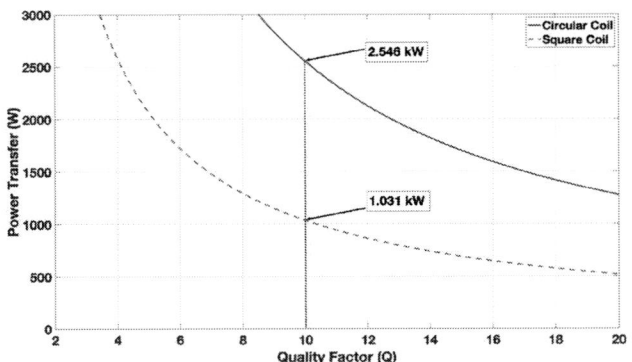

Fig. 8. Plot for power transfer vs quality factor

The variation of power transfer in both the coils due to the change in the quality factor is plotted as shown in Fig. 8, by by substituting eqn. (2) and eqn. (3) in eqn. (15). The power transferred to the receiving coil is plotted against the quality factor of 10. A quality factor of 10 is an acceptable value for a light load condition. The circular coil exhibits a higher power transfer value than the square coil at this test condition.

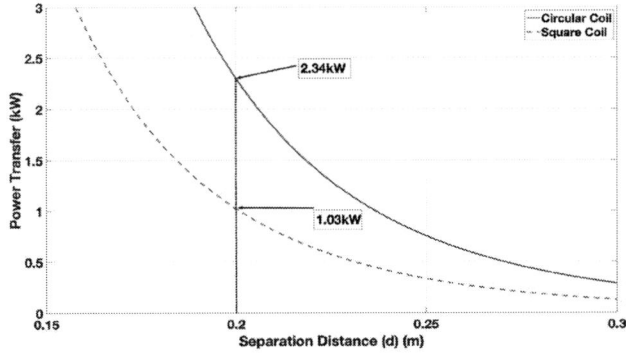

Fig. 9. Plot for Power Transfer vs ground clearance

For the ground clearance of 0.2 m, the power transfer has been plotted in Fig.9. The ground clearance value of 0.2 m is appropriate for lighter-load vehicles such as two wheeler EVs. The circular coil exhibits a higher power transfer value than the square coil at this test condition.

V. RESULT ANALYSIS

The coaxial circular coil has a better mutual inductance value for varying separation distance. The separation distance between the coils or ground clearance is taken such that general case two-wheeler EVs, passenger EVs, and electric buses are considered. For general vehicular transportation,

for minimal ground clearance, circular coils are better than square coils. The mutual inductance is better for circular than that of square. But as ground clearance increases for heavy vehicles, both the curves converge. Hence, it would be ideal to utilize square coils. The reason is that the coaxial square coils are better fit for the rectangular shape of the roads. Also, it has better space utilization. The better space utilization in rectangular roads give a slight edge when considering the charging of heavy vehicles (as the battery capacity would be higher). The conditions taken for both the coils are that the number of turns is 10, the radius of the coil is 0.1 m, and the side length of the square coil is 0.1 m. For the quality factor plot, the voltage is assumed to be 110 V.

From equation (2) and equation (3), the operating frequency is directly proportional to the quality factor which is in itself directly proportional to the efficiency. Hence, increase in operating frequency improves the efficiency but the INCIRP and SAE J2954 levels of operating frequencies must be maintained for safer operation. In Fig.8, the power transfer is plotted for both coils for varying quality factor. For the quality factor of 1 to 25, the curve is plotted. With an increase in Q, the circular coil still transfers more power than the square coil (square coil is less even initially because of the pi term in the denominator). As the Q factor increases, the coils become more saturated, so up until the quality factor of 10 (generally taken from 1 to 10), the circular coil holds up better than that of square coils. In Fig.9. power transfer with respect to ground clearance is plotted. Initially the power transfer remains nearly constant for both the coils, but as ground clearance increases, the power transferred drastically dips for the square coil. Hence, for greater the clearance, lesser is the power transfer (the medium of transfer is via air).

VI. CONCLUSION

The mathematical modeling of the coaxial circular and square coils has been carried out. The magnetic field at a point is determined. It is then integrated across the entire surface, to find the field across the surface. Consequently, the flux and flux linkage are found out. The value of the mutual inductance is determined based on the knowledge of the attained parameters. Furthermore, the importance of parameters and their impact on power transfer has been discussed. The power transfer efficiency and received power are found out. The mutual inductance for coaxial circular coil holds up better with variation and increase in coil distance. But as the distance between coils increases by more than 0.5 m, it is better to use a square coil. This is because the coaxial square coil provides better surface utilization for the rectangular roads. Hence, both the coils have their own benefits and drawbacks. One can evaluate the pros and cons judiciously and choose the coil based upon the implementation and use case.

REFERENCES

[1] X. Zhou et al., "The current research on electric vehicle," 2016 Chinese Control and Decision Conference (CCDC), Yinchuan, China, 2016, pp. 5190-5194, doi: 10.1109/CCDC.2016.7531925.

979-8-3315-3013-6/25 $31.00 © 2025 IEEE

[2] U. Fesli and M. B. Ozdemir, "Electric Vehicles: A Comprehensive Review of Technologies, Integration, Adoption, and Optimization," in IEEE Access, vol. 12, pp. 140908-140931, 2024, doi: 10.1109/ACCESS.2024.3469054.

[3] H. Zakaria, M. Hamid, E. M. Abdellatif and A. Imane, "Recent Advancements and Developments for Electric Vehicle Technology," 2019 International Conference of Computer Science and Renewable Energies (ICCSRE), Agadir, Morocco, 2019, pp. 1-6, doi: 10.1109/ICCSRE.2019.8807726

[4] R. Ahmad, V. Kumar, M. Bilal and S. Kumari, "Dynamic Wireless Power Transfer (DWPT) for Charging Application of Electric Vehicle," 2022 1st International Conference on Sustainable Technology for Power and Energy Systems (STPES), SRINAGAR, India, 2022, pp. 1-6, doi: 10.1109/STPES54845.2022.10006610.

[5] I. Aizpuru, E. Agirrezabala, M. Mazuela, U. Iraola, E. Oyarbide and C. Bernal, "Dynamic Wireless Power Transfer DWPT Time Domain model: xyz position and speed coupling effect," 2022 24th European Conference on Power Electronics and Applications (EPE'22 ECCE Europe), Hanover, Germany, 2022, pp. 1-9.

[6] C. -E. Lee, S. -F. Lin and Y. -C. Liu, "Continuous Dynamic Wireless Power Transfer for Circular Roadway with Optimal Load: Design and Analysis," 2023 IEEE/ASME International Conference on Advanced Intelligent Mechatronics (AIM), Seattle, WA, USA, 2023, pp. 1048-1054, doi: 10.1109/AIM46323.2023.10196270.

[7] Y. Dai, L. Ma, M. Huang and H. Su, "Modeling and Analysis Methods for the DWPT System Applicated in EVs Charging," 2018 IEEE PELS Workshop on Emerging Technologies: Wireless Power Transfer (Wow), Montreal, QC, Canada, 2018, pp. 1-6, doi: 10.1109/WoW.2018.8450910.

[8] L. Jinliang, D. Qijun, H. Wenshan and Z. Hong, "Research on quality factor of the coils in wireless power transfer system based on magnetic coupling resonance," 2017 IEEE PELS Workshop on Emerging Technologies: Wireless Power Transfer (WoW), Chongqing, China, 2017, pp. 123-127, doi: 10.1109/WoW.2017.7959378

[9] M. Kim, M. Jeong, M. Cardone and J. Choi, "Characterization of the Quality Factor in Spiral Coil Designs for High-Frequency Wireless Power Transfer Systems using Machine Learning," 2022 IEEE 23rd Workshop on Control and Modeling for Power Electronics (COMPEL), Tel Aviv, Israel, 2022, pp. 1-8, doi: 10.1109/COMPEL53829.2022.9830005.

[10] Y. Park, J. Kim and K. Na, "Resistance and Q-factor calculation of single and double layer coils with split circular pattern," 2020 IEEE Wireless Power Transfer Conference (WPTC), Seoul, Korea (South), 2020, pp. 304-307, doi: 10.1109/WPTC48563.2020.9295596

[11] F. Castelli-Dezza and M. M. Maglio, "A coil model for voltage distribution and electrical insulation analysis," 8th IEEE Symposium on Diagnostics for Electrical Machines, Power Electronics Drives, Bologna, Italy, 2011, pp. 444-450, doi: 10.1109/DEMPED.2011.6063661.

[12] J. P. K. Sampath, A. Alphones and D. M. Vilathgamuwa, "Coil optimization against misalignment for wireless power transfer," 2016 IEEE 2nd Annual Southern Power Electronics Conference (SPEC), Auckland, New Zealand, 2016, pp. 1-5, doi: 10.1109/SPEC.2016.7846159.

[13] K. N. Mude and M. T. Outeiro, "Coil misalignment analysis under different radius of coil and wire for Wireless Power Transfer System," IECON 2017 - 43rd Annual Conference of the IEEE Industrial Electronics Society, Beijing, China, 2017, pp. 5319-5323, doi: 10.1109/IECON.2017.8216921.

[14] T. Župan, Ž. Štih and B. Trkulja, "Fast and Precise Method for Inductance Calculation of Coaxial Circular Coils With Rectangular Cross Section Using the One-Dimensional Integration of Elementary Functions Applicable to Superconducting Magnets," in IEEE Transactions on Applied Superconductivity, vol. 24, no. 2, pp. 81-89, April 2014, Art no. 4901309, doi: 10.1109/TASC.2014.2301765.

[15] V. Pankrac, "Generalization of Relations for Calculating the Mutual Inductance of Coaxial Coils in Terms of Their Applicability to Non-Coaxial Coils," in IEEE Transactions on Magnetics, vol. 47, no. 11, pp. 4552-4563, Nov. 2011, doi: 10.1109/TMAG.2011.2148175.

[16] Fanpeng Kong, Yi Huang and L. Najafizadeh, "A coil misalignment compensation concept for wireless power transfer links in biomedical implants," 2015 IEEE Wireless Power Transfer Conference (WPTC), Boulder, CO, USA, 2015, pp. 1-4, doi: 10.1109/WPT.2015.7140152.

[17] M. Bajtos, R. Radil, L. Janoušek, M. Bajtos and N. Dang, "Numerical Simulations of Static Magnetic Fields with Mu-metal Cage Shielding," 2023 24th International Conference on Computational Problems of

Electrical Engineering (CPEE), Grybów, Poland, 2023, pp. 1-4, doi: 10.1109/CPEE59623.2023.10285311.

A Novel Multilevel Control Algorithm for Co-operative Energy Management in Interconnected DC Microgrids

Jithin K.
Department of Electrical Engineering
College of Engineering Trivandrum
APJ Abdul Kalam Tech. University
Thiruvananthapuram, Kerala, India
jithinkolampurath@gmail.com

Hari Kumar R.
Department of Electrical Engineering
College of Engineering Trivandrum
APJ Abdul Kalam Tech. University
Thiruvananthapuram, Kerala, India
harikumar@cet.ac.in

Mayadevi N.
Department of Electrical Engineering
College of Engineering Trivandrum
APJ Abdul Kalam Tech. University
Thiruvananthapuram, Kerala, India
maya@cet.ac.in

Abstract—Due to limited energy resources, autonomous DC microgrids (DCMGs) often face challenges in meeting system demand. Interconnecting these MGs into a cluster enables optimal resource utilization, improving overall reliability and stability. However, coordinating power flow among interconnected MGs while maintaining optimal performance remains a significant challenge. Advancements in cloud computing offer a powerful solution for managing distributed energy resources (DERs) across vast geographic areas. This paper presents a modular cloud-edge architecture for cooperative management of DERs in autonomous DCMG clusters. The proposed hierarchical framework consists of a cloud-based central controller that optimizes resource allocation across the cluster, a MG-level edge controller that prioritizes load management, and a local controller that ensures power-sharing with central setpoints. This multi-tiered approach enhances operational efficiency and grid resilience by enabling fast, localized responses while leveraging cloud-based intelligence for strategic decision-making. Performance evaluations demonstrate that the proposed control strategy improves renewable energy utilization, reduces dependency on storage, and ensures a reliable power supply to connected loads.

Index Terms—— cloud controller, cluster, DCMG, edge controller, interconnection, multi-level control, power management.

I. INTRODUCTION

Microgrids (MGs) are becoming increasingly popular for distributed power generation due to their potential to foster grid resilience and stability [1], [2]. They can be categorized into three main configurations: DC, AC, and hybrid MGs [3]. Among these, DCMGs have gained significant attention as a viable alternative to traditional AC microgrids (ACMGs), driven by rapid advancements in power electronics and their inherent advantages over ACMGs [4]. Current research focuses on developing applications for islanded DCMGs, emphasizing their off-grid, standalone, and self-sustaining capabilities. The development and implementation of islanded DCMGs have become a focal point in research, highlighting their potential as a fundamental component of future smart grid systems [5], [6].

Due to the intermittent nature of energy sources, storage plays a crucial role in ensuring the reliability of autonomous DCMGs. However, increasing storage capacity to enhance reliability also leads to higher maintenance requirements and costs. Additionally, despite having significant power generation potential, these sources may sometimes be underutilized. To address these challenges, integrating nearby MGs is essential [7].

Clustered MGs offer lower operational costs, enhanced reliability, and more efficient utilization of storage devices compared to independent MGs. By interconnecting DCMGs with a power link and optimizing energy storage, they can operate more reliably and dynamically without relying on the main grid. This concept is particularly beneficial for entities and consumers that cannot afford power outages, enabling the creation of self-sufficient energy islands capable of functioning even during grid-wide blackouts [8]. The interconnected MG approach was designed to mitigate power imbalances among different units, resulting in a highly resilient system that remains blackout-free and can source major portion of its required power from renewable energy sources, without requiring additional stabilization [9].

Effective control strategies that ensure proper power management among the multiple units within a microgrid cluster (MGC) are essential for the efficient and reliable operation of networked MGs. Notably, the MGC strategy closely resembles the cellular approach adopted in Europe for optimizing energy resource utilization [10]. Each unit within the cluster is equipped with a controller capable of monitoring and coordinating local supply and demand while also communicating with neighboring MGs. This interconnected operation maximizes the use of available renewable energy sources, ensuring efficient demand fulfillment.

Power exchange between MGs enhances resource utilization, reduces component stress and aging, improves system reliability and availability, lowers maintenance costs, and extends the lifespan of the distribution network [11]. Depending on the communication structure among MGs, the multi-level

979-8-3315-3013-6/25 $31.00 © 2025 IEEE

control framework for MGCs can be categorized into four types: hierarchical, distributed, decentralized, and centralized [12].

In the centralized approach, all available data from participating units is transmitted to a central control unit (CCU), which processes the information, makes decisions, and sends commands back to the managed devices via bi-directional communication links [13]. This method enables effective power exchange across tie-lines and ensures voltage control within clustered MGs. However, it requires high communication bandwidth and presents a potential single point of failure due to the concentration of data processing in one location. Additionally, the need for an advanced communication infrastructure limits the scalability of this approach for large-scale networked MGs [14].

In a decentralized model, local controllers independently validate control actions without requiring communication links, relying solely on local measurements within each MG. As a result, adding or removing a MG does not impact the operations of others within the system. To achieve effective coordination and control of networked MGs connected via tie-lines, a decentralized control technique was introduced in [15]. This method successfully regulates power flow between MGs within the cluster while maintaining DC-link voltage within acceptable limits under varying load conditions.

To address the limitations of both decentralized and centralized control approaches, the literature proposes a distributed control technique for DC-bus voltage regulation and power-sharing in DC MGCs. Unlike centralized control, which relies on a central unit, the distributed approach utilizes one-way communication networks, allowing MGs or DERs to make decisions based on their data as well as information from neighbors [16]. In [17], a distributed tertiary control method demonstrating cooperative behavior is discussed. This approach adjusts the power transfer between interconnected MGs and modifies each MG's voltage set point based on local load conditions. Although this method improves fault resilience and load sharing, it is not explored in this study due to potential tie-line power losses caused by unexpected terminal voltage differences, which may lead to circulating currents even when power sharing is unnecessary.

To enhance coordination among MGs and address control challenges arising from DER integration, several studies propose a hierarchical control framework [18]. This multi-tiered approach improves adaptability and efficiency in MG operation. It divides the MG control system into distinct levels, ensuring high reliability and smooth operation in both grid-connected and islanded modes. The primary control level manages initial power sharing, voltage regulation, and current control. Secondary control mitigates voltage deviations that may arise from the primary control layer. Tertiary control oversees optimal energy management tasks such as power flow coordination, economic dispatch, and resource optimization. However, the centralized control scheme is limited by its reliance on a central controller, high computational demands, and vulnerability to single-point failures. Similarly, decentralized control suffers from high operational costs and poor resilience in islanded modes due to costly point-to-point communication. Meanwhile, hierarchical control is highly effective in islanded MGs but struggles to accommodate sudden topological changes.

This study presents a novel cloud-based multi-level control strategy for interconnected DCMGs to overcome the limitations of existing approaches. The proposed method offers lower costs and reduced computational complexity while dynamically adapting to changes in the DCMG network. Additionally, it prioritizes customer needs during operation. The primary contributions of this work include:

- Development of a cloud-edge based hierarchical control algorithm for power regulation in interconnected DCMG architectures.
- Computation of optimal setpoints for all dispatchable sources by the MCC to mitigate mismatches between total generation and demand within the MGC, with edge control ensuring implementation.

The rest of the paper is structured as follows: Section II describes the energy management in DC MGC. The architecture of a cloud-based multilevel control in DC MGC is described in Section III. Section IV explains the proposed multilevel energy management algorithm. The algorithm is analyzed and simulated in Section V. Section VI compares the proposed energy management algorithm with the conventional techniques and Section VII brings the paper to a close.

II. ENERGY MANAGEMENT IN DC MGC

Interconnected DCMG clusters must be seamlessly coordinated to manage generation-load mismatches among participating MGs. An effective energy management strategy should be resilient to load disturbances and fluctuations in generation caused by the intermittent nature of renewable energy sources. Designing robust power-sharing algorithms is essential to prevent or mitigate power imbalances in DCMG clusters, ensuring optimal and reliable operation.

The integration of emerging technologies such as IoT, cloud computing, edge computing, 5G communication, and artificial intelligence (AI) can transform MGCs into intelligent, digital, and networked power systems. Cloud and edge computing-based interconnected MG architectures offer a sophisticated approach to energy management, combining traditional power grid infrastructure with advanced computing technologies. A schematic representation of a cloud-based DCMG cluster is shown in Fig. 1. Cloud computing enables the delivery of computing services—including storage, processing, and analytics—over the Internet. In the context of MGs, it provides significant advantages such as real-time control, monitoring, and data analytics. This facilitates real-time optimization, predictive maintenance, and remote management of MG assets. Meanwhile, edge computing processes data closer to the source of generation or consumption, enabling faster decision-making, enhanced security, and support for applications that require real-time responsiveness.

979-8-3315-3013-6/25 $31.00 © 2025 IEEE 554

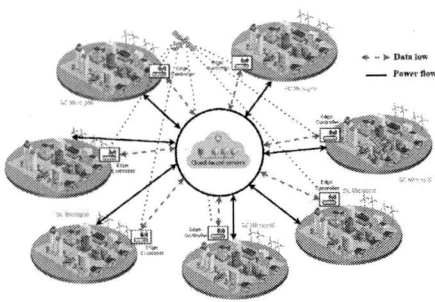

Fig. 1. Cloud-Based DC MGC

Cloud and edge-based hierarchical power-sharing control for interconnected DCMGs utilizes cloud computing infrastructure to coordinate and optimize MG operations. This approach enhances scalability, flexibility, and real-time monitoring and control capabilities. The structure of a DC MGC incorporating cloud and edge control is illustrated in Fig. 2.

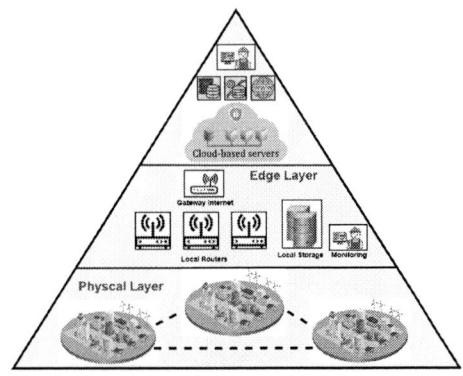

Fig. 2. Cloud-based multi-level control architecture

Each MG unit in the cluster is equipped with an edge controller responsible for collecting and processing data locally. These edge controllers communicate with the cloud, transmitting relevant time-stamped data. Decision-making occurs in the cloud, where data from each edge controller is analyzed to generate optimal setpoints for each MG unit. These setpoints are transmitted back to the edge controllers, which execute them to ensure precise and efficient power distribution within the MGC.

III. CLOUD-BASED MULTI-LEVEL CONTROL IN DC MGC

The multi-level control framework facilitates seamless collaboration between devices and the MG's edge. DC MGCs are integrated with an edge-computing service architecture designed for efficient energy management and real-time decision-making.

The bottom layer of this design consists of the MG's electrical components, including all DERs, which communicate by forming networks of electrical power flows. The middle layer comprises edge devices that continuously collect and analyze real-time operational data from DERs to assess the MG's state. Furthermore, it transmits the processed information to the cloud, enabling lightweight computation and localized decision-making.

The upper layer, the cloud, houses powerful computing resources, providing advanced analytics, optimization, and centralized decision-making for the interconnected DCMG cluster.

IV. PROPOSED CLOUD-BASED MULTI-LEVEL ENERGY MANAGEMENT ALGORITHM

The edge controller collects DER data from MG units, and computes total generation from both dispatchable and non-dispatchable sources, as well as total load. This data is transmitted to the cloud, along with the maximum capacity of each dispatchable source. The cloud platform then determines the appropriate control mode, such as generation control or load control, based on the real-time power balance equation.

During generation control, setpoints for fuel cells and batteries are determined by an optimization algorithm running in the cloud. The edge controller then executes the necessary actions based on the setpoints received from the cloud. The proposed algorithm is outlined in the following steps.

A. Algorithm

Edge Controller: Data Transmission to Cloud

Step-I:

- **Input**: Generation from dispatchable and non-dispatchable sources, fuel cell capacity, battery SoC, loads with priority.
- $P^i_{GND1}, P^i_{GND2} \ldots \ldots P^i_{GNDn}$ (generation from non-dispatchable sources in i^{th} MG)
- $P^i_{GD1}, P^i_{GD2}, \ldots \ldots P^i_{GDn}$ (available generation from dispatchable sources in i^{th} MG)
- $P^i_{L11}, P^i_{L12} \ldots \ldots P^i_{Lnk}$ (loads in i^{th} MG with k^{th} priority)

Step-II:

- Calculate $\Sigma\, T^i_{PGND}$ (Total non-dispatchable generation from i^{th} MG)
- Calculate $\Sigma\, T^i_{PGD}$ (Total dispatchable generation from i^{th} MG)
- Calculate $\Sigma\, T^i_{PLk}$ (Sum of kth priority load from i^{th} MG)

Step-III:

- Send data to the cloud (ΣT^i_{PGND}, $\Sigma\, T^i_{PGD}$, ΣT^i_{PLk}), fuel cell capacity, battery capacity based on SoC.

Cloud Controller

Step-I:

- Receive data from each MG and compute total load (ΣT_{PLk}) for each priority, total dispatchable generation (ΣT_{PGD}), and total non-dispatchable generation (ΣT_{PGND}), maximum capacity of dispatchable sources

$(\Sigma T_{PGD_{max}})$ from 'n' MG units in the cluster and line loss P_{Loss}.

- Calculate $\delta = (\Sigma T_{PL} + \Sigma P_{Loss}) - (\Sigma T_{PGD_{max}} + \Sigma T_{PGND})$
- **if** $\delta > 0$, Load control, Go to Step-II
- **else** $\delta < 0$, Generation control, Go to Step-III

Step-II: (Load control)

- Compute total load in 'n' MGs with priorities.
- Based on the value of 'δ', shed loads from the lowest priority and optimize if necessary.
- Send power to be shed to each edge controller.

Step-III: (Generation control)

- Calculate $\alpha = (\Sigma T_{PL} + \Sigma P_{Loss}) - (\Sigma T_{PGND})$
- **if** $\alpha > 0$, Optimize setpoints for battery and fuel cell
- **if** $\alpha < 0$, Optimize setpoint for Battery
- Send setpoints to the edge controllers

Edge Controller: Cloud Data Reception

- Receives setpoints or power to be shed
- **if** load control **then**
      Run "priority-based load management algorithm"
    else
      Send setpoints to dispatchable sources
- **End**

The term "α" in generation control represents the algebraic sum of the cluster's total load, total losses, and the power generated from non-dispatchable sources. It serves as the foundation for setpoint generation, guiding the operation of batteries and fuel cells. When "α" is positive, the dispatchable sources together with the non dispatchable sources can meet the demand and the algorithm generates setpoints for fuel cells and batteries to contribute to power generation within their nominal capacity limits. Conversely, when "α" is negative, indicating a surplus of generation, the algorithm sets the battery to charge. Thus the algorithm ensures maximum utilization of non-dispatchable sources, such as wind and photovoltaics.

B. Objective function (Generation control)

The generation control system determines the setpoints for dispatchable energy sources in real time. These setpoints are obtained through a cloud-based optimization algorithm, and the optimal values are then transmitted to the edge controllers of each MG unit. The optimization process utilizes the Particle Swarm Optimization (PSO) algorithm to optimize a cost-based objective function given by (??).

1) Cost Function for Battery:

$$C_B = \sum_{i=1}^{m}(C_{charge} * P_{Bat_i} - C_{discharge} * P_{Bat_i}) \quad (1)$$

The battery cost function is given by (1), where C_{charge} and $C_{discharge}$ are the costs during charging and discharging, respectively, and are taken as 20 \$/kWh [19]. P_{Bat_i} is the power delivered or absorbed by the ith battery.

2) Cost Function for Fuel Cell:

$$C_{FC} = \sum_{i=1}^{m}(a_i * P_{FC_i}^2) + (b_i * P_{FC_i}) + c_i \quad (2)$$

The cost function of the fuel cell is given by (2), where a_i, b_i, and c_i are the cost coefficients and are taken as a=0.0013, b=0.062, c=1.34 [20]. P_{FC_i} is the fuel cell power.

The final cost will be a sum of the cost functions of each dispatchable unit in the MGC and is given by (3).

$$Total\ Cost = \sum_{i=1}^{p}(C_B^i) + \sum_{j=1}^{q}(C_{FC}^j) \quad (3)$$

where p and q are the number of battery and fuel cell units in the cluster respectively.

Thus, the objective function (OF) is given by (4).

$$OF = min(Total\ Cost) \quad (4)$$

Subject to the constraints;

$$P_{Bat_{min}} \leq P_{Bat} \leq P_{Bat_{max}} \quad (5)$$

$$P_{FC_{min}} \leq P_{FC} \leq P_{FC_{max}} \quad (6)$$

V. Simulation and Validation of the Algorithm

The structure of the DC MGC used in this study is shown in Fig. 3. The cluster consists of four MG units interconnected by distribution lines. Each unit communicates with the cloud and is equipped with local edge controllers. After executing the optimization algorithm, the DERs within each unit are controlled based on the setpoints received from the cloud.

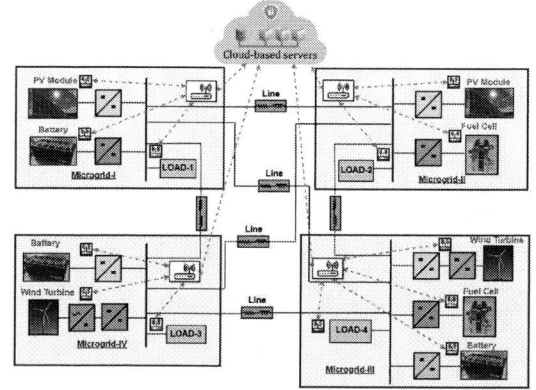

Fig. 3. DC MGC with multilevel control

A. System Specifications

The MG units in the cluster incorporate various energy sources, with MG-I integrating solar power and battery storage. MG-II integrates fuel cells with photovoltaic (PV) systems, while MG-III consists of fuel cells, batteries, and wind generators. Similarly, MG-IV is formed by combining batteries and wind generators. The specifications of each MG unit in the DC MGC cluster are detailed in Table I. The cluster is simulated in MATLAB/Simulink, and generation control is evaluated using the PSO algorithm with the objective function defined in (4).

TABLE I
SYSTEM SPECIFICATIONS

System	Sources
MG-I	PV:75 kW$_p$, Battery:150 Ah, 12 V, 80 nos.
MG-II	PV:75 kW$_p$, Fuel Cell:100 kW
MG-III	Wind:100 kW$_p$, Battery:150 Ah, 12 V, 80 nos, Fuel Cell:80 kW
MG-IV	Wind:40 kW$_p$, Battery:150 Ah, 12 V, 40 nos

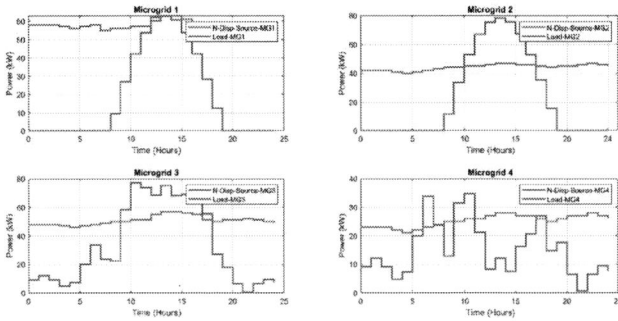

Fig. 4. Demand and dispatchable power generation in DC MGC

B. Results and Analysis

Fig. 4 illustrates the dispatchable source generation and total power demand for each MG in the cluster, with the corresponding data provided in Table II. A 24-hour profile of generation and demand is considered. In MGs I and II, PV serves as the non-dispatchable generation source, with power output determined based on irradiance data of the location. Similarly, in MGs III and IV, wind energy acts as the non-dispatchable source, with generation calculated using the wind speed profile. Additionally, each MG unit in the cluster operates with distinct load curves. The output of the cluster's generation control mode is depicted in Fig. 5.

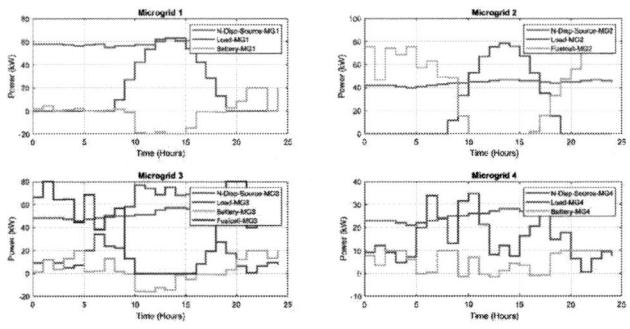

Fig. 5. DCMG cluster with cloud control

The results indicate that the power output from dispatchable sources decreases when wind or photovoltaic generation is

TABLE II
NON-DISPATCHABLE GENERATION AND DEMAND IN EACH MG UNIT

Time (hrs)	MG-I G (kW)	MG-I D (kW)	MG-II G (kW)	MG-II D (kW)	MG-III G (kW)	MG-III D (kW)	MG-IV G (kW)	MG-IV D (kW)
0	0	58	0	42	9.2	48	9.2	23
1	0	58	0	42	12.2	48	12.2	23
2	0	58	0	42	9.2	48	9.2	23
3	0	57	0	41	4.8	47	4.8	22
4	0	56	0	40	7.3	46	7.3	21
5	0	57	0	41	20	47	20	22
6	0	58	0	42	33.8	48	33.8	23
7	0	55	0	43	23.7	49	23.7	24
8	9.36	56	11.7	44	22.4	50	13	25
9	26.78	56	33.48	44	58.2	50	31.4	25
10	42.26	57	52.83	45	76.9	51	34.6	26
11	53.56	57	66.96	45	73.8	51	21.2	26
12	60.192	62	75.24	46	68.4	55	8.2	27
13	62.712	63	78.39	47	75	57	12.2	28
14	60.552	63	75.69	47	68.1	57	7.6	28
15	53.784	61	67.23	46	69.1	56	16.3	27
16	42.12	58	52.65	46	62.7	55	20.6	27
17	28.22	57	35.28	45	55.2	51	27	26
18	12.6	56	15.75	44	27.4	50	14.8	25
19	0	57	0	45	17.8	51	17.8	26
20	0	57	0	45	6.5	51	6.5	27
21	0	58	0	46	0.7	52	0.7	27
22	0	59	0	47	6.5	51	6.5	28
23	0	58	0	46	9.5	50	9.5	27
24	0	57	0	45	7.6	49	7.6	26

G-Generation, D-Demand

sufficient to meet the system's demand. Solar generation peaks between 12:00 and 15:00 hours, during which the cloud dispatches setpoints instructing the system's batteries to charge, as observed in MGs I and II. In MG III, which incorporates both a fuel cell and a battery, the setpoints optimize the power-sharing strategy between them. Likewise, when wind power reaches its peak and meets the demand in MGs III and IV, the fuel cell output is reduced, and the battery is charged during this period. Overall, the results confirm that the proposed algorithm ensures the optimal utilization of energy resources in the DC MGC.

VI. COMPARISON OF DIFFERENT CONTROL TECHNIQUES

Various control techniques used in DC MGCs are compared with the proposed multi-level control approach, as summarized in Table III. The key advantage of the proposed method over existing approaches is its ability to maximize the utilization of cluster resources efficiently. This approach is particularly effective for interconnected DC MGCs, as it can swiftly adapt to sudden variations in energy sources.

979-8-3315-3013-6/25 $31.00 © 2025 IEEE   557

Additionally, the proposed scheme is modular and the consumer demands are also taken into account. However, challenges such as data security and communication delays remain. Data security concerns can be mitigated by implementing advanced communication protocols similar to those used in smart grid architectures. Meanwhile, communication delays have minimal impact since the entire process operates at predefined intervals, reducing latency-related issues.

TABLE III
COMPARISON OF DIFFERENT CONTROL CONFIGURATIONS

Types	Advantages	Disadvantages
Centralized [21]	• Easy standardization and implementation • High reliability in islanded operation • Minimization of conflicts	• Dependence on the central controller • Heavy computation burden • Sensitive to single-point failure
Decentralized [22]	• Strong privacy protections • Strong plug-and-play functionality • Robust against single-point failures • High computation efficiency	• Low resiliency in islanded mode • High operation cost • Expensive point-to-point communication
Hierarchical [23]	• Preserve customer privacy • Less dependence on the central controller • High system redundancy	• Difficulty in managing sudden topological changes • Mostly used in independent MGs
proposed Cloud-based Hierarchical	• Optimal operation • Less computation complexity and cost • Adaptive to changes in case of interconnected DCMG • Modular	• Data security • Communication delay

VII. CONCLUSIONS

This study proposes an energy sharing control strategy for interconnected DC MGs. The performance of the proposed algorithm is validated using a clustered DCMG architecture and can be extended to multiple MGs, enhancing system modularity. These cluster maximizes resource use and raise overall stability and dependability. Future work will focus on MGs with priority-based load and the hardware implementation and validation of the proposed energy management algorithm.

REFERENCES

[1] M. VP *et al.*, "A novel system matrix building algorithm for stability analysis of interconnected dc microgrids." *International Journal of Circuit Theory & Applications*, vol. 52, no. 2, 2024.

[2] K. Jithin, N. Mayadevi, R. H. Kumar, and V. P. Mini, "Stability analysis of dc microgrid clusters based on step-by-step system matrix building algorithm," *Journal of Modern Power Systems and Clean Energy*, vol. 12, no. 3, pp. 900–912, 2024.

[3] B. Sahoo, S. K. Routray, and P. K. Rout, "Ac, dc, and hybrid control strategies for smart microgrid application: A review," *International Transactions on Electrical Energy Systems*, vol. 31, no. 1, p. e12683, 2021.

[4] F. Padhilah and K. Kim, "A centralized power flow control scheme of ev-connected dc microgrid to satisfy multi-objective problems under several constraints. sustainability 2021, 13, 8863," 2021.

[5] K. Jithin, N. Mayadevi, R. Hari Kumar, and P. Mini V, "The effect of multiple pv and battery penetration on stability of dc microgrid with single bus topology," in *2021 9th IEEE International Conference on Power Systems (ICPS)*, 2021, pp. 1–5.

[6] J. K., M. N., H. K. R., and M. V P., "Stability analysis of interconnected dc microgrid during disconnection of microgrid units," in *2024 Asia Pacific Conference on Innovation in Technology (APCIT)*, 2024, pp. 1–5.

[7] K. A. Khan and M. Khalid, "Improving the transient response of hybrid energy storage system for voltage stability in dc microgrids using an autonomous control strategy," *IEEE Access*, vol. 9, pp. 10 460–10 472, 2021.

[8] K. Jithin, P. P. Haridev, N. Mayadevi, R. P. Harikumar, and V. P. Mini, "A review on challenges in dc microgrid planning and implementation," *Journal of Modern Power Systems and Clean Energy*, vol. 11, no. 5, pp. 1375–1395, 2023.

[9] V. Vita, G. Fotis, C. Pavlatos, and V. Mladenov, "A new restoration strategy in microgrids after a blackout with priority in critical loads," *Sustainability*, vol. 15, no. 3, p. 1974, 2023.

[10] Z. H. A. Al-Tameemi, T. T. Lie, G. Foo, and F. Blaabjerg, "Control strategies of dc microgrids cluster: A comprehensive review," *Energies*, vol. 14, no. 22, p. 7569, 2021.

[11] M. E. Sezgin and M. Gol, "Distributed energy management and communication strategy for network of microgrids," *Electric Power Systems Research*, vol. 238, p. 111079, 2025.

[12] S. Ferahtia, H. Rezk, A. Olabi, H. Alhumade, H. S. Bamufleh, M. H. Doranehgard, and M. A. Abdelkareem, "Optimal techno-economic multi-level energy management of renewable-based dc microgrid for commercial buildings applications," *Applied Energy*, vol. 327, p. 120022, 2022.

[13] K. Jithin, R. Harikumar, N. Mayadevi, and V. P. Mini, "A centralized control algorithm for power management in interconnected dc microgrids," in *Proceedings of Symposium on Power Electronic and Renewable Energy Systems Control*, S. Mohapatro and J. Kimball, Eds. Singapore: Springer Singapore, 2021, pp. 451–461.

[14] M. G. Molina and P. E. Mercado, "Stabilization and control of tie-line power flow of microgrid including wind generation by distributed energy storage," *International journal of hydrogen energy*, vol. 35, no. 11, pp. 5827–5833, 2010.

[15] M. Nabatirad, R. Razzaghi, and B. Bahrani, "Decentralized energy management and voltage regulation in islanded dc microgrids," *IEEE Systems Journal*, vol. 16, no. 4, pp. 5835–5844, 2022.

[16] E. Espina, J. Llanos, C. Burgos-Mellado, R. Cárdenas-Dobson, M. Martínez-Gómez, and D. Sáez, "Distributed control strategies for microgrids: An overview," *IEEE Access*, vol. 8, pp. 193 412–193 448, 2020.

[17] F. Guo, Z. Lian, X. Zheng, C. Deng, C. Wen, and J. He, "Decentralized cluster-based distributed secondary control of large-scale dc microgrid cluster system," *IEEE Transactions on Sustainable Energy*, vol. 15, no. 3, pp. 1652–1662, 2024.

[18] A. Abhishek, A. Ranjan, S. Devassy, B. Kumar Verma, S. K. Ram, and A. K. Dhakar, "Review of hierarchical control strategies for dc microgrid," *IET Renewable Power Generation*, vol. 14, no. 10, pp. 1631–1640, 2020.

[19] L. Mauler, F. Duffner, W. G. Zeier, and J. Leker, "Battery cost forecasting: a review of methods and results with an outlook to 2050," *Energy & Environmental Science*, vol. 14, no. 9, pp. 4712–4739, 2021.

[20] K. Cabana-Jiménez, J. E. Candelo-Becerra, and V. Sousa Santos, "Comprehensive analysis of microgrids configurations and topologies," *Sustainability*, vol. 14, no. 3, p. 1056, 2022.

[21] R. Salas-Puente, S. Marzal, R. González-Medina, E. Figueres, and G. Garcera, "Experimental study of a centralized control strategy of a dc microgrid working in grid connected mode," *Energies*, vol. 10, no. 10, p. 1627, 2017.

[22] K.-H. Kim *et al.*, "Decentralized power management of dc microgrid based on adaptive droop control with constant voltage regulation," *IEEE Access*, vol. 10, pp. 129 490–129 504, 2022.

[23] M. Khushoo, A. Sharma, and G. Kaur, "Dc microgrid-a short review on control strategies," *Materials Today: Proceedings*, vol. 71, pp. 362–369, 2022.

Modelling and Performance Analysis of DFIG Integrated with LMF-PLL under Grid Abnormalities

Oinam Lotika Devi

Department of Electrical Engineering
Delhi Technological University
Delhi, India
olotika@gmail.com

Alka Singh

Department of Electrical Engineering
Delhi Technological University
Delhi, India
alkasingh.eed@gmail.com

Abstract - **Recently wind's energy has increased access in the electric grid that makes it vital to develop and test adequate acceptable electromechanical the wind turbine generator (WTG) models for analysing the effects of wind integration on the transient stability of the power system. A Doubly-Fed Induction Generator DFIG (asynchronous wound rotor turbine) integrated with Least Mean Fourth Phase Locked Loop (LMF-PLL) model is discussed in this paper. The modelling and the analysing of LMF-PLL integrated DFIG is carried out while considering different grid abnormalities like voltage dip (30%), phase angle jump (20°), frequency change (+3Hz), dc offset (10%) with polluted environment. Moreover, the performance of the integrated system is studied to observe its behaviour under varying rotational speed. The simulation performance of DFIG system integrated with the LMF-PLL under grid abnormalities is performed in Matlab/Simulink environment.**

Index Terms – Adaptive PLL, DFIG, Feedback Control System, LMF, Phase-Locked Loop, Renewable Energy.

I. INTRODUCTION

With increased greenhouse gas emission and decreasing non-renewable resource reserves, focus has shifted to renewable resources. Out of the assorted renewable energy sources available, wind energy is preferred due to its minimum environmental impact, low cost and high load demand [1-3]. The stator flux-oriented vector control (FoC) used for rotor side converter (RSC) decouples active and reactive powers. This causes ac machine to behaves as dc machine for analysing DFIG. The vector control also decouples the torque and flux components. Moreover, the grid side converter's (GSC) grid voltage-oriented vector control preserves nearly a resolute dc-link voltage providing reactive power support as required in compliance with grid regulation. Paper [4] discusses that DFIG improves the stability and regulation of the electrical system. DFIG's back-to-back converter connection to the grid demands phase locked loop (PLL) in order to lock quickly, accurate the grid voltage's phase angle, ensuring proper synchronization of inverter with the grid. Conventionally, with utility interface applications PLL algorithms are used in motor control to determine the electrical angular speed of the rotor [5–6]. However, the dynamic performance of wind turbines is affected by the PLL performance, as discussed in simulation results [7].

Hence the impact of PLL must be taken into account when analysing the dynamic behaviour of wind turbines. A detailed discussion of the LMF filter algorithm is done in [8].

The modelling of the DFIG with asynchronous wound rotor wind turbine has been discussed in [9-10]. Paper [11] uses the adaptive log filter based SOGI to create reference current for 1-ϕ grid following converter. In this paper, a model of 1kW, 320V, 50 Hz wind turbine of DFIG integrated with LMF-PLL is developed and tested. The dynamic performance of the developed system with LMF-PLL is simulated under different grid abnormalities and variable rotational speed. The simulation results show the transient response under grid abnormalities like volage dip (30%), frequency change(+3Hz), phase jump (20°), 10% dc offset, variable wind speed using MATLAB Simulink environment.

II. MODELLING A WIND ENERGY CONVERTER BASED ON DFIG INTEGRATED WITH LMF-PLL

The DFIG integrated with LMF-PLL based Wind Energy Converters (WECS) configuration is shown in Fig.1. A gearbox connects a wind turbine (WT) to the induction generator. The DFIG's rotor windings are connected to the grid via back-to-back converter. However, the stator windings directly connected.

A. Modelling of Wind Turbine and DFIG

The wind energy captured and converted to mechanical power, P_{mec} and torque, T_{mec} by rotor blades of WT is given by

$$P_{mec} = \frac{1}{2} \rho A v_w^{\ 3} C_p \tag{1}$$

$$T_{mec} = \frac{1}{2} \rho A R v_w^{\ 2} C_t \tag{2}$$

where A is the surface area of the turbine blade given in m² as $A = \pi R^2$, is the radius of turbine rotor, v_w is the wind speed in m/s, ρ is the air density in kg/m³, C_p and C_t denote the power coefficient and torque coefficient respectively. C_p is a function of tip speed ratio λ and pitch angle β. C_p is elated to C_t as $C_p = \lambda C_t$ and tip speed ratio is expressed as

$\lambda = \dfrac{R\omega_t}{v_w}$. Here ω_t is the rotational speed of the turbine and R

979-8-3315-3013-6/25 $31.00 © 2025 IEEE

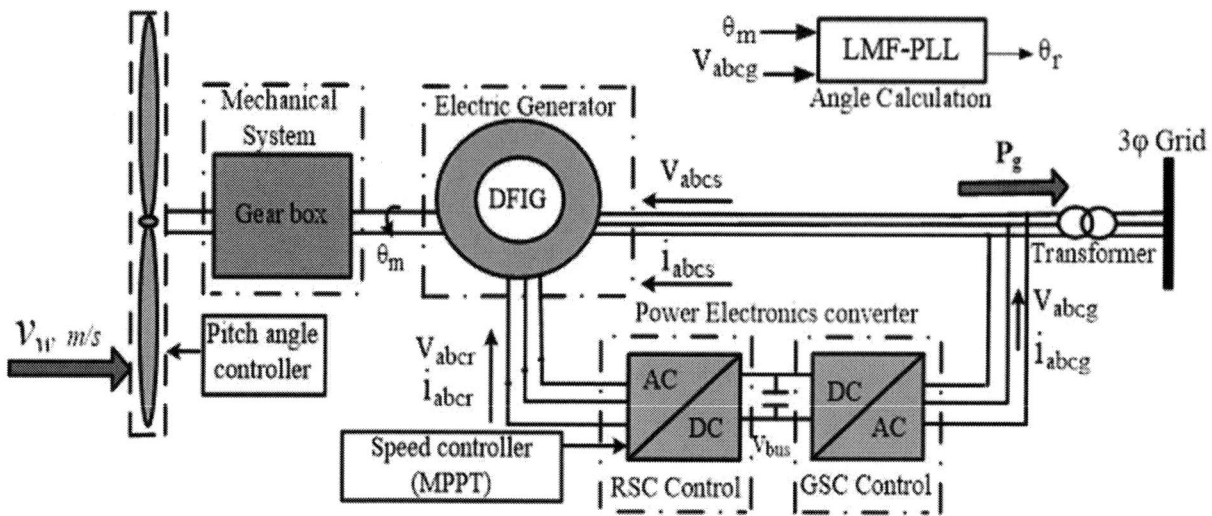

Fig. 1 Block diagram of DFIG integrated with LMF PLL WECS

being the length of the turbine blade. A 1kW, 320V, 2 pole machine is studied for this system The stator and rotor phase voltages ($\bar{v}_{st}$, $\bar{v}_r$) are related to current ($\bar{i}_{st}$ and $\bar{i}_r$), resistances (R_{st} and R_r) and fluxes in vector form and expressed

$$\bar{v}_{st} = \bar{i}_{st}R_{st} + \frac{d\bar{\psi}_{st}}{dt} \tag{3}$$

$$\bar{v}_r = \bar{i}_{rt}R_r + \frac{d\bar{\psi}_r}{dt} \tag{4}$$

When a balanced three-phase voltage of frequency "f" is provided to the three stator windings of DFIG, a stator's flux is induced in it. It moves with a synchronous speed "n_{syn}" given by the expression

$$n_{syn} = \frac{60 * f}{p} \tag{5}$$

where n_{syn} is synchronous speed in rpm and p is the number of pole pair.

The stator flux remains constant during stable condition. DFIG's stator flux ($\bar{\psi}_{st}$) and rotor flux ($\bar{\psi}_r$) in vector form can be expressed as

$$\bar{\psi}_{st} = L_{st}\bar{i}_{st} + L_m\bar{i}_r \tag{6}$$

$$\bar{\psi}_r = L_r\bar{i}_r + L_m\bar{i}_{st} \tag{7}$$

Here, L_{st} is the stator inductance, given as $L_{st} = L_{lst} + L_m$, L_r is rotor inductance, expressed as $L_r = L_{lr} + L_m$, and L_m is magnetising inductance, L_{lst} and L_{lr} are the leakage inductances of stator and rotor respectively.

In DFIG, the induced voltage in the rotor is determined by the stator flux and rotor's rotational speed rotational speed.

The induced rotor voltages and currents' angular frequencies are related as

$$\omega_r = \omega_{st} - \omega_m \tag{8a}$$

$$\omega_{slip} = \omega_{st} - p\omega_m \tag{8b}$$

Here ω_r is the angular speed of rotor winding (rad/s), ω_{st} is the angular speed of stator windings (rad/s), ω_m is the angular speed of the rotor (rad/s), ω_{slip} is the speed of rotor slip (rad/s) and p is the no. of pole pairs. The rotor current is synchronized with the slip angle (θ_{slip}) to control the rotor speed. Hence, the rotor currents can be controlled by aligning the control with the slip speed allowing both sub and super synchronous operation for DFIG

$$\theta_{slip} = \theta_{st} - \theta_m \tag{9}$$

The stator angle, θ_{st} is calculated using a PLL on the grid voltage and θ_m is the rotor mechanical angle. The Voltage differential equations of stator and rotor and current vectors of the generator at dq coordinates can be written as

$$\bar{v}_{stdq} = \bar{i}_{stdq}R_{st} + \frac{d\bar{\psi}_{stdq}}{dt} + j\omega_{st}\bar{\psi}_{stdq} \tag{10}$$

$$\bar{v}_{rdq} = \bar{i}_{rdq}R_r + \frac{d\bar{\psi}_{rdq}}{dt} + j(\omega_s - \omega_r)\bar{\psi}_{rdq} \tag{11}$$

where ω_{st} and ω_r are angular frequencies of stator and rotor.

B. Modelling of Grid Side Converter (GSC)

The grid and the GSC are connected by an LCL-filter. Active and reactive powers of grid are expressed as

$$P_g = \frac{3}{2}(U_d i_{gd} + U_q i_{gq}) \tag{12}$$

$$Q_g = \frac{3}{2}(U_q i_{gd} + U_d i_{gq}) \tag{13}$$

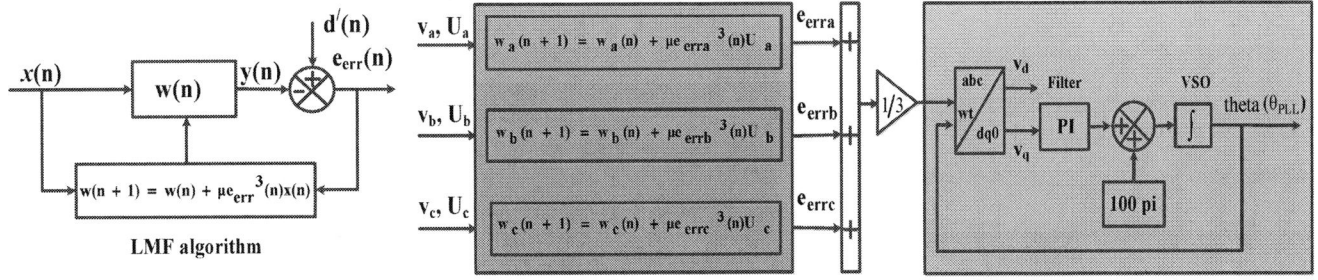

Figure 2(a) Block structure of LMF system	Figure 2(b) Block diagram of 3-phase LMF-PLL

where U_d and U_q, i_{gd} and i_{gq} are the dq components of grid voltage and currents respectively.

C. Modelling of Rotor Side Converter (RSC)

The RSC circuit independently controls the active and reactive powers to maximize the available power. RSC, the Stator Flux Orientation (SFO) scheme is chosen for DFIG power regulation and it can also be decoupled. Proportional integrator (PI) controllers are tuned to control the converter currents which controls both active and reactive powers.

The active and the reactive power of the stator is provided as

$$P_{st} = \frac{3}{2}(V_{std}i_{std} + V_{stq}i_{stq}) \qquad (14)$$

$$Q_{st} = \frac{3}{2}(V_{stq}i_{stq} + V_{std}i_{stq}) \qquad (15)$$

where V_{sd}, V_{sq} and i_{sd}, i_{sq} are the voltages and currents of dq components of the stator.

D. Least Mean Fourth Phase Locked Loop (LMF-PLL)

The Least Mean Fourth (LMF) structure is a popular adaptive method known for its quick convergence and minimal steady state. The general block structure of the LMF system and 3-phase LMF diagram are shown in Figure 2(a) and 2(b).

The LMF algorithm's weight update equation is depicted as [8]

$$W(n + 1) = W(n) + \mu e_{err}^3(n)X(n) \qquad (16)$$

where $W(n)$ is the current and $W(n + 1)$ is the succeeding weight vectors respectively, μ is a constant and regulates the stability and convergence rate, $e_{err}(n)$ is the available error while $X(n)$ is the input signal vector of the plant. The vector of adaptive weights for N samples is given as

$$W(n) = [w_1(n), w_2(n),......w_N(n)] \qquad (17)$$

The vector of the input data of N samples is given as

$$X(n) = [x(n), x(n - 1),......x(n - N + 1)]^T \qquad (18)$$

In Fig. 2(a) the adaptive LMF filter output is $y(n) = d' = e_{err}(n)$ where $e_{err}(n)$ is the error signal. $X(n)$ is the input signal, $d'(n)$ is the desired signal. In 3-phase LMF-PLL shown in Fig. 2(b) the LMF algorithm is used as the prefilter phase detector of the three-phase supply.

III. CONTROL OF RSC AND GSC

This section discusses DFIG based system control for power electronic converters. The RSC and GSC control loops form the two parts of the DFIG system's direct power control. The generator's rotor current, fluctuating fundamental frequency, and generator output voltage amplitude are controlled by the converters to keep under limit.

A. Rotor Side Converter (RSC) Control

Fig.3. demonstrates the RSC controls through stator-flux oriented frame. The unit template U_{rd}^* and U_{rq}^* are obtained from active and reactive powers (P^*, Q^*). The RSC independently adjust the generator's electromechanical torque and stator. Fig.4. shows that controller of current loop in RSC executes d-axis and q-axis voltages reference in response to reference and actual current error signal. Figure 5. depicts the angle calculation of slip, rotor and stator where the LMF-PLL has been used to obtain the synchronized frequency the phase angle of the grid.

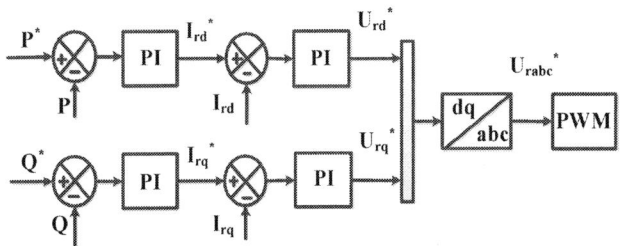

Fig. 3 Block diagram of RSC of DFIG integrated system

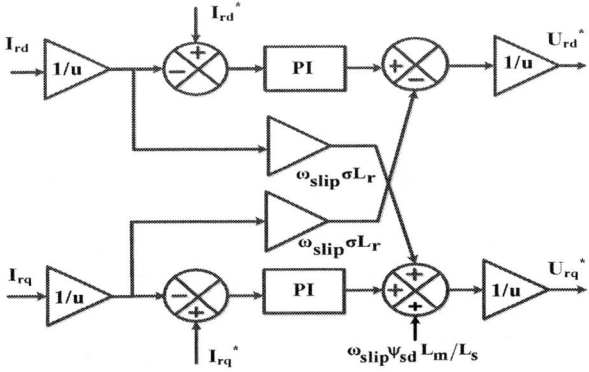

Fig. 4 Rotor current control in DFIG

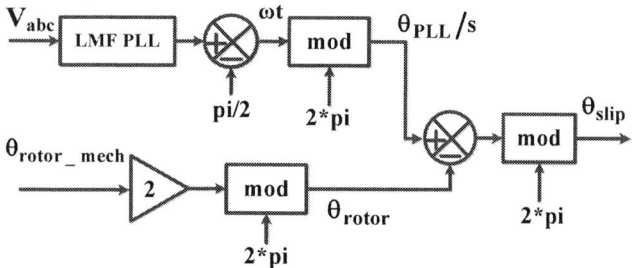

Fig. 5 Block diagram for angle calculation

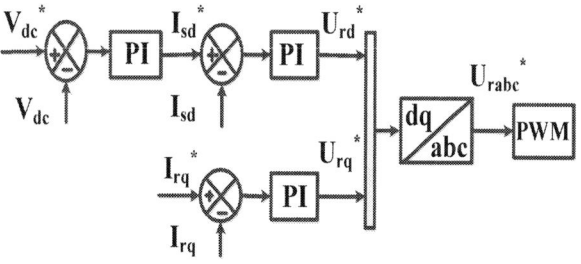

Fig. 6 Block diagram of control of GSC

B. Grid Side Converter (GSC) Control

The GSC controller maintains a steady dc voltage by controlling the reactive current component at the minimum. Figure 6. shows GSC control where q and d -axes loop are included for reactive power or grid voltage support control and dc-link voltage control, respectively and are subjected to signal processing techniques to keep high order harmonics out of the controllers.

IV. SIMULATION RESULTS

The simulation results showing the plots of dynamic performance of WES of 1kW, 320V, 50 Hz wind turbine integrated with LMF-PLL for phase angle, active and reactive power under various four abnormal grid cases: I. 30% voltage sag, II. +3Hz frequency change, III. 20° phase shift, IV. 10% dc offset are discussed in this Section. The comparison of stator voltage and sine theta of all the grid abnormalities is also carried out. The abnormalities are injected in the grid voltage for 0.35s to 0.65s time duration. The effect of this abnormalities is seen DFIG's performance under variable speed of wind. All the simulation results are obtained using MATLAB/Simulink.

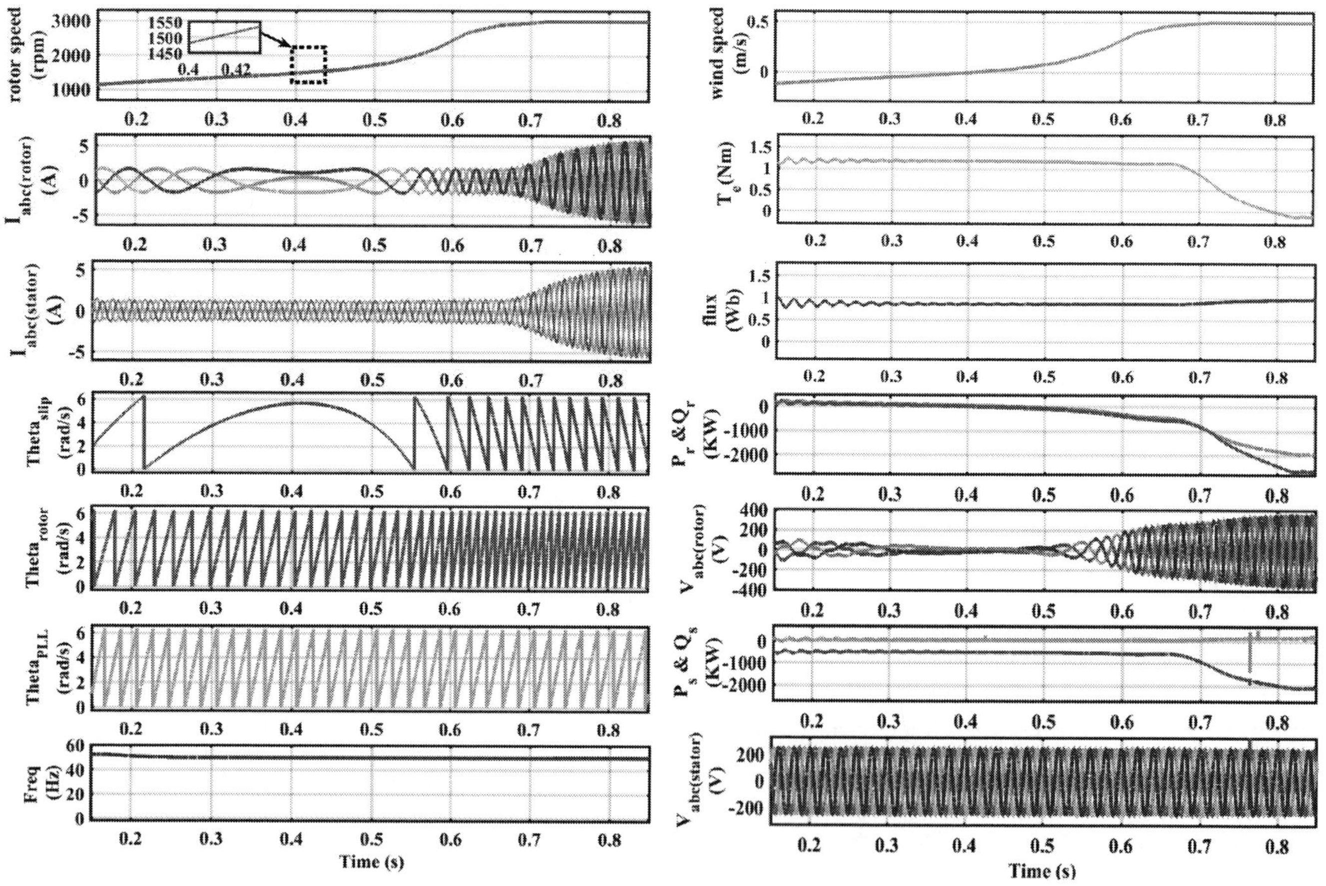

Figure 7 Simulated results of DFIG wind turbine under variable wind speed

979-8-3315-3013-6/25 $31.00 © 2025 IEEE

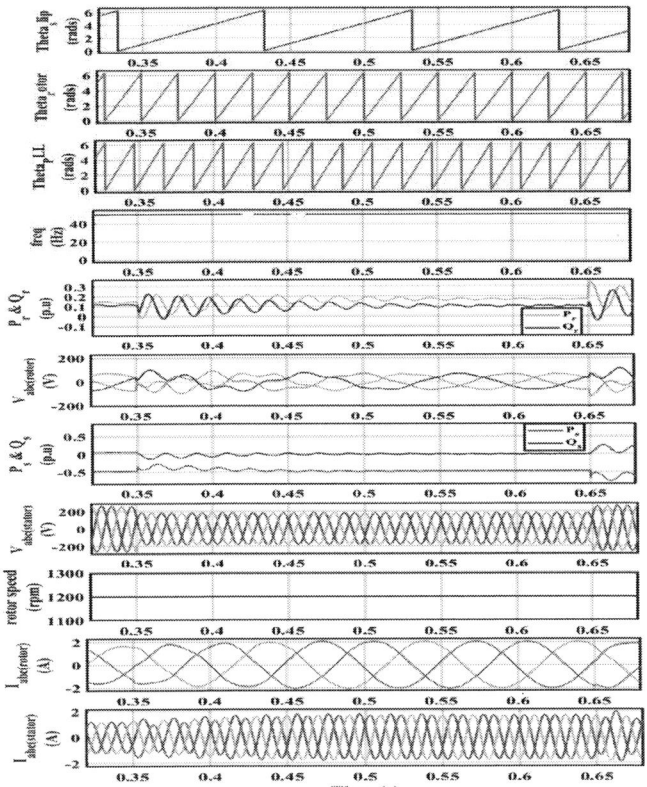

Fig.8 (a) Case I: Simulated results of DFIG under 30% voltage sag

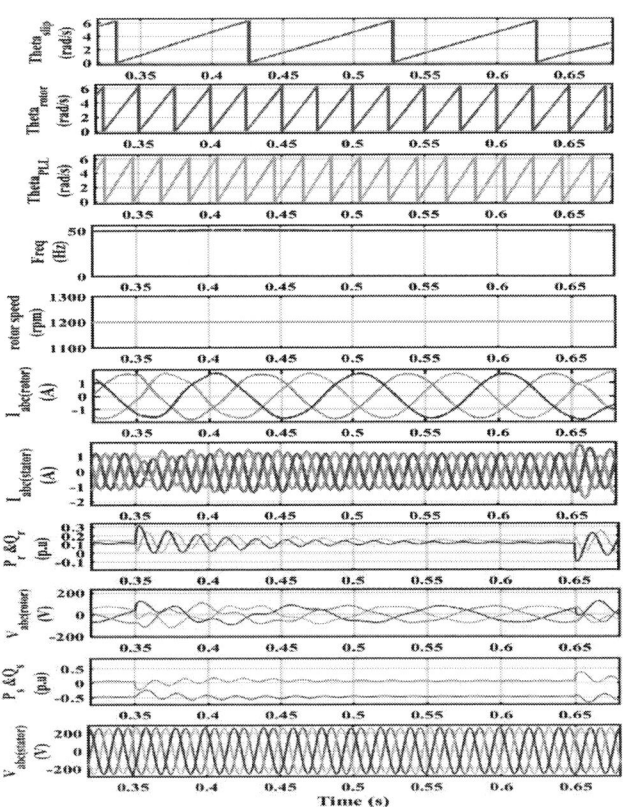

Fig. 8 (c) Case III: Simulated results for 20º phase shift in grid voltage

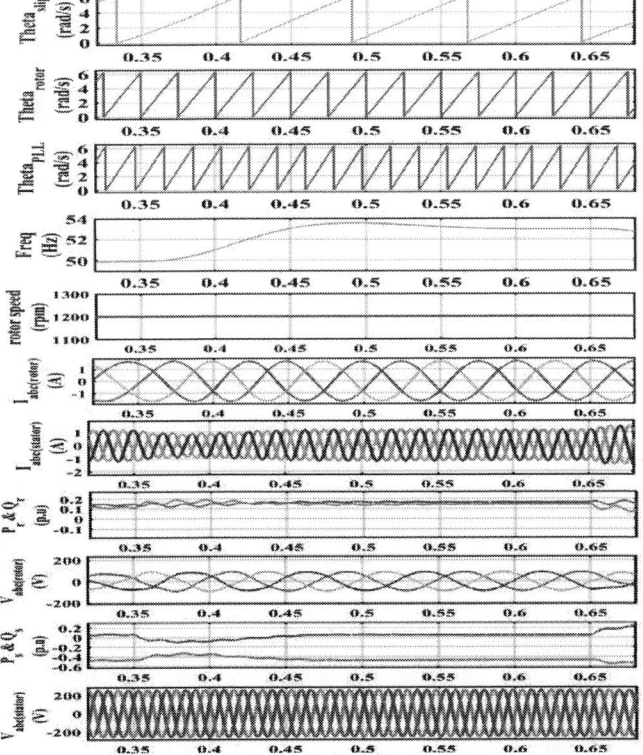

Fig. 8 (b) Case II: Simulated results for +3 Hz frequency change in grid voltage

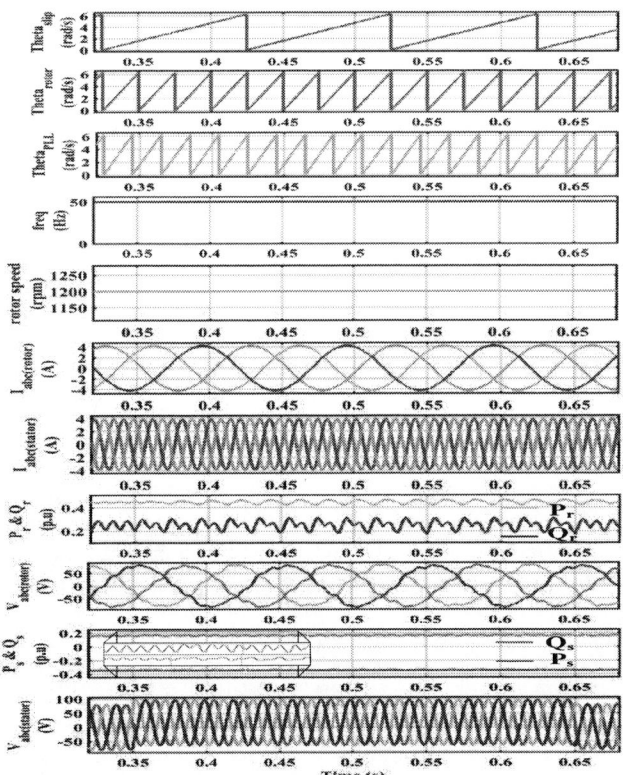

Fig. 8 (d) Case IV: Simulated results for 10% dc offset in grid voltage

979-8-3315-3013-6/25 $31.00 © 2025 IEEE

Fig.7 shows the simulation results of dynamic performance of DFIG under variable wind speed. It depicts results showing plots of phase angle (rad/s) of the slip and rotor phase angle (rad/s) and frequency (Hz) of LMF-PLL respectively, mechanical speed (rpm) of the rotor, three phase rotor and stator currents (A) and voltages (V), active and reactive powers (pu) of both rotor and stator, torques and flux of rotor, when the wind turbine is fed variable wind speed from 5m/s to 20m/s. The synchronous speed of the rotor is 1500rpm. The plot shows the phase sequence alterations in the rotor current of the DFIG during the wind's variable speed operation from sub synchronous to super synchronous mode of operation. It also shows that the flow of the power varies during this period.

Fig. 8 (a) shows the simulated results for Case I (under 30% voltage sag). It depicts the simulated results showing the plots of phase angle (rad/s) of the slip and rotor phase angle (rad/s) and frequency (Hz) of LMF PLL, mechanical speed (rpm) of the rotor, three phase rotor and stator currents (A) and voltages (V), active and reactive powers (pu) of both rotor and stator with 30% voltage sag grid abnormalities. It is observed that the active and reactive powers of stator (P_s and Q_s) have normal -0.5 and zero. Both have very less oscillations which dies down within 3 cycles. Active as well as reactive power of rotor (P_r and Q_r) at normal have 0.14. However, during abnormalities, they oscillate at 0.2 and sustain for longer time.

Fig. 8 (b) shows that P_s and Q_s shows small variation in their steady state value at t= 0.35s when frequency change of +3 Hz (Case II) is initiated and return back to their steady state values after t=0.45s. P_r and Q_r shows small oscillations at t=0.35s but settles downs at normal steady state values after t=0.43s.

Fig.8 (c) shows the Case III depicting for 20° phase change of grid abnormality. The steady state values of Ps and Qs are zero and -0.5 initially but after disturbance normal steady state are again obtained.

Fig. 8 (d) shows plots of Case IV for 10% dc offset grid abnormality. The P_s and Q_s at normal steady state have values of -0.34 and 0.17. They become normal after the disturbance from t=0.35s to t=0.65s. The rotor active and reactive powers depict 0.45 and 0.3 values at normal steady state. The disturbance dies down and comes to normalcy after t=0.65s.

V. CONCLUSIONS

In this paper, a Matlab model of DFIG is developed which is integrated with LMF-PLL at the rotor side for angle estimation. The modelled (RSC) converter controls the machine providing the active and reactive power. The vector control technique is employed to control the DFIG. Simulation results are studied for different test cases like (i) 30% voltage sag (ii)

+3Hz frequency change (iv)20° phase change and 20% dc offset and harmonics distortion in grid supply voltages. In all the cases studied the performance with the developed LMF PLL is satisfactory and the oscillations in active and reactive powers of stator and rotor settle down within four to five cycles. Further a new case with variable wind speed varying from 5m/s to 20m/sec case in DFIG is studied. It is observed that with increase in wind speed the rotor speed increases and all the three modes viz. sub synchronous, synchronous and hyper-synchronous mode of operation have been plotted. It is also observed that the slip angle, rotor current, powers changes accordingly as per the three modes of operations.

ACKNOWLEDGMENT

The authors are thankful to the DST for the FIST project SR/FST/ET-1/2022/1026.

References

[1] Ntanos, S.; Skordoulis, M.; Kyriakopoulos, G.; Arabatzis, G.; Chalikias, M.; Galatsidas, S.; Batzios, A.; Katsarou, A. Renewable Energy and Economic Growth: Evidence from European Countries. Sustainability 2018, 10, 2626.

[2] Sawant, M.; Thakare, S.; Rao, A.P.; Feijóo-Lorenzo, A.E.; Bokde, N.D. A Review on State-of-the-Art Reviews in Wind-Turbine and Wind-Farm-Related Topics. Energies 2021, 14, 2041.

[3] Javadi, M.A.; Ghomashi, H.; Taherinezhad, M.; Nazarahari, M.; Ghasemiasl, R. Comparison of Monte Carlo Simulation and Genetic Algorithm in OptimalWind Farm Layout Design in Manjil Site Based on Jensen Model. In Proceedings of the 2021 7th Iran Wind Energy Conference (IWEC2021), Shahrood, Iran, 17–18 May 2021.

[4] Liu, Chao & Tian, Xinshou & Chen, Kai & Su, Yuanyuan & Li, Yan. (2018). Effect of phase-locked loop on transient performance of wind turbines generator under voltage phase jump. The Journal of Engineering. 2019. 10.1049/joe.2018.8783.

[5] V. Blasko, J. C. Moreira, and T. A. Lipo, "A new field-oriented controller utilising spatial position measurement of rotor end ring current," in ProcPESC, 1989, pp. 295-299.

[6] F. Nozari, P. A. Mezs, A. L. Julian, C. Sun, and T. A. Lipo, "Sensorless synchronous motor drive for use on commercial transport rurplanes," IEEE Trans. Ind. Applicat., vol. 31, no. 4, July/Aug. 1995.

[7] Zhang, D., Wang, Y., Hu, J., et al.: 'Impacts of PLL on the DFIG-based WTG's electromechanical response under transient conditions: analysis and modelling', CSEE J. Power Energy Syst., 2016, 2, (2), pp. 30–39.

[8] E. Walach and B. Widrow, "The least mean fourth (LMF) adaptive algorithm and its family," IEEE Trans. Inform. Theory and its family," IEEE Trans. Inform. Theory, vol. IT-30, pp. 275–283, Aug. 1984.

[9] Haitham Abu-Rub, Mariusz Malinowski, Kamal Al-Haddad, "Power Electronics for Renewable Energy Systems, Transportation and Industrial Applications", ISBN: 978-1-118-63403-5, July 2014, Wiley-IEEE Press,

[10] Gonzalo Abad, Jesús López, Miguel A. Rodríguez, Luis Marroyo and Grzegorz Iwanski "Doubly Fed Indication Machine Modeling and Control for Wind Energy Generation", 2011.

[11] Kumar, Vineet & Singh, Alka & Chittora, Prakash. (2023). Control and Analysis of a Grid Following Converter using Adaptive Log Family based SOGI. 1-6. 10.1109/CERA59325.2023.10455350.

Aerodynamic Wing Flutter Control using IMU Sensor

1st Ganesh Ragava V
Electrical and Electronics Engineering
Vellore Institute of Technology
Chennai, Tamil Nadu, India
ganeshragava.v2022@vitstudent.ac.in

2nd M. A. Inayathullaah
Electrical and Electronics Engineering
Vellore Institute of Technology
Chennai, Tamil Nadu, India
inayathullah.a@vit.ac.in

3rd Gurucharan Gurunath
Electrical and Electronics Engineering
Vellore Institute of Technology
Chennai, Tamil Nadu, India
gurucharan.gurunath2022@vitstudent.ac.in

4th Ranganathan V
Electrical and Electronics Engineering
Vellore Institute of Technology
Chennai, Tamil Nadu, India
ranganathan.v2022@vitstudent.ac.in

5th Monish BKN
Electrical and Electronics Engineering
Vellore Institute of Technology
Chennai, Tamil Nadu, India
ranganathan.v2022@vitstudent.ac.in

Abstract—**This project intends to implement a prototype model of an aircraft wing's flap control to counteract forces during cruise. The wing design is based on NACA 2412 aerodynamic wings. The MPU 9250, which combines a 3-axis accelerometer, gyroscope, and magnetometer, is employed to monitor the orientation of the aircraft in real-time. A proportional controller logic is used to adjust the flaps on the wing in the opposite direction to counteract the forces, where the servo motor attached to the flaps readjusts itself. This solution aims to improve flight stability and aircraft performance, especially for autonomous flight systems.**
Keywords—Drone Stability, Flutter control, IMU Sensor.

I. INTRODUCTION

To counteract external forces from the wind during the cruise of the airplane, this project intends to implement flutter control to counteract the forces by controlling the flaps of the wings using a positive feedback and prevents the aircraft from turbulence. The use of an inertial measurement unit (IMU) controller causes the flaps to pitch up or down to counteract the direction of the wind. This can be very useful for stabilizing radio controlled drones. The project also implements a manual method of control by using a radio controller if the IMU stability fails as an backup. As per the work published, theoretical research into structural flutter originated in the 1950s, with experimental studies following later to validate this aeroelastic phenomenon. The choice of aerodynamic models proved pivotal in these investigations, as early analyses primarily relied on quasi-steady potential flow theory and linear potential flow theory. A notable example of flutter's consequences occurred in the 1990s when aeroelastic oscillations caused catastrophic structural failure in an F-117A stealth fighter during flight1. This instability mechanism remains a critical concern for aerospace engineering, as it compromises both airframe integrity and flight safety through complex interactions between aerodynamic forces and structural dynamics[1][2][3].

Accordingly, Aeroelastic flutter characteristics of aircraft structures are significantly important to investigate. The study of interaction between surface structures of an aircraft and the aerodynamic forces acting on structures is aeroelasticity [4]. One of the most typical aeroelastic problem encountered is Flutter control. The vibration amplitude of aircraft is gradually attenuated when the flight speed is low after it is being disturbed. The vibration amplitude remains unchanged when the flight speed is increased to a certain critical value, a limit cycle oscillation occurs on the aircraft structure that indicates that flutter has occurred. Flutter is one of the most common problems encountered in aeroelasticity [5][6]. The transfer of energy from unsteady aerodynamics associated with surrounding fluid to wing structure, causing in rapid change in behaviour is flutter. If the flutter can be controlled at cruise speeds, we can design lighter wings and, consequently, more efficient airplanes. It is in the aircraft designer's best interest to design innovative ways where flutter can be controlled without causing the structure to be too heavy. [7]. Wykes' development of the identically located accelerometer and force (ILAF) methodology aimed to enhance control system stability through aerodynamic damping forces proportional to linear velocity for modal stabilization. However, the approach's exclusive focus on damping mechanisms created inherent optimization constraints by limiting the range of adjustable parameters. Research literature demonstrates the counterintuitive phenomenon where introducing positive damping can induce instabilities in previously stable elastic systems. These findings reveal fundamental limitations in modal stabilization strategies for flutter suppression applications, particularly regarding flight condition adaptability and control law effectiveness. The combined evidence suggests conventional damping-based approaches require re-evaluation, with modern flutter suppression research now pursuing alternative strategies involving multi-input control architectures and adaptive systems[8][9][10].

979-8-3315-3013-6/25 $31.00 © 2025 IEEE

According to Ansari [11], The use of force jet and pulse width pulse frequency (PWPF) is recommended to control two-dimensional wing flutter. The PWPF modulator has the advantage of working similar to a linear mode of operation, low jet gas consumption and provide good accuracy when oscillations are encountered. Quasi-steady dynamic premises and compressible flow, as well as the thin airfoil theory is used in this scheme. The suggested method is to utilize piezoelectric materials and control surfaces. Control surfaces and smart structures such as piezoelectric (PZT) and electro rheological fluids (ERFs) materials are deployed for conventional methods of linear and nonlinear control of wing flutter. The two main methods to control the flutter of the aircraft structures are passive and active control [12]. By changing configuration of structures, the structural stiffness is increased by using passive flutter control, this results in reducing the deformations of the structures. However, changing configurations of structure is not always useful in the aerospace industry. With the development of active control technology, the aircraft design concepts have moved from passive control to active control. More control forces can be generated by controlling voltage applied on smart materials such as the piezoelectric materials, shape memory alloys and magneto rheological fluids through different control algorithms, and actively adjust aeroelastic stability of aircraft structures. The weight of the aircraft can be significantly reduced and also ensure that aircraft has good aeroelastic performance when active design concept is used. Aircraft morphing devices present cost and complexity challenges but address inherent design compromises across varying operational profiles. As an airfoil-level morphing mechanism, the morphing flap serves as an aerodynamic control surface that eliminates abrupt geometry transitions inherent in conventional systems. Traditional rigid-flap systems create discontinuous camber variations at hinge points, generating localized pressure differentials and premature flow detachment that significantly increase drag. Continuous curvature adaptation through gapless morphing surfaces maintains attached flow conditions while preserving wing skin integrity. Pioneering efforts in camber-morphing wing architectures date back to 1920s aviation engineering experiments, demonstrating early recognition of smooth airfoil transition benefits despite material and actuation limitations of the era. [13]. Sanders et al. further extensive studies into the aerodynamic and aeroelastic properties of a straight wing have highlighted that the implementation of a morphing flap could enhance the wing's maximum roll rate. [14]. This section of our paper explores active flutter control in aircraft using IMU-driven flap adjustments to counteract wind-induced turbulence, highlighting both traditional and modern adaptive techniques—such as smart materials and morphing flaps—for improved aeroelastic stability and flight efficiency. In Section II, we proceed with the 3D CAD Modelling of the wing which is done using the 3D modelling software, ONSHAPE. Continuing, in Section III the design of our proportional controller is done and the design calculations of the transfer function in question has been specified. In Section IV, the algorithm

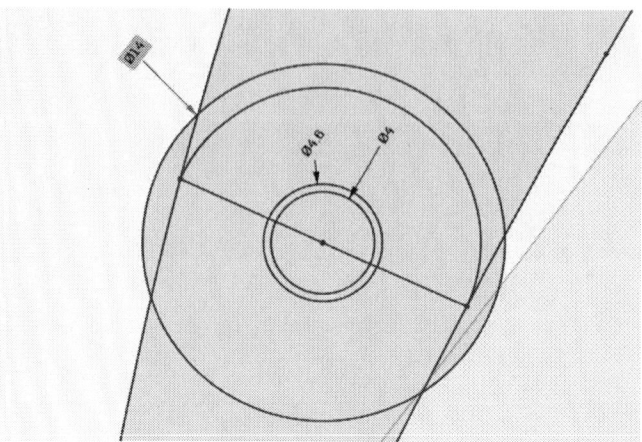

Fig. 1. Design of hinge

for our software coding of the overall system is specified in detail, step by step. In section V, our hardware implementation is mentioned along with the block diagram and pictures of the same. Concluding, the required stability of the system is achieved by counteracting the flutter wing with Proportional controller.

II. 3D CAD MODELLING OF THE WING

The 3D modelling software used to design the aerodynamic wing is ONSHAPE, it's an open-source tool that can directly be accessed online. The dimensions and the parameters to design the prototype was in accordance to cost efficiency and smaller scale. For the use of a low torque servo motor, the flap was designed in a way that it meets the torque requirement for movement. The dimensions and parameters of both the body and the flap are specified in relation to fig[2] and fig[3] . In reference to fig [4],[5] the side dimensions along with its pre-extrusion is shown. In fig [6], the extruded body and flap is shown as per the axis. Flight is determined by 4 fundamental forces: Lift, weight, drag and thrust. Flaps primarily opposes the force of drag, which is the opposing force that acts in the opposite direction of the aircraft. So, flaps oppose this force during slower take of speeds and where drag occurs during low speeds, so extending the flaps create more drag as byproduct of increased lift. At higher speed during cruise negative-flaps which reduces nose-down pitching moment of the aircraft. Fig[7] shows an extensive diagram of forces acting on the wing along with its components Fig 4.

III. PROPORTIONAL CONTROL DESIGN AND MODELING

This section presents a simplified control model for the servo-accelerometer system and the design of a proportional controller.

The system includes:

- A servo motor
- MPU9250 accelerometer (sensor for a_y, Y-axis acceleration)
- A proportional controller mapping error to servo angle

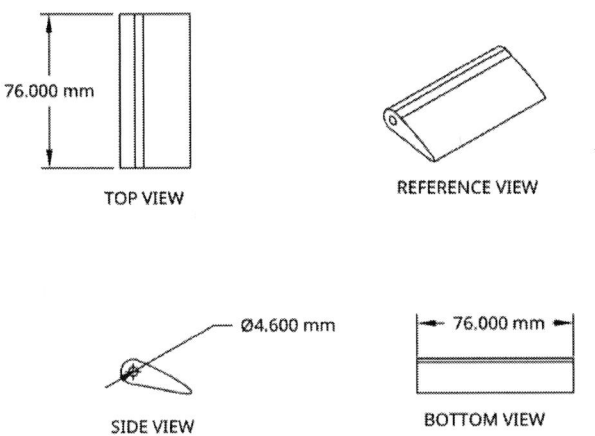

Fig. 2. 2D Dimensions of the Flap

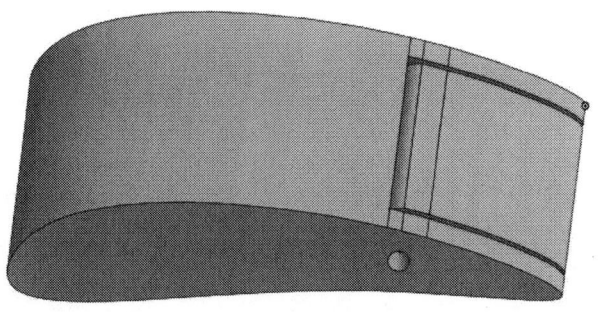

Fig. 3. CAD design of the wing

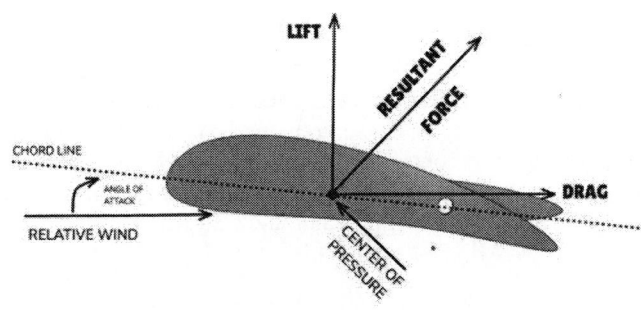

Fig. 4. Forces acting on the wing

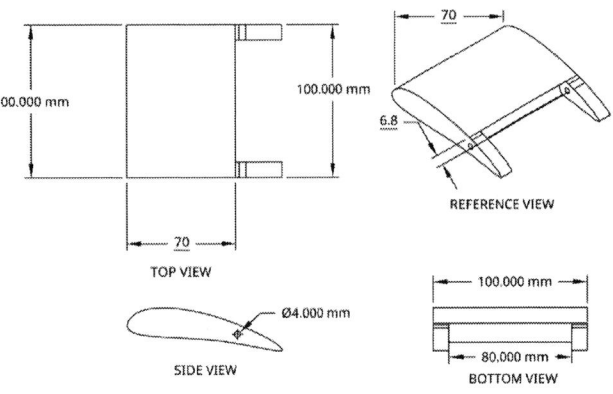

Fig. 5. Dimensions of the wing

The control goal is to minimize the deviation of a_y from a desired setpoint by adjusting servo orientation.

a) Assumptions:

- Servo dynamics modeled as a first-order system:

$$\frac{\theta(s)}{u(s)} = \frac{1}{\tau s + 1}$$

- Linear approximation of platform tilt:

$$a_y \approx g \cdot \theta \quad \text{(for small angles)}$$

with gain $k_\theta \approx 9.8$ m/s^2/rad

- System is stable and operates in linear regime

b) Plant Transfer Function: Combining the servo dynamics and tilt-to-acceleration relationship, the plant transfer function becomes:

$$G(s) = \frac{a_y(s)}{u(s)} = \frac{9.8}{\tau s + 1}$$

c) Proportional Controller: A proportional controller is given by:

$$u(t) = K_p \cdot e(t) \quad \Rightarrow \quad U(s) = K_p \cdot E(s)$$

The closed-loop transfer function becomes:

$$T(s) = \frac{9.8 K_p}{\tau s + 1 + 9.8 K_p}$$

Which yields a first-order closed-loop system with time constant:

$$\tau_{\text{cl}} = \frac{\tau}{1 + 9.8 K_p}$$

d) Gain Selection (K_p): Let the maximum acceleration error be ± 5 m/s^2, and the desired servo swing be $\pm 50°$ from center position:

$$K_p = \frac{50}{5} = 10 \quad \text{(deg/m/s}^2\text{)}$$

Thus, each 1 m/s^2 deviation results in $10°$ servo adjustment, which provides a responsive yet stable behavior.

Summary: With $K_p = 10$, the closed-loop system maintains stability and ensures the Y-axis acceleration converges rapidly to its setpoint.

IV. SOFTWARE CODING OF OVERALL SYSTEM

1. Connection is established between the MPU9250 (IMU) and the Arduino using I2C pins (SCL, SDA). The servo motor is attached to pin 3 of the Arduino.

2. MPU9250 is initialized using the `setup()` function and `mpu.begin()` method.

3. The accelerometer range is set to ±8G for handling moderate movements.

4. The gyroscope range is set to 500°/s for smoother angular velocity data.

5. The digital filter bandwidth of the MPU9250 is set to 21 Hz to reduce noise in the readings.

6. The servo motor is initialized using `servo.attach(3)`.

7. The sensor data is read using the `loop()` function.

8. The Y-axis acceleration data is mapped to a servo angle using the `map()` function.

9. Servo angle is written using `servo.write(value)`.

10. The Gyroscope data will give us the angular velocity in x, y, and z axes. The .temp will give the temperature data from the IMU.

11. The y-axis acceleration (a.acceleration.y) is extracted. The raw y-axis acceleration value (in units of m/s²) is mapped to a servo angle using the map() function:

12. The range of acceleration is assumed to be -10 to 10 m/s².This range is mapped to servo angles 0° to 180°.

13. The control effort u is calculated using the proportional term.

14. The servo initially starts at a neutral angle of 90 degree.

15. Constrain ensures that the value stays within a valid range of the servo angle (0-180) degree.

16. Servo Control: The calculated servo angle is written to the servo using servo.write(value). The servo motor physically moves to the corresponding angle based on the mapped value.

17. Data Output: The mapped servo angle is printed to the Serial Monitor using Serial.println(value) for debugging and observation.

18. Error is the difference between the desired acceleration (setpoint = 0) and the actual acceleration on the Y-axis.

19. Hardware components used.

a) arduino UNO- Microcontroller of the Project.

b) MPU 9250- gets the 3-axis value

c) Servo 9G-Manually adjusts the wing

d) NACA 2412 aerodynamic wings.

V. HARDWARE IMPLEMENTATION

The hardware setup were shown in Fig 5. To control an SG90 servo motor using an Arduino Uno and an MPU9250 sensor, it is connected to MPU9250 which is connected to the Arduino via I2C (SDA to A4 and SCL to A5) and power it using the 3.3V pin. The SG90 servo's signal pin is attached to Digital Pin 9 on the Arduino, and it is connected to VCC and GND pins to 5V and GND, respectively. The MPU9250 and servo libraries are used in the Arduino IDE.

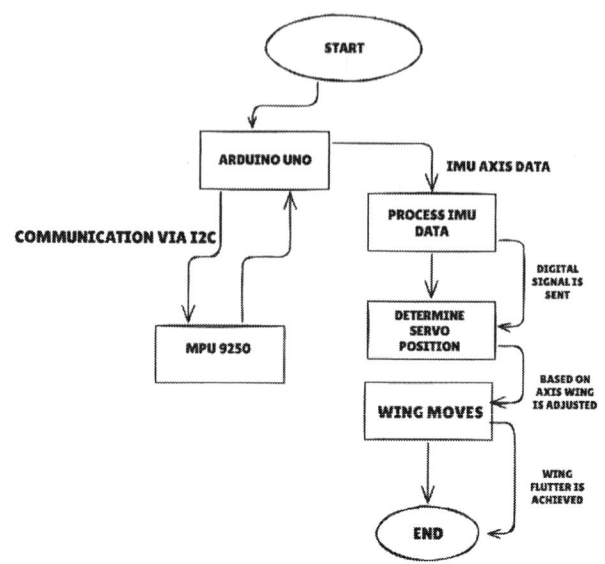

Fig. 6. Block Diagram

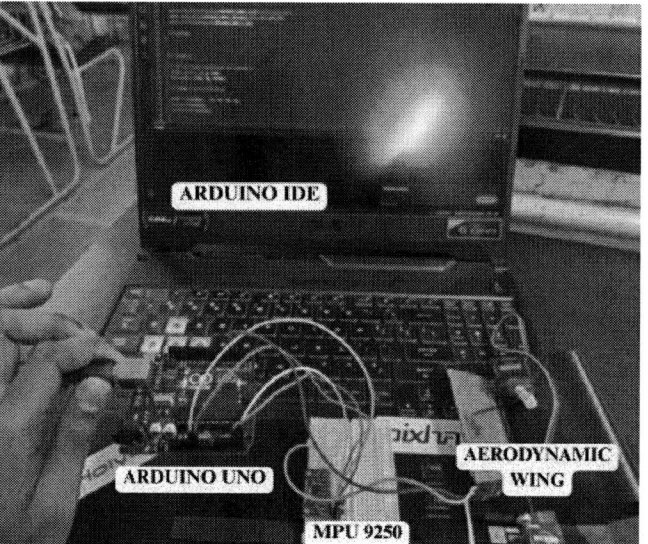

Fig. 7. Hardware implementation

In the code, the MPU9250 is initialized to read its pitch, roll, or yaw data, and map the desired axis data (like pitch) to a range suitable for the servo motor (0 to 180 degrees). the mapped value to control the servo's angle, causing the servo to tilt in response to the MPU9250's orientation. Finally, code is uploaded to the Arduino Uno, allowing the servo to move based on the MPU's detected tilt.

RESULTS AND DISCUSSIONS

The dynamic response of wing fluttering is basically how an aircraft wing behaves over time when it's hit with unpredictable, shifting air forces. These forces can

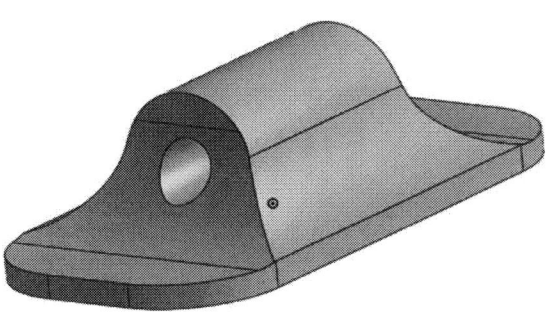

Fig. 8. Coupler

Fig. 9. Flap Down

Fig. 10. Flap up

trigger a dangerous condition known as flutter—an unstable vibration that comes from the way the air interacts with the wing's flexibility and movement. If the frequency of these aerodynamic forces matches the wing's natural tendency to vibrate—especially in bending or twisting—it can feed energy into the system, causing the wing to shake more and more with each cycle. The dynamic response helps us understand how the wing moves—how far it shifts, how fast it moves, and how quickly it accelerates—when these forces hit.

Stability is achieved when the IMU sensor receives input which is pitched out of its 180°, The digital pin from arduino provides the necessary control signals. It is converted to analog signal fed to servo motor for resetting the flaps to its original position and to achieve stability. In Fig. 6, the force is downwards of the aircraft, the flaps readjust upwards to counteract the force. Fig. 7, when there is force upwards of the aircraft the flaps readjusts itself upwards to counteract the force.

1. Setting up Add libraries for servo control (Servo), I2C communication (Wire), and MPU9250 (AdafruitMPU9250). Items: Make A (for the MPU9250 sensor) and a servo (for motor control).
2. Configuration Communication via Serial: To debug, start at 115200 baud. Servo Setup: Set the servo to 0° and attach it to pin 3. Configuration of the Sensor: Adjust the filter bandwidth to 21 Hz, the gyroscope range to ±500°/s, and the accelerometer range to ±8G. A 100 ms delay should be included for stabilizing.
3. Data Acquisition Use A.getEvent(a, g, temp) to fetch sensor readings. Extract Y-axis acceleration (a.acceleration.y) to represent sensor tilt.

4. Control by Proportion Charting: Connect 0° to 180° (servo range) linearly to -10 m/s2 to +10 m/s2 (input range) using control by proportion Value = map(value, -10, 10, 0, 180); is the copy code. K = Constant:K = Output Range. Input Range is equal to 180/20 = 9 K . Range of Output = 180/20 = 9. The servo moves by 9° for every 1 m/s 2 tilt.

5. Actuation of Servos Use the servo to move it to the mapped angle.write (value). For debugging, print the angle to the serial monitor 6. Remarks The tilt of the sensor is precisely reflected in real time by the servo. Responses are smooth and proportionate when K = 9. Although the system is stable, more sophisticated applications can benefit from additional tuning (such as PID control).

CONCLUSION

By controlling the flutter, Stability of the system is achieved and demonstrated. This can be used in RC enthusiast drone applications to control the cruising of the aircraft. The future plan is to implement a controller such as PID or IFC such that it could be used not only when cruising but also during takeoff or landing. Implementing the hardware inside the wings, makes the system more aerodynamic and streamlined

REFERENCES

[1] Bendiksen OO. Review of unsteady transonic aerodynamics: theory and applications. Prog Aerosp Sci. 2011.
[2] Garrick IE, Reed, III, WH. Historical development of aircraft flutter. J Aircr. 1981.
[3] Udrescu R. Effects of oscillating shock waves on the dynamics of fluttering panels. 19th AIAA Applied Aerodynamics Conference; 2001.
[4] Yuyang Chai, Wei Gao, Benjamin Ankay, Fengming Li, Chuanzeng Zhang. "Aeroelastic analysis and flutter control of wings and panels: A review", 2021.
[5] Karthik Palaniappan, Pradipta Sahu, Antony Jameson, Juan Jose Alonso. "Design of Adjoint-Based Laws for Wing Flutter Control", 2012.

979-8-3315-3013-6/25 $31.00 © 2025 IEEE

[6] David K. Schmidt, Brian P. Danowsky, Aditya Kotikalpudi, Julian Theis, Christopher D. Regan, Peter J. Seiler, Rakesh K. Kapania. "Modeling, Design, and Flight Testing of Three Flutter Controllers for a Flying-Wing Drone", 2020.

[7] Nissim E. Flutter Suppression Using Active Controls Based on the Concept of Aerodynamic Energy, 1971.

[8] Wykes, John H. Structural Dynamic Stability Augmentation and Gust Alleviation of Flexible Aircraft. AIAA Pap. No. 68-1067, Oct. 1968.

[9] Broadbent, E. G.; and Williams, Margaret. The Effect of Structural Damping on Binary Flutter. R.M. 3169, British A.R.C., 1960.

[10] Nissim, E. Effect of Linear Damping on Flutter Speed. Part I: Binary Systems. Aeronaut. Quart., vol. XVI, Pt. 2, May 1965, pp. 159-178.

[11] Ansari A.R., Novinzadeh A.R.B., "Designing a Control System for an Airplane Wing Flutter Employing Gas Actuators," Wiley Online Library, 2017.

[12] Gao Y., Li F., Zhang C., et al., "Aeroelastic analysis and flutter control of wings and panels: A review," Wiley Online Library, 2021.

[13] Ouyang Y., Gu Y., Kou X., Yang Z., "Active flutter suppression of wing with morphing flap," ScienceDirect, 2021.

[14] Sanders et al., Aerodynamic and aeroelastic characteristics of wings with conformal control surfaces for morphing aircraft: J. Aircr. (2003).

[15] Ranganathan V. (2025). "Wing Flap Control Code," Available at: https://github.com/ranga2012/Wing-Flap-Control-MPMC.

A Differential Current Unbalance Factor-Based Scheme for Detecting Shunt and High Impedance Faults in Microgrids

Chetan Anand
Dept. of Electrical Engg.
NIT Rourkela
India
uchetananand@gmail.com

Kunal Kumar
Dept. of Electrical Engg.
NIT Rourkela
India
kunalkumar.nitdurgapur@gmail.com

Susmita Kar
Dept. of Electrical Engg.
NIT Rourkela
India
karsusmita@nitrkl.ac.in

Abstract—With the growing incorporation of renewable energy sources (RES), microgrids (MGs) have emerged as a key solution, capable of operating in both grid-connected (GCM) and islanded modes (IM) to enhance reliability and flexibility. Due to the high penetration of Inverter-Interfaced Distributed Generators (IIDGs) such as PV and wind systems, fault current levels in microgrids (MGs) become significantly limited. This restricted fault current capability of IIDGs challenges conventional protection schemes, making fault detection and isolation more difficult. This paper proposes a novel Differential Change in Current Unbalance Factor (DCUF) scheme for detecting shunt faults (SFs) and high impedance faults (HIFs) in MGs. The scheme computes the Current Unbalance Factor (CUF) at both ends of the feeder between Buses M and N using positive and negative sequence currents under pre-fault (PRF) and post-fault (POF) conditions. A threshold-based approach ensures accurate fault detection while avoiding misclassification of MG routine operational disturbances. The proposed scheme is tested on a modified IEEE-13 bus system using MATLAB R2023b Simulink and obtained results under various SF and HIF scenarios confirm the method's effectiveness.

Keywords—Microgrid, Shunt Faults, HIFs, IIDGs

I. INTRODUCTION

In recent years, MGs have garnered significant attention for facilitating the incorporation of RESs into the power networks [1]. They are increasingly recognized as a practical and efficient solution for enhancing energy resilience and improving grid reliability. A key feature of microgrids is their ability to operate in either GCM or IM [2]. In GCM, microgrids remain synchronized with the utility grid, enabling power exchange and contributing to overall system stability [3]. In contrast, when operating in islanded mode, microgrids disconnect from the main grid and operate independently, relying solely on their local generation units and energy storage systems to meet load demands and faces several challenges [4].

Despite the numerous advantages offered by MGs, several technical challenges persist, notably in aspects related to protection. These challenges become more pronounced with the increasing integration of IIDGs, which exhibit distinct operational characteristics compared to conventional generation sources [5]. Developing an effective protection strategy for MGs with a high penetration of IIDGs presents significant challenges [6-8].

During GCM operation, the utility grid contributes substantial fault current, enabling conventional overcurrent relays (OCRs) to detect faults reliably. However, in IM, fault currents are considerably lower, particularly when most distributed generators are inverter based, due to the limited fault handling capability of voltage source inverters [6]. According to standards, IIDGs typically limit fault current to 1.5 times their rated output [7], in contrast to synchronous generators, which can supply 4 to 10 times their rated current during faults. As a result, traditional OCRs may fail to detect faults accurately or provide proper coordination in MGs dominated by IIDGs [8]. Numerous protection schemes for MGs operating in IM with inverter-based DGs are classified as either passive or active. Passive schemes do not interfere with DG control and rely on features like undervoltage relays with directional elements [9]. Recent passive protection methods for IM MGs with IIDGs utilize various techniques, including dq-axis voltage monitoring [10], and traveling wave-based fault detection and localization [11]. Other approaches involve fault classification using discrete wavelet transforms [12], wavelet decomposition combined with decision tree algorithms [13], and assessment of the phase shift between positive sequence voltage and current [14]. Additionally, a differential protection utilizes communication employing symmetrical current components and data mining has been proposed [15]. However, these methods commonly struggle with setting appropriate thresholds and accurately distinguishing between normal operational disturbance and fault conditions across different scenarios.

To address the limitations of passive methods, active protection schemes inject disturbances via the IIDG interface during faults to assist in detection and isolation. In [16], fifth harmonic currents with varying amplitudes are injected based on IIDG output impedance, allowing relays to differentiate between overload and fault scenarios. In [17], the authors introduced a protection method that detects the faulted phase by adjusting the IIDG's current sequence components in response to an asymmetrical fault. An active detection approach is introduced in [18] where IIDGs inject high-frequency components during fault conditions to assist in identifying the fault. Similarly, a protection scheme is introduced in [19] which harmonic currents are injected by the modified IIDG controller and detected by directional overcurrent relays for fault identification. In the proposed method [20], IIDGs are employed with voltage-restrained harmonic controllers that injecting three synthetic harmonic signals, which are then measured by local relays to ascertain

979-8-3315-3013-6/25 $31.00 © 2025 IEEE

the direction in which the fault has occurred. In [21], the IIDGs control injecting third harmonic components at the time of faults, enabling a definite-time harmonic directional relay to effectively coordinate and ensure reliable fault detection. In [22], enhanced current limiters were used to enable IIDGs to automatically inject inter harmonic currents after a fault, allowing an inter harmonic differential relay to recognize the fault and determine the corresponding faulty line. However, these schemes often rely on a separate fault detection mechanism to initiate disturbance injection, adding complexity and delay. Their effectiveness can also be limited by the use of harmonic signals, which may be distorted or blocked by transformer delta connections and nonlinear loads, undermining the reliability of protection. Moreover, current differential relays may fail to detect HIFs in such microgrids due to the low fault current not producing a sufficient differential signal.

In response to the existing gap in current research, this paper proposes a novel protection scheme i.e. Differential Change in Current Unbalance Factor (DCUF) designed for accurate detection of both SFs and HIFs in IM MGs with high penetration of IIDGs. The DCUF calculates the CUF at both ends of the feeder connecting Buses M and N by utilizing positive and negative sequence currents during PRF and POF conditions. The key contributions of this paper are as follows:

- A new fault detection metric, the DCUF, is proposed, which uses both positive and negative sequence current components at both ends of a feeder.

- The scheme is capable of detecting both SFs and HIFs.

- A threshold is set based on multiple operational transients, ensuring no false positives under normal conditions.

- The method is implemented and validated on a modified IEEE-13 Bus Test System with a high penetration of IIDGs using MATLAB R2023b Simulink.

II. Study Test System

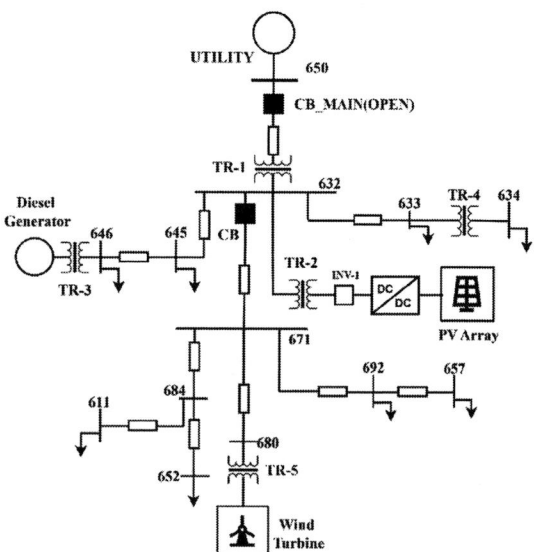

Fig. 1.   Modified IEEE-13 Bus Test System (BTS).

TABLE I.   Integrated Dgers Specification

IIDGs	Parameter	Symbolic name	Value
Solar PV	Rated max. Power	$P_m(PV)$	2 MW
	Open-Circuit Voltage	V_{oc}	34.4 V
	Short-Circuit Current	I_{sc}	7.72 A
	Operating voltage	V_{mp}	32.5 V
	Operating current	I_{mp}	7.47 A
Wind System	Rated max. Power	$P_m(wind)$	1 MW
	Speed of wind	W_s	13 m/sec

The efficacy of the proposed scheme is investigated on a modified IEEE-13 BTS as depicted in Fig.1. as utilized in [23]. The test system is modified by integrating two IIDGs: a solar PV unit at Bus 632 and a wind energy unit at Bus 680. Detailed configuration parameters for both DERs are listed in Table 1. The simulation and analysis of the proposed protection scheme are carried out in MATLAB R2023b Simulink platform. The IEEE-13 BTS represents an unbalanced distribution network operating at 4.16 kV, and the feeder between Buses 632 and 633 is selected for evaluating the performance of the scheme.

III. Proposed Scheme

A simplified two-bus network diagram is depicted in Fig. 2 to illustrate the proposed DCUF scheme. In this setup, Bus M is considered as Bus 632 hosts an IIDG-based solar PV system [24] that supplies power to a load connected at Bus N is considered as Bus 633 from modified IEEE1-13 BTS. A Current Transformer (CT1) and a Potential Transformer (PT1) are placed at the feeder end of Bus 632 and similarly, CT2 and PT2 are installed at the feeder end of Bus 633 to capture corresponding electrical quantities during PRF and POF states. These measurements form the basis for accurate fault detection using the proposed scheme.

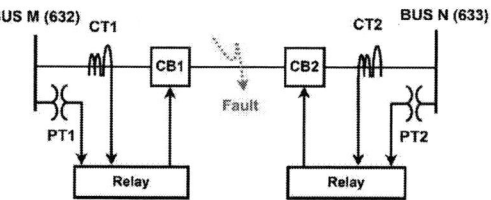

Fig. 2.   Two-bus system with fault and relay placement in the proposed DCUF scheme.

The proposed methodology utilizes the CUF at both ends of the feeder, which is derived from the positive sequence current (PSC) and negative sequence current (NSC) obtained using equation (1).

$$\begin{bmatrix} i_j^0 \\ i_j^+ \\ i_j^- \end{bmatrix} = \frac{1}{3}\begin{bmatrix} 1 & 1 & 1 \\ 1 & \alpha & \alpha^2 \\ 1 & \alpha^2 & \alpha \end{bmatrix}\begin{bmatrix} i_{r,j} \\ i_{y,j} \\ i_{b,j} \end{bmatrix} \quad (1)$$

The PSC and NSC can be obtained as

$$I_j^+ = \frac{1}{3}\left(I_{r,j} + \alpha I_{y,j} + \alpha^2 I_{b,j}\right) \quad (2)$$

$$I_j^- = \frac{1}{3}\left(I_{r,j} + \alpha^2 I_{y,j} + \alpha I_{b,j}\right) \quad (3)$$

Where, "j" represents the current that is obtained at Bus M or Bus N respectively. Additionally, " α " is defined as follows: $\alpha = 1\angle 120°$ and $\alpha^2 = 1\angle 240°$. "r", "y", " b" represents the three-phase system. The CUF is calculated as

$$CUF_{j_c} = \frac{I_{j_c}^-}{I_{j_c}^+} \qquad (4)$$

Here, "j" represents the CUF at Bus M or Bus N and "c" denotes the PRF and POF conditions. To calculate the DCUF, the CUF at both ends of the feeder is further analysed under PRF and POF conditions.

A. CUF considering PRF

At Bus M,

$$CUF_{M_{PRF}} = \frac{I_{M_{PRF}}^-}{I_{M_{PRF}}^+} = \frac{|I_{M_{PRF}}^-| \angle \delta_{M_{PRF}}^-}{|I_{M_{PRF}}^+| \angle \delta_{M_{PRF}}^+} \qquad (5)$$

$$CUF_{N_{PRF}} = \frac{I_{N_{PRF}}^-}{I_{N_{PRF}}^+} = \frac{|I_{N_{PRF}}^-| \angle \delta_{N_{PRF}}^-}{|I_{N_{PRF}}^+| \angle \delta_{M_{PRF}}^+} \qquad (6)$$

B. CUF considering POF

At Bus N,

$$CUF_{M_{POF}} = \frac{I_{M_{POF}}^-}{I_{M_{POF}}^+} = \frac{|I_{M_{POF}}^-| \angle \delta_{M_{POF}}^-}{|I_{M_{POF}}^+| \angle \delta_{M_{POF}}^+} \qquad (7)$$

$$CUF_{N_{POF}} = \frac{I_{N_{POF}}^-}{I_{N_{POF}}^+} = \frac{|I_{N_{POF}}^-| \angle \delta_{N_{POF}}^-}{|I_{N_{POF}}^+| \angle \delta_{M_{POF}}^+} \qquad (8)$$

The difference in CUF current at both bus end feeder is obtained as:

$$CUF_{d_{PRF}} = \left| CUF_{M_{PRF}} - CUF_{N_{PRF}} \right| \qquad (9)$$

$$CUF_{d_{POF}} = \left| CUF_{M_{POF}} - CUF_{N_{POF}} \right| \qquad (10)$$

The proposed scheme Differential Change in Current Unbalance Factor (DUCF) can be calculated as

$$DUCF = \left| CUF_{d_{POF}} - CUF_{d_{PRF}} \right| \qquad (11)$$

The positive-sequence equivalent (PSE) circuit of the two-bus system is illustrated in Fig. 3, where the fault is modeled using a voltage source V_F in series with fault impedance Z_F. This representation reflects the balanced behavior of the system under symmetrical conditions. The DG units are modeled as voltage or current sources depending on their type, rotating or inverter based, connected in parallel with their internal impedance. This configuration is essential for calculating the PSC of current at both ends of the feeder.

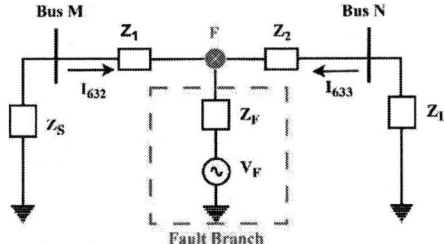

Fig. 3. PSE circuit of the system.

Correspondingly, the negative-sequence equivalent (NSE) circuit is shown in Fig. 4, where the fault branch is represented with a voltage source V_F in series with the same fault impedance Z_F. This model captures the system's

unbalanced condition during asymmetrical faults. The negative polarity of the voltage source reflects the phase inversion characteristic of negative-sequence components. Similar to the positive-sequence model, the DGs are represented by their respective current or voltage source models, enabling the evaluation of negative-sequence voltages and currents necessary for CUF and DCUF computation. The flowchart for proposed scheme is depicted in Fig.5.

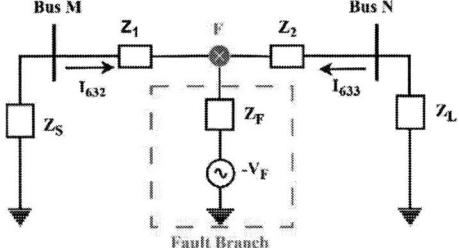

Fig. 4. NSE circuit of the system.

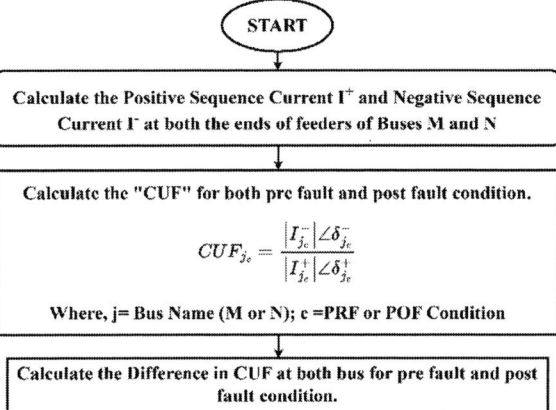

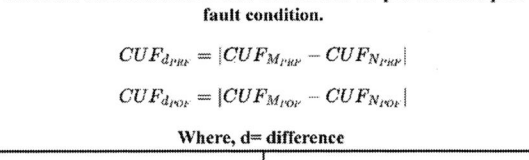

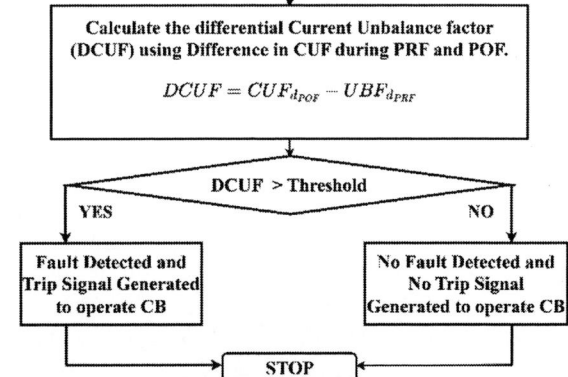

Fig. 5. Flowchart for Proposed DCUF scheme.

IV. SIMULATION STUDY

The proposed DCUF scheme is validated through extensive simulations carried out on a modified IEEE-13 BTS within the MATLAB R2023b Simulink environment. The simulations focus on evaluating the scheme's ability to detect SFs and HIFs in microgrid with high penetration of IIDGs,

979-8-3315-3013-6/25 $31.00 © 2025 IEEE

including a 1 MW solar PV at Bus 632 and a 2 MW wind unit at Bus 680. The test scenarios include both shunt faults (such as LG, and LLG), as well as HIFs with varying impedance levels, introduced at different locations across the feeder between Bus 632 and Bus 633. The proposed scheme calculates the CUF at both ends of the feeder under PRF and POF scenarios, and DCUF is derived accordingly. The two-diode model [24] is used to simulate HIF which is depicted in Fig.6. In the HIF model setup, resistances R_1 and R_2 are randomly adjusted within the range of 250 Ω to 1.1 k Ω, and the DC voltage sources V_1 and V_2 are varied between 1 V and 4.3 V to replicate fault current behavior influenced by grounding surfaces. The ideal diodes D_1 and D_2 are assigned resistance values of 0.01 Ω and 0.001 Ω, respectively.

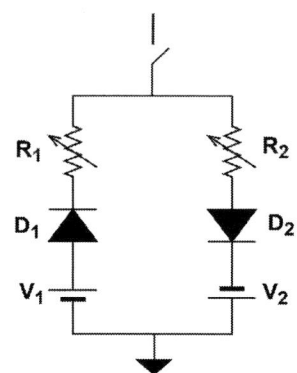

Fig. 6.   Two-diode HIF Model [25].

V. RESULT AND DISCUSSION

This section presents a comprehensive investigation of the proposed DCUF scheme, tested under a diverse fault scenario. The analysis focuses on verifying the scheme's accuracy, responsiveness, and robustness in detecting both SFs and HIF in different microgrid configuration with high integration of IIDGs.

A. Shunt Faults (SFs)

1) Results for Single Line to Ground (SLG) Fault

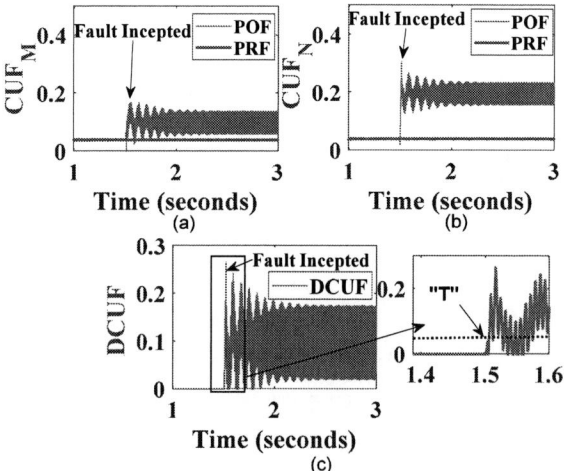

Fig.7.   Result for SLG Fault (a) CUF at Bus M (632) (b) CUF at Bus N (633) (c) DCUF

Figure 7 illustrates the system's response to a SLG fault introduced between phase b and ground on the feeder connecting Buses M (632) and N (633). The fault is created with a fault resistance of 0.001 Ω and is initiated at 1.5 seconds during the simulation. Figure 7 (a) presents the CUF at Bus M (632), while Figure 7 (b) illustrates the CUF at Bus N (633). In both plots, the blue curve denotes the PRF condition and the red curve corresponds to the POF condition for the SLG fault. It can be clearly observed that at 1.5 seconds, the red curve begins to deviate significantly from the blue curve, indicating the onset of the fault. Similarly, Figure 7 (c) illustrates the DCUF response, where a significant deviation is observed at 1.5 seconds. This deviation is evaluated against the predefined threshold value (T), depicted by the black plot. It is evident that the DCUF exceeds the threshold at this instant, thereby confirming the occurrence of the SLG fault.

2) Results for Double Line to Ground (LLG) Fault

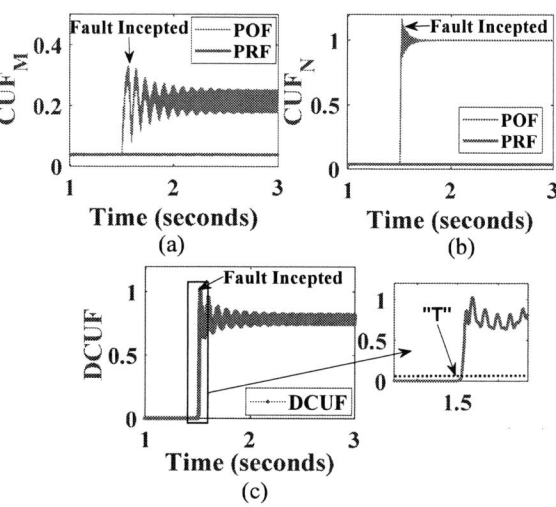

Fig. 8.   Result for LLG Fault; (a) CUF at Bus M (632) (b) CUF at Bus N (633) (c) DCUF

Figure 8 illustrates the system's response to a LLG fault introduced between phases r, phase y, and ground on the feeder connecting Buses M (632) and N (633). The fault is created with a fault resistance of 0.001 Ω and is initiated at 1.5 seconds during the simulation. Figure 8 (a) presents the CUF at Bus M (632), while Figure 8 (b) shows the CUF at Bus N (633). In both plots, the blue curve represents the PRF condition, and the red curve corresponds to the POF condition for the LLG fault. A distinct deviation of the red curve from the blue curve is observed at the fault inception time of 1.5 seconds, indicating the presence of an unbalanced fault. Similarly, Figure 8 (c) illustrates the DCUF response, where a significant deviation is observed at 1.5 seconds. This deviation is evaluated against the predefined threshold value (T), represented by the black plot. It is evident that the DCUF surpasses the threshold at this moment, thereby confirming the occurrence of the LLG fault.

To evaluate the effectiveness of the proposed scheme, HIFs are also considered, as they present unique detection challenges due to their low fault current magnitudes.

979-8-3315-3013-6/25 $31.00 © 2025 IEEE

B. HIF

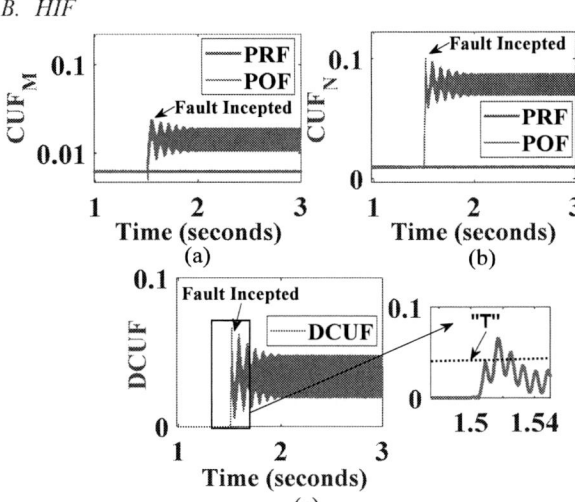

Fig. 9.   Result for HIF Fault; (a) CUF at Bus M (632) (b) CUF at Bus N (633) (c) DCUF

Figure 9 illustrates the system's response to a HIF introduced between the line connecting Buses M (632) and N (633), with a fault resistance of 200 Ω. The fault is initiated at 1.5 seconds during the simulation. Figure 9 (a) shows the CUF at Bus M (632), while Figure 9 (b) depicts the CUF at Bus N (633). In both plots, the blue curve represents the PRF condition, and the red curve indicates the POF condition. Although the fault exhibits high impedance, it leads to only a minor fluctuation in current magnitude, a distinct deviation of the red curve from the blue curve is still observed at 1.5 seconds, indicating the onset of the fault. Figure 9 (c) presents the DCUF, where a sharp increase in the red curve occurs at the moment of fault initiation. This deviation, when compared to the threshold value "T", exceeds the threshold, thereby confirming the presence of the HIF. These results demonstrate that the proposed DCUF-based scheme is efficient enough to detect even high impedance faults, an area where conventional protection schemes often fail, ensuring enhanced reliability and safety in IIDGs integrated MGs.

C. Threshold Value

To avoid false fault identification caused by normal microgrid operational behaviors in the proposed protection scheme, a threshold value "T" for the DCUF is carefully determined as per equation (12). IIDGs integrated MGs are subject to various transient events such as capacitor switching, sudden load variations including cold load start, generator outages, and line outages. While these events introduce momentary disturbances in current profiles, they do not constitute actual faults. To differentiate such non-fault transients from genuine fault conditions, each of these disturbance scenarios is simulated in the modified IEEE-13 BTS. The DCUF is computed for each case, and the maximum deviation observed among all scenarios is selected as the threshold value "T".

$$T = \max(DCUF_{non-fault\ transient}) + \delta \quad (12)$$

This threshold serves as a benchmark for comparison during fault detection. If the computed DCUF during a suspected event exceeds this threshold, the event is classified

as a fault; otherwise, it is treated as a normal disturbance. This threshold-based approach significantly enhances the robustness and selectivity of the proposed DCUF scheme, ensuring that spurious tripping is avoided while maintaining high sensitivity to both SFs and HIFs.

D. Variation in Fault Resistance

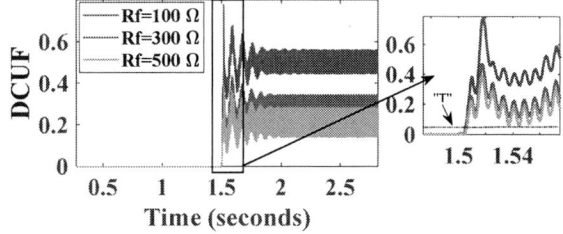

Fig. 10. DCUF at various fault resistance.

To validate the robustness of the proposed protection scheme under varying fault conditions, a series of simulations were conducted by introducing SLG faults between Buses M (632) and N (633), with fault resistance values ranging from 100 Ω to 500Ω as depicted in Figure 10. These high-resistance scenarios are critical in evaluating the ability of protection schemes to detect HIFs, which often generate low fault currents that may go undetected by conventional methods.

In each test case, the fault is initiated at 1.5 seconds, and the corresponding DCUF is computed and compared against the predefined threshold value "T". The DCUF consistently surpasses the threshold at each simulated fault resistance scenarios, confirming the proposed scheme's reliable detection of faults even under varying fault resistance.

VI. CONCLUSION

This paper introduced a novel protection scheme for inverter-dominated MGs based on the DCUF, aimed at accurately detecting both SFs and HIFs. By leveraging the PSC and NSC of current at both ends of a feeder, the method computes the CUF under PRF and POF conditions. The differential change in CUF effectively captures fault-induced asymmetries, and a carefully selected threshold ensures robustness against non-fault transients such as capacitor switching, load variations, and outages. Simulation studies conducted on a modified IEEE-13 BTS demonstrate that the proposed DCUF scheme shows strong capability in identifying a wide range of fault types, including SLG, LLG, and HIFs, even under low fault current conditions characteristic of IIDGs. The DCUF consistently exceeded the threshold during actual fault scenarios while remaining below it during normal operational disturbances, thereby validating both the sensitivity and selectivity of the scheme. Overall, the proposed DCUF scheme offers a simple yet reliable solution for enhancing microgrid protection, with no need for external signal injection or communication links, making it suitable for practical implementation in future decentralized power systems.

Due to space limitations, the paper does not include detailed comparison with existing protection techniques, extended simulations for all microgrid operational disturbances, or analysis of grid-connected to islanded mode transition. These aspects, along with the development of adaptive thresholding techniques and cyber-attack secure fault detection, are planned as part of future work to further establish the scheme's practicality and scalability.

REFERENCES

[1] K. Kumar, P. Kumar, and S. Kar, "A review of microgrid protection for addressing challenges and solutions," Renewable Energy Focus, vol. 49, Article no. 100572, June 2024, doi: 10.1016/j.ref.2024.100572.

[2] O. Dharmapandit, R. K. Patnaik, and P. K. Dash, "Detection, classification, and location of faults on grid-connected and islanded AC microgrid," *Int. Trans. Electr. Energy Syst.*, vol. 27, no. 12, p. e2431, 2017.

[3] M. Mehrasa, E. Pouresmaeil, B. N. Jørgensen, and J. P. Catalão, "A control plan for the stable operation of microgrids during grid-connected and islanded modes," *Electric Power Systems Research*, vol. 129, pp. 10–22, 2015.

[4] Chakraborty, S., Kumar, K. and Kar, S., 2024, July. Hierarchical Control Framework in Networked Microgrid. In 2024 IEEE International Conference on Smart Power Control and Renewable Energy (ICSPCRE) (pp. 1-5). IEEE.

[5] H. Yu, Z. Lu, Y. Liu, Y. Gao, and J. Wan, "Injected harmonic feature based protection scheme for active distribution networks with high proportion IIDGs," *IEEE Access*, 2024.

[6] M. N. Alam, S. Chakrabarti, and A. K. Pradhan, "Protection of networked microgrids using relays with multiple setting groups," *IEEE Transactions on Industrial Informatics*, vol. 18, no. 6, pp. 3713–3723, Jun. 2022.

[7] IEEE Standard 1547.2-2008, IEEE Application Guide for IEEE Std 1547™, IEEE Standard for Interconnecting Distributed Resources with Electric Power Systems, Apr. 2009.

[8] A. Hooshyar and R. Iravani, "Microgrid protection," *Proceedings of the IEEE*, vol. 105, no. 7, pp. 1332–1353, 2017.

[9] M. A. Zamani, T. S. Sidhu, and A. Yazdani, "A protection strategy and microprocessor-based relay for low-voltage microgrids," *IEEE Transactions on Power Delivery*, vol. 26, no. 3, pp. 1873–1883, 2011.

[10] H. Al-Nasseri, M. A. Redfern, and F. Li, "A voltage based protection for micro-grids containing power electronic converters," in *Proc. 2006 IEEE Power Engineering Society General Meeting*, 2006, pp. 7–7.

[11] X. Li, A. Dyśko, and G. M. Burt, "Traveling wave-based protection scheme for inverter-dominated microgrid using mathematical morphology," *IEEE Transactions on Smart Grid*, vol. 5, no. 5, pp. 2211–2218, 2014.

[12] A. H. N. Tajani, A. Bamshad, and N. Ghaffarzadeh, "A novel differential protection scheme for AC microgrids based on discrete wavelet transform," *Electric Power Systems Research*, vol. 220, Art. no. 109292, Jul. 2023.

[13] D. P. Mishra, S. R. Samantaray, and G. Joos, "A combined wavelet and data-mining based intelligent protection scheme for microgrid," *IEEE Transactions on Smart Grid*, vol. 7, no. 5, pp. 2295–2304, 2015.

[14] F. Zhang and L. Mu, "A fault detection method of microgrids with grid-connected inverter interfaced distributed generators based on the PQ control strategy," *IEEE Transactions on Smart Grid*, vol. 10, no. 5, pp. 4816–4826, 2018.

[15] E. Casagrande, W. L. Woon, H. H. Zeineldin, and D. Svetinovic, "A differential sequence component protection scheme for microgrids with inverter-based distributed generators," *IEEE Transactions on Smart Grid*, vol. 5, no. 1, pp. 29–37, 2013.

[16] Z. Chen, X. Pei, M. Yang, L. Peng, and P. Shi, "A novel protection scheme for inverter-interfaced microgrid (IIM) operated in islanded mode," IEEE Transactions on Power Electronics, vol. 33, no. 9, pp. 7684–7697, 2017.

[17] M. A. Azzouz, A. Hooshyar, and E. F. El-Saadany, "Resilience enhancement of microgrids with inverter-interfaced DGs by enabling faulty phase selection," *IEEE Transactions on Smart Grid*, vol. 9, no. 6, pp. 6578–6589, 2017.

[18] A. Soleimanisardoo, H. K. Karegar, and H. H. Zeineldin, "Differential frequency protection scheme based on off-nominal frequency injections for inverter-based islanded microgrids," *IEEE Transactions on Smart Grid*, vol. 10, no. 2, pp. 2107–2114, Mar. 2019.

[19] K. A. Saleh and A. Mehrizi-Sani, "Harmonic directional overcurrent relay for islanded microgrids with inverter-based DGs," *IEEE Systems Journal*, vol. 15, no. 2, pp. 2720–2731, 2020.

[20] K. Saleh, M. A. Allam, and A. Mehrizi-Sani, "Protection of inverter-based islanded microgrids via synthetic harmonic current pattern injection," IEEE Transactions on Power Delivery, vol. 36, no. 4, pp. 2434–2445, 2020.

[21] M. A. U. Khan, Q. Hong, A. Egea-Àlvarez, A. Dyśko, and C. Booth, "A communication-free active unit protection scheme for inverter dominated islanded microgrids," *International Journal of Electrical Power & Energy Systems*, vol. 142, Art. no. 108125, 2022.

[22] W. T. El-Sayed, E. F. El-Saadany, and H. H. Zeineldin, "Interharmonic differential relay with a soft current limiter for the protection of inverter-based islanded microgrids," *IEEE Transactions on Power Delivery*, vol. 36, no. 3, pp. 1349–1359, 2020.

[23] Kumar, K., Chakraborty, S., Kumar, P. and Kar, S., 2024, July. Blockchain-Based Defense Mechanisms for Mitigating Unnecessary Islanding in Microgrids Against Cyber-Attack. In *2024 IEEE International Conference on Smart Power Control and Renewable Energy (ICSPCRE)* (pp. 1-4). IEEE.

[24] Kumar, K., Kumar, P. and Bohre, A.K., 2023. Performance analysis of IC MPPT algorithm for applications of solar PV in DC microgrid. SN Computer Science, 4(5), p.579.

[25] Kumar, K. and Kar, S., 2024. An efficient protection scheme for microgrid using ROC of differential admittance angle. *Electric Power Systems Research*, 227, p.109969.

Enhanced Inertia Estimation in Microgrids using a Modified RoCoF Method

R. Venkatesh
School of Electrical and Computer Sciences
Indian Institute of Technology
Bhubaneswar, India
a22ee09006@iitbbs.ac.in

P.C. Sekhar
School of Electrical and Computer Sciences
Indian Institute of Technology
Bhubaneswar, India
pcsekhar@iitbbs.ac.in

C. N. Bhende
School of Electrical and Computer Sciences
Indian Institute of Technology
Bhubaneswar, India
cnb@iitbbs.ac.in

Abstract— The increasing integration of inverter-based resources alters the system's dynamic response, including inertia distribution. Accurate inertia estimation and its monitoring are crucial for ensuring stable operation in power grids with high renewable energy penetration. In this connection, this paper introduces an enhanced rate of change of frequency (RoCoF) based method for estimating inertia in a microgrid. The key attribute of the proposed method is its capability in estimating the inertia even without knowing the magnitude of disturbance, which is the main limitation of the conventional method. The effectiveness of the proposed approach is first validated on a system composed solely synchronous generators, followed by an evaluation in a hybrid energy system. The results are then compared with those obtained using conventional inertia estimation techniques. The findings demonstrate that the proposed method accurately estimates the system inertia, which is contributed by rotating/dynamic elements, however with a diverse generation mix without needing the disturbance magnitude.

Keywords— AC microgrid, Frequency stability, Inertia estimation, Photovoltaic system, Renewable energy sources.

I. INTRODUCTION

As concerns over climate change and global warming continue to rise, many nations are striving to increase the share of renewable energy in their electricity generation mix over the coming decade [1-2]. However, one of the key challenges associated with the large-scale integration of renewable energy sources is the significant reduction in system inertia [3-4] due to their power electronic converter interfaces. In traditional power systems, synchronous machines store kinetic energy in their rotating masses, which naturally counteracts frequency disturbances and helps stabilize the system during sudden fluctuations in generation or demand. This stored energy serves as a fundamental reference for designing frequency stability controllers and protection relays, ensuring the overall reliability of conventional power grids [5-7].

In contrast, most renewable energy sources, such as solar and wind power, connect to the grid through power electronic converters, which do not inherently provide the same level of inertial support. As the share of renewable generation increases, the system's overall inertia declines, making it more vulnerable to frequency instability [8-9]. A lower inertia grid experiences a steeper rate of change of frequency (RoCoF) and larger frequency deviations following disturbances. These conditions can activate RoCoF relays and under frequency load shedding mechanisms, potentially leading to the sequential tripping of generators and, in severe cases, complete grid failures [10]. According to predictions made for the UK electricity grid, system inertia will decrease to 30% by 2033–34 compared to the 2013–14 levels [11].

Given these challenges, accurately estimating system inertia under varying operating conditions has become increasingly important for maintaining grid stability. Inertia estimation enables grid operators to evaluate the system's ability to withstand disturbances, optimize frequency control strategies, and implement necessary countermeasures. As power systems transition toward greater renewable energy integration, ensuring reliable operation through effective inertia monitoring and management will be essential for preventing large-scale disruptions and maintaining a resilient energy infrastructure.

The conventional method for estimating power system inertia is based on the swing equation, which determines the inertia constant by analysing the frequency response following a known disturbance. This approach has been widely used in power system studies to assess stability and evaluate dynamic behaviour. In [12], the authors introduced a method for estimating the total inertia of the Great Britain (GB) power system by considering instantaneous power losses and regional variations in frequency transients, measured using synchronized phasor measurement units (PMUs). This approach divides the GB network into several regions based on constraint boundaries, with the inertia estimated at the regional level, thereafter aggregated for all regions to determine the overall system inertia. In [13], an algorithm was presented that estimates timing of a disturbance and the system inertia in a power system immediately following the disturbance using wide-area frequency and power measurements, allowing operators to continuously track inertia variations and enhance system stability. In [14], the authors proposed a closed-loop identification technique to estimate the equivalent inertia constant at a power system's connection bus using micro perturbations via multisine signals. Frequency and active power responses are analyzed to achieve real-time estimation with minimal system disruption and effectiveness of this approach is demonstrated through simulations and real-world system validation. In [15], an approach was introduced for estimating effective inertia using ambient frequency and active power signals from PMUs. Unlike conventional methods that rely solely on disturbances, this technique enables inertia estimation based on ambient data, allowing for near-continuous monitoring at intervals of minutes. Additionally, this method divides the system into multiple regions, estimating the inertia of each separately to provide a more detailed and dynamic assessment of the overall power network.

In [16], a switching Gaussian Markov model (SMGM) was proposed for dynamic inertia estimation, leveraging historical data on frequency and inertia variations. This model captures the complex dependencies between these parameters and represents different system operating conditions through a finite mixture model with a switching regime, improving the accuracy of inertia estimation under changing grid conditions. In [17], a robust swing equation-based technique was

---

979-8-3315-3013-6/25 $31.00 © 2025 IEEE

developed to estimate generator inertia during disturbances. This method was tested under various types of disturbances and noise conditions, demonstrating its reliability in practical applications.

Most of the studies on power system inertia estimation primarily utilize the conventional swing equation-based method. However, a key challenge with this approach is its accuracy, particularly when the disturbance size is unknown, making it less effective in such scenarios. Two major challenges that arise when estimating system inertia are determining the exact magnitude of the disturbance and the choice of an appropriate method used to calculate the RoCoF.

In this context, this paper introduces an improved inertia estimation approach based on a modified RoCoF method, applied to an AC microgrid test system. The system operates without speed governor and excitation controls, maintaining a constant mechanical input and fixed excitation to ensure precise inertia constant estimation. The actual variation in active power at the disturbance location is determined using the total active power generation profile, while RoCoF is derived through polynomial curve fitting applied to the system's frequency response.

The rest of the paper is organized as follows: Section II describes the test system, Section III covers methodology, Section IV presents results and discussion, and Section V concludes the paper.

II. DESCRIPTION OF TEST SYSTEM

In recent times, microgrids have gained significant attention as an effective solution for integrating distributed renewable energy sources, especially within medium and low-voltage power networks. A microgrid consists of a combination of generation units and electrical loads that can be managed as a unified system. It is designed to operate flexibly, either independently in islanded mode or in conjunction while connected to the main grid. This operational flexibility enhances the system's resilience, allowing it to maintain stability and supply power even during main grid disturbances or outages. Besides facilitating renewable energy integration, microgrids offer additional benefits such as delivering power to isolated or rural regions and enabling large power systems to be divided into smaller sections, reducing the likelihood of widespread blackouts. Furthermore, microgrids can help reduce greenhouse gas emissions by prioritizing clean energy sources and improving overall energy efficiency. Despite these advantages, the increased reliance on renewable energy within microgrids and their relatively small system size intensifies challenges related to reduced inertia and frequency instability.

A. Test System 1

The proposed methodology has been applied to a practical AC microgrid test system located in India. The test system 1 consists of seven buses, which are interconnected through five conventional synchronous generators, five transformers, and eight electrical loads. The microgrid has a total rated capacity of 95.8 MVA, while the combined active power demand of the loads is 36.1 MW. A clear overview of the system's configuration is presented in the single-line diagram shown in Fig. 1, and the detailed technical specifications of the synchronous generators used are listed in Table I.

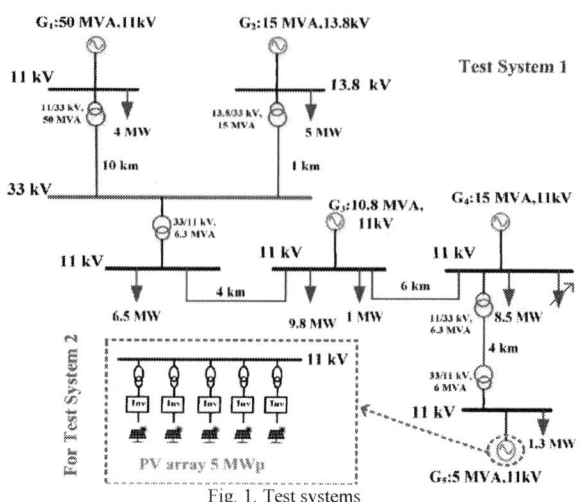

Fig. 1. Test systems

TABLE I. Test system 1 generator data

S.no	Synchronous Generator	Voltage (kV) (L-L)	Power (MVA)	Inertia constant (sec)
1	G_1	11	50	3.2
2	G_2	13.8	15	6
3	G_3	11	10.8	3.2
4	G_4	11	15	3.2
5	G_5	11	5	6

B. Test System 2

The proposed methodology is further evaluated on a modified version of the test system incorporating hybrid energy sources. In this configuration, Generator 5 from Fig. 1 is replaced with a 5 MWp photovoltaic (PV) array, thereby keeping the same generation capacity and load as same as that of Test system 1, as shown in Fig. 1. The updated generator and PV system specifications for the test system 2 are detailed in Table II.

TABLE II. Test system 2 generator data

S.no	Synchronous Generator	Voltage (kV) (L-L)	Power (MVA)	Inertia constant (sec)
1	G_1	11	50	3.2
2	G_2	13.8	15	6
3	G_3	11	10.8	3.2
4	G_4	11	15	3.2
5	PV	11	5	0

C. PV Array Unit

In Test System 2, the conventional generator G_5, which has a capacity of 5 MW in Test System 1, is replaced with a PV array system rated at 5 MWp. This PV system comprises 5 sub-units, each with a power generation capacity of 1 MWp. Each PV unit is integrated into the power network via a single-stage conversion process using an inverter, facilitating a direct connection to the 11 kV, 50 Hz bus, as depicted in Fig. 2. All five PV units are connected in parallel to the 11 kV bus through individual step-up transformers. Each PV unit operates at its maximum power point (MPP), generating 1 MWp at 580 V and 1666 A, as illustrated in Fig. 3. The PV modules output is supplied to an inverter, for grid integration. To maintain power quality and minimize harmonic distortions, the inverter output is processed through LC filters before being fed into the grid via the step-up transformer, as shown in Fig. 2. This configuration ensures efficient power transfer from the PV system to the utility grid while maintaining voltage and frequency stability.

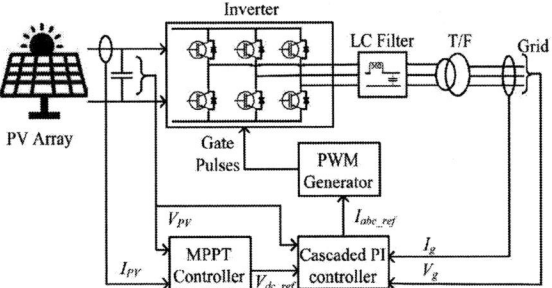

Fig. 2 Single stage grid connected PV unit

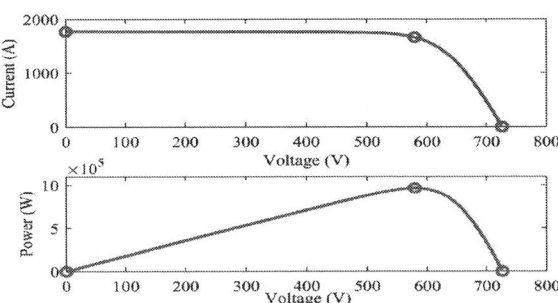

Fig. 3 *I-V* and *P-V* characteristics each PV array unit at 1000 W/m^2, 25^0 c

III. METHODOLOGY

This paper estimates the inertia of the selected test systems using two different approaches. The first one is the conventional swing equation method, and the second one is the proposed enhanced RoCoF based estimation method. A comprehensive explanation of both methodologies is provided in this section.

A. Conventional Method

A widely recognized approach for estimating power system inertia relies on the swing equation, which captures the dynamic interaction between mechanical and electrical power in a multi-machine network. This equation describes how the rotor speed of synchronous generators changes in response to any imbalance between mechanical input and electrical output power. The mathematical expression of this equation is given as

$$P_a = J\alpha \tag{1}$$

where P_a is accelerating power (W), J is moment of inertia (kg.m^2) and α is angular acceleration (rad/s^2). Further (1) can be rewritten as

$$P_m - p_e = \frac{2HS}{f_o}\frac{df}{dt} \tag{2}$$

where P_m and P_e are respective mechanical and electrical powers of generator, f_o is the steady state frequency of the system, f is measured frequency, S is apparent power and $\frac{df}{dt}$ is RoCoF (Hz/s), H is inertia constant of the system in seconds. From (2), H can be expressed as

$$H = \frac{p_m - p_e}{2S\frac{df}{dt}}f_o \tag{3}$$

rearranging (3), H can be given by

$$H = \frac{1}{2S}\left(\frac{f_o\,\Delta P}{\frac{df}{dt}}\right) = \frac{\Delta P}{2S\left(\frac{d\left(\frac{\Delta f}{f_o}\right)}{dt}\right)} \tag{4}$$

where, Δf denotes the deviation of measured frequency from f_o, ΔP represents magnitude of disturbance. The polynomial curve fitting is applied to ($\frac{\Delta f}{f_o}$) and the most suitable polynomial order can be mathematically formulated as

$$\left(\frac{\Delta f}{f_o}\right) = A_n t^n + A_{n-1}t^{n-1} + \ldots\ldots + A_1 t \tag{5}$$

The so obtained polynomial is used to map the coefficients of (5). By taking the time derivative of this polynomial expression, the coefficient A_1 is identified precisely at the moment the disturbance begins, which corresponds to the initial point of the disturbance, i.e., $t=0$. This coefficient is then directly utilized to calculate the inertia constant of the generator for the known size of disturbance using the formulated relationship, as follows.

$$H = \frac{\Delta P}{2SA_1} \tag{6}$$

The flow chart of the conventional estimation method is presented in Fig. 4.

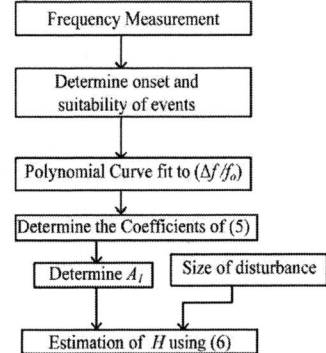

Fig.4 Conventional method for estimation of *H*

When the analysis is extended to a system consisting of N generators, the total or equivalent inertia of the entire system can be expressed as

$$H_{sys} = \frac{\sum_{i=1}^{N} S_i H_i}{\sum_{i=1}^{N} S_i} \tag{7}$$

In this formulation, S_i refers to the rated apparent power, and H_i denotes the inertia constant associated with the ith generator and overall system inertia constant is H_{sys}.

In large-scale power systems, frequency measurements taken from different parts of the network often show slight variations. These differences are mainly caused by oscillations between different generator groups. To address this challenge and improve the accuracy of inertia estimation, the concept of center of inertia (COI) frequency is used. The COI frequency provides a system-wide average frequency and is given by,

$$f_{COI} = \frac{\sum_{i=1}^{N} f_i H_i}{\sum_{i=1}^{N} H_i} \qquad (8)$$

Since the test system under consideration consists of multiple generators, the frequency f in (5) should be replaced by the f_{COI}. Similarly, the inertia constant H in (6) must be substituted with the system equivalent inertia H_{sys} for accurate inertia estimation.

B. RoCoF Method

Inertia estimation in power systems is predominantly carried out using the conventional swing equation-based method. However, this approach faces notable limitations, especially regarding its accuracy when the magnitude of the disturbance is unknown, which often restricts its applicability in real-world applications. To address the limitations of the conventional method, the RoCoF-based approach is introduced, incorporating an enhancement to improve estimation accuracy. This enhancement involves applying a curve-fitting process to the total generated power, allowing for a precise determination of the actual disturbance size (ΔP) while maximum value of RoCoF can be obtained from RoCoF plot. By refining the disturbance estimation, the overall accuracy of inertia calculation is significantly improved. The flowchart illustrating the RoCoF method is shown in Fig. 5. The inertia of the system can be calculated using (4) for unknown size of disturbance.

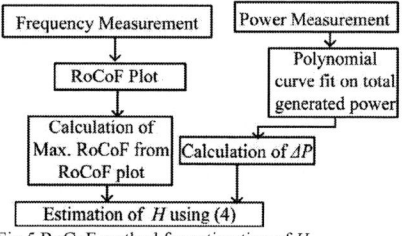

Fig.5 RoCoF method for estimation of H

C. Proposed Method

The conventional approach fails unless the disturbance magnitude is known. While the RoCoF-based method estimates the RoCoF using the system's transient response, however lacks a precise strategy for RoCoF calculation and the governor action is not excluded. To overcome these limitations, this paper proposes a modified RoCoF-based inertia estimation technique that enhances accuracy and applicability even in the case of unknown disturbance. In the proposed method, the actual change in active power at the disturbance point is obtained directly from the system's total active power generation profile, providing a more accurate disturbance measurement while considering inactive governor action.

In this method, the RoCoF is determined by fitting a suitable-order polynomial curve to the system's frequency response (Δf_{COI}). Once the optimal polynomial equation is obtained, its derivative with respect to time is calculated. This process allows for the accurate extraction of the coefficient associated with the time variable t at the instant when the disturbance occurs (at $t=0$), which is represents the system's first cycle response immediately after the disturbance. This extracted coefficient serves as a crucial parameter and is subsequently used to estimate the inertia constant of the system with improved precision. The inertia constant is then calculated using this improved method, as expressed in the following relation.

$$H_{Rocof} = \frac{\Delta P}{2S \left(\dfrac{d \left(\Delta f_{COI} \right)}{dt} \right)} f_o \qquad (9)$$

The flow chart of the proposed modified RoCoF estimation method is presented in Fig. 6.

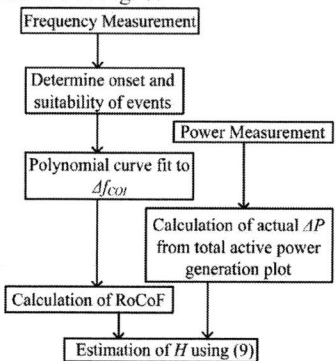

Fig.6 Modified RoCoF method for estimation of H

IV. RESULTS AND ANALYSIS

In this section, the inertia constants of both test systems are estimated, where one system consists only of conventional generation sources and the other system includes hybrid sources. The estimation is carried out using both the conventional swing equation-based method and the proposed modified RoCoF method. In both cases, the systems are operated without speed governor and excitation controls while maintaining a constant mechanical input and fixed excitation to ensure accurate and consistent inertia constant estimation.

A. Conventional Method

The inertia estimation with the conventional method is described in this subsection, for both the test systems to evaluate its effectiveness.

I. Test System 1

This test system consists of five synchronous generators, and the total system inertia is calculated as 3.78 seconds using (7). Additionally, f_{COI}, is determined using (8). Under steady-state conditions, the system operates at a nominal frequency of 50 Hz, as illustrated in Fig. 7.

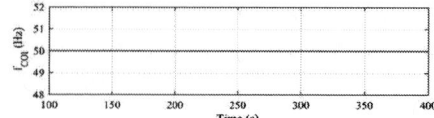

Fig. 7 f_{COI} of test system 1 under steady state

A step increase in load, ranging from 1% to 5% of the total load, is introduced at the 200th second as a known disturbance (ΔP). The system's frequency response is analyzed using polynomial curve fitting to $\Delta f/f_o$ and obtained a suitable-order polynomial equation. By differentiating this equation at the disturbance instant, the coefficient A_1 is determined and used in (6) to estimate the system inertia. The results obtained using the conventional method are summarized in Table III, and the curve fitting to $\Delta f/f_o$ for a 1% load increase (36.1 MW) is illustrated in Fig. 8.

TABLE III. Estimation of H for test system 1 using conventional method

S.no	% Load change	ΔP (MW)	H_{act} (sec)	H_{est} (sec)	% Error
1	1 %	0.361	3.78	3.77	-0.26 %
2	2 %	0.722	3.78	3.82	1.05 %
3	3 %	1.083	3.78	3.87	2.30 %
4	4 %	1.444	3.78	3.91	3.43 %
5	5 %	1.805	3.78	3.97	5.02 %

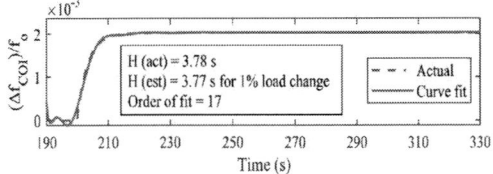

Fig. 8 Curve fitting to $\Delta f/f_o$ response for 1% load change at t = 200 s

II. Test System 2

This test system consists of four synchronous generators and one PV unit and the total system inertia is calculated as 3.47 seconds using (7). Additionally, f_{COI}, is determined using (8). Under steady-state conditions, the system operates at a nominal frequency of 50 Hz, as illustrated in Fig. 9.

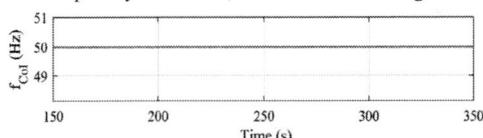

Fig. 9 f_{COI} of test system 2 under steady state

A step increase in load, ranging from 1% to 5% of the total load, is introduced at the 200th second as a known disturbance (ΔP). The results obtained using the conventional method for Test System 2 are summarized in Table IV, and the curve fitting to $\Delta f/f_o$ for a 1% load increase (36.1 MW) is illustrated in Fig. 10.

TABLE IV. Estimation of H for test system 2 using conventional method

S.no	% Load change	ΔP (MW)	H_{act} (sec)	H_{est} (sec)	% Error
1	1 %	0.361	3.47	3.57	2.8%
2	2 %	0.722	3.47	3.31	-4.6%
3	3 %	1.083	3.47	3.66	5.4%
4	4 %	1.444	3.47	3.34	-3.7%
5	5 %	1.805	3.47	3.59	3.4%

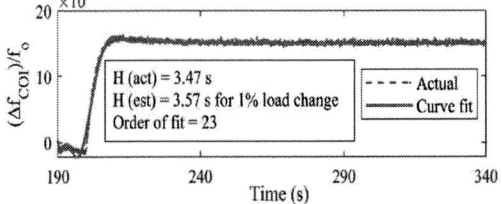

Fig. 10 Curve fitting to $\Delta f/f_o$ response for 1% load change at t = 200 s

B. Proposed Method

Similar to Subsection A, the proposed methodology is applied to both of the test systems to evaluate its effectiveness.

I. Test System 1

In this approach, the actual disturbance size (ΔP) is determined from the total active power generation profile as illustrated in Fig. 11. A load increase in steps of 1% to 5% of the total load is applied at the 200th second. The frequency response Δf_{COI} is then fitted with a suitable-order polynomial. The RoCoF is obtained by differentiating this polynomial at the disturbance instant. Finally, using (9) inertia constant of the system is estimated. The results from the proposed method are shown in Table V, and the curve fitting to Δf_{COI} for a 1% load increase (36.1 MW) is depicted in Fig. 12.

TABLE V. Estimation of H for test system 1 using proposed method

% Load change	ΔP (MW) Actual	ΔP (MW) measured	RoCoF (Hz/sec)	H_{act} (sec)	H_{est} (sec)	% Error
1 %	0.361	0.43	-0.0295	3.78	3.80	0.52 %
2 %	0.722	0.86	-0.0591	3.78	3.79	0.26 %
3 %	1.083	1.29	-0.0893	3.78	3.76	-0.52%
4 %	1.444	1.71	-0.1186	3.78	3.76	-0.52%
5 %	1.805	2.14	-0.1472	3.78	3.79	0.26%

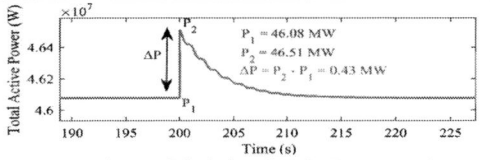

Fig. 11. Calculation of ΔP for Test system 1

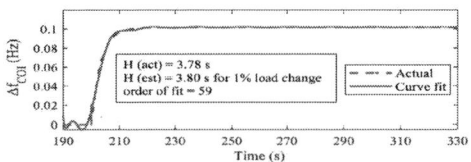

Fig. 12 Curve fitting to Δf_{COI} response for 1% load change at t = 200 s

II. Test System 2

In this approach, the actual disturbance size (ΔP) is determined from the total active power generation profile as illustrated in Fig. 13. A step load increase between 1% and 5% of the total load is applied at the 200th second, like ealier case studies. The results from the proposed method are shown in Table VI, and the curve fitting to Δf_{COI} for a 1% load increase (36.1 MW) is depicted in Fig. 14.

TABLE VI. Estimation of H for test system 2 using proposed method

% Load change	ΔP (MW) Actual	ΔP (MW) measured	RoCoF (Hz/sec)	H_{act} (sec)	H_{est} (sec)	% Error
1 %	0.361	0.39	-0.02950	3.47	3.44	-0.86%
2 %	0.722	0.80	-0.06076	3.47	3.43	-1.15%
3 %	1.083	1.15	-0.08548	3.47	3.51	1.15%
4 %	1.444	1.49	-0.11084	3.47	3.50	0.86%
5 %	1.805	1.88	-0.13872	3.47	3.53	1.72%

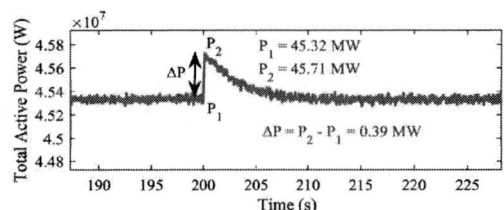

Fig. 13. Calculation of ΔP for Test system 2

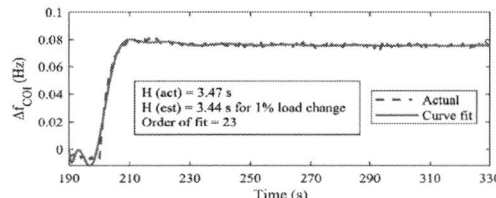

Fig. 14 Curve fitting to Δf_{COI} response for 1% load change at t = 200 s

The comparison between the conventional method and the proposed method for inertia estimation in both the Test Systems is summarized in Table VII.

TABLE VII. Comparison table

% Load change	% Error in			
	Test System 1		Test System 2	
	With Conventional method	With Proposed method	With Conventional method	With Proposed method
1	-0.26 %	0.52 %	2.8%	-0.86%
2	1.05 %	0.26 %	-4.6%	-1.15%
3	2.30 %	-0.52%	5.4%	1.15%
4	3.43 %	-0.52%	-3.7%	0.86%
5	5.02%	0.26%	3.4%	1.72%

VI. CONCLUSION

Two test systems, one consisting solely conventional synchronous generators and the other system having both the conventional and renewable sources are investigated for inertia estimations. Both the systems are operated without speed governor and excitation controls to ensure precise and consistent inertia estimation. The inertia is evaluated using both the conventional swing equation method and the proposed modified RoCoF-based method. The conventional method faces limitations, especially when the disturbance magnitude is unknown and in determining the RoCoF accurately while the RoCoF-based method considers disturbance magnitude, however lacks a precise strategy for RoCoF calculation, whereas the proposed method overcomes these limitations by calculating the actual disturbance from the total active power generation and obtaining RoCoF through polynomial curve fitting of the frequency response. Results reveal that the inertia of the system with conventional generators is 3.78 seconds, which reduces to 3.47 seconds when a synchronous generator of 5 MW is replaced by an equivalent PV unit. In all scenarios, the proposed method provides more accurate and reliable inertia estimation compared to the conventional approach even under unknown disturbances.

ACKNOWLEDGEMENT

This research work is supported in part by Central Power Research Institute through its RSoP with Ref. CPRI/R&D/TC/GDEC/2022.

REFERENCES

[1] I. Energy and C. Change, "World energy outlook—2017," in *Proc. OECD/IEA*, 2017, Paper 782.

[2] Lakshay and S. K. Jain, "A Review on Power System Inertia Estimation Techniques," *International Conference on Control, Automation, Power and Signal Processing (CAPS)*, Jabalpur, India, pp. 1-6, doi: 10.1109/CAPS52117.2021.9730607, 2021

[3] Gu, Huajie, Ruifeng Yan, and Tapan Kumar Saha, "Minimum synchronous inertia requirement of renewable power systems," *IEEE Transactions on Power Systems*, vol. 33, pp. 1533-1543, 2017

[4] R. I. Shupty and B. Chowdhury, "A Review of Inertia Estimation in Power Systems Using Measurement-Based Approaches," *North American Power Symposium (NAPS)*, Asheville, NC, USA, pp. 1-6, doi: 10.1109/NAPS58826.2023.10318733, 2023

[5] P.Tielens, "Operation and control of power systems with low synchronous inertia," Ph.D. dissertation, Katholieke Universiteit Leuven, Leuven, Belgium, 2017.

[6] R. Aljarrah, B. B. Fawaz, Q. Salem, M. Karimi, H. Marzooghi and R. Azizipanah-Abarghooee, "Issues and Challenges of Grid-Following Converters Interfacing Renewable Energy Sources in Low Inertia Systems: A Review," in *IEEE Access*, vol. 12, pp. 5534-5561, doi: 10.1109/ACCESS.2024.3349630, 2024

[7] A. Gupta, P. C. Sekhar, M. Z. Degefa, K. Jonatan R. A. and S. D'Arco, "Synchronization Controller for Seamless Interconnection of Mirogrids with Heterogeneous Sources," *22nd National Power Systems Conference (NPSC)*, New Delhi, India, pp. 344-349, doi: 10.1109/NPSC57038.2022.10069869, 2022

[8] Dowrah, Pardip, Kham Muan Lian, Jeush D. Sangma, Ferdinand Dkhar, and Shaik Affijulla. "Estimation techniques for power system inertia: A simulation-oriented review." *IEEE Mysore Sub Section International Conference (MysuruCon)*, pp. 36-41. IEEE, 2021.

[9] Li, Hongxin, Yibo Ding, Yi Lu, Jiapeng Li, Lijuan Fan, Huayi Wu, and Zhao Xu. "Inertia Analysis and Control of New Type Power System-An Overview." *International Conference on Power System Technology (PowerCon)*, pp. 1-5. IEEE, 2023.

[10] Phurailatpam, Chitaranjan, Zakir Hussain Rather, Behrooz Bahrani, and Suryanarayana Doolla. "Measurement-based estimation of inertia in AC microgrids." *IEEE Transactions on Sustainable Energy* 11, vol. 3, 2019

[11] Gayathri, K., and Manas Kumar Jena. "A practical approach to inertia distribution monitoring and impact of inertia distribution on oscillation baselining study for renewable penetrated power grid." *IEEE Systems Journal* 17, vol. 3, pp. 3593-3601, 2022

[12] Ashton, Phillip M., Christopher S. Saunders, Gareth A. Taylor, Alex M. Carter, and Martin E. Bradley. "Inertia estimation of the GB power system using synchrophasor measurements." *IEEE Transactions on Power Systems* 30, vol. 2, pp. 701-709, 2014

[13] Wall, Peter, and Vladimir Terzija. "Simultaneous estimation of the time of disturbance and inertia in power systems." *IEEE Transactions on Power Delivery* 29, vol. 4, pp. 2018-2031, 2014

[14] Zhang, Junbo, and Hanchen Xu. "Online identification of power system equivalent inertia constant." *IEEE Transactions on Industrial Electronics*, vol. 10, pp. 8098-8107, 2017

[15] Tuttelberg, Kaur, Jako Kilter, Douglas Wilson, and Kjetil Uhlen. "Estimation of power system inertia from ambient wide area measurements." *IEEE Transactions on Power Systems*, vol. 6, pp. 7249-7257, 2018

[16] Cao, Xue, Bruce Stephen, Ibrahim F. Abdulhadi, Campbell D. Booth, and Graeme M. Burt. "Switching Markov Gaussian models for dynamic power system inertia estimation." *IEEE Transactions on Power Systems* 31, vol. 5, pp. 3394-3403, 2015

[17] Wall, Peter, Francisco Gonzalez-Longatt, and Vladimir Terzija. "Estimation of generator inertia available during a disturbance." In *2012 IEEE Power and Energy Society General Meeting*, pp. 1-8. IEEE, 2012.

979-8-3315-3013-6/25 $31.00 © 2025 IEEE

Force Production in Permanent Magnet Synchronous Machine

1st Shahrukh
Department of Mechanical Engineering
Indian Institute of Technology Guwahati
Guwahati, India
shahrukh1995@iitg.ac.in

2nd Gopinath Sengupta
School of Energy Science and Engineering
Indian Institute of Technology Guwahati
Guwahati, India
g.sengupta@iitg.ac.in

3nd Karuna Kalita
Department of Mechanical Engineering
Indian Institute of Technology Guwahati
Guwahati, India
karuna.kalita@iitg.ac.in

Abstract—**Force production capability is integrated in permanent magnet synchronous machine, utilizing a stator winding configuration. The proposed three phase permanent magnet synchronous machine features a design where the same winding conductors can carry both torque producing and force producing currents. By introducing a bridge supply midway of the parallel path in each phase, additional $p \pm 1$ pole pairs are produced in a p pole pair machine. This leads to the production of electromagnetic force within the air gap. The electromagnetic forces are influenced by the current conditions. The bridge voltage between the parallel paths of each phase results in electromagnetic force generation in the air gap. The force producing capability of the machine is evaluated using the coupled field finite element analysis.**

Index Terms—**Permanent magnet synchronous machine, time stepping finite element analysis (TSFEM), force production, specialized stator winding.**

I. INTRODUCTION

Permanent magnet synchronous machines (PMSMs) exhibit high power density, simplified maintenance, and fast dynamic response compared to wound rotor machine [1]. The force production capability of permanent magnet synchronous machine can be explored for the development of bearingless permanent magnet synchronous machines. Ensuring stable rotor suspension during operation is one of the critical aspect of bearingless machine research. Such bearingless permanent magnet machines are particularly promising for high speed precision application such as advanced machine tools [2]. By integrating the magnetic levitation capability directly into the motor electromagnetic design, self bearing machines eliminate the reliance on mechanical or magnetic bearings. Accurate estimation of electromagnetic force in the air gap of the machine is important for the stable rotor support [3]. Bearingless machines can run at higher speed than the machine with mechanical or magnetic bearings due to reduced rotor length and utilize the machine magnetic flux as bias, significantly reduces the power required for rotor shaft levitation [4]. Independent

control of the stator coil currents enables the force regulation, with enhanced controlling capability by increasing the number of coils, demonstrated using six coils in a four pole induction machine [5]. Traditional magnetic bearings, while capable of position control in two axis, faces limitations in high speed applications and contribute to increased rotor length, conflicting with the mechanical strength [6]. Combining the torque producing and the levitation capability within a unified electromagnetic framework addresses these challenges. The operating principle of a self bearing machine is to generate the asymmetric air gap flux density in the air gap by injecting additional levitation currents into the winding [7]. A secondary winding is utilized to generate the additional levitation current. Although this approach simplifies control, it inherently results in higher power loss [8]. A single set winding configuration has also been proposed where the levitation current is superimposed onto the torque producing current within the same winding set [9]. Several bearingless motors have been proposed in past years. Magnetic levitation system employ the alternating currents and rotor position feedback to generate the stabilizing force, improving the life cycle and reducing maintenance [10]. A four pole main winding to produce torque and additional two pole control winding to produce the bearing force has been proposed in [6]. The levitation control force is generated through the interaction of the magnetic fields generated by the torque producing winding and suspension winding [11]. There exist a coupling between the torque producing and force control, which necessitates their decoupling to achieve stable performance of bearingless permanent magnet synchronous machines [12]. Several bearingless motor types have been explored, including bearingless induction machine, bearingless brushless DC machine and bearingless switched reluctance machine, the bearingless permanent magnet synchronous machine is prominent focus of research due to its high efficiency, high torque, power density and structrucal simplicity [13]. This highlights the permanent magnet synchronous machine's capability to produce precise electromagnetic forces utilizing

979-8-3315-3013-6/25 $31.00 © 2025 IEEE

single set of winding. The bridge configured winding enables electrical separation between the terminal for torque and those responsible for lateral force production. This approach eliminates the need for redundant winding, as the transverse force is genertaed using the existing stator windings [14].

This paper presents the force producing capability of 24 slots four pole PMSM equipped with Bridge configured winding. The application of bridge voltage across the parallel paths of each phase results in an uneven magnetic field distribution in air gap, enabling the lateral force generation. A coupled field circuit equation is solved using TSFEM to study the transverse force characteristics.

II. MODELLING OF FORCE PRODUCTION

A permanent magnet synchronous machine consists of essential components such as a rotor, stator, coil, permanent magnets, a spacer between the magnets, and a band to secure them during rotation. The proposed model is restricted to two dimensional. The two dimensional field equation for a PMSM is given as [15]

$$
\begin{aligned}
&\frac{\partial}{\partial x}\left[\frac{1}{\mu}\frac{\partial A}{\partial x}\right] + \frac{\partial}{\partial y}\left[\frac{1}{\mu}\frac{\partial A}{\partial y}\right] - \sigma\frac{\partial A}{\partial t} + \frac{N}{S}I \\
&= -\frac{\partial}{\partial x}\frac{1}{\mu}B_{ry} + \frac{\partial}{\partial y}\frac{1}{\mu}B_{rx}
\end{aligned}
\tag{1}
$$

where A is the magnetic vector potential, μ is the permeability of the material, σ is the conductivity of material, N is the number of turns of thin conductors, S is the total cross sectional area of the coil and B_{rx} and B_{ry} are the components of remanent flux density. The value of A is assumed to be zero at inner and outer boundaries of the machine due to negligible values at these boundaries. For a coil comprising of N turns of thin conductors with a cross section area of s, the circuit equation for the coil can then be represented as follows

$$
V = RI + L\frac{dI}{dt} + \frac{Nl}{S}\int_S \frac{\partial A}{\partial t}ds
\tag{2}
$$

where R is coil resistance, L is inductance, l is the length of the conductor and I is the coil current. The field and circuit equations are coupled together as expressed.

$$
\begin{bmatrix} P_1 & P_2 \\ 0 & P_4 \end{bmatrix}\begin{Bmatrix} A \\ I \end{Bmatrix} + \begin{bmatrix} Q_1 & 0 \\ Q_3 & Q_4 \end{bmatrix}\begin{Bmatrix} \frac{dA}{dt} \\ \frac{dI}{dt} \end{Bmatrix} = \begin{Bmatrix} B \\ V \end{Bmatrix}
\tag{3}
$$

Alternatively, equation (3) can be expressed as

$$
[P]\{X\} + [Q]\left\{\frac{dX}{dt}\right\} = \{F\}
\tag{4}
$$

where $[P]$ and $[Q]$ are the coefficient matrices of equation, and $[F]$ represents the input vector. The time discretization algorithm used here is Crank Nicholson method. When rotor is at the centre of the machine, the flux density distribution within the air gap is symmetric and no net electromagnetic force is produced. By supplying the bridge currents the $p \pm 1$ flux is generated and net radial force exerts on the rotor.

The electromagnetic field interaction within the air gap of a PMSM generates an electromagnetic force wave. The radial levitation force is predominately generated by the Maxwell stress tensor, a electromagnetic force that arises at the boundary of materials with different magnetic permeabilities. The determination of this force wave around the air gap utilizes the Maxwell stress method [16].

$$
\sigma_r = \frac{1}{2\mu_o}(B_r^2 - B_t^2) \text{ and } \sigma_t = \frac{1}{\mu_o}B_rB_t
\tag{5}
$$

where μ_0 represents the air gap permeability, while B_r and B_t are the air gap flux components along radial and tangential directions. The total electromagnetic force along the x and y axes can be expressed as:

$$
\begin{aligned}
F_x &= L\int_0^{2\pi r}(\sigma_r cos\theta - \sigma_t sin\theta)d\theta \\
F_y &= L\int_0^{2\pi r}(\sigma_r sin\theta + \sigma_t cos\theta)d\theta
\end{aligned}
\tag{6}
$$

where L is length of the stator and r is the air gap radial distance.

III. WINDING STRUCTURE

The machine parameters are summarized in Table I. Electrical machines equipped with dual winding systems typically feature a primary set of windings responsible for torque generation and a secondary set that produces lateral forces across the air gap [18]. However, these machines often face limitations in power ratings. To address this issue, advanced single-winding configurations, such as bridge-configured winding (BCW), have been implemented in permanent magnet motors [17]. Fig. 1 illustrates the stator winding layout of 24 slots PMSM. In the BCW configuration, the torque-producing current is divided into two parallel paths. Each path includes a power supply at its midpoint to facilitate the generation of lateral forces. Circuit diagram for bridge configured winding is shown in the Fig. 2. The current flowing through each phase is split into two parallel paths and the voltage supply responsible for the force production is connected midway the parallel paths of each phase. The flux generated by the bridge voltage is

TABLE I
MACHINE PARAMETERS

Parameter	Value
Number of stator slots	24
Axial length	100 mm
Stator internal diameter	95.50 mm
Stator external diameter	200 mm
Rotor internal diameter	42.40 mm
Rotor external diameter	91.50 mm
Air gap thickness	1.75 mm
Band thickness	0.25 mm
Supply frequency	50 Hz
Supply voltage	237.5 V
Speed	1500 RPM

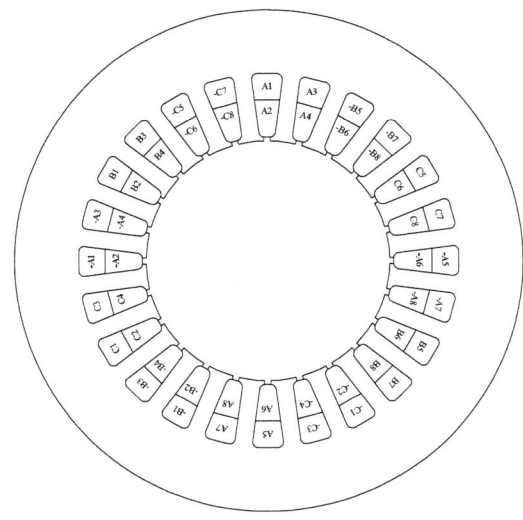

Fig. 1. Winding configuration.

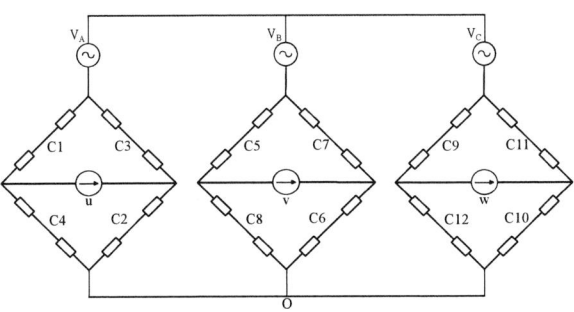

Fig. 2. Circuit diagram.

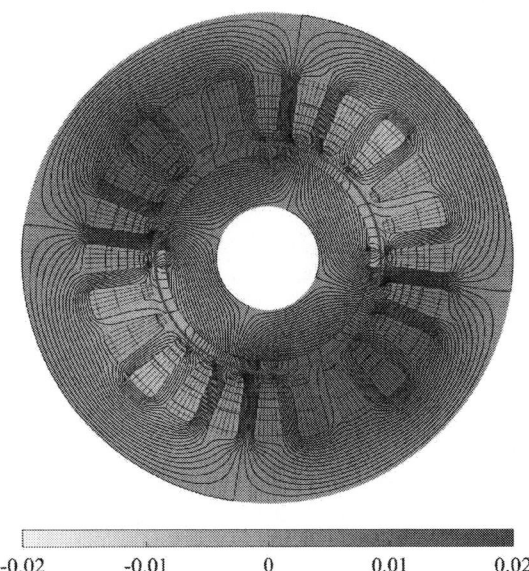

Fig. 3. Contour plot of the magnetic vector potential.

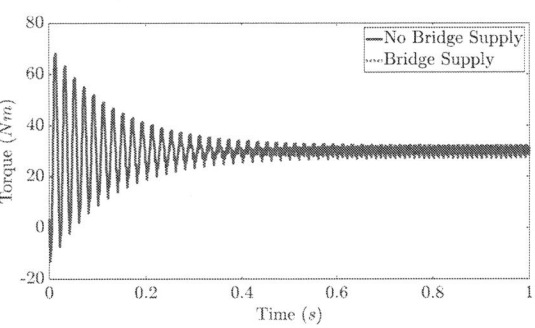

Fig. 4. Electromagnetic torque.

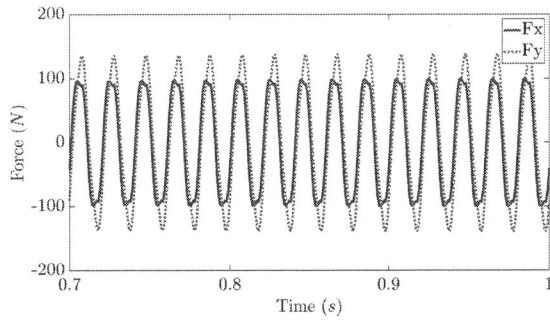

Fig. 5. Electromagnetic force due to bridge supply in phase A.

superimposed upon the magnetic flux produced by the motor or torque producing winding. The interaction between these fluxes results in variations in magnetic flux density in air gap. When the flux produced by the main supply align with the flux produced by the bridge supply, the flux density increases and a net radial force acts on the rotor [19].

IV. RESULT AND DISCUSSION

The implementation and simulations of this formulation were carried out in the MATLAB environment. The magnetic vector potential distribution across the machine's cross-sectional area at time $t = 1\ s$ is illustrated in Fig. 3. Maximum magnetic vector potential values are observed at the outer edges of the permanent magnets. Fig. 4 compares the torque production under the influence of bridge supply, revealing negligible impact on torque generation, maintaining an average torque output of $30.5\ Nm$ in both cases. The lateral force generated in x and y direction in shown in Fig. 5 when $1\ V$ bridge voltage is applied in phase A. The amplitude of the force in y direction is higher than force in x direction, reaching a peak value of $139\ N$ along y axis. This disparity in the force

production arises from the winding configuration. Fig. 6 shows the forces in x and y direction generated in the air gap when $1\ V$ bridge voltage is applied in phase B. Fig. 6 shows the force production a $1\ V$ bridge voltage is applied in phase B, the amplitude of force in y direction remains dominant over

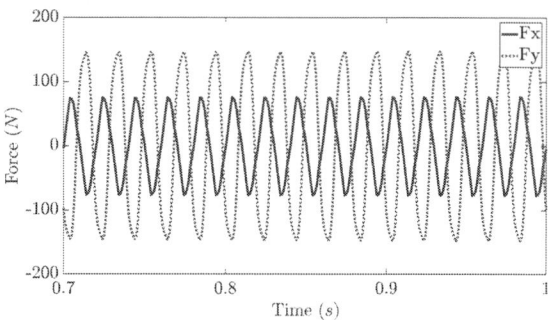

Fig. 6. Electromagnetic force due to bridge supply in phase B.

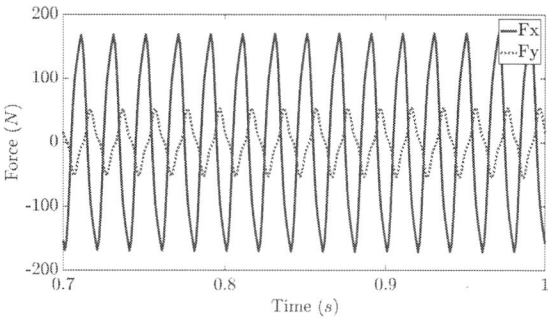

Fig. 7. Electromagnetic force due to bridge supply in phase C.

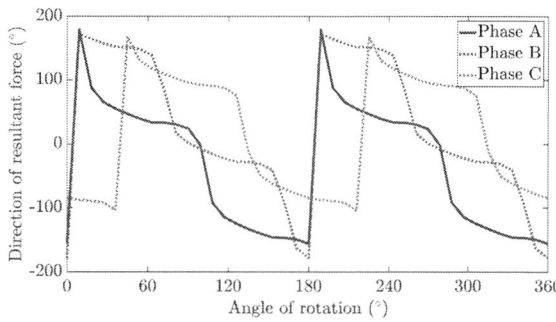

Fig. 8. Direction of resultant force.

the force in x direction, with a peak value of $148.5\ N$ along the y axis. However, the amplitude of force in x direction is marginally higher than that observed during the phase A bridge voltage application. When a $1\ V$ of bridge voltage is applied in phase C, the amplitude of force in x direction is significantly exceeds that in the y direction, with a peak value of $171.17\ N$ observed along x axis. The resultant force generated by the bridge supply in each phase independently exhibits periodic variation, repeating every half revolution, as depicted in Fig. 8, aligning with the behaviour of a four pole machine. This behaviour demonstrates the capability of bridge configured winding to force generation and potential utility in

application requiring the controllable force generation such as magnetic levitation or self bearing machine.

V. CONCLUSION

The analysis of permanent magnet synchronous machine equipped with bridge configured stator winding highlights its ability to produce the lateral force and dependence on the phase in which the bridge voltage is applied. When $1V$ bridge voltage is applied in phase A and B, the force in y direction dominates, with phase B showing slightly higher force in x direction compared to phase A. In contrast, bridge voltage in phase C results in force in x direction significantly higher than the force in y direction. This functionality highlights the bridge configured winding ability to produce the controllable forces, making it suitable for application such as self bearing machines and magnetic bearings.

REFERENCES

[1] L. Li, J. Hong, H. Wu, B. Kou, and R. Liu, "Direct and quadrature inductances measurement of the permanent magnetic linear synchronous machines," Energy Conversion and Management, vol. 52, no. 5, pp. 2282–2287, May 2011.

[2] H. Zhu, Q. Cheng, and C. Wang, "Modeling of bearingless permanent magnet synchronous motor based on mechanical to electrical coordinates transformation," Science in China Series E: Technological Sciences, vol. 52, no. 12, pp. 3736–3744, 2009, doi: 10.1007/s11431-009-0289-8.

[3] M. Ooshima, T. Kurokawa, M. Sakagami, A. Chiba, M. A. Rahman and T. Fukao, "An identification method of suspension force and magnetic unbalance pull force parameters in buried-type IPM bearingless motors," IEEE Power Engineering Society General Meeting, 2004., Denver, CO, USA, 2004, pp. 1276-1279 Vol.2, doi: 10.1109/PES.2004.1373063.

[4] C. Michioka, T. Sakamoto, O. Ichikawa, A. Chiba and T. Fukao, "A decoupling control method of reluctance type bearingless motors considering magnetic saturation," IAS '95. Conference Record of the 1995 IEEE Industry Applications Conference Thirtieth IAS Annual Meeting, Orlando, FL, USA, 1995, pp. 405-411 vol.1, doi: 10.1109/IAS.1995.530328.

[5] A. Ortiz Salazar, W. Dunford, R. Stephan and E. Watanabe, "A magnetic bearing system using capacitive sensors for position measurement," in IEEE Transactions on Magnetics, vol. 26, no. 5, pp. 2541-2543, Sept. 1990, doi: 10.1109/20.104791.

[6] A. Chiba, D. T. Power and M. A. Rahman, "Analysis of No-Load Characteristics of a Bearingless Induction Motor," in IEEE Transactions on Industry Applications, vol. 31, no. 1, pp. 77-83, January-February 1995, doi: 10.1109/28.363047.

[7] W. K. S. Khoo, R. L. Fittro, and S. D. Garvey, "AC polyphase selfbearing motors with a bridge configured winding," in 8th Int. Symp.Magnetic Bearings, Mito, Japan, Aug. 2002, pp. 47–52.

[8] A. Chiba, T. Deido, T. Fukao and M. A. Rahman, "An analysis of bearingless AC motors," in IEEE Transactions on Energy Conversion, vol. 9, no. 1, pp. 61-68, March 1994, doi: 10.1109/60.282477.

[9] Y. Okada, K. Dejima and T. Ohishi, "Analysis and comparison of PM synchronous motor and induction motor type magnetic bearings," in IEEE Transactions on Industry Applications, vol. 31, no. 5, pp. 1047-1053, Sept.-Oct. 1995, doi: 10.1109/28.464518.

[10] G. Yang, J. Chen, Y. Shen, Y. Zhou, A. Chiba and C. H. T. Lee, "Design and Analysis of Stator-Inset-PM Bearingless Slice Motor Utilizing DC Suspension Excitation," in IEEE Transactions on Industrial Electronics, doi: 10.1109/TIE.2025.3536634.

[11] Y. Lv, W. Zuo, X. Diao and H. Zhu, "Modeling and digital control system for bearingless permanent magnet synchronous motor based on magnetic energy equation," 2011 International Conference on Electrical Machines and Systems, Beijing, China, 2011, pp. 1-6, doi: 10.1109/ICEMS.2011.6073914.

[12] S. Zhang and F. L. Luo, "Direct Control of Radial Displacement for Bearingless Permanent-Magnet-Type Synchronous Motors," in IEEE Transactions on Industrial Electronics, vol. 56, no. 2, pp. 542-552, Feb. 2009, doi: 10.1109/TIE.2008.2003219.

[13] X. Sun, L. Chen and Z. Yang, "Overview of Bearingless Permanent-Magnet Synchronous Motors," in IEEE Transactions on Industrial Electronics, vol. 60, no. 12, pp. 5528-5538, Dec. 2013, doi: 10.1109/TIE.2012.2232253.

[14] W. K. S. Khoo, K. Kalita and S. D. Garvey, "Practical Implementation of the Bridge Configured Winding for Producing Controllable Transverse Forces in Electrical Machines," in IEEE Transactions on Magnetics, vol. 47, no. 6, pp. 1712-1718, June 2011, doi: 10.1109/TMAG.2011.2113377.

[15] Bastos, J.P.A., Sadowski, N. (2003). Electromagnetic Modeling by Finite Element Methods (1st ed.). CRC Press. https://doi.org/10.1201/9780203911174

[16] W. Zhu, S. Pekarek, B. Fahimi and B. J. Deken, "Investigation of Force Generation in a Permanent Magnet Synchronous Machine," in IEEE Transactions on Energy Conversion, vol. 22, no. 3, pp. 557-565, Sept. 2007, doi: 10.1109/TEC.2006.888034.

[17] W. K. S. Khoo, K. Kalita and S. D. Garvey, "Practical Implementation of the Bridge Configured Winding for Producing Controllable Transverse Forces in Electrical Machines," in IEEE Transactions on Magnetics, vol. 47, no. 6, pp. 1712-1718, June 2011, doi: 10.1109/TMAG.2011.2113377.

[18] G. Kumar, K. Kalita and K. Tammi, "Analysis of Bridge Currents and UMP of an Induction Machine With Bridge Configured Winding Using Coupled Field and Circuit Modeling," in IEEE Transactions on Magnetics, vol. 54, no. 9, pp. 1-16, Sept. 2018, Art no. 8104416, doi: 10.1109/TMAG.2018.2854666.

[19] L. Yu, Z. Zhang, Y. Shi and W. Lu, "Modeling and Analysis of Suspension Force of a New Bearingless Reluctance Machine With Independent DC Bias Winding," in IEEE Transactions on Magnetics, vol. 54, no. 11, pp. 1-5, Nov. 2018, Art no. 8204605, doi: 10.1109/TMAG.2018.2840325.

979-8-3315-3013-6/25 $31.00 © 2025 IEEE

Enhanced Half-Bridge Sub-module Modular Multilevel Converter and its Capacitor Voltage Balancing Control

Akshaya D. Bonde[1], Pradyumn Chaturvedi[2], Vijay B. Borghate[2]
[1]*Student Member, IEEE,* [2]*Senior Member, IEEE*
Visvesvaraya National Institute of Technology, Nagpur, India

Abstract—The Modular Multilevel Converter (MMC) is widely used in high-voltage applications due to its modularity, scalability, and superior power quality. This paper presents an Enhanced Half-Bridge Submodule (HBSM) MMC with a hybrid submodule structure, where 50% of HBSMs per arm operate at V_C and the rest at $2V_C$. This design reduces the number of submodules by 33.33%, leading to fewer switching devices, capacitors, and voltage sensors per arm. A capacitor voltage balancing algorithm is proposed to ensure stable operation. A comparative analysis with traditional HBSM and Full-Bridge Submodule (FBSM) MMCs is conducted. Simulation results in MATLAB/Simulink validate the proposed balancing strategy and demonstrate the Enhanced HBSM MMC's performance under various conditions.

Index Terms—Modular Multilevel Converter (MMC), Hybrid MMC, Pulse Width Modulation (PWM), Closed Loop Control, Sub-module Capacitor Voltage Balancing, Power Loss Analysis.

I. INTRODUCTION

Since their introduction, Modular Multilevel Converters (MMCs) have gained significant importance in power electronics by enabling high-power conversion with modular scalability [1]. MMCs are commonly employed in high-voltage direct current (HVDC) transmission, medium-voltage direct current (MVDC) grids [2], renewable energy integration, and motor drives applications, etc. The modular structure of MMC allows better multilevel voltage waveform generation, lowering the total harmonic distortion and thus improving power quality. Various modifications have been introduced to optimize MMC operation based on specific application requirements, incorporating different submodule (SM) topologies [3]. The conventional Half-Bridge Submodule (HBSM) MMC, as discussed in [4], has been widely adopted for its efficient operation and reduced conduction losses. Apart from HBSM topology, different SM topologies have been introduced in different literatures [3]. Different SM topologies [5] have been explored, including Full-Bridge (FB), Three-Level Cross-Connected (TLCC), Clamped Double (CD), Modified Switched Capacitor (MSC), etc. [6]- [7]. However, this increases control complexity due to a higher component count and intricate control requirements [8]- [9]. Apart from this, hybrid MMC configurations have also been proposed to balance performance and device count, such as a 50% HBSM and 50% FBSM per arm configuration

[10] or a 66.66% FBSM and 33.33% HBSM per arm structure [11]. However, these approaches introduce substantial control complexity. Among all the SM topologies, HBSM remains the simplest, offering lower control complexity and reduced semiconductor device count. Nonetheless, achieving high-resolution voltage waveforms with HBSM requires a large number of SMs, which can increase cost and implementation complexity [12].

To address these challenges while maintaining the simplicity of the HBSM topology, this paper introduces an enhanced HBSM MMC, which retains the fundamental structure of the traditional MMC. Therefore, the proposed design reduces the number of SMs without compromising output voltage levels by utilizing a voltage-scaling approach. Unlike the works that focus on Hybrid or Mixed-SM MMCs [13], this approach retains the simplicity of the conventional HBSM MMC while offering substantial cost and complexity reductions. In the proposed topology, 50% of the SMs in each arm operate with a SM capacitor voltage of V_C, while the remaining 50% operate at twice the SM capacitor voltage, i.e., $2V_C$. This voltage-scaling approach effectively reduces the total number of SMs required per phase by 33.33%, thus lowering the number of semiconductor switches, capacitor voltage sensors. This reduces the overall system complexity.

Maintaining capacitor voltage balance across the SMs is critical for stable MMC operation. Conventional sorting-based capacitor voltage balancing technique is developed in [14] for traditional HBSM MMC. Advanced model predictive control (MPC)-based methods [15] improve dynamic performance but introduce higher control complexity. Different SM capacitor voltage balancing approaches are reviewed in [16]- [17]. For enhanced HBSM MMC, capacitor voltage balancing becomes more challenging, as SMs operate at different voltage levels. The author has proposed a new voltage balancing technique for the proposed MMC configuration. A new SM capacitor voltage balancing algorithm needs to be developed, ensuring equalized capacitor voltages under both steady-state and dynamic operating conditions. The proposed algorithm dynamically regulates capacitor voltages, maintaining stability across all the SMs while effectively responding to load variations and modulation index changes. The pulse width modulation technique in [18] is extended for the proposed MMC configuration.

979-8-3315-3013-6/25 $31.00 © 2025 IEEE

Key Contributions of this work are: In enhanced HBSM MMC, there is a 33.33% reduction in no. of switching devices, no. of SM capacitors, no. of voltage sensors per arm compared to traditional HBSM MMC. Therefore, minimizes complexity and cost while maintaining the same output voltage waveform quality. Furthermore, the SM capacitor voltage balancing algorithm has less control complexity. It is easy to implement technique and offers reduced switching trasitions during SM capacitor voltage balancing. Its effectiveness has been validated through MATLAB/Simulink under varying modulation indices and various load transitions. The power loss analysis based on switching and conduction losses and its comparison with traditional HBSM MMC and traditional FBSM MMC has also been provided.

This paper is structured as follows: Section I briefly reviews the existing MMC configurations, SM topologies and capacitor voltage balancing strategies. Section II details the working principle of the proposed enhanced HBSM MMC. Section III introduces the proposed SM capacitor voltage balancing algorithm. Section IV provides simulation studies of both in MATLAB/Simulink along with the power loss analysis. Section V concludes the study. The proposed enhanced HBSM MMC offers a promising alternative to conventional MMC topologies by significantly reducing hardware complexity and cost while preserving performance.

II. Proposed Enhanced HBSM MMC

The conventional MMC topology consists of two arms/leg, forming a complete phase leg of the converter. Each arm comprises multiple cascaded HBSMs along with an arm inductor (L_{arm}), which serves to limit circulating currents within and between the converter legs. The HBSMs are capable of generating the two levels; 0 and $+V_C$. In conventional MMC configuration, for n ($=N+1$) discrete output phase voltage levels, $2N$ HBSMs/leg need to be incorporated, where N is the number of SMs per arm. The capacitor voltage of each SM must be regulated at $V_C=V_{dc}/N$. As the level increases, the number of SMs/arm increases resulting in more switch count, capacitors and their associated voltage sensors for its voltage balancing control.

The proposed enhanced HBSM MMC retains the fundamental structure of the conventional MMC but introduces a capacitor voltage distribution among SMs. Specifically, within each arm, 50% of the SMs ($N/2$SMs) maintain a capacitor voltage of V_C; while the other 50% SMs operate at $2V_C$. Fig. 1 shows the complete 3-ϕ n-level enhanced HBSM MMC. The SM capacitor voltage V_C can be calculated as $V_C=V_{dc}/(N+N/2)$. This results in output phase voltage levels to be improved from $n=N+1$ to $n=N+(N/2)+1$ i.e. $n=1.5N+1$.

$$V_{C_1} = \frac{V_{dc}}{1.5N}; V_{C_2} = 2 \times V_{C_1}; V_{phpk} = \pm\frac{V_{dc}}{2} \quad (1)$$

Compared to traditional HBSM MMC, the proposed enhanced HBSM MMC effectively reduces the total number of SMs required per phase by 33.33%, thus reducing the number of semiconducting switching devices by 33.33% while still achieving the same number of output voltage levels as a tradi-

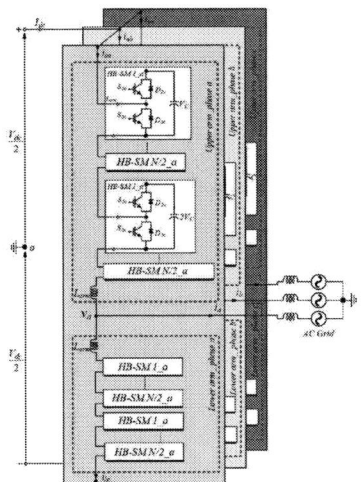

Fig. 1: 3-ϕ n-level Proposed enhanced HBSM MMC

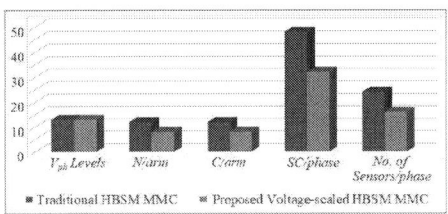

Fig. 2: Comparison of traditional HBSM MMC with proposed enhanced HBSM MMC based on V_{ph} Levels, N/arm, C/arm, switch count/phase (SC) and no. of Sensors/phase

tional HBSM MMC. The detailed comparison of the proposed enhanced HBSM MMC with traditional HBSM MMC [15] and with FBSM MMC [12] is tabulated in Table I (for better understanding, 13-level MMC example is considered). Table I also presents the % reduction in device count in proposed MMC structure compared to the traditional MMC structure. Consequently, as demonstarted in the barchart in Fig. 2, the number of capacitors and thus the voltage sensors required per phase is also reduced by 33.33%, leading to reduced complexity and lower implementation cost. The SM capacitor voltage (V_C) and peak ac output phase voltage (V_{phpk}) is given by Eq. (1).

TABLE I: Comparative analysis of different MMC configurations

Parameters (for 3-ϕ n-level)	HBSM MMC [4]	FBSM MMC [9]	Proposed Enhanced HBSM MMC
Input DC link voltage	V_{dc}	V_{dc}	V_{dc}
Apparent Power	S	S	S
Max. voltage stress per device	V_{dc}/N	V_{dc}/N	$2V_{dc}/N$
Output voltage levels, n	$N+1=13$	$N+1=13$	$1.5N+1=13$
No. of SMs/arm	$N=12$	$N=12$	$N=8$
Total no. of SMs	$6N=72$	$6N=72$	$6N=48$
Total no. of Capacitors	$6N=72$	$6N=72$	$6N=48$
Total no. of IGBTs	$12N=144$	$24N=288$	$12N=96$
Output voltage range	$\pm V_{dc}/2$	$\pm V_{dc}/2$	$\pm V_{dc}/2$

A. Modulation technique used for enhanced HBSM MMC

The single carrier PWM (SCPWM) technique in [18] is extended to generate the PWM pulses for proposed MMC configuration. Fig. 3 illustrates the modified SCPWM, wherein the modulating signal is initially amplified to $N+(N/2)$. The necessary adjustment of amplified sine signal involves employing remainder function, and is then compared with a carrier waveform having frequency f_{cr}. This comparison yields a pulsating signal ranging between 0 and 1, subsequently added to the floored signal of the amplified sine wave. The resulting waveform is the lower arm reference (U_l). By subtracting the resulting waveform from $1.5N$, the reference waveform for upper arm (U_u) is obtained. This reference waveforms are then used as a reference in voltage balancing algorithm, generating arm voltages (V_u & V_l) ranging from 0 to $1.5NV_C$. Therefore, the output phase voltage (V_{ph}) waveform ranges from $\pm V_{dc}/2$.

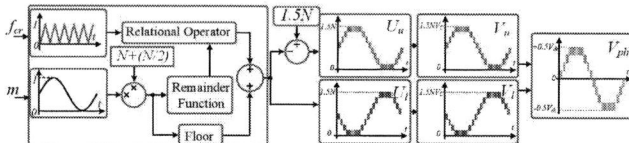

Fig. 3: Extended single carrier pulse width modulation (Extended SCPWM) technique

III. PROPOSED SM CAPACITOR VOLTAGE BALANCING

For stable operation of MMC, ensuring proper voltage balancing across the SM capacitors is critical. In traditional HBSM configuration, capacitor charges when the arm current is positive while discharging takes place when the arm current is negative. Unlike the traditional MMC, which requires uniform capacitor voltage regulation at V_C, the enhanced HBSM MMC necessitates distinct voltage balancing for two categories of SMs: one subset ($N/2$) is maintained at V_C, whereas the remaining ($N/2$) operates at $2V_C$. Since, some SMs are configured to V_C and some are to $2V_C$ at their SM terminals, direct switching for generating even and odd voltage levels can be achieved by sorting the SM capacitor voltages into two groups: one comprising SMs with V_C and the other with $2V_C$. A grouping-based strategy is proposed, offering a simpler and more practical implementation. This method relies on real-time sensing of SM capacitor voltages and arm currents. The proposed SM capacitor voltage balancing algorithm for proposed enhanced HBSM MMC is shown in flowchart Fig. 4. The SMs designated for V_C operation are grouped in matrix $[A]$, while those generating $2V_C$ are grouped in matrix $[B]$. For an MMC with N SMs, let H_1, H_2, ... $H_{N/2}$ represent the SMs generating V_C, grouped in $[A]$.

Similarly, $H_{N/2+1}$, $H_{N/2+2}$, ..., H_N denote the SMs generating $2V_C$, grouped in $[B]$. Further, two consecutive SMs from the $N/2$ SMs operating at V_C are paired forming X_1, X_2 .... $X_{N/2}$ and these pairs along with the SMs generating $2V_C$, are structured into matrix $[C]$. This grouping technique ensures improved voltage balancing. The three matrices are sorted and based on the reference arm voltage generated (U_u & U_l) using the PWM technique, the voltage balancing process

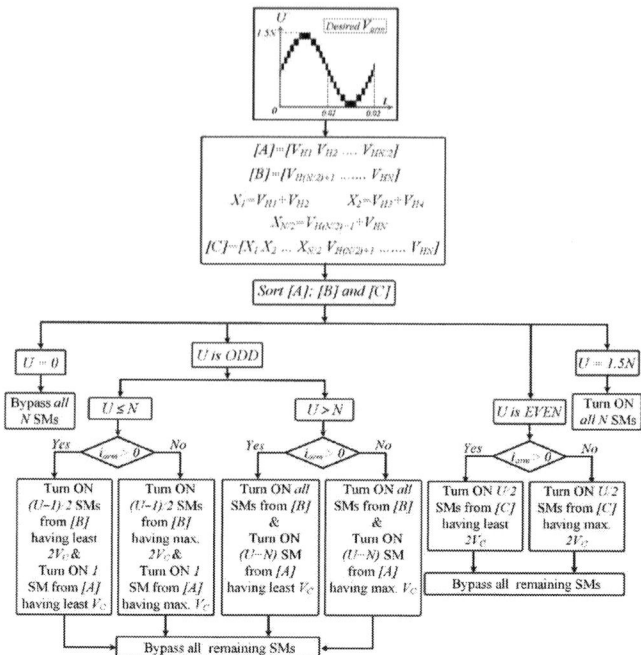

Fig. 4: Proposed SM capacitor voltage balancing algorithm

TABLE II: Manufacturer Ratings of SKM145GB066D [20] & SKM200GB12V [21] Modules

Device	Parameters	SKM145GB066D	SKM200GB12V
		[20] Values	[21] Values
IGBT	on-state resistance, R_S	8mΩ	7.60 mΩ
	collector emitter voltage, V_{CE0}	0.9V	0.98V
	energy required during turn on, E_{on}	8.5mJ	14mJ
	energy required during turn off, E_{off}	5.5mJ	22mJ
Diode	on-state resistance, R_D	4mΩ	6.8mΩ
	forward voltage, V_{F0}	1.0V	1.1V
	reverse recovery energy loss, E_{rr}	3.5mJ	13mJ

is categorized into five operational modes. The SM capacitor voltage balancing algorithm for both upper and lower arm is same. Thus, the generalized reference U instead of U_u and U_l is considered for further explanation. The control strategy is as follows: When $U=0$, all SMs are bypassed. When $U=1.5N$, $N/2$ SMs with V_C and $N/2$ SMs with $2V_C$ are inserted. For intermediate values of U, the insertion strategy depends on the parity of U and the polarity of arm current, i_{arm}. For even U, if the i_{arm} is positive, insert $U/2$ SMs or SM pairs from matrix $[C]$ with the lowest voltage of $2V_C$, and if the i_{arm} is negative, insert $U/2$ SMs (i.e. from $V_{H(N/2)+1}$ .... V_{HN}) or SM pairs (i.e. from X_1, X_2, ... $X_{N/2}$) from $[C]$ having maximum $2V_C$ and bypass the remaining SMs. For odd U with $U \leq N$, during positive i_{arm}, insert $U/2$ SMs from $[B]$ with least $2V_C$ and one SM from $[A]$ having minimum V_C. Similarly, during negative i_{arm}, insert $U/2$ SMs from $[B]$ with maximum $2V_C$ and one SM from $[A]$ having maximum V_C. For odd U, if $U>N$, during positive i_{arm}, insert all SMs from $[B]$ with $2V_C$ and $(U-N)$ SMs from $[A]$ with the lowest capacitor voltage V_C, and during negative i_{arm}, insert all SMs from $[B]$ with $2V_C$ and $(U-N)$ SMs from $[A]$ having maximum V_C. Bypass the remaining SMs. This approach ensures optimal voltage balancing while minimizing capacitor voltage ripples across the SMs.

IV. SIMULATION STUDIES

A 3-ϕ enhanced HBSM MMC system is modelled and simulated in MATLAB/Simulink environment. The simulation is performed for a DC link voltage of 6kV. In case of traditional HBSM MMC, the maximum voltage stress the semiconducting devices experience is V_{dc}/N. So, the number of SMs/arm are $N=12$, resulting the SM capacitor voltage of $V_{dc}/N=500$V. Therefore, the traditional HBSM MMC generates a 13-level ($n=N+1=13$) output phase voltage with the peak voltages at $\pm V_{dc}/2=\pm 3$kV.

However, in proposed enhanced HBSM MMC, the maximum voltage stress the semiconducting devices experience is $2V_{dc}/1.5N$. Thus, for this configuration, $N/2=4$ SMs are dedicated to generating a voltage of V_C, and $N/2=4$ SMs are are dedicated to generating a voltage of $2V_C$. Thus, the total SMs/arm are $N=8$. This arrangement increases the number of output voltage levels by 0.5 compared to the traditional HBSM MMC, resulting in a 13-level output phase voltage with peak voltage of $\pm V_{dc}/2=\pm 3$kV. The arm inductor, $L_{arm}=5$mH and the load of 0.846 lagging pf is connected at output terminals with fundamental frequency of 50Hz. The proposed SM capacitor voltage balancing algorithm (shown in Fig. 3) is employed to regulate the SM capacitor voltages within the 3-ϕ enhanced HBSM MMC (depicted in Fig. 2). The average switching transitions of switches in all HBSMs of upper and lower arm of traditional HBSM MMC with traditional voltage balancing and proposed enhanced HBSM MMC with proposed voltage balancing algorithm are presented in Table III.

A. Power Loss Analysis

Due to the uniform arm structure of MMC, power losses in each arm are identical, and SM currents remain consistent during normal operation. Only switching and conduction losses are considered here. Power loss for a single SM is calculated using an averaged approach from [19] and is extended to all SMs to determine the total power loss in the 3ϕ MMC. The necessary parameters for this calculation are taken from the Si-IGBT module SKM145GB066D (600V, 195A) datasheet in [20] and SKM200GB12V (1200V, 311A) datesheet in [21], tabulated in Table II.

Semiconductor devices consume energy during switching, leading to continuous losses in converter operation. In MMC, these losses depend on both the modulation and SM capacitor voltage balancing techniques. Switching losses are calculated using Eqs. (2) & (3).

$$P_S = \{2 \times (E_{on} + E_{off}) + 2 \times E_{rr}\}f_S \quad (2)$$

$$P_{S_{arm}} = N \times P_S ; P_S = 6 \times P_{S_{arm}} \quad (3)$$

where, P_S represents the switching losses of one SM. $P_{S_{arm}}$ and P_S represent the switching losses per arm and total switching losses of 3-ϕ Hybrid MMC, respectively.

Continuous current flow in semiconductor devices causes a voltage drop, leading to conduction losses. In an SM,

these losses result from all active IGBTs and diodes during operation.

$$P_{sw} = \{i_{arm_{avg}} \times V_{CE0}\} + \{i_{arm_{rms}}^2 \times R_S\} \quad (4)$$

$$P_d = \{i_{arm_{avg}} \times V_{F0}\} + \{i_{arm_{rms}}^2 \times R_D\} \quad (5)$$

$$P_C = (2 \times P_{sw}) + (2 \times P_d) \quad (6)$$

$$P_{C_{arm}} = N \times P_C ; P_C = 6 \times P_{C_{arm}} \quad (7)$$

$$P_{Loss} = P_S + P_C \quad (8)$$

The average ($i_{arm_{avg}}$) and RMS ($i_{arm_{rms}}$) arm currents determine conduction losses in MMC. Conduction losses in IGBT (P_{sw}) and diodes (P_d) are calculated using Eqs. (4) to (7), with total power loss (P_{Loss}) calculated by Eq. (8).

Switching losses are proportional to switching transition frequency, meaning fewer transitions reduce losses. For comparative analysis, a detailed comparison of the proposed MMC is conducted against traditional HBSM MMC and FBSM MMC, all operating for the same input dc link voltage. The results are summarised in Table III. Proposed enhanced HBSM MMC with proposed voltage balancing technique achieves a 25.36% reduction in average switching transitions (f_{st}), lowering switching losses (P_S) by 16.794%. A 33.031% decrease in conduction losses (P_C) lead to a 18.494% reduction in total power loss, improving efficiency by 1.215% compared to traditional HBSM MMC. (During power loss analysis, only major losses i.e. conduction and switching losses are considered.) Table IV details this comparison, highlighting the impact of proposed MMC configuration and the proposed SM capacitor voltage balancing algorithm.

B. Steady State & Dynamic Condition Analysis

To demonstrate the effectiveness of the proposed enhanced HBSM MMC and the associated proposed SM capacitor voltage balancing algorithm, both steady state and dynamic conditions are considered. The performance of the converter is analysed under different dynamic scenarios applied to the three phase MMC.

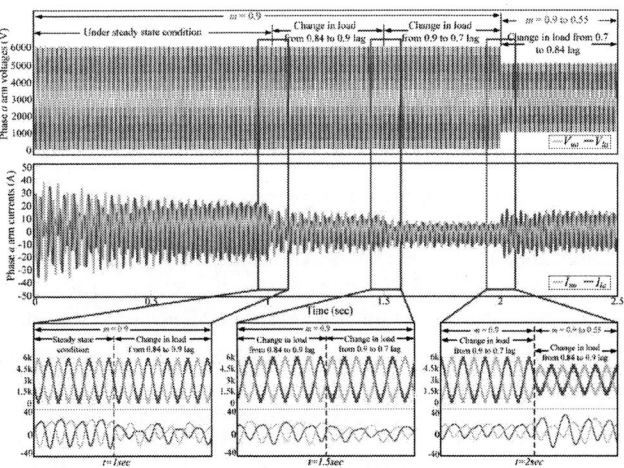

Fig. 5: Phase a arm voltages and currents & their zoomed views showcasing at $t=1$sec, $t=1.5$sec & $t=2$sec due to variations in load and/or m

TABLE III: Comparative analysis of simulation studies

MMC Configurations (for 13-level)	HBSM MMC [4]	FBSM MMC [9]	Proposed Voltage Scaled HBSM MMC
Voltage Balancing (VB) techq.	Traditional VB	Traditional VB	Proposed VB
N per arm	12	12	8
V_{dc}	6kV	6kV	6kV
V_C	500V	500V	500V, 1000V
total no. of current sensors	2	2	2
total no. of voltage sensors	**72**	72	**48**
$v_{u_{pk}}$ ($v_{u_{rms}}$)	5406V (3822V)	5407V (3822V)	5407V (3823V)
%THD in v_{arm}	11.90%	11.90%	11.91%
(i_{rms}) $i_{u_{rms}}$, $i_{l_{rms}}$	29.32A, 29.32A	21.2A, 21.2A	29.11A, 29.22A
V_{ph} range	±3kV	±3kV	±3kV
$V_{ph_{pk}}$ ($V_{ph_{rms}}$)	5.38kV (3.81kV)	5.37kV (3.79kV)	5.36kV (3.79kV)
%THD in V_{ph}	11.80%	11.71%	11.72%
$I_{ph_{pk}}$ ($I_{ph_{rms}}$)	82.81A (58.56A)	82.53A (58.36A)	82.49A (58.33A)
%THD in I_{ph}	0.58%	0.56%	0.55%
switching frequency, f_S	5kHz	5kHz	5kHz
avg. switching transition freq. f_{st_u}, f_{st_l}	4.145kHz, 4.142kHz	4.133kHz, 4.137kHz	3.082kHz, 3.109kHz
P_S per arm	1.722kW	3.471kW	1.432kW
Total P_S	**10.332kW**	20.83kW	**8.596kW**
P_C per arm	201.157W	402.314W	134.712W
Total P_C	**1.206kW**	2.413kW	**0.808kW**
Total Power loss, P_{Loss}	**11.538kW**	23.244kW	**9.405kW**
% Efficiency	**93.50%**	87.73%	**94.65%**

TABLE IV: Comparative analysis between Traditional & Proposed MMC Configuration

MMC	P_{dc} (kW)	%reduction in f_{st}	%Reduction in P_S	%Reduction in P_C	%Reduction in %P_{Loss}	Improvement in %Efficiency
Traditional HBSM MMC [4]	166.25	25.36%	16.794%	33.031%	18.494%	1.215%
Proposed Enhanced HBSM MMC	166.25					

At t=1sec, load is changed from a *pf* of 0.84 lag to 0.9 lag, with m=0.9. At t=1.5sec, the load is further modified from 0.9 lag to 0.7 lag with constant m=0.9. Subsequently, at t=2sec, the load is changed again from 0.7 lag back to 0.846 lag with a reduction in m from 0.9 to 0.55. The results in Fig. 5 illustrate that, under steady-state conditions, the upper and lower arm voltages of phase *a* vary between *0* and *1.5NV$_C$* i.e., 0 to 6kV. As m varies, the voltage amplitude decreases, and the number of levels reduces from 13 to 9. This behaviour is also observed in the other two phases. At t=0sec & during sudden load and/or m change, the initial arm currents exhibit significant variation & requires two to four cycles to be stable to achieve capacitor voltage equilibrium. Over time, the peak arm current stabilizes at approximately 22.5A. As the load changes, the effective resistance of the circuit shifts, causing adjustments in the arm current to maintain voltage balance. Fig. 4 shows the variations in arm voltages and currents for phase *a* during these load and m changes, along with detailed zoomed-in views for the specified time intervals.

Likewise, Fig. 6 presents the corresponding variations in the 3-ϕ output voltages and currents in response to load

and m changes, including zoomed views during the transition periods. It is evident that the output phase voltages fluctuate with changes in m. Under steady state conditions, the 13-level output voltages vary between ±3kV, while as m decreases from 0.9 to 0.55, the output voltage levels reduce to 9-level vary between ±2kV. These variations in m and load are reflected in the corresponding changes in the output currents. Under steady state conditions, the peak output current is 82.5A, whereas a reduction in peak current can be observed as the load and/or m changes. Finally, Fig. 7 displays the capacitor voltages of upper and lower arm SMs of phase *a*, where the voltages of SMs V_{H1} to V_{H4} SMs of both the arms are balanced at 500V, while the voltages of SMs V_{H5} to V_{H8} SMs are balanced at 1kV. The ripple content in SM capacitor voltages increases at the instant of load as well as m variations but remains within the acceptable limits as per *IEEE* standards. The proposed SM capacitor voltage balancing algorithm effectively manages the SM capacitor voltages even under unbalanced load conditions and m variations. This demonstrates the robust performance and reliability of the proposed SM capacitor voltage balancing algorithm.

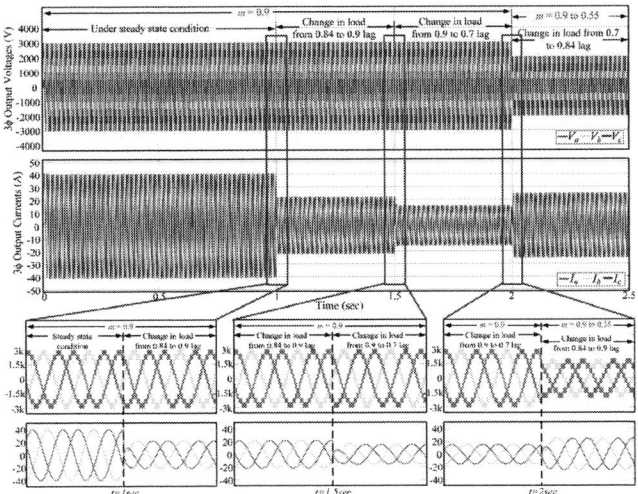

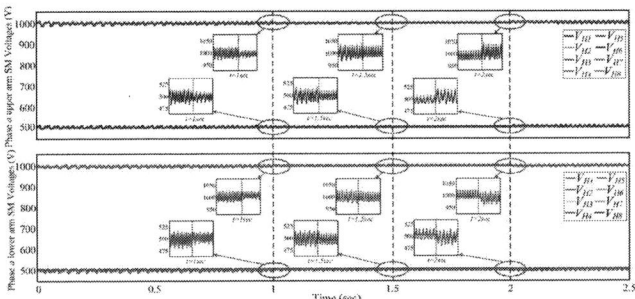

Fig. 6: 3-ϕ output voltages and currents & their zoomed views showcasing at t=1sec, t=1.5sec & t=2sec due to variations in load and/or m

Fig. 7: SM capacitor voltages of upper and lower arms of phase a and their zoomed views at load and/or m changing intervals at t=1sec, t=1.5sec & t=2sec

V. CONCLUSION

The proposed Enhanced HBSM MMC maintains the same operational performance as a conventional HBSM MMC while reducing the number of submodules, capacitors, and voltage sensors per arm by 33.33%. Its performance has been thoroughly assessed under steady-state and dynamic conditions. The developed capacitor voltage balancing algorithm ensures stable operation not only in steady-state but also during transients such as load variations and changes in m. MATLAB simulations at 6kV confirm the effectiveness of the balancing strategy, along with an analysis of power loss and dynamic performance. Additionally, the 33.33% reduction in switch count simplifies circuit design, lowers costs, and reduces conduction losses, thereby improving overall efficiency.

REFERENCES

[1] M. Kurtoğlu, F. Eroğlu, A. O. Arslan, and A. M. Vural, "Recent contributions and future prospects of the modular multilevel converters: A comprehensive review," Int. Trans. Electr. Energ. Syst., vol. 29, p. e2763, 2019

[2] G. Abeynayake, G. Li, T. Joseph, J. Liang and W. Ming, "Reliability and Cost-Oriented Analysis, Comparison and Selection of Multi-Level MVdc Converters," in IEEE Trans. on PD, vol. 36, no. 6, pp. 3945-3955, Dec. 2021

[3] S. Ali, Z. Ling, K. Tian and Z. Huang, "Recent Advancements in Submodule Topologies and Applications of MMC," in *IEEE Journal of Emerging and Selected Topics in PE*, vol. 9, no. 3, pp. 3407-3435, June 2021

[4] A. Lesnicar and R. Marquardt, "An innovative modular multilevel converter topology suitable for a wide power range," *IEEE Bologna Power Tech Conf. Proceedings*, Bologna, Italy, pp. 272 pp. vol.3,2003.

[5] Raju, M. N., Sreedevi, J., Mandi, R. P., and Meera, K. S., "Modular multilevel converters technology: a comprehensive study on its topologies, modelling, control and applications," IET PE, vol. 12, no. 2, pp. 149–169, 2019

[6] M. A. Perez, S. Bernet, J. Rodriguez, S. Kouro, R. Lizana, "Circuit Topologies, Modeling, Control Schemes & Applications of Modular Multilevel Converters," in *IEEE Trans. on Power Electronics*, vol. 30, no. 1, pp. 4-17, Jan. 2015

[7] A. D. Bonde, P. Chaturvedi and V. B. Borghate, "Capacitor Voltage Balancing of New Hybrid Modular Multilevel Converter Topology," 2023 IEEE 3rd International Conf. on Smart Tech. for Power, Energy and Control, pp. 1-6, 2023

[8] A. Dekka, B. Wu, R. L. Fuentes, M. Perez, and N. R. Zargari, "Evolution of Topologies, Modeling, Control Schemes, and Applications of Modular Multilevel Converters," *IEEE J. Emerg. Sel. Top. Power Electron.*, vol. 5, no. 4, pp. 1631–1656, Dec. 2017.

[9] G. P. Adam and I. E. Davidson, "Robust and generic control of full-bridge modular multilevel converter high-voltage dc transmission systems," IEEE Trans. Power Del., vol. 30, no. 6, pp. 2468–2476, Dec. 2015.

[10] A. D. Bonde, P. Chaturvedi, V. B. Borghate and S. K. Patro, "Reduced Switching Frequency (RSF) Voltage Balancing Technique for 21-Level Hybrid MMC," 2022 IEEE Int. Conf. on Power Electronics, Drives and Energy Systems, pp. 1-6, 2022

[11] R. Zeng, L. Xu, L. Yao and B. W. Williams, "Design and Operation of a Hybrid Modular Multilevel Converter," in IEEE Trans. on Power Electronics, vol. 30, no. 3, pp. 1137-1146, March 2015

[12] A. Farghly, A. Elserougi, A. Abdel-Khalik, and R. Hamdy, "A hybrid mixed-cells DC-DC modular multilevel converter with balanced arm energy and DC fault blocking capability for high voltage direct current systems," Int. J. Circ. Theor. Appl., vol. 52, no. 9, pp. 4556–4581, 2024

[13] R. Feldman et al., "A hybrid modular multilevel voltage source converter for HVDC power transmission," *IEEE Trans. Industrial Applications*, vol. 49, no. 4, pp. 1577–1588, Jul./Aug. 2013.

[14] J. Zhang, C. Jiang, and Y. Han, "Modular multilevel converter composite submodule topology and control," J. Eng., vol. 2019, pp. 2643–2648, 2019

[15] Y. Zhang, C. Luo, W. Yi, B. Luo, and Z. Cheng, "Model predictive control of modular multilevel converter based on fixed switch state set for offshore wind power," J. Eng., vol. 2023, p. e12259, 2023

[16] L. Angquist, A. Antonopoulos, D. Siemaszko, K. Ilves, M. Vasiladiotis and H. Nee, "Open-Loop Control of Modular Multilevel Converters Using Estimation of Stored Energy," in *IEEE Trans. on Industry Applications*, vol. 47, no. 6, pp. 2516-2524, Nov.-Dec. 2011

[17] M. A. Perez, S. Ceballos, G. Konstantinou, J. Pou and R. P. Aguilera, "Modular Multilevel Converters: Recent Achievements and Challenges," *IEEE Open Journal Industrial Electronics Society*, vol. 2, pp. 224–239, Feb. 2021.

[18] A. D. Bonde, P. Chaturvedi and V. B. Borghate, "Fault-Tolerant Operation of Hybrid Modular Multilevel Converter for MVDC Applications," *IEEE International Conference on Power Electronics, Drives and Energy Systems*, Mangalore, India, pp. 1-6, 2024.

[19] A. M. Shende, A. Gupta, S. K. Patro and M. A. Chaudhari, "Fault Tolerant Si-SiC based Hybrid Modular Multilevel Converter with Enhanced Efficiency," IEEE International Conference on Smart Tech. for Power, Energy and Control, India, pp. 1-6, 2021.

[20] Semikron International, "SKM200GB12V IGBT Module Datasheet," rev. 4, Mar. 2011.

[21] Semikron International, "SKM145GB066D IGBT Module Datasheet," rev. 02, Feb. 2014.

Parameter Identification and Controller Design of DAB Converter for Off-board EV Charging Applications

Aditya Kulkarni
Electrical Engineering Department
VNIT,Nagpur
Nagpur, India
iamadityakul123@gmail.com

Shashi Kumar Kondoju
Electrical Engineering Department
VNIT,Nagpur
Nagpur, India
Shashikumar.kondoju@gmail.com

Pradyumn Chaturvedi
Electrical Engineering Department
VNIT,Nagpur
Nagpur, India
pradyumn.c@eee.vnit.ac.in

Abstract—**Recent trends in Electric Vehicle (EV) charging infrastructure have shown development in the part of power electronic topology which is being used. Since most of the EVs are adopted to off-board charging mode because of the Reduced Vehicle Weight and Cost, Faster charging times, better thermal management, Grid integration and Smart charging there is a need for converter which can have High Power Capability. This paper deals with the design, modelling and the control of one prominent topology which is a Dual Active bridge (DAB) Converter. However, considering the charging and discharging mode of EVs which when considered for the bidirectional operation of converter in case of Vehicle-To-Grid (V2G) and Grid-To-Vehicle (G2V); the controller needs to be designed which can have high loop bandwidth, faster response time and better adoption to the uncertainties in the system such as sudden load variation, or transition from Constant Current (CC) to Constant Voltage (CV) mode. This paper addresses the issues and parameters which need to be considered while designing such controllers considering the Phase-Shift Modulation (PSM) technique. The performance of parameter identification and controller design is validated by simulation and hardware results.**

Keywords— **Dual Active Bridge, Single Phase Shift control, ZVS, Modelling, controller design.**

I. INTRODUCTION

There is now a greater need for effective, high-power charging infrastructure due to the quick uptake of electric cars (EVs). By removing the weight and size restrictions that come with on-board chargers, off-board EV chargers are especially beneficial for high-power charging applications. This enables more effective power conversion and thermal management [1]. For these chargers to operate safely, the power electronics converters must have galvanic isolation, high efficiency, and bidirectional power flow [2].
High efficiency, built-in soft switching, and bidirectional operation from vehicle to grid (V2G) and grid to vehicle (G2V) are only a few advantages of the DAB converter. Its modular and symmetrical design guarantees dependability and offers high power density. Galvanic isolation is another feature that makes the converter a popular option for applications with multiple inputs and outputs [3]. The DAB converter is the best option for subsequent-generation EV charging infrastructure. As shown in Fig. 1. the circuit diagram of DAB converter is described as follows:

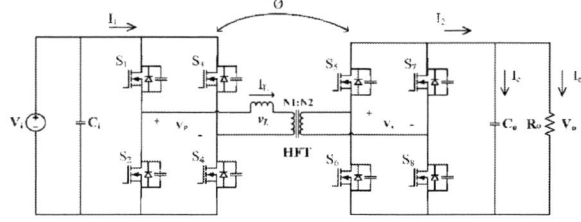

Fig. 1. Circuit diagram of Dual active Bridge converter.

A high-frequency transformer (HFT) connects two active bridges in the DAB converter's operating principle Fig.1.The secondary bridge delivers the output to the load, while the primary bridge receives the DC voltage. Phase shift modulation ($\emptyset$) is used to regulate the voltage between the transformer's primary and secondary sides. Fig.2. shows the matching waveforms of voltage and current, separated into four intervals (t_1, t_2, t_3, t_4).
V_i – Input DC voltage.
S_1, S_2, S_3, S_4 – Switches on primary side.
S_5, S_6, S_7, S_8 – Switches on secondary side.
V_p - Output voltage of primary bridge.
V_s -Output voltage of secondary bridge. (Voltage across secondary side of transformer).
I_L - Inductor current.
C_i, C_o - input and output capacitors respectively.
V_o - Output voltage (voltage across load).
Considering the system of 800/600V with switching frequency of 100kHz, and phase shift of 60° the following waveform shows nature of inductor current with phase shifted primary and secondary bridge voltages.

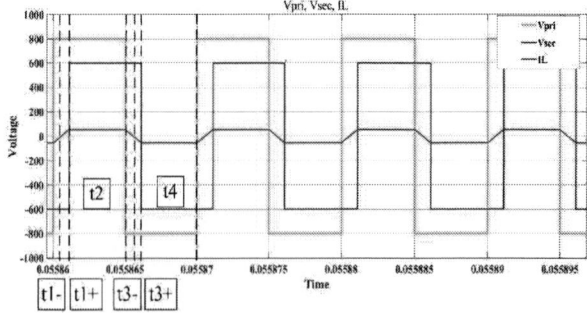

Fig. 2. Waveforms of Primary voltage, secondary voltage and inductor current for a divided into different time intervals.

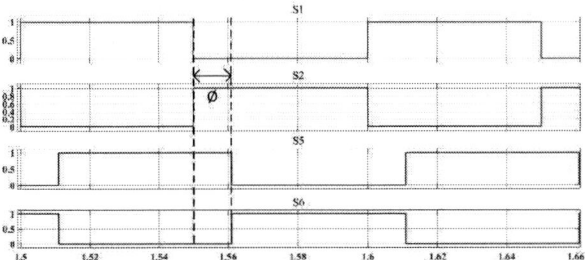

Fig. 3. Gate pulses for the switches of Dual Active Bridge converter

The current increases during the first interval as a result of the primary voltage (V_P) being more than the secondary voltage (V_s) which are being applied across the inductor. At all timing instances the voltage across inductor (V_L) can be defined as difference between primary and secondary bridge voltages which is v_p and V_s/n respectively. During interval t_1^+, secondary voltage starts becoming positive hence indicating a smaller slope then the previous interval (t_1^-), at instance t_2 both voltages are equal and hence the slope remains constant. Similar approach to analysis can be defined for timing intervals t_3 and t_4. The amount and direction of the current over the subsiquent intervals are controlled by the phase shifts of primary and seconday side voltages. The soft switching operation increases the converter's overall effieicney by lowering switching losses.

The Output power in case of Single Phase Shift (SPS) control technique is defined by,

$$P_{sps} = \frac{V_{in}V_0}{2nf_sL}D(1-D) \tag{1}$$

Where, D is called as phase shift ration and can be defined by change in phase shift ($\emptyset$) over half switching cycle ($T_s/2$).

$$D = \emptyset/T_s/2 = \emptyset/\pi$$

Along with traditional single phase shift control technique, as per literature various phase shift control techniques are used to control the output power flow.[3] [6]. Such as DPS (Dual Phase Shift) control, TPS (Triple Phase Shift) control and EPS (Extended Phase Shift) control for controlling the output power flow and to better manage the converter's operation This paper presents detailed modelling, parameter indetification and control design optimization of DAB converter.

The following sections of the paper are categorized as: section II gives brief analysis on the modelling of DAB converter and its control to output transfer functons for both CC-CV (Constant Current-Constant Voltage) modes of operations. Followed by section III which gives idea bout control architecture for EV charging systems using DAB with controller design approach. Sectiion IV summarizes the simulation results, related to modelling, design and control of the converter, with Section V gives the conclusion. Followed by references.

II. MODELLING OF DAB CONVERTER

Before designing the controller, parameters and implementing a proper control strategy for any power electronic converter it is always necessary to design a mathematical model of the converter based upon the passive or energy storing elements in the circuit. Which includes switches, states of passive elements such as inductor, capacitor etc. and the parasitic elements. To derive and determine the precise mathematical model of any converter or system, all the factors based upon their effect on the system needs to be considered. Based upon such approach, various modelling techniques are used and give required analysis on the performance of converter and its mathematical equivalent system for design applications [4].

Based upon such approaches, mentioned in [4] it is found that reduced order modelling [5] is advantageous compared to other modelling techniques such as Generalized average modelling [5] and Discrete time modelling [5] because of the low model complexity and better large signal and small signal accuracy than before mentioned modelling techniques. Implanting reduced order modelling in case of DAB converter, we consider both primary and secondary side bridges to be lossless, and as per [4] Since the inductor current (i_L), one of the state variables, is completely ac and has an average value of zero, simulating a DAB converter becomes more difficult. Ignoring the dynamics of i_L is one technique to model DAB. The properties of the current are described by the average values of the input and output currents throughout a single switching cycle (or half cycle). DAB is then reduced modelled to a first-order system.

As per circuit diagram of DAB converter shown in Fig. (1). it comprises of primary and secondary side bridges, with C_o as the DC output capacitor, L as the leakage inductance of transformer with turns ration $n:1$ which is also an energy storing element responsible for transferring power from primary side bridge to secondary side bridge or vice versa [7]. Vi and Vo are the input and output/load voltages with load resistance R_0. A SPS (Single Phase Shift) control technique is implemented for DAB modelling and controlling the DAB converter, where output voltage waveforms of both bridges are phase shifted by phase shift ($\emptyset$).

The voltage across the inductor (V_L) can be defined as

$$V_L = L\frac{di_L}{dt} = V_p - \frac{V_s}{n} \tag{2}$$

$$I_{2(avg)} = \frac{Q_{stored}}{n\frac{T_s}{2}} \tag{3}$$

$$I_b = \left(\frac{V_i + \frac{V_o}{n}}{L}\right) \times t_1 \tag{4}$$

$$I_a = \left(\frac{V_i + \frac{V_o}{n}}{L}\right) \times t_2 \tag{5}$$

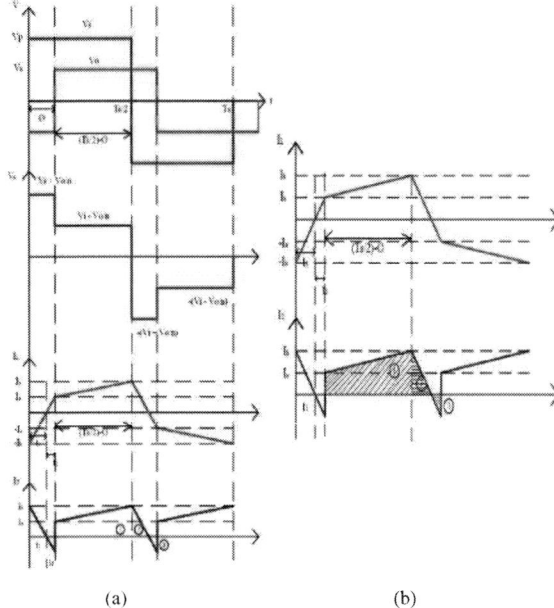

Fig. 4. (a) Phase shifted voltages across inductor and estimation of nature of inductor current and bridge output current from voltage across inductor (b) Inductor current and output current with respect to time instances t_1 and t_2.

$$\therefore I_a + I_b = \emptyset \times \left(\frac{V_i + \frac{V_o}{n}}{L} \right) \tag{6}$$

$$\therefore I_b - I_a = \left(\frac{T_s}{2} - \emptyset \right) \times \left(\frac{V_i + \frac{V_o}{n}}{L} \right) \tag{7}$$

Hence the area under $I_{2(avg)}$ curve which is also a rectified inductor current waveform can be defined as;

$$Q_{stored} = \frac{1}{2}(I_a + I_b)\left(\frac{T_s}{2} - \emptyset \right) + \frac{1}{2}t_1 I_b - \frac{1}{2}t_2 I_a \tag{8}$$

From (3)

$$I_{2(avg)} = \frac{Q_{stored}}{n\frac{T_s}{2}} = \frac{\left(\frac{V_i \emptyset \left(\frac{T_s}{2} - \emptyset \right)}{L} \right)}{n\frac{T_s}{2}} \tag{9}$$

$$= \frac{V_i D(1-D)}{2nLf_s}$$

Where, $D = \emptyset/\pi$

Hence, by doing small signal analysis with perturbation and linearization for average value of inductor current over a switching cycle we are able to derive the control to bridge output current ($I_{2(avg)}$) transfer function which can be defined (10)

$$\therefore G_{i_2\emptyset}(s) = \frac{V_i(1 - 2D_{nominal})}{2nLf_s} \tag{10}$$

The next step for modelling is to determine the control to output transfer function for DAB converter which can be calculated by,

Considering the output capacitance as a state variable, and averaging it over a switching cycle with respect to bridge output current (9) and it can be given as;

$$i_2(t)_{T_{s(avg)}} = C_o \frac{dV_o(t)_{T_{s(avg)}}}{dt} + \frac{1}{R_o}V_o(t)_{T_{s(avg)}} \tag{11}$$

$$\frac{V_i(t)_{T_{s(avg)}}\emptyset(\pi - \emptyset)}{2n\pi^2 f_s L} \tag{12}$$

$$= C_o \frac{dV_o(t)_{T_{s(avg)}}}{dt} + \frac{1}{R_o}V_o(t)_{T_{s(avg)}}$$

Substituting the value of $I_{2(avg)}$ calculated from equation (9) and neglecting the ac variations in the source or input side of DAB converter. After taking Laplace transform of the equation, the control to output voltage transfer function which is used for CV mode of charging in case of EV and can be given by,

$$G_{v\emptyset}(s) = \frac{(\pi - 2\emptyset_{nominal})}{2n\pi^2 f_s L C_o}\left(\frac{V_i}{s + \frac{1}{R_o C_o}} \right) \tag{13}$$

Similarly, the transfer function for control to output current which is used for CC mode of charging can be determined as

$$G_{i\emptyset}(s) = \frac{(\pi - 2\emptyset_{nominal})}{2n\pi^2 f_s L C_o R_0}\left(\frac{V_i}{s + \frac{1}{R_o C_o}} \right) \tag{14}$$

Considering EV charging system which is operating with CCCV (Constant Current Constant Voltage) mode, the system parameters as per the DC-DC fast EV charging system is considered [8] are given by;

TABLE I. PARAMETERS FOR SYSTEM DESIGN

Application	DC charging for EVs
Topology of choice	Dual Active Bridge (DAB)
Typical Power level	10-25kW
Target Power level	6.6kW
Target Vin	700-800V
Target Vout	250-600V
FETs to use	SiC FETs
Operating Frequency	>=100kHz

III. CONTROL OF DAB

Based upon the modelling and control to output transfer functions of DAB converter with respect to load voltage, load current and system parameters, control architecture for CC-CV mode of operation can be designed with suitable controller parameters.

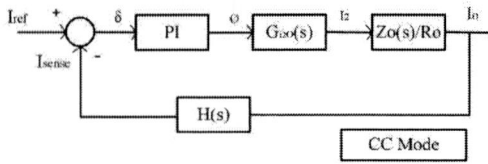

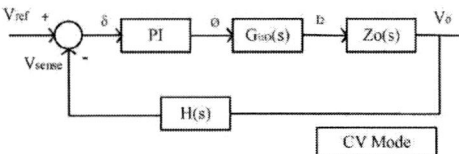

Fig. 5. Design of current and voltage loop for CC and CV mode of operation

Where both current loop and voltage loop are operated separately and can be cascaded based upon the SOC of the battery, hence whenever the drained battery starts to charge it will first operate in CC mode (CC loop) and once the certain desired level of SOC is reached it will change its mode from CC to CV and the second loop (CV loop) starts working.

To design PI controller, for both Voltage and Current loops, transfer functions of both systems are required. Which are obtained by substituting values from table (II). Since the variation in the resistive load is considered for getting constant load parameters, the transfer functions are derived with respect to 100% loading (6.6 kW, 600V) which is also the maximum case scenario for designing.

TABLE II.   DESIGN CONSTRAINTS FOR CONVERTER

P_{max}	10kW
P_o	6.6kW
V_i	800V
V_o	600V
N	1.33
L	80µH
C_0	1000µF
f_s	100kHz
$\emptyset_{nominal}$	38°

$$G_{V\emptyset}(s) = \frac{12309.20}{s + 20.5023} \quad (15)$$

$$G_{i\emptyset}(s) = \frac{224.2484}{s + 20.5023} \quad (16)$$

IV. DESIGNING OF CONTROLLER PARAMETERS

Considering a classical design approach for PI controller parameters, where a pole is to be placed at a particular frequency to obtain a desired crossover frequency from where the desired loop gain bandwidth and K_p, K_i parameters can be obtained.

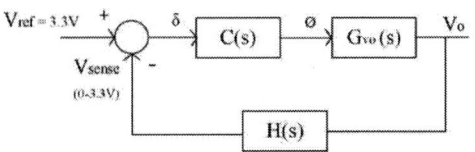

Fig. 6. Control system Block Diagram of voltage controlled mode of DAB converter

As per Fig. (8). The feedback sensor gain for the voltage sensor is considered and output has been step down to 3.3V range for standard microcontroller input.
Hence,

$$G_{V\emptyset}(s) = \frac{12309.20}{s + 20.5023} \quad (17)$$

$$H(s) = \frac{3.3}{600} = 0.0055 \quad (18)$$

$$C(s) = \left(\frac{K_p s + K_i}{s}\right) \quad (19)$$

Calculating the uncompensated loop gain (C(s)=1)

$$G_{L0}(s) = G_{v\emptyset}(s)H(s).1 = \frac{67.7006}{s + 20.5235} \quad (20)$$

$$G_{L0}(s) = \frac{T_0}{1 + \frac{s}{w_0}} = \frac{\frac{67.7006}{20.52}}{1 + \frac{s}{20.52}} = \frac{3.299}{1 + \frac{s}{20.52}} \quad (21)$$

The desired crossover frequency is taken as one twentieth of the switching frequency, hence for a 100kHz switching frequency;

$$f_c = 5000Hz$$

Hence, a high frequency loop gain can be determined by [7];

$$G_{c\infty} = \frac{f_c}{T_0 f_0} = \frac{5000}{5 \times 3.26} = 306.74 \quad (22)$$

Hence, a controller function can be given by,

$$C(s) = G_{c\infty}\left(1 + \frac{w_L}{s}\right) \quad (23)$$

Where, w_L is the frequency of pole which is to be placed to obtain a desired crossover frequency.
Hence the obtained loop gain transfer function is,

$$G_L(s) = G_{v\emptyset}(s)H(s)C(s) \quad (24)$$
$$= \frac{102.57(306.74s + 9.63 \times 10^5)}{s(s + 20.5235)}$$

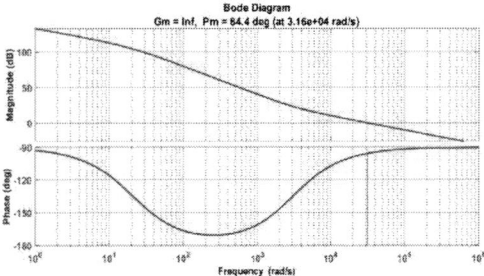

Fig. 7. Bode plot of compensated loop gain transfer function with calculated values of controller parameters

From the Fig. (8). it is evident that, the gain of the system is high (>-40dB) at switching frequency, which amplifies the noise present in the system. And thus, producing overshoots in the load voltage and load current.

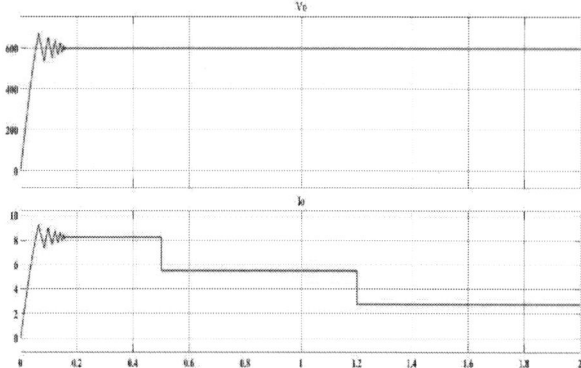

Fig. 8. Output voltage and output current waveforms of simulated results. With the calculated values of kp and ki.

As per Fig. (9). the load current changes as per different loading conditions, the calculated PI parameters causing initial noise and overshoots in the system before achieving steady state.Adjusting the PI controller parameters since the gain at switching frequency was more than minimum value (-40dB) tuning the PI parameters by adjusting the crossover frequency from 5000Hz to approximately 1kHz and the frequency of pole, the new obtained PI parameters not only mitigates the noise amplification issue but shows the response as stable system with a second order closed loop and both poles at left half (LHP).

Poles $= -133.33 \pm 139.87j$. with the gain at switching frequency reducing to -40dB.

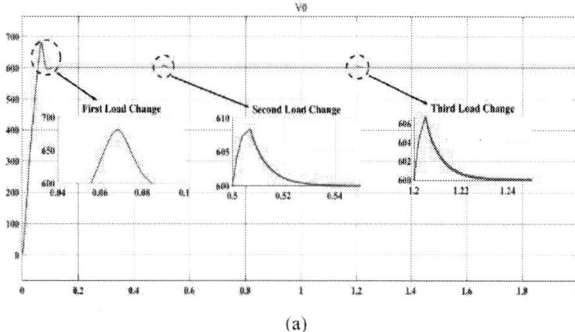

(a)

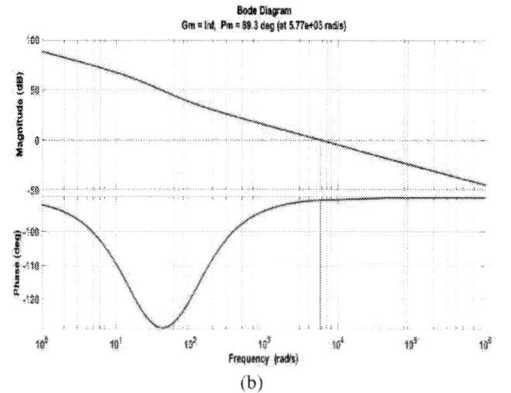

(b)

Fig. 9. (a).Reduction in noise/ripples in the load voltage after tuning the PI controller (b). Bode plot of compensated loop gain transfer function with calculated values of controller parameters

Similar approach can be implemented for finding the controller parameters of current loop, wherein from table (II) and equation (16) the transfer function for control to load current can be defined as

$$G_{i\emptyset}(s) = \frac{224.2484}{s + 20.5023} \qquad (25)$$

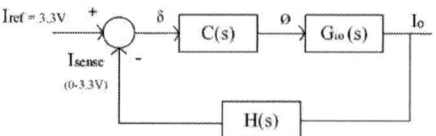

Fig. 10. Control system Block Diagram of current controlled mode of DAB converter

Similar to voltage sensor, considering current sensor feedback gain of

$$H(s) = \frac{3.3}{11.55} = 0.28571 \left(\frac{V}{A}\right) \qquad (26)$$

Uncompensated loop gain with (C(s)=1)

$$G_{L0}(s) = G_{v\emptyset}(s)H(s).1 = \frac{67.070}{s + 20.5235} \qquad (27)$$

Both equations for current loop and voltage loop are similar which proves the fact that same PI controller parameter can be used to control the load parameters.

Choosing $K_p = 90.002526$, $K_i = 7996.542315$ the loop gain at switching frequency is below the maximum limit of -40dB and stable system with a second order closed loop system with both poles at left half (LHP).

V. SIMULATION RESULTS

As per the identified new controller parameters, by changing the desired crossover frequency (from 5kHz to 1kHz) by pole placement technique the tuned PI parameters were able to reduce the noise/ripples in the system. Fig. (11) shows the CV mode of operation where the load voltage is maintained constantly at desired value of 600V with varying load current; by controlling the phase shift at each instance.

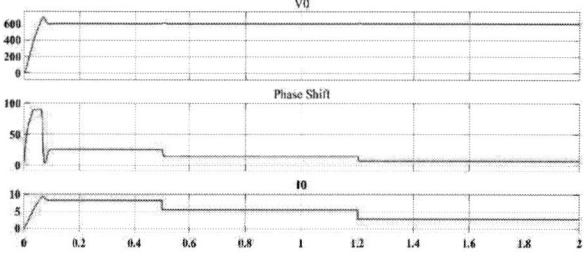

Fig. 11. Simulation waveforms of Constant load voltage (V_0) with varrying load current (I_0) and phase shift (Ø) in degrees

Fig.(12) shows the peaks in the inductor current when the load is being varried, with more finely tuned operation this peaks, and their settling time has been controlled with the new PI parameters and the faster response time of controller.

979-8-3315-3013-6/25 $31.00 © 2025 IEEE

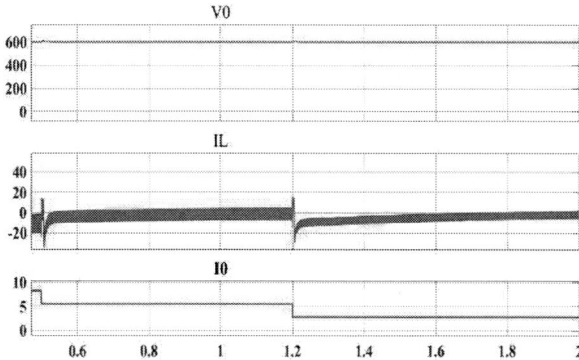

Fig. 12. Simulated waveforms of inductor current with respect to constant load voltage under varrying load current condition.

VI. HARDWARE RESULTS

The dual active bridge (DAB) converter's hardware configuration comprised a Silicon Carbide MOSFET (G3R75MT12D) with a rating of 1200V, 40A bidirectional power supply, Toshiba (TLP5214) gate driver integrated circuits, and a Texas Instruments (TMS320F28379D) digital signal processor for control.

The filter resistors of DSP are initialized as per the identified PI parameters, the voltage and current sensors at the feedback were configured with the ADC pin of the processor.

With the designed controller, and reducing the loop gain at switching frequency the converter was able to operate in Constant Current and Constant Voltage modes. Fig.13. shows the result at such instance where the input voltage of the converter was fixed at 100V, and the load voltage was being maintained constant at 75V with varying passive load from 100% loading (approx. 55Ω) to 50% loading (110Ω) the voltages across primary side and secondary side of the HFT showed less transient behavior and reduced ripples during the transitions in the load change.

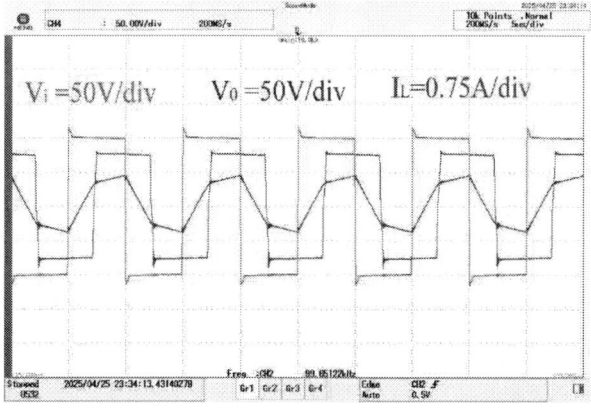

Fig. 13. Hardware obtained waveforms primary voltage, secondary voltage and inductor current.

VII. CONCLUSION

The research work focuses on the identification of the controller parameters of DAB Converter under CC-CV mode of operation. The reduced order modelling was done to derive the transfer functions for designing the controller and to maintain stable operation throughout load variations and transitions from one mode of operation to another. Controller parameters were designed, and it was observed that with the calculated controller parameters the load parameters (load voltage and current) had ripples at the output response. Thus, by changing the crossover frequency over a fix desired range of loop bandwidth, a particular frequency was selected at which the new PI parameters showed less ripple at the load. Earlier it was taken as one twentieth of switching frequency (5kHz0 and then modified to (1kHz). Though the response was little bit slower, the unwanted overshoots and ripples in the system were mitigated. With $K_p = 90.002526$, $K_i = 7996.542315$ the loop gain at switching frequency was able to drop down to the minimum limit of -40dB. The future work may include the balancing in the transients when system transitions from one mode to another, and implementation of controller parameters for Optimized Dual Phase Shift (DPS) or Triple Phase Shift (TPS) control technique.

REFERENCES

[1] M. N. Kheraluwala, R. W. Gascoigne, D. M. Divan and E. D. Baumann, "Performance characterization of a high-power dual active bridge DC-to-DC converter," in IEEE Transactions on Industry Applications, vol. 28, no. 6, pp. 1294-1301, Nov.-Dec. 1992, doi: 10.1109/28.175280.

[2] S. Patwa and M. Matcha, "An In-Depth Review of High-Performance DC-DC Converters for Electric Vehicle Charging Infrastructures," 2024 7th International Conference on Circuit Power and Computing Technologies (ICCPCT), Kollam, India, 2024, pp. 191-196, doi: 10.1109/ICCPCT61902.2024.10673045.

[3] H. Bai and C. Mi, "Eliminate Reactive Power and Increase System Efficiency of Isolated Bidirectional Dual-Active-Bridge DC–DC Converters Using Novel Dual Phase-Shift Control," in IEEE Transactions on Power Electronics, vol. 23, no. 6, pp. 2905-2914, Nov. 2008, doi: 10.1109/TPEL.2008.2005103.J. Clerk Maxwell, A Treatise on Electricity and Magnetism, 3rd ed., vol. 2. Oxford: Clarendon, 1892, pp.68–73.

[4] K. E. Lucas, D. J. Pagano and R. L. P. Medeiros, "Single Phase-Shift Control of DAB Converter using Robust Parametric Approach," 2019 IEEE 15th Brazilian Power Electronics Conference and 5th IEEE Southern Power Electronics Conference (COBEP/SPEC), Santos, Brazil, 2019, pp. 1-6, doi: 10.1109/COBEP/SPEC44138.2019.9065902.

[5] S. Shao et al., "Modeling and Advanced Control of Dual-Active-Bridge DC–DC Converters: A Review," in IEEE Transactions on Power Electronics, vol. 37, no. 2, pp. 1524-1547, Feb. 2022.

[6] M. Mishra and I. Sarkar, "EV Battery Charging using DAB DC-DC Converter with EPS and DPS modulations," 2023 IEEE International Students' Conference on Electrical, Electronics and Computer Science (SCEECS), Bhopal, India, 2023, pp. 1-6, doi: 10.1109/SCEECS57921.2023.10063090.

[7] A. Rodriguez, A. Vázquez, D. G. Lamar, M. M. Hernando and J. Sebastián, "Different Purpose Design Strategies and Techniques to Improve the Performance of a Dual Active Bridge With Phase-Shift Control," in IEEE Transactions on Power Electronics, vol. 30, no. 2, pp. 790-804, Feb. 2015, doi: 10.1109/TPEL.2014.2309853.

[8] V. M. Iyer, S. Gulur and S. Bhattacharya, "Small-Signal Stability Assessment and Active Stabilization of a Bidirectional Battery Charger," in IEEE Transactions on Industry Applications, vol. 55, no. 1, pp. 563-574, Jan.-Feb. 2019, doi: 10.1109/TIA.2018.2871101.

[9] Yu Yan, Yang Huang, RuiRui Chen, Hua Bai, "Building Common-Mode Analytical Model for Dual Active Bridge Incorporating With Different Modulation Strategies", IEEE Transactions on Power Electronics, vol-36, 2021.

Review of Conducted EMI in High Frequency based Traction Inverters

Mr. Kodeeswaran R
Dept. of Electrical and Electronics Engineering
National Institute of Technology Tiruchirappalli
Tamil Nadu, India
kodeeraja103@gmail.com

Dr. Sateesh Kumar Kuncham
Dept. of Electrical and Electronics Engineering
National Institute of Technology Tiruchirappalli
Tamil Nadu, India
sateesh@nitt.edu

Abstract—**The carrier wave modulation technique is considered a constructive approach to minimize conducted electromagnetic interference (EMI) and ease the stringent necessities for EMI filter design. This article briefed about various methods in reducing the conducted EMI for traction inverters. In addition, three different carrier signals namely triangular, sawtooth, and sinusoidal wave-based modulation schemes that are significantly used in traction inverters. Moreover, the analyzed strategies help to understand the behavior of the common mode and differential characteristics of the inverter and its selection. Additionally, the impact of total harmonic distortion, output voltage, and inverter's overall performance has been assessed using MATLAB simulations. The results show that the changed carrier wave can reduce conducted EMI in comparison to traditional carrier frequency PWM.**

Keywords— *Conducted electromagnetic interference (EMI), Pulse width modulation (PWM), Traction inverter*

TABLE. I.   NOMENCLATURE

Acronym	Explanations
CSFPWM	Constant Switching Frequency PWM
VSFPWM	Variable Switching Frequency PWM
RPWM	Random PWM
TWPWM	Triangle Wave PWM
STWPWM	Saw Tooth Wave PWM
SWPWM	Sinusoidal Wave PWM

I. INTRODUCTION

New power device technologies, such as gallium nitride (GaN) and silicon carbide (SiC), are being introduced to replace traditional silicon power devices in inverters. Generally, the wide band gap in semiconductor devices is working at higher switching frequencies to obtain the compact power converter system at higher efficiency. However, the operation at higher switching frequencies causes more transients in the voltage and current which leads to conductive and radiated EMI noise. Moreover, common mode (CM) noise is even adverse due to smaller parasitic capacitance impedance at higher operating frequencies. Thus, the systems with WBG devices essentially examined by EMI compliance in more rigorous than the low frequency devices. A Line Impedance Stabilization Network (LISN) is a device used in electromagnetic compatibility (EMC) and electromagnetic interference (EMI) testing. It functions as a low-pass filter positioned between the power source (AC or DC) and the equipment under test (EUT).

The LISN ensures a consistent line impedance, allowing for repeatable noise measurements. It also isolates noise from the power source to avoid interference with the testing process and provides a safe connection for sensitive measuring instruments like spectrum analyzers or EMI receivers. DC power is supplied to the inverter which has WBG devices via a linear impedance stabilization network (LISN) illustrated in Fig. 1. With the LISN network, EMI noises will be measured accurately. Fig. 1 illustrates the schematic of the LISN network.

Conductive electromagnetic interference (EMI) can be divided into two categories: differential-mode (DM) EMI and CM EMI, based on the various conductive paths through which the interference occurs.

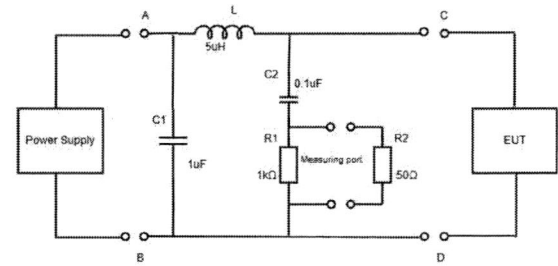

Fig. 1.   Inverter with LISN network

DM EMI results from voltage discrepancies between phases, which leads to conduction among the three phases. However, CM EMI occurs due to the common-mode voltage (CMV), which is the voltage potential relative to a shared reference point. The harmonics in the discrete output voltage generate low-frequency ripple currents. Furthermore, the fast-switching behavior of power devices generates high dV/dt, leading to significant currents for charging and discharging parasitic capacitances at high frequencies. With higher switching frequencies and increased DC bus voltage in the inverter, the resulting EMI is likely to pose greater challenges. Fig. 2 shows the current flow of common mode and differential mode noises.

979-8-3315-3013-6/25 $31.00 © 2025 IEEE

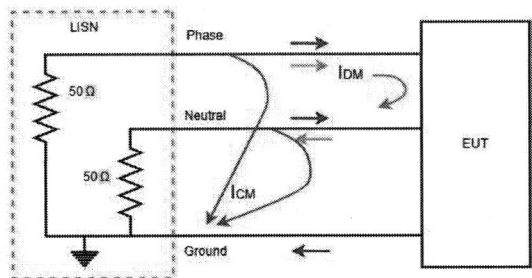

Fig. 2. Current flow of common mode & differential mode noises

$$Vcm = Va + Vb / 2 \qquad (1)$$
$$Vdm = Va - Vb / 2 \qquad (2)$$

where Va is the voltage at one input line, and Vb is the voltage at the other input line.

Common Mode Noise: This type of noise impacts both conductors in a circuit identically and moves in the same direction relative to a shared reference point (usually ground). It often arises from external electromagnetic interference (EMI) or poor grounding. Devices like common mode chokes are commonly used to reduce this noise.

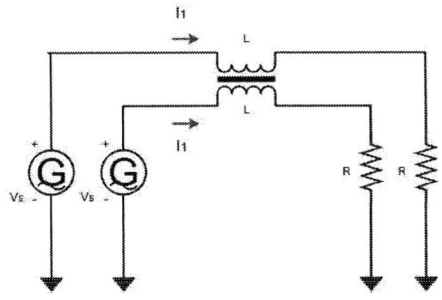

Fig. 3. Common mode choke

$$Z_{cm} = V_s/I_1 = jL_\omega + jM_\omega + R \approx j\omega(2L) \qquad (3)$$

where V_s is the voltage source, I_1 is the current flow and Z_{cm} is the common mode impedance.

Differential Mode Noise: This noise manifests as a disparity between two conductors in a circuit. Unlike common mode noise, it is unique to each line and is frequently caused by circuit-specific issues, such as ripple effects in a power supply.

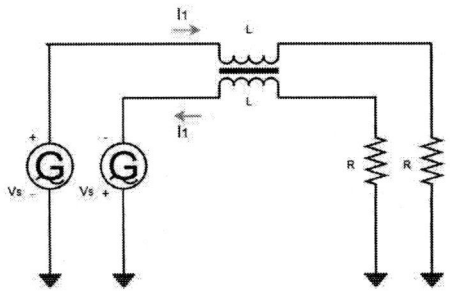

Fig. 4. Differential mode choke

$$Z_{dm} = V_s/I_1 = jL_\omega - jM_\omega + R \approx R \qquad (4)$$

where V_s is the voltage source, I_1 is the current flow and Z_{dm} is the differential mode impedance.

Fig. 3 and Fig. 4 are used to derive equations for the common mode and differential mode noise impedance. Numerous standards and regulations currently cover various sectors and electrotechnical fields including civil, military, and automotive industries. However, since this paper focuses on electric vehicles (EVs), only automotive-related standards and regulations are considered.

CISPR: The primary focus of the standard is to mitigate electromagnetic interference (EMI) that could disrupt vehicle communication systems, such as radios, navigation tools, or other communication devices, while also preventing interference with other electronic systems within or around the vehicle. Radiated emissions specify limits on the electromagnetic radiation emitted by equipment. Conducted emissions define restrictions on noise transmitted through power lines or cables. It sets limits and methods for measuring radio disturbances in the frequency range of 150 kHz to 5,925 MHz. Testing methods provide detailed procedures for testing emissions under controlled environments including specific test setups and equipment requirements. Electronic systems operate reliably across various electromagnetic environments without causing disruptive interference.

In power devices, the on-state and off-state operation results in high efficiency but also unavoidably produces dv/dt and di/dt. While pulse width modulation (PWM) strategies relying on the impulse equivalent principle are widely adopted in converters, the modulated voltage can only meet the intended target at the fundamental frequency. Despite this, the high-frequency band holds significant harmonic components that are responsible for generating electromagnetic interference (EMI) noise. Through conducted or radiated paths, high-frequency noise can spread, causing disruptions to the usual operation of power electronics converters and nearby systems. Fig. 5 shows the different methods for reducing conducted EMI.

II. ANALYSIS OF DIFFERENT EMI REDUCTION TECHNIQUES

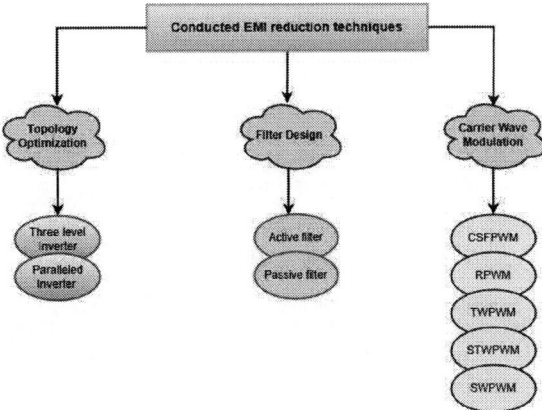

Fig. 5. Different EMI reduction techniques

979-8-3315-3013-6/25 $31.00 © 2025 IEEE

A. Topology Optimization:

Fig.6 and 7 illustrate the diagrams of the three-level and paralleled inverters, respectively. The three-level inverter offers more options for bypassing zero vectors while ensuring optimal use of the DC voltage. In reference [1], a three-level inverter is integrated with a four-leg inverter. The three-level inverter provides additional flexibility for avoiding zero vectors while still fully utilizing the DC voltage. The fourth leg not only assists in eliminating common-mode voltage (CMV) but also facilitates fault-tolerant designs for electric drives. In [2], the full potential of the fourth leg is realized. It is utilized to eliminate CMV during normal inverter operation, to implement fault-tolerant control in the event of IGBT faults, and to function as a boost converter when voltage sags occur.

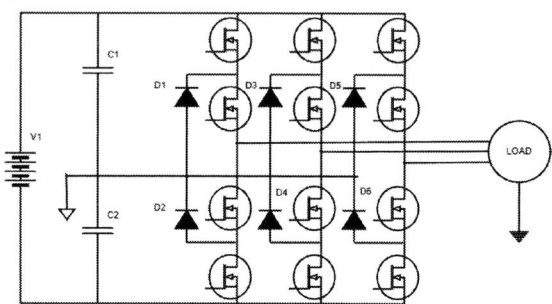

Fig. 6.   Three-level inverter

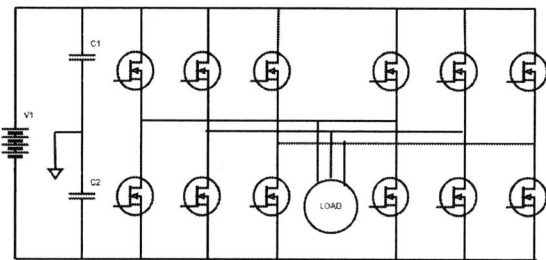

Fig. 7.   Paralleled Inverters

When it comes to EMI suppression, paralleled inverters offer greater flexibility than four-leg and three-level inverters in achieving Zero Common-Mode Voltage (ZCMV) PWM. Furthermore, by phase-shifting the carrier waves between two paralleled inverters, specific harmonics can be reduced. The harmonics that are minimized depend on the interleaving angle caused by the phase shift of the carriers. The effect of the interleaving angle between two inverter carrier signals on harmonic elimination is discussed in [3]. Moreover, in [4], carrier phase-shifting combined with discontinuous SVPWM is applied to a paralleled inverter to limit common-mode (CM) circulating currents. Although carrier phase-shifting can help reduce or eliminate certain harmonics, it does not fully eliminate CM voltage. The concept of paralleled space vectors is introduced in [5] and [6], where six paralleled space vectors are examined to achieve ZCMV PWM. In [7], a novel dual-segment three-phase permanent magnet (PM) machine is proposed, eliminating the need for inter-phase inductors

between the two inverters, thereby enhancing power density.

Although topologies have the potential to eliminate common-mode voltages, ZCMPWM may result in increased output ripple and harmonics, greater switching losses, and diminished bus voltage usage. Consequently, combining advanced modulation techniques with topological designs has become a key area of research in recent years. This integrated approach offers the potential for striking a balance between effective EMI reduction and enhanced inverter performance.

B. Filter Design:

Electromagnetic interference (EMI) suppression methods are typically implemented during the early stages of system design. However, in cases where EMI is insufficiently addressed at its origin or for systems that are already completed, EMI filters are often required to meet strict EMI regulations. Fig.8 depicts the schematic diagram of different passive filters. These filters aimed at combating conductive electromagnetic interference (EMI), specifically passive types, stand out as highly impactful and vital tools.

Placing such filters at the input end of inverters ensures superior feedback signal fidelity, while positioning them at the output side significantly minimizes the three-phase current ripple, which, in turn, elevates the motor drive system's overall efficiency. By creating resonant circuits along the EMI transmission pathways, these filters successfully redirect or block the interference currents. Employing multiple cascaded filter stages allows for comprehensive attenuation of EMI across diverse frequency bands, thereby enhancing insertion loss performance. Popular filter designs include configurations like LC, CL, CLC, and LCL. The selection of the appropriate configuration should be based on the impedance characteristics of both the source and the load, following the principle of impedance mismatch for optimal performance. To achieve an ideal filter design, it is essential to meticulously extract parasitic parameters and accurately model the high-frequency common-mode (CM) impedance.

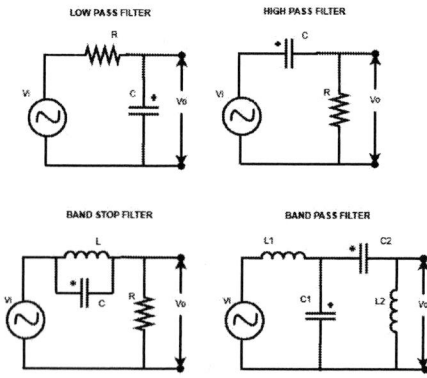

Fig. 8.   Types of Passive filters

C. Carrier Wave Modulation:

The study of the PWM spectrum originates from conventional sinusoidal PWM techniques, although the principles governing switching frequency variations differ across distinct PWM methodologies. To simplify implementation and enhance practicality, regular triangular carriers are employed alongside natural sampling. The recorded EMI spectrum is influenced by factors such as detection methods and bandwidth.

$$C = I_{RMS} / (2 * \pi * f_{SW} * \Delta V) \tag{5}$$

where I_{RMS} is the root mean square current, f_{SW} is the switching frequency.

While the mathematical derivations accounting for measurement complexities remains intricate and largely unexplored, the resulting formulas for the voltage spectrum are straightforward and clear, highlighting the effect of switching frequency on the spectrum. As a result, the voltage spectrum is utilized in place of the measured EMI spectrum for analytical purposes.

EMI enhancement at the source can be accomplished by applying variable switching frequency PWM (VSFPWM) and improving the switching transients of power devices. Modulation techniques have been applied to achieve VSFPWM to minimize EMI peaks at the switching frequency and its integer multiples. VSFPWM disperses the intense harmonics across a broader frequency range, effectively reducing EMI peaks. The switching frequency variation principle results in three main classifications: random PWM, periodic PWM, and programmed PWM. In random carrier frequency PWM (RPWM), the switching frequency is adjusted randomly, while in random pulse position PWM, the switching frequency remains constant, but the pulse position is varied randomly. Random PWM incorporates a random sequence into the signal that serves as the basis for modulation. In essence, periodic PWM provides a method to control power by modulating the duty cycle of a periodic signal, making it a versatile technique for various applications. Programmed PWM uses a pre-determined set of switching angles or patterns to generate the PWM signal. This contrasts with other PWM techniques that rely on a continuous comparison between a reference signal and a carrier wave.

a. Conventional CSFPWM:

In CSFPWM, harmonics appear at frequencies that combine the carrier and fundamental frequencies. Most of the harmonic energy gathers around the carrier harmonics, with their amplitude dropping quickly as the harmonic order increases. Although factors like bandwidth and detection methods can influence EMI testing, the outcomes of conducted EMI tests typically align with spectrum analysis results. EMI spikes found at carrier harmonics present considerable difficulties in designing EMI filters and positioning PWM.

$$C_{mn} = (2Vdc/\pi)(1/m)J_n(m*\pi/2*M)\sin[(m+n)*(\pi/2)] \tag{6}$$

where M is the modulation index, Vdc is the DC voltage, m is the carrier index variable, and n is the baseband index variable.

b. Random PWM:

Random PWM is considered the most traditional PWM modulation technique. By incorporating random fluctuations in the switching frequency, the autocorrelation function of a stationary random process is applied to determine the power spectral density. In [8], the presence of an impulse function in the power spectral density of CSFPWM indicates that the spectrum displays discrete features. In comparison, RPWM transforms the discrete spectrum into a continuous one, thereby mitigating EMI. The switching frequency fluctuates randomly in the time domain, with its statistical distribution resembling a uniform distribution.

As a result, the unpredictability of the switching frequency causes variability in the repeatability of conducted EMI noise. The peak value and associated frequency in the low-frequency range fluctuate with each EMI measurement for RPWM. Therefore, EMI filter designs must incorporate a sufficient margin to guarantee compliance with standards across all test conditions. While PWM modulation can reduce the emission levels of conducted EMI, the degree of EMI suppression depends on the properties of the switching frequency.

Among these techniques, RPWM proves to be the most efficient in minimizing conducted electromagnetic interference (EMI) compared to the other two methods. RPWM specifically involves the random alteration of the switching frequency. Nevertheless, the effectiveness of EMI reduction depends on the waveform's properties, and most of these approaches are not as efficient as RPWM in curbing EMI as discussed in [9]. Periodic PWM, however, is simple and flexible, allowing the adjustment of parameters like periodicity and consistency in the switching frequency waveform, thereby optimizing PWM efficiency. In [10], in waveform modulation, the switching frequency is directly controlled based on the switching rate, such as in triangle wave PWM (TWPWM) or sinusoidal wave PWM (SWPWM). Given the benefits and drawbacks of RPWM and periodic PWM, this paper discusses various switching carrier wave schemes tailored to reduce conducted EMI.

To begin with, when the distribution density at a particular frequency is significantly higher than that of other frequencies, prominent EMI spikes will appear in the spectrum. In [11], RPWM which utilizes uniform or normal distributions, and TWPWM based on an exponential distribution, are both effective methods for reducing EMI.

c. Periodic PWM:

The time-domain signals of the switching frequency for three common periodic PWM methods—TWPWM, STWPWM, and SWPWM are well documented in the literature.

1) TWPWM:

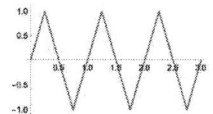

Fig. 9.   Triangular wave

A triangular wave is a type of non-sinusoidal waveform. It is characterized by a linear rise and fall, forming a pattern that alternates between a positive and a negative peak with a constant slope. Since the triangular wave has uniform distribution characteristics, it has been used as the carrier wave for the inverter. The reference signal and PWM integral signal will align at high frequencies, because of this nature triangle waves are used as carrier signals in most of the applications. Due to the linear and predictable characteristics of the triangular carrier wave, it allows for smooth and consistent modulation of the PWM signal.

$$f(t) = \{ \qquad 2A/T*t, \qquad\qquad 0<=t<T/2$$
$$\{ \qquad -2A/T*(t-T), \qquad T/2 <= t < T \qquad (7)$$

where A is the peak amplitude, T is the period it takes for one complete cycle and t is the time. The output voltage and THD are proven in applications for the uniform distribution of triangular carrier waves.

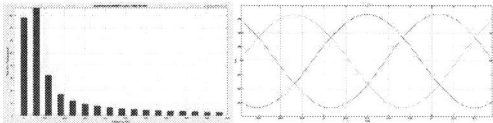

Fig. 10.   THD and Output Voltage

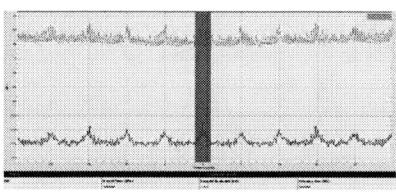

Fig. 11.   Common mode noise & Differential mode noise

2) STWPWM:

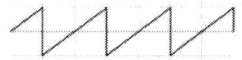

Fig. 12.   Sawtooth wave

A sawtooth wave is another type of non-sinusoidal waveform that has a distinctive, linear rise followed by a sudden drop (or vice versa), resembling the teeth of a saw, hence its name. The sawtooth wave's uniform distribution is a key factor in its role as a carrier wave in inverters. This property ensures consistent comparison with the reference sine wave for generating precise PWM signals. At high frequencies, the alignment between the reference signal and the triangular wave enhances accuracy in modulation, making it ideal for applications where reliable and efficient switching is critical. The linear, increasing slope of the sawtooth wave facilitates a straightforward comparison with the modulating signal.

$$y(t) = A\ (t - t0)\ /\ T,\ for\ t0 <= t <= t0 + T,$$
$$y(t) = 0,\ otherwise \qquad (8)$$

where A is the amplitude, T is the period it takes for one complete cycle, and t0 is the starting time of the ramp. The output voltage and THD are proven in applications for the uniform distribution of sawtooth carrier waves.

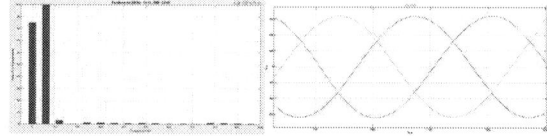

Fig. 13.   THD and Output Voltage

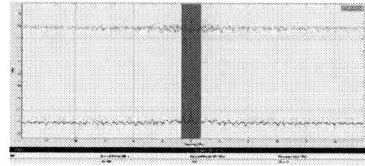

Fig. 14.   Common mode noise & Differential mode noise

3) SWPWM:

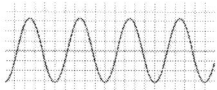

Fig. 15.   Sinusoidal wave

A sinusoidal wave is one of the most fundamental and commonly used waveforms in both physics and engineering. It represents smooth, periodic oscillations that are mathematically described by the sine function, which is a smooth and continuous curve. Since the sine wave does not have uniform distribution characteristics, mostly it won't be used as the carrier wave for the inverter. The reference signal and PWM integral signal will not align at high frequencies, because of this nature sine waves are not used as the carrier signals in most of the applications. Furthermore, sine waves maintain their shape when passing through linear systems, only changing in amplitude and phase, which simplifies the process of modulation and demodulation.

$$C(t) = A * sin(\omega t + \varphi) \qquad (9)$$

where A is the amplitude of the carrier wave, ω is the angular frequency of the carrier wave, t is time and φ is the phase shift of the carrier wave. The output voltage and THD are nearly matched with triangular and sawtooth carrier waves.

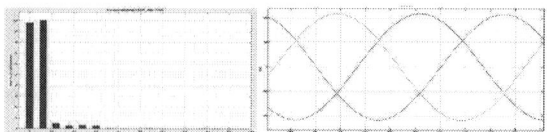

Fig. 16.   THD and Output Voltage

979-8-3315-3013-6/25 $31.00 © 2025 IEEE

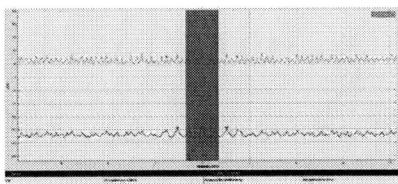

Fig. 17. Common mode noise & Differential mode noise

TABLE. II. COMPARISON OF RESULTS

Carrier Waveform	CM (dBm)	DM (dBm)	THD (%)
Triangular	-139	51.8	43.4
Sawtooth	-150	38.2	3.74
Sinusoidal	-146	40.9	7.19

III. CONCLUSION

This article combines the strengths of topology, filter design, and carrier PWM which proposes a different technique to reduce EMI in the inverter design. An overview of the topology and filter design is provided, enabling the strategy of reducing EMI frequencies. The article then presents experiments to implement three PWM modulation methods: Triangle, Sawtooth, and Sine waves. Finally, a detailed comparison of the three PWM techniques is carried out.

1. CSFPWM delivers the best overall performance, provided the common mode EMI is not considered.

2. While RPWM helps to reduce conducted EMI, generating a suitable random number is very challenging.

3. TWPWM has high total harmonic distortion, but the output voltage and common mode EMI are highly recommendable. Hence, TWPWM is widely used as the carrier wave in EV applications.

4. STWPWM has very low THD and to add upon the output voltage and common mode EMI are also good compared to TWPWM. So, we can use STWPWM as the carrier wave in automotive or commercial applications.

5. The impact of SWPWM on EMI reduction is not immediately apparent, and its performance in other areas is also not remarkable compared to TWPWM and STWPWM.

IV. REFERENCES

[1] R. Chen, J. Niu, H. Gui, Z. Zhang, F. Wang, L. M. Tolbert, B. J. Blalock, D. J. Costinett, and B. B. Choi, ``Investigation of fourth-leg for common-mode noise reduction in three-level neutral point clamped inverter fed motor drive," in Proc. IEEE Appl. Power Electron. Conf. Expo. (APEC), Mar. 2019, pp. 2582_2588, doi: 10.1109/APEC.2019. 8722319.

[2] P. Garg, S. Essakiappan, H. S. Krishnamoorthy, and P. N. Enjeti, ``A faulttolerant three-phase adjustable speed drive topology with active commonmode voltage suppression," IEEE Trans. Power Electron., vol. 30, no. 5, pp. 2828_2839, May 2015, doi: 10.1109/TPEL.2014.2361905.

[3] D. Zhang, F. Wang, R. Burgos, R. Lai, and D. Boroyevich, ``Impact of interleaving on AC passive components of paralleled three-phase voltagesource converters," IEEE Trans. Ind. Appl., vol. 46, no. 3, pp. 1042_1054, Mar. 2010, doi: 10.1109/TIA.2010.2045336.

[4] D. Zhang, F. Wang, R. Burgos, and D. Boroyevich, ``Common-mode circulating current control of paralleled interleaved three-phase two-level voltage-source converters with discontinuous space-vector modulation," IEEE Trans. Power Electron., vol. 26, no. 12, pp. 3925_3935, Dec. 2011, doi: 10.1109/TPEL.2011.2131681.

[5] D. Jiang and Z. Shen, ``Paralleled inverters with zero common-mode voltage," in Proc. IEEE Energy Convers. Congr. Expo. (ECCE), Sep. 2016, pp. 1_8, doi: 10.1109/ECCE.2016.7855330.

[6] D. Jiang, Z. Shen, and F. Wang, ``Common-mode voltage reduction for paralleled inverters," IEEE Trans. Power Electron., vol. 33, no. 5, pp. 3961_3974, May 2018, doi: 10.1109/TPEL.2017.2712369.

[7] Z. Shen, D. Jiang, T. Zou, and R. Qu, ``Dual-segment three-phase PMSM with dual inverters for leakage current and common-mode EMI reduction," IEEE Trans. Power Electron., vol. 34, no. 6, pp. 5606_5619, Jun. 2019, doi: 10.1109/TPEL.2018.2866338.

[8] H. Akagi and T. Hatada, ``Voltage balancing control for a three-level diode-clamped converter in a medium-voltage transformerless hybrid active _lter," IEEE Trans. Power Electron., vol. 24, no. 3, pp. 571_579, Mar. 2009, doi: 10.1109/TPEL.2009.2012528.

[9] H. Akagi and R. Kondo, ``A transformerless hybrid active _lter using a three-level pulsewidth modulation (PWM) converter for a mediumvoltage motor drive," IEEE Trans. Power Electron., vol. 25, no. 6, pp. 1365_1374, Jun. 2010, doi: 10.1109/TPEL.2009.2040002.

[10] C. Zhu and T. H. Hubing, ``An active cancellation circuit for reducing electrical noise from three-phase AC motor drivers," IEEE Trans. Electromagn. Compat., vol. 56, no. 1, pp. 60_66, Feb. 2014, doi: 10.1109/TEMC.2013.2267801.

[11] Y. Zhang, Q. Li, and D. Jiang, ``A motor CM impedance based transformerless active EMI _lter for DC-side common-mode EMI suppression in motor drive system," IEEE Trans. Power Electron., vol. 35, no. 10, pp. 10238_10248, Oct. 2020, doi: 10.1109/TPEL.2020.2980881.

[12] B. Narayanasamy and F. Luo, ``A survey of active EMI _lters for conducted EMI noise reduction in power electronic converters," IEEE Trans. Electromagn. Compat., vol. 61, no. 6, pp. 2040_2049, Dec. 2019, doi: 10.1109/TEMC.2019.295305.

[13] S. Ogasawara, H. Ayano, and H. Akagi, ``An active circuit for cancellation of common-mode voltage generated by a PWM inverter," IEEE Trans. Power Electron., vol. 13, no. 5, pp. 835_841, Sep. 1998, doi: 10.1109/63.712285.

[14] S. Ogasawara, H.Ayano, and H. Akagi, ``An active circuit for cancellation of common-mode voltage generated by a PWM inverter," in Proc. Rec. 28th Annu. IEEE Power Electron. Spec. Conf. Formerly Power Condition- ing Spec. Conf. 71st Power Process. Electron. Spec. Conf. (PESC), vol. 2, 1997, pp. 1547_1553, doi: 10.1109/PESC.1997.618067.

[15] F. Wang, A. K. Wallace, S. Dai, A. Von Jouanne, and H. Zhang, ``Multilevel inverter modulation schemes to eliminate common-mode voltages," IEEE Trans. Ind. Appl., vol. 36, no. 6, pp. 1645_1653, Nov. 2000, doi: 10.1109/28.887217.

[16] M. Duang-upra and Y. Kamsuwan, ``Three-segment switching sequences for a space-vector modulated three-level inverter to eliminate common-mode voltages," in Proc. 45th Annu. Conf. IEEE Ind. Electron. Soc. (IECON), Oct. 2019, pp. 2082_2087, doi: 10.1109/IECON.2019.8927324.

Axial Flux Permanent magnet motor for Drone Application

Madhav Yadav
Electrical Engineering Dept.
SECS, IIT Bhubaneswar
Jatni, Odisha, India
21ee01071@iitbbs.ac.in

K. L. Karthik Jandhyala
Electrical Engineering Dept.
SECS, IIT Bhubaneswar
Jatni, Odisha, India
21ee01074@iitbbs.ac.in

S. Rajesh
Electrical Engineering Dept.
SECS, IIT Bhubaneswar
Jatni, Odisha, India
23PD06005@iitbbs.ac.in

Dr. Ankit Dalal
Electrical Engineering Dept.
SECS, IIT Bhubaneswar
Jatni, Odisha, India
ankitdalal@iitbbs.ac.in

Abstract—**Axial flux motors offer higher torque and power density compared to radial flux motors. Thus, they are being extensively worked on for applications like Electric vehicles. Similarly, high torque and power density motors can greatly benefit drone applications, resulting in lighter drones and increased flight time. This paper compares the Axial Flux versus the Radial Flux Permanent Magnet Synchronous Motors (PMSM) for Drone applications. Performnace of motors, obtained from Finite Element Analysis (FEA) under geometrical constraints of outer diameter is compared. An analytical approach for the design of an Axial flux motor for optimal performance is also presented.**

Index Terms—**Axial flux motor, Drone application, PMSM, Torque density, power density.**

I. INTRODUCTION

Drone usage is greatly increasing across various market segments. Drones are being increasingly used not just in in critical services like unmanned wars and surveillance and but also in many commercial applications like agriculture, goods delivery, photography, etc. Critical performance parameters for drones are their flight time per charge and payload capacity. One of the most critical components affecting a drone's performance is its electric motor, which indirectly determines its efficiency, flight time, and payload capacity. Among different motor types, PMSM motors are preferred due to their high efficiency and power density [1]. PMSM motors can be classified into two categories based on magnetic flux direction: Axial Flux Permanent Magnet (AFPM) Motors and Radial Flux Permanent Magnet (RFPM) Motors. While outer rotor RFPM motors due to their high torque delivering capability are widely used in drones, AFPM can also be very suitable for drone application because of their small motor length, higher torque density, and improved power-to-weight ratio [2] [3].

The primary difference between these two motor types lies in the flow of magnetic flux. In RFPM motors, the flux moves radially from the center outward, giving them a cylindrical structure. In contrast, AFPM motors have a disc-like structure, with the magnetic flux flowing along the axis of rotation [4]. These design differences significantly affect motor torque, efficiency, and cooling performance. AFPM motors generally offer higher torque density and better efficiency,

whereas RFPM motors are preferred for their robustness, ease of manufacturing, and scalability [5].

For drone applications, an ideal motor should be lightweight, energy-efficient, and capable of delivering high power output [6]. While RFPM motors are commonly used due to their proven reliability, due to advancements in materials and manufacturing techniques, AFPM can be a suitable alternative to radial flux machines. Recent research has also focused on alternative motor designs, such as slotless lightweight motors, to enhance drone application performance [7]. High-speed AFPM motors have also been developed for specialized applications, offering improved power output and efficiency under demanding conditions [8].

This paper compares axial and radial flux PMSMs in terms of design, performance, and suitability for drone applications. A reference radial flux motor typically used for agricultural drones has been chosen as a reference. Keeping the outer diameter as a design constraint, an axial flux motor has been worked out analytically, and performance has been validated using Finite Element Analysis (FEA).

II. MOTORS FOR DRONE APPLICATIONS

A. Structures and Application Constrains

Table I compares radial and axial flux motors on various motor features important from an application viewpoint.

TABLE I
COMPARISON OF AXIAL FLUX AND RADIAL FLUX MOTORS

Feature	Axial Flux Motor	Radial Flux Motor
Magnetic Flux	Parallel to the shaft	Perpendicular to the shaft
Design Structure	Disc-shaped	Cylindrical
Power Density	Higher	Lower
Efficiency	Generally higher	Generally lower
Cooling	Challenging	Better
Manufacturing	Complex	Easier
Torque Density	Higher	Lower
Cost	High	Low
Rotor inertia	Low	High

B. Suitability to Drone application

An agricultural drone typically uses 4-6 motors to achieve the required thrust level. Thus, the motors' weight contributes significantly to the overall weight of the drone and affect thrust to weight ratio of the drone. Drones require motors that are lightweight, efficient, and capable of delivering high torque to ensure extended flight times and optimal performance. Axial flux motors provide significant advantages over radial flux motors due to their higher power-to-weight ratio and compact design. Their unique disc-shaped structure increases torque output, improving battery efficiency and enabling drones to carry heavier payloads without sacrificing speed or agility. This makes axial flux motors ideal for high-performance drone applications [9].

Despite some cooling challenges, axial flux motors excel in efficiency and torque density, directly improving a drone's flight time and operational range. Their superior energy conversion ensures lower energy losses, making them more suitable for applications that demand high endurance and reliability [6]. Furthermore, as advances in thermal management continue, axial flux motors are becoming increasingly viable for widespread drone use [10].

While still widely used, radial flux motors lack the same power efficiency and weight advantages as axial flux motors. Their bulkier design and lower torque density make them less suitable for next-generation drone applications prioritizing lightweight, high-power solutions. Therefore, axial flux motors stand out as the superior choice for drones, offering unmatched performance benefits [8].

Considering the motor requirements for agricultural drones, in this work, a commercially available radial flux motor has been chosen as a reference motor [12]. Fig. 1 shows the reference radial flux outer rotor motor, and Table II presents the motor specifications.

Fig. 1. Reference motor

TABLE II
REQUIRED SPECIFICATIONS OF THE DRONE MOTOR

Quantity	Value	Unit
Power	2.4	kW
Battery	46	V
Max Speed	6000	RPM
No of poles	28	-
Inner Diameter of Motor	30	mm
No of slots	24	-
KV rating	150	rpm/V

III. SIZING OF AXIAL FLUX MOTOR

Accurate motor design can be achieved with FEA tools. However, FEA takes significant amount of time especially for complex geometries like axial flux motor. As the flux does not flow in a 2-dimensional plane in an axial flux motor 3-D, FEA is essential for accurate results. To reduce the number of iteration of FEA analysis, an analytical approach for single stator single rotor AFPM motor is presented in this section.

In the single-sided machine with permanent magnet disc rotor, the armature winding is located on the stator core as shown in Fig. 2. The objective of the analytical approach is to determine the size of the motor and winding parameters, which are presented in detail in the following sub-sections.

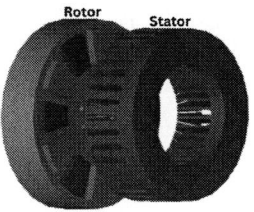

Fig. 2. Side view of single stator single rotor Axial Flux motor [11].

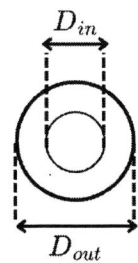

Fig. 3. Front view of single stator single rotor Axial Flux motor.

A. Determination of Inner and Outer Diameter Relation

Phase current can be calculated as :-

$$I_a = \frac{P_{out}}{m V_1 \eta \cos \phi} \tag{1}$$

where,
P_{out} : Desired Output Power
m : Number of phases
V_1 : Stator voltage per phase
η : Efficiency
$\cos \phi$: Motor power factor

Apparent Electromagnetic power in two stators can be written as :-

$$S_{elm} = m E_f I_a \tag{2}$$

$$= \frac{\pi^2 k_w n_s B_{mg} A_m (D_{in} + D_{out})(D_{in}^2 - D_{out}^2)}{16} \tag{3}$$

where,
E_f : Per phase back emf
n_s : Motor Speed in RPS
A_m : Average line current density
D_{out} : Outer diameter of the rotor as shown in Fig. 3
D_{in} : Inner diameter of the rotor as shown in Fig. 3
Apparent Electromagnetic power can be defined as:

$$S_{elm} = \frac{\epsilon P_{out}}{\eta \cos \phi} \tag{4}$$

where,
ϵ : Relative emf ratio and stator voltage per phase (E_f/V_1)

From Eq. 3 and Eq. 4, Relation between D_{out} and D_{in} can be extracted as :

$$\frac{\epsilon P_{out}}{\eta \cos \phi} = \frac{\pi^2 k_w n_s B_{mg} A_m (D_{in} + D_{out})(D_{in}^2 - D_{out}^2)}{16} \tag{5}$$

$$P_{out} = \frac{\pi^2 k_w n_s B_{mg} A_m \eta \cos \phi (D_{in} + D_{out})(D_{in}^2 - D_{out}^2)}{16\epsilon} \tag{6}$$

Torque for the motor can be defined as:

$$T_{out} = \frac{P_{out}}{2\pi n_s} \tag{7}$$

$$T_{out} = \frac{\pi k_w B_{mg} A_m \eta \cos \phi (D_{in} + D_{out})(D_{in}^2 - D_{out}^2)}{32\epsilon} \tag{8}$$

By taking D_{in} as constant for the particular application, Now the Eq.5 can be simplified as:

$$aD_{out}^3 + bD_{out}^2 + cD_{out} + d = 0 \tag{9}$$

where, cubic equation coefficients are :-

a = 1, b = D_{in}, c = $-D_{in}^2$, d = $-\left(D_{in}^3 + 16\alpha\right)$

where, α (Design Coefficient) is given as :-

$$\alpha = \frac{E P_{out} \times 10^3}{\pi^2 k_{w1} n_s \eta \cos\phi B_{mg} A_m} \tag{10}$$

B. Back Emf Calculations

Flux per pole (ϕ_f) can be calculated as :

$$\phi_f = \frac{B_{mg}(D_{out}^2 - D_{in}^2)}{4p} \tag{11}$$

where,
B_{mg} : Peak air gap flux density
p : Number of pole pairs

Back Emf (E_f) can be calculated as :-

$$E_f = \sqrt{2}\pi f N k_w \phi_f \tag{12}$$

where,
f : Supply frequency

N : Number of turns in series per phase
k_w : Winding factor
ϕ_f : Flux per pole
Using Eq 12, number of turns of the windig can b determined.

IV. RESULT AND DISCUSSION

The design parameter of the reference RFPM motor used in simulation are presented in Table III.

TABLE III
DESIGN PARAMETERS SIMULATION OF RADIAL FLUX MOTOR

Quantity	Value	Unit
Thrust	3.5-5.5	Kg
Efficiency	82.14-82.89	%
Recommended Battery Voltage	46	V
Outer Stator Diameter	61.53	mm
Inner Stator Diameter	38.25	mm
Outer Rotor Diameter	70	mm
Inner Rotor Diameter	64.04	mm
Stator stack length	18.5	mm
Rotor length	23.53	mm
Stator slots	24	-
Poles	28	-
Magnet width (M_w)	6.1	mm
Magnet length (M_l)	18.5	mm
Magnet depth (M_d)	1.1	mm

A FEA model show in Fig. 4 using ANSYS Maxwell is prepared with the dimensions mentioned in Table III and FEA analysis is performed.

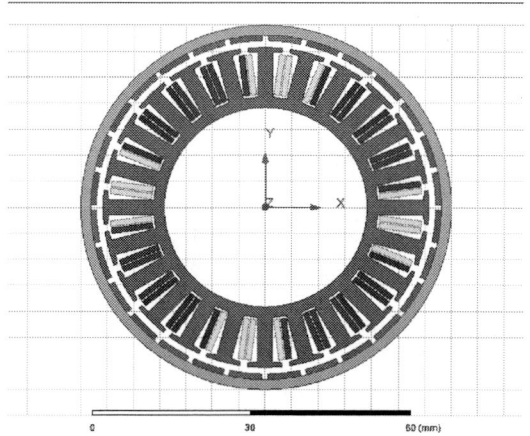

Fig. 4. FEA Model of Reference Motor

From above FEA model the back emf profile is obtained for 1000 RPM as shown in Fig. 5.
Using the Back emf result the voltage constant (K_E) is obtained as:

$$K_E = E_a/\omega = 5.73 \text{ V/kRPM}$$

where
E_a = Peak value of the back emf of the motor
ω = motor speed in kRPM.
Torque constant (K_T) can be calculated as,

979-8-3315-3013-6/25 $31.00 © 2025 IEEE

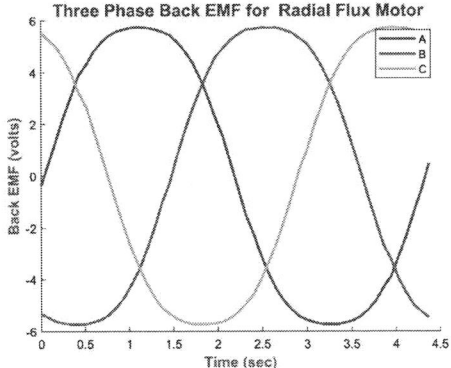

Fig. 5. Back Emf Profile of Reference Motor at 1000 rpm

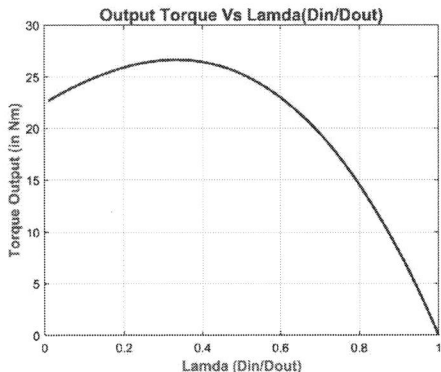

Fig. 6. Output Torque Vs D_{in}/D_{out}

$$K_T = K_E \frac{60}{2000\pi} = 0.05474 \text{ Nm/A}$$

A. Design and simulation of AFPM Motor

Sizing calculations for AFPM is performed as described in Section III-A. Motor specification and assumed parameters for analytical calculation are given in Tables II and IV. D_{out} can be calculated from the above motor specification using Eq.9,

$$D_{out} = 76.6 \text{ mm}.$$

TABLE IV
ASSUMED PARAMETERS FOR ANALYTICAL DESIGN

Quantity	Value	Unit
Efficiency*Powerfactor ($\eta \cos \phi$)	0.9	-
Magnetic Loading (B_{mg})	0.7	T
Line Current Density (Electric Loading) (A_m)	140000	A/m
Winding Factor (K_w)	0.85	-
$\epsilon(E/V)$	0.925	-
Parallel Conductor (a_w)	4	-
Parallel Current Paths (a_p)	1	-
Current Density (J_a)	4.5	A/mm^2
Maximum Flux Density in core (B_{max})	2.2	T
Slot Fill Factor	0.6	-

The number of turns in series per phase can be calculated as follows.

$$N = \frac{\pi(D_{in} + D_{out})A_m}{4\sqrt{2}mI_a} = 72 \quad (13)$$

Assuming $D_{out} = 76.6$ mm to be constant in Eq. 8, Equation can be simplified in terms of D_{in}/D_{out} ratio as follows :

$$T_{out} = \frac{\pi D_{out}^3 k_w B_{mg} A_m \eta \cos \phi (\frac{D_{in}}{D_{out}} + 1)((\frac{D_{in}}{D_{out}})^2 - 1)}{32\epsilon} \quad (14)$$

form above Eq. 14 the maximum Torrque Output is obtained at $\frac{D_{in}}{D_{out}} = 0.33$. As shown in the Fig. 6. Taking outer diameter as constraint we have taken D_{out} same as radial flux motor, $D_{out} = 70$ mm and $D_{in} =24.6$ mm is considered.

From the analytical design, an FEA model using ANSYS Maxwell is prepared for the axial flux motor as shown in Fig.

7 from the parameters mentioned in Tables II and IV. Stator and rotor models are shown separately in Fig. 8 and Fig. 9

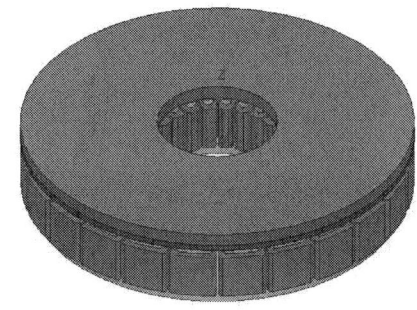

Fig. 7. FEA Model of Axial Flux Motor

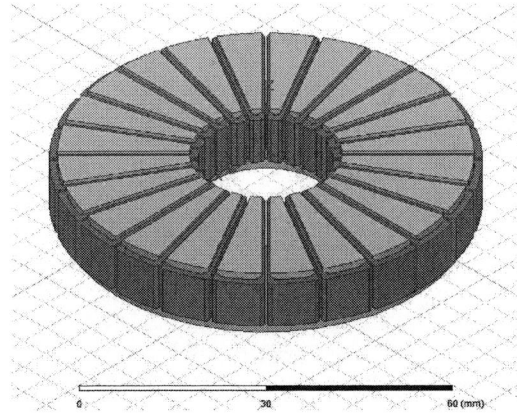

Fig. 8. FEA Model of Stator core with coils in Axial Flux motor

From the above FEA model the below back emf profile is obtained for 1000 RPM speed as shown in Fig. 10.

It can be observed from the Fig. 10, the waveform is not smooth, this is due to realtively coarse mesh chosen during 3-D FEA analysis to save computational time. However, the

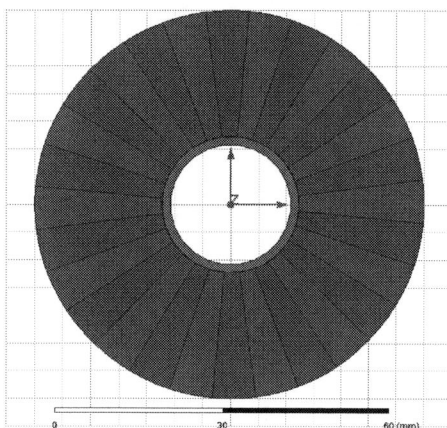

Fig. 9. FEA Model of Rotor core with PM's in Axial Flux motor

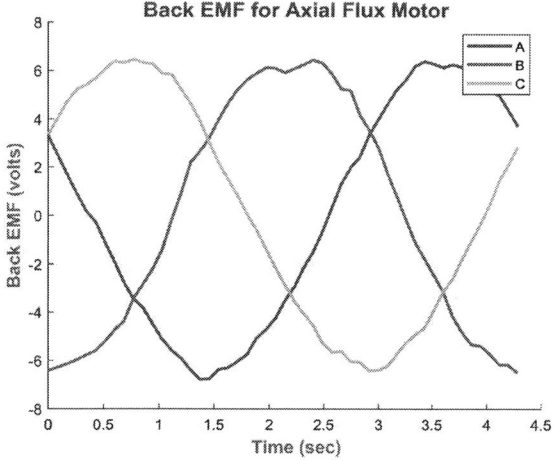

Fig. 10. Back Emf Profile of Axial Flux Motor at 1000 rpm

results is accurate enough to calculate the voltage constant of the motor. Using the Back emf the voltage constant (K_E) is obtained as:

$$K_E = E_a/\omega = 6.4 \frac{V}{kRPM}$$

where

E_a = Peak value of back emf of the motor

ω = motor speed in kRPM

Torque constant (K_T) can be calculated as,

$$K_T = K_E \frac{60}{2000\pi} = 0.06114 \frac{Nm}{A}$$

B. Comparison of RFPM and AFPM Motors

Table V shows the obtained parameters of both motors from FEA analysis by keeping outer diameter same for both motors. The positive percentage change shows the increase in parameter for axial flux motor and negative percentage change shows decrement. In the case of AFPM motor the back emf constant and the torque constant are observed to increase by 11.7%, So it is possible to get higher torque at lesser current as

TABLE V
RFPM Vs AFPM Motors

Parameter	RFPM	AFPM	Unit	% Change
K_E	5.73	6.4	V/kRPM	11.7%
K_T	0.05474	0.06114	Nm/A	11.7%
Motor Volume	90.55	60.035	cm^3	-33.7%
Magnet Volume	3.4756	3.7076	cm^3	6.675%
Copper Weight	72.4	117.8	g	62.7%

compared to radial flux motor, also their is decrease in motor volume by 33.7%, also the increase in magnet volume and copper weight is observed by 6.675% and 62.7% respectively.

V. CONCLUSIONS

This paper presents a comparison between AFPM and RFPM motors for drone application. An analytical approach for the design of the AFPM motor is presented along with this the performance of both motors is compared by FEA analysis using ANSYS maxwell. It is observed that for the same outer diameter the AFPM motor shows better performance with decrease in overall motor volume. It is also obseved that inspite of reduction of motor volume, the magnet volume and copper mass is more in AFPM compared to RFPM, indication efficient utilization of available volume. This results make AFPM Motors more suitable for drone application as compared to RFPM Motors. Further work on the Thermal and loading aspects of both these motors can be done for better comparison.

REFERENCES

[1] I. D. Chasiotis and Y. L. Karnavas, "A Study on Design and Optimization of High Power Density PMSM for Pod Propulsion System," 2018 XIII International Conference on Electrical Machines (ICEM), Alexandroupoli, Greece, 2018, pp. 534-540, doi: 10.1109/ICEL-MACH.2018.8506871.

[2] Deepak BBVL, Davinder Pal Singh, Sampath Kumar Kuppa, M Jayanthi Rao, Design optimization of drone BLDC motor for delivery service applications, Materials Today: Proceedings, 2023,

[3] J. F. Gieras, Axial Flux Permanent Magnet Brushless Machines, Springer, 2008.

[4] Hao, Z.; Ma, Y.; Wang, P.; Luo, G.; Chen, Y. A Review of Axial-Flux Permanent-Magnet Motors: Topological Structures, Design, Optimization and Control Techniques. Machines 2022, 10, 1178.

[5] Hendershot, J R, Jr., and TJE Miller, Design of Brushless Permanent-Magnet Motors (Oxford, 1995; online edn, Oxford Academic, 31 Oct. 2023).

[6] Ajay Vishwath N.C., Arvind R Yadav, Deep Mehta, Jatin Belani, Ravi Raj Chauhan, A guide to novice for proper selection of the components of drone for specific applications, Materials Today: Proceedings, Volume 65, Part 8, 2022, Pages 3617-3622, ISSN 2214-7853,

[7] M. V. Malarselvam, I. Monish, C. Carunaiselvane, "Dynamic Characteristic Analysis of Transitioning Radial to Axial Flux Motors in Electric Three-Wheeler Application," International Journal of Engineering Research and Technology (IJERT), vol. 9, no. 4, pp. 2278-0181, 2020.

[8] M. S. Islam, I. Husain, and R. Mikail, "Slotless lightweight motor for drone applications," 2017 IEEE Energy Conversion Congress and Exposition (ECCE), Cincinnati, OH, 2017, pp. 5041-5048, doi: 10.1109/ECCE.2017.8096851.

[9] S. Neethu, S. Pal, A. K. Wankhede, and B. G. Fernandes, "High performance axial flux permanent magnet synchronous motor for high-speed applications," IECON 2017 - 43rd Annual Conference of the IEEE Industrial Electronics Society, Beijing, China, 2017, pp. 5093-5098, doi: 10.1109/IECON.2017.8216880.

979-8-3315-3013-6/25 $31.00 © 2025 IEEE

[10] C. Jenkins et al., "Innovations in Axial Flux Permanent Magnet Motor Thermal Management for High Power Density Applications," in IEEE Transactions on Transportation Electrification, vol. 9, no. 3, pp. 4380-4405, Sept. 2023, doi: 10.1109/TTE.2023.3242698.

[11] A. Parviainen, "Design of Axial-Flux Permanent Magnet Low-Speed Machines and Performance Comparison Between Radial-Flux and Axial-Flux Machines," Ph.D. dissertation, Lappeenranta University of Technology, Finland, 2005.

[12] https://www.hobbywing.com/en/products/xrotor-x6-plus269

A High Frequency Switched Gallium Nitride Based Bi-directional EV Supply Equipment for Vehicle-to-Home Applications

Udaya Sagar V
Power Electronics Group
CDAC Thiruvananthapuram
Thiruvananthapuram, Kerala, India
udayasagar@cdac.in

Vishnu C V
Power Electronics Group
CDAC Thiruvananthapuram
Thiruvananthapuram, Kerala, India
vishnu.cv@cdac.in

Akshara S
Power Electronics Group
CDAC Thiruvananthapuram
Thiruvananthapuram, Kerala, India
aksharas@cdac.in

Jisha S R
Power Electronics Group
CDAC Thiruvananthapuram
Thiruvananthapuram, Kerala, India
jishasr@cdac.in

Amal S
Power Electronics Group
CDAC Thiruvananthapuram
Thiruvananthapuram, Kerala, India
amals@cdac.in

Dr. Chandrasekar V
Power Electronics Group
CDAC Thiruvananthapuram
Thiruvananthapuram, Kerala, India
vvcsekar@cdac.in

Abstract—The rapid growth of global electric vehicle (EV) adoption, fueled by advancements in power electronics and battery technologies, has led to an increasing demand for fast and efficient charging solutions. There is a rising research interest in compact EV supply Equipment(EVSE) with vehicle-to-grid (V2G), vehicle-to-home (V2H), and vehicle-to-load (V2L) capabilities, as these systems play a crucial role in improving grid resilience and energy flexibility.A need of high performance with improved power density has drifted the designs from Silicon (Si) to Gallium Nitride(GaN). This paper presents the design and implementation of a high-frequency bidirectional EV charger utilizing Gallium Nitride (GaN) power devices for improved efficiency and compactness. The system operates at 200 kHz and consists of two power conversion stages: a totem-pole power factor correction (PFC) converter, which rectifies the single-phase 230V AC input to 400V DC while achieving high power factor, and a dual active bridge (DAB) converter, which provides galvanic isolation and efficient DC-DC conversion to 48V DC. A novel control strategy is introduced to address zero-crossing challenges in the totem-pole PFC, ensuring stable operation and minimal harmonic distortion. The use of GaN devices enables high-frequency operation, reducing passive component value, size and enhancing power density. Experimental results validate the effectiveness of the proposed design, demonstrating high efficiency, stable operation, and reliable bidirectional power transfer. This work contributes to the development of compact, high-performance EV interfaced (V2H) systems that support grid stability and energy resilience in residential applications.

Index Terms—EV Supply Equipment (EVSE), Vehicle-to-Grid (V2G), Vehicle-to-Home (V2H), Gallium Nitride (GaN), Dual Active Bridge (DAB), Totem Pole PFC(TPPFC).

This research work is financially supported by Ministry of Electronics and Information Technology (MeitY), Govt. of India under NaMPET-III program

I. INTRODUCTION

The global pathway towards sustainable transportation has accelerated the adoption of electric vehicles (EVs), necessitating the development of robust and efficient charging infrastructure. Traditional unidirectional EV chargers only facilitate energy transfer from the grid to the vehicle battery system (G2V), limiting their role in modern energy ecosystems. In contrast, bidirectional EV chargers are gaining prominence, particularly for Vehicle-to-Home (V2H), Vehicle-to-Grid (V2G), and Vehicle-to-Load (V2L) applications. These technologies allow EV battery systems to function as distributed energy storage systems, optimizing energy utilization for power as well as transportation markets. V2H technology enables an EV battery to power a home, reducing dependence on the grid, lowering electricity costs during peak demand, and ensuring energy availability during grid failures or emergencies. Similarly, V2G technology allows bidirectional energy exchange between EVs and the grid, which helps in stabilizing voltage fluctuations, supporting renewable energy integration, and improve grid reliability. V2L expands this concept by enabling EVs to directly power external loads, offering flexible energy solutions for off-grid or remote applications. As a result, there is growing research interest in compact, high-efficiency bidirectional EVSE that can seamlessly integrate these functionalities and not limited to. Power Electronics community is stepping out from Silicon based to wide Band Gap (WBG) based converters for the performance and power density enhancements.

This paper presents the design and implementation of a high-frequency, GaN-based bidirectional EVSE that integrates a totem-pole power factor correction (TPPFC) converter and a dual active bridge (DAB) converter. The totem-pole PFC stage

979-8-3315-3013-6/25 $31.00 © 2025 IEEE

is selected for its higher efficiency over conventional AC-DC converters, leveraging GaN power devices to minimize conduction losses and achieve superior power factor correction. The DAB converter ensures efficient DC-DC conversion with galvanic isolation, facilitating stable power transfer between the 400V DC link and the 48V EV battery. A key challenge in totem-pole PFC design is the zero-crossing distortion, which affects performance and efficiency. This work introduces a novel control strategy to mitigate zero-crossing challenges, ensuring stable and low-distortion operation. The system was designed to operate at 200 kHz, which helped in reduction of passive components values and its mechanical sizes. This resulted in improved power density, and enhanced thermal performance. The proposed charger architecture is validated through experimental results, demonstrating high efficiency, stable operation, and reliable bidirectional power transfer with compactness. This study contributes to the development of next-generation EV charging solutions, supporting grid stability, energy resilience of power markets along with the sustainable transportation

II. LITERATURE REVIEW

Gallium Nitride (GaN) devices are transforming power electronics designs by offering superior performance compared to the traditional silicon-based. Their wide bandgap, high electron mobility, and low on-resistance enable high-frequency switching with minimal losses. GaN High Electron Mobility Transistors (HEMTs) are particularly well-suited for electric vehicle (EV) supply equipment higher power density and efficiency [1], [2] Operating at higher switching frequencies helps reduce the size of passive components like inductors and capacitors, making chargers more compact and lightweight. Studies have shown that GaN-based converters can achieve efficiencies better than 95% in bidirectional applications [3], [4]. However, high-frequency operation introduces challenges such as electromagnetic interference (EMI) and thermal management, which require advanced packaging and optimized gate drive circuits.The long-term reliability of GaN devices under high-frequency conditions remains an area of active research.

Dual Active Bridge (DAB) converters are commonly used in bidirectional EV chargers due to their galvanic isolation and soft-switching capabilities [6]. This topology supports efficient power transfer in both charging and discharging modes, making it ideal for vehicle-to-home (V2H) applications. Researchers are working on optimizing DAB control strategies to reduce circulating currents and improved efficiency. Modular converter designs are gaining popularity, offering scalability and adaptability to various battery voltages and power demands. These designs enhance power ratings and fault tolerance. Advanced control algorithms play a crucial role in managing bidirectional power flow, maintaining grid stability, and optimizing charging and discharging profiles. Active power factor correction (APFC) techniques helps to reduce harmonic distortion and ensure grid compliance.

In front end AC-DC stages of bidirectional EVSE, Totem-Pole PFC rectifiers leveraging GaN's high-frequency switching capabilities are under research. This bridge-less topology reduces conduction losses and improves efficiency compared to traditional boost PFC circuits. It is particularly effective at high power levels, benefiting significantly from the fast switching speeds of GaN devices [5], [7]. Additionally, advanced algorithms are being developed to optimize charging and discharging profiles and mitigate battery degradation effects.

Integrating bidirectional EV chargers into the grid requires careful management of power quality and stability. These chargers can introduce harmonic distortion and voltage fluctuations, impacting overall grid performance. Active filtering techniques and advanced control strategies help mitigate these issues. Vehicle-to-grid (V2G) functionality enables EVs to supply power back to the grid, supporting peak demand and frequency regulation. However, implementing V2G requires robust communication protocols and grid management systems. The development of smart grid technologies is essential for fully realizing V2G and V2H potential. Ensuring electromagnetic compatibility (EMC) is also critical, as the high switching frequencies of GaN devices can generate EMI, which must be managed through proper design and shielding.

Despite significant progress, challenges remain in the widespread adoption of GaN-based bidirectional EVSE. Cost remains a primary concern, as GaN devices are currently little expensive than silicon-based alternatives. Standardization and interoperability are also essential to ensure compatibility across different EV models and charging infrastructures. Future research need a focus on developing cost-effective solutions, advanced thermal management techniques, robust control strategies, and seamless smart grid integration.

III. DESIGN AND CONTROL STRATEGY

This section details the design of EVSE, along with the control strategies for better zero crossing in totem pole PFC and efficient bidirectional power flow. The system specification as mentioned in Table I

TABLE I
SPECIFICATIONS OF BI-DIRECTIONAL EV CHARGER

Parameter	Specifications
Input Voltage Range	180Vrms - 265Vrms
Nominal Input	230 Vrms, 50 Hz
Power Factor	0.99
Rated output voltage	48(44-54) V DC
Rated output Power	3.3 kW
Output current	70A DC (max)
Switching Frequency	> 200 kHz
Boost Inductor	118 uH
Converter Efficiency	> 94%
Communication Interface	CAN
Isolation	Galvanic Isolation
THD	3 %

A. Totem-Pole PFC Converter

The totem-pole PFC AC-DC converter is a bridgeless topology that eliminates the need for the diode bridge rectifier, significantly reducing conduction losses. It consists of two high-frequency GaN transistors operating at 200 kHz and two low-frequency Si-MOSFETs switching at the grid frequency (50 Hz). The boost inductor (L_{AC}) facilitates the energy transfer, while the output capacitor (Cout) maintains a stable 400V DC.

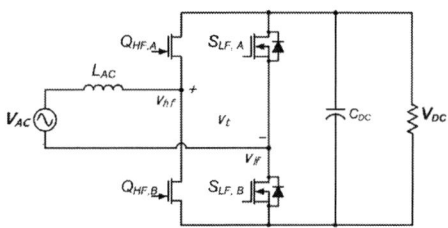

Fig. 1. Totem-Pole PFC

Fig. 2. GaN Halfbridge with Heatsink

During the positive half-cycle of the AC input, one of the GaN transistors is actively switching, while the corresponding low-frequency MOSFET remains ON, allowing current conduction. In the negative half-cycle, the second GaN transistor takes over, switching at high frequency, with the corresponding MOSFET conducting continuously. This asymmetric switching scheme ensures continuous conduction mode (CCM)operation and improves efficiency. The PFC inductors (L_{AC}) and DC bus capacitor (C_{DC}) of the TPPFC are designed using below equations (1) and (3) respectively.

$$L_{AC} = \frac{1}{\%\text{Ripple}} \cdot \frac{V_{ac}^2}{P_0} \left(1 - \frac{\sqrt{2} \cdot V_{ac}}{V_o} \right) \cdot T \quad (1)$$

$$I_{L,\max} = \frac{\sqrt{2} \cdot P_0}{V_{ac}} \cdot \left(1 + \frac{\%\text{Ripple}}{2} \right) \quad (2)$$

$$C_{DC} \geq \frac{2 \cdot P_0 \cdot t_{\text{hold}}}{V_o^2 - V_{o,\min}^2} \quad (3)$$

Based on the above design equations, the Boost inductor, Output capacitors and GaN FETs were selected. For high-frequency operation, GaN devices were selected over other devices. The devices selected for the high-voltage side is 650V, 30A GaN (MPN: LMG3522R050) transistors. Two such devices are fabricated in a single PCB to form a half-bridge. Silicon carbide (SiC) MOSFETs (MPN: SCT3022ALGC11) were used for low-frequency operation with a similar PCB construction. The loss analysis and estimation revealed that the majority of power loss (heat source) was observed in the magnetic components followed by HF GaN switching devices. The top-side cooled GaN device is attached with a heat sink and fan assembly for extracting the thermals as shown in Fig. 2.

B. Control Strategy in Totem Pole PFC

The AC input voltage zero-crossing in totem-pole Power Factor Correction (PFC) converters poses a significant challenge to achieving high efficiency and stable operation. Specifically, the abrupt polarity reversal of the AC input voltage

necessitates rapid switching transitions, which can induce substantial current spikes and exacerbate reverse recovery losses in semiconductor devices. This issue is caused by parasitic inductance within the circuit, which contributes to voltage overshoots and abnormal ringing during switching. To mitigate these effects, novel control strategies, including precise zero-crossing detection algorithms, soft-switching techniques are crucial. These techniques, coupled with optimized circuit layouts designed to minimize parasitic elements, are essential for ensuring stable operation and maximizing the efficiency of totem-pole PFC. The switching frequency was also finalized after making multiple comparison of switching losses at different frequencies. The state machine of the control algorithm is illustrated in Figure 3

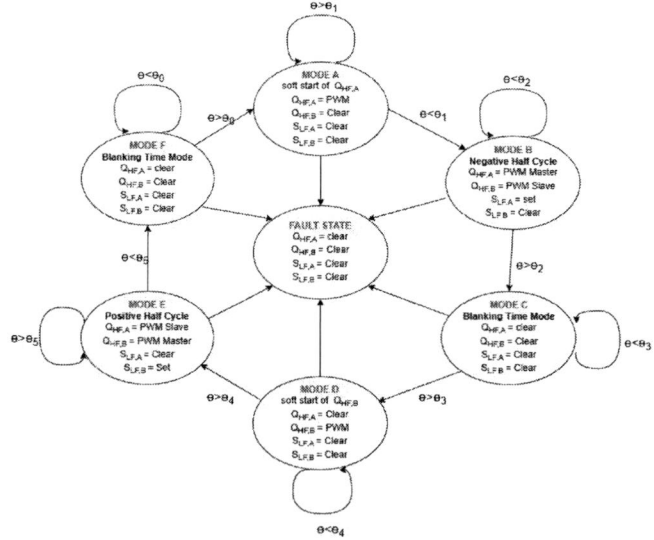

Fig. 3. State Machine of control strategy

With this applied soft start in each half cycle for the higher duty starting FET, and added a blanking time relating to the current. The PWM pulses with this control scheme generated as shown in Figure 4.

Blanking time and soft start are essential considerations

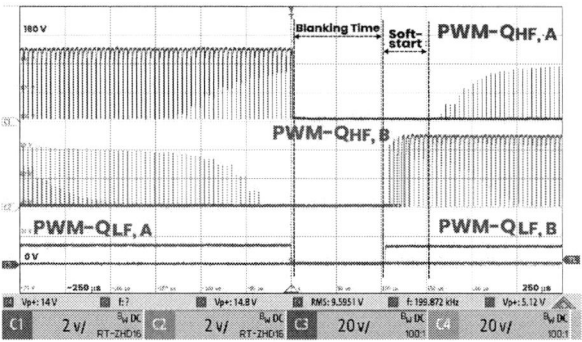

Fig. 4. Negative - Positive Zero crossing PWM

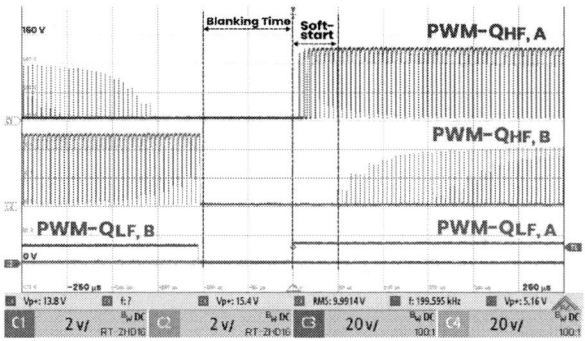

Fig. 5. Positive - Negative Zero crossing PWM

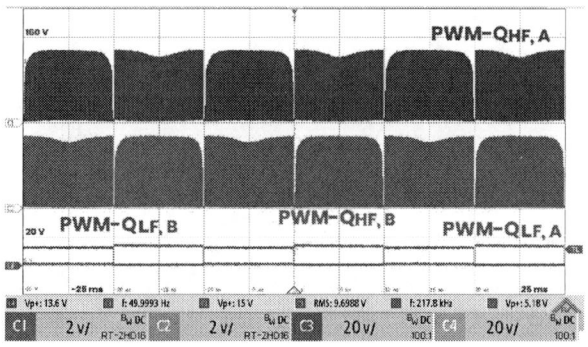

Fig. 6. Full PWM Sequence

for the operation of a Totem Pole Power Factor Correction (PFC) converter. Blanking time referred in this paper is a deliberate delay applied between turning OFF one switch and turning ON the complementary switch between the positive and negative half cycle of AC supply. This prevents shoot-through currents, which could otherwise lead to excessive losses or damage to the power switches. Optimizing blanking time is crucial to balancing efficiency and reliability. Soft start, on the other hand, is typically applied in high-duty cycle starting conditions to limit inrush current and avoid overstressing power components.

C. Dual Active Bridge

The Dual Active Bridge (DAB) converter is a high-frequency isolated DC-DC converter that facilitates bidirectional power flow between the 400V DC bus and EV battery. It consists of two full-bridge inverters, a high-frequency transformer, and a series inductor. The primary bridge converts the 400V DC bus from the Totem-Pole PFC stage into a high-frequency AC waveform, which is transferred through the high-frequency transformer to the secondary side. The secondary bridge rectifies the AC waveform to 48V DC for charging the EV battery.

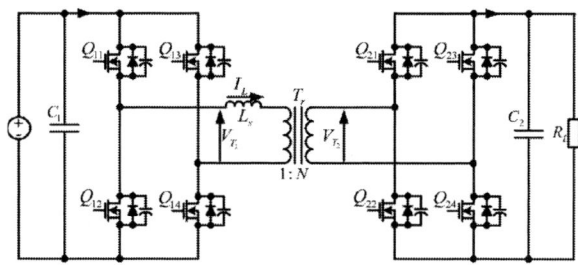

Fig. 7. Dual Active Bridge

The bidirectional nature of the converter enables power flow from the battery back to the grid or essential household loads when operating in V2H or V2G modes. The phase-shift control between the two full bridges regulates power flow and ensures soft switching conditions for improved efficiency.

$$P_{max} = \frac{V_1 V_2}{8 n f_{sw} L} \quad ; \quad \emptyset = \pm \frac{\pi}{2} \tag{4}$$

$$L_{max} = \frac{V_2 V_1}{2 n P_{max} f_{sw}} D_{max}(1 - D_{max}) \tag{5}$$

The maximum inductance (L_{max}) needed in a DAB converter's transformer for a given maximum power (P_{max})) is calculated with the above design equations. It depends on voltages, turns ratio, switching frequency, and duty cycle, ensuring efficient power transfer. Based on these equation,the designed DAB with a maximum D_{max} as 0.45.

D. Input Filtering and Protection

The input side of the Converter is filtered through a custom designed EMI filter for meeting the global standards. The Filter and Protection circuit comprises of a Common Mode Choke, Differential mode choke, X and Y capacitors with discharge resistors. In addition to the filter circuits SPDs and ESD protection devices are provided for the safety of the system as well as for user. The scheme of the Input Filtering and Protection circuit is shown in below Fig. 8.

E. Total System

The proposed bidirectional EVSE integrates a high-efficiency Totem-Pole PFC and a Dual Active Bridge (DAB) DC-DC converter, both utilizing GaN devices to achieve high power density and reduced switching losses at 200 kHz. The

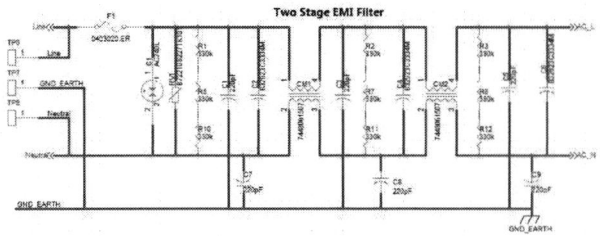

Fig. 8. Filter and Protection

Totem-Pole PFC stage efficiently converts single-phase AC 230Vrms to 400V DC, ensuring near-unity power factor. The DAB converter further steps down this 400V DC to 48V, providing galvanic isolation and bidirectional power transfer capability. This enables Vehicle-to-Home (V2H) functionality, allowing the EV battery to supply power to home loads during grid outages. The integrated scheme of the converter is shown in Fig. 9.

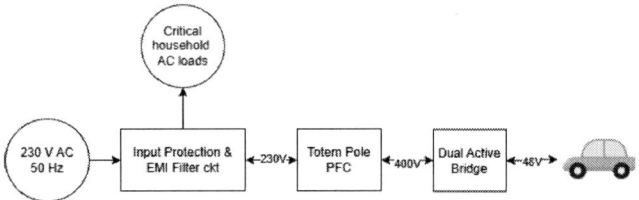

Fig. 9. Scheme of Integrated Motherboard

The control strategy ensures seamless mode transition between charging and discharging. The complete hardware implementation, including power stages, control circuits, and communication interfaces, is shown in Fig.10., depicting the system's compact design and high integration level.

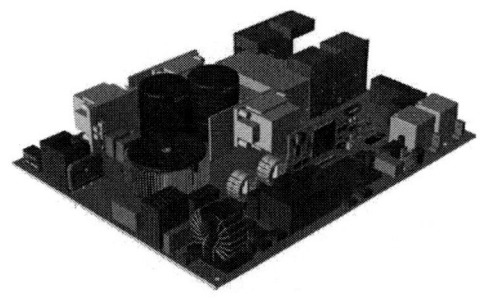

Fig. 10. 3D View of EV Charger Motherboard

IV. RESULT DISCUSSION

The experimental results validate the performance improvements achieved through the proposed control strategy and bidirectional operation of the GaN-based EV charger. The Totem-Pole PFC without the novel control exhibited significant input current distortion near the zero-crossing, leading to reduced power factor, as shown in Fig. 11. With the proposed control, the input current waveform closely follows the voltage waveform, achieving near-unity power factor and lower harmonic distortion (Fig. 12.). The expanded waveform

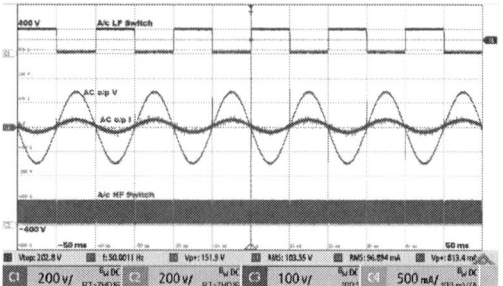

Fig. 11. System With typical controller

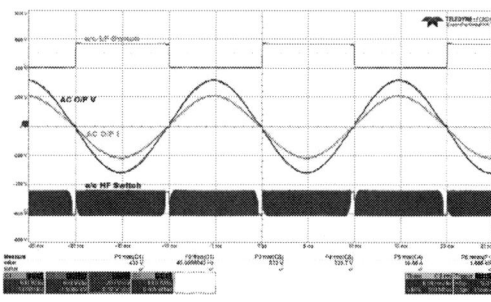

Fig. 12. System With modified Controller

during V2G at a higher power level with the effect of blanking time and soft start is shown in Fig. 13

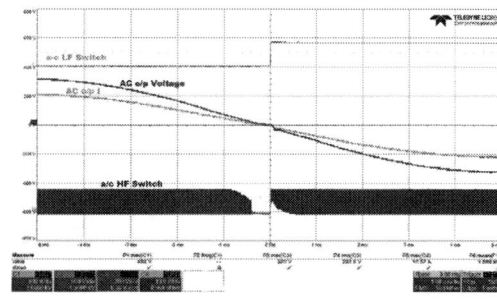

Fig. 13. O/P w/f (expanded view) with modified controller

These results confirm the high efficiency, improved power quality, and reliable bidirectional operation of the proposed system, making it a viable solution for compact, high-frequency bidirectional EVSE.The PFC front end converter achieved an efficiency of 98 % and DAB converter achieved an efficiency of 96% with an overall efficiency of the system as 94 %. The major losses were observed in magnetic components (Boost Inductor and DAB Transformers). The major challenges in this development is PCB layout design for minimized parasitic elements and tightly coupled magnetic components which requires special care right from design to fabrication. The price of GaN devices and the the unique production procedure have raised the system's overall cost, which limits user market demand temporarily.

V. CONCLUSION

This paper presents the design and implementation of a GaN-based bidirectional EV charger with high-frequency operation at 250 kHz, enabling grid-to-vehicle (G2V) and vehicle-to-home (V2H) functionalities. The system consists of a Totem-Pole PFC stage for efficient AC-DC conversion with power factor correction and a Dual Active Bridge (DAB) converter for galvanically isolated DC-DC conversion. A novel control strategy for the Totem-Pole PFC effectively eliminates zero-crossing distortions, ensuring a near-unity power factor and improved efficiency. Experimental results validate the system's high efficiency, stable voltage regulation, and effective bidirectional power transfer, demonstrating its potential as a compact and efficient EV charging solution. The proposed architecture supports sustainable energy integration by enhancing the flexibility and resilience of both electric vehicle charging and domestic power supply applications.

REFERENCES

[1] X. H. Yin, Y. Y. Song and H. S. Xu, "Research on Driving Technology of GaN HEMT and Application", 2016 International Symposium on Computer Consumer and Control (IS3C), pp. 837-840, 2016.

[2] B. Li, Q. Li, F. C. Lee, Z. Liu and Y. Yang, "A High-Efficiency High-Density Wide-Bandgap Device-Based Bidirectional On-Board Charger," in IEEE Journal of Emerging and Selected Topics in Power Electronics, vol. 6, no. 3, pp. 1627-1636, Sept. 2018.

[3] G. Liu, Y. Jang, M. M. Jovanović and J. Q. Zhang, "Implementation of a 3.3-kW DC–DC converter for EV on-board charger employing the series-resonant converter with reduced-frequency-range control", IEEE Trans. Power Electron., vol. 32, no. 6, pp. 4168-4184, Jun. 2017.

[4] B. Su, J. Zhang and Z. Lu, "Totem-pole boost bridgeless PFC rectifier with simple zero-current detection and full-range ZVS operating at the boundary of DCM/CCM", IEEE Trans. Power Electron., vol. 26, no. 2, pp. 427-435, Feb. 2011.

[5] Z. Liu, F. C. Lee, Q. Li and Y. Yang, "Design of GaN-Based MHz Totem-Pole PFC Rectifier," in IEEE Journal of Emerging and Selected Topics in Power Electronics, vol. 4, no. 3, pp. 799-807, Sept. 2016.

[6] Ganesan. P and K. Hatua, "Implementation of Vector control for Single Phase Dual Active Bridge to achieve ZVS and ZCS for Switching Loss Reduction," 2022 IEEE Energy Conversion Congress and Exposition (ECCE), Detroit, MI, USA, 2022

[7] Ravindranath Tagore Yadlapalli, Anuradha Kotapati, "Advancements in energy efficient GaN power devices and power modules for electric vehicle applications: a review", International Journal of Energy Research, April 2021

Reduced Switch Multilevel Inverter Topology for Electric Vehicle Powertrains

Shadab Murshid
Department of Electrical and Computer
Engineering
National *University* of Singapore
Singapore
shadab@nus.edu.sg

Prasanth Sundararajan
Department of Electrical and Computer
Engineering
National University of Singapore
Singapore
prsant.s@nus.edu.sg

Mrutyunjaya Sahani
Department of Electrical and Computer
Engineering
National University of Singapore
Singapore
mrutyunjayasahani@nus.edu.sg

Kolantla Dharani
Department of Electrical and Computer Engineering
National University of Singapore
Singapore
dharanik@nus.edu.sg

Sanjib Kumar Panda
Department of Electrical and Computer Engineering
National University of Singapore
Singapore
eleskp@nus.edu.sg

Abstract—This article explores a 7-level reduced switch multilevel inverter (RSMLI) topology for electric vehicle (EV) powertrain application. This work is aimed at enhancing efficiency and compactness, while minimizing the cost and complexity. Traditional multilevel inverters need a large number of power electronic switches, increasing the overall system size, switching losses, and control complexity. The proposed RSMLI topology significantly reduces the number of active switches and gate drivers without compromising output voltage quality or power delivery capability. This results in lower total harmonic distortion (THD), reduced conduction and switching losses, and improved overall system reliability. The inverter operates effectively across a wide range of operating conditions typically encountered in EVs, offering robust performance and fast dynamic response. Furthermore, the reduced switch count contributes to better thermal management, increased power density, and simpler control implementation. Simulation results confirm the proposed inverter's performance during different dynamic conditions, demonstrating superior efficiency and fast transient response. The compact nature of the design makes it highly suitable for integration in space-constrained EV applications. Comparative analysis with some of the recent multilevel inverters highlights the advantages on the basis of switch count, DC sources, power diodes, and capacitors used. The proposed RSMLI offers a promising solution for next-generation EV powertrains, promoting energy-efficient and sustainable transportation.

Keywords— Reduced Switch Multilevel Inverter, Self-Balancing Capacitors, Electric Vehicle, Powertrain

I. INTRODUCTION

Transportation electrification presents a promising solution to address the environmental challenges posed by greenhouse gas (GHG) emissions. In 2022, the transportation sector was responsible for approximately 28% of total U.S. GHG emissions, making it the largest contributor among all sectors [1]. Driven primarily by the rising demand for travel, this sector saw has witnessed an absolute increase in GHG emissions from 1990 to 2022 than any other sector. Switching from internal combustion engine vehicles to electric vehicles (EVs) offers a viable and impactful approach to significantly reduce these emissions. Beyond the environmental benefits, transportation electrification also brings several performance and operational advantages. These include better efficiency, improved acceleration, and reduced maintenance requirements, making EVs not only cleaner but also more cost-effective and reliable solution over time [2].

Using higher DC-link voltages in EVs offers several significant advantages [3]. Firstly, it enables extreme fast charging, greatly reducing the time needed to recharge the battery. Secondly, higher-voltage batteries allow for smaller and lighter cables, since lower current is required to deliver the same power. This contributes to overall vehicle weight reduction and improved packaging. Thirdly, operating at a higher voltage reduces motor current, leading to lower conduction losses and improved overall system efficiency. Additionally, manufacturing costs can be lowered due to reduced copper usage and simplified thermal management. The use of higher voltage also supports higher power density, allowing for more compact and efficient drivetrain designs [4]. Moreover, motor performance improves, as higher voltage operation enhances torque response and energy utilization.

While traditional two-level inverters have been widely used in EV powertrains, multilevel inverters (MLIs) are increasingly gaining attention as a preferred alternative. Their capability to produce high-quality output waveforms with lower total harmonic distortion (THD), reduced voltage stress on power devices, and improved overall efficiency makes them well-suited for high-power EVs, especially as the industry moves toward higher-voltage architectures. MLI topologies are generally divided into three primary categories: diode-clamped inverters, flying capacitor inverters, and cascaded H-bridge inverters. While these configurations offer improved output quality, they often require a larger count of switching devices and auxiliary components, such as clamping diodes or capacitors [5]. This not only increases the circuit complexity but also leads to higher overall cost, making them less suitable for cost-sensitive applications like EVs. Researchers have proposed various MLI topologies that utilize fewer switching devices while still achieving multiple voltage levels during recent years [6]. These reduced switch configurations aim to simplify the circuit design, lower costs, and enhance efficiency, making them more suitable for compact and high-performance applications such as EVs [7-18].

TABLE I. COMPARATIVE ANALYSIS OF 7-LEVEL MLI TOPOLOGIES

Type	N_{dcs}	N_{pes}	N_{sc}	N_{pd}	SR	CG	H-Bridge
[7]	1	10	2	1	No	Yes	No
[8]	1	9	2	0	No	No	No
[9]	1	10	1	0	Yes	No	No
[10]	1	12	4	0	Yes	No	No
[11]	1	9	3	2	No	Yes	No
[12]	2	8	4	2	Yes	No	Yes
[13]	1	10	3	2	Yes	No	No
[14]	1	10	3	2	Yes	No	No
[15]	1	11	3	3	Yes	No	Yes
[16]	1	8	2	1	No	No	Yes
[17]	1	10	2	0	No	No	No
[18]	2	14	2	0	Yes	No	No
Proposed	1	8	3	3	No	Yes	No

This work proposes a 7-level reduced switch multilevel inverter (RSMLI) topology for EV powertrain application. A comparative analysis of the presented 7-level RSMLI is conducted against several recently published 7-level multilevel inverter configurations to highlight its competitive advantages. Table I. provides a quantitative comparison of various topologies based on key component count parameters, including the number of DC sources (N_{dcs}), power electronic switches (N_{pes}), switching capacitors (N_{sc}) and power diodes (N_{pd}). In addition, a qualitative comparison is also presented, evaluating features such as voltage or current sensor requirement (SR), common ground (CG) configuration, and inclusion of an H-bridge on the load side.

From Table I., it is observed that most of the configurations use a single DC source, maintaining a compact and simplified structure. However, some designs, such as those in [12] and [18], use dual DC sources, which may complicate the control and increase the system's footprint. The proposed topology uses only one DC source, offering a favorable balance between simplicity and performance.

In terms of switch count, designs such as [10] and [18] employ up to 12 and 14 switches, respectively, increasing cost, complexity, and control effort. The proposed topology limits this to eight switches, representing a more efficient and cost-effective solution without compromising the output performance. Similarly, three switching capacitors and three diodes are used in the proposed design, which is comparable with or fewer than other configurations that often require up to four capacitors or three diodes.

The voltage or current sensor requirement is another important design factor. While some topologies, such as [9], [10], [12], and [13], require sensors for voltage or current monitoring and balancing, the proposed design operates without additional sensors, thereby simplifying control and reducing the risk of sensor failure or noise interference.

One of the standout features of the proposed inverter is the inclusion of a common ground configuration, which is present in only a few topologies like [7] and [11]. The CG setup is crucial as it bypasses the stray capacitance, thereby rejecting the leakage currents and significantly lowering the electromagnetic interference (EMI). This contributes to improved safety and reliability, especially in EV applications where EMI can be a concern.

Lastly, the proposed topology does not use H-bridge on load side, helping to reduce voltage stress for switches.

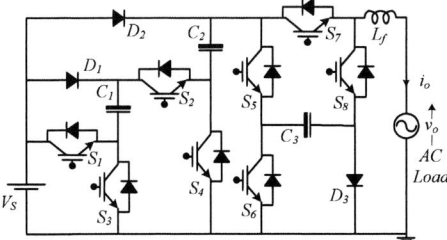

Fig. 1. Proposed 7-Level RSMLI

Overall, the proposed 7-level RSMLI strikes an optimal balance among device count, performance, and practical implementation requirements, making it a compelling candidate for cost-sensitive and compact power conversion systems.

II. PROPOSED RSMLI TOPOLOGY

The presented single source 7-level RSMLI is illustrated in Fig. 1. It comprises of eight power electronic switches (S_1-S_8), three switched capacitors (C_1, C_2 and C_3) and three diodes (D_1, D_2 and D_3). The topology is designed to address key challenges in multilevel inverter design while offering enhanced performance with a simplified structure.

The key contributions of the proposed RSMLI are summarized below:

1. Reduced number of power electronic switches are employed to produce a 7-level inverter voltage, which minimizes overall system complexity, cost, and switching losses.

2. All three capacitors are self-balanced, eliminating necessity for external sensors or complex control strategies. This not only simplifies the control algorithm but also reduces the risk of sensor-related issues such as noise sensitivity or failure.

3. The topology achieves a triple voltage gain using the self-balanced capacitors, making it suitable for applications where higher output voltage is required from a single low-voltage DC source.

4. H-bridge is not used at the output stage for polarity generation. This design choice significantly reduces the voltage stress across switches.

5. The presence of a CG configuration effectively eliminates leakage current and thereby minimizes EMI.

These features make the proposed RSMLI a promising and efficient solution for EV powertrain applications.

A. Operation

The conduction diagram showing operating states for the presented 7-level RSMLI is shown in Fig. 2. Table II illustrate the corresponding switching states used to generate various output voltage levels of the RSMLI. To achieve precise control of the power electronic devices, a level-shifted pulse width modulation (LSPWM) strategy is employed. In this method, multiple carrier waveforms are vertically stacked and compared against a single sinusoidal reference signal to produce the desired stepped multilevel output voltage. This modulation strategy is effective in controlling multilevel inverters due to its simplicity and good harmonic performance. For this work, the carrier waveform frequency is selected as 5 kHz, ensuring a balance between switching losses and output voltage quality.

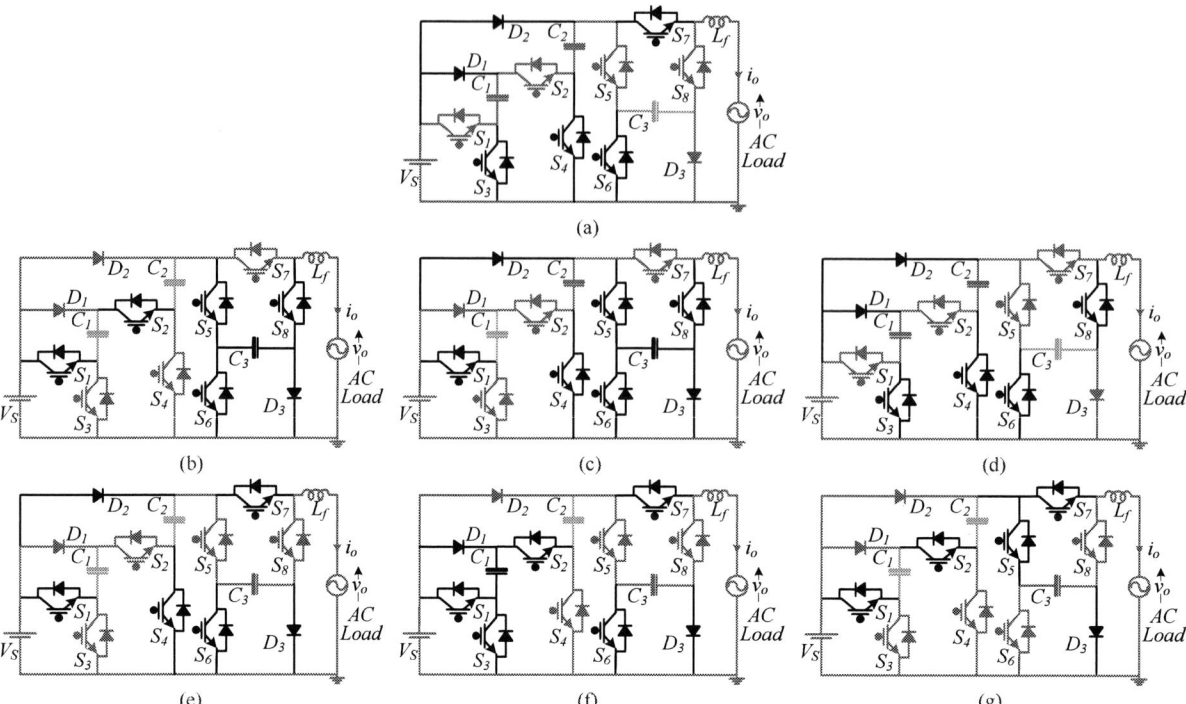

Fig. 2. Switching states for the proposed RSMLI

TABLE II. SWITCHING STATES FOR PROPOSED RSMLI

Level	Load Voltage	S_1	S_2	S_3	S_4	S_5	S_6	S_7	S_8	D_1	D_2	D_3	C_1	C_2	C_3	Fig.
0	0	1	1	0	0	1	0	0	1	NC	NC	C	NE	NE	C	2(a)
Pos 1	V_S	0	0	1	1	0	0	1	0	C	C	NC	CG	CG	NE	2(b)
Pos 2	V_S+V_{C2}	0	1	1	0	0	0	1	0	C	NC	NC	CG	DG	NE	2(c)
Pos 3	$V_S+V_{C1}+V_{C2}$	1	1	0	0	1	0	1	0	NC	NC	C	DG	DG	CG	2(d)
Neg 1	$V_{C3}-V_{C2}-V_{C1}$	0	1	1	0	1	0	0	1	C	NC	NC	CG	CG	DG	2(e)
Neg 2	$V_{C3}-V_{C2}$	0	0	0	1	1	0	0	1	NC	C	NC	NE	CG	DG	2(f)
Neg 3	V_{C3}	0	0	1	1	0	1	0	1	C	C	NC	CG	CG	DG	2(g)

1 = On; 0 = Off; C = Conducting; NC = Not Conducting; CG = Charging; DG = Discharging; NE = No Effect

B. Capacitor Design

The value of capacitors used MLI topology is primarily determined by their longest discharge duration (LDD) under peak loading conditions. In proposed 7-level RSMLI, the capacitor C_1 undergoes discharge during the inverter voltage levels of 0 and $+3V_{dc}$, while capacitor C_2 discharges at 0, $+2V_{dc}$ and $+3V_{dc}$. On the other hand, capacitor C_3 experiences discharge throughout the entire negative half-cycle, specifically at voltage levels $-V_{dc}$, $-2V_{dc}$ and $-3V_{dc}$. Integrating the peak load current (i_{lpeak}) through the capacitors during the LDD under yields the maximum discharge value (Q_{Ci}) in each cycle.

$$Q_{c_i} = \int_{LDD} i_{lpeak}(t)dt$$

The appropriate capacitance values *(C_i)* is determined by

$$C_i > {Q_{Ci}}/{\Delta V_{Ci}}$$

ΔV_{Ci} is the permissible capacitor voltage ripple. This ensures that voltage stability is maintained across the capacitors during operation, avoiding deep discharge.

III. RESULT AND DISCUSSION

Simulation validation are carried out using PLECS platform to evaluate the behaviour of the presented 7-Level RSMLI during steady-state and transient scenarios. The analysis is conducted for resistive (R) and resistive-inductive (RL) loads to gain a comprehensive understanding of the inverter's operation. Inverter output voltage and load current waveforms are examined, along-with their THD, to assess the quality of power delivery. Additionally, power losses across individual power electronic switches are analyzed to determine the RSMLI efficiency. The inverter's efficiency is further assessed under different loading conditions to gauge its overall performance. The findings from these simulations under various operating scenarios are discussed in the subsequent sections.

A. Steady State and Transient Behavior of Proposed RSMLI Toplogy for Resistive Loading

The behavior of the presented RSMLI configuration is analyzed under steady-state and transient scenarios for resistive loading with varying load values. The steady-state response of the inverter for resistive loads of 50 Ω and 100 Ω is illustrated

979-8-3315-3013-6/25 $31.00 © 2025 IEEE

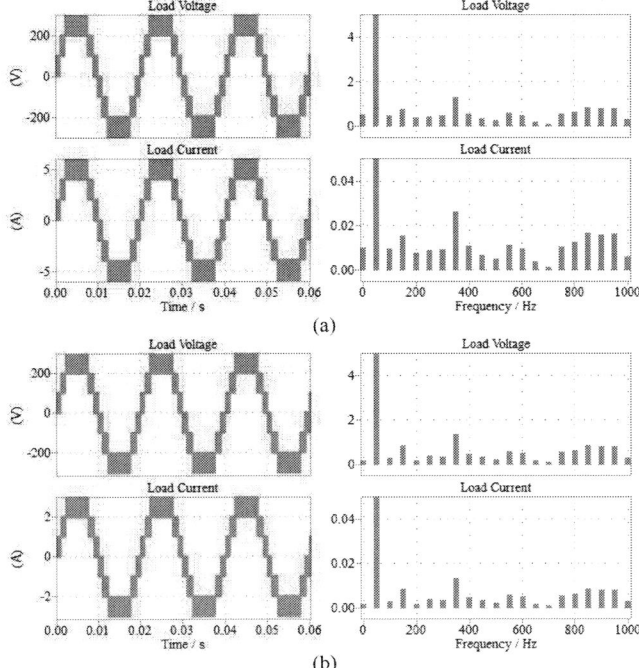

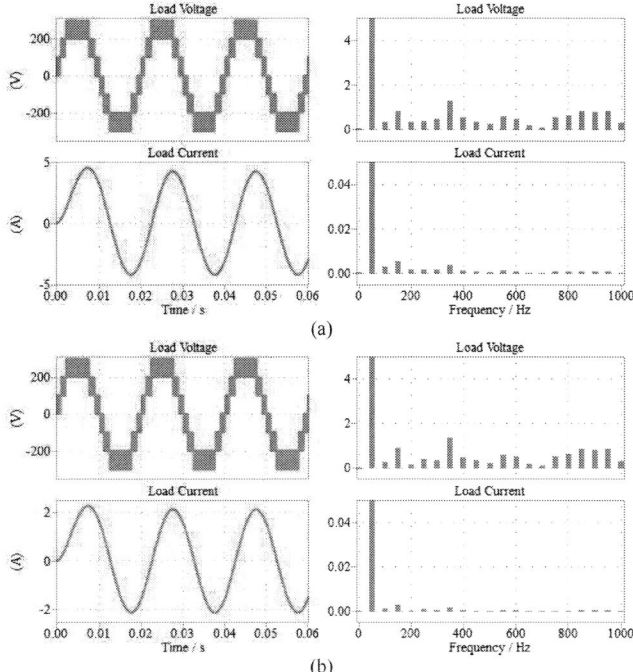

Fig. 3. Steady-state response of RSMLI for R loading: (a) R = 50 Ω and (b) R = 100 Ω

Fig. 5. Steady-state response of RSMLI for RL loading: (a) R= 50 Ω X_L= 50 Ω (b) R= 100 Ω X_L= 100 Ω

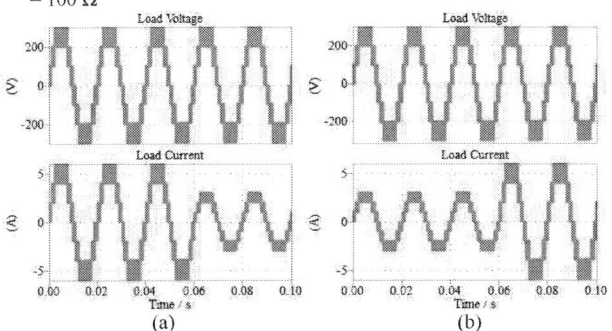

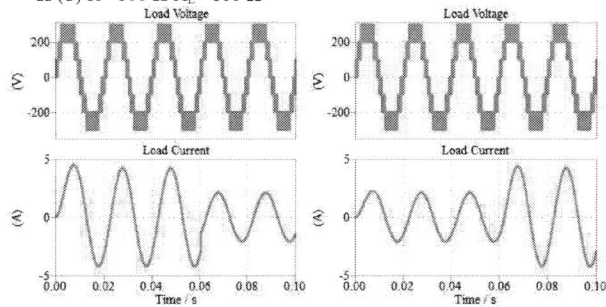

Fig. 4. Transient response of RSMLI for R load variation: (a) from R = 50 Ω to R = 100 Ω and (b) from R = 100 Ω to R = 50 Ω.

Fig. 6. Transient response of RSMLI for RL load variation: (a) from R= 50 Ω X_L= 50 Ω to R= 100 Ω X_L= 100 Ω and (b) from R= 100 Ω X_L= 100 Ω to 50 Ω X_L= 50 Ω.

respectively in Fig. 3(a) and Fig. 3(b). Additionally, transient response due to a load variation from 50 Ω to 100 Ω is illustrated in Fig. 4(a), while Fig. 4(b) presents the transient behavior for a change in resistance from 100 Ω to 50 Ω.

Fig. 3 confirm that for resistive loads, the inverter voltage and current are in-phase. Voltage waveform clearly exhibits a 7-level stepped output, and due to the absence of a filter, PWM switching patterns can also be observed in the inverter current waveform resulting in voltage and current THD around 18.4%.

During transient conditions, variations in inverter current are noticeable. As illustrated in Fig. 4(a), during the R load variation from 50 Ω to 100 Ω, the inverter current decreases accordingly, reducing to half its initial value. Conversely, in Fig. 4(b), as the R load is reduced from 100 Ω to 50 Ω, the inverter current doubles.

B. Steady State and Transient Behavior of Proposed RSMLI Toplogy for Resistive-Inductive Loading

The behaviour of the proposed RSMLI configuration is analyzed for RL loading under steady-state and transient condit-

ions. The steady-state response of the inverter for the load resistance (R) of 50 Ω with an inductive reactance (X_L) of 50 Ω is presented in Fig. 5(a), while the response for a RL load for R of 100 Ω and X_L of 100 Ω is illustrated in Fig. 5(b). The transient response during a load variation of R = 50 Ω, X_L = 50 Ω to R = 100 Ω, X_L = 100 Ω is illustrated in Fig. 6(a). Conversely, the transient response for RL load variation of R = 100 Ω, X_L = 100 Ω to R = 50 Ω, X_L = 50 Ω is depicted in Fig. 6(b).

From the results obtained in Fig. 5, it can be realized that the inverter current remains sinusoidal and smooth under resistive-inductive loading conditions. This is mainly attributed to the filtering effect of the inductance. Although the voltage THD is about 18.4%, the inductive reactance significantly reduces the current THD to approximately 0.28%.

During transient conditions, a reduction in inverter current occurs when the load resistance and inductance increases from R = 50 Ω, X_L = 50 Ω to R = 100 Ω, X_L = 100 Ω, as seen in Fig. 6(a). Conversely, an increase in load current is observed in Fig. 7(b) when the resistance and inductance decrease from R = 100 Ω, X_L = 100 Ω to R = 50 Ω, X_L = 50 Ω.

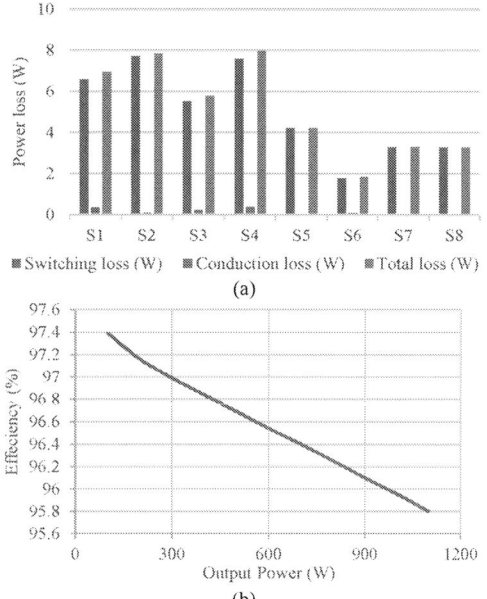

(a)

(b)

Fig. 7. Performance of the RSMLI: (a) power loss analysis and (b) efficiency analysis

Fig. 7 illustrates the system performance based on power loss and efficiency analysis. Switching, conduction, and total power losses for each switch at 1 kW load are depicted in Fig. 7(a). Meanwhile, the efficiency curve under varying load conditions is presented in Fig. 7(b). At a load of 100 W, the RSMLI achieves an efficiency of 97.4%, demonstrating its effectiveness in minimizing losses and maintaining high energy conversion efficiency.

IV. CONCLUSION

This paper introduces a 7-level RSMLI topology designed for EV powertrain applications. The proposed design has achieved a compact and efficient solution by minimizing the count of power electronic switches, gate drivers and passive components. The proposed inverter has delivered a 7-level output using only eight switches, three self-balanced capacitors, and three diodes, without requiring an H-bridge. The self-balancing capacitors have eliminated the requirement for voltage sensors and complex control, enhancing reliability and reducing system complexity. The proposed topology has also achieved triple voltage gain from a single low-voltage DC source, making it ideal for EV applications. Furthermore, the common ground configuration has effectively suppressed the leakage current and electromagnetic interference. Simulation results have demonstrated an excellent dynamic response, low THD and high efficiency under varying load conditions. Overall, the proposed RSMLI has offered a reliable, cost-effective, and high-performance solution for next-generation EV powertrains.

ACKNOWLEDGMENT

This research is supported by A*STAR, Singapore, under its "RIE2025 MTC IAF-PP: Development of High Performance Electric Traction Module (HiPe-ETraM)", Grant No. M22K4a0044.

REFERENCES

[1] "Fast Facts on Transportation Greenhouse Gas Emissions," United States Environmental Protection Agency, May 2024. [Available Online] https://www.epa.gov/greenvehicles/fast-facts-transportation-greenhouse-gas-emissions

[2] P. Sudararajan, D. Kolantla, J. Saha, S. Murshid, H. M. Kang and S. K. Panda, "Design of High-Density and High-Efficiency SiC-based Drivetrain Inverter," in *proc. 2024 IEEE 9th Southern Power Electronics Conference (SPEC)*, Brisbane, Australia, 2024.

[3] Poorfakhraei, M. Narimani and A. Emadi, "A Review of Multilevel Inverter Topologies in Electric Vehicles: Current Status and Future Trends," *IEEE Open J. Pow. Elect*, vol. 2, pp. 155-170, 2021.

[4] S. Murshid, P. Sundararajan, M. Sahani and S. K. Panda, "A 13-Level Switched Capacitior Boost MLI With Dual Configurability for EV Applications," in *proc. 2024 IEEE 9th Southern Power Electronics Conference (SPEC)*, Brisbane, Australia, 2024, pp. 1-4.

[5] S. Murshid, M. Tayyab, A. Sarwar, M. Tariq, A. Al-Durra and A. Tomar, "Self-Balanced Twenty Five Level Switched Capacitor Multilevel Inverter With Reduced Switch Count and Voltage Boosting Capability," *IEEE Trans. Ind. Appl.*, vol. 58, no. 2, pp. 2183-2194, March-April 2022.

[6] R. Barzegarkhoo, M. Forouzesh, S. S. Lee, F. Blaabjerg and Y. P. Siwakoti, "Switched-Capacitor Multilevel Inverters: A Comprehensive Review," *IEEE Trans. Pow. Elect.*, vol. 37, no. 9, pp. 11209-11243, Sept. 2022.

[7] A. Jakhar and N. Sandeep, "Switched-Capacitor-Based Seven-Level Boosting Inverter With Reduced Voltage Stress for Grid-Connected Photovoltaic Applications," *IEEE J. Emerg. and Sel. Top. Ind. Elect.*, [Early Access]

[8] V. S. K. Prasadarao, S. Peddapati and B. Kumar, "A Voltage-Boosting Seven-Level Switched Capacitor Multilevel Inverter With Reduced Device Count," *IEEE J. Emerg. and Sel. Top. Pow Elecs*, vol. 12, no. 1, pp. 743-753, Feb. 2024.

[9] S. Kumari, N. Sandeep and A. K. Verma, "T-Type Seven-Level Inverter With Triple Voltage-Boosting Gain," *IEEE J. Emerg. and Sel. Top. Ind. Elect.*, vol. 4, no. 3, pp. 899-906, July 2023.

[10] S. A. Saleh, "Maximum Resolution Based Method for Balancing Capacitor Voltages in 7-Level Single Phase Flying-Capacitor Wavelet Modulated Inverters," *IEEE Trans. Ind. Appl.*, vol. 59, no. 4, pp. 5019-5031, July-Aug. 2023.

[11] A. Srivastava and J. Seshadrinath, "A Single-Phase Seven-Level Triple Boost Inverter for Grid-Connected Transformerless PV Applications," *IEEE Trans. Ind. Elect.*, vol. 70, no. 9, pp. 9004-9015, Sept. 2023.

[12] P. Manoj, A. Kirubakaran and V. T. Somasekhar, "An Asymmetrical Dual Quasi-Z-Source Based 7-Level Inverter for PV Applications," *IEEE Trans. Ener. Conv.*, vol. 38, no. 2, pp. 1097-1107, June 2023.

[13] R. d. B. Cardoso, E. R. C. da Silva, L. R. Limongi and A. E. L. d. Costa, "A Seven-Level Inverter With Natural Balance and Boosting Capability," *IEEE Trans. Ind. Appl.*, vol. 59, no. 1, pp. 925-937, Jan.-Feb. 2023.

[14] S. K. Baksi and R. K. Behera, "A Reduced Switch Count Seven-Level Boost ANPC Based Grid Following Inverter Topology With Photovoltaic Integration," *IEEE Trans. Ind. Appl.*, vol. 59, no. 4, pp. 4238-4251, July-Aug. 2023.

[15] M. J. Sathik, M. F. Elmorshedy and D. J. Almakhles, "A New Boost Topology Seven-Level Inverter of High Voltage Gain Ability and Continuous Input Current With MPPT for PV Grid Integration," *IEEE Acc.*, vol. 11, pp. 139236-139248, 2023.

[16] P. S. V. Kishore, N. Jayaram, S. Jakkula, Y. R. Sankar, J. Rajesh and S. Halder, "A New Reduced Switch Seven-Level Triple Boost Switched Capacitor Based Inverter," *IEEE Acc.*, vol. 10, pp. 73931-73944, 2022.

[17] J. Zhao, Y. Chen, J. Zeng and J. Liu, "Low-Voltage Stress Seven-Level Inverter Based on Symmetrical Capacitors," *IEEE J. Emerg, and Sel Top. Pow. Electr*, vol. 10, no. 3, pp. 3033-3044, June 2022.

[18] V. Dargahi, A. K. Sadigh, R. R. Khorasani and J. Rodriguez, "Active Voltage Balancing Control of a Seven-Level Hybrid Multilevel Converter Topology," *IEEE Trans. Ind. Elect.*, vol. 69, no. 1, pp. 74-89, Jan. 2022.

Cyber Resilient Fully Distributed Secondary Control for Inverter Dominant AC microgrid

Phani Swecha Tadepalli
Dept. of Electrical Engineering
National Institute of Technology
Warangal, India
swechagsit@gmail.com

M. N. Alam
Dept. of Electrical Engineering
National Institute of Technology
Warangal, India
mnalam@nitw.ac.in

Deepak Pullaguram
Dept. of Electrical Engineering
Indian Institute of Technology Kharagpur
Kharagpur, India
drpullaguram@ee.iitkgp.ac.in

Abstract—A distributed consensus control strategy is preferred in the secondary control layer in AC microgrids above the primary droop control to regulate voltage and frequency. This control strategy involves information exchange among neighboring distributed generators(DGs) on a communication layer. This communication layer is prone to cyber intrusions, such as false data injection (FDI) attacks from malicious intruders that aim to disorient system control objectives. To deal with such attacks, this paper proposes a virtual layer-based distributed consensus control strategy that achieves resiliency against FDI attacks on secondary control layer of islanded AC microgrid. The proposed strategy achieves resiliency against sensor and communication FDI attacks without imposing any restrictions on the number of DGs attacked. The effectiveness of the proposed strategy is validated through simulation studies on a 4- DG test system under various FDI attack conditions.

Index Terms—Cyber-attack, false data injection (FDI), distributed generator (DG), distributed consensus control, resilient control, secondary control, virtual layer.

I. INTRODUCTION

Microgrids (MGs) have emerged as a viable solution to integrate renewable energy sources due to the rapid advancement of distributed power generation technologies. Compared to conventional power systems, MGs offer greater flexibility, efficiency, and environmental benefits. They can operate in grid-connected and islanded configurations and typically comprise distributed generators (DGs), energy storage systems, controllable loads, and power electronic converters. To ensure stable and reliable islanded operation, MGs generally adopt a hierarchical framework: primary, secondary, and tertiary control layers. Droop-based primary control enables decentralized power sharing, but introduces voltage and frequency deviations. Secondary control corrects these deviations for accurate frequency and voltage regulation. Tertiary control optimizes economic efficiency and power management. In contrast to traditional centralized control, AC microgrids using distributed control rely on peer-to-peer data exchange between DG units, which inherently reduces communication bandwidth, enhances scalability, and eliminates the risk of a single-point failure. Although decentralized secondary controllers are independent of communications, they suffer from poor coordination among distributed generators (DGs), resulting in suboptimal voltage

regulation and inaccurate current sharing, particularly under dynamic load conditions. This may lead to voltage imbalances, inefficient power distribution, and slower convergence during disturbances. Thus, distributed control offers an optimal trade-off between performance, scalability, and resilience requirements.

The reliance of distributed microgrids on communication and digital control systems makes them prone to cyberattacks that can disrupt normal operations, destabilize the system, and interrupt the power supply. The lack of global information in distributed cyber topologies will especially worsen control decisions in the presence of these attacks. False data injection (FDI) and denial of service (DOS) are two main types of attacks that commonly occur in the power industry that manipulate sensor data, controller inputs, and communication networks. While DoS attacks disrupt communication, hindering the timely operation of the network, FDI attacks manipulate the data exchanged between DGs, compromising the reliability of data.

This paper focuses on FDI attacks, which are highly deceptive in nature due to the difficulty in verifying data authentication. In the literature, two main approaches, data-driven and model-driven, have been developed to effectively counteract FDI attacks [1]. Data-driven approaches utilize historical input-output datasets combined with artificial intelligence (AI) techniques, such as neural networks [2], deep learning [3], reinforcement learning [4], and federated learning [5] to adaptive black-box models. Although these methods accurately identify attacks, they require significant computational resources and data collection, which limits their practical application. Although more accurate and robust against disturbances than model-based approaches, they remain emerging technologies, whereas model-based methods are considered more practical and reliable.

The authors in [6] proposed an observer-based control using confidence and trust factors to detect FDI attacks. This method detects and isolates attacked DGs to mitigate the propagation of the attack across the network. Instead of disconnecting the attacked DGs, [7] introduced a resilient observer method to mitigate FDI attacks by estimating malicious data injected into sensors or actuators. In [8], a resilient method is discussed that integrates sliding mode control and the communication

979-8-3315-3013-6/25 $31.00 © 2025 IEEE

link observer. A distributed resilient control technique against synchronization FDI attacks in AC microgrids is presented in [9]. The authors in [10] present a resilient control framework that employs an iterative observer-based approach to mitigate voltage and frequency deviations in the presence of an FDI attack. Although the method demonstrates effectiveness in counteracting such attacks, its performance depends on the number of iterations, leading to a high computational complexity. A hidden layer based resilient method has been discussed in [11] to effectively mitigate time-dependent and state-dependent FDI attacks. Similarly, a cross-layer control method to achieve resiliency against attacks in AC microgrids is discussed in [12]. These methods primarily address simple FDI attacks, neglecting sophisticated stealth attack strategies. In addition, these methods are based on the assumption that attackers cannot compromise all vulnerable points in the microgrid, which may not reflect real-world adversarial capabilities. To deal with stealthy state-dependent FDI attacks, a virtual layer based leader-follower distributed control is discussed in [13]. In [14], resilience is achieved against bounded and unbounded FDI attacks on the actuator and communication links. A recent study [15] introduced a resilient-by-design approach, developing a control strategy to counter strategic stealthy attacks while ensuring the privacy of DG measurements. However, this method relies on a state-dependent detection mechanism to identify destabilizing attacks. A proportional-integral observer based strategy is introduced in [16] to counter FDI attacks AC microgrids without requiring neighboring information.

This paper proposes an FDI- resilient dynamic average control strategy for islanded AC microgrids. The proposed strategy improves AC microgrid security by mitigating both strategic stealthy and destabilizing FDI attacks. It ensures a stable economic active and reactive power distribution while regulating the frequency and average voltage to nominal values. This method works effectively for both actuator and communication link attacks without imposing any restrictions on the number of attacked DGs.

The rest of this article is organized as follows: Section II details the configuration of the microgrid system considered in this study and its communication aspects. Section III discusses the proposed control strategy for seamless transition and power transfer. Section IV validates the proposed strategy through MATLAB/SIMULINK simulation studies. Section V provides the conclusion and future scope of the paper.

II. MICROGRID PRELIMINARIES AND PROBLEM FORMULATION

A. Modelling of Islanded AC Microgrid

This study investigates an islanded AC microgrid consisting of four parallel connected DGs through boost converters as illustrated in Fig. 1. The DGs are interconnected via tie lines to enable load sharing. Each DG exchanges its local measurements with neighboring units through a TCP/IP communication link, following a communication graph topology depicted by the blue lines in Fig. 1. Additionally, a virtual communication graph, represented by green dotted lines, is

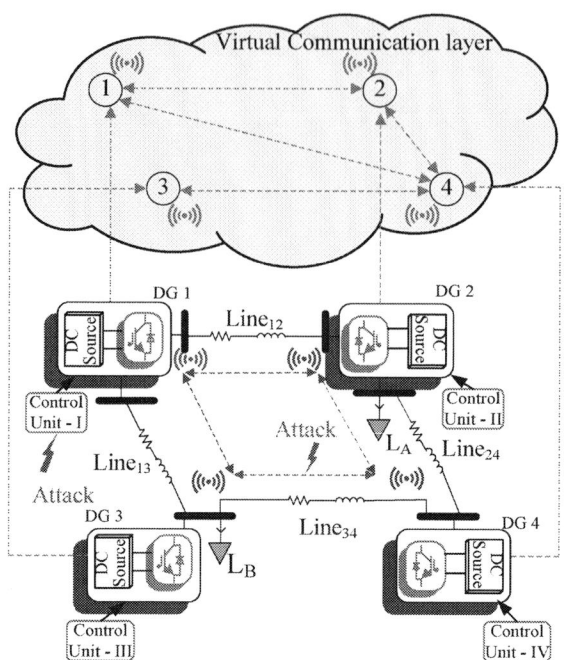

Fig. 1: Schematic of 4-DG AC microgrid system.

employed for control purposes and may differ from the actual physical communication network. Each DG is controlled by a hierarchical control strategy as discussed below.

At the primary control level, droop control is widely implemented, enforcing a linear correlation between frequency and active power as well as voltage and reactive power, which is mathematically expressed as,

$$\omega_i = \omega_{\text{nom}} - m_i P_i \qquad (1)$$

$$v_i = v_{\text{nom}} - n_i Q_i \qquad (2)$$

where, ω_i, v_i are the frequency and voltage, P_i, Q_i are the filtered active and reactive power injections, m_i, n_i are $\omega - P$ and $V - Q$ droop coefficients of ith DG, respectively, and ω_{nom}, v_{nom} are the nominal frequency and nominal voltage of the microgrid.

The primary droop control ensures decentralized operation while introducing frequency and voltage deviations, and the droop control alone may not guarantee proportional reactive power sharing among all connected distributed generators (DGs), due to line impedance mismatches and system disturbances. To address this, a distributed secondary control strategy is largely adopted, aiming to restore frequency, regulate the dynamic average voltage, and achieve accurate active and reactive power sharing. Further, in literature, many decentralized methods are proposed for the economic dispatch, which modifies, (1) as

$$\omega_i = \omega_{\text{nom}} - k_{\lambda_i} \lambda_i, \qquad (3)$$

$$\lambda_i = \frac{\partial C_i}{\partial P_i} \qquad (4)$$

979-8-3315-3013-6/25 $31.00 © 2025 IEEE

where $C_i(P_i) = a_i P_i^2 + b_i P_i + c_i$ is the active power quadratic cost functions of ith DG , and k_{λ_i} is the droop constant parameter that is chosen to have economic dispatch while maintaining stability. The a_i, b_i, and c_i are the cost coefficients. To have a cost-optimal operation, restore frequency, regulate the dynamic average voltage, and achieve proportional reactive power sharing, an economic droop-based secondary control is designed that feeds input to primary control, [17]:

$$\omega_i = \omega_{\text{nom}} - k_{\lambda_i} \lambda_i + \Delta \omega_i, \tag{5}$$

$$v_i = v_{\text{nom}} - n_i Q_i + \Delta v_i \tag{6}$$

where $\Delta \omega_i$ and Δv_i are the secondary control inputs to droop control. The frequency and economic dispatch secondary control loop are given by

$$\Delta \omega_i = \int u_{\omega_i} + \int u_{\lambda_i} \tag{7}$$

$$u_{\omega_i} = \sum_{j \in N_i} \alpha_{ij}(\omega_j - \omega_i) + \beta_i(\omega_{\text{nom}} - \omega_i) \tag{8}$$

$$u_{\lambda_i} = \sum_{j \in N_i} \alpha_{ij}(\lambda_j - \lambda_i) \tag{9}$$

The average voltage and proportional reactive power secondary control loop as

$$\Delta v_i = \int u_{v_i} + \int u_{Q_i} \tag{10}$$

$$u_{\bar{v}_i} = \beta_i(v_{\text{nom}} - \bar{v}_i) \tag{11}$$

$$\bar{v}_i = v_i + \int \sum_{j \in N_i} \alpha_{ij}(\bar{v}_j - \bar{v}_i) \tag{12}$$

$$u_{Q_i} = \sum_{j \in N_i} \alpha_{ij}(n_j Q_j - n_i Q_i) \tag{13}$$

This distributed secondary control strategy leverages local communication among DGs, avoiding reliance on a central controller. This distributed control strategy ensures robust microgrid operation, enhances resilience against disturbances, optimizes economic dispatch, and enables seamless plug-and-play capability for DGs. However, the reliance on distributed communication makes the system vulnerable to cyber threats, such as False Data Injection (FDI) attacks. Adversaries can manipulate exchanged data, leading to deviations in frequency, voltage regulation, and improper power sharing, and sometimes can lead to system instability. Thus, resilient distributed control mechanisms should be incorporated into the secondary control framework to enhance cybersecurity in the AC microgrids.

III. VIRTUAL LAYER BASED RESILIENT DISTRIBUTED CONTROL FOR AC MICROGRIDS

In this paper, a resilient controller proposed in [14] is adopted and modified to achieve cyber resiliency for AC mi-

crogrid systems against FDI attacks. The modified secondary control is given by

$$\Delta \omega_i = \int u_{\bar{\omega}_i} + \int u_{\bar{\lambda}_i}; \quad \Delta v_i = \int u_{\bar{v}_i} + \int u_{\bar{Q}_i} \tag{14}$$

$$u_{\bar{\omega}_i} = (\omega_{\text{nom}} - \bar{\omega}_i); \quad u_{\bar{\lambda}_i} = (\lambda_i - \bar{\lambda}_i) \tag{15}$$

$$u_{\bar{v}_i} = (v_{\text{nom}} - \bar{v}_i); \quad u_{\bar{Q}_i} = (Q_i - \bar{Q}_i) \tag{16}$$

The dynamics of the variables of form $(\bar{\cdot})$ in the proposed resilient controller for ith DG of the AC microgrid is given by

$$\dot{\bar{\omega}}_i = \alpha(\omega_i - \bar{\omega}_i) + \sum_{j \in \mathcal{N}_i} \beta\Big(\alpha(\bar{\omega}_j - \bar{\omega}_i) - (\sigma_{\omega_j} - \sigma_{\omega_i})\Big) + \dot{\omega}_i, \tag{17}$$

$$\dot{\sigma}_{\omega_i} = \alpha(\omega_i - \sigma_{\omega_i}) + \sum_{j \in \mathcal{N}_i} \beta\Big(\alpha(\sigma_{\omega_j} - \sigma_{\omega_i}) + (\bar{\omega}_j - \bar{\omega}_i)\Big) + \dot{\omega}_i, \tag{18}$$

Here in this paper, the dynamics are given only for $\dot{\bar{\omega}}_i$, but the same is valid for all other variables, $\dot{\bar{\lambda}}_i$, $\dot{\bar{v}}_i$, and $\dot{\bar{Q}}_i$

In the presence of communication $\delta_{\bar{\omega}_{ij}}(t)$, $\delta'_{\sigma_{ij}}(t)$ and actuator attacks $\delta_{u_{\bar{\omega}_i}}(t)$, $\delta'_{u_{\sigma_i}}(t)$, (17) and (18) are modified as,

$$\dot{\bar{\omega}}_i = \alpha(\omega_i - \bar{\omega}_i) + \sum_{j \in \mathcal{N}_i} \beta\Big(\alpha(\bar{\omega}_j - \bar{\omega}_i) - (\sigma_{\omega_j} - \sigma_{\omega_i})\Big) + \dot{\omega}_i$$
$$+ \delta_{u_{\bar{\omega}_i}}(t) + \delta_{\bar{\omega}_{ij}}(t) + \delta_{\sigma_{ij}}(t), \tag{19}$$

$$\dot{\sigma}_i = \alpha(\omega_i - \sigma_i) + \sum_{j \in \mathcal{N}_i} \beta\Big(\alpha(\sigma_j - \sigma_i) + (\bar{\omega}_j - \bar{\omega}_i)\Big) + \dot{\omega}_i$$
$$+ \delta'_{u_{\sigma_i}}(t) + \delta'_{\sigma_{ij}}(t) + \delta'_{\bar{\omega}_{ij}}(t), \tag{20}$$

The combined effect of communication and actuator attacks can be expressed as $\delta_i(t) \triangleq \delta_{u_{\bar{\omega}_i}}(t) + \delta_{\bar{\omega}_{ij}}(t) + \delta_{\sigma_{ij}}(t)$, $\delta'_i(t) \triangleq \delta'_{u_{\sigma_i}}(t) + \delta'_{\sigma_{ij}}(t) + \delta'_{\bar{\omega}_{ij}}(t)$. In this paper, it is assumed that both $\delta_i(t)$ and $\delta'_i(t)$ are bounded. To analyse the stability of the developed resilient controller, a error signal is defined as

$$e_{\omega_i} \triangleq \omega_i - \bar{\omega}_i, \quad e_{\sigma_i} \triangleq \omega_i - \sigma_i. \tag{21}$$

Taking derivative of the error e_{ω_i} and substituting $\dot{\bar{\omega}}_i$ from (19), we obtain

$$\begin{aligned}
\dot{e}_{\omega_i} &= \dot{\omega}_i - \dot{\bar{\omega}}_i \\
&= \dot{\omega}_i - \Big\{\alpha(\omega_i - \bar{\omega}_i) + \sum_{j \in \mathcal{N}_i} \beta\Big(\alpha(\bar{\omega}_j - \bar{\omega}_i) - (\sigma_{\omega_j} - \sigma_{\omega_i})\Big) \\
&\quad + \dot{\omega}_i + \delta_{u_{\bar{\omega}_i}}(t) + \delta_{\bar{\omega}_{ij}}(t) + \delta_{\sigma_{ij}}(t)\Big\} \\
&= -\alpha e_{\omega_i} - \sum_{j \in \mathcal{N}_i} \beta\Big(\alpha(\bar{\omega}_j - \bar{\omega}_i) - (\sigma_{\omega_j} - \sigma_{\omega_i})\Big) - \delta_i(t). \tag{22}
\end{aligned}$$

Similarly, differentiating e_{σ_i} and using (20) gives

$$\dot{e}_{\sigma_i} = -\alpha e_{\sigma_i} - \sum_{j \in \mathcal{N}_i} \beta\Big(\alpha(\sigma_j - \sigma_i) + (\bar{\omega}_j - \bar{\omega}_i)\Big) - \delta'_i(t) \tag{23}$$

A. Lyapunov Function Candidate

Considering a quadratic Lyapunov function is

$$V = \frac{1}{2} \sum_{i=1}^{N} \Big(e_{\omega_i}^2 + e_{\sigma_i}^2\Big), \tag{24}$$

or, in vector notation,

$$V = \frac{1}{2}\|e_\omega\|^2 + \frac{1}{2}\|e_\sigma\|^2,$$

where

$$e_\omega = \begin{bmatrix} e_{\omega_1} & \dots & e_{\omega_N} \end{bmatrix}^\top, \quad e_\sigma = \begin{bmatrix} e_{\sigma_1} & \dots & e_{\sigma_N} \end{bmatrix}^\top.$$

Differentiating V along the trajectories yields

$$\dot{V} = \sum_{i=1}^{N} \left(e_{\omega_i} \dot{e}_{\omega_i} + e_{\sigma_i} \dot{e}_{\sigma_i} \right).$$

Substitute the error dynamics from (22) and (23), we get

$$
\begin{aligned}
\dot{V} = & -\alpha \sum_{i=1}^{N} e_{\omega_i}^2 - \sum_{i=1}^{N} e_{\omega_i} \sum_{j \in \mathcal{N}_i} \beta \Big(\alpha(\bar{\omega}_j - \bar{\omega}_i) - (\sigma_{\omega_j} - \sigma_{\omega_i}) \Big) \\
& - \sum_{i=1}^{N} e_{\omega_i} \delta_i(t) - \alpha \sum_{i=1}^{N} e_{\sigma_i}^2 - \sum_{i=1}^{N} e_{\sigma_i} \delta_i'(t) \\
& - \sum_{i=1}^{N} e_{\sigma_i} \sum_{j \in \mathcal{N}_i} \beta \Big(\alpha(\sigma_j - \sigma_i) + (\bar{\omega}_j - \bar{\omega}_i) \Big).
\end{aligned}
$$

The coupling terms, involving sums over the neighbors, can be expressed compactly using the Laplacian matrix L of the communication graph. For a connected graph, L is positive semidefinite with one zero eigenvalue and all other eigenvalues positive. The cross terms appear in forms $-\beta \sum_{i=1}^{N} e_{\omega_i} \sum_{j \in \mathcal{N}_i} \alpha(\bar{\omega}_j - \bar{\omega}_i)$ or $\beta \sum_{i=1}^{N} e_{\omega_i} \sum_{j \in \mathcal{N}_i} (\sigma_{\omega_j} - \sigma_{\omega_i})$. Rearranging these sums in vector form, we get a quadratic expression as

$$-\beta \alpha\, e_\omega^T (L \otimes I) e_\omega \quad \text{and} \quad \beta\, e_\omega^T (L \otimes I) e_\sigma. \qquad (25)$$

Using Young's inequality:

$$ab \leq \frac{\epsilon}{2} a^2 + \frac{1}{2\epsilon} b^2, \quad \forall \epsilon > 0.$$

we can modify, (25), as

$$\beta \left| e_\omega^T (L \otimes I) e_\sigma \right| \leq \frac{\beta \lambda_{\max}(L)\epsilon}{2} \|e_\omega\|^2 + \frac{\beta \lambda_{\max}(L)}{2\epsilon} \|e_\sigma\|^2 \quad (26)$$

By choosing ϵ appropriately, these terms are bounded within limits and can be significantly lower than the terms $-\alpha \|e_\omega\|^2$ and $-\alpha \|e_\sigma\|^2$. Thus, the derivative $\dot{V}$ can expressed as

$$
\begin{aligned}
\dot{V} \leq & - \left(\alpha - \frac{\beta \lambda_{\max}(L)\epsilon}{2} \right) \|e_\omega\|^2 - \left(\alpha - \frac{\beta \lambda_{\max}(L)}{2\epsilon} \right) \|e_\sigma\|^2 \\
& + \delta_i + \delta'.
\end{aligned}
\qquad (27)
$$

In the absence of attack, for $\dot{V}$ to be negative definite, α and β are chosen as

$$\alpha > \frac{\beta \lambda_{\max}(L)\epsilon}{2} \quad \text{and} \quad \alpha > \frac{\beta \lambda_{\max}(L)}{2\epsilon}. \qquad (28)$$

Considering $\epsilon = 1$ leads to

$$\alpha > \frac{\beta \lambda_{\max}(L)}{2} \quad \Longrightarrow \quad \beta < \frac{2\alpha}{\lambda_{\max}(L)}.$$

Thus, with $\alpha > 0$, $0 < \beta < \frac{2\alpha}{\lambda_{\max}(L)}$, the proposed resilient controller is exponentially stable.

TABLE I: Test System data

Parameter	Value
V_{nom}	$400\ V$
Inverter rating	$3\,kVA$
Filter Parameters	$L_{f_i} = 3\,mH$ $R_f = 0.1$ $C_{f_i} = 1000\,\mu F$
Line parameters	$L_{12} = 0.318\,mH\ \ r_{12} = 0.23\,\Omega$ $L_{14} = 1.846\,mH\ \ r_{14} = 0.35\,\Omega$ $L_{34} = 1.846\,mH\ \ r_{34} = 1\,\Omega$ $L_{23} = 0.678\,mH\ \ r_{23} = 2.1\Omega$
Droop controller	$mp = 1.4e-3, n_q = 1.3e-3$ $a_1 = 0.05, b_1 = 1.2;$ $a_2 = 0.0025, b_2 = 1.2$ $a_3 = 0.004, b_3 = 1.2$ $a_4 = 0.004, b_4 = 1.2$
Secondary controller	$\alpha = 500, \beta = 1000$

The non-zero attack vectors, contribute additional bounded terms of the form

$$-\sum_{i=1}^{N} e_{\omega_i} \delta_i(t) - \sum_{i=1}^{N} e_{\sigma_i} \delta_i'(t).$$

As these disturbances are bounded, they can be upper bounded by

$$\left| \sum_{i=1}^{N} e_{\omega_i} \delta_i(t) \right| \leq \|e_\omega\|\, D,$$

and similarly for the e_{σ_i} terms. Consequently, the overall derivative inequality becomes

$$\dot{V}(t) \leq -c\, V(t) + D^*,$$

where $c > 0$ depends on α, β, and $\lambda_{\max}(L)$, and D^* depends on the disturbance bounds.

By standard Lyapunov arguments or the Input-to-State Stability (ISS) framework, the inequality

$$\dot{V}(t) \leq -c\, V(t) + D^*$$

implies that

$$V(t) \leq e^{-ct} V(0) + \frac{D^*}{c} \left(1 - e^{-ct} \right).$$

This shows that the state errors converge exponentially to a ball of radius proportional to D^*/c. Thus, having the resilient control is Uniform Ultimate Bounded (UUB). From the above analysis, it can be observed that the proposed resilient distributed control satisfies the UUB condition, and the error remains bounded. With large values of α and β gains the steady-state error during the attack can be reduced to near zero.

IV. SIMULATION RESULTS

A. AC Microgrid Operation with Conventional Hierarchical Control

The AC microgrid is initially operated with conventional hierarchical control [17] supplying a total load of (7+j0.7)

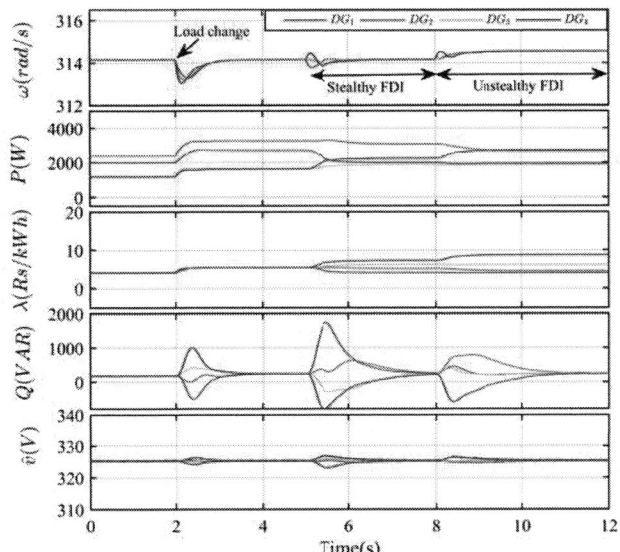

Fig. 2: Conventional hierarchical control under actuator FDI attack.

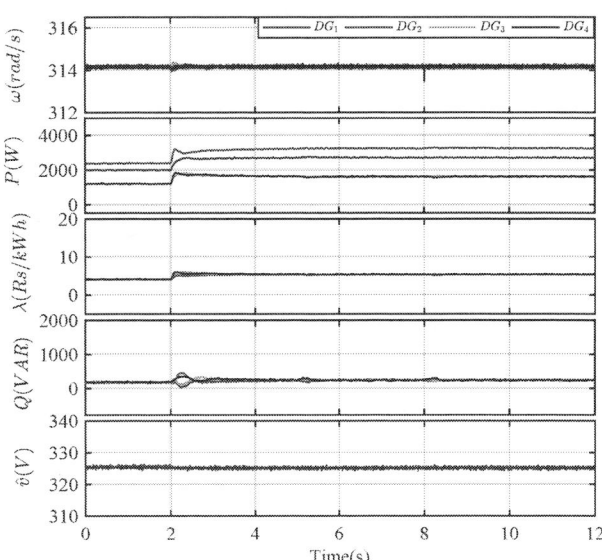

Fig. 3: Proposed resilient control under actuator FDI attack.

kVA. The system is at a steady state and acheives economic load-sharing while ensuring that frequency and average voltage are regulated at nominal values. At $t = 2\ s$ a load of (2.8+j0.2) kVA is added on DG_2. This increase in load is shared economically among all DGs as shown in Fig 2. At $t = 5\ s$ a stealthy FDI attack is performed by injecting a malicious data of $\omega = [3.1\ 0\ -3.1\ 0]\ rad/s$ into the actuator of DGs. Due the stealthy nature of injected attack no frequency deviations are observed, while power supplied by DGs increased imitating the case of load change to deceive the operator. At $t = 8\ s$ a destabilizing FDI attack is performed by changing the injected values to $\omega = [3.1\ 0\ 6.2\ 0]\ rad/s$ into the actuator of DGs. As a result of this attack, the frequency shifts from its nominal value and stabilizes at a new operating point of around $310\ rad/s$, while the incremental cost parameter λ also deviates from consensus, as illustrated in Fig. 2. The extent of deviation depends on the false data injected. However, the impact of bounded FDI on steady-state average voltage and reactive power remains minimal.

B. Proposed Resilient Control under Actuator Attack

This case study analyzes the performance of the proposed controller in an islanded AC microgrid under load variations and its resilience against actuator FDI attacks. Initially, the system operates as in the previous case. At $t = 2s$, the load at DG_2 is increased by (2.8+j0.2) kVA, which is economically distributed among the DGs, ensuring average voltage regulation, as shown in Fig. 3. At $t = 5s$, and $t = 8s$, a stealthy and unstealthy FDI attacks, respectively, as discussed in the previous case, are performed on the actuator of DGs by injecting malicious data into the frequency of the same magnitude as in the previous case. Almost instantaneously, the proposed controller restores powers to their pre-attack state. Additionally, the frequency ω at all DGs restores to

the nominal value of $314.15\ rad/s$, unlike the conventional controller.

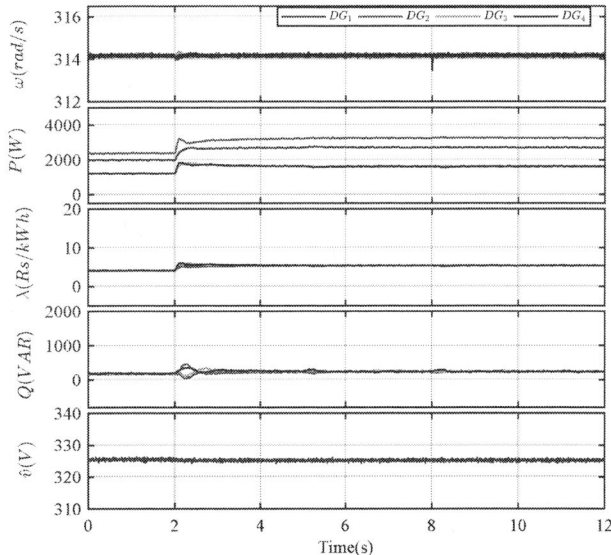

Fig. 4: Proposed resilient control under actuator and communication FDI attack.

C. Proposed controller under actuator and communication attacks

This case study assesses microgrid performance under state-dependent FDI attacks on actuator communication channels of DGs. At $t = 2s$ an FDI attack of $\dot{\omega} = F_1\omega + \omega_0$ is injected into all DG actuators and shared with neighboring agents. At the same time an FDI attack is also performed on the virtual

979-8-3315-3013-6/25 $31.00 © 2025 IEEE

layer as $\dot{\omega} = F_2\omega + \omega_0$. Here $F_1 = -\mathbf{I}$ and $F_2 = -2\mathbf{I}$ and $\mathbf{I}$ represents identity matrix. The proposed controller shows resiliency to such state-dependent attacks on both actual and virtual communication layers and restores average voltage and frequency as shown in Fig. 4.

V. CONCLUSION

This article presents a resilient distributed hierarchical control framework to defend against FDI attacks in islanded AC microgrids. This method utilizes a virtual layer to secure the integrity of the data exchanged. The proposed controller ensures frequency regulation and economic active power sharing even during the presence of FDI attacks. The approach accounts for potential manipulations in actuator/sensor data and data exchanged between neighboring DGs. It also works effectively when the virtual layer is also under FDI attack. This method does not impose any restrictions on the number of DGs attacked and shows robustness even when all distributed generation units are compromised. The effectiveness of the proposed strategy is validated through MATLAB/Simulink simulation results.

ACKNOWLEDGMENT

This work is partly supported by the NITW-Hitachi Energy Smart Electric Grid Laboratory.

REFERENCES

[1] P. S. Tadepalli and D. Pullaguram, "Distributed control microgrids: Cyber-attack models, impacts and remedial strategies," *IEEE Transactions on Signal and Information Processing over Networks*, vol. 8, pp. 1008–1023, 2022.

[2] P. S. Tadepalli, D. Pullaguram, and M. N. Alam, "Cyber-resilient strategy for dc microgrids against concurrent fdi and dos attacks," *IEEE Transactions on Industrial Informatics*, pp. 1–12, 2025.

[3] Y. Wang and B. C. Pal, "Destabilizing attack and robust defense for inverter-based microgrids by adversarial deep reinforcement learning," *IEEE Transactions on Smart Grid*, vol. 14, no. 6, pp. 4839–4850, 2023.

[4] A. J. Abianeh, Y. Wan, F. Ferdowsi, N. Mijatovic, and T. Dragičević, "Vulnerability identification and remediation of fdi attacks in islanded dc microgrids using multiagent reinforcement learning," *IEEE Transactions on Power Electronics*, vol. 37, no. 6, pp. 6359–6370, 2022.

[5] V. Veerasamy, L. P. M. I. Sampath, S. Singh, H. D. Nguyen, and H. B. Gooi, "Blockchain-based decentralized frequency control of microgrids using federated learning fractional-order recurrent neural network," *IEEE Transactions on Smart Grid*, vol. 15, no. 1, pp. 1089–1102, 2024.

[6] S. Abhinav, H. Modares, F. L. Lewis, F. Ferrese, and A. Davoudi, "Synchrony in networked microgrids under attacks," *IEEE Transactions on Smart Grid*, vol. 9, no. 6, pp. 6731–6741, 2018.

[7] M. Shi, X. Chen, M. Shahidehpour, Q. Zhou, and J. Wen, "Observer-based resilient integrated distributed control against cyberattacks on sensors and actuators in islanded ac microgrids," *IEEE Transactions on Smart Grid*, vol. 12, no. 3, pp. 1953–1963, 2021.

[8] A. J. Abianeh, M. M. Mardani, F. Ferdowsi, R. Gottumukkala, and T. Dragičević, "Cyber-resilient sliding-mode consensus secondary control scheme for islanded ac microgrids," *IEEE Transactions on Power Electronics*, vol. 37, no. 5, pp. 6074–6089, 2022.

[9] M. S. Sadabadi, S. Sahoo, and F. Blaabjerg, "A fully resilient cyber-secure synchronization strategy for ac microgrids," *IEEE Transactions on Power Electronics*, vol. 36, no. 12, pp. 13 372–13 378, 2021.

[10] H. Yang, C. Deng, X. Xie, and L. Ding, "Distributed resilient secondary control for ac microgrid under fdi attacks," *IEEE Transactions on Circuits and Systems II: Express Briefs*, vol. 70, no. 7, pp. 2570–2574, 2023.

[11] Y. Chen, D. Qi, H. Dong, C. Li, Z. Li, and J. Zhang, "A fdi attack-resilient distributed secondary control strategy for islanded microgrids," *IEEE Transactions on Smart Grid*, vol. 12, no. 3, pp. 1929–1938, 2021.

[12] Q. Zhou, M. Shahidehpour, A. Alabdulwahab, A. Abusorrah, L. Che, and X. Liu, "Cross-layer distributed control strategy for cyber resilient microgrids," *IEEE Transactions on Smart Grid*, vol. 12, no. 5, pp. 3705–3717, 2021.

[13] M. Jamali, M. S. Sadabadi, M. Davari, S. Sahoo, and F. Blaabjerg, "Resilient cooperative secondary control of islanded ac microgrids utilizing inverter-based resources against state-dependent false data injection attacks," *IEEE Transactions on Industrial Electronics*, vol. 71, no. 5, pp. 4719–4730, 2024.

[14] P. S. Tadepalli, D. Pullaguram, and M. N. Alam, "Resilient dynamic average secondary control for dc microgrids against fdi attacks," *IEEE Transactions on Industry Applications*, pp. 1–12, 2025.

[15] M. S. Sadabadi, "A resilient-by-design distributed control framework for cyber-physical dc microgrids," *IEEE Transactions on Control Systems Technology*, vol. 32, no. 2, pp. 625–636, 2024.

[16] Y. Jiang and Y. Yang, "A distributed proportional-integral observer-based hierarchical control for ac microgrids under fdi attacks," *IEEE Transactions on Industrial Electronics*, vol. 71, no. 12, pp. 15 780–15 792, 2024.

[17] D. Pullaguram, R. Rana, S. Mishra, and N. Senroy, "Fully distributed hierarchical control strategy for multi-inverter-based ac microgrids," *IET Renewable Power Generation*, vol. 14, no. 13, pp. 2468–2476, 2020.

Modeling and Analysis of 14-bus CIGRE Model

Bhanu Venkata Siva Saikiran Kodati
Department of Electrical Engineering
National Institute of Technolgy Warangal
Warangal, India
kb23eem3r03@student.nitw.ac.in

Mahamad Nabab Alam
Department of Electrical Engineering
National Institute of Technolgy Warangal
Warangal, India
mnalam@nitw.ac.in

Tadepalli Phani Swecha
Department of Electrical Engineering
National Institute of Technolgy Warangal
Warangal, India
tp720035@student.nitw.ac.in

Abstract—This paper provides a detailed dynamic analysis of the 14-bus medium-voltage distribution network coupled with several renewable and distributed energy resources, such as solar photovoltaic (PV) generators, wind turbines, battery energy storage systems (BESS), and proton exchange membrane fuel cells (PEMFC). The system is modeled in MATLAB/Simulink and includes elaborate modeling of each source of energy, as well as the use of decentralized control strategies on power flow and system performance. The high-rise issues or trends of the proposed framework encompass real-time load variation management, coordinated energy management, and voltage / frequency regulation under different conditions of operation. From the results of the simulation, it is evident that the integrated system facilitates the promotion of grid resilience, volatilities, and operational scenarios, and consequent intervention of transmission-related perturbations, proving to be robust, as well as adaptable to distribution networks today.

Index Terms—Battery energy storage system (BESS), distributed energy resources (DER), energy management, fuel cell, grid resilience, hybrid energy system, medium-voltage (MV) distribution network, renewable energy integration, solar photovoltaic (PV), wind energy.

I. INTRODUCTION

The shift towards greener energy sources is leading to more solar photovoltaic (PV), wind turbines, fuel cells, and batteries being installed around the world. However, making these new sources part of the current power distribution system is very difficult. In the past, MV distribution systems were developed for use with loads that did not interact and for one direction of power. As more people use inverter-based DERs, there are now more cases of voltage fluctuations, reduced inertia, frequency changes, and coordination complexity.

A mixture of different DERs, guided by effective control methods, provides an effective way to secure voltages and frequencies, save energy, and protect the system. The aim is to develop a flexible and distributed method to keep a modern MV network operating properly with real-time fluctuations in loads and generation.

Many studies have focused on how individual DERs are controlled and operated. Fuzzy MPC can be used for frequency regulation in AC microgrids, and PLL has improved PV coupling and stabilized the system in AC microgrids [7] [1]. A three-phase fuel cell that also has battery backup has been shown to enhance how energy is converted [8]. Other

researchers have worked on adaptive control for the rejection of disturbances in wind turbines [5] and the extraction of maximum power points in fuel cells [4].

Nevertheless, most studies focus on just a part of a hybrid energy system or on one kind of controller and do not give a complete outline of such systems. In addition, standard IEEE tests like the 14-bus network suit transmission applications and are not suitable for distribution-level DER use.

While different controls for a single DER have been investigated, there are few in-depth studies on:

- The way in which multiple DERs interact with each other in a hybrid system.
- Adding them to 14-bus systems developed by CIGRE,
- Managing the regulation of frequency and voltage by using distributed control during changes in the real-time load.

Nobody has systematically studied how a solar PV, wind, BESS and PEMFC system works inside the CIGRE MV benchmark, especially when factors are changing.

This research attempts to fill the gaps mentioned above through the contributions listed below.

- Designing a model for renewable energy with solar, wind turbines, PEM fuel cells, and batteries using the CIGRE 14-bus MV distribution network.
- Local voltage and frequency control is done using decentralized PID controllers with changing load frequencies.
- An in-depth study of the system with and without DERs is needed to understand whether the voltage remains stable, energy is balanced, and losses are reduced.
- The performance of the battery was checked by simulating it with MATLAB/Simulink, including its dynamic charge contributions and state-of-charge (SOC).

The framework supports the running of future distribution grids with many renewables that are safe and efficient.

This paper follows a specific structure that includes system modeling and component representation in Section II and simulation results in Section III before concluding in Section IV with findings and research directions.

II. SYSTEM MODELING

The modern power distribution networks require the integration of distributed energy resources (DERs) for increasing energy requirements and sustainability reasons and for

---

979-8-3315-3013-6/25 $31.00 © 2025 IEEE

the implementation of advanced renewable technology. The analysis examines solar photovoltaic (PV) with wind energy and fuel cells together with battery storage elements to study distribution grid management and performance using a 14-bus MV network design. The research evaluates the effects of DER on power system reliability and efficiency through combined modeling of regulatory behaviors with theoretical characteristics and operational performance.

A. Network Architecture and Power Flow Analysis

These networks that contain buses and feeders along with distributed generations and loads can be classified as distribution networks. The 14-bus medium-voltage network maintains two types of distributed generation units across its vital nodes. Dynamic analysis allows personnel to perform detailed power dynamic and energy management evaluations using the assessment tool. The special equation provides the mathematical basis required for the investigation of a power system.

$$P_{gen} = P_{load} + P_{loss} \tag{1}$$

$$P_{gen} = P_{PV} + P_{Wind} + P_{Grid} + P_{FC} + P_{Bat} \tag{2}$$

Where:

- P_{gen} is the total power generated by DERs,
- P_{load} represents the total demand across the system,
- P_{loss} accounts for power losses in transmission and distribution.

The equipment within the electric power network requires automatic response capabilities to achieve a precise energy supply balance during changing power demands. Grid networks function through the combination of energy storage systems and fuel cells as reliable backup protection to reduce supply disruptions and stabilize their control procedures.

B. Designing of inverter

A $3 - \phi$ inverter converts DC to AC using a bridge circuit and PWM control, with a DC-link controller maintaining a stable V_{dc} by balancing power flow is given by :

$$P_{dc} = P_{ac} + P_{loss} \tag{3}$$

A PI-based current controller regulates the inverter's output current in dq - frame, improving response and control accuracy. The transformation of $3 - \phi$ currents into dq components is given by the following:

$$\begin{bmatrix} i_d \\ i_q \end{bmatrix} = \frac{2}{3} \begin{bmatrix} \cos\theta & \cos(\theta - 120°) & \cos(\theta + 120°) \\ \sin\theta & \sin(\theta - 120°) & \sin(\theta + 120°) \end{bmatrix} \begin{bmatrix} i_a \\ i_b \\ i_c \end{bmatrix} \tag{4}$$

The voltage equations in the rotating frame are:

$$V_d = Ri_d + L\frac{di_d}{dt} - \omega Li_q \tag{5}$$

$$V_q = Ri_q + L\frac{di_q}{dt} + \omega Li_d \tag{6}$$

A PI controller regulates i_d and i_q by computing control voltages:

$$V_d^* = K_p(i_d^* - i_d) + K_i \int (i_d^* - i_d)dt + \omega Li_q \tag{7}$$

$$V_q^* = K_p(i_q^* - i_q) + K_i \int (i_q^* - i_q)dt - \omega Li_d \tag{8}$$

Power networks operate on automated controls that immediately respond to changes in system loads or electrical supply. A fuel cell combined with battery storage makes the system reliable by quickly responding to changes in demand and providing voltage support as needed. Real-time changes in solar and wind power are managed by MPPT and inverter-based voltage regulation strategies. Thanks to this approach, the system operates steadily, interruptions are reduced, and service is maintained as energy needs rise sustainably.

C. Solar Photovoltaic (PV)

The 14-bus medium-voltage (MV) power grid requires a deep solar power generation analysis to determine its network behavior. Photovoltaic (PV) technology transforms sunlight into electrical power by applying the photovoltaic effect to excite electrons within PV modules and produces ongoing electrical current. Monitoring solar power production dynamics helps power grid representatives better integrate renewable systems for efficient energy management and efficient operation of power networks.

$$P_{PV} = V_{PV} * I_{PV} \tag{9}$$

$$I_{PV} = I_{ph} - I_0 \left(e^{\frac{V_{PV} + I_{PV}R_s}{nkT}} - 1 \right) - \frac{V_{PV} + I_{PV}R_s}{R_{sh}} \tag{10}$$

The power generation of photovoltaic systems depends on the PV voltage and current through a single-diode model which accounts for series and shunt resistances during operation. The current output of the PV module results from the subtraction of two parallel loss components and the photogenerated current that includes series resistor current and shunt resistance voltage drop.

The solar power output depends on the amount of sunlight received and the operating temperature, which affects how high the system can generate power. Maximum Power Point Tracking controllers (MPPT), including Perturb and Observe (P&O) and Incremental Conductance (IncCond), automatically control the operating voltage to optimize energy conversion efficiency. The study implements the P&O MPPT controller to maximize the efficiency of solar power extraction.

The 14-bus distribution network enables the PV system to be connected through DC-AC inverters to specific buses that ensure power grid compatibility. The maintenance of the electricity grid becomes more reliable after adding PV energy while simultaneously lowering the greenhouse gas output and speeding up the growth of renewable energy technologies. Reliable system operation, together with optimal performance under real-world conditions, can be achieved, but requires accurate management approaches and sophisticated control systems.

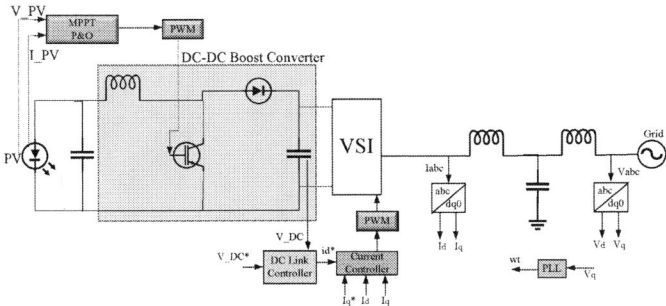

Fig. 1. Solar PV system connected to grid.

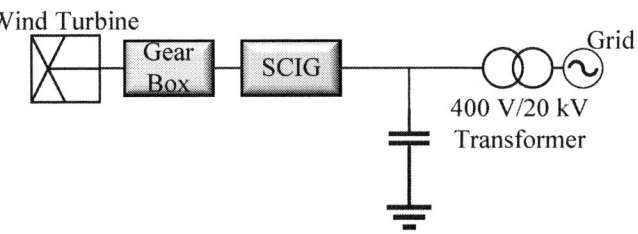

Fig. 2. Wind Energy connected to grid.

D. Wind Turbine

The generation of mechanical energy through wind turbines occurs by turning wind power into blade movement through kinetic energy. The available aerodynamic power P_{aer}. The mathematical formulation for power generation from turbine to rotor blades appears as stated in [5]:

$$P_{aer} = C_p(\lambda, \beta)P_w = C_p(\lambda, \beta)\frac{1}{2}\rho A_t w_t^3 \quad (11)$$

A wind turbine operates under the control of multiple essential operating parameters. The relationship between turbine blade rotation speed and wind speed is measured through the tip speed ratio (λ), yet the pitch angle (β) controls the blade orientation for maximum energy efficiency. The wind turbine receives its driving power from the wind kinetic power (P_ω) which produces the rotational speed of the wind turbine (WT) indicated by ω_t. The power output depends on the combination of air density (ρ) with the area swept by the turbine blade (A_t) because these factors determine the available wind energy for conversion directly.

The following expression provides the ratio of the turbine tip speed [5]:

$$\lambda = \frac{R\Omega_{tur}}{\omega_t} \quad (12)$$

The following expression represents the aerodynamic torque T_{aer} [5]:

$$T_{aer} = \frac{1}{2\Omega_{tur}}\rho A_t C_p(\lambda, \beta)\omega_t^3 \quad (13)$$

The mechanical torque (T_m) produced by the turbine is given by:

$$T_m = \frac{P_\omega}{\omega_r} \quad (14)$$

Through a gearbox, the wind turbine transmits its mechanical torque to an induction generator (IG) for efficient power conversion. The IG relies on slip for its operation and needs reactive power assistance to ensure a continuous connection to the grid. The dq reference frame voltage equations of the generator determine its performance through the dynamic of the stator and rotor for various operational conditions of the grid.

$$V_{ds} = R_s I_{ds} + \frac{d\psi_{ds}}{dt} - \omega\psi_{qs} \quad (15)$$

$$V_{qs} = R_s I_{qs} + \frac{d\psi_{qs}}{dt} + \omega_s\psi_{ds} \quad (16)$$

$$V_{d_r} = R_r I_{d_r} + \frac{d\psi_{dr}}{dt} - (\omega_s - \omega_r)\psi_{qr} \quad (17)$$

$$V_{qr} = R_r I_{qr} + \frac{d\psi_{qr}}{dt} + (\omega_s - \omega_r)\psi_{dr} \quad (18)$$

Where R_s, R_r are the resistances of the stator and rotor, I_{ds}, I_{qs}, I_{dr}, I_{qr} are the currents of the dq-axis and ψ_{ds}, $\psi_{qs}, \psi_{dr}, \psi_{qr}$, are the flux linkages.

E. Fuel Cell Modeling with MPPT

Fuel cells are crucial for the storage of electrochemical energy in renewable energy systems (REs), facilitating the conversion of stored hydrogen to electrical power. The process begins with the distillation of water, which is then fed into the electrolyzer. Here, each H_2O water molecule undergoes electrolysis, splitting into H_2 hydrogen and O_2 oxygen. The mathematical relationship between the electrical power output (P_e) and the rate of hydrogen generation in cubic meters per hour ($\dot{V}_{He}$), is expressed in equation (19) [6]:

$$\eta_e = \frac{P_e}{\dot{V}_{He}} = \frac{1}{3600} \cdot \frac{\rho_H \cdot e \cdot F \cdot U_{rev}}{\eta_I \cdot \eta_U \cdot M_r} \quad (19)$$

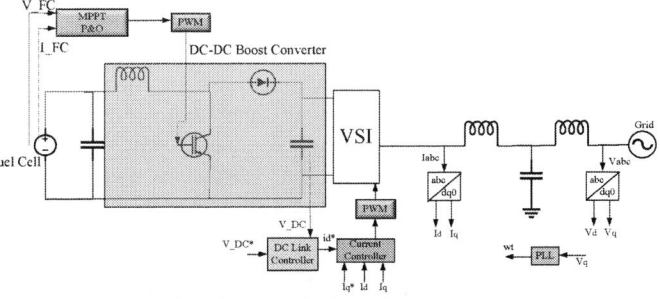

Fig. 3. Fuel Cell Power System connected to grid.

Where η_e, η_I, and η_U are electrical, current and voltaic efficiencies, respectively, ρ_H is the hydrogen density, e is the number of electrons implied in the reaction and equals 2 for the water splitting, F is Faraday's number, U_{rev} is the reversible cell voltage, and M_r is the relative molecular mass.

The PEM-FC stack composed of N_{cells} as a series connected cells, the stack voltage (V_{stack}) can be calculated as below (20):

979-8-3315-3013-6/25 $31.00 © 2025 IEEE

$$V_{stack} = N_{cells} \cdot (E_{oc} - V_{activation} - V_{Concentration} - V_{Ohmic}) \tag{20}$$

The variables in (20) are obtained using (21) to (23).

$$E = 1.299 - 0.85 \times 10^{-3}(T_{fc} - 298.15)$$
$$+ 4.3085 \times 10^{-5} T_{fc} \ln \left(P_{Hydrogen} \sqrt{P_{Oxygen}} \right) \tag{21}$$

Where T_{fuel} is the temperature of the fuel cell (K), and P_{Oxygen} and $P_{Hydrogen}$, respectively.

$$V_{activation} = -[\zeta_1 + \zeta_2 T_{fuel} + \zeta_3 T_{fuel} ln(C_{O_2})$$
$$+ \zeta_4 T_{fuel} ln(I_{fuel})] \tag{22}$$

where ζ_i ($i \in \{1, 2, 3, 4\}$) are empirical parameters, C_{O_2} is the concentration of O_2 (mol/cm^3).

$$V_{concentration} = -b \cdot ln \frac{J_{max} - J}{J_{max}} \tag{23}$$

where b is parametric coefficient, and J and J_max are actual and maximum density of current (A/cm^2), respectively.

$$V_{Ohmic} = I_{fuel}(R_m + R_c) \tag{24}$$

where R_m and R_c are resistances of the membrane and connections, respectively. P_{stack} is defined as below:

$$P_{stack} = V_{stack} \times I_{fuel} \tag{25}$$

For fuel cell (FC) MPPT technique is used to extract maximum power from FC. This MPPT algorithms continuously track the optimal operating V and I by adjusting the duty cycle of a DC-DC converter that interfaces the FC with the grid and load. This MPPT technique is similar to the MPPT used in PV.

F. Battery

A three-phase power grid requires an accurate model to represent power usage patterns together with grid synchronization and reliability performance during Battery Energy Storage System (BESS) integration. The Battery Energy Storage System serves several essential functions by distributing balanced energy while stabilizing power grids and improving electrical quality through its multiple support capabilities. The system operates as a two-way power supplier because it distributes electricity to the utility during times of high demand while taking in surplus energy released by the generation that exceeds consumption needs, thus maximizing both operational efficiency and reliability.

The battery power equation is given by:

$$P_{battery} = V_{battery} \times I_{battery} \tag{26}$$

The battery system controls charge and discharge activities through its state of charge system that shows battery capacity at the present moment. The SOC panel keeps refreshing

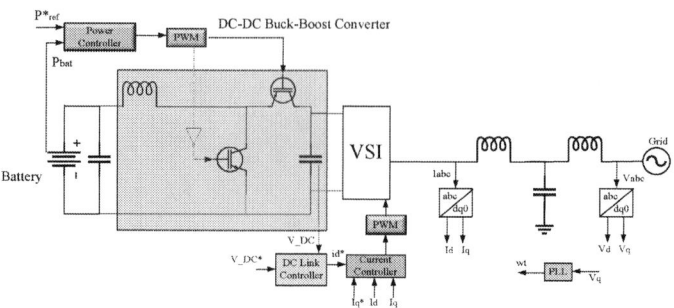

Fig. 4. Battery connected to grid.

its data throughout the operation. The SOC is dynamically updated as [13]:

$$SOC(t) = SOC(t - 1) + \frac{\eta}{E_{rated}} \int P_{battery} dt \tag{27}$$

$P_{battery}$ is +ve during discharging and −ve during charging.

To prevent overcharging and deep discharging, the SOC is typically maintained within operational limits:

$$SOC_{min} \leq SOC \leq SOC_{max} \tag{28}$$

The BESS connects to the three-phase grid through a power electronic converter to manage DC link voltage using a DC-DC converter and power convert AC using an inverter. The device uses PWM technology and needs a phase-locked loop (PLL) matching to remain synchronized with the grid power levels.

G. Coordination of DERs Under Transient Conditions

All the DERs are coordinated by using a decentralized control system. PV, wind turbine, battery, and fuel cell installations all have local controllers that are tailored to counter the distinctive dynamic behavior of each other and the nature of each of them in the grid. Inverter connections of all DERs include dq-frame PI control systems, and they are phase-locked to the grid by means of PLLs for constant voltage and frequency performance under steady and dynamic conditions.

When subject to sudden changes in load or idiosyncratic power supply, the battery energy storage system is agile to store and release energy in response, while the fuel cell system is able to cope with enduring deviations. The solar and fuel cell sources are controlled horizontally by MPPT controllers, and DC link voltage controllers help level the system's power consistency. With this integrated control scheme, the system can always supply power, maintain constant frequencies while in operations where changing shape rapidly can change, and regulate voltage.

III. SIMULATION AND RESULTS

A. Load Profile

The power distribution network load configuration represents the total number of feeders connected, but the simulation targets only one specific feeder. The detailed analysis of

979-8-3315-3013-6/25 $31.00 © 2025 IEEE

this feeder serves as the only study, while Fig. 5 shows the entire distribution pattern through network diagrams that show residential and commercial consumer load profiles daily.

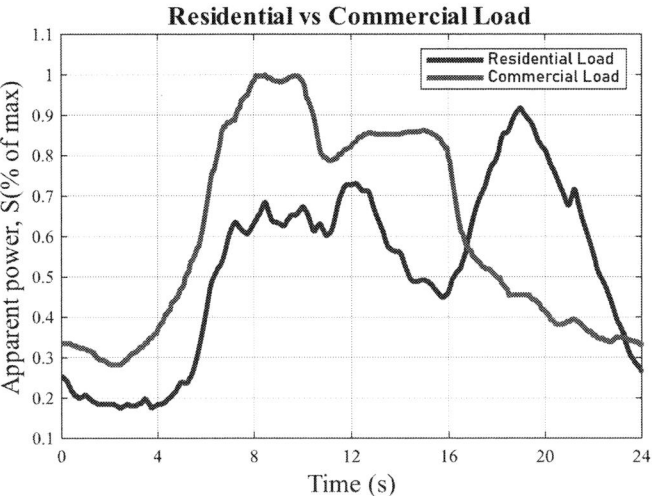

Fig. 5. Daily load profile of MV distribution network.

Fig. 5 and Fig. 6 show the network topology together with the system modeling, while Fig. 5 contains the daily load profiles. Using MATLAB/Simulink as the simulation platform, DERs were integrated at different locations within the framework. PV and wind turbine units operate as variable generation sources within the system but are dependent on deterministic fuel cells to support the reliability of the system. A high voltage primary transmission system operates at 220 kV, while the network topology that integrates DER units appears as depicted in Fig. 6.

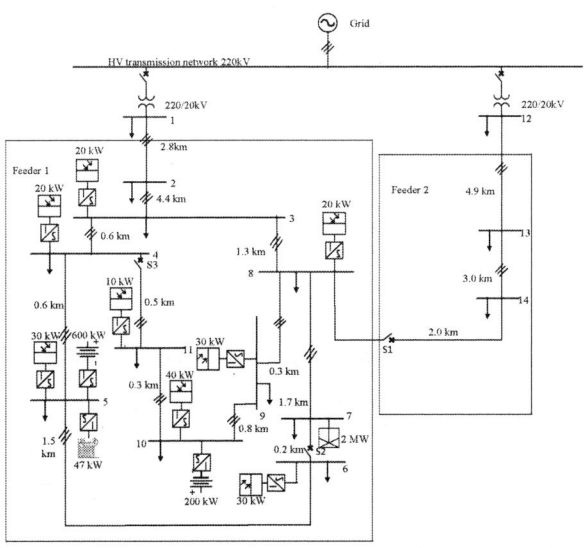

Fig. 6. System Architecture of the 14-bus distribution network.

Fig. 7 shows the generation profiles for each DER type:

wind turbine, solar (PV), battery, and fuel cell, respectively. During the test simulation, only one wind turbine with a rated power of 2 MW was connected to the network.

The proposed hybrid energy system is modeled and simulated in MATLAB/Simulink to evaluate its dynamic performance and operational reliability under different load conditions, and the performance metrics such as voltage regulation are analyzed. We also observed the variation in power supply and voltage profiles with DERs and without DERs.

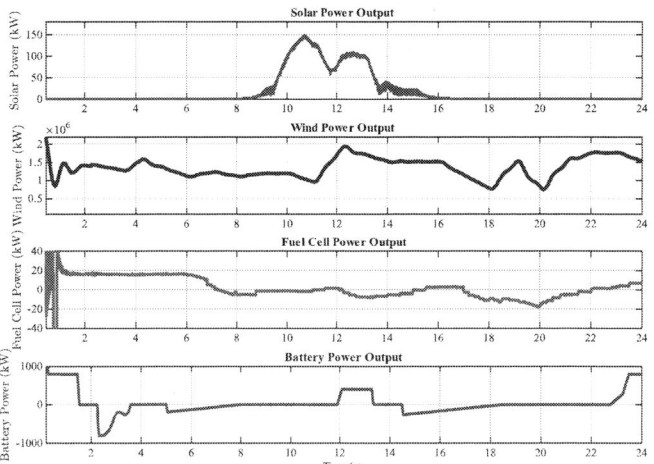

Fig. 7. Power flow of diverse resources in 14-bus distribution network.

The first sub-plot shows the total power obtained by PV arrays in the benchmark network, limited by sunlight. The second subplot shows that the wind power that varies with wind conditions throughout the whole simulation day was quite good. The third subplot indicates the net power of the fuel cell when providing the local load. The fuel cell runs at its rated maximum capacity for the entire time, and the power points in the plot indicate the residual output after satisfying the local demand. Positive values mean that the fuel cell sends excess power to the grid, while negative values mean that the local demand for the grid is more than the fuel cell can supply, and it obtains shortfalls through power importation from the grid. The fourth subplot gives the sum of the battery system outputs connected to the network which can either charge or inject power according to demand and helps avoid DER restrictions during bottlenecks. The positive value of the battery output indicates discharging, and the negative value indicates charging based on the SOC and the injected power on the bus.

The network can be enhanced by integration of the DER unit as shown in Fig. 8. In Fig. 8, the upper subplot depicts the power flow in line 2, with voltage profiles of node 3 and node 11 in the following subplots. The implementation of DER improves the voltage regulation of the system, while certain voltage levels exceed the acceptable limits.

To evaluate power performance, simulations assessed voltage levels at key buses (3 and 11) and power flow through major lines like Line 2. With DERs—PV, wind, fuel cells,

979-8-3315-3013-6/25 $31.00 © 2025 IEEE

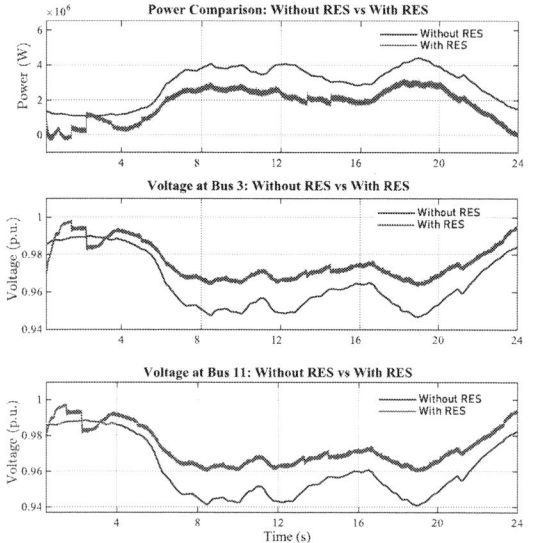

Fig. 8. Modification of voltage profiles due to DER units in 14-bus MV distribution network.

and batteries, the system showed reduced voltage fluctuations and better load support. Compared to conventional systems like PLL-based PV control [9] and single-stage fuel cell setups [1], the proposed hybrid system offers improved voltage regulation, lower losses, and better load balancing. Using inverter-based local control (MPPT, dq-PI, SOC), it ensures coordinated operation and enhanced grid resilience over earlier uncoordinated methods [7].

IV. CONCLUSION AND FUTURE SCOPE

This study presented the dynamic modeling, control and performance analysis of a 14-bus CIGRE medium-voltage distribution network integrated with hybrid DERs, including solar PV, wind turbines, BESS, and proton exchange membrane fuel cells (PEMFC).

Key performance metrics included voltage profile, active power contribution, system losses, and battery state of charge (SOC). The minimum voltages at Bus 3 and Bus 11 improved from 0.946 p.u. and 0.942 p.u. (without DERs) to 0.964 p.u. and 0.962 p.u. (with DERs), indicating an improvement of 1.9%–2.1% in voltage regulation and ensuring compliance with ±5% grid code requirements. DERs contributed up to 40% of total demand during peak solar hours, reducing reliance on upstream grid supply. System losses were reduced from 48 kW to 42 kW, marking a 12.5% improvement due to localized power injection.

The future work includes the use of HIL systems for real-time validation, use strategies such as PSO, GA and RL to optimize the controller and MPPT and conduct detailed fault analysis with coordinated protection schemes for better fault tolerance.

REFERENCES

[1] M. Jang, M. Ciobotaru and V. G. Agelidis, "A single-stage three-phase fuel cell system based on a boost inverter with a battery back-up

unit," 2012 Twenty-Seventh Annual IEEE Applied Power Electronics Conference and Exposition (APEC), Orlando, FL, USA, 2012, pp. 2032-2037, doi: 10.1109/APEC.2012.6166101.

[2] "Benchmark systems for network integration of renewable and distributed energy resources," Technical Brochure 575, CIGRE Task Force C6.04.02, 2014.

[3] M. Cucuzzella, S. Trip, A. Ferrara and J. Scherpen, "Cooperative Voltage Control in AC Microgrids," 2018 IEEE Conference on Decision and Control (CDC), Miami, FL, USA, 2018, pp. 6723-6728, doi: 10.1109/CDC.2018.8618898.

[4] Harrag, A. and Bahri, H. (2019), A Novel Single Sensor Variable Step Size Maximum Power Point Tracking for Proton Exchange Membrane Fuel Cell Power System. Fuel Cells, 19: 177-189. https://doi.org/10.1002/fuce.201800122.

[5] H. M. S. Sajid, M. B. Shafi, N. Malik, A. Muhammad and A. Amin, "Design of Three Phase Inverter System with LC filter," 2020 IEEE 23rd International Multitopic Conference (INMIC), Bahawalpur, Pakistan, 2020, pp. 1-5, doi: 10.1109/INMIC50486.2020.9318075.

[6] Laghridat, Hammadi, Essadki, Ahmed, Annoukoubi, Maha, Nasser, Tamou, A Novel Adaptive Active Disturbance Rejection Control Strategy to Improve the Stability and Robustness for a Wind Turbine Using a Doubly Fed Induction Generator, Journal of Electrical and Computer Engineering, 2020, 9847628, 14 pages, 2020. https://doi.org/10.1155/2020/9847628.

[7] Long, B., Liao, Y., Chong, K. T., Rodriguez, J., and Guerrero, J. M. (2021). Enhancement of Frequency Regulation in AC Microgrid: A Fuzzy-MPC Controlled Virtual Synchronous Generator. IEEE Transactions on Smart Grid, 12(4), 3138-3149. Article 9359668. https://doi.org/10.1109/TSG.2021.3060780.

[8] S. K. Jalan, B. C. Babu, S. K and G. Panda, "A Novel Phase Locked Loop based Control Strategy for a Three-phase Grid-tied Solar PV System," 2020 3rd International Conference on Energy, Power and Environment: Towards Clean Energy Technologies, Shillong, Meghalaya, India, 2021, pp. 1-6, doi: 10.1109/ICEPE50861.2021.9404371.

[9] S. Jahan, S. P. Biswas, S. Haq, M. R. Islam, M. A. P. Mahmud and A. Z. Kouzani, "An Advanced Control Scheme for Voltage Source Inverter Based Grid-Tied PV Systems," in IEEE Transactions on Applied Superconductivity, vol. 31, no. 8, pp. 1-5, Nov. 2021, Art no. 5401705, doi: 10.1109/TASC.2021.3094446.

[10] O. Plakhtii, V. Nerubatskyi, S. Mykhalkiv, D. Hordiienko, D. Shelest and I. Khomenko, "Research of Energy Characteristics of Three-Phase Voltage Source Inverters with Modified Pulse Width Modulation," 2021 IEEE 2nd KhPI Week on Advanced Technology (KhPIWeek), Kharkiv, Ukraine, 2021, pp. 422-427, doi: 10.1109/KhPIWeek53812.2021.9570071.

[11] A. Khalid, A. Stevenson and A. I. Sarwat, "Overview of Technical Specifications for Grid-Connected Microgrid Battery Energy Storage Systems," in IEEE Access, vol. 9, pp. 163554-163593, 2021, doi: 10.1109/ACCESS.2021.3132223.

[12] Magdi A. Mosa, Mariem Y. Yousef, Said M. El Masry, A.M Abdel Ghany, A.A. Ali, Frequency support of AC microgrid with high penetration of photovoltaic using super-capacitor, Sustainable Energy Technologies and Assessments, Volume 53, Part A, 2022, 102364, ISSN 2213-1388, https://doi.org/10.1016/j.seta.2022.102364.

[13] Y. Li and J. Wu, "Optimum Integration of Solar Energy With Battery Energy Storage Systems," in IEEE Transactions on Engineering Management, vol. 69, no. 3, pp. 697-707, June 2022, doi: 10.1109/TEM.2020.2971246.

[14] Agwa, A. M., Alanazi, T. I., Kraiem, H., Touti, E., Alanazi, A., & Alanazi, D. K. (2023). "MPPT of PEM Fuel Cell Using PI-PD Controller Based on Golden Jackal Optimization Algorithm. Biomimetics," 8(5), 426. ttps://doi.org/10.3390/biomimetics8050426

[15] Narendra Babu P., "Adaptive grid-connected inverter control schemes for power quality enrichment in microgrid systems: Past, present, and future perspectives," Electric Power Systems Research, Volume 230, 2024, 110288, ISSN 0378-7796, https://doi.org/10.1016/j.epsr.2024.110288.

Attention Based Bi-LSTM Bi-GRU model for Prediction of Electric Vehicle Charging Demand

Adithya R
Electrical Engineering
National Institute of Technology Calicut
Kozhikode, Kerala, India
adithya_b230759ee@nitc.ac.in

Harshan J
Electrical Engineering
National Institute of Technology Calicut
Kozhikode, Kerala, India
harshan_b230968ee@nitc.ac.in

S.N.Deepa
Associate Professor,EED
National Institute of Technology Calicut
Kozhikode, Kerala, India
sndeepa@nitc.ac.in

Abstract — The extensive progress and penetration of electric vehicles has led to the development of smart transportation systems, which moves the world towards a clean and sustainable future. The past few years have depicted the incremental volumes of electric vehicles (EVs) that have penetrated the market, leading to an increase in charging demand. As a result, it is necessary to predict the charging demand of EVs to minimize the burden on the electric grid setup and decrease the charging cost to consumers. This research study modelled a novel attention-based Bidirectional Long-Short Term Memory (Bi-LSTM) and Bidirectional Gated Recurrent Unit (Bi-GRU) to predict EV charging demand. The considered data for validating the proposed attention-based Bi-LSTM Bi-GRU (aBi-LSTM Bi-GRU) are time-series data, and the presence of Bi-LSTM and Bi-GRU overcomes the existence of vanishing and exploding gradients, as in the case of classic recurrent deep learning neural network models. The new modelled aBi-LSTM Bi-GRU was tested for its efficacy on an Electric Vehicle Charging dataset. Simulation experiments proved the superiority of the developed deep learning-based predictive model for the considered dataset over the other baseline models employed for comparison.

Keywords—Attention Mechanism, Charging Demand, Bidirectional LSTM, Bidirectional GRU, Prediction accuracy

I. INTRODUCTION

To combat the invariant climatic changes and minimize the emission of gases across the globe, the immediate solution that has come in the automobile sector is the utilization of electric vehicles. The extravagant evolution of EV technology has been an important factor in increasing the economy, employment opportunities, and power grids, and reducing emissions. EVs work on electric motors which replace traditional internal combustion engines that use fossil fuels. The power to the EVs comes from the battery or the solar panels within it and is charged by itself or through electricity from off-vehicle sources. Plugged-in EVs are widely manufactured and fall into plug-in hybrid and battery-powered electric vehicles. In plug-in hybrid EVs, the battery is powered by external power sources or onboard chargers. Battery-powered EVs use chemical energy stored in rechargeable lithium-based batteries and do not have internal combustion engines.

The utility of EVs has tremendously increased based on their extensive advantages: minimization of greenhouse gas emissions, health factors affected by air pollution, reduced diesel and petroleum needs, handling the energy consumed during the stationary period, reduced vehicle vibration, no requirement of gear box for torque variations, high power output, simple design, etc. In this scenario, the swift supply of millions of EVs across the universe has attained its highest utility rate. Robust growth is seen in electric car markets, wherein 14 million globally in 2023 and about 25% more in the year ending 2024 (Source: www.iea.org).

Considering the increased population size of vehicles, their charging requirements have become a serious concern among developers and the research community. Generally, EV charging requires a direct current supply (DC) to the EV battery, and because the grid power supply is alternating current due to transmission limitations, there is a need for a converter to provide DC power to the EV battery.

With respect to DC charging, the AC power is converted to DC externally and supplied to the battery without an on-board converter. This is typically done by charging stations or docks. Therefore, it is of high importance to recharge the battery source as and when needed for smooth and safe operation and running of EVs. The generalized power rating and charging models of an EV is as given in Table I.

TABLE I. EV POWER RATING

Charging Method	Power Rating	Supply
Normal Power Charging	Less than or equal to 8 kW	DC and AC
	8 kW to 22 kW	DC and AC
High Power Charging	22 kW to 48 kW	Only DC Supply
	48 kW to 200 kW	Only DC Supply

The increased electric power requirement and installation base for vehicle chargers has led to the need for charging demand prediction in EVs, which helps both the users and manufacturers and other utility and service providers to possess complete knowledge of the charging requirements based on the distance travelled and time incurred. The prediction of EV

979-8-3315-3013-6/25 $31.00 © 2025 IEEE

charging demand will help consumers to plan their distance to be travelled and to locate the existence of other charging stations in the near locations, as and when the charging battery is drained.

II. RELATED WORKS

Several studies have been conducted to predict the energy charging demand of electric vehicles employing different nonlinear programming algorithms, machine learning, and deep learning algorithms. A detailed literature study is conducted on the predictability of electric vehicle (EV) charging demand in this research study and is presented in this section.

Wang et al. (2023) [1] employed graph convolutional layers and gated recurrent units (GRUs) to extract spatio-temporal features in the observations and grouped regions based on graph embedding and point of interest (POI) data for EV charging demand prediction. Qu et al. (2024) [2] introduced large language models (LLMs) as EV charging demand predictors and formulated the prediction task into a text-to-text format. Feng et al. (2024) [3] an EV energy consumption prediction framework based on long short-term memory (LSTM) and Transformer models attaining a mean absolute percentage error (MAPE) of 6.7 % when SOC value got reduced from 90 to 40 for the considered dataset.

Cheng and Liu (2024) [4] developed a temporal convolutional network–discrete cosine transform-enhanced autotransformer for predicting charging loads at public EVs charging stations using the attained historical data. Kuang et al. (2024) [5] proposed a learning model for accurate EV charging demand prediction and reasonable pricing, enabling the integration of convolutional feature engineering, spatio-temporal dual attention mechanism and physics-informed neural models. Deng et al. (2024) [6] employed spatio-temporal fusion graph network (STFGN) to realize remaining useful life prediction in the charging of EVs. Wu et al. (2024) [7] developed a charging decision-making model based on fuzzy logic, by introducing a two-layer planning method. Zhang et al. (2024) [8] proposed a novel deep learning technique, weight fusion spatial–temporal graph convolutional network based EV charging load prediction framework to assess the impact of EVs on the multi-energy systems. The works in [9, 10] presented the applicability of machine learning models for EV charging demand prediction. Work on hybrid neuro-fuzzy prognostic framework, GRU models and also a multi-objective planning model has been modelled for EV charging demand and traffic-grid coupling [12-14].

Mekkaoui et al. (2025) [23] introduced the Probabilistic Dual-Adaptive Spatio-Temporal Graph Convolutional Network for forecasting of complex EV charging demands. Kuang et al. (2025) [24] proposed a variable selection network to adaptively learn dynamic auxiliary information and improve the Transformer encoder utilizing gated mechanisms for fluctuating charging time-series data and predicted EV charging demand. Shang et al. (2025) [25] developed a temporal GraphSAGE model adept at capturing spatiotemporal nuances and encompassed multiple prediction tasks of EV charging demand.

Siddiqui et al. (2025) [26] presented a novel DeepBoost approach for forecasting day-ahead EVs charging station load.

The literature review made on the development of predictive models for EV charging demand [1-26] has its own advantages and limitations. With millions of EVs flooding across the globe, it is time to consider, analyze, evaluate, and predict EV charging demand. Some of the commonly found limitations include scalability issues, premature and delayed convergence to the disparity among the datasets, stagnation, and global minimal occurrences. Hence, it is essential to have an effective and efficient predictive model of EV charging demand to facilitate providers and consumers and avoid the limitations of existing models. Considering this, the objective of this research work is to,

- Model a predictive time series model with the variant of recurrent deep learning models
- Design and development of attention based Bi-LSTM Bi-GRU Predictive model for EV charging demand
- Validation of the proposed model with baseline models.

The proposed model was trained and tested for its efficacy on the EV charging datasets.

III. PROPOSED METHODOLOGY

Effective and efficient models are needed to predict the EV charging demand (EVCD) and thereby facilitate the planning of charging requirements on both the user and supplier sides. In this study, an ensemble of the Bi-LSTM and Bi-GRU models was carried out, and this combined ensemble model operates based on the enabled attention mechanism. The attention mechanism has changed the way we operate using deep learning models. By using attention mechanism in sequence tasks we can make the model learn on contextual embeddings instead of general embeddings.

A. Attention Mechanism

A form of encoder-decoder architecture is the attention mechanism in a deep learning model that focuses on the specific section of the input during the task execution. In the attention mechanism, the weights are dynamically assigned pertaining to the various elements of the inputs based on their importance in the prediction process. By incorporating the attention mechanism, the model intends to select the highly relevant information to attend to and get processed by capturing the various relationships and dependencies that exist in the data. The attention mechanism is suitable for sequential and time-series data, such as EV datasets, as they enable the model to handle long-range dependencies.

The attention mechanism in the proposed study is as follows.

Step 1: Input Encoding – The EV time series data sequence is embedded and the data is transformed into a format so that it can be processed by the attention mechanism.

979-8-3315-3013-6/25 $31.00 © 2025 IEEE

Step 2: Query Vector Generation: With respect to the present state of the model, a query vector is generated, which indicates the information required by the model to extract or retrieve from the EV dataset.

Step 3: Key Value Creation: The inputs are split into key-value pairs, and these keys capture the information that can be utilized to determine the most relevant and important feature, with the key values comprising the actual data.

Step 4: Similarity Computation – For measuring the relevance and the compatibility between the query vector generated and key value created, a similarity metric is computed. This similarity metric can be – dot product, scaled dot product and cosine similarity. This similarity score can be given by,

$$Similarity\_Score\ (SS) = \begin{cases} h_q.y_i & Dot\ Product \\ (h_q)^T W y_i & General \\ v^T tanh\left(W\begin{bmatrix} h_q \\ y_i \end{bmatrix}\right) & Concate \end{cases} \quad (1)$$

In (1), 'h_q' is the encoder source hidden state at position 'q', 'y_i' is the encoder target hidden state at the position 'i', 'W' specifies the weight matrix and 'v' is the weight vector.

Step 5: *Attention weights calculation* – Similarity scores computed in (1) are applied on the 'softmax' activation to attain the attention weights. These weights specify the relevance of each key-value pair to obtain better input features of the complete data.

$Attention\_Weight\ (\alpha(q,i)) = softmax(Similarity\_Score(q,i))$

Step 6: Weighted Sum: The attention weights are multiplied by the respective values to generate a weighted sum. This step combines all the relevant information from the input data with respect to their importance as detected by the attention procedure.

$$d_t = \sum_{i=1}^{Q} \alpha(q,i)h_q \quad (2)$$

In (2), 'Q' specifies the total number of key-value pair.

Step 7: Context Vector – The weighted sum computed in (2), acts as the context vector, that represents the most relevant attended information from the input. This vector captures the relevant context for the present step.

Step 8: Integration with the proposed model: The context vector as given by (2) is combined with the model's present state, thereby providing additional data for the subsequent model layers.

Step 9: Iterative Procedure: Steps 2 – 8 are repeated for each time step and iterations proceed so that the attention mechanism dynamically concentrates on the relevant parts of the input data.

In the proposed procedure, the incorporation of the attention mechanism effectively captures the relevant dependencies and adaptively focus on the various element of the input, resulting in better performance for the EV charging demand prediction. Fig 1 shows the process of attention mechanism used in the proposed predictive model.

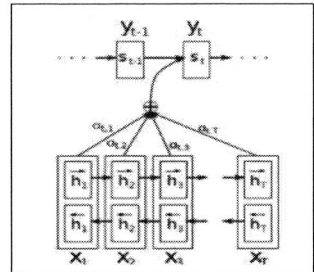

Fig 1. Attention mechanism

B. Bi-LSTM – Bi-GRU Model

In a long short-term memory (LSTM) network, the memory cells and gated mechanism activate the selection of forget data over time and retain data. This mechanism is intended to capture long-term dependencies.

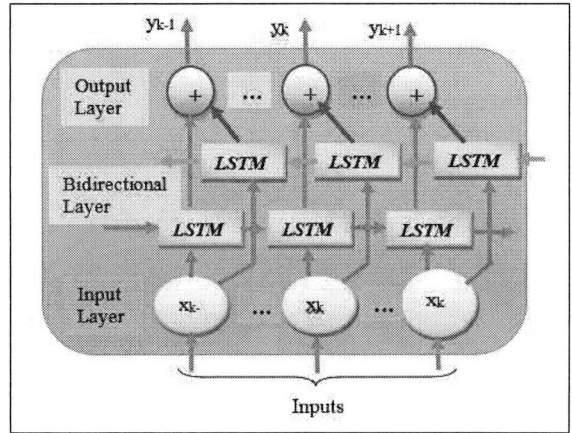

Fig 2. Architecture of the Bi-LSTM Model

Bi-LSTM model is designed to process the input data in both directions and with two LSTM layers. One LSTM layer processed the data in the backward direction, and the other LSTM layer processed the data in the forward direction. Each bidirectional layer tends to operate with its own hidden state and memory cells. In the forward step, Bi-LSTM captures the past information of the earlier time steps, and in the backward step, the future information pertaining to the upcoming time steps is captured. By activating in the forward and backward steps, Bi-LSTM captures the long-term dependencies in the input data. Fig 2 shows the architecture of the Bi-LSTM model. The computation of the Bi-LSTM Model is done using,

$$d_k = \gamma\left(w_{xi}^K x_k + w_{hi}^K h_{k-1} + b_i\right) \quad (3)$$
$$l_k = \gamma\left(w_{xf}^K x_k + w_{hf}^K h_{k-1} + b_f\right) \quad (4)$$
$$y_k = \gamma\left(w_{xy}^K x_k + w_{hy}^K h_{k-1} + b_y\right) \quad (5)$$
$$gt_k = tanh\left(w_{xg}^K x_k + w_{hg}^K h_{k-1} + b_g\right) \quad (6)$$
$$o_k = (f_k.o_{k-1} + i_k.gt_k) \quad (7)$$
$$h_k = \left(y_k.tanh(o_k)\right) \quad (8)$$

In (3) – (8), 'γ' is the sigmoidal function and '*tanh*' is the tangential activation that determines the output of each layer, '*h*' specifies the cell memory output, '*o*' the intermediate gate and '*d*','*l*' and '*o*' are input gate, forget gate and output gate respectively, '*k*' is the elapsed time steps, '*b*' is the bias-threshold levels, '*w*' represents the weights that exist among the interconnection of layers. The Bi-LSTM model operating in two directions stores the cell memory output and determines the final output by applying activations.

The Gated Recurrent Unit (GRU) model operates with a minimal number of gates compared to the LSTM model and thereby possesses simplified architecture, as shown in Fig 3. In the GRU process, the input and forget gates are controlled by one gate; thus, the input and forget gates are formulated as one gate. If $h_k=1$, the input gate is closed and new data can pass through the gate, and the forget gate opens; on the other hand, *if $h_k=0$*, the input gate opens to permit the new data and the forget gate is closed. The new input is combined with earlier memory cells to obtain the new state output using the reset gate.

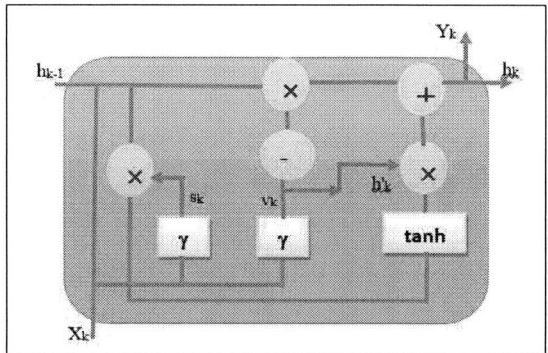

Fig 3. Architecture of the Individual GRU Model

The output of the GRU is attained using,

$$s_k = \gamma\left(w_{xr}^Q x_k + w_{hr}^Q h_{k-1} + b_r\right) \quad (9)$$
$$v_k = \gamma\left(w_{xu}^Q x_k + w_{ou}^Q h_{k-1} + b_u\right) \quad (10)$$
$$y_k = \left(v_k . y_{k-1} + (1 - v_k). y_{k-2}\right) \quad (11)$$

In (9) – (11), 's_k' is the reset gate, 'v_k' denotes the update gate, 'y_k' is the output gate, '*k*' specifies the time step. '*Q*' is length of complete data sequence, '*w*' are the weighted interconnections between layers and '*b*' is the bias level. Bi-GRU activates in the two directions – forward and backward with the mechanism in both the directions remaining the same.

C. Modelled new Attention based Bi-LSTM Bi-GRU Model

Combining deep learning is a compensation to the deep learning algorithms that are computationally expensive and numerous individual models are designed to learn from the same input sequences and result in an aggregation of output from which the final classification output is computed. Considering these merits of ensemble learning, this research study developed an ensemble bidirectional long short-term memory network – bi-directional gated recurrent units for performing EV charging demand prediction of benchmark

tweet datasets. The given model is novel in that no previous study has attempted to include both LSTM and GRU together using attention. This method works well because both the LSTM and GRU serve two different purposes. The LSTM helps in having the context of the long sequences while GRU is used to combine both the long- and short-range context. The attention mechanism is used as a way to give importance to features based on the context of the data given. By using attention, we are able to bring contextual awareness in the model. Fig 4 presents the configuration of the attention-based Bi-LSTM–Bi-GRU model.

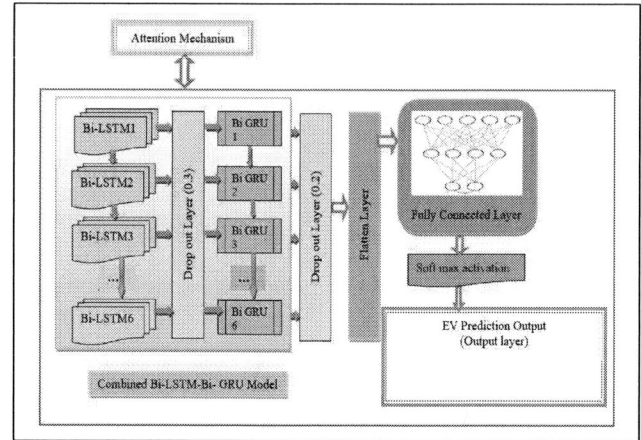

Fig 4. Proposed Attention based Bi-LSTM Bi-GRU Model

Bi-LSTM model operates in both directions so as to learn the discriminative data feature from the input sequence for each time steps. 20% of neurons are dropped out in the drop-out layer such that the generalization ability is maintained. 164 hidden neuron cells exist in the Bi-LSTM and Bi-GRU model, thereby the contextual features are extracted. In Bi-GRU processing, the dropout layer is executed with 10% neurons dropping out. Due to these operations, the over-fitting occurrences are completely avoided and flatten layer operated to compute a single column word feature vector. The last layer of the attention based Bi-LSTM-Bi-GRU is the dense fully connected layer comprising of 40 neurons and the output layer neurons is equal to 1 that is the charging demand prediction. The defined activations functions in the modelled new aBi-LSTM-Bi-GRU model are given by,

$$\gamma = \frac{1}{1+e^{-y}} \quad (12)$$
$$tanh = \frac{e^y - e^{-y}}{e^y + e^{-y}} \quad (13)$$
$$softmax\left(x_j\right) = \frac{e^{x_m}}{\sum_{k=1}^{N} e^{x_k}} \quad (14)$$

In (14), '*x*' is the input vector, '*k*' denotes the number of classes, *N* denotes the total number of samples and '*m*' specifies the sample number.

IV. RESULTS AND DISCUSSIONS

This section provides the performance metrics evaluated for the proposed attention-based Bi-LSTM-Bi-GRU model along with other deep learning (DL) models for the considered EV charging datasets.

A. Data Sets and its description

The dataset used for training, testing, and validating the proposed aBi-LSTM-Bi-GRU predictive model is a synthetically generated dataset with over 15000 days' worth of data, which is modelled using real-world data. Only a portion of this data is used in this study due to size limitations. This dataset is taken from Kaggle [27].

B. Dataset Processing

The given dataset is presented in its raw form. The dataset was processed to convert the given data into time sequences of 2 min, which is a suitable sequence length for both processing and computational efficiency. This is accompanied by the power drawn in that period of time derived from the KWh delivered. A total of 720476 data points were used, of which 72048 (10%) were used to train the model, and the remaining 6,48,426 (90%) were used to test the model. This ensured that there were sufficient data for testing the model to ensure fairness in the results.

C. Performance Model Evaluation

Model development and evaluation were conducted in Python (version 3.9.21) using the Anaconda platform. The TensorFlow library was employed for model construction, whereas packages such as NumPy, SciPy, matplotlib, and plotly facilitated the statistical analysis and data visualization. The model was trained on Huber Loss using the Adam Optimizer. Three standard statistical metrics were used to evaluate the effectiveness of the model: Mean Absolute Error (MAE), mean squared error (MSE), and root mean squared error (RMSE). The metrics are evaluated using,

$$MAE = \frac{1}{n}\sum_{i=1}^{n}|y_i - \hat{y}_i| \qquad (15)$$

$$MSE = \frac{1}{n}\sum_{i=1}^{n}(y_i - \hat{y}_i)^2 \qquad (16)$$

$$RMSE = \sqrt{\frac{1}{N}\sum_{q=1}^{N}\left(y_i - \hat{y}_i\right)^2} \qquad (17)$$

In (15), (16) and (17), the true value is y_i and the predicted value is $\hat{y}_i$.

TABLE II.   SAMPLE EV CHARGING DATASETS

Connection Time (Hours)	Charging Duration (Hours)	Energy Delivered (KWh)	Day Indicator
15.33	2.30	11.46	1
16.40	3.06	22.51	2
14.32	1.43	13.15	3
15.61	6.84	13.24	4
13.94	1.26	3.67	5
15.20	11.72	25.13	6
14.46	3.73	13.84	7

D. Results

The attention-based Bi-LSTM Bi-GRU model was simulated for the same EV charging datasets, and the results are shown in Table III. The dataset was divided into training and testing sets with 80% for training and 20% for testing. The main motive is to reduce the MAE, MSE, RMSE error metrics of the model's prediction and the actual values by iteratively optimizing the weights and biases of the model.

TABLE III.   RESULTS FOR PROPOSED ABI-LSTM BI-GRU MODEL

EV datasets	MAE	MSE	Prediction Accuracy
Gholizadeh, Nastaran (2024), "Electric Vehicle Charging Dataset"	0.00517	0.00007147	97.70%

Fig 5 confirms that the forecasted EV charging demand is on par with that of the actual charging demand of the EV charging station considered. The prediction accuracy evaluated during the convergence of the attention-based Bi-LSTM Bi-GRU predictor model was 97.70% with a minimal MAE of 0.00517.

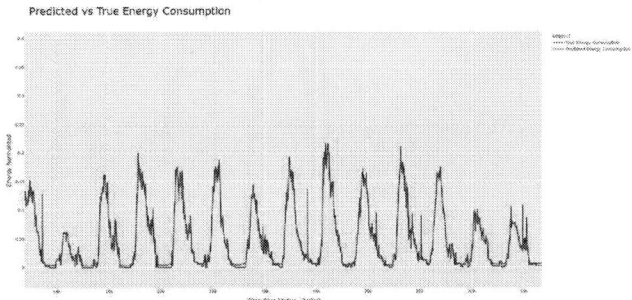

Fig 5. Proposed Attention based Bi-LSTM Bi-GRU Model

V. COMPARATIVE ANALYSIS

Table IV presents a comparative analysis of the developed models and the other baseline models.

TABLE IV.   COMPARATIVE ANALYSIS

DL Models	Prediction accuracy	MAE	MSE	RMSE
Zhang et al. (2024) [8]	88.16%	2.4578	2.5617×10^{-5}	0.3391
Khalid (2024) [12]	89.25%	1.9687	7.3123×10^{-3}	5.1082×10^{-03}
Alam et al. (2024) [13]	90.31%	1.8149	1.6500×10^{-4}	7.6632×10^{-03}
Wang et al. (2025) [22]	89.57%	0.8864	6.9172×10^{-6}	2.3109×10^{-04}
Siddiqui et al. (2025) [26]	91.55%	0.7148	8.2177×10^{-8}	8.5326×10^{-04}
Makaremi (2025) [20]	92.47%	0.3577	2.6618×10^{-9}	3.3614×10^{-05}
Proposed aBi-LSTM Bi-GRU	**97.70%**	**0.00517**	$\mathbf{7.147 \times 10^{-5}}$	$\mathbf{1.1497 \times 10^{-05}}$

979-8-3315-3013-6/25 $31.00 © 2025 IEEE

The results of this study are highly aligned and superior to those of previous studies published in this area [8, 12-13, 20, 22, 26]. The developed attention-based Bi-LSTM Bi-GRU resulted in better prediction accuracy and minimized error metrics. This is since the input data are perfectly captured with the attention mechanism and then are processed at the combined Bi-LSTM and Bi-GRU DL predictive model.

VI. CONCLUSION

In this study, a novel attention-based Bi-LSTM Bi-GRU predictive model was developed and applied to predict the electric vehicle charging demand. The devised predictive model combines the features of the attention mechanism, Bi-LSTM recurrent model with past information retained in the memory gates, Bi-GRU further activates to retrieve all the relevant information, and the end deep learning model for enhancing the depth of the architecture layers, which makes the prediction highly accurate. The simulation process was performed with the proposed predictor on the EV datasets, and the results proved their superiority over other existing methods of prediction of electric vehicle charging demand due to its accuracy. As a future extension of this research study, the proposed aBi-LSTM Bi-GRU predictive model should be applied to Indian EV Charging Datasets to facilitate the determination of the energy demand for Indian requirements.

REFERENCES

[1] Wang, S., Chen, A., Wang, P. and Zhuge, C., 2023. Predicting electric vehicle charging demand using a heterogeneous spatiotemporal graph convolutional network. Transportation Research Part C: Emerging Technologies, 153, p.104205.

[2] Qu, H., Li, H., You, L., Zhu, R., Yan, J., Santi, P., Ratti, C. and Yuen, C., 2024. ChatEV: Predicting electric vehicle charging demand as natural language processing. Transportation Research Part D: Transport and Environment, 136, p.104470.

[3] Feng, Z., Zhang, J., Jiang, H., Yao, X., Qian, Y. and Zhang, H., 2024. Energy consumption prediction strategy for electric vehicle based on LSTM-transformer framework. Energy, 302, p.131780.

[4] Cheng, F. and Liu, H., 2024. Multi-step electric vehicles charging loads forecasting: An autoformer variant with feature extraction, frequency enhancement, and error correction blocks. Applied Energy, 376, p.124308.

[5] Kuang, H., Qu, H., Deng, K. and Li, J., 2024. A physics-informed graph learning approach for citywide electric vehicle charging demand prediction and pricing. Applied Energy, 363, p.123059.

[6] Deng, S., Chen, Z., Lan, H., Yue, K., Huang, Z. and Li, W., 2024. Remaining useful life prediction with spatio-temporal graph transform and weakly supervised adversarial network: An application in power components. Energy, 313, p.133599.

[7] Wu, C., Wang, Y., Shi, Q. and Gao, S., 2024. A two-layer planning method for location and capacity determination of public electric vehicle charging stations. International Journal of Electrical Power & Energy Systems, 161, p.110205.

[8] Zhang, J., Cong, H., Zhou, H., Wang, Z., Wen, Z. and Zhang, X., 2024. Electric Vehicle Charging Load Prediction Based on Weight Fusion Spatial–Temporal Graph Convolutional Network. Energies (19961073), 17(19).

[9] Gunasekaran, R., Pareek, P.K., Gupta, S. and Shukla, A., 2024. Prediction of electric vehicle charging demand using enhanced gated recurrent units with RKOA based graph convolutional network. Discover Applied Sciences, 6(11), pp.1-18.

[10] Ostermann, A. and Haug, T., 2024. Probabilistic forecast of electric vehicle charging demand: analysis of different aggregation levels and energy procurement. Energy Informatics, 7(1), p.13.

[11] Akshay, K.C., Grace, G.H., Gunasekaran, K. and Samikannu, R., 2024. Power consumption prediction for electric vehicle charging stations and forecasting income. Scientific Reports, 14(1), p.6497.

[12] Khalid, M., 2024. Adaptive Neuro-fuzzy Inference System-Based Data-Driven Model for Optimal Recharging of Electric Vehicles and Cost Prediction in Energy Hubs. Arabian Journal for Science and Engineering, 49(12), pp.16477-16493.

[13] Alam, N., Rahman, M.A., Islam, M.R. and Hossain, M.J., 2024. Machine learning-based multi-variate forecasting of electric vehicle charging station demand. Electronics Letters, 60(23), p.e70104.

[14] He, B., Yang, B., Han, Y., Zhou, Y., Hu, Y., Shu, H., Su, S., Yang, J., Huang, Y., Li, J. and Jiang, L., 2024. Optimal EVCS planning via spatial-temporal distribution of charging demand forecasting and traffic-grid coupling. Energy, 313, p.133885.

[15] Kumar, R., Shakila, B. and Prakash, M., 2025. A novel hybrid deterministic temporal fusion transformer and dynamic recurrent neural network architecture for multi-horizon multi-variate electric vehicle energy consumption forecasting: mitigation pathway. International Journal of Information Technology, pp.1-19.

[16] Marzbani, F., Osman, A.H. and Hassan, M.S., 2025. Two-Stage Hybrid Feature Selection: Integrating ACO Algorithms with a Statistical Ensemble Technique for EV Demand Prediction. IEEE Transactions on Industry Applications.

[17] Tian, R., Wang, J., Sun, Z., Wu, J., Lu, X. and Chang, L., 2025. Multi-Scale Spatial-Temporal Graph Attention Network for Charging Station Load Prediction. IEEE Access.

[18] Zhang, H., Li, H., Zhou, Y. and Feng, D., 2024, September. Planning of Electric Vehicle Charging Stations Considering Inter-Station Interaction for Charging Demand Prediction. In Annual Conference of China Electrotechnical Society (pp. 667-674). Singapore: Springer Nature Singapore.

[19] Danish, S.M., Hameed, A., Ranjha, A., Srivastava, G. and Zhang, K., 2025. Block-FeDL: Electric Vehicle Charging Load Forecasting using Federated Learning and Blockchain. IEEE Transactions on Vehicular Technology, 74, pp.2048-2056.

[20] Makaremi, S., 2025. A multi-output deep learning model for energy demand and port availability forecasting in EV charging infrastructure. Energy, p.134582.

[21] Kene, R.O. and Olwal, T.O., 2025. Data-Driven Modeling of Electric Vehicle Charging Sessions Based on Machine Learning Techniques. World Electric Vehicle Journal, 16(2), p.107.

[22] Wang, S., Li, Y., Shao, C., Wang, P., Wang, A. and Zhuge, C., 2025. An adaptive spatio-temporal graph recurrent network for short-term electric vehicle charging demand prediction. Applied Energy, 383, p.125320.

[23] Mekkaoui, D.E., Midoun, M.A., Smaili, A., Feng, B., Talhaoui, M.Z. and Shen, Y., 2025. Probabilistic Dual-Adaptive Spatio-Temporal Graph Convolutional Networks for forecasting energy consumption dynamics of electric vehicle charging stations. Computers and Electrical Engineering, 122, p.109976.

[24] Kuang, H., Deng, K., You, L. and Li, J., 2025. Citywide electric vehicle charging demand prediction approach considering urban region and dynamic influences. Energy, p.135170.

[25] Shang, Y., Li, D., Li, Y. and Li, S., 2025. Explainable spatiotemporal multi-task learning for electric vehicle charging demand prediction. Applied Energy, 384, p.125460.

[26] Siddiqui, J., Ahmed, U., Amin, A., Alharbi, T., Alharbi, A., Aziz, I., Khan, A.R. and Mahmood, A., 2025. Electric Vehicle charging station load forecasting with an integrated DeepBoost approach. Alexandria Engineering Journal, 116, pp.331-341.

[27] Gholizadeh, Nastaran (2024), "Electric Vehicle Charging Dataset", Mendeley Data, V1, doi: 10.17632/5zrtmp7gwd.1

A Single Switch Structured Improved Quadratic Boost Converter for Renewable Energy Applications

Dharavath Anusha
Dept. of Electrical Eng,.
NIT Warangal, India
danusha94@gmail.com

Subbash Youvaraj
Dept. of Electrical Eng..
NIT Warangal, India
subbashguitar1@gmail.com

Ganesh Youvaraj
Dept. of Electrical Eng,.
NIT Warangal, India
ganeshyouvraj73@gmail.com

Harsh Bhanarkar
Dept. of Electrical Eng,.
NIT Warangal, India
harshbhanakar21@gmail.com

Bharath Marupatla
Dept. of Electrical Eng,.
NIT Warangal, India
bharathmaruputla465@gmail.com

Srinivasan Pradabane
Dept. of Electrical Eng,.
NIT Warangal, India
spradabane@nitw.ac.in

Abstract— **In the integration of renewable energy systems with DC grids, a DC-DC boost converter which produces a high gain output plays a crucial role. To meet this purpose, a non-isolated high-gain single switch based improved quadratic boost (SSIQB) converter is proposed in this paper. A single semiconductor switch, five diodes, three inductors, and two capacitors are utilized in the design of the proposed converter. This converter attains extremely high gain at lower duty ratios and high efficiency. The advantages include low switching stress and reduced current stress. The principle of operation and steady-state analysis in continuous conduction mode are studied in this paper. The parameter design and comparative analysis are included to show the features and how the proposed converter is better than the existing literature. The theoretical circuit analysis is validated in MATLAB/Simulink for a 500 W output power. The input voltage of the converter is set at 48V. It delivers an output voltage of 400 V at a load resistance of 320 Ω, attaining a voltage gain of 2/(1-D)² and operating with a duty ratio of 51.5%. The converter is observed to operate at a high level of performance, making it ideal for high-voltage and medium-power applications of renewable energy sources.**

Keywords— *DC-DC Converters, boost converter, high gain, step-up, low duty ratio, non-isolated converters*

I. INTRODUCTION

As technology is advancing ever so quickly, the use of DC-DC converters has been pivotal in high-power and renewable energy power systems. Generally, the converters used are categorized into two classes, i.e., isolated and non-isolated power converters. Transformers are used in isolated converters to electrically separate input and output. The cost of isolated converters rises as a result. Due to their reduced size, low switching losses, affordability, and relative superiority over isolated converters, non-isolated converters such as boost, buck, and buck-boost types are recommended.

The boost converter [1] plays a major role in modern applications, importantly in renewable energy and electric vehicles. In renewable energy applications, solar PV panel output is typically quite low compared to the rating of the electric DC-grid; a boost converter is employed to meet this demand. From a lower DC source to a higher output, it raises the voltage. As the equipment is manufactured to work in minimised tolerance to improve efficiency, it requires a power source with fewer ripples and high robustness. Often, the efficiency of DC-DC converters is limited due to parasitic components that are present in the circuit.

Many specialists have tackled the decrease in voltage gain in DC-DC converters by increasing the number of components in the basic circuit, and a number of strategies have been implemented to maximize profitability while considering costs [2]. A quadratic DC-DC boost converter is an advanced type of converter topology that achieves an increased voltage gain from the relation between the boosted output voltage and duty cycle. Quadratic converters employ a two-stage energy conversion method, in contrast to conventional converters that require large duty ratios to achieve noticeably high voltage. This design helps to minimize the size of inductors and capacitors while producing the required voltage gain. These converters are specifically useful in renewable energy systems. Many altered quadratic DC-DC boost converters of various sorts are put forward, which have the drawbacks of increased ripple input current and low voltage gain [3-9].

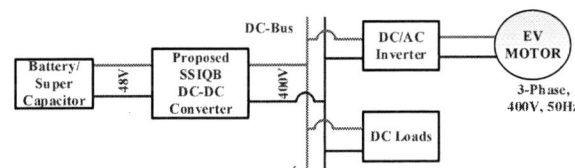

Fig. 1. Application of Proposed SSIQB DC-DC Converter.

A voltage gain of 2/(1-*D*) is developed in [3] which is an interleaved DC-DC converter. It attains maximum voltage gain of 10 with a duty ratio of 80%. It uses maximum number of power switches of 4 to attain minimum voltage gain. The equation of voltage gain derived is 2-*D*/(1-*D*)² in the paper [4] using 4 power switches and 2 diodes and attains maximum gain of 30. High-Gain DC-DC Converter for Hydrogen Fuel Cell Based Electric Vehicles discussed in [5] is seen to have a voltage gain of 2/(1-*D*)² using total components of 14. In [6] and [7], quadratic boost converters are proposed using 4 components including single switch with voltage gain of (1+*D*)/(1-*D*)². In [8], the SEPIC-based DC–DC converter which excludes coupled inductor for PV systems is proposed with high voltage gain of (1-*D*²+2*D*)/(1-*D*)² using two power switches. In [9] SEPIC DC-DC converters with a better voltage gain of 2*D*/(1-*D*)² with 14 number of components including single power switch is presented.

The application of proposed single switch based improved quadratic boost (SSIQB) converter is visualized in Fig.1. The high-gain step-up DC-DC converter configuration that is discussed in this paper reduces the current stress, achieves a voltage gain of $2/(1-D)^2$ and operates at a moderately lower duty ratio of 51.5%. The structure of the paper is as follows: The converter configuration and operating modes are shown in Section II, while Part III shows the simulation results. Section IV presents the summary.

II. PROPOSED SSIQB DC-DC CONVERTER

Fig. 2(a) depicts the proposed single switch based improved quadratic boost converter. The converter consists of single semiconductor power switch (Q_1), five diodes (D_1, D_2, D_3, D_4, and D_5), two intermediate capacitors and filter capacitor (C_1, C_2, and C_0), and three inductors (L_1, L_2, and L_3). Fig. 2(a) shows the active $2LCD$ network (a network consists of a pair of inductors, capacitor pair and a pair of diodes), on which the proposed converter design is based. This $2LCD$ network has been designed for efficient energy transfer through a step-by-step process. The initial energy stored in an inductor is transferred to and stored in a capacitor, which then, along with the input power source, transfers its energy to another inductor.

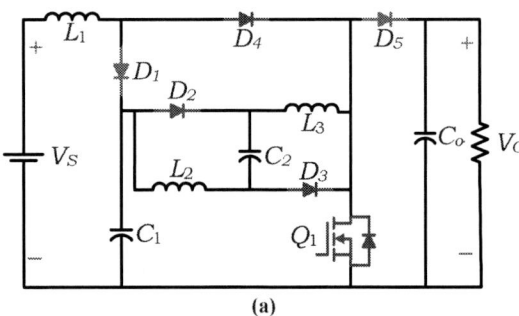

(a)

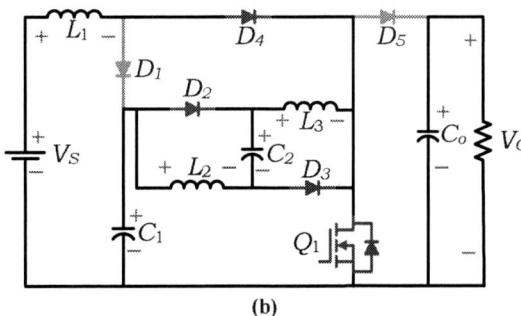

(b)

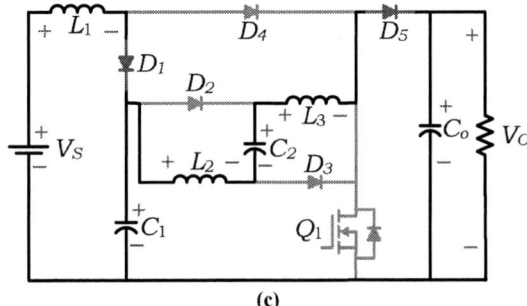

(c)

Fig. 2 (a) Proposed SSIQB converter (b) Mode-1 (c) Mode-2 operations.

III. OPERATION PRINCIPLE OF PROPOSE SSIQB CONVERTER

Fig. 2(b) and (c) shows how the SSIQB converter operates in Modes 1 and 2; at a duty ratio (D) of 51.5%. The circuit analysis equations in this paper are developed in continuous conduction mode under steady-state conditions. The operating intervals are determined by the ON and OFF states of the MOSFET (Q_1). The operational modes are explained in the following sections. The paper also includes analytical waveforms corresponding to both modes in Fig. 3.

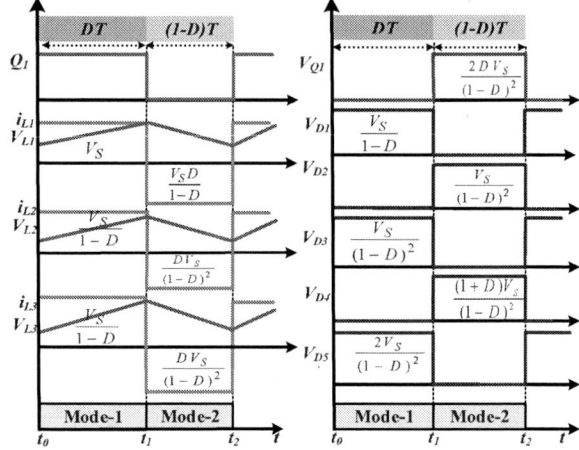

Fig. 3. Analytical Waveforms of SSIQB Converter.

A. Continuous Conduction Mode Operation

Mode 1 [t_0-t_1]: The switch (Q_1) is turned ON at time instant t_0 and inductor L_1 is charged directly from the source voltage V_S through diode D_4 and switch (Q_1) Simultaneously, in active network is $2LCD$, L_2 is charged using the energy stored in capacitor C_1 via diode D_2 and while L_3 is charged by the energy stored in capacitor C_2 through diode D_3. The equivalent circuit for Mode-1 operation is illustrated in the Fig. 2(b). By applying Kirchhoff's Voltage Law (KVL), the inductor voltage equations can be derived as follows:

$$V_{L1} = V_S \tag{1}$$

$$V_{L2} = V_{C1} \tag{2}$$

$$V_{L3} = V_{C2} \tag{3}$$

$$V_{L2} = V_{L3} = V_{C1} = V_{C2} \tag{4}$$

Mode 2 [t_1-t_2]: From t_1 (DT) to t_2 (T), the power switch (Q_1) is turned off in this state. During this time, diodes D_2, D_3, and D_4 remains off, while D_1, and D_5 conduct in reverse direction. In this phase, the inductors release their stored energy, which combines with the input voltage V_s to supply power to the load. At the same time, capacitors (C_1, and C_2) are charged using the energy stored in their corresponding inductors (L_1, L_2, and L_3).The equivalent circuit for this mode of operation is illustrated in the Fig 2(c). Using Kirchhoff's Voltage Law (KVL), the voltage across the inductors is expressed as:

$$V_{L1} = V_s - V_{C1} \tag{5}$$

$$V_{L2} + V_{L3} = 2V_{C1} - V_O \tag{6}$$

$$V_{L2} = V_{L3} = \frac{2V_{C1} - V_O}{2} \tag{7}$$

979-8-3315-3013-6/25 $31.00 © 2025 IEEE   642

B. Steady-State Analysis of Proposed Converter

a. Analysis of Gain

We know that the average voltage across all the inductors is zero according to the voltage-sec balance principle i.e.,

$$\int_0^{DT} V_{L1} + \int_{DT}^{T} V_{L1} = 0$$

Therefore, by considering the equations 1 and 5 for the inductor (L_1) we get voltage across capacitor C_1 that is V_{C1} in 8,

$$V_S D + (V_S - V_{C1})(1-D) = 0$$

$$V_{C1} = V_{C2} = \frac{V_S}{1-D} \tag{8}$$

Now considering the equations 2, and 7 for the inductor (L_2), we get the following result.

$$2V_{C1}D + \left(\frac{2V_{C1} - V_O}{2}\right)(1-D) = 0$$

$$V_{C1} = V_{C2} = \frac{(1-D)V_O}{2} \tag{9}$$

From the equations (8) and (9) we get the following result we get voltage gain equation in (10)

$$\frac{V_O}{V_S} = \frac{2}{(1-D)^2} \tag{10}$$

b. Stress Analysis

The voltage stress observed across various components in the converter has been theoretically calculated in terms of input voltage (V_S) and duty ratio (D) and are tabulated in the Table I.

TABLE I

Parameter	Voltage stress
Q_1	$\dfrac{2DV_S}{(1-D)^2}$
D_1	$\dfrac{V_S}{1-D}$
D_2	$\dfrac{V_s}{(1-D)^2}$
D_3	$\dfrac{V_S}{(1-D)^2}$
D_4	$\dfrac{(1+D)V_s}{(1-D)^2}$
D_5	$\dfrac{2V_S}{(1-D)^2}$

c. Parameter Design

The design values of the capacitors and inductors are calculated from voltage and current ripples as follows in (11) and (12) respectively:

$$C_1 = C_2 = \frac{Di_o}{(1-D)f_S \Delta V_C}; C_o = \frac{Di_o}{f_S}$$

$$L_1 = \frac{V_S D}{f_S \Delta i_{L1}}; L_2 = L_3 = \frac{V_S D}{(1-D)f_S \Delta i_L}$$

IV. COMPARATIVE ANALYSIS OF PROPOSED CONVERTER

Converters are intrinsically facing the limitation that cannot be at optimal in all design and operational aspects simultaneously. We compare the proposed converter against other high-gain converters, considering key aspects such as component count, voltage gain, and maximum voltage gain at duty ratio of 80%. The results, summarized in Table-II, highlight proposed SSIQB. In comparison with a boost converter of conventional type, the proposed converter has a voltage gain that is significantly higher. The converter maintains its efficiency at rated operating conditions while ensuring the switches operate under low voltage stress.

In terms of component count, the converters referenced in [3] and [4] utilize fewer components but provide a lower voltage gain the proposed converter. Hence, the converter that has been proposed has demonstrated a substantially higher voltage gain when compared with other converters listed in Table-II. The Duty ratio *vs* Voltage gain has also been compared with other existing converters to show the superiority of proposed converter in Fig. 4.

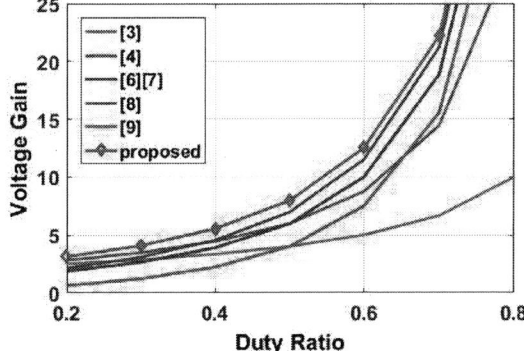

Fig. 4. Comparative Analysis plot of Duty Ratio *vs* Voltage Gain.

TABLE II

Topology	[3]	[4]	[5]	[6]	[7]	[8]	[9]	Proposed converter
N_Q	4	4	1	1	1	2	1	1
N_D	0	2	5	4	4	3	4	5
N_L	2	2	3	3	3	3	4	3
N_C	3	3	5	4	4	4	5	3
N_T	9	11	14	12	12	12	14	12
Gain in CCM	$\dfrac{2}{1-D}$	$\dfrac{2-D}{(1-D)^2}$	$\dfrac{2}{(1-D)^2}$	$\dfrac{1+D}{(1-D)^2}$	$\dfrac{1+D}{(1-D)^2}$	$\dfrac{1-D^2+2D}{(1-D)^2}$	$\dfrac{2D}{(1-D)^2}$	$\dfrac{2}{(1-D)^2}$
Gain at 80% duty Ratio	10	30	50	45	45	49	40	50
Input current	Pulsating	-	Pulsating	Pulsating	Pulsating	Pulsating	Pulsating	Pulsating
N_Q= Total Power Switches, N_D= Total Diodes, N_L= Total Inductors, N_C= Total Capacitors, N_T= Total Components								

V. SIMULATION RESULTS AND DISCUSSION

To evaluate the effectiveness of the proposed converter, a simulation was conducted using MATLAB/Simulink at an input voltage of 48 V, which gives an output of 400V, at a power of 500W. The specifications of inductors, capacitors and various other parameters are listed in the table III.

TABLE III

Parameter	Values
V_S	48 V
V_o	400 V
P_o	500 W
L_1, L_2, and L_3	2 mH
C_1, and C_2	100 µF
R	320 Ω

Fig. 5(a) shows the pulse of duty ratio D 51.5%, input voltage V_S is 48 V, the boosted output voltage V_o is 400 V and the output current(i_o) is 1.25 A. The inductor voltage V_{L1}, V_{L2} and V_{L3} are seen in Fig. 5 (b). The inductor currents i_{L1}, i_{L2} and i_{L3} is seen in Fig. 5 (c). Similarly, Fig. 5 (d) shows the capacitor voltages V_{C1} ~ 100 V, and V_{C2} ~ 100 V and diode voltages V_{D1} and V_{D5}. Fig. 6 shows the voltages across the power switch Q_1 and diodes D_2, D_3 and D_4 respectively.

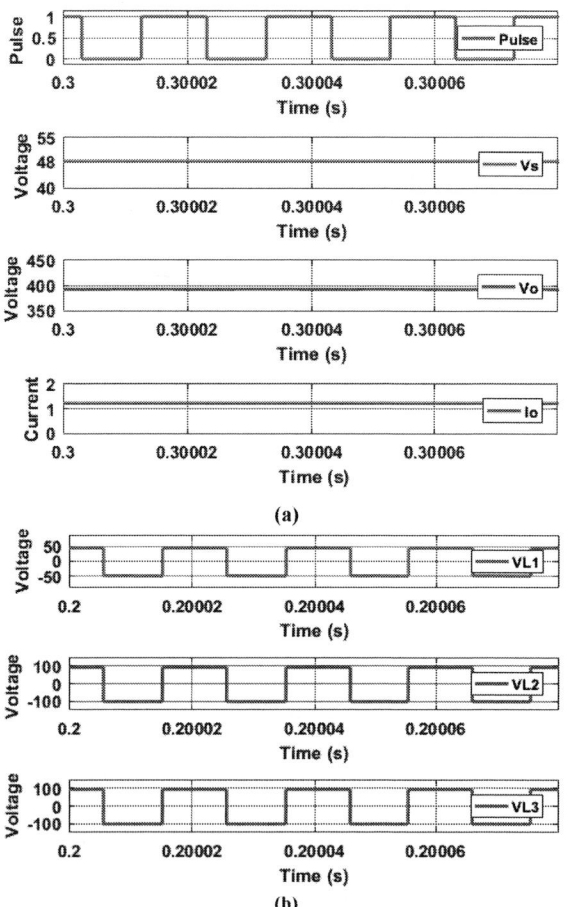

(a)

(b)

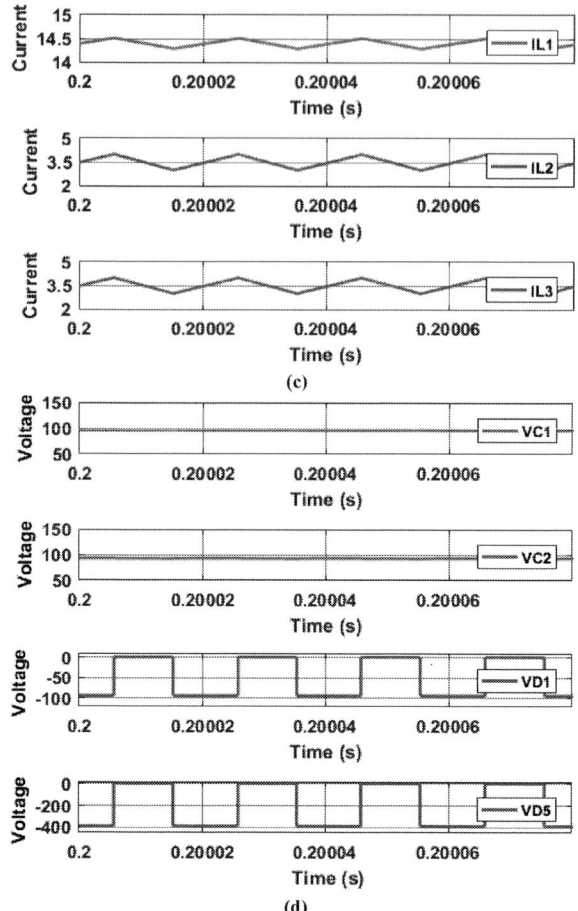

(c)

(d)

Fig. 5 Simulation results of proposed converter (a) Pulse, input voltage (V_S), output voltage (V_o) and output current (i_o) (b) Inductor voltages V_{L1}, V_{L2} and V_{L3} (c) Inductor currents i_{L1}, i_{L2}, and i_{L3} (d) Capacitor voltages V_{C1}, V_{C2}, and Diode voltage V_{D1}, V_{D5}.

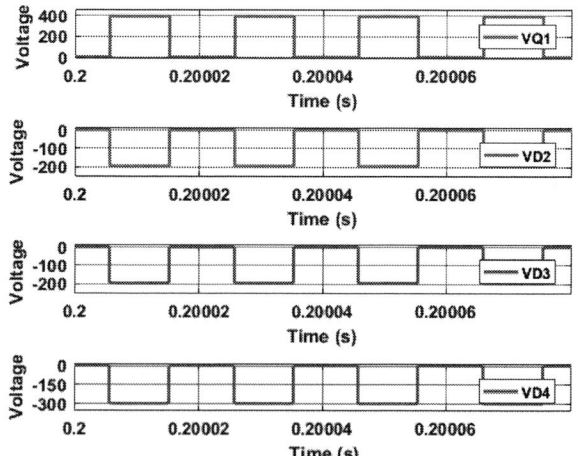

Fig. 6 Simulation Results of proposed converter voltages across switch voltage (V_{Q1}), V_{D2}, V_{D3}, and V_{D4}.

VI. CONCLUSION

The new DC-DC converter which produces a high-gain in voltage that is introduced in this paper utilizes a single power switch, five diodes, three inductors, and two capacitors. The developed topology achieves a voltage gain of $2/(1-D)^2$. Its performance is superior to recently developed converters, particularly at lower duty ratios. The advantages of the proposed converter are seen through a comparative analysis, it offers a higher voltage gain but simultaneously reduces voltage stresses on the switching components, thus enhancing overall efficiency and reliability. Using MATLAB/Simulink, simulations were conducted to validate the theoretical analysis. With an observed output power of 500 W, input voltage of 48 V, with a load resistance of 320 Ω, an output voltage of 400 V is delivered by the converter. Therefore, this proposed topology is highly effective for high-voltage, low-duty-cycle applications, making it a promising solution for micro-inverters and other high step-up power conversion applications.

REFERENCES

[1] Kumar, Avneet, Yi Wang, Xuewei Pan, Sajid Kamal, Xiaogang Xiong, Rui Hong Zhang, and Danyang Bao. "A high voltage gain DC–DC converter with common grounding for fuel cell vehicle." IEEE Transactions on Vehicular Technology 69, no. 8 (2020): 8290-8304.

[2] Karthikeyan, V., S. Kumaravel, and G. Gurukumar. "High step-up gain DC–DC converter with switched capacitor and regenerative boost configuration for solar PV applications." *IEEE Transactions on Circuits and Systems II: Express Briefs* 66, no. 12 (2019): 2022-2026.

[3] Mohurle, Divyani, and H. Nagendrappa. "Performance Analysis of Two Phase Interleaved Bidirectional DC-DC Converters for Electric Vehicle Application." In *2024 IEEE 4th International Conference on Sustainable Energy and Future Electric Transportation (SEFET)*, pp. 01-10. IEEE, 2024.

[4] Leelavathi, E., and Mahesh Kumar Mishra. "Design of High Gain Bidirectional DC-DC Converter with Improved Output Voltage." In *2024 International Conference on Electrical Electronics and Computing Technologies (ICEECT)*, vol. 1, pp. 1-6. IEEE, 2024.

[5] Kumar, Madhav, Akash Kumar, Jaya Kumari, Kaibalya Prasad Panda, Ritula Thakur, and Gayadhar Panda. "Development of a High-Gain DC-DC Converter for Hydrogen Fuel Cell Based Electric Vehicles." In *2024 6th International Conference on Energy, Power and Environment (ICEPE)*, pp. 1-5. IEEE, 2024.

[6] Leyva-Ramos, Jesús, Ricardo Mota-Varona, Ma Guadalupe Ortiz-Lopez, Luis Humberto Diaz-Saldierna, and Diego Langarica-Cordoba. "Control strategy of a quadratic boost converter with voltage multiplier cell for high-voltage gain." *IEEE Journal of Emerging and Selected Topics in Power Electronics* 5, no. 4 (2017): 1761-1770.

[7] Anusha, Dharavath, and Srinivasan Pradabane. "Analysis and Design of transformer less Single Switch based High Gain DC-DC Converter for Renewable Energy Applications." In *2023 11th National Power Electronics Conference (NPEC)*, pp. 1-6. IEEE, 2023.

[8] Heydari, Mojtaba, Hossein Khoramikia, and Alireza Fatemi. "High-voltage gain SEPIC-based DC–DC converter without coupled inductor for PV systems." *IET Power Electronics* 12, no. 8 (2019): 2118-2127.

[9] D. J. J. E. d. S. A. Murali, "Steady State Behavior of a SingleSwitch Non-isolated DC-DC SEPIC Converter Topology with Improved Static Voltage Gain," vol. 54, no. 3, pp. 445-452, 2021.

A Universal Input Single-Stage Bidirectional Charger for Light Electric Vehicles

Akash Kumar Swain
Department of Electrical Engineering
Indian Institute of Technology Bombay
Mumbai, India
akashks@iitb.ac.in

Vivek Agarwal
Department of Electrical Engineering
Indian Institute of Technology Bombay
Mumbai, India
agarwal@ee.iitb.ac.in

Abstract—This paper presents a universal input single-stage bidirectional charger for light electric vehicles (LEVs) utilizing a matrix converter. The proposed charger can operate with AC input from the grid and DC input from solar photovoltaic (PV) sources, enabling seamless integration with existing AC grids and renewable energy systems. The proposed architecture differs from traditional multi-stage chargers as it uses a direct ac-ac conversion method to mitigate the need for a bulky DC-link capacitor. Moreover, a bidirectional power flow capability is also possible, facilitating grid-to-vehicle (G2V) and vehicle-to-home (V2H) operations. The system is designed to charge a 48 V Li-ion battery pack, and the power level of the charger is 0.5 kW. Additionally, a robust control strategy is implemented to ensure stable battery charging. The converter is simulated in PLECS/Blockset, and the obtained outcomes validate the charger's performance for both AC and DC inputs. Furthermore, when operating with AC input, the charger achieves a unity power factor, ensuring compliance with grid standards while charging the battery.

Index Terms—Matrix converter, light electric vehicles, solar photovoltaic, DC-link capacitor, V2H.

I. INTRODUCTION

The rising global concern over carbon emissions and climate change has accelerated the transition to sustainable mobility solutions. Electric vehicles (EVs) have become a feasible option for traditional petroleum-based engines, thereby substantially decreasing greenhouse gas emissions. While much attention is given to high-power EVs, the growing adoption of light electric vehicles (LEVs), like e-rickshaws, e-bicycles, and electric two-wheelers, is gaining prominence in urban and commercial applications. LEVs have created a need for efficient, compact, and flexible charging solutions [1]. These vehicles typically operate with lower power requirements than electric cars; however, they demand reliable and versatile chargers that can operate under various input conditions, particularly in regions with inconsistent grid infrastructure or high penetration of renewable energy sources [2], [3]. Integrating AC and DC input capability into a single charger design enhances operational flexibility. It facilitates direct usage of energy from renewable sources, like solar photovoltaic (PV) systems, thereby reducing dependence on conventional grid

power. Conventional EV chargers often rely on two-stage power conversion, which increases system complexity, size, and cost [4]. Moreover, the two-stage topologies have less overall system efficiency as compared to the single-stage chargers.

Various single-stage charging topologies are investigated to address the challenges associated with the two-stage converters. These configurations eliminate the need for intermediate DC-link energy storage, reducing component count, cost, and conversion losses while enhancing system compactness. Among the prominent single-stage converters, isolated and non-isolated architectures have been investigated in [5], including totem-pole bridgeless rectifiers, matrix converters (MCs), active-clamped topologies, and interleaved structures [6]. While these solutions improve efficiency and power density, they often face trade-offs regarding control complexity, bidirectional operation, and the number of switches and components used.

The matrix converter-based single-stage charger is a promising alternative that effectively addresses these challenges. Unlike conventional topologies, the matrix converter enables the immediate transformation of utility frequency AC to high-frequency AC without requiring a bulky DC-link capacitor, significantly improving system reliability and lifespan. Additionally, it offers open-loop power factor correction (PFC) and power transfer between the vehicle and home in a bidirectional manner [7], [8].

This work presents a universal charging solution for LEVs that accommodates both AC and DC inputs, offering greater flexibility for diverse charging infrastructures. Unlike conventional chargers that rely on separate architectures for AC and DC charging, the proposed design integrates both functionalities within a single-stage topology, ensuring seamless charging operation based on the available power source.

The structure of the paper is outlined as follows. Section II presents the operation and control of the proposed universal charger during the charging process of LEVs during AC and DC input. Section III discusses the implementation of the charge controller, including control strategies to ensure effective and reliable charging. Section IV presents simulation results that demonstrate the effectiveness of the universal charger under different operating scenarios, including both

---

979-8-3315-3013-6/25 $31.00 © 2025 IEEE

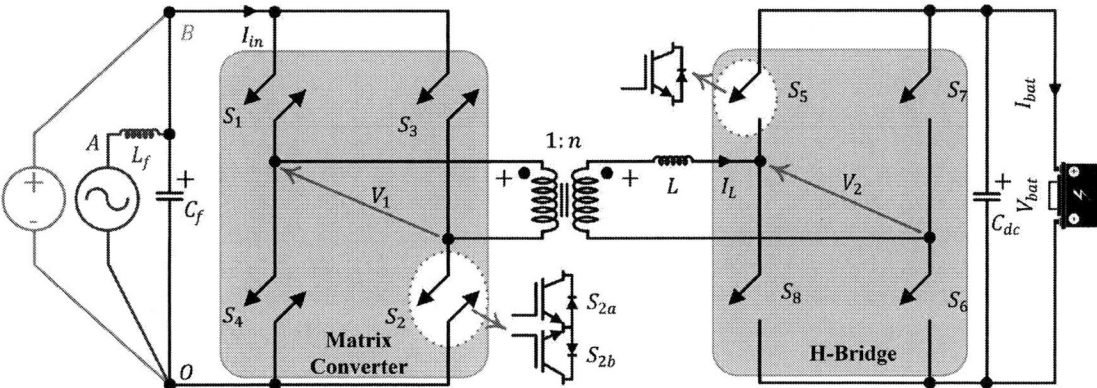

Fig. 1. Proposed topology for the universal LEV charger.

grid-to-vehicle (G2V) and vehicle-to-grid (V2G) modes. To summarize, Section V presents the concluding remarks of the study and outlines potential directions for future research.

II. OPERATION OF THE UNIVERSAL CHARGER FOR CHARGING THE LEVS

In Fig. 1, a schematic of the proposed universal charger is given. The proposed charger supports AC and DC inputs; however, only one input source can be provided at a time, i.e., either AC or DC can be connected during the charging operation. When the input to the converter is an AC source (connected across A and O), it works as an MC, whereas for a DC input (connected across B and O), it operates as a dual-active bridge (DAB) controlled through single-phase shift modulation.

A. Operation of the Converter with an AC Input

In this mode, the AC source is fed to the charger through an input filter inductor (L_f) and input filter capacitor (C_f) (Fig. 1). The bidirectional switches ($S_1 - S_4$) of the MC on the input side are modulated with a constant duty ratio (d_1) of 0.5, and the grid frequency AC voltage is converted to high-frequency AC. Let v_{ac} be the instantaneous value of the grid voltage and V_1 be the primary side voltage, then the relation between V_1 and v_{ac} can be expressed as:

$$V_1(t) = \begin{cases} +v_{ac}, & 0 \leq t < T_s/2 \\ -v_{ac}, & T_s/2 \leq t < T_s, \end{cases} \quad (1)$$

where T_s denotes the switching cycle corresponding to the switching frequency (F_s). Additionally, the active full bridge on the output side is operated with a duty cycle ($d_{2,1}$), and a phase difference in time (Δt) between the primary and the secondary side is regulated to control the amount of power transferred between the primary and secondary side [7], [8]. Let V_{bat} be the battery voltage corresponding to the converter's output voltage, and n be the turns ratio of the high-frequency transformer. Then, $d_{2,1}$ can be expressed as:

$$d_{2,1}(t) = n\frac{|v_{ac}|}{V_{bat}}. \quad (2)$$

Significantly, Δt is limited to guarantee that the pulse given to the H-bridge stays inside its specified $T_s/2$ time interval (Fig. 2). The phase shift ratio (δ) between the two sides is the ratio between the required phase difference in time and the maximum available phase difference, and the expression for δ is given by:

$$\delta = \frac{\Delta t}{0.25T_s}. \quad (3)$$

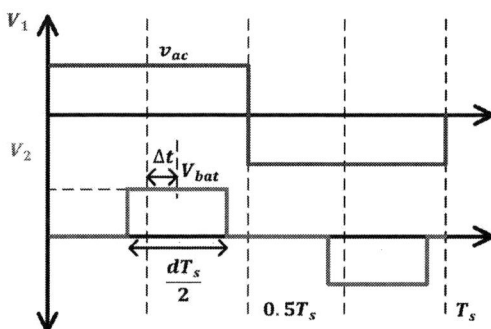

Fig. 2. Modulation of primary and secondary side voltages for AC input over a switching cycle.

The value of power that is transferred during a switching interval with a leakage inductance of L can be calculated as:

$$P_{sw}(t) = \frac{0.25n^2\hat{v}_{ac}^2\sin^2(wt)}{f_sL}\delta, \quad (4)$$

where $\hat{v}_{ac}$ is the peak value of v_{ac}. The average power during the entire grid cycle of the AC can be expressed as:

$$\bar{P}_{ac} = \frac{0.125n^2\hat{v}_{ac}^2}{f_sL}\delta. \quad (5)$$

Therefore, δ is directly proportional to the power delivered during a grid frequency AC cycle.

B. Operation of the Converter with a DC Input

In this operational mode, a DC source is connected to the proposed converter, as illustrated in Fig. 1. Notably, the filter inductor has no role, and this configuration allows the

979-8-3315-3013-6/25 $31.00 © 2025 IEEE

filter capacitor to function solely as a DC-link capacitor. The converter operates as a single-phase shifted DAB in this mode of operation [9]. The converter's input and output sides are operated with a fixed duty ratio of 0.5 each. On the primary side, the MC operates as an H-bridge. For the bidirectional switch S_1 to conduct, the current path is provided by the upper switch S_{1a}, and the body diode of the lower switch S_{2b}. Similarly, switches S_2–S_4 operate in the same manner, ensuring the matrix converter functions as an H-bridge. Coming to the secondary side, it is also modulated with 50% of the duty cycle; however, there exists a phase difference (ϕ) between the input and the output side, which accounts for the power transferred between the two sides (Fig. 3). The

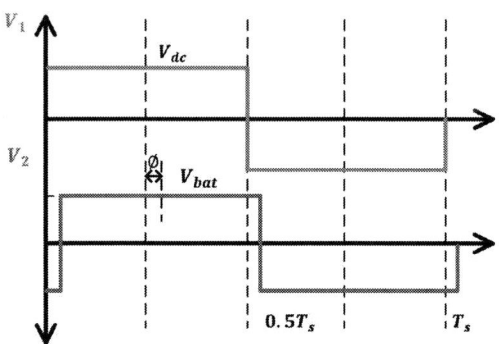

Fig. 3. Modulation of primary and secondary side voltages over a switching cycle for DC input.

general expression for the amount of power transferred in a DAB operating with a single-phase shifted modulation is given by:

$$P_{dc} = \frac{0.5 n V_{bat} V_{dc} \phi (\pi - |\phi|)}{\pi^2 f_s L}. \quad (6)$$

III. IMPLEMENTATION OF THE CHARGE CONTROLLER

Along with the primary and secondary side modulation schemes, regulating the phase difference between the two sides is essential to ensure efficient power transfer from the input side to the battery, thereby optimizing the charging process. As the charger supports both AC and DC voltage as input, depending on the input to the converter, the controller has to generate the corresponding phase difference between the two sides. In this section, two charge controllers for the proposed charger will be addressed.

As given in Fig. 4, a charge controller is used to regulate both the battery voltage and battery current (I_{bat}), ensuring efficient and reliable charging. The controller dynamically adjusts the modulation scheme based on the input voltage to the converter i.e., when an AC input is provided, the converter operates as a matrix converter, whereas for a DC input, it functions as a phase-shifted DAB. The controller processes the difference between the reference and instantaneous battery current and voltage, subsequently generating an appropriate phase shift to facilitate optimal power transfer. Notably, in matrix converter operation, the battery current exhibits a sinusoidal waveform with a constant average value and a

frequency twice that of the grid. This method reduces stress on the battery and improves the overall state of health of the battery [10], [11]. As a result, a sinusoidal ripple current (SRC) charging strategy is employed, followed by constant voltage (CV) charging to ensure effective battery utilization. Conversely, when the converter is supplied with a DC input, conventional CC and CV charging methods are adopted. This dual-mode capability enhances the flexibility of the proposed charger, enabling seamless integration with various power sources while maintaining efficient charging characteristics.

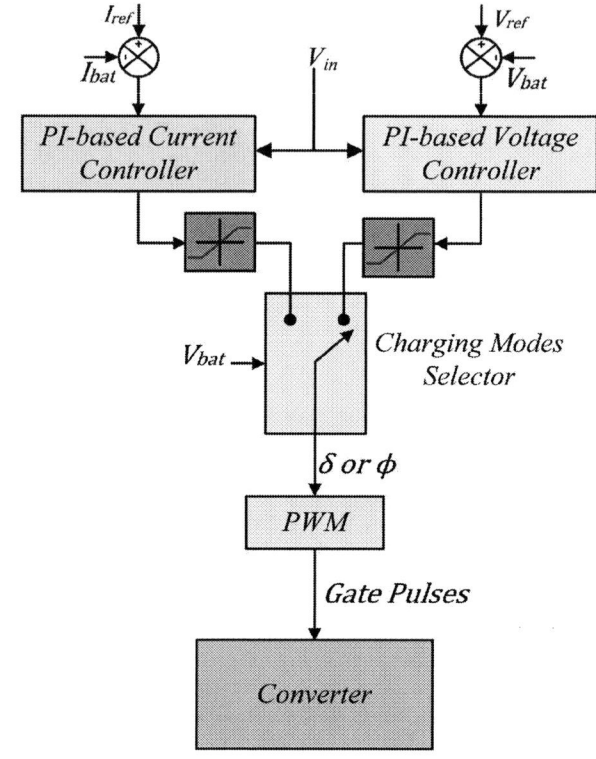

Fig. 4. Block diagram for the PI-based charge controller.

In the process of selecting the charging mode, the controller continuously monitors V_{bat} and I_{bat} to determine an appropriate mode for charging. Initially, the charger operates in constant current mode when a DC input is provided, ensuring a steady current supply to the battery. In the case of an AC input, the charger employs the SRC charging method, where the battery current follows a sinusoidal waveform with a constant average value. As the battery voltage gradually increases and reaches a predefined threshold value, the charging mode selector seamlessly transitions the operation from CC or SRC mode to CV mode. During CV mode, the controller maintains a fixed battery voltage while allowing the charging current to decrease progressively. This gradual reduction in current helps in preventing overcharging and enhances battery longevity. Additionally, the transition between charging modes requires proper control to ensure safe and efficient operation during the charging process. To achieve this, the charge controller dynamically adjusts the phase difference between the two

979-8-3315-3013-6/25 $31.00 © 2025 IEEE

sides of the charger, regulating the power flow accordingly. In this work, a well-tuned proportional-integral (PI) controller is designed and tuned using the Ziegler–Nichols tuning method, ensuring precise regulation of battery voltage and current.

IV. SIMULATION RESULTS

PLECS/Blockset is used to simulate the proposed universal charger. A 48 V Li-ion battery model is utilized in the simulation to charge and analyze the charging process of LEV. Table I provides the values of the various system parameters employed in the simulation. In the case of the AC input to the charger, the SRC-CV charging process is verified with PFC. Additionally, the CC-CV modes of charging are performed in case of a DC input to the charger, and the simulation outcomes confirm the efficacy of the charge controller.

TABLE I
PARAMETERS USED IN THE SIMULATION

Parameters	Values
Switching frequency (F_s)	10 kHz
AC voltage (v_{ac})	110 V, 50 Hz
DC voltage (V_{dc})	100 V
Turns ratio (n)	0.25
Nominal battery voltage (V_{bat})	48 V
Leakage inductance (L)	8 µH

The simulation results for the DC input case illustrate that the charging process initially operates in the CC mode, where a fixed current of 5 A is supplied to the battery. As the charging progresses, the battery voltage gradually increases, and the charge controller continuously monitors the voltage level. Once V_{bat} reaches 50 V, the mode selector seamlessly transitions the charger from CC to CV mode, as presented in Fig. 5.

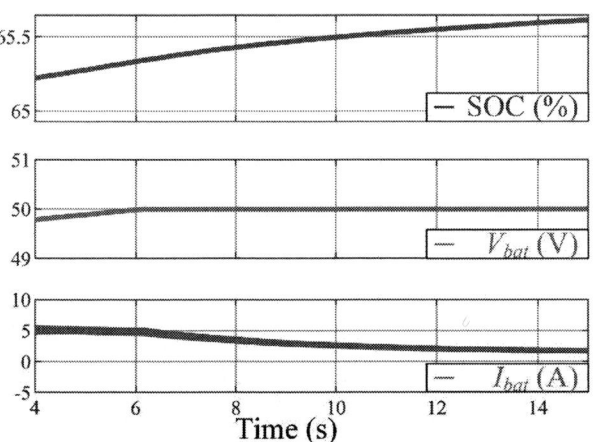

Fig. 5. Variation of battery parameters during CC-CV charging.

For the AC input case, the charger first operates in SRC mode, delivering an average charging current of 6.5 A. Similarly, once the battery voltage reaches 50 V, the mode selector changes the operation to CV mode (Fig. 6). Notably, the grid voltage and current remain in phase during AC operation,

achieving a unity power factor as depicted in Fig. 7. Moreover, Fig. 8 presents the waveforms of voltage and current during high-frequency switching of the MC and applied to the high-frequency transformer, thereby illustrating the single-stage conversion from utility-frequency AC to high-frequency AC.

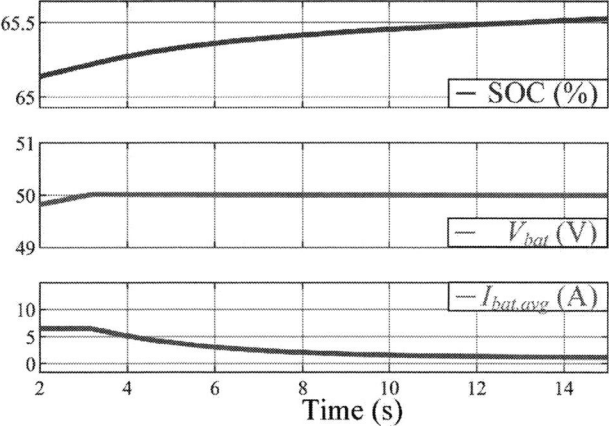

Fig. 6. Variation of battery parameters during SRC-CV charging.

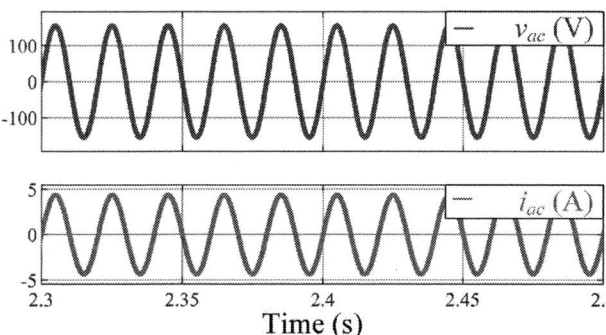

Fig. 7. Validation of unity power factor during an AC input.

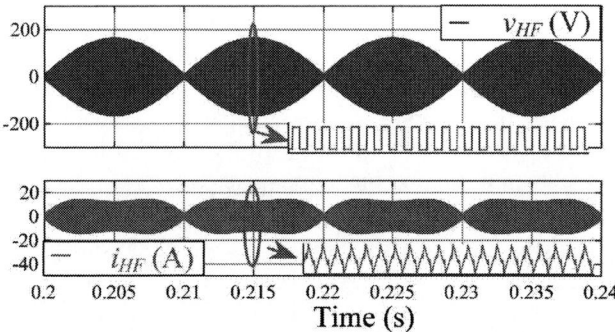

Fig. 8. Waveforms of voltage and current during high-frequency switching of the MC under AC input conditions.

To validate bidirectional power flow, V2G operation is simulated using the proposed universal charger. The corresponding grid voltage and current waveforms during this operation are given in Fig. 9, where the voltage and current are observed to be phase-shifted by 180 degrees, indicating reverse power flow.

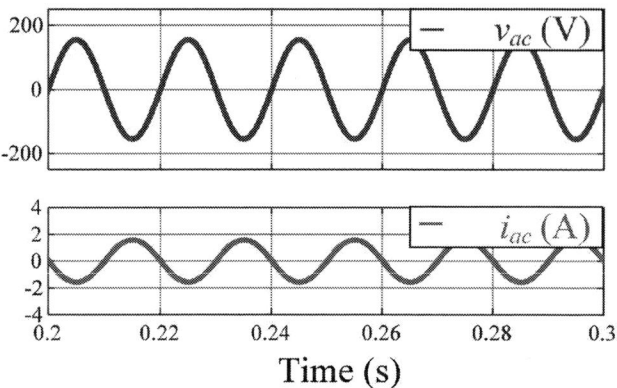

Fig. 9. Grid voltage and current during V2G mode of operation.

V. CONCLUSION

In this research work, a 0.5 kW single-stage universal charger is proposed utilizing a matrix converter for the LEV charging systems. The charger supports both AC and DC inputs, enabling seamless integration with various power sources, including operation in areas with frequent power outages in the grid and integration with solar PV systems for sustainable energy utilization. The charging of a 48 V Li-ion battery is simulated for both AC and DC inputs, demonstrating the effectiveness of the proposed charger in seamlessly transitioning between CC, SRC, and CV charging modes. The performance of the proposed charger is verified from the simulation outcomes for both AC and DC inputs. Notably, a unity power factor is achieved when the battery is charged from the electrical grid. Future work involves optimizing system parameters and implementing hardware for the proposed universal charger to validate the simulation results, including its performance under weak grid conditions. One key direction is the optimization of the system parameters to achieve power balance for both AC and DC input modes, resulting in balanced stress on the switching devices during any mode of operation. Furthermore, an extension of this proposed charger for large EVs can be explored by scaling the topology to support higher power levels.

REFERENCES

[1] S. K. Rastogi, A. Sankar, K. Manglik, S. K. Mishra and S. P. Mohanty, "Toward the Vision of All-Electric Vehicles in a Decade [Energy and Security]," *IEEE Consumer Electronics Magazine*, vol. 8, no. 2, pp. 103-107, March 2019.

[2] S. -G. Jeong, J. -M. Kwon and B. -H. Kwon, "High-Efficiency Bridgeless Single-Power-Conversion Battery Charger for Light Electric Vehicles," *IEEE Transactions on Industrial Electronics*, vol. 66, no. 1, pp. 215-222, Jan. 2019.,

[3] S. Ghosh and B. Singh, "A Multiport Charger for Light Electric Vehicles With Function of Powering Domestic Appliances," *IEEE Transactions on Consumer Electronics*, vol. 70, no. 1, pp. 308-317, Feb. 2024.

[4] R. P. Upputuri and B. Subudhi, "A Comprehensive Review and Performance Evaluation of Bidirectional Charger Topologies for V2G/G2V Operations in EV Applications," *IEEE Transactions on Transportation Electrification*, vol. 10, no. 1, pp. 583-595, March 2024.

[5] A. Ali, H. H. H. Mousa, M. F. Shaaban, M. A. Azzouz and A. S. A. Awad, "A Comprehensive Review on Charging Topologies and Power Electronic Converter Solutions for Electric Vehicles," *Journal of Modern Power Systems and Clean Energy*, vol. 12, no. 3, pp. 675-694, May 2024.

[6] G. M. Sahoo and V. Agarwal, "A Reduced Stage Half Bridge Based Isolated Interleaved Totempole Unidirectional AC-DC Converter," in *Proc. 2024 IEEE International Conference on Power Electronics, Drives and Energy Systems (PEDES)*, Mangalore, India, pp. 1-6, December 2024.

[7] N. D. Weise, G. Castelino, K. Basu and N. Mohan, "A Single-Stage Dual-Active-Bridge-Based Soft Switched AC–DC Converter With Open-Loop Power Factor Correction and Other Advanced Features," *IEEE Transactions on Power Electronics*, vol. 29, no. 8, pp. 4007-4016, Aug. 2014.

[8] G. Castelino, K. Basu, N. Weise and N. Mohan, "A bi-directional, isolated, single-stage, DAB-based AC-DC converter with open-loop power factor correction and other advanced features," in *Proc. of 2012 IEEE International Conference on Industrial Technology*, Athens, Greece, 2012, pp. 938-943.

[9] S. Wei, Z. Zhao, K. Li, L. Yuan and W. Wen, "Deadbeat Current Controller for Bidirectional Dual-Active-Bridge Converter Using an Enhanced SPS Modulation Method," *IEEE Transactions on Power Electronics*, vol. 36, no. 2, pp. 1274-1279, Feb. 2021.

[10] L. -R. Chen, S. -L. Wu, D. -T. Shieh and T. -R. Chen, "Sinusoidal-Ripple-Current Charging Strategy and Optimal Charging Frequency Study for Li-Ion Batteries," *IEEE Transactions on Industrial Electronics*, vol. 60, no. 1, pp. 88-97, Jan. 2013.

[11] Y. -D. Lee and S. -Y. Park, "Electrochemical State-Based Sinusoidal Ripple Current Charging Control," *IEEE Transactions on Power Electronics*, vol. 30, no. 8, pp. 4232-4243, Aug. 2015.

A Robust Data Driven FDI attack Detection Framework for Inverter based Microgrids

Suprabhath Sriranga Koduru
Department of Electrical Engineering
IIT(ISM) Dhanbad
Dhanbad

Venkata Siva Prasad Machina
Department of Electrical Engineering
IIT(ISM) Dhanbad
Dhanbad

Sreedhar Madichetty
Electronics and Computer Engineering
Ecole Centrale School of Engineering
Mahindra university,Hyderabad

Sukumar Mishra
Department of Electrical Engineering
IIT(ISM) Dhanbad
Dhanbad

Abstract—False Data Injection (FDI) attacks pose a major cybersecurity risk to power electronic systems, particularly inverter-based microgrids. These attacks compromise signals like current, voltage, and phase, causing incorrect power flow calculations, inverter instability, and system outages. This study presents a data-driven approach to detect FDI attacks using an attack classifier trained on data generated from a Model Predictive Control (MPC)-based current control scheme for a three-phase, two-level inverter. The dataset includes normal and attack-induced current deviations, enabling robust model training. The study addresses multiple FDI scenarios, including amplitude, offset, and phase manipulations. A multi-class classification approach is adopted, utilizing machine learning classifiers such as Support Vector Classifier (SVC), Decision Tree, Naive Bayes, Ensemble Boosting, and Neural Networks (NN). Classifier performance is evaluated using accuracy, precision, recall, and F1 score. Among the classifiers, the Neural Network outperforms the rest, achieving an overall accuracy of 99.6%, demonstrating superior detection for both covert and stealthy attacks. The use of an MPC-based control framework to generate training data adds novelty to the approach. The proposed methodology is implemented using MATLAB Simulink, highlighting the potential of neural networks in enhancing the cybersecurity of inverter-based microgrids.

Index Terms—Data-driven classification, False data injection attacks, Inverter-based microgrid, Model predictive control, Neural network classifier

I. INTRODUCTION

The growing integration of renewable energy sources (RES) like solar and wind in microgrids has made secure inverter control essential for stability [1]. Inverters convert DC from RES to AC power and maintain voltage, frequency, and power quality. Compromised inverter control can destabilize the microgrid, cause blackouts, and disrupt power sharing among distributed energy resources (DERs). Grid-forming inverters establish system voltage and frequency in islanded microgrids, while grid-following inverters synchronize with the main grid. Attacks targeting control logic can cause frequency deviations and voltage instability, leading to system collapse. Secure control systems with robust encryption and anomaly detection help maintain stability [2].

Inverters share active (P) and reactive (Q) power using droop control, but modern communication links increase vulnerability to attacks like denial-of-service (DoS) and data tampering. Inverters also filter harmonics from non-linear loads to maintain power quality. Cyberattacks introducing artificial harmonics can overload inverter control logic, causing system inefficiency. Fault ride-through (FRT) ensures inverter operation during faults, while black start capabilities allow microgrids to self-start. Attacks on FRT logic can cause inverter shutdowns, risking system-wide instability. Secure, consensus-based decision-making and redundant controls mitigate these risks. Cyber threats like false data injection (FDI) and malware can destabilize microgrids. Ensuring secure inverter control is vital for reliable, stable, and cyber-resilient microgrids [3].

A. False data injection attack

FDI attack on AC signals pose a critical threat to power grids and microgrids by compromising the integrity of measurement and control signals in key systems such as phasor measurement units (PMUs), smart meters, and SCADA systems. Given that AC signals have both magnitude and phase components, any tampering with these elements can disrupt essential operations like state estimation, load forecasting, and power flow control. Magnitude-based tampering methods include additive noise, multiplicative distortion, and signal clipping or saturation, each of which can distort signal accuracy and affect decision-making processes. Phase-based tampering involves phase shift insertion, where a phase shift (ϕ) is introduced by altering system delays, phase angle corruption, which modifies phase angle readings in PMU or smart meters, and signal injection, where a signal with a similar frequency but different phase is added to create distortions. Frequency-based tampering includes manipulating PMU sampling rates, injecting harmonics or inter-harmonics into the system, and interfering with control signals, leading to frequency fluctuations. Hybrid tampering simultaneously alters

979-8-3315-3013-6/25 $31.00 © 2025 IEEE

magnitude, phase, and frequency, often through coordinated attacks on multiple PMUs or sensors, making detection and mitigation more complex. Control signal tampering involves modifying instructions sent to inverters or controllers, and changing setpoints for voltage, frequency, and reactive power, thereby destabilizing the power system. Effective mitigation requires robust anomaly detection, secure communication protocols, and advanced AI-driven monitoring solutions [4].

B. Literature Review

MG vulnerabilities towards the communication delays and the denial of service attacks are discussed in [5], the cyber physical co-simulation framework is established to demonstrate the effect of attack. In [6], the researchers considered the black swan events on the secondary control of the MG, where the additive FDI attacks falsifies the meter readings. A small signal model of the islanded MG is used to design the attack detector.

FDI attack on real and reactive power setpoints in the 100% inverter-based microgrids is studied in [7], LSTM based mitigation methodology is used. FDI attack detection and mitigation strategy based on the classification of residues is proposed in [8], here the metered data is predicted with GRU-based time series prediction and compared with actual meter readings to classify the data. SVM-based FDI attack detection mechanism is developed in [9] to detect the voltage regulation attacks on the PV-based power distribution systems. The ANN-based observer is developed in [10] to estimate and mitigate the effect of FDI attack on the consensus-based secondary control of DC microgrid. FDI attacks on PMU units is addressed in [11], 1-D CNN model is used to detect the FDI attack. By following the above literature it can be understood that the data-driven methodologies are very efficient in detecting and mitigating the FDI attacks on different vulnerable points of the cyber-physical systems, especially microgrid systems. The majority of the work is performed in additive FDI attacks and DC microgrids. This research further improves the data-driven detection strategies by considering various signal parameters such as amplitude, phase, and offset.

II. METHODOLOGY

The intruder aims to destabilize the system by targeting vulnerable communication channels carrying sensor data. False Data Injection (FDI) attacks subtly alter feedback signals like voltage or current, misleading controllers while evading detection. In contrast, control references and actuator signals are more secure, making feedback manipulation the primary threat vector. To demonstrate the FDI attack's impact on the inverter, a robust current control mechanism is developed for the 3-phase 2L VSI using the model predictive control technique. Later, an FDI attack scenario is considered on one of the phase currents, and their impacts and detection methodologies are discussed in detail.

A. MPC based current control

MPC is superior compared to other existing controls such as hysteresis, SVM, and linear control. MPC is a nonlinear

controller that avoids the modulation stage, it has good steady-state and transient performances compared to its counterparts. On the negative side of it, the computational burden of this controller increases as the constraints and nonlinearities of the model increases. Fig. 1 denotes the three-phase 2L VSI with MPC and the data-driven attack classifier. Implementation of MPC involves three important steps, generation of reference currents, measurement of 3 phase currents, extrapolation of reference currents, prediction from the measured currents and the cost function minimization. Detailed mathematical formulation and implementation are given in [12].

1) Reference current generation: Generating reference currents for inverter current control is a fundamental aspect of ensuring stable and efficient operation in various applications like microgrids, renewable energy systems, and electric vehicle (EV) chargers. The methods of reference current generation differ depending on the control objectives, system requirements, and application scenarios. Grid-connected applications use, SRF theory, pq theory, PLL control and PR control for reference current generation. Whereas, islanded microgrid inverter uses, droop control, direct power control, MPC and virtual synchronous generator control for reference generation. Similarly, various methods are used for EV charging systems, and renewable energy applications use several other mechanisms.

In this article, a standalone inverter is considered for demonstrating an attack detection module, therefore user-defined reference currents are considered. Mathematical representation of the prediction of measured values and the prediction of reference currents are given in (1)-(4). The obtained measured ($\widehat{i}(t+1)$) and reference ($\widehat{i^*}(t+1)$) predictions are implemented on cost functions and the optimized switching instant is given to the digital I/O pins of the controller.

$$\widehat{i^*}(t+1) = 4i^*(t) - 6i^*(t-1) + 4i^*(t-2) - i^*(t-3) \quad (1)$$

$$\widehat{i}(t+1) = \left(1 - \frac{R_L T_s}{L}\right) i(t) + \frac{T_s}{L} \widehat{v}(t) \quad (2)$$

$$\widehat{v}(t) = \begin{bmatrix} 2/3 & -1/3 & -1/3 \\ -1/3 & 2/3 & -1/3 \\ -1/3 & -1/3 & 2/3 \end{bmatrix} \begin{bmatrix} p_a(t) \\ p_b(t) \\ p_c(t) \end{bmatrix} \quad (3)$$

$$C(t) = \left| \widehat{i^*}(t+1) - \widehat{i}(t+1) \right| \quad (4)$$

B. Formulation of FDI attacks

As discussed in the earlier section FDI attacks on an AC signal can be implemented on three parameters, amplitude, phase shift and offset. Attacker who targets to destabilize the control can choose any of these parameters depending on the level of unauthorized access he possess. Therefore, in this work these three parameters of the signal are considered as vulnerable and attack detection is designed. The attack on amplitude of the signal ($i_{fa}(t)$), phase of the signal ($i_{fp}(t)$) and the signal offset ($i_{fo}(t)$) are represented in (5)-(7) respectively. (8) represent the hybrid FDI attack ($i_{fh}(t)$) on the signal, where the attacker manipulates the multiple

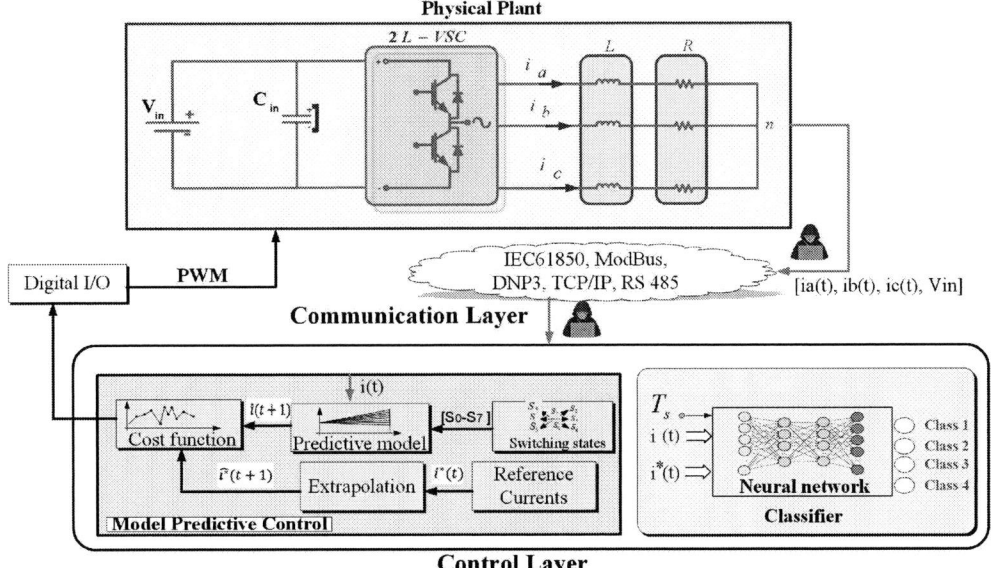

Fig. 1. Proposed classifier schematic of model predictive control based two level voltage source inverter

parameters to increase the impact. ε, ψ and β denotes the amplitude attack vector, phase attack vector and offset attack vector respectively.

$$i_{fa}(t) = (1 + \varepsilon) I_m sin (\omega t + \phi) \quad (5)$$

$$i_{fp}(t) = I_m sin (\omega t + \phi + \psi) \quad (6)$$

$$i_{fo}(t) = I_m sin (\omega t + \phi) + \beta \quad (7)$$

$$i_{fh}(t) = (1 + \varepsilon) I_m sin (\omega t + \phi + \psi) + \beta \quad (8)$$

III. DATA DRIVEN ATTACK CLASSIFICATION

Many data-driven classifiers are developed for identifying the anomalies and outliers in the signal data. Some of the advanced classifiers that can classify the data of nonlinear systems are support vector classifiers, decision trees, naive Bayes classifiers, ensemble classifiers, and neural network classifiers. Each of the above classifiers has its respective capabilities and limitations in their classification methodologies, which are chosen according to the requirement. Therefore, all the above-mentioned classifiers are implemented on the collected data to detect the presence of FDI attacks. Rather than designing separate classifiers (binary classification) to detect each attack, a single attack classifier is designed with multi-class classification.

A. ML based Classifiers

These classifiers follow traditional machine learning principles where the decision boundaries are explicitly derived from mathematical models and rules. The SVC aims to find a hyperplane that maximizes the margin between the classes. For multi-class classification, the one-vs-rest (OvR) or one-vs-one (OvO) approach is typically used in which each class is treated as a binary classification problem against the other classes.

(9) denotes the classification rule used, where the output of the classifier $f_j(x)$ is passed through (10) to attain the class assignment. y_i is the class label where $y_i \varepsilon -1, 1$, L_i is the lagrange multiplier and $K(x_i, x)$ is the kernel function used for SVC.

$$f_j(x) = \sum_{i=1}^{n} L_i y_i K(x_i, x) + \beta \quad (9)$$

$$\widehat{y} = arg \max_{j=1,2,3,4} f_j(x) \quad (10)$$

Decision tree classifiers use gini impurity index (g) to decide on the class the input belongs to as given in (11), p_j indicates the proportion of instances in class j.

$$g = 1 - \sum_{j=1}^{4} p_j^2 \quad (11)$$

For multi-class classification, Naive Bayes extends naturally, as we compute the posterior probability for each of the 4 classes and select the class with the highest probability. (12) shows the Bayesian formula for obtaining the probability of the occurrence and (13) indicates the classification rule.

$$P(y = j | x_1, x_2, ...x_n) = \frac{P(x_1, x_2, ...x_n | y = j) P(y = j)}{P(x_1, x_2, ...x_n)} \quad (12)$$

$$\widehat{y} = arg \max_{j=1,2,3,4} P(y = j) \prod_{i=1}^{n} P(x_i | y = j) \quad (13)$$

Gradient boosting also known as weighted sum is the ensemble learning technique used that predicts the probability of each of the 4 classes obtained as shown in (14). h_m^j indicates the prediction of mth weak learner of jth class.

$$f_j(x) = \sum_{m=1}^{M} \alpha_m h_m^j(x) \qquad (14)$$

B. Neural Network Classifiers

Neural network classifiers, inspired by the human brain, consist of layers of interconnected neurons. They use back-propagation and gradient descent for learning complex, non-linear relationships in data. These models excel in handling large datasets and unstructured data like images, audio, and text. Neural Network Classifiers: Consist of input, hidden, and output layers, where each layer performs nonlinear transformations of the data. They are highly effective for tasks like image classification, speech recognition, and deep learning-based time series forecasting. For multi-class classification, the neural network uses a softmax function at the output layer to produce probabilities for the 4 classes as shown in (15).

$$f_k(x) = \frac{e^{z_j}}{\sum_{k=1}^{4} e^{z_j}} \qquad (15)$$

C. Design of data-driven FDI attack classifier

The stepwise procedure followed to design the data-driven FDI attack classifier is given in Fig. 2. The initial step in the design process is dataset preparation. Dataset for the classifier is obtained from the MATLAB simulations. Various attack scenarios along with the normal operating conditions are implemented on the MPC controlled inverter. Multiple levels of ε, ψ and β are introduced in the normal signal and the attacked data is collected. It is assumed that the attacker targets the measured feedback signals. Therefore the training data for the classifier consists of measured current signals ($i(t)$), generated reference currents ($i^*(t)$) and time stamp (t). The proportion of the data samples generated with respect to each attack is tabulated in Table 1. Data obtained from the MPC and attack

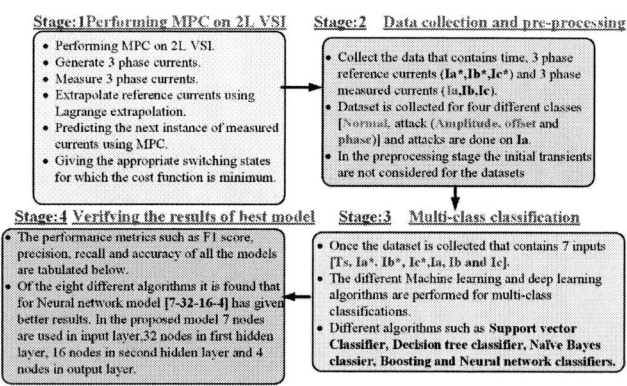

Fig. 2. Work flow for data-driven classification

scenarios is used for offline training of the multiple data-driven classifiers that are discussed above. Hyperparameter tuning and training parameter selection are performed to obtain the network architectures of each classifier. The training results of

TABLE I
SAMPLES COLLECTED FOR EACH CLASS

S.No	Class	Number of samples
1	Normal	499000
2	Amplitude	998900
3	Offset	499000
4	Phase	500000
5	Total	2496900

TABLE II
COMPARISON OF DIFFERENT EVALUATION METRICS FOR DIFFERENT CLASSIFIERS

Classifier	Precision	Recall	F1-score	Accuracy
Support vector machine	0.649	0.847	0.691	0.719
Decision tree	0.825	0.895	0.848	0.858
Naive Bayes	0.895	0.926	0.908	0.913
Boosting	0.988	0.988	0.988	0.99
Neural network	0.996	0.996	0.996	0.996

the classifiers are tabulated in Table 2, where the table consists of evaluation metrics such as accuracy, precision, recall, and F1 score. Based on the results obtained from the training, it can be understood that the ensemble classifier and the neural network classifier yield a better classification of the proposed FDI attacks.

IV. RESULTS

As a proof of concept, the proposed detection methodology is implemented on the 3-phase 2-Level VSI designed in the MATLAB Simulink platform with an input voltage of 520 V DC and the RL load of 10 Ω and 1 mH respectively. MPC-based inverter control is developed to achieve a robust control technique. Further, various attacks are introduced and the reliability of the designed classifier is examined. As the neural network classifier gives better performance in comparison of its counterparts, the NN classifier is considered for further examination.

A. NN Classifier Performance

Results from Table 2 show that the neural network attains an accuracy of 99.6%, the same is supported by the confusion matrix obtained from the training as shown in Fig. 3. Fig. 4 denotes the sample classifications during the NN classifier training. Fig. 4(a) denotes the correct classification samples of all the classes and Fig. 4(b) denotes the misclassified samples.

B. Normal operation

In this case, the normal operation of the MPC-inverter is examined. Fig. 5 denotes the accurate tracking of the model, where the reference currents are given as 10 A. Zoomed-in version indicates the overlapping of the reference and measured current with negligible error. The classifier performance is shown in dashboard created in MATLAB in Fig. 6, NN classifier correctly classifies it as normal condition.

Confusion matrix

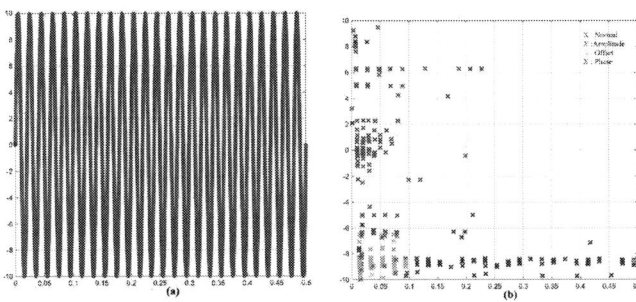

Fig. 3. Confusion matrix of NN classifier

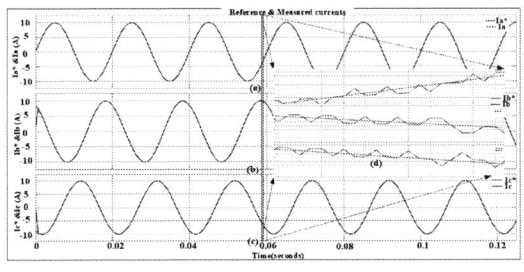

Fig. 4. (a) Correct classifications; (b) Misclassified samples

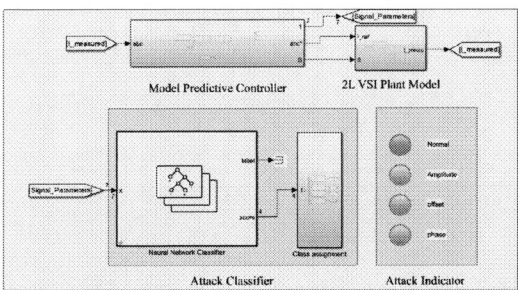

Fig. 5. a. Measured load currents and input reference currents (a) A phase; (b) B phase; (c) C phase and d. Zoomed version of currents

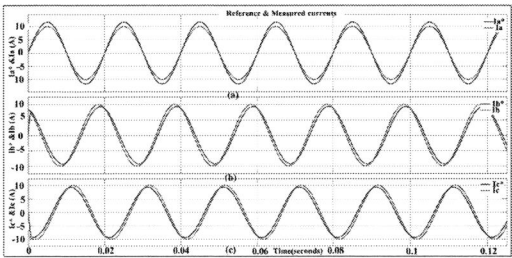

Fig. 6. Graphical user interface of proposed classifier algorithm whose output is normal class

C. Amplitude FDI attack

An amplitude of 1.5 A which is approximately 15 % of the measured signal is injected as an FDI attack. The response of the MPC-controlled inverter for this attack can be observed from Fig. 7, which shows a considerable difference in reference tracking. NN classifier classifies this as an amplitude attack as shown in Fig. 8(a).

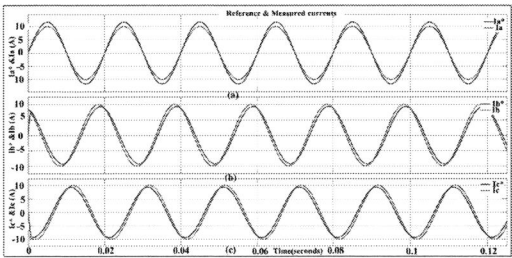

Fig. 7. Measured load currents and input reference currents during amplitude FDI attack (a) A phase; (b) B phase; (c) C phase

D. Offset

An offset of 5 A is introduced as FDI attack in this scenario and the response of MPC controller and the NN classifier can be observed in Fig. 9. It is evident that the offset attack produces a major tracking error and the classifier is able to detect the attack effectively and it will be observed on the GUI as shown in Fig. 8(b).

E. Phase

As part of creating a delay in the signal reaching the controller, an FDI attack of 30 degrees is introduced into the signal readings that causes a considerable delay of 1 ms. A conscious decision to introduce low values is taken to test the robustness and reliability of the NN classifier. Fig. 10 and Fig. 8(c) show the response of the MPC and NN classifier.

V. CONCLUSION

This article presented a comprehensive discussion on FDI attacks, emphasizing their significant impact on the integrity, stability, and security of power electronic systems, particularly in inverter-based microgrids. To address this challenge, a data-driven approach was adopted to design an FDI attack classifier. The training data for this classifier was generated using a MPC based current control scheme for a three-phase, two-level (2L) inverter. This control strategy provided a rich dataset capturing normal operating conditions and attack-induced deviations in the current measurements, enabling effective model training.

Various ML classifiers were evaluated for their ability to detect FDI attacks, including SVC, Decision Tree, Naive Bayes, Ensemble Boosting, and Neural Networks. Each classifier's performance was analyzed using key metrics such as accuracy, precision, recall, and F1 score. The comparative analysis revealed that, while traditional classifiers like Boosting and Naive Bayes performed reasonably well, the Neural Network classifier demonstrated superior detection capabilities. Further,

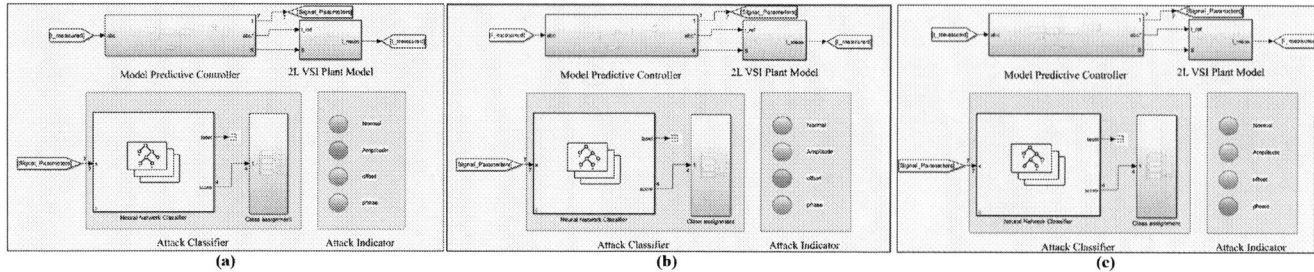

Fig. 8. Attack detection through Graphical user interface (a) Amplitude (b) Offset (c) Phase

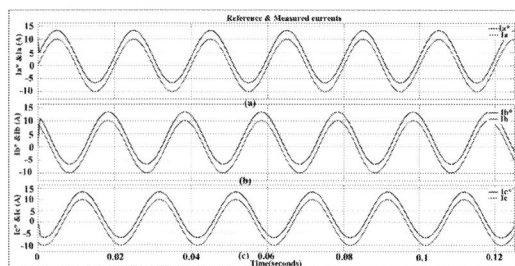

Fig. 9. Measured load currents and input reference currents during offset FDI attack (a) A phase; (b) B phase; (c) C phase

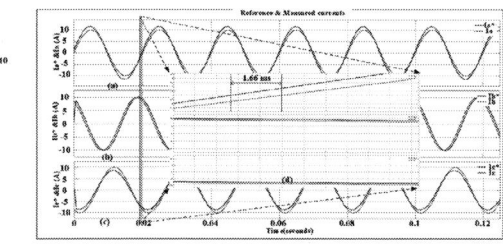

Fig. 10. Measured load currents and input reference currents during Phase FDI attack (a) A phase; (b) B phase; (c) C phase; d. Zoomed version

the study should be extended to higher dimensional systems such as grid connected inverters, EV charging systems, and renewable energy systems. The datasets from these realtime systems gives better insights and pose significant learning challenges for the classifiers. This research lays a solid foundation for future work on developing more generalized, scalable, and robust FDI detection systems that can be applied to broader microgrid and power system applications.

VI. ACKNOWLEDGEMENT

This work is supported by the Science and Engineering Research Board (SERB) under start-up research grant PDF/2023/001836 sponsored to Dr. Suprabhath Sriranga Koduru

REFERENCES

[1] Nguyen Duc Tuyen, Nguyen Sy Quan, Vo Ba Linh, Vu Van Tuyen, and Goro Fujita. A comprehensive review of cybersecurity in inverter-based smart power system amid the boom of renewable energy. *IEEE Access*, 10:35846–35875, 2022.

[2] Sai Nikhil Vodapally and Mohd Hasan Ali. Overview of intelligent inverters and associated cybersecurity issues for a grid-connected solar photovoltaic system. *Energies*, 16(16):5904, 2023.

[3] Yuanliang Li and Jun Yan. Cybersecurity of smart inverters in the smart grid: A survey. *IEEE Transactions on Power Electronics*, 38(2):2364–2383, 2022.

[4] BoHyun Ahn, Taesic Kim, Seerin Ahmad, Sudip K Mazumder, Jay Johnson, H Alan Mantooth, and Chris Farnell. An overview of cyber-resilient smart inverters based on practical attack models. *IEEE Transactions on Power Electronics*, 2023.

[5] Ola Ali, Tung-Lam Nguyen, and Osama A Mohammed. Assessment of cyber-physical inverter-based microgrid control performance under communication delay and cyber-attacks. *Applied Sciences*, 14(3):997, 2024.

[6] Ioannis Zografopoulos and Charalambos Konstantinou. Detection of malicious attacks in autonomous cyber-physical inverter-based microgrids. *IEEE Transactions on Industrial Informatics*, 18(9):5815–5826, 2021.

[7] Milad Beikbabaei, Mario Montano, Ali Mehrizi-Sani, and Chen-Ching Liu. Mitigating false data injection attacks on inverter set points in a 100% inverter-based microgrid. In *2024 IEEE Power & Energy Society Innovative Smart Grid Technologies Conference (ISGT)*, pages 1–5. IEEE, 2024.

[8] Wenxin Lei, Zhibo Pang, Hong Wen, Wenjing Hou, and Wen Han. Fdi attack detection at the edge of smart grids based on classification of predicted residuals. *IEEE Transactions on Industrial Informatics*, 18(12):9302–9311, 2022.

[9] Masoud Ahmadzadeh, Ahmadreza Abazari, and Mohsen Ghafouri. Detection of fdi attacks on voltage regulation of pv-integrated distribution grids using machine learning methods. In *2022 IEEE Electrical Power and Energy Conference (EPEC)*, pages 73–78. IEEE, 2022.

[10] Md Abu Taher, Mohd Tariq, and Arif I Sarwat. Enhancing security in islanded ac microgrid: Detecting and mitigating fdi attacks in secondary consensus control through ai-based method. In *2023 IEEE International Conference on Energy Technologies for Future Grids (ETFG)*, pages 1–6. IEEE, 2023.

[11] Saleh Almasabi, Zohaib Mushtaq, Nabeel Ahmed Khan, and Muhammad Irfan. Improving fdi detection for pmu state estimation using adversarial interventions and deep auto-encoder. *IEEE Access*, 2024.

[12] Machina Venkata Siva Prasad, Koduru Sriranga Suprabhath, Sreedhar Madichetty, Sukumar Mishra, and Abdelkader El Kamel. Design and implementation of model parameter independent robust current control scheme of three-phase inverter - a neural network-based classification approach. *CPSS Transactions on Power Electronics and Applications*, 9(2):166–174, 2024.

979-8-3315-3013-6/25 $31.00 © 2025 IEEE

An Interleaved DC-DC Converter for Low Voltage Bi-Polar DC Microgrid

Vipul Thakur
Department of Electrical Engineering
Delhi Technological University
Delhi, India
vipulthakur_23pes12@dtu.ac.in

Saurabh Mishra, *Senior Member, IEEE*
Department of Electrical Engineering
Delhi Technological University
Delhi, India
saurabhmishra@dtu.ac.in

Abstract— **This paper focuses on the design and analysis of a single input dual output (SIDO) non-isolated Boost-Luo interleaved (NBLI) DC-DC converter for low voltage bi-polar DC microgrid (LVBDCM), incorporating a voltage mode control (VMC). The presented converter integrates boost and negative Luo topologies. When compared to traditional converters of the same rating, the interleaving approach offers the advantages of lower current stress and less source current ripple. The converter that is being presented offers interleaving advantages in addition to multiple output voltage levels. In this regard, a NBLI converter is presented here. For a variety of applications, the converter can provide balanced or non-balanced output voltages, keeping output voltages steady even when loads are not balanced. On account of its voltage balancing capability, presented converter delivers DC-link voltage balancing for applications such as bipolar low-voltage DC (LVDC) distribution systems. This work comprises presented converter's designing, modelling and output voltage control. The efficacy of the presented system is verified in MATLAB/Simulink.**

Keywords— *Low Voltage Bi-polar DC Microgrid (LVBDCM), Continuous Conduction Mode (CCM), Interleaved Luo-Boost, Voltage Mode Control (VMC)*

I. INTRODUCTION

LVDC power systems have received growing interest in present times mainly in the growth of DC renewable power generation and consumption. This happens because of the advantages in terms of power transformation efficiency and adaptability of control. Low wattage rating renewable energy sources (RES) will be increasingly prevalent in domestic applications in future in a manner to match increasing energy needs and minimize the carbon emissions [1]. According to polarity, DC microgrids can be categorized as unipolar, bipolar, or homopolar. The most adaptable of these is the bipolar DC microgrid (BDCM). Three-wire DC distribution used by bipolar microgrid design consisted of a $+V_{dc}$ line, a neutral line, and a $-V_{dc}$ line. Key perk the method is providing that, output-side power converters could select input voltage from the voltage levels: $+V_{dc}$, $-V_{dc}$, or $2V_{dc}$. One of the primary characteristics of bipolar grids is that they reduce the voltage's magnitude in relation to ground, enabling larger loads to be connected to their full rating. By Taking an example for instance, here bipolar LVDC system having ±24 V we can utilized with either a 24 V or 48 V as voltage source, resulting in improving its flexibility. Thus, BDCM system offers pliability and increased efficiency as well as reliability [2]. High-power microgrids can now be adequately served by bipolar DC grids. This is mostly because to the higher power transmission capability of this design. The DC-DC converters, must need to have a monopolar input and bipolar output, required to connect any distributed generation system to bipolar DC grids. Being that the loads and sources are not evenly distributed between the two poles,

the bipolar DC microgrid is prone to voltage and current imbalances. To maintain equilibrium of these DC microgrids, particular power electronic-based solutions are therefore [1] needed [3]-[5]. Even though BDCM has many benefits, but linking RESs to BDCM is difficult because of the bipolar voltages, which must have to be proportionated for best functioning. Here, DC-DC converter that integrates the RESs with BDCM needs to be capable of high voltage gain. The RESs and BCDMs are interfaced using three-level boost (TLB) converters. TLBs' non-isolated structure allows them to efficiently proportionate output voltages and more effectiveness as well as affordability. Nevertheless, voltage gain of TLB converters is limited, and they lack an interleaved structure. Bipolar output voltages of several high gain non-isolated DC-DC converters are presented in order to counteract voltage gain constraints of TLB converters. Nonetheless, converter in [6] lacks interleaving, has high switching losses, and is unable to achieve perfect balance under all loading conditions, whereas converter in [9] having restricted functional limit in regard to duty cycle [6]-[9]. During integration with RESs or battery as source units, an interleaved structure is recommended because it lowers source current ripple, increasing RES MPPT performance and input power source's longevity. In [10], a Boost-Zeta Interleaved (NBZI) multiport, non-isolated converter is displayed. With a single input and dual outputs (SIDO), presented topology consists of a boost and zeta converter combination. The topology's principal objective is to produce regulated DC voltage, which qualifies it for LVBDCM. Here, converter is optimal for low voltage bi-polar DC grid but the presented converter has relatively low voltage gain. Although the presented converter has a relatively low voltage gain, it is ideal for low voltage bipolar DC grids. In [11], a Boost-Cuk derived converter for a bipolar DC microgrid with an interleaved topology is proposed and current control approach is employed.

As shown in Fig. 1, the bipolar LVDC is a DC bus structure with three wire DC lines that functions as a small DC distribution network for residences and utilities. It connects multiple DC sources of energy and loads to a distribution system with three voltage levels. The presented work introduces a SIDO NBLI type DC-DC converter. This topology is appropriate for LVBDCM applications and renewable power sources such as fuel cells and solar PV. In NBLI converter, the two outputs are designated for the Boost and Luo converter, respectively. Presented converter have steady source current with nominal ripples because of the implementation of interleaving technique along with considerable high gain. Objective of the presented converter is to increase the power handling capacity by creating multi ports at the output and provide high gain. Interleaving of

---

979-8-3315-3013-6/25 $31.00 © 2025 IEEE   657

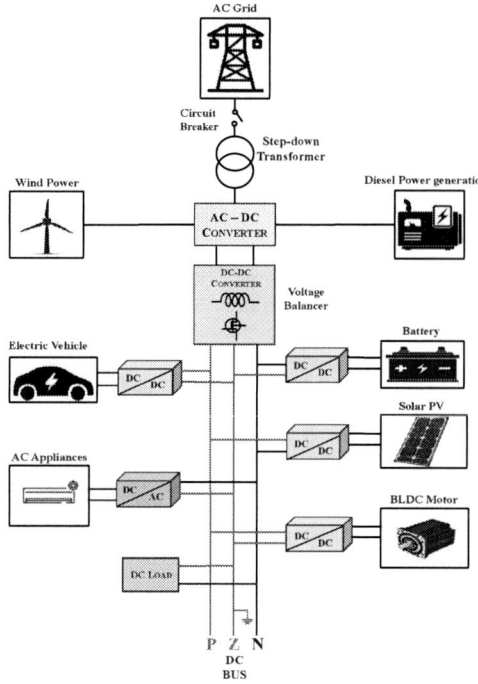

Fig. 1 Bipolar LVDC connected to grid

the converter, lowers the switches current rating, effective filter size, and current stress. As a result, a low voltage BDCM would benefit greatly from the presented topology. This converter operates as voltage balancer in BDCM. The following sections contain the remainder of this paper's content in the following order: The NBLI converter's schematic, design specifications, modes of operation, and circuit configuration are presented in Section II. Section III discusses voltage control schemes by employing PI controller and presenting dynamic modelling of converter. The outcomes of simulations results and discussion are included in section IV. Section V features the conclusion.

II. SCHEMATIC AND OPERATION OF NBLI DC-DC CONVERTER

In this section, configuration of circuit topology NBLI DC-DC converter, operating modes and specificizing the designing purpose of presented converter are comprised.

A. Circuit Configuration

The schematic representation for the NBLI DC-DC converter is shown in Fig. 2, which also shows the presented DC-DC converter configuration. 48V bus is linked to Boost converter's output, while 24V bus is linked to the Luo converter's output. Topology that is being presented combines a SIDO configuration with interleaving Boost-Luo converters. The input and output sides of the switching cells are connected in parallel and series, respectively. In the two-phase scenario, pulses for switching are delivered to S_L and S_B in an interleaved manner, means they are 180° degrees apart. When S_L is operating the S_B will be off and vice versa. V_{in} refers to voltage of input source; in this case, a battery is employed, and its voltage can range from 16 to 26 V. V_o is the overall DC bus voltage following by V_1, V_2 are the Boost and Luo output voltages. The Boost side boasts a load resistor R_{L1}, diode D_B, active switch S_B, and inductor L_B. The Luo side includes load

resistor R_{L2}, diode D_L, capacitors C_2, C_L, active switch S_L, and inductors L_{L1} and L_{L2}. As RESs can have low voltages between 18 and 26 V, negative Luo converter is selected to

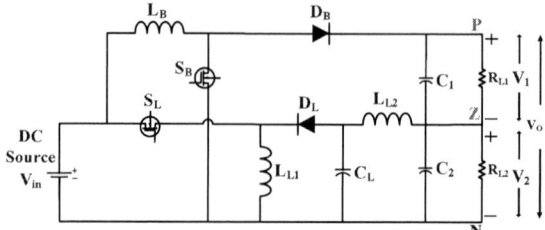

Fig. 2 Schematic of SIDO NBLI converter

interconnect with 24 V buses and a Boost to interconnect with 48 V buses. Luo converter is interleaved with Boost converter for its advantages, like low output voltage ripple, provides high efficiency and voltage gain with low conduction losses. In order to ensure CCM, the inductors are purposefully designed. Filter capacitors, resistive load linked within ports P-Z and Z-N as C_1, R_{L1} and C_2, R_{L2} respectively. Here, P-Z and Z-N are the poles i.e. positive (P), negative (N), neutral (Z) respectively.

B. Modes of Operation

Here, NBLI DC-DC converter's operating modes and a steady-state analysis of the converter in continuous conduction mode (CCM) are employed in this section. Ideal and lossless power components are assumed; parasitic components are not taken into account, and both converters have equal inductance and output capacitances values. Four CCM modes are presented based on the switching functions of S_B and S_L switches. These modes are represented in Table 1 as follows:

TABLE I CONDUCTING STATES OF SIDO NBLI CONVERTER

Conducting States S_B&S_L	Modes	L_B	L_{L1}	L_{L2}	C_L	C_1	C_2
00	Passive Mode 1	D	D	D	C	C	C
01	Active Mode 1	D	C	D	D	C	C
10	Active Mode 2	C	D	C	C	D	D
11	Passive Mode 2	C	C	C	D	D	D

❖ C- Charging, D- Discharging.

Representing voltage gains for CCM operation as per Fig. 2 are expressed as:

$$V_2 = \frac{V_{in} * d_1}{(1-d_1)} \tag{1}$$

$$V_0 = V_1 + V_2 = V_{in}\frac{1}{(1-d_2)} \tag{2}$$

here d_1, d_2 indicates duty cycles of switches S_L and S_B respectively.

$$M_B = \frac{V_0}{V_{in}} = \frac{1}{(1-d_2)} \tag{3}$$

$$M_L = \frac{V_2}{V_{in}} = \frac{d_1}{(1-d_1)} \tag{4}$$

M_B and M_L are gain of voltage for NBLI's Boost and Luo converters and V_{in} is source voltage respectively in CCM. Here, V_{C1} and V_{C2} are equal to V_1 and V_2.

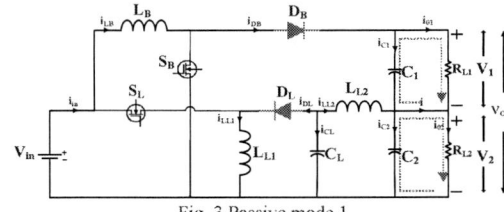
Fig. 3 Passive mode 1

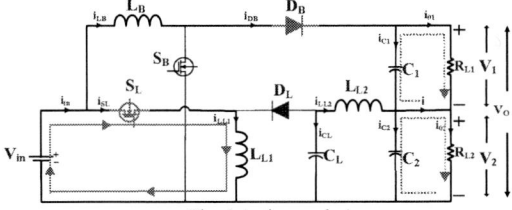
Fig. 4 Active mode 1

Fig. 5 Active mode 2

Fig. 6 Passive mode 2

Conduction Modes of SIDO NBLI converter: Fig. 3 Passive modes 1,
Fig.4 Active mode 1, Fig. 5 Active modes 2, Fig. 6 Passive mode 2

(1) Passive Mode 1: When S_L and S_B gated voltages are low, this mode starts. When neither of the active switches are conducting. R_{L1} is supplied with input source and L_b. L_{L1} will use its stored energy to charge C_L, and L_{L2} will release its energy to R_{L2}. Fig. 3 represents the circuit that goes with this mode.

$$V_{LB} = V_{in} - V_{C1} + V_{CL} - V_{C2} - V_{CL} \quad (5)$$

$$V_{LL1} = V_{CL} \quad (6)$$

$$V_{LL2} = V_{CL} - V_{C2} \quad (7)$$

$$V_0 = V_{C1} + V_{C2} \quad (8)$$

(2) Active Mode 1: For S_L, mode starts when the pulse for gating is low, and switch S_B, it is high. S_L is conducting when switch S_B is not conducting. Luo and Boost section share the input current. Supply input is charging L_{L1}, L_{L2} is discharging. Connected load (R_{L1}) is receiving energy from L_B in the Boost section. Fig. 4 represents circuit with this mode.

$$V_{LB} = V_{in} - V_{C1} + V_{LL2} - V_{CL} \quad (9)$$

$$V_{LL1} = V_{in} \quad (10)$$

$$V_{LL2} = V_{CL} - V_{C2} \quad (11)$$

$$V_0 = V_{C1} + V_{C2} \quad (12)$$

(3) Active Mode 2: When the gated voltage is high for S_B and low for S_L, this mode starts. S_B conducts as switch S_L aren't conducting. L_B is being charged by the input source, and C_1 discharges to load R_{L1}. L_{L2} is charging, L_{L1} is discharging through diode D_L. Fig. 5 depicts the circuit that goes with this mode.

$$V_{LB} = V_{in} \quad (13)$$

$$V_{LL1} = V_{CL} \quad (14)$$

$$V_{LL2} = V_{CL} - V_{C2} \quad (15)$$

$$V_0 = V_{C1} + V_{C2} \quad (16)$$

(4) Passive Mode 2: When gated voltages for S_L and SB are conducting, this mode starts. L_B and L_{L1} are being charged by the input source. C_L in the boost section provides the energy needed to load R_{L1}. In the Luo section, C_2 discharges and provides power to load R_{L2}. Fig. 6 displays the circuit that corresponds to this mode.

$$V_{LB} = V_{in} \quad (17)$$

$$V_{LL1} = V_{in} \quad (18)$$

$$V_{LL2} = V_{CL} - V_{C2} \quad (19)$$

$$V_0 = V_{C1} + V_{C2} \quad (20)$$

C. Design Specifications of NBLI converter

The desired current ripple can be achieved by selecting Boost's side inductance value, minimum values are calculated as indicated in expression as follows:

$$L_{B min.} \geq \frac{d_2 (1-d_2)^2 \times R_{L1}}{2 \times f_s} = \frac{0.5 \times (1-0.5)^2 \times 11.25}{2 \times 50 \times 10^3} = 16.75 \mu H \quad (21)$$

Similarly, Luo's inductors L_{L1} and L_{L2} can be expressed as shown

$$L_{L1 min.} \geq \frac{V_{in} \times d_1}{2 \times \Delta i_{LL1} \times f_s} = \frac{24 \times 0.5}{2 \times 1.9 \times 50 \times 10^3} = 63 \mu H \quad (22)$$

$$L_{L2 min.} \geq \frac{V_{in} \times (1-d_1)}{2 \times \Delta i_{LL2} \times f_s} = \frac{24 \times (1-0.5)}{2 \times 1.3 \times 50 \times 10^3} = 80 \mu H \quad (23)$$

Following is converter's minimum capacitance sizing, so that desired voltage ripple is achieved: (I = output current)

$$C_{1 min.} \geq \frac{d_2}{\frac{\Delta V_1}{V_1} \times R_{L1} \times f_s} = \frac{0.5}{\frac{0.1}{48} \times 11.52 \times 50 \times 10^3} = 416.6 \mu F \quad (24)$$

$$C_{2 min.} \geq \frac{I \times d_1}{\Delta V_{C2} \times f_s} = \frac{9.8 \times 0.5}{0.89 \times 50 \times 10^3} = 115.29 \mu F \quad (25)$$

$$C_{L\min.} \geq \frac{I \times d_1}{\Delta V_{CL} \times f_s} = \frac{9.8 \times 0.5}{1.3 \times 50 \times 10^3} = 75.38 \mu F \tag{26}$$

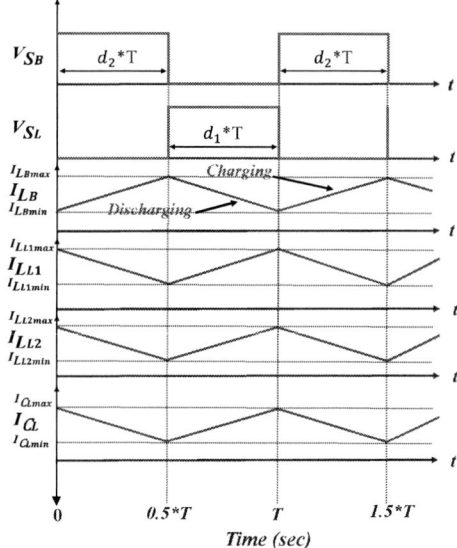

Fig. 7 Steady state waveform of SIDO NBLI converter in CCM

The range of duty cycles determines converter's operating modes. Here, d_1, $d_2 = 0.5$, operating modes are only active mode 1,2. Sole modes of operation for d_1, $d_2 < 0.5$ are active mode 1, 2 and passive mode 1. Additionally, sole modes of operation for d_1, $d_2 > 0.5$ are active mode 1, 2 and passive mode 2. As illustrated in Fig. 7, key waveforms of active switches' switching pulses, current of inductors and capacitor for d_1 and $d_2 = 0.5$ are assumed to be under CCM state.

III. DYNAMIC MODELLING AND CONVERTER CONTROL

This section comprises of the dynamic modelling and control implementation strategy for NBLI converter.

A. Dynamic Modelling

SIDO NBLI converter offers two operational modes. In the case of CCM, duty cycle regulation is employed to govern the input current and inductor currents at the input side. Corresponding dynamic model needs to be acquired in order to operate converter in each of the modes. Employs the state-space averaging method for getting converter's dynamic model. By taking into account a slight variation in the state variables (i_{LB}, V_1 for Boost, i_{LL1}, i_{LL2}, V_{CL}, V_2 for Luo) and duty cycles (d_1, d_2) of each section of converter, small signal analysis is carried out to produce a linearized AC model around a DC operation point as follows:

$$i_{Lm} = I_{Lm} + \hat{i}_{Lm} \tag{27}$$

$$v_{Cm} = V_{Cm} + \hat{v}_{Cm} \tag{28}$$

$$d_m = D_m + \hat{d}_m \tag{29}$$

In this case, m = b, L which is notation for Boost (b) and Luo (L) converter's representation. i_{Lm}, v_{Cm}, d_m are the current, voltage and duty cycle of NBLI converter which are subjected into perturbation for obtaining AC and DC model. In Eq. (30)-(35), A_m, B_m, C_m are average state space matrices for Boost and Luo converter, where A_m, B_m i.e. (A_{1m}, B_{1m}, A_{2m}, B_{2m}) are state and input matrix for active mode 1, 2.

$$A_b = A_{1b} \times d_2 + A_{2b} \times (1 - d_2) \tag{30}$$

$$B_b = B_{1b} \times d_2 + B_{2b} \times (1 - d_2) \tag{31}$$

$$C_b = C_{1b} \times d_2 + C_{2b} \times (1 - d_2) \tag{32}$$

$$A_L = A_{1L} \times d_1 + A_{2L} \times (1 - d_1) \tag{33}$$

$$B_L = B_{1L} \times d_1 + B_{2L} \times (1 - d_1) \tag{34}$$

$$C_L = C_{1L} \times d_1 + C_{2L} \times (1 - d_1) \tag{35}$$

The state vector X_m, in which m = b, L where, m representing Boost and Luo respectively. Here, steady state values of state vector are provided as follows:

$$X_b = \begin{bmatrix} I_{Lb} \\ V_O \end{bmatrix} = -A_b^{-1} \times B_b \times V_{in} \tag{36}$$

$$X_L = \begin{bmatrix} I_{LL1} \\ I_{LL2} \\ V_{CL} \\ V_2 \end{bmatrix} = -A_L^{-1} \times B_L \times V_{in} \tag{37}$$

Eq. (27), (28), and (29) are placed in the state-space averaged equations to produce the small signal model. Small signal and DC terms are then separated and further simplified. Then, laplace transformations to the small signal state space equations are used to formulate the resulting transfer functions, G(s) and T(s). The following are the expressions for the Boost and Luo converters i.e. G(s) and T(s) respectively:

$$G(s) = \frac{\hat{x}_b}{\hat{d}_2} = [sI - A_b]^{-1} \times \left[(A_{1b} - A_{2b}) X_b + (B_{1b} - B_{2b}) \times V_{in} \right] = \begin{bmatrix} G_{11} \\ G_{21} \end{bmatrix} \tag{38}$$

$$T(s) = \frac{\hat{x}_L}{\hat{d}_1} = [sI - A_L]^{-1} \times \left[(A_{1L} - A_{2L}) X_L + (B_{1L} - B_{2L}) \times V_{in} \right] = \begin{bmatrix} T_{11} \\ T_{21} \\ T_{31} \\ T_{41} \end{bmatrix} \tag{39}$$

$$\frac{\hat{V}_2}{\hat{d}_1} = \frac{\dfrac{v_{in}}{d_1 \times L_{L2} \times C_2} s^2 - \dfrac{d_1^2 V_{in}}{L_{L2} \times C_2 \times R_2 \times C_2 \times d_1^2} + \dfrac{V_{in}}{L_{L2} \times C_2 \times L_{L1} \times C_L}}{s^4 + \dfrac{s^3}{R_2 \times C_L} + s^2\left(\dfrac{d_1^2 V_{in}}{L_{L2} \times C_2} + \dfrac{1}{L_{L2} \times C_L} + \dfrac{d_1^2}{L_{L1} \times C_L}\right) + s\left(\dfrac{d_1^2}{R_2 \times C_L \times L_{L2} \times C_2}\right) + \dfrac{d_1^2}{L_{L1} \times C_L \times L_{L1} \times C_2}} \tag{40}$$

$$\frac{\hat{V}_0}{\hat{d}_2} = V_0 \left\{ \frac{(R_T) \times (1 - D_2) - L_B(s)}{(1 - D_2) \times (s^2 R_T \times L_B \times C_T + L_B(s) + R_T \times (1 - D_2)^2)} \right\} \tag{41}$$

Input and state matrices are represented by A and B in Eq. (38), (39). When one switch (S_{sw}) is on, matrices are A_{1m} and B_{1m} when other switch (S_{sw}) is off, the matrices are A_{2m} and B_{2m} which are input and state matrix for Boost and Luo converters in active mode 1 and 2. Eq. (40) and (41) represents transfer function T(s), G(s) for duty (d_1) to output voltage (V_2) and duty (d_2) to output voltage (V_o) respectively.

B. Voltage Control Scheme

An appropriate closed-loop control has been designed to control the NBLI converter's output voltages (V_0 and V_2). The transfer functions, G(s) and T(s) are used in voltage controller design for the converter that is being presented here. Table II contains a list of chosen parameters used in converter design.

TABLE II SPECIFICATIONS OF NBLI CONVERTER

Specifications	Symbols	Values	Units
Input Voltage	V_{in}	24	V
DC bus Voltage	V_0	48	V
Output Voltage (Luo side)	V_2	24	V
Rated Power	P_o	450	W
Frequency of Switching	f_s	50	kHz
Inductor	L_B, L_{L1}, L_{L2}	100	μH
Luo Capacitor	C_L	80	μF
Output capacitors	C_1, C_2	1000	μF
Load Resistances	R_1, R_2	11.52, 10	Ω
PI for V_0/d_2	K_p, K_i	0.0009978, 0.77	-
PI for V_2/d_1	K_p, K_i	0.0000044, 0.0799	-

Ziegler-Nichols tuning method is implemented for tuning transfer functions.

$$\frac{\hat{V}_0}{\hat{d}_2} = \frac{1.528 \times 10^4 s - 1.76 \times 10^8}{s^2 + 173.6s + 2 \times 10^6} \tag{42}$$

Lag compensator is designed, aims to improve the overall stability of converter, boost section (C_{boost}) as follows:

$$C_{boost} = \frac{-0.47107 * (s + 1845)}{s * (s + 3.5)} \tag{43}$$

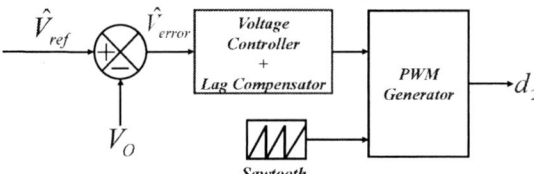

Fig. 8 Output Voltage V_0 Control

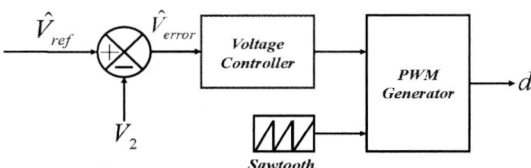

Fig. 9 Output Voltage V_2 Control

Applicable control-to-state $G_{21}(s)$ and $T_{41}(s)$ of NBLI's for output voltage control of V_0 and V_2 are obtained by replacing these parameters of table II in Eq. (40) and Eq. (41) are calculated. V_{0ref} and V_{2ref} are the reference voltages that are set to be desired output voltages respectively. The control signals are the error signals produced by comparing measured output voltages (V_o) and (V_2) with reference voltages. Voltage control is achieved by employing PI control method.

PI controller is implemented and tuned; its gains modified to prevent the output voltage from being impacted by dynamic conditions. Tuned controller eliminates error, which are the difference between V_o, V_2 and V_{0ref}, V_{2ref}. Fig. 8 and 9 display the control schematic for output voltage control.

IV. RESULTS AND DISCUSSION

This section IV contains simulation results of NBLI converter, a 450 W system demonstrated with presented DC-DC converter. System is validated in MATLAB/ Simulink. The converter's operation, here Boost side controls the 48 V (V_O) and Luo side controls the 24 V (V_2) output voltages. Performance of system is analyzed through waveforms of different parameters. Switches' voltage stress and current (V_{SB}), (V_{SL}), (I_{SB}), (I_{SL}) input current (I_{in}), Boost inductor currents (I_{LB}), Luo inductor currents (I_{LL1}), (I_{LL2}), diode current (I_{DB}), (I_{DL}), Input current I_{in} and Inductor Currents of Boost and Luo converters (I_{LB}), (I_{LL1}), (I_{LL2}), Input Voltage (V_{in}) and Output Currents (I_{01}), (I), (I_{02}) (A) are presented in figures from Fig. 10-14 respectively. Fig. 15 represents output voltages control of V_1 i.e. 72 V, $V_2 = 24$ V, and V_0 is 48 V. Voltage control for dynamic source voltage (V_{in}) is shown in this representation of output voltages V_1 (V), V_2 (V), and V_0 (V).

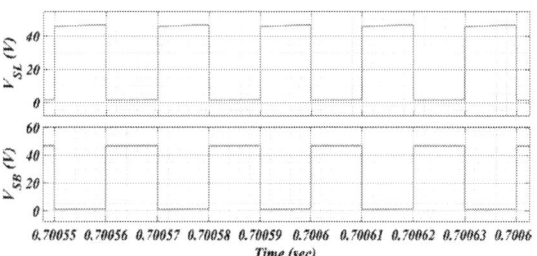

Fig. 10 Voltage Stress across switches V_{SL} (V), V_{SB} (V)

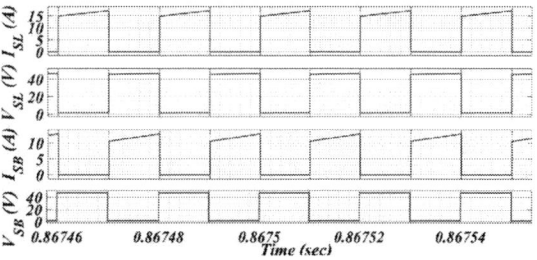

Fig. 11 Switches voltage stress V_{SL} (V), V_{SB} (V), Current I_{SL} (A), I_{SB} (A)

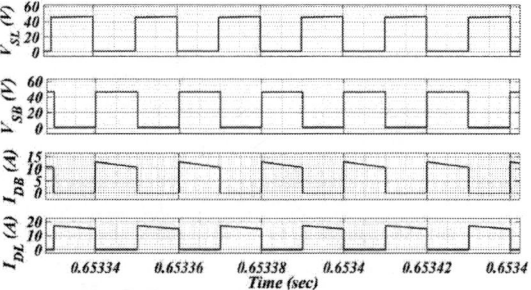

Fig. 12 Switch Voltage Stress (V_{SL}), (V_{SB}), Diode current (I_{DB}), (I_{DB})

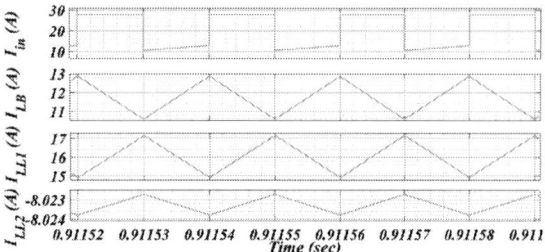

Fig. 13 Input current I_{in} (A) and Inductor Currents of Boost and Luo converters I_{LB} (A), I_{LL1} (A), I_{LL2} (A)

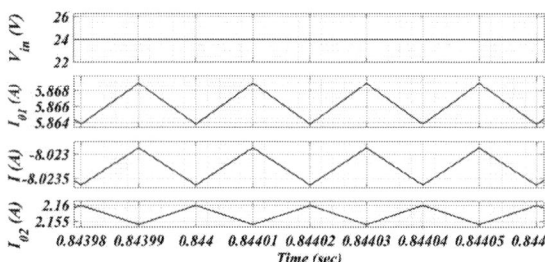

Fig. 14 Input Voltage V_{in} (V), and Output Currents I_{01} (A), I (A), I_{02} (A)

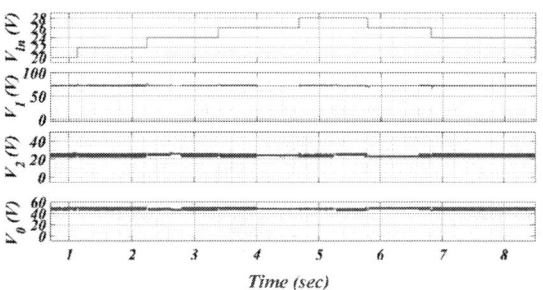

Fig. 15 Voltage Control of output voltages for dynamic source voltage

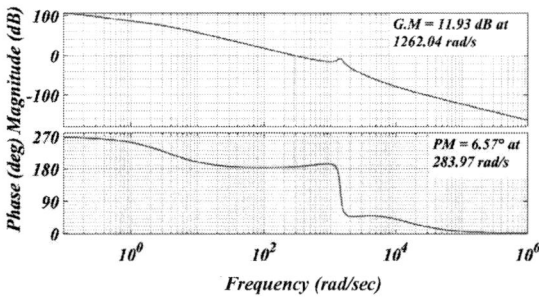

Fig. 16 Bode Plot of Compensated Boost section

In Fig. 16, Compensated bode plots of transfer functions GV_0/d_2 is presented. Employing the lag compensator for plot representing $G_{21}(s)$. For $G_{21}(s)$, phase margin (PM) is 6.57° at 283.97 rad/sec and gain margin (GM) is 119.93 dB at 1262.04 rad/sec.

V. CONCLUSION

A non-isolated SIDO NBLI DC-DC converter intended for BDCM operation, is presented in this paper. Due to interleaving of Boost and Luo converters, the input source current ripples are reduced. Presented converter is suitable for interfacing battery or PV as source to LVBDCM and providing multiple voltage levels V_1, V_2, and V_0 as output resulting in reducing requirement for several converters. This paper provides CCM's steady-state analysis in detail, design and analysis of converter, dynamic modelling, operating principle, VCM of converter and simulation results. Without voltage controller, output voltages are having alterations, for VCM PI controller approach is implemented to obtain constant intended output voltage levels. NBLI can provide bidirectional operation by substituting diodes (D_B and D_L) with MOSFETs/IGBTs as S_{B1} and S_{L1}. Body diodes of substituted switches function in manner of primary. Scope of improving the overall converter's stability by designing compensator, involving parasitic, losses and efficiency calculations, comparison of various interleaved topologies can be included. NBLI Converter's performance is validated and analysed by simulation results.

ACKNOWLEDGMENT

The authors convey their sincere thankfulness to the Electrical Engineering Department and Power Electronics Laboratory, Delhi Technological University.

REFERENCES

[1] Y. Ito, Y. Zhongqing and H. Akagi, "DC microgrid based distribution power generation system," *The 4th International Power Electronics and Motion Control Conference, 2004. IPEMC 2004*, Xi'an, China 2004.

[2] Q. Ren, Y. Han, M. Zhou, C. Yan, P. Yang and C. Wang, "Overview of Voltage Balancing Schemes in Bipolar DC Microgrids," *IEEE Transactions on Power Electronics*, vol. 40, no. 2, pp. 3469-3489, Feb. 2025.

[3] L. Li, K. -J. Li, K. Sun, Z. Liu and W. -J. Lee, "A Comparative Study on Voltage Level Standard for DC Residential Power Systems," *IEEE Transactions on Industry Applications*, vol. 58, no. 2, pp. 1446-1455, March-April 2022.

[4] L. E. Zubieta, "Are Microgrids the Future of Energy?: DC Microgrids from Concept to Demonstration to Deployment," *IEEE Electrification Magazine*, vol. 4, no. 2, pp. 37-44, June 2016.

[5] V. F. Pires, A. Cordeiro, C. Roncero-Clemente, S. Rivera and T. Dragičević, "DC–DC Converters for Bipolar Microgrid Voltage Balancing: A Comprehensive Review of Architectures and Topologies," *IEEE Journal of Emerging and Selected Topics in Power Electronics*, vol. 11, no. 1, pp. 981-998, Feb. 2023.

[6] T. Dragicevic, J. C. Vasquez, J. M. Guerrero and D. Skrlec, "Advanced LVDC Electrical Power Architectures and Microgrids: A step toward a new generation of power distribution networks," *IEEE Electrification Magazine*, vol. 2, no. 1, pp. 54-65, March 2014.

[7] H. -C. Chen and W. -J. Lin, "MPPT and Voltage Balancing Control With Sensing Only Inductor Current for Photovoltaic-Fed, Three-Level, Boost-Type Converters," *IEEE Transactions on Power Electronics*, vol. 29, no. 1, pp. 29-35, Jan. 2014.

[8] S. Rivera, R. Lizana F., S. Kouro, T. Dragičević and B. Wu, "Bipolar DC Power Conversion: State-of-the-Art and Emerging Technologies," *IEEE Journal of Emerging and Selected Topics in Power Electronics*, vol. 9, no. 2, pp. 1192-1204, April 2021.

[9] Q. Tian, G. Zhou, H. Li, Y. Yang and D. Zhou, "Symmetrical Bipolar Output Isolated Four-Port Converters Based on Center-Tapped Winding for Bipolar DC Bus Applications," *IEEE Transactions on Power Electronics*, vol. 37, no. 2, pp. 2338-2351, Feb. 2022

[10] V. Chapparya, A. Dey and S. P. Singh, "A Novel Non-Isolated Boost-Zeta Interleaved DC-DC Converter for Low Voltage Bipolar DC Micro-Grid Application," in *IEEE Transactions on Industry Applications*, vol. 59, no. 5, pp. 6182-6192, Sept.-Oct. 2023.

[11] S. Prasad, S. Mandal, P. Prabhakaran and A. D. D, "A Novel Boost Derived Input-Parallel Output-Series DC-DC Converter for Bipolar DC Microgrid," *2023 IEEE International Conference on Power Electronics, Smart Grid, and Renewable Energy (PESGRE)*, Trivandrum, India, 2023

A Novel Centralized Active Power Control Strategy for Multifrequency Microgrid with Energy Storage System

Sudeshna Mukherjee
Electrical Engineering
NIT Durgapur
Durgapur, India
sm.24ee1106@nitdgp.ac.in

Rajdip Dey
Electrical Engineering
NIT Durgapur
Durgapur, India
rdey.ee@nitdgp.ac.in

Tapas Kumar Saha
Electrical Engineering
NIT Durgapur
Durgapur, India
tapas.saha@ee.nitdgp.ac.in

Abstract—A Multifrequency microgrid (MFMG) is a type of Microgrid that operates at multiple frequencies simultaneously. Different frequency power maintains orthogonality and transmits simultaneously through the bus without mixing and different frequency voltages and currents are present on the Multi-Frequency Bus in an MFMG.

Due to the uncertainty of renewable sources, the energy storage system should be installed in MFMG. All converters used in MFMG are unidirectional. For higher reliability, an MFMG with an energy storage system is required, for which a bidirectional converter is needed. The power balancing technique for MFMG with ESS is not yet properly defined in any literature. This paper introduces a DC-to-MF converter as the fundamental component of a grid-forming converter in a Multi-frequency microgrid (MFMG). The proposed converter is analyzed in both grid feed and load scenarios. Its operation is examined through simulations conducted in MATLAB/Simulink, and the corresponding results are presented and discussed.

Index Terms—Multifrequency microgrid (MFMG), energy storage system (ESS), distributed energy resources (DER), grid feeding converter (GFE), grid forming converter(GF), and DC/AC+DC converter.

I. INTRODUCTION

A microgrid is a localized energy system that can function independently or alongside the conventional power grid. It comprises of distributed energy resources (DERs), including renewable sources like solar and wind power, Conventional generators (diesel, natural gas, etc.), energy storage system(batteries) and loads (such as buildings or industrial facilities).

AC, DC, and hybrid microgrids are the types of microgrids that are differentiated by the type of electric current they use for their operation. AC is the most common type of Microgrid as AC is the standard in most electrical grids and equipment. DC is less common in traditional power grids. It is used in many renewable energy systems(such as solar panels). A microgrid is a system that combines both AC and DC components to optimize energy use, reliability, and

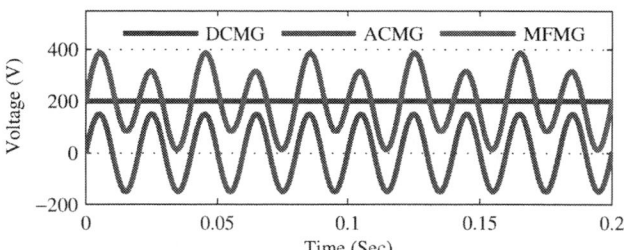

Fig. 1: Bus voltage of different Microgrids[7]

efficiency. It is called a hybrid microgrid. Microgrids offer several benefits compared to conventional transmission systems, including reduced transmission losses, improved load management, and seamless integration of distributed energy resources (DERs). However, AC microgrids face challenges related to control and stability, while DC microgrids face issues with protection and the need for new transmission infrastructure [6]. A multifrequency microgrid (MFMG) presents a viable solution by incorporating both AC and DC components, thereby combining their advantages and mitigating their drawbacks. With advances in power converter technology, MFMGs can be developed to enable the coexistence of multiple frequency components within a single conductor [7]. The design of a Multi-Frequency Microgrid (MFMG) can be achieved by applying the principles of superposition and orthogonal power flow theory [8]. Compared to conventional microgrids, MFMG offers improved efficiency and cost-effectiveness. The multifrequency concept holds significant potential for shaping future power markets.

Energy Storage Systems (ESS) play a crucial role in enhancing the reliability, stability, and efficiency of multi-frequency microgrids. As microgrids operate with multiple frequencies to accommodate diverse power generation and load requirements, ESS ensures a seamless power balance by mitigating fluctuations in supply and demand. A key feature of ESS in microgrids is the bidirectional power converter, which enables

979-8-3315-3013-6/25 $31.00 © 2025 IEEE

the battery to operate in both the charging and the discharging modes. During periods of excess power generation, the ESS stores surplus energy by charging the battery, whereas in times of high demand or frequency deviation, the stored energy is discharged back into the grid. This bidirectional energy flow improves grid stability, reduces dependency on conventional sources, and optimizes energy utilization.

The major contributions of the paper are designing a bidirectional DC/MF converter with filters and simulating the converter with ESS. The basic structure of MFMG with ESS is introduced and a novel active power balancing strategy is described. A 6 bus MFMG with ESS structure is simulated in Matlab Simulink and the power balancing strategy is verified with different results.

II. LITERATURE REVIEW

Very little literature is found on the multi-frequency microgrid. The use of multi-frequency systems in power transmission and converters is explored in this literature [1], highlighting their potential for efficient power management. Microgrids are described for the integration and efficiency of renewable energy [3]. Multifrequency power systems integrate renewable sources into smart grids, focusing on the quality and stability of power, as proposed in this literature [4] - a novel concept of power distribution for Multi-frequency microgrids, emphasizing efficient power flow and control. In [5], the concept of decoupling a microgrid into several independent power channels is presented. This architecture allows for flexible source-to-load power transfer using a shared transmission line. The study explores multifrequency power transfer within smart transformer-based grids, emphasizing enhancements in power quality and overall system efficiency. The approach is based on frequency-selective, orthogonal power transfer and leverages the superposition theorem to enable simultaneous, interference-free operation of multiple channels. A combined AC and DC power transmission system is introduced, highlighting its various applications and benefits. A newly proposed DC/DC+AC converter forms the basis for designing a single-phase MFMG architecture. The control techniques for this system are thoroughly explained, and simulations are conducted in both grid-connected and islanded modes to evaluate the performance of the grid-interactive converter.

All converters used in an MFMG are unidirectional. For higher reliability, an MFMG with an energy storage system is required, for which bidirectional converters are necessary. In the existing literature, the power balancing technique has not yet been properly defined for an MFMG with ESS. This paper proposes a bidirectional converter and its operation with an energy storage system to enable active power control of an MFMG with ESS, and the strategy is simulated using MATLAB Simulink.

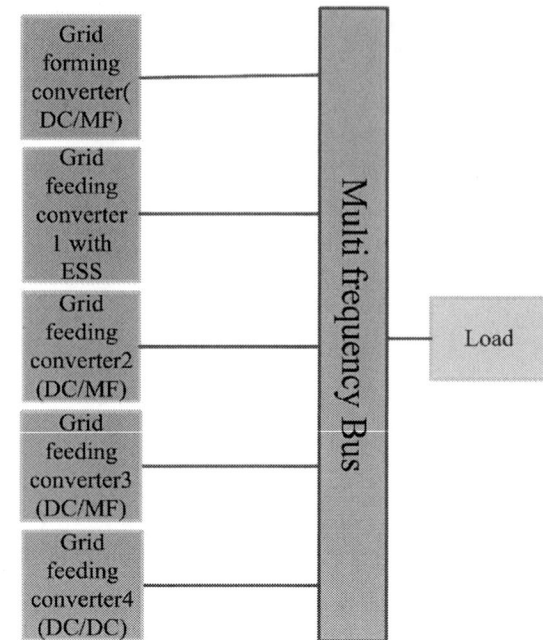

Fig. 2: Block diagram of MFMG with ESS

III. ARCHITECTURE FOR MULTIFREQUENCY MICROGRID WITH ESS

This section presents the structure of a multi-frequency microgrid with an energy storage system. The microgrid includes both AC and DC sources, various converters, buses, and loads. A proper architecture is essential to organize these components effectively. The connection of the Energy Storage System and the loads to the multifrequency bus are described in this section. The suggested architecture is presented in Fig. 2. In this study, three base frequencies are considered: 50 Hz, 25 Hz and 0 Hz. All energy sources are integrated into the Multifrequency bus through appropriate power electronic interfaces and are connected in parallel. The bus operates at 400 V with superimposed voltage levels of 230 V at 50 Hz and 150 V at 25 Hz.

A DC-to-MF converter plays a crucial role in the multifrequency microgrid. DC sources are connected to the bus through bidirectional converters, such as ESS-integrated DC/DC and DC/AC+DC converters. Moreover, resistive loads are interfaced with the MF bus. Based on the system architecture, the source-side DC/AC+DC converter operates in two distinct modes: grid-forming and grid-feeding. The grid-forming converter regulates the bus voltage and frequency, while the grid-feeding converter delivers power to the bus. This section provides a detailed explanation of the control strategies for the grid-forming converter, grid-feeding converter, and load-side regulation.

A. Grid forming converter

The GF converter is responsible for setting the voltage and frequency of the microgrid bus. It operates in parallel with

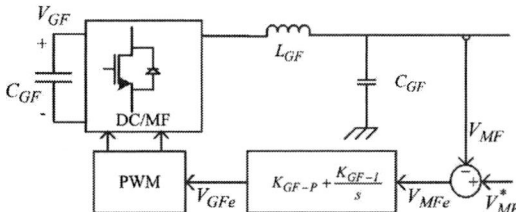

Fig. 3: Grid forming Control

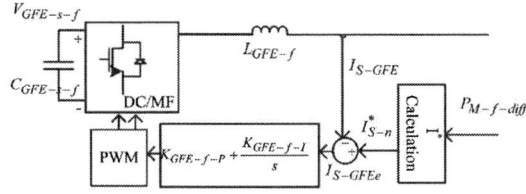

Fig. 4: Grid feeding Control

TABLE I: Simulation Parameters for MFMG

Parameters	Value
C,L	15.62uF,25mH
Kpgf , KIgf	25,0.5
Kpgf1, KIgf1	500,0.5
Kpgf2, KIgf2	500,0.5
Kpgf3, KIgf3	250,.25
Kpgf4, KIgf4	250,.25

all GFE converters and serves as the reference voltage and frequency of the converter. In this multi-frequency microgrid, a combined DC to AC+DC converter functions as the grid-forming converter. As a result, the bus voltage and frequency of the multi-frequency bus align with the output characteristics of the GF converter.

$$V*_{MF} = 400\,V_{DC} + 230\,V_{AC}\,@\,50\,Hz + 150\,V_{AC}\,@\,25\,Hz$$

B. Grid feeding converter

A grid-feeding converter operates as a current source, supplying power and current to the microgrid bus while maintaining a fixed reference voltage and frequency. These converters function in parallel, but require a GF converter to operate effectively. In this microgrid, there are four types of GFE converters: a bidirectional DC to AC+DC converter with an energy storage system (ESS), a DC/DC converter, and two DC to AC+DC converters.

C. Load

In the Simulink model, a resistive load was connected to observe the effect of sequentially connecting the resistor over time. Initially, at t = 0 s, a single 100 ohm resistor was connected to the circuit. After a time interval of t = 0.2 s, an additional 100 ohm was introduced into the circuit. The equivalent resistance of the circuit was observed to become 50 ohm after 0.2 s.

TABLE II: Battery Specifications

Parameter	Value
Voltage (V_{bat})	1000 V
Capacity (C_{bat})	50 Ah
State of Charge (SOC)	50 Percent

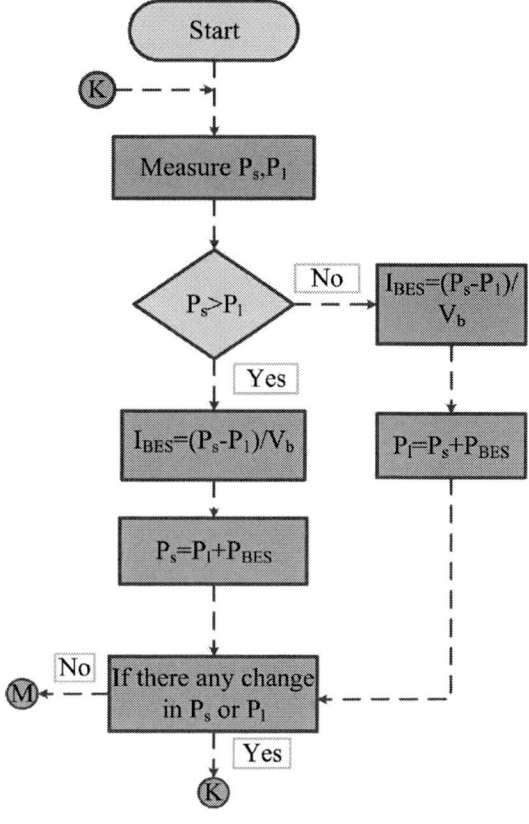

Fig. 5: Control flowchart for power balance between source and load using BESS.

IV. POWER BALANCING STRATEGY

In this study, a power balancing strategy is implemented for the Multifrequency Microgrid (MFMG) with an Energy Storage System (ESS) to ensure stable and reliable operation. The system consists of a Grid-Forming Converter, which maintains the bus voltage, four Grid-Feeding Converters that inject current into the multifrequency bus. The first grid feeding converter is a bidirectional converter connected to a battery storage system. The power-sharing mechanism ensures that all converters accurately track their respective reference currents to achieve proper load-sharing.

The control algorithm implemented to balance the source and load power is illustrated in the flow chart shown in Fig 5. The process begins with the measurement of source power (P_S) and load power (P_L). Based on the measured values, a decision is made to determine whether there is a surplus or deficit in power.

If the source power exceeds the load power ($P_S > P_L$), the excess power is directed towards charging the Battery Energy

Storage System (BESS). The charging current is calculated using:

$$I_{\text{BES}} = \frac{P_S - P_L}{V_b} \qquad (1)$$

where V_b is the DC bus voltage. The updated source power after accounting for BESS charging is given by:

$$P_S = P_L + P_{\text{BES}} \qquad (2)$$

In contrast, if the load power exceeds or equals the source power ($P_S \leq P_L$), the BESS compensates for the deficit by discharging. The discharge current is similarly computed as:

$$I_{\text{BES}} = \frac{P_S - P_L}{V_b} \qquad (3)$$

and the updated load power becomes:

$$P_L = P_S + P_{\text{BES}} \qquad (4)$$

The system continuously monitors for changes in either P_S or P_L. Upon detecting a variation, the algorithm re-initiates the measurement and decision loop to maintain system balance. If no change is observed, the system remains in the monitoring state.

Since the system operates with a purely resistive load, only active power is considered. The load current is manually calculated for different time intervals 0 to 0.2 seconds, 0.2 to 0.4 seconds, and 0.4 to 0.6 seconds shown in Table III.

TABLE III: Case study of power balancing

Time	0-0.2	0.2-0.4	0.4-0.6
Load	4A+2.3,50Hz+1.5A,25Hz	8A+4.6A,50Hz+3A,25 Hz	8A+4.6A,50Hz+3A,25Hz
GFC1	-3A	1.5A+2.3A,50Hz+2A,25Hz	2A+2.3A,50Hz+2A,25Hz
GFC2	4A+2.3A,50Hz	4A+2.3A,50Hz	3A+2.3A,50Hz
GFC3	2A+1.5A,25Hz	2A+1A,25Hz	2A+1A,25Hz
GFC4	1A	0.5A	1A

Based on these calculations, the reference currents for each Grid-Feeding Converter are set accordingly. Reference current values ensure that the converters proportionally share the load while maintaining system stability.

Each converter operates in current control mode, ensuring that its output current accurately follows the assigned reference. The Grid-Forming Converter maintains the bus voltage, while the Grid-Feeding Converters and the bidirectional battery-connected converter regulate current flow according to the set references.

The bidirectional converter connected to the ESS (battery) plays a crucial role in power balance. During periods of excess generation, the battery stores energy, and during load demand peaks, it discharges to maintain system stability. Power exchange with the ESS follows a predefined control strategy to prevent voltage fluctuations and ensure continuous power delivery.

The performance of the power balancing strategy is evaluated by verifying that all converters successfully track their reference currents and ensure that the bus voltage remains stable under varying load conditions and by confirming that the active power calculation at each instant aligns with the theoretical predictions.

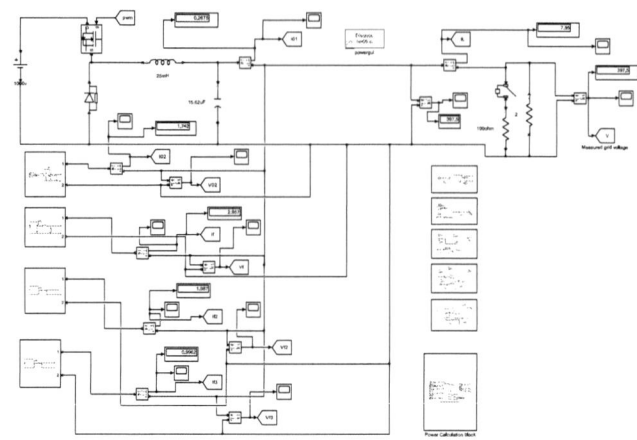

Fig. 6: Simulink diagram

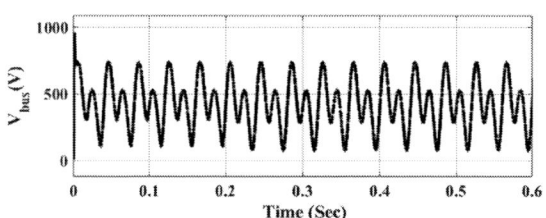

Fig. 7: Multifrequency bus voltage of MFMG

Simulation and/or experimental results validate the proposed power balance approach, demonstrating its effectiveness in maintaining a stable power distribution throughout the MFMG system.

V. SIMULATION RESULT

In this article, MFMG with ESS is introduced with a 6-bus premitive structure. The role of the grid-forming converter is played by the simulated DC to AC+DC converter. By using the PWM technique, the output voltage is controlled in the grid-forming converter. On the output side, an LC filter is connected. One 400V DC source and another source 230V,50Hz, and 150V, 25Hz are connected as a reference, which will produce an AC+DC voltage. Figure 6 shows the Simulink model diagram. In Fig. 7 the output voltage is presented. The output voltage follows the reference. So,the controller works properly.

During the operational state of the model, the circulating current in the grid-forming converter is much lower, indicating that the converter is functioning as expected shown in Fig. 8

Four types of grid feeding converters and loads are present in the MFMG structure. Grid feeding converter 1 is the ESS with a bidirectional converter. Grid feeding converter 2 is the dc to multi-frequency converter. Grid feeding converter 3 is also the dc to multi-frequency converter. Grid feeding

979-8-3315-3013-6/25 $31.00 © 2025 IEEE

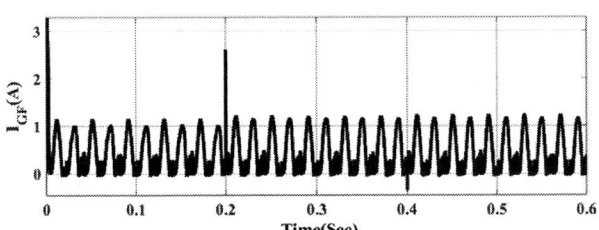

Fig. 8: Grid forming current

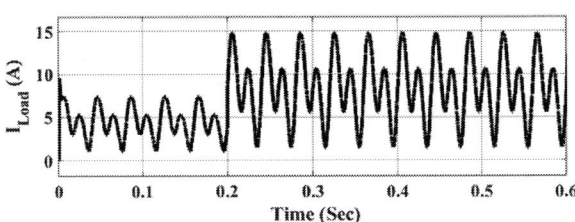

Fig. 9: Load Current

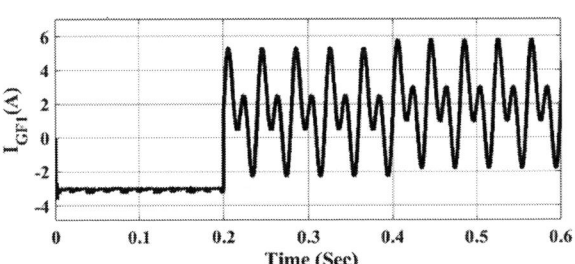

Fig. 10: Grid feeding Converter 1 current with ESS

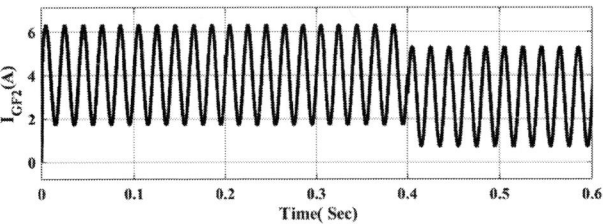

Fig. 11: Grid feeding Converter 2 current

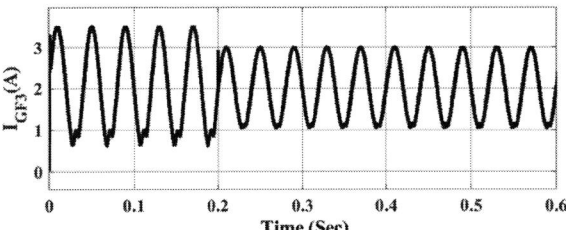

Fig. 12: Grid feeding Converter 3 current

converter 4 is the DC to DC converter. This multi-frequency bus is connected with a resistive load.

From fig 9 it is noticeable that the reference is being tracked by the output current of the converter. All reference currents at different times are shown in Table III.

The grid-feeding converter 1 operates as a bidirectional converter, facilitating both charging and discharging of the battery. During the time interval from 0 to 0.2 seconds, the battery charges from the multifrequency (MF) bus, storing energy. Subsequently, from 0.2 to 0.6 seconds, the converter switches to discharge mode, supply power back to the grid. This bidirectional operation ensures efficient energy management and grid support, aligning with dynamic power requirements. The waveform is shown in Fig. 10.

In Grid-Feeding Converter 2, the reference current is composed of both DC and AC components,with 50 Hz frequency. The converter employs a current control strategy to ensure that the output current accurately tracks the reference current shown in Fig. 11.

Grid-Feeding Converter 3 operates with a reference current consisting of both DC and AC components, where the AC component is derived from a 25 Hz source. The converter is designed to regulate its output current to precisely follow the reference signal, ensuring accurate power delivery. The effectiveness of this current tracking mechanism is validated through waveform analysis shown in Fig.12.

In Grid-Feeding Converter 4, the reference current is purely DC and varies over time. From 0 to 0.2 seconds, the reference current is set at 1 A DC. During the interval from 0.2 to 0.4 seconds, it decreases to 0.5 A DC. Finally, from 0.4 to 0.6 seconds, the reference current returns to 1 A DC. The converter ensures that the output current accurately tracks these variations, shown in fig 13.

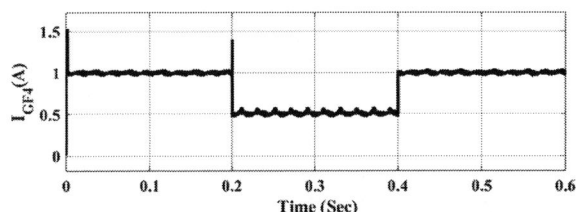

Fig. 13: Grid feeding Converter 4 current

The waveforms for battery voltage, state of charge (SOC), and current provide a comprehensive analysis of the performance of the Energy Storage System (ESS) in the proposed multifrequency microgrid. The battery voltage waveform demonstrates a stable profile during normal operation.These fluctuations remain within acceptable limits, indicating effective voltage regulation by the centralized control

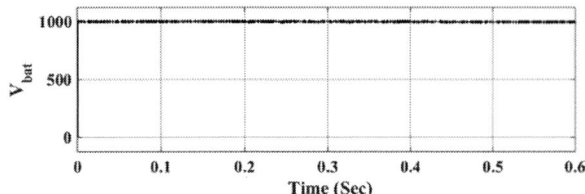

Fig. 14: Battery voltage

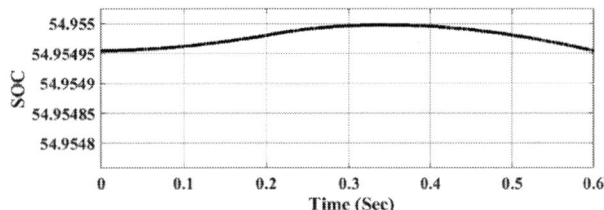

Fig. 15: State of Charge

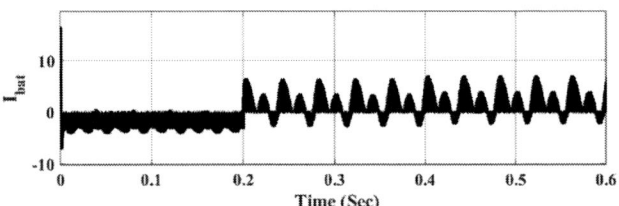

Fig. 16: Battery current

strategy. The voltage profile confirms that the ESS effectively maintains system stability while responding to dynamic load variations.

The SOC waveform reflects the charge-discharge cycle of the battery in response to power demand.The SOC remains within safe operating limits, validating the effectiveness of the proposed control technique shown in fig 15.

The battery current waveform further illustrates the dynamic operation of the ESS, showing (Fig. 16) distinct charging and discharging phases. Overall, the presented waveforms validate the efficiency of the proposed centralized control technique in managing ESS operation within a multi-frequency microgrid.

Total active power demand in the multi-frequency bus is 1.6 KW +0.529,50 Hz+0.223 KW, 25 HZ between 0-0.2 s, 3.2 kW +1.058 kW, 50 HZ +0.45 KW, 25 Hz between 0.2-0.4 s, 3.2 KW + 1.05 KW, 50 HZ +0.45 KW, 25 HZ between 0.4 -0.6 s. The MFMG is simulated using MATLAB Simulink to analyze the active power balance strategy. Variations in load power throughout the day create power imbalances at different time intervals. Since the microgrid operates independently without a grid connection, it must be self-sustaining. The achievement

of power balance requires effective control of all sources from the microgrid. The simulation results demonstrate that the proposed strategy successfully balances active power. All reference parameters are provided in Table I.

VI. CONCLUSION

This paper has presented a control technique for a Multi-frequency Microgrid (MFMG) integrated with an energy storage system (ESS), with stable operation. A 6 bus MFMG with ESS system has been simulated in Matlab Simulink. A detailed Centralized Control Strategy for MFMG with an Energy Storage System (ESS) has been effectively designed and validated. The proposed architecture of the Multifrequency Microgrid (MFMG) has been clearly defined with proper integration of multiple frequencies and power converters. Proper Active Power Balance has been achieved. Battery voltage, SOC (Stable within limits), and battery current (charge/discharge pattern) have confirmed the effectiveness of the control strategy. All converters accurately track reference currents. Future work may focus on different control techniques and the integration of more renewable energy sources to further enhance system resilience.

REFERENCES

[1] .H. Du, X. Zhang, Q. Sun and D. Ma, "Power Management Strategy of AC-DC Hybrid Microgrid in Island Mode," 2019 Chinese Control And Decision Conference (CCDC), Nanchang, China, 2019, pp. 2900-2905, doi: 10.1109/CCDC.2019.8833467.

[2] H. Han, X. Hou, J. Yang, J. Wu, M. Su and J. M. Guerrero, "Review of Power Sharing Control Strategies for Islanding Operation of AC Microgrids," in IEEE Transactions on Smart Grid, vol. 7, no. 1, pp. 200-215, Jan. 2016, doi: 10.1109/TSG.2015.2434849.

[3] H. Lotfi and A. Khodaei, "AC Versus DC Microgrid Planning," in IEEE Transactions on Smart Grid, vol. 8, no. 1, pp. 296-304, Jan. 2017, doi: 10.1109/TSG.2015.2457910.

[4] Multi-frequency power system for renewable source integration in smart grid Varun Chitransh, Mummadi Veerachary First published: 10 June 2019 https://doi.org/10.1049/iet-pel.2018.5101

[5] A. Alfergani, K. A. Alfaitori, A. Khalil and N. Buaossa, "Control strategies in AC microgrid: A brief review," 2018 9th International Renewable Energy Congress (IREC), Hammamet, Tunisia, 2018, pp. 1-6, doi: 10.1109/IREC.2018.8362575.

[6] S. Monesha, S. G. Kumar and M. Rivera, "Microgrid energy management and control: Technical review," 2016 IEEE International Conference on Automatica (ICA-ACCA), Curico, Chile, 2016, pp. 1-7, doi: 10.1109/ICA-ACCA.2016.7778452.

[7] R. Dey and S. Nath, "A new Power Distribution concept for Multifrequency Microgrid," 2021 IEEE 12th Energy Conversion Congress & Exposition - Asia (ECCE-Asia), Singapore, Singapore, 2021, pp. 491-496, doi: 10.1109/ECCE-Asia49820.2021.94790

[8] R. Dey and S. Nath, "Replacing silicon IGBTs with SiC IGBTs in medium voltage wind energy conversion systems," 2016 7th India International Conference on Power Electronics (IICPE), Patiala, India, 2016, pp. 1-6, doi: 10.1109/IICPE.2016.8079408.

[9] S. Brüske, G. De Carne and M. Liserre, "Multi-frequency power transfer in a smart transformer based distribution grid," IECON 2014 - 40th Annual Conference of the IEEE Industrial Electronics Society, Dallas,TX, USA, 2014, pp. 4325-4331, doi: 10.1109/IECON.2014.7049153.

[10] S. Akshatha, C. N. Arun, V. S. Abhijith and B. G. Fernandes, "A unified AC-DC microgrid architecture for distribution of AC and DC power on the same line," 2017 IEEE Applied Power Electronics Conference and Exposition (APEC), Tampa, FL, USA, 2017, pp. 430-433, doi: 10.1109/APEC.2017.7930729.

[11] R. Dey and S. Nath, "A New Active and Reactive Power Control Strategy for Multifrequency Microgrid in Islanded Mode," 2021 5th International Conference on Smart Grid and Smart Cities (ICSGSC), Tokyo, Japan, 2021, pp. 45-49, doi: 10.1109/ICSGSC52434.2021.9490493.

AUTHOR INDEX

Aarthi, A 236
Abhijith, G 469
Abhivarma, M. 475
Adithya, R. 635
Agarwal, Vivek 646
Aher, Prashant 336
Akshara, Kelli 487
Akshara, S. 612
Akshay, K 166, 372
Alam, Aftab 450
Alam, M. N. 623
Alam, Mahamad Nabab 629
Alam, Md Asif 535
Amal, S 107, 612
Amina, S 284
Anand, Chetan 571
Angel, K O 366
Anilkumar, Sandhya 342
Anirud, R S 306
Anusha, Dharavath 641
Aravind, G 376
Arya, Abhinav 388
Ayan, Syed 348
Babu, G. Suresh 272
Bandaru, Vamshi Krishna 394
Banik, Anindya 412
Barai, Mukti 535
Baruah, Ujjwal 541
Batta, Sai Sailu 432
Behera, Niranjan 388
Bhanarkar, Harsh 641
Bharadwaj, Ch Anil 272, 278
Bhaskar, M. S. 160
Bhende, C. N. 577
Bhowmick, Supratik 95
Bhowmik, Pritam 242
Boby, Rose Mary 119
Bonde, Akshaya D. 588
Borghate, Vijay B. 588
Borse, Harshada Dattatray 438
Bose, Indranil 342
Brahma, Bipul 481
Bulasara, Kavya Sree Sai 432
Castellazzi, Alberto 69
Chaganti, Srinivas Bhaskar 19
Chakka, Vijay Kumar 324
Chakrabarty, Supriya 388
Chakraborty, Aritra 107

Chakraborty, Chandan 95
Chander, Nikhil 348
Chandrasekar, V 469, 612
Chandrasekaran, Kandasamy 51
Chary, N Karthik 228
Chattopadhyay, Tishya 382
Chaturvedi, Pradyumn 588, 594
Chaudhari, Bhalchandra N 419
Chaudhary, Pramod 366
Chaudhuri, Sohini 101
Chauhan, Abhisha 266
Chejarla, Madhu Kishore Devara 312
Chittora, Prakash 406
Choudhury, Tanmoy Roy 547
Chowdhury, Dipanjana 511
Dalal, Ankit 200, 606
Das, Souvik 382
Dave, Rasesh 541
De, Dipankar 69, 131
Deepa, S. N. 635
Deepesh, Kv 354
Deo, Mayank P. 366
Deore, Rakesh 481
Desai, Nikhil K 493
Devabhaktuni, Swati 113, 125
Devanathan, V 228
Devarinti, Vani 125
Devi, Oinam Lotika 559
Dey, Rajdip 663
Dharani, Kolantla 618
Dharavath, Ramesh 236
Dhola, Hitesh 541
Divya, K 210
Dodda, Satishreddy 360
Dutta, Avismit 101
Edara, Shiva Prasad 394
Esha, Tangudu Sai 487
Fahnbulleh, Edwin Boima 76
Gadikota, Varun Reddy 528
Gajjar, Sandip 541
Gandhi, Manoj Leelachand 176
Ganesan, P. 469
Ganga, Ashok 487
Gautam, Jitendra 200
Gautam, Ravi Prakash 166
George, Kuruvachan K. 306
Ghatak, Sriparna Roy 76
Ghatakchoudhuri, Sumit 1

Ghose, T. 27
Ghosh, Joydeep 541
Ghosh, Shrestha 412
Ghoshal, Nabhomoni 499
Ghube, Aditya Purushottam 266
Girdharlal, Navle Sonal 206
Giridhar, A. V. 171
Gopi, Kandipalli 176
Goswami, Niranjanpuri 541
Goutham, Veerapu 432
Gowri, N. Vasantha 272, 278
Gupta, Deepak Kumar 222
Gupta, Nitish Kumar 528
Gupta, Rajesh 400, 426
Gupta, Rohit 426
Gurunath, Gurucharan 565
Harikrishnan, A. 131
Harshan, J 635
Inayathullaah, M. A. 565
Indhuja, L R 493
Jacob, Febin Koshy 342
Jagadanand, G 354
Jain, Jenis 324
Jandhyala, K. L. Karthik 606
Jisha, S R 612
Jithin, K 553
Kalita, Karuna 481, 522, 583
Kalla, Kanaka Raju 487
Kanagasabai, Lenin 13
Kannan, M 155
Kannan, S 228
Kar, Susmita 571
Karmakar, Subrata 511
Karn, Neha 462
Karri, Srujana 306
Kartikey, Umang 406
Katiyar, Trapti 33
Kaushal, Kaushlendra Kumar 206
Kazi, Wakar Hasan 166, 372
Keerthika, C 318
Kodati, Bhanu Venkata Siva Saikiran 629
Kodeeswaran, R 600
Koduru, Suprabhath Sriranga 651
Kolipaka, Spandana 113
Kondoju, Shashi Kumar 594
Kondreddi, Krishnaveni 278
Kotapati, Anuradha 394
Kothapalli, Nagaraju 260
Krishna, T. Murali 272, 278
Krishnaprasad, Mb 469
Kulkarni, Aditya 594
Kumar, Amit 33

Kumar, Brijesh 137
Kumar, C Santhosh 306
Kumar, Chandan 456
Kumar, Kunal 571
Kumar, R Hari 553
Kumar, R. M. Ram 522
Kumar, Ravinder 63
Kumar, Rohit 541
Kumar, S. Shiva 27
Kumar, V. Vignesh 493
Kumar, Varri Chandra Sekhar Pavan 131
Kumar, Vineet 406
Kumaravel, S. 210, 216, 342
Kumari, Aradhana 176
Kuncham, Sateesh Kumar 260, 444, 600
Kushwaha, Amarkumar 376
Lal, Vivek Nandan 248
Lekshmi, V 206
Lella, Venkatamaheshbabu 51
Machina, Venkata Siva Prasad 651
Madichetty, Sreedhar 651
Magar, Pradip M 366
Mahapatra, Subrat 330
Maiti, Suman 95
Makwana, Motibhai 541
Manju, R 469
Mankani, Ashok 107
Marupatla, Bharath 641
Matam, Sailaja Kumari 312
Mayadevi, N. 553
Mehta, Kush 541
Miriyala, Mahesh 432
Mishra, Jyotismita 302
Mishra, Manash Kumar 248
Mishra, Saurabh 657
Mishra, Shubham 462
Mishra, Sudhansu Kumar 63
Mishra, Sukumar 651
Misra, Himanshu 143
Mithun, T 149, 155
Modak, Prabhakar 160
Mohan, Prameeda 284
Mohapatra, Rachita 166
Mohapatro, Sankarsan 388
Monish, Bkn 565
Mukherjee, Sudeshna 663
Murshid, Shadab 618
Nag, Sunil P V 306
Nair, Supriya 541
Nandi, Sharmistha 76
Natarajan, Kumaresan 444
Naugraiya, Aditya 107

Nautiyal, Durgesh Chandra.................................324
Nayak, P Srinivasa Rao..............................149, 155
Nemani, Surya Teja.......................................306
Nikhitha, Matcha 487
Nutenki, Ratnakar 188, 194
Padole, Anshal S 419
Panchalai, V N 444
Panda, Anup Kumar 137
Panda, Aurobinda...101
Panda, Babita 76
Panda, Sanjib Kumar 7, 618
Panda, Swagat Kumar......................................89
Panigrahi, Bibhu Prasad..................................137
Panigrahi, Chinmoy Kumar 222
Panigrahy, Satyajit 511
Parmar, Darshan...541
Parmar, Prakash 541
Patel, Aman...266
Patel, Rishabh 182
Patil, Sanjaykumar 336
Pattnaik, Monalisa 475
Pennam, Akhilash 487
Peper, Pradeep K..45
Pippuri-Mäkeläinen, Jenni 522
Porpandiselvi, S 182
Praanesh, Kanala Srinivas 547
Pradabane, Srinivasan 641
Pradhan, Ashok Kumar 188, 194
Pradhan, K P 302
Pragya, Kumari Prakhar 450
Prakash, S. Lenin 318
Prasanthi, Pydisetti 487
Priya, Kumari...400
Priyadarshi, Neeraj 160
Pullaguram, Deepak Reddy 438
Pullaguram, Deepak 623
Puthusserry, Gireesh V 149
Radhakrishnan, Rahul 57, 119
Ragava, V Ganesh...565
Raghavendran, S. 318
Rahman, Shifa H..119
Rai, Dev 336
Rai, Pankaj 82
Rajeev, T 284
Rajesh, S. 606
Ramchand, Rijil 254
Ramesh, R. 444
Rane, Monish 419
Rane, Rhugved..336
Ranganathan, V 565
Rani, Kirlampalli Harija 39, 290
Ranjan, Rajeev 348

Rao, C. Srinivasa...505
Rao, D V Siva Krishna K 236
Rao, P. Maheswara 487
Rao, Yerramreddy Srinivasa 419
Rath, Dalija 296, 330, 499
Rathod, Tanmayee 336
Ray, Anjan Kumar.....................................101, 456
Reddy, G. Panduranga 505
Reddy, Pichali Reddykishore 45
Reddy, S Harsha Vardhan 372
Roul, Soumyasephalika 462
Routray, Aurobinda...................................188, 194
Roy, Molay...101
Sagar, V Udaya 612
Saha, Shantanu 412
Saha, Tapas Kumar 663
Sahani, Mrutyunjaya 7, 618
Sahoo, Biswajit 412
Sahoo, Rakesh 462
Sahu, Himanshu Sekhar 324
Sahu, Silpashree 69, 131
Samal, Mitali 222
Samal, Simanta Kumar....................................248
Samanta, Bikram Kumar 95
Samanta, Supriyo 57
Samanta, Susovon 296, 330, 499
Sandepudi, Srinivasa Rao.................................360
Saneep, K 149, 155
Sanith, Sl 469
Saranya, S T 242
Saravanan, B 318
Sarita, Kumari 82
Sasidharan, Nikhil 254, 354
Satpathy, Priya Ranjan 242
Saurabh, Kumar 107
Sawant, Rajendra R 419
Saxena, Vinayak 406
Seelam, Poornima 394
Sekhar, P. C. 577
Senapati, Dibyaranjan 45
Sengupta, Gopinath 522, 583
Shafeeque, K Muhammedali 254
Shahrukh, ...522, 583
Sharma, Akshat 143
Shastri, Sharankumar 517
Shekhar, Shashank 450
Shinde, Rushikesh U 366
Shishupal, ...1
Shreelakshmi, M P 366
Siddiqui, Mohammad Tahir 400, 426
Simhadri, Kumaraswamy 487
Simon, Sishaj P 149

Singaravel, M. M. Rajan 493
Singh, Abhinay Pratap 426
Singh, Alka 559
Singh, Bhim 517
Singh, Manoj Kumar 45
Singh, N P 541
Singh, Omkar 101
Singh, Pradeep 171
Singh, Prashant 166
Singh, Shivam 27
Singh, Shivendra Pratap 376
Singh, Shravan Kumar 348
Singh, Vipin Kumar 517
Soumiya, K 302
Sowmyasree, V 505
Sreekumar, V 45
Sriharitha, Saddikuti 200
Srivastava, Sachin 160
Subudhi, Bidyadhar 89
Subudhi, Sanjeet Kumar 462
Sudhakar, S 206
Suman, Santwana 33
Sundararajan, Prasanth 618
Sundararaju, K. 210
Sundareswaran, K 149, 155
Swain, Akash Kumar 646
Swecha, Tadepalli Phani 629
Syed, Abdul Mujeer 312
Tadepalli, Phani Swecha 438, 623
Tavhare, Sarika D. 342
Thakar, Aruna 541
Thakkar, Jayesh 33
Thakur, Vipul 657
Tharun, Meddi 107
Tikoo, Janyaa 336
Tiwari, Yogendra 406
Upadhyay, Dishang 541
Vadlamani, Venakata Sastry 45
Vakil, Gaurang 522
Varshath, S. Vinu 216
Veer, Karan 63
Venkatesaperumal, B. 493
Venkatesh, R. 577
Vinatha, U. 493
Vinopraba, T 228
Vishaal, S 318
Vishnu, C V 612
Vishwanathan, Neti 39, 290
Yadav, Madhav 606
Yadav, Ujjwal 176
Youvaraj, Ganesh 641
Youvaraj, Subbash 641

IEEE
445 Hoes Lane
Piscataway, NJ 08854-4141

ISBN 979-8-3315-3013-6